Forests, Water and People in the Humid Tropics
Past, Present and Future Hydrological Research for Integrated Land and Water Management

Forests, Water and People in the Humid Tropics is the most comprehensive review available of the hydrological and physiological functioning of tropical rainforests, the environmental impacts of their disturbance and conversion to other land uses, and optimum strategies for managing them.

The authors review existing guidelines for timber harvesting, land clearing and post-forest agriculture, and seek ways to enhance their application. The book also examines the possibilities of restoring the hydrological functioning of degraded areas. New techniques that may help researchers and managers to understand better the hydrological consequences of land management decisions are discussed. The editors have supplemented the individual contributions with invaluable overviews of the main sections and provide key pointers for future research.

This book brings together leading specialists in such diverse fields as tropical anthropology and human geography, environmental economics, climatology and meteorology, hydrology, geomorphology, plant and aquatic ecology, forestry and conservation agronomy. Specialists will find authenticated detail in chapters written by experts on a whole range of people–water–land use issues, and managers and practitioners will learn more about the implications of ongoing and planned forest conversion, while scientists and students will appreciate a unique review of the literature.

MIKE BONELL is Chief of the Hydrological Processes and Climate Section at the UNESCO Division of Water Sciences. He is the managing series editor of the International Hydrology Series, and is leading editor of *Hydrology and Water Management in the Humid Tropics* (1993; Cambridge University Press).

L. A. (SAMPURNO) BRUIJNZEEL is Senior Lecturer/Associate Professor of Eco-Hydrology at the Department of Hydrology and Geo-Environmental Sciences, Vrije Universiteit, Amsterdam. He is on the editorial board of the *Journal of Tropical Ecology*, *Hydrological Processes*, the *Encyclopedia of Forest Sciences (Forest Hydrology Section)*, and the *Journal of Land Use and Water Resources Research*.

INTERNATIONAL HYDROLOGY SERIES

The **International Hydrological Programme** (IHP) was established by the United Nations Educational, Scientific and Cultural Organization (UNESCO) in 1975 as the successor to the International Hydrological Decade. The long-term goal of the IHP is to advance our understanding of processes occurring in the water cycle and to integrate this knowledge into water resources management. The IHP is the only UN science and educational programme in the field of water resources, and one of its outputs has been a steady stream of technical and information documents aimed at water specialists and decision-makers.

The **International Hydrology Series** has been developed by the IHP in collaboration with Cambridge University Press as a major collection of research monographs, synthesis volumes and graduate texts on the subject of water. Authoritative and international in scope, the various books within the series all contribute to the aims of the IHP in improving scientific and technical knowledge of fresh-water processes, in providing research know-how and in stimulating the responsible management of water resources.

TITLES IN PRINT IN THE SERIES

M. Bonell, M. M. Hufschmidt and J. S. Gladwell *Hydrology and Water Management in the Humid Tropics: Hydrological Research Issues and Strategies for Water Management*

Z. W. Kundzewicz *New Uncertainty Concepts in Hydrology*

R. A. Feddes *Space and Time Scale Variability and Interdependencies in the Various Hydrological Processes*

J. Gibert, J. Mathieu and F. Fournier *Groundwater and Surface Water Ecotones: Biological and Hydrological Interactions and Management Options*

G. Dagan and S. Neuman *Subsurface Flow and Transport: A Stochastic Approach*

D. P. Loucks and J. S. Gladwell *Sustainability Criteria for Water Resource Systems*

J. C. van Dam *Impacts of Climate Change and Climate Variability on Hydrological Regimes*

J. J. Bogardi and Z. W. Kundzewicz *Risk, Reliability, Uncertainty and Robustness of Water Resources Systems*

G. Kaser and H. Osmaston *Tropical Glaciers*

I. A. Shiklomanov and John C. Rodda *World Water Resources at the Beginning of the Twenty-First Century*

A. S. Issar *Climate Changes during the Holocene and their Impact on Hydrological Systems*

INTERNATIONAL HYDROLOGY SERIES

Forests, Water and People in the Humid Tropics

Past, Present and Future Hydrological Research for Integrated Land and Water Management

Edited by

M. Bonell

UNESCO, Paris

L. A. Bruijnzeel

Vrije Universiteit, Amsterdam

CAMBRIDGE
UNIVERSITY PRESS

PUBLISHED BY THE PRESS SYNDICATE OF THE UNIVERSITY OF CAMBRIDGE
The Pitt Building, Trumpington Street, Cambridge, United Kingdom

CAMBRIDGE UNIVERSITY PRESS
The Edinburgh Building, Cambridge, CB2 2RU, UK
40 West 20th Street, New York, NY 10011–4211, USA
477 Williamstown Road, Port Melbourne, VIC 3207, Australia
Ruiz de Alarcón 13, 28014 Madrid, Spain
Dock House, The Waterfront, Cape Town 8001, South Africa

http://www.cambridge.org

First published 2004

Printed in the United Kingdom at the University Press, Cambridge

Typeface Times 9.5/13pt. *System* LaTeX 2ε [TB]

A catalog record for this book is available from the British Library

Library of Congress Cataloging in Publication data
Forests, water, and people in the humid Tropics : past, present, and future hydrological
research for intergrated land and water management / edited by M. Bonell and L. A. Bruijnzeel.
p. cm. – (International hydrology series)
Includes bibliographical references (p.)
ISBN 0 521 82953 4
1. Hydrology, Forest – Tropics. 2. Rain forests. I. Bonell, M. II. Bruijnzeel, Leendert Adriaan. III. Series.
GB840.F68 2004
333.75 – dc22 2003069746

ISBN 0 521 82953 4 hardback

Contents

Contributors

Aylward, B.
Deschutes Resources Conservancy, P.O. Box 1560, Bend, OR 97709, USA

Bago-Labis, J. P.
Heifer Project International/Philippines, Unit 907, South Center Tower, Madrigal Business Park, Alabang, Muntinlupa City 1771, Philippines

Barnes, C.
Climate and Agricultural Risk Unit, Agriculture and Food Sciences Program, Bureau of Rural Sciences, P.O. Box E11, Kingston ACT 2604 Canberra, Australia

Bidin, K.
School of Science and Technology, Universiti Malaysia Sabah, 88999, Kota Kinabalu, Malaysia

Bonell, M.
Hydrological Processes and Climate Section, Division of Water Sciences, UNESCO, 1 rue Miollis, 75732 Paris Cedex 15, France

Bruijnzeel, L. A.
Faculty of Earth and Life Sciences, Vrije Universiteit, Amsterdam, De Boelelaan 1085, 1081 HV Amsterdam, The Netherlands

Busby, A. L.
International Center for Aquaculture and Aquatic Environments, Department of Fisheries, Auburn University, Auburn, AL 36849, USA

Buttle, J. M.
Department of Geography, Trent University, Peterborough, Ontario, K9J 7B8, Canada

Callaghan, J.
Severe Weather Section, Bureau of Meteorology, G.P.O. Box 413, Brisbane, QLD 4000, Australia

Cassells, D. S.
The World Bank, Environment Department, 1818 H Street NW, Washington, DC 20433, USA

Cequiña, E. Y.
Central Mindanao University, Musuan, Bukidnon, Mindanao, Philippines

Chappell, N. A.
Centre for Research on Environmental Systems and Statistics, IENS, Lancaster University, Lancaster, LA1 4YQ, UK

Connolly, N. M.
Australian Centre for Tropical Freshwater Research, Rainforest Cooperative Research Centre, James Cook University, Townsville, LD Q4811, Australia

Connor, G.
Bureau of Meteorology, RAAF Base Garbutt, Townsville, QLD 4814, Australia

Costa, M. H.
Federal University of Viçosa, Brazil

Critchley, W. R. S.
CIS-Centre for International Cooperation/Faculty of Earth and Life Sciences, Vrije Universiteit, De Boelelaan 1105, 1081 HV Amsterdam, The Netherlands

Deutsch, W. G.
International Center for Aquaculture and Aquatic Environments, Department of Fisheries, Auburn University, Auburn, AL 36849, USA

Douglas, I.
School of Geography, University of Manchester, UK

Drigo, R.
Località Collina 5, I-53036 Poggibonsi, Siena, Italy

Fritsch, J. -M.
L'Institut de Recherche pour le Développement-LMTG, 38 rue des 36 Points, F-31400 Toulouse, France

Gash, J. H. C.
Centre for Ecology and Hydrology, Wallingford, OX10 8BB, UK

Grip, H.
Department of Forest Ecology, SLU, S-901 83 Umeå, Sweden

Guyot, J.-L.
L'Institut de Recherche pour le Développement-LMTG, 38 rue des 36 Points, F-31400 Toulouse, France

Hall, A. L.
LCSES, msn I-6-600, The World Bank, 1818 N Street, New Washington DC, 20433, USA

Hall, C. A. S.
College of Environmental Science and Forestry, State University of New York, Syracuse, NY 13210, USA

Hamilton, L. S.
East–West Center, 342 Bittersweet Lane, Charlotte, VT 05445, USA

Held, A. A.
CSIRO, Canberra, ACT, Australia

Hj Nik, A. R.
Forestry Research Institute of Malaysia, Kepong, 52109 Kuala Lumpur, Malaysia

Hölscher, D.
Institute of Silviculture, University of Göttingen, Buesgenweg 1, D-37077 Göttingen, Germany

Hooijer, A.
Department for River Basin Management, Delft Hydraulics, P.O. Box 177, 2600 MH Delft, The Netherlands

Jakeman, A. J.
Centre for Resource and Environmental Studies (CRES), The Australian National University, Canberra, ACT 0200, Australia

Kaimowitz, D.
Center for International Forest Research (CIFOR), PO Box 6596 JKPWB, Jakarta 10065, Indonesia

Kasran, B.
Forestry Research Institute of Malaysia, Kepong, 52109 Kuala Lumpur, Malaysia

Ko, J.-Y.
Coastal Ecology Institute, Louisiana State University, Baton Rouge, LA 70803, USA

Kundzewicz, Z. W.
Research Centre of Agricultural and Forest Environment, Polish Academy of Sciences, Bukowska 19, 60-809 Poznań, Poland *also* Potsdam Institute for Climate Impact Research, Potsdam, Germany

Lancaster, J. W.
Arup Water, 78 East Street, Leeds, LS9 8EE, UK

Mackensen, J.
Division of Policy Development and Law, United Nations Environmental Programme (UNEP), P.O. Box 30552, Nairobi, Kenya

Mahé, G.
L'Institut de Recherche pour le Développement – IRD-ex ORSTOM, 01 BP 182, Ouagadougou 01, Burkina Faso

Maley, J.
L'Institut de Recherche pour le Développement, BP 5045, F-34032 Montpellier Cedex 1, France

Malmer, A.
Department of Forest Ecology, Swedish University of Agricultural Science, SE-901 83 Umeå, Sweden

McDonnell, J. J.
Department of Forest Engineering, Oregon State University, Corvallis, OR, USA

Murdiyarso, D.
Center for International Forestry Research (CIFOR), Bogor, Indonesia

Ong, C. K.
Regional Land Management Unit, RELMA, International Centre for Research in Agroforestry, Nairobi, Kenya

Orprecio, J. L.
Heifer Project International/Philippines, Unit 907, South Center Tower, Madrigal Business Park, Alabang, Muntinlupa City 1771, Philippines

Pearson, R. G.
School of Tropical Biology, James Cook University, Townsville, QLD 4811, Australia

Planos-Gutierrez, E. O.
Instituto de Meteorología, Havana, Cuba

Proctor, J.
Department of Biological Sciences, University of Stirling, Stirling, FK9 4LA, UK

Roberts, J. M
Centre for Ecology and Hydrology, Wallingford, OX10 8BB, UK

Robson, A. J.
Centre for Ecology and Hydrology, Wallingford, OX10 8BB, UK

Rodriguez, E.
Jet Propulsion Laboratory, National Aeronautics and Space Administration, Pasadena, CA, USA

Scatena, F. N.
Department of Earth and Environmental Science, 240 South 33rd Street, University of Pennsylvania, Philadelphia, PA 19104, USA

Schellekens, J.
Faculty of Earth and Life Sciences, Vrije Universiteit, Amsterdam, De Boelelaan 1085, 1081 HV Amsterdam, The Netherlands

Schreider, S. Yu.
School of Mathematical and Geospatial Sciences, Royal University of Technology, Melbourne, Australia

Schweithelm, J.
Forest Mountain Consulting, Burlington, VT, USA

Scott, D. F.
FRBC Research Chair of Watershed Management, Okanagan University College, Kelowna, B.C., V1V 1V7, Canada

Serrão, E. A.
Embrapa Amazônia Oriental, Belém, Brazil

Servat, E.
L'Institut de Recherche pour le Développement, UMR Hydrosciences, BP 5045, F-34032 Montpellier Cedex 1, France

Sherlock, M. D.
Department of Geography, National University of Singapore, Singapore 117576, Malaysia

Tani, M.
Graduate School of Agriculture, Kyoto University, Kyoto, Japan

Thang Hooi Chiew
Forestry Department Peninsular Malaysia, Jalan Sultan Salahuddin, 50660 Kuala Lumpur, Malaysia

Thompson, I. S.
Department for International Development, Belém, Brazil

Tych, W.
Centre for Research on Environmental Systems and Statistics, IENS, Lancaster University, Lancaster, LA1 4YQ, UK

Van Noordwijk, M.
International Centre for Research in Agroforestry,
P.O Box 161, Bogor, Indonesia

Wallace, J. S.
CSIRO Land and Water, Townsville, QLD 4811, Australia

Young, A.
School of Environmental Sciences, University of East Anglia, Norwich, NR4 7TJ, UK

Yu, B.
Faculty of Environmental Sciences, Griffith University, Nathan, QLD 4111, Australia

Yusop, Z.
Institute of Environmental and Water Resource Management, Faculty of Civil Engineering, Universiti Teknologi Malaysia, 80990 Johor Bahru, Malaysia

Foreword

Management problems of water-source areas in developing countries show, within my experience, a characteristic pattern. For familiar ecological reasons, streamflow from forested hills supports the economic development of populations of the valleys and plains below. The protection of water source areas is therefore accepted, in principle, as necessary to national development. Such protection of remote areas is difficult to fund and to staff. The rapid growth of tropical populations has, however, resulted in large-scale invasion and destruction of upper-watershed forests by subsistence cultivators and graziers. Deterioration of streamflow regulation has become an all-too-familiar result, with regular flow replaced by flood flows and dwindling dry-season supply.

Authority resides in cities, but administration strong enough to protect these watershed forests must be resident in the hills. For the administrator, a posting to the remote hills is effectively a banishment to a life far from schools and other amenities as well as from opportunities for recognition and promotion. Thus although Forest Departments maintain their protective patrols by devoted staff, they are, in many countries, inadequately supported by the administration of the law.

Technical reports by hydrologists and land-use specialists, after making systematic surveys paid for by governments, have spelled out the critical importance of watershed protection, but the necessary following action has been neglected in at least a score of countries that I have been privileged to study. An important result of the compelling evidence described in this book will, I hope, be not only higher priority for funding the protection of watershed forests, but stronger interest in the more effective use of the funds provided.

Sir Charles Pereira

Preface

Although the areal extent of tropical rainforests has changed markedly through natural fluctuations in climate at a geological time scale, the rate of tropical forest harvesting and clearance during the second half of the twentieth century, has been unprecedented. Fuelled by the soaring demands for tropical hardwoods by 'northern' economies, timber harvesting relies heavily on the use of mechanised felling and extraction. This, in turn, has greatly disturbed the remaining vegetation, the soils and therefore the hydrological functioning of the forest. Further, the economic necessity for an adequate return on the capital invested in equipment, vehicles, roads and wood-processing mills makes it desirable to harvest all marketable logs during a single felling cycle, often at the cost of future growth. At the same time, traditional shifting cultivation practices of local communities have become unsustainable in many places due to the increased pressure on the land exerted by a growing population, resulting in gradual degradation or even total disappearance of closed forest. In addition to such 'unplanned' forest degradation and conversion there is an increasing trend towards planned, government-led conversions of tropical forest to apparently more profitable cattle ranching or commercial plantations.

The extensive disappearance of tropical forests during the last five decades has raised global alarm over the threats to climatic stability and the hydrological functioning of river basins posed by continued forest conversion, next to the well-being of forest dwellers and the conservation of biodiversity. Although the wave of publicity on rainforest conservation and related environmental issues has stimulated some changes, notably the development and testing of reduced-impact logging (RIL) techniques and timber certification schemes, their application is still the exception rather than the norm.

To discuss these issues, a symposium and workshop was organized jointly by the International Hydrological Programme (IHP) of UNESCO and the International Union of Forestry Research Organizations (IUFRO), which was hosted by Universiti Kebangsaan Malaysia, Kuala Lumpur, Malaysia between 30 July and 4 August 2000. The event, *Forest–Water–People in the Humid Tropics: Past, Present, and Future Hydrological Research for Integrated Land and Water Management*, provided a state-of-the-art overview of current knowledge on tropical forest hydrological functioning, the environmental impacts of forest disturbance and conversion, and the best ways to minimise these impacts. The meeting brought together some 94 people from 27 countries, representing a judicious mixture of senior professionals approaching the end of their research and management careers, and younger aspirants eager to follow in their footsteps. This book is based on contributions made to the Kuala Lumpur meeting, although several chapters dealing with specific topics not covered in detail by the symposium were added at a later stage.

Like the humid tropical environment it seeks to understand, tropical forest hydrology is changing. The relatively straightforward study of how water moves through forested catchments is rapidly giving way to a far wider approach embracing not just the physical aspects of water movement, but also how forest lands should be managed to optimise the environmental services and benefits they bring to all people living in, or downstream of forested catchments. Most importantly, the overriding need to alleviate poverty in many tropical countries requires the interfacing or even integration of the socio-economic, cultural and governance aspects when discussing forest–land–water management issues and seeking optimum solutions. The structure of the book reflects this importance.

The first global scientific programme devoted to hydrology and water resources, the UNESCO International Hydrology Decade (1965–74), provided an international impetus to the creation of long-term, hydrological data collection networks. In more recent times, however, there has been a progressive erosion of this long-term vision. Despite the threat of climate change, the need for long-term monitoring and research to address environmental and water resources management issues is no longer routine policy of most national governments, both within and outside of the humid tropics. Instead, there has been a drift towards funding short-term, high-visibility projects. The new UNESCO-led HELP (Hydrology for the Environment, Life and Policy) programme aims to promote just the type of integrated, interdisciplinary approach called for in this book.

There are some who argue that we know enough already and that there is little need for much more additional 'science'. Indeed, it is true that there is sufficient technical knowledge to minimise the adverse hydrological impacts associated with mechanised timber harvesting or land clearing and subsequent agricultural cropping. Thus the application of 'best management practices' is largely a matter of socio-economic acceptance and political will. At the same time, however, there are several important unanswered questions that require additional research. Two such issues that are of vital importance to the sustained livelihoods of countless upland farming communities and, indirectly, a great many more people living downstream, are:

(1) Will dry-season flows or even annual water yields decrease after clearing tropical montane headwater areas with cloud forest?
(2) Can the much reduced dry-season flows in heavily degraded areas be boosted, and if so, how?

Moreover, are we now in a position to predict the hydrological consequences of various management practices and land-use changes, including deforestation? Can we make these predictions in sufficient detail to be used by land users, managers or policy-makers wishing to avoid adverse hydrological consequences? And is the new hydrological knowledge uncovered by researchers being passed on to these stakeholders in a form they can?

We need to shift the emphasis back towards the longer-term vision necessary to solve the pressing environmental issues faced by tropical governments and their populations. This time, however, it is crucial that researchers involve local communities (who are often the de facto resource managers) and any non-governmental organisations representing them, as well as institution-based resource managers and policy-makers to help set the research agenda and translate the results of such research into concrete guidelines and tangible benefits.

We hope that this book will provide inspiration to all people involved in forest–land–water–people issues in the humid tropics and so contribute to a better management of precious natural resources to the benefit of people, animals, plants and their surroundings.

Acknowledgements

The production of a book of this scope and size involves the contributions and support from a great many people. In particular, it relies heavily on the goodwill of the peer community of the contributing authors, all of whom are experts in their field. All chapters of this book have been peer-reviewed internally and externally. Many have given their time to review and comment on the draft chapters and, on behalf of all contributors, we gratefully acknowledge their invaluable input. Known referees are:

Christian Brannstrom	Mark Lander	Fred Scatena
Nick Chappell	Christoph Leuschner	Jan Siebert
Will Critchley	Ian Littlewood	Murugesu Sivapalan
Oscar van Dam	Hua Lu	Mark Smith
Albert van Dijk	Ariel Lugo	Pradeep Tharakan
Horst Fölster	Anders Malmer	Chris Thorncroft
John Gash	Meine van Noordwijk	DesWalling
John Hayes	Paul Quinn	John Williams
Richard de Jeu	John Rodda	Maciej Zalewski
Timm Kroeger	Calvin Rose	

The editorial commitment to this venture has been substantial and was completed wholly within the editors' own time over a period of nearly three years. This required considerable personal support and understanding from family and friends and we express our deepest thanks to them all. In particular, Mike Bonell would like to make special mention of Kristof Koch, Marie-Camille Talayssat and Binnie Briffault of the UNESCO Division of Water Sciences and Daphne Mullett of the UNESCO Communication and Information Sector for their critical logistical support; and also to his family members, Catherine, Emma, Sarah and Bob for their sustained support and presence during this major editorial commitment. Sampurno Bruijnzeel extends particular thanks to Hester Dekker, Albert van Dijk, Linda and Larry Hamilton, Edi Purwanto, Ronald Vernimmen, Dorith van der Waerden, Maarten Waterloo and above all to Irene Sieverding for their invaluable support during times of illness. A similarly crucial role was played by our text editor, Celia Kirby, who kept everything together with her meticulous attention to detail, acting as liaison between editors and authors whenever required, and providing continuous support in all sorts of ways.

Our grateful thanks also go to all authors for their willingness to respond to comments and their patience in waiting for the book to appear. Such levels of co-operation from the authors have been remarkable and made the task of the editors in bringing this large project to a conclusion a lot easier. Part of the concluding stages of this editing was carried out whilst one of us (Bonell) had the privilege of residing within the monastery, L'Abbaye de St Pierre de Solesmes. Le Père Jobert is thanked for facilitating this most special experience.

In conclusion, we thank Dr Gerard Persoon for providing us with the beautiful cover photograph that captures the essence of this book in a nutshell and to Sir Charles Pereira, *éminence grise* of tropical hydrology, for his willingness to write the Foreword.

Symposium and Workshop

This book is based on contributions made at the joint UNESCO International Hydrological Programme (IHP) – International Union of Forestry Research Organization (IUFRO) Symposium and Workshop *Forest–Water–People in the Humid Tropics: Past, Present, and Future Hydrological Research for Integrated Land and Water Management*, hosted by Universiti Kebangsaan Malaysia, 30 July – 4 August 2000.

Our grateful thanks are due to all those – authors, delegates and organizers – whose efforts made the Kuala Lumpur event so successful. In particular, we wish to acknowledge that the Symposium would not have happened at all without the valiant efforts of six persons. Aminata Diaby and Nayla Naourfal of the UNESCO Division of Water Sciences, Paris, who provided more than two years of support in the preparations of the Symposium and Workshop at the international level as well as supporting the Technical Organizing Committee. Special mention must be made of Dr Mushrifah Idris (on secondment to UNESCO in 1999–2000 from the Universiti Kebangsaan), who unexpectedly appeared in Senior Editor's Office, just at the right time in May 1999, and offered the Universitit Kebangsaan as the suitable venue for the meeting, when previously all seemed lost. She quickly became the focal point of all local arrangements. The Vice-Chancellor of Universiti Kebangsaan Malaysia, Professor Anwar Ali, and the Deputy Vice-Chancellor, Professor Datuk Dr Zakri A. Hamid, kindly facilitated the symposium in support of Dr Mushrifah Idris. In addition, Mr Shamad Hussein, the Permanent Delegate from Malaysia to UNESCO in Paris until July 2000 also took a very close interest in the preparations of the symposium and on behalf of the government of Malaysia, provided the necessary support to the UNESCO IHP Secretariat and Mushrifah Idris.

The strategic directions taken by the Symposium were closely guided by the following members of the Technical Organizing Committee (TOC):

Dr Mike Bonell, UNESCO–Paris, Division of Water Sciences, France
Dr L. A. (Sampurno) Bruijnzeel, Faculty of Earth and Life Sciences, Vrije Universiteit, Netherlands
Dr Jean-Marie Fritsch, Institut de Recherches pour le Développement, Montpellier, France (recently on secondment to the World Meteorological Organization)
Dr John Gash, Centre for Ecology and Hydrology, Wallingford, UK
Dr Harald Grip, Swedish University of Agricultural Sciences, Umeå, Sweden (liaison with IUFRO)
Dr David Lamb, University of Queensland, Brisbane, Australia
Dr Jeffrey McDonnell, Oregon State University, Corvallis, USA
Dr Eduardo Planos Gutierrez, Cuban Meteorological Institute, Havana, Cuba.

The members of the TOC provided ideas and contacts, stimulating discussions, and helped with arrangements of funds for the meeting. In addition, Harald Grip, Jean-Marie Fritsch and John Gash who respectively provided hospitality during memorable preparatory meetings for the Symposium in Umeå (June 1998) and Montpellier (July 1999); and the concluding editorial meeting in Wallingford (February 2003).

All local organization was efficiently managed by the following members of the Local Organization Committee:

Datuk Professor Anwar Ali, Vice-Chancellor of Universiti Kebangsaan Malaysia
Datuk Hj Keizrul Abdullah, Director General, Department of Irrigation and Drainage, Malaysia
Dr Hj Mohd Nor Hj Mohd Desa, Director, Humid Tropics Centre, Kuala Lumpur, Malaysia
Prof Abdul Latiff Mohamad, Deputy Dean, Faculty of Science and Technology, Universiti Kebangsaan Malaysia
Dr Abdul Rahim Hj Nik, Division Director, Forest Research Institute Malaysia
Mr Mohan Nayer, Malaysian National Commission for UNESCO
Hj Baharuddin Kasran, Forest Research Institute, Malaysia
Mr Azman Hassan, Forest Research Institute, Malaysia
Mr W. Jayaweera, UNESCO, Kuala Lumpur office
Dr Mushrifah Idris, University Kebangsaan Malaysia (Symposium Collaborator).

In addition to the financial support from several sources in UNESCO (from the regular programme budget of the International Hydrological Programme of the Division of Water Sciences, Paris and the field offices – Montevideo, New Delhi, Nairobi and Jakarta; and separately a UNESCO Programme Participation Grant to Malaysia), we would like to express our grateful appreciation for additional sponsorship from several other sources:

The Australian Government's overseas aid program – AusAID
Center for International Forestry Research (CIFOR)
Cooperative Research Centre for Catchment Hydrology, Canberra, Australia (CRCCH)
Cooperative Research Centre for Tropical Rainforest Ecology and Management – Rainforest, Cairns, Australia (CRC)
Department for International Development, UK (DFID)
Forestry and Forest Products Research Institute, Japan
French Ministry of Foreign Affairs
International Center for Research in Agroforestry (ICRAF)
International Hydrological Programme (IHP) of UNESCO and Operational Hydrological Programme (OHP) of WMO
International Union of Forest Research Organizations (IUFRO)
Linköping Universitet, Sweden
National Committee of the Federal Republic of Germany for the
The Netherlands National Committee for IHP and HWRP
The Royal Society, UK
Svenska Institutet – The Swedish Institute, Sweden
The Swedish University of Agricultural Sciences
US National Committee on Scientific Hydrology (US IHP-NC)
Vrije Universiteit Amsterdam, The Netherlands

Introduction

This book reviews the current state of knowledge of forest hydrology and related land-water management issues in the humid tropics. As happened earlier in the related field of soil erosion and conservation, the days are long gone when land–water issues could be approached in a purely technical manner (cf. Hudson, 1971; Critchley, this volume), so much so that in a recent overview of responses to land degradation (Bridges *et al.*, 2001), the majority of chapters dealt with socio-economic, institutional and policy-related aspects rather than the physical aspects of soil erosion. In view of the importance of policy and governance aspects in environmental management, in particular the involvement of local communities and other resource managers, the present book also aims to bring together scientific, policy and management perspectives. Such perspectives address tropical forest–land–water management issues and concurrently also seek optimum solutions for the benefit of all interest groups involved. Of late, the term 'Blue Revolution' has been coined to describe the shift from the traditional technical approach to one that gives due consideration to socio-economic factors as well (Calder, 1999).

The contents of this book are based on contributions made to a joint UNESCO International Hydrological Programme (IHP) – International Union of Forestry Research Organisations (IUFRO) Symposium and Workshop *Forest–Water–People in the Humid Tropics : Past, Present, and Future Hydrological Research for Integrated Land and Water Management*, hosted by Universiti Kebangsaan Malaysia, Kuala Lumpur, Malaysia, 30 July – 4 August 2000. The Symposium was planned with the same structure as this book so that each were complementary, although a number of chapters were added after the meeting to achieve more complete coverage. The IHP-IUFRO Symposium originated from the UNESCO-IHP Humid Tropics Programme which was launched in 1990 as part of the Fourth Phase of the IHP (1990–1995). To mark the latter occasion, the book *Hydrology and Water Management in the Humid Tropics: Hydrological Research Issues and Strategies for Water Management* (Bonell *et al.*, eds. 1993) was published by Cambridge University Press, based on the First International Colloquium of the same title held in July 1989. In addition, the Humid Tropics Programme published *Hydrology of Moist Tropical Forests and Effects of Conversion : A State of Knowledge Review*, by L. A. Bruijnzeel, in 1990. More recently, UNESCO IHP Technical Document in Hydrology No. 52 *Hydrology and Water Management in the Humid Tropics* (Gladwell, ed., 2002) was published which included separate sections devoted to the hydrology of small islands and montane cloud forests, as well as sections on urban hydrology, groundwater and water quality issues. The current book complements and updates all these publications. It also marks the closure of the Humid Tropics Programme at the end of the Fifth phase of the IHP (1996–2001), and is a contribution to the new HELP (Hydrology for the Environment, Life and Policy) Programme within UNESCO (HELP Task Force, 2001; http://www.unesco.org/water/ihp/help) as part of IHP-VI (2002–2007).

At the First International Symposium on Forest Hydrology, held in 1965 at Pennsylvania State University (Sopper and Lull, 1967), only one contribution dealt explicitly with tropical land use hydrology (Pereira, 1967). Seen from this perspective, we have come a long way since then. As indicated however within the introductory paragraph, tropical forest hydrology is changing from the relatively limited study of how water and the transport of associated solid-debris and chemical species move through forested catchments, to how forest lands should be managed to maximise the environmental services and benefits they bring to the people living in, or downstream, of these forests. The tropics themselves are changing too: demographic and economic changes and, above all, the over-riding need to improve the livelihoods of the poorer strata in humid tropical societies all create massive pressure for both the exploitation and conversion of the remaining forest. Part of this quest for economic development concerns planned, government-based forest clearance and conversion to uses that are considered more profitable (e.g. large-scale cattle breeding, oil palm and cocoa plantations, irrigated rice cultivation in former wetlands). Elsewhere, closed forests are becoming degraded or disappear altogether as a result of continued, unplanned slash and burn activities by poor, land-hungry farmers (Drigo, this volume).

The structure of this book reflects the changing nature of tropical forest and land use hydrology although the emphasis is still on physical hydrology.

The book has five parts: Part I provides an overview of the current trends and perspectives on people–land–water issues in the humid tropics, where the use of the term 'humid tropics' is based on the criteria given by Chang and Lau (1993). Part I includes nine contributions which assess the rates, causes and patterns of land use change linked with policy within the broad dimensions of socio-economics, culture and governance. Particular emphasis is placed on the importance of incorporating local communities in the land use decision-making process. Part II presents an overview of the humid tropical, meteorological and climatic settings and outlines the biophysical aspects of tropical forest functioning through a systematic description of the principal hydrological, geomorphological and biogeochemical processes taking place in old-growth ('undisturbed') forest. Separate chapters are devoted to two special and hitherto under-researched rainforest types, swamp forest and montane cloud forest. The nine chapters making up Part II provide a baseline against which to not only evaluate the environmental impacts associated with forest disturbance (both natural and man-caused) and conversion, but also the changes accompanying forest recovery or reforestation and other rehabilitative measures such as agroforestry, as detailed in the eight chapters making up Part III. Next, the eight chapters of Part IV discuss the potential application in the tropics of a number of new tools for evaluating the biophysical impacts of land use change, including the transfer of technology and experience from more temperate latitudes. Examples include new sensors used in remote sensing, statistical techniques related to time series analysis, and several model approaches of varying complexity, some of which are particularly suited for use in data-poor areas such as the humid tropics. Consideration is also given to an assessment of the potential for using aquatic organisms as indicators of water quality. The four chapters constituting Part V present a critical appraisal of best management practices within the contexts of timber harvesting, land clearing and post-forest agricultural cropping. A concluding chapter provides a synthesis of the key issues emerging from the book, one of which is the overwhelming need for more integrated, multidisciplinary approaches in future tropical forest and land use hydrological research programmes. No chapter is devoted solely to groundwater, despite the fact that groundwater remains a neglected area of research in the humid tropics (see reviews by Foster and Chilton, 1993; Foster *et al.*, 2002). However, several contributions highlight the need for a better coupling of surface water–groundwater interactions in future assessments of the hydrological consequences of land use change. Examples where greater attention needs to be given to surface water–groundwater coupling include hillslope runoff generation (Bonell, this volume), nutrient retention in the riparian zone (Proctor, this volume), and the effects of reforestation on dry season flows (Scott *et al.*, this volume).

This book was conceived as a state-of-the-art record of tropical hydrological knowledge at the turn of the millennium. Several of the contributors commenced their careers during the first global programme devoted to hydrology and water resources, the UNESCO International Hydrology Decade (IHD), 1965–1974; and one of the aims of the Kuala Lumpur meeting, and this book, has been to capture their experiences to pass on to younger scientists. It is this younger generation who will have to take forward the recommendations made here for implementation in tropical forest-land-water management as well as addressing the associated research gaps. Their task is not made any easier, however, by the fact that there has a been decline of global hydrological monitoring networks especially in the humid tropics (Rodda, 1999). Moreover, at the national and international level there has been a progressive erosion of the longer-term vision that prevailed at the time of the IHD when the need for long-term monitoring and research to address environmental and water resource management concerns was still widely recognised by national governments (Bonell, 1999). Indeed, most of the hydrological data sets now proving so valuable for assessing the impacts of climatic variability, and global change in general, (e.g. the UNESCO IHP FRIEND project; Gustard and Cole, 2002) originate from the era of the IHD. Yet, during the initiation of the IHD, neither the notion of climate change nor global change were commonly part of the scientific vocabulary.

In more recent times, however, especially during the last decade, there has been a shift towards funding more short-term, high visibility international projects connected with water and climate. Partly, this reflects how science is managed nowadays in most developed countries where there is a need for 'products' over a one- to three-year economic cycle. As highlighted by Matsuura (2000), within this age of globalisation, we are also in an age of urgency, impatience and immediacy. Thus it would seem that international donors have become more inclined to sponsor high profile international meetings on water policy (Yamaguchi and Wesselink, 2000) rather than fund time-consuming technical-socio-cultural field studies. It is also pertinent to note that in some quarters there is a mistaken notion that 'we know enough science now'. On the one hand, this reflects the fact that many scientists are insensitive to relevant policy questions and usually preoccupied with their disciplinary orientation. At the same time, however, resource managers and policy makers also lose credibility because they lack an interest in incorporating new research results in their policies. The hydrological role of a good forest cover provides a case in point. Often, trees are planted in degraded areas within the context of massively funded watershed management projects, not only to arrest soil erosion and reservoir sedimentation but also in the expectation of restoring streamflow regimes (i.e. reduce 'floods' and enhance low flows; cf. Kaimowitz, this volume). Yet, the results of most hydrological research suggest a further lowering of stream discharges after reforestation, particularly during the dry season (Bruijnzeel, 1990, 1997; Scott *et al.*, this volume). Other solutions are needed, therefore, based on a sound understanding of the various processes governing the magnitude of dry season flows (Sandström, 1998; Bruijnzeel, 2005).

It is evident from the above and other examples given in this book that it is important to shift back towards a longer-term vision and maintain at least a number of longer-term experimental catchment projects. Such steps are imperative if we are to address adequately the impacts of such high-profile issues as climate variability and global change (Entekhabi *et al.*, 1999) but also other, less publicised issues, such as diminishing low flows, faced by tropical governments and their citizens. This time, however, it is essential to ensure the active involvement of major stakeholders outside scientific circles, notably local communities and institution-based resource managers, as well as government policy-makers, in helping to set the environmental research agenda (Bonell, 1999; Calder, 1999; HELP Task Force, 2001). This approach will improve the chances of research results becoming incorporated into national resource policy formulations and more specific guidelines for on-site land and forest management (Cassells and Bruijnzeel, Thang and Chappell, both this volume). In addition, the same approach will also aid the actual application of these guidelines, thereby providing such tangible benefits as improved agricultural production whilst maintaining water quality standards (Deutsch *et al.*, Critchley, both this volume).

Partly in response to economic pressures from funding bodies for more immediate 'products', coupled with the reduction in funding for longer-term field research signalled earlier, there has been a movement in hydrology and related sciences over the last two decades in favour of mathematical modelling and associated computer simulation. On the one hand, these developments have led to a greatly increased understanding of land surface – atmosphere interactions and the beginnings of an answer to the vexed question as to what extent the presence or absence of forest influences rainfall (Dolman *et al.*, 2004; Kabat *et al.*, 2004; cf. Costa, this volume). On the down side, however, the recent emphasis on modelling has also been at the expense of, rather than a complement to, field hydrological process studies (Philip, 1991; Klemeš, 1997; Shiklomanov, 2001). This book attempts to redress this imbalance by reporting on recent progress in both hydrological modelling and field studies in the humid tropics. Moreover, a fundamental message of the book is the need for a more integrated scientific approach to be adopted in future efforts which couple surface hydrology, groundwater, and ecohydrological aspects wherever required (cf. Sandström, 1998). Such an approach is advocated also within the HELP Programme (HELP Task Force, 2001; UNESCO-IAEA, 2002) to complement in-depth research along more traditional systematic disciplinary lines. Furthermore, strong emphasis is placed here on lateral fluxes of water, sediment and solutes at the small catchment scale (typically <10 km^2), i.e. the scale at which specific guidelines are usually applied in on-site land and forest management. In this regard, the technical aspects of this work complement the similarly ambitious overview by Kabat *et al.* (2003) which has a much greater focus on land surface – atmosphere interactions (notably exchanges of sensible and latent energy) and emphasises the impacts of global change on climate and the hydrological cycle from the regional up to the continental scale. The present book, however, also includes several chapters which 'bridge' these contrasting scales of scientific enquiry (notably those by Mahé *et al.* and Costa). Similarly, the present effort is also complementary to another end of millennium overview (Bridges *et al.*, 2001) which focuses on the extent, causes of, as well as possible remedies for land degradation. As indicated previously, the latter publication emphasises socio-economic, institutional and governance aspects rather than the more physical aspects of soil erosion. Moreover, the focus of attention of Bridges *et al.* (2001) is very much the farmer's field rather than downstream impacts of land degradation such as siltation of streams, irrigation works and reservoirs. Several chapters in the present book deal explicitly with erosion and sediment dynamics, both under undisturbed conditions and upon forest disturbance and conversion. In doing so, particular attention is paid to linking hillslope processes with the drainage network (i.e. streams) and the functioning of riparian buffer strips (see chapters by Douglas and Guyot; Chappell *et al.*; Tych, and Yu) whereas the chapter on soil and water management in humid tropical steeplands by Critchley typically acts as a 'bridge' to what might be called the 'land husbandry' community (Bridges *et al.*, 2001).

On a final note, this book also highlights the increasing tensions and inherent incompatibility between the goals of neo-classical economic development on the one hand and those of environmental sustainability on the other. Usually, the policy of rapid economic development is undertaken by way of a top-down type of governance and planning. This is shown by several contributors to be an important cause of degradation of tropical forests and associated land – water resources. The question is thus raised whether a new, 'biophysical' approach to economics is required, which focuses more on ways of increasing economic outputs for less material and energy inputs. Such an approach resembles the ecological economics vision which is becoming more widely accepted of late (Daley, 1998; Matthews, 2002). This new paradigm recognises that ecosystems are suppliers of all kinds of services to the economy and that these services represent a benefit (or a loss in the case of their disappearance) that needs to be incorporated into economic analysis before a decision is taken (cf. Hall and Ko, this volume; Aylward, this volume). Arguably, such an approach would also provide a much better framework for the incorporation and application of research outputs into forest and land management and policies (cf. Cassells and Bruijnzeel; Murdiyarso, both this volume).

We hope that this book will contribute towards the revival of some of the longer-term visionary values of the International Hydrological Decade within future national and international hydrological research agendas. We consider this vital if we are to adequately address the continued degradation of forest, land and water resources in the humid tropics. Such concepts are in line with the conclusions from the recent *World Summit on Sustainable Development* (WSSD, 2002) with its emphasis on a commitment to undertaking concrete actions and measures at all levels, and the recent 'Shiga Declaration on Forests and Water' (Shiga Declaration, 2002). Such world summits and declarations however, are of little use without commensurate action on the ground.

References

Bonell, M. (1999). Tropical forest hydrology and the role of the UNESCO International Hydrological Programme. *Hydrology and Earth System Sciences*, 3(4), 451–461.

Bonell, M., Hufschmidt, M. M. and Gladwell, J. S. (Eds.) (1993). *Hydrology and Water Management in the Humid Tropics*. Cambridge: Cambridge University Press, 590 p.

Bridges, E. M., Hannam, I. D., Oldeman, L. R., Penning de Vries, F. W. T., Scherr, S. J. and Sombatpanit, S. (2001). *Response to Land Degradation*. Science Publishers Inc., Enfield (NH), USA, 510 p.

Bruijnzeel, L. A. (1990). *Hydrology of Moist Tropical Forest and Effects of Conversion: a State of Knowledge Review*. Paris: UNESCO, and Amsterdam: Vrije Universiteit, 226 p.

Bruijnzeel, L. A. (1997). Hydrology of forest plantations in the tropics. In *Management of Soil, Nutrients and Water in Tropical Plantation Forests*, eds. E. K. S. Nambiar and A. G. Brown, pp. 125–167. Canberra – Bogor: ACIAR/CSIRO – CIFOR.

Bruijnzeel, L. A. (2004). Hydrological functions of tropical forests: not seeing the soil for the trees? *Agriculture, Ecosystems and Environment* (doi: 10.1016/j.agee.2004.01.015).

Calder, I. R. (1999). *The Blue Revolution*. London: Earthscan Publications, 192 p.

Chang, J.-H. and Lau, L. S. (1993). A definition of the humid tropics. In: *Hydrology and Water Management in the Humid Tropics – Hydrological Research Issues and Strategies for Water Management*, eds. M. Bonell, M. M. Hufschmidt and J. S. Gladwell, pp. 571–574. Paris – Cambridge: UNESCO – Cambridge University Press.

Daly, H. E. (1998). *Beyond Growth – The Economics of Sustainable Development*. Boston MA, USA: Beacon Press, 49 p.

Dolman, A. J., Van der Molen, M. K., Ter Maat, H. W. and Hutjes, R. W. A. (2004). The effects of forests on mesoscale atmospheric processes. In: *Forests at the Land–Atmosphere Interface*, eds. M. Mencucini, J. Grace, J. Moncrieff and K. McNaughton. CAB International, Wallingford, pp. 51–72.

Entekhabi, D., Asrar, G. R., Betts, A. K., Beven, K. J., Bras, R. L., Duffy, C. J., Dunne, T., Koster, R. D., Lettenmaier, D. P., McLaughlin, D. B., Shuttleworth, W. J., van Genuchten, M. T., Wei, M.-Y. and Wood, E. F. (1999). An agenda for land surface hydrology research and a call for the second International Hydrological Decade. *Bulletin of the American Meteorological Society* 80, 2043–2058.

Foster, S. S. D. and Chilton, P. J. (1993). Groundwater Systems in the Humid Tropics. In: *Hydrology and Water Management in the Humid Tropics – Hydrological Research Issues and Strategies for Water Management*, eds. M. Bonell, M. M. Hufschmidt and J. S. Gladwell, pp. 261–272. Paris – Cambridge: UNESCO – Cambridge University Press.

Foster, S., Smedley, P., and Candela, L. (2002). Groundwater quality in the humid tropics: an overview. In: *UNESCO IHP Technical Document in Hydrology* No 52. Paris: UNESCO, pp. 441–468.

Gladwell, J. S. (ed.) (2002). *Hydrology and Water Management in the Humid Tropics*, Proceedings of the Second International Colloquium Panama, Republic of Panama, 22–26 March 1999, *UNESCO IHP Technical Document In Hydrology* No 52. Paris: UNESCO, 487 p.

Gustard, A. and Cole, G. A. (eds.) (2002). *FRIEND – A Global Perspective 1998–2002*. Wallingford, UK – Paris: Centre for Ecology and Hydrology – UNESCO, 132 p.

HELP Task Force (2001). The design and implementation strategy of the HELP initiative. *UNESCO IHP Technical Document in Hydrology* No. 44. Paris: UNESCO, 67 p.

Hudson, N. W. (1971). *Soil Conservation*. London: Batsford, 320 p.

Kabat, P., Claussen, M., Dirmeyer, P. A., Gash, J. H. C., Bravo de Guenni, L., Meybeck, M., Pielke Sr, R. A., Vörösmarty, C., Hutjes, R. W. A. and Lutkemeier, S. (Eds.) (2004). *Vegetation, Water, Humans, and the Climate: a New Perspective on an Interactive System*. Heidelberg: Springer.

Klemes, V. (1997). Of carts and horses in hydrologic modelling. *Journal of Hydrologic Engineering*, ASCE 20, 43–49.

Matsuura, K. (2000). Address at the Forum Suisse de Politique Internationale, entitled *Tomorrow's UNESCO*, Geneva, Switzerland, 4 May 2000, UNESCO (DG/2000/19), 6 pp. Now available in: *A Year of Transition – Selected speeches – 15 November 1999–31 December 2000*. UNESCO Publications, *2002*. Part III: pp. 271–278.

Matthews, G. (2002). Hydrosolidarity – how to link to the economic development perspective. In Proceedings of the SIWI Seminar on Balancing Human Security and Ecological Security Interests in a Catchment – Towards Upstream / Downstream Hydrosolidarity. Stockholm, Sweden.

Pereira, H. C. (1967). Effect of land use on the water and energy budgets of tropical watersheds. In *International Symposium on Forest Hydrology*, eds. W. E. Sopper and H. W. Lull, pp. 435–450. Oxford: Pergamon Press.

Philip, J. R. (1991). Soils, natural science and models. Soil Science 151, 91–98.

Rodda, J. C. (1999). Measuring up to Water Resources Assessment. In: *Hydrometry – Principles and Practices* – 2nd edn., ed. R. W. Herschy, pp. 143–160. Chichester: John Wiley.

Rodda, J. C. (2001). Water under pressure. In: *Special Issue: Can Science and Society Avert the World Water Crisis in the 21st Century?*, Z. Kundzewicz (Ed.) *Hydrological Science Journal* 46, 835–839.

Sandström, K. (1998). Can forests 'provide' water: widespread myth or scientific reality? *Ambio* 27, 132–138.

Shiga Declaration (2002). Shiga Declaration on Forests and Water, 22 November 2002. In: *International Expert Meeting on Forests and Water, Shiga (Laforet Biwako), Japan, 20–22 November 2002, Proceedings of the Meeting*. Organised by the Ministry of Forestry, Agency of Agriculture, Forestry and Fisheries, the Government of Japan in collaboration with FAO, ITTO, UNESCO and the 3rd World Water Forum Secretariat, Available from: International Forestry Cooperation Office, Forestry Agency, Ministry of Agriculture, Forestry and Fisheries of Japan, 1-2-1 Kasumigaseki, Chiyoda-ku, Tokyo, 100–8952 Japan (*http://www.rinya.maff.go.jp/faw2002/iemfw-top.html*), Part 1, pp. 1–4.

Shiklomanov, I. (2001). Response to the award of the UNESCO-WMO-IAHS International Hydrology Prize. *IAHS Newsletter* 73, November 2001, pp. 4–5.

Sopper, W. E. and Lull, H. W. (1967). *International Symposium on Forest Hydrology*. Oxford: Pergamon Press, 813 p.

UNESCO-IAEA (2002). *Integration and application of hydrological and ecological processes as a tool for sustainable water resources management and society development*. Expert IHP-IAEA Workshop Warsaw, 3–7 July 2002, 17 pp. Available on http://www.unesco.org/water/ihp/help.

WSSD (World Summit on Sustainable Development) (2002). Plan of Implementation. Advance unedited text, 4 September 2002, Johannesburg, Republic of South Africa, 54 p.

Yamaguchi, A. and Wesselink, A. (2000). An overview of selected policy documents on water resources management that contributed to the design of HELP (Hydrology for the Environment, Life and Policy). *UNESCO IHP Technical Document In Hydrology No. 38*. Paris: Unesco, 46 p.

Part I
Current trends and perspectives on people–land use–water issues

SUMMARY

The extensive conversion of tropical forests to other land uses during, especially, the last four decades has raised global alarm on the threats posed by continued forest conversion to climatic stability and the hydrological functioning of river basins, next to the well-being of forest dwellers and the conservation of biodiversity. This part consists of nine chapters setting the scene for this book. It starts off with an account of the rates and underlying causes of deforestation in the three main tropical rainforest regions during the last two decades. This is followed by a critique of neo-classical market-based economics which are held responsible for stimulating environmental degradation. The next three chapters (Chapter 3–5) describe the adverse socio-economic consequences of large-scale planned forest conversions for forest dwellers and other poor strata in society in Latin America and South East Asia. After concluding that governments and donor organisations, whilst well aware of tropical environmental degradation, generally have no new ideas on how to mitigate the effects of adverse practices (Chapter 6), the final three chapters highlight ways of using economic theory to improve the negotiating position of upland farming communities and of actively involving these communities in the identification and solving of local environmental problems.

Setting the pan-tropical scene, Drigo discusses *how the latest FAO Tropical Forest Assessment and various related efforts (i.e. TREES II high resolution survey) to quantify the extent and rates of tropical deforestation* reveal a rather complex picture of a reduction in higher biomass densities during the last two decades, despite different definitions of various vegetation classes used in the respective surveys. A shift is noted in the rate of forest loss from the moist deciduous zone towards the wetter evergreen rainforest formations of late. The principal reasons for the observed changes in forest cover are either degradation or conversion to other land uses. The main driving force behind unplanned forest degradation is rural population pressure with its corresponding subsistence and energy demands. In contrast, planned activities result in immediate conversion of forest to other land uses within the framework of government-driven resettlement programmes, cattle ranching and permanent agriculture, as well as commercial plantations. Regional ecological and socio-economic settings are seen to determine many of the causes and factors influencing forest conversion. Africa differs from Latin America, for example, in that the process of forest conversion has been more progressive, i.e. a stepwise degradation to other land uses rather than outright forest clearance. On the other hand, Latin America, and the Amazon Basin in particular, depicts more the impacts of planned direct conversion of forest to large-scale cattle ranching and permanent agriculture. The highest rates of forest degradation, as well as forest conversion, occur in Asia, however, where the changes in forest cover reflect both the effect of population pressure (shifting cultivation) and centrally planned conversion to agriculture (e.g. irrigated rice in peat swamps) or commercial plantations (notably oil palm, some timber), with a gradual shift towards the latter during the last decade in particular.

Next, Hall and Ko *examine the economic efficiency of natural resource use in 12 tropical countries*. Focusing on the relationship between energy use and economic activity (expressed as gross domestic product, GDP), they find that greater use of energy (e.g. the use of energy-intensive fertilisers in agriculture) does not necessarily imply greater economic efficiency. Rather, there proves to be a strong positive correlation between GDP and energy use, and even more so with population growth. Although there is a positive correlation between the extent of forest conversion and economic growth, it is relatively weak and explained by the fact that most of the countries examined have already largely depleted their forest resources. In contrast, there is a strong positive linear relation between economic growth and water use. Consequently, 'market-based' economies do not lead to an efficient use of natural resources and, indeed, the imposition of the western model of economic sustainability is in direct opposition to the concept of environmental sustainability. The authors call for a new approach through 'biophysical' economics, which focuses more on increasing economic outputs for less energy or material inputs. Such new economics would take a more realistic position on available biophysical resources and economic possibilities and should therefore help to arrest the continued depletion of natural resources and ensure that basic needs (e.g. clean water) remain available to the populace. However, Hall and Ko emphasise that the issue of escalating populations should also be addressed. Otherwise, they argue, the persistent use of neo-classical economics is equivalent to 'an excuse to plunder'.

Serrao and Thompson, in their discussion of *the resource management situation in the Brazilian Amazon, provide an example of the environmental and socio-economic impacts of centrally planned (and therefore large-scale) forest conversion to (mostly) cattle ranches*. Resource management in Amazonia has a strong top–down character, with the federal and the state governments being by far the largest land managing bodies. In recent times the government has tried to establish more sustainable policies to correct for its earlier mismanagement (notably with respect to forest dwellers such as Indian Reserves) but at the same time such steps have come into conflict with the strong political desire for rapid economic development in the form of hydroelectric power generation (requiring the construction of huge dams), cattle ranching and, to a lesser extent, other forms of agriculture and commercial plantations. Ironically, an important psychological impediment to achieving environmental sustainability appears to be the sheer size of the Amazonian forest which not only gives the impression of inexhaustable abundance, but also tends to render the perceived environmental costs of forest conversion rather insignificant. Similarly, the existence of generally ample water resources in the area tends to negate the rather more short-lived, negative perception of (El Niño–Southern Oscillation-related) droughts which are often used to reinforce the environmental argument. Possible solutions centre around the need for strengthening institutional structures at the community level, coupled with economic incentives for reforestation as well as greater attention to socially and environmentally sustainable objectives, to balance the drive for economic development.

The adverse socio-economic and cultural impacts of forest disturbance and conversion on forest dwellers in South East Asia are highlighted by Schweithelm. The author notes that the activities of forest dwellers (including shifting cultivation) were sustainable at the population densities prevailing during the pre-colonial era. The forests provided food and shelter, as well as medicinal and trade products. Significantly, the institutional arrangements at the community level were strong then in terms of land tenure (ancestral claims) and legal rights (customary laws). The colonial era saw a shift towards legal control being taken over by Governments; a process which was accelerated once the emergent nations gained independence after World War II. Thus the forest dwellers' system of forest management came into direct conflict and became subordinate to national government policies of commercialisation (timber harvesting, clearing for plantations), at the expense of forest conservation. The dramatic shift from a subsistence-orientated (community-based) type of forest management to commercial (government-controlled) operations has not only resulted in the degradation of forests but also in water quality. Traditional institutional and legal structures have become radically weakened or lost altogether, causing mass migration to jobs in the market economy. Also, existing laws often do not recognise the legal land rights of forest dwellers. In recent times there has been an increasing recognition by governments that forest dwellers have rights in the remaining forests within the framework of a cash economy. Nonetheless in practice, governments have remained reluctant to transfer such rights back to the forest dwellers. Sadly enough even where ownership of the forest is retained, one consequence of forest dwellers suddenly becoming exposed to the cash economy is that the selling of timber concessions to logging companies, to maximise short-term financial returns, may still result in long-term environmental damage. The final chapters in this section explore ways of involving local communities to avoid such adverse practices whilst maintaining or even improving the people's livelihoods.

The next appraisal by A. Hall *of the fate of local populations in tropical forests builds on the above assessment by Schweithelm and confirms the adverse impacts of market-based economies signalled earlier* by C. Hall and Ko. The arrogance implied by the blanket application of 'western' economic thought to the forest dweller situation is still exhibited by many governments and international donor agencies, as well as by large-scale corporate and export-orientated enterprises, who tend to view forest peoples as an archaic legacy of a pre-industrial era. Further, forest people are considered to represent a hindrance to 'progress' (i.e. economic development) and national sovereignty with nothing to contribute to development and economic policy and thus expendable in the cause of modernisation. Hall highlights the commonalities here with Hardin's 'Tragedy of the Commons' hypothesis, that is, a progressive degradation of common property resources such as forests based on the assumption that individual, short-term, profit maximisation (the neo-classical economics of C. Hall and Ko) will always outweigh longer-term considerations of the collective good. However, since the 1990s there has been a shift in the policy of some national governments (e.g. Brazil) and donor agencies, towards recognising that for successful management of tropical forests, local communities must be integrated into policy formulation and implementation. Various well-publicised conflicts between forest dwellers and commercial interests have contributed to this shift. For example, in Brazil, rubber tappers and fishing communities successfully challenged commercial ranchers, loggers and fishing enterprises to secure territorial integrity of large areas of communal forests and waterways. The above change in policy also recognises that conservation and law enforcement within overall forest management are both expensive and labour intensive. Centralised 'command and control' methods usually result in a lack of human and financial capacities (or political will) to achieve these goals. Local populations, on the other hand, can offer the required institutional support based on their traditional knowledge, physical presence and customary systems of collective governance. Nonetheless, Hall too warns that the continued quest for economic development will put further pressure on the remaining tropical forests, thereby potentially jeopardising any recent advantages resulting from improved integration of local communities. He also cautions against the romanticised view that traditional groups have 'an inherent predisposition to conserving forest regardless', observing that the latter only applies as long as livelihoods are maintained and ecosystem carrying capacities are not exceeded. Further, once the equilibrium is disturbed, then the struggle to survive will inevitably outweigh environmental concerns.

In the next chapter, Kaimowitz *explores the myths and realities that surround the hydrological role of tropical forest in relation to watershed rehabilitation strategies followed by governments and major donors.* Particular attention is paid to the perceived ability of forest to control reservoir sedimentation, prevent downstream flooding and maintain dry season flows. Various examples from Central America are used to illustrate that, despite massive funding from governments and international donors, the positive impacts of recommended measures to control reservoir siltation (usually soil conservation schemes and reforestation) are usually meagre and at times perhaps even counterproductive. In some cases, this lack of a positive result is caused by the poor hydrological databases on which the evaluations were based, in others it also reflects a lack of understanding of the true role of forest in checking sediment production at the catchment scale. A case in point is provided by the massive landsliding associated with the extreme rainfall brought down by Hurricane Mitch in Honduras and Costa Rica. As explained later in the chapters by Douglas and Guyot (Part II) and Scatena *et al.* (Part III), the root system of a forest may provide some slope stabilisation with respect to shallow landslips but deeper mass movements are governed almost exclusively by geological and climatic factors. In other words, on certain geological substrates, extensive sediment production will still occur during excessive rains, regardless of whether they are covered with a well-developed rainforest or not. Furthermore, large-scale reforestation programmes recommended to reduce the risk of reservoir sedimentation, as for example in the Panama Basin, may even have an unintended adverse impact on dry-season flows. Whilst the traditional view maintained by most tropical foresters and donor agencies is that tree planting will lead to improved water yields, most experimental work suggests the contrary. The higher water use of forest vegetation compared to more shallow-rooted grass and crops may have a stronger (negative) effect on streamflow than the positive effect afforded by the higher infiltration rates commonly associated with forest, thereby creating a reduction in baseflows as the net result of forestation (see Scott *et al.*, Part III, for a detailed discussion of this complex subject). Despite the high degree of awareness of governments and donors of environmental degradation in the tropics, no 'new' ideas (i.e. based on current scientific understanding) on how to best mitigate the adverse effects of land-use change seem to have emerged. For example, Kaimowitz observes that follow-up proposals for catchment management in Central America after Hurricane Mitch focus again on 'the form of small-scale reforestation and soil conservation efforts implemented in the past and will probably have similar outcomes'. One may conclude that political expediency apparently still overrides scientific knowledge when it comes to implementing forest-land-water management policies in the tropics.

On a more positive economic note, the chapter by Aylward *represents a valiant attempt to 'translate' the hydrological consequences of land use change in economic terms.* In doing so, Aylward takes up known hydrological functions from existing scientific research (albeit undertaken at headwater down to slope and plot scales) rather than from some of the 'myths' described in the previous chapter. A set of basic mathematical equations is derived to describe the economic utility (in terms of production and consumption) of various hydrological functions at the catchment scale (notably water yield, flow regime, flooding, sediment yield and water quality) and changes therein due to changes in land use (called 'externalities'). In developing countries the greatest hydrological impact of land-use change on rural upland populations is usually connected with the production of food through subsistence farming. This may lead to various degrees of soil degradation and therefore diminished on-site productivity but also to enhanced surface runoff during intense rainfall, possibly to the extent that groundwater recharge becomes sufficiently impaired to cause a reduction in subsequent dry season flows. The latter, in turn, implies reduced availability of water for drinking, irrigation or hydropower generation whereas the increased peak flows during the wet season cause economic havoc downstream in the form of silted up streams, irrigation canals, reservoirs and ports (apart from direct lowland damage caused by floods). Examples of various attempts to mesh economics with hydrology in relation to land-use change are given for both tropical and non-tropical areas. Significantly, the much greater number of studies of the externalities associated with sedimentation (34 case studies) than with water quantity (13 case studies) suggests a greater ignorance of the biophysical impacts of land use change at the river basin scale. A further problem, common to most of the case studies, is an incompatibility in time scales between the short-term, economic policy-orientated work and the need for long-term monitoring to understand the hydrological system. Most of the case studies do not have a history of good hydrological measurement and evaluation, particularly with respect to water quality impacts. On the basis of his review Aylward suggests that watershed management policies should move away from the traditional emphasis on the protective values of forest because the hydrological externalities of land-use change are not necessarily always negative (e.g. increased water availabilty after conversion to well-managed grazing). Moreover, Aylward argues that, in economic terms, the production benefits of post-forest land use (from agricultural outputs, livestock, timber plantations) would have to decline significantly towards negative externalities of hydrological functions before rehabilitation could be considered economically justified, thereby indirectly confirming the concerns expressed by various authors earlier on in this section.

Turning next to *the governance aspect of water resources management in areas experiencing major land use change, Murdiyarso uses examples from Indonesia and the Mekong River basin to highlight the common mismatch in scale between public, land–water resources management policies at the national scale and the physical impacts of land-use change on water supply*. The latter is most significant at the local scale and directly affects communities. Governments, Murdiyarso argues, usually focus on the demand rather than the supply side of things and this easily leads to over-use of (limited) natural resouces. The bulk of forest conversion and land degradation in the humid tropics occurs generally in the uplands where many landless poor reside. Because the adverse environmental impacts of upland degradation (see previous chapter) often affect large downstream populations as well, Murdiyarso (like Aylward) proposes the introduction of a reward mechanism to transfer payments from (downstream) beneficiaries of environmental services (such as stable and high-quality streamflow) maintained by sound land husbandry practices to the providers in the uplands of these services. For this to be successful, however, greater decentralisation of decision-making down to the community level is needed through the involvement and capturing of local concerns within overall water–land resources policy formulation. Continued adjustments in governance to accommodate the above suggestions would be needed. In Indonesia, for example, recent decentralisation efforts have overestimated the capacity (human and technical) of local (district) governments to manage their natural resources sustainably. This capacity has not been correspondingly reinforced to meet the demands of decentralisation. Within the context of improved governance, Murdiyarso identifies the roles (often conflicting) of the three major stakeholder groups – scientists, policy-makers and resource managers (including local communities). It is felt that most scientists are insensitive to relevant policy questions and usually preoccupied with their disciplinary orientation. Consequently, research findings are often neglected in the public policy-making process. At the same time, policy-makers also lose credibility because they lack the scientific background or interest to utilise new research findings. To add to the complication, resources managers often follow their own business-orientated agenda. To reduce potential conflicts between different stakeholder groups, Murdiyarso calls for the development of decision support systems that are able to summarise quantitatively the detailed components of local concerns, thereby enhancing the negotiating position of poor local communities. Moreover, he emphasises the need for scientists to consider policy-relevant questions before a research agenda is designed. Equally, policy-makers and resource managers need to gain a better appreciation of the science to help focus research questions of practical relevance to policy formulation and environment (land–water–forest) resource management.

In response to the calls for greater community involvement made in several of the previous chapters, Deutsch *et al. provide a particularly inspiring example from the island of Mindanao, the Philippines, on how to involve community groups within a partnership of researchers, non-governmental organisations and government officials for assessing water quantity and quality within the 36 000 km*^{2} *Manupali River basin*. The results of a survey conducted by members of the communities themselves revealed a clear west-to-east pattern of progressive land degradation that was closely associated with increasing population pressure and corresponding changes in land use (notably from forest to fire-climax grassland and subsistence cropping). The steep environmental gradient towards progressively greater environmental degradation while moving eastwards, of which changes were well within living memory, provided a truly dynamic appreciation of the project's results by the local communities. 'Put simply,' say Deutsch *et al.*, 'a person in the middle of the watershed could "look west" to see where their environment had come from, and "look east" to see where it was going'. Although Deutsch *et al.* appreciate that their community-based surveys do not have the same rigour as scientific research (e.g. in the absence of automated recording and sampling equipment annual water and sediment yields are more than likely to be severely underestimated), the present approach does bring home the main points for environmental policy formulation and implementation of remedial measures. It may also assist in the formulation of working hypotheses for subsequent more rigorous research efforts. Most importantly, however, the involvement of local communities is shown to lead to a much better chance that research results actually have a policy impact. As such, it is heartening to note that the approach pioneered by Deutsch *et al.* is being adopted rapidly, not only elsewhere in the Philippines but also in mainland South East Asia and Latin America.

1 Trends and patterns of tropical land use change

R. Drigo
Formerly of FAO Forest Resources Assessment Programme

INTRODUCTION

Tropical regions have undergone dramatic land use changes in the last few decades. The myriad of changes that have, and still are, taking place are the effect of an equally large number of local causes and factors, highlighting a complexity that tends to defy easy generalisations. Major background driving forces can be recognised, however, such as rapid demographic expansion, with consequent booming demands for agricultural land and woodfuels, plus a change from a subsistence-based to a market-orientated economy, with the associated heavy pressure on natural resources to fuel economic growth and development or, in some regions, political instability and incurable conflicts.

From this complexity of tropical land use dynamics, one single resulting element has been recognized with considerable alarm over the past decades. This is the progressive depletion of natural tropical forests. It is in fact to assess the remaining area of forest and its rate of change, rather than to study the complexity of tropical land use dynamics per se, that large-scale studies have been carried out over tropical regions. From a hydrological point of view, the removal of forest cover causes important changes in runoff and sediment yields (cf. Grip *et al.*, this volume). Moreover, global concern over the fate of tropical forest has been coupled, especially during the last decade, with similar global concerns over the build-up of greenhouse gases in the atmosphere and the negative role that tropical deforestation plays in this regard (cf. Costa, this volume).

The key questions concerning the depletion of tropical forests that this chapter seeks to address include:

(1) How much forest is being lost?
This is a call for a quantification of the phenomenon and basically requires an estimation of annual rates of forest loss.

(2) Where? What are the 'hot fronts' of deforestation and degradation?
This is a call for a geographical definition of priority areas.

(3) How is forest depletion taking place? What are the processes, the cause-effect mechanisms and the trends?
This is a call for a deeper understanding of the phenomenon. Rather than a summary of the net effect, the need is to know the different forms of forest depletion (deforestation, degradation, fragmentation, etc.) and their relative importance; to know what happens to deforested lands; to know what are the processes associated with the land use change; to know the trends (deceleration, acceleration of deforestation and degradation rates).

In the last few years the understanding of the patterns and causes of land use change in the tropics has deepened considerably, highlighting the complexity of the processes involved, as well as their discontinuous and simultaneous character and, in summary, the extreme difficulty of developing the predictive models that are demanded so strongly by global change analysts (Lambin and Geist, 2001). In view of these limitations, in this chapter use is made of the information generated by large-scale scientific studies based on direct observations rather than predictive models and other more or less educated guesses. In line with this approach, the present analysis of patterns and trends in tropical land use change is based primarily on the information produced by the Forest Resource Assessment Programme (FRA) of the FAO and by the High Resolution Survey of the Humid Tropics carried out in the framework of the TREES II project (Achard *et al.*, 2002) of the Joint Research Centre of the European Commission. Of particular relevance in the analysis of processes influencing tropical land cover change and trends was the Remote Sensing Survey of tropical regions, carried out in the framework of FAO's FRA Programme during its most recent editions (FRA 1990 and FRA 2000; FAO, 1996, 2001a, 2003). These studies are all very recent and at the time of writing the available results were still rather 'crude' or available in summary format only with limited additional documentation, especially the TREES study and FAO

Forests, Water and People in the Humid Tropics, ed. M. Bonell and L. A. Bruijnzeel. Published by Cambridge University Press.

Remote Sensing Survey. In some cases this has reduced the depth of the comparative analysis.

Observing land cover, inferring land use

'Land use is characterised by the arrangements, activities and inputs people undertake in a certain land cover type to produce, change or maintain it' (FAO/UNEP 1999). The term 'land use' is used here to represent the complex system of relations that man establishes with the land. By its very nature, land use definition is a subjective matter and often based on contradictory judgement. It implies not only an objective status, such as a specific cropping system, but may also reflect future intentions (for example, . . . this land is classified as being under forestry use because the owner intends to plant it up, or because it is administered by the Department of Forests . . .). Moreover, multiple uses can overlap on the same piece of land, either at the same time, as is the case for agroforestry practices, or in different periods of the year, as in the case of seasonal rotation between agriculture and pasture. Similarly, to the human mind, land uses, and land use changes, represent a blend of physical and cultural, often political, elements. Knowledge of tropical land use changes over time would help to understand the underlying cause-effect mechanisms and formulate effective remedial policies, but land use changes are often subtle, thereby defying any attempt at objective measurement and leaving room for speculation.

'Land cover is the observed (bio)physical cover on the earth's surface' (FAO/UNEP 1999). Accordingly, 'land cover', which can be considered as the net effect of the land use (or uses) at a given moment, can be measured with some objectivity. Unlike land use classes, land cover classes do have measurable physiognomic or physical characters, such as vegetation height, tree crown and biomass density. As a result, land cover monitoring represents an indirect but more objective method to measure the effects of land use changes over time, whereby complex dynamics are consolidated into net land cover changes and biomass balances. In general, the reliability of the observed land cover changes increases with the time span of the study, since the effects of inter-annual dynamics or seasonal fluctuations are reduced over longer observation periods, while those of true changes are increased.

Practically all studies over large tropical regions have been limited to the observation of land cover types through remote sensing techniques. The quality and 'depth' of the information produced, however, varies enormously, depending on scale and resolution of the imagery, the number of land cover classes observed, measurement reliability and degree of consistency within time series. In this chapter the features and findings of the most consistent and up-to-date studies of tropical regions are reviewed to highlight the trends and patterns of tropical land cover change processes.

During the last 50 years, systematic country-by-country information on the state and change of tropical forests have been produced exclusively by the FAO whose reports, in spite of some criticism (e.g. Stokstadt, 2001), have been the main, and often the only, reference for discussion and analysis at regional and global levels. The FAO's assessment techniques have developed considerably since the initial 'questionnaire' approach that was used until the 1970s, which was based on the information provided by the governments themselves and therefore vulnerable to manipulation. In the 1980 Forest Resources Assessment, the questionnaire was replaced by collection and review, country-by-country, of original references such as national inventory reports, and complemented by expert opinions, especially concerning rates of change. In the 1970s and 1980s the FAO became the world leader in assisting tropical countries with their forest inventory programmes and it could benefit therefore from the advice of a unique pool of forest inventory experts (FAO, 1981a, b, c). This condition has degraded slowly since then, along with the reduction in the FAO's field inventory operations. During the 1990 Forest Resources Assessment (FRA 1990), the assessment of countries' forest areas was based on existing national and sub-national time series of forest area estimates that were integrated by modelling techniques based on ecological and demographic variables. The model, or adjustment function (Scotti, 1990) was used to adjust forest cover areas in standard reference years and assess rates of change only for countries without multiple observations. In addition, to complement country statistics, but also to respond to questions such as 'How do tropical forests change? How much forest is degraded, or fragmented? What are the causes of deforestation? What happens to deforested land?' a pan-tropical remote sensing survey was implemented with the aim of producing consistent, thematically and statistically sound results at the regional and pan-tropical level (FAO, 1996).

The most recent FAO effort, the Forest Resources Assessment 2000 (FRA 2000), continued to collect data per country to extend and improve the time series already available but discontinued the use of mathematical modelling to fill gaps in information. These were now filled by expert opinions, consultations or in-depth studies. The FRA of 2000 continued the remote sensing survey by re-visiting the sample sites on a third date to update change estimates and assess trends. The cited country statistics and the description of regional change processes and trends resulting from the last two global assessments by the FAO, represent the main source of information on the status of tropical forests. Their findings will be discussed in the following sections.

In addition to the FAO studies, the project TREES (Tropical Ecosystem Environment observation by Satellite) of the European Joint Research Centre (JRC) has recently completed an independent survey of humid tropical regions using high resolution satellite data (mainly Landsat) covering the period 1990–1997

Table 1.1. *Summary of country statistics of area and change of* natural forest *by tropical region as estimated by the FRAs of 1990 and 2000*

	Area of natural forest (million ha)				Annual change 1980–1990		Annual change post-1990	
Natural forest	1980	1990	1995	2000	Million ha	%	Million ha	%
America FRA 2000		940		893			−4.69	−0.50
America FRA 1990	983	918	889		−6.48	−0.66	−5.69	−0.62
Africa FRA 2000		684		629			−5.43	−0.79
Africa FRA 1990	556	514	495		−4.19	−0.75	−3.70	−0.72
Asia and Oceania FRA 2000		317		269			−4.81	−1.52
Asia and Oceania FRA 1990	348	308	290		−3.97	−1.14	−3.51	−1.14
All tropics FRA 2000		1941		1792			−14.93	−0.77
All tropics FRA 1990	1886	1740	1675		−14.64	−0.78	−12.90	−0.74

Sources: FAO (1997, 2001a).

(Achard *et al.*, 2002). It should be noted, however, that the results of the FAO and TREES remote sensing studies cannot be compared directly, since they refer to different portions of the tropics and to (slightly) different time periods. Nevertheless, the TREES and FAO surveys have many elements in common, including key methodological features and basic classification thresholds, which allow us to discuss their findings in a shared context.

MONITORING TROPICAL FORESTS

Country statistics of forest cover

The statistics of forest areas produced by FAO for tropical countries, based on existing time series, provide answers to the first main question: **HOW MUCH** FOREST IS BEING LOST? The latest statistics produced by FRA 2000 suggest that, for the period 1990–2000, the net annual world-wide deforestation rate is now about 9 million ha, compared to previous estimates of 11.3 million ha for the period 1990–1995 (FAO, 1997), and 13 million ha for the period 1980–1990 (FAO, 1995a). These recent findings give the impression that during the 1990s the pressure on the world's forests has reduced measurably. However, much of the apparent reduction in deforestation rate relates to the effect of new statistics published for China and former USSR states, where the forest area is estimated to have increased annually at the rate of 1.8 and 0.7 million ha respectively, while earlier estimates gave a cumulative loss for these countries of some 0.4 million ha (FAO, 1997).

From the statistics published for tropical countries, it appears that the annual loss of ***total forest*** (natural forest plus forest plantations) is 12.3 million ha, against the 12.6 million ha estimated previously for the period 1990–1995 (FAO, 1999a) This does not represent a significant reduction, even though the two statistics are not entirely comparable[1].

The results for ***natural forest*** are more striking. The change in natural forest area that can be deduced from the FRA 2000 statistics[2] implies that the net annual loss in tropical regions is 14.9 million ha (Table 1.1). Table 1.1 and Figure 1.1 summarise country statistics of state and change of tropical natural forests as estimated by the FRAs of 1990 (FAO, 1997) and 2000 (FAO, 2001a). The FRA 2000 estimates show important differences with respect to previous estimates by the FAO. The main discrepancies relate to Africa's forest area and the rates of change for Africa and Asia (higher rates obtained by FRA 2000) and Latin America (lower rate according to FRA 2000). The sudden dominant role of Africa in tropical deforestation should be further investigated, because the existing country data for this region are often very poor. Some of the highest deforestation rates (Sudan, Ivory Coast, Zambia, Zimbabwe, etc.) are not based on consistent time series but only on single-date statistics, often quite old, and expert opinions. Moreover, there seems to be little evidence of such high deforestation rates in Africa from the results of the independent surveys of tropical regions initiated by the FRA of 1990 and continued in 2000 and the TREES High Resolution Survey of the Humid Tropics. Both studies are presented in the following sections. In the presentation of FRA 2000 results, FAO acknowledged the various weaknesses in its estimates for African countries (FAO, 2001a), and applied some adjustment, limited to the global change estimates. This adjustment, however, did not modify the global

1 There is a significant difference in the total forest area of tropical regions as estimated by FRA 2000 and FRA 1990 (SOFO 97), because of new sources and a new approach in the standardisation of original country data. Total tropical forest area at year 1990 was estimated to be, respectively, 1981 and 1797 million ha, i.e. 41.4% and 37.5% of tropical land area.

2 FRA 2000 has not published separate statistics for natural forest but gave country-wise statistics for total forests only, which includes both natural forests and plantations, and, separately, country-wide plantation areas and planting rates. Natural forest areas and changes shown here were calculated by deducting from total forest the areas of plantation in the years 2000 and 1990.

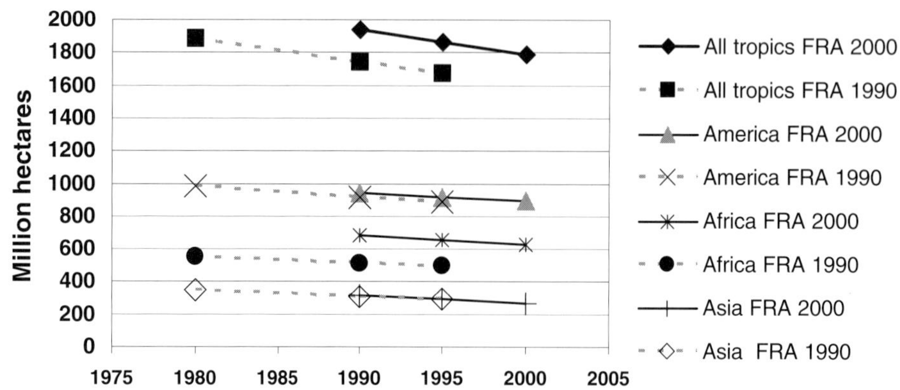

Figure 1.1 Summary of country statistics of natural forest area per tropical region as estimated by the FRA of 1990 (1980–1990–1995; FAO, 1997) and by the FRA of 2000 (1990–2000; deduced from FAO, 2001a).

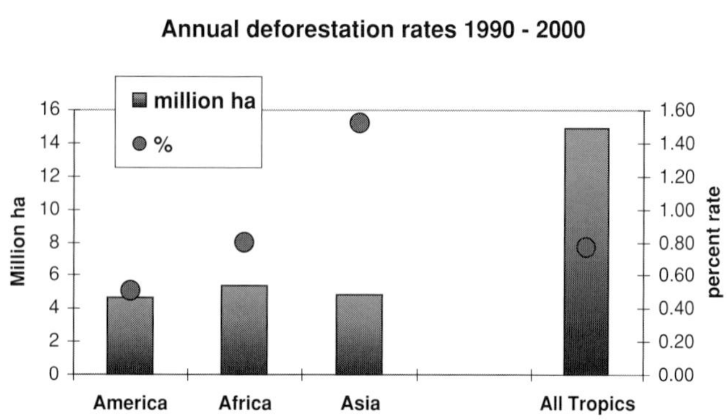

Figure 1.2 Annual deforestation rates during the period 1990–2000 in the three main tropical regions (deduced from FAO, 2001a).

change figures, since it included two calibrations, one reducing the global deforested area and one reducing the global planted area, of equal values and opposite signs. In another recent paper, FAO envisaged the need to apply some adjustment factor also to the published FRA 2000 country estimates (FAO 2001c).

Figure 1.2 summarises at the regional level the estimated annual rate of natural forest loss for the period 1990–2000 (FAO, 2001a), both in terms of area and percent of original forest areas. According to these estimates, the loss of natural forest in the three regions is similar in absolute terms but rather distinct in relative terms. Tropical Asia exhibits the highest pressure on forest resources, with an annual loss rate of 1.5%, followed at a distance by Africa (0.8%), and Latin America (0.5%).

Unfortunately, due to the non-comparability of country data produced by FRA 1990 and FRA 2000, the estimation of deforestation trends, i.e. acceleration or deceleration determined by the relation between the rates of the period 1980–1990 and those of the period 1990–2000, remains largely undetermined.

One cannot take the forest area at year 1980 from FRA 1990 and compare it to the areas estimated at years 1990 and 2000 by FRA 2000 because of the different database and estimation procedure. On the other hand, an analysis of deforestation trends is not possible solely on FRA 2000 country estimates, since the 2000 assessment covered the period 1990–2000 only and no revision was made to the estimates for 1980 based on the additional information available for that study (FAO, 2001a). With additional efforts, many of these new observations such as national surveys, and mapping, or simply new ways of harmonising earlier survey results could have been used to re-estimate the 1980 forest area and hence develop a consistent three-date time series suitable for trend analysis.

Next to the somewhat rigid country statistics, an important question concerns the ecological distribution of forests and forest change, i.e. which ecosystems are currently under the highest pressure at national, regional and global levels. Naturally, this has important ramifications in the context of biological diversity. To derive a tentative estimation of forest area and rate of depletion from an ecological perspective one must go back to the FRA of 1990, which produced data over the period 1980–1990 by broad ecological zones (FAO, 1993). These estimates were derived through cross-referencing ecological zone maps with the sub-national spatial and statistical databases of forest areas and change contained in the FAO's Forest Resources Information System (FORIS). The broad ecological zones that were distinguished included the Tropical Rainforest Zone, the Moist Deciduous Forest Zone, the Dry Deciduous and Very Dry Forest Zone, and the Hill and Montane Forest Zone, the latter one including the upland formations (above 600 m a.s.l.) from all rainfall zones. The results of this study, summarised in Figure 1.3, showed that in absolute terms deforestation was highest in the (lowland) Moist Deciduous Zone, while in relative terms the annual percent rate had its peak in the Hill and Montane Zone. This means that the forest cover was (and still is?) under the highest pressure in those areas, where the protective functions of forests are the most essential, such as in highlands and headwater areas (Hamilton and Bruijnzeel, 1997). The FRA of 2000 produced country statistics by ecological zones

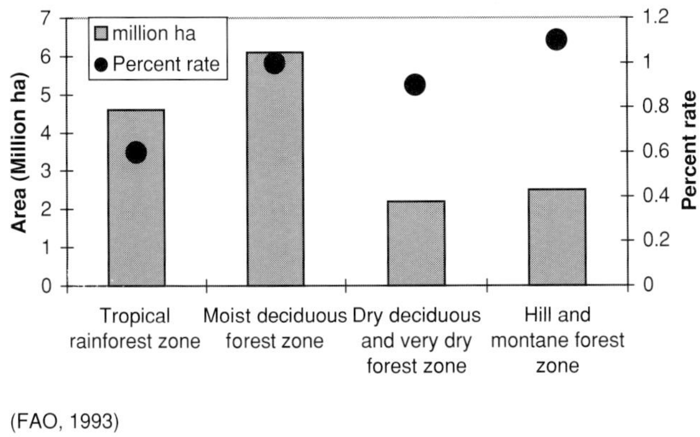

Figure 1.3 Annual deforestation rates in the tropics during 1980–1990 per forest formation. (FAO, 1993.)

limited to forest area. Unfortunately, it did not produce new statistics on forest change by ecological zone and therefore no update is available on this important aspect.

Hot spots of tropical deforestation as defined by the TREES project

An answer to the question **WHERE** IS DEFORESTATION TAKING PLACE? was recently provided, for the humid tropics, by the TREES project of the European Joint Research Centre (JRC), through the delineation of currently active fronts of deforestation (Achard *et al.*, 1998). The active-front areas, called Hot Spots, were located and described during an Expert Consultation Meeting, held at the JRC, Ispra, in November 1997. The main purpose of the meeting was to integrate the remote sensing approach of the project with first-hand field information from selected experts from the three main tropical forest regions of the world (South East Asia, Central Africa and Latin America). On the basis of their current understanding of deforestation trends in their region of expertise, the participants of the meeting described and outlined the main deforestation fronts that were active during the previous five years (1992–1997) and the areas considered to be at risk in the near future. The definition and delineation of Hot Spots based on expert opinion was conceived by Myers (1992) and further developed within the framework of the meeting. Figures 1.4a, b and c show the distribution of the Hot Spots in the three main tropical forest regions. A total of 111 Hot Spots were identified and delineated on the Humid Tropical Forest Map produced by the TREES Project (39 in Continental and Insular South East Asia, 19 in West and Central Africa, and 53 in Latin America). Each unit was accompanied by a summary description reporting, among other features, deforestation type and pattern, as well as the main causes, actors (initiators) and driving forces (Achard *et al.*, 1998).

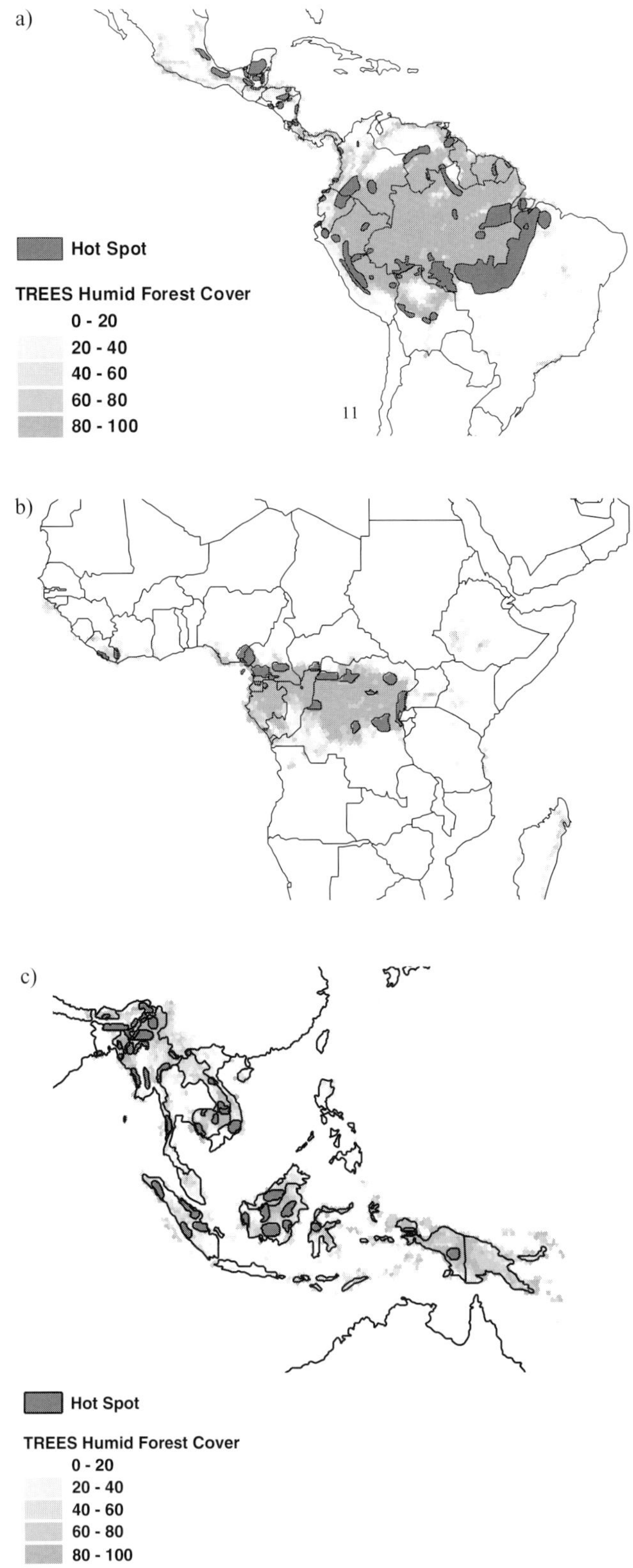

Figure 1.4 Location of deforestation Hot Spot areas in the three main humid tropical forest regions: (a) Central and South America, (b) West and Central Africa, and (c) South East Asia. (Source: Achard *et al.*, 1998.)

As shown in Figure 1.4, the hot spots lie along the edges of the humid forest blocks in South America and Central Africa, occupying the transition areas between moist and wet ecological zones. This distribution appears less clear in Southeast Tropical Asia, due to the fragmentation of the forest areas and/or to the complex mixture of ecological conditions.

Within the framework of the TREES Project, the Hot Spot map, after some adjustments and additions, was used as a stratification criterion for the optimal allocation of a statistical sample aimed at the study of tropical land cover changes, the results of which are discussed below. The TREES survey of humid tropical regions was based on an analysis of time series of high-resolution satellite imagery covering the period 1990–1997. In the TREES high-resolution survey, a higher sampling intensity was allocated to the Hot Spot areas, to enable a more efficient estimation of current land cover change rates.

Pan-tropical land cover monitoring (remote sensing component, RSS, of the FAO Forest Resources Assessments of 1990 and 2000)

The estimation of deforestation rates and the delineation of the main fronts, alone, offer little insight into the causes and mechanisms behind these phenomena, nor do they provide the understanding needed to define appropriate remedial measures. The question that remains unanswered is: **HOW** IS FOREST DEPLETION TAKING PLACE?

The first attempt to produce such insight at the pan-tropical level through objective measurements and sound statistical approach was carried out by a team of experts coordinated by Dr. K. D. Singh, within the framework of the Remote Sensing Component of the FAO Forest Resources Assessment 1990 (FAO, 1996). This Component was implemented with the specific objective of complementing the estimates based on existing information described above. Within the limits imposed by the tools available (i.e. satellite data with resolution ranging between 30 to 100 m interpreted at scale 1:250 000) and the detectable parameters (i.e. land cover classes), the survey was designed to address the following types of questions:

- How are the tropical forest resources changing?
- How much forest is being degraded or becoming fragmented?
- What is happening to deforested land?
- How are forest cover changes linked with ecological and socio-economic conditions?
- What are the causes of deforestation?

The study of land cover changes over large and diverse landscapes is not the simple task that one may think. Except for particularly simple conditions such as, for instance, the large and squared clearings that are common in the Brazilian Amazon, land cover changes are small elusive events whose reliable detection requires evaluation processes far more rigorous than normally accepted for conventional mapping purposes. The comparison of two land cover maps independently produced over the same areas will inevitably give differences that include true changes as well as other differences resulting from the different interpretation procedures adopted in each mapping process. Appendix 1.1 summarises the key features of the monitoring methodology developed for the FRA RSS and offers some considerations for the application of a similar approach to support hydrological research.

The FRA RSS study covered the entire tropical belt, following the standard FAO geo-political boundaries, including all ecological zones. To simplify data acquisition procedures and reduce costs, the sampling frame was constructed on the tessellation provided by the LANDSAT World Reference System 2 (WRS2), each unit covering approximately 3.2 million ha. The survey was based on a 10% sample, consisting of 117 randomly selected sampling units distributed per region as follows: 47 in Africa, 30 in Asia and 40 in Latin America (Figure 1.5). The 10% sample size was chosen to estimate forest cover at the global scale with a standard error of less than ±5% (Czaplewsky, 1991).

At each sample location, satellite images of the best quality and appropriate season, separated by an approximate ten-year interval, tentatively 1980 and 1990, were selected for examination. Two dates were studied during the FRA 1990 survey. Within the framework of FRA 2000, the 117 sample sites were re-visited at a third date, close to the year 2000, to assess current rates of change and the main trends with respect to the previous decade. To allow consistent aggregation of results of all sampling units at the regional, eco-regional and pan-tropical level, a standard classification system was adopted, based on simple physiognomic elements, as described in Table 1.2.

To overcome the constraints of a single, rigid definition of forest, the FAO study aggregated the above land cover classes to form three different definitions of forest and deforestation rates, ranging from a strict one that included only the class 'closed forest' to a much broader one that included also open, fragmented and degraded forest formations:

F1 = closed forest

F2 = closed + open forest + 2/3 fragmented forest

F3 = closed + open forest + fragmented forest + long fallow

Appendix 1.2 summarises the three definitions and the consequent categories of change adopted for the study. The 1990 study reported statistical results for each definition of forest, highlighting also the relation between the selected definition of forest and the implicit, consequent effect on the estimated rate of deforestation (FAO, 1996).

Table 1.2. *Land cover classification system applied in the remote sensing component of the FAO Forest Resources Assessment 1990*

Land cover classes	Average height	Canopy coverage	Description
Homogeneous			
Closed canopy forest	> 5 m	> 40%	Continuous tree formation of natural origin
Open canopy forest	> 5 m	10–40%	Continuous tree formation of natural origin
Shrubs	1–5 m	> 10%	Low woody vegetation of natural origin
Other land cover		< 10%	Land with woody vegetation below 10%
Plantations	> 1 m	(dense)	(synonymous with man-made woody vegetation) Forestry or agricultural plantation
Water			Sea, lakes, reservoirs, rivers, swamps
Composite			
Fragmented forest	(forest) > 5m	(forest)> 10%	Mosaic of forest and non-forest with forest fraction between 10% and 70% of total area (the estimated average of 33% is applied for the estimation of actual forest area)
Long fallow	Variable	Variable	Mosaic of mature forest, secondary forest, various stages of natural regrowth and cultivated areas with cultivated areas covering between 5% and 30% of total area
Short fallow	Variable	Variable	Mosaic of young secondary forest, various stages of natural regrowth and cultivated areas with cultivated areas covering between 30% and 50% of total area

Source: FAO (1996).

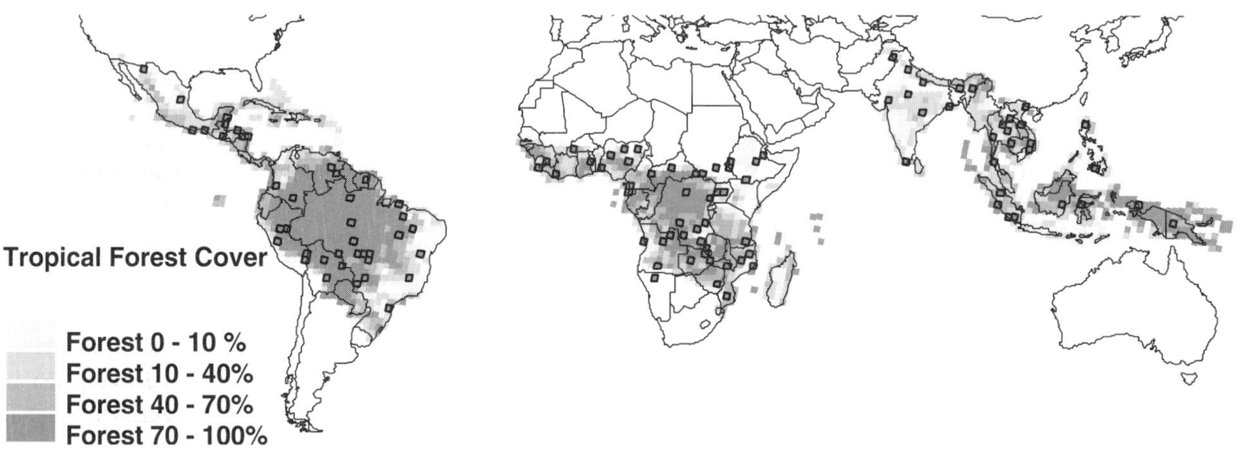

Figure 1.5 Sampling frame and selected sampling units of the FRA 1990 remote sensing survey (after FAO, 1996).

Figure 1.6 shows the estimated tropical forest cover in the benchmark years 1980, 1990 and 2000 for each definition of forest and the respective rates of deforestation[3]. Statistical analysis gave standard errors ranging between 4 and 5% for forest cover estimates (with the highest values obtained in the case of using forest definition F1 vs. slightly lower values for definitions F2 and F3), and SEs ranging between 14 and 17% for forest change estimates (with similar contrasts for the respective forest definitions). The precision achieved for the estimated change is much lower than for the forest cover estimates but, considering the 'event' character of change and the inevitable high variance, this result is more than acceptable. Figure 1.6 also shows an apparent decrease in deforestation rate with the broadening of the forest definition. This reflects the fact that the changes within the forest, when a

3 Mean deforestation rates are reported plus/minus standard errors. To define the 95% confidence limits, standard error intervals should be multiplied by 1.96.

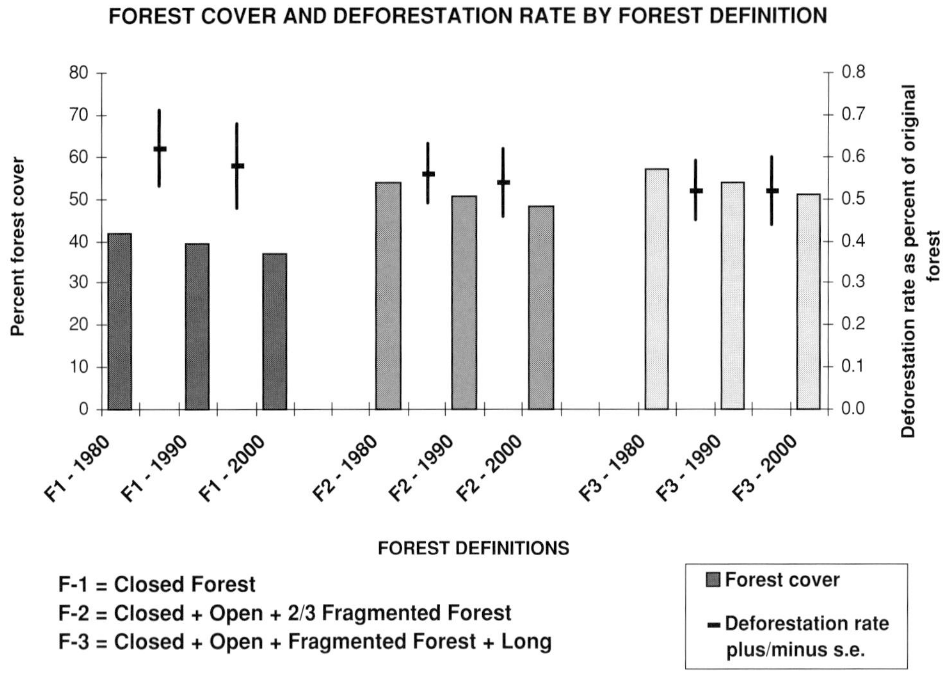

Figure 1.6 Estimated forest cover for three definitions of forest in 1980, 1990 and 2000 and the associated deforestation rates (plus and minus one standard error; after FAO, 2003). Note that all results are related to the surveyed area, which covered 63% of the total pan-tropical land area and some 87% of tropical forests (FAO 2001a). The survey excluded all non-forest areas such as desert zones and areas with negligible forest. Consequently, the values cannot be directly compared to estimates based on national data, which included all areas.

broad forest concept is used, are classified as degradation and fragmentation and not as full-blown deforestation.

A distinctive feature of the FAO FRA methodology is that it provided not only statistical results of forest cover and rates of change but also maps showing the spatial patterns and distribution of land cover changes and change matrices for each sample location. This enabled the estimation of class-to-class changes in land cover and forest categories between the two or three dates of interpretation at the sample, regional and pan-tropical scale, thus providing essential information for understanding the complex processes taking place, as well as their distinctive regional character.

Processes of land cover change and their trends at the pan-tropical, regional and eco-regional level

The satellite image interpretations carried out during the FRAs of 1990 and 2000 produced, for each sampling unit, two sequential transition matrices, referring to the periods between the three dates that were analysed. These sequential matrices were standardised to the common reporting periods 1980–1990–2000 on the basis of the individual annual class transition probabilities, following mathematical models specifically developed for the purpose (Rovainen, 1994; FAO, 2003). The standardised matrices were then statistically aggregated to represent the entire survey area as well as the three regions and main ecological zones. The resulting pan-tropical sequential change matrices, extrapolated to the entire surveyed area, are given in Table 1.3.

Much information may be derived from these matrices. To start with, one may compare area totals for the three sampling dates and thus assess the net change for each land cover class. The class *closed forest*, for instance, changed from 41.9% in 1980 to 39.3% in 1990 and 37% in 2000, thus representing a constant decrease but a small reduction in the absolute rate of change (79.4 million ha lost during the 1980s vs. 69.9 million ha lost during the 1990s). But more can be learned from examining the inner parts of the matrices. The matrices provide, *inter alia*, along the diagonal all areas that remained stable during the period under consideration and, away from the diagonal, all individual class-to-class transitions. The information on land cover dynamics contained in the change matrices can be represented efficiently, and in a more accessible manner, in the form of so-called woody biomass flux diagrams. The woody biomass flux diagram was conceived with the purpose of expressing better the magnitude of the land cover changes through the allocation of biomass densities to the individual land cover classes (FAO 1995b). By including the biomass perspective, one is able to visualise and better understand the change processes, and even assess their environmental impact through the release (or sequestration) of woody biomass related carbon. A nominal biomass value for each class thus permits the estimation

Table 1.3. *Pan-tropical area transition matrices for the periods 1980–1990 and 1990–2000*

Period 1: 1980–1990

Land cover classes in 1980	**State in 1980 (million ha)**	**Percentage of land area**	**Land cover classes in 1990**								
			Closed forest	Open forest	Long fallow	Frag-mented forest	Shrubs	Short fallow	Other land cover	Water	Planta-tions
Closed forest	**1284.6**	*41.9*	1200.4	6.3	9.5	11.3	1.7	15.1	35.5	2.1	2.7
Open forest	**317.4**	*10.3*	0.7	295.9	0.6	5.9	1.3	2.3	10.0	0.6	0.2
Long fallow	**73.0**	*2.4*	1.1	0.1	62.3	0.3	0.3	6.8	2.2	0.1	
Fragmented forest	**219.4**	*7.2*	0.7	0.8	0.2	197.5	0.8	3.9	14.8	0.4	0.2
Shrubs	**170.9**	*5.6*	0.2	0.1	0.2	0.1	149.9	0.3	19.2	0.6	0.3
Short fallow	**120.5**	*3.9*	1.1	0.4	1.3	0.7	0.3	109.2	7.2	0.2	0.2
Other land cover	**862.2**	*28.1*	0.8	1.0	0.3	1.6	1.6	1.2	853.6	1.4	0.9
Water	**4.0**	*0.1*	0.1	0.1		.1		.1	1.0	2.5	
Plantations	**16.1**	*0.5*	0.1					0.2	0.9		14.8
State in 1990	**3068.0**		**1205.1**	**304.5**	**74.4**	**217.5**	**155.9**	**139.0**	**944.4**	**7.8**	**19.3**
Percentage of land area			*39.3*	*9.9*	*2.4*	*7.1*	*5.1*	*4.5*	*30.8*	*0.3*	*0.6*

Period 2: 1990–2000

Land cover classes in 1990	**State in 1990 (million ha)**	**Percentage of land area**	**Land cover classes in 2000**								
			Closed forest	Open forest	Long fallow	Frag-mented forest	Shrubs	Short fallow	Other land cover	Water	Planta-tions
Closed forest	**1205.1**	*39.3*	1131.6	1.2	5.7	9.4	1.3	9.8	43.1	1.1	1.9
Open forest	**304.5**	*9.9*	0.2	287.3	0.5	6.8	0.7	2.2	6.6	0.1	
Long fallow	**74.4**	*2.4*	1.1	0.1	63.2	0.2		4.8	4.7		0.2
Fragmented forest	**217.5**	*7.1*	0.5	0.4	0.2	202.1	0.5	2.2	11.2	0.1	0.2
Shrubs	**155.9**	*5.1*	0.1	0.1		0.1	143.5	0.6	9.7	1.8	0.1
Short fallow	**139.0**	*4.5*	1.0	0.3	1.2	1.5	0.2	122.7	11.6	0.2	0.4
Other land cover	**944.4**	*30.8*	0.6	0.5	0.5	2.3	3.7	4.9	928.4	1.3	2.3
Water	**7.8**	*0.3*	0.2				0.8		1.2	5.6	
Plantations	**19.3**	*0.6*							1.1		18.0
State in 2000	**3068.0**		**1135.2**	**290.0**	**71.5**	**222.5**	**150.6**	**147.3**	**1017.6**	**10.2**	**23.2**
Percentage of land area			*37.0*	*9.5*	*2.3*	*7.3*	*4.9*	*4.8*	*33.2*	*0.3*	*0.8*

Note: the values along the diagonal (dark shade) represent stable areas; class losses are given along the rows and class gains are given along the columns; the areas with light shade represent transitions implying loss of biomass.
Source: FAO (2003).

of the biomass changes related to each class transition. The flux diagrams may be considered as some sort of 'signature', representing the dynamic character of a certain area over a certain period of time. As these diagrams may help to visualise the variety of such characters, Appendix 1.3 shows the flux diagrams from three locations from Africa, Latin America and Asia.

The woody biomass flux diagram in Figure 1.7, which combines the rates of change listed in Table 1.3 with estimated biomass values, is structured as follows:

- The *y*-axis, with its indicative biomass values, shows the order of the classes by their estimated biomass per hectare.
- The *x*-axis reports the areas of class-to-class transition, divided into *positive* and *negative* changes. The left side of the graph represents the lower-left part of the matrix, showing the positive class transitions (the arrow pointing upward indicates an increment in biomass), while the right-hand side of the graph represents the upper-right part of the matrix, showing the negative class transitions (the arrow pointing downward indicates a loss of biomass).

Each transition is defined by the area value on the *x*-axis and by the biomass value determined as the difference between the biomass values of the class of destination and the class of origin. Each transition is therefore represented by a rectangle, the area of which (area of change by biomass gradient) quantifies the total biomass gained or lost in a class-to-class transition (FAO, 1995b). The resulting pan-tropical diagrams in Figure 1.7 clearly show the complexity of the dynamics of the forest degradation and conversion processes and the main trends therein. (Appendix 1.4 provides a more detailed representation of the 1990–2000 diagram with an indication of the main transition types and causes of forest depletion).

It is interesting to analyse the inner parts of the matrices, which well represent the character and complexity of land cover dynamics. However, while comparing the two diagrams to assess main trends, one should be aware that statistical errors of change estimates are high, due to the uneven distribution of change events, and that, consequently, only few variations of individual class-to-class transitions have true statistical significance (FAO, 2003). The comments on the trends in class transitions that follow are based on the variations of larger size that are considered statistically significant.

The following features become apparent from Figure 1.7:

- negative changes are far more prevalent than positive ones;
- in both periods, ***closed forest*** has been by far the most common class of **origin** of land cover changes, and has been suffering the highest pressure;
- similarly, ***other land cover*** has been the most common class of **destination** in both periods (mainly cattle ranching and permanent agriculture in probably equal proportion), followed by ***short fallow*** (subsistence agriculture within the context of shifting cultivation) and ***fragmented forest***;
- the changes ***closed forest*** → ***other land cover*** and ***closed forest*** → ***short fallow***, are the two most frequent area transitions involving forest, as well as the ones that imply the largest amount of biomass loss;
- the fact that the class ***fragmented forest*** receives from the class ***closed forest*** and gives to the class ***other land cover*** (in similar amounts and in both periods) shows that this class represents an intermediate stage in the process of forest depletion;
- the biomass loss depends more on the biomass gradient than on the area of change; for example, in the transition ***shrubs*** → ***other land cover***, the biomass loss has been much less than in other transitions involving less area;
- in spite of an overall (small) reduction in the rates of loss of closed forest (see class totals in Table 1.3), the change ***closed forest*** → ***other land cover*** appears far more evident in the second diagram than in the first one. However, the increase in this transition is not evenly distributed and appears statistically significant only within the Rain Forest Zone (FAO 2003), where most of the changes occur;
- there is a significant reduction in the transitions from ***closed forest*** to ***short fallow*** and to ***long fallow***, indicating a lower relative influence of subsistence farming. This trend, combined with the point above, provides a perception of a process of radicalisation;
- during the first period, most of the new ***plantation*** areas (both agricultural and forestry ones) were created on previous ***closed forest*** area, therefore implying, if the average biomass values are accurate, a net loss of biomass;
- in the second period, more than half of the new ***plantations*** were established on previous ***other land cover***, which implies a weak but positive trend in the forestation of denuded lands (cf. Scott *et al.*, this volume).

In general, it appears that, in the context of a non-significant overall reduction, the conversion to subsistence farming associated with re-settlement programmes and traditional practices, as indicated by the transition ***closed forest*** to ***short fallow*** and to ***long fallow***, which represented a large share of the total change in the pre-1990 period, was reduced considerably during the last decade (Figure 1.7). This reduction, combined with the increased frequency in the transition from ***closed forest*** to ***other land cover***, which appears significant at least in the Tropical Rain Forest Zone, gives strong indications of an on-going process of radicalisation of the dynamics whereby the expansion of large-scale cattle ranching and permanent agriculture becomes more and more the dominant land use change associated with deforestation.

The pan-tropical flux diagrams of Figure 1.7 summarise the net land use dynamics associated with a variety of socio-economic

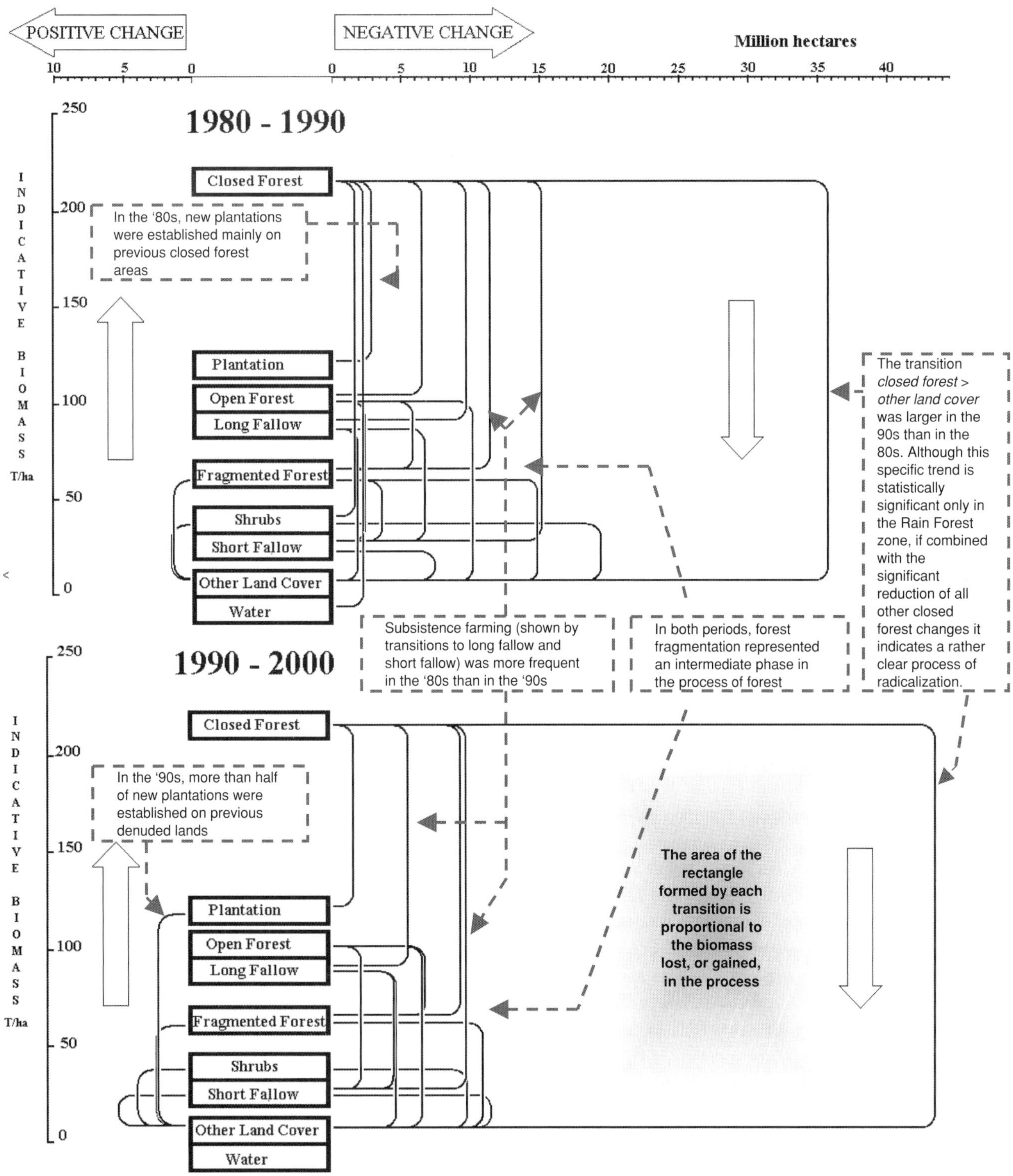

Figure 1.7 Pan-tropical flux diagrams of woody biomass for the periods 1980–1990 and 1990–2000 (elaboration of FAO, 2003). In order to focus on the most frequent and reliable class transitions and to improve the legibility of this and all following diagrams, the smallest transitions have been considered negligible and omitted from the diagrams. (See text for explanation.)

conditions and biophysical environments (the three local diagrams in Appendix 3 may help to visualise such variety). However, the pan-tropical diagrams (Figure 1.7) cannot represent the contributions of all these local dynamics to the global trend nor describe the characteristic processes of change for each region. These more local aspects, which are essential for understanding the cause-effect mechanisms threatening the forest resources in the respective regions, can be seen more clearly in similar flux diagrams summarising the results at the regional or eco-regional scale.

Regional character of processes governing land cover change

A synthesis of results derived from the regional transition matrices is presented in Table 1.4 and Figures 1.8–1.10. There are significant differences between these estimates and those presented in the previous section on the basis of country data. Part of the difference is due to the sampling universe adopted by the remote sensing survey, which excluded from sample selection all non-forest areas, such as deserts, and areas with negligible proportions of forest. The area actually surveyed covers 68% of the entire tropical area and includes approximately 90% of its forests (FAO, 1996). This fact, however, explains only part of the discrepancies. The differences between the remote sensing results shown here for Africa and the corresponding results based on the FRA 2000 country data appear to be less justified. The differences are very high indeed, in terms of both forest area and rates of change. In fact, on the basis of the country data, Africa appears as the region with the highest deforested area (5.4 million ha/year, as shown in Table 1.1 above) while according to the remote sensing survey this region appears to be the least deforested (2.1 million ha/year, as shown in Table 1.4 below). Considering the scarcity and generally low quality of African forest cover time series, the rate of change based on country data appears particularly weak, while that based on the interdependent interpretation of satellite time series appears more strongly based. Although there is always some concern on the representativity of the selected sample, the information produced by the remote sensing survey thus appears as the only robust analysis of forest change processes in Africa available so far.

The survey results further show that the highest pressure on the forest resources occurred in Asia, where the percent rate of change is highest (−8.2% over the last decade). In absolute terms, the largest area of change occurred in Latin America, with 41.4 million hectares of forest lost during the same period (Table 1.4a).

Although the trends observed for each sampling unit may be quite reliable in view of the consistency of the interdependent method of analysis, as discussed in Appendix 1.1, the regional and pan-tropical trends shown in Table 1.4 are not statistically significant in view of the relatively high standard errors of the estimated rates of change[4] and should be considered as indicative only. According to these indications, the pan-tropical deforestation trend reflects a rather stable situation, with only a slight reduction in annually deforested area but an almost equal value in percent change rates between decades, suggesting a relatively constant pressure on the remaining forests. At the regional level, Africa and Latin America present a slight reduction in deforestation rate while Asia shows a small increase. Interesting additional insights on the typologies of change and some indications on their trends may be obtained from an eco-regional analysis, as will be discussed further below.

The net forest degradation rates shown in Table 1.4b summarise the various positive and negative changes occurring within the forest area (see Appendix 1.2 for details). The qualitative changes that can be detected from Landsat data are rather limited, including only major changes in forest density (between closed and open conditions) and in the level of human disturbance (presence or absence of long fallow shifting cultivation). Other important modifications, positive or negative, such as change in species composition and forest structure, are not included in the FRA RSS results. However, based on the major physiognomic modifications mentioned above, the inferred degradation rates appear relatively small, with annual rates of some 0.08% (range 0.06–0.19) during the 1980s, diminishing to 0.04 (range 0.01–0.12) during the 1990s. It is interesting to note that the reduction in degradation rate has been observed in all three regions. As in the case of deforestation rates, the highest pressure is observed in Asia, where the relative degradation rate is several times higher than in the other two regions, mainly due to the large area of forest affected by long fallow shifting cultivation (see below).

REGIONAL CHANGE PROCESSES

In addition to statistics and rates it is interesting to analyse the dynamic processes of change taking place in the respective regions. These processes become more concrete when comparing the regional biomass flux diagrams for tropical Africa, Latin America and Asia (Figures 1.8–1.10). At this level of analysis one can differentiate more easily between typical change processes, and the specific cause-effect relationship can be better understood. During the FAO FRA RS survey it became evident that the flux diagrams represent a kind of signature and that the individual signatures belonged to quite distinct regional typologies. From Figures 1.8, 1.9 and 1.10 we can rapidly visualise the considerable differences in the changes in total area and biomass among the three regions. The specific regional characters resulting from the typologies of change and trends associated with the diverse socio-cultural settings, can be summarised as follows:

4 The standard errors of the deforestation rates range between ±15% of the mean at the pan-tropical level and ±20–25% at the regional level. (FAO, 2002).

Table 1.4. *(a) Regional forest cover (F3 definition) and associated rates of change in 1980, 1990 and 2000, and (b) net forest degradation[a] rates during 1980–90 and 1990–2000 as derived from the FAO*

	Land area studied (sampling frame) (million ha)	Forest area 1980	Change 1980–90 (million ha) (% rate)	Forest area 1990	Change 1990–2000 (million ha) (% rate)	Forest area 2000	Trend (as percent of 1980–90 rate)
(a) Regional forest cover							
Africa	1224	562	−23.5 (−4.3%)	539	−20.8 (−3.9%)	518	−7.6%
Latin America	1233	866	−44.8 (−5.3%)	822	−41.4 (−5.2%)	780	−2.6%
Asia	610	319	−23.3 (−7.6%)	295	−23.4 (−8.2%)	272	+8.7%
Total	3068	1748	−91.6 (−5.4%)	1656	−86.2 (−5.3%)	1570	−0.7%

	Forest area 1980 (million ha)	Annual forest degradation 1980–90		Forest area 1990 (million ha)	Annual forest degradation 1990–2000	
		Million ha	% rate		Million ha	% rate
(b) Net Forest degradation rates						
Africa	562	0.36	0.06	539	0.14	0.03
Latin America	866	0.50	0.06	822	0.11	0.01
Asia	319	0.62	0.19	295	0.35	0.12
Total	1748	1.45	0.08	1656	0.60	0.04

[a] See Appendix 1.2 for details on the definitions of forest and forest degradation.

Africa

In Africa, the observed processes of change appear to be distinguished by phases of progressive degradation, rather than outright deforestation, caused mainly by high rural and urban population pressure. Although Figures 1.8a and 1.8b differ somewhat in many small ways, it is evident that the process of forest depletion maintained the same typology, as characterised by a variety of relatively small changes, both in terms of area and biomass. The main thrust behind these processes has been rural population demands for land (subsistence farming, pastures) and wood (mainly fuelwood and, to a lower extent, timber and construction material).

The dominant transitions in land cover are:

- ***closed forest* → *short fallow***, which is the effect of small-scale subsistence farming whereby the fertility of the soil is regenerated during fallow periods of a few years, agronomic additives and fertilisers being inaccessible to poor farmers;
- the sequence ***closed forest* → *open forest* → *fragmented forest* → *other land cover*** clearly represents the various stages of forest depletion, and is an effect of rural and urban population needs for land and energy. Most probably, commercial logging triggered many of these processes, at least in the more productive forest zones, but logging is not a land cover class by itself and could not be detected in a consistent manner in this study.

Key factors behind the pressure on forests in Africa are (FAO, 2001a):

- rapid population growth, particularly that of urban population;
- poverty, slow economic development, inadequate economic policies;
- wars and conflicts (destruction of forests and infrastructures, refugee settlements and overall disincentive to international and national investments): During the last decade, Africa was afflicted by conflicts of various nature in Sierra Leone, Liberia, Chad, Ethiopia, Eritrea, Somalia, Sudan, Democratic Republic of Congo, Rwanda, Burundi, Angola and Zimbabwe.
- insecurity of land tenure (no clearly defined responsibility for management);
- desertification / climate change.

Among the direct causes of deforestation the following can be identified (FAO 2001a):

- poor farming practices (short fallow shifting cultivation)
- conversion to cash crop estates (Ivory Coast)
- mangroves being converted to rice fields or ponds for shrimp farming and cleared for woodfuels
- increased clearing and tree cutting for fuelwood and charcoal
- poor logging practices including over-exploitation

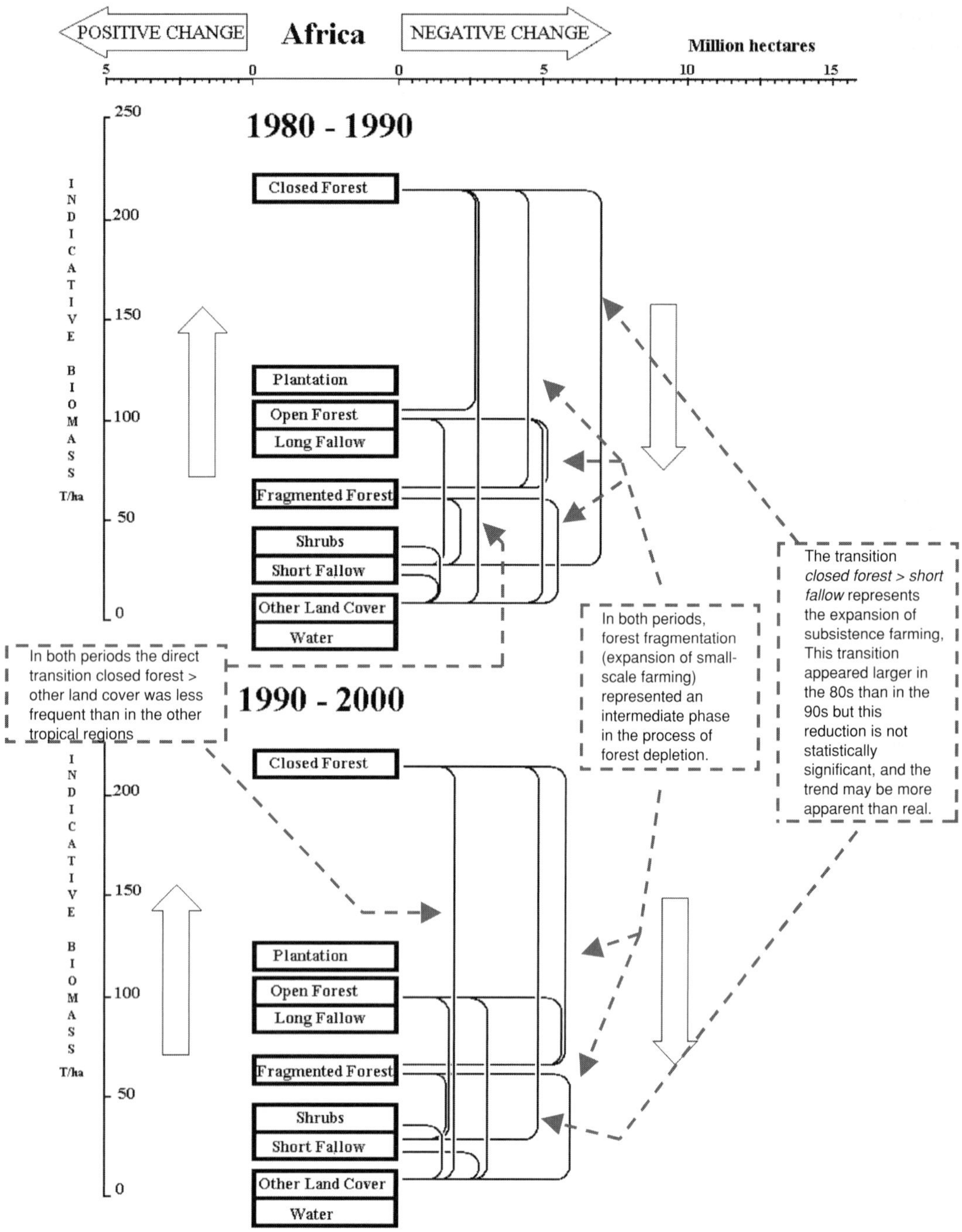

Figure 1.8 Woody biomass flux diagrams for Africa during 1980–1990 and 1990–2000 (elaboration of FAO, 2003).

- commercial logging as direct cause of degradation and indirect cause of full deforestation
- mining
- desertification in Sahelian countries.

Woodfuels (wood-based fuels such as fuelwood and charcoal) play an important but largely undisclosed role in the cause-effect mechanisms associated with deforestation and forest degradation in tropical regions, particularly in Africa. Following the major oil crisis of the 1970s, FAO produced a large-scale study on the status and prospects of fuelwood supplies in developing countries which predicted a dramatic deforestation driven by energy needs and an epochal fuelwood crisis by the year 2000 (FAO, 1983). The prediction proved wrong since the study was partly biased by the assumption that the fuelwood came mainly from forests, thereby underestimating the capacity of the agricultural sector to produce fuelwood as energy needs rise (Foley, 1987, Leach and Mearns, 1988, Dewees, 1989). The fact that the big crisis did not materialise conveyed a feeling – more than a proof – that there is no relationship between woodfuel needs and deforestation and forest degradation processes and that the woodfuel issue does not deserve much policy attention. Unfortunately, this opposite perspective seems equally biased, especially considering that the wood energy sector is largely informal and that woodfuel supplies, their sources and sustainability are poorly known and rarely studied, in spite of the paramount importance of this sector in tropical Africa[5]. An element of special concern is the fast rising demand for charcoal linked to the rapid increase in urbanisation (see Box 1.1).

Box 1.1 The charcoal issue

Recent global studies on the use of wood for energy (Broadhead, Bahdon, and Whiteman, 2001) projected, for the next 20 years in tropical Africa, a 74% rise in charcoal consumption (from 20 million t at year 2000, to some 34.7 million at year 2020) against a 22% rise in fuelwood consumption (from 400 million CUM in the year 2000, to some 488 million in the year 2020).

A review of recent national reports on wood energy supply and demand in Africa shows that the impact of charcoal-making on the remaining resources is considered to be extremely serious by many authors, and is often pointed out as a major cause of forest clearing in the countries concerned, even to the extent that charcoal-making is probably overtaking the practice of shifting cultivation in terms of its impact on natural resources (Drigo, 2001). Charcoal use is changing the relationship between household energy needs and wood resources in the region, transforming what was traditionally accepted as an all-time self-reliant practice (fuelwood gathering) into a vicious circle with potentially dramatic effects on the remaining natural forests and woodlands. Key aspects that contribute to the rapid increment of charcoal use in Africa and to its growing impact on forest resources include (Drigo, 2001):

- rapid urbanistion and a shift from fuelwood to charcoal by most urban dwellers;
- the shift from fuelwood to charcoal implies a doubling of the per capita wood demand as a result of the energy used in the carbonisation process and low transformation efficiency (needing up to twice the amount of wood for the same amount of end-use energy (FAO, 1999b));
- being strongly market orientated, charcoal-making opens up employment opportunities, promoting the law of profit with little attention, if any, to resource sustainability;
- charcoal-making is done almost exclusively with green wood from natural forests and woodlands, implying clearing operations and a generally high environmental impact, while fuelwood is more commonly a by-product of shifting cultivation practices and other land conversions and uses, or produced directly through energy plantations;
- charcoal production is economically convenient even at long distances from a market, thereby promoting intense exploitation of forests and wooded areas previously protected by their remoteness;
- the best charcoal quality comes from drier wood formations, where the regenerative capacity is lower, thereby potentially speeding up processes of desertification.

Unfortunately reliable estimates of the change in vegetation cover (forest and woodland) resulting from these practices are not available, leaving the issue simply as a vague threat or a subjective judgement. Of the overall weakness of existing information on woodfuel supply, the lack of data on charcoal is probably the most serious. The filling of this gap would deserve maximum efforts at local, national and international levels. Such an analysis of trends in charcoal demand should ideally be accompanied by adequate surveys of the associated land cover and biomass changes at local, sub-national and national levels.

A prime example of critical charcoal consumption and its impact on natural wood resources is the case of Madagascar. In this country charcoal provides only 11% of national household energy needs, but its impact on natural resources is far higher than that of fuelwood, which covers some 85% of household needs (less than 4% is covered by other fuels). This high impact of charcoal-making is due to the low carbonisation efficiency and to the fact that charcoal production takes place exclusively in forest zones while fuelwood comes mainly from non-forest areas[6].

The wood-based fuels issue is extremely important in the African context, and has far-reaching consequences, both for the socio-economic development and livelihoods of some two billion people that depend on woodfuels to satisfy their subsistence energy needs (FAO, 1995c) and, on the other hand, on natural forest resources, woodlands, and the environment at large. In less developed countries, bioenergy has crucial advantages over other energy sources as a tool for poverty reduction (Kartha and Leach, 2001) and the potential is certainly great, but sustainable resource management is equally crucial, if the benefits are to last for future generations as well.

5 Except for the five north African countries and South Africa, all African countries still depend heavily on wood to meet basic energy needs. In the various African regions, woodfuel share ranges from 61% to 86% of primary energy consumption, with a major part (74% to 97%) consumed by households. The management of woodfuel resources and demand should be considered a major issue in energy planning processes in Africa.

On the other hand, woodfuel consumption is a major contributor to total wood removal, accounting for around 92% of total African wood consumption and contributing to greenhouse gas emissions. Woodfuel use is therefore a major local and global environmental issue in Africa, and should be fully integrated into forestry planning and environmental protection processes. (FAO, 1999b).

6 Presentation by Mr Bertin Andriamanantsoa, Direction de l'Energie (MEM), at the national workshop 'Atelier de validation sur l'étude Pilote Bois Energie et Produits Forestiers Non Ligneaux', Antananarivo, 20–22 November 2001.

Latin America

A totally different typology of change dominates in **Latin America**. Here the single transition ***closed forest → other land cover***, which represents deforestation with the highest possible biomass loss resulting mainly from the direct conversion of the original forest to cattle ranching and permanent agriculture, was by far the most important change in both periods (Figure 1.9). The increment of this particular transition, although quite evident from the diagram, should be considered as indicative only since, due to the high variance of the sample in this region, it does not achieve statistical significance (FAO 2003).

Apart from this main type of transition and a few positive changes (from ***other land cover*** to ***short fallow***, ***shrubs*** or ***fragmented forest***) which represent regrowth of previously cleared forests, all other transitions showed a marked reduction during the 1990s (Figure 1.9b).

The second most frequent transition, ***shrubs → other land cover***, represents the large areas of Brazilian *caatinga* (steppe typical of the northeast regions of Brazil) and *cerrado* formations (tree and shrub savanna of south-west and central Brazil) that are being converted to cattle ranching. Most of these changes were the effect of policies and incentives to cattle ranching and other centrally planned operations on a comparatively large scale (large land ownership, energy schemes, resettlement and forest exploitation/conversion programmes), usually benefiting from consistent financial investment and heavy mechanisation (cf. Serrao, this volume). The estimated biomass loss associated with these transitions certainly is the highest anywhere in the tropics (cf. Figure 1.8 with Figures 1.9 and 1.7). The transition ***closed forest → short fallow*** as well as a number of other, less frequent, changes in land cover, represent the effects of high rural population pressure and small-scale farming, such as in the Amazon and Yucatan, and is often associated with resettlement programmes. A reduction in these types of changes, shown by the diagrams but not confirmed statistically,[7] may be due to several factors such as, for instance, changes in Brazilian policies regarding resettlement programmes (cf. Serrao, this volume). Resettlement programmes in the Brazilian Amazon during the 1970s and 1980s have been considered the main direct cause of forest depletion at the time (Fearnside, 1984) but after the initial colonisation phase, livestock production became subsidised, resulting in the proliferation of large-scale cattle ranching (Nepstad *et al.*, 1999). Nowadays it seems that the overwhelming majority of cleared forest is converted for cattle ranching rather than agriculture. It appears also that the impetus for the expansion of cattle ranching in the Brazilian Amazon currently comes largely from profit sources other than the sale of beef (including land speculation?; Fearnside, 2000; Serrao, this volume).

Other direct causes of forest loss in Latin America are mining and the construction of large hydroelectric schemes (Grainger, 1993). Outside the Brazilian Amazon, the complexity and variety of driving forces increases considerably. The most frequently acknowledged causes of forest depletion are demand for agricultural land, either directly due to high population pressure or induced by government policies. Examples include the credit programmes for agricultural production, which promoted forest conversion to agriculture in Nicaragua (FAO, 2001a), the expansion of beef and cash crop trade encouraged and promoted by government subsidies and colonisation schemes in Mexico and resettlement schemes supported by the Bolivian government in the foothills of the Andes (Achard *et al.*, 1998). Conflicts and related population migrations have also affected forest cover in Central America, causing forest expansion on abandoned fields during such conflicts and forest reduction around refugee camps and after repatriation (FAO, 2001a). Migrations and new settlements along the western edge of the Amazon Basin are the cause of large forest clearings in favour of a variety of different land uses (pastures, permanent and shifting agriculture, mining, etc.). In Colombia and Bolivia agricultural expansion is sometimes linked to drug production or to counter-measures undertaken by government authorities (Cavelier and Etter, 1995; Achard *et al.*, 1998). Agricultural expansion, mining, oil extraction and charcoal exploitation have also been indicated as being responsible for forest depletion in Venezuela's Orinoco Llanos and Delta (Achard *et al.*, 1998). Finally, expansion of urban areas is a common and widespread cause of deforestation and in the Caribbean the pressure on the forest is also linked to the uncontrolled expansion of the tourist industry (FAO, 2001a).

Asia

As shown in Table 1.4, tropical Asia presented the highest rates of deforestation and forest degradation among the three regions during the last two decades. These rapid changes were the effect of both high rural population pressure and centrally planned conversion programmes, as can be deduced from Figure 1.10ab, although the relative importance of the two components changed considerably over time. During the first decade both types of process were about equally important, resulting in considerable deforestation and forest degradation (Figure 1.10). Deforestation was represented by the conversion of the respective forest classes to ***other land cover*** and to ***short fallow shifting cultivation***. The former is largely the effect of centrally planned conversion programmes, mostly in the form of large resettlement schemes involving forest exploitation/conversion (particularly in Indonesia and Malaysia) and intensification of permanent agriculture on traditional shifting cultivation areas (South and South-east continental Asia). The second major conversion reflects the effect of high rural population pressure, and is represented by the expansion of subsistence farming into forest areas along logging roads and from

7 The reduction of the transition from **closed forest** to **short fallow** proved statistically significant at pantropical level but not at regional level (FAO 2003).

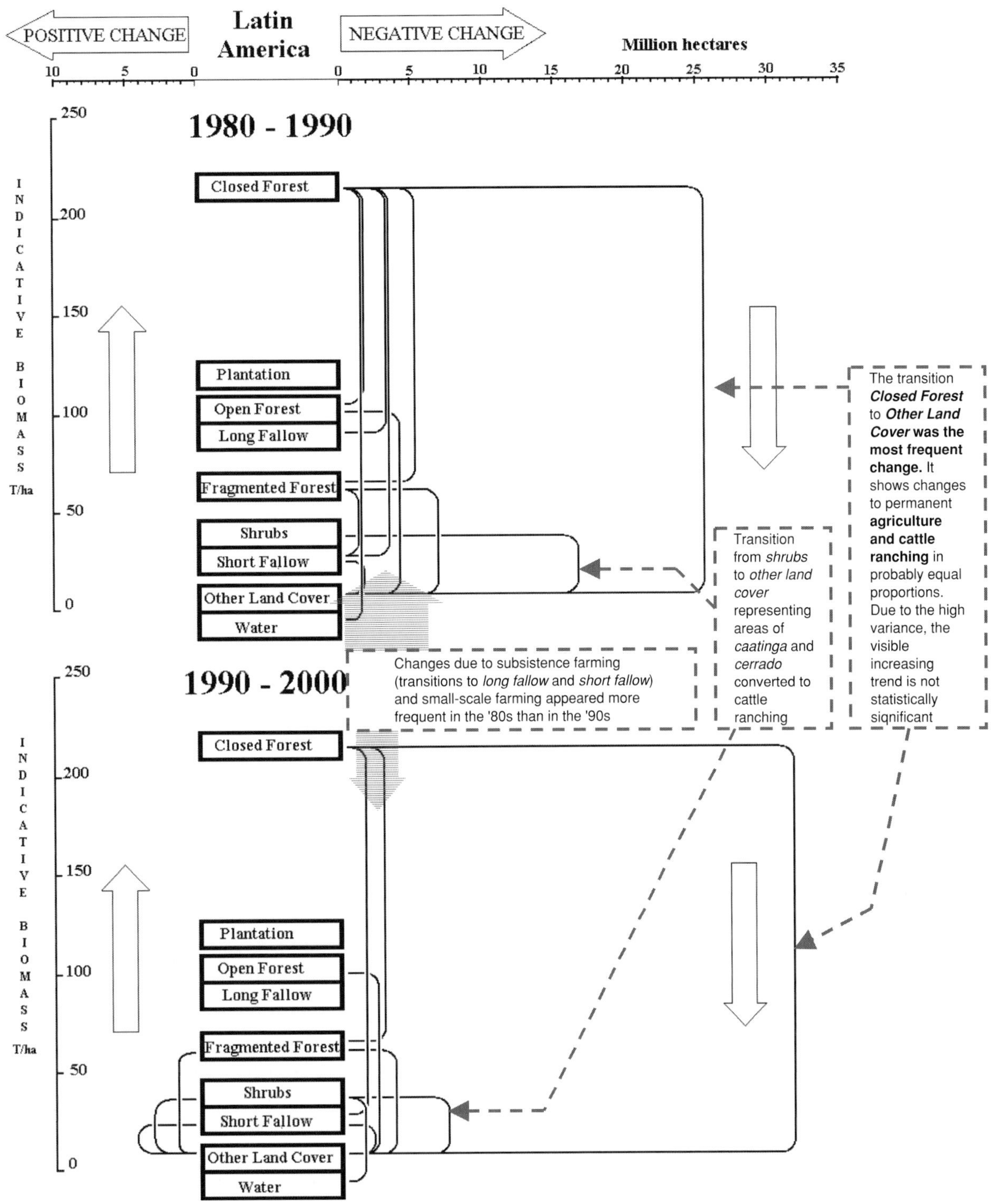

Figure 1.9 Latin America flux diagrams 1980–1990 and 1990–2000 (elaboration of FAO 2003). Note that the area scale (*X*-axis) of the Latin America diagram is more compressed than in Africa and Asia diagrams.

existing croplands. The process of forest **degradation** is represented mainly by the expansion of the area of forest affected by traditional shifting cultivation (long fallow shifting cultivation), that encroached on previously dense or undisturbed forest. An equal amount of the ***long fallow*** forest class was converted to ***short fallow*** and, a little less, to ***other land cover***, reflecting a sequence of progressive forest depletion as a direct effect of the growing population and related needs for farm land.

Shifting cultivation is by definition a cyclic form of land use. Traditionally, new areas under shifting cultivation were balanced by areas where cultivation was abandoned to revert to (secondary) forest conditions. This balance was lost a long time ago, thereby converting the cycle into a sequence of progressive degradation. The difference between the forest area going into long fallow (5.4 million ha) and the area of long fallow reverting to forest (0.8 million ha)[8] shows how unbalanced, and hence unsustainable, this originally sound practice has become (cf. Malmer *et al.*, this volume).

During the first decade the area covered by plantations in Asia, both forestry and agricultural plantations, increased significantly, although primarily at the expense of closed forest. Some 2.5 million hectares of closed forest were converted to plantations, mainly agricultural (oil palm, cocoa, rubber), representing the fourth most frequent transition observed (Figure 1.10). During the second decade the combination of population pressure and centralised conversions is still visible but in different proportions. As in the case of Latin America, there seems to be a process of radicalisation that favours land use changes associated with high-gradient transitions (read: clear-felling), although here as well the variations were not statistically significant. Another more solid difference with the previous decade is that in the 1990s almost half of the new plantation area was established on previously cleared lands This represents an important element counterbalancing, to a small extent at least, the negative general trend.

The main causes of forest depletion indicated for the deforestation hot spots of South and South East Asia (Achard *et al.*, 1998) are the following:

- Shifting cultivation is considered an ubiquitous cause of deforestation in this region, being associated with deforestation hot spots in NE India, Bangladesh, Myanmar, Lao PDR, Cambodia, Sumatera, Kalimantan, Sulawesi and Irian Jaya (cf. Figure 1.4c). A common process is the progressive reduction in the duration of the fallow period as caused by rapidly growing demographic pressure, as in the case of Bangladesh's Chittagong Hill Tracts, where the immigration of poor farmers from the overcrowded plains by far outnumber the original hill tribes.
- Intensive logging, mainly driven by the pulp industry, and conversion to large-scale agricultural plantations has been indicated as the main cause of deforestation in Indonesia. Examples of this process are the large-scale forest clearings in the lowlands of central Kalimantan for the national rice plantation programme (Rieley, 2001).
- Forest clearing for permanent cash crop agriculture of a small-medium scale was found to be the main cause of deforestation around the agricultural plains in Myanmar, along the Lao PDR-Cambodia boundary and in the Central Highlands of Vietnam. A mixture of illegal logging and forest clearing for shifting cultivation and small-scale cash crops is indicated as a diffuse cause of deforestation from NE India to Irian Jaya.
- Overexploitation of forest resources, i.e. logging above the productive capacity of the forests, is the main cause of forest degradation and fragmentation in many parts of NE India, Bhutan, Myanmar and Cambodia.
- Human-induced and natural habitat modifications, such as the reduced freshwater flow in the Indian and Bangladesh Sundarbans is considered the cause of species die-off and overall forest degradation in such areas (see also Hooijer, this volume).

SUMMARY OF REGIONAL PERSPECTIVE

Comparing the three regional situations, the dominance of the changes in Latin America becomes even more pronounced, as these combine the largest area of change with the highest biomass loss. In fact, the biomass gradient of Latin America's most frequent class transition (***closed forest*** $\longrightarrow$ ***other land cover***), represents the maximum biomass loss observed anywhere (cf. Figures 1.9 and 1.7).

The effects of **centrally planned operations** are evident in Latin America and in Asia but to a much lesser degree in Africa. The typical associated transitions include ***closed forest*** $\longrightarrow$ ***other land cover*** or, in Asia only, ***closed forest*** $\longrightarrow$ ***plantation***. Typical land uses related to these processes are: cattle ranching in the Brazilian Amazon, large resettlement and plantation programmes in South East Asia and, to a lesser degree, in West Africa. Comparing the 1980s and 1990s, there are strong indications that, except for Africa, these relatively high-investment and high-gradient transitions are taking the lead in recent years, thereby contributing to the radicalisation of land use change.

The other important component of the process, **rural population pressure**, is characterised by combinations of low-gradient transitions associated with subsistence and small-scale farming, such as long and short fallow shifting cultivation and processes of forest fragmentation and degradation. This component remains

8 The positive transition from ***long fallow*** to ***closed forest*** was estimated, (Asia 1980–1990), at 0.8 million hectares, an area too small to be represented in the regional biomass flux diagram in Figure 1.10.

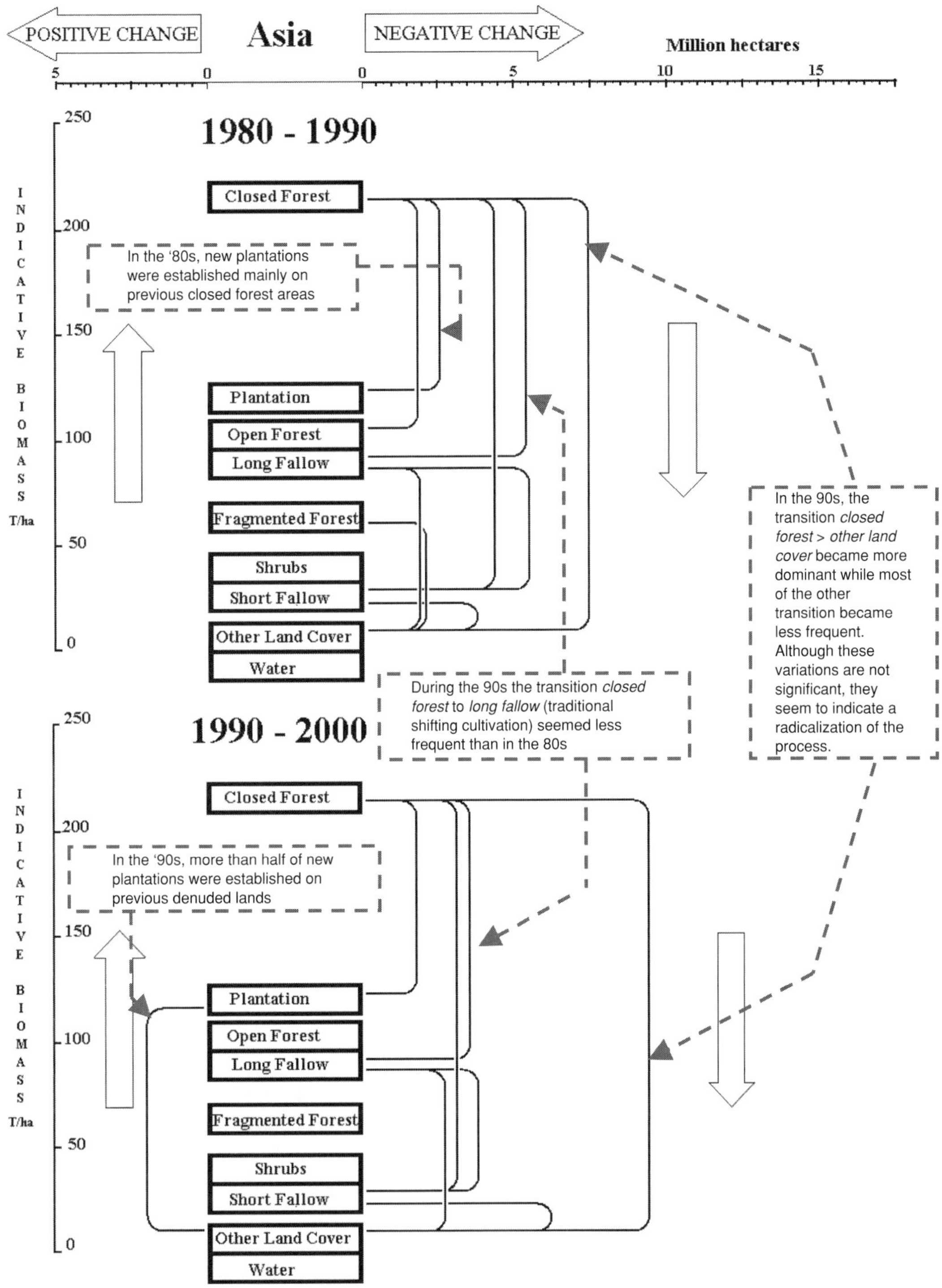

Figure 1.10 Asia flux diagrams 1980–1990 and 1990–2000 (elaboration of FAO 2003).

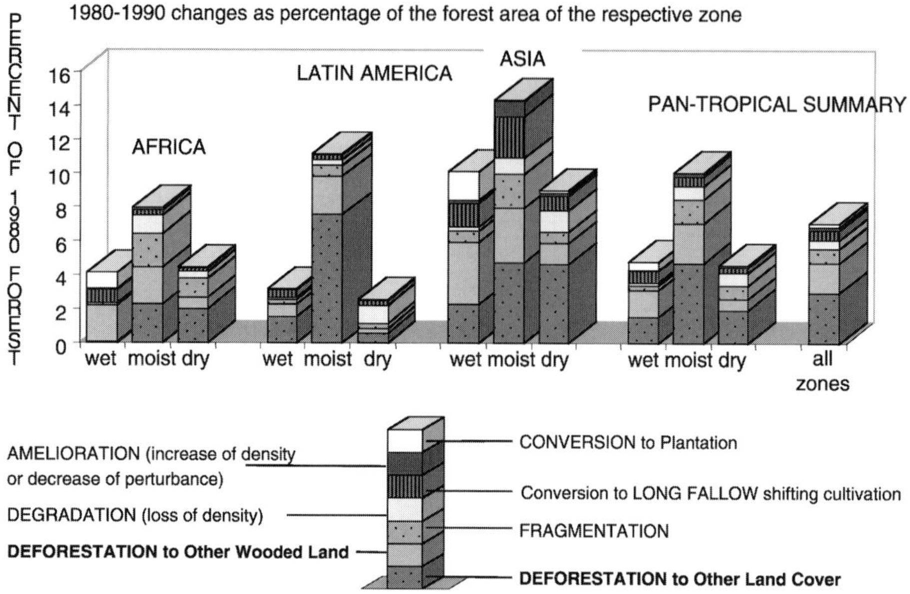

Figure 1.11 Changes in forest cover in 1980–1990 per eco-regional zone and at the pan-tropical level (after FAO, 1996).

dominant in Africa but it seems to be losing ground in Asia and Latin America (Figures 1.8–1.10).

Eco-regional distribution of forest change

Another interesting perspective is offered by the analysis of land cover change data at the eco-regional level. The FRA of 1990 reported on the distribution of major land cover changes in eco-regional terms, based on the FRA Remote Sensing Survey (FAO, 1996). The ecological parameters used were derived from the so-called Eco-floristic Zone Map of the Tropical Regions (FAO, 1988) which refers to the Holdridge Life Zone System (Holdridge, 1959). The final ecological zones that were adopted represent a simplification of the original classification and were defined on the basis of rainfall parameters, as follows:

Z1 = **Wet and very moist** (rainfall > 2000 mm)

Z2 = **Moist** (with short and long dry season) (rainfall 1000–2000 mm)

Z3 = **Sub-dry to very dry** (rainfall 200–1000 mm)

The information provided by an eco-regional analysis of change is particularly interesting, since it combines the regional character of deforestation processes and the effect of the ecological setting. However, in view of the limited number of sampling units studied for each ecological zone within each region, this eco-regional breakdown should be considered indicative only, and be used only to highlight major aspects.

Figure 1.11 describes the forms of change that occurred in the forest areas of the three main ecological zones and the relative amount of change expressed as a percentage of the respective original (1980) forest areas. Among major aspects, one can observe that:

- in all regions the forest resources of the moist tropical zone suffered the highest pressure, both in the form of full deforestation and forest degradation, with a relative rate of change which appears to be (at least) twice that observed for the wet and dry zones; however, from a carbon budget and biodiversity point of view, the effects of change in the wet and dry zone will be dramatically different even though the rates of change appear almost equal;
- the rate of change of Asia's forests is far higher than that of the other regions in all three ecological zones, although it appears that there is a diversity of type of change which includes significant proportions of amelioration (moist zone) and conversion to plantation (wet zone);
- considering only the two deforestation categories shown at the bottom of the stacked bars of Figure 1.11, which represent complete forest depletion to permanent agriculture or to short-fallow shifting cultivation, it appears that the highest relative rate (and even more so the absolute rate) occurs in the moist zone of Latin America.

Unfortunately, we do not know what happened in these particular zones after 1990, since the ecological subdivisions used in the analyses by the FRAs of 1990 and 2000 are different. Both studies divided the tropical region into three zones but used different thresholds. The FRA of 1990 used 1000 and 2000 mm of annual precipitation as thresholds, while the FRA of 2000 used basically the length of the dry period, with three and five dry months as

Table 1.5. *(a) Ecological zoning adopted by FAO's FRAs of 1990 and 2000 and the associated estimates of forest cover change (FAO, 1996, 2002)*[a]*; (b) changes per ecological zone as derived by a combination of the two studies (author's elaboration of 1.5 (a) data).*

(a) Ecological zoning

Rainfall		> 2000 mm	1000–2000 mm	200–1000 mm
Length of dry period		0–3 months	3–5 months	5–8 months
FRA 1990 zone	**All zones**	**Wet and very moist**	**Moist** (with short and long dry period)	**Sub-dry to very dry**
(sampling units) % of total forest		(26) 42.4%	(60) 43%	(29) 14%
Annual change 1980–90 10^6 ha* yr^{-1} % rate	 -9.36 -0.52%	 -2.76 -0.36%	 -5.91 -0.77%	 -0.69 -0.27%
FRA 2000 zone	**All zones**	**Rain forest**	**Moist deciduous forest**	**Dry forest**
(sampling units) % of total forest		(43) 57%	(44) 31%	(25) 12%
Annual change **1980–90** 10^6 ha* yr^{-1} % yr^{-1}	 -9.2 -0.54%	 -4.9 -0.50%	 -3.6 -0.69%	 -0.6 -0.32%
Annual change **90–2000** 10^6 ha* yr^{-1} % yr^{-1}	 -8.6 -0.53%	 -5.7 -0.61%	 -2.2 -0.44%	 -0.7 -0.37%

(FAO 1996, FAO 2003)

(b) Most active front of deforestation as deduced by combining the two studies above[b]

Rainfall		> 2000 mm		1000–2000 mm	200–1000 mm
Length of dry period		0–3 months		3–5 months	5–8 months
Percent of 1980 total forest	100%	42.4%	**14.6%**	43%	
Percent of 1980-90 total change	100%	30%	**24.6%**	45.5%	
Change 1980-90 10^6 ha* yr^{-1}	-9.23	-2.76	**~-2.27** (-2.14 /-2.4)	-4.2	
Change 1980-90 % yr^{-1}	-0.52%	-0.36%	**-0,89%**	-0.56%	

[a] The width of the columns in the table is not area proportional. They are meant to show that there are overlapping portions.

[b] FRA 90 and FRA 2000 gave slightly different estimates for forest area and change for the period 1980–90, as can be seen in Table 1.5a comparing FRA 1990 and FRA 2000 results for all zones. Such differences, which are due to the reduced area of overlap in the three-date time series used by FRA 2000, are very small but prevent the exact calculation of the values. In fact, the area of forest loss for the overlap zones (Moist in FRA 90 and Rainforest in FRA 2000) ranges between 2.14 and 2.4 million hectares per year, depending on the reference taken.

thresholds. Table 1.5 shows the relation between the ecological zoning adopted by the two studies. The main difference between the two lies in the width of the first zone. i.e. the 'Wet and Very Moist' zone distinguished in the FRA of 1990, and the 'Rain Forest Zone' of the FRA of 2000. The latter is much wider, and includes areas with rainfall as low as 1500 mm/yr (FAO, 2001b).

The fact that the recent trends in forest change in terms of the ecological subdivision employed in the FRA of 1990 are unknown is unfortunate, but the additional perspective provided by the new ecological subdivision counterbalances this to some extent. Paradoxically, the inconsistency between the two definitions may provide some additional insights into the ecological distribution of tropical deforestation since it highlights, through deductive reasoning, the distinct character of the sub-zone that 'moved' from the Moist zone (FRA 1990) to the Rain Forest zone (FRA 2000).

As shown in Figure 1.11 and Table 1.5a, the FRA of 1990 noted the highest rate of change in the Moist Zone, both at the regional and pan-tropical level. On the other hand, using the new zoning adopted by the FRA of 2000, the 1980–1990 rates of change of the Moist Deciduous Forest zone and Rain Forest zone are of comparable magnitude (−0.69% and −0.50% $year^{-1}$ respectively) (Table 1.5a). One should remember here that the differences due to the definitions are simply the results of different grouping, or post stratification, of the same sampling units. Given this, the wider Rain Forest zone can be divided into two sub-zones: the wetter one already described by FRA 1990 as 'Wet and Very Moist', and the remaining slightly drier sub-zone characterised, with some approximation, by an annual rainfall of 1500–2000 mm and a dry season of two-three months. As deduced from Table 1.5b, the latter sub-zone carried 14.6% of the total tropical forest area and as much as 24.6% of the entire tropical forest loss, with an annual change rate of −0.89%, which is far higher than the wetter sub-zone (−0.36%) and the remaining moist and dry zones (−0.56%, combined). It appears therefore that a relatively high rainfall and a short dry season characterised the hot fronts of tropical deforestation during the 1980s.

It seems also that these hot fronts are getting more intense and that they are fast moving towards the wetter cores of the remaining forest areas. This impression is confirmed by the acceleration of the deforestation rate in the Rain Forest zone, where the corresponding annual rate of change goes up from –0.50% (this figure can be replicated from Table 1.5b by weighting on the respective forest area) to −0.61%, which is one of the statistically most significant trends observed (FAO 2003). This acceleration is accompanied by the indication of a possible deceleration[9] of the deforestation rate in the Moist Deciduous Forest. Arguably, this shift of the deforestation front towards the wetter zones is the most significant trend observed in this study.

One important implication of this 'wet' shift of the deforestation front is that, due to the higher biomass densities of the forest formation being cleared and degraded, a higher per-hectare carbon emission can be assumed. In fact, from the carbon budget viewpoint, the increased biomass density of the forests currently under pressure may easily offset the effect of the slight reduction in deforestation rate shown by the trend analysis at the pan-tropical level (Table 1.4). These elements further strengthen the impression of a radicalisation of the processes of tropical deforestation as hinted at in the previous section.

In summary, it appears that socio-economic and cultural aspects, which tend to be more homogeneous within geographic regions, determine the nature of the change processes and more clearly indicate the underlying cause-effect mechanisms, while the ecological setting rather determines the intensity of change and reveals its environmental implications. In fact, the processes of change become clearer and more distinct when the perspective of analysis is regional, to the extent that the respective flux diagrams become a sort of 'signature' summarising the effect of the social, economic and cultural factors that distinguish the tropical regions of Africa, Latin America and Asia. Given this regional character, which defines also the magnitude of change specific to each region, the relative intensity of change seems to be well explained by the ecological perspective (Figure 1.11 and Tables 1.5a and b), which provides clear indication on where deforestation processes are more intense (hot spots) and on the direction of such hot deforestation fronts.

Having discussed the results of the global and regional surveys by the FAO's FRAs, let us now compare these with the recent findings of the high resolution survey of the humid tropics only by the TREES II project (Achard *et al.*, 2002).

MAIN RESULTS OF THE TREES II SURVEY OF DEFORESTATION IN THE HUMID TROPICS

The first set of global and regional results from the TREES high-resolution survey of humid tropical regions (Achard *et al.*, 2002) are summarised in Table 1.6.

In Table 1.7 the estimates of humid tropical forest area and net annual change as estimated by the TREES survey are compared with the corresponding FRA RSS results. As indicated earlier, the two surveys are not entirely comparable, as they refer to distinct geographic areas (humid tropics vs. all tropics, respectively) as well as time periods (1990–1997 vs. 1990–2000). Time inconsistencies are considered of minor relevance, as the bulk of the images actually used by both studies come from comparable years (around 1990–1997 for the TREES project and 1989–1997 for the FAO study) and the standard reference years are the result of mathematical extrapolations. More difficult to overcome is the difference in area surveyed, as forests and forest area changes are not evenly distributed. To facilitate the comparison, the two data sets are visually referred to the ecological zones adopted in the FRA of 2000. Other inconsistencies may arise from the definitions of forest adopted by the two studies. Although the physiognomic thresholds are compatible, the respective land cover classifications show considerable differences, especially concerning the number of classes distinguished. The FRA survey adopted a simple 10-class scheme for the interpretation as well as for the analysis (see Appendix 1.2) while TREES adopted a complex scheme structured hierarchically into four levels during the interpretation phase, which was simplified to a nine-class scheme for the analysis of results (Achard *et al.* 2002). A true comparison between the two studies can only be done on the basis of the detailed data set and after proper harmonisation in respect of surveyed area and definitions. Recognising these limitations, a first tentative comparison shows the following:

- Considering the relatively small portion of land in Africa covered by the TREES survey (Table 1.7), there is a good correlation between the two estimated rates of change. Both studies estimated an annual rate of change just below −0.4%, which strengthens the contention advanced earlier that the

9 The reduced rate of deforestation in the Moist Deciduous Forest, although quite consistent, was not considered statistically significant (FAO 2003).

Table 1.6. *Humid tropical forest cover estimates for the years 1990 and 1997 and estimates of mean annual change during the 1990–1997 period*

	Latin America (million ha)	Africa (million ha)	South East Asia (million ha)	Global (million ha)
Total study area	1155	337	446	1,937
Forest cover in 1990	669 ± 57	198 ± 13	283 ± 31	1,150 ± 54
Forest cover in 1997	653 ± 56	193 ± 13	270 ± 30	1,116 ± 53
Annual deforested area	2.5 ± 1.4	0.85 ± 0.30	2.5 ± 0.8	5.8 ± 1.4
rate of change	0.38%	0.43%	0.91%	0.52%
Annual regrowth area	0.28 ± 0.22	0.14 ± 0.11	0.53 ± 0.25	1.0 ± 0.32
rate of change	0.04%	0.07%	0.19%	0.08%
Net annual cover change	−2.2 ± 1.2	−0.71 ± 0.31	−2.0 ± 0.8	−4.9 ± 1.3
rate of change	0.33%	0.36%	0.71%	0.43%
Annual degraded area	0.83 ± 0.67	0.39 ± 0.19	1.1 ± 0.44	2.3 ± 0.71
rate of change	0.13%	0.21%	0.42%	0.20%

Notes: Sample figures were extrapolated linearly to the dates June 1990 and June 1997. Average observation dates are February 1991 and May 1997 for Latin America; February 1989 and March 1996 for Africa and May 1990 and June 1997 for South East Asia. Estimated ranges are at 95% confidence.
Source: After Achard *et al.* (2002).

Table 1.7. *Comparison of estimated annual changes in forest cover according to the TREES Survey of humid tropical forests and the FAO FRA pan-tropical Remote Sensing Survey (FRA RS). TREES reference period is 1990–1997; FRA RS reference period is 1990–2000. Shaded areas indicate (tentatively) the ecological zones covered by the surveys.*

		Survey area (million ha)	1990 Forest (million ha)	Annual forest area change (million ha ± 95% confidence interval) (Annual change rate as % of 1990 forest area)		
				Rain Forest zone	Moist Deciduous Forest zone	Dry Forest zone
AFRICA	FRA RS	1224	539	−2.07 ± 0.7 (−0.38 %)		
	TREES	337	198	−0.71 ± 0.3 (−0.36 %)		
ASIA	FRA RS	610	295	−2.33 ± 1.2 (−0.79 %)		
	TREES	446	283	−2.0 ± 0.8 (−0.71 %)		
LATIN AMERICA	FRA RS	1233	822	−4.18 ± 2.1 (−0.51 %)		
	TREES	1155	669	−2.2 ± 1.2 (−0.33 %)		
TOTAL	FRA RS	3068	1656	−8.57 ± 2.5 (−0.52 %)		
	TREES	1937	1150	−4.9 ± 1.3 (−0.43 %)		

Source: After FAO (2003) and Achard *et al.* (2002).

FRA Country Statistics estimates for Africa, with an estimated overall forest loss rate of –0.79% (see Table 1.1), are too high.

For Asia, the TREES project survey area was 73% of that covered by the FRA and included most of the region's forest area. Again, the estimated rates of change are similar (−0.79% for the FRA and –0.71% for TREES) and well within each other's confidence limit. Both estimates also agree with the FRA country-based estimates relative to total forest (natural forest and plantations) area[10], estimated at –2.4 million ha/yr or –0.78% (FAO, 2001a). The correlation with FRA country estimates is much poorer for natural forest only (i.e. plantations excluded), which indicate an annual loss of 4.8 million ha or –1.5% (Table 1.1). This considerable difference could be explained by the difficulty of differentiating between plantations and natural forest on satellite images, i.e. part of the area classified as 'natural' forest by the FRA RSS and TREES surveys is actually plantation, or by a certain overestimation of the plantation area and rates from country statistics. Most probably, the discrepancy is due to a mixture of both factors.

- Because the TREES and FRA RSS survey areas in Latin America are very similar (Table 1.7) one would expect the results of the two studies to be equally similar. For this particular region there was no significant difference in the rates of change obtained by the FRA RSS survey (−0.51%), and the FRA country data, which estimated annual rates of change of –0.50% for natural forest and –0.46% for total forest. Conversely, the difference with the rate of forest change estimated by the TREES project (−0.33%), appears considerable. The annual forest change of −2.2 million ha estimated by the TREES survey for the entire Latin American region appears rather low, considering that the results of the INPE PRODES Project already indicate an average deforestation of 1.6 million hectares within the Brazilian Amazon alone[11]. A large part of the discrepancy between FAO (both FRA RSS and country statistics data) and TREES deforestation estimates is due to the fact that the TREES project did not take account of changes occurring in seasonally deciduous forests, since the hot spot areas, which formed the basis for the stratification and sample selection criteria, were not delineated over this domain (Achard *et al.*, 2002). A good part of the changes estimated by FAO and INPE relate to these seasonally deciduous forests, which are common both within (INPE, 1997) and outside the Brazilian Amazon region. They also occur in Bolivia, Colombia, Peru and Venezuela. There are also some indications that these drier forest formations are disappearing more rapidly than the wetter formations. As shown in Figure 1.11, the dominance of changes in forest cover in the moist zone in Latin America is remarkable and much higher than in the other two regions. This factor may well explain the lower percentage rate of change derived by the TREES study.

  These regional considerations also go some way to explain the relatively low pan-tropical rate of change estimated by the TREES project, both in terms of area deforested and percent rate (Table 1.7). A more detailed comparative analysis, based on the full survey details, is still needed and may prove useful for a better understanding of tropical land use changes.
- Concerning the respective sampling errors and consequent confidence intervals, it appears that both studies achieved a similar precision, relatively high for forest area estimates and demonstrably lower for forest change estimates. Standard errors of forest cover estimates are rather low in both cases, at ±2% in the case of the TREES survey and ±4% in the case of the FRA RSS. The standard errors of the pan-tropical estimates of change range around ± 15% (FRA RSS) and ± 13% (TREES). The estimates of regional change by the FRA RSS and TREES studies also show similar standard errors: respectively, ±17% and 22% for Africa; ±26% and 20% for Asia; and ±26% and 28% for Latin America. The fact that standard errors are similar, in spite of TREES focus on deforestation risk, may be partly explained by the fact that TREES total survey area is smaller than FAO RSS (37% less) and that TREES sampling intensity is also smaller (6.5% instead of 10%).

An important conclusion is that the statistical variance of forest change estimates remains rather high, even when a more sophisticated change-orientated sample selection procedure is adopted, as in the case of the TREES survey. Obviously, much still needs to be done with respect to understanding and predicting tropical land use change. Arguably, the first step must be the full exploitation of the knowledge accumulated already by these two major initiatives.

CONCLUDING REMARKS

The aim of this chapter was to describe on-going changes and trends in tropical land cover on the basis of the most reliable and objective data available. To this end, the studies used in the present overview were: the FAO FRA 2000 Country Statistics (FRA CS),

10 Best correlation between FRA Country data and FRA RSS is for the F2 definition of forest, which estimates an annual change rate of –2.0 million ha or –0.84%.

11 Average annual deforestation estimated over the period 1988–1998 based on quasi-annual monitoring of the entire Legal Amazon area. Source: INPE web site, 2002.

the FAO pan-tropical FRA Remote Sensing Surveys of 1990 and 2000 (FRA RSS 1990 and 2000) and the TREES II high resolution survey. These studies gave different and sometimes contradictory pictures of tropical land use change, because of the different methods that were applied. To clarify the perception of current land cover dynamics and their recent trends, the present chapter tried to look for convergences and complementarities among their findings. However, these studies are all very recent and, at the time of writing, the available results were still rather 'crude' or available in summary format only, especially the TREES study and the FRA RSS 2000. Although this reduced in some cases the depth of the comparative analysis, the following may be concluded:

Land cover change rates

With respect to regional rates of land cover change, there seems to be general agreement by all three references on the rate of change in tropical Asia. In the case of tropical Latin America, there is substantial agreement between the FRA CS and the FRA RSS while the results of the TREES survey gave a much lower rate due, most probably, to the exclusion of seasonally deciduous forests from the analysis. In the case of Africa, there is good agreement between the FRA RSS and TREES data, but a substantial difference with the FRA CS results. The convergence of the remote sensing studies and other considerations, including FAO's own, support the idea that, due to a lack of reliable country statistics, the FRA CS-based estimate of the rate of forest change for Africa has been significantly overestimated.

Trends

The analysis of trends over the periods 1980–1990 and 1990–2000 derived from the FRA RSS study was interesting although rather constrained by the limited statistical significance of the variations reported.

The pan-tropical trend showed a non-significant decrease in the rate of forest change and provided rather strong indications of a process of radicalisation of the dynamics, as evidenced by comparatively higher frequencies of high-gradient changes (total clearing of closed forest), specifically in high-biomass zones, and lower frequencies of lower-gradient changes (expansion of shifting cultivation areas, forest fragmentation, forest degradation). At the regional level, the trends showed a non-significant change in the rates, with Africa and Latin America presenting a slight reduction in deforestation rate and Asia a small increase.

More significant are the differences and trends observed per ecological zone. Combining the different ecological zoning applied by FRA 1990 and FRA 2000, it appeared that the hottest fronts of tropical deforestation during the 1980s belonged to a relatively narrow sub zone with 1500–2000 mm annual rainfall and a short dry season. The trend over the two decades showed a marked move of the front of deforestation towards wetter forest formations. An important implication of this wet shift is that, due to the higher biomass densities of the forest formation being cleared and degraded, a higher per-hectare carbon emission can be assumed.

The combined effect of the 'wet shift' of the deforestation front and of the radicalisation of the processes has important implications on the pan-tropical carbon budget. In fact, from this viewpoint, the increased biomass density of the forests currently under pressure and the increased frequency of high-gradient transitions may heavily offset the effect of the slight reduction in deforestation rate shown by the trend analysis at the pan-tropical level.

CONCERNING THE REGIONAL CHARACTER OF CHANGE AND THE MAIN DRIVING FORCES

A comparative analysis of regional biomass flux diagrams proved most illustrative in displaying the processes influencing land cover change during the 1980s and 1990s. In Africa, the processes observed over both periods were distinguished by phases of progressive degradation, rather than outright deforestation, caused mainly by increasing rural and urban population pressure. In Latin America, the processes were dominated by the direct conversion of the original forest to cattle ranching and permanent agriculture, which represents deforestation with the highest possible biomass loss. Asia presented the highest rates of deforestation and forest degradation among the three regions. These rapid changes were the effect of both high rural population pressure and centrally planned conversion programmes, but there are reasonable indications that the relative importance of the two components changed over the two decades. During the first decade both types of process were about equally important, resulting in significant deforestation and forest degradation, while in the second decade forest conversion to permanent agriculture became more dominant and all other changes became less frequent. The main driving forces behind tropical land cover change are the subsistence needs of rural populations, expansion of commercial agriculture and animal breeding, resettlement programmes and large-scale plantation schemes, energy needs of urban populations . . . and many others, with timber logging often playing a catalytic role through the associated increased accessibility to remote forest areas.

METHODOLOGICAL CONSIDERATIONS

To achieve reliable change and trend estimations requires rigorous and extremely consistent methods of observation, as the item to be measured – change – is in general small and elusive. The methodology developed for the FRA RSS (FAO 1996), and specifically the interdependent interpretation of remote sensing time series, responds well to the task and is well suited and cost-effective also for local, intermediate scale applications (see Appendix 1.1). The key features of this approach, such as the thematic detail and the high spatial resolution, coupled with the very reasonable costs and the historical archive contained in satellite data, are conducive to the study and description of the processes of change at district, province, or, most relevant for hydrological studies, at the river basin scale. Such level of analysis may also help to establish a bridge between the often fragmented knowledge produced by on-site hydrological research and the broader, overall picture at the river basin scale; thus helping to highlight cause-effect mechanisms and to identify priority areas for action.

WHAT NEXT?

Speculations on the direct and underlying causes of tropical forest depletion remain indeterminate as they cannot be observed by the same tools used to observe land cover types and their modifications. Deforestation is the result of the complex interaction of many local factors, which defy easy generalisations. Economic models and hypotheses on the direct and underlying causes of deforestation have been produced in great numbers but these are often based on poor and/or local data, and countermeasures based on such generalisations may easily prove ineffective, if not counterproductive (Angelsen and Kaimowitz, 1999).

There is a strong need for the collection of objective and representative cause-effect data linked directly to objectively observed land use changes. Similarly important, from the climate change and carbon budget viewpoint, is to link the observed area changes to reliable biomass densities for all land cover classes and class transitions. It is therefore recommended that consistent investigations on these two aspects be promoted at all scales, from local to global, adopting as far as possible, compatible methods which will facilitate the integration of results. Therefore, in view of their compatible methodological features, a statistical integration of the FAO and TREES remote sensing studies should be undertaken to achieve a deeper and more robust analysis of tropical land cover changes.

APPENDIX 1.1 MONITORING TROPICAL LAND COVER CHANGE: KEY METHODOLOGICAL FEATURES

The study of large-scale land cover changes is no trivial task. The driving factors behind change are complex and often highly location-specific. Consequently, land cover changes are elusive events that are difficult to predict and that defy generalisations. The net result is that land cover change shows a high statistical variance: changes are often small, compared to many other conventional mapping items, and their estimation suffers enormously from less consistent estimation procedures that are commonly accepted for other more conventional purposes such as simple land cover mapping. These factors impose the use of rigorous methodologies when designing and implementing monitoring initiatives. In view of this, it is useful to highlight a few important methodological features on the basis of the experience gained by the large-scale studies reported in this chapter.

Key methodological features for the assessment of land cover change based on high-resolution satellite data include:

1. The **reliability** of the measurement of change depends primarily on the level of **coherence** in class delineation among all elements of the time series. The **visual interdependent interpretation procedure** developed in the framework of the FRA RSS study (FAO, 1991) secured the highest level of thematic and spatial consistency among the classifications of the series of images covering the study areas. A fundamental aspect of this interpretation procedure is that the class delineation of each image of the time series implies the consultation of all images of the series. This is an iterative process that eliminates the propagation of the kind of errors that are typical of independent image interpretation. To guarantee a thorough image-to-image comparison, this procedure includes also the re-delineation of all class boundaries on all images of the series, even where there are no pre-detected changes. This apparent redundancy is important since the re-delineation of detected changes only results in a systematic underestimation of total change. Such underestimation increases with the complexity and fragmentation of the area studied (FAO, 1991).

   The visual interpretation approach was considered most appropriate for the task, since it favours a critical and consistent interpretation of time-series data in spite of the common diversity of the individual images. The distinction between a real land cover change and the effect of temporary seasonal or meteorological factors is often subtle and in this the human brain is far more efficient and flexible than any numerical algorithm. Moreover, the visual approach proved more accessible to the interpreters whose main required competence was knowledge of specific field conditions, rather than remote sensing, GIS, or digital processing capacities. This procedure has been the most important element of the FRA RSS methodology since it reduced the error associated with the estimate of changes (FAO 1996) and enabled the production of consistent and highly informative change matrices.

   A similar procedure of interpretation was adopted in the TREES study, where it was adapted to the visual on-screen interpretation of digital data, which guaranteed highly consistent results (Drigo *et al.*, 2001) and considerably simplified the digital mapping process.

2. Other essential features that allow a more consistent evaluation of change are:
   - **Simple land cover classification schemes** based on distinct physiognomic classifiers that can be detected with acceptable confidence on remote sensing images. Given that a change is more reliable when there is a sharp contrast between the original land cover class and the final one (FAO, 1996), the presence of many classes with similar biomass densities separated by only small tonal differences may generate a cloud of low-reliability transitions, thereby enhancing the 'noise' in the resulting transition matrices.
   - Time series composed of **compatible satellite data**, with similar resolution or interpretability at the scale of interpretation.
   - **Common season** of image acquisition to limit to a minimum the chromatic variations linked to plant phenology.
   - **Clear interpretation responsibility**. The study in any given location must be carried out from A to Z by a single person with good knowledge of local field conditions, land uses, common practices, etc.
3. **Spatial and temporal scale aspects**. The study of land cover changes appears to be conducted most conveniently for intermediate scale strategic planning purposes, e.g. over entire provinces or catchment areas (river basins) of a few million hectares, and over suitably long time intervals, to become cost-effective. The methodology thus appears to become optimal at intermediate scales (ranging between 1:100 000 and 1:500 000) and over time intervals above five years. At more detailed levels, i.e. 1:50 000 and above, the analysis would become far more complex and expensive, since suitable historical satellite data would not be available, leaving as the sole alternative the use of historical aerial photographs, if accessible. Similarly, over very short time intervals the size of change would be too small to be detected with acceptable reliability.
4. **Cost**. The relatively low cost of this approach, if based on satellite data, is pertinent. Current pricing policies of remote sensing data, particularly that of the Landsat Programme, and the availability of rich data archives, make the study of land cover changes relatively inexpensive.

National/sub-national applications. The monitoring methodology based on satellite time series is suitable also for national and sub-national applications where it may provide essential information for the development of local models and scenarios to support territorial resource planning initiatives. The spatial resolution of the remote sensing data used and thematic detail of this approach, i.e. land unit classification and change matrix analysis, are also suitable for local applications, for instance to study and describe the processes of change in a district, province or, most relevant for hydrological studies, at the catchment level. Concerning the survey design for local monitoring studies, complete coverage is the obvious and most convenient approach. In addition, in a local monitoring study it would be easier to relate the observed land cover changes to other territorial features such as drainage pattern, slope and soil characters, settlements and infrastructural developments such as roads or dams, as well as taking into account socio-economic variations (both in space and time). Knowledge of the processes of change occurring in a certain area, as well as their impact and trends, adds enormously to simple statistics on available resources as derived from remote sensing and so facilitates the development of more realistic models and scenarios of land use change.

Research on forest hydrology is often location-specific, and the problem is to extrapolate from the plot or small catchment scale, to what might happen over a wider region or territory (Goudie, 1999). The overall picture may remain somewhat fragmented in this way, and priority issues and/or areas are accordingly difficult to define. Medium scale studies of land use change covering the entire territory of interest, and based on reliable and objective methods, may help to establish a bridge between the fragmented knowledge and the overall picture, thus helping to highlight cause-effect mechanisms and to identify priorities for action (see also Deutsch *et al.*, this volume).

Regional and global applications. Complete coverage for global or regional monitoring studies, although desirable in principle, remains an extremely demanding task, in terms of funding, timeliness and consistency of supervision. As discussed earlier, assessing changes in land cover is a delicate task and requires a highly consistent and intense analysis that would penalise the inevitable quantity/quality trade-off of a 'wall-to-wall' approach. Moreover, 100% coverage might even be unnecessary to meet the primary objectives of regional and global studies, which are to assess and describe patterns of change and trends at the corresponding scales. This is why the FAO and TREES studies adopted statistical sampling approaches for their global assessments.

Recently, the FRA RSS sampling approach was heavily disputed (Stokstad, 2001) on the basis of simulations done by Tucker and Townshend (2000), who concluded that 'because tropical deforestation is spatially concentrated, it is very improbable that an accurate estimate of deforestation by random sampling of Landsat scenes will be achieved'. However, Czaplewski (2002) using the same data set, clearly and convincingly rebutted these assertions, stating that 'FAO (FRA RSS) followed proper statistical principles for scientific inference with sampling. This allows construction of confidence intervals and tests of hypothesis, which help assure that the conclusions by the FAO are reasonable'.

The FRA RSS and TREES surveys showed that forest change has a much higher variance than forest state, with associated standard errors being ±14–15% for forest change estimates *vs.* ±2–4% for forest cover estimates. Even the more sophisticated stratification adopted by the TREES project, which was based on expert definition of high-risk areas (hot spots), could not improve sampling efficiency significantly. In view of these results, for a more precise estimation, it is recommended that the intensity of sampling be increased, maintaining at the same time a good representivity of all tropical forests. Sample selection probability should not vary too much, in order to avoid uncontrolled weights of unpredictable events.

APPENDIX 1.2 DEFINITIONS OF FOREST AND FOREST CHANGE

Definitions of ***forest*** applied in the remote sensing component of the FAO Forest Resources Assessment (FRA RSS) 1990 and 2000:

F1 = **closed forest**. Represents forest in the strictest sense, mostly dense, not fragmented nor (heavily) degraded.

F2 = **closed + open forest + 2/3 fragmented forest**. This definition is aimed at matching the concept of forest used in FORIS (Forest Resources Information System) by FAO in its periodic assessments based on existing information.

F3 = **closed + open forest + fragmented forest + long fallow**. This definition represents forest in its broadest sense, including all types and phases of degradation (but still with the connotation of forest). This definition of forest, that allows for the most detailed differentiation among changes, has been used in the analysis of change processes.

Key to change matrix analysis. Change categories relative to the F3 definition of forest and consequent definitions of gross and net deforestation and forest degradation (FAO 1996).

Interpretation classes at date 1			*Interpretation classes at date 2*									
COVER CATEGORIES			**Natural forest**				**Non-forest**				**Man-made woody v. plantations**
			Continuous forest				**Other wooded**		**Non-wooded**		
		Cover classes	**Closed forest**	**Open forest**	**Long fallow**	**Fragmented forest**	**Short fallow**	**Shrubs**	**Other land cover**	**Water**	
Natural forest	**Continuous natural forest**	**Closed forest**	-	Deg	[3]	2/3Def	Def	Def	Def	Def	Re/Cap
		Open forest	Am	-	Deg	2/3Def	Def	Def	Def	Def	Re/Ib
		Long fallow	[4]	Am	-	2/3Def	Def	De [1]	Def	Def	Re/Ib
	Fragmented forest		2/3Af	2/3Af	2/3Af	-	1/3Def	1/3Def	1/3Def	1/3Def	2/3Af/Ib
Non-forest	**Other wooded**	**Short fallow**	Af	Af	Af	1/3Af	-	Db	Db	Db	Af/Ib
		Shrubs	Af	Af	Af	1/3Af	Ib	-	Db	Db	Af/Ib
	Non-wooded	**Other land**	Af	[2]	Af	1/3Af	Ib	Ib	-	-	Af/Ib
		Water	Af	Af	Af	1/3Af	Ib	Ib	-	-	Af/Ib
Man-made woody v. plantations			-	Deg	Deg	2/3Def/Db	Def/Db	Def/Db	Def/Db	Def/Db	-

Gross deforestation = [1]

Net deforestation = [1] **minus** [2]

Net forest degradation = [3] **minus** [4]

Change categories:

Def	=	**Deforestation** of Continuous Natural Forest (from forest classes to non-forest classes)
2/3Def	=	**Fragmentation** of Continuous Natural Forest (partial deforestation, or loss of 2/3 of the actual forest)
1/3Def	=	**Deforestation** of Fragmented Forest (the actual loss of forest is estimated at 1/3 of the total area)
Deg	=	**Degradation** (decrease of density or increase of disturbance in forest classes)
Db	=	**Decrease** of non-forest woody **biomass**
Ib	=	**Increase** of non-forest woody **biomass**
Am	=	**Amelioration** (increase of density or decrease of disturbance in forest classes)
Af	=	**Afforestation** (from non-forest classes to forest classes or forest plantation)
1/3Af	=	**Partial afforestation** (from non-forest to fragmented forest)
2/3Af	=	**Partial afforestation** (from fragmented forest to Continuous Natural Forest)
Re	=	**Reforestation** (from forest classes to forest plantation)
Cap	=	**Conversion** (from closed forest to agricultural plantation)

APPENDIX 1.3
EXAMPLES OF LOCAL LAND COVER CHANGE PROCESSES

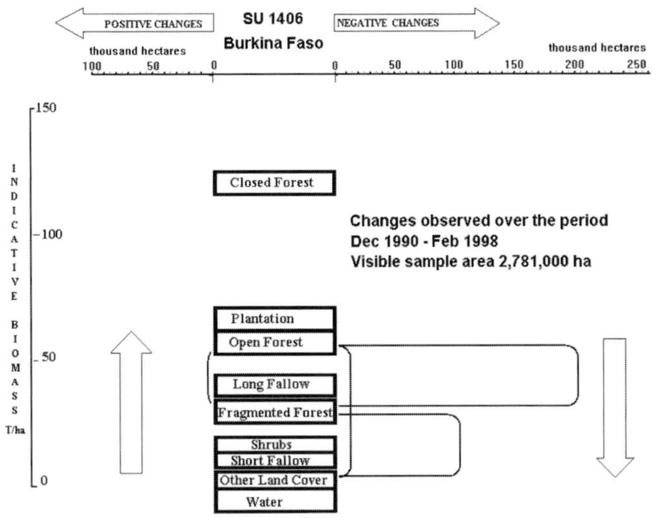

Figure 1.A1

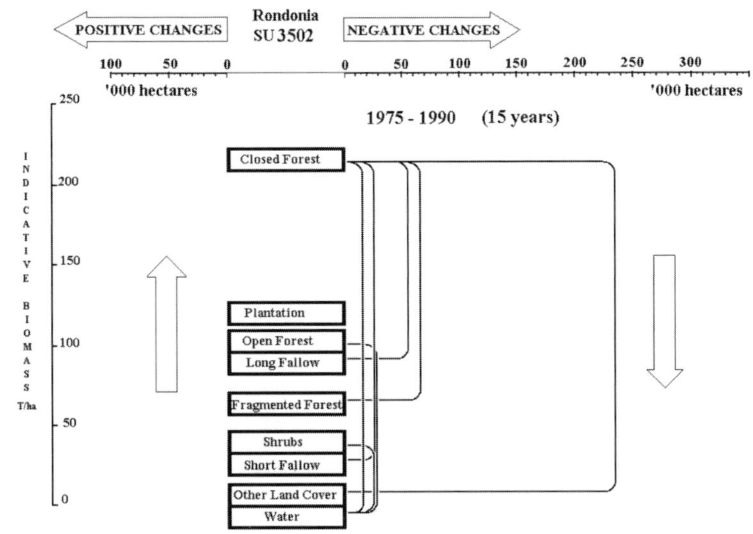

Figure 1.A2

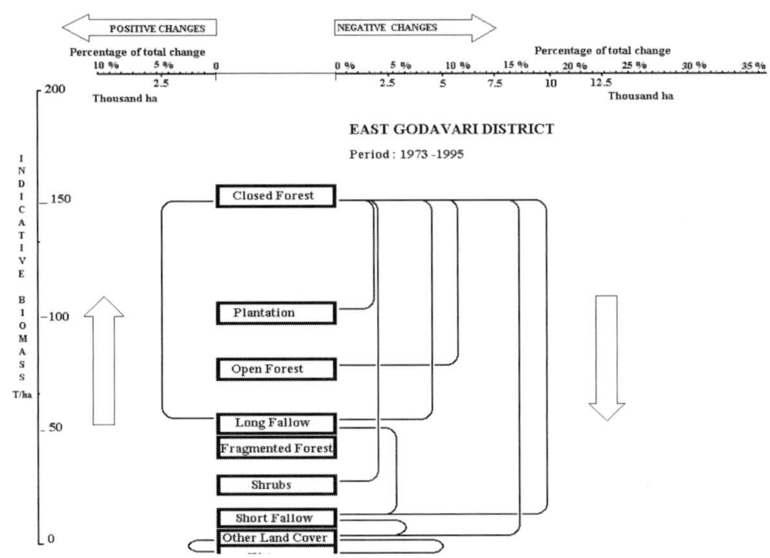

Figure 1.A3

The first diagram (Figure 1.A1) refers to an area of 2.8 million ha in western Burkina Faso, West Africa. The period covered was December 1990 – February 1998. The natural formations are mainly open forests belonging to the type *Sudanian woodland with abundant Isoberlinia* (White,1993). The change process was characterised by progressive fragmentation of the original open forest and final conversion to permanent agriculture.

The second diagram (Figure 1.A2) refers to an area of 2.8 million ha in the state of Rondonia, Brazil Amazon, along the river Guapore that borders Bolivia. The period covered is June 1975-August 1990. The change process is characterised by large-scale clearings, mainly for cattle ranching (here visible as the direct transition closed forest – other land cover), and fish-bone resettlement schemes, mainly represented by processes of fragmentation and long fallow-short fallow cultivations.

The third diagram (Figure 1.A3) refers to an area of 1 million ha in the district of East Godavari, Andhra Pradesh, India, over the period January 1973-January 1995. The process here is very complex, involving the expansion of short fallow subsistence farming and permanent agriculture on closed forest areas; various phases of forest degradation (closed to open forest and closed forest to shrubs); expansion of long fallow shifting cultivations in closed forest areas and regrowth of forest in previous long fallow areas, in a cycle that was common in the past but nowadays is very rare.

APPENDIX 1.4 PAN-TROPICAL WOODY BIOMASS FLUX DIAGRAM 1990–2000: MAIN TRANSITION TYPES AND CAUSES OF FOREST DEPLETION

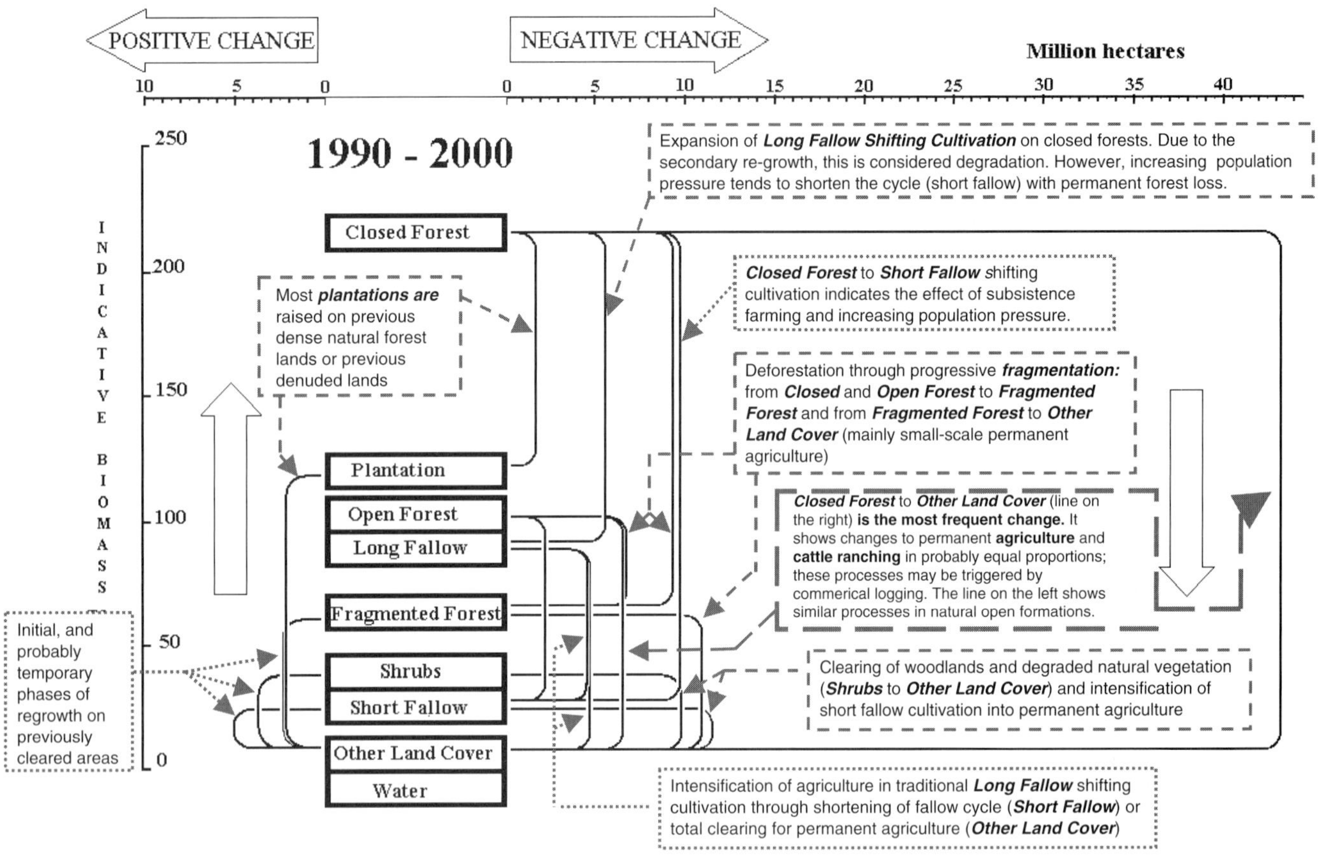

References

Achard, F., H. Eva, A. Glinni, P. Mayaoux, T. Richards, H. J. Stibig, 1998. Identification of deforestation hot spot areas in the humid tropics. TREES Publication Series B4, European Commission, Luxembourg, EUR 18079 EN.

Achard, F., H. Eva, H. J. Stibig, P. Mayaoux, J. Gallego, T. Richards, J.-P. Malingreau, 2002. Determination of Deforestation Rates of the World's Humid Tropical Forests. Science 297, 999 (2002).

Angelsen A., D. Kaimowitz, 1999. Rethinking the causes of deforestation: lessons from economic models. The World Bank Research Observer, vol.14, no.1 (February 1999), pp. 73–98.

Broadhead, J., Bahdon, J. and A. Whiteman. 2001. Woodfuel consumption modelling and results. Annex 2 in 'Past trends and future prospects for the utilization of wood for energy', Working Paper No: GFPOS/WP/05, Global Forest Products Outlook Study, FAO, Rome.

Cavelier, J. y Etter, A.1995. Deforestation of montane forests in Colombia as a result of illegal plantations of opium. In S. P. Churchill *et al.*, eds. Biodiversity and conservation of neotropical montane forests. Proceedings. Nueva York, The New York Botanical Garden, p. 541–550

Czaplewski, R. 1991. Analyses of alternative sample survey designs. FRA 1990 Project Document.

Czaplewski, R. L., 2002. Estimating Global Tropical Deforestation.

Dewees, P. A. 1989. *The Woodfuel Crisis Reconsidered: Observations on the Dynamics of Abundance and Scarcity*. World Development 17(8):1159–72.

D'Souza J. R., J. P. Malingreau, 1994. NOAA-AVHRR Studies of Vegetation Characteristics and Deforestation Mapping in the Amazon Basin. Remote Sensing Reviews, 10; pp 5 to 35.

Drigo, R. 1996. Survey of Pan-tropical Forest Resources Based on Multi-date High Resolution Satellite Data. Proceedings of the EUROSTAT Esquilino Seminar (27–29 November 1995), pp 111 to 141.

Drigo, R. 1999. Remote Sensing and Forest Monitoring in FRA 2000 and beyond. Proceedings of IUFRO Conference 'Remote Sensing and Forest Monitoring'. Rogow, Poland, 1–4 June 1999, pp 710 to 726. Document reported on http://www.fao.org/forestry/fo/fra/index.jsp

Drigo, R., A. Dell'Agnello, L. Peiser, V. Robiglio, 2001. Consistency assessment of the TREES-II high resolution exercise. Final report of JRC Contract n. AJ/08/2000. IAO, Firenze, Italy.

Drigo, R., 2001. Wood energy information in Africa. Working Document FOPW/01/4, FAO Project GCP/RAF/354/EC.

FAO, 1981a. Tropical Forest Resources Assessment project (in the framework of GEMS) – Forest resources of tropical Asia. Rome.

FAO, 1981b. Los recursos forestales de la America tropical. Rome.

FAO, 1981c. Forest resources of tropical Africa. Rome.

FAO, 1983. Fuelwood supplies in the developing countries. FAO Forestry Paper 42.

FAO, 1991. Monitoring Methodology – Procedures for the Interpretation and Compilation of High Resolution Satellite Data for Assessment of Forest Cover State and Change. Forest Resources Assessment 1990 Project Paper by R. Drigo.

FAO, 1993. Forest Resources Assessment 1990. Tropical Countries. Forestry Paper 112

FAO, 1995a. Forest Resources Assessment 1990. Global Synthesis. Forestry Paper 124

FAO, 1995b. Monitoring of Forest Resources at District Level Using Multi-date Satellite Data. Assistance to the Andhra Pradesh WB/GOI Forestry Project. Mission Report by R. Drigo.

FAO, 1995c. Forests, Fuels and the Future. Wood energy for sustainable development. Forestry Topics Report No. 5.

FAO, 1996. Survey of tropical forest cover and study of change processes. Forestry Paper 130.

FAO, 1997. State of the World's Forests 1997.

FAO, 1998. Estimation of recent deforestation rate in South America. Terminal Report by R. Drigo, Project GCP/RLA/131/EC.

FAO, 1999a. State of the World's Forests 1999.

FAO, 1999b. The role of wood energy in Africa. By Samir Amous FAO Working Paper FOPW/99/3.

FAO/UNEP 1999. Terminology for integrated resources planning and management. compiled and edited by Keya Choudhury and Louisa J. M. Jansen. Soil Resources, Management and Conservation Service. FAO Land and Water Development Division.

FAO, 2001a. Global Forest resources Assessment 2000. Main report. FAO Forestry Paper 140.

FAO, 2001b. Global Ecological Zoning for the Global Forest resources Assessment 2000. Final Report. FRA Working Paper 56.

FAO, 2001c. Comparison of forest area and forest area change estimates derived from FRA 1990 and FRA 2000. FRA Working Paper 59.

FAO, 2003. Pan-tropical survey of forest cover changes 1980–2000. FRA Working Paper 49.

Fearnside P. M. 1984. Brazil's Amazon settlement schemes: conflicting objectives and human carrying capacity. Habitat International 8. pp 45 to 61.

Foley, G. 1987. *Exaggerating the Sahelian woodfuel problem*? *Ambio* 16(6): 367–371.

Geist H., E. Lambin 2001. What drives tropical deforestation? A meta-analysis of proximate and underlying causes of deforestation based on subnational case study evidence. LUCC International Project Office.

Goudie A. S. 1999. The scientific significance of landuse and land-cover changes. University of Oxford (Development Office), UK. LUCC web site www.uni-bonn.de/ihdp/lucc/.

Grainger A. 1993. Rates of deforestation in the humid tropics: estimates and measurements. Geographical Journal 159. pp 33 to 44.

Hamilton, L. S., and Bruijnzeel, L. A. (1997). Mountain watersheds: integrating water, soils, gravity, vegetation, and people. Pp. 337–370 in Messerli, B., & Ives, J. D. (editors), *Mountains of the World. A Global Priority*. Parthenon Publishers, London.

INPE (Instituto Nacional de Pesquisa Espaciais, Brazil), 1997, Deforestation 1995–1997 Amazonia, INPE & IBAMA, Brazil.

INPE, 2000. Monitoring of the Brazilian Amazonian forest by satellite 1998–1999. INPE PRODES web site http://www.dpi.inpe.br:1910/col/dpi.inpe.br/banon/2000/09.12.17.24/doc/amz1998_1999/index_amz.htm

Jeanjean H., F. Achard, 1997. A new approach for tropical forest area monitoring using multiple resolution data. International Journal of Remote Sensing. 18. pp 2455 to 2461.

Kartha, S. and G. Leach, 2001. Using modern bioenergy to reduce rural poverty. Report to the Shell Foundation. London.

Lambin E. F., A. H. Strahler, 1994. Change-vector analysis: a tool to detect and categorize land-cover change processes using high temporal-resolution satellite data. Remote Sensing of Environment 48. pp 231 to 244.

Lambin E., 1994. Modelling Deforestation Processes – a Review. TREES Publication Series B1, EUR 15744, Loxembourg, European Commission.

Lambin, E. F., D. Ehrlich, 1997. The identification of tropical deforestation fronts at broad spatial scale. International Journal of Remote Sensing 18. pp. 3551 to 3568.

Lambin, E. F.; H. J. Geist,. 2001. Global land-use and land-cover change: what have we learned so far? Global Change Newsletter 46:27–20. www.igbp.kva.se.

Leach, G. and R. Mearns. 1988. *Beyond the Woodfuel Crisis: People, Land and Trees in Africa*. Earthscan Publications, London.

Marcoux, A., R. Drigo, 1999. Population Dynamics and the Assessment of Land Use Changes and Deforestation. Population Programme Service, FAO 1999. Document partly reported on http://www.fao.org/sd/WPdirect/WPan0030.htm

Nepstad *et al*., 1999. Large-scale impoverishment of Amazonian forests by logging and fire. Nature 398: 505–508 (R).

Rieley, J. O., 2001. Kalimantan's peatland disaster. Reported at http://www.insideindonesia.org/edit65/jack.htm

Rovainen, E. 1994. Estimates of tropical forest cover, deforestation and change matrices. Swedish University of Agricultural Sciences (SUAS), Sweden.

Scotti, R. 1990. Estimating and projecting forest area at global and local level: a step forward. FRA 1990 Project.

Stokstad, E. U.N. Report suggests slowed forest losses. (A Review of :) Science 291: 2294

Tucker, C. J., J. R. G. Townsend 2000. Strategies for monitoring tropical deforestation using satellite data. International Journal of Remote Sensing. Vol 21 pp. 1461–1471.

WRI, WCMC, WWF 1997. The Last Frontier Forests: Ecosystems and Economies on the Edge. By D. Bryant, D. Nielsen, L. Tangley.

2 The myth of efficiency through market economics: a biophysical analysis of tropical economies, especially with respect to energy, forests and water

C. A. S. Hall
State University of New York, Syracuse, USA
J.-Y. Ko
Louisiana State University, Baton Rouge, USA

INTRODUCTION

Tropical countries, in general, are changing much more rapidly than temperate ones. This is true with respect to population numbers, deforestation, economic growth (both positive and, occasionally, negative), influence of trade and, in general, various other aspects of globalisation (World Bank, 1998). At the same time, most tropical countries remain especially vulnerable to both natural and man-made disasters (Hurricane Mitch in Central America and the 1998 Asian economic 'meltdown' serve as ready examples). Within this context of uncertainty, 'sustainability' remains an obvious and highly desired goal for many, as is obvious in the promotional tourist literature of many tropical countries, such as Costa Rica. Similarly, one hears from various quarters the desirability of improving 'efficiency' and also the concept that with high levels of development, environmental improvements are not only possible but likely (e.g. the environmental Kuznets curve; see Rothman and de Bruyn, 1998). Often these are seen as important rationales for far-reaching programmes, such as the structural adjustment programmes implemented in many tropical countries by the World Bank and the International Monetary Fund (Lélé, 1991; Taylor, 1993) and even for large-scale conservation programmes (Goodland *et al.*, 1990).

The concept of sustainability

What would constitute this sustainability, if indeed it were able to be achieved? In fact, it turns out to be remarkably hard to characterise sustainability explicitly, despite thousands of references in the literature (i.e. a search for 'sustainable development' on the Amazon.com website turned up 1850 references!) Sustainable development is an extremely attractive concept and since it is rarely defined explicitly, it allows nearly anyone to define it in a way to have whatever one's cake is desired and, often, to eat it too.

There are at least nine basic definitions of sustainability (OTA, 1994). Most can be categorised as examples of three basic perspectives on sustainability, each of which are advocated by particular groups (Goodland and Daly, 1996). These are:

- economic sustainability, important to many of those focused on the material welfare of various groups,
- social or cultural sustainability, used especially by anthropologists and some others, and generally in reference to sustainability of cultures, and
- environmental sustainability, generally favoured by those concerned about resource depletion, deforestation, loss of biodiversity or the impacts of pollution.

The curious thing about these three concepts of sustainability is that they are often at variance with one another, that is, each one can be obtained only with at least some expense to one or both of the others. In addition, advocates of one perspective tend to be uninterested in, or oblivious to, the perspectives of others.

The concept of efficiency

Another important concept in these deliberations is that of efficiency. Efficiency – like motherhood – is golden, in that everyone is in favour of it. But the meaning of efficiency, like sustainability, is different in the minds of different beholders, and these differences have large implications for what it is we might be attempting to achieve and at what expense to other possible objectives. Most generally, efficiency is output over input, each of which can be defined in various ways.

Forests, Water and People in the Humid Tropics, ed. M. Bonell and L. A. Bruijnzeel. Published by Cambridge University Press.

There are three basic definitions of efficiency that are pertinent to our present interests:

- engineering efficiency, which often means energy output over energy input, such as mechanical work or electrical power out over steam or coal in for a steam engine
- neoclassical economic efficiency, which generally is perceived as low price per unit for a particular good or service (i.e. implying relatively small monetary inputs and hence efficient use of all resources used to generate that good or service) or more generally as a disparaging term used for the perceived lack of efficiency in some economic system or activity
- biophysical economic efficiency, somewhat of a hybrid, generally perceived as economic output over energy or material input.

The first and the third are normally given formally and mathematically, the second less commonly so, but it certainly could be represented as dollar cost of product over dollar cost of all inputs.

There is a widespread belief that efficiency has improved over time, especially recently, and that much greater improvements are possible in the near future if we put our minds and policies to that (e.g. Ausubel, 1996). In fact it is not hard to find examples of this occurring. For example, by some accounts the biophysical economic efficiency of the US economy has increased by some 24% since 1973 (Schipper and Howarth, 1990; Wernick *et al.*, 1996). An alternative perspective is based on the concepts of Howard Odum (e.g. Odum, 1971 and many others) that economic activity and its increment is nearly universally associated with energy use and with increases in energy use.

In this chapter we examine the biophysical economic efficiency, i.e. resources required per unit economic production, for a series of tropical countries to see if there is any indication of increase in efficiency by which we turn raw materials into economic wealth. Our focus is energy, forest products and water, because we believe these to be amongst the most fundamental and those that are likely to impact on water use and deterioration most directly. We use time series where possible and comparisons across nations. More formally, our hypothesis is: 'over time, technological progress has operated to decrease the resource intensity of economic activity'. We will accept this hypothesis if the ratio of inflation-corrected gross domestic product (GDP) per resource input, in tons or other physical units, increases over time for the nations examined. We will reject this hypothesis if there is a clear correlation (with a conservative r^2 of 0.5 or higher) of resource use and economic production, both across nations and for individual nations over time, or if resource use increases while economic production stagnates.

We have attempted to make this analysis as simple and straightforward as possible. Nevertheless, we must first acknowledge (although not necessarily deal with) some potential problems with such a simple analysis. The issues include boundaries (e.g. how far back do you follow energy inputs – do you include the energy required to, for example, support labour, etc?), convertibility (should you express purchasing power in terms of local buying power, where prices may be low, or as ability to purchase goods in dollars on the international market?) and comprehensiveness (e.g. should you include energy flows of nature, etc.?).

Some critics of international corporations argue that the way economic 'efficiency', i.e. low prices, is achieved is often by means that are injurious, socially or environmentally, such as low wages or not taking care of residual pollutants. Likewise there can be a problem with deriving biophysical economic efficiency since, for example, a car manufactured and sold in the United States is apt to have been built with steel (the particularly energy-demanding component) made in Brazil, South Korea or elsewhere. Since there is more value added from finishing the product than from generating the raw materials, this would make the United States look relatively energy-efficient and South Korea relatively energy-intensive (Ko, 2000). In reality, all analyses are to some degree incomplete. For example, should we include the pro-rated energy cost of manufacturing, or indeed developing, the steam engine or computer chip? How about the energy cost of obtaining the energy to run the engine? Probably the best way to deal with these issues is to acknowledge that there is no one best set of boundaries to work within, and thus to define the boundaries carefully and perhaps undertake analyses from several perspectives. But that is impossible on this scale of research, and so we simply assume that national boundaries and annual values (the criteria by which most of the data are maintained) are sufficient for our purposes.

METHODS

We chose four countries from Africa, Asia and Latin America to represent overall characteristics of demography, economic development and natural resource stocks of the tropical countries (Table 2.1). There are large differences in economic level among the nations. The World Bank (1997) has given estimates for 'natural capital' for all countries, which includes pasture land, crop land, timber resources, non-timber forest resources, protected areas and subsoil assets. The natural capital of Malaysia and Venezuela are high, due mostly to large petroleum reserves in these countries. Kenya, India and the Philippines have relatively low natural endowments. African countries have the highest recorded population growth rates for the last three decades. Asian countries are the most crowded regions among the three continents. Population growth in most of these countries has been higher than the world average for the 1990s. Thus we believe that we have chosen a suite of contrasting nations to examine our efficiency hypothesis. However, most of the selected countries have

Table 2.1. *A comparison of the 12 countries in relation to their economic output, demographic conditions and natural capital assets*

Countries	GDP per capita, 1995 (1987 US$)	Natural capital per capita (US$)	Population density, 1970 (per 1000 ha)	Population density, 1995 (per 1000 ha)	Population growth, 1990–5 (Percent)
Africa					
Kenya	369	1 730	20	47	2.91
Nigeria	354	N/a	60	121	3.00
Senegal	674	5 300	21	42	2.52
Zambia	286	5 490	6	11	2.24
Asia					
India	425	3 910	169	283	1.76
Malaysia	3 108	11 820	33	61	2.37
Philippines	637	2 730	125	226	2.20
Thailand	1 843	7 600	70	114	0.94
Latin America					
Brazil	2 054	7 060	11	19	1.44
Colombia	1 416	6 100	19	31	1.88
Costa Rica	1 885	7 860	34	67	2.41
Venezuela	2 627	20 820	12	24	2.27
World			27	42	1.48

Sources: World Bank (1997), World Resources Institute (1999).

had to implement neoliberal policies (e.g. structural adjustment programmes) to some degree in order to obtain aid from development banks, therefore we can empirically test the success of these policies in the generation of an efficient economy.

The main source of the time series data for our study was the World Resources Institute's (WRI) 1998–99 data set, which compiles environmental and resource information published by several international agencies, including the United Nations and the World Bank. Additionally, we downloaded agricultural data from the FAO Internet website.

The most comprehensive analysis of water use by various countries that we are aware of is found in Gleick (1998). He has reviewed the available data on water input (through rain, inflowing rivers and groundwater pumpage) for most of the nations of the world. Gleick also gives summaries of extraction of water for agriculture, industry, domestic purposes and total (Gleick, 1998). A similar analysis, with similar conclusions about the world distribution of per capita water and water shortages, was undertaken by Vorosmarty and Sahagian (2000). Unfortunately Gleick's data are generally for one year only, normally roughly 1990. Vorosmarty and Sahagian's projections, based on climate change and growth in human populations, are that whatever one's conclusions might be for 2000, the problems are likely to be much more severe into the future due, mostly, to human population growth. Thus we can assume that our conclusions based on Gleick's data are in some ways conservative (but see discussion).

Economic analysis

We used time series of GDP as an index of economic output for all countries. Gross domestic product figures presented in this study were corrected for inflation and different currencies by using the constant 1987 US$-based GDP, and we used total commercial energy consumption (in heat units, i.e. uncorrected for quality, so that more economically potent electricity gets the same rating per heat unit as coal) for the same countries. Quality corrections would probably make for stronger correlations, so in a sense this analysis is conservative.

Using these two data sets, we traced the pattern of per capita energy consumption vs. per capita GDP for the countries from 1970 to 1995. Finally, we calculated the energy efficiency of national economies by dividing constant dollar GDP by commercial energy use.

Agricultural analysis

We measured the fertiliser efficiency of cereal production as an index of agricultural efficiency. This was calculated by dividing the total cereal production by the nitrogen–phosphorus–potassium (NPK) fertiliser used on all cereals (barley, corn, rice, sorghum, wheat). This was not easy as no year-by-year numbers are kept on fertiliser use on specific crops. To estimate the fertiliser input to cereals, we used the following procedures:

(A) We calculated the ratio of fertiliser for 'total cereals' to 'total crops' for the only year available from *Fertilizer Use by Crop* (FAO, 1999), which was derived from questionnaires sent to national governments who returned the data (for example, the fertiliser consumption ratio of total cereals over total crops for the 'index' year for each country was 41.7% for Kenya in 1991, 9.35% for Malaysia in 1995, and 32.0% for Brazil in 1991).
(B) The total fertiliser used for all crops collectively, the yield of all cereals and the area harvested for both all cereals and all crops are available for each year from the agricultural statistics of FAO through the Internet website of FAO (http://www.fao.org).
(C) From A and B, we estimated fertiliser used for cereal production for each year by multiplying the total annual national fertiliser use by the ratio of fertiliser used on all cereals to total national fertiliser used for the index year, based on the assumption that the index year's ratio is applicable to the entire research period.
(D) From B and C, we estimated fertiliser input (kg ha^{-1}) for cereals by dividing the estimated fertiliser use for all cereals by the area harvested for all cereals.
(E) From B and D, we estimated fertiliser efficiency, a ratio of cereal production over fertiliser input, for each country during the research period.

Water analysis

The problem for our analysis is that Gleick's (1998) data are for various individual years, and there are no year-by-year data. Thus we were able to examine only the correlation of water use and economic activity (GDP) of different nations, where GDP was corrected for inflation and to the year of the availability of water data. We compared the sum of the quantities used in the categories 'domestic' and 'industrial' to values for inflation-corrected GDP for the same year.

Forest analysis

We were able to derive deforestation rates and extraction or exports of forest products for only some of the countries that we have considered. Our sources included United Nations Environmental Programme (UNEP, 2000) and Verissimo *et al.* (1997).

RESULTS

Contrary to our hypothesis, our results show no general pattern of increased efficiency (i.e. decreased resource use intensity per unit economic output) over time in any way and, more frequently, demonstrated linear increases in resource use with increased economic activity or, less frequently, a decreased efficiency, especially as intensity of use increases.

Energy use and economic activity

In general, there is a continuing pattern of increased energy use over time, similar to the increase in GDP (Figures 2.1 and 2.2), with the general exception that neither GDP nor commercial energy consumption has increased much in Africa during the period examined, and there was a significant drop in energy use in Zambia. Both economic activity and commercial energy consumption has increased, especially in Asian countries, as shown most markedly in Malaysia and Thailand (Figures 2.1 and 2.2).

Per capita energy consumption has decreased or increased only slightly for most countries, with the exception of Malaysia, Thailand and Venezuela where per capita energy use has increased significantly (Figure 2.3). Thus most of the increases in energy use are due simply to expanding populations.

There were in general very high correlations (often with r^2 from 0.8 to 0.99) for economic activity and energy use for each nation over time, although the correlations were not as strong in Africa. The correlations were even higher, and often reaching virtually 1.0, when economic activity was regressed against both human population levels and energy use (Table 2.2.)

In most countries, increases in energy use are matched fairly closely by population growth so there is little, if any, increase in energy use per capita (Figure 2.3). The same pattern is true for economic growth, i.e. growth in the economy is roughly the same as population growth so there has been little change in per capita inflation-corrected GDP (Figure 2.4). Per capita GDP has declined in Zambia, increased slightly in Kenya and has been steady in the other two African nations (Figure 2.4). The four Asian nations show a clear increase in both energy use and economic activity, with the increases strongest for Malaysia (Figure 2.4). Brazil, Colombia and Costa Rica also had relatively small increases in both energy use and economic activity, while Venezuela used more energy while economic activity decreased (Figure 2.4).

In general, there was a strong correlation between increases (or decreases) in economic activity and energy use, so that energy efficiency (GDP/energy) of national economies in many countries show little change over time. There are exceptions, and energy efficiency decreased especially for Nigeria and all Asian countries. Increases in efficiency occurred for Zambia (Figure 2.5). In sum, we did not see any consistent pattern of increasing energy efficiencies for 11 of the 12 countries. Zambia's increasing energy efficiency, the exception, was accompanied by economic depression.

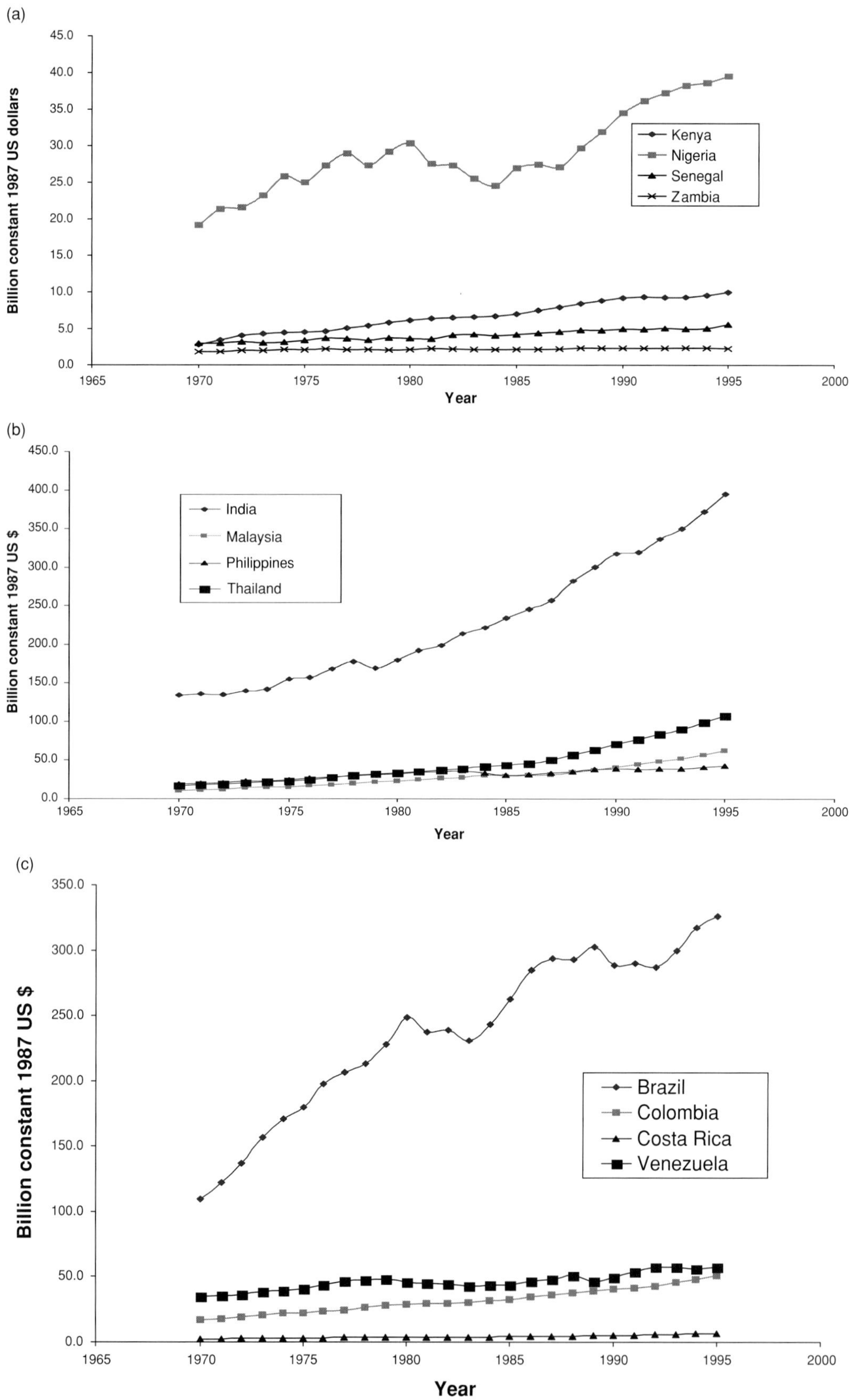

Figure 2.1 Gross domestic product, 1970–95. (a) Africa; (b) Asia; (c) Latin America.

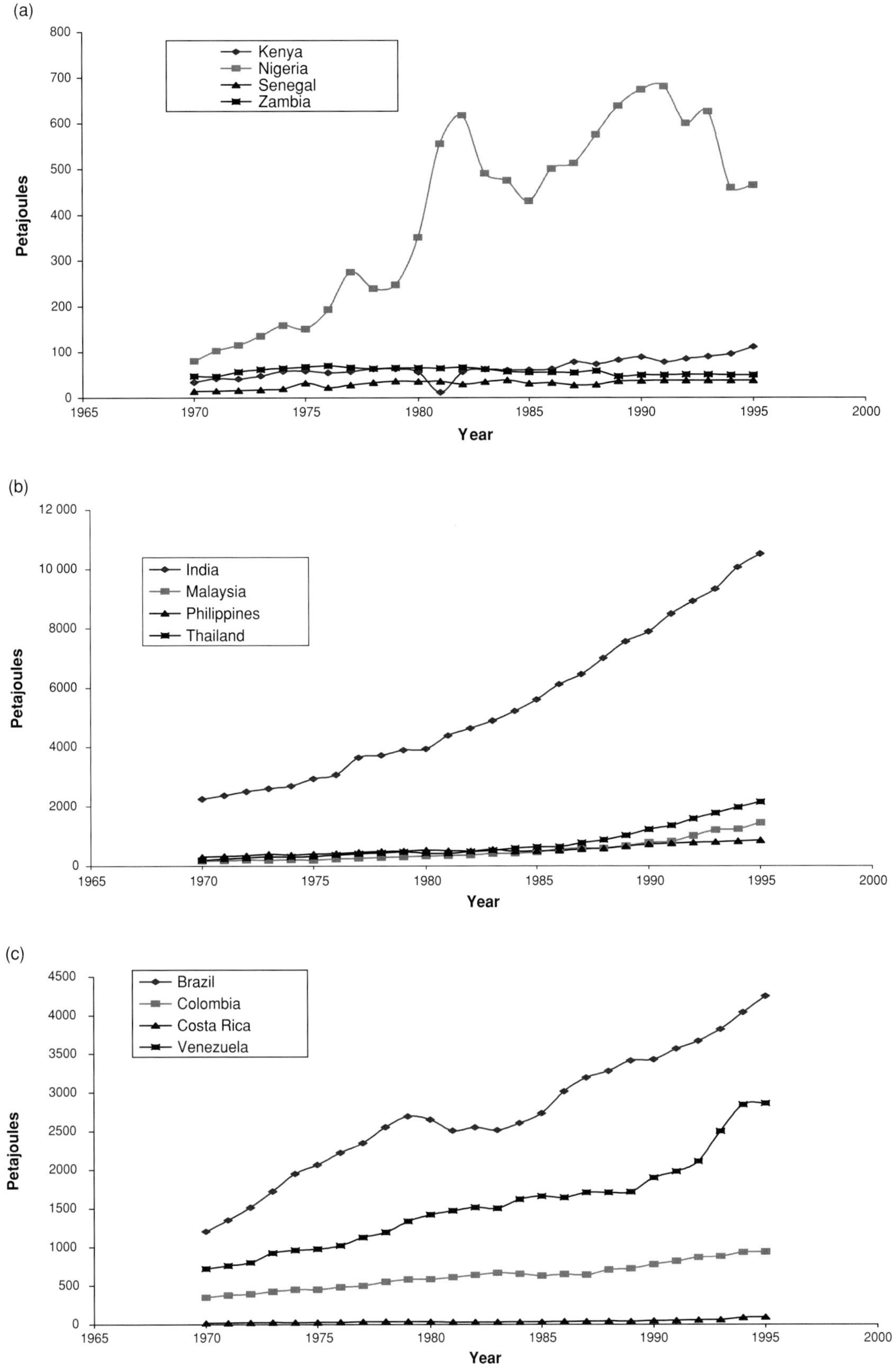

Figure 2.2 Annual energy use, 1970–95. (a) Africa; (b) Asia; (c) Latin America.

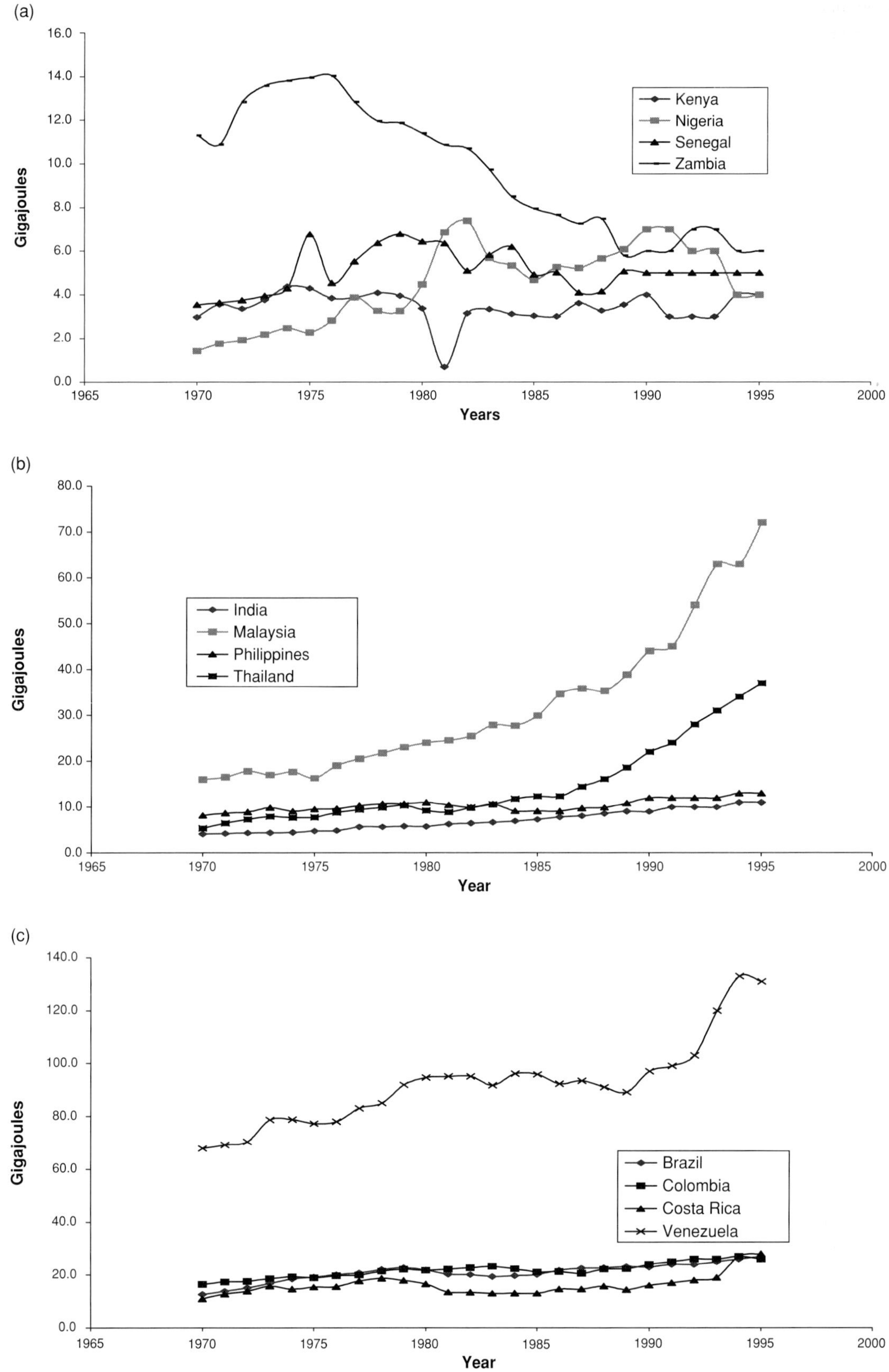

Figure 2.3 Commercial energy consumption per capita, 1970–95.
(a) Africa; (b) Asia; (c) Latin America.

Table 2.2. *Determinants of economic growth in the countries, 1970–95*

Country	Equation	Adjusted r^2	Durbin-Watson	Significance
A. The proposed equation: GDP = a*[ENGCONS] + b*[POP] + c				
Africa				
Kenya	GDP = −0.004 ENGCONS + 1.701 POP − 3352	0.9793	0.472	0.0001
Nigeria	GDP = −0.006 ENGCONS + 0.347 POP + 3754	0.8023	0.457	0.0001
Senegal	GDP = −0.007 ENGCONS + 0.621 POP + 569	0.9631	1.664	0.0001
Zambia	GDP = 0.007 ENGCONS + 0.130 POP + 1015	0.8214	1.194	0.0001
Asia				
India	GDP = 0.034 ENGCONS − 0.054 POP + 83934	0.9964	2.015	0.0001
Malaysia	GDP = 0.021 ENGCONS + 2.462 POP − 18785	0.9912	0.784	0.0001
Philippines	GDP = 0.015 ENGCONS + 0.458 POP − 475	0.8945	0.236	0.0001
Thailand	GDP = 0.037 ENGCONS + 0.941 POP − 26567	0.9971	0.753	0.0001
Latin America				
Brazil	GDP = 0.041 ENGCONS + 1.409 POP − 53837	0.9454	0.320	0.0001
Colombia	GDP = 0.015 ENGCONS + 1.567 POP − 22054	0.9913	0.522	0.0001
Costa Rica	GDP = 0.020 ENGCONS + 1.489 POP − 407	0.9778	0.834	0.0001
Venezuela	GDP = 0.005 ENGCONS + 1.005 POP + 22684	0.8268	0.607	0.0001
B. (Excluding population) The proposed equation: GDP = a*[ENGCONS] + b				
Africa				
Kenya	GDP = −0.082 ENGCONS + 1394	0.6391	0.876	0.0001
Nigeria	GDP = 0.019 ENGCONS + 21504	0.4361	0.247	0.0001
Senegal	GDP = −0.070 ENGCONS + 1977	0.4978	0.649	0.0001
Zambia	GDP = −0.003 ENGCONS + 2367	−0.022	0.301	0.5031
Asia				
India	GDP = 0.032 ENGCONS + 57313	0.9964	1.852	0.0001
Malaysia	GDP = 0.040 ENGCONS + 8573	0.9651	0.475	0.0001
Philippines	GDP = 0.040 ENGCONS + 9510	0.8681	0.212	0.0001
Thailand	GDP = 0.047 ENGCONS + 11284	0.9798	0.187	0.0001
Latin America				
Brazil	GDP = 0.074 ENGCONS + 35846	0.9368	0.244	0.0001
Colombia	GDP = 0.055 ENGCONS − 3014	0.9728	0.643	0.0001
Costa Rica	GDP = 0.056 ENGCONS + 1823	0.8290	0.469	0.0001
Venezuela	GDP = 79.53 ENGCONS− 2130022	0.8121	0.585	0.0001

Note: GDP, gross domestic product in constant 1987 US$ × 1 000 000; ENGCONS, total commercial energy consumption in terajoules; POP, total population × 1000.

Agricultural efficiency

The intensity of fertiliser use (i.e. kg ha^{-1}) increased in most countries throughout the study period except for Nigeria, Senegal and Venezuela, where there was a decline after roughly the 1970s or 1980s (Figure 2.6). Overall, the cereal output for each country over time did not increase as rapidly as fertiliser input, almost certainly in response to yield saturations. There was no evidence at all for any increase in the efficiency with which fertiliser was turned into food, and, at very high levels of application (e.g. Zambia in some years), strong evidence for a decrease in efficiency (Figures 2.7 and 2.8). This is probably also a general function of increasing expansion of land in agriculture over time, which tends to mean that land of increasingly poor quality is brought into production, generally lowering the average quality of land in production (e.g. Hall and Hall, 1993; Hall *et al.*, 1998).

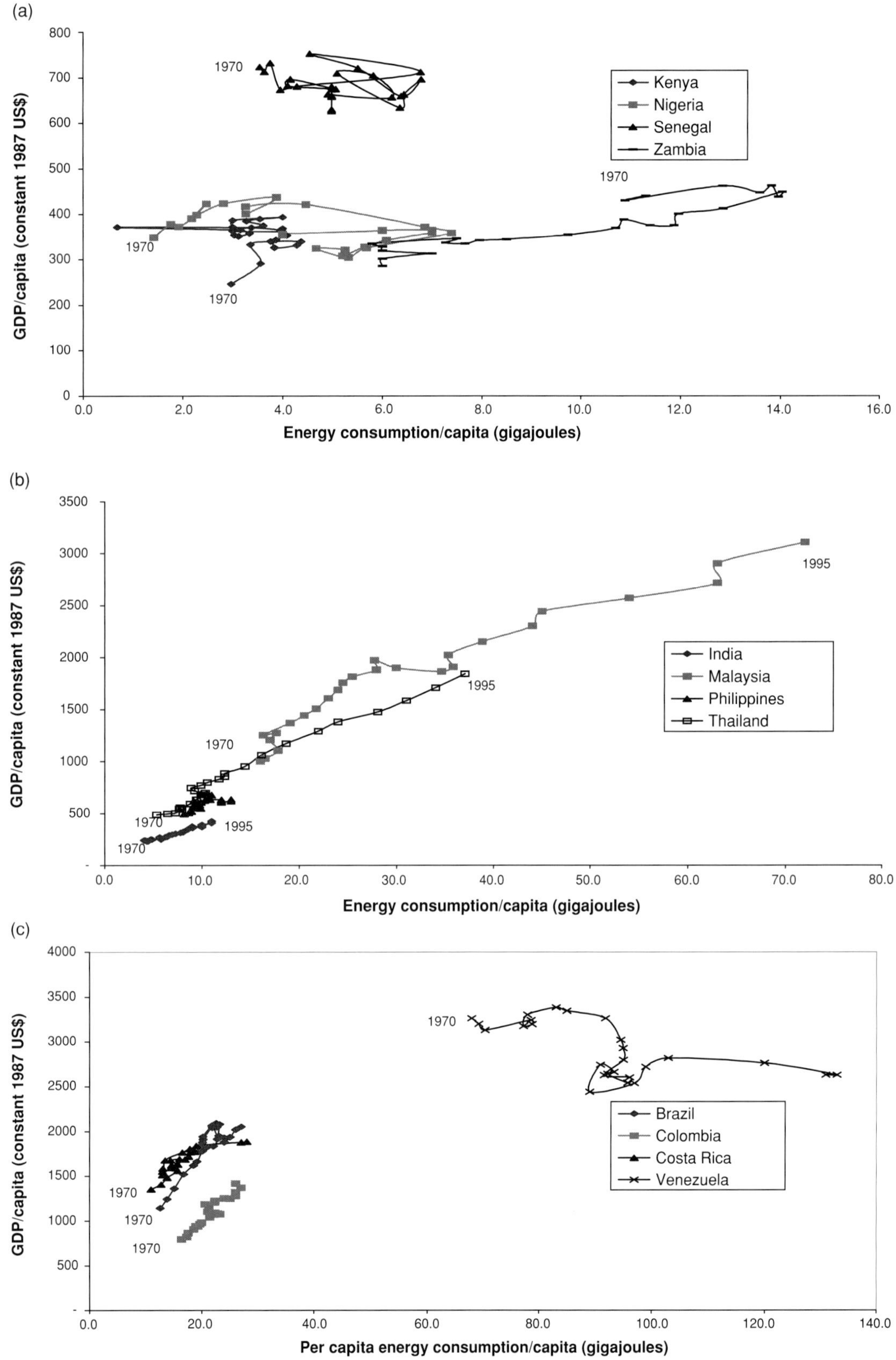

Figure 2.4 Per capita energy consumption vs. per capita GDP, 1970–95. (a) Africa; (b) Asia; (c) Latin America.

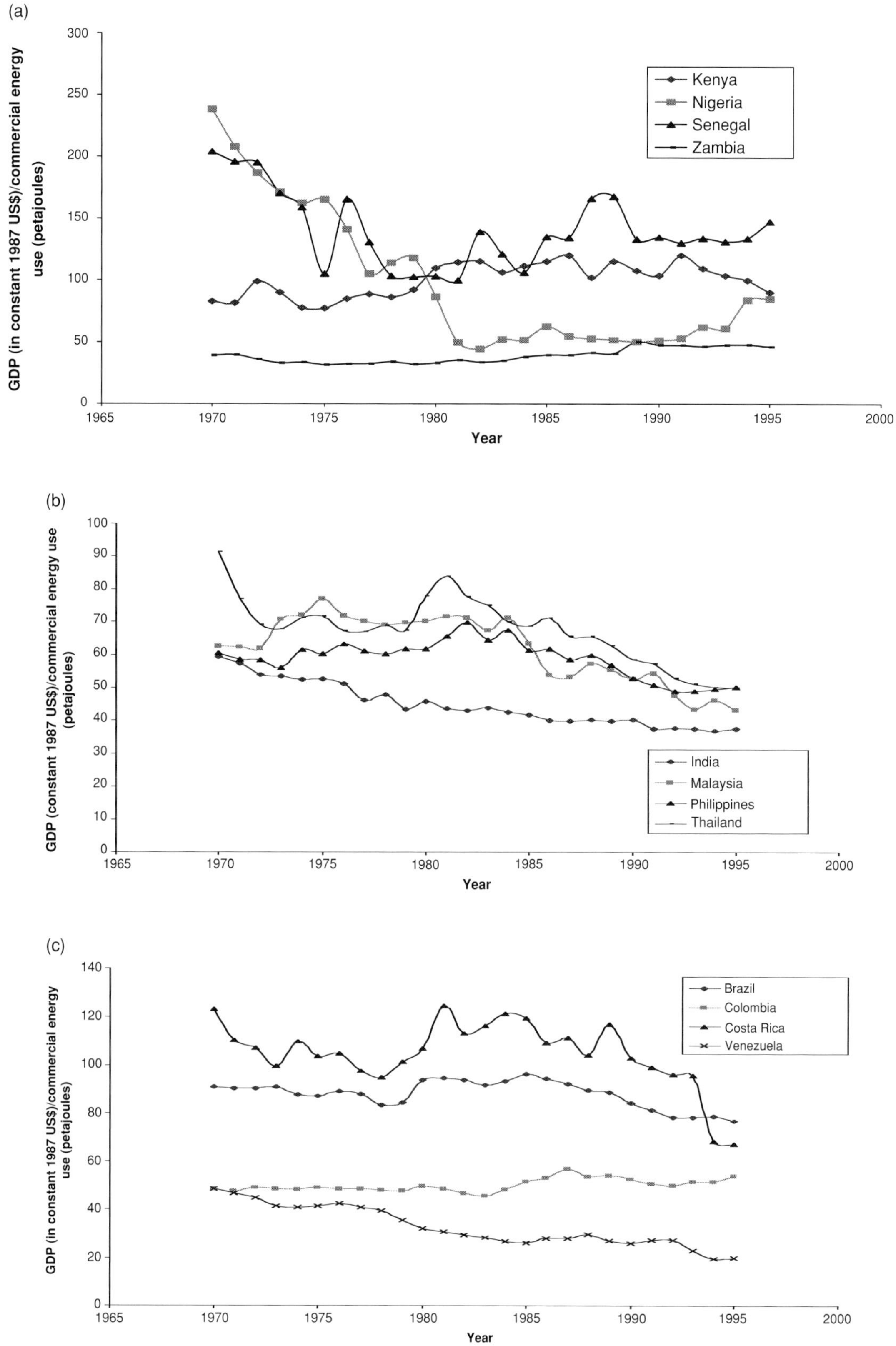

Figure 2.5 Energy efficiency of national economy, 1970–95. (a) Africa; (b) Asia; (c) Latin America.

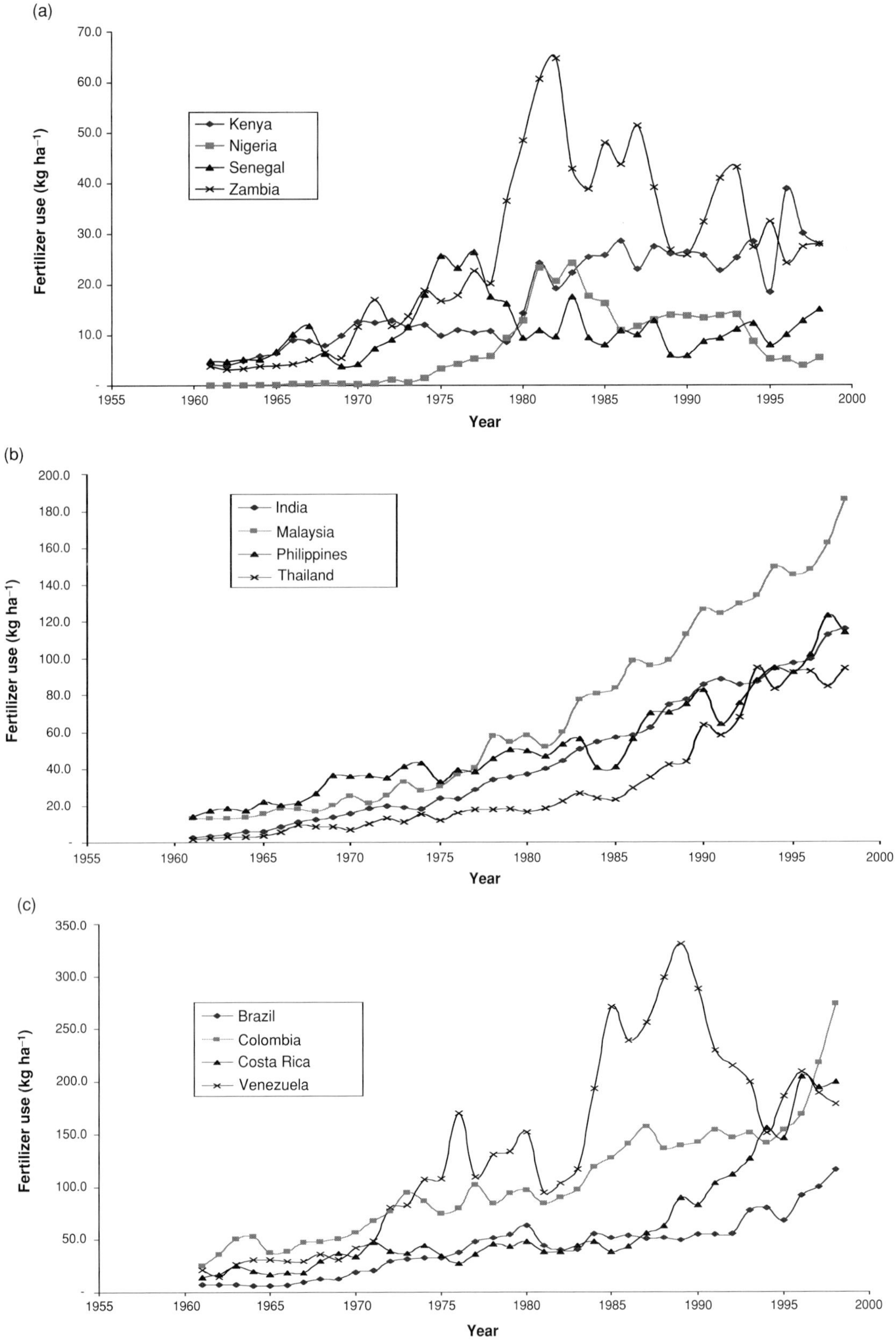

Figure 2.6 Ratio of total fertilizer use to cultivated area for total cereals, 1961–98. (a) Africa; (b) Asia; (c) Latin America.

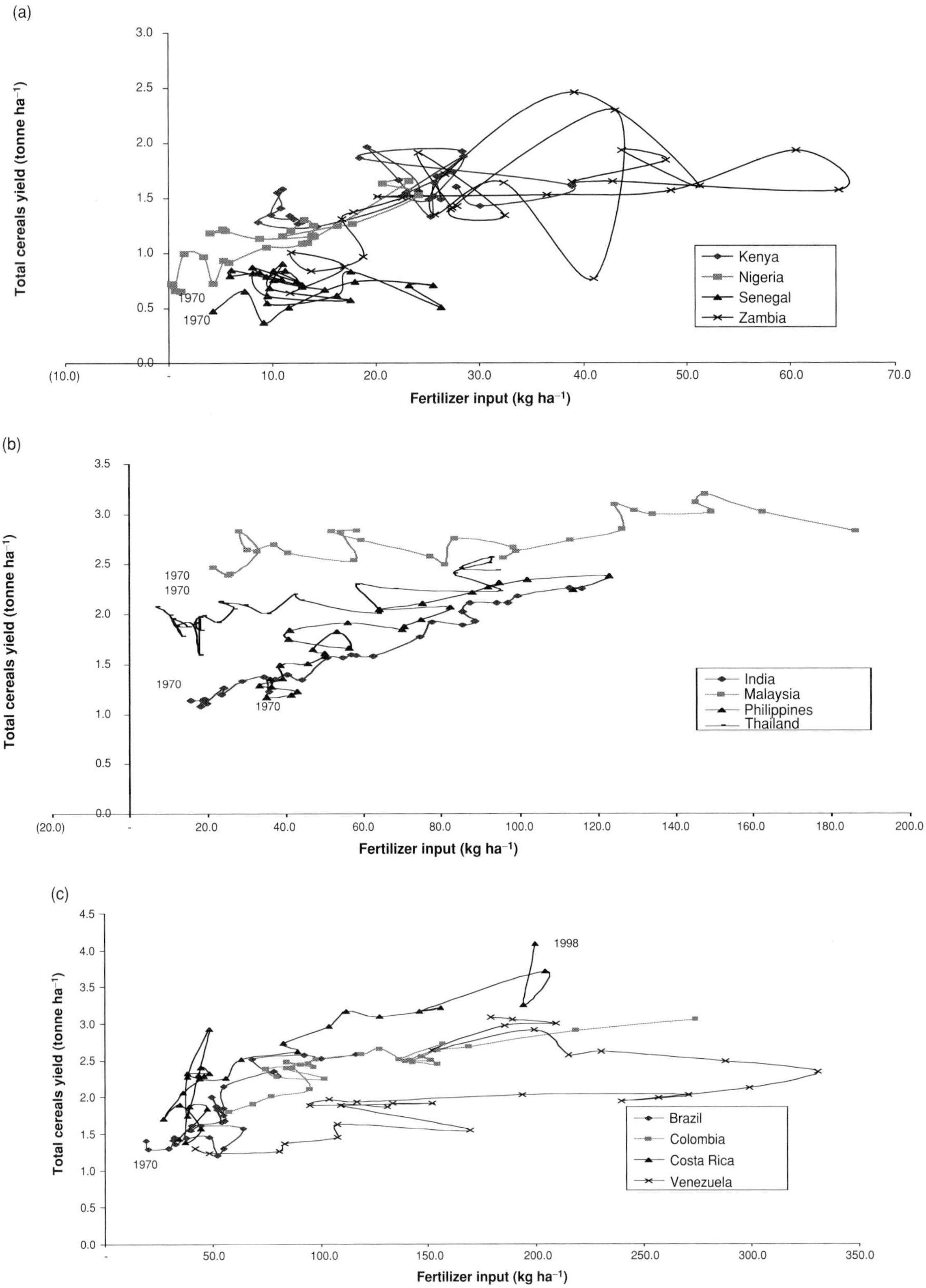

Figure 2.7 Fertilizer input vs. cereal yield, 1970–98. (a) Africa; (b) Asia; (c) Latin America.

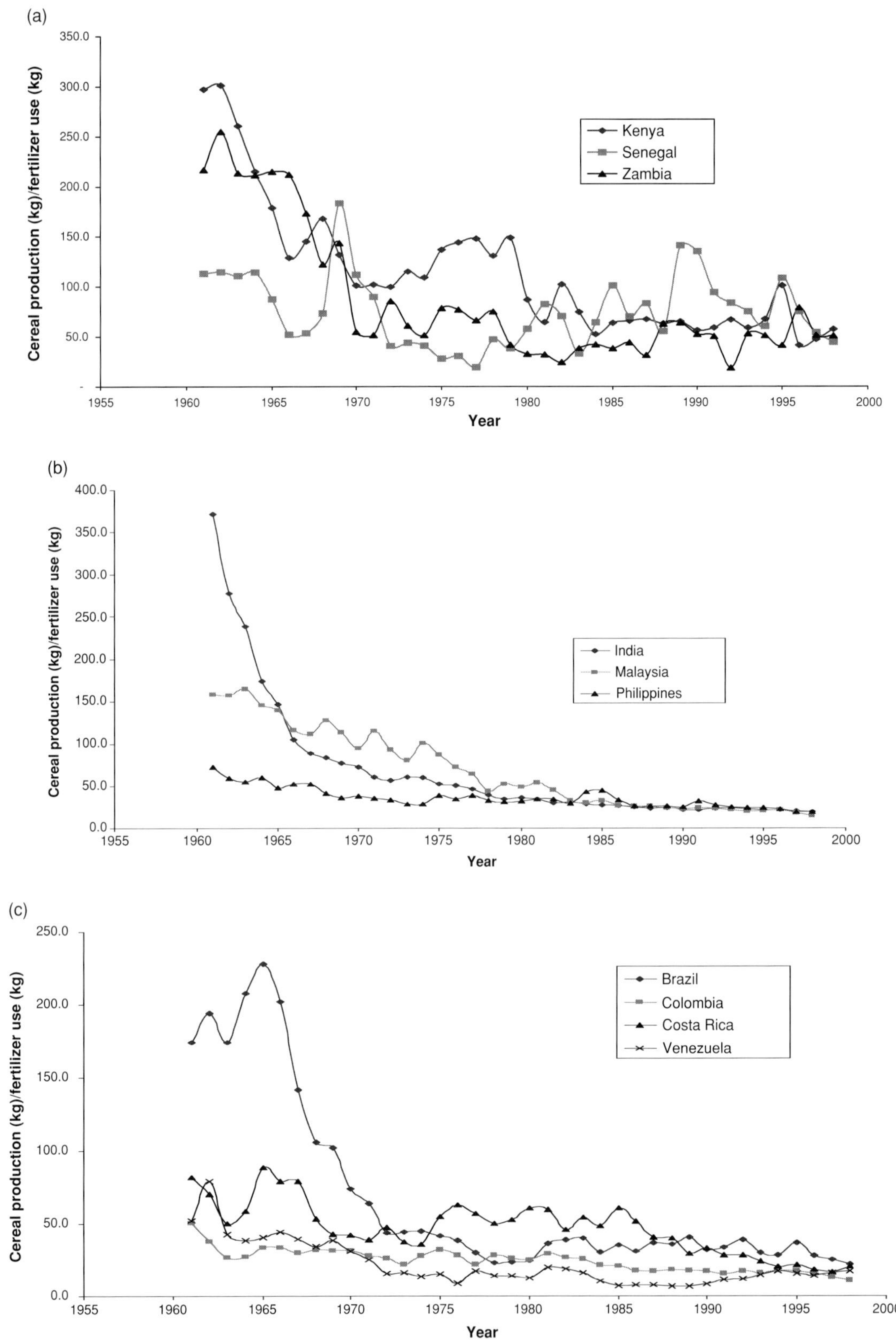

Figure 2.8 Fertilizer efficiency for cereal production, 1961–98.
(a) Africa; (b) Asia; (c) Latin America.

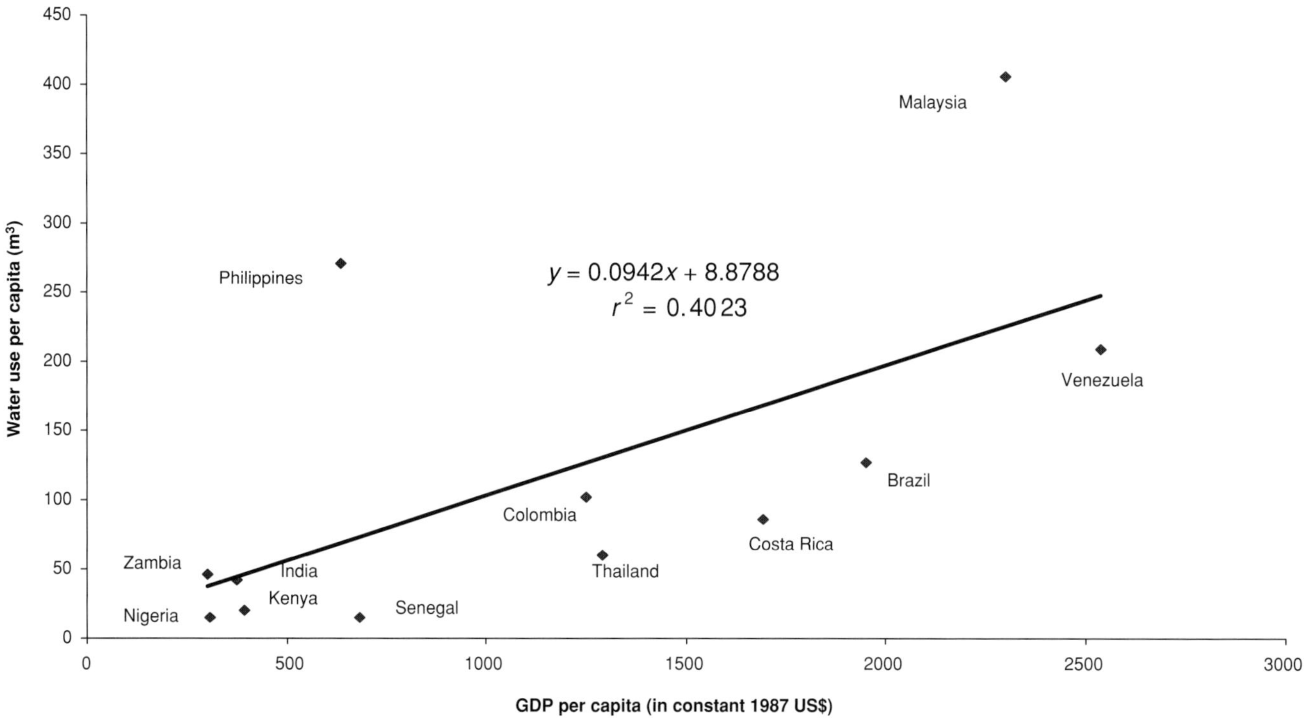

Figure 2.9 GDP per capita vs. water demand for domestic and industrial use per capita (Gleick, 1998). Data are for 1990, except Nigeria (1987), Senegal (1987) and Zambia (1994).

Water analysis

There is a positive linear relation between water use and economic growth among the 12 nations across the three continents of the Tropics (Figure 2.9). Poor countries and dry areas use less water, while relatively wealthy countries use more water. The Philippines and Malaysia, both relatively wet countries, stand out for using more water at a given level of GDP.

Deforestation and use of forest products

There is a positive but relatively weak correlation between deforestation and economic growth. In addition, there tends to be some relation between economic growth and the export of wood (in this case roundwood) (Figure 2.10). However, this weak relationship is influenced very much by the remaining area of forest land left to be exploited (e.g. Brazil) and by policy (e.g. Kenya and Malaysia) although these policies are of course influenced by the depletion of available timber (Figure 2.11).

DISCUSSION AND CONCLUSIONS

In Africa, as elsewhere, there is no increase in per capita wealth without an increase in per capita energy use, without implying which is the chicken and which is the egg. Nevertheless increasing energy use does not guarantee increasing wealth, as is clear from the case of Zambia. Thus increasing energy use appears to be a necessary but not sufficient component of increasing wealth. Another apparent condition is that energy use must increase more rapidly than population growth or else, as in Senegal or the Philippines, the population growth swallows any increase in economic activity and per capita wealth falls. These findings indicate that energy availability and population policy are far more important than fiscal or monetary policy for enhancing a nation's material wealth. But social/governmental stability or effectiveness is also required.

In India and the Philippines per capita economic activity barely increased despite enormous increases in energy use because of equally large population growth. In Malaysia, by contrast, energy use increased much more rapidly than population and an increase in per capita GDP occurred. This is probably related to Malaysia having become a leading nation in energy production despite its small population of 22 million. In Thailand a similar pattern occurred although not quite as strong. In Latin America most countries had small increases in per capita wealth and per capita energy use.

In sum, populations have increased steadily in most countries while periodically significant economic challenges, including oil shocks, debt and resultant International Monetary Fund

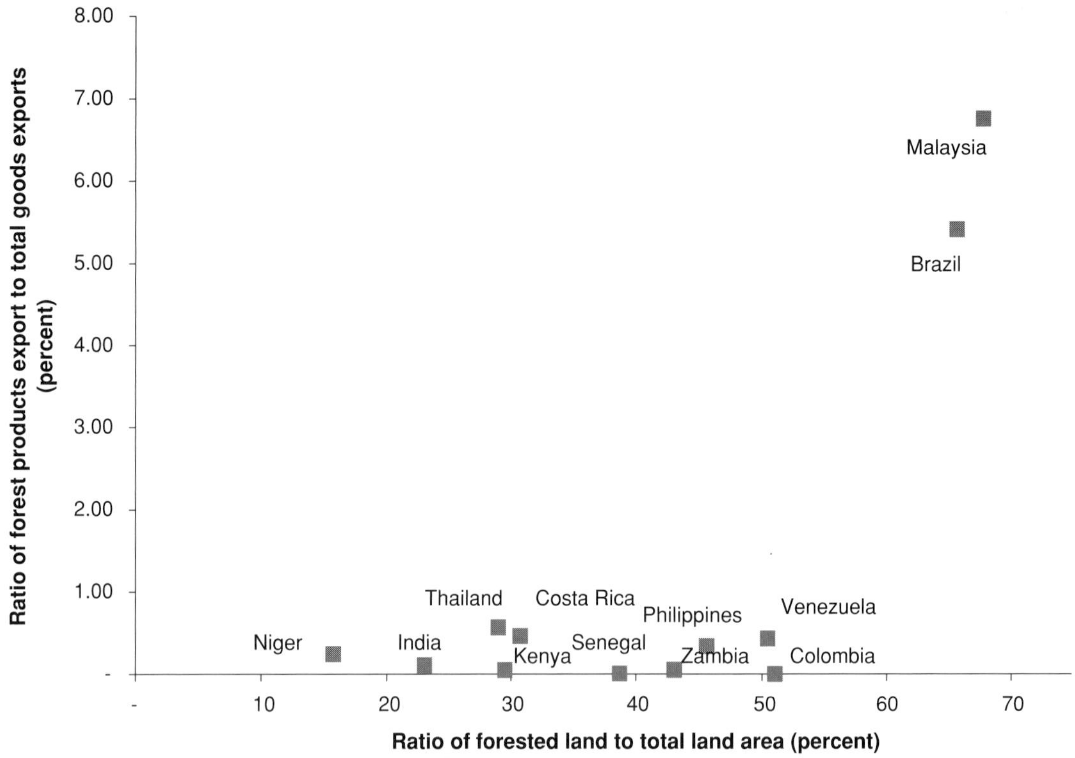

Figure 2.10 Ratio of forested land vs. ratio of forest products exports for 1994.

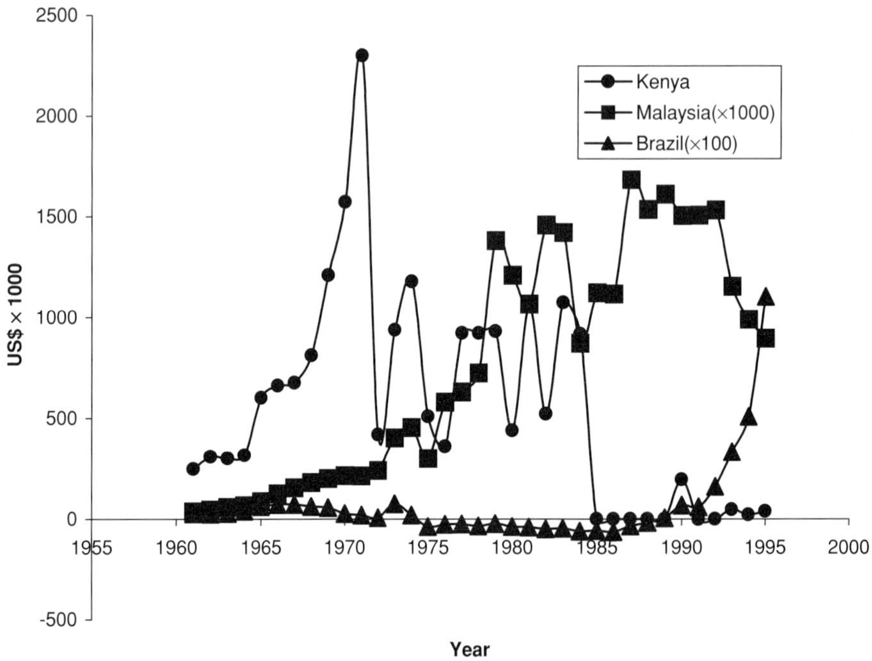

Figure 2.11 Net export of roundwood, 1961–95.

(IMF) impacts, have decreased economic activities and sometimes decreased energy consumption. Any reduced energy or fertiliser consumption appears to be driven not by technological developments leading to efficiency, as is commonly believed, but rather by economic constriction. Thus we found no clear pattern of decreasing energy, water or fertiliser use per unit of economic activity for any of the countries we have examined. The simplest summary is this: resource use expands at about the rate of economic growth, and different countries use resources in rough proportion to their economic activity. Hence our results do not lend any particular support to the idea that technology or some other factor is allowing economies to expand without impact on resource consumption or the environment, at least with respect to these resources that we have examined. If anything, these data show the contrary, that resource-use efficiency in many countries has *decreased*. They also show that if populations grow more rapidly than resources can be mobilised, there are no examples of per capita wealth increasing, at least within these countries examined. We conclude that an economy is largely a biophysical phenomenon, and economic theories or policies that do not take this into account are doomed to failure and will continue to exacerbate the dismal record of most of the developing world (see Hall (2000), Hall *et al.* and (2001) and LeClere and Hall (in press) for further analysis).

Implications for water management in the humid tropics

If there is to be development, and unless there is an enormous change from past patterns, more energy will be used, and more water will be used, both in general and specifically for energy projects. This argues, in agreement with Vorosmarty and Sahagian (2000), that the future of water availability and quality will depend more on 'global change' in human population and affluence than in, for example, possible climate change, although the latter may exacerbate or ameliorate the former. This phenomenon cannot be examined well without time series analysis of economic activity and water use for individual countries. Gleick is now undertaking such an analysis, with preliminary results showing two different patterns: many countries continue to use water in proportion to their economic activity while in others economic growth is continuing without a significant increase in water use.

The forest situation in Kenya, Malaysia and Brazil

Most of the forests in Africa have been reduced by increasing demands for wood products including roundwood. The consumption of forest products nearly doubled for the period 1970–94, but domestic production has been maintained only through increased logging with virtually no effective measures for sustainable forestry management (UNEP, 2000). The uncontrolled logging industry in Kenya was able to export roundwood while meeting domestic needs during the 1970s. However, it did so by exhausting its forest resources so that the country had become a net importer of roundwood by the 1990s. Unfortunately, the pressures on African forests are likely to continue due to the increasing populations in urban zones who need construction timber and roundwood.

Much of the primary forests in Asian nations have been depleted through serious deforestation, and many Asian nations have been trying to reduce deforestation by employing forest plantations (UNEP, 2000) or by restricting logging of the remaining natural forests, as has been done in Malaysia, for example.

In Brazil, 15 million ha of forest area disappeared in the period from 1988 to 1997 due to clearing for cropland and stock farming, construction of roads and other infrastructure development (see Drigo; Serrao and Thompson, both this volume). The deforestation increased between 1994 and 1995, with 2.9 million new hectares affected. The logging industry continues to expand without proper governmental planning or regulations (Verissimo *et al.*, 1997); roundwood exports from Brazil may well collapse in the near future as happened in Kenya.

The failure of development based on neoclassical economics to provide useful guidelines

The most important policy implication of this analysis (and see also Ko *et al.*, 1998; Hall, 2000; Tharakian *et al.*, 2001) is that we found no empirical justification at all for the commonly heard statement that neoclassical or market economics will lead to efficiency, at least if efficiency is defined as we do: the quantity of material, energy, etc. required to undertake a unit of economic production (see also Rothman and de Bruyn, 1998). Although there are examples of where engineering efforts have in fact increased the efficiency with which fuel is turned into mechanical work or even wealth, this does not necessarily lead to resource efficiency in the sense of using less, as the lower price may encourage increased use of the resource (this is known as Jevons' paradox after Jevons, 1865). Rather the contrary, neoclassical market economics lead to low prices (which are supposed to result in efficiency) through pushing prices to the floor, partly as a consequence of few buyers and many sellers (a situation analogous to Marx's arguments that the capitalist system pushes down the price – but not the value – of labour).

In many cases this is encouraged by the intervention of the wealthier nations. For example, the United States government encouraged simultaneously through external aid and the operation of the IMF the development of massive and highly industrialised banana production systems in both Costa Rica and Ecuador (Hernandez *et al.*, 2000). The net effect was large-scale deforestation, massive use of pesticides, a tremendous overproduction of bananas and, eventually, a sharp decline in their price, which has had a devastating effect on both economies and has led to the abandonment of many banana-growing fields and the devastation of entire communities. Similarly, coffee prices in 2001 were the lowest they had been in many decades, even when not corrected for inflation, devastating the revenues of tropical farmers and governments. In America, one can buy coffee for $3 or more per cup in upmarket coffee boutiques of which 5 or 10 cents relates to the coffee and of that, 1 or 2 cents goes to the grower. Somewhere in this chain there is a mockery of the free market deriving value, generating efficiency and solving economic problems.

The disconnect between economics and the resources upon which economics is based reaches a higher level in the premises of neoclassical economics itself. Hall *et al.* (2001) examine the fundamental model and premises of neoclassical economics and find they cannot be considered valid from a scientific viewpoint from at least three perspectives: the boundaries used for analysis are incorrect, the driving variables are incorrect and the fundamental theorems are presented as givens rather than as testable hypotheses. Indeed when they are tested, they are shown empirically to be incorrect at least as often as correct (reviewed in Hall, 1991).

This disconnect, in principle a theoretical matter, begins to have practical consequences when applied to issues of development. Montanye (1998) and Kroeger and Montanye (2000) have examined this issue in some detail for Costa Rica. For most of its existence Costa Rica was self sufficient in food while generating a relatively small amount of foreign exchange through the process of exporting high quality coffee. The Costa Rican government, which supported large and very popular programmes in health care and education (and having no army), borrowed only small amounts of money abroad from commercial banks.

In the 1970s several things happened to Costa Rica due in large part to increases in the international price of petroleum. First of all the price of petroleum and petroleum-derived products such as fertilisers, upon which Costa Rica had become heavily dependant, increased dramatically. But such inputs had become essential. This was due in part to the population growth that made adequate food production impossible without the use of inputs from the industrial countries and in part due to the advice and influence of US aid. The cost of growing crops increased dramatically relative to the price of agricultural exports. Costa Rica suddenly found itself much poorer but still quite dependant upon purchased inputs. At the same time, though, borrowing money became very cheap because of the large amount of 'petrodollars' (derived from oil-producing countries unable to spend all of their revenues) available in international banks. The interest rates, originally 2% per year, were so low that it made no sense either not to borrow or to repay the loans. But the rates suddenly became 10% per year and Costa Rica could not pay the interest let alone the principal. Costa Rica defaulted on its interest payments and thereafter could not get loans from normal commercial banks.

The Costa Ricans had to turn to the IMF. The IMF loaned Costa Rica new monies but as a requirement forced the government of Costa Rica to institute a series of 'reforms' (called 'structural adjustments') based on neoclassical economics. These changes reduced government expenditures for e.g. health and schooling, reduced tariffs and encouraged the growth of exports, including beef, bananas and 'non-traditional' crops such as cut flowers and macadamia nuts. Montanye (1998) and Kroeger and Montanye (2000) have examined whether or not the structural adjustment programmes met their own objectives and found that, in general, they did not, and that in addition they caused many destructive side effects in the general society and the environment. For example, massive deforestation was undertaken in a basically futile attempt to increase exported beef to make enough money to make interest payments. The elimination of trade barriers and the ending of Costa Rican government subsidies to their farmers required by the structural adjustment resulted in the virtual elimination of once-common maize farmers who were unable to compete with the cheap corn and wheat grown in the United States. But, as detailed in Kroeger and Montanye (2000), one of the reasons that the US corn is cheap is that all agriculture in the United States is heavily subsidied by the same United States government that is forcing Costa Rica to remove its own agricultural subsidies. These are but a few of the many inconsistencies and unproven economic policies thrust upon the Costa Rican people in the name of some ideologically derived aspect of neoclassical economics. What is not understood, however, is whether any other economic system could have done better. (The authors think that a biophysical approach would have allowed the recognition and avoidance of at least some the problems, and would have undermined the often inappropriate ideological basis for policy in the past.)

Neoclassical economics as an excuse for plunder

Finally there is a moral dimension by which (in our opinion) neoclassical economics often violates general standards of human decency. When the first author read Plutarch in college he was astonished to see who the men were that Plutarch thought great and hence should serve as an inspiration for others. The overwhelming preponderance of his great Greeks and Romans were military leaders who brought 'great glory, fame and riches' to their cities by the sack and desecration of other cities! The irony of this seems to have escaped Plutarch, but it would appear that great attention and glory was being paid to those who were basically thugs and common criminals, although perhaps not by the standards of their times. This perspective seems to have continued throughout the Middle Ages, as even the most casual trip through Europe will testify. Everywhere there are fortifications, and the local history is mostly about the jockeying for power and booty. The essence of much of history is told by the words of Coner Larkin, the main character in Leon Uris' novel *Trinity* who, when captured by the English, described to an English judge the devastation the British soldiers had laid upon Ireland. His summary was 'You're a bunch of damned hypocrites holding yourselves up to the world as the successors of the ancient democracies . . . All you're really in it for is the money.'

It is a mark of advances in civilisation that we no longer send in the troops to capture other people's wealth, steal slaves and so on. Cheap energy as well as moral outrage has made slaves unnecessary and inconvenient, so that the average American has the energy equivalent of 60 to 80 slaves to hew wood and haul water. Another question, however, is whether we are simply continuing that process without soldiers through the kind of international trade and economics that we espouse. These approaches influence enormously the lives of billions of people yet are not generally subject to government control, arbitration or even to much academic discussion, as governments and other institutions everywhere knuckle under to the enormous power of major international buyers. In other words, have military invasions, soldiers and colonialism been made unnecessary because we can now get the resources and energy that we need, or want, for an affluent life in the Western world more easily by 'free trade' than by conquest? Specifically, the concept that markets are 'good' because they allow the buyer to choose from competing suppliers, each of which is offering goods and services at the lowest possible price is, as partially argued above, certainly good for the buyer but not necessarily so for the supplier. It may also be that the treatment of other parts of the world as simply as supply depots for raw materials or as markets for manufactured products as is encouraged through market economics, dehumanises interactions with other cultures and may help lay the groundwork for events such as the terrorist attacks in New York City that took place while this text was being revised.

What might be an alternative is not so easily defined and we will be the first to admit that free trade and markets also bring benefits. Certainly there seems to be little reason or argument for returning to a planned economy or any form of Marxism, at least that has been tried to date. We have suggested the need to generate a 'biophysical economics' without giving any particular formulation for how economic decisions should be made (e.g. Hall, 2000; Hall *et al.*, 2001). Market mechanisms most probably remain good means for economics but are very poor devices for specifying the ends, as is the case in current practice. But it seems that given the continual depletion and destruction of the resource base and depletion of forests, soils, petroleum, fish, clean water, clean air and other basic resources, a completely new approach as to how we undertake economics is a first priority. And this new economics needs to have at its heart a discussion of population issues, the relation of total national wealth vs. per capita wealth, and the relation between biophysical resources and realistic economic possibilities.

Past neoclassical approaches to economic development are repudiated because they are based essentially on enhancing the dependency of those countries receiving aid on the developed western world and also the suppliers of oil. The failure of this approach was made clear inadvertently when the US closed its aid station in Costa Rica. The United States ambassador, J. Brian Atwood, celebrated 50 years of US aid to Costa Rica by saying that 50 years of foreign aid had helped to generate a peaceful, democratic, prosperous country. He also said that the \$1.7 billion of foreign aid to this small country (a mean of \$34 million a year) had been very good for the United States since Costa Rica had purchased more than \$2 billion dollars of imports from the United States per year during the mid 1990s, for example. In other words, it seems to us that the US aid was a very good investment, for by bringing 'modern' agriculture to Costa Rica, which has no fossil fuels, iron or heavy industrial base, it made it essential that Costa Rica purchase fuels, agrochemicals, tractors and their parts and so on from the developed world, including especially the United States. Costa Rica has 'paid for' these imports principally through tropical crops and through debt. The debt was increased terribly following the price increase of petroleum in the 1970s, and Costa Rica has not been able to retire that debt since then, although the principal has been paid back many times through extended interest payments.

The problems are getting worse again in the new millennium. Oil prices increased again for a while, the new hopes from a huge new INTEL factory and tourism are being dashed by the downturn in high-tech sales and stocks and the many cancellations of airplane flights and tourist hotel rooms. Meanwhile, the European countries have decided to purchase bananas only from their former colonies, devastating the Cost Rican banana industry. Coffee prices are the lowest in decades. Costa Rica is now left with even more people more dependent upon a fossil fuel-based world that is receding before them and an agricultural system once able to feed and employ the overwhelming majority of Costa Ricans but no longer able to do so. A reasonable question is whether the situation would be worse if there had been no development at all (probably) or better if we had had a different perspective on development that was based not on the neoclassical emphasis on growth and trade but rather on assessing and understanding the biophysical possibilities and limitations of the country's economy. Preferably, one that was tied to public discourse and policy discussion on, for example, desirable population levels.

Finally, the authors think that it is time to put to rest the term 'sustainable development', for it is not only an oxymoron but is also a serious impediment to resolving such of the world's resource problems as we can. Painting an activity or a nation green does little to avoid the essential issue that real economic development is costly in terms of resources. The use of the term 'sustainable development' mostly just makes people feel good when what they should be doing is undertaking hard and serious biophysical analysis. Science as an entity has contributed to the problems of today in that it has increased human survival without addressing the implications of that, or, in general, undertaking comprehensive, systems-based biophysical analysis of real economies (human-dominated ecosystems). At the same time the analysis of such systems has been left to social scientists or others without a proper understanding of thermodynamics and other aspects of biophysical reality. The net effect is that all of the wonderful advances brought about by science have mostly been lost on the developing world, the majority of which simply keeps getting poorer and poorer without even the benefits it once had of clean water, healthy air or some useable land.

References

Ausubel, J. H. (1996). Can technology spare the Earth? *American Scientist* **84**: 166–78.

Food and Agriculture Organization of the United Nations (FAO). (1999). *Fertilizer Use by Crop*, 4th edn. Rome, Italy: FAO.

Gleick, P. H. (1998). *The World's Water 1998–1999: The Biennial Report on Freshwater Resources*. Washingon, DC: Island Press.

Goodland, R. and H. Daly (1996). Environmental sustainability: universal and non-negotiable. *Ecological Applications* **6**: 1002–17.

Goodland, R. J. A., E. O. A. Asibey, J. C. Post *et al.* (1990). Tropical moist forest management: the urgency of transition to sustainability. *Environmental Conservation* **17**: 303–18.

Hall, C. A. S. (1991). An idiosynchratic assessment of the role of mathematical models in environmental sciences. *Environment International* **17**: 507–17.

(ed.) (2000). *Quantifying Sustainable Development: The Future of Tropical Economies*. San Diego, CA: Academic Press.

Hall, C. A. S. and M. Hall (1993). The efficiency of land and energy use in tropical economies and agriculture. *Agriculture, Ecosystems and Environment* **46**: 1–30.

Hall, C. A. S., J.-Y. Ko, C.-L. Lee *et al.* (1998). Ricardo lives: The inverse relation of resource exploitation intensity and efficiency in Costa Rican agriculture and its elation to sustainable development. In: S. Ulgiadi (ed.) *Advances in energy studies: Energy Flows in Ecology and Economy*. Rome, Italy: Musis.

Hall, C. A. S., D. Lindenberger, R. Kummel *et al.* (2001). The need to reintegrate the natual sciences with economics. *BioScience* **51**: 663–73.

Hernandez, C., S. G. Witter, C. A. S. Hall *et al.* (2000). The Costa Rica banana industry: can it be sustainable? In: C. A. S. Hall (ed.) *Quantifying Sustainable Development: The Future of Tropical Economies*, pp. 563–93. San Diego, CA: Academic Press.

Jevons, W. S. (1865). *The Coal Question: An Inquiry concerning the Progress of the Nations* (A. W. Flux (ed.) 1965). New York: A. M. Kelley.

Ko, J.-Y. (2000). An integrated assessment of energy and resource efficiency trends at regional, national, and international scales. Unpublished Ph.D. dissertation, State University of New York, College of Environmental Science and Forestry, Syracuse, NY.

Ko, J.-Y., C. A. S. Hall and L. G. L. Lemus (1998). Resource use rates and efficiency as indicators of regional sustainability: an examination of five countries. *Environmental Monitoring and Assessment* **51**: 571–93.

Kroeger, T. and D. R. Montanye (2000). An assessment of the effectiveness of structural adjustment policies in Costa Rica. In: C. A. S. Hall (ed.) *Quantifying Sustainable Development: The Future of Tropical Economies*, pp. 665–94. San Diego, CA: Academic Press.

LeClere, G. and C. A. S. Hall (in press). *Making Development Work: A New Role for Science*. Albuquerque, NM: University of New Mexico Press.

Lélé, S. M. (1991). Sustainable development: a critical review. *World Development* **19**: 607–21.

Montanye, D. R. (1998). Examining sustainability: an evaluation of U.S.AID policies for agricultural export-led growth in Costa Rica. Unpublished M.S. Thesis, State University of New York, College of Environmental Science and Forestry, Syracuse, NY.

Odum, H. (1971). *Environment, Power and Society*. New York: Wiley Interscience.

OTA (Office of Technology Assessment) US Congress (1994). *Perspectives of the Role of Science and Technology in Sustainable Development*, OTA-ENV-609. Washington, DC: Government Printing Office.

Rothman, D. and S. de Bruyn (1998). Special Issue: The Environmental Kuznets Curve. *Ecological Economics* 25.

Schipper, L. and R. B. Howarth (1990). United States energy use from 1973 to 1987: the impacts of improved efficiency. *Annual Review of Energy* **15**: 455–504.

Taylor, L. (1993). The world bank and the environment: the world development report 1992. *World Development* **21**: 869–81.

Tharakian, P., T. Kroeger and C. A. S. Hall (2001). Twenty-five years of indutrial development: a study of resource use rates and macro-efficiency indicators for five Asian countries. *Environmental Science and Policy* **4**: 319–32.

United Nations Environmental Programme (UNEP) (2000). *Global Environmental Outlook 2000*. Available online at http://www.unep.org/Geo2000/english/index.htm

Verissimo, A., C. S. Junior, S. Stone *et al.* (1997). Zoning of timber extraction in the Brazilian Amazon. *Conservation Biology* **12**: 128–36.

Vorosmarty, C. and D. Sahagian (2000). Anthropogenic disturbance of the terrestrial water cycle. *BioScience* **50**: 753–65.

Wernick, I. K., R. Herman, S. Govind *et al.* (1996). Materialization and dematerialization: measures and trends. *Daedalus* **125**: 171–98.

World Bank (1997). *Expanding the Measure of Wealth: Indicators of Environmentally Sustainable Development*. World Bank.

World Bank (1998). *World Development 1998–99: Knowledge for Development*. World Bank.

World Resources Institute (WRI) (1999). *A Guide to the Global Environment*. (CD data base). New York: Oxford University Press.

3 Impacts of land cover change in the Brazilian Amazon from a resource manager's perspective

E. A. Serrão
Embrapa Amazônia Oriental, Belém, Brazil

I. S. Thompson
DFID, Belém, Brazil

The Brazilian Amazon is conceived nowadays as a green ocean, containing one of the world's major river systems with a water course network of more than 6500 km and responsible for 20% of the world's river discharge to the oceans (Figure 3.1). The Amazon river system includes a large annually inundated floodplain, or *varzea* (Richey *et al*., 1990) which represents an important natural resource base for food and energy production to meet human needs.

The Amazon is presently home to about 20 million people, mainly distributed in large, medium and small size urban and rural developments along the roads and rivers and concentrated heavily in the eastern part of the region (Figure 3.2).

The area of the Brazilian Amazon extends for about 500 million ha (equivalent to about two-thirds the size of the continental United States of America) of which about 80% falls within the tropical forest zone. Deforestation is currently running at around 1.6 million ha per year and its distribution closely follows the road network as illustrated in Figure 3.3 (Alves, 1999).

Schneider *et al*. (2000), using data from the 1995–6 Agricultural Census, report on land use by rainfall zone in the Brazilian Amazon (Table 3.1). They observe that, of the area under agricultural use, pasture is the dominant system, representing nearly 80%.

The sheer size of the Amazon region gives it global importance but it is also important as one of the last great frontiers for 'modern man' where the natural vegetation is largely intact, representing a store of biodiversity and playing an important role in global processes such as climate, carbon and hydrological cycles. As such, it represents one of the last great opportunities for humankind to develop in harmony with nature.

The Brazilian government is planning large-scale investment in development projects in the Amazon region through its Avança Brasil (Advance Brazil) programme. One goal is to virtually double the extent of paved roads and to construct ports and waterways, railways and hydroelectric power plants. These sizeable developments have the potential to cause great adverse impact on the ecological and social balance of the region. Such a programme would certainly lead to increased deforestation. Forest clearing through cattle ranching and slash-and-burn agriculture would also cause large changes in biodiversity and to the water, nutrient, carbon and energy cycles. Besides the biological impoverishment of forest areas, the increase in fire occurrence and carbon emissions would result in unwanted changes in the hydrological system at the regional level and in an increase in Amazonia's contribution to global warming.

THE RESOURCE MANAGERS AND EXTERNAL FACTORS INFLUENCING THEM

It is an interesting exercise to reflect on the different perspectives of the major resource managers in this region. No specific studies exist and therefore what follows is largely a subjective analysis derived from the authors' experiences in a federal research organisation in the region whose clientele include those same resource managers.

Who are the resource managers in the Brazilian Amazon? Setting aside the community-level managers who are mainly family agriculturalists, perhaps the most obvious is the government. The federal government is responsible for the unallocated lands as well as the federally designated conservation areas (7.6%) including public forests. Through FUNAI (National Foundation for Indigenous Population) it is also responsible indirectly for the Indian reserves. State governments may also have similar conservation areas created at state level. At municipal level there is much less land resource management responsibility but there is a strong

Forests, Water and People in the Humid Tropics, ed. M. Bonell and L. A. Bruijnzeel. Published by Cambridge University Press.

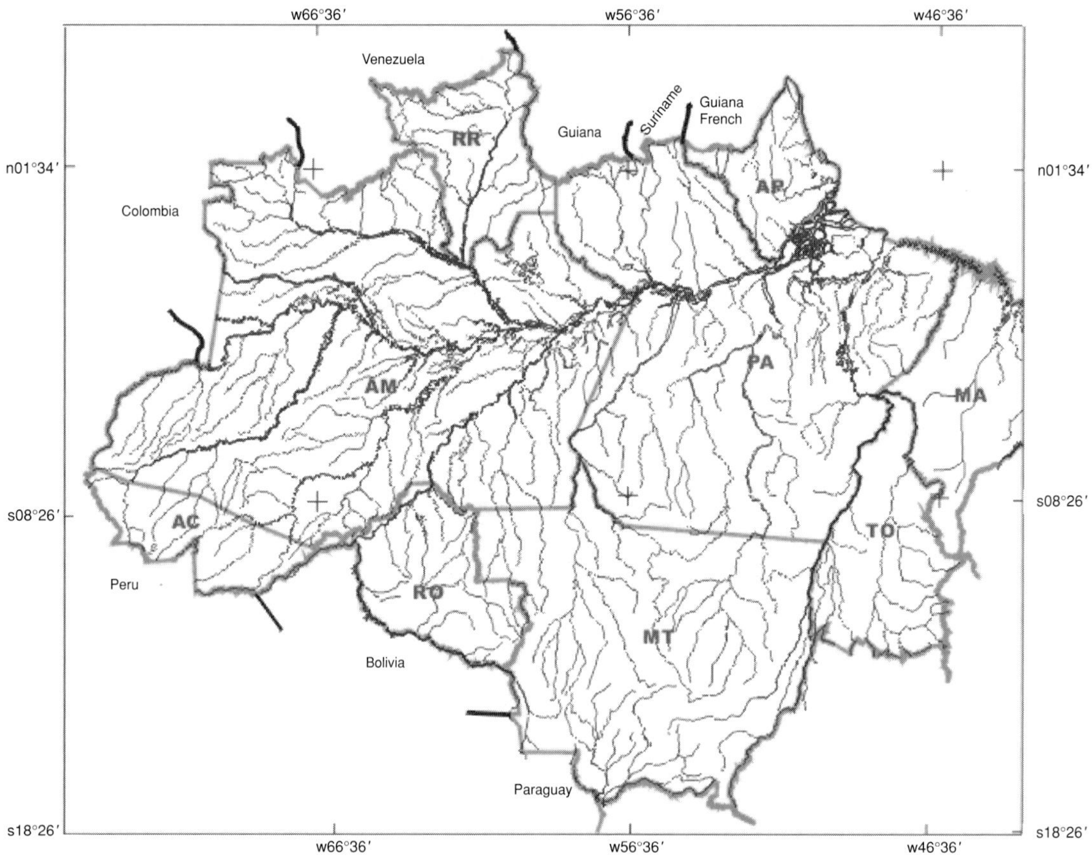

Figure 3.1 The hydrological system of the Brazilian Amazon.

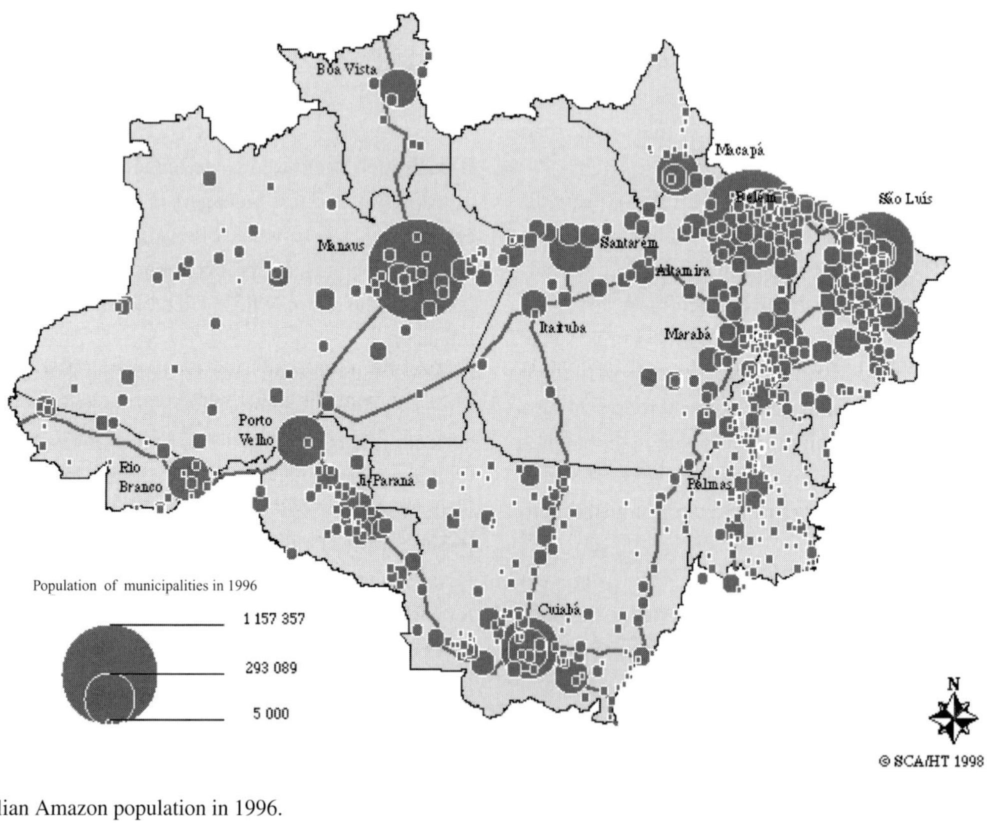

Figure 3.2 Brazilian Amazon population in 1996.

Table 3.1. *Land occupation by rainfall zone*

Rainfall	Land area (million ha)	Percent of total	Percent area in establishments	Percent area in agricultural use
Dry < 1800 mm	83.6	17	55.6	38.2
Transitional 1800–2200 mm	181.6	38	28.7	13.0
Humid > 2200 mm	219.4	45	7.5	3.2

Source: Adapted from Schneider.

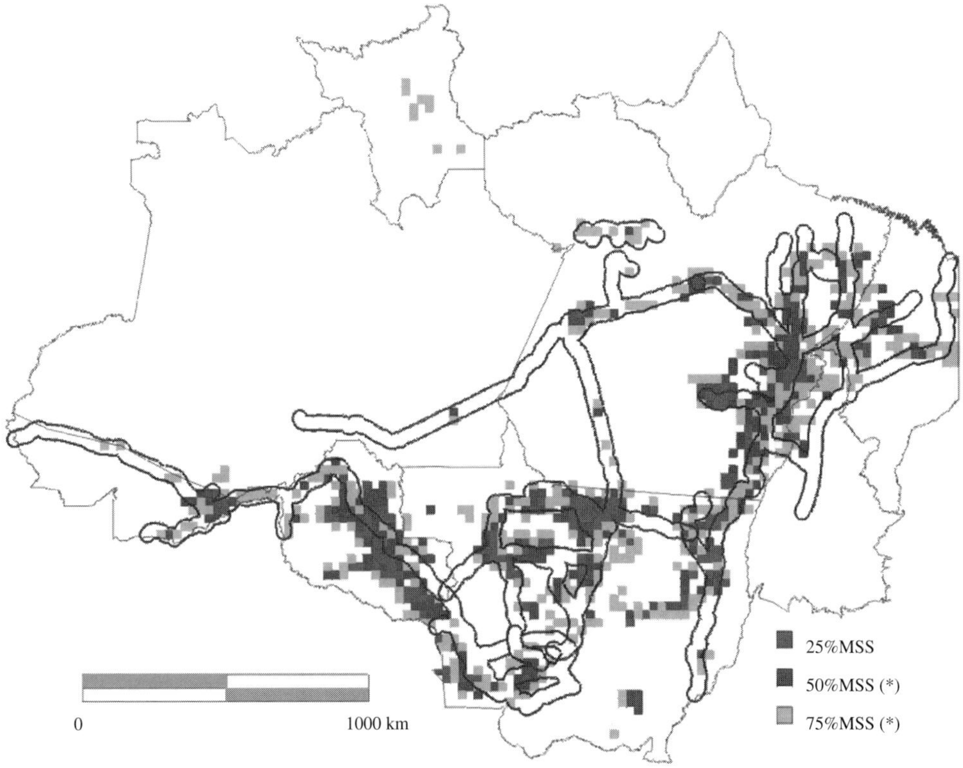

Figure 3.3 Distribution of deforestation hotspots and 25-km buffers around the western, eastern and central road networks.

tendency for decentralisation and municipalities are becoming involved in land tax collection.

Obviously the government as resource manager is influenced by government the policy-maker. However there may be a lack of integration and consistency, or at least a different perspective of the trade-offs between economic benefit and environmental cost between different branches of government. For example, the hydroelectric industry is a major factor in water and land use in the region. Brazil has some of the largest artificial lakes in the world, some of which are located in the Amazon region (e.g Tucuruí (2430 km^2) and Balbina (2360 km^2)) and there are a further 42 major dams under construction. The ten-year plan for expansion of Eletrobras (the energy parastatal) in Brazil for the period 1999–2008 projects an increase in the order of 65% in installed capacity from 156 new hydroelectric plants (Vainer, 2000).

The principal categories of private-sector managers are large ranchers, commodity crop developers, family-type subsistence farmers, and managers in settled and itinerant timber industries. Some effects on land use are indirect, for example the timber industry finances much forest logging and forest clearance although that industry is not directly responsible for the management of on-site timber extraction. The government has also contributed indirectly, for example through the financing of resettlement schemes which have led to forest clearance.

Private-sector resource managers are not immune to the policy-makers nor to communities. These form part of their

Table 3.2. *Relative environmental impacts of different land uses*

Type of use	Biodiversity	Soil	Water	Greenhouse gas emission
Non-timber forest products extraction	low	low	low	low
Logging	high	medium	medium-high	medium
Ranching	high	high	high	high
Slash-and-burn farming	medium-high	high	high	high
Hydroelectric power	high	medium	high	high
Mining	high	high	high	high
Ecotourism	low	low	low	low
Conservation units	none	none	none	none

decision-making environment. Official development programmes such as those administered by Superintendency for Development in the Amazon (SUDAM) have had and continue to exert a strong influence through incentives.

The trend in the financial sector towards green development is also beginning to stimulate the adoption of new management practices. Official and private banking agencies are beginning to create credit programmes specifically for green developments. The promotion of certification schemes to link market demand to production practices has been prerequisite to such developments. A Forest Stewardship Council (FSC) timber buyers group has been established in Brazil, the first in a developing country, and has challenged the timber industry to supply this new market demand. For example, the Tramontina Group with annual sales of US$500 million and one of the world's largest producers of wooden tool handles (Bihun, 1999), and the State Government of Acre, have become members.

In recent years, the Kyoto Protocol has increasingly been cited as a potential means to influence land use, making more conservationist land uses more financially attractive through the Clean Development Mechanism (CDM). On the other hand, the Brazilian government does not support the movement for preservation of existing forests to be included in this mechanism as such a step could be interpreted as a loss of sovereignty over the areas in question. In our view, fears over sovereignty loss can best be reconciled through more convincing, and better presented, evidence on the total sum of the ecological, economic and social costs and benefits of the alternative options.

A new Environmental Crimes Law sets out to influence land practices through regulation. It represents a real threat of major fines and custodial sentences for wrong-doers.

Community organisations at the local level are increasingly brought into land resource issues, e.g. local Watershed Basin Committees and non-government organisations (NGOs) such as Greenpeace at international level or Friends of the Earth, ISA (the Socio-environmental Institute) and IMAZON (Institute of Man and the Environment of Amazonia) at regional level are all increasingly influential through direct action or by influencing policy (advocacy).

Undoubtedly, large-scale land use change is the prerogative of the private sector. Properties of hundreds of thousands of hectares are not uncommon. Schneider *et al.* (2000) report that nearly half the land in the Amazon is in the 1% of holdings larger than 2000 ha. It is worth noting that for a manager to hold professional qualifications in the sector is the exception rather than the rule. The forest industry, for example, rarely employs graduate foresters. A lack of well-trained potential recruits is an easy explanation for this remarkable state of affairs. However, it is to be noted that even where external support for training staff has been provided, uptake by industry has been slow. An entrepreneurial philosophy of asset liquidation of an abundant resource may be at the root of this low investment in human capital.

Table 3.2 presents the authors' assessment of the relative impact of different actors on some key resource issues.

Given that there are increasingly strong external factors potentially influencing resource managers, let us now turn to their perceptions.

RESOURCE MANAGERS' PERCEPTIONS

Table 3.3 summarizes the level of attention paid to major impacts of land use change by different resource managers. While within each group there are exceptions we believe that this is a fair reflection of broad sensibilities.

Government

The government, as resource manager, seeks to establish models of sustainable use although there is tension between economic development and environmental impacts. This natural tension is exacerbated when there is strong segregation of responsibilities between ministries and little attention paid to 'joined-up' government. Plans for the development of the Amazon through the

Table 3.3. *Sensitivity to impacts by different resource managers*

Resource manager	Biodiversity	Water	Greenhouse gas emission
Government ministries			
Environment	high	medium	high
Science and Technology	high	medium	high
Transport	none	none	none
Agriculture	low	medium	medium
Foreign Affairs	medium	low	low
Private sector			
Industries	none	none	none
Farming	none	none	none
Logging	none	none	none
Companies seeking certification	low	low	low

government's new programme, Advance Brazil, have been much criticised as being in conflict with commitments to the environment such as in the Pilot Programme for the Protection of the Brazilian Tropical Forests (PPG7). Ambitious plans to link the Araguaia and Tocantins waterways to create waterways for grain transport are controversial and have led to charges of adulteration of the Environmntal Impact Assessment report. The government, through FUNAI, determines the scale of Indian reserves and the options for management within them. Even as demarcation of Indian reserves is proceeding, there is still no strong consensus on the long-term future of these lands.

On a smaller scale, government projects, such as Sustainable Management for Timber Production in the Tapajós National Forest, seek to develop new mechanisms for access to forest resources. Government, in partnership with NGOs, has had some success when seeking to decentralise management and promote public participation. For example a new category of conservation unit was established based on the successful model implemented at the Mamirauá Sustainable Development Reserve (Mamirauá, 2000) in Amazonas State. However, such promising initiatives are the exception, and public authorities have not been capable of establishing sound management of the lands under their responsibility, be it national production forests or nature conservation parks and reserves. The PPG7 is an ambitious programme to redress many of the institutional, technical and financial weaknesses but its slow progress illustrates the practical difficulties in achieving significant change.

The general political trend towards decentralisation also holds in land use, and control responsibilities are increasingly moving to State level where State environmental agencies are being strengthened and State legislation enacted.

Government supported research has sought to change land use practices and reduce the negative impacts of development through research on sustainable management of forests for timber, on alternatives to burning in secondary forest fallow systems, and on intensified livestock and agricultural production systems.

Private sector

Private-sector managers have tended to regard natural vegetation as a hindrance to be removed or as an endless free good to be cashed in, moving on when one locality is exhausted. Maintenance of forest as a land use has been regarded as an imposition instead of a valued option as seen in much of the current debate on the forest code. Indeed, rural politicians have introduced a proposal to reduce the legal reserve requirement in Amazonia from the current 80% to 20% claiming that it is inhibiting development, although the reserve area may be utilised under approved forest management plans. This was only avoided through a strong reaction by civil society organisations.

Impacts of land use change have generally been seen to be positive – economic and social development. The transformation of forest into pasture or perennial crops or grains is seen as an economic improvement and environmental costs are simply not perceived as a significant factor.

At specific points in time, fires in Pará and Roraima raise public concern over damage attributable to current land use practices, as has been the case in Asia; however, this is not linked to individual's management choices. Similarly, the El Niño phenomenon brought attention to the issue of water availability but again it is a phenomenon limited in time and distance in perception as a serious issue for the resource manager. In the year 2000, for example, the rainy season has been longer than usual in the Eastern Amazon, undermining efforts to raise awareness of trends. Researchers at IPAM (the Amazonian Environmental Research Institute) have calculated that probably 30–40% of the forests of the Brazilian Amazon are sensitive to small reductions in the amount of rainfall. They conclude that with an increase in the frequency and intensity of El Niño events, it will become more and more common for forests to dry out sufficiently so that they become flammable (IPAM, 2000). This has led to a political response in the form of a federally funded campaign (PROARCO) in the high-risk zone but there has not been a response from resource managers.

There is general public awareness over the loss of biodiversity in the Amazon region due to loss of natural habitat. However, the huge costs in quantifying the reality in terms of biodiversity loss weaken the potential impact, again particularly at the individual resource manager level. It has no effect on decision-making. IBAMA (the federal environmental agency) announced a new initiative recently to value national biodiversity with an initial estimated value of 4 trillion reais or five times Brazil's gross national product. There is more publicity given to biopiracy and access to resources by foreigners than to resource depletion.

It is economic controls and incentives that are now changing the perceptions of resource managers. Credit availability for reforestation and for more intensive pasture and agricultural uses such as grain and oil palm are beginning to change attitudes to new deforestation. Certification, which demands attention to social and environmental objectives as well as economic ones, will perhaps become the most powerful instrument for promoting sustainable management. Influential resource managers in the timber sector and in perennial crops such as palm hearts recognise the inevitability of certification or green labelling to respond to the general public concerns on adverse land use impacts. One illustration of this is that five years ago when looking for an industry partner to log an experimental forest area it took EMBRAPA (the Brazilian Agricultural Research Agency) three years to come to an agreement with a private company. Last year a private company was so insistent that it be included in another EMBRAPA forestry management research programme that it offered to finance all activities related to its participation. Again, in 1998, AIMEX (the Association of Timber Exporters of Pará) refused to participate in a FSC-sponsored meeting on certification but the editorial in the Association's magazine's in 1999 calls on its members to recognize that certification is here to stay.

OUTSTANDING QUESTIONS RELATED TO WATER RESOURCES

The Agenda 21 process in Brazil, managed by the Ministério do Meio Ambiente (MMA; Ministry for the Environment) (MMA, 2000) summarizes the current problems of water resource use in Brazil as:

- insufficient or inaccessible information to enable an adequate evaluation of the resources,
- non-existence of effective integrated management for multiple uses,
- insufficient legal base for decentralised management,
- inadequate soil management in agriculture,
- unjust distribution of social costs associated with intensive water use,
- incipient participation of civil society in management with over-reliance on government,
- water shortage for natural causes or due to intensive usage,
- a culture of abundance of water, and periodic flooding of large urban centres.

Kabat (LBA 2000), at the Large-Scale Biosphere-Atmosphere Experiment scientific conference in Amazonia, noted the need for an integrative approach to understanding the complex ecosystem of the Amazon region. He observed that changes in land use will certainly affect the global flux of key properties in the climate system. Water vapour and CO_2 fluxes, ground and cloud albedo, trace gas emissions, aerosol particles and radiation balance are amongst the properties changing fast about which there is little knowledge of the local or global implications. The programme adopts an Earth system perspective involving studies on physical climate, carbon storage and exchange, biogeochemistry, atmospheric chemistry, hydrology, land use and cover change, and human dimensions. The technical studies in this US$80 million programme, e.g. the development of a coupled ecological and hydrological model of the Amazon Basin (Coe *et al.*, LBA 2000) and in other initiatives such as the PPG7 and SHIFT (Studies on Human Impact in Forests and Floodplains in the Tropics) being conducted in the Amazon region with financial and technical support from the international community, are of fundamental importance for the advance of evidence-based policy-making.

However, allied to the outstanding technical demands, and equally important, is the need to discover effective means of involving resource managers in these issues; raising their awareness, seeking their participation in research, and interpreting results from their perspective. The effective involvement of all stakeholders is a complex and costly process but is fundamental if there is to be significant change in current practices and trends.

The Agenda 21 process offers a forum for multi-stakeholder involvement. One of the key means to involve different actors effectively is to provide objective methodologies to compare different values and discuss trade-offs for different choices. It is important to develop strategic impact assessment tools that can integrate analyses of environmental, social and economic costs and benefits and their risks, not only at the management unit level but at a regional scale, and so inform decision-making for resource managers and policy-makers alike. Strategic impact assessment tools need to be introduced through local institutional capacity development, and adapted and perfected in light of the region's characteristics.

CONCLUSIONS

Large private landowners and the government are the major resource managers in Amazonia. Despite a lack of specific studies on their perceptions of the impact of land use change, it is fair to say that they have generally low sensitivity to potential costs such as negative impacts on water resources, soils, biodiversity and climate change.

Within government, sustainable development initiatives tend to be small-scale and tensions exist within government sectors due to differing weights given to economic development gains and environmental losses. Sovereignty issues cloud analyses of international contributions. The way forward has to be firm and well-presented evidence, as support for policy development and improved governance through more joined-up government.

The private sector, with honourable exceptions, has little perception of the impacts of land use change at the management unit scale. The media has a tendency to report on disasters with short attention spans to the underlying issues that often involve cumulative, long-term change. It has, therefore, little lasting effect on resource managers' attitudes. Market instruments, such as certification, hold promise as tools of influence and merit official support.

Major efforts are under way to achieve an integrated understanding and a predictive ability of the complex processes that determine the impacts of land use changes in the Amazon, for example, the LBA programme. These technical advances must be linked to initiatives, such as Agenda 21, which provide a forum for all key stakeholders to participate in land use decisions. Perhaps most challenging of all is the introduction of strategic impact assessment methods and procedures with which to integrate the social, economic and environmental costs and benefits of the land-use alternatives and thus pave the way for evidence-based policy development.

References

Alves, D. S. (1999). An analysis of the geographical patterns of deforestation in Brazilian Amazônia in the 1991–1996 period. In *Proceedings of 48th Annual Conference of the Center for Latin American Studies*, Gainsville, FL, 23–26 March 1999.

Bihun, Y. (1999). Putting a new handle on a famous brand. *Timber and Wood Products International* 5 June 1999: 42–43.

LBA (2000). *Abstracts* of *1st LBA Scientific Conference*, Belém, Brazil, 26–30 June 2000.

Mamirauá (2000). Mamirauá Sustainable Development Reserve. http://www.pop-tete.rnp.br/mamirau.htm

MMA (2000). *Gestão dos Recursos Naturais: Subsídios à Elaboração da Agenda 21 Brasileira* (Maria do Carmo Lima Bezerra and Tânia Maria Tonelli Munhoz, General Coordinators). Brasília, Brazil: Ministério do Meio Ambiente.

Richey, J. E., Hedges, J. I., Devol, A. H. *et al.* 1990. Biogeochemistry of carbon in the Amazon river. *Limnol. Oceanogr.* 35: 352–71.

Schneider, R., Verissimo, A., Arima, E. *et al.* (2000). *Sustainable Forestry and the Changing Economics of Land: the Implications for Public Policy in the Legal Amazon*, draft World Bank report (cited with permission of authors). Brasília, Brazil.

Vainer, C. B. (2000). *Jornal no Brasil*, 4 April 2000.

4 Forest people and changing tropical forestland use in tropical Asia

J. Schweithelm
Forest Mountain Consulting, Burlington, USA

FOREST DWELLERS AND FORESTS

Until the latter part of the nineteenth century, most people living in the world's moist tropical forests were almost completely dependent on forestland and resources for food, shelter, medicine and trade products. Small groups typically claimed use rights over a specific area of forest and controlled access to agricultural plots and forest resources through customary law reinforced, in many cases, by religious restrictions. Most forest dwellers were isolated physically from the outside world by distance and natural barriers. They lived at or beyond the periphery of modern society and the market economy. The outside world had little interest in these people or forest resources, with the exception of a few high-value non-timber forest products (NTFPs) and timber species, some of which were traded for centuries.

The isolation that shaped interactions between forest dwellers and tropical forests began to be undermined when national governments and colonial powers extended their power into forested hinterlands; frequently accompanied by laws that shifted legal control of forestland and resources to the state (Poffenberger, 1999; Brookfield *et al.*, 1995). The super-imposition of formal tenure systems over traditional tenure had little initial impact on the lives of most forest dwellers because governments lacked the resources needed to enforce these laws and, in any case, there were few forest resources valuable enough to justify commercial exploitation. Early conflicts did occur where agricultural land was needed for commercial crops, such as tobacco plantations carved out of the forest in northern Sumatra in the late nineteenth century, or when forest products became widely commercialised, such as the *jelutung* resin tapped from *Dyera loweii* trees in southern Borneo in the first decade of the twentieth century (Potter, 1988) and the *Hevea brasiliensis* latex that created the nineteenth century rubber boom in the Amazon. Colonial governments in some cases seized forestland for timber production in areas accessible to transport or where high-value species, such as teak (*Tectona grandis*), could be grown. The spatial scale of these incursions was small in comparison to the vast areas of tropical forest that remained untouched by outsiders.

FORCES OF CHANGE

World War II marked a watershed in South East Asia in terms of control and use of forests (Brookfield *et al.*, 1995). Countries in the humid tropics of Africa and Latin America also began to change during this period, even though they were not directly involved in the war. Weakened colonial governments soon gave way to independent nations that saw forests as a source of capital for national development. Like their colonial predecessors, these governments believed that the forests were national resources and that the traditional uses of forest dwellers should be subordinate to public or commercial uses. Populations increased due to improved health care and sanitation, creating demand for agricultural land. Improved transportation infrastructure, market demand and technical advances made tropical timber harvesting economically viable (Poffenberger, 1999). These factors initiated and sustained massive changes in tropical forestland use that continue to the present.

The intensity and scale of forestland use in the humid tropics has changed greatly over the second half of the twentieth century, marked by a general shift from subsistence to commercial uses. The pace and nature of change has varied among countries and regions but has generally been driven by demand for timber and agricultural land. Governments have typically encouraged timber extraction and forest conversion to agriculture to support national economic development and to open up remote areas for settlement. Intensive timber harvests and forest conversion to plantations, pastures and annual cropping have altered ecological and hydrological processes, forest integrity and structure, and biological diversity. The recent epidemic of wildfires in Indonesia has been made possible by the degradation of forest integrity and accelerated forest conversion (Barber *et al.*, 2000).

Forests, Water and People in the Humid Tropics, ed. M. Bonell and L. A. Bruijnzeel. Published by Cambridge University Press.

The process of forest degradation and conversion is similar across the world's moist tropical forests. The process may start with increases in the intensity of forest resource extraction and swidden agriculture by local people, progressing through successive stages of intensive timber harvest, wildfire and conversion to permanent agriculture. Logging roads often open the forest to uncontrolled use by outsiders. Sometimes, logging and commercial conversion occur suddenly in rapid succession, with local people as bystanders and losers. The major islands of Indonesia illustrate the progressive stages in the forest conversion process. Aside from the upper slopes of Java's volcanoes, most of that island's forests were cleared or badly degraded by the beginning of the twentieth century. The degradation and conversion process in the Kali Kanto watershed of East Java is documented from the beginning in the mid-nineteenth century (Nibbering, 1988). The actors include early forest dwellers, agricultural settlers, the Dutch colonial government, coffee planters and the Japanese army during World War II. Most lowland forests on Sumatra are now nearing the end of the forest conversion process. Forests damaged by poor timber harvesting practices and further degraded by illegal logging and wildfire are being replaced by oil palm and pulp wood plantations. The forests of Kalimantan are about a decade behind Sumatra in this process. In contrast, most lowland forests in the vast province of Irian Jaya (also referred to as Papua) on the island of New Guinea have not yet been subjected to accelerated disturbance.

HOW DOES FOREST CHANGE AFFECT FOREST DWELLERS?

After millennia of experimentation and adaptation to climatic change, people living in the world's tropical forests developed sophisticated and relatively stable strategies to earn a livelihood from local forest resources, with increasing reliance on shifting agriculture among many groups (Hutterer, 1988; Meggers, 1988). Changes in forest quality, land use and market demand have forced many groups of forest dwellers to make unwanted changes in their livelihood strategies but have also provided new income opportunities in some cases. Collection of NTFPs for commercial sale has become more common and employment opportunities have temporarily or permanently attracted forest people to cities, timber concessions and commercial plantations. Communities or families are sometimes forced to relocate after losing their land to agricultural settlers or commercial firms.

Some forest dwellers have responded to market forces, voluntarily or involuntarily, by producing a narrow range of cash crops or collecting one or two forest products for sale, thereby foregoing the stability inherent in diverse traditional livelihood strategies. For example, groups living on the upper Barito River in Central Kalimantan lost the cultivated rattan gardens that formed the basis of their livelihoods to the Indonesia wildfires of 1997/98 (Barber *et al.*, 2000). Unlike the relative stability that characterised pre-industrial human interactions with tropical forests, these interactions now cover a spectrum from stable to highly unstable. The stability of community/forest interactions depends on the strength of traditional institutions and land tenure, the level of degradation of the forest, the degree of threat to forest resources, and the level of community dependence on these resources.

Some former forest dwellers have lost their connection with the forest in the wake of the advancing agricultural frontier, leaving them in an unfamiliar landscape. Others are on the frontier, trying to adjust their livelihood strategies to a forest that is depleted of resources and no longer under their control. Some families and communities have moved out of the forest, either voluntarily or at government insistence, to be closer to services, infrastructure, and jobs. The remote Apo Kayan watershed of East Kalimantan had 20 000 to 30 000 inhabitants in the 1930s and currently has about 10 000 people as the result of out-migration in recent decades (Eghenter, 1999).

OTHER GROUPS OF FOREST-DEPENDENT PEOPLES

The forces that drive forestland use change have also increased human interactions with tropical forests that do not fit the forest-dweller pattern described above. Other forest-dependent people can be broadly described as forest frontier settlers, forest-dependent agriculturalists, and forest product opportunists. These groups are not new but their numbers have greatly expanded in recent decades due to easier forest access and increased demand for forest products. Agricultural settlers moving into forest areas to establish farms are forest dwellers only in the sense that they live in an area where forest cover predominates. Their interaction with the forest is usually limited and their communities generally lack rules and institutions to regulate access to forest resources. The line between settlers and indigenous inhabitants is blurred when the former have settled in one place for multiple generations or in cases where the connection between forest dwellers and the forest is weakened.

Long-established agricultural communities often rely on forests and trees for critical resources. Humid and seasonally dry tropical agricultural landscapes are typically a mosaic of cropped fields and pastures interspersed with forest remnants, clusters of trees, and agro-forests. Agricultural communities are dependent on forest patches and scattered trees for fuel wood, fodder, construction materials and NTFPs. Because these resources arc scarce and valuable, agricultural communities commonly develop rules of resource access, although their forest management rights are seldom legally recognised except under community forest management agreements (Poffenberger, 1999).

Modern transportation has made it possible for people living far from forests to collect NTFPs and hunt wildlife for commercial sale. Booming markets for these products make this an attractive option for unemployed and landless people and for farmers during slack periods in the agricultural cycle. Opportunistic interactions increase when a forest management vacuum is created by weakened forest-dweller control over forestland that has not been replaced by effective government management. Many tropical forests are perceived to be open access resources as a result of this management vacuum (Fox, 1993; Peluso, 1993).

PERCEPTIONS OF FORESTLAND USE CHANGE AND IMPLICATIONS FOR THE FUTURE

Forest dweller perceptions of forestland use change largely depend on the gains and losses each family or community experiences in terms of resource access, lifestyle, government services and employment. Perceptions may vary among neighbouring communities and even among families in one community. Communities that have been displaced from their land or denied access to critical forest resources without compensation or options will be understandably bitter and angry. Other forest dwellers may see logging and forest conversion as beneficial to them if improved services and jobs follow. Forest-dependent agriculturalists and opportunists may share this positive view of forest disturbance if it affords them greater access to forestland and resources. Perceptions of change may also be influenced by past experiences with outsiders.

The dramatic changes of recent decades have generally weakened community social bonds and traditional institutions needed to adjust to change and to re-assert management rights. Past experience with unjust laws, heavy-handed officials and unscrupulous businessmen have left many communities mistrustful of outsiders and with little faith that land management rights will be returned to them. The process of returning forest management to communities requires building community trust, self-confidence and capability while changing the attitudes and incentives of government resource managers with respect to forest dwellers.

WHO SHOULD CONTROL THE FOREST?

Intensification and commercialisation of forest land use has changed the economic, legal, and cultural relationships between people and forests. Forest dwellers have had to accommodate formal land tenure laws imposed from above, greater integration into the cash economy, and outside cultural influences. Change has sometimes provided welcome economic and educational opportunities, but has usually been accompanied by loss of control over traditional land and forest resources, and has often put forest communities in conflict with government land managers, logging companies, commercial agricultural firms and agricultural settlers.

Traditional community-level institutions that evolved to regulate access to land and forest resources have weakened in the face of forest nationalisation, formal tenure systems, market demand for forest resources, and competition from outsiders for land and resources. Existing laws in many countries either do not legally recognise the traditional land-use rights of forest dwellers or accord greater legal standing to claims based on sedentary agriculture or extractive commercial uses. Government perceptions of community forest use rights have shifted since the immediate post-colonial period when these rights were largely denied, to the current situation in which most governments of tropical countries accept in principle that forest dwellers should have the right to participate in forest management. Policy and practice are starting to follow this perception as traditional forest users are beginning to be included as partners in sustainable forest management and forest conservation (Poffenberger *et al.* 1998; Poffenberger 1999). Cases in which full management authority has been returned to communities are far less common.

Efforts by governments and donors to work with communities on forest-based development began in the 1970s with social forestry projects designed to produce fuel and timber. These projects offered limited scope for community participation in management and have been replaced by community forestry laws that allow agricultural communities to manage forest remnants and forest areas according to their own needs and for their direct benefit.

Forest dwellers have been involved in protected area management in recent years with varying degrees of success. Communities living in and near production forests are demanding to be included in management and profit-sharing arrangements. Forest dweller land rights have become a major political issue in many countries, including the Philippines, where the government has created a process to legalise land claims based on ancestral tenure (Poffenberger, 1999). Indonesia's change of government in 1998 has allowed forest dwellers to express long-simmering resentment against timber companies and plantation firms that appropriated their lands and resources during the Suharto era (Barber *et al.*, 2000). Resentment repeatedly turned to protest in 1999 and 2000. Several timber concessions have been forced to suspend operations as a result of these clashes and others are now negotiating compensation and profit sharing agreements with communities.

NGOs and donor organisations promote the right of communities to manage nearby protected forests, often assuming that all forest communities have the capability and motivation to manage forests sustainably. Some social scientists have cautioned that the effectiveness of community-based forest management depends on the specific social, historical, political and economic environment that the community has experienced in the past and is currently experiencing (Eghenter, 2000a, 2000b; Brosius *et al.*, 1998). People/forest relationships vary over time and tend to change in response to economic and political conditions. Significant changes in people-forest relationships have been observed in Indonesia as a result of the economic crisis that began in 1997 (Sunderlin *et al.*, 2000) and major wildfire episodes in 1982/83 (Mayer, 1996) and 1997/98 (Barber *et al.*, 2000). Rural people have generally reacted to the destruction of forest resources and lost employment opportunities by seeking new resources to exploit or opening larger agricultural plots if they have sufficient labour to do so.

There is currently a wide spectrum of views on the appropriate level of community involvement in forest management and the degree of control over forest resources that should be legally vested in communities. Some governments see practical value in community forest management partnerships because they allow land managers to take advantage of indigenous ecological knowledge and reduce conflicts over land rights (A. Hall, this volume) while meeting domestic and international expectations of equitable treatment of indigenous people. Despite these advantages, community/government forest management partnerships are developing slowly in most countries due to lack of appropriate institutional mechanisms, and in many cases, mistrust and misunderstanding between the prospective partners. Even in countries where community forestry laws and procedures for implementing them exist, such as India and the Philippines, the pace of creating community forests is slowed by the cumbersome process of consultation and planning between communities and forestry officials.

Governments and international organisations are still discussing to what extent forest management rights should be transferred to communities and which communities are eligible to receive these rights. Most governments are reluctant to fully transfer forest ownership to communities because significant resources and power would be shifted away from the public sector, resulting in loss of revenues from forest product royalties and perhaps reduced investment in the forestry sector. The World Commission on Forests and Sustainable Development views community participation as a critical part of tropical forest management, but urges governments to clarify and enforce land tenure laws to protect the forests from invasion by agricultural settlers (Krishnaswamy, 1999).

Participants at a high level seminar on Indonesia's forests held in May, 2000, discussed efforts by Indonesia's new government to allow greater participation by indigenous peoples in forest management. They agreed that this policy change would not guarantee that the forests would be better protected. Uma Lélé, a World Bank official who attended the seminar, noted that 'there is a major gap between international expectations of how tropical forests must be managed and the expectations of local communities.' There is growing concern that communities may tend to manage forests to maximise short-term financial returns while ignoring environmental and social values. This concern is borne out by experience in Papua New Guinea, where clans that are empowered legally to manage ancestral forests have frequently made financially and ecologically short-sighted decisions to liquidate their forest capital (Filer with Sekhran, 1998). Many Indonesian forest communities, under the provisions of the 1999 revision of Indonesia's forestry law, are using their expanded forest management rights to accelerate commercial resource use and forest conversion.

FOREST HYDROLOGY AND FOREST PEOPLE

Forest dweller effects on hydrology

Indigenous forest dwellers and pioneer agricultural settlers in tropical forests are often cast as destroyers of the forest because of the perceived negative effects associated with their agricultural practices. Deforestation, waste of forest products, loss of biodiversity, increased incidence of wildfire, and changes in hydrology are commonly attributed to swidden agriculture and small-holder forest conversion. The effects of these agricultural practices on hydrological parameters attract less attention in the perhumid tropics than in the seasonally-dry tropics where dry season flows are critically important for lowland irrigation and water supply.

There is a long history of conflict between forest dwellers and government forest managers over swidden agriculture. During the colonial era, swiddening and annual burning of grasslands were generally viewed by colonial authorities as wasteful and destructive uses of forest resources and particularly detrimental to catchment management (Nibbering, 1988; Potter, 1988). Colonial anti-swiddening laws and attitudes persisted after independence and still exist in many tropical countries. Swidden agriculture has become more intensive in many areas over recent decades as fallow periods have been shortened or larger plots opened in response to population pressure and demand for cash crops. Economically marginalised or landless lowland farmers sometimes open temporary agricultural plots in nearby forests (for example, see Siebert and Belsky, 1985). These trends have made it difficult to draw the line between sustainable, long cycle shifting cultivation and less stable practices that often lead to forest conversion. Understanding the effects of the diverse practices described as shifting cultivation has been further complicated by the difficulty of isolating the effects of one type of forest land use within the overall context of forestland use change.

Studies by forest ecologists indicate that moist tropical forests are resilient and will recover from shifting cultivation if succession is allowed to proceed over a sufficient time period (Whitmore, 1998; Lawrence *et al.*, 1998; Schmidt-Vogt, 1998). Long cycle shifting cultivation results in a mosaic of forest patches at different stages of regeneration that mimic natural disturbance patterns. A review of scientific knowledge of the effects of various tropical land uses on hydrological parameters concluded that the effects of long cycle shifting cultivation are generally localised and of short duration and that total water yield is likely to be greater when part of a catchment is under shifting cultivation than would be the case if it were totally forested (Hamilton with King, 1983). Bruijnzeel (1990, 2004) concluded that water yield increases in proportion to forest clearance, with the greatest contribution to base flows that sustain dry season flow, assuming that infiltration is not decreased by soil disturbance.

Many scientists who have studied shifting cultivation under stable conditions have concluded that this agricultural system is well adapted to low fertility tropical forest soils and the diverse livelihood strategies of forest dwellers (for example, Fox, 1999; Colfer *et al.*, 1997; Peluso, 1993). Some observers argue that the negative attitudes of government officials and sedentary farmers toward swidden cultivators are based on a combination of factors that include ignorance, ethnic prejudice, and the desire to use the presumed negative effects of swiddening as a rationale to exert greater control over forest resources (see Dove, 1983 and 1985 with respect to Indonesia). Anti-swiddening attitudes have further marginalised forest dwellers in their societies and some countries have forced communities out of the forest to stop them from practising this form of agriculture.

Conflicts over the hydrological effects of swiddening continue, especially in the seasonally dry tropics. An on-going conflict in Chom Thong district of northern Thailand is based on the perception among lowland farmers and urban-based orchard owners that swiddening in hill forests is reducing water availability for irrigation. This conflict, which is part of a pattern of water and land use conflicts in northern Thailand, has made the ethnic minority upland forest dwellers the target of anger and violence, and become the subject of national debate in the late 1990s (Poffenberger, 1999). Laungaramsri (1999) contends that the scarcity of water is in fact due to increased water use by vastly expanded lowland orchards, exacerbated by reduced rainfall in recent years. In a similar case in Vietnam's Dak Lak Province, clearance of hill forest for coffee plantations is blamed by lowland farmers for water shortages (Funder, pers. comm.). The facts of the case are currently under investigation by a multi-disciplinary team that includes hydrologists and social scientists. In the Indonesian province of South Kalimantan, government officials have repeatedly attributed low dry season water levels in the reservoir of the Riam Kanan hydroelectric dam to grassland burning and swidden agriculture by watershed inhabitiants. Below-average rainfall in this relatively dry part of Borneo appears to be the primary cause (Schweithelm, 1987).

Water in the lives of forest dwellers

Water is an important part of the resource base for forest communities. It is needed for drinking, cooking, bathing and irrigation. Rivers and lakes provide habitat for fish and other aquatic food resources. Rivers are often the primary means of access and transport for forest communities. Forest people are attuned to seasonal and weather-related changes in hydrological parameters. Some groups recognise the relationship between forestland use change and changes in hydrological parameters. On the other hand, little research effort has been devoted to studying either how these changes affect the livelihood strategies of forest dwellers or the extent to which indigenous hydrological knowledge is translated into traditional rules that regulate community use of forestland for agriculture and forest products harvesting.

In recent decades, social scientists have investigated the factors that shape the livelihood strategies and resource management practices of forest dwellers (Peluso *et al.*, 1995) and how these are affected by forest disturbance and conversion (for example, Colfer *et al.*, 1997; Peluso, 1993). These studies have generally focused on the livelihood effects of decreased game and NTFP abundance and restricted access to forestland for agriculture, gathering and hunting. Reports of how disturbance-induced changes in forest hydrology affect forest dwellers are, for the most part, anecdotal and tend to focus on changes in fish yields. Changes in flow regimes and water quality take prominence only in cases where downstream uses are perceived to be affected adversely, such as the conflicts in Thailand, Vietnam and Indonesia described in the preceding section.

Some forest dwellers appear to have well-developed knowledge of forest hydrology. For example, some ethnic groups living in Kayan Mentarang National Park in East Kalimantan, Indonesia, include provisions in their traditional laws to restrict land use in the catchment of the village water supply and to forbid cutting trees along river banks, around salt springs, on steep slopes and atop hills (Eghenter, pers comm). The Kenyah people of East Kalimantan, for whom fish and aquatic organisms are important dietary items, acknowledge the relationship between forest cover and river water quality (Colfer *et al.*, 1997). These people have observed that upstream deforestation makes river water muddier, which they view as detrimental to their lives because it affects human health, makes river navigation more hazardous, reduces fish abundance and the aesthetic qualities of water, and causes more frequent and extreme flooding of riparian agricultural plots (Colfer, pers. comm.). It is likely that indigenous hydrological knowledge is more widespread than has been reported in the scientific literature. Some hydrological research activities now being conducted in the humid tropics under the auspices of the International Hydrological Programme of UNESCO include surveys of forest inhabitants to document their knowledge of local hydrology (Bonell, 1999).

INVOLVING FOREST PEOPLE IN HYDROLOGY

Why involve forest dwellers?

The underlying premise today is that the body of scientific knowledge in tropical forest hydrology contains important lessons for those who are charged with managing moist tropical forests and shaping forest policy and, furthermore, that these groups should be targeted by hydrological researchers as audiences for research findings. Targeting policy makers and forestland managers is relatively straightforward. The senior government officials responsible for shaping forest and land-use policy are a small and obvious group in most countries and the relationship between land use policies and forests have been studied widely. The legal authority and responsibilities of officials who manage various categories of forest are known and forest management issues are also generally well known.

The role of communities in forest management is much less clear, especially in view of rapidly changing forestland use, the diversity of ways that communities adapt to socio-economic change, and uncertainty over the pace and legal status of efforts to return forest management to community control. Amid this uncertainty, hydrologists, and others who are working to strengthen the scientific basis of tropical forest management, can choose to view forest dwellers as informants, audiences, partners, or simply as a complicating factor in research and forest management. Reasons why hydrologists should involve forest dwellers in their work are that:

- Forest communities that have ancestral claims to forestland have a moral, if not legal, right to participate in its management;
- Forest communities are *de facto* forest managers and may have a larger *de jure* role in forest management in the future, either as land owners or management partners;
- Forest dwellers have valuable, if not unique, long-term knowledge of local hydrology;
- Forest dwellers have an incentive to scientifically document the hydrological effects of their land use practices in cases where this is a point of conflict with outsiders and when forest disturbance affects their water use and livelihoods; and
- Research and donor organisations that shape the tropical forest research agenda and influence the opinions of governments and international bodies have put forest dweller rights and welfare squarely on the policy and research agenda, making it difficult for forest researchers to ignore these groups.

Engaging communities meaningfully requires skills that most physical scientists are not called upon to develop. To influence the thinking of forest dwellers directly, hydrologists must understand the basics of community dynamics and develop appropriate language and communication skills. Hydrologists currently have little professional incentive and few resources to develop these skills, nor do they routinely work with social scientists who can provide insights into forest communities. An alternative mechanism for involving communities in hydrology is to build partnerships with non-governmental organisations (NGOs) that have experience working with forest communities. Conservation, community forestry and community development NGOs can bring community relations skills to the partnership and may already have relationships based on mutual trust with target communities. Hydrologists must also provide some tangible benefit to their NGO partner, such as technical assistance, funding, or assistance with advocacy on behalf of the forest dwellers.

Barriers to involving forest dwellers in hydrology are costly to overcome, but the rationale for, and potential benefits of, their involvement will outweigh the costs in most instances. Developing working relationships will be much easier if the forest hydrology discipline endorses community involvement as a standard practice and provides incentives for hydrologists to build these relationships. (See also Deutsch *et al.*, this volume who provide a good example of community involvement.)

Potential roles for forest people

Forest hydrologists can choose to interact with forest-dwelling communities in a variety of ways, from one-way communication to more active engagement. Communities should be primary target audiences for scientific findings and management guidelines but can also be involved in hydrological research and forest management activities. Researchers are understandably reluctant to use scarce field time and research funds to bridge what is often a wide gap in culture, world-view and language, to develop meaningful dialogue with people living in their study areas. Such effort can be justified only if local people have relevant knowledge or can contribute to the research in other ways.

Forest dwellers are likely to have useful hydrological knowledge, especially with respect to infrequent, extreme events that hydrologists rarely witness. Forest dwellers can also conduct long-term data collection on behalf of researchers and are more likely to protect hydrological instruments if they are involved in the research. Perhaps most importantly, involvement in research gives the communities a sense of ownership of the results and a reason to support resulting land use prescriptions (Bonell, 1999). Conservation biologists that have involved forest dwellers in their research have gained valuable insights about long-term trends in plant and animal abundance and other aspects of local ecology. This cooperation usually leads to mutual understanding of the resource use perspectives of both parties and has sometimes led to long-term monitoring and management partnerships.

The first objective must be to determine what policy makers, managers and communities need to know about the relationship between hydrological parameters and land and water management practices. The second objective is to identify forest hydrology research results relevant to these information needs and determine how this information can be synthesised. The third objective is to use hydrological knowledge to formulate forest management guidelines (Parts III and IV take up these issues in detail). Addressing these objectives in a meaningful way requires involvement of communities and other stakeholders in an interactive dialogue to determine their information needs and communication style. Results of hydrological research often do not reach practitioners and communities in a form that is understandable and useful to them. Clear, unambiguous messages must be sent to these audiences about relationships between land use and hydrological parameters.

Guidelines are important, but may not bring about desired changes in practices if existing institutions, policies and procedures do not provide the means to translate guidelines into actions. Options for going beyond guidelines include:

- develop specific methods and approaches through case studies;
- identify ways that policies should be changed or institutions strengthened to facilitate implementation of guidelines; and
- build hydrology guidelines into other efforts to improve forest management, including forest certification, integrated conservation and development projects, and community forest management.

Efforts to identify knowledge gaps and research needs, with emphasis on underlying processes related to land-use planning and management practices, should address social, political and economic processes in addition to physical and ecological processes. Understanding how these processes actually affect forestland use and hydrology requires a close working relationship with communities and other forest stakeholder groups.

Strategies for engaging people and communities

As discussed above, forest hydrologists may wish to engage forest dwellers to: (1) provide hydrological information that will help them make better forestland use decisions; (2) tap their indigenous hydrological knowledge; and (3) make them participants in research and problem solving. Translating these goals into action can be achieved most efficiently by developing a strategy that fits partner communities and their environment by answering the following questions:

- What specific objectives should be achieved within designated time periods?
- What information, assistance and cooperation will be required of researchers and the communities to make collaboration work?
- What incentives do the communities and their members have to play their intended roles and are they capable of doing so?

The likelihood that efforts to work with a community on research or forest management will be successful depends primarily on the last point. The key internal factors that determine community capability are the strength of its leadership, social cohesiveness and institutions. Key external factors that determine capability and incentives are the nature and stability of the community's relationship with the forest and the local and national political, social and economic environment. Failing to appreciate differences in these internal and external factors can fatally flaw partnerships between forest dwellers and forest managers or researchers (Eghenter, 2000a, 2000b; Brosious *et al.*,1998). Communities must believe that they have a real stake in the forest, a recognized claim to valuable forest resources, and that their participation will help to maintain the productivity of their resources. Researchers must convince communities that there is a clear linkage between research and positive impacts on their lives. As Hall (this volume) points out, it is naive to expect that communities will blindly conserve forests if the forest is degraded, under high threat, or not clearly under community control.

Fortunately, much can be learned from efforts to involve communities in land and forest management for other purposes, particularly biodiversity conservation and community forestry. Each of these approaches has developed principles and methods for interacting with communities that are relevant to forest hydrology. Hydrology-based land use approaches, like integrated conservation and development approaches to protected area management, seek to change human behaviour in ways that improve forest management, but not usually to the immediate benefit of the community or its members. People are naturally less inclined to make these changes than to adopt changes that will benefit them directly and materially, such as is usually anticipated in community forestry. The benefits and costs associated with prescribed changes in land use must be clearly communicated to communities to win their support and prevent misunderstandings that could destroy the partnership.

Effective communication must be based on community characteristics and information needs, which will vary within and among communities. Key actors and relevant community institutions should be targeted specifically, and their roles, knowledge, interests and resources investigated along with their communication style. Researchers should explore community understanding of hydrology as a basis for dialogue. Knowledge of target audiences should be translated into a communication programme using appropriate languages and communication media and tailored to the needs and interests of each audience. Forest dwellers have repeatedly demonstrated that they can understand complex concepts related to their environment if communicated in a language and through mediums that are familiar to them. Many forest hydrology research findings are clearly relevant to communities but will not reach them unless mechanisms are put into place to ensure that communities become part of the communication network.

Hydrology research should involve communities where relevant and feasible. Forest dwellers are much more likely to accept and use research findings that they have participated in producing. Hydrologists may wish to follow the lead of community foresters and some conservation biologists in forging close working relationships with communities based on shared interests and objectives. Conducting joint research with social scientists can provide insights for both disciplines. To make positive changes in land use, researchers must understand the factors that are driving poor practices and how these affect hydrology.

Some community-related research topics that deserve attention from hydrologists and social scientists include:

- Forest-dweller knowledge of forest hydrology, especially extreme events, long term trends, and the relationship of hydrological parameters to different forms of forest disturbance and land use. This research should be conducted in several countries under varied physical and ecological conditions.
- Investigation of the hydrological characteristics of land uses practised by forest dwellers in comparison with disturbed forest and post-conversion uses.
- Work with social scientists and economists to quantify water-related costs to communities from forestland use change.

Forest management activities in partnership with communities provide a means to test and implement research results on larger spatial scales than is possible in experiments. Factors that affect the quality of community forest management can be observed and related to land use change. Community information and resource needs can be clearly identified. Forest management activities are expensive and time-consuming, and are therefore best pursued in partnership with projects or organisations that have a long-term field presence and relevant forest management interests. Hydrologists can take advantage of the expertise, relationships and logistical support that have been developed by their partner. Field activities may also be undertaken as part of long-term strategic alliances with organisations pursuing sustainable forest management and forest conservation on a regional or global basis.

CONCLUSIONS

Over a period of less than a century, forest dwellers have gone from being isolated and totally in control of their forest resource base to being under tenuous government control and part of the market economy with few, if any, legal rights over land and resources. Policy makers and forest managers are beginning to accept that communities have a role in forest management, but the extent of their role is controversial. Scientists who work in the tropical forests have the opportunity, if not the obligation, to broaden their research and increase its impact by involving forest dwellers as audiences and partners. This will enrich their research and help to establish the management role and credibility of communities. The urgent need to improve tropical forest management dictates that scientists not only create knowledge but also work to disseminate this knowledge and advocate its application at the policy level, in the field and in forest communities.

References

Barber, C. V. and Schweithelm, J. (2000) *Trial by Fire: Forest Fires and Forest Policy in Indonesia's Era of Crisis and Reform.* Washington, D.C.: World Resources Institute.

Bonell, M. (1999) Tropical forest hydrology and the role of the UNESCO International Hydrological Programme: some personal observations. *Hydrology and Earth System Sciences 3(4), 451–461.*

Brookfield, H., Potter, L. and Byron, Y. (1995) *In Place of the Forest: Environmental and Socio-economic Transformation in Borneo and the Eastern Malay Peninsula.* United Nations University Press, Tokyo, Japan.

Brosius, P. Lowenhaupt-Tsing, A. and Zerner, C. (1998) Representing communities: histories and politics of community-based natural resource management. *Society and Natural Resources*, 11, 157–68.

Bruijnzeel, L. A. (1990) *Hydrology of Moist Tropical Forests and Effects of Conversion: A State of Knowledge Review.* Paris, France:UNESCO International Hydrological Programme, Humid Tropics Programme.

Bruijnzeel, L. A. (2004) Tropical forests and environmental services: not seeing the soil for the trees? *Agriculture, Ecosystems and Environment*, doi:10.1016/J.agee.2004.01.015.

Colfer, C. J. with Peluso, N. and Chung, C. S. (1997) *Beyond Slash and Burn: Building on Indigenous Management of Borneo's Tropical Rain Forests.* Bronx, New York: The New York Botanical Gardens.

Dove, M. R. (1983) Theories of swidden agriculture and the political economy of ignorance. *Agroforestry Systems*, 1, 85–99.

Dove, M. R. (1985) The agroecological mythology of the Javanese and the political economy of Indonesia. *Indonesia*, 39, 1–36.

Eghenter, C. (1999) Migrants' practical reasonings: the social, political, and environmental determinants of long-distance migrations among the Kayan and Kenyah of the interior of Borneo. *Sojourn*, 14, 1–33.

Eghenter, C. (2000a) What is *Tana Ulen* good for? Considerations on indigenous forest management, conservation, and research in the interior of Indonesian Borneo. *Human Ecology*, 28(3), 331–357.

Eghenter, C. (2000b) Imagined models vs historical practices: considerations on *tana ulen* and community-based management of resources in the interior of Indonesian Borneo. In *Proceedings of the Conference on Community-Based Management* held in the Philippines, September 1998.

Filer, C. with Sekhran, N. (1998) *Loggers, Donors and Resource Owners: Papua New Guinea Country Study.* London: IIED.

Fox, J. (1993) Introduction. In *Legal Frameworks for Forest Management in Asia: Case Studies of Community/State Relations*, Occasional Paper No. 16, ed. J. Fox, pp. *ix–xix*. Honolulu, Hawaii, East-West Center Program on Environment.

Fox, J. (1999) Understanding a dynamic landscape: land use, land cover, and resource tenure in northeastern Cambodia. *Unpublished working paper*. East-West Center Program on Environment, Honolulu, Hawaii, USA.

Hamilton, L. S. with King, P. N. (1983) *Tropical Forested Watersheds: Hydrologic and Soils Responses to Major Uses and Conversions*. Boulder, Colorado: Westview Press.

Hutterer, K. L. (1988) The prehistory of the Asian rain forests. In *People of the Tropical Rain Forest*, ed, J. S. Denslow and C. Padoch, pp. 63–72. Berkeley, California: University of California Press.

Krishnaswamy, A. (1999) A global vision for forests in the 21st century. *Tropical Forest Update*, vol. 9, no. 4, 7–9.

Laungaramsri, P. (2000) The ambiguity of 'watershed': the politics of people and conservation in northern Thailand. A case study of the Chom Thong conflict. In *Indigenous Peoples and Protected Areas in South East Asia: From Principles to Practice*, ed. M. Colchester and C. Erni, pp. 108–33. Copenhagen: IWGIA.

Lawrence, D. Peart, D. and Leighton, M. (1998) The impact of shifting cultivation on a rainforest landscape in West Kalimantan: spatial and temporal dynamics. *Landscape Ecology*, 13, 135–148.

Mayer, J. H. (1996) Impact of the East Kalimantan fires of 1982–83 on village life, forest use, and on land use. In *Borneo in Transition: People, Forests, Conservation, and Development*, ed. C. Padoch and N. L. Peluso, pp. 187–218. Oxford: Oxford University Press.

Meggers, B. J. (1988) The prehistory of Amazonia. In *People of the Tropical Rain Forest*, ed. J. S. Denslow and C. Padoch, pp. 53–62. Berkeley, California: University of California Press.

Nibbering, J. W. (1988) Forest degradation and reforestation in a highland area of Java. In *Changing Tropical Forests: Historical Perspectives on Today's Challenges in Asia, Australasia, and Oceania*, ed. J. Dargavel, K. Dixon, and N. Semple, pp. 155–77. Canberra, Australia: Centre for Resource and Environmental Studies.

Peluso, N. L. (1993) *The Impact of Social and Environmental Change on Forest Management: A Case Study from West Kalimantan, Indonesia*, Community Forestry Case Study No. 8. Rome, FAO.

Peluso, N. L., Vandergeest, P. and Potter, L. (1995) Social aspects of forestry in South East Asia: a review of postwar trends in scholarly literature. *Journal of South East Asian Studies*, 26, 196–218.

Poffenberger, M., McGean, B. and Khare, A. (1998) Communities sustaining India's forests in the Twenty-first Century. In *Village Voices, Forest Choices: Joint Forest Management in India*, ed. M. Poffenberger and B. McGean, pp. 17–55. Delhi: Oxford University Press.

Poffenberger, M., ed. (1999) *Communities and Forest Management in South East Asia*. Gland, Switzerland: IUCN.

Potter, L. (1988) Indigenes and colonisers: Dutch forest policy in south and east Borneo (Kalimantan) 1900 to 1950. *Changing Tropical Forests: Historical Perspectives on Today's Challenges in Asia, Australasia, and Oceania*, ed. J. Dargavel, K. Dixon, and N. Semple, pp. 127–53. Canberra, Australia: Centre for Resource and Environmental Studies.

Schmidt-Vogt, D. (1998) Defining degradation: the impacts of swidden on forests in northern Thailand. *Mountain Research and Development*, 18, 135–149.

Schweithelm, J. (1987) The need for a method of land evaluation for watershed land use planning in the outer islands of Indonesia: a case study of Riam Kanan, Kalimantan. In *Proceedings of the International Workshop on Quantified Land Evaluation Procedures*. Washington, D.C. 28 April–2 May, 1986, ed. K. J. Beek, P. A. Burrough and D. E. McCormack, pp. 130–36. Enschede, The Netherlands, ITC.

Siebert, S. F. and Belsky, J. M. (1985) Some socioeconomic and environmental aspects of forest use by lowland farmers in Leyte, Philippines and their implications for agricultural development and forest management. *Philippine Quarterly of Culture and Society*, vol. 13, 282–96.

Sunderlin, W. D., Resosudarmo, I. A. P. and Angelsen, A. (2000) *The Effects of Indonesia's Economic Crisis on Small Farmers and Natural Forest Cover in the Outer Islands*. Bogor, Indonesia: CIFOR.

Whitmore, T. C. 1998. *An Introduction to Tropical Rain Forests*, 2nd edn. Oxford: Oxford University Press.

5 People in tropical forests: problem or solution?

A. L. Hall
The World Bank, Washington DC, USA

FOREST PEOPLES AS A PROBLEM?[1]

The treatment of tropical forest dwellers by development organisations has been mixed. Government and international agencies have tended to view forest peoples as, at best, an archaic legacy of a pre-industrial era who have little or nothing to contribute to development and environmental policy. At worst, such populations are perceived as a downright obstacle to progress, or even a threat to national sovereignty, that need to be dragged, screaming if necessary, into the 'modern' age. Throughout history, indigenous and other traditional groups have seen their forest resource base come under attack as geographical frontiers have been pushed back in the name of national integration and economic development. The strategies adopted have included highway construction, cattle ranching, small farmer settlement, export crop production, commercial logging and hydropower expansion. Nations such as Brazil, Indonesia and Malaysia, amongst others, have all pursued aggressive strategies of tropical forest occupation in which the needs of the native peoples themselves have been routinely ignored.

Until relatively recently, the social and environmental consequences of such policies were rarely given a second thought by planners and policy-makers. It is no exaggeration to say that forest peoples were (and to a large degree still are) considered expendable in the march towards modernisation and nationhood. Perhaps the clearest indicator of such pressures is deforestation. It is calculated that by 1988 the world had already lost 40% of its tropical forests (Park, 1992). The figure is undoubtedly much higher today given that some 14 million ha are destroyed every year (WCFSD, 1999). Brazilian Amazonia has lost 14% of its rainforest overall since the 1960s, although deforestation levels are three times this figure in some more intensively settled areas of the region. Furthermore, current detection methods seriously underestimate the true extent of forest loss (Hall, 1999). Figures for other tropical forest regions are much worse; Bangladesh, Haiti and India had all lost 100% of their primary rainforest by 1988 (Park, 1992). Asia as a whole retains just 6% of its original rainforest. By 1995, West Africa had lost three-quarters of its tropical moist forests, reaching 90% in some countries such as Nigeria (WCFSD, 1999).

In both environmental and social terms, the consequences of this rampant destruction have been dramatic. Many problems including soil erosion and degradation as well as loss of environmental services such as biodiversity maintenance, climate regulation and river basin management have been widely documented (Myers, 1984). However, until recently, the impacts on forest dwellers' livelihoods have, by and large, been given relatively little attention, especially in official circles. Forest populations have seen their traditional lands invaded by outsiders and the natural resource-base (terrestrial and aquatic) seriously undermined. Vast areas of common property have been enclosed by incoming farmers, agribusiness interests, loggers and land speculators (Hall, 2000c). More often than not these invasive strategies have been encouraged and heavily subsidised by national governments and international development institutions.

Since the 1960s in Brazil, for example, successive military and civilian governments have pursued an aggressive policy of frontier settlement in Amazonia. This has served a variety of development goals ranging from promoting regional economic development to strengthening national security and absorbing landless farmers from other regions of the country (Hall, 1989). All over the world, developers have promoted the myth of frontier zones as areas of unpopulated wilderness. This enabled two convenient assumptions to be made. First, new territories could be classified as a demographic void, thus removing the responsibility to consider the impacts of frontier occupation on indigenous populations. Second, by denying the presence of forest dwellers, frontier

1 'Forest peoples' are understood here as comprising groups, either indigenous or relatively recent migrants, who live within or adjacent to forests and whose livelihoods depend significantly upon the use of forest resources such as timber, tree products and fishing, etc.

Forests, Water and People in the Humid Tropics, ed. M. Bonell and L. A. Bruijnzeel. Published by Cambridge University Press.

zones could be categorised as open-access and ripe for exploitation by private commercial interests. Traditional common property regimes could then be ignored and indigenous rights usurped with a clear conscience (Colby, 1990; Gómez Pompa and Klaus, 1992). In other cases, the imposition of a European ideal of pristine wilderness incompatible with human occupation has legitimised the displacement of traditional communities to establish national parks and protected areas (Neumann, 1998).

The Amazon, which covers 60% of Brazil, was portrayed during the 1960s and 1970s by military governments as a vast, 'empty' space ripe for occupation and investment by modernising forces. In the words of one prominent policy-maker of the time, the aim was to 'inundate the Amazon forest with civilisation' (Schmink and Wood, 1992: 59). Traditional forest populations such as Amerindians and long-resident peasant farmers and extractivists of mixed descent (*caboclos*) were considered archaic remnants of a pre-capitalist era. They were seen as marginal to the development process, an inconvenient obstacle that was best removed or pushed out of sight – and out of mind. Reflecting the development priorities of the day, over US$5 billion was channelled to Amazon cattle ranchers in subsidies from 1971 to 1987, fuelling land speculation, deforestation and habitat destruction (Schneider, 1992). Major development schemes such as the Carajás iron-ore project and the Polonoroeste programme, both funded by the World Bank during the 1980s, served to aggravate demographic and socio-economic pressures on the rainforest and its peoples (Hall, 1989; Rich, 1994).

Yet over the same period almost no funds were allocated to assisting local forest populations, who have been obliged to organise in defence of their own interests. The struggle of the rubber tappers (*seringueiros*) during the 1980s is one of the best-known cases of successful resistance to land grabbing. This movement culminated in the death of leader Francisco 'Chico' Mendes in 1988 at the hands of cattle ranchers, leading eventually to the introduction in 1990 of the 'extractive reserve', a new policy instrument under Brazilian law designed to safeguard the livelihoods of forest dwellers (Hall, 1996, 1997a). Similarly, riverside communities in the middle and upper Amazon have been mobilising systematically since the 1970s. With the aid of the local church and NGOs, they have sought to protect their fishing grounds against incursions by large commercial boats using predatory techniques which threaten to deplete stocks and threaten people's sources of livelihood (Hall, 1997a; Goulding, *et al.*, 1996).

Amerindians have for centuries seen their land and humid forests systematically occupied by colonising forces of various kinds. In the Brazilian Amazon, numbers have fallen from several million at the time of the Conquest to just 200 000 today. Indigenous reserves and other protected areas cover some 25% of the region. However, they are poorly guarded and under constant pressure from illegal mining and logging operations (FOE, 1997). Sometimes there is official connivance but in most cases the state-level and federal environmental protection agencies are simply too ill-equipped to cope. In Ecuador and Peru, indigenous groups have been fighting a strong rearguard action to control the activities of oil companies in the Amazon jungle. Particularly in Ecuador, the uncontrolled dumping of toxic wastes, water pollution, forest loss and other ecological disturbances have had serious repercussions on indigenous people's forest-based livelihoods. Matters came to a head in 1993 when a class action lawsuit (Aguinda vs. Texaco), as yet unsettled, was filed in federal court in New York against the company on behalf of 30 000 settlers and Amerindians claiming US$1.5 billion in damages (Kimerling, 1991, 2000).

There are many other examples of tropical forest dwellers being harassed in the name of development and being obliged to defend their own interests in the face of official indifference to their plight. The controversial Chipko movement of the Garhwal Himalayas in India was formed to resist large-scale commercial felling in adjacent state-controlled forests. This resulted in stronger government environmental controls but also generated a popular backlash against what many now see as an overly restrictive approach that pays no heed to people's livelihood needs (Rangan, 1993). In Irian Jaya (Indonesian New Guinea) and Kalimantan (Borneo), tribal peoples have claimed customary rights to many areas that were designated by the government as prospective resettlement sites under Indonesia's Transmigration Programme and have come into conflict with the authorities (Rich,1994). In the case of Irian Jaya, these pressures have added to the conflicts generated by a long-standing guerrilla insurrection against the Indonesian annexation in 1969 of the western half of New Guinea (known locally as West Papua). In southern Cameroon, tribal groups have resorted to various forms of protest at the negative social and ecological consequences of large-scale commercial logging for export as unsustainable practices threaten the forest commons (Nguiffo, 1998). In Mexico, the *Zapatista* rebellion in Chiapas has been partly inspired by a popular rejection of the government's decision to privatise collectively owned farmlands and forests in *ejidos* (Stephen, 1998).

In all of the above examples, the interests of forest dwellers have proved largely incompatible with official development strategies. Planners have tended to view indigenous and other groups as an obstacle in the drive to exploit timber and mineral resources for rapid commercial profit or to expand the agricultural frontier. Politically, forest populations have been marginalised and have lacked parliamentary representation. Typically, they are despised and ostracised by society at large, regarded as inferior or 'exotic'. With neither political clout nor wider legitimacy, it is hardly surprising that their livelihood needs are generally ignored.

This low status is clearly reflected in mainstream environmental policy, which still largely views people as the enemies of nature. Thus, conservation strategies have been based primarily on a combination of preservation and criminalisation, otherwise known as the 'fences-and-fines' approach. Predicated on western concepts of centralised environmental control, this paradigm remains dominant and is based on two major ideas (Adams, 1990). First, the setting aside of areas for the preservation of Nature in protected units such as national parks, national forests, wildlife reserves and research stations, in which little or no anthropomorphic activity is permitted. Second, establishing environmental regulations together with systems of surveillance and punishment for those who abuse the law.

This narrow vision has been reinforced by neo-Malthusian ideas on the supposed incompatibility of Nature and human presence. According to the 'Tragedy of the Commons' hypothesis (Hardin, 1968), for example, there is an inevitable tendency towards the degradation of common property resources (CPRs) such as forests, based on the assumption that individual, short-term, profit-maximisation motives will always outweigh longer-term considerations of the collective good. However, it has been widely demonstrated that although this notion may be applicable to 'open access' CPRs where there are no controls, most traditional communities have in fact evolved their own informal and customary systems of collective resource governance (Bromley and Cernea, 1989). As will be discussed below, it is upon these systems which modern policy-makers and practitioners can build.

This mainstream approach to environmental management clearly has its merits and its place as part of any national policy framework. However, it has many drawbacks. Both conservation and law enforcement are expensive and highly labour-intensive activities. Most governments lack the human or financial capacity (and frequently the political will) to enforce such centralised policies, resulting in the increasingly rapid rates of tropical forest loss discussed in the opening paragraphs above. Furthermore, this approach is highly top-down and takes little or no account of the needs of forest dwelling populations, for whom neither fences nor fines are relevant. Conservation policies tend to exclude local groups or grant them minimal participation in management. Controls and laws penalise resource users, especially poor and weak subsistence groups, yet provide no incentive structures to encourage the adoption of non-destructive methods. Until quite recently, the notion that forest peoples could be involved directly and actively in the sustainable management of natural resources as part of official policy was not seriously considered.

FOREST PEOPLES AS A SOLUTION

It is now being recognised, especially since the early 1990s, that conventional nature conservation in the form of protected parks and similar areas is a valuable but limited solution to forest destruction. It is necessary but, on its own, an insufficient deterrent to deforestation and environmental degradation. Long-term conservation of tropical forests to minimise degradation, conserve biodiversity, maintain environmental services and strengthen people's livelihoods will depend increasingly on the integration of forest peoples into forest governance (Hall, 1997a, 1997b, 2000a; Schwartzman *et al.*, 2000). Participatory management is especially critical in the governance of common property resources, where trade-offs have to be negotiated between potentially conflicting objectives. Namely, individuals' propensity to engage in short-term profit maximisation through possibly destructive extraction on the one hand and, on the other, the need to conserve the resource base for public or collective benefit in the longer term. Environmental and livelihood concerns have thus converged. This realisation has sprung from two major sources.

Firstly, applied research has demonstrated quite unequivocally the potential value of local knowledge and community participation in natural resource management, including forests (IIED, 1994; Carney and Farrington, 1998; Wolvekamp, 1999; Haverkort and Hiemstra, 1999). This is discussed below in terms of key roles which forest communities may perform in the development effort: self-defence and vigilance, needs diagnosis and articulation, resource management and capacity-building and policy dialogue. Large numbers of people living in or near forests depend on forest products to varying degrees in a wide range of household survival strategies (Byron and Arnold, 1999). Secondly, as the examples cited above show, forest dwellers themselves have often taken the initiative to place their demands on the political agenda at national and international levels. Dissatisfied with the failure of official policies to address their livelihood needs, forest communities have increasingly given vent to their frustrations by taking direct action, actively resisting overt threats from commercial interests. Such initiatives have frequently involved an alliance of local groups, non-governmental organisations (NGOs), progressive government agencies and foreign donors. Due to a combination of growing technical sophistication and grassroots political pressure, therefore, planners and policy-makers at national and international levels have been obliged to acknowledge and respond to people's demands.

The growing legitimacy of, and need for, an integrated and participatory approach to forest management has also been reflected in international policy statements, ranging from the 1980 World Conservation Strategy, to the 1987 Brundtland Report and the 1992 Earth Summit. UNCED's 'Statement on Forest Principles', although non-legally binding, declares that, 'Governments should promote and provide opportunities for the participation of interested parties, including local communities and indigenous people . . . forest dwellers and women, in the development, implementation and planning of national forest policies' (UNCED, 1992: p. 292).

However, while general policy commitment has undoubtedly improved, there is less clarity about the specific ways in which forest groups may contribute to the governance process. The 'pessimist' might view local involvement as necessarily minimal, limited to filling a knowledge gap which outsiders are unable to fill, or perhaps engaging in token consultation as part of a predetermined agenda towards meeting objectives set by outside interests. The 'optimist', however, could perceive forest dwellers as forming the very foundation of management strategies, their traditional knowledge, governance capacity and decision-making powers being fundamental rather than a mere complement to conventional 'scientific' inputs. Clearly, however, a balance has to be struck. In re-designing forest management strategies, the danger of romanticising local people's potential contributions (as so often happens) has to be avoided and a realistic assessment made of their role and of the support needed to sustain these initiatives.

No developing country has embraced whole-heartedly the principle of community-based natural resource management. Yet there are many experiments under way which point to the potential value (and difficulties) of participatory approaches in terms of strengthening both environmentally sound economic development while supporting local populations. Projects for the joint management of national parks involving government, NGOs and local communities have been initiated in several countries including Costa Rica, Tanzania and Thailand (Wells and Brandon, 1992). Similarly, a community approach to wildlife management has been applied in Africa, in Ghana, Mali, Kenya and Zaire (IIED, 1994). Brazil's 'extractive reserves' for rubber tappers are a well-established policy innovation currently under implementation, while the concept is gradually being transferred to other groups such as fishing communities along the Amazon (Hall, 1997a, 2000c).

In one sense, these experiences are all unique, determined by the characteristics of each situation, the challenges and socio-political responses. Furthermore, these arrangements are invariably multi-institutional, in which the local population is but one of a series of key actors. Yet in spite of this apparent diversity, it is possible to identify a set of almost universal roles which forest populations may play in the process of participatory management and resource protection. These can be broadly classified as (1) self defence and vigilance, (2) community needs diagnosis and articulation, (3) long-term resource governance and capacity-building, and (4) broader policy dialogue.

Self-defence and vigilance

There is a strong argument that the best guardians of the forest are forest-dwellers themselves; that is, those who have a direct interest in forest conservation as a major factor in sustaining local livelihoods. The weakness of most environmental control agencies responsible for overseeing forest areas in developing countries renders them almost useless for detection and law-enforcement purposes. Thus, when external forces threaten, the first line of defence is in practice formed by the people on the ground. The history of western colonisation itself epitomises this struggle, as indigenous peoples in humid tropical forests in Latin America, Africa and Asia resisted European incursions. Today, pressures upon forest resources exerted by mining companies, logging interests, farming and commercial fisheries are an increasingly serious phenomenon which, in all too many instances, national governments are either unable or unwilling to discourage. Indeed, in the quest for 'development' at any cost, such investments (domestic and foreign) are often actively encouraged. Under these circumstances, forest communities have had little choice but to take matters into their own hands. They have both the motivation as well as the physical presence to form an effective barrier against rampant destruction.

As noted above, rubber tappers and fishing communities in Brazilian Amazonia have challenged successfully commercial ranchers, loggers and fishers in face-to-face confrontations (*empates*), securing the territorial integrity of large areas of common-pool forests and inland waterways, thus protecting many people's livelihoods (Hall, 1997a). Today, on the 'extractive reserves' which were set up as a result of the rubber tappers' movement, local people are employed as environmental officers to monitor forest use and report illegal entry to the government agency (IBAMA) so that appropriate action might be taken. Within comparable 'fishing reserves', floating guard posts equipped with short-wave radios, staffed by locals and placed at strategic entrance points also perform a vigilance function. In the tropical forests of Ecuador and Peru, indigenous groups have resisted oil companies *in situ* as well as through national and international campaigning, slowing down or stopping destructive exploration activities and obliging companies to adopt more environmentally sensitive practices (Kimerling, 1991, 2000). In Honduras during the 1970s, some 6000 families dependent on resin-tapping launched blockades and organised cooperatives to protect their livelihoods. 'Today, the villagers physically patrol the forest and limit access to loggers and agricultural encroachers' (Rich, 1994: 286).

The well-known Chipko Movement started in 1973 in India when women in the Garhwal Himalayas literally 'hugged' trees to protect them against logging contractors who had been granted official permission to deforest (Rangan, 1993). The long-standing nature of many such struggles is illustrated by Chipko, 'which is in reality a continuation of more than a century of rural revolts and peasant movements by Indian villagers against the enclosure and logging of common forest areas by state forestry agencies' (Rich, 1994: 285). In East Malaysia, the indigenous Dayaks have been engaged in physical and legal conflicts with officially authorised logging companies for adequate compensation in view of the extensive damage caused by timber extraction. Frustrated with the lack of government support and rejection of their requests to secure rights over communal forest areas, their direct action included a series of 'timber blockades' during 1987–88 (Hurst, 1990). In Papua New Guinea, tribal peoples have been relatively more successful in bargaining with logging companies since they received government support for their demands and timber concessions were restricted. However, they also were obliged to use physical confrontation and resistance to drive home their message (Hurst, 1990).

Needs diagnosis and articulation

Forest populations live typically at the spatial and political margins of society. They rarely enjoy any significant formal political representation nor are their local cultures respected. On the contrary, as already noted, they are often stigmatised as primitive vestiges of backward societies. Their livelihood needs are almost invariably ignored in mainstream policy-making, where concern for generating rapid profits for a few and strengthening national 'integration' generally outweigh all other considerations. However, within a conservation-and-development approach, it is essential to have a clear diagnosis and articulation of forest dwellers' felt needs and perceptions in the project/programme design and management process.

Thus, resource-users have a major role to play in ensuring that their requirements are placed firmly on the planning and policy agenda. If they do not force the issue and play an active part, it is far less likely that other institutions will be able to do so in an effective and appropriate manner. There is always the strong danger that outsiders' objectives will override the needs of local communities. This is essentially a political process and forest communities are usually in no position to progress very far on their own beyond the adoption of immediate defensive tactics such as those described above. Needs articulation in the longer term therefore requires an effective strategy of co-operation with a range of institutions through which such aims can be directed and supported. In most such initiatives, it is common to find a partnership involving grassroots organisations, non-governmental organisations (NGOs), progressive government bodies, foreign donors and occasionally the private sector.

NGOs in particular, by virtue of their independence from government control, often have a vital role to play in acting as a link and articulating forest people's needs, 'scaling up' their activities and placing grassroots interests on the political and policy agenda (Edwards and Hulme, 1992; Carney and Farrington, 1998). Most of the examples cited in the previous section have involved such multi-institutional arrangements. Increasingly, the private business sector will become involved as new markets for forest products are developed.

RESOURCE MANAGEMENT AND CAPACITY-BUILDING

It is the more dramatic and spectacular clashes of 'forest peoples versus outsiders' that attract the headlines and give hope to those in search of local demand-driven development models. However, involving traditional communities in long-term resource and environmental management successfully is a far more onerous challenge. This is particularly so in the case of common-pool resources such as forests, where individual needs have to be reconciled with those of the collective good to avoid a spiral of degradation and a possible 'tragedy of the commons'. Where customary systems of resource governance have broken down or are starting to disintegrate as a result of demographic, commercial, political and other pressures, modified systems must be put in their place. Ideally, these should build upon the existing capacities and knowledge of local populations as the hub of such new initiatives, complemented by technical, financial and political support from sympathetic outside institutions such as NGOs, international donors and progressive state organisations.

A successful outcome is far more likely when users have prior organisational experience in defending and managing their resources. In social science terms, the challenge is to strengthen or enhance existing 'social capital'. This term refers to networks of social relationships, norms and values that allow groups to meet their development objectives (Coleman, 1990). Like physical, human or financial capital, social capital may be viewed as a legitimate form of investment for development. Nowhere is this more so than in the case of common-pool resource management in the forest sector, where grassroots organisation and co-operation is so fundamental (Hall, forthcoming). Strengthening social capital is necessary in several key management areas; developing a common understanding of problems, building up mutual trust, encouraging organisational autonomy and setting economic incentives.

As far as developing a common understanding of problems is concerned, it is often assumed that such a perception is automatic. While there may be a superficial common interest, divisions within traditional populations may run deep along ethnic, class or caste lines. In many cases, forest dwellers in the humid tropics are spatially scattered over huge areas with very poor communications. Large or nuclear communities may not even exist, and settlements might typically comprise a few households. Resource-users may have little or no tradition of collective action apart from responding to immediate threats. At the same time, vertical ties of patron-clientage with political, religious and economic power holders could well undermine group collaboration (Leach *et al.*, 1997). Thus a major management task is to facilitate a better common understanding of the problems faced by the user group as a whole. As part and parcel of this process, establishing regular contacts amongst previously isolated groups and building up trust as the basis for co-operation is a huge challenge.

It is now accepted that, just as common-pool resource use does not necessarily lead to degradation, neither is it automatically conducive to self-governed regulation. 'Users will overuse the forest unless efforts are made to change one or more of the variables affecting perceived costs or benefits' (Ostrom, 1997: p. 6). One of these key variables is the extent to which forest groups are engaged effectively in the design and implementation of management systems. Ostrom (1990, 1997) has suggested a number of basic design principles governing such involvement, which are more likely to be conducive towards success than if such features are absent. An institutionalised system of regular meetings and local organisation which feeds directly into the decision-making process can help to ensure that these principles are followed with the interests of the resource-user group in mind.

These features include, for example: (1) mapping out clearly defined boundaries for common-pool resources and establishing ownership or usufruct rights together with rules for individual or household access, (2) effective and independent community involvement in collective-choice arrangements through appropriate decentralised and central organisational arrangements, (3) monitoring of resource conditions and resource user behaviour to avoid external or internal abuse and to ensure accountability, (4) setting up a system of graduated sanctions to punish offenders, (5) incorporating conflict-resolution mechanisms to settle problems arising amongst users, and (6) the existence of a supportive, wider policy and institutional environment.

One of the key areas of need that has to be effectively articulated and fed into project design and management is the whole question of economic incentives to encourage non-destructive forms of resource utilisation. Many outsiders, especially western intellectuals and 'radical greens', often fall into the trap of assuming that traditional groups have an inherent predisposition to conserve forests regardless of other considerations. In relatively undisturbed forest environments in which the carrying capacity of the land has not been exceeded and livelihoods are maintained, there is little or no reason for people to destroy the resource base. However, when this equilibrium is upset, for whatever reason, survival takes precedence. While it is therefore true that forest dwellers are more aware than most of their ecosystems' importance and fragility, it should not be forgotten that the struggle to survive invariably outweighs naked environmental concerns. If people are left with no option but to deforest in order to support themselves and their families, then that is what they will do. Indigenous and other traditional groups, as we have seen, often defend their forest resources quite literally to the death. Yet where coherent organisational or incentive structures are lacking to motivate such resistance, indigenous and other traditional groups may collaborate enthusiastically with commercial destroyers of the forest (through, for example, the sale of timber and mineral rights). Destabilising forces may range from demographic pressure due to natural increase or in-migration, encroachment by commercial interests such as logging, mining and fishing companies as well as market integration and exposure to consumer pressures.

Forest peoples commonly depend on a combination of extractive activities and small-scale farming to meet their livelihood needs, both in terms of satisfying immediate subsistence requirements and for generating additional income. However, world markets in basic items such as rubber, Brazil nuts and forest fruits are notoriously unstable. The development of synthetic substitutes has also led to gradual price declines, the classic example being rubber. Furthermore, most of the immediate trading in these commodities is dominated by local middlemen who (although they may provide other services to the community) tend to exercise purchasing monopolies, depriving producers of much potential profit. Thus, if access to forest resources has to be restricted in the name of conservation, or if existing production patterns are non-profitable for forest dwellers, it is necessary to complement and diversify people's income sources as an integral part of the environmental management process. This is critical for addressing immediate livelihood needs and will condition users' perceptions about the likely costs and benefits accruing to them as a result of their participation in collective management. It may also be a key issue in the quest for project self-sufficiency as beneficiaries are drawn into cost-sharing arrangements in order to reduce the dangers inherent in long-term dependence on foreign aid.

Much effort is now being devoted by development organisations to devising new schemes for forest peoples which offer an income flow but which do not, at least in theory, threaten the natural resource base. Some such projects are largely new concepts introduced from outside, while others may be firmly based on the use and adaptation of local knowledge. In the former category, the rapid expansion of so-called 'ecotourism' is a case in point. Tropical forests are becoming increasingly attractive destinations for adventure-seeking tourists. Countries such as Costa Rica and Brazil are investing heavily in forest-based infrastructure for tourism purposes, often in collaboration with local communities. Another example is that of nature-based tourism in the national parks of India, Zimbabwe and Indonesia. Tourism offers great potential for providing an alternative source of income and for benefiting the local economy but it carries potential dangers. These include environmental damage due to poor controls and the disproportionate appropriation of income by outsiders such as tour operators, leaving local populations marginalised (Wells and Brandon, 1992; IIED, 1994; Goodwin *et al.*, 1998).

Other revenue-generating options may be based more firmly on adapting local people's knowledge of their forest environment and its productive uses. Research shows that extractive products, far from being obsolete, have much economic potential in modern markets (Clüsener-Godt and Sachs, 1994). Agroforestry has also been hailed as a potentially lucrative activity. Agroforestry is based on combining limited short-cycle subsistence crop production (such as rice, beans and cassava) to meet immediate household food requirements) with the harvesting of perennial tree crops such as fruits and nuts (to generate a commercial profit). Agroforestry practices have long been common amongst indigenous groups in Amazonia, for example, as part of their forest management strategies (Anderson, 1990). Development agencies and local organisations are now adapting such ideas both to manage forested areas in a more sustainable manner and to recuperate degraded frontier areas through the planting of perennials. Thus, the ancient Amazonian practice of traditional agroforestry for subsistence purposes based on forest enrichment, managed fallows and home gardens has provided valuable technological knowledge for the development of modern, commercial agroforestry (Smith *et al.*, 1998; Smith 2000).

The variety of potentially marketable rainforest products is huge and increasingly well documented. Cleared areas may be ideal for combining local crops such as cocoa with new ones such as coffee and black pepper. Canopied forests may yield rubber, construction materials, wild fruits, fibres, medicines, dyes and handicraft materials. Many project-based experiences all over the tropics point to the potential enjoyed by agroforesty and extractive activities in providing economic incentives conducive to conservation (FOE, 1992; Plotkin and Famolare, 1992; Emperaire, 1996; Wolvekamp, 1999). In Indonesia, for example, 80% of rubber and resin production, along with 95% of fruits, comes from smallholder tree gardens (Küchli, 1997: 134).

Yet many problems that tend to raise production costs and reduce the competitiveness of new agroforestry products need to be addressed, including poor production and communications infrastructure, product perishability, limited markets and weak policy support. These problems may be surmountable but require new forms of investment from public and private sectors designed specifically for small, poorer producers who are commercially inexperienced. In some countries new incentive schemes are being introduced to stimulate such productive rainforest activities. In 1997, Brazil set up its PRODEX programme of subsidised credit for extractivism and agroforestry following a parliamentary campaign by congresswoman Marina Silva, herself the daughter of a rubber-tapper from the Amazonian state of Acre. Similarly, under the land reform programme in that country, the dedicated farmer credit scheme PROCERA now favours agroforestry activities which stabilise crop production, in an attempt to discourage the more commonly adopted slash-and-burn farming practices.

As described above, traditional knowledge and social capital are key elements underpinning community roles for forest management, ranging from self-defence to adapted production systems such as extractivism and agroforestry. However, social capital is also vital for catchment and river basin management (Pretty and Ward, 2001). There has been a significant expansion of micro-catchment management using resource-conserving practices implemented by community groups and associations. These have led to increased crop yields, improved groundwater recharge, increased tree cover and vegetation and microclimatic change as well as economic benefits for local areas. India alone, for example, is said to have 30 000 watershed and catchment groups in Rajasthan, Gujarat, Karnataka, Tamil Nadu, Maharashtra and Andhra Pradesh.

POLICY DIALOGUE

At the local level, community involvement in self-defence, needs articulation and management activities are all vital for project success. However, if such participation is to make a broader and enduring contribution to rainforest conservation and livelihood strengthening for forest people, practice must shape policy. In other words, action must extend beyond the project or programme level to influence the policy dialogue itself. The examples cited above, and many others, strongly suggest that there is growing enthusiasm and commitment at all levels to conservation-based, productive rainforest activities that offer a serious alternative to deforestation while supporting forest dwellers at the same time. This movement is multi-institutional and involves varying alliances of key actors such as grassroots organisations, NGOs, international donor organisations, progressive state agencies and the business sector. The momentum gained by these synergistic partnerships is bringing about incipient changes in national forest policies.

Brazil is a good example of such progressive change that belies the conventionally negative image often portrayed in the international media (Hall, 1997b, 2000a). The rubber-tappers' movement for security of land tenure led in 1990 to the introduction of an entirely new policy instrument, the 'extractive reserve'. Since 1993, this sector has received strong support from the US$350 million 'G7 Pilot Programme to Conserve the Brazilian Rainforest' (PPG7), set up in 1993. Presently, some 9% (around four million hectares) of Brazilian Amazonia's 45 million hectares in conservation units is protected under extractive reserves, both federal and state administered (Alves, 1996). There has been considerable pressure to extend this policy to other forest groups in the country. The movement by inland fishing communities in Mamirauá to protect their lakes resulted in a change of policy in the state of Amazonas and the introduction of the 'sustainable development reserve' concept. These initiatives have had a major influence upon extremely innovative legislation that was passed by Brazil's Congress in June 2000. The 'National System of Conservation Units' – SNUC, formally recognises the key role played by local communities in natural resource governance and redefines categories of protected area to allow for their formal participation in this process. It remains to be seen, of course, whether adequate funding and political support will be forthcoming to allow these principles to be applied widely. Yet this is a remarkable watershed in official thinking which lays a strong policy and legal foundation for constructing a new approach that recognises people as providers of solutions to forest degradation.

Community involvement in forest resource use and management has been scaled up significantly in Mexico, where over 70% of remaining forests are the common property of the rural population. This stems from successful projects within the Pilot Forestry Plan (PPF), assisted since 1986 by a bilateral technical agreement with Germany (Alatorre and Boege, 1998). Early experiences involving local communities, NGOs and innovative technical aid from Mexican and German agencies have been extended throughout 500 000 ha of *ejido* lands in Quintana Roo and Campeche states. During the late 1980s, an NGO network promoted a special programme to analyse and disseminate forestry experiences, leading to the creation in 1994 of the Mexican Civil Council for Sustainable Forestry (CCMSS). The CCMSS has since been able to influence national forest policy through its direct participation in government bodies such as the National Forest Technical Consultative Council. It is also involved in the process of introducing timber certification and at international level with initiatives to stimulate sustainable forestry. However, as Klooster (1998) points out, although community-based forestry has brought benefits to forest people, the monopolisation of profits by local elites, official corruption and timber smuggling remain serious problems in many areas which have constrained the effectiveness of people's participation.

In Thailand, pioneering project experiences have helped inform recent discussions on comprehensive policy reform in the forestry sector towards the introduction of sustainable management practices. During the 1970s and 80s, the Mae Moh village in northern Thailand developed a relatively successful social forestry model for teak harvesting. In the 1990s, the Royal Forest Department drew up a forest policy master plan that incorporated this 'forest village' concept that may involve up to 12 000 communities (Küchli, 1997). Village-based forest management initiatives in Nepal fed into wider debates which led to a new Forestry Act in 1978 and the Community Forestry Development Project. The Forestry Act of 1993 provides for the transfer of control over mountain forests to organised community user groups, and has resulted in notable progress in forest conservation (Küchli, 1997).

CONCLUSION

The past two decades have seen the beginnings of a transformation in the way that forest peoples are perceived. From inevitable destroyers of natural resources, driven by poverty, demographic pressure and individual greed, they are now seen as potential guardians of the forest and key providers of potential solutions in collaboration with other stakeholders. A growing number of project experiences, some driving progressive policy change, bears witness to this shift in thinking. At the same time, however, deforestation is proceeding apace throughout the world's tropical moist forests, driven largely by small farmer frontier settlement and, increasingly, commercial timber export interests and capital intensive agriculture. Ineffective systems of environmental monitoring and control, arguably, have little or no impact upon rates of forest loss, rendering vast areas of virgin forest virtually open access. Well-guarded conservation areas and community-protected zones may be the only exceptions to this rule. Valuable timber stands can be extracted at very low cost, generating huge profits for a handful of companies and their political allies.

Despite some progress in recent years, the policy environment in most developing countries is still generally not supportive of community forestry, providing few economic incentives to this sector, which remains heavily dependent on external funding. The financial subsidies which have for so long distorted the development process in favour of destructive commercial enterprises such as cattle ranching and logging have not yet been significantly re-directed towards more sustainable forms of resource use (de Moor and Calamai, 1997). Major public and private investments within forested regions into the development of transport, communications, agricultural and mining infrastructure will also stimulate new demographic and commercial pressures on natural resources. In Brazil, for example, under the *Avança Brasil* programme, the government envisages private and public investments of US$40 billion by 2007 for Amazonia in highway construction, hydropower, waterways, airports and telecommunications as well as other economic and social infrastructure. These are intended to integrate the region into Mercosul, a free-trade zone initially established by Brazil, Argentina, Uruguay and Paraguay but which also now includes Bolivia and Chile (Brazil, 1999). No official environmental impact assessment has yet been carried out on the potential consequences of *Avança Brasil*. However, detailed NGO research has predicted a rapid increase in the rate of Amazon deforestation over the coming decade, the spread of forest fires and a reduction in rainfall as well as much greater pressures on protected areas and loss of key environmental services such as carbon sequestration (IPAM, 2000). In 1970, just 4% of Brazilian Amazonia's original forest cover had been lost, but it has recently been predicted that by 2020 this figure could be between 25–72% (Laurance *et al.*, 2001).

The need to incorporate local populations into forest and environmental management is evident. In Brazilian Amazonia, for example, 40% of the area under 'conservation units' is classed as fully protected, although many such units are in practice inhabited and involve anthropomorphic activities. The remaining 60% are recognised officially as populated, suggesting the need to integrate communities into management strategies if these units are to be guarded effectively and local livelihoods assisted. Whether inside or outside officially protected areas, however, the urgent need to develop community-based forest resource management potential is becoming increasingly apparent. Yet the processes involved are still relatively poorly understood, both at the project level and in terms of the wider policy context. Further research in certain key areas could yield lessons that might help planners and practitioners to realise this potential. These might include the following, for example:

Socio-ecological hotspots

In the 1980s, British ecologist Norman Myers created the concept of biodiversity 'hotspots', arguing that a few hotspot ecosystems covering a small area accounted for a high percentage of global diversity. More recently, he and others have suggested specifically that 25 hotspots containing just 1.4% of the Earth's land surface contain 44% of all species of vascular plants and 35% of all species in four vertebrate groups (Myers *et al.*, 2000). Conservationists argue that scarce resources and preservation efforts should be concentrated in these hotspots in a 'silver bullet' strategy to slow down species loss most effectively (Conservation International, 2000).

In a similar vein, one could conceive of socio-ecological hotspots which combine on the one hand, key biodiversity and stocks of natural timber and water resources and on the other, local populations with a potential conservation and development role to play. The identification and mapping of especially vulnerable tropical moist forested areas in which local populations could thus be actively integrated into environmental protection and management strategies. In this strategy, the existence of social diversity would be seen as a prerequisite for the conservation of biodiversity and forest resources. This principle could be integrated systematically into legislation governing the management of populated conservation areas. This is happening, for example, in the case of Brazil's National System of Conservation Units (SNUC) mentioned above (Hall, 2000a). Within these areas, it is necessary to analyse the relationships between social groups and their environment, with particular reference to the relative weight of natural resources and other income sources within livelihood strategies (rural and urban). The roles of traditional knowledge and local social structures within this pattern should be analysed to explore the implications for designing conservation-and-development initiatives.

Economic mechanisms

Even if more effective local projects can be devised, their long-term success will depend upon bringing about a more supportive policy environment to make sustainable forest management more financially rewarding so that it might compete against the more profitable but destructive forms of forest exploitation. Research is needed into a range of proposed 'innovative incentive mechanisms' which could make sustainable forest and watershed management more attractive by 'internalising externalities'. That is, by incorporating non-market social and environmental costs and benefits into the financial returns of forest and watershed users (Richards and Moura Castro, 1999). These might include, for example:

(1) Internal and international transfer payments. Market-based instruments at the domestic level would transfer payments amongst stakeholders to offset non-market costs, together with innovative forest pricing mechanisms such as performance bonds. At the international level, options include debt-for-nature swaps, conservation trust funds and international timber trade taxes.
(2) Market or trade-based solutions on public good benefits. At the national level, sale of protection rights and ecotourism charges could be investigated. It has been estimated that forestry could offset up to 15% of the world's greenhouse gas emissions and provide capital for the sector. Thus, forest-based carbon offset trading linked to the Clean Development Mechanism of the Kyoto Protocol may become feasible. In Brazil, for example, slowing down deforestation is thought to offer the greatest potential for combating global warming compared with options such as plantation forestry and sustainable timber management (Fearnside, 1999). To the extent that social forestry could help reduce rates of forest loss, it could make a significant contribution towards carbon sequestration. Timber certification and bio-prospecting are other policy options at the international level.

Governance mechanisms

There is much optimism within development policy circles about the inherent capacities of forest populations to administer natural resources and maximise benefits effectively, whether for individual users and their families, for the community, for the wider public good and for the environment as a whole. Yet the inconvenient truth is that, notwithstanding those highly publicised instances of collective resistance, most forest groups are socially and politically fragmented, geographically isolated and lacking any tradition of collective action or organisation. More research is therefore needed into learning from successful experiences and designing appropriate mechanisms for strengthening grassroots-based governance. For example:

(1) Social capital and organisation. We need to know far more about existing forms of traditional social capital and the extent to which long-standing systems of social organisation amongst forest groups can be harnessed for the purposes of

resource governance. Gaps in traditional knowledge must be identified rather than ignored, and existing attributes complemented by capacity building in new skills for key areas such as needs diagnosis and articulation and day-to-day management.

(2) Property rights. It cannot be assumed that the establishment of particular forms of property rights, such as individual ownership for example, will lead automatically to rational exploitation and conservation of natural resources. Research into clearly defined and secure property rights is needed to set out appropriate legal arrangements, whether based on community usufruct or partial privatisation. These need to be married to appropriate economic incentives and systems of resource management.

(3) Self-sufficiency. Arguably, social or community forestry schemes should be subsidised by the state and international development organisations, using mechanisms such as those mentioned above, in view of the environmental and livelihood services that they provide. Yet in order that they be replicable, this has to be balanced with the need for financial self-sufficiency. Research is needed on appropriate institutional and economic arrangements that minimise the risk of over-dependence on external aid and maximise the likelihood of successful dissemination of social forestry models.

Yet whatever mechanisms are adopted, one thing is abundantly clear. The start of the third millennium is a critical period for forest and water resources all over the globe. Centralised, command-and-control policies have their role but this is limited in the context of growing human pressures which official agencies simply cannot deal with effectively. Civil society is playing an increasingly important part in the management of these strategic resources; civil society at all levels, from grassroots communities to NGOs and the private business sector. The sustainability of a viable resource base capable of meeting the needs both of present and future generations will become dependent upon these varied institutions reaching negotiated and workable agreements on how this base might be managed properly for the benefit of all.

References

Adams, W. (1990) *Green Development: Environment and Sustainability in the Third World*. Routledge, London.

Alatorre, G., and E. Boege (1998) Building Sustainable Farmer Forestry in Mexico, in Blauert and Zadeck, eds: 191–214.

Alves, K. (1996) Uma Visão Geral das Unidades de Conservação no Brasil, in A. Ramos and J. P. Capobianco, eds. *Unidades de Conservação no Brasil*, Instituto Socioambiental, São Paulo: 1–12.

Anderson, A. ed. (1990) *Alternatives to Deforestation: Steps Towards Sustainable Use of the Amazon Rainforest*. Columbia University Press, New York.

Blauert, J. and S. Zadeck, eds. (1998) *Mediating Sustainability: Growing Policy from the Grassroots*. Kumarian, West Hartford, Connecticut.

Brazil (1999) *Avança Brasil: Development Structures for Investment*. Brasilia.

Bromley, D. and M. Cernea (1989), *The Management of Common Property Natural Resources: Some Conceptual and Operational Fallacies*. World Bank Discussion Paper 57, World Bank, Washington, D.C.

Byron, N. and M. Arnold (1999) What Futures for the People of the Tropical Forests? *World Development*, 27 (5): 789–805.

Carney, D. and J. Farrington (1998) *Natural Resource Management and Institutional Change*. Routledge/ODI, London.

Clüsener-Godt, M., and I. Sachs, eds. (1994) *Extractivism in the Brazilian Amazon: Perspectives on Regional Development*. MAB Digest 18,UNESCO, Paris.

Colby, M. (1990) *Environmental Management in Development: The Evolution of Paradigms*. Discussion Paper 80, World Bank, Washington, D.C.

Coleman, J. (1990) *Foundations of Social Theory*. Harvard University Press, Cambridge, Mass.

Collinson, H. ed. (1996) *Green Guerrillas: Environmental Conflicts and Initiatives In Latin America and the Caribbean*. Latin America Bureau, London.

Conservation International (2000) *Hotspots*, Conservation International Foundation, Washington, D.C.

de Moor, A. and P. Calamai (1997) *Subsidizing Unsustainable Development*. Earth Council, Costa Rica/ Institute for Research on Public Expenditure, Netherlands.

Edwards, M. and D. Hulme (1992) *Making a Difference: NGOs and Development in a Changing World*. Earthscan, London.

Emperaire, L., ed. (1996) *La forêt en jeu: L'extractivisme en Amazonie central*. ORSTOM/UNESCO, Paris.

Fearnside, P. M. (1999) Forests and Global Warming Mitigation in Brazil: Opportunities in the Brazilian Forest Sector for Responses to Global Warming Under the 'Clean Development Mechanism'. *Biomass and Bioenergy*, 16: 171–189.

FOE (1992) *The Rainforest Harvest: Sustainable Strategies for Saving the Tropical Forests?* Friends of the Earth, London.

(1997) *Garimpagem Florestal*. Friends of the Earth, São Paulo.

Friedmann, J., and H. Rangan, eds. (1993) *In Defense of Livelihood: Comparative Studies on Environmental Action*. Kumarian, West Hartford, Connecticut.

Goldman, M. ed. (1998) *Privatizing Nature: Political Struggles for the Global Commons*. Pluto Press, London.

Gómez-Pompa, A. and A. Klaus (1992) Taming the Wilderness Myth, *BioScience*, 42 (4): 271–279.

Goodwin, H., I. Kent, K. Parker and M. Walpole (1998) *Tourism, Conservation and Sustainable Development*. International Institute for the Environment and Development, London.

Goulding, M., N. Smith and D. Mahar (1996) *Floods of Fortune*. Columbia University Press, New York.

Hall, A. (forthcoming) Enhancing Social Capital: Productive Conservation and Traditional Knowledge in the Brazilian Rainforest, in Posey, ed. (forthcoming)

(2000a) Environment and Development in Brazilian Amazonia: From Protectionism to Productive Conservation, in Hall (2000b): 99–114.

ed. (2000b) *Amazonia at the Crossroads: The Challenge of Sustainable Development*. Institute of Latin American Studies, University of London.

(2000c) Privatising the Commons: Liberalisation, Land and Livelihoods in Latin America, in W. Baer and J. Love, eds. *Liberalization and its Consequences: A Comparative Perspective on Latin America and Eastern Europe*. Edward Elgar, Cheltenham: 232–258.

(1999) Deforestation in Brazilian Amazonia: Trends, Causes and Policy Implications, *mimeo*.

(1997a) *Sustaining Amazonia: Grassroots Action for Productive Conservation*. Manchester University Press, Manchester.

(1997b) Peopling the Environment: A New Agenda for Research, Policy and Action in Brazilian Amazonia. *European Review of Latin American and Caribbean Studies*, 62, June: 9–31.

(1996) Did Chico Mendes Die in Vain? Brazilian Rubber Tappers in the 1990s, in Collinson, ed: 93–102.

(1989) *Developing Amazonia: Deforestation and Social Conflict in Brazil's Carajás Programme*. Manchester University Press, Manchester.

Hardin, G. (1968) The Tragedy of the Commons. *Science*, 162 (13): 1243–48.

Haverkort, B., and W. Hiemstra, eds. (1999) *Food for Thought: Ancient Visions and New Experiments of Rural People*. Compas, Leusden.

Hurst, P. (1990) *Rainforest Politics: Ecological Destruction in South East Asia*. Zed Press, London.
IIED (1994) *Whose Eden? An Overview of Community Approaches to Wildlife Management*. International Institute for the Environment and Development, London.
IPAM (2000) *Avança Brasil: Os Custos Ambientais para a Amazônia*. Belém.
Kimerling, J. (2000) Oil Development in Ecuador and Peru: Law, Politics and the Environment, in Hall (2000b), 73–96.
(1991) *Amazon Crude*. Natural Resources Defense Council, New York.
Klooster, D. (1998) Community-Based Forestry in Mexico: Can it Reverse Processes of Degradation? *mimeo*.
Küchli, C. (1997) *Forests of Hope: Stories of Regeneration*. Earthscan, London.
Laurance, W., M. Cochrane, S. Bergen, P. Fearnside, P. Delamônica, C. Barber, S. D'Angelo, and T. Fernandes (2001), The Future of the Amazon. *Science*, 291, 19 January: 438–439.
Leach, M., R. Mearns and I. Scoones, eds. (1997) Community-Based Sustainable Development: Consensus or Conflict? *IDS Bulletin*, 28 (4), October.
Myers, N. (1984) *The Primary Source: Tropical Forests and Our Future*. W. Norton, New York and London.
Myers, N., R. Mittermeier, C. Mittermeier, G. da Fonseca, and J. Kent (2000) Biodiversity hotspots for conservation priorities. *Nature*, 408, 24 February: 853–858.
Neumann, R. (1990) *Imposing Wilderness: Struggles over Livelihood Preservation and Nature Preservation in Africa*. University of California Press.
Nguiffo, S. A. (1998) In Defence of the Commons: Forest Battles in Southern Cameroon, in Goldman, ed: 102–119.
Ostrom, E. (1997) Self-Governance and Forest Resources. *Occasional Paper No. 20*, Centre for International Forestry Research (CIFOR), Jakarta.
(1990) *Governing the Commons: The Evolution of Institutions for Collective Action*. Cambridge University Press, New York.
Park, C. (1992) *Tropical Rainforests*. Routledge, London.
Plotkin, M. and L. Famolare, eds. (1992) *Sustainable Harvest and Marketing of Rain Forest Products*. Island Press, Washington, D.C.
Posey, D. ed. (forthcoming) *Human Impacts on Amazonia: The Role of Traditional Ecological Knowledge in Conservation and Development*. Columbia University Press, New York.
Pretty, J. and H. Ward (2001) Social Capital and the Environment, *World Development*. 29(2), February: 209–227.
Rangan, H. (1993) Romancing the Environment: Popular Environmental Action in The Garhwal Himalayas, in Friedmann and Rangan, eds: 155–181.
Rich, B. (1994) *Mortgaging the Earth*. Earthscan, London.
Richards, M. and P. Moura Castro (1999) Can Tropical Forestry be Made Profitable By Internalising the Externalities? *Natural Resource Perspectives*, 46, October, Overseas Development Institute, London.
Schmink, M. and C. Wood (1992) *Contested Frontiers in Amazonia*. Columbia University Press, New York.
Schneider, R. (1992) *Brazil: An Analysis of Environmental Problems in the Amazon*. Report No. 9104-BR, World Bank, Washington, D.C.
Schwartzman, S., A. Moreira and D. Nepstad (2000) Rethinking Tropical Forest Conservation: Perils in Parks. *Conservation Biology*, 14 (5), October: 1351–1357.
Smith, N. (2000) Agroforestry Developments and Prospects in the Brazilian Amazon, in Hall (2000b): 150–170.
Smith, N., J. Dubois, E. Lutz and C. Clement (1998) *Agroforestry Experiences in the Brazilian Amazon: Constraints and Opportunities*. Pilot Program to Conserve the Brazilian Rainforest, World Bank/Ministry of the Environment, Brasilia.
Stephen, L. (1998) Between NAFTA and Zapata: Responses to Restructuring the Commons in Chiapas and Oaxaca, Mexico, in Goldman, ed: 76–101.
UNCED (1992) *Agenda 21*. United Nations, New York.
Wells, M., and K. Brandon (1992) *People and Parks: Linking Protected Area Management with Local Communities*. World Bank/WWF/USAID, Washington D.C.
Wolvekamp, P., ed. (1999) *Forests for the Future*. Zed Press, London.
WCFSD (1999) *Our Forests, Our Future: Report of the World Commission on Forests and Sustainable Development*. Cambridge University Press, Cambridge.

6 Useful myths and intractable truths: the politics of the link between forests and water in Central America

D. Kaimowitz

Center for International Forest Research, Bogor, Indonesia

INTRODUCTION

In the final days of October 1998, Hurricane Mitch unleashed an apocalyptic rampage of floods and mudslides that wreaked havoc on Honduras, Nicaragua, Guatemala and El Salvador, causing 9000 deaths and US$6 billion in damage (Smyle, 1999; see also Bonell, Callaghan, and Connor, this volume). Once the floods subsided, people throughout the region began asking why the storm had sown such great destruction and how they could prevent future catastrophes. Press reports, public officials, environmentalists and international agencies claimed deforestation had greatly magnified the damage. To make the region less vulnerable to disasters they proposed greater support for reforestation, soil conservation and civil defence. 'Watershed management' and 'vulnerability' became watchwords. The agencies practically fell over one another to see who could invent more initiatives with those words in their titles.

Hurricane Mitch put watershed (river basin and/or catchment) degradation firmly on the Central American political landscape. Nevertheless, public concern about the problem had been growing steadily since the 1970s. News stories and consultant reports claiming that sediment was clogging up the region's dams, rivers and coasts had caused consternation in policy circles. Non-governmental organisations (NGOs), the media and others had convinced much of the public that deforestation had exacerbated seasonal water shortages by increasing surface runoff and reducing rainfall. Many agencies had set up reforestation, soil conservation and protected area projects in response to these concerns.

Recent interest in payments for environmental services has further fueled enthusiasm for catchment management. Over the past decade, the agricultural sector's political influence has greatly waned in Central America. Many of those associated with the sector see such payments as an opportunity to boost political support and funding for agriculture and forestry. They argue that urban consumers should pay farmers to protect the river basins that supply the cities' electricity and drinking water.

Public concern about basin degradation is well intentioned and well founded. However, myths and misunderstandings underlie much of the discussion about how forest cover relates to sedimentation, rainfall and water flows. Deforestation probably has only a slight effect on large-scale flooding and regional rainfall (Calder, 1999; Chomitz and Kumari, 1998). Sedimentation poses little medium-term threat to Central America's hydroelectric plants and the Panama Canal (OAS, 1992; Ordoñez, 1994). To the extent that sediment does constitute a problem, however, in many places road construction, urbanisation and other non-agricultural activities generate as much or more sediment as do agricultural activities (Enters, 2000; Nagle, Fahey and Lassoie, 1999). Forest clearing, if followed by land uses that prevent rainfall from percolating into the ground, increases runoff. That, in turn, may reduce dry-season water flows (see Scott *et al.*, this volume). But deforestation is as least as likely to have the opposite effect, since forests generally lose more water from evapotranspiration than shorter vegetation (Bosch and Hewlett, 1982; Bruijnzeel, 1990; Calder, 1999; Enters, 2000; Hamilton and King 1983).

The policies that development and environmental agencies are currently pursuing to mitigate catchment degradation are unlikely to achieve that goal. Most projects emphasise soil conservation and tree planting but pay scant attention to ensuring that farmers sustain those activities. Few projects select the locations for these efforts based on the potential off-site benefits and the areas involved are generally too small to have a significant impact at the landscape level (Nagle, Fahey, and Lassoie, 1999). The great difficulty in measuring the off-site effects of river basin projects and the pressure to respond to the immediate needs of local constituencies give project managers strong incentives to focus on on-site impacts, rather than the off-sites consequences (Aguedelo and Kaimowitz, 1987). The emphasis on soil erosion resulting from agricultural activities diverts attention from other sources of

Forests, Water and People in the Humid Tropics, ed. M. Bonell and L. A. Bruijnzeel. Published by Cambridge University Press.

erosion. Lack of clarity about whether the main objective is to provide off-farm benefits, increase agricultural productivity, produce forest products, or generate short-term employment often leads to efforts that do not achieve any of these goals effectively.

Despite all this, the myths about catchment degradation have yielded positive results. The concern about catchments has generated a favorable climate for addressing environmental issues. It has also provided a rationale for much needed development and conservation investments in rural areas that would not otherwise exist. Some soil conservation and reforestation efforts help farmers improve their incomes.

Alongside these useful myths, there are also intractable truths. Real off-farm catchment problems *do* exist. Even though sedimentation problems will not close the region's hydroelectric plants or the Panama Canal any time soon, the long-term off-site costs of soil erosion are probably substantial. In many instances, it may well be more cost-effective to prevent water pollution than to build expensive water treatment plants. We still do not know enough about the effects of land use changes on climate, water flows and sedimentation. But the simple fact that land use changes greatly alter existing ecological balances poses inherent risks and the precautionary principle makes it incumbent upon us to address them. Urbanisation, rising water consumption and soil compaction are depleting Central America's aquifers. However, no one has a good handle on these issues, much less a clear cost-effective solution for dealing with them. While the least risky solution might be to maintain natural forest cover, it is often too late for that or simply not feasible. Where that leaves us is not always very clear.

This chapter examines the policy debate related to the links between forests and water in Central America and the approaches that policymakers and others have used to address the perceived problems, with emphasis on the siltation of large reservoirs. It shows how political, institutional and technical factors have interacted to produce positive but sub-optimal results and offers suggestions for future initiatives. While the focus is on Central America, many of the arguments presented apply to other tropical regions. The next section provides a brief history of the debates surrounding catchment issues in Central America, followed by a summary of recent scientific literature on the biophysical and economic links between forests, climate, and water and sediment flows. Case histories of the El Cajón hydroelectric dam in Honduras, the Lempa River basin in El Salvador, the Panama Canal, and Hurricane Mitch, are then presented. These cases have many dimensions but we concentrate here exclusively on the aspects related to off-farm hydrological effects; there is no attempt to evaluate the projects involved, which may well be justified on other grounds. (Indeed, the 'useful myths' hypothesis suggests that is the case.) In addition, the main focus is on *forest* cover.

A QUARTER-CENTURY OF DEBATE

Concerns over soil erosion, sedimentation and the hydrological impacts of forest clearing in Central America go back to the 1920s, if not further (Alvarado, 1985). During the early 1970s, the Center for Research and Education on Tropical Agronomy (CATIE), the Food and Agriculture Organization of the United Nations (FAO) and the British Overseas Development Administration (ODA) promoted catchment management in several Central American countries (Mojica, 1975; UNDP / FAO, 1980; Wall, 1981). However, these initiatives failed to catch policymakers' imagination.

It took alarmist reports about the sedimentation of the Panama Canal Watershed and the region's main hydroelectric dams to put catchments on the political agenda. Wadsworth (1976) and Larson (1979) warned that degradation of the Panama Canal catchment could seriously affect the Canal's operations within a few decades. The Harza Engineering Company International (1976) claimed that sedimentation had reduced the original capacity of the El Salvador's '5 de Noviembre' dam by almost two-thirds. The United States Agency for International Development (USAID) issued reports asserting that siltation of the region's hydroelectric dams would greatly decrease their life span and cause hundreds of millions of dollars in damages (García, 1982; ROCAP, 1983). Two scientists from the Smithsonian Institute released a study purporting that deforestation had reduced rainfall in the Panama Canal Watershed and Northwest Costa Rica (Windsor and Rand, 1985). Thus, by the time Jeffrey Leonard published his influential assessment of natural resource degradation in Central America in 1987 it had become conventional wisdom that deforestation seriously endangered the region's energy supply and navigation routes and probably contributed to flooding and droughts.

These findings resonated among certain key international agencies and policymakers. Costa Rica, El Salvador, Guatemala, Honduras and Panama depend heavily on hydroelectric energy and hydroelectric dams account for a major share of their foreign debt. The Canal is central to Panama's economy and the reports on siltation problems were published about the same time that Panama and the United States were negotiating the future ownership of the Canal. The fact that US government agencies issued several of the more alarming reports and followed up by funding a regional catchment management project and several national projects, lent credibility to some of the sensationalist findings, thus causing even greater concern.

In this context, the Inter-American Development Bank (IDB) joined forces in 1988 with the Organization of American States (OAS) to formulate catchment management projects for areas near three of Central America's largest hydroelectric dams: Chixoy in Guatemala, El Cajón in Honduras, and Cerron Grande in El Salvador. IDB had funded a large share of the dams' construction and faced growing criticism for not protecting the catchments that housed them.

The team that formulated the three IDB catchment projects assumed initially that the projects' main objective would be to reduce dam siltation. Nevertheless, the project formulation studies concluded that dam sedimentation posed no real threat and that one could not justify the projects primarily on the basis of curtailing sediment flows (OAS, 1992). Apparently, this led the IDB to consider other justifications for the projects, which by then were already in the pipeline. By the time the IDB prepared the final projects, it was justifying them largely on their positive impacts on local communities, rather than on sediment control.

Even though the official project papers of the IDB's Chixoy, El Cajón and El Salvador projects stressed the on-farm benefits of soil conservation and crop diversification, rather than stemming sediment flows, many public officials and project personnel continued to view the latter as the projects' main objective. This contributed to the belief in these countries that soil conservation and tree planting play key roles in protecting urban consumers' energy supplies.

The decade of the 1990s witnessed the rapid expansion of non-government environmental activities. Foreign assistance agencies shifted support from public sector agencies to NGOs and gave the environment higher priority. Many NGOs wished to convince local farmers and communities that environmental problems affected their well-being directly and used catchment degradation as a case in point. These groups told farmers that if they cleared additional forest and failed to protect their soils, their water sources would dry up, their yields would decline and their crops would receive less rain. Numerous press reports echoed their message, which fit well with the popular perception that deforestation was drying up the region's rivers and streams.

The NGOs' messages found particularly fertile ground in Honduras after plummeting water levels in the reservoir of the El Cajón hydroelectric dam led to massive power outages in 1984 and almost caused economic collapse. The crisis caught the Honduran authorities unprepared and they had great difficulty explaining what had happened. That led the press and many government officials to speculate that a drought brought on by deforestation had caused the water shortages (Gellin, 1994; *El Heraldo*, 1993; Loker, 1995). Later, it came out that seepage in the dam itself was largely responsible for the problem (Mangurian, 1997). But by then the press had moved on to other stories.

In El Salvador, a national NGO called the Salvadoran Research Program on Development and the Environment (PRISMA) convinced many policymakers and opinion leaders that the country was on the verge of a serious water crisis. PRISMA's research showed that urban sprawl had negatively affected the recharge of the San Salvador aquifer and made it harder to meet the rising demand for water. Meanwhile, the contamination and siltation of the Lempa River made it much more costly to use the river as an alternative water source (Barry, 1994).

The link between drought and deforestation reappeared as a frequent topic of conversation and press reports in 1997, when the El Niño phenomenon caused strong droughts throughout the region. The droughts provided optimal conditions for forest fires to proliferate and the region experienced some of the worst fires in its history. Much of the public and the media attributed the drought and the fires to environmental degradation brought on by deforestation and logging.

None of these situations compares with the flurry of activity that followed in the wake of Hurricane Mitch, the worst hurricane ever to hit the Western Hemisphere. Overnight, everyone's attention shifted to the region's vulnerability to natural disasters, which many associated with the lack of forest cover. The World Bank, the IDB and many bilateral donors formulated a new generation of catchment management projects. Practically all the ministers of agriculture and environment added the topic to their list of priorities.

THE SCIENTIFIC VIEWPOINT

What exactly is catchment management? The term implies that someone manages land use at a scale larger than a farm to achieve collective benefits. Unless that is the objective, it makes little sense to focus on the catchment, rather than the farm, level (White, 1994). The question is what collective benefits concern us? Those mentioned most frequently include higher rainfall, flood control, greater dry season flow, landslide prevention, improved water quality, and reduced sedimentation of reservoirs, waterways and coastal zones.

Research on how deforestation affects rainfall remains inconclusive. Simulation models predict that massive deforestation will decrease rainfall in some areas and increase it in others (Bruijnzeel, 1990; Chomitz and Kumari 1998; Costa and Mahé *et al.*, both this volume). Scientists anticipate larger effects in regions where a large portion of the rain derives from evaporation within the region itself. This holds true for the Amazon but not so much so for regions such as Central America and South East Asia. Modellers have concentrated on simulating the effects of total forest conversion over very large areas. Whether the climate changes resulting from real changes in land use would be large enough to have major economic impacts remains uncertain.

Tropical montane cloud forests constitute a partial exception. These forests are known to intercept clouds or fog and channel some of the water to the forest floor as canopy drip. Thus, even though strictly speaking they may not affect rainfall, they do influence the amount of water that moves from clouds to the forest floor. As a result, removing cloud forests may well reduce the amount of water available for different purposes (Bruijnzeel, 1999; 2004 and also this volume).

To date, no compelling evidence has shown that land use changes involving only a few thousand kilometres, as typically occur in Central America, affect rainfall significantly.[1] Careful scrutiny of studies that make claims about land-use effects generally show that, in general, they used poor methodologies or that other plausible explanations may account for their findings (Bruijnzeel, 2004; Hart, 1992; Vargas, 1993). For the moment, one cannot discard the possibility that land use changes in Central America have caused rainfall to decline but most research suggests that is unlikely (Bruijnzeel, 2004; Calder, 1999).

Flood control provides a similar case. Research shows that land use affects the infiltration of water into the soils and changes in land use that compact the soil or diminish porosity will increase runoff and peak flows and, arguably, flooding. In addition, contrary to popular wisdom, the removal of tree cover tends to increase annual water yields, since more water evaporates from trees than from shorter crops (Bruijnzeel, 1990; Hamilton and King, 1983; Calder, 1999; Roberts *et al.*, this volume). This leaves more water that can contribute to flooding.

Nonetheless, these results hold mostly for small areas. At larger scales local effects average out and any storm long and intensive enough to cause major floods is likely to overwhelm the soil's capacity to absorb the rainfall early on. In such circumstances, land use is unlikely to affect greatly how much flooding occurs. Most studies of large-scale flooding find no relation between land use and flood intensity (Anderson Jr., Franca Ribeiro dos Santos and Diaz, 1993; Bruijnzeel, 2004; Calder, 1999; Chomitz and Kumari, 1998; Enters, 2000; see also discussion in Bonell and Scott *et al.*, both this volume). Hence, whether or not farmers deforest their catchments probably does not influence flooding intensity greatly in major floods such as those associated with Hurricane Mitch or the floods that regularly batter the lowlands of the north coast of Honduras.

The issue of dry season flows is less clear. On the one hand, forest vegetation usually reduces annual water yields, leaving less total water available (Finlayson, 1998; see also discussion in Chappell, Tych *et al.*; Grip *et al.*; Hölscher *et al.*, all this volume). On the other hand, any land use that improves water infiltration should help replenish groundwater reserves. Greater groundwater reserves imply more water available in the dry season. Whether the negative evapotransporation effect or the positive infiltration effect dominates depends largely on the rainfall regime, soil type and the land uses involved (Bruijnzeel, 2004; Calder, 1999). Young, rapidly growing tree plantations typically have higher evapotranspiration rates than mature forests. Burning, over-grazing and completely eliminating scrub vegetation typically reduce water infiltration. Certain soil conservation measures have the opposite effect. One cannot assume forest cover always leads to more water in the dry season. Indeed Finlayson (1998) claims that, 'the modern literature is unanimous in saying that most forests do not increase, but reduce dry-season flows'. It is particularly risky to assume that planting trees will re-establish or improve dry season flows in the medium-term. In the initial years, forest plantations have high evapotranspiration rates and poorly developed root systems. On the other hand, the evidence suggests that the conversion of tropical montane cloud forest to agricultural uses in Central America reduces dry season flows (Bruijnzeel, 1999).

Prior to Hurricane Mitch, natural resource specialists in Central America paid scant attention to landslides and mass wasting. Discussions of 'soil erosion' focused mostly on gradual and continuous soil loss (so-called 'sheet', 'laminar', or 'rill and interrill' erosion). However, Hurricane Mitch demonstrated the huge destructive potential of massive soil movements and the importance of episodic extreme events. Vegetation with deep and extensive root systems frequently provides greater soil stability, although landslides may occur nonetheless (Cassells *et. al.*, 1985; Smyle, 1999; see also Douglas and Guyot; Scatena, Planos-Gutierrez and Schellekens, all this volume).

Everyone agrees that sedimentation can adversely affect hydroelectric dams, waterways, irrigation systems and coastal zones. Marked differences exist, however, regarding the magnitude of the costs and whether any one can solve the problem cost-effectively. Aylward (1998) argues convincingly that previous studies have exaggerated the extent of sedimentation affecting hydroelectric reservoirs and the associated costs. However, Aylward's argument assumes that after countries degrade their existing catchments and allow their reservoirs to fill with silt, they will be able to obtain alternative sources of energy and drinking water at a reasonable price and will have the necessary resources to do so.[2] This is certainly possible, but should not be taken for granted. Aylward also ignores the advantages of using renewable energy sources that do not emit carbon, compared to alternatives based on fossil fuels. While Aylward's research takes into account most of the available evidence on the subject, great uncertainty remains about the extent

1 Gutierrez and Rapidal (1999) identified a significant decline in rainfall over a 100-year period in two locations in Nicaragua, but the authors do not attribute it to changes in land use. Fleming (1986) says precipitation in lowland Costa Rica fell and in higher areas it rose. Some intriguing new research suggests that smoke and dust can affect precipitation and rain droplet size (S. Bruijnzeel, personal communication, 2000; Rosenfeld, 1999). The practical implications of this for rainfall in Central America remain uncertain.

It is also widely acknowledged that cloud forests add to precipitation through fog or cloud deposition (Finlayson, 1998).

2 Traditional justifications for the use of discount rates assume that countries can find institutional mechanisms to make inter-generational transfers, per capita incomes rise over time, and all of the products and services involved have readily available substitutes (Portney and Weyant, 1999). It is not clear whether one can expect all these conditions to hold in the case at hand.

and sources of sedimentation, how sediment is distributed within reservoirs, and the role of extreme events, among other things. Nor has Aylward or anyone else analysed the costs of sedimentation on drinking water.

Even if one decides to curtail sediment flows, research suggests that tree plantations and mechanical soil conservation measures in agricultural fields are rarely the most cost-effective way to do so. Frequently, rural roads and construction activities and mass wasting contribute most to siltation and the channel system or flood plains may already store massive amounts of sediment awaiting transport to the reservoir (Nagle *et al.*, 1999). Soils often erode more under fast-growing tree plantations than under well-kept pastures or scrub vegetation (Calder, 1999). Most projects designed to control sedimentation from agricultural sources end up working mostly where farmers express the greatest interest in participating, rather than where the greatest sources of sediment are (Agudelo and Kaimowitz, 1997). In many fragile areas that have already lost their original forest cover, natural regeneration and fire control might be the most cost-effective means of reducing sediment flows, but few catchment management projects concentrate on those aspects.

To sum up, a good basic principle is that if the current land use provides the quantity and quality of water the population demands with an acceptable intra- and inter-annual distribution, any alteration will increase the risk of that situation changing. This is a strong argument for maintaining natural forest cover in many contexts. That being said, the evidence suggests that many of the claims about deforestation leading to reduced rainfall and dry season flows, greater flooding, and sediment flows that endanger dams and waterways in the medium term, are exaggerated. Long-term gradual sedimentation problems deserve serious attention even though using conventional cost-benefit approaches it might seem better to let them persist. But one needs to look to more creative and systematic approaches that have the clear objective of reducing sediment flows and focus more on rural roads, construction activities and natural regeneration.

EL CAJÓN IN HONDURAS

The Honduran government began building the El Cajón hydroelectric dam in 1979, with support from IDB and the World Bank. It was the largest construction project in Honduras' history. By the time it went on line in 1985 it had cost $800 million. In the early 1990s, the dam provided over 70% of all of Honduras' electricity and accounted for almost one-fifth of its foreign debt (Loker, 1995).

As part of the initial feasibility study in 1972, the Motor Columbus Company assessed the environmental issues related to the dam's construction and operation. That initial assessment proposed that the entire El Cajón catchment area (some 8000 square kilometres) be declared a forest reserve and that all grazing, logging and agricultural activities be halted to prevent sedimentation of the reservoir. Why the company recommended that remains unclear. Its own analysis suggested that sediment problems would not affect the dam over its anticipated fifty-year life span (IDB, 1990). In any case, the company had little to go on. The only information available was poor data on sediment loads collected over the previous eleven years.

Motor Columbus issued a second report in 1997, which was reviewed by a consultant for the Honduran government. That consultant concluded that given how little anyone knew about the true sediment transport and deposition levels, it would be wise to undertake a vigorous soil conservation and reforestation programme to minimise potential risks (IDB 1990).

During the 1980s, the Honduran government, the IDB, and the World Bank talked a lot about catchment management in El Cajón, but not much happened. The Honduran electrical company (ENEE) created and then disbanded an Ecology Division, responsible for catchment management, among other things. A few years later, it created a Watershed Management Department, which it also subsequently disbanded. No one collected much data on soil erosion or sedimentation. The IDB funded a commercial forestry project that covered much of the catchment. But that project did not focus on catchment management and never fully got off the ground. ENEE did set aside 336 square kilometres surrounding the reserve as a 'Forest Protection Zone' (IDB, 1990).

In 1980, Motor Columbus once again revised its sedimentation study. Unlike the first two reports, the new study concluded that most sediment would end up in the reservoir's live, rather than dead, storage area and thus affect energy generation almost immediately.[3] Even so, the Company argued that the cost of that loss would be negligible. Jennings and Cummins, two IDB consultants that reviewed the report in 1981, criticised it for not considering rapid population growth in the catchment and the potential for landslides. They argued that there was enough reason for concern to justify immediate action (OAS, 1990).

This was the situation in 1989 when the IDB began its joint programme with the OAS designed to formulate catchment management projects. No study had demonstrated that sediment flows in El Cajón posed a major threat. But several had raised doubts about the reports that claimed the opposite; and there was not much information to go on. NGOs and certain member governments were increasingly scrutinising the IDB's dam projects with

3 The live storage area of a reservoir is the volume that rests above the dam's outtake pipe and that stores water that dam managers can use for productive purposes. The dead storage area lies below the outtake pipe (Aylward, 1998).

regard to environmental and social issues, of which reservoir sedimentation was one.

Between 1989 and 1991, the OAS prepared a diagnosis of the El Cajón catchment and a proposal for how to manage it. Based on the previous reports and its own study that used a modified version of the Universal Soil Loss Equation (as outlined in Yu, this volume) to predict sediment flows, the team once again concluded that sedimentation did not threaten the dam significantly. The team proposed a project with four main components – soil conservation, forest development, protected areas, and studies. Its financial analysis centred entirely on the on-farm benefits of soil conservation and forestry activities. Its report lists 'protect the operation and maintenance of the El Cajón dam' as one of the project's three main objectives, but does not go into detail about how the project would promote that goal, beyond noting that soil conservation measures would reduce sedimentation (OAS, 1992).

Meanwhile, the IDB commissioned an ex-post environmental evaluation of its projects in El Cajón (IDB 1990). That evaluation ratified the previous studies' conclusion that sedimentation probably represented no major threat to El Cajón within its projected 50-year life span; it did, however, add several caveats and recommendations. It pointed out that no one had ever looked at the potential economic impact of sedimentation beyond the dam's official 50-year life span and noted that reducing sediment flows might increase the dam's life expectancy significantly. It underscored the urgency of measuring reservoir sediment accumulation through bathymetric / topographic studies and reiterated Jennings and Cummins' concerns about the possible effects of population growth and landslides in the catchment.

The IDB finally presented its proposed US$24.5 million 'Program for the Management of the Renewable Natural Resources in the Watershed of the El Cajón Reservoir' to its Executive Directors for approval in 1993. By that time any trace of its origins in the concern over dam sedimentation had largely vanished (except for the name). While the document drew heavily from the OAS study it dropped specific reference to sedimentation from the programme's formal objectives. It mentioned that the soil conservation component would reduce sedimentation and proposed to finance studies of slope stability and sediment flows, but these aspects did not feature prominently (IDB, 1993).

Given this history, one might have expected that when the El Cajón project got under way in 1996, project staff and the IDB would have justified their activities based on the direct effects on farmers' livelihoods. That was not the case. The project's public relations materials and presentations consistently stressed how important the El Cajón hydroelectric plant was to the national economy and implied that the project would help save the plant from impending destruction (AFE – COHDEFOR / IDB / ENEE, n.d.; *La Tribuna*, 1996; *Tiempo*, 1993). Conversations by the author with Honduras' Minister of Agriculture and local officials in the El Cajón catchment area leave little doubt that many Hondurans continued to believe the project's primary objective was to keep the reservoir from silting up. Presumably, project officials and IDB staff found it expedient to allow them to maintain that belief. The term 'catchment management' has a lot more appeal in Honduras these days than 'agricultural extension', 'commercial forestry', or 'protected areas'.

The project's actual operations have been much more in keeping with the formal loan proposal than with its public image. Its agricultural component has provided general technical assistance related to agricultural practices, with emphasis on soil conservation. The main forestry activities have been forest fire control and small-scale reforestation. The project pays farmers to participate in forestry activities, presumably because of the environmental services they provide, although no one has clearly specified what those services are. The project has only worked in part of the catchment that drains into the reservoir, and not necessarily the part that contributes the most sediment. Neither the agricultural nor the forestry activities affect areas large enough to make a discernible difference in sediment flows or hydrological services at the catchment level.

This would all be fine if sedimentation did not matter. But even at this late date that is still not clear. The slope stability studies financed by the project detected significant problems and proposed solutions, yet the project has no funds to implement those solutions. The project has never conducted the bathymetric / topographic studies proposed by the IDB's environmental evaluation and contemplated in the project proposal. If it does the studies now, the potentially large effects of Hurricane Mitch on sediment accumulation in the reservoir may confound the results. Moreover, even if it is true that sediment will cause only minor costs to the dam during the next fifty years, one cannot simply assume that fifty years from now Honduras will have enough funds and/or hydroelectric potential to replace El Cajón. Even less is known about how current land use may affect the catchment's hydrological regime. Nonetheless, one can safely say that the existing catchment management programme will have at best a marginal effect on land use or the hydrological regime.

THE LEMPA RIVER IN EL SALVADOR

El Salvador is a small, densely populated, country. The Lempa River is its principal lifeline. The river basin covers a large portion of the national territory, as well as parts of Guatemala and Honduras. The four hydroelectric plants along the river produce 70% of the country's electricity. In 1997, the river provided 30%

of San Salvador's drinking water (Government of El Salvador / OAS, 1994; Rosa *et al.*, 1999).

Concern about El Salvador's catchments goes back a long way. FAO began its first basin project in 1967. When the Harza company carried out the feasibility study for Cerron Grande, the largest hydroelectric dam on the Lempa River, in the early 1970s it found that siltation had greatly reduced the storage capacity of the '5 de noviembre' dam, which went on line in 1954. That caused some alarm in certain spheres and led the government and FAO to expand their catchment projects into the Lempa River area, with the explicit goal of increasing the life span of the hydroelectric dams. This was the first major programme that offered farmers incentives to adopt soil conservation measures and plant trees in the catchment. The government also requested support from the British government and began discussions with the IDB about a catchment management loan for the area (UNDP / FAO, 1980; Wall, 1981).

Despite its concerns about the '5 de noviembre' reservoir, the Harza evaluation concluded that siltation would not affect 'Cerron Grande' seriously for at least a century and that it would take 350 years before it put the dam out of operation. The British team did its own study of sedimentation from agriculture, roads, construction and river bank erosion along the Acelhuate River, one of the Lempa River's main sources of sediment, and corroborated the Harza findings. It also made the first serious attempt to calculate the on- and off-farm benefits of soil conservation mechanisms. That study concluded that the benefits from avoiding dam siltation were small, except perhaps on the steepest slopes, and that live barriers and hillside ditches were profitable investments for farmers, but bench terraces and stone walls were not (Wall, 1981) (note that Critchley, this volume, provides a comprehensive appraisal of improved land management optims in humid tropical steeplands).

These conclusions did not prevent the FAO from continuing to claim that only catchment management could keep the country from suffering 'grave consequences', such as a dramatic reduction in the life span of Cerron Grande. Nor did it stop the use of incentives to promote the same expensive soil conservation measures that the British study had found were not profitable (FAO, 1985). FAO justified its conclusions, in part, on a separate set of studies which it claimed showed sediment levels that would reduce the dam's economic life span to 60 years or less (Mojica, 1982).

Discussion of these issues died down during El Salvador's Civil War in the 1980s. Government agencies stopped collecting hydrological data and anti-government insurgents controlled much of the catchment. When the government requested that CATIE train people in catchment management in 1989, intense fighting in the capital forced the team to turn back at the airport (Ferran, 1993).

Things got going again in the early 1990s. CATIE and the Lempa River Hydroelectric Executive Commission (CEL) established a soil conservation project in the Las Cañas River basin, with funds from USAID. The project's main objectives were to improve farmers' productivity and reduce siltation of the Cerron Grande reservoir (Ferran, 1993). Meanwhile, the OAS began work on a catchment project for IDB, similar to those it had prepared for Chixoy and El Cajón.

Then in 1993, the United States Army Corps of Engineers released a bombshell. Based on bathymetric studies conducted by the Salvadoran government between 1988 and 1991, it asserted that sedimentation levels were at least six or seven times higher than previously estimated. It went on to say that if the sediment load continued at the existing rate the Cerron Grande dam would have a life span of only 30 to 50 years (US Army Corps of Engineers, 1993).

The OAS and IDB were in the midst of their study when the Corp of Engineers released its report. Based on the same method that the OAS team had used in El Cajón, they derived much lower sedimentation estimates than the Americans and which were not much different from the original Harza predictions. Nevertheless, the team's report noted that the great disparity in sedimentation estimates raised significant doubts. It calculated that based on the different sediment studies, estimates of the costs stemming from lower electricity production caused by sedimentation could range from anywhere between \$1 and \$7.8 million per year. To resolve the issue, the team proposed an on-going hydrological monitoring programme and a new bathymetric study. It also took sediment flows into account in selecting where the project would work and predicted that the project would reduce the total sedimentation of the Cerron Grande reservoir by 25% (Government of El Salvador / OAS, 1994b).

As in El Cajón, the OAS / IDB team concluded that the main benefits of the proposed catchment project would come from the improvements in on-farm incomes resulting from soil conservation and crop diversification. This led it to restrict the project's operations to locations where soil conservation measures could greatly increase agricultural productivity, i.e. in only moderately degraded areas.

The IDB commissioned a bathymetric study in early 1994 which reinforced the view that siltation was not a major problem. Based on the new data plus a review of the previous studies, its author concluded that the Army Corps of Engineers study had major methodological flaws and poor data (Ordoñez, 1994).

All the IDB / OAS studies showed that using soil conservation to reduce sedimentation would provide limited economic benefits. Nonetheless, the loan agreement for the 'El Salvador Environmental Program (PAES)' that the IDB and the Government of El Salvador signed in 1996 included \$11 million to finance incentives to induce farmers to adopt soil conservation measures, as part of its $36 million loan package. The agreement justifies those incentives based on the claim that soil conservation measures will have positive externalities, but fails to specify what those are (IDB, 1995).[4]

Meanwhile, beginning around 1994, PRISMA, the Salvadoran NGO, and the Salvadoran Development Foundation (FUSADES), an influential private sector think tank, began efforts to convince the country's opinion leaders of the dangers that catchment degradation posed. After PRISMA commissioned a review of the sedimentation debate by a leading Salvadoran specialist (Perdomo Lino, 1994), it avoided exaggerated claims about the effects of sedimentation on the life span of Cerron Grande. Instead, it stressed the negative consequences of paving over of large portions of San Salvador's aquifer, water pollution, how land-use change was influencing seasonal water flows, and the effects of sedimentation on water distribution and dam maintenance costs. Those issues had been raised in previous diagnosis and policy initiatives, but had played minor roles.

PRISMA established a compelling case that El Salvador had serious environmental problems related to water. To resolve those problems PRISMA called for payment for environmental services to restore catchment vegetation, urban zoning to ensure aquifer recharge, and greater regulation of industrial and domestic water pollution.

The current PAES programme funded by the IDB promotes basically the same soil conservation measures as the previous FAO programmes, despite studies that conclude such measures are often not profitable (Kaimowitz, 1993; Lutz *et al.*, 1994). If farmers do not perceive these activities as profitable, they will not maintain them. Thus, an evaluation of the FAO project in the same area where PAES currently operates found that only 40% of the 386 farmers interviewed who had implemented soil conservation measures had done anything to maintain them (MAG, 1992).

Even so, the discussion about catchment degradation in El Salvador has greatly increased awareness of the dangers of long-term gradual deterioration of the country's environment. It has also helped to justify the allocation of much needed funds for small farmers in marginal areas that probably would not have been forthcoming otherwise.

In this on-going saga, a bathymetric study by Harza released in 1999 has put an additional twist on the story. The study once again reaffirms that current sedimentation levels of Cerron Grande constitute no need for alarm. According to its figures, Cerron Grande has lost only 5% of its live storage space over the last 25 years and still has 172 years left to go before its managers have to dredge the reservoir or shut it down. However, it is argued that even though sedimentation does not constitute an immediate economic danger the government should take measures to keep it under control and notes that the '5 de noviembre' and '15 de setiembre' dams face greater problems. The surprise is that 48% of the sediment comes from Honduras. Anything that El Salvador does within its own territory will not affect that sediment flow (Harza Engineering Company International, 1999).

THE PANAMA CANAL

One hardly needs to stress the critical importance of the Panama Canal catchment. Each year the Panama Canal generates US$400–500 million in income. Panama City and Colon get their drinking water and some of their electricity from the catchment.

The first efforts to measure the siltation of the two artificial lakes that provide water to the Panama Canal, Alajuela and Gatún, date back to the 1920s. Between 1929 and 1931, engineers measured the sediment suspended in the Chagres River and concluded it would take 358 years before sediment reduced Lake Alajuela's water storage capacity by 25% (Alvarado, 1985). In the 1930s, the Panama Canal Commission (PCC) monitored the sediment passing through the reservoir's hydroelectric turbines. During the 1950s and 1970s, it sporadically evaluated sediment accumulation in the lakes.

As mentioned earlier, in the mid-1970s, reports by Wadsworth (1976) and Larson (1979) raised the spectre of sedimentation of Alajuela and Gatún potentially jeopardizing the canal. Wadsworth calculated that by the year 2000 Alajuela would lose 40% of its storage capacity. Larson said it would lose 20% (Alvarado, 1985). Both studies predicted that sediment would seriously affect canal operations, as well as water supplies and power production.

Wadsworth's report came out during the controversial Panama Canal Treaty negotiations. It caused such a stir that some of Panamanian President Omar Torrijos' advisors speculated that the only reason the US was willing to hand over the canal to Panama was because sedimentation would soon make the canal worthless.

USAID responded to the perceived sediment threat by funding the Panama Watershed Management Project. The United States was sensitive to claims that it had failed to pay sufficient attention to environmental issues in the catchment and wished to establish a good record of environmental stewardship when it turned over the Canal to Panama in 1999. Although the project also sought to conserve biodiversity, promote commercial forestry and encourage sustainable agriculture, its main objective was to provide adequate water supplies for canal operations and other uses by

4 Ricardo Quiroga, from IDB, subsequently clarified that the anticipated externalities included: conservation of local water sources, improved water quality, better conserved local habitats for flora and fauna, aesthetic and recreation benefits, reduced risks from slope instability, improved diets, and meeting energy needs more efficiently. He also mentions the value of education and 'cultural change' regarding resource management at the local level (personal communication, 2000).

controlling erosion. It was to achieve this through reforestation, soil conservation, pasture improvement and protected area management in locations selected for maximum impact on sediment flows (Associates for Rural Development, 1983).

The project's first four years of implementation were hardly encouraging. Only 3576 of a projected 10 500 hectares of trees were planted, and at a very high cost and in locations where the trees would have little impact on soil erosion. The project paid hired labourers to construct several hundred hectares of gully erosion control structures, but no one maintained them. The erosion control activities focused on Lake Gatún, even though studies had identified Lake Alajuela as having the greatest siltation problem. The project document envisioned a specific component to monitor soil erosion and water flows and set priorities for erosion control activities, but it never got off the ground (Associates for Rural Development, 1983).

In 1985, two new influential studies claimed that deforestation was creating serious sedimentation problems and affecting local rainfall patterns in the catchments that supply water to the Canal. The first, by Luis Alvarado, the Panama Canal Commission's leading expert on sediment flows, characterised the siltation of Lake Alajuela as critical (Alvarado, 1985). Alvarado gave credence to Larson and Wadsworth's estimates and attributed their results to large increases in forest clearing around Lake Alajuela since 1960. He predicted that if deforestation rates continued at their existing level, Lake Alajuela would lose 18% of its storage capacity by 2020. In the second report, Donald Windsor and Stanley Rand from the Smithsonian Tropical Research Institute claimed deforestation had caused annual rainfall in the Lake Gatún area to decline between 1925 and 1980 (Windsor and Rand, 1985).

The data Alvarado cited in 1985 came from renewed efforts by the PCC to measure sediment flows in response to the political concerns about sedimentation. Beginning in 1981, the PCC measured systematically the sediment transport at the mouths of the six largest rivers and prepared a detailed topographical map of sediment accumulation (Alvarado, 1985). The PCC found that as of 1983, Lake Alajuela had lost 4.7% of its storage capacity. Alvarado assumed most of this loss had occurred during the previous couple of years.

Initially, the PCC did not pay much attention to the Windsor and Rand study. Its engineers felt that even if the study were correct, a reduction in rainfall of the magnitude it reported would not affect the canal's operations for at least 50 years (Hart, 1992). Nevertheless, the study stimulated a series of sensationalist articles, which claimed that deforestation endangered the Canal. The PCC felt compelled to respond, so it asked its Engineering Division to investigate. The division produced two studies. Both concluded that rainfall had not declined significantly between 1914 and 1991 and that the most likely explanation for Windsor and Rand's findings was a change in how the Commission measured and recorded rainfall (Hart, 1992; Vargas, 1993).

Between 1986 and 1990, Panama's political crisis pushed these issues onto the backburner and USAID pulled out of the country. It returned in 1991 and almost immediately renewed its support for catchment activities. Together with the National Institute for Renewable Natural Resources, it sponsored a working group that prepared a strategic plan for the catchments that supply water to the Canal designed to reduce erosion to 10–12 tons ha^{-1} y^{-1} and to limit deforestation to 10 hectares per year (INRENARE, 1995).

In 1995, the same year that the working group presented its plan, Panama contracted Intercarib S.A. and Nathan Associates to prepare a separate plan for the catchment, concerned mostly with what to do with the properties in the Canal Zone that were supposed to revert to Panama. The Intercarib S.A. and Nathan Associates plan proposed to reduce the pasture area in the catchment from 142 000 hectares to 6 000 hectares and to create 70 000 hectares of commercial forest plantations. Its authors justified these measures by arguing that the area's topography and soils were unfit for pasture. The plan also called for actions to reduce siltation from urban sources and said that allowing existing land use trends to continue would lead to excessive erosion and sedimentation (ARI, n.d.) The authors calculated that the beneficial effects of erosion control on water supply for navigation that could be obtained by reforesting 100 000 hectares had a present value of \$9 per hectare (Aylward, 1998). More generally, they argued that even though the likelihood of environmental degradation impinging greatly on canal operations seemed remote, given what was at stake, it would be wise to take strong action.

In 1997, Panama's Legislative Assembly formally sanctioned the plan produced by Intercarib S.A. and Nathan Associates, giving it the weight of law. (It became known as Law 21.) Panama was under pressure to demonstrate it could administer the Panama Canal responsibly once the United States handed over control of the Canal's operations; having a detailed plan for managing the catchment was one way to show that.

During the El Niño phenomenon in 1997–8, the PCC had to impose draft restrictions on ships due to low water levels in lakes and reservoirs. For the first time the canal experienced a serious shortfall in its water supply. This once again renewed speculation about how land use was affecting the Canal's operations and showed that the PCC's conviction that water supply would not constrain canal operations for some time was at least partially unfounded (Gardner and Rojas, 1998).

Law 21 charged Panama's Ministry of Agriculture (MIDA) with implementing the proposal to convert pastures to forest plantations. To that end, MIDA sought support from the World Bank. The Bank, in turn, brought in a group of consultants to examine the issue. These consultants questioned whether the existing evidence justified using public funds to reforest pasture areas. They argued that more research was needed before it could be concluded that planting exotic tree species to reduce sedimentation made economic sense. They noted that the PCC had concluded that sedimentation was a smaller problem than believed previously and cited studies showing that sediment loads had declined thanks to the establishment of protected areas, greater control of forest clearing, and natural regeneration. They also pointed out that there was reason to doubt whether most of the sediment load came from agricultural activities and that poorly managed teak plantations could easily result in greater soil erosion than well managed pastures. Finally, they argued that a priori there was little reason to expect reforestation to have any beneficial effect on dry season stream flows (Aylward, 1999; Calder, 1999b; Gardner and Rojas, 1998).

The Panamanian authorities, USAID and local NGOs gave the consultants a mixed response. Some quietly admitted that sedimentation was not a major problem but expressed fears that admitting that openly would undermine support for conservation. It was felt that even if the sedimentation of Lake Alajuela and Gatún was a myth, it was a useful myth. Others expressed the position that lands with steep slopes and/or shallow soils should be under forest, independent of any economic consideration. Some argued that the transfer of Canal Zone properties to the Panamanian authorities would increase forest clearing and sedimentation from peri-urban sources, so the problem might quickly become more important.

The consultants' conclusion that surprised people in Panama the most was that forest cover would not necessarily improve dry season stream flow. Since the Panama Canal authorities typically have to discharge large amounts of water during the rainy season but occasionally face shortfalls during the dry season, this issue was potentially important. Officials from several government agencies and USAID pointed to one study that suggested that forest cover helped to regulate stream flow in the catchments that supply water for the Canal. They stressed that the consultants' arguments were based on research in regions with distinct soils and climates (D. Reese, personal communication, 1999). The consultants argued that the study suggesting forest cover in the catchment had a positive effect on dry season stream flows had been unable to rule out other possible explanations for its results and was based on only a couple of years of data. They also pointed out that the principles of forest hydrology applied equally to the Panama Canal catchment as to anywhere else (I. Calder and M. Rojas, personal communication, 1999).

Meanwhile, despite all evidence to the contrary, alarmist reports claiming that sedimentation poses a serious threat to Canal operations in the medium-term continued to appear in the media. An article in *La Prensa* commented that Panama's lakes were 'becoming filled with mud instead of water, which will seriously affect the Canal's water supply and its operations' (Esquivel, 1999). The Christian Science Monitor reported that 'if deforestation of the canal catchment continues, it could threaten the canal's operations' (Mitchell, 1997). CNN ran a story saying that 'without the forest, erosion and sedimentation would threaten the Canal's future' and expressed doubts about whether Panama's authorities could deal with the problem (Strieker, 1997). Another press report cited the Director of Watersheds and the Environment of Panama's Canal Ministry as saying that his department would support massive reforestation efforts to protect the Canal's water supply (Diaz, 2000).

Siltation in Lakes Alajuela and Gatún is almost certainly slower than these alarmist messages imply. Fast-growing tree plantations are probably not the best way to address whatever sedimentation problem exists and no one knows exactly how land use changes may affect dry season stream flow. However, even if cost-benefit analyses suggest it would not be profitable to control soil erosion, erosion may still pose a long-term threat. The Panamanian authorities are currently considering ways to meet the rising demand for water, including construction of a third artificial lake and a third set of locks. Such options would have huge economic, social and environmental costs. Any initiative that might delay the need for such investments could potentially have a high rate of return. The government's Canal Capacity Department claims to have analysed the available alternatives and concluded that water and soil conservation efforts could not make even a dent in the anticipated water shortfall but it has yet to release the evidence supporting that claim. Water pollution from industrial, agricultural and household activities pose real threats and sedimentation from urban and peri-urban sources could become more serious (Heckadón, Ibañez, and Condit, 1999).

HURRICANE MITCH

Within days after Hurricane Mitch swept through Central America in late October 1998, the media, academics and NGOs began blaming hillside deforestation for much of the destruction that it unleashed (Brown, 1998; DeWalt, 1998; *La Tribuna*, 1998; Marcus, 1998; *Tiempo*, 1999). They used two major arguments to support that idea. First, deforestation and subsequent soil compaction had reduced the soil's water retention capacity, which left more water to flood. Second, the removal of vegetative cover had made the slopes less stable and more prone to landslides and mass wasting. To avoid similar disaster in the future, they proposed massive reforestation efforts, greater restrictions on forest clearing and soil conservation measures, among other things.

Most multilateral and bilateral agencies shared these ideas. The IDB convened two major meetings of the 'Consultative Group for the Reconstruction and Transformation of Central America' in Washington and Stockholm to discuss what steps were needed to reconstruct the region in the wake of Hurricane Mitch. Central American governments, NGOs and international agencies attended those meetings and agreed that deforestation and poor catchment management had greatly aggravated the hurricane's negative impacts. Prominent themes included catchment management and land use planning (Kandel and Rosa, 1999). During 1999 and 2000, the IDB, the Swedish International Development Agency (SIDA), USAID, World Bank and others began formulating major catchment management projects in response to the perceived political support for such efforts, even though they remained uncertain about what exactly those projects should include.

Despite all the rhetoric, deforestation probably had little to do with how much water flooded during Hurricane Mitch. The hurricane took place well into the rainy season and most of the soils were already saturated or nearly so when the storm began. Those that were not quickly became so as the storm poured between 300 and 1900 mm of rain onto the hillsides for almost a week without stopping (Smyle, 1999; see discussion on Hurricane Mitch in Bonell *et al.*, this volume). Given what is known about the links between deforestation and flooding in large catchments, it seems almost certain that so much rain for so much time would have led to more or less the same amount of flooding whether or not forests covered the hillsides (see Bonell; Grip *et al.*; Scott *et al.*; all this volume).

The lack of forest cover and soil conservation measures probably did affect slope stability and soil erosion. One study that surveyed some 2,000 farmers in Guatemala, Honduras and Nicaragua and did field tests at a number of sites found that those farmers that practised soil conservation reported less damage as a result of Hurricane Mitch (pers. comm., E. Holt, 1999). Yet one should not exaggerate this point since major landslides and mass wasting caused most of the greatest damage. Land use and agricultural practices were practically irrelevant to soil movements of such great magnitude. It was geology, topography and climate (see Scatena *et al.*, this volume) that determined where these movements occurred, not how people managed the catchment.

Even though Hurricane Mitch greatly increased awareness of the dangers of environmental degradation, the discussions that followed have not come up with significant new ideas about what to do to mitigate it. Most catchment management proposals under discussion revolve around the same set of small-scale reforestation and soil conservation efforts implemented in the past and will probably have similar outcomes.

CONCLUSIONS

The slow, steady and diffuse degradation of Central America's hillsides has no easy solution. If one looks at the problem using traditional economic analysis, this degradation probably does not cause enough damage to justify major investments to curtail it. Most claims about huge medium-term costs that such degradation generates are exaggerated or unproven. Certain NGOs and international agencies have used these claims to justify their own institutional interests. The press has echoed their declarations out of a well-intentioned, but ill-informed, desire to protect the environment.

Nonetheless, serious problems *do* exist. A frog put into a pan of water whose temperature slowly rises will sit there until it dies. The changes in its environment are too slow for it to perceive, but deadly nonetheless. Current economic techniques assume that once the reservoirs, rivers, streams and coasts fill up with silt and all the topsoil erodes away a couple of hundred years from now, Central Americans will be able to find readily available substitutes. No one can guarantee that. Moreover, what people do not know can hurt them. Significant doubts remain about the effects of land use on dry season stream flow, aquifer recharge and climate, among other things. As Central Americans rapidly change their environments, they seriously increase the risk something will go wrong.

Traditional economic methods are poorly suited for assessing scenarios far into the future, where huge risks and uncertainties exist, the future of entire nations is at stake, and the public is poorly informed. Political processes that factor-in a wide variety of considerations and the interests of various constituencies and take into account the limited information available and the need for safeguards should be used to make decisions about such fundamental issues. Based on traditional cost-benefit analysis, the US government would probably never have established its National Park System. The American Endangered Species Act would undoubtedly never have been approved and the world's nations would not have adopted the United Nations Framework Convention on Climate Change.

The real question in the case of Central America catchments is what to do? We clearly need to move away from responding to immediate crises and exaggerated press reports and to take a longer-term positive approach based on careful analysis and monitoring. Sporadic short-term efforts to promote soil conservation and reforestation in individual plots selected on the basis of farmer interest are unlikely to have any discernible effect at the catchment level. They may not even increase farmers' yields or improve their incomes. They do provide needed investments and services to the rural areas, but at a high cost, with limited effectiveness, and little prospect of sustainability.

To the extent that payment for hydrological services implies a long-term commitment to land uses and agricultural practices that reflect environmental stewardship, it represents a step in the right direction, even if the specific services involved have not been fully demonstrated. But to be financially viable, it must promote low cost alternatives such as natural regeneration, forest conservation, commercially viable tree crops, and fire control. Managers must use clear criteria to select areas that have the greatest potential for providing environmental services and put in place or strengthen permanent local institutions to administer them. Ironically, the many sensationalist press reports about the dangers of environmental degradation may ultimately contribute to such long-term solutions.

Water quality and urban catchment issues deserve greater attention. In many cases, it may be cheaper to avoid water pollution than to treat the water once it has been contaminated; and forest cover can help maintain water quality. Similarly, road and housing construction activities that reduce aquifer recharge or generate major erosion problems deserve a place on policy agenda. Tropical montane cloud forests merit special consideration, since they do, in fact, seem to capture water and make it available to downstream users. Policymakers, NGOs and project managers urgently need to improve their understanding of under what conditions changes in land use will increase or decrease dry season stream flow and by how much; and to design appropriate policies accordingly.

Most catchment management projects give little priority to research and monitoring, even though these are essential to effective catchment management. Often, organisations use the results from those studies that are carried out to reaffirm their pre-existing positions, rather than to learn and to adapt their strategies. Others simply ignore them. Changing this will not be easy. Systematic learning has no real political or institutional constituency. Business-as-usual is much more politically expedient and few of the current decision-makers will be around to see the errors of their ways. That is the intractable truth.

References

Administración Foresal del Estado – Corporación Hondureña de Desarrollo Forestal, IDB, ENEE. (n.d.) 'Programa de Manejo de los Recursos Naturales Renovables de la Cuenca del Embalse el Cajón'. Tegucigalpa.

Agudelo, L. A. and D. Kaimowitz. (1997). 'Tecnología agrícola sostenible: Retos institucionales y metodológicos, dos estudios de caso en Colombia'. Serie Documentos de Discusión sobre Agricultura Sostenible y Recursos Naturales No. 3. San Jose, Costa Rica: IICA / GTZ.

Anderson Jr., R. J., N. Da Franca Ribeiro dos Santos, and H. F. Diaz. (1993). 'An Analysis of Flooding i the Parana / Paraguay River Basin'. LATEN Dissemination Note #5. Washington D.C.: World Bank. Associates for Rural Development Inc. (1983). 'Panama Watershed Management Evaluation Report', Burlington.

Autoridad de la Región Interoceánica (ARI). n.d. 'Plan Regional'. Panama.

Aylward, B. A. (1998). 'Economic Valuation of the Downsream Hydrological Effects of Land use Change: Large Hydroelectric Reservoirs'. Ph.D. dissertation. The Fletcher School of Law and Diplomacy.

Aylward, B. A. (1999). 'Panana Canal Watershed, Economic and Policy Aspects.' Mimeo.

Barry, D. (1994). 'El agua: Límite ambiental para el desarrollo futuro de El Salvador'. *PRISMA*. No. 5. January – March: 1–12.

Bosch, J. M. and J. D. Hewlett. (1982). 'A Review of Catchment Experiments to Determine the Effect of Vegetation Changes on Water Yield and Evapotranspiration'. *Journal of Hydrology*, 53: 3–23.

Brown, B. (1998). 'The Curse of Cut Trees'. *BBC*. November 17.

Bruijnzeel, S. (1990). *Hydrology of Moist Tropical Forests and Conversion: A State of Knowledge Review*. Paris: UNESCO International Hydrological Programme.

Bruijnzeel, S. (1999). 'Hydrology of Tropical Montane Cloud Forests: A Re-Evaluation', paper presented at the Second International Colloquium on Hydrology and Water Management in the Humid Tropics', Panama City, 22–24 March.

Bruijnzeel, S. (2004). 'Hydrological functions of tropical forests: Not Seeing the Soil for the Trees?' *Agriculture, Ecosystems, and Environment*, doi: 10.1016/j.agee.2004.01.015.

Calder, I. R. (1999). *The Blue Revolution, Land Use & Integrated Water Resources Management*. London: Earthscan Publications Ltd.

Calder, I. R. (1999b). 'Panama Canal Watershed, Hydrological Study Preparation'. Mimeo.

Cassells, D. S., M. Bonell, L. S. Hamilton, and D. A. Gilmour. (1985) 'The Protective Role of Tropical Forests: A State of Knowledge Review'. Paper presented at the Ninth World Forestry Congress, Mexico City.

Chomitz, K. M. and K. Kumari. (1998). 'The Domestic Benefits of Tropical Forests, A Critical Review'. World Bank Research Observer. Vol. 13, No. 3. February: 13–35.

DeWalt, B. (1998). 'Human Causes of a Natural Disaster'. *Pittsburgh Post Gazette*. November 22.

Diaz, I. (2000). 'Arboles maderables para reforestar cuenca del Canal'. *El Panama América*. June 11. *El Heraldo*. (1993). 'Anuncia subgerente de la Cohdefor: Destinaran Lps. 120 millones para proteger la cuenca de El Cajón.' March 10.

Enters, T. (2000). 'Methods for the Economic Assessment of the On- and Off-Site Impacts of Soils Erosion' 2nd Edition. International Board for Soil Research and Management. Bangkok: IBSRAM. Esquivel, E. 'La protección ecológica y ambiental del Canal de Panamá'. *La Prensa*. Panama. December 17: 79 A.

Ferran, F. I. (1993). 'Entre la guerra y la conservación: Estudio de caso de los antecedentes a la rehabilitación de la microcuenca del Rio Las Cañas, El Salvador'. Area de Manejo de Cuencas, Turrialba, CATIE. September.

Fleming, T. H. (1988). 'Secular Changes in Costa Rican Rainfall: Correlation with Elevation'. *Journal of Tropical Ecology* 2: 87–91.

Finlayson, W. (1998). 'Trees and Forests in the Upper Mahaweli Catchment: Their Effects on Water Yields and Sedimentaiton'. Colombo: Mahaweli Authority of Sri Lanka.

Gardner, B. and M. Rojas. (1998). 'Study of the Panama Canal Watershed'. Panama: Japan International Cooperation Agency.

Gellin, J. D. (1994). 'Trees Down, Lights Out in Honduras'. *Christian Science Monitor*. November 15.

García, L. (1982). 'Analysis of Watershed Management (El Salvador, Guatemala, Honduras)'. USAID Project #596–0000.6. Washington D.C.

Government of El Salvador. Ministry of Agriculture and Livestock (MAG) / Secretariat of the Environment (SEMA) / Organization of American States. Department of Regional Development and Environment. (1994). 'Programa Ambiental de El Salvador, Informe Final.' May.

Gutierrez, F., and B. Rapidel. (1999). 'Evolución de las precipitaciones en Nicaragua'. *Naturaleza* 16: 22–23.

Hamilton, L. S. with P. N. King. (1983). *Tropical Forested Watersheds, Hydrological and Soils Response to Major Uses or Conversions*. Boulder, Colorado: Westview Press.

Harza Engineering Company International. (1976). 'Estudio de factilibidad, Proyecto Hydroeléctrico Cerrón Grande'. San Salvador.

Harza Engineering Company International (1999). 'Estudio global de la sedimentación en la Cuenca del Rio Lempa, Resumen ejecutivo'. San Salvador, November.

Heckadón, S., R. Ibáñez, and R. Condit. (1999). *La Cuenca del Canal: Deforestación, contaminación y urbanización, Proyecto de monitoreo de la Cuenca del Canal de Panama (PMCC), Sumario ejecutivo del informe final*. Panama: STRI / USAID / ANAM.

Instituto Nacional de Recursos Naturales Renovables (INRENARE). 1995. 'La cuenca hidrográfica del Canal de Panamá: Prioridades y acciones para su manejo integral, Volumen I (Documento principal)', Panama, Comité Técnico Interinstitucional de la Cuena Hidrográfica del Canal de Panama – Proyecto Marena, February.

Inter-American Development Bank (IDB). IDB. (1990). 'Ex-Post Evaluation, El Cajón Hydroelectric Project Environmental Assessment, Honduras

(Loans 44/IC, 572/SF, 130/IC, and ATN/SF-2902). Washington D.C.: Office of the Controller, Operations Evaluation Department.

IDB. (1993). 'Honduras, Program for the Management of the Renewable Natural Resources in the Watershed of the El Cajón Reservoir (HO-0035), Loan Proposal'. Washington D.C.

IDB. (1995). 'El Salvador, Programa Ambiental de El Salvador (ES-0024), Propuesta de prestamo'. Washington D.C.

Kaimowitz, D. (1993). 'La experiencia de Centroamérica y la República Dominicana con proyectos de inversión que buscan sostenibilidad en las laderas'. Serie Documentos de Programa 40. San Jose, Costa Rica: Inter-American Institute for Cooperation in Agriculture.

Kandel, S. and H. Rosa. (1999). 'Después del Mitch: Temas y actores en la agenda de transformación de Centroamérica.' *PRISMA*. 36. San Salvador.

Larson, C. (1979). 'Erosion and Sediment Yield as Affected by Land Use and Slope in the Panama Canal Watershed.' *Proceedings of the II World Congress on Water and Resources. International Water Resources Association*. Mexico D. F., Part III: 1086–1095.

La Tribuna. (1998). 'Por que causo estragos el huracan Mitch?' December 3: 20-b.

La Tribuna. (1996). 'Programa de manejo de los recursos naturales renovables de la cuenca del Embalse el Cajón.' January 15: 83.

Leonard, H. J. (1985). *Natural Resources and Economic Development in Central America*. New Brunswisk, New Jersey: Transactions Books.

Loker, W. (1995). 'Social and Ecological Effects of the El Cajón Dam in Honduras'. Department of Sociology, Anthropology, and Social Work, Mississippi State University, mimeo.

Lutz, E., S. Pagiola, and C. Reiche (eds). (1994). 'Economic and Institutional Analyses of Soil Conservation Projects in Central America and the Caribbean.' World Bank Environment Paper #8. Washington D.C.: World Bank.

Mangurian, D. 'Honduras, Ingenuity Saves Dam'. *IDB America*. http://www.IDB.org/exr/IDB/1997/eng/8ds.htm

Marcus, D. L. (1998). 'Deforestation Worsened Mitch's Toll, Scientists Say'. *Boston Globe*. November 11.

Ministerio de Agricultura y Ganadería (MAG). (1992). 'Estudio de verificación de resultados del proyecto agroforestal desarrollado en Chalatenango: Evaluación de impacto'. Oficina sectoral de planificación agropecuaria. División de Seguimiento y Evaluación. San Salvador.

Mitchell, J. (1997). 'Water Woes: Deforestation Could Dry Up the Panama Canal'. *The Christian Science Monitor International*. October 23. http://www.csmonitor.com/durable/1997/10/23/intl/intl.2.html

Mojica, I. H. (1975). 'Mejoramiento y mantenimiento de cuencas hidrográficas'. CATIE. Departamento de Cuencas Hidrográficas'. Turrialba, Costa Rica.

Mojica, I. H. (1982). 'Las prácticas de conservación de suelos y el manejo de la Cuenca Norte del Embalse del Cerrón Grande'. San Salvador: UNDP / FAO. September.

Nagle, G. N., T. J. Fahey, and J. P. Lassoie. (1999). 'Management of Sedimentation in Tropical Watersheds'. *Environmental Management*, Vol. 23, No. 4: pp. 441–52.

Ordoñez, J. I. (1994). 'Análisis de la sedimentación en el embalse de Cerron Grande, República de El Salvador'. Santafé de Bogotá: IDB, January.

Organization of American States. (1992). *Honduras, Proyecto de manejo de los recursos naturales renovables de la Cuenca del Embalse el Cajón, Estudio de factibilidad*. Washington D.C. Department of Regional Development and the Environment.

Perdomo Lino, F. (1994). 'El suelo, la erosión y la sedimentación en El Salvador'. PRISMA. San Salvador. October.

Portney, P. R. and J. P. Weyant (eds.) (1999). *Discounting and Intergenerational Equity*. Washington D.C.: Resources for the Future.

Regional Office for Central America and Panama (ROCAP). (1983). 'AID Regional Tropical Watershed Management Paper'. 597–0106. USAID: Washington D.C.

Rosa, H., D. Herrador, M. González, N. Cuellar. (1999). 'El agro salvadoreño y su potencial como productor de servicios ambientales' *PRISMA*. 33.

Rosenfeld, D. (1999). TRMM Observed First Direct Evidence of Smoke from Forest Fires Inhibiting Rainfall. Geophysical Research Letters 26 (20): 3105–8.

Smyle, J. (1999). 'Disaster Mitigation and Vulnerability Reduction: Perspectives on the Prospects for Vetiver Grass Technology (VGT)'. Regional Unit for Technical Assistance (RUTA), World Bank, San Jose, Costa Rica, mimeo.

Strieker, G. (1997). 'Forests Along Panama Canal Face Uncertain Future'. *CNN World News*. November 18. http://www.cnn.com/WORLD/9711/18/panama.watershed/

Tiempo. (1993). '170 millones para proteger la cuenca en El Cajón.' April 30: 29

Tiempo. (1999). 'Advierten ecologistas, Si no se atiende el medio ambiente habrá estragos iguales a los del Mitch.' January 23: 16.

United Nations Development Program (UNDP) / FAO. (1980). 'Desarrollo forestal y ordenación de cuencas hidrográficas, El Salvador, resultados y recomendaciones del proyecto'. FO:DP/ELS/71/506 FO:DP/ELS/73/004. Informe terminal. Rome.

UNDP / FAO. (1985). 'Conservación y aprovechamiento de los recursos naturales renovables en cuencas hidrográficas del departamento de Chalatenango, El Salvador, resultados y recomendaciones del proyecto, FO:DP/ELS/78/004, informe terminal'. Rome.

United States Army Corps of Engineers. (1993). 'Sedimentation in the Rio Lempa Watershed – El Salvador, C. A.' Mobile, Alabama. March.

Wadsworth, Frank. (1976). 'Deforestation: Death to the Panama Canal.' United States Strategic Conference on Tropical Deforestation, US State Department and United States Agency for International Development. Washington: 22–4.

Wall, J. R. D. (editor) (1981). 'A Management Plan for the Acelhuate River Catchment, El Salvador: Soil Conservation, River Stabilization, and Water Pollution Control'. Land Resource Study 30. Land Resources Development Centre, Overseas Development Administration, Surrey, United Kingdom.

White, A. (1994) 'Collective Action for Watershed Management, Lessons from Hait'. Unpublished Ph.D. dissertation, University of Minnesota.

Windsor, D. M. and S. Rand. (1985). 'Cambios climáticos en los registros de lluvias en Panamá y Costa Rica', pp. 147–64, in Agonia de la naturaleza, ensayos sobre el costo ambiental del desarrollo panameño. S. Heckadon and J. Espinoza (editors). Panama: Instituto de Investigaciones Agropecuarias / Smithsonian Tropical Research Institute.

7 Land use, hydrological function and economic valuation

B. Aylward
Deschutes Resources Conservancy, Bend, USA

INTRODUCTION

Land use change affects economic activity both directly and indirectly. In the process of land colonisation that accompanies economic development and population growth, naturally occurring vegetation is typically affected in one of three ways: (1) available biomass and species are harvested and then left to regenerate before harvesting again, (2) the vegetation is simplified (in terms of its biological diversity) in order to increase production from selected species or (3) the existing vegetation is largely removed to make way for the production of domesticated species, the installation of infrastructure or urbanisation. The direct, and desired, impact of land use change under these circumstances is to raise the economic productivity of the land unit. Of course, many indirect (and perhaps unintentional) environmental impacts result as well. These impacts reflect the economic values attributed to natural vegetation and biogeophysical processes. Conversely, efforts to recuperate degraded lands or to protect natural ecosystems may forsake direct productive benefits in favour of fostering these indirect environmental values.

The loss of biodiversity and alteration of ecological processes accompanying the logging and conversion of forestland have captured the public imagination in the 1990s, with corresponding growth in research aimed at illustrating these indirect ecological and economic impacts (Perrings *et al.*, 1992; Barbier *et al.*, 1994). This chapter concerns itself with another type of environmental value: the impact of land use change on the hydrological cycle. Vegetation is an important variable in the hydrological cycle as it is the medium through which rainfall must pass to reach the soil and begin the journey back to the sea. Further, land use change invariably involves not just modification of land cover but alteration of soil surface and sub-surface conditions. The hydrological impacts that result from these changes are often grouped in terms of their impact on soils and changes in streamflow quality and quantity. The nature of these impacts on the economy can be summarised according to whether they feed back into the economic system through a reduction in on-site production (soils) or through a more distant, downstream impact on off-site production or consumption (streamflow quality and quantity).

To economists the theoretical implications of the on-site impacts of land use change are fairly straightforward. In a farming context, McConnell demonstrates that as long as farmers' objectives are consistent with society's objectives and social and private discount rates are identical, on-site losses of productivity due to soil erosion can be expected to follow an optimal path (McConnell, 1983). That is, soil would be 'used' over time so as to maximise the net present value of its contribution to production. The question of course is whether the assumptions of McConnell's model hold in the real world. As a result, considerable effort has been devoted to investigating policy, institutional and social imperfections that may lead to excessive rates of soil degradation (loss of soil depth or soil quality). Nevertheless, in the absence of serious imperfections, neoclassical economists are fairly sanguine about the ability of the market to provide a relatively efficient level of incentive for soil conservation (Crosson and Miranowski, 1982; Southgate, 1992; Lutz *et al.*, 1994).

In addition to the on-site impacts of soil degradation, a series of downstream hydrological impacts also accompany the disturbance of natural vegetation. Regardless of the perceived seriousness of the 'soil erosion problem,' economists and natural scientists have traditionally agreed that the downstream effects of land use change are potentially very serious (Crosson, 1984; Clark, 1985b; Pimentel *et al.*, 1995). This belief is based on the general perception that the hydrological impacts of land use change have unambiguously negative impacts on production and consumption and the suspicion that these impacts are often large in magnitude. As the effects are external to the land use decision-making process of landholders, the failure of the market to internalise these effects (externalities) is unquestioned. Consequently, this chapter uses the term 'hydrological externalities'

Forests, Water and People in the Humid Tropics, ed. M. Bonell and L. A. Bruijnzeel. Published by Cambridge University Press.

to refer to these downstream hydrological impacts of land use change.

This chapter examines the existing knowledge base with regard to the application of the tools of economic analysis to the valuation of these hydrological externalities of land use change, with an emphasis on the humid tropics. The objectives are to:

- specify the general theoretical linkages that govern the relationships between land use, hydrological function and downstream economic welfare;
- assess the existing empirical evidence in the economics literature regarding the significance of these hydrological externalities; and
- assess what a priori claims can be made regarding the direction and magnitude of the economic consequences of land use change and resulting downstream hydrological impacts.

Interest in the environmental benefits provided by forests and catchment management has never been greater (Johnson *et al.*, 2001). Investments in forest conservation and catchment management and the derivation of new regulations and market incentives in this regard are of increasing importance in both temperate and tropical zones. Thus, a systematic understanding of the relationships between upstream land use, hydrology and downstream economic activity, as well as practical methods for the quantitative evaluation of these linkages is required to guide project investments and policy-making.

Given the emphasis in other chapters of this book on the latest scientific findings in forest hydrology, the chapter begins with just a short and stylised summary of the biophysical impacts of land use change on hydrological function (sedimentation, water yield, seasonal flows, peakflows, etc.). This knowledge is used as a point of departure for a simple theoretical presentation of the linkages between land use, hydrology and individual utility. Hydrological services may enter into an individual's utility function directly through consumption, indirectly through the household production function or as factor inputs in production.

The types of economic impacts that can be expected to result from changes in hydrological services that are, in turn, related to changes in land use are then reviewed. The range of impacts that are caused by land use and subsequent hydrological change is amply demonstrated in the literature and the magnitude of these impacts is discussed. The ensuing section then discusses the general nature of these linkages between land use and hydrological externalities, drawing upon the empirical and theoretical ideas presented in the two previous sections. A final section summarises the findings of the chapter and presents recommendations for future research in this area.

LAND USE AND HYDROLOGY

As a means of introducing the hydrological issues and concepts employed, a brief overview of the hydrological impacts of land use change is provided, particularly as it relates to the case of the humid tropics.

Hydrological impacts of land use change

Disturbance of tropical forests can take many different forms, from light extraction of non-timber forest products through to wholesale conversion. Each type of initial intervention will have its own particular impacts on the pre-existing hydrological cycle (Hamilton and King, 1983). These hydrological impacts may be loosely grouped according to whether they are related primarily to water quality or water quantity. Under this typology erosion, sedimentation and nutrient outflow are grouped together under the heading of water quality impacts; and changes in water yield, seasonal flow, stormflow response, groundwater recharge and precipitation are considered as water quantity issues. Beginning with water quality and moving on to water quantity, the hydrological impacts of changes in land use and conversion of tropical forests can be summarised by compiling the general nature of these impacts as extracted from a number of authoritative reviews on the subject, including those in this volume (Hamilton and Pearce, 1986; Bruijnzeel, 1990; Calder, 1992; Bruijnzeel and Proctor, 1995; Bruijnzeel, 1997, 1998, 2002; Pielke *et al.*, 1999; and Bonell, Callaghan and Connor; Bonell *et al.*; Bruijnzeel; Chappell, Tych *et al.*; Grip *et al.*; Heil Costa; Scott *et al.*; this volume).

(1) Erosion increases with forest disturbance, at times dramatically, depending on the type and duration of the intervention.
(2) Increases in sedimentation rates are likely as a result of changes in vegetative cover and land use and will be determined by the kind of processes supplying and removing sediment prior to disturbance.
(3) Nutrient and chemical outflows following conversion generally increase as leaching of nutrients and chemicals is increased.
(4) Water yield is related inversely to forest cover, with the exception of upper montane cloud forests where horizontal precipitation may compensate for losses due to evapotranspiration.
(5) Seasonal flows, in particular dry season baseflow, may increase or decrease depending on the net effect of changes in evapotranspiration and infiltration.
(6) Peakflow may increase if hill-slope hydrological conditions lead to a shift from sub-surface to overland flows, although the effect is of decreasing importance as the distance from the site and the number of contributing tributaries in a river basin increase.

(7) Groundwater recharge is generally affected in a similar fashion to seasonal flows.
(8) Local precipitation is probably not significantly affected by changes in forest cover (at least up to a scale of 10 km). Exceptions are cloud forests (elevated cloud base following large-scale downslope forest clearance). At scales larger than 10 km there is evidence emerging from work in Florida, West Africa and large continental basins (such as the Amazon which is partially enclosed) that changes in forest cover can impact on the spatial and temporal variability of precipitation.

Finally, the authors cited above generally agree that in assessing the hydrological impact of land use changes it is important to consider not just the impacts of the initial intervention but also the impacts of the subsequent form of land use, as well as the type of management regime undertaken (Bosch and Hewlett, 1982; Hamilton and King, 1983; Bruijnzeel, 1990; Calder, 1999; Bruijnzeel, 2004).

LAND USE CHANGE, HYDROLOGY AND ECONOMIC WELFARE

A change in hydrological function as provoked by alteration of land use or land management practices will lead to changes in the downstream hydrological outputs associated with a given land unit. These outputs may be summarised generally as consisting of the streamflow over a given time period and the level of sediment and nutrient concentrations contained in this streamflow. The spatial and temporal point at which these outputs are evaluated will depend on the type and location of the affected economic activity. However, in general, a hydrological production function for a given site can be defined that relates land use, L, and a vector $\mathbf{Y}$ of other biophysical parameters to a vector of hydrological outputs, as follows:

$$\mathbf{H} = H(L, \mathbf{Y}) \tag{7.1}$$

The vector $\mathbf{H}$ then refers to the different hydrological outputs ($\mathbf{H} = h_1, \ldots, h_i, \ldots, h_m$) including sediment yield, annual water yield, peakflow, dry season baseflow, etc. Somewhat arbitrarily, L is defined such that an increase in L represents a change away from undisturbed natural forest (or vegetation) towards less vegetation and a more 'productive' land use. As noted above, the removal of forest cover tends to increase sediment yield, SY, as well as raising nutrient and chemical levels, FL. Similarly the effect of an 'increase' in land use is to raise annual water yield, WY, as well as peakflows, PF. The effect on dry season baseflow, BF, is indeterminate. Thus a majority of the relationships between land use and individual hydrological functions are increasing:

$$\frac{\partial \mathrm{SY}}{\partial L} > 0, \ \frac{\partial \mathrm{FL}}{\partial L} > 0, \ \frac{\partial \mathrm{WY}}{\partial L} > 0, \ \frac{\partial \mathrm{PF}}{\partial L} > 0, \ \frac{\partial \mathrm{BF}}{\partial L} < 0, \quad \text{or} \ \frac{\partial \mathrm{BF}}{\partial L} > 0.$$

However, given the existence of at least the possibility of one relationship that is decreasing (baseflow) no generalisation can be made about the net hydrological impact of a given change in land use in terms of first order effects. In any case, such a generalisation would have little meaning in practical terms as the direction of change of the hydrological function does not predetermine the direction of the accompanying change in economic welfare.

Three possibilities present themselves as to how the vector of hydrological outputs relates to utility (the economist's measure of well-being):

(1) **H** may enter directly into individual utility, for example if the degree of suspended sediment in surface waters affects the aesthetic pleasure derived by a recreationalist from sightseeing or hiking.
(2) **H** may be an input into the household production of utility-yielding goods and services, for example if poor quality of water drawn from a stream affects the health of people in the household.
(3) **H** may serve as a factor input in the production of a marketed good that in turn enters into the production of other marketed goods, household production or individual utility: for example if streamflow is used for hydroelectric power generation which in turn is consumed by businesses, households and individuals.

A simple theoretical presentation of each of these cases is presented below. In the discussion, an effort is made to identify the general type and nature and importance of downstream effects as they are felt through each medium in developed and developing economies (Freeman, 1993).

The approach taken in this chapter tends to focus on the ways in which land use affect hydrology and the ways that the resulting physico-chemical changes (in water, nutrients, sediment, etc.) feed into the economy. This is a very linear and straightforward approach to what is necessarily a complex and intertwined set of factors and events.

The same changes in land use and in hydrology may also affect economic activity through knock-on effects that are transmitted through changes in riparian zone and aquatic ecology. Changes in water quality and timing of water flow can have important ecological impacts that affect, for example, fish populations and those who depend on fish for their livelihood or income. At the same time changes in land use such as forest conversion or restoration can have direct impacts on these same riparian zones and aquatic ecosystems. Increases in light due to reduction in forest cover may lead to beneficial impacts on fish – at least up to some point (Zalewski, Thorpe, and Naiman, 2001). Even further, downslope riparian zones may play an important role in mitigating changes in water quality due to forest conversion upstream (Hubbard and Lowrance, 1996; Snyder *et al.*, 1998; Sheridan, Lowrance and Bosch, 1999).

Examination of these 'ecohydrology' impacts remains a relatively young science and the integration of these impacts into empirical work on economic valuation is a challenge for the future. It is important, however, to note that the addition of an ecohydrology perspective to the argument presented here would not change the outcome fundamentally – many of the studies that are emerging suggest that, a priori, ecosystem modification cannot be considered to be 'negative' and that ecosystems can indeed be managed in order to optimise the services they provide.

Hydrological outputs that enter directly into utility

As it is practically impossible for an upstream land user to prevent downstream users from enjoying or suffering (as the case may be) the consequences of upstream land use change, hydrological functions may be considered as non-exclusive in nature (Aylward and Fernández González, 1998). Absent regulation producers are unlikely to bear any downstream costs attributable to their upstream activities. Likewise, upstream 'producers' cannot capture any downstream benefits of their actions (or their restraint) by selling hydrological outputs in markets. This is not to preclude the possibility that property rights exist for these outputs further downstream. In many areas, for example, streamflow is appropriated under a system of private property rights. Deposited sediment may also be a marketable commodity once it is deposited. For example, in Thailand sediment dredged from rivers is subsequently resold (Enters, 1995). To the extent that these rights or products are then tradeable, these hydrological outputs may be marketable.

However, these cases involve the development of exclusivity, whether through institutional arrangements or investment in resource harvesting, only at the downstream end of the 'production' change. It remains the case that an upstream change in land use will alter the physical availability of the output regardless of any legal claim to the output, whether constituted as streamflow or sediment.[1] For this reason the vector of hydrological outputs may be assumed to enter into utility as a non-marketed good or service alongside a vector of marketed goods, **X**:

$$U = U(\mathbf{X}, \mathbf{H}) \tag{7.2}$$

where $U(\bullet)$ is a well-behaved and increasing individual utility function and **X** is composed of private good quantities ($\mathbf{X} = x_1, \ldots, x_j, \ldots, x_n$). The individual is then assumed to maximise utility subject to the budget constraint, where M equals money income and p refers to the prices of the marketed goods:

$$\sum_{j=1}^{n} p_j x_j \leq M \tag{7.3}$$

In developed economies, the principal manner in which change in hydrological function will affect utility directly, would be a change in water quality or quantity that directly affects aesthetic values. As in the example mentioned above, muddied waters may affect the attractiveness of a recreation or urban site, which then directly reduces the utility associated with the aesthetic aspect of the experience. There is also the possibility that people may hold existence values for the natural streamflow regime. For example, individuals may derive satisfaction or pleasure directly from the knowledge that free-flowing rivers continue to exist in their natural state, regardless of their past or planned future use of the river or its associated products and services. Donations to river conservation organisations are one example of how such existence values translate into willingness-to-pay for conservation. In developing economies it is more difficult to conceive of many instances where water quantity and water quality will simply be consumed directly by an individual, that is entered directly into the utility (or economic welfare) of the individual (Hearne, 1996). The exception may be the very poor where existence is literally 'hand to mouth'. In any event it is probable that hydrological outputs are more likely to enter directly as an input into household production processes in rural, developing households than in developed countries (or urban, developing households) where the household typically purchases basic services from public or private utilities.

Hydrological outputs as inputs to the household production

In the case of the household production function, utility of the household is assumed to be derived from a vector of final services, **Z**, that yield utility:

$$U = U(\mathbf{Z}) = u(z_1, \ldots, z_k, \ldots, z_0) \tag{7.4}$$

These final services are themselves produced by a technology that is common to all households and employ as inputs vectors of both marketed goods and non-marketed hydrological outputs:

$$z_k = z_k(\mathbf{X}, \mathbf{H}) \tag{7.5}$$

For example, changes in dry season baseflow or water quality (**H**) may affect the quantity of bottled water or the number of water

1 For an in-depth discussion of this topic and the possibility of a 'Coasian Bargain' wherein upstream and downstream parties may develop a voluntary arrangement that is in the interest of both parties see (Aylward and Fernández González, 1998) and for real-world examples see (N. Johnson *et al.*, 2001; Rojas and Aylward, Forthcoming).

filters (**X**) that are purchased by the household in providing drinking water (z_k, the utility-yielding service) for household members. Again, the budget constraint can be formulated as reflecting the need to spend less on the marketed goods than is available in money income. Thus, the household is assumed to maximise utility subject to the budget constraint, the level of **H** and the constraints implicit in Eqn 7.5.

In developed countries this model is applicable to certain cases of recreation. For instance, streamflow may be a factor along with canoes, equipment and other inputs in producing a household canoeing trip. Similarly, changes in water quality may affect riverine, estuarine or lacustrine ecological conditions, in turn affecting biomass and species composition of systems that are prized for fishing or diving. Stormflow and flooding are other examples where hydrological outputs may affect developed households directly. By and large household 'use' of water and other hydrological outputs is more often achieved through the purchase of marketed outputs produced by the state or the private sector, for example potable water for domestic use, electric power from hydroelectricity, food produced by irrigators and navigation from ferry services.

In developing countries, the use of water for recreation is likely to be limited to that by higher income or foreign recreationalists. Most probably, hydrological function more directly affects the rural household that 'uses' water for domestic and agricultural use, waterways for navigation, and waterpower as an energy source. Thus, streamflow and water quality may serve as inputs (along with other marketed or non-marketed inputs of labour and capital) into the preparation of food and drink, subsistence farming, transport of produce to market, the accomplishment of repetitive and small-scale mechanical tasks. In developing countries then the bulk of rural populations will experience the hydrological impact of land use change through the household production function.

Hydrological outputs as factor inputs into production

The vector of hydrological outputs can also appear directly in the production function along with other factor inputs. Production of the marketed good, x, then depends on the production function as follows:[2]

$$x = x(k, w, \dots, \mathbf{H}) \tag{7.6}$$

Production is assumed initially to be an increasing function of capital, k, and labour, w, so that an additional unit of each will yield an increase in x. Typically, production is assumed to be an increasing function of the environmental service. As formulated in the case of **H**, this may not be strictly true. An increase in water yield may be beneficial while an increase in sediment yield may not improve production. For example, an increase in streamflow (as a result of forest conversion) may be assumed to have a positive impact on production in the case of hydroelectric power generation. Meanwhile, an increase in sediment delivery may lower production, other things being equal – e.g. holding expenditure on dredging constant. Given that the hydrological functions and their economic impacts will be site specific, it is not possible, *a priori*, to draw any generalisation about which effect will predominate.

Change in hydrology will thus alter both the cost curve for x as well as the demand for inputs of capital, k, and labour, w. Given factor prices, p, the cost function is:

$$C = C(p^w, p^k, x, \mathbf{H}) \tag{7.7}$$

The producer is assumed to minimise cost and the impacts of a change in **H** are felt by consumers (as prices change) or by producers in the input markets (as demand, and hence prices, for capital and labour inputs change).

As suggested above, the analysis of economic consequences of changes in land use and hydrology for developed countries will often draw on this formulation of the problem, particularly as it relates to impacts on hydroelectric power production, domestic water treatment and supply, and industrial water supply. The same goes for developing countries where urban households, industrial concerns and commercial farmers purchase water-related products from public/private utilities and state agencies.

DOWNSTREAM ECONOMIC IMPACTS OF CHANGES IN HYDROLOGICAL FUNCTION

A number of the points typically held as conventional wisdom regarding the downstream impacts of changes in hydrological function require a re-assessment. To this end the empirical literature on the economic valuation of hydrological externalities is reviewed and critiqued below. This leads to a series of conclusions regarding the direction and magnitude of these externalities to the extent possible.

The conventional wisdom emerging from the literature holds that forest conversion (or 'deforestation' as it is often called in developing countries or clear-cutting in developed countries), leads to large costs in terms of losses in on-site productivity and costly sedimentation of downstream hydropower, water supply and irrigation facilities. In addition, conventional wisdom holds that the forest attracts rainfall and acts as a sponge, soaking up and storing excess water for use at later times, thus providing benefits in terms of increased water supply, flood reduction,

2 Following on the tradition of 'bioeconomic' modelling, such a production function could be called a 'hydroeconomic' production function. However, in order to avoid confusion this function is simply referred to as an 'economic' production function to distinguish it from the 'hydrological' production functions that model the land use-hydrology relationship.

improved navigation and dry season flow to agriculture and other productive activities. Although these views seem to be shared across developed and developing regions, they are often emphasised in humid areas of the tropics where 'rainforests' are the dominant natural vegetation type.

There exists another strand of conventional wisdom, which concerns ecological systems that receive less rainfall, oftentimes including ecosystems where forests are not the native vegetation. Conventional wisdom emphasises the negative effects of the choice of agricultural production technology on hydrological function rather than questioning the choice of land use per se. In this context, the debate over the severity of the erosion problem and its economic impact on productivity is complemented by the debate over the relative magnitude of the off-site costs of erosion and other surface and sub-surface water quality impacts of agricultural land use (some of which may result indirectly from the need to fertilise eroded and degraded soils). While most of the evidence comes from North America, the issue clearly applies in other regions. Although the evidence is far from conclusive, many analysts have suggested that these off-site impacts may be at least as important as the on-site costs.

Another issue receiving increased attention in the North American context is the growing evidence that the over-appropriation and abstraction of instream flows for irrigation, urban and industrial use is having increasingly negative impacts on recreation and fish stocks. According to this view, an increase in streamflow would restore these use and existence values. The implicit suggestions being that altering land use and land management practices so as to increase streamflow would have the same effect as reducing water abstraction for agricultural, domestic and industrial uses.

The earlier discussion of the hydrological impact of land use change noted that the conventional wisdom regarding the relationship between forest conversion (and reforestation) and water yield, seasonal flows, flooding and precipitation is often at odds with the scientific understanding, particularly in the tropics (Hamilton and King, 1983; Bruijnzeel, 2004). Much however remains to be learned in this regard as many of the existing studies have been undertaken at small scales (less than 10 km^2) in headwater basins and over relatively short durations, making accurate extrapolation and 'upscaling' difficult (Bonell, pers. com.). Moreover, the net economic effect of land use change in a given circumstance will depend not only on the land use and hydrological function relationship but also the direction of the relationship between hydrological change and economic welfare. Accurate identification, quantification and valuation of the hydrological externalities associated with land use change are complicated further by the need to consider both a range of potential changes in hydrological function and a series of potential economic impacts that may be associated with a given hydrological function.

Below, a review of the available literature on these topics is undertaken with four objectives in mind. The first objective is to demonstrate the range of economic activities that may be affected by change in hydrological functions. The second objective is to give the reader an idea of the degree to which these impacts have been explored in both developed and developing countries. The third objective is to summarise what this research has to say about the relative magnitude and importance of these downstream effects, as well as noting the direction (positive or negative) of the externalities identified. As will be shown, there are considerable gaps and misinterpretations in the literature. Thus, the final objective, which is taken up in the next section, is to suggest the extent to which the direction of the individual impacts can be generalised as increasing or decreasing with respect to land use.

Prior to turning to the empirical literature it is worth stating that there are a large number of techniques available for use in the valuation of non-marketed environmental goods and services. Many authors have surveyed the use of these methods in determining the user cost of soil erosion (Pierce *et al.*, 1983; Stocking, 1984; Bishop, 1992; Olson, Lal, and Norton, 1994; Barbier and Bishop, 1995; Bishop, 1995; Barbier, 1998). Less frequent in the literature are surveys that include methods for use in valuing downstream changes in hydrological function (Gregersen *et al.*, 1987; De Graaff, 1996; Aylward, 1998; Enters, 1998). For example, Gregersen *et al.* (1987) investigate systematically different aspects of hydrological function (including downstream effects) and suggest appropriate valuation techniques. The techniques they consider, while perhaps still the most applicable, represent only a small subset of currently available techniques. Aylward (1998) provides a more recent survey of valuation methods and identifies those applicable to the valuation of hydrological externalities.

Valuation of water quality impacts

The literature on water quality impacts is fairly well spread out over developed and developing countries (see Table 7.1). The lack of cited studies from European countries does not indicate that they do not exist, rather it probably reflects the reliance in this review on English language sources, primarily those from the United States. At the same time, applied work in natural resource and environmental economics has a longer history in United States universities than in their European counterparts.

The bulk of the literature on water quality impacts in both developed and developing countries surrounds the off-site effects of erosion, otherwise referred to as 'sedimentation.' This literature is reviewed first before assessing what material is available regarding the effects of nutrient and chemical outflows.

Studies of externalities associated with sedimentation are found in the literature on tropical moist forests and temperate agricultural

Table 7.1. *Summary of valuation literature on water quality*

Region	Country	Source[a]
Africa	Cameroon	Ruitenbeek (1990)
	Morocco	Brooks *et al.* (1982)
Latin America	Chile	Alvarez *et al.* (1996)
	Costa Rica	Quesada-Mateo (1979); Duisberg (1980); Rodríguez (1989); CCT and CINPE (1995); Aylward (1998)
	Dominican Republic	Veloz *et al.* (1985); Santos (1992); Ledesma (1996)
	Ecuador	Southgate and Macke (1989)
Asia	Indonesia	Magrath and Arens (1989); De Graaff (1996)
	Lao PDR	White (1994)
	Malaysia	Mohd Shahwahid *et al.* (1997)
	Panama	Intercarib S.A. and Nathan Associates (1996)
	Philippines	Briones (1986); Cruz *et al.* (1988); Hodgson and Dixon (1988)
	Sri Lanka	Gunatilake and Gopalakrishnan (1999)
	Thailand	Johnson and Kolavalli (1984); Enters (1995)
North America	Canada	Fox and Dickson (1990)
	United States of America	Guntermann *et al.* (1975); Kim (1984); Clark (1985a); Duda (1985); Forster and Abrahim (1985); Crowder (1987); Forster *et al.* (1987); Holmes (1988); Ralston and Park (1989); Hitzhusen (1992); Pimentel *et al.* (1995)

[a] These studies include a number that are summary studies in the sense that they report on results obtained by other researchers.

production systems. The specific economic activities examined and type of values estimated by these studies are summarised below:[3]

(1) The loss of hydroelectric power generation due to sedimentation of reservoirs (Aylward, 1998; Briones, 1986; Cruz, Francisco and Conway, 1988; De Graaff, 1996; Duisberg, 1980; Gunatilake and Gopalakrishnan, 1999; Ledesma, 1996; Magrath and Arens, 1989; Quesada-Mateo, 1979; Rodríguez, 1989; Santos, 1992; Southgate and Macke, 1989; Veloz *et al.*, 1985).

(2) The loss of irrigation production due to sedimentation of reservoirs (Briones, 1986; Brooks *et al.*, 1982; Cruz, Francisco and Conway, 1988; De Graaff, 1996; Magrath and Arens, 1989).

(3) The loss of flood control benefits due to sedimentation of reservoirs (De Graaff, 1996).

(4) The increase in operation and maintenance costs incurred by sedimentation of drainage ditches and irrigation canals (Alvarez *et al.*, 1996; Brooks *et al.*, 1982; Forster and Abrahim, 1985; Fox and Dickson, 1990; Gunatilake and Gopalakrishnan,1999; Kim, 1984; Magrath and Arens, 1989).

(5) The increase in dredging and maintenance costs associated with sedimentation of hydroelectric reservoirs (Rodríguez, 1989; Southgate and Macke, 1989).

(6) The increase in costs of water treatment associated with sedimentation (CCT and CINPE, 1995; Forster *et al.*, 1987; Fox and Dickson, 1990; Gunatilake and Gopalakrishnan, 1999; Holmes, 1988).

(7) The increasing dredging costs associated with harbour siltation (Magrath and Arens, 1989).

(8) The loss in production due to the effects of sedimentation on subsistence or commercial fisheries (Hodgson and Dixon, 1988; Gunatilake and Gopalakrishnan, 1999; Johnson, 1984; Ruitenbeek, 1990).

(9) The loss of tourism revenues or recreational benefits (including fishing) following sedimentation of water systems (Fox and Dickson, 1990; Hodgson and Dixon, 1988; Ralston and Park, 1989).

(10) The loss of hydroelectric power production and increased dredging costs associated with sedimentation of settling ponds (Mohd Shahwahid *et al.*, 1997)

(11) The loss of navigation opportunities associated with sedimentation of water supply reservoirs used to supply water to canal locks (Intercarib S.A. and Nathan Associates, 1996).

In the most comprehensive examination of the off-site costs of erosion in the United States to date, Clark (1985a) identifies the

3 Studies that merely present the results of other studies or aggregate them are not included in this list.

full range of economic impacts that eroding soils may cause. Of these impacts, a number are missing from the list above including: impact of sediment on biological systems, lake clean-up, damage caused by sediment in floods and damage caused to productive activities and consumption by residual sedimentation in end use water supplies. Thus, even a single hydrological output, sedimentation, may cause an enormous number of external effects.

The results of these studies confirm the intuition that in general utility will be a decreasing function of sedimentation and, consequently, that utility will be a decreasing function of land use. In other words, land use change that increasingly modifies natural vegetation can be expected to produce negative hydrological externalities. A dissenting voice on this topic is that of Enters (1995) who cautions that sedimentation may also confer benefits and not just costs on society. This claim is based on the author's observation that illegal dredging of deposited sediment in the Ping River, Thailand, demonstrates positive externalities associated with sedimentation. It has also been noted that erosion and sediment transport lead to increased soil fertility on footslopes (van Noordwijk *et al.*, 1998; Malmer *et al.* this volume). Still, these benefits are likely to simply reduce the net negative effect of sediment rather than suggesting that sedimentation impacts are positive on net.

These observations are complemented by noting that in many river systems (e.g. the Nile, the Senegal, the Mekong) natural flooding and sedimentation historically played vital roles in the renewal of soil fertility in floodplain and recession agriculture systems, as well as the renewal of geomorphological processes in delta ecosystems. The loss of these downstream services due to the construction of dams or their confinement to river channels by levees has now led to interest in the possibility of re-establishing natural flood regimes and instream flows artificially so as to restore the benefits of sedimentation. At a larger, basin-scale then, the issue of costs and benefits of natural and accelerated erosion and sedimentation requires a careful assessment.

A number of the studies demonstrate significant external effects. For the United States, Clark (1985a) gathers related research on practically every conceivable off-site impact of eroding soils and provides a nationwide estimate of the annual monetary damage caused by soil erosion of US$6.1 billion (in 1985). Even so, Clark concludes that this figure may be severely under-estimated as the impact of erosion on biological systems and subsequently on economic production and consumption is not included. At the same time it should be acknowledged that Clark includes in his analysis the effects of 'erosion-associated' contaminants. In other words, the figures relate to water quality more generally, not simply the effects of soil erosion, and include the effects of pesticides and fertilisers that are used in agricultural production. This of course goes beyond the scope of the hydrological externalities envisioned in this chapter where the concern is with nutrient and chemical outflows related to a change in vegetation accompanying a change in land use.

Nonetheless, Clark's estimates serve the purpose of dramatising the potential magnitude of the off-site damage caused by soil erosion. Clark's compilation also suggests that the literature on the topic as reported on in this chapter is but a representative sample of a much larger literature. However, it must be acknowledged that the quality of a majority of the studies drawn upon by Clark and, indeed, of those gathered for this chapter, is mediocre. Holmes (1988) summarises this criticism by stating that the Clark (1985a) study 'is based to a large degree on *ad hoc* interpretation of a widely divergent group of studies.' The majority of these studies rely on simple damage function estimates of changes in costs or revenues, with no consideration of optimising behaviour on the part of consumers and producers as reflected in supply and demand curves.

Interestingly, Holmes' (1988) more sophisticated study of the nationwide costs of soil erosion to the water treatment industry produces a range of US$35 million to US$661 million per year. This range is close to that provided by Clark (1985a) of from US$50 to US$500 million, even though Holmes' best estimate of US$353 million is three times larger than Clark's best estimate of US$100 million. At the same time, it must be acknowledged that despite the sophistication in methods, the large range obtained by Holmes indicates continued uncertainty over the true magnitude of these sorts of damage estimates.

Clearly much work remains to be done in refining such estimates. In particular, one difficulty of many of these studies is that they simply measure existing damage levels and do not consider to what extent these damages could be mitigated by alternative land uses or production technologies. Nor do they subsequently assess the trade-off between alternatives and the existing situation. This may be an important point as even improved technologies will produce some erosion and sedimentation. Of course, oftentimes an understanding of how damage relates to different sediment levels is missing from the studies as well, making it difficult to understand the form of the relationship and how it might be altered by partial reductions in sedimentation rates. The application of a damage function approach that evaluates the choice between the option to undertake conservation and postpone the decision may be worth investigating in this regard (Walker, 1982).

In sum, it is likely that substantial off-site damages are caused by soil erosion due to agricultural production in the United States and similar areas around the world. Whether the claim is accurate that these damages are as big as, if not larger than, the on-farm impacts is probably a moot point, given that the estimates of on-farm losses are just as debatable as the off-site losses on methodological grounds. For example, Crosson (1995) elegantly rebuts the exaggerated claims made by Pimentel *et al.* (1995) regarding on-site productivity losses due to soil erosion. What is probably more important to evaluate is whether off-site damages are important enough to merit action, a point that is often disregarded by the literature. To be fair, however, it may be difficult to generalise due to the site-specific nature of the biophysical and economic relationships involved.

In tropical regions, many of the studies are more explicit in targeting land use per se as the cause of hydrological externalities, particularly the conversion of tropical forests to other uses. A number of these studies even go so far as to include damage estimates into cost-benefit analyses in order to demonstrate the need for changes in policies affecting land use or to justify conservation projects. For example, in Ruitenbeek's valuation of the Korup Project in Cameroon, the benefits from erosion control were estimated to be almost half of the direct conservation benefits of conserving the forest, benefits which outweighed the sum of the direct and opportunity costs of conservation (Ruitenbeek, 1990). Santos (1992), Southgate and Macke (1989), and Veloz *et al.* (1985) all suggest that sedimentation will have significant effects on hydroelectric power plants in Latin America and the Caribbean.

Nevertheless, there is an additional series of studies demonstrating that the externalities associated with sedimentation are often not terribly large or important. In the Philippines, the effect of sedimentation derived from the conversion of large areas to open grasslands in the Magat Basin on the length of life of the reservoir downstream was valued at 0.10 pesos ha^{-1} yr^{-1}, or under one US cent per hectare per year (Cruz *et al.*, 1988). Meanwhile the benefits of erosion control through reforestation in the Panama Canal Zone comes to a present value of just \$9 ha^{-1} in terms of its effect on storage reservoirs and water supply for navigation (Intercarib S.A. and Nathan Associates, 1996).

In Arenal, Costa Rica, the present value of the cost of sedimentation from pasture (as opposed to reforestation) in terms of lost hydroelectric production ranged from US\$35 to US\$75 ha^{-1} (Aylward, 1998). The Arenal study is unusual in that it employed a formal model of the impact of sedimentation on both the dead and live storage areas of the reservoir, enabling it to separate out the differential effects on these areas. Given the large dead storage relative to sediment inflow for this particular reservoir the effect of sedimentation on dead storage produced benefits, not costs, in the case of Arenal as the sediment effectively displaces water upwards into the live storage during dry periods. Arenal is an interannual regulation reservoir and thus during a series of dry years in which the reservoir does not fill but is gradually drawn-down, the sediment occupying the dead storage effectively makes additional water available for power generation (Aylward, 1998).

In Malaysia, a simulation of the effect of logging on downstream run-of-stream hydroelectric power and treated water production indicated that a programme of reduced impact logging would have essentially no effect on water supply and would lead to only a minimal disturbance of hydropower generation through sedimentation of the settling ponds (Mohd Shahwahid *et al.*, 1997). In other words, the gains from logging could easily compensate for the losses incurred by the hydroelectricity producer due to sedimentation. Finally, in Sri Lanka a comparison of measures for preventing or mitigating the impact of sedimentation on the Mahaweli reservoirs suggested that the costs of the measures outweighed their potential benefit (Gunatilake and Gopalakrishnan, 1999).

In sum, the results are mixed on the magnitude of the economic impact of sedimentation as caused by the conversion and modification of tropical forests. Such a conclusion is not counter intuitive as it is logical to expect that site-specific characteristics such as geology and climate, drainage area and topography, type and size of reservoir or other infrastructure, and demand for end use goods and services, will determine the magnitude of these effects in particular cases. In addition, it must be said that many of these studies present only fairly crude estimates, just as in the case with the studies from developed countries.

Turning briefly to water quality issues beyond merely the off-site effect of erosion, no studies were found in the developing country literature that specifically assess the downstream externalities associated with nutrient or chemical outflows associated with land use change (though see chapters by Proctor, Connolly and Pearson, this volume, for more on the biogeochemical impacts). In a developed country context, there are of course many studies of the economic damage caused by poor water quality (Bouwes, 1979; Epp and Al-Ani, 1979; Young, 1984; Ribaudo, Young and Shortle, 1986; Lant and Mullens, 1991). Typically these studies are not linked to land use in specific geographical areas, nor do they evaluate damage that is directly and only related to land use change. Oftentimes the measure of water quality that can actually be evaluated (as perceived by recreationalists, for example) is extremely crude (i.e. water quality is good or bad), so that associating the measure of damage with a particular type of non-point source pollution is impossible. These are precisely the 'erosion-associated' contaminants surveyed by Clark. Clearly these (gross) impacts are important and perhaps particularly so in the case of the biological impacts that Clark does not estimate. The extent to which they are associated with land use per se and not simply the prevalence of pesticide and fertiliser use as part of a production technology package is difficult to assess.

Valuation of water quantity impacts

The external effects of land use change on streamflow levels will affect four types of hydrological outputs: (1) annual water yield, (2) seasonal flows, (3) peakflow and (4) groundwater levels (Gregersen *et al.*, 1987). These outputs will in turn affect a host of different economic activities, including most of those affected by water quality changes. An increase in water yield or baseflow

Table 7.2. *Summary of valuation literature on water quantity*

Region	Country	Function valued	Source
Latin America	Bolivia	Flood control Groundwater recharge	Richards (1997)
	Costa Rica	Dry season flow Peak flows	Quesada-Mateo (1979)
		Annual water yield Dry season flow[a]	Aylward (1998)
	Guatemala	Dry season flow	Brown *et al.* (1996)
	Panama	Dry season flow	Intercarib S.A. and Nathan Associates (1996)
Africa	Cameroon	Flood control	Ruitenbeek (1990)
	South Africa	Annual water yield	de Wit *et al.* (2000)
		Annual water yield	de Wit *et al.*, forthcoming
Asia	China	Annual water yield	Guo *et al.* (2001)
	Indonesia	Annual baseflow	Pattanayak and Kramer (2001a, b)
	Malaysia	Dry season flow	Kumari (1995)
	Thailand	Dry season flow	Vincent and Kaosa-ard (1995)
Temperate countries	Australia	Annual water yield	Creedy and Wurzbacher (2001)
	United Kingdom	Annual water yield	Barrow *et al.* (1986)
	United States of America	Annual water yield	Kim (1984)

[a] Sensitivity analysis only.

will change reservoir storage and irrigation capacity leading to changes in water supply for hydropower, irrigation, navigation and recreation. Similarly, changes in water yield and baseflow may affect these activities directly in the absence of hydrostorage capacity in the system. Changes in peakflows are felt principally through a change in localised flood frequency and can damage infrastructure (bridges, culverts, roads, embankments) and agriculture (sedimentation of crop land with coarse material), as well as putting homes and lives at risk. Changes in the groundwater table in upland areas will influence directly spring discharges used for local water supply and have downstream impacts on the productivity of local biological systems (such as wetlands) that may provide recreational or preservation benefits, as well as affecting downstream agricultural and other productive systems.

The methods that may be applied in valuing such external effects are essentially no different than those in the case of water quality. Nonetheless the literature on this topic is scanty in comparison to that on water quality effects. Just 13 studies were found in comparison to the 34 studies of sediment. The countries for which such studies were found are listed in Table 7.2.

Of the studies that examined the off-site costs of sedimentation, only five considered the attendant issue of water quantity (Aylward, 1998; Intercarib S.A. and Nathan Associates, 1996; Kim, 1984; Quesada-Mateo, 1979; Ruitenbeek, 1990). Indeed, such impacts were rarely, if ever, even identified and listed in qualitative terms. Whether this is due to a suspicion that the magnitude of the changes is insignificant or simply represents an ignorance of the biophysical impacts of land use change on water yield is unclear. As an indication that this situation is changing, 12 of the 16 studies were published since 1995. Interestingly, seven of the studies considered water quantity issues but did not raise the issue of water quality (Barrow *et al.*, 1986; Brown *et al.*, 1996; Guo *et al.*, 2001; Pattanayak and Kramer, 2001a, b; Richards, 1997; Vincent *et al.*, 1995).

An additional avenue of research, primarily in a developed country context, concerns the valuation of increases in instream flows. A number of studies have examined the recreation, fishery and hydroelectric power benefits that would be gained by restoring instream flows in the western United States (Daubert and Young, 1981; Narayanan, 1986; Ward, 1987; Johnson and Adams, 1988; Brown, Taylor, and Shelby, 1992; Duffield, Neher, and Brown, 1992). Once again, these studies are not linked directly to land use, but could be used to indicate the economic benefits associated with land use change that subsequently alters streamflow.

ANNUAL WATER YIELD

Of the seven studies on annual water yield reviewed here, five suggest that catchment protection values are negative, i.e. that utility is increasing as a function of land use. In the earliest study of this nature, Kim (1984) simulates the increase in annual water yield associated with a change in land use management from no grazing to grazing in the Lucky Hills catchment of southeastern Arizona.

Based on a review of the literature Kim (1984) assumes a 30% increase in water yield under grazing over a simulated 50-year rainfall cycle (based on climatic records). Under the additional assumption that all the extra water would be used for irrigated agriculture and employing a US$1.2 m^{-3} value for irrigation water based on studies from the region, Kim calculates the net present value over the 50 years to be US$342 at a 7% discount rate. Unfortunately, it is not clear if this is the catchment total or a per acre figure. Assuming the former this comes out to a little over US$7 ha^{-1} for the 44-hectare catchment. When Kim adds in the costs of excavating the sediment settling ponds (US$1068) and the benefits of animal weight gain (US$740), the net present value of the returns to the land use management change are barely positive at $14 or about US$0.25 ha^{-1}.

A study of the effects of afforestation on hydroelectricity generation in the Maentwrog catchment in Wales and forty-one catchments in Scotland by Barrow *et al.* (1986) indicates that the increased evaporation under reforestation (in comparison with grazing) lead financially marginal sites (for forestry) to become financially sub-marginal once hydropower losses were included into the analysis. While there was some variation in results depending on site conditions, the example clearly shows the negative impact on productivity associated with afforestation in a hydroelectric catchment.

A study in Arenal, Costa Rica, confirmed the results obtained by Barrow *et al.* (1986) by showing that water yield losses due to reforestation of pasture areas may lead to large efficiency losses in downstream hydroelectric power production (Aylward, 1998). The externalities associated with water yield effects were calculated to be one order of magnitude greater than those associated with the sedimentation costs (as already referred to above). Best estimates for both cloud and non-cloud forest areas suggested positive present values in the range of US$250 to US$1100 ha^{-1} for pasture. Sensitivity analysis showed that the values will be reduced to two-thirds of these figures with higher discount rates and in the event that all the water yield gain under pasture were to arrive during the wet season (instead of being received proportionately across wet and dry seasons). The values may also rise to almost US$5000 ha^{-1} if dry periods lengthen or occur early in the seventy-year simulation period. Further sensitivity analysis examined what would be the economic outcome if reforestation resulted in net gains in dry season flow, in spite of the expected overall losses in total annual water yield. A switching value (where the value of total hydrological externalities go to zero) was obtained only when all of the annual water yield gain *and* an amount equal to an additional 50% of this amount was redistributed to arrive during the wet season (when water is less valuable for power generation). When the analysis of livestock productivity was incorporated into a cost-benefit analysis of land use options, strong synergies between livestock production and hydroelectric power generation in the catchment were demonstrated (Aylward and Echeverria, 2001).

The South African study by De Wit *et al.* (2000) examines issues related to the catchment management charge (approximately US$1 ha^{-1} yr^{-1}) that is to be levied on forestry activities as Stream Flow Reduction Activities under existing legislation. Combining information from detailed hydrological studies of the effect of forestry on evapotranspiration, the authors calculate that forestry consumes 7% of South Africa's water (see also Scott *et al.*, this volume). Collation of macroeconomic data on value added in forestry suggests that the value added per cubic metre in forestry is low (2.8 Rand or about US$0.50) but still higher than irrigation. De Wit *et al.* (2000) use an input-output model to confirm that due to the existence of higher value uses for water (than forestry) such changes lead to economy-wide gains in output. In a related study De Wit (forthcoming) calculates the present value cost of water consumed by black wattle (*Acacia mearnsii*) in South Africa as US$1.4 billion using information on the difference between streamflow and value added of black wattle as versus alternative land uses.

In a study of ecological services in Victoria, Australia, a counter-example to the trend shown above is provided by Creedy and Wurzbacher (2001). In this case the authors are assessing the effect of harvesting old-growth Eucalypt forest. These forests have the unique property that they transpire very little water. Thus, the effect of harvesting and allowing regrowth will lead to a decline in annual water yield, not an increase as would be otherwise expected (Vertessy *et al.*, 1998). Creedy and Wurzbacher (2001) do not provide explicit value estimates in per hectare terms. However, they do show that given the projected costs of alternative sources of water to the public utility, incorporating the loss of water benefits alongside the wood benefits of logging leads to an infinite length of the optimal rotation. In other words logging is not worth the costs it incurs in terms of forgone water supply.

In examining the value of ecosystem services in Xingshan County of Hubei Province, (north-eastern) China, a study by Guo *et al.* (2001) purports to value the water conservation value of forests in terms of 'hydrological flow regulation' and 'water retention and storage'. However, all the figures employed in the study are annual, thus it can only be concluded that this is a study of annual water yield. Unfortunately, the authors' definition of forest hydrological function is confused, leaving out transpiration and defining canopy interception as one of the elements of rainwater 'conserved' by a forest ecosystem. The authors' empirical analysis concludes that, in comparison with a scenario of forest conversion to shrub and grass, the forest alternative 'conserves' such large amounts of water that 42% of the value of downstream hydroelectric production is due to the conservation of forest. This study only serves to illustrate how inadequate hydrological analysis and simplistic applications of economic valuation can lead to gross exaggerations of hydrological externalities (see also Cheng, 1999).

FLOOD CONTROL

The remainder of the literature that was surveyed portrays utility as a decreasing function of land use. Ruitenbeek (1990) estimates the flood control benefits to be generated by protecting forested catchments in Korup National Park in Cameroon. Ruitenbeek's calculation is based on the share of local income that would be lost in a flood event multiplied by the percentage of cleared forest area in the Park. As reported by Bonell (this volume), the hydrological literature (subject to the scale constraints of experimentation mentioned earlier) does not support definitively the contention that land use change would lead to changes in flood frequency or magnitude at the scale suggested by Ruitenbeek, and thus the results must be regarded as suspect until proven otherwise.

Richards (1997) examines the potential benefits of a flood control programme in the Taquina catchment in the Bolivian highlands. The approach taken is more data intensive than that by Ruitenbeek, insofar as the costs of damage from a recent flood are actually gathered to motivate the damage cost estimate. Assumptions regarding flood frequency and intensity are then made under the 'with' and 'without' project cases, accounting for a gradual phase-in of project benefits. Straight multiplication is then used to arrive at yearly flood control benefits as the difference between the 'with' and 'without' project scenarios. By year five, the nominal flood control benefits outweigh the project costs by a ratio of 3:1.[4] While the benefits of flood control appear quite large, it is not clear to what degree they are a response to land use change in terms of on-farm soil conservation technologies as opposed to the effect of hydraulic works and infrastructure located in gullies and stream courses.

Interestingly, neither of the two studies mentioned above attempts to apply the standard methodology for evaluating flood damages as recounted by Gregersen *et al.* (1987). Under this methodology flood frequency curves (the probability that a given instantaneous streamflow level or stage height will be exceeded) are developed for the 'with' and 'without' project scenarios. A damage function is then developed that relates peakflow levels to damage costs. A practical difficulty in applying this technique in developing economies is the poor availability of historical data on the damages of past flood events. This problem is exacerbated by rapid urbanisation, industrialisation and population growth, which make the relationship between peakflow levels and damage costs unreliable over time. A further limitation of flood frequency analysis from a hydrological standpoint is that it rests on the assumption of stationarity: the analysis ignores changes in river and stream discharge linked to climate variability and land use change over time.

Quesada-Mateo (1979) develops a deterministic simulation model that enables the user to determine the maximum amount of firm (reliable) power that could be produced from a hydroelectric reservoir. What makes the model interesting is that it incorporates explicitly the effect of the accumulation of sediment in the live storage, as well as a change in stormflow regime. To the detriment of the analysis, the author assumes erroneously that the removal of forest cover will lead to an increase in peak flow during the wet season and a decrease in baseflow during the dry season. While the first assumption is likely to be correct, the latter does not necessarily follow.

DRY SEASON FLOW AND GROUNDWATER STORAGE: HYDROLOGICAL ANALYSIS

Eight studies were found that attempt to quantify the purported benefits provided by forest cover in terms of enhanced groundwater storage and subsequent dry season baseflow. All but the Quesada-Mateo (1979) study (reviewed above) are recent in origin and most of the studies suffer from the same problem, namely difficulty with the direction and magnitude of the land use and hydrological relationship. As irrigated agriculture and navigation will clearly benefit from an increase in dry season baseflow there is little doubt that the relationship between the hydrological outputs (dry season baseflow) and economic activities is increasing. However, if the direction or magnitude of the land use and hydrology relationship is misstated, the overall conclusions of the studies regarding the hydrological externalities would be erroneous. As this concern is central to the interpretation of the results obtained by these studies, the hydrological analyses are explored below at some length.

In the Sierra de las Minas Biosphere Reserve of Guatemala a comparison between dry season baseflow in a forested and a partially cleared catchment was used to estimate the percentage increase in baseflow associated with a forested catchment (Brown *et al.* 1996). Unfortunately, study limitations implied that only four months of dry season data from 1996 were compared. As the two catchments were not calibrated prior to the change in land use it is not possible to rule out the possibility that the effect observed is a result of some other situational variable and not land use. For example, the forested catchment faces southeast and sits at an altitude of 1900–2400 metres. The cleared catchment faces southwest, is located some ten kilometres to the west of the forested catchment and sits at an altitude of 1400–2120 metres. The forested catchment is known to be a cloud forest area and the study concerned reports on the capture of horizontal precipitation during the dry season in this catchment. Given the lack of calibration the higher level of baseflow in the forested catchment may simply be attributable to climatic conditions such as the presence of cloud forest moisture or rainfall levels and not only

4 The study does not give the present value of flood control but only the project internal rate of return.

to conversion of the other catchment.[5] Bruijnzeel (this volume) also notes that the two catchments are of different size, which may also affect baseflow levels. To make matters even more difficult, the cleared catchment is not in the basin in which the impact of baseflow changes is valued, while the forested catchment is within one of these basins.

Brown *et al.* (1996) also note that high values for the capture of fog moisture were observed only in an elevation zone that occupied a very slight percent of the total catchment area and that the lower catchment was well below this zone. Despite the intuition, then, that the existence of forest will serve to strip moisture from clouds in the dry season thus adding to dry season baseflow as compared to a scenario in which forest conversion occurs, the simulations undertaken in the study are not very well supported by the hydrological analysis.

The study of the Panama Canal Basin relies on a similar 'paired' catchment analysis that does not have an experimental basis (i.e. calibration followed by treatment) (Intercarib and Nathan Associates, 1996). Nevertheless, the data are more convincing as the monthly streamflow for six forested and cleared catchments (three each) compared are based on 21 years of data. The data reveal that monthly streamflow measured as a percent of total precipitation is less responsive in the case of the forested catchments. The authors use this information to substantiate the claim that land that remains in forest stores a larger amount of water going into the dry season. This capacity is then available to refill the dams that release their stored water in the dry months, thereby augmenting reduced streamflow during these months.

Once again, the potential existence of confounding variables has not been ruled out in the analysis. Further, as annual water yield from a cleared catchment can be expected to rise, even a lowering in monthly streamflow in percentage terms during the dry season does not rule out an increase in streamflow in absolute terms. In this regard it is worth noting that the Intercarib study ignores the potential decrease in water yield that would presumably result from reforesting the cleared areas of the Canal Basin. Thus, the study emphasises one type of hydrological change and ignores another, in addition to falling short of providing firm evidence of the hydrological effect that is subsequently included in the valuation exercise.

The analysis of the Mae Teng Basin in Thailand by Vincent *et al.* (1995) resolves a number of the issues encountered above by employing historical data on streamflow and precipitation. By analysing data from periods before and during the period of land use change the authors strengthen their case further. The authors use regression analysis to demonstrate that:

- no change in streamflow is observed prior to land use change (1952–1972);
- dry season streamflow is reduced during the period in which land use change occurs (1972–1991);
- climatological factors do not explain the reduction in water yield.

The land use change that took place in Mae Teng during the 1972 to 1991 period consisted of both an increase in irrigated agriculture and an expansion of pine forestry plantations. As both of these activities can be expected to increase water use, the authors conclude that land use change has indeed led to the reduction in water yield, particularly during the dry season. Unfortunately, the authors are unable to define clearly to what extent the conversion of land to agriculture, the use of water in irrigation or the growth of pine plantations were responsible for the observed decrease in streamflow.

Pattanayak and Kramer (2001a, b) value 'drought mitigation' in a large number of catchments that lie below the Ruteng Park, on the island of Flores in eastern Indonesia. In the longer of the two papers, the authors estimate an explicit hydro-economic model of how changes in baseflow lead to changes in profits received by farmers from crops (Pattanayak and Kramer, 2001b). In the other paper, the authors explore what farmers would be willing to pay to obtain 'drought mitigation' services from forest areas in the Park (Pattanayak and Kramer, 2001a).

The authors cite a number of sources as providing evidence that forest in the Park plays a role in drought mitigation, with one consultancy report explicitly cited as claiming higher dry season baseflow under forest. And clearly it seems logical that more water in the dry season would increase farm productivity and, indeed, the willingness-to-pay survey confirms this expectation (Pattanayak and Kramer, 2001a). The hydrological portion of the model, however, weakens the meaningfulness of the hydroeconomic analysis.

First, the authors actually do not include dry season baseflow in the model, but rather total annual baseflow. That agricultural production is related to total water availability is not in question, however it seemed that the intent of the paper was to get at the marginal benefit associated with increased flows when they presumably matter most, that is during dry periods.

A second difficulty encountered by the authors, however, concerns their effort to develop a quantitative linkage between forest cover and baseflow. The authors estimate a cross-sectional regression equation using data from 37 catchments and a series of explanatory variables, amongst them three for forest cover: area of forest cover, percent of forest cover, and the square of percent of forest cover. As the squared term produces a negative coefficient, the end result is that the simulation of increases in forest cover in the catchments leads to a mixture of expected losses and gains in farmers' profits as a result of increases in forest cover (Pattanayak and Kramer, 2001b). The study illustrates the importance of multidisciplinary cooperation as poor theoretical formulation and

5 The authors also do not provide data on yearly rainfall totals in the two catchments, but indicate that rainfall levels will vary with elevation and that at high elevations precipitation may vary greatly within short distances.

execution of the hydrological portion of this study undermines an otherwise excellent economic analysis.

Richards (1997) values the aquifer recharge benefits of the same Bolivian soil conservation programme mentioned above. Apparently, the intuition is that the project will increase infiltration but without the project infiltration rates will fall. There appears to be some confusion, however, as the author first misrepresents the direction of water quantity effects as found in the literature and then states that with the project 'runoff would be reduced by 15–25%' (Richards, 1997:26). By year 50 the author calculates that aquifer recharge would be 80% higher with the project than without the project. Further, although the benefits of aquifer recharge under the project are considerable, there is no discussion of seasonality of runoff or water storage and thus it is not clear how the change in aquifer recharge is translated into water supply benefits.

The last of the studies is a valuation of the hydrological function provided by peat swamp forest in Malaysia by Kumari (1995). Unfortunately, insufficient detail of the hydrological basis for the analysis is provided in the paper to provide an informed content and thus cannot be analysed further here. Interestingly, however, the paper does refer to a controversy over the role of forests in the production of dry season padi rice.

The studies reviewed above demonstrate the difficulty of developing convincing hydrological analyses of the linkages between specific land uses and dry season flows. This is particularly acute when the study site does not have a history of hydrological measurement or evaluation and points to the difficulty of undertaking short-term policy-oriented studies where long-term hydrological research or calibration of process-based models to local conditions is probably necessary to guarantee the reliability of results.

DRY SEASON FLOW AND GROUNDWATER STORAGE: ECONOMIC ANALYSIS

The decline of sophistication in the economic modelling conducted for these studies also varies tremendously. In the Guatemalan, Indonesian and Thai studies detailed econometric analyses of agricultural production are used to estimate the change in revenue that would be associated with changes in flows. In the Thai case it is not possible to link the significant (roughly 50%) loss in revenues to a particular causal factor. In the Indonesian case, the weakness of the hydrological analysis undermines the results provided by a full hydro-economic model.

In the Guatemalan study only aggregate figures are provided, not estimates of the loss in irrigated agriculture in dollars per hectare or net present value terms. Employing data from the report, however, the loss of revenue accruing to the area in the two catchments that would be cleared under the simulation can be calculated to be US\$7.5 ha^{-1} yr^{-1} and US\$47 ha^{-1} yr^{-1}.[6] If the effect is assumed to continue indefinitely and the money flows are converted into present value terms at a 10% discount rate, the figures may be multiplied tenfold to obtain approximate present values of from US\$75 to US\$470 ha^{-1}. Such values would be low to respectable values for land of presumably marginal productive potential in the region. Thus, should the claimed hydrological effect be substantiated, the authors have demonstrated a significant hydrological externality of forest conversion in these Guatemalan catchments.

In the Bolivian case, the economic methodology employed is fairly simple. Unit values for water are multiplied by the changes in aquifer storage (Richards, 1997). Again, this linkage is not well demonstrated but as presented is significant.

In the Panamanian case, the valuation hinges on the prospects for developing a third set of locks in the Canal, at which point the current water storage capability would not be sufficient (Intercarib and Nathan Associates, 1996). The benefits of water storage offered by 132 000 ha of existing forest are estimated to be an additional 1500 m^3 ha^{-1} yr^{-1} based on the hydrological analysis. The costs of building additional capacity are US\$0.185 m^{-3}. The study reports water storage benefits for these existing forest areas as US\$277 ha^{-3} in present value terms. The same figure is calculated for the water storage benefits of reforesting an additional 100 000 ha in the Canal Basin.

The study apparently uses the Polestar software to generate different scenarios for how land use determines water and sediment inflows to the dams and water supply to the system of locks is modelled over a 60-year planning horizon. According to results presented in the study, there is an anticipated water shortage only if the third set of locks is built, an event projected for the year 2020. Unfortunately, it is not possible to come close to the per hectare calculation using a 10% discount rate (the exact discount rate employed is not cited in the document). It is however, possible to calculate the US\$36 million present value attributable to the 132 000 ha of existing forest, by simply multiplying the number of hectares by the annual water storage figure and the per unit cost of building the new dam. However, assuming that the new dam would not need to be built until 2020, the present value of such a figure would be more in the region of US\$3 million than US\$36 million.[7] Further, it has been estimated recently that sedimentation levels in the Canal Basin have dropped back to background levels given that land use has stabilised in the last decade (Stallard, 1997). In all likelihood then the hydrological benefits of engaging in massive reforestation of the Panama Canal Basin due to both water storage and erosion control are substantially overstated, if they exist at all.

Whether as a result of questions regarding the hydrological assumptions or modelling, or the economic interpretation of these

6 In Brown *et al.* (1996), on page ii the percentage of remaining forest area that is cleared under the simulation is presented and on pages 69 for the Jones catchment and page 80 for the Hato catchment the remaining forest areas can be derived from land use and area data.

7 Current intentions in Panama greatly exceed such marginal changes with plans to build a series of three dams in order to double the water supply to the Canal by approximately 2010.

relationships, the results of the Bolivian, Guatemalan, Indonesian and Panamanian studies examined above must be regarded as highly questionable. The cautious stance taken by the Thai study simply reflects the inherent difficulties in undertaking such an integrated hydrological and economic analysis of dry season flows.

THE DIRECTION OF HYDROLOGICAL EXTERNALITIES

The effects of changes in hydrological outputs on economic consumption and production will vary with different types of hydrological function and types of economic activities. For instance, an additional unit of baseflow into an irrigation scheme during the dry season will lead to additional output by raising water availability during a critical period. If baseflow is an increasing function of land use then the relationship between land use and agricultural production will be increasing. On the other hand a rise in sedimentation of the irrigation canals will be associated with either a loss in production as the sediment impairs the ability of the canal to deliver water or an increase in, for example, labour expended on dredging. In this case then, production will be a decreasing function of land use.

In general, an increase in sedimentation, nutrification or leaching can be expected to impact negatively on the profits from activities such as irrigation, hydroelectric power generation, water treatment and navigation. Similarly, the effects of increases in these outputs on developing country households may be negative. However, it is at least conceivable that on occasion they may also have positive elements, as in the case in South East Asia where sediment is actually harvested (Enters, 1995; van Noordwijk, 1998). The augmentation of natural processes of renewing soil fertility cannot be assumed to be negative. In addition, it should be noted that there is no general intuition that requires a given change in chemical or nutrient outflows to have a negative impact on the household. Much will depend on how ideal the starting point is with respect to desired water quality characteristics and what thresholds or discontinuities exist in the relationship. Finally, it is reasonably clear that reduction in water quality of waterways and lakes has a negative impact on recreation opportunities. In other words, the conventional wisdom with regard to the sign of the water quality effect is likely to be correct, though questions remain regarding the magnitude of the problem.

The case with the different measures of water quantity is much less certain and will depend on the hydrological functions that are germane to the production technology and end use demand. For example, an increase in land use that leads to soil compaction and an increase in peakflows will affect profits adversely from a run-of-stream hydroelectric plant, whilst having no effect on an annual storage reservoir used for irrigation, hydroelectricity or navigation control. An increase in annual water yield may raise profits for a large hydroelectric reservoir that stores water interannually while having little to no impact on a downstream water treatment plant that is fed from such a reservoir. In other words, profits (and eventually utility) may be either an increasing or decreasing function of these hydrological outputs and of land use itself. This result is clearly at odds with the conventional wisdom on the effects of changes in water quantity on productive activities.

The situation with regard to consumptive values of water quantity in developed countries is somewhat clearer. On the one hand, in cases where streamflow is already greatly diminished or altered (for example due to abstraction, dams or levees), the benefits to recreation activities of increases in these flows are clear. However, the restoration of original vegetation cover in the catchment may only provoke a worsening of the situation if it means replacing shallow-rooted vegetation (crops) with deep-rooted vegetation (forest). A further consideration is that the extent to which developed country consumers actually are aware of the nature of original streamflow conditions is debatable, given the large modifications and extractions already made to most waterways in developed countries. Thus, although a change back to the original land use would alter the *status quo*, it is not clear that such a change would produce perceived improvement in aesthetic values. In other words the direct effects of land use change on utility as experienced through hydrological functions may not be terribly large, nor may utility necessarily be a decreasing function of land use for these functions. Again, much will depend on the severity of the problem posed by current streamflow and hydrological conditions at the site.

An added difficulty to the process of unravelling the implication of downstream hydrological change is that a single hydrological output may affect a series of productive or consumptive activities. A study in the Philippines demonstrates that logging of a coastal catchment may lead to an increase in sedimentation of a coral reef downstream (Hodgson and Dixon, 1988). This sedimentation subsequently has negative effects on both coral cover (biomass production) and coral diversity. As coral cover and diversity are assumed implicitly by the authors to enter into an ecotourism production function, the knock-on effect of the change in hydrology is negative. At the same time the loss in coral cover has a negative impact on the biological production function for fish in the area. Fish in turn are a key input in the fishing production function, which is also affected adversely by the logging and subsequent change in catchment hydrology.

This example demonstrates the need to clearly specify the intricate relationships that may exist between the outputs of the hydrological production function and their subsequent impacts on economic production functions. This impact may occur directly, as inputs into economic production functions, or indirectly, as inputs affecting other biophysical production functions that subsequently produce another level of outputs that in turn enter an economic production function (i.e. turbidity impacts on fish that are the object of fisheries production). It is also the case that a single economic production function may be affected in different ways by a number of hydrological outputs that are linked to a given land use change.

In sum, although hydrological function is more often than not an increasing function of land use (interpreting an increase in land use as modification of original vegetation and intensification of land use), there may also be cases where it is a decreasing function of land use. On the other hand, utility (whether affected directly or indirectly) may be either an increasing or decreasing function of hydrological function. Increases in land use that lead to an increase in sedimentation, nutrification and leaching will generally be related negatively to utility. Similarly, increases in peakflows that lead to increased and localised flooding may affect utility negatively. However, increases in land use that lead to increase in downstream annual water or increased dry season baseflow will be related positively to utility. Thus, while in many cases utility will be a decreasing function of land use it will by no means be the rule.

As a single land-use change will cause a series of changes in utility it is clear that the net impact will reflect the trade-off between the different functions and value changes that result. Added to the complexity of understanding the net result is that an individual hydrological output (for example sediment) may affect a number of economic activities (for example recreation and hydropower production) and a given activity (for example hydropower production) may be affected by changes in a number of hydrological outputs (for example sediment, water yield and dry season flow).

Thus, given the nature of hydrological function and the range of economic activities that depend on this function, it will not be possible to generalise regarding the sign of hydrological externalities. A reduction in the intensity of land 'use' (i.e. reforestation of pasture) may lead to a decrease in sedimentation, subsequently improving water storage capacity for hydroelectric production. At the same time, however, the increase in forest cover may also lead to a decrease in water yield thereby decreasing water inflow to the reservoir. Aylward (1998) traces out these linkages in providing a formal model linking land use to hydropower generation for the case of large hydroelectric reservoirs. The model illustrates the effect of a change in land use on discharge from the reservoir, power production and, hence, the marginal opportunity costs of power generation. As both streamflow and sediment yield functions are increasing (i.e. increase with forest conversion), but have opposing effects on discharge, it cannot be assumed that forest conversion will not be unambiguously positive or negative. Summarising the results from Aylward (1998) as cited earlier, Box 7.1 provides an example of how the perception of the direction (and magnitude) of hydrological externalities can vary as additional hydrological components are incorporated into the valuation exercise.

Box 7.1 Direction of hydrological externalities: an example from Arenal, Costa Rica

Aylward (1998) valued the hydrological consequences of land use practices in the Rio Chiquito catchment in Arenal, Costa Rica. Lake Arenal is an enlarged natural lake found at the headwaters of the Arenal river that is of national importance to Costa Rica for the production of electricity. Given that annual inflow is normally less than the live storage capacity, technical production levels are limited by water availability rather than by production capacity (362 MW) and the facility is operated as an interannual storage reservoir to buffer the national grid during dry years.

Located on the Atlantic side of the continental divide in Costa Rica, the Rio Chiquito catchment occupies 8900 ha and makes up approximately one-fourth of the entire drainage system for Lake Arenal. The dominant characteristics of the catchment are steep slopes with abrupt ridges and valleys with 90% of the area on slopes greater than 25%. Elevation ranges from lake level at 545 m up to 1800 m. Four of the Holdridge life zones are present in Rio Chiquito including Wet Premontane Forest, Premontane Rainforest, Lower Montane Rainforest and Wet Lower Montane Forest. The main categories of land cover in the Rio Chiquito catchment are pasture and forest with minimal amounts of agriculture. Between 1960 and 1992, pasture areas more than quadrupled, while the area under primary forest was cut in half, from almost 80% to just under 40%.

Three aspects of the change in hydrological function expected from pasture areas (as opposed to reforestation) were valued explicitly in the analysis, based on the scientific literature and available hydrological studies and data for the site. These included the effect of sediment on the dead and live storage volumes of Lake Arenal and the impact of a change in water yield on water inflows. In addition, sensitivity analysis explored what would be the effect if pasture and livestock contributed to a loss of dry season flow through soil compaction and a reduction in infiltration opportunities.

Table 7.3 presents the value figures for one of the land use units included in the analysis (Wet Premontane Forest) in a sequential format. What the table seeks to demonstrate is that the conclusion as to what is the direction of the externalities – positive or negative – will depend on which effects are valued explicitly. If the dead storage impacts of sedimentation is the only effect valued then the hydrological externalities of pasture would be seen to be positive. If the impacts on live storage are included then the conclusion would be that pasture represents a loss of benefits had under forest. However, once the effects of the gain in water yield on the interannual hydropower facility are included it can be seen that these dominate the costs of sedimentation, once again leading to the conclusion that the hydrological impacts are positive in economic terms. Finally, the sensitivity analysis showed that if a strong loss of dry season flows was associated with pasture, then it is possible (though remote) that the net effect in terms of hydrological externalities could go to zero – i.e. neither positive nor negative. While the hydrological basis for the valuation exercise could be improved, the example demonstrates the potential for obtaining false results if the range of hydrological impacts considered is restricted arbitrarily.

Table 7.3. *Valuation of hydrological externalities provided by pasture (as opposed to reforestation), Arenal, Costa Rica*

	Sedimentation effect		Water effect:	Total hydrological externalities	
Values taken into account	Loss of dead storage	Loss of live storage	change in water inflow	Direction	Magnitude
1. Dead storage impact on sedimentation only	$6 ha^{-1}			Positive	$6 ha^{-1}
2. Both dead and live storage impacts of sedimentation	$6 ha^{-1}	($80 ha^{-1})		Negative	($74 ha^{-1})
3. Sedimentation impacts plus reduction in annual water yield	$6 ha^{-1}	($80 ha^{-1})	$1149 ha^{-1}	Positive	$1075 ha^{-1}
4. Sensitivity analysis of a loss in dry season flow under pasture (i.e. all water yield gained under pasture accrue during wet season and 50% again of the amount of the water yield gain is redistributed from dry to wet season)	$6 ha^{-1}	($80 ha^{-1})	$74	Neutral	$0 ha^{-1}

In reality, then, there will often be a number of hydrological functions (sedimentation, water yield, water regulation, etc.) that need to be considered in determining the net impact (direct or indirect) upon a range of affected economic activities. Thus, the general statement that forest provides soil and water conservation benefits, or catchment protection benefits, is disingenuous in implying uni-directional effects, i.e. benefits only.

CONCLUSIONS

The findings from research into land use–hydrological interactions suggest that the reduction or conversion of natural vegetation accompanying land use change is likely to increase downstream sediment levels and lead to higher nutrient and chemical outflows. The empirical literature on this topic supports the conventional wisdom that the end result will be a decrease in economic welfare due to a myriad of downstream effects on production by enterprises, the household production function and consumption by individuals. Although the general direction of the effect of land use change on water quality can be surmised, there remain legitimate questions as to whether the literature available conveys accurately the magnitude of these damages. In particular, conventional wisdom that such adverse water quality effects must always be of disastrous proportions and merit immediate attention across the board is probably flawed as the economic valuation studies reviewed in this chapter demonstrate that the magnitude of the effects will likely vary according to the economic and biophysical characteristics and conditions of the site.

With regard to the effects of land use change on water quantity variables, the review of the hydrological literature reveals that the conventional wisdom that forests 'conserve' water and act like a 'sponge' persists in the face of a good deal of empirical evidence of cases where this does not apply. The literature on forest hydrology reveals that a reduction in normal vegetation levels will likely increase annual water yield and may either raise or lower dry season baseflow. Intensification of land use that involves substantial soil compaction, will certainly lead to an increase in the flood potential. Where such compaction is small in area (as is commonly the case), however, this effect will be localised and will not extend to the basin scale. Finally, there is evidence emerging that forest cover could have a direct relationship with precipitation at scales greater than 10 km^2 and certainly at the scale of the Amazon (see Pielke *et al.* (1999), as well as Costa; and Bonell, Callaghan and Connor, this volume). Thus the relationship between land use and these hydrological variables is mixed, with some positive and some negative effects and others for which there is no generalisation.

Changes in water quantity will affect a large range of productive and consumptive activities, often affecting the same activities influenced by sedimentation. Interestingly, however, few of the empirical studies of sedimentation have also considered water quantity effects. In forest areas, land use change may lead to major changes in rates of evapotranspiration and so it would appear indispensable to combine both aspects into the analysis of externalities. This concern may be less pressing in temperate grassland areas; however, the study by Kim (1984) suggests that even in a drier grassland environment the choice of land management technique may have a large impact on water yield.

It is also the case that many of the studies appearing in the literature are either extremely simplistic or flawed in their formulation or implementation, limiting the reliability of their results and at worst leading to the confusion of positive and negative externalities. As observed by Aylward (1998) there is also a large methodological gap between the rudimentary valuations provided in the externalities literature (reviewed here) and the complex dynamic optimisation models employed in the design and operation literature (as found for example in *Water Resources Research* and *Journal of Water Resources Planning and Management*). In this regard, it is worth noting that the failure to make a connection between these two larger sets of literature is mutual. The optimisation of reservoir operation is not mentioned in the literature on economic valuation of watershed management. While sedimentation and land use are occasionally mentioned in the operations literature, issues of land use and water quantity are effectively ignored (Howard, 2000).

The range of empirical studies reviewed in this chapter reveals a heavy emphasis on the economic evaluation of sedimentation impacts with only a few studies examining water quantity and water quality (excluding sedimentation) impacts. Given that water quality and water quantity impacts may affect the same consumptive or productive activity, the exclusion of water quantity impacts from consideration implies that much of the literature is incomplete.

Combining the analysis of hydrological effects and economic effects, a discussion of the sign (or direction) of the different impacts confirms that in most cases land use change (away from natural vegetation) will affect economic welfare negatively through its impact on water quality. However, it cannot be argued a priori that all water quantity impacts will have a similarly negative economic outcome. Review of the empirical evidence on sedimentation impacts also suggests that these impacts may often be of limited economic consequence. Meanwhile, empirical studies of water quantity impacts often either misinterpret the direction of hydrological change (based on erroneous conventional wisdom) or rely on questionable hydrological and economic assumptions to demonstrate negative impacts.

Thus, the principal conclusion of the chapter is that both theory and empirical evidence suggest that it would be incorrect to assume that the hydrological externalities resulting from land use change are necessarily negative. As a result it may be time to reconsider the conventional wisdom that land use change away from natural vegetative states must always impair catchment protection values – when these are narrowly defined as hydrological in nature. Clearly, in any comprehensive assessment of land use choices such hydrological externalities would be but one criterion amongst many, contributing to the decision process alongside biodiversity, timber and other values.

Notwithstanding the larger decision framework, this chapter has shown that on theoretical grounds the case can be made that, a priori, the net outcome of the effect of land use on the different hydrological functions is indeterminate. Empirically, the existing literature cannot be taken as evidence that in practice the net effect is typically negative as most studies are either incomplete or unreliable. The small but growing number of studies reviewed here sustain the theoretical conclusion that there will be cases where the net result of the hydrological impacts of forest clearance (away from natural vegetation) may lead to increases in economic welfare or produce only trivial losses in welfare. In such cases the initial production benefits (e.g. timber, livestock, agricultural outputs) of the subsequent land use would have to decline significantly (or be negative), before basin rehabilitation would be warranted.

This analysis argues for more emphasis on ‘catchment management’ as opposed to ‘catchment protection’. In the case where existing old-growth forest is under threat there will be a host of other goods and services that need to be included in any evaluation of land use alternatives. When combined with the prevalence of uncertainty regarding the direction or magnitude of hydrological externalities associated with potential land use change (due to their site-specific biophysical and economic nature) a risk-averse posture is likely to be the prudent approach. Water quality impacts can be expected to be negative and water quantity impacts ambiguous, so it is natural to prefer to realise the other goods and services provided by forests and avoid potentially negative impacts on hydrological function.

Thus, the analysis in this chapter should not be taken as providing support to relaxing the protection and management afforded to forest areas. It does, however, point out that in developing markets for environmental services, and in this case for hydrological or catchment services, critical analysis is needed to avoid blithely assuming that downstream water users should be willing to pay large amounts of money for hydrological ‘services’ provided by intact forest. It would be unfortunate indeed if large transfers of this kind led to little or a perverse impact and it was clear that no effort had been made to even consult the available scientific and economic literature in designing the scheme.

Further, the decision to take a risk-averse approach has implications for the case of reforestation of already converted lands. In this case, the goods and services already being supplied will be the crop or livestock values of the land. The downside risk of reforesting will be the potential for negative impacts on dry season baseflow. Clearly, the extent of land degradation and the downstream effects on water quality are a prime consideration, but the point is that risk aversion works against land use change here as well. This is particularly true where land-use change involves significant socio-economic dislocation or entails large up-front costs in terms of retraining of rural workers and the direct costs of reforestation. In order to authorise large expenditures of this nature, decision-makers should logically place the burden on the proponent to present a clear and well-reasoned assessment of the potential hydrological benefits and risks, as well as the costs of such programmes, the losses in agricultural values and gains in timber and non-timber forest product values.

Of course, where reforestation leads to carbon sequestration these values must also be included and considered. Viewed from a macro-perspective, the potential for reforestation to reduce water availability raises the larger issue of the trade-off between two equally critical sets of ecological services. As carbon is a service of global value and hydrological services of local or national value, the act of balancing these two sets of demands or needs may well evolve into an important policy issue.

Future research priorities for the valuation of hydrological externalities will revolve primarily around efforts to encourage multidisciplinary work. It is likely that effort needs to be devoted not to the development of new methods, per se, but rather that an investment must be made in determining how models and methods applied in each area can be joined into a comprehensive approach to the problem. Given the complexity of the interactions involved, the investigation of hydrological externalities is likely to require participation by experts in land use/productive systems, forest hydrologists, engineers and economists. While economists are conversing increasingly with hydrologists, engineers tend to be left out of the equation and land use aspects are simply assumed.

The literature review suggests that water quantity and water quality impacts are largely under-researched and that there is great scope for expanding our understanding of the relationships between the different variables. Additional case studies and more general theoretical work would greatly assist in the development of a clear set of rules of thumb and shortcuts that could contribute to better project and policy formulation. In this regard, a fundamental question to which hydrologists need to respond is whether, to what extent, and under what conditions, it is possible to develop reliable predictive models for land use and hydrology interactions in the absence of calibrated datasets for catchments. As noted, much of the policy-oriented studies are short-term when compared to long- or medium-term hydrological data collection and research. Nor is it possible to guarantee that catchments that are to be the subject of policy or project interventions are those that have historically been metered.

There are many reasons, some more or less obvious, for advocating increased stakeholder participation in research programmes – whether academic or applied (Deutsch *et al.*, this volume). Two central objectives of stakeholder involvement are to ensure that the research responds to local conditions and concerns and to increase the likelihood of the practical uptake of research results in actual practice. Stakeholders will include both those who live and work in the catchments as well as those who benefit at a distance from the services provided by water resources. Policy-makers and technical staff of relevant agencies and utilities are also an important set of stakeholders.

The degree of involvement of stakeholders will vary with the objectives and content of the research. For applied work that is aimed at policy or project development, stakeholders should be consulted and involved in the project on a continual basis, from assisting in the identification and prioritisation of research topics and sub-themes through to the dissemination, outreach and policy/project formulation phases. For basic research, stakeholder consultation will likely be more punctual, but nevertheless should be used to ensure that the research design addresses local concerns and issues where feasible. The time and money costs of participation will vary, but it is important that they be provided for explicitly in project budgets and time schedules. In particular, it is important to avoid under-budgeting resources for outreach and communication of research results.

From an economic standpoint the overriding concern has to do with the economic contribution that such research can make to local and national development goals. If it is true that such research can greatly improve the productivity and efficiency with which catchment resources are managed, then there is no better way to ensure that such research is funded than by providing assistance to the actors that will actually reap these benefits (or avoid the costs of poor management/investments). While water is a public good, the distribution of the benefits of improved catchment management may often be localised in a particular region. Achieving stakeholder buy-in to a research programme will thus not only increase the likelihood that the research will lead ultimately to welfare improvement but may open up new partnerships and funding avenues for researchers.

References

Alvarez, M., Aylward, B., and Echeverría, J. (1996). *Management of Natural Forest for Off-Site Hydrological Benefits in the Magallenes National Reserve.* Santiago, Chile: Corporación Nacional Forestal, Environmental Resources Management, Ltd., British Council.

Aylward, B. (1998). *Economic Valuation of the Downstream Hydrological Effects of Land Use Change: Large Hydroelectric Reservoirs.* Unpublished Ph.D. diss., Fletcher School of Law and Diplomacy, Medford, MA.

Aylward, B., and Echeverria, J. (2001). Synergies between Livestock Production and Hydrological Function in Arenal, Costa Rica. *Environment and Development Economics*, *6*(3), 359–82.

Aylward, B., and Fernández González, A. (1998). Incentives and Institutional Arrangements for Watershed Management: A Case Study of Arenal, Costa Rica, *CREED Working Paper Series*. London: International Institute for Environment and Development and Institute for Environmental Studies, Free University, Amsterdam.

Barbier, E. B. (1998). Natural Capital and the Economics of Environment and Development. In E. B. Barbier (Ed.), *The Economics of Environment and Development*. Cheltenham, UK: Edward Elgar.

Barbier, E. B., and Bishop, J. (1995). Economic Values and Incentives Affecting Soil and Water Conservation in Developing Countries. *Journal of Soil and Water Conservation*, *50*(2), 133–37.

Barbier, E. B., Burgess, J. C., and Folke, C. (1994). *Paradise Lost? The Ecological Economics of Biodiversity*. London: Earthscan.

Barrow, P., Hinsley, A., and Price, C. (1986). The Effect of Afforestation on Hydroelectricity Generation. A Quantitative Assessment. *Land Use Policy*(4), 141–51.

Bishop, J. (1992). Economic Analysis of Soil Degradation, *LEEC Gatekeeper Series* (Vol. 92–01, pp. 13). London: International Institute for Environment and Development.

(1995). The Economics of Soil Degradation: An Illustration of the Change in Productivity Approach to Valuation in Mali and Malawi, *LEEC Discussion Paper* (Vol. 95–02, pp. 79). London: International Institute for Environment and Development.

Bosch, J. M., and Hewlett, J. D. (1982). A Review of Catchment Experiments to Determine the Effect of Vegetation Changes on Water Yield and Evapotranspiration. *Journal of Hydrology*, *55*, 3–23.

Bouwes, N. W. (1979). Procedures in Estimating Benefits of Water Quality Change. *American Journal of Agricultural Economics*, *61*(3), 535–39.

Briones, N. D. (1986). Estimating Erosion Costs: A Philippine Case Study in the Lower Agno River Watershed. In K. W. Easter and J. A. Dixon and M. M. Hufschmidt (Eds.), *Watershed Resources Management: An Integrated Framework with Studies from Asia and the Pacific* (pp. 205–18). Boulder: Westview Press.

Brooks, K. N., Gregersen, H. M., Berglund, E. R., and Tayaa, M. (1982). Economic Evaluation of Watershed Projects: An Overview Methodology and Application. *Water Resources Bulletin*, *18*(2), 245–50.

Brown, M., de la Roca, I., Vallejo, A., Ford, G., Casey, J., Aguilar, B., and Haacker, R. (1996). *A Valuation Analysis of the Role of Cloud Forests in Watershed Protection: Sierra de las Minas Biosphere Reserve, Guatemala and Cusuco National Park, Honduras*. Guatemala: RARE, Defensores de la Naturaleza and Fundación Ecologista Hector Rodrigo Pastor Fasquelle.

Brown, T. C., Taylor, J. G., and Shelby, B. (1992). Assessing the Direct Effects of Streamflow on Recreation: A Literature Review. *Water Resources Bulletin*, *27*(6), 979–89.

Bruijnzeel, L. A. (1990). *Hydrology of Moist Tropical Forests and Effects of Conversion: A State of Knowledge Review*. Paris: International Hydrological Programme of the United Nations Educational, Scientific and Cultural Organization.

Bruijnzeel, L. A. (1997). Hydrology of Forest Plantations in the Tropics. In E. K. S. Nambiar and A. G. Brown (Eds.), *Management of Soil, Nutrients and Water in Tropical Plantation Forests* (pp. 125–67). Canberra/Bogor: ACIAR/CSIRO/CIFOR.

(1998). Soil Chemical Changes after Tropical Forest Disturbance and Conversion: The Hydrological Perspective. In A. Schulte and D. Ruhyat (Eds.), *Soils of Tropical Forest Ecosystems. Characteristics, Ecology and Management* (pp. 45–61). Berlin: Springer.

(2004). Tropical forests and environmental services: not seeing the soil for the trees? *Agriculture, Ecology and Environment*. doi: 10.1016/j.agee.2004.01.015.

Bruijnzeel, L. A., and Proctor, J. (1995). Hydrology and Biogeochemistry of Tropical Montane Cloud Forest: What Do We Really Know? In L. S. Hamilton and J. O. Juvik and F. N. Scatena (Eds.), *Tropical Montane Cloud Forests* (Vol. Springer Ecol. Studies, pp. 38–78). New York: Springer-Verlag.

Calder, I. R. (1992). The Hydrological Impact of Land-Use Change (with Special Reference to Afforestation and Deforestation), *Proceedings of the Natural Resources and Engineering Advisers Conference on Priorities for Water Resources Allocation and Management, Southampton, July 1992* (pp. 91–101). London: United Kingdom Overseas Development Administration.

(1999). *The Blue Revolution: Land Use and Integrated Water Resources Management*. London: Earthscan.

CCT, and CINPE. (1995). *Valoración Económico Ecológica del Agua*. San José: Centro Científico Tropical and Centro Internacional en Política Económica para el Desarrollo Sostenible.

Cheng, G. W. (1999). Forest Change: Hydrological Effects in the Upper Yangtze River Valley. *Ambio*, *28*, 457–59.

Clark, E. H. (1985a). *Eroding Soils: The Off-Farm Impacts*. Washington, DC: The Conservation Foundation.

(1985b). The Off-Site Costs of Soil Erosion. *Journal of Soil and Water Conservation*, *40*(1), 19–22.

Clark, R. (1996). Methodologies for the Economic Analysis of Soil Erosion and Conservation, *CSERGE Working Paper* (Vol. GEC 96-13). Norwich: University of East Anglia.

Creedy, J., and Wurzbacher, A. D. (2001). The Economic Value of a Forested Catchment with Timber, Water and Carbon Sequestration Benefits. *Ecological Economics*, *38*(1), 71–83.

Crosson, P. (1984). New Perspectives on Soil Conservation Policy. *Journal of Soil and Water Conservation*, *39*(4), 222–25.

(1995). Soil Erosion and its On-Farm Productivity Consequences: What Do We Know?, *Discussion Paper* (Vol. 95-29, pp. 17). Washington: Resources for the Future.

Crosson, P., and Miranowski, J. (1982). Soil Protection: Why, by Whom, and for Whom? *Journal of Soil and Water Conservation*, *37*(1), 27–28.

Crowder, B. M. (1987). Economic Costs of Reservoir Sedimentation: A Regional Approach to Estimating Cropland Erosion Damages. *Journal of Soil and Water Conservation*, *42*(3), 194–97.

Cruz, W. D., Francisco, H., and Conway, Z. (1988). The On-Site and Downstream Costs of Soil Erosion in the Magat and Pantabangan Watersheds. *Journal of Philippine Development*, *15*(26), 85–111.

Daubert, J., and Young, R. (1981). Recreational Demands for Maintaining Instream Flows: A Contingent Valuation Approach. *American Journal of Agricultural Economics*, *63*(4), 666–75.

De Graaff, J. (1996). *The Price of Soil Erosion: An Economic Evaluation of Soil Conservation and Watershed Development* (Vol. 14). Wageningen: Wageningen Agricultural University.

de Wit, M., Crookes, D. J., and van Wilgen, B. W. (forthcoming). Conflicts of Interest in Environmental Management: Estimating the Costs and Benefits of Black Wattle (Acacia mearnsii) in South Africa. *Biological Invasions*, 32(2), 167–178.

de Wit, M., Crookies, D., Scott, D., Visser, W., and Netshiluvhi, T. (2000). *Environmental Impacts of the Forestry Sector in South Africa with Specific Reference to Water Resources* (ENV-P-C 99016, Project: JN716). Pretoria: Division of Water, Environment and Forestry Technology, CSIR; PO Box 395, Pretoria 0001.

Duda, A. M. (1985). Environmental and Economic Damage Caused by Sediment from Agricultural Nonpoint Sources. *Water Resource Bulletin*, *21*(2), 225–34.

Duffield, J. W., Neher, C. J., and Brown, T. C. (1992). Recreation Benefits of Instream Flow: Application to Montana's Big Hole and Bitterroot Rivers. *Water Resources Research*, *28*(9), 2169–81.

Duisberg, P. (1980). Erosión y Conservación de Suelos. In CCT (Ed.), *Estudio Ecológico Integral de las Zonas de Afectación del Proyecto Arenal* (pp. Anexo 2:1-65). San José: Centro Científico Tropical.

Enters, T. (1995). The Economics of Land Degradation and Resources Conservation in Northern Thailand. In J. Rigg (Ed.), *Counting the Costs: Economic Growth and Environmental Change in Thailand* (pp. 90–110). Singapore: Institute of South East Asian Studies.

(1998). Methods for the Economic Assessment of the On- and Off-site Impacts of Soil Erosion, *Issues in Sustainable Land Management* (Vol. 2, pp. 60). Bangkok: International Board for Soil Research and Management.

Epp, D. J., and Al-Ani, K. S. (1979). The Effect of Water Quality on Rural Non-Farm Residential Property Values. *American Journal of Agricultural Economics*, *61*, 529–34.

Forster, D. L., and Abrahim, G. (1985). Sediment Deposits in Drainage Ditches: A Cropland Externality. *Journal of Soil and Water Conservation*, *40*(1), 141–43.

Forster, D. L., Bardos, C. P., and Southgate, D. (1987). Soil Erosion and Water Treatment Costs. *Journal of Soil and Water Conservation*, *42*(1), 349–52.

Fox, G., and Dickson, E. (1990). The Economics of Erosion and Sediment Control in Southwestern Ontario. *Canadian Journal of Agricultural Economics*, *38*, 23–44.

Freeman, A. M. (1993). *The Measurement of Environmental and Resource Values*. Washington DC: Resources for the Future.

Gregersen, H. M., Brooks, K. N., Dixon, J. A., and Hamilton, L. S. (1987). Guidelines for Economic Appraisal of Watershed Management Projects, *FAO Conservation Guide* (Vol. 16, pp. 144). Rome: United Nations Food and Agriculture Organization.

Gunatilake, H. M., and Gopalakrishnan, C. (1999). The Economics of Reservoir Sedimentation: A Case Study of Mahaweli Reserviors in Sri Lanka. *Water Resources Development*, *15*(4), 511–26.

Guntermann, K. L., Lee, M. T., and Swanson, E. R. (1975). The Off-site Sediment Damage Function in Selected Illinois Watersheds. *Journal of Soil and Water Conservation*, *30*, 219–25.

Guo, Z., Xiao, X., Gan, Y., and Zheng, Y. (2001). Ecosystem Functions, Services and their Values – a Case Study in Xingshan County of China. *Ecological Economics*, *38*, 141–54.

Hamilton, L. S., with King, P. N. (1983). *Tropical Forested Watersheds: Hydrologic and Soils Response to Major Uses or Conversions*. Boulder: Westview Press.

Hamilton, L. S., and Pearce, A. (1986). Biophysical Aspects in Watershed Management. In C. W. Howe (Ed.), *Watershed Resources Management: An Integrated Framework with Studies from Asia and the Pacific* (1st ed., Vol. 10, pp. 33–49). Boulder: Westview Press.

Hearne, R. R. (1996). A Review of Economic Appraisal of Environmental Goods and Services: With a Focus on Developing Countries, *Environmental Economics Programme Discussion Paper* (Vol. 96-03, pp. 13). London: International Institute for Environment and Development.

Hitzhusen, F. J. (1992). The Economics of Sustainable Agriculture: Adding a Downstream Perspective. *Journal of Sustainable Agriculture*, *2*(2), 75–89.

Hodgson, G., and Dixon, J. A. (1988). Logging Versus Fisheries and Tourism in Palawan: An Environmental and Economic Analysis, *Occasional Paper* (Vol. 7, pp. 95). Honolulu: East-West Center.

Holmes, T. P. (1988). The Offsite Impact of Soil Erosion on the Water Treatment Industry. *Land Economics*, *64*(4), 356–66.

Howard, C. D. (2000). Operations, Monitoring and Decomissioning of Dams, *WCD Thematic Review* (Vol. IV.5). Cape Town: World Commission on Dams.

Hubbard, R. K., and Lowrance, R. (1996). Solute Transport and Filtering Through a Riparian Forest. *Transactions of the ASAE*, *39*(2), 477–88.

Intercarib S. A, and Nathan Associates. (1996). *Manejo Ambiental, Aspectos Institucionales, Económicos y Financieros: Volumen 2 de 2, Estudios Sectoriales*. Panamá: Intercarib S.A. y Nathan Associates, Inc.

Johnson, N., White, A., and Perrot-Maitre, D. (2001). *Developing Markets for Water Services from Forests: Issues and Lessons for Innovators*. Washington, DC: Forest Trends, World Resources Institute, The Katoomba Group.

Johnson, N. S., and Adams, R. M. (1988). Benefits of Increased Streamflow: The Case of the John Day River Steelhead Fishery. *Water Resources Research*, *24*(11), 1839–46.

Johnson, S. H., and Kolavalli, S. (1984). Physical and Economic Impacts of Sedimentation on Fishing Activities: Nam Pong Basin, Northeast Thailand. *Water International*, *9*(4), 185–88.

Kim, I.-H. (1984). *An Economic Analysis of Watershed Practices: Impact of Grazing on Watershed (Erosion, Sedimentation, Valuation)*. Unpublished Ph.D. diss., Utah State University.

Kumari, K. (1995). An Environmental and Economic Assesment of Forest Management Options: A Case Study in Malaysia, *Environmental Economics Series, Environment Department Papers* (Vol. 26, pp. 46). Washington, DC: World Bank.

Lant, C. L., and Mullens, J. B. (1991). Lake and River Quality for Recreation Management and Contingent Valuation. *Water Resources Bulletin*, *27*(3), 453–60.

Ledesma, H. R. (1996). *Effects of Reservoir Sedimentation and the Economics of Watershed Management: Case Study of Aguacate Dam, Dominican Republic*. Unpublished Ph.D. diss., University of Florida.

Lutz, E., Pagiola, S., and Reiche, C. (1994). The Costs and Benefits of Soil Conservation: The Farmer's Viewpoint. *The World Bank Research Observer*, *9*(2), 273–95.

Magrath, W., and Arens, P. (1989). The Costs of Soil Erosion on Java: A Natural Resource Accounting Approach, *Environment Department Working Paper* (Vol. 18, pp. 60). Washington DC: World Bank.

McConnell, K. (1983). An Economic Model of Soil Conservation. *American Journal of Agricultural Economics*, *65*(1), 83–89.

Mohd Shahwahid, H. O., Awang Noor, A. G., Abdul Rahim, N., Zulkifli, Y., and Ragame, U. (1997). Economic Benefits of Watershed Protection and Trade-off with Timber Production: A Case Study in Malaysia, *EEPSEA Research Report Series*. Singapore: Economy and Environment Program for South East Asia.

Narayanan, R. (1986). Evaluation of Recreational Benefits of Instream Flows. *Journal of Leisure Research*, *18*(2), 116–28.

Olson, K. R., Lal, R., and Norton, L. D. (1994). Evaluation of Methods to Study Soil Erosion-Productivity Relationships. *Journal of Soil and Water Conservation*, *49*(6), 586–90.

Pattanayak, S. K., and Kramer, R. A. (2001a). Pricing Ecological Services: Willingness to Pay for Drought Mitigation from Watershed Protection in Eastern Indonesia. *Water Resources Research*, *37*(3), 771–78.

(2001b). Worth of Watersheds: A Producer Surplus Approach for Valuing Drought Mitigation in Eastern Indonesia. *Environment and Development Economics*, *6*, 123–46.

Perrings, C., Folke, C., and Maler, K.-G. (1992). The Ecology and Economics of Biodiversity Loss: The Research Agenda. *Ambio*, *21*(3), 201–11.

Pielke, R. A., Walko, R. L., Steyaert, L. T., Vidale, P. L., Liston, G. E., Lyons, W. A., and Chase, T. N. (1999). The Influence of Anthropogenic Landscape Changes on Weather in South Florida. *Am. Meteorol. Soc.*, (127), 1663–72.

Pierce, F. J., Larson, W. E., Dowdy, R. H., and Graham, W. P. (1983). Productivity of Soils: Assessing Long-Term Changes Due to Erosion. *Journal of Soil and Water Conservation*, *38*(1), 39–44.

Pimentel, D., Harvey, C., Resosudarmo, P., Sinclair, K., Kurz, D., McNair, M., Crist, S., Shpritz, L., Fitton, L., Saffouri, R., and Blair, R. (1995). Environmental and Economic Costs of Soil Erosion and Conservation Benefits. *Science*, *267*(24 February), 1117–23.

Quesada-Mateo, C. A. (1979). *Effect of Reservoir Sedimentation and Stream Flow Modification on Firm Power Generation*. Unpublished Ph.D. diss., Colorado State University, Fort Collins, CO.

Ralston, S. N., and Park, W. M. (1989). Estimation of Potential Reductions in Recreational Benefits due to Sedimentation. *Water Resources Bulletin*, *25*(6), 1259–65.

Ribaudo, M. O., Young, C. E., and Shortle, J. S. (1986). Impacts of Water Quality Improvement on Site Visitation: A Probabilistic Modeling Approach. *Water Resources Bulletin*, *22*(4), 559–63.

Richards, M. (1997). The Potential for Economic Valuation of Watershed Protection in Mountainous Areas: A Case Study from Bolivia. *Mountain Research and Development*, *17*(1), 19–30.

Rodríguez, R. (1989). *Impactos del Uso de la Tierra en la Alteración del Régimen de Caudales, la Erosión y Sedimentación de la Cuenca Superior del río Reventazón y los Efectos Económicos en el Proyecto Hidroeléctrico de Cachí, Costa Rica*. Unpublished MSc thesis, Centro Agronómico Tropical de Investigación y Eneñanza, Turrialba.

Rojas, M., and Aylward, B. (Forthcoming). Mechanisms for Upstream-Downstream Cooperation in Rural Watersheds: The Case of a La Esperanza: A Small, Private, Hydropower Producer in Costa Rica. In B. Kiersch (Ed.), *Watershed-Based Integrated Land and Water Management: Valuation of Land Use Impacts on Water Resources and Mechanisms for Upstream-Downstream Cooperation in Rural Watersheds*. Rome: FAO.

Ruitenbeek, H. J. (1990). *Economic Analysis of Tropical Forest Conservation Initiatives: Examples from West Africa*. Surrey, UK: World Wide Fund for Nature.

Santos, B. A. (1992). *Cost Benefit Analysis of Soil Erosion Control: The Case of Plan Sierra (Dominican Republic)*. Unpublished Ph.D. diss., University of California, Berkeley.

Sheridan, J. M., Lowrance, R., and Bosch, D. D. (1999). Management Effects on Runoff and Sediment Transport in Riparian Forest Buffers. *Transactions of the ASAE*, *42*(1), 55–64.

Snyder, N. J., Mostaghimi, D. F., Berry, D. F., Reneau, R. B., Hong, S., McClellan, P. W., and E. P., S. (1998). Impact of Riparian Forest Buffers on Agricultural Nonpoint Source Pollution. *Journal of the American Water Resources Association*, *34*(2), 385–95.

Southgate, D. (1992). The Rationality of Land Degradation in Latin America: Some Lessons from the Ecuadorian Andes, *LEEC Gatekeeper Series* (Vol. 92-04, pp. 9). London: International Institute for Environment and Development.

Southgate, D., and Macke, R. (1989). The Downstream Benefits of Soil Conservation in Third World Hydroelectric Watersheds. *Land Economics*, *65*(1), 38–48.

Stallard, R. F. (1997). Detailed Report of the Soil and Hydrology Component, *Informe Anual del Program de Monitoreo del Canal*. Panamá: Programa del Monitoreo del Canal.

Stocking, M. A. (1984). *Erosion and Soil Productivity: A Review*. Norwich: Overseas Development Group, University of East Anglia.

van Noordwijk, M., and al, e. (1998). Erosion and sedimentation as multiscale, fractal processes: implication for models, experiments and the real world. In F. W. T. Penning de Vries (Ed.), *Soil Erosion at Multiple Scales*. Wallingford: CABI Publishing.

Veloz, A., Southgate, D., Hitzhusen, F., and Macgregor, R. (1985). The Economics of Erosion Control in a Subtropical Watershed: A Dominican Case. *Land Economics*, *61*(2), 145–55.

Vertessy, R. A., Watson, F., O'Sullivan, S., Davis, S., Campbell, R., Benyon, R., and Haydon, S. (1998). *Predicting Water Yield from Mountain Ash Forest Catchments*. Canberra: CRCCH.

Vincent, J. R., and Kaosa-ard, M. (1995). Policy Brief: The Economics of Watershed Protection: A Case Study of the Mae Teng River. Cambridge: Harvard Institute for International Development.

Walker, D. J. (1982). A Damage Function to Evaluate Erosion Control Economics. *American Journal of Agricultural Economics*, *64*(11), 690–98.

Ward, F. A. (1987). Economics of Water Allocation to Instream Uses in a Fully Appropriated River Basin: Evidence from a New Mexico Wild River. *Water Resources Research*, *23*(3), 381–92.

White, W. C. (1994). *The Study of Sustainable Hydrological Development in Lao PDR*. Unpublished Ph.D. diss., Fletcher School of Law and Diplomacy, Tufts University, Medford, MA.

Young, C. E. (1984). Perceived Water Quality and the Value of Seasonal Homes. *Water Resources Bulletin*, *20*(2), 163–66.

Zalewski, M., Thorpe, J. E., and Naiman, R. J. (2001). Fish and Riparian Ecotones: A Hypothesis. *Ecohydrology and Hydrobiology*, *1*(1–2), 11–24.

8 Water resources management policy responses to land cover change in South East Asian river basins

D. Murdiyarso[1]
Bogor Agricultural University, Bogor, Indonesia

INTRODUCTION

Water supply is usually taken for granted in the humid tropics and perceived as stable, i.e. it is assumed that water availability will not change over a relatively long period. It is also often believed that water is an unlimited resource, hence water-related policies (if any) are seldom reviewed and adjusted. In reality water resources in the humid tropics are getting more and more depleted in terms of both quantity and quality (Bruijnzeel, 1990, 1998). The economic losses due to water shortage or excess result mainly from a lack of adjustment and responses in the public policy-making process regarding the use of resources. Water scarcity in particular and, hence, insecurity is a growing issue in areas where population pressure and rates of environmental degradation are high. Moreover, water scarcity is increasingly causing political and social tensions between upland and lowland communities and between neighbouring districts or countries or other political and administrative boundaries. Conflicts between the riparian states in the Lower Mekong Basin and within the Greater Mekong Subregion are classic examples. Controversy over the impacts of navigation and hydroelectric power projects is widespread through the region because three-quarters of the population of the Lower Mekong Basin, mainly farmers and fishermen, earn their living by utilising natural resources directly. Therefore an integrated management plan that involves the various stakeholders is necessary to avoid conflicts (Dubash *et al.*, 2001).

One of the most important causes of changes in water supply (streamflow) is land use or land cover change (Bruijnzeel, 1990, 2002). In many parts of the tropics the driving forces of land use/cover change as identified by Turner *et al.* (1995) are occurring over a range of scales and are multi-sectoral. They range in scale from a plot or production unit to a village or a catchment, and even to the regional scale which encompasses large river basins. On the one hand land use change may be driven by land development policies originated by the government, such as large-scale forest conversion for agricultural or plantation development (food crops, oil palm and timber plantations). Other examples include timber market and pricing policies, and land allocation and tenure systems. Moreover, poverty, which is caused by low productivity of the land and lack of access to markets, also drives conversions of forested areas using unsustainable techniques (see Drigo, this volume). These include slash-and-burn clearance, steep-slope cultivation without adequate soil and water conservation, and intensive cropping systems that enhance the depletion of soil fertility. Needless to say, the situation is aggravated by a high rate of population growth.

Unfortunately, global concern about deforestation has seen a binary classification of forested versus non-forested areas without considering sufficiently the land use options in between (Murdiyarso *et al.*, 2002). Appropriate policy responses, therefore, require a better understanding of land use decisions by the stakeholders, ranging from national policies to local collective decisions. The effects of forest conversions on water yield, soil and nutrient losses in the humid tropics have been well documented (Bruijnzeel, 1990, 1998, 2004). Adjusting water resource policy should, therefore, also be closely linked to land use policies and decisions.

This chapter reviews several cases at the river basin scale where biophysical and socio-economic aspects are assessed and public policy-making processes are involved, with special reference to South East Asia. Examples from regional, national and local scales demonstrate the importance of a proper planning phase and at the same time conflict resolution during the implementation phase. It is emphasised that policy responses regarding land use options are needed at all scales and phases of management.

1 Former Deputy Minister of Environment, The Republic of Indonesia.

Forests, Water and People in the Humid Tropics, ed. M. Bonell and L. A. Bruijnzeel. Published by Cambridge University Press.

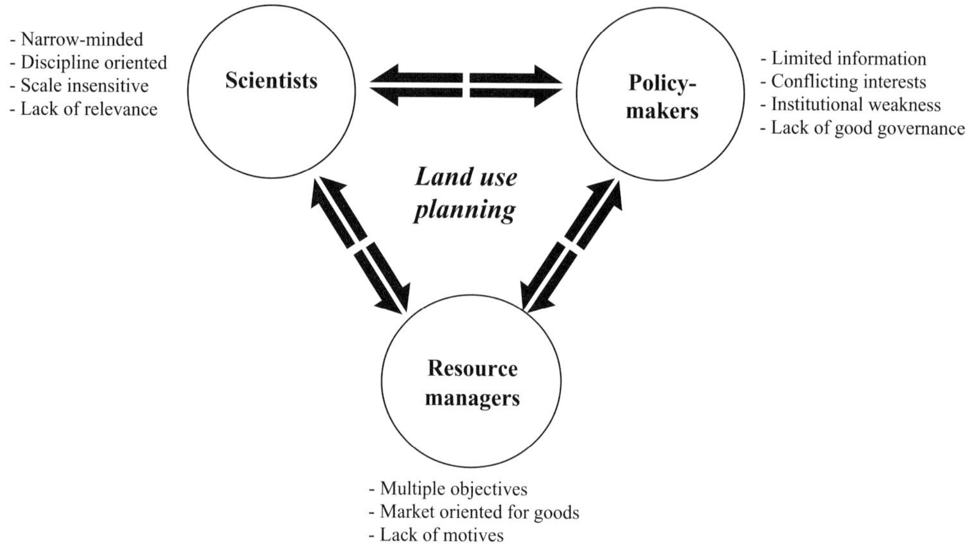

Figure 8.1 Identification of common weaknesses of major stakeholders in land use planning that potentially widen the gaps. The broken-and-split arrows indicate how much needs to be done to narrow the gaps.

LAND USE PLANNING AND WATERSHED MANAGEMENT

Land use planning in the context of river basin management may occur at different scales according to different objectives. However, there are common themes: among others, to address divergent viewpoints, reduce conflicting interests, and improve communication and understanding among stakeholders regarding development outcomes, social justice and environmental integrity.

The three major stakeholder groups identified here are scientists, policy-makers and resource managers. Each of these normally have their own agenda in solving the problems but often the various approaches do not necessarily complement each other. Scientists are usually preoccupied with their disciplinary orientation and are often insensitive to the scale and scope of the problems, so that their efforts often lack relevance to public policy development. Likewise, the policy-makers often lose their credibility since the public policy they generate lacks the scientific background that has to be tapped from the scientific community. In addition, in many parts of the developing world institutional weakness due to lack of capacity is very common and this contributes to the difficulty of implementing, monitoring and evaluating any policies. On the other hand, the resource managers often simply follow their own business-oriented agenda. Their motives are short-term benefits without consideration of the potential of long-term sustainable development objectives.

Clearly, in the initial stage of land use planning it has to be guaranteed that gaps between the three different groups are narrowed so that they can come together with a better understanding of each other's problems and are well prepared to communicate. Figure 8.1 indicates the common weaknesses of each side that may cause the widening of the signalled gap between the three interest groups.

Most scientists are not aware of relevant policy questions. Consequently, their excellent and important research findings are often neglected in the public policy-making process. Policy-relevant research questions need to be communicated and discussed before a research agenda is designed. With such an approach it can be expected that results are communicable to policy-makers and appropriate water resource managers. Efforts have been made to communicate the effects of conversion on the hydrology of moist tropical forest (see e.g. Bruijnzeel, 1990). In this way, policy-makers would have the opportunity to appreciate the importance of science so that they can provide more focused research questions that are relevant to policy formulation and the design of the associated legal instruments. Resource managers can help to define practical questions. They can also offer vital links between science and the community. This new approach of narrowing the gap between these three stakeholder groups is also the philosophy behind the new Hydrology for the Environment, Life and Policy (HELP) programme led by UNESCO (UNESCO, 2001).

The issue of scales should be addressed comprehensively, knowing that biophysical and non-biophysical processes behave differently. The scale issues may have social, economic and legal implications that are relevant to stakeholder groups. At the small scale, where government is largely decentralised, there is a growing involvement by members of the community in assessing the sustainability of natural resources. To a large extent they are the true managers whose participation may be enhanced in the research–development continuum. The potential success of such

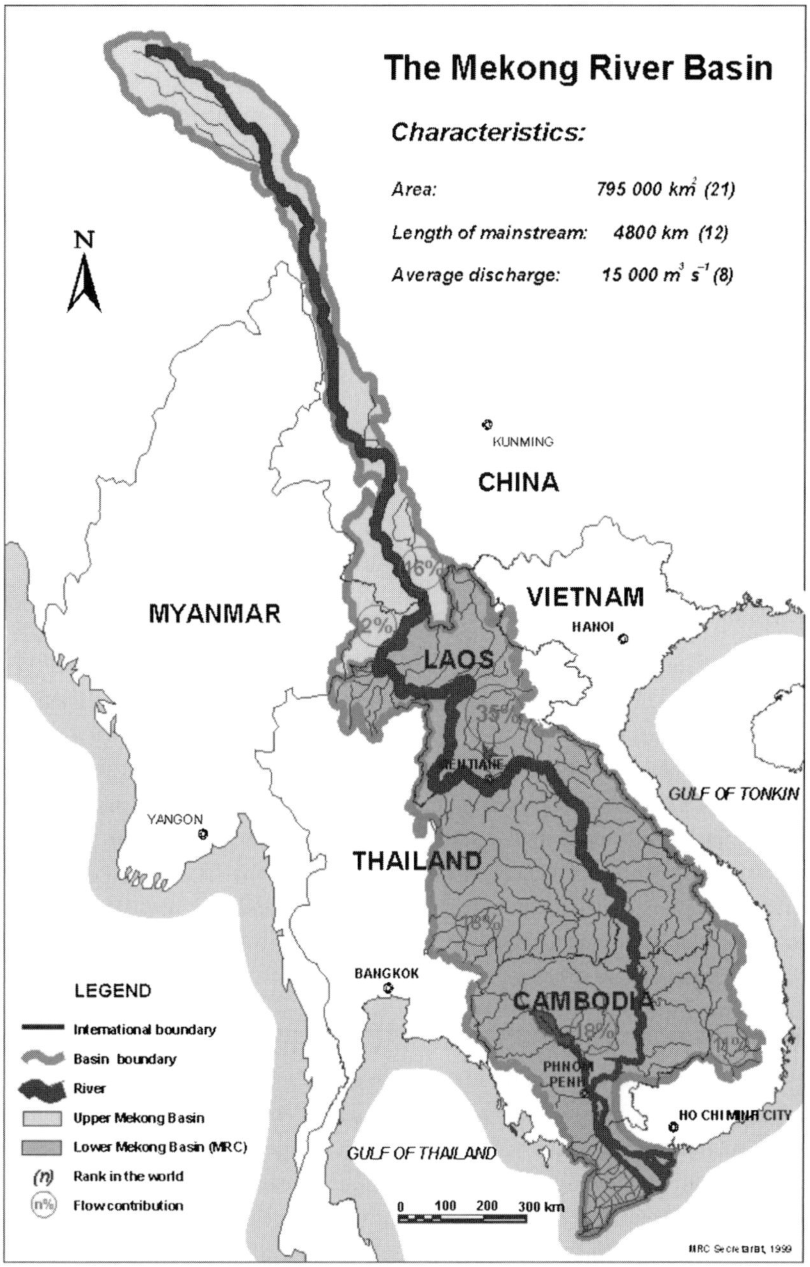

Figure 8.2 The Mekong River Basin. (From Mekong River Commission, 2001.)

an approach has been demonstrated by Deutsch *et al.* (this volume) in community-based water quality assessments in the Philippines. The strength of this approach is that the findings which are credible and scientifically sound may be implemented directly without going through complicated bureaucratic channels. While the start-up of the Philippine collaborative project was relatively slow compared with a more traditional research approach (that does not involve the local community), initial results nevertheless indicate that projects planned with the involvement of communities have a higher chance of project sustainability than those carried out by scientific communities or government officials or communities in isolation.

The following section describes the planning and management processes at regional, national and local scales for South East Asia.

Regional scale

The Mekong River drains an area of 795 000 km^2 (Figure 8.2), of which 606 000 km^2 is in the Lower Mekong Basin (LMB),

Table 8.1. *Approximate catchment area and streamflow distribution of the Mekong River in the riparian countries*

Country	Catchment area (km^2 (%))	Average discharge ($m^3\ s^{-1}$ (%))
China	165 000	2 410
	(21)	(16)
Myanmar	24 000	300
	(3)	(2)
Laos	202 000	5 270
	(25)	(35)
Thailand	184 000	2 560
	(23)	(18)
Cambodia	155 000	2 860
	(20)	(18)
Vietnam	65 000	1 660
	(8)	(11)
Total	795 000	15 060

Source: Saysanovongphet (1996).

comprising almost the whole territories of Laos and Cambodia, one-third of Thailand and two-fifths of Vietnam (Table 8.1). The LMB entered into an international agreement in 1995 and formed the Mekong River Commission (MRC). One of the purposes of the Agreement is to promote the sustainable utilisation, management and conservation of water and related resources for the well-being of all riparian states (Phanrajsavong, 1996). This may be taken as an example of basin management at the regional scale where national policy-making is highly political since it has to be integrated with trans-boundary concerns. Land use planning is one of the critical issues in the upland parts of the basin as this will have effects on the development opportunities and conservation efforts downstream, particularly in the domains of hydroelectric power, navigation, irrigation and flood control and freshwater fishery.

Although the two countries in the upper reaches of the Mekong River, Myanmar and China, are not members of the MRC, they are involved in the programme called the Greater Mekong Subregion (GMS) which was developed in 1992 and funded by the Asian Development Bank. In April 2000, China (Yunnan Province), Myanmar, Thailand and Laos signed an agreement on commercial navigation in the Lancang–Mekong River. Based on this, dredging, destruction of rapids and islets, as well as construction of (large) dams and ports were made possible. Environmental activists are very concerned about any impacts that might be caused by the completion of two Chinese dams in the Upper Mekong and the six more that are to follow. However, the MRC cannot take effective measures against China since the country is not a member of the commission.

In Laos, land use change is usually associated with shifting cultivation, considered as the major cause of deforestation, soil erosion and declining agricultural productivity. To stabilise shifting cultivation by the year 2000, the government introduced a land allocation programme. If the programme was strictly implemented each family would receive 5 ha of land (4 ha for annual crops and 1 ha for trees and other perennial crops) which is less than half of the current shifting cultivation area per family (Fisher, 1996). One of the principles of the programme is that the fallow period should be less than three years. This means that the majority of the proposed land package would enhance soil erosion and water scarcity as the cropping cycle becomes more intensive (see Malmer *et al.*, this volume). Thus forest encroachment may well be stabilised but land degradation and productivity will be worsened. The choice between unsustainable land use by shifting cultivators on the one hand and this rather unrealistic yet intentionally introduced government programme on the other represents a true dilemma. In this case the public policy-making process does not necessarily lead to a solution of the problem.

With such a large volume of streamflow (see Table 8.1) Laos could also generate as much as 13 000 MW of electricity if managed properly, which is more than 70% of the hydroelectric power potential in the entire LMB (Phanrajsavong, 1996). It has also been predicted that by 2020 Laos would still need less than 1000 MW. This would mean that Laos could sell its electricity to neighbouring countries such as Thailand, which has been predicted to consume more than 50 000 MW by 2020 (Phanrajsavong, 1996). The key factor is how much of this hydroelectric power potential would be constrained by enhanced sediment transfer within the rivers from the land allocation programme referred to above.

Another concern for Laos is that it is a land-locked country. Creating new trade routes through river navigation which are capable of transporting large amounts of freight is obviously in its interest. Therefore, the recent commercial navigation agreement with China, Thailand and Myanmar cited above was most welcome. However, it has not been well documented just how much the agreement would cost in terms of the destruction of natural beauties (and therefore loss of revenues from tourism) offered by the waterfall, shoals and rapids in historic spots like the old capital of Laos, Luang Prabang; the changes in water regime due to river works that will affect downstream freshwater fisheries; and so on (Saysanovongphet, 1996).

As with the case for Laos, all Cambodian territories are within the LMB. These include the very fragile wetland ecosystem around the Great Lake (Tonle Sap) where freshwater fisheries are economically important. The productivity and dynamics of this lake ecosystem depends very much on the water level in the Mekong. It has been recognised that the prime limitation to develop the fishery sector is the seasonally uneven distribution of water yield. The flow of the Mekong at Kratie, Cambodia, before entering the delta areas ranges between 1800 $m^3\ s^{-1}$

and 52 000 $m^3 s^{-1}$ (Phanrajsavong, 1996). This large variation is caused mainly by climatic variability linked with monsoons; it is unlikely that at this scale land use change plays much of a role. In the Nam Pong catchment, northeast Thailand, no significant effect on water yield of sustained forest clearance over almost 40 years could be demonstrated (Wilk *et al.*, 2001). Even in the much smaller catchment (100 km^2) of the Cikapundung, West Java, Susetyo and Murdiyarso (1992) demonstrated that the effect of a double-CO_2 climate on water yield was more pronounced than changing the land use from the existing conditions into more conservation-orientated patterns. It was shown that the increased rainfall predicted by a doubling of the CO_2 concentrations would double water yield, while the ideal land use scenario only increased flows by up to 10%.

According to Quang (1996) the area of the Mekong Basin in Vietnam consists of three main regions: the Central Highlands (11 450 km^2), the Dien Bien area (1392 km^2), and the Mekong Delta (39 000 km^2). Each of these has its own potential and constraints in developing water resources. The Central Highlands area still has a relatively good forest cover and the development potential for hydroelectric power is promising. The construction of dams is expected to regulate runoff better since 80% of the streamflow is concentrated in the rainy season. The intermediate Dien Bien area has largely been deforested, causing even more seasonally uneven river discharges that hamper the development of irrigation schemes. The ethnic minority group that dwells in this area does not practise shifting cultivation but subsistence agriculture with relatively low productivity. In contrast, the Mekong Delta is the rice bowl of Vietnam and accounts for most of the national rice production, with an annual yield of around 13 million tonnes. The Delta is also a productive fishing ground with relatively well-maintained mangrove forests (260 000 ha).

In addition to the problems in the Upper Mekong Basin mentioned earlier, a number of conflicts are reported over the use of water resources within the LMB related to community development and environmental issues. These include the construction of the Theun Hinboun dam in Laos, a 240-MW hydroelectric scheme on the Se San River and the Yali Falls dam in Vietnam (Burton, 2000). There was no conflict associated with land use, which indicates that water resource management at this scale is mostly seen from the demand side. Therefore, the challenge is to find ways to manage the development in the different parts of the basin so that the benefits are shared equally and harm to the environment is minimised. To this end the role of science is very strategic when it comes to producing a large-scale baseline dataset. The use of remote sensing and geographic information system (GIS) technologies to establish and provide land use and other biophysical data for river basin models that have predictive capabilities would be beneficial in the assessment of development projects, regardless of political boundaries.

National scale

River basin management in archipelagic countries like Indonesia has its own challenges. In the early 1990s most 'critical' river basins in Indonesia were prioritised in terms of planning and river basin management. The assessment of a basin's status and water resources problems, and the planning of remedial measures such as soil conservation campaigns, were usually centralised at the Ministry of Forestry and/or Public Works in Jakarta. Traditionally, the ratio of maximum and minimum annual streamflow (Q_{max}/Q_{min}) has been used by Indonesian engineers as an indicator of such biophysical parameters related to runoff as infiltration and water retention, and to classify the level of a catchment's priority for rehabilitation (Ministry of Forestry, 1990, 1991, 1992). It may be misleading, however, to use such a single indicator to compare large basins without knowing their geological characteristics, as these determine catchment groundwater reserves or deep leakage (Hardjono, 1980; Bruijnzeel, 1989). At the next level, the regional offices coordinate the work themselves with relevant agencies at the local level to implement the plans. Among others treated this way were river basins such as the Saddang in South Sulawesi, the Solo in Central and East Java, and the Batanghari in the Jambi Province, Sumatera (Ministry of Forestry, 1990, 1991, 1992). Nowadays, in the more decentralised government of Indonesia, the design of a general management plan is translated into a Technical Work Plan, which is meant to provide technical guidance for the District Governments to implement the plan. Figure 8.3 shows the top–down approach that is used to produce work plans for the District Governments. The implementation of the plan has been decentralised and should ideally be carried out using a bottom–up approach by the District Governments.

The maps and manuals produced in the process are normally large to medium in scale (1 : 250 000 to 1 : 50 000) and cover an area of up to 40 000 km^2. The maps show the location and severity of problems, using indices of, *inter alia*, rainfall erosivity, soil erodibility and erosion hazard. Figure 8.4 shows a map of erosion hazard of Batanghari river basin, Sumatera, produced by the central government as a management tool. The district or local government would then use the map to design the Technical Work Plan which is implemented annually. The hazard indices indicate the ratio between potential annual soil loss estimated using the universal soil loss equation (USLE) (Wischmeier and Smith, 1978; see Yu, this volume) and the tolerable soil loss which varies with soil depth (Hammer, 1988). It was realised that the uncertainties associated with both figures remain high. Since there is no practical alternative, this approach will be used and the use of a modelling approach is still a long way off (see Yu, this volume). These are among the limitations and problems faced by local governments trying to obtain plausible results on which to base their campaigns.

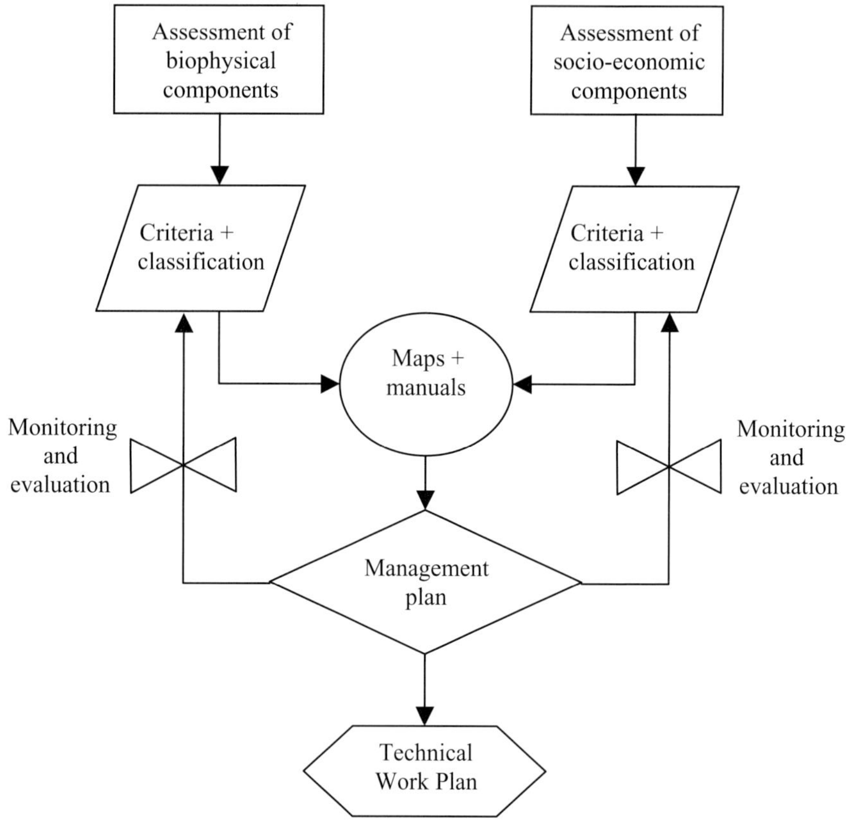

Figure 8.3 Diagrammatic representation of the top–down approach to produce Technical Work Plans for catchment management at the District level in Indonesia. The local government is to implement the plan ideally using a bottom–up approach.

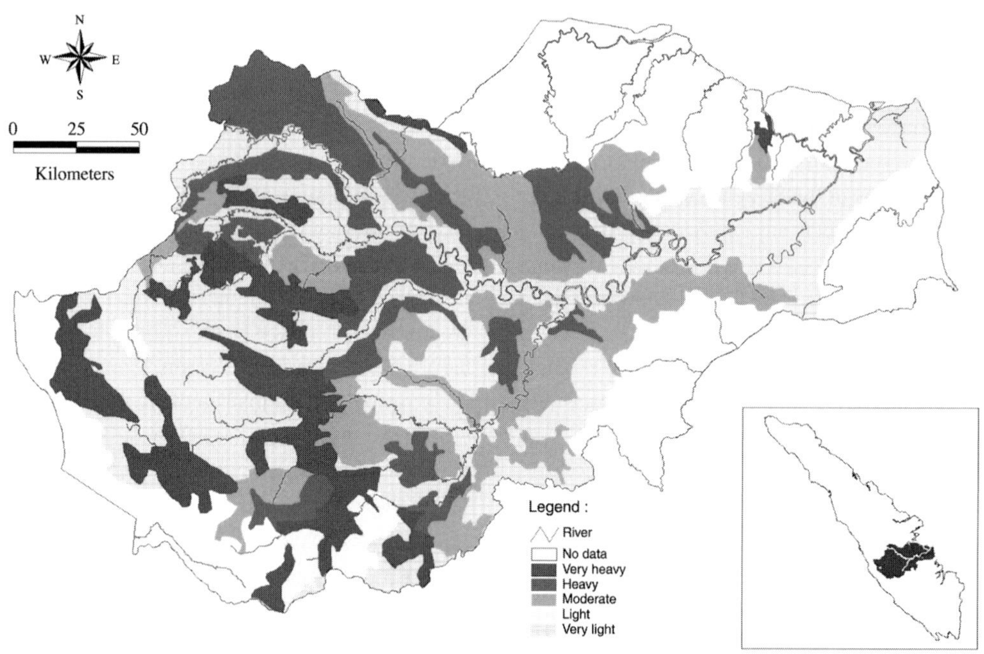

Figure 8.4 Map of erosion hazard of Batanghari river basin, Sumatera, calculated as ratio between potential annual soil loss, based on the universal soil loss equation (Wischmeier and Smith, 1978) and the so-called tolerable soil loss (Hammer, 1988). The map is a typical macro-mangement tool produced by central government before being translated into a Technical Work Plan by district or local government. (Redrawn from Ministry of Forestry (1992)).

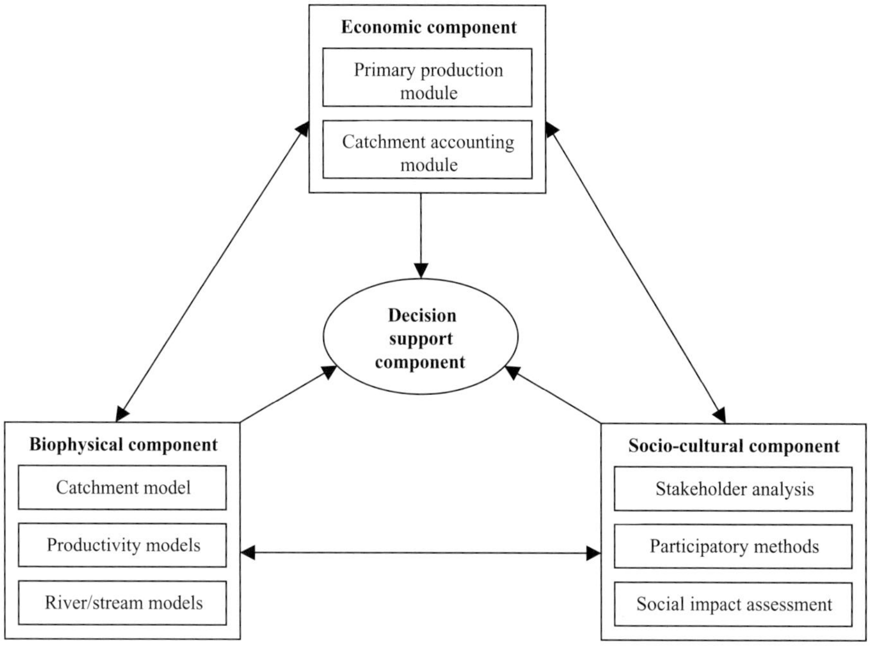

Figure 8.5 Linkages between project components to produce decision support components in the Upper Chao Phraya headwaters, northern Thailand. (From Jakeman *et al.*, 1997.)

Information on socio-economic conditions, such as job opportunities, income and demographic profiles are presented by district in tabular form. The general conservation prescription as described in the manuals may be categorised as the structural or engineering approach (gully plugs, terracing) and the vegetational approach (filter strips, optimal cover) (see Critchley, this volume). These conservation projects are financed by the central government from a limited budget which needs to be approved annually. However, the problems can hardly be tackled at all given the current lack of resources. Ultimately, central government is responsible for monitoring the implementation of the Technical Work Plan and should provide an evaluation for further improvements. Whilst the manuals indicate the actions to be taken and the procedures that need to be followed (Figure 8.3), the availability of technical solutions does not guarantee the success of the projects due to various socio-economic reasons (Purwanto, 1999; Critchley, this volume).

In the Upper Chao Phraya headwaters, northern Thailand, water resource assessment and management have been carried out in an integrated manner with strong scientific support since 1997 (Jakeman *et al.*, 1997). Confidence in the approach increased when modelling tools were employed to allow the prediction of the natural streamflow and the remaining discharge after irrigation diversion (Schreider and Jakeman, this volume). Assessments were made of three main aspects: biophysical, socio-cultural and economic, including their components. One of the main outcomes of the exercise has been a decision support system (DSS) developed by a multidisciplinary team (Figure 8.5). This is one of the rare positive examples in the developing world where the decision-making process is based on scientific findings. The catchment accounting or assessment was carried out at the local level and it was realised that most local governments have the capacity to do this (Jakeman *et al.*, 1997). The team considered that the major constraint in disseminating the outcomes was the need to communicate them to various stakeholder groups, especially the group of land managers represented by the local people or local farmers. In fact these groups are the most important indirect users of the outcomes of the studies (Mackey *et al.*, 1997).

For Indonesian forests, one of the greatest challenges is weak governance during the present time of transition. Good governance should enhance transparency, accountability, partnership and law enforcement. The recent decentralisation efforts in Indonesia have overestimated the capacity of local governments to manage their natural resources sustainably. In this regard local governments even lack the capacity to coordinate with neighbouring districts because they are used to a more centralised system (Jepson *et al.*, 2001).

Local scale

The assessment of water resources problems at the local scale is the most likely to involve diverse stakeholders and actors, including smallholder farmers. In a situation where decision-making processes are decentralised it is high time to support the

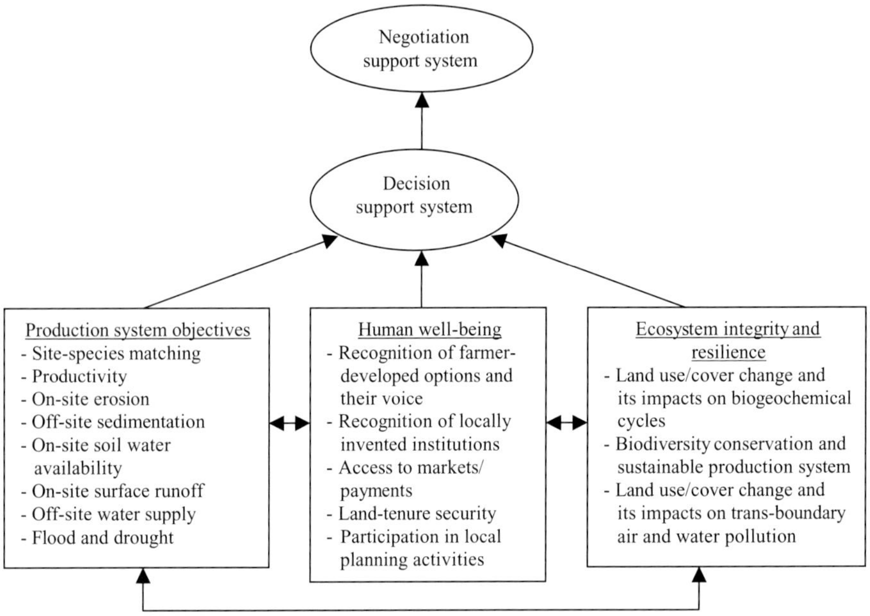

Figure 8.6 Proposed process for the evolution of a locally developed negotiation support system (NSS) from scientifically based decision support system (DSS). (Modified from Van Noordwijk *et al.*, 2001.)

processes at the local scale. In many cases local governments usually do not have the capacity to develop detailed spatially distributed development plans for the sustainable use of natural resources. The institutions and human resources are often weak. In the meantime, the central government usually provides only macro-management plans and indicative work plans. Starting from this point, as shown in Figure 8.6, the bottom–up process that involves a wide range of stakeholders should be engaged to develop a local decision support system (DSS). It is equally important to promote the evolution of a local DSS into a negotiation support system (NSS) as proposed by Van Noordwijk *et al.* (2001) on which a range of performance indicators should be collectively identified by the stakeholders as being relevant in the landscape.

Although smallholder farmers rarely plan water resources as such, one can start facilitating individual decisions and turning them into more collective decisions by screening and justifying them from a scientific point of view. Lessons learnt from community-based water resource assessment and valuation systems developed in the Philippines (see Deutsch *et al.*, this volume) and in Latin America (see Aylward, this volume) may be adopted. This is one way of implementing good environmental governance at the local scale. A DSS should accommodate stakeholders' concerns, including the sustainable production of their farms and their personal well-being as members of society at the local and national scale. The DSS should also be able to summarise quantitatively the detailed components of local concerns (see example in Figure 8.6). It is realised that decision-making at the farm or village level may have some national and trans-boundary or even global implications (Tomich *et al.*, 2000; Murdiyarso *et al.*, 2002).

The DSS/NSS approach will only be applicable if local participation is facilitated and locally developed options are recognised. The roles of outside actors, including scientists and higher-education institutions, to increase the credibility of analysis would be beneficial. Such analyses should be based on high-quality data and information covering both biophysical and non-biophysical components. Information delivery networks should be designed to bridge the gap with the less technologically skilled stakeholders (Mackey *et al.*, 1997).

The major Jakarta flood of early 2002 was an example of a complex chain of causes and effects involving water resources management at the local scale. Subsequent analyses and commentaries blamed local auothorities without credible evidence of the true causes, and therefore no meaningful solution was suggested. Four main causes that may be considered to have affected the event include: extraordinarily high rainfall accompanied by a high tide in the same period, the poor drainage system of the city and spatial mismanagement of increasingly populated uplands. To reduce the devastating effects of similar events in the future, a multi-stakeholder approach should be taken, as demonstrated by Jakeman *et al.* (1997) and Mackey *et al.* (1997).

DRIVING FORCES OF LAND USE/COVER CHANGE

South East Asia has long been the home of tropical forests but when development was accelerated in the 1970s the forests were heavily extracted to earn revenue in the form of foreign currencies. In the 1980s the rates of deforestation ranged between 60 000 ha yr^{-1} (Cambodia) and 600 000 ha yr^{-1} (Indonesia). Twenty years later the Indonesian deforestation rate became uncontrollable, amounting to 1 600 000 ha yr^{-1} (MoFEC, 1997), leaving the area under closed forest cover at 93 million ha compared to 120 million ha in 1980 (UNEP/DEWA, 2001). According to the most recent Forest Resource Assessment in 2001, carried out by the FAO, the annual loss of forest in Asia is currently more than 4 million ha (Drigo, this volume). The main cause of tropical forest depletion is the flawed public policy associated with their utilisation. In Indonesia, the Ministry of Forestry Decree No. 682/Kpts/Um/8/1981 legitimised forest conversions in the early 1980s at over 30 million ha. Data on actual rates of deforestation are usually available only a few years later. With the current uncertainties in forest governance, the deforestation rate in Indonesia within the next five years may well reach 3 million ha yr^{-1}. The growing demands for wood and timber have largely initiated deforestation in Indonesia, both for export and local forest industries, including pulp and paper. This will usually be followed by land conversions for pulp and oil palm plantations. Industrial demand for wood in Indonesia is around 80 million m^3 yr^{-1} while the officially reported harvest is only 21.4 million m^3 yr^{-1} (World Bank, 2000). This large shortfall has resulted in widespread over-cutting and illegal logging. Of the 120 million ha officially designated forest land in Indonesia, almost half (55 million ha) is considered to be production forest. However, in practice, the area available for production is much smaller. There are only a few lowland forest patches left in Sumatera and these will probably disappear in the very near future. Five years ago the actual area of production forest was 10 million ha, while the area allocated for conversion was 2 million ha yr^{-1}. The demand for oil palm plantation has exceeded the allocated area, causing a deficit of 8 million ha which will potentially lead to encroachment into production forest. Even protected wetlands and peatland forests are seriously threatened (see Hooijer and Hamilton, both this volume). In Kalimantan, the 14 million ha of conversion forest allocated in 1981 has completely gone (Manurung, 2000). In addition, demands for new lands to support settlements and agricultural expansion have also contributed to a great deal of forest loss in the country (Kartodihardjo and Supriono, 2000).

The loss of tropical forests leads to a decline of job opportunities and sources of income. At the same time environmental functions may become severely disturbed. These are amongst the local and national concerns that merit further quantification. Meanwhile there are also global concerns that are affected by forest loss, such as loss of carbon stocks, changes in biogeochemical cycles through emission of greenhouse gases (GHGs), and loss of biodiversity. Table 8.2 demonstrates the complementarities and trade-offs of land-use decisions following 'deforestatation' for wider environmental services.

Although water yield in small catchments is more directly affected by land use change than in larger basins (Bruijnzeel, 2004), developing a sound water resource policy at such a small scale is almost a secondary objective for most decision-makers. At a larger scale that involves several districts or states, water resources issues may be addressed more directly. Figure 8.7 shows the cascade of land use decisions that represent the different levels of scales and stakeholders involved. At the household and local scale, sometimes even at a national scale, land use decisions still dominate the process. The benefits of land use change usually flow in one direction of the cascade, from the lower level to the higher ones. In South East Asia generally, there is as yet no market or reward mechanism to transfer payments from the beneficiaries to the environmental service providers (e.g. from urban lowlanders to upland farmers or forest dwellers) as is beginning to be developed, legalised and implemented in the Philippines (Deutsch *et al.*, this volume) and Costa Rica (Aylward, this volume).

To reverse the driving forces of deforestation and land degradation, the process has to be initiated in the uplands where many landless poor reside who could benefit from the resources present within their areas. Sustainable use of these resources has to be rewarded or paid for in a real sense and transfer mechanisms have to be devised in a market setting (Aylward, this volume). Conflicting interest groups at different levels within decision-making bodies (local, national and global) have to come to some form of agreement on how such payments may be arranged. The payments may be used to improve the use of natural resources, increase the capacity of community organisation, increase social status or even secure land ownership (Jensen, 2002). In the Philippines, upland communities collaborating in basin management could be paid or compensated in terms of wages for services rendered, provision of tree planting materials, skills training, technical assistance and land tenure security (Arocena-Francisco, 2002).

POLICY RESPONSES

Three-quarters of the population of the LMB (mainly farmers and fishermen) earn their living by utilising the natural resources base. This is why it is so important to take fully into consideration how the environmental, economic and social changes brought by

Table 8.2. *Land use options in Jambi Province, Sumatera, and the related environmental goods and services provided from the perspectives of local, national and global concerns*

	Local/national concerns				Global concerns				
						Green house gases emission ($m^{-2}\ hr^{-1}$)			
Land use options	Returns to labour (US$ per day[a])	Number of people supported	Agronomic sustainability	Watershed functions	Carbon stocks (Mg C ha^{-1})	CO_2 (mg C)	CH_4 (μg C)	N_2O (μg N)	Plant species richness
Natural forest	0	0	Low	Low	250	67	–20	0.7	120
Managed forest	4.58	0.2	Low	Low	175	–	–	–	100
Logged-over forest	8.33	17	Medium	Medium	150	102	–18	5	90
Complex agroforest	1.67	59	Medium	Low	116	–	–34	2	90
Simple agroforest	2.50	80	Medium	Medium	103	–	–	–	60
Jungle rubber	1.25	71	Medium	Low	97	74	–13	0.2	25
Oil-palm plantation	4.17	58	Medium	Medium	91	–	–16	55	25
Food crop	1.67	54	Low	High	39	46	–11	2	15

[a] The exchange rate was Rp 2400 for US$1.

Source: Tomich *et al.* (2001) and Murdiyarso *et al.* (2002).

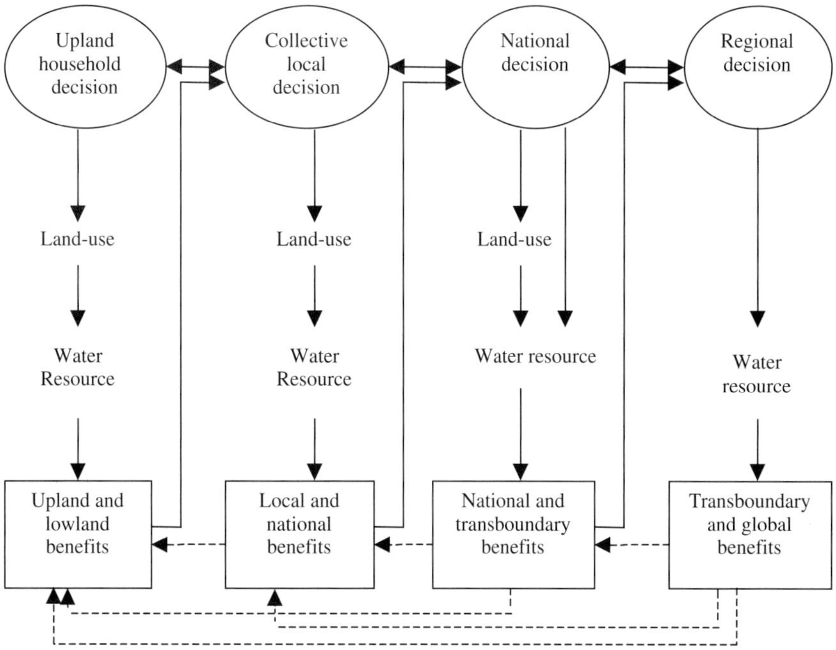

Figure 8.7 Multi-level land use decision cascade that will indirectly affect water resource management. The broken arrows indicate the transfer of payments from higher level beneficiaries to lower level environmental service providers still has to take place.

regional policies, such as trade and transportation, will affect their livelihoods. To achieve this, the policies, standards and knowledge that have been developed and established through hard-won international treaties are indispensable. Most of the issues are associated with the development of (large) dams, not only in the upper parts but also in the lower parts of the basin. Also, it is expected that the associated land use changes over such a large scale may eventually affect water yield and streamflow regime. Therefore, notification of major river works and any changes in land and water use, and the application of international standard environmental impact studies, are among the major 'rules' governing water resource planning between countries sharing international waters. In a setting such as the Mekong River Basin, the risk of benefiting one sector at the expense of others is real. It is critical that universally accepted planning and resource-sharing arrangements are adhered to, and are seen as fair by all parties. Participation of non-parties may be encouraged to recognise the existing agreement in a wider perspective and discuss the decision-making process for Mekong-related natural resource planning.

In such a complex socio-political domain it is important to establish credible planning and policy responses by and within the MRC through adequate and fair representation, independence, transparency and inclusiveness, as suggested by Dubash *et al.* (2001). Representation should emphasise individual capacity rather than formal institutional representatives and the burden of legitimacy should be put on personal and professional reputation. Knowing the nature of the agreement this is a real challenge to achieve. One of the practical elements in maintaining independence is diversifying sources of funding. Besides the contributions from the member countries, the LMB is supported by multiple sources of funding, including the governments of Japan, Switzerland, Sweden, Denmark and Australia, as well as the World Bank and the Asian Development Bank. Therefore, the LMB should be able to seek a balance with donors and stakeholders. Transparency is perhaps the weakest area that the LMB is facing since broad participation in the decision-making process, and no mechanisms have been established to adequately acknowledge public contributions. As a result, the inclusiveness that may be interpreted as a sense of belonging for all stakeholders is still a long way off. This may be gradually achieved and enhanced by reaching out to previously unheard voices (Deutsch *et al.*, this volume), regular provision of updates on the progress of the work programme and direction, and holding public hearings to accept general submissions of diverse viewpoints (Dubash *et al.*, 2001).

It is suggested here that to address public concerns on water resources properly, policy responses should be made and reviewed regularly, depending on the spatial and temporal scale of concerns as summarised in Table 8.3. One response may be appropriate for one concern in one temporal and/or spatial scale but not for others. Concerns may be growing but they should have definitive temporal and spatial dimensions.

Table 8.3. *Public concerns that require policy responses on water resources at different spatial and temporal scales*

	Spatial scale		
Temporal scale	Farm/plot	Landscape/catchment	River basin
Seasonal	Water availability for crops and fishery	Fluctuation of water yield (Q_{max}/Q_{min})	Upstream–downstream conflicts on water supply and demand
Annual	Water supply for sustainable crop and fish yields	Excess and limited water supply and demand	Supply and demand for industries and navigation
Decadal	Water security	Engineering costs	Regional economic development

CONCLUDING REMARKS

Although the effects of land use change on water resources are more pronounced at the small catchment scale, sound water resources policies are rarely made at that same scale. In such situations top–down planning is still a useful approach to implement macro-management plans. Macro-policy adjustments may be carried out based on the feedback mechanism of monitoring and evaluation. However, it has to be followed by bottom–up planning which is capable of capturing local concerns such as improvement of local crop and fish productivity, upland dweller well-being and improvement in water and general environment quality.

The driving forces of land use change are largely associated with government policies on large-scale land development and land management. To a lesser extent, poverty and high rates of population growth aggravate the situation. The impacts on water resources may be reversed by internalising (i.e. assigning an economic price or attaching a price tag) to the causes in water resource management, supported by appropriate legal instruments and institutional arrangements, and sound data upon which to base all these.

Providing an arrangement for payment mechanisms in a real market setting could potentially reduce conflicts of interests. Uplanders providing environmental services such as stable streamflow may get the rewards from the beneficiaries in the lowlands. Proper compensations must be based on a sound understanding of the roles and impacts of land-use change vis-à-vis inherent, natural climate variability (and change) on streamflow. The role of the scientific community is to provide insight in hydrological processes and above all in distinguishing between natural and man-made effects. It is also an urgent task to evolve a scientifically established decision support system into a negotiation support system. This will enhance the bargaining power of the upland community in connection with the conservation of the upland resources.

References

Arocena-Francisco, H. (2002). Environmental service 'payments': experience, constraints and potential in the Philippines. In: *Regional Workshop on Developing Mechanisms for Rewarding the Upland Poor in Asia for the Environmental Services They Provide*, Bogor, Indonesia, 6–8 February 2002.

Bruijnzeel, L. A. (1989). (De)forestation and dry-season flow in the tropics: a closer look. *Journal of Tropical Forest Science* **1**: 229–243.

(1990). *Hydrology of Moist Tropical Forest and Effects of Conversion: a State of Knowledge Review*. Paris, France: UNESCO.

(1998). Soil chemical responses to tropical forest disturbance and conversion: the hydrological connection. In: A. Schulte and D. Ruhiyat (eds.) *Tropical Forest Soils and Their Management*, pp. 45–61. Berlin, Germany: Springer Verlag.

(2004). Hydrological functions of tropical forest: not seeing the soil for the trees? *Agriculture, Ecology and Environment*, doi:10.1006/j.agee.2009.01.015.

Burton, R. (2000). Mekong basin dams claim lives, causes poverty, bank warned. Australia Vietnam Science-Technology Link. Available online at http://ens.lycos.com/ens/jun2000/2000L-06-27-01.html.

Dubash, N., Dupar, M., Kothari, S. *et al.* (2001). *A Watershed in Global Governance? An Independent Assessment of World Commission on Dams*. World Resources Institute.

Fisher, B. (1996). Shifting cultivation in Laos: is the government's policy realistic? In: B. Stensholt (ed.) *Development Dilemmas in the Mekong Subregion*. Clayton, Australia: Monash Asia Institute. University of Monash.

Hammer, W. I. (1988). *Second Soil Conservation Consultant Report*, AGOF/INS/78/006, Technical Note no. 10. Bogor, Indonesia: Center for Soil Reseach.

Hardjono, H. W. (1980). Influence of a permanent vegetation cover on streamflow. In: *Proceedings of the Seminar on Watershed Management, Development and Hydrology*, Surakarta, Indonesia, 3–5 June 1980, pp. 280–297 (in Indonesian).

Jakeman, T., Ross, H., Wong, F. *et al.* (1997). Integrated water resource assessment and management for sustainable resource management in Northern Thailand. In: *Proceedings International Congress on Modelling and Simulation*, Hobart, Tasmania.

Jensen, C. (2002). Development assistance to upland communities in the Philippines. In: *Regional Workshop on Developing Mechanisms for Rewarding the Upland Poor in Asia for the Environmental Services They Provide*, Bogor, Indonesia, 6–8 February 2002.

Jepson, P., Jarvie, J. K., MacKinnon, K. *et al.* (2001). The end for Indonesia's lowland forests?. *Science* **292**: 5518.

Kartodihardjo, K. and Supriono, A. (2000). *Effect of Sectoral Development on Conversion and Degradation of Natural Forests: Cases in Timber and Estate Crop Plantations in Indonesia*, CIFOR Occasional Paper no. 26 (in Indonesian). Bogor, Indonesia: CIFOR.

MoFEC (1997). *Forest Inventory and Mapping Programme*. Jakarta, Indonesia: Ministry of Forestry and Estate Crops.

Mackey, B., Trisophon, K., Ekasingh, M. *et al.* (1997). A decision support system for integrated water resources assessment and management: a case study of the Upper Chao Phraya Headwaters, Northern Thailand.

In: *Proceedings International Congress on Modelling and Simulation*, Hobart, Tasmania.

Manurung, E. G. T. (2000). Why forest conversions must be stopped? In: *Proceedings of Workshop on Natural Forest Logging Moratorium and the Closure of Financially Unhealthy Timber Industries* (in Indonesian) Jakarta, Indonesia: Ministry of Forestry.

Ministry of Forestry (1990). *Integrated Watershed Management Plan for Saddang River*. Jakarta, Indonesia: Ministry of Forestry.

Ministry of Forestry (1991). *Integrated Watershed Management Plan for Solo River*. Jakarta, Indonesia: Ministry of Forestry.

Ministry of Forestry (1992). *Integrated Watershed Management Plan for Batanghari River*. Jakarta, Indonesia: Ministry of Forestry.

Murdiyarso, D., van Noordwijk, M., Wasrin, U. R. *et al.* (2002). Environmental benefits and sustainable land-use options in Jambi transect, Sumatra. *Journal of Vegetation Science* **13**: 429–38.

Phanrajsavong, C. (1996). Hydropower development in the Lower Mekong Basin. In: B. Stensholt (ed.) *Development Dilemmas in the Mekong Subregion*. Clayton, Australia: Monash Asia Institute, University of Monash.

Purwanto, E. (1999). Erosion, sediment delivery and soil conservation in an upland agricultural catchment in West Java, Indonesia. Unpublished Ph.D. thesis, Vrije Universiteit, Amsterdam, The Netherlands.

Quang, N. N. (1996). The Mekong basin development: Vietnam's concerns. In: B. Stensholt (ed.) *Development Dilemmas in the Mekong Subregion*. Clayton, Australia: Monash Asia Institute, University of Monash.

Saysanovongphet, S. (1996). Mekong development management: views of the Lao PDR. In: B. Stensholt (ed.) *Development Dilemmas in the Mekong Subregion*. Clayton, Australia: Monash Asia Institute, University of Monash.

Susetyo, B. and Murdiyarso, D. (1992). Simulation model of hydrological processes based on climate and land use change. *Bulletin of Agrometeorology* **1**: 34–45 (in Indonesian).

Tomich, T. P., van Noordwijk, M., Budidarsono, S. *et al.* (2000). Agricultural intensification, deforestation and the environment: assessing trade-offs in Sumatra, Indonesia. In: D. R. Lee and C. B. Barrett (eds.) *Tradeoffs or Synergies?: Agricultural Intensification, Economic Development and the Environment*, pp. 221–244. Wallingford, UK: CAB International.

Turner, B. L., II, Skole, D., Sanderson, S. *et al.* (1995). *Land-Use and Land-Cover Change*, vol. I, IHDP Report no. 7–IGBP Report no. 35. Stockholm, Sweden: IHDP.

UNEP/DEWA (2001). *An Assessment of the Status of the World's Remaining Closed Forests*, Division of Early Warning and Assessment no. TR.01–2. Nairobi, Kenya: UNEP.

UNESCO (2001). *The Design and Implementation Strategy of the* HELP *Initiative*, Technical Document in Hydrology no. 44. Paris, France: UNESCO.

Van Noordwijk, M., Tomich, T. P. and Verbist, B. (2001). Negotiation support models for integrated natural resource management in tropical forest margins. *Conservation Ecology* **5**(2): 21–42.

Wilk, J., Andersson, L. and Plermkamon, V. (2001). Hydrological impacts of forest conversion to agriculture in a large river basin in northeast Thailand. *Hydrological Processes* **15**: 2729–48.

Wischmeier, W. H. and Smith, D. D. (1978). *Predicting rainfall Erosion Loss: A Guide to Conservation Planning*, USDA Agricultural Handbook, no. 537. Washington, DC: Government Printing Office.

World Bank (2000). *Indonesia: Environmental and Natural Resource Management in a Time of Transition*. Washington, DC: World Bank.

9 Community-based hydrological and water quality assessments in Mindanao, Philippines

W. G. Deutsch and A. L. Busby
International Center for Aquaculture and Aquatic Environments, Auburn University, Auburn, USA
J. L. Orprecio and J. P. Bago-Labis
Heifer Project International, Muntinlupa City, Philippines
E. Y. Cequiña
Central Mindanao University, Mindanao, Philippines

INTRODUCTION

Philippine water issues

In spite of the fact that the Philippines is water rich, with nearly 5000 cubic metres per capita of renewable water resources, there is a national crisis regarding conservation of a dwindling supply of high quality water. This has led to presidential decrees and other legislative action at the federal level, including Senate Bill No. 1082 which is designed to institute 'a comprehensive water development act thereby revising and consolidating all the laws governing the appropriation, utilisation, exploitation, conservation, development and management of water resources, creating the National Water Commission' (Policy Forum, 1997).

Water quality of both coastal marine and inland freshwater environments of the Philippines is threatened by soil erosion and sedimentation, excess nutrient runoff and bacterial contamination. These types of pollutants often come from broad areas of both rural and urban land (usually classified as polluted runoff or non-point source pollution). Although polluted runoff is the most common source of water degradation in the Philippines and worldwide, it is much more difficult to control than pollution from specific sources.

As in most parts of the developing world, there is a limit to what government can do to protect and conserve water because of a lack of personnel, equipment and finances. This is especially true in remote, rural areas where rates of natural resource loss generally exceed local governments' attempts to remedy environmental problems. In particular, specific information of water conditions needed to establish management strategies is generally lacking.

Regardless of governmental resources, many of the current environmental problems are not solvable by government regulation alone. Citizens need to become aware of the issues and take an active part in finding solutions. They have the greatest vested interest in conserving local water supplies and, potentially, a greater capacity than that of the government to measure conditions, identify specific problems and decide upon a proper course of action.

Of pressing need are practical, environmental indicators that local communities can use to determine trends in their natural resources and evaluate the appropriateness of their collective actions. The degree of virtually irreparable damage to upland forests, streams and coastlines, and rates at which degradation continue, underscores the urgency needed for addressing seriously catchment management at the local level.

Decentralisation and potential for local environmental management and policy

Since the era of President Ferdinand Marcos and the 1986 revolution, the Philippines has moved squarely in the direction of decentralised authority, including decentralised natural resource management. The 'people power' democratisation process saw the flourishing of non-governmental organisations (NGOs) eager to play an active role in the country's development, and the Philippines now has one of the highest numbers of NGOs in the world. An evaluation of this transition (Jutkowitz *et al.*, 1997) indicated that '. . . the Philippines has made significant progress in establishing legal guidelines for greater local government autonomy, for more responsive and accountable local government, and for broader participation by civil society at the local level.'

Forests, Water and People in the Humid Tropics, ed. M. Bonell and L. A. Bruijnzeel. Published by Cambridge University Press.

The Philippine government reversed the centralised political power and governance primarily through the enactment of the Local Government Code in 1991. The code includes the following provisions (Jutkowitz *et al.*, 1997):

(1) Devolves power and authority to deliver services to local government units and calls for health, agriculture, environment, infrastructure and social welfare services to be run by barangays (municipal subunits).
(2) Provides for quarterly distribution of internal revenue allotments to local government units from national revenue collected, using a formula based on government level and population (such allotments may be used for natural resource management and protection).
(3) Mandates participation of government-accredited NGOs (nonprofit organisations) and peoples' organisations (community-based membership organisations) in local government council deliberations.
(4) Authorises local initiatives and referenda to allow registered voters to propose, enact, repeal, or amend ordinances directly at the local government level.

Continued decentralisation of authority over the last 8–10 years has provided a foundation for community-based environmental management and policy. At the local level, municipal mayors are being mandated by federal and provincial governments to develop natural resource management plans that usually include plans for water. Although not always the case, many mayors and local government units are becoming more receptive to cost-effective ways by which they may obtain information to formulate municipal policies of environmental protection and restoration.

Both the need for natural resource management alternatives and the political climate in the Philippines made a new research programme, called the Sustainable Agriculture and Natural Resource Management, Collaborative Research Support Program (SANREM CRSP), relevant and timely. The programme is funded by the US Agency for International Development, through the University of Georgia, and has been implemented from 1992 to the present by a consortium of international (primarily US) and Philippine-based universities, governmental agencies and non-governmental organisations. The goal of SANREM is to conduct drainage basin or catchment-scale, participatory research involving farmers and other stakeholders, to elucidate linkages between land use, environmental quality and overall sustainability of the ecological and social system (Foglia, 1995; Cason, 1999).

Study area

The ecosystem under investigation in the Philippines is the 36 000 ha Manupali River drainage basin in central Bukidnon Province of the southern island of Mindanao. The northern, larger portion of the basin, where most programme activity takes place, is in the Municipality of Lantapan. Elevations range from 2,900 m at the top of Mt. Kitanglad in a national park, to about 300 m in the lowlands where the Manupali River flows into the larger Pulangi River (Figure 9.1). Soils of the higher elevations of this volcanic slope are primarily Ultisols and Inceptisols whereas soils of the footslopes and alluvial terraces are mainly Oxisols (Poudel and West, 1999). The landscape of the Manupali catchment encompasses agro-ecological zones of upland forests, agroforestry buffer zone, vegetables, corn, sugar-cane and lowland rice, all transected by several streams.

Rainfall varies from about 2000 to 3000 mm yr^{-1} (Table 9.1). Greatly reduced rainfall during the El Niño event that began in November 1997 resulted in two to three crop failures and significant hardship to local communities.

The population of Lantapan is about 40 000 and is made up of a diverse mix of ethnic groups. The indigenous tribe, called the Tala-andig, claims ancestral rights to much of the land in the Manupali uplands and groups of Dumagat, Igorot and Ilocano settlers have been migrating to the area from other parts of the country over the last several decades. The population growth rate from 1960 to 1995 was more than 4%, almost double the national average, primarily because of in-migration. The population of Lantapan is relatively young (about 40% are 0–14 years) and without out-migration, the municipality will continue to grow well beyond the 21st century (Paunlagui and Suminguit, 2001). Other social, economic and agroecological aspects of Lantapan may be found in Coxhead and Buenavista (2001).

METHODS

A participatory research approach

Several sectors of the local community, including agriculture, business, government, religion and education, were involved with researchers and community organisers in an initial, catchment or basin-scale evaluation of the site called a Participatory Landscape/Lifescape Appraisal or PLLA (Bellows *et al.*, 1995). During this six-week analysis, scientists, development workers and local residents worked as a team to interview people of the Lantapan community and become aware of perceptions and environmental concerns. This appraisal was vital to the development of the framework plan, which led to the water indicators research.

Several interdisciplinary research projects on soil, water and biodiversity were designed within the larger SANREM/Philippines programme, based on information gained during the PLLA and the interests of collaborative research teams. 'Priming activities' led by NGO partners helped the community to understand and feel comfortable working with researchers throughout

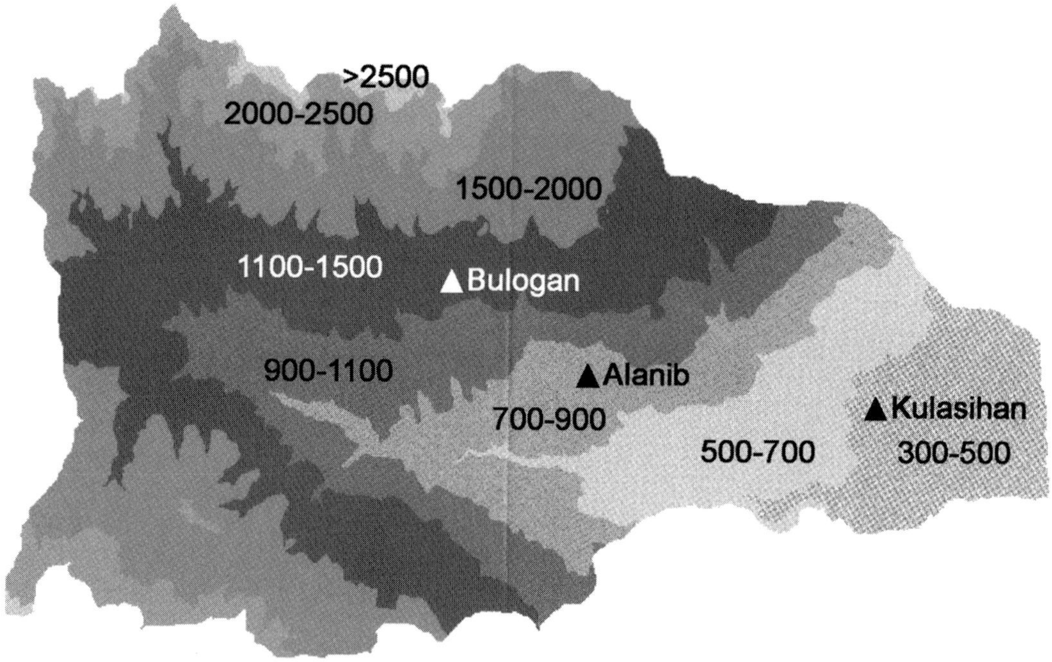

Figure 9.1 Elevation map (m) of the Manupali River watershed with approximate locations (triangles) of the Bulogan, Alanib and Kulasihan automated weather stations (based on a 1 : 50 000 scale US Army and Bureau of Soils map series, UTM coordinates).

Table 9.1. *Total annual rainfall (cm) at three weather stations in the Manupali River watershed, 1994–2000*

	Weather station		
Year	Bulogan	Alanib[a]	Kulasihan[b]
1994	226	208	165
1995	293	301	280
1996	235	229	273
1997	222	265	209
1998	196	151	169
1999	331	188	257
2000	300	286	254

[a] No data were available from the Alanib station in January 1998 and June–August and October 1999.

[b] No data were available from the Kulasihan station from September to December 1994.

the period of developing the research plan. For example, a study tour was organised and led by a local educator and project partner to enable several local farmers to travel from their upland communities through various portions of a large river valley and to the sea (some for the first time). This helped the farmers to understand more clearly certain biophysical and social linkages between their land and the downstream areas that the researchers intended to study.

As one component of the SANREM research plan, a project led by Auburn University and Heifer Project International focused on local water quality assessment and management. The overall goal of the project was to foster the development of community-based water monitoring groups, and to collect credible water quality and quantity data that lead to environmental and policy improvements. This was accomplished primarily by conducting a series of workshops and field exercises to train interested community groups in the evaluation of water quality using portable test kits and other basic analytical tools.

The approach of this project was to develop and test specific water quality indicators that were appropriate for natural resource management by community volunteers and the local government unit. In that regard, the following criteria were established for each indicator:

(1) Scientifically valid methods, for credible qualitative and quantitative information.
(2) Relevant to the community, for their endorsement and participation in data collection.
(3) Practical and relatively inexpensive, for sustainable use and applications using locally available materials.

Many of the methods used were modelled after those developed in Alabama Water Watch, a citizen volunteer, water quality monitoring programme involving about 70 community groups that is now under way in the US (Deutsch *et al.*, 1998). Filipino partners of this SANREM project who were educators and

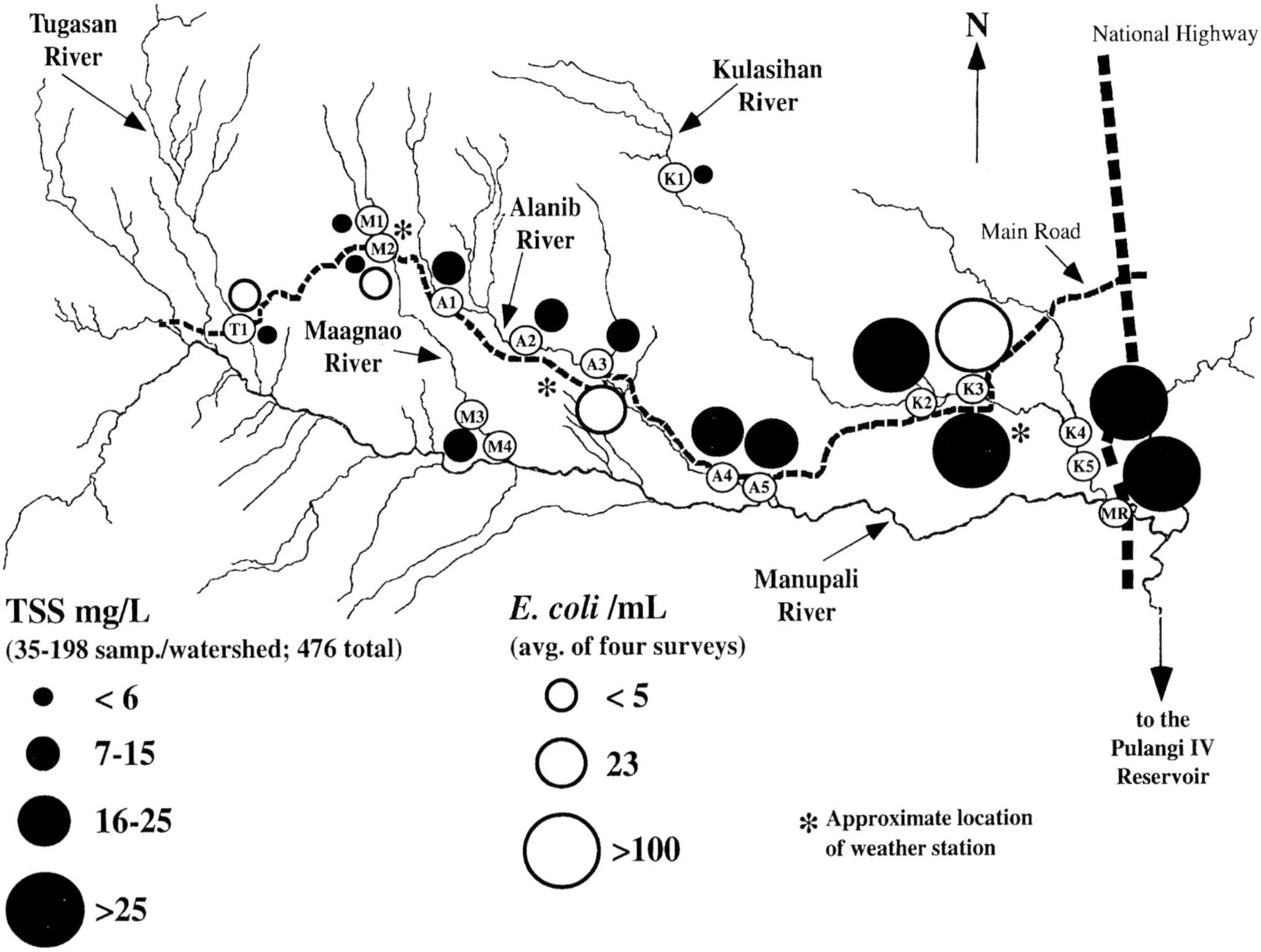

Figure 9.2 The Manupali River catchment, with total suspended solids (TSS) and *E. coli* bacteria concentrations in four subcatchments, August 1995–July 1996.

community developers helped customise the workshops and sampling techniques to the local situation. Community volunteers who attended workshops and began monitoring water included farmers, teachers, members of certain women's organisations and some members of the local government unit. Monitoring results were disseminated to community members, educators and local policy makers through oral presentations and written reports. After several months of involvement in the project, the core group of water monitors proceeded, in Filipino fashion, to form a people's organisation and incorporate as an officially recognised NGO in 1995 (Deutsch *et al.*, 2001a). The group's name is Tigbantay Wahig, or 'water watchers' in the local Binukid dialect.

The project researchers and volunteer water monitors selected 16 sampling sites on four main tributaries of the Manupali River (Figure 9.2). Sites were chosen that were generally accessible and representative of the diverse portions of the overall landscape, including subcatchments of varying degrees of forest cover, agricultural land and population. A 'menu' of possible water quality indicators was made available to the monitors in the workshops. These included the physicochemical parameters of temperature, pH, alkalinity, hardness (primarily concentrations of calcium and magnesium), nitrates, phosphates, dissolved oxygen, turbidity, total suspended solids and stream discharge. Biological parameters included biotic indices of stream macroinvertebrates and measurements of *Escherichia coli* and other coliform bacteria concentrations (Deutsch and Orprecio, 2000; Deutsch *et al.*, 2001b).

Although pesticides in water were a concern among the local community, analyses were logistically and technically difficult and cost prohibitive. Alternatively, techniques of rapid biological assessment of water quality using stream macroinvertebrates were introduced to serve as biotic indicators of pesticides. Many invertebrates are sensitive to pesticides and significant shifts in their community structure may occur if this type of pollutant is present.

After several months of testing water for the 8–10 parameters presented in the training workshops, the data began to suggest that the relatively few parameters related to soil erosion, disrupted stream flows and bacterial contamination were the most productive to pursue as indicators. For example, biological indicators using stream invertebrates required considerable taxonomic expertise that seemed too academic for community members (see Connolly and Pearson, this volume). Both the citizen monitors and researchers concurred that it was more practical to pursue in-depth study and application of a limited number of promising indicators. The following is a summary of the rationale and methodology related to these community-based water indicators.

SOIL EROSION AND SUSPENDED SOLIDS

Because the community of Lantapan is primarily agrarian, measurements of erosion and sedimentation were particularly relevant to volunteer monitors. Farmers generally understood that soil loss usually meant a reduction in the fertility of their fields, with accompanying reduction in crop production. Farmers of lowland rice clearly realised the negative impacts of upland soil erosion because the irrigation canals had become heavily silted and as a system of water conveyance were only about 25% efficient.

Further downstream, the Manupali River flows into the Pulangi River, which is impounded to create a series of hydroelectric generating stations. Interviews and information gathering activities of the PLLA revealed that the Pulangi IV reservoir was silting at an alarming rate of nearly one metre per year at the dam, and that the reservoir capacity had been reduced by about 50% by sedimentation. This also contributed to premature wear of hydropower turbines and frequent power outages or 'brown outs.'

One indicator of soil erosion was the measurement of total suspended solids, or TSS. A relatively simple and inexpensive technique was adapted in which a known volume (usually 1 litre) of stream water was filtered for calculation of mg l^{-1} suspended solids (see also Douglas and Guyot, this volume). The plastic filtering apparatus was lightweight and easily portable. The glass fiber filters (6 μm pore size) were prepared and weighed before and after sampling using a Metler balance (nearest 0.1 mg) at a nearby university. Each filter was contained in a small, plastic Petri dish that was labelled and taped shut. Pre-weighed filters in their individual dishes were brought to the catchment in batches of about 50–100 for monitors to sample TSS.

In the field, a filter was placed on a stage between two chambers of the apparatus and a hand pump was used to create a vacuum in the lower chamber that drew sample water through thte filter. Sampled filters were air-dried in a relatively dust-free environment and returned to the university for oven drying, weighing and data recording. Solids usually dried hard and remained on filters during transport to the laboratory. Dislodged pieces of dried solids remained in the plastic dishes and could be added to the filter during post-weighing.

TSS measurements were taken once or twice monthly during base flow conditions at the main road bridge crossing of each of the four subcatchment streams. Additional TSS measurements were made at each stream at 30-minute intervals during selected rainfall events. Rainfall data were collected from three weather stations in the catchment and obtained electronically from researchers at Central Mindanao University who maintained the stations and the weather database.

ALTERED STREAMFLOWS

Typically, stream discharge measurements are made by researchers using expensive installations such as concrete or metal weirs, flumes and gauging stations. Such methods are usually impractical for rural communities using their own resources, so low-tech methods were developed and adapted for use by the volunteer water monitors in Lantapan.

Stream velocity and discharge measurements were made with locally available materials, including rope, measuring sticks and a float. A cross-sectional map of each of the four streams was made at the main bridges, using the regular, concrete sides of the revetment wall under the bridge as boundaries when possible. A rope was stretched perpendicularly across the stream between two fixed points and stream depth was determined at 1 m intervals along the rope. Measurements of stream width and depth were used to draft cross-sectional maps and calculate area.

Another rope of known length was stretched parallel to the stream bank to mark the distance that a floating orange would travel while being timed. Multiple measurements of the time required to float a known distance in different parts of the stream were used to determine average surface current velocity. Mean surface velocities were multiplied by a standard factor of 0.8 to estimate stream current velocity.

Together, the cross-sectional area of the stream (square metres) and its current velocity (metres per second) were used to estimate stream discharge (cubic metres per second). For comparisons among subcatchments, discharges were measured on the same day of each month, and were normalised by subcatchment area (expressed as specific discharge in liters per second per hectare, $l\,s^{-1}\,ha^{-1}$).

SEDIMENT YIELD

Stream discharge estimates were not only valuable in understanding seasonal hydrographs of the four subcatchments, but were also used to estimate sediment yield. Concurrent TSS (weight per volume) and discharge (volume per time) measurements were made monthly and used to calculate instantaneous sediment yields (expressed as weight per time and normalised for catchment area, $mg\,s^{-1}\,ha^{-1}$).

BACTERIAL CONTAMINATION OF WATER

Levels of potentially harmful bacteria in streams, wells and piped drinking water were of primary concern to many citizens of Lantapan because of obvious public health risks and personal experiences of illness. As with related memories of community members regarding stream degradation from pesticides and silt, older adults recounted how they freely drank from streams in the past at places where they knew they would become ill today.

Evaluation of water for bacteria in the community had been infrequent, and the tests that were done occasionally by the Department of Health or the Barangay Health Workers detected only the presence or absence of fecal coliforms without determining a concentration value. As with all other techniques and indicators to be developed for practical use, bacteriological monitoring methods were chosen and adapted based on simplicity, accuracy and low expense.

A relatively new technique of measuring concentration of *E. coli* and other coliform bacteria was used for the monitoring (Deutsch and Busby, 1999). With this method, a 1 ml sample of water is collected using a sterile, plastic pipette and squirted into a 10 ml bottle of sterile, liquid medium. The medium (with color indicators for coliform bacteria, including *E. coli*) containing the water sample is poured onto a sterile, plastic dish which has a layer of calcium ions on the bottom that causes the liquid medium to solidify in about 20 minutes.

Incubation of sample plates at ambient tropical temperature was sufficient to grow the bacterial colonies for enumeration in about 30–36 hours. A simple incubator made from a foam box with a lightbulb is preferable because of the ability to maintain a temperature of 35–37 °C for up to 48 hours. No sterilizers or glassware were needed for this technique and necessary supplies (which cost less than US $2 per sample) could be easily transported to remote areas for sampling scores of sites per day. Following the incubation period, bacterial colonies of *E. coli* and other coliforms were enumerated with the naked eye based on their colour (purple and pink, respectively).

Each colony on the plate after incubation represented a single bacterial cell collected from the stream, so concentrations could be determined and compared with health standards. Four bacteriological surveys of the four major tributaries of the Manupali River were conducted in different seasons throughout 1995–96. In addition to the surveys of surface waters, the municipal drinking water system was tested from spring sources to several taps (faucets) throughout many barangays. Bacteriological tests were also made at selected households and included water storage tanks and drinking utensils.

DEMOGRAPHICS AND LAND USE

The human population in each of the four subcatchments of Lantapan was estimated from data in the 1990 Census of Population and Housing conducted by the National Statistics Office (NSO). Census data were gathered by the NSO for each of 14 barangays (villages) of the Municipality (Paunlagui and Suminguit, 2001). Subcatchment population estimates were made by determining which barangays (or portions of barangays) were in a given subcatchment. Land use data were obtained from Li (1994) and in personal communications with Dr. Allan Dela Cruz, SEARCA, the Philippines.

RESULTS AND DISCUSSION

The significant findings of the community-based water monitoring project were related to each of six indicators, as follows:

Indicator 1: Community perceptions, memories and experience

The first discussions between community members and researchers during the Participatory Landscape/Lifescape Appraisal revealed that residents were commonly concerned with water contaminants, such as pesticides and pathogens, in addition to soil erosion and sedimentation of streams and irrigation canals. Some farmers did not water their livestock in streams during rainfall events, citing loss or illness of animals from pesticide runoff. Public health records, although scanty, indicated a higher than average infant mortality and morbidity rate in the community and many common ailments were caused by waterborne pathogens.

Besides water quality concerns, some residents questioned or lamented the fact that some streams were no longer maintaining regular flows, but were cycling through seasonal flood and drought. Memories of stable stream flow and clean water were within the last few decades. The community believed that flash floods were increasingly common in the eastern part of the Manupali basin, resulting in severe soil erosion, crop loss and occasional loss of livestock or human life. Overall, the pattern of degradation experienced by the community was typical of that in upland landscapes of the Philippines and in many other parts of the world.

The indigenous people of the Tala-andig tribe had distinctive perceptions of environmental problems that were important to consider. The overarching worldview of the tribe was that spirits of water, air, forests and other natural and human phenomena were to be respected, and that lack of respect led to natural disasters. For example, one Tala-andig man explained that a recent flash flood that killed a young girl of the tribe resulted from outsiders who came to the forest and were loud and irreverent. The view was that water came suddenly from the ground, independent of rainfall, as a judgment.

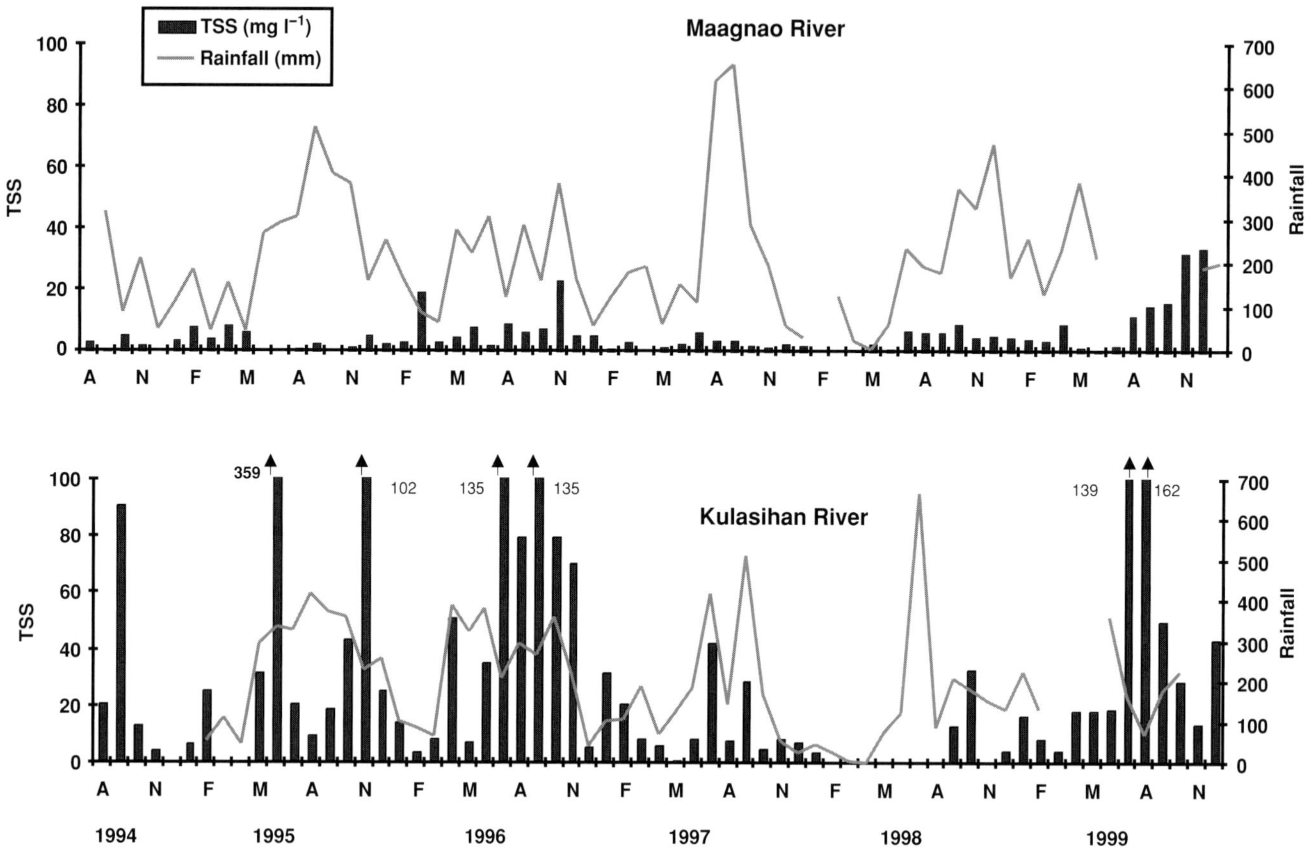

Figure 9.3 Average monthly total suspended solids (TSS) and total monthly rainfall in the Maagnao and Kulasihan rivers, 1994–9.

To help reconcile differing cultural views of the environment and raise awareness of the tribal way of life, the Tala-andig leadership invited researchers and development workers to a several-hour 'ritual of understanding' in the tribal centre. Subsequently, researchers and community members interested in studying water quality and quantity obtained the permission of the tribe to enter and sample the water of the streams. Modern testing methods for determining water quality merged with an ancient, tribal spirituality of water. In one instance, a rice offering in a banana leaf was left to the water spirit by a Tala-andig man who had just measured various chemical and biological parameters of a stream as part of the Tigbantay Wahig monitoring group.

Community perceptions became an important part of the research design and implementation. Factors to be carefully considered included how much time community members had to volunteer for water monitoring, how relevant an environmental variable was to their daily life and how important the data were to making a positive change. These factors had to be balanced continually with research needs, budgets and project evaluations from funders and the scientific community. The project prioritised community desires for water monitoring, even if it meant sacrificing data needs perceived by the researchers, because long-term, local participation and action for natural resource management were the ultimate goals. In turn, the community was often willing to volunteer valuable time and monitor some less relevant variables on 'good faith' that they were important to the research needs.

Indicator 2: Soil erosion and suspended solids

After the community had collected several hundred TSS samples throughout the subcatchments of the Manupali River valley, this indicator of soil erosion began to reveal patterns of degradation that went beyond the simple observations of 'clear' and 'muddy' water in various streams. The monitoring data strongly suggested differences in erosion rates at the subcatchment level, with sharp increases in TSS concentrations from the western to eastern subcatchments (Figures 9.2 and 9.3).

When correlated with rainfall data collected from three local weather stations (Figure 9.1), it became clear that seasonal difference in TSS occurred in each subcatchment (Figure 9.3). These differences were probably caused by natural factors, such as changing rainfall frequency and intensity in rainy and dry seasons,

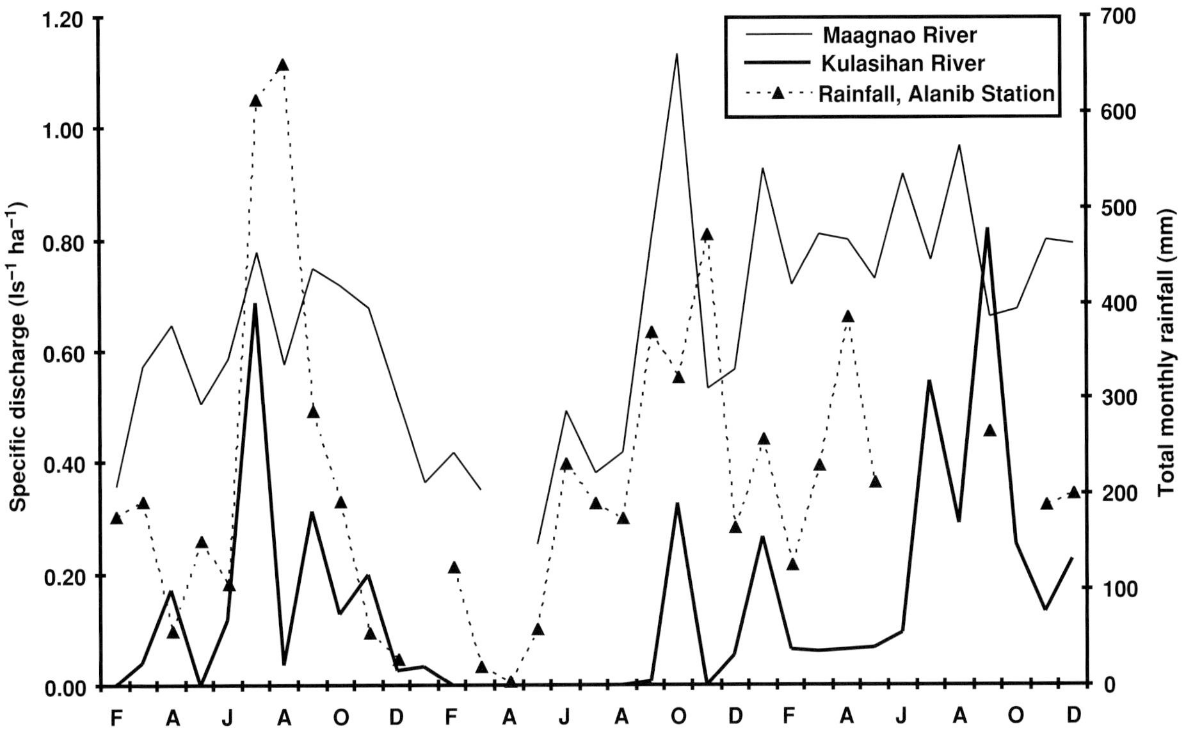

Figure 9.4 Specific discharge (measured monthly on the same day) and total monthly rainfall in the Maagnao and Kulasihan rivers, 1997–9.

as well as cropping patterns. Sometimes, a combination of these natural and human induced changes greatly increased TSS in streams, such as when farmers ploughed and exposed large areas of bare soil for planting just prior to heavy spring (May–June) rains.

The TSS measurements made during base flow were useful for relative comparisons among streams and over time, but were an underestimate of the greatly increased erosion rates during strong storms. Recognising this fact, the monitors offered to measure TSS more frequently, just before and during selected rainfall events in each subcatchment. Results were sometimes dramatic, and in one case, TSS increased by 1000-fold within a two-hour period of a heavy rain, to reach about 18 kg of soil in each cubic metre of water. To communicate this fact better to farmers and other community members, such a rate of erosion was likened to the weight of a sack of seed corn in each unit volume of water that approximated the size of a small desk. The TSS indicator became an increasingly important way for the Lantapan residents to quantify environmental change and lay the foundation for local action and policy changes.

Indicator 3: Altered streamflows

The measurements of stream discharge provided an indicator of subcatchment stability and seasonal hydrological patterns.

Table 9.2. *Average specific discharge, range and coefficient of variation (CV as %) of four tributaries of the Manupali River (measured monthly on the same day), February 1997– December 1999*

Stream	Average discharge and range (l s^{-1} ha^{-1})	CV
Tugasan	0.30 (0.04–0.64)	51
Maagnao	0.65 (0.26–1.14)	31
Alanib	0.13 (0.01–0.31)	60
Kulasihan	0.14 (0.00–0.82)	138

Monthly discharge measurements taken in each of the four streams on the same day were used to produce hydrographs that indicated distinct subcatchment differences. For example, the Kulasihan River had a much greater range of flow during an annual cycle than that of the Maagnao River, in spite of relatively similar rainfall patterns (Tables 9.1 and 9.2; Figure 9.4).

Of particular note was the response of the subcatchments to the severe El Niño drought which began in November 1997. The Kulasihan River was most affected by the drought and had no surface flow from March through August 1998 (Figure 9.4). This caused considerable hardship for local residents who depended on the river for washing, watering livestock and, in some cases,

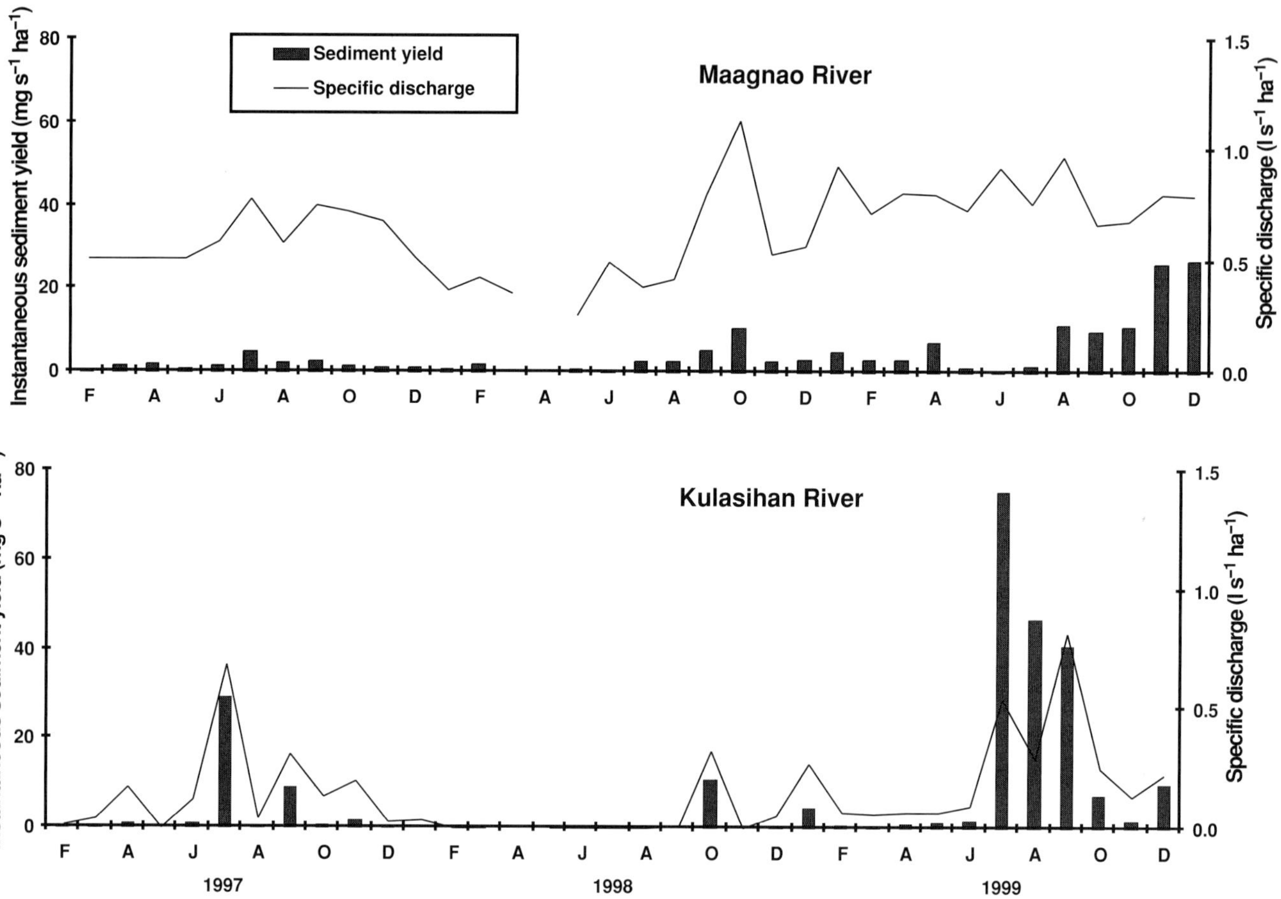

Figure 9.5 Sediment yield estimates and specific discharge (measured monthly on the same day) in the Maagnao and Kulasihan rivers, 1997–9. No discharge data were collected for the Maagnao River in April 1998; there was no surface flow in the Kulasihan River in May 1997, February–August 1998, and November 1998.

gathering household drinking water. Throughout this period, western subcatchments like the Maagnao River maintained relatively stable flows.

The instability or 'flashiness' of the Kulasihan River, indicated by its abrupt flooding and drought cycle (Figure 9.4), has intensified over the last several years and is becoming a serious problem for the local municipality. After a flash flood in July 1997, a new culvert system needed to be constructed in which three additional concrete tubes were installed to convey floodwaters and prevent washout and blockage of the main access road. Such problems for local government, along with property damage and loss of crops and soil from flooding, underscore the importance of the stream discharge indicator as an early alert to catchment disruptions.

The coefficient of variation (CV) in monthly streamflows from 1997 to 1999 ranged from 31% in the Maagnao River to 138% in the Kulasihan River, and provided a simple, additional indicator of watershed stability (Table 9.2).

Indicator 4: Sediment yield

As expected from TSS observations, sediment yield from the Kulasihan subcatchment generally exceeded that from the Maagnao (Figure 9.5) and other subcatchments. Some of the greatest contrast in sediment yield among the subcatchments occurred during July of 1997 and 1999 when the Kulasihan River flooded. At these times, sediment yield was about 6–40 times greater in the Kulasihan than in the Maagnao. The often dramatic difference between the two subcatchments was almost certainly because the Kulasihan is more deforested and populated, with larger areas of bare soil.

A disturbing trend that warrants further research and community action is the increase in sediment yield throughout the Manupali River basin over the last 3–4 years. Increases in sediment yield in the relatively pristine Maagnao subcatchment, from usually less than 5 mg s^{-1} ha^{-1} to more than 20 mg s^{-1} ha^{-1} (Figure 9.5), suggest increased human settlement and land disturbance. A recent population census indicated that some of the barangays in the Maagnao subcatchment are the fastest growing in the Municipality. This growth, coupled with relatively steep slopes, make this subcatchment particularly vulnerable to soil erosion and downstream sedimentation.

Indicator 5: Bacterial contamination of water

Bacteriological results were strikingly similar to the pattern observed for TSS, sediment yield and flow variability at these same locations, and reinforced the conclusion that degradation was occurring in a west-to-east gradient across the landscape (Figure 9.6). According to World Health Organization and US Environmental Protection Agency standards, bacterial concentrations in the Tugasan and Maagnao rivers were generally safe for human 'whole body contact' whereas those in the Alanib and Kulasihan rivers exceeded that safety standard typically 10 to 50 times.

A useful strategy taken by the project was to trace the quality of drinking water 'from source to mouth' by sampling for bacteria at springs, taps, water containers, water storage tanks and vessels in the homes. The simplicity and speed of analysing these components of a system on site has facilitated the identification of 'weak links' that need attention. In the case of the piping system for conveying water to communities, strategic sampling can determine if the contamination is coming from a main line, secondary line or local tap. This can speed up the repair process and be very cost-effective for local government units.

Bacteriological surveys of the municipal drinking water system revealed that water from several taps had become contaminated with *E. coli* because of breaks in the pipes and seepage into them from contaminated soils and water. Community members indicated that some contamination problems were because of the old system of pipes that needed replacement. Others noted that some immigrants who settle on uplands that are far from community standpipes tap into municipal water lines illegally by breaking the pipe and splicing a hose into it. This break then becomes a site for seepage and contamination downslope.

Whereas the initial participants in water quality training workshops and monitoring were predominantly young men, bacteriological monitoring generated much interest among women and girls. It is believed that this parameter was of particular interest to women because of its direct tie to family health, especially that of infants and children. It also may have been more relevant than other parameters because the measurement was made from community faucets and public springs that had a close connection to household affairs and daily chores. Strong involvement from the Federation of Lantapan Women's' Organisation and other women of the community added a new dimension to community-based water quality indicators and their applications. Overall, the concentration of coliform bacteria has become an important indicator of water quality, used by diverse sectors of the community.

Indicator 6: Demographics and land use

The community-based indicators of TSS, specific discharge, sediment yield and *E. coli* concentrations within the four subcatchments were compared with both demographic and land cover patterns determined from the government census and remote sensing data (Li, 1994) to understand better the linkages between land use and environmental quality. This comparison revealed a clear, yet disturbing, pattern. The progressive decrease in forest cover and increase in cleared land from west to east across the Manupali River valley were correlated generally with the patterns of water quality degradation that the community monitors had detected (Figure 9.6).

The overall results of this project indicated that the community-based indicators (Table 9.3) might be very important for describing landscape-scale trends. For example, abrupt increases in TSS occurred when subcatchment forest cover dropped below 30% and agricultural land made up more than 50% (Figure 9.6). Knowing such thresholds of unsustainable soil erosion by using an indicator like TSS could be of great value to natural resource managers and policy makers.

In the case of Lantapan, the western two subcatchments may be even more vulnerable to severe erosion with deforestation than the eastern two subcatchments because their average slope of about 20% is much greater (Figure 9.6). It is also noteworthy that about 75% of the population of Lantapan live in the two eastern subcatchments (although human density in all subcatchments is similar, ranging from 0.9–1.8 person per hectare). Much larger populations in the Alanib and Kulasihan subcatchments, including many more houses and roads, certainly contributed to the sharply elevated levels of soil erosion, *E. coli* concentrations and other measures of water-related problems.

FROM INDICATORS TO POLICY

As the water quality information increased through greater community involvement and monitoring activity, the environmental indicators helped to explain and communicate the changes in land and water that were occurring. It seemed clear that the 'price of development' in Lantapan under current technologies and land use strategies included streams that were silt laden, contaminated with bacteria and unstable in their seasonal flow.

What made the emerging scenarios of development and degradation more stark was that this rather extreme environmental

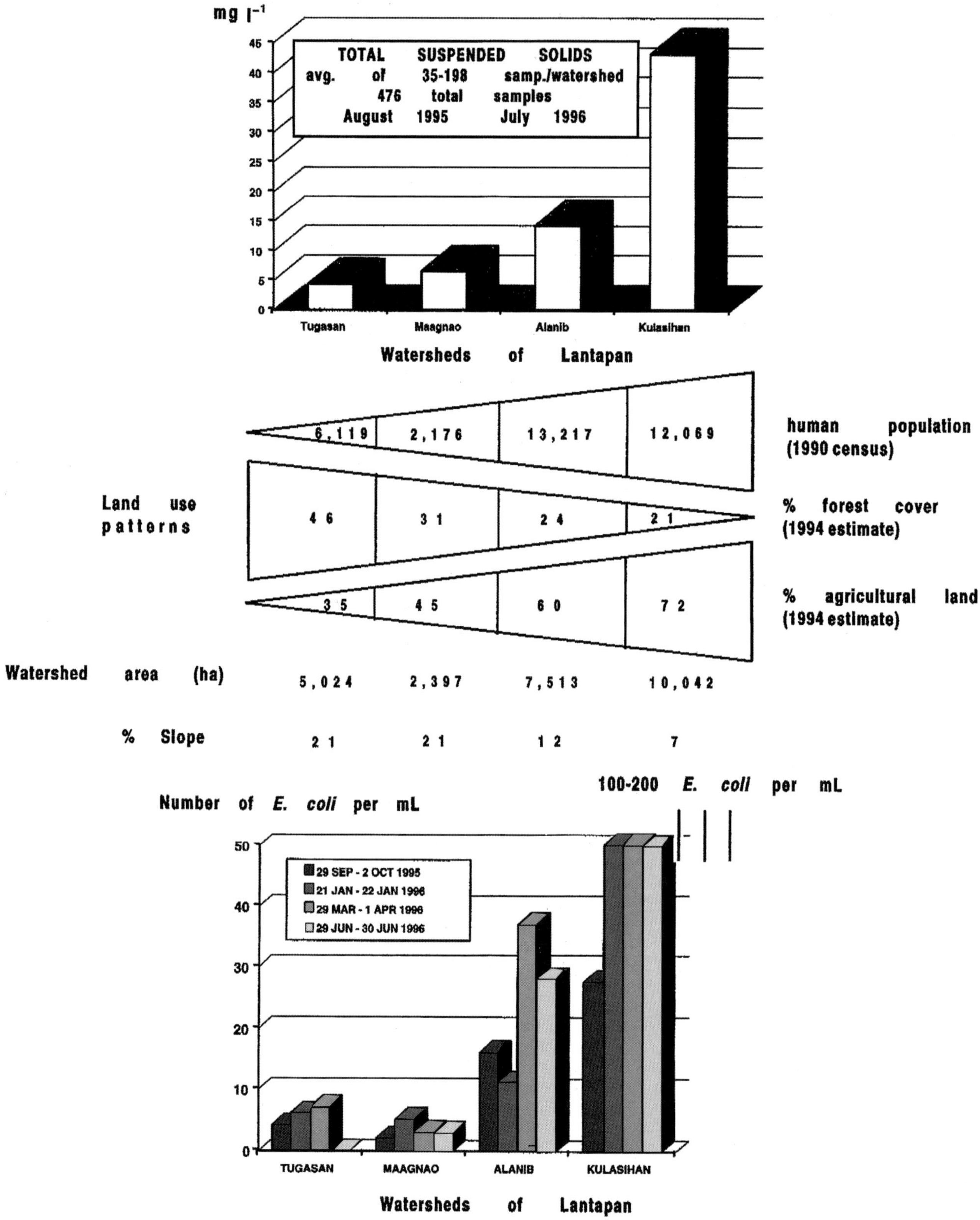

Figure 9.6 Total suspended solids, land use patterns and concentrations of *E. coli* bacteria in four subcatchments of the Manupali River, August 1995–July 1996.

Table 9.3. *Summary of community-based water quality indicators*

Issue/problem	Indicator	Unit of measure
General Environmental degradation	Community perceptions, memories, experiences	Anecdotal, or questionnaires/surveys
Soil erosion	Suspended soils in water	TSS, mg l^{-1}
	Instantaneous sediment yield	kg s^{-1} ha^{-1}
Disrupted streamflow	Specific discharge	l s^{-1} ha^{-1} (monthly measurement)
	Flow variability	Coefficient of variation (comparisons: time, space)
Bacterial contamination	Coliform concentration	Number of colonies per millilitre of water (*E. coli* and other coliforms)

gradient occurred in one, medium-sized river valley, and the changes were well within human memory. Community members did not have to envision a hypothetical, pristine or highly degraded watershed or a centuries-long process. They had seen for themselves by monitoring the dynamic ecosystems of the Manupali River valley in which they lived, and many were beginning to understand the consequences of land use decisions on a landscape scale.

To increase public understanding and action, the catchment information was popularised as a 'Walk Through Time.' Subcatchments of the west, including the Tugasan and Maagnao rivers, represented relatively natural conditions of the past whereas the Alanib and Kulasihan watersheds of the east illustrated the environmental costs of using traditional technologies to clear land for agriculture, homes and roads over the last few decades (Figure 9.6). Put simply, a person in the middle of the catchment could 'look west' to see where their environment had come from, and 'look east' to see where it was going. This basic way to use indicators to describe environmental change and suggest human causes and responses contributed to policy debates and decisions.

A municipal natural resource management plan

The environmental information collected in the water monitoring project was provided by the Tigbantay Wahig and other members of the research team to representatives of the local government unit in a variety of forms. At the invitation of the mayor of Lantapan, a summary of the research findings was presented orally with visual aids to the municipal council. This prompted the local government to incorporate community-based water testing and some of the research findings and recommendations into their Natural Resource Management Plan (NRMDP, 1998). The plan is well under way, and begins with the following statement:

> The Natural Resource Management and Development Plan (NRMDP) of Lantapan is a practical, not wishful action plan that presents practical intervention to the critical conditions of the natural resources. This has led to the identification of 'hot spots' or fragile areas that need immediate attention before it will be totally degraded over the next few years. The NRMDP evolved from a strong, participatory planning and collaboration of various sector groups in the community and the local legislators that comprised the Municipal Natural Resource Management Council (MNRC), together with different stakeholders from concerned government agencies at the provincial level . . . The plan will likely become a development model or template for natural resource management and environmental planning to other municipalities in the province of Bukidnon.

Among the many 'implementable actions' of the plan is a strategy to improve water quality, quantity and distribution. Key activities within this strategy involve continuous water quality monitoring and the expansion of membership of the Tigbantay Wahig group through the organisation of community chapters. Such a strategy represents a major step toward the practical application of community-based water quality indicators by a local government unit of the Philippines.

Of extra significance was the recent mayoral appointment of the president of the Tigbantay Wahig to the newly formed Natural Resource Management Council of the municipality. This created a direct link between the water monitors and government policy makers, and was in accord with the trend toward greater citizen participation in governance, provided for in the new Local Government Code.

Other effects on policy

In addition to the actions taken by the municipal government, the water quality project has affected decisions and policies of certain barangays and the local school system. In one recent case, a barangay leader in Lantapan was interested in tapping some mountain springs to convey drinking water to several household of the barangay. She requested the services of the water monitors to determine the bacterial level of the water prior to making the final decision of installing the pipes. The tests revealed that some of the springs had unsafe levels of coliform bacteria, and this type of information was obviously useful in choosing alternative water sources, saving government funds and minimising the risk of waterborne disease.

Through presentations to schools and involvement of teachers and their students in the water monitoring activities, young people are becoming more aware of environmental indicators and their meaning. Some of the elementary students of Lantapan are now being taught which of the rivers of their municipality are clean and which are polluted (Mrs. Natividad Durias, Head Teacher, Alanib Elementary School, pers. comm.). Beyond awareness of the environmental problems, some of the school students and their teachers have begun restoration activities including tree plantings on riverbanks to prevent bank erosion and downstream sedimentation.

The initially informal way of extending the information of water quality indicators to schools has become more systematised through discussions with representatives of the school district and the Philippine Department of Education, Culture and Sports (DECS). The Secretary of DECS has endorsed the overall SANREM programme and has requested that additional steps be taken to enhance outreach and environmental education in schools. Additionally, the water research findings are being used in various courses of the local university, through a faculty partner in the project.

OUTLOOK AND DIRECTIONS

Largely through the practical development and application of water quality indicators, the local government and community have acknowledged increasingly the advantages of having an ongoing, citizen water quality monitoring programme. Regular dissemination of the water information in a variety of forms and to different audiences has done much to convince policy makers and the public of the value of water assessment using simple indicators.

The NRMDP of Lantapan is still in a formative stage, and much remains to be done to have a clear policy that results in specific conservation measures. National and local elections result in changes of leadership, from President to mayor, that have profound effects on the way NRM planning is conducted. In the meantime, citizen participation in monitoring and restoration activities is increasing and will, hopefully, ensure that elected officials continue to implement their much-needed plan.

Factors for successful use of indicators in policy

An evaluation of the project suggests that two key factors combined to create a strong potential for water quality indicators to have lasting policy impacts.

1. PERCEIVED NEED AND RECEPTIVITY OF THE COMMUNITY AND LOCAL GOVERNMENT UNIT

The landscape of Lantapan shows obvious signs of degradation that have resulted in a general concern among local residents. As in most rural settings, daily life and well-being depends upon reliable sources of clean water and productive soil without the luxury of expensive inputs and treatments. Lawrence *et al.* (1996) found that farmers of the upland Philippines were 'more articulate' about environmental problems than those in the lowlands, and that this pattern also occurred in Bangladesh and India. They attributed this to the fact that farmers are most aware of issues that affect them directly, and that soil erosion and increasingly unreliable or scarce water supplies (often attributed to deforestation) are upland farmers' principal agricultural problems over recent decades. As a result, there is a strong consensus among academics, development workers and farmers regarding the problems. The Lantapan project supports this observation, and found that many in the community have an interest in environmental integrity that carries a sense of urgency and goes far beyond academic interests.

As authority in natural resource management is decentralised and 'people power' flourishes in the Philippines, municipal and provincial planning and policy is focused increasingly on a long-term, sustainable course. The status of NGOs is probably higher in the Philippines than in most Asian countries, and they often interact well with government (Lawrence *et al.*, 1996). The ability of Lantapan citizens to enter the political process as accepted stakeholders will be vital. The formation of the Tigbantay Wahig group, with the mentoring and backstopping by an established, filipino NGO partner in the project (Heifer Project International) will sustain the development and practical application of environmental indicators in the Manupali basin.

2. PARTICIPATORY RESEARCH WITH FOCUS ON INDICATORS AND POLICY

The SANREM programme in general and the water quality project in particular provided financial resources and expertise that complemented the community interests and political climate of the Philippines. A natural resource research programme that stressed inter-sectoral collaboration, community participation and a landscape scale approach fits well with the predisposition of the local residents and the new Local Government Code. An added emphasis on environmental indicators (Bellows, 1995), and on-site coordination of the programme and water project enhanced this synergy.

Development of a 'menu' of practical, low-tech water indicators (Table 9.3) gave the community options for exploring their local environment and identifying areas needing conservation and restoration. This process was facilitated in Lantapan by adapting techniques that were developed previously in other contexts of citizen monitoring (Deutsch *et al.*, 1998). The Philippines experience, in turn, led to further refinement of methods and indicators for applications in other places, including improvements to US programmes. The physical features of water make it conducive to measuring a variety of important parameters using simple tests with colour-changing chemical indicators (colorimetric methods) and inexpensive equipment. Additionally, the hands-on activities of environmental monitoring are a tremendous motivation for community participation, awareness and action.

Future needs and applications of indicators

As the process of affecting public policy using community-based water quality indicators has progressed over the last few years, three major needs for further research and applications of findings have emerged.

1. TEST AND COMPARE COMMUNITY-BASED INDICATORS WITH THOSE OF RESEARCHERS

Much more needs to be learned about the value of community-based water indicators, and their application to scientific knowledge and natural resource management. It is conceded that these indicators lack the precision of more sophisticated tests that are commonly used by researchers. Moreover, the typical sampling times, locations and frequencies of citizen monitors often miss rare but significant events affecting water quality, such as strong storms or pollution spills. In the case of the Lantapan project, TSS values collected near the stream surface may be lower than those near the stream bed, and stream discharge and *E. coli* concentration measurements probably did not capture the extremes of an annual cycle (as would be detected by continuous monitoring equipment).

The lack of precision and possible bias that may stem from community-based monitoring techniques must be weighed against the advantages of simplicity, mobility, cost-effectiveness and local relevance. An underlying question is how useful such measurements and derived indicators are for environmental managers and policy-makers (such as the local government unit of Lantapan). What are the limits and constraints of the community-based approach, and does it capture enough information to be consistently valuable for environmental assessment and policy recommendations? Specific answers to these questions are important to pursue and would require side-by-side studies using different levels of analyses. Several years of conducting such side-by-side studies through the Alabama Water Watch programme indicates that good training and careful monitoring by community groups produces data that are comparable to those of researchers and governmental agencies.

Some of the community-based indicators were similar to indicators developed by research organisations, and raise intriguing questions of comparability. For example, the Lantapan water monitoring study found that abrupt increases in TSS occurred when forest cover dropped below 30% (Figure 9.6). A threshold of 30% minimum cover before severe environmental degradation occurs was also determined for upland tropical forests (Pereira, 1989).

2. APPLY INDICATORS TO RESTORATION ACTIVITIES

After seven years of research in Lantapan, the community and local government unit is more receptive to incorporating environmental indicators into specific action plans to restore degraded areas or 'hot spots' within the landscape. The water quality indicators have the potential to not only identify these areas more quickly but also to be a useful tool for evaluating the effectiveness of restoration activities. For example, concentrations of *E. coli* in piped water may be used to identify specific areas of the public drinking water system needing repair, and to make a stronger case in municipal grant proposals for federal aid to do extensive pipe replacements. This indicator of water safety may also be used to evaluate quickly the existing mountain spring sources of public water, as was begun already by one barangay leader of Lantapan. Such strategies are in accordance with national policies which 'are likely to under-emphasise new water supply projects and focus instead on changes leading to more efficient utilisation and management of water resources' (Rola, 1997).

Environmental protection policies in Lantapan will probably also include recommendations on soil, water and biodiversity conservation measures (as outlined in Part V of this book) such as the establishment of streamside (riparian) zones, selected reforestation, ravine restoration and contour farming. A variety of simple indicators could help guide this process.

3. EXTEND DEVELOPMENT AND USE OF INDICATORS BEYOND LANTAPAN

Several initiatives are in progress to disseminate the methodology and significant findings of the research programme, including environmental indicators, beyond Lantapan. The Provincial Planning and Development Office (PPDO) of Bukidnon has facilitated a forum in which the water project and key indicators have been presented to policy-makers and planners in the 15 other municipalities of the province. The PPDO also maintains records of how municipalities use the internal revenue allotment from the federal government, and they plan to work with and encourage them to apply portions of the allotment to natural resource management (Mr Antonio Sumbalan, PPDO of Bukidnon, pers. comm.).

Additional outreach activities have included presentations regarding community-based water monitoring and indicators to scores of college and university faculty at a national seminar and workshop on environmental education and management at

Central Mindanao University. The level of response and enthusiasm toward the water quality indicators suggested that significant impacts on water management could be promoted throughout the country via university researchers.

Because of strong programme partnerships within the Philippine national government, the approach of local environmental management that has begun in Lantapan can be extended formally throughout the country. Already, the establishment of indicators of sustainability from research in Lantapan has contributed to the implementation of the Philippine Agenda 21 (Dr William Dar, former Director, Philippine Council for Agriculture, Forestry and Environmental Resource Research and Development, pers. comm.). The strategy of the SANREM programme in coming years is to continue this process of extension throughout South East Asia and in other regions of the world.

Although the impacts of the Tigbantay Wahig's work is yet to find its full potential in Lantapan, it continues to grow and has attracted considerable interest among other Municipalities in the Philippines. Study tours of local government representatives from Sarangani Province (southern Mindanao) led to the start of a similar, community-based water monitoring effort in the Municipality of Maitum in 1999. Importantly, this was done with the local government's initiative and financial resources (which included the purchase of the test kits). A similar programme, requested by the Governor of the Province of Bohol, began in 2001.

The model of community-based watershed assessment and management has also been extended via the SANREM programme to Ecuador, and was adopted by about 40 Quechua (native American) communities in the Canton of Cotacachi. Through the Lantapan project partner, Heifer Project International (HPI), the model was introduced to the upper Yangtze River catchments in China, with plans to extend into several HPI country programmes in the Mekong River Basin. The Christian Children's Fund (CCF) of Brasil financed the implementation of a similar project in Minas Gerais State, with plans to expand to other parts of the country and possibly other CCF country projects.

The strong interest in locally led, water quality/quantity monitoring among non-governmental organisations and governmental agencies has recently led to the development of an Auburn University-based network of community groups called, 'Global Water Watch.' This network will consist of groups around the world that adopt voluntarily the techniques and quality assurance protocols developed in Alabama and Lantapan. Local trainers and university personnel will provide the necessary workshops and technical support to respond to community needs, and Internet web sites of each group will be linked for accessing group information and data. Adopting verified, standardised protocols and sharing information across sites and regions has been motivational for all groups.

Lessons learned

A major strength of collaboration in participatory environmental indicator research is that development and extension of information and community action are occurring simultaneously. Instead of a traditional model of conducting the research in isolation from the local community and then trying to extend the significant findings to them through such things as technology transfer and the media, the citizens, community organisers and scientists have learned together. The start-up of this collaborative project was relatively slow compared with research that does not involve the community, but initial results indicate that the potential for lasting benefits and project sustainability are much higher than if attempted by a community, NGO, university or government agency in isolation.

In the search for the elusive 'indicators of sustainability,' it is important to factor in social as well as biophysical criteria. Particularly in participatory research, perceived relevance of a given environmental variable to the needs and aspirations of the community is essential for it to be monitored consistently by local people. This project demonstrated that drinking water quality, especially with regards to bacterial contamination and infant health, was of utmost concern. In that sense, bacteriological monitoring became one of the key 'entry points' for introducing community-based monitoring. As volunteer monitors gain skill and confidence in gathering data and understanding its implications for themselves and their families, they are more likely to extend their concern to the community and the greater catchment issues. This involves expansion of individual perceptions in both space and time, from a farm to a landscape, and from this year to the next decade and beyond. If the environmental monitoring group's motivation, camaraderie and recruitment are maintained, they may become a significant catalyst for renewed community service and positive change on a grand scale. As in the case of this project, such enthusiasm and success quickly gains the attention of others, and increased awareness, outreach and regional spread becomes a natural process.

Most scientists are aware that excellent and important research findings often go under-utilised because they do not enter the political process. Instead, the data remain in professional journals and away from meaningful action. The type of information needed by policy makers for natural resource management planning should be science-based but need not necessarily meet all the requirements of the scientific community with regard to precision and rigour. This is especially true in catchments that are degrading rapidly, with severe consequences for the local community. In these situations, application of partly understood conservation practices, with full community involvement, may be far better than waiting for a 'complete' scientific understanding.

Glover (1995) noted that rigorous research requires a clear definition of a problem and the variables to be measured, but the objectives of government policies and programmes tend to be loosely defined and sometimes contradictory. He added that, 'In the research domain, there is no single recipe for policy impact. Luck and persistence, along with good science, are vital ingredients.' The case study of Lantapan suggests that when science and persistence directed toward natural resource management comes from within the community, there is a much greater probability of policy impact.

References

Bellows, B. (ed.) (1995). *Proceedings of the Indicators of Sustainability Conference and Workshop, August 1–5, 1994*. SANREM Research Report 1-95. Athens, Georgia: University of Georgia.

Bellows, B., Buenavista, G. and Ticsay-Rusco, M. (eds.) (1995). *Participatory landscape lifescape appraisal, vol. 1 The Manupali watershed, Province of Bukidnon, the Philippines*. SANREM CRSP Philippines: The practice and the process. SANREM Research Report 2-95. Athens, Georgia: University of Georgia.

Cason, K. (ed.) (1999). *Choosing a Sustainable Future*. SANREM CRSP 1999 Annual Report. Sustainable Agriculture and Nature Resource Management Collaborative Research Support Programme. Athens, Georgia: University of Georgia.

Coxhead, I. and Buenavista, G. (eds.) (2001). *Seeking Sustainability: Challenges of Agricultural Development and Environmental Management in a Philippine Watershed*. Philippine Council for Agriculture, Forestry and Natural Resources Research and Development, Department of Science and Technology: Los Baños, Laguna.

Deutsch, W., Busby, A., Winter, W., Mullen, M. and Hurley, P. (1998). *Alabama Water Watch, The First Five Years*. Research and Development Series 42, International Center for Aquaculture and Aquatic Environments. Auburn, Alabama: Auburn University.

Deutsch, W. G. and Busby A. L. (1999). *Quality Assurance Plan for Bacteriological Monitoring for Alabama Water Watch*. Auburn, Alabama: Auburn University.

Deutsch, W. G. and Orprecio, J. L. (2000). Formation, Potential and Challenges of a Citizen Volunteer Water Quality Monitoring Group in Mindanao, Philippines. In *Cultivating Community Capital for Sustainable Natural Resource Management. Experiences from the SANREM CRSP*. ed. K. Cason, 62 pp. Sustainable Agriculture and Natural Resource Management Collaborative Research Support Programme.

Deutsch, W. G., Orprecio, J. L. and Bago-Labis, J. P. (2001). Community-based water quality monitoring: the Tigbantay Wahig Experience. In *Seeking Sustainability: Challenges of Agricultural Development and Environmental Management in a Philippine Watershed*. ed. I. Coxhead and G. Buenavista. Philippine Council for Agriculture, Forestry and Natural Resources Research and Development, Department of Science and Technology: Los Baños, Laguna.

Deutsch, W. G., Orprecio, J. L., Busby, A. L., Bago-Labis, J. P., and Cequiña, E. Y. (2001). Community-based Water Quality Monitoring: From Data Collection to Sustainable Management of Water Resources. In *Seeking Sustainability: Challenges of Agricultural Development and Environmental Management in a Philippine Watershed*. ed. I. Coxhead and G. Buenavista. Philippine Council for Agriculture, Forestry and Natural Resources Research and Development, Department of Science and Technology: Los Baños, Laguna.

Glover, D. (1995). *Policy researchers and policy makers: never the twain shall meet? Where research meets policy*. 4–6. IDRC Reports.

Folgia, K. (1995). *Choosing a Sustainable Future*. SANREM CRSP Annual Report. Sustainable Agriculture and Nature Resource Management Collaborative Research Support Programme. Athens, Georgia: University of Georgia.

Jutkowitz, J., Stout, R. and Lippman, H. (1997). *Democratic local governance in the Philippines*. Impact Evaluation. PN-ABY-235, (1). United States Agency for International Development.

Lawrence, A., Garforth, C., Dagoy, S., Go, A., Go, S., Hossain, A., Kashem, M., Krishna, K., Naika, V. and Vasanthakumar, J. (1996). Agricultural extension, the environment and sustainability: research in Bangladesh, India and the Philippines. *Agricultural Research and Extension Newsletter*, 33, 15–22.

Li, B. (1994). The impact assessment of land use change in the watershed area using remote sensing and GIS: A case study of Manupali Watershed, the Philippines. A thesis proposal submitted for Master of Engineering, School of Environment, Resources and Development, Asian Institute of Technology, Bangkok, Thailand. 119 pages.

NRMDP (1998). *Natural Resource Management and Development Plan 1998–2002 (Five Year Indicative Plan)*. Province of Bukidnon, Philippines: Municipality of Lantapan.

Paunlagui, M. M. and Suminguit, V. (2001). Demographic development of Lantapan. In *Seeking Sustainability: Challenges of Agricultural Development and Environmental Management in a Philippine Watershed*. ed. I. Coxhead and G. Buenavista. Philippine Council for Agriculture, Forestry and Natural Resources Research and Development, Department of Science and Technology: Los Baños, Laguna.

Pereira, H. C. (1989). *Policy and Practice in the Management of Tropical Watersheds*. Westview Press, Boulder, Colorado.

Policy Forum (1997). In *Center for Policy and Development Studies*, vol. 12, no. 4, p. 7. Los Baños, Philippines: University of the Philippines Los Baños.

Poudel, D. D. and West, L. T. (1999). Soil development and fertility characteristics of a volcanic slope in Mindanao, the Philippines. *Soil Science Society of America Journal*, 63:1258–1273.

Rola, A. (1997). Water and food security. In *Center for Policy and Development Studies, vol 12, no. 4, 5–7*. Los Baños, Philippines: University of the Philippines Los Baños.

Part II
Hydrological processes in undisturbed forests

SUMMARY

Callaghan and Bonell *introduce the main features of the tropical atmospheric circulation as a step towards linking different synoptic-scale, rain-producing phenomena, their associated rainfall characteristics and the subsequent impacts on runoff hydrology*.

Within the monsoon regions, three systems of convergence are identified, the northern monsoon shearline, the southern monsoon shearline and the maximum cloud zone associated with the monsoon westerlies in the vicinity of the equator. The most active monsoon shearline (otherwise known as the monsoon trough) is identified with the summer hemisphere. It is along this system that tropical cyclones often develop in response to convergent, opposing equatorial westerlies and trade wind easterlies, coupled with sea surface temperatures in excess of 26 °C. Low latitude tropical cyclones can, however, occur more rarely within 5° of the equator, despite the common belief that the Earth's deflection (Coriolis) force is too weak in this zone for these storms to form. An alternative explanation is the short duration of the year when the monsoon trough is resident near the equator so that the chances for tropical cyclones to form are much reduced. It is noted that the zone of deepest convection and persistent cloud is associated with the maximum cloud zone of the equatorial westerlies due to the convergence of inter-hemispheric airstreams. Activity waxes and wanes in this sector in response to the varying strengths of the inter-hemispheric trade wind systems and the eastward propagation of a Kelvin wave known as the Madden–Julian Oscillation.

Outside of the monsoon regions, rain-producing activity is associated with a linear feature, known as the zonal trough in the easterlies, which separates the more tangential convergence of the northeast and southeast trade winds. The prevailing cooler sea surface temperatures, the absence of convergence of 'opposing' winds and the near-equatorial location of this trough, all militate against tropical cyclone development. Convection within cloud clusters is the main rainfall source.

Callaghan and Bonell emphasise the lack of a monsoon circulation, i.e. cross-equatorial air flow, within the Amazon Basin. The tapering southwards of the South American land mass, the unusual topographic setting of the Amazon Basin combined with a distorting heating effect, all set up a synoptic climatology quite distinct from the rest of the humid tropics. A common synoptic-scale cause of summer rainfall within the Amazon is the northward penetration of former cold fronts of southern hemisphere origin as part of the South Atlantic Convergence Zone (and more occasionally from the northern hemisphere). The South Atlantic Convergence Zone is part of a family of tropical–extratropical cloud bands that occur at selected points around the tropics in both hemispheres. They act as 'conduits' for the transfer of surplus energy out of the tropics into the higher latitudes. Moreover, there are preferred seasons when these cloud bands are more active. Such activity however is not seasonally consistent within and between inter-hemispheric regions.

Finally, Callaghan and Bonell provide an overview of climatological features connected with inter-annual, decadal and longer term (>70 years) variability which account for rainfall being non-stationary across the humid tropics. Particular focus is given to the El-Niño–Southern Oscillation (ENSO) phenomenon, West African interdecadal variability and monsoon variability. Special attention is given to the occurrence of anomalous equatorial westerly gales which propagate eastwards in the equatorial western Pacific. Such circumstances 'push' warm water eastwards which could be linked with the onset of El Niño events.

Bonell, Callaghan and Connor provide meteorological details and rainfall characteristics of various rain-producing systems at both the synoptic (>2000 km length scale) and mesoscale (2–2000 km length scale). During the survey of tropical cyclones and synoptic-scale easterly perturbations, the role of upper winds in providing a diffluent (outflow) environment for the intensification and steering of these perturbations is strongly emphasised. In comparison to the temperate latitudes of the northern hemisphere, there is a minimal network of upper air monitoring stations in the humid tropics which poses a difficulty in forecasting and gaining a better understanding of such cyclonic and easterly perturbation dynamics.

Tropical cyclones are by far the most frequent cause of the flood-producing rainfall events. Spectacular daily as well as short-term rainfall totals are recorded especially when the forward movement of these vortices is slow or near stationary. Examples are given that show the important interactions between the upper winds, the steering mechanisms and orography linked with the areas of highest rainfall. The much-published Hurricane Mitch belongs to this category of slow moving high rainfall producing events which was further aggravated by orographic uplift, thus resulting in the devastating floods and land slides. Overall, the inner eye wall of tropical cyclones contributes at least 25% of total rainfall from the whole system. It can however attain more than 30% depending on the mean strength of this inner core. Thus there is temporal as well as spatial variability of rainfall across these tropical vortices which has ramifications on the runoff generation process and preferred areas for flooding.

Because of their destructive properties, considerable focus of research attention has been on tropical cyclones. More common, however, are rainfalls originating from disturbances in the surface tropical easterlies and monsoon westerlies at both the synoptic and mesoscale. Whilst storm event totals are much lower than in tropical cyclones, high equivalent hourly intensities can be recorded over shorter time increments. This account highlights the complexity in origin and structure of these disturbances, especially easterly perturbations.

Of particular relevance to process hydrology is the appreciation of the temporal and spatial variability of the convective and stratiform components of mesoscale cloud systems (MCSs). There is at least one order of magnitude difference in the short-term rain intensities and rain totals in favour of the convective portion of MCSs. However, it is significant that the stratiform area of coverage within rainfields expands progressively at the expense of decaying convective cells. Thus the highest short-term rainfall intensities usually only occupy the smallest proportion of grid cells, except within the inner core of tropical cyclones.

As part of the global regionalisation of rainfall characteristics, an assessment of the differences in rainfall frequency–magnitude–duration and classification of rainfall characteristics at the global scale shows that the simple distinction between cyclone-prone vis-à-vis non-cyclone-prone (convective) regions is too simplistic. Tropical cyclone-prone areas in general produce the highest daily rain totals in comparison to the near-equatorial, convective regions although there are exceptions.

The need to take a synoptic climatology perspective when regionalising rainfall even at the mesoscale is emphasised. Work in northeast Australia has shown how changes in the spatial and temporal occurrence of preferred rainfall areas takes place in response to a corresponding change in the surface and upper wind patterns and their interactions with topography for different synoptic meteorological systems.

Thus process hydrology needs to take a more dynamic perspective to include the characteristics of rain fields such as the movement and changing life cycles of MCSs (convective, stratiform rain). Further, to take into account the impact of different wind circulations at the mesoscale and their interaction with topography in producing different preferred spatial areas of rainfall.

Mahé et al. provide a paleoecology and paleoclimatic history of tropical forests from the lower Cretacious to the Holocene eras. Contrary to the common perception that dense tropical forests were the most stable ecosystems over geological time, recent advances in methodologies for reconstructing past climates connected with plate tectonics – palynology, anthracology and paleoclimatology – have all shown that the spatial and temporal distribution of these forests have undergone profound changes. Paleoclimatic changes from 70 000 years BP to the Holocene in particular are relatively well documented because of access to more paleo-data. During the Last Glacial Maximum (20 000–15 000 years BP), a severe reduction in the global area of dense tropical forests occurred in response to global cooling, increasing aridity and a much weaker monsoon circulation. Average temperature across the tropics reduced by about 4 °C. At this time the increase in desiccation caused an extension of the savannah at the expense of forest.

With the recovery of global temperatures and enhanced rainfall, the beginning of the Holocene around 10 000 years BP coincided with the last phase of the maximum expansion of rainforests. Even so, climatic fluctuations during the mid to late Holocene and the drying of the climate facilitated expansion of savannahs in the eastern part of the Amazon forest and parts of central Africa. The causal influences of sea surface temperature fluctuations rather than the actions of humans in accounting for the corresponding spatial changes in dense forests compared with open savannah is emphasised.

Mahé et al. also introduce the current debate on whether the ongoing dramatic changes in land use are having a major influence on climate from the regional to continental scale. With horizontal temperature gradients being weak in the tropics, the atmosphere is very sensitive within the vertical plane to changes within the radiation energy budget over the land and ocean. The sensitive parameters are albedo, temperature, humidity and vegetation type linked with the vertical exchange of sensible heat and water vapour.

The biogeophysical feedbacks, which include the surface–albedo feedback and various hydrological feedbacks, directly affect near-surface energy, moisture and momentum fluxes as a result of changes in land use and associated changes in albedo, roughness and leaf areas. The surface–albedo feedback is particularly sensitive to conversion of forests to selected (but not all) land use types. The albedo of dense forests is much lower than that of bare ground so there is more latent energy available for evaporation and its recycling as rainfall in proximity to forests. Towards the margins of dense forests, the desert–albedo feedback is particularly relevant in sub-Saharan Africa. As the surface albedo increases, more solar radiation is reflected back into space which thus reduces the net radiation available for evaporation and transpiration. Moreover, the enhanced longwave (back) radiation from less vegetated areas encourages radiative cooling of the overlying upper atmosphere. This radiative cooling is compensated for by enhanced subsidience through adiabatic sinking and heating of the air (known as the Charney mechanism). Hydrological feedbacks associated with a reduction in transpiration from forest clearance is also inferred because forests have deeper roots and larger areas. This reduction in transpiration could reduce local rainfall and therefore militate against forest regrowth.

The difficulty of separating the impacts of human-induced land use change from natural climatic fluctuations on regional rainfall and runoff is strongly emphasised by Mahé *et al.* Various phenomena connected with the ocean–atmosphere linkage are highlighted as being influential on rainfall variability over tropical north Africa. Interestingly, since the 1970s, ENSO events have had a stronger impact in this region, i.e. weak rains in the Sahel region after an ENSO warm phase and heavy rains after an ENSO cold phase. Sea surface temperature anomalies associated with the tropical Atlantic ocean also encourage an approximate ten-year cycle of rainfall variability, especially in west Africa.

Overall, at least 50% of observed rainfall variability in west Africa can be directly attributed to sea surface temperature variations. There still remains considerable uncertainty on what proportion of rainfall variability is influenced by changes in terrestrial vegetation (surface conditions)–climate interactions resulting from anthropogenic activities. Runoff variability was noted to be greater than rainfall variability in the recent sustained drought of the 1970s which is attributed to decreasing groundwater storage. Whilst annual monthly and low flows have decreased, on an event basis, runoff coefficients and peak discharges can be higher (especially in the Sahel–Sudan areas). Such increases in event runoff are in response to changes in land cover from forest to agriculture associated with reduction in surface infiltration capacities and higher soil moisture content during the wet season.

A key component of the biogeophysical feedbacks linked with terrestrial–climate interactions is evaporation. Roberts et al. *elaborate on the various micrometeorological, physiological and hydropedological controls that influence evaporation processes of tropical forests under wet canopy (interception) and dry canopy (transpiration) conditions.* Most experimental catchment studies do not provide robust values of evaporation for several reasons. There are difficulties in defining catchment-wide rainfall inputs (as highlighted by Bonell *et al.*), the possible lack of coincidence between topographic and groundwater divides causing interbasin transfers of groundwater and the difficulty of establishing topographical catchment boundaries in areas of low relief. Consequently, the evaporation term of the water balance equation is the repository for cumulative errors when calculated in the form $E =$ precipitation $-$ runoff $\pm$ changes in soil moisture storage $\pm$ changes in groundwater storage.

The alternative methods for evaporation measurement are the use of eddy correlation flux measurements (micro-meteorological) from flux towers, physiological studies (sap flow measurements) and interception–throughfall measurements at both the stand scale and within canopy plot scale. Thus there has had to be a reduction in the scale of measurements to gain a better understanding of evaporation processes, although eddy correlation measurements are likely to integrate sensible heat and water vapour contribution from a larger area. The measurement of evaporation (interception–throughfall) during storm events is technically more problematic as outlined by Roberts *et al.*, who appropriately provide a more detailed evaluation of the weaknesses of different experimental methodologies.

Accepting errors associated with the sampling of above-canopy precipitation (gross precipitation) and below-canopy throughfall (net precipitation), Roberts *et al.* note that interception (wet canopy evaporation) losses are comparatively small in lowland tropical forests compared with temperate forests. This is due to the nature of convective storms in the tropics which are generally of high intensity but short in duration, especially in equatorial, continental areas such as the Amazon Basin. Typically, within equatorial continental or continental edge sites in the humid tropics, between 12% and 18% of rainfall is intercepted and evaporated directly, compared with around 30% for humid temperate forests. In contrast, maritime sites show higher interception in the range of 18% to 52% of gross precipitation. These higher evaporation losses are attributed to additional fluxes of horizontal advected energy originating upwind from an adjacent warm sea. Nonetheless, the failure of existing evaporation models to reproduce these measured high interception losses highlights a major lack of understanding of detailed processes connected with wet canopy losses in these maritime, tropical forests.

Having considered 'inputs' and above-surface 'losses', the next step in the water cycle as it functions in the humid tropics is to look at runoff. Bonell's chapter on storm runoff generation processes in tropical forests is a thorough and in-depth study of this massive topic. Compared with the humid temperate forests, there is a much greater diversity to be found in hillslope and headwater basin response patterns. Overland flow (mostly of the 'saturation excess', SOF type) is also much more frequent within rainforest-covered Acrisol landscapes where there is a marked decline in permeability with depth. The higher prevailing rain intensities of the humid tropics more easily cause subsurface stormflow and associated perched water tables to emerge at the soil surface which subsequently leads to saturation (saturation-excess) overland flow. This mechanism contributes to the more highly responsive stream hydrographs and downstream flooding during major storm events, especially within tropical-cyclone prone regions. There remain insufficient data to generalise storm runoff responses from other soilscapes such as Ferrasols. Some Ferrasols, such as those within the central Amazon Basin, favour predominantly vertical flowpaths during storms in consequence of their much higher soil permeabilities and which are maintained to greater depths. Other Ferrasols studied in South East Asia favour subsurface stormflow (SSF) at various depths within the soil profile. Acrisols and Ferrasols combined cover only about 60% of the humid tropics which means that other end-members of pedo-hydrological functioning still need to be identified. Examples from West Africa identify infiltration-excess overland flow (Hortonian overland flow, HOF) as a dominant pathway in soils with shallow hardened layers (e.g. Plinthosols). This pathway type occurs more extensively in tropical semi-arid landscapes where the soil surface is less protected by vegetation. It also occurs in the humid tropics following forest conversion where surface soils are compacted or not protected by a closed vegetation cover.

The coupling of hydrometric-hydrochemistry (environmental tracers) methodologies have detected significant contributions to the storm hydrograph from pre-existing, 'old' water, as noted elsewhere in humid temperate studies. A precise understanding of the mechanisms responsible for such 'old' water contributions remain uncertain, and in this context, it is a common research challenge not unique to the humid tropics. Current evidence also suggests however, that subsurface transfer of water from hillslopes into organised drainage occurs at a much faster rate than indicated from point measurements of permeability. In line with Chappell *et al.* (Chapter 31, later in this volume), Bonell emphasises that hydraulic conductivity is a very sensitive parameter in process-hydrology modelling.

Hydrometric-hydrochemistry studies highlight the much stronger connectivity between permanent groundwater and hillslope hydrology than was appreciated previously. Through preferential flow in macropores, recharge to deeper groundwater bodies can take place during major storm events. The resulting steepening of water table gradients enables significant contributions of groundwater to the storm hydrograph from this source. Thus future hillsope hydrology studies need to be integrated with a detailed hydrogeological component.

On the other hand, there is evidence from environmental tracer studies for much larger contributions of 'new' (rainwater) water to storm hydrographs (compared to humid temperate areas), especially during the flood-producing rains of tropical-cyclone-prone areas. The implication here is that forests are not 'sponges' with an infinite capacity to absorb rainwater as commonly believed. Under extreme rainfalls, floods can occur from tropical forests as well. Of course, any change in the surficial, soil hydraulic properties following forest clearance can, for certain soils, cause a dramatic shift in the dominant storm flowpaths, especially where vertical percolation is reduced. More detailed hillslope hydrology studies are required to address this issue.

The need for greater coupling of hillslope hydrology with surface water–groundwater processes in the riparian and hyporheic zone is highlighted. Traditional hillslope hydrology experimental studies have given little attention to these zones. Apart from their coupling role of water transfer from hillslope to organised drainage, the same areas have important biogeochemical functions as part of nutrient cycling (see Proctor, Chapter 16) which need to be linked to the hillslope hydrology processes. Also, because of field logistics, there is a shortage of hillslope hydrology data during flood-producing (extreme) events. The long-term goal should be to couple hillslope hydrology with radar imagery of transient rain cells to detect temporal changes in the dominant runoff pathways with corresponding within-storm temporal variations of rainfall intensities. Such linking would have much more practical value when linked with flood forecasting in addition to the further development of land/water management guidelines.

Runoff generation leads naturally to a consideration of erosion and sediment yield. Douglas and Guyot *assess these factors in two stages: the fundamental continental and major river basin scale and the local catchment scale identified with experimental catchments.* The diversity of tectonics, lithology, climate, relief and vegetation in producing wide variations in sediment yield is emphasised. Although their account focuses on solid-debris and dissolved load transport from tropical forests at the small scale, other land covers are included in this overview at the larger river basin scales.

Douglas and Guyot identify the weathering environment as transport-limited and weathering-limited erosion regimes. In tectonically active areas such as the headwaters of basins in Papua New Guinea, the younger rocks weather and break up more rapidly. Transport processes such as landslides and soil creep work faster than weathering processes on the steep slopes so that sediment yields within rivers are weathering-limited. In contrast, the low relief of older landscapes such as the Guyana shield of the Amazon Basin encourages thick weathering profiles where transport-limited erosion occurs. Thus the highest sediment yields in the humid tropics originate from tectonically active areas such as the mountains of New Guinea (and Taiwan just outside the humid tropics) which experience mean annual sediment yields in the order of 10^4 t km^{-2} yr^{-1}. By contrast, the lowest sediment yields (in the order of 10 to 10^2 t km^{-2} yr^{-1}) occur over old landscapes or sedimentary basins of low relief such as the Congo Basin in Africa, and the central and lower Amazon and Orinoco Basins of South America. Consequently river basins which do not have headwaters with considerable relief, such as the Congo and

Negro, and corresponding large lowland areas, produce the lowest sediment yields.

Prolific rainfall associated with tropical cyclones in conjunction with high relief also lead to high sediment yields arising from landslides. During hurricane events, SSF (and presumably SOF) produce debris flows, debris slides and slumps up to several hundred metres long that contribute to the high event sediment yields. Thus landscapes do not have to be disturbed for landslides to occur. In extreme situations, where a combination of high relief, high tectonic uplift rates and intense rainfall occurs, these can lead to maximum daily sediment discharges which exceed the average annual load of rivers. The cyclone-prone drainage basins of Taiwan provide such examples.

An ongoing research programme in Sabah (Danum Valley), Malaysia, illustrates various erosion processes which take place at the small scale. These include localised erosion from stemflow at the base of tree boles, the role of termites and other soil fauna as sediment sources, and the sensitivity of streamhead hollows to water table fluctuations and runoff production, and therefore being a major source of sediment. Linkages with the dominant pathways of hillslope runoff are made but in general this review highlights the poor coupling between hillslope hydrology and erosion process studies. Sediment and dissolved local transport by SSF in macropores and pipes has seldom been observed, let alone measured yet this process is probably a major source of supply of both solid debris and solutes.

A pertinent conclusion from this review of small-scale studies is that sediment originates from several discrete localities within catchments (soil fauna, streamhead hollows, mass movements, subsurface pipes, woody debris dam collapse); rather than from general surface erosion. Such discrete sediment sources make it irrelevant to use slope maps derived from topographic maps for the prediction of soil loss and sediment yields using the soil loss equation approach. Moreover, sediment transfer in tropical forests is event-driven by major storms occurring only a few times a year. In between, there are long periods of inactivity when sediment remains in various forms of temporary storage until the next large storm event.

Bedload measurements in undisturbed tropical streams are rare and are a research need. Moreover, depending on the lithology, a considerable proportion of bedload material can be lifted and transported as suspended load (notably coarse sand-sized or finer material). Thus estimates of bedload contributions to total sediment load are presently crude and range from 20% to 70%.

Douglas and Guyot also provide a summary of dissolved loads in selected tropical rivers. As they note, while climate, lithology and tectonics control the dissolved loads of tropical rivers, the persistence of high rainfall and temperatures encourages high weathering rates of rock materials. These circumstances lead to potentially high dissolved loads in comparison with the low suspended load discharge from tropical forests. As with sediment loads, lithology is a major control over dissolved load export from small tropical forested basins. Granite tends to favour greater dissolved load transport than suspended sediment whereas the reverse applies to metamorphic and sedimentary rocks.

Proctor provides an overview of the rainforest nutrient cycle which has important practical applications in connection with reforestation of degraded lands and the rehabilitation of disturbed forests in general. Despite the need for a comprehensive baseline understanding, beyond the influx of rainwater, no study has succeeded in accurately quantifying the pools and fluxes within the remainder of the rainforest nutrient cycle. There are several reasons. Many rainforest formations are caused by soils which are distinct chemically and hydrologically; this requires a sound understanding of variability of fertility within as well as between forest formations. The most widespread formation (lowland evergreen rainforest) encompasses a wide range of soils which are poorly defined from a nutrient status perspective. At the plot or sub-hillslope scale, there are practical difficulties associated with sampling of the deeper nutrient pools of soil. The lower boundary conditions of nutrient extraction by trees is fundamentally linked with the need for a good understanding of the spatial organisation of root networks in the vertical. As with evaporation processes, the dearth of such knowledge is a barrier to closing the nutrient cycle.

An in-depth appraisal of the use of various soil analyses linked with the nutrient supply to trees is provided. This includes the proliferation of nutrient-addition experiments to clarify limiting nutrient factors for tree growth. No generalisations can be made. Selected rainforest species respond to nutrient additions at different stages of growth. Specific mention is made of soil acidity and the relative roles of H^+ and Al^{3+} ions connected with nutrient supply. The role of roots and mycorrhizal symbiosis in nutrient uptake is highlighted as one of the prime research needs.

One of the most original contributions to this book is the description of a peat swamp study in Sarawak by Hooijer. Little is known about the hydrology of wetland forests because their environmental circumstances make access and accurate monitoring extremely difficult. Paradoxically, the most difficult measurements to make are surface discharge and the delineating of catchment area which in traditional non-wetland basins are the most easy to quantify. By contrast, the persistent high water tables and limited depth of the unsaturated zone make it possible to determine forest transpiration and surface evaporation more confidently, as well as groundwater seepage and the storage coefficient of the soil from a single diurnal water table record.

With surface gradients usually below 0.5 m km^{-1}, it is difficult to delineate catchment boundaries, even more so during storm events when such boundaries can shift. A dense network of elevation measurements using lasers, coupled with groundwater levels taken from transects across the study area, were used to confirm that the shape of the groundwater body coincided with the peat surface. The probable maximum and minimum catchment areas were then estimated. Even so, the diffuse pattern of surface runoff through the microtopography of hollows and hillocks makes it impractical to gain acceptable runoff estimates at high flows. In this context, there are some parallels with erosion process modelling as the same characteristic of anastomising flow patterns creates difficulties (see Yu, Chapter 33).

For stream discharges between high and low flows, Hooijer used 'coherent acoustic Doppler flow profiling sensors' along a cross-section through the main discharge channel. Two sets of discharge measurements were proposed, of which one is upstream of tidal influences. The flows measured at the two stations were then combined after filtering out the tidal influences at the lower station.

Changes in groundwater storage were estimated from a single water level record in the centre of the catchment, after previously showing that the latter was comparable to the average of the water level fluctuations from the wells along a transect. Subsurface flow (<0.1m depth) and groundwater flow (>0.1m depth) were estimated from water table drawdown data linked with saturated hydraulic conductivities, K_{sat} and storage coefficient estimates. Total evaporation (transpiration plus evaporation) from this tropical peatswamp forest was found to be not too dissimilar from other lowland tropical forest types, i.e. 1550 mm yr^{-1}. Groundwater seepage from water table drawdown is estimated in the early morning hours (when E_t is negligible). E_t can then be estimated from the daily, diurnal water table drawdown after discounting for groundwater seepage and if the storage coefficient is known. Thus surface flow (~950 mm yr^{-1}) is calculated as the difference between rainfall and total evaporation, groundwater flow and subsurface flow combined.

These estimates were then cross-checked with the use of a simple linear reservoir model (see Barnes and Bonell, Chapter 29) based on depression storage (surface), subsurface storage (upper 0.1 m depth) and groundwater storage (deeper than 0.1 m depth). The different storage coefficients and permeabilities are incorporated for the latter two storages. Net rainfall and total evaporation are respectively the inputs and outputs to the model. Outflow rates from the three storages were the previous calibration from one year of discharge and water level data. The catchment area from the model was found to resemble closely the areas from topographic estimates by fitting long-term discharges to observed totals.

On the basis of this modelling, only about 3% of rainfall is discharged from open water storage: this immediately suggests that 3% of the catchment is permanently inundated. It takes a few days for shallow subsurface flow (<0.1 m depth) to drain the upper peat layer. The model confirms that the deeper groundwater (>0.1 m depth) flows are only a small component of overall discharge.

These findings have some management implications. The slow drainage of the upper layer subsurface flow supported by the ancillary small groundwater contributions confirm to some extent that wetlands have the ability to sustain river baseflow (delayed flow). Once the water table declines into the lower, less permeable layer, Hooijer's modelling suggests the 'low flow maintenance' function is equivalent to only 0.2 mm day^{-1} during the dry season which is not as quantitatively significant as commonly thought. Under extreme drought, peatswamp outflow may cease altogether. Thus whilst the hydrological service (see Aylward, Chapter 7) performed by sustainable swamp management may not be as significant as expected, the alternatives such as subsidence, salinisation, acidification and fires from swamp conversion are economically unfeasible.

The other finding of Hooijer is that the 'sponge' function of these swamp forests has a limited upper capacity of storage due to the persistent high water tables. Once this capacity is exceeded, these peat swamps 'efficiently' discharge excess water towards organised drainage so that they do not 'drown'. This excess water may still enter the main channel slowly due to obstructions created by backwatering effects and storage in the (depressional) floodplains fringing the peat swamp forest domes. These effects create an apparent 'delayed peat swamp response' function during flooding, i.e. attenuation of flood peaks.

The global tropical forest assessment of Drigo has already indicated that upland forests have been especially vulnerable to conversion during the last two to three decades. *Amongst tropical montane forests, so-called tropical montane cloud forests (TMCF) are set apart because they are exposed to (various degrees of) fog and low cloud. In the final chapter of this Part, which marks a transition to the next Part on forest disturbance,* Bruijnzeel *assesses their hydrological functioning and the consequences of their conversion to pasture and agricultural cropping.* Depending on elevation, degree of exposure and thus cloud incidence, three groups of TMCFs are distinguished: lower montane cloud forest, upper montane cloud forest and sub-alpine cloud forest. Stunted, low-elevation 'elfin' cloud forest constitutes another important, azonal type of cloud forest.

A key function of TMCFs is their ability to capture cloud water droplets, otherwise known as 'cloud stripping', 'occult precipitation', 'horizontal interception' or 'cloud water interception', CW. These CW amounts add to the normal precipitation inputs and so are a gain to the water balance. The annual accumulation of CW is dependent on several functions such as frequency of cloud incidence (enveloping the TMCF), exposure (windward or leeward) and structural as well as floristic features of the forest to capture cloud droplets.

Bruijnzeel's account highlights the difficulties in securing reliable data from these forests. In particular, there are technical problems of gaining good estimates of CW from the present generation of 'fog gauges', because they do not mimic the complexities of a live forest canopy. Moreover, there remains considerable debate on the most suitable type of fog gauge (e.g. wire mesh cylinder, wire harp type, louvered screen, poly propylene), and the means to distinguish between CW and wind-driven rain. An alternative is to measure throughfall plus stemflow (net rainfall) beneath the forest canopy and at the same time gross rainfall above the canopy or in an adjacent clearing. In the absence of cloud, rainfall interception (re-evaporation from the forest canopy) shows a net loss to the forest.

Accepting the above technical difficulties, CW additions to the input side of the water balance can commonly range up to 20% of mean annual precipitation across the three types of TMCF. Extreme CW additions in excess of 150% of mean annual precipitation have been recorded. Despite the recognised uncertainties of these results, there is an emerging consensus that CW interception reaches its peak during the dry season which underlines the importance of TMCF in sustaining dry season stream flows.

Due to the protection of remaining TMCFs, there have been no controlled experiments using paired catchments whereby direct comparisons of the water balance can be made before one of the catchments forest cover is converted to pasture or agriculture. Under uncontrolled conditions, there have been several reports of diminished dry season flows in areas that have experienced a considerable reduction in montane forest cover. It remains unclear, however, whether such reductions are due to the loss of cloud stripping (CW), diminished rainfall (as part of inherent climate variability), reduced infiltration and water retention capacities or even increased diversions of stream flow for irrigation. Another consideration concerns recent evidence for more elevated cloud base levels during the dry season in Central America and the Caribbean which reduces the potential for cloud stripping. Apart from climatic variability (and possible change), extensive forest conversion in the lowlands and foothills adjacent to TMCF areas implies a reduction in 'local' water vapour recycling from evapotranspiration and a converse increase in surface temperature (see Mahé *et al.*). The impact of this 'local' change in the terrestrial–atmospheric exchange of energy and water vapour could also contribute to the noted reductions in dry season stream flow.

In conclusion, Bruijnzeel calls for more systematic observations of CW interception and net precipitation along elevational gradients according to an internationally accepted (standard) measuring protocol. Such transects would contribute towards addressing the differing impacts of forest clearance on dry season stream flow between lower montane rainforest, lower montane cloud forest and upper montane cloud forest. Further, whilst controlled catchment studies cannot be undertaken, combined hydrological process work which includes hillslope and groundwater hydrology (as well as rainfall cloud interception, and water uptake and stream flow) at specific sites should address the question of conversion of TMCF to vegetable cropping or pasture on potential reductions in dry season flows. So far, no hillslope hydrology and associated detailed infiltration measurements have been undertaken in TMCF studies, i.e. there is a total lack of hydrological process work associated with stream flow dynamics and land use change.

10 An overview of the meteorology and climatology of the humid tropics

J. Callaghan
Bureau of Meteorology, Brisbane, Australia
M. Bonell
UNESCO, Paris, France

INTRODUCTION

Because of the positive net radiation received in the tropics, this energy is the driver of the hydrological cycle, as is reflected in the frequency of some of the highest rainfall intensities (by global standards) found across the duration spectrum. There remains, however, considerable spatial and temporal variability in rainfall across the humid tropics. Such variability is partly a consequence of the different synoptic-scale, rain-producing meteorological phenomena which occur in this climatic region. Moreover, the link between synoptic climatology/rainfall characteristics/storm runoff hydrology, for example, is insufficiently represented within the hydrological literature, especially that pertaining to tropical forest hydrology.

Consequently, it will be necessary to go into some detail both within this chapter as we introduce the main features of the tropical atmosphere circulation and also in the subsequent one, where the focus is on particular synoptic- and meso-scale rain-producing systems, in an attempt to highlight the important linkage between synoptic climatology and rainfall characteristics. Later, varying responses in the storm runoff hydrology of tropical forests will be cross-referenced with material presented here.

Within the meteorological and climatological literature, there is no consensus on the terminology used to describe the various meteorological systems affecting the tropics. Commonly, many such rain-producing systems are 'lumped' under the phenomenon, the intertropical convergence zone (ITCZ). As this chapter (and the one following) will outline, the term ITCZ incorporates several phenomena between the synoptic (say 10° latitude by 40° longitude, Davidson *et al.*, 1983) and mesoscale (length scale, 2–2000 km; Orlanski, 1975). This chapter therefore provides a synoptic climatology overview of the general atmospheric circulation of the tropics and defines various meteorological systems that operate in both the lower and upper atmosphere. Subsequently, various aspects of the Walker and monsoon circulation, tropical-extratropical cloud bands and the penetration of 'cold surges' into the humid tropics will be presented. Attention will then be given to the intra-annual variability (e.g. 30–60 day Madden-Julian Oscillation), interannual variations (e.g. the El Niño-Southern Oscillation, ENSO, and the role of specific synoptic scale weather systems in triggering interannual variations and interdecadal longer term variability.

A glossary of scientific terms, which are possibly unfamiliar to some readers, is included at the end of this chapter.

OVERVIEW OF THE ATMOSPHERIC CIRCULATION IN THE TROPICS

Low-level circulation

A detailed overview of the atmospheric circulation in the tropics is given in Manton and Bonell (1993) including the Hadley and Walker circulations and more specifically, a description of the various synoptic-scale meteorological systems that have a bearing on rainfall generation. The background large-scale flow for the tropics is what is known as the Hadley circulation, which drives the low-level, easterly trade winds and in which heat is transported from the tropics to higher latitudes by the meridional return flow associated with the upper westerlies. Within this broad scale flow are several low-level and upper atmospheric troughs and ridges that need more rigorous definition than is commonly found in climatological texts (Sadler and Harris, 1970; Manton and Bonell, 1993). Such remarks apply particularly to the *surface* axis of the Hadley circulation, a thermal low pressure trough between the respective subtropical high pressure belts in the northern and southern hemispheres (Figure 10.1).

Forests, Water and People in the Humid Tropics, ed. M. Bonell and L. A. Bruijnzeel. Published by Cambridge University Press.

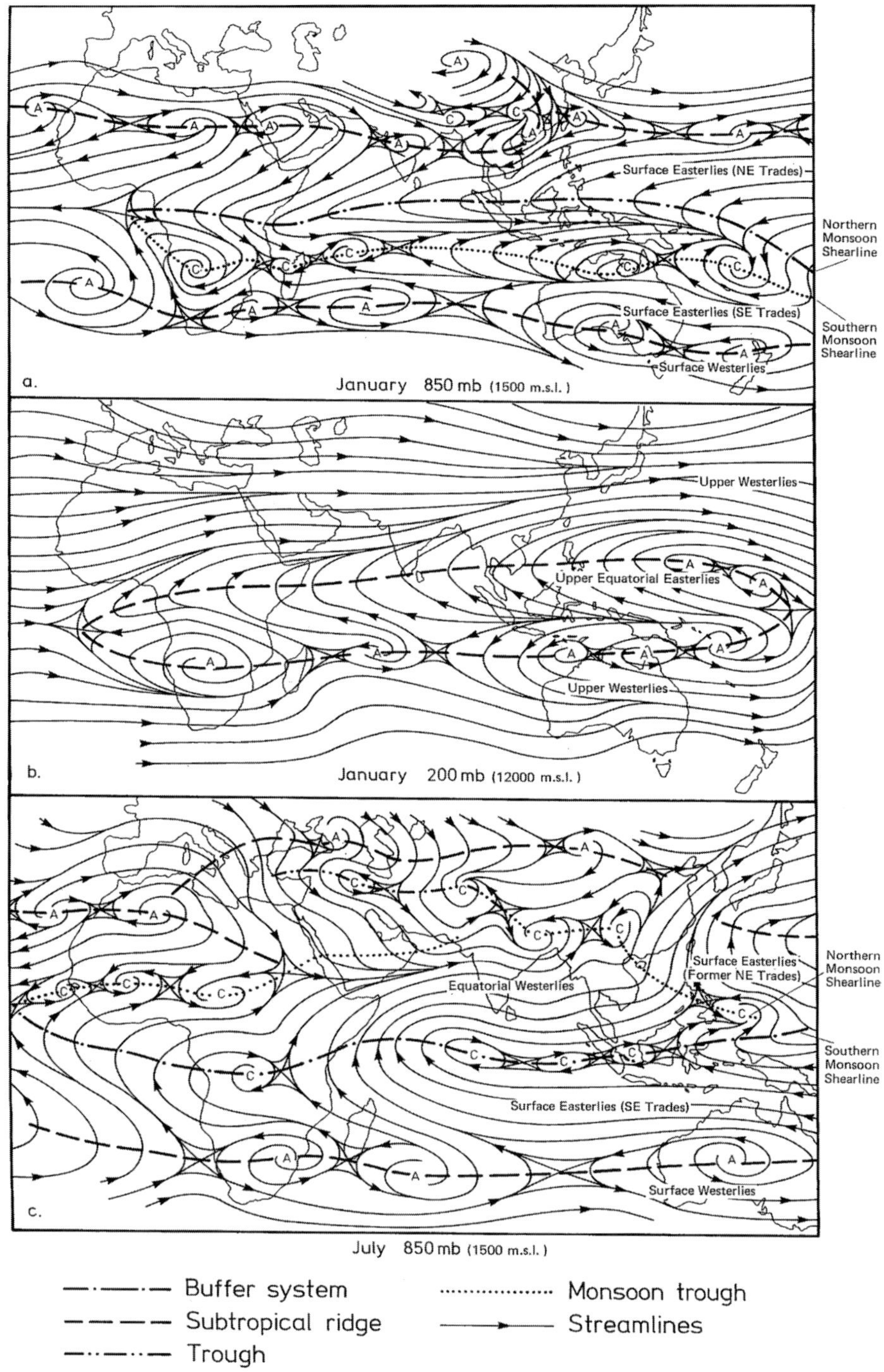

Figure 10.1 Mean 850 hPa streamline analysis for January (a) and July (c), and mean 200 hPa streamline analysis for January (b). (adapted from Sadler and Harris, 1970, and reproduced in Bonell *et al.*, 1991.)

The zonal, near-surface thermal trough is referred to in the literature as the equatorial trough, near-equatorial trough, equatorial front, equatorial convergence zone, intertropical front and, most commonly, the intertropical convergence zone (ITCZ) (Sadler and Harris, 1970). None of these terms will be used in this chapter.

As noted by Sadler and Harris, it is incorrect to identify this 'convergence zone' as a zone of maximum cloudiness. In fact the *maximum cloud zone* (known as the MCZ) occurs on the equatorial side of the thermal trough of low pressure within the *equatorial westerlies* as shown in Figure 10.1 (Davidson *et al.*, 1983; McBride, 1983) and is a feature of all monsoon areas such as the eastern Atlantic, for example Sadler (1974). The thermal low pressure trough is referred to as the *monsoon trough* (Sadler and Harris, 1970) or more specifically the **monsoon shear line**

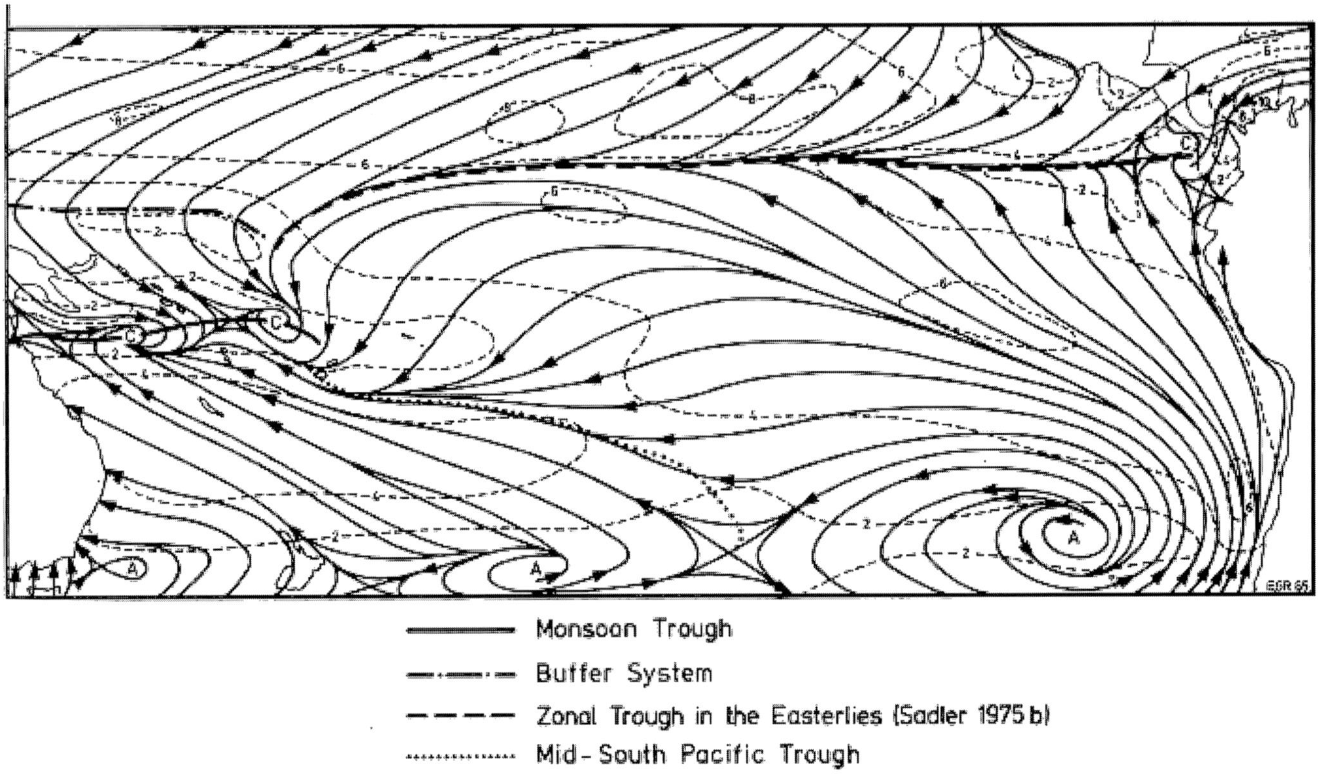

Figure 10.2 Mean low level streamlines for February (period 1900 to 1979) over the Pacific Ocean. (Source: Sadler *et al.*, 1987 and reproduced by Bonell *et al.*, 1991.)

(after R. Falls in McAlpine *et al.*, 1983) and is a zonal pressure trough which separates the *surface trade wind easterlies* and *the equatorial westerlies*.

A factor encouraging the synoptic scale MCZ of deep convection and cloudiness is the convergence of low level westerlies originating respectively from the surface easterlies of the northern and southern hemispheres. Considering only zonal (i.e. along the west to east axis), the latter trade winds when they cross the Equator change from an easterly direction to a westerly direction in response to the Earth's rotational (Coriolis) force. It is within the MCZ that extensive cloud clusters form, which will be elaborated upon in the next Chapter as *mesoscale convective complexes (MCCs)* or *mesoscale convective systems (MCSs)*.

There are two monsoon shear lines in regions of strong monsoon flow, identified geographically by hemisphere as the northern and southern monsoon shear lines; these systems separate the respective trade wind easterlies from the opposing equatorial westerlies. The monsoon shear line in the summer hemisphere is the more active and is the thermally-induced low pressure belt or monsoon trough identified by Sadler and Harris (1970). Satellite imagery shows that cloudiness is not necessarily persistent along this shear line (monsoon trough). Large cloud clusters, commonly associated with vortices of low pressure cells, are separated by relatively cloud-free areas. Some of these vortices develop into tropical cyclones over the tropical oceans. The corresponding winter hemisphere monsoon shear line is the less active and represents the wind-turning mechanism from easterly trades to equatorial westerlies driven by the earth's rotational deflection force (Coriolis force). Sadler and Harris (1970) referred to this shear line as the *buffer system* but during the transitional months (April, May, November, December) both shear lines can be equally active in the Indian ocean and the western Pacific, as evidenced by the observed pairing of tropical cyclones astride the Equator. Consequently, the term buffer system is less appropriate during these months if it is understood to be the less active of the two monsoon shear lines. Consequently, the terms *northern monsoon shear line* and *southern monsoon shear line* are preferred (Bonell *et al.*, 1991). It should be noted that during the winter months, 'out of season' tropical cyclones (following R. Falls in McAlpine *et al.*, 1983) can still occur along the northern monsoon shear line over the north west Pacific Ocean and Bay of Bengal (Sadler, 1967).

Outside the preferred regions of monsoon flow such as the central and western parts of the north Atlantic, and the eastern Pacific Ocean (east of the International Date Line) during the Southern hemisphere summer (Figure 10.2), the basic surface wind flow pattern is a more tangential convergence of northeast and southeast

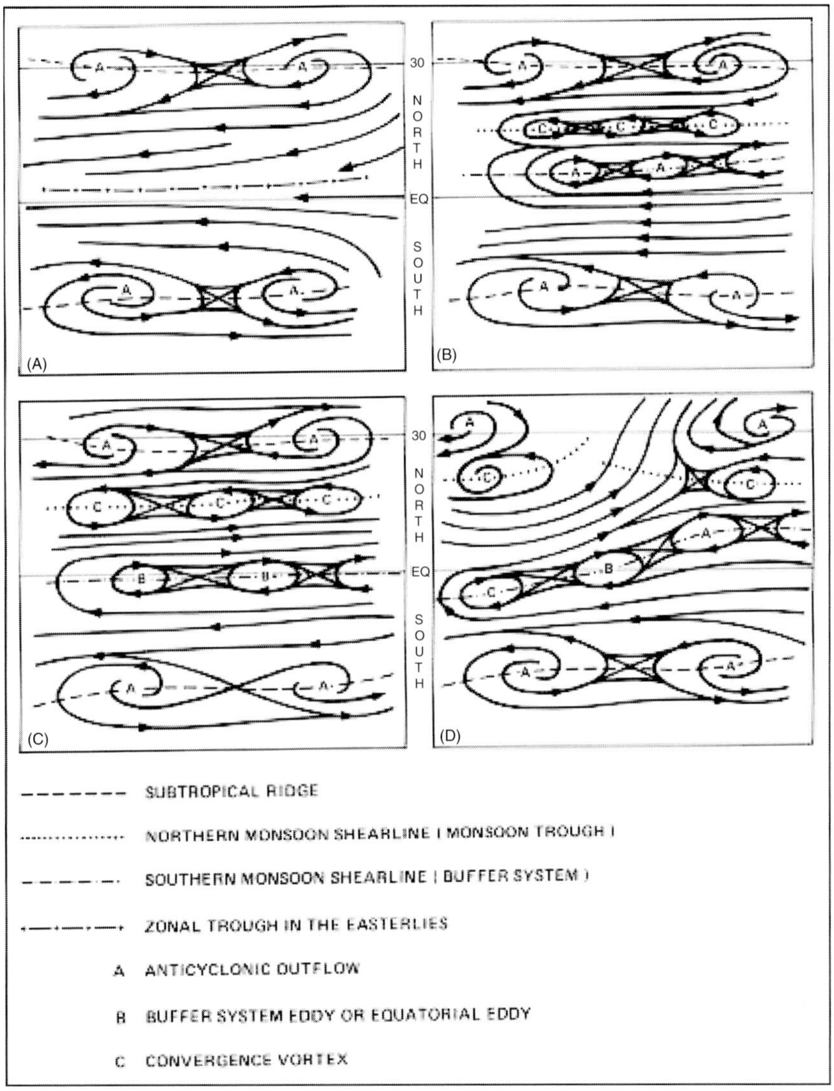

Figure 10.3 Schematic of the low-level flow for an evolving monsoon circulation in the Northern hemisphere summer and the various synoptic systems. Part D is typical of the low-level flow pattern between the longitudes of India and the Philippines. (Adapted from Sadler and Harris, 1970, and reproduced from Manton and Bonell, 1993.)

trade winds into a feature termed *the zonal trough in the easterlies (ZTE)*, (Sadler 1975b). The cloudiness and rainfall structure of the ZTE is complex (see Sadler, 1975b; Ramage *et al.*, 1979; Manton and Bonell, 1993) but this system is not associated with persistent, deep convection, as found within the monsoon regions. The ZTE is found over the ocean where there are little data and consequently the internal structure of the ZTE has seldom been investigated and is poorly understood. Ramage *et al.* (1979) detected alternate strips of convergence and divergence rather than a single strip of convergence in the central Pacific. Significantly, convection along the ZTE is much less deep than in the monsoon regions due to the presence of dry upper equatorial westerlies (Sadler, 1975a) and an Equator-ward extension of a trade wind type inversion across the ZTE (Ramage *et al.*, 1979). Cloud and rain development along the ZTE of the central Pacific is mostly an 'orographic' phenomenon of shallow uplift (Ramage *et al.*, 1979). The complex structure of the ZTE means that locating the MCZ in relation to the surface trough is difficult because of difficulty in determining the exact trough position within the broad and weak-pressure zone (e.g. mid-north Atlantic Ocean, Sadler, 1975b). During El Niño-Southern Oscillation (ENSO) events, the ZTE of the central Pacific Ocean is replaced by the eastward expansion of a monsoon circulation. During the Northern hemisphere summer, the ZTE is replaced seasonally by a monsoon pattern over the eastern Pacific Ocean. This circulation occasionally penetrates as far east as the southwest Caribbean Sea and is the cause of the heavier rain events associated with more organised, tropical vortices.

Figure 10.3 is a schematic diagram of the low level flow for an evolving monsoon circulation in the Northern hemisphere summer and the various synoptic systems.

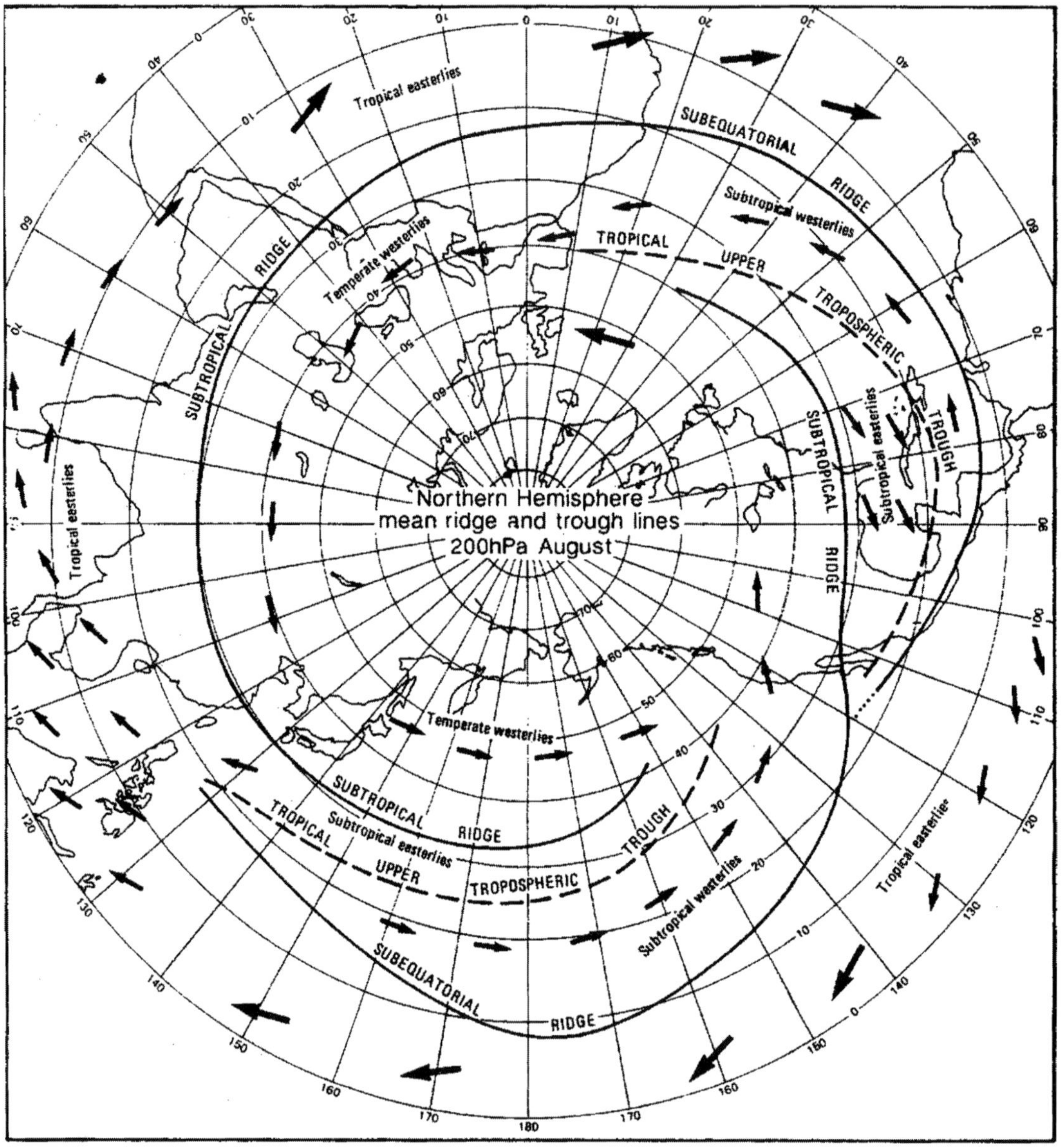

Figure 10.4 The location of troughs, ridges and major currents at 200 hPa during August. (Source: Sadler, 1975a.)

Features of the upper atmospheric circulation

The classic work and system definitions of Sadler (1975a) is a principal source for outlining the upper atmospheric circulation. Within the monsoon regions (Figures 10.1 and 10.4), the low-level equatorial westerlies are overlain at higher levels (commonly at elevations above the 500 hPa level *c.* 6000 m a.s.l.) by *upper tropical easterlies (or upper equatorial easterlies)*. These upper easterlies are deflected eventually by the Coriolis effect through the *upper subtropical ridge* overlying the monsoon shear lines and join directly either the upper temperate westerlies or the subtropical westerlies (which subsequently join the temperate westerlies) associated with the *Tropical Upper Tropospheric Trough* (TUTT).

Either of these westerly channels provide the conduits for the export of surplus, latent energy from the tropics to the higher latitudes, especially to the winter hemisphere. The same channels also provide ideal outflow or vents for intensifying tropical cyclones when in the presence of either a trough in the temperate westerlies (e.g. north-east Australian region, Holland, 1984) (Figure 10.5a) or in the proximity of TUTT (mid-north Atlantic or Caribbean sea (Sadler, 1976, Figure 10.5b). In contrast, the unidirectional upper airstream in Figure 10.5c is less favourable for venting of the upper anticyclonic flow of a vortex, and favours the southern sector only. These circumstances partly account for the decay of vortices, emerged previously from the monsoon region

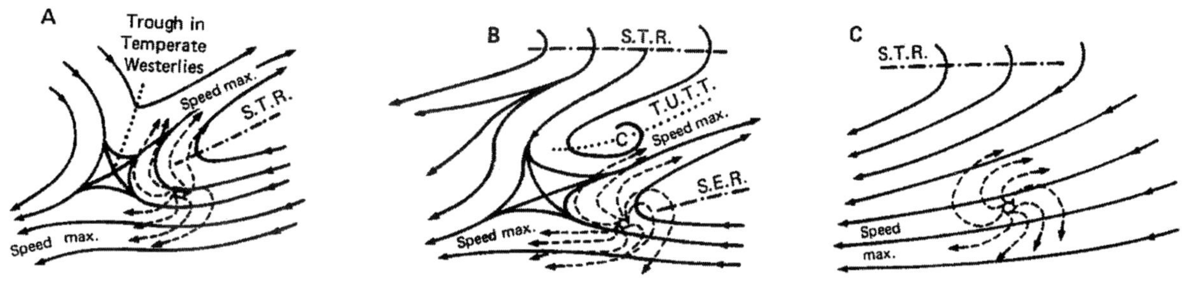

Figure 10.5 Schematic of storm flow outflow interaction (dashed lines) with the larger scale upper tropospheric circulation (solid lines). STR is the subtropical ridge; SER the sub-equatorial ridge. TUTT, the tropical upper tropospheric trough. (Adapted from Sadler, 1976a.)

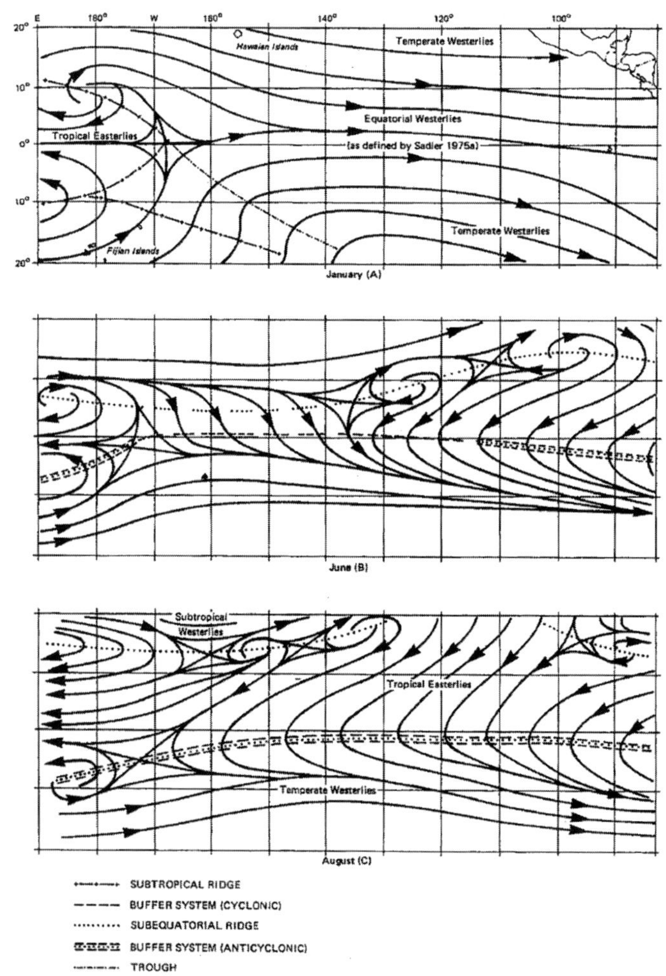

Figure 10.6 A schematic of the 200 hPa ridge development of the Northern hemisphere over the eastern Pacific and the concurrent establishment of a buffer system in the equatorial region as part of the Hadley circulation. (Source: Sadler, 1975a.)

of west Africa, within the mid-north Atlantic ocean during the summer (we will elaborate on this topic in the next chapter).

Figure 10.1b shows a classic monsoonal pattern, with an upper anticyclone over the monsoon trough redistributing air aloft as tropical (or equatorial) easterlies to the north and circumpolar, temperate westerlies to the south. The monsoon trough shear line slopes Equator-ward with height; but during very active phases of the monsoon, the depth of the lower equatorial westerlies can extend up to the 200 hPa level 13 000 m a.s.l. (Hendon *et al.*, 1989) before being replaced by upper tropical easterlies. This is much higher than the traditionally accepted position of 500 hPa level (McAlpine *et al.*, 1983).

During the Northern hemisphere winter when the surface circulation favours the development of the ZTE over the eastern Pacific Ocean and west-central Atlantic Ocean, the prevailing *upper equatorial westerlies* (Sadler, 1975a) merge with the temperate westerlies at higher latitudes. Thus overall, the upper atmospheric circulation is much less complicated than that over the monsoon regions or during the Northern hemisphere summer for the easternmost section of the tropical north Pacific (Figure 10.6). During the same period of the year (i.e. the Northern hemisphere winter) these upper equatorial westerlies envelop the northern Amazon basin (Kousky, 1979, Kousky and Ferreira, 1981, Molion, 1993).

The importance of TUTT systems needs further elaboration. As indicated in Figure 10.4, these are located over the western north Pacific and north Atlantic Oceans in the Northern hemisphere summer. A third preferred area for TUTT occurrence is over the south-east, south Pacific Ocean during the summer. These systems are capable of developing their own vortices, which on occasions can extend down to the surface (Figure 10.7) and are thus a source of rain-producing perturbations in the surface easterlies. Some of these perturbations subsequently form into tropical cyclones (about 10% of the annual total of these storms) in both the north Atlantic and Pacific Oceans (Sadler, 1976). The same systems can commonly be mistaken (Sadler, 1975b) for 'easterly waves' following Riehl's model (Riehl, 1954; 1979). Moreover, the complex upper wind pattern over the western parts of the north Atlantic (Figure 10.4) provides favourable conditions for the re-intensification of vortices of north African origin. This applies especially when a TUTT extends into the Hispanic area of the Caribbean (Sadler, 1975a; Sadler, 1976) and so provides

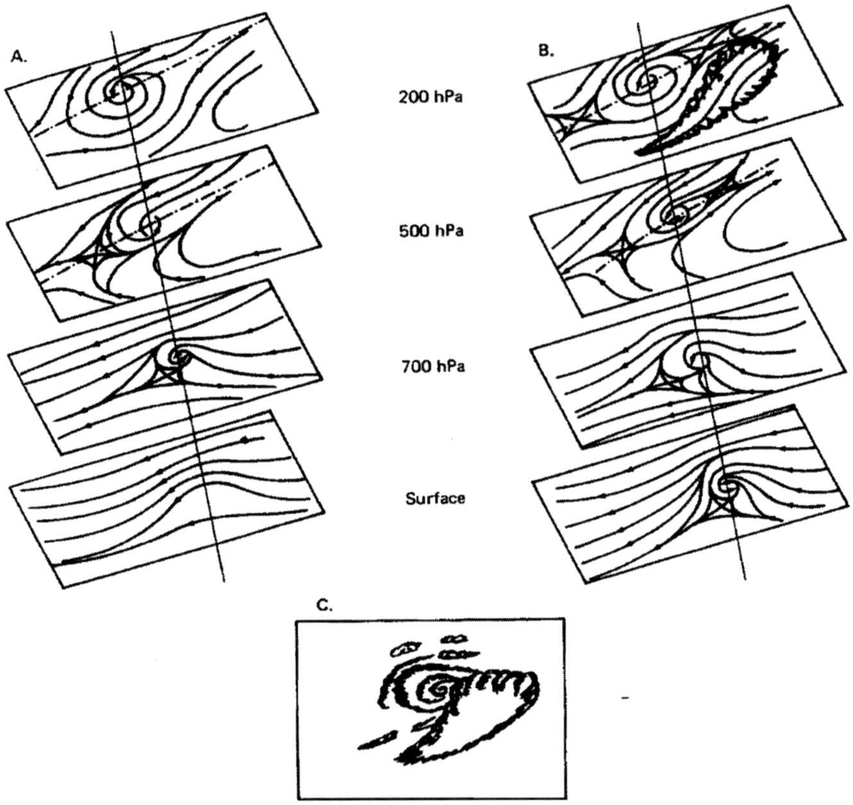

Figure 10.7 Three-dimensional structure of the Tropical Upper Tropospheric Trough (TUTT) systems and the plan view of a commonly observed cloud system. In (A) the vortex has penetrated through the 700 hPa level and shows in the surface level as an induced trough, whereas in (B) the penetration is to the surface as a vortex in the trade wind easterlies. (C) is a schematic plan view of the cloud system of a moderate-to-strong cell in the central Pacific having a southeast slope with decreasing altitude. Whilst the view is under (B), it is equally appropriate to (A). (Source: Sadler, 1967.)

a favourable upper outflow (venting) environment for tropical cyclone development. Figure 10.8 shows the corresponding seasonal shift in tropical cyclone formation over a 20-year period with the formation of these cyclones in general following the migration of the monsoon trough. However, as noted earlier, in the January to March period some cyclones form over the Bay of Bengal and in the northwest Pacific. The formation of tropical cyclones in the north Atlantic without a monsoon trough will be discussed in more detail in the next chapter. Tropical cyclones also form in the south Pacific east of the date line, as will be shown later.

During the 1920s, while scientists in South America were busy documenting the local effects of El Niño, Sir Gilbert Walker was studying monsoons in India. Walker, who was the head of the Indian Meteorological Service, had been asked in 1904 to try to understand how to predict the vagaries of India's monsoons after an 1899 famine caused by monsoon failure. As he sorted through world weather records, he recognised some patterns of rainfall in South America and associated them with changes in ocean temperatures. He also found a connection between barometer readings at stations on the eastern and western sides of the Pacific, at Tahiti and Darwin. He noticed that when pressure rises in the east, it usually falls in the west, and vice versa. He coined the term Southern Oscillation to dramatise the ups and downs in this east–west seesaw effect. He also realised that, under certain barometric conditions, Asian monsoon seasons were often linked to drought in Australia, Indonesia, India and parts of Africa, and mild winters in western Canada.

He was the first person to claim a connection between monsoons in India and unusually mild winters in Canada. In a modern rendition of 'the world is flat' scenario, he was criticised for suggesting that climatic conditions over such widely separated regions of the globe could be linked. Walker conceded that he could not prove his theory, but predicted that whatever was causing the connection in weather patterns would become clear once wind patterns above ground level, which were not being observed routinely at that time, were thrown into the equation. He was right.

The Walker circulation was later formally named by Bjerknes for two circulation cells in the equatorial atmosphere, one over the Pacific and one over the Indian Ocean. Schematically, these are longitudinal cells where, on one side of the ocean, convection and

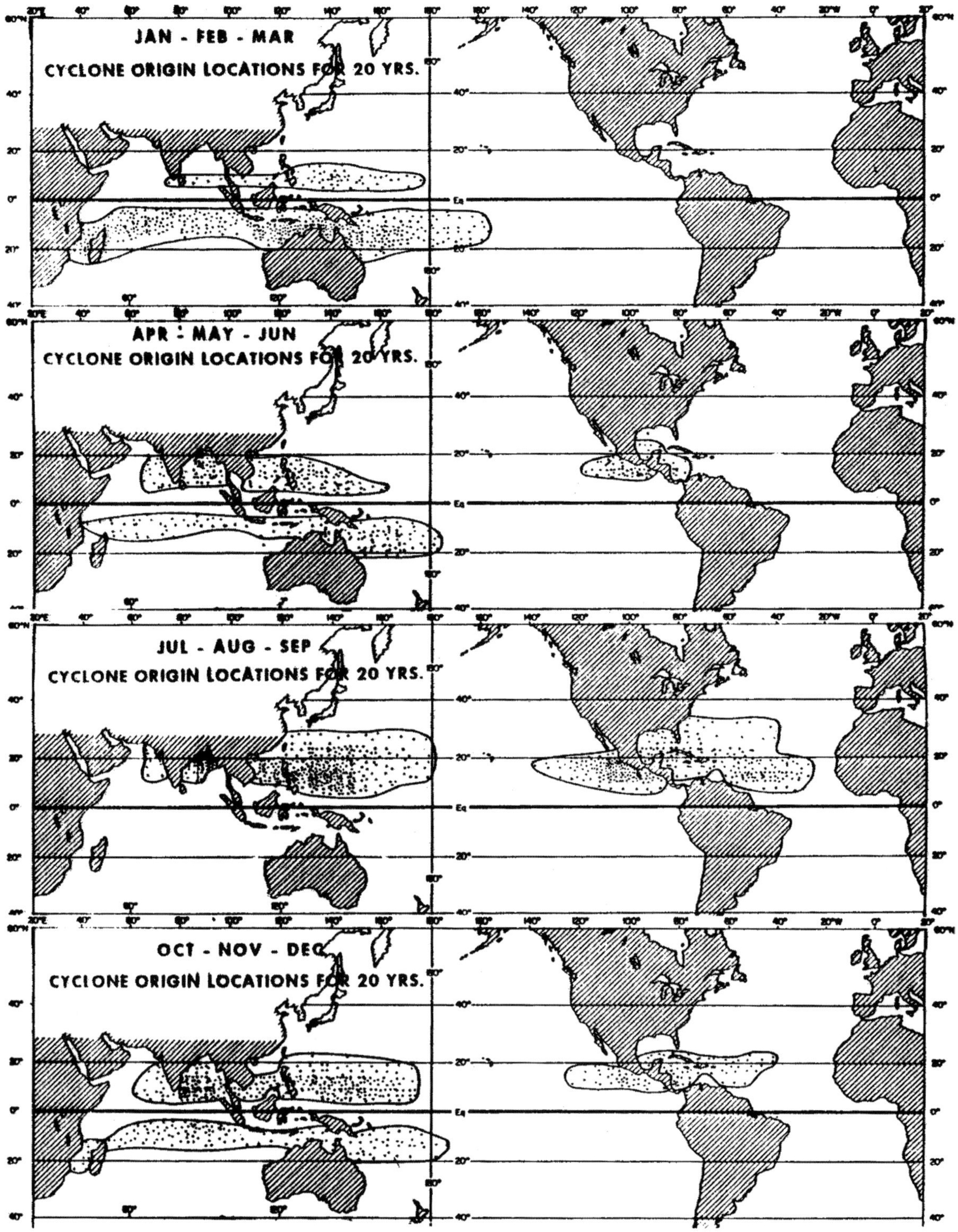

Figure 10.8 Seasonal distribution of the formation positions of tropical cyclones over a 20-year period. (Source: Gray, 1975.)

the associated release of latent heat in the air above lifts isobaric surfaces upward in the upper troposphere and creates a high pressure region there. The lack or lesser degree of the same process on the other side of the ocean results in lower pressure there, and a longitudinal pressure gradient is established which, being on the equator, cannot be balanced by the Coriolis force. Thus a direct zonal circulation is driven in the equatorial plane with countervailing winds at the surface and in the upper troposphere, and with concomitant rising and sinking branches on the appropriate sides of the ocean.

The normal Walker circulation in the Pacific consists of air rising over Indonesia, west winds in the upper troposphere, sinking air off the west coast of South America, and east winds near the surface. A reversed but weaker Walker circulation (and an enhanced Hadley circulation) occurs during ENSO years. In the Indian Ocean the circulation cell proceeds in the opposite sense (to the normal Pacific Walker cell), with sinking air over cold waters off the Somali coast and a low level acceleration from west to east along the Equator in the lower atmosphere.

LOW-LATITUDE TROPICAL CYCLONES

Many meteorologists were surprised when tropical storm Vamei reached typhoon intensity slightly equatorward of 2 degrees north during December 2001 in the north-west Pacific sector, since one gains the impression from the bulk of the meteorological literature that tropical cyclones cannot form within 5 latitude degrees of the equator, because the Coriolis force tends towards zero at the equator so inhibiting circular motion. Brunt (1969) questioned this limiting effect of a weak Coriolis force for tropical cyclone formation. He thought that an important limiting factor on cyclogenesis may be the short time the monsoon trough spends in these low latitude regions.

The inspiration for the study of Brunt was tropical cyclone Annie which developed near 6 deg S and caused widespread damage and loss of life in the maritime provinces of Papua New Guinea in November 1967 and his paper could have hardly been more prophetic. During 1970, typhoon Kate (Holliday and Thompson, 1986) at 4.5N 131E reached a minimum sea level pressure of 938 hPa (as reported from a reconaissance flight). Kate went on to make landfall in the Davao Gulf of Mindanao (usually a typhoon-free zone) and caused 631 deaths. In Papua New Guinea two more areally small but intense cyclones affected the region, Hannah in May 1972 and Adel in May 1993, which both developed near 5.5 deg S. During April 1973, a tropical cyclone with an eye evident on infrared satellite imagery at 0140 UTC 29/4/1973 near 8S 121.5E, was responsible for the loss of 1500 fishermen at sea, sunk a ship with the loss of 21 and caused 53 deaths on the Indonesian Island of Flores.

Brunt listed several other destructive low latitude tropical cyclones between 1778 and 1960 based on both historical and instrumental records. He noted that the occurrences of such low latitude storms within the Indonesian area and the PNG area were bimodal; November/December as the monsoon trough passed south on its way down to Australia, and April/May as it was retreating back to the Northern hemisphere.

MONSOONS

'Monsoon' was derived from the Arabic word *Mausim*, meaning season, and referred originally to the winds of the Arabian Sea which blow for about six months from the northeast and then six months from the southwest. With greater understanding of the monsoon circulation the definition has been broadened to include almost all the phenomena associated with the annual weather cycle within the tropical and sub-tropical continents of Asia, Africa and Australia and the adjacent seas. There is also a monsoon-like circulation extending into the eastern Pacific from Central America. The fundamental driving force is the result of the differences in the annual temperature trends over land and sea. This gives rise to a pressure gradient which drives the winds from high to low pressure, combined with the effects of the rotation of the earth (Coriolis force) which deflects the winds to the left (right) in the Southern (Northern) hemisphere. The British scientists Edmund Halley and George Hadley recognised this around the beginning of the eighteenth century. Nevertheless, the existence of the monsoon was known in ancient times and an Arab pilot with a strong knowledge of the monsoons guided Vasco da Gama from East Africa to India in 1498.

The Indian monsoon

At the equator, westerly winds constantly occur near the surface in the Indian region throughout the year. Surface easterlies reach only to 20°N in February. Late in March the sun is over the equator and moving north, bringing with it atmospheric instability due to surface heating, and the associated convective clouds and rain. India is particularly prone to rapid surface heating in April as the Himalayan mountain chain to the north prevents any incursions of cold air.

During May the Tibetian Plateau acts as an elevated heat source and radiates heat that is transmitted readily to the atmosphere above. This heating warms up the atmospheric column aloft by several degrees Celsius per day. Lower-layer air in the surrounding area of the plateau is then drawn towards the plateau. The north to south gradient of the upper-tropospheric temperature to the south of the plateau is reversed gradually in favour of the development of tropical upper easterlies. All these provide suitable conditions

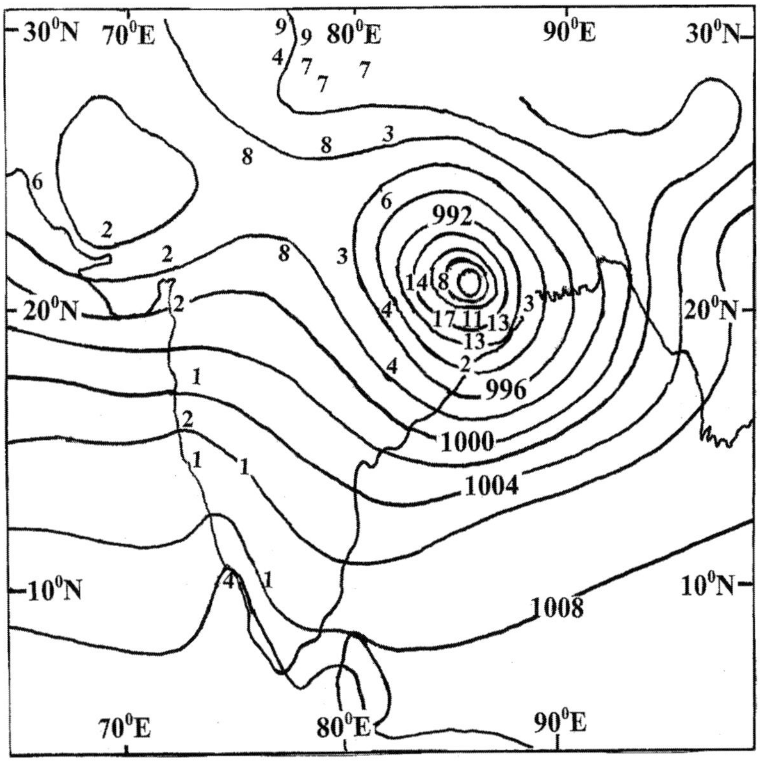

Figure 10.9 Mean sea level (MSL) pressure distribution (hPa) and 24 hour rainfall plots (cm) for a typical Bay of Bengal depression. (Source: Fein and Stephens, 1987.)

for the seasonal transition of the circulation in East Asia and the onset of a monsoon. On many occasions the onset of the surface south west monsoon is preceded by a sudden increase in wind speed in the upper easterly jet stream some 1500 km further north.

During June the upper easterly jet becomes firmly established between the 150 hPa and 100 hPa levels. This upper jet reaches its greatest speed at its normal position along 15°N from China through India. The position of the upper easterly jet controls the location of the monsoonal rains. The heavy rain occurs in the south and southwestern sectors of the maximum wind speed zone of the jet. The surface flow is a strong moist southwesterly, however, which brings heavy squally rain that is the *burst* of the monsoon. Most spectacular cloud and rain occur against the Western Ghats, the mountain ranges lying along the west coast of India. The windward slopes extensively receive from 2000 to 5000 mm of rain in the monsoon season. Monsoon seasonal rain over much of India, however, is limited to 400 to 500 mm. Nevertheless, topography produces some extraordinary totals. For example, over the Ganges Valley, the low level winds are deflected southeasterly by the Himalayas and Cherrapunji, on the southern slopes of the Khasi Hills (north of the Ganges Delta) receives on average 2730 mm of rain in July. In July 1861 the total for the month reached 9300 mm. Over the 4-day period 12–15 September 1974, Cherrapunji recorded 3721 mm of rain (Bureau of Meteorology, 1994).

Monsoon depressions

A good part of Indian monsoon rainfall is generated by westward moving depressions which form along the northern monsoon shear line which at this season is positioned across the Bay of Bengal. On average, two or three depressions are observed in the monsoon months of July and August and the horizontal diameter of these systems is around 1000 km (Das, 1987). The lifetime of a depression is about one week and usually they move towards the west-northwest for the first three or four days. After this, they either re-curve to the north or continue on in the same direction. Figure 10.9 shows the mean sea level (MSL) pressure (hPa) and 24 hour rainfall (cm) distribution associated with a Bay depression. The heavier rainfall tends to be concentrated to the south and southwest of the centre of the system. The resulting rainfall from these depressions is generally distributed more evenly than in June.

Mostly, these systems do not stay long enough over water to develop into tropical cyclones. However, some do become weak tropical cyclones before making landfall and bring disastrous flooding with them. During August, persistent cloudiness and decreasing solar radiation causes temperatures to fall and the southwesterly low level winds decrease; by September dry, cool northerly winds flow around the highlands into northwestern India. The upper easterly jet moves south and the low level southwesterly winds also retreat. Northerly flow at low levels spreads over India during October and the northeasterly winter monsoon takes shape. This causes an October to December rainy season for the extreme southeast of the Deccan (including Madras) and Sri Lanka.

Indian tropical cyclones

Although only about 7% of the global tropical cyclones occur over the northern Indian Ocean, they are the most deadly. The shallow warm waters of the Bay of Bengal, the flat coastal terrain and the funneling shape of parts of the coast can lead to devastating loss of life and property from storm surges produced by cyclones of even moderate intensity. The Buckerganj cyclone of 1876 and the Bhola cyclone of 1970 each killed more than 200 000 people. In 1991 more than 100 000 people died as a result of a storm surge in Bangladesh.

On average, five to six cyclones form in the basin per season and five times as many form in the Bay of Bengal than in the Arabian Sea. Nevertheless, the Arabian Sea storms can still be disastrous. A cyclone crossed the coast between Mumbai and Karachi in June 1998. There were reports of 1063 deaths in Gujarat State and relief workers indicated that up to 14 000 people disappeared without a trace. There are two periods of tropical cyclone activity. The first occurs from April to June, before the monsoon season when the northern monsoon shear line moves up onto the Asian continent. The second and most active period occurs from September to December after the northern monsoon shear line moves back over the Bay of Bengal.

The Asian–Australian monsoon

The substantial water masses between Asia and Australia have a moderating effect on temperatures in the troposphere and weaken the boreal summer monsoon. The northern limit of this summer monsoon is around 25°N. As a result monsoon rains occur (generally north of 10°N) in June and also in late August and September with weak highs often breaking up the monsoon in July.

At the global scale, the northern Asian winter monsoon constitutes one of the most energetic circulations which results in a concentrated area of deep convection over the 'maritime continent' (centred over the Indonesian archipelago of Ramage, 1968) (see Bonell *et al.*, this volume). Its influence can extend to other parts of the globe (Lau and Chang, 1987) due to the transfer to

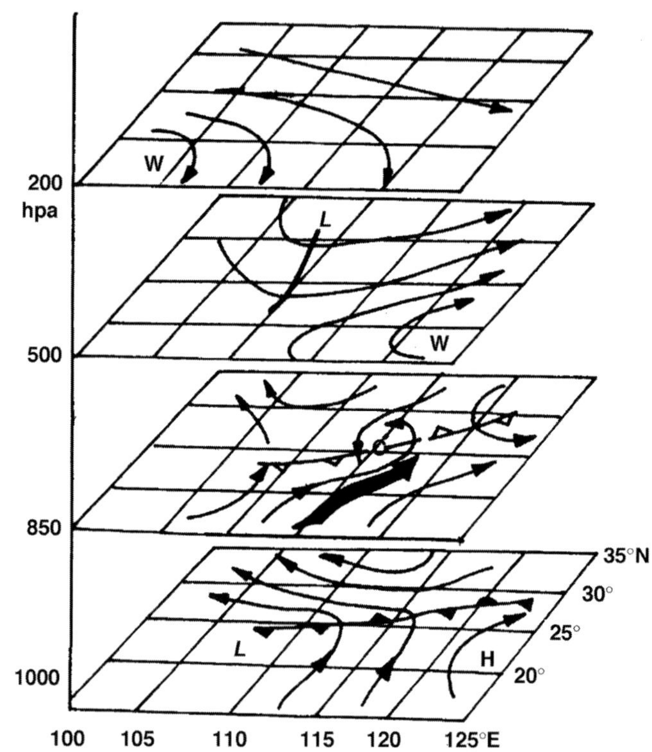

Figure 10.10 Schematic flow patterns during heavy rainfall events in South China in May and June. (Source: Chang and Krishnamurti, 1987, after Huang, 1982.)

latent energy via the upper atmospheric circulation across the Pacific Ocean towards higher latitudes in both hemispheres. At low levels in South China and the Philippines, trade winds prevail from October to April, strengthened by the outflow from the intense stationary Siberian high. Their disappearance and replacement by opposite southwesterly winds in the May to September period is the essence of the monsoon. On the larger islands, the slopes facing the prevailing winds get the most rain.

In China a quasi-stationary belt of heavy rainfall migrates northwards with the summer monsoon. For southern China the heavy rain occurs mainly in May, in Taiwan from May 15 to June 15 and in the Yangtze Valley from about 10 June to 10 July when the term *Mei-Yu* (plum rains) is officially used. During the same period the *Baiu* rains occur over Japan. The May rain in southern China is caused by an interaction between the summer monsoon winds and the mid-latitude flow from the north. The rainfall develops along a quasi-stationary front, with most of the heavy rain falling on the warm side of the front. Huge stationary, persistent thunderstorms are initiated by the diverging upper flow overlying converging low level flow. Figure 10.10 is a schematic diagram (from Huang, 1982) showing the flow pattern at different levels during these events. The winds veer with height (warm air advection Northern hemisphere) from low level northerlies to upper level diffluent westerlies. It can be shown (Figure 2 in Hoskins *et al.*, 1978) that warm air advection denotes motion upwards where the warm air advection is maximised due to the presence of a jet in the layer. There is usually a low-level southwesterly jet at 850 hPa increasing the likelihood of forced ascent in the area. The jet is an important factor in the formation of heavy rainfall, with 38 of the 40 cases from 1971 to 1978 occurring when a low-level jet stream was present in the southwest monsoon. In southern China, most of the non-typhoon rainfalls that cause severe floods occur in late May and early June. At Haifeng (23°N 115.3°E) on 31 May 1977, 884 mm of rain was recorded in 24 h and at Yanjiang (21.9°N 111°E) on 29 May 1973, 850 mm of rain was recorded in 24 h. Both these places are located near the coast, which is orientated almost west to east so that the low-level southerly flow is directly onshore. A second period of heavy rain occurs in southern China anywhere between mid-summer and autumn and is associated with typhoons.

In Vietnam and Thailand the boreal summer monsoon brings plentiful but not extraordinary rain from May to October. November to February is the cool dry season and March to April is the hot dry season. Along the east coasts and eastern slopes more rain is brought by the boreal winter monsoon. The Asian winter monsoon brings much rain to countries south of latitude 10°N. In Indonesia the wettest months are December in Sumatra and January elsewhere. The wettest month at Khoto Baru on the northeast coast of Malaysia is December while Singapore's wettest months are January and December. Thus, the heavy rain contracts eastwards over the boreal winter. In Indonesia, however, rainfall patterns can be localised. In Java, for example, there are two major climatic regions at sea level, with an equatorial west with no dry season and a monsoon east with extreme drought in August and September.

Australia, with its relative large size and compact shape, has relatively simple monsoon patterns, with the north coast subject to a northwesterly monsoon between November and April. However, this rain-bearing monsoon is often held offshore and is most likely to move overland during January and February. The trades can often extend up to the north coast throughout the austral summer, pushing the monsoon trough northwards.

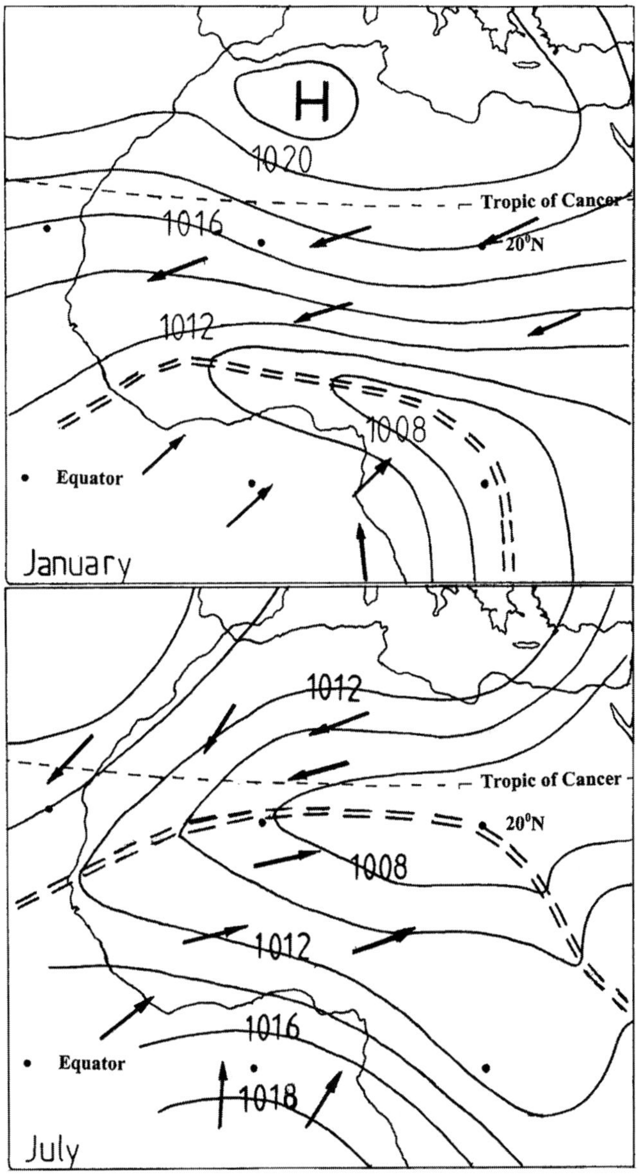

Figure 10.11 Average MSL pressure distribution (hPa), low level flow (bold arrows) and monsoon trough (parallel dashed lines) for January and July. (Source: Encyclopaedia Britannica.)

The West African monsoon

The main characteristics of this monsoon have been known for more than 200 years. The southwest monsoon flows as a shallow humid layer of surface air overlain by the primary northeast trade wind, which blows from the Sahara and the Sahel and is a deep dry dusty stream (Figure 10.11). During the cool season, the surface northeasterly is known as the Harmattan, which is a dry gusty wind, cool at night and scorchingly hot by day. The West African monsoon is thus an alternation of the southwesterly wind and the Harmattan at the surface. As seen in Figure 10.11, the monsoon belt alternates between 9°N and 20°N throughout the year.

The normal drought conditions existing north of 20°N (the Sahara Desert) become shorter as one heads towards the Equator. At 12°N the drought lasts about half the year; it disappears at 8°N. Further south, a different and lighter drought begins to appear in the high sun months when the monsoon southwesterly winds are strongest and relatively drier air arrives from the Southern hemisphere. The advancing fringe of the monsoon is mostly too shallow to produce much convection and, in general, thunderstorm activity occurs up to 400 km behind the front where the moist air is deeper. These storms form in squall lines aligned roughly north–south and will be discussed in the following chapter.

The American monsoon

In Central America a true monsoon cycle occurs over a small area facing the Pacific, between 5°N and 12°N. It undergoes a complete seasonal reversal of the winds and the rainfall regime is typically monsoon, with a dry winter and wet summer. The rainy season begins earlier in the south (May) and later in the north, coming at the end of June in southern Mexico. The result is a rainy season that lasts for three months in southern Mexico increasing to six to seven months in Costa Rica.

LARGE-SCALE CLOUD BANDS

The South Pacific Convergence Zone (SPCZ)

Cloud bands extending from the tropics to the mid-latitudes are persistent in preferred meridional locations (see Kuhnel, 1989), especially during the respective Northern and Southern hemisphere summers. These bands act as conduits for the concentrated transport of latent heat and moisture into higher latitudes. In the process, these poleward excursions of tropical moisture often trigger extensive, and often intense, precipitation (Kuhnel, 1989).

Satellite imagery shows three persistent cloud-bands in the Southern hemisphere, the SPCZ, which extends from the Solomon Islands southeastward into the mid-latitude Pacific. This feature has a marked control over the precipitation regime of many South Pacific islands. Over South America, the South Atlantic Convergence Zone (SACZ) drains excess latent heat and moisture from the Amazon Basin across sub-tropical southeast Brazil. The cloud band connecting the Congo Basin with south-east Africa is less persistent but remains a significant transitory feature which contributed to the flooding in Mozambique in February/March 2000. Wright (1997) also described transitory, tropical/extra-tropical cloud-bands, which affect Australia during the April-October cool season. There are two types, i.e. 'oceanic' cloud bands and 'continental' cloud bands. The oceanic group originated west of 120°E from the Indian Ocean and provided 70–90% of cool season rain in north-western Australia (parts of which are in the wet/dry tropics defined by Chang and Lau, 1993). These oceanic cloud bands are most active between April and July, but decline quickly thereafter to be replaced by continental interior (continental) cloud bands which increase. The eastward movement of both oceanic and continental cloud bands makes a significant contribution to the cool season rainfall of tropical northeast Australia. Over the perhumid, northeast Queensland tropics, these upper disturbances provide the most organised rainfall rather than the low-level SE Trades (Bonell and Gilmour, 1980). At the synoptic scale, there is a marked inflow of advected low to mid-level moisture from the Coral Sea and northern Arafura Sea (including the Gulf of Carpentaria) which sustains the activity of these upper troughs.

In the Northern hemisphere cloud bands extend from both the tropical Atlantic and tropical Pacific northeastward during the winter 'cool' season. Manton and Bonell (1993) have already reviewed the Pacific-North America (PNA) pattern of Horel and Wallace (1981). Significant tropical higher latitude connections also occur during the Indian monsoon, as will be discussed later.

The origins of the SPCZ, a semi-permanent feature, are still not clear (Matthews *et al.*, 1996) and cannot be explained by a zone of maximum sea surface temperature (SST) alone. The NW–SE axis of the SPCZ produces a direct connection with the southern monsoon shear line over Papua New Guinea. Along this axis, summer tropical cyclones often develop embedded within the SPCZ and move along this system during their southeastward movement, especially during the opening and closing stages of the tropical cyclone season. Transient troughs in the mid-latitude westerlies also interact with the SPCZ during their penetration into the tropics.

Over the annual cycle, Basher and Zheng (1998) noted that the SPCZ is the most active during January to March, with central parts of the zone producing monthly rainfalls in excess of 400 mm. In February-March, the SPCZ also displaces a few degrees further south, thus increasing rainfall over the Tonga-Fiji area, and with a converse decrease over Samoa. Subsequently, April–June is the transition to dry season conditions with a rapid decline in SPCZ rainfall, with totals in excess of 200 mm occupying only about 15% of the SPCZ axis. During July to September the SPCZ is relatively inactive, with monthly rainfall in excess of 200 mm confined to the northwest South Pacific, that is, in the vicinity of Papua New Guinea and the Solomon Islands. From October to December, the higher monthly rainfall area propagates eastward, with maxima in the Samoan region which corresponds with the rapid broadening and extension of the SPCZ.

The 'splitting' of convective activity between the southern monsoon shearline/SPCZ over Papua New Guinea produces a dry, wedge-shaped region (see Falkland and Brunel, 1993) of low-level divergent easterlies further east lying between these two zones.

South Atlantic Convergence Zone (SACZ)

The South Atlantic Convergence Zone (SACZ), like the SPCZ, is a semi-permanent feature aligned northwest–southeast. It is associated with convergent warm, moist air from the southwestern flank of the South Atlantic High and, further west, strong northwesterly flow originating from the Amazon Basin down the eastern flanks of the Andes (Figure 10.12). The latter airstream had previously entered the Amazon Basin north of about 15°S as tropical easterlies. Precipitation activity along the SACZ is enhanced by the strengthening and southward movement of the upper anticyclone (300–200 hPa level) over the Andes, termed the Bolivian high

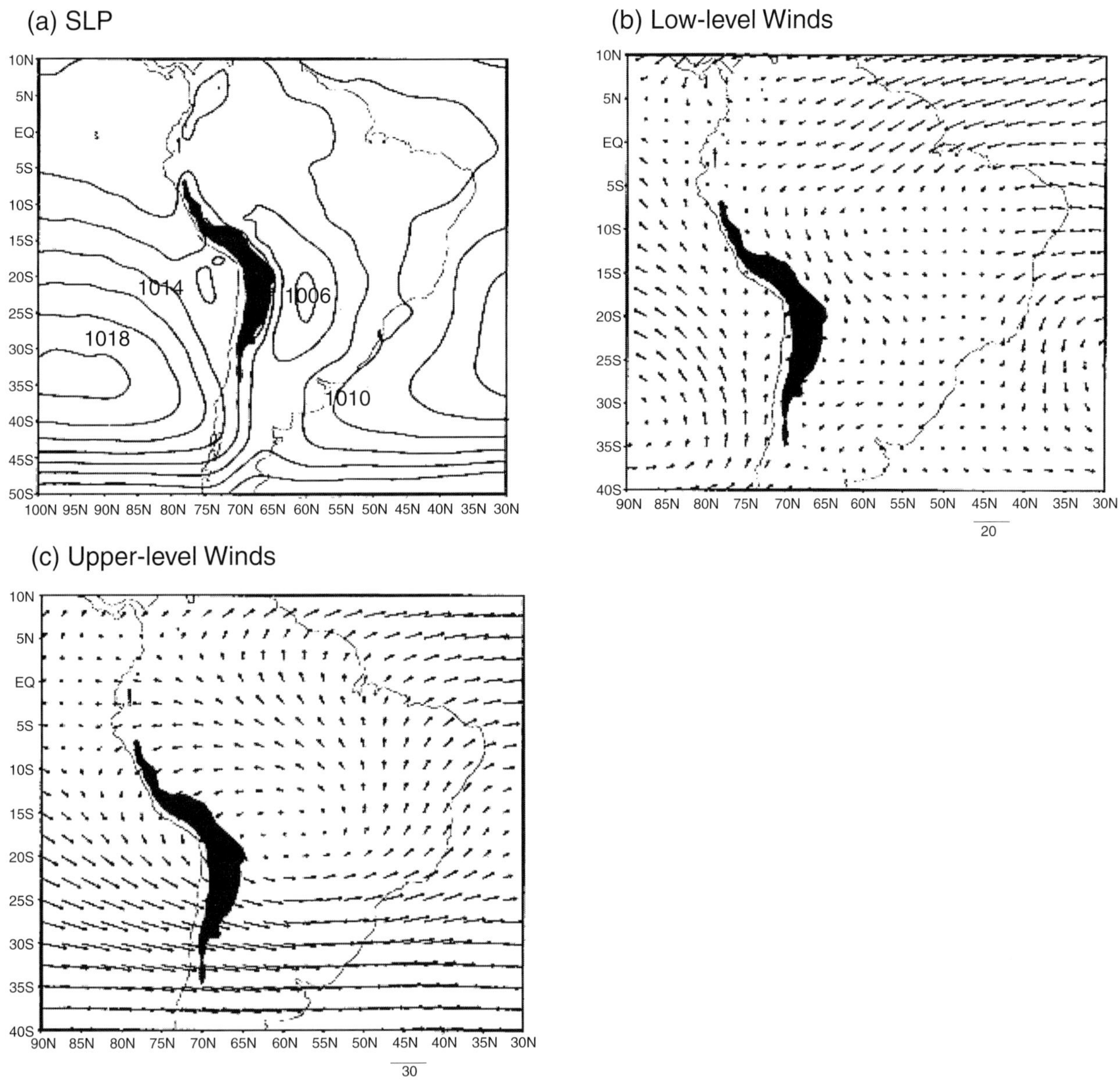

Figure 10.12 Austral summer (DJF) mean climatological fields of (a) SLP (contour interval is 2 hPa), (b) low-level (925–850 hPa) winds, and (c) upper-level (300–200 hPa) winds. (Source: Garreaud and Wallace, 1998.)

(Lenters and Cook, 1999). The SACZ (cf. SPCZ) is more active in the summer (austral) months, with Lenters and Cook (1999) identifying three anomalous circulation patterns that affected precipitation over the tropics at this time. Some of these anomalies are the result of the SACZ interacting with mid-latitude fronts and associated cyclonic circulations.

Kousky (1979), and later Molion (1993), identified a link between the convective cloudiness pattern over South America with the northward penetration of mid-latitude air through near decaying cold fronts, east of the Andes, which have the capability of penetrating the Amazon Basin. These disturbances are most prominent in December to February. They enhance convective activity in the southern Amazon Basin, to merge with and strengthen the SACZ. During this season, classifying such phenomena as 'cold fronts' (at the surface) is questionable (Garreaud and Wallace, 1998) but satellite imagery identifies them as a NW–SE oriented band of enhanced convection which is propagating equatorward. To the rear is an area of suppressed convection. The composites of convection anomalies shown as an index (Garreaud and Wallace, 1998) (Figure 10.13) clearly show a displacement of enhanced convection that moves from the mid-latitudes (40°–35°S) into the tropics (as far as 5°S) over five days,

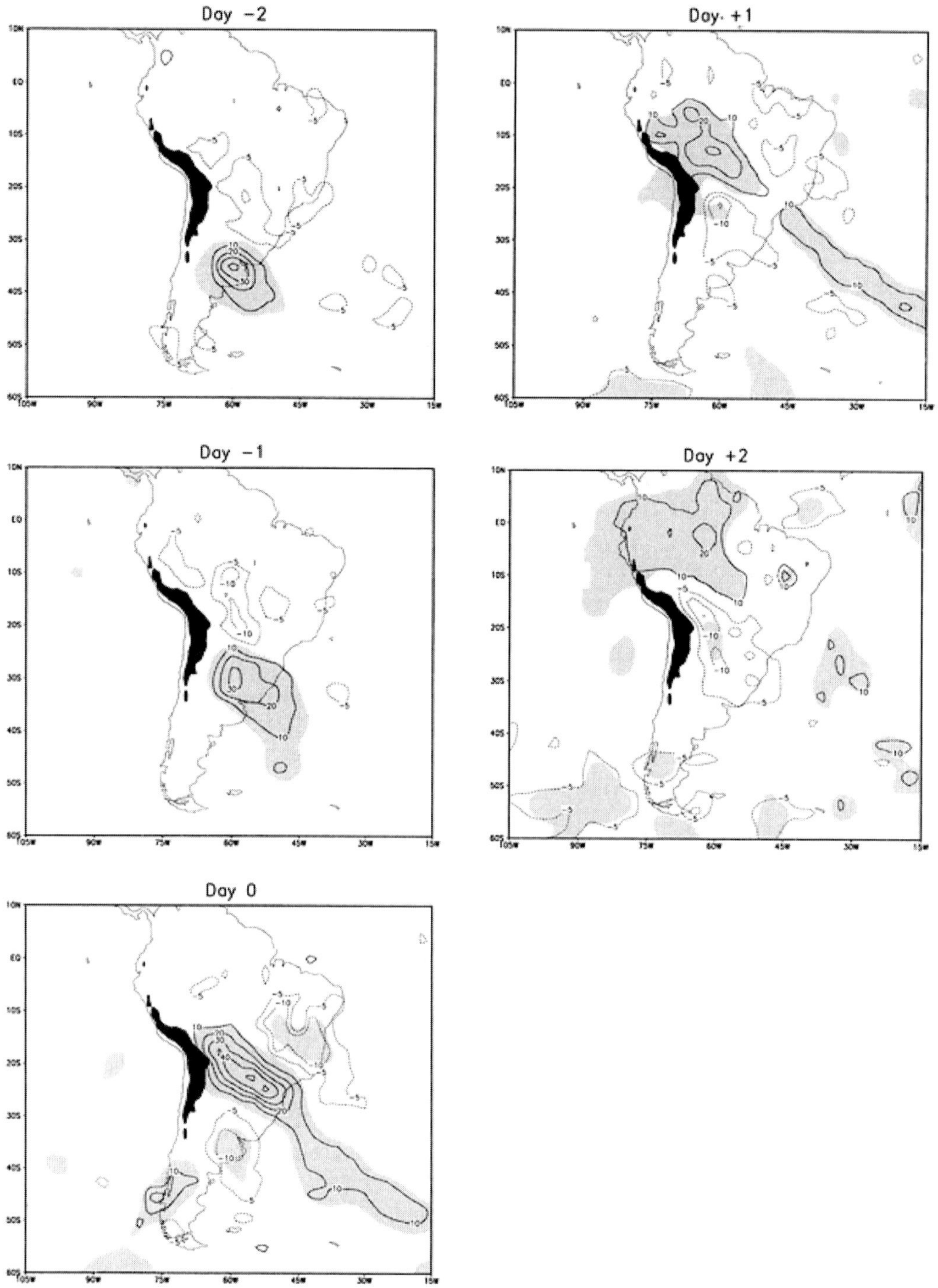

Figure 10.13 Convective index composite anomalies from day −2 to day +2. Contour interval is 10 W m^{-2} for positive anomalies (solid contours) and 5 W m^{-2} for negative anomalies (dotted contours). Black area indicates terrain elevations in excess of 2000 m. Light shading indicates the 95% significance level. (Source: Garreaud and Wallace, 1998.)

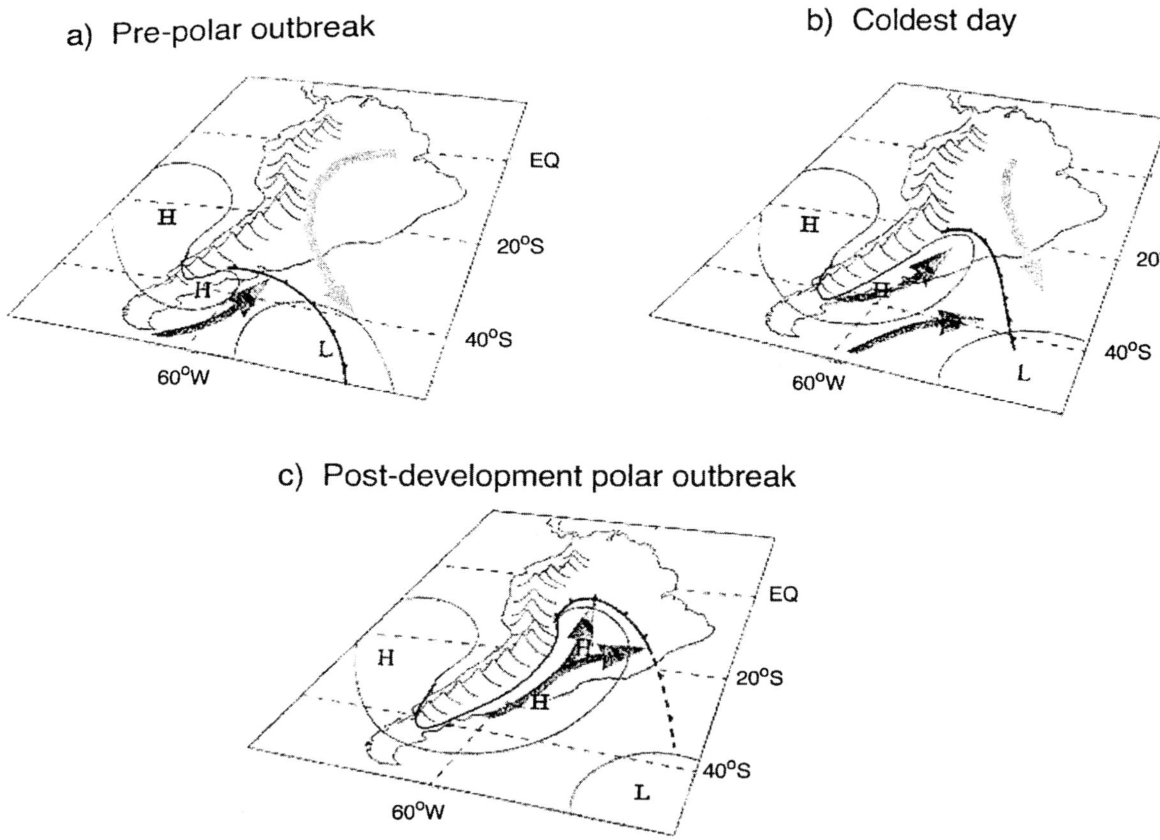

Figure 10.14 Conceptual model of wintertime polar outbreaks in South America (adapted from Marengo *et al.*, 1997; Garreaud, 2000.) (Source: Marengo and Rogers, 2001.)

at a mean speed of ~10 ms^{-1}. This band has longitudinal and transverse scales of 3000 × 800 km. Such equatorward migrations recur between seven and 12 days: they account for about 30% of total summertime precipitation over the west side of the Amazon Basin and ~20% over the northeast coast of South America. The upper Bolivian high has a strong influence on day +1 (Figure 10.13) when it covers a large fraction of the tropical domain and facilitates divergent outflow aloft for the deep convection over the southern and central parts of the Amazon basin. Such latent heat release strengthens the Bolivian upper high and is also possibly a cause of its existence. On days +1 and +2 drier southerly winds penetrate much of the western part of the Amazon basin in response to a transient, cold anticyclone over southern Argentina and northward displacement of the continental low pressure system along the eastern Andean flanks. The equatorward propagation of cooler air subsequently disappears due to the modification of this air mass over the warmer central part of South America.

Mid-latitude incursions to the east of the Andes occur throughout the year. These have more vigorous temperature falls and pressure rises (twice as large as in the summer season) associated with the southerly wind to the rear of these equatorward propagating fronts during the winter period (see description of Marengo *et al.*, 1997). Marengo and Rogers (2001) presented a conceptual model of wintertime polar outbreaks in South America, as shown in Figure 10.14. Paradoxically, minimal convectivity activity occurs (see Figure 10.16 in Garreaud and Wallace, 1998, p. 2729) during these winter episodes in comparison with other geographic regions where similar phenomena occur, e.g. Caribbean, South China Sea (Schultz *et al.*, 1997).

Among the precipitation anomalies described by Lenters and Cook (1999) is enhanced precipitation over the central Andes (day 0 in Figure 10.13) as a result of the westward shift of the intense, active SACZ and extra-tropical cyclonic activity southeast of central Andes. The SACZ separates warm, moist, low level inflow from the northwest from southerlies to the west of the extra-tropical cyclone. During this situation, dry conditions prevail over the Amazon Basin (Figure 10.13, Day 0).

The second anomalous precipitation pattern (Lenters and Cook, 1999) corresponds with Day +1 in Figure 10.13, except that a cold-cored subtropical low has developed off the coast of southern Brazil. This feature intensifies precipitation along the SACZ (including increased central Andean rainfall) in response to marked convergence of mid-latitude southerlies with low-level warm, humid northwesterly flow along the eastern flanks of the north-central Andes. A band of convergent westerlies occurs on the northern side of the sub-tropical low.

The third anomalous precipitation type of Lenters and Cook (1999) does not fit in with the composite model of Garreaud and Wallace (1998). There is a westward enhancement of the South Atlantic high, such that a separate anomalous high cell is located over south-central South America (Uruguay – southern Brazil). This results in a broad low-level easterly flow on to the central Andes, which feeds moisture from the eastern Amazon and South Atlantic, resulting in high precipitation. Further east, the outflow from the high cell has a more southerly component, which results in anomalously dry conditions over the Amazon Basin. A westward shift in the SACZ over the enhanced precipitation area is suggested, but in these circumstances the SACZ does not have the usual diagonal orientation: if anything, it is more north–south along the western flank of the high cell.

Cold surges into the humid tropics

In the preceding section we showed that whilst southerly 'cold' surges in the winter season marked by clearly definable cold fronts commonly occurred east of the Andes, their equatorward extensions were not significant rainfall producers in comparison to the summer season. Such variability in activity demands that more attention is paid to understanding the fate of mid-latitude cold fronts as they approach the tropical latitudes. In general, tropical atmosphere–cold front interactions are poorly understood but the penetration of mid-latitude air has long been recognised as a source for strengthening the trade winds and subsequently the activity of the zonal trough in the easterlies (ZTE) or monsoon shearline (e.g. Henry, 1979, 1980; Riehl, 1954).

The North American and East Asian continental areas (e.g. Lau and Chang, 1987; Schultz *et al.*, 1997) present the opportunity for the build-up and southward penetration of very marked cold surges in response to very low temperatures associated with cold continental anticyclones. In Central America, the Caribbean and even northern Venezuela. for example, such cold surges are common following the passage of an intense extra-tropical cyclone east of the Rocky Sierra Madre Mountains; these phenomena are termed the CACS (Central American Cold Surges).

Schultz *et al.* (1997) provided a case study of a CACS that occurred in March 1993 where the interactions between synoptic-scale air flow and mesoscale phenomena, such as topographic channelling, were complex. Thus at the mesoscale, the structure and southward penetration of the cold surge bore no similarities with those of classic cold fronts. There was a series of pre-frontal cloud bands which moved either side of the Sierra Madre at speeds much faster than the advective wind speed. Further above the surface (*c.* 900 hPa), the surge had a 'tipped-forward frontal structure', with the cold air advancing much quicker at this level. Schultz *et al.* (1997) put forward several hypotheses for explaining both the 'tipped forward' structure, as for example, surface friction and the impacts of adiabatic ascent–turbulent mixing in the lower atmospheric levels under a low subsidence inversion (850 hPa). In addition, various principles of the hydraulics of fluid flow were used to explain the presurge cloud bands such as pressure jumps, bores linked with gravity flow (the gravity current model). However, none of these theories explained the dynamics of the CAC boundary comprehensively.

In the context of the broader global debate of cold front penetration into the tropics, Schultz *et al.* (1997), did provide evidence to support the hypothesis that cold fronts in the tropics lose their frontal properties and become shear lines (density discontinuity) so that the pressure trough separates from the temperature gradient and, as previously mentioned, the front arrives first above the surface. For example, in eastern Mexico during the March 1993 event, the lowest pressure did not coincide with the onset of strong pressure rises, a rapid temperature drop and the strongest northerlies. Rather, the lowest pressure preceded these features by several hours and 150–200 km, thus implying that the leading edge of the surge reformed 'in prefrontal air' as if it were a pressure jump propagating along an elevated inversion within the prefrontal air. On the other hand, only one ground station could provide any data to support the possible existence of this mechanism (Schultz *et al.*, 1997).

The impact of the cold surge on precipitation was highly variable. Maximum rain totals in Central America occurred on the east- and north-facing slopes with a local maximum of 79 mm in Honduras. Short-term intensities were low, and the Honduran rain amount emanated from a two-day period of nimbostratus drizzle. Further south, smaller amounts were recorded – usually 12 mm or less – but there were streamflow rises of 10–20 times pre-storm levels in Costa Rica. During the later stages of the surge, Colombia recorded a station maximum of 99 mm and 119 mm from a squall line in Cuba (Schultz *et al.*, 1997). The largest totals in most cases were caused by orographically enhanced precipitation.

INTRA-ANNUAL VARIATIONS: THE MADDEN–JULIAN OSCILLATION

Kelvin waves

A Kelvin wave is a kind of wave in a rotating system that relies on gravity for its restorative force and is called a boundary wave because it transmits along the boundary of the fluid. The amplitude of the wave decreases exponentially with distance from the boundary (or wall). In the case of inertial gravity waves, which

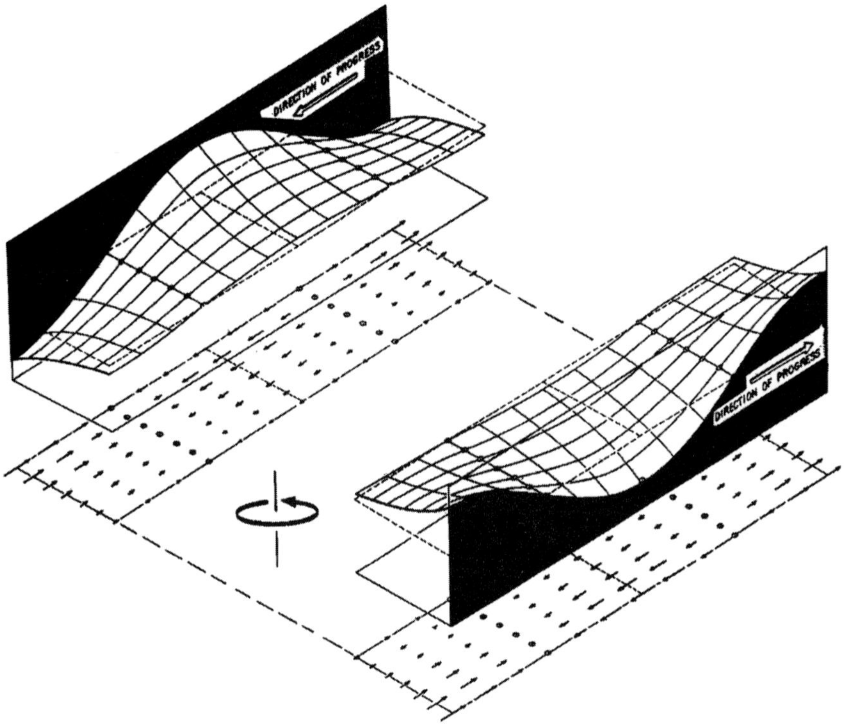

Figure 10.15 Northern hemisphere coastally trapped Kelvin waves in the ocean on opposite sides of a channel that is wide compared with the Rossby radius. In each vertical plane, which is parallel to the coast, the currents (shown by the arrows) are entirely within the plane and are exactly the same as those for a long gravity wave in a non-rotating channel. However, the surface elevation varies exponentially with distance from the coast in order to give geostrophic balance. This means Kelvin waves move with the coast on the right in the Northern Hemisphere and on their left in the Southern Hemisphere. (Source: Gill, 1982.)

are another kind of gravity wave, the Coriolis force forms part of the restorative force whereas Kelvin waves use the Coriolis forces to push against the wall and not as a restorative force. Figure 10.15 is a schematic example of a coastally trapped oceanic Kelvin wave. The movement of the particles of the fluid in a Kelvin wave is always parallel to the wall, so the Coriolis force, which acts on them, balances out with the pressure gradient perpendicular to the wall. In other words, Coriolis force does not act parallel to the wall, so the period of a Kelvin wave becomes the same as that of a stationary system. An important property of the Equator is that it acts as a wave-guide to disturbances in the atmosphere and ocean. These disturbances are trapped in the vicinity of the Equator. The simplest wave that illustrates this property is a Kelvin wave (Gill, 1982). The wave guide properties of the Equator come from the change in sign of the Coriolis force across the Equator.

Madden–Julian oscillation (MJO)

At mid-latitudes the weather is governed largely by the upper-tropospheric Rossby waves, which are controlled by the rotation of the Earth. These waves interact with surface conditions in a process called baroclinic instability. No such dominant wave motion exists in the tropics and as a consequence the weather is less predictable for the one-to-ten day period. Until recently, it was believed that tropical weather variations on time scales of less than a year were essentially random.

Madden and Julian (1971) discovered a 40–50 day oscillation when analysing zonal wind anomalies in the tropical Pacific, which became known as the Madden Julian oscillation (MJO) (reviewed and shown in Figure 10.11 (a) in Manton and Bonell, 1993). Further studies showed the MJO, also referred to as the 30–60 day or 40–50 day oscillation, to be the main fluctuation that caused weather variations in the tropics over periods of less than a year. Equatorially trapped waves (Kelvin and Rossby waves) are the driving mechanism for the MJO. These waves occur in the entire troposphere from 30°N to 30°S, mainly in the Eastern hemisphere. The MJO exhibits the mixed Kelvin-Rossby wave structure over the eastern hemisphere, but over the western hemisphere it shows a Kelvin wave structure only. The MJO affects the entire tropical troposphere but is most evident in the Indian and western Pacific Oceans. The MJO involves variations in wind, sea surface temperature (SST), cloudiness and rainfall. Because most tropical rainfall is convective, and convective cloud tops are

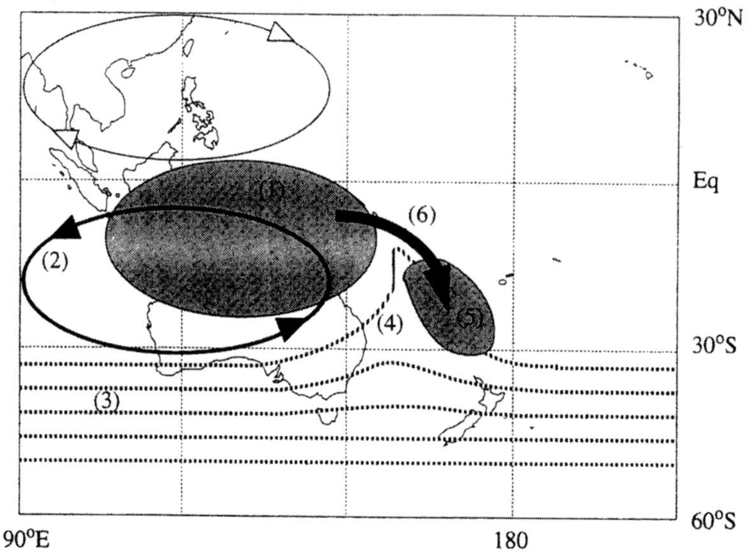

Figure 10.16 Schematic of hypothesised mechanism for the development of convection along the South Pacific Convergence Zone (SPCZ) during a Madden–Julian oscillation (MJO). Convection over Indonesia (1) associated with the passage of a MJO leads to an upper tropospheric anticyclone (2). Poleward of the anticyclone there is a large potential-vorticity (PV) gradient, associated with the subtropical jet and the tropopause (3). Equatorward advection of high PV air on the eastern flank of the anticyclone leads to an upper tropospheric trough (4), which induces deep ascent to the east (5). This region of deep ascent, to the southeast of Indonesia, is over the SPCZ, an area susceptible to deep convection. Hence strongly enhanced convection can be triggered by the deep ascent, and convection develops from Indonesia in the SPCZ (6). (Source: Matthews *et al*., 1996.)

very cold (emitting little long-wave radiation), the MJO is most obvious in the variation in outgoing long-wave radiation (OLR), as measured by an infrared sensor on a satellite.

These OLR anomalies in the eastern hemisphere propagate to the east at around 5 m s^{-1}. The OLR signal in the Western hemisphere is weaker, and the recurrence interval for the eastward propagating OLR anomalies in the eastern hemisphere is about 30 to 60 days. How exactly the anomaly propagates from the dateline to Africa (i.e. through the western hemisphere) is not well understood. It appears that near the dateline a weak Kelvin wave propagates eastward and poleward at a speed exceeding 10 m s^{-1}.

Associated with the propagation of convective anomalies, the MJO has impacts on the global circulation. The MJO affects the intensity and break periods of the Asian and Australian monsoons. Wet spells in the Australian monsoon occur about 40 days apart. The oscillation is stronger in the northern hemisphere winter. It is also in this season that the negative OLR anomalies are most likely to propagate along the Equator from the Indian Ocean to the central Pacific Ocean. In the northern hemisphere summer, many of the anomalies veer away from the tropics before they make it to the central Pacific.

Matthews *et al.* (1996) presented a mechanism for linking Rossby waves along the SPCZ with Kelvin gravity waves of the MJO in equatorial latitudes based on an intensive study in March/April 1988. Figure 10.16 presents a conceptual summary of the mechanism. The large-scale convection over the Indonesian archipelago (influenced by the MJO) forces a Rossby wave propagation along the SPCZ which is facilitated by large potential vorticity (PV) gradients of an upper anticyclone, in combination with an upper trough. Subsequently, convection continues eastwards along the Zonal Trough in the Easterlies (ZTE), in accordance with the eastward translation of the MJO, as well as southeastwards along the SPCZ.

The occurrence of Kelvin waves in equatorial latitudes associated with the MJO and the spawning of Rossby waves meridionally towards the outer tropics is also a significant feature of the Indian monsoon (Maloney and Hartmann, 1998; Krishnamurti *et al*., 1997). The regulation of the Indian monsoon between *bursts* and *break* periods is strongly phase locked with MJO events (Madden and Julian, 1994). Through composites (see Maloney and Hartmann, 1998, for details of their construction), the eastward progression of the MJO spawned respective Rossby wave perturbations in phase 5 and 6 (Figure 10.17). These Rossby wave circulations propagate into both the northern and southern hemispheres both poleward and westward away from the equatorial convective area. This takes the heavy precipitation area towards the Indian sub-continent and SE Asia. Elsewhere Krishnamurti *et al.* (1997) show that these 'Rossby wave trains' move subsequently into the north Pacific Ocean area.

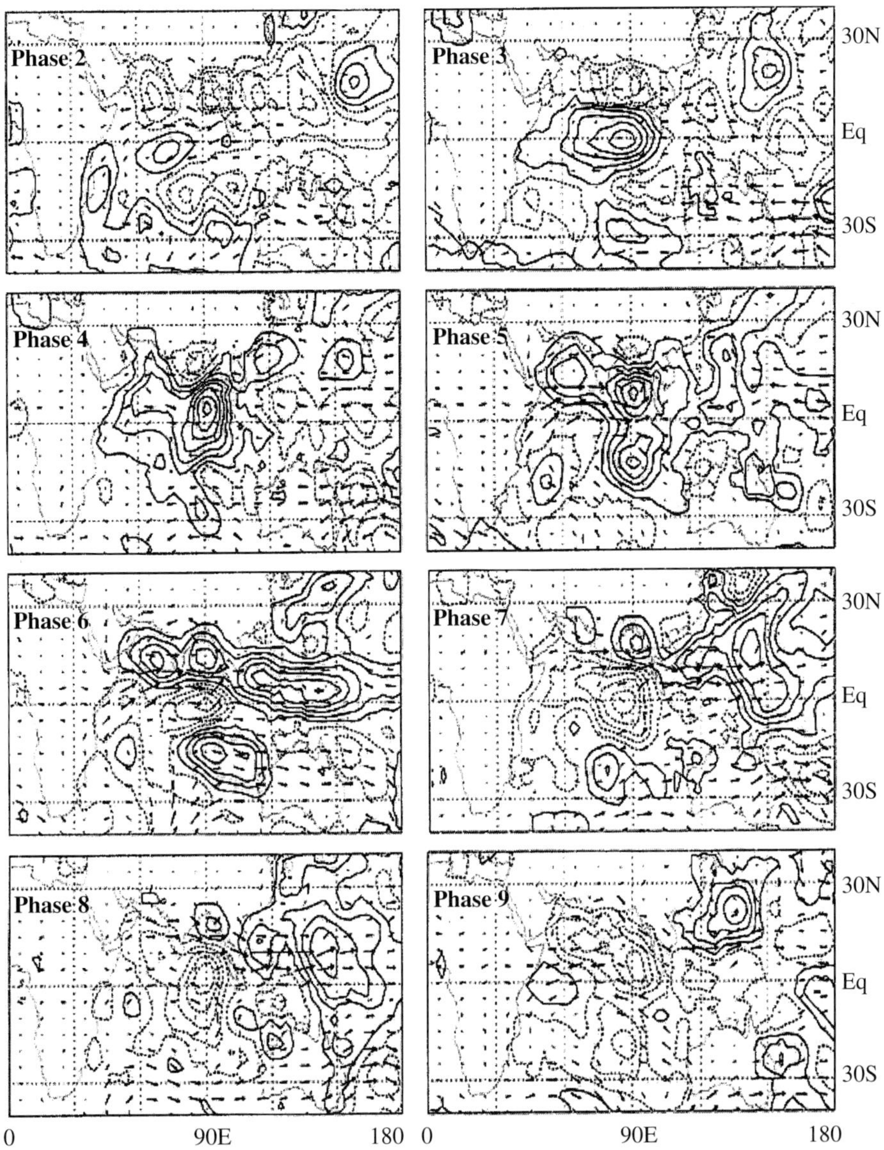

Figure 10.17 May–October 850-mb wind anomalies and microwave sounding unit (MSU) precipitation anomalies as a function of phase for 1979–95. Wind vectors significant at the 90% level are plotted in black and non-significant vectors are gray. Contours are every 0.6 mm day^{-1} starting at 0.3 mm day^{-1}. Maximum vectors are 3.3 m s^{-1}. Dashed contours are negative. (Source: Maloney and Hartmann, 1998.)

Following the spawning of Indian Ocean Rossby waves, the 850 hPa westerly flow (which had previously replaced easterly flow) 'dries out' the equatorial region of the Indian Ocean, thus terminating convection. The centre of convective activity (Phase 7–9, Figure 10.17) then shifts eastwards into the west Pacific with further Rossby wave disturbances (i.e. tropical cyclones and depressions) spawned in a manner similar to the Indian Ocean, for example, in the SPCZ (Krishnamurti *et al.*, 1997). Two other features need mention. Positive water vapour anomalies in the low-level convergent easterlies (in accordance with equatorial Kelvin wave theory) to the east of the convection area precede the positive precipitation anomalies. The equatorial westerlies, which are strengthened by the formation of Rossby waves, are divergent to the west of the principal convective area (in accordance with equatorial Kelvin wave theory; Maloney and Hartmann, 1998).

INTERANNUAL VARIATIONS

The El Niño/Southern Oscillation (ENSO) in Australia

Variations in Pacific Ocean temperatures are often associated with ENSO, the most important coupled ocean-atmosphere phenomenon causing global climate variability on interannual time scales. A useful measure of ENSO activity is the Southern

Oscillation Index (SOI) which is defined here as ten times the normalised difference in monthly pressure anomaly between Tahiti and Darwin.

The Southern Oscillation Index (SOI) is defined by

$$SOI = 10 \times [dP(Tahiti) - dP(Darwin)]/SD \quad (10.1)$$

where dP(Tahiti) = Tahiti monthly pressure anomaly (monthly mean minus 1882–1985 mean, averaging 3-hourly observations); dP(Darwin) = Darwin monthly pressure anomaly (monthly mean minus 1882–1985 mean, averaging 0900 hr, 1500 hr observations); SD = monthly standard deviation of the difference.

At the negative end of the SOI lies El Niño, where tropical waters around Australia often have relatively cool temperatures while waters over the equatorial Eastern Pacific are anomalously warm. El Niño is often associated with drought in Australia (e.g. Nicholl, 1992; Allan, 1991).

At the positive end of the SOI spectrum lies La Niña, where tropical waters around Australia have relatively warm temperatures while waters over the equatorial eastern Pacific are anomalously cool. La Niña is associated with increased rainfall in Australia.

The Tropical Cyclone Coastal Impacts Program (TCCIP) was launched in Australia in 1994 to help focus research attention and resources on the problem of increased hazard levels and vulnerability of coastal communities from tropical cyclone impacts. As part of the TCCIP, the record of tropical cyclones in eastern Australia has been reviewed. This review follows improvements that have occurred in the knowledge of tropical cyclone structure since the original case studies were constructed. Additionally, since the creation of the Regional Severe Weather Sections in the Bureau of Meteorology in 1988, dedicated staffing resources are available to create tropical cyclone case histories and review past case histories.

In all, between 1878 and 2000, 179 tropical cyclones are known to have had an impact over eastern Australia. Below is an analysis of the relationship between their rate of incidence with the SOI. The SOI figure used is the three months mean, centred on the month of the occurrence of the event.

75 impacts occurred when the SOI was greater than 5
40 impacts occurred when the SOI was between zero and 5
26 impacts occurred when the SOI was between zero and −5
38 impacts occurred when the SOI was less than −5

This analysis highlights the fact that there is a strong relationship between eastern Australian tropical cyclone impacts and the SOI, with almost twice as many impacts during La Niña ($SOI > 5$) than during El Niño ($SOI < -5$).

South Pacific Ocean

In the cold (La Niña) phase of ENSO, the zonal trough in the easterlies (ZTE) replaces the southern monsoon shearline near 180°E (Sadler *et al.*, 1987). For the South Pacific islands, mean monthly rainfalls are spatially highly variable, ranging from 30 to 450 mm (see Figures 3 and 4 in Basher and Zheng, 1998; Falkland and Brunel, 1993).

Also, these islands are particularly vulnerable to the interannual variability of rainfall arising from movements from the negative (warm) phase of ENSO to corresponding positive (cold) phases. The impact of ENSO on the interannual rainfall of Kiribati, for example, is provided by Falkland *et al.* (1991) and Falkland and Brunel (1993), where annual rainfall ranged between just less than 200 mm to about 1800 mm over the period 1950–1982. Significantly, the strongly negative ENSO phases all produce above average annual rainfalls which coincide with the eastward progression of the equatorial low level westerlies and associated monsoon shearlines (Manton and Bonell, 1993). Basher and Zheng (1998) observed that during such ENSO phases, composites reveal more uniform rainfall fields with weaker meridional gradients and a less developed equatorial dry zone. In contrast, during the La Niña (positive ENSO phase) the meridional rainfall gradients are much steeper and the equatorial dry zone, east of 180°, becomes extensive. Thus during negative ENSO phases, Basher and Zheng (1998) noted that the annual 200 mm isohyet shifts eastward by nearly 3000 km. During positive ENSO phases, this trend is reversed and the 50 mm isohyet moves westward by 3000 km in association with the enlargement of the equatorial dry zone. Basher and Zheng (in Figure 4, 1998) show the impact of ENSO phases on three–monthly rainfall total for the southwest Pacific.

Basher and Zheng (1995) also highlighted a marked change in the distribution of tropical cyclones between warm and cold Pacific episodes. During La Niña episodes, tropical cyclones tend to be confined to the area west of the date line as in Figure 10.8. During El Niño, however, they are distributed right across the Pacific and occur well east of the date line. For example during the strong negative phase of 1982–1983, Sadler (1984) (reviewed in Manton and Bonell, 1993) observed the most easterly occurrence on record of a cyclone storm which took place at 8°S, 113°W. In addition, five storms traversed the Society Islands within 500 km of Tahiti where only nine storms had occurred between 1939 and 1982.

Northwest Pacific–Asia

Chen and Weng (1998a, b) showed that the points of development and frequency of tropical synoptic-scale disturbances exhibits interannual variation in the northwest Pacific. In warm summers (El Niño conditions) the occurrence of tropical synoptic waves is confined to the region south of 15°N, but stretched longitudinally eastward across the International Date Line. In contrast, in cold summers (La Niña conditions) the occurrence of tropical synoptic disturbances is concentrated in the region west of 150°E when the north–south extent is enlarged but the longitudinal spread is reduced.

Residual lows of tropical disturbances from the South China Sea – western tropical Pacific region may propagate westward over northern Indochina to reach the Bay of Bengal. It is possible for these residual lows to revive over the Bay of Bengal and to form monsoon depressions. During cold summers, the maximum occurrence and frequency of tropical synoptic disturbances shifts westward. The interannual variation in the westward propagation of residual lows from the western tropical Pacific may also be a contributing factor in the interannual variation of Indian monsoon rainfall.

The active tropical cyclone season in the northwest Pacific covers the boreal summer and autumn. Chen and Weng (1998a, b) noted the following characteristics and interannual differences in the summer locations of tropical cyclones:

(1) the climatological latitudinal location of the monsoon trough is 15°N;
(2) the eastern end of the climatological monsoon trough is near 150°E;
(3) interannually, the monsoon trough undergoes a north–south migration as well as a longitudinal variation;
(4) during cold (La Niña) summers the monsoon trough exhibits a northward migration across 15°N and westward retreat across 150°E;
(5) during warm (El Niño) summers the monsoon trough exhibits a southward migration across 15°N and an eastward extension across 150°E;
(6) these interannual variations in the monsoon trough result in the enhancement (reduction) of tropical cyclone genesis frequency north of 15°N and west of 150°E during cold (warm) summers.

During autumn, the South China Sea – western tropical Pacific monsoon diminishes, but the monsoon trough still exists between 10° and 15°N. Unlike the summer season, the autumn monsoon trough does not show a pronounced north–south migration. However, the monsoon trough undergoes a longitudinal variation in accordance with the interannual variation of the SST in the central and eastern equatorial Pacific:

(1) the monsoon trough extends (retreats) eastward (westward) across 150°E during warm (cold) autumns;
(2) the tropical cyclone genesis frequency west of 150°E is suppressed (enhanced) accordingly in warm (cold) autumns.

Asian monsoon variability

The Asian summer monsoon is a major feature of the general circulation that dominates the Eastern hemisphere for over one-third of the year, with its influence extending to many regions remote from South East Asia. For example, there is increasing evidence to suggest that the arid regions of north Africa and the dry summers of the eastern Mediterranean are a direct consequence of the Asian summer monsoon (Rodwell and Hoskins, 1996). The monsoon displays substantial interannual variability which can have profound social and economic consequences. This variability is related intimately with the phase of ENSO (e.g. Rasmusson and Carpenter, 1983; Webster and Yang, 1992.

Indian rainfall

Many researchers have studied the way the Indian monsoon rainfall (IMR) varies on the interannual time scale (Sikka, 1980; Pant and Parthasarathy, 1981; Rasmusson and Carpenter. 1983; Shukla and Paolino, 1983; Parthasarathy and Pant 1985; Shukla, 1987). If we consider the Niño-3 (150°W–90°W and 5°N–5°S) SST anomaly as an index of the ENSO variability, the long-term (based on 127 years) correlation between the seasonal anomalies of IMR and Niño-3 SST is 0.63, significant at 99.9% level. This correlation represents a tendency of the Indian monsoon to be below normal when the eastern Pacific is warm.

One half of the Indian monsoon rainfall is brought to this subcontinent by monsoon depressions. Chen and Weng (1999) showed that over a period of 16 summers (1979–94), only five of 96 monsoon depressions were formed by *in situ* genesis in the Bay of Bengal. The large majority of monsoon depressions (91 of 96, 95%) developed from the re-genesis of the westward-propagating residual lows from the east. Saha *et al.* (1981) traced their residual lows to tropical cyclones and weak disturbances generated indirectly in association with tropical cyclones and by land genesis over Indochina. In addition to these three types of residual lows, there exist two other types derived from 12–24-day monsoon lows and equatorial waves. However, only two monsoon depressions over the 16 summers (1979–94) were linked to equatorial waves.

A clear interannual variation emerges from the genesis of monsoon depressions and frequency. Their enhancement (reduction) is, on average, about one-third of the climatological level during the cold (warm) summers. The large majority of monsoon depressions, 77 out of 91 (85%), analysed in this study are formed by residual lows related to tropical cyclones and 12–24-day monsoon lows. The occurrence frequencies of tropical cyclones and 12–24-day monsoon lows combined in the South China Sea (SCS) region exhibit an interannual variation tendency highly correlated (with a correlation of 0.93) with that of monsoon depressions formed by the residual lows of these weather disturbances. The interannual variation of monsoon depression formation is attributed primarily to the interannual variation of tropical cyclones and 12–24-day monsoon lows combined over the western tropical Pacific–SCS region. The latter interannual variation is regulated by the interannual variations of the summer circulation over the western tropical Pacific-SCS region in response to SST variations in the central and eastern equatorial Pacific.

Thailand

It has been observed by the Meteorological Department of Thailand that the dry–wet conditions in Indochina are highly correlated with the occurrence and frequency of westward-propagating tropical cyclones and other strong synoptic disturbances from the South China Sea region (Chen and Yoon, 2000). These disturbances extend further westward towards Indochina during cold (La Niña) summers, adding to the rainfall over Indochina.

Asian winter monsoon

Wang *et al.* (2000) focussed on major El Niño (1957/58, 1965, 1972, 1982/83, 1991/92, 1997/98) and La Niña (1970/71, 1973/74, 1975/76, 1984/85, and 1988/89) episodes for the period from 1950 to 1998. The El Niño (La Niña) composite showed weaker (stronger) than normal northeasterly winter monsoon along the East Asian coast and a warmer (cooler) than normal winter in that region.

THE ENSO EFFECT ON GLOBAL RAINFALL

The following images were provided by the NOAA-CIRES Climate Diagnostics Center, Boulder Colorado from their web site at: http://www.cdc.noaa.gov/

The years used for ENSO composites:

(1) Warm events: 1877 1880 1884 1891 1896 1899 1902 1904 1911 1913 1918 1923 1925 1930 1932 1939 1951 1953 1963 1965 1969 1972 1976 1982 1986.
(2) Cold events: 1886 1889 1892 1903 1906 1908 1916 1920 1924 1928 1931 1938 1942 1949 1954 1964 1970 1973 1975 1988.

Summer and winter global rainfall for both hemispheres for the year after the El Niño or La Niña begins are compared in Plates 1 and 2. The rainfall anomalies can be seen to change sign in many places over the tropics between the El Niño and the La Niña composites.

ROLE OF SYNOPTIC SCALE WEATHER SYSTEMS IN THE ONSET OF EL NIÑO

The forecast of the 1997–98 El Niño, one of the strongest on record, was not predicted as well as had been hoped (Pearce, 1997; McPhaden, 1999). It was thought that an El Niño could be predicted a year in advance. However, from van Oldenburgh (2000), most predictions (Stockdale *et al.*, 1998; Ji *et al.*, 1996, forecasts are available online at http://nic.fb4.noaa.gov; Schneider *et al.*, 1999, forecasts available online at http://www.iges.org/ellfb; Kleeman *et al.*, 1995, forecasts available online at http://www.bom.gov.au/bmrc/mrlr/r2k/climfcn3.htm) only started to indicate a weak event six months ahead of time. One reason for this may be that El Niño depends not only on slow internal factors but also on external noise in the form of weather events in the western Pacific.

The 1997/1998 ENSO event developed more rapidly than any other since sufficient instrumental records have been available to monitor such phenomena. The reason it developed so quickly may be linked to the influence tropical cyclones played in its initiation, as will be outlined below.

Equatorial westerly gales

Earlier in the season (preceding the 1997/1998 ENSO) a strong equatorial westerly wind episode occurred north of New Guinea. From 20 December 1996 to 26 December 1996 the wind data for the 850 hPa level, obtained from the European Centre for Medium Range Weather Forecasting computer model output (EC), showed an area of westerly winds with speeds >15 m s^{-1} north of Papua New Guinea (Figure 10.18). This area was between the typhoon *Fern* in the Northern hemisphere and a very active monsoon trough south of the Equator containing two developing tropical cyclones, *Phil* (Gulf of Carpentaria) and *Fergus* (Coral Sea).

Later in the season there was a much more intense belt of westerly gales along the Equator. At 1200 UTC on 4 March 1997 (Figure 10.19) an active monsoon trough extended east northeastwards from northern Australia towards tropical cyclone *Gavin*. By 1200 UTC on 6 March 1997, tropical cyclone *Justin* had absorbed much of the monsoon trough in the Australian region and was developing off the Queensland coast. At 1200 UTC 8 March 1997, a large area of monsoon gales extended from Papua New Guinea across to the date line north of *Justin* and *Gavin*. *Gavin* then moved down towards New Zealand and the monsoon winds eased a little. However, a new cyclone (*Hina*) was developing at that time north of Fiji and by 1200 UTC 14 March 1997 *Justin* moved up towards the southeastern tip of Papua New Guinea.

The following comments were added to the Satellite Bulletin issued by the Brisbane Tropical Cyclone Warning Centre (BTCWC) at 0515 UTC 14 March 1997:

> Of interest is the strong near-equatorial pressure gradient between Irian Jaya and this low, producing a fetch of 4000 km of near gale force winds directed towards Tuvalu. Models are forecasting these winds to further increase. Dangerous swell conditions are expected to worsen over the next few days.

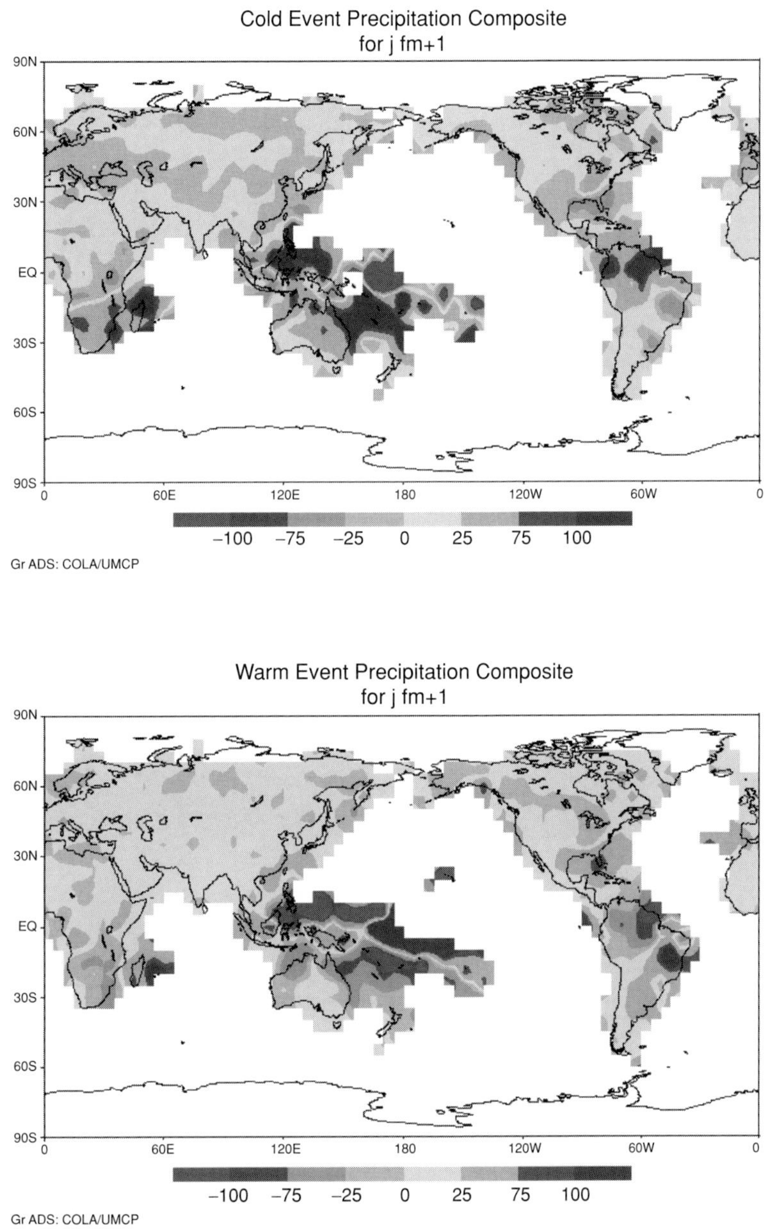

Plate 1 Global rainfall anomalies January/February/March for the years following the onset of La Niña (top) and El Niño (bottom). Units are mm/month. (Source: NOAA-CIRES Climate Diagnostics Center: http://www.cdc.noaa.gov/) (See also colour plate section.)

The low referred to the developing *Hina*. The vast area of gales attracted numerous ship observations. The analysed ECMWF wind data at the 850 hPa (lower right panel in Figure 10.19) gave an excellent portrayal of the near-equatorial MSL wind fields associated with this event, except that the actual MSL winds were a little stronger than indicated along the Equator.

Kelvin waves and El Niño

Rasmusson and Carpenter (1982) found significant westerly wind anomalies occurring over the equatorial western Pacific region in the onset phase of El Niño events over the period 1950–1976. An important property of the Equator is that it acts as a wave-guide to disturbances in the atmosphere and ocean. These disturbances are trapped in the vicinity of the Equator. The simplest wave that illustrates this property is the previously described Kelvin wave (Gill, 1982). Equatorial Kelvin waves over the ocean can be initiated by strong westerly winds blowing along the Equator for up to a week. Typically, these waves would take about two months to travel from the warm waters near Papua New Guinea to South America. The Kelvin waves have two effects: they generate anomalous east-west currents and they depress the thermocline. The waves move

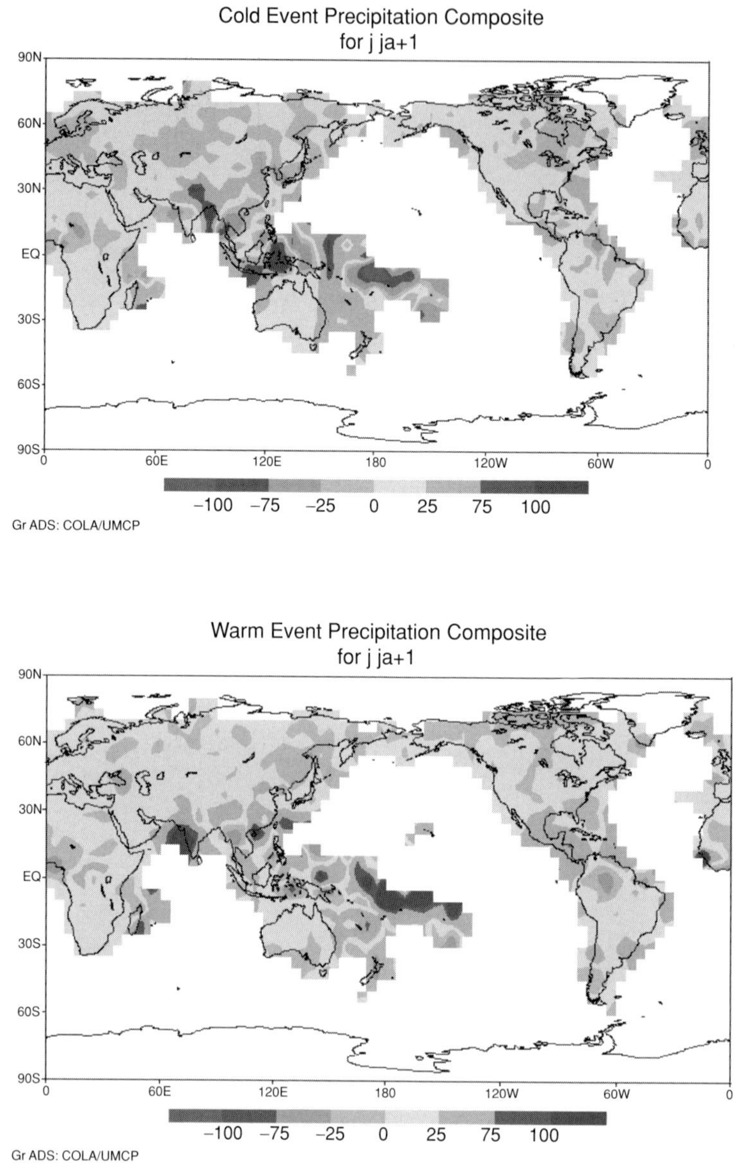

Plate 2 Global rainfall anomalies June/July/August for the years following the onset of La Niña (top) and El Niño (bottom). Units are mm/month. (Source: NOAA-CIRES Climate Diagnostics Center: http://www.cdc.noaa.gov/) (See also colour plate section.)

warm water eastwards, allowing it to accumulate in the central Pacific and also depress the thermocline in the vicinity of the South American coast helping to warm the sea surface temperature (SST) there. It has been speculated that a possible mechanism to produce westerly wind anomalies may be tropical cyclones on either side of the Equator (twin cyclones) in the western-central Pacific (Keen, 1982).

The Kelvin wave signatures in ocean temperature analyses leading into the 1997/1998 El Niño are shown in Plate 3. Initially, a wave following the December 1996 event travelled across the equatorial Pacific, depressing the thermocline in its wake. The evidence of this wave in Plate 3 is the orange to red band moving across the Pacific during January/February 1997 and signifying a lowering of the 20 °C isotherm (core of the thermocline). A much stronger Kelvin wave is evident in Figure 10.19 associated with the massive westerly gale event involving tropical cyclone *Justin*. As the central and eastern equatorial Pacific warmed up, a series of un-seasonal tropical cyclones and lows developed, mostly east of the International Date Line in the South Pacific. This began with tropical cyclone *Keli* in June, tropical lows in September and October, tropical cyclones *Martin, Nute, Osea* in November and *Pam* in December. All this activity plus the associated westerly wind events as shown in Plate 3, accounts for the large Kelvin wave activity in the latter half of 1997.

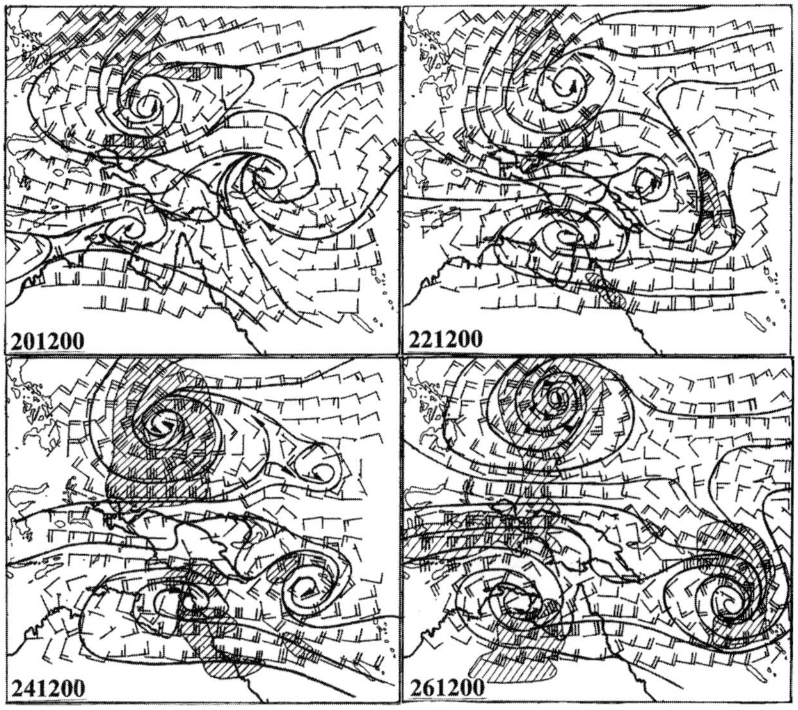

Figure 10.18 Sequence of 850 hPa streamlines and wind plots every 48 hr from 1200 UTC 20 December 1996 (201200) to 1200 UTC 26 December 1996 (261200). Hatched area denotes area where wind speeds exceed 15 m s^{-1}. The domain in each panel extends eastwards to 180° (date line) and the equator is located where wind plots change to the Northern hemisphere plotting convention. The plotting convention used is where half barbs/barbs/flags denote wind speeds of 2.5 m s^{-1} (5 knots)/5 m s^{-1} (10 knots)/25 m s^{-1} (50 knots) respectively.

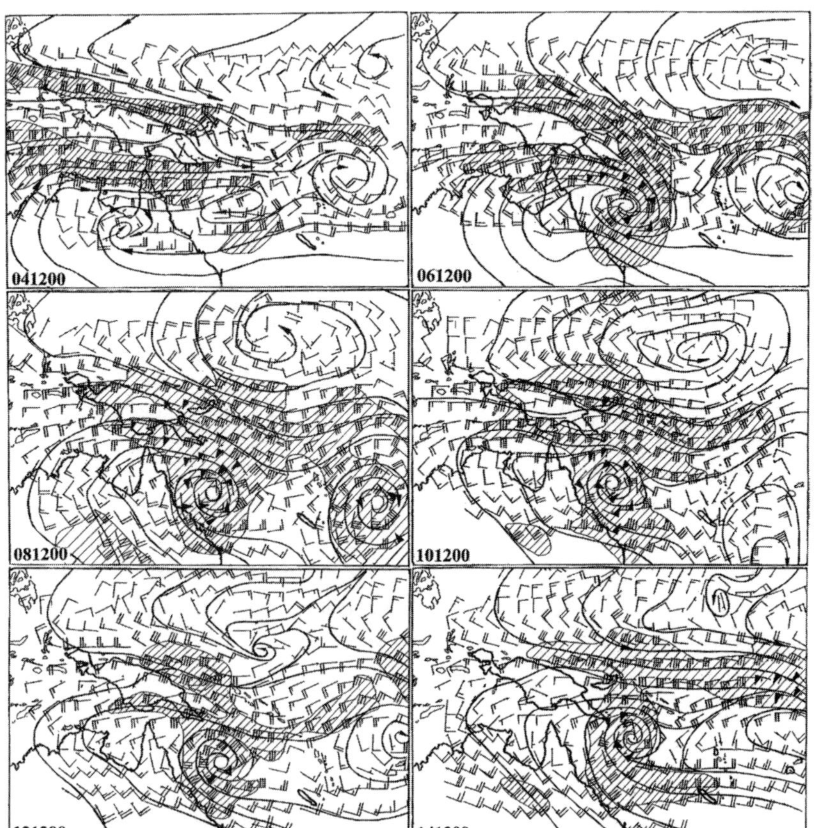

Figure 10.19 As in Fig. 10.18 but period is 1200 UTC 4 March 1997 (041200) to 1200 UTC 14 March 1997 (141200).

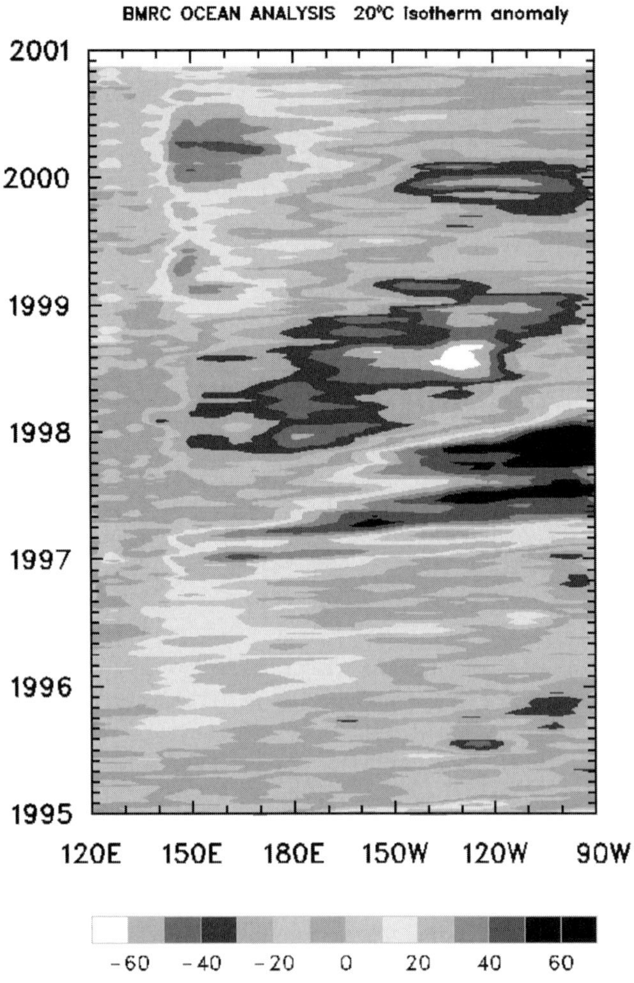

Plate 3 Hovmoller (time/longitude) diagram of the anomaly of the 20 °C depth (metres) for period 1995 to 2000. (Source: Australian Bureau of Meteorology.) (See also colour plate section.)

At the end of January 1997 the Southern Oscillation Index (SOI) was +4.1 and the SST in the equatorial Pacific reflected a weak La Niña pattern with weak cool anomalies in the eastern Pacific. Similarly, at the end of February 1997 the SOI was +13.1 and a weak La Niña SST pattern was still evident. During March 1997 warm anomalies began to appear near the South American coast. Over succeeding months the equatorial warm anomalies continued to develop over the central and eastern Pacific and by July the characteristic El Niño pattern had evolved.

Vertical sections of temperature anomalies across the equatorial Pacific are examined in Plate 4 to show the eastward traverse of warm deep anomalies after the March 1997 cyclone event leading to the onset of the El Niño.

Comparison with the 1982/1983 El Niño

The development of the almost equally strong 1982/1983 El Niño was more gradual. A twin cyclone event occurred near Papua New Guinea in late March/early April 1982 (*Odessa* (NH), *Bernie* (SH)). The associated equatorial gales were limited in horizontal extent and no strong Kelvin wave signature was evident in the ocean temperature analyses. Another equatorial westerly wind episode occurred in June 1982. However, the strongest Kelvin wave signature was observed in November 1982 in the central Pacific. This was associated with Hurricane *Iwa* and an unnamed circulation in the Southern hemisphere.

INTERDECADAL VARIABILITY

West African interdecadal variability

Elsewhere in west Africa, considerable attention has been given to the causal factors of the Sahelian drought which are mostly unconnected with the ENSO phenomenon. For example, the persistent dry period since the 1960s arises partly from sea surface temperature (SST) anomalies between two ocean groups. Marked warming in the Southern hemisphere (plus the Indian Ocean) oceans relative to the Northern hemisphere produces a strong negative correlation between the resulting SST anomaly and rainfall anomalies for the west African Sahel (see discussions by Folland *et al.*, 1991; Hulme, 1992). Mahe (1993) showed that the Sahel dry (wet) periods were linked with:

(1) a stronger (weaker) African easterly jet (at 500–700 hPa level);
(2) a weaker (stronger) tropical easterly jet (at 100–200 hPa level) (Sadler, 1975a);
(3) smaller (greater) amount of water vapour in equatorial areas and inside the African easterly jet;
(4) positive (negative) equatorial upwelling SST anomaly (inside grid points latitude 2°N and 2°S and longitude 8°W and 12°W);
(5) and a westward (eastward) position of the centre of the St. Helena high pressure system over the south Atlantic Ocean.

Thus, attributing recent population pressure, combined with bad land management causing landscape degradation (commonly identified with the term 'desertification', Williams and Balling, 1994) as the prime factor responsible for climate and hydrological change (through lower rainfalls), is too simplistic. In the case of Africa, there is increasing evidence that annual rainfall amounts are non-stationary and are an inherent feature of the Sahel and other regions for time scales ranging from several decades to centuries (Nicholson, 1978, 1989; Hubert and Carbonnel, 1987; see review by Sircoulon *et al.*, 1998). Moreover, a statistical analysis of representative rainfall data sets in the western Sahel indicate that there would seem to be an *abrupt* change in the annual rainfall averages, followed by an apparent 'stationarity' before the next well defined 'jump' in the trend of mean precipitation.

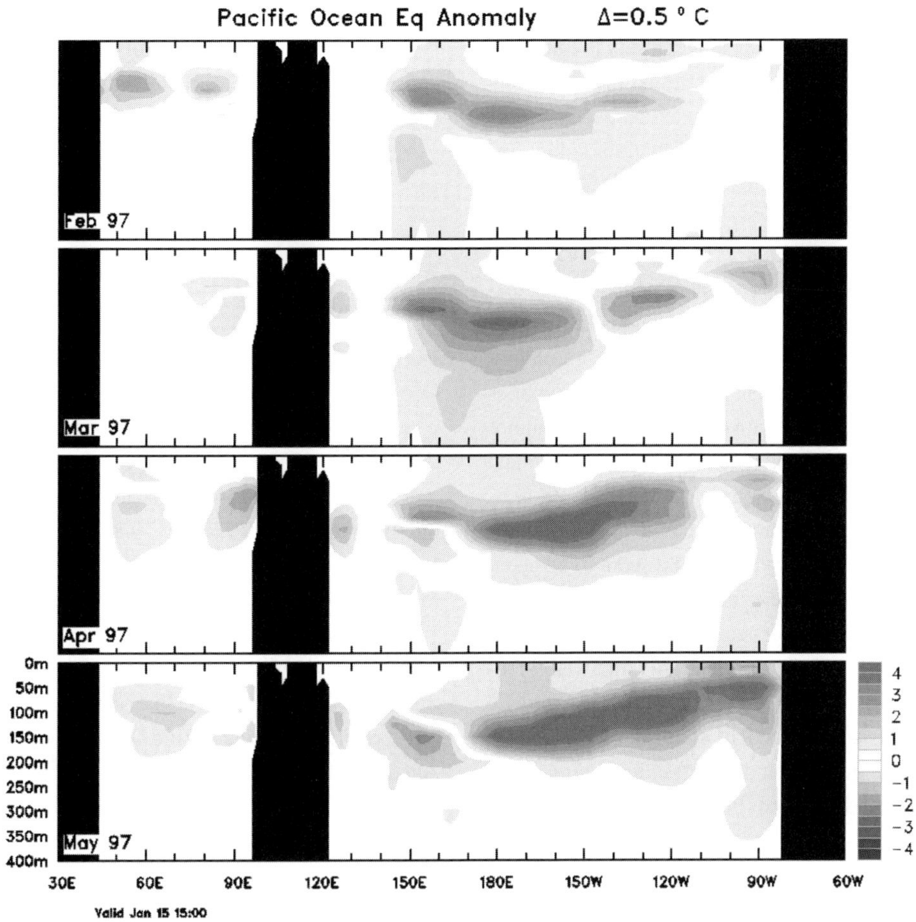

Plate 4 Sequence of the 150 m depth averaged temperature anomalies (K) for period February 1997 to May 1997. (Source: Australian Bureau of Meteorology.) (See also colour plate section.)

That is, there are two stable precipitation regimes represented by a 'humid' and a 'dry' state which are separated by a sharp transition (Demaree and Nicolis, 1990).

In the last decade, considerable attention has been given to the impacts of land use change on the adjustment of the energy balance by way of a change in the local surface fluxes of available energy, sensible heat and evaporation, all of which have the potential to affect the amount and distribution of rainfall. For example, modelling by Gong and Eltahir (1996) predicted that 27% of the precipitation in the west African region (including the humid coastal regions) originated from the recycling of local precipitation.

Although the HAPEX-Sahel Field Experiment was undertaken further north in the semi-arid Sahelian region, the recent conclusions from the experiment (Goutorbe *et al.*, 1994), need mention because there may be potential implications for the humid tropics (especially the wet/dry tropics sub-region cited by Chang and Lau, 1993). Gash *et al.* (1997) noted that there was substantially *less* evaporation from millet crops than from two natural vegetation types, fallow savanna and tiger bush. With increasing population pressure, the area of millet has increased, with the prospect of a reduction in the atmospheric moisture content at a regional scale and consequential decrease in rainfall. Conversely, there is the counterbalancing effect that the greater sensible heat generated will enhance convection and increase precipitation (Dolman *et al.*, 1997). On the other hand, during periods of weak convection when evaporation contributes more strongly to the potential energy of such events (Polcher, 1995), then the reductions in evaporation reported by Gash *et al.* (1997) (through the conversion of savanna to agricultural millet production) could exacerbate any decline in the frequency and amounts of rain per event. Such comments apply especially to the opening and closing stages of the rainy season when weak convection events are dominant (Dolman *et al.*, 1997). During the core period of the wet season, however, larger-scale moisture convergence and internal energy conversion associated with synoptic-scale meteorological systems (for example the *northern monsoon shearline* referred to below) become more important, which makes the role of surface evaporation relatively small at this time (Dolman *et al.*, 1997). Some support for this hypothesis emerges from the long-term rainfall analysis of Le Barbe and Lebel (1997), who noted that the mean event rainfall has been rather constant during the core period of the wet season. In contrast, there have been greater fluctuations in the mean event rainfall near the commencement and termination of the rainy season.

Thus, when considering the west African Sahelian region, one has to view the contributory causes of climatic and rainfall variability across different scales as being highly complex. Further, the conclusions from HAPEX-Sahel caution strongly against taking a single cause-effect perspective (Dolman *et al.*, 1997). Nonetheless, it would seem that the synoptic scale provides the overall climatic setting which will be favourable for either 'humid' or 'dry' states. Such remarks are in the context of SST anomalies and the corresponding adjustment in the structure of the atmospheric circulation, the latter of which affects both the depth and inland penetration of the moisture supply from the south-west monsoon. At the mesoscale, the complex interactions described between climate and the local hydrology would seem to be capable of reinforcing the synoptic-scale effects under certain circumstances. For example, there is evidence emerging of mesoscale processes connected with land cover (especially for weak convection in the opening and closing phases of the wet season (Le Barbe and Lebel, 1997)) and antecedent wetness (Taylor *et al.*, 1997) contributing to the generation of rainfall, and its resulting complex spatial and temporal variability, over the Sahel (Gash *et al.*, 1997; Dolman *et al.*, 1997).

It is significant that coastal, humid west Africa (along the Gulf of Guinea) was also affected by a reduction in rain amounts and number of rain days in parallel with the Sahelian drought of the 1970s and 1980s, although the effect was not spread evenly (Servat *et al.*, 1997; Patural *et al.*, 1997). Côte d'Ivoire and Nigeria were the most affected, Benin to a lesser extent and Togo and Ghana were the least affected (Patural *et al.*, 1997). As shown by Opuku-Ankomah and Cordery (1994), the correlation relationship between SSTs and rainfall in Ghana (south of the Sahel) is opposite in sign from that of the Sahel seasonal (July–September) rainfall-SST relationship. The explanation may be the adjustment in the synoptic climatology of west Africa during periods of Sahelian drought, as indicated above, whereby the *northern monsoon shear line* and associated rainfall producing perturbations are displaced further south over selected coastal areas (including Ghana), or further offshore in the case of other countries, in line with the analysis by Patural *et al.* (1997). Such displacement is also linked with positive SST anomalies to the south of 10°N over the eastern, equatorial Atlantic ocean (see discussion in Opuku-Ankomah and Cordery, 1994).

When considering possible terrestrial-atmosphere interactions in the more humid regions, a statistical analysis of long-term rainfall series undertaken by van Rompaey (1995) indicated that there was no significant difference in the decadal variations in rainfall between still predominantly forested Liberia, compared with Côte d'Ivoire which has experienced extensive deforestation. He attributed such 'deforestation-independent variability' to SST anomalies on the lines of Folland *et al.* (1991) linked with the adjustments in the atmospheric circulation mentioned above.

Mahé *et al.* (this volume) will focus on west African climatic variability in more detail.

Interdecadal changes in the Sea Surface Temperature (SST) patterns

Wang (1995) described long term changes in the tropical Pacific Ocean SST patterns since 1950 which have affected the manner in which El Niño events evolve. Over recent decades the SST long-term pattern changed around 1977 in the Pacific, which resulted in warming along the Equator and eastern Pacific and cooling in the southwest Pacific. Figure 10.20 illustrates how the SOI altered from a tendency to remain positive (negative) in the 20 years prior to (after) January 1977.

Power *et al.* 1999 highlighted the potential importance of a long-term cycle of rising and falling sea surface temperatures in the Pacific Ocean called the Inter-decadal Pacific Oscillation (IPO) (Also referred to as the Pacific Decadal Oscillation or PDO) for the Australian climate. While El Niño and La Niña are generally year-to-year events, the IPO has been shown to last decades – 10, 20 or even 30 years. When the IPO either warms or cools the central Pacific, it alters the impact of El Niño and La Niña. For example, over the period from 1949 to 1998, the four strongest La Niña (El Niño) episodes occurred before (after) 1977, corresponding with negative (positive) IPO indices.

North Pacific

Several recent studies (Allan, 1993; Allan *et al.*, 1995; Kachi and Nitta, 1997; Zhang *et al.*, 1997) have shown that El Niño – Southern Oscillation (ENSO) is part of an interdecadal variability. Zhang *et al.* (1997) showed that the interdecadal variability in the tropical Pacific is strongly related to the interdecadal variability in the North Pacific that was discussed in earlier studies (Tanimoto *et al.*, 1993; Kawamura, 1994; Kachi and Nitta, 1997). They also showed that the structure of the interdecadal variability is very similar to the interannual El Niño – Southern Oscillation (ENSO) mode. The ENSO-like interdecadal variability may be considered as a coupled ocean – atmosphere mode of oscillation for which different coupled mechanisms have been proposed (Latif *et al.*, 1997; Latif, 1998; Knutson and Manabe, 1998).

Indian monsoon

The Indian summer monsoon is known to have gone through alternating epochs of above-normal and below-normal conditions,

Figure 10.20 Southern Oscillation Index (SOI) cumulative monthly values January 1957 to February 1998. To obtain the cumulative SOI the monthly value of the SOI is continually added to generate the curve, starting from zero at the beginning of the period under examination. Over periods when the SOI is positive more than it is negative, the curve will rise inexorably. This curve shows how the earth's climate changed in January 1977, the curve falling much more than it rises since the SOI has been negative for much of the period.

each lasting about three decades (Krishnamurthy and Goswami, 2000). Kripalani and Kulkarni (1997) showed that there are more El Niño–related droughts in the decades when the Indian monsoon is generally below normal than during the decades when the Indian monsoon is generally above normal. Based on the record of Indian monsoon rainfall from 1871 to 1997 and extrapolating the trend of the interdecadal variability, Krishnamurthy and Goswami (2000) suggested that the Indian monsoon might be at the beginning of a period where above normal rainfall may persist over the next 20 or 30 years or even longer.

During a warm eastern Pacific phase of the interdecadal SST variation, the regional Hadley (near-meridional circulation in the troposphere resulting from strong heating at the Equator) circulation associated with El Niño reinforces the prevailing anomalous interdecadal Hadley circulation, while that associated with La Niña opposes the prevailing interdecadal Hadley circulation. Therefore, during the warm phase of the interdecadal oscillation, El Niño events are expected to be related strongly to monsoon droughts, while La Niña events may not have a significant relationship. On the other hand, in the cold eastern Pacific phase of the interdecadal oscillation, La Niña events are more likely to be related to monsoon floods while El Niño events are unlikely to have a significant relationship with the Indian monsoon. Thus, there is a fundamental reason why the monsoon–ENSO relationship is not very strong on the interannual timescale. In either phase of the interdecadal oscillation, only one phase of the interannual variation reinforces the local Hadley circulation while the other phase almost cancels the interdecadal Hadley circulation. Whenever the interdecadal and the interannual Hadley circulations reinforce each other, the coupled mode (or boundary forcing) is strong and may overcome the effects of the internal dynamics. On the other hand, when the interdecadal and interannual Hadley circulations oppose each other in the monsoon region, internal processes may govern the state of the Indian monsoon.

North Atlantic hurricane frequency

Landsea *et al.* (1994) showed that hurricane occurrence is greater in the Atlantic over multidecadal periods when La Niña dominates than over similar periods when El Niño dominates.

Longer-term variations

Recent NASA research has indicated that there may be a second, much longer, PDO cycle that lasts about 70 years. Yi Chao, an oceanographer at NASA's Jet Propulsion Laboratory in Pasadena, California, and colleagues Michael Ghil and James McWilliams of the University of California, Los Angeles, found evidence of the PDO's two-part structure in a study of the past 92-year record of sea-surface temperatures in the North and South Pacific.

CONCLUSIONS

Three systems of convergence have been identified within the envelope of the monsoon regions, i.e. the northern monsoon shear line, southern monsoon shear line and the maximum cloud zone (MCZ). There is a seasonal progression of activity along the respective monsoon shear lines, with the most active favouring the one located in the summer hemisphere corresponding to the zonal thermal low pressure trough (otherwise known as the monsoon trough).

The convergent, opposing equatorial westerlies and trade wind easterlies in the lower level of the atmosphere favour highly the development of vortices, some of which develop further into tropical cyclones (Sadler, 1967; McBride and Keenan, 1982). During the transitional months of April–May and November–December; both monsoon shear lines can be similarly active because the above convergence mechanism is occurring. Conversely, the winter hemisphere monsoon shear line is the less active because it is more 'a wind turning zone' from easterlies to westerlies due to the Coriolis effect. Nonetheless, weak vortices can develop due to this wind-backing mechanism, some of which are capable of developing into 'out of season' tropical cyclones in the north west Pacific ocean, and occasionally in the Bay of Bengal, especially if the shear line is positioned north of 5°N (or much more rarely south of 5°S in the south-west South Pacific) (Sadler, 1967). Away from these latitudes, the Coriolis force favours more organised, circular motion.

The MCZ is an area of deep convection associated with the equatorial westerlies in monsoon regions, which arise in part from the convergence of inter-hemispheric airstreams on passing through the respective shear lines (McBride and Keenan, 1982). Deep convection (as mesoscale cloud systems, MCSs) waxes and wanes coinciding with the varying strength of trade wind airstreams into the equatorial regions (Davidson, 1984; Love, 1985) and the eastward propagation of the Madden-Julien Oscillation (MJO) as a Kelvin wave within equatorial latitudes, especially from east Africa to the east Pacific Ocean. The MJO, in contrast, has minimal impact across the Atlantic Ocean and most of Latin America except for an eastward extension of the MJO over the Caribbean Sea during strong ENSO warm phases.

Significantly, the MCZ only 'spills over' the monsoon trough coinciding with the development of deep low pressure systems along the more active of the monsoon shear lines. Elsewhere along both the northern and southern monsoon shear lines, comparatively cloud-free conditions can occur.

Outside of the monsoon regions over the oceans, in the absence of continental heating effects, the Zonal Trough in the Easterlies (ZTE) occurs, as represented by the more gentle, tangential convergence of north east and south east trade winds. Weak, stationary and ephemeral vortices may develop along the ZTE but the absence of a mechanism of 'opposing' convergence of winds, cf. the monsoon trough, the lower Sea Surface Temperatures (SSTs), the potential shearing of convective cloud cells by the upper equatorial westerlies and the near-equatorial location of the ZTE (where the Coriolis force is weakest) all militate against any tropical cyclonic development (Sadler, 1967; Ramage *et al.*, 1979).

Variations in the monsoon characteristics across regions have been reviewed. Interestingly, South America, notably the Amazon basin, is not within the monsoon region. Moreover, the distorting heating effects of this basin and the occurrence of other meteorological systems such as the northward penetration of former cold fronts of southern hemisphere origin (and more occasionally from the Northern hemisphere) provide a different synoptic climatology. These aspects will be described in the following Chapter and see also Molion (1993) for specific details of Amazon basin synoptic climatology. In the latter reference it is shown by satellite imagery that the ZTE is maintained over the Atlantic Ocean only, off the north-east coast of South America.

Emphasis was also given to tropical-extratropical cloud bands which act as preferred geographic points (i.e. conduits) for surplus energy transfer out of the tropics into the higher latitudes. The same systems are also the dominant sources of rainfall and the Amazon basin is one of several examples where these systems prevail, especially in the summer. Some of these cloud bands also coincide with the penetration and decay of austral 'cold surges' from higher latitudes.

Finally, the profound impacts of inter-annual (El Niño – Southern Oscillation and decadal variability (e.g. in West Africa) on tropical climatology and rainfall variability need to be considered within the context of non-stationarity in the various components of the hydrological cycle. The same non-stationarity is reflected in the readjustment of the geographic position of the meteorological systems described. The propagation of the monsoon region meteorological systems (including tropical cyclones) further east of the International Date Line, over the South Pacific Ocean during warm ENSO phases, is an example (Sadler, 1984, reviewed in Manton and Bonell, 1993). The apparent failure of climate models to predict the disastrous 1997/1998 El Niño highlights the role that synoptic scale weather events can sometimes play in triggering ENSO events.

APPENDIX 10.1 GLOSSARY OF TERMS USED IN THIS CHAPTER AND THE FOLLOWING CHAPTER

Adiabatic An adiabatic process (thermodynamic) is one in which heat does not enter or leave the system. (Greek, *a* not, and *diabaino* pass through.)

Because the atmosphere is compressible and pressure varies with height, adiabatic processes play a fundamental role in meteorology. Thus, if a parcel of air rises it expands against its lower environmental pressure; the work done by the parcel in so expanding is at the expense of its internal energy and its temperature falls, despite the fact that no heat leaves the parcel. Conversely, the internal energy of a falling parcel is increased and its temperature raised, as a result of the work done. Observation shows that such processes determine, to a large extent, the vertical temperature distribution within the troposphere. It also supports the view that, to a first approximation, it is justifiable to treat the vertically moving, individual masses of air of indefinite size (termed 'parcels') as closed systems which move through the environment without unduly disturbing it or exchanging heat with it. Various non-adiabatic processes such as condensation, evaporation, radiation and turbulent mixing also operate to produce temperature changes in the free atmosphere but their effects are generally negligible in comparison with those caused by appreciable vertical motion.

Advection Advection like convection refers to the process of transfer (of an air mass property) by virtue of motion. In meteorology the term is however, applied to signify horizontal motion only.

Baroclinic A baroclinic atmosphere is one in which surfaces of pressure and density (or specific volume) intersect at some level or levels. The atmosphere is always, to some extent, baroclinic. Strong baroclinicity implies the presence of large horizontal temperature gradients and thus of large wind vector changes with height. Surfaces of pressure and density (or specific volume) coincide at all levels. The concept of barotropy, though idealised, gives a useful first approximation in some types of atmospheric problem. The contrasting atmospheric state is the baroclinic.

Convection Here convection in atmospheric processes refers to currents, which can be set up by low level convergence either by heating effects (such as solar radiation) or from colliding air masses. Such convection currents primarily move vertically and account for many atmospheric phenomena, such as clouds and thunderstorms.

Convergence and divergence The term diffluence (confluence) refers to the separation (joining) of adjacent streamlines in the direction of flow. If this diffluent (confluent) flow is not decreasing (increasing) in speed downstream it is also divergent (convergent) which means mass is depleting (increasing) per unit volume.

Coriolis force The deflection (to the right in the Northern, left in the Southern, hemisphere) and acceleration relative to the Earth's surface, caused by the Earth's rotation.

Hadley circulation Net radiation heating of the surface in the tropics leads to widespread low static stability of the atmosphere there. Evaporation from the warm tropical oceans assures a large supply of water vapour in the air. There is, consequently, a large scale, persistent band of organised convection throughout the tropics called the Inter Tropical Convergence Zone (ITCZ). The ITCZ extends along a long band and can often be traced around the whole globe.

If the effects of the earth's rotation are neglected, the Hadley circulation comprises a high-level poleward flow from heat source to heat sink. This is in response to a horizontal pressure gradient (at high level, pressure is greatest above the heat source since pressure decreases less rapidly with height in the warmer air column). A compensating low level flow then occurs towards the heat source.

Meridional and zonal The term meridional means the direction along a geographic meridian and the word zonal refers to the west to east direction.

Microwave 85 GHz data In additional to conventional meteorological observations this study has made use of newly available remote sensing data using microwave radiation. Over the last two years, tropical cyclone forecasters and researchers around the globe received a large incremental advance in data concerning the rain band structure of tropical cyclones. The data can be downloaded from the Navy Research Laboratory Monterey (California) tropical cyclone home page on the worldwide web. The URL for this site is: http://kauai.nrlmry.navy.mil/sat-bin/tc_home

Observations of tropical cyclones can be obtained from satellite data in remote locations over the oceans. The most continuous data are for visible and infrared imagery from geostationary satellites. Data at microwave frequencies from polar-orbiting satellites, however, are more directly related to precipitation than are those from visible and IR channels. The upwelling radiation at these microwave frequencies can therefore be used to assess structure of the tropical cyclone's precipitation regions.

The 85-GHz microwave channel is sensitive to precipitation-sized ice particles, which scatter the upwelling radiation and reduce the brightness temperature. This result is termed an 'ice-scattering signature,' as the depressed brightness temperature indicates the presence of precipitation-sized ice aloft. A low 85-GHz brightness temperature can therefore imply increased convection and precipitation.

The Defense Meteorological Satellite Program (DMSP) F14, F13, F11 and F10 satellites carry Special Sensor Microwave/Imager (SSM/I) in sun synchronous orbits. Independently, McGaughey *et al.* (1996), using high-resolution data from the Advanced Microwave Precipitation Radiometer, derived 225 K as a threshold for tropical oceanic convection. Mohr *et al.* (1996) noted a dramatic increase in lightning flash rates in continental convection having PCT below 200 K.

The Tropical Rainfall Measuring Mission (TRMM) satellite is a joint project between the United States (under the leadership of NASA's Goddard Space Flight Center) and Japan (under the leadership of the National Space Development Agency). This was the first spacecraft designed to monitor rain over the tropics and was successfully launched from Tanegashima, Japan, on November 27, 1997. It placed in low Earth orbit the first precipitation radar to be flown in space, along with a 9-channel SSM/I-like passive microwave imager.

Polarization Corrected Temperature (PCT) To differentiate between low brightness temperatures due to ice scattering and those due to the low surface emissivity of the ocean (especially evident in the horizontally polarised channel), Spencer *et al.* (1989) define a polarisation corrected temperature (PCT) as

$$\mathrm{PCT} = 1.818\mathrm{TBV} - 0.818\mathrm{TBH}$$

where TBV is the brightness temperature for the vertically polarised channel and TBH is that of the horizontally polarised channel at 85 GHz. Spencer *et al.* (1989) found that the PCT range of 250–260 K is generally a threshold below which precipitating systems are found, with 250 K roughly corresponding to a moderate rain rate (3 mm h^{-1}). Spencer *et al.* and the Goddard scattering algorithm (GSCAT, described by Adler *et al.*, 1994) basically utilise a linear relationship between 85-GHz brightness temperature depression and rain rate. Mohr and Zipser (1996a,b) used the existence of a 225 K PCT (10 mm h^{-1}) to indicate the presence of cumulonimbus convection. Independently, McGaughey *et al.* (1996), using high-resolution data from the Advanced Microwave Precipitation Radiometer, derived 225 K as a threshold for tropical oceanic convection. Indeed, 225 K pixels at the satellite scale invariably feature inhomogeneities with lower brightness temperatures in the high resolution data, indicative of deep convection. Mohr *et al.* (1996) noted a dramatic increase in lightning flash rates in continental convection having PCT below 200 K.

In summary:

250 K PCT is considered an indicator of moderate rain;
225 K PCT is considered an indicator of deep convection;
and lower PCTs are considered indicators of more intense convection.

Potential temperature That which a given sample of air would attain if transferred in a dry adiabatic process to the standard pressure, 1000 hPa.

Potential vorticity In adiabatic motion of a column of air, the quotient of the absolute vorticity (ζ_a) of the air column to the pressure difference between the top and bottom of the column (Δp) is constant (potential vorticity theorem), i.e. the value of the absolute vorticity of a column which corresponds to a standard value of Δp (say, 50 mb) is termed the potential (absolute) vorticity.

Stability A system which is subjected to a small disturbing impulse, is said to be in stable, neutral or unstable equilibrium, according to whether it returns to its original position, remains in its disturbed position, or moves farther from its original position, respectively, when the disturbing influence is removed.

Inertial stability A type of dynamic stability associated with the Earth's rotation, in which an air particle, embedded in a wind flow along a circle of latitude, tends to return to this latitude on being subjected to a small displacement from it. Inertia instability arises when such a displacement results in an acceleration of the particle away from its original latitude.

The condition for stability is that the Coriolis parameter (or Earth vorticity $2\Omega \sin \phi$) should exceed the northward (Northern hemisphere). increase of the geostrophic west-wind component, i.e. $2\Omega \sin \phi > \partial u_g/\partial y$ (Northern hemisphere). Similarly, instability arises when $2\Omega \sin \phi < \partial u_g/\partial y$ (Northern hemisphere).

The inertial stability condition in this zonal flow is simply that the absolute vorticity be positive (negative) in the Northern (Southern) hemisphere.

Static stability Investigation of the static stability of the atmosphere is made most simply by the parcel method, in which an assessment is made of changes of kinetic energy of a test parcel of air, displaced vertically and adiabatically with respect to its environment. The environment is termed stable, neutral, or unstable as the kinetic energy of the parcel decreases, remains constant or increases, respectively.

(i) Absolute stability exists if the environment fall in temperature with height (lapse rate) is less than the SALR.
(ii) Absolute instability exists if the lapse rate is greater than the DALR
(iii) Conditional instability exists if the lapse rate is between the DALR and SALR.

Static temperature profiles are calculated according to the following expression:

$$T_s = T + (g/C_p)z + (L/C_p)Q_{vs},$$

where T, Q_{vs}, g, C_p, L are respectively the temperature at height z, the saturated water vapour mixing ratio, the gravity acceleration, the specific heat for air and the latent heat (Dhonneur, 1978 cited and used by Asselin de Beauville, 1995, in the next chapter).

Tropical cyclone A non-frontal synoptic cyclonic rotational low pressure system of tropical origin in which ten-minute mean winds of at least gale force 63 km h^{-1} occur, the belt of maximum winds being in the vicinity of the system's centre, (Australian Bureau of Meteorology (1978) *Australian Tropical Cyclone Forecasting Manual*. Australian Bureau of Meteorology, Melbourne).

The World Meteorological Organization definition of a tropical cyclone is: *A non-frontal cyclone of synoptic scale developing over tropical waters and having a definite organised wind circulation with average wind of gale force (34 knots* or *63 km/h) or more surrounding the centre.*

Tropical low or depression Usually refers to an area of low pressure, which can be enclosed by a closed isobar (2 hPa spacing) in the tropics.

Vorticity The vorticity at a point in a fluid is a vector, which is twice the local rate of rotation of a fluid element. The component of the vorticity in any direction is the circulation per unit area of the fluid in a plane normal to that direction. The dimensions are s^{-1}.

Vorticity is a three-dimensional property of the field of motion of a fluid. In large-scale motion in the atmosphere the vorticity component of chief significance is that which occurs in the horizontal plane (i.e. rotation about the vertical axis); the other components are, however, significant in some dynamical problems. In vector notation the vorticity of a velocity vector **V** is written as curl **V** or rot **V** or $\nabla \times \mathbf{V}$.

In 'solid rotation' of angular velocity ω the vorticity is 2ω. At latitude ϕ, where the angular velocity of the earth about the vertical axis is $\Omega \sin \phi$, the earth has a vorticity about this axis of $2\Omega \sin \phi$ which is cyclonic in sense. Air partakes of the vorticity of the Earth appropriate to its latitude, in addition to any relative vorticity it may possess. Thus, at latitude ϕ, absolute vorticity is given by

$$\zeta_a = \zeta + 2\Omega \sin \phi \text{ (or relative vorticity plus Earth's vorticity).}$$

In the Northern hemisphere relative vorticity in a cyclonic sense is reckoned positive, in an anticyclonic sense negative.

References

Adler, R. F., G. J. Huffman, and P. R. Keehn, 1994. Global tropical rain estimates from microwave-adjusted geosynchronous IR data. *Remote Sens. Rev.*, 11, 125–152.

Allan, R. J., 1991: In: Teleconnections linking worldwide climate anomalies. Glantz, M, *et al.* (Eds), Cambridge Uni. Press, Cambridge, UK, 73–120 pp.

Allan, R. J., 1993: Historical fluctuations in ENSO and teleconnection structure since 1879: Near global patterns. *Quat. Australasia*, 11, 17–27.

Allan, J. A. Lindesay, and C. J. C. Reason, 1995. Multidecadal variability in the climate system over the Indian Ocean region during austral summer. *J. Climate*, 8, 1853–1873.

Basher, R. E. and Zheng, X. 1995. Tropical Cyclones in the Southwest Pacific: Spatial Patterns and Relationships to Southern Oscillation and Sea Surface Temperatures. *Jnl of Climate, 8, 1249–1260.*

Basher, R. E. and Zheng, X. (1998) Mapping Rainfall Fields and their ENSO Variation in Data-Sparse Tropical South-West Pacific Ocean Region. *Int. J. Climatol. 18*: 237–251.

Bonell, M. (1991) Progress and future research needs in water catchment conservation within the wet tropical coast of NE Queensland. In: *Tropical Rainforest Research in Australia: Present Status and Furure Directions for the Institute for Tropical Rainforest Studies*, Proceedings of a Workshop held in Townsville, Australia, 4–6 May, 1990. Institute for Tropical Rainforest Studies, James Cook University, Townsville, pp. 59–86.

Bonell, M. and Gilmour, D. A. (1980) Variations in short-term rainfall intensity in relation to synoptic climatological aspects of the humid tropical northeast Queensland coast. *Singapore J. Trop. Geog. 1(2)*: 16–30.

Brunt, A. T., 1969: Low latitude Cyclones. *Aust. Met. Mag.*, 17, 67–90.

Bureau of Meteorology 1994. *The Estimation of Probable Maximum Precipitation: Generalised Short-Duration*. Melbourne, Australia.

Chang, C-P. and Krishnamurti, T. N. (eds). 1987. *Monsoon Meteorology*. Oxford Monographs on Geology and Geophysics No 7. Oxford University Press. 544 pp.

Chang, J-H and Lau, L. S.: 1993, A definition of the humid tropics. In: *Hydrology and Water Management in the Humid Tropics – Hydrological Research Issues and Strategies for Water Management*, Bonell, M., Hufschmidt, M. M., and Gladwell, J. S., eds., UNESCO – Cambridge University Press, Cambridge, pp. 571–574.

Chen, S. S. and Houze, Jr. R. A. (1997) Diurnal Variation and life-cycle of deep convective systems over the tropical Pacific warm pool. *Int. J. Climatol.* 123: 357–388.

Chen, T.-C. and Weng, S.-P. 1998a. Interannual Variation of the Summer Synoptic Scale Disturbance Activity in the Western Tropical Pacific. *Monthly Weather Review*: Vol. 126, 1725–1733.

Chen, T.-C. and Weng, S.-P. 1998b. Interannual Variation in the Tropical Cyclone Formation over the Western North Pacific. *Monthly Weather Review*: Vol. 126, No. 4, pp. 1080–1090.

Chen, T.-C. and Weng, S.-P. 1999. Interannual and Intraseasonal Variations in Monsoon Depressions and Their Westward-Propagating Predecessors. *Monthly Weather Review*: Vol. 127, No. 6, pp. 1005–1020.

Chen, T.-C and Yoon, J-h. 2000. Interannual Variation in Indochina Summer Monsoon Rainfall: Possible Mechanism. *Journal of Climate*: Vol. 13, No. 11, pp. 1979–1986.

Das, P. K. 1987. Short and Long-Range Monsoon Prediction in India. *Monsoons* J. S. Fein and P. L. Stephens, published by John Wiley and sons Ney York 632 pages.

Davidson, N. E., McBride, J. L. and McAvaney, B. J. (1983) The onset of the Australian monsoon during winter MONEX: Synoptic aspects. *Mon. Weath. Rev.* 117: 1181–1192.

Demarre, G. R. and Nicolis, C.: 1990. 'Onset of Sahelian drought viewed as a Fluctuation-Induced Transition', *Q.J.R. Meteorol. Soc.*, **116**, 221–238.

Dolman, A. J., Gash, J. H. C., Goutorbe, J-P., Kerr, Y., Lebel, T., Prince, S. D. and Strickler, J. N. M.: 1997, 'The role of the land surface in Sahelian climate: HAPEX-Sahel results and future research needs', *J. Hydrol.* **188–189**, 1067–1079.

Dunne, T. and Black, R. D. (1970a). An experimental investigation of runoff production in permeable soils. *Wat. Resour. Res.*, 6, 478–490.

Dunne, T. and Black, R. D. (1970b). Partial area contributions to storm runoff in a small New England watershed. *Wat. Resour. Res., 6*, 1296–1311.

Falkland, A. C., Custodio, E., Diaz, Arenas, A. and Simler, L. (1991). *Hydrology and water resources of small islands: a practical guide*. Studies and reports in hydrology, No. 49. UNESCO, Paris.

Falkland, A. C. and Brunel, J. P. (1993). *Review of Hydrology and Water Resources of Humid Tropical Islands*. (In) Hydrology and Water Management in the Humid Tropics: Bonell, M., Hufschmidt, M. M. and Gladwell, J. S. 10: 135–166.

Fein, J. S. and Stephens, P. L. (eds), 1987. *Monsoons*. John Wiley, Canada. 632 pp.

Folland, C. K., Owen, J. A., Ward, M. N. and Colman, A. W.: 1991. 'Prediction of seasonal rainfall in the Sahel region using empirical and dynamical methods', *Journal of Forecasting* **10**, 21–56.

Garraud, R. D. 2000. Cold air incursions over the subtropical and tropical South America: structure and dynamics. *Mon. Weath. Rev.* 128, 2544–2559.

Garreaud, R. D. and Wallace, J. M. 1998. Summertime Incursions of Midlatitude Air into Subtropical and Tropical South America. *Monthly Weather Review*. 126, 2713–2733.

Gash, J. H. C., Kabat, P., Monteny, B. A., Amadou, M., Bessemoulin, P., Billing, H., Blyth, E. M., deBruin, H. A. R., Elbers, J. A., Friborg, T., Harrison, G., Holwill, C. J., Lloyd, C. R., Lhomme, J.-P., Moncrieff, J. B., Puech, D., Soegaard, H., Taupin, J. D., Tuzet, A. and Verhoef, A.: 1997. 'The variability of evaporation during the HAPEX-Sahel: Intensive Observation Period', *J.Hydrol.* **188–189**, 385–399.

Gill, A. E. 1982. *Atmosphere-Ocean Dynamics*. International Geophysics Series, Vol. 30, Academic Press, London, 662 pages.

Gong, C. and Eltahir, E.: 1996, 'Sources of moisture for rainfall in west Africa', *Water Resour. Res.*, **32**, 3115–3121.

Goutorbe, J-P., Lebel, T., Tinga, A., Bessemoulin, P., Brouwer, J., Dolman, A. J., Engman, E. T., Gash, J. H. C., Hoepffner, M., Kabat, P., Kerr, Y. H., Monteny, B., Prince, S., Said, F., Sellers, P., Wallace, J. S.: 1994, 'HAPEX-Sahel: a large-scale study of land-atmosphere interactions in the semi-arid tropics', *Ann. Geophysicae* **12**, 53–64.

Gray, W. M. 1975. Tropical cyclone genesis. Dept of Atmos Sci., paper No 323, Colorado State University.

Hendon, H., Davidson, N. E. and Gunn, B. (1989). Austrailian summer monsoon onset during AMEX 1987. *Mon. Weath. Rev. 117*: 2554–2574.

Henry, W. K. (1979). Some aspects of the fate of cold fronts in the Gulf of Mexico. *Mon. Weath. Rev. 107*: 1078–1082.

Henry, W. K. (1980) Three late spring fronts in the Carribean. *Nat. Wea. Dig. 5(1)*: 18–24.

Holland, G. J. (1984). On the climatology and structure of tropical cyclones in the Australian/south-west Pacific region: (a) Data and tropical storms. *Aust. Met. Mag 32*: 1–15; (b) Hurricanes, 17–31 (C) Major hurricanes, 33–46.

Holliday, C. R. and Thompson, A. H. (1986). An Unusual Near-Equatorial Typhoon. *Mon. Weath. Rev.*, 114(12), 2674.

Horel, J. D. and Wallace, J. M. (1981) Planetary-scale atmospheric phenomena associated with the Southern Oscillation. *Mon. Weath. Rev. 109*: 813–829.

Horton, R. E. (1933). The role of infiltration in the hydrological cycle. *Trans. Am. Geophys. Union*, 14, 446–460.

Horton, R. E. (1945). Erosional development of streams and their drainage basins: hydrophysical approach to quantitative morphology. *Bull. Geolog. Soc. Am.*, 56, 275–370.

Hoskins, B. J., I., Draghici and H. C. Davies. 1978. A new look at the Omega equation. *Quarterly Journal or the Royal Met. Soc.* 104, 31–38.

Huang, S. S. ed. (1982). Heavy rainfall in South China in pre-typhoon season. Guangdon Institute of Tropical Meteorology, (in Chinese) Guangdon.

Hubert, P. and Carbonnel, J. P.: 1987, 'Approche statistique de l'aridification de l'Afrique de l'Ouest', *J. Hydrol.* **95**, 165–183.

Hulme, M.: 1992, 'Rainfall changes in Africa: 1931–1960 to 1961–1990', *Int. J. Climatol.* **12**, 685–699.

Ji, M., A. Leetmaa, and V. E. Kousky, 1996: Coupled model predictions of ENSO during the 1980s and 1990s at the National Centers for Environmental Prediction. *J. Climate*, 9, 3105–3120.

Kachi, M., and T. Nitta, 1997: Decadal variations of the global ocean–atmosphere system. *J. Meteor. Soc. Japan*, 75, 657–675.

Kawamura, R., 1994: A rotated EOF analysis of global sea surface temperature variability with interannual and interdecadal time scale. *J. Phys. Oceanogr.*, 24, 707–715.

Keen, R. A. 1982. The role of cross equatorial cyclone pairs in the Southern Oscillation, *Mon. Wea. Rev.*, 110, 1405–1416.

Kleeman, R., and A. M. Moore, and N. R. Smith, 1995: Assimilation of subsurface thermal data into a simple ocean model for the initialization of an intermediate tropical coupled ocean–atmosphere forecast model. *Mon. Wea. Rev.*, 123, 3103–3114.

Knutson, T. R., and S. Manabe, 1998: Model assessment of decadal variability and trends in the tropical Pacific Ocean. *J. Climate*, 11, 2273–2296.

Kousky, V. E. (1979) Frontal influences on northeast Brazil. *Mon. Weath. Rev.* 107: 1140–1153.

Kousky, V. E. and Ferreira, J. (1981) Frontal influences on northeast Brazil: Their spatial distribution, origins and effects *Mon. Weath. Rev. 109*, 1999–2008.

Kripalani, R. H., and A. Kulkarni, 1997: Climatic impact of El Niño/La Niña on the Indian monsoon: A new perspective. *Weather*, 52, 39–46.

Krishnamurti, T. N., Wagner, C. P., Cartwright, T. J. and Oosterhof, D. (1997) Wave Trains Excited by Cross-Equatorial Passage of the Monsoon Annual Cycle. *Mon. Weath. Rev. 125*: 2709–2715.

Krishnamurthy, V. and Goswami, B. N. 2000. Indian Monsoon–ENSO Relationship on Interdecadal Timescale. *Journal of Climate*: Vol. 13, No. 3, pp. 579–595.

Krishnamurti, T. N. (1979) *Tropical Meteorology. Compendium of Meteorology II*, WMO-No. 364, World Meteorological Organization, 428 pp.

Kuhnel, I. (1989) Tropical-extratropical cloudband activity climatology based on satellite data. *Int. J. Climatol. 9* : 441–463. Press: 161–202.

Landsea, C. W., Gray, W. M., Mielke, Jr. P. W., and Berry, K. J. 1994. Seasonal forecasting of Atlantic hurricane activity. *Weather*. 49, 273–283.

Latif, M., 1998: Dynamics of interdecadal variability in coupled ocean–atmosphere models. *J. Climate*, 11, 602–624.

Latif, R. Kleeman, and C. Eckert, 1997: Greenhouse warming, decadal variability, or El Niño? An attempt to understand the anomalous 1990s. *J. Climate*, 10, 2221–2239.

Lau, K.-M. and C.-P. Chang 1987. Planetary Scale Aspects of the Winter Monsoon and Atmospheric Teleconnections. *Monsoon Meteorology* edited by C.-P. Chang and T. N. Krishnamurti. Published by Oxford University Press UK, 544 pages.

Le Barbe, L. and Lebel, T.: 1997, 'Rainfall climatology of the HAPEX-Sahel during the years 1950–1990', *J. Hydrol*, 188–189, 43–73.

Lenters, J. D. and Cook, K. H. (1999) Summertime Precipitation Variability over South America: Role of the Large-Scale Circulation. *Mon. Weath. Rev.* 127: 409–431.

Love, G. (1985a) Cross-equatorial influence of winter hemisphere subtropical cold surges. *Mon. Weath. Rev. 113(9)*: 1487–1498.

Love, G. (1985b) Cross-equatorial interactions during tropical cyclongenesis. *Mon. Weath. Rev. 113(9)*: 1499–1509.

Madden, R. A. and Julian, P. R. (1971) Detection of a 40–50 day oscillation in the zonal wind in the tropical Pacific. *J. Atmos. Sci. 28*: 702–708.

Madden, R. A. and Julian, P. R. (1994) Observations of the 40–50-day tropical oscillation – A review. *Mon. Weath. Rev. 122*: 814–837.

Mahe, G.: 1993, *Les Ecoulements Fluviaux sur la Facade Atlantique de l'Afrique – Etude des Elements de Bilan Hydrique et Variabilite Interannuelle, Analyse de Situations Hydroclimatiques Moyennes et Extremes,* **Editions de l'ORSTOM**, Institut Francais de Recherche Scientifique Pour le Developpement en Cooperation, Collection **ETUDES et THESES**, Paris 1993, 438 pp.

Maloney, E. D. and Hartmann, D. L. (1998) Frictional Moisture Convergence in a Composite Life Cycle of the Madden-Julian Oscillation. *Jnl. Of Climate. 11*: 2387–2403.

Manton, M. J. and Bonell, M. (1993) *Climate and Rainfall Variability in the Humid Tropics*. (In) Hydrology and Water Management in the Humid Tropics: Bonell, M., Hufschmidt, M. M. and Gladwell, J. S. *2*: 13–33.

Marengo, J. A. and Rogers, V. C. (2001). *Polar Outbreaks in the Americas: Assessment and Impacts during modern and past climates*. *In*: Interhemispheric Climate Linkages, Markgraf, V. (ed), Academic Press, San Diego, pp. 31–51.

Marengo, J., Carnejo, A., Saymurty, P., Nobre, C. and Sea, W. (1997) Cold surges in tropical and extratropical South America: The strong event in June 1994. *Mon. Weath. Rev. 125*: 2759–2786.

Matthews, A. J., Hoskins, B. J., Slingo, J. M. and Blackburn, M. (1996) Development of convection along the SPCZ within a Madden-Julian oscillation. *Q.J.R. Meteorol. Soc. 122*: 669–688.

McAlpine, J., Keig, G. and Falls, R. (1983) *Climate of Papua New Guinnea,* CSIRO-ANU Press, 11–38.

McBride, J. L. and Keenan, T. D. (1982) Climatology of tropical cyclone genesis in the Austrailian region. *J. Climatol. 2*: 13–33.

McGaughey, G., E. J. Zipser, R. W. Spencer, and R. E. Hood, (1996) : High-resolution passive microwave observations of convective systems over the Tropical Pacific Ocean. *J. Appl. Meteor.*, 35, 1921–1947.

McPhaden, M. J., 1999: Genesis and evolution of the 1997–98 El Niño. *Science*, 283, 950–954.

Mohr, K. I., and E. J. Zipser, 1996a: Defining mesoscale convective systems by their 85-GHz ice scattering signature. *Bull. Amer. Meteor. Soc.*, 77, 1179–1189.

______, and ______, 1996b: Mesoscale convective systems defined by their 85-GHz ice scattering signature: Size and intensity comparison over tropical oceans and continents. *Mon. Wea. Rev.*, 124, 2417–2437.

______, E. R. Toracinta, E. J. Zipser, and R. E. Orville, 1996: A comparison of WSR-88D reflectivities, SSM/I brightness temperatures, and lightning for mesoscale convective systems in Texas. Part II: SSM/I brightness temperatures and lightning. *J. Appl. Meteor.*, 35, 919–931.

Molion, L. C. B.: 1993, Amazonia rainfall and its variability. In: *Hydrology and Water Management in the Humid Tropics – Hydrological Research Issues and Strategies for Water Management*, Bonell, M., Hufschmidt, M. M. and Gladwell, J. S., eds., Cambridge, pp. 99–111.

Nicholls, N., 1992: Historical El Niño/Southern Osscillation variability in the Australian region. In Diaz, H. F. and Margraf, V. (Eds), El Niño, historical and paleoclimate aspects of the Southern Oscillation. Cambridge Uni. Press, Cambridge, UK 151–174 pp.

Nicholson, S. E.: 1978, 'Climatic variations in the Sahel and other African regions during the past five centuries', *J. Arid Environments* **1**, 3–24.

Nicholson, S. E.: 1989, 'Long-term changes in African rainfall', *Weather* **44**, 46–56.

Opoku-Ankomah, Y. and Cordery, I.: 1994, 'Atlantic sea surface temperatures and rainfall variabilty in Ghana' *J. Climate,* 7, 551–558.

Orlanski, I. (1975) A relatipnal subdivision of scales for atmosphere processes. *Bull. Amer. Met. Soc.* 56: 527–530.

Pant, G. B., and B. Parthasarathy, 1981: Some aspects of an association between the Southern Oscillation and Indian summer monsoon. *Arch. Met. Gaffes. Biokl.*, 1329, 245–252.

Parthasarathy, B., and G. B. Pant, 1985: Seasonal relationship between Indian summer monsoon rainfall and Southern Oscillation. *J. Climatol.*, 5, 369–378.

Patural., J. E., Servat, E., Kouame, B., Lubes, H., Ouedraogo, M. and Masson, J. M.: 1997 'Climatic variability in humid Africa along the Gulf of Guinea Part II: an integrated regional approach', *J. Hydrol.* **191**, 16–36.

Pearce, F., 1997: Sneaky El Niño outwits weather forecasters. *New Sci.*, 31 May, 6.

Polcher, J.: 1995 'Sensitivity of tropical convection to land surface processes', *J. Atmos. Sci.*, **52**, 3143–3161.

Power, S., Casey, T., Folland, C., Colman, A., Mehta, V. 1999. Inter-decadal modulation of the impact of ENSO on Australia *Climate Dynamics* 15, 319–324.

Ramage, C. S. (1968). Role of a 'maritime continent' in the atmospheric circulation. *Mon. Weath. Rev.* 96, 365–370.

Ramage, C. S., Khalsa, S. J. S. and Meisner, B. N. (1979) *The Central Pacific Near – Equatorial Convergence Zone,* UHMET 79-11, Department of Meteorology, University of Hawaii.

Ramage, C. S., and Schroeder, T. A. (1999) Trade wind rainfall atop Mount Waialeale, Kauai. *Mon. Weath. Rev.*, *127*: 2217–2226.

Rasmusson, E. M. and Carpenter, T. H. 1982. Variations in Tropical Sea Surface temperatures and surface wind fields associated with the Southern Oscillation/El Nino. *Mon Wea.Rev.*, 354–384.

Rasmusson, E. M., and T. H. Carpenter, 1983: The relationship between eastern equatorial Pacific sea surface temperature and rainfall over India and Sri Lanka. *Mon. Wea. Rev.*, 111, 517–528.

Riehl, H. (1954) *Tropical meteorology*. McGraw-Hill, New York.

Riehl, H. (1979) *Climate and Weather in the Tropics*. Academic Press.

Rodwell, M. J., and B. J. Hoskins, 1996: Monsoons and the dynamics of deserts. *Quart. J. Roy. Meteor. Soc.*, 122, 1385–1404.

Saha, K., F. Sanders, and J. Shukla, 1981: Westward propagating predecessors of monsoon depressions. *Mon. Wea. Rev.*, 109, 330–343.

Sadler, J. C. (1967) On the origin of tropical vortices. *Proc. Of working panel on tropical dynamic meteorology*, Monterey, California, Navy Weather Research Facility, Report 12-1167-132, 39-75. Available from the department of Meteorology, University of Hawaii.

Sadler, J. C. (1974) The monsoon circulation and cloudiness over the GATE area. Dept. of Meteorolgy, University of Hawaii. (Also presented at the Eighth Technical Conference of Hurricans and Tropical Meteorology, Miami, Florida, 14–17 May 1973.)

Sadler, J. C. (1975a) *The upper tropospheric circulation over the global tropics*, Dept. of Meteorolgy, University of Hawaii.

Sadler, J. C. (1975b) The monsoon circulation and cloudiness over the GATE area. *Mon. Weath. Rev. 104*: 369–387.

Sadler, J. C. 1976: A role of the tropical tropospheric trough in early season typhoon development. *Mon. Wea. Rev.*, 104, 1266–1278.

Sadler, J. C. (1984) The anomalous tropical cyclones in the Pacific during the 1982–1983 El-Nino. Postprints volume, *15th Conf. Hurricans and Tropical Meteorology,* January 9–13, 1984, Miami, Florida, American Meteorological Society, Boston, Mass., 51–55.

Sadler, J. C., Lander, M. A., Hori, A. M. and Oda, L. K. (1987) *Tropical Marine Climatic Atlas, Volume I* (Indian Ocean and Atlantic Ocean) and *Volume II* (Pacific Ocean), Department of Meteorology, University of Hawaii.

Sadler, J. C. and Harris, B. E. (1970) *The mean tropospheric circulation and cloudiness over southeast Asia and neighbouring areas*, Hawaii Institute of Geophysics, University of Hawaii. Available from the Department of Meteorology, University of Hawaii.

Schneider, E. K., B. Huang, Z. Zhu, D. G. DeWitt, J. L. Kinter III, B. Kirtman, and J. Shukla, 1999: Ocean data assimilation, initialization, and predictions of ENSO with a coupled GCM. *Mon. Wea. Rev.*, 127, 1187–1207.

Schultz, D., Bracken, W. E., Bosart, L. F., Hakim, G. J. Bedrick, M. A., Dickinson, M. J. and Tyle, K. R. 1997. The 1993 superstorm cold surge: Frontal structure, gap flow and tropical impact. *Monthly Weather Review*, 125, 5–39.

Servat, E., Patural., J. E., Lubes, H., Kouame, B., Ouedraogo, M. and Masson, J. M. 1997: 'Climate variability in humid Africa along the Gulf of Guinea Part I: detailed analysis of the phenomenon in Cote d'Ivoire', *J. Hydrol.* **191**, 1–15.

Shukla, J., 1987: Interannual variability of monsoons. Monsoons, J. S. Fein and P. L. Stephens, Eds., New York, Wiley, pp. 399–464.

Shukla, J., and D. A. Paolino, 1983: The Southern Oscillation and long-range forecasting of the summer monsoon rainfall over India. *Mon. Wea. Rev.*, 111, 1830–1837.

Sikka, D. R., 1980: Some aspects of the large scale fluctuations of summer monsoon rainfall over India in relation to fluctuations in the planetary and regional scale circulation parameters. Proc. Ind. Acad. Sci. (Earth and Planet. Sci.), 89, 179–195.

Sircoulon, J., Lebel, T. and Arnell, N. W.: 1998, 'Assessment of the impacts of climate variablilty and change on the hydrology of Africa', In: *Impact of Climate Change and Climate Variability on Hydrologic Regimes*, van Dam, J. C., ed., Cambridge University Press – UNESCO International Hydrology Series, in press.

Spencer, R. W., H. M. Goodman, and R. E. Hood, 1989: Precipitation retrieval over land and ocean with the SSM/I: Identification and characteristics of the scattering signal. *J. Atmos. Oceanic Technol.*, 6, 254–273.

Stockdale, T. N., D. L. T. Anderson, J. O. S. Alves, and M. A. Balmaseda, 1998: Global seasonal rainfall forecasts using a coupled ocean–atmosphere model. *Nature*, 392, 370–373.

Tanimoto, Y., N. Iwasaka, K. Hanawa, and Y. Toba, 1993: Characteristic variations of sea surface temperature with multiple time scales in the North Pacific. *J. Climate*, 6, 1153–1160.

Taylor, C. M., Said, F. and Lebel, T. 1997 Interactions between the land surface and mesoscale rainfall variability during HAPEX-Sahel, *Mon. Weath. Rev.* 125, 2211–2227.

van Oldenborgh, G. J. 2000. What Caused the Onset of the 1997–98 El Niño? *Monthly Weather Review*: Vol. 128, No. 7, pp. 2601–2607.

van Rompaey, R. S. A. R.: 1995, 'Climate change and deforestation in west Africa: a space-time trend analysis of rainfall series from Cote d'Ivoire and Liberia', In: *Climate Change Research: Evaluation and Policy Implications,* Zwerver, S., van Rompaey, R. S. A. R., Kok, M. T. J. and Berk, M. M., eds., Elsevier, Amsterdam, pp. 417–420.

Wang, B. 1995. Interdecadal Changes in El Nino Onset in the Last Four Decades. *J. Climate, 8*, 267–285.

Wang, B., Wu, R., and Fu, X., 2000. Pacific–East Asian Teleconnection: How Does ENSO Affect East Asian Climate? *J. Climate*, 13, 1517–1536.

Webster, P. J., and S. Yang, 1992: Monsoon and ENSO: Selectively interactive systems. *Quart. J.Roy. Meteor. Soc.,* 118, 877–926.

Williams, M. A. J. and Balling, R. C.: 1994, *Interactions of Desertification and Climate*, WMO/UNEP, 230pp.

W.M.O. 1995. Global Perspective on Tropical Cyclones. World Meteorological Organization Report No. TCP-38 289 pages.

Wright, W. J. (1997) Tropical-Extratropical Cloudbands and Australian Rainfall: I. Climatology. *Int. J. Climatol.* 17: 807–829.

Zhang, Y., J. M. Wallace, and D. S. Battisti, 1997: ENSO-like interdecadal variability: 1900–93. *J. Climate*, 10, 1004–1020.

11 Synoptic and mesoscale rain producing systems in the humid tropics

M. Bonell
UNESCO, Paris, France
J. Callaghan
Bureau of Meteorology, Brisbane, Australia
G. Connor
Bureau of Meteorology, Townsville, Australia

INTRODUCTION

A key facet of the hydrology and climatology of the humid tropics is the occurrence of more persistent high-rainfall intensities compared with those occurring over the higher latitudes. The equivalent hourly intensities of short-term rainfalls of, for example, over one minute, are commonly one or two orders of magnitude higher than those experienced in humid temperate areas (Bonell, 1993). Thus, the magnitude of rainfall is one of the principal drivers in accounting for the much wider range of preferred pathways of storm runoff in tropical forests (see Bonell, this volume). Less well appreciated in the literature is a possible linkage between preferred pathways of storm runoff and the spatial and temporal variability of different rain-producing meteorological systems. For example, there is a substantial difference in the range of rainfall intensity-frequency-duration characteristics identified with tropical cyclone-prone areas (e.g. northeast Queensland) as against tropical islands where trade wind 'stream' showers are more persistent (e.g. the Hawaiian Islands). Moreover, within the outer tropics, there can be a significant seasonal change in synoptic-scale meteorology systems and corresponding rainfall characteristics (notably rain intensities). In northeast Queensland, the dominant pathways of storm runoff in tropical forest show dramatic changes over a few months in response to a change from a monsoon regime of persistent tropical depressions (and cyclones) through to a regime dominated by upper troughs, and finally, stream showers associated with the SE trades (Bonell and Gilmour, 1980; Bonell *et al.*, 1991). Elsewhere in this volume, Scatena *et al.* provide rainfall intensity-duration data for different storm types (cyclonic, non-cyclonic, frontal) in the northern Caribbean. In common with northeast Queensland, there is a progressive seasonal decline in rainfall intensities towards the cool season in the northern Caribbean. Thus, process hydrologists need to have a better appreciation of different rain-producing systems in the humid tropics, their resulting different rainfall intensity regimes and their contribution to the diverse range of runoff generation responses that occur.

Definitions and descriptions of the synoptic scale features of the tropical atmospheric circulation were given in the previous chapter (Callaghan and Bonell). Here we present a more detailed overview of various perturbations at different scales, initially by assessing the characteristics of tropical cyclones using a series of case studies to highlight their genesis, intensification, steering mechanisms and subsequent decay. These storms, in conjunction with moisture uplift produced by topography, are associated with some of the most extreme rainfalls experienced in the tropics.

Perturbations embedded within the surface tropical easterlies, at both the synoptic and mesoscale, are still not understood comprehensively and yet they are a significant rain-producing mechanism. Thus, considerable attention is given to their origins and westward propagation, placing strong emphasis on West African and Caribbean examples. Some of the easterly perturbations observed in the Caribbean are thought to originate from the African Sahel. Consequently, we will also incorporate material from outside the humid tropics when referring to the African Sahelian region in terms of the dynamics of easterly perturbations and mesoscale convective systems, and the associated rainfall characteristics. Moreover, the descriptions of these phenomena are also thought to be relevant further south to the wet/dry tropics zone (Chang and Lau, 1993), of West Africa (Lebel *et al.*, 1998; Taylor *et al.*, 2000).

The nature of Mesoscale Convective Systems (MCSs) within the surface easterlies, the Zonal Trough in the Easterlies (ZTE) and the synoptic features of the monsoon circulation need to be described because their internal structure (connective, stratiform)

Forests, Water and People in the Humid Tropics, ed. M. Bonell and L. A. Bruijnzeel. Published by Cambridge University Press.

and varied diurnal cycles have a major effect on terrestrial hydrology. Various studies of MCSs over both oceanic and continental areas will be outlined, with much more detail provided for the tropical northwest Pacific, Fiji, northern Australia and the Amazon basin. In addition, a considerable proportion of the outer tropics is occupied by the trade wind belt and we also show how much rain is thus accounted for by the passage of 'stream' showers and orographic uplift over islands and coastal hinterlands, rather than from well-organised, synoptic-scale perturbations.

The concluding sections of this chapter are devoted to various aspects of tropical rainfall, such as rainfall frequency-intensity-duration, impacts of orographic uplift and the spatial organisation of rainfall. The latter will also include the possible impacts of land-use conversion on the reorganisation of mesoscale rainfall.

Appendix 11.1 provides a listing of acronyms commonly used in this chapter. The Glossary of Terms (Appendix 10.1) presented in the previous chapter is also relevant here.

TROPICAL CYCLONES

Most surface low pressure systems over tropical oceans do not develop into tropical cyclones even when the low level atmospheric circulation is favourable (e.g. a monsoon shearline) and sea surface temperature (SST) exceeds 26 °C. A key control is the need for an upper tropospheric environment of low wind shear and diffluent outflow to evacuate low level convergent inflow efficiently. Box 11.1 provides a case study for the rapid formation and intensification of tropical cyclone *Susan* over the South Pacific Ocean in response to a favourable upper troposphere environment.

Current debate on the mechanisms connected with the formation and intensification of tropical cyclones

Many forecasters would view the above rapid intensification of *Susan* (Box 11.1) to have resulted because it was in a low vertical wind shear environment due to its location north of a deep layered ridge and there were suitable upper wind currents present in the vicinity of the system to provide good upper outflow channels from the storm. There is vigorous debate among tropical cyclone forecasters and researchers, however, regarding the role of upper outflow channels and mid-latitude upper troughs in the intensification of tropical cyclones (McRae, 1956; Sadler, 1978; Holland and Merill, 1984; Chen and Gray, 1986).

There is further debate on the role of SST warm anomalies in tropical cyclone intensification. The SST exceeding 26 degrees Celsius has been shown to be a necessary, though insufficient, condition for tropical cyclogenesis (Palmen, 1948). Thin layers of warm SST in the paths of developing storms may be mixed with underlying cool layers and therefore not contribute to the development of the system. Deep warm oceanic mixed layers offer reservoirs of high heat content for the continued development and intensification of tropical cyclones (Shay, 1988).

Box 11.1 The rapid formation and intensification of a very intense tropical cyclone in the South Pacific Ocean

The large scale wind analyses at 200 hPa (constructed from ECMWF wind data) leading up to the genesis of tropical cyclone *Susan* (the eastern cyclone) are shown in Figure 11.1. There was strongly divergent flow above the general area where *Susan* developed as the upper wind currents branched between flow into the westerlies in both hemispheres and flow into the near equatorial easterlies. A southerly wind current flow began to extend directly from the cyclone into the Northern hemisphere. Then over the remaining period covered in Figure 11.1, the upper outflow became focussed above *Susan*. From the MSL sequence (Figure 11.2) we see initially two low pressure centres located in a large broad monsoon trough over the southwest Pacific Region. Major deepening in the monsoon trough over a wide area of the South Pacific occurred after 0000 UTC 2 January 1998.

Closer examination at 1200 UTC 31 December 1997 (top left panel in Figure 11.1) indicates there was a strong upper outflow centre in the extreme northeast of the domain. The main deep heavy convective rain area was located about 5 latitude degrees south of this outflow zone and near the eastern MSL low. By 1200 UTC 1 January 1998 (top right panel in Figure 11.1), diffluent upper flow developed over the deep convection and MSL low. Twenty-four hours later (middle left panel), a strong upper outflow centre developed over the MSL low, which occurred immediately after the initial intensification of this MSL feature. Coincident with the formation of this upper outflow centre was the development of a 200 hPa ridge between Vanuatu and Fiji. This ridge was a deep system, which can be verified by the MSL pressure rises south of *Susan* by 0000 UTC 3 January 1998. The development of this ridge helped establish a strong east-southeasterly upper wind current from *Susan* towards Papua New Guinea. This same deep layered ridge may have also been an important factor in the intensification of the cyclone, because it provided a low vertical wind shear environment (i.e. small change in wind velocity with height).

The most rapid intensification occurred around 0000 UTC 4 January 1998 and the central pressure dropped 60 hPa in the following 24 hours. Wind and height contour charts constructed from actual observations (Figure 11.3) show a 200 hPa trough passing well to the south of *Susan* around the time when the most rapid intensification was taking place (centre and lower right panels in Figure 11.3). The upper outflow from *Susan* into the westerlies increased as this trough system weakened the upper portion of the ridge south of the cyclone.

The 700 hPa sequence during the period of rapid intensification in Figure 11.3 shows that, as the trough system passed to the south, a large break formed in the Subtropical Ridge (STR) at low to middle levels south of *Susan*. The cyclone was turned southwest then southward towards this weakness in the ridge.

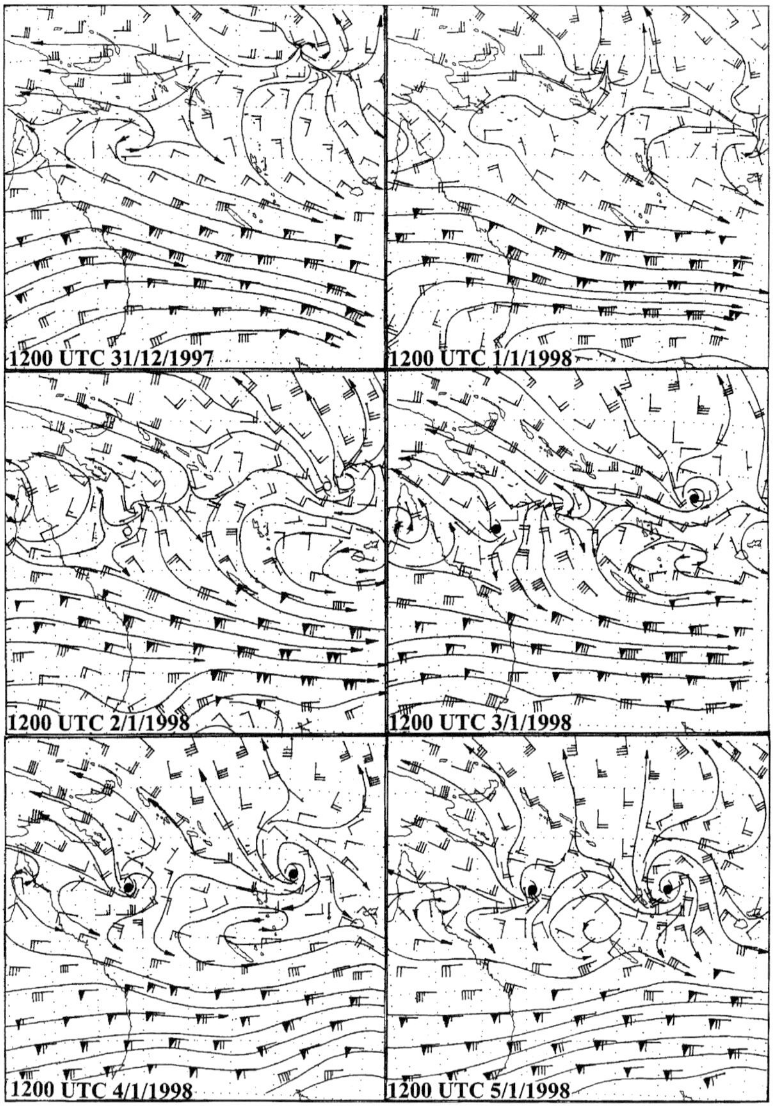

Figure 11.1 Streamlines at 200 hPa drawn from ECMWF analysed winds 1200 UTC 31 December 1997 to 1200 UTC 5 January 1998. Unfilled (filled) cyclone symbol denotes pre-cyclone disturbance (tropical cyclone) position.

SST anomalies appeared to play a secondary role in the intense development of *Susan*: the cyclone moved over weakly positive (warm) anomalies (less than 10 cm) as it intensified. The actual SST over the oceans where it developed was about 1 degree Celsius above normal.

Common upper wind pattern associated with the genesis and intensification of tropical cyclones

There is one upper wind pattern that is commonly associated with tropical cyclogenesis around the globe. In the Southern (Northern) hemisphere, the disturbance intensifies under diffluent east to northeasterly winds at 200 hPa west of an upper anticyclone. The northeasterly (southeasterly) winds turn into the mid-latitude westerly flow through a weakness in the upper ridge which extends westward from the upper anticyclone. Ritchie and Holland (1999) found this to be the most common pattern leading to tropical cyclogenesis in the western North Pacific (42% of all cases studied). Bracken and Bosart (2000) also found it to be the most common pattern associated with tropical cyclogenesis in the North Atlantic. An example of the development of a tropical cyclone over the Bay of Bengal (Box 11.2) also shows that the diffluent zone, located west of a developing upper anticyclone, proved a favourable environment for the rapid intensification of the storm.

Rapid intensification of tropical cyclones adjacent to the east coasts of continents

A common feature of the more severe tropical cyclones is their rapid intensification just to the east of landmasses. Hurricanes

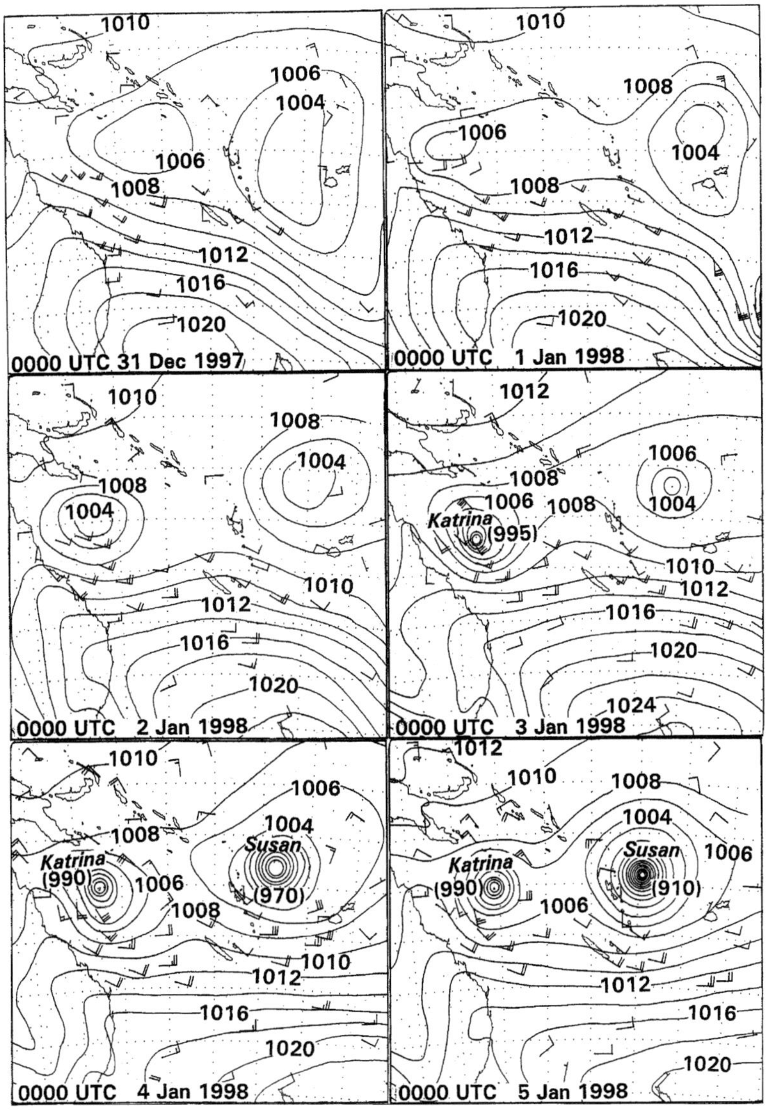

Figure 11.2 MSL pressure distribution (hPa) and selected wind observations 0000 UTC 31 December 1997 to 0000 UTC 5 January 1998.

Andrew (1992) and Mitch (1998) (Box 11.3) are recent examples. The optimal location of such storms in relation to either a near-stationary trough over a continent or transient troughs in the upper westerlies are key factors responsible for the rapid intensification.

Summary of the characteristics of rapid intensification

We have investigated tropical cyclones that underwent very rapid intensification and identified the upper wind patterns that were conducive to this process. The upper outflow from these intensifying cyclones merged readily with these upper wind patterns to form outflow channels on both the Equatorward and poleward sides of the cyclones. The cyclones were all found to be interacting with upper troughs embedded in the westerlies of the respective Northern and Southern hemispheres. The amplitude of these troughs and their distance from the cyclone appeared crucial to the intensification process.

The rapid intensification occurred under low vertical wind shear zones, associated with deep layered coherent weather systems. For example, when cyclones move south or south-southwest, bursting through the subtropical ridge in the Southern hemisphere, the deep layered ridge was weakened by an upper trough system which exerted its influence down to low levels (thus weakening the ridge coherence down to low levels). This trough system remained suitably distant from the cyclone so as not to bring strong

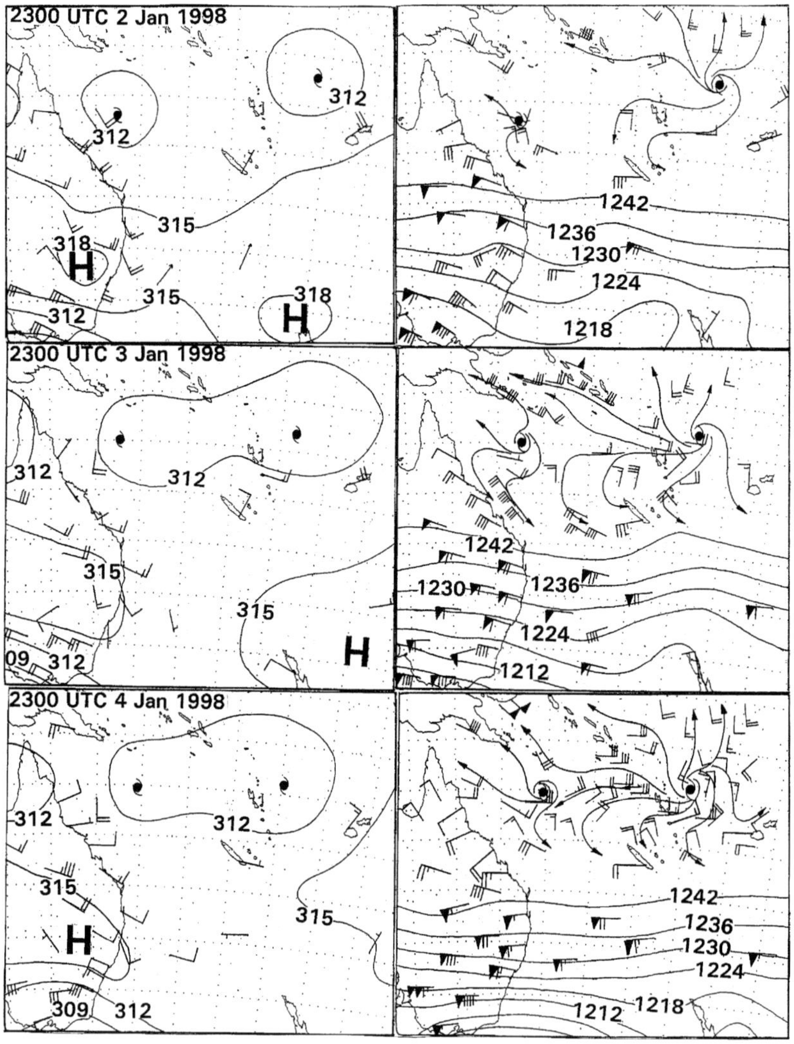

Figure 11.3 Height contours (decametres) and wind observations at 700 hPa (left panels) and 200 hPa (right panels) from 2300 UTC 2 January 1998 to 2300 UTC 4 January 1998.

shearing upper winds close to the cyclone and thus weaken the storm.

Tropical cyclones moving in a general westerly direction were steered by deep-layered persistent subtropical ridges. They interacted with upper trough systems which remained either well to the west of the cyclones over a continental land mass or well to the east in the transient systems, e.g. Hurricane Mitch. The trough systems were close enough to provide an access for an upper wind outflow into the mid-latitude westerlies, but far enough removed from the cyclone so as not to cause shearing (which would weaken the storm) from strong upper winds. Cyclones moving in general north-easterly (northern hemisphere) or south-easterly (southern hemisphere) directions were steered by deep layered trough systems which weakened the subtropical ridge at low levels and thus did not block the progress of the intensifying cyclones.

EXTREME TROPICAL CYCLONE RAINFALL EVENTS

By their very nature, tropical cyclones are very efficient systems in converting atmospheric water vapour to rainfall through rapid low-level convergence and subsequent uplift within these storms. Consequently, they are prolific rain producers and account for some of the highest recorded rainfalls by both tropical and as well as global standards. It is important to recognise, however, that for different meteorological reasons there are preferred quadrants of more intense precipitation.

Box 11.2 Indian tropical cyclone

A recent disastrous example was tropical cyclone, code numbered 5B, which struck the east coast of India in October 1999 and was thought to have caused more than 10 000 deaths in Orissa Province. Wind data at 200 hPa (Figure 11.4) from the ECMWF numerical model output, covers the rapid intensification of the cyclone which occurred in the diffluent zone between flow into the mid-latitude westerlies and flow into the tropical easterlies. The diffluent zone was located west of a developing upper anticyclone. A deep trough system well to the north in the westerlies was approaching the longitudes of the cyclone. This was weakening the upper ridge to the north of the cyclone allowing the upper outflow from the cyclone to flow into the mid-latitude westerlies.

SSM/I and TRMM images (Plate 5) show a very large open eye had formed at 1142 UTC 27 October 1999 (around the time of the middle panel in Figure 11.4). At 1410 UTC 26 October 1999 an SSM/I image (not shown) showed the main convective feature was a large curved rain-band. This implies that the cyclone was rapidly intensifying at 1142 UTC 27 October 1999. At 1543 UTC 28 October 1999 the TRMM image (lower panel Plate 5) shows a tightly coiled compact eye had formed indicating it was a very intense cyclone. Therefore, over the 48-hour period covered by Plate 5, the system developed from a tropical system with a curved rain band into a very intense hurricane with intense convection surrounding a tight compact eye.

Box 11.3 Amplifying transient upper trough to the east of an intensifying cyclone, Hurricane Mitch

The influence of a transient trough in the upper westerlies is illustrated by Hurricane Mitch. The period of rapid intensification of this hurricane is examined in Figure 11.5 through an analysis of the upper winds at the 100 hPa to 250 hPa layers. These winds were derived from water vapour imagery and supplied by the Cooperative Institute for Meteorological Satellite Studies at the University of Wisconsin (CIMSS). The height contours were drawn from NCEP/NCAR re-analyses data. Throughout the period of intensification, the major upper trough system was located over longitudes to the east of Mitch.

In the top panel, Mitch was a weak system with a central pressure of 997 hPa. An upper anticyclone centre lay to the northwest of Mitch over the Gulf of Mexico and a weaker upper ridge was located east and south of the hurricane. By 0000 UTC 25 October 1998 (centre panel), the upper ridging and the ridge axis lay north and east of Mitch. This synoptic environment further gave the storm access to more outflow on its northwestern flank. By then, a *good outflow current had developed into the upper westerlies where a major trough was passing to the northeast.* A strong Equatorial upper outflow channel had also developed and Mitch had subsequently deepened to 973 hPa.

By 0000 UTC 25 October 1998, Mitch had deepened to 924 hPa, with the upper winds showing a good outflow structure into the easterlies (westerlies) to the south (north) of the hurricane. This occurred as the major anticyclone centre in the upper ridge moved to the northeast of the hurricane. Mitch continued to deepen and the lowest central pressure of 906 hPa (calculated from aircraft reconnaissance data) was reached at 1800 UTC 26 October 1998. Over this period Mitch was at the category 5 stage (10 min average wind speeds > 57 m s^{-1} or 110 Knots).

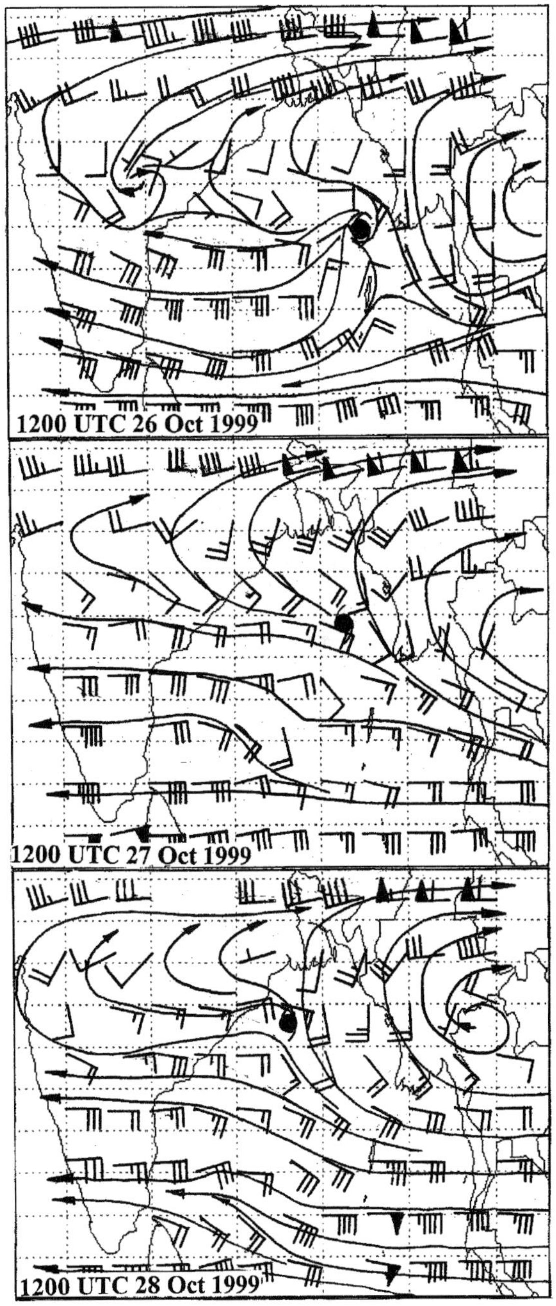

Figure 11.4 Streamlines at 200 hPa drawn from ECMWF analysed winds 1200 UTC 26 October 1999 to 1200 UTC 28 October 1999.

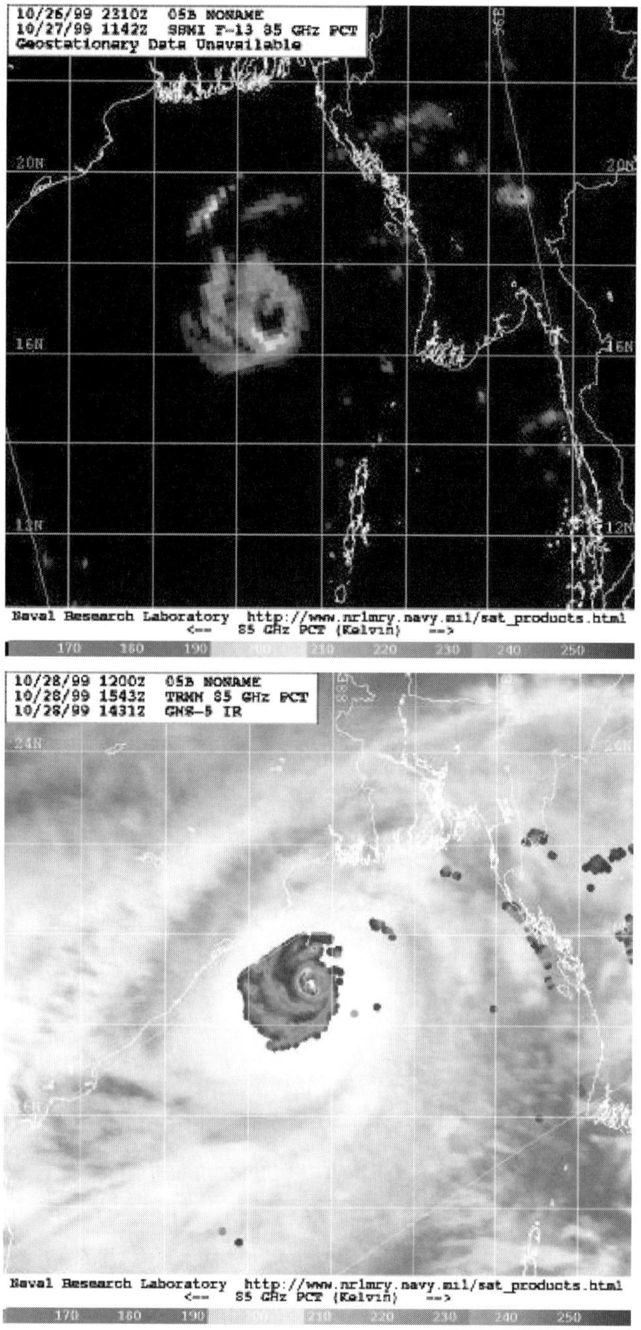

Plate 5 SSM/I image (top) 1142 UTC 27 October 1999 (top) and TRMM image (bottom) 1543 UTC 28 October 1999. (Source: US Naval Research Laboratory, Monterey, CA, USA http://www.nrlmry.navy.mil/tc-bin/tc.home) (See also colour plate section.)

For example, in the previous chapter we showed that the rain structure of Indian monsoon depressions has persistent heavier rainfall in the Equator-ward quadrants away from the cyclone centre. The southern hemisphere example of tropical cylone *Les*

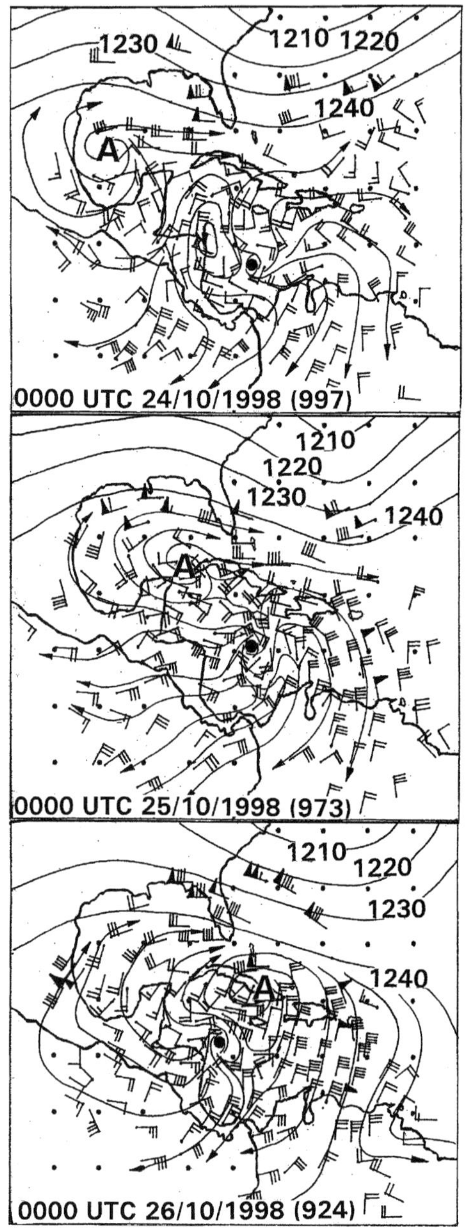

Figure 11.5 Height contours (decametres) and wind observations at 200 hPa from 0000 UTC 24 October 1998 to 0000 UTC 26 October 1998. (See text for sources of data.)

(Box 11.4) also shows preferred heavier rainfall to the north and west of the storm centre, i.e. equatorward. An additional feature of tropical cyclones is that they can be slow moving or even become near-stationary on making landfall, which further accentuates the amount of rainfall.

The rain structure of tropical cyclones and depressions

There are many examples around the globe where higher latitude weather systems influence the tropical weather: in South East Asia

Box 11.4 Tropical cyclone *Les*

Les was rapidly developing in the Gulf of Carpentaria along the north Australian coast when it made landfall in the Northern Territory. It then crossed the Top End of the Northern Territory as an intense rain depression causing record flooding in the Katherine and Daly Rivers. The main rain gauge at Katherine recorded 221 mm in the 24 hours to 2300 UTC 25 January 1998 and 160 mm for the period ending 2300 UTC 26 January 1998. The Katherine River peaked at 20.5 metres, exceeding the previous record of 19.3 metres set during 1957.

The rain areas are delineated in the sequence of SSM/I imagery (Plate 6). The green areas show rain, the yellow shows heavy rain, while the red indicates the areas of heaviest precipitation. These rain bands were located to the north and west of the centre of the cyclone which is similar to the rain structure of the Indian monsoon depressions where the heaviest rain was also equatorward from the cyclone centre, as described in the previous chapter.

Like the Indian monsoon systems, Les lay under the upper easterlies (Figure 11.6) as it crossed the Top End of the Northern Territory. The upper wind pattern over Les showed diffluent and accelerating (divergent) flow over the rain areas between the easterlies and a south to southeasterly wind current to the north of Australia.

(Chang *et al.*, 1979, Lau *et al.*, 1983; and Wu Chan, 1997), in tropical South America (Lenters and Cook, 1999; Garreaud and Wallace, 1998) while Matthews and Kiladis (1999) describe a tropical-extratropical interaction in the central Pacific. We show here how extreme rainfall occurs when upper troughs extend into the northeast Australian tropics and their impact on poleward moving tropical lows. The extreme rainfall occurs where the wind direction backs with height associated with warm air advection; these directions are confined to the north-east quadrant below 500 hPa. In the Northern hemisphere, the equivalent wind structure would be veering winds with height confined to the southeast quadrant. More details of this warm air advection–vertical ascent mechanism are described in Hoskins *et al.* (1978; Figure 11.2) and Holton (1979).

Box 11.5 provides a case study of a tropical low (ex-tropical cyclone Sid) which was steered by a deep upper trough system from the Gulf of Carpentaria to the Townsville region of north east Queensland.

Common synoptic patterns between the south west Pacific and south west Indian Ocean during record rainfall events (northeast Queensland and La Réunion Island)

North-east Queensland and La Réunion both have elevated rain recording stations, at Mount Bellenden Ker (1555 m a.s.l.) and

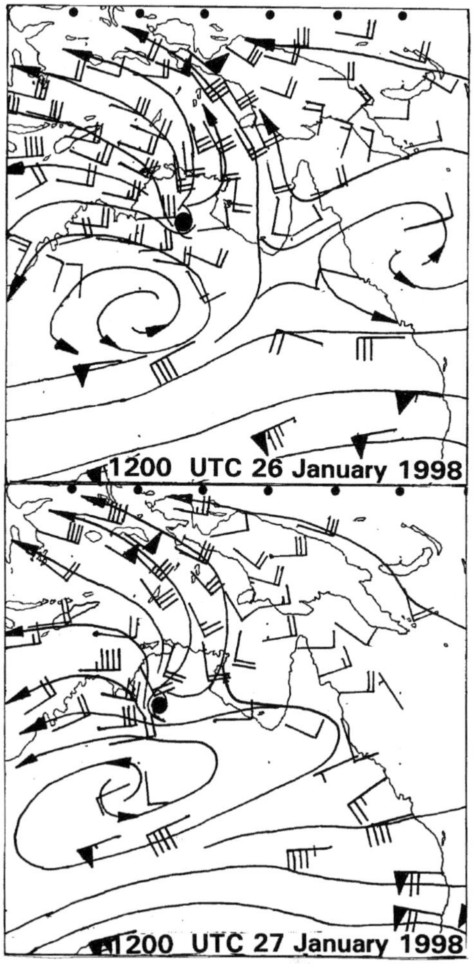

Figure 11.6 Streamlines and wind plots (conventional observations and satellite derived winds) at 200 hPa for 1200 UTC 26 January 1998 to 1200 UTC 27 January 1998.

Baril (1600 m a.s.l), respectively. Both stations have similar meteorological and topographical settings which are highlighted by separate examples of synoptic events.

In January 1979, ex-tropical cyclone Peter hovered off the north-east Queensland coast after previously developing in the Gulf of Carpentaria and transiting eastwards across Cape York Peninsula before entering the Coral Sea. Several Australian record rainfalls were broken at Mt. Bellenden Ker Top Station where 1947 mm was recorded in the 48-hour period ending 2300 UTC January 1979.

In the south-west Indian Ocean, a similar synoptic situation occurred in February 1993. Several world records were broken when a slow moving tropical low developed to the northwest of La Réunion Island. Over the 48-hour period close to the period covered by Figure 11.9, Baril, on La Réunion Island recorded 3000.5 mm of rain (Barcelo *et al.*, 1997). The island can be

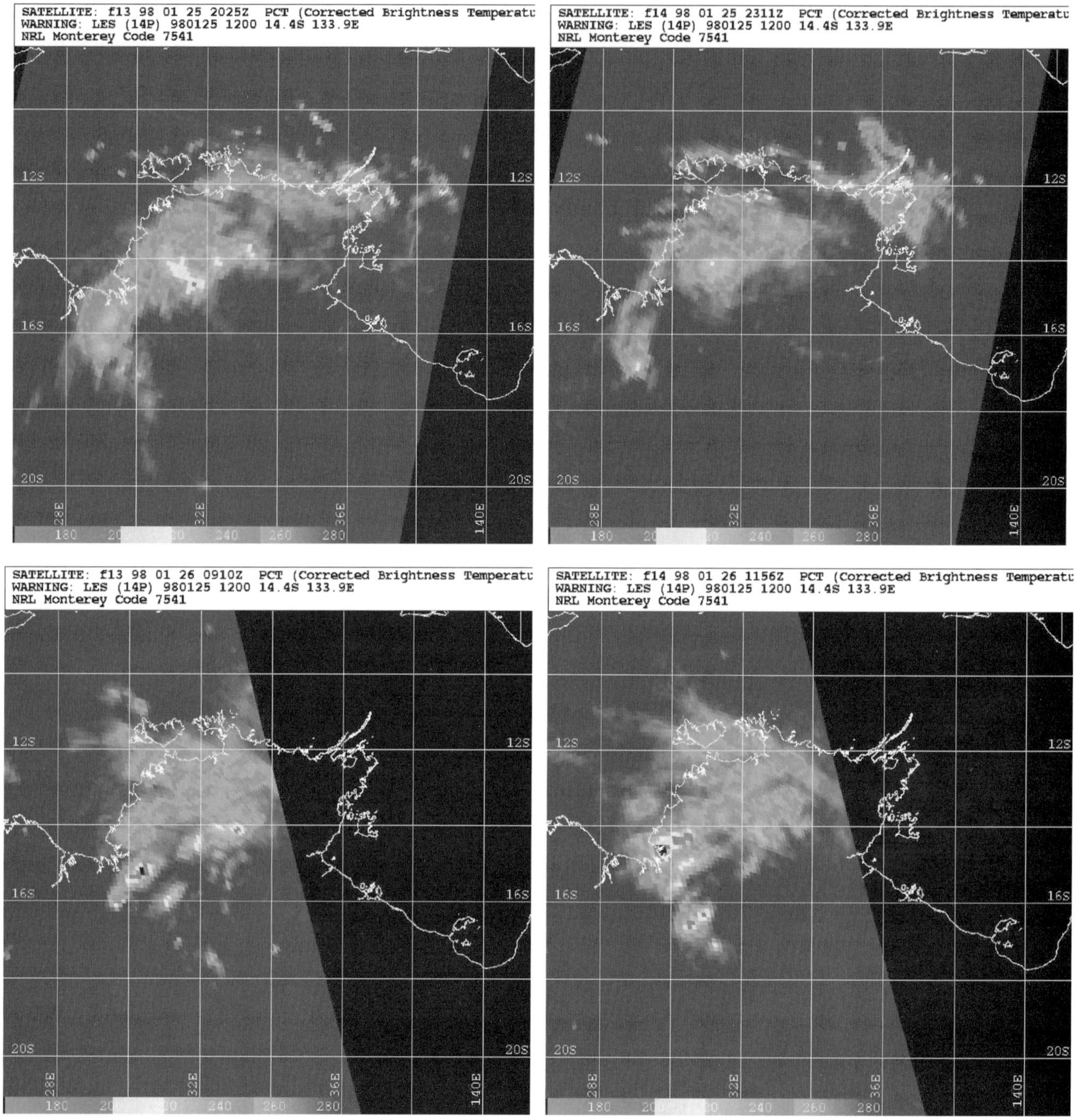

Plate 6 SSM/I imagery at 2025 UTC 25 January 1998 (top left), 2311 UTC 25 January 1998 (top right), 0910 UTC 26 January 1998 (bottom left) and 1156 UTC 26 January 1998 (bottom right). (Source: US Naval Research Laboratory, Monterey, CA, USA http://www.nrlmry.navy.mil/tc-bin/tc.home) (See also colour plate section.)

seen to be located in a similar meteorological environment to Bellenden Ker, i.e. in the confluent zone south of the monsoon trough. Other similarities are the large slow-moving MSL high south of the heavy rain area and the trough to the southeast breaking the ridge at 500 hPa, so that middle level steering is weak, resulting in slow movement of the low which was developing over Madagascar. The analyses in Figure 11.9 were from the NCEP/NCAR re-analyses. The MSL pattern at 0000 UTC 1 March 1993 was similar to the MSL pattern in Figure 11.2 of Barcelo *et al.* (1997) except that that their low had a central pressure of 1000 hPa against 1007 hPa for the NCEP/NCAR low. We have found that this is typical with these re-analyses in that they

Box 11.5 The 1998 Townsville floods

At 2300 UTC 8 January 1998, a trough at the 200 hPa level (Figure 11.7 top panel) lay from southeast Australia up into the tropics with strong upper northwesterly winds over most of Queensland. By 2300 UTC 9 January 1998, the trough had moved westward (Figure 11.7 lower panel) and strong northwesterly winds continued to blow over Queensland at high levels. The trough was reflected at lower levels at 2300 UTC 8 January 1998, however, it became less defined there 24 hours later indicating that it had weakened at these lower levels as the upper trough moved west. Also by 2300 UTC 9 January 1998, heights had risen over southern Queensland and New South Wales at 850 hPa and 700 hPa as strong ridging was developing in the Tasman Sea.

Movement of the low

The deep trough system initially provided the steering mechanism for the low, moving it southeastward across Cape York Peninsula towards Townsville. As the trough weakened by 2300 UTC 9 January 1998, the steering was also weakened, particularly in the lower parts of the troposphere where the developing ridge blocked the path of the low. At 500 hPa and 200 hPa, 2300 UTC 9 January 1998, however, the trough was still evident west of the low and north to northwesterly winds blew across the low at these levels. This had the effect of advecting the rain areas when they developed towards the southern side of the low.

MSL development

At 0500 UTC 9 January 1998 (top left, Figure 11.8), the low was overland west of Cairns with the dashed line marking the location of the monsoon trough which crossed the coast just to the north of Cooktown. By 1100 UTC, the monsoon trough was near Cairns and rapidly moving southwards to a band of developing strong winds. By 1700 UTC 9 January 1998, the monsoon trough merged with this band of strong winds and the heavy rains began to develop in the zone just south of the monsoon trough. This area is typically confluent in the low levels where the moist flow of monsoon origin meets the southeast flow from a poleward anticyclone. During monsoon situations along the eastern coast of Queensland the heaviest rain is always recorded just to the south of the area where the monsoon trough crosses the coast. The low then remained very slow moving just to the north of Townsville for the 24 hours after 2300 UTC 9 January 1998. This kept the Townsville area in the zone of heavy rain over this period.

Rainfall

Thus intense rain fell in the Townsville area during the 24 hour period ending 2300 UTC 10 January 1998. Table 11.1 shows the rainfall depths duration at the Townsville Meteorological Office during this period. The distribution of the 24 hour rainfall totals in the Townsville area exceeded 200 mm over a very large area.

Upper wind vertical profile

As indicated above, extreme heavy rain events in Queensland have in the past been found to be associated with winds which back in direction with height. The direction from which these winds blow is generally confined to the northeast quadrant. For example, the heavy flood rains in southeast Queensland in 1992 (Bureau of Meteorology, 1992) were associated with winds which backed with height from low level easterlies to middle level northerlies. The heaviest flood rains in southeast Queensland in May 1996 (Bureau of Meteorology, 1996) were also associated with low level easterly winds which backed with height to middle level northeasterlies. When concerning the evolution of the vertical wind profile at Townsville up to 0500 UTC 10 January 1998 (the latter time was just before the heaviest rain began to fall at Townsville), the winds increased in speed at lower levels over the 30-hour period and kept a backing with height profile. In time, the wind directions below 500 hPa became confined to the northeast quadrant. Additionally, the winds at the lowest levels turned north of easterly, which gave these low level winds a larger normal component on to the coast near Townsville.

Table 11.1. *Rainfall depths and duration at Townsville Meteorological Office*

Duration	Rainfall (mm)	Period ending
6 minutes	17.0	0943 UTC 10 January 1998
12 minutes	33.4	0949 UTC 10 January 1998
18 minutes	49.4	0950 UTC 10 January 1998
30 minutes	75.4	0957 UTC 10 January 1998
1 hour	131.0	1016 UTC 10 January 1998
2 hours	212.0	1049 UTC 10 January 1998
3 hours	253.0	1125 UTC 10 January 1998
6 hours	361.0	1423 UTC 10 January 1998
12 hours	482.0	2013 UTC 10 January 1998
24 hours	564.0	0008 UTC 11 January 1998

portray the synoptic scale features well but do not resolve the intense inner core of deep low pressure systems.

Comparison of two severe tropical cyclones in Fiji of contrasting vertical wind and rain structure

Despite the strong emphasis on the role of the warm air advection-vertical ascent mechanism, this feature is not always present in tropical cyclones. For example, in late 1992 and early 1993, two tropical cyclones of similar intensities passed over the Fiji group. The second, tropical cyclone Kina, produced prolonged heavy rain and extensive flooding. The heaviest rainfall was recorded at Monasavu on the highlands of the main island of Vitu Levu.

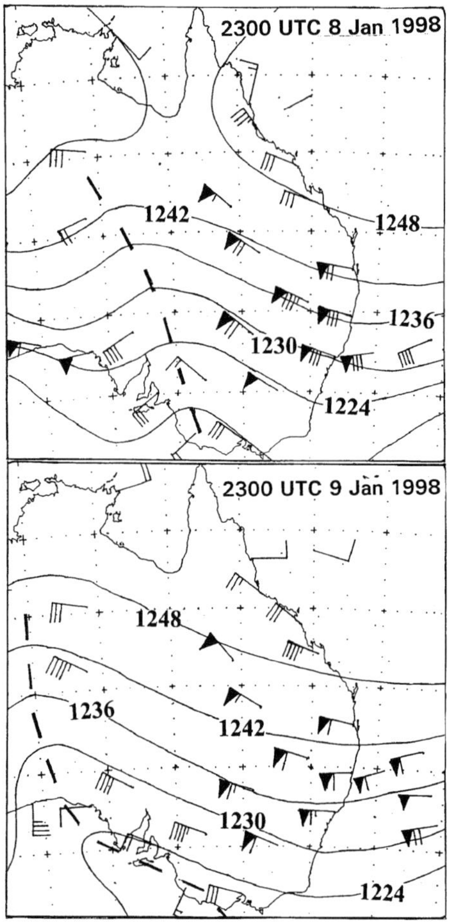

Figure 11.7 Height contours (decametres) and wind observations at 200 hPa for 2300 UTC 8 January 1998 (top) and 2300 UTC 9 January 1998 (bottom).

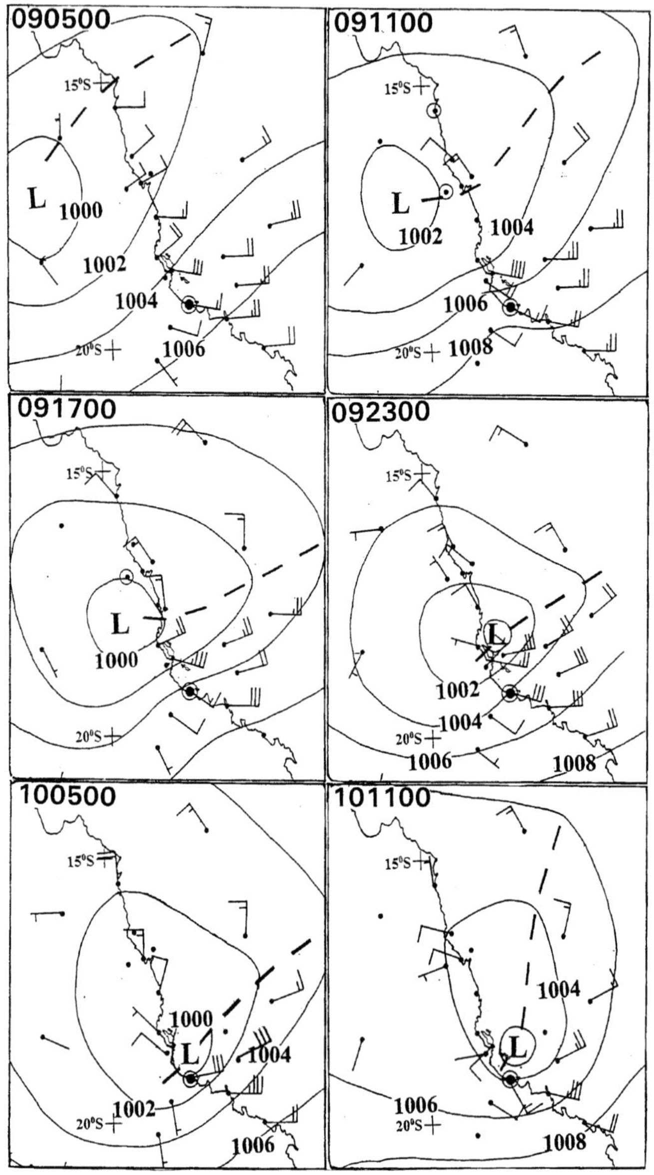

Figure 11.8 MSL pressure distribution (hPa) and wind observations over tropical northeast Australia from 0500 UTC 9 January 1998 (denoted by date time group on panel 090500) to 1100 UTC 10 January 1998 (101100).

Total rainfall over the period 27 December 1992 (2100 UTC) – 2 January 1993 (2100 UTC) amounted to 1007 mm, of which 550mm occurred in the final 24 hours to 2100 UTC, 2 January 1993. Three weeks earlier, the first cyclone, Joni, was of similar intensity but much smaller in size. The associated rainfall was less intense than Kina, with flooding confined to areas near the eye of the cyclone.

To understand these contrasting rainfall-flooding characteristics between storms of similar intensity, one must examine the synoptic meteorological environments around the cyclones. These environments were quite different, as shown by Figure 11.10. As Kina approached Fiji, the upper winds at Nadi indicated the now familiar pattern of wind direction mostly backing with height (turning anticlockwise), that is, the warm air advection-vertical ascent mechanism which accounted for the flood-producing rains. In contrast, the winds at Nadi as Joni approached showed a tendency to veer (turn clockwise with height) which is associated with subsiding flow in the Southern hemisphere. Consequently Joni produced less rainfall and associated flooding than Kina.

The synoptic meteorological controls of Hurricane Mitch: an example of a slow moving system which produced high rainfalls

We described previously the intensification of Mitch: now let us discuss the linkage between the trajectory of this storm and the high rainfalls it produced.

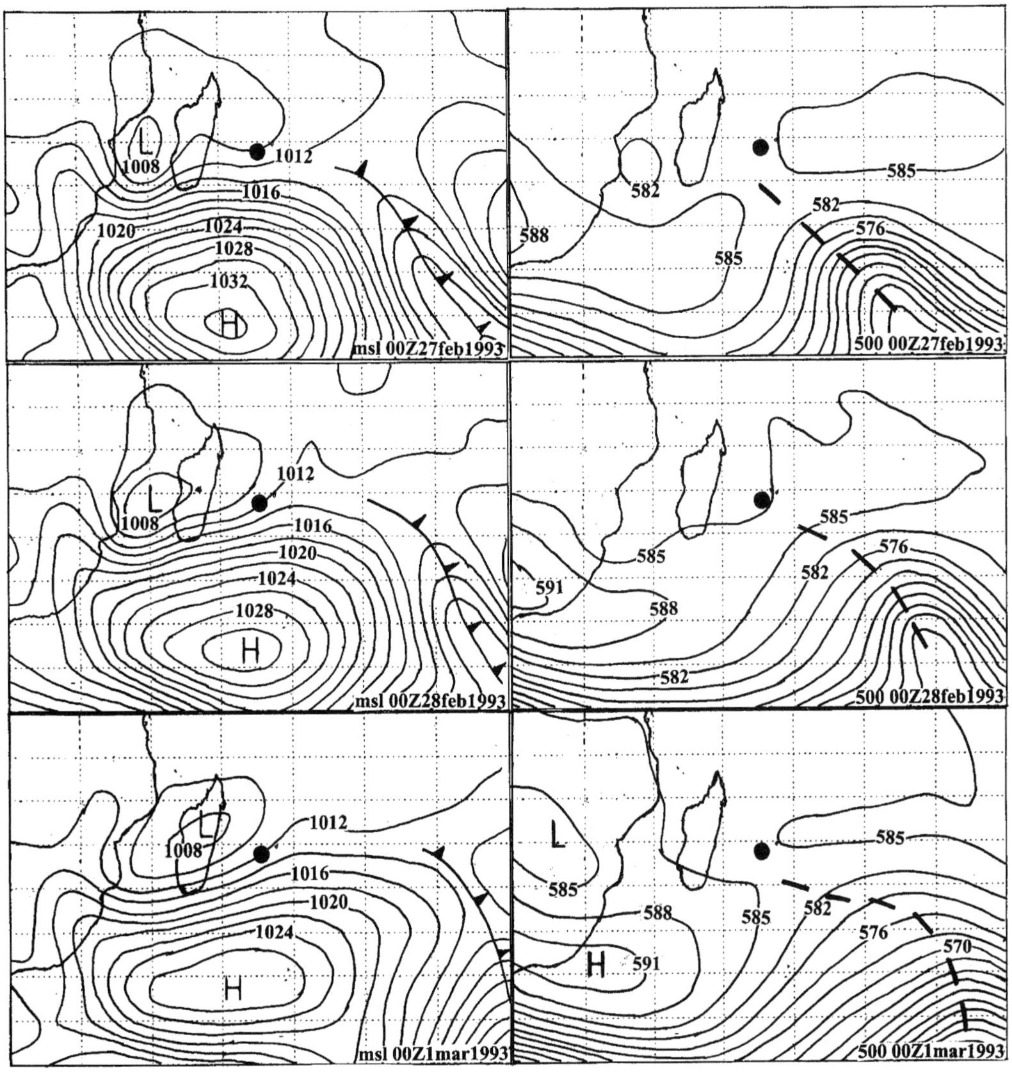

Figure 11.9 MSL pressure distribution (hPa and left frames) and 500 hPa height contours (decametres and right frames) from 0000 UTC 27 February 1993 to 0000 UTC 1 March 1993.

Mitch was slow-moving, leading up to landfall at 1200 UTC 29 October 1998 130 km east of La Ceiba on the North Coast of Honduras. The estimated central pressure of Mitch rose to 77 hPa from 0000 UTC 27 October 1998 until landfall. Mitch then moved slowly southward, then south-west-ward and west-ward over Honduras, weakening to a tropical storm by 0600 UTC 30 October 1998 and to a tropical depression by 1800 UTC 31 October 1998. The overall motion was slow (Figure 11.11a), less than 7.5 km h^{-1}, for a week. This resulted in extremely heavy rainfall, estimated at over 900 mm, primarily over Honduras and Nicaragua. The estimated total storm rainfall is shown in Figure 11.11b.

The deep mean layer winds between 850 hPa and 300 hPa are the best charts to gain an appreciation of the contribution to the steering of a tropical cyclone from the environment around it. Neumann (1979) demonstrated that the middle levels (around 500 hPa) are the best single levels to use in the absence of deep mean layer field data. The 500 hPa sequence (from NCEP/NCAR re-analyses data) in Figure 11.12 show Mitch moving westward from 0000 UTC 26 October 1998 to 0000 UTC 27 October 1998. It was being steered by a large and deep high pressure ridge, visible to the north of the hurricane at the 500 hPa level. A trough to the northwest of the hurricane was, however, slowly eroding the eastern end of this ridge. Thus, from 0000 UTC 28 October 1998, the main system in the ridge at 500 hPa was a high over Mexico. Therefore, the storm was cradled in a ridge which was strongest on the western flank of the storm. This environment would tend to steer the storm slowly south or southwest. From Figure 11.12, it can be seen that Mitch did indeed move slowly south and then

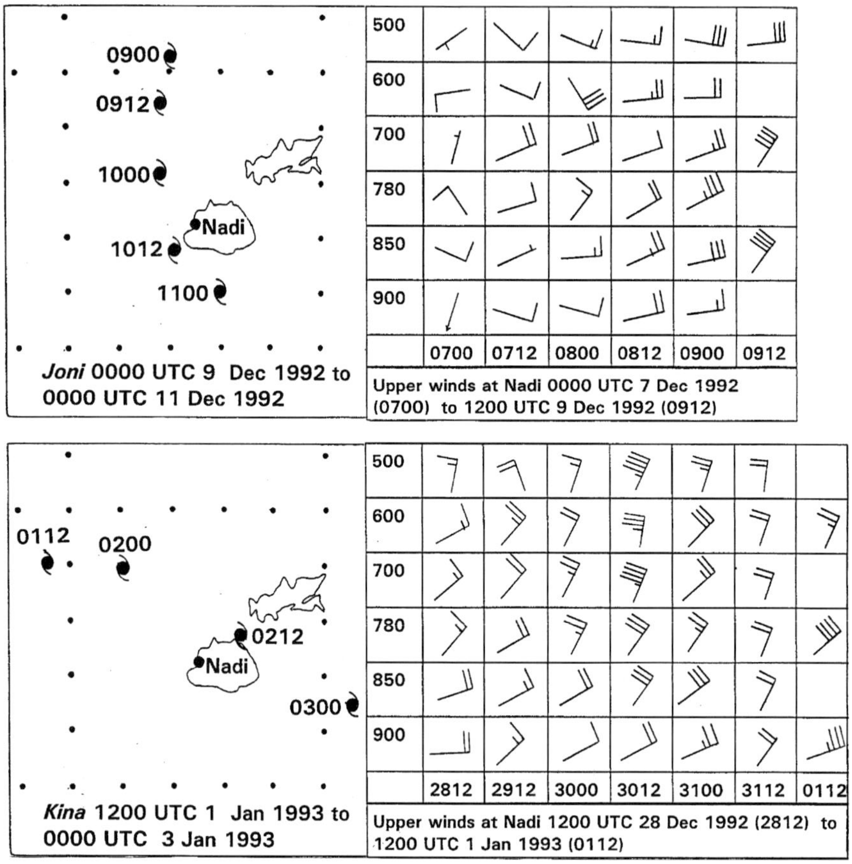

Figure 11.10 Tracks of tropical cyclone Joni (top) and Kina (bottom) together with corresponding time series of upper winds from Nadi.

south-west over Honduras. Subsequently, the cyclone remained cradled by the ridge which by 0000 UTC 31 October 1998 (not shown) strengthened to the northeast of Mitch as the trough system moved away out into the Atlantic. Thereafter, the hurricane resumed a more westward track. In summary, the interaction between the large scale ridge to the north of the hurricane and the trough in the Atlantic resulted in a slow tortuous track over Central America.

In a later section, we shall elaborate on satellite estimates of rainfall during this storm; and the rainfall intensities as measured by a surviving autographic recorder but there are two important aspects of the rain characteristics to be discussed here. First, and surprisingly, the highest recorded rainfall over the period from 25 to 31 October 1998 (912 mm) was from Choluteca near the Pacific Coast in Honduras. The maximum 24-hour total there was 467 mm. The highest report from the north coast of Honduras (where landfall occurred and the heaviest totals would normally be expected) was at La Ceiba where 877 mm was recorded from the 25 to 31 October and 24-hour totals reached 284 mm.

Second, the location of Choluteca, close to the Casito Volcano in Nicaragua, was the scene of a major disaster. Intense, near-stationary rain bands between 0157 UTC 29 October 1998 and 0025 UTC 31 October 1998 produced the exceptional rainfall near the Pacific Coast. The crater lake atop the dormant volcano filled and parts of the wall collapsed. The resulting massive mud flows covered an area 16 by 8 km. At least four villages were totally buried. Over 2000 of the dead were from the areas around the volcano.

The development of the near-stationary rain bands along the Pacific Coast is evident in the SSM/I sequence in Plate 7. The progressive destruction of the circular inner core of Mitch can be seen when compared with the near perfect circular eye-wall evident at 0951 UTC 26 October 1998 (Plate 8).

Climatological aspects of tropical cyclones linked with rainfall

Brunt (1966) observed that weaker tropical cyclones, or systems which had weakened to tropical lows, were more prolific producers of rain than those associated with more severe systems. The previous case studies of tropical cyclones substantiate some of Brunt's observations. Bonell and Gilmour (1980) provided

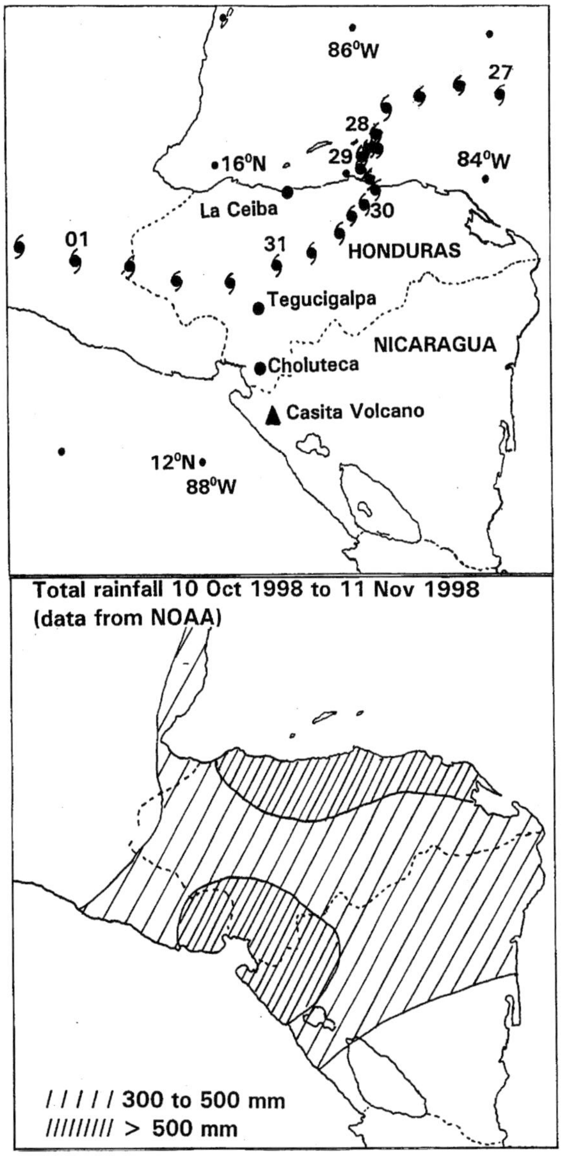

Figure 11.11 (a) (Top) Track of hurricane Mitch from 0000 UTC 27 October 1998 to 0600 UTC 1 November 1998. Cyclone symbols denote the position of Mitch every 6 hours with the date marked alongside the 0000 UTC position. (b) (Lower frame) Total storm rainfall derived from NOAA National Hurricane Center web site. (http://www.nhc.noaa.gov/index.shtm)

short-term rainfall intensity data for a weakening tropical cyclone Keith in January 1977, a category 1 storm (mean winds around the centre were only just above tropical cyclonic strength, 63 km h^{-1}) by the time it passed the Babinda experimental tropical rainforest catchments. Total storm rain amounted to 423 mm, with maximum equivalent hourly intensities ranging from 91.8–96.00 mm h^{-1} equivalent over durations from 15 mins up to 1 hour. By contrast, in December 1990 tropical cyclone Joy (a category 4 storm), which was near-stationary for more than 24 hours and whose well-developed eye was within 80 kms of the same catchments, did not produce short-term rainfalls much above 30 mm h^{-1} equivalent hourly intensity although total rainfall amounted to 530 mm.

A recent study by Cerveny and Newman (2000) provides a more comprehensive analysis of rainfall contribution across the spectrum of tropical cyclone intensities. These writers assessed the linkages between tropical cyclones and rainfall through the analysis of 877 tropical cyclones associated with the Atlantic and north Pacific basins over the period 1979 to 1995.

Several interesting features emerged. The centre grid (inner eye wall) contributes about 26.3% ± 0.2% of the total rainfall associated with the entire tropical cyclone. Of even more interest, a U-shaped pattern emerges when the ratio of the centre grid cell rainfall to the average tropical cyclonic rainfall is compared with maximum surface wind speed (in 10 knot categories). The inner core (centre grid cell) provides more rainfall (in excess of 30% total rainfall) for both weak storms (central mean winds < 30 knots, 15.4 m s^{-1}) and strong storms (central mean winds > 120 knots, 61.8 m s^{-1}). Thus the dynamics of precipitation production is different between the inner core and the outer spiral bands; even though overall there is highly significant relationship between the daily amount of tropical cyclone precipitation over the nine grid cells and the daily maximum windspeed of these vortices.

Evident from the analysis of Cerveny and Newman (2000) is the greater contribution of rainfall from the inner core of weak storms (i.e. tropical lows). These circumstances will become more geographically extensive during the weakening stage of tropical cyclones (especially after landfall), in line with Brunt's (1966) earlier observations, and was also shown in the case of Hurricane Mitch.

PERTURBATIONS IN THE EASTERLIES

During the northern hemisphere summer (May to October), a steady sequence of meridional disturbances within the trade wind easterlies of the north tropical Atlantic (and the tropical north west Pacific), known as *easterly waves*, has long been a focus of attention. The easterly wave model of Riehl (1954) has been widely cited within the climatological literature (Sumner, 1988). Manton and Bonell (1993) cautioned against the acceptance of easterly wave model as the only one occurring within the trade easterlies and in support strongly emphasised Sadler's alternative mechanisms (1967; 1975a, b; 1978). For example, penetration of TUTT vortices towards the surface (on the lines of Figure 10.7 in the previous chapter) at 700 hPa can either induce a trough at the surface or even a vortex within the trade wind easterlies. Subsequent

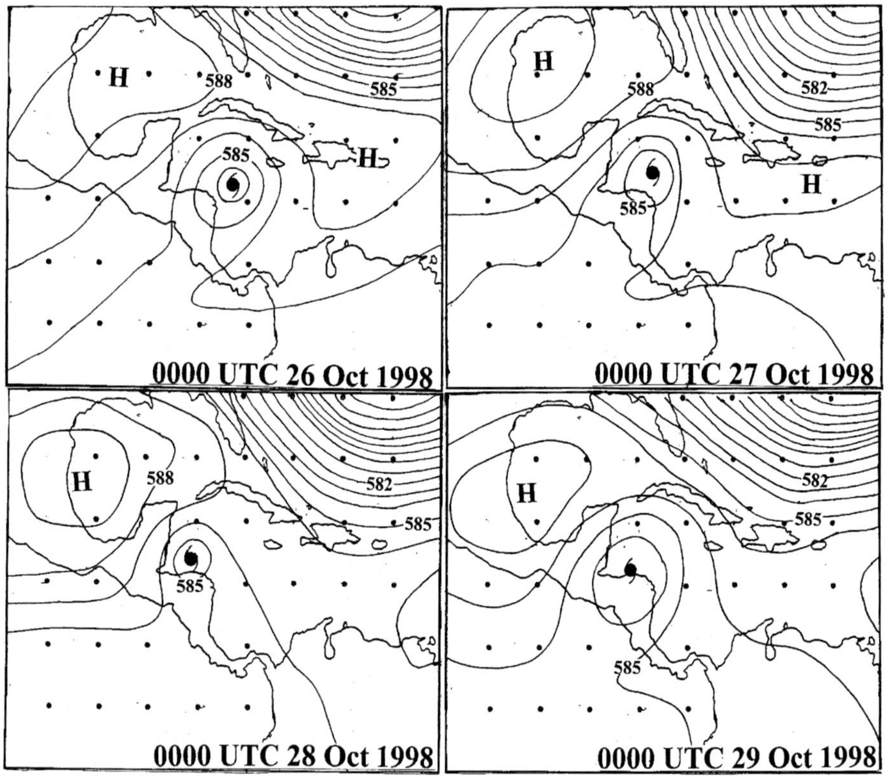

Figure 11.12 Height contours (decametres and from NCEP/NCAR re-analyses) at 500 hPa from 0000 UTC 26 October 1998 to 0000 UTC 29 October 1998.

westward movement of these TUTT–origin disturbances can be mistakenly identified as easterly waves.

Surprisingly, there has been little reappraisal of easterly disturbances since the early balanced study of Merrit (1964), in connection with the Caribbean. Even from the limited database then available, Merritt observed that the classic Riehl 'waves in the easterlies' model was '. . . smaller in scale than indicated in previous studies and has a smaller frequency of occurrence'. More appropriately, Merrit preferred the expression 'easterly perturbations' and identified five distinctly different cloud distributions, with those most frequently observed being related to a closed cyclonic circulation in the mid-troposphere. There was little in that study which disputed Sadler's later ideas. Merritt also made other pertinent observations, such as that '. . . distortion of Riehl's original concepts has been the prime source of this controversy and confusion'. In addition, the same writer noted that part of the problem would seem to be many meteorologists 'forcing' all easterly disturbances into the classic easterly wave model into their '. . . analysis whenever any perturbation less intense than a tropical cyclone is detected in the tropical easterlies'.

Since the review of Manton and Bonell (1993), the traditional focus of attention has been on the genesis and propagation of easterly perturbations over the African continent (commonly labelled African Easterly Waves, AEW, but we will refer to them as African Easterly Perturbations, AEP) during the Northern hemisphere summer. More recent, mostly theoretical and modelling work, has provided better insights into the dynamics of such perturbations and their interannual variability (Thorncroft and Hoskins, 1994a, b; Thorncroft, 1995; Thorncroft and Rowell, 1998; Pytharoulis and Thorncroft, 1999). Such work has subsequently been supplemented by a 20-year review of AEP activity over the African continent and north Atlantic (Thorncroft and Hodges, 2001). The ability of easterly perturbations to progress across the Atlantic Ocean and trigger tropical cyclogenesis in both the Caribbean and eastern Pacific Ocean also continues to be a focal point of attention.

The west African region is of great interest because 'easterly perturbations' occur as westward moving disturbances occurring, on average, every three to five days (Burpee, 1972); and with two preferred regions of development on either side of the African Easterly Jet (AEJ) at 600 hPa. The first is a more northerly track of westward moving troughs originating between 18 °N and 25 °N and 10 °W and somewhere between 15 °E and 30 °E downwind of the Hoggar (Ahaggar) mountains in the Sahara. The second is between 8 °N and 15 °N, and 0 and 10 °E in the Maximum Cloud Zone as defined by Callaghan and Bonell in the previous chapter.

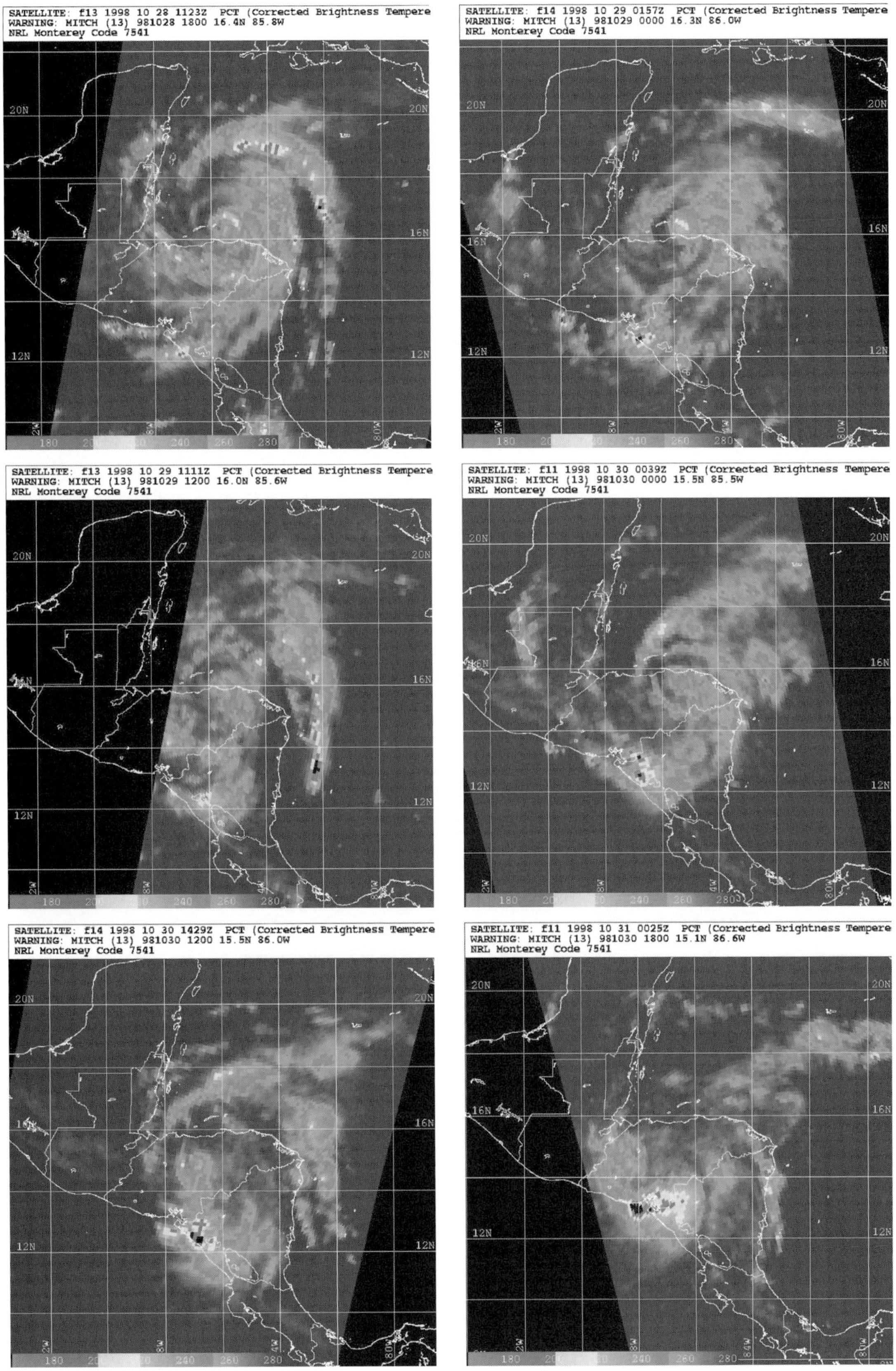

Plate 7 The rain-bands of hurricane *Mitch* (SSM/I) imagery from 1123 UTC 28 October 1998 to 0025 UTC 31 October 1998. (Source: NCEP/NCAR re-analyses data from http://wesley.wwb.noaa.gov/reanalysis.html) (See also colour plate section.)

Plate 8 TRMM image of hurricane Mitch at 0951 UTC 26 October 1998. (Source: US Naval Research Laboratory, Monterey, CA, USA http://www.nrlmry. navy.mil/tc-bin/tc.home) (See also colour plate section.)

Little is known about the exact mechanisms which trigger AEPs.

The decaying process of these perturbations can be partly attributed to their passage over the relatively cool, east Atlantic Ocean. In addition, Sadler (1967, 1975, a, b; reviewed and cited above by Manton and Bonell, 1993) explained the decaying process on the movement of the perturbations out of the low level monsoon shear zone of opposing convergent winds (NE/SW) into the zonal trough in the easterlies (ZTE). Thus, many of these easterly perturbations (but not all) noted in the north Atlantic Ocean originated from low level vortices embedded on the northern monsoon shearline further east over west Africa at the surface (located immediately north of the 600 hPa AEJ) between the opposing NE trades and Equatorial westerlies. Such vortices are a characteristic feature of monsoon shearlines, as outlined in the previous chapter.

A climatological reassessment of African easterly perturbations, 1979–1998

Following earlier theoretical work, Thorncroft and Hodges (2001) used ECMWF analyses to develop a 20-year climatology of AEP activity based on the automatic tracking of vorticity centres. This automatic tracking detected only systems that had closed vorticity contours with values $\geq 0.5 \times 10^{-5}\ s^{-1}$, which were more appropriate to the stronger, more coherent systems relevant to tropical cyclogenesis over the north Atlantic.

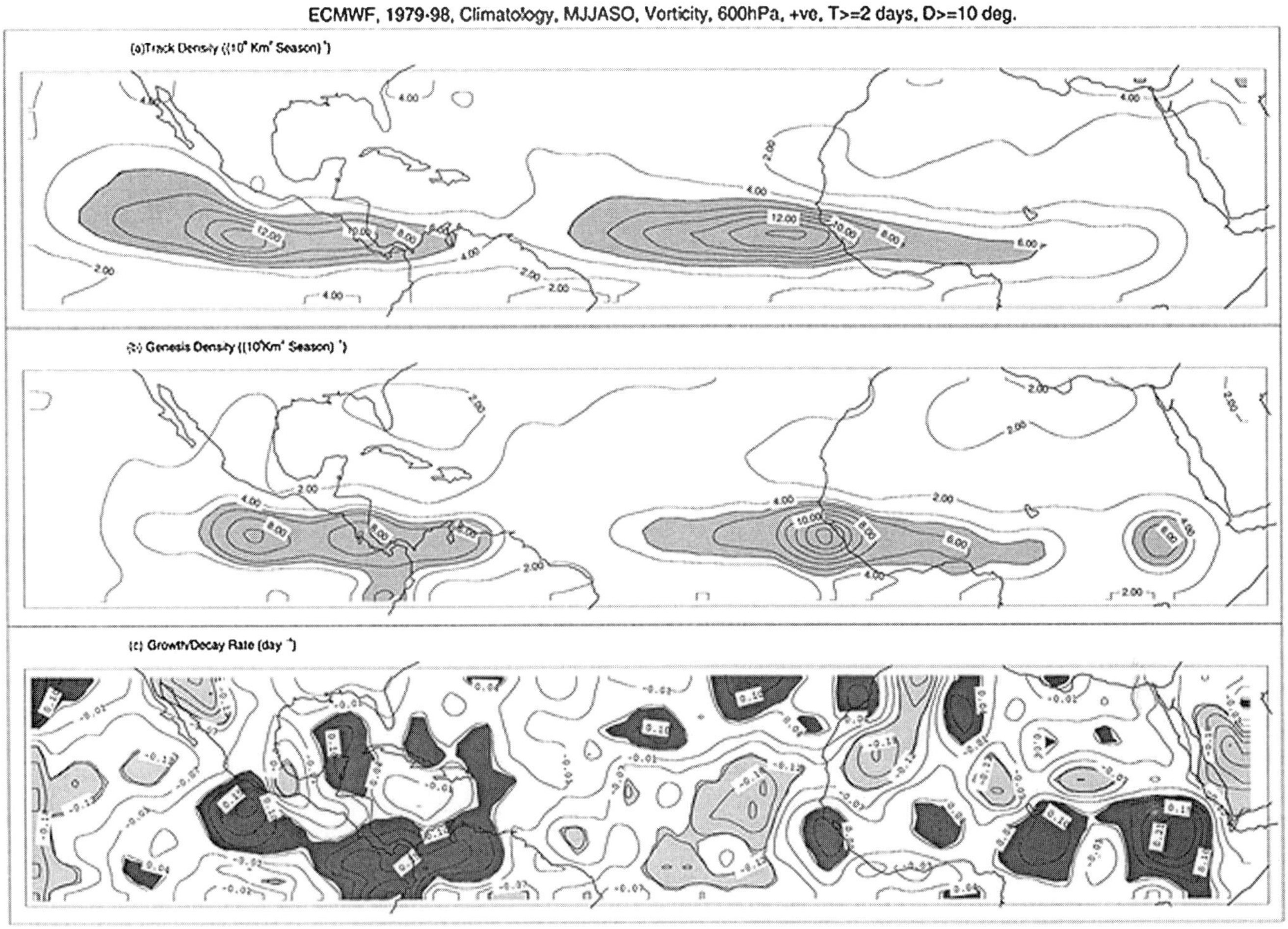

Figure 11.13 Climatological tracking statistics at 600 mb based on the ERA data (197993) and the ECMWF analyses (199498). (a) Track density scaled to number density per unit area (106 km^2) per season (MJJASO), shading for values greater than 6. (b) Genesis density per unit area (106 km^2) per season (MJJASO), shading for values greater than 5. (c) Growth and decay rates in units of per day, shading for values greater than 0.05 and less than −0.1. (Source: Thorncroft and Hodges, 2001.)

A preliminary evaluation of AEP tracks in the extremely active year of 1995 for tropical cyclones (19 named storms) (Landsea *et al*., 1998) compared with 1994 (relatively inactive with only seven named storms) established significant differences between the principal AEP tracks over the north Atlantic compared with those over north Africa. At 850 hPa, the tracks over the north Atlantic ocean are similar to those at the 600 hPa level. Over the west African coastal region, the principal AEP track was Equator-ward of 15 °N at the 600 hPa level which subsequently extended westwards into the Atlantic. *Significantly, many of the tracks recurved poleward before reaching the Carribean*. In contrast, the main AEP storm track was poleward of 15 °N over north Africa at the 850 hPa level being associated with the low-level temperature gradient in proximity to the Sahara-Sahelian zone (Thorncroft and Hodges, 2001).

A subsequent synthesis of the climatology of AEP tracking statistics at 600 hPa and 850 hPa levels provide additional insights into several features of these easterly perturbations outlined earlier. These statistics are based on track density, genesis density and growth and decay rates which are summarised for the 600 hPa level in Figure 11.13 and for the 850 hPa level in Figure 11.14.

The track density at 600 hPa level shows two peak epicentres, one immediately downstream of the west African coastline around 20 °W and a second downstream of Central America in the eastern north Pacific. Activity over the African continent is weaker. The main axis of the 600 hPa level density is near 10–15 °N, commencing over east Africa before extending westwards towards Venezuela and Colombia, and then into the eastern north Pacific.

Despite the diminishing track density in the western north Atlantic (Figure 11.13), this feature does not discourage the school of thought that these AEPs can stimulate cyclogenesis in the Caribbean and eastern north Pacific. Moreover, the spatial

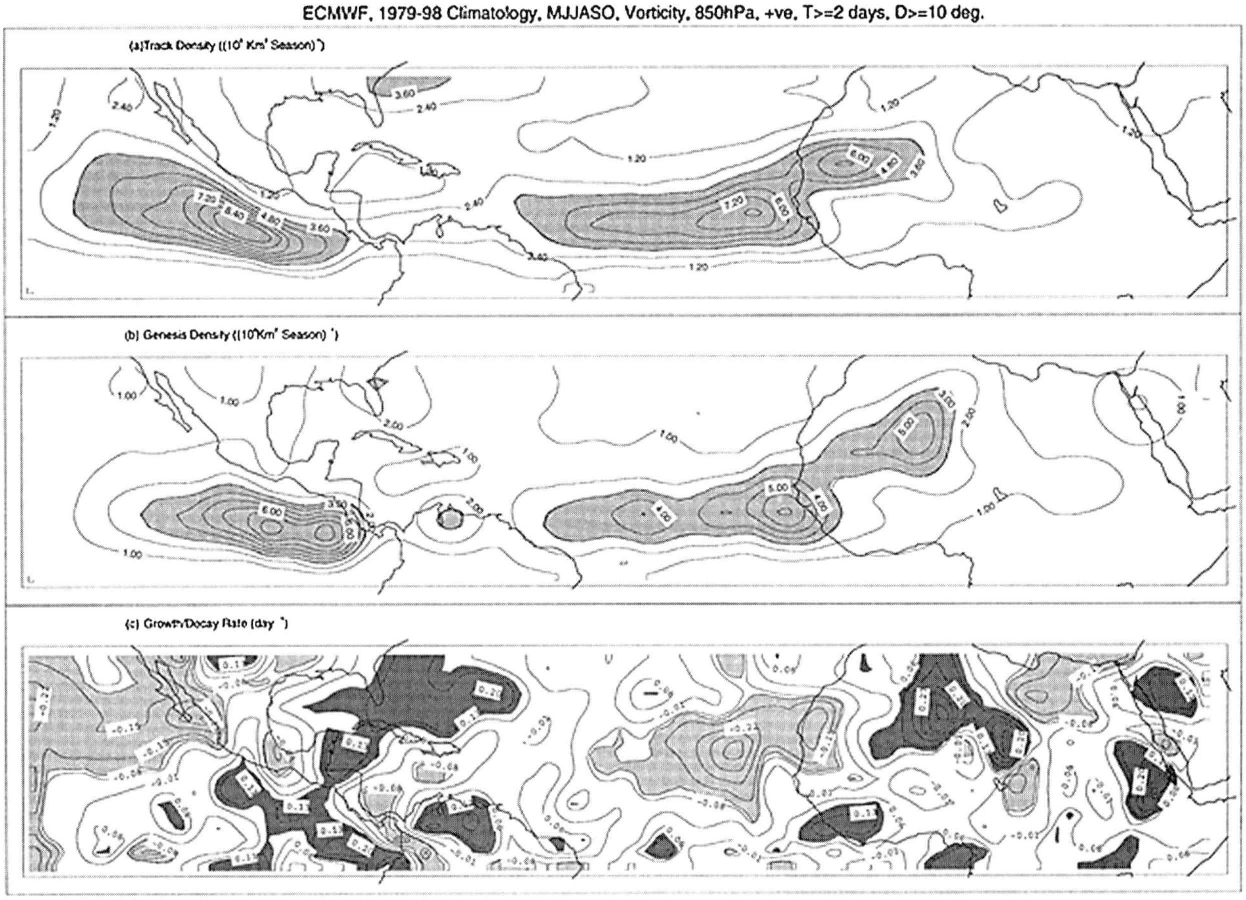

Figure 11.14 Climatological tracking statistics at 850 mb based on the ERA data (197993) and the ECMWF analyses (199498). (a) Track density scaled to number density per unit area (106 km^2) per season (MJJASO), shading for values greater than 3.6. (b) Genesis density per unit area (106 km^2) per season (MJJASO), shading for values greater than 5. (c) Growth and decay rates in units of per day, shading for values greater than 0.1 and less than −0.1. (Source: Thorncroft and Hodges, 2001.)

organisation of the track density, and the growth and decay rates are in line with the Sadler (1967) conceptual model (as shown in Figure 11.15) associated with the movement westwards and corresponding decay of vortices originating from the northern monsoon shearline over the tropical, east-north Atlantic.

Significantly, the genesis regions at 600 hPa conform with the storm track density. Also interesting is the prominent peak (10 °N, 35 °E) on the western side of the Ethiopian highlands. The main peak is located over the sea, just off the west African coast, following a re-emergence of genesis upstream west of about 20 °E. In line with Carlson (1969a, b), decay over the central-north Atlantic is extensive, partly due to colder SSTs there. In addition, the potential vorticity sign reversals (discussed by Molinari *et al.*, 1997 and reviewed shortly) weaken in the same region which encourages the AEPs to also weaken by Rossby wave dispersion. Further west, over northern Venezuela and Colombia, a genesis peak is depicted which is also a region identified by Molinari *et al.* (1997) as a potential vorticity sign reversal region. The latter writers thus argue that AEPs are able to be reinvigorated prior to their continued westward movement into the eastern north Pacific.

The track density at 850 hPa level (Figure 11.14) shows major differences over the African continent. The dominant track is poleward of 15 °N and commences further west than the 600 hPa.

The corresponding genesis density for the 850 hPa level shows two peaks. The first peak polewards of 15 °N over land is just downstream of the Hoggar mountains (25 °N, 10 °E) indicating a possible orographic influence to supplement the low-level temperature gradient. The second peak, like that at the 600 hPa level, is just off the west African coast. When the growth and decay rates are examined, significantly, these low-level perturbations start to dissipate towards the coast. Thorncroft and Hodges (2001) thus question whether many many of these low level AEPs generated on the poleward side of the AEJ participate in cyclogenesis over the north Atlantic. Rather, the 850 hPa level perturbations over the north Atlantic are generated separately off the coast in association with the westward movement of the 600 hPa disturbances

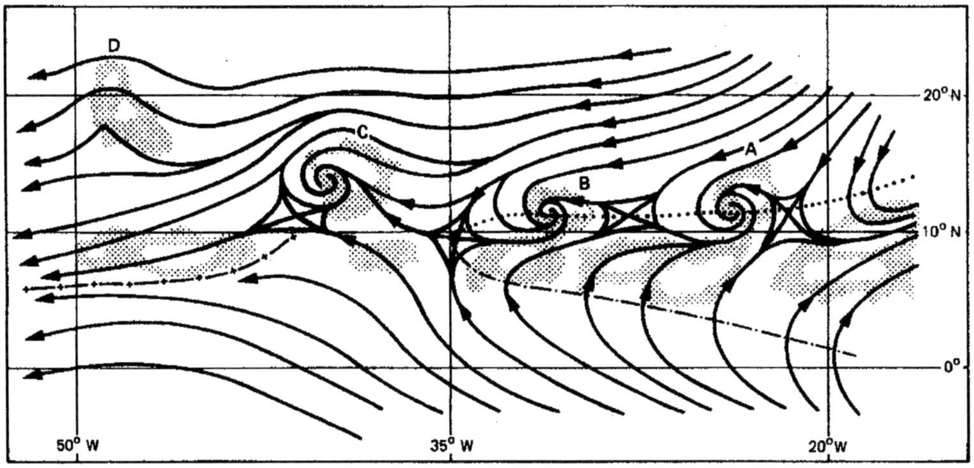

Figure 11.15 A schematic model of the low-level cyclones in the north Atlantic. The model depicts either a chain of cyclones (A to D) or the life history of one cyclone. The major satellite-observed cloud systems are stippled and open areas within the stippling represents the deeper, more convective cloud groups. Note the wind streamlines have been modified near 50 °W to conform with later information presented in Figure 1 of Sadler (1975b). (Modified from Sadler, 1967.)

from over the land Equatorward of the AEJ. From thereon, there is just one storm track with the 600 hPa and 850 hPa level activity co-located over the north Atlantic Ocean. Thorncroft and Hodges (2001) conceded however, that during the active hurricane season of 1995, a few low level perturbations from the poleward side of the AEJ were detected, moving southwestward offshore to join the main north Atlantic track. Thus the tracking Equatorward of some of these poleward low level disturbances can still occur. In addition, these northern low level disturbances often show coherence with the 600 hPa waves more Equatorward, thus indicating a complicated, multi-centred easterly perturbation structure as well as facilitating possible later co-location over the north Atlantic.

The major genesis area off the coast is of special interest and was attributed by Thorncroft and Hodges (2001) to the enhancement of instability due to latent heat release near 10 °N, 10 °W. Moreover, Figure 11.14 shows at the 850 hPa level a situation of weak growth near 7 °N which stretches from the African coast to the Caribbean and so permits the continued westward progression of these low level disturbances. In contrast, at the 600 hPa level there is a broad region of decay.

This analysis of Thorncroft and Hodges (2001) provides a spatial framework for appreciating the complex genesis and tracking of these African easterly perturbations. New light is shed on the varying importance of the Equatorward vis-à-vis poleward tracks over the African continent in terms of their contribution towards the north Atlantic track density and the ability of such perturbations, sometimes with recurvature polewards, to attain the Caribbean region.

What remains missing is an updated synoptic climatological perspective (including the use of GATE data) along the lines of Sadler (1967; 1975b) to better appreciate how the characteristics of these easterly perturbations fit within the synoptic scale wind streamline analysis and the corresponding monsoon shearlines, and zonal trough in the easterlies. Moreover, as conceded by Thorncroft and Hodges (2001), the weaker vorticity amplitudes or multicentred nature of the AEWs over the African continent require complementary methods for detecting these.

Interannual variability of African easterly perturbations

Until the GCM modelling study of Thorncroft and Rowell (1998), there had been scant mention of interannual variability of African easterly perturbation activity. Over the period 1967–1991, Avila and Pasch (1992) noted that the average number of 'waves' moving westwards off the west African coast in one year averaged 59, with extremes of 49 in 1980 and 76 in 1990. A link between this inter-annual variability of 'waves' and the variability in tropical cyclogenesis over the Atlantic has been long suggested by several authors (e.g. Burpee, 1972, 1974; Reed *et al.*, 1988 a, b; Avila and Pasch, 1992; Gray *et al.*, 1994).

The possible links between topographic relief and the genesis of African easterly perturbations

The significance of the Hoggar mountains in the initiation of easterly perturbations was suggested by the GCM modelling study of Thorncroft and Rowell (1998), through grid point correlations with local zonal wind 'wave' activity in tropical north west Africa.

The later climatological study of Thorncroft and Hodges (2001) pinpointed the Ethiopian Highlands as an orographic trigger point for African easterly perturbations on the southern, Equatorward track of the AEPs. After a subsequent temporal weakness of these perturbations downstream (from the Ethiopian Highlands) (Figure 11.13), interactions further west with other topographic features such as the Jos Plateau and the Air Mountains (west of 20 °E) has the potential to reinvigorate the genesis density of such disturbances along this southern track.

Rainfall

Using power spectra analysis within their modelling study, Thorncroft and Rowell (1998) inferred that easterly perturbations affect daily rainfall over the more humid Guinea Coast (4 °N–8.75 °N) and West Sudan (wet/dry, 8.75 °N–11.25 °N) regions of West Africa. Peaks in the power spectra were established at the 'easterly wave' time scales. The West Sahel (11.25 °N–18.75 °N), however, did not show peaks in the power spectra which suggests that whilst easterly perturbations are present and produce rainfall further south, the same disturbances are not always associated with rainfall in the more northerly zone because the atmosphere is too dry. Later, Thorncroft and Hodges (2001) noted that the variability of African easterly perturbations does not have a straightforward relationship with west Sahelian rainfall variability. In fact, in some active AEP years (e.g. 1985), the west Sahel was comparatively dry.

Easterly perturbations over the Caribbean and eastern Pacific

Early work (Riehl, 1954; Frank, 1970; Carlson, 1969a, b; Burpee, 1972) put forward evidence for the westward propagation of some African 'waves' which can remain as 'debris' to reach the western Atlantic, the Caribbean and the east Pacific, and so regenerate into tropical storms. During a study of air mass characteristics linked with rainfall over mainly Guadeloupe, Asselin de Beauville (1995) presented satellite imagery of 'tropical wave' examples, some of which fit the aforementioned 'debris' category. Later, Shapiro (1986) argued that strong African waves with substantial convection are capable of maintaining their structure on moving across the north Atlantic and even into the eastern Pacific Ocean. On the other hand, there is plentiful evidence of the weakening of such perturbations on passing from the African continent over the cooler central, north Atlantic (Carlson, 1969a, b; Sadler, 1967; Manton and Bonell, 1993) and into a less favourable zone of low-level convergence (Sadler, 1967; 1975).

Using the summer of 1991 as a case study, Molinari *et al.* (1997) put forward the spatial distribution of reversals in the meridional potential vorticity (PV) gradient (at the 310-K isentropic surface near 700 hPa level), coupled with the Madden-Julian oscillation (MJO), as factors enabling easterly perturbations to cross the west Atlantic, Caribbean, and into the east Pacific. Molinari *et al.* (1997) noted three regions where time-averaged PV **decreased** northward *viz* the Caribbean sea between Hispaniola and the South American coast; over the east Pacific just west of Central America; and a marginally unstable region over west Africa and the east.

In a subsequent case study of Hurricane Hernan (1996) in the east Pacific, Molinari *et al.* (2000) suggested that components from several of the theories for cyclogenesis in this region were detected in this one case study. These included the tracking back to north Africa of the origin of the 700 hPa level disturbance; strong interactions of airflow with the Central American mountains; a dynamically unstable background state and a resultant surge in the SW monsoon at the 1000 hPa level.

Precipitation in tropical easterly perturbations: a case study over the Lesser Antilles archipelago (Guadeloupe)

Asselin de Beauville (1995) presented precipitation data for the 1991 summer rainy season, principally for Guadeloupe. Following Betts (1974), Asselin de Beauville (1995) classified perturbations into three categories over the flat part of Guadeloupe:

(1) no rain
(2) rain between 1 mm and 5 mm
(3) rain exceeding 5 mm.

Rain amounts will obviously be higher over the mountainous parts of the island, but the rain gauge network is sparse in these higher relief areas. In line with the rest of the Caribbean region, the recorded mean precipitation amounts fall within the range 0 and 40 mm corresponding to easterly perturbations (Asselin de Beauville, 1995); with a maximum rainfall of 26 mm over Basse-Terre (Guadeloupe) in contrast to only 2 mm over Grande-Terre (Guadeloupe) for the same event. Such rainfall variability is typical during the passage of these disturbances over the Caribbean islands.

The zero or low precipitation amounts in the first category of classified perturbations is attributed to a strong wind shear of westerlies above 3000 m (the upper Equatorial westerlies of Sadler (1975a) outlined by Callaghan and Bonell in the previous chapter) and a weak, low level easterly flow of mean relative humidity below 50%. The perturbations in categories (2) and (3) above have a relative humidity in excess of 50%, within the 0–3000 m layer (below the trade wind inversion). Further, these categores have static temperature profiles (defined by Asselin de Beauville, 1995, p. 165) which are high (~330 °K) within the 0–1500 m atmospheric layer. Such high values are associated with higher latent heat release from moderate to deep convection. Moreover, the enhancement or weakening of the easterly perturbations studied was associated with the state of the relative humidity above the trade wind inversion (3000–10000 m).

Easterly perturbations in the north-west Pacific

The work of Chen and Weng (1998a, b) provides a good summary of the interannual variability of the origins of tropical-synoptic scale disturbances in the north-west tropical Pacific and the corresponding adjustments in the wind and pressure fields at 850 hPa.

A later review of TOGA results will show that there are *westward* propagations of easterly perturbations embedded within *eastward* moving convection associated with the MJO. Elsewhere, Molinari *et al.* (2000) suggested that some of the factors which influenced the development of Hurricane Hernan (1996) might also recur in the west north Pacific as well, especially if easterly perturbations are active upstream of a monsoon trough.

Perturbations in the easterlies over the Coral Sea

Whilst perturbations in the easterlies of the southwest Pacific occur, especially during cold (La Niña) ENSO phases, they have not been described in the same detail as for other regions such as the Caribbean; nor do they necessarily have the same structure. Lyons and Bonell (1992) presented rainfall data for the Townsville area for an easterly perturbation event in March 1989 (the 1988–89 wet season also coincided with a La Niña phase). We now provide more recent case studies of perturbations over the Coral Sea in the form of an amplifying trough in the easterlies (Box 11.6) and tropical cyclone *Tessi* (Box 11.7).

Box 11.6 Intensifying trough in the easterlies over the Coral Sea

At MSL, a trough in the easterlies south of the eastern tip of Papua New Guinea at 1100 UTC 15 March 2000 (top panel in Figure 11.16). Over the next 12 hours this trough sharpened and moved towards the coast and the ridge along the coast to the south strengthened. By 0800 UTC 16 March 2000 a low formed as the system reached the coast. Notice the lack of any monsoon westerly flow. A very similar sequence occurred at 700 hPa and 500 hPa with a trough in the easterlies forming a closed low at landfall. Middle level steering towards the coast was always evident with a strong 700/500-hPa ridge maintained south of the disturbance.

At 200 hPa (Figure 11.17) an upper anticyclone moved towards the coast following the disturbance. The low developed under an upper diffluent zone northeast of the upper anticyclone centre. Over the interior of Queensland, an upper northwesterly jet lay to the east of a relaxing upper trough over Central Australia. The middle to upper level pattern was similar to situations where severe cyclones strike the East Coast. That is, a weakening upper trough over Central Australia while the cyclone is steered onto the coast by a strong middle level ridge.

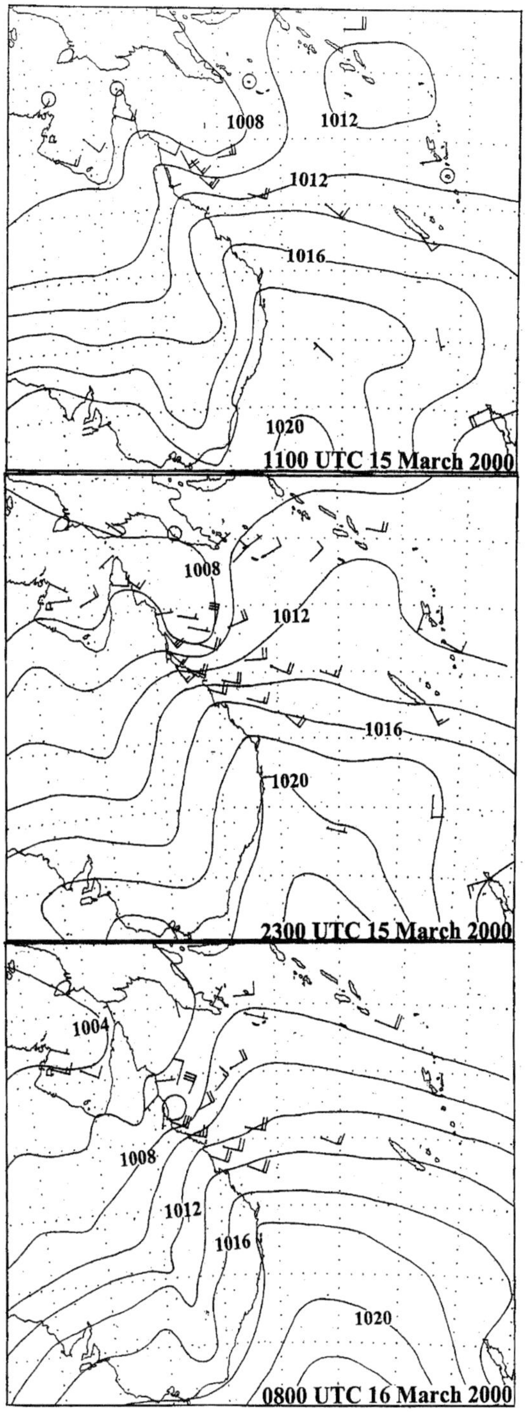

Figure 11.16 MSL pressure distribution (hPa) for 1100 UTC 15 March 2000 (top), 2300 UTC 15 March 2000 (centre) and 0800 UTC 16 March 2000 (bottom).

The first ever recorded tropical cyclone in the South Atlantic

At the time of going to press, there is considerable on-going discussion concerning the first ever recorded tropical cyclone in

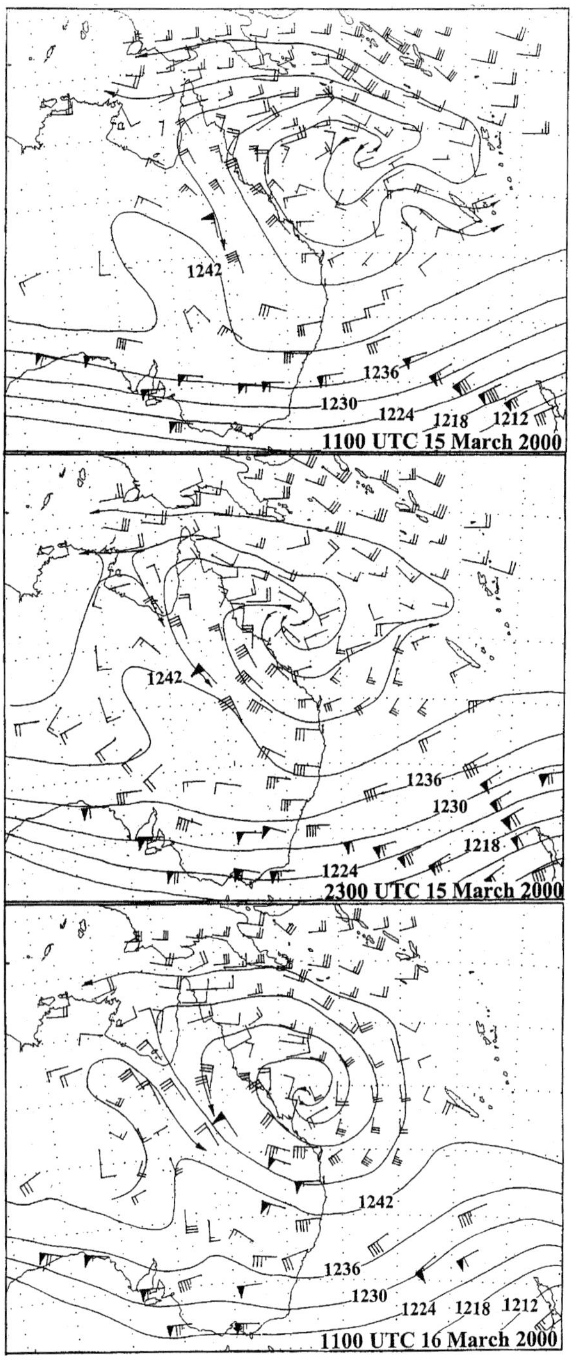

Figure 11.17 200 hPa height contours (decametres), streamlines and winds observations (large plots conventional observations and small plots computer analysed winds) for 1100 UTC 15 March 2000 (top), 2300 UTC 15 March 2000 (centre) and 1100 UTC 16 March 2000 (bottom).

Box 11.7 The development of tropical cyclone Tessi within an easterly trough over the Coral Sea

Tropical cyclone Tessi underwent extremely rapid development near the Australian coast as it passed close to Townsville, Australia's largest tropical population centre. It was a weak tropical cyclone at 1800 UTC 1 April 2000 and almost reached hurricane force intensity near the coast 21 hours later. An automatic weather station (AWS) on Magnetic Island (15 km northeast of the Townsville Meteorological Office) recorded 10-minute average wind speeds of up to 59 knots).

At 1000 UTC 31 March 2000, the 850-hPa analysis (Figure 11.18) shows the position of the lowest MSL pressure by the unfilled cyclone symbol. The system was a trough in the easterlies orientated northwest to southeast *with no monsoon westerlies to the north*. This trough extended up at least to 500 hPa and provided an environment with weak vertical shear. The trough at low levels had been located in the region for nearly a week and originally developed to the east of an amplifying upper trough over eastern Australia. The winds at 200 hPa (lower panels Figure 11.18) show at 1100 UTC 31 March 2000 the developing low lay in the diffluent area west of an upper anticyclone with a weak trough passing to the south. As mentioned previously, this upper wind pattern is often associated with tropical cyclogenesis globally. Over the next 12 hours, the trough rapidly developed into a closed low with a good upper outflow pattern evident at 200 hPa. The upper pattern leading up to landfall (Figure 11.19) shows initially (left panel) that the wind over eastern Australia turned westerly; and the upper outflow to the west and south of the cyclone weakened. Then, just before landfall (right panel Figure 11.19), the upper outflow, particularly on the southern side where winds to the southwest of the cyclone had become more northwesterly.

The MSL sequence (Figure 11.20) shows the developing tropical cyclone, *with an absence of any monsoon flow*, moving towards the northeastern Australian coast (top panel). By 2300 UTC 1 April (centre panel in Figure 11.48), the cyclone had intensified and note how the 1008 hPa isobar decreased in radius (indicating rising pressures around the system). At 2000 UTC 2 April 2000 (lower panel) it was a very small 'midget' cyclone with the pressure continuing to rise around it.

Figure 11.21 shows the MSL sequence illustrating the cyclone passing to the north of Townsville between 1500 UTC and 1600 UTC 2 April 2000. The maximum wind speed was reported from the Magnetic Island AWS at 1600 UTC. Notice that the intense pressure gradient extended onto the coast near Rollingstone just after 1600 UTC. This coincided with the arrival of the destructive winds. The vortex decreased in size between 1800 UTC and 2000 UTC with the most intense pressure gradient over a thinly populated section of the coast and over the sea.

the South Atlantic which developed off the coast of Brazil in the last week of March 2004. The storm (locally known as 'Catarina') crossed the coast of the Brazilian state of Santa Catarina on 28 March 2004 with winds of near 144 km h^{-1} (90 mph), as estimated by the US National Hurricane Center (http://www.metoffice.com/sec2/sec2cyclone/catarina.html). Significantly, this storm developed outside a monsoon region, and in turn, in the absence of a

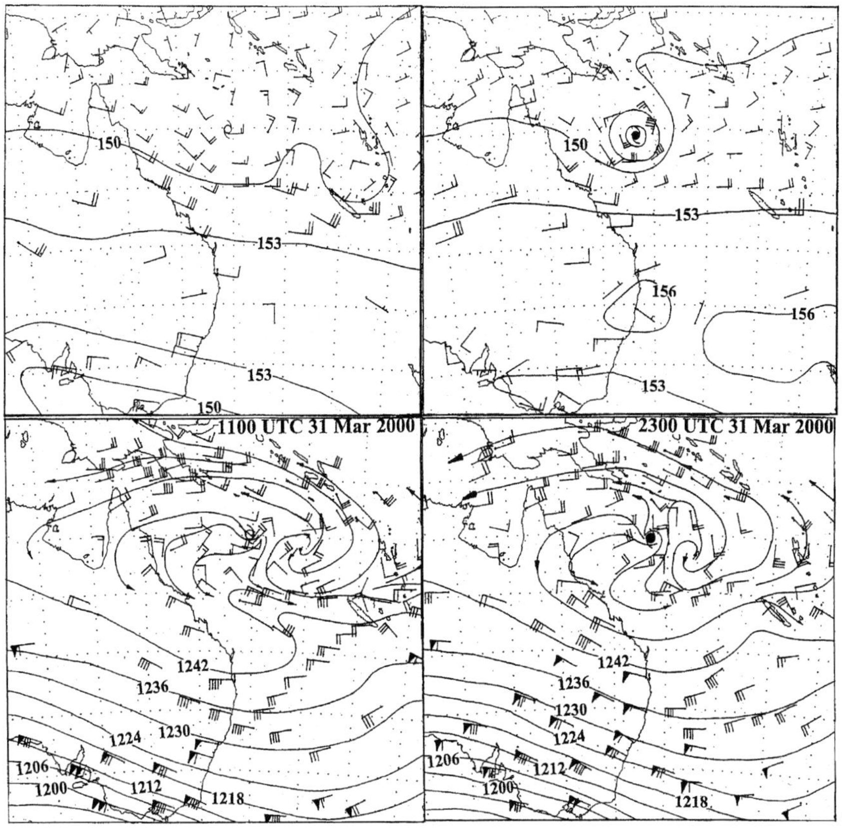

Figure 11.18 Height contours (decametres) and winds observations (large plots conventional observations and small plots computer analysed winds) at 850 hPa (top) for 1100 UTC 31 March 2000 (left) and 2300 UTC 31 March 2000 (right). Height contours (decametres), streamlines and winds observations (conventional and satellite derived winds) at 200 hPa (lower panels) for 1100 UTC 31 March 2000 (left) and 2300 UTC 31 March 2000 (right).

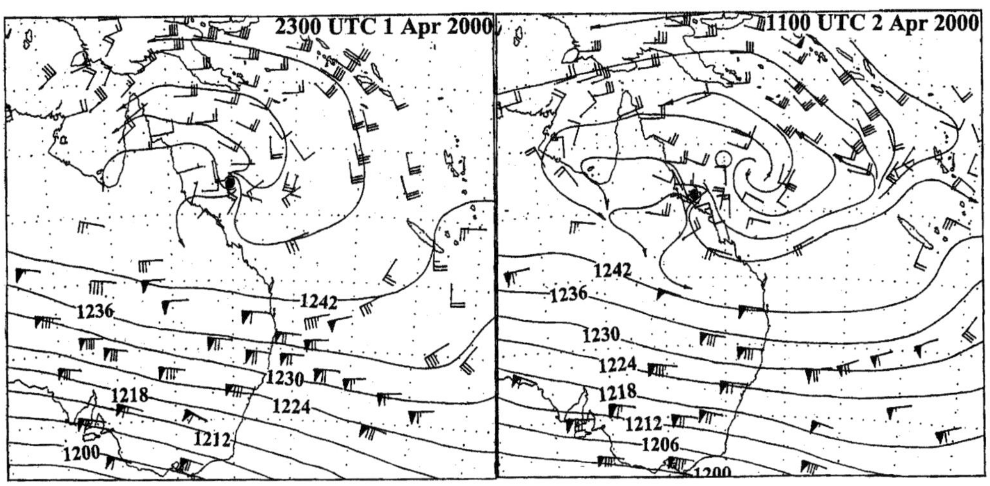

Figure 11.19 Height contours (decametres), streamlines and wind plots (observations and satellite derived winds) at 200 hPa for 2300 UTC 1 April 2000 (left) and 1100 UTC 2 April 2000 (right).

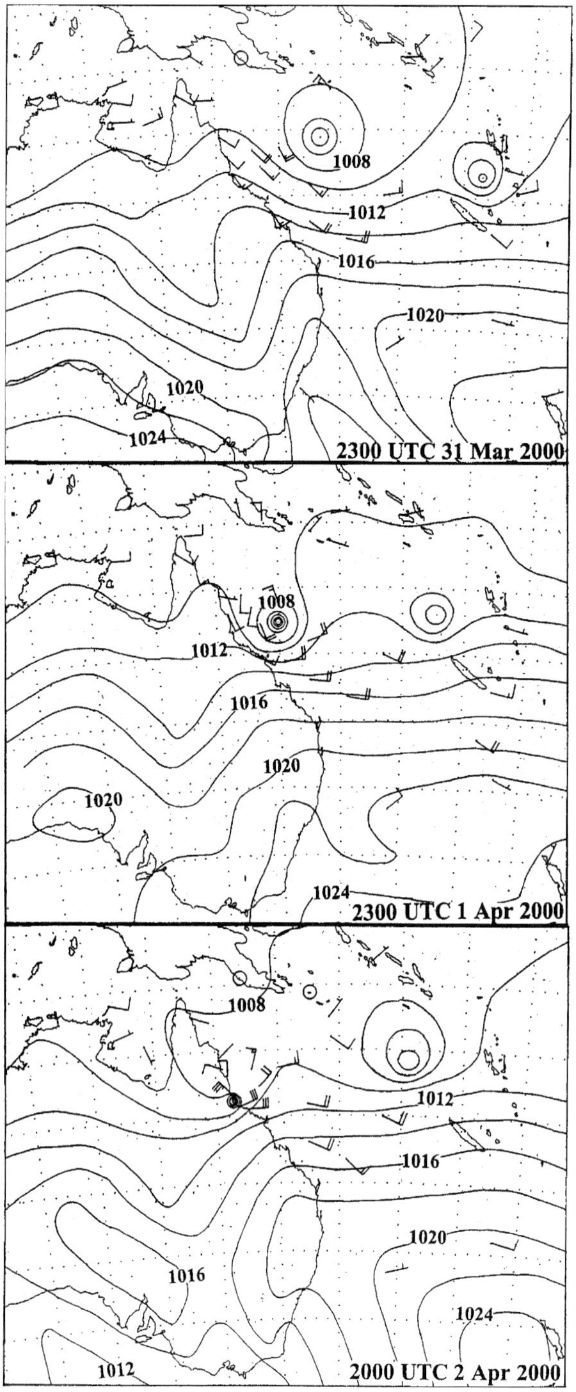

Figure 11.20 Pressure distribution (hPa) and selected wind plots for 2300 UTC 31 March 2000 (top), 2300 UTC 1 April 2000 (centre) and 2000 UTC 2 April 2000 (bottom).

low level monsoon shearline (which does not occur over the tropical south west Atlantic). Further the prevailing upper, equatorial westerlies normally are able to shear any such latent storms.

Using the UK model analyses, discussion and associated charts are to be found on http://www.bom.gov.au/bmrc/clfor/cfstaff/jmb/synoptic.html. The cyclone initially developed from mid latitude baroclinoc processes and then moved across the spectrum towards a tropical type system. It initially formed as an extra-tropical low where the source of the MSL pressure falls came from warm upper air associated with a tropopause undulation (Hirschberg & Fritsch 1991a, b, c; 1993a, b; 1994) or a PV anomaly (see Hoskins *et al.*, 1985). The undulation then weakened but left favourable upper (200 hPa) warm air over the cyclone, which by then was vertically stacked and therefore in a low vertical wind shear environment. Then convection near the centre enhanced the warm air at upper levels over the system leading to further pressure falls. The SSTs were normal to slightly below normal (24–25 °C) in the region where the storm underwent tropical transition and these SSTs are 1–2 °C below optimal values for TC formation.

Catarina was also straddling an increasing low to middle tropospheric, thermal gradient between a warm thermal high over land to its southwest and a cold 700–500 hPa cold low near and north-east of the centre. Such thermal gradients are observed in both extra-tropical and tropical cyclones though with the former the gradients are much stronger and therefore the tilt of the system is larger than a tropical cyclone. These thermal gradients provide dipoles of warm and cold advection across the system and may be critical in helping form the convection. As we showed earlier, warm air advection in the tropics is associated with very heavy rainfall.

Using quasi-geostrophic and semi-geostrophic theory Hoskins *et al.* (2003) sought to find the missing link between the PV and vertical motion dynamical perspectives on mid latitude cyclogenesis. They identified a vertical velocity term which may link the upper PV with the development of the thermal pattern such as we seeing at 700 hPa in the vicinity of intensifying tropical cyclones. Black *et al.* (2002) found 850/500 hPa vertical wind shear in the core of very intense US hurricanes. They found this shear tended to be consistent with the synoptically analysed environmental flow.

These described characteristics of 'Catarina' have also been observed in hybrid examples over the Coral Sea. In such cases, the Bureau of Meteorology, Australia, would operationally assign the status, tropical cyclone. So for Catarina it would be considered a tropical cyclone (Callaghan, pers. comm.).

North Atlantic patterns for tropical cyclogenesis from troughs in the easterlies

The synoptic-scale flow during tropical cyclogenesis in the North Atlantic basin has been examined (Bracken and Bosart, 2000) using storm-centred composites. Composites are created using only wind data at the 900 hPa and 200-hPa levels. The results suggest that two different large-scale upper-tropospheric flow patterns

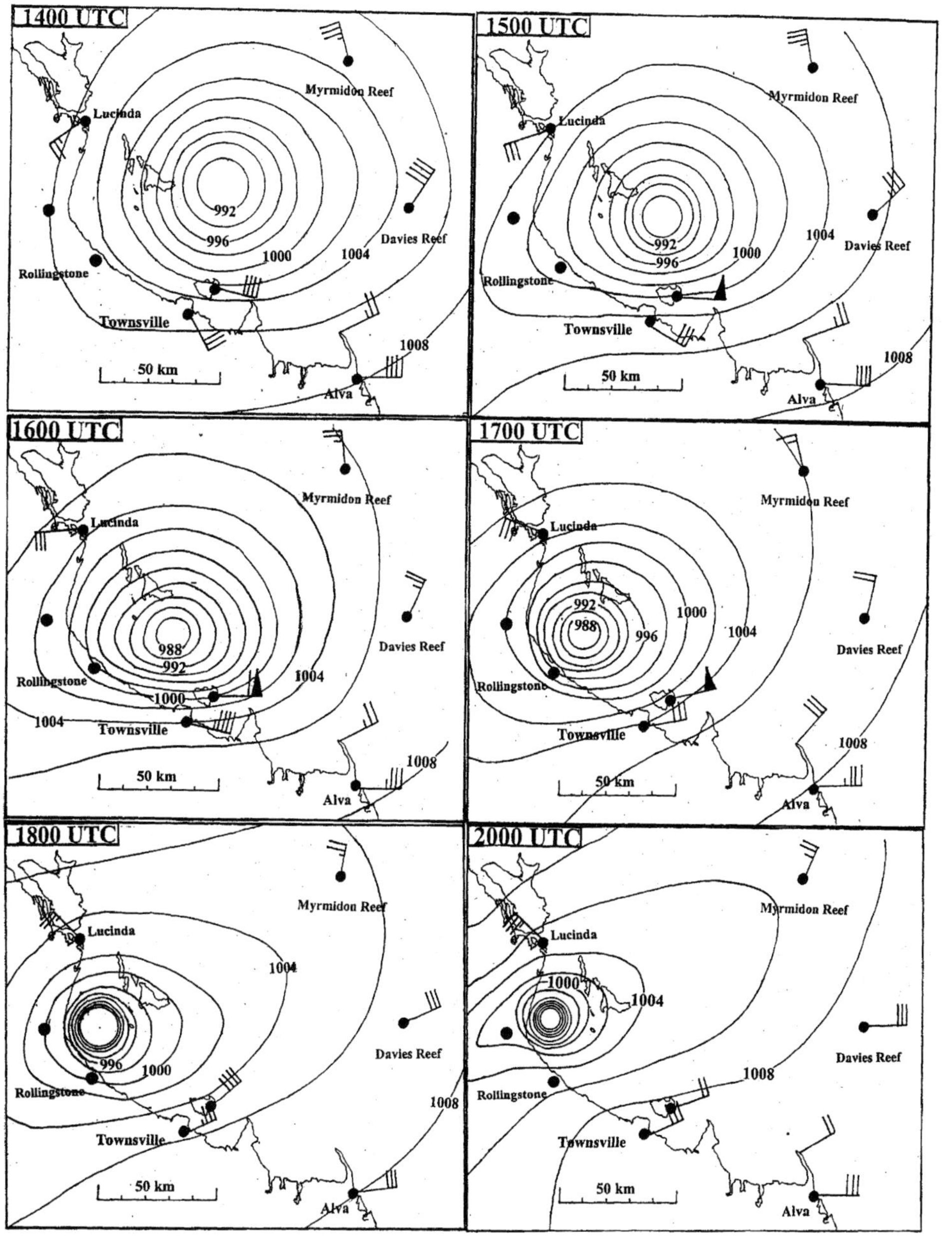

Figure 11.21 Pressure distribution (hPa) and selected wind plots in the Townsville area from 1400 UTC 2 April 2000 (top) to 2000 UTC 2 April 2000.

are most commonly observed during genesis over the basin. One flow pattern is characterised by an upper-tropospheric trough-ridge couplet and is most commonly observed in the Bahamas region. The low-level cyclonic vorticity maximum in the Bahamas composite is located beneath the poleward flow east (west) of the upper-level trough (ridge) (Figure 11.22). The second flow pattern is commonly observed in the Cape Verde region and is characterised by an upper-tropospheric ridge axis poleward of the low-level cyclonic vorticity maximum.

MESOSCALE CONVECTIVE COMPLEXES (MCCs) AND MESOSCALE CONVECTIVE SYSTEMS (MCSs) IN THE TROPICS

Definitions and theoretical considerations

As observed by Orlanski (1975) the term mesoscale defines all of the intermediate atmospheric states between macroscale (i.e. synoptic scale) (length scale > 2000 km) and microscale (length

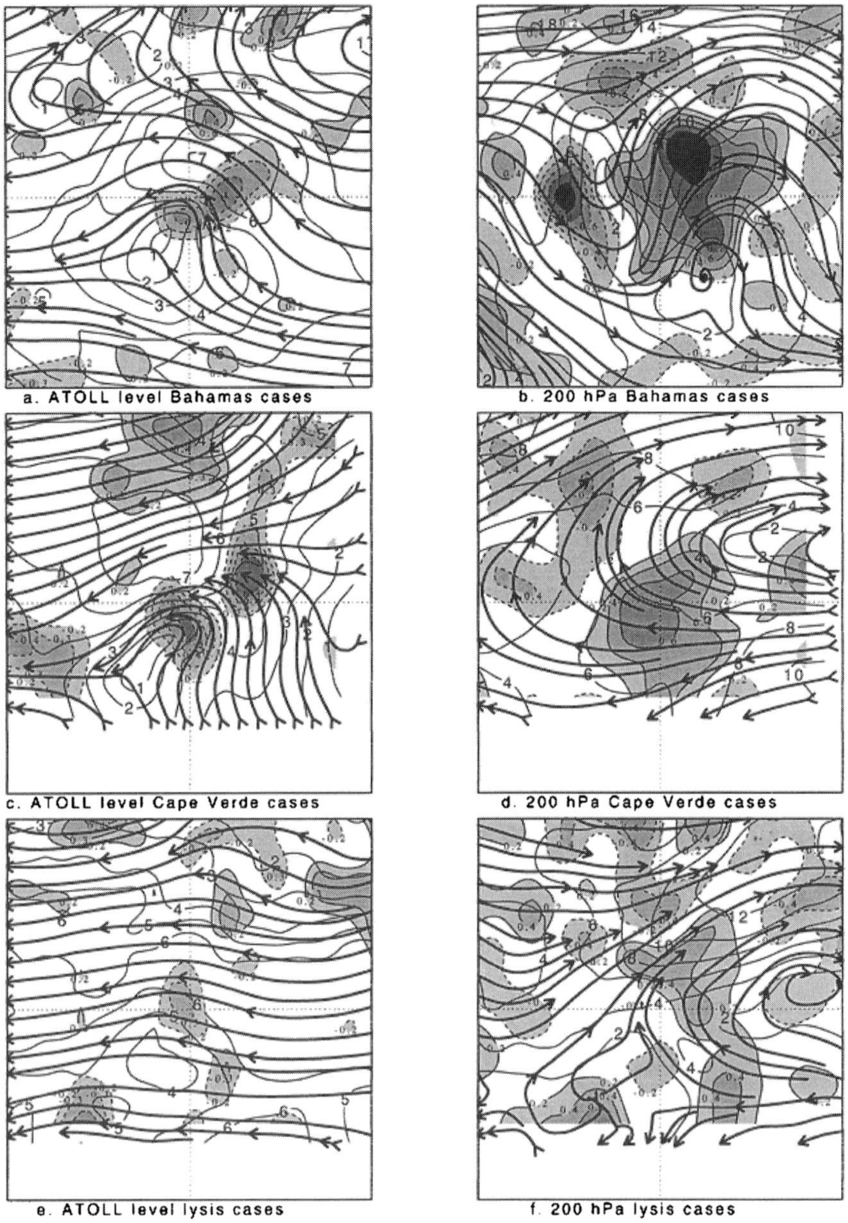

Figure 11.22 Composite flow streamlines (thin solid), isotachs (every 1 m s^{-1}; thick solid), and divergence (every 0.2×10^{-5} s^{-1} in all panels except every 0.1×10^{-5} s^{-1} in e; greater than 0.2 and less than 0.2 shaded with positive values surrounded by thin solid lines and negative values surrounded by thin dashed lines) at the ATOLL level for (a) Bahamas subregion, (c) Cape Verde subregion, and (e) lysis subregion; and the 200-hPa level for (b) Bahamas subregion, (d) Cape Verde subregion, and (f) lysis subregion. Thin dotted lines cross at the center of the composite depression. Composite domain approximately 3000 km × 3000 km. Blank areas in analyses are regions where data from a storm were missing. (Source: Bracken and Bosart, 2000.)

scale < 2 km). Thus Orlanski (1975) divided up mesoscale phenomena by length and time scales into meso-τ scale (2–20 km), e.g. thunderstorms (mins to hours duration), meso-β scale (20–200 km), e.g. squall lines, cloud clusters (hours to approximately one day), and meso-α scale (200–2000 km), e.g. fronts, hurricanes, in order to distinguish atmospheric processes operating at different spatial and temporal scales. This classification has been adopted by many writers in meteorology and climatology (e.g. Atkinson, 1981), but in some cases with modifications for specific meteorological circumstances (e.g. Maddox, 1980; see next paragraph). The meso-α scale, which includes 'fronts' and 'hurricane' phenomena, is considered by some groups in the scientific community to be on the synoptic scale whilst many others consider them to be on an intermediate scale between macroscale and mesoscale

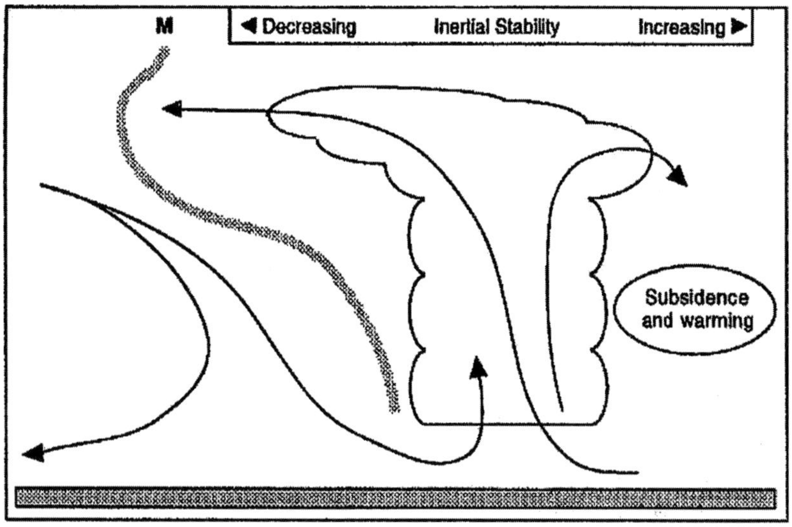

Figure 11.23 Cross-section in the y–z plane showing the conceptual model of trajectories of air parcels in inertially stable and unstable regions. In the inertially stable region (right side), outflow material descends in the near environment, resulting in drying and warming. The region of inertial instability (left side) permits meridional accelerations of outflow material. The thick, shaded line is a momentum (M) surface typical of inertially unstable regions. The vertical shear vector is directed into the page. (After Blanchard *et al.*, 1998.)

(Orlanski, 1975). The inclusion of the meso-α scale, for example, takes into account a number of meso-β scale convective components within the cloud shield of a larger weather system such as within hurricanes (Maddox, 1980) or perturbations in the tropical easterlies. In this Chapter, a discussion of mesoscale phenomena will encompass all three sub-divisions of Orlanski (1975).

Maddox (1980) coined the term *mesoscale convective complexes* (MCCs) and was referring to the development of nocturnal sub-synoptic-scale storm clusters which occur frequently over the central United States. In his definition, these storm clusters are characterised as having length and time scales associated with the meso-α scale, 250–2500 km with duration $\geq$6 h. The central United States is at great advantage to study MCCs because there is an unparalleled concentration of surface meteorological instruments, rain radars, wind profilers and additional aircraft data collection over intensive periods. Thus the most comprehensive understanding of MCCs is based on that experience where well-defined frontal systems associated with the mid-latitudes occur. Nonetheless, from the work of Cotton *et al.* (1989) onwards, several studies have provided evidence of the complexity and varied range of types of convective processes and structures. In the tropics, the more general term *mesoscale convective systems* (MCSs) is used in the literature (e.g. Houze, 1989; Machado *et al.*, 1998; Cifelli and Rutledge, 1998) and will be the term used here.

In the Sahel, Lebel *et al.* (1997) indicate that this is a 'privileged' region for the formation of MCSs and their evolution into MCCs. In the EPSAT-Niger data set covering a 16 000 km^2 area, MCSs were identified as storms having produced rainfall over more than 30% of the study area, that is, at least 5000 km^2. The definition for the evolution of Sahelian MCSs into MCCs was considered to be 'a large spatial extension (area not given but thought to be in the order of 10 000 km^2), high rainfall efficiency, steady displacement (westwards) and the presence of a stratiform region at the rear of the system, are the characteristics of MCCs' (Lebel *et al.*, 1998). These authors noted that the majority of MCCs are non-squall clusters, but over a four-year study period (1990–1993) squall lines produced almost half (44%) of the MCCs rainfall. As mentioned later, the main features of Sahelian MCCs are 'identical' for the area extending south to at least 10 °N in the humid tropics.

The development of MCSs are linked with larger-scale forcing from synoptic meteorological systems (Houze, 1989), supplemented by surface features such as topography, elevated heat sources (e.g. areas of higher sea surface temperatures, SSTs) and other localised mesoscale forcing. Thus, when examples from the tropics are considered, inevitably there will be some overlap with discussion elsewhere on various other disturbances at the synoptic-scale. Based on USA experience, Blanchard *et al.* (1998) tested a conceptual model explaining the upscale development of MCSs and their longevity. Central to the model was the observation that MCC/MCS development occurred in areas with weak inertial stability or inertial instability. In addition, symmetric instability is required which can occur in deep convective systems where there is sufficient CAPE (Convective Available Potential Energy). Figure 11.23 provides a pictorial representation of the process. The geostrophic momentum surface, *M* in Figure 11.23, can be advected further away by the acceleration of the outflow air parcels (Elassien, 1951) which therefore enables the inertially

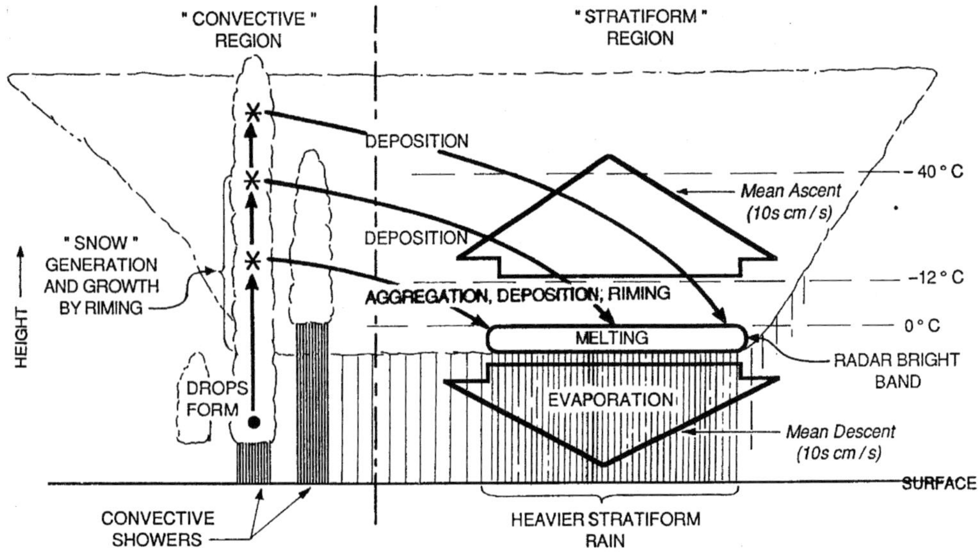

Figure 11.24 Schematic diagram of the precipitation mechanisms in a tropical cloud system. Solid arrows indicate particle trajectories. (Adapted from Houze, 1989.)

stable region to become increasingly less stable, thus enhancing the low level positive feedback process.

Studies of MCSs in the tropics

The intensive field experiments of GATE (Global Atmospheric Research Programme, Atlantic Tropical Experiement) in 1974, WMONEX (Winter Monsoon Experiment) in 1978, and EMEX (the Equatorial Mesoscale Experiment), as part of the Australian Monsoon Experiment (AMEX) 1986–87, enabled writers such as Houze (1989) to develop an idealised conceptual model of the cloud and precipitation structure of tropical MCSs. In the mature phase, the conceptual model of Houze identifies the heaviest rainfall with deep convective towers (Figure 11.24), grading to lighter rainfall associated with a stratiform region of rainfall extending over a horizontal distance of 100–200 km. Within this stratiform region is an area of heavier rain where the convectively generated snow particles reach the 0 °C level after their passage through the stratiform cloud. Significantly, Houze concluded that the broad features of his conceptual model were applicable to a wide variety of MCSs, including Equatorial cloud clusters, Bay of Bengal depressions and hurricanes. However, the horizontal arrangement of convective and stratiform precipitation varied between different MCS.

More recently, the TOGA COARE (**T**ropical **O**cean and **G**lobal **A**tmospheric **C**oupled **O**cean-**A**tmospheric **R**esponse **E**xperiment) (Webster and Lukas, 1992) obtained an extensive, four-month data set of observations, November 1992–February 1993. TOGA was preceded by the DUNDEE (**D**own **U**nder **D**oppler and **E**lectricity **E**xperiment) near Darwin, Australia, over the 1989–90 and 1990–91 wet seasons (Cifelli and Rutledge, 1994, 1998). Both these field campaigns have extended our understanding of MCSs. In addition, there have been a large number of satellite studies of MCSs using mostly infrared/visible radiance images from geostationary satellites. To balance the more recent concentration of mostly oceanic studies in the western Pacific, Machado *et al.* (1998) undertook a satellite coverage of MCSs in both Central and South America, in advance of the LBA experiment (**L**arge Scale **B**iosphere-**A**tmosphere Experiment in Amazonia), to provide a comparison of tropical convection over land with that over oceans.

The structure of MCSs

Of significance to hydrology is the identification of those parts of MCSs which are convective (with higher rain intensities) as against those parts which are stratiform (with associated lower rainfall intensities). The spatial and temporal variability of these convective and stratiform rainfields contribute to temporally (within-storm event) varying dominant runoff pathways within selected soils, especially those of the 'Acrisol'-type end-member (see Bonell, this volume).

Two methods are mostly used for separating the convective from stratiform parts of MCSs, namely IR (infrared) satellite data and radar detection. Yuter and Houze (1998) employed a combination of both methods in TOGA-COARE (due to availability of airborne radar) but rely mostly on IR data. Machado *et al.* (1998) depended solely on IR data whereas Cifelli and Rutledge (1998) used a combination of radar and very high frequency wind profiler data. As noted by Yuter and Houze (1998), a definition of convective activity identified from radar data is fundamentally more robust than an IR-based definition of convective activity because of ambiguities in IR temperature interpretation. One problem is that relying on IR data alone can be deceptive because, depending on the stage of storm evolution, the coldest parts of the cloud may not lie over the deepest convective centres. Radar offers the advantage of distinguishing regions with distinct dynamical and microphysical properties; it can thus detect lower altitude precipitation associated with storm dynamics that can not be discerned from cloud tops, as seen by satellite.

Radar data are only available during intensive field experiments over oceans like TOGA-COARE (i.e. airborne radar) or in coastal areas from ground-based stations like the DUNDEE experiment (Cifelli and Rutledge, 1994), which means that strong reliance is placed on IR data. There has been extensive discussion on which cloud top temperatures to use as indicators of precipitation, and the distribution of occurrence of convective and stratiform precipitation (reviewed in Mapes and Houze, 1993, Yuter and Houze, 1998; Machado *et al.*, 1998). In addition, the determination of probabilities of the occurrence of precipitation and spatial and temporal occurrence of convective and stratiform types differ, depending on the grid scale of IR data used.

Figure 11.25 provides a simplified 2-D projection of precipitation structure phase space for cold clouds at a grid scale of 240 × 240 km in TOGA-COARE. This projection is subdivided into the active inter-seasonal oscillation (ISO) (the 30–60 day Madden–Julian oscillation) and suppressed phases of the ISO. During the active ISO phases, spatially large deep convection occurs within a generally eastward-propagating ensemble of cloud clusters. During the suppressed phases of the ISO, convective activity continues but on smaller temporal and spatial scales, and large cloud clusters are notably absent.

In Figure 11.25(a) there is a large phase space which incorporates both active and suppressed phases of the ISO; in addition, this phase space includes a broad range of main IR temperature and precipitation-area size. Larger precipitation areas with IR temperatures of cloud tops (<235 K) occur predominantly in the active phases of ISO. The large regions of empty space imply there were no samples. On the other hand, the breadth of the shaded area does not provide a good indicator of precipitation occurrence by area nor by ISO phase.

Figures 11.25(b) and (c) provide a better understanding of the importance of rain by precipitation area and by per cent of total rain that originates from the stratiform structure of MCSs. Most pertinent, during the active phases of ISO, larger precipitation areas (>20% of area) with very cold cloud tops (<235 K mean IR temperature) are >75% stratiform by area and have stratiform rain fractions >70%. *The proportion of grid cells occupied by convective cell activity is never more than 30% of a fine grid cell (24 × 24 km) within a ~240 km coarse resolution. Thus, the highest rain intensities are spatially and temporally much smaller from these convective cells than the lower rain intensities from the stratiform region.* What ensures the larger contribution of stratiform cloud by area and by rainfall is, paradoxically, the *sustainability* of convective activity over a region. As each convective cell collapses, it evolves into stratiform cloud and develops precipitation characteristics which have a much longer life cycle than a convective cell. Thus the area of the stratiform cloud region continues to expand as each cell completes its convective phase and so adds to the stratiform area.

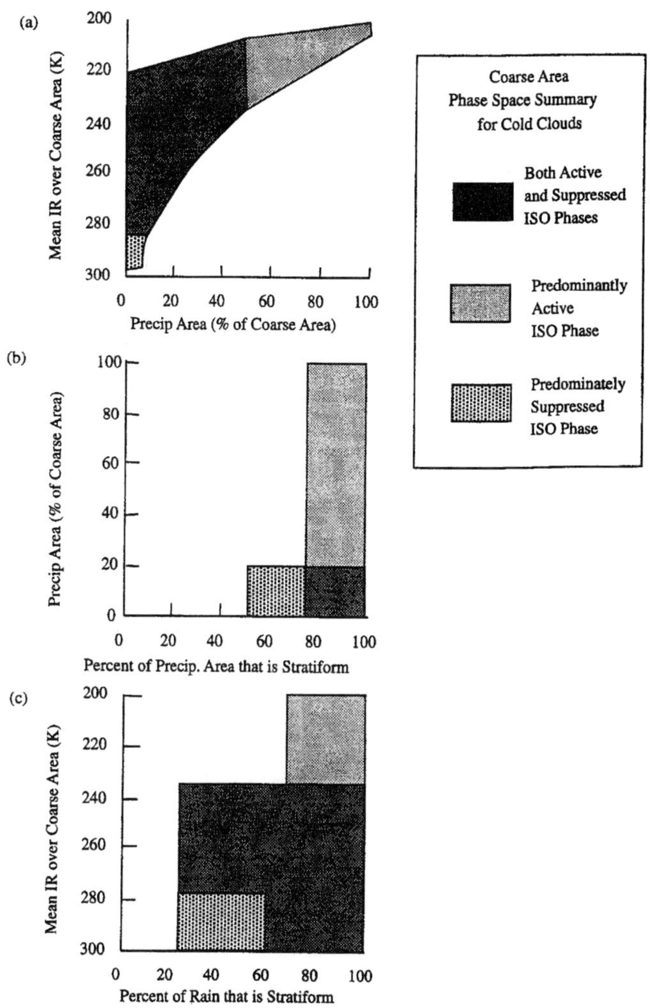

Figure 11.25 Simplified 2-D projections of a n-dimensional precipitation structure phase space for cold clouds at coarse resolution (~240 km). (a) precipitation area versus mean infrared (IR) temperature, (b) stratiform area fraction versus precipitation area, and (c) stratiform rain fraction versus mean IR temperature. (After Yuter and Houze, 1998.)

A separate modelling study by Raymond (1994) concluded that convective heating and associated precipitation can occur in

Table 11.2. *Rainfall statistics for the six break-period events using all available rain gauges from the Darwin network*

Event date (yymmdd)	Number of gauges available	Rain		Rain intensity		
		Convective (%)	Stratiform (%)	Convective (cm d^{-1})	Stratiform (cm d^{-1})	Maximum (mm h^{-1})[a]
891205	22	79	21	36.0	6.1	147.6
900118	20	84	16	47.5	19.9	194.6
900122	20	87	13	27.7	5.9	70.8
900128	24	74	26	34.1	6.5	153.6
901121	15	90	10	39.3	4.9	90.0
901215	18	79	21	17.7	6.1	54.0
Average	20	82	18	33.7	8.2	118.4

[a] Maximum for any reporting gauge during the observation period based on rainfall acuumulation over a 10-minute period. The rain traces were subdivided into convective and stratiform components using the partitioning algorithm described in the text.
Source: Ciffelli and Rutledge (1998).

regions with moderate subsidence in situations where stratiform precipitation is suppressed. The convective heating is capable of opposing moderate subsidence from a larger-scale atmospheric circulation pattern. On the other hand, in unstable regions of general uplift, stratiform rain is more extensive by area and by overall fraction of total precipitation. During TOGA-COARE, stratiform fractions varied widely during suppressed phases of the ISO due to the smaller size of the precipitation areas.

Raymond's (1994) modelling work also highlighted the importance of the evaporation of precipitating stratiform rain in the setting up of heating gradients for enhancing instability and uplift in the upper troposphere to maintain MCS development. On the other hand, horizontal temperate gradients at the ocean-boundary layer interface leads to maximum (convective) ascent at low levels.

Significantly, even allowing for different life cycle effects, the stratiform-dominated MCSs of oceanic origin within the 'active' phase (NW monsoon) in the DUNDEE experiment (Cifelli and Rutledge, 1994) showed the most significant and strongest updrafts in the upper troposphere whereas the lower troposphere convective updrafts were weaker than in the 'break' (south-easterlies) squall lines. In contrast, the most pronounced updraughts occurred in the lower troposphere (up to 4 km) during the 'break' regime where the MCSs were triggered by heating and intense continental convection (i.e. squall lines) from low-level south-easterly flow passing over the top end of Australia. There were, however, secondary deeper updraughts in the middle to upper atmosphere behind the squall line. Within the stratiform area, vertical air motions were nearly identical between the 'break' and monsoon MCSs, with a similar small region of upward motion generally restricted to the upper troposphere occurring in both the break MCSs as with those of the monsoon. Overall, in the upper troposphere, convective updraughts were much stronger in monsoon MCSs.

The work of Raymond (1994) has also another implication in terms of maximum upward motion at lower levels being linked with larger SST gradients. Where sea surface temperature gradients are weak over a large 'pool' of warm water, such as in the 'maritime continent' (centred on the Indonesian archipelago), the heating maximum due to deep convection occurs consistently at higher elevations when compared with the eastern Atlantic, where temperature gradients are more significant (Raymond, 1994). Thus, the resulting deeper convection over the maritime continent gives a greater potential for higher rain intensities from the convective portion of MCSs.

The results of the DUNDEE experiment provide an interesting contrast to the structure of MCSs and associated rain characteristics. In total, 13 tropical MCSs were analysed over two wet seasons (1989–90, 1990–91) which consisted of six MCSs connected with the 'break' regime. The latter were characterised by a leading 'squall' line of convection with intense precipitation and a trailing stratiform region with light rainfall. By contrast, the remaining seven MCSs of the north-west monsoon regime were relatively unorganised clusters of areally more extensive stratiform cloud. Each cluster contained embedded, linear convective bands which moved on-shore with the monsoon flow. Within about an hour, the monsoon convective elements had passed through a complete life cycle and subsequently decayed into the stratiform region of precipitation.

Tables 11.2 and 11.3 provide a summary of the rainfall statistics for the respective 'break' and monsoon events. Significantly, the contribution of stratiform rain to the total MCS rainfall remains the smallest proportion, especially in the break (range 10% to

Table 11.3. *Same as Table 11.2 except for the seven monsoon events*

	Number of	Rain		Rain intensity		
Event date (yymmdd)	gauges available	Convective (%)	Stratiform (%)	Convective (cm d^{-1})	Stratiform (cm d^{-1})	Maximum (mm h^{-1})[a]
900110	24	82	18	24.0	7.2	81.6
900112	26	77	23	17.5	5.5	88.8
901210	21	70	30	20.1	7.6	98.4
901212	22	63	37	22.1	14.1	110.4
910109	19	65	35	25.3	13.6	100.8
910129	18	68	32	15.1	4.9	46.8
910130	17	75	25	22.5	8.5	64.8
Average	21	71	29	20.9	8.8	84.5

[a] Maximum for any reporting gauge during the observation period based on rainfall accumulation over a 10-minute period.

Source: After Ciffelli and Rutledge (1998).

26%, 18% average). For monsoon events, the stratiform contribution is somewhat higher than the break fraction and ranged from 18% to 37% (29% average). On the other hand, the monsoon stratiform fraction is smaller than other estimates of tropical oceanic stratiform precipitation (40–60% e.g. Churchill and Houze, 1984; Houze and Rappaport, 1984; see review of Yuter and Houze, 1998). Cifelli and Rutledge (1998) suggested that some of the differences in these estimates could be technical due to the use in previous studies of a single $Z-R$ (radar reflectivity – rain rate) relationship. Such relationships for convective and stratiform rain are different and require the use of two calibrations (as was used in Cifelli and Rutledge, 1998). Consequently, previous studies are likely to have over-estimated the contribution of stratiform rain.

When concerning rain intensities, the 'break' MCSs produced values about 40% larger than the corresponding maximum rainfall intensity on average (Tables 11.2 and 11.3). This finding is consistent with the observation of squall lines from intense convection over land that form during the 'break'. Absolute maximum recorded hourly intensities were also 1.5 to nearly two times higher in favour of the 'break' period when considering the respective three highest values. *Moreover, the stratiform intensities are generally an order of magnitude lower than the corresponding convective portion of each MCS.* Figure 11.26 shows that, with one exception, there is a distinct 2-D clustering of 'break' vs. 'monsoon' intensity characteristics which highlights the contrasting origins of land and oceanic MCSs.

Box 11.8 The MCS near Fiji 18–20 January 1999

At 0000 UTC 18 January 1999, a trough extended southeastwards from tropical cyclone *Dani* to a tropical low located southwest of the Fiji group with a north to northwest low level flow across the Islands. In the six hours to 1800 UTC 18 January 1999, 237 mm of rain was recorded at Nadi in thunderstorms; and in the 24 hours to 0600 UTC 19 January 1999, 451 mm of rain was recorded.

The pattern at MSL at 0000 UTC 18 January 1999 (Figure 11.27) was characterised by northwesterly flow over Fiji with a trough extending along a line from *Dani* over to the southwest of Fiji. Over the period ending 1800 UTC 18 January 1999 (period of heaviest rain), surface winds increased in strength as the major trough deepened and a smaller scale complex trough formed over Fiji. Upper wind, around the 200 hPa level, mainly derived from satellite data shortly after at 0000 UTC 19 January 1999 (Figure 11.28), show a strongly diffluent wind pattern over the Fiji Islands between the westerlies to the south and the southerlies to the north.

Satellite imagery showed very deep convection over the Islands. At 1939 UTC 17 January cold convective cloud tops lay to the north and east of Viti Levu. At 1531 UTC 18 January 1999 (in Figure 11.29 and the time of the heaviest rain), Nadi lay on the south to southwestern edge of an area of cold cloud tops (colder than −80 °C). That is, the heaviest rain was in this zone and the upper cold cloud was being advected towards the north-northeast by the upper winds.

Widespread severe flooding and damage occurred in the western parts of Vitu Levu with destruction to infrastructure, livestock and crops. Six people were killed and six were still missing after the event.

Example of a severe MCS near Fiji 18–20 January 1999

In contrast to the findings of the above DUNDEE experiment, more severe MCSs can occur over oceans embedded within westerly monsoonal flow. We provide an example in Box 11.8 based on an event over Fiji in the southwest Pacific. Rainfalls up to 451 mm in 24 hours were recorded from a complex trough system embedded within north-westerly monsoon flow. The upper wind

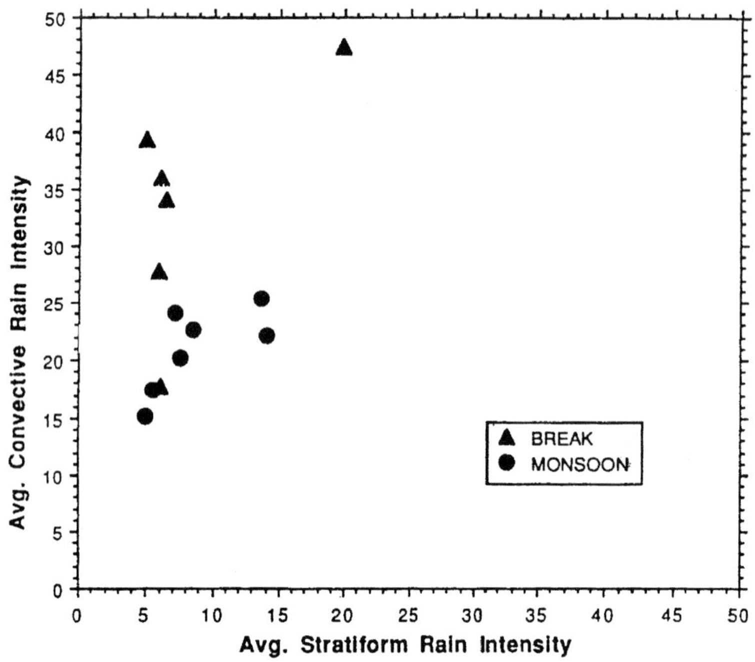

Figure 11.26 Average convective rain intensity as a function of average stratiform rain intensity (cm d^{-1}) for each of the mesoscale convective systems. The rainfall statistics were derived from the Darwin area rain-gauge network and are summarized in Tables 11.2 and 11.3. (After Cifelli and Rutledge, 1998.)

pattern at 200 hPa was markedly diffluent which enabled convective cloud tops to penetrate to this level with temperature less than –80 °C. The most intense rain was in this zone.

The diurnal march of the MCS life cycle

Manton and Bonell (1993) suggested two groupings of MCS (and related rainfall) diurnal activity. Generally, for land areas, rainfall maxima occur between the mid-afternoon and the nocturnal decay period, associated with the rapid build-up of convective heating, and nocturnal to mid-morning maxima centred on the early morning hours for oceanic areas. Nonetheless, the remarks that '. . . there is an inadequate understanding of the dynamics of mesoscale atmospheric–ocean–topography interactions in terms of diurnal cloud and rainfall within the humid tropics compared with recent progress in higher latitudes' (Manton and Bonell, p. 21) still applies. The TOGA-COARE campaign has since provided some new understanding (e.g. Chen and Houze, 1997) over oceanic areas. Garreaud and Wallace (1997) also analysed nine years of IR satellite data and documented a diurnal march of convective cloudiness over the tropical and subtropical Americas.

South America

Analyses by Garreaud and Wallace (1997) reconfirmed our understanding of the prevalence of deep cumulus convection over land areas during the respective Boreal (Northern hemisphere) and Austral (Southern hemisphere) summers. Garreaud and Wallace based their analysis on GOES (Geostationary Operational Environmental Satellite) satellite data. The results are presented in Figures 11.30 and 11.31. The coldest clouds (190–235 K cloud top temperature) reach their peak amplitudes by 15.00 h Local Standard Time (LST) and are associated with the strongest convective updrafts. Thereafter, there is a gradual decaying process with warmer clouds (~250 K) attaining their peak frequency around midnight. Continental areas that exhibit the highest frequency of convective cloudiness, such as those in South America, tend to be aligned parallel to certain coastlines (some linked with the penetration of afternoon sea breezes) and the main topographic barriers (e.g. the Sierra Madre Occidental and the Andes Mountains linked with orographic uplift and mountain-valley winds).

During the Austral summer (Dec–Feb), maximum convective cloudiness over central and the southern part of the Amazonia is organised into two parallel bands, aligned north-west to south-east of greater than 2000 km in length and 400 km in width (b2 and b3, in Figures 11.30 and 11.31) in the absence of significant topography (local terrain is rather flat). The peak in convective cloudiness (and rainfall) occurs in the late afternoon or early evening. There are additional bands of maximum convective cloudiness along the subtropical Andes and along the north east coast of the Amazon basin (b2 and b4 in Figures 11.30 and 11.31).

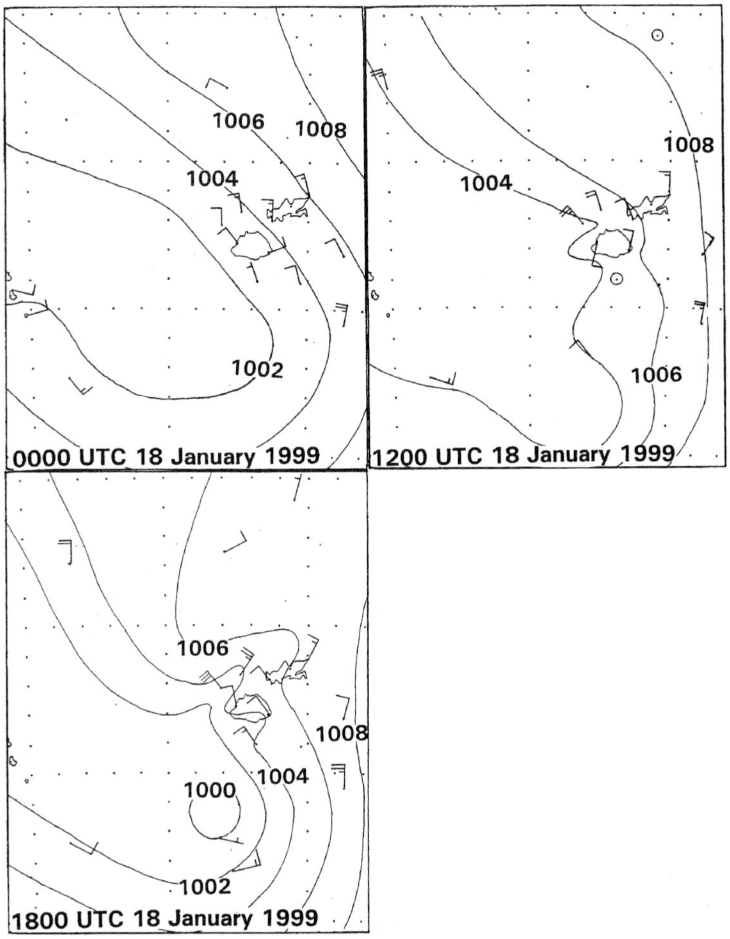

Figure 11.27 MSL pressure distribution (hPa) for 0000 UTC 18 January 1999 (top left), 1200 UTC 18 January 1999 (top right) and 1800 UTC 18 January 1999 (lower left).

The bands b4 and b3/b2 in Figures 11.30 and 11.31 have a linkage to the westward passage of Amazon Coastal Squall Lines (ACSLs) (Garstang *et al.* 1994). The 300 km-wide b1 band, just inland from the coast, is characterised by nearly convective cloud-free conditions from mid-night to noon, then, frequencies of cold clouds increasing to a maximum in excess of 30% around 18.00 h LST (the *Intensification phase* of Garstang *et al.*, 1994). There is also a weak maxima offshore around 14.00h LST which propagates onshore in association with the sea-breeze front (*coastal genesis phase* of Garstang *et al.*, 1994). *At maturation* (near midnight LST), an ACSL can reach a total length of 3000 km, and shows up on satellite images usually as a discontinuous line or arc of discrete clusters of cells rather than a continuous line of cells. Commonly, only 30–40% of an ACSL has active deep convection (Garstang *et al.*, 1994). Subsequently, the ACSLs enter a *weakening* phase with warmer cloud top temperatures and a decreasing ACSL band width. In some cases, the ACSLs reduce their forward speed and weaken as they approach the confluence of major rivers near Manaus. A *regeneration phase* of ACSLs (Garstang *et al.*, 1994) commonly occurs west of Manaus during the late afternoon/early evening of the second day in response to maximum diurnal heating. This process contributes to the central Amazonia maximum convective cloudiness band approximately 1500 km inland from the coast and about 300 km inside the Amazon basin, as noted by Garreaud and Wallace (1997) (see Figures 11.30 b2 and b3). Further westwards, the weakening process is repeated and the individual clusters become more ragged and eventually lose their identity. The forward phase speeds of ACSLs are in the order of 50–60 km h^{-1}, with some ACSLs maintaining their structure to arrive at the westernmost boundaries of the Amazon basin 24–48 hr after their genesis along the northeastern coast.

The conceptual model of the flow structure for a mature Amazon Coastal Squall Line is shown, as proposed by Garstang *et al.*, 1994), in Figure 11.32. There is a marked mesoscale rear-to-front inflow which introduces downdrafts immediately to the rear of the main updraft with a strong inflow of cold, dry air at 750 hPa.

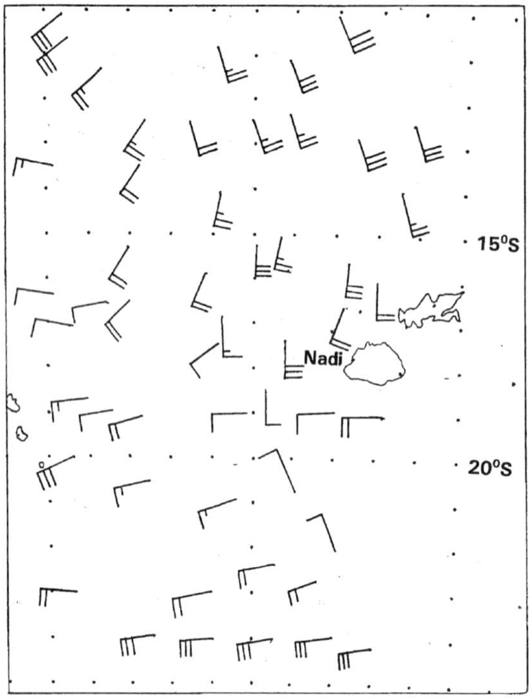

Figure 11.28 Upper winds around the 200 hPa level chiefly derived from satellite data for 0000 UTC 19 January 1999.

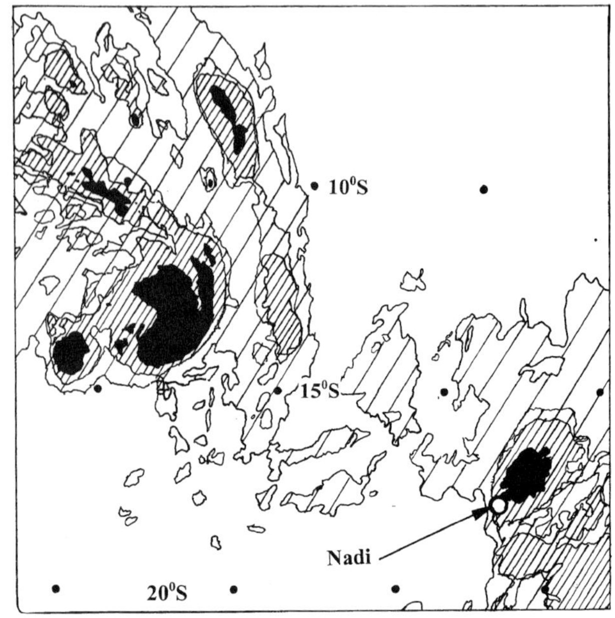

Figure 11.29 Analysis of infrared satellite imagery at 1531 UTC 18 January 1999, diagonally hatched areas denote cloud tops colder than −30 °C, −63 °C and −70 °C with the colder zones marked by increased hatching. Black areas denote cloud tops colder than −80 °C.

There is also an inflow near 600 hPa that reduces buoyancy of the convective updraft in the midlevels. Thus, a near horizontal tilt exists before a deep concentrated core of tropospheric ascent which marks the *leading edge convection (LEC)*. The secondary ascent towards the rear of the ACSL, above 500 hPa, corresponds with a *trailing stratiform region (TSR)*, and the vertical uplift is an order of magnitude smaller than in the LEC. This double-updraft structure is typical of MCS in the tropics associated with squall lines and is a variant of the Houze (1989) model. Ahead of the LEC squall line, new convective elements (cumulus towers) often develop in the prestorm region. Garstang *et al.* (1994) cite measured total rainfall and duration and maximum rain rate for four stations which formed part of the PAM *(Portable Automated Mesonet)* measurement system, just north-east of Manaus in central Amazonia for two ACSL events. An arbitrary rain rate of 0.25 mm min^{-1} was used to separate the convective from the stratiform rain (using a 0.25 mm resolution tipping bucket). The fraction of total rain which was convective vis-à-vis stratiform is unfortunately not cited for each station. Nonetheless, at one station (Embrapa) (Figure 11.6 in Garstang *et al.*, 1994) it is evident that the convective fraction dominates the total of 33 mm rain recorded. The maximum 1 min intensity was 1.7 mm (102 mm hr^{-1}). For the remaining stations, two others recorded maximum rain rates (mm min^{-1}) of 1 (2F1) and 1.5 (Carapana) with respective totals of 16 and 34 mm. The fourth station (Ducke) lay completely beneath the stratiform shield of the ACSL and only recorded 1 mm in total. In the second event, maximum rain rates were smaller and ranged from 0.3 to 1.3 mm min^{-1}, and the total rain recorded was 1–15 mm. Moreover, in the second event the duration was longer, ranging between 1.0–8.0 hrs in contrast to 0.25 to 3.0 hr in the first event. The latter was clearly the more intense, and indicates the spatial and temporal variability of convective activity in ACSLs.

This work by Garstang *et al.* (1994) was part of the Amazon Boundary Layer Experiment (ABLE 2B), which was an intensive mesoscale measurement campaign from 1 April to 14 May 1987 (Greco *et al.*, 1990). Over the preceding period, these workers classified three main modes of precipitation of which the ACSLs (which these writers termed *Coastal Occurring Systems COS*) was the first (Figure 11.33(a)). They also identified a second mode which contributes towards our understanding of the observed banding (b2 and b3) of Garreaud and Wallace (1997) and were named *Basin Occurring Systems (BOS)* (Figure 11.33(b)). These mesoscale to synoptic-scale systems, of the order of 1000 to 1 000 000 km^2, form mainly to the north and east of Manaus. Their forward speed during westward propagation ranges from 10 to 40 km h^{-1}. The third group, termed *Locally Occurring Systems* (LOS), are much smaller in area (<1000 km^2) and generate within the Manaus region (Figure 11.33(c)). These 'local' convective storms usually have a life cycle of an hour.

An assessment of the contribution of each of the three systems and their occurrence over the diurnal cycle provides an insight into

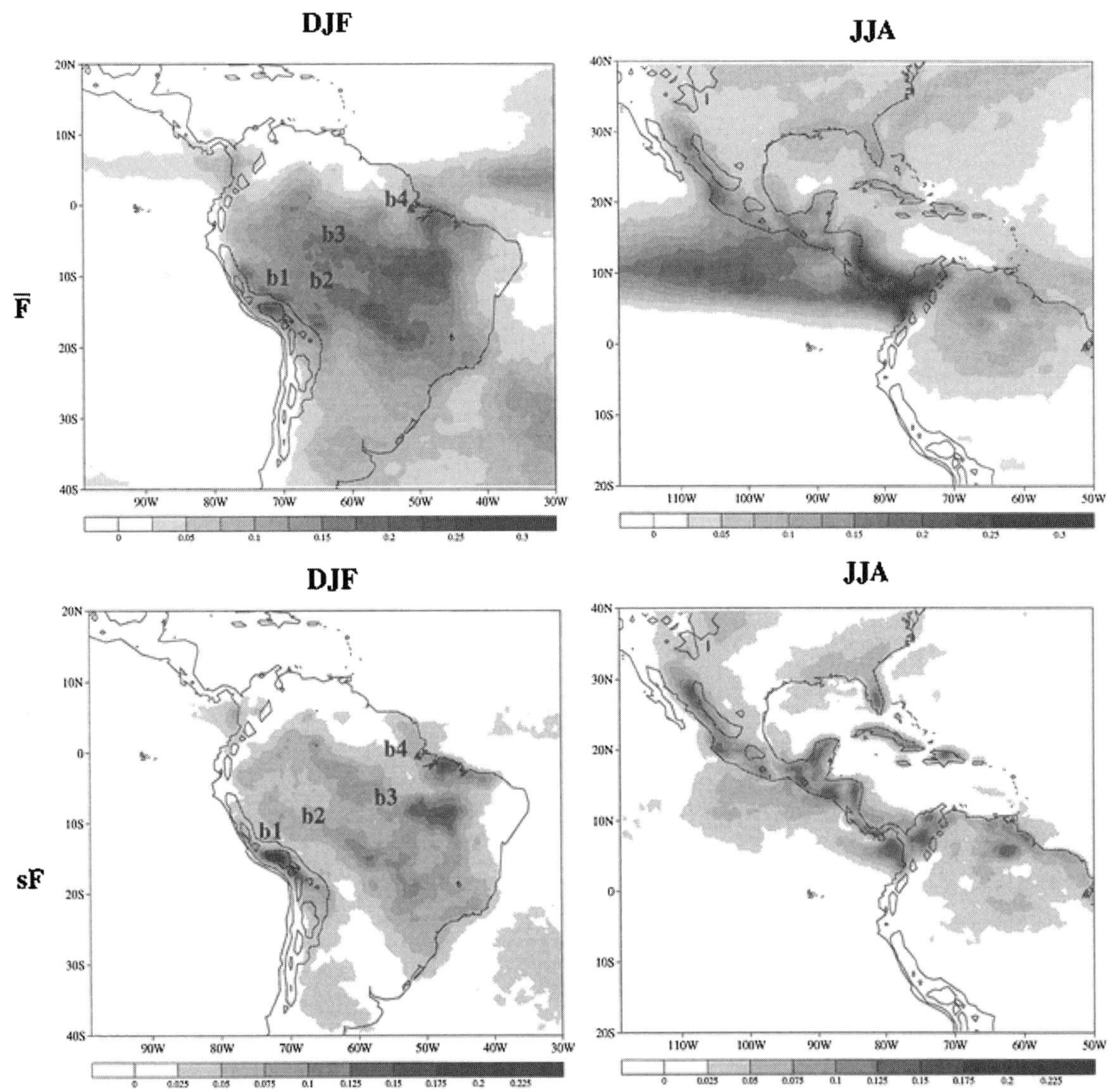

Figure 11.30 Upper panels: daily fractional cold (Tb ≤ 235 K) cloud coverage (denoted as F̄) during December–January–February (DJF) and June–July–August (JJA). Lower panels: As in upper panels but for the seasonal standard deviation (denoted as sF) of the 3-h fractional cold cloud coverage. The 2000 m and 4000 m topographic contours are also indicated. Labels b1, b2, b3, and b4 refer to bands of maximum summertime convective cloudiness over South America. (Source: Garreaud and Wallace, 1997.)

a continental humid tropics regime. Figures 11.34, 11.35 and 11.36 provide a summary of the number of events and the amounts at the four PAM sites for each of the three modes of precipitation. Each has a very distinct pattern. The COS generally affected the PAM network between 1400–1900 UTC (1000–1500 hr LST) following their previous development, up to 24 hours earlier, along the coastal hinterland. The peak rainfall amounts occurred between 1600–1800 UTC (1200–1400 hr local time) which take advantage of the diurnal heating; this peak was followed by an abrupt decrease after 1800 UTC (1400 LST). This rapid decrease was thought to be a result of strong down draughting at the rear of the ACSL (Figure 11.32).

The BOS systems affected the PAM network earlier than the COS, and preferentially affected the mesoscale network 1000–1400 UTC (0600–1000 LTC) after previously generating to the north east of the Amazon Basin on the previous afternoon and evening. Subsequent diurnal heating enables a secondary peak between 1800–2400 UTC. The diurnal heating cycle is strongly evident in Figure 11.33 for LOS with most rainfall occurring in the late afternoon and early evening (1800–0100 UTC).

Despite the frequent occurrence of LOS, 82% of the total rainfall was produced by the well-organised convection of COS and BOS, with only 18% of the PAM network rainfall identified solely with LOS days (Table 11.4). Thus, whilst LOS occurred the most

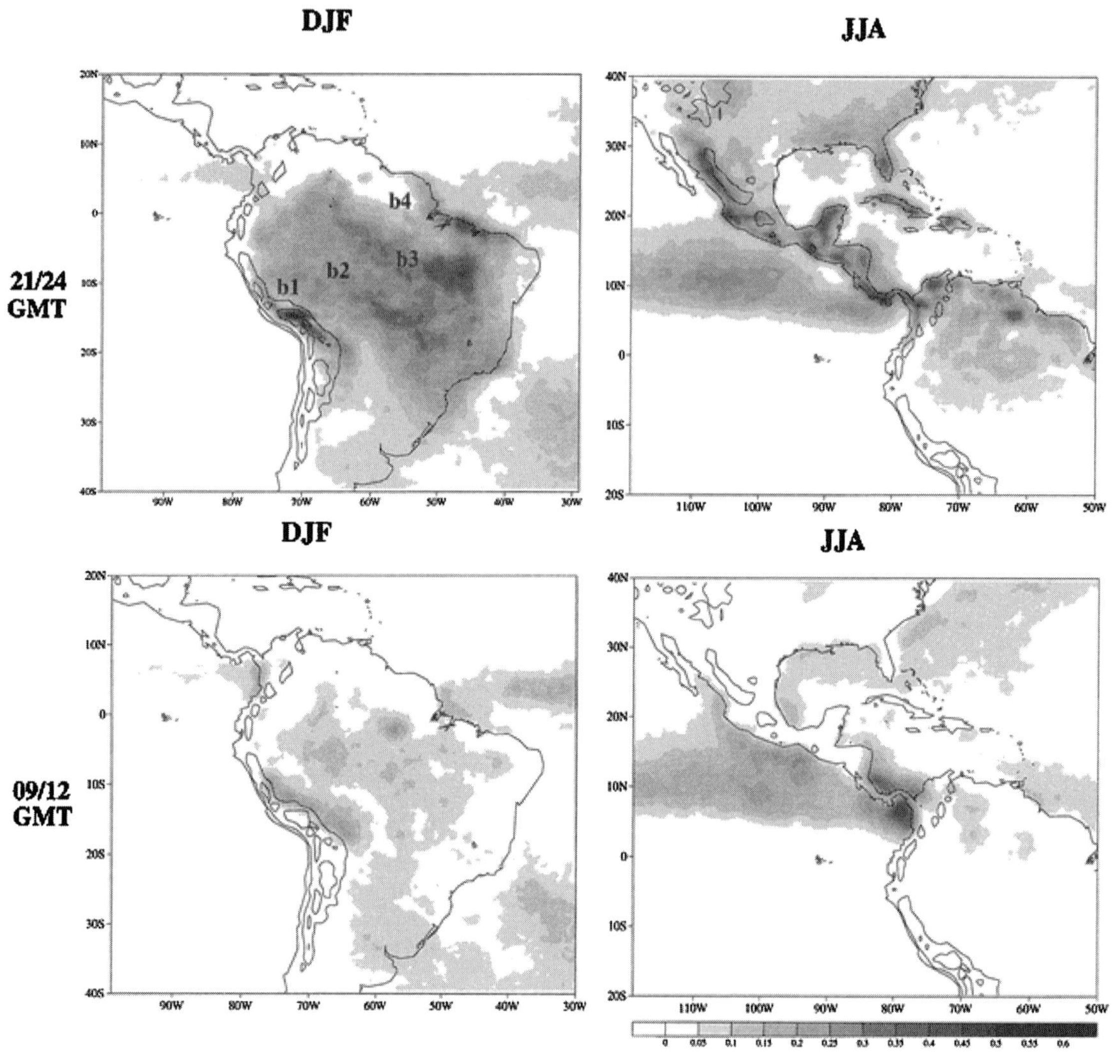

Figure 11.31 Upper panels: fractional cold (Tb $\leq$ 235 K) cloud coverage at 2100–2400 UTC (evening conditions) during December–January–February (DJF) and June–July–August (JJA). Lower panels: As in upper panels but for 0900–1200 UTC (morning conditions). The 2000 m and 4000 m topographic contours are also indicated. Labels b1, b2, b3, and b4 refer to bands of maximum summertime convective cloudiness over South America. (Source: Garreaud and Wallace, 1997.)

frequently (42% of the time during the ABLE2B campaign), it produced the least amount of rain. The remarks of Greco *et al.* (1990, p 17014) are highly pertinent here in stating '. . . like most other regions of the world, Amazon Basin rainfall is dominated by organized synoptic–scale rain bearing systems. Despite the apparently strong diurnal signal observed by satellites over the Amazon Basin, only a small fraction of the total rainfall is derived from local convection.'

Table 11.4 indicates that the average per cent contribution of rain over PAM network from COS and BOS were almost identical (*c*. 41%). Greco *et al.* (1990) also assessed the kinematic-rainfall relationships (Table 11.5) linked with the Houze (1989) model. The BOS produce the highest daily average rain (63.1 mm) in response to stronger surface convergence, duration of convergence, and mass inflow when compared to COS. The reason for these larger values is that BOS systems moved slower and took longer to pass over the PAM network. The weakness of LOS in all parameters (Table 11.4) is also evident.

Based on the intensive rainfall measurement campaign in Rondonia, December–February 1999, as part of the *LBA WET-AMC (Large-scale Biosphere-Atmosphere experiment in Amazonia Wet Season Mesoscale Campaign)*, Marengo *et al.* (2001) describe the diurnal variations of rainfall for a low level 'easterly' and 'westerly' regime. The westerly regime is related to an enhanced South Atlantic Convergence Zone (SACZ, see Callaghan and Bonell, this volume) and a corresponding intense and/or wide Low Level Jet (LLJ) from the northwest ahead of the SACZ. The easterly regime is related to easterly propagating

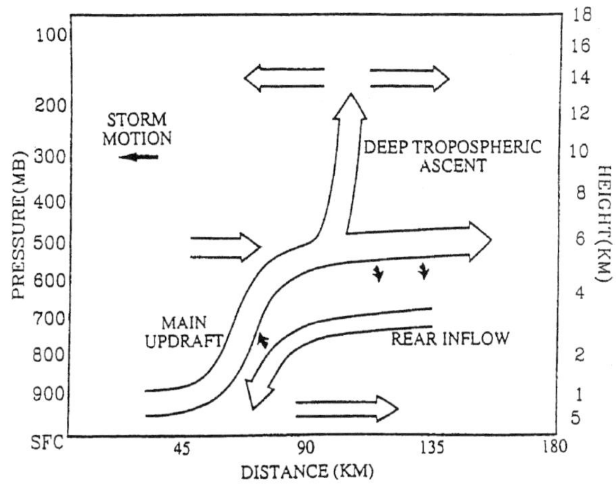

Figure 11.32 Conceptional model of the flow structure for a mature ACSL constructed from a combination of vertical velocity and storm relative flow calculations and two-dimensional cloud model simulations. (Source: Garstang *et al.*, 1994.)

systems (the COS and BOS described above) with possible weakened or less frequent LLJs and a suppressed SACZ.

Differences in the magnitude of diurnal rainfall and its timing were attributed by Marengo *et al.* to the proportion of convective and stratiform clouds connected with both easterly and westerly regimes. Convective development is much stronger, leading to more intense rainfall with an afternoon peak under the easterly regime. The westerly regime is also characterised by an afternoon peak but the smaller proportion of convection embedded within more extensive stratiform clouds leads to rainfall amounts almost 50% less than in the easterly regime. Significantly, the easterly and westerly regimes show secondary peaks in the early morning (0200 LST) and evening (2100 LST), both attributed to re-developing stratiform cloud.

Middle America and the Caribbean

During the Boreal summer (June to August) the spatial and temporal variability of convective (Figures 11.30 and 11.31) activity is complex. By and large, however, coastal shapes (e.g. Gulf of Panama) and mountain ranges (e.g. Yucatan Peninsula and the western slopes of Sierra Madre Occidental) are significant controls over diurnal patterns of convection. The Florida peninsula and the major Caribbean islands also have a prominent diurnal cycle. Over the land areas, Garreaud and Wallace (1997) revealed the strong influences of diurnal heating (and associated orographic and sea-breeze influences) in coastal hinterlands. These regions are characterised by a rapid increase of convective cloud build-up in the afternoon and then a gradual decay through the evening into the early morning hours. Significantly, peak convective cloudiness in the Boreal summer occurs about an

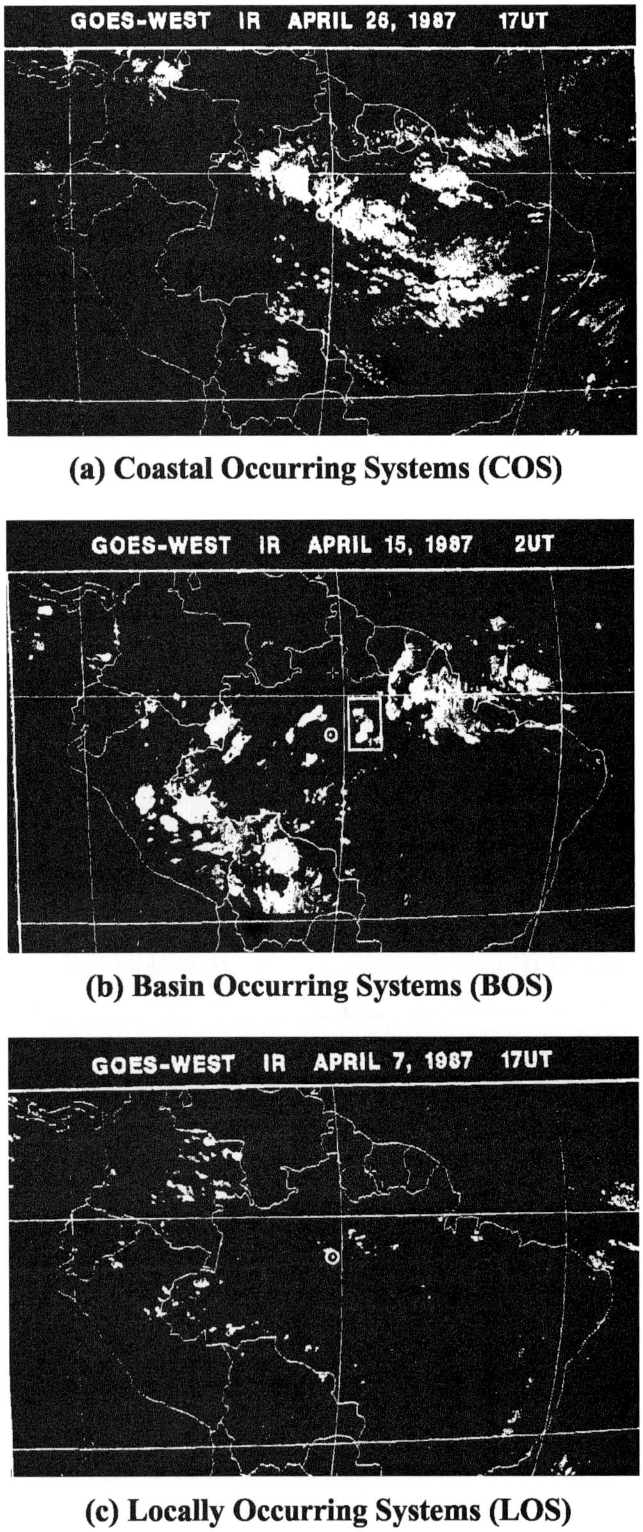

Figure 11.33 (a) Coastal Occurring Systems (COS) (b) Basin Occurring Systems (BOS) (c) Locally Occurring Systems (LOS). (Source: Greco *et al.*, 1990.)

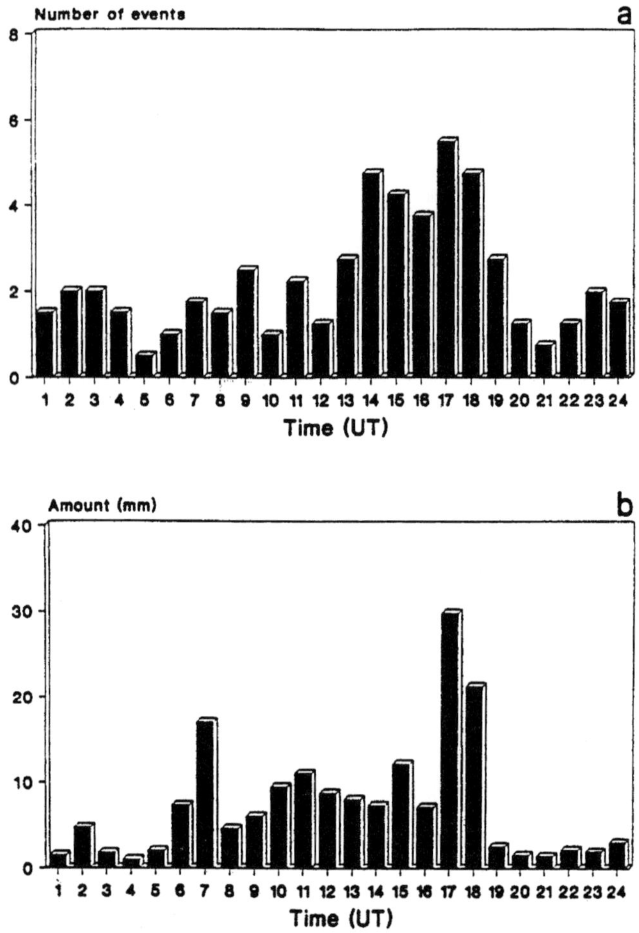

Figure 11.34 Hourly distribution of averaged PAM network rainfall (a) events and (b) amounts for COS. (Source: Greco *et al.*, 1990.)

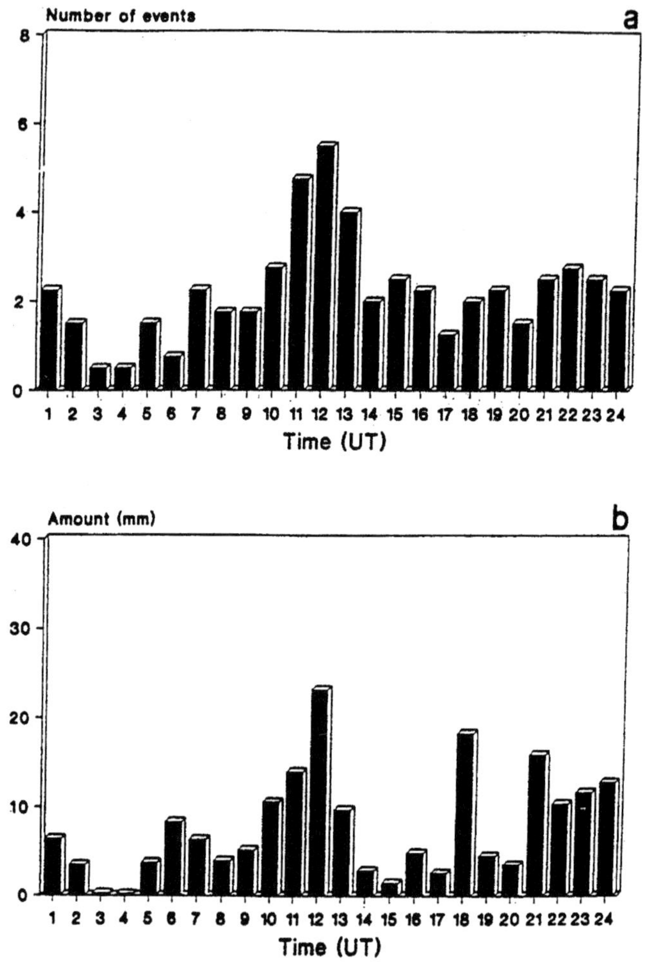

Figure 11.35 Hourly distribution of averaged PAM network rainfall (a) events and (b) amounts for BOS. (Source: Greco *et al.*, 1990.)

hour or two later than in the Austral summer over South America. Otherwise, mean convective cloudiness and late afternoon / early evening convective cloudiness show remarkably similar spatial signatures over both South America and Central America and the Caribbean.

The oceans adjacent to Central and South America

Two regimes in the diurnal march of convection can be identified. The first group concerns the near-coastal areas which exhibit a marked diurnal variability in convection and rain activity, especially along coastlines with prominent concavities, for example, the Gulf of Panama, both the Pacific and the Caribbean coasts of Central America and the north-east coast of South America (March–May). Convection typically attains its maximum near noon and a minimum around midnight although the exact mechanisms responsible for this diurnal pattern are not totally clear.

The second group of convective activity is identified with synoptic-scale phenomena such as the zonal trough in the easterlies over the Atlantic ocean (Austral summer) and the northern monsoon shearline in the eastern Pacific (Boreal summer). These systems exhibit very small sF standard deviations ($<3\%$), indicative of a weak 'diurnal march' in favour of early morning dawn maxima.

The trajectories of individual MCS, tracked by Machado *et al.* (1998) in the Austral summer show the effect of ACSLs (COS) and BOS in the Amazon basin in Figures 11.37(c) and (d). Movement around the upper Bolivian high is also evident. MCS activity in the tropics moves northwards during the Boreal summer (Figure 11.37) into Central America, the Caribbean, and northern South America. The movement of MCS, westwards off the Central American coast and also across northern South America are particularly evident. In general, the MCSs are much smaller (maximum radii about 170 km) in tropical America in comparison with those observed in the western Pacific (maximum

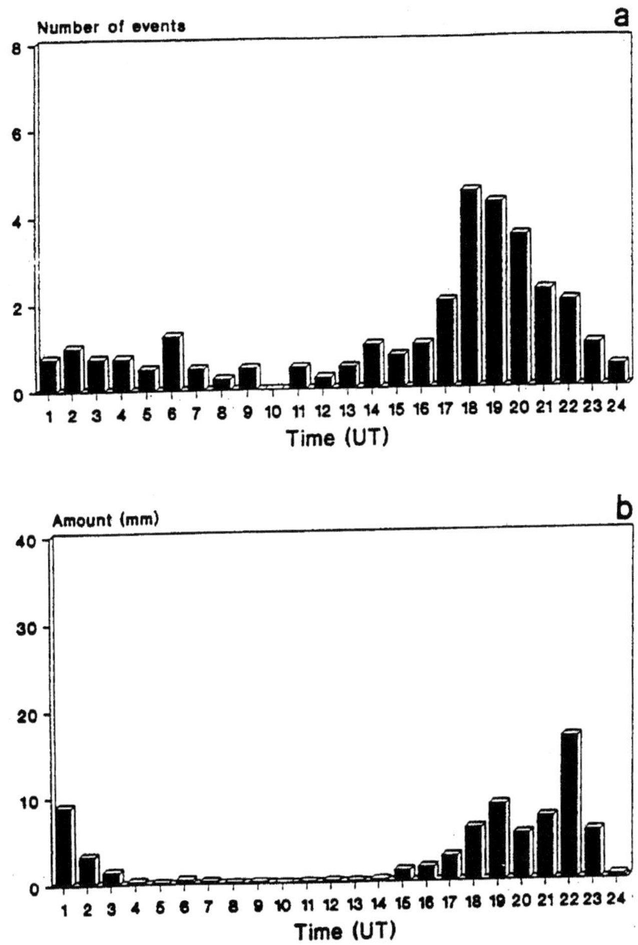

Figure 11.36 Hourly distribution of averaged PAM network rainfall (a) events and (b) amounts for LOS. (Source: Greco *et al.*, 1990.)

radii up to 500 km) (Chen and Houze, 1997; Mape and Houze, 1993).

An example of the diurnal evolution of tropical island convection within the maritime continent

As emphasised in the previous chapter, the 'maritime continent' of Ramage (1968) is one of the epicentres for global heat release which drives both the Walker and Hadley circulations. Aside from the monsoonal systems and associated tropical cyclones, diurnal convection across the Malaysian peninsula, Indonesian Archipalago, northern Australia and Papua New Guinea at relatively fixed locations is a principal contributor to latent heat release. As part of the *Maritime Continent Thunderstorm Experiment (MCTEX)*, an intensive study of the evolution of diurnal convection over the Tiwi Islands (Bathurst and Melville Islands) just to the north of Darwin, Australia, was undertaken in November 1995 (Saito *et al.*, 2001). The topography of the islands are relatively shallow and mostly less than 100 m a.s.l. Field measurements were compared with numerical simulations using the Japanese Meteorological Institute non-hydrostatic models (MRI NHM) at two grid scales (2.5 km and 1 km). Box 11.9 provides meteorological and rainfall details.

Box 11.9 Evolution of diurnal convection

Figure 11.38 presents a schematic representation of a conceptual model which depicts the three main stages leading to convection. Initially, a Sea Breeze Front (SBF) triggered by the temperature difference between sea and land propagates more rapidly from the windward coast. Upward motion at the SBF is less than 1 m s^{-1} in the early stages which does not attain the Local Condensation Level (LCL). Later at the condensation stage (Figure 11.38a) shallow convective cells (Rayleigh-Benard cells) develop. In addition, Horizontal Convective Roll (HCR) clouds develop upwind of the islands, so that clouds over the island develop preferentially at the cross points of SBF and HCRs leading to enhanced convection. At this stage the depth of the sea breeze is about 500 m and sea/land temperature difference of 2 °C occurs. Upwards motion associated within the horizontal convection cells reaches around 3–4 m s^{-1}, but there is mostly an absence of precipitation.

At the precipitation stage (Figure 11.38) upward motion attains 7–8 m s^{-1} and precipitation commences due to the convergence of the respective lee and tail–wind SBFs into the principal convergence zone. Weak outdrafting from the precipitating clouds commences to suppress the Rayleigh-Bernard convection cells.

Later (Figure 11.38c) there is a merging of clouds along an east-west line coinciding with an 'explosive' increase in convective activity. Updrafts exceed 20 m s^{-1}, but the deep moist layer present in this region means downdrafts are comparatively weak in the absence of dry air. Significantly, cloud tops attain the tropopause (*c.* 16000 m a.s.l) and extensive anvil overcast develops. Subsequently, convective activity decays as solar heating decreases towards evening.

Figure 11.39 shows the time sequence of convective activity using the 1 km grid NHM model. The 'explosive' convection occurs about 1400 h LST (330 min in Figure 11.39b) with the maximum updrafts up to 29 m s^{-1}, and cloud tops in excess of 16 km. At about 1600 LST convection enters the decaying stage. During the merging (explosive convection) stage very high instantaneous rain intensities are detected of up to 180 mm h^{-1}, which coincide with the maximum updraft velocities. Significantly, a separate simulation which excluded topographic variations over the island still gave average rain rates not significantly different from those reported in Figure 11.39b.

This study shows that diurnal convective activity is not only just controlled by convective available potential energy (CAPE). A low wind shear environment coupled with high atmosphere moisture are also conducive to marked diurnal activity. Local circulations linked with initial horizontal convection are also important.

Table 11.4. *Amounts (mm) and relative percent contribution of the rain produced by the three types of convective systems, which influence the mesoscale network.*[a]

Station	Basin Occurring Systems (BOS)	Coastal Occurring Systems (COS)	Locally Occurring Systems (LOS)	Total
Ducke	114.53	117.90	59.69	292.12
Percent of total	39.2	40.4	20.4	
Embrapa	210.31	235.42	72.55	518.28
Percent of total	40.6	45.4	14.0	
Carapña	238.27	204.98	89.16	532.41
Percent of total	44.8	38.5	16.7	
ZF-1	171.74	183.17	91.83	446.74
Percent of total	38.4	41.0	20.6	
Experimental average percent contribution	40.7	41.33	17.93	
Percent contribution of number of experiment days	30	28	42	

[a] Values given for all four PAM stations.
Source: After Greco *et al.* (1990).

Table 11.5. *Daily averaged surface kinematic conditions for the four different rain producing systems*[a]

	Locally Occurring Systems (LOS)	Coastal Occurring Systems (COS)	Basin Occurring Systems (BOS)
Convergence, s^{-1}	-1.1×10^{-5}	-1.3×10^{-5}	-1.4×10^{-5}
Duration of convergence, hours	6.5	8.0	9.5
Vertical velocity, cm s^{-1}	+0.8	+1.0	+1.1
Mass inflow, g	$1.9 \times 10^{+14}$	$2.7 \times 10^{+14}$	$3.6 \times 10^{+14}$
Rain, mm	12.8	59.4	63.1

[a] All values are average daily values for each rain system.
Source: After Greco *et al.* (1990).

DIURNAL VARIABILITY OF OCEANS (TOGA)

MCSs over the TOGA area

The overview of diurnal variability over oceans in the Americas has highlighted a different oceanic cycle; namely, that away from coasts, the diurnal variation is relatively weak. On the other hand, a comprehensive understanding of the mechanism(s) responsible for such variability has been poorly understood until the recent TOGA campaign. Common explanations for the nocturnal enhancement of precipitation are attributed to long wave (LW) radiation. For example, differential cooling between clear and cloudy regions was put forward as the mechanism for generating secondary circulations, which, in turn, enhance convective activity (Gray and Jacobson, 1977). As there is little LW cooling in the cloud interior, enhanced subsidence in the 'cooler' clear air leads to convergence into the cloud region. A second, widely-favoured, mechanism involves nocturnal cloud-top cooling from LW radiation which destabilises cloud (especially stratiform cloud) and enhances convective overturning, and hence precipitation. The increased differential of temperature between cooler cloud tops and a warming cloud base fed by a warm ocean is the catalytic

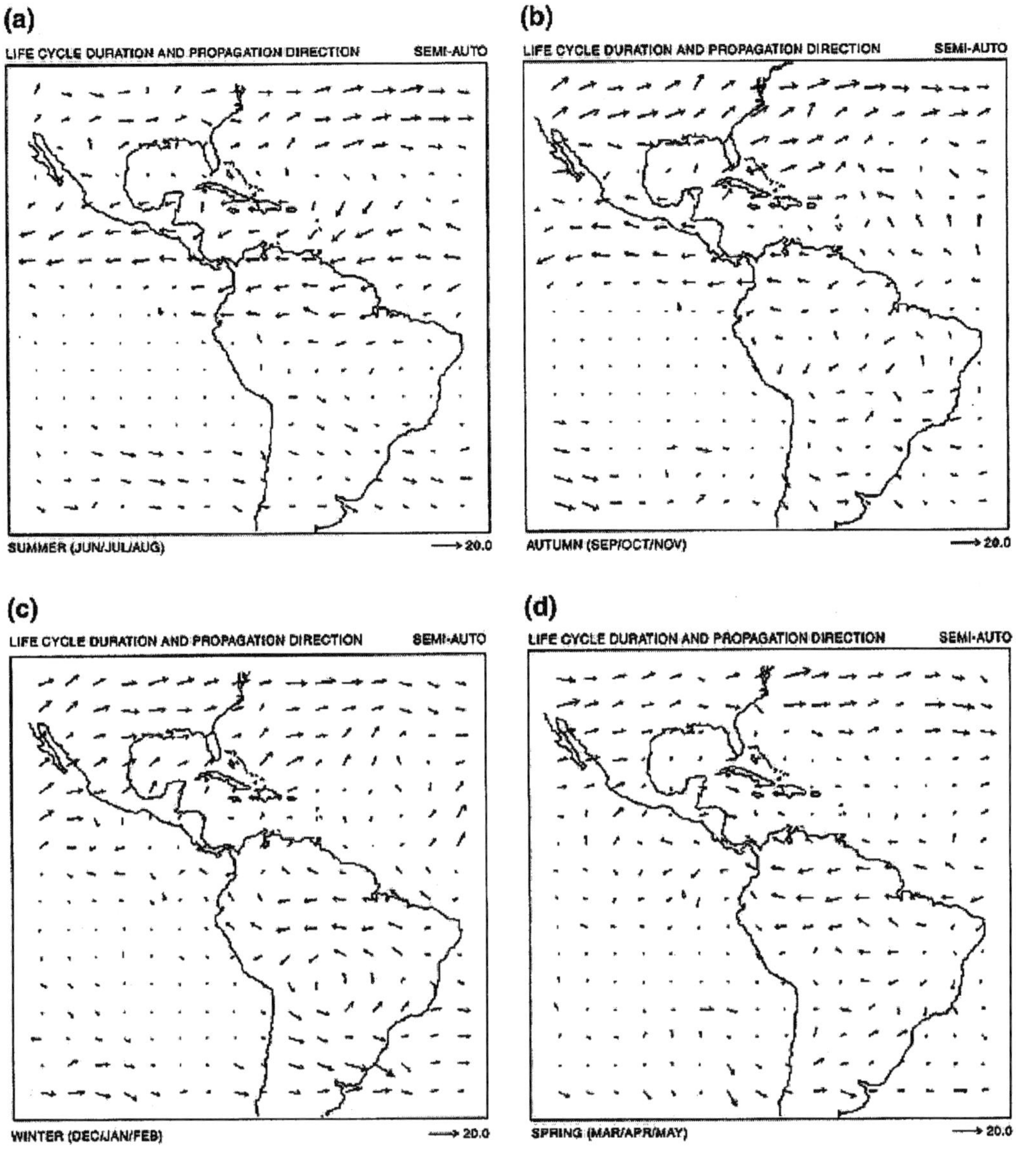

Figure 11.37 Average lifetime (in hours, arrow length) and average direction of propagation (arrow direction) of convective systems during boreal (a) summer, (b) autumn, (c) winter, and (d) spring (July 1987–June 1988). (Source: Marchado *et al.*, 1998.)

point for increased convection (Webster and Stephens, 1980). Such postulated mechanisms are very persuasive, and provide an explanation with satellite observations and island data of very cold cloud tops over tropical oceans which occur most frequently between 0300–0600 LST (Gray and Jacobson, 1977; Janowiak *et al.*, 1994). Modelling (e.g. Dudhia, 1989) has also emphasised the role of LW radiative cooling which, when excluded, reduces net rainfall in excess of 30%.

Through the use of a two-dimensional cloud-resolving model, Dharssi *et al.* (1997) simulated three mechanisms: (i) differential cooling between cloudy and clear regions; (ii) cloud-top cooling/cloud base warming (both referred to above) and the inclusion of a third (iii) domain-wide cooling. Domain-wide cooling involves cooling which extends to the surface and thus encourages enhanced vertical fluxes from a warm ocean surface. Destabilisation results from this increased convective activity, which also causes a corresponding differential between equivalent potential temperature (θ_e) in the mid-troposphere. Thus, there is also a scale difference between the first two mechanisms and the third, the latter functioning across a much broader scale. Dharssi *et al.*'s sensitivity experiments showed that the domain-wide cooling mechanism was of primary importance whereas the other two mechanisms (whilst still contributing) were of minor importance in terms of their overall impact.

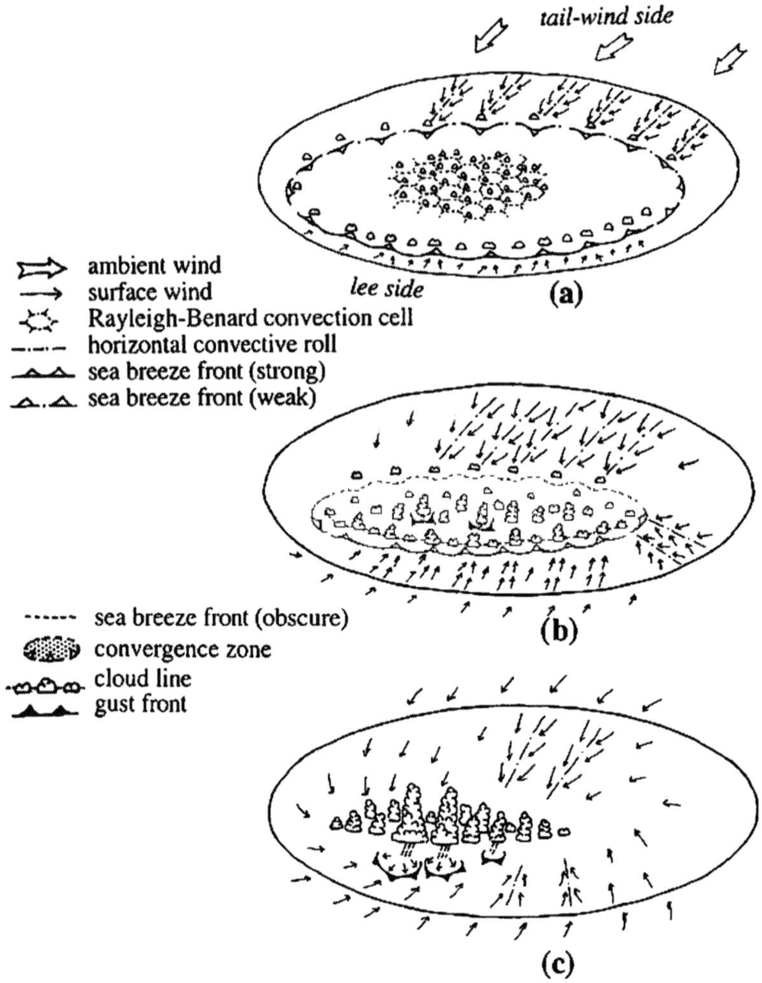

Figure 11.38 Schematic representation of a conceptual model showing stages in the diurnal evolution of tropical island convection. (a) Condensation stage, (b) precipitation stage, (c) merging stage. (Source: Saito *et al.*, 2001.)

Chen and Houze (1997) provided new insights into the oceanic diurnal cycle based on TOGA-COARE. Several features emerged from this important study. When referring to the LW radiative cooling effect, Chen and Houze emphasise that diurnal variability is the product of a complex surface–cloud–radiation–Equatorial inertio–gravity waves interaction and not just a cloud-radiation interaction alone. Further, the diurnal cycle needs to be coupled with the life-cycle of individual cloud systems for both the suppressed and convectively-active intra-seasonal oscillation (ISO) phases (the MJO). The diurnal response is very different between these two phases.

Thus, during convectively–suppressed phases of the ISO small cloud systems (horizontal dimension <80 km) with spatially limited 'cold' tops (<208 K) and small clouds with cloud tops warmer than 208 K, dominated the TOGA area. Significantly, the trigger for this cloud population was diurnal heating of the ocean surface and overlying atmospheric surface layer so that these small cloud systems formed almost exclusively in the afternoon. Moreover, their life cycle was usually 1–3 hours so that the maximum size of cloud cover occurred in the afternoon. Thus, the diurnal behaviour of these small cloud systems during suppressed phases of the ISO were similar to that which occurs over land masses.

During active phases of the ISO, spatially large convective systems (super-convective systems >300 km in horizontal dimension) with more extensive 'cold' cloud tops (<208 K) occur. In line with the 'suppressed' phase small convective systems, these super convective systems generally formed in the afternoon due to the same causal factors. On the other hand, the life cycles of these super convective systems were in the order of one day. Consequently, they attained maturity, with the most extensive cold cloud tops, through the night to sunrise period. Thereafter, solar absorption results in the very cold cloud top area (<208 K) to

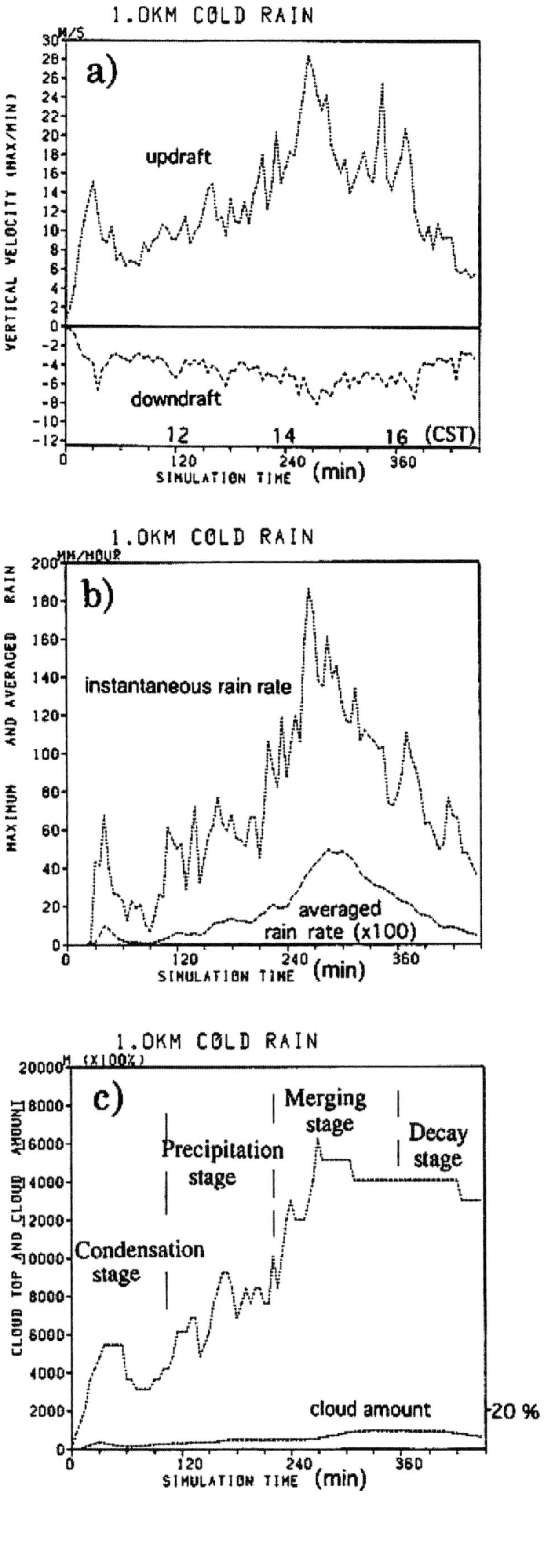

Figure 11.39 Time sequence summarising convective activtity as simulated by 1-km NHM (a) Maximum updraft (upper) and maximum downdraft (lower). (b). Maximum instantaneous surface rain intensity (dotted line) and the averaged rain rate (broken line) over a rectangular modelled domain centred on Tiwi Islands. Averaged rain rate is multiplied hundred times. (c) Maximum cloud top height and cloud amount (%) over the modelled domain. Note that the simulation time has a 90-min lag. (Source: Saito *et al.*, 2001.)

diminish, but overall the area of cloud continues to expand but with warmer tops. Thus, the period of most active convective rain occurs overnight but stratiform rain persists throughout and its dominance is enhanced on the second day during the decaying process.

The longevity of these super convective systems also results in an interesting spatial and temporal pattern in their occurrence. Essentially, the persistence of the decaying cycle on the second day results in the cloud canopy partially shading the ocean surface from the sun. The net result is a marked spatial (as well as temporal) variability which Chen and Houze (1997) termed *diurnal dancing* so that, for a given location, the large convective systems occur on every alternate day.

The above 'diurnal dancing' also has strong connections with the two-day disturbances observed by Haertl and Johnson (1998) to pass through the TOGA IFA (Intensive Flux Array) during active phases of the ISO. These two-day disturbances are a cluster of MCSs which propagate in various directions but with the cluster as a whole moving westward at about 16 m s^{-1} (average for TOGA period). They characteristically generate close to the equator at about 170 °E and dissipate near 140 °E within a larger, active convective envelope, which is moving *eastward*, associated with the eastward movement of the MJO. Thus, there are westward-moving MCSs within a larger eastward-moving central envelope of convection. Each two-day MCS cluster originates to the east of where its predecessor originated, so that the net result is that the centre of convection moves slowly eastward in conformity with the MJO convective envelope. Integrated with this broad-scale dynamics, Chen and Houze (1997) proposed that the 'diurnal dancing' of MCSs trigger and then phase-lock with westward-propagating inertio-gravity waves (Matsuno, 1966) of similar two-day periodicity (Figures 11.40(a) and (b)). Interestingly, such disturbances had been identified previously as 'easterly waves' (e.g, Reed and Recker, 1971). It is clear that they do not conform to the same structure (Roux, 1998; Haertel and Johnson, 1998), as previously described by Thorncroft and Hoskins (1994a, b) and Thorncroft (1995). However Haertel and Johnson (1998) noted the existence of a weak easterly jet at 500 hPa which could contribute to the organisation of the MCS convection.

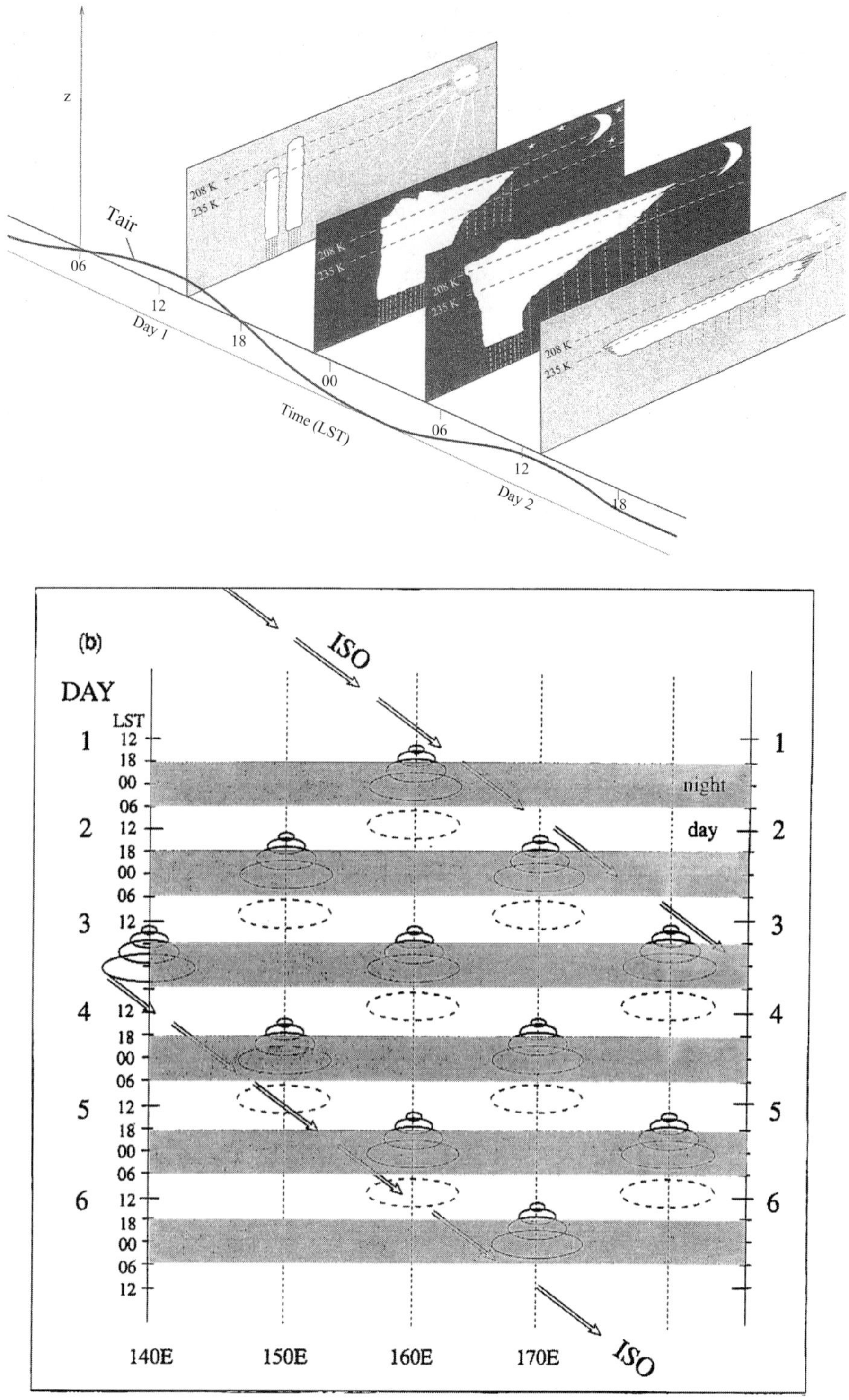

Figure 11.40 (a) Schematic of the 2-day cycle of surface–cloud–radiation interaction for large convective systems. T_{air} is the surface air temperature. (b) Schematic of the time–longitude variation of cold tops of large convective systems during the active phases of the intra-seasonal oscillation (ISO). Spatial variation of the large convective systems within the envelope of the ISO caused by the diurnal behaviour of these large systems alone (solid circles <208 K, dashed circles ~235–260 K). The smallest circle represents the early stage of the system (first afternoon cross-section in (a)), the largest circle represents the mature stage with the maximum areal extent of cold cloud tops (pre-dawn cross-section in (a)), and the dashed circles represent the warmer cloud deck (last cross-section in (a)). (c) The westward-propagating speed of '2-day' equatorial inertia-gravity waves is added to the westward displacement of the starting location of a new convective system. The pattern in (b) is the sum of the effects of the ISO, 2 day waves, and the diurnal cycle of large convective systems. This superposition accounts for much of the observed cold cloudiness patterns seen in satellite infrared data. (Source: Chen and Houze, 1997.)

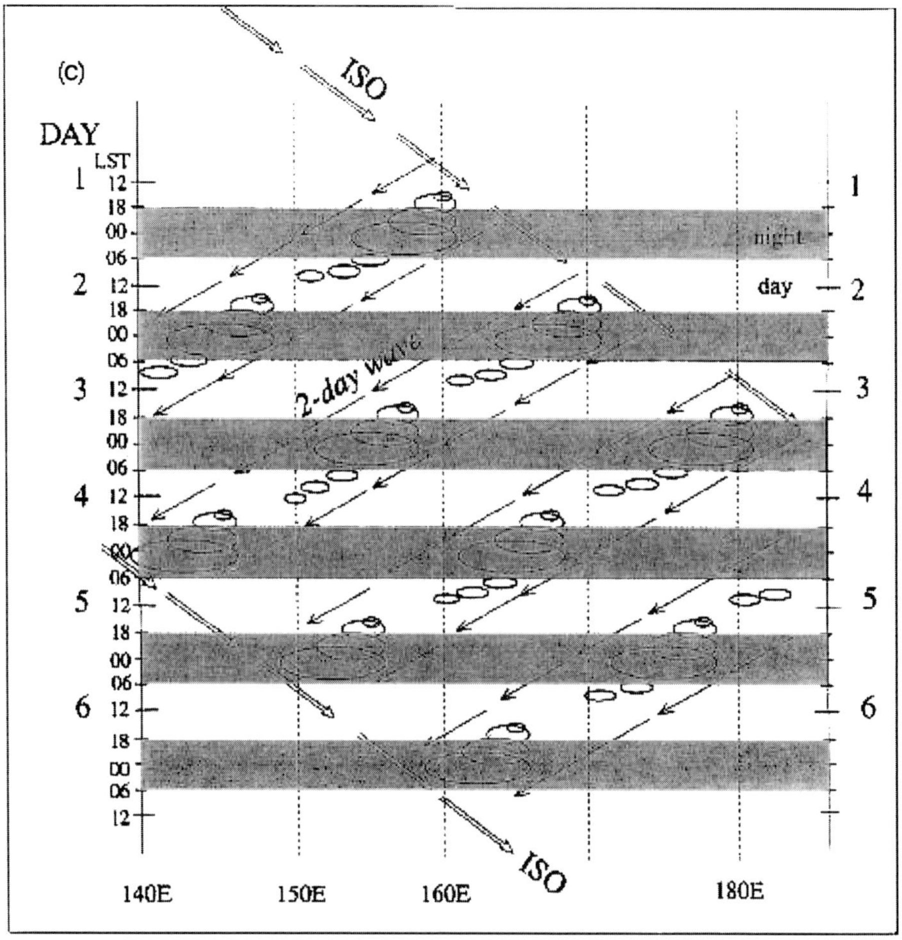

Figure 11.40 (*cont.*)

The diurnal cycle of surface rainfall in TOGA–COARE over the ocean

Five rain gauges attached to moored buoys (Janowiak *et al.*, 1994) were located as a transect along the equator which included sampling both the inner Intense Flux Array (IFA) and the western equatorial section of the Large-Scale Array (LSA) (see Figure 11.1 of Chen and Houze, 1997).

A composite (Haertel and Johnson, 1998) of the two-day disturbances showed substantial rainfall over the IFA with a rain peak (rain rate of 1.8 mm hr^{-1}) occurring six hours prior to the passage of the disturbance. Overall, 480 mm of rain accumulated over the IFA, but as these measurements are 24-hour block averages, actual rain intensities are likely to be much higher over shorter durations.

Results from Janowiak *et al.* (1994) provide more insight into the rainfall characteristics linked with diurnal activity. Three categories of three-hourly rainfall amounts were used: *highest* >44 mm which accounted for 10% of three-hourly amounts, *medium* (4–44 mm), 58%, *and light* – the lowest 32% (1–3 mm). Significantly, all of these categories showed a diurnal maximum frequency in the early morning (0030–0600 LST). The highest category showed the greatest preference for the pre-dawn maximum period, and was consistent with the pre-dawn diurnal maximum of very cold cloud top temperatures (<208 K) (Chen and Houze, 1997). In line with many tropical rain stations, about 10–15% of heaviest rain events contributed about 50% of the total precipitation (Janowiak *et al.*, 1994).

The observed diurnal cycle of precipitation (of ~25–100%) in favour of nocturnal hours (Janowiak *et al.*, 1994) remains within the range of modelling simulations by Dudhia (1989) (only 36% net rainfall when LW radiation excluded), but remains much larger as against the simulations of Tao *et al.* (1996, cited by Chen and Houze, 1997) and Dharssi *et al.* (1997). Tao *et al.* (1996) simulated a ~0–7% difference in rainfall between daytime (short wave SW and LW radiation) and nightline (LW radiation only). Dharssi *et al.* (1997) simulated a maximum 23% difference between LW radiation and no radiation. Chen and Houze (1997) therefore argue that the cloud / LW radiation interaction hypothesis is insufficient, especially when linked with the hypothesis of solar SW absorption suppressing convection during the daytime. Convection is most active during daylight hours after noon in the suppressed phase of the ISO. The prime controls, as we have noted, are large-scale (ISO-MJO) and mesoscale (two-day disturbances) dynamic and thermodynamic processes.

Diurnal variations over West Africa

Shinoda *et al.* (1999) analysed the diurnal variations of three-hourly rainfall for selected stations in Niger. Although this study was undertaken outside the humid tropics, it does present an interesting link between diurnal variability and the westward propagation of MCSs associated with the easterly perturbations. The described characteristics may have similar applications further south, for example, in the wet/dry humid tropics. At Niamey in Niger, the long-term climatology showed two peaks of three-hourly rainfall amounts. The primary peak occurred in the early morning (0300–0600h LST) with a secondary peak in the afternoon (1500–1800 LST). A comparison of composites for wet and dry years showed that these diurnal cycle characteristics persisted, but the amounts of rain respectively increased and decreased. Most pertinent, the most significant reduction of rain coincided with the peak diurnal period, 0300 h–0600 h LST, during dry years.

The early morning peak is interesting because previous studies, based primarily on satellite data, indicate that convective clouds (and rain) attain their maxima in spatial extent during the late afternoon and early evening over the west African land area (e.g. Duvel and Kandel, 1985). Solar heating has a strong influence on convective activity and the triggering of squall lines. On the other hand, the double diurnal maxima noted at Niamey are also reported (Griffiths, 1972; Ojo, 1977) for selected stations in northern Nigeria located in the humid (wet/dry) tropics of Chang and Lau (1993).

Shinoda *et al.* (1999) detected that stations close to the Air Mountains and Jos Plateau, where squall lines are triggered, showed the primary diurnal peak in the late afternoon (1500 h–1800h LST) or early evening (1800 h–2100 h LST). Further west, this primary peak was progressively delayed in association with the westward migration of MCSs and their subsequent gradual decay. For the more organised MCSs, peak diurnal activity could be delayed as late as 0900 h–1200 h LST on the next day during their westward propagation.

As with the findings in TOGA, Shinoda *et al.* (1999) highlight the need for a comprehensive understanding of the dynamics of MCSs, and their propagation, to interpret more precisely the causal factors for varying diurnal activity of rain over space.

MESOSCALE RAINFALL PRODUCING SYSTEMS ASSOCIATED WITH TRADE WINDS

One needs to recall that a considerable area of the humid tropics is occupied by the trade wind belt for a large part of the year. Thus, for large areas of the higher tropics a significant proportion of annual rainfall over oceanic islands and exposed coasts of continents is dominated by the interaction of the trade winds with coastal geography. Commonly, rainfall emanates from the passage of 'stream' showers and orographic uplift over islands and coastal hinterlands in the absence of well-organised, synoptic-scale perturbations. Thus attention needs to be given to these trade wind phenomena.

Topography is one of the main factors that attenuates the rainfall in this regime (Lyons, 1982). Tropical coastal locations that present significant relief to the trade winds can have annual precipitation totals in excess of 4 m. The general principles underlying the forcing mechanisms for trade wind precipitation at an exposed coastal location are well understood. As the moist onshore trade flow passes over the coastline, a combination of frictional drag, orographic and thermal effects create a zone of low-level convergence which enhances the development of convection within the flow (Crowe, 1971). However, the complexity of quantifying these effects at the mesoscale remains as a challenge to meteorologists. Rainfall regimes such as the trade winds, where the smaller scale convective processes dominate, present a far greater forecast problem than regimes where rainfall can be attributed to synoptic scale systems (Glahn, 1985; Olsen *et al.*, 1995).

Trade wind layer depth and moisture

A range of studies has noted the importance of the subsidence inversion height in regulating tropical convection. Chen and Feng (1995) found that the trade wind layer depth can explain a considerable amount of rainfall variance for windward locations in Hawaii where the maritime character of the prevailing trade flow ensures adequate moisture beneath the inversion. For environments where dry continental air streams can influence the trade wind layer, such as in Australia, Connor and Bonell (1998) could not establish a clear precipitation prediction signal from this factor alone. To explain the properties of the prevailing air mass more fully, a suite of factors such as the strength of the inversion, or the degree of potential instability below the inversion, must also be considered. For example, there is a general association of a low inversion height with low rainfall where a significant warm layer, above the trade wind layer, will inhibit deeper convection triggered from the surface. The inversion will usually decouple the moist layer from transient dynamic features in the upper troposphere that may add to upper divergence. However, heavy rainfalls can occur on days of shallow trade wind layer depth where a lifting mechanism is sufficient to break through the inversion and release large amounts of convective available potential energy (CAPE).

Topography, trade wind strength and atmospheric stability

Takehashi (1986) has related trade wind rainfall on a windward location with the wind strength in the middle of a trade wind layer and Connor and Bonell (1998) note that onshore wind components are dominant in influencing precipitation. There is a distinct topographical influence as high rainfall regions generally correspond to locations that present an orographic obstacle to the moist trade flow (Giambelluca *et al.*, 1986). This mechanism is obviously enhanced with a strong flow incident to the orography, where the terrain-induced uplift acts as a persistent trigger for rainfall.

For the north-east coast of Australia, atmospheric stability parameters generally display a poor correlation with trade wind rainfall when considered individually. Stronger links with precipitation were displayed for dynamic parameters for different mean pressure levels of the atmosphere such as onshore wind components, wind magnitude and bearing, stream confluence components and backing angle with height (Connor, 1999; Connor and Bonell, 1998). However, the stability of the lower atmosphere is an important consideration when studying the air flow characteristics and the dynamic blocking effect that a coastline presents to the trade flow. In the Hawaiian context, on occasions when the trade inversion is dominant, Leopold (1949), Smolarkiewicz *et al.* (1988), and Rasmussen *et al.* (1989) note that vertical suppression of trade flow by the inversion forces most of the flow around the orography. However, a portion of the flow is 'dynamically blocked' by the island and forced downward along the windward slopes forming a shallow density current that opposes the prevailing trade winds. Carbone *et al.* (1995) suggest an alternative view and describe the flow reversal as influenced primarily by evaporative cooling from orographic precipitation.

Diurnal modulation of rainfall in the trade winds over coasts and islands

Earlier discussions on rainfall variability over oceans and over West Africa indicate that the diurnal variation of rainfall presents a dominant precipitation signal. This signal is also clearly apparent for rainfall attributed to the trade winds in Hawaii and north-eastern Australia (Janowiak *et al.*, 1994; Connor and Woodcock, 2000) in tropical areas where baroclinic disturbances are less influential and the diurnal variation of rainfall presents a dominating precipitation signal. In trade wind zones, the diurnal behaviour of precipitation can be broadly summarised according to the air mass characteristics at a specific location. In regions where a continental air mass prevails, most precipitation occurs in the afternoon or evening. For oceanic regions, most of the rainfall occurs at night or in the early morning. Coastal regions can fall into either category, depending on the prevailing weather regime or season.

Dependence on the synoptic setting

Trade winds are subject to local fluctuations in the low-level wind and moisture fields. The amount of rainfall in an area is also affected by stability changes and the transition of dynamic features. Austin *et al.* (1996) found that trade flow patterns and cloud fields are greatly modified by synoptic scale influences. The synoptic setting also has an effect on the complex interactions between the trade flow, orography, and the diurnal modulating mechanisms described previously. Recent studies (e.g. Kodama and Barnes, 1997; Ramage and Schroeder, 1999; Connor and Bonell, 1998) provide evidence that it is unwise to use a single forecast parameter to explain rainfall variability in the trade regime.

Perturbations in trade wind flow at the mesoscale

Perturbations in trade wind flow can be grouped into those disturbances which have developed well offshore over the ocean and those which originate from the interaction between the prevailing trade wind flow and local land and sea breezes. The latter low-level, air flow reversals are associated with the diurnal cycle of heating and cooling over islands and coastal areas of larger land masses.

Less is known of the detailed dynamics of perturbations over the ocean because there are no data; however, the *Hawaiian Rainband Project (HaRP)*, July-August 1990 provided an opportunity to evaluate the organisation of disturbances at the 30–100 km scale using aircraft and radar measurements. Two case studies, just east of the island of Hawaii, were described by Raymond and Lewis (1995) when the trade-wind inversion was higher than normal, up to 650 hPa.

In the first case study, a vortex with a radar 'eye' of about 5 km diameter was detected in a large, amorphous mass of cloud (Figure 11.41). This disturbance was located in the lower trade wind flow of 3.5–4 km depth; and above this layer was a westerly wind shear and a region of lower humidity. The strongest low level winds and precipitation rates were on the northern side of the vortex.

In the second example, of 27 July 1990, although embedded in relatively strong NE trades, the wind on the west side of north-south line of convergence was westerly at the surface and southerly on the east side (Figure 11.42).

These two case studies present a contrast in terms of development. The reflectivity values of the 20 July storm suggested that the system was largely in a late phase of development with a stratiform structure, whereas the case study of July 27 was still in

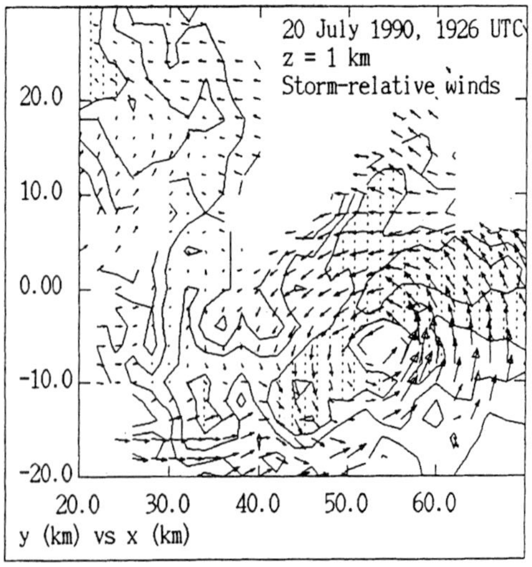

Figure 11.41 Storm-relative horizontal winds and reflectivity of vortex on 20 July 1990 at 1926 h and $z = 1$ km. The contour interval on reflectivity is 5 dBZ with reflectivity exceeding 20 dBZ hatched. The scale for wind vectors is 2.5 m s^{-1} km^{-1}. Data have been thinned to 2 km resolution for clarity. (Source: Raymond and Lewis, 1995.)

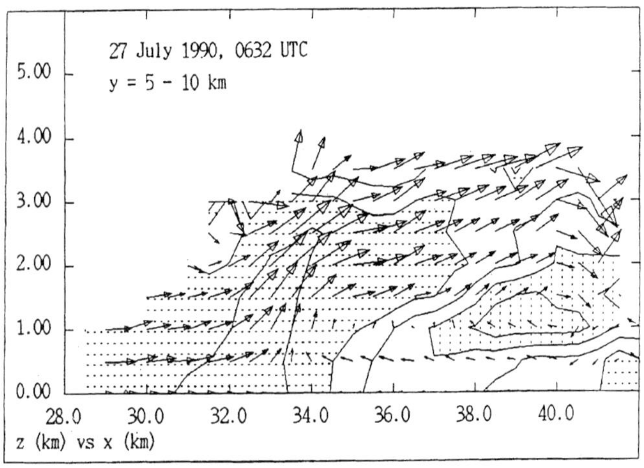

Figure 11.42 Storm-relative wind components in an east-west section averaged over $5 \leq y \leq 10$ km for the 27 July 1990 case. The southerly relative wind component is contoured at intervals of 1 m s^{-1}, with horizontal hatching showing winds less than 2 m s^{-1} and vertical hatching showing winds exceeding 4 m s^{-1}. The vectors show winds in the plane of the figure with a scale of 5 m s^{-1} km^{-1}. (Source: Raymond and Lewis, 1995.)

its early life as a vigorously growing convective system, with little stratiform component in the life cycle of convective-stratiform cloud development of the Houze (1989) model.

Acknowledging that the two case studies are not typical of ambient trade wind conditions (Raymond and Lewis, 1995) where cloud bases are normally higher and the trade wind inversion much lower (here the cloud bases were 500–600 m lower than normal and the marine layer deeper than the usual 850 hPa), the work nonetheless highlights the complexity at the mesoscale of trade-wind rain producing systems which can occur when the trades are less stable.

ASPECTS OF TROPICAL RAINFALL: INTENSITY–FREQUENCY–DURATION, AND MORE ON TOPOGRAPHIC INTERACTIONS

Previous work by Bonell and Balek, considered several aspects of rainfall: the spatial organisation of tropical rainfall; estimation of rainfall using remote sensing; the observed structure of MCSs; stochastic rainfall modeling; scaling linked with MCSs; and rainfall intensities. That review of Bonell with Balek (1993) still provides essential background information although updates on several of these topics have already been covered in preceding sections. The main focus from here on will be to provide more detail on various tropical rainfall characteritics and the spatial organisation of rainfall in general, linked with cyclonic and non-cyclonic areas of the humid tropics.

Apart from difference in soil orders (see discussion in Bonell, this volume), the magnitude of rainfall is one of the driving forces in differentiating between the response of hydrological processes across the humid tropics. It is also a principal cause for a much greater range of dominant runoff pathways observed in hillslope hydrology, both temporally and spatially, during tropical storm events compared with non-tropical areas. Equivalent hourly intensities of maximum 1- to 6-min rain amounts are commonly one or two orders of magnitude higher than those measured in the humid temperate latitudes.

Thus greater attention needs to be given to coupling runoff generation processes (hillslope hydrology) with synoptic climatology-rainfall characteristics. Hillslope hydrology studies can be placed into two groups: those undertaken in tropical cyclonic regions and against those in non-cyclone regions, to ascertain if there are any commonalities in dominant runoff pathways (Bonell, 1998). A reasonable expectation is that overland flow (principally of the saturation-excess type, see Bonell, this volume) would be more common in the cyclonic regions because of higher rainfalls, especially over durations >6 hours. As Bonell indicates, however, saturation overland flow can occur anywhere in the humid tropics if short-term rainfall intensities are relatively high, coupled with *less* permeable soils to compensate for lower, long duration rainfalls in non-tropical cyclonic regions. An important step towards our understanding of hydrological processes, especially runoff generation, is then the knowledge of short-term rainfall from point measurements over periods of 6 minutes (but ideally 1 minute) to 24 hours.

The synthesis of tropical rainfall characteristics by Jackson (1989) showed that there is a lack of rainfall frequency-intensity-duration (RFID) information for the humid tropics. Consequently, one cannot yet ascertain comprehensively what are the differences (as well as the similarities) in rainfall characteristics between the various rain producing systems linked with synoptic climatology. A starting point is to compare and contrast a selection of the limited RFID available for both tropical cyclone-prone and non-cyclonic regions.

Rainfall intensity–frequency–duration

During his comprehensive review of tropical rainfall characteristics, Jackson (1989 p. 110) suggested that 'for short time periods, say up to one hour, thunderstorms may produce the heaviest falls. Since the intensity of these systems is likely to be greatest where there is greatest supply of moist air, highest values are likely in the general area of the Equatorial trough' (the ZTE or the MCZ of the monsoon westerlies). Jackson then continued (p. 112) '. . . for longer time periods, high totals will be associated with major organised disturbances (e.g, tropical cyclones) – especially, a slow moving, declining system'. Since hurricanes, in particular, do not commonly occur within about 5° of the Equator (the previous chapter in this book, by Callaghan and Bonell, indicated exceptions), it follows that maximum totals are found in higher latitudes. Despite the paucity of RFID information, Table 11.6A provides some support to Jackson's (1989) latter remarks when concerning the longer time periods.

RFID data for varying periods of record, time intervals and return periods are presented for cyclonic (La Réunion, Australia) and non-cyclonic stations (east Africa, central Africa, Malaysia) in the humid tropics. The inclusion of two La Réunion stations provides an example of extremes in RFID on a global scale since La Réunion has had some of the world's greatest observed point rainfalls due to the frequent visitation of tropical cyclones in combination with orographic uplift. Cilaos is located near the centre of La Réunion where orographic uplift is particularly pronounced. Gillot represents a coastal station.

When comparing 24-hr rainfall for different return periods, the much higher totals identified with La Réunion (Cilaos, Gillot) and Babinda (North and South Creek), north-east Queensland, conform with Jackson's (1989) observations of much higher rain amounts for longer durations in cyclone-prone areas. On the other hand, Kota Bharu in north-east Peninsular Malaysia is an exception for a near-Equatorial station where 24-hr totals are within the range of those for Gillot and Babinda. Even the 3-hr rainfall amounts are not too dissimilar from Gillot for return periods of up to 20 years. Clearly, monsoonal influences, especially the Northern hemisphere NE monsoon, have a major impact on some of the east coast Peninsular Malaysian stations. In contrast, the east African stations (6 hr) Kinshasa (24 hr) and Kuala Kubu (Salengor) (24 hr) have much lower rainfall amounts. In this regard, Kuala Kubu (inland Selangor on the western side of the Malaysian Peninsula) provides an interesting contrast to Kota Bharu in also having much lower 24-hr rain totals.

On the other hand, for durations of 1 hour or less, the trend for higher magnitudes of rainfall in near-Equatorial stations (Jackson, 1989) is not conclusive in Table 11.6A. For example, the 30-min amounts at Rochambeau (French Guyana), the Malaysian stations and Kinshasa, are close to those presented for Babinda and Gillot. For the selected east African stations, the amount of rainfall is much less. Unfortunately, available data for less than 30 minutes is not consistent across RFIDs by temporal resolution in order to provide careful interpretation. Such information is not even available for the La Réunion stations. Most other stations provide rain amounts for 15 mins, with the exceptions being Babinda (6-min increments) and Kinshasha (10 and 20 mins). Nonetheless, if one considers a return period of once in two years and a duration of 15 mins (12 and 18 mins for Babinda, 10 and 20 mins for Kinshasha), Mbeya and Mombasa have marginally lower rain amounts compared with the remaining stations. Thus for these stations a separation between cyclonic vis-à-vis convective (non-cyclonic) is not apparent.

A clear distinction between cyclonic and non-cyclonic regions may be more evident if short-term rain amounts over much smaller temporal resolutions (say 1 to 6 mins) were available. In this context, it is significant that Rochambeau (non-cyclonic) has a higher of amount of rain over 6 minutes compared with Babinda.

High equivalent hourly intensities over short durations have been reported elsewhere, as in the Amazon basin (non-cyclonic) for specific rain events during intensive measurement campaigns (Hodnett *et al.*, 1997) and the ABLE 2B experiment. In both cases, maximim 1- to 5-min equivalent hourly intensities exceeded 100 mm h^{-1} for specific events. In fact, Hodnett *et al.* noted four events with average intensities exceeding 100 mm h^{-1} for 30 minutes.

Unpublished instantaneous rainfall intensity data (Koichiro, pers. comm.) for a storm event observed in Lambir Hills National Park, Sarawak, Malaysia, provides further evidence for the high short-term rainfall amounts that can be experienced. Peak equivalent hourly intensities (over less than 1 min increments) were in excess of 150 mm h^{-1}, and in one instance greater than 200 mm h^{-1}. These data highlight the high magnitude of short-term rain intensities which can occur in near Equatorial stations (non-cyclonic) of the maritime continent. Significantly, in the same event (Koichiro, pers. comm.) the average storm intensity (over a duration of 50 min) was an order of magnitude lower, at 45 mm h^{-1}. This feature highlights the need for rainfall data of fine

Table 11.6A. *Rainfall (mm) for various durations and return periods over different periods of record*

Tropical cyclone prone areas

North and South Greek experimental catchments (Wyvuri Holding) Babinda, NE Queensland (Australia): 1971–83 (after Bonell, 1991)

Return period (years)	Duration						
	6 min	12 min	18 min	30 min	60 min	3 h	24 h
2	*12.7*	*23.1*	*29.3*	*42.2*	*67.5*	*120.9*	*335.8*
7	*20.5*	*28.3*	*40.3*	*58.5*	*90.7*	*209.3*	*497.0*
14	*24.7*	*32.0*	*45.5*	*64.5*	*103.9*	*244.0*	*660.9*

Cilaos (La Réunion): 1962–1980 (source: MétéoFrance, 2000)

Return period (years)	Duration						
	30 min	45 min	60 min	90 min	3 h	6 h	24 h
2	*40*	*51.8*	*61*	*82.5*	*120*	*174*	*408*
5	*64*	*81.8*	*96*	*129*	*189*	*270*	*648*
10	*80*	*102*	*119*	*160.5*	*231*	*342*	*864*

Gillot (La Réunion): 28 years record (period unspecified) (source: MétéoFrance, 2000)

Return period (years)	Duration						
	30 min	45 min	60 min	90 min	3 h	6 h	24 h
2	*26.0*	*34.5*	*43.0*	*55.5*	*81*	*102*	*168*
5	*34.5*	*47.3*	*57.0*	*78.0*	*135*	*144*	*240*
10	*40.0*	*54.8*	*69.0*	*90.0*	*138*	*174*	*312*
20	*45.0*	*62.3*	*79.0*	*106.5*	*159*	*216*	*384*

Non-tropical cyclone affected areas

South America

Rochambeau (French Guyana): 1955–81 (after Service Météorologique de la Guyane, cited in Fritsch, 1992)

Return period (years)	Duration								
	6 min	15 min	30 min	60 min	2 h	3 h	6 h	12 h	24 h
2									
$mm\ h^{-1}$	*150*	*90*	*68*	*47*	*30*	*22*	*15*	*9.5*	*6*
mm	*15*	*23*	*34*	*47*	*60*	*66*	*90*	*114*	*144*
10									
mm^{-1}	*215*	*132*	*105*	*70*	*45*	*35*	*26*	*14.5*	*8.2*
mm	*22*	*33*	*53*	*70*	*90*	*105*	*156*	*174*	*196*

Africa

Kinshasa (Lower Congo, Central Africa): period of record not given (after Griffiths, 1972)

Return period (years)	Duration									
	10 min	20 min	30 min	40 min	50 min	60 min	70 min	80 min	90 min	24 h
2	*23.3*	*37.5*	*46.5*	*56.2*	*62.0*	*66.1*	*67.3*	*69.5*	*69.8*	
10	*30.6*	*49.3*	*61.2*	*74.4*	*82.4*	*87.9*	*89.5*	*92.5*	*92.7*	*117*
25										*132*
50										*143*

(*cont.*)

Table 11.6A. *(cont.)*

Mombasa (Kenya, East Africa): 1946–62 (after Taylor and Lawes, 1971)

Return period (years)	Duration 15 min	30 min	1 h	3 h	6 h
2	*19*	*29*	*39*	*56*	*73*
5	*25*	*40*	*50*	*77*	*99*
10	*28*	*46*	*57*	*91*	*116*
20	*31*	*53*	*64*	*104*	*133*
50	*35*	*61*	*73*	*121*	*155*

Dar-es-Salaam (Tanzania, East Africa): period record not given (after Taylor and Lawes, 1971)

Return period (years)	Duration 15 min	30 min	1 h	3 h	6 h
2	*23*	*35*	*46*	*56*	*66*
5	*28*	*40*	*54*	*70*	*89*
10	*32*	*44*	*59*	*80*	*104*
20	*35*	*47*	*64*	*89*	*119*
50	*39*	*52*	*72*	*100*	*138*

Mbeya (Tanzania, East Africa): 1932–58 (less 1949, 1950, 1955) (after Taylor and Lawes, 1971)

Return period (years)	Duration 15 min	30 min	1 h	3 h
2	*18*	*25*	*32*	*40*
5	*24*	*32*	*42*	*52*
10	*28*	*37*	*49*	*60*

Table 11.6B. *Rainfall characteristics for the stations Agumbe and Bhagamandala, Western Ghats, Karnataka State, India*

Station	Year	Annual rain (mm)	Number of rain events	Number of rainy hours	Average rain per day (mm day^{-1})	Maximum daily rain (mm)	Average hourly rain (mm h^{-1})	Maximum hourly rain (mm h^{-1})	Average 15 min rain depth (mm) and equivalent mm h^{-1}	Maximum 15 min rain depth (mm) and equivalent mm h^{-1}
Agumbe	1992	7558	33	392	6.8	256.0	5.43	50.0	2.37 (9.48)	25.0 (100.00)
Bhagamandala	1992	5485	61	402	4.0	304.8	3.91	60.0	1.74 (6.96)	28.0 (112.00)
	1986	4758	34	214	5.5	252.5	3.92	31.5	1.65 (6.60)	25.0 (100.00)

Source: Adapted from Putty *et al.* (2000).

Table 11.6C. *Percentage contribution of falls exceeding various 15-min intensities and percentage time for which they last*

		Contribution (x%) and duration of occurrence (y%) of falls exceeding									
		24 mm h^{-1}		32 mm h^{-1}		40 mm h^{-1}		60 mm h^{-1}		80 mm h^{-1}	
Station	Year	x	y	x	y	x	y	x	y	x	y
Agumbe	1992	41	11	28	6.0	19	3.8	6.2	1.3	2.0	1.00
Bhagamandala	1992	34	7	22	3.6	13	1.8	4.5	0.6	2.2	0.40
	1986	25	5	13	2.0	8	1.2	1.8	0.1	1.3	0.07

Source: Adapted from Putty *et al.* (2000).

temporal resolution to be of more practical value for comparative purposes in rainfall climatology as well as for the study of runoff generation processes.

In contrast, Elsenbeer *et al.* (1994) in western Amazonia (see Table 14.10, Bonell, this volume) report rainfall intensities up to two orders of magnitude lower than for cyclone-prone rain stations. This work, along with evidence from Sabah, Malaysia (Bidin, 2001), led Douglas and Guyot (this volume) to remark that the reports of high rainfall intensities in the near-Equatorial zone could be exaggerated.

Tables 11.6A and 11.6C presents the rainfall characteristics for two stations (Agumbe and Bhagamandala) in the Western Ghats of India for the year 1992 (and 1986 also for Bhagamandala) from Putty *et al.* (2000). According to these writers, the rainfall of 1992 represents a comparatively high intensity year and the annual rainfall corresponds to a recurrence interval of two years: the 24-hour and 1-hour maximum depths are of five-year frequency. Rainfall in 1986, in contrast, has an annual rainfall and rainfall intensities of a return period of less than two years (Putty *et al.*, 2000).

The Western Ghats have a range of relief in excess of 1000 m which border the west coast of the Indian peninsula. Putty *et al.* (2000) focused on the Western Ghats between 11 °N and 15 °N in their evaluation of the rainfall characteristics. Annual rainfall in this section exceeds 2000 mm along the coastal ridges of which more than 85% occurs during the south-west monsoon between June and September. Significantly, the high rainfall arises principally from orographic forcing of the SW stream, with little cyclonic activity.

Agumbe (Table 11.6B) records the highest known rainfall in south India (>7500 mm) whereas Bhagamandala, along the western Ghats escarpment, has an annual rainfall of about 5600 mm which is towards the mid-range of rain stations in the area (Putty *et al.*, 2000). Average daily rainfall is comparatively high, with maximum daily falls in excess of 250 mm (comparable with tropical cyclone-prone regions). In contrast, the average and maximum equivalent hourly intensities over 15 minutes are comparatively weak. Moreover, the duration and contribution of 15 min rain depths to annual rainfall in excess of 60 mm h^{-1} are also comparatively small (Table 11.6C). Thus, despite the marked concentration of rain in a few months and the high daily rain totals (similar to tropical cyclonic areas), short-term intensities are comparatively weak when compared with tropical cyclone prone regions. This is an environment of persistent rain events of long duration but paradoxically with low short-term intensities, most of which occur within a few concentrated spells of four to ten days over the SW monsoon season. For example, more than 50% of the annual rainfall occurs in only about 15% of the rainy days, but the duration of rainfall is in excess of 20 hours per day during concentrated spells (Putty *et al.*, 2000).

It can be thus concluded that the Western Ghats rainfall characteristics of the SW monsoon are quite distinct from other non-tropical cyclone affected regions. On the one hand, daily rain totals are comparable with tropical-cyclone prone regions but on the other, the much weaker short-term rain intensities paradoxically are not. The latter is in part due to the lack of additional convective forcing in the absence of, for example, persistent tropical vortices. The northern monsoon shearline is much further north of the Karnataka portion of the Western Ghats during the peak of the summer monsoon (see Callaghan and Bonell, this volume, e.g. Figure 10.9).

Extreme rainfalls linked with topographic interactions

The almost complete absence of rain data from high mountainous areas (Berndtsson and Niemczynowicz, 1988) is a major gap both for climatological and hydrological applications. It is from such areas that floods originate and data scarcity makes accurate flood forecasting more difficult with, at best, rainfall-radar information available only for limited areas.

A COMPARISON BETWEEN NORTHEAST QUEENSLAND AND LA RÉUNION

Bonell with Balek (1993) highlighted some extreme rainfalls from the telecommunications station at Mt. Bellenden Ker Top (1561 m.a.s.l) south of Cairns (and immediately north-west of Babinda) in north-east Queensland. The station has a mean annual rainfall in excess of 8000 mm (8065 mm, 1974–89 inclusive). Also, we highlighted earlier how a combination of orographic uplift and a near-stationary tropical low (ex-tropical cyclone *Peter*), north-east of Cairns, provided a record 48-hr rainfall total for Australia of 1947 mm in January 1979. In the same month there was the highest recorded weekly fall (3847 mm), and highest monthly fall (5387 mm) (Hall, 1984). Short-term data over 1 minute periods were initiated from 1984 by Bonell (although there are breaks in the records and a comprehensive analysis is still required). Bonell (1991) provided information during the landfall of tropical cyclone 'Ivor' to the north of Cooktown. Daily totals of 482.5 mm and 182.0 mm were recorded on 19 and 20 March 1990, respectively. During that period, consecutive one-minute rain amounts showed sustained equivalent hourly intensities between 30–60 mm h^{-1}, with peaks up to 150 mm h^{-1}. It is the persistence of such rainfall totals and short-term equivalent intensities which result in extensive saturation – excess overland flow (and even infiltration – excess) occurring on the slopes of these forested mountains (Bonell, 2004). During the January 1979 event, 'a sheet of overland flow up to 15cm depth' was observed flowing out of the forest at one point at the base of Mt Bellenden Ker (Australian Telecom, 1979, pers. comm. to Bonell). Such observations dispel the commonly held myth in the humid tropics that forests

Table 11.7. *Baril 1600 rain gauge tip rates during the heaviest 6-day, 2-day and 6 hour rainfall periods February–March 1993*

Tip number per minute	Rainfall over 6 days 0710 LST27 Feb 1993 0710 LST5 March 1993 (8640 min, 9570 tips)		Rainfall over 2 days 2033 LST27 Feb 1993 2033 LST 1 March 1993 (2880 min, 6000 tips)		Rainfall over 6 h 1835 LST28 Feb 1993 0035 LST 1 March 1993 (360 min, 1336 tips)	
0	3840	44.4	393	13.6	3	0.8
1	2465	28.5	960	33.3	54	15.0
2	1104	12.8	594	20.6	53	14.7
3	573	6.6	381	13.2	57	15.8
4	340	3.9	265	9.2	75	20.8
5	170	2.0	149	5.2	50	13.9
6	89	1.0	82	2.8	35	9.7
7	41	0.5	39	1.4	22	6.1
8	15	0.2	14	0.5	8	2.2
9	3	0.0	3	0.1	3	0.8
Total	8640 min	100%	2880 min	100%	360 min	100%

Source: After Barcelo *et al.* (1997).

'act as sponges' (see Bonell, Bruijnzeel *et al.*, this volume, for discussion).

La Réunion (and one could also suggest parts of Mauritius and Madagascar as well) is noted for more persistent high rainfalls than northeast Queensland because these islands are located within the persistent tracks of tropical cyclones in the south–west Indian Ocean; the existence of sharp topographic gradients and the maritime location are also influential factors. The summer monsoon meteorology between the south-west Indian Ocean and north-east Queensland is otherwise very similar (as highlighted previously in Figure 11.9). Table 11.6A for Cilaos, La Réunion, shows large magnitudes of rain in absolute terms across all the time intervals for various return periods.

Barcelo *et al.* (1997) provide a recent example of persistent high rainfall along a rain station transect (the Baril transect), south eastern La Réunion (Figure 11.1 of Barcelo *et al.* (1997)). Significantly, all four gauges were of the tipping bucket type (capacity 0.5 mm) to enable short-term rainfall amounts (and equivalent hourly intensities) to be measured. In addition, Météo-France complete the rain gauge network with several other rain gauges with a 6-min measuring frequency and 0.2 mm sensitivity.

As we described earlier (Figure 11.9), during the period 27 February–5 March 1993, La Réunion was affected by a weak (in terms of cyclonic winds) tropical vortex (called 'Hutelle') embedded on a very active southern monsoon shearline extending east from Madagascar. As Barcelo *et al.* (1997, p. 3343) observed '. . . rainfall totals measured at an altitude of 1600 m on the south-east slope of Piton de la Fournaise, were globally unprecedented on 2-day (3000 mm), 3-day (3587 mm), 4-day (3982 mm), 5-day (4608 mm), 6-day (4785 mm) and 7-day (4834 mm) timescales'. Whilst these longer term totals are impressive, Barcelo *et al.* (1997, p. 3343) indicated very short-term rainfall was in contrast of only 'moderate' intensity over the six days of observation. The maximum intensity recorded was 4.5 mm min^{-1} (270 mm h^{-1}) but by and large it was '. . . the persistence of such moderate rains that generated the exceptional rain amounts'. These remarks have much in common with the characteristics of rainfalls experienced in north-east Queensland.

For the heaviest rainfall over six hours, two days and six days, the respective modal peaks in rain gauge bucket tips per minute for Baril 1600 were 2 mm min^{-1} (20.8% frequency), 0.5 mm min^{-1} (33.3% frequency) and 0 mm min^{-1} (44.4% frequency). The latter shows that there was a large proportion of the six days with no rain occurring. The second highest frequency (28.5%) had an intensity of 0.5 mm min $^{-1}$ (see Table 11.7). Elsewhere, the Météo-France operated gauges reported new local records, and exceeded 1000 mm over the six-day period, even at stations located near sea level. The highest rainfalls (>3000 mm) occurred within the maximum rainfall zone, between 1300 and 2000 m a.s.l. Above 2000 m, rainfalls in excess of 2000 mm occurred in a topographic zone which normally is relatively dry, being persistently above the trade wind inversion (Table 11.8).

Barcelo *et al.* (1997) remarked that whilst the above period produced impressive rainfall records, the frequency of flood-producing rains in La Réunion remains high. These writers cited previous record-breaking events, in January 1980 (tropical cyclone *Hyacinthe*), and other meteorological systems not connected with tropical cyclones which produced rainfall totals of 2121 mm (26 February–3 March 1994) and 2845 mm (1–5 May 1995) within the Flanc Est (east slope) area, all within the

Table 11.8. *Daily rainfall (mm and extracted from 0700 to 0700 LST) from 27 February 1993 to 4 March 1993 for ten stations on the Piton de la Fournaise massif*

Station	27 Feb 1993	28 Feb 1993	1 March 1993	2 March 1993	3 March 1993	4 March 1993	Total
Baril 1600	1254.5	1582.0	750.0	237.0	738.0	175.0	4581.5
Baril 1200	905.0	1061.5	484.5	145.0	472.0	109.5	3177.5
Baril 900	848.5	1082.5	404.5	109.5	422.5	78.5	2946.0
Baril 650	708.0	881.0	289.0	93.0	357.0	70.5	2398.5
Baril 130	388.4	460.8	123.2	48.8	205.8	87.2	1314.2
Basse-Vallée 1200	974.0	1122.0	478.0	105.0	400.5	75.0	3154.5
His Ste-Rose (860 m)	650.0	600.0	514.0	131.0	510.0	564.5	2969.5
Bellecombe (2250 m)	661.5	506.0	427.0	102.0	389.0	115.0	2300.5
Le Tremblet (120 m)	422.5	179.0	92.0	17.5	137.0	376.0	1224.0
Soufrière (2490 m)	688.0	>526.0	>199.0	>77.0	>407.0	169.0	>2066.5

Source: After Barcelo *et al.* (1997).

maximum rainfall zone (1300–2000 m). Moreover, the Baril 1600 gauge, over 672 operating days (8 February 1993–14 May 1995), exceeded 136 mm 10% of the time, 228 mm 5% of the time and 642 mm 1% of the time. On three occasions, 24-hr totals exceeded 1000 mm. Over the complete year, 27 February 1993–26 February 1994, the Baril 1600 rain gauge received over 18 000 mm of rain, 2600 mm higher than the previous La Réunion record (Barcelo *et al.*, 1997).

Most pertinent, these rainfall amounts were a record because of the recent establishment of a measurement network in an area remote from human settlement, and where rainfall had not been measured previously. The easterly aspect of the upper east slopes of la Fournaise volcano favour pronounced orographic uplift of an east-southeast flux. In addition, this location is commonly in a zone of convergence (heavy rain area), sandwiched between a tropical vortex to the north and an anticyclone further south; on very much the same lines as was described for the tropical cyclone Peter case study in connection with Bellanden Ker. The MSL high is once again slow moving and the trough to the south east is breaking the ridge at 500 hPa so that middle level steering is weak; thus causing the tropical low to be slow moving.

HURRICANE MITCH (OCTOBER 1998)

The heavy rains from Hurricane Mitch over Central America between 28 October and 1 November 1998 captured global attention because of the extensive flooding and mudslides that resulted; and an estimated death toll of 9086 lives with another 9191 people missing (Ferraro *et al.*, 1999). It is appropriate to place this event within the global context of extreme events.

Previously we described the genesis and causes of the slow, tortuous track followed by this storm (for a time a severe category 5) before it moved slowly over Honduras and Nicaragua after making landfall about 110 km east of La Ceiba. This system gained more notoriety with the exceptional rains that it produced by regional standards, due to the hurricane's slow movement, orography and mesoscale interactions. The extensive east-west mountain range, with peaks approaching 3050 metres, covers this part of central America and thus contributed to the high rainfall totals by orographic uplift.

On an hourly temporal basis, there were two areas of concentrated rainfall (as outlined in Figure 11.11b). The first area was centred on the west coast of Nicaragua, the scene of the most severe flooding; and the second area along the north coast of Honduras where Mitch made landfall. Interestingly, whilst rainfall along the northern Honduras coast was the more intense, it did not persist as long as that over Nicaragua.

The pertinent question arises: were these rainfalls exceptional for the Caribbean region and by global standards? Hellin *et al.* (1999), based on an autographic rain gauge in southern Honduras, commented that the measured totals over a range of time intervals, were not that exceptional for hurricanes and tropical storms in the Caribbean basin. A surviving rain gauge (13.28 °N, 87.07 °W) was located on a hill crest (100 m a.s.l.) in the foothills of Cerro Guanacause (1007 m a.s.l.). Nearby flooding and landslides were extensive, coinciding with two periods of maximum short-term intensities. Over more than a four-day period (1800 hr LST 27 October–2100 hr 31 October 1998), 896 mm of rain was measured. The most intense and prolonged period occurred over less than two days (1500 hr LST 29 October–0700 hr 31 October 1998) when 698 mm rain fell. Within this period, two maximum rainfalls occurred over six hours in coincidently the same time of day, i.e. 186 mm and 245 mm, 1600–2200 hr on 29 and 30 October respectively. At that time, maximum intensities ranged from 138 mm h^{-1} (2 minute period) to 58.4 mm h^{-1} (60 minute period). These figures are comparable with those commonly measured in La Réunion and north-east Queensland. It was during these high

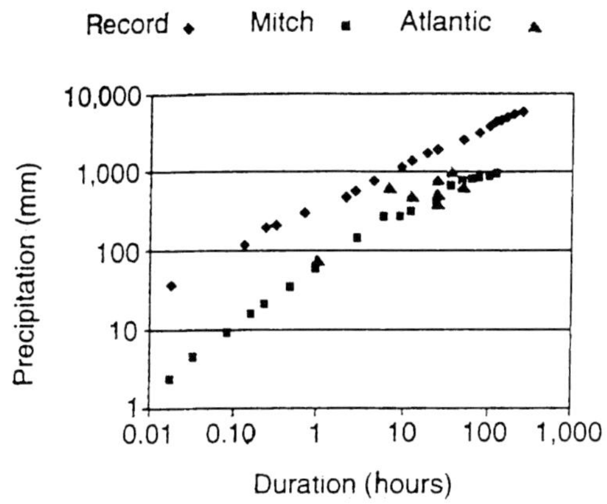

Figure 11.43 Plot of maximum rainfall amounts (squares) against duration from a rain gauge in southern Honduras during hurricane/tropical storm Mitch ($y = 51.64x^{0.674}$, $R^2 = 0.99$). Also plotted are updated rainfall events (diamonds) for different durations that define the curve of maximum potential rainfall ($y = 353.07x^{0.519}$, $R^2 = 0.99$) and data (triangles) from recent major Atlantic hurricanes and tropical storms. (Source: Hellin *et al.*, 1999.)

rainfalls and equivalent hourly intensities that approximately 20% of landscape was affected by landslides and major flooding in the Choluteca river occurred.

When plotting the maximum rainfall amounts for Mitch, (Figure 11.43 square symbols), as against updated record rainfall events (diamond symbols) and recent data (triangles) for major Atlantic and Caribbean cyclonic systems, it is evident that Mitch is well below record rainfall events, and comparable to between 1 and 100 hours for tropical storms in the Atlantic basin. It is pertinent that the maximum potential rainfall curve is strongly influenced by La Réunion measurements where topographic forcing is more pronounced than was experienced in Nicaragua and Honduras (Hellin *et al.*, 1999). As Hellin *et al.* (p. 316) remarked 'the data suggest that extensive damage in Honduras and Nicaragua was accentuated by several factors: the storm struck at the end of the rainy season when the soil was saturated, resulting in catastrophic flooding and landslides; agricultural extension caused by land pressures had left many hillsides denuded; and the population was ill-prepared, . . .' because the landfall of the storm had been expected further north.

It is not unreasonable to suggest that the runoff generation and erosion processes reported elsewhere in high rainfall areas (see Bonell, this volume) where saturation excess overland flow is frequently extensive (in both undisturbed/disturbed areas) also occurred during Hurricane Mitch. Infiltration excess overland flow is also likely to have been extensive, especially during the periods of maximum rain intensities, over human-impacted, agricultural landscapes. There is great value then in transposing the research findings from more cyclone-prone areas to comparably less affected areas of the humid tropics so that forest-land-water management practices are developed to accommodate the infrequent, but devastating, events like Mitch.

THE SPATIAL ORGANISATION OF TROPICAL RAINFALL

Overview of methodologies

Using various rainfall parameters developed from a spatial network of rain stations, commonly based on a temporal resolution of 24 hours, a logical step is to establish regions of similar precipitation characteristics. Commonly known as regionalisation methods (e.g. Sumner, 1988), the standard approach is to develop *precipitation affinity areas* (or regions of coherent precipitation). Elementary linkage analysis (based on spatial correlation between individual rain stations) is rather limited however, because it only considers the highest correlation coefficients in the two-dimensional plane. On the other hand, antecedent spatial correlation provides a useful insight into the spatial organisation of MCSs and individual rain cells.

With the availability of more powerful computers and more user-friendly software, the use of eigentechniques (principal components analysis (PCA), common factor analysis (CFA), empirical orthogonal functions (EOF)) has become more prevalent. Being multidimensional, these methods have the advantage of determining more than one characteristic that influences the spatial organisation of rainfall. Nonetheless, despite undertaking a rigorous exploratory analysis, there will always remain a degree of subjectivity in the interpretation of different combinations of methods.

An extension of the use of eigentechniques is to import the loadings into a clustering strategy (e.g. Williams, 1976; Everitt, 1980) for classification of rain stations into groups. Lyons and Bonell (1994), for example, used the PC loadings derived from the *Harris-Kaiser II* BTB rotation as inputs to the Ward clustering strategy for regionalisation of the total wet season record. Once again, exploratory analysis is required before a decision can be made on the selection of the most suitable clustering strategy.

Before reviewing selected results from the use of the above methods, it is pertinent to emphasise that even the most comprehensive syntheses of global data sets by Jackson (1986, 1988) and, most recently, Jackson and Weinand (1994, 1995), lack a synoptic climatology-rain producing systems dimension in the interpretation of their results. Just outside the wet/dry region of the humid tropics (Chang and Lau, 1993), the French EPSAT-NIGER (Estimation des Précipitations SATellite – expérience NIGER) Project in the Sahel of west Africa is the most comprehensive study in terms of data bases (93 gauges over a study area of 16 000 km^2 at different spatial and temporal scales) supported by IR METEOSAT satellite data and a C-band weather radar system

Table 11.9. *Summary of the characteristics of each of the 13 synoptic rain-producing types that occurred over the Townsville region during the 1988–1989 wet season*

Number	Type	Characteristics
1	Type A	Shallow, moist east to south-easterly winds associated with a strong ridge of high pressure along the Queensland coast.
2	Type B	A ridge of high pressure off the Queensland coast which is directing more moist, low-level north to north-easterly winds.
	Type C	Dominated by a mid-level (500 hPa) trough in the temperate westerlies which is progressing eastwards across the region. Three subtypes are identified on the basis of surface wind vectors:
3	Subtype C1	Upper trough accompanied by east to south-easterly winds similar to Type A.
4	Subtype C2	Upper trough accompanied by north to north-easterly winds similar to Type B.
5	Subtype C3	Upper trough associated with a surface low pressure system located to the south-east of Townsville.
6	Type D	Deep south westerlies (up to 200 hPa level) and a marked mid-level trough over the western Coral Sea. This circulation can be characterised by high dew-point levels from the surface up to 500 hPa and is often a precursor to the onset of the summer monsoon phase.
	Type E	The monsoon trough. Three subtypes identified on the basis of the monsoon trough position and in one case, complex interaction with a disturbance in deep easterlies:
7	Subtype E1	Deep, moist south to south-easterlies associated with a vortex embedded within monsoon trough located to the north of Townsville.
8	Subtype E2	Deep, moist north to north-westerlies associated with an active monsoon trough to the south of the Townsville region.
9	Subtype E3	A perturbation in deep easterlies which subsequently interacted with the monsoon north-westerlies to develop a weak vortex and reform the monsoon trough to the north of the study area (Cooktown area).
	Type F	Tropical cyclone. Three subtypes are identified on the basis of position of tropical cyclone:
10	Subtype F1	Landfall of a tropical cyclone, south of Townsville, which approached the coast on a south-westerly track.
11	Subtype F2	Low-level convergence between moist north-easterlies associated with a high pressure ridge off the coast and a decaying tropical cyclone, south-west of Townsville.
12	Subtype F3	Deep cyclonic circulation from a very severe system in the central, east Coral Sea.
13	Type G	The penetration into the tropics of a surface trough embedded in the temperate westerlies.

Source: After Lyons and Bonell (1992).

(Thauvin and Lebel, 1991; Lebel *et al.*, 1997; 1998). EPSAT-NIGER also adopted a dynamic, process-oriented perspective. We will briefly highlight a few aspects of this work in the context of Sahelian MCCs and MCSs. Lebel *et al.* (1998, p. 1713) remarked that the main features of MCCs rainfall, identified within the Sahelian region, were considered to be also almost identical with MCCs which occur further south to the 10 °N parallel in the Soudanian region (wet/dry region of the humid tropics of Chang and Lau, 1993).

Selected studies

Rainfall data sets in the Queensland work were subdivided by first establishing the basic synoptic climatological circulation types based on near-surface (Sumner and Bonell, 1986), and the later inclusion of upper tropospheric (Lyons and Bonell, 1992), rain-producing phenomena. This required extensive consultation of daily streamline charts. For Queensland, Sumner and Bonell (1986) adopted eight near-surface principal circulation types plus two additional variants of the easterly flow category. The well-developed Walker circulation (Lockwood, 1984) and positive ENSO phase associated with the 1988–89 wet season (Bureau of Meteorology, 1989) resulted in a wide range of synoptic influences which affected the Townsville region (Lyons and Bonell, 1992). Thus, 13 distinct synoptic weather types were identified (Table 11.9), also summarised in Figure 11.5 of Lyons and Bonell, 1992 which were used to stratify the rainfall records prior to analyses.

Apart from the daily rainfall, simple rainfall indices were selected to indicate the magnitude of rainfall activity associated with each circulation type by Sumner and Bonell (1986, 1988) and Lyons and Bonell (1992). The first, (mean rainfall per wet day) provided an index of the amount of rainfall activity on days when rain occurs. The second, (wet/dry index) is a ratio between the number of wet days and the number of dry days and is effectively an indicator of the likelihood of rain, regardless of amount. Sumner and Bonell (1986) also added a third index to assess extreme values by using the highest total daily rainfall depths attained during a four-year period as an index of likely severity of individual daily events. Each of these indices were mapped for each circulation type, and provided an important step towards understanding the changing spatial patterns and identifying their likely causal factors. Lyons and Bonell (1992, 1994) also used radar imagery and the mapping of the spatial organisation of specific events to support their interpretations. All these basic approaches were an essential step in identifying the underlying causal factors responsible for each PC component and loading distributions, plus the groupings of rain stations formed from the clustering strategy of Ward (1963) during regionalisation.

Within north Queensland, Sumner and Bonell (1986) determined that orographic uplift of the easterlies along the coast could be as significant in rainfall production, if not more so, than the MCZ of the equatorial westerlies to the north of the southern monsoon shearline over central and northern Cape York Peninsula. This applies especially when there is a major vortex embedded in the monsoon trough. During the occurrence of well organised low pressure systems (tropical depressions, tropical cyclones) some of the highest magnitude rainfalls are produced. Nonetheless, from both types of circulation, the bulk of the rainfall, and its highest incidence, occur again down the east coast. Feeder bands advecting moisture from the Coral Sea, coupled with orography, occur for example when the disturbance is over the Gulf of Carpentaria (Type 6 – Sumner and Bonell, 1986). Otherwise, the overriding influence of local topography and exposure to moist air flows emerged as an important determinant. Furthermore, orography and exposure to the prevailing SE trade winds is particularly important in the absence of pronounced synoptic perturbations.

At the mesoscale, Lyons and Bonell (1992, 1994) reinforced emphasis on the influence of topography and exposure to moisture sources as the major control of rainfall distribution and amount. Specific case studies were presented to highlight the spatial patterns of the indices, mean rainfall per wet day and wet/dry ratios (see Figures 11.6a–d in Lyons and Bonell, 1992). For the prevailing low-level easterly flow, the highest mean daily values are found on the exposed eastern facing slopes of principal orographic features On the other hand, there are other means whereby topography controls precipitation.

The upper level steering of thunderstorm cells are commonly deflected either side of a principal orographic feature, Mt Stuart in this case (Figure 11.44). In addition, in situations of deep instability, orographic enhancement of precipitation can then travel a significant distance downstream before it decays. Figure 11.45 shows the orographic enhancement as a deep, moist northerly flow passes over Great Palm Island producing rain intensities in excess of 100 mm h^{-1}. The rain band then continues downstream to impact the Townsville region. Lyons and Bonell (1994, Figure 11.9) present another example for SE flow.

A final role of topography as a generator of precipitation is the interesting interaction of low level and upper level winds. The outer monsoon regions are characterised by a horizontal wind shear between opposing low level and upper level winds. In this case between low level south-east to north easterlies, and upper south-west to north-westerlies. (Sadler, 1975a; Sadler *et al.*, 1987). Thus, topography initially causes uplift of the low level easterlies on the windward (exposed) slopes. The resultant uplift (cloud) then interacts with the steering influence of the upper winds causing precipitation to apparently move in the opposite direction to the low level flow. Thus, the area affected extends 'upstream' to the low level flow in response to the reversal in vertical flow. Figure 11.46 shows the precipitation pattern controlled by a marked shear at 850 hPa between surface NE and upper SW winds.

Thus, by taking a synoptic climatology perspective, these Queensland studies have highlighted the changing spatial organization of rain in response to changes in the exposure and orographic uplift; and the interaction with upper level winds in terms of the steering of rain cells and MCSs.

Elsewhere, Lebel *et al.* (1998) take a more dynamic approach when concerning the westward propagation of MCSs in the Sahelian region. They present a disaggregation model to reproduce the dynamics of Sahelian MCSs both in space and time. The spatial disaggregation of event rainfall is undertaken using a turning band algorithm (see Lebel *et al.*, 1998, p. 1715). The simulation of the dynamics and displacement of the MCS westwards is achieved by imposing a typical hyetogram which reproduces the existence of a convective front and stratiform trail based on observations in the EPSAT network. Figure 11.47 shows the simulation in time steps of 20 minutes of rainfields over 5 mins. In terms of size, shape and width of the convective front, there is good agreement. Lebel *et al.* (1998) also tested their simulations (12, 4 and 2 km) using rainfields observed at different scales, e.g. 12 × 24 km, 1 gauge per 20 km^2. The best simulations with the observed rain records were at the same scale of measurement (12 km). More complex, spatial structures become evident however as the scale of simulation is progressively reduced to 2 km. *If the latter are transposed to an assessment of dominant runoff pathways (hillslope hydrology) for example, the resultant spatial and temporal response could potentially be equally complex* (Figure 11.48).

A classification of global tropical rainfall stations

The preceding discussion has focused on case studies for evaluating the spatial and temporal organisation as a step towards regionalisation. The lack of a high density of rain gauging stations within the humid tropics (Jackson, 1989), which can also supply detailed RFID information, has been a major impediment

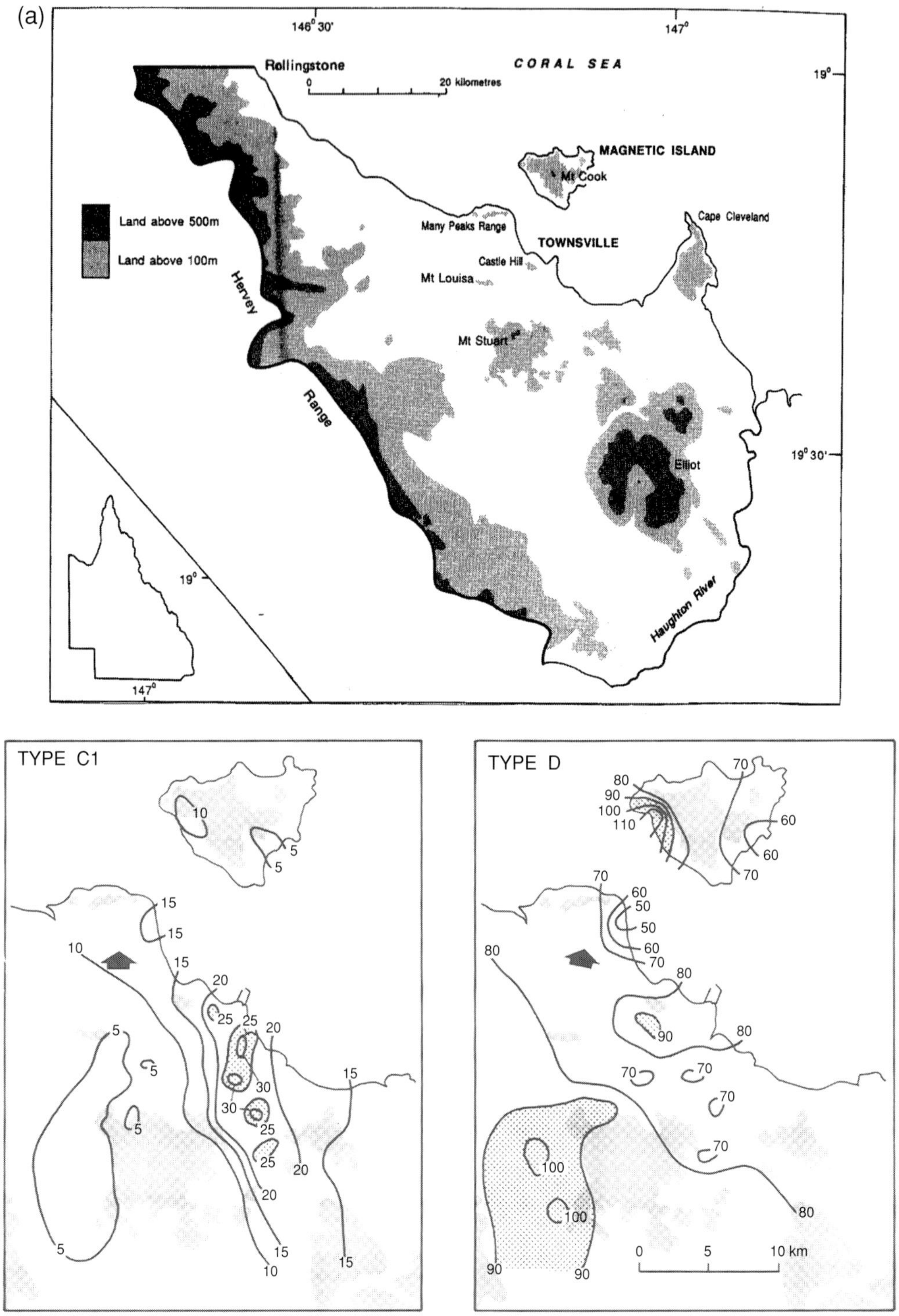

Figure 11.44 (a) Topography of the Townsville area. (b) Daily rainfall (24-h totals) over the Townsville area associated with weather types C1 (26 October 1988) and D (29 December 1988) (see Table 11.9). The trajectory of the cells is on either side of a principal topographic feature (Mt Stuart). Note alignment of heaviest rainfall parallel to suggested steering wind (arrows indicate suggested steering wind direction). (Source: Lyons and Bonell, 1992.)

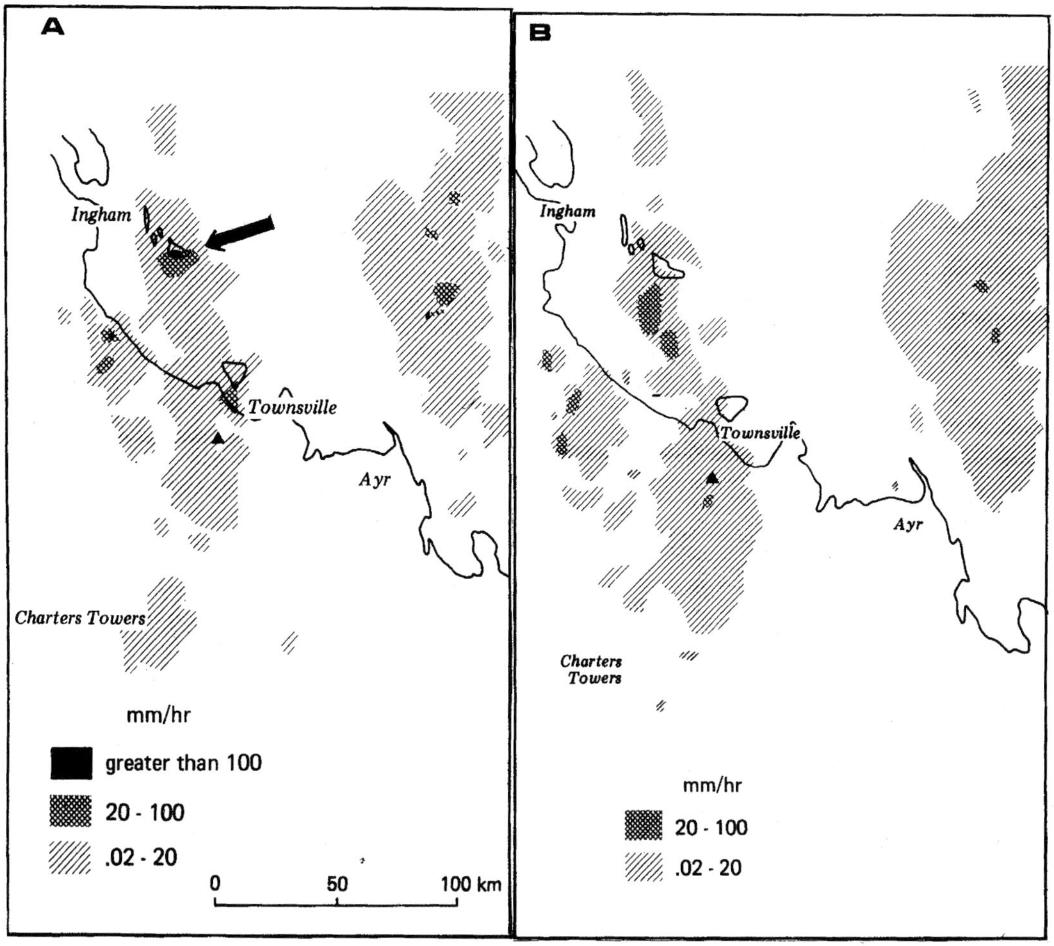

Figure 11.45 Radar reflectivity for southward moving rain-band associated with ex-tropical *Aivu* on 5 April 1989. (a) Orographic enhancement of rainfall over Great Palm Island (arrowed) at 0010 UTC. (b) Down-streaming of precipitation from Palm Island at 0030 UTC, with some indication of precipitation occurring downwind of Mount Stuart and on the coast to the NW of Mt Stuart. (Source: Lyons and Bonell, 1992.)

towards developing a global classification of tropical rainfall stations. The work of Jackson and co-workers (e.g. Jackson, 1986, 1988; Jackson and Weinand, 1994, 1995), however, provides an important step towards this objective and the pertinent findings need to be highlighted.

A variety of different regression models (i.e. arithmetic, power, semi-logarithmic, logarithmic) were used between the respective dependent variables (average number of rain days, mean daily rainfall intensities) and monthly rainfall. The global coverage included 28 stations, of which 24 occupied the three sub-regions of the humid tropics of Chang and Lau (1993). Of particular interest here is the mean daily intensity (MDI) parameter of Jackson (1986). MDI was derived from dividing monthly rainfall averages by the average number of rain days.

Residuals from the best fit regression relationships were then examined by individual stations, the rainfall regions of Jackson (1986) and also by geographic region. Significantly, the two Australian stations (Darwin, Daly Waters, both in the Northern Territory) showed consistently positive deviations (higher mean daily falls) in residuals for all months – similar to the wet/dry tropics of Central Africa, e.g. Keno (Nigeria). In contrast, South America tended to be the opposite with a dominance of negative deviations. In the case of rain days, both central Africa and Australia had consistently only negative deviations (fewer rain days); whereas South America was dominated by positive deviations. This led Jackson (1988) to analyse 51 stations in northern Australia (some of which were outside the humid tropics region of Chang and Lau, 1993) focusing on the summer wet season, December-March. He determined that the characteristics for Darwin and Daly Waters applied to the majority of these northern Australian stations. Rainfall is much more concentrated, with fewer rain days and higher mean daily intensities than would

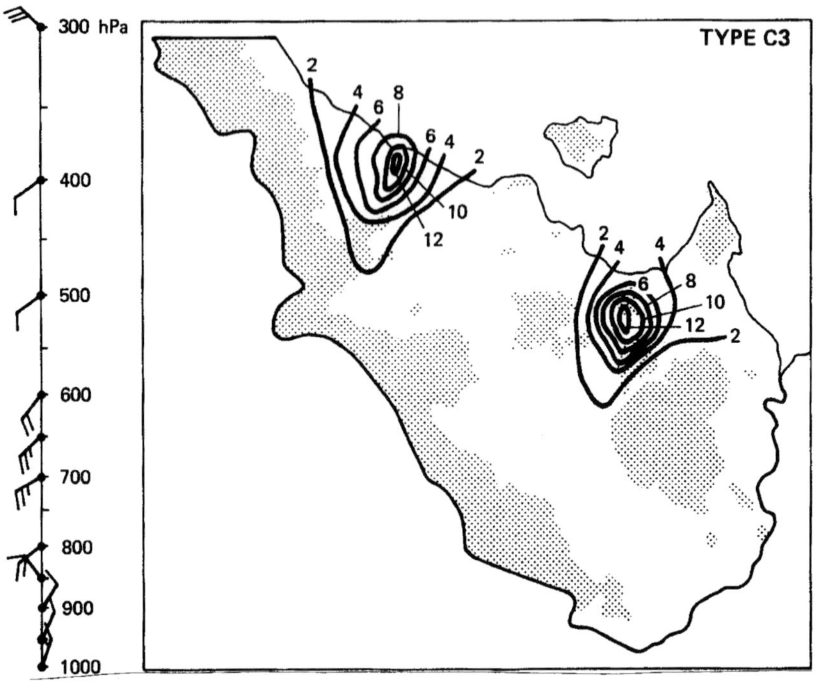

Figure 11.46 Daily isohyetal map for weather types C3 (see Table 11.11) at 2300 UTC 19 December 1988 and Townsville winds at 0500 UTC 19 December 1988. (Source: Lyons and Bonell, 1992.)

be predicted from monthly totals and the worldwide regressions. Jackson (1988) attributed the persistence of southern monsoon shearline lows and tropical cyclones as a principal explanation for the regression deviations. He also noted that San Salvador, Kingston and Rangoon tended to show a pattern of deviations similar to that of Australia; and there was a common tropical cyclonic influence (Jackson, 1986). In contrast, the lack of tropical cyclonic influences in South America could account for the tendency of lower MDI and higher rain days values than monthly totals predicted there (Jackson, 1986). However, the fact that the area showing statistically the closest agreement with Australian deviation patterns was central Africa (e.g. Keno, Nigeria; and Bulawayo, Zimbabwe (just outside the humid tropics), albeit represented by only these two stations, cautions ***against*** the simple division into cyclonic and non-cyclonic regions. Jackson (1988) highlighted other factors such as orography (topographic influences) to explain some of these positive MDI deviations. In the case of Keno, we have already established the persistency of well-organised easterly perturbations, and associated MCSs, which characterise the summer season in west Africa. The triggering influences of high relief, such as the Jos Plateau, which act in combination with the westward propagation of the easterly trough lines have also been outlined. Such characteristics support Jackson's (1988) above cautionary remarks.

More recent work (Jackson and Weinand, 1994, 1995), used 34 rainfall variables in the following groups: (a) general seasonal or annual characteristics (5 variables); (b) shorter period characteristics – wet season months (months >50 mm) (25 variables) and shorter period characteristics – dry season months (months <50 mm) (4 variables). MDI is one of the 25 variables in the latter group. As with all tropical rainfall studies, the availability of data controlled the statistical sample sizes. In total, 87 rain stations were selected for varying periods and lengths of record. Not all of the stations are in the humid tropics, but the sample includes a broad range of continental and maritime humid tropics stations; cyclonic and non cyclonic; and the three sub-regions of the humid tropics of Chang and Lau (1993).

Fifteen groups were developed based on the characteristics identified (Table 11.1, p. 279) in Jackson and Weinand (1994) who adopted a combined principal components and clustering strategy. Those groups relevant to the humid tropics, and shown in Figures 11.49(a)–(d), are Groups 1, 2, 4, 6, 7, 8, 10, 11 and 15. Group 14 refers to the four most northerly stations in Nigeria, in locations bordering the wet/dry region and the Sahel. This area is dominated by the described characteristics of MCSs in this region. Pertinent features of the other groups are:

Group 1 All stations have coastal locations and characterised by heavy rainfall. They include both cyclonic and non-cyclonic affected areas.

Group 2 Mostly inland stations with low average annual rainfall at locations on the edge of the humid tropics region.

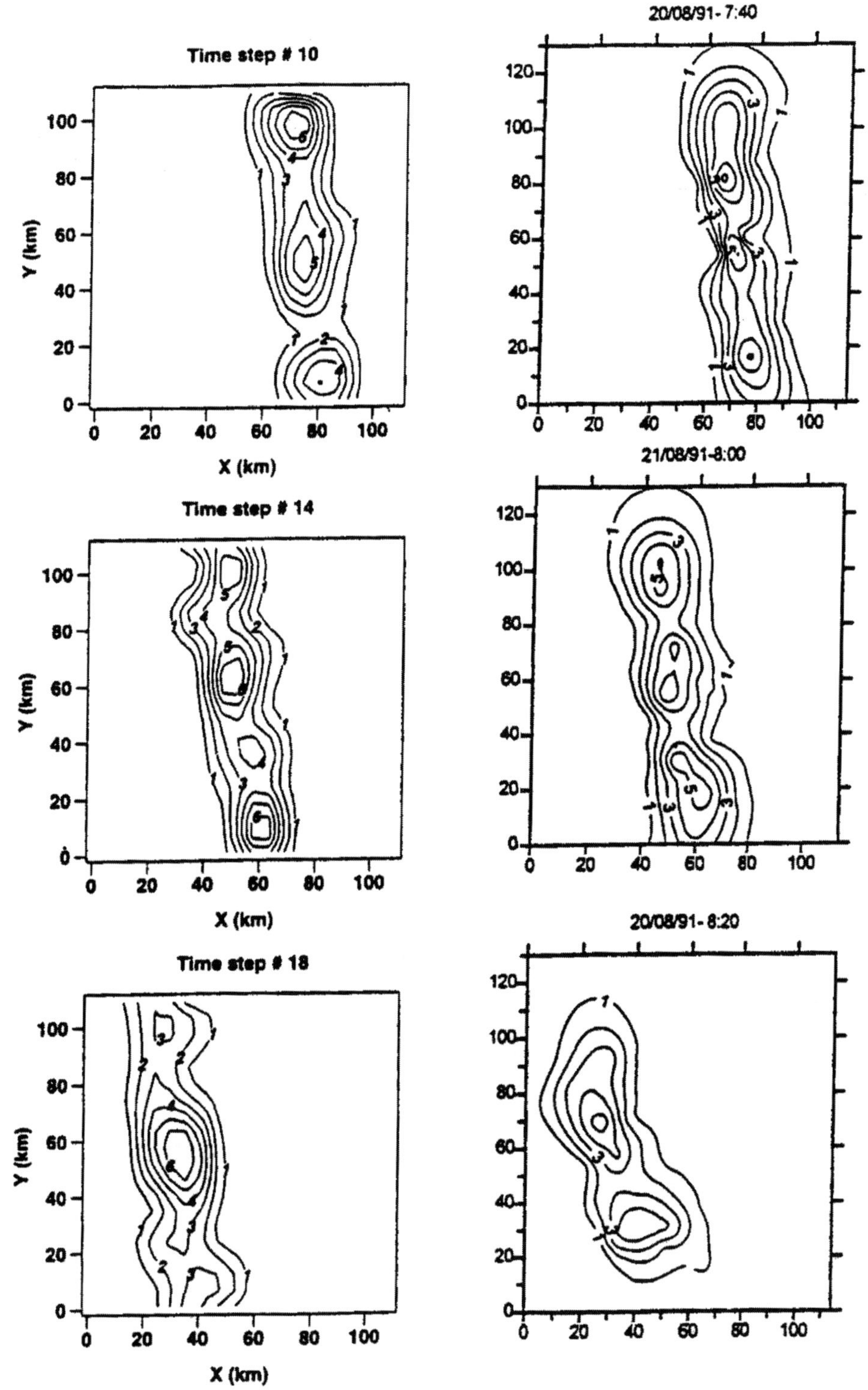

Figure 11.47 Comparison of simulated (left) and observed (right) 5 min rain fields at 20 min intervals for the 20 August 1991 squall line. (Source: Lebel *et al.*, 1998.)

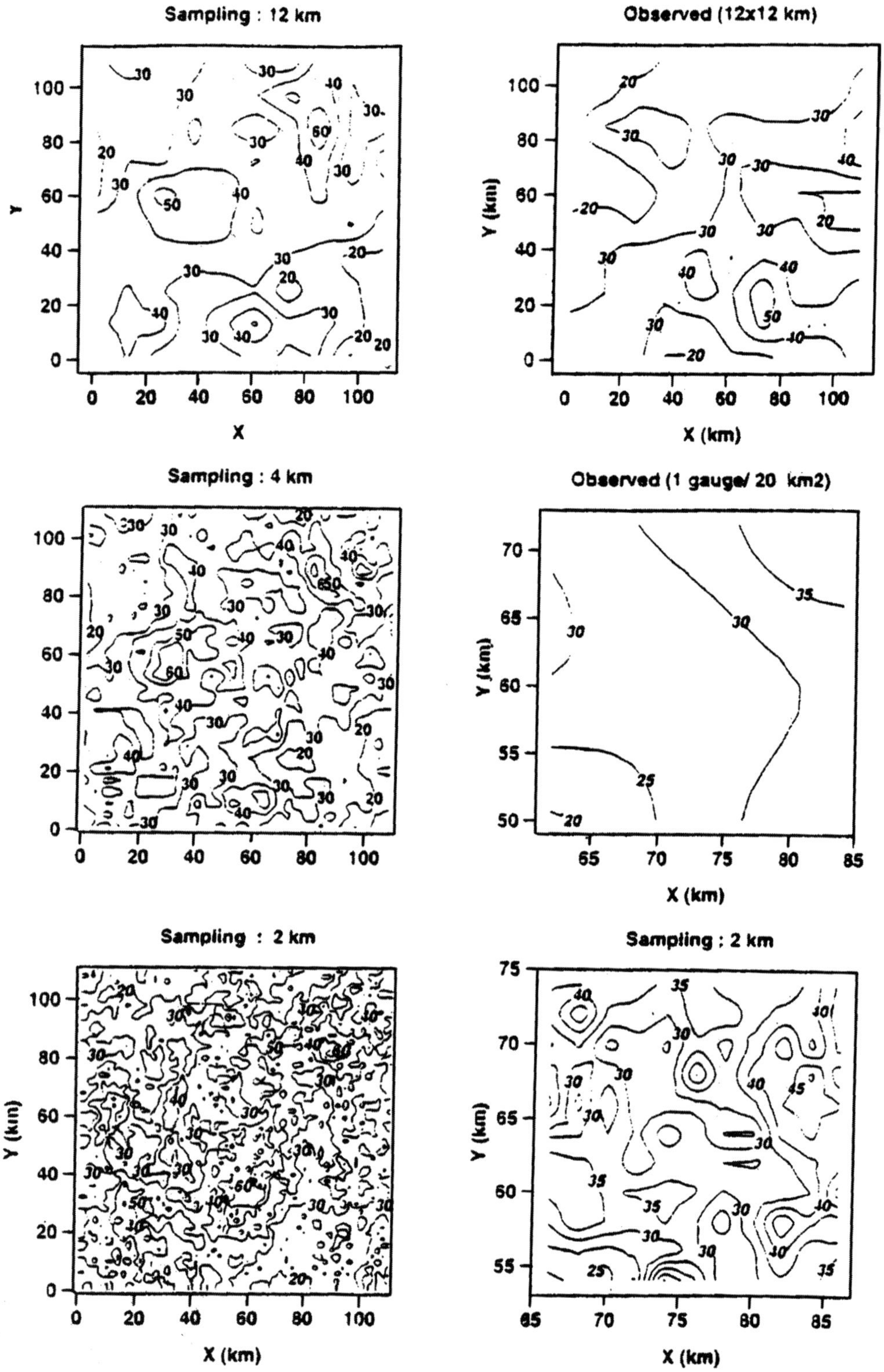

Figure 11.48 Simulated storm rain fields for the 20 August 1991, squall line, at increasing sample frequency from top to bottom (left). The corresponding observed rain field over the whole domain (top) and observed (middle) and simulated (bottom) zooms over the target area (right). (Source: Lebel *et al.*, 1998.)

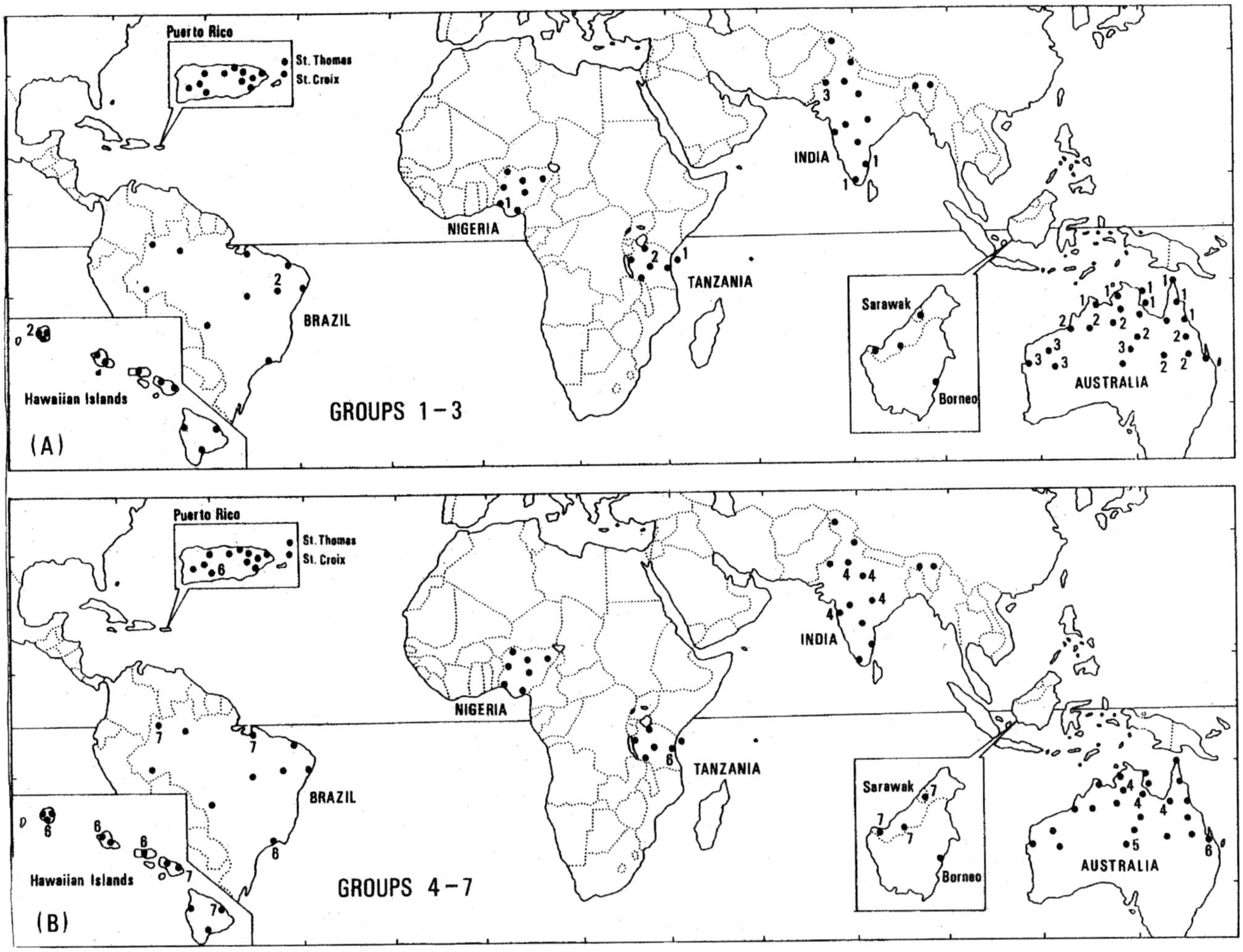

Figure 11.49 A classification of global tropical rainfall stations: spatial distribution of 15 groups following classification using principal components analysis and a clustering strategy. (Source: Jackson and Weinand, 1994.)

There is an absence of long wet spells of low rainfall, and interchanging wet/dry spells. Large rainfalls by event are relatively few.

Group 4 The Australian stations and Mumbai (India) are typical of the monsoon regions. They are characterised by heavy rainfall, and also intervening long dry spells.

Group 6 Only the stations in Hawaii and the south coast of Puerto Rico are inside the humid tropics. There is an absence of heavy falls but with long dry spells and dry days in the wet season. But moderate to large rainfalls can occur in the dry season.

Group 7 There is an interesting inclusion of low-latitude, non cyclonic stations in Brazil, Sarawak and two east coast (exposed to NE trades) Hawaiian stations. This group has large daily and monthly falls, with an alternating pattern of short wet/dry spells. However, long wet spells of light rain are also a characteristic.

Group 8 Most of the Puerto Rico stations plus single stations in Hawaii and Borneo occupy this group. Mostly short wet and dry spells alternate, and there is an absence of large dry season falls.

Group 10 Three Brazilian, two eastern Nigerian and two stations in north-east India are included; and are characterised by heavy rainfalls and large monthly totals (Table I, Jackson and Weinand (1994)). Longer wet spells of light rain can also occur. It is interesting that both non-monsoonal (Brazil) and monsoonal regions are included in this group.

Group 11 The Caribbean island stations of St. Thomas and St. Croix, one station in south-west Brazil, and one on Oahu, Hawaii, and two stations in north-west Tanzania. In all cases,

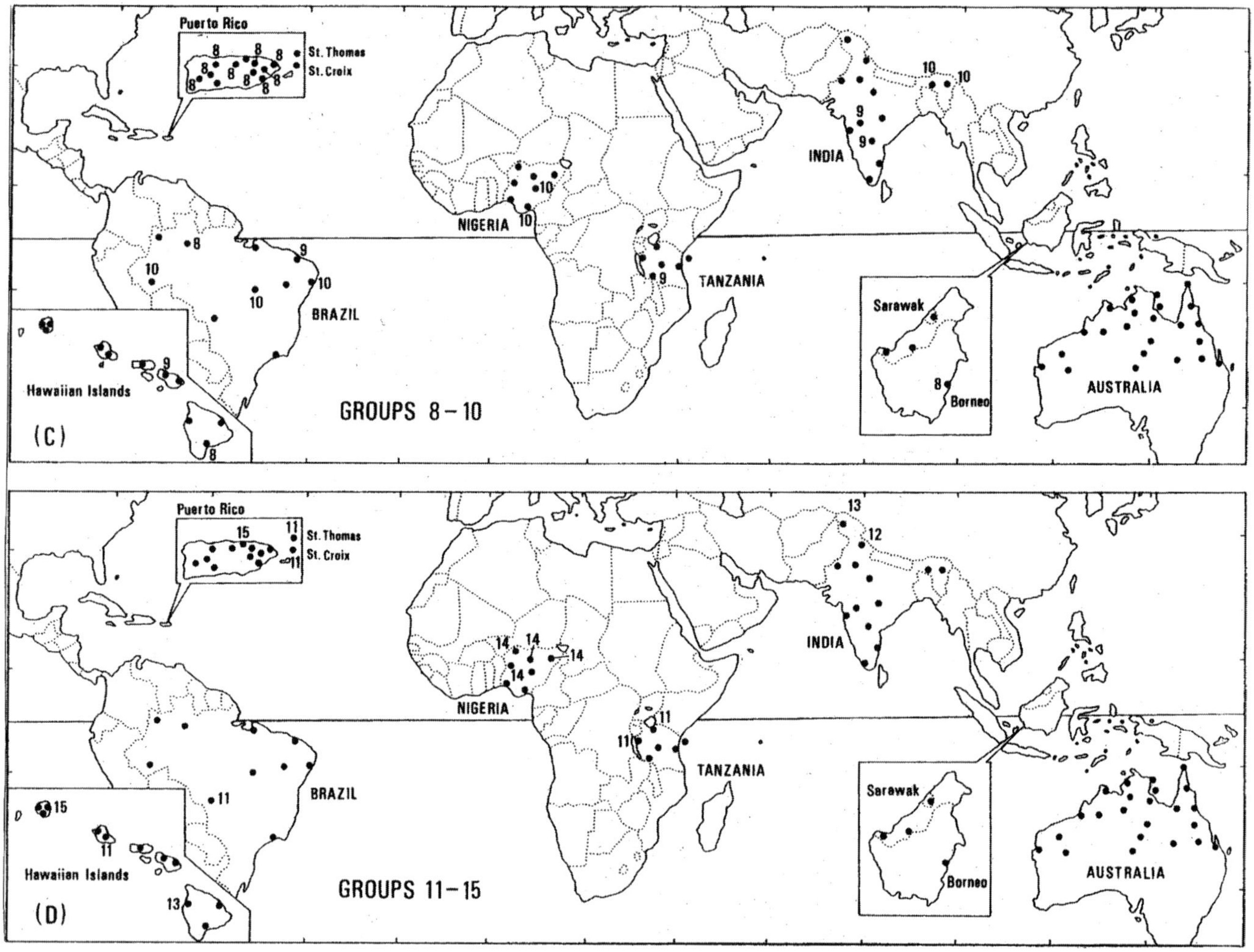

Figure 11.49 (*cont.*)

there is a relative absence of heavy rainfall and alternating short wet and dry spells.

Group 15 This includes one station on the north coast of Puerto Rico and another on the east coast of the Kanai island, Hawaii. Mostly affected by longer spells of light rain and in absence of heavy falls in both wet and dry seasons.

Although Jackson and Weinand (1994) stated that a climatological interpretation of the groups was not an immediate objective, it is evident from these groupings that degree of coherence does not always include stations within a geographic entity, e.g. the Amazon Basin. In addition, the islands of Hawaii and to a lesser extent, Puerto Rico, show representation in a few groups which highlights the climatological gradients linked with varying exposure to the prevailing winds. Moreover, there is no clear distinction between cyclonic and non-cyclonic regions (e.g. Group 1, Group 8). Thus, concerning rainfall characteristics, the identification of corresponding regions of spatial hydrological coherence in terms of processes (e.g. runoff generation) is also likely to have the same degree of complexity.

POSSIBLE IMPACTS OF LAND USE CONVERSION ON THE SPATIAL AND TEMPORAL ORGANISATION OF PRECIPITATION

Precipitation recycling

The possible impacts of land use conversion on climate are highlighted elsewhere in this volume (Roberts *et al.*; Mahé *et al.*; Costa, Grip *et al.*). We will now elaborate on this issue, linked with potential changes in the spatial and temporal organisation of precipitation at the mesoscale and regional scale.

Recycling refers to how much evaporation in an area contributes to the same precipitation in the same area and is expressed as a precipitation recycling ratio, ρ. Saleti *et al.*, 1979, based on the isotopic gradient of precipitation from different stations in the Amazon basin, claimed that up to 50% of forest evaporation could be re-cycled precipitation. The Amazon basin is physiographically a semi-enclosed entity, except for the east which allows the entry of prevailing moist, easterly surface winds from the Atlantic Ocean (Molion, 1993). However, the use of isotopes per se are indicators only and do not provide quantitative estimates. Moreover, the atmosphere of the Amazon basin is not a closed system and there is a significant 'leakage', especially to the south-southwest along the foothills of the Andes (see discussion on the SACZ, Callaghan and Bonell, this volume and cf. Marengo *et al.*, 2001).

Subsequent modelling estimates of recycled precipitation by Eltahir and Bras (1994) and Trenberth (1999) (using the model of Brubaker *et al.*, 1994) are lower than those of Salati *et al.* (1979). Eltahir and Bras (1994), using a rectangular 2.5° latitude by 2.5° longitude grid and a ECMWF data set, provide estimates of recycled precipitation between 25 to 35% in the Amazon basin, with minima of $<10\%$ near the Amazon river estuary and a small area $>50\%$ in the south-west at the foothills of the Andes. In the latter case, this coincides with the longest trajectory of the easterly flux. Trenberth (1999) also noted $\rho > 20\%$ in the southern part of the Amazon basin, based on a modelling resolution of 500 km. For the complete Amazon basin of 2750 km in length, the annual average recycling is about 34% (Trenberth, 1999). This result is not incompatible with the earlier estimate of Eltahir and Bras (1994), despite some differences in the modelling approach (see discussion in Eltahir and Bras, 1994, pp. 863–866; Trenberth, 1999, pp. 1372–1374).

Elsewhere, modelling by Gong and Eltahir (1996) for the west African region predicted 27% of the precipitation to be derived from local precipitation. Thus implications of land use change such as forest conversion, and corresponding changes in the 'local' supply of water vapour for precipitation, has triggered numerous investigations. These are not exclusive to the humid tropics. For example, model simulations by Zheng and Eltahir (1998) argue that the meridional conditions of the land surface, as characterised by vegetation cover and soil moisture, play a significant role in the dynamics of the west African monsoon and rainfall variability. The resulting reduction in surface net radiation and total heat flux from the surface is considered to be capable of reducing dramatically the monsoon circulation, with land use conversions along the humid, southern coast being the most sensitive region.

On the other hand, Trenberth (1999) noted that the fractions of moisture flowing through a region which is precipitated out exceed 40% over the areas of the tropical African monsoon; the 'maritime continent' of SE Asia; the southern part of the Amazon basin; and the Western Ghats of India. These high fractions coincide with the strong low-level maritime transports (i.e. synoptic-scale moisture advection) associated with the seasonal movement of the precipitation fields.

GCM simulations of the Amazon basin

From the foregoing, the key is the impact of land use change on *changes* in the partitioning of net radiation into sensible and latent turbulent heat flux and ground heat conduction. For example, when more short wave radiation is diverted into sensible heat flux (e.g. dry soils, sparse vegetation cover) then the convective boundary layer, the CBL (otherwise known as the planetary boundary layer, PBL) in the lower atmosphere deepens in response to greater turbulent kinetic energy. On the other hand, transpiration from vegetation presents a feedback which encourages cloud development and possible rain recycling. Modification of the vegetation albedo (i.e. shortwave reflection coefficient), whereby a larger fraction of short-wave radiation is reflected back into space, also affects the surface energy balance. For example, in the case of the Amazon basin, there is a general consensus from GCM studies that there will be a reduction in rainfall, soil moisture availability and evaporation for the simulation of *imposed wholescale conversion* from forest to pasture. The pasture has a much higher albedo and is shallower rooted so, in the latter case, is less able to extract deeper, subsurface soil moisture. Moreover, there is a reduction in 'roughness' and surface area of vegetation which reduces wet canopy evaporation. Conversely, complete removal of the Amazonian forest increases surface temperatures. Thus one of the latest GCM simulations by Lean and Rowntree (1997) indicated that complete removal of the Amazonian forest produced area-mean changes that are in general agreement with trends predicted in earlier GCM studies, *viz*, decreases in evaporation of 0.76 mm day^{-1}(18%) and rainfall of 0.27 mm day^{-1} (4%) and a rise in surface temperature of 2.3 °C. On the other hand, the reduction in precipitation is much less than earlier predictions (see Lean and Rowntree, 1997; Lean *et al.*, 1996; Bonell, 1998) because the rise in temperature causes a feedback of increased moisture convergence from the Atlantic Ocean, for example. Elsewhere in this volume Costa gives more detail on the large-scale hydrological impacts of forest conversion with particular reference to the Amazon basin.

Effect of surface heterogeneity of land cover

Spatial variations in land surface cover (and properties) cause, in turn, horizontal variations in the surface energy budget. Such variability can generate mesoscale atmospheric circulations that have a bearing on rainfall from the land to the regional scale. It is generally accepted that at spatial (length) scales of 5 km or smaller, surface inhomogeneities have less impact on local climate due to the 'homogeneisation' of turbulent fluxes. When the spatial scale is about 10 km or greater, such surface inhomogeneities of differential energy balances have greater potential to impact on the PBL and can develop their own mesoscale circulations (Pielke *et al.*, 1988).

There is a general consensus emerging that mesoscale land surface variability can influence the amount of precipitation and its spatial organisation (see review of Pielke *et al.*, 1998). Factors involved in increasing rainfall include changes in energy budget, frictional effects, changes in horizontal convergence and associated vertical velocities. Other causes include more rapid evaporation from intercepted rainfall over vegetation (especially forests), rather than over soil, which causes a change in the mixing ratio of the boundary layer. Thus, any local wind circulations that concentrate water vapour from say transpiration or wet canopy evaporation, favour the formation of deep cumulus clouds in a deeper PBL.

There have been no detailed studies within the humid tropics of the impacts of surface heterogeneity of land cover on rainfall although such work is part of the current LBA experiment (Concise Experimental Plan, LBA, 1996). Elsewhere, Pielke *et al.* (1999) inferred potential changes in the organisation and intensity of precipitation over southern Florida between 1900 and 1973, and also 1993 during the months July and August. They noted that significant land use change had occurred over this period, with extensive everglades cover being replaced by urban and agricultural land. Simulated temperature and precipitation based on these land use changes suggested that average maximum temperature had increased by about 0.4 °C for the 1973 landscape, and by 0.7 °C for the 1993 landscape, compared to the model run for a 1900 landscape. Conversely, there was a 9% decrease in rainfall averaged over south Florida with the 1973 landscape; and 11% decrease with the 1993 landscape as compared to the 1900 landscape.

The key factor is that much of the summer rainfall depends on 'local' evaporation from the everglades. The extensive land use conversion has thus accentuated additional inherent, precipitation (climatic) variability, with the net result that as much as 11% less average rainfall has occurred compared to simulations where the landscape had been left undisturbed.

In the Sahel, Taylor and co-workers (Taylor *et al.*, 1997; Taylor and Lebel, 1998) observed a positive feedback between the land surface and rainfall in semi-arid conditions. Bearing in mind the acknowledged potential similarities in meteorology of MCSs with the wet/dry zone of the humid tropics of west Africa and the fairly similar landscapes and open vegetation cover, these findings may also have some application to the more humid region further south.

For the 1992 summer wet season, a rainfall gradient of 270 mm over a 9 km section had developed in an area not normally favoured by high rainfall. Taylor *et al.* (1997) cited a specific example where the antecedent specific humidity over a savanna site was 1g kg^{-1} more moist (and 0.2 °C cooler) than over an adjacent tiger bush site. The following passage of a gust front produced rainfall rates in excess of 60 mm h^{-1} over the savanna compared to ~20 mm h^{-1} over the tiger bush.

The above positive feedback mechanism has the following implications. *Towards the end of the wet season*, and at spatial scales of about 10 km and time scales of up to 40 days, rainfall variability *would seem more sensitive to antecedent rainfall patterns* than to land use type or state of land degradation. It is feasible that such characteristics could also occur over the outer edges of the humid tropics, especially in seasons of below-average rainfall and low relief (such as parts of northern Australia as well as west Africa). At larger scales, it remains unclear how and if this positive feedback mechanism affects precipitation.

The nature of the land surface cover and antecedent moisture also has implications for the maintenance of easterly perturbations during their progression westwards over the more sparse, moisture-limited vegetation typical of the Sahelian region, compared with the better watered, more dense vegetation further south. Modelling by Taylor *et al.* (2000) noted that the surface land cover and antecedent moisture has a major influence on the PBL depth and potential temperature, with the depth of the moist layer *being deeper and warmer* over sparse compared to dense vegetation within the pressure ridge (24 hours ahead of the passage of the easterly trough). The larger sensible heat and long wave heating for sparse vegetation causes the depth of the PBL to increase, thus enabling a near-surface parcel of air to more likely reach the lifting, condensation level. A consequence is the higher CAPE over sparse vegetation, thus triggering deep convection *sooner* over sparse vegetation compared with more dense vegetation. As a result, the subsequent passage of an easterly perturbation does not require necessarily the additional forcing of daytime heating to be convectively very active from the previous build-up of large conditional instabilities to support travelling easterly perturbations. The consequences are that the thermodynamic environment over the Sahelian vegetation suppresses daytime rainfall for several days due to sparse, moisture-limited vegetation. On the other hand, these environmental circumstances enhance the development of a deeper PBL and a progressive build-up of specific humidity. With the passage of an easterly perturbation, the antecedent deeper PBL and deep moist layer contribute relatively large amounts of rainfall to the total as squall lines.

Further south towards the humid tropics of west Africa, well-watered, dense vegetation provides a higher evaporative fraction, and enables more frequent, short-lived, daytime convection rain to be triggered. Thus, the relative contribution of squall lines to the total rainfall over a wet season tends to reduce over the more dense vegetation further south (Taylor *et al.*, 2000) which is supported by other observational evidence (Omotosho, 1985; Laing and Fritsch, 1993). Moreover, heavy rainfall ahead of an easterly trough only develops during daylight hours due to the requirement of additional forcing from diurnal (afternoon) heating (Taylor *et al.*, 2000). As indicated above, this is not a requirement for the Sahelian thermodynamic environment whereby modelling by Taylor *et al.* (2000) showed a rainfall maximum up to six hours earlier during the passage of an easterly trough, and with a sharper peak, in response to the large CAPE.

The implications of modelling work of Taylor *et al.* (2000) are that the nature of the surface land cover and antecedent moisture have a significant influence on the maintenance of easterly trough squall lines. The same factors affect synoptic meteorological influences such as the location of the African Easterly Jet, and the position of the monsoon trough and associated circulation.

CONCLUSIONS

Compared with the temperate latitudes of the Northern hemisphere, there is a minimal network of upper air monitoring stations in the humid tropics which thus poses a difficulty in forecasting and gaining a better understanding of storm dynamics. Moreover, a review of tropical cyclones reveals the frequent flood-producing rainfall which results from these well-organised vortices are most pronounced, especially when their forward movement is slow or near-stationary. The much publicised Hurricane Mitch belongs to this category of rain-producing systems but it was a rare event for Central America. As indicated elsewhere, north-east Australia and the islands of the southwest Indian Ocean (as well as south/southeast Asia) commonly experience extreme rainfalls from these slow moving synoptic systems.

We have also emphasised the importance of the advection mechanism of warm and very moist air within tropical cyclones which is vertically uplifted by backing winds, on progression from the surface to higher levels of the atmosphere. This mechanism contributes to extreme rainfall and is a persistent feature, especially in less intense tropical depressions. In summary, the highest, say 24-hr, rain amounts are mostly (but not completely as shown by the Western Ghats) associated with the tropical vortices of the monsoon regions.

There has been considerable research focus on tropical cyclones but the overview of easterly perturbations, at both the synoptic-scale and mesoscale (for example, in the trade wind flow), highlights their complexity in origins and structure. Most attention has been given to African easterly 'waves' but even here more field measurements are required to elaborate on modelling work. The westward propagation of these African easterly perturbations in terms of rain-intensity structure and interaction with different land surface and antecedent wetness has considerable ramifications for storm runoff hydrology. Further west, there still remains considerable debate on whether these African easterly perturbations have the ability to propagate over the north Atlantic and intensify over the Caribbean before moving into the tropical north-east Pacific. Examples cited from the Amazon basin, Guadeloupe and Niger highlight that total rain amounts from these easterly perturbations are much less than from tropical cyclones. Short-term rain intensities can however be high.

Elsewhere, during the description of TOGA, it is clear that the easterly perturbations in the north-west tropical Pacific are not comparable with those over west Africa and the north Atlantic. The role of the *eastward* propagating MJO in being one of the triggers of these *westward* moving perturbations is of particular interest. Even more, the description of tropical disturbances and cyclones, which originate from troughs in the surface easterlies of the south-west Pacific and which are totally disconnected from the monsoon shearline further north, are of particular relevance. Previous work by McBride and Keenan (1982) attributed most (~95%) vortices as originating from along the southern monsoon shearline. It is possible that these tropical disturbances within the surface easterlies recur mostly during La Niña phases of the Southern Oscillation, as previous studies in this region had noted an absence of such perturbations (e.g. McBride, 1983). The description of the HaRP (Hawaiian Rainband Project: Raymond and Lewis, 1995; Austin *et al.*, 1996) and the northeast Queensland study of Connor and Bonell (1998) also emphasises the need for a better understanding of the dynamics of mesoscale rain areas embedded within the trade winds. Under less stable conditions, the HaRP detected the existence of a mesoscale vortex with an eye detected by radar embedded within the easterly flow in one case study. Statistical analyses of trade wind rainfall also highlight the need to consider several dynamic parameters related to wind flow as well as atmosphere stability parameters.

Thus overall, the earlier remarks of Manton and Bonell (1993, p. 25) when they stated 'that our understanding of disturbances in the easterlies is far from complete, and that further research is urgently required . . .' still remain valid. Such disturbances are largely not responsible for the rainfall extremes of the tropics, compared with the 'flood-producing' rains of tropical cyclones. Nevertheless, easterly disturbances account for a considerable proportion of annual rainfall, especially within the trade wind belt, west Africa, and even parts of the Amazon basin. What is surprising is the absence of a detailed field and satellite study since GATE (GARP Atlantic Tropical Experiment, where GARP represents the Global Atmospheric Research Programme) of the 1970s to address the genesis and dynamics of easterly perturbations. What is urgently needed is an updated combined satellite imagery–synoptic analysis, with particular attention to the west, north Atlantic; the Caribbean sea and the east north Pacific. The West African monsoon experiment (AMMA, 2002) and the current five-year experiment (1999–2004) in the eastern Pacific, EPIC (Eastern Pacific Investigation of Climate Processes) (Cronin *et al.*, 2002) will rectify this deficiency in field studies.

Of particular relevance to process hydrology is the appreciation of the temporal and spatial variability of convective and stratiform components of MCSs. Moreover, the stratiform area of coverage within rainfields expands progressively at the expense of the decaying convective cells. Thus, the highest short-term rainfall intensities usually occupy only the smallest proportion of grid cells, except under the optimal development of well organised rain-producing systems, e.g. tropical cyclones.

A particularly challenging area is the potential impact of land use change such as forest conversion on the energy balance for scales upwards from 10 km^2 and the resultant potential reorganisation of MCSs.

Finally, the global regionalisation of rainfall on the lines of Jackson and Weinand (1994), needs to take a more synoptic climatology perspective, as was demonstrated by Lyons and Bonell in a northeast Queensland study (1992, 1994). The advent of a new generation of satellites for remote rain measurement, coupled with continued progress in a better understanding of synoptic and mesoscale tropical meteorology, may facilitate this objective over the next decade.

APPENDIX 11.1

Acronyms

a.s.l.	Above sea level
ABLE 2B	Amazon Boundary Layer Experiment (1 April–14 May 1987)
ACSL	Amazon Coastal Squall Lines
AE	Auto-Estimates rainfall detection using GOES IR satellite data
AEJ	African Easterly Jet (north Africa)
AMEX	Australian Monsoon Experiment (1986–1987)
BOS	Basin Occurring Systems (referring to the Amazon basin)
CAPE	Convective Available Potential Energy
CBL	Convective Boundary Layer
CFA	Common Factor Analysis
COS	Coastal Occurring Systems (referring to the Amazon basin)
DUNDEE	Down Under Doppler and Electricity Experiment (1989–1990 and 1990–1991 Australian summer monsoon seasons)
ECMWF	European Centre for Medium-Range Weather Forecasts
EMEX	Equatorial Monsoon Experiment (1986–1987)
ENSO	El Niño – Southern Oscillation
EOF	Empirical Orthogonal Functions
EPIC	Eastern Pacific Investigation of Climate processes in the coupled ocean atmospheric system (Cronin *et al.*, 2002). Further information is available from http://www.pmel.noaa.gov/tao/epic/ and http://www.atmos.washington.edu/gcg/EPIC/
EPSAT–NIGER	Estimation des Précipitations SATellite-expérience NIGER
ERA	ECMWF Reanalysis Dataset
GATE	Global Atmospheric Research Programme, Atlantic Tropical Experiment, 1974
GCM	Atmospheric General Circulation Model
GMSRA	GOES Multispectral Rain Algorithm
GOES	Geostationary Operational Environmental Satellite
HaRP	Hawaiian Rainband Project (July–August 1990)
IR	Infrared Satellite data
ISO	Inter-Seasonal Oscillation (the 30–60 day Madden-Julian Oscillation, MJO)
LEC	Leading edge convection (of squall line, Amazon basin)
LFC	Level of Free Convection
LOS	Locally Occurring Systems (referring to the Amazon basin)
LBA	Large Scale Biosphere–Atmosphere Experiment in Amazonia
LST	Local Standard Time
LW	Long Wave (radiation)
MCCs	Mesoscale Convective Complexes
MCSs	Mesoscale Convective Systems
MCZ	Maximum Cloud Zone
MDI	Mean Daily Intensity
METOSAT	Geostationary METOrogical Satellite (European Space Agency)
MJO	Madden–Julian Oscillation
NACL	North Australia Cloud Line
NCAR	The National Center for Atmospheric Research
NCEP	The National Center for Environmental Prediction
PAM	Portable Automated Mesonet (central Amazonia, near Manaus)
PBL	Planetary Boundary Layer
PC	Principal Component model
PCA	Principal Components Analysis
PV	Potential Vorticity (see glossary of selected terms in Chapter 10)
RFID	Rainfall Frequency-Intensity-Duration
RPCA	Rotated Principal Components Analysis
SACZ	South Atlantic Convergence Zone
SSM/I	Special Sensor Microwave Images (satellite)
STR	Subtropical Ridge (upper atmosphere)
TOGA IFA	TOGA Intensive Flux Array
TOGA–COARE	Tropical Ocean and Global Atmospheric – Coupled Ocean-Atmospheric Response Experiment (November 1992–February 1993)
TSR	Trailing Stratiform region (of squall line, Amazon basin)

TUTT	Tropical Upper Tropospheric Trough
UTC	Coordinated Universal Time (equal to Greenwich Mean Time)
WMONEX	Winter Monsoon Experiment, 1978
ZTE	Zonal Trough in the Easterlies

References

AMMA (African Monsoon Multidisciplinary Analyses) (2002) *On African Monsoon and its components*, CNES, CNRS/INSU, IRD, Météo-France, version February 2002, 52 pp + 14 Figures (available from web site: http://medias.obs-mip.fr/amma).

Asselin de Beauville, C. (1995) Air mass characteristics as related to tropical wave precipitation over Lesser Antilles archipelago. *Atmospheric Research 36*: 157–170.

Atkinson, B. W. (1981). Mesoscale Atmosphere Circulations, Academic Press, London.

Austin, G. R., Rauber, R. M., Ochs III, H. T. and Miller, L. J. (1996) Trade-wind clouds and Hawaiian rainbands. *Mon. Weath. Rev.*, *124*: 2126–2151.

Avila, L. A. and Pasch, R. J. (1992). Atlantic tropical systems of 1991. *Mon. Weath. Rev. 120*: 2688–2696.

Barcelo, A., Robert, R. and Coudray, J. (1997). A Major Rainfall Event: The 27 February–5 March 1993 Rains on the Southeastern Slope of Piton de la Fournaise Massif (Reunion Island, Southwest Indian Ocean). *Mon. Weath. Rev. 125*: 3341–3346.

Berndtsson, R. and Niemczynowicz, J. (1988). Spatial and temporal scales in rainfall analysis – some aspects and future perspectives. *J. Hydrol. 100*: 293–313.

Betts, A. (1974). Thermodynamic classification of tropical convective soundings. *Mon. Weath. Rev.*, 102: 760–764.

Bidin, K. (2001). *Spatio-temporal variability in rainfall and wet-canopy evaporation within a small catchment recovering from selective tropical forestry*. Unpublished PhD thesis, University of Lancaster, Lancaster, UK pp 201.

Black, M. L., Gamache, J. F., Marks, F. D., Jr., Samsury, C. E. and Willoughby, H. E. 2002. Eastern Pacific hurricanes Jimena of 1991 and Olivia of 1994: the effect of vertical shear on structure and intensity. *Mon. Weath. Rev. 130*(9): 2291–2312.

Blanchard, D. O., Cotton, W. R. and Brown, J. M. (1998). Mesoscale Circulation Growth under Conditions of Weak Inertial Instability. *Mon. Weath. Rev. 126*: 118–140.

Bonell, M. (1991). Progress and future research needs in water catchment conservation within the wet tropical coast of NE Queensland. In: *Tropical Rainforest Research in Australia: Present Status and Furure Directions for the Institute for Tropical Rainforest Studies*, Proceedings of a Workshop held in Townsville, Australia, 4–6 May, 1990 Institute for Tropical Rainforest Studies, James Cook University, Townsville, pp 59–86.

Bonell, M. (1993). Progress in the understanding of runoff generation dynamics in forests. *J. Hydrol.*, *150*: 217–275.

Bonell, M. (1998). Possible impacts of climate variability and change on tropical forest hydrology. *Climatic change 39*: 215–272. Based on the WWF Conference on the Potential Impacts of Climate Change on Tropical Forest Ecosystems, Puerto Rico, 24–28 April 1995.

Bonell, M. (2004) Selected Issues on Mountain Hydrology of the Humid Tropics. *Conference on Forestry and Forest Products 1999 Series: Forestry and Land Use Perspectives*, 31 March–1 April 1999, Kuala Lumpur, Malaysia, 25 pp. To be published as a UNESCO Technical Document in Hydrology (UNESCO Jakarta Office), in press.

Bonell, M. with Balek, J. (1993). Recent Scientific Developments and Research Needs in Hydrological Processes of The Humid Tropics. (In) *Hydrology and Water Management in the Humid Tropics*: Bonell, M., Hufschmidt, M. M. and Gladwell, J. 4: 167–260.

Bonell, M. and Gilmour, D. A. (1980). Variations in short-term rainfall intensity in relation to synoptic climatological aspects of the humid tropical north-east Queensland coast. *Singapore J. Trop. Geog. 1(2)*: 16–30.

Bonell, M., Gilmour, D. A. and Cassels, D. S. (1991). The links between synoptic climatology and the runoff response of rainforest catchments on the wet tropical coast of north-eastern Queensland. In: P. A. Kershaw and G. Werran, (eds.) *Australian National Rain Forests Study Report* 2: 27–62, Australian Heritage Commission, Canberra, Australia.

Bracken, W. E. and L. F. Bosart., (2000). The role of synoptic-scale flow during tropical cyclogenesis over the North Atlantic Ocean. *Mon. Wea. Rev.*, *128*: 353–376.

Brunt, A. T. (1966). Rainfall associated with tropical cyclones in the northeast Australian regions. *Aust. Met. Mag.*, *14*: 85–109.

Bureau of Meteorology (1989). *Darwin Tropical Diagnostic Statement, April 1989*, 8(4), Bureau of Meteorology Research Section, Darwin.

Burpee, R. W. (1972). The origin and structure of easterly waves in the lower troposphere of North Africa. *J. Atmos. Sci. 29*: 77–90.

Burpee, R. W. (1974). Characteristics of the North African easterly waves during the summers of 1968 and 1969. *J. Atmos. Sci. 31*: 1556–1570.

Carbone, R. E., Cooper, W. A., and Lee, W. C. (1995). On the forcing of flow reversal along the windward slopes of Hawaii. *Mon. Weath. Rev.*, *123*: 3466–3480.

Carlson, T. N. (1969a) Synoptic histories of three African disturbances that developed into Atlantic hurricanes. *Mon. Weath. Rev. 97*: 256–276.

Carlson, T. N. (1969b). Some remarks on African disturbances and their progress over the tropical Atlantic. *Mon. Weath. Rev. 97*: 716–726.

Cerveny, R. S. and Newman, L. E., (2000). Climatological Relationships between Tropical Cyclones and Rainfall. *Mon. Weath. Rev.*, *128*: 3329–3336.

Chang, C.-P., Erickson, J. E. and Lau, K. M. (1979). Northeasterly cold surges and near – equatorial disturbances over the winter MONEX area during December 1974. Part 1 synoptic aspects. *Mon. Weath. Rev.*, 107, 812–829.

Chang, J.-H. and Lau, L. S. (1993). Appendix: A definition of the Humid Tropics. (In) *Hydrology and Water Management in the Humid Tropics*: Bonell, M., Hufschmidt, M. M. and Gladwell, J. S. pp. 571–574.

Chen, L. and Gray, W. M. (1986). Global view of the upper level outflow patterns associated with tropical cyclone intensity change during FGGE. *Department of Atmospheric Science Paper No. 392*, Colorado State University, Fort Collins, CO, 80523, 126pp.

Chen, Y.-L., and Feng, J. (1995). The influences of inversion height on precipitation and airflow over the island of Hawaii. *Mon. Weath. Rev.*, *123*: 1660–1676.

Chen, S. S. and Houze, Jr., R. A. (1997). Diurnal Variation and life-cycle of deep convective systems over the tropical Pacific warm pool. *Int. J. Climatol.* 123: 357–388.

Churchill, D.D and Houze, Jr., R. A. (1984). Development and structure of winter monsoon cloud clusters on 10 December 1978. *J. Atmos. Sci. 41*: 933–960.

Cifelli, R. and Rutledge, S. A. (1994). Vertical motion structure in maritime continent mesoscale convective systems: Results from a 50 HRz profiler. *J. Atmos. Sci. 51*: 2631–2652.

Cifelli, R. and Rutledge, S. A. (1998). Vertical motion, diabatic heating, and rainfall characteristics in north Australia convective systems. *Q. J. R. Meteorol. Soc. 124*: 1133–1162.

Concise Experimental Plan, LBA: (1996). *The Large Scale Biosphere – Atmosphere Experiment in Amazonia (LBA)*, Compiled by the LBA Science Planning Group, 40pp. Available from LBA project offices:- *Centro de Previsao de Tempo e Estudos Climaticos, Institutio Nacional de Pesquisas Espaciais (INPE), 12630–000 Cachoeira Paulista, SP, Brazil*; (EU-LBA office) *DLO Winand Staring Centre for Integrated Land, Soil and Water Research, PO Box 125, NL-6700AC Wageningen, The Netherlands and* (North American LBA-office) *NASA, 300 E Street SW, DC 20546 Washington, USA*.

Connor, G. J. (1999). An exploratory study into synoptic classification techniques for tropical precipitation forecasting. *Aust. Met. Mag.*, *48*: 241–253.

Connor, G. J., and Bonell, M. (1998). Air mass and dynamic parameters affecting trade wind precipitation on the northeast Queensland tropical coast. *Int. J. Climatol.*, *18*: 1357–1372.

Connor, G. J. and Woodcock, F. (2000). The application of synoptic stratification to precipitation forecasting in the trade wind regime. *Weath. Forecasting*, *15*: 276–297.

Cotton, W. R., Lin, M.-S., Mc Anelly, R. L. and Tremback, J. J. (1989). A composite model of mesoscale convective complexes. *Mon. Weath. Rev.*, *117*: 765–783.

Cronin, M. F., Bond, N., Fairall, C., Hare, J., McPhaden, M. J. and Weller, R. A. (2002). Enhanced oceanic and atmospheric monitoring underway in eastern pacific. *EOS*, 83(19), 7 May 2002, pp. 205, 210–211.

Crowe, P. R. (1971). *Concepts in Climatology*, Longmans, London, 589 pp.

Dharssi, I., Kershaw, R. and Tao, W. K. (1997). Sensitivity of a simulated tropical squall line to long-wave radiation. *Q.J.R. Meteorol. Soc. 123*: 187–206.

Dudhia, J. (1989). Numerical study of convection observed during the Winter Monsoon Experiment using a mesoscale two-dimensionnal model. *J. Atmos. Sci. 46*: 3077–3107.

Duvel, J. P. and Kandel, R. S. (1985). Regional-scale diurnal variations of outgoing infrared radiation observed by METEOSAT. *J. Climate Appl. Meteorol. 24*: 335–349.

Elassien, A. (1951). Slow thermally or frictionally controlled meridional circulation in acircular vortex. *Astrophys. Norv. 5*: 19–60.

Elsenbeer, H., Cassel, D. K. and Zuniga, L. (1994). Throughfall in the terra firme forest of western Amazonia. *J. Hydrol.(NZ)*, 32(2): 30–44.

Eltahir, E. A. B. and Bras, R. L. (1994). Precipitation Recycling in the Amazon Basin. *Quart. J. Roy. Meteorol. Soc.*, *120*: 861–880.

Everitt, B. (1980). *Cluster analysis*. 2nd Edition, Social Science Research Council. Heinemann, London, 136pp.

Ferraro, R., Vicente, G., Ba, M., Gruber, A., Scofield, R., Li, Q. and Weldon, R. (1999). Satellites Techniques Yield Insight into Devastating Rainfall from Hurricane Mitch. *EOS, Trans. Am. Geophy. Un. 80, No 43*: 505, 512–514.

Frank, N. L. (1970). Atlantic tropical systems of 1969. *Mon. Weath. Rev. 98*: 307–314.

Garreaud, R. D. and Wallace, J. M. (1997). The Diurnal March of Convective Cloudiness over the Americas. *Mon. Weath. Rev. 125*: 3157–3171.

Garreaud, R. D. and Wallace, J. M. (1998). Summertime Incursions of Midlatitude Air into Subtropical and Tropical South America. *Mon. Weath. Rev.*, *126*: 2713–2733.

Garstang, M., Massie, Jr. H. L., Halverson, J., Greco, S. and Scala, J. (1994). Amazon Coastal Squall Lines. Part I: Structure and Kinematics. *Mon. Weath. Rev. 122*: 608–622.

Giambelluca, T. W., Nullet, M. A. and Schroeder, T. A. (1986). *Rainfall Atlas of Hawaii*, Dept. of Land and Natural Resources. State of Hawaii, 267 pp.

Glahn, H. R. (1985). Yes, precipitation forecasts have improved. *Bull. Am. Met. Soc.*, *66*: 820–830.

Gong, C. and Eltahir, E.: (1996). Sources of moisture for rainfall in west Africa, *Wat. Resour. Res.*, 32: 3115–3121.

Gray, W. and Jacobson, R. W. (1977). Diurnal variation of deep cumulus convection. *Mon. Weath. Rev. 105*: 1171–1188.

Gray, W. M., Landsea, C. W., Mielke, Jr. P. W. and Berry, K. J. (1994). Predicting Atlantic basin seasonal tropical cyclone activity by June 1. *Weath. Forecasting 9*: 103–115.

Greco, S., Swap, R., Garstang, M., Ulanski, S., Shipham, M., Harriss, R. C., Talbot, R., Andreae, M. O. and Artaxo, P. (1990). Rainfall and Surface Kinematic Conditions Over Central Amazonia During ABLE 2B. *J. Geophys. Res. 95*: 17,001–17,014.

Griffiths, J. F. (1972). *Climates of Africa*, World Survey of Climatology 10, Elsevier, Amsterdam, 604 pp.

Haertel, P. T. and Johnson, R. H. (1998). Two-day disturbances in the equatorial western Pacific. *Q. J. R. Meteorol. Soc. 124*: 615–636.

Hall, A. J. (1984). Hydrology in tropical Australia and Papua New Guinea. *Hydrol. Sci. J. 29*: 399–423.

Hellin, J., Haight, M. and Marks, F. (1999). Rainfall characteristics of hurricane Mitch. *Nature*, *399*: 316.

Hirschberg, P. A., and Fritsch, J. M. (1991a). Tropopause undulations and the development of extratropical cyclones. Part I: Overview and observations from a cyclone event. *Mon. Weath. Rev. 119*: 496–517.

(1991b). Tropopause undulations and the development of extratropical cyclones. Part II: Diagnostic analysis and conceptual modal. *Mon. Weath. Rev. 119*: 518–550.

(1993a). A study of the development of extratropical cyclones with an analytic model. Part I: The effects of stratospheric structure. *J. Atmos. Sci. 50*: 311–307.

(1993b). On understanding height tendency. *Mon. Weath. Rev. 121*: 2646–2661.

(1994). A study of the development of extratropical cyclones with an analytic model. Part II: Sensitivity to tropospheric structure and analysis of height tendency dynamics. *Mon. Weath. Rev. 122*: 2312–2330.

Hodnett, M. G., Vendrame, I., De O. Marques Filho, A., Oyama, M. D. and Tomasella, J., (1997). Soil water storage and groundwater behaviour in a catenary sequence beneath forest in central Amazonia: I Comparisons between plateau, slope and valley floor, *Hydrol. Earth Sys. Sc.*, 1(2), 265–277.

Holland, G. J. and Merrill R. T. (1984). On the dynamics of tropical cyclone structure change. *Q. J. R. Met. Soc.*, 110: 723–745.

Holton, J. R. (1979). *An Introduction to Dynamic Meteorology*, Academic Press, New York, 309pp.

Hoskins, B. J., Draghici I. and H. C. Davies (1978). A new look at the Omega equation. *Q. J. R. Met. Soc. 104*: 31–38.

Hoskins, B. L., McIntyre, M. E., and Robertson, A. W. (1985). On the use and significance of isentropic potential vorticity maps. *Q. J. R. Meteorol. Soc. 111*: 877–946.

Hoskins, B. J., Pedder, M. and Wyn Jones, D. 2003. The omega equation and Potential Vorticity. *Q. J. R. Meteorol. Soc. 129*: 3277–3303.

Houze, R. A. (1989). Observed structure of mesoscale covective systems and implications for large-scale heating. *Q. J. R. Met. Soc. 115*: 425–461.

Houze, R. A. and Rappaport, E. N. (1984). Air motions and precipitation structure of an early summer squall line over the eastern tropical Atlantic. *J. Atmos. Sci. 41*: 553–574.

Jackson, I. J. (1986). Relationships between raindays, mean daily intensity and monthly rainfall in the tropics. *J. Climatol. 6*: 117–134.

Jackson, I. J. (1988). Daily rainfall over Northern Australia: deviations from the world pattern. *J. Climatol. 8*: 463–476.

Jackson, I. J. (1989). *Climate, Water and Agriculture in the Tropics*, 2nd edn., Longman, London.

Jackson, I. J. and Weinand, H. (1994). Towards a classification of tropical rainfall stations. *Int. J. Climatol. 14*: 263–286.

Jackson, I. J. and Weinand, H. (1995). Classification of tropical rainfall stations: a comparison of clustering techniques. *Int. J. Climatol. 15*: 985–994.

Janowiak, J. E., Arkin, P. A. and Morrissey, M. (1994). An examination of the diurnal cycle in oceanic tropical rainfall using satellite and *in situ* data. *Mon. Weath. Rev. 122*: 2296–2311.

Kodama, K. and Barnes, G. M. (1997). Heavy rain events over the south-facing slopes of Hawaii: Attendant conditions. *Weath. Forecasting, 12*: 347–367.

Laing, A. G. and Fritsch J. M. (1993). Mesoscale convective complexes in Africa, *Mon. Weath. Rev.*, *121*: 2254–2263, 1993.

Landsea, C. W., Gray, W. M., Mielke, Jr. P. W. and Berry, K. J. (1998). Seasonal forecasting of Atlantic hurricane activity. *Weather 49*: 273–283.

Lau, N.-P., Chang, C.-P. And Chan, P. H. (1983). Short-term planetary scale interactions over the tropics and mid latitudes. Part II: Winter MONEX periods. *Mon. Weath. Rev.*, *111*: 1372–1388.

Lean, J. and Rowntree, P. R. (1997). Understanding the sensitivity of a GCM simulation of Amazonian deforestation to the specification of vegetation and soil characteristics. *J. Climate*, *10*: 1216–1235.

Lean, J., Bunton, C. B., Nobre, C. A. and Rowntree, P. R.: (1996). The simulated impact of Amazonian deforestation on climate using measured ABRACOS vegetation characteristics. In: *Amazonian Deforestation and Climate*, Gash, J. H. C., Nobre, C. A., Roberts, J. M. and Victoria, R. L., eds., Wiley, Chichester, pp. 549–576.

Lebel, T., Taupin, J. D. and D'Amato, N. (1997). Rainfall monitoring during HAPEX-Sahel. 1. General conditions and climatology. *J. Hydrol. 188–189*: 74–96.

Lebel, T., Braud, I. and Creutin, J.-D. (1998). A space-time rainfall disaggregation model adapted to Sahelian mesoscale convective complexes. *Wat. Resour. Res. 34*: 1711–1726.

Lenters, J. D. and Cook, K. H. (1999). Summertime Precipitation Variability over South America: Role of the Large-Scale Circulation. *Mon. Weath. Rev. 127*: 409–431.

Leopold, L. B. (1949). The interaction of trade wind and sea breeze, Hawaii. *J. Meteorol.*, *6*: 312–320.

Lockwood, J. G. (1984). The Southern Oscillation and El Niño. *Progress in Physical Geography 8(1)*: 102–110.

Lyons, S. W. (1982). Empirical orthogonal function analysis of Hawaiian rainfall. *J. Appl. Meteorol.*, *21*: 1713–1729.

Lyons, W. F. and Bonell, M. (1992). Daily, meso-scale rainfall in the tropical wet/dry climate of the Townsville area, North-East Queensland during the 1988–1989 wet season: synoptic-scale airflow considerations. *Int. J. Climatol. 12*: 655–684.

Lyons, W. F. and Bonell, M. (1994). Regionalization of daily mesoscale rainfall in the tropical wet/dry climate of the Townsville area of North-East Queensland during the 1988–1989 wet season. *Int. J. Climatol. 14*: 135–163.

Machado, L. A. T., Rossow, W. B., Guedes, R. L. and Walker, A. W. (1998). Life cycle variations of mesoscale convective systems over the Americas. *Mon. Weath. Rev. 126*: 1631–1654.

Maddox, R. A. (1980). Mesoscale convective complexes. *Bull. Amer. Meteor. Soc., 61*: 1374–1387.

Manton, M. J. and Bonell, M. (1993). Climate and Rainfall Variability in the Humid Tropics. (In) *Hydrology and Water Management in the Humid Tropics:* Bonell, M., Hufschmidt, M. M. and Gladwell, J. S. 2: 13–33.

Mapes, B. E. and Houze, Jr. R. A. (1993). Cloud clusters and superclusters over the oceanic warm pool. *Mon. Weath. Rev. 121*: 1398–1415.

Marengo, J., Fisch, G. F., Vendrame, I., Cervantes Dias, P. G., Morales. C. A, (2001). Diurnal variability of rainfall in Southwest Amazonia during the LBA-WET AMC campaign of the summer of 1999. *Unpublished manuscript*.

Matsuno, T. (1966). Quasi-geostrophic motions in the equatorial area. *J. Meteorol. Soc. Japan 44*: 25–42.

Matthews, A. J. and Kiladis, G. N. (1999). The Tropical-Extratropical Interaction between High-Frequency Transients and the Madden-Julian Oscillation. *Mon. Weath. Rev. 127*: 661–677.

McBride, J. L. (1983). Satellite observation of the southern hemisphere monsoon during winter MONEX, *Tellus, 35A (3)*: 189–197.

McBride, J. L. and Keenan, T. D., (1982). Climatology of tropical cyclone genesis in the Australian region. *J. Climatol., 2*: 13–33.

McRae, J. N. (1956). The formation and development of tropical cyclones during the 1955–56 summer in Australia. *Proceedings of the Tropical Cyclone Symposium Brisbane 1956*, Bureau of Meteorology, Australia, 233–262.

Merritt, E. S. (1964). Easterly waves and perturbations, a reappraisal. *J. Applied Met.* 4: 367–382.

Molinari, J., Knight, D., Dickinson, M., Vollaro, D. and Skubis, S. (1997). Potential Vorticity, Easterly Waves, and Eastern Pacific Tropical Cyclogenesis. *Mon. Weath. Rev. 125*: 2699–2708.

Molinari, J., Vollaro, D., Skubis, S and Dickinson, M. (2000). Origins and Mechanisms of Eastern Pacific Tropical Cyclogenesis: A Case Study. *Mon. Weath. Rev.*, 128: 125–139.

Molion, L. C. B.: (1993). Amazonia rainfall and its variability. In: *Hydrology and Water Management in the Humid Tropics – Hydrological Research Issues and Strategies for Water Management*, Bonell, M., Hufschmidt, M. M. and Gladwell, J. S., eds., Cambridge, pp. 99–111.

Neumann, C. J. (1979). On the use of deep layer mean geopotential height fields in statistical prediction of tropical cyclone motion. 6th Conference on Probability and Statistics in Atmospheric Sciences, *Amer. Met. Society*, Boston 32–38.

Observatoire Réunionnais de l'EAU (1994) Le cyclone tropical Colina et les évènements hydrologiques majeurs du 1er trimestre 1993. *Etude hydrologique générale*.

Ojo, O. (1977). *The Climate of West Africa*, Heinemann, London, 219 pp.

Olsen, D. A., Junker, N. W., and Korty, B. (1995). Evaluation of 33 years of quantitative precipitation forecasting at NMC. *Weath. Forecasting, 10*: 498–511.

Omotosho, J. B. (1985). The separate contributions of line squalls, thunderstorms and the monsoon to the total rainfall in Nigeria, *J. Climatol., 5*: 543–552.

Orlanski, I. (1975). A relatipnal subdivision of scales for atmosphere processes. *Bull. Amer. Met. Soc. 56*: 527–530.

Palmen. E. (1948). On the formation and structure of tropical cyclones. *Geophysics*, 3, 26–38.

Pielke, R. A., Avissar, R., Raupach, M., Dolman, A. J., Zeng, X. and Denning, A. S. (1998). Interactions between the atmosphere and terrestrial ecosystems: influence on weather and climate. *Global Change Biology 4*: 461–475.

Pielke, R. A., Walko, R. L., Steyaert, L. T., Vidale, P. L., Liston, G. E., Lyons, W. A. and Chase, T. N. (1999). The Influence of Anthropogenic Landscape Changes on Weather in South Florida. *Am. Meteorol. Soc. 127*: 1663–1672.

Putty, M. R., Prasad, V. S. R. K. and Ramaswamy, R., (2000). A study of the rainfall intensity pattern in western Ghats, Karnataka. *In*: Proc. Workshop on *Watershed Development in Western Ghats Region of India*, 28–29 February 2000, Varadan, K. M. (Ed.), Centre for Water Resources Development and Management, Kozhikode-673 571, Kerala.: pp 44–51.

Pytharoulis, I. and Thorncroft, C. D, (1999). The low-level structure of African easterly waves in 1995. *Mon. Weath. Rev. 127*: 2266–2280.

Ramage, C. S., (1968). Role of a 'Maritime Continent' in the atmospheric circulation. *Mon. Weath. Rev. 96*: 365–370.

Ramage, C. S., and Schroeder, T. A. (1999). Trade wind rainfall atop Mount Waialeale, Kauai. *Mon. Weath. Rev. 127*: 2217–2226.

Rasmussen, R. M., Smolarkiewicz, P. K., and Warner, J. (1989). On the dynamics of Hawaiian cloud bands: Comparison of model results with observations and island climatology. *J. Atmos. Sci. 46*: 1589–1608.

Raymond, D. J. (1994). Convective processes and tropical atmospheric circulations. *Q. J. R. Meteorol. Soc. 120*: 1431–1455.

Raymond, D. J. and Lewis, S. A. (1995). Rotating convective disturbances in the trades. *Q. J. R. Meteorol. Soc. 121*: 271–299.

Reed, R. J. and Recker, E. E. (1971). Structure and properties of synoptic wave disturbance in the equatorial western Pacific. *J. Atmos. Sci. 28*: 1117–1133.

Reed, R. J., Hollingsworth, A., Heckley, W. and Delsol, F. (1988a). An evaluation of the performance of the ECMWF forecasting operational system in analyzing and forecasting easterly wave disturbances over Africa and the tropical Atlantic. *Mon. Weath. Rev. 116*: 824–865.

Reed, R. J., Klinker, E. and Hollingsworth, A. (1988b). The structure and characteristics of African easterly wave disturbances as determined from the ECMWF operational analysis/forecast system. *Meteor. Atmos. Phys. 38*: 22–33.

Riehl, H. (1954). *Tropical meteorology*. McGraw-Hill, New York.

Ritchie, E. A. and Holland, G. H. (1999). Large-ScalePatterns Associated with Tropical Cyclogenesis in the Western Pacific. *Mon. Weath. Rev. 127*: 2027–2043.

Roux, F. (1998). The oceanic mesoscale convective system observed with airborne Doppler radars on 9 February 1993 during TOGA-COARE: Structure, evolution and budgets. *Q. J. R. Meteorol. Soc. 124*: 585–614.

Sadler, J. C. (1967). On the origin of tropical vortices. *Proc. Of working panel on tropical dynamic meteorology*, Monterey, California, Navy Weather Research Facility, Report 12-1167-132, 39–75. Available from the department of Meteorology, University of Hawaii.

Sadler, J. C. (1975a). *The upper tropospheric circulation over the global tropics*, Dept. of Meteorolgy, University of Hawaii.

Sadler, J. C. (1975b). The monsoon circulation and cloudiness over the GATE area. *Mon. Weath. Rev. 104*: 369–387.

Sadler, J. C. (1978). Mid-season typhoon development and intensity changes and the Tropical Upper Tropospheric Trough. *Mon. Weath. Rev. 106*: 1137–1152.

Sadler, J. C., Lander, M. A., Hori, A. M. and Oda, L. K. (1987). *Tropical Marine Climatic Atlas, Volume I* (Indian Ocean and Atlantic Ocean) and *Volume II* (Pacific Ocean), Department of Meteorology, University of Hawaii.

Saito, K., Keenan, T., Holland, G. and Puri, K. (2001). Numerical Simulation of the Diurnal Evolution of Tropical Island Convection over the Maritime Continent. *Mon. Weath. Rev. 129*: 378–400.

Salati, E., Dallo'Olio, A., Marsui, E. and Gat, J. R. (1979). Recycling of Water in the Amazon Basin: An Isotope Study. *Wat. Resour. Res. 15*: 1250–1258.

Shapiro, L. J. (1986). The three-dimensionnal structure of synoptic-scale disturbances over the tropical Atlantic. *Mon. Weath. Rev. 114*: 1876–1891.

Shay, L. K. (1988). Ocean Processes excited by tropical cyclones. Rapporteur Reports of the Fourth WMO International Workshop on Tropical Cyclones (IWTC-IV), Haikou, China.

Shinoda, M., Okatani, T. and Saloum, M. (1999). Diurnal variations of rainfall over Niger in the West African Sahel: a comparison between wet and drought Years. *Int. J. Climatol. 19*: 81–94.

Smolarkiewicz, P. K., Rasmussen, R. M., and Clark, T. L. (1988). On the dynamics of Hawaiian cloud bands: Island forcing. *J. Atmos. Sci., 45*: 1872–1905.

Sumner, G. N. (1988). *Precipitation – Processes and Analyses*. Wiley, Chichester.

Sumner, G. N. and Bonell, M. (1986). Circulation and daily rainfall in the north Queensland wet season 1979–1982. *J. Climatol. 6*: 531–549.

Sumner, G. N. and Bonell, M. (1988). Variation in the spatial organisation of daily rainfall during the north Queensland wet season, 1979–82. *Theor. Appl. Climatol. 39*: 59–72.

Takehashi, T. (1986). Wind shear effects on water accumulation and rain duration in Hawaiian warm clouds. *J. Meteorol. Soc. Japan, 64*: 575–584.

Tao, W.-K., Lang, S., Simpson, J., Sui, C.-H., Ferrier, B. and Chou, M.-D. (1996). Mechanisms of cloud-radiation interaction in the tropical latitudes. *J. Atmos. Sci. 53*: 2624–2651.

Taylor, C. M. and Lebel, T. (1998). Observational Evidence of Persistent Convective-Scale Rainfall Patterns. *Mon. Weath. Rev. 126*: 1597–1607.

Taylor, C. M., Said, F. and Lebel, T. (1997). Interactions between the Land Surface and Mesoscale Rainfall Variability during HAPEX-Sahel. *Mon. Weath. Rev. 125*: 2211–2227.

Taylor, C. M., Clark, D. B. and Cox, P. M. (2000). Land surface controls on west African rainfall in a GCM. In: World Meteorological Organisation (2000) *Proceedings of the workshop on west African monsoon variability and predictability (WAMAP)*. Geneva, WMO Tropical Meteorology Research Programme Report Series *63*: 43–46.

Thauvin, V. and Lebel, T. (1991). International workshop on precipitation measurement IV: EPSAT-NIGER- Study of rainfall over the Sahel at small time steps using a dense network of recording raingauges. *Hydrol. Proc. 5*: 251–260.

Thorncroft, C. D. (1995). An idealized study of African easterly waves. III: More realistic basic states. *Q. J. R. Meteorol. Soc. 121*: 1589–1614.

Thorncroft, C. D. and Hoskins, B. J. (1994a) An idealized study of African easterly waves. I: A linear view. *Q. J. R. Meteorol. Soc. 120*: 953–982.

Thorncroft, C. D. and Hoskins, B. J. (1994b). An idealized study of African easterly waves. II: A nonlinear view. *Q. J. R. Meteorol. Soc. 120*: 983–1015.

Thorncroft, C. D. and Rowell, D. P. (1998). Interannual Variability of African Wave Activity in a General Circulation Model. *Int. J. Climatol. 18*: 1305–1323.

Thorncroft, C. D. and Hodges, K. (2001). African Easterly Wave Variability and Its Realtionship to Atlantic Tropical Cyclone Activity. *J. Clim., 14*: 1166–1179.

Trenberth, K. E. (1999). Atmospheric Moisture Recycling: Role of Advection and Local Evaporation. *Am. Meteorol. Soc. 12*: 1368–1381.

Ward, J. H. (1963). Hierarchical grouping to optimize an objective function. *Am. Sci. Assoc. J. 58*: 236–244.

Webster, P. J. and Lukas, R. (1992). TOGA COARE: The coupled ocean-atmosphere response experiment. *Bull. Am. Meteorol. Soc. 73*: 1377–1416.

Webster, P. J. and Stephens, G. L. (1980). Tropical upper-tropospheric extended clouds: Inferences from Winter MONEX. *J. Atmos. Sci. 37*: 1521–1541.

Williams, W. T. (ed.) (1976). *Pattern Analysis in Agricultural Science*. CSIRO and Elsevier, Melbourne.

Wu, M. C., and Chan, J. C. L. (1997). Upper-Level features associated with winter monsoon surges over South China. *Mon. Weath. Rev. 125*: 317–340.

Yuter, S. E. and Houze, Jr., R. A. (1998). The natural variability of precipitating clouds over the western pacific warm pool. *Q. J. R. Meteorol. Soc. 124*: 53–99.

Zheng, X. and Eltahir, E. A. B. (1998). The role of vegetation in the dynamics of west African monsoons. *J. Climate 11*: 2078–2096.

12 Climatic variability in the tropics

G. Mahé, E. Servat and J. Maley
Institut de Recherche pour le Développement, Montpellier, France

ANALYSIS OF THE HISTORY OF FORESTS AND ASSOCIATED CLIMATES

Previously, dense tropical forests were considered to be the most stable ecosystems on the planet, and their exceptional richness has often been associated specifically with their resistance to past climate changes. Because of recent advances in paleoecology, however, it has been shown that dense forests, such as those in Africa and in Amazonia, have in fact undergone profound changes in response to global climatic changes.

The history of dense forests and their dynamics can be reconstructed by the study of fossils such as pollens or – much rarer – wood or carbon, within specific disciplines such as palynology, paleo-botany or anthracology. Reconstitution of paleo-vegetation is one method, among others, of reconstructing the paleo-climates of past eras. Nevertheless, even though there has been much progress in studying the history of intertropical forests and their associated climates, several problems remain, linked mainly to the small amount of data available. This introduces subjectivity in describing the succession of paleo-environments.

Nearly all trees in dense tropical forests are angiosperms. It would be logical to begin the history of these forests at the time when angiosperms evolved, i.e. during the lower Cretaceous era through to the Barremian and the Aptian eras, around 120 million BP (Maley, 1996a). Up until then, gymnosperms were the dominant plant form but by the end of the Cretaceous, dense tropical forests had become made up almost entirely of angiosperms. Pollen data have shown that the dense forests in Africa and South America were then quite similar and characterised by a great number of palm trees (Maley, 1990; 1996a). Palm tree species have remained abundant in South American forests while becoming relatively rare in African forests. It is since the upper Eocene (around 40 million BP) in particular, that the floral composition of these forests has begun to resemble their current state (Maley, 1996a).

Palynology is the study of pollen. Pollen is of microscopic size and is spread in great numbers. The study of fossilised pollen from lakeside sediments or peat bogs is often used to reconstruct the history of vegetation (Bonnefille, 1993). It is first necessary to extract sediment core samples. The various levels of a sample are dated using radiocarbon techniques. The pollen is then extracted using a physical chemistry procedure and then subject to microscopic examination. The range of pollen types can thus be established and statistical techniques used to give an idea of the changes over time in the regional distribution of the plant varieties that lived for thousands of years around the lakes or peat bogs under study (Jolly *et al.*, 1998).

Carbonisation preserves the fine structure of wood and carbonised wood can then be identified by botanists because the ligneous species have a specific anatomical constitution according to the make-up and relative quantity of cellular elements (Tardy, 1998). Anthracology, the determination of fossil woods coming from radiocarbon dated soil samples, allows the identification of specific taxa from flower trails. It also allows these taxa to be placed in the chronological context and so permit the reconstruction of the vegetation dynamics and the history of the plants (Vernet *et al.*, 1994).

Reconstructing past climates

The most spectacular advances have been made within the last 20 years in the polar and oceanic regions (Jouzel *et al.*, 1994). It is now possible to:

(1) determine the surface temperature variation of both the polar ice caps and Greenland by analysing oxygen isotopes from deep ice core samples which date back over 400 000 years (Raynaud *et al.*, 2000),
(2) measure the quantity of CO_2 in air bubbles trapped in ice samples and thus reconstitute the CO_2 component of the earth's

Forests, Water and People in the Humid Tropics, ed. M. Bonell and L. A. Bruijnzeel. Published by Cambridge University Press.

atmosphere in parallel with the temperature variation over time (Raynaud *et al.*, 2000),

(3) determine ocean temperatures through analysis of foraminifer plankton taken from marine core samples. These results show that major climatic variations are global in nature and follow patterns linked to variations in the Earth's orbit.

Because of the complexity of the global climatic system, these data cannot be extrapolated to the continents where studies to determine the nature of past climates are concentrated on areas around lakes and river headwaters and on the analysis of pollen samples.

There are several other difficulties in reconstructing the history of intertropical forests and associated climates. The continental climatic dynamics can be deduced only by comparing many pollen diagrams, accurately dated and studied in detailed resolution, in other words within a densely sampled study area (Bonnefille, 1986; Maley and Brenac, 1998; Jolly *et al.*, 1998). In Africa, this requirement causes some problems as, for example, through (i) the irregular distribution of test sites because when there is only one site available for a very large region, it is difficult to extrapolate; and (ii) the irregular time distribution of data because there is much more information available for the last 20 000 years than there is for the Tertiary era (Maley, 1996a).

The irregularity of data in both space and time makes it difficult to propose a precise distribution scheme for the various types of ecosystems in forests of the past. Nevertheless, the data we have for the past 20 000 years shows that the main changes in the composition and distribution of tropical forests were synchronised with the major variations in the global climate; this leads us to think that this has been the case for millions of years.

The origin and history of intertropical forests

It would be difficult to use the current distribution of the major intertropical forest regions as a reference point from which to relate the fluctuations of forest boundaries over the past millions of years. For the current study, a starting point could be when the continents that contain the forests of South East Asia, South America and Africa attained their present proportions (Figure 12.1).

The tropical rainforests of South East Asia are, geographically speaking, relatively spread out, in particular, on numerous islands in Indonesia and New Guinea. It has been shown that these forests were at the outermost limits of their extension during the beginning of the Tertiary Period some time during the last 65 million years when they reached as far as Japan and China. Over the last 25 million years they have been shrinking progressively from their northern and southern boundaries (Heaney, 1991).

In South America, the presence of a tropical forest closely resembling what we know today has existed since the Oligocene era, or about 36 million years ago. At this time the north–south boundaries of this paleoforest stretched much further, bearing witness to a hotter and more humid climate than exists today. The range and diversity of this vegetation seem to have been greatest during the Miocene era (Van der Hammen, 1991). Between 10 and 2 million BP (end of the Pliocene), the gradual formation of the Andes mountain chain reordered the continental drainage directions and gave the Amazon Basin its western facing contour. The morphogenesis of the South American continent, therefore, separated the forest into two zones (Choco and Magdalena Valley) and changed the zones of heavy rainfall in the western Amazon. During the upper Miocene era the forest diminished progressively, affected by the cooling of the global climate.

In Africa, from the end of the Cretaceous to the Eocene, because of the shift of the African plate, the tropical forest extended well into what is now the Sahara desert and the northern Sudanian savannahs. It was from the beginning of the Miocene especially that the forest attained its current position around the Gulf of Guinea (Maley, 1996a). The Ethiopian plateau was covered by forests from the Eocene until the Miocene period (Bonnefille, 1993). In eastern Africa, to the east of the Rift, the savannahs began to expand in the Oligocene period (Hamilton and Taylor, 1991; Harris, 1993).

The historical and climatic framework of African tropical forests from the end of the Tertiary period to the Quaternary period

Starting in the Miocene period, the major variations in tropical forests can be interpreted in a global context of temperature variations and in particular, of cooling phases which were marked by the extension of the polar ice caps. (Maley, 1996a).

Between 15 and 10 million BP, at the end of the Tertiary period, following the increase in the Antarctic ice mass, the climate became drier and cooler, with the ascent of the polar fronts and, progressively, a pattern of seasonal climates alternating from dry to humid. This aridification would have had a direct impact on the African vegetation which opened up and dried out (Bonnefille, 1993). A more humid period occurred between 8 and 6.5 million BP, associated with a new forest extension. Then, towards 5 million BP, there came a significant expansion of the Antarctic ice mass which led to a drier period in tropical Africa and was accompanied by a new period of savannah expansion (Bonnefille, 1993). Once again, the climate became more temperate with oscillations between dry and humid periods but with less of an overall effect than in the preceding period, and a lot less effect than in the period that was to follow, starting around 2.5 million BP.

The study of pollen deposits from the Niger River delta (Morley and Richards, 1993) as well as the East African sequences (Bonnefille, 1993) show a new and significant expansion of

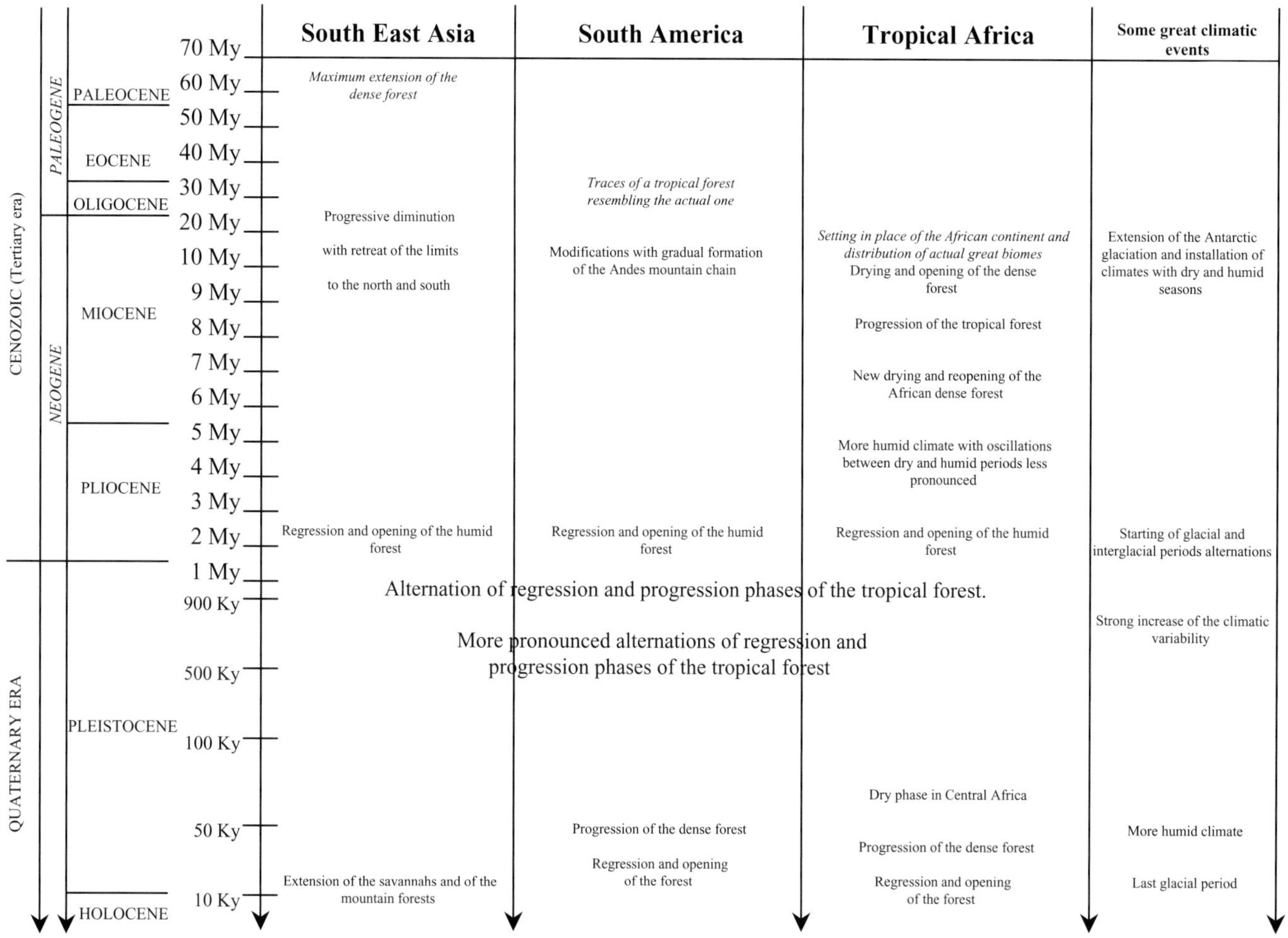

Figure 12.1 Tropical forests and associated climates from the Lower Tertiary era to the beginning of the Quaternary era. My, million years; Ky, thousand years.

savannahs around 2.5 million BP, i.e. at the beginning of the alternating ice ages and interglacial periods. This major change occurred simultaneously with another important event: the first major ice age, which was marked by the extension of the polar ice caps into the Arctic region of the northern hemisphere and, at the same time, by new glacial development in the Antarctic (Maley, 1996a). Then came a progressive increase in the magnitude of glacial variation, marked by two principal phases: the first occurred between 2.5 million BP and 800 000 BP, and was characterised by ice age/interglacial cycles of about 40 000 years; the second takes us up to the current era and is characterised by dominant cycles of about 100 000 years. These cycles are controlled by the main parameters of the Earth's orbit, as shown by Milankovitch (Mc Intyre *et al.*, 1989). This alternation of ice age and interglacial periods that began with an arid phase about 2.5 million BP would have a major impact on the forest ecosystems of the entire tropical zone.

The ice ages controlled the water level of the ocean; for example, the sea level was lower by 120 m around the year 18 000 BP, during the last glacial maximum. This entailed a concurrent change in the extent of evaporating water surface. At the same time, global temperatures rose or fell (depending on which phase we are in, interglacial or ice age). This synchronous variation in water surface and temperature affects the quantity of water vapour in the air and leads to an increase or decrease in precipitation. On the continents, the resulting decrease (increase) in rainfall leads to an expansion (reduction) of the savannas and open areas or forests, depending on the conditions.

The periods of maximum forest expansion, comparable to the current situation, began towards the beginning of the Holocene at about 9500 BP; and corresponded to the warmest climatic phases in which ice masses of both polar ice caps were reduced. The pollen study of marine core samples taken from the Gulf of Guinea show that the preceding phase of the African forest's maximum expansion occurred between 130 000 and 115 000 BP during the Isotopic stage 5e (Dupont *et al.*, 2000).

For about 800 000 years, it seemed, therefore, that the length of interglacial stages associated with the great expansion of dense forests corresponded to approximately 10% of the time while the remaining 90% corresponded to the ice ages linked to the expansion of savannah areas (Maley, 1996a).

In tropical Africa, there are no pollen data for before 30 000–40 000 BP. On the other hand, in South America, in the Andes near Bogota, Columbia, there are long continuous pollen records dating back to the upper Pliocene (Hooghiemstra and Van der Hammen, 1998). In South East Asia, 400 000 years old pollen records for the extreme south of China have been published (Zheng and Lei, 1999). In these two records, the fluctuations of forest vegetation and open areas were also in phase with global climatic changes.

Although the study of the history of intertropical forests and associated climates during the Tertiary and the major part of the Quaternary is handicapped by a lack of paleo-data, the upper Quaternary, particularly from the last ice age, is relatively well documented.

From all tropical regions there is evidence that the Last Glacial Maximum was both drier and cooler than the present climate. In some areas the desiccation is more obvious, and in others the cooling. There is clear evidence of depressions of forests limits, which may be interpreted as a reduction of mean annual temperature. According to Lezine (1998), during the Glacial-Interglacial transition at 15 000 then at 10 000 BP the regime was dominated by meridional exchanges. The Holocene is, to the contrary, characterised by low continent–ocean exchanges, with a dominance of the Atlantic monsoon fluxes.

Concerning the current forested regions of central Africa, pedological, geological and archaeological data show that between 70 000 and 40 000 BP this region was relatively dry and subject to intense erosion, leading to the first generation of stone lines that are found frequently at the base of the soil. A second generation of stone lines appears in this period at the limit between the Pleistocene and Holocene around 11 500 years BP, coinciding with the first large increase in rains in the regions which were, until then, still dominated by open space vegetation (Maley, 1996b).

At the global level, the maximum cooling period occurred between 20 000 and 15 000 BP, characterised by major expansion of the polar ice cap into the northern parts of North America and Europe. In response to this global cooling, monsoons reduced dramatically which entailed a severe reduction in the area of forests. Such reductions resulted in nothing more than a series of isolated forested areas, not far from the coast of the Gulf of Guinea and some others near the centre of the Congo basin (riverine forests) and at the foot of the mountains of the African Rift (Maley, 1996a; 1997).

With regard to the second generation of stone lines mentioned above, it is interesting to observe that the driest phase, which was synchronous with the last major ice age, was not the most erosive phase, probably because the reduced rainfall was not of the erosive type (Maley, 1996b). It was calculated that the average temperature dropped by close to 4 °C in one sector of the mountainous zone of the East African Rift (Bonnefille, 1991; 1993); in Barombi Mbo, in west Cameroon and in the region of Lake Bosumtwi in Ghana, there was an estimated temperature drop of about 3 °C, based on the lower altitude of certain mountainous taxa (Maley, 1991).

In South East Asia, West Pacific, New Guinea and Australia, the general pattern is one that exhibits lowland rainforest in the Holocene, but some variation from this at the Last Glacial Maximum. Lower montane elements in some places suggest that the lowland climate at that time was both drier and cooler (Flenley, 1998).

In Africa, the amount of mean annual temperature cooling is of 4±2 °C, though a reduction of 5–8 °C is still required by snow-line and forest limit data (Bonnefille *et al.*, 1990) if the latter are to be explained by temperature alone (Flenley, 1998).

In tropical Latin-America there is some evidence for desiccation during the Late Pleistocene in some places. In north-east Brazil there are indications of an extension of the savannah between 22 000 and 11 000 BP, at the expense of forest. The Amazon forest could therefore have been divided into blocks at the Last Glacial Maximum (Flenley, 1998).

The end of this last dry period initiated the beginning of a phase of forest regrowth, which, at the start of the Holocene, achieved its optimum range over the equatorial zone, including South East Asia and the Amazon.

The amount of such reduction is, however, of the order of 6–10 °C, at least in South East Asia and the Western Pacific, and in Latin America.

The maximum forest extension during the Holocene: The chronological lag between the African and the Amazonian rainforests

The start of the Holocene around 10 000 BP coincided almost exactly with the last phase of the maximum expansion of the rainforests in all the equatorial zones (Servant *et al.*, 1993). In South America, the history of the eastern part of the Amazon forest recorded at Carajas shows that after a first expansion between 10 000 and 8000 BP, the forest diminished considerably until about 4000 BP, with the driest period occurring between 6000 and 5000 BP (Siffedine *et al.*, 1994). This drying of the climate created favourable conditions for recurring forest fires, including in French Guyana, until around 4000 BP (Tardy, 1998). The level of Lake Titticaca in the Andes was also relatively low in the middle

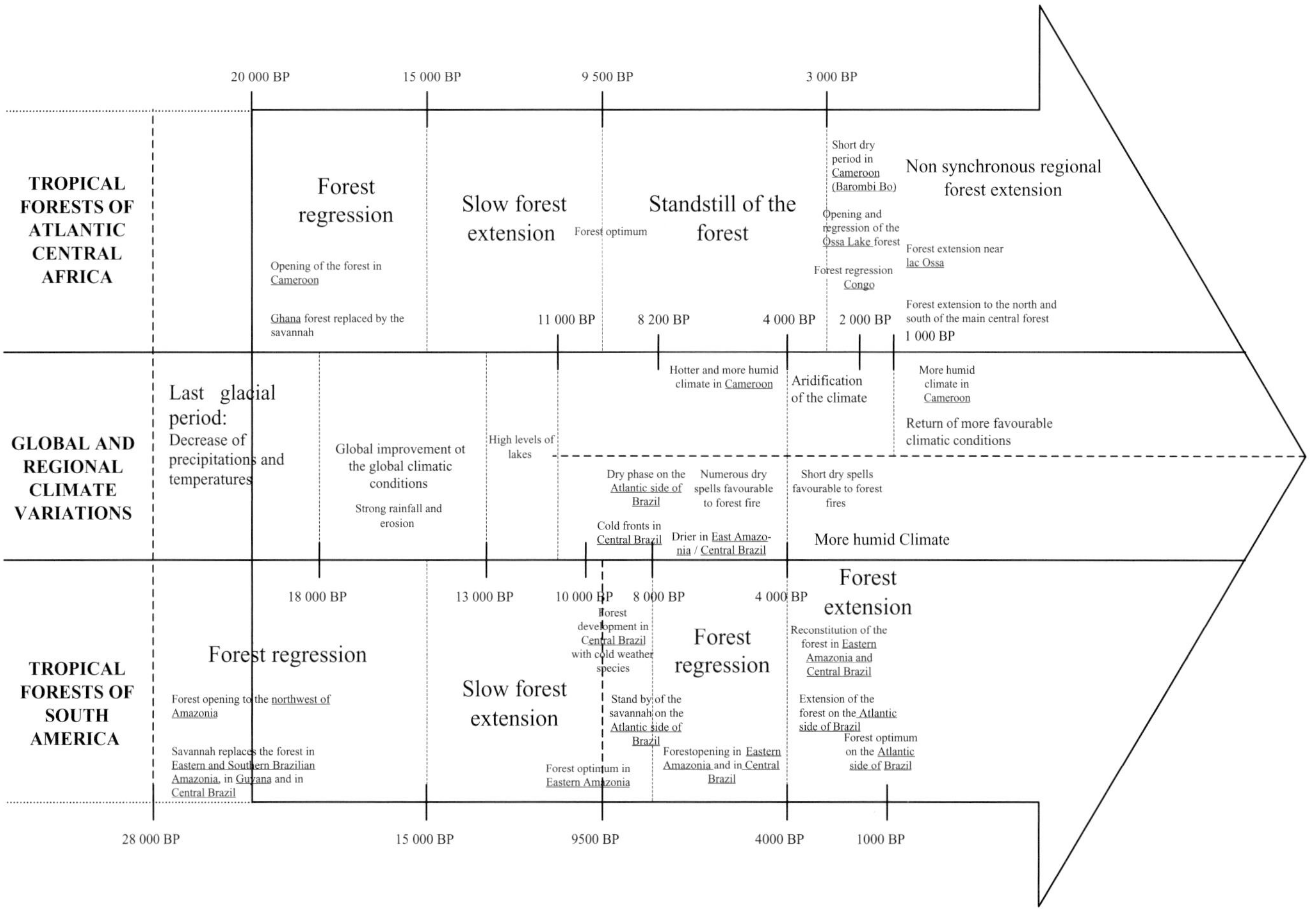

Figure 12.2 Changes in South American and Atlantic Central African tropical forests over the past 20 000 years. (From Vincens *et al.*, 1996.)

Holocene and then, towards 3800 BP, it rose abruptly by about 20 m to reach roughly its current level (Martin *et al.*, 1993). Shortly after 4000 BP a new phase of forest expansion began in Carajas and Guyana which lasted until modern times.

The situation was very different in central Africa. The expansion phase of forests began around 9500 BP in western Africa (Lake Bosumtwi) and central Africa (Lake Barombi Mbo) (Maley, 1991; Maley and Brenac, 1998). In western Africa the forest expanded continuously until the present (Maley, 1991) but in central Africa there was a major interruption around 2800 BP in southern Cameroon and western Congo (Maley and Brenac, 1998; Maley *et al.*, 2000; Vincens *et al.*, 2000). Extremely dry conditions were present in these regions between 2800 and 2000 BP, facilitating the expansion of savannahs and open spaces. At the same time there was a significant increase in pioneer taxa that would allow a rapid recovery of forests starting in about 2000 BP. According to the sites studied, the forest recovery and the succession of forest formations were not synchronous. For example, in the region of Lake Ossa near Edea, it was not until around 800 BP that the evergreen forest, rich in Caesalpiniaceae, dominated again (Reynaud-Farrera *et al.*, 1996). The configuration of the contemporaneous different forest types is largely the result of this former perturbation. Differences between the history of the South American and African forests are presented in Figure 12.2.

The maximum extension of the African forest seems to have been synchronous with a sudden rise in the sea surface temperatures of the Gulf of Guinea (Maley, 1997). Monsoons pick up moisture from the eastern Atlantic and this rise in water temperature has the effect of increasing sharply the water vapour pressure and ultimately increasing rainfall in the neighbouring continent (see next section). The variations occurring in South America, especially the expansion of open areas during the middle Holocene, were also connected with variations in the sea surface temperatures which include such phenomena as El Niño (Martin *et al.*, 1993).

The opposite behaviour between the western (Lake Bosumtwi in Ghana) and central parts (Lake Barombi Mbo in western

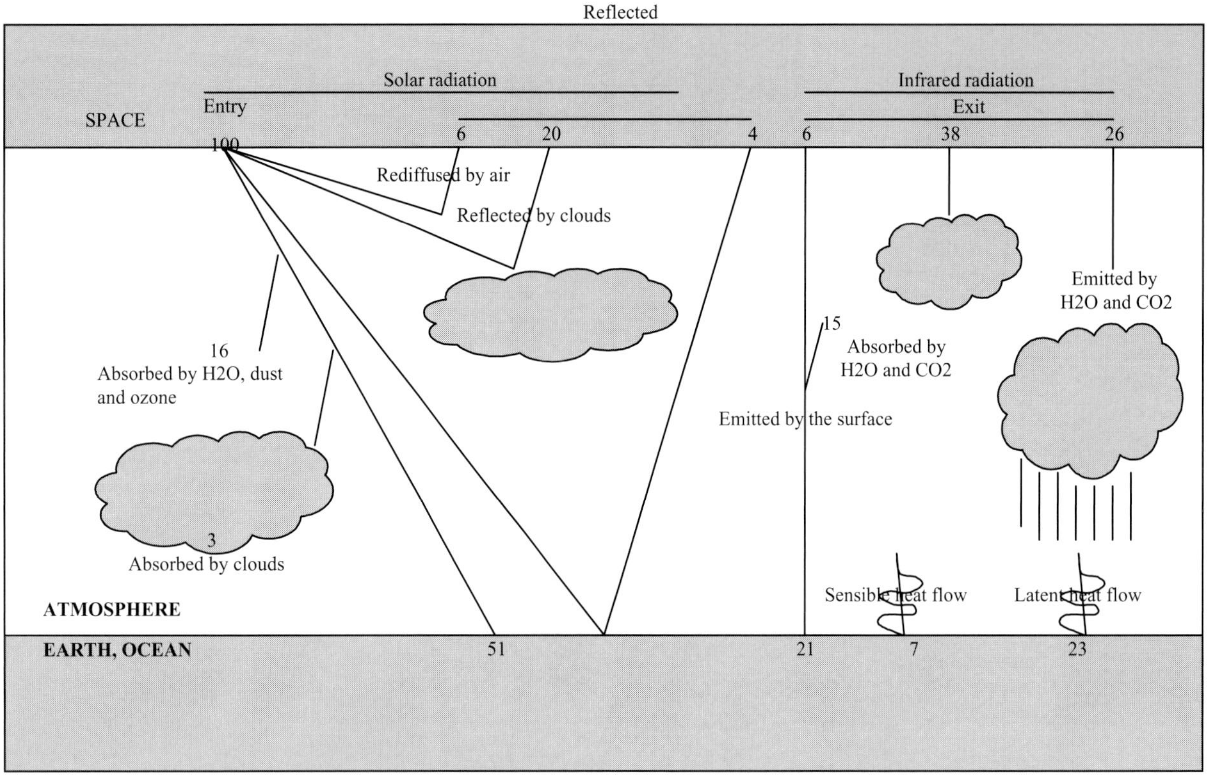

Figure 12.3 Average and overall atmospheric energy. (From Polcher, 1994.)

Cameroon) of the forest domain is to be linked to sea surface temperature variations and also to the particular monsoon features in the south Cameroon and Gabon areas (Maley and Brenac, 1998; Maley *et al.*, 2000).

For the period between 2800 and 2000 BP, rather than speaking of an 'arid phase', the term 'climatic pejoration' is used. This is because this particular climatic phase appears to have resulted from an accentuation of the seasonality due to a shortening of the annual rainy season and, at the same time, to an increase in disturbance lines with cumuliform clouds. This last feature is deduced from the heavy erosion which characterised this period (Maley, 1997). These large paleo-environmental variations cannot be attributed to the actions of Man but rather to global changes in the climate (Schwartz *et al.*, 2000). It would even seem that the perturbations which occurred between 2800 and 2000 BP could have been the cause, or one of the main causes, of the Bantu migration through central Africa (Schwartz, 1992).

RELATIONSHIP BETWEEN FORESTS AND CLIMATE VARIABILITY

It is accepted that vegetation depends essentially on the climate. Its geographical distribution and its seasonal behaviour are influenced largely by rainfall, water being the main component of plants (Bigot, 1997). However, it has also been shown that forest, within the earth-ocean-atmosphere system, has a significant impact on climate, stemming from its use of energy from the global system and its involvement in the water cycle.

Earth-ocean-atmosphere system energy

Earth radiation budget (ERB)

The earth-ocean-atmosphere system receives its energy from the sun in the form of short wave radiation. The sun's rays are reflected back into space in an infrared frequency bandwidth to maintain equilibrium of the radiation budget (Figure 12.3).

The sun's rays interact differently with the Earth's atmosphere and with its surface (Polcher, 1994): (i) the atmosphere reflects a significant part of this radiation (26%) mostly because of clouds, and absorbs only a small amount (19%), and (ii) on the contrary, the surface of the Earth reflects a small amount (4%) and absorbs a large portion (51%) of the short wave radiation received.

The Earth's surface therefore absorbs over half the sun's rays and, to ensure its equilibrium, the Earth reflects this energy in two different forms: a layer of moisture between the surface and the atmosphere engenders a flow of latent heat (linked to evaporation). A flow of sensible heat ensures the diffusion of heat all along this thermal layer between the Earth and its atmosphere (Polcher, 1994).

The heat dispersion that warms the lower atmosphere depends on climatic conditions but even more so on the surface conditions. The continental land surfaces are likely to have an immediate influence on the ERB (Fontaine and Janicot, 1993).

With the horizontal temperature gradients being weak in the Tropics, the atmosphere is very sensitive to land and ocean surface conditions (relief, albedo, temperature, humidity, vegetation), which influence the distribution and the intensity of heat sources and heat sinks, for which the atmospheric response is principally by vertical advection throughout the entire layer of the troposphere (Fontaine *et al.*, 1998a; 1998b). These vertical movements develop most notably within the deep convection systems which, with the associated atmospheric instability, are capable of creating rain clouds and completing condensation of atmospheric water vapour in the convective systems (Bigot, 1997). Such deep convection is responsible for the majority of rainfall in tropical Africa and Amazonia. Contrary to conditions in the middle latitudes, the variations of surface conditions strongly influence vertical movement, with significant repercussions on diverging circulation systems such as those of Hadley and Walker (Fontaine *et al.*, 1998b).

The forest system, because of its great propensity for solar energy absorption and its capacity for evaporation, plays the role of an enormous energy converter: it absorbs more solar energy than any other plant surface. It uses this energy to limit heating and to vaporise water that its root system extracts from the soil (Monteny, 1987). The exchanges of energy that it maintains with the atmosphere influence the physical air mass parameters of the atmosphere layer closest to the Earth (Monteny *et al.*, 1996). This role is linked to several properties, which Polcher (1994) catalogues as three characteristics that determine the sensitivity of the climate to surface processes:

(1) The density of the forest system is such that the albedo is very weak compared to that of bare ground. Land clearing increases the portion of bare ground exposed to the sun's rays and therefore increases albedo.
(2) The high rate of evaporation, comparable to that of oceans, is one of the main characteristics of forests whose leaf density allows them to intercept and re-evaporate a large part of rainfall. The root systems of trees allow them to extract water from a greater portion of the soil than could be done in any other surface system.
(3) The surface variation caused by the different heights of trees within a forest enhances the aerodynamic roughness. This increases turbulence, which is favourable to triggering precipitation.

The two key variables seem to be albedo and soil moisture content. The two are closely linked since wet soil, whether or not covered with vegetation, has a lower albedo and greater evaporation capacity than the same soil when bare and dry (Fontaine and Janicot, 1993). The degree of surface variation is a more delicate parameter to study since it has been shown that on a smooth surface such as a pasture, the presence of a few isolated trees generates more turbulence than a whole forest. Therefore, we will focus on the analysis of the first two variables, the albedo (determining the capacity of forests to absorb solar energy) and the involvement of evaporation in the water cycle.

The role of albedo in plant-atmosphere interaction

The threat of forest removal and eventually the destruction of all tropical forests has led climatologists to investigate the climatic impact of such changes to the Earth's surface (Gash and Shuttleworth, 1991), all the more so because computer simulations have affirmed the sensitivity of the climate to the surface processes (Polcher, 1994).

Historically, albedo is the first variable considered to be connected to recent deforestation. Charney (1975) proposed a now famous mechanism that illustrates the retroaction linked to albedo and its influence on the regional climate. Charney's work showed that an increase in albedo, after a reduction in surface vegetation left a greater amount of exposed bare ground, brought about a decrease in net ground back radiation (i.e. the preponderant term of the energy balance for the study of heat and mass exchange (Monteny *et al.*, 1996)) and thus a decrease in the sum of sensible and latent heat flows. Because of this, a column of atmospheric air would be cooled and this heat loss would be compensated for adiabatically by subsidence (Fontaine and Janicot, 1993) and thus contribute to a reduction in precipitation. In turn, the latter will react positively on the first part of Charney's mechanism by decreasing the amount of vegetation on the surface (Figure 12.4).

Charney's work (1975) gave rise to numerous criticisms: for example, that his mechanism did not in any way incorporate the role of water and especially the humidity of the Earth's surface. This role is essential in tropical regions covered by dense forest, however, where the forest is closely integrated in the water cycle. A diminishing albedo and therefore a higher loss of solar energy at the surface means a significant reduction in energy available for evapotranspiration, thus disrupting the recycling of water and which in turn causes a decrease in the water vapour contribution from the continental land masses (Brou Yao, 1997).

The role of water and the involvement of the tropical forest in the water cycle

Forests are key regions for interactions between the biosphere and the atmosphere. Indeed, it is through evapotranspiration that vegetation recycles moisture locally and influences the regional distribution of precipitation (Bigot, 1997; Bonell, Callaghan and Connor, this volume). In the monsoon regions that characterise a large part of the tropical zone, the quantities of water precipitated

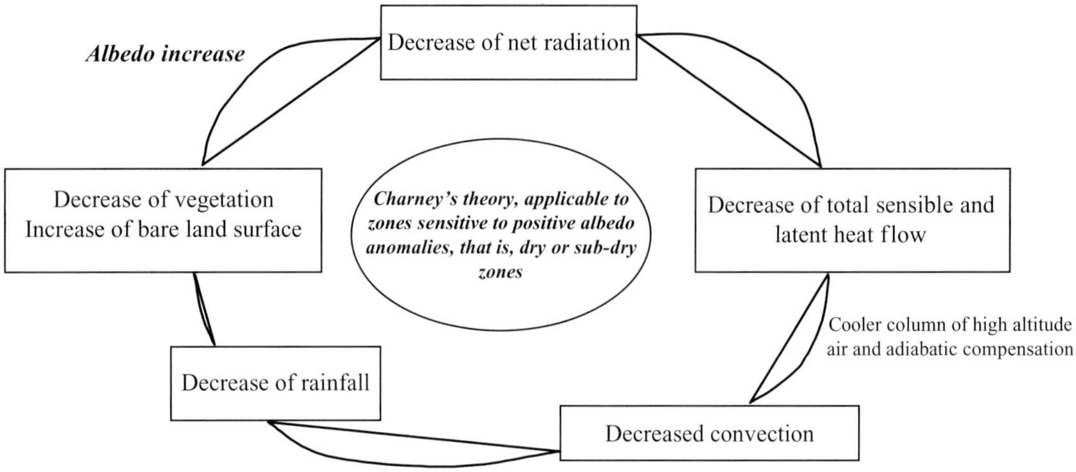

Figure 12.4 Charney's theory (1975) on retroaction of vegetation in dry and sub-dry regions (Polcher, 1994; Fontaine and Janicot, 1993).

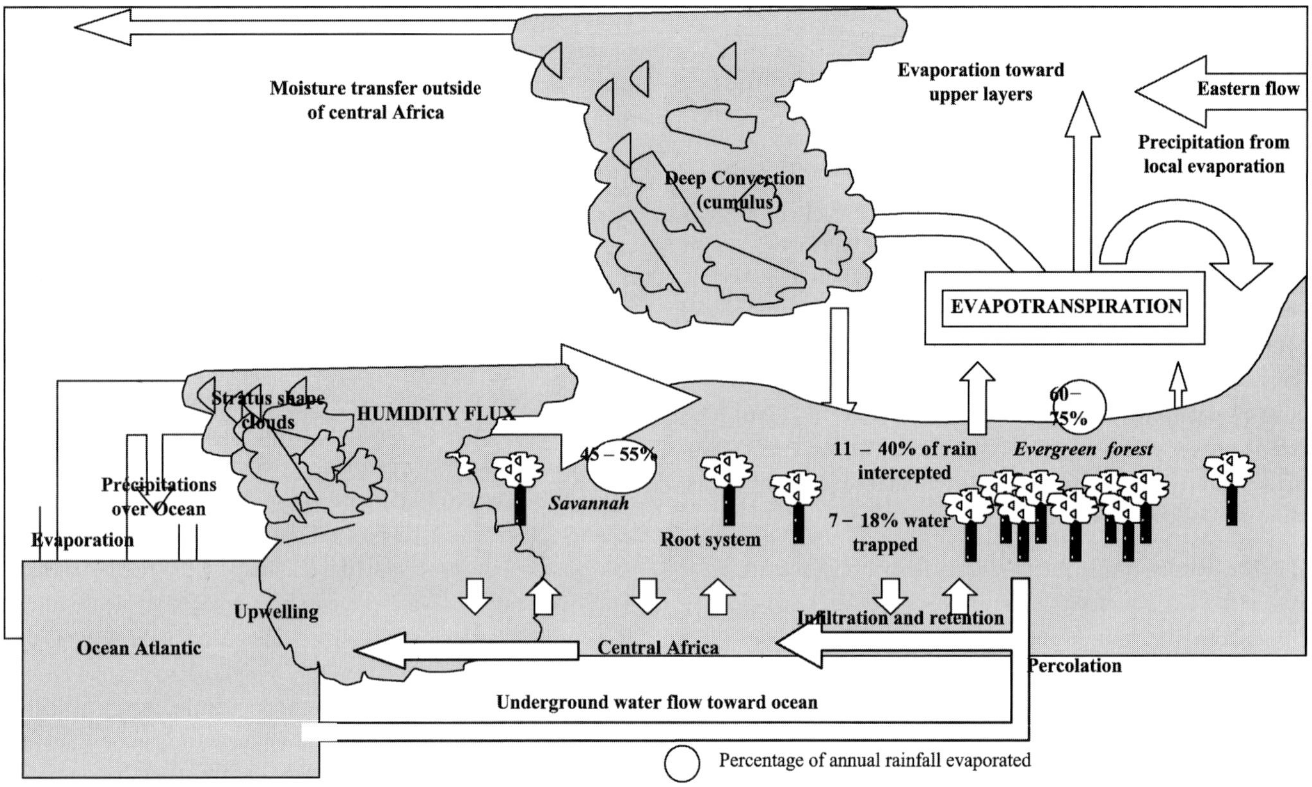

Figure 12.5 Main interactions between the water cycle and the ocean–atmosphere–forest interface. (From Bigot, 1997.)

on the continent come from the condensation of water vapour accumulated in the mass of air as it passes over the ocean. It has been shown that for central Africa a large part of the moisture transfer into the atmosphere, which is generated by evapotranspiration, contributes to the formation of cloud systems (Bigot, 1997) (see Figure 12.5) and that the rainfall associated with these convective systems depends not only on the monsoon flow but also on the recycling of moisture by the forest (Cadet and Nnoli, 1987).

Monteny (1987) described the importance of this water recycling in supplying water vapour to this monsoon flow through the study of favourable conditions for the creation of dense African

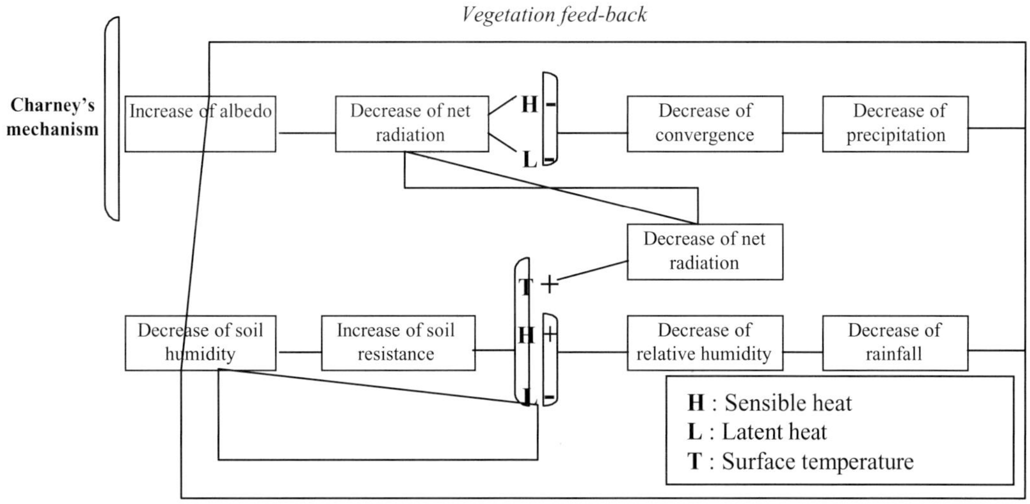

Figure 12.6 Retroaction of soil humidity on Charney's mechanism as presented by Mylne and Rowntree (1992) and Polcher (1994).

forests, conditions which permit water recycling and movement towards regions further to the north. Brou Yao (1997), after Monteny (1987), writes that the tropical rainforests of the southern Ivory Coast inject into the atmosphere the equivalent of 55–75% of the annual precipitation. His work on the Ivory Coast underscored the important role of the forest concerning potential rainfall. This role is twofold, since the forest system is both a receiver of precipitation (especially monsoon rains); and a generator of rainfall by means of evapotranspiration and more local processes (Bigot, 1997). By contrast, Gong and Eltahir (1996) gave a lower estimate (simulated) of recycled precipitation over West Africa of 27%. This raises the question of our understanding of the precipitation process and its recycling over West Africa, a feature which is still very difficult to measure and to take into account correctly in models.

The variation in available moisture at the surface is added to the effects of changes in albedo in the sense that it has a direct link with the radiation budget. Charney *et al.* (1977) reformulated his theory on the increase in albedo and the retroaction of vegetation; the new theory included the role of water and especially changes in the Bowen ratio which establishes the relationship between sensible and latent heat flows (sensible heat flow density divided by latent heat flow density). The results of Charney *et al.* (1977) were similar to those formulated by the 1975 theory. Subsequently Mylne and Rowntree (1992) proposed a mechanism which linked Charney's process to ground moisture (Figure 12.6). Essentially, a decrease in precipitation induced by an increase of albedo (Charney's mechanism) also brings about a decrease in ground moisture. This surface drying leads to a reduction in evaporation through an increase in soil resistance. Bowen's ratio is also lowered since the latent heat flow is reduced which means lower air humidity and, consequently, less precipitation. The sensible heat flow and the soil temperature rise because of this surface humidity decrease, all of which leads to a decrease in net radiation and an amplification of Charney's mechanism (Mylne and Rowntree, 1992). This leads to a reduction in the total of heat flows (sensible and latent), leading to a diminished contrast between land and ocean. Conversely, dense forest increases this net radiation at the surface by the combined effect of weak albedo and high surface humidity, and increases the sum of sensible and latent heat flows. These flows supply the humid static energy (HSE) in the boundary layer and reinforce the HSE layer between Earth and ocean which is the driving force behind the circulation of monsoons (Zheng and Eltahir, 1998).

Highlighting the forest system's contribution to the regional climate leads naturally to a concern about the possible impacts of the current trend in massive deforestation, along with the increased agricultural production and the acceleration of land clearing activities. The impact of such actions is extremely difficult to evaluate and gives rise to considerable uncertainty because the cause and effect links between the forest and climate are not yet clearly understood (Brou Yao, 1997).

Evaluating the climatic impact of forest conversion: modelling global terrestrial vegetation–climate interactions

Even if it is accepted that decreasing forestlands affects, in theory, the climate (Salati and Nobre, 1991), it is difficult to evaluate the impact of massive deforestation on the climate. We are thus obliged to use climate simulation models (Bigot, 1997). But large-scale modelling is not without its problems, given the significant uncertainty that exists concerning the relationships between climate and vegetation. This topic is likely to be debated thoroughly within the next few years, as the resolution of global and regional circulation models is improved.

According to Bigot (1997) the inadequate understanding of the relationships between climate and vegetation is due principally to an incomplete knowledge of average rain fields in tropical forest regions (mainly in Africa), seasonal vegetation activity, and the variability of these two items as a function of climatic anomalies. Accurate parameterisation of the physical processes taking place within vegetation, i.e. resistance to heat and water vapour transfer between soil and atmosphere, amount of surface variation, albedo, interception of a portion of precipitation, are indispensable to all realistic modelling and require multiple field campaigns to gather specific measurements (Fontaine and Janicot, 1993).

Predictions by global atmospheric models are highly sensitive to prescribed large-scale changes in vegetation cover, such as removal of tropical forests (Henderson-Sellers *et al.*, 1993; Polcher and Laval, 1994; Zheng and Eltahir, 1997; cf. Costa, this volume). The majority of recent forest clearance simulations by GCMs indicate what a considerable influence the disappearance of forest cover would have, eventually, on tropical regions. Polcher (1994) expands on Charney's hypothesis by showing that a reduction in sensible heat flow, which could be caused by forest conversion, seems to affect the number of convective events and thus to reduce local precipitation. Zheng and Eltahir (1998) support the hypothesis of the dramatic influence of forest clearance on the north coast of the Gulf of Guinea, which would lead to reduced rainfall and a weakening of moisture convergence in all West Africa. But simulation results from climatic models still yield contradictory conclusions and some of these simulations show that the complete disappearance of the Amazonian forest, while having significant consequences on albedo, would bring about only a slight decrease in local precipitation or evapotranspiration (Bonnefille, 1993; Bonell, Callaghan and Connor, this volume; Costa, this volume).

For validating the hypothesis, Foley *et al.* (1994) suggested the investigation of past environments such as the climate of the early to middle Holocene, some 6000–9000 years ago, for which strong differences in global vegetation pattern are amply documented. According to Claussen (2001), most researchers do not agree on the relative importance of biospheric feedbacks on climate. Moreover, Claussen (1994) discovered the possibility of multiple equilibria in the three-dimensional atmosphere–vegetation system, which seems to be specific to the subtropics and particularly to North Africa, as the high-latitude climate–vegetation system is much more stable (Levis *et al.*, 1999). Two solutions seem to be possible: the arid, present-day climate and a humid solution resembling more that of the mid-Holocene, some 6000 years ago, with a Sahara greener than today. Simulations with mid-Holocene vegetation yield only one solution, the green Sahara (Claussen and Gayler, 1997), while for the Last Glacial Maximum (LGM) some 21 000 years ago, two solutions exist (Kubatzki and Claussen, 1998). Brovkin *et al.* (1998) show that for the present-day climate, the green equilibrium is less probable than the desert equilibrium, and this explains the existence of the Sahara desert as it is today. The difference in albedo between desert and vegetation cover appears to be the main parameter that controls an existence of multiple stable states. Claussen (1997), Claussen and Gayler (1997) and Claussen *et al.* (1998) explain this positive feedback by an interaction between the high albedo of Saharan sand deserts and the atmospheric circulation as hypothesised by Charney (1975), whilst Texier *et al.* (1997) suggest an additional feedback between sea-surface temperatures and land-surface changes. Claussen *et al.* (1998) found that the velocity potential patterns, which indicate divergence and convergence of large scale atmospheric flow, differ between arid and humid solutions mainly in the tropical and subtropical regions. It appears that the Hadley-Walker circulation shifts slightly to the west. For the mid-Holocene boreal summer, the large-scale atmospheric flow is already close to the humid mode, even if one prescribes present-day land surface conditions. This is caused by differences in insolation: in the mid-Holocene boreal summer, the Northern Hemisphere received up to 40 $\mathrm{W m^{-2}}$ more energy than today, due to a change in the ellipsoid orbit of the Earth around the sun, thereby strengthening the African monsoon (Kutzbach and Guetter, 1986). During the LGM, insolation was quite close to present-day conditions.

After using a coupled atmosphere-vegetation-ocean model, Ganopolski *et al.* (1998) conclude that the biospheric feedback dominates in the subtropics, while SST adds only a little. Claussen *et al.* (1999) clearly show that subtle changes in the seasonal ellipsoid orbital forcing triggered changes in the North African climate. Such changes were then strongly amplified by biogeophysical feedbacks in this region, leading to a rather fast desertification within a few centuries, starting around 5500 years ago. This seems to be in agreement with palaeogeological reconstructions (Petit-Maire and Guo, 1996).

De Noblet *et al.* (2000) compared two GCMs (LMD 5.3 and ECHAM 3) coupled asynchronously to an equilibrium biogeography model to give steady-state simulations of climate and vegetation 6000 years ago, including biogeophysical feedback. They found surprisingly different results of simulation of climate and vegetation for 6000 years ago, neither GCM being fully realistic and both being unaffected by the choice of green or modern initial conditions, due to inadequate strength in the tropical summer circulation in the GCMs. Such results highlight the importance of correct simulation of atmospheric circulation features for the sensitivity of climate models to changes in radiative forcing, particularly for regional climates where atmospheric changes are amplified by biosphere–atmosphere feedbacks.

For Claussen *et al.* (2001) biogeochemical and biogeophysical processes do not operate independently and, being triggered by large-scale land cover changes, oppose each other on the global scale. Tropical forest conversion tends to warm the planet because the increase in atmospheric CO_2 and hence, atmospheric radiation, outweighs the biogeophysical effects. Thus the sensitivity of the tropical climate to forest conversion remains open to debate. The change in surface energy transfer exchange between latent energy and sensible heat flows, linked to this degradation of forest cover, could have an impact on precipitation but at the present time we are not able to evaluate its magnitude (Polcher, 1994). Indeed, even if 50–70% of the western African forest lands have been transformed into agricultural use or left fallow, the magnitude of this transformation does not seem to have yet affected the regional climate significantly (Bonnefille, 1993).

Elsewhere, according to Bonnefille (1993), the recent 'desertification' has resulted more from a change in rainfall distribution than from a reduction in the total amount of rainfall. This leads the author to deduce that the regional climatic variations that have been highlighted by the historical data are not only the result of human activities but may also be attributed to a large degree to *natural climatic fluctuations*.

CLIMATIC VARIABILITY IN TROPICAL FOREST REGIONS

Climatic variability is defined as being the distribution of climatic elements around their average values calculated over 30 years; this natural variability is an intrinsic characteristic of climate (Janicot, 1995). A major objective of climatology is to understand the variations in the succession of average states of the atmosphere and to forecast these climate variations, particularly in the tropics, where climatic variability has a profound impact on the lives of the people there and on the evolution of ecosystems (Fontaine *et al.*, 1998b).

Climatic variability and impact on rainfall and runoff in West and Central Africa

Bonell (1998) showed that only a small section of the hydrological community was studying climate change and its linkages with hydrology. This is mainly due to uncertainties in climate projections, insufficient knowledge of land-cover changes, and the inadequacies in the resolution of GCMs (too coarse) and the small scale of actual development problems in hydrology and water resources. Nevertheless, significant results have been obtained on rainfall and runoff variability and their links with climatic variability, on the basis of decades of measurements and studies in most of the Western and Central African countries (Mahé and Olivry, 1995; Bricquet *et al.* 1997; L'Hote and Mahé, 1996; Servat *et al.*, 1998).

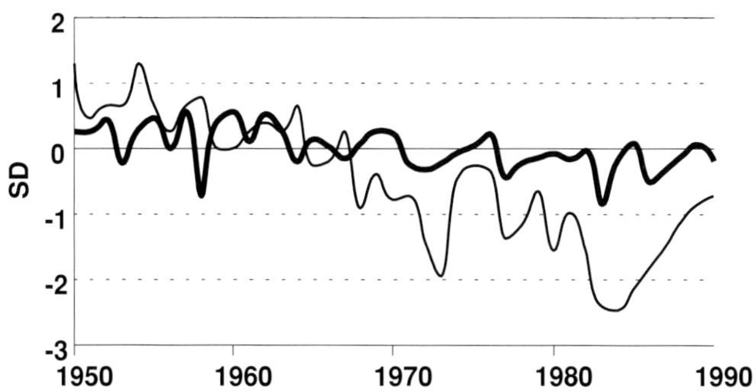

Figure 12.7 Regional rainfall interannual variability (standard deviation) over the period 1951–89. Bold line: humid Africa; thin line: dry Africa. Humid Africa is taken south of 8° N, including western coast of the Gulf of Guinea and the Zaire Basin. Dry Africa is north of 8° N from Senegal to Chad.

Mainly annual and monthly time series from the beginning of the century have been analysed for hundreds of raingauge stations over regional areas (Wotling *et al.*, 1995; Servat *et al.*, 1997; Paturel *et al.*, 1998; Mahé *et al.*, 2001) and these show that the recent drought period affected not only the Sahel but also that there was a decrease in rainfall over the more humid parts of tropical and equatorial Africa along the Gulf of Guinea. Statistical tests for detection of discontinuities in time series have been applied (Hubert *et al.*, 1998), showing that the discontinuities in rainfall time series were often observed around the year 1970, with some regional variability (Figure 12.7).

Runoff series have been also studied: annual and monthly flows, and low flows (Servat and Sakho, 1995; Aka *et al.*, 1996; Servat *et al.*, 1997; Mahé and Olivry, 1999; Laraque *et al.*, 2001) have also declined since 1970. Seasonal floods have changed: in tropical West African basins their magnitude is lower and their rise and decrease is more rapid since 1970, which is partly due to an increase in the recession coefficients (Bricquet *et al.*, 1997; Mahé *et al.*, 2000). In equatorial Africa one major impact of climatic variability on river regimes is to shorten and reduce the magnitude of the boreal spring flood since 1970, as is particularly evident in the case of the Ogooue and the Kouilou rivers (Mahé *et al.*, 1990; 2000) (Figure 12.8). These studies showed that the variability in runoff is greater than that for rainfall, mainly in the case of tropical and Sahelian rivers, and is most likely due to a decrease in groundwater levels. The same observations apply also for many equatorial rivers, except for special cases where the aquifer plays a major buffering role (Laraque *et al.*, 2001).

Table 12.1 shows rainfall and runoff decadal variability over eight regional areas. Over the period 1971–1989, the rainfall decrease was stronger (−25%) in north-western West Africa (Senegal, Guinea, Mauritania). Over Central West African areas

Table 12.1. *Decadal variations of precipitation (P) and runoff (Q) over eight regional areas in West and Central Africa: percentage deviation from the 1951–89 average.*

		1951–60	1961–70	1971–80	1981–89	Cumulative 1971–89[a]
Senegal/Gambia Rivers	P	+23.0	+13.0	−8.5	−16.5	−25.0
North Guinea: Rivers Corubal, Konkoure	Q	+32.6	+23.6	−24.1	−35.7	−59.8
South Guinea, Sierra Leone	P	+10.3	+5.2	−3.5	−13.3	−26.8
and Liberia Rivers	Q	+19.6	+15.7	−9.3	−28.8	−38.1
Niger River	P	+11.3	+3.1	−4.2	−11.2	−15.4
	Q	+14.8	+13.4	−8.7	−21.5	−30.2
North Coast of Gulf of	P	+9.3	+4.6	−5.5	−9.4	−14.9
Guinea: Ivory Coast, Ghana, Togo, Benin	Q	+23.4	+21.8	−18.4	−29.9	−48.3
Coastal rivers of Nigeria,	P	+3.1	+7.4	−1.4	−9.6	−11.0
Central Cameroon: Mungo, Wouri, Sanaga	Q	+10.5	+12.6	−9.3	−15.3	−24.3
Angola, incl. Cubango and	P	+2.6	+8.3	−5.2	−6.1	−11.3
Cunene Rivers, and except Zaire River Basin	Q	+1.2	+8.7	−6.9	−4.0	−10.9
South Cameroon: Nyong	P	+1.7	+3.6	−3.2	−1.4	−4.6
and Ntem Rivers Gabon/ Congo: KouiLou/Ogooue/ Nyanga Rivers	Q	+1.2	+11.5	−6.9	−3.9	−10.8
Zaire/Congo River	P	+1.3	+3.2	−2.9	−0.6	−3.5
	Q	−4.0	+14.7	−1.8	9.9	−11.7

[a] The righthand column gives the cumulative deviation over the last two dry decades 1971–89.

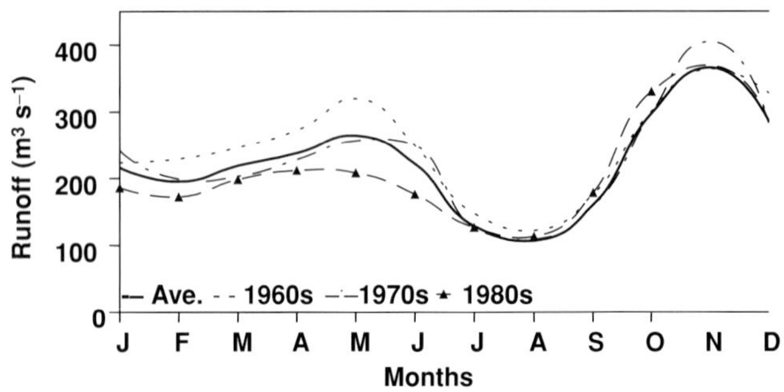

Figure 12.8 Decadal variability of the monthly flows (runoff, m^3s^{-1}) of the Kouilou Basin, Republic Congo Brazzaville.

the rainfall decrease was smaller (−15%). The rainfall deficit diminished towards the Equator: −11% over Cameroon, −4% over Gabon and Congo, and −3% over the Zaire River basin. The runoff decrease also diminished from the Sahel to the Equator: from −60% for the Senegal region, to −11% for the Zaire River. In addition, good relationships have been shown for the interannual variability in rainfall, low flows and groundwater levels for the Bani tropical river in Mali (Mahé *et al.*, 2000).

Adaptations of river flow simulation models have been developed, mostly at the monthly time step but sometimes at lower time scales (daily or decadal) (Servat *et al.*, 1990; Servat and Dezetter, 1991; 1993; Paturel *et al.*, 1995), which helped in the study of the sensitivity of each water cycle component of tropical basins. A recent study (Mahé *et al.*, 2002) showed that the river flows of a Sahelian basin have increased significantly since changes in land cover (increased deforested areas for agriculture) took place and despite the reduction in rainfall. Simulations of river flows were improved by modifying the annual values of the soil water holding capacity, with respect to the satellite-observed land cover changes during the period.

In the Ivory Coast, Brou Yao *et al.* (1998) observed that the coffee and cocoa producing areas shifted following the rainfall changes, and that the linked deforestation might have had a significant impact on local precipitations.

The diminution in rainfall since 1970 has been well described. It concerns the whole of West Africa, and Central Africa to a lesser extent. Runoff diminution is amplified by the groundwater level decrease, which causes less groundwater to participate in the surface runoff during the flood, and which also causes an increase in the speed of the flood recession (Mahé *et al.*, 1998). The changes in land-cover seem to have a great impact on the hydrological cycle and the rainfall-runoff relationships. The impact of the forest clearance on the rainfall/runoff relationships seems to be dependent on the type of climate/vegetation system. In Sahelo-Sudanian areas, the forest clearance, associated with an increase of agricultural activities, induces a rapid destructuring of the top layer of the soil such that infiltrability decreases and runoff coefficients increase, even during a dry phase. In equatorial and tropical humid areas, the major impact of the forest conversion is to reduce the local evapotranspiration, thus reducing the total amount of available water vapour for precipitation recycling.

Sensitivity of the tropical climate to surface conditions

In the lower latitudes, the seasonal variation in temperatures is slight. The organisation of average tropical and subtropical climates is therefore most often a function of total rainfall and seasonal rhythms creating alternate dry and wet seasons, and thus reflecting the activity and variations of the water cycle in the atmosphere. This water cycle activity is expressed in terms of evapotranspiration, water vapour flows and precipitation (Fontaine *et al.*, 1998a). The surface conditions, especially the plant cover of dense rainforests and the upper layer of oceans, have a significant effect on the atmospheric water cycle and also on the vertical movements within the tropical atmosphere.

The space-time variability of climate depends principally on the interaction between the surface conditions (temperature, albedo, humidity) and the atmosphere: the linkage manifested by wind pressure and sensible and latent heat flows. Recent research focuses on the heat content of the mixed global ocean layer, in other words, the upper layer subjected to the action of surface winds and in which the thermal gradients are weak (Fontaine *et al.*, 1998a).

The ocean–atmosphere linkage

Covering 70% of the Earth's surface, the oceans constitute its largest water reservoir and an important reservoir of energy in tropical latitudes. Moreover, and contrary to the continents, oceans possess a strong thermal inertia, which makes the sea surface temperature (SST) the most influential variable on the atmosphere and at the same time, an indicator of land climatic variability (IPCC, 1996).

As noted by Bigot (1997), the three essential parameters controlling the ocean–atmosphere linkage are:

(1) SSTs (thermodynamic estimation of the ocean–atmosphere interface),
(2) precipitation (physical estimation of one element of the atmospheric water cycle),
(3) wind (dynamic estimation of changes in atmospheric movement).

Even if the ocean's annual thermal amplitude is slight, due to its weak specific heat, the linkage is intensified by the release of latent heat. This energy release influences surface winds, which then change upwellings, the ocean drift currents, and thus the SSTs (Bigot, 1997).

SST variations determine significant atmospheric responses (Fontaine *et al.*, 1998b) and, moreover, the ocean's strong thermal inertia defines the characteristic time steps of the ocean–atmosphere linkage, and accordingly, those of low latitude climatic variability (Fontaine *et al.*, 1998a). This SST influence is also found seasonally, where the monsoon pattern is its most obvious manifestation (Fontaine, 1991; Fontaine *et al.*, 1991).

Interannual and multiannual scale: the Southern Oscillation

One of the most important expressions of the ocean–atmosphere linkage is the phenomenon of the El Niño–Southern Oscillation (ENSO) in the equatorial Pacific (Folland *et al.*, 1998; Fontaine *et al.*, 1998a; Bigot, 1997). Although the consideration of oceans in the climatic equation is rather recent (since only the beginning of the 1980s), the fact is that oceans are now the object of innovative research and have a special place in the global scientific arena, being the main element of the earth climate and playing a major role in the earth's climatic variability. This is attributable to the increased intensity of ENSO related events, in particular, that of 1982–1983. Unusual in its intensity and singularity, this event was an 'extraordinary catalyst' for research and permitted scientists to confirm ENSO as an important element in the ocean–atmosphere linkage on an interannual and multiannual basis (Fontaine and Janicot, 1993).

ENSO is deemed to be the dominant cause of interannual and long term climate variability in the world (Trenberth, 1997; Callaghan and Bonell, this volume), particularly in the tropics where it influences the rainfall in many countries such as in India, eastern, southern and north-eastern Africa as well as in Australia (Janicot, 1999). Indeed, few tropical regions are not affected by ENSO (Fontaine and Janicot, 1993) and this phenomenon has a considerable and continuous impact on temperature ranges and precipitation, not only around the Pacific and Indian Ocean but also in the tropical Atlantic regions where ENSO has been associated with SST and trade wind variability (Hastenrath *et al.*, 1987; Nicholson and Kim, 1997; Fontaine *et al.*, 1998a).

The relationship between ENSO and interannual climate variability in Central Africa was well documented by Bigot (1997) who associated ENSO with a decrease in cloud convection and precipitation, caused mainly by a slowing of western winds that transport water vapour. Other studies which came to the same conclusions were done in Guinean Africa (Janicot and Fontaine, 1997), tropical Amazonia (Fu *et al.*, 1999), Uruguay and Brazil (Diaz *et al.*, 1998), and indicate a notable reduction in rainfall in tropical forest regions during the warm phase of the Southern Oscillation.

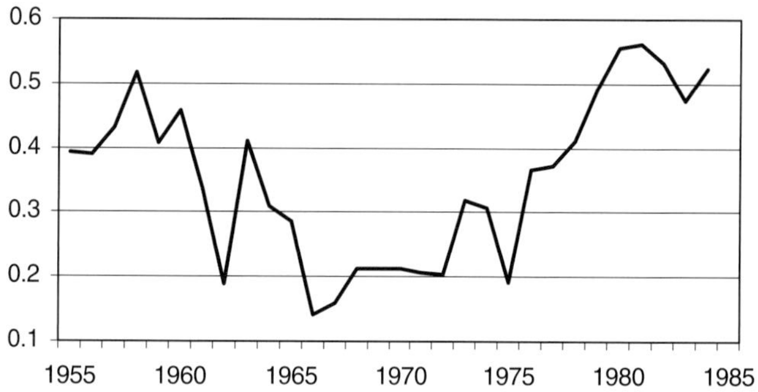

Figure 12.9 Evolution of correlation between ENSO index and Sahelian rainfall during the period 1955–84 (From Janicot *et al.*, 1996.)

THE EVOLUTION OF ENSO EVENTS AND THEIR REPERCUSSIONS IN THE TROPICS SINCE 1970

Since the 1970s, the intensity and frequency of ENSO events have changed, as well as their impacts on tropical regions. Indeed, we can observe that, starting around 1975, there was an increase in the magnitude of positive anomalies in the Southern Oscillation Index (SOI), occurring with increasing frequency in relation to global warming. Similarly, a number of studies have shown an evolution in the correlations between anomalies of SST fields in the equatorial Pacific and precipitation in tropical regions (Nicholson and Entekhabi, 1986; Moron *et al.*, 1995; Janicot *et al.*, 1996; Semazzi, 1996).

In central Africa, we can see an increase in the amplitude of annual rainfall during ENSO events after the 1970s, notably in 1977, 1982 and 1987 (Bigot, 1997). A study of the precipitation impact in western Africa of ENSO events created some controversy because of the high interannual variability in the evolution of correlation of the SOI and western African rainfall (Figure 12.9). Although these correlations were not significant before the 1970s, they have become important in the last 25 years, indicating correctly weak rains in the Sahel region after an ENSO warm phase and heavy rains after an ENSO cold phase (Janicot *et al.*, 1996).

This evolution in relationships between ENSO events and Sahelian climate variability can be explained, first by an observed increase in ENSO's interannual variability, then by an increase in the number of major warm events over the past 20 years and lastly, by the positive correlation between rainfall and the ten-year global SST variations.

The relationships between the Southern Oscillation and precipitation in tropical regions tends to show the links between SST variations in the Pacific and those recorded in the Atlantic Ocean. During an ENSO event, there is a positive correlation in SST anomalies in the north of the tropical Atlantic with a delay of a few months (Uvo *et al.*, 1998). There are also some complex interrelations, not yet well understood, between sea surface temperature variability of the Atlantic and Pacific oceans and rainfall in tropical Africa (Bigot, 1997; Janicot and Fontaine, 1997) as well as in tropical South America (Fu *et al.*, 1999; Diaz *et al.*, 1998).

The particular geometry of the Atlantic Ocean, which is very wide near the Equator and narrower in high latitudes, introduces other phenomena linked to ocean dynamics and gives special importance to meridional thermal gradients. Tropical Amazonia, like western and central Africa situated on the Atlantic perimeter, is subject to the double influence of the Atlantic and the Pacific which affect both the Hadley and Walker circulations (Fontaine *et al.*, 1998a). Recent studies have also showed signs of multi-annual and decadal variations specific to SST anomalies in the Atlantic Ocean (Janicot, 1999; Fontaine *et al.*, 1998b), separate from the impact of ENSO.

An approximate ten-year cycle: the tropical Atlantic

Numerous studies (Bigot, 1997; Fontaine *et al.*, 1998a, 1998b, 1999) have distinguished two regional modes of SST variability: in the north Atlantic and in the southern equatorial Atlantic. The natural variation of these two modes induces almost ten-year meridional gradient fluctuations in the tropical Atlantic SST anomalies which dominate one portion of the long-term surface thermal structure of this ocean (Fontaine *et al.*, 1998b).

The Atlantic thermal dipole has a significant impact on the variability of precipitation over neighbouring continents. It is associated with the atmospheric circulation in the equatorial and south Atlantic and has a major influence on the positioning of the monsoon shearline (see Callaghan and Bonell, this volume), which modulates a large part of tropical precipitation and defines the seasonal rain cycle in regions such as western Africa. When the tropical north Atlantic is abnormally warm and the south Atlantic abnormally cool, the northern monsoon shearline tends to migrate farther to the north during the rainy season. Conversely, when the tropical north Atlantic is cooler and equatorial and south Atlantic are warmer, the southern monsoon shearline is the most active (Janicot, 1999).

This modulation and meridional movement of the monsoon shearlines influences precipitation in tropical countries that have monsoon seasons. In western Africa, weakening of south and equatorial Atlantic circulation, characterised by a weak St Helene's anticyclone – sometimes shifted towards Brazil – plus a warming of south Atlantic waters and reduced equatorial and coastal upwellings, results in a particularly dry season in the Sahel (Fontaine, 1991; Mahé and Citeau, 1991; Mahé, 1993). Similarly, studies on central Africa by Bigot (1997) affirmed a major influence of this meridional Atlantic thermal gradient on annual rainfall in the Congo and Gabon, via its influence on the convergence zones' meridional position and the associated conveyance of moisture over central Africa. Wotling *et al.* (1995), found good relationships between West African rainfall variability and Atlantic SSTs. More recently, a strong oceanic influence on Sahelian rainfall was also found during global (Koster *et al.*, 2000) and regional (Dolman *et al.*, in press) atmospheric modelling exercises.

Moreover, this decadal variability of the SST has amplified ENSO's impact in different areas of the globe by increasing the correlation between precipitation and ENSO events in central Africa. The same phenomenon has also affected the cumulative rainfall in tropical Amazon's rainy season (Fu *et al.*, 1999). However, the relationship between the Pacific and Atlantic Oceans' SSTs are still fuzzy because on the one hand, the cause and effect links remain difficult to discern (Diaz *et al.*, 1998) and on the other, because the Atlantic falls into an intrinsic mode of SST variability such as is more usually identified with multi-decadal intervals (Fontaine and Janicot, 1993).

The multi-decadal scale: the inter-hemisphere reversal

On a larger area scale, studies based on SSTs recorded all over the globe during the past 20 years show that the thermal structure of the surface is dominated by a third type of variability. This is a multi-decadal type of variability that affects the gradient between the two hemispheres and, at the same time, the zonal gradients at the Equator through the relative warming of the southern and equatorial basins as well as the Indian Ocean (Fontaine *et al.*, 1998a). This mode of low frequency variability describes a structure that associates warm (cold) anomalies in southern hemisphere oceans and the Indian Ocean with cold (warm) SST anomalies in the north Atlantic and north Pacific oceans (Fontaine and Janicot, 1993). This mode of variability, dominated by a slow inter-hemispheric reversal, is independent statistically from the two other regional modes described above and defines fluctuations in the Atlantic meridian thermal gradients in cycles of almost ten years (Fontaine *et al.*, 1998b).

When the large, dipolar, multi-decadal system experienced warm anomalies in the south or cold anomalies as far as 30 °N, a correlation was made with dry periods in western Africa (Folland *et al.*, 1986; Fontaine *et al.*, 1998b; Fontaine and Janicot, 1993). The large-scale low frequency oceanic influence is then associated with other regional fluctuations, which thus explains in large part the variability in rainfall over western Africa (Fontaine *et al.*, 1998b). Studies in Guinean Africa (Janicot and Fontaine, 1997) and central Africa (Bigot, 1997), linked the general influence of tropical forest rainfall to the change in SSTs since 1945. Such changes corresponded to ocean warming in the Southern hemisphere and the Indian Ocean, and simultaneous cooling of the oceans in the Northern hemisphere. The effect of this trend is to strengthen the inter-hemispheric gradient in the Boreal winter which is conversely reduced in the Boreal summer (Janicot and Fontaine, 1997). There is a corresponding significant impact on the seasonal northerly migration of the northern monsoon shearline, which has caused a slow reversal of inter-hemisphere SST anomalies in the world's oceans since the 1970s. As a result, all of Africa is often subject to the same fluctuations in precipitation anomalies, despite some modifications from regional anomalies (Mahé, 1993; Mahé and Olivry, 1995; Bigot, 1997).

The recent climatic variability observed during the last 30 years leads us to focus on more recent studies that have dealt with trend detection in climatic variables.

Recent fluctuations in climate: the trend towards increasing temperatures and decreasing rainfall over West Africa

Since the end of the 19th century, i.e. the end of the Little Ice Age, the Earth's climate has been affected by a large-scale warming trend that has resulted in an average increase in global air temperature of 0.5 °C (Janicot, 1995). However, this trend has been evolving at different rates in the two hemispheres over the last few decades, with the northern hemisphere warming more slowly than the southern hemisphere (Figure 12.10). Consequently, air temperatures as well as surface water temperatures are lower in the northern hemisphere.

This temperature change differential between the two hemispheres may be compared by using the bias in recorded cumulative rainfall in intertropical regions of the northern hemisphere since the 1960s (Figure 12.11). A dry period is especially pronounced over the last 30 years in Guinean and central Africa as well as in the Sahelian region, where dependence on rain-fed agriculture by the population makes it all the more dramatic (Janicot and Fontaine, 1997; Bigot, 1997; Brou Yao *et al.*, 1998). A temporal drying phase of a tropical climate also has a major impact on forest regions where rainfall is a critical factor for the growth of vegetation (Janicot, 1995), which is further increased by the intensification of human activities causing land degradation. Even if it seems clear that such anthropogenic disturbances have aggravated

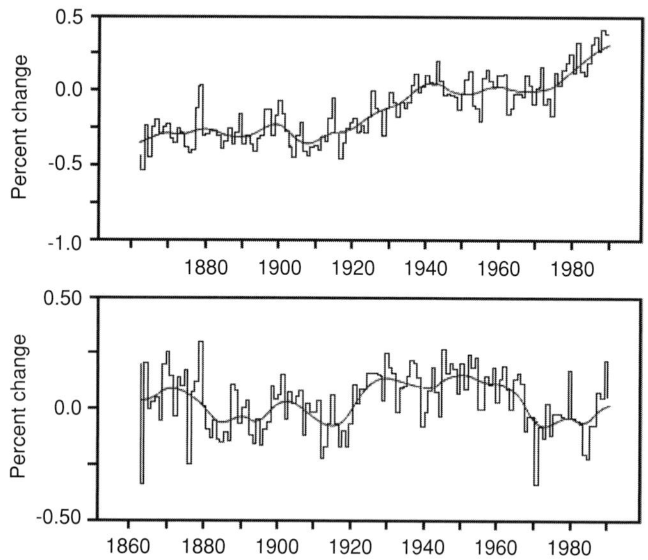

Figure 12.10 Changes in air temperature differentials since the middle of the nineteenth century. (From Janicot, 1995.)

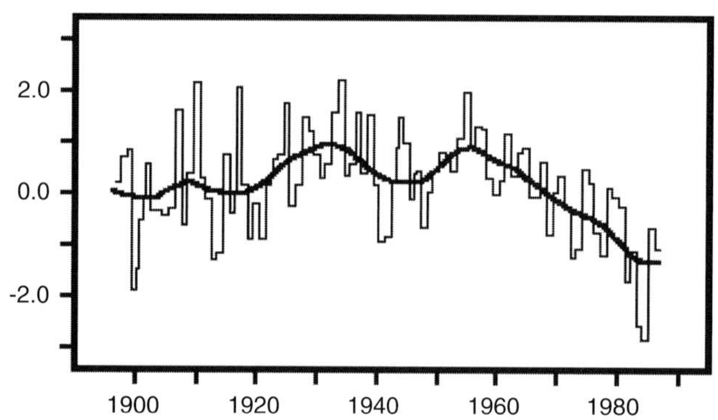

Figure 12.11 Change in cumulative rainfall in intertropical regions of the northern hemisphere, Janicot (1995).

this global climatic trend, it is very difficult to separate the relative contributions of natural variations and human activities to long-term climatic variability. Shukla (1998) demonstrated that tropical atmospheric flow paths and rainfall, especially over the open ocean and in the more 'maritime' tropics, are so strongly determined by the underlying SST that they show little sensitivity to changes in initial conditions of the atmosphere (humid or dry). Shukla further hypothesised that the latitudinal dependence of the rotational force of the earth and solar heating together produced the unique structure of the large-scale tropical motion field, such that, for a given boundary condition of SST, the atmosphere is stable with respect to internal changes. As a result, it is now quite possible, once an ENSO event has begun, to predict its growth and maturation for the following 6–9 months (Shukla, 1998). Furthermore, global-scale simulations by Koster *et al.* (2000) showed that land and ocean processes have rather different domains of influence in the world. The amplification of precipitation variance by land-atmosphere feedbacks appears to be more important outside of the (tropical) regions that are affected most by SST. In other words, the impact of land cover on the precipitation signal is expected to be muted in regions with a large oceanic contribution, such as South East Asia and the Pacific, West Africa, the Caribbean side of Central America and north-western South America (Koster *et al.*, 2000).

The variations in thermal structure of the ocean's surface explain the variability of tropical climate to a large degree, especially regarding precipitation. Over 41% of rainfall variation in central Africa is due to deep water convection according to Bigot (1997); between 25 and 40% variation in rainfall in Guinea (Janicot and Fontaine, 1997), and 50% of the yearly variability in Sahelian rainfall (Fontaine and Janicot, 1993). Nevertheless, more than half of the rainfall in these tropical regions is not explained by SST variations. The increase in anthropogenic activities over the last 50 years could have contributed significantly to the current climatic trend.

Indeed, the increase in carbon dioxide, relative to increased industrial activities, is often evoked as a cause of global warming (Houghton, 1994; Le Houerou, 1993; Duglas, 1993). But for the extensive tropical forest regions, such as in central and western Africa and the tropical Amazon, recent studies have focused more on the impact of deforestation on climate fluctuation. In these regions sensitive to surface conditions where the atmospheric humidity has land mass origins (Fontaine, 1991), an increase in drylands would provoke warmer air temperatures. This is firstly because there is a greater warming of air by heat transfer and secondly, because the reduction of forest cover which would naturally absorb carbon dioxide in the atmosphere, contributes to the increase in greenhouse gases in the atmosphere (Janicot, 1995; Myers, 1991; Myers and Goreau, 1991; Keller *et al.*, 1991). However, it is pertinent to note that in a recent simulation of interdecadal climate variability in the Sahel, Zeng *et al.* (1999) found that incorporation of land-surface characteristics (albedo, soil moisture status) in a coupled atmosphere-ocean circulation model did not improve the correlation between observed and predicted year-to-year rainfall variability, but substantial improvement was obtained for interdecadal (>10 yr) rainfall variability. The authors ascribed the lack of influence exerted by land-surface feedbacks at the short term to the existence of a phase lag between the occurrence of rainfall and the adjustment (recovery) of the vegetation.

Recent work by Brou Yao *et al.* (1998) on rainforests in Ivory Coast underscores the similar relationship between rainfall and forest surface area: a marked increase in albedo after 1970 and massive forest conversion, while the decrease in forest cover has reduced the land-based component of the water cycle and therefore diminished the amount of water that is recycled in the atmosphere.

All of these factors seem to have contributed to the reduced rainfall of the 1970s. Nevertheless, there is still significant uncertainty as to the relative contributions of anthropogenic effects and natural variations in climate to the current trend of global warming and decreased rainfall (Janicot, 1995).

CONCLUSIONS

Recent findings in paleoclimatology, coupled with reconstructions of the past interactions between the climate and the vegetation cover in low latitudes has stimulated recent research to evaluate the connections between surface processes and the regional climate. Schemes for these surface processes are now being integrated within General Circulation Models (GCMs). Most of the results of these simulations support the palaeoclimatic evidence of the major role of the vegetation cover on GCM simulations of climatic variability. Difficulties remain, however, in realistic representation of the biosphere dynamics of the climatic modelling, due to the inadequate representation of the summer tropical atmospheric circulation in GCMs, and to their too large resolution. There remains also an uncertainty about biogeochemical and biogeophysical process interactions, which seem to oppose each other at the global scale, with recent studies indicating that the atmospheric feedback on climate may be more important at a global scale than the biogeophysical feedback.

There are some clear examples of strong hydrological impact of forest clearance on runoff. For example, the diminution of rainfall and runoff since 1970 concerns the whole of West and Central Africa. The changes in land-cover seem to have a great impact on the hydrological cycle and the rainfall-runoff relationships. The impact of forest clearance on these rainfall/runoff relationships seems to be dependent on the type of climate/vegetation system. In Sahelo-Sudanian areas, the forest clearance associated with an increase in agricultural activities, induces a rapid destructuring of the top layer of the soil with a resulting decrease in infiltrability and an increase in runoff coefficients. But in more humid tropical and equatorial areas, such a correspondence is not observed. In equatorial humid areas the major impact of the forest conversion is to reduce the local evapotranspiration, thus reducing the total amount of available water vapour through local recycling for monsoon rainfall. Due to lack of direct measurements, it is very difficult to estimate the impacts of a massive forest conversion on climate dynamics as well as on evapotranspiration.

Numerous uncertainties still remain about the knowledge of vegetation-climate relationships and about the part that anthropogenic forcing plays in the current climatic evolution at the global scale. Nevertheless, the human impact on local surface conditions is obvious, at least at the regional scale of forest conversion where hydrological changes are clearly connected with an increase in agricultural activities.

References

Aka A. A., Servat E., Paturel J. E., Kouame B., Lubes H., Masson J. M. 1996. Analysis of the temporal variability of runoff in Ivory Coast: statistical approach and phenomena characterization. Hydrological Sciences Journal, 41, 6, 959–970.

Bigot S. 1997. Les précipitations et la convection profonde en Afrique Centrale: Cycle saisonnier, variabilité interannuelle et impact sur la végétation. Thèse de docorat de l'université de Bourgogne, 283 p.

Bonell M. 1998. Possible impacts of climate variability and change on tropical forest hydrology. Climatic Change, 39, 215–272.

Bonnefille R. 1993. Afrique, paléoclimats et déforestation. Sécheresse, 4, 4, 221–231.

Bonnefille R. 1986. Modifications des végétations intertropicales et changements climatiques. In: Changements globaux en Afrique durant le Quaternaire passé-présent-futur 1986, INQUA-ASEQUA Symposium International, Dakar, Avril 1986, Ed. ORSTOM, coll. Travaux et Documents no. 197, Paris.

Bonnefille R., Roeland J. C., Guiot J. 1990. Temperature and rainfall estimates for the past 40,000 years in equatorial Africa. Nature, 346, 347–349.

Bonnefille R. 1991. Températures et Précipitations des derniers 40 000 ans en Afrique Equatoriale. Sécheresse, 4, 2, 279.

Bricquet J. P., Bamba F., Mahé G., Toure M., Olivry J. C. 1997. Water resource variations of the Atlantic river basins of Africa: the long term effect of rain shortage. Journal of Water Science, 3, 321–337.

Brou Yao T. 1997. Analyse et dynamique de la pluviométrie dans le sud forestier ivoirien: recherche de corrélations entre les variables climatiques et les variables liées aux actions anthropiques. Thèse de doctorat de l'Université de Cocody, 211 p.

Brou Yao T., Servat E., Paturel J. E. 1998. Contribution à l'analyse des inter-relations entre activités humaines et variabilité climatique: cas du sud-est forestier ivoirien. Comptes Rendus de l'Académie des Sciences de Paris, Serie Iia, 327, 833–838.

Brovkin V., Claussen M., Petoukhov V. Ganopolski A. 1998. On the stability of the atmosphere-vegetation system in the Sahara/Sahel region. J. Geophys. Res. 103, D24, 31.613–31.624.

Cadet D. L., Nnoli N. O. 1987. Water vapour transport over Africa and the Atlantic ocean during summer 1979. Quart. J. Roy. Met. Sci., 113, 581–602.

Charney J. 1975. Dynamics of deserts and droughts in the Sahel. Quat. J. of Roy. Meteo. Soc., 101, 193–202.

Charney J., Quirk W., Chow S., Kornfield J. 1977. A comparative study of the effects of albedo change on drought in semi-arid regions. Journal of Atmospheric Science, 34, 1366–1385.

Claussen M. 1994. On coupling global biome models with climate models. Climate Research, 4, 203–221.

Claussen M. 1997. Modeling bio-geophysical feedback in the African and Indian monsoon region. Climate Dynamics, 13, 247–257.

Claussen M., Gayler V. 1997. The greening of the Sahara during the mid-Holocene: results of an interactive atmosphere-biome model. Global Ecology and Biogeography Letters, 6, 369–377.

Claussen M., Brovkin V., Ganopolski A., Kubatzki C., Petoukov V. 1998. Modelling global terrestrial vegetation-climate interaction. Phil. Trans. R. Soc. Lond. B., 353, 53–63.

Claussen M., Kubatzki C., Brovkin V., Ganopolski A., Hoelmann P., Pachur H. J. 1999. Simulation of an abrupt change in Saharan vegetation in the mid-Holocene. Geophysical Research Letters, 26, 14, 2037–2040.

Claussen M. 2001. Biogeophysical feedbacks and the dynamics of climate. In: Global Biogeochemical Cycles in the Climate System, Academic Press Ed., 61–71.

Claussen M., Brovkin V., Ganopolski A. 2001. Biogeophysical versus biogeochemical feedbacks of large-scale land cover change. Geophysical Research Letters, 28, 6, 1011–1014.

Diaz A. F., Stduzinski C. D., Mechoso C. R. 1998. Relationships between precipitation anomalies in Uruguay and Southern Brazil and sea surface temperatures in the Pacific and Atlantic Oceans. Journal of Climate, 11, 251–271.

Dolman, A. J., Van der Molen, M. K., Ter Maat, H. W. and Hutjes, R. W. A. (2003). The effects of forests on mesoscale atmospheric processes. In: *Forest and the Atmosphere*, eds. M. Mencucini and J. Grace (in press).

Duglas J. 1993. Effet de serre et activités humaines. Sécheresse, 4, 4, 211–220.

Dupont L. M., Jahns S., Marret F., Ning S. 2000. Vegetation change in equatorial West Africa: time-slices for the last 150 ka. Palaeogeogr., Palaeoclim., Palaeoecology, 155, 95–122.

Flenley J. R. 1998. Tropical forest under the climates of the last 30,000 years. Climatic Change, 39, 177–197.

Folland C. K., Palmer T. N., Parker D. E. 1986. Sahel rainfall and worldwide sea temperature. Nature, 320, 602–607.

Folland C. K., Parker D. E., Colman A. W., Washington R. 1998. Large Scale Modes of Ocean Surface Temperature since the late nineteenth century. In: Climate Research Technical Note no. 81, Hadley Center for Climate Prediction and Research, Meteorological Office, London, 38 p.

Fontaine B. 1991. Variations pluviométriques et connexions climatiques: l'exemple des aires de mousson indienne et ouest-africaine. Sécheresse, 4, 2, 259–264.

Fontaine B., Janicot S. 1993. L'évolution des idées sur la variabilité interannuelle récente des précipitations en Afrique de l'Ouest. La Météorologie, 1, 28–53.

Fontaine B., Janicot S., Durand A. 1991. La variabilité climatique en Afrique. Sécheresse, 4, 2, 280–281.

Fontaine B., Janicot S., Moron V., Roucou P., Trzaska S. 1998a. Anomalies de températures de surface de la mer et précipitations tropicales. La Météorologie, 23, 14–35.

Fontaine B., Trzaska S., Janicot S. 1998b. Evolution of the relationships between near global and atlantic sea surface temperatures modes and the rainy season in West Africa: statistical analysis and sensitivity experiment. Climate Dynamics, 14, 451–473.

Fontaine B., Janicot S., Roucou P. 1999. Coupled Ocean-Atmosphere Surface Variability and its Climate Impacts in the Tropical Atlantic Region. Climate Dynamics, 15, 451–473.

Fu R., Zhu B., Dickinson R. E. 1999. How do atmosphere and land surface influence seasonal changes of convection in the Tropical Amazon. Journal of Climate, 12, 1306–1321.

Ganopolski A., Kubatzki C., Claussen M., Brovkin V., Petoukhov V. 1998. The influence of vegetation-atmosphere-ocean interaction on climate during the mid-Holocene. Science 280, 1916–1919.

Gash J. H. C., Shuttleworth W. J. 1991. Tropical Deforestation: Albedo and the Surface-Energy Balance. Climatic Change, 19, 1–2, 123–133.

Gong C., Eltahir E. 1996. Sources of moisture for rainfall in West Africa. Water Resources Research, 32, 10, 3115–3121.

Hamilton A. C., Taylor D. 1991. History of climate and forests in tropical Africa during the last 8 millions years. Climatic Change, 19, 65–78.

Hastenrath S., de Castro L. C., Aceituno P. 1987. The Southern Oscillation in the Tropical Atlantic Sector. Beitr. Phys. Atmosph., Vol. 60, No 4: 447–763.

Harris J. 1993. Ecosystem structure and growth of the African savanna. Global and Planetary Change, 8, 231–248.

Heaney L. R. 1991. A synopsis of climatic and vegetational change in South East Asia. Climatic Change, 19, 53–61.

Hooghiemstra H., Van Der Hammen T. 1998. Neogene and Quaternary development of the neotropical rainforest: the forest refugia hypothesis, and a literature overview. Earth-Science Review, 44, 147–183.

Houghton R. A. 1991. Tropical Deforestation and Atmospheric Carbon Dioxide. Climatic Change, 19, 1–2, 99–117.

Hubert P., Servat E., Paturel J. E., Kouame B., Bendjoudi H., Carbonnel J. P., Lubes-Niel H. 1998. La procédure de segmentation, dix ans après. In: Water resources variability in Africa during the XXth century (Servat E., Hughes D., Fritsch J. M. and Hulme M. Eds), IAHS pub. no. 252, 267–273.

IPCC 1996. Climate Change 1995 – The Science of Climate Change. In: Contribution of Working Group 1 to the second assessment report of the Intergovernmental Panel on Climate Change, Houghton J. T., Filho Meira L. G., Callander B. A., Harris N., Katenbberg A., Maskell K. Ed. OMM/UNEP, Cambridge University Press, Cambridge, 572 p.

Janicot S. 1995. Le point sur la désertification et les changements climatiques. La Météorologie, 4, 1–15.

Janicot S. 1999. Atlantique Tropical et Climat Africain. In: Colloque PNEDC, 13–14 Décembre 1999, Météo France, Toulouse, France.

Janicot S., Moron V., Fontaine B., 1996. Sahel droughts and ENSO dynamics. Geophysical Research Letters, Vol. 23, No 5: 515–518.

Janicot S., Fontaine B. 1997. Evolution saisonnière des corrélations entre précipitations en Afrique guinéenne et températures de surface de la mer (1945–1994). Comptes Rendus de l'Académie des Sciences de Paris, Serie IIa, 324, 785–792.

Janicot S., Moron V., Fontaine B. 1996. Sahel droughts and ENSO dynamics. Geophysical Research Letters, 23, 515–518.

Jolly D., Prentice I. C., Bonnefille R. and 30 authors 1998. Biome reconstruction from pollen and plant macrofossil data for Africa and the Arabian peninsula at 0 and 6000 years. J. of Biogeography, 25, 1007–1027.

Jouzel J., Lorius C., Raynaud D. 1994. Climat du futur: le témoignage du passé. Natures, Sciences, Sociètés, 2, 4, 298–308.

Keller M., Jacob D. J., Wofsy S. C., Harriss R. C. 1991. Effects of tropical deforestation on global and regional atmospheric chemistry. Climatic Change, 19, 139–157.

Koster, R. D., Suarez, M. J. and Heiser, M. 2000. Variance and predictability of precipitation at seasonal to inter-annual timescales. *Journal of Hydrometeorology* 1, 26–46.

Kubatzki C., Claussen M. 1998. Simulation of the global bio-geophysical interactions during the Last Glacial Maximum. Climate Dynamics, 14, 461–471.

Kutzbach J. E., Guetter P. J. 1986. The influence of changing orbital parameters and surface boundary conditions on climate simulations for the past 18,000 years. J. Atmos. Sci., 43, 1726–1759.

Laraque A., Mahé G., Orange D., Marieu B. 2001. Spatiotemporal variations in hydrological regimes within Central Africa during the XXth century. J. Hydrology, 245, 104–117.

Le Houerou H. N. 1993. Changements climatiques et désertisation. Sécheresse, 4, 95–111.

Lezine A. M. 1998. Pollen records of past climate changes in West Africa since the last glacial maximum. In: A. S. Issar and N. Brown (eds), Water, environment and society in times of climatic change, Kluwer Academic Publishers, 295–317.

Levis S., Foley J. A., Brovkin V., Pollard D. 1999. On the stability of the high-latitude vegetation-climate system in a coupled atmosphere-biosphere model. Global Ecol. Biogeography, 8, 6, 489–500.

L'Hote Y., Mahé G. 1996. Map of mean annual rainfall for West and Central Africa (1951–1989), 1/6000000. Collection des cartes ORSTOM. ORSTOM (IRD) Ed.

Mahé G., Lerique J., Olivry J. C. 1990. Le fleuve Ogooue au Gabon. Reconstitution des débits manquants et mise en évidence de variations climatiques à l'équateur. Hydrologie Continentale, Ed. ORSTOM, 5, 2, 105–124.

Mahé G. 1993. Les Ecoulements Fluviaux sur la Façade Atlantique de l'Afrique: Etude des Elements du Bilan Hydrique et Variabilité Interannuelle, Analyse des Situations Hydroclimatiques moyennes et extrêmes. Thèse de doctorat, Ed. ORSTOM, coll. Etudes et Thèses, Paris, 438 p.

Mahé G., Citeau J. 1993. Interactions between Ocean-atmosphere-continent in Africa, related to the Atlantic monsoon flow. General pattern and the 1984 case study. Veille Climatique Satellitaire Ed. ORSTOM-METEO FRANCE, 44, 34–54.

Mahé G., Olivry J. C. 1995. Variation des précipitations et des écoulements en Afrique de l'Ouest et Centrale de 1951 à 1989. Sécheresse, 6, 109–117.

Mahé G., Dessouassi R., Cissoko B., Olivry J. C. 1998. Comparaison des fluctuations interannuelles de piezometrie, precipitation et debit sur le bassin versant du Bani a Douna. In: Water resources variability in Africa during the XXth century (Servat E., Hughes D., Fritsch J. M. and Hulme M. Eds), IAHS pub. no. 252, 289–295.

Mahé G., Olivry J. C. 1999. Assessment of freshwater yields to the ocean along the intertropical Atlantic coast of Africa (1951–1989). C. R. Acad. Sci. Paris, serie IIa, 328, 621–626.

Mahé G., Olivry J. C., Dessouassi R., Orange D., Bamba F., Servat E. 2000. Relations eaux de surface-eaux souterraines d'une rivière tropicale au Mali. C. R. Acad. Sci Paris, serie IIa, 330, 689–692.

Mahé G., L'Hote Y., Olivry J. C., Wotling G. 2001. Trends and discontinuities in regional rainfall of West and Central Africa: 1951–1989. Hydrological Sciences Journal, 46, 2, 211–226.

Mahé G., Dray A., Paturel J. E., Cres A., Kone F., Manga M., Cres F. N., Djoukam J., Maiga A. H., Ouedraogo M., Conway D., Servat E. 2002. Climatic and anthropogenic impacts on the flow regime of the Nakambe

River in Burkina. FRIEND 4th International Conference, Cape Town, South Africa, 2002, IAHS Pub. 274, 69–76.

Maley J. 1990. L'histoire récente de la forêt dense humide africaine: essai sur le dynamisme de quelques formations forestières. In: Paysages quaternaires de l'Afrique Centrale Atlantique, 367–389.

Maley J. 1991. The African rainforest vegetation and paleoenvironments during late Quaternary. Climatic Change, 19, 79–98.

Maley J. 1996a. The African rainforest: main characteristics of changes in vegetation and climate from the upper Cretaceous to the Quaternary. Proceed. R. Soc. Edinburg, Biol. Sc., 104B, 31–73.

Maley J. 1996b. Fluctuations majeures de la forêt dense humide africaine au cours des 20 derniers millénaires. In: L'alimentation en forêt tropicale: interactions bioculturelles et perspectives de développement vol.1. Paris, UNESCO, 55–76.

Maley J. 1997. Middle to late Holocene changes in tropical Africa and other continents: paleomonsoon and sea surface temperature variation. NATO ASI series, 1, 611–639.

Maley J., Brenac P. 1998. Vegetation dynamics, Paleoenvironments and Climatic Changes in the forests of Western Cameroon during the last 28 000 years B. P. Review of Paleobotany and Palynology, 99, 157–187.

Maley J., Brenac P., Bigot S., Moron V. 2000. Variations de la végétation et des paléoenvironnements en forêt dense africaine au cours de l'Holocène. Impact de la variation des températures marines. In: Dynamique à long terme des Ecosystèmes forestiers intertropicaux, M. Servant and S. Servant-Vildary eds., Mémoire UNESCO, Paris, 205–220.

Martin L., Fournier M., Mourguiart P., Sifeddine A., Turcq B., Absy M. L., Flexor J. M. 1993. Southern Oscillation signal in South American paleoclimatic data of the last 7000 years. Quat. Res., 39, 338–346.

Monteny B. A. 1987. Contribution à l'étude des interactions végétation-atmosphère en milieu tropical humide. Thèse de doctorat de l'Université d'Orsay, 176 p.

Monteny B. A., Lhomme J. P., Chebhouni A., Amadou M., Troufleau D., Brunel J. P., Bessemoulin P., Sicot M., Galle S., Lloyd C., Said F. 1996. Les Interactions Surface Continentale-Atmosphère en Milieu Sahélien. Passage de l'Echelle Stationnelle à l'Echelle Régionale. In: Interactions Surface Continentale-Atmosphère: l'Expérience HAPEX-Sahel, Actes des Xe Journées Hydrologiques de Montpellier, 13–14 Septembre 1994, Ed. ORSTOM, Ed. Scient.: Hoepffner M., Lebel T., Montény B., 287–303.

Morley R. J., Richards K. 1993. Gramineae cuticle: a key indicator of Late Cenozoic climatic change in the Niger delta. Rev. of Palaeobotany and Palynology, 77, 119–127.

Moron V., Bigot S., Roucou P., 1995. Rainfall variability in subequatorial america and africa and relationships with the main sea-surface temperature modes (1951–1990). Int. J. Climatol., Vol. 15: 1297–1322.

Myers N. 1991. Tropical Forests: Present status and Future Outlook. Climatic Change, 19, 1–2, 3–31.

Myers N., Goreau T. J. 1991. Tropical forests and the greenhouse effect: a management response. Climatic Change, 19, 215–225.

Mylne M., Rowntree P. 1992. Modelling the effects of albedo change associated with tropical deforestation. Climatic Change, 21, 317–343.

Nicholson S. E., Entekhabi D. 1986. Rainfall Variability in Equatorial and Southern Africa: Relationships with Sea Surface Temperatures along the Southwestern Coast of Africa. Journal of Climate and Applied Meteorology, Vol. 26, No 5: 561–578.

Nicholson S. E., Kim J. 1997. The Relationship of the El Niño-Southern Oscillation to African Rainfall. Int. J. Climatol., Vol 17: 117–135.

De Noblet-Ducoudre N., Claussen M., Prentice C. 2000. Mid-Holocene greening of the Sahara: firstresults of the GAIM 6000 year BP Experiment with two asynchronously coupled atmosphere/biome models. Climate Dynamics, 16, 643–659.

Paturel J. E., Servat E., Vassiliadis A. 1995. Sensitivity of conceptual rainfall-runoff algorithms to errors in input data – case of the GR2M model. J. Hydrology, 168, 111–125.

Paturel J. E., Servat E., Delattre M. O., Lubes-Niel H. 1998. Analyse de séries pluviométriques de longue durée en Afrique de l'Ouest et Centrale non sahélienne dans un contexte de variabilité climatique. Hydrological Sciences Journal, 43, 6, 937–946.

Petit-Maire N., Guo Z. 1996. Mise en évidence de variations climatiques holocenes rapides, en phase dans les deserts actuels de Chine et du Nord de l'Afrique. Sciences de la Terre et des Planetes, 322, 847–851.

Polcher J. 1994. Etude de la sensibilité du climat tropical à la déforestation. Thèse de doctorat de l'Université de Paris VI, 185 p.

Raynaud D., Barnola J. M., Chappellaz J., Blunier T., Indermuhle A., Stauffer B. 2000. The ice record of greenhouse gases: a view in the context of future changes. Quat.Sc.Rev., 19, 9–17.

Reynaud-Farrera I., Maley J., Wirrmann D. 1996. Végétation et climat dans les forêts du Sud-Est Cameroun depuis 4770 ans B.P.: Analyse pollinique des sédiments du Lac Ossa. Comptes Rendus de l'Académie des Sciences de Paris, Serie IIa, 322, 749–755.

Salati E., Nobre C. A. 1991. Possible Climatic Impacts of Tropical Deforestation. Climatic Change, 19, 1–2, 177–195.

Schwartz D. 1992. Assèchement climatique vers 3000 B.P. et expansion Bantu en Afrique Centrale Atlantique: quelques réflexions. Bull Soc Géol France, 163, 352–361.

Schwartz D., Elenga H., Vincens A., Bertaux J., Mariotti A., Achoundong G., Alexandre A., Belingard C., Girardin C., Guillet B., Maley J., De Namur C., Reynaud-Farrera I., Youta-Happi J. 2000. Origine et évolution des savanes des marges forestières en Afrique centrale Atlantique (Cameroun, Gabon, Congo): approche aux échelles millénaires et séculaires. In: Dynamique à long terme des Ecosystèmes forestiers intertropicaux, M. Servant and S. Servant-Vildary eds., Mémoire UNESCO, Paris, 325–338.

Semazzi F. H. M., 1996. A GCM Study of the Teleconnections between the Continental Climate of Africa and Global Sea Surface Temperature Anomalies. J. of Climate, Vol. 9, No 10: 2480–2497.

Servat E., Lapetite J. M., Bader J. C., Boyer J. F. 1990. Satellite data transmission and hydrological forecasting in the fight against onchocerciasis in West Africa. J. Hydrology, 117, 187–198.

Servat E., Dezetter A. 1991. Selection of calibration objective functions in the context of rainfall-runoff modelling in a sudanese savannah area. Hydrological Sciences Journal, 36, 4, 307–330.

Servat E., Dezetter A. 1993. Rainfall-runoff modelling and water resources assessment in northwestern Ivory Coast. Tentative extension to ungauged catchments. J. Hydrology, 148, 231–248.

Servat E., Sakho M. 1995. Instability of water resources and management of a planned water system in non-Sahelian West Africa. Hydrological Sciences Journal, 40, 2, 217–230.

Servat E., Paturel J. E., Lubes-Niel H., Kouame B., Travaglio M., Marieu B. 1997. De la diminution des écoulements en Afrique de l'Ouest et Centrale. C. R. Acad. Sci. Paris, série IIa, 325, 679–682.

Servat E., Paturel J. E., Lubes H., Kouame B., Ouedraogo M., Masson J. M. 1997. Climatic variability in humid Africa along the Gulf of Guinea. Part I: detailed analysis of the phenomenom in Cote d'Ivoire. Journal of Hydrology, 191, 1–15.

Servat E., Paturel J. E., Kouame B., Travaglio M., Ouedraogo M., Boyer J. F., Lubes-Niel H., Fritsh J. M., Masson J. M., Marieu B. 1998. Identification, caractérisation et conséquences d'une variabilité hydrologique en Afrique de l'Ouest et Centrale. In: Water resources variability in Africa during the XXth century (Servat E., Hughes D., Fritsch J. M. and Hulme M. Eds), IAHS pub. no. 252, 323–337.

Servant M., Maley J., Turcq B., Absy M. L., Brenac P., Fournier M., Ledru M. P. 1993. Tropical forests changes during the late Quaternary in African and South American lowlands. Global and Planetary Change, 7, 25–40.

Shukla, J. 1998. Predictability in the midst of chaos. *Science*, 282, 728–731.

Sifeddine A., Frohlich F., Fournier M., *et al.* 1994. La sédimentation lacustre indicateur de changements des paléoenvironnements au cours des 30 000 dernières années (Carajas, Amazonie, Brésil). Comptes Rrendus de l'Académie des Sciences de Paris, Serie IIa, 318, 1645–1652.

Tardy C. 1998. Paléoincendies naturels, feux anthropiques et environnements forestiers de Guyane Française du Tardiglaciaire à l'Holocène récent. Thèse de docorat de l'Univ. Montpellier-II, 343 p.

Texier D., de Noblet N., Harrison S. P., Haxeltine A., Jolly D., Joussaume S., Laarif F., Prentice I. C., Tarasov P. 1997. Quantifying the role of biosphere-atmosphere feedbacks in climate change: coupled model simulations for 6000 years BP and comparison with palaeodata for northern Eurasia and Northern Africa. Climate Dynamics, 13, 865–882.

Trenberth K. E. 1997. The definition of El Niño. Bull of the Amer Meteo Soc, 78, 2771–2777.

Uvo C. B., Repelli C. A., Zebiak S. E., Kushnir Y. 1998. The relationships between Tropical and Atlantic sea surface temperature and the North-East Brazil monthly precipitation. Journal of Climate, 11, 551–562.

Van der Hammen T. 1991. Paleoecological background: Neotropics. Climatic Change, 19, 37–47.

Vernet J. L., Wengler L., Solari M. E., Ceccantini G., Fournier M., Ledru M. P., Soubies F. 1994. Feux, climats et végétations au Brésil central durant l'Holocène. C. R. Acad. Sci. Paris, Serie IIa, 319, 1391–1397.

Vincens A., Elenga H., Reynaud-Farrera I., Schwartz D., Alexandre A., Bertaux J., Mariotti A., Martin L., Meunier J. D., Nguetsop F., Servant M., Servant-Vildary S., Wirrmann D. 2000. Réponse des forêts aux changements du climat en Afrique Atlantique équatoriale durant les derniers 4000 ans et héritage sur les paysages végétaux actuels. in Dynamique à long terme des Ecosystèmes forestiers intertropicaux, M. Servant and S. Servant-Vildary eds., Mémoire UNESCO, Paris, 381–387.

Vincens A., Schwartz D., Bertaux J., Elenga H., Denamur C. 1998. Late Holocene climatic changes in Western Equatorial Africa inferred from pollen from Lake Sinnda, Southern Congo. Quat Res., 50, 34–45.

Wotling G., Mahé G., L'Hote Y., LeBarbe L., 1995. Analyzing the space-time variability of annual rainfall linked to the African monsoon using regional vectors. Veille Climatique Satellitaire, ED. ORSTOM-MeteoFrance, 52, 58–73.

Zeng, N., Neelin, J. D., Lau, K. M. and Tucker, C. J. 1999. Enhancement of interdecadal climate variability in the Sahel by vegetation interaction. *Science* 286, 1537–1540.

Zheng X., Eltahir E. 1997. The response to deforestation and desertification in a model of West African monsoons. Geophys. Res. Lett., 24, 155–158.

Zheng X., Eltahir E. A. B. 1998. The role of vegetation in the dynamics of West African monsoons. Journal of Climate, 11, 2078–2097.

Zheng X., Lei Z. Q. 1999. A 400,000 year record of vegetational and climatic changes from a volcanic basin, Leizhou Peninsula, southern China. Palaeogeogr., Palaeoclimat., Palaeoecol., 145, 339–362.

13 Controls on evaporation in lowland tropical rainforest

J. M. Roberts and J. H. C. Gash
Centre for Ecology and Hydrology, Wallingford, UK
M. Tani
Kyoto University, Japan
L. A. Bruijnzeel
Vrije Universiteit, Amsterdam, The Netherlands

INTRODUCTION

Although there has been substantial deforestation in recent decades, lowland rainforests still constitute an important fraction of land cover in the tropical regions (cf. Drigo, this volume). There are around 1800 million hectares of natural forest cover in the tropics, about 37% of the land area. Of this forest cover around 1000 million hectares are lowland rainforest (FAO, 2001). The bulk of lowland rainforest is in the Equatorial Regions, in the Amazon basin, South East Asia and Central Africa (Richards, 1996).

In the past two decades there has been a sustained interest in the factors that control water fluxes from tropical rainforest. The motivations for that interest are many. It has long been realised that the hydrology of vegetation has important links with the partitioning of available energy at the surface of the Earth and the impact of land cover changes on surface climate. There is also a need to understand the influence of global changes in climate on water resources and the behaviour of large areas of tropical vegetation will be important to this understanding. Detailed knowledge and understanding of aspects of tropical forest hydrology is key to a reliable prediction of the effects of rainforest management on the amount and timing of streamflow.

Catchment studies have considerable value in enabling direct comparisons of the effects of changes in land cover on the amount and the timing of streamflow to be made. However, because very little information emerges on the contribution of different soil and vegetation types in the landscape or of the contribution of different evaporation mechanisms, the results of catchment studies are difficult to use in a deterministic way for prediction other than simply in a statistical manner, unless they are accompanied by process studies (Bruijnzeel, 1996). There are also various methodological constraints associated with catchment studies, notably, the proper estimation of areal rainfall inputs due to spatially variable convective rainfall or steep terrain (Chappell *et al.*, 2001), problems associated with defining the catchment boundary in areas of low relief or swamps (cf. Hooijer, this volume), and in some tropical catchments with deep permeable soils the possibility of inter-basin transfer (Bruijnzeel, 1990).

Evaporation from vegetation comprises two very different processes: transpiration of water taken up from the soil by roots and evaporation of rainfall intercepted by the forest canopy. Providing a full understanding of the functional behaviour of tropical rainforests requires that the two processes are examined separately. Studies of the different evaporation processes, rainfall interception and transpiration, can be studied at a range of scales. Inevitably at the largest spatial scale the information tends to be for the whole forest and little understanding emerges about the behaviour of forest growing on different topographical elements of the landscape, the contribution of different vertical components in the rainforest or species differences at the plot scale. The information provided at the smallest scale, i.e. the tree or the leaf, provides the most insight into the functioning at that scale but the scaling up to plot or landscape will require considerable additional effort.

Despite the large amount of effort that has been directed, in various ways, at measuring evaporation losses from tropical forests and understanding the controlling processes, our knowledge and understanding is still fragmentary and incomplete. Therefore the capacity to predict the hydrological and climatological consequences of rainforest disturbance or conversion to other land uses is restricted. The purpose of this chapter is to examine the data available from a range of scales and evaluate the state of our

Forests, Water and People in the Humid Tropics, ed. M. Bonell and L. A. Bruijnzeel. Published by Cambridge University Press.

current knowledge. Data have been drawn from a range of tropical lowland forest studies in the old and new world tropics. In this volume tropical montane forests are discussed separately in the chapter by Bruijnzeel whereas swamp forests are dealt with in the chapter by Hooijer.

Tropical forests contain about half of the carbon present as biomass in the world's terrestrial ecosystems, suggesting a key role for these forests in the global carbon balance (Grace *et al.*, 1999). There is considerable interest in the fate of the carbon sinks in tropical forests in scenarios for future global climate. The functioning of the forests in terms of carbon metabolism is closely linked to their hydrological functioning. Important predictions have been made that global climate change associated with increased atmospheric CO_2 levels, will lead to increased drought in the Amazon Basin (e.g. Cox *et al.*, 2000; White *et al.*, 2000). Unfortunately, at present such predictions are based on sparse information on the response not only to temperature increases but also to severe seasonal soil water deficits.

In this chapter we examine the recent information on the control of evaporation losses from tropical rainforests by evaporation of intercepted rainfall and transpiration losses of water taken up from the soil. Key objectives are to identify consistent similarities and differences in the rates and controls of evaporation from tropical rainforests that have been studied and to establish in what geographical areas or at what scales we lack data. We also examine how well we can answer arguably the most important question at present: how will evaporation from lowland tropical forest respond to global climate change?

HYDROLOGICAL STUDIES IN TROPICAL RAINFORESTS

Evaporation estimates (rainfall minus streamflow) from hydrological studies in lowland tropical rainforests were reviewed by Bruijnzeel (1990). Figure 13.1 summarises data collected in Latin America, Africa and South East Asia. Calder (1999) believes that the net radiation equivalent of evaporation at these sites was in the range 1500–1550 mm. This would mean that only in dry years or at the Guma, Sierra Leone site that has a five-month dry season, was evaporation substantially less than the net radiation supply (cf. Ledger, 1975). Zhang *et al.* (2001) have proposed a general function to describe the relationship of annual forest evaporation with annual rainfall, including humid tropical conditions. This relationship is also shown in Figure 13.1 but the function is a poor fit to the field data collated by Bruijnzeel. As indicated earlier, there are various technical difficulties in measuring evaporation from rainfall less streamflow from tropical forest catchments. As such, there may be some question about the robustness of forest evaporation estimates from catchment studies. In addition, there

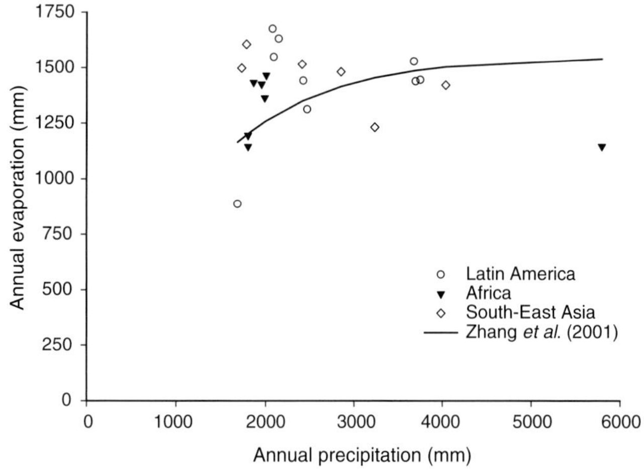

Figure 13.1 Annual precipitation and evaporation for lowland rainforests in Latin America, Africa and South East Asia (after Bruijnzeel, 1990). The solid curve represents the empirical relationship proposed by Zhang *et al.* (2001).

are difficulties in exploiting the information in a predictive way to evaluate the effects of land cover changes on streamflow or to gauge the likely impact of the influence of climate change. As will be argued below, the answer to this is believed to be a judicious combination of catchment-based and hydrological process work.

Most of the values for lowland rainforest evaporation shown in Figure 13.1 have also included studies of rainfall interception loss. Estimates of annual values of transpiration calculated as the difference between total evaporation and interception give values ranging from 885 to 1285 mm yr^{-1} with an average of 1045 mm (Bruijnzeel, 1990). However, these estimates are likely to be crude because of the shortcomings of catchment water budget-based estimates of total evaporation already referred to. There are also problems of adequately representing forest interception losses in tropical rainforests because of sampling difficulties (Lloyd and Marques Filho, 1988; see below).

The highest values of forest transpiration are observed in lowland tropical rainforest which do not have a marked seasonality. Transpiration can often exceed 1000 mm yr^{-1} and in many cases transpiration can account for substantially more than half of the annual rainfall. Transpiration losses from these forests can reach 3.5–4.0 mm day^{-1} with over 70% of the available radiative energy being used in transpiration daily (Shuttleworth, 1988; Tani *et al.*, 2003). There is far less information available for tropical lowland rainforests which experience a marked seasonality. Annual transpiration for these forests has been estimated to fall between 500–600 mm yr^{-1} (Bruijnzeel, 1990).

Hydrological information derived from tropical rainforest catchments is valuable as a baseline if land use changes need to be compared on the same or an adjacent catchment, but the data are often site specific and there are the methodological constraints already mentioned. A better understanding of the functioning of tropical rainforest under both wet and dry canopy conditions has come from process studies carried out at the stand or within canopy scale. It is from such studies that physical and physiological models may be derived that can be used to evaluate scenarios of land cover or global climate change. It would be rewarding if the results of catchment studies could be used to validate these models (cf. Bruijnzeel, 1996).

Evaporation from leaves and canopies

Water is lost from leaves and canopies in two ways. Firstly there is transpiration. In the transpiration process water is taken up from the soil by the roots, passes up the tree as the transpiration stream or sap flow and passes as water vapour from the stomatal pores in the leaf surfaces into the atmosphere. The second way by which water is evaporated is through the interception of rainfall by the canopy and the evaporation of this water from wetted leaves, branches and trunks both during and after the rain event. This second evaporation process is termed, in full, rainfall interception loss, interception loss or, simply, interception. There is of course also direct evaporation from the forest floor/soil. However, in tropical forests the dense canopy means that solar radiation reaching the forest floor can be as little as 1% of that above the forest (Shuttleworth *et al.*, 1984b). Consequently, the contribution from the forest floor to total forest evaporation is small ~35–70 mm yr^{-1} (e.g. Jordan and Heuveldop, 1981; Roche, 1982).

There are some common features between transpiration and interception but there is also an important difference. In both cases there are three prerequisites. Firstly, an input of energy is necessary to sustain evaporation. Secondly; the atmosphere must have an affinity for water, i.e. there must be a humidity gradient from the vegetation surface to the atmosphere. Thirdly, water vapour must transfer from the leaves/canopies into the atmosphere through conductances (reciprocal of resistances) associated with the boundary-layers of leaves and the canopy space. In the cases of all conductances high values are associated with rapid, efficient transfer while low conductances are associated with slow rates and reduced losses. The important difference between the transpiration and interception process is the involvement of the stomata in the transpiration process while interception loss from leaves and branches occurs directly from their surfaces. The degree of opening of the stomata constitutes a further conductance for water vapour which is called the stomatal conductance.

While the evaporation of intercepted rainfall is largely a physical process, the involvement of the stomata makes transpiration a combination of physiological and physical processes. Stomatal conductance is influenced by a wide range of external factors, particularly levels of solar radiation, temperature, air humidity deficit and carbon dioxide concentration. There are also internal factors in the leaf that are known to control stomatal conductance. These factors include the leaf water potential that reflects the readiness with which leaves are supplied with water from the soil, chemical messages from the roots which are also influenced by soil drying (e.g. Wilkinson and Davies, 2002) and even the rates of transpiration themselves (e.g. Monteith, 1995).

The most realistic description of evaporation from leaves and canopies is given by the Monteith version of the Penman Equation (Monteith, 1965), hereafter referred to as the PM Equation. When the canopy is dry the flux of water vapour, transpiration (λE_t), from the canopy surface can be expressed as:

$$\lambda E_t = \frac{\Delta A + \rho c_p D g_a}{\Delta + c_p(1 + g_a/g_c)/\lambda} \tag{13.1}$$

where A is the available radiative energy, c_p the specific heat capacity of air at constant pressure, E_t the transpiration, D the air vapour pressure deficit, γ the psychrometric constant, Δ the slope of the saturation vapour pressure curve, λ the latent heat of vaporisation of water and ρ the density of air. In the case of a complete canopy, g_a is the aerodynamic conductance that constitutes the control of vapour transfer from the foliage surfaces into the free atmosphere. The symbol g_c, the surface (or canopy) conductance, represents the combined influence of the stomata for all the canopy foliage. The PM equation (13.1) can be used for individual leaves, in which case g_b substitutes for g_a and is the boundary-layer conductance of an individual leaf and g_s, which replaces g_c, is the stomatal conductance of a single leaf.

Given all the inputs on the right hand side of Eqn 13.1, transpiration from a forest or leaf can be calculated. This type of approach is often used to predict transpiration for forest areas for which there are only meteorological data, with conductances being substituted from information provided from other studies where values are already available for forests or other vegetation.

When the canopy is wetted by rain, evaporation of the intercepted rainfall is dominated by the aerodynamic conductance (or the boundary layer conductance in the case of an individual leaf) and the PM equation simplifies to a form that omits the conductance associated with leaf or canopy physiology (g_s or g_c) and becomes (Monteith, 1965):

$$\lambda E = \frac{\Delta A + \rho c_p D g_a}{\Delta + c_p/\lambda} \tag{13.2}$$

The control of transpiration at the leaf level

In many rainforests the height and density of the foliage means that there are considerable changes in most micro-climatic parameters such as solar radiation, air temperature, water vapour concentration, wind speed and carbon dioxide concentrations downwards through the canopy compared to their values above the canopy

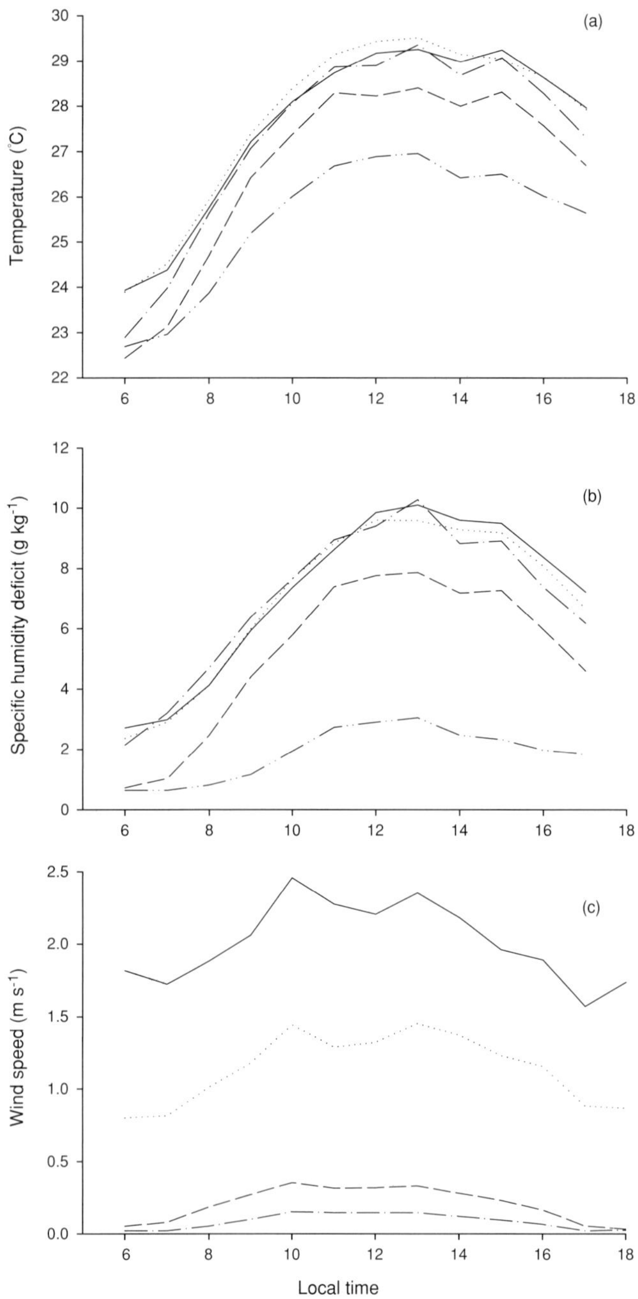

Figure 13.2 Variation in (a) air temperature, (b) specific humidity deficit and (c) wind speed at 45m, — ; 36m, ... ; 23m, — . — ;13m, — — — ; and 1.5m, — .. — .. — at the Reserva Ducke, Manaus. Values are averages of data taken throughout July 1984. (After Roberts *et al.*, 1990.)

(Schulz, 1960 – Suriname; Kira *et al.*, 1969 – Cambodia; Aoki *et al.*, 1978 – Malaysia; Shuttleworth *et al.*, 1985 – Brazil). There is usually an almost exponential decline in solar radiation (R_s) and photosynthetically active radiation (PAR) down through the canopy (e.g. Carswell *et al.*, 2000). Figure 13.2 shows the typical diurnal variation in micro-climatic variables above and down through the forest canopy in the Reserva Ducke, Manaus, Brazil. At midday the differences in air temperature above the canopy to that at the forest floor can be as much as 7–8 °C while air specific humidity deficit (D) can differ by up to 8 g kg^{-1}. In the day time period wind speeds above the canopy can typically be 3 to 4 m s^{-1}, declining to values one tenth or less than this at the floor of the forest (Shuttleworth *et al.*, 1985).

One consequence of the gradients in microclimate, particularly that of radiation, is that there is a substantial vertical differentiation of foliage properties and physiological capacity. Figure 13.3 (Carswell *et al.*, 2000) shows the vertical variation in leaf nitrogen and specific leaf area through the canopy at the Reserva Cuieiras (some 30 km from the Reserva Ducke). McWilliam *et al.* (1996) working in the Reserva Jaru, Ji-Paraná (SW Amazon) showed a similar strong positive relationship between leaf photosynthesis and stomatal conductance through the depth of the forest canopy as found by Carswell *et al.* (2000) at Reserva Cuieirias. The expected strong links between leaf photosynthesis and leaf nitrogen contents and the strong positive relationship between photosynthesis and stomatal conductance offers some optimism that canopy functions can be scaled up from information on leaf nitrogen content (e.g. Kull and Kruijt, 1999). In a temperate forest (with only two tree canopy species), Roberts *et al.* (1999) have shown that through residual nitrogen contents, specific leaf area or other leaf properties, leaf litter can be used to estimate the relative amounts of foliage at different canopy positions. It remains to be seen to what extent similar work may be developed in the much more diverse tropical forest canopies.

Physiological studies in rainforests have been used to determine the sources of transpiration through the forest canopy by examining stomatal conductance and its control in various parts of the canopy profile. Figure 13.4 (Roberts *et al.*, 1990) shows the average diurnal fluctuations in leaf stomatal conductance (g_s) at different levels throughout the canopy. There are characteristics of these data that have emerged as common features, at least in the relatively few studies where measurements have been made through the whole canopy height profile of the rainforest (e.g. Kira *et al.*, 1969; McWilliam *et al.*, 1996; Sá *et al.*, 1996; Carswell *et al.*, 2000). Maximum values are observed at the top of the canopy and there is a more or less steady decline down through the canopy. Sharp peaks in g_s can be observed, particularly in foliage at the uppermost levels around the mid-morning period. Lower in the canopy these peaks become less pronounced while they are absent in foliage at the base of the canopy.

Studies at the leaf or tree level have attempted to identify the major environmental controls of g_s but unfortunately this has only been possible in a correlative way by seeking to find which environmental factors account for the most variation in g_s. In common with similar studies in temperate forests, direct manipulative studies have been rare. Studies in which this approach has been adopted

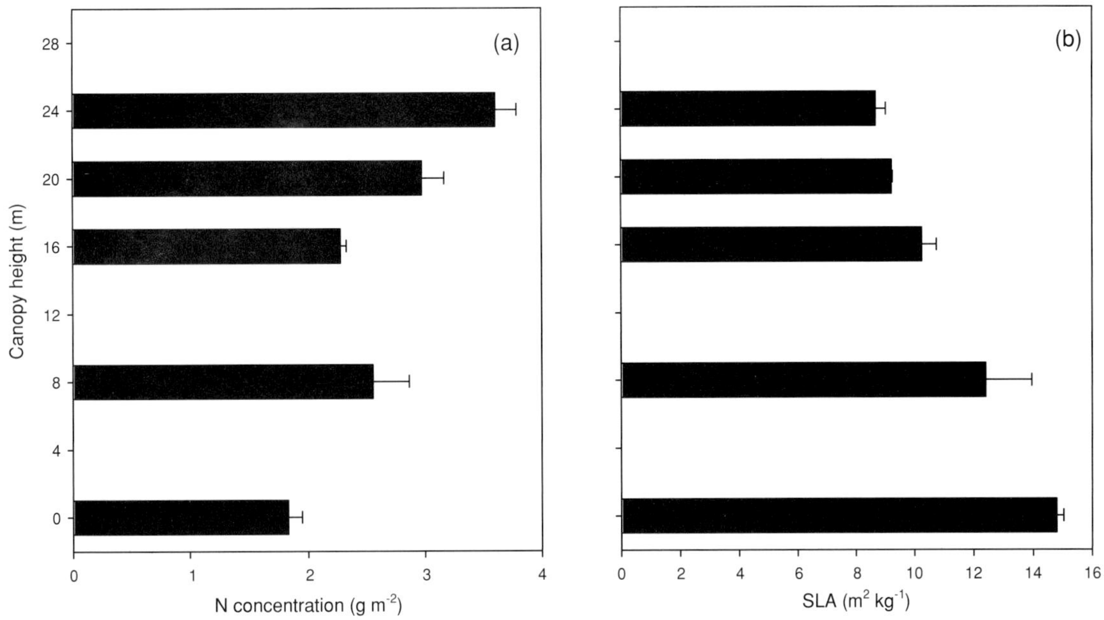

Figure 13.3 (a) Foliar nitrogen concentration on a leaf area basis for each of five canopy layers at the rainforest at Cuieiras, Manaus, Brazil. (b) Vertical variation in specific leaf area. (After Carswell *et al.*, 2000.)

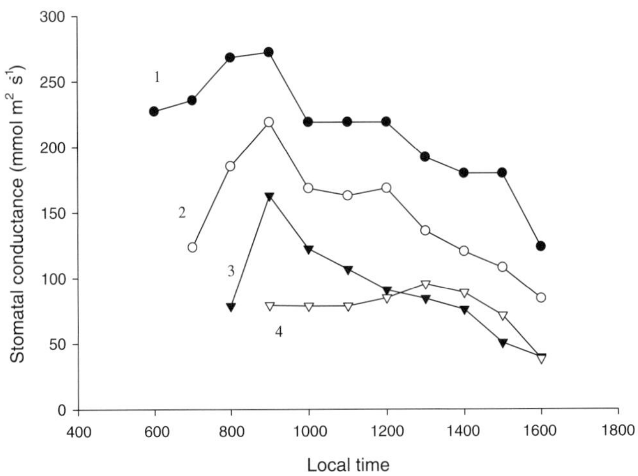

Figure 13.4 Diurnal variation in stomatal conductance of selected species around the sampling tower at Reserva Ducke, Manaus. (1) *Piptadenia suaveolens* at 33 m; (2) *Licania micrantha* at 25.6 m; (3) *Naucleopsis glabra* at 17 m; (4) *Scheelea* spp. at 3 m. (From Roberts *et al.*, 1990.)

have been made largely in forests in Amazonia but some information has also come from other regions. Illumination measured either as levels of R_s or *PAR* above or through the canopy, and D (or the gradient in D between the leaf and the air), have emerged as the most important environmental factors linked to fluctuations in g_s. Roberts *et al.* (1996) compared the linear regressions of g_s against D for a range of radiation classes using data taken from three forest sites (Reserva Jaru, Ji-Paraná; Reserva Ducke, Manaus and the Reserva Vale do Rio Doce, Marabá) in the Brazilian Amazon. In all cases, although the upper canopies had the highest values of g_s, these declined more sharply in response to increasing D than at lower canopy levels. This behaviour parallels the comparison made of the surface conductances at Jaru and Ducke which are shown later in this chapter (Figures 13.9 and 13.8 respectively). Physiological studies throughout rainforest canopies at sites through the Brazilian Amazon have not shown a major role for seasonal fluctuations in soil moisture as a control of g_s.

The relation between g_s and soil water content has been derived in different rainforests either from observations during the course of routine studies or by the imposition of drought. Thus far, no clear patterns seem to be emerging. At the Reserva Ducke, Roberts *et al.* (1990) found somewhat lower g_s of some species in the upper canopy in the dry season but it was also possible that leaf age had an effect. This possibility has also been considered by Meinzer *et al.* (1993) who studied the behaviour of g_s of *Anacardium excelsum* in a rainforest in Panama. In the wet season the maximum value of g_s was 300 mmol m^{-2} s^{-1} vs. 90 mmol m^{-2} s^{-1} in the dry season. To a large extent the differences in g_s could be related to differences in D but Meinzer *et al.* (1993) reported that differences in leaf age could also have contributed to the lower g_s in the dry season. At Reserva Jaru, Ji-Paraná, McWilliam *et al.* (1996) found little influence of low soil water content on g_s but one emergent tree became completely leafless in the dry season.

Because of reductions in wind speed down through the canopy space there is also a decline in the values of leaf boundary layer conductance (g_b) from the values obtained at the top of a rainforest

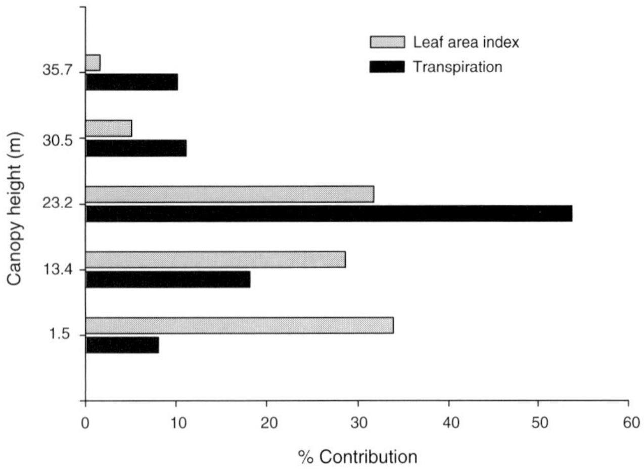

Figure 13.5 The percentage contribution to total leaf area index and canopy transpiration from each of five different canopy layers at Reserva Ducke, Manaus. (After Roberts *et al.*, 1996.)

canopy to those measured lower down. In parallel to studies of g_s made through the canopy profile at the Reserva Ducke referred to above, Roberts *et al.* (1990) also estimated g_b from weight losses of wetted leaf replicas. At the top of the canopy the average value of g_b was around 1400 mmol m^{-2} s^{-1} whereas at ground level it was only 300. At any level in the canopy the value of g_s is only 20–30% of the value of g_b. The PM equation reveals that these relative magnitudes of g_s and g_b will mean that when the canopy is dry the control of transpiration is dominated by the much lower stomatal conductance, g_s.

There are several important consequences of the profiles of microclimate and leaf physiology to the functioning of the forest canopy as a whole. Studies by Kira *et al.* (1969) in Pasoh Forest, Malaysia showed that although there are substantial amounts of leaf area distributed through the vertical profile of the forest, the relative contribution is greater from foliage in the upper parts of the canopy compared to that below. This observation was confirmed for the forest at the Reserva Ducke, Manaus, Brazil, by Roberts *et al.* (1993) using the CLATTER[1] model (Figure 13.5). This model uses data on the climate within the canopy, along with variations with canopy position of stomatal and boundary layer conductances and leaf area index in a multi-layer (bottom up) PM Equation to estimate canopy transpiration. The variation in the environmental variables constituting evaporative demand and the magnitudes of leaf conductance through the forest canopy both contribute to the different relative contributions from foliage from separate vertical canopy positions.

The control of forest transpiration at the stand level

The physical and physiological influences of vegetation within the PM framework are through the available energy term (A), the aerodynamic conductance (g_a) and the canopy conductance (g_c).

Table 13.1. *Albedo values for tropical lowland rainforests*

Albedo	Location	Reference
0.120	Nigeria	Oguntoyinbo, 1970
0.130	Thailand	Pinker *et al.*, 1980b
0.125	Manaus, Brazil	Shuttleworth *et al.*, 1984b
0.121	Manaus, Brazil	Culf *et al.*, 1995
0.149	Marabá, Brazil	Culf *et al.*, 1995
0.133	Ji-Paraná, Brazil	Culf *et al.*, 1995
0.120	Pasoh, Malaysia	Tani *et al.*, 2002
0.120	Bisley, Puerto Rico	Van der Molen, 2002
0.130[a]	Sabana, Puerto Rico	Van der Molen, 2002

[a] Coastal wetland forest; albedo rising to ~0.15 after flush of new leaves and persisting for next 6 months.

In addition to the various climatic variables which influence transpiration, the physical and physiological features of the vegetation which limit water loss are:

(1) the solar reflection coefficient or albedo which determines the amounts of net radiation available at the vegetation surface,
(2) the aerodynamic roughness, which can be directly related to g_a (see Equation 13.4 below), controls the turbulent exchange of water vapour from the vegetation into the atmosphere, and
(3) the surface conductance through which the forest canopy is able to control water loss by changes in stomatal conductance in each of the leaves in response to environmental factors such as solar radiation and air humidity deficit, levels of soil moisture and even rates of transpiration themselves.

ALBEDO

An important factor that controls the amount of energy available for evaporation is the reflection coefficient for solar radiation of the vegetation, the surface albedo (α). Lowland tropical rainforests tend to have values of α that are much lower than most other vegetation types (crops, grassland), which is a consequence of their high leaf area index distributed over tall, deep canopies. Such canopies are particularly effective in trapping solar radiation. Table 13.1 summarises the average α values for a range of lowland tropical rainforests. There is diurnal and seasonal variation in α which is mostly a consequence of changes in sun angle. Culf *et al.* (1995) showed that there was also a seasonal fluctuation in α at all three rainforest sites studied in the Brazilian Amazon in the ABRACOS project (Anglo-Brazilian Amazonian Climate Observation Study) (Gash *et al.*, 1996). The maximum values of α were observed at the same time as the driest soil moisture conditions. Given that leaf area index at all the sites remained high, Culf *et al.* (1995) concluded that the increase in α was associated

1 **CLATTER** – Canopy **L**ayer **A**nd **T**otal **T**ranspiration **E**stimation **R**outine.

with increased dehydration of the foliage that would change their optical properties. Some support for this contention comes from the fact that of all the values given in Table 13.1 the highest annual average α was observed at the site in the eastern Amazon, Marabá, which had the longest dry season and the largest seasonal development of soil water deficit. In the dry season at Marabá there would be an advantage to survival in reducing radiation capture, thereby limiting transpiration losses. On the other hand, Van der Molen (2002) observed an increase in albedo of a coastal *Pterocarpus officinalis* wetland forest in Eastern Puerto Rico (from 0.12–0.13 to 0.14–0.16) after a period of heavy rainfall which was followed by a new flush of leaves.

AERODYNAMIC ROUGHNESS

The specification of the aerodynamic interaction of tropical forest with the atmosphere requires that values for the aerodynamic roughness, represented by the forest's roughness length z_0, and the zero plane displacement d, are provided. The idealised relationship between wind speed, u, at a height z above the forest is assumed to have the form:

$$u = u_* / k \ln[(z - d)/z_0] \qquad (13.3)$$

where u_* is the friction velocity and k is von Kármán's constant (0.41). Values of d and z_0 are generally determined by fitting the equation to measured wind profiles under near-neutral conditions of atmospheric stability (Thom, 1975). Measurements of wind profiles through and above tropical forest have been made only rarely. Estimates for the Reserva Ducke, near Manaus, based on the work of Molion and Moore (1983) yielded an average value of $d = 30.1$ m ($0.86h$ where h is the height of the vegetation) and $z_0 = 2.1$ m ($0.06h$). From a similar analysis at the Reserva Jaru, Ji-Paraná, Rondonia in south-western Amazonia, Wright *et al.* (1996b) came to the conclusion that d and z_0 for that forest were similar to the values derived for the Reserva Ducke. Mean values of d and z_0 at the Sakaerat forest in monsoonal northern Thailand were 27.58 m ($0.79h$) and 4.69 ($0.13h$) (Thompson and Pinker, 1975) but there were substantial seasonal differences. In January, when wind speeds were low the mean values of d and z_0 were 29.53 m ($0.84h$) and 0.83 ($0.023h$). Wind speeds were much higher in the other periods (June and September) in Thompson and Pinker's studies. In June, d and z_0 were 27.16 m ($0.78h$) and 5.62 ($0.16h$), while in September the values were 28.09 m ($0.80h$) and 3.61 ($0.10h$).

The relationship between aerodynamic conductance g_a, and wind speed, u, the roughness length (z_0) and zero plane displacement (d) of the vegetation, is given in Eqn 13.4:

$$g_a = \frac{k^2 u}{\{\ln[(z - d)/z_0]\}^2} \qquad (13.4)$$

where all the variables are as indicated previously.

It is possible to compare the relative importance of the two conductances governing evaporation – the aerodynamic conductance g_a and the surface or canopy conductance g_c for a few rainforest sites. The estimate of g_a for the forest at Ducke at a wind speed of 3 m s^{-1} is 250 mm s^{-1} or (expressed in the commonly used units for g_c) 5.25 mol m^{-2} s^{-1}. In this case g_a exceeds the maximum values of g_c for the site by around five times (see Figure 13.6 below). Sensitivity studies using the PM equation to estimate transpiration, however, show that the magnitude and fluctuations in g_a contribute only trivially to differences in transpiration when g_c and g_a differ in magnitude of the order shown above. However, as will be shown later, the magnitude of g_a becomes crucial in determining rates of evaporation during rainfall.

SURFACE CONDUCTANCE

Studies of energy balance and turbulence were made above a semi-deciduous (monsoonal) forest in Thailand by Pinker and her colleagues (Pinker *et al.*, 1980a) but comprehensive studies of evaporation fluxes from tropical rainforests were not made until the Anglo-Brazilian collaboration in the central Amazon near Manaus within the framework of the ARME project (Amazonian Regional Micro-meteorological Experiment; Shuttleworth, 1988) and continued throughout Amazonia in ABRACOS (Gash *et al.*, 1996), TIGER (Grace *et al.*, 1995) and subsequent studies (Mahli *et al.*, 1998; 2002). The Large Scale Biosphere-Atmosphere Experiment in Amazonia (LBA) is an international research initiative led by Brazil. LBA is designed to create the new knowledge needed to understand the climatological, ecological, biogeochemical and hydrological functioning of Amazonia, the impact of land use change on these functions, and the interactions between Amazonia and the Earth system. Within LBA (http://www-eosdis.ornl.gov/lba_cptec/indexi.html) which started in 1996 and will run until 2003, considerable enhancement of the evaporation flux studies established in ARME, ABRACOS, TIGER as well as in other studies is being pursued throughout Amazonia. In addition, there are basin-wide studies of physical climate, hydrology, carbon fluxes and biogeochemical cycling. Eddy correlation studies of H_2O and CO_2 fluxes have also been carried out in rainforest at Pasoh in Peninsular Malaysia by Ohtani *et al.* (1997) and Tani *et al.* (2003). Similar techniques have also been used recently in a coastal wetland forest by Van der Molen (2002) and in a study comparing various rainforest types along an elevational gradient (by F. Holwerda), both in eastern Puerto Rico. Granier *et al.* (1996) measured the transpiration of dominant and co-dominant species in the rainforest at Paracou and St. Elie in French Guyana using sap flow techniques. The latter measurements were made in the dry season that extends from August to November. Basic information about the sites used in the various micro-meteorological and sap flow studies of stand transpiration is given in Table 13.2.

Table 13.2. *Details of the lowland rainforest sites used for transpiration measurements*

Site	Latitude	Longitude	Annual rainfall (mm)	Soil	Height (m)	Leaf area index	Method[a]	References
Pasoh Reserve, Negiri Sembilan, Malaysia	2° 58′ N	102° 18′ E	1800	Ultisol	35–40	6.5	BR	Tani *et al.*, 2002
Reserva Ducke, Manaus, Brazil	2° 57′ S	59° 57′ W	2500	Oxisol	30–35	6.0	EC	Shuttleworth *et al.*, 1984a; Shuttleworth, 1988
Reserva Cuieiras, Manaus, Brazil	2° 35′ S	60° 06′ W	1500	Oxisol	30–35	5.0–6.0	EC	Mahli (2002)
Reserva Jaru, Ji-Paraná, Brazil	10° 05′ S	61° 56′ W	1500	Oxisol	30–35	4.5	EC	Wright *et al.*, 1996a
Kourou, French Guyana	5° 12′ N	53° W	2300–3200	Ultisol	30–35	8.6	SF	Granier *et al.*, 1996
Bisley, Puerto Rico	18° 18′ N	65° 50′ W	3500	Ultisol	20–25	6.0–7.0	EC	Van der Molen, 2002

[a] BR, Bowen ratio; EC, eddy correlation; SF, sap flow.

Diurnal trends

Surface conductance has been calculated by many researchers from the latent heat flux or transpiration sap flux measurements referred to above. An inverted form of the PM equation is used for this with measured or estimated weather variables.

$$\frac{1}{g_c} = (\Delta\lambda\beta/c_p - 1)/g_a + \rho D/E \qquad (13.5)$$

where β is the Bowen ratio (the ratio of sensible heat to evaporation).

In this form, given meteorological information and aerodynamic conductance (usually measured or estimated using wind speed/vegetation height relationships along the lines indicated in the previous section), it is possible to deconstruct transpiration values to yield estimates of the surface conductance, g_c. The surface conductance values obtained in this way can be used to explore relationships with controlling variables such as climate, soil moisture or leaf area index.

The maximum surface conductances calculated from latent heat flux data or from sap flow studies at the stand level differ between studies. Figure 13.6 shows the diurnal trend in average hourly values of surface conductance calculated in this way for six out of the seven forests listed in Table 13.2. Three of those sites are in Amazonia, Brazil (Reserva Jaru, Ji-Paraná; Reserva Ducke, Manaus and Reserva Cuieiras, Manaus). Figure 13.6 also shows data for the cited studies at Pasoh, Malaysia and at St Elie, French Guyana (Kourou). In all cases except Cuieiras, the highest values occur in the mid-morning with a decline through the rest of the day. Mean maximum values are highest at Jaru, with maximum values on individual days reaching 2 mol m^{-2} s^{-1} (50 mm s^{-1}). At Jaru we also see the biggest diurnal reduction in g_c during the day. The maximum daily values of g_c at the other sites are far more modest and do not exceed 0.8 mol m^{-2} s^{-1}. The magnitude of g_c and its diurnal range were similar at Ducke, Cuieiras and Pasoh. However, the diurnal behaviour differs at the two neighbouring sites in central Amazonas. At Reserva Ducke there is a steady diurnal decline following a maximum value in the mid-morning whereas at Cuieiras the values remain steady during the day or even increase slightly.

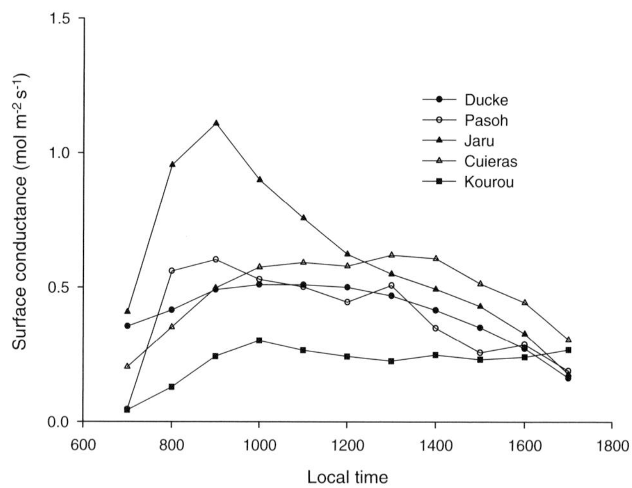

Figure 13.6 Average diurnal trends in surface conductance, g_c, derived for five lowland tropical rainforests.

Surface conductance calculated from the sap flow data collected in the Kourou rainforest in French Guyana by Granier *et al.* (1996) showed maximum values of g_c up to 0.7 mol m^{-2} s^{-1} but generally the values were much lower; the average maximum over 19 bright days being less than 0.32 mol m^{-2} s^{-1}, which constitutes the

Table 13.3. *Leaf area index values for lowland tropical rainforests*

Location	*Leaf area index*	*Reference*
Reserva Ducke, Manaus, Central Amazon	5.7	McWilliam *et al.*, 1993
Reserva Cuieiras, Manaus, Central Amazon	5.0–6.0	Mahli *et al.*, 2002
Marabá, Eastern Amazon	5.2	Roberts *et al.*, 1996
Ji-Paraná, South West Amazon	4.4	Roberts *et al.*, 1996
Ji-Paraná, South West Amazon	4.3	Meir, 1996
El Verde, Puerto Rico[a]	6.6	Jordan, 1969, 1971; Odum, 1970
Rio Negro/Branco, Venezuela	6.9	Williams *et al.*, 1972
San Carlos, Venezuela[b]	5.2	Jordan and Uhl, 1978
San Carlos, Venezuela	7.5	Saldarriaga, 1985
San Carlos, Venezuela[c]	5.1	Klinge and Herrera, 1983
Kourou, French Guyana	8.2	Granier *et al.*, 1996
Pasoh, Malaysia	6.5–7.0	Tani *et al.*, 2002; Kato *et al.*, 1978
Khao Chong, Thailand	~11.0	Kira *et al.*, 1964, 1967; Ogawa *et al.*, 1965
Cambodia	7.4	Kira *et al.*, 1969
Darien, Panama	10.6–22.4	Golley *et al.*, 1971, 1975
Abidjan, Ivory Coast	3.2	Muller and Nielsen, 1965
Abidjan, Ivory Coast	8.0–10.0	Bernhardt-Reversat *et al.*, 1978
Luanza, Zaire	3.5	Malaisse, 1981

[a] Lower montane.
[b] Low elevation rainforest on oxisols.
[c] Heath forest on spodosols.

lowest maximum g_c in any of the studies we have examined. Also, following the peak value the decline in g_c during the following daylight hours is the smallest in the Kourou forest (Figure 13.6). It is possible that transpiration and surface conductance for this forest are underestimated. There is a conversion factor (K) used in calculating sap flux from heat probe output proposed by Granier (1985, 1987) and presumably used at Kourou. Working in a beech stand in the UK, Roberts *et al.* (2001) showed that K determined in calibrations for trees from this forest exceeded Granier's empirical value substantially. Using the empirical value, estimates of transpiration would be nearly half of the estimate using K determined for beech from the stand. Using K determined for trees at the site and with scaling up to the stand level, Roberts *et al.* (2001) showed that for all dry days in the leafy periods, transpiration from sap flow agreed with measurements from eddy correlation to within 4% in one year and to within 10% in a second. For a lowland rainforest at Surumoni, Venezuela, Szarzynski and Anhuf (2001) found an excellent qualitative agreement between sap flux density of trees of three species and micro-meteorological estimates of transpiration. The authors acknowledged that quantitative comparison of transpiration from the sap flow method and micro-meteorological determinations requires the determination of total sap flow; i.e. the product of sap flux density and sapwood cross section.

The differences in maximum g_c observed at the different forest sites in Figure 13.6 cannot be explained solely by differences in leaf area index, L^*. In fact, L^* as measured at Kourou (8.2; Granier *et al.*, 1996) is associated with the lowest g_c, while conversely, at Jaru, which has the highest maximum g_c, Roberts *et al.* (1996) estimated L^* at 4.6, the lowest value of the three ABRACOS sites.

Measuring L^* of a tropical rainforest by destructive sampling requires a considerable effort. Despite this, there are numerous reports of the total leaf area index (L^*) for a range of tropical forests although, as with stomatal conductance (g_s) determinations, reports come predominantly from studies in South and Central America with less data from Asia and Africa (Table 13.3). Apart from a very high value ($L^* = 22$) for a riverine forest in Panama given by Golley *et al.* (1971, 1975), most L^* values fall between 4 and 8, whereas in forests in South America most reports even give an L^* of 6 or below. McWilliam *et al.* (1993) suggested the relatively low nutrient status of the soils, particularly in the Amazon Basin, as an important factor for the low leaf area indices of these forests. However, within the Amazon Basin itself, the highest values are observed in those areas with more rain and with lower seasonal variation in rainfall, such as at the Reserva Ducke, Manaus ($L^* = 5.7$) compared to sites with less rainfall and with a more pronounced seasonal rainfall distribution (Reserva Jaru, Ji-Paraná, Rondonia, $L^* = 4.4$; Reserva Vale do Rio Doce, Marabá, Para, $L^* = 5.2$). Later we show that when L^* is around 6 as at the Reserva Ducke, the lower canopy foliage contributes very little to forest transpiration. This result is not unique:

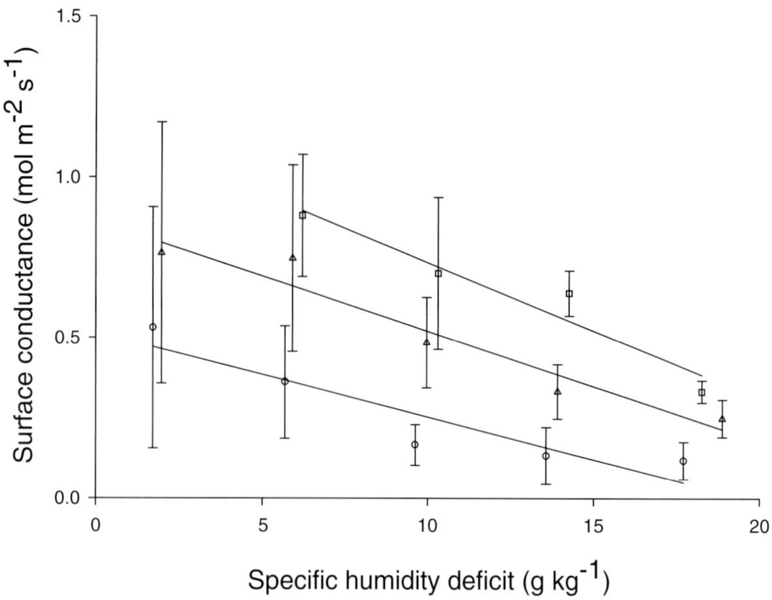

Figure 13.7 Relationship between surface conductance and specific humidity deficit as a function of solar radiation (R_s) at Pasoh forest, Peninsular Malaysia. □, $R_s \geq 800$ W m^{-2}; △, $R_s > 800$ W m^{-2} to ≥ 400 W m^{-2}; and ○, Rs > 400 W m^{-2}. (After Tani *et al.*, 2002.)

in the forest in south western Cambodia, Kira *et al.* (1969) found that the upper third of the canopy accounted for only 27% of the total L^* of 7.4 but was responsible for more than half of the photosynthetic capacity. Therefore, determinations of total L^* do not give any indications of the relative effectiveness of the different layers of foliage but this is possible with further measurements of the changes with height in the canopy of specific leaf area or leaf nitrogen (see Figure 13.3 and associated text).

Relationships of surface conductance with solar radiation and air humidity deficit

Previous analysis of the surface conductance data from the Reserva Ducke (Shuttleworth, 1988; Dolman *et al.*, 1991) emphasised the importance of levels of air humidity deficit (D) and solar radiation (R_s) in providing the major influence on g_c of this particular rainforest. Sufficient detailed information for R_s is not available for Kourou and Reserva Cuieiras to evaluate the type of function proposed by Dolman *et al.* (1991) for those particular sites but the function can be seen to fit well the data of Tani *et al.* (2003) from Pasoh (Figure 13.7). The diagram also shows the influence of R_s and D on surface conductance, g_c.

The relationship of g_c with D which is shown for the Reserva Ducke, Manaus, in Figure 13.8 and the Reserva Jaru, Ji-Parana, in Figure 13.9. The relative lack of influence of soil moisture in different seasons at these sites will be dealt with in the following section but at both sites there is a marked decline in g_c as D increases. An important difference between the studies at Ji-Paraná and Reserva Ducke is the rate of change in g_c with D. At Ji-Paraná the decrease in g_c for a change in D of 20 g kg^{-1} is almost double that observed at the Reserva Ducke. Although maximum g_c is much higher for the forest at Ji-Paraná than for the forests at Manaus (Figure 13.6) the differences in D might be a significant consideration and may mean that transpiration differs less than might be expected from the differences in maximum g_c alone. Figure 13.10 shows the frequency distribution of D at Ji-Paraná in 1992 and at Manaus in 1985. There is a higher percentage of hours with D below 10 g kg^{-1} at Manaus but D above 10 g kg^{-1}, is more frequent at Ji-Paraná. This difference between sites is a general phenomenon, not a specific effect of the years when the experiments were conducted (Culf *et al.*, 1996). At Kourou, Granier *et al.* (1996) also found that there was a strong negative relationship of g_c with D but the sensitivity was considerably less than that found in the data from Ji-Paraná and somewhat less than at the other Amazonian sites.

The negative relationship of g_c with D is not unexpected. It has been observed many times before in a wide variety of vegetations and has been reported both in terms of responses at the canopy and at the leaf level (Roberts, 1983). The decline of leaf stomatal conductance, g_s with D has been reported for several lowland tropical rainforest sites (e.g. Roberts *et al.*, 1990; McWilliam *et al.*, 1996) and discussed earlier. Despite at least three decades of plant physiological study, the functional relationship between stomatal closure and increasing dryness of the atmosphere is still not well understood. Nevertheless, the response has important hydrological and ecological implications. It means that when atmospheric demand is highest there is compensatory stomatal closure with the result that, on a daily basis, transpiration rates remain modest

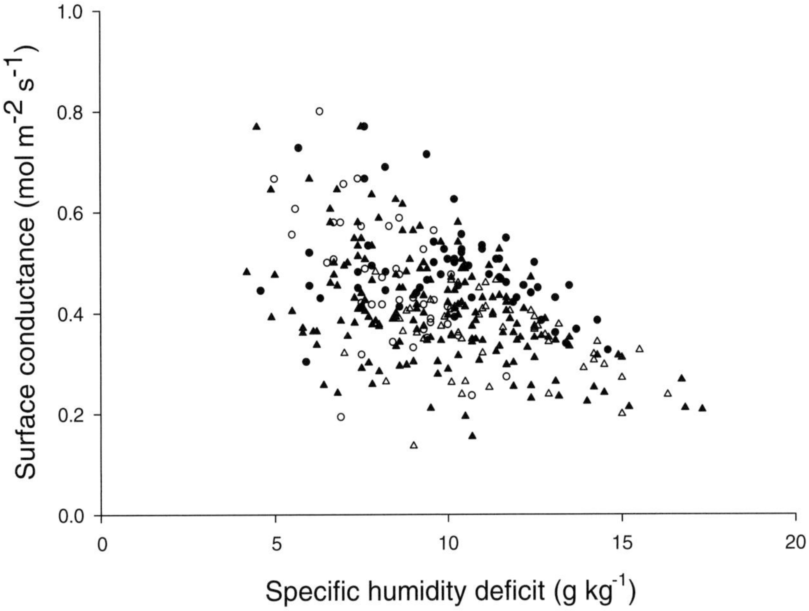

Figure 13.8 Surface conductance versus air humidity deficit at the Reserva Ducke, Manaus. •, in March/April 1985; △, May/June 1985; ○, August/September 1985; and ▲, September 1983.

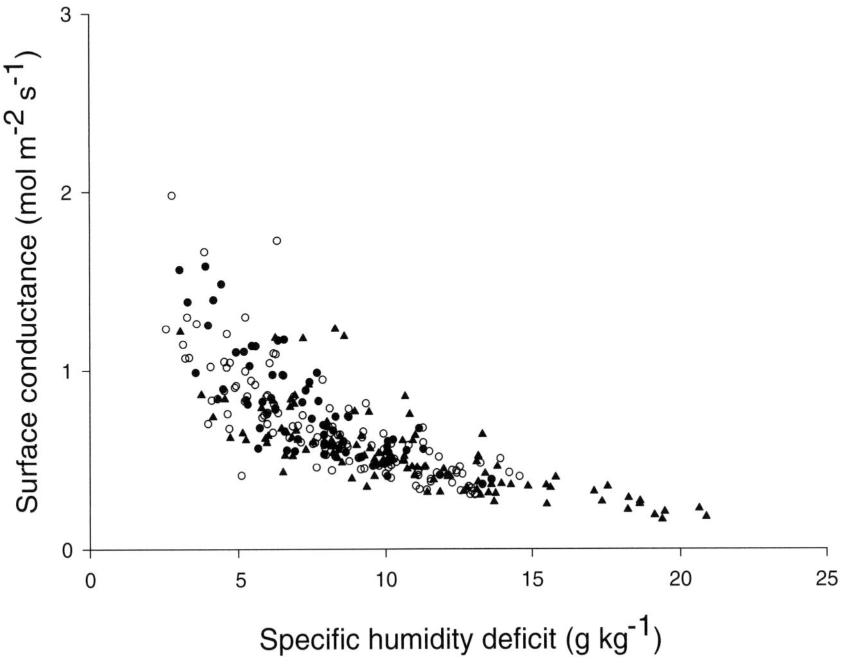

Figure 13.9 The relationship between air humidity deficit and surface conductance at Reserva Jaru, Ji-Paraná in the wet (•), intermediate (○) and dry (▲) seasons.

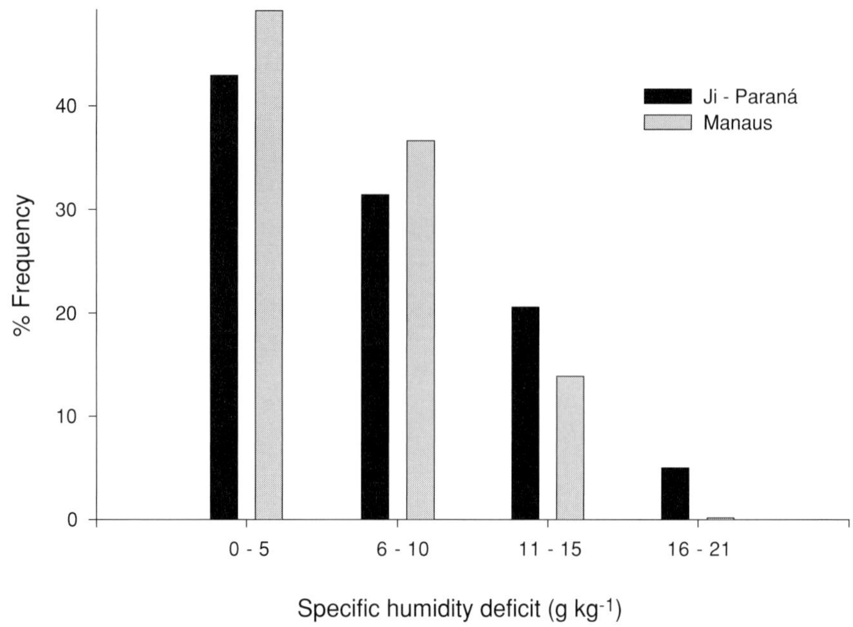

Figure 13.10 The frequency of occurrence of hours with differing air humidity deficit at the Reserva Ducke, Manaus and Reserva Jaru, Ji-Paraná.

and below potential rates determined by climatic conditions. Granier *et al.* (1996) found at Kourou that the typical daily rates of transpiration on bright days were only around 3.8 mm day^{-1}, and similar to those reported from Ducke (3.66 mm day^{-1}) by Shuttleworth *et al.* (1984a) and Shuttleworth (1988). It is not unreasonable to view these modest daily rates of transpiration as a form of rationing operating well in advance of limitations on transpiration due to soil water deficits (cf. Roberts, 1983).

Relationships of surface conductance and transpiration with soil moisture

Figure 13.8 indicates g_c plotted against D for groups of data collected at the Ducke Forest when R_s exceeded 600 W m^{-2} in four different seasonal periods differing in soil moisture conditions. The periods with the lowest levels of soil moisture are those from September 1983 and August to September, 1985. For a particular level of D there is a substantial amount of variation in g_c but there is little indication of systematic variation in g_c associated with the differences in season and the associated differences in soil water content. Figure 13.9 shows a similar graphical approach for g_c data from Reserva Jaru. In this case data are shown for three consecutive periods with increasing soil moisture deficits. The change in available soil water content from the wet season to the dry season is much greater at Ji-Paraná (Figure 13.9) compared to Manaus (Figure 13.8) but again there is no systematic difference in g_c in the different seasons for a particular radiation class and level of D. On the basis of this information it is tempting to conclude that soil water stress is not likely to be a major factor determining forest water uptake in Amazonia. However there was a marked response of g_c to reductions in soil water content at the Reserva Cuieiras, also near Manaus (Mahli *et al.*, 2002). Figure 13.11 shows a substantial lowering in g_c of the forest at Reserva Cuieirias, even at rather small reductions in soil water content. The difference in responses to soil moisture observed at Reserva Ducke and Reserva Cuieiras, which are only about 30 km apart and are thought to have much in common in terms of climate, soils and topography, is surprising. Results from the forest at Kourou (French Guyana) indicate that there is a reduction in transpiration in response to lowering of soil water availability. Granier *et al.* (1996) showed that in common with the data from the Amazon the ratio of actual transpiration to the potential was normally around 0.75. However, the ratio showed a decline down to around 0.60–0.65 during a month-long dry period.

Differences in response of the forest to reductions in soil moisture have also been observed in other parts of the Amazon. Detailed micro-meteorological studies were not made at Marabá, the ABRACOS site in the eastern Amazon with the longest dry season and with groundwater not within 20 metres of the soil surface. However, soil moisture studies (Hodnett *et al.*, 1996) showed that the rate of depletion of soil moisture did not decline substantially throughout the four-month long dry periods at the Marabá site. The implication of this is that the rates of evaporation were unchanged while soil water content reduced significantly. However, there is some contrasting information that soil moisture levels may play a

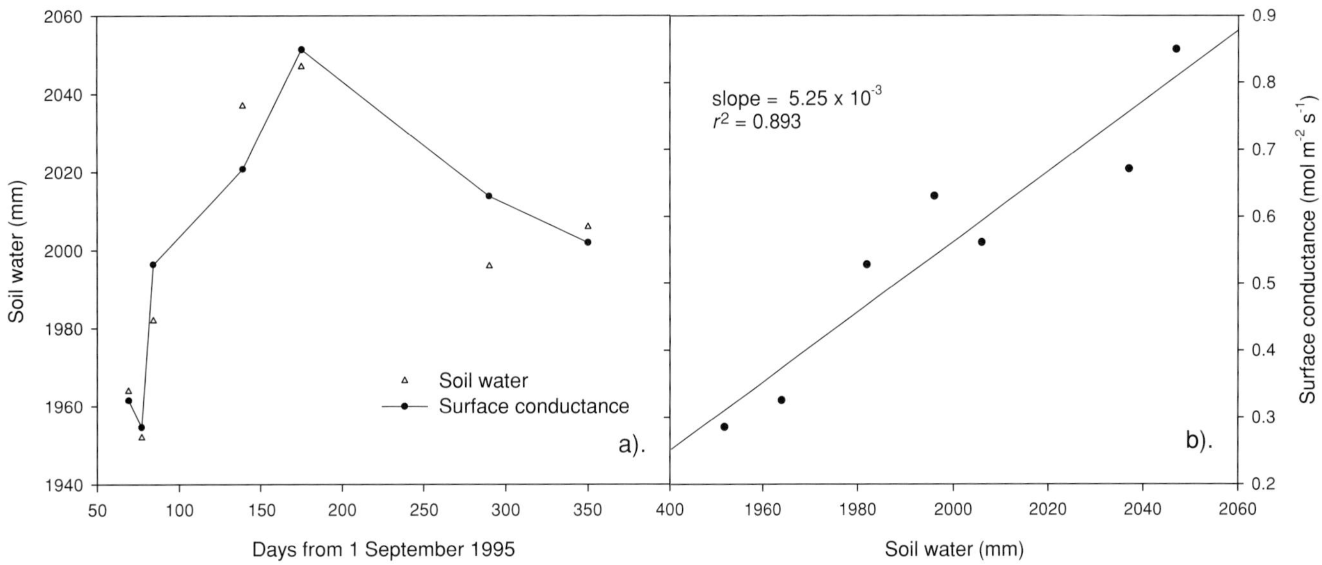

Figure 13.11 (a) Surface conductance and soil water storage (down to 4 m) on selected days after 1 September 1995. (b) Surface conductance plotted against soil water storage. Data obtained at the Reserva Cuieiras, Manaus. (After Mahli *et al.*, 2002.)

bigger part in controlling transpiration losses from tropical rainforest in the same region. The influence of reduced rainfall and dry soils on transpiration may be greater in El Niño years. On these occasions, in the eastern Amazon, complete recharge of soil moisture stores does not always occur if the previous wet season does not have adequate rainfall (Jipp *et al.*, 1998). The apparent reduction in forest evaporation in conditions of substantial soil water deficit that has been implied by Jipp *et al.* (1998) was derived from measurements of changes in soil water content down to a depth of 8 m. Nepstad *et al.* (1994) found occasional roots well below 10 m at the same site though.

The micro-meteorological data available for Pasoh, Peninsular Malaysia, do not allow an equally detailed investigation of effects of soil moisture on surface conductance as was possible for the Ducke and Jaru forests. However, based on the ratio of latent heat flux to available energy during dry spells, Tani *et al.* (2003) believe that soil moisture effects are small despite the relatively low annual rainfall at Pasoh (1800 mm). This probably reflects the good water holding capacity of the clayey soils at Pasoh (cf. Leigh, 1978). Similarly, from (short-term) soil moisture measurements in mature secondary rainforest in West Java (an area with high annual rainfall and a moderately developed dry season), Calder *et al.* (1986) concluded that the naturally-occurring maximum soil moisture deficit was only around 30 mm, and not considered likely to limit forest transpiration substantially. In Eastern Borneo, Bruijnzeel *et al.* (1993) reported large increases in the amounts of litter on the forest floor of a series of rainforests along an altitudinal transect following an ENSO event. Interestingly, no elevated litterfall occurred above 700 m, the general level of the cloud base.

Information from Africa about responses of rainforest to reduced soil moisture is sparse. Information has been reported for a high elevation site in Kenya with deep volcanic soils and a well-developed dry season of four months. On the basis of soil moisture measurements down to a depth of 4.5 m Eeles (1979) demonstrated that the ratio of actual to potential water uptake gradually decreased as the soil water content declined, despite the fact that Kerfoot (1962) found roots down to a depth of 6 m in this forest. One may conclude tentatively from these observations that these deep roots are apparently not numerous enough to fully prevent drought stress. Rather they may represent an adaptive mechanism aimed primarily at survival during extended dry periods.

No consistent pattern seems to be emerging from the studies of the response of lowland rainforest to reductions in soil water availability. It is possible that some of the variation in behaviour observed between sites in the same region can be explained by the fact that individual studies took place in years with markedly different rainfall. Nevertheless, there are several unresolved issues and we are not well equipped to predict the influence of an increasingly dry climate because of global climate change. There are important questions for which we require answers. At what level does the lack of available soil water become critical to tree survival? Does this differ for different species? What is the role of deep roots during drought events? Are drought effects more severe for forest on plateau sites compared to valley bottoms? Is short-term research based around intensive measurement campaigns the best way to study the influences on water use, photosynthesis and growth of lowland tropical rainforest of droughts with return periods at the decade level or more? Some of these questions are considered further in the last part of this chapter.

Modelling forest transpiration

The first attempts to provide a model of lowland rainforest transpiration came from Shuttleworth (1988). Shuttleworth calculated hourly values of g_c for daylight periods using the measured evaporation flux in dry periods in the inverted form of the PM equation (Eqn 13.5). Plots of the values showed consistent diurnal trends with a steep rise in g_c to a peak in mid-morning, associated with the increase in solar radiation, followed by a steady decline throughout the rest of the day in parallel with the rise in air humidity deficit, D, or perhaps the increase in water potential of the trees. Shuttleworth derived a quadratic equation to describe the mean diurnal trend in g_c (mm s^{-1}) and it took the form:

$$g_c(t) = 12.17 - 0.531(t - 12) - 0.223(t - 12)^2 \qquad (13.6)$$

where t is the time of day, in hours, beginning at midnight. Dolman *et al.* (1991) compared three models of varying complexity for describing the behaviour of surface conductance at Ducke for use in the PM equation against all the available data. The simplest model used a single daily value of g_c based on a mean value of eight days' data available from Shuttleworth *et al.* (1984a). A model of this type could estimate average evaporation accurately but would require the mean g_c value to be derived for most of the data, not just a few days. A second model used a quadratic description of g_c based on a diurnal fluctuation, essentially Eqn 13.5. Although empirical in nature, this model worked well because much of the diurnal variation in g_c is linked closely to the diurnal variation in two important variables, namely solar radiation and air humidity deficit, which change in a diurnal fashion. This model fittted the Reserva Ducke data well but is unlikely to be useful for situations widely different from the circumstances under which the empirical function was derived.

The third, most complex, model proposed by Dolman *et al.* (1991) is based on a Jarvis (1976) type approach that involves parameters for specific humidity deficit (D), solar radiation (R_s), air temperature (T) and soil moisture deficit (θ):

$$g_c = a_1 f(D) f(R_s) f(T) f(\delta\theta) \qquad (13.7a)$$

where the individual response functions are given by:

$$f(D) = \exp(-a_2 D) \qquad (13.7b)$$

$$f(T) = [(T - T_l)(T_h - T)^t]/[(a_3 - T_l)(T_h - a_3)^x] \qquad (13.7c)$$

$$f(R_s) = [(R_s/(a_4 + R_s)]/[1000/(1000 + a_4)] \qquad (13.7d)$$

with the exponent, ^{x}, given by $(T_h - a_3)/(a_3 - T_l)$, and T_h and T_l representing maximum and minimum temperatures respectively; a_1 represents the maximum surface conductance under non-limiting conditions, and a_2, a_3 and a_4 are constants.

A soil moisture function was not included however because of the low sensitivity for this factor demonstrated for the Reserva Ducke data (cf. Figure 13.8). The optimised parameters in the surface conductance model produced by Dolman *et al.* were very similar to those published for the so-called Simple Biosphere Model (SiB) by Sellers *et al.* (1989) and Shuttleworth (1989) to predict the effects of large-scale forest conversion in the Amazon (cf. Heil Costa, this volume).

Wright *et al.* (1996a) found that the Ducke-based parameterisation of the surface conductance model proposed by Dolman *et al.* (1991) was not adequate for the prediction of g_c as determined for the forest at Ji-Paraná, Rondonia. The model did not reproduce the high g_c values observed at this site (compare the values for the Reserva Ducke, Manaus and the Reserva Jaru, Ji Paraná in Figure 13.6). Furthermore, the already cited substantial reduction in g_c with lowering of soil water content at the Reserva Cuieiras (Figure 13.11) indicates that the model parameterisation would also need to be adjusted to describe seasonal changes in g_c fully at the Reserva Cuieiras. There is therefore still a need to develop the model so as to predict surface conductance well over a range of sites and while encompassing the different responses due to fluctuations in soil moisture.

RAINFALL INTERCEPTION LOSSES

Measuring rainfall interception in tropical rainforest

A proportion of the rain falling on a forest is intercepted by the canopy and evaporated directly back into the atmosphere without reaching the ground. This interception loss is measured conventionally as the difference between the incident (gross) rainfall, and the sum of the throughfall and stemflow (net rainfall) measured on the forest floor. Rainfall in the lowland tropics is generally characterised by short duration, high intensity storms and the interception loss as a percentage of rainfall is generally rather small. It is important therefore that net precipitation is measured with maximum accuracy, because when measured in the conventional way interception is a small difference between two large numbers. The size of the problem can be seen by considering a typical water balance, with errors assumed to be 3% in both gross and net rainfall. With a typical annual rainfall of 2000 (±60) mm and a net rainfall of 1700 (±50) mm (adding the errors quadratically) gives an interception loss of 300 (±80) mm, i.e. an error of ±26% in the evaporation. It may be possible to achieve lower errors, but if the rainfall is not measured above the canopy but in a clearing some distance away, or if the throughfall and stemflow are not sampled adequately, the error may easily be more. Before discussing published interception values for tropical rainforests we will examine various ways of how good interception data may be obtained.

Gross rainfall is best measured in the same location as the net rainfall measurements, but above the forest canopy. The height of rainforest canopies makes this difficult and expensive. Rainfall from convective tropical storms can vary considerably, even over small distances, and the alternative of measuring rainfall in a clearing may give results which are not representative, if the clearing is any distance from the net rainfall site. Unfortunately, measurements of rainfall made above a forest canopy are liable to greater error than those made at the surface, particularly if the topography is not flat, or there are strong winds. The errors in above-canopy rainfall measurements are still the subject of active research, but measurements are probably best made just above the canopy. Measurements made too close to the canopy might be influenced by splash off the trees. Alternatively, a rain gauge placed at an extended height above the forest might be subjected to high wind speeds and the increased turbulence around the gauge will reduce the catch. However, this approach derives from concerns usually associated with the need to measure above-canopy rainfall in north-west Europe or in mountainous areas where wind speeds and turbulence are higher than in lowland tropical rainforests.

The large number of species in lowland tropical rainforest, and the size of the dominant trees, means that special sampling techniques are required to obtain a representative sample of net precipitation. Lloyd and Marques Filho (1988) have analysed the errors involved in measuring throughfall with an array of gauges in the Reserva Ducke near Manaus. To ensure sampling the throughfall beneath a sufficiently large number of dominant trees, Lloyd and Marques Filho used a 100 m by 4 m grid with 36 gauges moved to new random positions each week. The throughfall was found to be highly variable with large leaves often either sheltering the gauges or concentrating the throughfall into the gauges via so-called drip points. Lloyd and Marques Filho (1988) ascribed their much reduced interception estimate (9%) as obtained with this roving gauge technique compared with an earlier study in the same forest using 20 fixed gauges (19.8%; Franken *et al.*, 1982) to a more representative inclusion of drip points. The latter phenomenon often gave measured point throughfall greater than the gross rainfall, as shown in Figure 13.12 (Lloyd and Marques Filho, 1988). A similar result was found by Jetten (1996) in two forests in neighbouring Guyana and by Ubarana (1996) at sites elsewhere in the Brazilian Amazon. By contrast, recent work by F. Holwerda in a rainforest in Eastern Puerto Rico revealed no systematic difference in median or average throughfall when using 30 roving gauges or 30 fixed gauges, although both estimates of average throughfall were distinctly higher than suggested by a set of 20 gauges used in a long-term study in the same forest. In addition, the standard error of the mean throughfall estimate was much reduced in the case of roving gauges. Interestingly, the roving gauge arrangement did not sample more drip points than the fixed gauge arrangement as found earlier in Amazonia (F. Holwerda, pers. comm.).

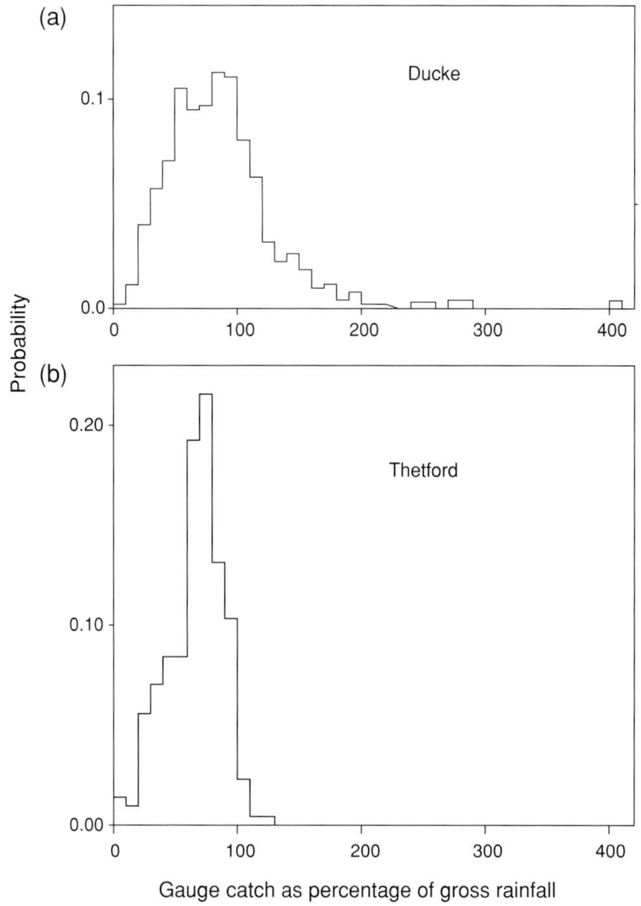

Figure 13.12 Probability distribution of throughfall gauge catch expressed as a percentage of gross rainfall for (a) rainforest, Reserva Ducke, Manaus and (b) temperate coniferous forest, Thetford Forest, UK. (After Lloyd and Marques Filho, 1988.)

The accuracy of the throughfall sample thus increases with the number of gauges used and the number of times they are moved. Some workers (e.g. Calder *et al.*, 1986; Calder *et al.*, 1996) preferred to use large plastic sheet net rainfall gauges, which collect a 100 per cent sample over an area of some 20 m^2. However, while this method overcomes the problem of the small scale variability and works well in plantation forest, the variability at the scale of dominant tree separation (10 to 20 m or more, see below) in species-rich rainforests may be missed unless more than one sheet is used, as throughfall can be sampled under only relatively few dominant trees. There are also practical difficulties in dealing with the large volumes of water that result from heavy storms, and in maintaining the plastic sheet free from leaks – insects and rodents eat the plastic – and because snakes like to live underneath it, access can be dangerous.

The spatial variability problem was recently examined by Loescher *et al.* (2002) in a lowland forest in northern Costa Rica (La Selva). Thirty-six funnel-type collectors were placed on radial transects extending away from the centre of a 160 × 160 m plot

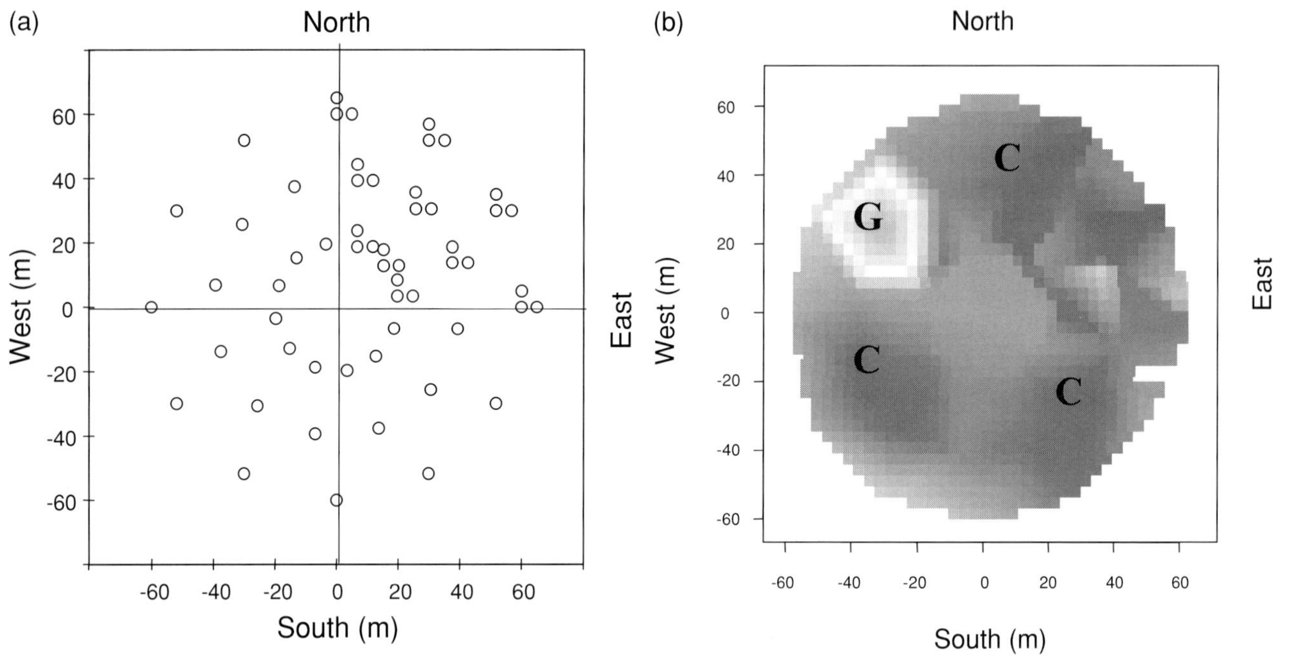

Figure 13.13 (a) Radial arrangement of throughfall collectors. (b) Variogram depicting mean throughfall volumes. [G] signifies a gap was present with a horizontal distance >30 m, and [C] depicts specific tree crowns with minimum horizontal distance of 40 m. Lighter shading, () indicates areas where throughfall << bulk precipitation; white, areas where throughfall < bulk precipitation; and dark shading, (), where throughfall ≈ bulk precipitation. (After Loescher *et al.*, 2002.)

(divided in four 80 × 80 m quadrants) at 20, 40 and 60 m from the centre. Each transect was repeated at 30°. To test for spatial autocorrelation in the throughfall measurements, an additional 20 gauges were laid out 5 m N and 5 m E of all the collectors in the NE quadrant (see Figure 13.13a). Measurements were made on an event basis during five weeks and the collectors were not moved between sampling occasions. So-called semi-variograms were constructed from which the 'range distance', i.e. the distance between gauges beyond which their respective estimates can be considered spatially independent, may be derived (see Loescher *et al.* 2002 for details). In addition, leaf area above each gauge was estimated from light measurements in an attempt to explain the high spatial variability in throughfall. The results were rather revealing: (i) only measurements made with collectors spaced as much as 43 m apart were statistically independent; (ii) the relationship between percentage canopy cover and gauge catch was weak (coefficient of determination of only 0.11, at $p < 0.02$); and (iii) the coefficient of variation of the throughfall stabilised (at 25%) after using 15 gauges, with the use of adding extra gauges not contributing further to the precision of the mean. Loescher *et al.* (2002) suggested that the 'range distance' derived in their study (43 m) corresponded with the typical size of emergent tree crowns or gaps (cf. Figure 13.13b). Brouwer (1996) used a similar statistical approach during a very short-term study in Guyana and derived a range distance of only 7 m. It is possible that the contrast in range distances reported for the two studies reflects the sampling interval (20 m vs. 10 m) between gauges rather than actual differences in canopy structure. More work is needed.

Although the regular relocation of throughfall gauges may improve the accuracy of the mean, it should be noted that an important assumption underlying the roving gauge approach is that (apart from random errors) only spatial differences in canopy structure determine the observed variation in throughfall (Lloyd and Marques Filho, 1988). In other words, *temporal* changes in canopy characteristics are considered negligible. Brouwer (1996) put this contention to the test by examining the degree of correlation between throughfall quantities in each of 20 fixed gauges on successive sampling dates. The average correlation coefficient decreased steadily from *c.* 0.7 (one sampling interval, i.e. a few days to a week) to *c.* 0.6 (interval of four months) and to less than 0.3 (interval of nine months). Interestingly, correlations with earlier or later measurements were especially poor during the period of maximum leaf litter production (Brouwer, 1996). It must be concluded that, even under more or less per-humid equatorial conditions, the rainforest canopy structure cannot be considered as simply constant in time. Because even two consecutive measurements showed a correlation coefficient of *c.* 0.7, canopy structure roughly accounted for *c.* 50% of the observed variability. The remaining variability must be ascribed to random variation and differences in rainfall or evaporative conditions during different storms (Brouwer, 1996).

Relative to the throughfall, stemflow has been found to be a small component of the water balance of most lowland rainforests – typically less than 2% of the gross rainfall (e.g. Lloyd *et al.*, 1988; Hutjes *et al.*, 1990; Jetten, 1996, Ubarana, 1996). However, some palms have very high stemflow. Much higher stemflow fractions have been reported for some montane rainforests (Bruijnzeel, this volume).

Other, more direct, methods of measuring the evaporation of intercepted rainfall have been attempted, for example the micrometeorological, Bowen ratio method (Stewart, 1977). Alternatively, various methods have been developed of measuring the change in the amount of water on the canopy, for example by weighing cut branches (Rutter, 1967); straingauges applied to branches *in situ* (Hancock and Crowther, 1979); and gamma ray (Calder and Wright, 1986) and microwave (Bouten *et al.*, 1991) attenuation. But none of these methods have been taken up as a generally applicable technique, and all would be difficult to implement in the tropical rainforest environment. However, Mizutani *et al.* (1997) and Gash *et al.* (1999) have shown recently that fast response, ultrasonic anemometers, with good water-shedding properties and spike-suppressing, on-line software, can function normally during rainfall. This allows the possibility that the eddy correlation method can be used to measure evaporation during rainfall directly in the turbulent boundary layer above the vegetation. Ideally, the eddy correlation method uses a fast response anemometer to measure vertical windspeed fluctuations and a fast response hygrometer to measure simultaneous humidity fluctuations. It is not yet established whether any currently available hygrometer will work during rainfall, but sensible heat flux can be obtained by correlating vertical windspeed with temperature. If the sonic anemometer's measurement of the speed of sound is used to give temperature, the problem of the rain causing a conventional thermometer to act as a wet bulb is then avoided. Gash *et al.* (1999) have used the method to measure sensible heat flux and then derive the evaporation as the residual in the energy balance equation. This development is important. A wealth of data is currently being collected on the rainforest carbon balance at a number of sites in Amazonia (e.g. Grace *et al.*, 1995; Mahli *et al.*, 1998). These sites are part of the Large Scale Biosphere-Atmosphere Experiment in Amazonia (LBA), which has an objective of measuring the long-term carbon balance using the eddy correlation technique. These sites also, by default, measure sensible heat flux and the data could be used to provide estimates of evaporation from wet rainforest canopies. The method does not give a complete measurement of interception because when the canopy is partially wet it cannot distinguish the evaporation of intercepted rain from transpiration. However these measurements of evaporation during saturated canopy conditions should have an important role in constraining and validating models.

Modelling rainforest rainfall interception

Rutter *et al.* (1971) first modelled forest rainfall interception with a model that recognised that the process was driven primarily by evaporation from the wetted canopy. The model has a canopy store that is filled by rainfall and emptied by evaporation and throughfall drainage. The evaporation is calculated from the PM equation, with the surface conductance (g_c) eliminated (Eqn 13.2) – the appropriate situation for a wet surface (Monteith, 1965).

There is a family of interception models based on the Rutter model, the simplest of which is the storm-based, analytical model of Gash (1979), which was later modified by Gash *et al.* (1995) to give improved boundary conditions and take account of forest sparseness, and later still by Van Dijk and Bruijnzeel (2001a and b) to allow the model to be applied to vegetation types with changing leaf area index through time (e.g. in forest gaps). The analytical model has the advantage that, as long as the rainfall climate is such that the assumption of one storm per day is justified, it can be run from daily rainfall records. The analytical model makes it explicit that the interception loss is controlled primarily by the mean evaporation rate during rainfall, $\overline{E}$, and the size of the canopy store, S.

When $\overline{E}$ is estimated from the PM equation and meteorological variables measured above the forest it has been found to be a remarkably conservative parameter. From a study of four forests in Great Britain, Gash *et al.* (1980) suggested a value of 0.22 mm hr^{-1} might be suitable for the whole country and (provided account was taken of the forest sparseness) similar values (0.17 mm hr^{-1}) were obtained for pine forest in south-west France (Gash *et al.*, 1995), and 0.32 and 0.20 mm hr^{-1} for pine and eucalypt plantation forest respectively, in Portugal (Valente *et al.*, 1997). For rainforest in central Amazonia, Lloyd *et al.* (1988) found a similar value with $\overline{E} = 0.21$ mm hr^{-1}, whereas Hutjes *et al.* (1990) in West Africa and Calder *et al.* (1986) in West Java derived values of 0.34 and 0.17 mm hr^{-1} respectively. As such, it would seem that during tropical rainfall the driving meteorological variables are similar to those in temperate latitudes: radiation is low, saturated vapour pressure deficit is small and windspeeds are similar – the higher temperatures in the tropics have little effect.

There is often confusion over the definition of the canopy storage parameter, S, as used in the family of Rutter-type models. The original definition is that S is the minimum amount of water that the canopy holds when just saturated at the end of a storm. Higher amounts of water may be held temporarily on the canopy during a storm, but at the end of a storm the canopy store drains rapidly to a depth of water, S. Jetten (1996) lists values of S found for tropical rainforest. The values are typically equivalent to a rainfall depth of just less than 1 mm – perhaps less than might be expected given the high leaf area of many rainforests (cf. Table 13.3). The low storage values are usually attributed to the good water-shedding properties

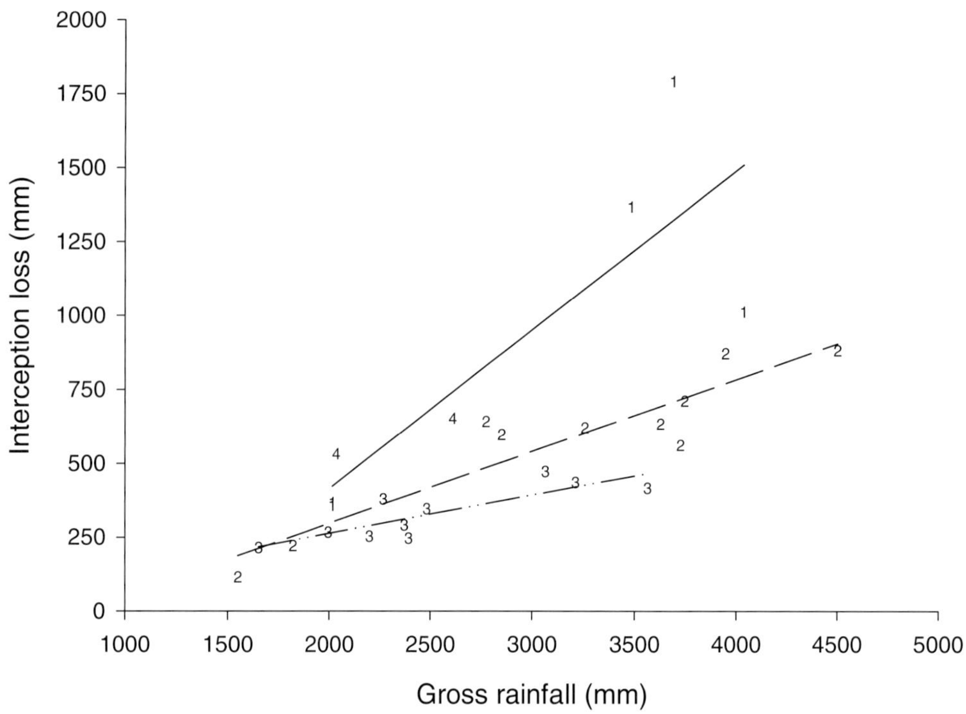

Figure 13.14 Interception loss plotted against gross rainfall for lowland tropical rainforest sites. Island sites are shown as '1', fitted regression (——) (data from Gilmour (1975); Schellekens *et al.* (2000); Waterloo *et al.* (1999)). Continental sites '2' (– –) (data from Abdul Rahim *et al.* (1995); Burghouts *et al.* (1998); Calder *et al.* (1986); Chappell *et al.* (2001); Dykes (1997); Hölscher *et al.* (1997); Malmer (1992) and personal communication; Richardson (1982); Roche (1982) and Sinun *et al.* (1992)). Mid-continental sites as '3' (— . . —) (data from Asdak *et al.* (1998); Collinet *et al.* (1984); Elsenbeer *et al.* (1994); Hutjes *et al.* (1990); Jetten (1994); Kuraji and Paul (1994); Shuttleworth (1988) and Tobón-Marin *et al.* (2000)). Also shown as '4' are data for two temperate coastal sites (data from Pearce and Rowe (1979); Rowe (1979); Calder (1990)). (Modified after Schellekens *et al.*, 2000.)

of rainforest vegetation. However, it may be more realistic for *S* to be treated as a variable rather than a fixed parameter and several models have been developed around this argument. For example, Calder (1986) and Hall (1992) have developed a stochastic model in which the canopy store varies with drop size and therefore rainfall intensity, while Jetten (1996) used a multi-layer canopy approach to increase the amount of water that can be accommodated by the canopy during large or high intensity storms. Both these models gave better fits to the observed throughfall – particularly within storms. However, although the CASCADE model of Jetten (1996) did improve predictions of throughfall amounts during very large storms in a Puerto Rican rainforest compared to the Rutter model, within-storm patterns were poorly represented (Schellekens *et al.*, 1999). It would seem that, apart from the Rutter *et al.* (1971) and Gash (1979) models and their derivatives, no model has been taken up as a generally applicable technique. For most hydrological purposes the analytical model of Gash *et al.* (1995) should give acceptable results, however. Nevertheless, a comparative modelling exercise is long overdue – several of these more complex models should be compared against independent datasets (cf. Schellekens *et al.*, 1999).

Interception from maritime sites

All the above discussion relates mostly to equatorial continental or continental edge rainforest sites in Latin America or Africa, and much less to the 'maritime continent' of South East Asia. However there is increasing evidence (e.g. Schellekens *et al.*, 2000) that forests on small islands, for example in South East Asia, the Pacific or the Caribbean and, as indeed observed also for coastal sites in temperate regions, may have much higher rates of interception. Schellekens *et al.* (2000) presented the relationships between gross rainfall and evaporation of intercepted rainfall for tropical forest and plantations at coastal and island sites (Type 1), continental edge, equatorial sites (Type 2), continental, equatorial (Type 3) and also for native woodland and plantations at coastal sites in temperate latitudes (Type 4). Figure 13.14 extends the analysis of Schellekens *et al.* (2000) by including further data for Type 2 sites from Brunei (Dykes, 1997); Sabah (Sinun *et al.*, 1992; Burghouts *et al.*, 1998; Chappell *et al.*, 2001) and Type 3 sites in central Kalimantan (Asdak *et al.*, 1998); the Colombian Amazon (Tobón-Marin *et al.*, 2000); the Brazilian Amazon (Ubarana, 1996) and Guyana (Jetten, 1996). Calder *et al.* (1986) recorded 21 per cent of the gross rainfall being intercepted by a mature secondary forest on Java (a Type 2 or continental edge site according to the legend employed in Figure 13.14). Expressed over a year this amount is equivalent to the measured, annual net radiational input to the site. This result was discussed by Shuttleworth (1989) who argued that this could not occur over continents because there would then be insufficient sensible heat to provide the convection necessary for generating rainfall, postulating a negative feedback effect acting to reduce rainfall and thus interception under non-maritime conditions. The measurements by Dykes (1997) in Brunei (again, a Type 2 site) gave an interception loss of 18%. The Gash *et al.* (1995) model predicted the measured interception loss only when the derived value of $\overline{E} = 0.75$ mm hr^{-1} was used. Similar high values of $\overline{E}$ were needed by Bruijnzeel and Wiersum (1987) and Waterloo *et al.* (1999) working in plantation forests in Java and Fiji, respectively. Recently Schellekens *et al.* (1999 and 2000) working on the island of Puerto Rico have reported the exceptionally high interception loss of 52%. Interception at this particular site showed a distinct seasonal variation, estimates ranging from *c.* 50% during summer months to about 30% in winter (Schellekens *et al.*, 2000). This latter work was supported by a catchment water balance study that required evaporation of intercepted rainfall of this amount to close the catchment water budget.

In the Calder *et al.* (1986) study on Java the Rutter model underestimated the evaporation by about half when driven by the measured meteorology and the best result was obtained with a model that effectively increased the canopy store and the aerodynamic conductance. Similarly, the Schellekens *et al.* study on Puerto Rico found $\overline{E}$ estimated from the measured meteorology to be 0.11 mm hr^{-1}, but an increase of an order of magnitude was needed to obtain agreement between model estimate and the rainfall interception data. Agreement could only otherwise be reached by increasing the canopy store to a value in excess of 5 mm. However, unlike the increase in $\overline{E}$, raising the value of S did not result in a better fit between predicted and measured throughfall data. In particular, the use of a large storage value led to an overestimation of throughfall during large storms and vice versa during small storms (Schellekens *et al.*, 1999). If the possibility of such large values of the canopy store are discounted therefore, then all this work suggests that the use of the PM equation is in some way inappropriate[2]. However, the equation has a sound basis: i.e. the surface energy balance and the flux diffusion equations. These can only be inappropriate if the assumption of horizontal homogeneity is not valid, i.e. if there is horizontal advection of energy (Thom, 1975). Indeed, it has been hypothesised that the high interception losses from the Puerto Rican and Fijian sites are at least partly the result of energy being advected from the warm sea upwind of these islands (Schellekens *et al.*, 1999; Waterloo *et al.*, 1999). This argument implies a horizontal gradient of temperature in the direction of the mean wind. The equation to describe this horizontal flux divergence is given by Thom (1975) as:

$$H_x = \int_0^{z_r} \frac{\partial \rho c_p u T}{\partial x} dz \qquad (13.8)$$

where H_x is the divergence of the horizontal sensible heat flux (i.e. advected energy) in the direction of the mean horizontal wind, u, between the surface and the reference height, z_r, and ρ and c_p are the density and specific heat of air respectively. H_x could be deduced from observations of the horizontal temperature gradient. Thom (1975) calculated that a flux divergence of 100 W m^{-2} would result from a temperature gradient of 1 K per 100 m. The evaporation rates are apparently enhanced by a flux of this size, so (if the hypothesis of advection is true) it should be possible to measure H_x. Alternatively H_x could be measured directly using the eddy correlation method, as the divergence of the forward diffusion ($\rho C_p u'T'$), where u' and T' are the instantaneous deviations of the horizontal wind and temperature from their means). Measurements of this type are technically feasible and may provide a way of resolving these apparently anomalous results.

Certainly, there is a major lack of understanding in the process of rainfall interception in maritime, tropical forests. The reasons for the failure of the models to reproduce the measurements must be established, preferably by experiment and observation, and new knowledge incorporated into the models so that they can describe the hydrology of these forested island catchments adequately.

FUTURE RESEARCH NEEDS IN LOWLAND TROPICAL RAINFORESTS

In the past two decades hydrological research in lowland rainforests has concentrated on two main aspects. Firstly, the evaporation characteristics of rainforest and vegetation that would replace rainforest after human intervention have been studied. The behaviour of forests has been the main focus of attention here. The second important area of research has been the examination of key properties of soils and subsurface layers and how changes in them, e.g. following forest clearance, influence the dynamics of streamflow. This subject has been dealt with comprehensively by Bruijnzeel (1990, 1996; see also the chapter by Grip *et al.*, this volume). Although the forest evaporation research has revealed a great deal, there are still gaps. The geographical coverage of this effort has focused largely on forests in South America, less on South East Asia and to some extent Australia but there has

2 Later work in the same forest by F. Holwerda and using many more (up to 60) throughfall gauges than the 20 employed by Schellekens *et al.* (1999, 2000) gave somewhat lower interception values, even after accounting for seasonal differences. As such, the degree of underestimation of wet canopy evaporation by the Penman-Monteith model will be somewhat less than inferred by Schellekens *et al.*

been very little effort dedicated to rainforests in Central and West Africa. Although smaller in extent than forests in Amazonia, the geographical location of the forests of central Africa will mean that they also have an important influence on global and regional climate (cf. Mahé *et al.*, this volume).

Recent initiatives such as the Large Scale Biosphere-Atmosphere Experiment in Amazonia (LBA) based largely in Brazil, emphasise the bias which has developed towards research, particularly in micro-meteorology and plant physiology, towards the lowland rainforests of Latin America and the Caribbean. This means that the level of knowledge and understanding of tropical forests of other regions, particularly in Africa, remains very low. An associated, very important, benefit to the location of a research initiative in particular countries or regions is the opportunity for training and development of young scientists from the host countries. The scientific outputs from experimental programmes such as ARME and ABRACOS are substantial. Of equal importance was the involvement of a great number of young researchers who developed the experiments with and studied alongside experienced scientists. The development of knowledge, experience and, perhaps most importantly, confidence has enabled many of the Brazilian collaborators in ARME and ABRACOS subsequently to initiate their own research programmes. The lack of substantial research in rainforests in some geographical regions, particularly Africa, means that, as well as a slow increase in scientific knowledge and understanding of the forests there, it is likely that the skills and experience of young, local scientists will also be slow to develop.

The long-term influence of human interventions, burning of fossil fuels and deforestation have led to increases in atmospheric CO_2, which is generally regarded as a key force in global climate change (IPCC, 2001). Predictions of changes in global climate during the coming century, such as increased temperatures, are considered likely to increase the frequency and intensity of droughts in tropical rainforest areas, e.g. Amazonia (Hulme and Viner, 1998). The disappearance of tropical rainforest from the Amazon Basin (Cox *et al.*, 2000) and from northern South America, southern equatorial Africa and to a lesser extent in South East Asia (White *et al.*, 2000) is even offered as a potential consequence of these increased droughts. Unfortunately, our knowledge of the response of rainforests to reductions in soil water availability is still sketchy, so a sound judgement cannot be made on the plausibility of global circulation models predictions of the demise of rainforest.

Extended droughts do not mean per se that tropical forest cannot exist. For example, in the eastern parts of the Amazon there is a dry season of three to four months and Nepstad *et al.* (1994) found that tree roots could extract water from a depth of 18 metres, thereby allowing the forest to survive even though this was not sufficient to avoid major drops in transpiration rates (Jipp *et al.*, 1998). In Sierra Leone, lowland evergreen rainforest is found on deeply weathered substrates under strongly seasonal climatic conditions (five months dry season, Ledger, 1975). Bonell *et al.* (1998) report that rainforest in North-East Australia also survives a long dry season by virtue of access to groundwater reserves. In the Western Ghats, India annual rainfall ranges from 2000 to 6000 mm, the majority of which falls in as few as three to four months. Bonell (1999) has speculated whether the trees survive the long dry season (> six months) because deep tap roots can exploit groundwater. We do not know how readily trees are able to extend their root systems downwards to cope with the development of quasi-permanent drier conditions as a result of climate change.

There are a number of important questions remaining about deep roots in lowland rainforests. The numbers of studies indicating the presence of deep roots are rather few. How widely occurring are deep roots below tropical forest? Are they an intrinsic genetic feature of most tropical tree species or are they stimulated to be produced in response to drought episodes? It is important to quantify water uptake by deep roots under dry soil water conditions. To what extent can deep roots fully supply the transpiration requirements of a forest when upper soil horizons are completely depleted of water? Studies in the cerrado, savannah vegetation south of the tropical rainforest in Brazil, by Meinzer *et al.* (1999b) indicated that although there was an extensive deep root system, it had a low hydraulic conductance, and did not exploit the deep soil water resources that were available. This would probably mean that the contribution of the deep root system might be limited to providing a limited amount of water for survival in extended dry periods. We need studies such as those carried out in the cerrado to be made in tropical rainforest.

Severe droughts followed by fire and complete loss of tree cover are considered to have occurred in the past in several of the regions where lowland tropical rainforest still persists today. There is the evidence of charcoal layers in the soil e.g. in the San Carlos de Rio Negro region of Venezuela (e.g. Saldarriaga, 1985) and also evidence obtained from forest structure and species composition in Sabah (e.g. Walsh and Newbery, 1999). It is probable that widespread fires do not occur while the forest remains evergreen but unfortunately we have no direct experience of the severity of drought that needs to occur to result in wholesale leaf loss and probably tree death also.

There are a number of ways that trees can avoid drought. This chapter has given a number of examples of where forest transpiration is limited because g_c is negatively correlated with increasing D: in fact, the two may be functionally related. The g_c versus D relationship will mean that on a daily basis the amount of water removed as transpiration will normally be restricted to around 0.75 of the atmospheric demand. Obviously, for transpiration to be maintained, this amount of water has to be reliably supplied from the soil. Walsh and Newbery (1999) have discussed how trees can deal with air embolisms (cavitation) which occur in the conducting tissues when transpiration from the canopy exceeds soil water supply. In the case of very severe droughts, complete dysfunction of the hydraulic system of some or all species may occur. If this situation persisted even for less than a few days, death of the trees would ensue. What we know little about for lowland tropical species is under what conditions does this occur and how frequently. One approach to evaluate the frequency of conditions that influence stomatal closure and photosynthesis that has begun to be exploited for species in temperate woodlands is the variation in carbon isotope discrimination in the annual rings of the xylem of the tree (e.g. Duquesnay *et al.*, 1998). Unfortunately the approach needs some development before it can be applied to tropical trees because they lack the distinct annual growth rings required for dating purposes.

An approach that is being used to examine the effects of increased drought on tropical forest has been to establish rain-out experiments. Adjacent droughted and control plots (100 m × 100 m) have been established by Daniel Nepstad and his colleagues (Woods Hole Laboratory, USA) in the Tapajos National Forest, near Santarem, Brazil. These plots will be used to compare the response of trees to intense drought. A similar approach has been adopted by John Grace and his team from Edinburgh University, at Caxiuana, west of Belem. The conclusions of these studies are awaited with interest.

The effects of drought on rainforest evaporation from studies that have measured latent heat fluxes directly have been described. The results are to some extent conflicting and a clear picture has yet to emerge. Even in the same region of the central Amazon around Manaus, studies by Shuttleworth (1988) showed little influence of soil water deficits on transpiration while a strong influence was observed by Mahli *et al.* (2002). Both of these studies involved the eddy correlation approach and were located on plateau areas. As Mahli *et al.* point out, there is a need to determine if small scale differences exist in plant and soil hydraulic properties, even for the same part of the toposequence (i.e. plateau sites), e.g. due to differences in soil texture.

One of the drawbacks of most hydrological research initiatives in lowland tropical forests is that they are usually fairly short in duration. This short-term nature is often determined because the research is linked to a studentship or a particular research call. There are no places where there are extended measurements from which the impact of severe, but infrequent droughts on forest transpiration, carbon fixation and growth can be evaluated. It could be argued that some of the resources directed towards very detailed, short-term process studies might be better directed towards less detailed but extended studies which focus on the effect of severe soil water deficits on forest function.

A typical feature of many tropical forest landscapes is that they are highly variable in elevation, with plateau, slope and valley bottom areas occurring within close proximity. The existence of these different topographical elements with very different seasonal water supply probably means that current GCM predictions of the disappearance of tropical forests are too simplistic (cf. Costa, this volume). Some parts of the landscape (e.g. footslopes and valley bottoms) will probably remain well supplied with water, even though rainfall might be drastically reduced. Up to the present there is no systematic study of the physiological responses of forests in different parts of the toposequence to seasonal fluctuations in rainfall and soil water content. Virtually no reliable information is available on the water use of swamp forests, whether seasonally or permanently inundated (Hooijer, this volume).

The simplest method for measuring evaporation from tropical forests that has emerged in the last 20 years is eddy correlation. This method can provide an areal average of water vapour (and CO_2) fluxes. In a detailed study at the Reserva Ducke, central Amazon, good agreement was achieved between eddy correlation measurements of transpiration and estimates made from physiological studies (Shuttleworth, 1988; Roberts *et al.*, 1993). However, it is possible that the agreement is only fortuitous. There is considerable scope for variation in evaporation over a unit of the landscape within the sample footprint of the eddy correlation method that will not be captured at a single tower site used for physiological sampling.

Eddy correlation flux measurement systems are likely to integrate contributions from the vegetation from all the toposequences. As noted above, it is very likely that the forests occurring on plateaus, slopes and valley bottoms will behave very differently during protracted rainless periods but the eddy correlation systems might not discriminate this behaviour. This reinforces the need to determine the physiological responses of trees in different parts of the landscape, e.g. slope elements, and to investigate variation at the plot scale requires measurements to be made on individual trees. One of the ways forward in sampling the transpiration of trees throughout a catchment which both overcomes the averaging problem of the eddy correlation approach and the time-consuming work from towers required for leaf gas exchange studies, is the use of sap flow methods. Nevertheless, there are still some technical difficulties to be overcome. There is good evidence (Roberts *et al.*, 2001) that the empirical calibration factor proposed by Granier (1985, 1987) for the type of sap flow gauges pioneered by him is not universal and might need to be determined for individual species. Furthermore, providing sufficient power will always be a problem and there is also a need to have assured ways to determine the area of sapwood in a range of tropical species and a good prospect for determining the variation in sap flow velocity across the sapwood area.

There are a number of studies in temperate forests which show that various physiological functions of forests such as photosynthesis and transpiration decline with forest age (Yoder *et al.*, 1994; Ryan *et al.*, 1997; Watson *et al.*, 1999; see also the chapter on tropical tree plantations by Scott *et al.*, this volume). Because the amount of foliage maintained by the tree also declines with age the canopy interception may also fall. Watson *et al.* (1999) showed for *Eucalyptus regnans* forests in Victoria, Australia that new regrowth following wildfire has a higher transpiration and interception loss that declines steadily as the forest matures. The trees comprising these forests are long-lived (>250 years), so the impact on streamflow of changes with age of the hydrological functioning of the woodlands on the catchment can have a substantial effect. In a tropical rainforest the oldest trees will be the scattered emergents and those comprising the upper layer of the main canopy (Richards, 1996; Whitmore, 1998). Two factors, at least, will contribute to higher water stress of the emergents and the upper canopy trees compared to trees lower in the canopy. A decrease in the hydraulic conductance has been shown for taller older trees (e.g. Hubbard *et al.*, 1999) but also their canopies will experience a much more demanding atmosphere because of high radiation, temperature and air humidity deficit (cf. Figure 13.2). The imposition of more intense droughts associated with future global climate change may mean that a more rapid turnover of individuals is a consequence. It would be valuable to know what degree of xylem dysfunction occurs in upper canopy rainforest trees in drought situations.

There are other processes occurring at the plot scale that might predispose some trees to the effects of drought rather than others. Working with the rainforest in French Guiana, Bonal *et al.* (2000) found that one species, *Eperua falcata*, was able to exploit water from depths below 3 m while another, *Dicorynia guianensis*, depended on water in the surface layers of the soil. *Eperua* was thought to have advantages under occasional severe moisture stress. At Barro Colorado Island, Panama, there is typically a pronounced dry season. Using stable isotope techniques, Meinzer *et al.* (1999a) examined the soil water extraction patterns of a number of species during the four-month dry season. They found that there was a clear vertical separation of the zones in the soil from which different trees obtained water. It was also shown that species that were able to tap increasingly deep sources of soil water as the dry season progressed also showed the smallest seasonal variability in leaf fall. Understorey species and herbs on the other hand which were less deeply rooted experienced drought stress (Jackson *et al.*, 1995). Van Dam (2001) observed a significant autocorrelation between rainfall and leaf litter production in forest in Guyana whereas leaf fall quadrupled during ENSO events (cf. Bruijnzeel *et al.*, 1993). There seems to be a pressing need for these types of integrated studies to be repeated in a wide range of tropical forest types (cf. Proctor, this volume). However, a full understanding of the evaporation characteristics of tropical rainforests will only be achieved by employing studies at a range of scales from the leaf and tree up to the landscape/catchment scale.

It is very clear (e.g. Figure 13.14) that rainfall interception increases greatly from continental sites to those at continental edges or on islands. It would be valuable to measure the sources of energy supporting the high evaporation rates observed in the case of small islands. This might be done micro-meteorologically as we have proposed earlier. In this context it is interesting to reflect how little effort has been devoted to direct measurements of interception loss and its control at the leaf/canopy level since the pioneering work of Jack Rutter who examined interception losses by measuring the weight loss of individual branches (Rutter, 1967). In contrast, very many studies of stomatal conductance, transpiration and photosynthesis measured on individual leaves and twigs with porometers and portable infra-red gas analysers have added considerably to the insight that has come from larger scale measurements.

Another area of study which needs development is the interaction between landscape elements. These may be elements which differ in vegetation or perhaps topography or both. Although we may have a reasonable knowledge of the hydrological behaviour of different vegetation types, our understanding of how they interact in the landscape is only rudimentary (see also the final part of the chapter by Malmer *et al.*, this volume). As an example, it is important to recognise the potential importance of lowland rainforest evaporation to climatic conditions in adjacent montane systems. Lawton *et al.* (2001) used satellite imagery to show that in the dry season, deforested areas of the Caribbean lowlands of Costa Rica remain relatively cloud-free while forested regions have well-developed clouds. The authors also used mesoscale atmospheric simulations to estimate the cloud base height in a simulation using broadleaf evergreen forest and also one using short grass. In the forest run the cloud base height was substantially less than observed over the pasture. From their satellite imagery studies and modelling, Lawton *et al.* suggested that changes in land cover in tropical lowlands have serious impacts on the ecosystems in adjacent mountains. Similarly, Van der Molen (2002) predicted raised cloud condensation levels after converting coastal wetland forest to pasture in NE Puerto Rico, despite the fact that the pasture evaporated more water than the forest. The higher sensible heat flux above the forest resulted in a stronger land-sea interaction and this in turn caused more air to move upslope and form clouds (see also the chapter on montane forests by Bruijnzeel, this volume).

Over the past two decades there has been a large amount of investigation and a substantial increase in the understanding about factors controlling the evaporation from lowland tropical rainforest. A major incentive for this research has been the need to parameterise global circulation models and to understand better the factors controlling the partitioning of solar energy into sensible and latent heat fluxes by different land covers. To a significant extent, these requirements have been achieved. However, there are still a number of issues for which we lack information, understanding and the ability to predict. Examples are the enhanced evaporation of intercepted rainfall at island and continental edge sites and the interaction between different vegetations in the landscape. Nevertheless the priority for research is undoubtedly the response of tropical lowland rainforest to climate change, particularly the increased occurrence and severity of droughts. To succeed in acquiring a better understanding of the likely responses of tropical lowland rainforests to less reliable supplies of water, a different research emphasis will be probably required than the one used to quantify energy partitioning. There will also be a need to adjust the time and space scales over which studies are conducted. Hitherto, evaporation studies in lowland tropical rainforests, particularly those involving micro-meteorological measurements, comprised fairly short-term experiments, perhaps one to two years, and studied fluxes at the km^2 scale. The occurrence of drought episodes is usually greater than this time scale but the impact of drought on the forest is likely to be observed at smaller space scales than 1 km^2.

There will be a need to concentrate more on studies below ground. We need to know more about root distributions and the ability of the root system, and especially deep roots, to support transpiration during extended dry periods. Lowland tropical rainforest usually occurs in landscapes where topography varies a great deal over short distances. The topographical position is intimately related to the vertical depth to groundwater and streams. It is likely therefore that the response of forest in a particular region to rainfall deficits will depend on the position of the forest in the toposequence. The micrometeorological approaches used to measure evaporation flux and its control at the km^2 scale will be inappropriate to distinguish the responses of tropical forest growing at different positions in a slope sequence. We look to techniques such as sap flow measurements to provide the type of information we need about the responses of forest to variation in the supply of soil moisture. There are difficulties in sustaining such instrumentation over long periods so the approach is not ideal for long-term (i.e. several years) monitoring of the responses of forests at different topographical positions in the landscape. Arguably, we lack a methodology to measure tree water use and tree water stress that can be deployed for long periods on individuals throughout catchments that have varying topography. Equally we lack a methodology equivalent to the sap flow approach, in the scale at which it operates, that can deliver information about individual tree carbon accumulation. Up to the present, only the humble increment girth band approaches this ideal instrument. It can give information at short time scales about water stress but can also signal the long-term influence of fluctuations in soil water on tree productivity. The need for new technological advances is important. While modelling initiatives will have an important part to play in exploring the impact of future climate scenarios on the functioning of lowland tropical rainforests, there are still many questions that will only be answered by direct measurements.

APPENDIX 13.1 LIST OF SYMBOLS

A	available energy
D	air specific humidity deficit
E_i	interception loss
E_t	transpiration
$\bar{E}$	mean evaporation rate during rainfall
H_x	horizontal sensible heat flux, advection
K	empirical Granier factor
L^*	leaf area index
PAR	photosynthetically active radiation
R_s	solar radiation
S	canopy capacity for rainfall storage
T	air temperature
T'	deviation from mean temperature
c_p	specific heat at constant pressure
d	zero plane displacement
g_a	aerodynamic conductance
g_b	leaf boundary layer conductance
g_c	surface or canopy conductance
g_s	stomatal conductance
h	vegetation height
k	von Karman's constant (0.41)
t	time of day
u	horizontal wind speed
u_*	friction velocity
u'	deviation from mean horizontal wind speed
x	exponent (Eqn 13.7c)
z	height
z_0	zero plane displacement
z_r	reference height
α	albedo, the solar radiation reflection coefficient
β	Bowen ratio
γ	psychrometric constant
λ	latent heat of vaporisation of water
ρ	air density

References

Abdul Rahim, N., Saifiddin, S. and Zulkifi, Y. (1995). Water balance and hydrological characteristics of forested watersheds in Peninsular Malaysia. In: *Second International Study Conference on GEWEX in Asia and GAME*. Pp. 6–10, Pattaya, Thailand.

Aoki, M., Yabuki, K. and Koyama, H. (1978). Micrometeorology of Pasoh forest. *Malayan Nature Journal*, 30: 149–159.

Bernhardt-Reversat, F., Huttel, C. and Lemee, G. (1978). La forêt sempervirente de basse Côte d'Ivoire. In: *Problemes d'Ecologie: Structure et Fonctionnement des Ecosystemes Terrestres*. (M. Lamotte and F. Bouliere (eds). Masson, Paris. Pps. 32–345.

Bonal, D., Atger, C., Barigah, T. S., Ferhi, A., Guehl, J-M. and Ferry, B. (2000). Water acquisition patterns of two wet tropical canopy tree species of French Guiana as inferred from $H_2{}^{18}O$ extraction profiles. *Annals of Forest Science*, 57: 717–724.

Bonell, M (1999). Tropical forest hydrology and the role of the UNESCO International Hydrological Programme. *Hydrology and Earth System Sciences*, 3: 451–461.

Bonell, M., Barnes, C. J., Grant, C. R., Howard, A. and Burns, J. (1998). High rainfall, response-dominated catchments: A comparative study of experiments in tropical NorthEast Queensland with temperate New Zealand. In: *Isotope Tracers in Catchment Hydrology*, (eds. C. Kendall and J. J. McDonnell), Elsevier, Amsterdam. Pp. 347–390.

Bouten, W., Swart, P. J. F. and de Water, E. (1991). Microwave transmission, a new tool in forest hydrological research. *Journal of Hydrology*, 124: 119–130.

Brouwer, L. C. (1996). Nutrient Cycling in Pristine and Logged Tropical Rain Forest. Tropenbos-Guyana Series, 1, Tropenbos-Guyana Programme, Georgetown, Guyana. 224 pp.

Bruijnzeel, L. A. (1989). (De)forestation and dry season flow in the humid tropics: a closer look. *Journal of Tropical Forest Science*, 1: 229–243.

Bruijnzeel, L. A. (1990). *Hydrology of Moist Tropical Forests and Effects of Conversion: a State of Knowledge Review*. IHP–UNESCO Humid Tropical Programme, Paris, 224 pp.

Bruijnzeel, L. A. (1996). Predicting the hydrological impacts of land cover transformation in the humid tropics: the need for integrated research. In: *Amazonian Deforestation and Climate*, Edited by J. H. C. Gash, C. A. Nobre, J. M. Roberts and R. L. Victoria, John Wiley, Chichester, UK. Pp. 15–55.

Bruijnzeel, L. A., Waterloo, M. J., Proctor, J., Kuiters, A. T. and Kotterink, B. (1993). Hydrological observations in montane rainforests on Gunung Silam, Sabah, Malaysia, with special reference to the 'Massenerhebung' effect. *Journal of Ecology*, 81: 145–167.

Bruijnzeel, L. A. and Wiersum, K. F. (1987). Rainfall interception by a young *Acacia auriculiformis* (A. Cunn) plantation forest in west Java, Indonesia: Application of Gash's Analytical Model. *Hydrological Processes*, 1: 309–319.

Calder, I. R. (1986). A stochastic model of rainfall interception. *Journal of Hydrology*, 89: 65–71.

Calder, I. R. (1990). *Evaporation in the Uplands*. John Wiley, Chichester, UK. 148 pp.

Calder, I. R. (1999). *The Blue Revolution: Land Use and Integrated Water Resources Management*. Earthscan Publications Ltd., London.192 pp.

Calder, I. R., Hall, R. L., Rosier, P. T. W., Bastable, H. G. and Prasanna, K. T. (1996). Dependence of rainfall interception on drop size: 2. Experimental determination of the wetting functions and two-layer stochastic model parameters for five tropical tree species. *Journal of Hydrology*, 185: 379–388.

Calder, I. R. and Wright, I. R. (1986). Gamma ray attenuation studies of interception from Sitka Spruce: some evidence for an additional transport mechanism. *Water Resources Research*, 22: 409–417.

Calder, I. R., Wright, I. R. and Murdiyarso, D. (1986) A study of evaporation from tropical rainforest – west Java. *Journal of Hydrology*, 89: 13–31.

Carswell, F. E., Meir, P., Wandelli, E. V., Bonates, L. C. M., Kruijt, B., Barbosa, E. M., Nobre, A. D., Grace, J. and Jarvis, P. G. (2000). Photosynthetic capacity in a central Amazonian rainforest. *Tree Physiology*, 20: 179–186.

Chappell, N. A., Bidin, K. and Tych, W. (2001). Modelling rainfall and canopy controls on net-precipitation beneath selectively-logged tropical forest. *Plant Ecology*, 153: 215–229.

Collinet, J., Monteny, B and Poyaud, B. (1984). Le milieu physique, in *Recherche et Aménagement en Milieu Forestier Tropical Humide: le Projet Taï de Côte d'Ivoire*, Notes Techniques du MAB no. 15, pp. 35–58, UNESCO, Paris.

Cox, P. M., Betts, P. M., Jones, C. D., Spall, S. A. and Totterdell, I. J. (2000). Acceleration of global warming due to carbon-cycle feedbacks in a coupled climate model. *Nature*, 408: 184–187.

Culf, A. D., Esteves, J. L., Marques, A. and da Rocha, H. R. (1996). Radiation, temperature and humidity over forest and pasture in Amazonia. In: *Amazonian Deforestation and Climate*, edited by J. H. C. Gash, C. A. Nobre, J. M. Roberts and R. L. Victoria, John Wiley, Chichester, UK. Pp. 175–191.

Culf, A. D., Fisch, G. and Hodnett, M. G. (1995). The albedo of Amazonian forest and ranchland. *Journal of Climate*, 8: 1544–1554.

Dolman, A. J., Gash, J. H. C., Roberts, J. M. and Shuttleworth, W. J. (1991). Stomatal and surface conductance of tropical rainforest. *Agricultural and Forest Meteorology*, 54: 303–318.

Duquesnay, A., Bréda, N., Stievenard, M. and Dupouey, J.-L. (1998). Changes in tree-ring δ ^{13}C and water-use efficiency of beech (*Fagus sylvatica* L.) in north-eastern France during the past century. *Plant, Cell and Environment*, 21: 565–572.

Dykes, A. P. (1997). Rainfall interception from a lowland tropical rainforest in Brunei. *Journal of Hydrology*, 200: 260–279.

Eeles, C. W. O. (1979). Soil moisture deficits under montane rainforest and tea. *East African Agricultural and Forestry Journal*, 43: 128–138.

Elsenbeer, H., Cassel, D. K. and Zuñiga, L. (1994). Throughfall in the Terra Firme forest of western Amazonia. *Journal of Hydrology (New Zealand)*, 32: 30–44.

FAO (2001). *State of the Worlds Forests 2001*. Food and Agriculture Organization of the United Nations, Rome, Italy. 181 pp.

Franken W., Leopoldo P. R., Matsui, E. and Ribeiro, M. de N. G. (1982). Estudo da interceptação da áqua de chuva em cobertura floestal amazônica do tipo terra firme. *Acta Amazonica*, 12: 327–331.

Gash, J. H. C. (1979). An analytical model of rainfall interception in forests. *Quarterly Journal of the Royal Meteorological Society*, 105: 43–55.

Gash, J. H. C., Lloyd, C. R. and Lachaud, G. (1995). Estimating sparse forest rainfall interception with an analytical model. *Journal of Hydrology*, 170: 79–86.

Gash, J. H. C., Nobre, C. A., Roberts, J. M. and Victoria, R. L. (Eds.). (1996). *Amazonian Deforestation and Climate*. John Wiley, Chichester, UK. 611 pp.

Gash, J. H. C., Valente, F. and David, J. S. (1999). Estimates and measurements of evaporation from wet, sparse pine forest in Portugal. *Agricultural and Forest Meteorology*, 94: 149–158.

Gash, J. H. C., Wright, I. R. and Lloyd, C. R. (1980). Comparative estimates of interception loss from three coniferous forests in Great Britain. *Journal of Hydrology*, 48: 89–105.

Gilmour, D. A. (1975). Catchment water balance studies on the wet tropical coast of North Queensland. *Unpublished Ph.D. Thesis*, James Cook University, Townsville, Australia.

Golley, F. B., McGinnis, J. T. and Clements, R. G. (1971). La biomassa y la estructura mineral de algunos bosques de Darién, Panama. *Turrialba*, 21: 189–196.

Golley, F. B., McGinnis, J. T., Clements, R. G., Child, G. I. and Duever, M. J. (1975). *Mineral cycling in a tropical moist ecosystem*. University of Georgia Press, Athens, Georgia.

Grace, J. (1999). Environmental controls of gas exchange in tropical rainforests. In: *Physiological Plant Ecology*, edited by M. C. Press, J.D. Scholes and M. G. Baker, Blackwell Science, Oxford, UK. Pp. 367–389.

Grace, J., Lloyd, J., McIntyre, J, Miranda, A., Meir, P., Miranda, H., Moncrieff, J., Massheder, J., Wright, I. and Gash, J. (1995). Fluxes of carbon dioxide and water vapour over an undisturbed tropical forest in south-west Amazonia. *Global Change Biology*, 1: 1–12.

Granier, A. 1985. Une nouvelle méthode pour la mesure de flux de sève brute dans le tronc des arbres. *Annales de Sciences Forestières*, 42: 193–200.

Granier, A. 1987. Evaluation of tranpiration in a Douglas-fir stand by means of sap flow measurements. *Tree Physiology*, 3: 309–320.

Granier, A., Huc, R. and Barigah, S. T. (1996). Transpiration of natural rainforest and its dependence on climatic factors. *Agricultural and Forest Meteorology*, 78: 19–29.

Hall, R. L. (1992). An improved numerical implementation of Calder's stochastic model of rainfall interception – a note. *Journal of Hydrology*, 140: 389–392.

Hancock, N. H. and Crowther, J. M. (1979). A technique for the direct measurement of water storage on a forest canopy. *Journal of Hydrology*, 41: 105–122.

Hodnett, M. G., Oyama, M. D., Tomasella, J. and Marques Filho, A. de O. (1996). Comparisons of long-term soil water storage behaviour under pasture and forest in three areas of Amazonia. In: *Amazonian Deforestation and Climate*, Edited by J. H. C. Gash, C. A. Nobre, J. M. Roberts and R. L. Victoria, John Wiley, Chichester, UK. Pp. 57–77.

Hölscher, D., Sá, T. D. de A., Bastos, T. X., Denich, M. and Fölster, H. (1997). Evaporation from young secondary vegetation in eastern Amazonia. *Journal of Hydrology*, 193: 293–305.

Hubbard, R. M., Bond, B. J. and Ryan, M. G. (1999). Evidence that hydraulic conductance limits photosynthesis in old *Pinus ponderosa* trees. *Tree Physiology*, 19: 165–172.

Hutjes, R. W. A., Wierda, A. and Veen, A. W. L. (1990). Rainfall interception in the Tai Forest, Ivory Coast: application of two simulation models to a humid tropical system. *Journal of Hydrology*, 114: 259–275.

Hulme, M. and Viner, D. (1998). A climate change scenario for the tropics. *Climatic Change*, 39: 145–176.

IPCC. (2001). *Climate Change 2001: The Scientific Basis*. Edited by J.T Houghton, Y. Ding, D. J. Griggs, M. Noguer, P. J. van der Linden, X. Dai, K. Maskell and C. A. Johnson. Working Group 1 Report to the Intergovernmental Panel on Climate Change, Third Assessment, Cambridge University Press, Cambridge. 881 pp.

Jackson, P., Cavalier, J., Goldstein, G. and Meinzer, F. C. (1995). Partitioning of water resources among plants of a lowland tropical forest. *Oecologia*, 101: 197–203.

Jarvis, P. G. (1976). The interpretation of the variations in leaf water potential and stomatal conductance found in canopies in the field. *Philosophical Transactions of the Royal Society*, B 273: 593–610.

Jetten, V. G. (1994). *Modelling the effects of logging on the water balance of a tropical rainforest*. Unpublished Ph.D. thesis, University of Utrecht, Utrecht, The Netherlands.

Jetten, V. G. (1996). Interception of tropical rainforest: performance of a canopy water balance model. *Hydrological Processes*, 10: 671–685.

Jipp, P. H., Nepstad, D. C., Cassell, D. K. and de Carvalho, C. R. (1998). Deep soil moisture storage and transpiration in forests and pastures of seasonally-dry Amazonia. *Climatic Change*, 39: 395–412.

Jordan, C. F. (1969). Derivation of leaf area index from light quality on the forest floor. *Ecology* 50: 663–666.

Jordan, C. F. (1971). Productivity of a tropical forest and its relation to a world pattern of energy storage. *Journal of Ecology* 59: 127–142.

Jordan, C. F. and Heuveldop, J. (1981). The water budget of an Amazonian rainforest. *Acta Amazonica* 11: 87–92.

Jordan, C. F. and Uhl, C. (1978). Biomass of a 'Terra Firme' forest of the Amazon basin. *Oecologia Plantarum* 13: 387–400.

Kato, R., Tadaki, Y. and Ogawa, H. (1978). Plant biomass and growth increment studies in Pasoh Forest. *Malayan Nature Journal*, 30: 211–245.

Kerfoot, O. (1962). Root systems of forest trees, shade trees and tea bushes. *East African Agricultural and Forestry Journal*, 27: (Special Issue).

Kira, T. H., Ogawa, H., Yoda, K. A. and Ogino, K. (1964). Primary production by a tropical rainforest of southern Thailand. *Botanical Magazine of Tokyo* 77: 428–429.

Kira, T. H., Ogawa, H., Yoda, K. A. and Ogino, K. (1967). Comparative ecological studies on three main types of vegetation in Thailand. IV. Dry matter production with special reference to the Khao Chong rainforest. *Nature and Life in SE Asia* 5: 149–174.

Kira, T. H., Shinozaki, K. and Hozumi, K. (1969). Structure of forest canopies as related to their primary productivity. *Plant and Cell Physiology*, 10: 129–142.

Klinge, H. and Herrera, R. (1983). Phytomass structure of natural plant communities on spodosols in southern Venezuela: the tall Amazon caatinga forest. *Vegetatio*, 53: 65–84.

Kull, O. and Kruijt, B. (1999). Acclimation of photosynthesis to light – a mechanistic approach. *Functional Ecology*, 13: 24–36.

Kuraji, K. and Paul, L. L. (1994). Effects of rainfall interception on water balance of two tropical rainforest catchments, Sabah, Malaysia, in *Proceedings of the International Symposium on Forest Hydrology*, 291–298. Tokyo, Japan.

Lawton, R. O., Nair, U. S., Pielke Sr., R. A. and Welch, R. M. (2001). Climatic impact of tropical lowland deforestation on nearby montane cloud forests. *Science*, 294: 584–587.

Ledger, D. C. (1975). The water balance of an exceptionally wet catchment area in West Africa. *Journal of Hydrology*, 24:207–214.

Leigh, E. G. (1978). Slope hydrology and denudation in the Pasoh Forest Reserve. 1. Surface wash; Experimental techniques and some preliminary results. *Malayan Nature Journal*, 30: 179–197.

Lloyd, C. R., Gash, J. H. C., Shuttleworth, W. J. and Marques, A. de O. (1988). The measurement and modelling of rainfall interception by Amazonian rainforest. *Agricultural and Forest Meteorology*, 43: 277–294.

Lloyd, C. R. and Marques Filho, A. de O. (1988). Spatial variability of throughfall and stemflow measurements in Amazonian rainforest. *Agricultural and Forest Meteorology*, 42: 63–73.

Loescher, H. W., Powers, J. S. and Oberbauer, S. F. (2002). Spatial variation of throughfall volume in an old-growth tropical wet forest, Costa Rica. *Journal of Tropical Ecology*, 18: 397–407.

Mahli, Y., Nobre, A. D., Grace, J., Kruijt, B., Pereira, M. G. P., Culf, A. D. and Scott, S. (1998). Carbon dioxide transfer over a central Amazonian rainforest. *Journal of Geophysical Research-Atmospheres*, 103: D24, 31593–31612.

Mahli, Y., Pegoraro, E., Nobre, A. D., Pereira, M. G. P., Grace, J., Culf, A. D. and Clement, R. (2002). The energy and water dynamics of a central Amazonian rainforest. *Journal of Geophysical Research-Atmospheres*, (in press).

Malaisse, F. (1981). In: *Dynamic Properties of Forest Ecosystems* (D. E. Reichle, ed.). Cambridge University Press, Cambridge.

Malmer, A. (1993). *Dynamics of Hydrology and nutrient losses as response to establishment of forest plantation: A case study on tropical forest land in Sabah, Malaysia*. Unpublished Ph.D. Thesis, Swedish University of Agricultural Sciences, Umeå, Sweden.

McWilliam, A-L. C., Roberts, J. M., Cabral, O. M. R., Leitao, M. V. B. R., da Costa, A. C. L., Maitelli, G. T. and Zamparoni, C. A. G. P. (1993). Leaf area index and above-ground biomass of terra firme rainforest. *Functional Ecology*, 7: 310–317.

McWilliam, A-L. C., Cabral, O. M. R., Gomes, B. L., Esteves, J. L. and Roberts, J. M. (1996). Forest and pasture leaf-gas exchange in south-west Amazonia. In: *Amazonian Deforestation and Climate*, Edited by J. H. C. Gash, C. A. Nobre, J. M. Roberts and R. L. Victoria, John Wiley, Chichester, UK. Pp. 265–286.

Meinzer, F. C., Andrade, J. L., Goldstein, G., Holbrook, N. M., Cavalier, J. and Wright, S. J. (1999a). Partitioning of soil water among canopy trees in a seasonally dry tropical forest. *Oecologia*, 121: 293–301.

Meinzer, F. C., Goldstein, G., Franco, A. C., Bustamante, M., Igler, E., Jackson, P., Caldas, L. and Rundel, P. W. (1999b). Atmospheric and hydraulic limitations on transpiration in Brazilian cerrado woody species. Functional Ecology, 13: 273–282.

Meinzer, F. C., Goldstein, G., Holbrook, N. M., Jackson, P. and Cavalier, J. (1993). Stomatal and environmental control of transpiration in a lowland tropical forest tree. *Plant, Cell and Environment*, 16: 429–436.

Meir, P. W. (1996). *The exchange of carbon dioxide in a tropical forest*. Unpublished Ph.D. thesis, University of Edinburgh. 208 pp.

Mizutani, K., Yamanoi, K., Ikeda, T. and Watanabe, T. (1997). Applicability of the eddy correlation method to measure sensible heat transfer to forest under rainfall conditions. *Agricultural and Forest Meteorology*, 86: 193–203.

Molion, L. C. B and Moore, C. J. (1983). Estimating the zero-plane displacement for tall vegetation using a mass conservation method. *Boundary-Layer Meteorology*, 26: 115–125.

Monteith, J. L. (1965). Evaporation and environment. *Symposium of the Society for Experimental Biology*, 19: 205–234.

Monteith, J. L. (1995). A reinterpretation of stomatal responses to humidity. *Plant, Cell and Environment*, 18: 357–364.

Müller, D. and Nielsen, J. (1965). Production brute, perte par respiration et production nette dand la foret ombrophile tropicale. *Forst. ForsVaes. Danm.*, 29: 60–160.

Nepstad, D. C., de Carvalho, C. R., Davidson, E. A., Jipp, P. H., Lefebvre, P. A., Negreiros, G. H. D., da Silva, E., Stone, T. A., Trumbore, S. E. and Veira, S. (1994). The role of deep roots in the hydrological and carbon cycles of Amazonian forests and pastures. *Nature*, 372: 666–669.

Odum, H. T. (1970). Summary: an emerging view of the ecological system at El Verde. *A tropical rainforest* (eds. H. T. Odum and R. F. Pogeon). Division of Technical Information, US Atomic Energy Commission, Washington DC.

Ogawa, H., Yoda, K. A., Ogino, K. and Kira, T. H. (1965). Comparative ecological studies on three main types of vegetation in Thailand. II. Plant biomass. *Nature and Life in SE Asia* 4: 49–80.

Oguntoyinbo, J. S. (1970). Reflection coefficient of natural vegetation, crops and urban surfaces in Nigeria. *Quarterly Journal of the Royal Meteorological Society*, 95: 430–441.

Ohtani, Y., Okano, M., Tani, M., Yamanoi, K., Watanabe, T., Yasuda, Y. and Nik, A. R. (1997). Energy and CO_2 fluxes above a tropical rainforest in Peninsular Malaysia – Under-estimation of eddy correlation fluxes during low wind speed conditions. *Journal of Agricultural Meteorology*, 52: 453–456.

Pearce, A. J. and Rowe, L. K. (1979). Forest management effects on interception, evaporation and water yield. *Journal of Hydrology (New Zealand)*, 18: 73–87.

Pinker, R. T., Thompson, O. E. and Eck, T. F. (1980a). The energy balance of a tropical evergreen forest. *Journal of Applied Meteorology*, 19: 1341–1350.

Pinker, R. T., Thompson, O. E. and Eck, T. F. (1980b). The albedo of a tropical evergreen forest. *Quarterly Journal of the Royal Meteorological Society*, 106: 551–558.

Richards, P. W. (1996). *The Tropical Rain Forest*. 2nd Edition. Cambridge University Press, Cambridge, UK. 575 pp.

Richardson, J. H. (1982). Some implications of tropical forest replacement in Jamaica. *Z. Geomorph. N.F.*, 44: 107–118.

Roberts, J. M. (1983). Forest transpiration: a conservative hydrological process? *Journal of Hydrology*, 66: 133–141.

Roberts, J. M., Cabral, O. M. R. and de Aguiar, L. F. (1990). Stomatal and boundary-layer conductances in an Amazonian terra firme rainforest. *Journal of Applied Ecology*, 27: 336–353.

Roberts, J. M., Cabral, O. M. R., Fisch, G., Molion, L. C. B., Moore, C. J., and Shuttleworth, W. J. (1993). Transpiration from an Amazonian rainforest calculated from stomatal conductance measurements. *Agricultural and Forest Meteorology*, 65: 175–196.

Roberts, J. M., Cabral, O. M. R., da Costa, J. P., McWilliam, A-L. C. and Sá, T. D. de A. (1996). An overview of the leaf area index and physiological measurements during ABRACOS. In: *Amazonian Deforestation and Climate*, Edited by J. H. C. Gash, C. A. Nobre, J. M. Roberts and R. L. Victoria, John Wiley, Chichester, UK. Pp. 287–306.

Roberts, J. M., Hopkins, R. and Morecroft, M. D. (1999). Towards a predictive description of forest canopies from litter properties. *Functional Ecology*, 13: 265–272.

Roberts, J. M., Rosier, P. T. W. and Smith, D. M. (2001). *The effects of afforestation on chalk groundwater resources*. Final Report to UK Department of Environment Transport and the Regions. 81 pp.

Roche, M. A. (1982). Evapotranspiration réelle de la foret amazonienne en Guyana. *Cahiers Orstom, Série Hydrologie*, 19: 37–44.

Rowe, L. K. (1979). Rainfall interception by a beech-podocarp-hardwood forestnear Reefton, North Westland, New Zealand. *Journal of Hydrology (New Zealand)*, 18: 63–72.

Rutter, A. J. (1967). An analysis of evaporation from a stand of Scots Pine. In: Sopper, W. E. and Lull, H. W. (Eds.), *International Symposium on Forest Hydrology*, Pergamon Press, Oxford.

Rutter, A. J., Kershaw, K. A., Robins, P. C. and Morton, A. J. (1971). A predictive model of rainfall interception in forests, I. Derivation of the model from observations in a stand of Corsican pine. *Agricultural Meteorology*, 9: 367–384.

Ryan, M. G., Binkley, D. and Fownes, J. H. (1997). Age-related decline in forest productivity: Pattern and Process. *Advances in Ecological Research*, 27: 213–262.

Sá, T. D. de A., da Costa, J. de P. R. and Roberts, J. M. (1996). Forest and pasture conductances in southern Pará, Amazonia. In: *Amazonian Deforestation and Climate*, Edited by J. H. C. Gash, C. A. Nobre, J. M. Roberts and R. L. Victoria, John Wiley, Chichester, UK. Pp. 287–306.

Saldarriaga, J. G. (1985). *Forest succession in the upper Rio Negro of Colombia and Venezuela*. Unpublished Ph.D. thesis, University of Tennessee, Knoxville, USA.

Schellekens, J., Scatena, F. N., Bruijnzeel, L. A. and Wickel, A. J. (1999). Modelling rainfall interception by a lowland tropical rainforest in north eastern Puerto Rico. *Journal of Hydrology*, 225: 168–184.

Schellekens, J., Bruijnzeel, L. A., Scatena, F. N., Holwerda, F. and Bink, N. J. (2000). Evaporation from a tropical rainforest, Luquillo Experimental Forest, Puerto Rico. *Water Resources Research*, 36: 2183–2196.

Schulz, J. P. (1960). Ecological Studies on the rainforest of northern Suriname. In: *The vegetation of Suriname, 2*, Ed by I. A. Hulster and J. Lanjouw, Van Eedenfonds, Amsterdam for Bot. Mus. and Herb., Rijksuniversitet, Utrecht.

Sellers, P. J., Shuttleworth, W. J., Dorman, J. L., Dalcher, A. and Roberts, J. (1989). Calibrating the simple biosphere model for Amazonian tropical forest using field and remote sensing data. Part I. Average calibration with field data. *Journal of Applied Meteorology*, 28:728–759.

Shuttleworth, W. J. (1988). Evaporation from Amazonian forest. *Proceedings of the Royal Society of London*, B 233: 321–346.

Shuttleworth, W. J. (1989). Micrometeorology of temperate and tropical forest. *Philosophical Transactions of the Royal Society, London*, B 324: 299–334.

Shuttleworth, W. J., Gash, J. H. C., Lloyd, C. R., Moore, C. J., Roberts, J. M., Marques, A., Fisch, G., Silva, V., Ribeiro, M. N. G., Molion, L. C. B., Sá, L de A., Nobre, C. A., Cabral, O. M. R., Patel, S. and de Moraes, J. C. (1984a). Eddy correlation measurements of energy partition for Amazonian forest. *Quarterly Journal of Royal Meteorological Society*, 110: 1143–1162.

Shuttleworth, W. J., Gash, J. H. C., Lloyd, C. R., Moore, C. J., Roberts, J. M., Marques, A., Fisch, G., Silva, V., Ribeiro, M. N. G., Molion, L. C. B., Sá, L de A., Nobre, C. A., Cabral, O. M. R., Patel, S. and de Moraes, J. C. (1984b). Observations of radiation exchange above and below Amazonian forest. *Quarterly Journal of Royal Meteorological Society*, 110: 1163–1169.

Shuttleworth, W. J., Gash, J. H. C., Lloyd, C. R., Moore, C. J., Roberts, J. M., Marques, A., Fisch, G., Silva, V., Ribeiro, M. N. G., Molion, L. C. B., Sá, L de A., Nobre, C. A., Cabral, O. M. R., Patel, S. and de Moraes, J. C. (1985). Daily variations of temperature and humidity within and above Amazonian forest. *Weather*, 40: 102–108.

Stewart, J. B. (1977). Evaporation from the wet canopy of a pine forest. *Water Resources Research*, 13: 915–921.

Szarzynski, J. and Anhuf, D. (2001). Micrometeorological conditions and canopy energy exchanges of a neotropical rainforest (Surumoni-Crane Project, Venezuela). *Plant Ecology*, 153: 231–239.

Tani, M., Abdul Rahim, N., Ohtani, Y., Yasuda, Y., Mohd Mohd, S., Baharudin, K., Takanashi, S., Noguchi, S., Zulkifli, Y. and Watanabe, T. (2003). Characteristics of energy flux and surface conductance estimated from the meteorological monitoring above a tropical rainforest in Peninsular Malaysia. In: *Pasoh: Ecology of a Lowland Tropical Rain Forest in South East Asia*. (Ed. by T. Okuda, K. Niiyama and S. C. Thomas). Springer, Tokyo.

Thom, A. S. (1975). Momentum, mass and heat exchange. In: *Vegetation and the Atmosphere*, Vol 1, ed. J. L. Monteith, Academic Press, London, 57–109.

Thompson, O. E. and Pinker, R. T. (1975). Wind and temperature profile characteristics in a tropical evergreen forest in Thailand. *Tellus*, 27: 562–573.

Tobón-Marin, C., Bouten, W. and Sevink, J. (2000). Gross rainfall and its partitioning into throughfall, stemflow and evaporation of intercepted water in four forest ecosystems in western Amazonia. *Journal of Hydrology*, 237: 40–57.

Ubarana, V. N. (1996). Observation and modelling of rainfall interception loss in two experimental sites in Amazonian forest. In: *Amazonian Deforestation and Climate*, eds. J. H. C. Gash, C. A. Nobre, J. M. Roberts and R. L. Victoria, J Wiley, Chichester, 151–162.

Valente, F., David, J. S. and Gash, J. H. C. (1997). Modelling interception loss for two sparse eucalypt and pine forests in central Portugal using reformulated Rutter and Gash analytical models. *Journal of Hydrology*, 190: 141–162.

Van Dam, O. (2001). *Forest filled with gaps: Effects of gap size on water and nutrient cycling in tropical rainforest. A study in Guyana*. Tropenbos-Guyana Series, 10, Tropenbos-Guyana Programme, Georgetown, Guyana. 208 pp.

Van der Molen, M. K. (2002). *Meteorological impacts of land use changes in the maritime tropics*. Ph.D. Thesis, Vrije Universiteit, Amsterdam, The Netherlands. 262 pp.

Van Dijk, A. I. J. M. and Bruijnzeel, L. A. (2001a). Modelling rainfall interception by vegetation of variable density using an adapted analytical model. Part 1. Model description. *Journal of Hydrology*, 247: 230–238.

Van Dijk, A. I. J. M. and Bruijnzeel, L. A. (2001b). Modelling rainfall interception by vegetation of variable density using an adapted analytical model. Part 2. Model validation for a tropical upland mixed cropping system. *Journal of Hydrology*, 247: 239–262.

Walsh, R. P. D. and Newbery, D. M. (1999). The ecoclimatology of Danum, Sabah, in the context of the world's rainforest regions, with particular reference to dry periods and their impact. *Philosophical Transactions of the Royal Society, London*, Series B, 354: 1869–1883.

Waterloo, M. J., Bruijnzeel, L. A., Vugts, H. F. and Rawaqa, T. T. (1999). Evaporation from *Pinus caribaea* plantations on former grassland soils under maritime tropical conditions. *Water Resources Research*, 35: 2133–2144.

Watson, F. G. R., Vertessy, R. A. and Grayson, R. B. (1999). Large-scale modelling of forest hydrological processes and their long-term effect on water yield. *Hydrological Processes*, 13: 689–700.

White, A., Cannell, M. G. R. and Friend, A. D. (2000). CO_2 stabilization, climate change and the terrestrial carbon sink. *Global Change Biology*, 6: 817–833.

Whitmore, T. C. (1998). An Introduction to Tropical rainforests, 2nd Edition. Oxford University Press, Oxford. 282 pp.

Wilkinson, S. and Davies, W. J. (2002). ABA-based chemical signalling: the co-ordination of responses to stress in plants. *Plant, Cell and Environment*, 25: 195–210.

Williams, W. A., Loomis, R. S., and Alvim, P. de T. (1972). Environments of evergreen rainforest on the lower Rio Negro, Brazil. *Tropical Ecology*, 13: 65–78.

Wright, I. R., Gash, J. H. C., da Rocha, H. R. and Roberts, J. M. (1996a). Modelling surface conductance for Amazonian pasture and forest. In: *Amazonian Deforestation and Climate*, Edited by J. H. C. Gash, C. A. Nobre, J. M. Roberts and R. L. Victoria, John Wiley, Chichester, UK. Pp. 437–458.

Wright, I. R., Nobre, C. A., Tomasella, J., da Rocha, H. R., Roberts, J. M., Vertamatti, E., Culf, A. D., Alvala, R. C. S., Hodnett, M. G. and Ubarana, V. N. (1996b). Toward a GCM surface parameterization of Amazonia. In: *Amazonian Deforestation and Climate*, Edited by J. H. C. Gash, C. A. Nobre, J. M. Roberts and R. L. Victoria, John Wiley, Chichester, UK. Pp. 473–504.

Yoder, B. J., Ryan, M. G., Waring, R. H., Schoettle, A. W. and Kaufmann, M. R. (1994). Evidence of reduced photosynthetic rates in old trees. *Forest Science*, 40: 513–527.

Zhang, L., Dawes, W. R. and Walker, G. R. (2001). Response of mean annual evapotranspiration to vegetation changes at catchment scale. *Water Resources Research*, 37: 701–708.

14 Runoff generation in tropical forests

M. Bonell

UNESCO, Paris, France

INTRODUCTION

The nature of the soil surface is the key factor in deciding how rainfall will infiltrate and move through the soil, i.e. whether water will move downwards or sideways. Surface soil hydraulic properties control the rate of entry (i.e. infiltration) but, if unimpeded vertically, incoming water will move through the regolith as percolation to reach the water table. More commonly, however, there is a reduction in the permeability in the upper soil horizons at various points because of the presence of more impervious soil layers. These deflect water laterally, either at the surface (as *infiltration excess (Hortonian) overland flow*, *HOF* (Horton, 1933; 1945)) or subsurface (as *subsurface stormflow*, *SSF*, or *interflow*) (Chorley, 1978). This SSF can emerge at the surface as *return flow* and combine with precipitation falling on saturated soils to produce saturation (or *saturation-excess*) overland flow, *SOF*. This is also known as the Dunne mechanism (Dunne and Black, 1970a, b).

As highlighted by Bonell and Williams (1989), the soil hydraulic properties of 'undisturbed' tropical landscapes tend to be in equilibrium with the prevailing rainfall characteristics (notably short-term rain intensities). Thus in closed tropical forest, HOF is not generally favoured (exceptions will be outlined later) because the dense root mat and the incorporation of soil organic matter in the topmost soil layers encourage very high infiltration rates. Annual erosion rates from closed tropical forests at the headwater catchment scale are thus small in comparison with disturbed landscapes (see Douglas and Guyot; Chappell, Tych *et al.*, this volume). Where SOF prevails, however, significant intra-catchment re-distribution of sediment may occur on hillslopes where it is trapped by large lateral roots (Ruxton, 1967). 'Pristine' forests also promote a positive feedback by encouraging vertical percolation to comparatively deep groundwater bodies of large capacity in selected (but not all) geological formations (Ogunkoya *et al.*, 1984; Foster, 1993; Foster and Chilton, 1993; Foster *et al.*, 2002; Bonell, 1998a). These groundwater bodies act as reservoirs for sustaining the transpiration demands of forests during seasonal or interannual drought as well as sustaining 'dry weather' stream discharge. The same subsurface water bodies may also participate in the generation of storm runoff if the surface-groundwater hydrology is well coupled (Bonell *et al.*, 1998) within selected environments.

The adverse impacts from natural disasters (such as landslides emanating from rainfall-induced natural hazards as, for example, in Venezuela in December, 1999 (Larsen *et al.*, 2001; see Scatena *et al.*, Douglas and Guyot, this volume) or more extensively from human activities associated with either forest conversion, fire or shifting cultivation, may all cause a shift towards a disequilibrium state (Bonell and Williams, 1989). The surface soil hydraulic properties (notably infiltration rates) are particularly sensitive to any of the preceding perturbations which may cause a dramatic shift in the amount of rainfall partitioned between lateral and vertical pathways of stormwater transfer. Overland flow occurrence is enhanced, with corresponding dramatic increases in erosion rates, depletion of nutrient stores and degradation of within-stream water quality. Several other chapters in this book (notably in Section 3) will elaborate on these various adverse impacts.

A process-oriented perspective, as encapsulated in the hillslope hydrology discussed here, is essential for a better understanding of the consequences of land-use change on the hydrology and associated nutrient-erosion dynamics at the catchment scale. Through a detailed review of case studies, this chapter will introduce early work on hillslope hydrology undertaken in the humid tropics and also the fundamental environmental characteristics that control the runoff generation process. For definitions of terminology used throughout this Chapter, refer to the *Glossary of Terms* compiled by Chorley (1978) in the benchmark publication *Hillslope Hydrology* (Kirkby, 1978). The work is supplemented by an additional Glossary of more recent terms by Chappell, Tych *et al.*, elsewhere in this volume. When comparing studies, where possible, soil classifications based on FAO-UNESCO (1974, 1988) will be supplemented by the use of the USDA (Soil Survey Staff, 1975) soil taxonomy in some cases. For the Australian soilscapes

Forests, Water and People in the Humid Tropics, ed. M. Bonell and L. A. Bruijnzeel. Published by Cambridge University Press.

(Stace *et al.*, 1968; Northcote, 1979; Murtha *et al.*, 1996) the Northcote (1979) system of soil classification is preferred because other international classifications do not describe tropical Australian soils adequately. Isbell (1996, 2002) subsequently used the Northcote system as a basis for the development of an Australian soil classification which has been endorsed as the official national system (McKenzie *et al.*, 2004).

A diagnostic for inferring preferred stormwater pathways is the use of point measurements of soil hydraulic properties linked with prevailing rainfall characteristics. There are, however, inherent weaknesses, both in the experimental methodologies and in the representativeness of point measurements for upscaling (i.e. at larger scale). Initially, a succinct overview of such limitations will be provided, followed by a more detailed comparison of the broad spectrum of hillslope hydrology responses across the humid tropics then highlighting the importance of riparian zones and deeper groundwater in the runoff generation process. The limited testing of spatial hydrological models for runoff procedure is also assessed briefly in the context of their appropriate application elsewhere in the humid tropics. Ultimately, the controversial issue of whether unsubstantiated (in a scientific context) reports of forest conversion enhance the frequency of floods will be evaluated in the context of existing hillslope hydrology work in the humid tropics.

EARLY WORK UP TO 1980

With the exceptions of the Babinda study in north-east Queensland (Bonell and Gilmour, 1978; Bonell *et al.*, 1979; Gilmour and Bonell, 1979; Gilmour *et al.*, 1980) and within the Reserva Ducke in Central Amazonia (Nortcliff *et al.*, 1979), there had been little research activity on hillslope hydrology under tropical forests compared with the rigorous research approach undertaken in the humid temperate regions (e.g. Dunne and Black, 1970a, b; Harr, 1977; Whipkey, 1965, 1969). Any reference to runoff generation was usually made by geomorphologists when considering landscape denudation processes (e.g. Ruxton, 1967; Douglas, 1973; Thomas, 1973; Lewis, 1976; Leigh, 1978a, b; Peh, 1978; Walsh, 1980). Thus for the most part, there are only qualitative observations available although some manually collected runoff plot data were presented for the Pasoh Forest Reserve (Leigh, 1978a, b; Peh, 1978).

Perhaps the most commonly held belief typical of that time was summarised by Thomas (1973). He referred to the high surface infiltrability (Hillel, 1980) of the deep tropical soils which prevented the occurrence of overland flow except on steep slopes. Thus Thomas (1973) favoured the concept of slower storm recharge to streams by subsurface stormflow within the deep regolith. Whilst concurring with the idea of dominance of subsurface stormflow, both Douglas (1973) and Ruxton (1967) observed what they called 'surface wash' (i.e. overland flow) in respectively north-east Queensland and northern Papua New Guinea. These researchers attributed the concentration of stemflow at the base of tree boles as being capable of exceeding the infiltration rate, thus allowing overland flow to occur during moderate rainfall events. Significantly, Herwitz (1986) later provided quantitative evidence for this process in the high rainfall environment of Mt. Bellenden Ker, north-east Queensland. According to Ruxton (1967), overland flow was also initiated by sealing of the soil surface caused by raindrop and throughfall water drop impacts but no supporting infiltration data were provided. Elsewhere, reports on the occurrence of overland flow were confined to valley bottoms and channel head hollows (e.g. Boulet *et al.*, 1979; Leigh, 1978a, b) and even then measured quantities were small (Peh, 1978).

At that time, there was no appreciation of the spatial and temporal organisation of the preferred pathways of runoff delivery within tropical forests, unlike in humid temperate areas where major advances had already been made (reviewed by Dunne, 1978; 1983) within the framework of the variable source area concept of runoff generation[1] (Cappus, 1960; Hewlett, 1961a, b; Hewlett and Hibbert, 1967).

FUNDAMENTAL ENVIRONMENTAL CHARACTERISTICS WHICH CONTROL THE RUNOFF GENERATION PROCESS

Rainfall

As indicated elsewhere in this volume (Bonell *et al.*) short term, maximum rainfall intensities by storm event and daily rain totals are commonly an order of magnitude higher in the humid tropics than those which are experienced in humid temperate areas. Consequently, a principal 'driver' of the runoff generation process is the much broader range of rainfall application rates linked with different rain-producing systems. This point is strongly emphasised by Bonell and Gilmour (1978) and Gilmour *et al.* (1980), based on the early hillslope hydrology work in north-east Queenland. Thus the more extensive and frequent occurrence of saturation overland flow in north-east Queensland tropical rainforest was attributed to the prevailing well-organised perturbations associated with the southern monsoon shearline during the summer wet season (Manton and Bonell, 1993; Bonell with Balek, 1993). High short-term rainfall intensities (>60 mm h^{-1}) as well as high daily totals (>100 mm) are a persistent feature between December and April (Bonell *et al.*, 1991).

Most hillslope hydrology studies undertaken in the humid tropics have placed insufficient emphasis on synoptic climatology-rainfall characteristics (intensity-frequency-duration) criteria

1 See Hewlett (1974) for a succinct commentary on the development of the variable source area concept of runoff generation.

(Bonell, 1998a). In part, this can be attributed to surface permeabilities of rainforest soils being able to accept the prevailing rainfalls, for example as found in central Amazonia (Nortcliff and Thornes, 1981). More pertinently, the period of hillslope hydrology monitoring has been either too short or absent to be able to gauge the impacts of the extreme rain events capable of producing occasional devastating floods (even outside tropical-cyclone prone areas), such as occurred in southern Thailand in November 1988 (reported in Bonell with Balek, 1993, p. 228) and the December 1999 flood of north-east Venezuela (Larsen *et al.*, 2001).

Soil hydraulic properties

Based on humid temperate work (Whipkey, 1965, 1969; Dunne, 1978) there was early recognition of the importance of surface and subsoil hydraulic properties (i.e. the impact of soil anisotropy and heterogeneity) in controlling the preferred stormflow pathways in tropical forests (Bonell and Gilmour, 1978; Walsh, 1980). Several studies have since presented field saturated hydraulic conductivity, K^* (Bouwer, 1966, based on *in situ* methods and is usually lower than K_s determined under sorption conditions in the laboratory[2]) or saturated hydraulic conductivity, K_{sat} (laboratory-tested soil cores, pumping tests) which have highlighted the changes in permeability with depth and the water retention properties of the soil as being the prime 'controls' on the preferred pathways of storm runoff (e.g. Bonell *et al.*, 1981, 1983b; Nortcliff and Thornes, 1981; Elsenbeer and Cassel, 1990, 1991; Elsenbeer and Lack, 1996a b; Elsenbeer *et al.*, 1999; Tomasella and Hodnett, 1996; Hodnett *et al.*, 1997a, b). Accepting a lack of standardisation in the determination of K^* and K_{sat}, the limitations of small-scale measurements (i.e. the small volume of samples tested and the depths of the soil profile tested, all of which affect precise inter-site comparisons), a persistent feature reported from hillslope hydrology studies is the decline in K^* (or K_{sat}) with depth. The depth of the A – B soil horizon interface and the soil/weathered bedrock boundaries in particular are important controls on the dominant pathways of hillslope storm runoff generation because it is at these points where there is usually a discontinuity in transmissivity. The very high surficial infiltration rates are encouraged by the proliferation of macropores (see Table 1 in Beven and Germann, 1982 for definitions) arising from the extensive biological activity and the incorporation of organic matter associated with forest (Bonell, 1998a, b). Soil hydraulic conductivity measurement and the implications of soil anisotropy will be elaborated on below.

The role of topography

The role of topography such as hillslope hollows or convergent headwater areas in encouraging the convergence of lateral soil water movement and the resulting subsurface stormflow as well as saturation overland flow is well recognised (Freeze, 1972; Anderson and Burt, 1978) as being universal to both humid tropical as well as humid temperate areas (Bonell with Balek, 1993). Such controls form the basis of digital terrain models for runoff procedure (e.g. TOPOG, O'Loughlin, 1986; TOPMODEL, Beven *et al.*, 1995) developed originally for more shallow soils of the humid temperate regions. These models will be reassessed shortly for application to the deeply weathered regoliths found in some parts of the humid tropics.

THE LIMITATIONS OF POINT MEASUREMENTS AND *IN SITU* PARAMETERISATION OF SOIL AND ROCK PROPERTIES LINKED WITH HILLSLOPE HYDROLOGY

In the following sections there will be considerable emphasis on the vertical changes in soil and weathered rock hydraulic conductivity linked with rainfall characteristics to ascertain the preferred pathways of stormflow within hillslopes. Of all the soil hydraulic properties, saturated hydraulic conductivity, K_s, is now widely acknowledged as being one of the most sensitive parameters in physically-based modelling of hillslope hydrology (e.g. Freeze 1972; Beven *et al.*, 1995) and also in the control of subsurface stormwater transfer and the behaviour of soil water within hillslopes (e.g. Chappell and Ternan, 1992; O'Loughlin, 1986). The limitations of using point measurements of K_{sat} (or K^*) however must be acknowledged despite the reliance that is placed on such data during subsequent comparisons of hillslope hydrology studies in the humid tropics.

Essentially, there are inherent weakness in either the methodologies for determination of K_{sat} (or K^*) of soil hydraulic properties or with the limited volumes of soil of weathered rock sampled, or a combination of both (Bonell with Balek, 1993). A principal issue is the representation of macropores and pipes (using the specifications of Table 1 in Beven and Germann, 1982) which are now recognised as a principal means for rapid transfer of subsurface stormflow. The occurrence and importance of this pathway has also been highlighted through the use of environmental tracers (e.g. Bazemore *et al.*, 1994; McDonnell, 1990; Peters *et al.*, 1995; Wilson *et al.*, 1991a, b).

Controlled experiments using the sprinkled application of tracers (e.g. Bronswijk *et al.*, 1995; Lange *et al.*, 1996) demonstrated that only a small proportion of total porosity participates actively in stormwater transfer and at a much faster rate than suggested from the determination of field-saturated hydraulic conductivity, K^*, and total porosity (*cf.* Hubbert, 1940). From bromide tracer studies in heavy clay soil, Bronswijk *et al.* (1995), for example, noted bromide transport in three domains, namely large

2 Bouwer (1966) showed that K^* may be as low as $0.5\ K_s$.

continuous pores (fast preferential flow paths), mesopores (relatively slow preferential pathways) and matric flow. Bromide transport in the first domain amounted to only a few per cent of the total, yet was the most important route for mobile water. Elsewhere, Lange *et al.* (1996) using LiBr as a tracer, estimated that less than 15% of the estimated total soil water in a small forested sub-catchment was mobile water. The principal problem is that knowledge of these subsurface networks of lateral transfer, most notably in pipes, is almost nil (Jones, 1990). By contrast, most reports of K^* or K_{sat} associated with the following descriptions of hillslope hydrology studies in tropical forest, are biased towards the representation of soil matric hydraulic properties only. This means that the application of these K^* or K_{sat} data to larger scales (i.e. hillslope scale or larger) in the interpretation of the runoff generation process – and in particular as inputs to subsequent applications of models – remains open to challenge.

Considerable use has been made of the laboratory determination of K_{sat} from several sources:

- soil cores (Boersma, 1965)
- ring infiltrometers (e.g. Talsma, 1969)
- constant head well permeameter (Guelph permeameter) for K^* determination above a water table (Talsma and Hallam, 1980; Reynolds *et al.*, 1983, 1985) (the successor to the 'shallow-well pump in' technique of Boersma (1965) and Bouwer and Jackson (1974))

and in situations where a shallow water table is present (reviewed in Boersma, 1965; Bouwer and Jackson, 1974; Hendrickx, 1990; Jenssen, 1990):

- auger hole (Kirkham and van Bevel, 1948; van Beers, 1958)
- piezometer (Luthin and Kirkham, 1949)
- tube pumping tests (Reeve and Kirkham, 1951) in situations where a shallow water table is present (reviewed in Boersma, 1965; Bouwer and Jackson, 1974; Hendrickx, 1990; Jenssen, 1990).

Less use has been made so far of the disc permeameter (Perroux and White, 1988). One of the few reports of its application in hillslope hydrology concerns an on-going study in the Western Ghats, India (Purandara *et al.*, 2004). Details of all these various methodologies are reported in the references cited and in the review of Bonell with Balek (1993, pp. 203–206). Subsequent work (as for example Purandara *et al.*, 2004, unpublished data; Buttle and House, 1997; Davis *et al.*, 1999; Sherlock *et al.*, 2000) continues to focus on this issue of measurement uncertainty of soil hydraulic properties, including field saturated hydraulic conductivity. In this regard, many of the problems outlined in the 1950s during a comparison of laboratory and field methods of K_{sat} determination by writers such as Hvorslev (1951), Reeve and Kirkham (1951) and Kirkham (1955) are still valid; all of which have subsequently been encapsulated in the concepts of *repetitive unit and representative elementary volume* (REV) (Bear, 1979) and *representative elementary area* (REA) (Wood *et al.*, 1988; Bloschl and Sivapalan, 1995). All these concepts infer the need for field measurements to incorporate much larger volumes or areas and are in line with the call by Youngs (1983) for more attention to be given to measurements of the 'bulk properties of the whole system'. Using a cascade system of troughs in open eucalypt woodland (whose length and spacing far exceeded the repetitive unit of the vegetation mosaic), Williams and Bonell (1988) showed that K^* estimates from permanently inserted infiltration rings exceeded by 2–7 times those from the troughs using a simple continuity model. More recently, Brooks *et al.* (2004) compared small core and Guelph permeameter measurements of K_{sat} in A, B and E horizons with hillslope-scale K_{sat} measurements. The latter exceeded the core estimates by 13.7, 4.1 and 3.2 times respectively. The intensity of macroporosity and the role of the soil and plant biology probably accounts for these contrasting findings. Williams and Bonell (1988) highlighted the preferential channelling (short-circuiting) of ponded infiltration at the base of spinifex tussocks and the individual ant holes enclosed by the infiltration rings. The low density of spinifex and the high rain intensities by event experienced in central-north Queensland temporally compacted and sealed the intervening bare soil to reduce K^*. In contrast, Brooks *et al.* (2004) noted abundant macropores, especially in the A horizon, including animal burrows and back-filled channels of tree roots. Such characteristics enhanced K_{sat} with increasing scale. The fundamental message from such work is to measure K^* at the sub-hillslope to hillslope scale. Moreover, the current gap between small-scale K_{sat} (or K^*) and hillslope scale estimates in heterogenous models *is a measurement problem, not a scaling problem* (Williams and Bonell, 1988; Bonell, 1998b; Brooks *et al.*, 2004). Such concerns will contribute towards the limited success in applications of digital terrain models for runoff simulations outlined later.

Two further field methodologies by Chappell *et al.* (1998) and Chestnut and McDowell (2000), both undertaken in the humid tropics and which provide a basis for better representation of K^* and K_{sat} respectively up to the hillslope scale, will be outlined later. Elsewhere in this volume Chappell, Bidin *et al.*, provide more details on the former technique.

Linked with the uncertainty of parameterisation of K^* (or K_{sat}) is the increasing attention being given to the concept of pressure waves passing through porous media which travel at velocities much higher than Darcian velocity (Beven, 1981, 1982a, b, 1989; Germann, 1990a, b; Germann and Di Pietro, 1999; Smith, 1977; Torres *et al.*, 1998; Lancaster, 2000; Rasmussen *et al.*, 2000). During the course of presenting evidence of water level responses to storm events in wells, piezometers and pressure-transducers (matric potentials) and independently the dominance of 'old' water in chemohydrograph separations (Kendall and MacDonnell, 1998, ed.), a common observation is the much faster responses in hydraulic potentials and considerable displacement of 'pre-event' soil water or groundwater than would be anticipated from measured K^* and the corresponding Darcian velocities. The precise role of pressure waves leading to displacement of pre-existing old water vis-à-vis the physical propagation of subsurface water by preferential flow (both of which can occurr during transient saturation of porous media within storm events) thus remains *the* key research issue in hillslope hydrology (Bonell, 1998b).

There are reports from field experiments of the possible occurrence of both mechanisms, for example at Walker Branch (Mulholland, 1993) and at Coos Bay, Oregon (Torres *et al.*, 1998). Moreover, kinematic waves may not be the only conceptual model (Beven, 1989a, b; Germann, 1990a, b; Germann and Di Pietro, 1999). Rasmussen *et al.* (2000) put forward evidence from laboratory experiments for diffusion-dominated soil water pressure waves whose velocities initially exceeded kinematic wave celerities (velocities) in the more shallow depths of intact saprolite columns. Subsequently, these pressure wave velocities eventually graded with depth towards those conforming with kinematic wave theory. In contrast, Germann and Di Pietro (1999) present theory (remaining within the kinematic wave approach) and lysimetry experimental data to show wave acceleration with depth up to at least 1metre depth below the surface following infiltration. These writers attribute such accelerations to the 'tearing off' of water films from pre-existing soil water adsorbed on to individual soil aggregates. The water films become more freely moving, especially along macropores, such that momentum dissipation overrides the attraction of capillary potential. Thus the vertical translation of these 'momentum dissipation' waves may continue for at least one metre or more if soil structure and momentum permit.

Significantly, the experimental hydrology community at large have not considered in detail the role of pressure waves (Bonell, 1998b). McDonnell (1990, p. 2829) inferred its possible existence in augmenting the displacement of old water in pipe flow but no field evidence was put forward, despite a comparatively sophisticated database. Elsewhere, during the course of using sprinkler applications to simulate stormwater transfer in the Oregon Coast range, Torres *et al.* (1998) noted that the pressure head advanced through the soil profile on average 15 times faster than water and wetting front velocities. When a natural rain event was superimposed on the near-zero matric potential field, this caused 'spike' increases in piezometers (lag 1.7–2.5h) which were attributed to the translation of a pressure wave causing the rapid effusion of old stored water (Torres *et al.*, 1998, p. 1873). Significantly, no analysis based on the hydrometric evidence was put forward to confirm this explanation. A critical attribute central to the argument for the occurrence of pressure waves is the properties of the soil-water retention curve. Torres *et al.* (1998) indicated that the soil water-matric potential retention curve θ (Ψ) shows a rapid change in soil water content (Figure 3 in Torres *et al.*, 1998, p. 1870) for small changes in potential. Thus the unsaturated zone, the saturated zone and stream discharge become coupled more closely so that any sudden increase in rainfall intensities are capable of generating '. . . rapidly advancing pressure waves that induced slight changes in head gradient and very large changes in hydraulic conductivity' (Torres *et al.*, 1998, p. 1878). It is suggested that this mechanism is capable of delivering significant volumes of stored soil water.

It is clear from the preceding discussion that even if K^* (or K_{sat}) is parameterised better at larger scales, such measurements still do not necessarily account for the higher celerities associated with the potential occurrence of pressure waves. *Such remarks have a significant bearing on the parameterization and associated future developments in process-based hillslope hydrology modelling linked with subsurface stormwater transfer.* The same remarks should also be borne in mind during interpretations of potential preferred pathways of stormflow on the basis of soil hydraulic properties and rainfall characteristics alone or during descriptions of various responses in hydrometric or hydrochemistry data. As Loague and Kyriakidis (1997, p. 2883) remarked '. . . the scale at which the spatial and temporal variability of various near-surface soil hydraulic properties can be represented in process-based hydrologic response simulation has proven to be a delimiter' in hillslope hydrology. They continue '. . . that the Achilles heel of large-scale hydrologic simulation with process-based models is the characterisation of small-scale variability in near-surface soil hydraulic property information at larger spatial and temporal scales'.

COMPARATIVE STUDIES OF HILLSLOPE HYDROLOGY IN THE HUMID TROPICS

Elsenbeer and Vertessy (2000) presented the first attempt at providing a conceptual framework for comparing and contrasting preferred stormflow pathways across the humid tropics. These writers observed that it is the soilscape properties and their interaction with prevailing rainfall intensities which control hydrological response patterns. A wide spectrum of response patterns for hillslopes and drainage basins were envisaged but detailed knowledge of this spectrum remains indeterminate because existing field studies are too few in number. As a first step towards documenting this spectrum, a conceptual framework for stormflow generation and flowpaths, as presented in Figure 14.1, was developed.

Soil anisotropy and heterogeneity, as expressed by changes in K^* (or K_{sat}) with depth, may be a decisive control on all storm pathways. In the case of an impeding layer at depth (such as the soil 'B' horizon or soil-bedrock interface), subsurface stormflow occurs which can also emerge at the surface as return flow and produce saturation-excess overland flow. If the soil anisotropy control is at the surface, then infiltration-excess overland flow occurs. Thus the

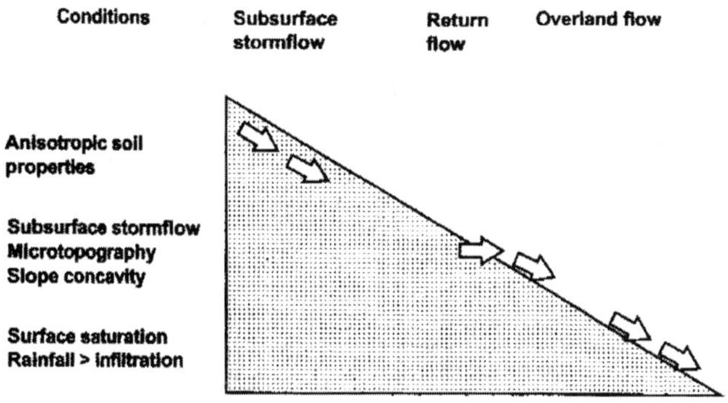

Figure 14.1 A conceptual framework of hillslope runoff-generating mechanisms. (After Elsenbeer and Vertessy, 2000.)

interaction of soil hydraulic properties (including antecedent soil moisture) with the prevailing rainfall characteristics (intensity, amount, frequency) play a fundamental role. As observed by Elsenbeer and Vertessy (2000), a principal exception to soil anisotropic control concerns low relief landscapes and pronounced valley floors where significant vertical amplitudes in the position of the water table occur. In these cases, it is a geomorphological control in operation and saturation-excess overland flow occurs when the water table emerges at the surface.

As part of formalising the approach for comparative studies, Elsenbeer and Vertessy (2000) suggested that there were three stages in a field approach for the characterisation of a catchment's hydrological response pattern.

Stage 1 (S1). A survey of soil hydraulic conductivity should be undertaken at various depths, either at fixed intervals or per soil horizon, in conjunction with compilation of soil physical and catchment properties and hydrometeorological information (especially rainfall intensity-frequency-duration). A field assessment of the geomorphology and depth to the water table should also be included. The field component at this stage principally requires a soil hydraulic conductivity survey supplemented by a soil profile description, either in a generalised form or for specific soil pit profiles.

Stage 2 (S2). The second stage requires the installation of equipment to monitor permanent and temporal saturated zones by wells and piezometers and of the energy status of soil water by means of tensiometers. Most of the literature reports manual measurements but recent advances in technology now enable the monitoring of temporal changes in hydraulic (using wells and piezometers) and matric potentials. This is desirable for a more comprehensive understanding of within-storm flowpaths and the mechanisms of stormflow generation.

Stage 2 will confirm or contradict the inferences from Stage 1 through the identification of transient positive pressure heads for the development of subsurface stormflow; also, whether saturated zones (transient or permanent) emerge at the surface to generate saturation-excess overland flow.

Stage 3 (S3) involves the investment of the necessary infrastructure to link the spatial and temporal variations in storm flow pathways with the runoff hydrograph. This hydrometric approach ideally should be complemented later by hydrochemistry (environmental tracers) studies to provide new insights into the origins and residence times of various contributing sources to the storm hydrograph.

It is desirable that a detailed hydrometric investigation precedes the hydrochemistry approaches for the development of the appropriate hypotheses and subsequent interpretation. The use of a combined hydrometric-hydrochemistry approach in the humid tropics is very limited in comparison with the more extensive investigations that have taken place in the humid temperate latitudes (e.g. see review of Generaux and Hooper, 1998 and Buttle and McDonnell, this volume).

On the basis of the above framework, Elsenbeer and Vertessy (2000) provided a comparative spectrum of expected and documented flowpaths in humid tropical forest ecosystems based on the successive stages of experimentation listed above. These writers categorised the outcome according to the dominance of expected (stage 1 or stage 2 evidence) or documented (stage 3 evidence) flowpaths; this provided the end-members of the spectrum, that is, *predominantly* lateral vis-à-vis vertical. The operative word 'predominantly' is important and refers to 'widespread' versus 'localised' and, in both a spatial and temporal sense, to more stormflow generating events activated in one rather than other pathways. Thus in the case of 'predominantly vertical pathways' it does not mean that saturation overland flow (SOF) is excluded; the latter can occur in valley floors (even though stage 3 evidence is not available). The Reserva Ducke (Nortcliff *et al.*, 1979, Nortcliff and Thornes, 1989), Fazenda Dimona (Hodnett *et al.*, 1997 a, b) Bukit Tarek, (Noguchi *et al.*, 1997a, b) and Mgera (Lørup, 1998) studies are in this group.

The original Table 1 of Elsenbeer and Vertessy (2000) has been modified (which is shown here also as Table 14.1) so that the predominantly lateral pathways are subdivided further into four subgroups, namely subsurface stormflow (SSF) at the soil-bedrock interface, subsurface stormflow (SSF) within the soil (e.g. at a soil horizon boundary), saturation overland flow (SOF) and infiltration-excess overland flow or Hortonian overland flow (HOF). The humid temperate examples of Elsenbeer and Vertessy have also been excluded. On the other hand, new studies have been inserted, from Hodnett *et al.* (1997a, b) (Amazonia); Chappell *et al.* (1998) (Baru, Sabah, Malaysia); Elsenbeer and Cassel, 1990, 1991; Elsenbeer and Lack, 1996b; Elsenbeer and Vertessy, 2000 (La Cuenca, Western Amazonia, Peru); Elsenbeer *et al.* (1999) (Rancho Grande, Rondonia); Dykes and Thornes (2000) (Kuala Belalong, Brunei); Schellekens (2000) (Bisley II,

Table 14.1. *The spectrum of expected and documented flowpaths in tropical forest ecosystems. S1, S2 and S3 refer to the stages of a field approach to hydrological processes as outlined in the text, and indicate the strength of the evidence available in favour of the given pathway categories: S1, soil physical and hydrometeorological information; S2, information on soil moisture or energy status of soil water; S3, actual flowpath information*

Predominantly vertical pathways	Predominantly lateral pathways (infiltration-excess overland flow)	Predominantly lateral pathways (saturation-excess overland flow)	Predominantly lateral pathways (subsurface stormflow within the soil)	Predominantly lateral pathways (subsurface stormflow at the soil-bedrock interface)
Reserva Ducke,[a] S1+S2	Tai Forest,[f] S1	South Creek,[g] S1 to S3	Danum Valley (Sungai) (Barn Barat),[k] S1+S2(*)	Kuala Belalong,[r] (S1+S2)
Bukit Tarek,[b] S1+S2		Barro Colorado,[h] S2	Rancho Grande,[l] S1	
Fazenda Dimona,[c] S1+S2		La Cuenca,[i] S1 to S3	Bisley II,[m] (S1+2)*	
Mgera,[d,e]		Mabura[e,j]	Dodmane,[n] S1	
			Kannike, [e,o]	
			Ife,[e,p]	
			ECEREX,[e,q]	

[a] Central Amazonia, Brazil (Nortcliff *et al.*, 1979; Nortcliff and Thornes, 1989).
[b] Peninsular Malaysia (Noguchi *et al.*, 1997a, b).
[c] Central Amazonia, Brazil (Hodnett *et al.*, 1995; Tomasella and Hodnett, 1996; Hodnett *et al.*, 1997a, b).
[d] Southern Tanzania Highlands (Lørup, 1998).
[e] Field methodology did not fit in precisely with stages 1 to 3 classification of Elsenbeer and Vertessy (2000).
[f] Ivory Coast (Wierda *et al.*, 1989).
[g] Northeast Queensland, Australia (Bonell and Gilmour, 1978; Bonell *et al.*, 1981, 1991; Gilmour *et al.*, 1980).
[h] Panama (Dietrich *et al.*, 1982).
[i] Western Amazonia, Peru (Elsenbeer and Cassel, 1990, 1991; Elsenbeer and Lack, 1996a, b; Elsenbeer and Vertessy, 2000).
[j] Tropenbos site, Guyana (less permeable Ferrasols on steep slopes only) (Jetten, 1994).
[k] Ulu Segama, Sabah, Malaysia (Chappell *et al.*, 1998).
[l] Rondonia, Brazil (Elsenbeer *et al.*, 1999).
[m] Luquillo, Puerto Rico (Schellekens, 2000).
[n] Uttar Kannada, Western Ghats, India (Purandara *et al.*, 2004, unpublished data).
[o] Talakaveri, Western Ghats (Putty and Prasad, 2000a).
[p] South-west Nigeria (Jeje *et al.*, 1986).
[q] ECologie, ERosion, EXperimentation, west of Sinnamary, French Guyana (Fritsch, 1992).
[r] Batu Apoi Forest Reserve, Brunei (Dykes and Thornes, 2000).
Source: Adapted from Elsenbeer and Vertessy (2000, Table 1) with extensive modifications.

Luquillo, Puerto Rico) and Purandara *et al.* (2003, unpublished data, Dodmane, Western Ghats, India). In addition, the studies of Jeje *et al.* (1986) (Ife, Nigeria), Jetten (1994) (Mabura, Tropenbos site, Guyana), Putty and Prasad (2000a, b) (Kannike, Western Ghats, India) are included even though the field methodology does not conform precisely with stages 1 to 3 of Elsenbeer and Vertessy (2000).

In those environments which favour vertical flow, soil anisotropy is poorly developed due to high K^* compared with prevailing rain intensities. In the case of Reserva Ducke (Nortcliff and Thornes, 1984) at a depth of 0.7 m ($K^* = 22$ mm h^{-1}) and below 1.1 m depth ($K^* = 17$ mm h^{-1}) at Fazenda Dimona (Hodnett *et al.*, 1997a, b), there is a decline in K^* compared with the more permeable upper horizons. The order of magnitude of K^* at these deeper soil layers still remains comparatively high, however, in contrast to the 'lateral flow' group. Nonetheless, Bonell (1993) had suggested that these layers may still act as 'a throttle layer' and lead to deep lateral SSF during heavy rainfall. Hodnett *et al.* (1997a, p. 274) did not discount this layer as 'a potential interflow route under conditions of prolonged and intense rainfall. However, saturated conditions are unlikely to persist for long because of vertical drainage through the underlying layer. As a result, interflow will be a very transient process.' In contrast, those environments which contain a marked soil anisotropy differ in their preferred lateral flowpath response for a given rainfall regime, depending on the depth of the low K^*. Thus SSF at the soil-bedrock interface at Kuala Belalong, Brunei, (Dykes and Thornes, 2000) with highly permeable upper soil horizons, contrasts with more shallow

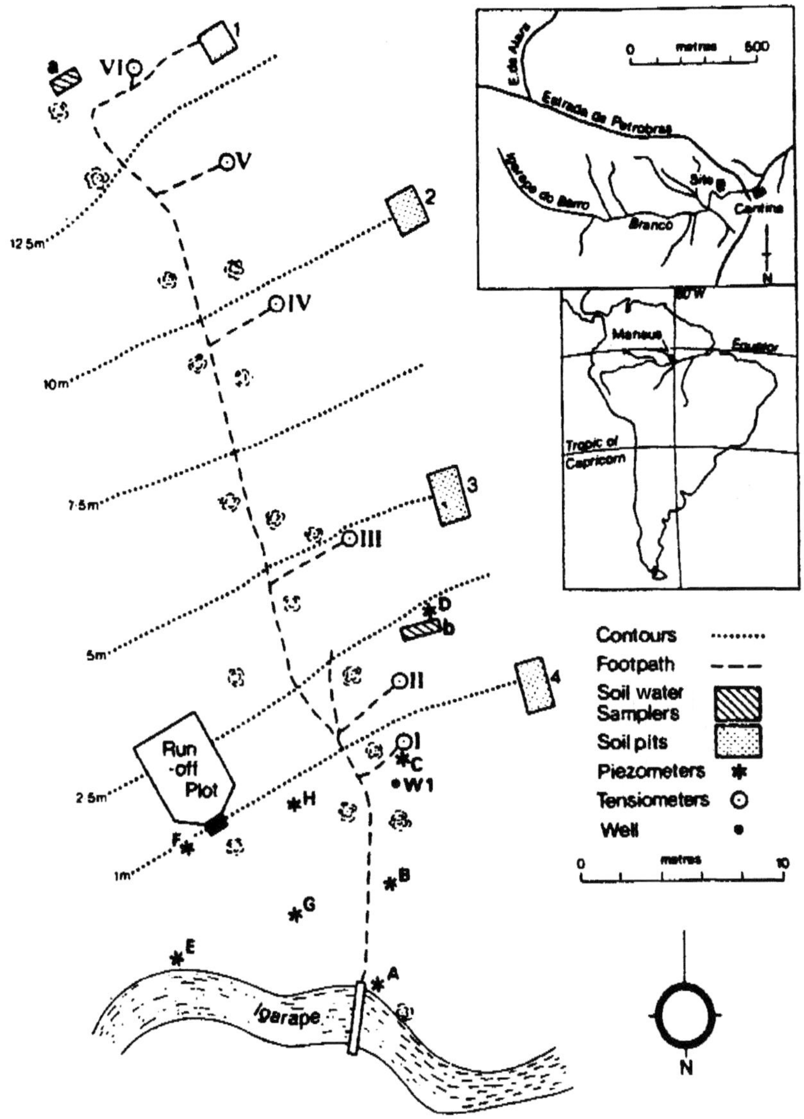

Figure 14.2 Map showing the installation of equipment at Reserva Ducke. (After Nortcliff *et al.*, 1979.)

SSF within the soil at several sites (Table 14.1). Overland flow can also be prevalent (in contrast to humid temperate forest sites, Elsenbeer and Vertessy, 2000), and is principally of the saturation-excess (saturation overland flow), SOF, type. But, more rarely, the infiltration-excess (overland flow) mechanism is reported where soil surface K^* is exceptionally low such as in the Tai Forest of Côte d'Ivoire (Ivory Coast).

Table 14.1 thus presents a much more diverse range of preferred stormflow pathways than those reported from humid temperate studies (e.g. Kirkby, 1978; Anderson and Burt, 1990; Elsenbeer and Vertessy, 2000). Moreover, this diversity of responses can potentially occur within the same geographical region (Table 14.1). Elsenbeer and co-workers (Elsenbeer and Lack, 1996; Elsenbeer *et al.*, 1999, for example), make the point that too much emphasis has been placed on the results from central Amazonia studies (e.g. Nortcliff and Thornes, 1981, where preferred vertical pathways dominate) as typical of Amazonia, when subsequent work in western Amazonia (Elsenbeer and Cassel, 1990, 1991) and Rondonia (Elsenbeer *et al.*, 1999) shows that predominantly lateral pathways are much more prevalent.

The overall spectrum of runoff responses as presented in Table 14.1 and the mechanisms of storm rainfall generation postulated must still be regarded as tentative. As Elsenbeer and Vertessy (2000) noted, several sites still require the implementation of stage 3 experimentation to confirm the expected predominant pathways. Most important, the existing suite of experimental sites are too limited in number, and do not include all the geo-ecological controls and rainfall regimes. Thus there is a likelihood of – as yet unidentified – new 'hillslope hydrology environments' with distinct processes of runoff generation.

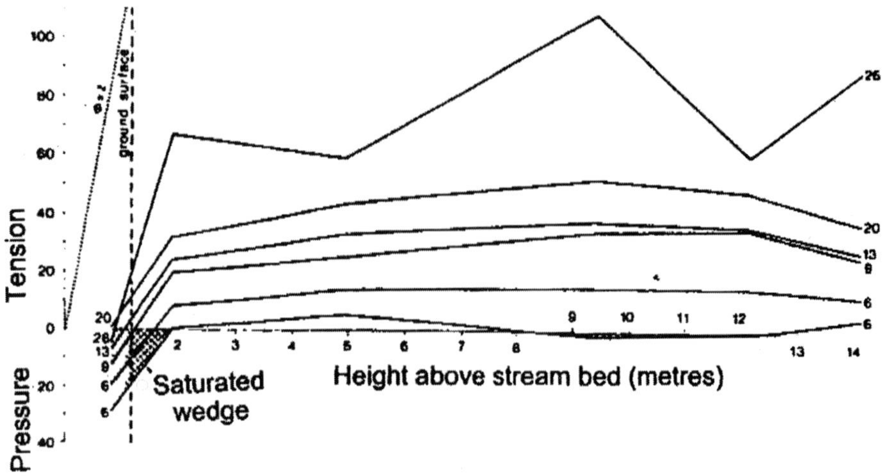

Figure 14.3 The distribution of tension over the hillslope and through time at a depth of 30 cm after the storm (5–6 May 1977) over the Reserva Ducke. (After: Nortcliff *et al.*, 1979.)

HYDROLOGICAL PROCESSES

Predominantly vertical pathways

Reserva Ducke, Brazil

One of the first hillslope hydrology studies undertaken in the humid tropics was reported in a series of papers by Nortcliff and co-workers (Nortcliff *et al.*, 1979; Nortcliff and Thornes, 1981, 1988, 1989) within the Reserva Ducke near Manaus (annual rainfall = 2442 mm; 1966–1992, Hodnett *et al.*, 1997a, b) based on an experimental design shown in Figure 14.2. The reported K_{sat} (from laboratory testing of soil cores) of the prevailing Ferralsols (Oxisols, USDA) soils appear to be too high by an order of magnitude (Elsenbeer and Lack, 1996b) when compared with subsequent *in situ* K^* determinations in Oxisols at the nearby Fazenda Dimona site (Tomasella and Hodnett, 1996). Comparisons between the Babinda, north east Queensland and Reserva Ducke also assumed the above order of magnitude overestimation and made the appropriate correction (Bonell, 1993; Bonell with Balek, 1993), but even so, the subsoil permeability at Reserva Ducke still remains high and thus facilitates predominantly vertical percolation (K^*; 921 mmh^{-1}, 0–0.15 m; 157 mmh^{-1}, 0.15–0.60 m; 61 mmh^{-1}, 0.60–0.90 m; 22 mmh^{-1}, 0.90–1.15 m).

Nortcliff *et al.* (1979) noted an absence of overland flow on the slopes of the Reserva Ducke (but, in contrast, Franken, 1979, did observe overland flow) and, on the basis of a sampled storm, noted positive matric potentials in tensiometers only at the foot of the slope. Elsewhere along the slope transect there was a preponderance of near vertical fluxes which indicated little lateral movement. The only exception was site VI at the top of the transect where a lateritic horizon at 0.8–0.9 m impeded vertical percolation which caused lower, negative matric potentials. Figures 14.3 and 14.4 highlight the position of the saturated, lower slope wedge and the overall trend of vertical fluxes. Nortcliff and co-workers concluded that water supplied to the river must originate entirely from groundwater (from within the hillslope) which subsequently emerges at the surface over the riparian zones. Thus the hillslope hydrology of the mid to upper slopes is decoupled from the storm runoff generation process except through the provision of vertical fluxes to the deeper saturated zone and subsequent movement of this groundwater towards the floodplain (although this was not measured).

To account for the rapid stream hydrograph responses (Franken, 1979), Nortcliff and Thornes (1989) suggested that rapid drainage (mostly vertical) occurred through a macropore system leading to a near-instantaneous response of subsurface stormflow (at the foot of the slope), saturated floodplain storage and almost immediate floodplain overland flow (Nortcliff and Thornes, 1988). No direct measurements of runoff generation over the floodplains were made, but a comparison of the spatial extent of saturation based on field mapping between optimal wet (April) and dry (August) seasons was provided (Nortcliff and Thornes, 1988).

With the aid of soil solute concentrations, Nortcliff and Thornes (1989) supported the bi-phasic nature of soil water drainage (Beven and Germann, 1981; Haria *et al.*, 1994; Lange *et al.*, 1996; Philip, 1986) whereby the rapid drainage response is through the macropore system, with a much slower response through the micropore system. The latter leads to '. . . the consequent limitations of the Darcian approach to water and solute transfer' (Nortcliff and Thornes, 1989, p. 44). These writers were among the first in the humid tropics literature to highlight this bi-phasic nature of soil water drainage as an important issue which will be re-iterated when more recent studies are reviewed.

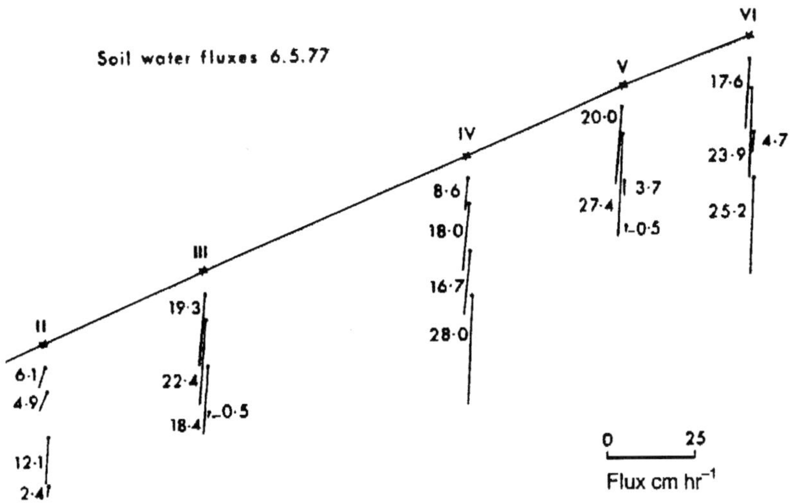

Figure 14.4 Resultant fluxes on 6 May. (After Nortcliff *et al.*, 1979.)

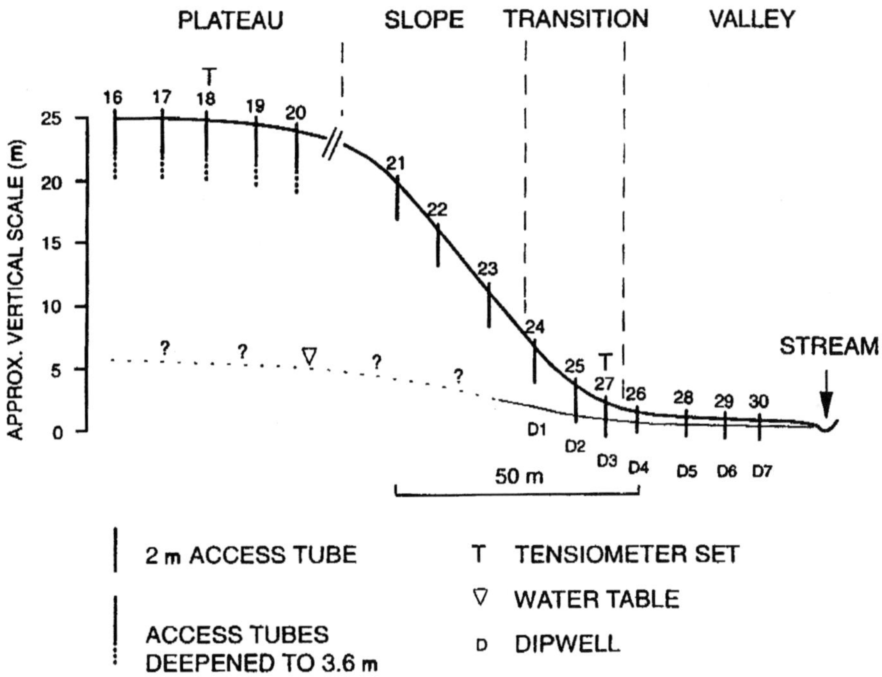

Figure 14.5 Schematic cross-section showing form of hillslope and position of instruments within the forest adjacent to the Fazenda Dimona cattle ranch, north of Manaus. (After Hodnett *et al.*, 1997a.)

Fazenda Dimona, Brazil

Later, Hodnett and co-workers were able to provide more insight into the above conclusions from the Reserva Ducke as part of the ABRACOS project (Anglo-Brazilian Amazonian Climate Observation Study) (Hodnett *et al.*, 1995; 1996a, b; 1997a, b; Tomasella and Hodnett, 1996). The principal focus of their study was to support contiguous micrometeorological research by determining the availability of soil water and groundwater storage to meet the transpiration requirements of forest and land converted from forest to pasture (see Roberts *et al.*, this volume). Nonetheless, significant data were obtained on seasonal variations in unsaturated/saturated zone hydraulic potential profiles and corresponding changes in the water table profile along a slope (Figure 14.5) transect (part of the S2 stage of Elsenbeer and Vertessy, 2000). The K^* within the adjacent pasture, and other soil hydraulic properties, were also presented by Tomasella and Hodnett (1996) (but not for the forest). Thus the S1 stage of Elsenbeer and Vertessy (2000) is technically not met (in the forest), but the Tomasella and Hodnett (1996)

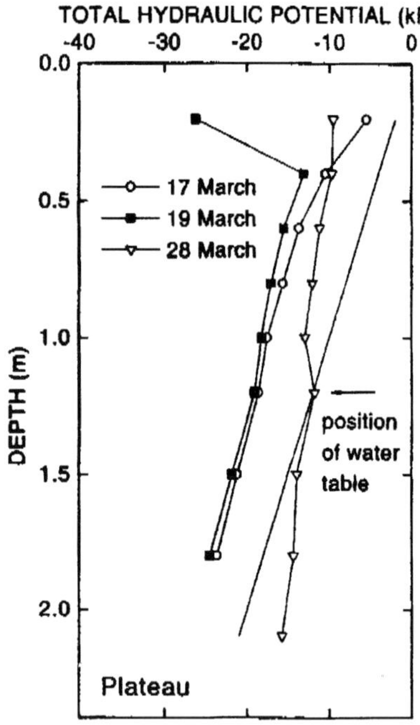

Figure 14.6 Plateau profiles of total hydraulic potential during the 1991 wet season. (After Hodnett *et al.*, 1997a.)

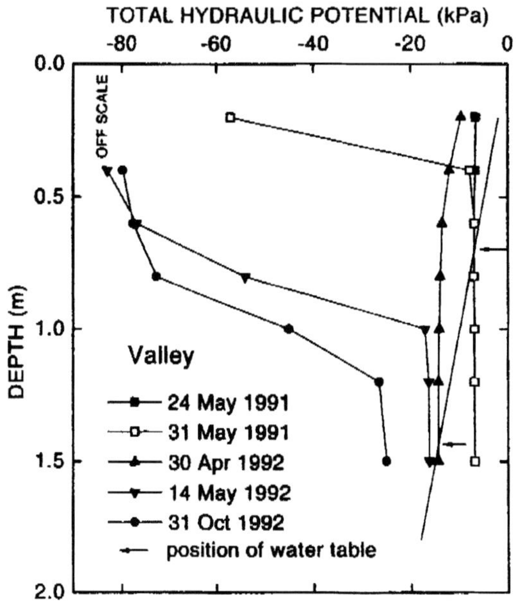

Figure 14.7 Valley profiles of total hydraulic potential in the 1990–1 and 1991–2 wet seasons, and the 1992 dry season. (After Hodnett *et al.*, 1997a.)

approach to field and laboratory determination of soil hydraulic properties was comprehensive (e.g. ring permeameter, Perroux and White, 1988; instantaneous profile method, Hillel *et al.*, 1972). The subsoil K^* provided an indication of corresponding estimates beneath the surface horizons of the adjacent forest. The soils are also similar to those of the Reserva Ducke being clayey Xanthic Ferrasols (haplic acrothox, Oxisols USDA Soil Taxonomy) derived from the unconsolidated sediments of the Tertiary Barreiras formation.

Tomasella and Hodnett (1996) determined high K^* at 0.3 m depth (97 mm h^{-1}) with a decline to 17 mm h^{-1} at 1.05 m depth. They considered that these high values were the result of intense macroporosity, particularly between 0.4–1.1 m depth, in line with the observations of Nortcliff and Thornes (1989). A consequence is the very rapid decrease in unsaturated hydraulic conductivity as a result of these macropores emptying between 0 and –3 KPa matric potential. Significantly, Tomasella and Hodnett (1996) observed that these macropore effects decreased with depth and became negligible below 1.3 m. Nonetheless, hydraulic potential profiles for the plateau (Figure 14.6) and the valley (Figure 14.7) immediately after rain show a downward potential gradient (Figure 14.6, e.g. 17 March; Figure 14.7, e.g. 30 April 1992). Otherwise, a reversal of gradients occurred towards the surface in between events in response to transpiration. During the course of the study, saturated conditions were noted only once on the plateau (28 March) in response to 146 mm of rain. In the valley, the position of the water table is marked (arrow) (Figure 14.6). At other times the water table was below 1.5 m depth because of a below-average wet season in consequence of a negative ENSO phase (see Callaghan and Bonell, this volume).

The dominant pathway for vertical transmission of deep drainage as noted previously by Nortcliff and co-workers (Nortcliff *et al.*, 1979; Nortcliff and Thornes, 1981) is confirmed beneath the plateau and slope (Hodnett *et al.*, 1997a, b). Such deep drainage (groundwater recharge) occurs however only after the profile has attained optimal conditions of wetness during the wet season (whereby a downward hydraulic potential occurs throughout the profile and the unsaturated hydraulic conductivity has increased). The depth of the water table within the valley floor is significantly affected by the amount of groundwater discharge from beneath the plateau and slope to the stream. Hodnett *et al.* (1997b) noted a minimal seasonal change in water storage (about 50 mm) which resulted in a consequent narrow range in depth to the water table of between 0.1–0.8 m (except during the below-average rainfall year of 1992). Thus the antecedent conditions for the development of saturation overland flow over the valley floors exist, although Hodnett *et al.* (1997a, b) did not measure this flow pathway directly. The seasonal change in the water table (Figure 14.8) shows that this feature attained the surface at D4 for a short period at Day 523. Also of interest is that the water table showed different behaviour between the early wet season compared with the late wet season. Water table gradients reversed temporarily with movement *into* the slope following rainfall (Figure 14.9b) with a maximum 15-minute intensity of 138

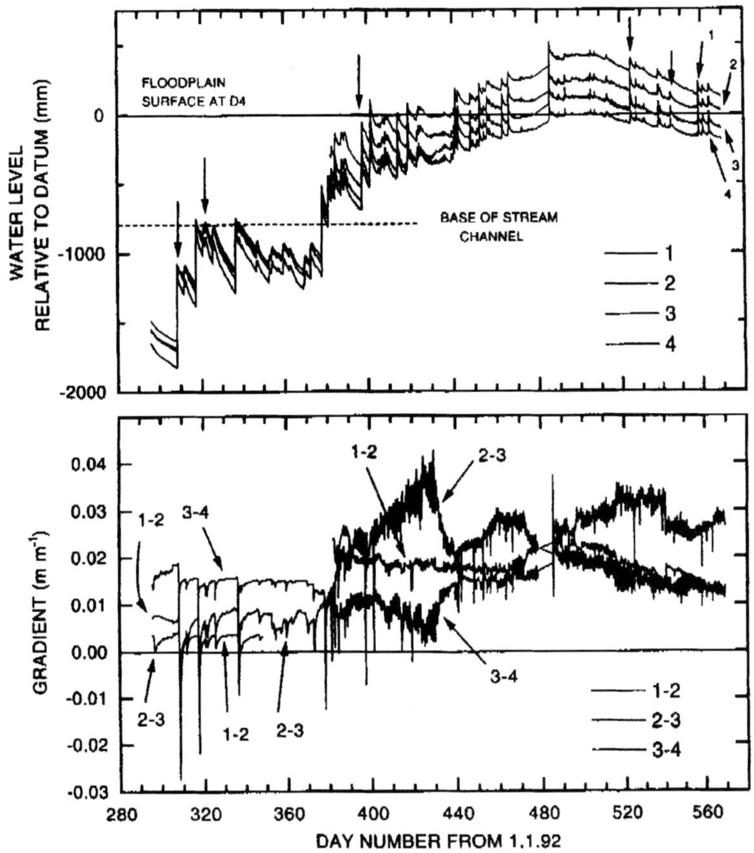

Figure 14.8 Water level data from four dipwells logged at 10-minute intervals (21 Oct 1992 to 22 July 1993). Also shown are the water table gradients between the dipwells. (After Hodnett *et al.*, 1997b.)

mm h^{-1}. Later in the wet season, this gradient reversal characteristic terminated because the antecedent water table was above the level of the floodplain on the upper slopes (Figure 14.10b). Paralleling these changes is the decreasing time lag in response of the well water levels to rainfall, which becomes almost instantaneous near the commencement of the storm.

Thus the findings from Hodnett and co-workers tend to confirm the earlier conclusions of Nortcliff and Thornes (1984, 1989) that the water table within the valley floor is maintained by groundwater recharge from the adjacent plateau and slope which provides the basis for the generation of SOF over the riparian zone. Bearing in mind the arbitrary nature of storm hydrograph separation, the quickflow response ratios (QRR, percentage of storm rainfall leaving a basin as quickflow) as determined by Nortcliff and Thornes (1984, 1988) and for the late wet season for Reserva Ducke (Barro Bronco) by Leopoldo *et al.* (1995), were about 5% in both studies (from the occurence of SOF) which apply to conditions where the water table is close to the surface in the valley bottom. When the water table is lower due to below-average rainfall (as was recorded by Hodnett *et al.*, 1997a, b), Hodnett *et al.* (1997b) suggested SOF would not be so prevalent, thus reducing the percentage of quickflow. For example, Leopoldo *et al.* (1995) noted lower QRR of 2–3% in the early wet season when the water table was deeper.

In summary, this is an environment in which the most important process is the subsurface movement of groundwater from recharge beneath the plateau and slope areas. Such movement then enters the stream through a perennial groundwater body in the valley bottoms. Leopoldo *et al.* (1995), for example, attributed 91% of total annual runoff from a 1.3 km^2 catchment to baseflow. The QRR for these central Amazonia studies are, in consequence, at the very low end of the spectrum of runoff generation activity by humid tropical standards, and can be attributed to restricted SOF occurrence in the valley bottoms, following the classic field model of Dunne and Black (1970a, b). For the most part, the *upper profile of the adjacent slopes and plateau are decoupled from the runoff generation process* except for those areas acting as a conduit for vertical recharge to the deep groundwater body. Elsewhere, Proctor (this volume) highlights the evolutionary

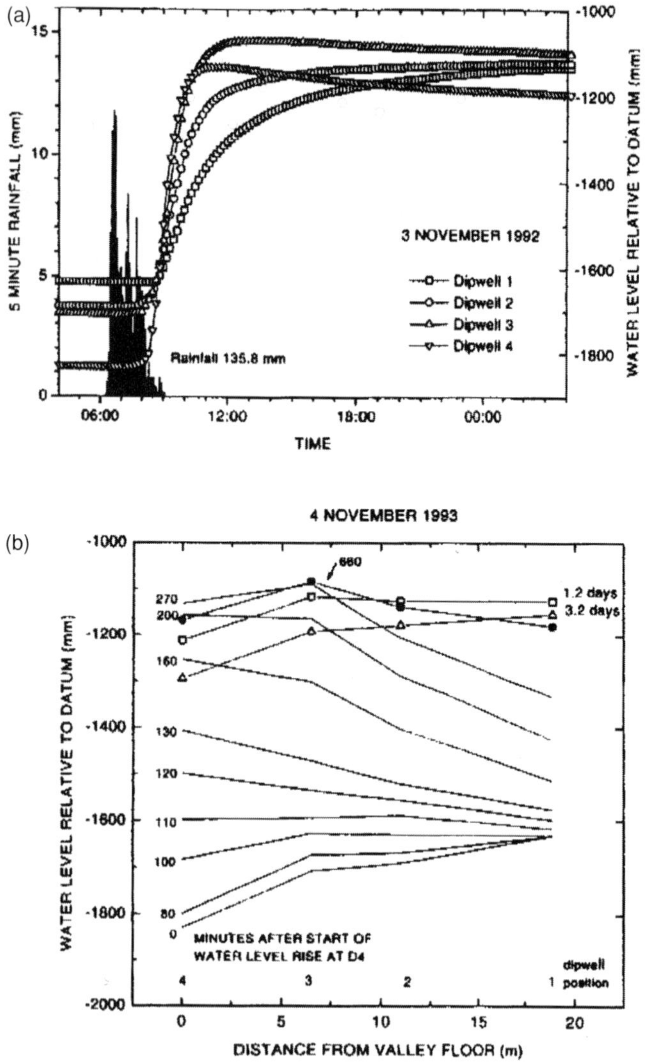

Figure 14.9 (a) Ten-minute water level data for a storm on 3 Nov 1992 (rainfall = 135 mm). (b) Cross-section showing water table response during and after the storm. (After Hodnett *et al.*, 1997b.)

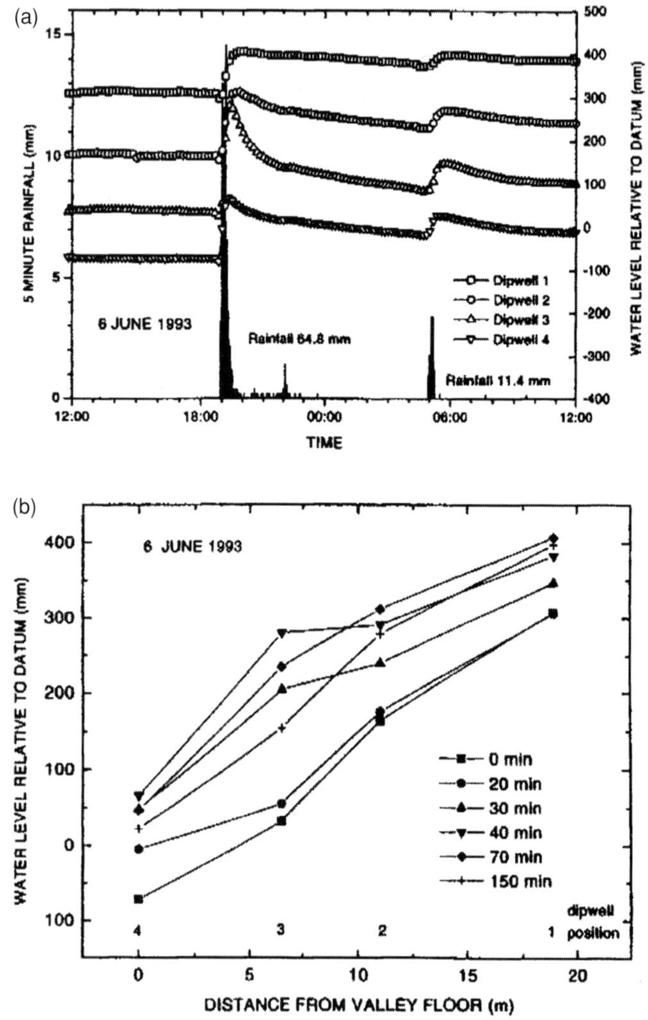

Figure 14.10 (a) Ten-minute water level data from a storm on 6 June 1993 (rainfall = 57 mm in 50 minutes, total 64.8 mm). (b) Cross-section showing water table response during and after the event. (After Hodnett *et al.*, 1997b.)

progression upslope of podzols at the expense of the Ferralsols in the central Amazonia landscape such as Reserva Ducke. The former soils encourage SOF whilst the latter Ferrasols are dominated by vertical percolation.

An exception to the above conclusion is the zone of high saturated hydraulic conductivity, especially between 0.4–1.05 m on the plateau and slope, as measured by Tomasella and Hodnett (1996). Hodnett *et al.* (1997a) measured a minimum in soil water profiles within this horizon, and could not rule out the occurrence of interflow (subsurface stormflow, *SSF*) during and immediately after major storms, as a mechanism for transporting water rapidly towards the floodplain. Certainly, the very high short term rain intensities reported by Hodnett *et al.* (1997a) for six events (5 min intensities exceeding 100 mm h^{-1}) and four events with average intensities exceeding 100 mm h^{-1} for 30 minutes, makes deeper SSF likely at such times above approximately 1 m depth. These observations fit within the scenario suggested by Bonell (1993). However, despite these much higher short-term rain intensities compared with 'cyclone-prone' regions (see Bonell *et al.*, 1991; Bonell *et al.*, this volume), the corresponding durations are much shorter (see Figures 14.9a and 14.10a). Hodnett *et al.* (1997b) thus suggesting that the impact of within-slope SSF on the lower floodplain hydrology would be limited to the occurrence of very large storms and/or very wet antecedent conditions. The latter circumstances would allow interflow to move in significant quantities through the slope. Similar reasoning was given for the generation and re-distribution of infiltration-excess overland flow in drier tropical climates (Bonell and Williams, 1986;

Williams and Bonell, 1988; Dunne *et al.*, 1991; Van de Giesen *et al.*, 2000). In these drier environments, only during the larger storms was overland flow able to exceed runon (infiltration) and thus contribute significantly to organised drainage.

Earlier hydrograph work by Nortcliff and Thornes (1984) determined a 6 hour rapid rise and fall in discharge to individual storm events (attributed to SOF from the valley floodplain), followed by a two-stage recession over the following 20 hours prior to stream discharge returning to the pre-storm rate. It is conceivable that the first stage of the recession could in part be the delay in interflow contributions from upslope due to the longer transit distances in comparison with SOF (Hodnett *et al.*, 1997b).

Igarape Mote, Brazil

Elsewhere in Central Amazonia, Lesack (1993a, b) reported the hydrograph characteristics of 173 storms (2870 mm in total) from a 23.4 ha sub-basin of the Igarape Mote stream system (near the Solimoes River). The environmental conditions (topography, soils, vegetation) are similar to those of the more detailed hillslope hydrology studies above (Reserva Ducke, Fazenda Dimona). In common with the Barro Bronco (Leopoldo *et al.*, 1995), the baseflow component amounted to ~95% of total streamflow (1650 mm) giving 88 mm as quickflow (5%). The QRR of a subset of 47 storms ranged from 0.5% to 4.0%. The volumes of streamflow covered 1.5 orders of magnitude (9 to 0.3 mm), and the quickflow volumes covered 3 orders of magnitude (4 to 0.004 mm) while the corresponding rainfall by event was 100 to 1 mm. Lesack (1993b) thus calculated a volume-weighted record of QRR as being 2.8%.

No supporting hillslope hydrology studies were undertaken in Igarape Mote following the model of Elsenbeer and Vertessy (2000). The work of Lesack (1993b) provides additional insights, however. The deeply weathered Ferralsols (Oxisols) are also highly transmissive to at least 4 m depth. Through the use of the method of Lee and Cherry (1979), indirect estimates of K_{sat} along a slope transect of piezometers show persistent high K_{sat} of 23.8 to 133.9 mm h^{-1} (1 m depth), 16.6 to 81 mm h^{-1} (2 m depth), and 38.5 to 134.3 mm h^{-1} (4 m depth). In common with the other central Amazonia studies, bedrock was not attained. Estimates of subsurface (groundwater leakage) losses were minimal however, representing 2.5% of non-evaporative flow from the basin; thus the derived water balance was considered credible (including the QRR estimates) (Lesack, 1993b). Moreover, Lesack (1993b) related the occurrence of SOF within the valley bottoms to the rainfall characteristics. He suggested that the lower QRR could be attributed to direct channel precipitation during frequent small storms. It was only during the occurrence of the less frequent category of storms of higher magnitude (which produced >1 mm of runoff) that SOF enhanced the QRR. *This less frequent but larger volume storm group applies to only 20% of the total number of storm events. The same group does, however, account for 75% of the rainfall volume and 80% of the quickflow volume; and is probably associated with the more organised rain-producing systems in the Amazon basin described elsewhere* (e.g. Garstang *et al.*, 1994; Bonell *et al.*, this volume).

Bukit Tarek, Malaysia

The work of Noguchi *et al.* (1997a, b) in the Bukit Tarek Experimental basin (annual rainfall 2414 mm, 1992–1994) of Peninsular Malaysia (and elsewhere by Sherlock *et al.*, 1995) also details the dominant role of vertical pathways to groundwater as a fundamental component of the storm runoff generation process. There are some differences with central Amazonia in that under optimal conditions of catchment wetness, shallow SSF was detected in the Bukit Tarek study. This flow pathway thus makes a more significant contribution to the quickflow component.

Previous water balance work (Abdul Rahim, 1988; Law and Cheong, 1987) also indicated low annual runoff coefficients (6 to 17%) for the nearby Sungai Takim experimental catchment undisturbed forest, with quickflow only accounting for 21% of the annual runoff component. In comparison with the Central Amazonia, however, these estimates are higher.

Noguchi *et al.* (1997a) undertook a systematic catchment survey (S1 of Elsenbeer and Vertessy, 2000) of soil physical and soil hydraulic properties of the Bukit Tarek experimental basin. This survey was supplemented by a dye test (methylene blue and diluted white liquid paint) sprinkled over small plots using rain simulators. The determined K_{sat} values (geometric means) ranged from 1466 mm h^{-1} at 0.1 m depth to 169 mm h^{-1} at 0.8 m depth, and this trend in high transmissivity persisted from ridge top to the base of a slope transect (Figure 14.11). In comparison with many other global K_{sat}-depth measurements, Bukit Tarek has one of the most permeable sub-soils in common with the Reserva Ducke, Fazenda Dimona and an additional Indonesian basin (Bukit Soeharto) (see Figure 4, Noguchi *et al.*, 1997a). Thus SOF was not considered a dominant pathway but the occurrence of SSF could not be discounted.

The prevalence of macropores was shown by the large changes in volumetric water content at matric potentials less than 0.3 m at all measured depths between 0.1–0.8 m. Of particular interest was that the dye test showed evidence for lateral deflection of vertical percolation between the organic-rich surface soil horizon and the B layers. The same dye test showed the influence of both decayed and living roots in encouraging preferential flow paths in both vertical and lateral (downslope) directions. The manipulation of the annulus of live tree roots by preferential flow was observed qualitatively in the Babinda study (Bonell, 1993) and discussed briefly in Bonell with Balek (1993, p. 200) in the context of the review of Kozlowski (1981, p. 111). This review noted that few data existed for tree roots in comparison with herbaceous plants.

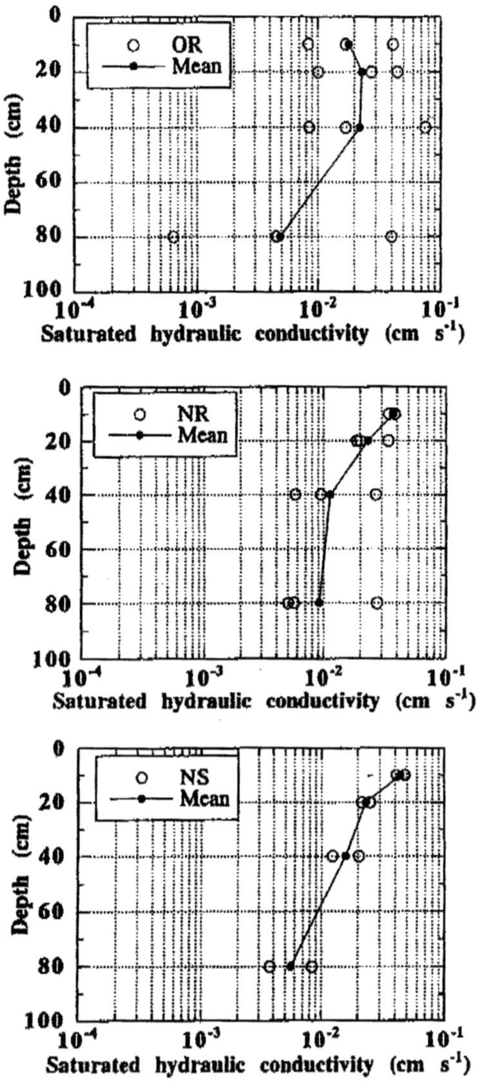

Figure 14.11 The relationship between depth and saturated hydraulic conductivities (K_s) and the geometric mean. (After Noguchi *et al.*, 1997a.)

The work of Noguchi et al. (1997a) is thus one of the first to provide experimental evidence for annular rootflow in a tropical forest. Noguchi *et al.* (1997b) also highlighted the association of decayed roots (macropores) within termite nests, based on the dye test. However, the role of termite nests in controlling storm subsurface pathways remained inconclusive in their study, except to point out that these termite nests are likely to have a different hydraulic conductivity relative to the soil around them.

Subsequent experimentation (S2 of Elsenbeer and Vertessy, 2000) determined that stormflow generation depended strongly on antecedent wetness, as represented by the initial runoff rate (Noguchi *et al.*, 1997b), based on a slope transect (Figure 14.12). During dry periods when marked negative matric potentials prevailed, streamflow responded quickly to rain events but declined quickly after rain stopped (Figure 14.13). As the soil became wetter, storm hydrograph recessions become more gradual, as shown in September 1994 (Figure 14.13). The corresponding hydraulic potential profiles at all sites along the slope transect converged towards a hydraulic gradient of 1.0 with increasing soil wetness. These profiles (Figure 14.14) and those for optimal wetness (Figure 14.15) bear similarities with the central Amazonia work described by Hodnett *et al.* (1997a, b). In both cases downward flow of soil water (recharge to groundwater) prevails.

During wet conditions, matric potentials are also only marginally negative, with positive pressures recorded at 1.6 m depth after storm events; and also at 0.1 m depth at the lower slope sites T2 and T3 during storms (not shown in Figure 14.15) (Noguchi *et al.*, 1997b). These results suggest that shallow SSF between 0.1–0.2 m depth can occur on the lower slopes (which was supported by the dye test), even though the K_{sat} values exceed the prevailing rainfall intensities (Noguchi *et al.*, 1997a). Thus during wet conditions, the hillslope hydrology is more likely coupled by SSF to stormflow within the riparian and stream channel areas, than noted earlier in the central Amazonia studies, cf. Hodnett *et al.* (1997a, b). The same SSF could also contribute to the more gradual hydrograph recessions. Nonetheless, the dominant preferred pathway remains vertical in the form of persistent downward soil water fluxes for recharging deeper groundwater (unfortunately groundwater was not monitored). *In line with the central Amazonia description, it is this groundwater discharge into the riparian and stream channel areas that has a significant influence on storm hydrograph characteristics.*

By contrast, the rapid hydrograph recessions under dry conditions suggest most rain is retained within the soil, as is borne out by the prevailing negative matric potentials. At such times, groundwater contributions to the hydrograph are reduced and SSF is not operating. Moreover, when rain by event is less than 30 mm, less than 10% of the rainfall appeared as quickflow; the latter was not correlated with soil moisture conditions along the slope transect. Noguchi *et al.* (1997b) therefore suggested that under dry conditions, quickflow originated mainly from the riparian (SOF) and stream channel areas although no direct measurements were made, for example of SOF generation over the riparian zone cf. central Amazonia (Nortcliff *et al.*, 1979; Hodnett *et al.*, 1997a, b).

Three other aspects of this work need mention. First, the sensitivity of quickflow (stormflow) to antecedent soil moisture is inferred by the three antecedent initial runoff rates shown in Figure 14.16. When the initial runoff rate was $\geq$0.1 mm h^{-1} all rainfall except about 30 mm contributed to quickflow, thus giving QRR in excess of 10%, and up to 50% for event rain up to about 60 mm. Only the small volumes of quickflow ($\leq$10% of rainfall) occurred when the initial runoff rate was <0.1 mm h^{-1}.

Second, Noguchi *et al.* (1997b) established no distinct differences in the soil moisture profiles along the slope which suggests

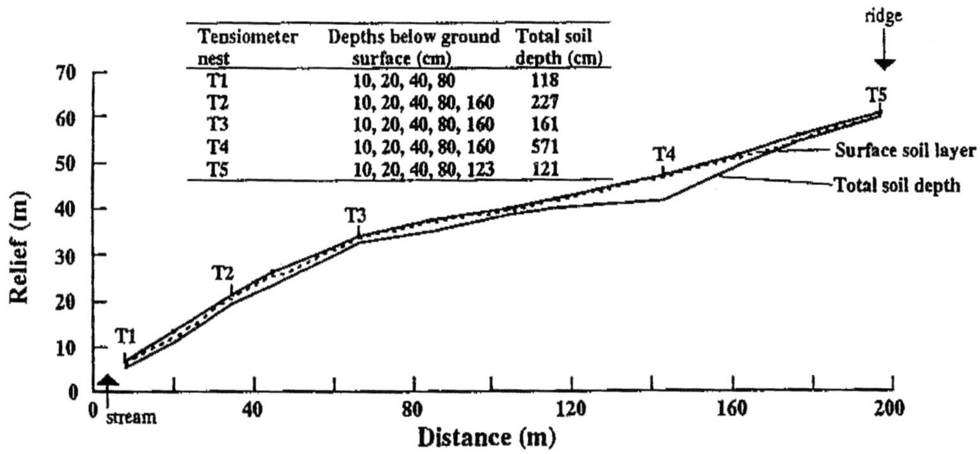

Figure 14.12 Soil depth along the longitudinal axis of an experimental slope, the location of sites and depths of tensiometer nests. (After Noguchi *et al.*, 1997b.)

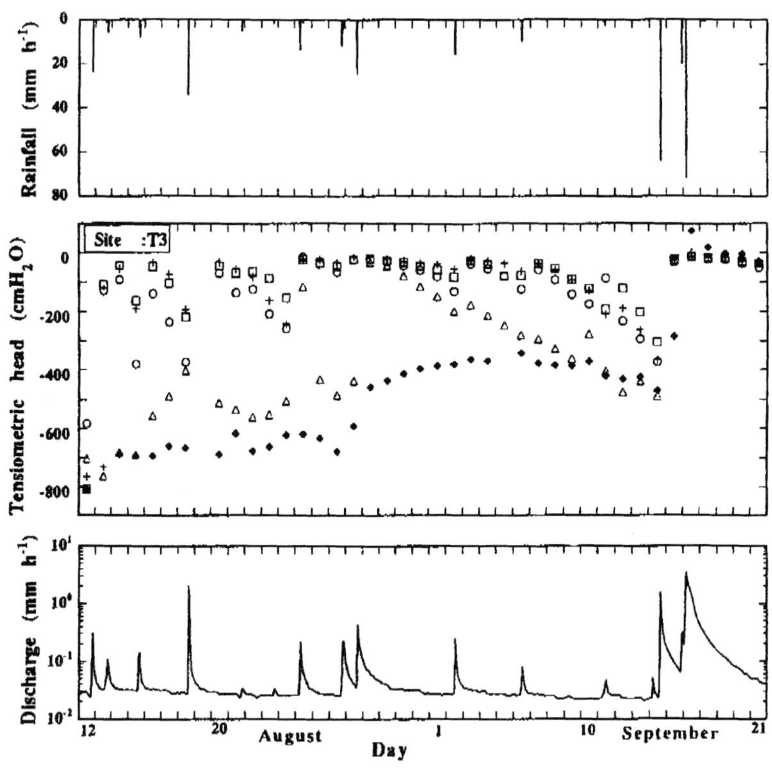

Figure 14.13 Hyetograph, hydrograph and changes in tensiometric heads at T3 through dry condition to wet condition. Tensiometric head: O, 10 cm; □, 20 cm; +, 40 cm; △, 80 cm; ♦, 160 cm. (After Noguchi *et al.*, 1997b.)

micro-topography and soil depth rather than hillslope scale topography is more influential. The ridge top site (T5) was the exception, and was generally drier. *At the hillslope scale, these findings may have implications for the application of spatial models (discussed later) that assume a strong topographic control on soil moisture. Clearly, there are other considerations apart from topography that control the spatial organisation of soil moisture.*

Third, the earlier work of Jeje *et al.* (1986) in south-west Nigeria measured lateral, sub-surface pathways within the shallow, upper organic layer to support Noguchi *et al.* (1997a) dye experimental results. Thus the potential importance of this flow vector to the storm hydrograph (Noguchi *et al.*, 1997b) is highlighted. More details of the Jeje *et al.* (1986) work will be provided later.

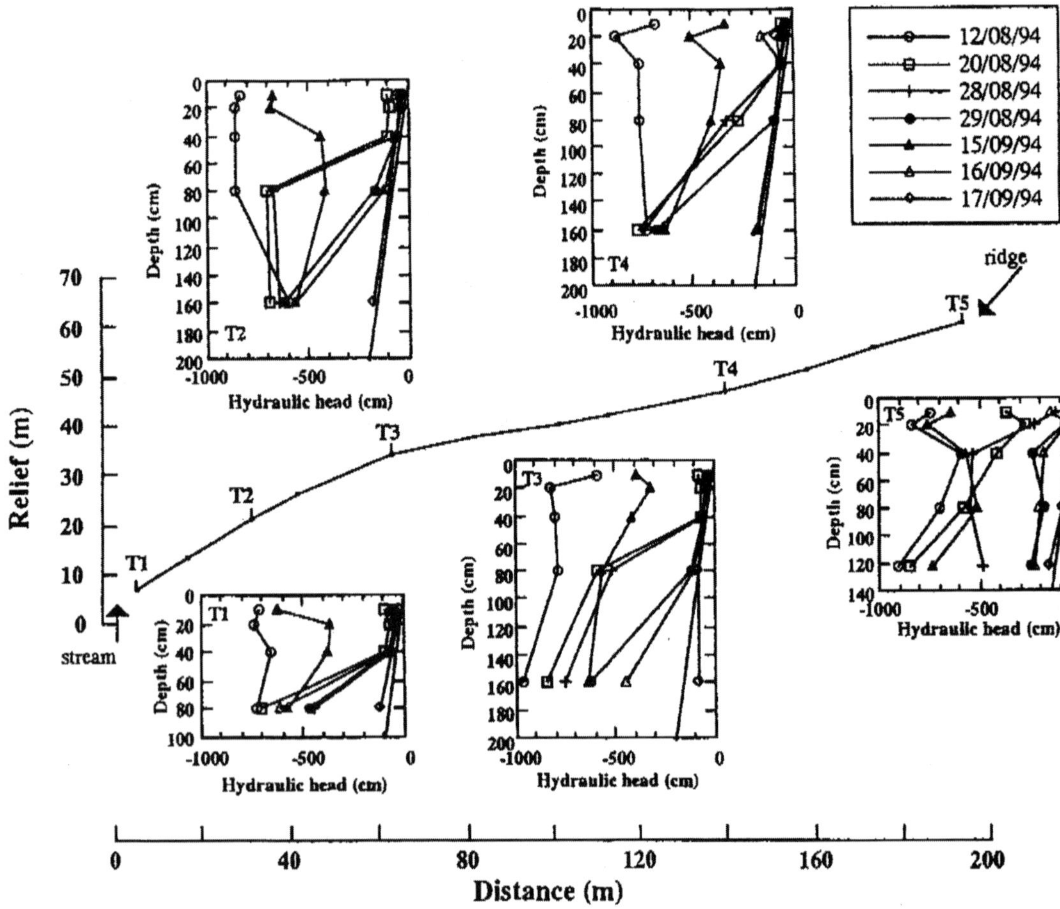

Figure 14.14 Time variation of the hydraulic head profiles on the experimental slope through dry condition to wet condition. (After Noguchi *et al.*, 1997b.)

Southern Tanzanian highland catchments

Whilst east Africa was the first location in the humid tropics to establish experimental catchment studies (Blackie *et al.*, 1979), there has been no detailed hillslope hydrology research undertaken so far (the work of Sandstrom, 1995 in Tanzania was outside the humid tropics). A recent water balance study by Lørup (1998), however, provides some insight into some of the pertinent storm runoff hydrology controls in three southern Tanzanian highland catchments (*Cultivated*: Grendavaki, 4.48 km^2, Muhur, 4.87 km^2, *Forest*: Mgera, 5.16 km^2). Annual rainfall in the three basins remains at the low end of the spectrum (1154 to 1329 mm, 1993–1996) by humid tropical standards. The corresponding annual runoff coefficients for the forest (0.22) and two cultivated catchments (0.33 each) are also modest, despite average slope gradients of 30–30%. A significant influence is the deeply weathered metamorphic bedrock, especially on the middle and upper slopes, where the soils (Humic Acrisols, FAO, Ultisols, USDA) are in excess of 8 m depth. These soils are characterised by a high clay content, 40–60% in the B-horizon and usually lower in the A horizon (25–45%). On the other hand, they also have a high sand content (34–42%) which enhance the permeability whereas the silt content is low (5–15%). Thus most of the soils are texturally classified as clay or sandy clay. By contrast, in the lower parts of the slope and in the valley bottom the regolith is much shallower, 0.5–3 metres deep. It is this regolith-bedrock interface which constitutes the main groundwater reservoir. Infiltration rates on the slopes are very high and are not conducive to shallow or surface runoff pathways, except during extremely wet periods (Lørup, 1998). Thus vertical fluxes prevail on the hillslopes away from the lower slope sections and riparian zones.

In line with studies in the Central Amazonia, Lørup (1998) placed strong emphasis on the role of the riparian zone in storm runoff generation. The annual quickflow percentage of annual rainfall was consistent in all catchments over the three-year period, and ranged from 3.2 to 4.8%. These percentages were closely related to the extent of the wetland (riparian) area rather than land use in each catchment, thus suggesting that the major part of the quickflow originates from rain falling directly on the riparian

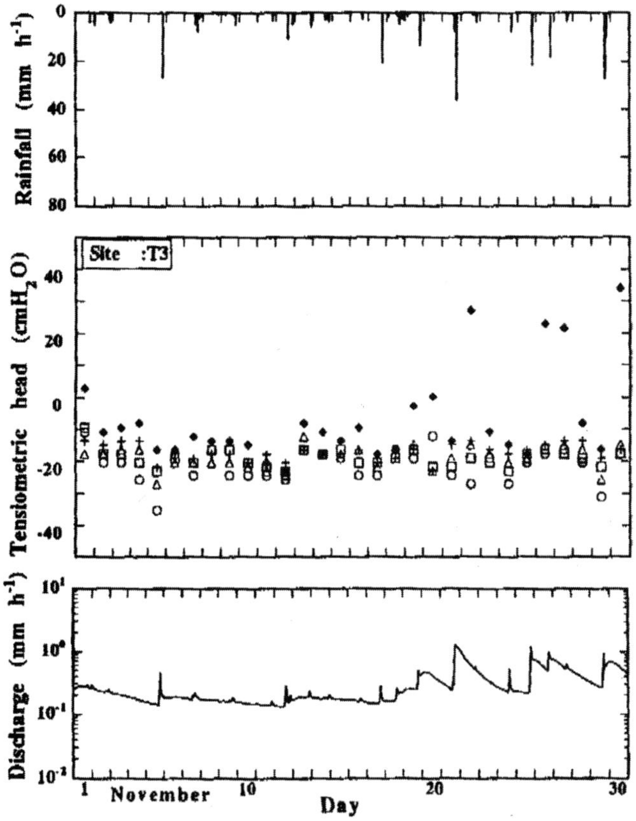

Figure 14.15 Hyetograph, hydrograph and changes in tensiometric heads at T3 during wet condition. Tensiometric head: O, 10 cm; □, 20 cm; +, 40 cm; △, 80 cm; ♦, 160 cm. (After Noguchi *et al.*, 1997b.)

area. End Member Mixing Analysis used by Knudby (1996) also suggested that approximately two-thirds of storm runoff originated from the wetlands. The remaining part originated primarily from groundwater. Any overland flow contribution from the valley slopes remained less than 1%.

Following the approach of Hewlett *et al.* (1977), the quick flow response ratio (QF/P), and time-to-peak discharge were, in turn, statistically regressed against four independent variables, namely total rainfall intensity, I_{30}, accumulated rainfall since the start of the wet season and antecedent rain over the previous five days. Thirty selected storms were used. The most pertinent result was the lack of statistical significance for the correlation and regression coefficients between the quickflow response ratio and maximum rain intensity, *irrespective of land use*. As Lørup (1998) noted, this result is particularly interesting for the cultivated catchments, and strengthens the notion of the storm runoff volume originating from the riparian area.

Lørup thus remarked (1998, p. 642) 'if surface runoff from the slopes had constituted a major part of the storm runoff, it would have been expected that the quickflow response ratio would have increased with increasing rainfall intensity'. Peak discharge, however, had a higher level of statistical significance (0.1%) against

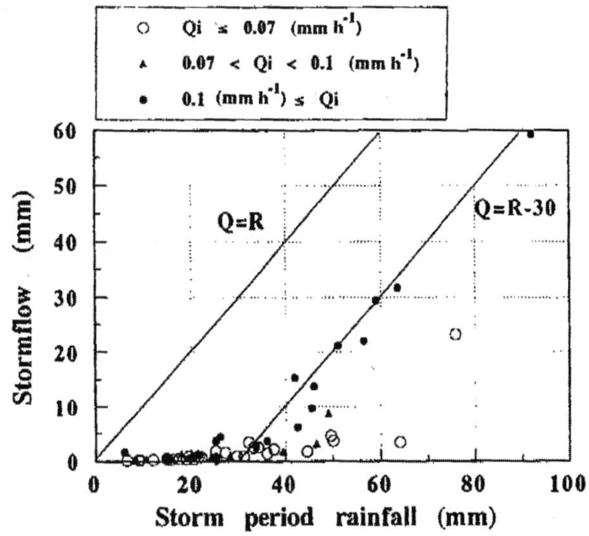

Figure 14.16 Relationship between storm flow and storm period rainfall. (After Noguchi *et al.*, 1997b.)

I_{30} in the two cultivated catchments in contrast to the forest (5%). A similar trend was observed when concerning antecedent five days rain with peak discharge.

One can conclude that with highly permeable hillslopes, a comparatively low rainfall-frequency-duration (see Bonell *et al.*, this volume) and in the absence of cyclonic-type rain systems, the east African highlands present a low responsive hydrological environment with vertical pathways and the riparian zones playing a dominant role. Discharge of deeper groundwater into organised surface drainage via the riparian zones is also an important mechanism here. The stream hydrograph characteristics of the earlier African catchment experiments in Kenya (Kericho) and Tanzania (Mbeya) (Blackie *et al.*, 1979; reviewed in Bonell with Balek, 1993, p. 227) would seem to have a similar stormflow hydrology to that described by Lørup (1998).

Predominantly lateral pathways

SUBSURFACE STORMFLOW AT THE SOIL–BEDROCK INTERFACE

The potential for significant contributions to the hydrograph of deeper SSF at the soil-bedrock interface where there is a reduction in hydraulic conductivity, has long been recognised since Weyman (1973) reported this mechanism in the Mendip Hills of the UK. In the case of Weyman's study, the regolith is shallow (typical of previously glaciated humid temperate environments) so this phenomenom can be observed more easily (e.g. Peters *et al.*, 1995, a Canadian Shield basin). Subsequent work elsewhere in humid temperate environments with deeper regoliths have also reported soil-bedrock SSF, for example in New Zealand (*Maimai*,

McDonnell, 1990) and the USA (*Walker Branch*; Wilson *et al.*, 1991a, b; Mulholland, 1993; *Panola*; McDonnell *et al.*, 1996).

The deeper regoliths associated with some parts of the tropics means that the majority of hillslope hydrology studies (using wells and piezometers) are not able to penetrate deep enough to the soil-bedrock interface. The principal obstacles are the difficult field logistics working in tropical forests and costs. Consequently, existing reports of SSF at the soil-bedrock interface are sparse for the humid tropics. It is likely however that the occurrence of this mechanism is much more extensive than is presently credited within the literature.

There were early reports by Williams and Coventry (1979) of lateral flow below open eucalypt woodlands in north-eastern Australia in a contrasting deeply weathered sandstone of associated yellow earths soils (0.4–2.5 m depth) and red earths soils (5–36 m + depth) (Stace *et al.*, 1968) in an environment similar to the West African Sahel. A saturated zone developed during the summer wet season within the yellow earths at the soil-bedrock interface. A weak hydraulic gradient then induced movement towards the deeper, bedrock topography under the red earths. In this particular case, the low topography and the corresponding drainage network is too poorly organised for such lateral flow to have a major effect on the ephemeral storm hydrographs. Nonetheless, in similar environments of more contrasting surface and subsurface bedrock topography, this SSF mechanism would become more important. Significantly, the work of Williams and Coventry (1979) is one of the few tropical studies to include wells and piezometers that penetrated the deeply weathered regoliths. Elsewhere in north-eastern Australia, recent hydrometric-hydrochemisty work in the Babinda study (Bonell *et al.*, 1998) highlights the greater importance of groundwater contributions to the storm hydrograph than previously credited: this issue will be referred to later.

Kuala Belalong, Brunei

More recently, one of the few hillslope hydrology studies to highlight the soil-bedrock SSF mechanism is the work of Dykes and Thornes (2000) in Brunei (Kuala Belalong Field Studies Centre). The topography is 'steep and sharp' which contributes towards a shallow clay regolith (Orthic Acrisols), derived from a dense argillaceous and impermeable shale, which rarely exceeds 3 m in depth. Although a survey of soil hydraulic (and physical-chemical properties) was undertaken using soil pits to collect cores, the provision of specific details by depth and soil horizon (on the lines of Elsenbeer and Vertessy (2000), S1 stage) were not provided by Dykes and Thornes (2000). Their work was supplemented by spot single-ring infiltration tests on intact forest floor and excavated subsoil surfaces at 0.2–0.3 m depth.

Laboratory-determined K_{sat} from intact bulk samples collected from soil pits were reported to be from between 27.4–468 mm h^{-1} with an average *c.* 180 mm h^{-1}. This high value was attributed to extensive macroporosity existing within the regolith. The corresponding soil moisture retention curves also provided evidence of macroporosity. Apart from the influence of soil biological factors, Dykes and Thornes (2000) also attributed this macroporosity to aggregation and micro-aggregation of the clay particles by iron oxides and the presence of goethite. As these samples were taken to within 0.1 m of the unweathered rock from within the soil pits, the range and mean K_{sat} cited above is then presumably an aggregate for the shallow clay regolith (although this is not clear in Dykes and Thornes, 2000). The surface mean infiltration capacity of the forest floor was reported to be 793 mm h^{-1} (due to the incorporation of organic matter and macropores) whose value far exceeds the highest 5-minute rain intensity recorded of 175.2 mm h^{-1} (Dykes and Thornes, 2000) to permit (infiltration-excess) overland flow. Consequently, the clay regolith was considered transmissive to vertical percolation as far as the shale bedrock. No field measurements of K_{sat} of this bedrock were made but a separate finite simulation of the regional groundwater hydrology (as reported by Dykes and Thornes, 2000) indicated that the bedrock is 'effectively impermeable' compared to the upper soil regolith.

With the aid of a tensiometer network along a slope transect, the hydrological responses to rainfall events in both the dry and wet seasons were compared. In some respects, there were similarities with the earlier reports of Noguchi *et al.* (1997a, b). Rapid differential wetting in the vertical (bypass flow in macropores) was noted during 'dry' soil states, but such flow did not penetrate the regolith and influence the hillslope hydrology. On the other hand, once the soil profile was wetted up, downward vertical percolation occurred within the clay regolith as far as the bedrock. At such times, positive matric potentials were observed, especially in the deeper tensiometers (at 1.2 m depth). An example of maximum saturation conditions is shown in Figure 14.17 for a storm of 87 mm (4 May 1992). Positive pressures (i.e. negative tensions) developed rapidly in most tensiometers. The perched water table over the bedrock surface is subsequently dissipated by return flow commonly through pipes and insect burrows at the base of the slope. Moreover, the return flow was observed to commence below tensiometer nest E (Figure 14.18) within minutes of the onset of high intensity rainfall. Consequently, drainage of the slope is rapid due to the effectiveness of SSF within the perched water body (i.e. beneath the perched water table).

In common with other studies, Dykes and Thornes (2000) presented data to show that the tensiometers were very responsive to wet season storms leading to extremely rapid saturation (e.g. Figure 14.18). They attributed this phenomenon to conversion of a 'capillary fringe' to a true saturation state (e.g. Abdul and Gillham, 1989) due to the prevailing low antecedent matric potentials (also discussed later for the Babinda and La Cuenca studies). Dykes and Thornes (2000) also conceded that bypass flow

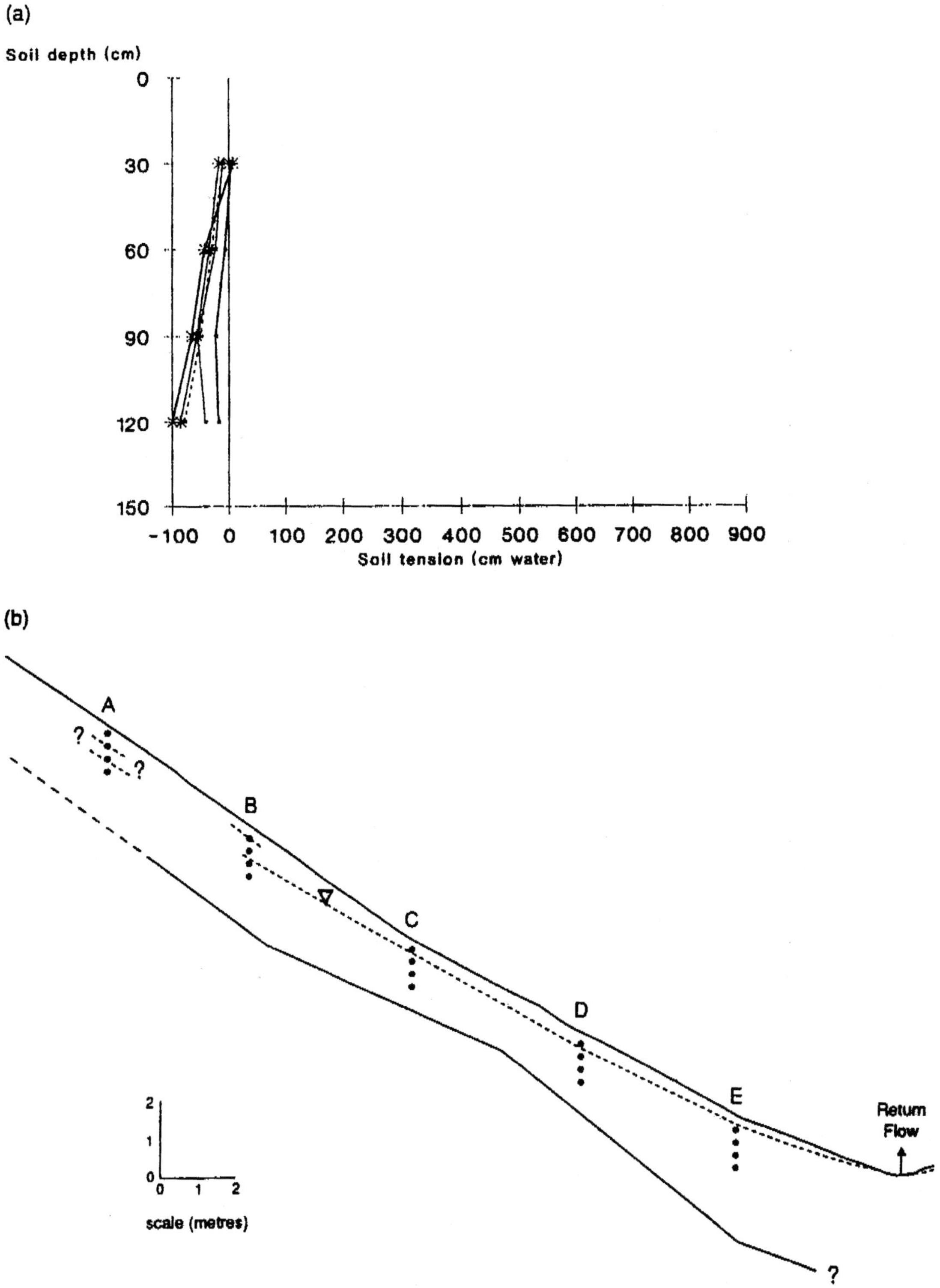

Figure 14.17 Maximum saturation conditions during the storm of 4 May 1992. (a) Soil tension variations with depth at tensiometer nests A–E at 15.05 h. (b) Estimated position of perched water table within the instrumented regolith profile at 15.05 h, as derived from tensiometer readings. The positions of the tensiometers are shown. The base of the profile is drawn according to the depths of the adjacent soil pits. Readings from nests A and B were inconclusive. (After Dykes and Thornes, 2000.)

Table 14.2. *Sungai Esu discharge and storm runoff data*

	Storm 1	Storm 2
Date	24 April 1993	4 May 1993
Pre-storm discharge (m^3 s^{-1})	0.04	0.11
Equivalent reduction in soil moisture (mm day^{-1})	3.1	8.4
Corresponding equivalent Sg. Belalong discharge (m^3 s^{-1})	5.4 (e.g. 24 Jan. 1992)	14.6 (e.g. 18 Nov. 1991)
Soil tensions on the corresponding date (cm H_2O)	4–58 (av. 16.0)	0–29 (av. 11.6)
Rainfall depth (estimated) (mm)	81	32
Rainfall duration (min)	75	80
Time from start of rain to hydrograph rise (min)	15	15
Time from start of rain to peak discharge (min)	70	51
Peak discharge (m^3 s^{-1})	3.21	0.96
Assumed baseflow (m^3 s^{-1})	0.10	0.11
Total storm runoff (m^3)	34450	14450
Equivalent runoff depth (mm)	30.9	13.0
Percent of total storm rainfall	38	41
Percent of estimated throughfall	47	50

Source: Dykes and Thornes (2000).

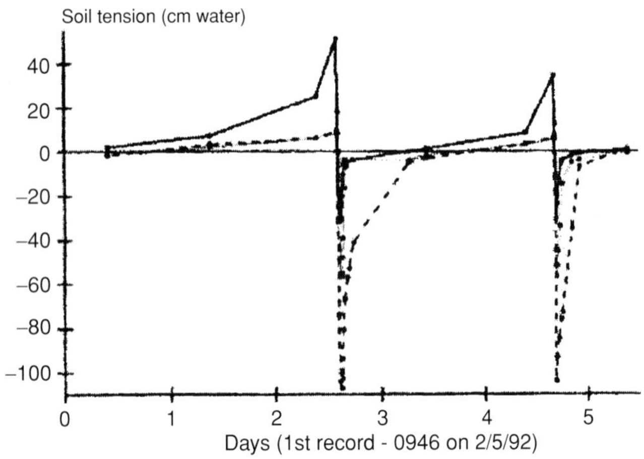

Figure 14.18 Variations in calculated hydraulic head at each of three tensiometer cups in response to the storms of 4 and 6 May 1992. Solid line, 60 cm depth at nest C; broken line, 90 cm depth at nest D; dotted line, 120 cm depth at nest E. (After Dykes and Thornes, 2000.)

(preferential flow) within the extensive macropore networks of their clay regolith could propagate stormwater fast enough to produce the same effect. This issue will be taken up later.

Whilst no storm hydrographs were available for the Sungai Belalong which hosted the hillslope study outlined above, Dykes and Thornes (2000) provided examples of two wet season storms at the nearby Sungai Esu (112 km^2) (the Sungai Belalong displayed similar hydrograph characteristics to Sungai Esu) to show the effectiveness of deep SSF over bedrock in combination with return flow. The storm hydrographs (Figure 14.19) are highly responsive, based on the antecedent conditions and storm characteristics, as shown in Table 14.2, with relatively high QRR.

SUBSURFACE STORMFLOW WITHIN THE SOIL

With the exception of the Rondonia (Elsenbeer *et al.*, 1999) and Western Ghats work (Purandara *et al.*, 2004, unpublished data), none of the studies outlined in this section (see Table 14.1) systematically follow Stages 1 to 3 of Elsenbeer and Vertessy (2000). Moreover, the former remain at the S1 stage.

Sungai Baru Barat, Sabah, Malaysia

The work of Chappell *et al.* (1998) in evaluating laboratory and field methodologies for K_{sat} estimation was undertaken in a Haplic Acrisol at Sungai Baru Barat (annual rainfall 2778 mm, 1986–1996) in Sabah where SSF occurs, including soil piping. Artificial application of spray-irrigation rates of 100–200 mm h^{-1} as part of the pulse wave test did not generate either infiltration-excess or saturation-excess overland flow outside the sprayed areas (1 m by 0.5 m). Although the vertical distribution of K_{sat} from cores is presented (Figure 14.20), these are not compared with prevailing rain intensities by Chappell *et al.* (1998) (i.e. Stage 1, Elsenbeer and Vertessy, 2000) but are much greater than the 10

(a)

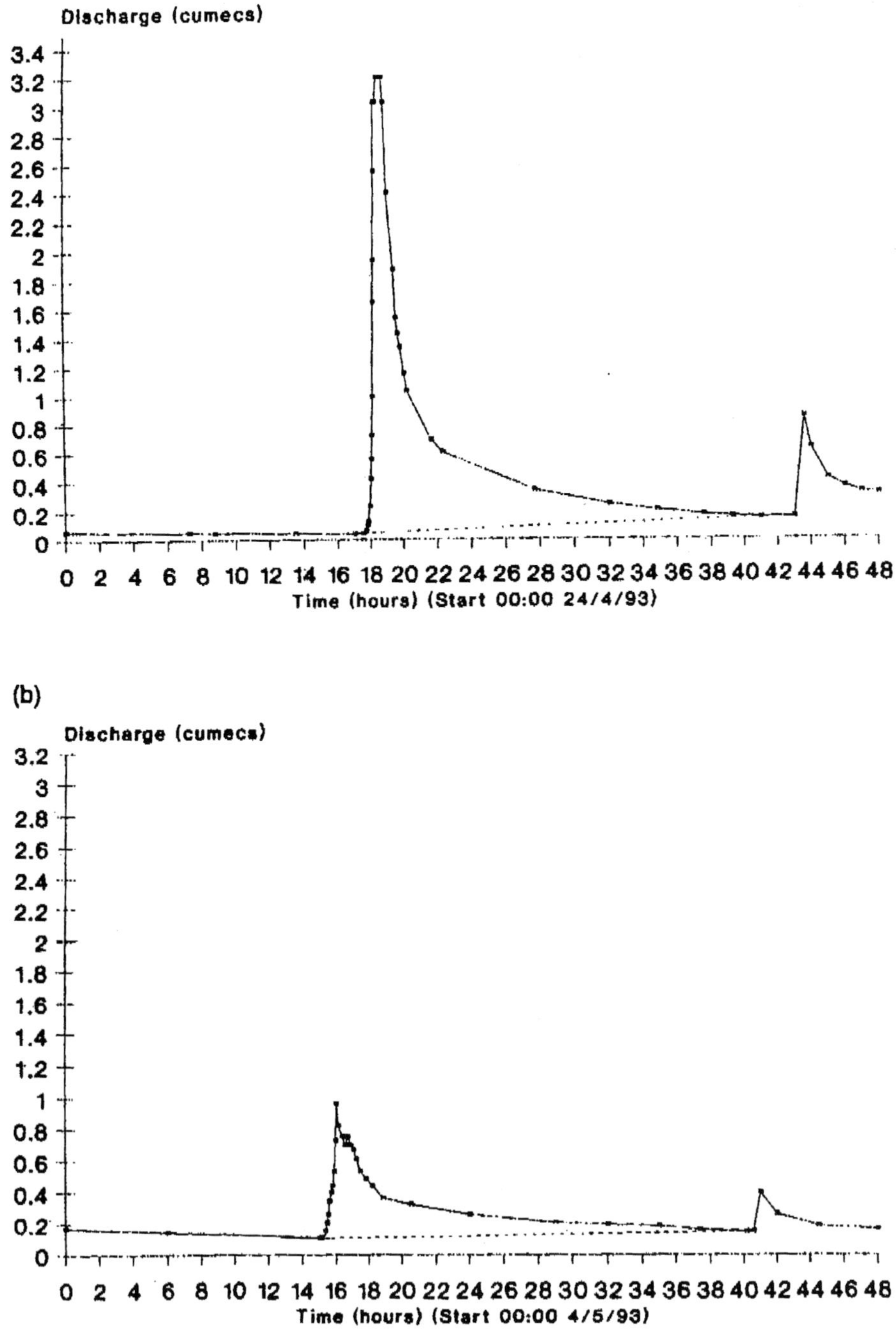

Figure 14.19 Storm hydrograph from Sungai Esu detailed in Table 14.2: (a) 24–25 April 1993; (b) 4–5 May 1993. (After Dykes and Thornes, 2000.)

mm h^{-1} equivalent rainfall intensity (using the 5 min. integrated data) monitored later at the site (Bidin, 2001). Even so, this is one of the few studies to show the vertical distribution of K_{sat} down to the bedrock. The impact of the argillation process in reducing permeability is also evident (Figure 14.20) through the A and B horizons. The K_{sat} reduces further from B to C (weathered rock) to R (solid rock, (mudstones and sandstones)) in response to sustained *in situ* chemical weathering (this decline in vertical permeability contrasts with the results of Dykes and Thornes, 2000). The geometric mean of these core scale measurements (for the entire vertical distribution) was 5.94 mm h^{-1}. Corresponding estimates of lateral block permeability derived from the hillslope

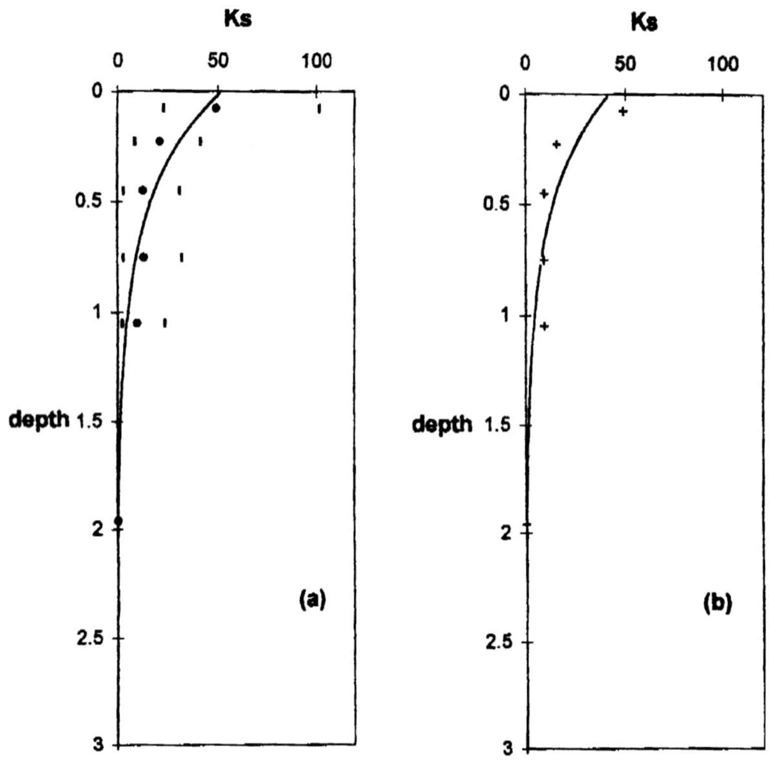

Figure 14.20 Vertical distribution of core-scale permeability data, with layer-specific centroid values represented by (a) the median (•), and (b) the geometric mean (+). Permeability has units of $\times 10^{-6}$ m s^{-1} and sampling depth is in metres. The bars in (a) represent the interquartiles range (Wrinkler and Hays, 1975). (After Chappell *et al.*, 1998.)

pulse-wave experiment were an order of magnitude higher, ranging from 32 mm h^{-1} to 53 mm h^{-1} depending on whether one selects the time from water injection to steady state response (V_s in Chappell *et al.*, 1998) or the centroid time between injection and steady-state response. These higher estimates are a result of the injected water plume incorporating the influence of pipe flow (not included in soil cores) and/or pressure wave effects. Thus SSF is able to be propagated more effectively downslope and produce the observed return flow (Chappell *et al.*, 1998) than otherwise suggested from core K_{sat} estimates. The fundamental difference between this study and that of Dykes and Thornes (2000), is the marked vertical change in K_{sat} with depth which causes the preferred position of SSF to differ. Otherwise the outcome would have seemed similar, with return flow into the channel bed being a principal contributor to storm hydrograph responses. Chappell, Bidin *et al.*, elsewhere in this volume provides more detail of this work.

Rondonia, Brazil

Work undertaken by Elsenbeer *et al.* in Rondonia (Stage 1) provides an interesting contrast to previous studies reported for central Amazonia. Elsenbeer *et al.* (1999) remarked appropriately that, when concerning the Amazon basin, there had been 'over-reliance' on predominantly vertical pathways linked with Tertiary Barreiras geological formation. Data presented for soils of similar taxonomic classification to those of central Amazonia (Ferralsols, FAO; Oxisols, Soil Survey Staff, 1975) but derived from Pre-cambrian basement rocks showed a marked vertical change in K^* with depth under forest as well as pasture and a small teak plantation (Figure 14.21). Disregarding for the moment the low K^* at the forest surface, the decrease in K^* at a shallow depth is in marked contrast to the reports from the Barreiras formation of central Amazonia (Nortcliff and Thornes, 1989; Elsenbeer and Lack, 1996a, b; Hodnett *et al.*, 1997a, b). Further, this marked decline means that the subsoil K^* are two to three orders of magnitude lower than those reported for central Amazonia. When compared with the rainfall reference lines of one-hour intensities (24 and 15 mm h^{-1}) with return intervals of 10 and 30 times per year, it is evident that a perched water table can develop above 0.3 m depth which encourages a *strong lateral* SSF component (*cf.* Reserva Ducke) of hillslope flowpaths in this soilscape. Under sustained high rain intensities, SOF is likely to be generated which would be independent of land cover due to the marked anisotropic profile of K_{sat}.

The cause of the low K^* at the forest surface is interesting and was attributed to hydrophobicity effects by Elsenbeer *et al.*

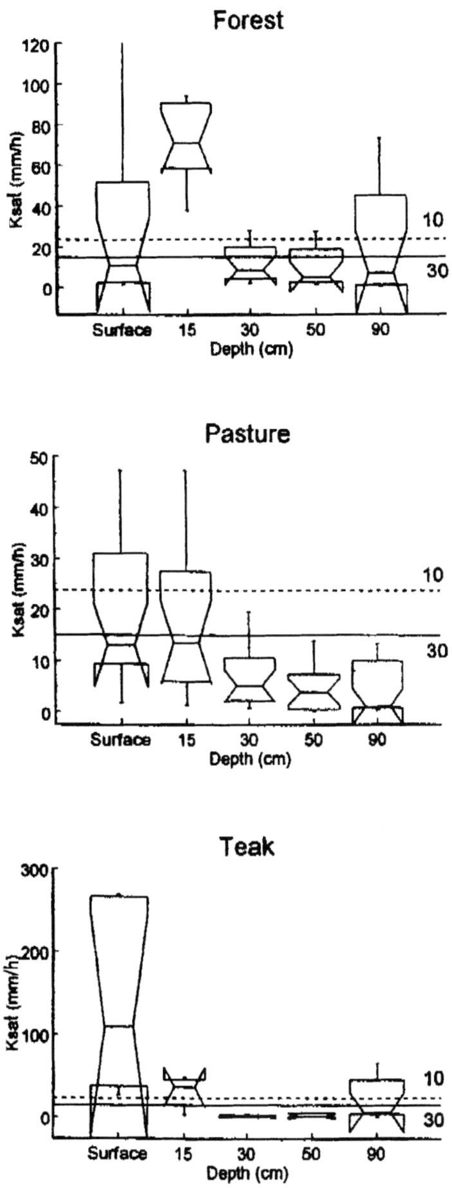

Figure 14.21 Diagram to show K_{sat} as a function of depth under forest, pasture and teak. The reference lines labelled '10' (dashed) and '30' (solid) indicate the 1-hour rainfall intensities (24 and 15 mm h^{-1}) that have a recurrence interval of 10 and 30 times per year, respectively. (After Elsenbeer *et al.*, 1999.)

(1999) arising from sustained drought linked with a negative phase of ENSO (see Callaghan and Bonell, this volume). Independent rainfall simulator results had yielded infiltration capacities an order of magnitude higher (146 mm h^{-1} and 181 mm h^{-1}). In addition, during a high intensity rainstorm, the widespread ponding observed in adjacent pasture did not occur under the forest. These features led Elsenbeer *et al.* (1999) to consider the measured K^* to be grossly underestimated and the presented surface data 'suspect'. Whilst the Elsenbeer *et al.* (1999) reasoning is supported from the above evidence and other global data sets, their suggestion of hydrophobicity effects (transitory though they may be) could occur more frequently than presently credited by these writers, especially in more strongly negative phases of ENSO. The impact of hydrophobicity on the runoff generation process in the Rondonian forest should thus not be discounted as necessarily an aberration of this study. For example, there are reports of the occurrence of infiltration-excess overland flow from south-east Australian eucalypt forests which occur temporarily due to soil hydrophobicity after summer drought (Topalidis and Curtis, 1982; Burch *et al.*, 1987, 1989).

South-west Nigeria studies

Elsewhere, one of the few studies to monitor SSF directly is the work of Jeje *et al.* (1986) in Nigeria, although a rigorous survey of soil hydraulic properties (S1 stage) had previously not been made. This study was undertaken in south-west Nigeria along select slopes under forest (Jeje *et al.*, 1986) and mixed cultivation and scattered forest (Ogunkoya *et al.* 2000, 2003) underlain by pre-Cambrian granite gneisses which supported SSF as the prime delivery mechanism. Jeje *et al.* (1986) followed an experimental design very similar to that of Whipkey (1965) except lateral seepage was monitored in the present study under natural rainfall (rather than using a rainfall simulator) within a 10% forested slope. No soil classification was provided but their hydrological behaviour was in line with that described later by Elsenbeer (2001) as part of the 'Acrisol' End-Member group.

Four soil layers were identified, namely, 0–500 mm clay loam (K_{sat} ~400 mm h^{-1}), 500–900 mm, sandy clay, K_{sat} ~3.5 m h^{-1}; 900–1200 mm, sandy clay loam, K_{sat} ~25 m h^{-1} and 1200–1800 mm, sandy clay loam, K_{sat} ~4 mm h^{-1}. The preferred stormflow pathways of SSF were within the shallow, organic layer, 0–30 mm (c.f. Noguchi *et al.*, 1997a, b) and the deeper 500–900 mm layer above the impeding lower horizons, >900 mm depth. Significantly, the largest contributions of seepage were from the subsurface 0–30 mm horizon where the lateral conductance (K_{sat}) of the organic layer was far in excess of the main 30–500 mm horizon. Over the sampling period, total rainfall was 924 mm from 106 rainy days. By contrast, the total amount of seepage was only 67.7 mm from 29 seepage days. Thus the total seepage response was confined to the double maxima, seasonal rain pattern.

The temporal occurrence of seepage showed that the 500–900 mm layer responded first before the 0–30 mm layer. Smaller flows (due to lower K_{sat}) were monitored from the deeper horizons (900–1200 mm, 1200–1800 mm) towards the end of the wet season (see Table 14.3) when soil moisture storage was at its highest.

Table 14.3. *Timing and amount of seepages from the soil horizons and the API and rainfall intensities*

Date	API (mm)	Rainfall intensity ($mm\ h^{-1}$)	Seepage 0–30 mm	30–500 mm	500–900 mm	900–1200 mm	1200–1800 mm	Total
4/6/82	92.62	12.80			13.30			13.30
6/6/82	84.29				1.10			1.10
7/6/82	75.86				1.00			1.00
10/6/82	70.70	6.25			0.30			0.30
11/6/82	63.63				1.60			1.60
13/6/82	53.34	10.80			0.20			0.20
20/6/82	39.55		3.60	0.40	0.40			4.40
21/6/82	39.40	3.80	0.10	0.10	0.10			0.30
22/6/82	54.46	9.50	0.70		1.00			1.70
24/6/82	58.17	5.56	0.10		0.80			0.90
29/6/82	58.80	3.10	7.00					7.00
2/7/82	59.36	5.50	3.00					3.00
3/7/82	63.43	5.50	2.20					2.20
7/7/82	53.62		0.70	0.20				0.90
12/7/82	35.38	3.40			0.40			0.40
13/7/82	31.85		0.70					0.70
17/7/86	44.75	6.00	0.20		0.20			0.40
18/8/82	14.39	2.76		0.30	0.70			1.00
29/8/82	14.66	4.80		0.30	0.40			0.70
6/9/82	30.07	3.75	3.00			0.20		0.20
15/9/82	46.69	3.98	5.00					5.00
24/9/82	41.08		0.80		0.05	0.40		1.25
26/9/82	42.27		1.20		0.05	0.04		1.29
1/10/82	43.74	7.80	0.90					0.90
3/10/82	69.43	5.93	8.50		2.50	1.30	0.50	12.80
11/10/82	60.64		1.10			0.04		1.14
14/10/82	61.10	1.88			0.03			0.03
25/10/82	41.01		0.40		0.05			0.45
1/11/82	52.08		0.30		0.20			0.50
Total			39.50	1.30	24.38	1.98	0.50	67.66

Source: Jeje *et al.* (1986).

An assessment of rainfall intensity frequency-distribution in relation to K_{sat} explains why no overland flow occurred. Only seven events exceeded 20 mm h^{-1} (7% by frequency distribution) with the maximum recorded being 52.2 mm h^{-1} while most of the 'heavy' intensity rainfall had values between 20 and 30 mm h^{-1} (Jeje *et al.*, 1986). These values are an order of magnitude lower than 30–500 mm soil horizon K_{sat}. Significantly, none of these heavy intensity rainfalls produced seepage because of the inherent, low soil moisture contents associated with the opening stages of the wet season when the highest intensity rain events were recorded (Jeje *et al.*, 1986).

There was a close correspondence between seepage and soil moisture content. For example, when soil moisture was highest at 200 mm depth, seepage from the 0–300 mm layer was also correspondingly higher than from the 500–900 mm layer. Under optimal soil moisture wetness, there was a statisticaly significant relationship between amount of rain (>10 mm) and amount of seepage. In contrast, there was no statistical relationship between rainfall intensity and cumulative daily seepage, cf. Bonell and Gilmour (1978).

A subsequent assessment of the spatial and temporal variability of soil moisture (Ogunkoya *et al.*, 2000, 2003) along a hillslope transect (103 m) in the nearby Agbogbo catchment (0.44 km^2, mean annual rainfall, 1474 mm) provides a further insight into the previous interpretations of Jeje *et al.* (1986). Of interest, the clay content and moisture retention properties increase upslope along the toposequence from the footslope to the upper rectilinear slope. A consequence is that despite the existence of a gradient of increasing moisture downslope, *more water is required to raise the moisture status to saturation level in the sandy soils within the footslope zone. Such circumstances are counter to conventional wisdom of trends in the development of saturation within hillslope hydrology*. Thus the spatial and temporal distribution of soil moisture and soil physical properties do not facilitate the development of saturation to the surface and SOF. The runoff hydrology depends on contributions from lateral subsurface drainage within preferred horizons (e.g. 600–900 mm depth) of the footslope which are also encouraged by the incised nature of the stream channel (Freeze, 1972). Thus a dynamic surface source area of runoff generation could not be defined.

Further upslope, the preferred runoff pathway was SSF within the 450–600 mm horizon in the mid-slope section where lighter textured soil overlay a more compact horizon (Ogunkoya *et al.*, 1995a, b). Surface infiltration rates were in excess of 800 mm h^{-1} which precluded overland flow except from the infiltration-excess type from base rock surfaces on the upper slope associated with the convex inselberg rock faces (Ogunkoya *et al.*, 2003). SSF was most likely to occur during periods of heavy rainfall within the rainy season at midslope. On the other hand, the regular occurrence of saturation within the 600–900 mm horizon at the footslope was considered to be connected hydraulically with the common occurrence of rapid streamflow responses during the rainy season. Time lags were approximately 30 minutes between peak rainfall and peak streamflow.

The above small-scale sites were underlain by either coarse grained granite gneisses (Agbogo catchment) or variously magnetised gneiss and schists (i.e. study slope of Jeje *et al.*, 1986), which produce soils belonging to the Luvisol-Acrisol (FAO) or Alfisol-Ultisol (USDA) complex. A series of papers by Ogunkoya and co-workers (Adejuwon *et al.*, 1983; Ogunkoya *et al.*, 1984; Ogunkoya, 1988) have emphasised the diversity of hydrological response patterns within south-west Nigeria, and such patterns need to be considered to place these smaller scale studies in perspective.

Fifteen third-order basins were selected within the Upper Owena to include the diverse geology associated with the Precambrian Basement Complex suite, geomorphic attributes and land. Figure 14.22a and b and Tables 14.4 and 14.5 provide the various attributes of these basins. Compared with cyclone-prone parts of the humid tropics, annual rainfall is low and maximum daily rainfall modest in total. Nonetheless these figures are typical of the Gulf of Gunean coastal hinterland (Ayibotele, 1993).

The variation in amount of annual rainfall, however, does not account for variations in runoff amongst these 15 basins (Adejuwon *et al.*, 1983; Ogunkoya *et al.*, 1984). Further, with the exception of the Opapa basin, annual runoff coefficients are low with many basins less than 10%. Another interesting feature is that despite the highly responsive nature to storms during the wet season, several basins have no daily flow (101–262 days) centred on the dry season (Figure 14.23 and Table 14.6). The recession constants (K) range from 0.64 to 0.98 with high values depicting basins with large groundwater storage and a slow release of this storage during the recession period; the low K values imply a sharp decline from peak to baseflow, low groundwater contribution to streamflow and low groundwater storage in such basins (Table 14.6).

A classification of the above basins (Figure 14.24) from the use of principal components and cluster analyses produced five groups (Ogunkoya, 1988). The overriding influence central to these groupings was the varied nature of the geology. These basins, which incorporate quartzitic rocks, encourage substantial groundwater storage from percolation through the extensive fissures and joints. The groundwater aquifers promote low discharge variability and sustained delayed flow (e.g. Okun, Opapa, Alura, Ohoo). In contrast, the granitic gneisses, amphibolites, schists and the associated clayey saprolites are poorly jointed. Consequently, the runoff hydrology is dominated by quickflow and, where the relief is low, also higher soil moisture retention. The variability index of runoff (Table 14.6) is much higher and the lack of groundwater storage results in sustained periods of no stream discharge. Further, most of the basins on these rocks have low relief and are forested.

Two points emerge from this larger scale study. The small-scale hillslope hydrology is more representative of the second, broad group of basins above. *In addition, the important role of groundwater in the runoff hydrology, associated with the quartzitic rocks, is highlighted.* This point will recur throughout this review. Under these latter circumstances the hillslope hydrology is likely to be different from the previous descriptions of Jeje *et al.* (1986) and Ogunkoya *et al.* (2000, 2003).

Bisley II catchment, Luquillo, Puerto Rico

Schellekens (2000) presents some data from Puerto Rico to suggest that shallow SSF is the dominant pathway in the Bisley II catchment (6.4 ha, annual rainfall 3530 mm) within a dissected mountainous terrain of the Luquillo Experimental Forest (Scatena, 1989). The upper 0.8–1.0 metres of the solum are clayey in texture, classified as strongly leached Acrisols (Ultisols, USDA) which have developed from thick-bedded tuffaceous sandstones and indurated siltstones (Scatena, 1989).

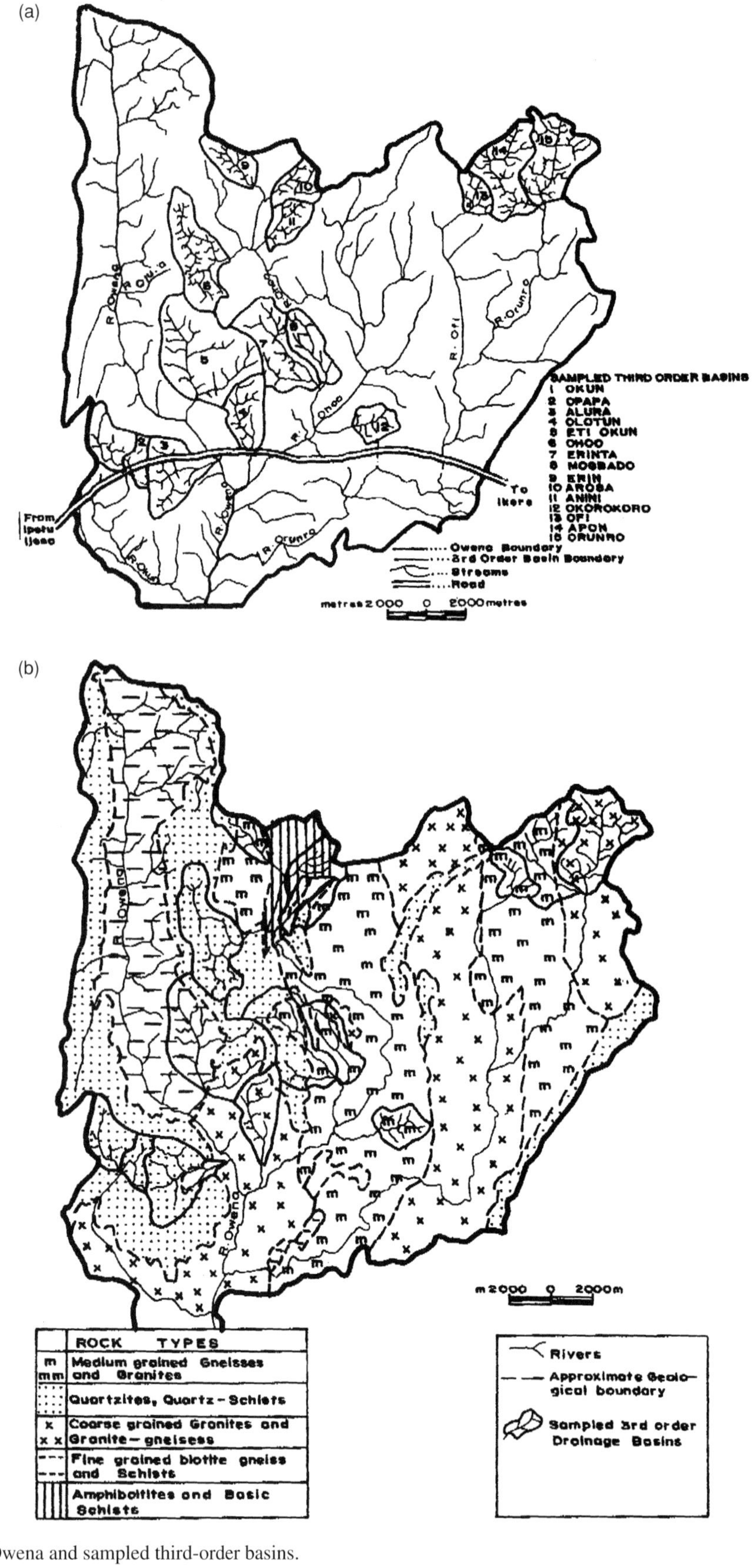

Figure 14.22 The upper Owena and sampled third-order basins. (a) Drainage composition. (b) Geology. (After Ogunkoya, 1988.)

Table 14.4. *Physiographic and land use attributes of the basins*

Basin number and name	Percentage of area of basin underlain by[a]			Area (km^2)	Drainage density (km^{-1})	Relief ratio	Percentage of area of basin covered by		Total annual rainfall	Maximum weekly rainfall (mm)	Total runoff (mm)	Annual runoff coefficient
	Qz	Gg	Amph				Forests	Farms				
1. Okun	96	3	1	5.0	1.3	0.08	47	53	1719	168.4	387.0	22.5
2. Opapa	96	3	1	2.0	2.5	0.14	41	59	1719	168.4	680.1	39.6
3. Alura	78	21	1	9.3	1.5	0.08	78	22	1444	132.3	189.0	13.1
4. Olotun	8	91	1	6.1	1.8	0.05	69	31	1412	114.3	28.4	2.0
5. Etiokun	36	63	1	18.8	1.3	0.07	86	14	1425	151.2	31.1	2.2
6. Ohoo	96	3	1	10.4	2.0	0.06	14	86	938	55.3	164.9	17.6
7. Erinta	58	41	1	10.6	1.5	0.05	62	38	1438	132.7	118.2	8.2
8. Mogbado	13	86	1	5.0	1.6	0.04	62	38	1438	132.7	58.0	4.0
9. Erin	33	63	4	3.2	2.5	0.07	26	74	927	55.1	135.7	14.6
10. Arosa	3	12	85	3.9	2.3	0.07	52	48	950	54.0	10.9	1.1
11. Anini	29	26	45	4.1	22.8	0.04	23	77	965	53.3	29.4	3.0
12. Okorokoro	1	98	1	3.3	2.2	0.01	64	36	1472	113.7	14.6	1.0
13. Ofi	1	98	1	3.3	2.8	0.05	75	25	1458	88.6	46.0	3.2
14. Apon	3	96	1	8.9	2.2	0.02	87	13	1532	192.4	66.1	4.3
15. Orunro	1	96	1	10.5	2.1	0.03	79	21	1532	192.4	88.2	5.8

[a]Qz, quartzitic rocks; Gg, granitic gneisses; Amph, amphibolites.

Source: Compilation after Ogunkoya, O. O., Adejuwon, J. O. and Jeje, L. K., 1984.

Table 14.5. *Hydrological response parameters of the fifteen third-order basins*

Basin number and name	$Q_{\bar{A}}$[a] ($l\ s^{-1}\ km^{-2}$)	Q_{90}[a] ($l\ s^{-1}\ km^{-2}$)	Q_{50}[a] ($l\ s^{-1}\ km^{-2}$)	Q_{10}[a] ($l\ s^{-1}\ km^{-2}$)	Variability index	K = recession constant	Total annual runoff (mm)	Total dry season runoff (mm)	Total dry season runoff as a percentage of total annual runoff	Runoff coefficient	Number of days without flow
1. Okun	12.2	3.0	11.6	24.8	0.44	0.94	387	82.3	21.3	22.5	0
2. Opapa	21.5	6.5	21.0	33.0	0.29	0.98	680	293.8	43.2	39.6	0
3. Alura	6.0	2.0	3.8	11.1	0.36	0.95	189	44.5	23.5	13.1	0
4. Olotun	1.0	0	0	4.1	1.92	0.81	28	2.2	7.7	2.0	192
5. Etiokun	1.0	0	0.4	2.7	0.70	0.90	31	2.5	8.0	2.2	106
6. Ohoo	5.2	3.4	4.3	9.8	0.27	0.98	165	51.6	31.3	17.6	0
7. Erinta	3.8	1.4	2.1	10.0	0.46	0.96	118	25.5	21.6	8.2	0
8. Mogbado	1.8	0	0	6.6	0.67	0.87	58	3.1	5.4	4.0	225
9. Erin	4.4	1.3	4.4	6.6	0.45	0.96	136	60.3	44.4	14.6	0
10. Arosa	0.3	0	0	1.0	0.25	0.88	10	1.3	12.2	1.1	262
11. Anini	1.0	0.2	0.5	2.4	0.62	0.93	29	5.2	17.7	3.0	0
12. Okorokoro	0.6	0	0	1.5	2.12	0.64	15	0.1	0.2	1.0	197
13. Ofi	1.5	0	0	6.7	1.44	0.82	46	0.2	0.3	3.2	147
14. Apon	2.0	0	0.3	7.9	0.77	0.92	66	3.5	5.3	4.3	143
15. Orunro	2.8	0	0.6	10.6	0.79	0.92	88	5.7	6.5	5.8	101

[a] $Q_{\bar{A}}$, mean daily discharge; Q_{90}, Q_{50}, Q_{10}, daily mean discharge equaled or exceeded 90%, 50% and 10% of the time.
Source: After Ogunkoya (1988).

Table 14.6. *Variability and recession indices of discharge*

Rivers	Variability index	Number of days without flow	*K* recession constant	Q_0[a]	Q_t[a]	*t*, time of recession (days)
Okun	0.44	0	0.94	698	1	99
Opapa	0.29	0	0.98	113	31	68
Alura	0.36	0	0.95	668	19	79
Olotun	1.92	192	0.81	520	0.02	61
Etiokun	0.70	106	0.90	754	0.6	69
Ohoo	0.27	0	0.98	405	35	99
Erinta	0.46	0	0.96	592	15	99
Mogbado	0.67	225	0.87	261	2	36
Erin	0.45	0	0.96	410	9	88
Arosa	0.25	262	0.88	14	1	21
Anini	0.62	0	0.93	87	0.8	65
Okorokoro	2.12	197	0.64	281	0.001	28
Ofi	1.44	147	0.82	192	0.03	46
Apon	0.77	143	0.92	201	1	73
Orunro	0.79	101	0.92	315	2	66

[a] Q_0, discharge at beginning of recession ($l\ s^{-1}$); Q_t discharge at time *t* after Q_0 ($l\ s^{-1}$).
Source: After Adejuwon *et al.* (1983).

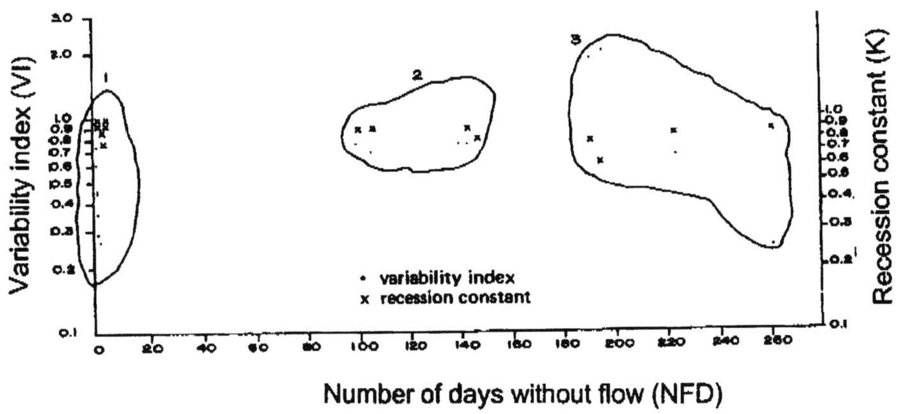

Figure 14.23 Groupings of the 15 selected third-order basins of the Upper Owena, Nigeria, based on the indices of discharge: variability (VI), recession constant (K) and number of days when no observable discharge occurred (NFD). (After Adejuwon *et al.*, 1983.)

Essentially, the work of Schellekens (2000) was undertaken at the S1 stage, but with ancillary data included as part of S2 and S3. *Significantly, whilst the basin is within the belt of 'perturbations in the easterlies' and occasional tropical cyclones (see Callaghan and Bonell, this volume) the QRR ratios are low and ranged from 0.02 to 0.28, with an average of 0.09 during the study.* Taking into account the very high interception losses of about 50% in this environment, the above QRR values double, with an average of 0.18. Recorded rainfall intensities are also comparatively low. For example, during a 66-day period the average intensity of 80 events was 3.0 mm h^{-1}, the average storm size was 10.7 mm and the average duration was 3 hr 43 min. The maximum storm of 227.5 mm was delivered over nearly 21 hours of which more than 120 mm fell in 3 hours. It was the latter storm that produced the highest QRR of 0.28 (0.56 of net rainfall). Thus despite the steep terrain (more than half of the catchment has slopes greater than 45%) these low QRR suggest that, based on the findings of other studies, overland flow is not extensive.

An innovative feature of Schellekens (2000) is that it is the first study in the humid tropics to present the results of a geophysical survey (Figure 14.25) and provides a valuable insight into the possible mechanisms of runoff generation. *In situ* weathering has proceeded for such a long time that the weathering front is 60 m deep in places. The most transmissive layer (of low

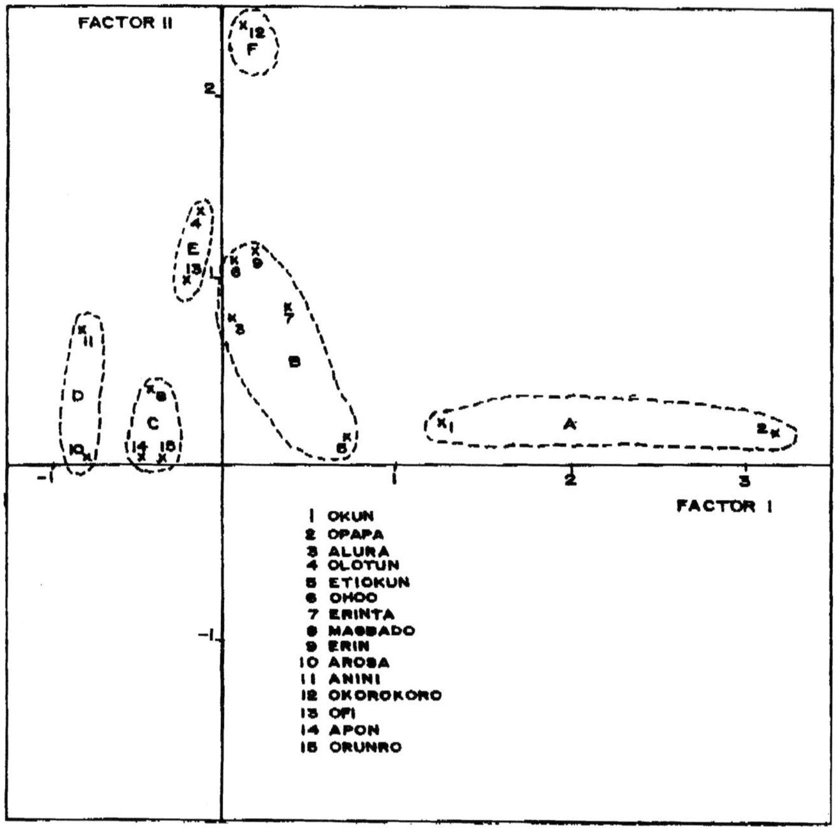

Figure 14.24 Grouping pattern of the 15 basins. (After Ogunkoya, 1988.)

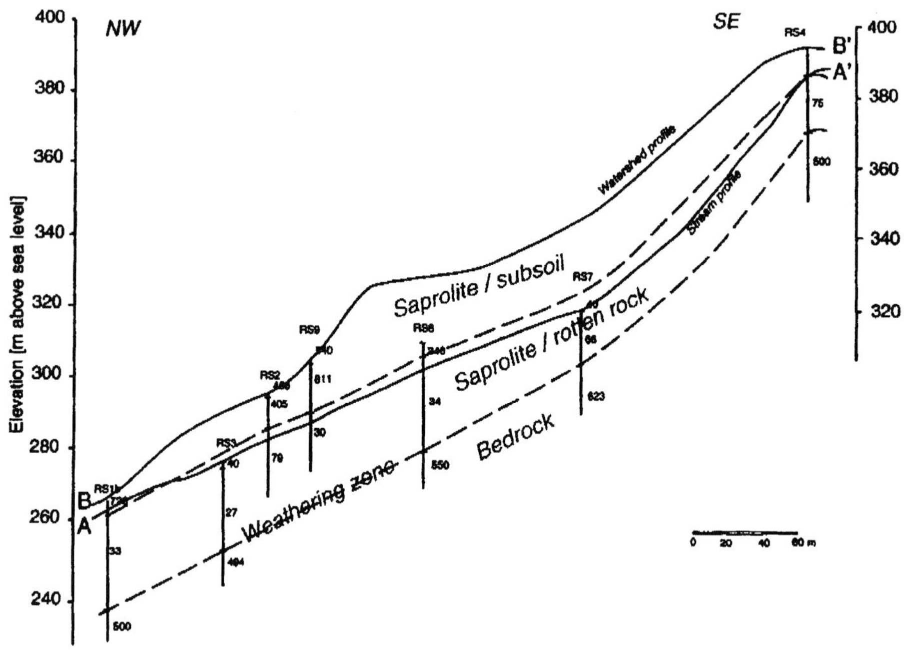

Figure 14.25 NW–SE profile of the subsurface of the Bisley II catchment showing the locations of Schlumberger resistivity soundings (indicated by vertical bars with their location code; the numbers along the bars are resistivities in Ohmm) and the interpreted layered structure as indicated by the dashed lines. The lower dashed line separates unweathered bedrock (bottom layer of high resistivity values) from a zone with low resistivity values, the rotten rock part of the saprolite; the contact between these zones indicates the zone of active weathering. The upper dashed line separates the low resistivity zone from the highly leached upper part of the saprolite, the subsoil. (After Schellekens, 2000.)

resistivity) occurs as a thin top layer (0.1–0.5 m thick) of mean K^*, 262.5 mm h^{-1} (based on the reverse auger hole method of van Beers, 1958) which grades down to 29.2 mm h^{-1} away from the surficial macropore networks. The C horizon consists of a 5–40 m thick layer of nutrient-poor, highly weathered, saprolite material of medium to high resistivity values. The K_{sat} determinations in the uppermost layer of this material show a low mean permeability of 1.7 mm h^{-1}. Other estimates by McDowell *et al.* (1992) in another part of the Bisley catchment reported a topsoil K_{sat} range of 42–833 m h^{-1} reflecting the surficial macroporosity. Well level recovery after pumping for the B and C horizons was in the K^* range of 0.38–3.8 mm h^{-1} which also points towards a relatively impermeable subsoil. The subsequent layer of less weathered saprolite material is 20–25 metres thick and is of lower resistivity (and presumably more transmissive). Significantly, the zone of active weathering has attained the local base of the stream channel. A consequence of this extensive and thick porous medium, of low permeability, is that the soil moisture is relatively low and no groundwater was found down to a depth of 8 m in boreholes. In effect, this layer insulates the lower stratigraphy from participating in the runoff generation process in the short term.

There was some independent support for the insulating role of the less weathered saprolite from collected hydrochemical data. The silicon Si^{4+} concentrations from runoff sources such as a gully (return flow), shallow groundwater in piezometers and soil matric water did not resemble the much higher Si^{4+} concentrations in stream water during baseflow conditions. These high Si^{4+} concentrations were considered to be the result of protracted percolation through the slowly 'permeable' saprolite of low resistivity, prior to lateral movement at the saprolite-fresh bedrock interface and entry into organised drainage, the latter intersecting this substrate. Schellekens (2000) thus argued that SSF in the topmost layer was the predominant pathway which emerged as return flow on the lower slopes. Direct measurements of this flow vector were not made but were inferred from a soil water modelling exercise into which field data could only fit with the inclusion of substantial lateral fluxes of soil water. In addition, the hydrograph of a tributary gully had close similarities with modelled amounts of lateral drainage.

Further supporting evidence of the inferred runoff generation process was provided by hydrochemical data and the application of the three-component separation model (De Walle *et al.*, 1988), although interpretations from the latter application are less reliable in the absence of a parallel or preceeding detailed hydrometric study, *cf.* La Cuenca, Babinda. For example, the various combinations of chemical constituent pairs (e.g. Si-K, K-Ca, Cl-Ca, Cl-Mg) indicated that there was more than one choice of possible end members. Particularly inconsistent were the estimates of proportional contributions to the storm hydrograph of soil water and return flow. Such difficulties may have been resolved had a detailed hillslope hydrometric experiment been undertaken *a priori* to better understand the system and establish the appropriate hypotheses. Nevertheless Schellekens (2000) concluded that:

(1) for the largest event (227.5 mm), the chemohydrograph is dominated by a different fast flow path (most likely SOF) than that found in most smaller events;
(2) soil water contributions to quickflow are much smaller than previously envisaged, when using return flow as an end member;
(3) two fast flow paths were identified, i.e. shallow SSF through macropores and SOF, both of which emerge as principal contributors to quickflow. The exact contribution of SOF could not be determined because it was not sampled directly, but the hydrochemistry analyses suggest that SOF becomes progressively more important in larger events (e.g. 227.5 mm) *cf.* the Babinda study. SOF was visually observed in the field during the largest storm, and at other times (Schellekens, 2000).

Western Ghats, India

Within the Western Ghats of Karnataka State, India, recent papers by Putty and Prasad (2000a, b) suggest that pipeflow and return flow is the dominant pathway in evergreen forests; although no detailed measurements at either the S1 or S2 stage (Elsenbeer and Vertessy, 2000) were presented to substantiate this reasoning. No detailed soil descriptions were provided but soil thickness extends up to 20 m on well-vegetated slopes which originated from a Precambrian formation with gneiss and intrusive granites being the principal rock types. Soils in the surface layer are usually sandy loams (Putty and Prasad, 2000b).

Putty and Prasad (2000a) depended on measured infiltration rates and the results from the installation of runoff collectors within trenches, plus a large number of vertical faces of soil exposed at road cuttings near valley bottoms (Putty and Prasad, 2000a). No overland flow was observed on the slopes, although the occurrence of SOF in the riparian valley bottom areas was still considered very important following the Dunne and Black mechanism (Dunne and Black, 1970a, b). The lack of overland flow was attributed to the very high infiltration capacities of the forests (61–780 mm h^{-1}) which far exceed the maximum short-term intensities (*c.* 60 mm h^{-1}). Moreover, no SSF was collected from the soil matrix. Most SSF emerged from pipes varying in diameter from a few centimetres to more than one metre and which drained more than 12 m of soil mantle on the forested slopes. Significantly, Putty and Prasad (2000b) consider that *dynamic subsurface saturated zones* exist which may contribute substantial quantities of quickflow via pipeflow in addition to SOF from contributing areas riparian to stream channels. Elsewhere in the humid temperate forests of the Oregon Coast Range, *a subsurface variable source area contribution to runoff generation* has also been identified (Anderson *et al.*, 1997; Montgomery *et al.*, 1997; Torres *et al.*, 1998; Montgomery and Dietrich, 2002). At the exit of these pipes, return flow (Putty and Prasad, 2000a, referred to this mechanism as 'pipe overland flow') was considered to be the principal mechanism for surface runoff because piezometer water levels were more than 0.8 m deep (although no specific details of the piezometer responses were given) within the lower slope, riparian zones. They also observed that return flow did not occur extensively on the slopes within a small forested basin (8 ha, annual rainfall 6750 mm) because the transmissivity of the forest soils are so high that pipe outflow does not easily induce saturation of the soil. The storm hydrograph was thus considered to be an integration of contributions from return flow ('pipe overland flow') and SOF, pipeflow and delayed flow. Subsequently, Putty and Prasad (2000b) incorporated a pipeflow component along with SOF during the course of modelling rainfall-runoff satisfactorily within seven drainage basins which ranged from 4.5 to 600 km^2 in the Western Ghats. These basins incorporated the spectrum of forest types and other land covers associated with the Western Ghats escarpment.

A current survey within the Western Ghats of soil hydraulic properties (disc and Guelph permeameters) across different land covers in the Uttar Kannada district of Karnataka by Purandara *et al.* (2004, unpublished data, National Institute of Hydrology/Karnataka Forests Dept/UNESCO project) provides more data at the S1 stage to complement Putty and Prasad (2000a, b). Within an evergreen forest (Dodmane) of light textured soil, log mean K^* measurements taken at the surface (62 mm h^{-1}), 0.1 m depth (38 mm h^{-1}) and 0.45–0.60 depth (76.6 mm h^{-1}) confirm the high transmissivities of the upper evergreen forest mantle in this region. At 1.5 m depth however, K^* declines to 7.5 mm h^{-1}, thus the occurrence of SSF at this depth cannot be ruled out.

In agreement with Putty and Prasad (2000a, b), the analysis of rainfall-frequency-duration for selected stations within the Uttar Kannada region by Purandara *et al.* (2004, unpublished data) also shows that short-term rain intensities are comparatively low for a high annual rainfall area. Computed one-hourly rainfalls are within the range between *c.* 30–70 mm h^{-1} for progressive return periods from 2–100 years respectively. Thus despite annual rainfall exceeding 6000 mm along the Western Ghats escarpment, of which more than 90% is confined to the four monsoon months, an average number of rainy days of 120–140 per year, as well as high daily values recorded (see discussion of rainfall characteristics of the Western Ghats region by Bonell *et al.*, this volume), short term intensities are, by contrast, remarkably low (Putty and Prasad, 2000b). As Putty and Prasad (2000b, p. 216) noted, a major proportion '. . . of rainfall is contributed by four to five spells each lasting 8–10 days but relatively moderate rainfall intensities result because of the very long duration of events. Thus 15-minute intensities seldom exceed 80 mm h^{-1} and contribute about 2% of the annual rainfall, while hourly intensities of 60 mm h^{-1} contribute less than 1% of the annual rainfall.' These comparatively low application rates (cf. cyclone-prone areas, Bonell *et al.*, this volume) in combination with transmissive upper soil mantles within evergreen and semi-evergreen forests indicate that SSF (including pipeflow) is the more likely dominant storm pathway.

SATURATION OVERLAND FLOW

With short-term rain intensities at least one order of magnitude higher compared with those of humid temperate environments, coupled with a rapid decline in K^* in subsoils, leads to the frequent occurence of SOF in selected tropical forests. Reports of persistent SOF occurrence have emerged from both cyclone-prone and some non-cyclonic (convective) areas in northeast Australia (Babinda) and Peru (La Cuenca) respectively. Further, as both studies have followed the systematic S1 to S3 approach of Elsenbeer and Vertessy (2000), including complementary hydrometric and hydrochemical studies, these projects provide an opportunity for a more detailed comparison.

South Creek, northeast Australia

The Babinda study initially launched a prolonged hydrometric phase, assessing the water balance of undisturbed and forest converted to pasture using the traditional paired catchment study approach (Gilmour, 1975, 1977), set within the wet tropics of northeast Queensland (Figure 14.26), namely South Creek (undisturbed, 25.7 ha) and North Creek, (disturbed, 18.3 ha of which 12.8 ha was previously logged and cleared, 1971–1973). The highly responsive storm hydrographs, high quickflow response ratios and statistically weak differences in peak hydrograph discharges between converted forest and the control catchment led to the initiation of a hillslope-hydrometric campaign (Bonell and Gilmour, 1978; Bonell *et al.*, 1979, 1981, 1983a, b; 1987; Gilmour *et al.*, 1980) which was supplemented by tritiated water tracing experiments (Bonell *et al.*, 1982, 1983a, 1984). Subsequently, a complementary hydrochemistry phase was undertaken using both environmental isotopes (Bonell *et al.*, 1998) and chemical species (Elsenbeer *et al.*, 1994a; 1995a).

Figure 14.26 shows the experimental design used for the combined hydrometric-environmental isotope (deuterium) campaign (Bonell *et al.*, 1998). Other sources for the earlier hydrometric experimental plans using runoff troughs and *in situ* determination of soil hydraulic properties can be found in Bonell *et al.* (1981, 1987) and Gilmour *et al.* (1980). Some of the experimental plans for the tritiated water tracing and the use of chemical species (non-isotopes) as environmental tracers will be discussed later (Elsenbeer *et al.*, 1994a, b; 1995a).

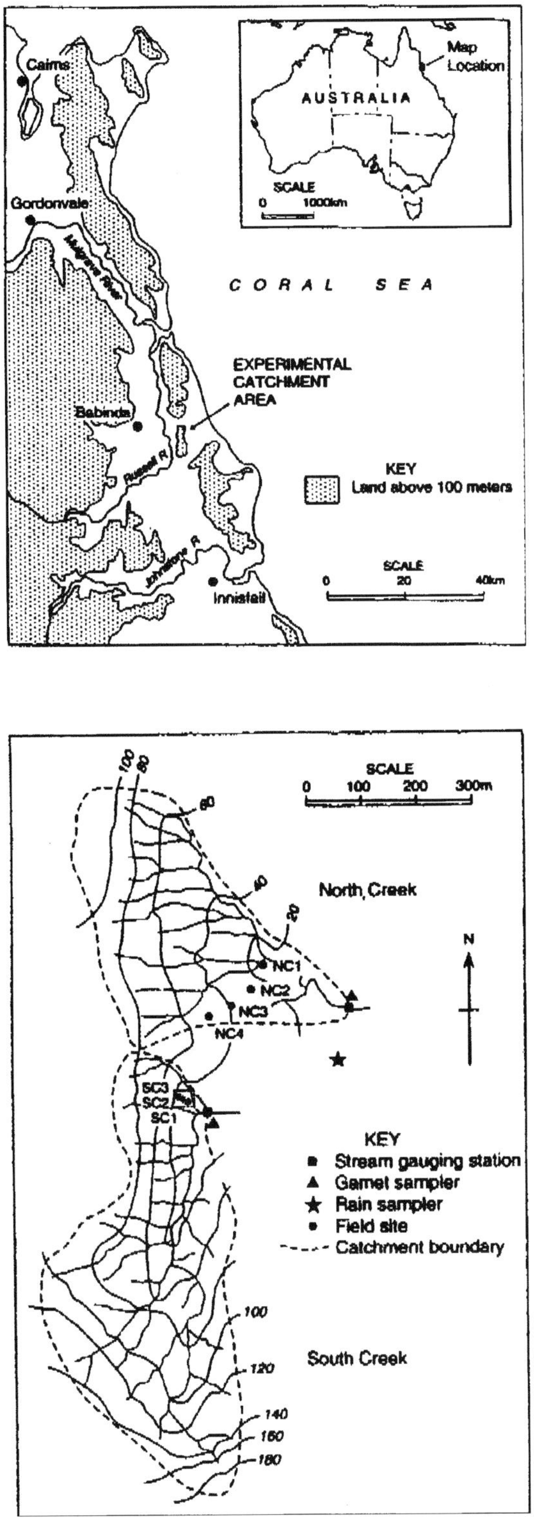

Figure 14.26 (a) The physical setting and location of the paired catchment study near Babinda. (b) The experimental plan in North and South Creek. (After Bonell *et al.*, 1998.)

The underlying geology of the catchments consists of basic metamorphic rocks which belong to the Babalangee Amphibolite of De Keyser (1964). The resulting soils are complex and incorporate haplothox and tropeptic haplothox of the Soil Survey Staff (1975) or the Bingel, Galmara and Bicton series of CSIRO (Murtha *et al.*, 1996; Red Kandosol, Isbell, 2002). Elsenbeer *et al.* (1995a) generalised the steep side slopes of South Creek as Cambisols with Ferralsols (Inceptisols with Oxisols, USDA) prevailing on the small interfluve areas. These soils are dominated by kaolin-dominated silty clay loam to clay soils which may continue to 6 m depth.

A significant feature of the Babalangee Amphibolite is that it exhibits a strong and complete schistosity with a coarse grained texture (Arnold and Fawckner, 1980) so that the kind and degree of deformation (i.e. schistosity) may affect the dominant hydrological pathways. For example, on the upper slopes of South Creek the schistosity of the weathered rock is preserved in sections of the deep soil profiles, notably at the headwaters of first-order streams. Such areas provide localised 'conduits' of high soil hydraulic conductivities of up to two orders of magnitude greater than the surrounding soil matrix which in turn facilitates preferential soil water (or groundwater) movement within the differentially weathered schist (Bonell *et al.*, 1984). These exposures of schistosed material are represented by 'seepage' areas which persist during the wetter parts of the year (see plate 2.4 of Gilmour *et al.*, 1980) and contribute to maintenance of surface discharge in the first-order streams. The exact role of these 'seepage' areas within the runoff generation process could not be ascertained, despite a hydraulic conductivity survey across both South and North Creek (Bonell *et al.*, 1987, 1991). Most of the field tests of K^* undertaken in the catchments were in the same order of magnitude as those determined for the runoff and artificial tracing plots. Nonetheless the 75 m grid used in this K^* survey may have been too coarse in scale to detect the potential subsurface flow contributions of this schistosed material to the later stages of the storm hydrograph. The occurrence of pipeflow has also been noted in South Creek but qualitative observations suggest that the network of pipes is less dense than that observed in the La Cuenca study (Elsenbeer and Lack, 1996a, b).

The Babinda study was located within a high rainfall environment (annual rainfall 4009 mm, 1970–1983) with a marked concentration (63% annual rainfall) occurring between December and March in association with the 'monsoon' season when tropical cyclonic perturbations occur. Analyses of rainfall and runoff characteristics over the period January 1974 to May 1981 were undertaken which are highly relevant to the ensuing process hydrology descriptions and also mirror the seasonal change in synoptic climatology and associated hillslope hydrology (Howard, 1993). As part of this analysis, the quickflow component of the storm hydrograph was subdivided into the following categories, *viz*, 2–5 mm,

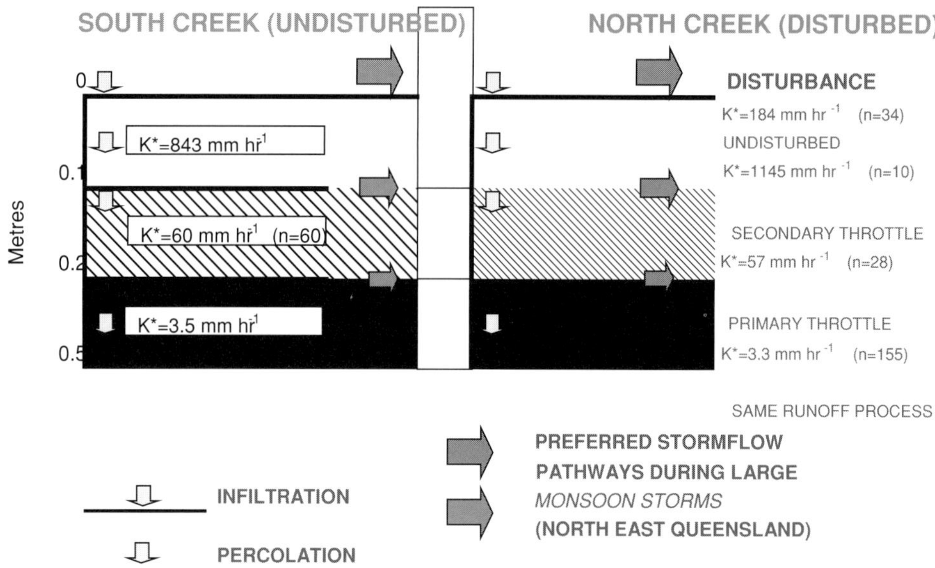

Figure 14.27 Dominant pathways of storm runoff in the Babinda catchment, linked with field saturated hydraulic conductivity, K^*. (After Bonell, 1991.)

5–10 mm, 10–50 mm, 50–100 mm, and greater than 100 mm, using the hydrograph separation technique of Lyne and Hollick (1979). During the monsoon season, the average maximum 6-min rain intensities by event I_6 were in the range 14–82 mm h^{-1} across the preceding spectrum of quickflow classes. Total rainfall by event, in excess of 250 mm, is common (Howard, 1993). The rainfall intensity-frequency-duration for Babinda, reported elsewhere in this volume (see Bonell *et al.*, this volume) refers to this season.

Subsequently, rainfall intensities (and corresponding quickflow amounts) relax in the 'post-monsoon season' (April-June) which accounts for about 21.5% of the annual rainfall. The corresponding average maximum I_6 values by event are in the range 14–27 mm h^{-1} (for the quickflow categories 2–5 mm, 5–10 mm and >10 mm), with the maximum being 119 mm h^{-1}. The average total rainfall by event also declines and for the three quickflow categories (2–5 mm, 5–10 mm, >10 mm) was in the range 7.6–41.5 mm (Howard, 1993). There is a further marginal reduction in average I_6 in the 'winter' dry season (July–September) (15 and 25 mm h^{-1} for the respective quickflow categories 1–5 mm and >5 mm), but the average total rainfall by event remains similar to the post-monsoon season being 10.5 and 43.2 mm for these respective quickflow categories (Howard, 1993). During October through to December occasionally 'pre-monsoon' events occur to break the protracted dry season. At such times shorter-term intensities (I_6) can attain monsoon levels up to 110 mm h^{-1} or occasionally higher (Gilmour *et al.*, 1980).

Thus four meteorological (rain-type) seasons distinguish this study, namely monsoon, post-monsoon, winter 'dry' season and pre-monsoon (Bonell and Gilmour, 1980), each of which impresses a temporal change in the hillslope hydrology. There is also a progressive adjustment in the quickflow response ratios, QRR (total quickflow/total precipitation, QF/P). During the monsoon season, for quickflow volumes in excess of 250 mm per event, the median QRR is 57% for the undisturbed forest catchment and 55.5% for the disturbed forest basin. Overall, the QRR is 50.5% (forest) and 54.5% (disturbed) for storms with quickflow exceeding 100 mm for the larger monsoon season storms. There is a progessive reduction in median QRR for smaller quickflow volumes during smaller rain events of the monsoon season (e.g. for the undisturbed forest, QF 2–5 mm, 7.05%; 5–10 mm, 12%; 10–50 mm, 25%; 50–100 mm, 25%). Quickflow volumes in the subsequent 'post-monsoon' and 'dry' season reduce in line with diminishing rainfall intensities, so that in the undisturbed forest the median QRR does not exceed 22% as, for example, for quickflow volumes in excess of 10 mm during the post-monsoon season. For smaller quickflow volumes of less than 10 mm, the QRR are in the same order of magnitude as those for the smaller events of the monsoon season (Howard, 1993).

The runoff generation process for the Babinda catchments using data from the hydraulic conductivity survey is summarised in Figure 14.27. The high prevailing rainfalls of the monsoon season exceed the transmission capacity of the subsoil (away from the organic-dominated, surficial layer). Moreover, the persistent near-positive matric potentials until the commencement of the winter 'dry' season (Figure 14.28) results in the frequent occurrence of SOF when short-term rainfall intensities exceed the combined

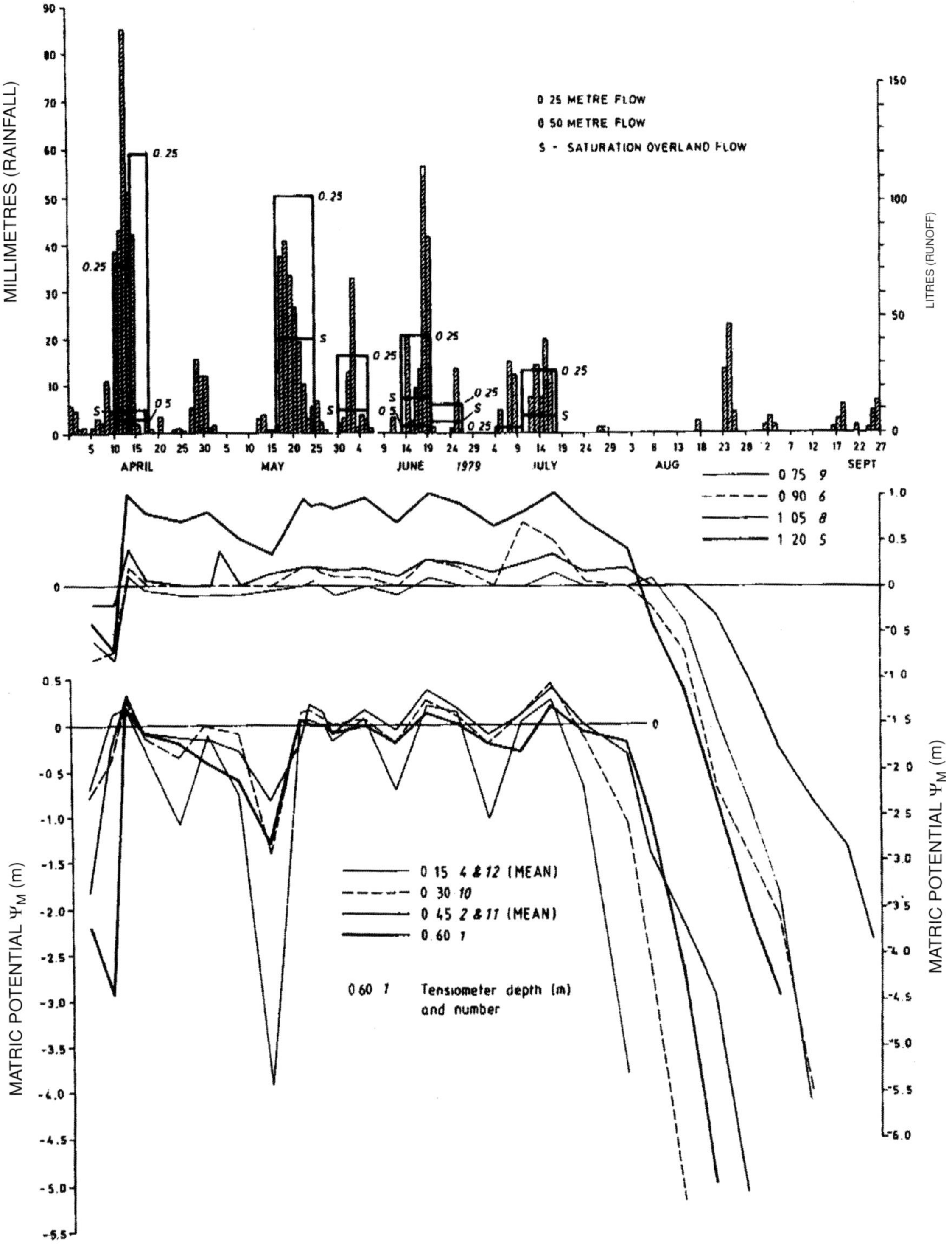

Figure 14.28 The rainfall, hillslope runoff and matric potential (at lower slope tracing site 1b; see Bonell *et al.*, 1981, 1983s for location) for the period April–September 1979 during the post-monsoon and the opening stages of the winter 'dry' season, 1979. (After Bonell *et al.*, 1983a.)

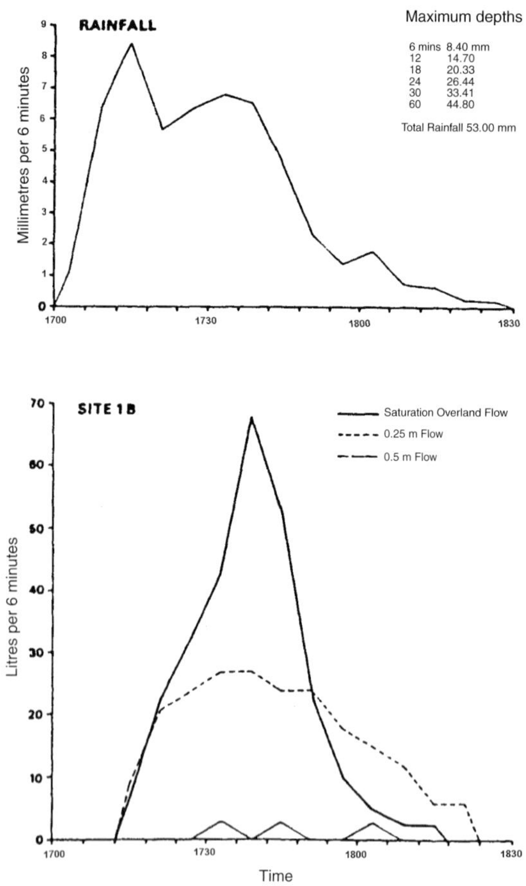

Figure 14.29 The summer response of site 1b to a summer monsoon storm. (After Bonell *et al.*, 1981.)

threshold of the subsoil K^* and the available water storage capacity of the upper transmissive layer (0–0.2 m depth).

Unbounded runoff plot responses (Figure 14.29) show that SOF is concentrated near the rainfall intensity peaks. As short-term rainfall intensities decline, SSF prevails as is particularly apparent from the post-monsoon season onwards where Inceptisols dominate on the lower slopes (sites 1A and 1B, Figure 14.30). Towards the dry season, the dominant pathway is SSF, with the exception of the Ferralsols (Oxisols) on the interfluves where SOF can still occur (site 2, Figure 14.31). Total runoff plot contributions are minimal, however, in comparison with the monsoon and post-monsoon seasons.

Whilst the contribution of pre-monsoon storms to the storm hydrograph on an annual basis is negligible, it is significant that the dominant pathway changes from SOF to SSF, even though short-term rainfall intensities can be high (Figure 14.32). A combination of large negative matric potentials and enhanced macroporosity, resulting from the dry state of the kaolin-dominated soils, permits the occurrence of 'short-circuiting' (Bouma and Dekker, 1978) or rapid by-pass flow (Foster and Smith-Carrington, 1980).

The application of time series methods to selected monsoon (Bonell *et al.*, 1979) and post-monsoon storms (Bonell *et al.*, 1981) showed the high responsiveness of South Creek. Using 6-min time increments, maximum cross-correlation between rainfall and SOF was about 6–12 mins, 12 min for SSF between the surface and 0.25 m depth, and 24 min for stream discharge.

The preceding hydrometric approach enabled the appropriate hypotheses to be made and provided the means to interpret a complementary hydrochemistry campaign which was undertaken in the early 1990s. The use of the environmental isotope deuterium (Bonell *et al.*, 1998) and several environmental chemical species (Elsenbeer *et al.*, 1994a, 1995a) confirmed the domination of 'new' or 'event' water during high rainfall intensity events of the monsoon season. Through the application of a modelling approach to storms of different magnitude, Barnes and Bonell (this volume) show that the contributions of 'new' water is highly sensitive to rainfall intensity. As the latter declines temporally during the monsoon season (or more persistently during the post-monsoon season), the volumes of 'new' water decrease in line with a corresponding decrease in SOF occurrence, as was shown from the hillslope runoff plots.

Elsenbeer *et al.* (1994a, 1995a) were also able to demonstrate the sensitivity to rainfall intensity of 'new' water contributions within the chemohydrograph on the basis of two contrasting events (high-intensity, low-intensity). These storms were separated by only five days during February 1993 (see Table 14.7), with the lower intensity event associated with a higher antecedent catchments wetness. In addition to sampling rainfall, streamflow, soil water and groundwater using the experimental design outlined in Bonell *et al.* (1998), supplementary grab samples were taken from both an incised concentrated flow-line and an ephemeral flow-line without a defined channel, to assess the contributions of 'new' and 'old' water to overland flow (Elsenbeer *et al.*, 1994a). Additional spot checks were made in several poorly defined rills. The chemical composition of 'old' water was determined from delayed flow samples collected during February 1993.

Whilst the rainfall characteristics of the two events differed considerably, the stormflow chemistry in South Creek was characterised by a sharp decrease in Ca^{2+}, Mg^{2+}, Na^{+}, Si^{4+}, Cl^{-}, Ec, ANC, alkalinity and total inorganic carbon. In contrast, pH remained nearly constant with discharge whereas K^{+} increased. Figures 14.33 and 14.34 illustrate the temporal changes for Ca, Mg, Na and SiO_2 concentrations for both events. The gully A (Figure 14.35) is the ephemeral flow-line, supplemented by additional mean concentrations in other ephemeral gullies C, D and E. The gully B is the incised, intermittent gully where sampling took place near its confluence with South Creek.

Significantly, the temporal variations in the chemical species for event 1 in Gully A (and C, D, and E) closely match those of the stormflow pattern for South Creek. The shallow nature of

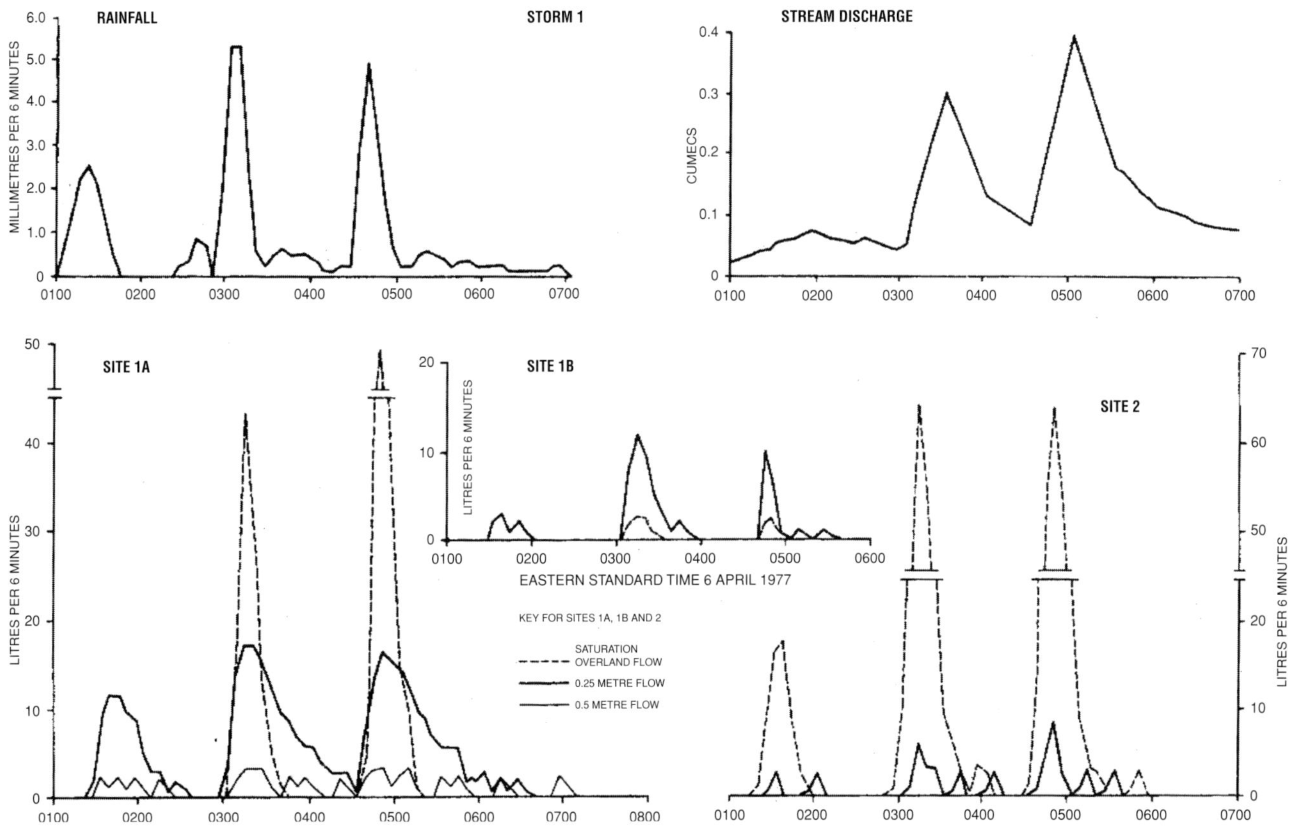

Figure 14.30 The continuous record for rainfall, saturation overland flow, subsurface flow and stream discharge for storm 1, 6 April, 1977. (After Gilmour *et al.*, 1980.)

these ephemeral gullies means that they are recipients of principally SOF (supplemented by exfiltration SSF) and thus 'new' water. In consequence, the stormflow hydrograph for the high intensity event is dominated by 'new' water. Whilst Gully B also behaved like South Creek, the 'new' water signal was more dampened because the incision is much greater than in gully A so that the contribution from SSF is more enhanced. In contrast, event 2 shows that the initial 'new' water signal in gully A was subsequently overwhelmed by a different signal which was attributed to the rapid drainage of the perched water table by SSF. Thus, despite higher antecedent moisture conditions prior to event 2, the contributions of new water are less *because of the overall lower rainfall intensities and corresponding peak discharge* in South Creek (404 l s^{-1} as against 1841 l s^{-1} for event 1). It should be noted that in event 2 the arithmetic mean flow of gullies C, D and E are less closely aligned with gully A, presumably because these gullies are much less incised and so with lower rain intensities, SSF is more prevalent in gully A.

The work described above was extended by Elsenbeer *et al.* (1995a) to achieve chemohydrograph separation through the adaptation of end-member mixing analysis (EMMA) to accommodate overland flow. The mixing analysis was based on potassium (K^+) and the acid-neutralising capacity (ANC) because they provided the best separation of the potential sources. Initially, soil-water samples taken from depths 0.3, 0.6 and 1.2 m were used, narrowed down subsequently to data from 0.6 m. The principal challenge was to distinguish between true SOF generated at the surface (event water) with high concentrations of K^+ and low ANC following unchannelled (non-incised) pathways, from SOF more mixed with SSF in incised pathways which conversely would produce a lower K^+ signal (concentration) and higher ANC. Use was made of data from Gully C (OF2) to represent SOF in a non-incised pathway and from Gully A (OF1) which had a degree of incision (but not on the scale of Gully B), to represent the incorporation of some contributions from SSF (see Figure 14.36).

Figures 14.36 and 14.37 show the event and pre-event basin chemistry for both storms. For event 1, the end-member SOF is poorly characterised because it is too low in K^+. On the basis of previous discussion the idealised overland flow end-member (Of_{id}) must be more aligned with peak stormflow in South Creek. Consequently, a triangle formed by HGW, SW60 and Of_{id} (based on the highest stormflow K^+ concentration) encompassed nearly

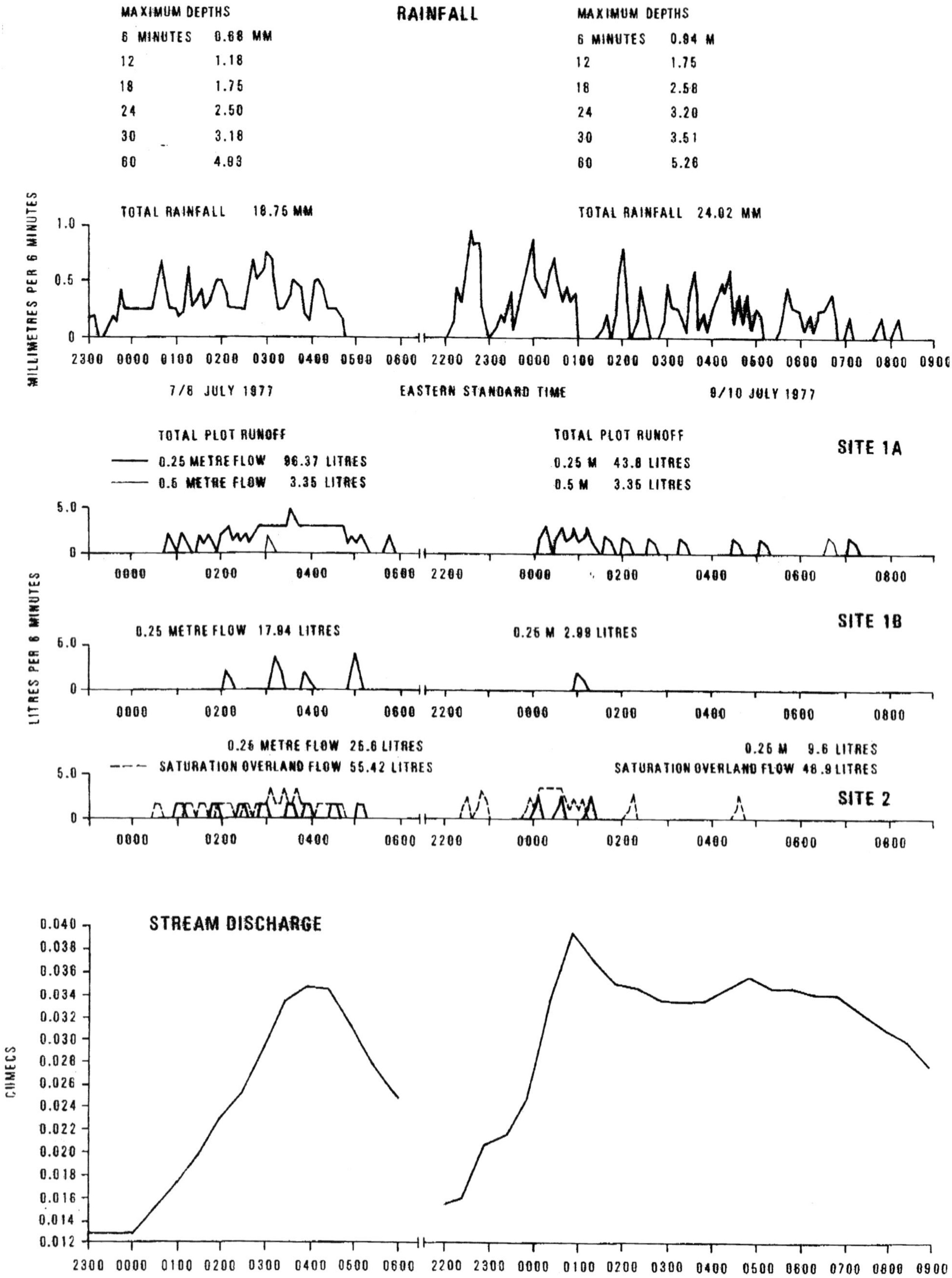

Figure 14.31 The continuous record for rainfall, saturation overland flow, subsurface flow and stream discharge for winter storms on 7/8 and 9/10 July, 1977. (After Gilmour *et al.*, 1980.)

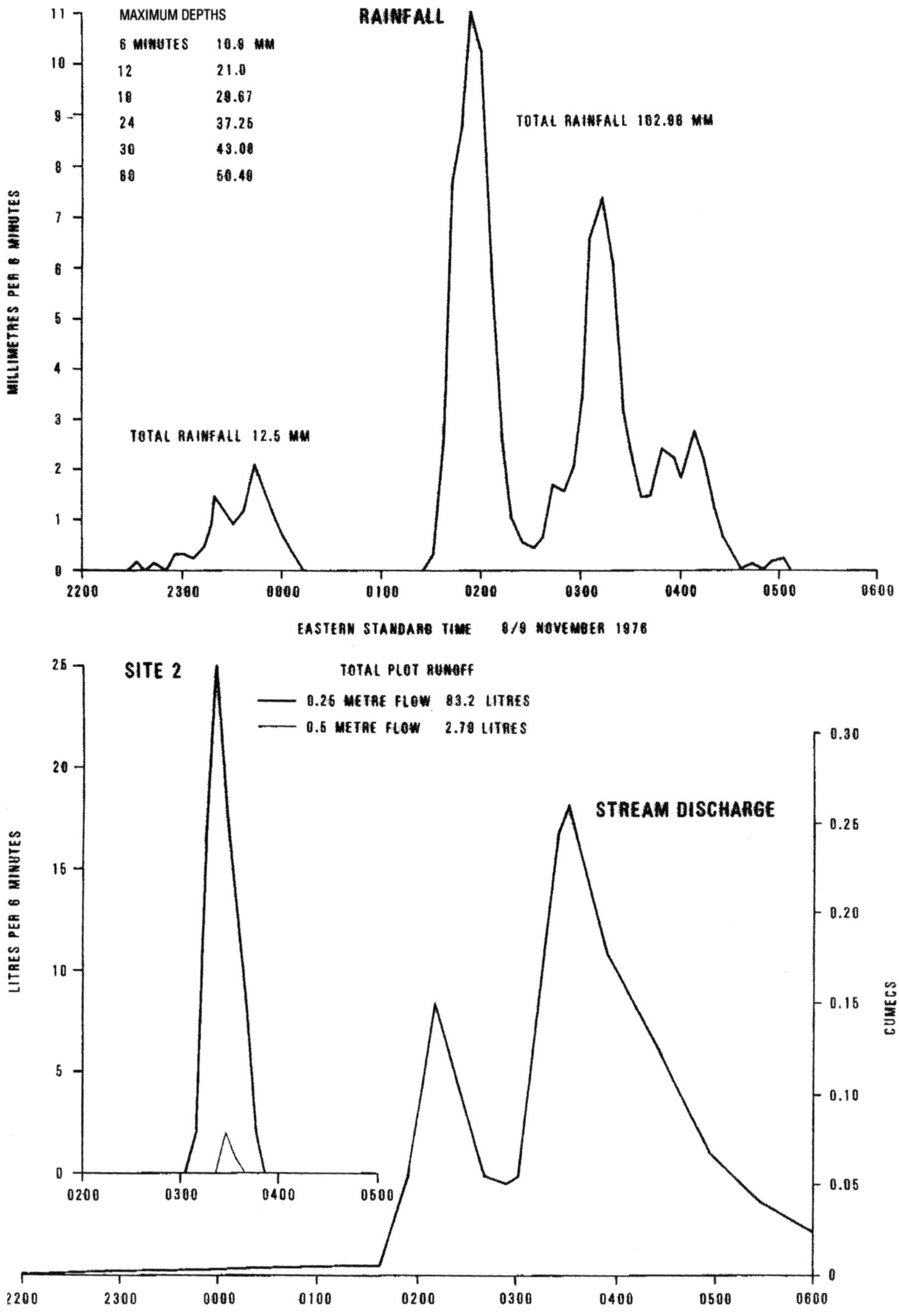

Figure 14.32 The continuous record for rainfall, subsurface flow and stream discharge for pre-monsoon 'transitional' storm on 8/9 November, 1976. (After Gilmour *et al.*, 1980.)

Table 14.7. *Precipitation characteristics of the events of 18 and 23 February 1993*

Variable	Event 1	Event 2
Magnitude, mm	177.7	44.2
Duration, hours	18.1	3.0
I_6max,[a] mm h^{-1}	90.0	63.0
I_{10}max,[a] mm h^{-1}	79.8	60.0
I_{30}max,[a] mm h^{-1}	57.6	45.0
I_{60}max,[a] mm h^{-1}	50.7	27.8
I_{120}max,[a] mm h^{-1}	45.5	17.5

[a] These are maximum intensities, with the subscripts referring to the time period in minutes over which they were evaluated.
Source: Elsenbeer *et al.* (1995a).

all the streamflow and overland flow from OF1 (Gully A) much better. It is evident that the selection of the soil water source is less critical. The same procedure was followed for event 2 and, with the use of Of_{id}, all samples are incorporated in the triangle.

Subsequently, a three-component mixing model was used based on the contributing sources HGW, SW 60 and SOF (representing both OF2 and Of_{id}) (Figures 14.38 and 14.39). Comparisons can be made between the use of inferred 'idealised' SOF composition (Of_{id}) and OF2. Although the former yields a much higher contribution of soil water at the expense of SOF, overall the patterns are similar from the use of either representative of the surface pathway. Consequently Of_{id} is selected.

When considering the bottom panels of Figures 14.38 and 14.39, the contribution of SOF reaches almost 80% near peak flow, whilst the concurrent groundwater contribution is nearly extinguished. The soil water contribution reaches a maximum (*c.* 40%) midway on the hydrograph recession limb and is nearly equal to the groundwater contribution. For event 2, the contribution of SOF does not exceed 60% and that of HGW does not fall below 25%, in contrast to event 1. Most significantly, soil water contribution is very pronounced on the rising limb of the storm hydrograph.

Thus, overland flow sampled from within an incised flow-line such as Gully A (OF1), whilst a fast pathway, is composed of both event and pre-event water and not event *per se* as commonly assumed. By pooling hillslope groundwater and soil water contributions to represent SSF of pre-event water, Elsenbeer *et al.* (1995a) then produced a separation between SOF and SSF. Predictably, SOF dominated event 1 whereas the contribution of SOF for the low rain intensity event 2 was much less. These findings are in line with the isotope modelling reported elsewhere (Barnes and Bonell, this volume). Further, within event 2, pre-event subsurface sources were mobilised more quickly to contribute to the rising limb as SSF in line with the earlier hillslope runoff (hydrometric) data. The fact that checking this analysis using SW30 and SW120 produced similar results to SW60 raises the issue as to whether the whole vadose zone contributes to stormflow or the use of mixing analysis alone does not define precisely all the contributing sources (due to the weak chemical differentiation between the soil water end-members). The latter still remains an issue under debate. Bonell *et al.* (1998) identified up to five sources contributing to stormflow, including those identified by EMMA and the three-component mixing model. The earlier hydrometric hillslope runoff study suggested that most SSF takes place in the top 0.25 m whilst volumes from lower depths (at 0.5 m and 1.0 m) were minimal (Bonell and Gilmour, 1978). This supports the concept of SW30 as a source of stormflow. On the other hand, upward fluxes of soil water during storms as detected by tensiometry and tritiated water tracing experiments, shortly to be outlined, indicate the complexity of subsurface soil water movement and thus add to this uncertainty.

The significant contributions of soil water to the storm hydrograph are established nonetheless, and the proportions outlined by Elsenbeer *et al.* (1995a) are in line with those from humid temperate forests in the south-eastern USA reported by Mulholland (1993) (Walker Branch) and Bazemore *et al.* (1994) (Shaver Hollow). The principal differentiating factor is the very much higher rainfall intensities, especially identified with event 1 (a typical monsoon storm), which ensures the domination of SOF.

Whilst the subsequent hydrochemistry campaign confirmed the general understanding of the runoff generation process based on the previous hydrometric studies, the position with regard to the role of the permanent groundwater is still evolving. Initially, the water table was monitored in two wells near an upper slope runoff plot (site 2, Bonell and Gilmour, 1978). The profusion of rocks on the lower slopes prevented hand augering to access the water table in the more incised part of South Creek. Figure 14.40 shows the seasonal change in the water table over two years based on weekly measurements. These spot measurements had suggested that the water table remained too deep to participate in the runoff generation process. The measurement regime included a period of heavy, persistent rain in February 1977 when water levels remained below 1m depth. In contrast, piezometers located at 0.5 m, 1.0 m, 2.0 m and 3.0 m depths remained dry for the most part, except during storms when the upper cavities at 0.5 m and 1.0 m depth were particularly responsive (Figure 14.41), with maximum cross-correlation (lag response) between rainfall and piezometer responses (using 6 min time units) ranging between 18 and 54 minutes for the events shown in Figure 14.41. The 0.5 m piezometer lag responses (24–36 mins) were on par with, or slower than, the stream discharge responses. Moreover, the deeper 1 m cavities had longer time lags (up to 54 mins). In addition, statistical analyses produced no evidence to support the hypothesis that the initial and maximal storm response times were faster at

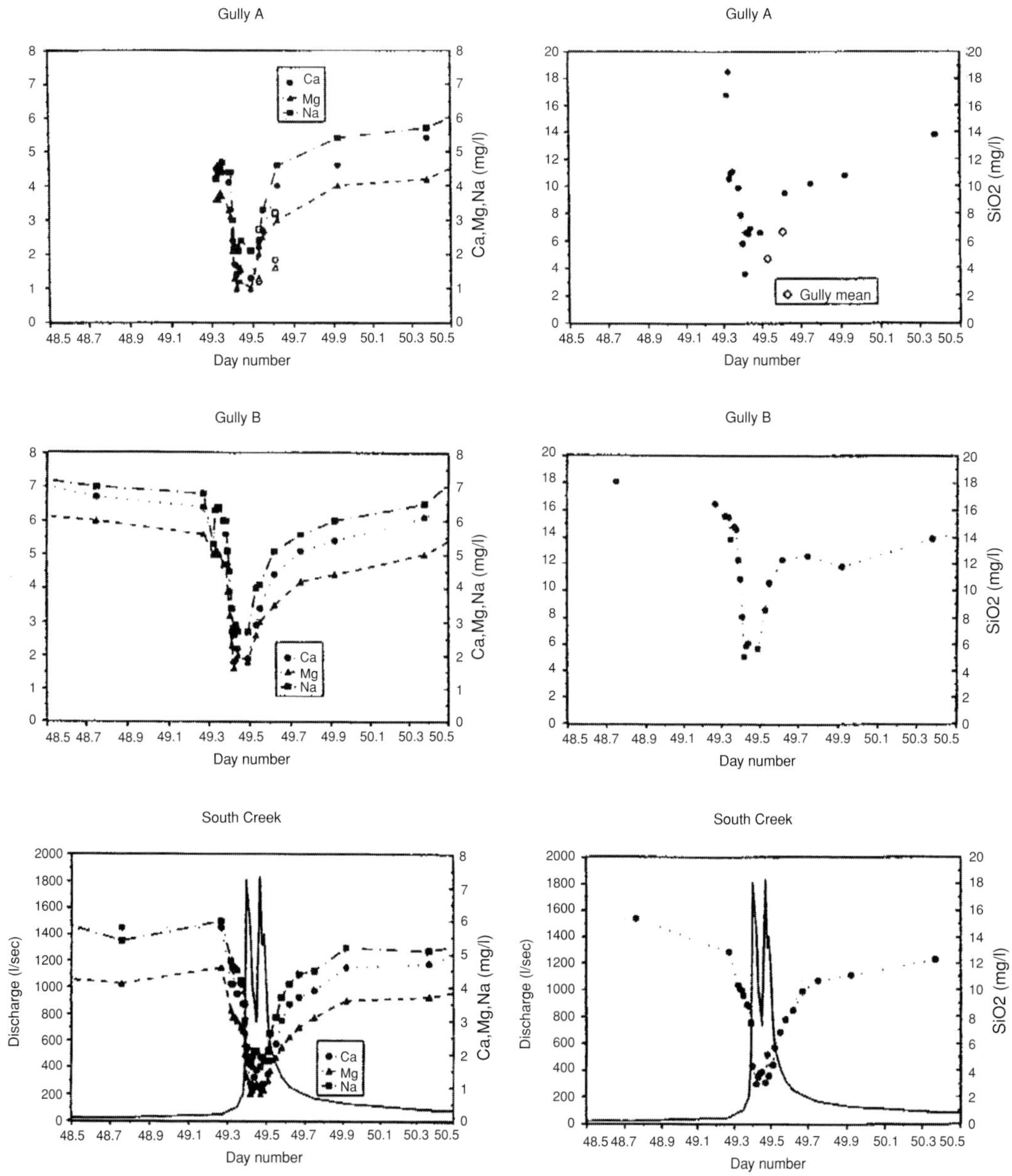

Figure 14.33 South Creek discharge (solid line, bottom), calcium, magnesium, sodium and silica concentrations in gully A (top), gully B (middle) and South Creek (bottom) during event 1. The open symbol in the top panels represents the mean concentration of gullies C, D and E. (After Elsenbeer *et al.*, 1994a.)

the bottom of the slope than at the top. Thus at this sub-hillslope scale, the preceding lag responses suggest that near-instantaneous saturation occurs, irrespective of topographic position (Gilmour *et al.*, 1980; Bonell *et al.*, 1981).

The focus of attention in the early 1980s centred otherwise on the mechanisms of vertical recharge rather than surface-deeper subsoil water interactions. A series of artificial tracing experiments using tritiated water coupled with measured matric potentials from tensiometry did suggest appreciable lateral downslope movement, as well as vertical movement, by a combination of interstitial piston flow (Foster and Smith-Carrington, 1980) and preferential flow (Beven and Germann, 1981) occurring simultaneously. A line injection of tritiated water at 0.2 m depth (near the termination of rapid SSF) was monitored both vertically and laterally by soil water extractors installed 2 m apart (lower slope runoff plot 1b, in Bonell and Gilmour, 1978; Bonell *et al.*, 1983a) and 0.30 m apart (a new plot, north of upper slope former runoff plot 2 (Figure 14.42a); Bonell *et al.*, 1982; 1984). The high initial activity in the more shallow soil water extractions and the centre of mass (calculated using Zimmerman *et al.*, 1967) on 3 March 1980 (Table 14.8) were indicative of preferential flow. The progressively slower, apparently vertical movement (as shown by the centre of

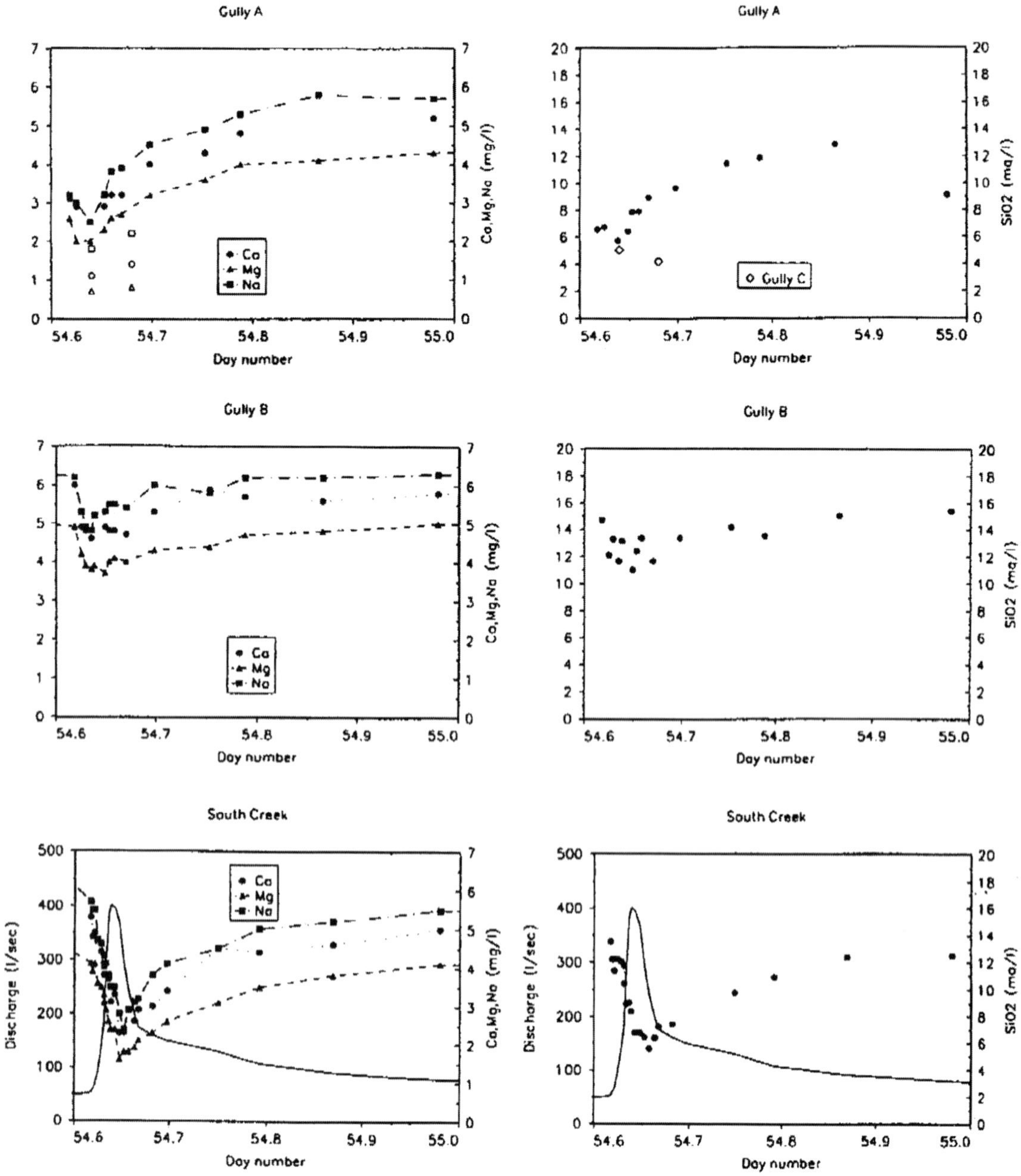

Figure 14.34 South Creek discharge (solid line, bottom), calcium, magnesium, sodium and silica concentrations in gully A (top), gully B (middle) and South Creek (bottom) during event 2. (After Elsenbeer *et al.*, 1994a.)

mass at subsequent times), is the result of slower interstitial piston flow. An additional cause for the slow vertical translation of the tritiated pulse over the experimental time period is the occurrence of upward movement of soil water at both tracing sites in different parts of the subsoil during storm events (Figure 14.42b). Within the framework of this complex three-dimensional flow pattern, the convergence of soil water from downward percolation with a counter upward movement provided the basis for lateral movement into or out of the cross-section. At that time though (in the early 1980s), it was not clear whether this was either an on-site (localised) phenomenon or that this counter upward movement mechanism occurred more extensively within this hydrological system. Moreover it was then uncertain what were the implications on the runoff process of this counter upward movement.

Another interesting feature was that similar amounts of lateral recharge (assuming the piston flow model of Zimmerman *et al.*, 1967) were required under prevailing monsoonal rainfalls to 'push' soil water downslope and cause a peak response at 0.35 m and 0.50 depths on the lowest soil water transect (Table 14.9), despite differences in distance between the points of

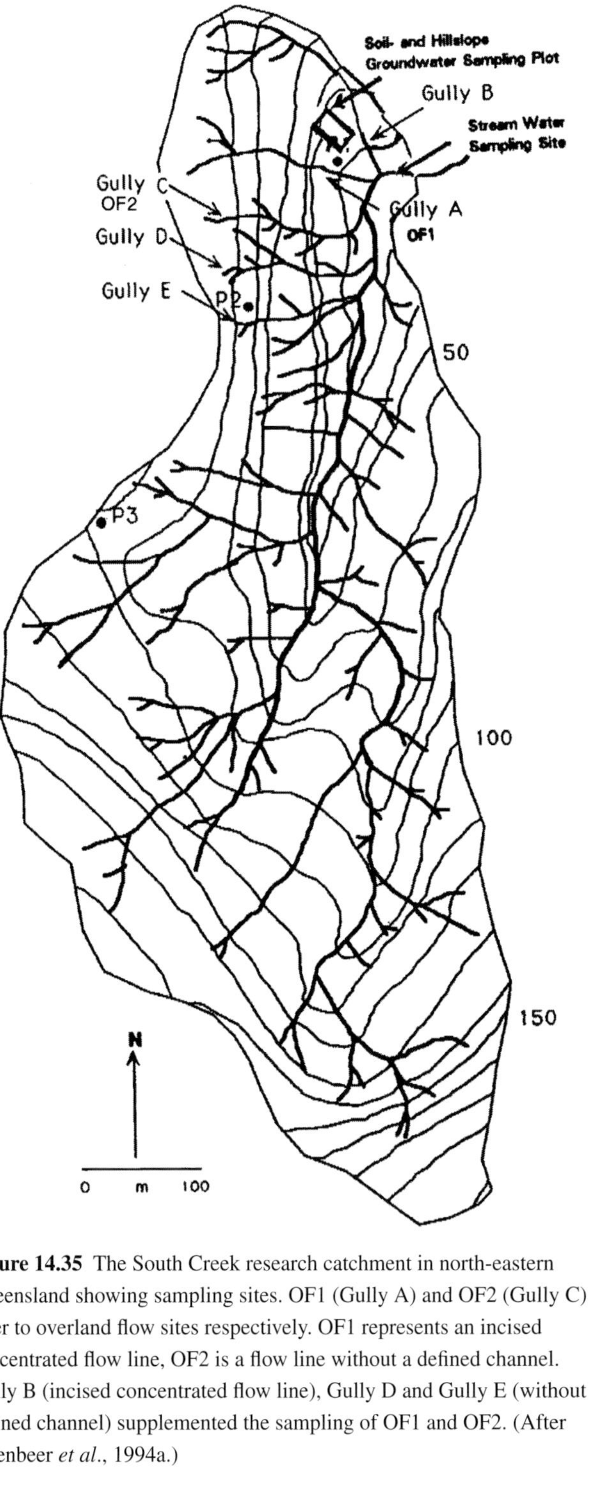

Figure 14.35 The South Creek research catchment in north-eastern Queensland showing sampling sites. OF1 (Gully A) and OF2 (Gully C) refer to overland flow sites respectively. OF1 represents an incised concentrated flow line, OF2 is a flow line without a defined channel. Gully B (incised concentrated flow line), Gully D and Gully E (without defined channel) supplemented the sampling of OF1 and OF2. (After Elsenbeer *et al.*, 1994a.)

injection, i.e. 2.0 m on site 1 (incised, lower slope of South Creek) and 0.6 m distance on the upper slope site 2. Greater amounts of recharge were required at 1.50 m depth for site 1, however, as would be expected.

The above tracing experiments, albeit undertaken at a very small scale, pointed towards a much more complex pattern of subsurface soil water movement below the most active SSF layer and above the permanent groundwater table than had been appreciated earlier. The detection of preferential flow below 0.2 m depth also inferred that the subsoil 'impeding' layer could still be breached. Further, the persistency of tritiated water below 1m depth inferred long soil water transit times.

The ability of deeper subsoil water to return to the surface layers (through the reversal of vertical hydraulic potential gradients, see also Figure 14.5 in Bonell *et al.*, 1983a) and thus participate in the storm runoff process, was not detected from these experiments. On the other hand, the earlier responses of a shallow piezometer to a tropical cyclone had suggested that strong upward movement and contributions of subsoil water to both SSF and SOF by exfiltration could be plausible (Figure 14.43) (Bonell *et al.*, 1982).

A subsequent hydrometric-environmental isotope study using deuterium verified the 'leaky' nature of the impeding subsoil in some parts of the soil profile whereas other parts conformed with the previous conceptual model of runoff generation. Moreover, it was also established during this campaign that the surface hydrology was much more connected to the permanent groundwater during major storm events than previously conceived (Bonell *et al.*, 1998). Figure 14.44 shows the changes in deuterium concentration of soil water during optimal wet season conditions. The sites SC1 and SC2 show only small changes in isotopic concentration at below 0.45 and 0.90 m depths respectively, despite 905 mm of rain falling in the period 6 February–5 March 1991. In contrast, soil water isotopic concentrations show temporal variability throughout the record at all depths at sites SC3 and NC4. The relative lack of variation with depth of isotopic concentrations for a particular time at NC2 suggest that the soil profile readily saturates during intense rainfall and drains slowly thereafter. This is consistent with adjacent piezometer responses (Figure 14.45) which show the rapid attainment of saturation after the commencement of a monsoon storm of 259.8 mm; the penetration of the 'impeding' subsoil with a progressive time lag with depth in the initial piezometer responses; and the compounding effect of upward soil water movement (hydraulic potential reversal) between 0.60 m and 0.30 m depth. Detection of the latter phenomenon supports the earlier observations from the tritiated water tracing experiments (Bonell *et al.*, 1998).

The connectivity of the water table with the surface hydrology was shown by a comparison of deuterium concentrations (Figure 14.46) in selected wells of both North and South Creek, with background stream samples taken from the respective catchments. Despite the fact that well samples were only taken in between storm events, the temporal variations in deuterium concentrations suggest the penetration of 'new' water to depths in excess of 3 m by preferential flow. Moreover, the differences between isotope concentrations in different wells and the

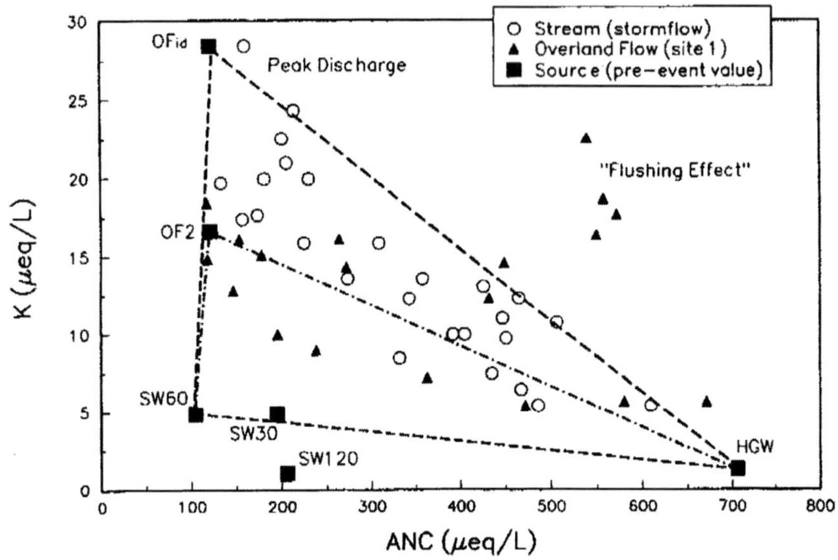

Figure 14.36 K-ANC mixing plot for event 1 (18 Feb 1993). The sources ('end-members') expressed as the median of three sites, are hillslope groundwater (HGW) and soil water from depths 1.2 (SW120), 0.6 (SW60) and 0.3 m (SW30; saturation overland flow is the mean of two observations at OF2 or the deduced concentration (Of_{id}); accordingly, there are two mixing regions. (After Elsenbeer *et al.*, 1995a.)

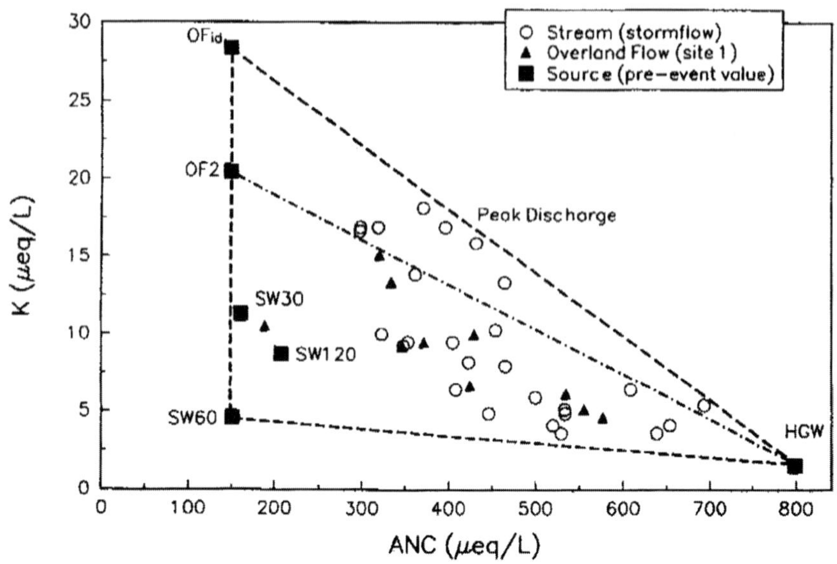

Figure 14.37 K-ANC mixing plot for event 2 (23 Feb 1993). The sources ('end-members') expressed as the median of three sites, are hillslope groundwater (HGW) and soil water from depths 1.2 (SW120), 0.6 (SW60) and 0.3 m (SW30; saturation overland flow is the mean of two observations at OF2 or the deduced concentration (Of_{id}); accordingly, there are two mixing regions. (After Elsenbeer *et al.*, 1995a.)

associated stream indicates a complex groundwater body, with different areas of the catchment contributing groundwater of significantly different isotopic concentrations and residence times (as defined by Zuber and Malszewski, 2001) to streamflow.

The complexity of the groundwater response is shown by the records of wells NCC, NCD, SCA and SCB during the February 16 storm (Figure 14.47). The pipe depths ranged between 6.9 to 7.6 m in NCC and NCD, but these remained insufficient to maintain contact with the permanent water table for long periods because of their elevated topographic position in North Creek. In contrast, SCA and SCB were continuously connected with a permanent water table and were shallower in depth (1.9–4.86 m depth) because of their lower elevated positions. All well pipes were forced into undersized auger holes (Luthin and

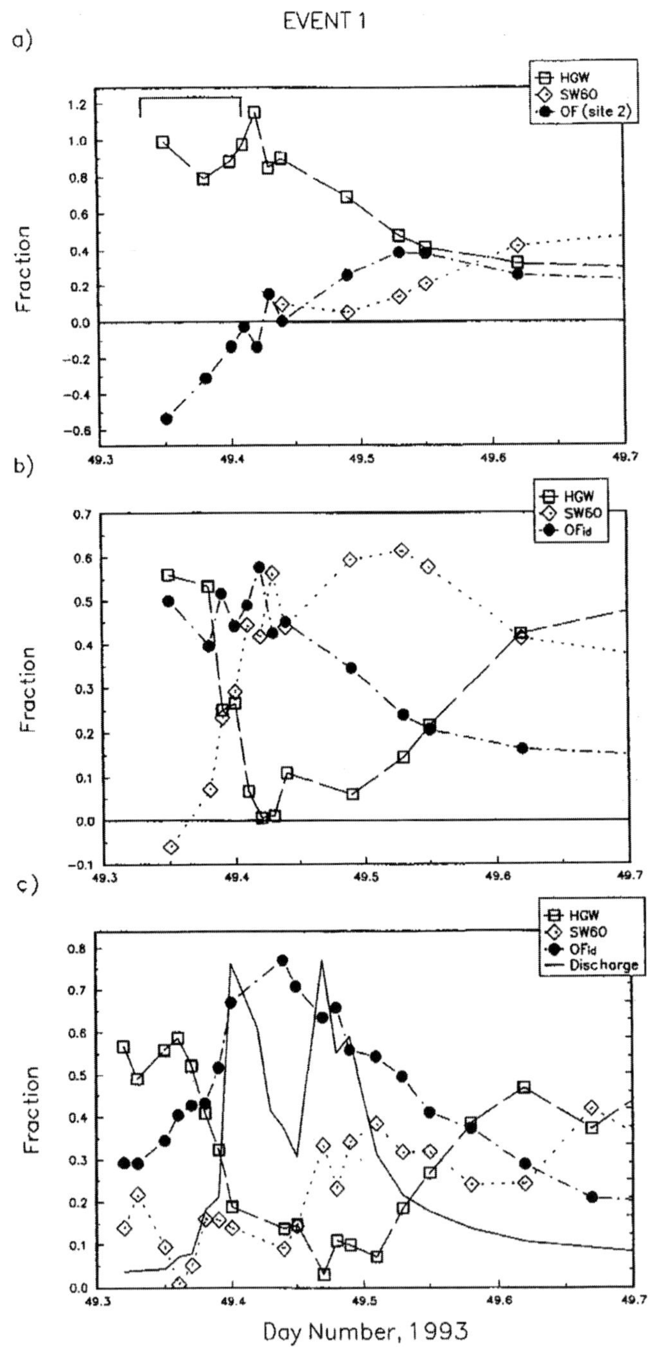

Figure 14.38 The time-dependent contribution of sources, expressed as fraction of overland flow and stream flow, respectively, for event 1. (a) Overland flow, with OF (site 2) (see Figure 14.35) representing saturation overland flow. (b) Overland flow, with Of_{id} representing saturation overland flow. (c) Stream flow. (After Elsenbeer *et al.*, 1995a.)

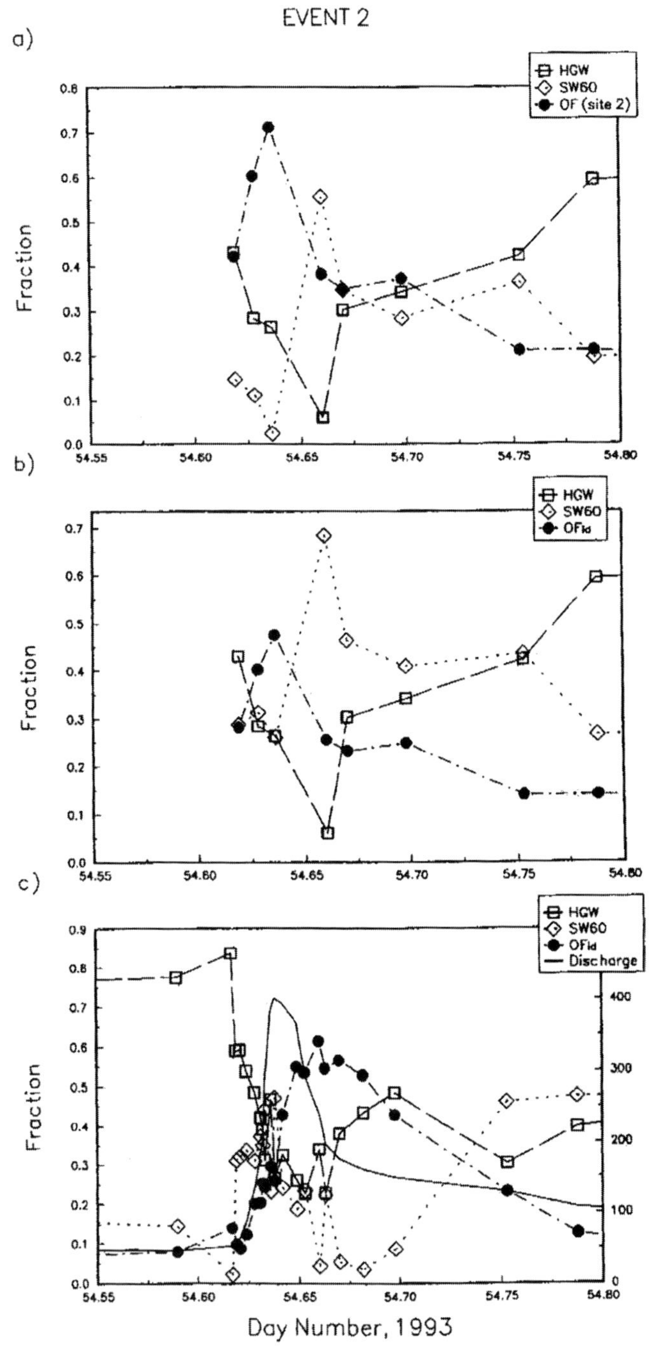

Figure 14.39 The time-dependent contribution of sources, expressed as fraction of overland flow and stream flow, respectively, for event 2. (a) Overland flow, with OF (site 2) (see Figure 14.35) representing saturation overland flow. (b) Overland flow, with Of_{id} representing saturation overland flow. (c) Stream flow. (After Elsenbeer *et al.*, 1995a.)

Kirkham, 1949) and sealed with bentonite and concrete in the top 1 metre.

NCC, SCA and SCB are responsive to the February 16 storm. In contrast, NCD shows only a weak response but subsequently there is a progressive increase in head until it peaks six days later following additional rainfall. These contrasting responses suggest very different pathways, with the first three sites connected to the surface hydrology. NCD more nearly follows the earlier hydrometric study conceptual model of an impeding layer to vertical percolation during storms.

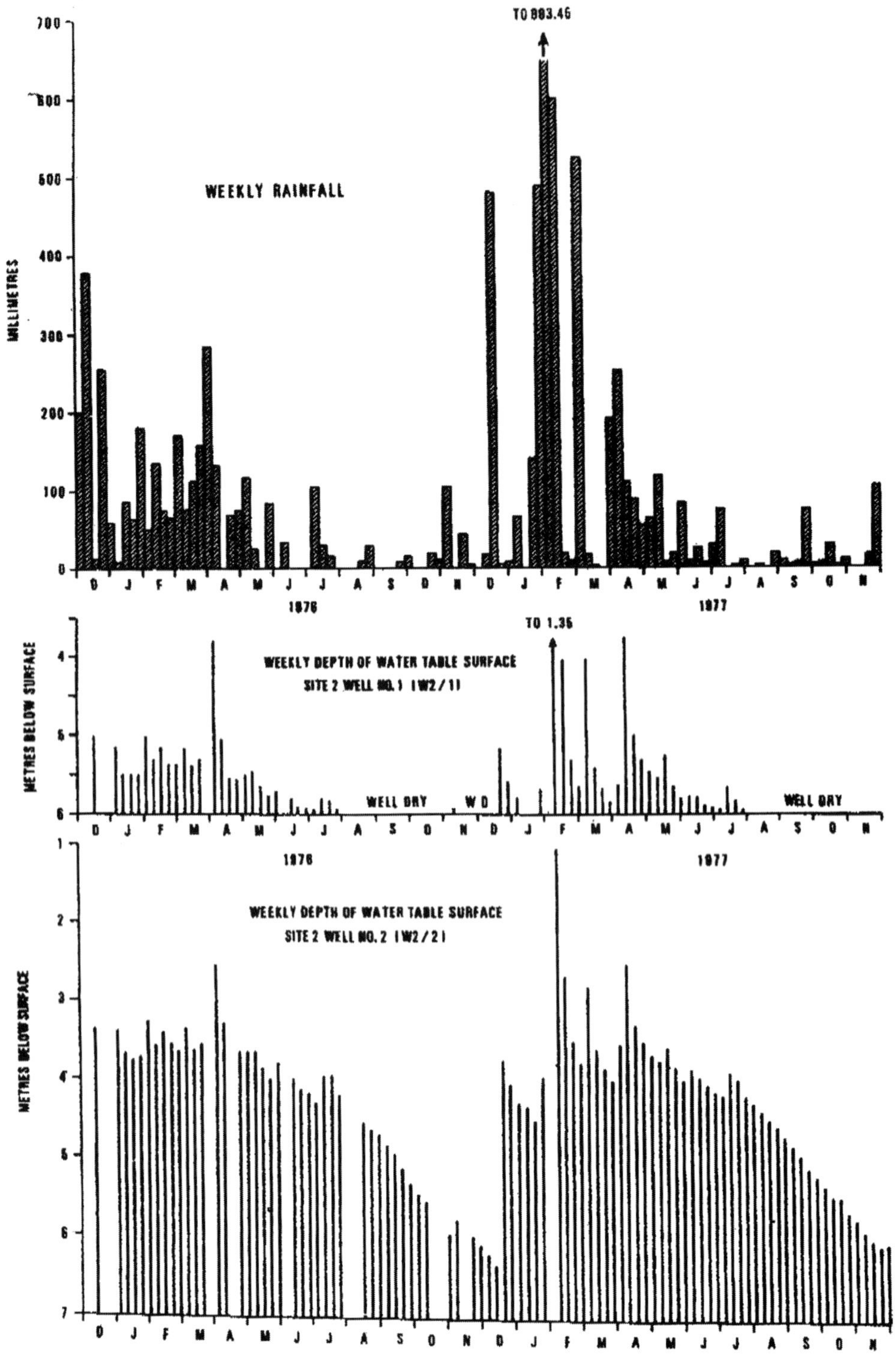

Figure 14.40 The well hydrographs for site 2 wells 2/1 and 2/2 for the period 1 Dec 1975–30 Nov 1977. (After Gilmour *et al.*, 1980.)

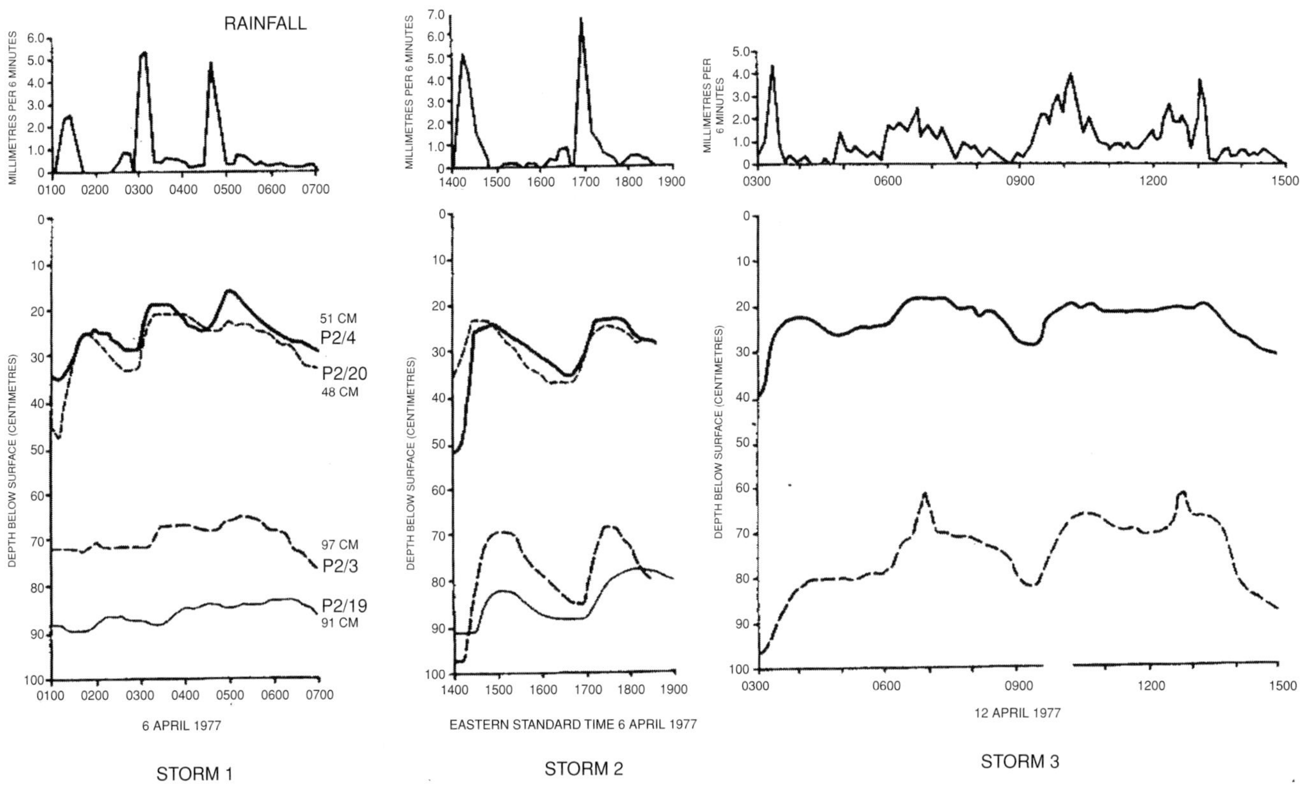

Figure 14.41 The continuous record of selected site 2 piezometers during storms 1 and 2, 6 April, and storm 3, 12 April 1977. (After Gilmour *et al.*, 1980.)

Some clues concerning the roles of possible groundwater contributions to the storm hydrograph were provided by crude water table profiles for selected events from the South Creek wells (Figure 14.48). Data logger problems precluded complete reconstruction of the profiles for the 'wet' period. Nonetheless, there is an indication that the water table 'pivots' in the vicinity of SCA in response to rapid and direct infiltration, and shows a progressive increase in hydraulic gradient with increasing storm size and antecedent streamflow (another indicator of antecedent catchment wetness). Furthermore, the maximum hydraulic gradient occurs near or soon after the stream hydrograph peak which implies the potential for significant groundwater contributions to stormflow at this time. Thus the protracted recovery to pre-storm background levels of the deuterium concentration profiles of both North and South Creek streamflow (following the hydrograph peaks) could in part be the contributions of groundwater of different residence times as well as the draining of more shallow SSF from the hillslopes (Bonell *et al.*, 1998).

The availability of improved technology (data loggers) and the complementary application of hydrochemistry studies has thus provided a much better insight into the runoff generation process. More important, there has been a significant modification of the earlier hydrometric study conceptual model to highlight the connectivity of the surface hydrology with permanent groundwater, and the potential for significant groundwater contributions to the storm hydrograph. These are shown in the revised conceptual model (Figure 14.49) which has many features in common with the corresponding three-component model of hydrological flowpaths as summarised by Mulholland (see Figure 1 in Mulholland, 1993) for the Walker Branch watershed (Wilson *et al.*, 1991a, b). The principal difference is that the overland flow is minimal in Walker Branch as a result of the very high infiltration capacities of the surface soils and lower prevailing rainfall intensities in eastern Tennessee, USA. Both conceptual models show SSF from perched saturation, and the contribution of a deeper saturated zone overlying a saturated bedrock zone (with exchange of water between these stores) to streamflow. The Babinda conceptual model also allows for a mixing zone of 'new' water inputs at the water table surface on the lines of Anderson and Burt (1982) with older water below, the latter of which has a capacity of at least 3 m equivalent depth of rainfall on the basis of modelling (Barnes and Bonell, 1996; Bonell *et al.*, 1998) (see discussion by Barnes and Bonell, this volume). At the end of the 1991 wet season, for example, deuterium concentrations of background streamflow samples were within 5 parts per million of their initial values, even after more than 3000 mm of rainfall; they showed no significant

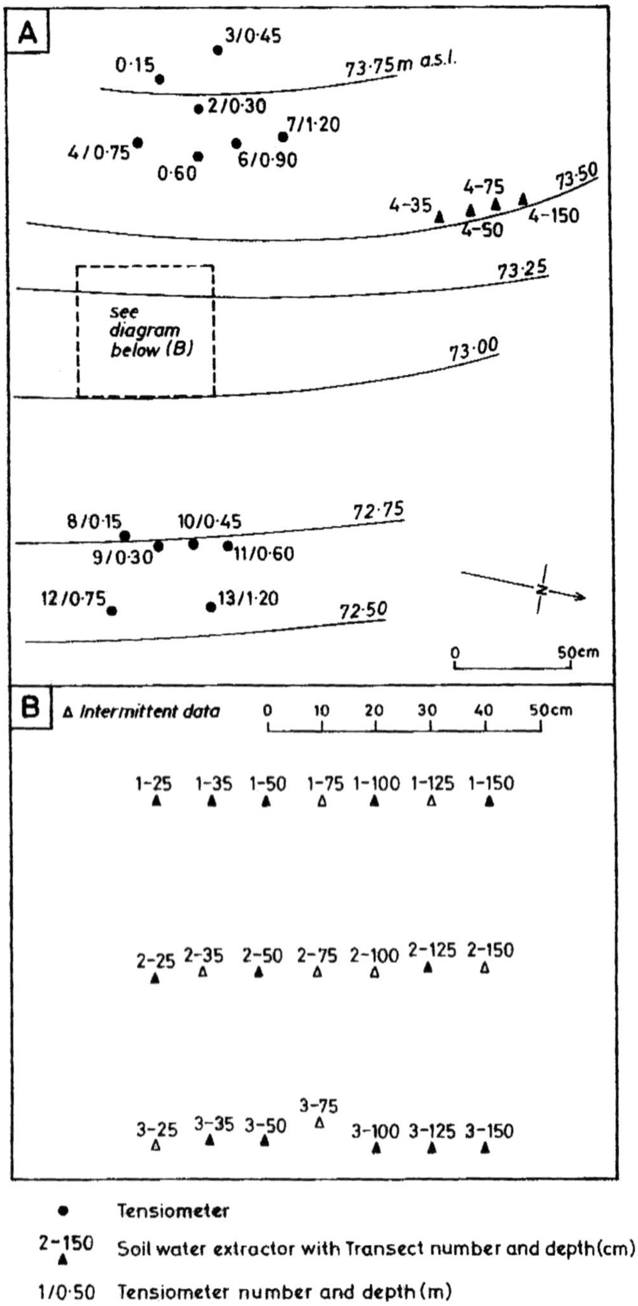

Figure 14.42 (a) The experimental design of the upper slope tracing site 2 (for location see Bonell *et al.*, 1982.) (b) The equipotential and flow lines on 14.03.80 at tracing site 2. The hydraulic potential, Φ is given as $\Phi = \Psi_M + z$ where z is the elevation above sea level (upward is positive) and Ψ_M is the matric potential at a particular point. (After Bonell *et al.*, 1984.)

change at all as a result of the February 16 event of 259.8 mm described above. These observations further suggest the existence of active storages of deeper groundwater (in excess of 3 m depth of water equivalent) which is well-buffered from 'new. water inputs (Bonell *et al.*, 1998).

The resulting near-constant deuterium concentrations in the delayed flow component of streamflow are likely to be the result of different water particle velocities and ages beneath the water table which appear to be 'well-mixed water' at the point of effluence of groundwater into surface drainage on the lines proposed by Raats (1978). Otherwise the notion of perfectly mixed water is a contentious issue. As Maloszewski and Zuber (1998, p. 297) remarked '. . . we insist that the term 'perfect mixing' is unfortunate for groundwater systems in which prefect mixing can never occur. Mixing usually takes place at the outlets, i.e. drainage areas, springs or pumping wells.'

Wet tropical coast region of north-east Queensland, Australia

A valid criticism of concentrated work in experimental catchments is the problem of extrapolation of research findings to other 'similar' environments. In response, a field survey of soil hydraulic properties, notably K^*, was undertaken within the tropical forests of the wet tropical coast region of north east Queensland by Bonell *et al.* (1983b). In total, 13 additional sites were included to cover additional parent materials (granite, basalt), great soil groups (following the CSIRO Australian classification, Stace *et al.*, 1968; Northcote, 1979), slope angles and annual rainfall, and these were compared with the runoff plot hydrometric sites 1a and 2 (Bonell *et al.*, 1981) in South Creek (see detailed description in Table 1 in Bonell *et al.*, 1983b) (Figure 14.50). Reconnaissance soil investigations in the area had indicated that the granites and metamorphics develop both red and yellow soils (Isbell *et al.*, 1968; Red Kandosol or Red Dermosol, Yellow Dermosol, most profile descriptions are Acidic, Dystrophic: Isbell, 2002)[3] or Acrisols-Alisols (Inceptisol, Ultisol-Inceptisol, USDA). An attempt was made to sample the climatic/pedology spectrum by selecting both red and yellow soil phases of the granites and metamorphics at both the drier and wetter end of the range. Two basalt sites of Krasnozems (Stace *et al.*, 1968) (Dystric Nitosols; Oxisols, USDA; Red Ferrosol: Isbell, 2002) were also chosen at each end of the annual rainfall spectrum. An additional site in a granitic colluvial material was sampled because of the common occurrence of such areas on the eastern edge of the coastal ranges.

The work highlighted the persistence of intense biological activity in the top 0.1 m which makes all the great soil groups capable of accepting the prevailing rain intensities. Similar conditions also apply to the 0.1–0.2 m layer despite a decline in the log mean K^*. Paradoxically, both Babinda catchment runoff plots (sites 1 and 2) and a yellow podzolic (site10) had a lower log mean K^* at 0.1–0.2 m depth which indicated that some impedance to vertical fluxes during monsoon storms was more common. Below 0.2 m depth most soil profiles operated as 'impeding' layers to

3 Neil McKenzie, CSIRO Land and Water, Canberra, is thanked for providing the updated Isbell (2002) Australian Soil Classification (i.e. abbreviated name of soils) for the mentioned Australian soils.

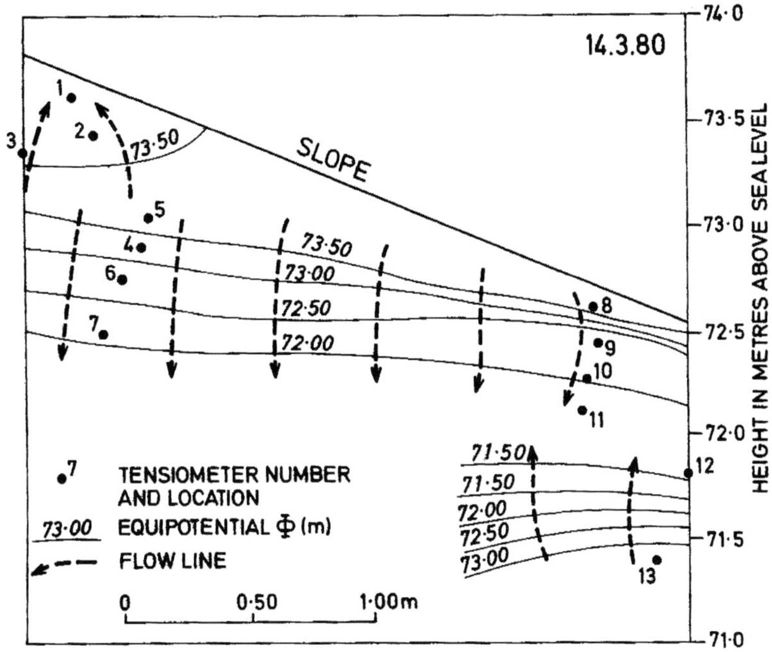

Figure 14.42 *(cont.)*

Table 14.8. *The vertical displacement of tracer along the injection line and recharge on selected dates*

Date	Site	Centre of mass, z, (m)	Average volumetric water content, $\bar{\theta}$	Accumulated inputs from beginning of experiment: Rainfall (mm)	Throughfall[a] (mm)	Apparent vertical recharge, R_v: Recharge (mm)	Percentage of throughfall[a]
3.3.80	1	0.595	0.405			159.97	725.82[b]
				34.90	22.04		
	2	0.270	0.390			27.30	121.88[b]
26.3.80	1	0.815	0.464			285.36	71.35
				574.40	399.95		
	2	0.635	0.515			224.03	56.01
14.5.80	1	1.060	0.457			393.02	56.89
				892.80	690.89		
	2	0.780	0.525			304.50	49.92
26.6.80	1	1.140	0.483			454.02	54.08
				1237.50	839.50		
	2	0.920	0.545			392.40	46.74

[a] Throughfall calculated from daily rainfall by equation $y = 0.7 + 0.725x$ where y = throughfall (mm), x = rainfall (mm) (Gilmour, 1975).

[b] These high values are the result of short-circuiting.

Source: After Bonell *et al.* (1984).

Table 14.9. *Lateral recharge from injection line to peak response in selected extractors on lowest transect*

Extractor depth (site no.)	Date of peak response	Accumulated rainfall (mm)	Accumulated throughfall (mm)	Average volumetric water content, $\bar{\theta}^{a}$	Lateral recharge R_c, accumulated throughfall $\times \bar{\theta}$ (mm)
0.35 (1)	12.3.80	279.4	193.49	0.445	86.10
0.35 (2)	14.3.80	282.0	194.45	0.480	93.34
0.50 (1)	9.4.80	603.0	417.67	0.403	168.32
0.50 (2)	19.3.80	433.1	301.32	0.535	161.20
1.50 (1)	9.4.80	603.0	417.67	0.450	187.95
1.50 (2)	12.3.80	279.4	193.49	0.540	104.48

[a] Average volumetric water content on the given date in a unit area of profile between 0.20 m on top transect and the depth of each extractor on lowest transect.
Source: After Bonell *et al.* (1984).

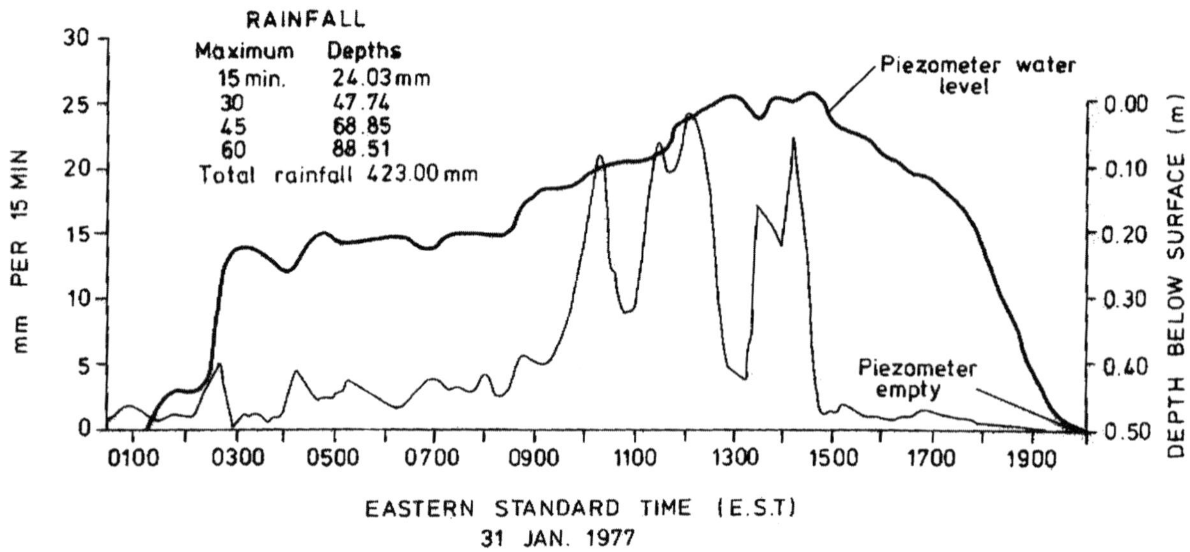

Figure 14.43 The continuous record for a piezometer set 0.5 m below ground level on former runoff plot, site 2, during tropical cyclone Keith. (After Bonell *et al.*, 1982.)

prevailing rainfall intensities. The principal exception was the basalt (Dystric Nitosol, Oxisol: USDA; Red Ferrosol: Isbell, 2002) at site 12 which remained relatively transmissive ($K^{*} \sim 57$ mm h^{-1}) down to 0.5 m depth before declining thereafter ($K^{*} \sim 5$ mm h^{-1}, 0.5–1.0 m depth).

The application of the Least Significant Difference test (LSD) and subsequently cluster analyses enabled the log mean K^{*} of the two impeding layers (0.2–0.5 m, 0.5–1.0 m) to produce a classification. A scatter plot of log mean K^{*} (Figure 14.51) shows a nearly continuous distribution, grading from predominantly yellow soils (Yellow Dermosol: Isbell, 2002) at one end of the spectrum, through a cluster of three red soils (sites 3 and 7, Red Dermosol, site 9 Red Kandosol: Isbell, 2002), into the higher hydraulic conductivity sites of the basalt (Red Ferrosol), colluvium (Melacic Dystrophic Red Kandosol) and South Creek runoff plot 1a (Red Kandosol: all classifications Isbell, 2002). The within South Creek differences are also emphasised with runoff plot 2 (the more prolific producer of SOF, Gilmour *et al.*, 1980) being located in the least permeable group. Subsequent application of up to four clustering strategies consistently produced the same three groups (Figure 14.52). The resulting dendograms highlight the marked difference between the lower permeable group 1 and the remainder. In the former, one can surmise that SOF will continue to prevail during any temporary decline

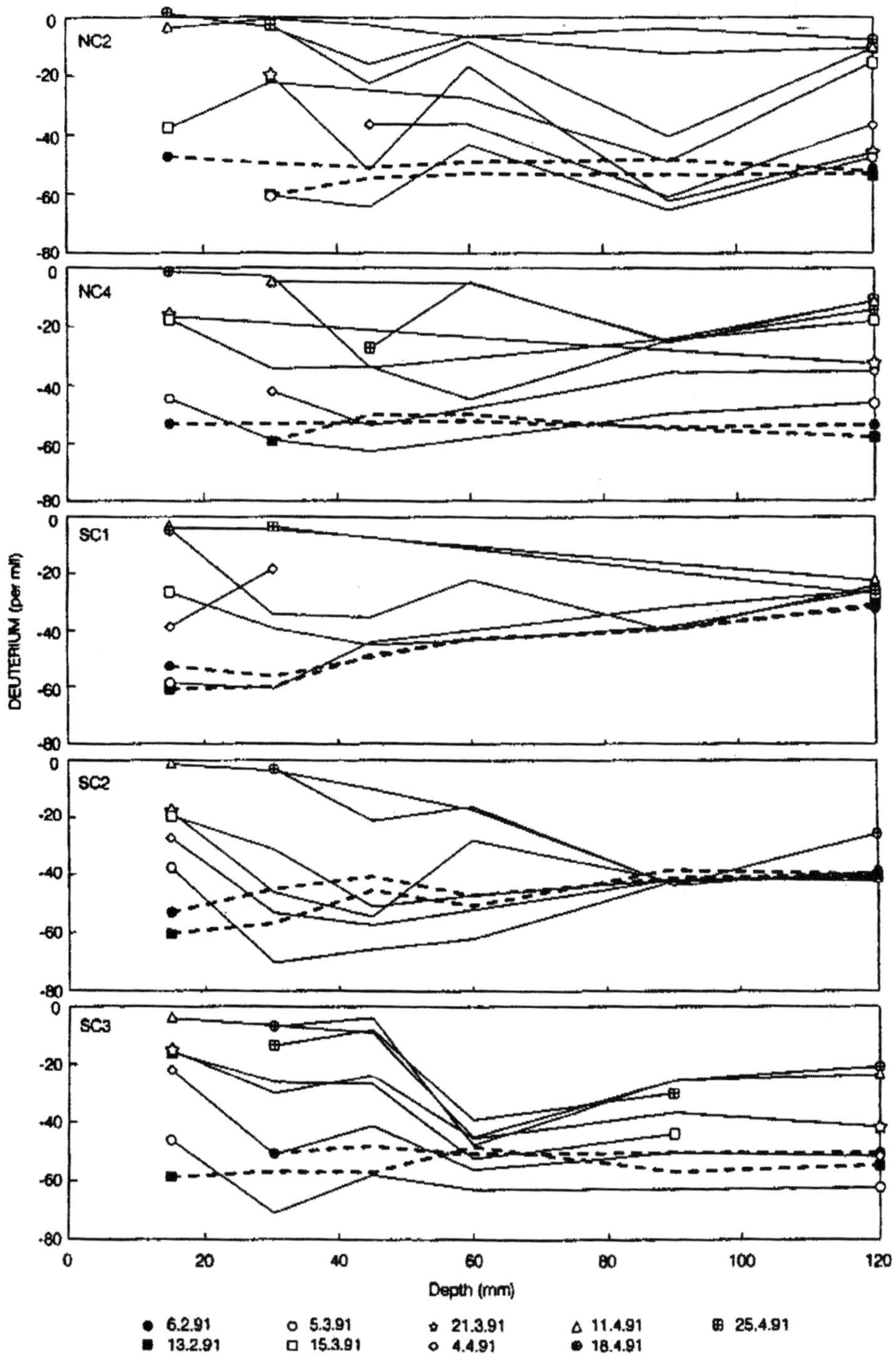

Figure 14.44 Isotopic data from soil lysimeters from North and South Creek catchments. (After Bonell *et al.*, 1998.)

in short-term rainfall intensities of summer monsoon storms, or more generally during the post-monsoon season on the basis of measured responses at South Creek runoff plot 2. On the other hand, SSF becomes more important at sites in group III except at the highest rain intensities when SOF will dominate in line with South Creek runoff plot 1a (site 1) (Bonell *et al.*, 1981). Sites in group II might take an intermediate position in the runoff process.

A later soil hydraulic conductivity survey undertaken on the slopes of Mt Bellenden Ker (immediately west of the Babinda catchments, Figure 14.50) by Herwitz (1986) also confirmed the same vertical trends in K^* in soils of granitic origin. This area

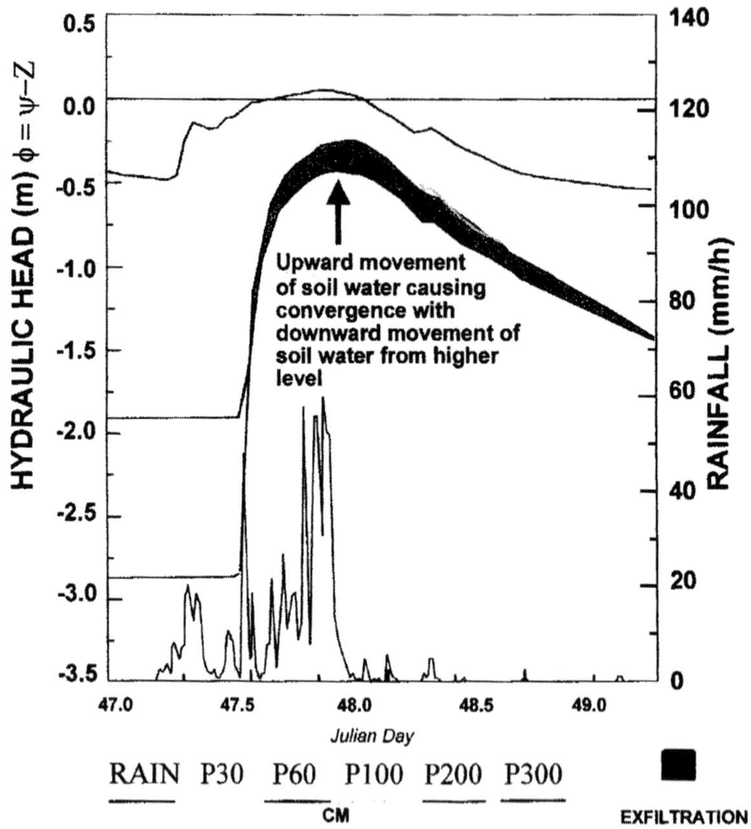

Figure 14.45 The response of the piezometers at NC2 to a monsoon storm event, 16 February 1991. (After Bonell *et al.*, 1998.)

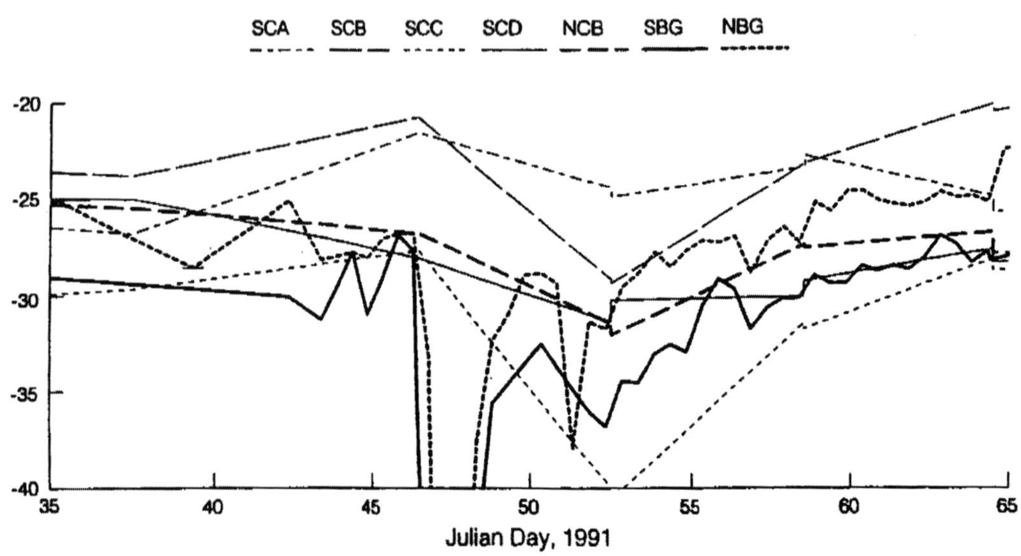

Figure 14.46 Isotopic data for well samples taken during the period 4 Feb–6 March (JD 35–65) 1991, for South Creek (SCA, SCB, SCC and SCD) and North Creek (NCB). Also shown are background stream samples for North Creek (NBG) and South Creek (SBG) for the same period. (After Bonell *et al.*, 1998.)

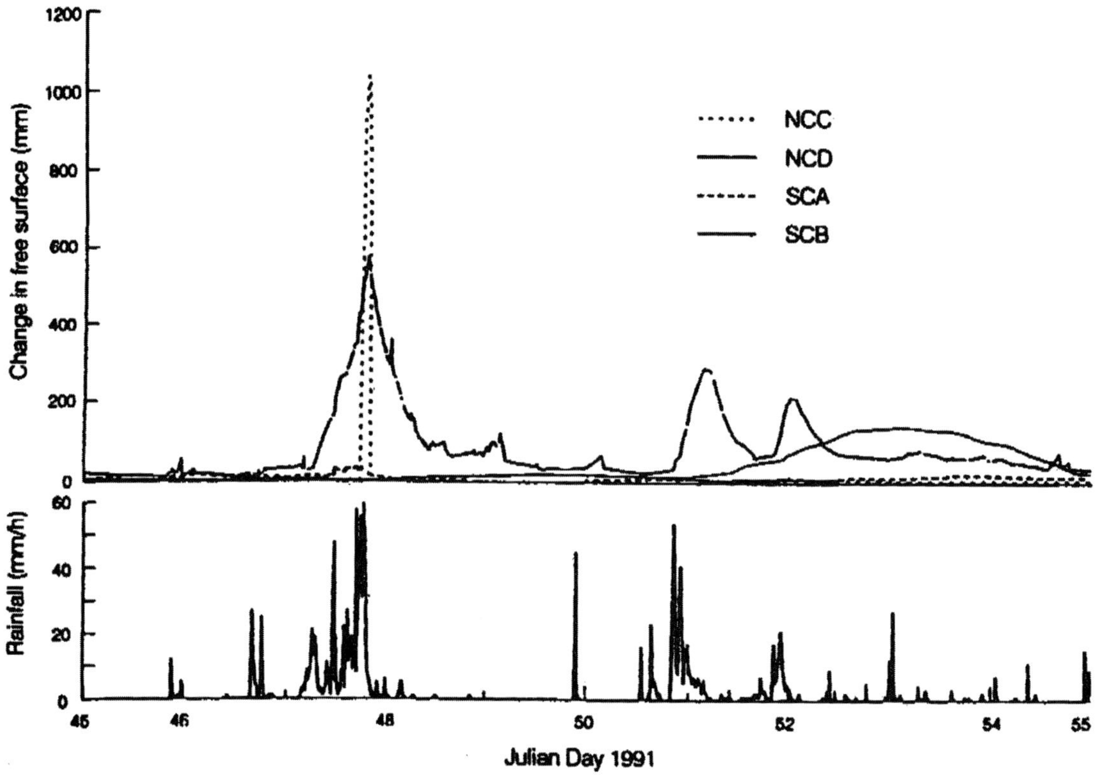

Figure 14.47 The change in depth to water table with rain between 14–25 Feb 1991 for wells NCC, NCD, SCA and SCB. (After Bonell *et al.*, 1998.)

experiences even higher annual rainfall (~ 6570 mm) than the sites considered earlier by Bonell *et al.* (1983b). Herwitz (1986) emphasised the morphological role of the associated upland forests in modifying the runoff generation process by funnelling very high volumes of stemflow with fluxes as high as 314 mm min^{-1}, when the corresponding rain intensity is 2 mm min^{-1}. Consequently, 'localised' infiltration-excess overland flow at the base of the trees of the montane forest add to the prevailing SOF (Herwitz, 1986).

Near to site 12 (Figure 14.51), Prove (1991) measured extensive SOF during monsoon events over basalt (Dystric Nitosols: Oxisols, USDA; krasnozems: Stace *et al.*, 1968; Red Ferrosol: Isbell, 2002) in former forest land converted to sugar cane. The surficial ploughed layer provided a much higher K^* than the layer below, in common with the vertical changes in K^* of rainforest soils. Thus the runoff generation process was the same.

La Cuenca, Peru

Elsewhere in western Amazonia, Peru, the work of Elsenbeer and collaborators (Elsenbeer and Cassel, 1990, 1991; Elsenbeer *et al.*, 1992; Elsenbeer *et al.*, 1994b, 1995b, 1996; Elsenbeer and Lack, 1996a, b; Elsenbeer and Vertessy, 2000) in a first-order tropical rainforest catchment (La Cuenca, Figure 14.53), provides a contrast with the preceding Babinda study. Elsenbeer and co-workers also highlight the prevalence of SOF compared with the dominant vertical pathways associated with central Amazonia. Moreover, through undertaking an intensive study at such a small scale (0.75 ha), this study has brought to attention additional facets of hillslope hydrology insufficiently emphasised elsewhere; namely the difficulty in differentiating between HOF, SOF, SSF and return flow (RF) and the less dominant role of topography in the spatial and temporal organisation of overland flow at this small scale, cf. Noguchi *et al.* (1997b).

The topographic setting and experimental design of La Cuenca is presented in Figure 14.53. Acrisol-Alisol (Ultisols, USDA) prevail from tertiary 'Red Beds' (sandstones, siltstones and shales) which grade into Cambisols (Inceptisols, USDA) on the steeper side slopes near the valley floor, cf. Babinda. An interesting feature is the existence of an intricate network of pipes which makes pipeflow an important contributor to the runoff generation process (see Elsenbeer and Lack, 1996b for photographic details).

Mean annual rainfall is about 3300 mm; for the study period 1987 (3190 mm) and 1988 (2750 mm) totals were somewhat lower. Of more interest are the rainfall characteristics by event as presented by Elsenbeer *et al.* (1994b) (Table 14.10) for an 18-month period. Elsenbeer *et al.* (1994b) noted that the largest event was a single burst of 120.7 mm; and six events larger than 70.3 mm

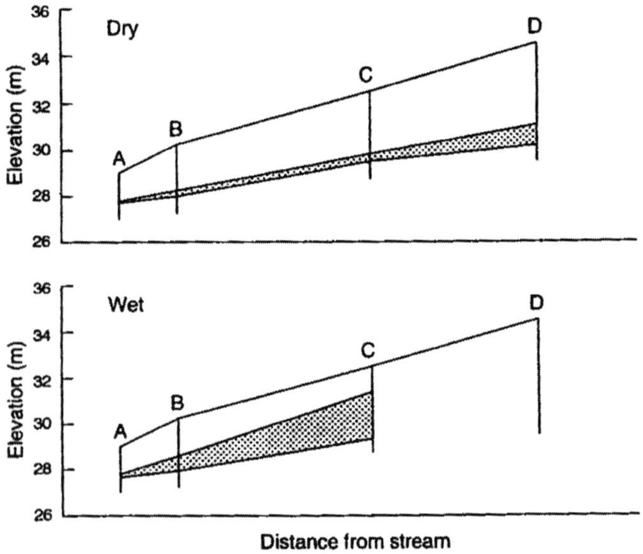

Figure 14.48 Changes in the position of the water along the cross-section of the South Creek hillslope monitoring system during the periods 17 Jan–17 Mar 1990 (a 'dry' period) and 8 Nov–8 Dec 89 (a 'wet' period). the upper line shows the soil surface while the vertical lines indicate the depth of penetration of the wells. The shaded area indicates the range of water table movement in the respective period. Note that in the 'wet' period, hydraulic gradients are more than doubled. (After Bonell *et al.*, 1998.)

could not be sampled because of the limited capacity of the containers connected with an associated throughfall study. Otherwise, the monitored 214 events covered the complete range of recorded rainfall intensities during the La Cuenca throughfall study period. These rainfall characteristics *are up to two orders of magnitude lower* than the monsoon rain characteristics of Babinda (when SOF is also most prevalent). By contrast, the short-term intensities for the post-monsoon and dry season events in Babinda have more in common with those of La Cuenca when SSF is the more dominant pathway in the Babinda study (Howard, 1993). In terms of rainfall characteristics *per se*, the differences between a cyclone-prone and non-cyclone (convective clusters) meteorological environment is thus clear (see Bonell *et al.*, this volume). Nonetheless, SOF is very prevalent in La Cuenca. During a comparison between the two studies, Bonell with Balek (1993, p. 218) observed that the much lower rainfall intensities in La Cuenca are compensated by the impeding soil layer (0.1–0.2 m depth) being more shallow than in the Babinda study and the corresponding K^* for La Cuenca below 0.3m depth, an order of magnitude lower than in South Creek. Elsenbeer and Vertessy (2000) subsequently formalised the preceding observation through the presentation of Figure 14.54 which links soil hydraulic conductivity and rainfall intensity of 26 selected events (Table 14.11). As Elsenbeer and Vertessy (2000, p. 2373) observed '. . . a hydrological discontinuity (thus favouring SOF and SSF) at a depth of about 0.1–0.2 m is the salient feature controlling runoff generation in this environment, given the prevailing rainfall characteristics'. However, in common with the Babinda work, Elsenbeer and Vertessy (2000) noted that this vertical discontinuity was spatially discontinuous, on the evidence of numerous outlying data points in Figure 14.54. Thus, once again, a connection between near-surface flowpaths and deep soil or groundwater is implied. Also in common with several other hillslope hydrology studies in the humid tropics, positive matric potentials persist (reflecting the existence of a perched water table with the hydrological discontinuity) cf. Babinda, which only departs from this pattern during the dry season (Figure 14.55).

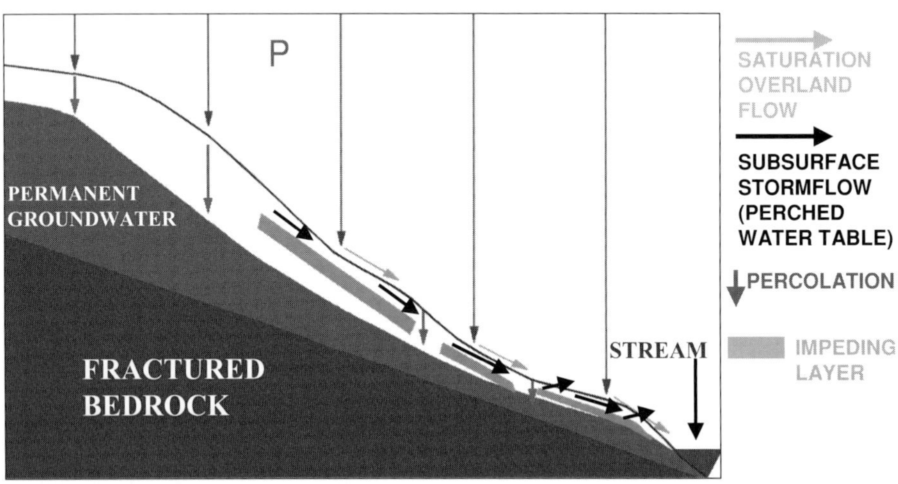

Figure 14.49 Revised conceptual model of storm water transfer in the South Creek. (After Barnes and Bonell, unpublished data.)

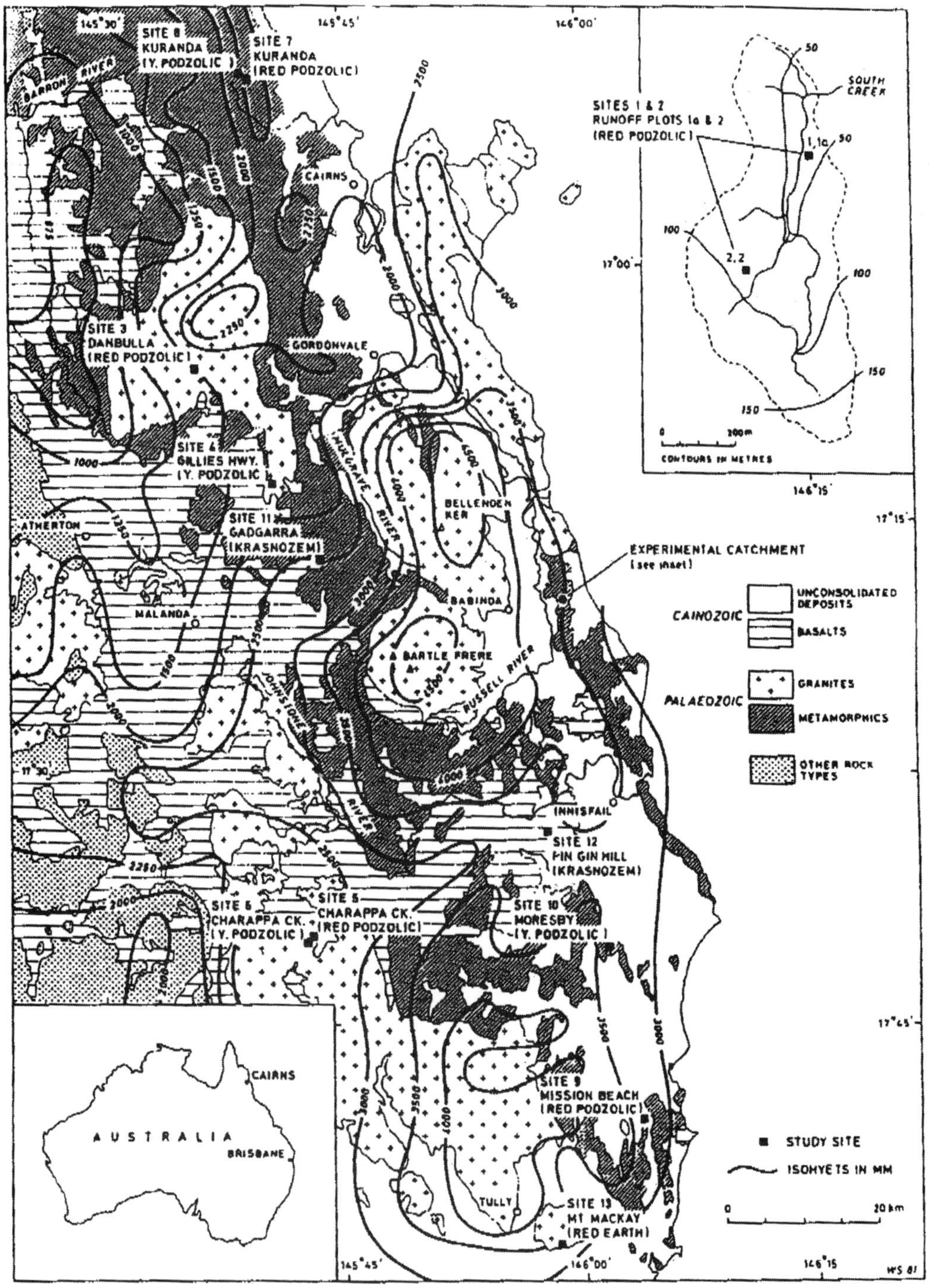

Figure 14.50 The location of the study sites in relation to the geology and 30-year mean annual rainfall (1926–1955) of the wet tropical hinterland. (After Bonell *et al.*, 1983b.)

The spatial and temporal extent of overland flow during the study period is summarised in Figure 14.56. Several features of these hydrometric observations need highlighting. *The accepted linkage between topography (i.e. depressions) and frequency of occurrence of overland flow only partly holds*. Overland flow is persistent even in topographically divergent areas. Moreover, some downslope positions are less favourable to overland flow occurrence over other slope positions. Elsenbeer and Cassel (1991) provided a rationale for this apparent inconsistency. In addition to SOF occurrence, they showed that maximum rainfall

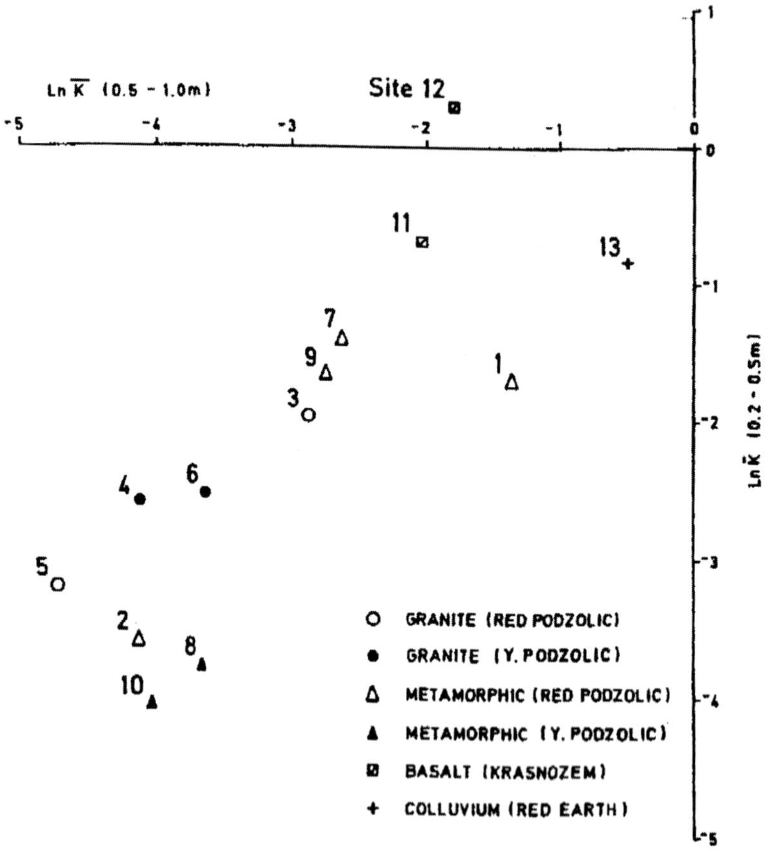

Figure 14.51 The scatter plot of the $\log_e K$ means of the impeding layers at the study sites of the wet tropical hinterland. (After Bonell *et al.*, 1983b.)

intensities exceed the surface K_{sat} occasionally in some places to produce Hortonian (infiltration-excess) overland flow *whose occurrence is independent of topography*. Stratified sampling however, indicated that HOF was particularly favoured on the steep valley side slopes in the western half of the catchment where either the soils are poorly developed or there is a missing A horizon combined with a patchy and sparse litter layer. In addition, monitored return flow from pipe exits (e.g. pipes 3 and 4) is not necessarily dependent on topographic position. As noted by Elsenbeer and Lack (1996b, p. 947), 'the outstanding feature at La Cuenca is the numerous pipe outlets that produce perhaps most of the overland flow in this environment and certainly explain its prevailing occurrence as concentrated flow lines' rather than sheetflow (cf. the 'pipe overland flow' of Putty and Prasad, 2000a). Figure 14.56 indicates that one out of three precipitation events produced overland flow (SOF and, on a more limited scale of occurrence, HOF) at 50% or more of sites. Examples of overland flow hydrographs from sites 1–3 (S1 to S3 in Figure 14.57 are presented for a typical wet and dry season storm, respectively. The overland flow at sites 1 and 3 are fed by outflow from pipes (pipes 3 and 6 in Figure 14.53) and thus qualify as return flow. The hydrograph for S2 (rainy season event) differs because it is generated by SOF which is notably absent in the dry season storm. Some pipe outflow hydrographs (e.g. P1 in Figure 14.58) were not so responsive but show a delayed and dampened response in comparison with those of sites 1 and 3 shown above as well as S1 and S2 for the event shown in Figure 14.58. Thus some pipes have networks that are draining the shallow perched water table above the soil hydraulic conductivity discontinuity whereas others are accessing much deeper subsurface water. With the dominance in this environment of return flow (RF) from near-surface sources, separating out the contributions of RF, SOF, HOF and SSF to overland flow thus becomes becomes a matter of semantics.

A parallel recession analysis by Elsenbeer and Vertessy (2000) (Figure 14.59) showed that there was negligible differences between the hillslope sites, with median recession constants ranging between 6 to 12 min. These constants were smaller than the first recession constant of streamflow which was still identified with the surficial contributing pathways. The low magnitude of the hillslope constants for S1 and S3 are as much proxies for pipeflow and represent the rapid drainage of the perched water table above

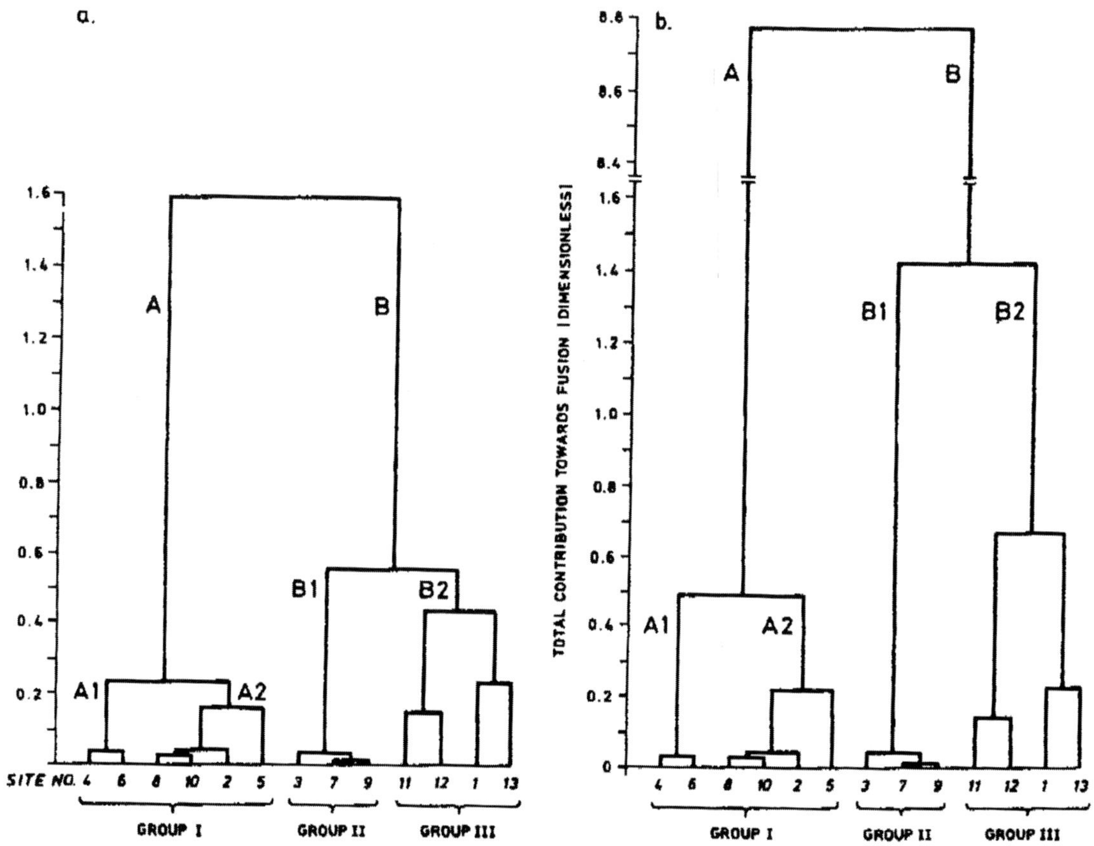

Figure 14.52 The dendogram hierarchy ((a) Sokal and Michener (1958), (b) Burr (1970)), based on the classification of the field hydraulic conductivity of the impeding layers (0.2–0.5 m, 0.5–1.0 m) of tropical rainforest soils in north-east Queensland. (After Bonell *et al.*, 1983b.)

the soil hydraulic conductivity discontinuity. Thus the contributing store is too small to allow for longer recession constants. The limited sample of P1 also suggests a small recession constant, despite the delayed and dampened response.

A complementary hydrochemistry study was presented in detail by Elsenbeer *et al.* (1995b) which used five environmental tracers (hydrogen, calcium, magnesium, potassium and silica) to distinguish 'fast' flow paths, influenced mainly by the biological subsystem from 'slow' flow paths in the geochemical subsystem. The sampling approach followed is summarised in Table 14.12.

Following a comparison of environmental tracers by hydrological compartments and flow paths (outlined in Table 14.12), some rationalisation was then undertaken by using silica to represent all those solutes whose streamflow concentration generally decreases with increasing discharge (calcium, magnesium, silica) and potassium whose concentration can tend to increase with discharge due to its origins from biomass. Potassium can also be independent of discharge from flushing during the rainy season i.e. 'flat' K^+ stormflow chemographs. Thus, through the use of silica and potassium, Elsenbeer and Lack (1996a) showed that overland flow does explain the storm hydrograph signal in many cases. On the other hand, due to the flushing effect, K^+ is not an entirely reliable indicator for overland flow. An unambiguous distinction of flowpaths and contributing sources was better provided by the K/SiO_2 ratio (Elsenbeer *et al.*, 1995b). Fast pathways interacting with the biological subsystem are characterised by a high K/Si ratio, while the ratio is reversed in slow pathways interacting with the geological subsystem because SiO_2 originates from the weathering zone.

An example of the use of the K/SiO_2 is presented in Figure 14.60. The signature of OF is quite distinct from a soil water source (SW) and groundwater (FGW, seep). In the rainy season example, the chemohydrograph approaches the K/SiO_2 concentration of overland flow, thus confirming the contribution of this flow pathway to the storm hydrograph from the parallel hydrometric work. In the dry season case study, the K/SiO_2 ratio of streamflow appears to exceed the corresponding OF ratio. This is probably due to 'noise' from the 'flushing effect' of K^+ after a period of low antecedent moisture conditions, and temporal variability in the chemical composition of OF during the storm which has been masked by taking an average concentration. Because of this time-dependency of environmental tracers, linked with fast

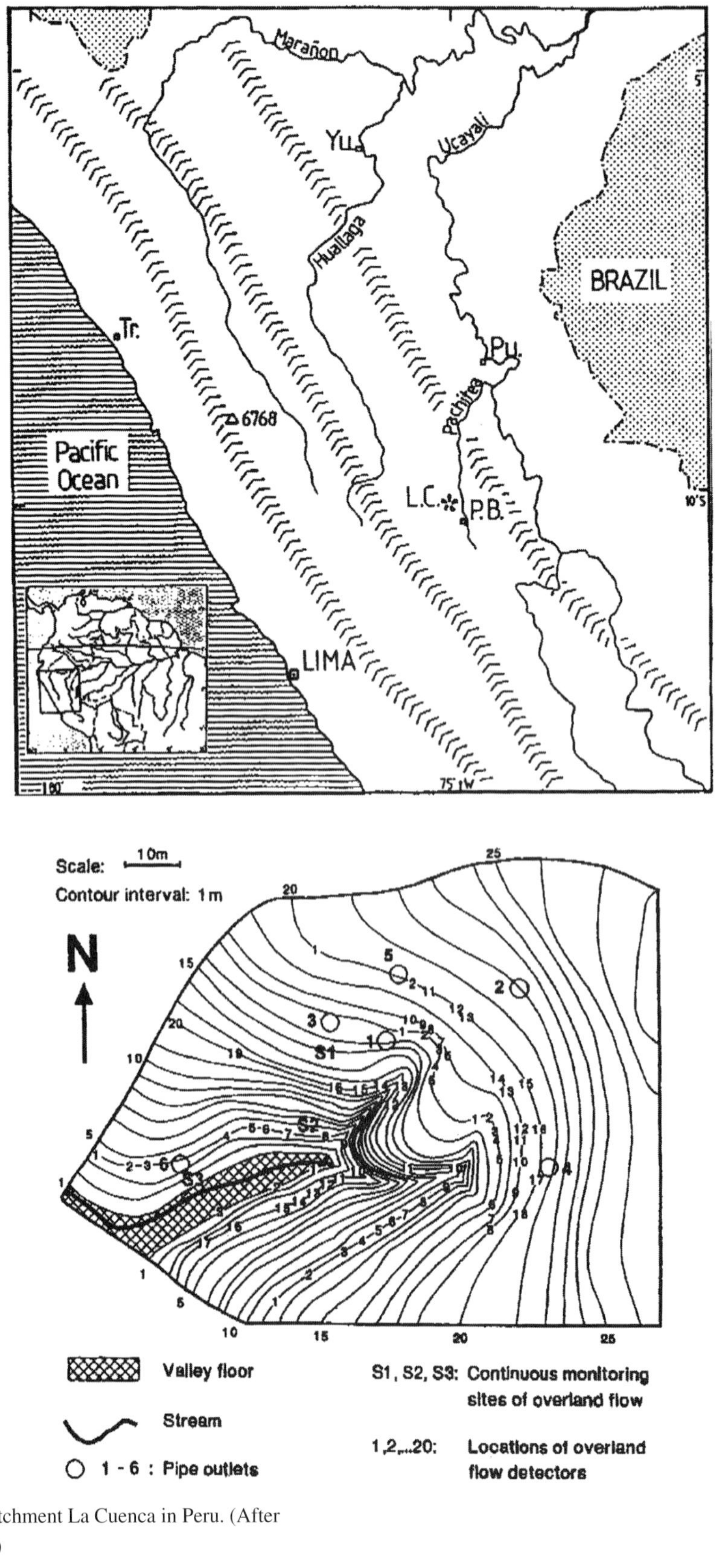

Figure 14.53 The research catchment La Cuenca in Peru. (After Elsenbeer and Vertessy, 2000.)

Table 14.10. *La Cuenca research catchment: summary of descriptive statistics of rainfall variables*

Variable	Units	Events	Median	Maximum	Minimum
Magnitude	mm	214	8.7	70.3	0.3
I_{60} max	mm h^{-1}	170	6.7	50.0	0.5
I_{30} max	mm h^{-1}	198	9.1	66.0	0.4
I_{15} max	mm h^{-1}	211	12.4	76.0	0.8
I_5 max	mm h^{-1}	214	19.2	96.0	1.2
R	mm h^{-1}	214	3.2	31.1	0.4
Duration	min	214	170	960	10

Source: After Elsenbeer *et al.* (1994b).

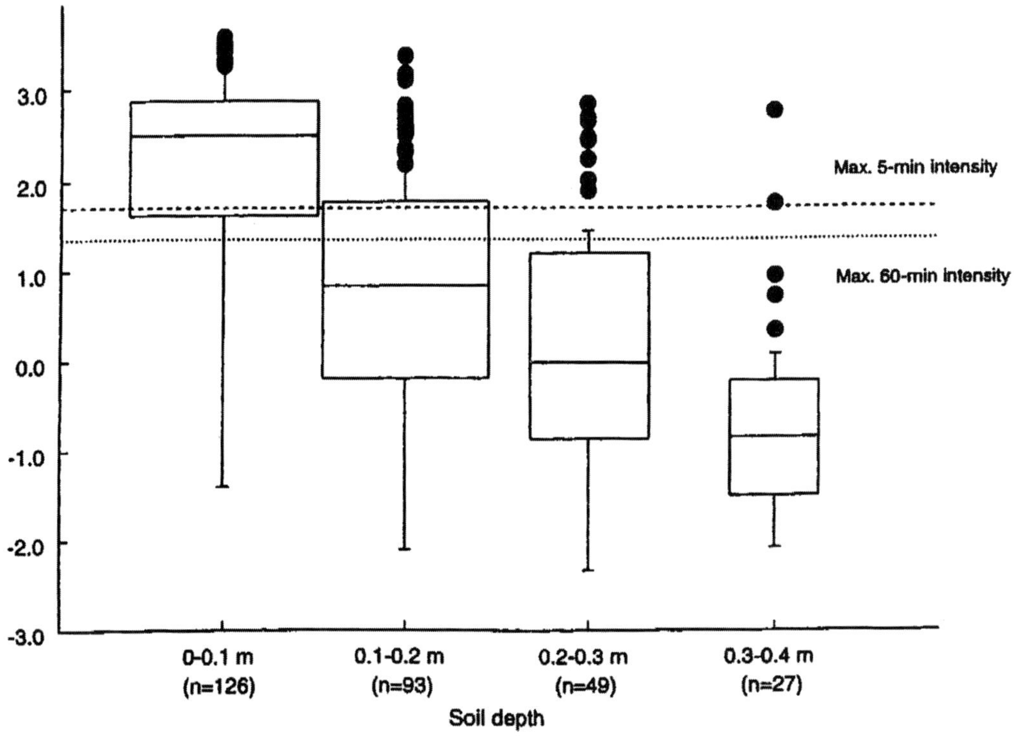

Figure 14.54 Soil hydraulic conductivity and rainfall intensity. The width of each box is proportional to the sample size (in parenthesis), the depth represents the interquartile range, the horizontal bar the median, and the solid circles those data points with a distance from the upper quartile larger than 1–5 times the interquartile range. For better visualisation, the data are presented in logarithmic form. The rainfall intensity lines are the medians of the 26 storms considered in this study. (After Elsenbeer and Vertessy, 2000.)

flowpaths, future sampling schemes need to take such temporal variability into account (Elsenbeer and Lack, 1996a).

Overall, the La Cuenca study emphasises the point that had a complementary hydrometric study not been undertaken to establish the dominant overland flow pathway, then correct interpretation of the hydrochemistry linked to hydrological pathways in this environment would have been more difficult. Conversely, the hydrochemical approach was necessary to confirm the dominant role of overland flow with respect to coupling the hillslope hydrology with the storm hydrograph. Furthermore, the distinction between overland flow and pipeflow (in the context of return flow) is shown to be meaningless from a hydrochemistry perspective. Elsenbeer *et al.* (1995b), for example, showed no significant differences between the SiO_2 concentrations of overland flow and pipeflow.

INFILTRATION-EXCESS OVERLAND FLOW

The preceding survey has already highlighted the comparatively restricted occurrence of Hortonian (infiltration-excess) overland flow in the tropical forests of west Africa (Dubreuil, 1985),

Table 14.11. *Characteristics of the 26 selected rainfall events*

Variable	Minimum	Median	Mean	Maximum
Amount (mm)	13.9	32.1	34.4	77.3
Maximum 5-min intensity (mm h^{-1})	8.2	51.6	54.6	85.1
Maximum 60-min intensity (mm h^{-1})	3.1	22.8	21.0	47.3
Duration (min)	75	210	426	1665

Source: After Elsenbeer and Vertessy. (2000).

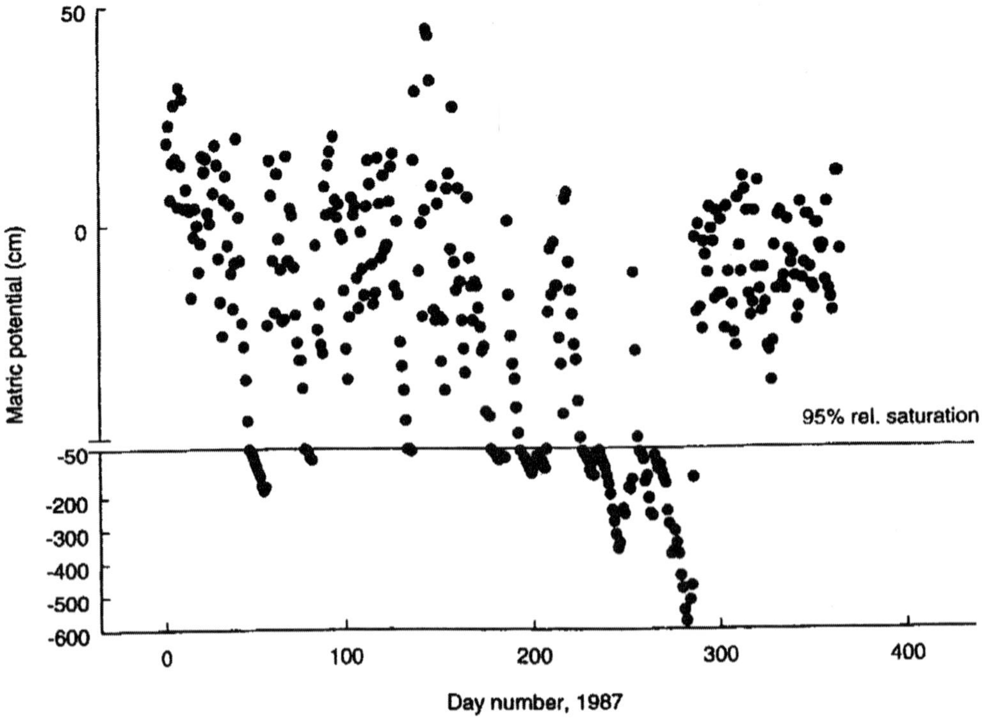

Figure 14.55 Time series of matric potential at a depth of 0.3 m. Data points represent the median of nine tensiometer readings from sites S1 ($n = 4$) and S3 ($n = 5$). See Figure 14.53 for location. (After Elsenbeer and Vertessy, 2000.)

western Amazonia (Elsenbeer and Cassel, 1991), Borneo Island (Malmer, 1993; Chappell *et al.*, 1999) and northeast Queensland (Herwitz, 1986a, b). In the case of west Africa, Dubreuil (1985) suggested that HOF occurred, especially over lateritic-type soils (Acrisol-Alisol), although detailed measurements were lacking in his overview. Chevallier and Planchon (1993) subsequently provided more details of HOF within the Booro-Borotou catchment over soils which are ferruginous and have a massive structure. In the case of western Amazonia (La Cuenca), Elsenbeer and Cassel (1991) noted that a positive feedback loop could be inferred for the steeper valley side slopes. Thus this positive feedback loop takes the form, erosion → low surface infiltration rates → infiltration excess overland flow → erosion. Thus in the La Cuenca study, HOF seems restricted in spatial occurrence principally to these less permeable soils on steep slopes. By contrast, the supplementary impact of forest morphology which leads to concentrated stemflow at the base of tree boles provided the hydropedological environment for localised HOF. In the high rainfall area of Mt. Bellenden Ker, northeast Queensland (Herwitz, 1986), this flow vector added to the prevailing SOF.

Elsewhere, in the forest of Guyana, Jetten (1994) reported overland flow occurrence (HOF was inferred) which varied between 3% to 15% of rainfall for slope angles from 25% to 60% respectively over the less permeable Ferralsols soils. Based on flow velocity measurements using the NaCl 'salt-pulse' method, Jetten considered that most of this HOF (but not all) would subsequently

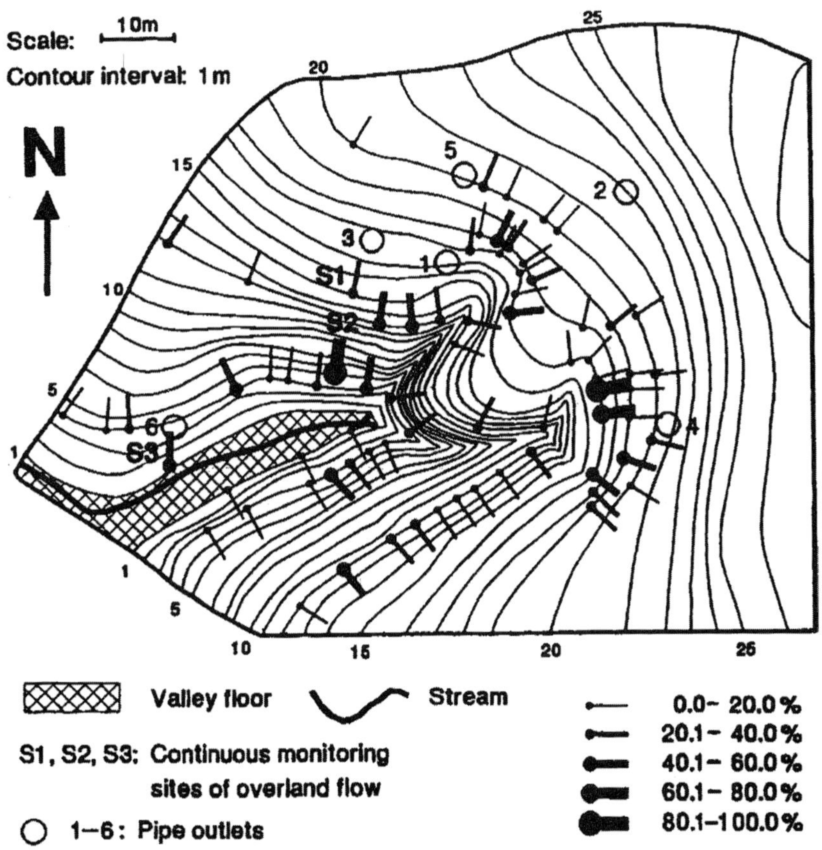

Figure 14.56 The spatial and temporal frequency of overland flow at La Cuenca. The head of each overland flow symbol marks the position of a runoff detector; the thickness reflects the temporal frequency of overland flow as a percentage of 187 monitored events. (After Elsenbeer and Lack, 1996a.)

infiltrate over the lower floodplain before it was able to attain organised drainage, i.e. a stream. The quickflow component of discharge was attributed more to SOF emanating from the high groundwater table near the stream channel although HOF would contribute to the occurrence of a high groundwater table as well. In common with the preceding studies of Herwitz (1986) and Elsenbeer and Cassel (1991), the occurrence of HOF in Guyana was associated with near, or at, soil saturation (Jetten, 1994). In general, however, HOF is temporally and spatially restricted in occurrence. For example, Malmer (1993) reported that 2.9% of rainfall in his period of study in Sabah, Malaysia, occurred as infiltration-excess overland flow over clays (Acrisols), but it was restricted to the occasional high magnitude storms.

Tai Forest, Côte d'Ivoire

Exceptions to this notion of restricted occurrence are the studies of Casenave *et al.* (1984) and Wierda *et al.* (1989) in the Tai Forest of Côte d'Ivoire. Casenave *et al.* (1984) had suggested that overland flow and natural erosion occurs in the Tai Forest. Later, Wierda *et al.* (1989) applied the Mein and Larson's (1973) extension of the Green–Ampt model in the Tai National Park rainforest with reasonable success (reviewed in Bonell with Balek, 1993, p. 201). Of particular interest was the measured low saturated infiltration rates (approximately K^*) which ranged from 7 to 12 mm h^{-1} along a catenary slope. These infiltration rates declined from the upper to the down slope (grading from Plinthic Acrisols on the upper slope through to Xanthic Ferralsols, mid-slope, and imperfectly drained Ferric/Clayey Acrisols on the down-slope segment), in line with the trend towards finer-textured soils at the footslope, cf. Ogunkoya *et al.*, 2000, 2003 (the valley bottom water-logged soils were not included in the study). No direct measurements of hillslope storm runoff were made, but when the infiltration rates were compared to rainfall data, it was concluded that HOF frequently occurred, especially on the middle and lower slopes sections where K^* was lower. Thus it is the low, surface K^* (by forest standards) rather than exceptionally high short-term rain intensities in this Côte d'Ivoire study (which are comparatively moderate in west Africa by global standards, see Bonell *et al.*, this volume; Jeje *et al.*, 1986; Ogunkoya *et al.*, 1984) which encourages this hillslope response (Bonell, 1993).

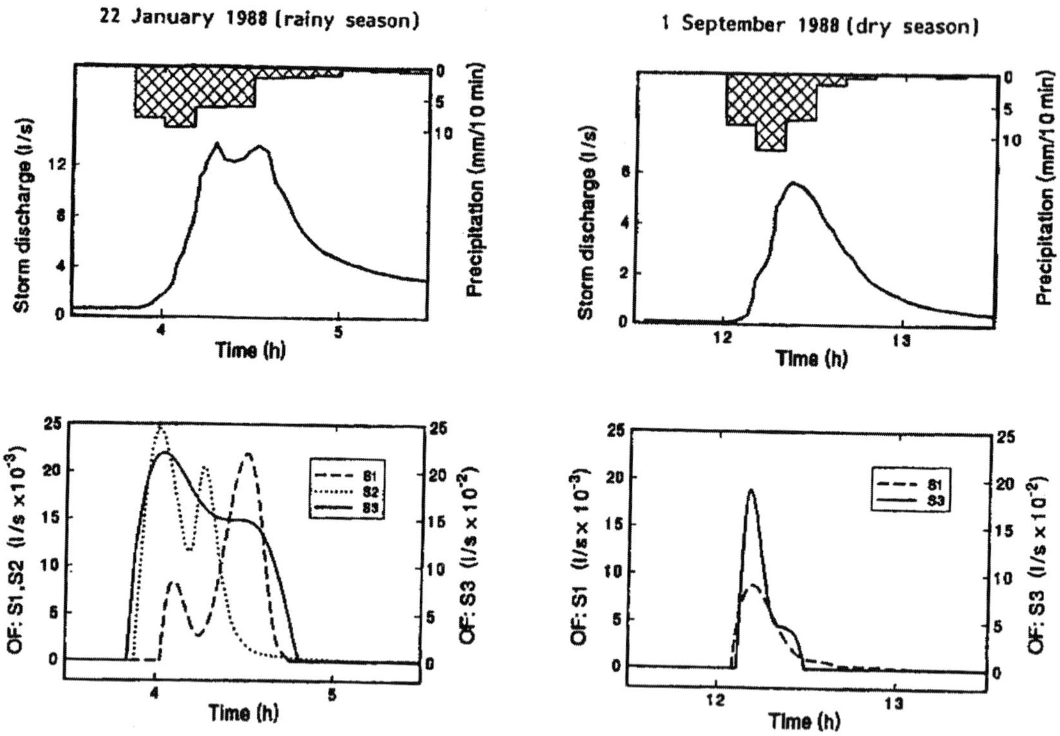

Figure 14.57 Hydrometric characteristics of the events of 22 Jan and 1 Sept 1988. Top panels: hyetographs and stream flow hydrographs; bottom panels: overland flow (OF) hydrographs from sites S1 and S2 (left *y*-axis) and S3 (right *y*-axis). (After Elsenbeer and Lack, 1996a.)

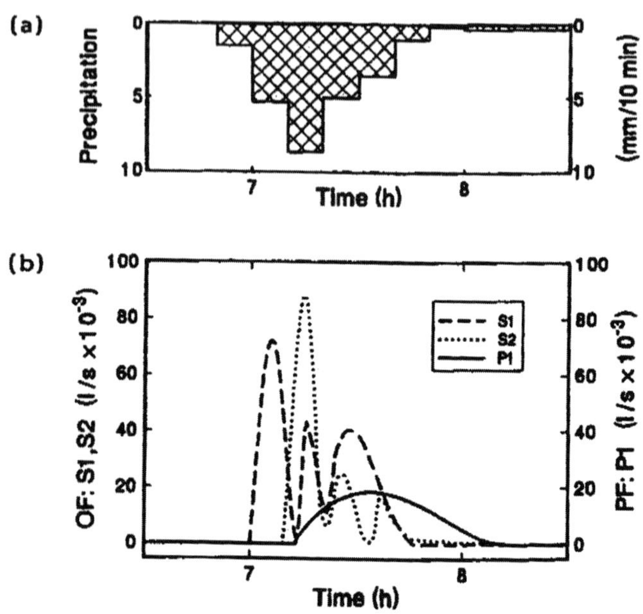

Figure 14.58 Hydrometric characteristics of the event of 6 May 1987: (a) hyetograph; (b) overland flow (OF) hydrographs from sites S1 and S2 (left *y*-axis) and pipeflow (PF) hydrograph from site P1 (right *y*-axis). (After Elsenbeer and Lack, 1996b.)

THE IMPACT OF COMPLEX GEOLOGY AND SOILS ON DOMINANT STORMFLOW PATHWAYS

Only a limited number of studies have highlighted the complexity of the runoff generation process at smaller scales in response to the complex, spatial organisation of geological strata and associated soils. None of these studies however, has undertaken a rigorous experimental strategy on the lines S1 to S3 (Elsenbeer and Vertessy, 2000). It is evident however, that in at least one study (Chevallier and Planchon, 1993), the prospect of being able to generalise a dominant pathway is more difficult.

Booru-Borotou, Côte d'Ivoire

During an overview of principally ORSTOM[4] work, Dubreuil (1985) had suggested that a wide range of runoff pathways occurred in west African landscapes. A later review of forest hydrology research in Francophone west Africa by Adokpo Migan (2000) also suggested a range of complex runoff responses. The

4 ORSTOM: Office de la Recherche Scientifique et Technique Outre Mer but now known as IRD, L'Institut Française de Recherche Scientifique par le Développement en Coopération.

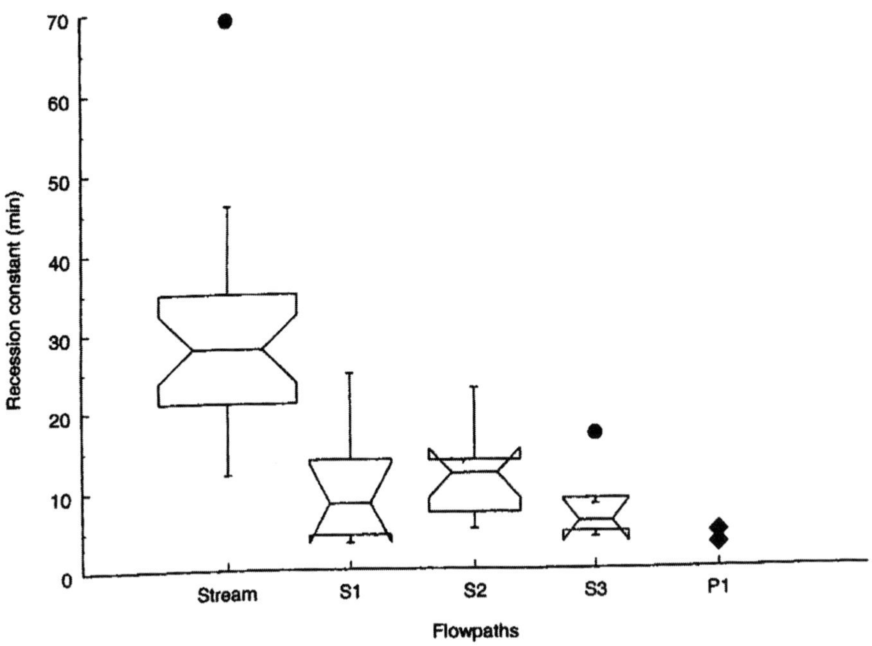

Figure 14.59 Recession constants of stream flow (first constant), overland flow (S1, S2 and S3) and pipeflow (P1). (After Elsenbeer and Vertessy, 2000.)

Table 14.12. *Sampling approach*

Compartment	Code	Frequency	Sample size
Precipitation	P	event-based	62[a]
Throughfall	T	event-based	55[a]
Overland flow	OF1	event-based	30[a]
Overland flow	OF2	event-based	119[a]
Pipe flow	PF	event-based	58[a]
Soil water at 0.3 m	SW3	fixed-interval	119[b]
Soil water at 0.6 m	SW6	fixed-interval	123[b]
Soil water at 0.9 m	SW9	fixed-interval	126[b]
Floodplain groundwater	FGW	fixed-interval	168[b]
Hillslope groundwater	Seep	fixed-interval	267[b]
Stream water, base flow	BF	fixed-interval	276[b]
Stream water, storm flow	SF	within event	30

[a] Number of events.
[b] Number of daily samples.
Source: After Elsenbeer *et al.* (1999b).

study by Chevallier and Planchon (1993) within Côte d'Ivoire however provides more intricate detail. Located in the north-west of Côte d'Ivoire, on the border between the wet-dry/subhumid regions, the Booro-Borotou catchment study (1.36 km^2, annual rainfall 1360 mm, within the wet/dry zone of Chang and Lau, 1993) sampled the complexity of geomorphic, vegetation and land use, and pedogenic features, typical of west African landscapes. All these characteristics have a consequential bearing on the resulting complex hillslope hydrology response (Chevallier and Planchon, 1993). Although no soil classifications were provided by Chevallier and Planchon (1993), three different soil systems corresponding to the geomorphological organisation were identified over the magmatitic gneiss geology:

- Ferralithic upstream, where the soils are red and permeable with high clay and gravel contents. SSF is the dominant pathway.
- Ferruginous at midslope where the structure is massive, often compacted and which the soils are ochre, with much gravel. This is an area which generates HOF.
- The soils are hydromorphic on the lower slopes and the valley bottom which forms a large sandy reservoir containing a fluctuating water table. This system is the source of both SSF and SOF.

The experimental approach followed by Chevallier and Planchon (1993) was somewhat unconventional by comparison with other studies, in that the work relied partly on rainfall simulators and surface runoff traps (as indicators of runoff); *in situ* soil hydraulic properties were not presented. Over a three-year period, 30 major rainfall-runoff events were recorded. Total rainfall ranged from 14.4 to 82.7 mm, with corresponding maximum 30-min intensities, 11.4 to 95.7 mm h^{-1}. Quickflow volumes generally remained low, however, and only exceeded 2 mm for three events, despite the occurrence of infiltration-excess

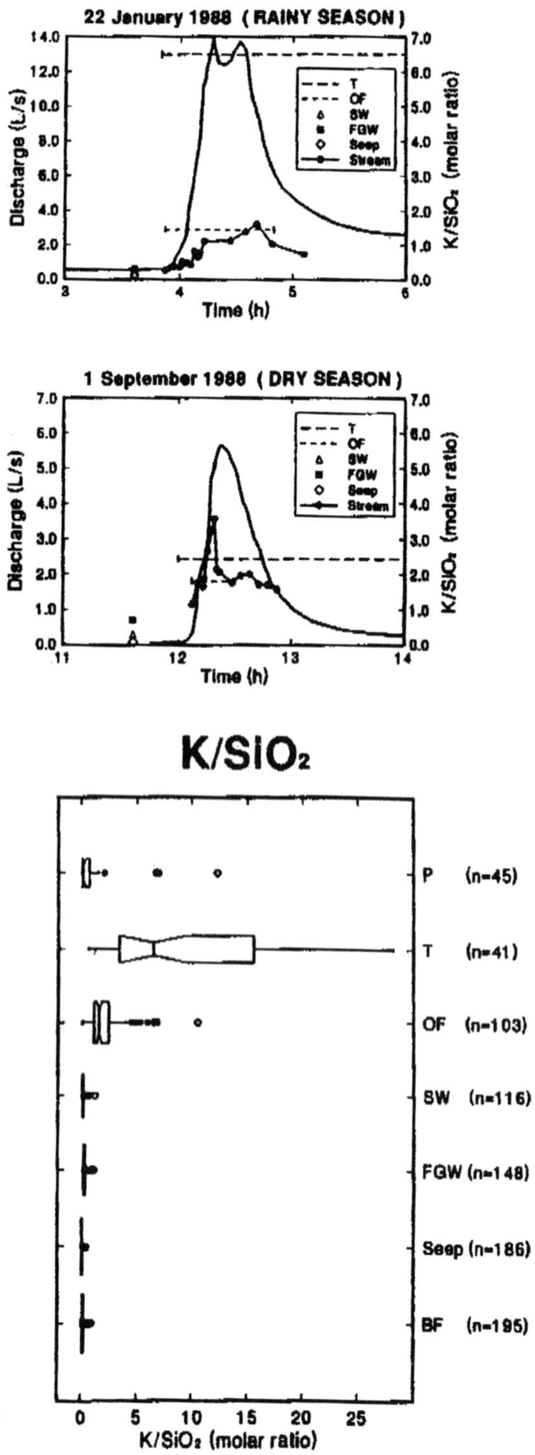

Figure 14.60 The chemistry of the K/SiO_2 ratio at La Cuenca. Top panel: discharged (solid line) and storm flow chemographs during two events. The length of the throughfall (T) and overland flow (OF) lines show the duration of rainfall and overland flow, respectively. Bottom panel: boxplot characterisation of compartments and flowpaths covering the whole study period. P, event precipitation; T, throughfall; OF, overland flow; SW, soil water; FGW, floodplain groundwater; Seep, hillslope groundwater; BF, baseflow. Sample sizes in parentheses. The crossbars in the boxes indicate the sample medians, the notches their 95% confidence intervals. The length of a box represents the sample interquartile range. (After Elsenbeer and Lack, 1996a.)

overland flow over crusted soil in selected cultivated areas (7–8% of catchment) in the higher parts of the basin as well as over the midslope area. The latter corresponds with soils which are ferruginous and have a massive structure. By contrast, infiltrability is high on the upper plateau and lower slopes in the lower riverine forest and in the valley bottoms. A significant proportion of the infiltration–excess overland flow re-infiltrates both downslope and downstream in small gullies from mid-hillslope sections. On exceptional occasions, it seems that, based on dye tracing experiments, this overland flow can attain the catchment outlet. Consequently this redistribution (re-infiltration) of overland flows, for the most part, accounts for the low quickflow responses. Neutron probe data also showed both surface saturation within a clay-sand upper layer and lateral subsurface flow redistribution from upslope within in a lower sandy layer. Thus water tables in the sandy hillslope bottom (riverine) areas acted as a repository for redistributed moisture upslope and were considered to play the most significant role in the runoff generation process. Piezometric levels were highly responsive, which indicated the prime contributing source being lateral subsurface flow into the stream; possibly supplemented by saturation-excess overland flow. Thus, despite the small size of Booro-Borotou, '. . . flow may take all possible paths . . .' and in various combinations, that is '. . . surface runoff in the sense of Horton (1933), overland flow (saturation-excess) from variable contributing areas, baseflow supplied by the aquifer, or different types of rapid internal flows (i.e. macropores)' (Chevallier and Planchon, 1993, p. 189).

M'bé, Côte d'Ivoire

Chevallier and Planchon (1993) had reported double-peak storm hydrographs with the second peak corresponding to the highest rise of the water table to the surface, although the specific hydrograph details and mechanisms were not provided, cf. Anderson and Burt (1978). In a subsequent study, also in Côte d'Ivoire (but further south than the Booro-Boroton experimental basin), Masiyandima *et al.* (2003) presented storm hydrograph analyses supported by a piezometer network within the M'bé experimental catchment (130 ha, annual rainfall 967 mm, 1992–2000). The catchment was only partly forested, principally on the mid- to upper slopes and upland (the remainder being natural grass fallow and rain-fed or irrigated rice). Nonetheless, the heterogenous physiography and hydropedology were the principal controls of storm runoff mechanisms, and so are relevant to the current discussion.

Rainstorms originate principally from easterly perturbations (see Bonell *et al.*, this volume) being of short duration (<30 min), high short-term intensities (data not reported) but comparatively moderate rain totals (18–33 mm), for three events outlined by Masiyandima *et al.* (2003). Topography is gentle (0–4%), but like the Booro-Boroton, the hydropedology is complex across a typical cross-section, with four physiographic-pedology units. The lowest, valley bottom area (slope <1%) has a perched water table overlying an impermeable clay layer close to the surface. During the wet season, this perched water table is coupled with the deeper regional groundwater table. Saturation excess overland flow (SOF) can contribute up to 43% of the strom hydrograph from this valley bottom, with a clear link between total rain and the percentage of SOF. It is this source that accounts for the first hydrograph peak. The second hydrograph peak was delayed from minutes to hours, and was dependent on the depth of the perched water table immediately upslope within an adjacent hydrographic zone (width 25–50 mm, slope 1–4%). The closer the perched water table was to the surface preceding a storm, so the time difference between the first and second peak contracted. In other optimal conditions of antecedent wetness, the two peaks would merge. Masiyandima *et al.* (2003) considered that the main SSF contributions to the second peak emanate from this hydromorphic zone where the overlying soil has a conductivity of ~400 mm h^{-1}.

The remaining physiographic-pedalogical unit, viz. the mid and upper slopes (slope 2–4%) and upland (slope 0–20%) also favour SSF (and localised HOF over exposed impermeable iron pan (known as *cnirasse*)); but Masiyandima *et al.* (2003) considered that for short timescales, such flow pathways and groundwater from these higher physiographic-pedological units had minimal effect on changes in the streamflow in the valley bottom, cf. Chevallier and Planchon (1993).

A principal message of this work is that the surface water–groundwater is well coupled during storms within the valley bottom and hydromorphic zone; and such coupling is responsible for runoff generation mechanisms and the double-peaked storm hydrograph responses.

ECEREX, French Guyana

Elsewhere, in the ECEREX (Ecology-Erosion-Experimentation) within the French Guyana (annual rainfall ~ 3300 mm), ten elementary research basins with drainage areas ranging from 1 to 2 ha were established from 1975 onwards to evaluate the impacts of different forest conversions (Sarrailh, 1990; Fritsch, 1990; see the detailed review of ECEREX in Grip *et al.*, this volume). Two of these basins, catchment A (1.3 ha) under pasture and the control catchment B (1.6 ha) (Figure 14.61) (tropical moist forest) were investigated by a French research team in the early 1990s using tensiometry and environmental tracers for chemohydrograph separation (Bariac *et al.*, 1995a, b; Millet, 1996) and the application of TOPMODEL (Molicova *et al.*, 1997).

The Guyanan soils, bordering the Amazon basin, are highly complex. They are composed of several layers of contrasting hydraulic conductivities about 25 m above the bedrock (micaschiste sprinkled by pegmatite veins). To summarise, two families of soils were identified by Boulet (1981) during a pedological survey of the ECEREX basins:

- One group of soils with vertical drainage behaviour which show a very homogeneous structure and high transmissivity with deep percolation to more than 2 m depth.
- The other group of lateral drainage soils also have a high surface permeability, but there is a marked decline in hydraulic conductivity with depth. A compact clayey horizon limits vertical percolation and can constitute the base of perched water table development. In this soil group, Boulet (1981) suspected that the dynamics of water was superficial and parallel to the slope.

A parallel root survey by Humbel (1978) reported that on average the lateral drainage soils contained 87% of the weight of roots between 0 and 0.20 m depth, whilst the proportion is only 66% for soil with vertical drainage. For 0.20–0.60 m depths, the respective proportions of root weight are 9% for lateral drainage and 25% for vertical drainage soils. Below 1 m depth, there are almost no roots in lateral drainage soils, while 5% of the total weight are still present in vertical drainage profiles. In addition, the number of large roots (10–40 mm diameter) was three times higher in the 0–0.20 m layer of the lateral drainage soils. Thus the existence of abundant macropore flows along the lateral drainage slopes (of 15–30% topographical gradients) to facilitate the occurrence of SSF (at a minimum SSF and even SOF) are in place.

A schema of the different catenary stages of pedological evolution on the Bonidoro schists within the ECEREX basins are presented in Figure 14.62. The catchment B (control) is typified by stage II and catchment A by stage III (Sarrailh, 1992). The dominant flowpaths are inferred from the pedological survey of Boulet (1981), although runoff production from the surficial top 0.05 m was investigated by a 40 m × 10 m runoff plot experiment close to the boundaries of the forest control experiment on a lateral drainage soil type. A comparison of runoff of catchment B and the plot was carried out over three years (Sarrailh, 1990; Fritsch, 1992). The production of the catchment B was slightly higher than the runoff plot, but the cumulated difference was only between 14%–17% over the three years. Thus Fritsch (1992) inferred that the dominant storm pathway was close to the surface in line with Boulet (1981). In catchment B (control), the soils with deep and free vertical drainage occupy the upper parts (about 10% of the soil area) (Molicova *et al.*, 1997); Haplic Acrisols occupy the lower two-thirds of transects, with Haplic Ferrasols dominating

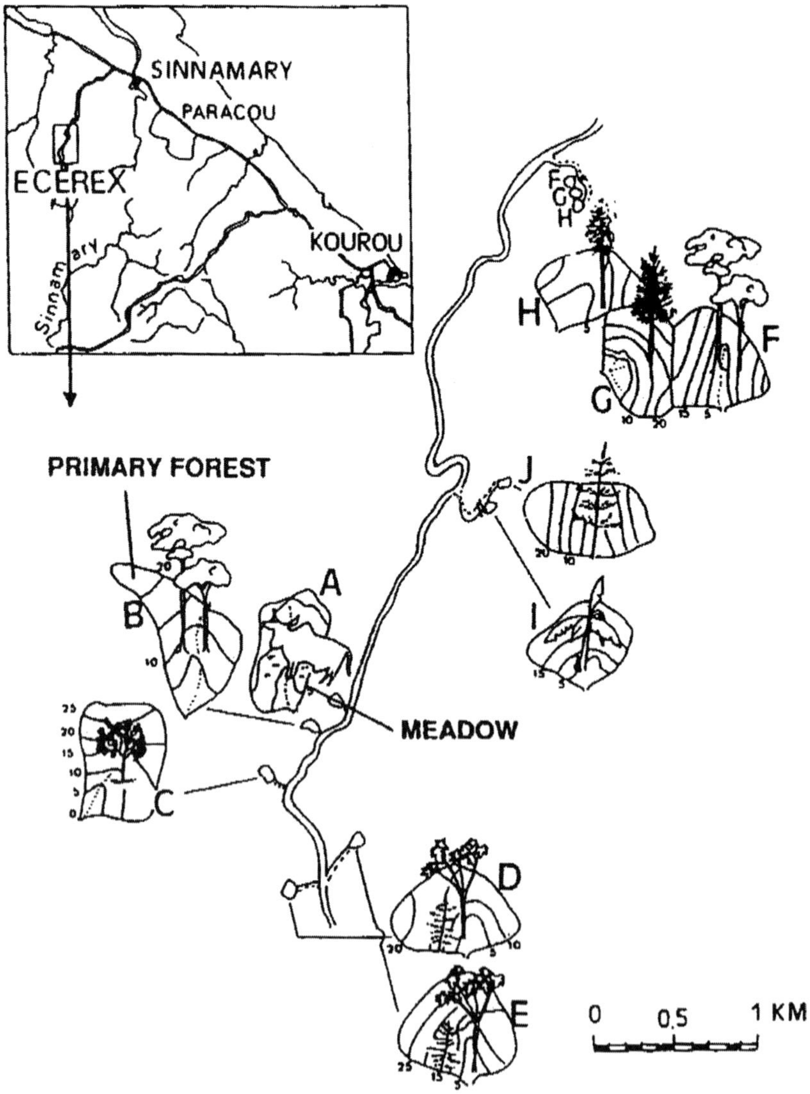

Figure 14.61 Location of the two experimental catchments (basins 'A' and 'B') in the vicinity of Sinnamary (ECEREX operation, Saint Elie's track, French Guiana). (After Bariac *et al.*, 1995a.)

the remaining upper transect. Catchment A (converted to pasture) is monopolised by lateral drainage soils (100% of the soil area). In the control catchment A, the more superficial lateral drainage zone is situated on the midslope and downslope (90% of the total soil area). Although no soil hydraulic conductivity *in situ* measurements were undertaken (S1 stage of Elsenbeer and Vertessy, 2000), soil core K_{sat} estimates taken from soil pits were in the range 36–360 mm h^{-1} for the superficial horizons (0–0.40 m depth) and 3.6–36 mm h^{-1} (below 0.40 m depth) (Molicova *et al.*, 1997).

Molicova *et al.* (1997) assessed the runoff generation process for two storms through the interpretation of tensiometer measurements taken during the course of these events. A complex pattern of perched water table development over time and space was noted. Whilst SSF was the dominant pathway, tensiometry data indicated that SOF occurs though its spatial occurrence is discontinuous, especially over the better drained upper slopes. A consequence is that SOF is disconnected from the lower percoline zone. In addition, the development of SOF was near-instantaneous following the commencement of the two events, in line with the near-zero antecedent matric potentials. What makes this particular experiment unique is that a combination of the complex hydropedology and small-scale of these research basins results in no delayed flow within organised surface drainage in between events. The deeper, regional groundwater is disconnected from the hillslope hydrology during storms (Sarrailh, 1990; Fritsch, 1992). More detail will be given later on the spatial and temporal occurrence of saturation

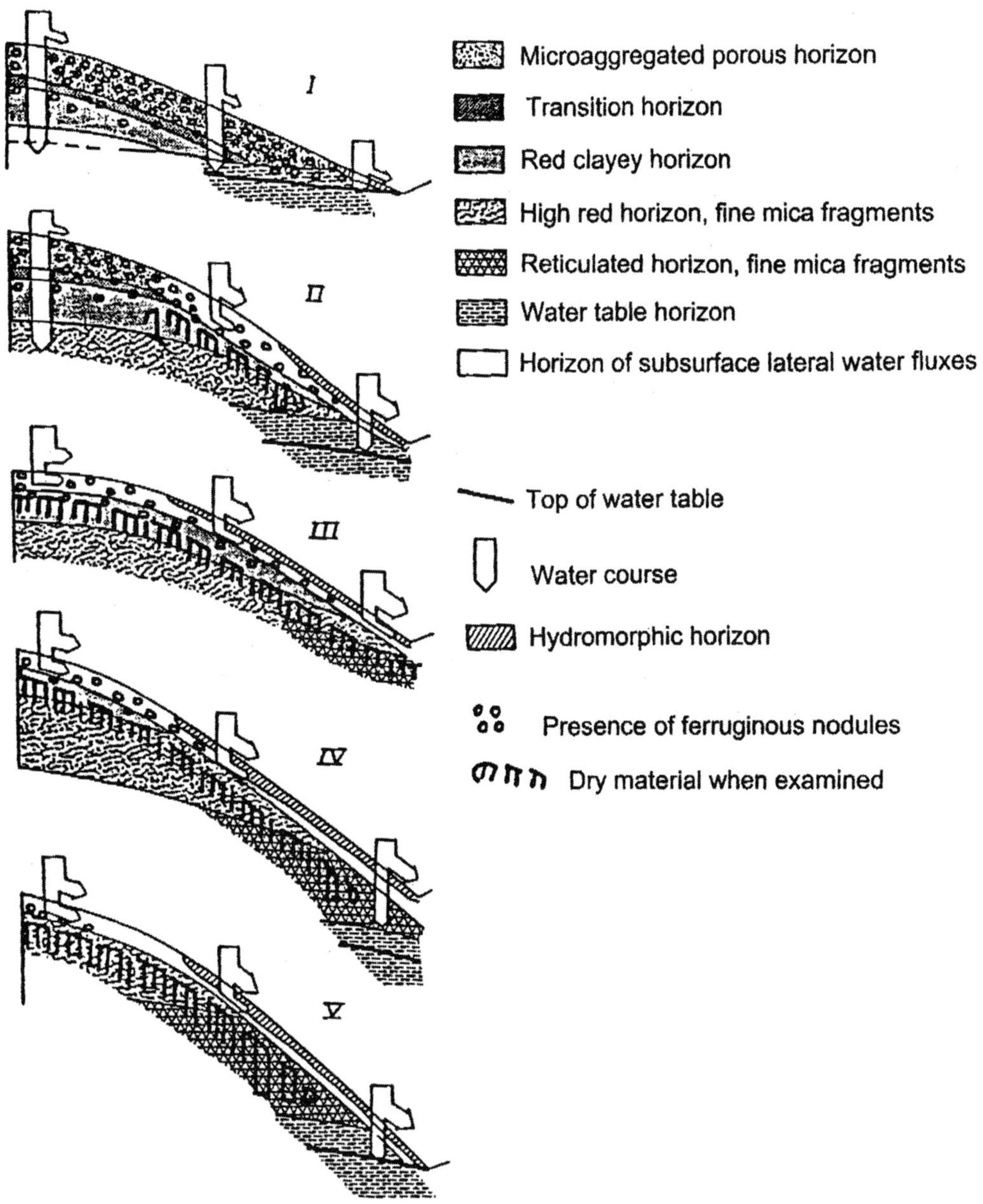

Figure 14.62 Schema of the evolution of pedological transects over Bonidoro schists. (After Boulet, 1979.)

during storms, as presented by Molicova *et al.* (1997), during a discussion of the application of digital terrain models for runoff procedure.

Although no comprehensive hydrometric study had been undertaken previously, an isotopic-hydrochemical tracer study was carried out on both the control basin (catchment B) and the converted catchment A occupied by pasture (Bariac *et al.*, 1995a, b; Millet, 1996). Chloride and potassium were used as natural chemical species and ^{18}O as the environmental isotope tracer. There are no minerals containing chloride in these basins, the only source being precipitation and throughfall. A preceding survey by Grimaldi (1988) and Grimaldi *et al.* (1994), showed that chlorides were more concentrated in 'old' waters in the soil than in the current stormwater in the environment. In agreement with Elsenbeer *et al.* (1995b), the major source of potassium added to throughfall is deemed to be from vegetation. Thus a high concentration of potassium is a strong indicator of 'new' water (Grimaldi, 1988).

Chemohydrograph separation was based on the notion of three reservoirs (subsequently modified by Millet, 1996, from that previously reported in Bariac *et al.*, 1995a, b), i.e. 'superficial' with a signature of throughfall (i.e. event water), 'intermediate'

Table 14.13. *Characteristics of some rainstorms from the ECEREX study*

				Origin of stormflow stream water (percent of total flow)		
Flood	Rainfall (mm)	Total runoff (percent of rainfall)	Maximum discharge (l s^{-1})	superficial (rainfall)	intermediate (soil matrix) (<40 cm)	deep (water table)
1	56.7	28	92	32	54	14
2	64.0	43	86.2	21	47	32
3	21.8	1	6.5	68	32	–

Source: Bonell and Fritsch (1997).

composed of soil matrix pre-event water (above 0.40 m depth) and a third component identified as 'deep' water. The ultimate basis of division into three contributing components was the assemblage of soil survey evidence; tensiometry evidence of the vadose zone (Molicova *et al.*, 1997) and specific results from the hydrochemical and isotopic experiment. The fact that Millet (1996) 'displaced' upwards the earlier classification of reservoirs adopted by Bariac *et al.* (1995a, b) (where the occurrence of SOF was not envisaged by the latter writers), was more in agreement with the subsequent hydrometric approach taken by Molicova *et al.* (1997). These modifications of reservoir sources by Millet (1996) reiterate that there is a need for a detailed hydrometric study to precede a combined hydrochemical-isotopic study.

The characteristics of three rainstorms are shown in Table 14.13 for catchment B, based on a three-component model (Bariac *et al.*, 1995b) on similar lines to that reported by Ogunkoya and Jenkins (1993) and Hinton *et al.* (1994). Elsewhere Molicova *et al.* (1997) report the hydrometric characteristics of the two larger storms.

The initial two storms are highly responsive, in contrast to the third event. Perched water table development was identified for the two larger storms only. Whilst the proportion of 'new' water is significant (from 21–68%), it is interesting that the higher the magnitude of the storm, the lower the proportion of this new water in the hydrograph. Such findings contrast markedly with the cyclone-prone Babinda environment (Bonell *et al.*, 1998) where rainfall intensities are much higher. Moreover, the event water noted in the ECEREX study is associated with the rising segment of the hydrograph (emanating from SOF and shallow SSF in the lateral drainage soils), while the contribution of soil matric water is dominating near the hydrograph peak and during the recession. In contrast, the catchment A (pasture) contributed a much smaller proportion of event water to the storm hydrograph, despite the total runoff as a percent of rainfall (41%) being higher. This event water also occurred around the hydrograph peak (Bariac *et al.*, 1995a). The upper soil layer (0–0.40 m depth) was the largest contributor, in line with the dominant lateral drainage soils which occupy this basin completely.

During a review by Bonell and Fritsch (1997) of this ECEREX work and those elsewhere in La Cuenca and Babinda, a warning is given against generalising the above results on the basis of a few storms. For the control catchment B in ECEREX, for example, 771 storm events were identified during seven years of continuous monitoring, with a maximum and minimum number of events per year being 128 and 65 (Fritsch, 1992). Moreover, 9% of total stormflow is generated during individual storms with less than 2.5 m runoff equivalent (*cf.* storm 3 in Table 14.13), 24% during storms which comprised between 2.5 and 10 mm, and 50% of stormwater was produced by storms in the range of 10–50 mm of runoff (cf. storms 1 and 2 in Table 14.13) (Fritsch, 1992). The mechanisms for runoff generation across this classification must vary. Thus the tracer experiments reported above are limited, both in magnitude and in time, and therefore do not sample comprehensively the range of hydrometeorological conditions of the ECEREX environment (Bonell and Fritsch, 1997).

A CONCEPTUAL FRAMEWORK OF HILLSLOPE HYDROLOGY RESPONSES LINKED WITH TROPICAL RAINFOREST SOIL LANDSCAPES

The preceding overview of dominant storm pathways includes some mention of the associated reference soil groups (FAO, 1974) in which these flow vectors occur. Elsenbeer (2001) formalised the linkage between tropical forest soilscapes and the hillslope hydrology response. Two groups of research sites were considered. The first were 'prediction' sites which conformed with the S1 stage of Elsenbeer and Vertessy (2000) except at some locations rainfall data were not available. Their basis for the grouping was detailed records of vertical changes in saturated soil hydraulic conductivity together with adequate information to permit a soil classification. These sites are summarised in Figure 14.63. As acknowledged by Elsenbeer (2001), the six sites do not show a clear picture. The Reserva Ducke (RD) (Nortcliff and Thornes,

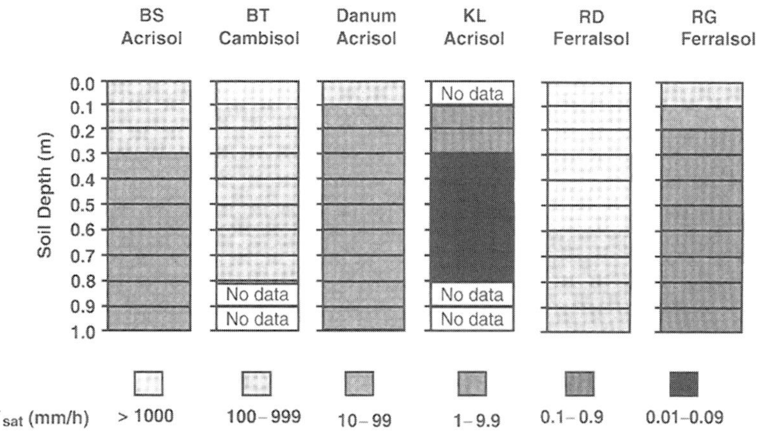

Figure 14.63 K_{sat} as a function of depth at the 'prediction' sites Bukit Soeharto (BS), Bukit Tarek (BT), Danum, Kiani Lestari (KL), Reserva Ducke (RD) and Rancho Grande (RG). Thick lines indicate the approximate position of impeding layers, defined as K_{sat} <6 mm h^{-1}. (After Elsenbeer, 2001.)

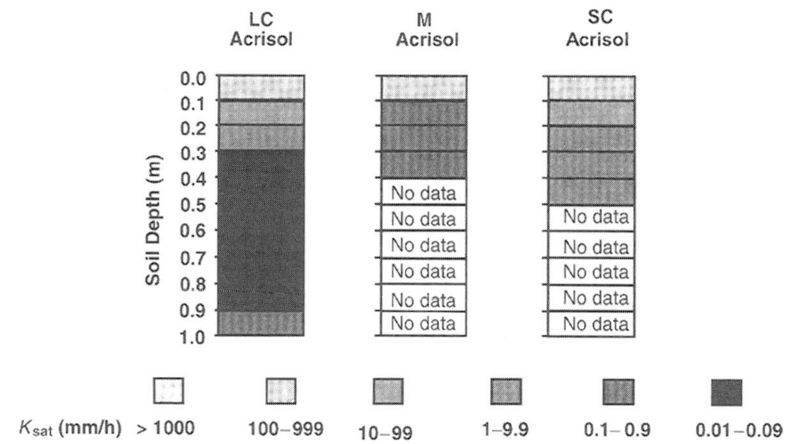

Figure 14.64 K_{sat} as a function of depth at the 'test' sites La Cuenca (LC) and South Creek (SC). Thick lines indicate the approximate position of impeding layers, defined as K_{sat} <6 mm h^{-1}. (After Elsenbeer, 2001.)

1989) has the weakest anisotropy associated with a Ferralsol; and the strongest anisotropy with the Kiani Lestari (KL) (Wenzel *et al.*, 1998) site identified with an Acrisol. The Rancho Grande (RG) Ferralsol (Elsenbeer *et al.*, 1999) has a more pronounced K_{sat} anisotropy than other Acrisols (BS, Ohta and Effendi, 1992; Danum, Chappell *et al.*, 1998). Aside from topographically and seasonally favoured areas such as riparian zones, predominantly vertical storm pathways are suggested for Bukit Soeharto (BS), Bukit Tarek (BT) (Noguchi *et al.*, 1997a, b) and Reserva Ducke (RD) based on Figure 14.63. The most likely lateral near-surface flow component is inferred for Kiani Lestari (KL) which could potentially include SSF, SOF, and/or return flow, RF. Lateral SSF was suggested at Danum (at 1m. depth) and at Rancho Grande, RG (from 0.2m depth onwards), with RF possible at both sites.

The second group of sites were known as 'test' sites (La Cuenca, Peru, Elsenbeer and Lack, 1996b; Mendolong, Sabah, Malaysia, Malmer, 1996; South Creek, Australia, Bonell and Gilmour, 1978) because all had been documented in detail at the S3 stage of experimentation (Elsenbeer and Vertessy, 2000). These are shown in Figure 14.64 and the more detailed hydrological responses available provide a basis for the postulated responses at the former 'prediction' sites, as well as their integration to produce a spectrum of hydrological flowpaths in tropical rainforests (Figure 14.65). Whilst the Ferralsol sites (RD, RG) are not consistent in dominant storm pathways, the Acrisols are consistent, with a dominant lateral, near-surface flow component. So, at one end of the spectrum is an 'Acrisol'-type end member and at the other are the 'Ferralsol-type' end-members, as represented by the Reserva Ducke (RD), of dominant vertical pathways. This cluster includes functionally

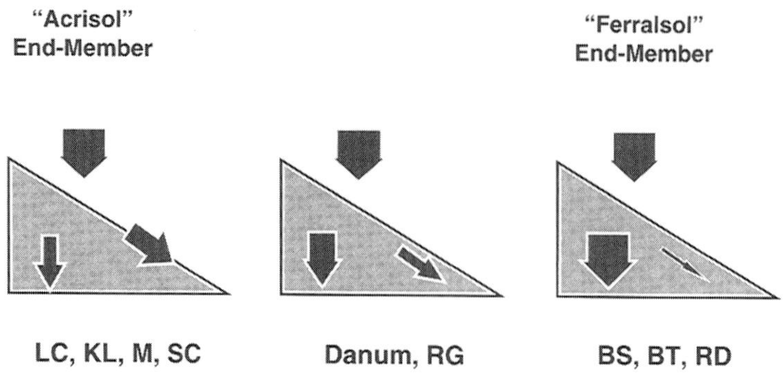

Figure 14.65 The spectrum of hydrological flowpaths in tropical rainforests. The arrows indicate the idealised partitioning of rainfall. The gradient of the lateral component is from strong and near-surface at the left to weak and subsurface at the right. LC, La Cuenca; KL, Kiani Lestari; M, Mendolong; SC, South Creek; RG, Rancho Grande; BS, Bukit Soeharto; BT, Bukit Tarek; RD, Reserva Ducke. (After Elsenbeer, 2001.)

similar, but soil taxonomically different, Bukit Soeharto (BS) and Bukit Tarek (BT) sites. The remaining Danum (Acrisol) and RG sites (Ferralsol) have a modest lateral subsurface component. Figure 14.65 represents an average situation using a median, event-based, 'global tropical' $I_{60\,max}$ of 6 mm h^{-1}. For extreme events, I_{60max} would move the intermediate sites Danum and RG towards the 'Acrisol' End-Member group. Furthermore '. . . favourable topographic conditions are likely to add a strong lateral surface component to runoff generation regardless of dominant soil type' (Elsenbeer, 2001, p. 1757).

Despite the uncertainties mentioned above, there seems to be a plausible association between rainforest-covered Acrisol landscapes and overland flow occurrence as a potential pathway. The Ferralsols require further documentation before the Reserva Ducke (RD) site can be considered the archetypal 'Ferralsol-type' slope end member of predominatly vertical flowpaths. Even here, SOF in the lower riparian zones is inferred, underlain by Podsols, Fluvisols or Gleysols within the Ferralsol landscape (Nortcliff *et al.*, 1979; Franken, 1979; Lesack, 1993). Thus overland flow is an important mechanism of runoff generation, especially in connection with Acrisols which occur in about one-third of the humid tropics (Kauffman *et al.*, 1998).

As Elsenbeer (2001) remarked, additional field testing on the lines of stages S1 to S3 (Elsenbeer and Vertessy, 2000) are required before the current hypothesis that overland flow is fundamental to the runoff generation mechanism in forested Acrisol but not in Ferralsol landscapes can be confirmed. In addition, Acrisols and Ferralsols combined cover only about 60% of the humid tropics (Kauffman *et al.*, 1998) which means that additional extreme end-members, in terms of pedo-hydrological functioning, may still need to be identified. For example, the West African studies in Côte d'Ivoire (Tai Forest Study, Casenave *et al.*, 1984; Wierda *et al.*, 1989; Booro-Borotou, Chevallier and Planchon, 1993) identified HOF as a dominant pathway in soils with a shallow hardened layer (e.g. Plinthosols). Such soils require more detailed research to ascertain whether they should be placed in the 'Acrisol-type' category or they represent a new, more extreme, member.

In the meantime, it is encouraging that additional studies outlined in this review linked with Acrisols (e.g. West Africa, Jeje *et al.*, 1986; Puerto Rico, Schellekens, 2000; French Guyana, Molicova *et al.*, 1997; Brunei, Dykes and Thornes, 2000) fit within the hydrological behaviour of the 'Acrisol-type' End Member group of Elsenbeer (2001). A possible exception is the Mgera catchment in the southern Tanzanian highlands (Lørup, 1998).

THE ROLE OF RIPARIAN ZONES IN THE RUNOFF GENERATION PROCESS

Despite considerable emphasis on the significance of riparian zones as a principal source of SOF within environments which otherwise favour more vertical pathways (e.g. central Amazonia, Malaysia, Southern Highlands, Tanzania), no detailed studies of the runoff generation dynamics of such zones have been reported. Direct measurements of SOF are lacking, and this storm pathway has been inferred from hydrograph analyses and stream water chemistry.

On the other hand, some relevant hydrological work has been undertaken in support of a better understanding of carbon (C) and nitrogen (N) dynamics within both riparian and hyporheic zones of tropical forests. Such reports have emerged outside the traditional sources of hillslope hydrology literature (e.g. Bowden *et al.*, 1992; Chestnut and McDowell, 2000; McClain and Elsenbeer, 2001; McClain *et al.*, 1994, 1997; McDowell *et al.*, 1992). These riparian and hyporheic zones (where stream water and subsurface water actively interact; e.g. Bencala, 2000) are areas of rapid N and C transformation; hyporheic sediments in particular seem capable of immobilising a significant proportion of N and C fluxes from adjacent hillslopes before their entry into stream water. It is beyond the scope of this chapter to outline the dynamics of N and C removal or retention by the near-stream or hyporheic zones (see Proctor, Connolly and Pearson, this volume) but the latter requires an integrated hydrological appreciation of stormflow pathways and groundwater-surface water interactions.

A key contribution from the biogeochemical research community has been an assessment of subsurface water movement through the riparian and hyporheic zones, presently lacking in hillslope hydrology studies. To illustrate these contributions, the work in the Luquillo Experimental Forest, Puerto Rico, is used as an example because it links with the earlier description of hillslope hydrology work of Schellekens (2000).

McDowell *et al.* (1992) had earlier compared two sub-basins (Bisley, Icacos) of the Luquillo Experiment Forest for nitrate, ammonium, dissolved oxygen in stream water and shallow groundwater. Differences in speciation and concentration of nitrogen within the riparian zone groundwater appeared to be controlled by corresponding differences in geology and hydrogeology (Figure 14.66). At the Icacos site, after infiltrating through oxic sandy red clay and saprolite on the upper slopes, the resulting relatively deep SSF passes through a deep layer of highly conductive coarse sand towards the stream, below the rooting zone of riparian vegetation. Under these circumstances the processes of mineralisation, nitrification, plant uptake and denitrification are spatially segregated so that upslope, nitrates dominate whereas ammonium levels prevail in the riparian area. In contrast, at the Bisley site, dense clays retard deep percolation following rapid infiltration upslope (as described by Schellekens, 2000). Thus the more shallow SSF has a trajectory which has to pass through a more variably oxidised zone as well as the rooting zone. As a result, there is a co-existence of mineralisation, nitrification, plant uptake and denitrification. Consequently there are no consistent trends in the spatial distribution of nitrate and ammonium, and instead there is the co-existence of both ions across the slope transect and generally lower concentrations (McDowell *et al.*, 1992).

Chestnut and McDowell (2000) extended this investigation through an intensive study of the hydrological and chemical characteristics along a 100 m reach of a sandy-bottom tributary of the Rio Icacos using the experimental design shown in Figure 14.67. Transects of riparian and hyporheic wells were installed to detect groundwater movement through the floodplain and its outwelling within stream. The hyporheic wells comprised three sets of vertically nested wells (at 0.1, 0.3, 0.5 and 0.8 m below the stream bed surface) on the centre, left and right sides of the stream. This enabled the construction of groundwater flow nets (Figure 14.67) and the calculation of vertical as well as lateral hydraulic gradients. For the hyporheic wells, the difference in stream and subsurface hydraulic head was measured as the difference between the respective stream and water table levels. Saturated hydraulic conductivities were measured using auger hole pumping tests (Hvorslev, 1951; Boersma, 1965; Freeze and Cherry, 1979). These measurements enabled both lateral and vertical fluxes to be calculated using the Darcy equation and then compared with the cumulative groundwater discharge during baseflow conditions (as measured by dilution gauging with sodium bromide) over the 100 m reach. Some of the results are summarised in Table 14.14 and Figure 14.68. From this work it was determined that an additional $\sim$1.5 l s^{-1} of groundwater was supplied to the 100 m stream transect.

Chestnut and McDowell (2000) compared the calculated vertical hydraulic gradients using the cumulative groundwater input Q_{GW} ($\sim$ 1.5 l s^{-1}) and determined K_{sat} in Darcy's equation with the corresponding average vertical hydraulic gradients (VHG) from the hyporheic wells. The average VHG necessary to account for Q_{GW} was small, 0.02 (range –0.06 to +0.07 cm cm^{-1}) (see Table 14.14), thus inferring limited upwelling or downwelling within the hyporheic zone. In this particular case study, the steep hydraulic gradients between the riparian zone and the stream (a consequence of the steep topography) was by far the more important in generating a near-constant, diffuse outwelling of groundwater into the stream under baseflow conditions. It was this hydrological vector that provided the basis for significant N and C retention or loss in the near-stream zone as stream concentrations were remarkably near constant along the 100 m reach, despite this effluent groundwater.

Chestnut and McDowell (2000) thus showed the surface-groundwater interactions which one can conclude must be well coupled within the near-stream zone during storm events. Chestnut and McDowell (2000) *also provide an experimental design for providing an aggregate left-and-right bank estimate of* K_{sat}, *although these writers do not highlight this important contribution.* With the field determination of hydraulic gradients (both vertical and lateral) and Q_{GW}, one can back-calculate K_{sat} in Darcy's equation for the hyporheic zone initially and subsequently for the near-stream zone, providing that the hyporheic downwelling or upwelling is very limited.

DIGITAL TERRAIN MODELS FOR RUNOFF SIMULATION

When considering the practical application of hillslope hydrology, spatial hydrological models based on digital terrain models (Moore *et al.*, 1991) offer a landscape perspective for assessing the spatial and temporal variability of soil moisture and runoff production. In the context of land and water management, these spatial hydrological models can assist in determining the areas most

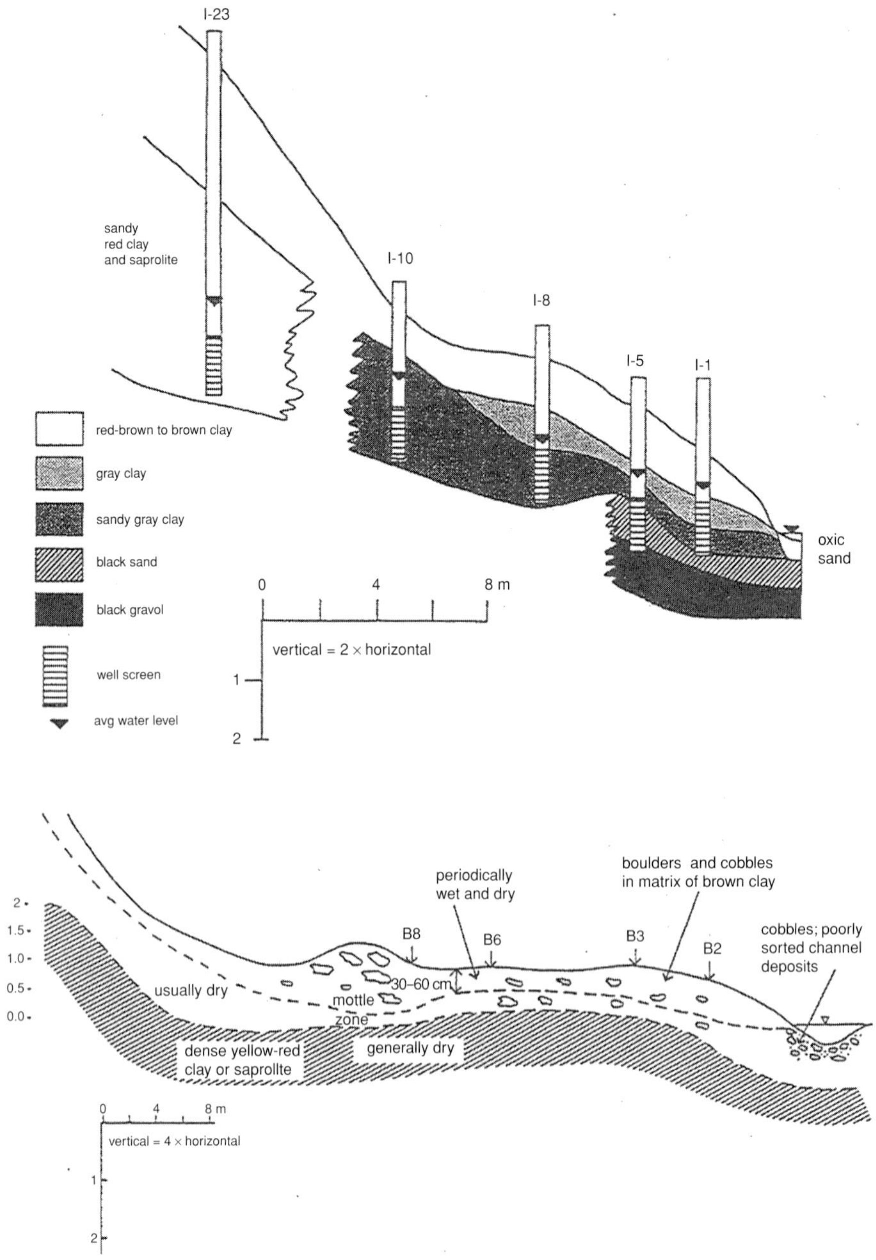

Figure 14.66 Soil profiles in a transect across the Icacos (upper panel) and Bisley (lower panel) well fields, showing well locations and depths. For Icacos, average elevation of the water table is also included. Note differences in vertical exaggeration. (After McDowell *et al.*, 1992.)

vulnerable to land degradation from disturbance as well as providing guidance on the most appropriate species for reforestation, based on soil moisture and groundwater requirements (Bonell, 1998b).

Spatial hydrological models have evolved progressively following early pioneering work by Kirkby (1975), Beven and Kirkby (1979) and O'Loughlin (1981), which resulted in the initial presentation of TOPMODEL (Beven *et al.*, 1984), TOPOG (O'Loughlin, 1986) and later THALES (Moore *et al.*, 1991; see Barnes and Bonell, this volume). These models have continued to be refined through their application and testing in a wide variety of environments. When considering their potential application to the humid tropics, Bonell with Balek (1993) incorporated a detailed review of TOPMODEL and TOPOG. Since then, the literature has expanded considerably, especially in relation to TOPMODEL, for example as recently presented in a Special Issue of *Hydrological Processes* (vol. 11, 1997) (Beven, 1997) and elsewhere (Beven *et al.*, 1995; Gunter *et al.*, 1999). In addition, the previous steady-state component of TOPOG (O'Loughlin, 1986; O'Loughlin, 1990a, b) has now been extended to include an unsteady component based on

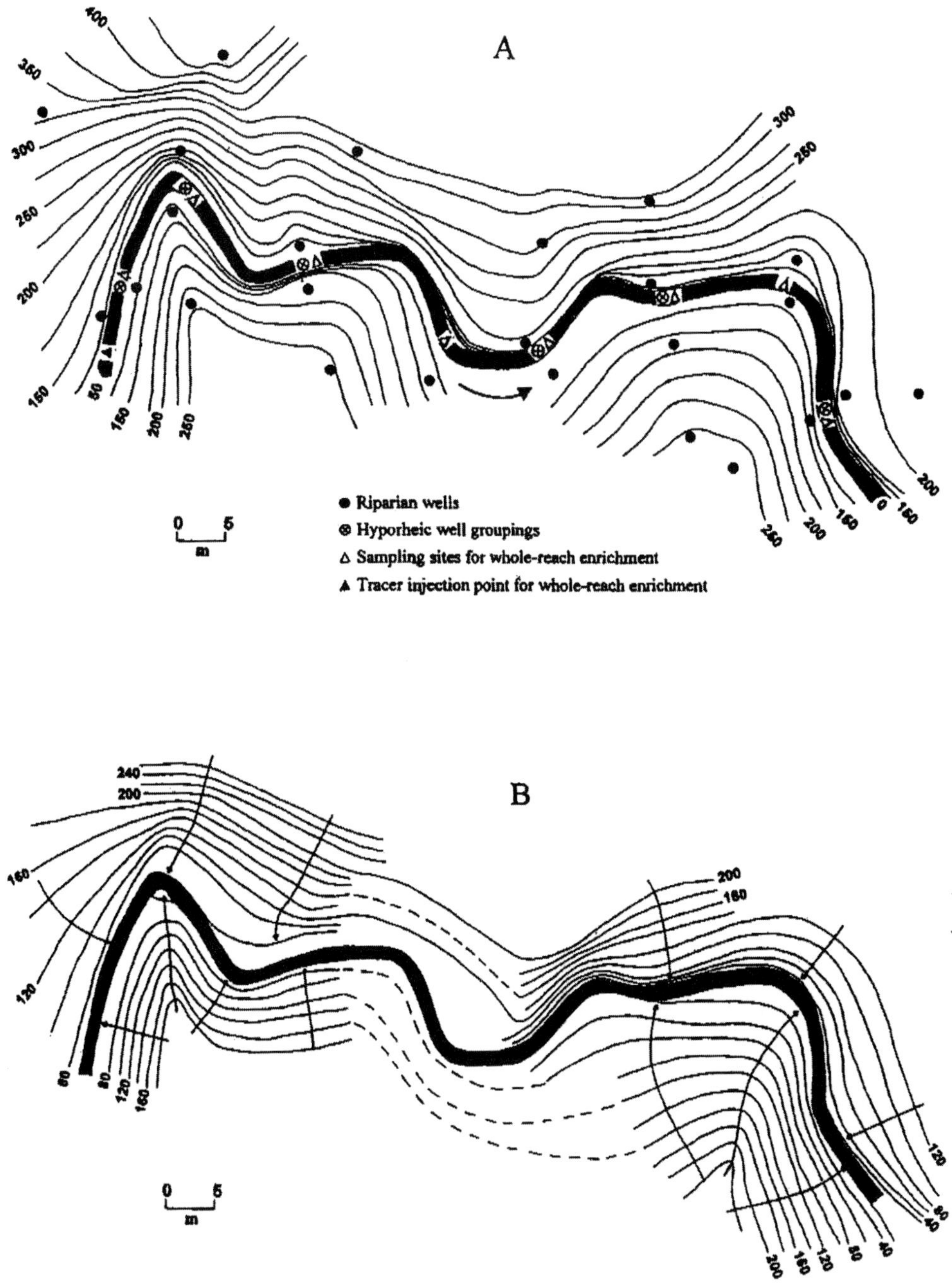

Figure 14.67 (A) Topographic map of 100-m study reach along a tributary of the Rio Icacos. Scale = 1:350, topographic contour interval = 25 cm, elevation change from upstream to downstream = 50 cm. Bold arrow indicates direction of stream flow. (B) Groundwater flownet for the study. Scale = 1:350, water table contour interval = 20 cm. Streamlines showing the direction of groundwater flow are represented by arrows. (After Chestnut and McDowell, 2000.)

several versions of TOPOG-DYNAMIC (Vertessy *et al.*, 1993, 1996; Dawes *et al.*, 1997). Elsewhere, Chappell *et al.*, b, (this volume) provide a more comprehensive overview of these recent developments; consequently, the present discussion will focus on the limited field testing of these spatial models in selected experimental catchments in the humid tropics. First, the assumptions implicit in these spatial hydrological models will be evaluated and second, case studies involving the respective testing of TOPMODEL (by Molicova *et al.*, 1997) and TOPOG (by Vertessy and Elsenbeer, 1999 and later Schellekens, 2000) will be considered.

Digital terrain models for runoff generation are based on the notion that ultimately it is landscape morphology that controls storm water flow pathways and the post-storm redistribution of

Table 14.14. *Hydrological data for the 100-m study reach along a tributary of the Rio Icacos*

Variable[a]	Value
Q_{gw} (L / s)	1.1–1.9
Q_{stream} (L / s)	12.8–15.0
Percent Q_{gw} contributing to Q_{stream}	8–15
Stream gradient (percent)	0.50
K of streambed (cm s^{-1})	1.0×10^{-2}
Vertical hydraulic gradient of streambed (cm cm^{-1})	0.003 (0.02)

[a] Discharge (Q) data are reported as a range for three whole-reach tracer experiments. Q_{gw}, groundwater discharge. Hydraulic conductivity (K) is reported as a geometric mean based on data from hyporheic wells ($n = 35$). Vertical hydraulic gradient is reported as a mean ($\pm$ 1 SD) based on data from hyporheic wells ($n = 70$).
Source: After Chestnut and McDowell (2000).

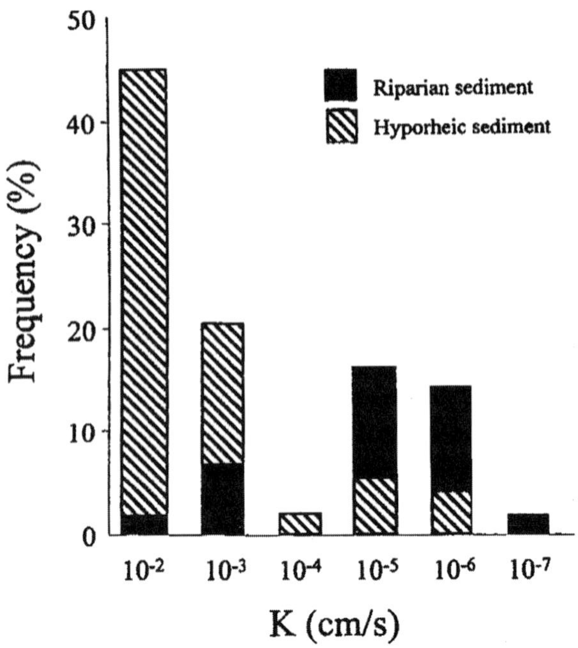

Figure 14.68 Frequency distribution of hydraulic conductivity (K) for riparian and hyporheic sediments based on bail-mode measurements of wells. (After Chestnut and McDowell, 2000.)

soil moisture. There is extensive support for such morphological controls, both theoretically (e.g. Zaslavsky and Sinai, 1981) and experimentally (e.g. Hewlett and Hibbert, 1963) as lateral flow occurs widely, even though the conceptual model of delivery mechanisms of hillslope runoff varies across environments (Beven 1986; see review of Bonell, 1993) and even within the humid tropics, as outlined earlier. Implicit in this morphological control is the existence of a subsoil layer impeding vertical percolation which therefore encourages a shallow water table and lateral subsurface stormflow. Significantly, the TOPMODEL school (Beven and Kirkby, 1979, Beven *et al.*, 1984) were influenced by the shallow soils associated with glacially-affected landscapes of the Pleistocene in the UK while the origins of TOPOG were encouraged by the hydrology of duplex soils which occur extensively within south-east Australia (O'Loughlin, 1981, 1986) and continue to remain a focus in the field-testing of the assumptions of spatial hydrological models (Grayson *et al.*, 1997; Western *et al.*, 1999). Thus it is assumed, from humid temperate experience, that the soil is of finite depth and that the hydraulic conductivity is negligible at the base of the profile. Therefore, with the aid of digital elevation models, the topography is then divided into flow strips to calculate a *topographic* index (the higher indices indicate the potential for wetter areas) in TOPMODEL (Beven *et al.*, 1995) or a *wetness index* in TOPOG (O'Loughlin, 1986). Both indices effectively determine the areas most susceptible to soil waterlogging and indicate the preferred areas for runoff generation.

The initiators of TOPOG (O'Loughlin, 1990a,b) and TOPMODEL (Beven *et al.*, 1995) have acknowledged the restrictive assumptions and limitations of these models. They work best where there is a water table which is quasi-parallel to the topographic surface, with the water table depth controlled mostly by topography (and not the soil-bedrock topography, McDonnell *et al.*, 1996). Further, the soil transmissivity decreases either exponentially (as assumed in TOPMODEL, Beven *et al.*, 1995) or linearly with depth as assumed in TOPOG, (O'Loughlin, 1986) for sharp soil A-B horizons (e.g. duplex soils). The latter is preferred by Ambroise *et al.* (1996) for use in TOPMODEL.

Because of the depth of soil and weathered rock in the profiles of the humid tropics, they present a much more complex hydrological environment in terms of the spatial and temporal development of perched water tables during storms (Molicova *et al.*, 1997; Bonell *et al.*, 1998) and in the occurrence of pipeflow (Elsenbeer and Cassel, 1990, 1991; Elsenbeer and Lack, 1996). For example, in the control catchment B of ECEREX in French Guyana, Molicova *et al.* (1997) showed the impact of the previously highlighted complex catenal succession of soils (Figure 14.69) and a corresponding complex spatial and temporal development of perched water tables during storms within the superficial layers (<1.5 m depth) (Figure 14.70). Of even more interest, diametric examples of wetting front propagation were highlighted during one storm event which produced both saturation build-up upwards from a lower boundary (e.g. B4 in Figure 14.70, as conceptualised in TOPMODEL) and conversely, a downwards development and progression of saturation from the soil surface (e.g. B3 in Figure 14.70 and B1 (not shown)) on the lower slope section. This complex response indicated a lack of spatial correlation of hydrodynamic conditions across the slope and violates the TOPMODEL

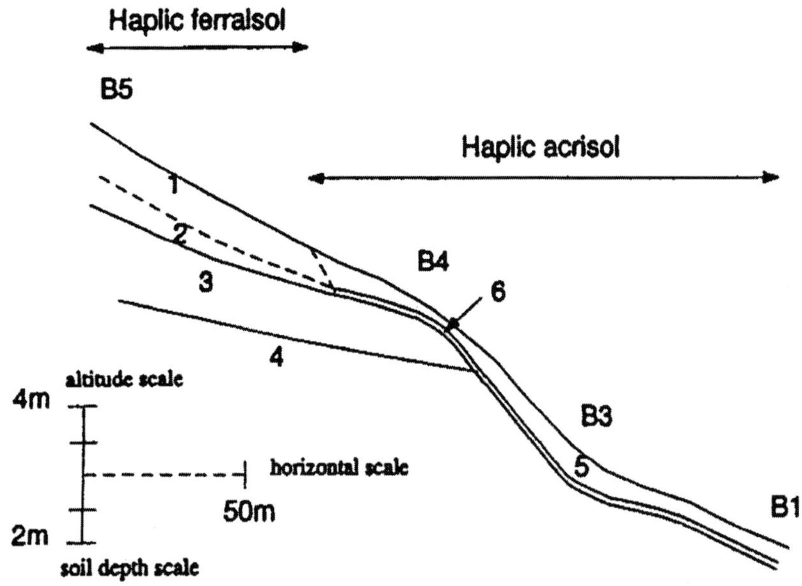

Figure 14.69 Schematic, pedological transect of catchment B in ECEREX (after M. Grimaldi, unpublished data): (1) reddish brown clayey microaggregated horizon:clayey microaggregated horison:ferralic horizon; (2) transition ferralic horizon; (3) red clayey weathering horizon; (4) dark red clayey-silty massive saprolite; (5) yellowish brown sandy clayey horizon:eluvic horizon; (6) yellowish brown clayey horizon:argic horizon. (After Molicova *et al.*, 1997.)

assumption of identical hydrological behaviour of catchment pixels, which are scaled only by the topographic index distribution (Molicova *et al.*, 1997).

Elsewhere, in the Babinda catchment study, earlier discussion had highlighted the conceptual model of runoff generation whereby a shallow 'impeding' soil layer (in *hydraulic* terms) existed in response to the large inputs of percolation arising from the prevailing summer monsoon rainfalls. Emphasis has been made, however, of the highly heterogeneous spatial nature of vertical percolation through the 'impeding' layer via preferential flow and the possible occurrence of pressure waves. The role of deeper groundwater in the storm runoff process is also apparent through a better connection with the surface hydrology than previously thought. *None of these phenomena is explicitly parameterised in the current generation of distributed or semi-distributed, spatial hydrological models*. Even more of interest, the dynamics of the spatial and temporal extent of runoff production areas in both the French Guyana and north-east Queensland catchment studies do not follow the standard framework for humid temperate areas as presented in spatial hydrological models. Thus the notion of these saturated areas (co-axial with organised drainage and riparian zones) slowly expanding and contracting from lower slopes over time by different stormflow delivery mechanisms – depending on the storm occurrence (Beven, 1986) – is not necessarily the only variant of the variable source area conceptual model. For example, Molicova *et al.* (1997) noted that saturation-excess overland flow occurred first on the *upper slopes* which is completely inverse to the classic concept of the development of saturation overland flow taking place initially on the low slopes (Figure 14.70). In the Babinda catchments, Cassells *et al.* (1985, and later corroborated by Bonell *et al.*, 1998) indicated that when antecedent soil wetness is large (under optimal wet season conditions), the intense rainfall exceeds the surficial, unsaturated zone storage capacity of the soil profile of tropical rainforest and *is independent of any downslope subsurface stormflow*. This mechanism is capable of inducing near-instantaneous areas of saturation from transient, perched water tables over a large area of the catchment. Neither of the above mechanisms and spatial organisation of saturation overland flow obey the existing assumptions of spatial hydrological models. Elsewhere, Noguchi *et al.* (1997b) and Elsenbeer and Cassel (1991) had already indicated that microtopography, soil depth and pipe exit point rather than hillslope scale topography were more influential in the spatial organisation of soil moisture and overland flow.

Initial attempts at the testing of spatial hydrological models in tropical rainforest were presented by Molicova *et al.* (1997) (ECEREX Catchment B, TOPMODEL) and Vertessy and Elsenbeer (1999) (La Cuenca, TOPOG). In both cases the catchment areas were very small (1.5 ha, 0.75 ha respectively). Interestingly, the mean annual rainfall of ~3300 mm of both studies were comparable, but the synoptic climatology rain producing systems are dissimilar in the respective maritime and mountain environments.

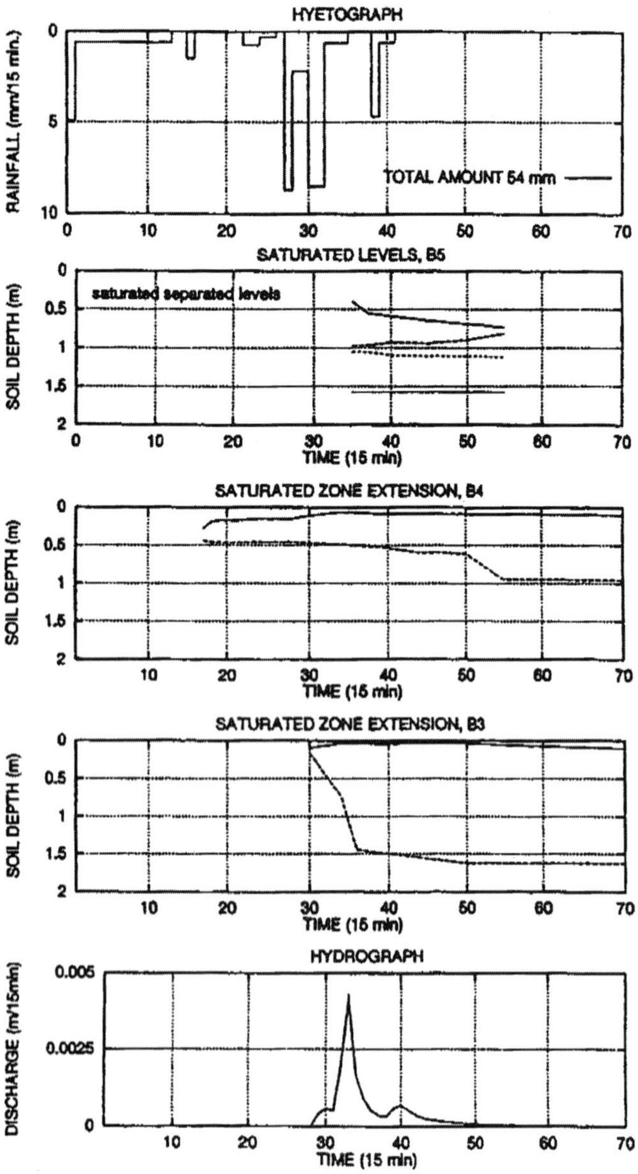

Figure 14.70 Spatial and temporal saturation extension along the pedological transect of catchment B in ECEREX (see Figure 14.69) during the event of 24 May 1992. (After Molicova *et al.*, 1997.)

In the French Guyana study, all input parameters were permitted to vary freely and only two storm events of sufficient size for the testing of TOPMODEL were measured over four field campaigns. Within slope, the interpretation of results were supported by the aforementioned tensiometer network along a slope transect, a throughfall gauge network (31 gauges distributed at random along a 100 m section) and permeability measurements from soil cores. By contrast, Vertessy and Elsenbeer (1999) were able to draw upon a much larger data base which included 34 storm events, measured saturated hydraulic conductivity involving 740 undisturbed small cores (Elsenbeer *et al.*, 1992), and measured overland flow occurrence at 72 detector sites which sampled three distinct land units. Vertessy and Elsenbeer (1999) used the SBM (Soil Bucket Model) version of TOPOG-DYNAMIC for their simulations (see Chappell *et al.*, b, this volume) based on a more rigorous approach using eight modelling sets which allowed parameterisation for K_{sat}, a scaling parameter (Vertessy and Elsenbeer, 1999, p. 276), Manning's roughness coefficient and the spatial representation of K_{sat} to operate in different combinations (i.e. lumped, distributed). Some of the principal conclusions of these analyses will be presented here.

The authors of both of the above case studies acknowledge that 'good' simulations of storm hydrograph with observed data do not infer good model performance in terms of a realistic representation of the internal spatial and temporal dynamics of runoff generation. On the other hand, Molicova *et al.* (1997) were able to 'successfully' simulate the storm hydrograph for one of the two storms, and corroborate from field evidence that the preferred delivery mechanism was subsurface stormflow, as indicated from TOPMODEL simulations. Nevertheless, the results – even for the successful simulations (Figure 14.71) – showed two weaknesses. The initial peaks of the multi-peaked hydrographs had been missed which suggested a weakness in the TOPMODEL root zone/evaporation procedure. The root zone reservoir, as represented by the SRMAX parameter, was a 'black box' in this application Another weakness was the small over-prediction in the hydrographs which were attributed to possible deep leakage and *uncertainty in the spatial and temporal variability of the precipitation data. The latter explained why the second storm could not be simulated initially* (Figure 14.72). With the aid of the 31-gauge, throughfall transect, it was established that the rainfall had not been spatially uniform, despite the small catchment size (~1.5 ha) (Figure 14.73). On the contrary, the rain amounts *increased upslope which also coincided with the more preferred areas for saturation – excess overland flow development (B4 in Figure 14.69).* Thus the average depth of the 31-gauge network could not be used as a unit rainfall rate in TOPMODEL, and subjective manual calibration of the throughfall transect results had to be undertaken to fit the hydrograph. Thus, despite the criticism of Vertessy and Elsenbeer (1999, p. 2184) that for 'successful' application of TOPMODEL, Molicova *et al.* (1997) '. . . still had to modulate the rainfall record for one of the two storms to obtain a credible rainfall prediction for it', the more pertinent point missed by the former writers, was *the extreme sensitivity* of spatial hydrological models *to erroneous inputs* (e.g. Franchini *et al.*, 1996, Shah *et al.*, 1996), as represented by the spatial and temporal resolution of the rainfall/throughfall data. *The implications of these findings for larger catchments is even more disturbing where the impacts of spatial and temporal heterogeneity in tropical rain fields will be even more pronounced* (Bonell *et al.*, this volume). The solution is the incorporation of more intensive throughfall/rainfall gauge

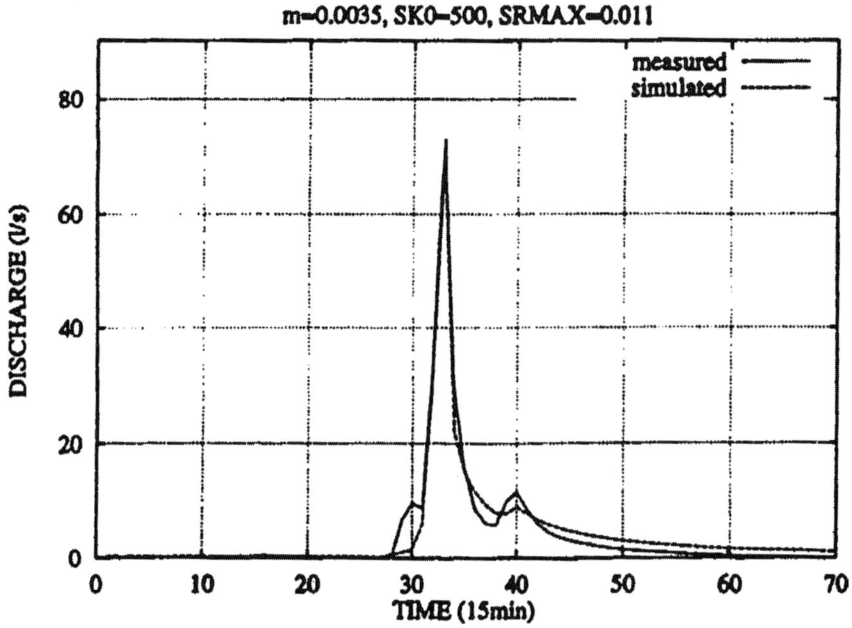

Figure 14.71 TOPMODEL simulation for the 24 May 1992 storm flow in catchment B, ECEREX. (After Molicova *et al.*, 1997.)

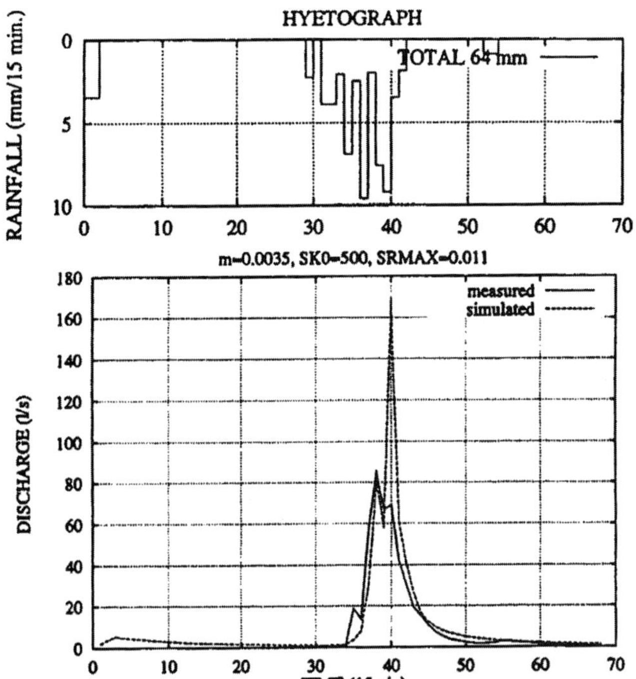

Figure 14.72 TOPMODEL simulation for 15 May 1993 storm flow without any correction account for rainfall spatial variability. (After Molicova *et al.*, 1997.)

networks linked with radar rain detection but, in more remote locations, even this suggestion is prohibitive on grounds of cost alone.

For modelling data sets which were calibrated on one storm and constrained using different representations of K_{sat} (as against input parameters which were varied freely), the TOPOG-SIB provided the best hydrograph simulations for the La Cuenca study (Vertessy and Elsenbeer, 1999) for only about half of the events. An inadequate representation of antecedent soil moisture conditions, due to insufficient knowledge of pre-storm soil moisture patterns, was thought to be responsible for some of the poor simulations. However, simulations undertaken in circumstances where successive storm events were too temporally close to permit adequate catchment drainage (which violated the assumption of 'well-drained'), predicted hydrograph peaks were fitted very well for only two events. Vertessy and Elsenbeer (1999) thus concluded that errors in the model representation of initial conditions could not be responsible for all the poor predictions.

One of the principal criticisms of spatial hydrological models is that the 'good' model simulations of storm hydrographs do not imply a realistic representation of within-catchment (pixel) soil moisture and runoff dynamics (see discussion in Grayson *et al.*, 1992). Outside the humid temperate latitudes (e.g. Grayson *et al.*, 1997; Western *et al.*, 1999; Woods *et al.*, 2001), the work of Vertessy and Elsenbeer (1999) is the only one so far in the humid tropics to compare field observations of overland flow occurrence with the internal predictions of a spatial hydrological model, i.e. TOPOG-SBM. One of the model data sets (K_{sat} values pooled in a single catchment population, random allocation of cumulative frequency of log-transformed values across the whole catchment) generated the most physically realistic pattern

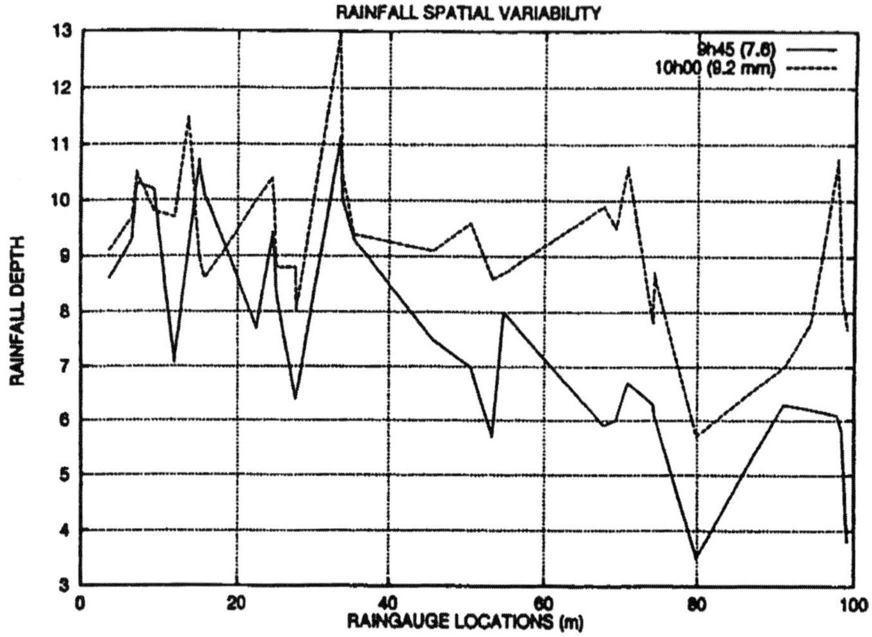

Figure 14.73 The spatial variability of the throughfall gauge record which explains the inconsistency of the rainfall-runoff data in Figure 14.72. The corresponding two 15 minutes throughfall rain gauge records linked with the third rainfall peak (finishing at time step 40) (see Figure 14.72) along a 100 m transect in catchment B. The throughfall inputs are increasing upslope so that the higher inputs correspond with the B4 zone in Figure 14.69. (After Molicova *et al.*, 1997.)

of overland flow generation. In these circumstances, the volume of runoff was well predicted but the hydrograph peak and time to rise were underpredicted and overpredicted, respectively. Rather than underestimating the saturated area, Vertessy and Elsenbeer (1999) attributed the Manning's roughness coefficient ($n = 0.7$) as being too high, thus reducing overland flow contributions. Any decrease in the latter value, however, had the adverse impact of over-steepening hydrograph recessions.

A major obstacle to Vertessy and Elsenbeer (1999) having more success in their simulations is probably their inability to represent the rapid transfer of subsurface stormflow by pipes, both spatially and volumetrically, in the absence of this information. As these writers acknowledge, such information would allow for a reduction in Manning's roughness coefficient (i.e. reduce the model's reliance on fast overland flow from surface saturation). *The fact that pipeflow is a critical component in this study, in retrospect, raises the issue of whether the importing of soil core K_{sat} estimates into a spatial modelling exercise was sufficient and even realistic*. Soil cores are biased towards the soil matrix hydraulic properties which, in comparison with the role of pipes, are passive in subsurface stormwater transfer (see Chappell, Bidin *et al.*, this volume). As noted earlier, estimates from cores are also unrepresentative in the volumes measured. The same criticisms apply to the work of Molicova *et al.* (1997). The latter writers found that the optimised hydraulic conductivity from TOPMODEL simulations were three orders of magnitude higher than that corresponding to K_{sat} estimates from soil cores.

There has been extensive discussion, especially in relation to TOPMODEL, on the calibration experiences in several studies which give much higher values of transmissivity than were previously indicated from small–scale measurements (see reviews of Beven *et al.*, 1995; Beven, 1997). In part, this is attributed to the hydraulic conductivity being sensitive to the grid size of the DTM, and tends to increase with increasing grid dimensions (Saulnier *et al.*, 1997). Alternatives to the exponential transmissivity function with depth (see Ambroise *et al.*, 1996) have also been put forward to better 'capture' deeper vertical percolation in humid tropics soils as well as possibly much deeper rapid subsurface flow pathways (Molicova *et al.*, 1997; Vertessy and Elsenbeer, 1999). Further, Molicova *et al.* (1997) noted some of the parameters imported into TOPMODEL are interchangeable. For example, a small *m* (*m* is the catchment baseflow recession parameter which characterises the soil storage-catchment discharge relationship) can be part compensated by a high transmissivity. As the catchment baseflow recession parameter is one of the most sensitive in TOPMODEL theory, particular focus of attention was given by Molicova *et al.* (1997) to their estimates of *m* as a source of error, and the resulting compensatory high transmissivity, to explain their high efficiency of simulations (Molicova *et al.*, 1997). The latter could be a consequence of the fortuitous interaction between these two parameters. Estimates of m from baseflow recession curves were not possible in the ECEREX study because there was no permanent (dry weather) flow in the catchment and, as an alternative, m derived from storm hydrograph recession curves corresponding to 'quickflow' had to be used (Molicova *et al.*, 1997).

Schellekens (2000) subsequently tested the TOPOG model linked with the Bisley II catchment (6.4 ha) in Puerto Rico which is a larger basin than the preceding two case studies; it also presents a more dissected terrain of much steeper slopes and a different hydroclimatic regime. Four modes of operation of TOPOG were tested, based on daily and 5-minute time steps for the respective TOPOG_DYNAMIC and TOPOG_SBM versions. Of more relevance here is a review of the results from the 5-minute time steps for 16 selected storms. These storms ranged from 3 to 78 mm in total rainfall with the associated streamflow totals ranging from 0.7 mm to 63.8 mm. Significantly, the largest event (227.5 mm) monitored during the study of Schellekens (2000) was not incorporated in this testing programme. Most of the 16 selected events produced multi-peaked hydrographs with time lags in the order of 114 min (minimum 55 min, maximum 350 min). A Manning roughness coefficient of 0.3 was used throughout the modelling, cf. 0.7, Vertessy and Elsenbeer (1999).

Despite differences between the two versions of TOPOG in terms of parameterisation, the heavily parameterised TOPOG_DYNAMIC did not show much improvement over the much simpler TOPOG_SBM. Model efficiencies (using the Nash and Sutcliffe, 1970, criterion) for individual storms ranged from 0.9 (a good fit) to values less than 0.2 (a poor fit). In one event (net precipitation 4.2 mm after correction for interception) even gave a negative value. The poorest results (in model efficiency terms) were linked with the smallest storms where absolute differences between actual and modelled storm hydrographs were small. In addition, modelled maximum discharges were generally underestimated whereas they were significantly overestimated for the smallest storms. Moreover, discharges at the end of the storm (including the recessions) were often poorly predicted, with observed values being grossly over-predicted for about 50% of the storms.

Despite its greater sophistication, results obtained from the TOPOG_DYNAMIC model were in general similar to those from the much simpler TOPOG_SBM version. A critical exception was the largest event (net precipitation 77.9 mm, total streamflow 63.8 mm) which '. . . caused TOPOG_DYNAMIC to crash. The severe numerical instability that was caused by the combination of the high net precipitation intensities (up to 65 mm h^{-1}) that occurred during this major storm and a clayey soil that rapidly became saturated during the process which resulted in very high mass balance errors while solving the Richards equation. Predicted discharges even became zero during the peak of the storm and remained that low until the end of the simulation' (Schellekens, 2000, p. 109). As described elsewhere (Bonell *et al.*, this volume), the preceding largest event is not exceptional by humid tropical standards; and it is storms of this magnitude or greater which are of more practical interest in terms of the various impacts of floods on society and land-forest management practice. Consequently, it is important to assess possible reasons as to why TOPOG_DYNAMIC did not work.

Schellekens (2000) placed strong emphasis on the lack of a macropore flow algorithm in TOPOG, and some of the noted shortcomings could be attributed to its absence. As observed earlier, SSF is the dominant pathway in this basin and macropore flow is the principal conveying mechanism. Furthermore, the stability must be questioned of the widely used water retention-hydraulic conductivity functions such as van Genuchten (1980) (as used by Schellekens in his study) and others (Brooks and Corey, 1964; Campbell, 1985), and also the Richards equation itself during transient saturation due to the complicating effects of macropores. As noted by van Genuchten and Leij (1992, p. 8) '. . . these phenomena lead to apparent discontinuities in the hydraulic properties near saturation that are not easily captured with existing equations. A broader issue is the validity of the Richards equation for conditions involving macropore and preferential flow . . .' In addition, the earlier criticism directed towards the work of Vertessy and Elsenbeer (1999) and Molicova *et al.* (1997) in importing either K_{sat} or K^* estimates into these spatial hydrological models, which are unrepresentative of macropore or pipeflow effects at larger scales, also applies to this Puerto Rico study. Schellekens (2000) used K^* estimates in TOPOG_DYNAMIC using a methodology comparable with the CHWP, and so such estimates will not capture the dynamics of SSF and thus introduce error into this modelling exercise. During the most intense storm, SOF was also expected to occur (see earlier discussion) and a failure to capture the spatial and temporal extent of SSF would also affect the model, accounting for such SOF contributions on the steep slopes away from the riparian zones. Furthermore, the mechanism of near-instantaneous saturation without the need for significant upslope SSF contributions, as pertaining to the Babinda study, cannot be discounted as having occurred also during this large storm event. For reasons given earlier in this section, this mechanism would violate one of the assumptions in TOPOG.

The above remarks go some way to explaining why Schellekens failed to model the largest event, and even raises the issue whether in these circumstances the simpler version, TOPOG_SBM, should always be the preferred option. In terms of representing physical reality however, the latter model is not very accurate. For example, Schellekens (2000) observed that there is no concept of a perched water table (relevant to shallow SSF) included in TOPOG_SBM. Rather, this model only allows for lateral flow out of the saturated part of elements, and these elements can only be saturated from the bottom upwards. Thus lateral flow in the early stages of a storm will occur only from those elements which are already saturated, and not from newly-developed perched water tables. In contrast, the soil moisture accounting scheme in TOPOG_DYNAMIC does allow for saturation at the base of a thin, highly conductive top layer (without the need to first saturate the rest of the soil profile from the bottom up) (Schellekens, 2000). One then moves into a circular argument whereby the instability of the Richards' equation near saturation and lack of macropore algorithm to represent a fast lateral path (shallow SSF) loses the advantage of TOPOG_DYNAMIC representing physical reality better. Until we really can represent such physical processes, there is a case for remaining with the more simple version of TOPOG. The most pertinent observation is that Schellekens (2000) had not been able to follow the S1 to S3 methodology of Elsenbeer and Vertessy (2000) systematically prior to applying TOPOG. Had this been achieved, then such better understanding of the runoff dynamics of the basin would likely have influenced the selection of the appropriate modelling strategy. As Schellekens (2000, p. 115) conceded '. . . the present results underline the importance of a good conceptual understanding of a catchment's behaviour before undertaking a hydrological modelling exercise (cf. Vertessy and Elsenbeer, 1999) . . .' and he goes on to state '. . . it is sobering to realise that the availability of a substantial amount of data alone is not necessarily enough'.

Probably the most critical issue, and external to these spatial hydrological models, is the inadequate representation of K^* at the hillslope scale or higher, from existing *in situ* methodologies (which originate from the laboratory scale). As already highlighted, evidence suggests that only a small percentage of total catchment porosity participates in the 'rapid' subsurface water transfer both laterally and vertically, during and immediately after storms. Thus, existing field parameterisations of K^* are biased towards the soil matrix, and not at the appropriate scale to include these rapid subsurface pathways. The catchment 'effective' hydraulic conductivity is much higher, as was shown by Bazemore *et al.* (1994) based on a crude Darcian representation of an entire catchment. Bonell (1998a) reported a similar order of magnitude of 'effective' hydraulic conductivity for South Creek (to that reported by Bazemore *et al.*, 1994) using the same method of Bazemore *et al.* (1994). So despite the aforementioned flaws in TOPMODEL, the fact that high transmissivities are consistently reported from the existing suite of high efficiency simulations using TOPMODEL in various catchments (e.g. Beven, 1997) lends credibility to these reports and reinforces the preceding comments of inadequate field methodologies for determining K^* at the appropriate scale of interest (Chappell and Ternan, 1992; Chappell *et al.*, b, this volume). This led Davis *et al.* (1999) to suggest that modelling (such as TOPOG) should place more emphasis 'on the most complete set of physical input parameters' which are more easily measurable, and then treat K_{sat} as a calibration parameter.

THE ROLE OF 'DEEP' GROUNDWATER

The preceding overview of studies in south-west Nigeria had highlighted the important role of groundwater in the runoff hydrology from basins underlain by quartzitic rocks. As Foster *et al.* (2002) more recently remarked, the humid tropics incorporate a wide range of groundwater systems by geological build (Foster and Chilton, 1993), and most of these systems are characterised by shallow groundwater tables directly connected to the surface hydrology. Thus '. . . aquifers tend to fill up rapidly in the wet season with the water table virtually reaching the land surface . . . so that. . . . further excess rainfall is then rejected because of the absence of storage space and will lead to overland flow' (Foster, 1993, p. 447). It is interesting that there is commonality here in the perspective of a hydrogeologist (Foster, 1993) with the hillslope hydrology reported in the current work. Nonetheless, despite recognition of this connectivity between the surface hydrology with hydrogeology, globally the number of studies which have parameterised these groundwater systems are few in number (Foster, 1993; Foster and Chilton, 1993) and when concerning groundwater-hillslope hydrology interactions, such perspectives are even more limited (Bonell *et al.*, 1998; Dykes and Thornes, 2000; Elsenbeer and Vertessy, 2000; Chestnut and McDowell, 2000; Masiyandima *et al.*, 2003).

An output from the preceding overview of various hillslope hydrology studies, however, has emphasised the potential greater role of 'deeper' groundwater in the runoff generation process, e.g. the Babinda study (Bonell *et al.*, 1998). Earlier during the testing of TOPMODEL by Quinn *et al.* (1991) (reviewed in Bonell with Balek, 1993, p. 233) using information from the Booro-Borotou catchment (1.36 km^2) in Côte d'Ivoire, the impact of more deeply weathered soil and rock encourages a deeper water table which affects the assumption of the latter being quasi-parallel to the surface. A concept of a *reference level* was introduced to cater for the deviations (and contributions) in the water table from the soil surface but the lack of detailed hydrogeological information militates against the application of a reference level in most tropical environments (Bonell, 1993).

The above considerations are even more complex when one considers the work of Cosandey and De Oliveira (1996) based on a catchment study in subtropical south-east Brazil. They determined that the *subsurface paleotopography*, rather than the surface topography, was the controlling influence on the spatial and temporal development of saturated zones (Figure 14.74). Thus in the case of TOPMODEL, the topographic index would be better

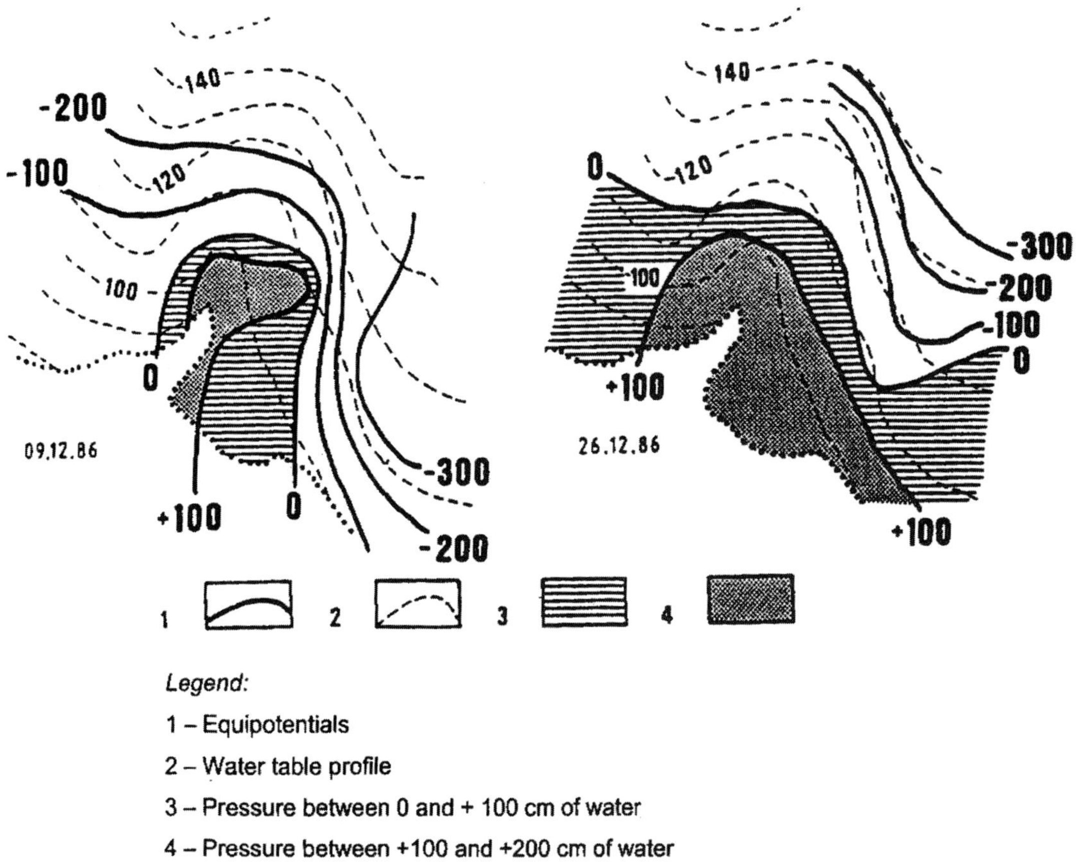

Figure 14.74 Extension of the saturation zone on the slope (redrawn following Fernandes, 1990). (After Cosandey and de Oliveira, 1996.)

based on the subsurface paleotopography (see also Freer *et al.*, 1997).

One can conclude that in humid tropical environments which have deeply weathered regoliths, the distinction between the role of topography vis-à-vis bedrock topography in terms of saturated zones, and their control on subsurface stormflow pathways, is poorly understood (Bonell, 1998b). The application of spatial hydrological models in these circumstances (which do not include the function and contributions of deeper groundwater to the storm hydrographs), must be treated with considerable caution in the absence of a preceding geophysical and hydrogeological survey. Current work outside the tropics in the North Island of New Zealand (the MAhurangi River Variability EXperiment, MARVEX, Woods *et al.*, 2001) in clay loam soils (*c.* 1 metre depth) also emphasises these points. There is little correlation between topographic position and soil moisture in the top 300 mm in comparison with the duplex soils of south-east Australia (Grayson *et al.*, 1997) and '. . . points to the hydrological responses being controlled by deeper groundwater processes or to (deeper) lateral processes on the hillslopes being strongly influenced by preferred flow pathways' (Woods *et al.*, 2001, section 3.3). These deeply weathered regoliths also have ramifications on the use of spatial hydrological models as part of either a reforestation or afforestation strategy of degraded lands in the humid tropics (see review of Bonell and Molicova, 2003).

A CONTROVERSIAL ISSUE: DOES STORMFLOW INCREASE AND DELAYED FLOW DECREASE FOLLOWING FOREST CONVERSION?

Contrary to controlled experimental catchments both in the humid tropics as well as the humid temperate latitudes, there are reports (but unverified) that forest conversion results in an increase in the quickflow component of the storm hydrograph and conversely, a decrease in delayed flow (see review by Bonell with Balek, 1993, pp. 224–228, Bonell, 1998b, pp. 242–246). As noted elsewhere (Bruijnzeel, 1989; Pereira, 1991; Bonell, 1999) existing controlled experiments are not representative of the long-term degraded lands where human impacts have been sustained for decades into centuries; either the converted forest was replaced by an alternative crop, or the experiment continued only for a short period after tree fellings. Based on the hypotheses of Bruijnzeel (1989), Bonell (1993, 1998a) encouraged a hillslope hydrology perspective to resolve this debate The focal point concerns the impact of forest clearance on the surficial, soil hydraulic properties. Following Bruijnzeel (1989):

- If surface infiltration characteristics are reasonably maintained following forest conversion, because of stable soil aggregates, immediate soil conservation measures or low rainfall erosivity, then percolation to groundwater will be maintained. Thus the effect of reduced evaporation after clearing will result in the widely reported dry weather flow (delayed flow) increases from controlled experiments.
- Conversely, if the surficial, soil hydraulic properties discourage vertical percolation after forest removal then overland flow occurrence is enhanced which results in the quickflow component of the hydrograph also increasing. The reduced percolation reduces the storage capacity of groundwater which, in turn, reduces dry season, delayed flow.

There has been lack of either *in situ* measurements of soil hydraulic properties or controlled experimental catchments in the humid tropics to evaluate the above hypotheses (see also Scott *et al.* this volume); although such preliminary work is on-going, in the form of a UNESCO IHP project (implemented by the Karnataka Forest Department/National Institute of Hydrology) within the Western Ghats forests and degraded landscapes of Karnataka State, India. The most conclusive evidence originates from forests of semi-arid environments (see review of Bonell, 1998b) such as the recent presentation of results from Tanzania (Sandstrom, 1995; 1996; 1998). Sandstrom (1995) established that an increase in frequency of floods since the mid-1940s was caused by accelerated forest clearance and not climatic change (or variability). He emphasised the loss of macropore networks, due to raindrop and anthropogenically-induced compaction which cause a reduction in surficial infiltration rates, especially on *the steeper slopes of fine-textured soils*. The latter reduced groundwater recharge via percolation, and led to an increase in infiltration–excess overland flow. Thus great importance was attached to greater relief (which prevents redistribution of overland flow and subsequent infiltration), coupled with the role of macropores in finer-textured soils '. . . which are present in forest soils but are lost after deforestation and land degradation' (Sandstrom, 1995, p. 12, paper VI). Of particular interest in this study was the linkage between the proportion of 'old/new' water in storm hydrographs (using environmental tracers) with the presence or absence of macropores (Sandstrom, 1996). Sandstrom (1996) established that the storm hydrographs in the degraded catchment consisted mostly of 'new' water, irrespective of storm magnitude, due to the absence of macropore networks. The reduction in opportunities for the 'discharge' of 'old' water from the degraded catchment, due to loss of contributions of subsurface stormflow pathways, was the prime explanation. In contrast, the forested catchment emitted 'old' water, although significantly the storm hydrographs progressively incorporated greater volumes of 'new' water as rain intensity increased (Sandstrom, 1996), similar to the Babinda catchment study (Bonell *et al.*, 1998).

Other studies have noted the loss of macropores following forest disturbance (e.g. Waterloo, 1994, Jetten, 1994), although Jetten (1994) in Guyana highlighted exceptions to this conclusion, depending on soil type. In this regard, the foregoing reviews of studies in the central Amazon basin are relevant here. Based on qualitative observations in the Reserva Ducke (Bonell, 1993, Figure 1) had suggested that any form of disturbance seems to cause the surface soil fabric to collapse (possibly due to a change in surface charge, Uehara, 1995), thus inducing depression storage of water even on walking footpaths, whereas in the adjacent forest surface permeabilities are high and no overland flow occurs. Bonell (1993), therefore put forward the hypothesis that environments where the translation to the surface of an impeding layer (originally located at significant depth) from forest conversion *could encourage a dramatic change in the dominant stormflow pathways*, and lead to the frequent occurrence of infiltration–excess overland flow (which previously had not existed). Thus the quickflow component of the storm hydrograph would increase significantly in volume. Factors influential to this hypothesis are the intensity of surface compaction, associated with land management, e.g. overgrazing with cattle, coupled with the effects of raindrop compaction.

Based on the ABRACOS campaign in central Amazonia, Tomasella and Hodnett (1996) presented a reduced K^* (66 mm h^{-1}) at the surface (from cattle trampling) from a figure previously in excess of 800 mm h^{-1} found in the forest. Further, as reported earlier, the soil profile below was highly permeable to 1.1 m depth (K^*, 97 mm h^{-1}), prior to a marked decline at that depth (K^*, 17 mm h^{-1}, below 1.1 m depth). These trends in K^* with depth lend support to the above hypothesis of Bonell (1993). In contrast to the central Amazonia studies, Elsenbeer *et al.* (1999) in Rondonia showed previously that K^* decreases abruptly with depth under forest and teak in Ferralsols or Oxisols (Soil Survey Staff, 1975), but more gradually under pasture. *Further, the decrease in K^* with depth under pasture is at odds with the more universally accepted 'inflection' in K^* with depth whereby there is a reduction due to compaction at the surface, then an increase in the subsoil followed by decrease at lower depths.* Nonetheless the earlier comparison of one-hour rainfall intensities (24 and 15 mm h^{-1}) with return intervals of 10 and 30 times per year (Elsenbeer *et al.*, 1999) demonstrated that ponding, and Hortonian (infiltration-excess) overland flow, occurs quite frequently over pasture. The marked anisotropic profiles also favour perched water tables, SSF and SOF within all three land covers. The work of Elsenbeer *et al.* (1999) also presents new evidence for re-affirming the sensitivity of GCMs to changes in surface soil hydraulic properties (Lean *et al.*, 1996, see review of Bonell (1998b)) arising from imposed, wholescale forest conversion in the Amazon basin.

Elsewhere, the review of the ECEREX by Grip *et al.* (this volume) highlighted that the catchments with free-draining soils proved much more sensitive to disturbance by forest conversion than catchments with impeded drainage and marshy bottoms. In line with Bonell (1993), Grip *et al.* attributed a shift towards the surface of dominant runoff pathways during rainfall being due to changes in surface hydraulic properties of these free-draining soils (notably from compaction by machinery used in the forest conversion process). Thus the change was from previously predominantly vertical percolation to more rapid surficial routes (especially infiltration-excess overland flow).

A subsequent K^* survey by Godsey and Elsenbeer (2002) of Acrisols-Alisols (Ultisols) under secondary forest of less than five years old in Rondonia showed that the permeability down to 0.125 m depth continued to reflect the previous intensive land uses (i.e. grazing and banana-cocoa plantation). In particular, impeded drainage and lateral flow (and possibly SOF) were surmised from the very low K^* measured at 0.125 m depth under former intensive grazed pasture. This led Godsey and Elsenbeer (2002) to conclude that for the topsoil, the K^* values of the former land uses could be assumed for this recent secondary forest regrowth. At greater depths (>0.20 m), the measured K^* seems more influenced by pedological factors rather than land use. So for this particular environment, the subsoil K^* (and by inference the dominant stormflow pathways) are not significantly different from the primary forest.

Elsewhere, (Scott *et al.*, this volume) provide a more detailed water balance perspective on the hydrological impacts of reforestation–afforestation of degraded lands.

ACHIEVEMENTS AND RESEARCH GAPS

Since the detailed overview of process hydrology needs (including hillslope hydrology) in the humid tropics by Bonell with Balek (1993), there has been a significant expansion in case studies to enhance our understanding of runoff generation. These comments apply especially to the Equatorial region of the humid tropics which are not influenced by tropical cyclones and where total rainfalls are comparatively lower; work undertaken in the Western Ghats, India and Puerto Rico are the exceptions. One refers here to new data, for example, from Malaysia (Noguchi *et al.*, 1997a, b; Chappell *et al.*, 1998; Dykes and Thornes, 2000), India (Putty and Prasad, 2000a, b; Purandara *et al.*, 2004, unpublished data), Tanzania (Lørup, 1998), Nigeria (Ogunkoya *et al.*, 2003, 2004), Puerto Rico (Schellekens, 2000), Guyana (Jetten, 1994), French Guyana (Molicova *et al.*, 1997) and the Amazon basin (Tomasella and Hodnett, 1996; Hodnett *et al.*, 1997a, b; Elsenbeer *et al.*, 1999).

Furthermore, recent publications have built on previous work in Peru (Elsenbeer *et al.*, 1995b; Elsenbeer and Lack, 1996; Vertessy and Elsenbeer, 1999; Elsenbeer and Vertessy, 2000) and north east Queensland (Elsenbeer *et al.*, 1994; Elsenbeer *et al.*, 1995a; Barnes and Bonell, 1996; Bonell *et al.*, 1998; Barnes and Bonell, this volume) which have incorporated a more comprehensive process understanding through taking a combined hydrometric-hydrochemistry approach, and the testing of assumptions implicit in spatial hydrological models.

This progress has enabled the initation of a conceptual framework (Elsenbeer and Vertessy, 2000; Elsenbeer, 2001) for the global comparison of hillslope hydrology under tropical forest and the early development of a spectrum of dominant flowpaths based on two soil orders or reference groups (i.e., Acrisol, Ferrasol) of the FAO (1974). The hypothesis (Elsenbeer, 2001) that overland flow is potentially an important mechanism of runoff generation within tropical forests (cf. humid temperate forests) of Acrisol landscapes, but not necessarily in Ferralsol landscapes, provides a basis for further field testing.

Nonetheless, in comparison with the proliferation of literature identified with the humid temperate areas (e.g. Bazemore *et al.*, 1994; Bonell, 1993; 1998a; Buttle, 1998; Mulholland, 1993), our spatial understanding of hillslope hydrology in the humid tropics continues to remain limited and to be too site-specific to apply globally to the degradation of land and water resources issues of this region. In addition, there has been little progress in our understanding of the hillslope hydrology of degraded lands linked with the adoption of rehabilitation measures, despite the earlier calls of Bonell (1993) to develop such research initiatives. The current Karnataka study in the Western Ghats of India, albeit still preliminary in scope, remains an exception. Further, our knowledge of hillslope hydrology in mountain cloud forests is absent (Bonell, 2004; see Bruijnzeel, this volume).

In using the conceptual frameworks of Elsenbeer and Vertessy (2000) and later Elsenbeer (2001) for comparison of hillslope hydrology studies, one is obliged to acknowledge that this approach (despite being the first of its kind for tropical forests) still takes a 'static' perspective in describing dominant storm pathways. A more dynamic perspective of runoff generation also needs to be encouraged whereby the dominant pathways of runoff can potentially change within a storm, especially in Acrisol landscapes, in response to temporal variations in rain intensity. Earlier contributions in this volume (Bonell *et al.*, this volume) place strong emphasis on the intricate linkage between synoptic climatology – different rain-producing systems – rain fields (the spatial and temporal occurrence of rain) – preferred pathways of hillslope runoff. The ultimate goal is to be able to predict the potentially changing dominant pathways and contributions of runoff during the transient passage of specific rain fields over a catchment on similar lines to the strategy adopted in a 46.6 km^2 basin of the MARVEX project in New Zealand (Woods *et al.*, 2001).

When comparing radar imagery (150 m grid resolution) of transient rain cells (1–5 km across) with data from a ground truth network of rain gauges and stream gauging stations, a lag of 1 hour between time-averaged rainfall and runoff was determined by Woods *et al.* (2001) to be the most sensitive after an evaluation of smaller temporal resolutions down to two minutes. The higher rain intensities of the humid tropics encourages a more diverse suite of dominant runoff pathways compared with the humid temperate forests, especially in environments with shallow 'impeding' layers associated with a rapid decline in K^* with depth (e.g. north-east Queensland, Bonell *et al.*, 1983; Rondonia, Brazil, Elsenbeer *et al.*, 1999). As Woods *et al.* (2001) remarked '. . . it is possible that the two-minute detail is superfluous to an understanding of storm response for the monitored catchments. However, for smaller catchments, or runoff processes that depend strongly on 'instantaneous' rain intensity, this detail could be significant'. One could hypothesise that, during the larger rain events, there is a spatial and temporal alternation of dominant runoff, i.e. overland flow (saturation-excess or infiltration-excess) and subsurface stormflow in correspondence with the increase and then relaxation of short-term rain intensities. As shown in Figures 14.29 and 14.30, Gilmour *et al.* (1980) presented evidence for such alternation in dominant runoff pathways at specific sites linked with the storm hydrograph response. In attaining this long-term goal, hillslope hydrology would have much more practical value when linked with flood forecasting, in addition to existing land-water management applications.

Floods in the humid tropics (e.g. Hurricane Mitch in Nicaragua and Honduras, October 1998; Venezuela, December 1999; Mozambique, February 2000) and their associated impacts are going to generate increased publicity, in line with the escalation in population and corresponding intensification of socio-economic pressures in this region. With a few exceptions (e.g. the Babinda and La Cuenca studies), the very difficult field logistics of 'capturing' hillslope hydrology data during such extreme events has remained illusive. For the most part, the hillslope hydrology as reported here is based on moderate rain events of higher frequency of occurrence. *They are not representative of extreme events, which thus remains a research need.* It is not unreasonable, however, to propose that the *widespread* saturation-excess (and corresponding subsurface stormflow) and infiltration-excess overland flow mechanisms, as reported for north-east Queensland (Bonell *et al.*, 1991, 1998; Prove, 1991; Elsenbeer *et al.*, 1995a) were similar to those mechanisms which triggered the devastating floods and landslides during, for example, Hurricane Mitch (October 1998), (see, Bonell *et al.*, this volume), Venezuela (December 1999) (Larsen *et al.*, 2001), and Thailand (November, 1988), (Rao, 1998) reported in Bonell with Balek, 1993, p. 228).

Elsewhere, Bonell with Balek (1993) provided a comprehensive overview of existing gaps in our understanding of processes connected with the unsaturated zone, soil hydraulic properties measurement and groundwater linked with runoff generation processes. That same work also refered to environmental tracer applications and applications in digital terrain models of runoff procedure. In that overview, considerable resort had to be made to developments undertaken in higher latitudes, which were then either projected into the humid tropics or highlighted as research needs. Moreover, the universality of hydrological principles means that the research issues identified at higher latitudes apply as much to the humid tropics.As identified in the opening paragraphs of this section, significant progress had been made. By and large, however, the research issues highlighted in 1993 (Bonell with Balek) still remain. These include measurement of soil hydraulic properties (including the problematic representation of pipes and macropores) at the hillslope (or larger) scale for inputting into spatial hydrological models (Bonell, 1998a). Several studies in Table 1 (e.g. La Cuenca, Peru; Bukit Tarek, Peninsular Malaysia; South Creek, Babinda; Danum Valley, east Malaysia; Bisley, Puerto Rico; Kannike, India; ECEREX, French Guyana; Kuala Belalong, Brunei Darusaalem) have strongly emphasised the role of macropores and pipes in the hillslope hydrology of tropical forests.

In addition, the link between rain fields and improved representation of soil hydraulic properties (e.g. Chappell *et al.*, 1998; Chappell *et al.*, b, this volume) for establishing the temporal and spatial changes in dominant pathways of storm runoff continues to be a major research objective.

Elsewhere, in common with humid temperate research, greater attention needs to be directed towards a better understanding of the mechanisms contributing 'old' water to the storm hydrograph from both the unsaturated and saturated zones, as part of combined hydrometric-hydrochemistry studies. This includes new experimental approaches to detect displacement mechanisms (e.g. pressure waves) linked with different pathways (e.g. Raats, 1978), mean residence times of water (turnover time) and transit times (mean tracer age) (Zuber and Maloszewski, 2001). So far, hillslope hydrology (in humid temperate latitude work) has focused on the use of lumped parameter models (see review of Zuber and Maloszewski, 2001) in connection with the more rapid pathways identified with SSF (e.g. Stewart and McDonnell, 1991). The coupling of more shallow subsurface water with the mean residence time of deeper groundwater has been notably absent, with the exception of work in the Black Forest, Germany (Uhlenbrook, 1999; Uhlenbrook *et al.*, 2002) and earlier at Plynlimon, UK

(Robson *et al.*, 1995). Significantly, in the lumped parameter modelling strategy referred to above, there has also been a surprising lack of testing of the assumptions within experimental basins. In part this has been due to the lack of an integrated hillslope hydrology-hydrogeology approach in field experiments. As part of this integrated experimentation, the determination of mean tracer ages using chlorofluorocarbons (CFCs) should be used (Cook *et al.*, 1995; Oster *et al.*, 1996; Szabo *et al.*, 1996) to complement tritium (and helium, see Solomon *et al.*, 1998) for the necessary dating of water samples extracted from both soil and weathered rock.

The preceding remarks are linked with complementary hydrometric evidence. For example, under conditions of high antecedent soil wetness, continuous records of well levels (Hodnett *et al.*, 1997a, b), matric potentials as measured by pressure transducers (Bonell *et al.*, 1998; Dykes and Thornes, 2000) or manually (Elsenbeer and Cassel, 1991; Molicova *et al.*, 1997; Sherlock, 1997), and piezometers (Bidin, 1995; Bonell *et al.*, 1998), have all presented evidence of small time lags (i.e. highly responsive) between rainfall and either matric or hydraulic potentials within both the unsaturated and saturated zones of hillslopes. Closer attention needs to be given in ascertaining which are the dominant mechanisms (or combinations thereof) responsible for such high responsiveness if we are to make significant progress in modelling through the incorporation of such processes.

In the above context, the potential role of pressure waves (displacement mechanism) vis-à-vis preferential flow (or a combination of both) have already been reviewed. Other mechanisms include the transmissivity feedback concept of Bishop (1991), whereby a saturated layer enters a domain of increasingly higher hydraulic conductivity arising from the proliferation of roots which encourages more rapid preferential flow or macropore flow in the downslope direction. The high order of magnitude of 'effective hydraulic conductivity', as calculated in some studies (Chappell and Binley, 1992; Chappell *et al.*, 1998; Bazemore *et al.*, 1994); and subsequently by Bonell, 1998b for South Creek using Bazemore *et al.* methodology) was attributed in part to the transmissivity feedback mechanism. Elsewhere Dykes and Thornes (2000) inferred the rapid conversion of small negative matric potentials of the capillary fringe (Abdul and Gillham, 1984, 1989) into positive matric potentials to permit much faster transmissions of pre-existing soil water. Evidence presented here certainly confirms the rapid conversion of small negative matric potentials into transient, positive pressures. On the other hand, the extension of the capillary fringe mechanism to a corresponding groundwater ridging mechanism (Abdul and Gillham, 1989), thus causing significant displacement of groundwater, still remains a highly contentious issue (see review of Bonell, 1998a). One is referred to subsequent exchanges by McDonnell and Buttle (1998) and Gillham and Jayatilaka (1998) on this topic, where the former writers placed more emphasis on the role of rapid preferential flow via macropores in causing dramatic recoveries in near-surface water tables.

It is clear, however, from the recent hydrometric-hydrochemistry studies in Babinda (North and South Creek) and La Cuenca that the surface water-groundwater hydrology is much more tightly coupled in the storm runoff generation process than previously credited in the humid tropics literature. Recent hydrometric work in central Amazonia (Fazenda Dimona, Kuala Belalong in Table 14.1), the biogeochemical-hydrology approach of Chestnut and McDowell (2000) Côte d'Ivoire (the M'be) (Masiyandima *et al.*, 2003), also present supporting evidence for this coupling: indeed, the groundwater **is** a dominant storm runoff source in the central Amazonia hillslope hydrology studies. Thus future research needs to incorporate a detailed hydrogeological component. Such steps increase costs however, both in the need for additional financial as well as human resources. A consequence may mean fewer research sites but more concentrated effort. In this way experimental hydrology might contribute better towards the needs of physically-based models.

Central to those environments where vertical pathways are predominant (Table 14.1) is the supplementary contribution of SOF to quickflow from riparian zones. Direct measurements of this supplementary process are currently lacking and, in general, greater attention needs to be given to a comprehensive understanding of runoff generation processes within riparian zones.

Finally, the limited testing of spatial hydrological models in the humid tropics (Molicova *et al.*, 1997; Vertessy and Elsenbeer, 1999; Schellekens, 2000) have also strongly highlighted the need for better parameterisation of soil hydraulic properties, e.g. K_{sat} and the 'capturing' in these models of many of the processes identified here as research needs.

References

Abdul, Rahim, N., 1988. Water yields changes after forest conversion to agricultural land use in peninsular Malaysia. *J. Trop. Forest Sci.* 1: 67–84.

Abdul, A. S. and Gillham, R. W., 1984. Field studies of the effects of the capillary fringe on streamflow generation. *Wat. Resour. Res.*, *10*: 691–698.

Abdul, A. S. and Gillham, R. W., 1989. Field studies of the effects of the capillary fringe on streamflow generation. *J. Hydrol.*, *112*: 1–18.

Adejuwon, J. O., Jeje, L. K., and Ogunkoya, O. O., 1983. Hydrological response patterns of some third order streams on the Basement Complex of southwestern Nigeria. *Hydrol. Sci. J.*, *28*: 377–389.

Adokpo Migan, S., 2000. The impact of forest conversion on hydrology: A synthesis of French work in west Africa and Madagascar. *IHP–VI Technical Documents in Hydrology*, No 36 *Unesco – Paris. 76pp.*

Ambroise, B., Beven, K. and Freer, J., 1996. Toward a generalization of the TOPMODEL concepts: Topographic indices of hydrological similarity. *Wat. Resour. Res.*, *32*: 2135–2145.

Anderson, M. G. and Burt, T. P., 1978. The role of topography in controlling throughflow generation. *Earth Surf. Proc.*, *3*: 331–344.

Anderson, M. G. and Burt, T. P., 1982. The contribution of throughflow to storm runoff: an evaluation of a chemical mixing model. *Earth Surf. Proc. and Landf.*, *7*: 565–574.

Anderson, M. G. and Burt, T. P., 1990. *Process Sudies in Hillslope Hydrology*. Wiley, Chichester, UK. 539pp.

Anderson, S. P., Dietrich, W. E., Montgomery, D. R., Torres, R., Conrad, M. E. and Loague, K., 1997. Subsurface flow paths in a steep unchanneled catchment. *Wat. Resour. Res.*, 33, 2637–2653.

Arnold, G. O. and Fawckner, J. F., 1980. The Broken River and Hodgkinson Provinces. *In*: *The Geology and Geophysics of North-East Australia*, R. A. Henderson and B. Stephenson (eds), Geology Society Australia Incorporated, Queensland Division. 175–189.

Ayibotele, N. B., 1993. Regional Hydrology and Water Resources in the African Humid Tropics. In: *Hydrology and Water Management in the Humid Tropics – Hydrological Research Issues and Strategies for Water Management*, Bonell, M., Hufschmidt, M. M. and Gladwell, J. S., (eds), UNESCO-Cambridge University Press, Cambridge, UK. 112–134.

Bariac, T., Millet, A., Ladouche, B., Mathieu, R., Grimaldi, C., Grimaldi, M., Hubert, P., Molicova, H., Bruckler, L., Bertuzzi, P., Boulegue, J., Brunet, Y., Tournebize, R. and Granier, A., 1995a. Stream hydrograph separation on two small Guianese catchments. In: *Tracer Technologies for Hydrological Systems, IAHS Publication No 229*: 193–209.

Bariac, T., Millet, A., Ladouche, B., Mathieu, R., Grimaldi, C., Grimaldi, M., Hubert, P., Molicova, H., Bruckler, L., Bertuzzi, P., Boulegue, J., Brunet, Y., Tournebize, R. and Granier, A., 1995b. La dècomposition géochimique de l'hydrogramme de rive: le rôle du sol (les bassins versants de la piste Saite-Elie, Guyane). *Comité National Français de Géodésie et de Géophysique, Rapport Quadriennal 1991–1994*, juillet 1995. 271–282.

Barnes, C. J. and Bonell, M., 1996 Application of unit hydrograph techniques to solute transport in catchments, *Hydrol. Proc.*, 10: 793–802.

Bazemore, D. E., Eshleman, K. N. and Hollenbeck, K. J., 1994. The role of soil water in stormflow generation in a forested headwater catchment: synthesis of natural tracer and hydrometric evidence. *J. Hydrol.*, *162*: 47–75.

Bear, J., 1979. *Hydraulics of Groundwater*. McGraw-Hill, New York, N.Y. 544pp.

Bencala, K. E., 2000. Hyporheic zone hydrological processes. *Hydrol. Proc.*, 14: 2797–2798.

Beven, K. J., 1981. Kinematic subsurface stormflow. *Wat. Resour. Res.*, 17: 1419–1424.

Beven, K. J., 1982a. On subsurface stormflow: an analysis of response times. *Hydrol. Sci. Bull.*, 4: 505–521.

Beven, K. J., 1982b. On subsurface stormflow: predictions with simple kinematic theory for saturated and unsaturated flows. *Wat. Resour. Res.*, 18: 1627–1633.

Beven, K. J., 1986. Runoff production and flood frequency in catchments of order *n*: an alternative approach. In: *Scales Problems in Hydrology*, V. K. Gupta, I. Rodriguez-Iturbe and E. F. Wood (eds), Reidel, Dordrecht, The Netherlands. 107–131.

Beven, K. J., 1989. Changing ideas in hydrology – the case of physically-based models. *J. Hydrol.*, 105: 157–172.

Beven, K. J. 1989. Interflow. In: *Unsatured Flow in Hydrologic Modelling*, H. J. Morel-Seytoux (ed.), Kluwer, Germany. 191–219.

Beven, K. J., 1997. TOPMODEL: A Critique. *Hydrol. Process.*, *11*: 1–17.

Beven, K. J. and Kirkby, M. J., 1979. A physically-based variable contribution area model of basin hydrology. *Hydrol. Sci. Bull.*, 24: 43–69.

Beven, K. J. and Germann, P. F., 1981. Water flow in soil macropores II. A combined flow model. *J. Soil Sci.*, 32: 15–29.

Beven, K. J. and Germann, P. F., 1982. Macropores and water flow in soils. *Wat. Resour. Res.*, *18*: 1311–1325.

Beven, K. J., Kirkby, M. J., Schofield, N. and Tagg, A. F., 1984. Testing a physically-based flood forecasting model (TOPMODEL) for three UK catchments. *J. Hydrol.*, *69*: 119–143.

Beven, K., Lamb, R., Quinn, P., Romanowicz, R. and Freer, J., 1995, *TOPMODEL* (Chapter 18). In: *Computer Models of Watershed Hydrology*, V. P. Singh (ed.), Water Resources Publications, Boulder, USA. 627–668.

Bidin, K., 1995. *Subsurface flow controls of runoff in a Bornean natural rainforest*. Unpublished MSc. Thesis, University of Manchester, UK. 128pp.

Bidin, K., 2001. *Spatio-temporal variability in rainfall and wet-canopy evaporation within a small catchment recovering from selective tropical forestry*. Unpublished Ph.D. thesis, University of Lancaster, UK. 201 pp.

Bishop, K. H., 1991. *Episodic increases in stream acidity, catchment flow pathways and hydrograph separation*. Unpublished Ph.D. Thesis, University of Cambridge, UK. 246pp.

Blackie, R., Edwards, K. A. and Clarke, R. T. (Compiled), 1979. Hydrological Research in East Africa. *E. Afr. Agric. For. J.*, *43*, *Special Issue: 1–313*.

Bloschl, G. and Sivapalan, M., 1995. Scale issues in hydrological modelling: A review. *Hydrol. Process.*, *9*: 251–290.

Boersma, L., 1965. Field measurement of hydraulic conductivity above a water table. In: *Methods of Soil Analysis*, C. A. Black (ed.), Agronomy 9, American Society of Agronomy, Madison, Wisconsin, 234–252.

Bonell, M., 1993. Progress in the understanding of runoff generation dynamics in forests. *J. Hydrol.*, 150: 217–275.

Bonell, M., 1998a. Possible impacts of climate variability and change on tropical forest hydrology. (Based on the WWF Conference on the Potential Impacts of Climate Change on Tropical Forest Ecosystems, Puerto Rico, 24–28 April 1995.) *Climatic Change*, 39: 215–272.

Bonell, M., 1998b. Challenges in runoff generation research in forests from the hillslope to headwater drainage basin scale. *J. Amer. Water Resour. Assn.*, 34: 765–785.

Bonell, M., 1999. Tropical forest hydrology and the role of the UNESCO International Hydrological programme. *Hydrol. Earth Syst. Sci.*, 3: 451–461.

Bonell, M., 2004. Selected Issues in Mountain Hydrology of the Humid Tropics. Keynote address at the Joint FRIM (Forest Research Institute Malaysia) – WWF (World Wildlife Fund, Malaysia) Conference on Forestry and Forest Products Research 1999 Series, Kuala Lumpur, 31-March-1 April 1999. *Water: Forestry and Land Use Perspectives. UNESCO Technical Documents in Hydrology*. (Abdul Rahim, M., ed.)

Bonell, M. and Gilmour, D. A., 1978. The development of overland flow in a tropical rainforest catchment. *J. Hydrol.*, 39: 365–382.

Bonell, M. and Gilmour, D. A., 1980. Variations in short-term rainfall intensity in relation to synoptic climatological aspects of the humid tropical north-east Queensland coast. *Singapore J. Trop. Geog.*, 1: 16–30.

Bonell, M. and Williams, J., 1986a. The two parameters of the Philip infiltration equation: Their properties and the spatial and temporal heterogeneity in ared earth of tropical semiarid Queensland. *J. Hydrol.*, 87: 9–31.

Bonell, M. and Williams, J., 1986b. The generation and redistribution of overland flow in a massive oxic soil in a eucalypt woodland within the semiarid tropics in north Australia. *Hydrol. Proc.*, 1: 31–46.

Bonell, M. and Williams, J., 1989. Reply to P. I. A. Kinnells' Comments on 'The generation and Redistribution of Overland Flow on a Massive Oxic Soil in a Eucalypt Woodland within the Semiarid Tropics of North Australia' Mike Bonell and John Williams. *Hydrol. Proc.*, 3: 97–100.

Bonell, M. with Balek, J., 1993, Recent scientific developments and research needs in hydrological processes of the humid tropics. In: *Hydrology and Water Management in the Humid Tropics – Hydrological Research Issues and Strategies for Water Management*, M. Bonell, M. M. Hufschmidt and J. S. Gladwell (eds), UNESCO-Cambridge University Press, Cambridge, UK. 167–260.

Bonell, M. and Fristch, J. M., 1997. Combining hydrometric-hydrochemistry methods: a challenge for advancing runoff generation process research. *IAHS publication No*. 244: 165–184.

Bonell, M. and Molicova, H., 2003. A hydrological approach to tropical reforestation: Controversial issues and possible applications of topographic-wetness models. *J. Trop. Forest. Sci.*, 15: 411–431.

Bonell, M., Gilmour, D. A. and Sinclair, D. F., 1979. A statistical method for modelling the fate of rainfall in a tropical rainforest catchment. *J. Hydrol.*, 42: 251–257.

Bonell, M., Gilmour, D. A. and Sinclair, D. F., 1981, Soil hydraulic properties and their effect on surface and subsurface water transfer in a tropical rainforest catchment. *Hydrol.Sci. Bull.*, 26, 1–18.

Bonell, M., Gilmour, D. A. and Cassells, D. S., 1982. Vertical and lateral soil water movement in a tropical rainforest catchment. *First National Symposium on Forest Hydrology, Instn Engineers, Australia, National Conference Publication* No. 82–6: 30–38.

Bonell, M., Gilmour, D. A. and Cassells, D. S., 1983a, Vertical soil water movement in a tropical rainforest catchment in northeast Queensland. *Earth Surf. Process. Landf.*, 8: 253–272.

Bonell, M., Gilmour, D. A. and Cassells, D. S., 1983b. A preliminary survey of the hydraulic properties of rainforest soils in tropical north-east Queensland and their implication for the runoff process. *In: Rainfall Simulation, Runoff and Soil Erosion*, J. de Ploey (ed.). *Catena* Suppl. 4: 57–78.

Bonell, M., Gilmour, D. A. and Cassels, D. S., 1984. Tritiated water movement in clay soils of a small catchment under tropical rainforest in North-East

Queensland. Proc. ISSS Symposium on water and solute movement in heavy clay soils, J. Bouma and P. A. C. Raats (eds). *ILRI Publication* 37: 197–204.

Bonell, M., Cassels, D. S. and Gilmour, D. A., 1987. Spatial variations in soil hydraulic properties under tropical rainforest in north-eastern Australia. In: *Proc. Int. Conf. On Infiltration Development and Application*, Yu-Si Fok (ed.). Water Resour. Res. Center, Univ. of Hawaii, Manoa. 153–165.

Bonell, M., Gilmour, D. A. and Cassells, D. S., 1991. The links between synoptic climatology and the runoff response of rainforest catchments on the wet tropical coast of north-eastern Queensland. In: *The Rainforest Legacy – Australian National Rainforests Study Report Vol. 2*, P. A. Kershaw and G. Werran (eds). Australian Heritage Commission, Canberra, Australia. 27–62.

Bonell, M., Barnes, C. J., Grant, C. R., Howard, A. and Burns, J., 1998. High rainfall response-dominated catchments: A comparative study of experiments in tropical north-east Queensland with temperate New Zealand. In: *Isotope Tracers in Catchment Hydrology*, C. Kendall and J. J. McDonnell (eds), Elsevier. 347–390.

Boulet, R., 1981. Etude ptographie. In: *Le projet ECEREX – compte rendu des journées*, 4–8 March 1983, GERDAT, INRA, MUSEUM, ORSTOM, 23–52.

Boulet, R., Brugière, J. M. and Humbel, F. X., 1979. Relations entre organisation des sols et dynamique de l'eau en Guyane Française: conséquences agronomiques d'une évolution déterminée par un déséquilibre d'origine principalement téctonique. *Sc. Sol.*, 13–18.

Bouma, J. and Dekker, L. W., 1978. A case study on infiltration into dry clay soil, I: Morphological observations. *Geoderma*, 20: 24–40.

Bouwer, H., 1966. Rapid field measurements of air entry value and hydraulic conductivity of soils as significant parameters in flow system analysis. *Wat. Resour. Res.*, 2: 729–738.

Bouwer, H. and Jackson, R. D., 1974. Determining soil properties. In: *Drainage for Agriculture*. J. van Schilfgaarde (ed.), Agronomy 17, American Society of Agronomy, Madison, Wisconsin, USA. 611–616.

Bowden, W. B., McDowell, W. H., Asbury, C. E. and Finley, A. M., 1992. Riparian nitrogen dynamics in two geomorphologically distinct tropical rainforest watersheds: nitrous oxide fluxes. *Biogeoch.*, 18: 77–99.

Bronswijk, J. J. B., Hamminga, W. and Oostindie, K., 1995. Field-scale solute transport in a heavy clay soil. *Wat. Resour. Res.*, 31: 517–526.

Brooks, R. H. and Corey, A. T., 1964. Hydraulic properties of porous media, *Hydrology Paper* 3, Colorado State University, Fort Collins, CO, USA. 27pp.

Brooks, E. S., Boll, J. and McDaniel, P. S. 2004. A hillslope-scale experiment to measure later saturated hydraulic conductivity. *Water Resour. Res.*, 40, WO4208. doi 10.1029/2003WR002858, 2004.

Bruijnzeel, L. A., 1989. (De)forestation and dry season flow in the tropics: A closer look, *J. Trop. Forest Sc.*, 1: 229–243.

Burch, G. J., Bath, R. K., Moore, I.D. and O'Loughlin, E. M., 1987. Comparative hydrologic behaviour of forsted and cleared catchments in south-eastern Australia. *J. Hydrol.*, *90*: 19–42.

Burch, G. J., Moore, I.D. and Burns, J., 1989. Soil hydrophobic effects on infiltration and catchment runoff. *Hydrol. Proc.*, *3*: 211–222.

Burr, E. J., 1970. Cluster sorting with mixed character types II. Fusion strategies. *Austral. Computer J.*, 1: 97–99.

Buttle, J. M., 1998. Fundamentals of small catchment hydrology. In: *Isotope Tracers in Catchment Hydrology*, C. Kendall, C. and J. McDonnell (eds). 1–50.

Buttle, J. M. and House, D. A., 1997. Spatial variability of saturated hydraulic conductivity in shallow macroporous soils in a forested basin. *J. Hydrol.*, 203: 127–142.

Campbell, G. S., 1985. *Soil Physics with BASIC*, Elsevier, New York, USA. 73–97.

Cappus, P., 1960. Etude des lois de l'écoulement. Application au calcul et à la prévision des débits, bassin expérimental d'Alrance (Investigation of the laws of flow. Application to the computation and prediction of discharge, the Alrance experimental basin). *La Houille Blanche* A: 493–520.

Casenave, A., Flory, J., Mahieux, A. and Simon, J. M., 1984. *Etude hydrologique des bassins de Tai*, Campagne 1981, ORSTOM, Centre d'Adiopodocimé, Côte d'Ivoire.

Cassells, D. S., Gilmour, D. A. and Bonell, M., 1985. Catchment response and watershed management in tropical rainforests in north-east Australia. *Forest Ecol. Manage.*, *10:* 155–175.

Chang, J-H and Lau, L. S., 1993. A definition of the humid tropics. In: *Hydrology and Water Management in the Humid Tropics – Hydrological Research Issues and Strategies for Water Management*, M. Bonell, M. M. Hufschmidt and J. S. Gladwell (eds). UNESCO – Cambridge University Press, Cambridge, UK. 571–574.

Chappell, N. A. and Binley, A. M., 1992. The impact of rainforest disturbance upon near-surface groundwater flow: modelling of hillslope flow experiments. *Annales Geophysicae, 10*: 330.

Chappell, N. A. and Ternan, J. L., 1992. Flow-path dimensionality and hydrologic modelling. *Hydrol. Proc.*, 6: 327–345.

Chappell, N. A., Franks, S. W. and Larenus, J., 1998. Multi-scale permeability estimation for tropical catchment, *Hydrol. Proc.*, 12, 1507–1523.

Chappell, N. A., McKenna, P., Bidin, K., Douglas, I. and Walsh, R. P. D., 1999. Parsimonious modelling of water and suspended-sediment flux from nested-catchments affected by selective tropical forestry. *Phil. Trans. Royal Soc. London, Series B*, 354: 1831–1846.

Chestnut, T. J. and McDowell, W. H., 2000. C and N dynamics in the riparian and hyporheic zones of a tropical stream, Luquillo Mountains, Puerto Rico. *J. N. Amer. Benthol. Soc.*, 19: 199–214.

Chevallier, P. and Planchon, O., 1993. Hydrological processes in a small severe humid savanna basin (Ivory Coast). *J. Hydrol.*, 151: 173–191.

Chorley, R. J., 1978. Glossary of terms. In: *Hillslope Hydrology*, M. J. Kirkby (ed.), Wiley, Chichester, UK. 365–375.

Cook, P. G., Salomon, D. K., Plummer, L. N., Busenberg, E. and Schiff, S. L., 1995. Chlorofluorocarbons as tracers of groundwater transport processes in a shallow, silty sand aquifer. *Wat. Resour. Res.*, 31: 425–434.

Cosandey, C. and de Oliveira, M., 1996. Surfaces saturées, surfaces contributives: localisaton et extension dans l'espace du bassin versant. *Hydrol. Sci. J.*, 41: 751–760.

Davis, S. H., Vertessy, R. A. and Silberstein, R. P., 1999. The sensitivity of a catchment model to soil hydraulic properties obtained by using different measurement techniques. *Hydrol. Proc.*, 13: 677–688.

Dawes, W. R., Zhang, L., Hatton, T. J., Reece, P. H., Beale, G. T. H. and Packer., I., 1997. Evaluation of a distributed parameter ecohydrological model (TOPOG_IRM) on a small cropping rotation catchment. *J. Hydrol.*, 191: 64–86.

De Keyser, F., 1964. *Explanatory Notes on the Innisfail*, Qld. Sheet SE 55-G, Department of National Development, Bureau of Mineral Resources, Geology and Geophysics, Commonwealth of Australia. 30pp.

De Walle, D. R., Swistock, B. R. and Sharpe, W. E., 1988. Three-component tracer model for stormflow on a small Applachian forested catchment. *J. Hydrol.* 104: 301–310.

Dietrich, W. E., Windsor, D. M. and Dunne, T. 1982. Geology, climate and hydrology of Barro Colorado Island. In: The Ecology of a Tropical Forest Seasonal Rhythms and Long-term Changes, E. G. Leigh, A. S. Rand and D. M. Windsor (eds). Smithsonian Institute Press, Washington, D.C. 21–46.

Douglas, L. 1973. Rate of denudation in selected small catchments in eastern Australia. *Occasional Paper in Geography No. 21*, University of Hull, UK. 127 pp.

Dubreuil, P. L., 1985. Review of field observations of runoff-generation in the tropics. *J. Hydrol.*, 80: 237–264.

Dunne, T., 1978. Field studies of hillslope processes. In: *Hillslope Hydrology*, M. J. Kirkby (ed.) Wiley, Chichester, UK. 227–294.

Dunne, T., 1983. Relation of field studies and modelling in the prediction of storm runoff. *J. Hydrol.*, 65: 25–48.

Dunne, T. and Black, R. D., 1970a. An experimental investigation of runoff production in permeable soils. *Wat. Resour. Res.*, 6: 179–191.

Dunne, T. and Black, R. D., 1970b. Partial area contributions to storm runoff in a small New England watershed. *Wat. Resour. Res.*, 6: 1269–1311.

Dunne, T., Zhang, W. and Aubry, B., 1991. Effects of rainfall, vegetation, and microtopography on infiltration and runoff. *Wat. Resour. Res.*, 27: 2271–2285.

Dykes, A. P. and Thornes, J. B., 2000. Hillslope hydrology in tropical rainforest steepland in Brunci. *Hydrol. Proc.*, 14: 215–235.

Elsenbeer, H., 2001. Hydrologic flowpaths in tropical rainforest soilscapes-a review. *Hydrol. Proc.*, 15: 1751–1759.

Elsenbeer, H. and Cassel, D. K., 1990. Surficial processes in the rainforest of western Amazonia. In: *Research Needs and Applications to Reduce Erosion and Sedimentation in Tropical Steeplands*, R. R. Zimmer, C. L. O'Loughlin and L. S. Hamilton (eds). *Int. Assoc. Hydrol. Sci. Publ. No. 192*, 289–297.

Elsenbeer, H. and Cassel, D. K., 1991. The mechanisms of overland flow generation in a small catchment in western Amazonia. In: *Water Management of the Amazon Basin Symp.*, B.P.F. and Fernandez-Jauregui, C. A., (eds), Proc. Manaus Symp., Aug. 1990, Braga, UNESCO (ROSTLAC), Montevideo, Uruguay. 275–288.

Elsenbeer, H. and Lack, A., 1996a. Hydrometric and hydrochemical evidence for fast flowpaths at La Cuenca, Western Amazonia. *J. Hydrol.*, 180: 237–250.

Elsenbeer, H. and Lack, A., 1996b. Hydrological pathways and water chemistry in Amazonian rainforests. In: *Advances in Hillslope Processes, Volume 2*, M. G. Anderson and S. M. Brooks (eds), Wiley, Chichester, UK. 939–959.

Elsenbeer, H. and Vertessy, R. A., 2000. Stormflow generation and flowpath characteristics in an Amazonian rainforest catchment. *Hydrol. Proc.*, 14: 2367–2381.

Elsenbeer, H., Cassel, K. and Castro, J., 1992. Spatial analysis of soil hydraulic conductivity in a tropical rainforest catchment. *Wat. Resour. Res.*, 28: 3201–3214.

Elsenbeer, H., West, A. and Bonell, M., 1994, Hydrologic pathways and stormflow hydrochemistry at South Creek, northeast Queensland. *J. Hydrol.*, 162: 1–21.

Elsenbeer, H., Cassel, D. K. and Zuniga, L., 1994. Throughfall in the Terra Firme Forest of Western Amazonia. *J. Hydrol.(NZ)*, 32: 30–34.

Elsenbeer, H., Lorieri, D. and Bonell, M., 1995a. Mixing model approaches to estimate storm flow sources in an overland flow-dominated tropical rainforest catchment. *Wat. Resour. Res.* 31: 2267–2278.

Elsenbeer, H., Lack, A. and Cassel, K., 1995b. Chemical fingerprints of hydrological compartments and flow paths at La Cuenca, western Amazonia. *Wat. Resour. Res.* 31: 3051–3058.

Elsenbeer, H., Lack, A. and Cassel, K., 1996. The Stormflow chemistry at La Cuenca, western Amazonia. *Intercienca,*, 21: 133–139.

Elsenbeer, H., Bradley, E., Newton, Dunne, T, and Moraes, J. M., 1999. Soil hydraulic conductivities of latosols under pasture, forest and teak in Rondonia, Brazil. *Hydrol. Proc.*, 13: 1417–1422.

FAO-UNESCO, 1974. FAO-UNESCO Soil map of the world. Vol. 1, legend, UNESCO Press, Paris. Later revised as FAO-UNESCO (1988) Soil map of the world. Revised legend. *World Soil Resources Report 60*, FAO, Rome.

Fernandes, N. F., 1990. Hidrologia superficial e propriedaes fisico-mecanicas dos ≪Complexo de rampa≫ Bananal (SP), tese de maestrado. Universitade federal do Rio de Janeiro, Instituto de Geosciencas, Rio de Janeiro, Brazil.

Foster, S. S. D., 1993. Groundwater conditions and problems characteristic of the humid tropics. In: *Hydrology of Warm Humid Regions*, J. S. Gladwell (ed.). *IAHS Publication No. 216*, 433–449.

Foster, S. S. D. and Smith-Carrington, A., 1980. The interpretation of tritium in the chalk unsaturated zone. *J. Hydrol.*, 46: 343–364.

Foster, S. S. D. and Chilton, P. J., 1993. Groundwater Systems in the Humid Tropics. In: *Hydrology and Water Management in the Humid Tropics – Hydrological Research Issues and Strategies for Water Management*, M. Bonell, M. M Hufschmidt and J. S. Gladwell, J. S. (eds), UNESCO – Cambridge University Press, Cambridge, UK. 261–272.

Foster, S., Smedley, P. and Candela, L., 2002. Groundwater quality in the humid tropics: an overview. In: *IHP Technical Document in Hydrology No 52*: UNESCO – Paris. 441–468.

Franchini, M., Wendling, J., Obled, C. and Todini, E., 1996. Physical interpretation and sensitivity analysis of the TOPMODEL. *J. Hydrol.* 175: 293–338.

Franken, W., 1979. Untersuchungen im Einzugsgibiet des zentralamazonishen Urwaldbaches 'Barro Branco' auf der 'terra firme'. 1. Abflussverhalten des Baches. *Amazoniana,* 6: 459–466.

Freer, J., McDonnell, J., Beven, K. J., Brammer, D., Burns, D., Hooper, R. P. and Kendall, C., 1997. Topographic controls on subsurface storm flow at the hillslope scale for two hydrologically distinct small catchments. *Hydrol. Proc., 11*: 1347–1352.

Freeze, R. A., 1972. Role of subsurface flow in generating surface runoff, 2. Upstream source areas. Wat. Resour. Res., 8: 1272–1283.

Freeze, R. A. and Cherry, J. A., 1979. *Groundwater*, Prentice-Hall, Englewood Cliffs, New Jersey, USA. 604pp.

Fritsch, J.-M., 1990. The effects of manual and mechanized clear-cut of the Amazonian moist forest and of alternative land uses of small experimental watersheds, *PhD Thesis*, Université des Sciences et Techniques du Languedoc, Montpellier, 391 pp. [in French].

Fritsch, J-M., 1992. Les effets du défrichement de la forêt amazonienne et de la mise en culture sur l'hydrologie de petits bassins versants: Opération ECEREX en Guyane Française. *Editions de l'ORSTOM, Institut Français de Recherche Scientifique pour le Développement en Coopération, Collection Etudes et Thèses, Paris. 392pp.*

Garstang, M., Massie, Jr. H. L., Halverson, J., Greco, S. and Scala, J., 1994. Amazon Coastal Squall Lines. Part I: Structure and Kinematics. *Mon. Weath. Rev.*, 122: 608–622.

Generaux, D. P. and Hooper, R. P., 1998. Oxygen and hydrogen isotopes in rainfall-runoff studies. In: *Isotope Tracers in Catchment Hydrology*, C. Kendall and J. J. McDonnell (eds), Elsevier, Amsterdam. 319–346.

Germann, P. F., 1990a. Macropores and hydrologic hillslope processes. In: *Process Studies in Hillslope Hydrology*, M. G. Anderson and T. P. Burt (eds.), John Wiley and Sons, Chichester, UK. 327–363.

Germann, P. F., 1990b. Preferential Flow and the Generation of Runoff. 1. Boundary Layer Flow Theory. *Wat. Resour. Res.*, 26: 3055–3063.

Germann, P. F. and Di Pietro, L., 1999. Scales and dimensions of momentum dissipation during preferential flow in soils. *Wat. Resour. Res.*, 35: 1443–1454.

Gillham, R. W. and Jayatilaka, C. J., 1996. A deterministic-empirical model of the effect of the capillary fringe on near-stream area runoff 1. Description of the model. *J. Hydrol.*, 184: 299–315.

Gillham, R. W. and Jayatilaka, C. J., 1998. Response to Comment by Jeffrey J. McDonnell and James Buttle on 'A deterministic-empirical model of the effect of the capillary fringe on near-stream area runoff. 1. Description of the model' (Journal of Hydrology Vol. 184 (1996) 299–315). *J. Hydrol.*, 207: 286–289.

Gilmour, D. A., 1975. *Catchment Water Balance Studies on the Wet Tropical Coast of North Queensland.* Unpublished Ph.D. Thesis, Dept. of Geography, James Cook Univ. of North Queensland, Townsville, Australia. 254 pp.

Gilmour, D. A., 1977. Effects of logging and clearing on water yield and quality in a high rainfall zone of north-east Queensland. In: *The Hydrology of Northern Australia* (Proc. Brisbane Symp. June 1977), Inst. Engrs. Austral., Canberra, A. C. T., Nat. Conf. Publ., 77/5. 156–160.

Gilmour, D. A. and Bonell, M., 1979. Runoff processes in tropical rainforests with special reference to study in north-east Australia. In: *Geographical Approches to Fluvial processes*, A. F. Pitty (ed.), Geo Books, Norwich, UK. 73–92.

Gilmour, D. A., Bonell, M. and Sinclair, D. F., 1980. An investigation of storm drainage processes in a tropical rainforest catchment, Australian Government Publ. Service, Canberra, A. C. T. *Australian Water Resources Council Tech. Paper 56.* 93pp.

Godsey, S. and Elsenbeer, H., 2002. The soil hydrologic response to forest regrowth: A case study from southwestern Amazonia. *Hydrol. Proc.*, 16: 1519–1522.

Grayson, R. B., Moore, I.D. and McMahon, T. A., 1992. Physically based hydrologic modelling, 2. Is the concept realistic? *Wat. Resour. Res.*, 26: 2659–2666.

Grayson, R. B., Western, A. W, Chiew, F. H. S. and Bloschl, G., 1997. Preferred states in spatial soil moisture patterns: Local and non local controls. *Wat. Resour. Res.*, 33: 2897–2908.

Grimaldi, C., 1988. Origine de la composition chimique des eaux superficielles en milieu tropical humide: Exemple de deux petits bassins versants sous forêt en Guyane Française. *Sci. G. Bull. 41* Strasbourg, France: 247–262.

Grimaldi, C., Fritsch, J. M. and Boulet, R., 1994. Composition chimique des eaux de nappe et évolution d'un matériau ferralitique en présence du système muscovite-kaolinite-quartz. *Compte-Rendu de l'Académie des Sciences de Paris. Série III.* 319: 1383–1389.

Guntner, A., Uhlenbrook, J., Seibert, J. and Leibundgut, Ch., 1999. Multi-criterial validation of TOPMODEL in a mountainous catchment. *Hydrol. Proc., 13*: 1603–1620.

Haria, A. H., Johnson, A. C., Bell, J. P. and Batchelor, C. H., 1994. Water movement and isoproturon behaviour in a drained heavy clay soil: 1. Preferential flow process. *J. Hydrol.*, 163: 20–216.

Harr, R. D., 1977. Water flux in soil and subsoil on a steep forested slope. *J. Hydrol.*, 33: 37–58.

Hendrickx, J. M. H., 1990. Determination of soil hydraulic properties. In: *Process Studies in Hillslope Hydrology*, M. G. Anderson and T. P. Burt (eds). Wiley, Chichester, UK. 43–92.

Herwitz, S. R., 1986. Infiltration-excess caused by stemflow in a cyclone-prone tropical rainforest, *Earth Surf. Proc. Landf.*, 11: 401–412.

Hewlett, J. D., 1961a. *Soil moisture as a source of base flow from steep mountain watersheds*. US Dept. Agric. Forest Ser., Southeastern Forest Experiment Station, Asheville, North Carolina, Station Paper No. 132. 11pp.

Hewlett, J. D., 1961b. *Watershed management*. US Dept. Agric. Forest Ser., Southeastern Forest Experiment Station, Asheville, North Carolina, Report for 1961. 61–66.

Hewlett, J. D. and Hibbert, A. R., 1963. Moisture and energy conditions within a sloping soil mass during drainage. *J. Geoph. Res.*, 68: 1081–1087.

Hewlett, J. D., 1974. Comments on letters relating to 'The role of subsurface flow in generating surface runoff, 2. Upstream source areas,' by R. A. Freeze. *Water Resour. Res.*, 10: 605–607.

Hewlett, J. D. and Hibbert, A. R., 1967. Factors affecting the response of small watersheds to precipitation in humid areas. In: *International Symp. on Forest Hydrology*, W. E. Sopper and H. W. Lull (eds), Pergamon, Oxford, UK. 275–290.

Hewlett, J. D., Forston, J. C. and Cunningham, G. B., 1977 : The effect of rainfall intensity on stormflow and peak discharge from forest land. *Wat. Resour. Res.*, 13: 259–266.

Hillel, D., 1980. *Fundamentals of Soil Physics*. Academic press, New York. USA. 413pp.

Hillel, D., Krentos, J. D. and Stylianou, Y., 1972. Procedure and test of an internal drainage method for measuring soil hydraulic characteristics in situ. *Soil Sci.*, 114: 395–400.

Hinton, M. J., Schiff, S. L. and English, M. C., 1994. Examining the contributions of glacial till water to storm runoff using two- and three-componenthydrograph separations. *Wat. Resour. Res.*, 30: 983–993.

Hodnett, M. G., Da Silva, L. P., Da Rocha, H. R. and Cruz Senna, R., 1995. Seasonal soil water storage changes beneath central Amazonian rainforest and pasture. *J. Hydrol.*, 170 : 233–254.

Hodnett, M. G., Oyama, M. D., Tomasella, J. and Marques Filho, A de O., 1996a. Comparisons of long-term soil water storage behaviour under pasture and forest in three areas of Amazonia. In: *Amazonian Deforestation and Climate*, J. Gash, C. A. Nobre, J. M. Roberts and R. L. Victoria (eds), Wiley, Chichester, UK. 57–77.

Hodnett, M. G., Oyama, M. D., Tomasella, J. and Marques Filho, A de O., 1996b. Deep soil water uptake by forest and pasture in central Amazonia: predictions from long-term daily rainfall using a simple water balance model. In: *Amazonian Deforestation and Climate*, J. Gash, C. A. Nobre, J. M. Roberts and R. L. Victoria (eds), Wiley, Chichester, UK. 79–99.

Hodnett, M. G., Vendrame, I., De O. Marques Filho, A., Oyama, M. D. and Tomasella, J., 1997a. Soil water storage and groundwater behaviour in a catenary sequence beneath forest in central Amazonia: I Comparisons between plateau, slope and valley floor. *Hydrol. Earth Sys. Sci.*, 2: 265–278.

Hodnett, M. G., Vendrome, I., de O. Marques Filho, A., Oyama, M. D. and Tomasella, J., 1997b. Soil water storage and groundwater behaviour in a catenary sequence beneath forest in central Amazonia II. Floodplain water table behaviour and implications for streamflow generation, *Hydrol. Earth Sys. Sci.*, 2: 279–290.

Horton, R. E., 1933. The role of infiltration in the hydrological cycle. *Trans. Amer. Geophys. Union 14*: 446–460.

Horton, R. E., 1945. Erosional development of streams and their drainage basins: Hydrological approach to quantitative morphology. *Bull. Geol. Soc. Amer.*, 56: 275–370.

Howard, A. J., 1993. The effect of rainfall intensity on stormflow peak flow within the humid tropics of Northeastern Queensland. Unpublished honours thesis, Dept. of Geography, James Cook Univ. of North Queensland, 141pp.

Hubbert, M. K., 1940. The theory of groundwater motion. *J. Geol.*, 48: 785–944.

Humbel, F. X., 1978. Caractérisation par des méthodes physiques, hydriques et d'enracinement des sols de Guyane Française à dynamique de l'eau superficielle. *Sci. Sol.*, 2: 83–93.

Hvorslev, M. J., 1951. Time lag and soil permeability in groundwater observations. *US Army Corps of Engineers, Waterways Experimental Station Bulletin 36*. US Army Corps of Engineers, Vicksburg, Mississipi, USA. 50pp.

Isbell, R. F. 2002. *Australian Soil Classification*, revd edn, Australian Soil and Land Survey Handbooks Series, vol. 4. Melbourne: CSIRO Publishing.

Isbell, R. F., Webb, A. A. and Murtha, G. G., 1968. *Atlas of Australian Soils. Sheet 7, North Queensland with explanatory data*. C.S.I.R.O and Melbourne University Press, Melbourne, Australia. 99pp.

Jeje, L. K., Ogundoya, O. O. and Uyi, O. E., 1986. Subsurface flow from a forested slope in Ife area of southwestern Nigeria. *Hydrol. Sci. J.*, 31: 584–594.

Jenssen, P. D., 1990. Methods for measuring the saturated hydraulic conductivity of tills. *Nordic Hydrol.*, 21: 95–106.

Jetten, V. G., 1994. Modelling the effects of logging on the water balance of a tropical rainforest: A study in Guyana, *Tropenbos Series* 6, Universiteit Utrecht, Department of Physical Geography, Faculty of Geographical Sciences, P.O. Box 80.115, 3508 TC Utrecht, The Netherlands/Tropenbos, Wageningen, The Netherlands. 196pp.

Jones, J. A. A., 1990. Piping effects in humid lands. In: *Groundwater Geomorphology*, C. G. Higgins and Coates, D. R., Boulder (eds): Geological Society of America Special Paper 252: 111–138.

Kauffmann, S., Sombroek, W. and Mantel, S., 1998. Soils of rainforest. Characterization and major constraints of dominant forest soils in the humid tropics. In: *Soils of Tropical Forest Ecosystems*, A. Schulte and D. Ruheyat (eds), Springer, Berlin, Germany. 9–20.

Kendall, C. and McDonnell, J. J., 1998. *Isotope Tracers in Catchment Hydrology*, Elsevier, Amsterdam, The Netherlands. 839 pp.

Kirkby, M. J., 1975. Hydrograph modelling strategies. In: *Processes in Physical and Human Geography* R. Peel, R. Chisholm. and P. Haggett (eds), Heinemann, Oxford, UK. 69–90.

Kirkby, M. J. (ed.), 1978, *Hillslope Hydrology*. Wiley, Chichester, UK. 389pp.

Kirkham, D., 1955. Measurement of the hydraulic conductivity of soil in place. Symposium on permeability of soils. *American Society for Testing Materials, Special Technical Publication* 163: 80–97.

Kirkham, D. and Van Bavel, C. H. M., 1948. Theory of seepage into auger holes. *Soil Sci. Soc. Amer., Proc.*, 13: 75–89.

Knudby, C. J., 1996. Runoff generation from hillslopes in the southern Tanzanian highlands. *M.Sc. Thesis*, Dept. of Hydrodynamics and Water Resources, Tech. Univ. of Denmark. 147pp.

Kozlowski, T. T. (ed.), 1981. *Water Deficits and Plant Growth, Vol. VI (Woody plant Communities)*. Academic Press, New York, USA. 672pp.

Lancaster, J., 2000. Multi-scale estimation of effective permeability within the Greenholes Beck catchment. Unpublished Ph.D Thesis, University of Lancaster, UK. 389pp.

Lange, H., Lischeid, G., Hoch, R. and Hauhs, M., 1996. Water flow paths and residence times in a small headwater catchment at Gardsjon, Sweden, during steady state storm flow conditions. *Wat. Resour. Res.*, 32: 1689–1698.

Larsen, C., Wieczorek, G. F., Eaton, L. S., Morgan, B. A. and Torres-Sierra, H., 2001. Venezuelan debris flow and flash flood disaster of 1999 studied. *EOS*, 82, No. 47: 572–573.

Law, K. F. and Cheong, C. W., 1987. Effects of land use changes on the hydrological characteristics of Sungai Tekam experimental basin. Paper presented at UNESCO/MRP Workshop on *Impact on Operations in Natural and Plantation Forest on Conservation of Soil and Water Resources*, 23–26 June 1987, Universiti Pertanian, Malaysia.

Lean, J., Bunton, C. B., Nobre, C. A. and Rowntree, P. R., 1996. The simulated impact of Amazonian deforestation on climate using measured ABRACOS vegetation characteristics. In: *Amazonian Deforestation and Climate*, J. Gash, C. A. Nobre, J. M. Roberts and R. L. Victoria (eds), Wiley, Chichester, UK. 549–576.

Lee, D. R. and Cherry, J. A., 1979. A field exercise on groundwater flow using seepage meters and minipiezometers. *J. Geol. Educ.*, 27: 6–10.

Leigh, C. H., 1978a. Slope hydrology and denudation in the Pasoh Forest Reserve. I- Surface wash: experimental techniques and some preliminary results. *Malay. Nat.* J., 30: 179–97.

Leigh, C. H., 1978b. Slope hydrology and denudation in the Pasoh Forest Reserve. II-Throughflow: experimental techniques and some preliminary results. *Malay. Nat.* J., 30: 199–210.

Leopoldo, P. R., Franken, W. K. and Villa Nova, N. A., 1995. Real evapotranspiration and transpiration through a tropical rainforest in central Amazonia as estimated by the water balance method. *For. Ecol. Manag.*, 73: 185–195.

Lesack, L. F. W., 1993a. Export of Nutrients and Major Ionic Solutes From a Rain Forest Catchment in the Central Amazon Basin. *Wat. Resour. Res.* 29: 743–758.

Lesack, L. F. W., 1993b. Water Balance and Hydrologic Characteristics of a Rain Forest Catchment in the Central Amazon Basin. *Wat. Resour. Res.* 29: 759–773.

Lewis, L. A., 1976. Soil movement in the tropics – a general model. Z. *Geomorph.*, N. F., Suppl. – Bd. 25, 132–44.

Loague, K. and Kyriakidis, P. C., 1997. Spatial and temporal variability in the R-5 infiltration data set: Déjà vu and rainfall-runoff simulations. *Wat. Resour. Res.*, 33: 2883–2895.

Lørup, J. K., 1998. Effects of land use on the water resources in the southern Tanzania highlands, Ph.D. Thesis, Institute of Hydrodynamics and Water Resources, Technical University of Denmark. 201pp.

Luthin, J. N. and Kirkham, D., 1949. A piezometer method for measuring permeability of soil in sites below a water table. *Soil Sci.*, 68: 349–358.

Lyne, V. D. and Hollick, M., 1979. Stochastic time-varying rainfall-runoff modelling. In: *Proc. Hydrology and Water Resources Symposium, Perth*. Inst. Engrs, Canberra, Australia. 89–92.

Malmer, A., 1993. Dynamics of hydrology and nutriend losses as response to establishment of forest plantation: a case study on tropical rainforest land in Sabah, Malaysia. Dissertation, Department of Forest Ecology, Swedish University of Agricultural Sciences, Umea. 54pp + 6 supporting manuscripts.

Malmer, A., 1996. Hydrological effects and nutrient losses of forest plantation establishment on tropical rainforest land in Sabah, Malaysia. *J. Hydrol.*, 174: 129–148.

Maloswewski, P. and Zuber, A., 1998. A general lumoed parameter model for the interpretation of tracer data and transit time calculation in hydrologic systems. (Comments on Amin, I. E. and Campana, M. E., 1996. A general lumped parameter model for the interpretation of tracer data and transit time calculation in hydrologic systems. *J. Hydrol.* 179:1–21). *J. Hydrol.*, 204: 297–300.

Manton, M. J. and Bonell, M., 1993. Climate and rainfall variability in the humid tropics. In: *Hydrology and Water Management in the Humid Tropics – Hydrological Research Issues and Strategies for Water Management*, M. Bonell, M. M Hufschmidt and J. S.Gladwell (eds), UNESCO-Cambridge University Press, Cambridge, UK. 13–33.

Masiyandima, M. C., van de Giesen, N., Diatta, S. *et al.*, 2003. The hydrology of inland valleys in the sub-humid zone of West Africa: rainfall–runoff processes in the M'bé experimental watershed. *Hydro. Process.*, 17: 1213–1225.

McClain, M. E. and Elsenbeer, H., 2001. Terrestrial inputs to Amazon streams and internal biogeochemical processing. In: *Biogeochemistry of the Amazon basin*, M. E. McClain. R. Victoria and J. E. Richey (eds), Oxford University Press, UK. 185–208.

McClain, M. E., Richey, J. E. and Pimentel, T. P., 1994. Groundwater nitrogen dynamics at the terrestrial-lotic interface of a small catchment in the Central Amazon Basin. *Biogeoch.*, 27: 113–127.

McClain, M. E., Richey, J. E., Brandes, J. A. and Pimentel, T. P., 1997. Dissolved organic matter and terrestrial-lotic linkages in the central Amazon basin of Brazil. *Glob. Biogeoch. Cycles* 11: 295–311.

McDonnell, J. J., 1990. A rationale for old water discharge through macropores in a steep, humid catchment. *Wat. Resour. Res.*, 26: 2821–2831.

McDonnell, J. J. and Buttle, J. M., 1998. Comment on 'A deterministic-empirical model of the effect of the capillary-fringe on near-stream area runoff.1. Description of the model' by Jayatilaka, C. J. and Gillham, R. W., *Journal of Hydrology, (1996) Vol. 184*: 299–315. *J. Hydrol.*, 207: 280–285.

McDonnell, J. J., Freer, J., Hooper, R., Kendall, C., Burns, D., Beven, K. and Peters, J., 1996. New method developed for studying flow on hillslopes. EOS, Transactions, *Am. Geophys. Union 77*, No. 47: 465–472.

McDowell, W. H., Bowden, W. B. and Asbury, C. E., 1992. Riparian nitrogen dynamics in two geomorphologically distinct tropical rainforest watersheds: subsurface solute patterns. *Bioch.* 18: 53–75.

McKenzie, N., Jacquier, D., Isbell, R. F. and Brown, K. 2004. *Australian Soils and Landscapes: An Illustrated Compendium*, Melbourne: CSIRO Publishing, 432pp.

Mein, R. G. and Larson, C. L., 1973, Modelling infiltration during a steady rain. *Wat. Resour. Res.*, 9: 384–394.

Millet, A., 1996. Bilan hydrique et grandes forêts tropicales. Ph.D. Thesis, Mem. Sc. Terre Univ. P. et M. Curie, Paris 6, France. 199pp.

Molicova, H., Bonell, M., Grimaldi, M. and Hubert, P., 1997. Using TOPMODEL towards identifying and modelling the hydrological patterns within a headwater humid tropical catchment. *Hydrol. Proc.*, 11: 1169–1196.

Montgomery, D. R. and Dietrich, W. E., 2002. Runoff generation in a steep, soil-mantled landscape. *Wat. Resour. Res.*, 38: 7–1 to 7–8.

Montgomery, D. R., Dietrich, W. E., Torres, S. P., Anderson, S. P., Heffner, J. T. and Loague, K., 1997. Piezometric response of a steep unchanneled valley to natural and applied rainfall. *Wat. Resour. Res.*, 33: 91–109.

Moore, I. D., Grayson, R. B. and Ladson, A. R., 1991. Digital terrain modelling: A review of hydrological, geomorphological and biological applications. *Hydrol. Process.*, 5: 3–30.

Mulholland, P. J., 1993. Hydrometric and stream chemistry evidence of three storm flowpaths in Walker Branch Watershed. *J. Hydrol.*, *151*: 291–316.

Murtha, G. G., Cannon, M. G. and Smith, C. D., 1996. Soils of the Babinda-Cairns area, north Queensland. *CSIRO Div. Soils, Div. Report*, No 123. 79 pp.

Nash, J. E. and Sutcliffe, J. V. 1970. River flow forecasting through conceptual models, part 1 – a discussion of principles. *J. Hydrol.*, 10: 282–290.

Noguchi, S., Abdul Rahim, N., Kasran, B., Tani, M. Sammori, T. and Morisada, K., 1997a. Soil physical properties and preferential flow pathways in tropical rainforest, Bukit Tarek, Peninsular Malaysia. *J. For. Res.*, 2: 115–120.

Noguchi, S., Abdul Rahim, N., Yusop, Z., Tani, M. and Sammori, T., 1997b. Rainfall-runoff responses and roles of soil moisture variations to the response in tropical rainforest, Bukit Tarek, Peninsular Malaysia. *J. For. Res.* 2: 125–132.

Nortcliff, S. and Thornes, J. B., 1981, Seasonal variations in the hydrology of a small forested catchment near Manaus, Amazonia, and the implications for its management. In: *Tropical Agricultural Hydrology – Watershed Management and Land Use*, R. Lal and E. W. Russell (eds), Wiley, Chichester, UK. 37–57.

Nortcliff, S. and Thornes, J. B., 1984. Floodplain response of a small tropical stream. In: *Catchment Experiments in Fluvial Hydrology*, T. P. Burt and D. E. Walling (eds), Geo-Books, Norwich, UK. 73–85.

Nortcliff, S. and Thornes, J. B., 1988. The dynamics of a tropical floodplain environment with reference to forest ecology. *J. Biogeog.*, 15: 49–59.

Nortcliff, S. and Thornes, J. B., 1989. Variations in soil nutrients in relation to soil moisture status in a tropical forested ecosystem. *Mineral Nutrients and Tropical Forest and Savenna*, J. Proctor (ed.). Publ. No. 9, British Ecological Society, Blackwell, Oxford, UK. 43–54.

Nortcliff, S., Thornes, J. B. and Waylen, M. J., 1979. Tropical Forest Systems: A Hydrological Approach. *Amazonia, VI*, 4: 557–568.

Northcote, K. H., 1979. *A Factual Key for the Recognition of Australian Soils*. 4th Edition, Rellim Technical Publications, Glenside, South Australia, 124 pp.

Ogunkoya, O. O., 1988. Towards a delimitation of southwestern Nigeria into hydrological regions. *J. Hydrol.*, 99: 165–177.

Ogunkoya, O. O. and Jenkins, A., 1993. Analysis of storm hydrograph and flow pathways using a three-component hydrograph separation model. *J. Hydrol.*, 97: 23–32.

Ogunkoya, O. O., Adejuwon, J. O. and Jeje, L. K., 1984. Runoff response to basin parameters in southwestern Nigeria. *J. Hydrol.*, 72: 667–84.

Ogunkoya, O. O., Jeje, L. K. and Aina, P. O., 2000. Spatial variability of soil moisture in a small catchment in western Nigeria. Nigerian George. J. (n.s.), 3/4: 183–198.

Ogunkoya, O. O., Aina, P. O. and Nurudeen, O., 2003. Temporal variability of soil water in a small catchment in Ile, Ife, southwestern Nigeria. *J. Mining Hydrol.*, 39: 53–59.

Ohta, S. and Effendi, S., 1992. Ultisols of 'lowland Dipterocarp forest' in East Kalimantan, Indonesia. I. Morphology and physical properties. *Soil Science and Plant Nutrition*, *38*: 197–206.

O'Loughlin, E. M., 1981. Saturation regions in catchments and their relations to soil and topographic properties. *J. Hydrol.*, 53: 229–246.

O'Loughlin, E. M., 1986. Prediction of surface saturation zones in natural catchments by topographic analysis. *Wat. Resour. Res.*, 22: 794–804.

O'Loughlin, E. M. 1990a. Perspectives on hillslope research. In: *Process Studies in Hillslope Hydrology*, M. G. Anderson and T. P. Burt (eds), Wiley, Chichester, UK. 501–516.

O'Loughlin, E. M., 1990b. Modelling soil water status in complex terrain. *Agric. Forest Meteorol.*, 50: 23–38.

Oster, H., Sonntage, C. and Münnich, K. O., 1996. Groundwater age dating with chlorofluorocarbons. *Wat. Resour. Res.*, 32: 2989–3001.

Peh, C. J., 1978. Rates sediment transport by surface wash in three forested areas of peninsular Malaysia. Department of Geography, University of Malya, Kuala Lumpur Occasional Paper No. 13, 130pp.

Pereira, H. C., 1991. The role of forestry in the management of tropical watershed. In: *Forests, A Heritage for the Future*, J. Parde and G. Blanchard (eds). Proc. 10th World Forestry Congress, Paris, Sept 1991 Revue Forestière Française, Hors Série No. 3 (Proc. 3), ENGREF, F-54042 Nancy Cedex, 139–150 (English), 151–160 (French), 161–170 (Spanish).

Perroux, K. M. and White, I., 1988. Designs for disc permeameters. *Soil Sci. Amer. J.*, 52: 1205–1215.

Peters, D. L., Buttle, J. M., Taylor, J. M. and LaZerte, B. D., 1995. Runoff production in a forested, shallow soil, Canadian Shield basin. *Wat. Resources Res.* 31: 1291–1304.

Philip, J. R., 1986. Steady infiltration from spheroidal cavities in isotropic and anisotropic soils. *Wat. Resour. Res.*, 22: 1874–1880.

Prove, B. G., 1991. A study of the Hydrological and Erosional Processes under Sugar cane Culture on the Wet Tropical Coast of North Eastern Australia. Unpublished PhD Thesis, James Cook University of North Queensland. 273pp.

Putty, M. R. Y. and Prasad, R., 2000a. Understanding runoff processes using a watershed model – a case study in the Western Ghats in South India. *J. Hydrol.*, 228: 215–227.

Putty, M. R. Y. and Prasad, R., 2000b. Runoff processes in headwater catchments – an experimental study in Western Ghats, South India. *J. Hydrol.*, 235: 63–71.

Quinn, P. F., Beven, K. J., Chevallier, P. and Planchon, O., 1991. The prediction of hillslope flow paths for distributed hydrological modelling using digital terrain models. *Hydrol. Process.*, 5: 59–79.

Raats, P. A. C., 1978. Convective Transport of solutes by steady flows. II Specific flow problems. *Agric. Water Manage.* 1: 219–232

Rasmussen, T. C., Baldwin, R. H., Dowd, J. F. and Williams, A. G., 2000. Tracer vs pressure wave velocities through unsaturated saprolite. *Soil Sci. Soc. Am. J.*, 64: 75–85.

Reeve, R. C. and Kirkham, D., 1951. Soil anistropy and some field methods for measuring permeability. *Trans. Am. Geophys. Un.*, 32: 582–596.

Reynolds, W. D., Elrick, D. E. and Topp, G. C., 1983. A reexamination of the constant head well permeameter method for measuring saturated hydraulic conductivty above the water table. *Soil Sci.*, 136: 250–268.

Reynolds, W. D., Elrick, D. E., and Clothier, B. E., 1985. The constant head well permeameter: Effect of unsaturated flow. *Soil Sci.*, 139: 172–180.

Robson, A. J., Neal, C. and Beven, K. J., 1995. Linking mixing techniques to a hydrological framework – an upland application. In: *Solute Modelling in Catchment Systems*, S. Trudgill (ed.), Wiley, Chichester, UK. 347–370.

Ruxton, B. P., 1967. Slopewash under Mature Primary Rainforest in Northern Papua. In: *Ladform Studies from Australia and New Guinea*, J. N. Jennings and J. A. Mabbutt (eds), Australian National University Press, Canberra, A.C.T. 85–94.

Sandstrom, K., 1995. Forests and water – friends or foes? Hydrological implications of deforestation and land degradation in semi-arid Tanzania. Ph.D. thesis, Linkoping Studies in Arts and Science 120, Linkoping University, S-581 83 Linkoping, Sweden. 69pp + 6 supporting manuscripts.

Sandstrom, K., 1996. Hydrochemical deciphering of streamflow generation in semi-arid East Africa. *Hydrol. Process.*, 10: 703–720.

Sandstrom, K., 1998. Can forests ‘provide’ water: Widespread myth or scientific reality?, *Ambio*, 27: 132–138.

Sarrailh, J. M., 1990. *Mise en valeur de l’écosystème forestier guyanais (Opération ECEREX)*. INRA, CTFT, Paris. 273 pp.

Saulnier, G. M., Obled, C. and Beven, K., 1997. Analytical compensation between DTM grid resolution and effective values of saturated hydraulic conductivity within a TOPMODEL framework. *Hydrol. Process.*, 11: 1331–1346.

Scatena, F. N., 1989. An introduction to the physiography and history of the Bilsey Experimental Watersheds in the Luquillo Mountains of Puerto Rico. *Gen. Tech. Rep.* SO-72, USDA, Forest Service, Southern Forest Experiment Station, New Orleans, USA. 22pp.

Schellekens, J., 2000. Hydrological processes in a humid rainforest: A combined experimental and modelling approach. *Vrije Universiteit*, Amsterdam, The Netherlands. 158pp.

Shah, S. M. S., O’Connell, P. E. and Hosking, J. R. M., 1996. Modelling the effects of spatial variability in rainfall on catchment response2. Experiments with distributed and lumped models. *J. Hydrol.*, 43: 45–65.

Sherlock, M. D., 1997. Plot-scale hydrometric and tracer characterisation of soil water flow in two tropical rainforest catchments in South East Asia. Unpublished Ph.D thesis, University of Lancaster, UK. 352pp.

Sherlock, M. D., Chappell, N. A. and Greer, T., 1995. Tracer and Darcy-based identification of subsurface flow, Bukit Timah forest, Singapore. *Singapore J. Tropic. Geogr.*, 16: 197–215.

Sherlock, M. D., Chappell, N. A. and McDonnell, J. J., 2000. Effects of experimental uncertainty on the calculation of hillslope flow paths. *Hydrol. Process.*, 14: 2457–2471.

Smith, R. E., 1977. Approximate soil water movement by kinematic characteristics. *Soil. Sci. Soc. Amer. J.*, 47: 3–8.

Sokal, R. R. and Michener, C. D., 1958. A statistical method for evaluating systematic relationships. University of Kansas, *Scientific Bulletin, 38*: 1409–1438.

Soil Survey Staff, 1975. *Soil Taxonomy*. USDA Handbook 436. US Government Printing Office, Washington, DC, USA. 754pp.

Solomon, D. K., Cook, P. G. and Sanford, W. E. 1998. Dissolved gases in subsurface hydrology. In: *Isotope Tracers in Catchment Hydrology*, C. Kendall and J. J. McDonnell (eds.), Elsevier, Amsterdam, 291–318.

Stace, H. C. T, Hubble, G. D., Brewer, R., Northcote, K. H., Sleeman, J. R., Mulcahy, M. J. and Hallsworth, E. G., 1968. *A Handbook of Australian Soils*. Rellim Technical Publications, Glenside, South Australia. 435pp.

Stewart, M. K. and McDonnell, J. J., 1991. Modeling base flow soil water residence times from deuterium concentrations. *Wat. Resour. Res.*, 27: 2681–2693.

Szabo, Z., Rice, D. E., Plummer, L. N., Busenberg, E., Drenkard, S. and Schlosser, P., 1996. Age dating of shallow groundwater with chlorofluorocarbons, tritium/helium 3, and flow path analysis, southern New Jersey coastal plain. *Wat. Resour. Res.*, 32: 1023–1038.

Talsma, T., 1969. In situ measurement of sorptivity. *Austral. J. Soil Res.*, 7: 269–276.

Talsma, T. and Hallam, P. M., 1980. Hydraulic conductivity measurement of forest catchments. *Austral. J. Soils Res.*, 18: 139–148.

Thomas, M., 1973. Landforms in Equatorial Forest Areas. In: *The Unquiet landscape*, D. Brunsden and J. C. Doornkamp (eds.), David and Charles, Newton Abbott, UK. 141–146.

Tomasella, J. and Hodnett, M. G., 1996. Soil hydraulic properties of an oxisol under pasture in central Amazonia. In: *Amazonian Deforestation and Climate*, J. H. C. Gash, C. A. Nobre, J. M. Roberts and R. L. Victoria, R. L., (eds.), Wiley, Chichester, UK. 101–124.

Topalidis, S. and Curtis, A. A., 1982. The effect of antecedent soil water conditions and rainfall variations on runoff generation in a small eucalypt catchment. In: *The First National Symposium on Forest Hydrology*, E. M. O’Loughlin and K. J. Bren (eds.), Melbourne. Vic. Inst. Eng., Aust., Canberra, A. C. T. Australia. 43–49.

Torres, R., Dietrich, W. E., Montgomery, D. R., Anderson., S. P., Conrad, M. E. and Loague, K., 1998. Unsatured zone processes and the hydrologic response of a steep, unchanneled catchment. *J. Hydrol.*, 57: 1865–1879.

Uehara, G., 1995. Management of Isoelectric Soils of the Humid Tropics. In: *Soil Management and the Greenhouse Effect*, R. Lal, J. Kimble, E. Levine and B. A. Stewart (eds.), CRC Lewis Publishers, Boca Raton, FL, USA. 271–278.

Uhlenbrook, S., 1999. Untersuchung und Modellierung der Abflußbildung in einem mesoskaligen Einzugsgebiet (Examination and modelling of runoff generation in a mesoscaled basin). PhD Thesis, Institute of Hydrology, University of Freiburg, Germany. 201pp.

Uhlenbrook, S., Frey, M., Leibundgut, C. and Maloszewski, P., 2002. Hydrograph separations in a mesoscale mountainous basin at event and seasonal time scales. *Wat. Resour. Res.*, 38: 31–1 to 31–14.

van Beers, W. F. J., 1958. The auger hole method, a field measurement of the hydraulic conductivity of soil below the water table. *Int. Inst. Land Reclam. Improv. Bull. 1*, Wageningen, The Netherlands. 32 pp.

van Genuchten, M.Th., 1980. A closed-form equation for the predicting the hydraulic conductivity of unsaturated soils. *Soil Sci. Soc. Amer. J.*, 44: 892–898.

van Genuchten, M. Th. and Leij, F. J., 1992. On estimating the hydraulic properties of unsaturated soils. In: *Indirect methods for estimating hydraulic properties of unsaturated soils*, M. Th. van Genuchten, F. J. Leij and L. J.

Lund (eds.), Proc. Int. Workshop, Riverside, California, October 11–13, 1989, Univ. California, CA 92521, USA. 1–14.

van de Giesen, N. C., Stomph, T. J. and de Ridder, N., 2000. Scale effects of Hortonian overland flow and rainfall-runoff dynamics in a West African catena landscape. *Hydrol. Proc.*, 14: 165–175.

Vertessy, R. A., T. J. Hatton, O'Shaughnessy, P. J. and Jayasuriya, M. D. A., 1993. Predicting water yield from a mountain ash forest catchment using a terrain analysis-based catchment model. *J. Hydrol.*, 150: 665–700.

Vertessy, R. A., Hatton, T. J., Benyon, R. J. and Dawes, W. R., 1996. Long term growth and water balance predictions for a mountain ash (E. regnans) forest subject to clearfelling and regeneration. *Tree Physiol.*, 16: 221–232.

Vertessy, R. A. and Elsenbeer, H., 1999. Distributed modeling of storm flow generation in an Amazonian rainforest catchment: Effects of model parameterization. *Wat. Resour. Res.*, 35: 2173–2187.

Walsh, R. P. D., 1980. Runoff processes and models in the humid tropics, *Z. Geomorphol.* N. F. Suppl.-Bd. 36, 176–202.

Waterloo, M. J., 1994. Water and nutrient dynamics of *Pinus caribaea* plantation forests on former grassland soils in southwest Viti Levu, Fiji. Thesis, Vrije Universiteit Amsterdam, The Netherlands. 478 pp.

Wenzel, W. W., Unterfrauner, H, Schulte, A., Ruhiyat, D., Simorangkir, D., Kuraz, V., Brandstetter, A. and Blum, W. E. H., 1998. Hydrology of acrisols beneath Dipterocarp forests and plantations in East Kalimantan, Indonesia. In: *Soils of Tropical Forest Ecosystems*, A. Schulte and D. Ruhiyat (eds.), Springer, Berlin, Germany. 62.72.

Western, A. W., Grayson, R. B. and Green, T. R., 1999. The Tarrawarra project: High resolution spatial measurement, modelling and analyis of soil moisture and hydrological response. *Hydrol. Process.*, 13: 633–652.

Weyman, D. R., 1973. Measurements of the downslope flow of water in a soil. *J. Hydrol.*, 20: 267–288.

Whipkey, R. Z., 1965. Subsurface storm flow from forested slopes. *Int. Assoc. Sci. Hydro. Bull.*, 10: 74–85.

Whipkey, R. Z., 1969. Storm runoff from forested catchments by subsurface routes. In: *Floods and their Computation, Studies and Reports in Hydrology*, Leningrad, UNESCO-IASH-WMO, 3. 773–779.

Wierda, A., Veen, A. W. L. and Hutjes, R. W. A., 1989. Infiltration at the Tai rainforest (Ivory Coast): measurements and modelling. *Hydrol. Proc.*, 3: 371–382.

Williams, J. and Bonell, M., 1988. The influence of scale of measurement on the spatial and temporal variability of the Philip infiltration parameters – an experimental study in an Australian savannah woodland. *J. Hydrol.*, 104: 33–51.

Williams, J., and Coventry, R. J., 1979. The contrasting soil hydrology of red and yellow earths in a landscape of low relief. In: *The hydrology of areas of low precipitation., Int. Assoc. of Hydrol. Sci. Publ. no. 128*: 385–395.

Wilson, G. V., Jardine, P. M., Luxmoore, R. J., Zelazny, L. W. and Todd, D. E., 1991a. Hydrogeochemical processes controlling subsurface transport form an upper Walker Branch Watershed during storm events. 1. Hydrologic transport processes. *J. Hydrol.*, 123: 297–316.

Wilson, G. V., Jardine, P. M., Luxmoore, R. J., Zelazny, L. W. and Todd, D. E., 1991b. Hydrogeochemical processes controlling subsurface transport form an upper Walker Branch Watershed during storm events. 2. Solute transport processes. *J. Hydrol.*, 123: 317–336.

Wood, E. F., Sivapalan, M., Beven, K. J. and Band, L., 1998. Effects of spatial variability and scale with implications to hydrologic modelling. *J. Hydrol.*, 102: 29–47.

Woods, R. A., Grayson, R. B., Western, A. W., Duncan, M. J., Wilson, D. J., Young, R. I., Ibbit, R. P., Henderson, R. D. and McMahon, T. A., 2001. Experimental design and initial results from the Mahurangi river variability experiment: MARVEX. In: *Land Surface Hydrology, Meteorology and Climate: Observations and Modelling*, V. Lakhsmi, J. D. Albertson. and J. Schaake (eds). *Water Resources Monograph*, AGU. 201–213.

Wrinkler, R. L. and Hays, W. L., 1975. *Statistics: probability, inference and decision*, 2nd Edition, Holt, Rinehart and Winston, New York, USA. 889pp.

Youngs, E. G., 1983. Soil physical theory and heterogeneity. *Agric. Wat. Manage.*, 6: 145–159.

Zaslavsky, D. and Sinai, G., 1981. Surface hydrology: I. Explanation of phenomena, II. Distribution of raindrops, III. Causes of lateral flow, IV. Flow in sloping layered soil, V. In-surface transient flow. *Proc. Am. Soc. Civil Engrs. J. Hydraulics Div. HY 1*: 1–93.

Zimmermann, U., Munnich, K. O. and Roether, W., 1967. Downward movement of soil moisture traced by means of hydrogen isotopes. In: *Isotope techniques in the hydrological cycle*, G. E. Stout (ed.), American Geophys. Un., Geophysical Monograph Series no. 1, Washington, DC, USA. 28–36.

Zuber, A. and Maloszewski, P., 2001. Lumped parameter models. *In*: *Environmental isotopes in the hydrological cycle, Principles and applications.* W. G. Mook (ed.), IHP-V, *Technical Documents in Hydrology, No 39, Vol. VI*, UNESCO, Paris. 5–35.

15 Erosion and sediment yield in the humid tropics

I. Douglas
University of Manchester, UK
J. L. Guyot
L'Institut de Recherche pour le Développement, Toulouse, France

INTRODUCTION

Under natural conditions, sediment loads in rivers are primarily a function of climate and geology. Climate is expressed through the amount, seasonality and intensity of precipitation. Geology is expressed through the surface rocks of the catchment area and the tectonic style of the terrain over which the catchment has developed. The influence of these interactions between climate and geology will be discussed here at both the fundamental continental and major river basin scale and the local catchment scale. The continental scale analysis considers variations in the fundamental drivers of sediment delivery: the supply of erosive energy and the resistance to erosion. The local catchment scale analysis examines how flows and activities within the ecosystems produced by these climatic and geologic drivers influence the sources and delivery of sediment.

INFLUENCE OF TECTONICS AND LITHOLOGY

Although tropical rainforests cover many parts of the Equatorial regions, they extend over great geological diversity, with wide variations in sediment yield. Where tectonics are active, earth movements frequently trigger landslides and thus supply large quantities of sediments to rivers. Active volcanic areas also create large sediment supplies, particularly through ejected ash and its mobilisation in lahars. Lahar flows can contain sediment concentrations as high as 66% by volume and peak discharges of several hundred m^3 s^{-1} (Major *et al.*, 1996; Pierson *et al.*, 1992). Such tectonic controls have usually been considered as azonal phenomena but in the humid tropics their influence is increased in power by the violence of the climate and the efficient biogeochemical cycling in rainforest ecosystems (see Proctor, this volume).

The lithologic controls are more subtle than the tectonic controls but relate closely to the erodibility of rocks, particularly their behaviour when exposed in river banks and channel beds or by landsliding. Much of the research literature gives detailed descriptions of deeply weathered granitic rocks and the characteristics of the sediment produced by their breakdown. There is also a considerable literature on sandstone and limestone landforms but far less on the highly prevalent mudrocks that dominate large areas of sedimentary rock sequences in the tropics, particularly in Borneo and adjacent areas. Where rocks are highly dispersible and tend to disintegrate in contact with water, or when abraded by gravels being transported across them, sediment yields tend to be high.

It is thus possible to develop a schema of how sediment yields will vary under natural conditions in the humid tropics (Table 15.1). These help to explain the broad inter-regional differences in sediment yields in the humid tropics. The highest sediment yields in the humid tropics occur in tectonically active areas, such as the mountains of New Guinea and Taiwan, experiencing sediment yields of the order of 10 000 t km^2 y^{-1} (Pickup *et al.*, 1981; Shimen Reservoir Authority, 1975). The lowest sediment yields are on old land surfaces or sedimentary basins of low relief. Areas like the Congo Basin in Africa have sediment yields of the order of 100 t km^2 y^{-1} (Milliman and Meade, 1983).

INFLUENCE OF RAINFALL

Tropical rainforests feature climatic diversity, particularly rainfall depth/area/duration characteristics (see Bonell *et al.*, this volume). Seasonality of rainfall varies as distance from the Equator increases. Occurrence of tropical cyclones (hurricanes or typhoons) greatly affects patterns of sediment yields. Across areas of similar geology and relief, larger channels and higher sediment yields occur in cyclone-affected river basins. Precipitation

Forests, Water and People in the Humid Tropics, ed. M. Bonell and L. A. Bruijnzeel. Published by Cambridge University Press.

Table 15.1. *Types and characteristics of humid tropical rivers*

Tectonic setting	Major channel characteristics	Dominant sediment source	Estimated erosion rate (range of annual sediment yield) (km^{-2} y^{-1})	General depth of soil on slope	Example	Reference
Active plate margin Rift, or half graben edge	Braided channels abundant gravel	Mass movement often triggered by earthquake	Up to 10 000	Thin	Markham River, Papua New Guinea	Löffler, 1977
Active volcanic areas of recent lava flows	Braided channel	Volcanic debris unstable ash deposits	Up to 10 000	Skeletal	Toto Amarillo, Costa Rica	Kasel, 1985
Tectonically active mountain areas	Deep gorge sections where river usually occupies whole valley floor in flood; some braiding, but lateral movement of river restricted	Valley wall failure, land sliding associated with seismic activity	7000–10 000	Thin	Upper Fly River, Papua New Guinea	Pickup, 1984
Late Tertiary tectonic activity and weak mud rocks	Wide channel with gravel bare, frequent undercut banks and grassed flood deposits	Bank erosion in main channel and tributaries; erosion by saturated overland flow in streamhead hollows	About 1000	Around 1 m	Segama River, Sabah, Borneo	Douglas *et al.*, 1992
Passive margin with relief of 2000 m in equatorial climate	In upland, boulder-strewn channels with mature trees right up to water's edge; Abundant quartz sand between boulders	Continuation of boulders, surface wash on slopes, bank erosion	50–100	Up to 30 m	Gombak, Malaysia	Douglas, 1968
Passive margin with relief of 2000 m in tropical cyclone zone	In upland, boulder-strewn or rock-cut channels, with areas of exposed rock or boulders at low flows; channel capacities much greater than previous case	Disintegration of boulders, surface wash on slopes, bank erosion	50–200	Up to 30 m	Babinda and Behana Creeks, North Queensland	Douglas, 1967
Ancient craton with relief of 2000 m	In upland, channels guided by ancient structural lineaments, resistant angular blocks and some derived gravel; monsoonal climates produce large broad lowland valleys	Little sediment supply except through river action on rock of channel wall	Up to 50	Up to 30 m	Mahaweli Ganya System, Sri Lanka	Bremer, 1981
Ancient craton with erosion surface	Wide sandy channels, little incision; occasional rock bars that suffer little erosion	Wash from etchplain surface	Up to 30	Up to 30 m	Zaire, Africa	Büdel, 1965, 1981
Sedimentary basin on ancient craton	Wide anastomosing channels, often with legacies of past phases of fluvial erosion	Little locally derived sediment, except from bank erosion	Up to 30	Up to 30 m	Amazon basin	Garner, 1974; Baker, 1978

intensity and duration directly influence detachment of soil and stream discharge and thus the supply and transport of material by streams.

The characteristics of tropical rainfall are often given great attention in discussing erosion and sediment sources in tropical rainforest areas. In Hong Kong, rain drops in the open at intensities of less than 20 mm h^{-1} have a median drop size of less than 2.25 mm, but at 60 mm h^{-1} it is close to 3 mm and larger than 3.25 mm for intensities over 90 mm h^{-1} (Jayawardena and Rezaur, 2000). Kinetic energy ranged from 780 J m^{-2} in a storm of 119.63 mm h^{-1} intensity to 30 in a storm of 12.31 mm h^{-1} intensity.

The frequency of high intensity rainfalls in the tropics may be exaggerated; in the Danum Valley area of Sabah, 5-minute intensities rarely exceed 100 mm h^{-1} and most 5-minute intensities are below 100 mm h^{-1} (Bidin, 2001). In the period May 1995 to April 1998, 5-minute intensities of over 100 mm h^{-1} occurred on only 26 occasions and on only two of these days was an intensity of over 50 mm h^{-1} sustained for over 25 minutes.

The forest canopy modifies raindrop characteristics. One argument is that as drops accumulate on leaves, they coalesce as they run towards the leaf tip and much larger drops fall from leaves to the ground than originally fell in the rain above the forest canopy. In tall forest, this can allow large drops to reach their terminal velocity and so have a greater kinetic energy than did the rain in the open. On the other hand, undisturbed forest canopies are extremely variable, with many forest patches having young saplings with leaves within a few metres of the ground and gaps where pioneer and primary species are competing and the density of foliage within 2 m of the ground surface is high. Under such circumstances, high-energy raindrops do not reach the ground.

Studies of rainforest interception bear out this heterogeneity of the forest canopy and the variation in the catch of sub-canopy rainfall collectors (Lloyd and Marques, 1988; Herwitz and Slye, 1992). Between 13 and 28% of sub-canopy gauges in undisturbed forest at Danum Valley caught more rainfall than gauges in the open (Wong, 1991; Bidin, 2001). Such heterogeneity of catch may influence patterns of soil detachment and erosion on the forest floor.

El Niño and La Niña events produce significant responses in the humid tropics that affect sediment yield markedly. During the 1982–83 El Niño event in central Amazonia, larger precipitation than normal occurred at the stations of Parintins, Prainha and Oriximina (90.0 mm), while at the other stations around 45.0 mm more occurred. However, during the 1997–98 El Niño event, precipitation deficits of the order of 150.0 mm less than normal occurred in the months of June and July 1997 (Mendes and Cohen, 2000).

Fifteen years' hydrological investigations at Danum Valley, Sabah, have provided dramatic evidence of the effects of extremes of drought, as in the 1992 and 1997 ENSO events, and of flood, as in January 1996. From May 1997 to April 1998, only two high 5-minute intensity events occurred at the KM63 gauge site at Danum Valley (Table 15.2), compared to nine in the 1995–96 water year (Bidin, 2001). In April 1992, the undisturbed forest W8S5 catchment stream dried out completely for the whole 1.5 km defined channel above the gauging station, but after the exceptional rains of January 1996, integrated concentrated flow occurred from within a few metres of the catchment divide, a total flow length to the gauge of some 1.8 km. Inevitably under such conditions, erosion is episodic. Sediment is discharged in pulses driven by storm events, collapse of debris dams and occasional landslips. The effects of the 1992 drought year and of individual extreme events are also borne out in the pattern of sediment yield from the Ulu Segama catchment. 1992 contributes little to the cumulative sediment yield, whereas a few large storm events contribute almost 50% of all the sediment discharged over the period (Figure 15.1).

Table 15.2. *Number of storms in the water year at Danum Valley's KM 63 gauge reaching an intensity of 50 mm h^{-1} for a given duration*

	Duration in minutes			
	10	15	20	25
1995–96	9			1
1996–97	5			1
1997–98	2			

Source: Based on data in Bidin (2001).

Sediment discharge by humid tropical rivers is thus largely an event-driven process. All rivers, even the largest, show seasonal and storm period variations in the relationship between sediment concentration and discharge. For example, concentrations of suspended sediment in the Rio Orinoco and the Amazon are greater during rising stages than at equal water discharges during the falling stages, as the Beni River clearly shows (Figure 15.2).

PATTERNS OF SEDIMENT YIELD IN LARGE BASINS AND THROUGHOUT THE HUMID TROPICS

In large river basins with considerable relief, at least in their headwaters, such as the Amazon and Orinoco, the sediment yields are high. In such large catchments erosion and sediment yield are a complex function of topography (tectonics), lithology and climate, as contrasts in sediment yields in the Bolivian upper Amazon indicate (Guyot and Hérail, 2000) (Figure 15.3). For example, the Beni River sediment yields of about 3000 t km^{-2} y^{-1} are similar to those that prevail in New Guinea. The rapid uplift of New Guinea has created mountains whose peaks locally exceed 4000 m. Combined

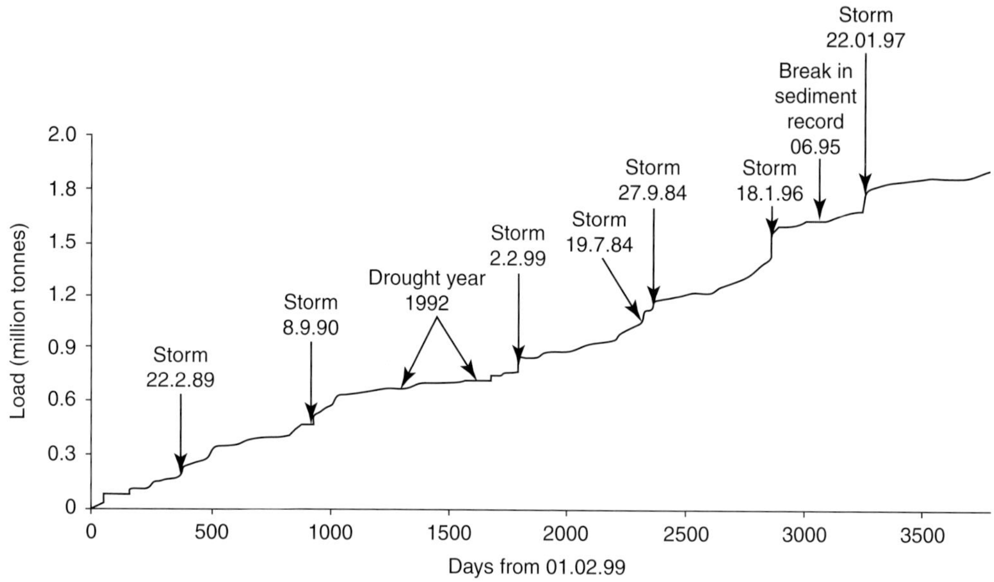

Figure 15.1 Cumulative curve of daily suspended sediment discharge in the Ulu Segama River, at the Danum Valley Field Centre, Sabah (Catchment area km^2; mean discharge m^{-3} s^{-1}; mean sediment yield t km^{-2} y^{-1}. (Based on data supplied by I. Douglas.)

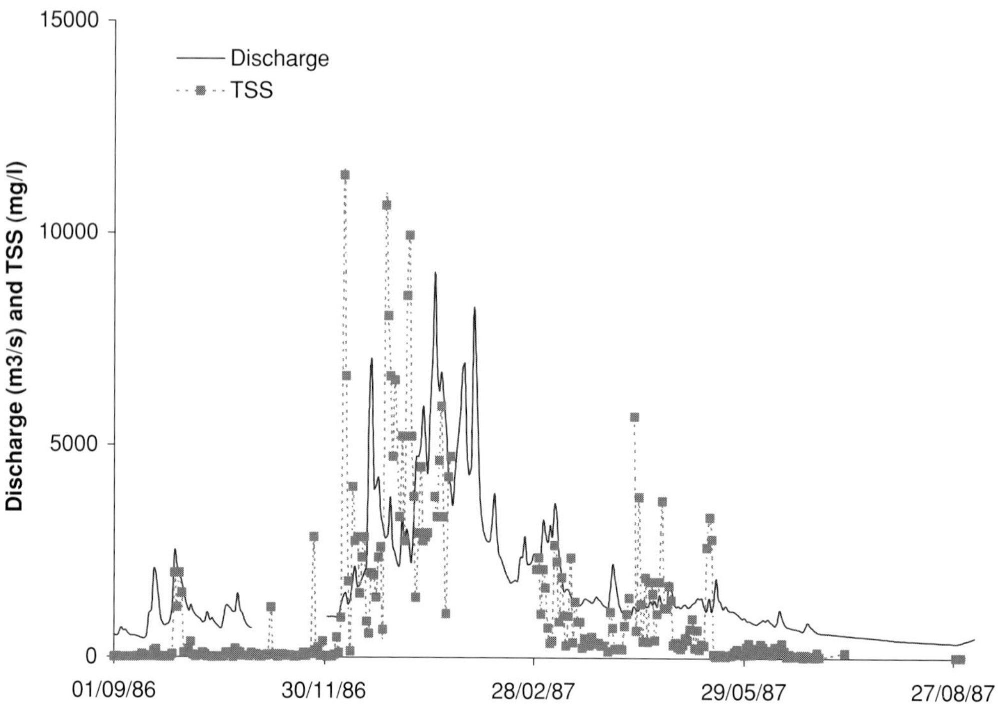

Figure 15.2 Sediment concentrations and discharge September 1986 to August 1987 of the Beni Tiver at Rurrenabaque, Amazonia. Catchment area 67.500 km^2; mean discharge 2200 m^{-3} s^{-1}; mean sediment yield 3200 t km^{-2} y^{-1}. (Data supplied by J.-L. Guyot.)

with the high rainfall, these conditions result in annual sediment yields exceeding 1000 t km^{-2} (Figure 15.4). Large episodic sediment delivery to river channels occurs due to landsliding induced by intense rain and earthquakes. Before catchment disturbance, the combined sediment discharge of the Ok Tedi and the Fly where they join was about 840 t km^{-2} yr^{-1} (Pickup *et al.*, 1981; Pickup, 1984). The Strickland, which penetrates farther into the fold and thrust belt, and drains twice the drainage area, discharges about 2000 t km^{-2} yr^{-1}. The total sediment discharge from rivers draining New Guinea is estimated to be about 1.5 times that of

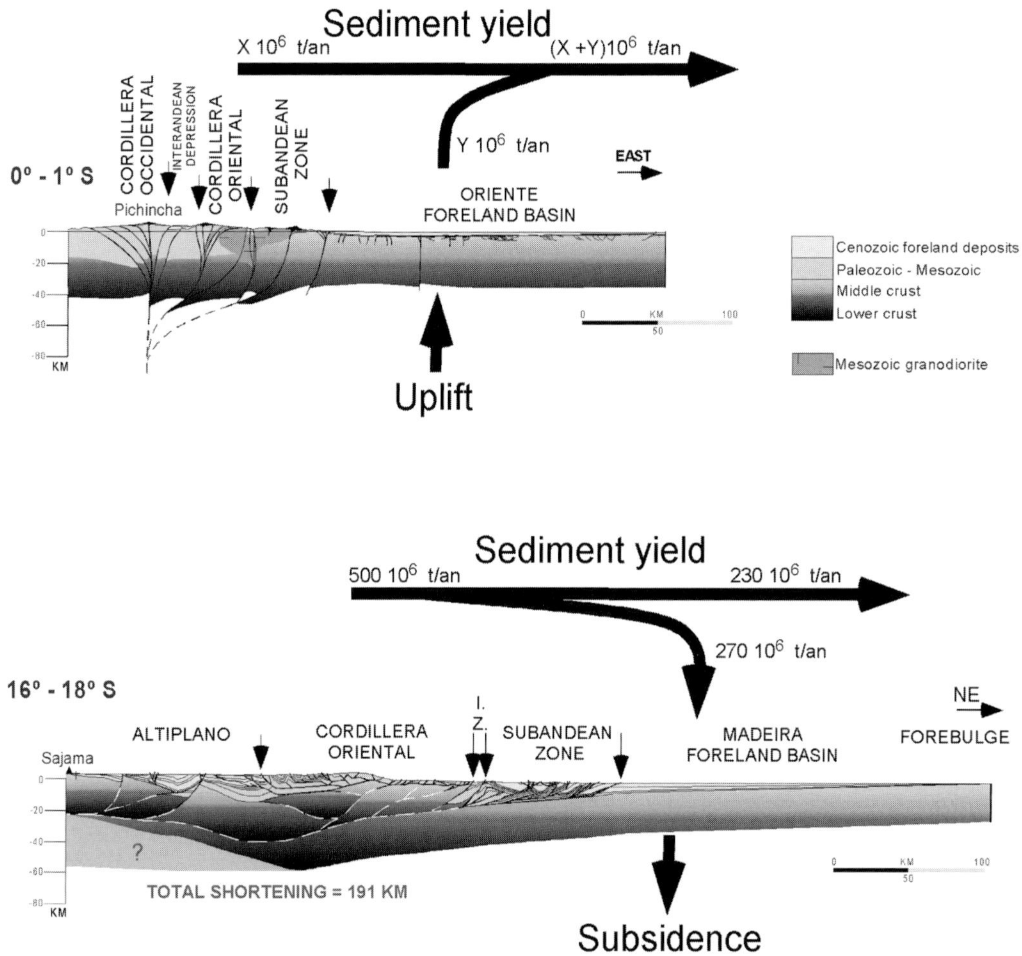

Figure 15.3 Sediment yields in relation to tectonics and lithology in the Andean foreland of Western Amazonia. (Diagram by J.-L. Guyot.)

the Amazon River, although the island has only one-eighth the drainage area.

In the steeplands of eastern Taiwan, local relief of 1000 m or more and high tectonic uplift rates combine with a vigorous climate (average annual precipitation is 3000 mm) to cause a basin-wide denudation rate of around 5.2 mm yr^{-1}. Instantaneous rates of landsliding vary to the extent that the maximum daily sediment discharge observed on the Hualien Chi may exceed the average annual load of this river. Maximum daily sediment discharges measured on the Ma-An Chi (136 km^2) and Wan-Li Chi (242 km^2) are 5000 and 4200 t km^{-2} day^{-1}, respectively.

These Andean, New Guinea and Taiwanese examples reinforce Milliman and Syvitski's conclusion (1992) that small steep mountainous basins are responsible for a large part of the total sediment supply to the world's oceans. A disproportionate amount of this supply is provided by the islands of the Indo-Pacific archipelago, the six high-standing East Indies islands alone supplying about 25% of the fluvial sediment discharged to the global ocean (Milliman *et al.*, 1999).

However, further downstream in major catchments, beyond the tectonically active high relief zones, erosion under stable forest in the humid tropics is spatially discrete (only a few localities being sediment sources). Where major rivers are not dominated by mountainous headwaters and have large lowland areas, sediment yields are low, as in the Congo and Negro (Table 15.3). On St. John, US Virgin Islands, where landslides and debris flows are relatively rare, typical sediment yields from relatively undisturbed catchments are 20 to 40 t km^{-2} y^{-1} (MacDonald *et al.*, 2001).

Not all tropical rivers rise in well-watered headwaters. Some flow from dry areas into the tropics, as does the Mekong which receives only 18% of its runoff from the steep slopes of the eastern edge of the Tibetan plateau but some 33% from the high rainfall areas of the mountains of Laos. Thus influences on sediment yields vary greatly within large river systems.

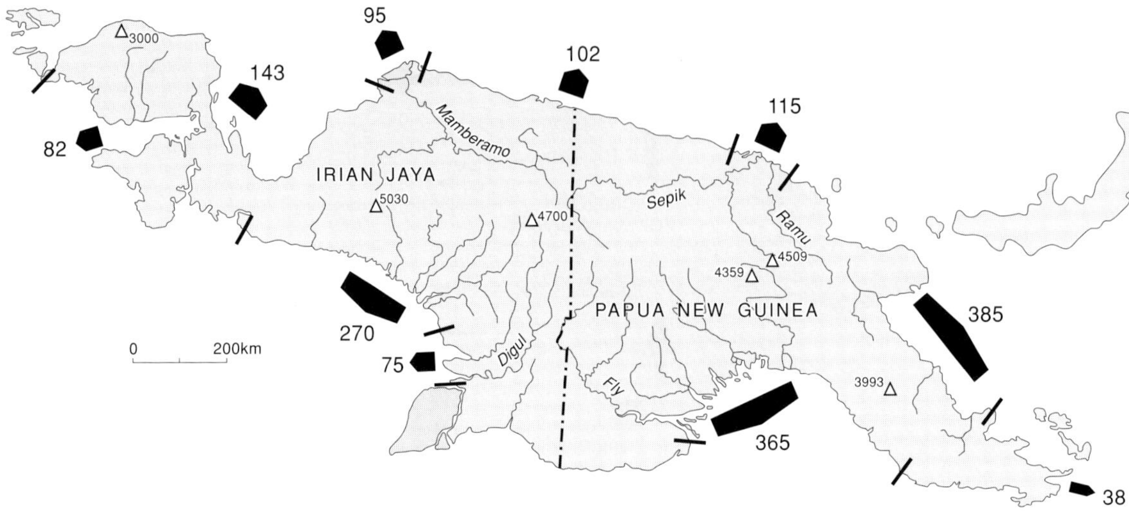

Figure 15.4 Sediment budgets for the island of New Guinea. (After Milliman, 1995.)

When rivers like the Mekong and the Amazon traverse tropical lowlands their sediment transfer becomes dependent upon hydrological regime (climate) and river channel geometry (tectonics). Complex depositional systems develop through time and flood plain lake systems and seasonally inundated (flooded) forests can become sediment traps.

EROSION PROCESSES IN SMALL CATCHMENTS

Physical phenomena and the activities of forest organisms lead to two kinds of response to extreme events in the undisturbed forest:

(1) Much more erosion and larger changes to the land surface than would be expected by the magnitudes of the water flows involved.
(2) The occurrence of changes due to biological and pedogenetic processes not linearly related to the magnitudes of the events concerned.

Undisturbed tropical forest ecosystems have high turnover in biogeochemical cycles, most of the nutrient pool being held in the plant biomass, with rapid recycling of nutrients between plants and the soil. Litterfall is a key element in this nutrient recycling but the litter layer on the forest floor is variable, both in time and space. In places such as the Middle Caqueta (Colombian Amazonia) a thick litter layer with abundant fine roots over highly weathered, nutrient-poor, mineral soils plays an important role in the forest water regime (Marin *et al.*, 2000). Other humid tropical forests have thin ground cover with several bare patches.

In a tropical rainforest, hydrological processes are affected by both physical phenomena and the activities of forest organisms, and are therefore subject to both hydrometeorological and biotic temporal and spatial controls. Treefall and debris dam formation and collapse, the fluvial transport of coarse woody debris, the actions of termites and other burrowing organisms in creating subsurface pathways for water, and of deer and other quadrupeds in creating surface routes for soil erosion and water flow, are all parts of the intimate links between ecological and hydrological processes.

Some of the patches of bare soil occurring irregularly on catchment slopes in tropical rainforests are created by the activities of termites and other insects. Other bare patches are created by the way stemflow washing down tree trunks sweeps a small area downslope of the tree base clear of plant debris. Animals create further bare patches: hollows with steep, easily eroded back walls are made by rhinoceros and wild pigs; deer and elephant create tracks down which concentrated flow occurs. Tree fall also creates considerable areas of bare earth, especially when a whole mature tree is uprooted. None of these sources is long-lasting. After a few months they may be covered with fallen leaves and other debris, or regrowth of creepers and saplings has begun to protect them from raindrop impact.

In the Ulu Segama area of Sabah, Malaysian Borneo, the effects of stemflow illustrate this point well. In March 1987, along a 0.5 km transect along a ridge crest, 80% of a sample of 70 trees had bare patches on their downslope sides, while 20% had no clear signs of any surface wash adjacent to the trees. A year later, such bare patches were less apparent, over 50% of the trees being surrounded by leaf litter (Spencer *et al.*, 1990). By 1991, the proportion with bare patches had declined to 20%. However, it is not

Table 15.3. *Dissolved and solid loads of major tropical rivers*

River		Basin area (1000 km^2)	Runoff (l s^{-1} km^{-2})	Dissolved load (t km^{-2} yr^{-1})	Solid load estimates[a] (I) (t km^{-2} yr^{-1})	(II) (t km^{-2} yr^{-1})	(III) (t km^{-2} yr^{-1})	Reference
Congo	Africa	4000	9.8	12	13	20.5		Hanson (1999)
								Lerman and Meybeck (1988)
								Milliman and Syvitski (1992)
								Ramesh and Subramanian (1993)
								Restrepo and Kjerfve (2000)
Niger	Africa	1125	5.5	9	60	4.5		
Zambezi	Africa	1340	5.3	12	75			
Amazon	S. America	6300	28	46	143	190	204.2	
Madeira	S. America	1380	23	42	457			
Magdalena	S. America	240	31	117	1000	560		
Maranon	S. America	407	27	90	250			
Negro	S. America	755	60	10	10			
Orinoco	S. America	950	31.6	52	91	150	220.8	
Ucayali	S. America	400	24	136	307			
Brahmaputra	Asia	580	33	130	1370			
Ganges	Asia	975	12	78	537			
Godavari	Asia	313	9.34	53.3	543			
Krishna	Asia	251	4.09	46.4	16			
Irrawaddy	Asia	430	31.2	211	700	833		
Mekong	Asia	795	23	75	435	362		

[a] Alternative estimates are provided to show the range of results reported in the literature. Periods and intensity of sampling of sediment loads vary greatly. However, these estimates support the general ranges of values used in Table 15.1.

known whether large amounts of stemflow, during extreme rainfalls, temporarily wash leaf litter away from the downslope sides of trunks. Stemflow varies greatly from tree to tree and particularly between species, *Malotus wrayi*, of the Euphorbiaceae family, producing up to four times the stemflow of other trees, with about 15% of rainfall emerging as stemflow under such trees, compared with 2 to 4% elsewhere (Wong, 1991).

Roles of termites and other insects in erosion processes

The activity of soil macrofauna (e.g. earthworms, termites) is critical to the maintenance of favourable soil structure and mineral cycling. But such insects also influence runoff, infiltration and soil detachability. The clearest influence of termites on water movement is the creation of macropores at the soil surface into which water can move. These macropores communicate with galleries below the soil, creating a high infiltration capacity. Although infiltration increases with termite activity, at least 30 foraging holes per square metre appear to be necessary for the effect to be significant (Léonard and Rajot, 2001).

Where relatively sandy topsoils are found over clayey subsoils, earthworms may stimulate the formation of the sandy surface soils by producing clay-rich worm casts that are susceptible to erosion, as was found in an undisturbed forested basin in south-western Ivory Coast. The worm casts were easily disintegrated by rain splash and overland flow and cast material contributed to the 0.12 kg m^{-2} yr^{-1} of highly organic suspended sediment removed by surface water drainage (Nooren *et al.*, 1995).

Soil faunal activity plays an important role in preparing sediment for erosion at Danum Valley in Sabah. In an experiment where one 20 m^2 plot had ground cover material removed and the other was left in its natural state, the undisturbed plot had a higher sediment yield (293 as opposed to 229 kg ha^{-1} y^{-1}) and had many more bare earth pillars and tubes made by termites or cicadas. Larger animals contribute to the sediment supply through their trampling of parts of the forest, particularly the deer whose tracks crossing streams are lines of preferential water and soil movement. Despite these catchment surface sources of sediment, there is rarely any overland flow discharging directly to streams in these natural forests, overland flow accounting for only about 5% of the total runoff at Danum Valley (Anderson and Spencer, 1991; Sinun, 1991; cf. Bonell, this volume).

OVERLAND FLOW, SUBSURFACE FLOW, CHANNEL HEAD DYNAMICS AND SEDIMENT SUPPLY

Overland flow can be a powerful agent for the transportation of sediment to streams in tropical forests during the heaviest rain events. The interactions between soil physical and rainfall characteristics determine exactly which flow paths are activated in response to a given event. Vertical movement of soil water prevails between rain events and during most small events, but under heavy and prolonged rain, the horizontal transfer of water develops, producing lateral flow. Changes in soil hydraulic conductivity with depth play a decisive role in determining the flowpaths along which water moves.

In experiments on vegetated hillslope plots on St. John, US Virgin Islands, the only two storms that produced runoff from the plots were hurricanes Bertha and Hortense, the rainfalls from which have a recurrence interval of 1 to 1.5 times per year (MacDonald *et al.*, 2001). The majority of the runoff on St John is generated by saturation overland flow (SOF) during the largest storm events. While SOF is more frequent in convergence zones, such as swales and toeslopes, it was also observed on a relatively planar and moderately steep forested hillslope on the north-west coast of St John. Subsurface storm flow (SSF) is probably a major contributor to the development of SOF in downslope and convergent areas. On 24° to 43° slopes in the Luquillo Experimental Forest, Puerto Rico, monthly surface runoff was only 0.2 to 0.5 percent of monthly rainfall (Larsen *et al.*, 1999).

In the western Amazon in Peru, a sharp decrease in soil hydraulic conductivity with depth and high rainfall intensity and frequency favour rapid near-surface flowpaths, mainly in the form of saturation excess overland flow and return flow (Elsenbeer and Vertessy, 2000). At this site, a perched water table is rapidly created at a depth around 30 to 50 cm which coincides with the presence of a number of pipes. The saturation overland flow is discontinuous and overlaps in operation with pipeflow. There is thus a link between overland flow and return flow in pipes.

In the clayey soils of eastern Puerto Rico, soil surface values of K_{sat} varied widely from 0.4 to 300 mm h^{-1}, partly as a function of the methods used. However, K_{sat} decreased rapidly below 50 cm depth. Macropores in the soil profile influence the infiltration rate (Schellekens, 2000). Weathered granite on slopes of up to 30° in southern China had K_{sat} values of around 110 mm h^{-1} while those of 16° had K_{sat} values of 220 mm h^{-1} (Kimoto *et al.*, 1999). On the granite of Singapore's Bukit Timah Reserve, K_{sat} decreased from a surface 1800 to 2500 mm h^{-1} to 0.5 to 17 cm h^{-1} at 50 cm depth (Sherlock *et al.*, 2000). However, this work identified several sources of error in K_{sat} measurements and argued that different combinations of flow paths could arise in different environments. In Brunei (Dykes and Thornes, 2000), the infiltration capacity of the surface soil is 500 to 1000 mm h^{-1}; that of the exposed mineral subsoil is an order of magnitude less, with a K_{sat} of around 180 mm h^{-1} at a depth of 150 cm. There was no indication that K_{sat} reduced with depth except very near the bedrock interface.

The general, but not universal, high hydraulic conductivities in the surface layer of the forest floor produce rapid movement of surface water into the soil. The rapid decrease in K_{sat} down the soil profile and the presence of many macropores in the upper part of the profile allows SSF to form rapidly. This moves rapidly downslope to converge at stream-head hollows, also termed colluvium-filled bedrock depressions (Crozier *et al.*, 1990) or zero-order drainage basins (Slaymaker, 1991). The colluvium in these hollows is often subject to the development of major pipes, which may partially collapse leaving holes in the hollow floor. In those holes, pipes can be seen entering from upslope and continuing downslope. These may enlarge and become totally open channels. Much sediment movement down pipes occurs during this process.

In the hollows at the heads of zero-order streams in the Danum Valley area, the response of the water table to rainfall is rapid, almost nine times faster than K_{sat} values suggest, depending on rainfall intensities and antecedent water table level. If initially less than 0.7 m below the zero-order channel, the water table reaches its maximum height of water table in less than two hours from the onset of rain (Bidin *et al.*, 1993). Diurnal fluctuations of the water table are correlated with daily hours of sunshine when the water table is between 0.5 and 0.9 m below the channel floor, reflecting transpiration by canopy trees. While subsurface flow occurs immediately after storm rains begin, surface flow in the zero-order channel occurred on only 22 times in the study year. The generation of surface runoff by the rise of the water table to the channel floor indicates the importance of the saturated zone and the filling of macropores as the water table rises in the overall down-valley discharge of infiltrating water.

Stream head hollows are not only highly sensitive to water table fluctuations but are also a major source of sediment. When the water table rises above the channel bed, initiating flow in the ephemeral hollow, water begins to seep out of the back wall and sides of the hollow, producing small-scale slumping of fragments of rock and clay no more than 3 cm in diameter down the sides of the hollow into the channel. Macropore outflows during large storms may entrain channel sediment and contribute to the storm suspended sediment load. Thus a combination of mechanisms exist which favour higher sediment supply and transport in wet periods.

These observations emphasise how hydrometeorological variations and geomorphic processes in the forest are regulated by rainfall and subsurface water. However, rare events produce a disproportionately large amount of geomorphic work, erosion and deposition, in relation to the rainfall volumes involved, as erosion bridge measurements at Danum Valley, Sabah, illustrate. Detailed erosion and deposition measurements at six channel head locations in the undisturbed primary forest in the W8S5 catchment from July 1990 to July 1997 show a complex year to year pattern, with erosion dominating in 1990–1992, deposition in the relatively drier years 1992–1994 and a combination of the two (erosion on the slopes and in the channel cross-sections, but deposition at the channel heads) in 1994–1995. The year 1995–6, which included the extreme rainstorm of 19th January 1996, was characterised by the highest erosion at all three categories of site in the entire record. Erosion was recorded at 12 of the 13 slope sites, averaging 3.84 mm at six of the seven channel head sites, where it averaged 4.64 mm, and at six of the eight cross-sections, where it averaged 6.90 mm. The next year, 1996–7, saw relatively little change, with examples of both erosion and deposition. The high erosion rates in 1995–6 clearly reflect the rarity and magnitude of the 19 January 1996 storm event that produced a flushing effect, scouring available sediment from all elements of the channel head area. Erosion rates on undisturbed forest slope sites averaged 0.85 mm y^{-1} between 1990 and 1997, a rate exceeding the 0.28–0.32 mm y^{-1} measured in the Rupununi, Guyana, but lower than figures of 1.5–3.5 mm y^{-1} for the Ivory Coast (Rougerie, 1960), 4.0–4.7 mm y^{-1} for the Bukit Timah Reserve, Singapore (Chatterjea, 1989), 2.61 mm y^{-1}) for the Pasoh Reserve, peninsular Malaysia (Leigh, 1972) and 7.6 mm y^{-1} in the wet interior of Dominica (Walsh, 1993).

The shallow subsurface lateral movement of water is thus a major runoff pathway. Links to the channel by macropores and pipes may thus be highly significant routes for sediment supply to channels (Baillie, 1974). Observations of sediment discharge from macropores in undisturbed rainforests have so far been restricted to incidental sightings, but new work on this phenomenon is under way.

LANDSLIDING AND MASS MOVEMENTS

Landslides are a natural phenomenon on steep forested slopes in many parts of the humid tropics (Larsen and Torres-Sánchez, 1998) but slopes underlain by certain rock types are more susceptible to slipping than others. Naturally occurring landslides may be more common than casual observation suggests. Some 120 major landslide scars were recorded by Day (1980) in the Gunung Mulu National Park, Sarawak, Malaysia, some multiple scars having a combined length of 250 m and a width of 200 m. In the vicinity of Gunung Api, three slide scars covered 60 000 m^2 and another eight separate scars covered 70% of a slope area. Day (1980) considered heavy storms to have been the main trigger for a number of landslides observed in the Gunung Mulu National Park in Sarawak. Increases in pore water pressure during heavy storms or when drainage is impeded, cause failure.

Most of the slides observed in the Gunung Mulu National Park were situated on the lower slope sections although landslides are clearly associated with steep slopes. Of 21 landslides examined at Gunung Mulu, all were located on slopes in excess of 40° and 18 on slopes in excess of 50° (Day, 1980). In nearby Brunei, at Ulu Temburong, landsliding contributes about 16.5% of the annual erosion rate (Dykes, 1995). The weathered and fractured slaty shales of the Mulu Formation are clearly prone to slipping, with most of the slides occurring on slopes parallel to the dip of the rocks. In Peninsular Malaysia, landslides occur on slopes underlain by granitic rocks, as on the slopes of the Main Range of Peninsular Malaysia during the exceptional rainfalls of late December 1926 and January 1971.

Major storms of November 20–23, 1988 produced 885 mm of rain in four days, triggering large-scale landsliding in the Nakhon Si Thammarat Range of Phipun District in southern Thailand. In a catchment of 92 km^2, debris avalanches and debris flows transported approximately 107 t km^{-2}. The minimum annual erosion rate in this area with this sort of landsliding every 150–300 years would be 356–712 t km^{-2} yr^{-1} (Harper, 1995). Further landsliding occurred at least twice in 1997 in southern Thailand, in association with heavy rains in August and November.

In New Guinea, landsliding is the dominant source of fluvial sediment, Simonett (1960) having demonstrated how tectonic activity triggers frequent sediment releases to channels. In Hawaii, rockfalls have occurred solely as a result of the weathering and erosion of steep slopes, as in the Sacred Falls landslide (Jibson and Baum, 1999). Slides in the weathered mantle are widespread, often leaving bright scars on hillsides for months. A good example is a slide that occurred in Olokele Canyon on Kauai in October, 1981. The slide face was about 300 metres wide and about 800 metres high.

In Puerto Rico, the igneous rocks in the Upland Province weather rapidly to form a deep, predominantly coarse-grained saprolitic soil mantle. When saturated, this saprolite produces debris slides, debris flows, and slumps ranging from a few to several hundred metres long. In May 1985, severe storms in west-central Puerto Rico triggered hundreds of such debris slides and debris flows, which choked streams, blocked roads, and destroyed homes and other structures (Jibson 1987a, b). Hurricanes David and Frederic in 1979 produced extreme rainfalls in north-eastern Puerto Rico that triggered several debris slides up to 750 metres wide and as much as 25 metres deep on slopes of deeply weathered intrusive igneous rocks. Rock falls from natural slopes often occur in the igneous and sedimentary rocks of the Upland province. In Puerto Rico, landslide-producing storms occur at an average rate of 1.2 per year (Larsen and Simon, 1993). Generally, short-duration, high intensity storms trigger shallow soil slips and debris flows, while long-duration, low intensity rainfalls produce larger, deeper debris avalanches and slumps (see also Scatena *et al.*, this volume).

Landslides, like lahars, supply sediment to slope foot areas, channel heads and stream channels, where it may partially block, or even divert, the stream. Much of this added sediment may not be moved further until subsequent large rain events erode material and carry it further downstream.

Within-channel sediment sources

Within-channel sediment sources are probably more important than slope erosion for the immediate responses of streams to storm rainfall. In the easily fragmented mud rocks of the Kuamut formation in the Ulu Segama area of Sabah, break-up of weak rocks in stream beds is rapid, with boulders of quartz being the main resistant material in the channel. Behind the boulders and beneath a veneer of coarse quartzitic gravel is a stock of sand-sized material awaiting transport. Lateral bars of silt and sand often exist at tributary junctions where the water from the tributary has been backed up by a higher level flow in the main stream. Elsewhere, lateral bars of sediment occur one or two metres above the level of the minor, low flow channel.

In addition to the accumulated sediments from past erosive events, active bank erosion supplies new sediment. In many places banks are actively being undercut by streams, with freshly exposed weathered rock and soil appearing after most high storm runoff events. Bank erosion often undermines riparian trees which fall across streams and lead to the build-up of debris dams. The channel then may migrate laterally to avoid the fallen vegetation, continuing to erode its bank and eventually undermining another tree. This process of bank erosion is the most obvious immediate source of sediment for downstream transport.

Regulation of sediment discharge by coarse woody debris

Log jams and other accumulations of large woody debris (LWD) are well known as a factor in sediment yield from studies in the forests of the Pacific Northwest of the USA (Keller and Tally, 1979; Marston, 1982; Swanson *et al.*, 1967; Orme, 1990) but play an even more dynamic role in tropical rainforests (Spencer *et al.*, 1990). The tropical dynamism results from the greater frequency of high energy events, intense rainfalls and high per unit area discharges, and from the rate of decomposition of fallen plant material.

Two mechanisms are involved in the elimination of woody debris dams: extreme discharges which float logs upward and the *in situ* rotting of major tree trunks blocking the channel. These two types of event exert a major control on suspended sediment yields of small streams. The surveys of LWD in the undisturbed W8S5 stream at Danum Valley (Figure 15.5) reveal periods of debris dam build-up followed by episodes of debris dam removal. For example, the survey of March 1988 (3.88 in Figure 15.5) shows five debris dams along the channel. In August of the same year (8.88 on Figure 15.5), only three dams remained. A few weeks

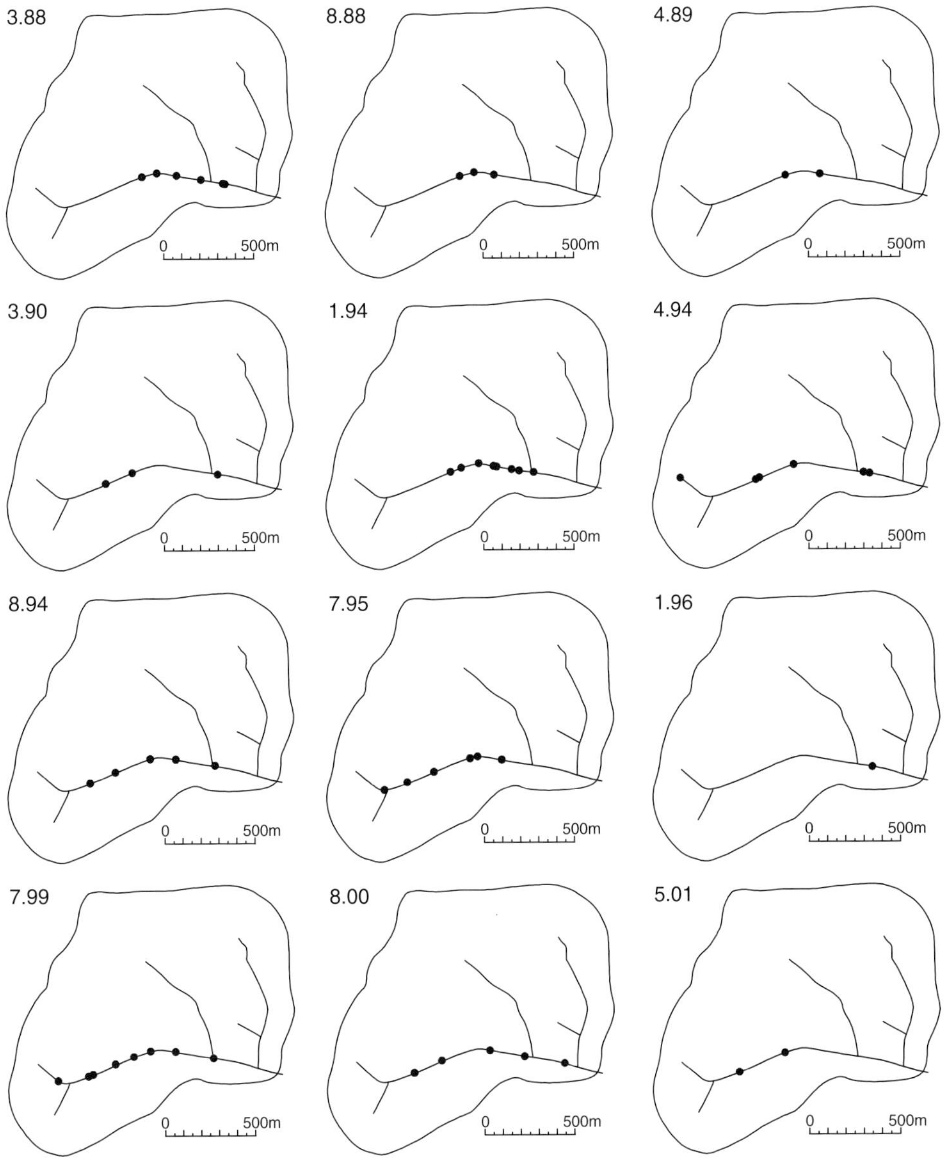

Figure 15.5 Surveys of debris dams in W8S5 a 1.5 km^2 tributary of the Ulu Segama, Sabah. (Based on data supplied by I. Douglas.)

after the survey, on March 29th, this 1.7 km^2 catchment experienced a peak instantaneous discharge of 29 m^3 s^{-1} that moved most of the LWD in the channel, except for one particularly large 1.8 m diameter log that had remained intact.

The large log formed a dam, with an accumulation of gravel on the upstream side that had built up to be flush with the top of the log, so that effectively the dam acted as a waterfall in the stream. In such a situation, with the stream flow passing over the log, there is little chance for the debris dam to be floated or pushed out of the way by a high flow. The April 1989 survey found this dam still in place (the downstream of the two dams on 4.89 in Figure 15.5). However, by the March 1990 survey (3.90) the dam had disappeared. Examination of the suspended sediment concentration record showed an exceptionally high instantaneous concentration (for W8S5) of 5020 mg l^{-1} on 20 November 1989 during the falling stage of a storm whose peak instantaneous discharge was 4.24 m^3 s^{-1}, a much lower peak discharge than that of 29 March 1988. Field examination of the site of that debris dam in March 1990 revealed a large area of erosion of the right bank of the stream immediately downstream. The log was still embedded in the left bank, but the stream had eroded laterally to the right, having breached the decomposed log. In this case, the biological decay of the log had regulated the release of the accumulated gravel and trapped fine silt-clay sized material, so producing the high suspended sediment concentration of 20th November 1989.

Other episodes of debris dam removal are revealed by the surveys of January 1996 and August 2000 (1.96 and 8.00 respectively). The January 1996 survey was made two days after the largest 24-hour rainfall (178 mm) since records began at the Danum Valley Field Centre in 1985. The high flow had washed out all the dams shown in the July 1995 survey (7.95 on Figure 15.5) and had created a single new dam, relatively low down in the catchment. Similarly, the positions of dams in July 1999 and August 2000 differ (7.99 and 8.00 respectively) because another extreme event (243 mm in three days, 29–31 January 2000, made up of two days of just over 50 mm each and 139 mm on the third day) eliminated most of the pre-existing LWD accumulations.

The undisturbed W8S5 catchment has thus had alternating phases of debris dam building and removal. On average, dams last 12 months, although several individual logs have remained in fixed positions for over ten years. These are usually logs bridging the channel, normally dry and thus suffering relatively little decomposition.

Removal of a dam, or the development of a water pathway beneath a restraining log does not always lead to the immediate total removal of the gravel that had accumulated upstream of the obstruction. Small remnant gravel terraces where there had once been debris dams are often found in the channel. The natural forest channel is thus in a highly dynamic state. Wood is entrained, accumulated and dispersed within 8 to 20 months. Long-lasting debris dams are rare. The operation of biotic processes is thus a major factor in the sediment discharge of small tropical rainforest streams. Biological processes of this type introduce an element of non-linearity in the hydraulics of sediment transport, contributing to the difficulty of predicting sediment yield as a function of runoff, or even of rainfall.

The role of lateral bars of fine sediment

Among the readily available sediment sources, lateral bars of fine sediment play an important role in many natural streams in the humid tropics. Fine material may be washed downslope by SOF or SSF, or may be transferred by lateral mechanical eluviation to accumulate at the stream margin until a storm event flushes it a little further downstream. Sediment washed out from natural pipes may be a major contributor to these lateral bars. In places sediment is carried to the channel margin by flow along animal tracks crossing streams. Just as the slumping of undercut banks adds new sediment to the channel, the sediment supplied to lateral bars accumulates until there is sufficient streamflow energy and volume to remove it.

Throughout the length of small streams in tropical rainforests, the overhanging vegetation supplies litter and plant material to the channel. This is not only incorporated in the channel bar sediments, but is readily removed once streams start to rise. Thus early on during the rising stage of flow from an intense storm, the stream flushes a large amount of organic matter downstream and sediment concentrations rise rapidly as the fine sediment is entrained. Characteristically, rising stage sediment concentrations are greater than those of the falling stage. However, while the overall pattern of the W8S5 loops is clockwise, the falling stages show many fluctuations in concentration, perhaps influenced by variations in runoff, but possibly due to the flushing of mixtures of sediment and organic matter out of ephemeral headwater channels that are usually dry, but are favoured zones of activity by mammals, worms, cicadas and termites.

Variation in storm period-sediment yield with antecedent conditions

Storm period sediment yields vary greatly, according to antecedent conditions, sediment availability and actual patterns of rainfall intensity during individual storms. Three storms in 1993–94, each producing about 27 mm rainfall (with similar durations, rainfall totals and intensity patterns) over the Ulu Segama small catchment illustrate this varied response. The peak suspended sediment concentration on 24 March 1994 was over five times that on 8 June 1993 and the total suspended sediment load was 15 times greater. Antecedent conditions may explain the differences between the storms, with over five times more rain falling in the ten days prior to 22 March 1994 than before 8 June 1993.

The higher outputs of suspended sediment on 14 February and 22 March may arise from three causes:

(1) Saturation of much of the soil before the day of the storm, thus converting more rainfall to concentrated overland flow which locally eroded more soil and transported it to the stream.
(2) Mobilisation of sediment already in the stream channel by the higher discharge.
(3) Detachment and displacement of sediment by storms in the days before the storm, giving more available sediment to be entrained by the higher stream discharges.

The sediment yields in W8S5 appear to be much higher than those reported for other parts of Malaysia (Table 15.4). Two reasons contribute to this. The mudstones of the Kuamut Formation are much more erodible than the weathered granites and partially metamorphosed sediments of the areas studied in peninsular Malaysia, and also probably more erodible than the orthic acrisols

Table 15.4. *Sediment yields for small tropical rainforest catchments in Malaysia*

Catchment name	Catchment area (km^2)	Sediment yield ($t\ km^{-2}\ y^{r-1}$)	Reference
Sg. Telom, Cameron Highlands	77.0	53	Shallow, 1956
Sg. Mupor, Johor	21.8	41	Leigh, 1973
Sg. Gombak, Selangor	140.0	97	Douglas, 1975
Sg. Kalangan, Mt Kinabalu, Sabah	2.5	15	Sinun and Douglas, 1997
W8S5, Ulu Segama, Sabah	1.1	312	Douglas *et al.*, 1992

of the Sipitang area of Sabah. However, many other investigators acknowledge that their sampling procedures do not always measure suspended sediment concentrations in streams during all storm runoff events. Furthermore, extensive sampling strategies may miss storm peaks; we have demonstrated in this study the rapidity of rise-to-peak flow in these catchments. Thus calculations may underestimate sediment loads during the all-important extreme events.

CONCLUSIONS

All rivers, even the largest, show seasonal and storm period variations in the relationship between sediment concentration and discharge. For example, concentrations of suspended sediment in Río Orinoco, the Amazon and the Mississippi are all greater during rising stages than at equal water discharges during falling stages (Meade *et al.*, 1983; Schmidt, 1972; Robbins, 1977).

Physical phenomena and the activities of forest organisms lead to two kinds of response to extreme events in the undisturbed forest: (1) much more erosion and larger changes to the land surface than would be expected by the magnitudes of the water flows involved and (2) the occurrence of changes due to biological and pedogenetic processes not linearly related to the magnitudes of the events concerned.

Far from being a steady, regulated, water-conserving environment, the tropical forest is an event-driven hydrological system (quasi-ephemeral in nature), subject to long periods of relative inactivity and sudden dramatic changes during rare large storms.

The bulk of the sediment transport by major rivers occurs in a few days (few weeks) in any year. A 1-in-10 year storm event may remove more sediment in 24 hours than the river carries in a dry year.

The lessons from the natural forest suggest that variations in lithology, tectonics and climate produce a wide variety of erosion conditions that need to be understood before attempting to model or predict the likely effects of human activity on erosion and sedimentation.

APPENDIX 15.1 CHARACTERISTICS OF THE WORLD'S EIGHT LARGEST TROPICAL RIVERS

River	Drainage area ($10^3 \times km^2$)	Water discharge		Sediment discharge	
		$m^3\ s^{-1}$	$km^3\ yr^{-1}$	$t\ km^2yr^{-1}$	$10^3t \times yr^{-1}$
Amazon	6 150	200 000	6 300	146	900 000
Zaire (Congo)	3 820	40 000	1 250	11.25	43 000
Orinoco	990	34 880	1 100	212	210 000
Ganges–Brahmaputra	1 480	30 790	971	1 128	1 670 000
Yangtze	1 940	28 540	900	246	478 000
Mekong	790	14 900	470	202	160 000
Parana–La Plata	2 830	14 900	470	32.5	92 000
Niger	1 113			4.49	5 000

References

Anderson, J. M. and Spencer, T. 1991 Carbon, nutrient and water balances of tropical forest ecosystems subject to disturbance: management implications and research proposals. *MAB Digest 7*, 1–95. UNESCO, Paris.

Baillie, I. C. 1974 Piping as an erosion process in the uplands of Sarawak. *Journal of Tropical Geography*, **41**, 9–15.

Baker, V. R. 1978 Adjustments of fluvial systems to climate and source terrain in tropical and subtropical environments. In: *Fluvial Sedimentology*, Miall. A. D. (ed.) 7 (Can. Soc. Petrol. Coal.), 211–230.

Bidin, K. 2001 Spatio-temporal variability in rainfall and wet-canopy evaporation within a small catchment recovering from selective tropical forestry. Unpublished Ph.D. thesis, University of Lancaster.

Bidin, K., Douglas, I and Greer, T. 1993 Dynamic response of subsurface water levels in a zero-order tropical rainforest basin, Sabah, Malaysia. *International Association of Hydrological Sciences Publication, 216*, 491–496.

Bremer, H. 1981 Reliefformen und reliefbildendo Prorzesse in Sri Lanka. *Relief, Boden, Palaoklima (Zur Morphogenesea in den feuchten Tropan)*, 7–183.

Büdel, J. 1965 Dia Relieftypen dar FINchenapUl-Zone SUd-Indiens Ostabfall Derken gegen Madras. *Colloqium Geograficum Bonn 8*, 100pp.

1981 *Kiima Geomorphologie* (2 Auflage), Borntraeger, Berlin.

Chatterjea, K. 1989 Surface wash in rainforest of Singapore. *Singapore Journal of Tropical Geography*. **10**, 95–109.

Day, M. J. 1980 Landslides in the Gunong Mulu National Park. *Geographical Journal*, **146**, 7–13.

Douglas, I. 1973 No details – appears in Table 15.4.

Douglas, I. 1967 Erosion of granite terrain under tropical rainforest in Australia, Malaysia and Singapore. *International Association of Hydrological Sciences Publication, 75*, 31–39.

1968 Erosion in the Sungai Gombak catchment, Selangor, Malaysia, *Journal of Tropical Geography, 26*, 1–16.

1975 The impact of urbanisation of river systems. *Proceedings International Geographical Union Regional Conference, Palmerston North, New Zealand, December 1974*, 307–317.

Douglas, I., Spencer, T., Greer, T., *et al.* 1992 The impact of selective commercial logging on stream hydrology, chemistry and sediment loads in the Ulu Segama rainforest, Sabah. *Philosophical Transactions Royal Society London*, B, 335, 397–406.

Douglas, I. *et al.*, 1992. No details – appears in Table 15.4.

Dykes, A. P. 1995 Regional denudation by landslides in the tropical rainforest of Temburong District, Brunei. *International Association of Geomorphologists South East Asia Conference Singapore, 18–23 June 1995, Programme with abstracts*, 39.=

Dykes, A. P. and Thornes, J. B. 2000 Hillslope hydrology in tropical rainforest steeplands in Brunei. *Hydrological Processes* 14 215–235.

Elsenbeer, H. and Vertessy, R. A. 2000 Stormflow generation and flowpath characteristics in an Amazonian rainforest catchment. Hydrological Processes, 14, 2367–2382.

Garner, N. F. 1974 *The Origin of Landscapes*. Oxford Uni. Press, New York, 736pp.

Guyot, J. L. and Hérail, G. 2000 L'érosion actuelle de la Cordillère Oriental des Andes boliviennes. 18ème Réunion des Sciences de la Terre, RST 2000, Paris 17–20 Avril, 200, 154.

Hanson, L. 1999 *Deltas*. Available online at http://www.salem.mass.edu/~lhanson/gls214/gls214_deltas.html

Harden, P. O. and Sundborg, A. 1992 *The Lower Mekong Basin: Suspended Sediment Transport and Sedimentation Problems*. A Report Submitted to the Mekong Secretariat.

Harper, S. B. 1995 Landslides in southern Thailand: contributing factors, sedimentological characteristics and denudation impact. *International Association of Geomorphologists South East Asia Conference Singapore, 18–23 June 1995, Programme with abstracts*, 44.

Herwitz, R. S. and Slye, R. E. 1992 Spatial variability in the interception of inclined rainforest by a tropical rainforest canopy. *Selbyana*, 13, 62–71.

Jabatan Alam Sekitar (1996) *Guidelines for Prevention and Control of Soil Erosion and Siltation in Malaysia*. Ministry of Science, Technology and Environment, Malaysia, Kuala Lumpur.

Jayawardena, A. W. and Rezaur, R. B. 2000 Drop size distribution and kinetic energy load of rainstorms in Hong Kong. Hydrological Processes. 14, 1069–1082.

Jibson, R. W., 1987a. Debris flows triggered by a tropical storm in Puerto Rico. In A. P. Schultz and C. S. Southworth (eds.), Collected papers on Eastern North American Landslides, p. 9–10. US Geological Survey Circular 1008.

Jibson, R. W., 1987b. Landslide hazards of Puerto Rico. In W. W. Hays and P. L. Gori (eds.), Proc. Assocessment of Geologic Hazards and Risk in Puerto Rico, p. 183–188, US Geological Survey Open-File Report 87–008.

Jibson, R. W. and Baum, R. L. 1999 Assessment of landslide hazards in Kaluanui and Maakua Gulches, Oahu, Hawaii, following the 9th May 1999 Sacred Falls Landslide. US Geological Survey Open-File Report 99–364, http://geology.cr.usgs.gov/pub/open-file-reports/ofr-99-0364/index.html

Kasel, R. W. 1985 Tropical fluvial geomorphology. *In* Pitty, A. F. (ed.) *Themes in Geomorphology*. Croom Helm, London, 102–121.

Keller, E. A.; Tally, T. 1979. Effects of large organic debris on channel form and fluvial processes in the coastal redwood environment. In: Rhodes, D. D.; Williams, G. P., eds. Adjustments in the fluvial system. Dubuque, IA: Kendell Hunt Publications; 168–198.

Kimoto, A., Uchida, T., Asano, Y., Mizuyama, T. and Li, C. 1999 Surface runoff generation of devastated weathered granite mountains – Comparison between Dahoi experimental basin in the southern part of China and Jakujo Rachidari Experimental Basin in Tanakami Mountains. *Journal of the Japan Society of Erosion Control Engineering*, 51 (6) 13–19.

Krishnaswamy, J., Richter, D. D., Halpin, P. M., *et al.* 2001 Spatial patterns of suspended sediment yields in a humid tropical watershed in Costa Rica. *Hydrological Processes*, 15, 2237–2257.

Larsen, M. C. and Simon, A. 1993 Rainfall-threshold conditions for landslides in a humid-tropical system. Geografiska Annaler, 75A, 13–23.

Larsen, M. C. and Torres-Sánchez, A. J. 1998 The frequency and distribution of recent landslides in three montane tropical regions of Puerto Rico. *Geomorphology*, **24**, 309–331.

Larsen, M. C., Torres-Sánchez, A. J. and Concepción, I. M., 1999 Slopewash, surface runoff, and fine litter transport in forest and landslide scars in humid-tropical steeplands, Luquillo Experimental Forest, Puerto Rico. Earth Surface Processes, 24, 481–502.

Leigh, C. H. 1982 Sediment transport by surface wash and throughflow at the Pasoh Forest Reserve, Negri Sembilan, Peninsular Malaysia. *Geografiska Annaler*, 64A, 171–179.

Léonard, J. and J. L. Rajot 2001 Influence of termites on runoff and infiltration: quantification and analysis Geoderma. **104** 17–40.

Léonard, J., Perrier, E. and Rajot, J. L. (in press). Biological macropores effect on runoff and infiltration: a combined experimental and modelling approach. *Agriculture, Ecosystems and Environment* (special issue GCTE Focus 3 conference: Soil Erosion, Biology and Organic Matter).

Lerman, A. and Meybeck, M. 1988 *Physical and Chemical Weathering in Geochemical Cycles*, Kluwer, Dordrecht.

Lloyd, C. R. and Marques, A de O., 1988 Spatial variability of throughfall and stemflow measurement in Amazonian rainforest. *Agricultural and Forest Meteorology*, 42, 63–73.

Loffler, K. 1977 *Geomorphology of Papua New Guinea*, ANU Press, Canberra, 196pp.

MacDonald, L. H., Sampson, R. W. and Andersonb, D. M. 2001 Runoff and road erosion at the plot and road segment scales, St John, US Virgin Islands. *Earth Surface Processes and Landforms*, 26, 251–272.

Major, J. J., Janda, R. J. and Arturo S. Daag, A. S. 1996 Watershed Disturbance and Lahars on the East Side of Mount Pinatubo During the mid-June 1991 Eruptions. in Newhall, C. G. and Punongbayan, R. S. (eds) *Fire and mud: eruptions and lahars of Mount Pinatubo, Philippines*. Philippine Institute of Volcanology and Seismology, Quezon City, and University of Washington Press, Seattle and London, 765–800. http://pubs.usgs.gov/pinatubo/major

Marin, C. T., Bouten, I. W. and Dekker, S. 2000 *Forest floor water dynamics and root water uptake in four forest ecosystems in northwest Amazonia. Journal of Hydrology* 237, 169–183.

Mendes, D. and Cohen, J. 2000 Impact of the phenomenon El Niño on the regime of precipitation in the area of the middle Amazon. *American Meteorological Society Annual Meeting Abstracts*, http://ams.confex.com/ams/annual2000/11global/abstracts/10528.htm

Milliman, J. D., Farnsworth, K. L. and Albertin, C. S. 1999 Flux and fate of fluvial sediments leaving large islands in the East Indies. *Journal of Sea Research*, 41:97–107.

Milliman, J. D. and Meade, R. H. 1983 World-wide delivery of river sediment to the oceans. *Journal of Geology*, 91, 1–21.
Milliman, J. D. and Syvitski, J. P. M. 1992 Geomorphic/tectonic of sediment transport to the ocean: the importance of small mountainous rivers. *Journal of Geology*, 100, 525–544.
Nooren, C. A. M., N. van Breemen, J. J. Stoorvogel and A. G. Jongmans 1995 The role of earthworms in the formation of sandy surface soils in a tropical forest in Ivory Coast *Geoderma*, 65, 135–148.
Olivry, J. C. 1988 Transports solides en suspension au Cameroun. *International Association of Hydrological Sciences Publication*, 122, 134–141.
Ongwenyi, G. S., Kitheka, J. U. and Nyagaga, J. M. 1993 The impact of hydrological and land use processes on the quality of water in the Gucha catchment, southwestern Nigeria. *International Association of Hydrological Sciences Publication*, 216, 79–85.
Orme, A. R. 1990 Recurrence of debris production under coniferous forest, Cascade Foothills, northwest United States. *In* Thornes, J. B. (ed) *Vegetation and erosion: Processes and environments*, Wiley, Chichester, 67–84.
Pickup, G. 1984 Geomorphology of tropical rivers I. Landforms, hydrology and sedimentation in the Fly and Lower Furori, Papua New Guinea. *Catena Supplement 5*, 1–17.
Pickup, G., R. J. Higgins, and R. F. Warner. 1981. Erosion and sediment yeild in Fly River drainage basins, PNG. Erosion and Sediment Transport in Pacific Rim Steeplands, I.A.H.S. Publication No. 132 (Christchurch, NZ):438–457.
Pickup, G. 1984. Landforms, hydrology and sedimentation in the Fly and lower Purari, PNG. In: A. P. Schick (ed.), Channel Processes–Water, Sediment, Catchment Controls. Braunschweig, Catena Supplement 5:1–17.
Pierson, T. C., Janda, R. J., Umbal, J. V., and Daag, A. S., 1992, Immediate and long-term hazards from lahars and excess sedimentation in rivers draining Mount Pinatubo, Philippines: US Geological Survey Water-Resources Investigation Report 92-4039, 35 pp.
Ramesh, R. and Subramanian, V. 1993 Geochemical characteristics of the major tropical rivers of India. *International Association of Hydrological Sciences Publication*, 216, 157–164.
Restrepo, J. D. and Kjerfve, B. 2000 Magdalena river: interannual variability (1975–1995) and revised water discharge and sediment load estimates. *Journal of Hydrology*, 235, 137–149.
Rougerie, G. 1960 Le façonnement actuel des modèles en Côte d'Ivoire forestière. *Mem. Inst. Française Afrique Noire 58*, Dakar.
Schellekens, J. 2000 Hydrological Processes in a humid tropical rainforest: a combined experimental and modelling approach. Proefschrift, Vrije Universiteit Amsterdam.
Schmidt GW 1972 Chemical properties of some waters in the tropical rainforest region of central-Amazonia along the new road Manaus-Caracarai Amazoniana 3: 199–207.
Seyler, P., Olivry, J. C., and Sigha Nkamdjou, L. 1993 Hydrogeochemistry of the Ngoko River, Cameroon: chemical balances in a rainforest equatorial basin. *International Association of Hydrological Sciences Publication*, 216, 87–105.
Shallow, P. G. D. 1956 *River Flow in the Cameron Highlands*. Malaysia Hydroelectric Technical Memorandum No. 3.
Sherlock, M. D., Chappell, N. A. and McDonnell, J. J. 2000 Effects of experimental uncertainty on the calculation of hillslope flow paths. *Hydrological Processes*, 14, 2457–2471.
Shimen Reservoir Authority 1975 *Shimen Reservoir catchment management work report*. Taipei: The Authority (in Chinese).
Sinun, W. 1991 *Hillslope hydrology, hydrogeomorphology and hydrochemistry of an equatorial lowland rainforest, Danum Valley, Sabah, Malaysia*. Unpublished M.Sc. thesis, University of Manchester.
Sinun, W. and Douglas, I., 1997 Geomorphic and hydrologic response to humid tropical steepland montane forest disturbance on Gunong Kinabalu, Sabah, Malaysia. In Webb, B. W. (ed.) *Erosion and Sediment Yield: Global and Regional Perspectives. Poster Report Booklet, International Association of Hydrological Sciences, Exeter*, 100–102. (A/149.)
Slaymaker, O. 1991 Mountain geomorphology: a theoretical framework for measurement programmes. *Catena*, *18*, 427–437.
Spencer, T., Douglas, I., Greer, T., Sinun, W. 1990 Vegetation and fluvial geomorphic processes in South-east Asian tropical rainforests. *In* Thornes, J. B. (ed.) *Vegetation and Erosion: processes and environments*, Wiley, Chichester, 451–469.
Swanson, F. J., Lienkamper, G. W., Sedell, J. R. History, physical effects and management implications of large organic debris in western Oregon streams. *USDA Forest Service, General Technical Report*, PNW-56, 1–15. 1967.
Walsh, R. P. D. 1993 Problems of the climatic geomorphological approach with reference to drainage density, chemical denudation and slopewash in the humid tropics. *Würzburger Geographische Arbeiten* 87, 221–239.
Wang, H. and Yao, L.-X. 2000 Soil and water loss in the Lancang River-Mekong River Watershed (in Yunnan section, China) and its control measures. *Journal of Environmental Sciences (Beijing)*, 12, 90–97.
Wong, W. M., 1991 *Interception and stenflow in tropical rainforest*. Unpublished M.Sc. thesis, University of Manchester.

16 Rainforest mineral nutrition: the 'black box' and a glimpse inside it

J. Proctor
University of Stirling, UK

INTRODUCTION

This chapter deals mainly with undisturbed tropical rainforests and its aims are threefold:

- to summarise the current knowledge of selected topics in rainforest mineral nutrition;
- to emphasise the links between hydrology, pedogenesis and rainforest mineral nutrition; and
- to identify the most important gaps in knowledge and to suggest the most promising lines of research.

The rainforest nutrient cycle

Nutrients enter the ecosystem with rain, deposition of dust and aerosols, by fixation by microorganisms (in the case of N) above and below ground, and (except for N) by weathering of the substratum (Figure 16.1). The major above-ground pool of nutrients is formed in the tree boles and large branches. There is a flow of nutrients from these and the many other components of the above-ground pool of nutrients to the forest floor in small and large litterfall and in throughfall and stemflow of rainwater, enriched by nutrients from leaves and bark. The nutrients in dead organic matter are released gradually by decomposition, mediated by soil animals and microorganisms. Decomposition can involve immobilisation of nutrients as well as their release. In extreme cases the immobilisation may involve a conversion of the litter to stable organic matter which holds nutrients indefinitely. Such occurs in those forest types that have very wet or acid conditions (such as heath forests, peat swamp forests and upper montane cloud forests) which are described by Whitmore (1984). Nutrients are taken up from the soil by roots (probably usually in association with mycorrhizal fungi) which provide a living below-ground pool and which transfer them back to the canopy. The roots also release nutrients to the soil as secretions and by the death and decomposition of their parts. Permanent loss of nutrients occurs through erosion, fires, loss in drainage water and, in the case of N, by abiotic or microbial denitrification. Phosphorus, in particular, may effectively leave the system by conversion into insoluble forms within the soil.

The links between tree mineral nutrition and hydrology are numerous. There is the input to the forest floor of rainfall, throughfall and stemflow of which the chemical composition varies depending on the climate (seasonality, annual rainfall total), the vegetation, on sources of chemical constituents (e.g. volcanoes, the oceans), and on the rates of chemical transformations in the atmosphere (Forti and Neal, 1992).

Water is the solute in which ions, arriving at the top of the forest floor or top of the rooted zone, the principal source of mineral nutrients for plants, are transported to the roots. The ion supply by the soil solution involves a contribution from diffusion and mass flow (Nye and Tinker, 1977). Mass flow has two components, the contribution made by the movement of water to plants by transpiration and the less well researched movement of solutions as wetting fronts in soils (Lambers, Chapin and Pons, 1998). The relative importance of mass flow and diffusion depends on the ions and soil in question and its physical state (the moisture content is particularly important). A reduction in transpiration will reduce mass flow while conversely, a reduction in soil water content will reduce the permeability of roots to nutrients, change the nutrient concentrations in the soil solution, and decrease the diffusion coefficients of ions in the soil (Sands and Mulligan, 1990).

The effect of soil drying on nutrient uptake will vary from ion to ion and the different root and soil combinations. Since a plant functions as an integrated whole and not as a collection of isolated organs (e.g. Russell, 1974) there are important consequences for the relationships between water and nutrients. The amount of plant growth which goes into the root system will be balanced by that of the above-ground parts. This in turn will influence water and nutrient requirements and use, depending on whether water or nutrients

Forests, Water and People in the Humid Tropics, ed. M. Bonell and L. A. Bruijnzeel. Published by Cambridge University Press.

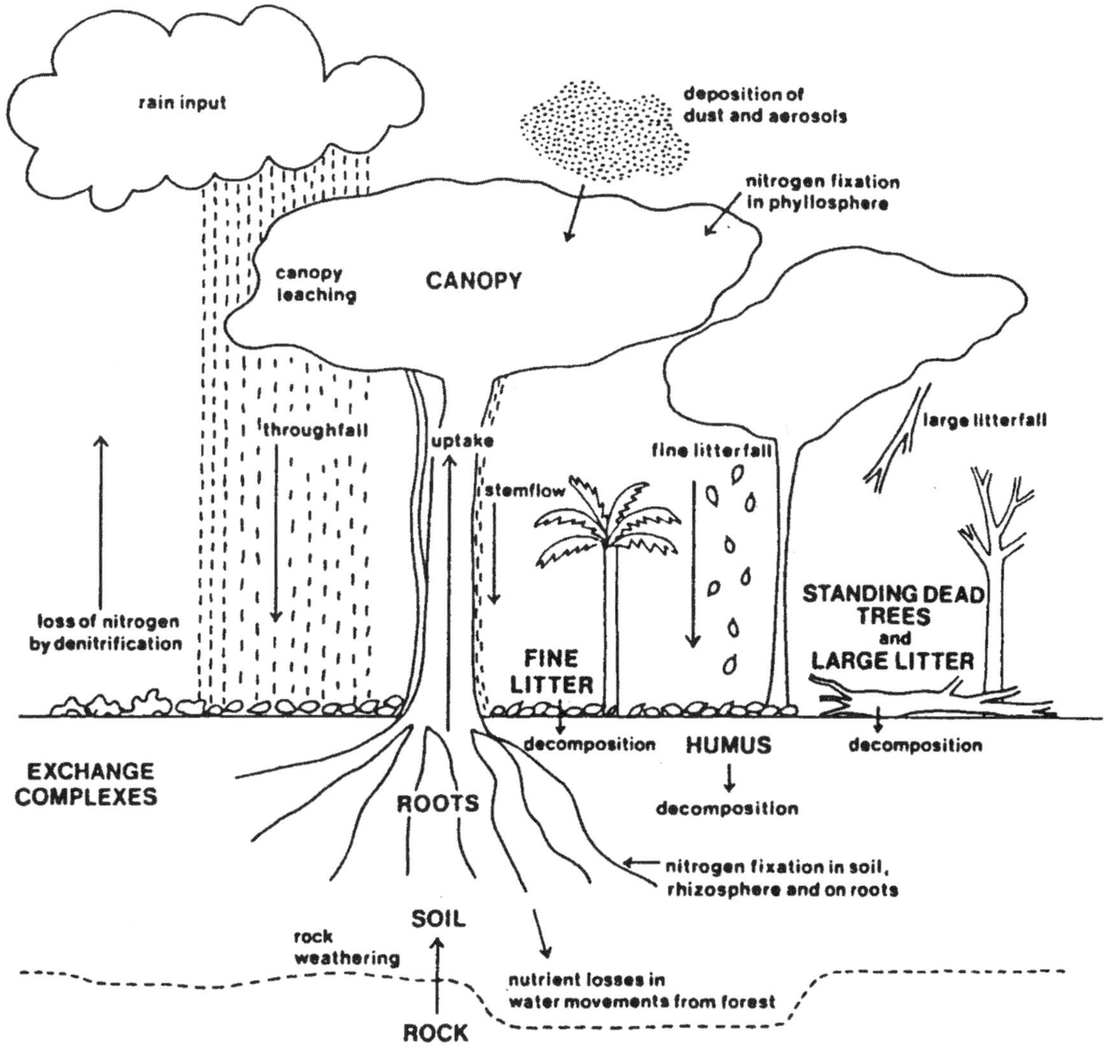

Figure 16.1 The nutrient cycle in moist tropical forest.

or light are limiting growth. A nutrient deficit has been reported to decrease the hydraulic conductance of root systems and hence an enhanced supply of nutrients could change tree water relations by increasing it (Minshall, 1975). A case of direct involvement of a nutrient in plant water relations is that of the K^+ ion, which is well known to be important in osmotic adjustment and turgor maintenance (Mengel and Arneke, 1982).

There is an interacting involvement of micro-organisms, vegetation, water and other abiotic factors which has been discussed in some detail by, amongst others, Coleman, Reid and Cole (1983). The balance between limitation to growth by N and P may be influenced by the soil moisture regime (Lloyd and Pigott, 1967) because the rate of N-mineralisation (and hence supply of this essential element) is dependent on soil water status (Marrs *et al.*, 1991).

Plants determine the rate of weathering in terrestrial ecosystems since they are a critical component in the hydrological cycle, both locally and regionally, influencing the magnitude of the fluxes of water percolating through the soil and producing major weathering agents including CO_2, organic acids, and ligands (Drever, 1994). In rainforests a considerable proportion of rainfall may be subject to evapotranspiration (cf. Roberts *et al.*, this volume) and hence made unavailable for leaching. A dense plant cover also reduces surface erosion (Bach, Wierenga and Ward, 1986) and thus losses via superficial pathways while at the same time maximising rainfall infiltration (cf. the chapter on runoff generation mechanisms by Bonell, this volume).

Quantification of pools and fluxes

The task of understanding rainforest nutrient cycling involves measuring the nutrients in the different pools and their fluxes into and out of the system and between the pools (Figure 16.1). It is true to say that beyond the influx of nutrients in rainwater, no study has succeeded in quantifying accurately any of the fluxes and pools. The pool sizes of nutrients in litter have been estimated satisfactorily only for the smaller surface-litter fractions and those in the below-ground and larger litter fractions are virtually unknown (e.g. Proctor, 1983, 1984; Poels, 1987; Burghouts *et al.*, 1998).

Steinhardt (1979) has perhaps come closest to good estimates in his work on a Venezuelan (montane) forest and has at least provided probabilities for his data. It has to be acknowledged, however, that all attempts to estimate the dry weights and nutrient contents of samples of natural tropical forests (for example, see Table 1 in Proctor, 1987; Scatena *et al.*, 1993) have been flawed for five main reasons:

(1) The difficulty of sampling adequately a structurally and floristically diverse forest to get a representative estimate of the above-ground biomass, let alone that of roots.
(2) The huge range of nutrient concentrations among species or even between different individuals of the same species.
(3) The likelihood that most of the nutrients in a sample will be in one or a few large trees which occur rarely in the rest of the forest.
(4) Problems associated with deciding the depth of soil sampling and of obtaining roots for chemical analyses.
(5) The wide range of soils on which rainforests occur and the difficulties associated with soil analyses.

It is unlikely that improved estimates of pools and fluxes will be made in the foreseeable future because the cost would be very high, the labour tedious, and the subject area is unfashionable. Moreover, the existing nutrient pool size data, although poor, can be used to test some important ideas. For example, the widespread view that rainforest nutrients are largely contained in the above-ground biomass rather than in the soil (cf. Proctor, 1987) is still conventional wisdom on tropical forests. This is linked with notions about poor forest recovery following disturbance and tree removals which allegedly reduce the ecosystem nutrients so that their dearth limits tree re-growth. Data collated by Proctor (1987) suggest that the forest which comes closest to the conventional view is that on very nutrient-poor soil investigated by Klinge (1976) in central Amazonia, which had a high proportion of total K, Ca and Mg in the above-ground biomass (estimated at 450 t ha^{-1}). However, in many cases a high proportion of nutrients lies below the ground. This is always so for N and often for P, but raises the question of the true availability of these elements in the soil to plants. Again, notwithstanding the limitations in the data, some pool size estimates have been used successfully when assessing nutrient losses caused by logging operations (Bruijnzeel, 1998). The studies which are available suggest that nutrient removal as harvested timber generally exceeds those associated with enhanced leaching, both after selective logging and clear-cutting operations. The best estimates are that after commercial logging, the time required for the replenishment of nutrients (Ca, Mg, K) by rainwater and other natural sources (e.g. weathering) is of the order of 30–60 years. Shorter recovery periods are associated with light selective logging (Bruijnzeel, 1998; see also the chapter on selective logging impacts by Chappell *et al.*, this volume). The key question in assessing whether such nutrient losses are important in the long-term regeneration of the forest is that of the nutrient requirements of the trees, which remains substantially unanswered (e.g. Proctor, 1995).

ATMOSPHERIC INPUTS OF NUTRIENTS

There are now good compilations of data on atmospheric inputs of nutrients to tropical forests following Bruijnzeel (1989b), Veneklaas (1990), Waterloo *et al.* (1997) and Hafkenscheid (2000). Many of these data are included in Table 16.1 and illustrate the great variation in atmospheric nutrient additions among sites. The causes of this variation include the amounts and seasonality of rainfall and site proximity to the sea or volcanic activity. Waterloo *et al.* (1997) have demonstrated the large contributions to overall atmospheric nutrient addition that can be made by extreme events, such as the passage of the occasional hurricane (cf. Scatena *et al.*, this volume). Throughfall chemistry is a topic where there is a particularly close link between plant physiology and hydrology.

Many studies of throughfall quality in tropical lowland forests have reported a net enrichment from the canopy regardless of soil fertility (e.g. Vitousek and Sanford, 1986; Forti and Neal, 1992) although its true extent is not easy to measure because it depends among other things on an accurate assessment of throughfall volumes (Bruijnzeel, 1989a; see also Roberts *et al.*, this volume). It is now clear that changes in rainfall chemical composition as the water moves through the canopy are governed by a considerable number of processes, including dry deposition, wash off, leaching from within leaves, and uptake by epiphylls. Brouwer (1996) has provided a detailed discussion of how these processes can be influenced by species composition, size, age, and form and nutritional status of the forest as well as the size and intensity of rain storms, and seasonal influences. The nutrients have a specific behaviour. Potassium is well known to leach easily, and P, Na, Ca, Mg, Mn and S are also reported to increase after passage through the canopy, but to a lesser extent and not so consistently. The scleromorphic (thick, waxy and tough) leaves of many trees growing on nutrient-poor sites may be an adaptation to restrict nutrient loss (e.g. Sobrado and Medina, 1980) but there are other explanations, as discussed later. In tropical rainforests the acidity of throughfall is usually less than that of incoming rainfall (Parker, 1983).

The situation in montane tropical forests with their often high epiphytic biomass is different and epiphytes may have an important role in patterns of nutrient cycling. The epiphyte contribution to nutrient cycling will depend on their mass (reported range

Table 16.1. *Annual fluxes of nutrients (kg ha^{-1} yr^{-1}) in rainfall on a range of tropical forests*

Location	Annual rainfall (mm)	N	P	K	Na	Ca	Mg	References
Queensland A	8631	–	–	8.1	158	17.1	19.4	Brasell and Gilmour (1980), Bellenden Kerr site
Puerto Rico A	5000	4.2	0.01	27	247	47	30	Asbury *et al.* (1994)
Java	4670	15.4	1.2	9.6	13.3	9.8	4.0	Bruijnzeel (1989b)
Puerto Rico B	3750	–	–	18.2	57.2	21.8	4.9	Jordan *et al.* (1972)
Panama	3510	7.3	0.7	13.5	63.5	27.9	4.1	Cavelier *et al.* (1997)
Costa Rica	3191	3.4	0.05	3.0	20.5	5.8	2.4	Hendry *et al.* (1984)
Jamaica	3060	–	<0.13	<8.34	20.6	<8.96	<2.01	Hafkenscheid (2000)
Costa Rica	2820	5.0	0.1	2.5	5.8	1.4	1.0	Clark *et al.* (1998)
Amazonia A	2545	10.3	0.1	–	–	3.8	3.2	Anonymous (1972), Manaus
Queensland B	2520	–	–	4.5	20.8	2.3	2.9	Brasell and Gilmour (1980), Gadgarra site
Taiwan	2420	–	–	7.3	8.8	17.7	3.0	King and Yang (1984)
Malaysia	2380	13.5	–	6.4	22.9	4.2	0.7	Manokaran (1980)
Colombia	2115	18.3	0.72	7.9	24.1	10.1	3.2	Veneklaas (1990) 2550 m site, N value for NH_4-N only
Fiji A	2113	3.6	<0.5	<5.6	–	<1.8	<3.0	Waterloo *et al.* (1997), Tulasewa site
Amazonia B	2050	–	0.1	3.4	9.3	–	–	Franken and Leopoldo (1984), Reserva Ducke (26 km from Manaus)
Fiji B	1904	5.2	<0.2	<4.2	–	<2.6	<5.4	Waterloo *et al.* (1997), Koromani site
Ghana	1850	14	0.4	17.5	–	12.7	11.3	Nye (1961)
Fiji C	1796	9.1	<0.2	<3.2	–	3.3	<4.3	Waterloo *et al.* (1997), Korokula site
Queensland C	1650	60	<0.2	3.4	50	3.2	5.9	Westman (1978)
Venezuela	1575	9.9	1.1	2.6	3.3	5.6	5.2	Steinhardt (1979)
Colombia	1453	11.2	0.48	6.9	15.9	7.3	2.5	Veneklaas (1990) 3370 m site, N value for NH_4-N only

Source: Modified from Bruijnzeel (1989b); Waterloo *et al.* (1997); Hafkenscheid (2000).

2.3 t ha^{-1} in stunted ridgetop forest in Jamaica (Tanner, 1980) to 44 t ha^{-1} in a very mossy cloud forest in Colombia (Hofstede, Wolf and Benzing 1993), their position on the tree, and the prevailing climatic conditions. They are particularly effective at absorbing nutrients from mist and hence can be expected to make a substantial contribution to the nutrient input of montane cloud forests (Asbury *et al.*, 1994; Clark *et al.*, 1998). Other sources of nutrients (albeit the proportions are as yet unquantified) include decomposition from the host tree bark, litterfall and throughfall, excrement from canopy dwelling animals, debris carried by ants and termites from the forest floor, and free-living and symbiotic N-fixers. It has been suggested (Nadkarni, 1986) that epiphytes may act as 'nutrient capacitors' by buffering the pulses of nutrients that arrive in dry spells, and releasing them only under heavy rainfall conditions during wet spells.

There now seems to be little doubt that some elements in rain and fog can be absorbed by the canopy of montane rainforests as suggested by Cavelier *et al.* (1997) for sulphur (SO_4-S) and to a lesser extent for N for an epiphyte-rich (but not mossy) Lower Montane Forest in Panama. Clark *et al.* (1998) found in a very mossy montane cloud forest in Costa Rica that the canopy retained H^+ (92%), NO_3^- (80%) and NH_4^+ (61%). The N seemed to be retained by the epiphytes. Hafkenscheid (2000) also reported substantial retention of NO_3^- in two Jamaican montane forests, despite a relatively small epiphytic biomass. It remains unknown to what extent nutrient capture by epiphytes can contribute to the mineral nutrition of the host tree by foliar absorption. There are recent indications (e.g. Burgess, Dubinsky, and Dawson, 2001) that fog water (and presumably the nutrients within it) can be absorbed by redwood trees in the coastal zone of California, even to the extent that the sapflow is reversed and internal water contents are replenished (perhaps even those in the soil). It should be borne in mind, however, that overall climatic conditions on wet tropical mountains are much rainier than those in coastal California and the need for foliar absorption of mist water may be correspondingly smaller.

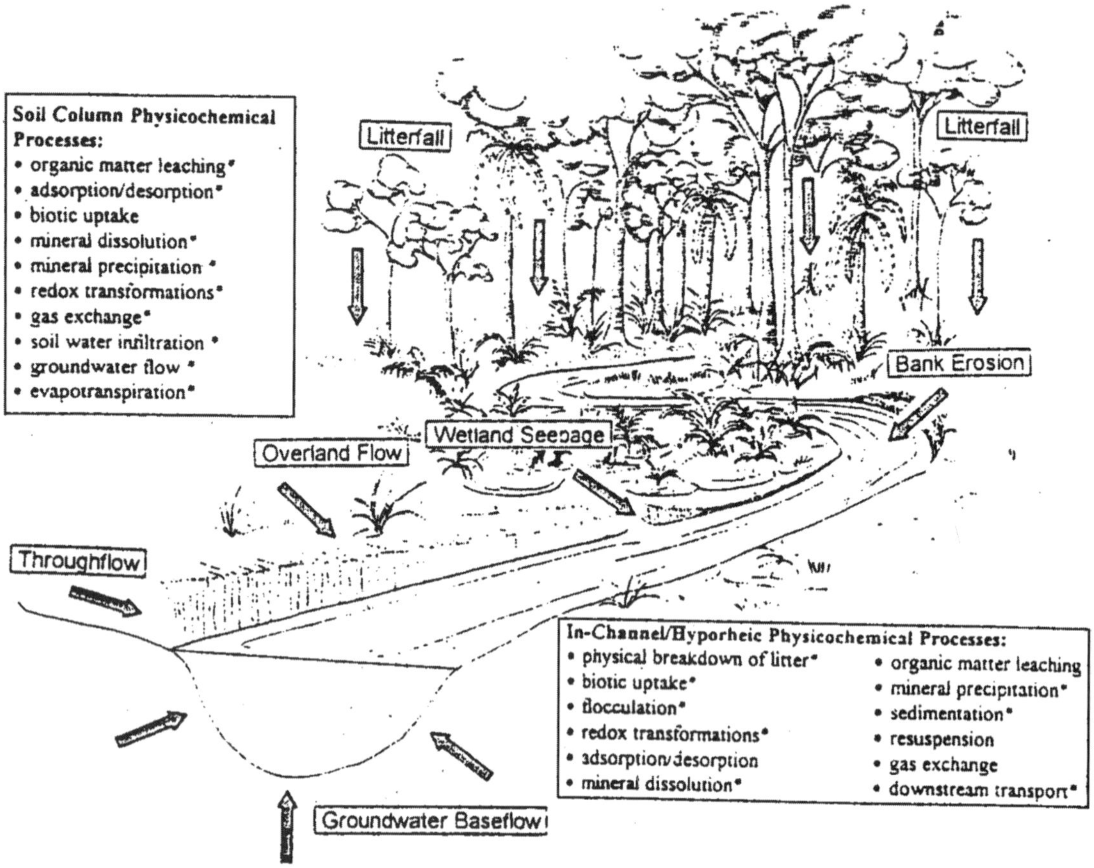

Figure 16.2 The major pathways linking terrestrial and stream ecosystems in the Amazon Basin and the main physicochemical processes operating in each ecosystem. Processes identified by an asterisk have been examined in Amazon stream system although the number of these studies and diversity of terrain types examined is generally very limited.

NUTRIENT LOSSES IN DRAINAGE

Bruijnzeel (1991, 1998) has reviewed the overall inputs and outputs of nutrients via waterbound pathways in rainforests. He has discussed the specific difficulties and limitations of the main methods commonly used to evaluate nutrient losses via leaching at the plot and catchment level. Sollins and Radulovich (1987) and Lesack (1993) stressed the importance of proper quantification of nutrient losses via macropore flow and there must always be concern about the representativeness of soil water concentrations obtained by ceramic cup suction lysimetry, not just in terms of the strength of the suction applied (Nortcliff and Thornes, 1989) but also the material of the cup (Zimmermann, Price and Montgomery, 1978). Nortcliff and Thornes (1989) raised the issue of different nutrient concentrations in soil macropores and soil micropores – the biphasic system of soil water. The larger the pores the shorter will be the water residence time and the smaller the loss of nutrients as they will pass through the macropores largely unused. Samples of water using suction lysimetry will be of micropore water which will have had time to equilibrate with the surrounding soil and is much more likely to represent the soil solution from which the roots access the nutrients. Considerable progress in quantifying nutrient losses by leaching at the plot level has been made recently by Klinge *et al.* (2001) who obtained plausible estimates by using fast recording tensiometers at different depths in combination with a soil water simulation model to obtain rates of drainage. They also used specially designed suction lysimeters capable of extracting soil water at ambient suction levels rather than at an arbitrary pre-set value as is usually done.

Attempts to measure nutrient losses from forests at the catchment scale by measuring concentrations in streams are difficult to interpret because the soils in the riparian zone or even the organic matter within the stream itself may have a dominant influence on the composition of the stream water for some elements, notably N. Pathways linking the two systems are shown in Figure 16.2. McClain and Elsenbeer (2001) have discussed four of the specific pathways by which nutrients enter a stream. There are direct inputs to the channel from litterfall, mass wasting, precipitation and throughfall. The second of these includes landslides and bank failures and can be quantitatively large as well as sporadic, depending on geological and climatic settings (cf. Douglas and Guyot,

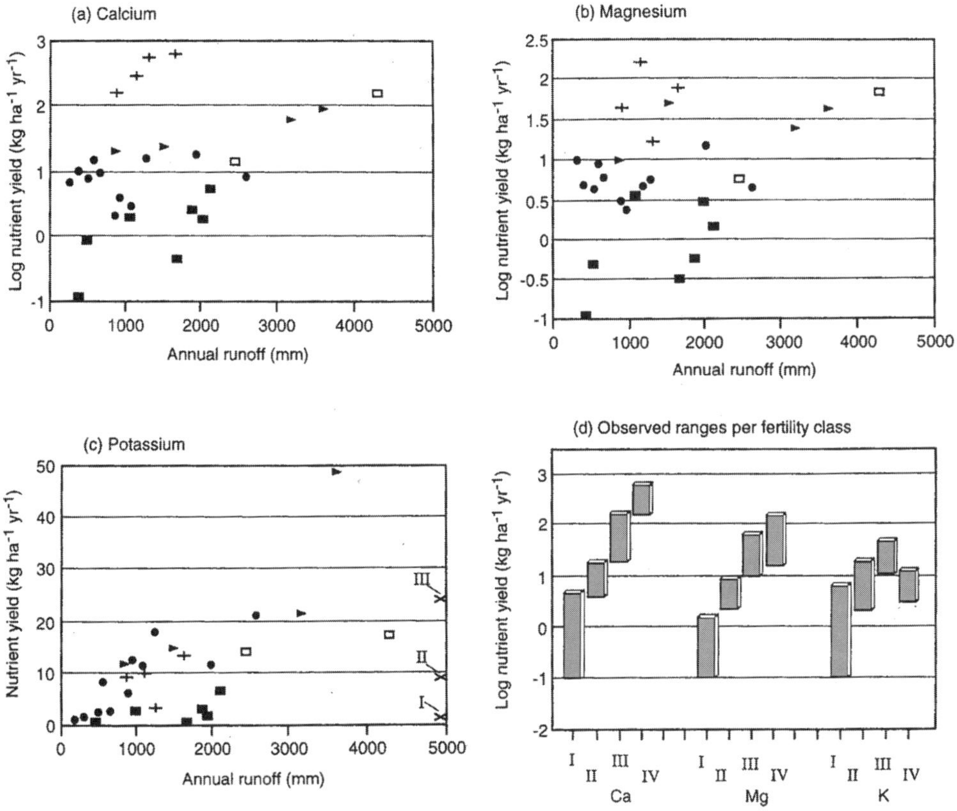

Figure 16.3 Solute losses of (a) calcium, (b) magnesium and (c) potassium (ka ha^{-1} yr^{-1}) *vs.* amounts of streamflow and the ranges in output per soil fertility class (I–V) for 28 tropical forest ecosystems (updated from Bruijnzeel, 1991). ■ Group I, soils of very low fertility (spodosols, oxisols); ● Group, II, moderately infertile soils (ultisols, some oxisols); ▸ Group III, moderately fertile soils (inceptisols); and + Group IV, soils of high fertility (mollisols, vertic soils); □ large catchment; × group average (potasssium only, Figure 16.3c).

this volume). The relative importance of processes in the riparian zone compared with those occurring in the stream itself will vary from area to area; the former will tend to predominate in the upper reaches of river systems since there will be more riparian habitat per volume of water in the smaller tributaries. The remaining pathways are overland flow, lateral subsurface flow and groundwater flow into the stream. All of these are hydrological flowpaths, subject to many interactions among precipitation, soil structure, topography, lithology and vegetation (as is discussed by Bonell *et al.*, this volume). When the nutrients enter the stream they are subject to an unknown extent of chemical and biological activity (McClain and Elsenbeer, 2001). Brandes, McClain and Pimentel (1996), using $\delta^{15}N$ to determine the source of inorganic N in the stream water of a small Amazonian catchment, found that groundwater DIN (dissolved inorganic N) was not the primary source. Their data suggested that remineralisation of organic N within the stream itself may be a major source of stream DIN, and that the majority of DIN entering the stream via groundwater flowpaths is removed at the riparian-stream interface. Further evidence that the riparian zone is important in C or N retention or loss was provided by Chestnut and McDowell (2000) for a stream in the Luquillo Mountains, Puerto Rico. All these observations point to the difficulties of interpreting ecosystem nutrient losses (especially N) in stream water. To this should be added losses in eroded sediment, vegetation and floating litter for which very little quantitative data are available (Brinkman, 1983, 1985; Malmer, 1993). Despite the limitations of the data, Bruijnzeel (1991, 1998) detected a general pattern of increasingly large nutrient losses in streamflow per volume of flow with increasing site fertility, as inferred from general soil classification information (Figure 16.3).

The quantification of soil mineral weathering is generally weak in rainforest areas although stable isotopes, for example the $^{87}Sr/^{86}Sr$ quotients which can be used as proxies for Ca and carry the signature of the parent material, offer promise (Blum *et al.*, 1994). The crucial question remains as to what extent the nutrients released by rock weathering are available to the root network and to what extent they are just carried off via deep drainage (Bruijnzeel, 1991). Baillie (1989) has pointed out the two extreme situations which can occur in rainforests: those with deep, highly-weathered, nutrient-poor soils where the nutrients are concentrated near the surface and where the nutrient losses are balanced by inputs from rainfall; and shallower, more nutrient-rich soils where there is nutrient input from rock weathering and where the nutrients are less tightly cycled.

SOILS AND THE SUPPLY OF NUTRIENTS

Rainforest and soil classifications

One of the most widely used schemes of rainforest classification is that described by Whitmore (1984). This includes 13 tropical rainforest formations, nine of which are lowland forests of which no less than eight are on distinctive soils: heath forest, forest over limestone, forest over ultrabasic rocks, beach vegetation, mangrove forest, brackish water forest, peat swamp forest, swamp forest (including the 'freshwater' and 'seasonal' variants). 'Forest over ultrabasic rocks' is probably an untenable formation because the forest physiognomy and floristic variation are so great, corresponding to large but poorly defined lithological variation (Proctor, 1995). 'Forest over limestone' may be an untenable category for similar reasons. Nevertheless, at this level of forest classification, soil type (with distinct nutrition and hydrology) is clearly causal for many types of rainforest. The correlation is much less clear for Whitmore's (1984) lowland evergreen rainforest, by far the most widespread of the formations, which can occur on a wide range of 'zonal' soils in the broad ultisol and oxisol groups and some 'azonal' soils also.

Rainforest soils are very diverse and soil classification is a first step to categorise their features. The Soil Taxonomy (Soil Survey Staff, 1975) has much information on soil mineralogy, organic matter content and charge chemistry, all of which are important for an understanding of nutrient dynamics (Sollins, 1989). However, conventional soil classifications have many limitations in the information they can provide (Baillie, 1989). For example, fertility levels implied by conventional soil classifications give little information on nutrient supply to trees (Baillie, 1989) and offer little beyond broad indications of soil fertility. Examples of the use of these broad classifications are by Vitousek (1984) and Vitousek and Sanford (1986) in the context of rainforest litter production and Bruijnzeel (1991, 1998) in the context of the relationship between soil class and nutrient losses in drainage water.

Walker and Syers (1976) proposed a theory of pedogenesis to predict changes in nutrient limitation to plant growth during soil development. At first, N would be limiting because N is not present in rocks and must come from atmospheric sources. Later on, rock-derived elements such as P are lost or immobilised and will become relatively more limiting. This idea is supported by some circumstantial evidence and certainly applies in some instances in temperate areas where one notes, for example, the high incidence of N-fixing trees developing in the wake of glaciers (Crocker and Major, 1955). Its applicability to the tropics is less clear but it must be admitted that there is some supporting evidence, best seen in the unusual volcanic situation that prevails in Hawaii. For example, Vitousek *et al.* (1993) showed a trend for Hawaiian montane volcanic soils with a switch from N limitation on young soils to P limitation on old soils. Vitousek (1984) and Vitousek and Sanford (1986) had hypothesised from analyses of leaf litterfall that total productivity of lowland tropical rainforest on oxisols and ultisols was commonly limited by the availability of P. However, plant response in many nutrient addition experiments has failed to support the Walker and Syers (1976) premise.

Because the lowland evergreen rainforest formation is so diverse and can occur on many types of soil which are undefined from the point of view of nutrient supply, it is axiomatic that each example has to be studied separately and that wide generalisations are still some way ahead.

Soil analyses

Soil analyses are a frequently used tool in understanding rainforest mineral nutrition but it is important to be aware of the limitations. For example, Silver *et al.* (1994) found that exchangeable Ca, Mg, and K were significantly lower in soils extracted fresh with NH_4Cl solutions than from soils which were dried and ground prior to extraction with KCl or $NaHCO_3$ solutions. It remains an arbitrary decision whether soils should be analysed fresh or after drying and grinding, as is usually the case. There is an increasing use of ion exchange resins to characterise the availability of soil nutrients in ecological-hydrological studies (e.g. Crews *et al.*, 1995; Yavitt and Wright, 1996) because they extract the ions carried by soil-water fluxes over a period of time. Although it makes for difficult comparisons with the results of other methods of soil analysis, the use of these resins should be encouraged.

Conventional analytical extractants are designed by agriculturalists to be relevant for the growth of annual crops and their application to the growth of long-lived trees is guesswork (Wild, 1989). There is evidence (e.g. Bolan *et al.*, 1994) that low molecular weight organic acids released by plant roots are able to increase the solubility of mineral P but the time it takes to do this in the field is not known. If there is mineral matter present around the roots it may supplement the trees' P supply. If solubilisation is occurring at great depth as part of the weathering process then P may be released directly into the drainage water and may miss the roots altogether. The arbitrariness of the decision of the depth to which soils should be sampled is a major problem in understanding rainforest nutrient cycles. A facile presumption might be that the samples should extend as far as the tree rooting depth but this itself is difficult to estimate. At a site in eastern Brazil with deep alluvial soils, functional roots have been reported to reach 18 m below the soil surface (Nepstad *et al.*, 1994). There is some doubt about the wider applicability of Nepstad *et al.*'s (1994) observations and many rainforest areas have bedrock much closer to the surface, depending on the geological and tectonic setting. Stable continental areas (e.g. Amazonia, central Africa) of low relief and little erosion often have deep profiles of soil and regolith whereas tectonically more active steepland areas (e.g. much of South East Asia) with more intense erosion usually have much shallower profiles (Burnham, 1989; Douglas and Guyot, this volume). Moreover, the seasonal rainfall at Nepstad *et al.*'s (1994) site may have induced deeper root growth. Nevertheless, roots probably always exploit most of the soil depth range that is available to them. For example, Brouwer (1996) reported roots at 305 cm depth, just above the cemented B-horizon in a giant podsol, and at 430 cm depth at the fresh bedrock in another soil in perhumid Guyana.

It might be possible to use suction lysimetry at a range of depths to sample the soil water and hence gain some knowledge of nutrient supply but fine root and soil nutrient concentration changes with depth may, in any case, be a poor predictor of root efficacy in nutrient uptake. Figure 16.4 shows that the rate of ion uptake by a particular section of root depends on signals given by the plant as a whole and is unrelated to any measurable part of that piece of root system (Drew and Nye, 1969). This means that it is impossible to determine the proportion of nutrient supplied relative to a particular depth of the root system.

The availability of soil N is dependent entirely on its release by microbial activity from the soil organic matter and the methods of analysis for plant-available N have often involved incubation techniques in which the release of N from organic matter is measured over a period of 7–14 days. The high insolubility of phosphates means that there are special difficulties associated with P analyses. The most intensive effort to analyse soil P supply has been made by Newbery *et al.* (1997) who followed the fate of several fractions of soil P over three years in Korup, Cameroon, using a method developed from that of Hedley *et al.* (1982). They found marked temporal fluctuations in the different fractions but there was a poor match between the labile soil fractions and litterfall P with time. The authors interpreted this as showing that their sampling occurred over only part of a longer phenological cycle involving droughts, fruiting, and the activity of ectomycorrhizal fungi (Newbery *et al.* 1997).

Overall, it has to be admitted that soil analyses have been of limited use in understanding rainforest nutrient cycling. This has been vividly illustrated by Scott *et al.* (1992) who compared the tree litterfall chemical concentrations from two sites (La Selva, Costa Rica and Maracá Island, Brazil) which differed greatly in their soil nutrient concentrations (Table 16.2). The soils at La Selva were much richer in all nutrients (except for dilute acid-extractable P). Yet the forest there produced lower quantities of non-woody small litterfall with lower concentrations of P, K and

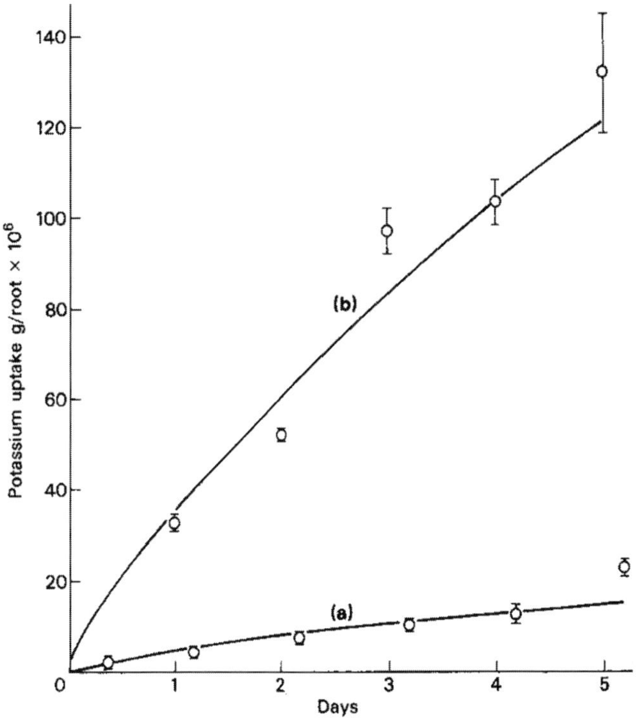

Figure 16.4 Plant demand and the absorption of potassium by young plants of ryegrass (*Lolium multiflorum*). Absorption by a 1 cm segment of the root system to which labelled potassium was provided is compared when: (a) the remainder of the whole root system received the same concentration of unlabelled potassium; (b) potassium was supplied only to the labelled segment. (After: Drew and Nye, 1969.)

Mg and not markedly higher N and Ca concentrations compared with Maracá. Explanations for the Maracá–La Selva anomaly may lie in the two-fold higher rainfall at the La Selva site or (and more likely) that the Maracá trees were obtaining soil water and nutrients at greater depth than that at which the soil samples were taken.

MINERAL AND ORGANIC SOILS: TWO TYPES OF ACIDITY

The parent materials of mineral soils usually contain Al, and Al^{3+} or its hydrated forms are the dominant cations of most mineral soils in the humid tropics. The hydrated forms of the Al^{3+} ion dissociate to act as weak acids – hence their role in creating soil acidity. Al facilitates the immobilisation of P, is antagonistic to the uptake of basic cations, and is moderately toxic in its own right to a wide range of plants (Baillie, 1996). Experiments using plants from temperate areas have long shown the phenomenon of tolerance to Al toxicity (e.g. Clarkson, 1966). Kidd and Proctor (2000b) found differential adaptation to Al^{3+} among temperate *Betula pubescens* races from a wide range of soils. Godbold *et al.*

Table 16.2. *Soil chemical data for 0–10-cm deep samples in a volcanic clay soil (dystropept) at La Selva, Costa Rica (Marrs* et al., *1988) and a very sandy drift-derived soil (grossarenic plinthic paleudultat) Maracá, Brazil (Thompson* et al., *1992) and Marrs* et al. *(1991) for N mineralisation (wet season data) together with mass and chemical composition of non-woody litterfall at La Selva (Heaney and Proctor, 1989) and Maracá (Scott* et al., *1992)*

	Site	
Variable	La Selva	Maracá
Soil data		
pH_{H_2O}	4.2	4.9
N mineralization rate		
(μg N g^{-1} 30 day^{-1})	81	8.5
P concentration		
(μg g^{-1})		
Total	900	61
Extractable	2.2	5.1
Exchangeable cations		
(m-equiv kg^{-1})		
K^+	2	0.7
Na^+	1	0.02
Ca^{2+}	12	2.3
Mg^{2+}	3	1.8
Litterfall data		
Mass (t ha^{-1} $year^{-1}$)	7.6	7.9
Concentration (mg g^{-1})		
N	14	13
P	0.54	0.72
K	2.0	5.5
Ca	8.3	7.0
Mg	1.7	2.7

(1988) have suggested that deficiencies in nutrient cations develop in plants because Al^{3+} ions compete with Mg^{2+} and Ca^{2+} for exchange sites in the cell walls of the root cortex. Al^{3+} ions have two notable beneficial effects: they buffer the pH at around 4.0 (Fitter and Hay, 2002), and they ameliorate the toxic effects of H^+ ions (Kinraide, 1993). Al solubility is controlled by several factors besides pH including dissociation/formation reactions with kaolinite and gibbsite, $Al(OH)_3$, and organic acids. Al^{3+} toxicity can be ameliorated in many ways- by NH_4^+ ions, by organic acids that chelate the Al in non-toxic forms, by SiO_4^{4-} (Kidd and Proctor, 2001b), by Ca^{2+}, and even H^+ ions (Barcelo *et al.*, 1993). The Al^{3+} is usually regarded as the most toxic although other ionic species of Al can be taken up by plants. There is at least one instance of Al (presumably supplied as $Al(OH)_4^-$) being accumulated in leaf dry matter concentrations of up to 2300 μg g^{-1} in an unidentified species of *Memecylon* (Melastomataceae) occurring on soils with a pH of *c*.7 on Mount Bloomfield in Palawan, the Philippines (Proctor *et al.*, 2000). Tyler (1994) showed that leaf concentrations of Al growing on limestone soils (pH 8) were not very different from those found in plants on acid silicate soils, despite the significantly lower soil concentrations. Since the aluminate ion has some structural similarity to $H_2PO_4^-/H(PO_4)^{2-}$, Tyler suggested aluminate may be taken up by the same mechanism as phosphate.

Organic soils in the lowland tropics occur principally in the Amazon Basin of which they reputedly occupy 6% of the Brazilian part (Whitmore, 1984) and are also found in Amazonian Colombia, Ecuador, Peru and Venezuela. They occur on the Orinoco system and in Suriname and Guyana. In Asia they have a smaller extent and are best developed in Borneo in the heath forests and peat swamps. Their occurrence in Africa seems poorly documented and needs further investigation. Organic soils are widespread in montane forests. They have developed a thick layer of organic matter because of slower decomposition caused by low temperatures in high montane situations, waterlogging in cloud forests and peat swamp forests, and extreme acidity in heath forests. There is often a sharp boundary between the organic and mineral horizons (Whitmore, 1984).

In organic soils the acidity is caused primarily by H^+ ions because they often have little Al in them (Fitter and Hay, 2002). Soils with an excess of H^+ over Al^{3+} are likely to be very acid (pH around 3.0) and to be buffered probably by some organic acids. The plants on them will have N supplied mainly in the form of NH_4^+ ions. We can expect such plants to be small leaved (to increase stomatal control over water loss), and to be xeromorphic (to reduce the mass flow delivery of H^+ ions). Moreover they are likely to have low tissue N concentrations since both NH_4^+ uptake and symbiotic N_2 fixation result in the extrusion of H^+ ions which would make the soil solution even more acid (Marschner, 1995). H^+ ions may be very toxic because of the metabolic costs of maintaining the cytoplasmic pH between 7.0 and 7.5 by proton pumps located in the tonoplast and the plasma membranes of all cells. These proton pumps will consume much more metabolic energy when the external pH is very low (Marschner, 1991).

Proctor (1999) showed that a Bornean hill rice variety which grew well on moderately acid (and presumably potentially Al-toxic) clay soils from an old (*c*.50 years) shifting cultivation fallow was unable to produce roots on a very acid organic heath forest soil unless it had been fertilised with $CaCO_3$. The only likely toxin in the acid soil was H^+ since Al is not abundant in these waterlogged sandy soils. Luizão (1996) obtained similar results in his bio-assay work on Brazilian heath forests.

Kidd and Proctor (2000a) showed that a race of the temperate grass species *Holcus lanatus*, collected from an organic soil, was more tolerant to H^+ ions, confirming that there are likely to be specific adaptations for H^+ toxicity. This remains to be demonstrated for a tropical species, although there is some circumstantial evidence that this is the case. Proctor (1999) reported preliminary experiments with seedlings from Barito Ulu, Central Kalimantan, which were consistent with the view that H^+ excluded non-adapted species from the extremely acid and poorly buffered heath forest soils.

A further feature of organic soils is the presence of large amounts of polyphenols. These may have an important role in regulating soil processes and there have been suggestions (Rice and Pancholy, 1973) that they tend to inhibit nitrification as well as decomposition and nutrient cycling (e.g. Kuiters, 1990). However, Smith *et al.* (1998) found that litter N and polyphenol concentrations were not related to rates of mass-loss in decomposition in an Amazonian forest. Moreover, R. H. Marrs and J. Proctor (unpublished) in Proctor (1999) found evidence of nitrification in highly organic heath forest soils in Kalimantan that were presumably rich in polyphenols. Phenolic acids are widely believed to reduce plant growth (e.g. Fitter and Hay, 2002) and the drainage waters of heath forest areas are characteristically tea-coloured by these compounds (the phenomenon of blackwater rivers). However, when added to soils or tested hydroponically at concentrations found in soil solutions, Kidd and Proctor (2000a) found that they had beneficial effects on the growth of *Holcus lanatus* but only at pH 3.5 and not at pH 6.5. Previous workers who have found that phenolic acids inhibited growth appear to have used unrealistically high concentrations of phenolic acids in their experiments. Northup *et al.* (1995) have explained the increasing stunting of conifer forest on a Californian marine terrace soil acidity gradient, as caused by increasing polyphenols in the system. Proctor and Kidd's (2000a) paper throws this interpretation into question and there seems little evidence that the stunting of the conifers is caused by anything other than high soil acidity (pH as low as 2 in the litter layer of the smallest forest).

It is tempting to ascribe the similarities (e.g. dense tree crowns, a predominance of microphylls (i.e. small-leaved trees), dense hard wood (Whitmore, 1984) among heath forests, upper montane forests or cloud forests and the central parts of domed peat swamps, to the influence of H^+ toxicity associated with high soil organic matter. The most detailed work so far (Hafkenscheid, 2000) on upper montane forests in Jamaica has lent some support to the H^+ toxicity hypothesis (Table 16.3). The forest stature is ranked in the order of both H^+ and Al^{3+}. Although Hafkenscheid (2000) found relatively low Al concentrations in the leaves and fresh litter, he decided that (when assessing the likelihood that H^+ or Al^{3+} was limiting growth in these forests) 'I could not eliminate one of these, so both explanations are plausible in Jamaica' (R. L. L. J. Hafkenscheid, pers. comm.).

One would expect that upper montane forest soils would be much more variable than those under heath forests because of the greater range in climatic conditions and bedrock types compared with lowland conditions and sandy substrates only. However, both Proctor (1999) and Hafkenscheid (2000) agree that soil toxicity is a major cause of the distinctive features of heath forests and upper montane forests. It remains to be seen whether the causes of the toxicity are the same or reflect subtle interactions among Al^{3+}, H^+, NH_4^+, Si, and phenolic and other organic acids.

Table 16.3. *The stature and soil moisture concentrations of H^+ and Al^{3+} in the Ah horizons from four upper montane forest sites in Jamaica (from Hafkenscheid, 2000)*

Forest type	Canopy height (m) and depth of water samplers (cm)	H^+ (μmol l^{-1})	Al^{3+} (μmol l^{-1})
Well-developed on mull humus	12–17, 0–14	11.22	1.14
Poorly developed on mull humus	7–12, 0–14	15.74	1.46
Moderately developed on mor humus	5–8, 0–5	81.37	14.52
Stunted on mor humus	5–7, 0–9	132.50	19.48

Source: Hafkenscheid (2000).

It must be emphasised that because of the difficulties with conventional soil pH measurements (i.e. those involving the insertion of an electrode into a soil/water mix), these may not be a reliable indicator of soil acidity and are unsatisfactory for detemining if the soils are acid because of Al^{3+} or H^+. Such measurements are affected by drying and storage between sampling and preparation of the suspension and by the choice of the solution used to prepare the suspension (Courchesne, Savoie and Dufresne, 1995). Soils buffer pH through a combination of ion exchange of the clays, sesquioxides, organic matter, $CaCO_3$, and various hydrated Al ions, thereby retaining equilibrium between the potential and active acidity. Measurements of the concentrations of H^+ and Al^{3+} in the soil solution (vacuum tube lysimetry) or KCl extracts are necessary to determine which ion is dominant.

Nitrogen supply

It is clear that the free-living soil microbial population plays a key role in mineralising soil organic matter and supplying nutrients to plants (e.g. Coleman *et al.*, 1983; Fitter and Hay, 2002). A wide range of environmental conditions will affect soil microbes and soil water and the effects on soil oxygen supply are known to be important (e.g. Silver *et al.*, 1999). Although the mineralisation of organic matter will eventually release all the mineral nutrients within it, I do not intend to review the vast subject of rainforest decomposition and mineralisation for which Villela and Proctor (2002) and Luizão *et al.* (1998) have provided some recent examples. Instead, the focus will be on aspects of N supply because N is required in large quantities by all plants and is the only major nutrient which has an almost entirely organic pool.

The first stage of inorganic N mineralisation from organic matter is the ammonification process by which a wide range of micro-organisms yield NH_4^+ and OH^- ions. The former are oxidised, by specialist chemotrophic micro-organisms which have more precise environmental requirements than the ammonifiers, to yield first NO_2^- (which usually has a transient existence) and then NO_3^- in a process called nitrification during which H^+ ions are released (Robertson, 1989). During the ammonification, N may be immobilised by microorganisms and hence ammonification is often a rate-limiting stage in N supply. It has been demonstrated (e.g. by Jones and Richards, 1977) that nitrifiers are poor competitors for NH_4^+ ions and this is reflected in the critical C: N ratio (between 30: 1 and 20:1; Fitter and Hay, 2002) above which no N is released for higher plants.

The proximal controls on nitrification are the supply of NH_4^+ and soil O_2 that are themselves dependent on the quantity of water in the soil. There is a converse process called denitrification where the nitrate is reduced to N_2 (and OH^- ions are released) which occurs under conditions of high NO_3^- supply and low supply of O_2. Nitrification and denitrification are two of the potentially most important regulators of mineral N retention in tropical soils because by their influence on acidity they may regulate the status of the variable-charge soils. Where there is a cation exchange capacity, NH_4^+ will be retained: where there is an anion exchange capacity NO_3^- will be retained; while at the point of zero charge there will be no retention of mineral N. The influence on acidification can be balanced to an unknown extent by plant uptake. If N is taken up primarily as NH_4^+ then the plant releases H^+ ions to maintain electro-neutrality. If NO_3^- is the N source then the pH of the soil solution may increase because the plant can take up H^+ with the NO_3^- to maintain the balance.

There have been recent advances in the understanding of the controlling factors for nitrification and denitrification (from work on the soil N supply in a Dutch coniferous forest on an acid soil) but much remains speculative and one awaits similar intensive techniques being used in rainforests (Laverman *et al.*, 2000a, b, 2001).

Nitrification varies greatly in undisturbed forests probably partly because of real differences between soils but also because of a large temporal and spatial heterogeneity. Recorded values of nitrification rates from primary forest range from < 0.1 – > 6 μg g^{-1} day^{-1} NO_3^- N. (de Rham, 1970; Tanner, 1977; Lamb, 1980; Vitousek *et al.*, 1983; Chandler, 1985, Robertson, 1984). Marrs *et al.* (1988) showed the decreasing N supplying power of soil with increasing altitude on Volcán Barva, a wet tropical mountain in Costa Rica. At *c.* 100 m the nitrification rate was 1.6 μg g^{-1} day^{-1}, at 2600 m the rate was 0.21 μg g^{-1} day^{-1}.

Whilst nitrification is clearly an important part of the nitrogen supply, its converse process – dentrification of NH_4^+ to N_2 gas – can be a cause of considerable N-loss to the whole system. Field measurements of denitrification in tropical forest soils are technically difficult but Robertson and Tiedje (1988) measured denitrification in four primary forest sites in Costa Rica using the acetylene incubation technique on intact soil cores. Over a 14-month period, average rates extrapolated to an annual flux of 7.6–21 kg ha^{-1} $year^{-1}$ N gas.

The importance of non-symbiotic fixation in tropical forests remains unknown but is almost certainly small. A qualified step towards estimating the quantities of N fixed by legumes in tropical forests has been made by Roggy *et al.* (1999). They used the natural ^{15}N abundance method on legume trees from rainforests on oxisols and spodosols in French Guyana. On the spodosol, the available soil N was isotopically similar to that of fixed-N and no reliable estimate of N-fixation was possible. On the oxisols, estimates were possible and, extrapolating from the sampled trees by calculating their proportional contribution to the ecosystem, Roggy *et al.* (1999) arrived at a fixed N value of 7 kg ha^{-1} y^{-1}. Sprent (1995) had argued previously that N-fixation by tropical rainforest trees is likely to be small and the work of Roggy *et al.* (1999) in the legume-rich forest supports this view. However, Roggy *et al.* (1999) admitted that the $\delta^{15}N$ method (combined with biomass assessments) yields only approximate estimates of N_2 fixation. Moreover, the legitimacy of using $\delta^{15}N$ to estimate amounts of N_2-fixation has been questioned by Handley and Scrimgeour (1997). They found no consistent patterns in $\delta^{15}N$ values in woody plants in an old Scottish field that could be ascribed unequivocally to N_2-fixation and showed that $\delta^{15}N$ values could vary for a number of reasons, including differences in translocation of stored N within the plant, differences in N uptake, and differences in N mineralisation. All of these processes confound the use of $\delta^{15}N$ values to assess the extent of N-fixation and may be involved in Roggy *et al.*'s failure to provide N_2-fixation estimates for the forest on the spodosol.

There is a further question about possible advantages of N-fixation, bearing in mind that N is often not in short supply and that the energetic cost of fixation is higher than for assimilation into amino acids and proteins (Sprent and Raven, 1985) even when NO_3^- ions are the N source. (In order to participate in assembling these compounds the NO_3^- ions have to be reduced, at an energetic cost, to NH_4^+.)

Climax communities are characterised by low rates of nitrification and it has been held (Smirnoff *et al.*, 1984; Smirnoff and Stewart, 1985) that woody plants appear to take up inorganic N mainly in the form of NH_4^+. NH_4^+ assimilation is associated with lower energy requirements, a fact which may be of importance for understorey plants severely limited by light. Shade-bearing species did not show NR (nitrate reductase) induction after exposure to additional NO_3^-. Freeden *et al.* (1991) obtained evidence of a greater reliance on NH_4^+ by shade-tolerant species within the genus *Piper* in Mexico. Stewart *et al.* (1992) looked at nitrate reductase (NR) activity in a range of tropical species and found that NO_3^- was not the main source of N. However, most species had the potential to use NO_3^- because NR activity could be induced upon addition of NO_3^-.

Soil heterogeneity

An important factor which is crucial to understanding nutrient cycling is soil heterogeneity.

SPATIAL HETEROGENEITY

Vertical heterogeneity in soil chemical and physical characteristics is obvious, particularly so in soils where the horizons are clearly demarcated. One extreme component of heterogeneity is the presence of animal nests. One of the pits dug by Nepstad *et al.* (1994) in alluvial soils in eastern Brazil had cut through a leaf-cutter ants' nest 5 m below the surface, well beyond the depth of conventional soil analyses and likely to contribute substantially to the mineral nutrition of the surrounding trees. The importance to the forest overall of these animal nests remains unquantified but will be a function of the number of nests per hectare.

The horizontal component of heterogeneity has been shown many times: examples include the work of Silver *et al.* (1994) in Puerto Rico, Burghouts, van Stralen and Bruijnzeel (1998) in lowland dipterocarp forest in Sabah and Nooren *et al.* (1995) in the Ivory Coast.

Catenary changes with topography influence soil properties. Silver *et al.* (1994) found that soil nutrient availability decreased with sample depth and that several soil properties varied in a simple catenary model across the landscape. Exchangeable base cation concentrations and pH increased along a gradient from ridge tops to riparian valleys, while soil organic matter and exchangeable Fe decreased along this gradient. Nutrient availability in the upper catena appeared to be controlled primarily by biotic processes, particularly the accumulation of organic matter. Periodic flooding and impeded drainage in the lower catena resulted in a more heterogeneous environment. Burghouts *et al.* (1998) demonstrated a trend of increased Ca in topsoil and litterfall from ridgetops to footslopes in the Danum whereas Nooren *et al.* (1995) observed a reverse trend at Taï, despite a concurrent downslope increase in the number of nutrient-rich worm casts.

The reality of small-scale soil heterogeneity for tree growth has been demonstrated by L. Nagy (pers. comm.) who grew seedlings of *Vatica vinosa* in soil collected from three different sample sites, about 10 m apart under a single mother tree, in a lowland evergreen rainforest from Barito Ulu, Central Kalimantan. He found that the mean leaf relative growth rates were ($cm^2\ cm^{-2}\ 210\ d^{-1}$) 4.330 (s.d. 2.537), 1.816 (s.d. 0.815), and 0.849 (s.d. 0.155). The first and last values were significantly different at the 1% level.

TEMPORAL HETEROGENEITY

Temporal heterogeneity has been shown many times for rainforest soils although Yavitt and Wright (1996) lamented that 'we still have woefully inadequate knowledge about temporal variation in soil nutrient levels and the processes and mechanisms controlling the variability'. Van Dam (2001) has pointed out that there may be considerable ecosystem nutrient supply effects associated with the lengthy dry spells during 'El Niño' years. Grimaldi *et al.* (1992) showed that seasonal variations in Amazonian precipitation cause physical modifications to the soil (e.g. in relation to water flushes and filling of the soil pores), biological modifications (e.g. decomposition rates and soil organism population fluctuations) and chemical modifications (fluctuating nutrient concentrations). Grimaldi *et al.* (1992) concluded that to disregard these 'nutrient pulses' in forest management 'would be as prejudicial as to ignore the rhythm of "samba" in the "Carnival"! Luizão *et al.* (1998) showed similar wet and dry season values for the concentrations of several nutrients but there were higher values for N mineralisation and nitrification in the wet season. Where seasonal dryness occurs this will limit microbial activity and decomposition of soil organic matter and nutrient release from detritus will tend to be lower in the dry than in the wet season. Roy and Singh (1995), working in a seasonal forest in India, found that wetting of dry soil disrupted the osmotic balance of soil micro-organisms causing nutrient release to the soil. It should be mentioned that relatively high rates of dry-season leaf litter decomposition have been recorded for *Peltogyne gracilipes* on Maracá Island, Brazil (Villela and Proctor, 2002) and they have argued that the high concentrations of magnesium released during it is a potential cause of the dominance of this species at this particular site.

Yavitt and Wright (1996) reported the temporal changes that occurred with seasons on Barro Colorado Island, Panama, and included a dry-season irrigation experiment. In general, they observed less temporal variation than had been observed in more seasonal forest in India (length of dry season > 6 months) (Roy and Singh, 1995). Yavitt and Wright (1996) suggested that the length and severity of the dry season plays a critical role in nutrient cycling. It seems as though the heavy clay soil on Barro Colorado Island stays wet enough during the four-month dry season to permit movement within the soil of ions as measured using ion-exchange resins. Irrigation had little effect on nutrient accumulation rates by the resins but did affect the timing in temporal variation in K, Na, and inorganic N and P. In the mineral soil, inorganic N and P pools and N transformation rates showed virtually no temporal variation and no response to irrigation. The authors concluded that temporal patterns disappear with depth in the mineral soil because the nutrient efflux from the forest floor litter is 'mitigated in the mineral soil possibly because of immobilization by plants and microbes'. There was a strong correlation between NO_3^-, NH_4^+ and Mg^{2+}, and Ca^{2+} release from fallen litter and their accumulation in resin-exchange bags placed on the soil surface. Irrigation enhanced decomposition (Wieder and Wright, 1995) and changed its timing but had no corresponding effects on litterfall. Decomposition rates during the dry season were 1.48 times higher in the irrigated plots than in control plots but were only 0.88 of those in control plots in the wet season.

SOIL HETEROGENEITY AND TREE SPECIES DISTRIBUTION

Attempts to relate tree distribution solely to soil nutrient concentrations on the scale of 100–400 m^2 have not been successful except where there are gross differences between the soils, for example, the juxtaposition of ultisols and spodosols (i.e. clayey and sandy soils) or different geological substrata, e.g. outcrops of limestone (Proctor *et al.*, 1983) or ultramafic (ultrabasic) rocks (Proctor *et al.*, 2000). In practice, such an approach has many problems (which are more fully discussed later): the tree roots are as widespread as the crown and hence will integrate forest soil conditions, including gaps which show a wide range in areas depending on the forest type. Luizão *et al.* (1998) showed that on Maracá Island tree falls often leave no gaps in the canopy whereas at La Selva, Costa Rica, tree falls are typically associated with gaps of *c.* 1450 m^2. Spatial and temporal heterogeneity will confound single-time sampling. Finally, without experiments, one can rarely be certain that correlated soil chemical properties are causal for tree distribution – they may merely reflect the influence of the tree on the soil. By contrast, hillslope hydrological patterns determined by topographical positions have long been known to have an influence on tree species distributions, mostly through their influence on soil water status and composition. Ridgetops are often more prone to drought stress than valley bottoms which tend to be wetter and, as long as there are no major changes in geological substrate involved, often more nutrient rich as well (Baillie, 1996).

Oliveira-Filho *et al.* (1994) looked at the effects of soil and topography on the distribution of tree species in a tropical riverine forest in south-eastern Brazil. They found strong correlations between tree species distribution and soil P on the one hand and between tree species distribution and topography on the other. They claimed that spatial heterogeneity in the environment can be an important factor promoting the co-existence of tree species in a forest community. Topographic features associated with differential groundwater regimes or with soil properties or both have also been found to be strongly correlated with tree species distribution in gallery forests of central Brazil (Oliviera-Filho *et al.*, 1990; Ratter, 1980) and of the Chaco region of Argentina (Adámoli *et al.*, 1991).

Austin and Greig-Smith (1968) investigated within-site floristic heterogeneity in a 1.8 ha site in the Sepilok Forest Reserve, Sabah. They divided the area into 20 m × 20 m plots and used the most abundant species (25% of the total number of species) in a principal components analysis ordination of the vegetation and showed that the first axis of the ordination was associated with the topographic gradient from ridge to valley. Ashton (1976) working at the Pasoh Forest reserve, Peninsular Malaysia, used association analysis to show differences in floristic composition between 'hillside, alluvium and lower hillside/undulating land'. The clearest demonstration of the relationship between vegetation and topographic position is perhaps that of Franco and Dezzeo (1994) for the San Carlos area in Venezuelan Amazonia. They distinguished five forest types, the first three of which occurred on well-drained clay soils and the remaining two on sandy valley fills. The first forest, called 'Complex Forest' is the equivalent of the lowland evergreen rainforest formation of Whitmore (1984); this is followed by 'Yevaro Forest' on the same formation but dominated by the legumes *Eperua purpurea* and *Micrandra spruceana*; and 'Guaco Forest', again on the same formation but dominated by the legume *Monopteryx uaucu*. There is 'Tall Caatinga', which is the equivalent of the heath forest formation of Whitmore (1984); and 'Tall Bana' (or Low Caatinga), which is a shorter facies of heath forest. Franco and Dezzeo (1994) measured daily variations in the soil water tension at different positions above the groundwater table for 450 days and were able to relate the different forest types to a soil catena along which the following four factors seemed to be the most influential (cf. Proctor, 1999): the degree of saturation in the very wet months; the degree and frequency of water deficit periods; the water retention capacity of the soil; and the root distribution in the soil. It is axiomatic that the different species composition, caused by the differences in topography, groundwater regimes and soil changes, will have reciprocally different influences on soil mineral nutrients and hydrology.

A further feature of soil heterogeneity concerns the importance of gaps. Marrs *et al.* (1991) and Luizão *et al.* (1998) used artificial gaps of a range of sizes from 40–2500 m^2 and unfelled forest controls at their field site on Maracá Island, northern Brazil. They found higher maximum and lower minimum air temperatures in gaps but there was little consistent effect of gap size on soil chemistry, including soil N mineralisation and nitrification and on several other ecological processes. For example, from a litter bag study they concluded that decomposition and nutrient release were not influenced in a consistent way by gap size. It may be that small areas of high nutrient release can be caused by the local additions of large quantities of litter even though process rates remain unaltered. Luizão *et al.* (1998) also found that at least on Maracá there is no support for the hypothesis that the heterogeneity associated with gaps will involve an interaction between enhanced nutrient supply and increased light as is often assumed (Denslow, 1980, Riddoch *et al.*, 1991). Natural treefalls on Maracá normally do not result in substantial gaps because trees fall through the crowns of standing trees and remain covered by them. As such, there is no 'chablis' effect (for example, as described by Jacobs, 1988) causing an increase in light reaching the forest floor. An intensive study of artificial gaps has been made recently by van Dam (2001) in Guyana. He observed a reduction of up to 95% in litterfall and hence also the nutrients within it in the centre of large gaps compared with closed forest. Furthermore and in contrast to Luizão *et al.* (1998), in Amazonia he found that decomposition rates of the litter fractions and N-mineralisation rates were slower in the gaps. Evidence of nutrient stress limiting regeneration was lacking, however. Van Dam (2001) also showed increased nutrient leaching and soil acidification (with Al concentrations up to 19-fold higher) with increasing gap size and hypothesised (albeit without experimental evidence) that these could 'seriously depress regeneration in gaps larger than 400 m^2'.

THE UPTAKE OF NUTRIENTS BY TREES

Roots and mycorrhizas

Information on tree roots in tropical forests remains inadequate although there are several estimates of root biomass, e.g. Lawson *et al.* (1970), Klinge (1973), Huttel (1975), and Green (1992). There is a frequently made but as yet unproven assumption that a surface root mat development is an important mechanism to enhance nutrient conservation in forest ecosystems on poor soils where the roots or their mycorrhizal derivatives absorb directly nutrients released from decomposing organic matter (Herrera *et al.*, 1978; Jordan and Herrera, 1981; Khiewtam and Ramakrishnan, 1993; Newbery *et al.*, 1997, Went and Stark, 1968a, b). However, an equally likely explanation is that the surface root mat exists as a means of allowing nutrient uptake whilst avoiding the soil toxins or competition from soil micro-organisms in the very acid or waterlogged situations (where high C:N ratios prevail) where the root mat is best developed. For example, on nutrient-poor (but not very acid) sandy soils of Maracá Island with nutrient-rich litterfall, no root mat exists (Thompson *et al.*, 1992a). Luizão (1994) studied bagged decomposing *Clitoria* leaves from a central Amazonian rain forest on clay soils and used the same substrate in a nearby heath forest on a sandy soil. Both forests had a well developed root mat. She found that disturbance (where the bags were lifted at weekly intervals to prevent ingrowth of roots) had no effect on the rate of N or P release. It is suggested that in this case direct nutrient uptake from the decomposing leaves must be negligible.

Mycorrhizas are associations between plant roots and fungi that are held to be beneficial to the former in the uptake of nutrients, water, or in the production of growth substances. The fungal component receives carbohydrate from the host. Recent summaries about their ecology are to be found in Fitter and Hay (2002) and Lambers *et al.* (1998). They enhance the symbiotic plants' below-ground absorbing surface and for endomycorrhizas (in which there is intra-cellular penetration of the host cells by the fungi) this is held to be the main mechanism by which mycorrhizal plants increase nutrient uptake, especially that of the poorly mobile phosphate ion. For ectomycorrhizas (which envelop the root in a fungal sheath but have only intercellular penetration) there are other potential aids to enhanced nutrient uptake such as the excretion of organic acids and hydrolytic enzymes. There seems little doubt that in the laboratory ectomycorrhizal fungi are able to produce enzymes and utilise forms of organic N and P (e.g. amino acids, peptides, proteins, chitin and nucleic acids) and transfer them or their N and P into the host plant which would otherwise have no access to them (Chalot and Brun, 1998; Antibus *et al.*, 1997). The key question regarding the ecological functioning of mycorrhizas is one of cost and benefit. Does the supply of carbohydrates by the host justify the benefits conferred by the fungal partner? In fact, conflicting evidence abounds concerning the functioning of rainforest mycorrhizas.

There have been a large number of pot experiments with mycorrhizas but few in the field, and virtually none in natural rainforest communities. Even the pot experiments have given variable results which are not easy to summarise. Lee and Alexander (1994) grew the dipterocarp species *Hopea helferi* and *H. odorata* in a sandy loam soil (obtained from a 40-yr-old plantation of *Dryobalanops aromatica* and *Dipterocarpus* spp.) to which more sand was added. The soil was sterilised and all except the control treatment were fertilised with N, Ca, Mg, S, B, Co, Cu, Fe, Mn, Mo, and Zn. The experimental soils were additionally fertilised with P and K in various combinations. In spite of the fact that Ca had been added, the seedlings grown in the amended treatments showed, at least initially, Ca-deficiency symptoms. The symptoms were worse in non-mycorrhizal plants that had not been inoculated by the fungal partner. Ectomycorrhizal infection increased the Ca concentration and relieved the symptoms. *Hopea odorata* showed the classic mycorrhizal response (Harley and Smith, 1983) of increased shoot P concentration and total dry weight to the same or greater extent as those of uninfected plants grown in soil to which P had been added (Figure 16.5a). Lee and Alexander (1994) claimed that 'this is the first direct experimental evidence

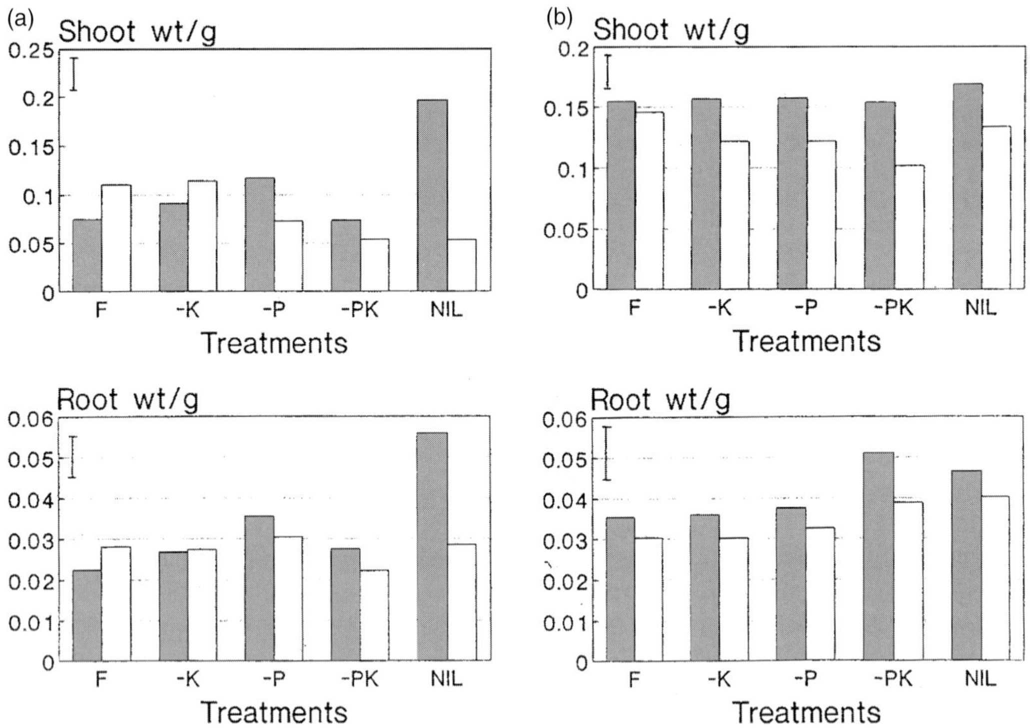

Figure 16.5 (a) Shoot and root dry weights of *Hopea odorata* plants after 100 days growth in nutrient amended soil with P and K additions (F), with P but no K addition (−K), with K but not P addition (P), without P or K addition (−PK), and in unamended soil (NIL). (b) Author to supply caption in proof. Shaded bars, ectomycorrhizal; unshaded bars, non-mycorrhizal. Bar is pooled LSD for comparison between treatment means within the mycorrhizal or non-mycorrhizal series.

that ectomycorrhizal infection can improve P uptake and growth of a dipterocarp species'. It is striking however that the response to mycorrhizal infection was greater than the response to P addition. *Hopea helferi*, by contrast, showed a positive growth response to ectomycorrhizal infection but no growth response to nutrient treatments (Figure 16.5b). It was concluded that P availability was not limiting growth in *H. helferi* in the experiment. Other investigators have found similar positive responses to ectomycorrhizal infection but no response to added nutrients. Alexander *et al.* (1992) found that *Intsia palembanica*, an ectomycorrhizal legume, also responded better to infection than to P addition. Smits (1994) came to a similar conclusion in that the dipterocarps he examined on ultisols in East Kalimantan would not grow well without mycorrhizas and that these could not be substituted by fertilisers. He favoured the view that the mycorrhizas produced a growth substance required by the dipterocarps.

There is still much research to be done on mycorrhizas before they can be accepted unequivocally as a necessary component of rainforest nutrient cycling. Newbery *et al.* (1997) have provided some circumstantial evidence for the importance of ectomycorrhizas in the rainforests of Korup, Cameroon, but their work was done with soils which were not very low in total P at least. For example, the mean total P in the mineral layer (of which the depth varied) of plots in their 'transect P' was 119 $\mu g\ g^{-1}$ in those with few ectomycorrizal trees and 192 $\mu g\ g^{-1}$ in those with many ectomycorrizal trees. Both values are substantially higher than in the rainforest of Maracá Island, Brazil, which had a mean total P concentration in the undisturbed plots of 78 $\mu g\ g^{-1}$ (0–10 cm depth in the mineral soil) and where no special mycorrhizal mechanisms were invoked (Thompson *et al.*, 1992a; Scott *et al.*, 1992). Experimental glasshouse work on two species from Korup has indicated that the mycorrhizal influence may be subtle. For example, Moyersoen *et al.* (1998) found that P uptake by seedlings of *Tetraberlinia moreliana* was correlated with ectomycorrhizal colonisation but there was no such correlation with seedling growth. It has been observed that ectomycorrhizal fungi from Korup show strong phosphatase (enzymes capable of breaking down organic matter and releasing its phosphate) activity in pure culture (I. J. Alexander, unpublished data).

The need for field experiments is paramount and experiments should be carried out as they have been in temperate grasslands (e.g. Fitter, 1986) using benomyl, a fungicide which inhibits endomycorrhizas. Unfortunately no fungicide is currently available which has a corresponding effect on ectomycorrhizas. It is possible that mycorrhizas may be involved in drought-P interactions since the diffusion coefficients of this element are reduced disproportionately in drying soils (Nye and Tinker, 1977) and diffusion along ectomycorrhizal hyphae may be much faster. There is evidence from grasslands that mycorrhizas may help drought tolerance (Fitter, 1986).

The mineral nutrient requirements of trees

TISSUE NUTRIENT CONCENTRATIONS

Rainforest trees vary enormously in their mineral composition between species, individuals, and within individuals (e.g. Masunaga *et al.*, 1997, 1998a,b). Proctor *et al.* (2000) have shown the great range in element concentrations that occur in leaves of rainforest trees, even when they are growing on the same sites. Their data from a forest on graywacke in the Philippines showed a 3-fold range for N, a 6-fold range for P, 13-fold range for Ca, and 11-fold for Mg. In spite of the large area covered by the roots of a single tree and the reduced effect of soil spatial heterogeneity, equally large contrasts occur in foliar concentrations between individuals of the same species (Tanner, 1977, 1985; Hafkenscheid, 2000). Some of the causes of the different nutrient concentrations may be because trees have different proportions of woody tissues but the question can probably only be resolved by experiments at the molecular level. Attempts at short-cuts in tree-nutrient studies have been unrewarding. Masunaga *et al.* (1998a) attempted to show correlations between bark and foliar samples in the modest hope that bark samples might be substituted for leaves which are much more difficult to collect. Unfortunately the correlations were not strong for P, Ca, and Mg (N was not analysed) and hence the use of bark samples to assess the nutrient concentrations in trees is not feasible. Quantification of nutrients in timber is difficult because the nutrient composition varies along the length of a tree and also radially (Poels, 1987; cf. Waterloo, 1994) so there is no alternative to felling the trees and analysing sections along their length. In the forest, most of the nutrients of the trees are in wood and will be concentrated in the boles of very large individual trees which are impossible to sample. The difficulties of working with roots are even greater.

Plant analyses, except in the case of some plantation species showing symptoms of an apparent nutrient deficiency, will always be a poor guide to the quantification of essential nutrients since plants readily take up nutrients in excess of their immediate requirements (Lambers *et al.* 1998). There have been attempts to use foliar N/P quotients to assess which of these two nutrients is more limiting but this is unlikely to be a simple matter. For example, Proctor *et al.* (2000) found a 3.5-fold variation in N/P quotients in 36 tree species from a forest on graywacke in the Philippines.

The nutrient requirement may vary with the supply of other nutrients and it is of relevance that the concern about Ca losses during logging operations (Nykvist, 1998) has failed to recognise that reasonably productive forest can grow on soils with very low Ca concentrations (Proctor *et al.*, 1983). Ca may only be required as a micronutrient if the soil solution concentrations of other potentially toxic elements, for example Mg, are low (Wyn Jones and Lunt, 1967).

Nutrient addition experiments

A direct approach to the question of nutrient requirements comes from the nutrient addition experiments which have proliferated over the last 15 years or so.

POT AND SMALL-SCALE FIELD EXPERIMENTS

Denslow *et al.* (1987) grew six species of shrubs (*Piper* and *Miconia* species, including pioneers and shade bearers) and one large pioneer herb (*Phytolacca rivinioides*) in a Costa Rican lowland rainforest soil (probably an inceptisol) developed on basalt. The *Phytolacca* responded positively to P (three-fold increase in dry matter production), spectacularly so to a 'complete nutrient' addition (*c.* 40-fold increase) but not to N alone (*c.* 20% decrease). The shrubs showed no significant response to P or N+P fertilisation but had a *c.* two- to five-fold increase in growth on a 'complete nutrient' addition. Concentrations of N or P in leaves were always higher in the +N, +P, or +NP treatments than in the controls or 'complete nutrient additions'. Denslow *et al.* (1987) pointed out that 'N and P enrichment of tissues in the absence of a strong growth response is further indication that the growth of these species is additionally limited by availability of other nutrients'. A further point of interest in this work is that the growth response to added nutrients was not correlated with pioneer or shade-bearing strategies. Although the response of the herbaceous *Phytolacca* was characteristic of the ruderal (fast-growing, light-demanding weedy herb) growth strategy (Grime, 1977; Chapin, 1980) the light-demanding pioneer shrubs did not respond more strongly to nutrients than did the shade-bearing species. Several workers have observed no response to P fertilisation in pot experiments (for example, Turner, Brown and Newton, 1993, and Burslem, Grubb and Turner, 1995).

Sometimes there has been a large response of tree seedlings to added nutrients, for example, Nussbaum *et al.* (1995). They added a mixed NPK fertiliser to a badly degraded and nutrient-poor subsoil from a log landing in Sabah. There was a positive response by all species (the pioneers *Macaranga hypoleuca*, and *M. gigantea*, and the shade-bearing dipterocarps *Dryobalanops lanceolata*, and *Shorea leprosula*) (Figure 16.6). Burslem, Turner and Grubb (1994) found that the non-mycorrhizal seedlings of the early successional shrub *Melastoma malabathricum* showed, on a Singaporean infertile ultisol, a primary limitation by P; under conditions of increased P supply, any other macronutrient and a mixture of micronutrients became limiting to plant growth.

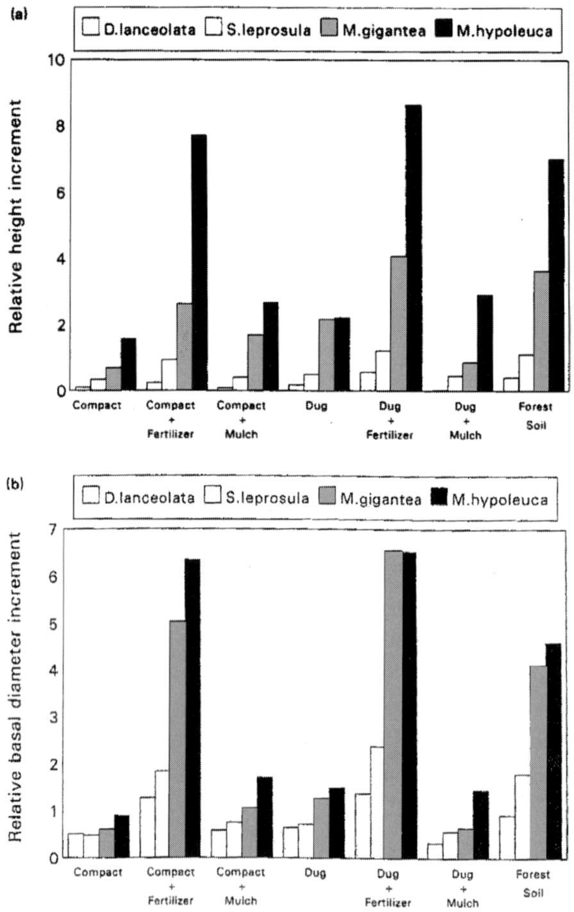

Figure 16.6 Mean growth of seedlings of four tree species (*Dryobalanops lanceolata, Shorea leprosula, Macaranga gigantea* and *Macaranga hypoleuca*) six months after planting in each of seven soil treatments ($n = 12$). (After Nussbaum *et al.*, 1995.)

Burslem *et al.* (1995) went on to test if the availability of P limits the growth of mycorrhizal shade-tolerant tree seedlings when grown in the same infertile ultisol as that used in the *Melastoma* bioassays. The seedlings used in the experiments were *Antidesma cuspidatum, Calophyllum tetrapterum, Dipterocarpus kunstleri,* and *Garcinia scortechinii*. All four show some shade tolerance as seedlings. Only seedlings of *Antidesma* showed increased growth response to nutrient supply and a later experiment confirmed that the limiting nutrient was not P but likely to be Mg (Burslem, Grubb and Turner, 1996). In the same experiment the dipterocarps *Hopea griffithii* and *Vatica maingayi* showed no statistically significant response to Mg addition although earlier work by Burslem *et al.* (1994) had suggested that Mg may limit the establishment of dipterocarp species in the ultisol from Bukit Timah, Singapore. The evidence of limitation by Mg^{2+} is supported by the seedling work of Gunatilleke *et al.* (1997) in Sri Lanka and brings to mind the suggestion of Baillie and Ashton (1983) that this element may be an important component of rainforest tree species composition.

Fetcher *et al.* (1996) demonstrated positive responses to N and P addition by two pioneer species planted in hurricane gaps in Puerto Rico. The two non-pioneer species they planted responded to N but not to P. The authors commented that neither pioneer nor non-pioneer species appeared to be particularly well adapted to colonise exposed parent material, for growth was very slow in the absence of added nutrients: this is a surprising result since one would expect pioneers to be adapted to gaps. The explanation may be that other factors such as compact subsoil or active erosion are limiting recolonisation (see also Dalling and Tanner, 1995). Ferns, including *Gleichenia* spp., seem to be much more widespread colonisers of landslides and road sides in Puerto Rico (L. A. Bruijnzeel, pers. comm.).

PLOT EXPERIMENTS

Other workers, mainly in montane forests, have fertilised forest plots and measured the girth increments of the trees and in some cases the mass and chemical composition of the litterfall. The montane forest fertilisation work has been critically reviewed by Tanner *et al.* (1998) who have collated a large number of instances of individual positive responses of tree growth to added N or P or both. Not all species in the system may be nutrient limited and even within a species, different individuals may be limited because of light and water rather than by nutrients. The only published nutrient addition work relating to lowland rainforest is that of Mirmanto *et al.* (1999) in central Kalimantan. They found during five years a significant increase in litterfall but only in the +NP treatment, a significant increase in P cycled (again only in the +NP treatment) but no significant effects on overall mean girth increment. The contrasting results of Mirmanto *et al.* (1999) on girth increment may reflect a genuine difference caused by the lowland forest compared with the montane forests investigated by other workers. It is possible that the relatively short time between measurements (five years) has been insufficient for significant effects to occur although one might expect a much more rapid response in the lowland rainforest compared with the montane rainforest in view of the higher temperatures and faster process rates. However, some of these processes, such as higher respiratory rates and potentially enhanced competition, may have negative impacts on growth.

ROOT INGROWTH EXPERIMENTS

Fine roots often proliferate in nutrient-rich microsites (Raich *et al.*, 1994) and there have been several attempts to assess nutrient limitation from measurements of root growth in mesh bags placed on or in the soil and containing inert media to which nutrients have been added. The best known of these experiments was by Cuevas and Medina (1988) who found no response to major nutrient addition in roots in growth-bag experiments made in short and tall heath forest on podzols and lowland evergreen rainforest sites on clayey soils in the Venezuelan Amazon, despite the fact that both soil types are among the least fertile of any soils in the tropics. They did find a positive response to $+NH_4Cl$ in their tall heath forest, and to the +P and $+CaCl_2$ in their lowland evergreen rainforest but, curiously, only when the bags were placed on the soil surface, not when they were in the soil. Raich *et al.* (1994) found a positive response to N addition in a *Metrosideros* -dominated forest at 1175 m altitude in Hawaii and were able to calibrate their results with those of a fertiliser experiment at the same site. In Central Amazonia, Luizão (1994) used the technique in two types of lowland heath forest on very acid sandy soils where there was a positive response to added $CaCO_3$ and $CaCl_2$, and in lowland evergreen rainforest where there was no statistically significant response to nutrient addition (despite the fact that the soil was extremely infertile). Proctor (1999), in Central Kalimantan, compared root ingrowth in two types of heath forests on podsols, a lowland evergreen rainforest (LERF1) which shared some intermediate characterisics with those of heath forests but which was on an ultisol, and a more widespread type of lowland evergreen rainforest (LERF2) (which resembled those plots in Mirmanto *et al.*'s (1999) plot fertilisation experiment). The results were very variable and he found no significant effects of N, P, or N+ P in the heath forests, a significant positive effect of P and N + P fertilisation in LERF1 and no significant effects in LERF2. The results agree with those of Mirmanto *et al.* (1999) who found no positive growth effect of P addition in the plot fertilisation experiment just discussed. It should be mentioned that the theoretical basis of root-ingrowth experiments needs questioning since roots may not necessarily proliferate in nutrient-rich sites and the plant may allocate its resources to root growth in parts of the soil where nutrients are less readily obtained.

PLANT COMPETITION

The effect of root competition on nutrient uptake was demonstrated by Coomes (1995) working in heath forests at La Esmeralda, southern Venezuela. He dug trenches between saplings and adult trees and found that there was an increase in growth rate, a 9.5% increase in foliar N concentration and a 17% increase in P concentration in the saplings after trenching. The application of N fertiliser to trenched saplings in gaps led to further increases in growth.

Yap (1998) trenched with and without fertiliser application treatments and effectively provided evidence of root competition in the logged-over forests in his experiments in Sabah. Fertiliser and trenching treatments had taller and bigger seedlings than the smaller-sized seedlings of the control or the fertiliser-only treatment. Competition could depress seedling growth, as shown in a study on the importance of nutrient limitation, soil moisture and site stability on regenerating species on landslide slopes by Dalling and Tanner (1995) in Jamaica. Nutrients limited growth of all three of their experimental species on open landslides but by contrast growth of the same species in the understorey of the adjacent forest (which had more nutrient-rich soils) was limited by photosynthetically-active radiation. In some cases they actually showed reduced seedling growth following fertiliser application in the understorey which they attributed to the effects of shade and root competition from mature trees.

PHOTOSYNTHETIC RESPONSES TO N AND P SUPPLY

There has been much work involving physiological experiments on tropical tree leaves and their photosynthetic responses to increases in N and P concentrations (e.g. Cromer *et al.*, 1993). Usually a positive response to both elements has been found which has continued until they cease to be limiting. Physiological work has established the strong correlations that exist among species between photosynthesis and leaf N concentrations (Lüttge, 1997). Reich *et al.* (1994) examined this relationship in plants from five habitat types at San Carlos in Venezuelan Amazonia (cf. Figure 16.4): cultivated ground, early secondary successional and primary lowland evergreen rainforest, and large and small stature heath forest. The authors showed that the relationship between photosynthetic capacity and leaf N varies in ecologically patterned ways: photosynthesis increased with increasing foliar N concentrations most markedly in the species from the disturbed sites; the species from the heath forest sites showed an intermediate response; and those from lowland evergreen rainforest showed the least photosynthetic response to increasing leaf N.

CONCLUSIONS ON EXPERIMENTAL WORK

Ingrowth experiments do offer an approach to assessing nutrient limitation but the results in any one experiment have been very variable, and sometimes difficult to interpret.

The interpretation of nutrient addition experiments in pots and on a small scale in the field is not straightforward for a number of reasons. First, forests are highly heterogeneous not only in soil conditions but also in light and water supply. Second, trees from infertile sites may not respond to added nutrients by increased growth rates because they are naturally slow-growing. Third, carefully controlled experiments with large tree species are not practicable unless seedlings are used and the nutrient requirements may well change during the life of the tree. Fourth, nutrients may only become limiting when there is adequate water and light which implies that, in theory, nutrients could limit the growth of one individual of a species but not that of another. There is some evidence (Burslem *et al.*, 1996) that rainforest seedlings (from Singapore) may be limited by nutrient shortage even when they are in deep shade.

Plot experiments are superficially attractive in that they provide a direct test for hypotheses on limiting nutrients at the ecosystem level. In practice they are expensive, labour intensive, do not yield meaningful results for several years, may be difficult to interpret in species-rich forests, and may alter the vegetation being studied.

The work on the relationship between photosynthesis and foliar nutrient concentrations (e.g. Reich *et al.*, 1994) needs comment: the assumed limiting-N relationship with photosynthesis may be less clear when leaf N is expressed on an area rather than on a mass basis (Reich and Walters, 1994) and there is no guarantee of correspondence between actual soil nutrient supply and the composition of culture solutions used in the experiments. These comments imply that the photosynthesis work falls short of strongly supporting arguments on which nutrients might be limiting in the field. The same applies to work on respiration, for example that by Raaimakers *et al.* (1995) in Guyana and Meir, Grace and Miranda (2001) in Brazil, who showed that very low leaf P concentrations in lowland rainforest tree leaves seemed to constrain leaf respiration more strongly than N.

The results from all these experiments leave one with the unsatisfactory conclusion that under some conditions some rainforest species at some stages of their growth will respond to certain nutrient additions. An equally unhelpful conclusion applies to plot experiments if we adopt the definition of nutrient limitation used by Tanner *et al.* (1998): 'limitation is shown if the rate of an ecosystem process is increased by the addition of that nutrient'.

Nutrient redistribution within the plant

It is widely accepted that one way of conserving nutrients under infertile conditions is their withdrawal from fresh leaves before abscission. The question of adaptive withdrawal is not fully resolved, however. Killingbeck (1996) has stressed the differences between the concepts of resorption efficiency (proportion of nutrient in a senesced leaf relative to that in a mature leaf) and resorption proficiency (the nutrient concentration in senesced leaves relative to the least concentrations which are biochemically possible). He claims that 'resorption is one of the most important of all strategies employed by plants to conserve nutrients and consequently influences processes as varied as competition, nutrient uptake, and productivity'. Because most nutrients are directly or indirectly dependent on nutrients available in plant tissues deposited as litter, and because falling leaves account for about 70% of all above-ground litter, it is clear that resorption is a key process in most if not all ecosystems. Killingbeck made the important point that not only is there a large inter-specific difference in resorption efficiency (Scott *et al.*, 1992) but also that for the same individual the resorption can vary hugely from year to year. Other workers (e.g. Veneklaas, 1990) have claimed that there is a general trend for nutrient withdrawal (particularly P) to be greater when soil supplies of this element are low. Because of problems with interpreting soil analyses, ensuring that adequate attention is paid to temporal aspects of sampling and to correct weighting of individual species' contributions, it is difficult to prove that this is the case. The recent work of Hafkenscheid (2000) lends little support to Veneklaas's (1990) premise but again shows that large amounts of nutrients are withdrawn before abscission.

TREES, MINERAL WEATHERING AND PEDOGENESIS?

For European coniferous forests Jongmans *et al.* (1997), van Breemen *et al.* (2000), and Landeweert *et al.* (2001) have provided evidence that the ectomycorrhizal mycelium is able to penetrate and colonise mineral microsites which are innaccessible to plant roots. They reported that fungal hyphae physically penetrate aluminosilicate minerals via strongly complexing acid exudates and create pores that may further enhance the weathering rates of minerals. They have found that weatherable minerals in all podzol surface soils and shallow granitic rock under European coniferous forests are criss-crossed by numerous tubular open pores, 3–10 μm in width. They suggest that these pores were formed by complex-forming, low molecular weight organic acids exuded by or formed in association with mycorrhizal fungi. Dissolved products could be translocated to the host plant roots, bypassing the soil solution which often has toxic Al^{3+} concentrations in podzols and bypassing competition for nutrient uptake by other organisms. The importance of this type of weathering for the nutrition of rainforest trees will depend on the distribution of roots at the weathering front and their nutrient contributions relative to those nearer the soil surface.

Si is well known as a key element in pedology and although it is not accepted as a generally essential nutrient for all plants, it does have a role in their biology (Epstein, 1994; Raven, 1983). The soil solution contains silicic acid, H_4SiO_4, at 0.1–0.6 mM-concentrations similar to those of K, Ca and other major nutrients. Plants can take up Si in large quantities. Leaf Si concentrations show a wide range e.g. 23–17900 $\mu g\ g^{-1}$ from leaves of different species growing on graywacke at the base of Mount Bloomfield, Palawan, the Philippines (Proctor *et al.*, 2000). Much higher values have been reported and high leaf Si concentrations seem to be a feature of the Dipterocarpaceae (Gautam-Basak and Proctor, 1984). The plant Si is mainly in the form of microscopic opaline particles (phytoliths) but some must be dissolved and biologically active. Kidd and Proctor (2001b) have shown that Si can ameliorate Al^{3+} toxicity in the temperate grass species *Holcus lanatus*, and that at high concentrations (>2500 μM) Si can act as a toxin in its own right.

Once the parent material's silica is dissolved it may be leached directly from the system, particularly if the weathering front is at depth. In shallower soils much of it may enter the biological cycle again where it may take part in secondary clay (kaolinite) formation, form pedogenic opaline minerals, or enter plants (and be cycled or immobilised). Biogenic silica varies in solubility and where it is more soluble than primary silicates it will suppress the rate of their weathering. Where the solubility of biogenic silica is lower than other soil silicates then the phytolith material will act as a Si sink and remove it from the biological cycle (Alexandre *et al.*, 1997, Kelly *et al.*, 1998).

In a comparison of soils under pasture and forest in Hawaii, Kelly *et al.* (1998) found that the changes in compartmentalisation of Si when forests are converted to pasture are substantial. The total Si content of both forests and pastures was similar but the ratio of oxalate-extractable Si to total Si was much higher under forest soils. The authors suggested that grasses, which produce more phytoliths per unit weight than trees, had stabilised a greater proportion of soil Si by producing phytoliths. This biological transformation of the more soluble Si to a less soluble form could have a marked effect on the weathering of soils and the total biogeochemical cycle of these systems.

Further work involving Si uptake by plants and mineral weathering has been made by Lucas *et al.* (1993, 1996). They observed two unusual features of Amazonian soils. First, the apparent stability of kaolinite in topsoil horizons where thermodynamic considerations predict gibbsite and, secondly, the frequent association of ferralitic soils (probably Oxisols, Soil Survey Staff, 1975) with podzols. These observations led to the investigation of why podzols or Oxisols develop in the same area from the same parent material. The plateau Oxisols soils of the Manaus area consist of a 3 to 8-m thick kaolinitic clay overlying a more gibbsitic 3-m thick nodular or bauxitic horizon. The kaolinites are renewed at each level of the profile, suffering a permanent disequilibrium by precipitation-dissolution. As the balance of these opposite processes preserves a kaolinitic material, this latter is in dynamic equilibrium. The Si input on the topsoil (40–50 kg ha^{-1} yr^{-1}) by vegetation (mainly in litterfall) is about four times greater than the Si leached out of the system (11 kg ha^{-1} yr^{-1}). The conclusion is that the dynamic equilibrium of the kaolinitic material in the upper horizons is sustained by biological activity. The surface of the Oxisol plateau soils remains perfectly flat as it slowly sinks through downward movement of Al (and other elements) which is conserved in gibbsite nodules lower down or in the production of new kaolinite. This is combined with a slow loss of Si in the vertically moving drainage water. In contrast, where valleys form, the drainage becomes inclined toward the axis of the valley. The drainage on the convex slopes becomes more rapid as one approaches the stream and some Al from the upper layers in organo-metallic compounds is now exported to rivers instead of deeper layers. As a result, the kaolin is destroyed faster than it is replaced, and the valley deepens. Podzols are formed which progress upslope at the expense of the Oxisols as the slope develops and become vegetated with increasingly small heath forest.

The Manaus case is a good example of how soil genesis in Equatorial areas can be greatly controlled by biological activity. The Oxisols soils are in a dynamic equilibrium sustained by the forest cycling of elements. The precipitation-dissolution kinetics of most secondary minerals are relatively fast. Thus, those reflect the present physico-chemical conditions in the profile level where they are found. The podzols result from an alteration in this dynamic, when the organo-metallic compounds are leached out of the system before their microbial degradation. This depends on the water dynamics in the system and on the type of microbiological activity.

CONCLUDING REMARKS

The one certainty among the topics discussed here is that rainforest mineral nutrition and hydrology are inextricably linked and that despite this there has been very limited collaboration between the two disciplines. The quest for generalisations based on existing data is problematic for the many reasons discussed. However, some progress has been made. For example, Bruijnzeel (1998) was able to draw useful conclusions about rainforest nutrient losses via drainage as a function of general soil type. Since I last reviewed this subject in 1987 there have been tangible improvements in techniques such as the use of stable isotopes, and a wide extension of experimental work, notably in the field of fertilisation. The latter is still in its infancy but one hopes will progress to the point where the mineral requirements of mature trees can be assessed. There has been a growing realisation of the importance of formerly underconsidered factors such as the influence of within-stream processes on stream nutrient composition, the redistribution of nutrients within plants, the role of plant competition, the biological involvement in Si movements, and the role of H^+ ions in heath forests and montane forests.

Scott *et al.* (1992) suggested that the rapid cycling of nutrients by litterfall, throughfall, and decomposition was the means by which the trees could be well supplied with nutrients on nutrient-poor soils. Burghouts *et al.* (1998) and Newbery *et al.* (1997) have made similar suggestions but the vital question remains as to what extent the trees are obtaining nutrients from below, say, 1 m soil depth or are tapping groundwater of unknown nutrient concentration. The 'La Selva-Maracá Anomaly' (Table 16.2) might be explained if the trees were using one or both of such nutrient sources. We have no evidence at present that this is the case but the solution to this important question about a key link between rainforest hydrology and mineral nutrients might be resolved by using stable isotope techniques, sapflow gauges in roots, systematic sampling with depth of water contents and composition all the way down to the weathering front. There have been some commendable attempts to model nutrient and moisture cycling in tropical forests, notably by Boersma, van Schaik and Hogeweg (1991) and Noij *et al.* (1993) but at present these are not generally accepted because of the complexity of the real-life situation. Such models should be developed using the improving data sets.

The view of Bruijnzeel (1998) that work on rainforest hydrology and nutrient cycling should be carried on in detail at a few selected sites has been the thrust of this chapter since some key issues can only be resolved by teams of ecologists, hydrologists, tree physiologists and pedologists working in collaboration. There is still a need for more superficial studies, however, since they are necessary to indicate where more substantial studies might be most rewarding. One is mindful of Hobbs (2001): "The joy of being an ecologist is that ecosystems are infinitely variable and do differ in different parts of the world. We need generalised frameworks to organise our thinking, but we also need the detail because that's our bread and butter, and it's what we are trying to conserve in our rapidly changing world".

References

Adámoli, J., Sennhauser, E., Acero, J. M. and Rescia, A. (1991). Stress and disturbance: vegetation dynamics in the dry Chaco region of Argentina. *Savanna Ecology and Management. Australian Perspectives and Intercontinental Comparisons*, ed. P. Werner, pp. 147–156. Oxford: Blackwell Scientific Publications.

Alexander, I. J., Norani, A. and Lee, S. S. (1992). The role of mycorrhizas in the regeneration of some Malaysian forest trees. *Philosophical Transactions of the Royal Society of London B*, 335, 379–388.

Alexandre, A., Meunier, J.-D., Colin, F. and Koud, J. M. (1997). Plant impact on the biogeochemical cycle of silicon and related weathering processes. *Geochimica et Cosmochimica Acta*, 61, 677–682.

Anonymous (1972). Regenwasseranalysen aus Zentralamazonien, ausgefuhrt in Manaus, Amazonia, Brasilien, von Dr. Harald Ungemach. *Amazoniana*, 3, 186–198.

Antibus, R. K. *et al.* (1997). Root surface phosphatase activities and uptake of 32P-labelled inositol phosphate in field collected gray birch and red maple roots. *Mycorrhiza*, 7, 39–46.

Asbury, A. E., McDowell, R., Trinidad-Pizarro, R. and Berrios, S. (1994). Solute deposition from cloud water to the canopy of a Puerto Rican montane forest. *Atmospheric Environment*, 28, 1773–1780.

Ashton, P. S. (1976). Mixed dipterocarp forest and its variation with habitat in the Malayan lowlands: a re-evaluation at Pasoh. *Malaysian Forester* 39, 56–72.

Austin, M. P., and Greig-Smith, P. (1968). The application of quantitative methods to vegetation survey. II. Some methodological problems of data from rainforest. *Journal of Ecology*, 56, 827–844.

Bach, L. B., Wierenga, P. J., and Ward, T. J. (1986). Estimation of the Philip infiltration parameters from rainfall simulation data. *Soil Science Society of America Journal*, 50, 1319–1323.

Baillie, I. C. (1989). Soil characteristics and classification in relation to the mineral nutrition of tropical wooded ecosystems. In *Mineral Nutrients in Tropical Forest Ecosystems* ed. J. Proctor, pp., 15–26. Oxford: Blackwell Scientific Publications.

Baillie, I. C. and Ashton, P. S. (1983). Some aspects of the nutrient cycle in mixed dipterocarp forests in Sarawak. In *Tropical Rain Forest: Ecology and Management*, ed. S. L. Sutton, T. C. Whitmore, and A. C. Chadwick, pp. 347–356. Oxford: Blackwell Scientific Publications.

Baillie, I. C. (1996). Soils of the humid tropics. In *The Tropical Rain Forest* (ed. P. W. Richards), pp. 256–286. Cambridge: Cambridge University Press.

Barcelo, J., Guevara, P., and Poschenrieder, C. (1993). Silicon amelioration of aluminium toxicity in teosinte (*Zea mays* L. ssp. *mexicana*. *Plant and Soil*, 154, 249–255.

Boersma, M., van Schaik, C. P. and Hogeweg, P. (1991). Nutrient gradients and spatial structure in tropical forests: a model study. *Ecological Modelling*, 55, 219–240.

Blum, J. D., Erel, Y. and Brown, K. (1994). $^{87}Sr/^{86}Sr$ ratios of Sierra Nevada stream waters: implications for relative mineral weathering rates. *Geochimica and Cosmochimica Acta*, 58, 5019–5025.

Bolan, N. S., Naidu, R., Mahimairaja, S., and Baskaran, S. (1994). Influence of low-molecular weight organic acids on the solubilization of phosphates. *Biology and Fertility of Soils*, 18, 311–319.

Brandes, J. A., McClain, M. E., and Pimentel, T. P. (1996). ^{15}N evidence for the origin and cycling of inorganic nitrogen in a small Amazonian catchment. *Biogeochemistry*, 34, 45–56.

Brasell, H. M., and Gilmour, D. A. (1980). The cation composition of precipitation at four sites in far north Queensland. *Australian Journal of Ecology*, 5, 397–405.

Brinkmann, W. L. F. (1983). Nutrient balance of a central Amazonian rainforest: comparison of natural and man-managed ecosystems. *Hydrology of Humid Tropical Regions with Particular Reference to the Hydrological Effects of Agriculture and Forestry Practice*, IAHS Publication No. 140, pp. 153–163.

Brinkmann, W. L. F. (1985). Studies on hydrogeochemistry of a tropical lowland forest system. *Geojournal*, 11, 89–101.

Brouwer (1996). *Nutrient Cycling in Pristine and Logged Tropical Rain Forest. A Study in Guyana*. The Netherlands: University of Utrecht.

Bruijnzeel, L. A. (1989). Nutrient content of bulk precipitation in south-central Java, Indonesia. *Journal of Tropical Ecology*, 5, 187–202.

Bruijnzeel, L. A. (1989b). Nutrient cycling in moist tropical forests: the hydrological framework. In *Mineral Nutrients in Tropical Forest and Savanna Ecosystems*, ed. J. Proctor, pp. 383–415. Oxford: Blackwell Scientific Publications.

Bruijnzeel, L. A. (1991). Nutrient input-output budgets of tropical forest ecosystems: a review. *Journal of Tropical Ecology*, 7: 1–23.

Bruijnzeel, L. A. (1998). Soil chemical changes after tropical forest disturbance and conversion: the hydrological perspective. In *Soils of Tropical Forest Ecosystems*, ed. A. Schulte and D. Ruhiyat, pp. 45–61. Berlin: Springer.

Burgess, S. S. O., Dubinsky, E. and Dawson, T. E. (2001). The role of fog in the ecology and water relations of coast redwood Sequoia sempervirens. In R. S. Schemenauer and H. A. Puxbaum (Eds). Proceedings of the 2nd International Conference on Fog and Fog Collection, 15–20 July 2001, St. John's, Canada, pp. 121–124.

Burghouts, T. B. A., Van Stralen, N. M. and Bruijnzeel, L. A. (1998). Spatial heterogeneity of element and litter turnover in a Bornean rainforest. *Journal of Tropical Ecology*, 14, 477–506.

Burnham, C. P. (1989). Pedological processes and nutrient supply from parent material in tropical soils. In *Mineral Nutrients in Tropical Forest and Savanna Ecosystems*, ed. J. Proctor, pp. 27–41. Oxford: Blackwell Scientific Publications.

Burslem, D. F. R. P., Grubb, P. J., and Turner, I. M. (1995). Responses to nutrient addition among shade-tolerant tree seedlings of lowland tropical forest in Singapore. *Journal of Ecology* 83, 113–122.

Burslem, D. F. R. P., Grubb, P. J., and Turner, I. M. (1996). Responses to simulated drought and elevated nutrient supply among shade-tolerant tree seedlings of lowland tropical forest in Singapore. *Biotropica*, 28, 636–648.

Burslem, D. F. R. P., Turner, I. M., and Grubb, P. J. (1994). Mineral nutrient status of coastal hill dipterocarp forest and *Adinandra* belukar in Singapore: bioassays of nutrient limitation. *Journal of Tropical Ecology*, 10, 579–599.

Cavelier, J., Jaramillo, M., Solis, D., and de Léon., D. (1997). Water balance and nutrient inputs in bulk precipitation in tropical montane cloud forest In Panama. *Journal of Hydrology*, 193, 83–96.

Chalot, M. and Brun, A. (1998). Physiology of organic nitrogen acquisition by ectomycorrhizal fungi and ectomycorrhizas. *FEMS Microbiological Reviews*, 2, 21–44.

Chandler, G. (1985). Mineralization and nitrification in three Malaysian forest soils. *Soil Biology and Biochemistry*, 17, 347–353.

Chapin, F. S. (1980). The mineral nutrition of wild plants. *Annual Review of Ecology and Systematics*, 11, 233–260.

Chestnut, T. J. and McDowell, W. H. (2000). C and N dynamics in the riparian and hyporheic zones of a tropical stream, Luquillo Mountains, Puerto Rico. *Journal of the North American Benthological Society*, 19, 199–214.

Clark, K. L., Nadkarni, N. M., Schaeffer, D. and Gholz, H. L. (1998). Atmospheric deposition and net retention of ions by the canopy in a tropical montane forest, Monteverde, Costa Rica. *Journal of Tropical Ecology*, 14, 27–45.

Clarkson (1966). Aluminium tolerance in species within the genus *Agrostis*. *Journal of Ecology*, 53, 427–435.

Coleman, D. C., Reid, C. P. P. and Cole, C. V. (1983). Biological strategies of nutrient cycling in soil systems. *Advances in Ecological Research*, 13, 1–55.

Coomes, D. A., (1995). *The Effects of Root Competition on Saplings and Seedlings in Amazonian Caatinga Forest in Southern Venezuela*. Ph.D. thesis, University of Cambridge, U.K.

Courchesne, F., Savoie, S., and Dufresne, A. (1995). Effects of air-drying on the measurement of soil pH in acidic forest soils of Quebec, Canada. *Soil Science*, 160, 56–67.

Crews, T. E., Kitayama, K., Fownes, J. H., Riley, R. H., Herbert, D. A., Mueller-Dombois, D., and Vitousek, P. M. (1995). Changes in soil phosphorus and ecosystem dynamics across a long chronosequence in Hawaii. *Ecology*, 76, 1407–1424.

Crocker, R. L. and Major, J. (1955). Soil development in relation to vegetation and surface age, Glacier Bay, Alaska. *Journal of Ecology*, 43, 427–448.

Cromer, R. N., Cameron, D. N., Rance, S. J., Ryan, P., and Brown, M. (1993a). Response to nutrients in *Eucalyptus grandis*. 1. Biomass accumulation. *Forest Ecology and Management*, 62, 211–230.

Cromer, R. N., Cameron, D. N., Rance, S. J., Ryan, P., and Brown, M. (1993b). Response to nutrients in *Eucalyptus grandis*. 2. Nitrogen accumulation. *Forest Ecology and Management*, 62, 231–243.

Cromer, R. N., Kriedemann, P. E., Sands, P. J., and Stewart, L. G. (1993c). Leaf growth and photosynthetic response to nitrogen and phosphorus in seedling trees of *Gmelina arborea*. *Australian Journal of Plant Physiology*, 20, 83–98.

Cuevas, E. and Medina, E. (1988). Nutrient dynamics within Amazonian forest. II Fine root growth, nutrient availability and leaf litter decomposition. *Oecologia*, 76, 222–235.

Dalling, J. W., and Tanner, E. V. J. (1995). An experimental study of regeneration on landslides in montane rainforest in Jamaica. *Journal of Ecology*, 83, 55–64.

Denslow, J. S. (1980). Patterns of plant species diversity during succession under different disturbance regimes. *Oecologia*, 46, 18–21.

Denslow, J. S., Vitousek, P. M., and Schultz, J. C. (1987). Bioassays of nutrient limitation in a tropical rainforest soil. *Oecologia*, 74, 370–376.

De Rham, P. (1970). L'azote dans quelques fôrets, savanes et terrains de culture d'Afrique tropicale humide. *Veröffentlichen des Geobotanischen Institutes der Eidgenössische Technische Hochschule, Stiftung Rübel in Zurich*, 45.

Drever, J. I. (1994). The effect of land plants on weathering rates of silicate minerals. *Geochimica and Cosmochimica Acta*, 58, 2325–2332.

Drew, M. C. and Nye, P. H. (1969). The supply of nutrient ions by diffusion to plant roots in the soil. II. The effect of root hairs on the uptake of potassium by roots of rye-grass (*Lolium multiflorum*). *Plant and Soil*, 31, 407–424.

Epstein, E. (1994). The anomaly of silicon in plant biology. *Proceedings of the National Academy of Sciences*, 91, 11–17.

Evans, J. R. (1989). Photosynthesis and nitrogen relationships in leaves of C_3 plants. *Oecologia*, 78, 9–19.

Fetcher, N., Haines, B. L., Cordero, R. A., Lodge, D. J., Walker, L. R., Fernandez, D. S., and Lawrence, W. T. (1996). Responses of tropical plants to nutrients and light on a landslide in Puerto Rico. *Journal of Ecology*, 84, 331–341.

Fitter, A. H. (1986). Effect of benomyl on leaf phosphorus concentration in alpine grasslands: a test of mycorrhizal benefit. *New Phytologist*, 103, 767–776.

Fitter, A. H. and Hay, R. K. M. (2002). *Environmental Physiology of Plants*. 3rd edition. London: Academic Press.

Forti, M. C. and Neal, C. (1992). Hydrochemical cycles in tropical rainforests: an overview with emphasis on Central Amazonia. *Journal of Hydrology*, 134, 103–115.

Franco, W. and Dezzeo, N. (1994). Soils and soil water regime in the terra firme-caatinga forest complex near San Carlos de Rio Negro, State of Amazonia, Venezuela. *Interciencia*, 19, 305–316.

Franken, W. and Leopoldo, P. R. (1984). Hydrology of catchment areas of Central-Amazonian forest streams. In *The Amazon.Limnology and Landscape Ecology of a Mighty Tropical River and its Basin*, ed. H. Sioli, pp. 501–509. Dordrecht: W. Junk.

Freeden, A. L., Griffin, K, and Field, C. B. (1991). Effects of light quantity and quality and soil nitrogen status on nitrate reductase activity in rain forest species of the genus *Piper*. *Oecologia*, 85, 441–446.

Gautam-Basak, M., and Proctor, J. (1983). Micronutrients, silicon and ash in leaf litterfall from forests in Gunung Mulu National Park, Sarawak. *Malaysian Forester* 46, 224–232.

Gilman, G. P., Sinclair, D. F., Knowlton, R., and Keys, M. G., 1985. The effect on some soil chemical properties of the selective logging of a north Queensland rainforest. *Forest Ecology and Management*, 12, 195–214.

Godbold, D. L., Fritz, E., and Hutterman, A. (1988). Aluminium toxicity and forest decline. *Proceedings of the National Academy of Sciences* U.S.A., 85, 3888–3892.

Green, J. J. (1992). *Fine Root Dynamics in a Bornean Rain Forest*. Unpublished Ph.D. thesis, University of Stirling, U.K.

Grimaldi, C., Luizão, R. C. and Chauvel, A. (1992). Hydric and biogeochemical pulses in the functioning of soils in Amazonian forest ecosystems. In *Second International Symposium on Environmental Studies of Tropical Rainforests- Forest 92 Proceedings Volume*, ed. Anon., pp. 170–171. Rio de Janeiro: Brazilian Society for the Environment.

Grime, J. P. (1977). Evidence for the existence of three primary strategies in plants and its relevance to ecological and evolutionary theory. *American Naturalist*, 111, 1169–1194.

Gunatilleke, C. V. S., Gunatilleke, I. A. U. N., Perera, G. A. D., Burslem, D. F. R. P., Ashton, P. M. S., and Ashton, P. S. (1997). Responses to nutrient addition among seedlings of eight closely related species of *Shorea* in Sri Lanka. *Journal of Ecology*, 85, 301–311.

Hafkenscheid, L. L. R. (2000). *Hydrology and Biogeochemistry of Tropical Montane Rain Forests of Contrasting Stature in the Blue Mountains, Jamaica*. Amsterdam: Free University.

Handley, L. and Scrimgeour, C. M. (1997). Terrestrial plant ecology and $\delta^{15}N$ natural abundance: thepresent limits to interpretation for uncultivated systems with original data from a Scottish old field. *Advances in Ecological Research*, 27, 133–212.

Harley, J. and Smith, S. E. (1983). *Mycorrhizal Symbiosis*. London: Academic Press.

Heaney, A. and Proctor, J. (1989). Chemical elements in litter in forests on Volcán Barva, Costa Rica. In *Mineral Nutrients in Tropical Forest and Savanna Ecosystems*, ed. J. Proctor, pp. 255–271. Oxford: Blackwell Scientific Publications.

Hedley, M. J., Stewart, J. W. B., and Chauhan, B. S. (1982). Changes in inorganic and organic soil phosphorus fractions induced by cultivation practices and by laboratory incubations. *Soil Science Society of America Journal*, 46, 970–976.

Hendry, C. D., Berish, C. W. and Edgerton, E. S. (1984). Precipitation chemistry at Turrialba, Costa Rica. *Water Resources Research*, 20, 1677–1684.

Herrera, R., Merida, T., Stark, N. *et al.* (1978). Direct phosphorus transfer from leaf litter to roots. *Naturwissenschaften*, 65, 208–209.

Hobbs, R. J. (2001). From our southern correspondent. *Bulletin of the British Ecological Society*, 32, 32–33.

Hofstede, R. M., Wolf, J. H. D., and Benzing, D. H. (1993) Epiphytic biomass and nutrient status of a Colombian upper montane rain forest. *Selbyana*, 14, 37–45.

Huttel, C. (1975). Root distribution and biomass in three Ivory Coast rainforest plots. In *Tropical Ecological Systems* ed. F. B. Golley and E. Medina, pp. 123–130. Berlin: Springer.

Jacobs, M. (1988). *The Tropical Rain Forest. A First Encounter*. London: Springer.

Jongmans, A. G., Breemen, N. V., Lundstrom, U., Finlay, R. D., van Hees, P. A. W., Giesler, R., Melkereud, P.-A., Olsson, M., Srinavasan, M., and Unestam, T. (1997). Rock-eating fungi: a true case of mineral plant nutrition? *Nature*, 389, 682–683.

Jones, J. M. and Richards, B. N. (1977). Effect of reforestation on turnover of N-15 labelled ammonium plus nitrate in relation to changes in soil microflora. *Soil Biology and Biochemistry*, 9, 383–392.

Jordan, C. F. and Herrera, R. (1981). Tropical rainforests: are nutrients really critical? *American Naturalist*, 117, 167–180.

Jordan, C. F., Kline, J. R., and Saszcer, D. S. (1972). Relative stability of mineral cycles in forest ecosystems. *American Naturalist*, 106, 237–254.

Kelly, E. F., Chadwick, O. A. and Hilsinki, T. E. (1998). The effect of plants on mineral weathering. *Biogeochemistry*, 42, 21–53.

Khiewtam, R. S. and Ramakrishnan, P. S. (1993). Litter and fine root dynamics of a relict sacred grove forest at Cherrapunji in north-eastern India. *Forest Ecology and Management*, 60, 327–344.

Kidd, P. S. and Proctor, J. (2000a). The growth response of ecotypes of *Holcus lanatus* from different soil types in northwestern Europe to phenolic acids. *Plant Biology* 2, 335–343.

Kidd, P. S. and Proctor, J. (2000b). Effects of aluminium on the growth and mineral composition of *Betula pendula* Roth. *Journal of Experimental Botany*, 51, 1057–1066.

Kidd, P. S. and Proctor, J. (2001a). Why plants grow poorly on very acid soils: are ecologists missing the obvious? *Journal of Experimental Botany*, 52, 791–799.

Kidd, P. S. and Proctor, J. (2001b). The growth response of ecotypes of *Holcus lanatus* L. from different soil types in northwestern Europe to aluminium and silicon. *Fresenius Environmental Bulletin*, 10, 590–594.

Killingbeck, K. T. (1996). Nutrients in senesced leaves: keys to the search for potential resorption and resorption proficiency. *Ecology*, 77, 1716–1727.

King, H. B. and Yang, B. Y. (1984). Precipitation and stream water chemistry in Pi-Lu-Chi watersheds. January, 1981–December, 1982. *Bulletin of the Taiwan Forestry Research Institute, Taipei*. No. 427. (In Chinese with English summary).

Kinraide, T. B. (1993). Aluminium enhancement of plant growth in acid rooting media. A case of reciprocal alleviation of toxicity by two toxic cations. *Physiologia Plantarum*, 88, 619–625.

Klinge, H. (1973). Root mass estimation in lowland tropical rainforests of central Amazonia, Brazil. I. Fine root masses of a pale yellow latosol and a giant humus podzol. *Tropical Ecology*, 14, 29–38.

Klinge, H. (1976). Bilanzierung von Hauptnahrstoffen in okosystem tropischer Regenwald (Manaus)-vorlaufige Daten. *Biogeographica*, 7, 59–76.

Klinge, R., Schmidt, J. and Fölster, H. (2001). Simulation of water drainage of a rainforest and forest conversion plots using a soil water model, *Journal of Hydrology*, 246, 82–95.

Kuiters, A. T. (1990). Role of phenolic substances from decomposing forest litter in plant-soil interactions. *Acta Botanica Neerlandica*, 39, 329–348.

Lamb, D. (1980). Soil nitrogen mineralization in a secondary forest succession. *Oecologia*, 47, 257–263.

Lambers,H., Chapin, F. S., and Pons, T. L. (1998). *Plant Physiolgical Ecology*. New York: Springer.

Landeweert, R., Hoffland, E., Finlay, R. D., Kuyper, T. W., and van Breemen, N. (2001). Linking plants to rocks: ectomycorrhizal fungi mobilize nutrients from minerals. *Trends in Ecology and Evolution*, 16, 248–254.

Laverman, A. M., Zoomer, H. R., Engelbrecht, D., Berg, M. P., van Straalen, N. M., van Verseveld, H. W., and Verhoef, H. A. (2000a). Soil layer specific variability in net nitrification and denitrification in an acid coniferous forest. *Biology and Fertility of Soils*, 32, 427–434.

Laverman, A. M., Zoomer, H. R., van Verseveld, H. W., and Verhoef, H. A. (2000b). Temporal and spatial variations of nitrogen transformations in an acid coniferous forest. *Soil Biology and Biochemistry*, 32, 1661–1670.

Laverman, A. M., Zoomer, H. R., and Verhoef, H. A. (2001). The effect of oxygen, pH and organic carbon on soil-layer specific denitrifying capacity in acid coniferous forest. *Soil Biology and Biochemistry*, 33, 683–687.

Lawson, G. W., Armstrong-Mensah, K. O., and Hall, J. B. (1970). A catena in tropical moist semi-deciduous forest near Kade, Ghana. *Journal of Ecology*, 58, 371–398.

Lee, S. S. and Alexander, I. J. (1994). The response of seedlings of two dipterocarp species to nutrient additions and ectomycorrhizal infection. *Plant and Soil*, 163, 299–306.

Lesack, L. F. W. (1993). Export of nutrients and major ionic solutes from a rainforest catchment in the Central Amazon basin. *Water Resources Research*, 29, 743–758.

Lloyd, P. S. and Pigott, C. D. (1967). The influence of soil conditions on the course of succession in southern England. *Journal of Ecology*, 55, 137–146.

Lucas, Y., Nahon, D., Cornu, S., and Eyrolle, F. (1996). Genèse et fonctionnement des sols en milieu équatorial. *Comptes Rendues de l' Académie des Sciences*, 322, lla, 1–16.

Luizão, F. J. (1996). *Ecological studies in Contrasting Forest Types in Central Amazonia*. Unpublished Ph.D. thesis, University of Stirling, U.K.

Luizão, F. J., Proctor, J., Thompson, J., Luizão, R. C. C., Marrs, R. H., Scott, D. A., and Viana, V. (1998). Rain forest on Maracá Island, Roraima, Brazil: soil and litter process response to artificial gaps. *Forest Ecology and Management*, 102, 291–303.

Luizão, R. C. C. (1994). *Soil Biological Studies in Contrasting Types of Vegetation in Central Amazonian Forests*. Unpublished Ph.D. thesis, University of Stirling, U.K.

Lüttge, U. (1997). *Physiological Ecology of Tropical Plants*. London: Springer.

Malmer, A. (1993). *Dynamics of Hydrology and Nutrient Losses as Response to Establishment of Forest Plantation. A Case Study on Tropical Rainforest land in Sabah, Malaysia*. Lund, Sweden: Department of Forest Ecology, Swedish University of Agricultural Sciences.

Manokaran, N. (1980). The nutrient contents of precipitation, throughfall and stemflow in a lowland tropical rainforest in peninsular Malaysia. *Malaysian Forester*, 43, 266–289.

Marrs, R. H., Proctor, J., Heaney, A., and Mountford, M. D. (1988). Changes in soils, nitrogen mineralization and nitrification along an altitudinal transect in tropical rainforest in Costa Rica. *Journal of Ecology*, 58, 297–299.

Marrs, R. H., Thompson, J., Scott, D., and Proctor, J. (1991). Nitrogen mineralization and nitrification in 'terra firme' forest and savanna soils on Ilha de Maracá, Roraima, Brazil. *Journal of Tropical Ecology*, 7, 123–137.

Marschner, H. (1991). Mechanism of adaptation of plants to acid soils. *Plant and Soil*, 134, 1–20.

Marschner, H. (1995). *Mineral Nutrition of Higher Plants*, 2nd edn. London: Academic Press.

Masunaga, T., Kubota, D., Hotta, M., Shinmura, Y., and Wakatsuki, T. (1998a). Distribution characteristics of mineral elements in trees of tropical rainforest, West Sumatra, Indonesia. In *Soils of Tropical Forest Ecosystems*, ed. A. Schulte and D. Ruhiyat, pp. 168–174. Berlin: Springer.

Masunaga, T., Kubota, D., Hotta, M., and Wakatsuki, T. (1997). Nutritional characteristics of mineral elements in leaves of tree species in tropical rainforest, west Sumatra, Indonesia. *Soil Science and Plant Nutrition*, 43, 405–418.

Masunaga, T., Kubota, D., Hotta, M., and Wakatsuki, T. (1998b). Nutritional characteristics of mineral elements in leaves of tree species in tropical rainforest, west Sumatra, Indonesia. *Soil Science and Plant Nutrition*, 44, 315–329.

McClain, M. E. and Elsenbeer, H. (2001). Terestrial inputs to Amazon streams and internal biogeochemical processing. In *The Biogeochemistry of the Amazon Basin and its Role in a Changing World*, ed. M. E. McClain, R. L. Victoria, and J. E. Richey, pp. 185–208. Oxford: Oxford University Press.

Meir, P., Grace, J., and Miranda, A. C. (2001). Leaf respiration in two tropical rainforests: constraints on physiology by phosphorus, nitrogen and temperature. *Functional Ecology*, 15, 378–387.

Mengel, K., and Arneke, W. W. (1982). Effect of potassium on the water potential, the pressure potential, the osmotic potential and cell elongation in leaves of *Phaseolus vulgaris*. *Physiologia Plantarum*, 54, 402–408.

Minshall, W. H. (1975). Stimulation of transpiration by nitrogenous materials. *Canadian Journal of Botany*, 53, 1259–1263.

Mirmanto, E., Proctor, J., Green, J., Nagy, L., and Suriantata. (1999). Effects of nitrogen and phosphorus fertilization in a lowland evergreen rainforest. *Philosophical Transactions of the Royal Society B*, 354, 1825–1829.

Moyersoen, B., Alexander, I. J., and Fitter, A. H. (1998). Phosphorus nutrition of ectomycorrhizal and arbuscular mycorrhizal tree seedlings from a lowland tropical rainforest in Korup National Park, Cameroon. *Journal of Tropical Ecology*, 14, 47–61.

Nadkarni, N. M. (1986). The nutritional effects of epiphytes on host trees with special reference to alteration of precipitation chemistry. *Selbyana*, 9, 44–51.

Nepstad, D. C., De Carvalho, C. R., Davison, E. A., Jipp, P. H., Lefebvre, P. A., Negreiros, G. H., da Silva, E. D., Stone, T. A., Trumbore, S. E., and Vieira, S. (1994). The role of deep roots in the hydrological and carbon cycles of Amazonian forests and pastures. *Nature*, 372, 666–669.

Newbery, D. M., Alexander, I. J., and Rother, J. A. (1997). Phosphorus dynamics in a lowland African rainforest; the influence of ectomycorrhizal trees. *Ecological Monographs*, 67, 367–409.

Noij, I. G. A. M., Janssen, B. H., Wesselink, L. G., *et al.* (1993). *Modelling Nutrient and Moisture Cycling in Tropical Forests*. Wageningen, The Netherlands: Tropenbos Foundation.

Nooren, C. A. M., van Breemen, N., Stoorvogel, J. J., *et al.* (1995). The role of earthworms in the formation of sandy surface soils in a tropical forest in Ivory Coast. *Geoderma*, 65, 135–148.

Nortcliff, S. and Thornes, J. B. (1989). Variations in soil nutrients in relation to soil moisture status in a tropical forested ecosystem. In *Mineral Nutrients in Tropical Forest and Savanna Ecosystems*, ed. J. Proctor, pp. 43–54. Oxford: Blackwell Scientific Publications.

Northup, R. R., Dahlgren, R. A. and Yu, Z. (1995). Intraspecific variation of conifer phenolic concentration on a marine terrace acidity gradient: a new interpretation. *Plant and Soil*, 171, 255–262.

Nussbaum, R., Anderson, J., and Spencer, T. (1995). Factors limiting the growth of indigenous tree seedlings planted on degraded rainforest soils in Sabah, Malaysia. *Forest Ecology and Management*, 74, 149–159.

Nye, P. H. (1961). Organic matter and nutrient cycles under moist tropical forest. *Plant and Soil*, 13, 334–346.

Nye, P. H. and Tinker, P. B. (1977). *Solute Movements in the Root-Soil System*. Oxford: Blackwell Scientific Publications.

Nykvist, N. (1998). Logging can cause a serious lack of calcium in tropical rainforest ecosystems: an example from Sabah, Malaysia. In *Soils of Tropical Forest Ecosystems*, ed. A. Schulte and D. Ruhiyat, pp. 87–91. Berlin: Springer.

Oliveira-Filho, A. T., Vilela, E. A., Carvalho, D. A., and Gavilanes, M. L. (1994). Effects of soils and topography on the distribution of tree species in a tropical riverine forest in south-eastern Brazil. *Journal of Tropical Ecology*, 10, 483–588.

Oliveira-Filho, A. T., Ratter, J. A. and Shepherd, G. J. (1990). Floristic composition and community structure of a central Brazilian gallery forest. *Flora*, 184, 103–117.

Parker, G. G. (1983). Throughfall and stemflow in the forest nutrient cycle. *Advances in Ecological Research*, 13, 57–133

Poels, R. L. H. (1987). *Soils, Water and Nutrients in a Forest Ecosystem in Suriname*. Wageningen, The Netherlands: Agricultural University.

Proctor, J. (1983). Tropical forest litterfall. I. Problems of data comparison. In *Tropical Rain Forest: Ecology and Management*, eds. S. L. Sutton, T. C. Whitmore and A. C. Chadwick, pp. 267–273. Oxford: Blackwell Scientific Publications.

Proctor, J. (1984). Tropical Forest litterfall. II. The data set. In *Tropical Rain Forest: The Leeds Symposium*, eds. A. C. Chadwick and S. L. Sutton, pp. 83–113. Leeds: Leeds Philosophical and Literary Society.

Proctor, J. (1987). Nutrient cycling in primary and old secondary rainforests. *Applied Geography*, 7, 135–152.

Proctor, J. (1995). Rainforests and their soils. In *Ecology, Conservation, and Management of South-East Asian Rainforests*, ed. R. B. Primack and T. E. Lovejoy, T. E. (eds) pp. 87–104. London: Yale University Press.

Proctor, J. (1999). Heath forests and acid soils. *Botanical Journal of Scotland*, 51, 1–14.

Proctor, J., Baker, A. J. M., Bruijnzeel, L. A., van Balgooy, M. M. J., Fairweather, G. M., and Madulid, D. A. (2000). Foliar chemistry and leaf herbivory on Mount Bloomfield, Palawan, Philippines. *Botanical Journal of Scotland*, 52, 79–89.

Raaimakers, D., Boot, R. G. A., Dijkstra, P., Pot, S., and Pons, T. (1995). Photosynthetic rates in relation to leaf phosphorus content in pioneer versus climax tropical rainforest trees. *Oecologia*, 102, 120–125.

Raich, J. W., Riley, R. H., and Vitousek, P. M. (1994). Use of root-ingrowth cores to assess nutrient limitations in forest ecosystems. *Canadian Journal of Forest Research*, 24, 2135–2138.

Ratter, J. A. (1980). Notes on the Vegetation of Fazenda Agua Limpa (Brasilia, DF, Brazil). Edinburgh: Royal Botanic Gardens.

Raven, J. A. (1983). The transport and function of silicon in plants. *Biological Reviews*, 58, 179–207.

Reich, P. B., Walters, M. B., Ellsworth, D. S. *et al.* (1994). Photosynthesis – nitrogen relations in Amazonian tree species. I. Patterns among species and communities. *Oecologia* 97, 62–72.

Rice, E. I. and Pancholy, S. K. (1973). Inhibition of nitrification by climax ecosystems II. Additional evidence and possible role of tannins. *American Journal of Botany*, 60, 691–702.

Riddoch, I., Lehto, T., and Grace, J. (1991). Photosynthesis of tropical tree seedlings in relation to light and nutrient supply. *New Phytologist*, 137–147.

Robertson, G. P. (1984). Nitrification and and nitrogen mineralization in a lowland rainforest succession in Costa Rica. *Oecologia*, 61, 99–104.

Robertson, G. P. (1989). Nitrification and denitrification in humid tropical ecosystems: potential controls of nitrogen retention. In *Mineral Nutrients in Tropical Forest and Savanna Ecosystems*, ed. J. Proctor, pp. 55–69. Oxford: Blackwell Scientific Publications

Robertson, G. P. and Tiedje, J. M. (1988). Denitrification in lowland tropical rainforest. *Nature*, 336, 756–759.

Roggy, J. C., Prevost, M. F., Garbaye, J. and Domenach, A. M. (1999). Nitrogen cycling in the tropical rainforest of French Guiana: comparison of two sites with contrasting soil types using $\delta^{15}N$. *Journal of Tropical Ecology*, 15, 1–22.

Roy, S. and Singh, J. S. (1995). Seasonal and spatial dynamics of plant-available N and P pools and N-mineralization in relation to fine roots in dry tropical forest habitat. *Soil Biology and Biochemistry* 27, 33–40.

Russell, R. S. (1974). *Plant Root Systems*. London: McGraw-Hill.

Sands, R. and Mulligan, D. R. (1990). Water and nutrient dynamics and tree growth. *Forest Ecology and Management*, 30, 91–111.

Scatena, F. N., Silver, W., Siccama, T., Johnson, A. and Sanchez, M. J. (1993). Biomass and nutrient content of the Bisley experimental watersheds, Luquillo experimental forest, Puerto Rico, before and after Hurricane Hugo, 1989. *Biotropica*, 25, 15–27.

Scott, D. A., Proctor, J. and Thompson, J. (1992). Ecological studies on a lowland evergreen rainforest on Maracá Island, Roraima, Brazil. II. Litter and nutrient cycling. *Journal of Ecology*, 80, 705–717.

Silver, W., Lugo, A. E. and Keller, M. (1999). Soil oxygen availability and biogeochemistry along rainfall and topographic gradients in upland wet tropical forest soils. *Biogeochemistry*, 44, 301–328.

Silver, W., Scatena, F. N., Johnson, A. H., Siccama, T. G., and Sanchez, M. J. (1994). Nutrient availability in wet montane forest: spatial patterns and methodological considerations. *Plant and Soil*, 164, 129–145.

Smirnoff, N. and Stewart, G. R. (1985). Nitrate assimilation and translocation by higher plants: comparative physiology and ecological consequences. *Physiologia Plantarum*, 64, 133–140.

Smirnoff, N., Todd, P. and Stewart, G. R. (1984). The occurrence of nitrate reduction in the leaves of woody plants. *Annals of Botany*, 54, 363–374.

Smith, C. K., Gholz, H. L., and De Assis Oliveira, F. (1998). Fine litter chemistry, early-stage decay, and nitrogen dynamics under plantations and primary forest in lowland Amazonia. *Soil Biology and Biochemistry*, 30, 2159–2169.

Smits, W. T. M. (1994). *Dipterocarpaceae: Mycorrhizae and Regeneration*. The Netherlands, Wageningen: The Tropenbos Foundation.

Sobrado, M. and Medina, E. (1980). General morphology, anatomical structure and nutrient content of sclerophyllous leaves of the 'Bana' vegetation of Amazonia. *Oecologia*, 45, 341–345.

Soil Survey Staff (1975). *Soil Taxonomy: A Basic System of Soil Classification for Making and Interpreting Soil Surveys*, Agricultural Handbooks no. 436. Washington, D.C.: United States Department of Agriculture.

Sollins, P. (1989). Factors affecting nutrient cycling in tropical soils. In *Mineral Nutrients in Tropical Forest and Savanna Ecosystems*, ed. J. Proctor, pp. 85–95. Oxford: Blackwell Scientific Publications.

Sollins, P. (1998). Factors influencing species composition in tropical lowland rainforest: does soil matter? *Ecology*, 79, 23–30.

Sollins, P. and Radulovich, R. (1988). Effects of soil physical structure on solute transport in a weathered tropical soil. *Soil Science Society of America Journal*, 52, 1168–1173.

Sprent, J. I. (1995). Legume trees and shrubs in the tropics: N_2 fixation in perspective. *Soil Biology and Biochemistry*, 27, 401–407.

Sprent, J. I. and Raven, J. I. (1985). Evolution of nitrogen fixing symbioses. *Proceedings of the Royal Society of Edinburgh*, 85B, 215–237.

Steinhardt, U. (1979). Untersuchungen über den Wasser- und Nährstoffhaushalt eines andinen Wolkenwoldes in Venezuela. *Göttinger Bodenkundliche Berichte* no. 56.

Stewart, G. R., Joly, C. A., and Smirnoff, N. (1992). Partitioning of organic nitrogen assimilation between roots and shoots of cerrado and forest trees of contrasting plant communities of South East Brazil. *Oecologia*, 91, 511–517.

Tanner, E. V. J. (1977). *Mineral Cycling in Montane Rain Forests in Jamaica*. Unpublished Ph.D. thesis, University of Cambridge.

Tanner, E. V. J. (1980). Studies on the biomass and productivity in a series of montane rain forests in Jamaica. *Journal of Ecology*, 68, 573–588.

Tanner, E. V. J. (1985). Jomaican montane forests: nutrient capital and cost of growth. *Journal of Ecology*, 73, 553–568.

Tanner, E. V. J., Vitousek, P. M. and Cuevas, E. (1998). Experimental investigation of nutrient limitation of forest growth on wet tropical mountains. *Ecology*, 79, 10–22.

Thompson, J., Proctor, J., Viana, V., Milliken, W., Ratter, J. A., and Scott, D. A. (1992). Ecological studies on a lowland evergreen rainforest on Maracá Island, Roraima, Brazil. 1. Physical environment, forest structure and leaf chemistry. *Journal of Ecology*, 80, 689–703.

Turner, I. M., Brown, N. D., and Newton, A. C. (1993). Effect of fertilizer application on dipterocarp seedling growth and mycorrhizal infection. *Forest Ecology and Management*, 57, 329–337.

Tyler, G. (1994). Plant uptake of aluminium from calcareous soils. Experientia, 50, 701–703.

Van Breemen, N., Finlay, R., Lundstrom, U., Jongmans, A. G., Giesler, R., and Olsson, M. (2000). Mycorrhizal weathering: a true case of mineral plant nutrition. *Biogeochemistry* 49: 53–67.

Van Dam, O. (2001). *Forest Filled with Gaps*. The Netherlands: University of Utrecht.

Van Schaik, C. P. and Mirmanto, E. (1985). Spatial variation in the structure and litterfall of a Sumatran rainforest. *Biotropica*, 17, 196–205.

Veneklaas, E. J.(1990). *Rainfall Interception and Aboveground Nutrient Fluxes in Colombian Montane Tropical Rain Forests*. The Netherlands: Utrecht, Rijksuniversitheit.

Villela, D. M. and Proctor, J. (2002). Leaf litter decomposition and monodominance in the *Peltogyne* forest of Maracá Island, Brazil. *Biotropica*, 34, 334–347.

Vitousek, P. M. (1984). Litterfall, nutrient cycling, and nutrient limitation in tropical forests. *Ecology*, 65, 285–298.

Vitousek, P. M. and Sanford, R. L. (1986). Nutrient cycling in moist tropical forest. *Annual Review of Ecology and Systematics*, 17, 137–167.

Vitousek, P. M., Walker, L. R., Whiteaker, l. D., and Matson, P. A. (1993). Nutrient limitations to plant growth during primary succession in Hawaii Volcanoes National Park. *Biogeochemistry*, 23, 197–215.

Walker, T. W. and Syers, J. K. (1976). The fate of phosphorus during pedogenesis. *Geoderma*, 15, 1–19.

Waterloo, M. J. (1994). *Water and Nutrient Dynamics of* Pinus caribea *Plantation Forests on Former Grassland Soils in Southwest Viti Levu, Fiji*. Ph.D. thesis, The Netherlands: Vrije Universität, Amsterdam.

Waterloo, M. J., Schelleken, J., Bruijnzeel, L. A., *et al.* (1997). Chemistry of bulk precipitation in southwestern Viti Levu, Fiji. *Journal of Tropical Ecology*, 13, 427–447.

Went, F. W. and Stark, N. (1968a) Mycorrhiza. *BioScience*, 18, 1035–1039.

Went, F. W. and Stark, N. (1968b). The biological and mechanical role of the soil fungi. *Proceedings of the National Academy of Sciences of the USA*, 60, 497–504.

Westman, W. E. (1978). Inputs and cycling of mineral nutrients in a coastal subtropical eucalypt forest. *Journal of Ecology*, 66, 513–531.

Whitmore, T. C. (1984). *Tropical Rain Forests of the Far East*. Oxford: Clarendon Press.

Wieder, W. and Wright, S. J. (1995). Tropical forest litter dynamics and dry season irrigation on Barro Colorado Island, Panama. *Ecology*, 76, 1971–1979.

Wild, A. (1989). Mineral nutrients in tropical ecosystems: a soil scientist's view. In *Mineral Nutrients in Tropical Forest and Savanna Ecosystems*, ed. J. Procter, pp. 441–456. Oxford: Blackwell Scientific Publications.

Wyn Jones, R. G. and Lunt, O. R. (1967). The function of calcium in plants. *Botanical Reviews*, 33, 407–426.

Yap, S. W. (1998). *Large Scale Enrichment Planting of Dipterocarps in Logged-Over Forests in Sabah, Malaysia: Potential Role of Fertiliser Application*. Unpublished Ph.D thesis, University of London.

Yavitt, J. B. and Wright, S. J. (1996). Temporal patterns of soil nutrients in a Panamanian moist forest revealed by ion-exchange resin and experimental irrigation. *Plant and Soil*, 183, 117–129.

Zimmerman, C. F., Price, M. T. and Montgomery, J. R. (1978). A comparison of ceramic cup and teflon in situ samplers for nutrient pore water determinations. *Estuarine, Coastal and Marine Sciences*, 6, 93–97.

17 Hydrology of tropical wetland forests: recent research results from Sarawak peatswamps

A. Hooijer
Delft Hydraulics, Delft, The Netherlands

INTRODUCTION

Freshwater wetlands are distributed widely throughout the tropical areas of Africa (340 000 km^2; Pajunen, 1996) and South America (430 000 km^2; Lappailanen and Zurek, 1996). Many of these wetlands are covered with forest (Lugo *et al.*, 1990). However, South East Asia has more wetland forest than any other part of the world: here, huge expanses of so-called 'peatswamp' forest can be found – often in association with other wetland forest types such as sago, melaleuca, nipah palm and mangrove. In Indonesia alone, the original extent of peatswamp forest exceeded 200 000 km^2 (Rieley *et al.*, 1996), while Malaysia had 25 000 km^2 (Ambak and Chye, 1996). Significant peatland areas also exist in other Asian countries. In fact, the total wetland area of Asia as a whole is estimated at no less than 2 268 000 km^2, of which 1 119 000 km^2 is peatland (including non-forested and non-tropical peatlands; Lappailanen, 1996). In contrast, the peatland area in Africa and South America is estimated at only 58 000 and 102 000 km^2, respectively. There is a significant expanse of wetland forest in the southern parts of the United States too but only some 10% of this is peatland (Malterer, 1996).

The peatswamps of South East Asia, with peat deposits frequently over 3 metres deep and often covered with lush rainforest, could be considered as the most 'typical' type of tropical wetland forest because here ecological functioning and wetland hydrology are truly interdependent (Page *et al.*, 1999) (Box 17.1). Indeed, descriptions of the development of peatswamps (Anshari *et al.*, 2001; Staub and Esterle, 1994) suggest that they represent a 'climax-phase' in wetland development, requiring relatively stable hydrological conditions over a long period. The present chapter will focus specifically on this wetland forest type, to show in what ways the hydrology of tropical wetland forest can differ from that of other tropical forest types, and to explore the implications of these differences on research and management. Although the extent of peatland forest in South East Asia is unique, there are similar peatland systems elsewhere (e.g. in Panama; Phillips *et al.*, 1997b), and hydrologically there are many similarities with tropical wetland forest systems on mineral soils (e.g. the cypress swamps of Florida; Riekerk and Korhnak, 2000).

Forests and peatswamps: a valuable combination

Why would we want to think specifically of the hydrology of tropical peatswamp forests? Luckily, there are better arguments than just that of it being different from other hydrological systems and relatively unknown. Though they do not yet enjoy the kind of publicity that tropical rainforest receives in general, there is increasing international awareness of the economic and ecological values of peatswamps (e.g. Page and Rieley, 1998; Davie and Samardja, 1997) and peatlands in general (e.g. Immirzi, 1997). The following specific functions are often used as arguments for peatswamp conservation:

- If well-managed, they can be a sustainable source of high-quality timber and crops like sago and fruits – as they have been for centuries (Kumari, 1996).
- As the dominant ecosystem within their range, they are the only environment in which many specialised species can survive.
- Their inaccessibility to man means that many species survived here which are less specialised and have become rare or extinct in other environments.
- They are needed for the functioning of many other associated ecosystems, which in turn harbour ecological and economic riches – e.g. coastal mangroves, which are essential for fisheries.
- They are often credited with maintaining river flows during dry periods and are therefore important for downstream water supply and irrigation – especially in coastal areas, where fresh water is scarce (Phillips *et al.*, 1997a).

Forests, Water and People in the Humid Tropics, ed. M. Bonell and L. A. Bruijnzeel. Published by Cambridge University Press.

- They are also credited with attenuating flood peaks, thereby preventing flood damage to downstream areas (Ramadasan *et al.*, 1999).
- To scientists, it is of interest that peatswamp forests represent a type of ecosystem that is similar to those which produced the Earth's coal deposits, millions of years ago (Staub and Esterle, 1994). Also, they preserve a wealth of information on more recent paleoclimates in their peat, and thus on climate change (Anshari *et al.*, 2001).
- The acidic blackwater produced by peatswamps in Sarawak is crucial for maintaining water quality well beyond the swamp boundaries: e.g. it causes flocculation of fine sediments when mixed with river water from mineral areas (Staub and Esterle, 1993), thus partly controlling the coastal sediment balance.
- Peatswamps store an enormous amount of carbon and therefore play a role in global climate control. This carbon is released upon drainage or burning (Wösten *et al.*, 1997).
- Last but not least: careful management is the only sustainable way to utilise peatswamp areas. Any use that disrupts their hydrological system through the associated drainage will result in progressive subsidence of the soil surface (Wösten *et al.*, 1997), with a range of associated problems. In the case of coastal peatswamps, subsidence may cause seawater intrusion, complicating agriculture and habitation unless very expensive coastal protection (and non-gravity drainage) is implemented. Ultimately, subsidence can also result in exposure of the marine clays underneath the peat, which in some cases causes acid sulphate problems.

Box 17.1 General characteristics of Sarawak peatswamps

Location and morphology

In the Malaysian state of Sarawak, peat deposits are mostly concentrated in the deltas of three main rivers: the Baram, Rajang and Lupar. The peat surface is convex ('dome-shaped'), with steeper slopes towards the edges. Peat depths are generally around 3 metres or more – depths up to 25 metres have been reported. Individual peat bodies here may range from 50 to 1000 km^2. Despite the fact that these are often elongated rather than having the 'ideal' round dome shape, they may be more than 25 km across. Over 7000 km^2 of peat exceeding 1 m in depth is found in and around the Rajang Delta of Sarawak alone. The swamps are separated by wide, meandering branches of the Rajang river and drained by narrower 'blackwater' streams. The peatswamps have progressively invaded former mangrove areas over the last 5000 years, and are therefore usually underlain by marine clays (Staub and Esterle, 1993, 1994). Marine influence in the coastal swamps continues to date, through tidal fluctuations in river water levels that extend up to 120 km inland in the larger rivers, and through periodic intrusions of saline water into swamp streams when low flow conditions coincide with very high tide.

Genesis and vegetation

Peatswamps occur where organic matter accumulates due to permanent waterlogging and the associated anaerobic and acidic conditions (pH is generally between 3 and 5). They are usually found where rainfall is high and runoff and internal drainage are impeded, often in low-gradient coastal zones. Plant remains accumulate here to form thick blankets of peat, which eventually become dome-shaped and create their own radial drainage pattern. Anderson (1964) found that the centres of peatswamps may be elevated 4 metres (near the coast) or even 9 (inland) metres above adjacent channels; peat depths up to 25 metres have been found under the highest (and oldest) domes. Surface slopes rarely exceed 1.5 m km^{-1} and are usually less than 0.5 m km^{-1} near the top of the peat domes and at the lower fringes of the slopes. (Staub and Esterle, 1993; Anderson, 1964, 1983). In the climax-stage of peatswamp development (which takes thousands of years; Anderson, 1964; Anshari *et al.*, 2001), the only input of water into these domed peat bodies is from rainfall and the swamps become increasingly nutrient-poor. In Sarawak, this results in clear vegetation-gradients across peat domes: from fringe through slope to centre. Anderson (1964, 1983) distinguished 6 zones (simplified to 3 here) in the Baram area on the basis of floristics, species diversity, tree density and tree size. These differences can be explained solely by hydrological flows:

- At the ***fringe*** of peatswamps, nutrients in runoff accumulate and there may be some influence from river water. Flooding by river water extends a few hundreds of metres inland at most (Bruenig, 1990), so the extent of this forest type is often limited. Here, 'topogenous' peats support a species-rich, structurally complex forest up to 50 metres tall. As water tables can frequently be metres above the surface here, specific adaptations like buttress roots are most commonly seen in this zone. The species *Shorea albida* is present but rarely dominant.
- Higher up the ***slope*** of the dome, dense, uniform forests of *Shorea albida* forest are found. These may range from up to 70 metres near the fringe to 30 metres closest to the centre.
- The 'ombrogenous' peats in the ***centre*** of fully developed domes support a forest of stunted trees between 3 and 12 metres high, more similar to savannah or heath forest than to 'normal' rainforest. The lack of nutrients is also indicated by the presence of *Sphagnum* mosses, similar to those which dominate temperate bogs. It should be noted that this forest type was not found in the study area reported upon in this chapter.

While it is not hard to see that peatswamps perform important economic functions, little research has been carried out to establish the extent to which each function is actually performed by a particular tropical wetland forest type. Clearly, different swamp types in different regions perform different functions; some of these functions may even be mutually exclusive. By unjustifiably claiming they may perform most of the above functions in most cases, the argument for sustainable management may actually become less convincing. This may contribute to the fact that the alleged benefits are often insufficiently considered in management decisions, and peatswamps are still considered 'worthless' in much of South East Asia (Muhamad, 2001).

The ecological and economic functions of peatswamp forests – and other wetland forest types – are closely linked to their unusual hydrological characteristics: the high water table, elevated peat surface and considerable water retention capacity (cf. Box 17.1). Therefore, there is a distinct need to study the hydrology of tropical wetland forests in more detail – if only to provide a sound scientific basis for management decisions.

Threats to peatswamp survival in South East Asia

Improved management of peatswamp forests is no luxury. Despite rising awareness of the need for sustainable forest and wetland management, and the fact that peatswamp areas are generally not densely populated, they are under very serious threat from logging, fires and drainage schemes for the establishment of rice and oil palm plantations. Often, these developments go hand-in-hand. In ten years time, the peatswamp forest area in Indonesia has been reduced from over 25 to 17 million hectares, while the area in Malaysia has dropped from 2.5 to 1 million hectares (Rieley *et al.*, 1997; Rieley 2001).

Logging and fires have severe implications in any forest environment (cf. the respective chapters by Chappell *et al.* and Malmer *et al.*, this volume), but peatswamps are exceptionally vulnerable due to two unique characteristics:

(1) Their internal hydrology makes them very sensitive to drying out by drainage, which will greatly diminish the flora and fauna they can sustain.
(2) Their deep organic soils are flammable when dry, producing fires which are impossible to extinguish and can keep burning for months until the next rainy season arrives.

Once activated, these processes are self-enhancing: if even only the edge of a swamp dome is drained, the change in surface gradient will result in progressive drainage far beyond the intended drainage area. This process is accelerated once fires take hold: internal drainage increases, more peat dries out, etc.

Peatswamp reclamation occurs often at the local scale (Abe, 1997), but some very large development projects may have caused the most damage recently. A controversial example of peatswamp mismanagement is the 'Mega Rice Project' in Central Kalimantan, Indonesia (Muhamad, 2001; Rieley, 1999, 2001). In 1995, over 1 million hectares of peatswamp were deforested and drained for conversion to rice plantations. Within a few years, it was evident that rice production was not feasible in most of the area. Instead, fires destroyed the dried-out peat and what was left of the vegetation, thereby bringing the project to a halt: what was productive rainforest had been transformed rapidly into a wasteland, without economic benefits. Unfortunately, the 'Mega Rice Project' was not an isolated incident: analysis of remote sensing images suggests that the 1997 fires alone destroyed over 2 000 000 ha in Indonesia, affecting mainly peatswamp forests that were already degraded by land clearance or logging (Legg and Laumonier, 1997). In fact, the peatswamp area affected in 1997 may even be far larger: analysis of radar images puts the forest area with 'burn scars' in East-Kalimantan alone at 5.2 million hectares (Ruecker and Siefert, 2000). These fires were so huge that much of South East Asia was covered in a smoke haze for months, and the air quality in the region deteriorated severely.

Filling the gaps

Tropical wetland forests may form a relatively little-studied environment, but they combine elements of two better known environments: wetlands and tropical 'dry-land' rainforests. Hydrological studies in tropical wetland forests can benefit especially from information on other types of wetlands, e.g. temperate peat bogs, which have been described extensively (e.g. Ingram, 1983; Gilman, 1994). As there is no lack of information on peatswamp ecology (Anderson, 1983; Bruenig, 1990; Phillips, 1998; Page *et al.*, 1999), the absence of hydrological studies may be caused by the simple fact that most components of the water balance are very difficult to measure in these areas. In this chapter, information from a few sources is presented: (i) a single but comprehensive hydrological study carried out in Sarawak peatswamps during 1995–1997, (ii) hydrological studies in somewhat similar environments elsewhere, and (iii) non-hydrological studies in Sarawak. The aim is not to provide a full overview of the hydrology of tropical wetland forests but rather to discuss some unique features and methodologies that can be used in developing a better understanding of how to manage this special environment sustainably.

PEATSWAMP HYDROLOGY

Introduction

The hydrology of peatswamp forests is exceptional for the following reasons:

- The defining characteristic of peatswamp hydrology is the groundwater table: not only is it close to the soil surface for most of the time but one could almost say that it *is* the soil surface, as the peat consists of over 85% of water. The association between the water table and the surface of peatlands

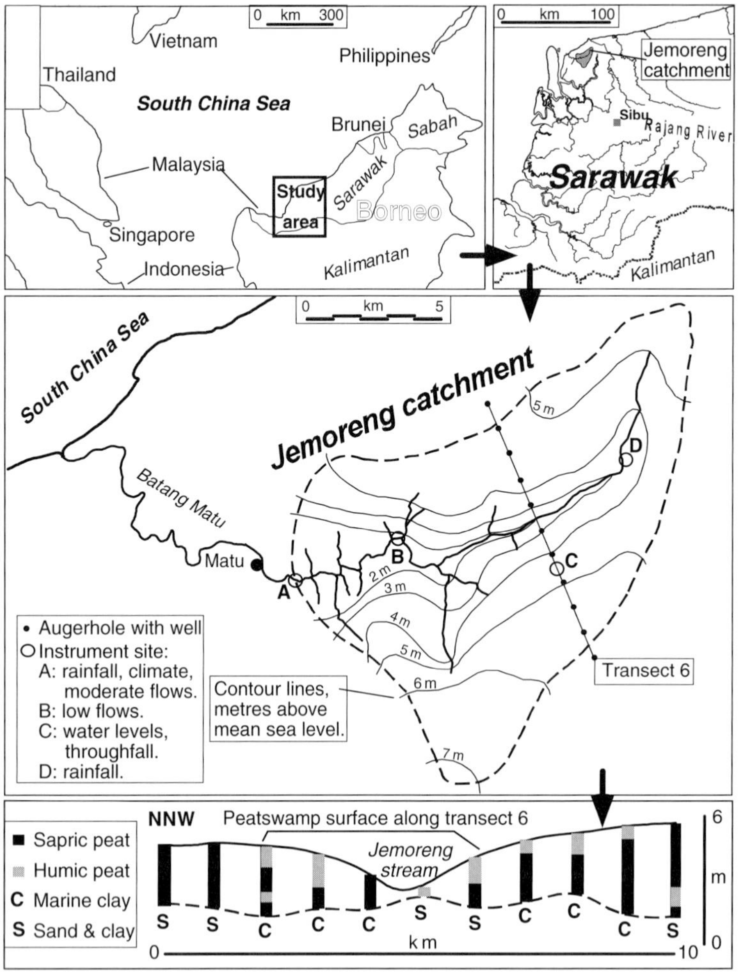

Figure 17.1 Location, topography and geology of the Jemoreng study catchment. Elevations are relative to mean sea level.

is so close that the surface of temperate bogs is known to oscillate seasonally (Ingram, 1983). In tropical peatswamp forests, too, it is clear that the shape of the swamp surface follows that of the 'groundwater dome' (Anderson 1983; see Figure 17.2A below). This implies that drainage will result in immediate lowering of the soil surface (Wösten *et al.*, 1997). Knowledge of the water table regime is therefore the basis of understanding peatswamp functioning, and water table management is crucial for minimising the impact of land use changes.

- In peatswamps, perhaps more than in any other environment, abiotic and biotic factors are completely interdependent (Page *et al.*, 1999). The morphology, substratum and hydrology are shaped by the vegetation (through accumulation of organic material), and this vegetation can only exist in the environment it has helped to create. Therefore, the slightest change in vegetation will have an immediate effect on the morphology and hydrology of the swamps, while at the same time any change in hydrology will change the vegetation.
- The peatswamp water balance is difficult to quantify. Because of the degree of uncertainty in so many hydrological variables, the water balance of peatswamps (as in other wetland types; Labaugh, 1986; Gilvear and Bradley, 2000) can only be determined accurately if an attempt is made to measure all water balance components in the field, and over a long period (Hooijer *et al.*, 1997).
- Hydrological monitoring in peatswamps requires the use of some unusual data collection techniques. Some of these techniques are worth discussing here because the difficulties in data collection, even more than the complexity of the hydrological system itself, determine the limitations and possibilities for hydrological studies and water balance modelling in this environment.

In the following sections, these points will be discussed in some detail, using the results of a recent study of the Jemoreng peatswamp catchment in Sarawak, Malaysia (Figure 17.1; Boxes 17.2 and 17.3).

Box 17.2 A study of the hydrology of peatswamp forests in Sarawak

The study reported here was carried out for the Sarawak State Government, represented by the Sarawak Water Resources Council (with 'Jabatan Kerja Raya', JKR, as implementing agency), as part of the development of a Water Resources Master Plan for Sarawak. More specifically, the goal of the study was to assess the water yields during drought periods from 10 peatswamp catchments. Because of the lack of existing information, this required an extensive monitoring scheme and modelling. The study and field instrumentation started by mid-1995; the results presented here are based on data collected from 11 January 1996 until 10 May 1997 (a period of 497 days) in the Jemoreng catchment, which was studied the most intensively. The results of more limited studies in the other 9 catchments support those for Jemoreng (Hooijer *et al.*, 1997).

Study set-up

The hydrological study of the Jemoreng catchment consisted of the following main elements:

(1) A topographic and geological survey was carried out along a grid of 0.5 × 1 km.

(2) An extensive monitoring network was set up (and maintained for 2 years). The following variables were monitored, using data loggers unless indicated otherwise (see Figure 17.1 for locations):
- *Weather parameters*: near station A, for calculation of Penman potential evaporation.
- *Rainfall*: at stations A and C, near the outlet and inside the catchment.
- *Throughfall*: four gauges were installed below the canopy, and moved once.
- *Catchment discharge*: with acoustic instruments at stations A and B.
- *Channel water levels*: at stations A and B, using pressure transducers.
- *Ground water level*: in 55 wells along transects across the swamp. These wells were anchored in the 'solid' mineral substratum at several metres depth:

  In one of these wells (station C, along transect 6), water levels were monitored hourly, with a pressure transducer.

  Water levels in another 12 wells along transect 6 were monitored at 2-weekly intervals.

  Water levels in the remaining wells were monitored every few months.

(3) Data were analysed and interpreted taking into account knowledge available on tropical rainforests and temperate peatswamps.

(4) A wetland water balance model was developed on the basis of the analysis of the hydrological system, and used to provide long-term records of water levels and stream discharges for use in water resources planning.

Box 17.3 Description of a typical peatswamp: the Jemoreng study catchment

Location and land use

The Jemoreng catchment is situated in the north-eastern part of the Rajang delta (Figure 17.1), a sparsely populated area north of the regional capital Sibu. It is 123 km^2 in size. The land use of the catchment is representative of many other peatswamps in the area: analysis of a Landsat image for 1996 (Wallace, 1997) showed that 81% of the Jemoreng catchment was covered with dense secondary forest, with 12% of the area being recently logged or 'jungle garden', and 4% sago plantations. Agriculture covers 5% of the area, and no human settlements are located within the catchment boundaries.

In this part of Sarawak, logging practices appear to be adapted specifically to the peatswamp environment. Logs are placed on sleds, pulled with cables over the soft wetland soil to a makeshift railway line, and transported by light trains. These railway lines are removed and rebuilt at 1 km intervals. This system appears to allow relatively rapid regrowth of a forest cover; Landsat images show that several of the logging patterns (abandoned railway lines) are overlapping, which suggests that much of the Jemoreng forest has been selectively logged several times in the last century. The continuing density of the canopy bears testimony to the resilience of this forest type, but it is assumed that the canopy of the forest studied here is considerably lower than that of the natural peatswamp forests described by Anderson (1964).

Hydro-physiography

Peat depths in the Jemoreng catchment are generally between 2 and 5 metres (see inset Figure 17.1) and the gradient of the peat surface is below 0.5 m km^{-1} in over 60% of the catchment area, and rarely more than 1 m km^{-1} though steeper slopes occur around the edges and near streams. Though the overall peat surface appears flat, it is irregular at the microscale with protruding 'buttress' tree roots and numerous rivulets and shallow ponds. The stream gradient in the Jemoreng catchment is also very low: 1.5 m over 14 km (along the stream) between stations A and D (Figure 17.1). The peat is underlain by impermeable marine clay in most of the area; deposits of fine sand are found locally. The catchment is situated relatively close to the South China Sea and tidal influence on stream water levels is pronounced. Due to the lack of roads, the very soft soils and the dense vegetation, access into the catchment is only possible by boat along the main stream and over elevated wooden walkways constructed along transects for this study.

Climate and study period

Rainfall rates in the Sarawak coastal region are high: 3275 mm yr^{-1} on average at Matu (see Figure 17.1 for location) between 1957 and 1994. There is a distinct seasonality with rainfall in the wettest and driest months (January and June) being almost double and half the annual monthly average, at 531 and 148 mm $month^{-1}$ respectively

(1957–1994 rainfall data). Extensive periods without significant rainfall can occur from May to August, and have a severe impact on streamflows. Annual rainfall during the study period was close to the long-term average, at 3386 mm yr^{-1}. The average 24h temperature in the swamps was 25.4 degrees Celsius, the average 24h humidity was as high as 92% and average 24h windspeeds above the canopy only 0.11 m s^{-1} (Hooijer *et al.*, 1997).

Water balance studies in wetlands

As for non-wetland catchments, the water balance for peatswamps can be written as: $P = \text{ET} + Q + \Delta S$ (where P = rainfall, ET = evapotranspiration, Q = discharge (drainage) and ΔS = change in catchment storage, including groundwater storage). However, wetlands differ from other catchments in the accuracy with which most of these components can be determined:

- In most hydrological studies, rainfall, discharge and catchment area can be determined with relatively high accuracy. Evapotranspiration is usually determined indirectly from meteorological data or by difference (between P and Q), while the changes in catchment storage (mostly groundwater storage) are almost always determined by difference as they are rarely measured directly (Ward and Robinson, 1990).
- The situation in wetlands is often the opposite: discharge and catchment area are extremely difficult to measure (and often only with a considerable error margin) while evapotranspiration and catchment storage can be determined directly and accurately from water table information. Moreover, accurate rainfall records can be hard to obtain in extensive and unpopulated wetlands. Therefore, application of a 'standard' water balance approach to wetlands often results in significant errors (Labaugh, 1986). To avoid such errors, unconventional data collection methods are needed.

In the Jemoreng study catchment in Sarawak, it has been possible to determine all main water balance components (apart from rainfall interception and surface runoff) independently, through collection and analysis of field data (Hooijer and Sivapalan, 1995; Hooijer *et al.*, 1997, Wong *et al.*, 1997). The data collection programme is outlined in Box 17.2.

WATER TABLE STUDIES: AN IMPORTANT TOOL IN WETLAND HYDROLOGY

The analysis of fluctuations in water levels is a widely applied component in studies of wetland hydrology. In peat bogs and floodplains it has been used for the determination of evapotranspiration, soil moisture retention characteristics and groundwater seepage rates (e.g. Dolan *et al.*, 1984; Laine, 1984; Hooijer, 1996). In lakes it is used routinely to determine changes in water storage over large areas. In the current study such techniques have been used successfully for all these purposes. The applicability of these methods is unique to wetlands because the following assumptions can be made:

- Due to the limited depth of the unsaturated zone (generally less than 0.3 m), there is very little delay in the response of the water table to changes in storage. Water table fluctuations can therefore be linked directly to individual rainfall and evaporation events.
- Due to the high rates of capillary rise in peat soils it may be assumed that moisture content in the unsaturated zone remains close to field capacity (except during extreme drought events); changes in storage (ΔS) are therefore proportional to changes in water level (ΔL) (Ingram, 1983; Dolan *et al.*, 1984; Laine, 1984; Hooijer, 1996). Based on this principle, it is possible to determine forest transpiration plus surface evaporation rates (Et), groundwater seepage rates (G) and the storage coefficient of the soil (Sf) from a single diurnal water table record. The principle of this approach is demonstrated in Figure 17.2C.
- Considering the relative geological, topographical and botanical uniformity within peatswamps, it is not surprising that water table fluctuations and other hydrological characteristics were also found to be quite uniform, as illustrated in Figures 17.2A and 17.2B. Point data for ΔS, Et and soil moisture storage coefficients can therefore often be applied (with a few corrections) to the entire catchment area – such simple extrapolation is not possible in most hydrological studies in non-wetland areas.

CATCHMENT BOUNDARY DELINEATION

The normally relatively simple task of defining catchment boundaries is challenging in peatswamps, as in most wetlands. Not only is it difficult to define the water divide when surface gradients are generally under 0.5 m km^{-1}, but even the concept of a single constant catchment area does not fully apply here. Catchment boundaries may shift rapidly with extreme rainfall events (and in some cases with tidal events), they will change gradually as the swamp develops, and they can also change progressively due to peat subsidence caused by drainage – even when drainage activities take place well away from the catchment boundaries. Moreover, even relatively minor artificial depressions like logging tracks are likely to lead water across the natural catchment boundaries. While modern techniques like digital elevation model generation using remote sensing data are increasingly useful for catchment area delineation, they cannot be used in low-gradient areas with a dense canopy, like the peatswamps (Werner *et al.*, 2000; see also Held, this volume). Therefore, determination of the

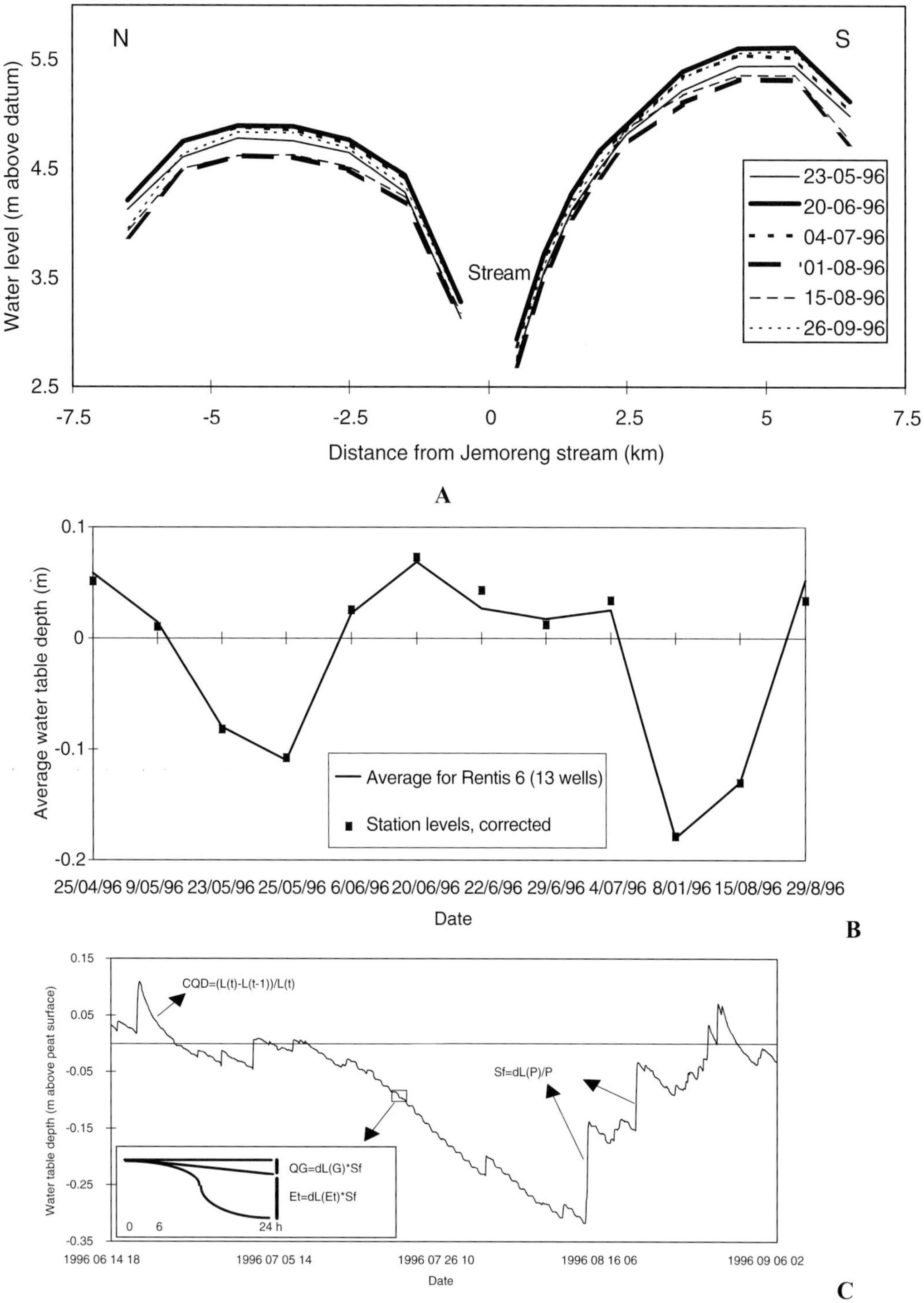

Figure 17.2 Fluctuations and uniformity in water table levels in the Jemoreng peatswamp, and the use of water level information in hydrological studies. (A) Typical fluctuation pattern of the water table in the Jemoreng peatswamp, along transect 6 (Figure 17.1). Note that water levels usually fluctuate uniformly, and by less then 0.3 m above or below the peat surface (approximately the water level at 23-05-96).
(B) Average water levels (manually monitored) for all 13 wells along the same transect 6, compared to those in a single well (water level monitoring station C; Figure 17.1). Note the very small difference between the two, confirming the high degree of spatial uniformity in water level fluctuations. (C) Water table fluctuations at water table monitoring station C in the Jemoreng catchment were used to determine the following: the storage coefficient (*Sf*) is determined from the rapid response of the groundwater table to rainfall; rates of groundwater seepage (*QG*) and evapotranspiration (E_t) are determined from diurnal fluctuations in periods without rainfall; the fraction of water standing in depressions on the surface that is discharged by surface runoff is determined from the daily drawdown rate of the water level when it is above the peat surface (corrected for drawdown due to evapotranspiration).

catchment boundaries does not only require highly accurate (and recent) topographic data (e.g. Loftin *et al.*, 2000), but it is also necessary to verify the catchment area through the water balance (e.g. Riekerk and Korhnak, 2000).

To delineate the boundaries of the Jemoreng study catchment, elevation measurements were made at over 300 points (using laser instruments) at 0.5 km intervals along the 14 transects (1 km apart) covering the entire catchment. In addition, water level readings in 55 wells along four transects were used to confirm that the shape of the groundwater body coincided with that of the peat surface (Figure 17.2A). As the catchment boundaries are located on the flat summit of the peat domes, they can only be delineated with a margin of uncertainty in the order of 5% to 10% at best. Without such detailed data the error could easily exceed 50% – as is common in wetlands (e.g. Riekerk and Korhnak, 2000). In addition, the boundaries may vary somewhat with the position of the water table and the location of intense rainstorms. Therefore, the probable catchment boundaries had to be estimated for a range of different conditions. The probable maximum catchment area was determined as 135.8 km^2, and the probable minimum area as 123.5 km^2. The minimum estimate was confirmed by water balance modelling (see model discussion below).

STREAM DISCHARGE

Discharge measurements in peatswamps are complicated by a number of factors:

- Peatswamps are characterised by their ***diffuse drainage pattern***: runoff flows (or trickles) through a micro-topography of hummocks and hollows. Only near the fringes of a peat dome will runoff concentrate in recognisable 'blackwater' channels, but even here discharges cannot be measured during floods because part of the stormflow leaves the catchment outside of the stream channel, often inundating an area hundreds of metres wide. In the Jemoreng catchment, the highest stream discharge that could be measured accurately was only about one-third of the highest daily peak outflow during the study period (as simulated using a water balance model; cf. Figure 17.6 below).
- ***Tidal influence*** and other backwater effects can also be a problem in these low-gradient areas. In the case of the Jemoreng catchment, the stream water level fluctuates by up to 2 metres at the outlet (station A, Figure 17.1) and flow direction reverses four times a day. During low- to mid-flow conditions, tidal flow is dominant and only a small part of the total flow in the channel consists of actual catchment discharge. Conventional rating curve methods cannot be applied in these conditions, and flows must therefore be measured directly.

In the Jemoreng catchment, mid-range stream discharges were monitored using three 'coherent acoustic Doppler flow profiling sensors' along a cross-section through the channel (station A, Figure 17.1). Each of these instruments continuously measured a velocity profile between the riverbed and the surface, as well as water depth and flow direction. In this way, it was possible to separate tidal flows from actual catchment discharges for most of the time, but not during extreme conditions:

- During drought periods, catchment discharge forms only a few percent of total channel flows.
- During peak flow events, part of the catchment outflow occurs outside of the channel and hence cannot be measured using equipment installed within the channels alone.

To improve discharge measurements during dry periods, flows were also monitored at a location further upstream (station B in Figure 17.1), beyond the reach of tidal influences. The data from the two stations (A and B) were then combined in a single flow record by inserting data from station B only for the periods with the lowest flows (increasing measured flows at station B by 14% to correct for the smaller catchment area upstream of this station). The end-result of this approach was an accurate record of flows between 0.5 and 15 mm d^{-1}, with a data coverage of 85.1% over the study period. It should be noted that many brief data gaps occurred rather than a few long ones. These are due not only to water levels and flows frequently being too high and in some cases too low (in combination with extreme tidal events) for accurate measurement, but also to periodic failure of monitoring instruments subjected to extreme conditions (Hooijer *et al.*, 1997).

EVAPOTRANSPIRATION

Temperate wetlands can have evapotranspiration rates equal to or in excess of potential evaporation rates (Ingram, 1983), and this was also found to be the case in the Jemoreng peatswamp. The following methodology was followed to establish this:

(1) Actual evapotranspiration (E_t, i.e. tree water uptake plus soil evaporation but excluding rainfall interception) rates were determined from analysis of a continuous record of the position of the groundwater table in a representative part of the forest (monitored at station C, Figure 17.1), using the basic equation $E_t = \Delta L(E_t) * Sf$. The diurnal drawdown of the water table (ΔL, Figure 17.2C) is largely caused by loss of water through evapotranspiration, which can be determined accurately if the drawdown due to groundwater seepage (ΔLG) and the storage coefficient (Sf) of the peat soil are known. The latter variables can be determined from the same water table record: groundwater seepage from the drawdown

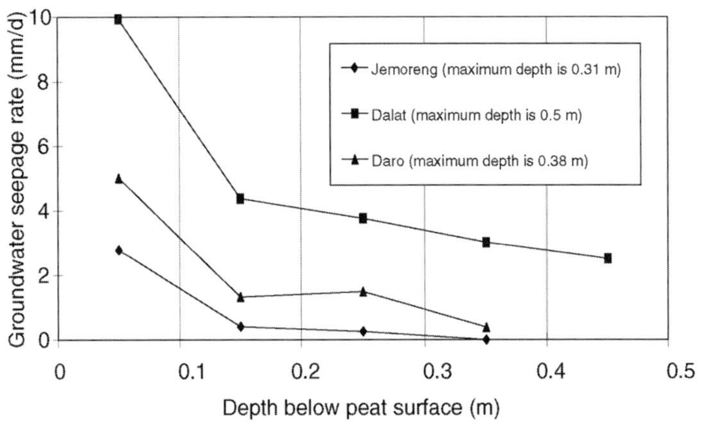

Figure 17.3 Seepage rates at the Jemoreng, Dalat and Daro water level stations, as determined from water table drawdown between 0:00 and 6:00 am. The seepage rates in the Jemoreng catchment were calculated using storage coefficients of 0.71 at depths to 0.1 m and 0.29 at greater depths (see text). Seepage rates decrease with water table depth at all stations. In all swamps, the sharpest decrease occurs when the water table drops below approximately 0.1 m depth and the subsurface flow system ceases to be active.

between midnight and 6:00 am (when E_t can be assumed to be negligible) and *Sf* from the water table response during intense rainfall events (Figure 17.2C). This method has been used successfully in several other wetland types (e.g. Dolan *et al.*, 1984; Laine, 1984; Hooijer, 1996). Groundwater seepage was found to be negligible (<0.1 mm d^{-1}) when the groundwater table was more than 0.1 m below the peat surface (Figure 17.3). An *Sf* value of 0.71 was found for peat depths less that 0.1 m, while a value of 0.29 applied at greater depths. Using these results, E_t was then determined as: $E_t = L*0.29$ for rainless days when the water table was at least 0.1 m below the surface (Figure 17.2C).

(2) Penman potential evapotranspiration (*PET*) was determined using hourly records of temperature, humidity, windspeed and radiation.

(3) *PET* and E_t proved to be very close: ET = *1.05*ET* (Figure 17.4). Using this relation, a record of daily actual evapotranspiration was generated. Average daily E_t (excluding rainfall interception) was 2.94 mm (annual total 1073 mm).

(4) Rainfall interception could not be determined with sufficient accuracy using the set-up of only four throughfall gauges (cf. Roberts *et al.*, this volume). Instead, an average interception fraction for lowland rainforests in Borneo of 14% of *P* (Dykes, 1997; Asdak *et al.*, 1998) was used to complete the water balance. Annual E_i is therefore estimated at 476 mm for 3400 mm of rainfall (Box 17.3).

Total evaporative losses ($E_t + E_i$) from the Sarawak peatswamp over the study period are therefore estimated at *c.* 1550 mm/y. This is reasonably close to the average of *c.* 1460 mm y^{-1} derived by Bruijnzeel (1990) for non-wetland lowland rainforests in South East Asia, even more so when considering the rather wide confidence limits that are attached to both estimates. Van der Molen (2002) estimated E_t for a coastal (brackwater) swamp forest in Puerto Rico at *c.* 1220 mm y^{-1} and rainfall interception *Ei* at *c.* 240 mm y^{-1} (14% of P), giving an annual total ET of *c.* 1460 mm. It appears that evapotranspiration in tropical peatswamp forests is not too dissimilar from that in other tropical lowland forest types (cf. Roberts *et al.*, this volume).

CHANGES IN CATCHMENT STORAGE

In peatswamps, unlike in most other environments, it is sometimes possible to obtain a record of changes in catchment water storage from a single water level record, In the case of the Jemoreng catchment, this record was collected at station C in the centre of the catchment (Figure 17.1). Water storage is a function of water level; therefore, if the storage coefficient of the peat (*Sf*) is known and a water level record representative for the entire catchment is available, it is possible to determine a record of daily water storage changes.

Figure 17.2B shows that the average of the water level fluctuations in 13 wells (2-weekly measurements) along transect 6 ($L_{transect}$) was almost identical to water levels at water level monitoring station C alone. A close relationship was also found between the station water level and average water levels for 44 wells (irregular measurements), distributed over the entire catchment. A slightly modified water level measured at a single station is therefore considered representative for the catchment ($L_{catchment} = 0.035 + 1.026*L_{station}$; $r^2 = 0.99$). Changes in catchment water storage can thus be calculated from water level fluctuations monitored at a single site, using an *Sf* of 0.71 just below the surface and of 0.29 at greater depths (see above).

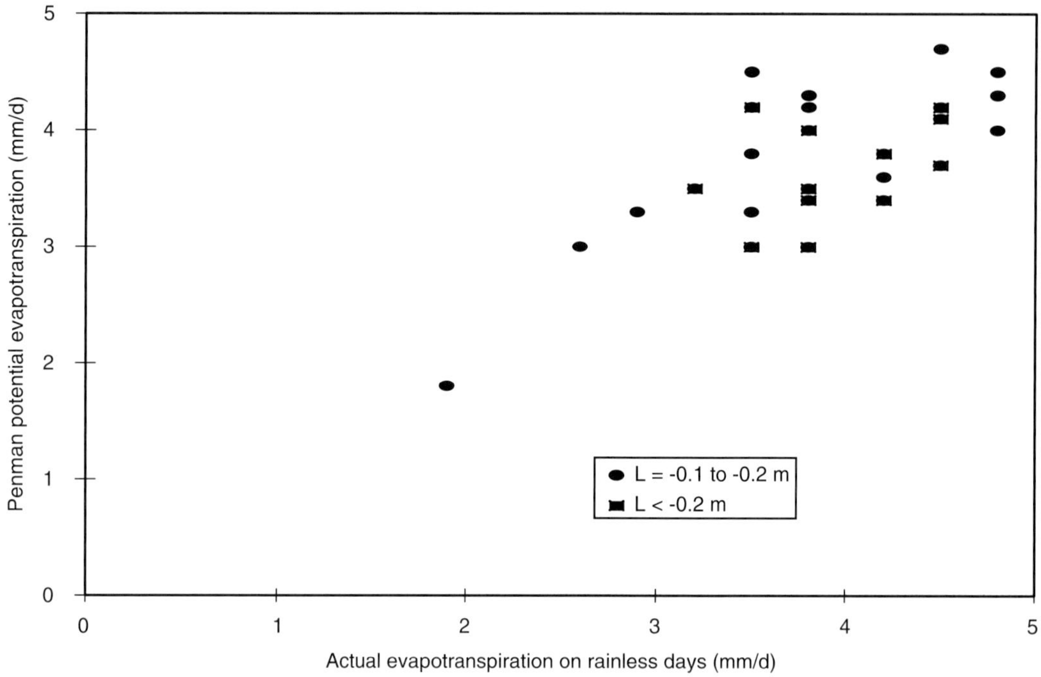

Figure 17.4 Actual evapotranspiration (*Et*) at the Jemoreng water table station. Determined from diurnal water table fluctuations in rainless periods when the water level was at least 10 cm below the soil surface – versus Penman potential evaporation (ET), calculated from climatic data monitored at Matu WTP. *Et*/ET = 1.05; $r = 0.7$.

RUNOFF FLOW PATHS

Groundwater flow, QG (at depths below 0.1 m; Figure 17.5): The vegetation in most of the peatswamp receives moisture and nutrients exclusively from rainwater due to the lack of external groundwater inputs (Anderson, 1964, 1983; Box 17.1). Peat permeability is inversely related to the degree of humification, which is moderate to high for the Sarawak peatswamps (see inset Figure 17.1). Peat permeabilities are therefore moderate to low, typically in the order of 1 m d^{-1} at most but often as low as 10^{-3} m d^{-1} (Ingram, 1983). In temperate bogs these low permeabilities, in combination with the limited aquifer depth and catchment gradient, restrict groundwater flows to very low rates (Bay, 1969). Tropical peatswamps may be similar to other peatlands in this respect: in the case of the Jemoreng peatswamp catchment, it was shown that the water table drawdown due to groundwater percolation rates was less than 0.3 mm d^{-1} (when water tables were below 0.1 m below the peat surface; Figure 17.3). On an annual basis, this implies that groundwater flow towards the stream is 100 mm yr^{-1} at the most. For comparison, Figure 17.3 also shows percolation rates in two smaller peatswamp catchments where gradients are higher due to artificial drainage; it is clear that the drainage causes groundwater flows to be higher there.

Subsurface flow, QS (at depths between 0 and 0.1 m; Figure 17.5): considering the low groundwater flow rates and high rainfall rates in peatswamps, it is no surprise that most runoff occurs very close to the peat surface, i.e. just below or above it. Figure 17.3 shows that water table drawdown due to subsurface flows is typically in the order of 3 mm d^{-1} when the water table is between the peat surface and 0.1 m depth, which it is for most of the time (Figures 17.2B and 17.6). This means that flows through this zone, corrected for the time the water table is below 0.1 m depth (18% of the time during the study period) is about 800 mm yr^{-1}.

Surface flow, QD: As in other wetlands, flows over the surface are an important flow component in peatswamps. As this is a water balance component that could not be measured independently in the Jemoreng study, it was determined from the peatswamp water balance, by difference:

$$\begin{aligned} \textit{Rainfall} &= \textit{Evapotranspiration} + \textit{Groundwater flow} \\ &\quad + \textit{Subsurface flow} + \textit{Surface flow}. \\ 3400 &= 1550 + 100 + 800 + \textit{Surface flow} \\ &\Rightarrow \text{Surface flow} \approx 950\ mm\ yr^{-1} \end{aligned}$$

Water balance modelling approach

For the Jemoreng study area, a wetland water balance model was developed for two purposes: (A) as an aid to understand the relation between water levels and flow mechanisms within the

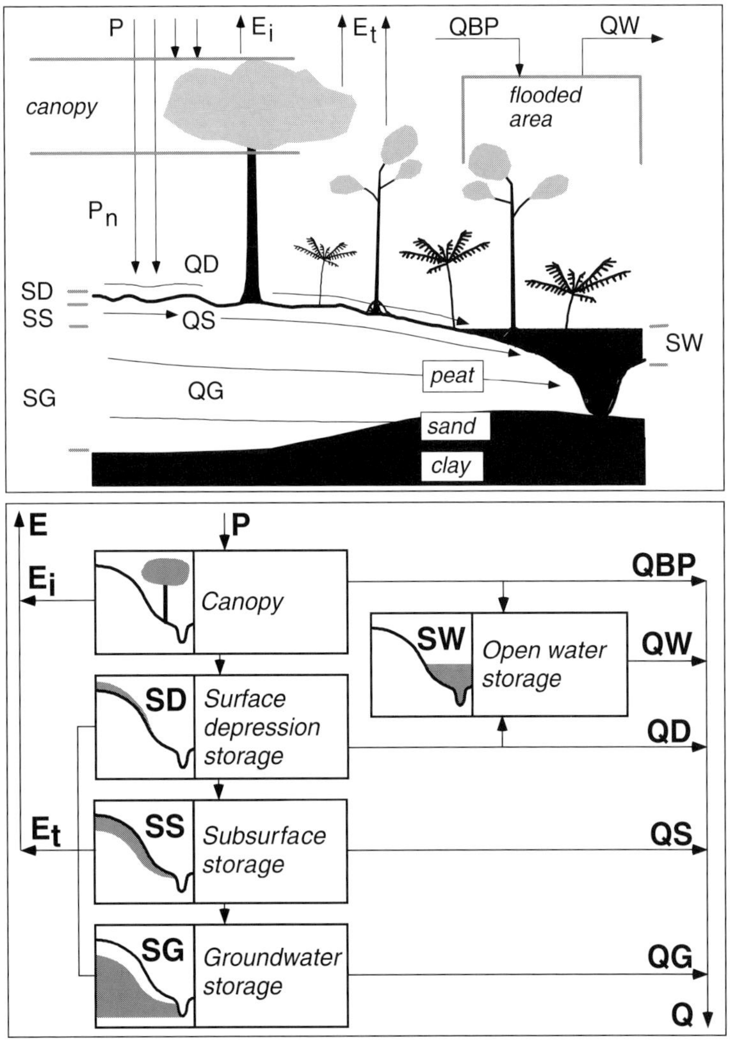

Figure 17.5 Schematic representation of water stores and flows on a peatswamp.

swamps, and (B) to simulate long-term flows in extreme conditions for which no observed flow data could be collected. The model will not be described here in detail, but some model characteristics and results that are relevant to hydrological studies in peatswamp forest will be discussed (see Hooijer *et al.*, 1997 and Phillips *et al.*, 1997a for more details).

From the description of the hydrology of peatlands, it will be clear that modelling their water balance is complicated by the fact that flow processes are very different from those found in other hydrological systems. This has led to the development of specific hydrological models for wetlands (e.g. Mansell *et al.*, 2000; Wilsnack, 2001). Because the use of model parameters that are valid in other environments is usually not feasible in a peatswamp context, models must often be developed and calibrated 'from scratch'. To make matters worse, calibration using discharge data is not always possible, as these are very hard to acquire – even over a limited flow range. Peak and very low flows proved impossible to measure in the case of the Jemoreng study area or any of the other Sarawak peatswamps studied.

On the bright side, it has been possible to measure changes in catchment storage in peatswamps with some accuracy, so it is actually possible to calibrate the water balance with both discharge and water storage (or water level, which is directly related to both). The model for the Jemoreng catchment was built using this principle. This conceptual, non-distributed, reservoir model operates at a daily timestep. It was developed with simplicity and transparency in mind; model stores should represent actual peatswamp water stores as much as possible, and model fluxes should be interpretable in terms of actual runoff mechanisms (Hooijer and Sivapalan, 1995; Hooijer *et al.*, 1997).

Essential in the modelling approach is the use of water level as the state variable on which all discharges depend. The difference

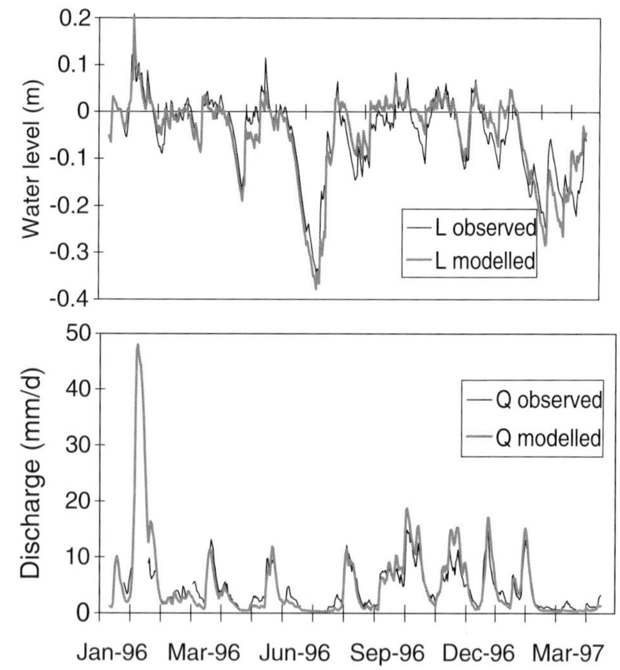

Figure 17.6 Jemoreng peatswamp catchment. Simulated discharges and water levels were fitted to observed values through optimisation of a limited set of model coefficients that could not be determined independently, from field measurements. The *coefficient of efficiency* (R^2; Nash and Sutcliffe, 1970) was used as the optimisation criterion for both discharges and water levels, and both were optimised to achieve the best overall model performance, with a higher weighting for discharges: $R^2_{\text{mode el}} = \frac{2R^2_Q + R^2_L}{3}$

with most other reservoir models is the possibility to calibrate it by optimising modelled versus observed catchment water levels, and thus optimising actual changes in water storage. This ensures that model simulations represent the hydrological conditions in the catchment more accurately than would be the case when only discharge data were used for optimisation. Thus, the predictive value of simulations for extreme conditions, which may not be encountered during the calibration period, will also be higher.

MODEL OUTLINE

The model set-up (stores and fluxes) is shown schematically in Figure 17.5. Records of net rainfall and evapotranspiration are input to the model. Part of the net rainfall falls on low-lying permanently inundated or saturated areas along streams, and thus 'bypasses' the hydrological system of the peat-domes: *open water storage (SW)*. All flows from the peat dome are also collected in this reservoir, before entering the streams.

Most of the rain falls on the peat-dome itself, in which three layered water reservoirs (storages *SD, SS and SG*) are distinguished. Each reservoir must be filled before the reservoir above it can contain (and discharge) water. The outflow rate from each reservoir (*QD, QS and QG*) is linearly proportional to the amount of storage at each time step (Figure 17.5).

- *Depression storage* (*SD*) is located above the swamp surface, where rainfall is retained in ponds and rivulets when the peat is fully saturated. Due to the low gradients and considerable roughness of the peat surface, some of this water may be stored for several days before the outflow from this store (*QD*) reaches a channel; it was found that 'time to peak' was usually between 1 and 2 days in the Jemoreng catchment.
- *Subsurface storage* (*SS*) occupies the upper 10 cm of the peat deposit and is transitional between the standing water above and the solid peat below. This zone is in fact a dense litter layer which has some characteristics in common with the 'acrotelm' surface peat layer typical for temperate bogs (Ingram, 1983). This layer has a high permeability and storage coefficient ($Sf = 0.71$). The depth of 0.1 metre was determined from water table records that show that both discharge rates and the storage coefficient drop sharply below this depth. The characteristics of this layer are crucial for peatswamp hydrology, as the water table is located within this zone most of the time (Figure 17.6).
- *Groundwater storage* (*SG*) takes place at greater depths in the peat dome. While the considerable depth (2 to 5 m) and storage coefficient of the peat ($Sf = 0.29$) ensure a large water storage capacity, only a fraction of this is available to baseflow due to the low permeabilities of the aquifer and the high evapotranspiration of the vegetation.

MODELLING RESULTS

The water balance model was calibrated using 1 year of discharge and water level data for the Jemoreng catchment (though discharge data were not available for 14.9% of the time). From Figure 17.6 it is evident that good fits between simulated and observed data were obtained both for discharges and for catchment water levels, with the coefficient of determination for discharges R^2_Q of 0.91 and for water levels R^2_L of 0.83.

Verifying the delineated catchment area through model optimisation

Calibration of the water balance model is only possible if the catchment area is known, as observed discharges need be converted from volumes ($m^3\ d^{-1}$) to specific discharges ($mm\ d^{-1}$). The difficulties encountered in delineating catchment areas in peatswamps have been mentioned earlier. Fortunately, it proved possible to estimate the catchment area from model results, by fitting simulated long-term discharges to observed totals. This approach is feasible in this study because all other water components (*P, Et,* ΔS) are inputs to the model. It was found that the catchment area derived using this method was 119 km^2, which is very close to the 'minimum probable area' of 123 km^2 delineated from topographic information (see above). A catchment area of 123 km^2 was therefore used in all further modelling.

The extent of the permanently inundated area
Model results indicate that 3% of rainfall is discharged from the 'open water storage' (*SW*) immediately, regardless of the initial storage conditions in the swamp. This suggests that 3% of the catchment is permanently inundated, or at least saturated, which corresponds with a zone of 250 m on average around the main Jemoreng stream. In reality, permanently inundated areas are likely to occur in 'patches'. During wet periods, the inundated area is much larger than 3%, however.

Modelling the source of peatswamp discharge
The results of the Jemoreng water balance model further indicate that most rainfall that is not discharged as surface runoff (*QD*) from the peat dome within a day, runs off within a few days as subsurface flow (*QS*) through the highly permeable upper 10 cm of peat. The model confirms the finding that groundwater flows at depths greater than 10 cm (*QD*) only form a small component of overall discharge. Water levels in most of the catchment area are just below the peat surface most of the time (Figures 17.2A and 17.6) and only rarely exceed 0.03 m above the surface. Most rainwater is stored for less than one day above the surface as depression storage *SD*, as it runs off rapidly into the open water storage (*SW*) in low-lying areas along streams. Storage in these flooded areas controls the characteristics of the catchment discharge. Backwatering is common during periods of high rainfall, due to the very low gradients within the catchment and the even lower gradient between the catchment outlet and the sea. Floodwater is therefore released only slowly from the open water storage, accounting for the 'sluggish' storm flow response that is sometimes attributed to retention on the peat surface or in the peat dome itself (the 'peatswamp sponge' concept).

Modelling peatswamp discharge during droughts
Despite the fact that groundwater flows are only a small part of the water balance, peatswamps have a reputation for maintaining minimum river flows by producing steady baseflows – and this claim is confirmed to some extent. A 38-year record of daily rainfall was used to simulate long-term discharges from the Jemoreng catchment, as well as water levels. The aim was to determine long-term flow characteristics of peatswamp hydrology, such as the lowest water yield and water level. It was shown that baseflows (from the groundwater reservoir) ceased on two occasions in this period, when the water table dropped to 0.7 m below the peat surface. The ecological literature confirms that serious droughts may occur naturally during the development of peatswamps, which must have some capacity to regenerate after such events (Bruenig, 1990). In most years, however, the lowest baseflows are not below 0.2 mm d^{-1} and the water table does not drop below 0.5 m (Hooijer *et al.*, 1997).

CONCLUDING REMARKS

Hydrological functions of peatswamps

Peatlands are often considered to perform a number of hydrological functions with economic value (Page and Rieley, 1998; Immirzi, 1997). The current study allows an assessment of two of these functions: maintenance of low flows in streams and attenuation of peak flows. The difficulty with assessing such benefits is that they are relative: do we compare peatswamp functions with those of other environments or with a situation after peatswamp drainage?

Taking the example of the 'sponge' function that would help flood storage: peatswamps act as 'sponges' only to store enough water to maintain their shape and vegetation. The peat surface can indeed store rainwater efficiently, but once they are saturated they are equally efficient in discharging excess water rapidly towards streams, in order not to 'drown'. The water may enter the main channel slowly, creating the impression of a 'delayed peatswamp response', but this is not due to storage on the peat dome, but rather to backwatering effects and storage in the depressional floodplains fringing the peat domes (cf. Box 17.1). Of course, larger peat domes than the one studied here may store water for a longer time. Also, it could be argued that peat domes and 'fringing' floodplains are part of the same hydrological system and could not exist without each other, but it may be important to make the distinction for understanding the system as a whole, and for practical management purposes.

Similarly, peatswamp hydrology is 'aimed' at storing water during droughts, not at releasing it to maintain baseflows, to the detriment of peatswamp vegetation. Surface flows cease shortly after storms and groundwater outflows are limited, at 100 mm yr^{-1} in the case of the Jemoreng swamp catchment. Therefore, peatswamps may perform a 'low flow maintenance' function, but a contribution of 0.2 mm d^{-1} during the dry season cannot be called spectacular compared to water systems where the groundwater component is more important. In fact, the present study showed that peatswamp outflow may cease altogether during (rare) extreme drought events. However, peatswamps with shallow peat deposits overlaying sand deposits have been reported to produce higher baseflows than 'typical' peatswamps like Jemoreng (Hooijer *et al.*, 1997).

The findings above do not mean that peatswamps do not perform valuable hydrological functions, but it is important to understand which functions are really performed by a specific swamp, if the case for sustainable management is to be strengthened. However, the most convincing economic argument for sustainable peatswamp management which maintains the peatswamp hydrology, may be that other types of management produce unproductive wastelands in the long term. Subsidence, salinisation, acidification, fires and other problems often render alternative uses economically unfeasible within decades. Although this was not the subject of this study, it would appear that many peatswamps are economically more productive (in the long term) when managed for low-yield timber production in a sustainable way – which would of course exclude drainage and clear-cutting – than when reclaimed for agriculture (Kumari, 1996; Page and Rieley, 1998). Experiences in other types of wetland forests also suggest that sustainable exploitation for timber requires knowledge of the hydrological system and a very cautious approach (e.g. Sun *et al.*, 2001).

Specific problems and opportunities in studies of peatswamp hydrology

The collection of good hydrological data is the chief problem when studying peatswamps, and the methodologies developed to achieve this were an important output of the present study. Forested peatswamps do not only pose a number of unique challenges to hydrologists, but also some unique opportunities for research. The most important are:

(1) Groundwater levels are so high that peatswamps are practically inaccessible. However, these high water tables also create some unique opportunities for research. Being so close to the soil surface, peatswamp water levels react to the smallest input or output. This makes it possible to determine several water balance components ('actual' evapotranspiration and groundwater seepage) using high-definition groundwater level records. Furthermore, groundwater levels also fluctuate uniformly throughout the (small) catchment. This allows determination of changes in storage from changes in groundwater level as measured at only a few locations, or even at a single site.

(2) The topography is so flat, and even variable in time, that accurately delineating a catchment boundary is often impossible. However, it has been shown that a water balance model may be used successfully to optimise the catchment area as delineated through topographic surveys.

(3) In coastal areas, water levels in the low-gradient peatswamp streams may be tidally influenced for long distances inland, rendering discharge monitoring using rating curves impossible. However, with continuous discharge measurements using acoustic Doppler instruments, it was possible to distinguish 'true' catchment outflow from tidal flows (Wong *et al.*, 1997; Hooijer *et al.*, 1997).

Water balance modelling is not usually part of wetland research, due to the considerable data requirements. However, it proved a critical part of the present study, where it served several purposes:

(1) Improved understanding of pathways for water flow within peatswamps.

(2) Filling of gaps in the monitored discharge record.

(3) Simulation of extreme discharges and water levels, that could not be monitored.

(4) Estimation of 'optimum' catchment size.

The results of the modelling were remarkably good in many respects: the coefficient of determination for discharge was 0.91 and the 'optimised' catchment area was very close to the 'minimum probable catchment area' delineated in the field. It is believed that these results show the potential of water balance modelling in tropical wetland forests in other parts of the world.

The importance of water table management in peatswamps and other tropical wetlands

Peatswamps share their basic hydrological characteristics with most other tropical wetland forests: high water tables, low gradients and a high degree of interdependence between hydrology and vegetation. Therefore, the basic conclusions on the hydrology of peatswamps are also likely to apply to other types of tropical wetland forest. The dominant feature of peatswamp forests, i.e. their high water table, is also their weakest point: any change in water level will have significant effects on the hydrology and ecology of the swamps. Because almost all runoff occurs just above or just below the surface, even limited drainage (e.g. via logging tracks) will be very damaging. Finally, most peatswamp functions (ecological and economic ones) listed in the introduction depend primarily on their 'wetland' characteristics. The conclusion is therefore inescapable that the basis for any sustainable use of peatswamps or other forested wetlands must be sound water level management, for which hydrological understanding is needed.

Acknowledgement

Financial support for this challenging field study was provided by the Sarawak Water Resources Council, with technical support provided by Konsultant (Kuching), the Geological Survey of Malaysia and Montgomery Watson Consultants.

References

Abe, K., 1997. Cari rezeki, numpang, siap: the reclamation process of peat swamp forest in Riau. South East Asian Studies Kyoto, Vol. 34, No. 4: 622–632.

Ambak, K. and Chye, L. A. 1996. Peat in Malaysia. In: Global peat resources. Ed.: Lappalainen, E. International Peat Society, Jyskä, Finland. 183–187.

Anderson, J. 1983. The tropical peat swamps of Western Malaysia. In: Gore, A. J. P. (ed.): Mires: swamp, bog, fen and moor, Elsevier, Amsterdam. 181–199.

Anderson, J. 1964. Structure and development of the peat swamps in Sarawak and Brunei. *Journal of Tropical Geography*, 18: 7–16.

Anshari, G., Kershaw, A. P., van der Kaars, S. 2001. A late Pleistocene and Holocene pollen and charcoal record from peat swamp forest, Lake Sentarum Wildlife Reserve, West Kalimantan, Indonesia. In: Quaternary environmental change in the Indonesian Basin. Eds.: Dam, R., van der Kaars, S. Palaeogeography – Palaeoclimatology – Palaeoecology 171: 213–228.

Asdak, C., Jarvis, P. G. and van Gardingen, P. V. 1998. Rainfall interception loss in unlogged and logged forest areas of Central Kalimantan, Indonesia. *Journal of Hydrology*, 206: 237–244.

Bay, R. R. 1969. Runoff from small peatland watersheds. *Journal of Hydrology* 9. 90–102.

Bruenig, E. F. 1990. Oligotrophic forested wetlands in Borneo. In: Forested wetlands – Ecosystems of the world no. 15. Eds: Lugo, A. E., Brinson, M. M., Brown, S. Elsevier, Amsterdam. 299–334.

Bruijnzeel, L. A. 1990. Hydrology of moist tropical forests and effects conversion: a state of knowledge review. Unesco Humid Tropics Programme.

Davie, J., Samardja, E. 1997. The protection of forested coastal wetlands in southern Sumatra: a regional strategy for integrating conservation and development. *Pacific Conservation Biology*, Vol. 3, No. 4: 366–378.

Dolan, T. J., Hermann, A. J., Bayley, S. E. and Zoltek, J. 1984. Evaporanspiration of a Florida, USA, freshwater wetland. *Journal of Hydrology* 74: 355–371.

Dykes, A. P. 1997. Rainfall interception from a lowland tropical rainforest in Brunei. *Journal of Hydrology*, 200: 260–279.

Gilman, K. 1997. Bog hydrology: the water budget of the mire surface. In: Conserving Peatlands, edited by: Parkyn, L, Stoneman, R. E. and Ingram, H. A. P. CAB International, Wallingford. 132–138.

Gilman, K. 1994. *Hydrology and wetland conservation*. Wiley, London.

Gilvear, D. J. and Bradley, C. 2000. Hydrological monitoring and surveillance for wetland conservation and management; a UK perspective. Phys. Chem. Earth (B), Vol. 25, No. 7–8: 571–588.

Hooijer, A. and Sivapalan, M. 1995. Review and fieldwork planning for water balance modelling in Sarawak peatswamps. Research Report WP-1037-AH, Centre for Water Research, University, University of Western Australia.

Hooijer, A. 1996. Floodplain hydrology – an ecologically oriented study of the Shannon Callows, Ireland. Published PhD Thesis, Vrije Universiteit Amsterdam, The Netherlands.

Hooijer, A., Phillips, R. L., Pattiaratchi, C. B., Sivapalan, M. 1997. Sarawak Water Resources Study: Modelling Research Studies. Research Report WP-1274-AH, Centre for Water Research, University, University of Western Australia. Summary: http://www.cwr.uwa.edu.au/cwr/publications/reports/1274.html

Immirzi, P. 1997. Wasted Assets or Wasting Assets: Peatland Functions and values. In: Conserving Peatlands, ed.: Parkyn, L, Stoneman, R. E. and Ingram, H. A. P. CAB International, Wallingford, UK. 147–157.

Ingram, H. A. P. 1983. Hydrology. In: Mires: swamp, bog, fen and moor. Elsevier, Amsterdam. 67–158.

Kumari, K. 1996. Sustainable forest management: myth or reality? Exploring the prospects for Malaysia. Ambio Vol. 25. No. 7: 459–467.

Labaugh, J. W. 1986. Wetland ecosystem studies from a hydrologic perspective. Water Resources Bulletin, Vol. 22, No. 1: 1–10.

Laine, J. 1984. Estimation of evapotranspiration from peatlands by means of daily water table hydrographs. Publication 5. Dept. Peatl, For., Un. Helsinki.

Lappailanen, E. (ed.) 1996. Global Peat Resources. International Peat Society, Jyskä, Finland.

Lappailanen, E. and Zurek, S. 1996. Peatlands in Central and South America. In: Global Peat Resources. Ed.: Lappalainen, E. International Peat Society, Jyskä, Finland. 279–282.

Legg, C. A., Laumonier, Y. 1997. Fires in Indonesia, 1997: a remote sensing perspective. Ambio, Vol. 28, No. 6: 479–485.

Loftin, C. S., Rasberry, W., Kitchens and W. M. 2000. Development of a grid-cell topographic surface for Okefenokee Swamp, Georgia. WETLANDS, Vol. 20, No. 3: 487–499.

Lugo, A. E., Brown, S. and Brinson, M. M. (eds.) 1990. Forested Wetlands – Ecosystems of the World no. 15. Elsevier, Amsterdam.

Malterer, T. J., 1996, Peat resources of the United States. In: Global Peat Resources. Ed.: Lappalainen, E. International Peat Society, Jyskä, Finland. 253–260.

Mansell, R. S., Bloom, S. A., Sun, G. 2000. A model for wetland hydrology: description and validation. Soil Science, Vol. 165, No. 5: 384–397.

Muhamad, N. Z. 2001. Management of tropical peatlands in Indonesia: Mega Reclamation Project in Central Kalimantan. Unpublished Report for the School of Geography, University of Nottingham, UK. At: http://www.geocities.com/kopitubrok/Report2.html

Page, S. E., Rieley, J. O. Shotyk, O. W., Weiss, D. 1999. Interdependence of peat and vegetation in a tropical peat swamp forest. Phil. Trans. R. Soc. Lond. B (1999). No. 354: 1885–1897.

Page, S. E., Rieley, J. O., 1998, Tropical peatlands; a review of their natural resource functions, with particular reference to South East Asia. International Peat Journal, Vol. 8.

Pajunen, H. 1996. Peatlands in Africa. In: Global Peat Resources. Ed.: Lappalainen, E. International Peat Society, Jyskä, Finland. 213–225.

Parkyn, L, Stoneman, R. E. and Ingram, H. A. P. (eds.) 1997, Conserving Peatlands. CAB International, Wallingford.

Phillips, R. L., Hooijer, A., Sivapalan, M., Pattiaratchi, C. B. 1997a. Preliminary results from a modelling study of catchment yield and saline intrusion in the Sarawak coastal peat wetlands. Proceedings of the Conference on Recent Advances in Soft Soil Engineering, 1997, Kuching, Malaysia. 499–510.

Phillips, S., Rouse, G. E. Bustin, R. M. 1997b. Vegetation zones and diagnostic pollen profiles of a coastal pet swamp, Bocas del Toro, Panama. In: Quaternary environmental change in the Indonesian Basin. Eds.: Dam, R., van der Kaars, S. Palaeogeography – Palaeoclimatology – Palaeoecology, Vol. 171, No. 3–4: 301–338. Phillips, V. D. 1998. Peatswamp ecology and sustainable development in Borneo. Biodiversity and Conservation, Vol. 7, No. 5: 651–671.

Ramadasan, K., Abdullah, D. M., Koay, L. 1999. The influence of highland forests, wetlands on floods in Malaysia. In: Proceedings of Intl. Workshop on Flood Forecasting in Tropical Regions, 1999, Kuala Lumpur, Malaysia.

Riekerk, H. and Korhnak, L. V. 2000. The hydrology of cypress wetlands in Florida pine flatwoods. WETLANDS, Vol. 20, No. 3: 448–460.

Rieley, J. O. 2001. Kalimantan's peatland disaster. Unpublished Report: http://www.insideindonesia.org/edit65/jack.htm

Rieley, J. O. 1999. Borneo's chainsaw massacre. The Guardian Online, 18 February 1999.

Rieley, J. O., Page, S. E., Shepherd, P. A. 1997. Tropical bog forests of South East Asia. In: Conserving Peatlands, ed.: Parkyn, L, Stoneman, R. E. and Ingram, H. A. P. CAB International, Wallingford, UK. 35–34.

Rieley, J. O. Page, S. E., Setiadi, B. 1996. Distribution of peatlands in Indonesia. In: Global peat resources. Ed.: Lappalainen, E. International Peat Society, Jyskä, Finland. 169–177.

Ruecker, G. and Siegert, F. 2000. Burn scar mapping and fire damage assessment using ERS-2 SAR images in East Kalimantan, Indonesia. IAPRS, Amsterdam, Vol. XXXIII.

Staub, J. R. and Esterle, J. S. 1993. Provenance and sediment dispersal in the Rajang river delta / coastal flood plain, Sarawak, East Malaysia. Sedimentary Geology, 85: 191–201.

Staub, J. R. and Esterle, J. S. 1994. Peat-accumulating depositional systems of Sarawak, East Malaysia. Sedimentary Geology, 89: 91–106.

Sun, G., McNulty, S. G., Shepard, J. P., Amatya, D. M., Riekerk, H., Comerford, N. B., Skaggs, W., Swift, L. 2001. Effects of timber management on the hydrology of wetland forests in the southern United States. Forest Ecology and Management, Vol. 143, No. 1–3: 227–236.

Van der Molen, M. K., 2002. Meteorological impacts of land use change in the maritime tropics. Ph.D. thesis, Vrije Universiteit, Amsterdam, The Netherlands, 262 pp.

Wallace, J. 1997. Classification of Landcover for Water Catchment Areas in Sarawak using Landsat TM Imagery. CSIRO, Perth, Australia.

Ward, R. C. and Robinson, M. A., 1990. Principles of Hydrology, 2nd edition, McGraw-Hill, Maidenhead, UK.

Wilsnack, M. M., Welter, D. E., Montoya, A. M. Restrepo, J. I., Obeysekera, J. 2001. Simulating flow in regional wetlands with the modflow wetlands package. Journal of the American Water Resources Association, Vol. 37, No 3: 655–674.

Werner, C. L., Wiesman, A., Siegert, F., Kuntz, S. 2000. JERS INSAR DEM generation for Borneo. Proceedings of the International Geoscience and Remote Sensing Symposium (IGARSS 2000), Honolulu, USA, Vol. 5: 2248–2250.

Wong, M., Malone, D., Tite, I., Hayes, J. 1997. Monitoring of hydrological processes in the Sarawak Coastal peat wetlands. Proceedings of the Conference on Recent Advances in Soft Soil Engineering, 1997, Kuching, Malaysia. 484–498.

Wösten, J. H. M., Ismail, A. B., van Wijk, A. L. M. 1997. Peat subsidence and its practical implications: a case study in Malaysia. Geoderma 78: 25–36.

18 Tropical montane cloud forest: a unique hydrological case[1]

L. A. Bruijnzeel
Vrije Universiteit, Amsterdam, The Netherlands

INTRODUCTION

The paper by Zadroga (1981) on the hydrological significance of tropical montane cloud forests (TMCF) in northern Costa Rica probably marks the start of the enhanced interest in these remarkable forests although the importance of fog deposition on vegetation surfaces as an extra source of moisture has been acknowledged for a long time (see Kerfoot's 1968 review of early literature).

Arguably, this increased interest is in no small measure due to the unstinting efforts of one man, Professor Lawrence S. Hamilton, who recognised the far-reaching implications of Zadroga's preliminary work and who kept stressing the hydrological and ecological importance of TMCF on numerous occasions. Hamilton's efforts culminated in the organisation of the First International Symposium on Tropical Montane Cloud Forests, held in San Juan, Puerto Rico, from 31 May until 5 June 1993 (Hamilton, Juvik and Scatena, 1995), and the launching of *'A Campaign for Cloud Forests'* by the World Conservation Union (IUCN) in 1995 (Hamilton, 1995a). The hydrological and biogeochemical evidence on TMCF was reviewed in detail at the Puerto Rico Symposium by Bruijnzeel and Proctor (1995). These authors stressed how little is actually known about the hydrological functioning of different types of montane forests exposed to varying degrees of cloud impaction; the role of epiphytes in cloud water interception and retention; cloud forest carbon dynamics and the factors limiting their growth; and, above all, the uncertainty surrounding the water use of different types of TMCF and the effect of their conversion to pasture or vegetable cropping on downstream water yield. Bruijnzeel and Proctor (1995) also called for the establishment of a pan-tropical network linking the more data-rich TMCF research sites where these important questions could be addressed in an integrated manner.

The nomenclature of montane forests, including TMCF, is confusing. Stadtmüller (1987) listed at least 35 different names that have been used to typify 'cloud forest'. Therefore, before reviewing the results of hydrological research in TMCF (with emphasis on post-1993 work, i.e. published or initiated after the Puerto Rico Symposium), a simple classification of TMCF types is proposed to allow for hydrological distinctions between the different forest types. In addition, background information is provided on the chief controls governing TMCF occurrence. Finally, the chapter identifies the chief remaining research questions with suggestions for where and how these might be addressed.

TROPICAL MONTANE CLOUD FORESTS: DEFINITIONS AND OCCURRENCE

With increasing elevation on wet tropical mountains, distinct changes occur in forest appearance and structure. At first, these changes are gradual. The tall and often buttressed trees of the multi-storeyed *lowland rainforest* (main canopy height 25–45 m, with emergents up to 60 m), gradually give way to *lower montane forest*. With a mean canopy height of up to 35 m in the lower part of the montane zone and emergent trees as high as 45 m, lower montane forest can still be quite impressive. Yet, with two rather than three main canopy layers, the structure of lower montane forest is simpler than that of lowland forest. Also, the large buttresses and climbers that are so abundant in the lowland forest all but disappear while epiphytes (orchids, ferns, bromeliads) become more numerous on branches and stems with increasing elevation (Whitmore, 1998). The change from lowland to lower montane forest seems to be controlled largely by temperature as it is normally observed at the elevation where the average minimum temperature drops below 18 °C. At this threshold many lowland tree species are displaced by a floristically different assemblage of montane species (Kitayama, 1992). On large equatorial inland mountains this transition usually occurs at an altitude of

1 Largely based on Bruijnzeel (2001/2002a) and updated in April 2003.

Forests, Water and People in the Humid Tropics, ed. M. Bonell and L. A. Bruijnzeel. Published by Cambridge University Press.

1200–1500 m but it may occur at much lower elevations on small outlying island mountains and away from the Equator (see also below). As elevation increases further, the trees not only become gradually smaller but also more 'mossy' (changing from *c.* 10% to 25–50% moss cover on the stems). There is usually a very clear change from relatively tall (15–35 m) lower montane forest to distinctly shorter-statured (2–20 m) and much more mossy (70–80% bryophytic cover) *upper montane forest* (Frahm and Gradstein, 1991). Although at this point the two forest types are not separated by a distinct thermal threshold, there can be little doubt that the transition from lower to upper montane forest coincides with the level where cloud condensation becomes most persistent (Grubb and Whitmore, 1966). On large mountains in equatorial regions away from the ocean this typically occurs at elevations of 2000–3000 m but incipient and intermittent cloud formation is often observed from *c.* 1200 m upwards, *i.e.* roughly at the bottom end of the lower montane zone. On small oceanic island mountains, however, the change from lower to upper montane-looking forest may occur at much lower altitudes (down to less than 500 m above sea level) (Van Steenis, 1972). Mosses also start to cover rocks and fallen trunks on the soil surface in the upper montane forest zone. With increasing elevation and exposure to wind-driven fog and rain, tree stems become increasingly crooked and gnarled, and bamboos often replace palms as dominant undergrowth species (Kappelle, 1995). The eerie impression of this tangled mass, wet with fog and glistening in the morning sun, has given rise to names like 'elfin' forest or 'fairy' forest for the more dwarfed forms of these upper montane forests (Stadtmüller, 1987).

A third major change in vegetation composition and structure typically occurs at the elevation where the average maximum temperature falls below 10 °C . Here the upper montane forest gives way to still smaller-statured (1.5–9 m) and more species-poor *subalpine forest* (or scrub) (Kitayama, 1992). This forest type is characterised not only by its low stature and gnarled appearance but also by even tinier leaves, and a comparative absence of epiphytes. Mosses usually remain abundant, however, confirming that cloud incidence is still a paramount feature (Frahm and Gradstein, 1991). On large equatorial mountains the transition to subalpine forest is generally observed at elevations between 2800 and 3200 m. As such, this type of forest is encountered only on the highest mountains, mostly in Latin America and Papua New Guinea, where it may extend to *c.* 3900 m (Whitmore, 1998).

It follows from the preceding descriptions that most lower montane, and all upper montane and subalpine forests, are subject to various degrees of cloud incidence. As indicated earlier, definitions, names and classification of the respective vegetation complexes are myriad, as well as overlapping and, at times, contradictory (Stadtmüller, 1987). Bruijnzeel and Hamilton (2000) proposed to distinguish the following forest types that become increasingly mossy with elevation:

- lower montane forest (tall forest little affected by low cloud but rich in epiphytes, particularly in the upper reaches of the zone);
- lower montane cloud forest (25–50% moss cover on stems);
- upper montane cloud forest (70–80% epiphyte cover on stems); and
- subalpine cloud forest.

In doing so, the widely adopted broad definition of cloud forests as 'forests that are frequently covered in cloud or mist' (Stadtmüller, 1987; Hamilton *et al.*, 1995) is included whilst at the same time recognising the important influence of temperature and humidity on montane forest zonation. However, a fifth and more or less 'a-zonal' cloud forest type should be added, low-elevation dwarf (or 'elfin') cloud forest (see below).

The large variation in elevation at which one forest formation may replace another is caused by several factors. For example, the transition from lower to upper montane forest is mainly governed by the level of persistent cloud condensation (Grubb and Whitmore, 1966). Cloud formation, in turn, is determined by the moisture content and temperature of the atmosphere. Naturally, the more humid the air, the sooner it will condense upon being cooled during uplift. With increasing distance from the ocean the air tends to be drier. As such, it will take longer to cool to its condensation point and the associated cloud base will be higher. Likewise, for a given moisture content, the condensation point is reached more rapidly for cool air than for warm air. Thus, at greater distance from the Equator, the average temperature, and thus the altitude at which condensation occurs, will be lower (Nullet and Juvik, 1994). Superimposed on these global atmospheric moisture and temperature gradients are the more local effects of sea surface temperatures and currents, the size of a mountain and its orientation and exposure to the prevailing winds, as well as local topographic factors (Stadtmüller, 1987). It goes almost without saying that sea surface temperatures influence the temperature of the air overhead and thus the 'starting point' for cooling. Also, where warm, humid ocean air is blown over a comparatively cold sea surface, a low-lying layer of persistent coastal fog tends to develop. Well-known examples are the fog-ridden west coast of California where tall redwood forests thrive in an otherwise sub-humid climate (Dawson, 1998), and the coastal hills of Chile and Perú, where, under conditions approaching zero rainfall, forest groves are able to survive solely on water stripped from the fog by the trees themselves (Aravena, Suzuki and Pollastri, 1989).

The occurrence of low-statured mossy, upper montane-looking forest at low elevations on small, isolated coastal mountains has

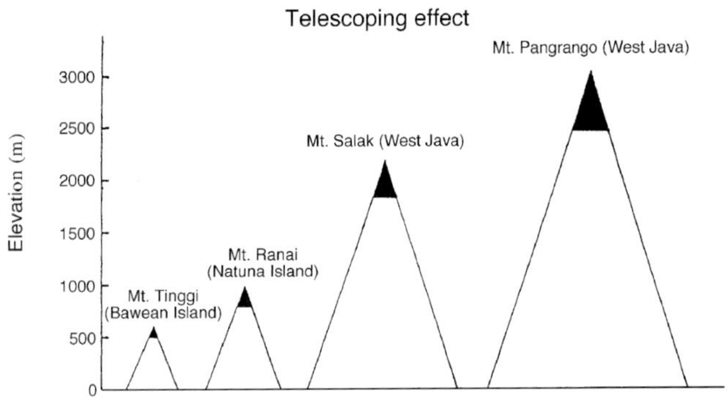

Figure 18.1 The telescoping effect of vegetation zonation on differently sized mountains. (After Van Steenis, 1972.)

puzzled scientists for a long time. This phenomenon is commonly referred to as the 'mass elevation' or 'telescoping' effect (Van Steenis, 1972; Whitmore, 1998) (Figure 18.1). The sheer mass of large mountains exposed to intense radiation during cloudless periods is believed to raise the temperature of the overlying air, thus enabling plants to extend their altitudinal range. Whilst this may be true for the largest mountain ranges it is not a probable explanation for mountains of intermediate size on which the effect is also observed. Instead, the contraction of vegetation zones on many small coastal mountains must be ascribed to the high humidity of the oceanic air promoting cloud formation at (very) low elevations rather than to a steeper temperature lapse rate with elevation associated with small mountains. Further support for this comes from the observation that the effect is most pronounced in areas with high rainfall and thus high atmospheric humidity (Van Steenis, 1972; Bruijnzeel *et al.*, 1993).

Whilst the cloud base on small islands is often observed at an elevation of 600–800 m, dwarf cloud forests reach their lowermost occurrence on coastal slopes exposed to both high rainfall and persistent wind-driven cloud. Examples from the equatorial zone include Mount Payung near the western tip of Java and Mount Finkol on Kosrae island (Micronesia) where dwarf forests are found as low as 400–500 m (Hommel, 1987; Merlin and Juvik, 1995). An even more extreme case comes from the island of Gau in the Fiji archipelago where the combination of high precipitation and strong winds has led to the occurrence of a wind-pruned dwarf cloud forest at an altitude of only 300–600 m above sea level (Watling and Gillison, 1995).

These examples illustrate the importance of site exposure. Generally, the lower limits of mossy forest of any kind (upper montane, subalpine, or dwarf cloud forest at low elevation) on drier and more protected leeward slopes lie well above those on windward slopes. In extreme cases, such as in the Colombian Andes, the difference in elevation may be as much as 600 m (Stadtmüller, 1987). Also, leeward forests tend to be better developed than their more exposed windward counterparts at the same elevation. For example, in the Monteverde Cloud Forest Preserve, northern Costa Rica, the trees of 'leeward cloud forest' are 25–30 m tall *v.* 15–20 m in nearby floristically similar 'windward cloud forest'. Moreover, towards the exposed crests of the windward slopes the height of the vegetation decreases further to 3–10 m along an altitudinal gradient of only 30–50 m (Lawton and Dryer, 1980; cf. Weaver (1995) for similar contrasts in Puerto Rico).

Although the stunted appearance of low-elevation dwarf cloud forests resembles that of the transition from high-elevation upper montane to subalpine cloud forests at first sight, the two differ in several important respects. At low elevations, the leaves are much larger and the floristic composition is very different (Grubb, 1974). Also, the degree of moss cover on the ground (but not the vegetation) is generally much less pronounced at lower altitudes (Frahm and Gradstein, 1991). Lastly, the temperatures and thus overall evaporative demand to which the forests are exposed are (much) higher at lower elevations (Nullet and Juvik, 1994). The soils of upper montane and dwarf cloud forests (regardless of elevation) are typically very wet and, in extreme cases, persistently close to saturation. As a result, decomposition of organic matter is slow and topsoils become peaty and acid (Bruijnzeel and Proctor, 1995). Recent work in the Blue Mountains of Jamaica suggests that the most stunted upper montane cloud forests suffer from toxic levels of aluminium in their soils which, in turn, affect nutrient uptake by the trees and a host of other forest ecological processes (see Hafkenscheid (2000) for details). At the other end of the scale, the very tall (up to 50 m) montane oak forests found at high elevations (up to 3000 m) on the large inland mountain massifs of Latin America (Kappelle, 1995) and Papua New Guinea (Hyndman and Menzies, 1990) more than likely reflect a fortunate combination of slightly warmer and drier air (due to the 'mass elevation' effect, distance to the sea and topographic protection) and the presence of well-drained soils in which the toxic conditions described by Hafkenscheid (2000) for the wettest localities do not easily develop.

So far, the focus has been on the climatic gradients and other factors governing the elevation of the cloud *base*. Another climatological phenomenon, which influences the vertical temperature profile of the air and the *top level* of cloud formation, is the so-called 'trade wind inversion'. As part of a large-scale atmospheric circulation pattern (the Hadley cell), heated air rises to great elevation in the equatorial zone, flowing poleward and eastward at upper atmospheric levels and descending in a broad belt in the outer tropics and subtropics from where it returns to the Equator. This subsidence reaches its maximum expression at the oceanic subtropical high-pressure centres and along the eastern margins of the oceanic basins. As the air descends and warms up again, it forms a temperature inversion that separates the moist layer of surface air (that is being cooled while rising) from the drier descending air above. The inversion forms a tilted

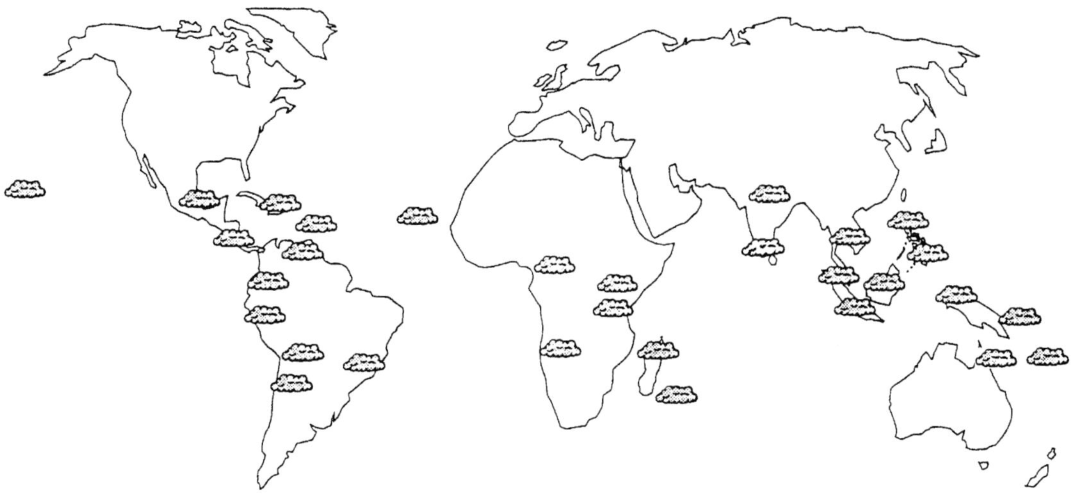

Figure 18.2 Generalized occurrence of montane cloud forests in the (sub)tropics. (Adapted from Hamilton *et al.*, 1995.)

three-dimensional surface, generally rising towards the Equator and from east to west across the oceans. Over the eastern Pacific Ocean, the inversion is found at only a few hundred metres above sea level, e.g. off the coast of southern California, rising to about 2000 m near Hawai'i and dissipating in the equatorial western Pacific (Nullet and Juvik, 1994). The low elevations at which the inversion occurs on mountains situated away from the Equator may well be another reason why the vegetation zonation tends to become compressed on smaller mountains (Stadtmüller, 1987). The consequences of the trade wind inversion for the occurrence of the upper boundary of montane cloud forest are profound. For instance, at 1900–2000 m on the extremely wet windward slopes on islands in the Hawai'ian archipelago, the montane cloud forest suddenly gives way to dry sub-alpine scrub because the clouds (which generally deliver more than 6000 mm of rain per year below the inversion layer) are prevented from moving upward by the presence of the temperature inversion (Kitayama and Müller-Dombois, 1992). One of the best-known examples of the trade wind inversion and its effect on vegetation zonation comes from the Canary Islands. Situated between 27 and 29 degrees north, a daily 'sea of clouds' develops between 750 m and 1500 m which sustains evergreen Canarian laurel forests in an otherwise rather arid environment (Ohsawa, Wildpret and del Arco, 1999).

As a result of the various climatic and topographic gradients described in the previous paragraphs, concentrations of montane cloud forests in the tropical and subtropical parts of the world occur approximately as shown on the generalised map in Figure 18.2. Further details on TMCF distribution can be found in Hamilton *et al.* (1995) and a draft directory of TMCF sites has been published by Aldrich *et al.* (WCMC, 1997).

HYDROLOGICAL PROCESSES IN TROPICAL MONTANE CLOUD FORESTS

Rainfall and cloud interception

One of the most important aspects in which cloud forests differ from montane forests that are not affected much by fog (low cloud) concerns the deposition of cloud water onto the vegetation. Whilst the hydrological and ecological importance of this extra input of moisture is widely recognised, its quantification is notoriously difficult (Kerfoot, 1968). Two approaches are usually followed: (i) the use of 'fog' gauges, of which there are many types, and (ii) a comparison of amounts of canopy drip as measured inside the forest with amounts of rainfall measured in the open. Both methods are fraught with difficulties of measurement and interpretation of the results. Therefore, before presenting recent research results obtained with either method, the chief limitations of the two approaches are discussed below.

Fog gauges

The inherent problem of fog gauges is that no gauge, whether of the 'wire mesh cylinder' (Grünow) type (Russell, 1984), the 'wire harp' type (Goodman, 1985), the 'louvered-screen' type (Juvik and Ekern, 1978), or the more recently proposed poly-propylene 'standard' fog collector of Schemenauer and Cereceda (1994), can mimic the complexities of a forest canopy. Also, each forest represents a more or less unique situation that defies standardisation. Therefore, fog gauges can only be used as comparative instruments (e.g. for site climatic characterisation) and, provided they are protected against direct rainfall and equipped with a recording mechanism, for the evaluation of the timing and frequency of occurrence of fog. Where concurrent information on wind speed is available as well, the liquid water content of the fog may also be evaluated from measured volumes of fog water (Padilla *et al.*, 1996). However, apart from its intrinsic trapping efficiency, the catch of a fog gauge is highly dependent on its position with respect to the ground and nearby obstacles. It has been recommended to install gauges at a 'standard' height of 2 m (Schemenauer and Cereceda, 1994) or 3 m (Juvik and Ekern, 1978). Often, however, studies using fog gauges in the tropics have not specified gauge height or position, rendering interpretation of the results more difficult, even more so when placed above an aerodynamically rough surface such as a forest canopy (Bruijnzeel and Proctor, 1995).

A major problem of interpretation associated with most fog gauges concerns the distinction between cloud water and wind-driven rain (Hafkenscheid, 2000; Cavelier, Solis and Jamarillo, 1996). Adding a protective cover to keep out vertical rain may further complicate things as different amounts of wind-driven rain will be included in the catch depending on wind speeds and the inclination of the rain drops (Sharon, 1980; cf. Juvik and Nullet, 1995a). For particularly windy and exposed conditions, Daube *et al.* (1987) proposed the use of a wire harp collector enclosed in a rain-proof box in which air flow is restricted by two baffles. The front baffle causes the passing air to accelerate and project heavy raindrops against the rear baffle where they are drained away. The lighter fog particles continue on and impact against the collecting harp. This type of fog collector has been used successfully above a fog-ridden lower montane forest in southern Queensland, Australia, by Hutley *et al.* (1997). Elsewhere in Queensland, Herwitz and Slye (1992) used a more theoretical approach to evaluate the angle of wind-driven rain as a function of wind speed and drop size (based in turn on rainfall intensity; Sharon, 1980).

There has been some debate as to what is the most suitable type of fog gauge under the windy and rainy conditions that prevail on many tropical mountains (Juvik and Nullet, 1995a; Schemenauer and Cereceda, 1995). Metal louvered screen collectors have been shown to drain their catch (both rain and cloud water) more efficiently than wire mesh screens (Juvik and Ekern, 1978), whereas cylindrical designs are considered superior to two-dimensional screens in terms of presenting the same silhouette and catchment surface configuration regardless of wind direction (Juvik and Nullet, 1995a). On the other hand, the catching surface of cylindrical gauges is generally much smaller than that of a large screen such as that proposed by Schemenauer and Cereceda (1994). The latter may thus generate measurable deposition rates when fog liquid water contents are low or winds are light (Schemenauer and Cereceda, 1995). There is a need to test the relative performance of the various gauge types under typical cloud forest conditions against concurrent measurements of visibility or cloud liquid water content (Burkard *et al.*, 2002). Therefore, the results of two such ongoing comparative experiments at East Peak in the Luquillo Mountains of Puerto Rico and in the windy Tilarán range near Monteverde, Costa Rica (F. Holwerda and K. F. A. Frumau, pers. comm.) are awaited with interest.

Measurement of net precipitation

Subtracting amounts of throughfall (T_f) plus stemflow (S_f) (together making up net rainfall) as measured below the forest canopy from gross rainfall measured above the forest or in a nearby clearing (P_g), gives the amount of precipitation intercepted by the canopy and evaporated back to the atmosphere during and shortly after the event. This process is usually referred to as rainfall interception (E_i) or wet canopy evaporation and implies a *net loss of water* to the forest:

$$E_i = P_g - (T_f + S_f) \qquad (18.1)$$

where the terms are as defined above and expressed in mm of water per time period. Where fog or cloud only is present, a similar process of cloud interception (CW) may be expressed as:

$$\mathrm{CW} = E_{i_{cw}} + T_f + S_f \qquad (18.2)$$

However, because neither the actual amounts of cloud interception (CW) nor those evaporated from the wetted vegetation ($E_{i_{cw}}$) are easily quantifiable in a direct manner, a more practical approach is to measure net precipitation and equate the amount to *net* cloud interception CW_{net}:

$$\mathrm{CW}_{net} = T_f + S_f \qquad (18.3)$$

where the term 'cloud interception' implies a *net gain of water* to the ecosystem. In the more complex case of rainfall plus cloud incidence, separate knowledge of the total evaporation from the vegetation wetted by both rain and fog (ΣE_i) would be required to solve the wet canopy water budget equation for CW:

$$P_g + \mathrm{CW} = \Sigma E_i + T_f + S_f \qquad (18.4)$$

Solving Eqn (18.4) under the climatic conditions prevailing at many cloud forest sites is not easy for a number of reasons. Firstly, depending on wind speeds and rainfall intensities, P_g can be severely underestimated because of unaccounted wind-driven rain missed by a standard rain gauge. Although various corrections have been proposed for this phenomenon (e.g. Sharon, 1980; Yang *et al.*, 1998) there is the added complication afforded by live forest canopies in which emergent trees sticking out of the main canopy tend to catch inclined rainfall (and CW) more efficiently than their more sheltered neighbours, thereby increasing the overall catch to a level that exceeds amounts of conventionally measured rainfall in the open (Herwitz and Slye, 1992). Thirdly, because net precipitation under these conditions often exceeds P_g, it is not possible to estimate ΣEi in a manner analogous to Eqn 18.1. Some investigators have therefore used the wet canopy version of the Penman-Monteith equation to approximate ΣEi although this may result in an underestimation due to problems with advected energy (see discussion in Roberts *et al.,* this volume). Furthermore, for the proper quantification of net precipitation ($T_f + S_f$) large numbers of throughfall gauges (>20–30) are usually needed to account for the high spatial variability of rainforest canopies. In addition, it is advisable to apply a 'roving' gauge technique that is considered to include 'drip' points (where rain or fog drip becomes concentrated because of peculiarities in the configuration of the trees) in a more representative manner than a fixed gauge arrangement would do. Although amounts of throughfall sampled in this way in lowland rainforest have been shown to be significantly higher than when a fixed gauge network is used (Lloyd and Marques, 1988), the roving gauge technique has been little used so far in TMCF (Bruijnzeel and Proctor, 1995) and published results may therefore represent underestimates (cf. Hafkenscheid *et al.*, 2002). On the other hand, regularly relocating one's throughfall gauges easily causes major disturbance to the often fragile soils of TMCF, particularly on steep slopes. As such, the use of a large number of gauges in a fixed spatial arrangement has been preferred by some investigators, particularly under seasonal conditions in which the assumption of a temporally non-variant canopy that underlies the roving approach no longer holds (Köhler, 2002; cf. Brouwer, 1996). Given the very high stemflow proportions generated in the more stunted cloud forest types with their multiple-stemmed and crooked trees, adequate attention should also be paid to the stemflow sampling design.

The classic approach to evaluate contributions by cloud water to forests by simply comparing amounts of net and gross precipitation for events with and without fog ignores potential contrasts in rates of wet canopy evaporation under the respective conditions (Kashiyama, 1956; Harr, 1982). Furthermore, given the high spatial variability in net precipitation already referred to, this method only works well if cloud water contributions are substantial and temporally well-defined or if the confidence intervals for the net precipitation estimates are narrow through the use of a sufficiently large number of throughfall and stemflow gauges. For example, both Hafkenscheid (2000) and Schellekens *et al.* (1998) reported that regression equations linking gross and net precipitation in TMCF in Jamaica and Puerto Rico, respectively, did not differ significantly for events with and without fog, thus rendering the approach meaningless from the statistical point of view in these particular cases, possibly because an insufficient number of (roving) gauges was used (10–12); (Table 18.2).

ALTERNATIVE APPROACHES

In view of the above-mentioned difficulties with the more traditional approaches various alternative methods have been advanced but these too have met with variable success. Exploiting the fact that concentrations of sodium and chloride in cloud water are generally much higher than in rainfall (Asbury *et al.*, 1994; Clark *et al.*, 1998), Hafkenscheid, Bruijnzeel and De Jeu (1998) attempted to evaluate the contribution of cloud water to net precipitation in two upper montane cloud forests in Jamaica of varying exposure using a *sodium mass balance approach*:

$$(P_g \times C_{P_g}) + (\mathrm{CW} \times C_{\mathrm{cw}}) = (T_f \times C_{T_f}) + (S_f \times C_{S_f}) \quad (18.5)$$

in which C denotes the concentration of Na (or any other suitable constituent) in the respective components. Whilst a reasonable estimate was obtained for the most exposed forest, an unexpectedly high cloud water input was derived for the less exposed forest, suggesting that application of the chemical mass balance approach may be less than straightforward in complex mountainous terrain, possibly due to spatial variations in dry deposition. A similar approach makes use of the difference in isotopic composition of rain and fog water (fog often being enriched in the heavier isotopes ^{2}H and ^{18}O relative to rainfall in the same region; Ingraham and Matthews, 1988, 1990; Scholl *et al.*, 2002). Dawson (1998) and Te Linde *et al.* (2001) applied this *isotope mass balance technique* to quantify contributions by CW to the water budgets of a redwood forest in California and elfin cloud forest in Puerto Rico, respectively. In the latter case, Eqn 18.4 was then used to derive a plausible estimate for ΣE_i (4.4% of P_g) where the simple subtraction of net precipitation from gross rainfall had previously yielded a negative value of 7.5%.

Of late, various studies have attempted to measure fog deposition onto forest canopies in the temperate zone directly through the use of a so-called eddy covariance set-up in which a three-dimensional sonic anemometer (measuring turbulence) is combined with an active high speed cloud particle spectrometer (measuring fog liquid water content) (Kowalski *et al.*, 1997; Eugster *et al.*, 2001; Burkard *et al.*, 2002). Although promising, the technique is not without problems, among others because of flux divergences related to uncertainties in the magnitude of advected air streams and what has been termed the 'local net source and sink term' (i.e. condensation and evaporation of fog droplets; Burkard *et al.*, 2002). The first application of the eddy covariance technique in a tropical montane cloud forest setting (Puerto Rico) suggested contributions by CW to be much smaller than those by wind-driven rain (F. Holwerda, pers. comm.).[2]

An alternative experimental process-based approach to the evaluation of CW has been followed by M. Mulligan and A. J. Jarvis (pers. comm., February 2000) who *monitored the changes in weight of a known mass of living mossy epiphytes* suspended below the canopy of a TMCF in Colombia over an extended period

2 A similar experiment was initiated by the University of Bern and the Vrije Universiteit Amsterdam in windward lower montane cloud forest in the Tilarán range of northern Costa Rica in February 2003 (R. Burkard and K. F. A. Frumau, pers. comm.).

(weeks). Their alternative 'cloud trap' was protected against rainfall and extended over the first 5 m above the forest floor. No fog drip was recorded from this device, suggesting that either most of the intercepted cloud water was evaporated again or that absorbed amounts were too low to generate drip. Rates of both processes were comparable. A different result might have been obtained if the interceptor had been allowed to be wetted by throughfall or if it had been placed at a more exposed position higher up in the canopy. Chang, Lai and Wu (2002) followed a similar approach in a montane coniferous forest at a particularly foggy location in Taiwan using much smaller moss samples in a series of wetting and drying experiments under field conditions. By multiplying the average rate of fog absorption for various mosses times the estimated epiphytic biomass a stand-scale deposition rate of 0.17 mm h^{-1} (mosses only) was derived. Although the fog stripping capacity of individual conifer leaves was about half that of the mosses (0.30 vs. 0.63 g H_2O g^{-1} dry weight h^{-1}) the corresponding leaf biomass was about 20 times larger. As such, the fog stripping capacity of the canopy as a whole was considered to be closer to *c.* 2 mm h^{-1}. Unexpectedly, little change in weight occurred during moss exposure on a sunny day. With average daily fog durations of 4.7–11 h (depending on the time of year) overall fog deposition at this site can be expected to be considerable, therefore (Chang *et al.*, 2002). Köhler (2002), on the other hand, reported a distinct seasonal variation in moss and epiphyte water content in an upper montane forest in Costa Rica subject to comparatively little fog, indicating that evaporation may be important under certain conditions. Values ranged from *c.* 405% at the height of the wet season to less than 25% in the dry season. D. Hölscher *et al.* (2004) were able to reproduce this seasonal variation in epiphyte water contents using a simple running water balance model linked to an adapted analytical model of rainfall interception (Van Dijk and Bruijnzeel, 2001). Out of a total annual interception of 28%, 6% was predicted to be contributed by the epiphytes (D. Hölscher *et al.*, unpublished).

Finally, considerable progress has also been made during the last decade in the estimation of cloud water deposition in complex terrain using *physically-based models* (Joslin, Mueller and Wolfe, 1990; Mueller, 1991; Mueller, Joslin and Wolfe, 1991; Walmsley, Schemenauer and Bridgman, 1996; Walmsley, Burrows and Schemenauer, 1999). Such models include assumptions about the shape and spacing of the trees, their fog water collection efficiency, the frequency of fog, and the vertical rate of change of the liquid water content within ground-based clouds. Topographical data are used as a forcing function in wind flow models to derive a spatially explicit representation of the wind field. Although the application of such advanced models has given promising results for the estimation of cloud water deposition onto montane coniferous forest in Canada (Walmsley *et al.*, 1999), virtually none of the required input data is presently available for TMCF environments. Clearly, the application of physically-based models to remote tropical mountain sites remains a major challenge for some years to come. In the meantime, a 'hybrid' approach in which (some) physical modelling is combined with empirically derived estimates of fog characteristics, such as employed successfully to evaluate fog water contributions to catchment water budgets in the Maritime Provinces of Canada by Yin and Arp (1994), may constitute a suitable alternative that is worth exploring in a tropical montane context.

Results of post-1993 rainfall and cloud interception studies in TMCF

Measurements made with various types of fog gauges in areas with TMCF as reviewed by Bruijnzeel and Proctor (1995) suggested typical cloud water deposition rates of 1–2 mm/day (range 0.2–4.0 mm/day), with a tendency towards lower values during the dry season and with increasing distance to the ocean. Several studies employing fog gauges or measuring net precipitation in TMCF environments have been published since 1993, the results of which are summarised in Tables 18.1 (cloud interception data) and Table 18.2 (overall interception data).

At 0.3–2.43 mm d^{-1}, the majority of the post-1993 results for cloud water interception fall within the previously established range. Minimum values of *c.* 0.3 mm d^{-1} have been derived for forests in Honduras and Venezuela during rather dry periods whereas a maximum of 6.3 mm d^{-1} (or 2300 mm $year^{-1}$) has been claimed for an exposed site at 1100 m on the Pacific-Caribbean water divide as part of a transect study in western Panama (Cavelier *et al.*, 1996). There are strong indications that the latter figure is unrealistically high. Firstly, it is based on measurements with uncovered Grünow-type fog gauges, the poor performance of which has been hinted at already. Secondly, the rainfall at this windy site is reported as only 1495 mm $year^{-1}$ whereas annual totals at similar elevations on either side of the main divide were consistently above 3600 mm (Cavelier *et al.*, 1996), suggesting severe underestimation of rainfall and thus overestimation of the fog input at this site. Finally, Cavelier *et al.* (1997) obtained a rather low throughfall fraction (63% or *c.* 2200 mm $year^{-1}$) for lower montane rainforest at a similar elevation (1200 m) in the same area (Table 18.2). Adding the 2300 mm of allegedly intercepted cloud water to the 3500 mm of rain received annually by this forest would suggest a total precipitation input of *c.* 5800 mm, of which *c.* 3600 mm (5800 minus 2200 mm of throughfall) would then be required to have been lost through evaporation from the wet canopy ('rainfall interception'). As will be shown below, reported *total* evaporative losses (i.e. both wet and dry canopy evaporation) from lower montane rainforests do not exceed 1380 mm $year^{-1}$. Nevertheless, although the claim of excessively large cloud water inputs in western Panama by Cavelier *et al.* (1996)

Table 18.1. *Post-1993 studies of cloud water interception (CW) in tropical montane environments; figures represent annual averages unless indicated otherwise*

Location	Elevation (m)	Forest type[a]	Mean annual precipitation (mm)	Cloud water interception		Remarks	Reference
				mm d^{-1}	percent[b]		
Australia	1000	LMCF	1350	0.94	35	CW equal to 'excess' T_F	Hutley *et al.* (1997)
Costa Rica	1500	LMCF	2520	2.43	28	Artificial foliage collector 1410 cm^2 at 17 m height	Clark *et al.* (1998)
Costa Rica	1500	LMCF	3300	0.53	6	CW equal to 'excess' T_F[c]	Fallas (2002)
		fragment LMCF		1.25	14		
		secondary LMCF		0.70	8		
Guatemala	2400	UMCF	2500	0.72	13	CW equal to 'excess' T_F	Brown *et al.* (1996)
	2750	UMCF		1.65d	281d	Dry season (Jan–March)	
Guatemala	2550	UMCF	2500	0.64	8	Dry season (Jan–March)	Holder (1998)
				1.30d	53d	Dry season (Nov–March)	
Hawai'i	2600	SA(C)F	<500	0.61[d]	38	Louvered fog gauge (3m)	Juvik and Nullet (1995b)
Honduras	900–1400	LMCF	4200	0.3d	6d	CW equal to 'excess' *TF*; Dry season (Jan–May)	Brown *et al.* (1996)
	1500	LMCF	2500	0.37d	12d	Dry season (Jan–March)	
Jamaica	1850	UMCF	2850	1.84	22	Grünow fog gauge above	Hafkenscheid (2000)
	1825			1.0	12	forest canopy	
	1850			0.53	6	Covered gauge in clearing	
	1810			0.53	6	Net precipitation method[e]	
Panama	1100	LMF	>3600[f]	6.30[f]	154	Grünow fog gauges	Cavelier *et al.* (1996)
	1250	LMF	5700	1.23	8		
Puerto Rico	1015	ECF	4500	1.33	7	Net precipitation method[a]	Schellekens *et al.* (1998)
Venezuela	2300	LMCF	3000	0.29	7	'Standard' collector (5m); 7 months (rather dry)	Ataroff (1998)

[a] LM(C)F, lower montane (cloud) forest; UMCF, upper montane cloud forest; SA(C)F, (dry) subalpine (cloud) forest; ECF, low-elevation dwarf cloud forest.

[b] Expressed as percentage of associated rainfall.

[c] Adding the 320 mm yr^{-1} of rainfall intercepted by a nearby LMF only seasonally affected by cloud (Fallas, 2002), would raise these values to: 1.40, 2.13 and 1.58 mm d^{-1}, and 15%, 23% and 17%, respectively.

[d] Expressed as mm/event (0.27 mm per calendar day).

[e] Minimum value due to exclusion of fog deposition during and shortly after rain (Hafkenscheid, 2002).

[f] See text for explanation of this excessively high value.

must thus be ascribed to instrumental error and misinterpretation of the data and is more likely to represent wind-driven rain, substantial (short-term) additions of cloud water (up to 5 mm d^{-1}) have been observed occasionally on rainless days at the highest elevations in the Sierra de las Minas, Guatemala (Brown *et al.*, 1996; Holder, 1998) (Figure 18.3c below).

Typical ranges of net precipitation determined for the respective types of montane forests as reviewed by Bruijnzeel and Proctor (1995) were: (i) 67–81% (average 75%; $n = 9$) for lower montane forest not affected much by cloud; (ii) 80–101% (average 88%; $n = 4$) for lower montane cloud forest; and (iii) 81–179% (average 112%; $n = 10$) for upper montane and low-elevation dwarf cloud forests. These averages for the respective forest types do not change much after incorporating the results of post-1993 studies (Table 18.2). The extent of the change, however, sometimes depends on the forest type to which each 'new' study is assigned and this presents difficulties in several cases. For example, throughfall fractions obtained for 'leeward' lower montane cloud forests in Costa Rica and, especially, Venezuela are so low (65% and 55%, respectively) (Table 18.2) that these forests effectively behave like LMRF not affected by cloud. This is partly a result of the protected location of the two sites but possibly also reflects their high epiphyte loading (Clark *et al.*, 1998; Ataroff, 1998; cf. Cavelier *et al.*, 1997). Epiphytes, especially

Table 18.2. *Post-1993 studies of throughfall* (T_f)*, stemflow* (S_f) *and apparent rainfall interception* (E_i) *fractions in tropical montane environments*

Location	Elevation (m)	Forest type[a]	T_f	S_f	E_i[b] (% of *P*)	Remarks	Reference
Australia	1000	LMCF	90	2	8	25 fixed troughs; 10–21 day sampling interval	Hutley *et al.* (1997)
Costa Rica	1500	LMCF	65	–	35	20 fixed standard gauges; 1–3 day sampling interval	Clark *et al.* (1998)
Costa Rica	1500	LM(C)F	90/131d[c]	–	10	10 fixed troughs; daily	Fallas (2002)
		LMCF	106/175d	–	−6		
		Remnant LMCF	114/174d	–	−14	4 fixed troughs, daily	
		Secondary LMCF	108/135d	–	−8		
Ecuador	1900–2300	LMCF	57	1	42	10 fixed gauges	Wilcke *et al.* (2001a)
Guatemala	2200	LM(C)F	81	<1	18	3–6 fixed gauges with	Brown *et al.* (1996)
	2400	L/UMCF	113	1–2	−15	large (52 cm dia) funnels; 4–7 day intervals	
	2750	UMCF	281d	2	−183		
Guatemala	2550	UMCF	108	–	−8	58 fixed gauges, weekly;	Holder (1998)
			100w	–	(0)	Rainy season (April–Oct)	
			153d	–	−53	Dry season (Nov–March)	
Hawai'i	2600	SA(C)F	75	–	25	2 fixed recording troughs;	Juvik and Nullet (1995b)
Honduras	900–1400	LMCF	95	–	5	3–6 fixed large diameter	Brown *et al.* (1996)
	idem		106d[d]	–	−6	gauges; 4–7 day intervals	
	1500	LMCF	111d[e]	–	−11		
Jamaica	1810	UMCF	73	12	14	1 recording trough + 12	Hafkenscheid (2000)
	1825	UMCF	60[f]	18	22	roving gauges (3–4 days)	
Panama	1200	LMRF	63	<1	37	50 fixed troughs; daily	Cavelier *et al.* (1997)
Puerto Rico	1015	ECF	89	(5)[g]	6	3 recording gutters + 10 roving gauges (1–3 days)	Schellekens *et al.* (1998)
Venezuela	2300	LMCF[h]	55	<<1	45	6 fixed trough gauges; weekly sampling	Ataroff (1998), Ataroff and Rada (2000)

[a] LM(C)F, lower montane (cloud) forest; UMCF, upper montane cloud forest; SA(C)F, (dry) sub alpine forest; LMRF, lower montane rainforest not affected significantly by cloud; ECF, low-elevation dwarf cloud forest.

[b] $E_i = P - (T_f + S_f)$, apparent interception, including ungauged contributions by cloud water.

[c] Dry season: December–April.

[d] Dry season: January–May.

[e] Dry season: January–March.

[f] Underestimate due to insufficient number of gauges (see Hafkenscheid *et al.* (2002) for details).

[g] Based on data from Weaver (1972).

[h] Transitional to UMCF (Ataroff and Rada, 2000).

'tank' bromeliads and moss 'balls', are known to have very high *potential* moisture storage capacities and to release their water rather slowly (Pócs, 1980; Nadkarni, 1984; Veneklaas *et al.*, 1990; Richardson *et al.*, 2000). On the other hand, *effective* storage capacities will depend strongly on actual water contents which are known to vary rapidly depending on variations in precipitation and evaporation (Köhler, 2002). More work is needed along the lines indicated by Chang *et al.* (2002), Köhler (2002) and Hölscher *et al.* (2004) to solve this question.

The remaining new studies in undisturbed lower montane cloud forests listed in Table 18.2 (n = 4; i.e. leaving the two transitional Guatemalan forests at 2200 and 2400 m aside as they rather resemble LMRF and UMCF, respectively) produced results that are in agreement with previously established values for this type of forest. Adding these new results to the existing dataset for LMCF raised the average net precipitation fraction slightly from 88% to 92% (n = 8). Instead, incorporating the 'anomalous' results for the Costa Rican and Venezuelan sites as well would not only lower the 'old' average value from 88% to 85% but particularly extend the bottom end of the reported range for LMCF (from 80% to 55%). Alternatively, assigning these two studies plus the Guatemalan forest at 2200 m and the Panamanian forest to the class of LMRF not affected much by low cloud would bring the associated average value down from 75% to 71%. Adding the new results obtained for upper montane and elfin cloud forests in Guatemala, Jamaica and Puerto Rico ($n = 5$) also lowers the associated average value of net precipitation slightly (from 112% to 109%). However, this is entirely due to the inclusion of the rather low throughfall figure obtained for the forest at 1810 m in Jamaica (Table 18.2).[3] The very high stemflow fractions obtained for the two Jamaican forests are noteworthy and were ascribed to the presence of multiple-stemmed and crooked trees (Hafkenscheid *et al.*, 2002). Similarly high values have been reported for stemflow in 10–15 year-old leeward secondary montane forest in Costa Rica which showed both a very high stem density and a high frequency of multiple-stemmed trees (Köhler, 2002).

Given the rapid conversion of TMCF to other land uses (mostly pasture; Bruijnzeel and Hamilton, 2000), the finding of enhanced net precipitation in small primary and secondary forest fragments compared to nearby tall undisturbed windward lower montane cloud forest in northern Costa Rica (Fallas, 2002; study no. 3 in Table 18.2) is potentially of great interest. More work is needed to confirm these preliminary (because based on few, fixed gauges) results.[4]

Summarising, the extended dataset for net precipitation in tropical montane forests suggests a steady increase in average values from lower montane forest (*c.* 70%), through lower montane cloud forest (*c.* 90%) to upper montane and low-elevation dwarf cloud forests (*c.* 110%). More work is needed to elucidate the precise role of epiphytes in the interception process, as evidenced by the very low throughfall fractions obtained for some cloud forests despite considerable inputs of cloud water (e.g. as in Costa Rica and Venezuela, Tables 18.1 and 18.2). Future work could profitably combine several of the approaches outlined in the previous sections, notably the use of stable isotopes, electronic field monitoring of epiphytic water content, and interception modelling. Furthermore, amounts of net precipitation generated in cloud forest fragments deserve more study as well.

Most of the data collated in Tables 18.1 and 18.2 concern annual averages. However, in many areas (e.g. Central America) cloud interception is a highly seasonal process that assumes its greatest importance during the dry season. As such, a cloud forest with an overall net precipitation figure well below 100% may still experience a much higher value during particular times of the year. A case in point is the Sierra de las Minas, Guatemala, where in the LMCF zone at 2200 m average throughfall is 81% (Table 18.2). At the height of the rainy season (August, September) relative throughfall drops to about 65% but during the dry season (January–March) it exceeds incident rainfall (Figure 18.3a). At 2400 m (transition zone to UMCF), cloud interception is important year-round but again reaches its peak during the dry season (Figure 18.3b; cf. the seasonal contrast observed by study no. 6 in the same area; Tables 18.1 and 18.2). Cloud interception is still more pronounced at 2750 m (UMCF zone). In the period January–March 1996, throughfall exceeded rainfall by 147 mm (181%; study no. 5 in Table 18.2), with the excess reaching maximum values of as much as 40–50 mm over 3–4 day periods (Figure 18.3c). Such findings contradict the suggestion by Vogelmann (1973) that absolute amounts of fog incidence in eastern Mexico decreased with elevation within the cloud belt and during the dry season compared to the rainy season. However, Vogelmann's contention was based on measurements made with cylindrical wire-mesh fog gauges, the limitations of which have been stressed already. It is more than likely therefore that these early measurements at least partly reflect effects of wind-driven rain, notably during the rainy season (cf. Cavelier *et al.*, 1996). More importantly, the findings from Guatemala underline the importance of TMCF for sustained dry season flows (cf. Zadroga, 1981). We will come back to this important point in the section on TMCF and water yield.

There is a need for additional studies like that of Brown *et al.* (1996) to document the changes in net precipitation with elevation, slope aspect and season in different regions.

Transpiration and total forest water use

In their review of pre-1993 work on TMCF water use (both total evapotranspiration, E_T and transpiration, E_t), Bruijnzeel and Proctor (1995) had to draw mostly on catchment and site water balance studies. In addition, in the absence of direct estimates of E_t, only approximate values – obtained by subtracting apparent interception totals E_i from E_T – could be presented. Apart from the general limitations of the water budget technique for the evaluation of E_T (see Bruijnzeel (1990) for a discussion in a tropical context) there is the added complication in the case of TMCF that unmeasured inputs via the interception of cloud water and wind-driven rain will lead to correspondingly lower values of E_T. As such, the estimates of E_T for cloud-affected forests cited below represent *apparent* values only and care should be taken when comparing them with values derived by non-water budget based methods.

3 The results for the nearby forest at 1825 m are excluded from the present analysis because these are considered underestimates due to the use of an insufficient number of throughfall gauges.

4 Recent ongoing work in the same area seems to confirm the findings of Fallas (2002) (K. F. A. Frumau, pers. comm.).

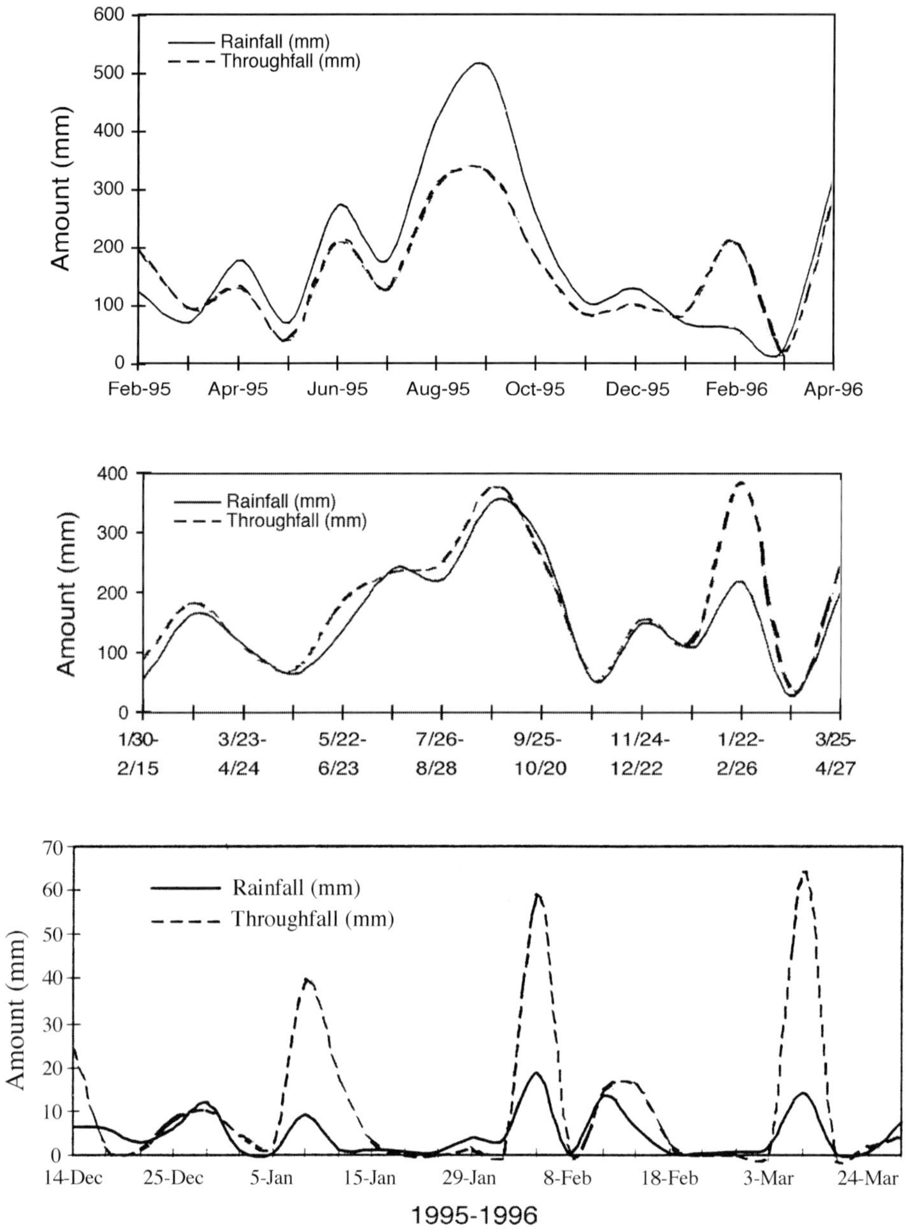

Figure 18.3 Seasonal rainfall and throughfall patterns with elevation in the Sierra de las Minas, Guatemala. (a) 2200 m, lower montane cloud forests; (b) 2400 m, transition from lower to upper montane cloud forest; (c) 2750 m, upper montane cloud forest, dry season only (after Brown *et al.*, 1996).

The pre-1993 dataset on E_T for tropical montane forests can be summarised as follows: (i) equatorial lower montane forest with negligible fog incidence: 1155–1380 mm year^{-1} (average 1265 mm year^{-1}; $n = 7$); (ii) lower montane cloud forest with moderate fog incidence: 980 mm year^{-1} ($n = 1$); upper montane cloud forest with high fog incidence: 310–390 mm year^{-1} ($n = 3$). Corresponding 'guesstimates' for E_t are: (i) 510–830 mm year^{-1}; (ii) 675 mm year^{-1}; and (iii) 250–285 mm year^{-1}, respectively (see Bruijnzeel and Proctor (1995) for details). Thus, whilst information on total water use (E_T) of cloud forests of any type is admittedly scarce, the data for upper montane cloud forest seem at least consistent. Conversely, there is considerable uncertainty about the water use of lower montane cloud forest. The only estimate available (980 mm year^{-1} for a tall forest at 2300 m in Venezuela) is based on energy budget calculations which involved numerous assumptions (Steinhardt, 1979). Bruijnzeel *et al.* (1993) derived

Table 18.3. *Post-1993 water budget studies in tropical montane cloud forest environments (values rounded off to the nearest 5 mm)*

Location	Elevation (m)	Forest type[a]	Mean annual precipitation	E_T (mm yr^{-1})	E_t	Reference
Australia[b]	1000	LM(C)F	1350	1260	845	Hutley *et al.* (1997)
Jamaica[c]	1810	UMCF[h]	2850	1050	620	Hafkenscheid *et al.* (2002)
	1825	UMCF*		890[j]	510	
Venezuela[d]	2350	L/UMCF[h]	3125	2310	560	Ataroff and Rada (2000)
Puerto Rico[e]	780	UMCF[h]	4450	1045	–	Van der Molen (2002)
	1015	ECF[i]	4450	705	–	
Puerto Rico[f]	1015	ECF[i]	4450	1145	–	Garciá-Martino *et al.* (1996)
Puerto Rico[g]	1015	ECF[i]	4450	435	170	Holwerda (1997), Schellekens *et al.* (1998)

[a] LMCF, lower montane cloud forest; UMCF, upper montane cloud forest; ECF, low-elevation dwarf cloud forest.
[b] E_t evaluated from soil water budget (net precipitation vs. change in soil water storage); $E_T = E_i + E_t$,
[c] E_t estimated via energy-balance temperature fluctuation (EBTF) method for nearby secondary forest; scaling up to forest plots according to relative values of leaf area index; values must be considered approximate; $E_T = E_t + E_i$.
[d] E_t via portable gas exchange measurements; $E_T = E_t + E_i$.
[e] E_T estimated with Bowen ratio energy balance method; no distinction made between wet and dry canopy conditions, and therefore no separate estimate for E_t.
[f] E_T evaluated as the difference between average rainfall (with 10% cloud water added) and runoff, both estimated from regression equations against elevation.
[g] E_t via EBTF method; $E_T = E_t + E_i$, with $E_i = 6\%$.
[h] Relatively tall-statured forest.
[i] Stunted ridge top forest.
[j] Value lowered by 300 mm by the present writer to account for underestimation of throughfall and corresponding overestimation of E_i (Hafkenscheid *et al.*, 2002).

an E_T of 695 mm year^{-1} for a 'stunted lower montane cloud forest' at 870 m on a coastal mountain in East Malaysia. However, their estimate was based on short-term site water budget calculations for a particularly dry period and is therefore not included in the present analysis.

Despite the great need for additional information on TMCF water use signalled at the 1993 Puerto Rico Symposium, comparatively little new evidence has become available since then. The results of several recent studies in three different types of cloud forest are summarised in Table 18.3.

Table 18.3 shows that all new estimates are considerably higher than those derived previously for the respective forest types. For example, at 1260 mm, the annual E_T of a tall lower montane forest subject to 'frequent, low intensity rainfall associated with low cloud' in Queensland, Australia is very close to the average value of E_T for lower montane forests *not* affected by fog and low cloud cited earlier (1265 mm year^{-1}). Similarly, E_t for this forest (845 mm year^{-1}) approximates the maximum value inferred by Bruijnzeel (1990) for non-cloud forests (830 mm). However, the investigators stressed that their evaporation estimates should be seen as maximum values because the observations concerned a small plot dominated by a single emergent tree. If a larger plot had been monitored, with additional species, openings in the canopy and individuals with less exposed crowns, the result might well have been lower (Hutley *et al.*, 1997). At first sight, the E_T values estimated for two upper montane cloud forests of contrasting stature in the Blue Mountains of Jamaica (1050 and 890 mm year^{-1}, Hafkenscheid *et al.*, 2002) also seem to be closer to the single value reported earlier for tall *lower* montane cloud forest (980 mm year^{-1}; Steinhardt, 1979) rather than to the 310–390 mm year^{-1} derived by Bruijnzeel and Proctor (1995) for upper montane cloud forests proper. However, the latter water budget based estimates would need to be raised by the corresponding amounts of *CW* to be directly comparable with the present micro-meteorological estimates. In the absence of measured values for *CW* for these other sites, Bruijnzeel (1990) tentatively added amounts measured at comparable nearby locations, obtaining approximate E_T values of 570–775 mm year^{-1}. These are more in line with the presently found estimates but still lower. The cloud water incidence experienced by the Jamaican forests is relatively low, however (Table 18.1), which may go some way towards explaining their somewhat higher E_t. Further support for the Jamaican values comes from soil water depletion patterns measured at the same sites (Hafkenscheid *et al.*, 2002) as well as from the almost identical value derived for a comparable forest type in Puerto Rico (Van der Molen, 2002). Albedoes (0.11–0.14) and net radiation fractions of incoming short-wave radiation for these forests were comparable to values derived for lowland forest (Hafkenscheid, 2000; Van der Molen, 2002; Shuttleworth, 1988), suggesting that any contrasts in cloud forest water uptake should rather reflect differences in stomatal behaviour or leaf area index (see below). The estimated total evaporation from a tall cloud forest in the transitional zone from lower to upper montane forest in Venezuela reported by Ataroff and Rada (2000) is excessively high. Although based on comparatively short-term measurements of gas exchange, the estimate for *Et* seems plausible at 560 mm year^{-1}. The very high value for total E_T is largely due to the comparatively even higher wet canopy evaporation component (1750 mm year^{-1}) to which the authors added a further 215 mm of litter evaporation, thereby bringing the total annual evapotranspiration figure to as much as 2525 mm. Since available radiation totals in this otherwise 'very humid and cloudy environment' almost certainly will not be sufficient to sustain such high rates of evaporation it cannot be excluded that the author's throughfall sampling design (Table 18.2) is in need of intensification (cf. Hafkenscheid *et al.*, 2002).

The three estimates for E_T of low-elevation dwarf cloud forest in Puerto Rico that have become available in recent years are very different at 435–1145 mm year^{-1} (Table 18.3). The highest of these estimates must be considered suspect for two reasons. Firstly, it is based on the subtraction of an average runoff figure from an average rainfall figure (though corrected for unmeasured cloud water inputs), both of which were estimated from regressions linking rainfall/runoff to elevation (Garciá-Martino *et al.*, 1996). Secondly, it is much higher than the value established by Holwerda (1997) for the reference open-water evaporation total for the dwarf forest zone (670 mm year^{-1}). Because inclusion of ungauged additions to the water budget by wind-driven rain would have increased estimated E_T even more, it is more than likely that this excessively high value is caused by catchment leakage problems. Assuming the contribution to overall E_T by wet canopy evaporation from the elfin forest to be 5 ± 1% of the rainfall (Schellekens *et al.*, 1998; Te Linde *et al.*, 2001), suggests the average transpiration rate to vary between 0.47 (Holwerda, 1997) and 1.2 mm d^{-1} (Van der Molen, 2002). Given these rather contrasting findings, the results of an ongoing study in the same forest by F. Holwerda using more sophisticated equipment are awaited with interest.

Despite these recent additions to the literature on TMCF water use, it must be concluded that our knowledge remains fragmentary and at times contradictory. Further work is urgently needed (see also the next section). Nevertheless, it can also be concluded that differences in actual water use by lower and upper

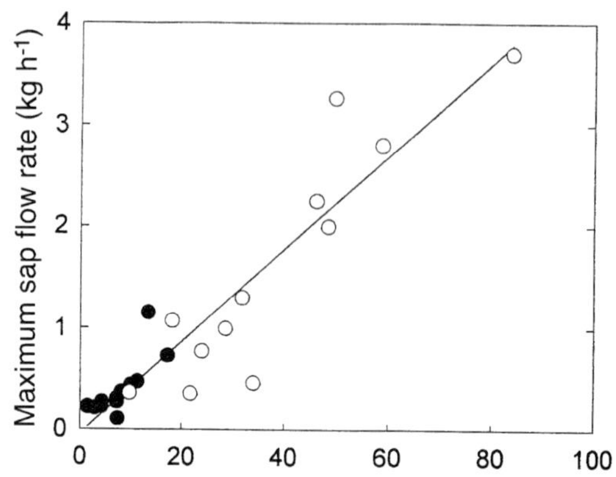

Figure 18.4 Relationship between maximum sapflow rate and total leaf area from level, waterlogged sites (closed circles) and from sloped, better drained sites (open circles) in TMCF in Hawai'i. (After Santiago *et al.* (2000); reproduced with permission.)

montane cloud forests are much smaller than inferred initially from catchment- or site water budget calculations. In theory, future studies of cloud forest water uptake could make use of a variety of plant physiological techniques (cf. Dawson, 1998; Smith and Allen, 1996; Wulschleger, Meinzer and Vertessy, 1998). A recent study that employed sapflow gauges to study transpiration in montane cloud forest in Hawai'í is that reported by Santiago *et al.* (2000). Their measurements demonstrated a clear dependence of (maximum) instantaneous transpiration rates on tree leaf surface areas that, in turn, were governed by site drainage conditions. Stand leaf area index, and therefore E_t, was much lower on waterlogged, level sites compared to better drained, sloping sites (Figure 18.4). However, transpiration rates were often too low for monitoring by the sapflow method. Similar problems were encountered in Jamaican cloud forest (R. L. L. J. Hafkenscheid, pers. comm.). It would seem, therefore, that there may be scope for alternative approaches, such as the use of stable isotopes (Dawson, 1998).

TROPICAL MONTANE CLOUD FORESTS AND WATER YIELD

Due to the combination of added moisture inputs from cloud water interception (Tables 18.1 and 18.2) and somewhat lower water use (Table 18.3), water yields for a given amount of rainfall from cloud forested headwater areas tend to be higher than those emanating from montane forests not affected by fog and low cloud. Similarly, flows from cloud forest areas tend to be more stable during extended periods of low rainfall. Therefore, fears have been expressed that the conversion of TMCF to other land uses could result in significant declines in overall and dry season flows (Zadroga, 1981; Brown *et al.*, 1996).

The original extent of TMCF worldwide was given as about 50 million ha (Persson, 1974). Although this estimate was probably somewhat higher than reality (Hamilton, 1995b) and no accurate information is available as to how much might now remain, there can be little doubt that cloud forests are disappearing rapidly. In Central America and the Caribbean, LaBastille and Pool (1978) considered as early as the 1970s that cloud forests were declining faster than any other forest type. Similarly, it has been estimated that some 90% of the cloud forests of the northern Andes of Colombia has been lost, mostly to pastures and agricultural fields (Doumenge *et al.*, 1995). The causes of cloud forest disappearance and degradation are myriad but worldwide the greatest loss comes from its conversion to grazing land, especially in seasonally dry climates. Other, regionally important causes include conversion to temperate vegetable cropping and harvesting of wood for charcoal production, timber harvesting, mining, unsustainable harvesting levels of non-wood products (e.g. orchids and bromeliads), recreation and eco-tourism, introduction of alien species, and the establishment of an ever increasing number of telecommunication installations on cloud forested mountain tops (Doumenge *et al.*, 1995; Hamilton, 1995b; Bruijnzeel and Hamilton, 2000). Finally, as discussed more fully below, there is increasing evidence that TMCFs are also threatened by global warming.

Where tropical forests of any kind are replaced by annual cropping or grazing there are bound to be profound changes in the area's hydrology (Bruijnzeel, 1990). The beneficial effects on soil aggregate stability and water intake capacity afforded by the high organic matter content and abundant faunal activity of forest soils may linger for a year or two after clearing. However, exposure of the soil surface to the elements generally leads to a rapid decline thereafter, particularly if fire was used during the clearing operation (Lal, 1987). An additional aspect in densely populated agricultural steep lands is that considerable areas may become permanently occupied by compacted surfaces, such as houses, yards, trails and roads (Ziegler and Giambelluca, 1997). In areas with heavy grazing pressure, soil infiltration capacities suffer further from compaction by trampling cattle (Gilmour, Bonell and Cassells, 1987). As a result, conversion of forest to annual cropping or grazing is almost inevitably followed by increases in amounts of surface runoff (Bruijnzeel, 1990; cf. Grip *et al.*, this volume).

A second consequence of forest clearing relates to the associated changes in net rainfall – no longer are there trees to intercept rainfall or fog. Neither, of course, are (low) levels of forest water use (transpiration) maintained. Whilst annual crops and grass also intercept rainfall and cloud water, and take up water from the soil, the associated amounts are (much) smaller than for forest due to the generally larger total leaf surface area and deeper root systems of forests compared to crops or grass (Calder, 1998). Thus, the clearing of a montane forest that does not experience appreciable inputs of cloud water (i.e. LMRF) will result in an increase in the total volume of streamflow, typically by 100–400 mm year^{-1} (depending on rainfall; Bruijnzeel, 1990). In theory, the extra amount of moisture available in the soil due to the reduction in E_i and E_t after converting (non-cloud) forest should permit an increase in baseflow levels – *as long as soil infiltration capacity is maintained*. In practice, however, the degeneration of the soil's infiltration capacity after forest removal is often such that the potential gain in soil water afforded by the reduced water use after clearing is more than offset by increases in overland flow and peak runoff during the wet season, with diminished streamflow during the dry season as the result (Bruijnzeel, 1989, 2004).

The risk of reduced dry season flows following forest clearance becomes even more serious in the case of clearing TMCF. The extra inputs of water to the forest ecosystem afforded by cloud interception during the dry season can be substantial, particularly at exposed locations (Tables 18.1 and 18.2). Also, such extra additions assume particular importance during periods of low rainfall (Figure 18.3c). Whilst the cloud stripping ability of solitary or groups of trees that have been left standing remains more or less intact and could even be enhanced due to greater exposure to passing fog (cf. study no. 3 in Table 18.2; Fallas, 2002; Weathers *et al.*, 1995), it would surely disappear altogether in the case of a wholesale conversion to vegetable cropping or grazing (Zadroga, 1981). The eventual effect on streamflow will depend on the relative proportions of the catchment that were occupied by the respective types of cloud forest. For example, exposed ridge top forests may intercept large amounts of cloud water but their spatial extent is limited (Brown *et al.*, 1996).

In recent years, diminished dry season flows have been reported for various parts of the tropics that experienced a considerable reduction in montane forest cover, including Costa Rica's Monteverde area (Pounds, Fogden and Campbell, 1999), Honduras' Cusuco National Park and Guatemala's Sierra de las Minas (Brown *et al.*, 1996), eastern Mexico (I. Garciá, pers. comm.) and (possibly) Flores, eastern Indonesia (Patthanayak and Kramer, 2004). However, it is not clear to what extent these reductions in flow are primarily the result of the loss of the fog stripping capacity of the former forest, or of diminished rainfall related to larger-scale climatic fluctuations (cf. Mahé *et al.*, this volume), reduced infiltration and water retention capacities of the soil due to erosion after forest clearance or, in some cases, increased diversions for irrigation. For instance, the two catchment pairs in Honduras and Guatemala for which Brown *et al.* (1996) derived a

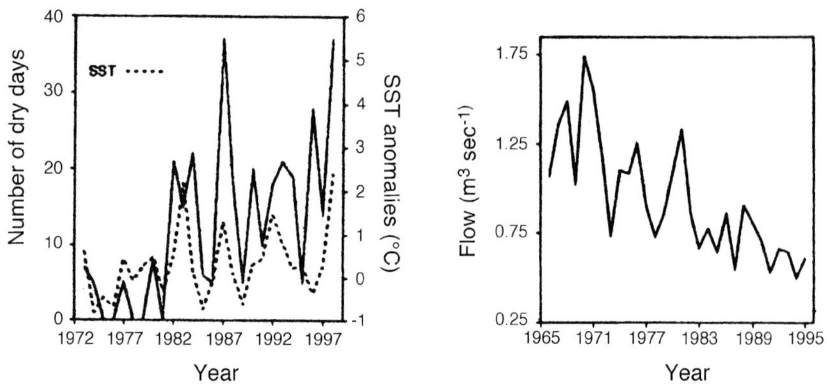

Figure 18.5 Trends for (a) sea surface temperature anomalies and number of rainless days, and (b) minimum streamflows in north-west Costa Rica since the 1970s. (Adapted from Pounds *et al.*, 1999.)

50% reduction in dry season flow after conversion to vegetable cropping, were rather different in size and elevational range, and therefore in their exposure to fog and rainfall (Table 18.1). A more convincing, albeit non-tropical, case was provided by Ingwersen (1985) who observed a (modest) decline in summer flows after a 25% patch clearcut operation in the same catchment in the Pacific Northwest region of the US for which Harr (1982) had inferred an annual contribution by fog of *c.* 880 mm. The effect disappeared after five to six years. Because forest cutting in the Pacific Northwest is normally associated with strong increases in water yield (Harr, 1983), this anomalous result was ascribed to an initial loss of fog stripping upon timber harvesting, followed by a gradual recovery during regrowth. Interestingly, the effect was less pronounced in an adjacent (but more sheltered) catchment and it could not be excluded that some of the condensation not realised in the more exposed catchment was 'passed on' to the other catchment (Ingwersen, 1985; cf. Fallas, 2002). Little is known of the soil physical changes accompanying cloud forest conversion to pasture. Duisberg-Waldenberg (1980) did not detect consistent changes in infiltration capacity in grazing land in northern Costa Rica some five to seven years after forest clearance. More recently, however, substantial reductions were observed in 30-year-old pasture in the same area, with the degree of the reduction being commensurate with the frequency of cattle passing. Infiltration-excess overland flow was observed regularly on hillslope contour cattle trails which were estimated to make up at least 15% of the total area (C. Tobón, pers. comm.). Such findings suggest that it may take a number of years before soil degradation reaches a critical level and surface hydrological processes are altered sufficiently to affect streamflow regimes (cf. Sandström, 1998). Identifying the precise cause(s) of the observed decreases in dry season flows and finding ways of restoring them should be given very high research priority in the years to come (cf. Scott *et al.*, this volume; Bruijnzeel, 2004).

On a related note, there is increasing evidence that TMCFs are also threatened by regional and global warming of the atmosphere. The latter tends to raise the average level of the cloud condensation level (Scatena 1998; Still, Foster and Schneider, 1999; Lawton *et al.*, 2001). Apart from adverse hydrological consequences such as diminished opportunities for cloud water interception, a lifting of the cloud base is bound to produce important ecological changes as well. The organisms living in TMCFs are finely attuned to the rather extreme climatic and soil conditions prevailing in these already stressed ecosystems (Benzing, 1998; Pounds *et al.*, 1999; Hafkenscheid, 2000). One of the best-documented cases in this respect is provided by the Monteverde Cloud Forest Preserve in northern Costa Rica (Pounds *et al.*, 1999). Here, a decrease in fog frequency since the mid 1970s has been inferred using the number of days with no measurable precipitation as an index of fog frequency. The most extreme decreases within an overall downward trend occurred in 1983, 1987, 1994 and 1998, which appeared to be correlated with higher sea surface temperatures (Figure 18.5a). Anoline lizard populations in the area have declined in association with this pattern, with major population crashes in 1987, 1994 and 1998. Currently, 20 out of 50 species of frogs and toads, including the spectacular and locally endemic golden toad, have disappeared. At the same time, species from drier, lower elevations are invading and becoming residents (Pounds *et al.*, 1999).

The evidence presented in Figure 18.5 suggests a worrying relationship between global warming and drying of the air on the one hand, and reduced streamflows on the other in the case of northwest Costa Rica (Fleming, 1986). It could be argued that these data pertain to an area that is rather protected from the moisture-bearing trade winds coming from the Caribbean and that one should therefore be careful not to generalise such findings to 'all' cloud forest situations. For example, simulation studies of the effects of global warming on rainfall patterns predict a distinct rise in some cloud

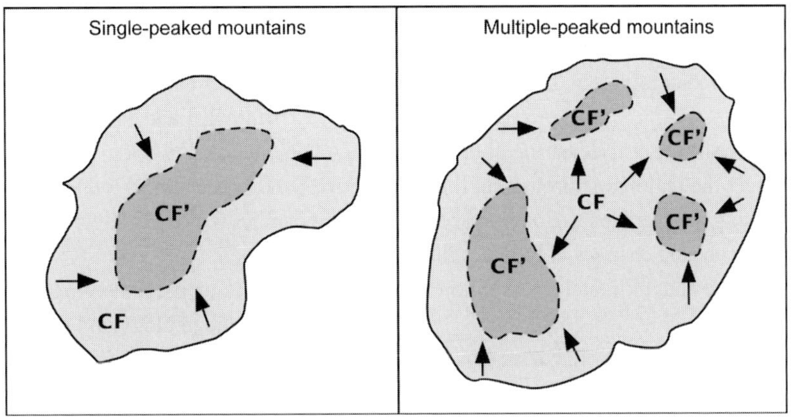

Figure 18.6 Possible changes in spatial cloud forest distribution in response to a rise in the cloud condensation level on (a) a single peak and (b) a mountain with several peaks. (After Sperling (2000); reproduced with permission.)

forest areas, such as the Pacific slopes of the Andes in southwest Colombia (Mulligan, 2000). Nevertheless, recent observational evidence reported by Lawton *et al.* (2001) suggests that widespread deforestation in the Atlantic lowlands of northern Costa Rica does indeed reduce cloud formation during the dry season, with possible consequences for cloud water interception in the adjacent uplands. The authors were able to reproduce the observed differences in cloud formation using a meso-scale atmospheric circulation model although values applied to parameterise soil water status below forest and pasture have been subject to criticism (Bruijnzeel, 2004). The current measurements of variations in surface climate, soil water content and the height of the cloud base by Lawton and collaborators (pers. comm.) in the area should resolve the issue before too long, however (see also Foster, 2001). Further evidence of the importance of land cover in the lowlands on the height of the cloud base comes from Puerto Rico. Here, the average cloud condensation level was lifted temporarily by several hundred metres after Hurricane Hugo had effectively defoliated the forests on the lower slopes of the Luquillo Mountains in September 1989. The resulting drop in forest water use caused a significant rise in the temperature of the overlying air and thus in the average position of the cloud base. Interestingly, the effect disappeared in a few months after the leaves had grown back again (Scatena and Larsen, 1991; see also photographs in Bruijnzeel and Hamilton, 2000). In the same area, Scatena (1998) interpreted the presence of isolated stands of large and very old (>600 years) Colorado trees (*Cyrilla racemiflora*) at elevations well below the current cloud base that experience relatively low rainfall (<3000 mm year^{-1}) as evidence of a gradual upward shift in vegetation zonation over the past several centuries. *Cyrilla* is currently a dominant tree in areas above the cloud base (> 600 m) and is most common where mean annual rainfall exceeds 4000 mm. Similarly, Brown *et al.* (1996) reported the occurrence of pockets of mossy cloud forest below the current average cloud base in Honduras. There is a need for more systematic research linking such empirical evidence to records of current and subrecent climatic change (cf. Scatena, 1998).

On single mountains, a lifting of the average cloud condensation level will result in the gradual shrinking of the cloud-affected zone (Figure 18.6a). On multiple-peaked mountains, however, the effect may be not only that, but one of increased habitat fragmentation as well (Figure 18.6b), adding a further difficulty to the chances of survival of the remaining species (Sperling, 2000). Much more research is needed to confirm and extend the results obtained so far at Monteverde (Pounds *et al.*, 1999) and, to a lesser extent, Puerto Rico (Scatena, 1998). Apart from amphibians (Pounds *et al.*, 1999), the epiphyte communities living in the more exposed parts of TMCF canopies might prove equally suited to detecting changes in climatic conditions, rainfall and cloud water chemistry, and possibly enhanced ozone and UV-B levels (Lugo and Scatena, 1992; Benzing, 1998; cf. Gordon, Herrera and Hutchinson, 1994).

PUTTING CLOUD FORESTS ON THE HYDROLOGICAL RESEARCH AGENDA

Having reviewed the post-1993 evidence on the hydrological functioning of TMCF, what are the primary remaining gaps in knowledge? Arguably, the most urgent research questions in relation to the perceived 'added' hydrological value of TMCF over other montane forests are:

- Does conversion of TMCF to vegetable cropping or pasture indeed lead to reductions in dry season flows, or even total water yield? And if so, is this mainly because of the loss of the cloud interception function, or does the reduction in dry season flow rather reflect a deterioration in infiltration opportunities after deforestation?

- Do changes in (dry season) flow after forest clearance differ for the respective types of montane forest (i.e. LMRF, LMCF, and UMCF) and thus local climatology (e.g. distance to the ocean, exposure)?
- To what extent does global warming or the clearing of forests *below* the cloud belts affect the regional hydrological function of TMCF through raising the average level of cloud condensation? What are the associated changes in fog stripping opportunities, forest water use and, ultimately, streamflow?

Furthermore, most hydrological studies in TMCF have concentrated on quantifying net precipitation (crown drip). Whilst these have indicated important differences between contrasting forest types and topographic situations, published results differ widely in their reliability and comparisons between sites and forests are, therefore, rather difficult to make (Bruijnzeel and Proctor, 1995) (cf. Tables 18.1 and 18.2). There is a distinct need for more systematic observations of cloud water interception and net precipitation along elevational gradients according to an internationally accepted (standard) measuring protocol. This would require the adoption of a robust standard cloud water collection device that may be used for comparative purposes (site characterisation; cf. Schemenauer and Cereceda, 1994, 1995; Juvik and Nullet, 1995a; Daube *et al.*, 1987). Net precipitation measurements should involve the use of a sufficient number of roving throughfall gauges to allow for an adequate representation of 'drip points' (cf. Lloyd and Marques, 1988; Hafkenscheid *et al.*, 2002). In addition, the potential contrast in concentrations of stable isotopes in rain and cloud water to evaluate cloud water contributions to overall net precipitation totals (cf. Dawson, 1998) deserves further exploration in a TMCF context. The same holds for the further development and testing of physically-based cloud water deposition models on tropical mountains (cf. Joslin *et al.*, 1990; Mueller, 1991; Yin and Arp, 1994; Walmsley *et al.*, 1996).

Finally, before a sound understanding can be obtained of the influence of TMCF on streamflow amounts, reliable information is urgently needed on the water uptake rates (transpiration) of these forests. Such information is lacking almost entirely at present (Table 18.3; Bruijnzeel and Proctor, 1995). There are no published studies that have combined hydrological process work (rainfall and cloud interception, water uptake) and streamflow dynamics in any TMCF environment, although such work was recently initiated by a consortium led by the Vrije Universiteit Amsterdam in northern Costa Rica (Bruijnzeel, 2002).

Having defined the chief hydrological research needs in TMCF in the previous sections, where could these be addressed best? Arguably, the most cost-effective approach would be to identify sites with ongoing work (hydrological and/or ecological) and plan additional observations and experiments as part of a network that covers the range of environmental conditions encountered in the pan-tropical cloud forest belt (Figure 18.2). A preliminary inventory of current climatological and hydrological research efforts in TMCF (Bruijnzeel and Proctor, 1995; this chapter) establishes: (i) A notable lack of work in African, and to a lesser extent, Asian TMCF; and (ii) an almost total absence of studies linking streamflow dynamics with hydrological process work or land use change (see also Bonell, *in press*).

On the basis of previously executed or ongoing work, representativity of site geology and climate, as well as logistical considerations, the sites listed in Table 18.4 may be considered to be the most promising for inclusion in such a pan-tropical TMCF research network. To answer the questions raised earlier with respect to the effect on water yield of converting TMCF to other land uses would ideally require setting up a paired catchment experiment in which flows from a forested control catchment are compared against flows from a cleared catchment *after initial intercalibration of the two areas in the undisturbed state* (Hewlett and Fortson, 1983). 'Direct' comparisons of streamflow emanating from forested and cleared catchments may easily give biased results due to potential differences in ungauged, subterranean water transfers into or out of the catchments, especially in the kind of volcanic terrain prevailing at almost all of the sites listed in Table 18.4 (cf. Bruijnzeel, 1990; Brown *et al.*, 1996; Garciá-Martino *et al.*, 1996). However, because much of the remaining forest in these areas is officially protected, experimental clearing within the context of a paired catchment is almost certainly not feasible. The other option then is to compare streamflows from catchments with contrasting land uses whilst accounting for differences in deep leakage and within-basin processes (interception of rainfall and cloud, transpiration, soil water depletion, deep drainage). It goes without saying that any inferences for streamflow made on the basis of (short-term) measurements of forest and pasture water use will reflect the quality of such measurements. For example, based on rather short-term measurements of gas exchange Ataroff and Rada (2000) claimed an increase in soil water uptake of as much as 1510 mm year^{-1} after converting cloud forest in the transition zone from lower to upper montane forest to pasture. Even though their grassland was said to exhibit active growth, the cited value seems excessively high. Adding the interception loss of 625 mm brings the total evaporation to the very high value of 2690 mm year^{-1}. Further work seems necessary which could usefully employ energy budget or lysimetric measurements to back up the plant physiological observations.

Some of the more suitable sites for such an experiment include Mt Kinabalu, Malaysia, and Sierra de las Minas, Guatemala (forest clearance for vegetable cropping); and Monteverde, Costa Rica and the Cauca area, Colombia (conversion to pasture). The effects of timber harvesting (oaks and *Podocarpus*) and forest clearing for orchards or grassland under somewhat drier climatic conditions

Table 18.4. *Key research sites in tropical montane cloud forests*

Site	Elevation range (m)	Mean annual precipitation (mm)	Forest types[a]	Reference
'Maritime' tropics				
Mt Kinabalu, Sabah, Malaysia	600–3400	2600–4100	LRF – SACF	Kitayama (1992), Frahm (1990a, 1990b)
Hawai'i islands	500–3200	<7000	LMRF-SACF	Kitayama and Mueller-Dombois (1992), Juvik and Ekern (1978), Juvik and Nullet (1995b), Santiago *et al.* (2000), Nullet and Juvik (1997)
Luquillo Mnts, Puerto Rico	600–1050	<4500	LMCF-ECF	Weaver (1972, 1995), Schellekens *et al.* (1998), Baynton (1968, 1969), Van der Molen (2002), Scatena (1998), Vugts and Bruijnzeel (1999)
Blue Mnts, Jamaica	1500–2265	2600–4000	LMRF-UMCF	Tanner (1977), Kapos and Tanner (1985), Hafkenscheid (2000); Hafkenscheid *et al.* (2002)
'Continental' tropics				
Cauca, Southwest Colombia	1375–2895	>5000	LMCF-UMCF	M. Mulligan and collaborators (http://www.kcl.ac.uk/herb)
San Francisco, South Ecuador	1900–2300	2200	LM(C)F	Wilcke *et al.* (2001a, 2001b), Bussmann (2001), Schrumpf *et al.* (2001)
Merida, Venezuela	2200–3000	<2950	LMCF-UMCF	Ataroff (1998), Ataroff and Rada (2000)
Monteverde, Costa Rica	1200–1850	2500	LMCF	Lawton and Dryer (1980), Zadroga (1981), Clark *et al.* (1998), Nadkarni (1984), Fallas (2002), Still *et al.* (1999), Pounds *et al.* (1999), Lawton *et al.* (2001), Bruijnzeel (2002)
Talamanca, Costa Rica	2000–3300	<6300	LMCF-UMCF	Kappelle (1995), Hölscher *et al.* (2002, 2004), Köhler (2002), Dohrenwend (1979)
Sierra de las Minas, Guatemala	1400–2700	2500	LMRF-UMCF	Brown *et al.* (1996), Holder (1998)

[a] LM(C)F, lower montane (cloud) forest; UMCF, upper montane cloud forest; SA(C)F, (dry) sub alpine forest; LMRF, lower montane rainforest not affected significantly by cloud; ECF, low-elevation dwarf cloud forest.

may be studied in the Talamanca area, Costa Rica (Kappelle, 1995) or in the Merida area, Venezuela (Ataroff and Rada, 2000).

In view of the ongoing monitoring of amphibian and bird populations (Pounds *et al.*, 1999), bryophyte communities (cf. Nadkarni, 1984), and climate change modelling efforts (Still *et al.*, 1999; Lawton *et al.*, 2001) in the Monteverde area, this site must rank as the prime location for continued long-term observations of ecological changes due to climate change. In addition, it was recently chosen as the site for a process-based evaluation of the hydrological impacts of cloud forest conversion to pasture[5].

At several of the sites listed in Table 18.4 (e.g. Malaysia, Hawai'i, Puerto Rico), observations of climatic variables along the elevational gradient have been made whereas in others (e.g. Guatemala, Puerto Rico) preliminary estimates of net precipitation vs. elevation are available as well. It would be of great interest to both regional water resource planning and TMCF conservation efforts to also initiate gradient studies of forest water and energy budgets (including transpiration) at key sites. We seem to have reached a crucial point where additional hydrological information is required if true progress in the promotion of the water values of TMCF – and therefore their chances of being afforded adequate protection – is to be made. This, together with the rapid disappearance of TMCF in many areas (Doumenge *et al.*, 1995) is why it is 'decision time' for cloud forests (Bruijnzeel and Hamilton, 2000).

It is encouraging to note, therefore, that in 1999 IUCN, WWF International, the World Conservation Monitoring Centre (Cambridge, UK) and UNESCO-IHP joined hands to form the

5 See Bruijnzeel (2002) for details.

'Tropical Montane Cloud Forest Initiative'. The objectives of the Initiative include: 'the building and strengthening of networks of TMCF conservation and research organisations around the globe' and 'to increase recognition and resources for cloud forest conservation around the world, emphasizing their role in maintaining water catchments and biodiversity'. For further details, see: http://www.unep-wcmc.org/forest/cloudforest/english/homepage.htm. The Steering Committee of the Initiative brings together representatives from the founding organisations, scientists actively working in TMCF, and NGO representatives from Asia, Latin America and Africa. In June 2000, a 40-page document entitled: '*Decision time for cloud forests*' (Bruijnzeel and Hamilton, 2000) was published by the Initiative to help raise awareness among a non-scientific audience.[6]

A concerted effort is needed now to put tropical montane cloud forests firmly on the agendas of tropical hydrologists, conservationists, resource managers, donor agencies and policy makers. The chances of achieving this have never been better than today but time is running out in some areas. One can only hope that the recently initiated 'payment for environmental services' schemes installed in some Latin American countries and considered by others (Calvo, 2000; see Aylward, this volume, for details) will help to turn the tide. At the same time policy-makers would do well to remember that such schemes need to be based on sound hydrological information. Otherwise they may well suffer loss of credibility with lowlanders being charged for environmental services (such as a reliable supply of good quality water) that cannot be met in practice.

References

Aravena, R., Suzuki, O., and A. Pollastri, A., 1989. Coastal fog and its relation to ground-water in the IV region of northern Chile. *Chemical Geology* 79: 83–91.

Asbury, C. E., McDowell, W. H., Trinidad-Pizarro, R., and R. Berrios, 1994. Solute deposition from cloud water to the canopy of a Puerto Rican montane forest. *Atmospheric Environment* 28: 1773–1780.

Ataroff, M., 1998. Importance of cloud water in Venezuelan Andean cloud forest water dynamics. In: *First International Conference on Fog and Fog Collection*, Schemenauer, R. S. and H. A. Bridgman (eds). ICRC, Ottawa, pp. 25–28.

Ataroff, M. and Rada, F., 2000. Deforestation impact on water dynamics in a Venezuelean Andean cloud forest. *Ambio* 29: 440–444.

Baynton, H. W., 1968. The ecology of an elfin forest in Puerto Rico, 2. The microclimate of Pico del Oeste. *Journal of the Arnold Arboretum* 49: 419–430.

Baynton, H. W., 1969. The ecology of an elfin forest in Puerto Rico, 3. Hilltop and forest influences on the microclimate of Pico del Oeste. *Journal of the Arnold Arboretum* 50: 80–92.

Benzing, D. H., 1998. Vulnerabilities of tropical forests to climate change: the significance of resident epiphytes. *Climatic Change* 39: 519–540.

Bonell, M. (in press). Selected issues in mountain hydrology of the humid tropics. In *Water: Forestry and Land Use Perspectives. UNESCO Technical Documents in Hydrology* No. 5x. Paris: UNESCO.

Brouwer, L. C., 1996. Nutrient Cycling in Pristine and Logged Tropical Rain Forest. Tropenbos-Guyana Series, 1, Tropenbos-Guyana Programme, Georgetown, Guyana. 224 p.

Brown, M. B., de la Roca, I., Vallejo, A., Ford, G., Casey, J., Aguilar, B. and R. Haacker, 1996). *A Valuation analysis of the Role of Cloud Forests in Watershed Protection. Sierra de las Minas Biosphere Reserve, Guatemala and Cusuco N.P., Honduras.* RARE Center for Tropical Conservation, Philadelphia, USA

Bruijnzeel, L. A., 1989. (De)forestation and dry season flow in the humid tropics: a closer look. *Journal of Tropical Forest Science* 1: 229–243.

Bruijnzeel, L. A., 1990. *Hydrology of Moist Tropical Forests and Effects* converting tropical montane cloud forest to pasture, with initial *of Conversion: a State of Knowledge Review*. IHP-UNESCO Humid Tropical Programme, Paris, 224 pp.

Bruijnzeel, L. A., 2001. Hydrology of tropical montane cloud forests: a reassessment. *Land Use and Water Resources Research* 1: 1.1–1.18. [Also published under the same title as: Bruijnzeel (2002a) in *Hydrology and Water Management in the Humid Tropics*, (ed). J. S. Gladwell, pp. 353–383. *UNESCO Technical Documents in Hydrology* No. 5x. Paris: UNESCO.

Bruijnzeel, L. A., 2002. Hydrological impacts of reference to northern Costa Rica. Project Memorandum Form R7991, Forestry Research Programme, Natural Resources International, Aylesford, UK, 60 pp.

Bruijnzeel, L. A., 2004. Hydrological functions of tropical forests: not seeing the soil for the trees? *Agriculture, Ecosystems and Environment* doi: 10.1016/j.agee.2004.011.015.

Bruijnzeel, L. A. and L. S. Hamilton, 2000. Decision Time for Cloud Forests. *IHP Humid Tropics Programme Series* no. 13, IHP-UNESCO, Paris, IUCN-NL, Amsterdam and WWF International, Gland, 44 pp.

Bruijnzeel. L. A. and J. Proctor, 1995. Hydrology and biogeochemistry of tropical montane cloud forests: what do we really know? In: *Tropical Montane Cloud Forests. Ecological Studies* 110, Hamilton, L. S., Juvik, J. O. and F. N. Scatena (eds). Springer Verlag, New York, pp. 38–78.

Bruijnzeel, L. A., Waterloo, M. J., Proctor, J., Kuiters, A. T. and B. Kotterink, 1993. Hydrological observations in montane rainforests on Gunung Silam, Sabah, Malaysia, with special reference to the '*Massenerhebung*' effect. *Journal of Ecology* 81: 145–167.

Burkard, R., Eugster, W., Wrezinsky, T. and Klemm, O., (2002). Vertical divergence of fogwater fluxes above a spruce forest. *Atmospheric Research* 64: 133–145.

Bussmann, R. W., 2001. The montane forests of Reserva Biológica San Francisco (Zamora-Chinchipe), Ecuador. *Die Erde* 132: 9–25.

Calder, I. R., 1998. Water use by forests: limits and controls. *Tree Physiology* 18: 625–631.

Calvo, J. C., 1986. An evaluation of Thornthwaite's water balance technique in predicting stream runoff in Costa Rica. *Hydrological Sciences Journal* 31: 51–60.

Calvo, J. C. (2000). Case study: payments for watershed services in Costa Rica. In: M. F. Price and N. Butt (Eds.), *Forests in Sustainable Mountain Development. A State-of-Knowledge Report for 2000*, pp. 428–429, CABI, Wallingford, UK.

Cavelier, J., Jaramillo, M., Solis, D. and D. De Leon, 1997. Water balance and nutrient inputs in bulk precipitation in tropical montane cloud forest in Panama. *Journal of Hydrology* 193: 83–96.

Cavelier, J., Solis, D. and M. A. Jaramillo, 1996. Fog interception in montane forests across the Central Cordillera of Panama. *Journal of Tropical Ecology* 12: 357–369.

Chang, S. C., Lai, I. L., and Wu, J. T. (2002). Estimation of fog deposition on epiphytic bryophytes in a subtropical montane forest ecosystem in northeastern Taiwan. *Atmospheric Research* 64: 159–167.

Clark, K. L., Nadkarni, N. M., Schaeffer, D. and H. L. Gholz, 1998. Atmospheric deposition and net retention of ions by the canopy in a tropical montane forest, Monteverde, Costa Rica. *Journal of Tropical Ecology* 14: 27–45.

Daube, B., Kimball, K. D., Lamar, P. A. and K. C. Weathers, 1987. Two new ground-level cloud water sampler designs which reduce rain contamination. *Atmospheric Environment* 21: 893–900.

Dawson, T. E., 1998. Fog in the California redwood forest: ecosystem inputs and use by plants. *Oecologia* 117: 476–485.

Dohrenwend, R. E., 1979. Hydrologic behavior at the top of a tropical mountain. *Michigan Technological University Forestry Center Research Note no. 29*, 14 pp.

6 A Spanish translation appeared in 2001. Both documents are freely available as long as stocks last from the Humid Tropics Programme, UNESCO-IHP, Paris, or from the author.

Doumenge, C., Gilmour, D. A., Ruiz Perez, M. and J. Blockhus, 1995. Tropical montane cloud forests: conservation status and management issues. In:. *Tropical Montane Cloud Forests. Ecological Studies* 110, L. S. Hamilton, J. O. Juvik, J. O. and F. N. Scatena (eds). Springer Verlag, New York, pp. 24–37.

Duisberg-Waldenberg, P. (1980). Erosion y conservacion de suelos. In: Centro Cientifico Tropical. *Estudio Ecológico Integral de las Zonas de Afectación del Proyecto Arenal.* Centro Cientifico Tropical, San José, Costa Rica, pp. 2-1-2-65.

Eugster, W., Burkard, R., Klemm, O. and Wrezinsky, T. (2001). Fog deposition measurements with the eddy covariance method. In: *Second International Conference on Fog and Fog Collection*, R. S. Schemenauer and H. Puxbaum (eds). ICRC, Ottawa, pp. 193–196.

Fallas, J., 2002. Net precipitation patterns in undisturd and fragmented Costa Rican cloud forests. In: J. S. Gladwell (Ed.), *Proceedings of the Second International Colloquium on Hydrology and Water Management.* CATHALAC, Panamá City, pp. 389–398.

Fleming, T. H., 1986. Secular changes in Costa Rican rainfall: correlation with elevation. *Journal of Tropical Ecology* 2: 87–91.

Foster, P., 2001. The potential negative impacts of global climate change on ropical montane cloud forests. *Earth-Science Review* 55: 73–106.

Frahm, J. P., 1990a. The ecology of epiphytic bryophytes on M. Kinabalu, Sabah (Malaysia). *Nova Hedwigia* 51: 121–132.

Frahm, J. P., 1990b. The altitudinal zonation of bryophytes on M. Kinabalu. *Nova Hedwigia* 51: 133–149.

Frahm, J. P. and S. R. Gradstein, 1991. An altitudinal zonation of tropical rainforests using bryophytes. *Journal of Biogeography* 18: 669–676.

Garciá-Martino, A. R., Warner, G. S., Scatena, F. N. and D. L. Civco, 1996. Rainfall, runoff and elevation relationships in the Luquillo Mountains of Puerto Rico. *Caribbean Journal of Science* 32: 413–424.

Gilmour, D. A., Bonell, M. and D. S. Cassells, 1987. The effects of forestation on soil hydraulic properties in the Middle Hills of Nepal: A preliminary assessment. *Mountain Research and Development* 7: 239–249.

Goodman, J., 1985. The collection of fog drip. *Water Resources Research* 21: 392–394.

Gordon, C. A., Herrera, R. and T. C. Hutchinson, 1994. Studies of fog events at two cloud forests near Caracas, Venezuela-II. Chemistry of fog. *Atmospheric Environment* 28: 323–337.

Grubb, P. J., 1974. Factors controlling the distribution of forest types on tropical mountains; new facts and a new perspective. In:. Altitudinal Zonation in Malesia, J. R. Flenley (ed). *University of Hull, Department of Geography Miscellaneous Series* no. 16, 1–25.

Grubb, P. J. and T. C. Whitmore, 1966. A comparison of montane and low land rainforest in Ecua dor. II. The climate and its effects on the distribution and physiognomy of the forests. *Journal of Ecology* 54: 303–333.

Hafkenscheid, R. L. L. J., 2000. *Hydrology and Biogeochemistry of Montane Rain Forests of Contrasting Stature in the Blue Mountains of Jamaica.* PhD thesis, Vrije Universiteit, Amsterdam.

Hafkenscheid, R. L. L. J., Bruijnzeel, L. A. and R. A. M. De Jeu, 1998. Estimates of fog interception by montane rainforest in the Blue mountains of Jamaica. In: *First International Conference on Fog and Fog Collection,* Schemenauer, R. S. and H. A. Bridgman (eds). ICRC, Ottawa, pp. 33–36.

Hafkenscheid, R. L. L. J., Bruijnzeel, L. A., R. A. M. De Jeu and N. J. Bink (2002). Water budgets of two upper montane rainforests of contrasting stature in the Blue mountains, Jamaica. In: J. S. Gladwell (Ed.), *Proceedings of the Second International Colloquium on Hydrology and Water Management.* CATHALAC, Panamá City, Panamá, pp. 399–424.

Herwitz, S. R., and Slye, R. E. (1992). Spatial variability in the interception of inclined rainfall by a tropical rainforest canopy. *Selbyana* 13: 62–71.

Hewlett, J. D. and J. C. Fortson, 1983. The paired catchment experiment. In: *Forest Water Quality*, J. D. Hewlett (ed). University of Georgia, School of Forest Resources, Athens, Ga., pp. 11–14 (77).

Hommel, P. F. M., 1987. *Landscape Ecology of Ujung Kulon (West Java, Indonesia).* PhD Thesis, Wageningen Agricultural University, Wageningen, The Netherlands.

Hamilton, L. S., 1995a. *A Campaign for Cloud Forests.* International Union for the Conservation of Nature, Gland, Switzerland.

Hamilton, L. S., 1995b. Montane cloud forest conservation and research: a synopsis. *Mountain Research and Development* 15: 259–266.

Hamilton, L. S., Juvik, J. O. and F. N. Scatena (eds), 1995. *Tropical Montane Cloud Forests. Ecological Studies* 110, Springer Verlag, New York, 407 pp.

Harr, R. D., 1982. Fog drip in the Bull Run municipal watershed, Oregon. *Water Resources Bulletin* 18: 785–789.

Harr, R. D., 1983. Potential for augmenting water yield through forest practices in western Washington and western Oregon. *Water Resources Bulletin* 19: 383–393.

Holder, C. D., 1998. Fog precipitation in the Sierra de las Minas Biosphere Reserve, Guatemala. In: *First International Conference on Fog and Fog Collection*, Schemenauer, R. S. and H. A. Bridgman (eds). ICRC, Ottawa, pp. 101–103.

(2003). Fog precipitiation in the Sierra de las Minas Reserve, Guatemala. *Hydrological Processes*, 17, 2001–2010.

Hölscher, D., Köhler, L., Leuschner, C. and Kappelle, M. (2002). Nutrient fluxes in stemflow and throughfall in three successional stages of an upper montane rainforest in Costa Rica. *Journal of Tropical Ecology* 19: 557–565.

Hölscher, D., Köhler, L., van Dijk, A. I. J. M., and Bruijnzeel, L. A. 2004 The importance of epiphytes to total rainfall interception by a tropical montane rainforest in Costa Rica. *Journal of Hydrology*, 292, 308–322.

Holwerda, F., 1997. A Study of Evaporation from Lowland and Montane Tropical Rain Forest in the Luquillo Mountains, Puerto Rico. MSc Thesis, Vrije Universiteit, Amsterdam.

Hutley, L. B., Doley, D., Yates, D. J. and A. Boonsaner, 1997. Water balance of an Australian subtropical rainforest at altitude: the ecological and physiological significance of intercepted cloud and fog. *Australian Journal of Botany* 45: 311–329.

Hyndman, D. C. and J. I. Menzies, 1990. Rainforests of the Ok Tedi headwaters, New Guinea: an ecological analysis. *Journal of Biogeography* 17: 241–273.

Ingraham, N. L. and R. A. Matthews, 1988. Fog drip as a source of groundwater recharge in northern Kenya. *Water Resources Research* 24: 1406–1410.

Ingraham, N. L. and R. A. Matthews, 1990. A stable isotopic study of fog: the Point Reyes Peninsula, California, USA *Chemical Geology* 80: 281–290.

Ingwersen, J. B., 1985. Fog drip, water yield, and timber harvesting in the Bull Run municipal watershed, Oregon. *Water Resources Bulletin* 21: 469–473.

Joslin, J. D., Mueller, S. F. and M. H. Wolfe, 1990. Tests of models of cloudwater deposition to forest canopies using artificial and living collectors. *Atmospheric Environment* 24A: 3007–3019.

Juvik, J. O. and P. C. Ekern, 1978. *A Climatology of Mountain Fog on Mauna Loa, Hawai'i Island. Technical Report* no. 118, Water Resources Research Center, Honolulu, Hawai'i, 63 pp.

Juvik, J. O. and D. Nullet, 1995a. Comments on 'a proposed standard fog collector for use in high elevation regions'. *Journal of Applied Meteorology* 34: 2108–2110.

Juvik, J. O. and D. Nullet, 1995b. Relationships between rainfall, cloudwater interception and canopy throughfall in a Hawaiian montane forest. In: *Tropical Montane Cloud Forests. Ecological Studies* 110, Hamilton, L. S., Juvik, J. O. and F. N. Scatena (eds). Springer Verlag, New York, pp. 165–182.

Kapos, V. and E. V. J. Tanner, 1985. Water relations of Jamaican upper montane rainforest trees. *Ecology* 66: 241–250.

Kappelle, M., 1995. *Ecology of Mature and Recovering Talamancan Montane Quercus Forests, Costa Rica.* PhD thesis, University of Amsterdam, Amsterdam.

Kashiyama, T., 1956. Decrease of sea-fog density by a model shelterbelt. *Proceedings of the 12th Congress of the International Union of Forestry Research Organizations Oxford*, November 1956.

Kerfoot, O., 1968. Mist precipitation on vegetation. *Forestry Abstracts* 29: 8–20.

Kitayama, K., 1992. An altitudinal transect study of the vegetation on Mount Kinabalu, Borneo. *Vegetatio* 102: 149–171.

Kitayama, K. and D. Müller-Dombois, 1992. Vegetation of the wet windward slope of Haleakala, Maui, Hawai'i. *Pacific Science* 46: 197–220.

Köhler, L. (2002). Die Bedeutung der Epiphyten im ökosystemaren Wasser- und Nährstoffumsatz verschiedener Altersstadien eines Bergregenwaldes in Costa Rica. PhD thesis, University of Göttingen, Göttingen, Germany, 131 pp.

Kowalski, A. S., Anthoni, P. M., Vong, R. J., Delany, A. C. and MacLean, G. D. (1997). Deployment and evaluation of a system for ground-based measurement of cloud liquid water turbulent fluxes. *Journal of Atmospheric and Oceanic Technology* 14: 468–479.

Lal, R., 1987. *Tropical Ecology and Physical Edaphology*. Wiley and Sons, New York.

Lawton, R. and V. Dryer, 1980. The vegetation of the Monteverde cloud forest reserve. *Brenesia* 18: 101–116.

Lawton, R. O., Nair, U. S., Pielke Sr., R. A. and Welch, R. M. (2001). Climatic impact of tropical lowland deforestation on nearby montane cloud forests. *Science* 294: 584–587.

Lloyd, C. R. and A. de O. F. Marques, 1988. Spatial variability of throughfall and stemflow measurements in Amazonian rainforest. *Forest and Agricultural Meteorology* 42: 63–73.

Lugo, A. E. and F. N. Scatena, 1992. Epiphytes and climate change research in the Caribbean: a proposal. *Selbyana* 13: 123–130.

Merlin, M. D. and J. O. Juvik, 1995. Montane cloud forests in the tropical Pacific: Some aspects of their floristics, biogeography, ecology and conservation. In: *Tropical Montane Cloud Forests. Ecological Studies* 110, Hamilton, L. S., Juvik, J. O. and F. N. Scatena (eds). Springer Verlag, New York, pp. 234–253.

Mueller, S. F., 1991. Estimating cloud water deposition to subalpine spruce-fir forests. I. Modifications to an existing model. *Atmospheric Environment* 25A: 1093–104.

Mueller, S. F., Joslin, J. D. and M. H. Wolfe, 1991. Estimating cloud water deposition to subalpine spruce-fir forests. II. Model testing. *Atmospheric Environment* 25A: 1105–1122.

Mulligan, M., 2000. Downscaled climate cyhange scenaria for Colombia and their hydrological consequences *Advances in Environmental Monitoring and Modelling* 1, 3–40.

Nadkarni, N. M., 1984. Epiphyte biomass and nutrient capital of a neotropical elfin forest. *Biotropica* 16: 249–256.

Nullet, D. A. and J. O. Juvik, 1994. Generalised mountain evaporation profiles for tropical and subtropical latitudes. *Singapore Journal of Tropical Geography* 15: 17–24.

Nullet, D. and J. O. Juvik, 1997. Measured altitudinal profiles of UV-B irradiance in Hawai'i. *Physical Geography* 18: 335–345.

Ohsawa, M., Wildpret, W. and M. del Arco, 1999. *Anaga Cloud Forest: A Com-parative Study on Evergreen Broad-leaved Forests and Trees of the Canary Islands and Japan*. Laboratory of Ecology, Chiba University, Chiba, Japan.

Padilla, H., Belmont, R., Torres, M. C., Garcia, R. and A. Baez, 1996. A field comparison of cloud water collectors in a mountainous region under low wind speed conditions in eastern Mexico. *Atmosfera* 9: 189–199.

Pattanayak, S. and R. Kramer, 2004. Valuing watershed services: concepts and empirics from South East Asia. *Agriculture, Ecosystems and Environment* (in press).

Persson, R., 1974. *World Forest Resources*. Royal College of Forestry, Stockholm, Sweden.

Pócs, T., 1980. The epiphytic biomass and its effect on the water balance of two rainforest types in the uluguru Mountains (Tanzania, East Africa). *Acta Botanica Academia Scientifica Hungarica* 26: 143–167.

Pounds, J. A., Fogden, M. P. A. and J. H. Campbell, 1999. Biological response to climate change on a tropical mountain. *Nature* 398: 611–615.

Richardson, B. A., Richardson, M. J., Scatena, F. N. and W. H. McDowell, 2000. Effects of nutrient availability and other elevational changes on bromeliad populations and their invertebrate communities in a humid tropical forest in Puerto Rico. *Journal of Tropical Ecology* 16: 167–188.

Russell, S., 1984. Measurement of mist precipitation. *Techniques Notebook, The Bryological Times* 25: 4.

Sandström, K. (1998). Can forests 'provide' water: widespread myth or scientific reality? *Ambio* 27, 132–138.

Santiago, L. S., Goldstein, G., Meinzer, F. C., Fownes, J. H. and D. Müller-Dombois, 2000. Transpiration and forest structure in relation to soil waterlogging in a Hawaiian montane cloud forest. *Tree Physiology* 20: 673–681.

Scatena, F. N., 1998. An assessment of climatic change in the Luquillo Mountains of Puerto Rico. In: *Third International Symposium on Water Resources*, American Water Resources Association, pp. 193–198.

Scatena, F. N. and M. Larsen, 1991. Physical aspects of hurricane Hugo in Puerto Rico. *Biotropica* 23: 317–323.

Schellekens, J., Bruijnzeel, L. A., Wickel, A. J., Scatena, F. N. and W. L. Silver, 1998. Interception of horizontal precipitation by elfin cloud forest in the Luquillo Mountains, eastern Puerto Rico. In: R. S. Schemenauer and H. A. Bridgman, eds., *First International Conference on Fog and Fog Collection*, ICRC, Ottawa, pp. 29–32.

Schemenauer, R. S. and P. Cereceda, 1994. A proposed standard fog collector for use in high elevation regions. *Journal of Applied Meteorology* 33: 1313–1322.

Schemenauer, R. S. and P. Cereceda, 1995. Reply to comments by Juvik and Nullet (1995). *Journal of Applied Meteorology* 34: 2111–2112.

Schol, M. A., Gingerich, S. A., and Tribble, G. W., 2002. The influence of microclimates and fog on interpretation of regional hydrology using $\delta^{18}O$ and δD: East Maui, Hawaii. *Journal of Hydrology* 264: 170–184.

Schrumpf, M., Guggenberger, G., Valarezo, C. and Zech, W., 2001. Tropical montane rainforest soils. Development and nutrient status along an altitudinal gradient in the south Ecuadorian Andes. *Die Erde* 132: 43–59.

Sharon, D. (1980). The distribution of hydrologically effective rainfall incident on sloping ground. *Journal of Hydrology* 46: 165–188.

Smith, D. M. and S. J. Allen, 1996. Measurement of sap flow in plant stems. *Journal of Experimental Botany* 47: 183–1844.

Sperling, F., 2000. Tropical montane cloud forests – Ecosystems under the threat of climate change. Unpublished report, World Conservation Monitoring Centre, Cambridge, UK, 10 pp.

Stadtmüller, T., 1987. *Cloud Forests in the Humid Tropics. A Bibliographic Review*. The United Nations University, Tokyo and Centro Agronomico Tropical de Investigacion y Ensenanza, Turrialba, Costa Rica, 81 pp.

Steinhardt, U., 1979. Untersuchungen über den Wasser- und Nährstoffhaushalt eines andinen Wolkenwaldes in Venezuela. *Göttinger Bodenkundliche Berichte* 56: 1–185.

Still, C. J., Foster, P. N. and S. H. Schneider, 1999. Simulating the effects of climate change on tropical montane cloud forests. *Nature* 398: 608–610.

Tanner, E. V. J., 1977. Four montane rainforests of Jamaica: a quantitative characterization of the floristics, the soils and the foliar mineral levels, and a discussion of the interrelations. *Journal of Ecology* 65: 883–918.

Te Linde, A. H., Bruijnzeel, L. A., Groen, J., Scatena, F. N., and Meijer, H. A. J. (2001). Stable isotopes in rainfall and fog in the Luquillo Mountains, eastern Puerto Rico: a preliminary study. In: *Second International Conference on Fog and Fog Collection*, R. S. Schemenauer and H. Puxbaum (eds). ICRC, Ottawa, pp. 181–184.

Van der Molen, M. K. (2002). Meteorological impacts of land use change in the maritime tropics. PhD thesis, Vrije Universiteit, amsterdam, The Netherlands, 262 pp.

Van Dijk, A. I. J. M. and Bruijnzeel, L. A. (2001). Modelling rainfall interception by vegetation of variable density using an adapted analytical model. 1. Model description. *Journal of Hydrology* 247: 230–238.

Van Steenis, C. G. G. J., 1972. *The Mountain Flora of Java*. E. J. Brill, Leiden, The Netherlands.

Veneklaas, E. J., Zagt, R. J., Van Leerdam, A., Van Ek, R., Broekhoven, A. J. and M. Van Genderen, 1990. Hydrological properties of the epiphyte mass of a montane tropical rainforest. *Vegetatio* 89: 183–192.

Vogelmann, H. W., 1973. Fog precipitation in the cloud forests of Eastern Mexico. *BioScience* 23: 96–100.

Walmsley, J. L., Burrows, W. R. and R. S. Schemenauer, 1999. The use of routine observations to calculate liquid water content in summertime high-elevation fog. *Journal of Applied Meteorology*, 38: 369–384.

Walmsley, J. L., Schemenauer, R. S. and H. A. Bridgman, 1996. A method for Estimating the hydrologic input from fog in mountainous terrain. *Journal of Applied Meteorology* 35: 2237–2249.

Watling, D. and A. N. Gillison, 1995. Endangered species in low elevation cloud forest on Gau island, Fiji. In: *Tropical Montane Cloud Forests. Ecological Studies* 110, Hamilton, L. S., Juvik, J. O. and F. N. Scatena (eds). Springer Verlag, New York, pp. 333–342.

Weathers, K. C., Lovett, G. M. and Likens, G. E., 1995. Cloud deposition to a spruce forest edge. *Atmospheric Environment* 29: 665–672.

Weaver, P. L., 1972. Cloud moisture interception in the Luquillo Mountains of Puerto Rico. *Caribbean Journal of Science* 12: 129–144.

Weaver, P. L., 1995. The Colorado and dwarf forests of Puerto Rico's Luquillo Mountains. In: *Tropical Forests: Management and Ecology*, Lugo, A. E. and C. Lowe (eds). *Springer Ecological Studies* 112, pp. 109–141.

Whitmore, T. C., 1998. *An Introduction to Tropical Rain Forests (2nd Edition)*. Clarendon Press, Oxford.

Wilcke, W., Yasin, S., Valarezo, C. and Zech, W., 2001a. Changes in water quality during the passage through a tropical montane rainforest in Ecuador. *Biogeochemistry* 55: 45–72.

Wilcke, W., Yasin, S., Valarezo, C. and Zech, W., 2001b. Nutrient budget of three microcatchments under tropical montane rainforest in Ecuador – preliminary results. *Die Erde* 132: 61–74.

World Conservation Monitoring Centre, 1997. *A Global Directory of Tropical Montane Cloud Forests (draft)*. M. Aldrich *et al.* (eds). WCMC, Cambridge, UK

Yang, D., Goodison, B. E., Ishida, S. and Benson, C. S. (1998). Adjustment of daily precipitation data at 10 climate stations in Alaska: Application of World Meteorological Organization intercomparison results. *Water Resources Research* 34: 241–256.

Yin, X. W. and P. A. Arp, 1994. Fog contributions to the water budget of forested watersheds in the Canadian Maritime Provinces: a generalized algorithm for low elevations. *Atmosphere-Ocean* 32: 553–566.

Zadroga, F., 1981. The hydrological importance of a montane cloud forest area of Costa Rica. In: *Tropical Agricultural Hydrology*, Lal, R. and E. W. Russell (eds). J. Wiley, New York, pp. 59–73.

Ziegler, A. D. and T. W. Giambelluca, 1997. Importance of rural roads as source areas for runoff in mountainous areas of northern Thailand. *Journal of Hydrology* 196: 204–229.

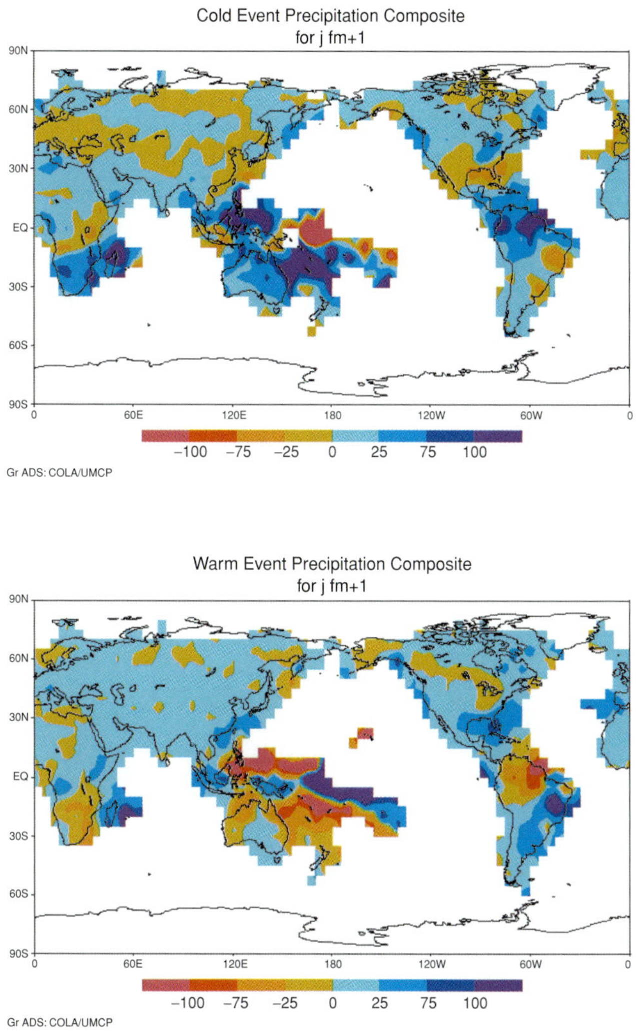

Plate 1 Global rainfall anomalies January/February/March for the years following the onset of La Niña (top) and El Niño (bottom). Units are mm/month. (Source: NOAA-CIRES Climate Diagnostics Center: http://www.cdc.noaa.gov/)

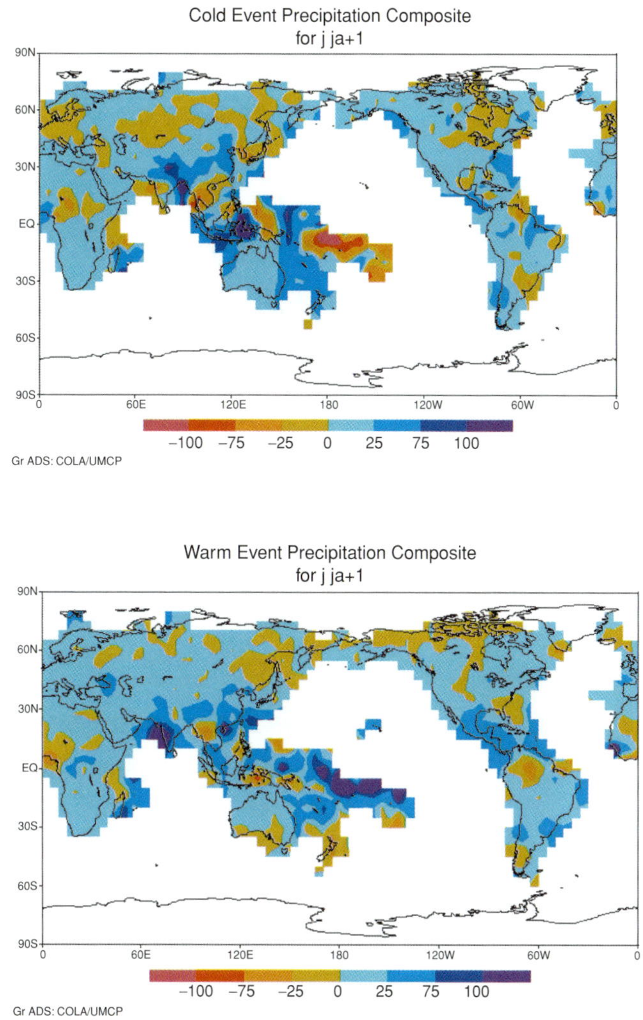

Plate 2 Global rainfall anomalies June/July/August for the years following the onset of La Niña (top) and El Niño (bottom). Units are mm/month. (Source: NOAA-CIRES Climate Diagnostics Center: http://www.cdc.noaa.gov/)

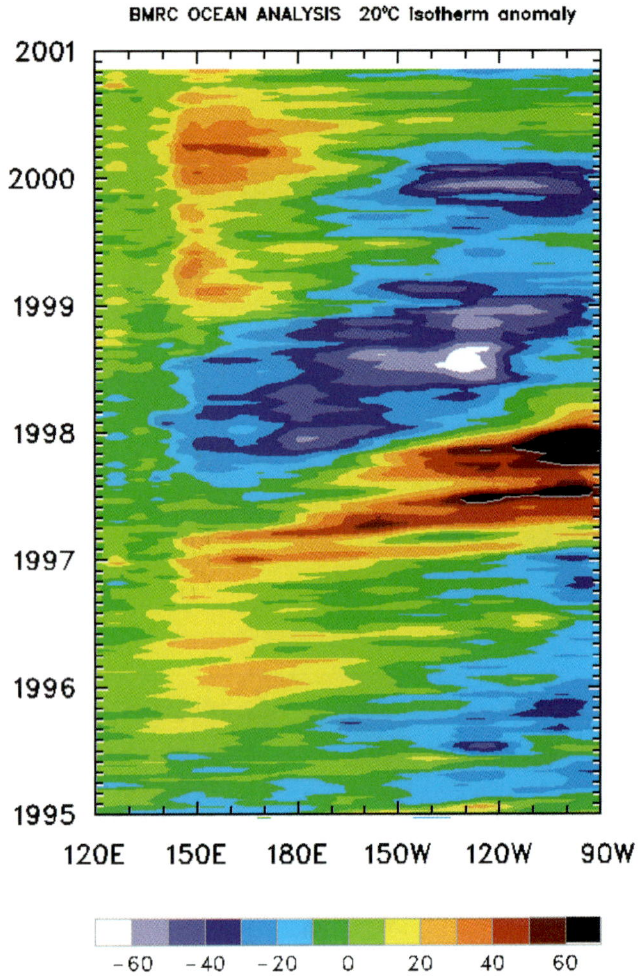

Plate 3 Hovmoller (time/longitude) diagram of the anomaly of the 20 °C depth (metres) for period 1995 to 2000. (Source: Australian Bureau of Meteorology.)

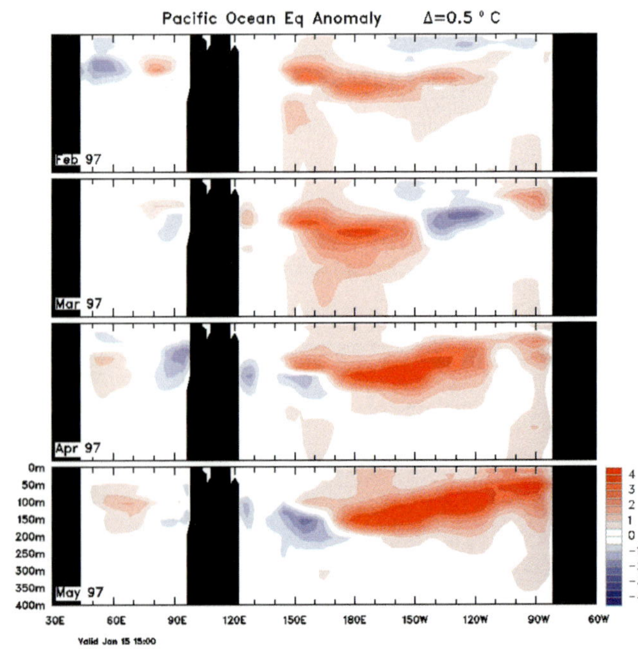

Plate 4 Sequence of the 150 m depth averaged temperature anomalies (K) for period February 1997 to May 1997. (Source: Australian Bureau of Meteorology.)

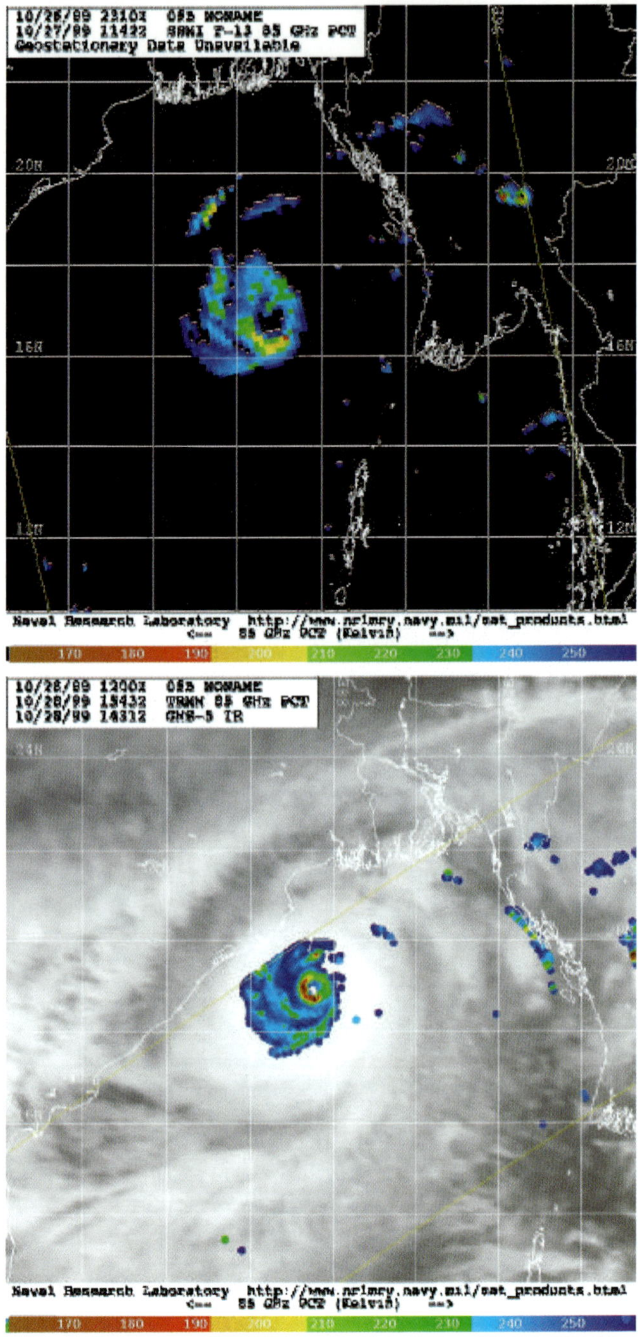

Plate 5 SSM/I image (top) 1142 UTC 27 October 1999 (top) and TRMM image (bottom) 1543 UTC 28 October 1999. (Source: US Naval Research Laboratory, Monterey, CA, USA http://www.nrlmry.navy.mil/tc-bin/tc.home)

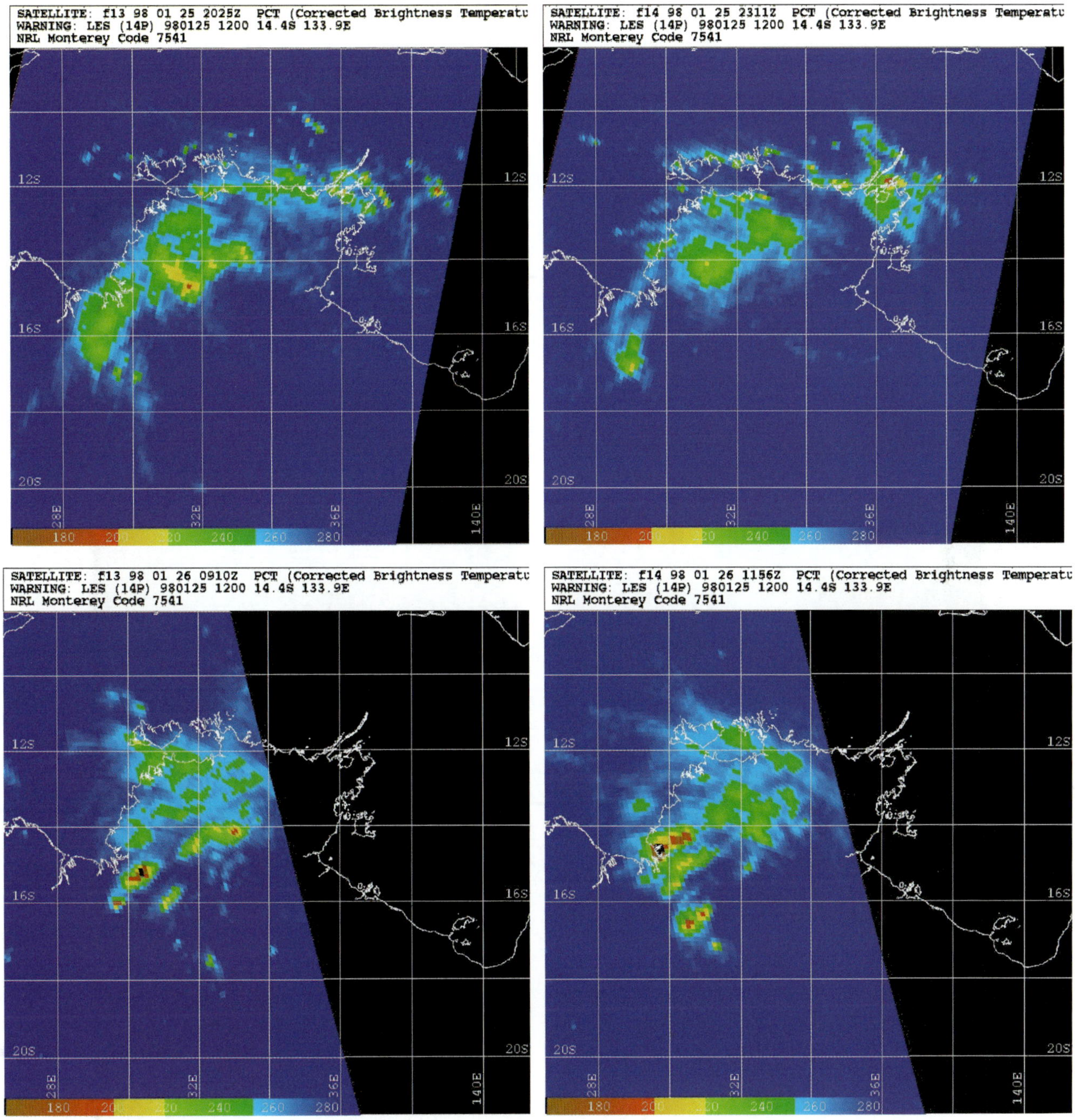

Plate 6 SSM/I imagery at 2025 UTC 25 January 1998 (top left), 2311 UTC 25 January 1998 (top right), 0910 UTC 26 January 1998 (bottom left) and 1156 UTC 26 January 1998 (bottom right). (Source: US Naval Research Laboratory, Monterey, CA, USA http://www.nrlmry.navy.mil/tc-bin/tc.home)

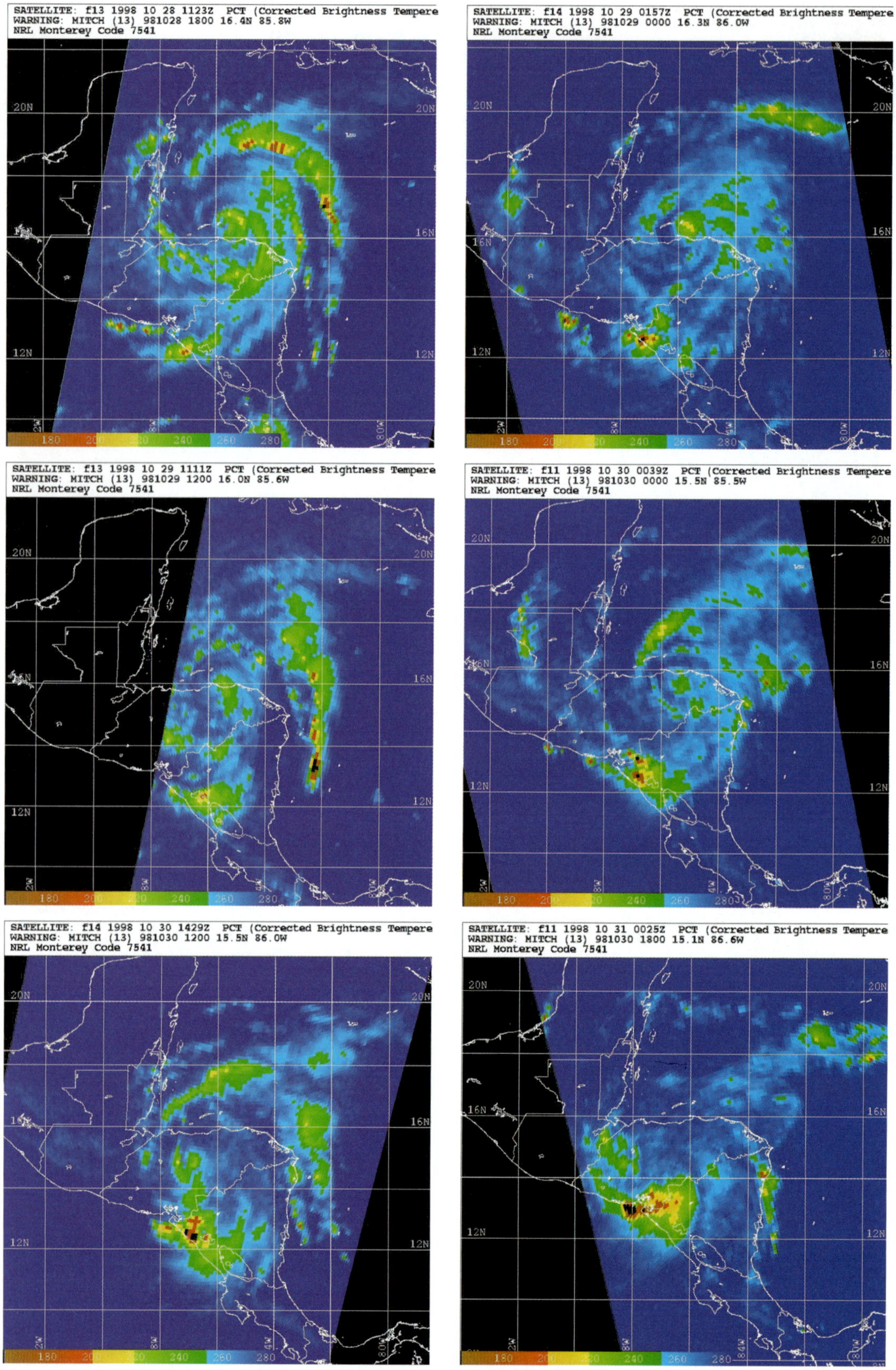

Plate 7 The rain-bands of hurricane *Mitch* (SSM/I) imagery from 1123 UTC 28 October 1998 to 0025 UTC 31 October 1998. (Source: NCEP/NCAR re-analyses data from http://wesley.wwb.noaa.gov/reanalysis.html)

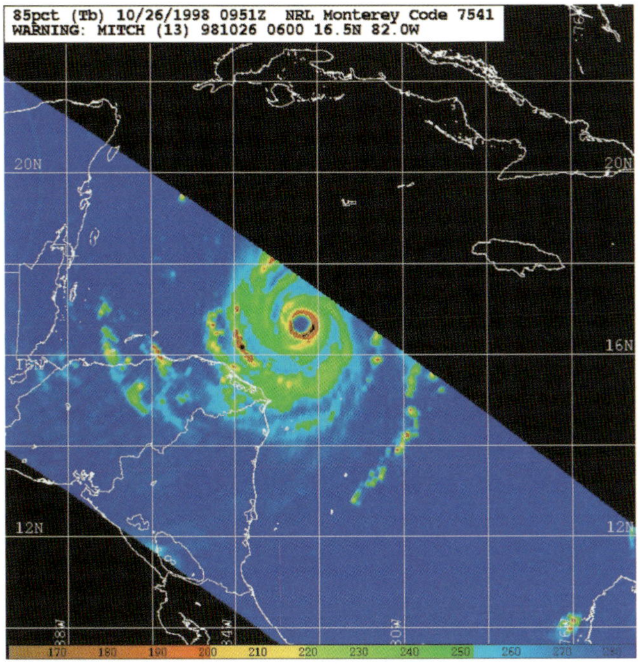

Plate 8 TRMM image of hurricane Mitch at 0951 UTC 26 October 1998. (Source: US Naval Research Laboratory, Monterey, CA, USA http://www.nrlmry.navy.mil/tc-bin/tc.home)

Plate 9 Three-dimensional representation of a 126-band hyperspectral image acquired by the 'Hymap' airborne sensor over a 1.5 km × 1.5 km area surrounding Cape Tribulation in Australia. In this reprsentation, each 3 m resolution image pixel contains the geographic, spatial dimensions in the *x*- and *y*-dimension, while the spectral reflectance information, from 400 to 2500 nm, is shown in the third, *z*-dimension. The brighter yellow and red colours represent higher reflectance and dark blue colours represent low reflectance areas in the specific wavelegths. Notice the sharp changes in reflectance between land-based and water-based areas. (After Held, unpublished data.)

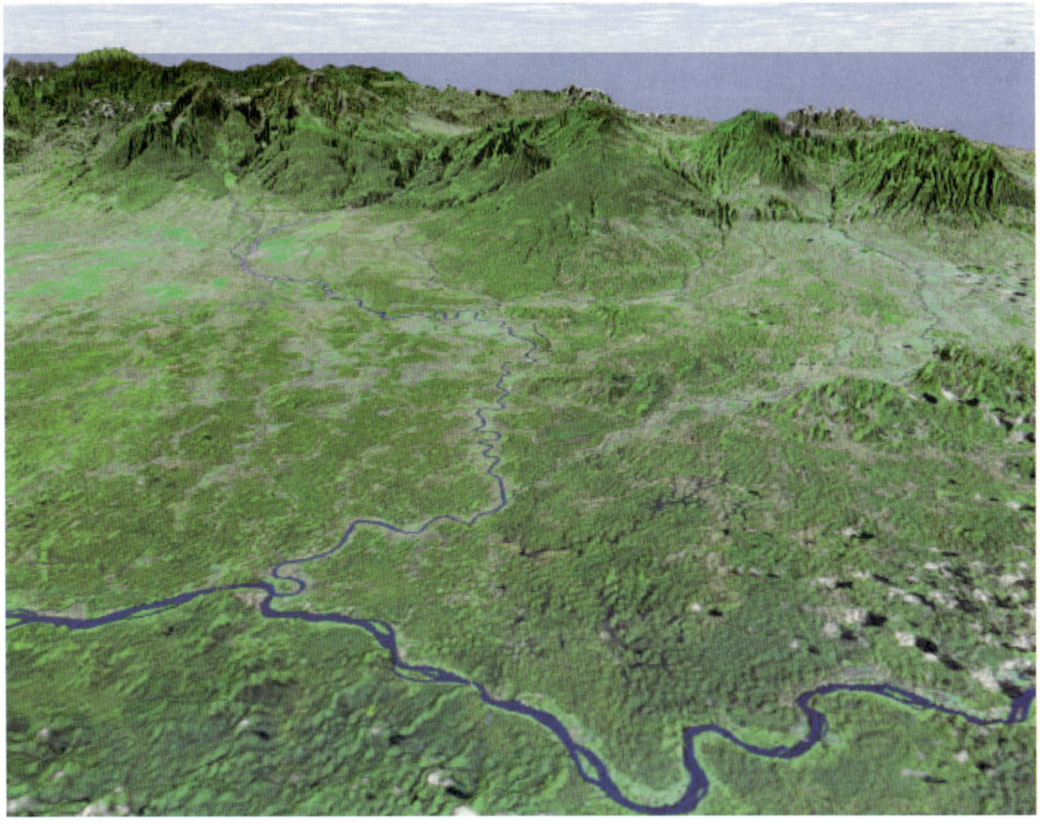

Plate 10 Perspective view of the northern coastal plain of Costa Rica and the Rio San Juan, combining a Landsat TM image with SRTM-determined topography, exaggerated about six times vertically. Both datasets were acquired in February, 2000. (from NASA/JPL/NIMA/USGS.)

Plate 11 Age map of a 275 km^2 area of regenerating rainforest near Manaus, Brazil, as derived from time-series of 30-m resolution Landsat sensor data. (After Lucas *et al.*, 2001.)

Plate 12 Sample terrain classification of a 1,140 km^2 rainforest area in the Amazon using JERS-1 (satellite radar) data processed to a 90 m pixel resolution. Water and flood areas are classified blue, bare soil and sand areas are yellow, whereas forest is green. (After Siqueria *et al.*, 1997.)

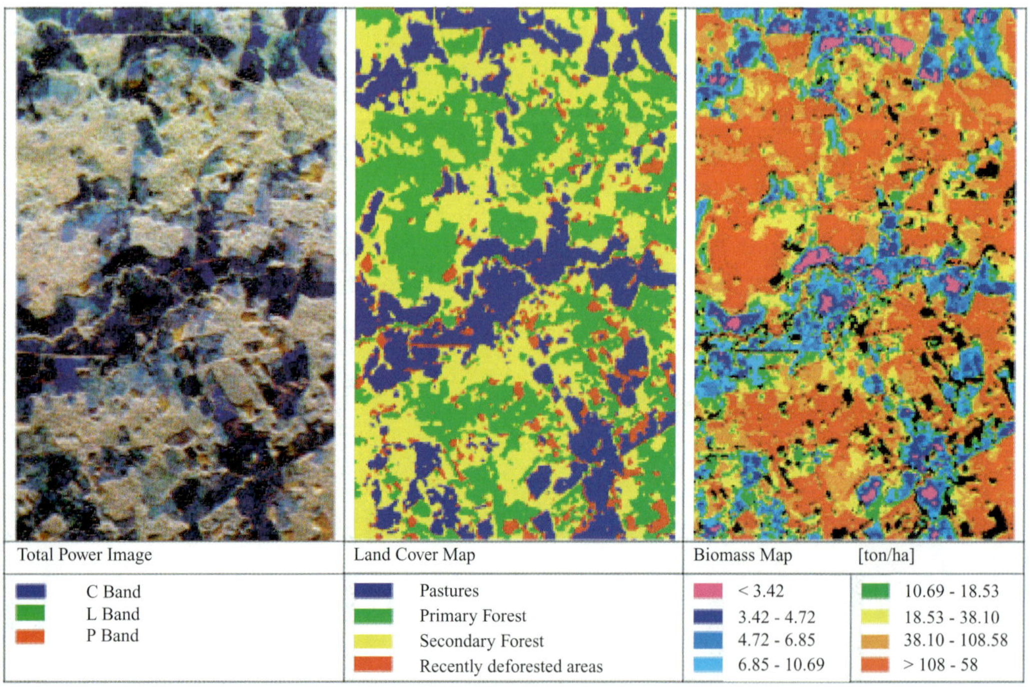

Plate 13 (a) AirSAR Total Power image, (b) land cover map and (c) biomass map, for a 4×7 km^2 section of a 40×8 km^2 area mapped in Guaviare, Colombia. In the biomass map the areas of recently cut forest are masked (black). (After Hoekman and Quiñones, 2000.)

Plate 14 (Left) High spatial resolution (1m pixel) true colour image of a tropical canopy collected with the casi airborne imager (cf. Table 27.1) over rainforests in Queensland, Australia. (Right) Image analysis product in the form of polygons, corresponding to delineated crowns. Different colour polygons (reproduced have in grays) represent crowns of different spectral characteristics (After Held *et al.*, 2001.)

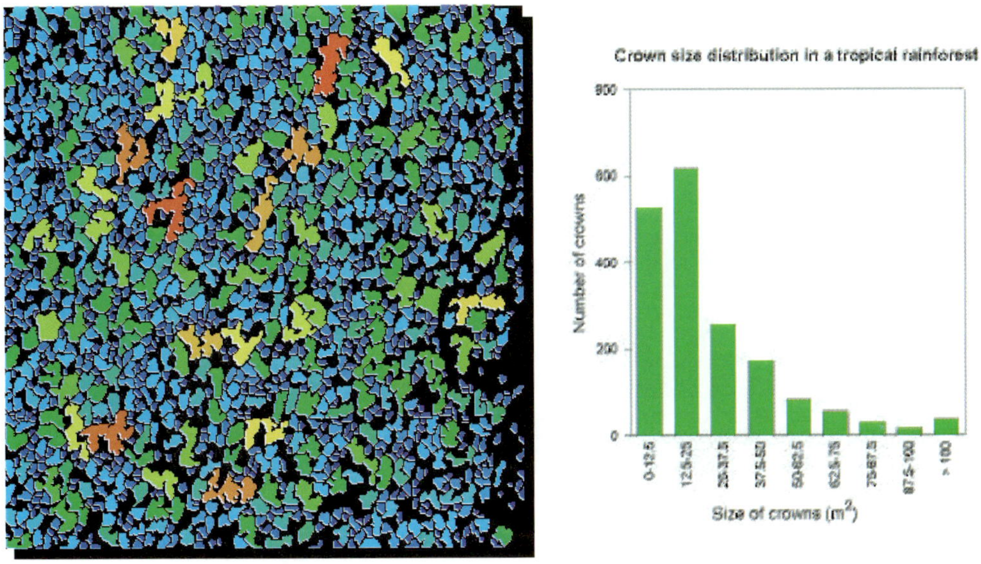

Plate 15 Automatic, computer assisted crown mapping image and crown-size distribution, derived from a high resolution forest canopy image. (After Held and Billings, 1998.)

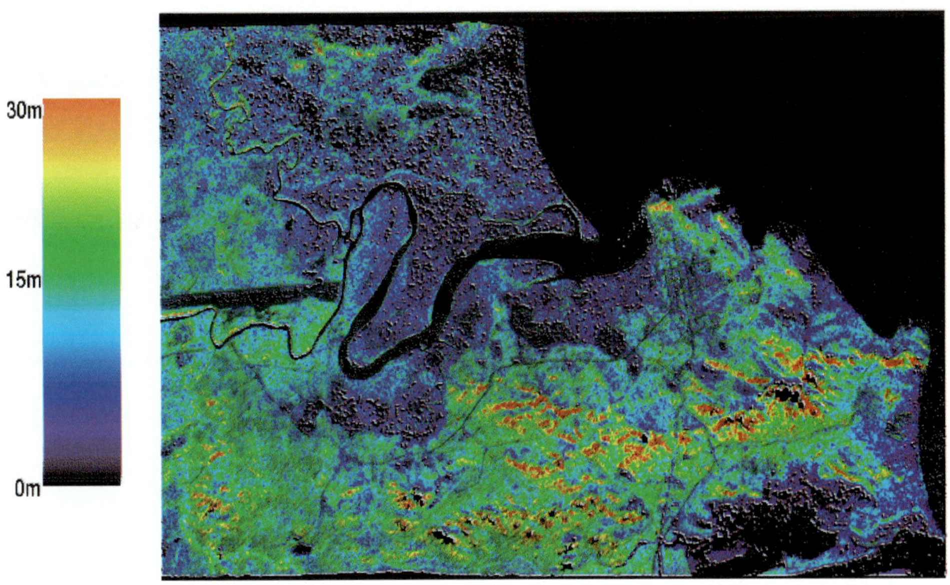

Plate 16 Estimated canopy height in the Endeavour River estuary, Queensland, Australia using JPL/NASA C-band interferometric radar data. (After Rodriguez *et al.*, 1996.)

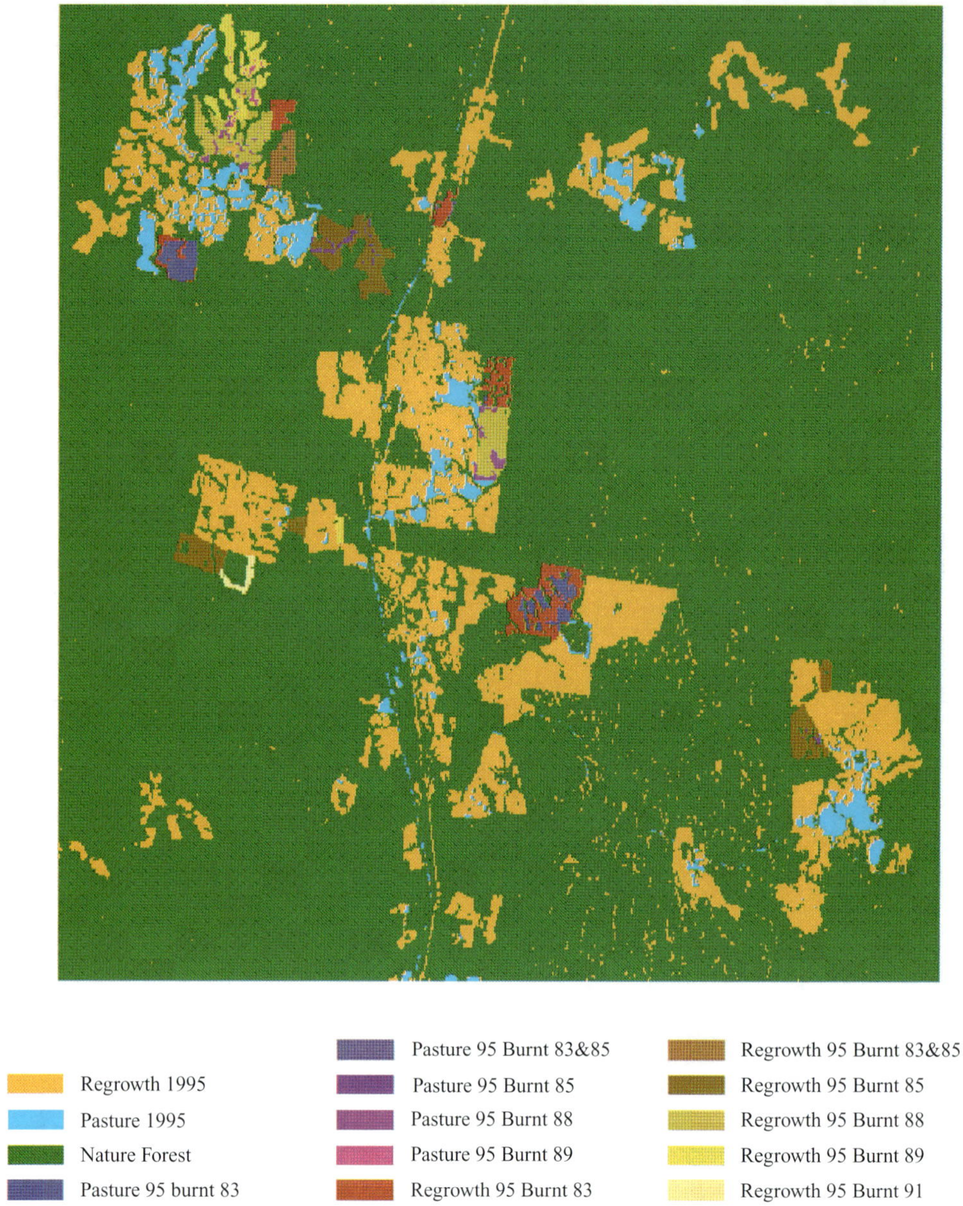

Plate 17 The fire history of a 275 km^2 area of rainforest near Manaus, Brazil, as derived from time-series comparisons of mapped fire scars within 80 m resolution Landsat MSS and 30 m Landsat TM data. (After Lucas *et al*., 2001.)

Part III
Forest disturbance, conversion and recovery

SUMMARY

This part contains eight chapters, the first three of which deal with the soil and water impacts of various kinds of forest disturbances, in order of increasing intensity. These are then followed by two chapters discussing the impacts of forest clearing and burning for the establishment of other land uses (pasture, cropping, plantations) at the small catchment, and river basin scale, respectively. The final three chapters focus on the changes in hydrology and soil nutrient reserves associated with forest regeneration, tree planting and the mixed crop-tree systems collectively known as agroforestry.

It is commonly overlooked that humid tropical forests are subject to a range of natural disturbances, including treefalls, landslides, hurricanes, floods, droughts, fires, volcanic eruptions, earthquakes and, in coastal areas, tsunamis. With the exception of treefalls and landslides, all of these have a significant impact on hydrological functioning and nutrient cycling at the catchment scale, with the geomorphology of an area often affected too. Scatena, Planos-Gutierrez and Schellekens provide an overview of the *hydrological impacts of the principal natural disturbances occurring in humid tropical forest*, emphasising that most forests experience disturbance-generating rainfalls at least once every decade. Rainfalls of *c*.200 mm day^{-1} may cause treefall gaps, landslides (especially in steep, tectonically-active areas) and localised flooding. Events in the order of 400–500 mm of rain day^{-1} (commonly associated with hurricanes) can cause widespread landscape modification, mainly through landsliding. Extreme event rainfalls in excess of 1000 mm generally trigger extensive mass movements in steeplands, regardless of the presence or absence of a forest cover. In general, intense short-duration rainfalls tend to cause shallow soil slips and debris flows whereas longer duration, lower-intensity events produce larger, deeper debris avalanches and slumps. Apart from excess rainfall, mass movements are also associated with tectonic activity and various examples are given of volcanic eruption and earthquake-triggered landslides. Although most landslides are covered again with vegetation within a few years, it still takes many decades before biomass levels are comparable with those of adjacent 'undisturbed' forest. Tree mortality and gap formation have an inherent, local impact on such hydrological processes as throughfall and drainage but their overall contribution to catchment hydrology is usually negligible. The strong winds associated with tropical cyclones change this situation radically, however, with extensive uprooting of trees and massive defoliation causing strong temporary reductions in forest evaporation. An important message of this chapter is that the disturbance connected with prolonged droughts (commonly associated with severe El Niño Southern Oscillation phases, ENSO) is generally underestimated. The reduced streamflows during such droughts decrease the aeration and self-cleansing capacity of surface waters, with adverse consequences for overall water quality. Also, the forest fires that often coincide with ENSO events temporarily enhance nutrient concentrations in stream water as well as the transfer of sediment to streams during subsequent rain events. Overall, many tropical forest species need disturbed conditions for their regeneration. However, there remains considerable uncertainty regarding the resilience of tropical forests when the effects of natural disturbances become aggravated by anthropogenic disturbances (see below) or climate change.

Next, the *impacts of selective logging* on water, nutrient and sediment flows in tropical rainforests are discussed by Chappell *et al.*, (Chapter 20). The authors begin by stressing how the very limited number of catchment-scale studies undertaken to investigate such impacts effectively means that the subject is still surrounded with considerable uncertainty. The main focus of attention is the Bukit Berembun timber harvesting experiment in Peninsular Malaysia where one catchment was kept as a forested control, a second was selectively-logged by unsupervised methods and a third by 'supervised' or 'reduced-impact-logging' (RIL) procedures. Statistical and parametrically-parsimonious conceptual modelling (PPCM) (see also Part IV) are applied to the rainfall-runoff behaviour of the three catchments and the results are interpreted in terms of (inferred) flow paths on the hillslopes. The magnitude of the changes in catchment-scale exports of water, nutrients and sediment imposed by logging are also assessed and compared with results from east Malaysia, Suriname and Guyana. From the evidence of the limited data available, it seems that catchment water yield is not affected by low-intensity logging (up to 20 m^3 ha^{-1}) but increases by *c.* 40–70% after medium-intensity logging (40–60 m^3 ha^{-1}), depending whether conventional or RIL extraction techniques are used. The increase in water yield is attributed to a reduction in evaporation due to the selective removal of large emergent trees. However, the evidence with regard to the duration of the period of enhanced flows is rather conflicting, with a return to pre-harvesting levels being observed in some places but not in others. The reasons for this discrepancy are as yet unclear. River responsiveness to storms (peak flows, quickflows) increases only marginally at low to intermediate harvesting intensities and the effect is generally short-lived. Nutrient exports via stream flows at Berembun increased by a factor of 1 to 6 in the year of harvesting, depending on the element under consideration. Nitrate concentrations returned to background levels within 6 months. Other nutrients (phosphorus, potassium, calcium and magnesium) took between 3 and 5 years. The rapid recovery of nutrient levels in stream flow suggests a high ecosystem resilience and is ascribed to biological uptake rapidly utilising the additional supply of nutrients from decomposing logging debris. Parallel increases in stream suspended sediment flow ranged by a factor of 2 to 50, though again there remains considerable uncertainty as to what are the reasons for this range. Part of the uncertainty concerns the effect of natural climatic cycles (such as ENSO) on river flows. Chappell *et al.* therefore call for more studies at the landscape scale (0.1–50 km^2) and for observations covering 10–30 year periods, arguing that quantification of the true environmental impacts of timber harvesting at varying intensities can only be achieved after such cyclic effects and the associated changes in ecosystem behaviour have been filtered out.

Taking the intensity of forest disturbance one step further, the *impacts of forest fires and shifting cultivation* (slash-and-burn) on hydrology, sediment transport and nutrient losses are assessed by Malmer, van Noordwijk and Bruijnzeel. From the outset, the authors emphasise the scarcity of field data related to this topic. A distinction is made between 'forest fires' due to natural causes and the uncontrollable fires started by the use of fire during forest clearance, which are termed 'wild fires'. There is a lack of hard data on the causes of natural forest fires (lightning, ENSO), but in recent times, major wild fires in South East Asia and the Amazon have occurred repeatedly during ENSO-related droughts. Increasingly intensive slash and burn agriculture, logging and forest conversion to other land uses all increase the risk of ignition and occurrence of wildfires during prolonged dry spells. Generally, young secondary forests and recently logged forests are the most vulnerable due to the large amounts of decomposing slash present. As for the hydrological impacts of fire, Malmer *et al.* suggest, on the basis of (modelling) studies of large-scale forest conversion (see also Costa, below), that it cannot be excluded that large-scale degradation of former forestland to fire climax grasslands may have an adverse effect on rainfall patterns and amount due to the associated changes in energy budget and surface roughness, particularly under non-maritime climatic conditions (see also Bonell *et al.*, Mahé *et al.*, Grip *et al.*). Where forest vegetation is able to recover from fire, however, so does evaporation (see also Hölscher *et al.*, below). In the absence of mechanised methods, the manual felling of trees as part of shifting cultivation does not create too much surface disruption but soil bulk density and pore size distribution can be radically changed by excess heating during the burning of vegetation piles. Soil hydrophobicity, and with it widespread gullying, may also occur after intense wild fires. As for shifting cultivation, most studies at the farmer-field scale report large increases in overland flow (and erosion) due to the already cited soil impacts but, at the catchment scale, the typical mosaic of open, newly burned fields and patches of secondary vegetation in various stages of regeneration generally does not have dramatic effects on stormflows (an exception is where new fields dominate the overall lay-out in small catchments). As long as catchment-wide groundwater replenishment is not impaired too much, therefore, adverse effects on low flows will remain limited. Solute leaching losses can be enhanced considerably after burning but virtually no studies of the moderating effect of riparian buffer zones in controlling nutrient leaching to streams have been undertaken. In conclusion, there is a distinct need for more research to quantify hydrological processes and the corresponding rates and pathways of sediment and nutrient transfers associated with fire. In the meantime, Malmer *et al.* provide a conceptual framework for such studies in the context of (i) traditional shifting cultivation, (ii) forest fire, (iii) intensified shifting cultivation, and (iv) wild fires in logged forests and young secondary forests.

In the most intensive form of tropical forest disturbance, i.e. *forest conversion to other usages (e.g. pasture, cropping, plantations)*, mechanised methods of extracting timber and the deliberate use of fire are combined. The environmental impacts of forest clearance are reviewed by Grip, Fritsch and Bruijnzeel. Conversion includes the complete removal of the original forest cover (usually employing heavy tracked or wheeled machinery), a brief phase with slash lying on the ground, which is followed by some form of site preparation that normally includes rearranging and burning the slash, and the planting and development of the new vegetation cover. From a soil and water perspective, the most dramatic changes occur immediately after forest felling and burning of slash, when much of the soil lies exposed to the elements and no new cover has established yet. In addition, much of the topsoil has become compacted to some degree through repeated machine passage, thereby destroying the macro-pores that maintained the infiltration capacity under forested conditions. Furthermore, an extensive network of roads and other compacted surfaces is usually in place now. The reduced evaporation from bare soil and very young vegetation, plus the cited adverse effects on soil infiltration capacity, combine to produce large increases in annual water yield (reported range: 125–825 mm yr^{-1}, depending on rainfall and degree of surface disturbance), stormflow volumes (typically by 50–200%) and sediment yield. During the stabilisation phase, many of these initial effects decline with time, usually within 4–5 years, but some changes are permanent. Examples include the higher water yields and stormflows under pasture and cropping. Post-clearing erosion 'hot spots' typically include roads and roadsides, disturbed streambanks and hollow-log bridges. Many tropical soils have inherent low fertility, but forest conversion does not necessarily decrease the fertility status of the soil further. It is difficult to generalise the results obtained by the very few forest clearance experiments that have been undertaken in the tropics to date. However, Grip *et al.* suggest that parameterisation of the observed phenomena and conceptualising them into simple, practical models may be a step forward. Despite limitations in the data, a considerable amount of knowledge has been accumulated during the last 10–20 years and the database is increasing rapidly, particularly for vegetation characteristics and water use, albeit at a much less rapid pace in terms of catchment experiments.

Following up on the previous chapter dealing with more local effects of forest clearing, Costa presents an assessment of the *impacts of large-scale tropical forest conversion* based on a combined hydrological and atmospheric model approach, focusing on the Amazon basin. Such a coupled approach is required because at the length scales under consideration (10^2–10^4 km) various feedback mechanisms between the land surface and the atmosphere connected with the exchange of terrestrial atmospheric energy and water vapour transfer come into play which are not apparent at the small scale. Stated differently, the upscaling of results from small experimental catchment studies has to be modified to incorporate the atmospheric feedbacks operating at the larger scale. Costa demonstrates this elegantly by first simulating the current Amazonian climate under fully forested conditions without allowing for large-scale atmospheric moisture convergence. Replacing the forest by pasture then causes a decrease in evaporation and an increase in runoff, in line with results expected from small-scale catchment clearing studies. However, after incorporating the atmospheric moisture convergence into the simulation, forest conversion to pasture causes a decrease in precipitation, which induces drier topsoils and, in turn, a further decrease in evaporation, such that the average runoff rate changes from the previously obtained increase of 0.5 mm day^{-1} to a *decrease* of 0.1 mm day^{-1}. The significance of moisture re-cycling in the Amazon basin is highlighted and Costa makes the important point that, whilst the annual average proportion of water vapour being of local origin is about 30 per cent under forested conditions, this figure can rise to nearly 100 per cent during the dry season. This buffering effect of the forest is particularly important during inter-annual and decadal decreases in input of water vapour from the Atlantic. This stability in evaporation during dry periods is attributed to the deep root system of the Amazonian forest which is capable of accessing the estimated 3000 mm of water stored in the deep regoliths prevailing in the area. Indeed, closure of the dry season water balance for Amazonian forest is not possible without allowing uptake of soil water stored at depths greater than 2 m. Costa concludes that there is now enough evidence of the intimate connection between tropical rainforests and climate. Thus, large-scale man-induced disturbances of this equilibrium can lead to important adverse changes in climate at corresponding spatial scales.

As old-growth forests continue to disappear, secondary forests are becoming an increasingly important element in humid tropical landscapes. Hölscher, Mackensen and Roberts review those features of *forest recovery* which are pertinent to ecosystem nutrient reserves and the hydrological cycle. Of particular importance is the question whether the effects of converting old-growth forest to agricultural land are reversed during forest recovery and, if so, to what extent and how rapidly? The impact of moderate and heavy pasture use on vegetation recovery is shown to be more severe than that of shifting cultivation. The same holds for the recovery of topsoil hydraulic conductivities. It may take several centuries for regenerating forests to regain their original composition while the corresponding recovery of above-ground phytomass has been estimated to range between less than 100 years on fertile soils to *c.* 200 years on relatively poor soils. Information on the development of root systems under secondary forest is rare but there is evidence that 60-year-old forests have root masses still below those of old-growth forests. On the other hand, the leaf area index of recovering forests can reach values (~5) of old-growth forests within the time-scale of years to 1–2 decades. Ecosystem nutrient reserves during vegetation recovery are closely related to the rate of biomass accumulation. The rapid recovery of leaf area and canopy height of secondary forests results in rainfall interception, albedo and aerodynamic conductance values that approach the equivalent values of old-growth forests from a decade upwards. In addition, pioneer species show very high stomatal conductances to transpiration compared to late-successional species, so much so that the water use of young (3–5 years) vegetation may even exceed that of the original forest. Although experimental evidence to date is limited, the effect may be enhanced further by positive heat advection towards neighbouring secondary forest patches from recently cleared areas experiencing limited cooling by evaporation. Another potentially important but little researched factor in this regard may be the increased turbulence associated with the sudden changes in vegetation height that are typical of fragmented landscapes. Aside from these advection and small patch effects, it is only in the later stages of secondary forest succession that there is a similarity in water use by old growth forest and regrowth forest.

Scott, Bruijnzeel and Mackensen address the challenging question of the environmental impacts *of reforestation of grasslands and degraded lands* in the tropics. Although forest plantations are expanding across the tropics and subtropics at a current rate of *c*. 2 million ha yr^{-1}, this is far less than the rate at which old-growth forests are disappearing (cf. Drigo, Part I). The plantations consist mostly of eucalypts, acacias and pines which provide an important source of industrial timber and paper pulp. Their value for soil protection and rehabilitation and, indirectly, for the conservation of overall biodiversity by reducing the pressure on the remaining natural forest is emphasised. The authors make a distinction between reforestation (i.e. establishing trees in areas that were once under forest) and afforestation (planting trees in areas whose climate does not support natural climax forest). In addition, Scott *et al.* distinguish between the forestation of catchments in good hydrological condition (where soil hydraulic properties still favour infiltration) and catchments with severely degraded soils, poor infiltration and thus rampant overland flow and erosion. Most examples cited refer to the former condition as there is very little firm quantitative information on the hydrological behaviour of degraded catchments. As such, the authors are forced to base their judgement of the possible hydrological consequences of tree planting in severely degraded areas on various inferences, analogies and expert opinion. On the basis of the limited available evidence, Scott *et al.* present an overall assessment of changes in annual water yield, low flows and peak flows as a function of plantation age, tree species and soil conditions. The close link between tree productivity and water use is highlighted. For young, rapidly growing plantations, the trend is one of higher water uptake compared to more mature forest (>25–30 years) of lower vigour. Because rainfall interception does not change much after canopy closure, the changes in annual water yield with plantation age tend to follow those observed for water uptake. The storm flow (quickflow) component of the stream hydrograph hinges on the status of the fixed upper soil water storage capacity (accessed by roots) which can be exceeded during heavy rainfall events. At such times, truly large peak flows ('floods') can occur. Otherwise, well-established tree plantings gradually delay and reduce stormflows, reflecting improved surface infiltration and the larger soil water deficits associated with active growth. However, the authors infer that chances of improving dry season flows in degraded catchments through forestation are slight because the gains afforded by improved infiltration are outweighed by the much higher water use of the trees. Experimental evidence for this contention is lacking, however. Concerning changes in soil nutrient reserves following forestation of degraded land, a general improvement in nutrient status is expected as the trees act as 'nutrient pumps', bringing nutrients from deeper layers to the surface via their litterfall, which would not be available to shallower-rooted shrubs and crops. On the other hand, replacement of native forest by plantations generally reduces fertility stepwise with / upon each successive rotation. The chapter concludes with a lengthy list for further research work, of which the evaluation of the socially important changes in low flows after forestation of degraded areas for various combinations of climate, infiltration, soil depth and tree species should arguably receive the highest priority.

In the final chapter of this part, Wallace, Young and Ong make the important point that most, if not all, future increases in food and wood production for the growing population in the tropics will have to be achieved from existing land and water resources. Consequently, considerable effort is needed to improve the efficiency with which land and water are currently used. One option is the use of various *agroforestry systems*. Agroforestry combines woody perennials (trees, shrubs) and herbaceous plants (crops, grasses) or livestock in various spatial arrangements and rotations. Up to 20 major agroforestry systems have been identified, of which six are suited to the humid tropics. Key benefits of agroforestry claimed in the literature include: improved soil conservation and maintenance of soil fertility; as well as water conservation and/or more efficient use of water. As such, agroforestry is widely thought of as contributing towards environmental sustainability. The soil conservation value of agroforestry systems depends on their ability to maintain a good ground cover of litter. Multi-strata and plantation crop combinations are the most efficient in this respect because of the associated density of plants. Whilst contour hedgerow systems may also reduce overland flow and erosion significantly on slopes up to 30 per cent, this particular agroforestry system has not been widely accepted by farmers. The maintenance requirements of hedgerows and competition for space, light and nutrients are the reasons commonly given (see also Critchley, Part V). Model studies have shown that the intensity of nutrient recycling under well-managed agroforestry can approach levels similar to those in natural forest. An important function of the tree component is to act as 'nutrient pumps' as described in the previous section on plantations. A central issue of this chapter is the search for evidence whether and to what extent agroforestry can enhance water conservation, particularly under subhumid conditions. An important concept in this regard is rainfall use efficiency, i.e. can more rain be used directly by the plants whilst minimising losses via soil evaporation, overland flow and drainage? The authors use a recent detailed water balance study of an agroforestry system within a sub-humid part of Kenya to examine this further, employing a one-dimensional model. Whilst the trees did increase rainfall interception losses compared to those from agricultural crops, this was compensated by a reduction in soil evaporation under semi-arid conditions, but only partially so under more humid conditions. Amounts of overland flow, soil water and deep drainage were all reduced by adding trees under both climatic regimes, suggesting that the higher water uptake by the trees compared to crops did increase the rainfall use efficiency of the system.

19 Natural disturbances and the hydrology of humid tropical forests

F. N. Scatena
University of Pennsylvania, Philadelphia, USA
E. O. Planos-Gutierrez
Instituto de Meteorología, Cuba
J. Schellekens
Vrije Universiteit, Amsterdam, The Netherlands

INTRODUCTION

Humid tropical forests are highly dynamic ecosystems that are affected by a wide array of environmental processes and disturbances (Figure 19.1). Quantifying the magnitude, frequency, and impacts of natural disturbances is essential for designing hydraulic structures, developing water management strategies, and distinguishing between natural variation and man-made influences.

A disturbance can be defined as any discrete event that transfers mass and energy from one part of a system to another in a manner that disrupts ecosystem, community, or population structure and changes resource availability or the physical environment (see White and Pickett 1985 for a detailed discussion). Natural disturbances can be driven by both external factors – for example, hurricanes, meteor impacts – and the biological properties of the system such as senescence and pathogens. The natural disturbances specified by the United Nations in the International Decade of Natural Disaster Reduction were earthquakes, windstorms, tsunamis, floods, landslides, volcanic eruptions, wildfires, grasshopper and locust infestations, drought and desertification (Board on Natural Disasters, 1999). Additional natural disturbances known to affect the hydrology of humid tropical forests are tree falls, pathogens, exotic invasions and meteor impacts.

Quantifying the effects of disturbances on landform morphology and ecosystem development have been major themes in geomorphology and ecology (Wolman and Miller, 1960, Connell, 1978). This approach has led to the paradigm that landscapes are structured by the processes acting upon them (O'Neill *et al.*, 1986; Urban *et al.*, 1987; Scatena, 1995). It is now generally recognised that the ability of a disturbance to affect the morphology of a landscape or the structure of an ecosystem depends on: (1) the type of disturbance (e.g., flood, fire, landslide, biologic, anthropogenic); (2) the force exerted (e.g. wind velocity and duration, rainfall magnitude and intensity, earthquake magnitude); (3) the ecosystem component that is impacted directly (e.g. soil, biomass, leaf area); (4) the area affected and the spatial distribution of impacts; (5) the return period or frequency of the event; (6) the condition of the system at the time of the disturbance (e.g. structure, regeneration phase, time since last disturbance); and (7) the magnitude of the constructive or restorative processes that occur between disturbances.

ATMOSPHERIC SYSTEMS

The humid tropics have some of the largest and most intense rainfall events in the world (Table 19.1). They are also areas where annual variability in precipitation is commonly greater than seasonal variability and a relatively high proportion of annual rainfall is delivered in rainfall events that exceed 100 mm day^{-1}. Moreover, rainfall events with intensities of 25 mm hr^{-1} or more commonly account for more than 30% of annual precipitation in tropical sites but less than 5% in many temperate sites (Walsh, 1997). The most common disturbance-generating weather systems can be grouped broadly into six basic meteorological systems (Planos, 1999; Callaghan and Bonell, this volume):

- Equatorial troughs
- Cyclones
- Inter-tropical systems
- Cold surges and extra-tropical fronts
- Monsoons
- Inter-annual oscillations.

Forests, Water and People in the Humid Tropics, ed. M. Bonell and L. A. Bruijnzeel. Published by Cambridge University Press.

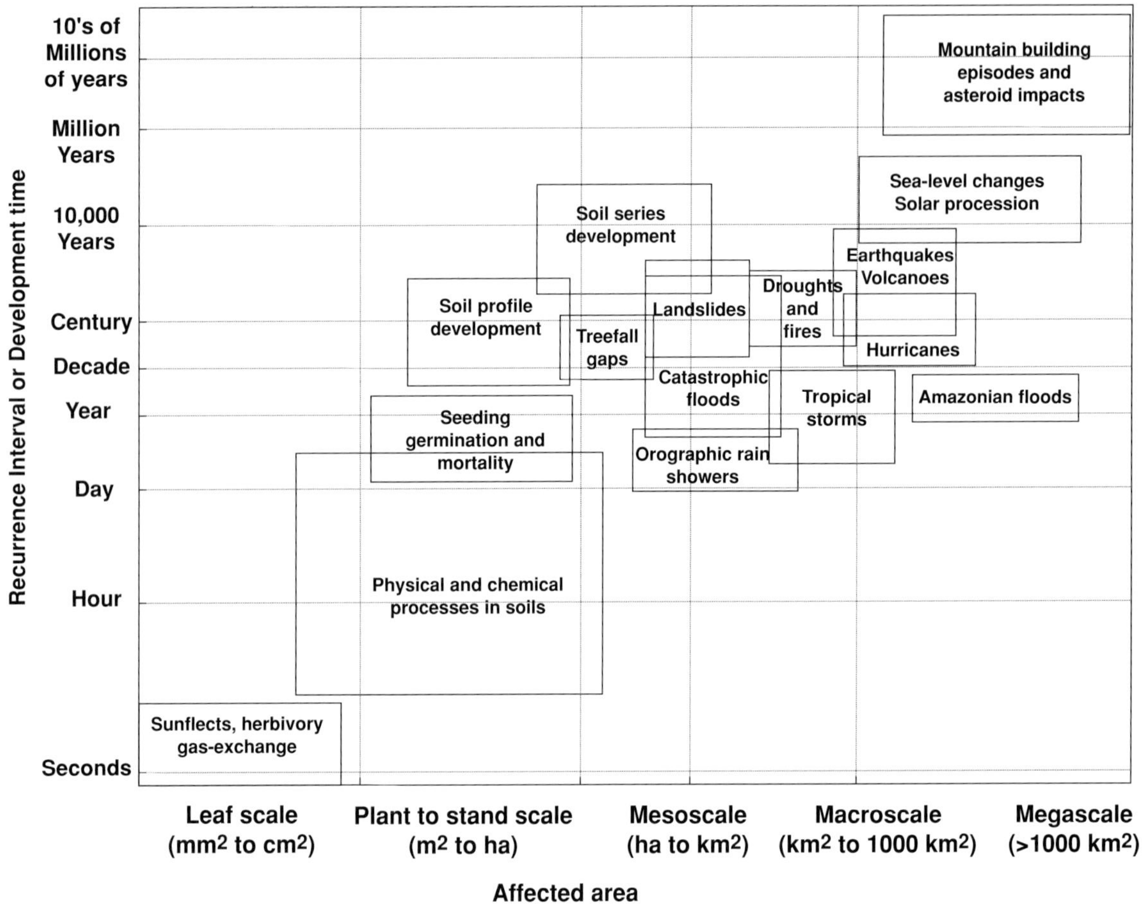

Figure 19.1 Spatial and temporal scales of common natural disturbances affecting humid tropical forests in the American tropics.

The equatorial troughs

The idealised, time-averaged, large-scale atmospheric system for the tropics is the system of Hadley circulation cells on either side of the Equator that transport heat toward the poles through low-level easterly trade winds and upper-level westerlies (see Callaghan and Bonell, this volume). The transition between these cells and the associated northeast and southeast trade winds takes place within a relatively narrow belt located several degrees north of the Equator that is most commonly called the inter-tropical convergence zone (ITCZ). This zone is also referred to as the Equatorial trough, near-Equatorial trough, Equatorial front, the inter-tropical front and is actually discontinuous in space and time and consists of combinations of cloud clusters, low pressure systems, upper level systems and near-surface thermal troughs (see Callaghan and Bonell, this volume, for more precise definitions of different rain-producing systems associated with the ITCZ). These systems have a range of structure and characteristics that can produce collectively the largest and most intense rains in the tropics. For example, in parts of Venezuela during December 1999, ITCZ-related rains brought twice the annual average rainfall in the first 16 days of the month and 911 mm in three subsequent days (Table 19.1). These catastrophic rainfalls resulted in widespread landsliding and flooding that ultimately affected over 400 000 people and destroyed completely 26 000 houses (Pan American Health Organization, 1999).

The broadly defined ITCZ migrates through series of daily advances and retreats but follows a seasonal path where it moves towards the equator between January and March and moves away between June and July (Coen, 1983). The magnitude and location of this seasonal migration has been linked to droughts in the African Sahel, India, Zimbabwe, South Africa, Central America and the Chaco region of South America (Lauer, 1989). Changes in the location and intensity of the ITCZ have also been implicated in late Quaternary climate change in Africa and South America (Servant *et al.*, 1993). These changes include the mid-Holocene dry period that appears to have affected many humid tropical forests approximately 5000 years BP, plus changes in the

Table 19.1. *Examples of extreme rainfall events in the humid tropics*

Total rainfall (mm/event)	Average mm day^{-1} during event	Location	Dates	Source
Hurricanes				
5678	568	La Réunion	18–27 Jan 1980	Landsea, 1999
3240	1080	La Réunion	24–27 Jan 1980	Landsea, 1999
2467	1233	La Réunion	8–10 Apr 1958	Landsea, 1999
2287	327	Cinchona, Jamaica	4–11 Nov 1909	Gupta, 1988
2025	405	Tacajo, Cuba	3–8 Oct 1963	Gupta, 1988
1825	1825	La Réunion	7–8 Jan 1966	Landsea, 1999
1524	762	Silver Hill, Jamaica	5–7 Oct 1963	Gupta, 1988
1248	1248	Pai Shih, Taiwan	10–11 Sept 1963	Gupta, 1988
1168	1168	Baguio, Philippines	14–15 July 1911	Gupta, 1988
1400–800	266–466	Nicaragua	29–31 Oct 1998	Ferraro *et al.*, 1999
Monsoon				
3388	484	Cherrapunji, India	9–16 June 1876	Gupta, 1988
3213	536	Cherrapunji, India	24–30 June 1932	Gupta, 1988
1036	1036	Cherrapunji, India	14 June 1876	Gupta, 1988
493–468	493–468	Various sites in India		Gupta, 1988
Frontal systems and inter-tropical convergence zone				
2789	930	Bowden Peninsula, Jamaica	22–25 Jan 1960	Gupta, 1988
1109	1109	Bowden Peninsula, Jamaica	23 Jan 1960	Gupta, 1988
977	977	Bowden Peninsula, Jamaica	22 Jan 1960	Gupta, 1988
911	304	Maiquetia, Venezuela	14–16 Dec 1999	Pan American Health Organization, 1999
867	867	Cienfuegos, Cuba	1 June 1996	Planos, 1999

magnitude and frequency of ENSO events (see Mahé *et al.*, this volume).

Cyclones

Cyclones are atmospheric disturbances characterised by masses of air that rapidly rotate around a low-pressure centre (Table 19.2). Historically, they have been the most costly type of natural disaster (Diaz and Pulwarty, 1997, Board on Natural Disasters, 1999). During the 1900s alone, 23 Atlantic hurricanes each caused damage in excess of $US 1 billion (inflation-adjusted) (NOAA, 1999). The long-term estimates of economic losses caused when Hurricane Andrew struck South Florida in 1992 exceed US$ 30 billion. When Hurricane Mitch crossed over Central America in 1998 over 1.5 million people were affected (Pan American Health Organization, 1999; see also Bonell *et al.*, this volume).

Approximately 82 tropical cyclones occur in a typical year (Table 19.3). Because they tend to develop north and south of the ITCZ and travel poleward, many equatorial humid tropical forests are unaffected by hurricanes (Figure 19.2). They are very important disturbances in the Caribbean, the southwest Indian Ocean (Madagascar, Mauritius, La Réunion), the northern Philippines, Taiwan, parts of Indo-China, Indonesia, the southwest Pacific islands and tropical Queensland (Walsh, 1997). They are rare to non-existent over the humid tropical forests of South America and Africa and a large part of those in Malaysia and Indonesia.

Table 19.2. *Classification of cyclones by maximum wind speed*

	Maximum velocity (km hr^{-1})
Tropical depression	35–65
Tropical storm	63–117
Hurricane	> 118

Hurricanes are the most intense type of cyclones and are classified by barometric pressure, storm surge and destructive potential based on wind velocity (Table 19.4). Hurricane-force winds (e.g. >118 km hr^{-1}) typically extend 40 to 50 km from the eye in small hurricanes and up to 250 km for large hurricanes. Tropical storm force winds (63–117 km hr^{-1}) can extend as far as 500 km from the centre of large hurricanes. As a general rule in the northern hemisphere, the hurricane's right side relative to the direction of motion (the left side in the southern hemisphere) has the greatest winds and rain and is therefore the most destructive (Anthes, 1982).

Table 19.3. *Mean annual cyclone frequency in tropical regions*

Region	Storms per year	Hurricanes per year	All cyclones	Percent of total	Principal humid tropical forests affected
North Indian Ocean	3.5	2.2	5.7	7	Andaman Islands
North Atlantic/Caribbean	4.2	5.2	9.4	11.5	Caribbean
Southwest Indian Ocean	7.4	3.8	11.2	13.7	Mauritius, La Réunion, Madagascar
Southwest Pacific	10.9	3.8	14.8	18.1	Queensland, Fiji, Solomons, Vanuatu
Eastern North Pacific	9.3	5.8	15.2	18.6	None
Western North Pacific	7.5	17.8	25.3	31.0	Philippines, Taiwan, South China, Northern Borneo
Total	42.8	38.6	81.6	100	

Source: Adapted from Walsh (1997) and Crutcher and Quayle (1974).

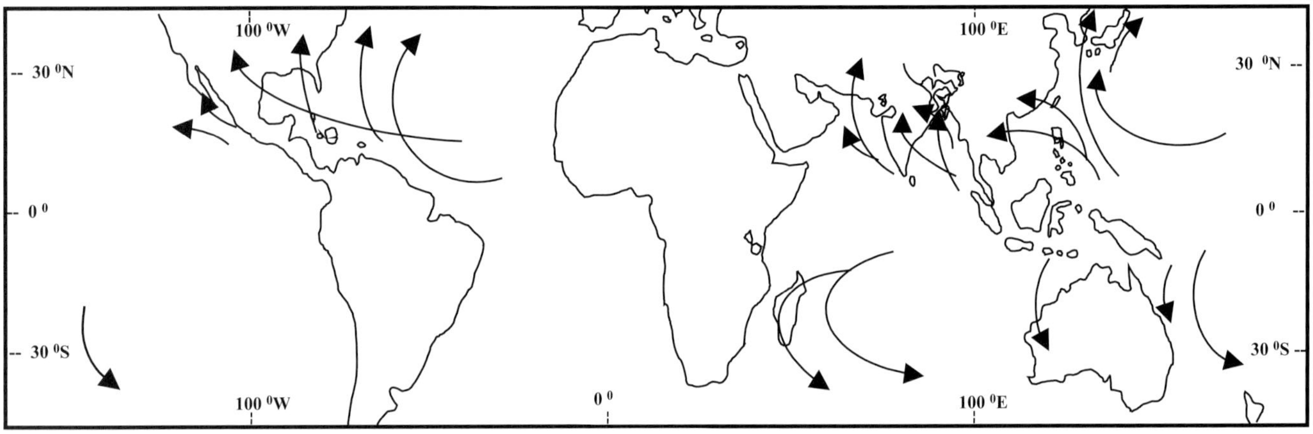

Figure 19.2 Common tracks of tropical hurricanes and cyclones. (After NOAA 1999.)

Table 19.4. *Central pressure, sustained winds, storm surge and relative destruction potential of winds by Saffir–Simpson storm categories*

Saffir–Simpson category	Central pressure (mb)	Maximum sustained winds (m s^{-1})	Storm surge (m)	Relative destruction
1	980	33–42	1.0–1.7	1 (Minimal)
2	965–979	43–49	1.8–2.6	10 (Moderate)
3	945–964	50–58	2.7–3.8	50 (Extensive)
4	920–944	59–69	3.9–5.6	100 (Extreme)
5	<920	>69	>5.6	250 (Catastrophic)

Source: After Gray *et al.* (1997).

The frequency of cyclones varies with season (Table 19.5), decade (Figure 19.3), and local physiography (Table 19.6). Multi-decadal variation in cyclone activity has been linked to variations in thermohaline oceanic circulation, global sea surface temperatures, West African monsoons, African droughts and ENSO events (Gray *et al.*, 1997; see discussion in Callaghan and Bonell, this volume). Locally, frequency can vary with physiography (e.g. coastal plains, interior mountains) and the location of the land masses with respect to storm trajectories. For example, hurricane frequencies vary by more than a factor of two between the western and central regions of the island of Cuba (Table 19.6).

There is no simple relationship between hurricane intensity, total rainfall, or destructive power (Planos, 1999; Callaghan and Bonell, this volume). Locally, many hurricanes are classified as

Table 19.5. *Monthly distribution of Atlantic cyclones in the Northern Caribbean between 1890 and 1990*

	June	July	Aug	Sept	Oct	Nov	Other	Total
Percent of total	6	8	25	34	20	5	2	100
Total	50	64	206	288	171	38	15	832
Maximum recorded per month	3	4	7	7	6	2	2	2

Source: After Planos (1999).

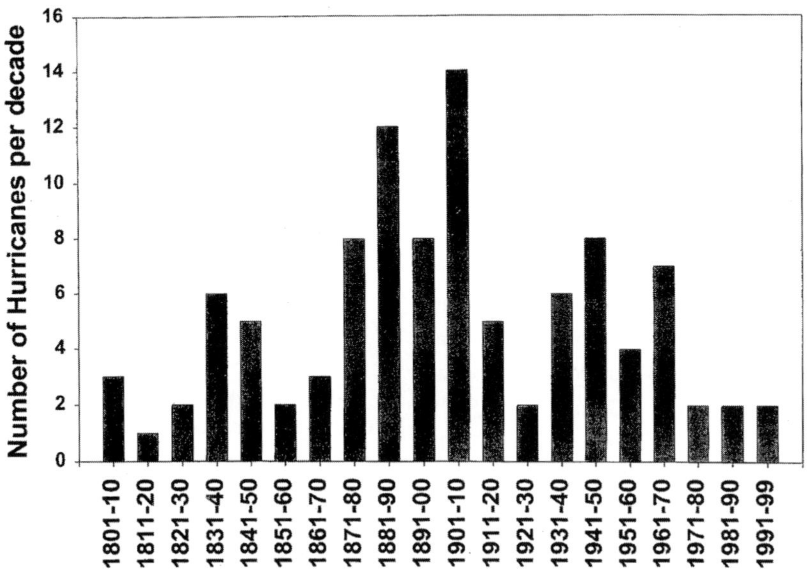

Figure 19.3 The number of hurricanes directly affecting the island of Cuba by decade between 1800 and 2000. (After Perez Suarez, 1999.)

either wind or rain dominated storms. In general, the total amount of rain that falls in a given area depends on: (a) the intake of humid air into the circulating system; (b) the velocity of the winds within the hurricane; (c) the forward velocity of the eye and the length of time the hurricane affects a particular area directly; and (d) the position of the storm and site relative to the ocean. While the typical hurricane brings between 150 to 300 mm of rain (NOAA, 1999), event totals over 1500 mm are not uncommon (Table 19.1). On average, rainfalls within 222 km of the eye are of the order of 100 mm day^{-1} while rainfall between 222 and 444 km of the eye average 30 to 40 mm day^{-1} (Anthes, 1982). Tornadoes and spiral bands with intense convection and rainfall can also extend several hundred kilometres from the eye.

Over the next 100 years or so, the magnitude and possibly the frequency and spatial extent of hurricanes is expected to increase as a consequence of a CO_2-warmed climate (Emanuel, 1987, 1997; Knutson *et al.*, 1998). Unfortunately, the magnitudes of these increases are uncertain. However, high resolution computer simulations of 51 western Pacific storms indicate wind speeds will increase by 3 to 7 m s^{-1} (5 to 12%), with a 2.2 °C increase in sea surface temperature (Knutson *et al.*, 1998). Simulations of tropical forest response to changes in hurricane frequency indicate that a range of forest compositions can occur with different hurricane regimes (O'Brien *et al.*, 1992). In general, a decrease in hurricane frequency will result in more mature forest containing large trees while an increase in frequency will result in shorter and younger forests with abundant pioneers species and few mature forest trees.

Table 19.6. *Return periods (in years) by storm type for the major physiographic regions of Cuba, based on data between 1800 and 2000*

Region	Tropical depression	Tropical storm	Hurricane
Western (Occidental)	1.3	2.4	2.6
Central	3.0	5.6	6.3
Eastern (Oriental)	2.3	4.5	4.8

Source: After Planos (1999).

Inter-tropical systems

This group of atmospheric systems includes those that originate and generally remain within the tropics, such as orographic and meso-scale convective systems. These systems are most common in the summer months and can produce 100 mm of rain in 24 hours or less (Lauer, 1989). The most intense rains occur over relatively small areas (e.g. ha to km^2) and rainfalls greater than 200 mm per event are known such as the 719 mm which fell in 24 hours over Havana, Cuba, on June 19, 1982.

One of the most common disturbance-generating systems in this group are easterly and westerly perturbations. These synoptic-scale low pressure systems form within the trade wind belt and can produce several days of heavy convectional showers (Wallace and Hobbs, 1977). These perturbations, commonly but imprecisely called waves (Bonell *et al.*, this volume, discuss the characteristics of these perturbations in some detail), are frequent during the summer throughout the humid tropics but are most common in maritime regions like the Caribbean, South Pacific and the southwest Indian Ocean. They also bring the largest proportion of summer rain in West Africa where they are associated with winds as high as 20 m s^{-1} and event rainfalls of over 100 mm (Lauer, 1989).

Orographic uplift can also result in intense, and generally localised, rainfalls. These rains are most important where coastal mountains are aligned with the normal prevailing winds. Because of orographic related rains, the zone of maximum rainfall on tropical mountains is generally between 1000–1400 m (Walsh, 1997). In Costa Rica and much of Central America the rainfall maximum occurs at about half the average elevation of the mountain peaks, or about 1000 m (Coen, 1983).

Extra-tropical frontal systems

While summer rainfall in the humid tropics comes typically from cyclones and the inter-tropical systems, winter rainfall usually comes in organised frontal systems that originate in extra-tropical areas (Lauer, 1989). These fronts are part of the Large Scale Cloud Bands of Callaghan and Bonell (this volume) and are regions of strong thermal contrast and intense rainfalls. Individual fronts can be several hundred kilometres in length and 10 to 100 kilometres in width and are important in the interchange of temperature and humidity between tropical and mid-latitude regions, especially in the sub-tropics and parts of tropical North America, the South Pacific, the South Atlantic, South Africa and Australia. Cool tropical climates during the late Quaternary glacial period and glacial-interglacial transition have also been linked to an increase in the frequency of polar air masses reaching the tropics (Servant *et al.*, 1993).

Of these systems, cold and stationary fronts typically produce the largest disturbances. Rains associated with these systems are usually of low intensity, last for several days, and produce less than 150 mm per event. Nevertheless, intense rainfalls can be associated with frontal systems (Table 19.1) and landslides and flooding are common when intense downpours follow several days of persistent, soil-saturating rain. The steady and continuous winter rains of Central America called 'Temporales' are caused by frontal systems that encounter mountains, rise, cool and generate rain (Coen, 1983).

Monsoons

Monsoon rainfall regimes incorporate northern and southern monsoon shearlines and equatorial westerlies (Callaghan and Bonell, this volume) and the complex interactions of thermal heating of land masses and the seasonal migration of the trade winds (Walsh, 1997; Lauer, 1989). These systems have distinct wet and dry seasons and are well developed on the Indian and the Asian continent. However, they are also common in humid tropical forests in northern Australia, the Guinea coast of Africa and in a small area of Central America facing the Pacific.

The monsoon rainfall regime has two major traits: (a) they are responsible for some of the worlds highest annual rainfalls; and (b) rainfall is so seasonal that 75% or more of the annual rainfall falls in three or four months of the year. In lowland areas most of this rain comes with tropical waves (perturbations), depressions and cyclones (Walsh, 1997). In mountainous areas orographic uplift plays an important role. Maximum daily rainfalls during the wet monsoon are commonly over 100 mm (Table 19.1) but it is the continuous, widespread, multi-day character of these rainfall events that often causes widespread flooding and slope failures. For example, in the Brahmaputra plains of India, one or two days of monsoonal rain can cover 20 000 km^2 with over 200 mm of rain and several thousands of km^2 with over 500 mm of rain (Raghavendra, 1982).

Inter-annual oscillations

On inter-annual time scales the interaction of the oceans and atmosphere are considered to be the principal cause of global climate variability (see Callaghan and Bonell, this volume). The most widely known system, the El Niño/Southern Oscillation (ENSO), refers to the large-scale warming of the tropical Pacific ocean and the coupled oscillations in atmospheric surface pressure (Neelin and Latif, 1998). Other systems, such as the North Atlantic Oscillation have also been identified and related to regional and global climatic variability (see also Mahé *et al.*, this volume).

Stratigraphic evidence from Ecuador suggests that between 15 000 and 7000 years BP, ENSO had a periodicity greater than or equal to 15 years (Rodbell *et al.*, 1999). The modern ENSO periodicity of 2 to 8.5 years, with an average occurrence of

4 years, was established about 5000 calendar years ago. This phenomenon has been linked to: (1) increased global hurricane activity; (2) increased droughts and fires in Africa, Amazonia, Indonesia and Australia; and (3) intense flooding in Chile, Ecuador and the United States. The phenology and growth of tropical forests, the impacts of logging and deforestation, insect outbreaks and epidemics in East Africa, Caribbean atmospheric dust, and the carbon dynamics of tropical oceans have also been correlated with ENSO events. Nevertheless, the local impacts of ENSO events can be quite variable. In Borneo many of the dominant canopy tree species disperse seed within a 1–2 month period during ENSO events (Curran *et al.*, 1999). However, during the severe 1983 ENSO, high monthly temperatures and rainfall variability in the Kibale Forest of Uganda affected the leaf flush and phenology adversely in 73% of the species studies (Struhsaker, 1997).

COMMON NATURAL DISTURBANCES IN HUMID TROPICAL FORESTS

Most natural disturbances affect tropical forests by modifying vegetation and/or soil cover directly. Both the initial impacts and subsequent changes in the cycling of water and nutrients vary through the character of the disturbance and the local site conditions. The most common types of natural disturbances affecting humid tropical forests are tree mortality, earth movements, floods, hurricanes, droughts and fires. Mega-events associated with volcanic eruptions, earthquakes, tsunamis and meteor impacts, although less frequent, can also be locally and globally important.

Tree mortality and tree-fall gaps

Mortality is a complex process that occurs over many spatial and temporal scales and may be caused by natural and anthropogenically induced biotic and abiotic factors. Mortality events may affect individual trees because of 'background mortality' or become large-scale 'catastrophic mortality events' (Lugo and Scatena, 1996). Background mortality is typically associated with senescence, competition and succession while catastrophic mortality events usually occur when a forest is impacted mechanically or chemically by an external force. However, catastrophic stand-level diebacks of indigenous montane rainforests in Hawaii, for example, can occur as a relatively continuous process acting over hectares to square kilometres that cannot be reduced to a specific biotic disease or abiotic stress (Mueller-Dombois, 1987). When expressed as a percent of stems or biomass per year, background mortality is typically less than 3 percent per year while catastrophic events are greater than 3 percent per year. The median value of background mortality in 68 pan-tropical moist, wet and rainforest stands was 1.6 percent per year and is similar to those reported from temperate and boreal forests (Lugo and Scatena, 1996).

Table 19.7. *Canopy turnover periods of tree-fall and hurricanes induced gaps for some Neotropical forests*

Location	Years	Reference
Bisley, Puerto Rico		Scatena and Lugo, 1995
Hurricane induced	57	
Tree-fall gap induced	165	
Barro Colorado, Panama	62–159	Foster and Brokaw, 1982
La Selva, Costa Rica	79–137	Hartshorn, 1990
Tierra firme, Amazonia	100	Uhl and Murphy, 1981
Central America	62–155	Brokaw, 1985
Los Tuxtlas, Mexico	61–138	Bongers *et al.*, 1988

The creation of canopy gaps by individual or multiple tree falls is a common process involved in maintaining the structure and diversity of tropical forests (Denslow, 1987; Whitmore, 1984). The size of tree-fall gaps can range considerably but generally lies between 50 and 100 m^2 (Brokaw, 1985; Hartshorn, 1990). The rate of gap formation in mature tropical forests is typically around 1 gap ha^{-1} yr^{-1} while the canopy turnover periods by gaps range from 50 to 200 years (Table 19.7). In any area, the size and frequency of gaps can vary with topography, aspect, soil type and forest age. However, in most humid tropical forests, less than 1 to 2% of the forest area has open, recent gaps at any instant in time.

Once created, tree-fall gaps have a different microclimate and hydrology than the surrounding closed-canopy forest. The magnitude of the differences is related positively to gap size and negatively to gap age. The four most common changes are: (a) increase and shift in the spectral composition of solar radiation reaching the forest floor; (b) increases in maximum, means, and ranges of air and soil temperature, especially on sunny days; (c) reduction in daytime relative humidities and increased saturation deficit, especially on sunny days; (d) an increase in local wind speed and turbulence (Schulz, 1960; Walsh, 1997).

Although solar radiation and air temperatures are higher in gaps than in the adjacent forest, dry season surface soil (0–20 cm) moisture is typically greater in both temperate and tropical forest gaps (Vitousek and Denslow, 1987; Becker *et al.*, 1988; Ostertag, 1998). After a 21-day dry period, surface soil moisture (0–20 cm) in 18 recent gaps in the lowland forest of La Selva, Costa Rica, was 6.6% greater than in adjacent forests (Ostertag, 1998). Fine root length and biomass were also significantly lower in these gaps compared to the adjacent forest. However, the magnitude of change in below-ground biomass may be dependent on soil fertility and unrelated to gap size, openness or age. Higher gap soil moisture levels have been attributed to: (1) concentration of precipitation from the drip line of tree crowns that surround the gap; (2) greater canopy throughfall; and (3) reduced root density and plant uptake (Becker *et al.*, 1988).

The pit and mound topography that accompanies uprooted trees can also influence catchment hydrology by changing flow paths and water storage. In general, tree fall pits increase surface storage and promote the development of subsurface pipes and macropores. The percentage of tree-fall gaps created by uprooted trees relative to snapped trees typically ranges between 20 and 50% and may be greater in steepland than lowland tropical forests (Putz, 1983; Scatena and Lugo, 1995). Tree throws contribute between 2.5 and 15% of hillslope sediment erosion in steep forested catchments in Puerto Rico (Larsen, 1997; see also Douglas and Guyot, this volume). In nearby agricultural and suburban catchments their contributions were less than 5%. The pit and mound features caused by tree uproots typically occupy less than 0.1% of the ground surface of humid tropical forests but as much as 60% in some temperate environments (Putz, 1983). These differences are due in part to the dynamic surficial processes in tropical forests that act to remove rather than to preserve the pit and mound features. Nevertheless, both pits and mounds can be important microhabitats for the reproduction of certain tropical plants (Putz, 1983; Walker, 2000).

Experimental manipulations and comparative sampling of tree fall gaps and adjacent neo-tropical forests indicate that soil water concentrations of some nutrients can increase following tree fall gaps that are larger than 0.5 ha (Parker, 1985; Vitousek and Denslow, 1987; Uhl *et al.*, 1988; Silver *et al.*, 1996). Single-tree gaps smaller than 0.02 ha typically have less effect on soil water chemistry. In both large and small gaps, environmental changes are generally isolated to the disturbed area and last for less than one year.

Impacts of tree fall gaps on the hydrological cycle and water supply

The hydrological processes impacted by tree mortality are those associated with the removal of leaf area and the shift in water cycling from mature canopy vegetation to understorey vegetation and regeneration. Although the structural form and composition of a gap can often be recognised for decades, their influence on local or watershed scale hydrology is much shorter. In most gaps, the disturbed soil is covered with herbaceous vegetation after a few months and the opening begins to fill up (Hoelscher *et al.*, this volume). Canopy throughfall in the centre of a recent single-tree gap can be 30 to 50% higher than that of the adjacent forest (Scatena, 1990). However, within a year it approaches that of the adjacent forest as the understorey and adjacent canopy develop. Because gaps account for only a small fraction of a catchment at any time, their contribution to overall catchment throughfall is relatively small (e.g. 3% in the Bisley watersheds of Puerto Rico, Scatena, 1990). Likewise, there is little evidence to suggest that these small spatial scale changes cause significant changes in downstream water quality (Uhl *et al.*, 1988).

In summary, tree mortality and gaps can have measurable impacts on local, stand scale, hydrological processes. However, since the environmental changes that do occur are limited both spatially and temporally, individual tree falls have relatively little short-term impacts on watershed scale hydrological processes. Nevertheless, they contribute to baseline conditions and are a essential process in forest regeneration and soil development.

Mass earth movements

Mass movements of earth are a common landform scale disturbance in most, if not all, upland humid tropical forests and are commonly associated with changes in land use or land cover. The velocity of downslope movement can range from the continuous downslope creep of soil profiles that occurs on the order of mm/yr. (Lewis, 1974) to debris flows that move tens of kilometres per hour. Rapidly moving failures are triggered by heavy rainfalls and seismic activity but are often exacerbated by land use change. In many areas they are often the most severe natural disturbances because they modify the landscape to the extent that forest recovery must start from primary succession on bedrock or subsoil.

Landslides, debris flows and avalanches claim hundreds of lives per year and are chronic disturbances in many tropical and subtropical areas. Moreover, mass movements typically cause at least one catastrophic disaster in the tropics each decade. In 1970, earthquake-triggered debris avalanches in Peru killed an estimated 17 000 people and buried a whole city under 5 m of mud and debris (Costa, 1984). In 1985, a moderate size eruption of the Nevado del Ruiz volcano in Colombia triggered a flow of snow, ice and earth that ultimately buried about 23 000 people. In 1998, Hurricane Mitch caused mud and rock slides that buried five Nicaraguan communities (Showstack, 1999). One of these debris avalanches was caused by the collapse of a dormant volcano. This failure was initially about 200 000 m^3 in size and grew at least ten-fold as it moved downslope.

Neither the temporal nor spatial occurrence of landslides are random. At the regional scale, landslides are much more common in mountainous areas than coastal lowlands and in tectonically active areas compared to tectonically passive landscapes. They also occur preferentially along fault zones and fractured areas that are related to mountain building processes (Ahmad *et al.*, 1993b). Landslides associated with a 1976 earthquake in Panama covered approximately 12% of a 450 km^2 area (Garwood *et al.*, 1979). In this area, earthquake-triggered landslides are expected to denude about 2% of the forest per century, a rate that is similar to tree fall gaps. In tectonically active Papua New Guinea, 8 to 16% of

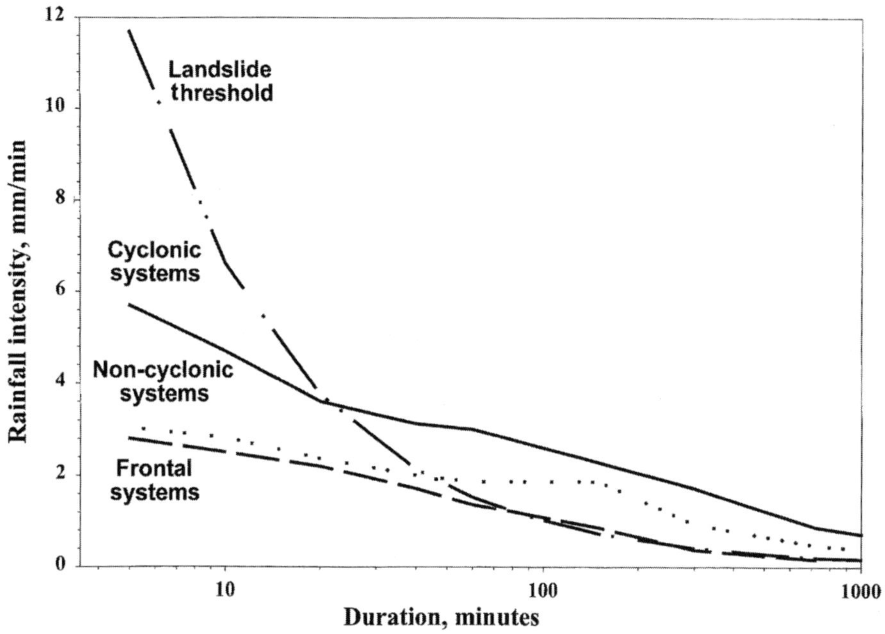

Figure 19.4 Landslide threshold curve and rainfall intensities by storm type for the northern Caribbean. (After Larsen and Simon (1993) and Planos (1999).)

the forest each century may be denuded by earthquake generated landslides (see also Douglas and Guyot, this volume).

In mature humid tropical forests, landslide frequencies and the rate of revegetation have been related to bedrock geology, elevation and mean annual rainfall (Myster *et al.*, 1997; Larsen and Parks, 1997; Larsen and Torres-Sanchez, 1998). Within areas of similar geology and mean annual rainfall, mass wasting is 5 to 8 times more frequent along roads and is most common on hillslopes that: (1) have been modified anthropogenically; (2) have slopes greater 12 degrees; and (3) face the prevailing trade winds.

The occurrence of landslides is also related to the duration and intensity of rainfall events. In general, intense short duration events create shallow soil slips and debris flows while long-duration, low intensity events produce larger, deeper debris avalanches and slumps. In humid sub-tropical South Africa, the number of annual and wet season landslides was correlated positively with the annual maximum 30-day rainfall and negatively correlated with the minimum monthly wet season rainfall (Garland and Olivier, 1993). On the relatively wet island of Puerto Rico, landslide-producing storms occur at an average rate of 1.2 per year (Larsen and Simon, 1993). Storms with duration of 10 h or less typically required an intensity of nearly 14 mm h^{-1} to trigger landslides while events of 100 hr or more trigger landslides with an average intensity of 2–3 mm h^{-1}. Comparison of the Puerto Rican landslide threshold relationship with rainfall-intensities from the nearby island of Cuba indicates that all common atmospheric systems can produce landslide-generating storms (Figure 19.4).

Impacts of massive earth movements on hydrological cycle and water supply

At the scale of individual mass movements, impacts on the hydrological cycle include those associated with the removal of vegetation and the upper soil profile. Moreover, localised increases in runoff and sediment discharge occur until the landslide scars are vegetated. Although it is not uncommon for failures in geologically unstable situations to remain active for decades or even centuries, most landslides are covered with vegetation in a few years. In humid tropical environments landslides are covered with herbaceous vegetation typically within one or two years, have closed canopies of woody vegetation in less than 20 years, and above-ground biomass of the adjacent forest after several decades (Guariguata, 1990; Zarin and Johnson, 1995a and b; Myster and Fernandez, 1995; Walker *et al.*, 1996).

Slope wash and surface runoff from landslide scars also tend to approach that of adjacent forest after a few years. Annual slopewash collected in Gerlach troughs on recent landslide scars was 100 to 349 g m^2 yr^{-1} initially but decreased to 3 to 4 g m^2 yr^{-1} after four years (Larsen *et al.*, 1999). Monthly surface runoff was less than 1% in both landslides and adjacent forest. However, there was considerable variation in the amount of runoff between scars

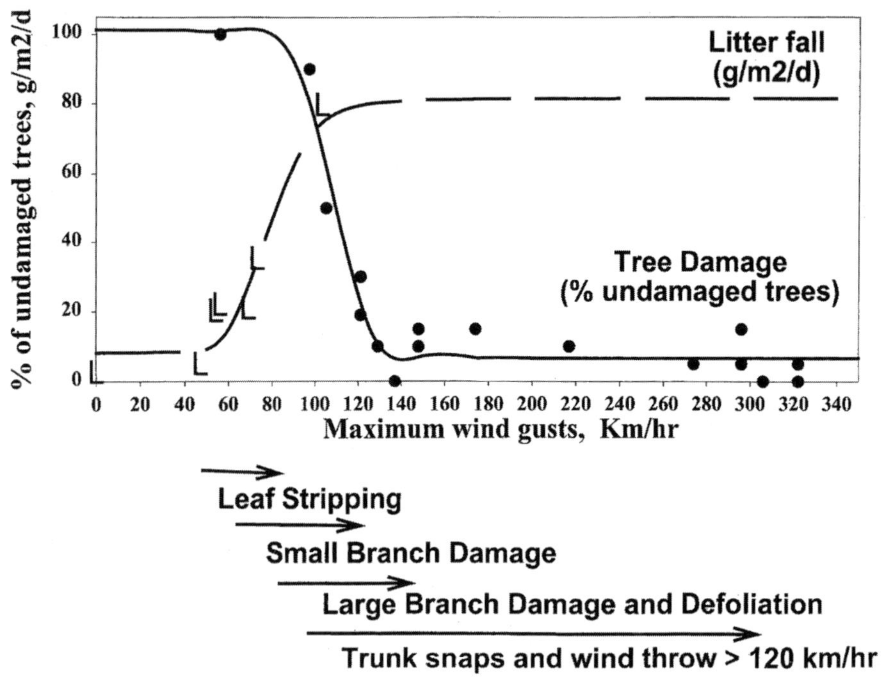

Figure 19.5 Maximum hurricane wind gusts, tree damage (•), and litterfall (L) for Puerto Rican Forests. (Developed from Francis and Gillespie (1993), Scatena *et al.* (1991), and Scatena (unpublished data).)

in different forest types and the ratio of landslide runoff to runoff in the adjacent forest ranged from 1 to 4.

At the catchment scale, landslides can be major sources of stream sediment in upland humid tropical environments (see Douglas and Guyot, this volume). They also disrupt roads and water conveyance systems and can be so chronic that roads and pipelines need to be relocated or face continual maintenance (Ahmad *et al.*, 1993a; Olander *et al.*, 1998). Landslides can also cause catastrophic dam-break floods. When a 240 M m^3 rock slide fell into the 150 M m^3 Vaiont reservoir in Italy, the displaced water overtopped the dam and created a flood wave that killed 2600 people downstream (Morris and Fan, 1997).

Landslides are also known to create dams and lakes when they cross rivers and water courses. In 1992, a 3 M m^3 rock slide dammed the Rio Toro river in Costa Rica (Mora *et al.*, 1993). The vibrations generated by the rapidly moving landslide were recorded in seismographs as far as 110 km away and the debris eventually created a 75 m long, 600 m wide dam with a maximum height of 100 m. The lake impounded by the blockage was 1200 m long, had a maximum depth of 52 m, a volume of 0.5 million m^3 and took three days to fill. However, because the dam material was blocky and pervious, large volumes of water infiltrated and created internal piping that collapsed within in a month. Another landslide dam in Ecuador attained a length of 2.6 km (Plaza-Nieto *et al.*, 1990). It required seven days for bulldozers to construct a spillway and an additional 26 days until stream flows returned to normal.

HURRICANES

Hurricanes, the most destructive type of cyclone (see Tables 19.2, 19.3 and 19.4), affect different parts of the landscape in different ways. At the scale of individual trees, winds in excess of about 100 to 130 km hr^{-1} kill or damage trees lethally within a few hours (Figure 19.5). Winds in excess of 60 km hr^{-1} cause large scale defoliation and litterfall. At this scale, the extent of the damage is related to species, tree morphology, age, size, health and rooting conditions. In general, fast-growing low-density woods are more susceptible to wind damage than high-density, later successional species (Putz *et al.*, 1983; Zimmerman *et al.*, 1994).

At the landform scale, variations in wind damage result from differences in exposure and the modification of wind velocity caused by the landforms themselves. For example, valleys orientated parallel to the direction of dominant winds will often receive more damage than nearby valleys that are perpendicular to the hurricane force winds. Defoliation and the associated transfer of nutrients from the canopy to the forest floor can also cause major shifts in nutrient cycling pathways at the stand and landform scale (Lodge *et al.*, 1994; Scatena *et al.*, 1996).

At the scale of individual mountains, the spatial pattern of damage can be complex and is correlated strongly with aspect relative to the prevailing winds and forest type (Walker *et al.*, 1991). In general, damage is spatially uniform in low-lying, gentle rolling landscape and more complex in highly dissected mountainous terrain. At the regional scale, the configuration of coastlines and mountains relative to the storm track determines how a particular storm will weaken when it crosses land (Boose *et al.*, 1994). Factors controlling forest damage at this scale include gradients in wind velocity that are related to the size and intensity of the hurricane and large topographic features.

At all scales, hurricanes create patches of survivors and new regeneration that undergo changes in structure and composition that can last for decades (Crow, 1980; Weaver, 1986). In the dry evergreen forests of Sri Lanka these changes include a greater susceptibility to overgrazing and mortality by primates (Dittus, 1985). In the Yucatan Peninsula of Mexico, fires commonly follow hurricanes and can cause more mortality than the initial hurricane (Whigham *et al.*, 1991). Changes in the composition and structure of populations of forest organisms and food webs have also been observed (Walker *et al.*, 1996). Nevertheless, hurricanes may be unable to erase the signature of past land use on the species composition at the stand scale, as the composition of post-hurricane regeneration can be related directly to the prior land use (Zimmerman *et al.*, 1995). In some stands, forest composition reflects the prior composition of shade coffee plantations even after 100 years of abandonment and the direct impacts of several hurricanes (Garcia-Montiel and Scatena, 1994).

Impacts of hurricanes to hydrological cycles and water supplies

The extensive defoliation and tree mortality associated with hurricanes can have considerable impacts on both water and nutrient cycles. The initial impacts to hydrological cycles are related typically to the reduced evapotranspiration associated with defoliation. Impacts to nutrient cycles and stream water chemistry are related to the large nutrient pool that is transferred from live biomass to necromass. Unlike the small-scale isolated changes associated with tree falls and landslides, hurricanes can disturb hundreds of square kilometres of adjacent forest and influence watershed scale hydrological processes for years or decades.

In the Luquillo Mountains of Puerto Rico, Hurricane Hugo completely defoliated two adjacent catchments and reduced the above-ground biomass by 50% (Scatena *et al.*, 1993; Scatena *et al.*, 1996). The magnitude and storm response of stream flow returned to pre-storm conditions within a year. Canopy throughfall recovered within two years, and forest litter fall recovered in about five years. The hurricane also increased the concentrations of NO_3, NH_4, dissolved organic N, base cations, Cl and SiO_2 in shallow groundwater within five months of the hurricane (McDowell *et al.*, 1996). The largest relative change in concentration occurred for K, which remained elevated for at least 5.5 years after the hurricane. Most other solutes had returned to background levels within 1–2 years of the storm.

Stream water chemistry can also change following the passage of hurricanes (Waterloo, 1994; Schaefer *et al.*, 2000). In a hurricane-damaged pine plantation in Fiji, 10% of the boles were damaged and the stream water concentrations of Na, K, Mg, Ca and Cl increased significantly in the weeks following the storm (Waterloo, 1994). Sulphate decreased and ammonium and nitrate were unchanged. In watersheds severely damaged by Hurricane Hugo in the Luquillo mountains of Puerto Rico, K, nitrate-N, and ammonium-N remained elevated for up to two years and increased by 90%, 191% and 98% in the first year, respectively (Schaefer *et al.*, 2000). However, the increased exports in the first two years were equivalent to only 1–4% of the nutrients in the hurricane-derived litter, reflecting how effective this forest is at retaining and recycling nutrients following severe natural disturbances.

In addition to changes in the internal cycling of water and nutrients, hurricanes and their associated flooding commonly cause considerable damage to water supply infrastructure. For example, when Hurricane Mitch passed over Honduras in 1998, 80% of the water supply systems were damaged (Pan American Health Organization, 1999). Pools of stagnant flood water also created massive breeding grounds for insects carrying malaria and dengue. While storm discharges can be damaging and post-hurricane stream water concentrations elevated, suspended sediment concentrations during hurricanes may be lower than non-hurricane storms with similar discharges (Gellis, 1993). Apparently, the defoliation by the hurricane-force winds creates debris dams on hillslopes and in small channels that act to trap sediment and reduce suspended sediment concentrations. Nevertheless, because high stream flow can last for several days, the total sediment transport associated with the passage of a hurricane can be significant and very detrimental to reservoir storage. Statistical analysis indicates that the explicit inclusion of hurricane sediment transport to a reservoir in the Philippines can reduce the predicted reservoir life from 139 years to 50 years (White, 1990).

STORM SURGES AND TSUNAMIS

Storm surges are large domes of water that sweep across the coastline with hurricanes or tropical storms (Table 19.8). Tsunamis are tidal waves caused by volcanic eruptions, earthquakes, and submarine landslides and are typically larger than storm surges. When either hit populated coastal areas they are often responsible for the greatest damage to life and property associated with a particular event and they are especially destructive when they coincide with high tide.

Storm waves form ahead of and along the margins of hurricanes and other storms as they cross the ocean. Wave heights vary from a few tens of centimetres to 12–14 m (Table 19.8). Both the magnitude and frequency of storm waves is influenced strongly by local physiography and can vary considerably over small areas. For example, the Gulf of Tokyo had five large hurricane surges over a 50-year period while there were eight waves in 35 overlapping years in the nearby Gulf of Suruga (Nalivkin, 1983). Likewise,

Table 19.8. *Selected examples of large hurricane-related storm surges*

Location	Year	Height (m)	Human deaths	Comments
Bay of Bengal				
River Hooghly	1737	12	30 000	20 000 boats destroyed
Calcutta	1864	12	50 000	
Calcutta	1876	3–9	100 000	150 000 post-flood deaths in epidemics
Bhola, Hatiya, and Ramgati Islands	1960	3.5 to 6.6	7 591	
Pacific Ocean				
North Vietnam	1881		300 000	
Marshall Islands	1905	14		
Shantou, China	1922	2.5–7.5	60 000	
Fiji	1910	10.5		
Hawaii	1960	6–9		
Australia	1899	12–13	307	Damaged 120 km of coastal vegetation
Australia				
Nagoya, Japan	1959	5.2	5 500	Flooded 350 km^2
Gotsen Iva, Japan	1956	10		
Gulf of Osaka, Japan	1934	4.58	2 900	Flooded 100 km^2
Atlantic				
Gulf Coast, Hurricanes Carol and Edna	1953	12–13		
New England Hurricane	1938	3–4		
South Carolina, Hurricane Hugo	1989	5–8		Increased sea salts in soil water
North Sea, England, Holland, Belgium	1953	9–10	3 000	Water penetrated 65 km inland
St Petersburg	1924	3.69		Additional surges occurred in 1777, 1824, 1955

Source: After Nalivkin (1983) and Blood *et al.* (1991).

the Gulf of Toyama experienced nearly ten times the frequency of the Gulf of Tokyo.

Between 1992 and 1998 eight major tsunamis impacted coastal areas around the world (Tapin, 1999). Seventy-five percent of these occurred in the tropics or subtropics, including Nicaragua in 1992, Flores (Indonesia) 1992, East Java (Indonesia) 1994, Mindoro (Philippines) 1994, Chimbote (Peru) 1996, and Papua New Guinea, 1998. While tsunamis occur in all large water bodies, the Pacific rim countries are particularly vulnerable. Between 1888 and 1998, six large tsunamis occurred along the northern coast of Papua New Guinea (Tapin, 1999). The 1998 tsunami impacted a 25 km stretch of coast with wave heights of 10 to 15 m and killed over 2500 people. In the northern Caribbean earthquakes generated devastating tsunamis in 1867, 1918, and 1946 (Dillon and Brink, 1999).

Impact of storm surges and tsunamis on the hydrological cycle and water supply

While the impacts of storm surges and tsunamis can be devastating and extend to tens of kilometres inland, their impacts are limited to coastal zones and coastal forests. Impacts range from catastrophic stand mortality to the covering of the herbaceous understorey by fine sediment. They can also have prolonged effects on coastal water tables and soil water dynamics. When Hurricane Hugo struck the South Carolina Coast in 1989, a 5 to 8 m storm surge inundated coastal forests with over 3 m of standing salt water (Blood *et al.*, 1991). Low-lying areas were flooded by up to a metre of water for about two weeks and approximately 1.9 cm of sea water entered the saturated soil matrix. Soil solution salinity increased greatly and the total ionic content of soil water increased from 1.6 to 3.4 meq l^{-1} prior to the storm to 227 to 778 meq l^{-1} two days after. The Na input from the storm was 50 times greater than the annual atmospheric input and the average ammonium-N concentrations in the Bh horizon increased 115 times in the three months following the storm.

Floods and fluvial processes

The streams and rivers that drain humid tropical forests are important sources of water, food, transportation and recreation. Waterways and river dynamics may also influence the distribution of forest species and have a long-term role on speciation and forest diversity (Rasanen *et al.*, 1987; Salo *et al.*, 1986). Their riparian forests can also have distinct plant communities, disturbance regimes, edaphic conditions, soils and nutrient cycles (Shreve, 1914; Davis and Richards, 1933; Foster, 1990;

McDowell *et al.*, 1992; Scatena and Lugo, 1995; Richards, 1996). These riparian forests also play an important role in maintaining aquatic habitat and are important sources of timber, fibre and agricultural soils. Therefore, there are very few humid tropical riparian forests that have not been disturbed by human activity directly.

Two general types of flood disturbances are commonly distinguished in the humid tropics: (1) seasonal inundation-type floods where extensive areas are covered with lake-like water for extended periods (e.g. weeks to months) each year; and (2) event floods which are relatively short duration (e.g. hours to days) events with high velocity stream flows. The distribution of flood related plant communities has been further grouped into areas with: (1) occasional flooding, like floodplains and terrace habitats; (2) annual short-term flooding; (3) annual long-term flooding; and (4) annual submersion by flood waters, like oxbow lakes (Junk, 1989; Lamotte, 1990; Foster, 1990; Kalliola *et al.*, 1991; Thompson *et al.*, 1992). It is also common to distinguish permanently waterlogged swamp forests from seasonally flooded marsh forests (Richards, 1996).

Seasonal inundation-type flooding is most common in lowland areas along passive continental margins. The most extensive areas of relatively undisturbed seasonally flooded forests are along the Amazon, Orinoco and Paraná rivers where large areas are flooded every year with up to 15 m of water. In the Amazon basin, varzea type forests that are flooded during annual high water periods occupy over 55 000 km^2 (Richards, 1996). In the Peruvian Amazon, 12% of the forests are undergoing primary succession on recent riverine deposits (Salo *et al.*, 1986). Seasonally flooded forests also occur in West Africa, in the Zaire basin, Borneo, Sumatra and New Guinea (Richards, 1996). In many areas of the eastern tropics, notably Indonesia, the original swamp forests have been largely cleared to irrigate rice (see Hooijer, this volume).

Event-type floods are short duration disturbances (e.g. hours to days). However, they can alter channel morphology and floodplain vegetation drastically and leave imprints on the landscape that can persist for decades, especially in streams where floods transport and deposit coarse bed load sediment (Gupta, 1988). While large floods can occur in all parts of the humid tropics, Gupta considers the following areas to have the geological and physiographic conditions necessary to maintain the morphology of large floods events:

(1) Rivers valleys of East Asia, especially Taiwan and the Philippines
(2) Upland areas of Vietnam, Sumatra, Java and Burma
(3) Humid areas of the Indian subcontinent
(4) Madagascar and neighbouring parts of coastal East Africa
(5) North and northeast Australia
(6) Caribbean basin and Central America highlands.

Regression models indicate that peak flood discharges in forested humid tropical watersheds are influenced by both climatic and morphologic factors. Drainage area, mean annual rainfall, the maximum two-year 24 hour rainfall, the length of the main channel and the total length of tributaries have been positively related to peak flood discharge and annual peak discharges (Rivera-Ramirez, 1999; Ramos-Gines, 1999). Likewise, depth-to-bedrock and watershed shape factors have been negatively correlated with peak discharge in forested watersheds.

Seasonal differences in precipitation and stream flow can also influence channel morphology and flood-related disturbances. In Jamaica, the morphology of streams draining areas with similar annual rainfall and geology varies with the seasonal distribution of rainfall (Gupta, 1975). Areas that receive rainfall throughout the year have meandering-like channels while areas with a strongly seasonal rainfall tend to have more braided channels. Likewise, sandy bedded streams in the tropical monsoon environments of India often have braided stream channel morphologies during the dry season and meandering channels during the wet season (Gupta and Dutt, 1989).

Impacts of floods on hydrological cycle and water supply

In general, seasonal inundation-type flooding has large influences on the distribution of forest types, aquatic organisms and trace gas emissions. Most of these impacts are related to biogeochemical changes in soil chemistry that occurs with standing water. However, because these floods are annual and relatively predictable, their influence on infrastructure and water supplies is generally low.

The high streamflows and stream-power associated with event-type flooding commonly causes local catastrophic disturbances. and can damage water distribution systems significantly. Event-type floods are responsible for most reservoir sedimentation, as 50% or more of the annual sediment load is often transported in streamflows that occur only 1% of the time (Morris and Fan, 1997). Over 90% of the annual sediment discharge occurs typically in less than 10% of the time and large events can transport decades worth of average annual sediment.

While flood disturbances affect riparian forests, riparian environments also influence stream water chemistry and aquatic habitats. Surface soils in humid tropical riparian environments are typically saturated and have intensive microbial activity. Consequently they have a high potential to reduce and remove nitrogen from soil and groundwater before it enters the stream channel (Bowden *et al.*, 1992; McDowell *et al.*, 1992; Proctor, this volume). Moreover, because much of the shallow subsurface storm runoff from upland areas passes through saturated riparian areas before it enters the stream channel, sediment trapping and biogeochemical transformations in riparian environments can have a disproportionate influence on stream water chemistry. These influences, combined with the importance of riparian vegetation to aquatic organisms, have been the impetus for the establishment of riparian buffer zones in many areas.

Both event and seasonal inundation-type flooding can play important roles in structuring tropical aquatic communities. Event-type floods and droughts have been shown to cause significant invertebrate mortality in streams in Malaysia, Ghana, Hong Kong, India, Ecuador, tropical Australia, the Andean piedmont of Venezuela and the Caribbean (Flecker and Feifarek, 1994). Significant positive relationships also exist between invertebrate densities and the number of days since the previous major rainstorm. However, the response of coexisting aquatic species can vary depending on their habitat requirements, populations, and life history traits (Covich *et al.*, 1996; Connolly and Pearson, this volume).

DROUGHTS AND FIRES

Prolonged droughts and fires are an important, but often underestimated, disturbance in humid tropical forests (Brunig, 1969; Wright, 1992; Walsh and Newbery, 1999; Malmer *et al.*, this volume). Paleoclimatic evidence from Africa, Asia, Amazonia, Central America and the Caribbean indicate that most humid tropical forests have experienced numerous fires and extended droughts during the past 10 000 years (Sanford *et al.*, 1985; Hodel *et al.*, 1991; Servant *et al.*, 1993; Guilderson *et al.*, 1994; Mahé *et al.*, this volume). Currently, as much as 200 000 km^2 of the Brazilian Amazon may burn each year (Cochrane and Schulze, 1999) and accidental fires may have affected nearly 50% of the remaining un-cleared Amazonian forests (Cochrane *et al.*, 1999). Furthermore, model simulations of future CO_2-induced climate change indicate that large areas of the humid and seasonal tropics will experience decreases in soil moisture and increases in dry season length (Emanuel, 1987, 1997; Hulme and Viner, 1995; Knutson *et al.*, 1998). These increases imply that fires and droughts in these forests will also become more common.

At the continental scale, the seasonality and the frequency of droughts varies with global oscillations (e.g. ENSO, NAO) and the relative position of particular geographic location to the trade winds and ocean. In general, maritime locations that receive precipitation from daily and orographic processes have a lower occurrence of droughts than inland areas that receive their moisture from large-scale global weather systems. Nevertheless, all humid tropical forests experience extended rainless periods. There is no satisfactory way of defining an ecological drought because (1) forests differ in their susceptibility to drought; (2) water shortages vary with soil characteristics, slope and topography; (3) different components of the forest experience water shortages at different points with epiphytes and ground herbs being more susceptible than trees (Walsh and Newbery, 1999). In the lowlands of Sarawak, 30-day periods with less than 60 to 100 mm of rain are considered to be of sufficient duration to be of ecological significance (Brunig, 1969). Based on the assumption that monthly potential transpiration of lowland rainfall forests is 100 mm m^{-1}, 100 mm m^{-1} has been widely used to define 'wet' and 'dry' months. Drought intensity has also been ascertained by summing the amounts by which the rainfall of each month in a dry month sequence falls below 100 mm (Walsh and Newbery, 1999). In most humid tropical forests, Cumulative Rainfall Deficits (CRD) of between 5 to 10% of mean annual precipitation (MAP) are common on annual and decadal time scales while CRDs greater than 15% are rare events (Figure 19.6). Available data from Borneo suggests that within the same region there can be considerable spatial variation in the frequency of low intensity droughts (e.g. CRDs $<$ 10% MAP) but convergence in the frequency of intense droughts (e.g. CRDs $>$ 15% MAP).

At the scale of forest stands, short-term (e.g. weeks to months) dry periods have been linked to declines in the abundance of common forest lizards, spiders and exotic earthworms (Reagan and Waide, 1996; Zou and Gonzalez, 1997). Short-term droughts and wet and drying cycles can also stimulate microbial biomass growth, enhance microbial nitrogen immobilisation, and affect detrital food chains (Lodge *et al.*, 1994). Prolonged droughts have also been linked to Anoline lizard decline at the Monteverde cloud forest in Costa Rica (Pounds *et al.*, 1999). After three days without rain in the Luquillo Mountains of Puerto Rico, the abundant tree frogs have empty stomachs because of the lack of insects that are normally borne in wet forest litter (Figure 19.7). One week without rain or canopy throughfall occurs once nearly every year and will result in the wilting of herbaceous vegetation in open areas like gaps and roadways. Once every 10 to 20 years there are enough consecutive rainless days that small headwater streams dry up, aquatic habitat becomes limiting and canopy vegetation increases litterfall and/or flower.

Prolonged droughts have been linked to the massive mortality of rainforest epiphytes, heavy mortality and flowering of trees, mortality of moisture loving species, increased fires and grassland expansion (Richards, 1996; Whitmore, 1984; Hubbell and Foster, 1990). However, the relationship between drought and tree mortality is often obscured by fire and the resultant increase in growth and productivity of some tree species. A severe drought in 1972–73 is reported to have killed about 50% of the vegetation on rock faces near the summit of Mt Kinabalu Sabah (Lowry *et al.*, 1973; Smith, 1979). During the 1982–83 ENSO related drought in Borneo annual rainfall was 36 to 44% of the long-term average and over 30 000 km^2 of forested lands and peat swamp were severely damaged by the drought and drought-linked fire (Woods, 1989). During this drought vegetation on steeper slopes was more affected than that on gentler slopes and areas that had been logged had greater mortality than unlogged areas. While there is a general pattern for areas with deeper soils and therefore more soil moisture

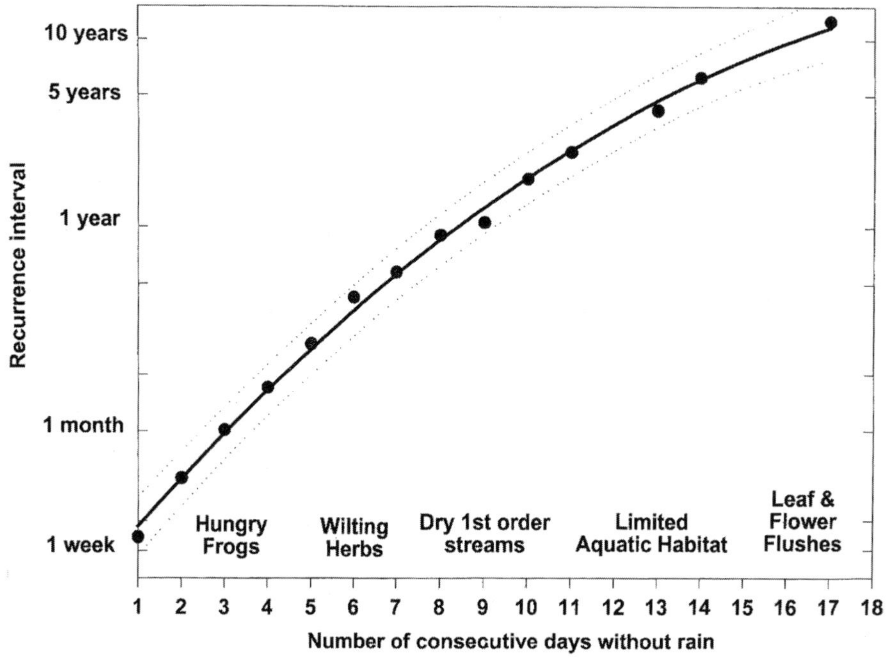

Figure 19.6 Number of drought periods per century and cumulative rainfall deficits as percent of mean annual precipitation for various humid tropical forests. (After Walsh and Newbery, 1999.)

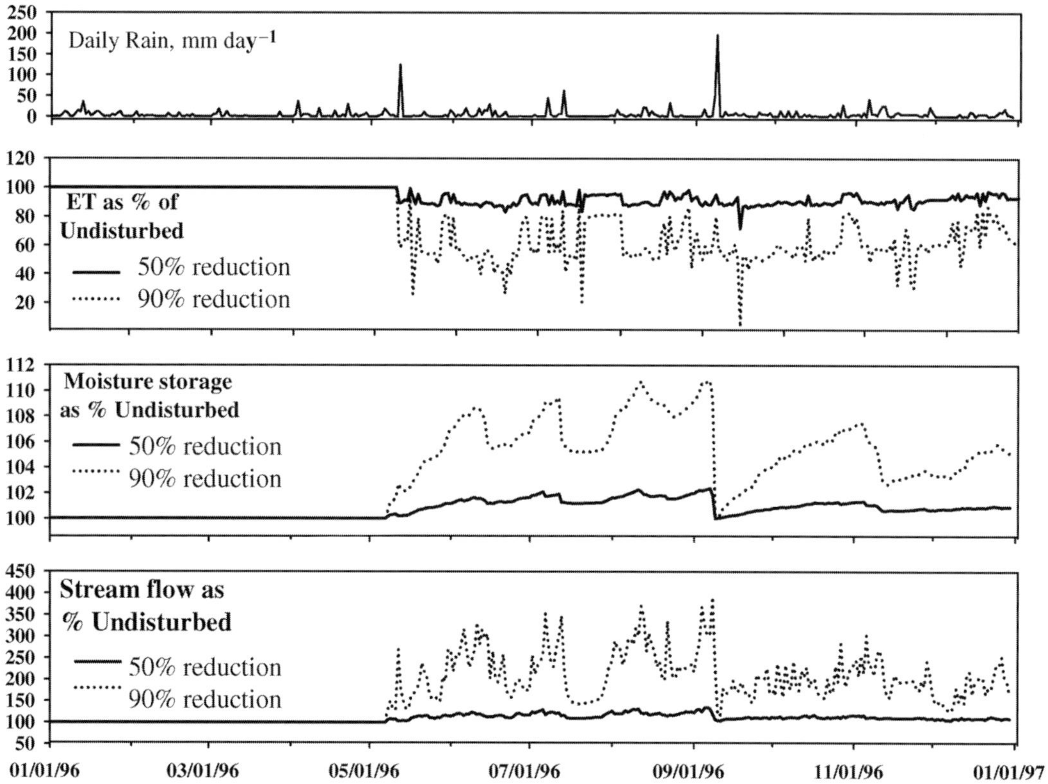

Figure 19.7 Recurrence interval and ecosystem response for number of consecutive days without rain in the Tabonuco Forest of the Luquillo Mountains in Puerto Rico.

Table 19.9. *Estimates of annual biomass burning (Tg yr^{-1}) in the tropics*

	Seiler and Crutzen, 1980	Hao *et al.*, 1990	Crutzen and Andreae, 1990	Andreae, 1991	Hao and Liu, 1994
Deforestation	550–880	280–560	440–1560	420	510
Shifting cultivation	900–2500	750–940	1110–2220	840	1310
Savannah fires	480–1900	3690	670–3560	3690	2670
Fuel wood	620	680	670–1330	1260	620
Agricultural residues	710	660	1110–1780	1360	280
Total, Tg yr^{-1}	3260–6610	6060–6530	4000–10450	7570	5390

Source: Barbosa *et al.* (1999).

reserves to have less mortality, there is less consistency in the relationship between tree size and drought-related mortality. Greater mortality of smaller trees during droughts has been reported in the forests of Sabah (Woods, 1989), Sarawak (Nakagawa *et al.*, 2000), Costa Rica and Peru (Hartshorn, 1990). In Panama and in other areas larger trees are apparently more susceptible to drought-related mortality (Condit *et al.*, 1995).

Estimates of the annual burning of tropical biomass from natural and anthropogenic causes are large and reflect the importance of fire to tropical ecosystems (Barbosa *et al.* 1999). The vulnerability of tropical landscapes to burning is related to population density, vegetal type and land use. Recent estimates of the annual biomass burned in the tropics indicate that between 33 and 61% comes from deforestation, shifting cultivation or fuel wood consumption (Table 19.9). Recent analysis from Amazonia indicates that accidental fires may cause more deforestation than intentional clearing (Cochrane and Schulze, 1999). While human-induced fires may be the dominant cause of modern fires, the widespread occurrences of pre-historic soil charcoal indicates that fires are a common natural disturbance (Sanford *et al.*, 1985; Goldammer, 1999). In Indonesia and Amazonia drought-related fires increase during El Niño events and are probably part of the long-term natural disturbance regime (Neelin and Latif, 1998; Walsh and Newbery, 1999). Nevertheless, the evolutionary adaptations of humid and wet tropical forests to fire are uncertain and tree mortality is expected to be high even during light ground fires (Cochrane and Schulze, 1999).

Susceptibility to forest fires can increase following hurricanes (Whigham *et al.*, 1991) and after logging and clearing (Woods, 1989; Cochrane *et al.*, 1999). Although both hurricanes and logging can decrease transpiration and thereby increase soil moisture, the increase in fuel wood supplies apparently overrides any increase in soil moisture, especially in areas with distinct dry seasons. In all areas, the impacts and frequencies of fires may be greatest in dense secondary forests and areas with abundant woody debris. In the humid coastal plain forests of north-eastern Queensland, the size of host trees and the distribution of epiphytes are related to fire frequency (Bartareau and Skull, 1994). In this area, larger trees and more abundant epiphytes occur in areas with fire frequencies between 10 and 20 years. In the eastern Amazon, burn forests can have substantial variation in forest structure and fire damage over distances of less than 50 m (Cochrane and Schulze, 1999). Burning substantially increases the susceptibility to additional fires and at least 50% of all previously burned areas are predicted to become flammable with 16 consecutive rainless days. While undisturbed Amazonian forest is less susceptible to fires than logged forests, fires may occur after multiple years of low rainfall or during extreme droughts.

Lightning-induced fires have been considered an important natural disturbance in humid tropical forests in Sarawak, the Solomon islands and Costa Rica (Whitmore, 1984; Horn, 1991) and in dry tropical forests in Australia, Southern Africa, South East Asia and Central America (Middleton, 1997). Lightning can disturb forests by creating canopy gaps or by starting fires. In *Shorea albida* peat swamp forest in Sarawak, between 0.2 and 3% of individual plots consisted of lightning-induced gaps and gap formation by lightning was comparable to that caused by wind-throw (Brunig, 1969, as reported in Walsh, 1997). In a dry forest of Costa Rica a lightning strike created a burned area of 5.4 hectares (Middleton *et al.*, 1997).

Influence of droughts and fires on the hydrological cycle and water supply

Droughts not only result in a direct loss of water for municipal and industrial needs, they can also lower the quality of surface water supplies. Moreover, reduced flows can decrease the aeration and self cleansing capacity of rivers and increase the concentration of pesticides, insecticides, and other point and non-point pollutants. Therefore, it is often necessary to increase or vary chemical additives in water treatment plants during droughts to reduce risks to public health (Pan American Health Organization, 1998). In many areas, the reduction in surface water supplies during droughts is mitigated by increasing groundwater withdrawals temporarily.

Nevertheless, the expected increase in droughts and fires caused by local and global change may be particularly troublesome for many humid tropical areas that already have stressed municipal water systems.

When fires occur they cause temporary increases in the concentrations of nutrients and sediments in stream waters and trigger forest succession (Malmer *et al.*, this volume). In experimentally burned forest plots in Venezuela, stream water concentrations of magnesium, nitrate, calcium and potassium increased for one to two years following burning (Uhl *et al.*, 1982; Uhl and Jordan, 1984) and are similar in magnitude to the increases observed following hurricanes (Schaefer *et al.*, 2000). Windblown ash can also contaminate drinking water sources and cause damage to water treatment plants by clogging filters and treatment tanks (Pan American Health Organization, 1998).

Volcanic eruptions and earthquakes

Both volcanic activity and earthquakes are common disturbances in many humid tropical areas. There are over 290 active volcanoes in the tropics and subtropics (volcano.und.nodak.edu, 1999) of which approximately 80% are in humid tropical life zones. In Central America alone there are 80 geologically young volcanoes (Cunningham *et al.*, 1984). On an event basis, volcanic eruptions and earthquakes are the most costly type of natural disaster (Board on Natural Disasters 1999) and individual eruptions commonly cause thousands of fatalities. Their impacts are especially important in tropical regions, where as much as 80% of the human fatalities associated with major volcanoes in the past 500 years have occurred (Table 19.10).

The impacts of earthquakes on tropical forests include local landslides and tree-uproots (Garwood *et al.*, 1979). Impacts associated with volcanic eruptions generally cover a larger area and can range from light ash falls to the complete covering of the landscape with thick accumulations of rock (Whittaker and Walden, 1992). Vegetation can also be affected by volcanically induced lightning strikes, blast waves and gas poisoning. Because of repeated eruptions and ash-falls that interrupt the recolonisation of vegetation, their impacts can be prolonged.

Revegetation following earthquake-induced landslides is presumably similar to that following rainfall induced landslides. In contrast, revegetation on hard-rock lava flows can take decades and may occur at a slower rate than with any other type of natural disturbance. Following the 1883 eruption of the Krakatau islands in Indonesia the first plants were blue-green algae and grasses (van Leeuwen, 1936; Bush *et al.*, 1995). After 25 years a light grassy woodland thicket had established and by 50 years the islands were completely forested (Bush *et al.*, 1995). Similar rates of colonisation and revegetation have been observed on other tropical lava flows and ash deposits (Richards, 1996). The first plants to establish on solidified lava are usually bird or wind dispersed ferns and herbs. In areas disturbed by ash accumulations and blast waves, the invading species are usually those that are common on abandoned cultivated land. Recent studies have also demonstrated that the seeds of some species can survive burial by volcanic ash for about 60 years (Whittaker *et al.*, 1995).

The vegetation recovery in upper elevation tropical montane volcanic sites can take much longer than at nearby lower elevations sites and often begins with moss and lichen (Beard, 1945; Sheridan, 1991; Richards, 1996). Ten years after the Soufrière volcano erupted on the Caribbean island of St. Vincent, vines, herbs and a few bushes were established on the upper elevations. After 30 years there was a light forest canopy at lower elevations but no higher plants had established in the upper elevations of the mountain. Canopy closure in upper elevation blast areas on volcanoes in Papua New Guinea and Krakatau can also take decades and remain species-poor for considerably longer periods (Richards, 1996).

In addition to stand-level effects on forest regeneration, volcanic eruptions can also cause regional and global impacts (Sadler and Grattan, 1999). Locally and regionally they can increase the magnitude and frequency of lightning strikes and the acidity of rainfall. Globally, volcanic activity has been linked to ENSO events, global temperature decreases and glacier expansion, Holocene climate change, tree growth and mass extinctions (Sadler and Grattan, 1999; Olsen, 1999). Most of these linkages are related to the potential of eruptions to depress global temperatures by increasing the amount of upper atmospheric aerosols and thereby decreasing the available solar energy. However, none of these linkages are well understood and the direction and magnitude of responses to historic eruptions have varied considerably between regions, especially when they have occurred in conjunction with ENSO events.

Impacts of volcanoes and earthquakes on the hydrological cycle and water supplies

Based on the rate of regeneration of forest cover, the direct impacts of volcanic disturbances on stand-level hydrological cycles can last for decades in lower elevation areas and possibly centuries in upper montane sites. These tectonic processes can also modify landform morphology, subsurface storm flow paths and stream water chemistry. In Costa Rica, stream reaches associated with geothermal inputs from Pleistocene lava flows have high solute, phosphorus-rich waters (Pringle *et al.*, 1990; Pringle, 1991). Adjacent streams and reaches that do not receive geothermal inputs are relatively solute-poor and have different biotic assemblages.

The impact of volcanic activity on water supplies can range from short-term contamination to total destruction. The contamination of drinking water sources by volcanoes is typically caused by deposits of ash or toxic fluids (Pan American Health Organization,

Table 19.10. *Fatalities associated with some notable volcanic disasters between 1500 and 1989*

			Primary cause of death				
Volcano	Country	Year	Pyroclastic flows	Debris flows	Lava flows	Post-eruption starvation	Tsunami
Kelut	Indonesia	1586		10 000			
Vesuvius	Italy	1631			18 000		
Etna	Italy	1669			10 000		
Merapi	Indonesia	1672	300				
Awu	Indonesia	1711		3 200			
Oshima	Japan	1741					1 480
Cotopaxi	Ecuador	1741		1 000			
Papadajan	Indonesia	1772	2 960				
Lakagígar	Iceland	1783				9 340	
Asama	Japan	1783	1 150				
Unzen	Japan	1792					15 190
Mayon	Philippines	1814	1 200				
Tambora	Indonesia	1815	12 000			80 000	
Galunggung	Indonesia	1822		4 000			
Nevado del Ruíz	Colombia	1845		1 000			
Awu	Indonesia	1856		3 000			
Cotopaxi	Ecuador	1877		1 000			
Krakatau	Indonesia	1883					36 420
Awu	Indonesia	1892		1 530			
Soufrière	St Vincent	1902	1 560				
Mont Pelée	Martinique	1902	29 000				
Santa Maria	Guatemala	1902	6 000				
Taal	Philippines	1911	1 330				
Kelut	Indonesia	1919		5 110			
Merapi	Indonesia	1951	1 300				
Lamington	Papua New Guinea	1951	2 940				
Hibok-Hibok	Philippines	1951	500				
Agung	Indonesia	1963	1 900				
Mount St Helens	USA	1980	60				
El Chichón	Mexico	1982	>2 000				
Nevado del Ruíz	Colombia	1985		>22 000			

Source: Adapted from Tilling (1989).

1998). Deposits of fine volcanic ash can also damage water treatment plants by clogging filters and treatment tanks. The most common impact of earthquakes on water resources is the physical damage they bring to water distribution systems. This damage can be caused by direct ground motion or landslides and typically includes the breaking of water and sewage distribution networks. Seismic activity or landslides falling into reservoirs can also generate large waves that overtop and / or damage dams. Rapid fluctuations in well water levels are often observed during earthquakes and changes in groundwater hydrology have been reported after earthquakes (Meinzer, 1942). These changes include: changing well water levels; changes in spring flow and well discharge; changes in groundwater temperature; emission of hydrogen sulphide and other gases; change in mineral content of water; and the creation of new springs.

Meteor impacts

Catastrophic meteor impacts are events that can have local, continental and global scale consequences (Toon *et al.*, 1997). Moreover, a meteor impact in the Caribbean basin has been implicated in the global Cretaceous–Tertiary mass extinction and may have generated a free-standing wave over 500 m high (Hildebrand and Boynton, 1990; Florentin *et al.*, 1991). Catastrophic disturbances associated with meteor impacts include earthquakes, blast waves, tsunamis and fires. Other effects, like high-altitude particles from

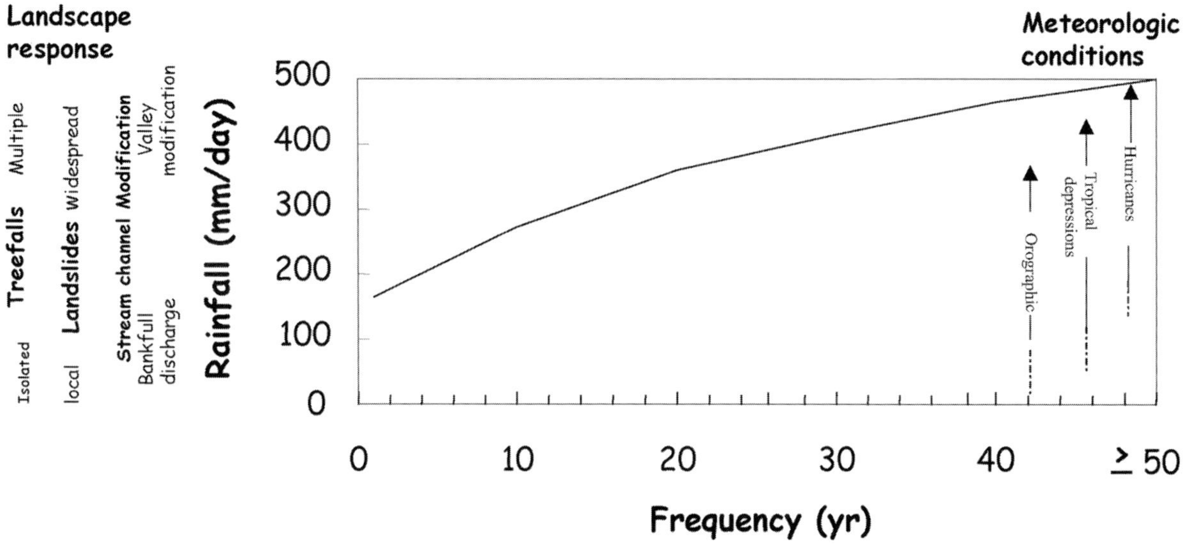

Figure 19.8 Daily rainfall for 1996 and model simulations of evapotranspiration, moisture storage and streamflow as percentage of undisturbed conditions for a 50% and 90% reduction in leaf area. Based on TOPOG dynamic hydrological model of the Bisley II watersheds, Luquillo Experimental Forest, Puerto Rico.

dust, smoke and acid rain, may have longer term impacts on global climate and biota.

Environmental perturbations associated with impact frequencies in the order of 60 000 years are predicted to be important over areas on the order of 10^4or 10^5 km^2 (Toon *et al.*, 1997). If an impact of this frequency fell within an ocean, one kilometre of coastal plain along the margins of the entire ocean basin could be flooded. Impacts with recurrence intervals of two million years or greater would have global effects of much larger magnitudes. While these are clearly rare events, when it is considered that the physical features of much of Latin America have been present since the Tertiary (1.8 to 65 million years BP), large areas of ancestral humid tropical forests in Latin America may have been affected a dozen times by global scale impacts and hundreds of times by impact-generated tsunamis. The impacts of these events on forest ecosystems or hydrological cycles are unknown but, given their magnitudes, should be substantial.

MODELLING

As the above discussion demonstrates, the impacts of disturbances on the hydrology of humid tropical forest is quite variable and complex. Not only can impacts differ with the type and magnitude of disturbance, they can also differ with initial and subsequent conditions. To demonstrate the magnitude of initial, short-term (e.g. months) responses to disturbance-induced changes in leaf area, watershed specific simulations (see Figure 19.8) were made using the TOPOG dynamic catchment hydrological watershed model which operated at daily timesteps (O'Loughlin, 1986; Vertessy *et al.*, 1993; Bonell, this volume, Chappell, Bidin *et al.*, this volume). The model was calibrated to actual conditions that occurred in the forested Bisley II watershed of Puerto Rico during 1996 (Schellekens, 1999; Schellekens *et al.*, 2000). The element network used for these simulations incorporated a 2.5 m contour interval and consisted of 2876 elements with an average element size of 23.55 m^2. The average slope of all the elements in the catchment is 0.75 m m^{-1} (0.54 m m^{-1} when weighted for element size). Four different soil units, closely related to topographic position and slope morphology, were used in the model: ridgetop soils, hillslope soils, valley bottom soils and patches of rocky stony land (see Scatena, 1989 for a full description of the soil types).

Three scenarios were modelled: (a) undisturbed forest during 1996; (b) a reduction in leaf area index (LAI) of 50% after 125 days, and (c) a 90% reduction in LAI after 125 days (Figure 19.8). Transpiration was estimated within the model using the Penman-Monteith formula. In scenarios (a) and (b), rainfall interception was estimated using the Gash model (Schellekens *et al.*, 1999). In scenario (c), interception was estimated to be 25% of precipitation, a value similar to that observed in the watershed following defoliation in response to the passage of Hurricane Hugo (Scatena *et al.*, 1996). These simulations reflect the initial, short-term (e.g. < 1 year) impacts before forest vegetation begins to recover.

The simulations indicate that a decrease in LAI results in greater streamflow and moisture storage and decreases in evapotranspiration (Figure 19.8). The differences between undisturbed and disturbed conditions increased significantly when the reduction in LAI was greater than 50% (Table 19.11). Moreover, a 50% reduction in LAI resulted in a 12% increase in runoff while a 90%

Table 19.11. *Summary of simulation results for the Bisley watersheds in Northeastern Puerto Rico with different levels of canopy cover*

		240-day post-disturbance period		
	Entire year, 100% canopy	100% canopy	50% canopy reduction	90% canopy reduction
Net precipitation, mm	1975	1481	1563	2630
Discharge, mm	1295	995	1113	2291
(% undisturbed)			(112)	(230)
Soil evaporation, mm	6.7	4.2	9.2	24.3
(% undisturbed)			(219)	(578)
Transpiration, mm	662	443	396	247
(% undisturbed)			(89)	(56)
Catchment saturation, %	9.7	10.6	11.4	17.2
(range)	(2.3–78.6)	(4.8–78.6)	(5.1–80.8)	(6.4–97.9)
Moisture storage, mm	551	560	566	591
(range)	(483–620)	(513–620)	(524–621)	(531–621)

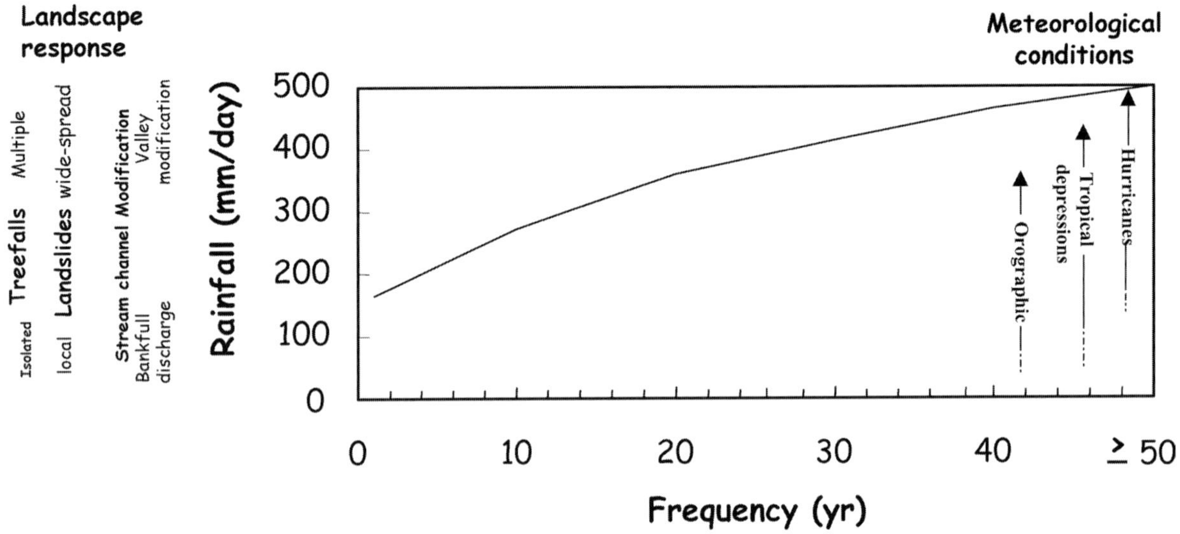

Figure 19.9 Rainfall intensity and general ecosystem response for montane humid tropical forests. (After Lugo and Scatena, 1995.)

reduction in LAI resulted in a 130% increase. Simulated increases in soil moisture and catchment saturation after a 90% reduction in LAI were similar to those observed in recent tree fall gaps (Becker *et al.*, 1988; Ostertag, 1998).

During larger rainstorm events, the differences between disturbed and undisturbed conditions tended to decrease, especially when daily rainfall was greater than 200 to 300 mm day^{-1} (Figure 19.8). This suggests that during large rainfall events, like many of those listed in Table 19.1, the influence of different types of forest cover on the hydrological processes decreases (Bruijnzeel, 1990). For the Bisley II catchment modelled here, the rapid runoff response to precipitation is governed mostly by net precipitation rather than by antecedent soil moisture (Schellekens, 2000). This behaviour was thought to be a result of the clayey soils that have a rapid decrease in hydraulic conductivity with depth, the abundant macropores in the upper soil, and the high frequency of rain that maintains soil moisture.

CONCLUSION

Humid tropical forests are affected by a range of potential disturbances. Nearly all types of weather systems that affect humid tropical forests can produce intense, landscape-modifying rainfalls.

While the response of any particular landscape is complex, most humid tropical forests experience disturbance-generating rainfalls at least once every decade. Synthesis of observations from a number of upland tropical forests indicates that daily rainfalls of the order of 200 mm day^{-1} will tend to cause tree fall gaps, landslides and localised flooding (Figure 19.9). Events of the order of 400 to 500 mm day^{-1} or more can cause widespread modification to vegetation, hillslopes and stream channels. The recovery of hydrological cycles following these atmospherically-induced events varies from a few months to several decades, depending on the type and extent of damage.

Numerous ecological studies have shown that many of the plant and animal species in the humid tropics depend on some type of disturbance during their life cycle. Moreover, many species need disturbed conditions for regeneration and are well adapted to the natural disturbance regimes in a region. Nevertheless, there is considerable uncertainty regarding the resilience of these species and ecosystems when natural disturbances interact with anthropogenic disturbances and/or climate change, especially when populations are isolated and ecosystems are fragmented. The synergistic interactions of fires and logging roads, floods and land cover, landslides and roads indicate that the effects of natural and anthropogenic disturbances can be additive and result in complex positive feedbacks. Understanding these synergistic interactions and how the frequencies and impacts of disturbances will change during periods of land use and climate change is a major research need and challenge for the future.

References

Ahmad R., Earle A. H., Hugues P., Maharaj R., and Robinson E. 1993a. Landslide Damage to the Boar River Water supply pipeline, Bromley Hill Jamaica: Case Study of a Landslide Caused by Hurricane Gilbert (1988). *Engineering Geology Bulletin* **47**: 59–70.

Ahmad R., Scatena F. N., and Gupta A. 1993b. Morphology and Sedimentation in Caribbean Montane streams: examples from Jamaica and Puerto Rico. *Sedimentary Geology* **85**: 157–169.

Anthes R. A. 1982. Tropical Cyclones. Their Evolution, Structure, and Effect. American Meteorological Society. *Meteorological Monographs* **19**(41) pp. 208.

Barbosa P. M., Stroppiana D., Gregoire J., and Cardoso Pereira J. M. 1999. An assessment of vegetation fire in Africa (1981–1991): Burned areas, burned biomass, and atmospheric emissions. *Global Biogeochemical Cycles* **13**(4): 933–950.

Bartareau T., Skull S. 1994. The Effects of Past Fire Regimes on the Structural Characteristics of Coastal Plain *Melaleuca viridiflora* Sol. Ex Gaert. Woodlands and the Distribution Patterns of Epiphytes (*Dendrobium canaliculatum R. Br., Dischidia nummularia R. Br.*) in Northeastern Queensland. *Biotropica* **26**(2): 118–123

Becker P., Rabenold P. E., Idol J. R., and Smith A. P. 1988. Water potential gradients for gaps and slopes in a Panamanian tropical moist forest dry season. *Journal of Tropical Ecology* **4**: 173–184.

Beard J. S. 1945. Progress of Plant Succession on the Soufrière of St. Vincent. *Journal of Ecology* **33**(1): 1–9.

Blood E. R., Anderson P., Smith P. A., Nybro C., and Ginsberg K. A. 1991. Effects of Hurricane Hugo on Coastal Soil Solution Chemistry in South Carolina. *Biotropica* **23**(4a): 348–355.

Board on Natural Disasters. 1999. Natural Hazards and Policy. Mitigation Emerges as Major Strategy for Reducing Losses Caused by Natural Disasters. *Science* **284**: 1943–1947.

Bongers F., Popma J., Moave del Castillo J., Carabias J., 1988. Structure and floristic composition of a lowland rainforest of Los Tuxtlas, Mexico. *Vegetatio* **74**: 55–80

Boose E. R., Foster D. R., and Fluet M. 1994. Hurricane impacts to tropical and temperate forest landscapes. *Ecological Monographs* **64**(4): 396–400.

Bowden W. B., McDowell W. H., Asbury C. E., and Finley A. M. 1992. Riparian nitrogen dynamics in two geomorphically distinct tropical rainforest watersheds: nitrous oxide fluxes. *Biogeochemistry* **18**: 77–99.

Brokaw N. V. L. 1985. Treefalls, regrowth and community structure in tropical forests. In: S. T. A. Pickett and P. S. White (Eds.) *The Ecology of Natural Disturbances and Patch Dynamics*. Academic Press, New York. 53–69p.

Bruijnzeel L. A. 1990. *Hydrology of Moist Tropical Forests and Effects on Conversion: A State-of-Knowledge Review*. IHP-UNESCO Humid Tropical Programme, Paris.

Brunig E. F. 1969. On the Seasonality of Droughts in the Lowlands of Sarawak (Borneo). *Erdkunde* **23**(2): 127–133.

Bush M. B., Whittaker R. J., and Partomihardjo T. 1995. Colonization and Succession on Krakatau: An Analysis of the Guild of Vining Plants. *Biotropica* **27**(3): 355–372.

Cochrane M. A., Alencar A., Schulze M. D., Souza C. M., Neostad D. C., Lefebvre P., and Davidson E. A. 1999. Positive Feedbacks in the Fire Dynamics of Closed Canopy Tropical Forest. *Science* **284**: 1832–1834.

Cochrane M. A., and Schulze M. D. 1999. Fire as a recurrent event in tropical forests of the eastern Amazon: Effects in Forest Structure, Biomass, and Species Composition. *Biotropica* **31**(1): 2–16.

Coen E. 1983. Climate. In D. H. Janzen (ed.) *Costa Rican Natural History* pp. 35–46. University of Chicago Press. Chicago.

Condit R., Hubbell S. P., Foster R. B., 1995. Mortality rates of 205 tropical tree and shrub species and the impact of a severe drought. *Ecological Monographs* **65**: 419–439.

Connell J. H. 1978. Diversity in tropical rainforest and coral reefs. *Science* **199**: 1302–1210.

Costa J. E. 1984. Physical Geomorphology of Debris Flows. In *Developments and Applications of Geomorphology*. Edited by J. E. Costa and P. J. Fleisher. Springer-Verlag-Berlin Heidelberg. pp. 268–317.

Covich A. P., Crowl T. A., Johnson S. L., and Pyron M. 1996. Distribution and abundance of tropical freshwater shrimp along a stream corridor: response to disturbance. *Biotropica* **28**(4a): 482–492.

Crow T. R. 1980. A rainforest chronicle: a 30-year record of change in structure and composition at El Verde, Puerto Rico. *Biotropica* **12**: 42–55.

Crutcher H. L., Quayle R. G., 1974. *Mariners' Worldwide Climate Guide to Tropical Storms at Sea*. US Dept. Commerce, NOAA/EDS. National Climate Center, Ashville, North Carolina.

Cunningham C. G., Fary R. W., Guffanti M., Laura D., Lee M. P., Master C. D., Miller R. L., Quinones F., Pebbles R. W., Reinemund J. A., and Russ D. P. 1984. Earth and Water Resources and Hazards in Central America. *US Geological Survey Circular* **925**: 1–40.

Curran L. M., Caniago I., Paoli G. D., Astianti D., Kusneti M., Leighton M., Nirarita C. E., and Haeruman H. 1999. Impact of El Niño and logging on canopy tree recruitment in Borneo. *Science* **286**: 2184–2185.

Davis T. A., and Richards P. W. 1933. The vegetation of Moraballi Creek, British Guiana: An ecological study of a limited area of tropical rainforest. Part 1. *Journal of Ecology* **21**: 350–384.

Denslow J. S. 1987. Tropical treefall gaps and tree species diversity. *Annual Review of Ecology and Systematics* **18**: 431–451.

Diaz H. F., and Pulwarty R. S. 1997. *Hurricanes, Climate and Socioeconomic Impacts*. Springer-Verlag Heidelberg.

Dillon W., and U. ten Brink. 1999. Seismic and tsunami hazards in Northeast Caribbean. *EoS* **80**(28): 309–310.

Dittus W. P. J. 1985. The influence of leaf monkeys on their feeding trees in a cyclone-disturbed environment. *Biotropica* **17**: 100–106.

Emanuel K. A. 1987. The dependence of hurricane intensity on tropical cyclones. *Nature* **326**: 483–485.

Emanuel K. A. 1997. Climate variations and hurricane activity: some theoretical issues. In Diaz H. F. and Pulwarty R. S. (Eds.) *Hurricanes, Climate and Socioeconomic Impacts*. Springer-Verlag Heidelberg.

Ferraro R., Vicente G., Ba M., Gruber A., Scofield R., Li Q., and Weldon R. 1999. Satellite techniques yield insight into devastating rainfall from Hurricane Mitch. *EoS* **80**(3): 505.

Flecker A. S., and Feifarek B. 1994. Disturbance and the temporal variability of invertebrate assemblages in two Andean streams. *Freshwater Biology* **31**: 131–142.

Florentin J. M., Maurrasse R., and Gautam S. 1991. Impacts, tsunamis, and the Haitian Cretaceous-Tertiary boundary layer. *Science* **252**: 1690–1692.

Foster R. B. 1990. The Floristic composition of the Rio Manu Floodplain forest. In *Four Neotropical Rainforests* (A. H. Gentry, ed.) Yale University Press, New Haven, pp. 99–111.

Foster R. B., Brokaw N. V. L., 1982. Structure and history of the vegetation on Barro Colorado Island. In Leigh E. G. Jr., Rand A. S., and Windsor, D. M. (eds.) *The Ecology of a Tropical Forest: Seasonal Rhythms and Long-term Changes*. Smithsonian Institution Press, Washington D.C., pp. 67–81.

Francis J. K., Gillespie A. J. R., 1993. Relating gust speed to tree damage in Hurricane Hugo, 1989. *Journal of Arboriculture* **19**(6): 368–373.

Garcia-Montiel D., and Scatena F. N. 1994. The effect of human activity on the structure and composition of a tropical forest in Puerto Rico. *Forest Ecology and Management* **63**: 57–78.

Garwood N., Janos D. P., and Brokaw N. 1979. Earthquake-caused landslides: a major disturbance to tropical forests. *Science* **205**: 997–999.

Garland G. G., and Olivier M. J. 1993. Predicting landslides from rainfall in a humid, subtropical region. *Geomorphology* **8**: 165–173.

Gellis A. 1993. The effects of Hurricane Hugo on suspended sediment loads, Lago Loiza Basin, Puerto Rico. *Earth Surface Processes and Landforms* **18**: 505–517.

Goldammer J. G. 1999. Forests on fire. *Science* **284**: 1782–1783.

Gray W. M., Sheaffer J. D., and Landsea C. W. 1997. Climate trends associated with multidecadal variability of Atlantic hurricane activity. In Diaz H. F., and Pulwarty R. S. (eds.) *Hurricanes, Climate and Socioeconomic Impacts*. Springer-Verlag, Berlin Heidelberg, pp. 15–53.

Guariguata M. R. 1990. Landslide disturbances and forest regeneration in the Upper Luquillo Mountains of Puerto Rico. *Journal of Ecology* **78**: 814–832.

Guilderson T. P., Fairbanks R. G., and Rubenstone J. L. 1994. Tropical temperature variations since 20,000 years ago: modulating inter-hemispheric climatic change. *Science* **263**: 663–664.

Gupta A. 1975. Stream characteristics in eastern Jamaica, an environment of seasonal flow and large floods. *American Journal of Science* **275**: 825–847.

1988. Large floods as geomorphic events in the humid tropics. In Baker V. R., Kochel R. C., and Patton P. C. (eds.) *Flood Geomorphology*. John Wiley and Sons. New York, pp. 301–315.

Gupta A., and Dutt A. 1989. The Auranga: description of a tropical monsoon river. Zeitschrift Geomorphologie, N.F. **33**(1): 73–92.

Hartshorn G. S. 1990. An overview of neotropical forest dynamics. In A. H. Gentry (ed.) *Four Neotropical Rainforests*. Yale University Press, New Haven. pp. 585–599.

Hildebrand A. R., and Boynton W. V. 1990. Proximal Cretaceous–Tertiary boundary impact deposits in the Caribbean. *Science* **248**: 843–846.

Hodel, D. A., Curtis J. H., Jones G. A., Higuera-Gundy A., Brenner M., Binford M. W., and Dorsey K. T. 1991. Reconstruction of Caribbean climate change over the past 10, 500 years. *Nature* **352**: 790–793.

Horn S. P. 1991. Fire history and fire ecology in the Costa Rican Paramos. In *Fire and the Environment: Ecological and Cultural Perspectives*, Proceedings of an International Symposium. Gen. Tech. Rep., SE-69. Asheville NC, USDA Forest Service, Southern Forest Experiment Station, pp. 289–296.

Hubbell S. P., and Foster R. B. 1990. Structure, dynamics and the equilibrium status of old-growth forest on Barro Colorado Island. In A. H. Gentry (ed.) *Four Neotropical Rainforests*. Yale University Press, New Haven, pp. 522–541.

Hulme M., and Viner D. 1995. *A Climate Change Scenario for Assessing the Impact of Climate Change on Tropical Rain Forests*. A Report for the World Wildlife Service by the Climatic Research Unit, School of Environmental Sciences. University of East Anglia, Norwich, UK.

Junk W. J. 1989. Flood tolerances and tree distributions in Amazonian floodplains. In L. B. Nieelsen, I. C. Nielsen, and H. Balslev (eds.) *Tropical Forests: Botanical Dynamics, Speciation, and Diversity*. Academic Press, London, pp. 47–64.

Kalliola R., Saol J., Puhakka M., and Rajasilta M. 1991. New site formation and colonizing vegetation in primary succession on the Western Amazon floodplains. *Journal of Ecology* **79**: 877–901.

Knutson T. R., Tuleya R. E., and Kurihara Y. 1998. Simulated increase of hurricane intensities in a CO_2-warmed climate. *Science* **279**: 1018–1020.

Lamotte S. 1990. Fluvial dynamics and succession inn the Lower Ucayali River Basin, Peruvian Amazonia. *Forest Ecology and Management* **33/34**: 141–156.

Larsen M. C. 1997. Tropical geomorphology and geomorphic work: a study of geomorphic processes and sediment and water budgets in montane humid tropical forested and developed watersheds, Puerto Rico. Ph.D. thesis, University of Colorado.

Larsen M. C., and Simon A. 1993. A rainfall intensity–duration threshold for landslides in a humid-tropical environment, Puerto Rico. *Geografiska Annaler* **75A**: 1–2: 13–23.

Larsen M. C., and Torres-Sanchez A. J. 1998. The frequency distribution of recent landslides in three montane tropical regions of Puerto Rico. *Geomorphology* **24**: 309–331.

Larsen M. C., and Parks J. E. 1997. How wide is a road? The association of roads and mass wasting in a forested montane environment. *Earth Surface Processes and Landforms* **22**: 835–848.

Larsen M. C., Torres-Sanchez A. J., and Concepción I. M. 1999. Slopewash, surface runoff and fine-litter transport in forest and landslide scars in Humid-Tropical Steeplands, Luquillo Experimental Forest, Puerto Rico. *Earth Surface Processes and Landforms* **2**: 481–502.

Lauer W. 1989. Climate and weather. In Leith H. and Werger M. J. A. (eds.) *Tropical Rain Forest Ecosystems*. Amsterdam, Elsevier, p. 7–54.

Lewis, L. A. 1974. Slow movement of earth under tropical rainforest conditions. *Geology* **2**: 9–10.

Lodge D. J., McDowell W. H., and McSwiney C. P. 1994. The importance of nutrient pluses in tropical forests. *Trends in Ecology and Evolution* **9**(10): 384–387.

Lowry J. B., Lee D. W., Stone B. C., 1973. Effect of drough on Mount Kinabalu. *Malayan Nature Journal* **26**: 178–179.

Lugo A. E., and Scatena F. N. 1996. Background and catastrophic tree mortality in tropical moist, wet, and rainforest. *Biotropica* **28**(4a): 585–599.

McDowell, W. H., Bowden W. B., and Asbury C. E. 1992. Riparian nitrogen dynamics in two geomorphologically distinct tropical rainforest watersheds: subsurface solute dynamics. *Biogeochemistry* **18**: 53–75.

McDowell W. H., McSwiney C. P., and Bowden W. B. 1996. Effects of hurricane disturbance on groundwater chemistry and riparian function in a tropical rainforest. *Biotropica* **28**(4a): 577–584.

Meinzer O. E. 1942. *Hydrology*. New York, Dover Publications Inc.

Middleton B. A., Sanchez-Rojas E., Suedmeyes B., and Michels A. 1997. Fire in a tropical dry forest of Central America: a natural part of the disturbance regime? *Biotropica* **29**(4): 515–517.

Mora S., Madrigal C. Estrada J., and Schuster R. L. 1993. The 1992 Rio Toro Landslide Dam, Costa Rica. *Landslide News* August 1993: 19–22.

Morris G. L., and Fan J. 1997. *Reservoir Sedimentation Handbook*. New York, McGraw-Hill.

Mueller-Dombois D. 1987. Natural dieback in forests. *BioScience* **37**: 575–583.

Myster R. W., Thomlison J. R., and Larsen M. C. 1997. Predicting landslide vegetation in patches on landscape gradients in Puerto Rico. *Landscape Ecology* **12**: 299–307.

Myster R. W., and Fernandez D. S. 1995. Spatial gradients and patch structure on two Puerto Rican landslides. *Biotropica* **27**: 149–159.

Nalivkin D. V. 1983. *Hurricanes, Storms, and Tornadoes*. Rotterdam, A. A. Balkema.

Nakagawa M., Tanaka K., Nakashizuka T., Ohkubo T., Kato T., Maeda T., Sato K., Miguchi H., Nagamasu H., Ogino K., Teo. S. Hamia A. A., Lee H. S. 2000. Impact of severe drough assocuiated with the 1997–1998 El Niño in a tropical forest in Sarawak. *Journal of Tropical Ecology* **16**: 355–367.

Neelin J. D., and Latif M. 1998. El Niño Dynamics. *Physics Today*: December 1998, 32–36.

NOAA 1999. *Hurricane Basics*. http://www.nhc.noaa.gov.

O'Brien S. T., Hayden B. P., and Shugart H. H. 1992. Global climatic change, hurricanes, and a tropical forest. *Climate Change* **22**: 175–190.

Olander L. P., Scatena F. N., and Silver W. L. 1998. Impacts of disturbance initiated by road construction in a subtropical cloud forest in the Luquillo

Experimental Forest, Puerto Rico. *Forest Ecology and Management* **109**: 33–49.

O'Loughlin, E. M. 1986. Prediction of surface saturation zones in natural catchment by topographic analysis. *Water Resources Research* **22**: 794–804.

Olsen P. E. 1999. Giant lava flows, mass extinctions, and mantle plumes. *Science* **284**: 604–605.

O'Neill R. V., DeAngelis D. L., Waide J. B., and Allen T. F. H. 1986. *A Hierarchical Concept of Ecosystems*. Princeton, NJ: Princeton University Press.

Ostertag R. 1998. Belowground effects of canopy gaps in a tropical wet forest. *Ecology* **79**(4): 1294–1304.

Pan American Health Organization 1998. *Natural Disasters Mitigation in Drinking Water and Sewage Systems: Guidelines for Vulnerability Analysis*. Disaster Mitigation Series. 110 pages.

Pan American Health Organization 1999. *The Devastating Path of Hurricane Mitch in Central America. A Supplement to Disasters: Preparedness and Mitigation in the Americas*. Newsletter no. 1. http://www.PAHO.org.

Parker G. G. 1985. The effect of disturbance on water and solute budgets on a hillslope tropical rainforest in northeastern Costa Rica. Ph.D. thesis, University of Georgia, Athens.

Perez Suarez R., 1999. *Informe Cientifico del Proyecto de los Ciclones Tropicales de Cuba, Su Variabilidad y Su Posible Vinculacion con Los Cambios Climaticos*. La Habana, Cuba, Cuban Instituto de Meteorologia.

Planos E. 1999. *Analisis Hidrologico de las Grandes Precipitraciones: Suprovincia Fisico-Geografica Cuba Occidental y Central*. La Habana, Cuba, Investigacion del Instituto de Meteorologia.

Plaza-Nieto G., Yepes H., and Schuster R. L. 1990. Landslide dam on the Pisque River, northern Ecuador. *Landslide News*: July 1990, 2–4.

Pounds A., Fogden M. P. L., Campbell J. H., 1999. Biological response to climate change on a tropical mountain. *Nature* **398**: 611–615.

Pringle C. M., Trisha F. J., and Browder G. 1990. Spatial variation in basic chemistry of streams draining a volcanic landscape of Costa Rica's Caribbean slope. *Hydrobiologia* **206**: 73–85.

Pringle C. M. 1991. Geothermally modified waters surface at La Selva Biological Station, Costa Rica: volcanic processes introduced chemical discontinuities into lowland tropical streams. *Biotropica* **24**(4b): 523–529.

Putz F. E. 1983. Treefall pits and mounds, buried seeds, and the importance of soil disturbance to pioneer trees on Barro Colorado Island, Panama. *Ecology* **64**: 1069–1074.

Putz, F. E., Coley P. D., Lu K., Montalvo A., and Aiello A. 1983. Uprooting and snapping of trees: Structural determinants and ecological consequences. *Canadian Journal of Forestry Research* **13**: 1011–1022.

Raghavendra V. K., 1982. A case study of rain storm in the eastern mountanious catchments of Brahmaputra leading to highest ever recorded flood level at Dibrugarh. In *Proceedings of International Symposium on Hydrological Aspects of Mountainous Watersheds*, School of Hydrology, University of Roorkee, Uttar Pradesh, India, pp. I-39 to I-44.

Ramos-Gines O. 1999. *Estimation of Magnitude and Frequency of Floods for Streams in Puerto Rico: New Empirical Models*. US Geological Survey Water-Resources Investigations Report 99–4142. Washington, DC: US Government Printing Office.

Rasanen M. E., Salo J. S., and Kalliola R. J. 1987. Fluvial perturbance in the western Amazon Basin: regulation by long-term, sub-Andean tectonics. *Science* **238**: 1398–1401.

Reagan D. P., and Waide R. B. 1996. *The Food Web of a Tropical Rain Forest*. Chicago, IL, University of Chicago Press.

Richards P. W. 1996. *The Tropical Rain Forest: An Ecological Study*, 2nd edn. Cambridge, UK: Cambridge University Press.

Rivera-Ramirez H. D. 1999. Peak discharge prediction, recession analysis, and evaluation of streamflow response to rainfall for watersheds in the Luquillo Experimental Forest, Puerto Rico. M.Sc. thesis, University of Connecticut, Storres, CT.

Rodbell D. T., Seltzer G. O., Anderson D. M., Abbott M. B., Enfield D. B., and Newman J. H. 1999. A 15,000 year record of El Niño-driven alluviation in southwestern Ecuador. *Science* **283**: 516–520.

Sadler J. P., and Grattan J. P. 1999. Volcanoes as agents of past environmental change. *Global and Planetary Change* **21**: 181–196.

Salo J., Kalliola R., Hakkinen I., Makinen Y., Niemela P., Puhakka K., and Coley P. D. 1986. River dynamics and the diversity of Amazon lowland forests. *Nature* **322**: 254–258.

Sanford R. L., Saldarriaga J., Clark K. E., Uhl C., and Herrera R. 1985. Amazon rain-forest fires. *Science* **227**: 53–55.

Scatena, F. N. 1990 Watershed scale rainfall interception on two forested watersheds in the Luquillo Mountains of Puerto Rico. *Journal of Hydrology* **113**: 89–102

Scatena F. N. 1995. Relative scales of time and effectiveness of watershed processes in a tropical montane rainforest of Puerto Rico. *American Geophysical Union Geophysical Monographs* **89**: 103–111.

Scatena, F. N., Silver W., Siccama T., Johnson A. J., Sanchez M. J., 1993. Biomass and nutrient content of the Bisley experimental watersheds, Luquillo Experimental Forest, Puerto Rico, before and after Hurricane Hugo, 1989. *Biotropica* **25**(1): 15–27.

Scatena, F. N., and Lugo A. E. 1995. Geomorphology, disturbance, and the soil and vegetation of two subtropical wet steepland watersheds of Puerto Rico. *Geomorphology* **13**(1–3): 199–213.

Scatena F. N., Moya S., Estrada C., and Chinea J. D. 1996. The first five years in the reorganization of aboveground biomass and nutrient use following Hurricane Hugo in the Bisley Experimental Watersheds, Luquillo Experimental Forest, Puerto Rico. *Biotropica* **28**(4a): 424–441.

Schaefer D. A., McDowell W. H., Scatena F. N., and Asbury C. E. 2000. Effects of hurricane disturbance on stream water concentrations and fluxes in eight watersheds of the Luquillo Experimental Forest, Puerto Rico. *Journal of Tropical Ecology* **16**: 189–207.

Schellekens J. 2000. Hydrological processes in a humid tropical rainforest: a combined experimental and modelling approach. Ph.D. thesis, Faculty of Earth Sciences, Vrije Universiteit, Amsterdam, The Netherlands.

Schellekens J., Scatena F. N., Bruijnzeel L. A., and Wickel A. J. 1999. Modeling rainfall interception by a lowland tropical rainforest in northeastern Puerto Rico. *Journal of Hydrology* **225**: 168–184.

Schulz XYZ 1960. Ecological studies of the rainforest in Northern Suriname. In De Hulster, I. A., and Lanjouw J. (eds.) *The Vegetation of Suriname*, vol. 2. Utrecht, The Netherlands, Botanical Museum Ryksuniversitet.

Servant M., Maley J., Turcq B., Absy M., Bernac P., Fournier M., and Ledru M. 1993. Tropical forest changes during the late Quaternary in African and South American lowlands. *Global and Planetary Change* **7**: 25–40.

Sheridan R. P. 1991. Nitrogenase activity by Hapalosiphon flexuosus associated with Sphagnum erythrocalx mats in the cloud forest on the Volcano La Soufrière, French West Indies. *Biotropica* **23**(2): 134–140.

Showstach R. 1999. Hazards and geologic controls of stratovolcano collapse. *EoS* **80**(46): 549–550.

Shreve F. 1914. *A Montane Rain-Forest: A Contribution to the Physiological Plant Geography of Jamaica*. Washington, D.C., Carnegie Institution.

Silver W. L., Scatena F. N., Johnson A. H., Siccama T. G., and Watt F. 1996. At what temporal scales does disturbance affect belowground nutrient pools? *Biotropica* **28**(4a): 441–457.

Smith J. M. B., 1979. Vegetation recovery from drought on Mt. Kinabalu. *Malayan Nature Journal* **32**: 341–342.

Struhsaker T. T. 1997. *Ecology of an African Rainforest: Logging in Kibale and the Conflict between Conservation and Exploitation*. Gainesville, FL, University of Florida Press.

Tapin D. R. 1999. Sediment slump likely caused 1998 Papua New Guinea tsunami. EoS **80**(30): xx–xx.

Thompson J., Proctor J., Viana V., Milliken W., Ratter J. A., and Scott D. A. 1992. Ecological Studies on a lowland evergreen rainforest on maraca Island, Roraima, Brazil. 1. Physical environment, forest structure and leaf chemistry. *Journal of Ecology* **80**(4): 705–717.

Toon O. B., Zahnle K., Morrison D., Turco R. P., and Covey C. 1997. Environmental perturbations caused by the impacts of asteroids and comets. *Reviews of Geophysics* **35**(1): 41–78.

Uhl C., Clark K., Dezzeo N., and Maquirrino P. 1988. Vegetation dynamics in Amazonian treefall gaps. *Ecology* **69**: 751–763.

Uhl C., Jordan C. F., Clark K., Clark H., and Herrera R. 1982. Ecosystem recovery in Amazon caatinga forest after cutting, cutting and burning, and bulldozer clearing treatments. *Oikos* **38**: 313–320.

Uhl, C. and Murphy, P. G., 1981. Composition, structure and regeneration of a tierra firme forest in the Amazon Basin of Venezuela. *Tropical Ecology* **22**: 219–237.

Uhl C., and Jordan C. F. 1984. Succession and nutrient dynamics following forest cutting and burning in Amazonia. *Ecology* **65**: 1476–1490.

Urban, D. L., O'Neill R. V., Shugart H. H., 1987. Landscape ecology. *BioScience* **37**(2): 119–127.
van Leeuwen W. M. 1936. *Krakatau 1883–1933, Part A, Botany*. Leiden, The Netherlands, E. J. Brill.
Vertessy, R. A., Hatton, T. J., O'Shaughnessy, P. J., Jayasuriya, M. D. A. 1993. Predicting water yield from a mountain ash (*E. regnans*) forest subject to clearfelling and regeneration. *Journal of Hydrology* **150**: 665–700.
Vitousek P. M., and Denslow J. S. 1987. Differences in extractable phosphorus among soils of La Selva Biological Station, Costa Rica. *Biotropica* **19**: 167–170.
Walker L. R., 2000. Seedling and sapling dynamics of treefall pits in Puerto Rico. *Biotropica* **32**: 262–275.
Walker L. R., Brokaw N. V. L., Lodge D. L., and Waide R. B. 1991. Ecosystem, plant, and animal responses to hurricanes in the Caribbean. *Biotropica* Special Issue **23**(4a): 521.
Walker L. R., Zarin D. J., Fetcher N., Myster R. W., and Johnson A. H. 1996. Ecosystem development and plant succession on landslides in the Caribbean. *Biotropica* **28**(4a): 566–576.
Wallace J. M., and Hobbs P. V. 1977. *Atmospheric Science*. New York, Academic Press.
Walsh R. P. D. 1997. Climate. In P. W. Richards (ed.) *The Tropical Rain Forest*, 2nd edn. Cambridge, UK, Cambridge University Press, pp. 159–205.
Walsh R. P. D, Newbery D. M., 1999. The ecoclimatology of Danum, Sabah in the context of the world's rainforest regions, with particular reference to dry periods and their impact. *Philosophical Transactions of the Royal Society of London* **354**: 1869–1883.
Waterloo M. J. 1994. Water and nutrient dynamics of *Pinus caribea* plantation forests on former grassland soils in southwest Viti Levu, Fiji. Ph.D. thesis, Vrije Universiteit, Amsterdam, Netherlands.
Weaver P. L. 1986. Hurricane damage and recovery in the montane forests of the Luquillo Mountains of Puerto Rico. *Caribbean Journal of Science* **22**: 53–70.
Whigham D. F., Olmsted I., Cabrera Cano E., and Harmon M. E. 1991. The impact of Hurricane Gilbert on Trees, litterfall, and woody debris in a dry tropical forest in the northeastern Yucatan Peninsula. *Biotropica* **23**(4a): 434–442.
White P. S., and Pickett S. T. A. 1985. Natural Disturbances and Patch Dynamics: An Introduction. In Pickett S. T. A. and White P. S. (eds.) *The Ecology of Natural Disturbances and Patch Dynamics*. San Diego, CA, Academic Press, pp. 3–13.
White S. M. 1990. The influence of tropical cyclones as soil eroding and sediment transporting events: an example from the Philippines. *International Association of Hydrological Sciences Publication* **192**: 259–269.
Whitmore T. C. 1984. *Tropical Rain Forests of the Far East*, 2nd edn. Oxford, UK, Clarendon Press.
Whittaker R. J., and Walden J. 1992. Post-1883 ash fall on Panjang and Sertung and its ecological impact. *GeoJournal* **28**(2): 153–171.
Whittaker R. J., Partomihardjo T., and Riswan S. 1995. Surface and buried seed banks from Krakatau, Indonesia: implications for the sterilization hypothesis. *Biotropica* **27**(3): 346–354.
Wolman, M. G., and Miller J. P. 1960. Magnitude and frequency of forces in geomorphic processes. *Journal of Geology* **68**: 54–74.
Woods P. 1989. Effects of logging, drought, and fire on the structure and composition of tropical forests in Sabah, Malaysia. *Biotropica* **21**: 290–298.
Wright S. J., 1992. Seasonal drought, soil fertility and the species density of tropical forest plant communities. *Trends in Ecology and Evolution* **7**: 260–263.
Zarin D. J., and Johnson A. H. 1995a. Base saturation, nutrient cation, and organic matter increases during early pedogenesis on landslide scars in the Luquillo Experimental Forest, Puerto Rico. *Geoderma* **65**: 317–330.
Zarin, D. J., and Johnson A. H. 1995b. Nutrient accumulation during primary succession in a montane tropical forest, Puerto Rico. *Journal of the Soil Science Society of America* **59**: 1444–1452.
Zimmerman J. K., Everham E. M., Waide R. B., Lodge D. J., Taylor C. M., and Brokaw N. V. L. 1994. Response of tree species to hurricane winds in subtropical wet forest in Puerto Rico: implications for tropical tree life histories. *Journal of Ecology* **82**: 911–922.
Zimmerman J. K., Aide T. M., Rosario M., Serrano M., and Herrera L. 1995. Effects of land management and a recent hurricane on forest structure and composition in the Luquillo Experimental Forest, Puerto Rico. *Forest Ecology and Management* **77**: 65–76.
Zou X., and Gonzalez G. 1997. Changes in earthworm density and community structure during secondary succession in abandoned tropical pastures. *Soil Biology and Biochemistry* **29**: 627–629.

20 Spatially significant effects of selective tropical forestry on water, nutrient and sediment flows: a modelling-supported review

N. A. Chappell
Lancaster University, UK

W. Tych
Lancaster University, UK

Z. Yusop
Universiti Teknologi, Malaysia

N. A. Rahim
Forestry Research Institute of Malaysia, Kuala Lumpur, Malaysia

B. Kasran
Forestry Research Institute of Malaysia, Malaysia

INTRODUCTION

Selective forestry is a set of commercial forestry practices that involves the selective removal of particular trees within an 'annual logging coupe' of forest (Conway, 1982). Selective harvesting within 'natural forests' (i.e. those forests that have not been clear-felled for non-forest uses or converted to plantation or agro-forestry) covers a very wide range of practices, including high-lead and tractor yarding, harvesting of only large, commercial trees, protection of riparian vegetation along rivers and protection of forest on very steep hills. As a consequence, the intensities of the impacts on the water environment (i.e. water, nutrient and sediment systems) are expected to be very varied. Some of these impacts can be profound. One of the most significant environmental impacts of all types of forestry operations within the humid tropics is accelerated soil erosion (Bruijnzeel, 1992). The resultant input of sediments into rivers leads to damage to fish populations (Martin-Smith, 1998), reduced quality of water supplies, reductions in channel capacity which affects flood risk and boat traffic (Sheffield *et al.*, 1995), and the inundation of offshore corals (MacDonald *et al.*, 2001).

Development of selective harvesting techniques when applied to natural forests in the tropics are currently being focused on so-called 'Reduced-Impact-Logging' (RIL) or 'closely supervised' methods which aim to improve the 'sustainability' of timber production and reduce wider environmental damage. These RIL procedures include optimising skid trail networks, given a knowledge of the exact location of each tree to be felled, minimising stream crossings, minimising 'skid-trail' earthworks, maintaining canopy-cover over skid trails, construction of water-bars on unused haulage roads and critically, careful supervision of all forestry operations (Abdul Rahim *et al.*, 1997; Pinard *et al.*, 1995; van der Hout, 1999). Additionally, where natural forests have been logged and regeneration has been poor, then 'enrichment planting' of commercial (and non-commercial) trees is beginning to be used as part of selective forestry management (Adjers *et al.*, 1995; Kobayashi *et al.*, 2001).

Research into the water-related impacts of natural forest management is important throughout tropical regions given (i) the economic importance of such forestry, (ii) the desire to identify the least damaging forestry practices, and (iii) the large areal extent of natural forest. For example, within South and South East Asia, natural forests (managed and undisturbed) currently cover 25% of the land area while within the continents of South America and Africa they occupy 47 and 12%, respectively (Iremonger *et al.*, 1997).

Forests, Water and People in the Humid Tropics, ed. M. Bonell and L. A. Bruijnzeel. Published by Cambridge University Press.

Scale and variability

Selective forestry is managed by dividing the forestry concession into annual harvesting coupes, typically 10–50 km^2 in area. At this scale, processes controlling the pathways of water, nutrients and sediments are highly heterogeneous. Tropical water pathways are spatially variable, in part, because subsurface flow often emerges only in near-stream areas, some slope sections have areas of extensive and highly conductive soil piping (Chappell *et al.*, 1998; Jones, 1990), and permeability variations are observed along the soil catena (Chappell and Ternan, 1992; see also Bonell, this volume). Tropical nutrient pathways will be variable because of the heterogeneity of (i) the controlling water-paths and catenal changes in soil chemistry (Whitmore and Burnham, 1969; Dixon, 1986), (ii) weathering rates of different parent rocks, and (iii) the rate of nutrient release/uptake with different vegetation associations. Tropical erosion and sediment delivery will also vary as a result of patterns in the water-paths but also because of changes in the soil stability resulting from changes in the underlying rock (Rahman, 1993), local topography (Larsen and Torres-Sanchez, 1998) or root-anchoring properties of different plant species (Collison and Anderson, 1996). Superimposed on this pattern of natural processes, selective logging will generate new local patterns of land-use including zones of surfaced haulage roads (with slope cuts), skid trails, highlead foci, lightly-impacted forest and protection forest. New patterns of water-related processes, arising from an interaction of their natural distribution and the new land-use patterns, will then result. Clearly, this inherent heterogeneity means that statistically meaningful changes to the rates and distribution of water processes need to be assessed over scales that capture a distribution of the natural and anthropogenically-induced landforms. The scale of the 'experimental catchment' of perhaps 0.1–50 km^2 has been seen by hydrologists and geomorphologists as this fundamental scale of integration (Gregory and Walling, 1973). While the impacts of any intensity of land-use change can be observed at the local or individual landform scale, clearly, it is those land-use practices that *can be demonstrated* to have very significant impacts over 'experimental catchment' scales, which we could call the 'landscape scale', where improvements to the practices would give real economic and environmental benefits. Further, we need to demonstrate that there is indeed a measurable impact at these landscape scales before we need to consider the complex physical and chemical processes that might have led to the change.

Catchment-scale studies

Very few catchment studies have examined the effect of selective forestry on water-related processes (Bruijnzeel, 1989a; 1990; 1992; 1996; Abdul Rahim and Harding, 1992), making it very difficult to separate the individual effects of climatic regime, geological setting and forest type from those of specific selective forestry practices. Even with the results of each catchment experiment being taken only as a 'case study with a set of known environmental variables', the quality of the individual data-sets remains paramount with such limited numbers of studies. Ironically, the need to utilise only high quality data-sets (Bruijnzeel, 1991; 1996) limits further the choice of published results for detailed examination.

The 'quality' of the catchment data-sets and interpretations will be dependent on a number of factors:

(1) The first factor is accuracy of the water, nutrient or sediment variables measured at specific sampling locations. For example, Bruijnzeel (1989a; 1991) has questioned the evapotranspiration and nutrient budgets of many tropical catchment studies because the authors have assumed implicitly that no subsurface water or nutrients cross (surface-defined) catchment divides. Similarly, Douglas *et al.* (1992) note that where tropical sediment budgets have not been derived from data collected by automatic storm-sampling equipment, highflow concentrations will be characterised inadequately and budgets under-estimated.

(2) The second factor is the inability to separate land-use effects from those caused by natural trends and cycles in the local climate (see Mahé *et al.*, Callaghan and Bonell, both this volume). This effect is dependent, in part, on the number of years of record that can be examined before and after the land-use change. The study of Subba Rao *et al.* (1985) on the effect of plantation thinning in Rajpur Forest (North India) is good in this respect, as it utilises nine years of rainfall-runoff data from the pre-thinning period. Despite the length of their records, however, the impact of the plantation thinning on fortnightly water-yield was not observable.

While differences between a manipulated catchment and a nearby 'control' (un-manipulated) catchment can be identified through natural climatic fluctuations, the relative magnitude of the anthropogenic impacts may, however, depend on whether they took place during a wet or a dry period (Abdul Rahim and Harding, 1992; Douglas *et al.*, 1999). Thus use of 'paired catchments' does not remove all of the effects of climate dynamics.

(3) The third factor is a difference in size of the manipulated and control catchment. Experimental catchments with similar rainfall, soil/rock type and vegetation, but small differences in size, perhaps less than a factor ten, may have a different balance of surface and subsurface flow processes (Chappell *et al.*, 1999a). This will affect the sensitivity of the catchment to land-use change, with a catchment having greater proportions of surface flow being more sensitive to terrain modifications.

(4) The fourth factor is the danger of interpolating the effects for selectively-managed forest from the behavioural range derived from (a) catchments in an undisturbed state, and (b) after clearfelling. This is because the selective removal of trees under a selective management system may not give the same effect as partial clearfelling (i.e. removal of large patches of forest) even where the same regional timber yields are obtained. There are several reasons for this. First, the selective removal of large, commercial trees leaves younger trees, saplings and seedlings that rapidly take advantage of the new micro-climate generated. This means that in many areas, the ground surface is soon protected from impacting rain-drops, which may increase infiltration and reduce erosion (Douglas *et al.*, 1995). Further, some colonising vegetation may have greater transpiration and wet-canopy evaporation rates in comparison to areas completely cleared of vegetation (Swank *et al.*, 1988; Bidin, 2001; Chappell *et al.*, 2001; Restom and Nepstad, 2001; see also Roberts *et al.* and other chapters in Part III of this volume). A second important issue why selective forestry should not be considered as 'partial clearfelling' relates to the haulage road network. With selective forestry operations in natural forest the road network is constructed with the 'consideration' (though not complete assurance) of access during the next phase of selective logging perhaps 30–60 years ahead. As a result, it is often considered better to built roads close to ridge tops rather than in the wetter valley floors. In contrast, clearfelling may only require access during the single clearance process and hence not necessitate the construction of roads with any longevity.

(5) The fifth factor is the length of observations of the impact. Most studies on the impact of selective forestry are carried out over only the one to three years of the haulage road construction, harvesting operations and immediate post-logging phase. These studies clearly allow only limited assessment of the persistence of the impacts, an issue critical in assessing the 'environmental sustainability' of forestry operations. Issues that are important in this respect include: (i) the re-vegetation of skid trails and consequent reduction in overland flow and erosion (Douglas *et al.*, 1995), (ii) accelerated growth of vines and pioneer trees with different wet-canopy evaporation (Chappell *et al.*, 2001) or transpiration rates (Becker, 1996; Eschenbach *et al.*, 1998; Davies, 1998), and (iii) persistent instability of road cuts (Chappell *et al.*, 1999a).

(6) The sixth factor is the quality of the records detailing the type of forestry practices adopted, timber yields extracted and the spatial distribution of the extraction systems (i.e. skid trails, log landings and highlead foci). Newly established catchment studies have benefited from recent moves towards 'Reduced Impact Logging' (RIL) and 'certified' harvesting systems that has meant that forest management agencies and companies have improved the quality of their forest management records.

Given these issues, we seek here to assimilate the results of studies of selective forestry impacts on catchment-scale water, nutrient and sediment flows in tropical natural forests, and also to examine the value of data-based modelling in the assimilation of the most reliable case studies.

WATER FLOWS

Selective forestry may affect the pathways of precipitation (e.g. rainwater and fog-drip) from the tropical rainforest canopy and terrain to the river or back to the atmosphere as evapotranspiration.

Evapotranspiration, catchment water-balance and water yield

An integral component of the selective logging of natural forests is the construction and use of lorry haulage roads (sometimes stone-surfaced), 'skid trails' (i.e. the tracks used by logging tractors), and 'log-landing areas' where timber is loaded on to the lorries. If these surfaces are impacted by frequent use, vegetation re-growth may be inhibited locally. Rates of total evaporation (evapotranspiration) from these bare earth surfaces are likely to be less than those from the vegetated surfaces that they replace. Additionally, 'highlead yarding', where timber is dragged from all directions to a central mast (Conway, 1982), leaves patches (perhaps 20–50 m in diameter) of shrubs (e.g. Zingiberaceae), herbs and sprawlers distributed about the natural forest (Bidin, 2001; Chappell *et al.*, 2001). This ground vegetation may have transpiration rates less than high forest with its greater rooting depth and leaf area index (Roberts *et al.*, 1993; Roberts *et al.*; Scott *et al.*, both this volume). Recent measurement and modelling work in Guyanan rainforest, South America (van Dam, 2001), has indeed suggested that the selective removal of climax trees leads to a reduced total evaporation in the immediate area.

The effect of clearfelling experimental catchments covered by most types of tropical natural forests is well attested – complete clearance results in greater riverflow or 'water yield' (Oyebande, 1988; Bruijnzeel, 1990; 1996; 2001). The possible exception to this is montane forest with its high precipitation input as fog-drip, though the data are too limited for generalisation (Bruijnzeel, 1996; 2001; Hafkenscheid, 2000; see Bruijnzeel, this volume). As selective logging generates smaller gaps in the forest, where new growth of pioneer trees, accelerated growth of younger and smaller commercial trees and/or vine growth takes place, the marked increases in water yield seen with clearfelling climax trees may be partly offset by the rapid growth of water-demanding pioneer trees and vines. Restom and Nepstad (2001) and Eschenbach *et al.*

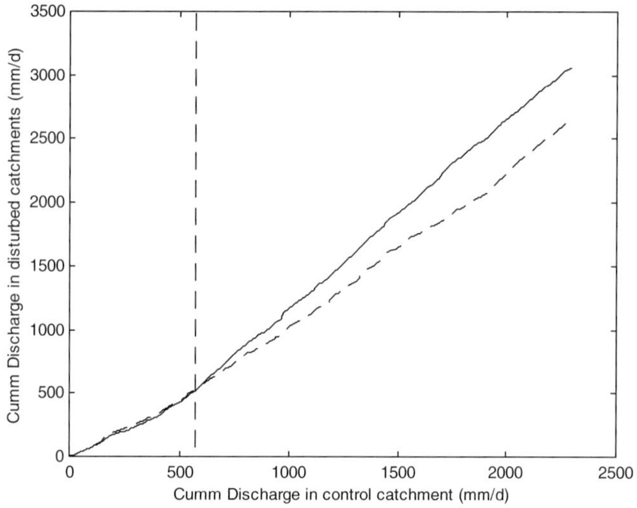

Figure 20.1 The double-mass curve of riverflows for the conventionally-logged catchment versus the control catchment (black line), and RIL catchment versus the control catchment (broken line), Bukit Berembun, Negri Sembilan, Malaysia. The vertical dashed line indicates the start of forestry activities (July 1983) within the catchments.

(1998) have noted high rates of transpiration from vines in secondary Amazonian forest and pioneer trees in East Malaysian rainforest, respectively.

The most reliable catchment water-balance study that captures the effects of selective logging of tropical natural forests is that undertaken within the Bukit Berembun catchments in Peninsular Malaysia. Three catchments (2°46′ N, 102°6′ E) were installed within 'lowland dipterocarp rainforest' on Acrisol soils derived from a weathered granite geology (Abdul Rahim, 1990; Abdul Rahim and Zulkifli, 1994). One catchment was kept as a natural control ('C2'), one was selectively-harvested by 'unsupervised' methods ('C1') and one selectively-harvested by 'supervised' or 'Reduced-Impact-Logging' (RIL) procedures ('C3'). The 'San-Tai-Wong' logging method (which involves the use of both tracked-skidders and winch lorries) was used within both harvested catchments. The data-series derived from these catchments were chosen for analysis, given that high quality riverflow data, collected at 120° V-notch weirs, is available for a pre-logging phase and a relatively long record for the logging and 'terrain recovery' phase.

The double-mass curve (Searcy and Hardison, 1960) of the two selectively-logged, Bukit Berembun river catchments *versus* the control catchment shows an increased water-yield from the beginning of the road construction and harvesting in July 1983 through to the end of intensive monitoring in 1989 (Figure 20.1). The catchment with the greater harvesting intensity (i.e. 40% timber extraction rather than 33%; Abdul Rahim and Harding, 1992) resulted in a greater increase in water-yield. This selective-logging produced a gradual increase in water-yield over the initial harvesting period, compared with the step change in the double-mass-plot observed for the clearfelled Sungai Tekam 'Catchment A', also in Peninsular Malaysia (see Figure 20.5 in Abdul Rahim, 1988). Interestingly, the greater water-yields following selective forestry persisted over the six years of intensive monitoring after the harvesting phase (Figure 20.1). Bidin (2001) examining catchment water yield of a selectively managed forest in East Malaysia from one to eight years post-disturbance, could not detect a change in the water balance with forest recovery (above the changes associated with climate dynamics). Over the whole post-logging period, average water-yield (1984–1989) of the commercially-logged, Bukit Berembun catchment increased 1.47-fold (+129 mm) and the RIL-catchment 1.24-fold (+66 mm) compared with the undisturbed, control catchment.

In some contrast to the Bukit Berembun study, Jetten (1994) noted no significant change in catchment water balance following light selective harvesting of Guyanan rainforest of South America. While not being directly applicable to the impacts of selective forestry on water balance in tropical natural forests, the light logging (20 m^3 ha^{-1}) catchment study of Gilmour (1977) in Queensland, Australia, and the 20% thinning of a *Shorea* plantation covering a catchment in northeast India (Subba Rao *et al.*, 1985) similarly failed to note observable changes in the water balance following timber harvesting. Failure to observe water balance changes directly after timber extraction may have been explained by accelerated growth of existing trees within the new light environment, offsetting the effect of the removal of climax trees.

The relatively small increase in water yield within Bukit Berembun might be explained by a catchment-average reduction in the transpiration and/or wet canopy evaporation. The work of van Dam (2001) in Guyana shows that in the gaps created by selective harvesting, transpiration may be reduced in comparison with that of the original climax trees. Thus reduced transpiration from the greater number of canopy gaps could explain the increase in water yield following selective felling. Other plot-based work within selectively-managed forest in Indonesian Borneo suggests that catchment water-yields might be increased by reduced rates of wet-canopy evaporation (or 'interception-loss') when climax trees are removed (Asdak *et al.*, 1998). A more recent study conducted in similar selectively-managed forest in neighbouring Malaysian Borneo has, however, shown that rates of wet-canopy evaporation can be greater in highly damaged patches of rainforest compared with that in remnants of climax forest (Bidin, 2001; Chappell *et al.*, 2001). Clearly, new studies undertaken through the first cycle of selective logging of natural forests are needed to apportion catchment-scale evapotranspiration more definitively into the components of transpiration and wet-canopy evaporation.

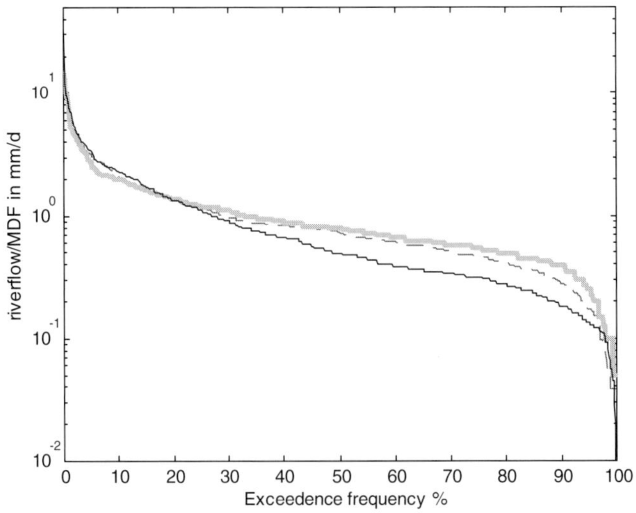

Figure 20.2 The flow-duration curves using data for the 1981–2 period for the control catchment (grey line), the conventionally-logged catchment (broken black line), and the RIL catchment (solid black line). Bukit Berembun riverflow is normalised by the mean daily flow (MDF).

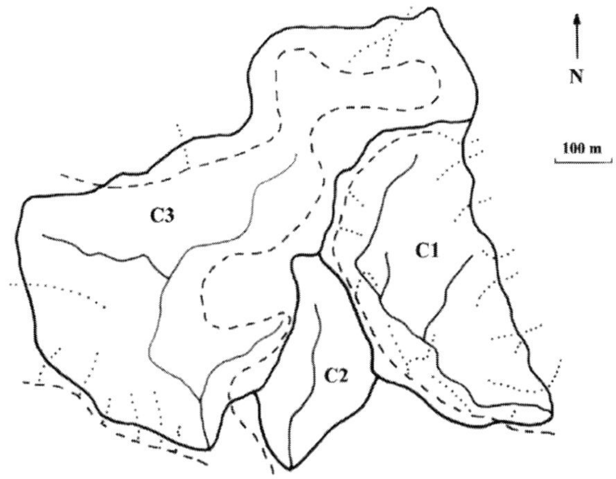

Figure 20.3 A map of the Bukit Berembun C1, C2 and C3 catchments in Peninsular Malaysia, showing catchment divides (wide, solid lines), streams (narrow, solid lines), timber haulage roads (broken lines) and skidder trails (dotted lines).

Flow-paths and rainfall-runoff behaviour

The dynamic characteristics of a river are integrated measures of the varied responses of complex water-pathways draining to that river. If there are substantial changes in catchment water-paths resulting from forestry operations, then they will be reflected in (i) the riverflow dynamics and (ii) the mathematical relationship between the incoming rainfall and outgoing riverflow.

The construction and use of forestry haulage roads and skid trails has been demonstrated to have a *local impact* on rates of infiltration into Humic Acrisols in Costa Rica (Spaans *et al.*, 1990), Haplic Acrisols (Malmer and Grip, 1990) and Alisols (Van der Plas and Bruijnzeel, 1993) in East Malaysia, and Ferralsols in Guyana (Jetten *et al.*, 1993). This has then had a local impact on the amounts of water travelling over the ground surface as 'infiltration-excess overland flow' (Baharuddin, 1995; see Bonell, this volume). The key issue is whether these impacts, which may be significant at certain points within a catchment or forest-logging-coupe, are extensive enough to impact at the landscape-scale itself and thereby affect the river behaviour.

The range of flows within a river can be characterised graphically by estimation of the flow-duration curve or FDC (Searcy, 1959). A river's flashiness can then be estimated from the slope of a specified segment of the FDC. If we examine the data for the Bukit Berembun catchments again – prior to any forestry activities (1981–2) – the three Bukit Berembun catchments have different natural flow regimes (Figure 20.2). The C3 catchment was much more flashy prior to RIL logging compared with the other two catchments. The smallest of the catchments, the 4.6 ha C2, had the greatest lowflows per unit catchment area (in relative and absolute terms) and the largest, 30.8 ha C3, had the least lowflow. This is not expected, as water that has percolated to a significant depth has more likelihood of returning to the ground surface before, or 'upstream' of, a river gauging structure as catchment size increases up to 1 km^2 (Chappell *et al.*, 1999a). As the smallest catchment (C2) is on only the lower slopes of the hill that all three catchments occupy (Figure 20.3), it may be that deep (and slow) preferential flow (within the weathered granite) from the other catchments is feeding C2 rather than their own lower catchments.

Following selective harvesting, both conventional and RIL logging techniques produced an increase in the highflows as expressed in the Q10 statistic – the riverflow equalled or exceeded for 10% of the time (Table 20.1). In the case of the conventional logging (C1 catchment), Q10 flows were increased by 1.43-fold directly after harvesting, at a time when the control catchment highflows reduced slightly. These effects are larger than those observed by the studies of Subba Rao *et al.* (1985) and Gilmour (1977), though the studies are not directly comparable. Subba Rao *et al.* (1985) monitored a 9% increase in peakflow in the first year of a 20% thinning of a *Shorea* plantation catchment in north India, while Gilmour (1977) failed to observe any change in peakflows during the light logging catchment study in Queensland, Australia.

The ratio of the riverflow observed for at least 30% of the time to that observed for more than 70% of the time – the Q30/Q70 statistic, has been used as a measure to characterise the 'flashiness' of river regimes within temperate catchments (Ward, 1981). The greater the value of the statistic, the greater the flashiness of the river behaviour. Table 20.2 shows these statistics calculated from the FDCs for the Bukit Berembun riverflows (normalised by mean daily flow) (Figure 20.4). The undisturbed, control catchment (C2) maintained a very similar Q30/Q70 statistic throughout

Table 20.1. *Daily riverflow (mm/d) equalled or exceeded for 10% of the time (Q_10 statistic describing highflows) within the Bukit Berembun experimental catchments*

	C1 Unsupervised logging (13.3 ha)	C2 Control (4.6 ha)	C3 Reduced-impact logging (RIL) (30.8 ha)
Pre logging (1981–2)	1.4562	1.5224	1.6001
Logging and recovery (1984–5)	2.0802	1.4285	1.7489
relative change from 1981–2 period	Increase 1.430-fold	Reduce 0.938-fold	Increase 1.090-fold
Logging and recovery (1984–5)	1.5224	1.6001	1.9706
relative change from 1981–2 period	Increase 1.045-fold	Increase 1.051-fold	Increase 1.232-fold

Table 20.2. Q_{30}/Q_{70} *River flashiness statistics for the Bukit Berembun experimental catchments; riverflows in the FDC are normalised by the respective mean daily flows (MDF)*

	C1 Unsupervised logging (13.3 ha)	C2 Control (4.6 ha)	C3 Reduced-impact logging (RIL) (30.8 ha)
Pre logging (1981–2)	1.9815[a]	1.9568[a]	2.6905[a]
difference to 'control'	0.0247	–	0.7337
direction of difference	Same	–	Much more flashy
Logging and recovery (1984–5)	2.2039	1.9545	2.4851
absolute change[b]	0.2214	0.0023	0.2054
direction of change	Slightly more flashy	Same	Slightly less flashy
Logging and recovery (1984–8)	1.9390	1.9524	2.1783
absolute change[b]	0.0425	0.0044	0.5122
direction of change	Same	Same	Much less flashy

[a] A greater value indicates an increase in 'riverflow flashiness' (i.e., a steeper curve).

[b] Absolute value of the Q_{30}/Q_{70} statistic for 1981–2 minus that for either 1984–5 or 1984–8.

the whole study period. The conventionally logged C1 catchment became slightly more flashy immediately after harvesting activities but then recovered to the pre-logging condition within a few years (Figure 20.4b). In some contrast, the river regime within the RIL logged catchment (C3) became and remained more damped (Figure 20.4c) in comparison with the pre-disturbance condition.

The contrasting behaviour of the two selective-logging systems is interesting. The increase in lowflows relative to highflows following selective-logging of the C3 catchment may be attributable to reduced transpiration which takes place throughout inter-storm periods. While absolute values of lowflow have increased with logging in the C1 catchment, the overall response may have become more flashy because of greater increases in the higher flows. This may have been because of the greater lengths of indurated or compacted road surfaces with the conventionally logged C1 catchment compared with the supervised RIL harvesting in C3. The length of haulage roads and skid trails in the conventionally logged C1 is 0.14 km ha^{-1} but 0.10 km ha^{-1} within RIL (Abdul Rahim and Harding, 1992). This may have led to greater quantities of rapid flowing surface flows on road surfaces or in roadside drains that would give slightly more peaked river hydrographs (see e.g. Macdonald *et al.* (2001) work in the US Virgin Islands). With vegetation re-growth on skid trails and unsurfaced haulage roads, surface flow would reduce, as demonstrated in the East Malaysian plot studies of Douglas *et al.* (1995), thus reducing storm peaks. The slightly more flashy behaviour of the conventionally logged C1 catchment could also be explained by a marked reduction in wet-canopy evaporation which then allowed much more water to enter the catchment system during storms (cf. Asdak *et al.*, 1998). With vegetation recovery, rates of wet-canopy evaporation would then increase again, damping storm behaviour within rivers. While this second explanation is consistent with the results of C1, it would not be totally consistent with those of C3, so we suggest that the mechanism of small changes to the quantities of surface flow is the more likely explanation.

Interpretation of flow-duration curves is somewhat dependent on the particular sections of the FDC that are used to derive the statistics. Dynamic modelling could be used to derive the

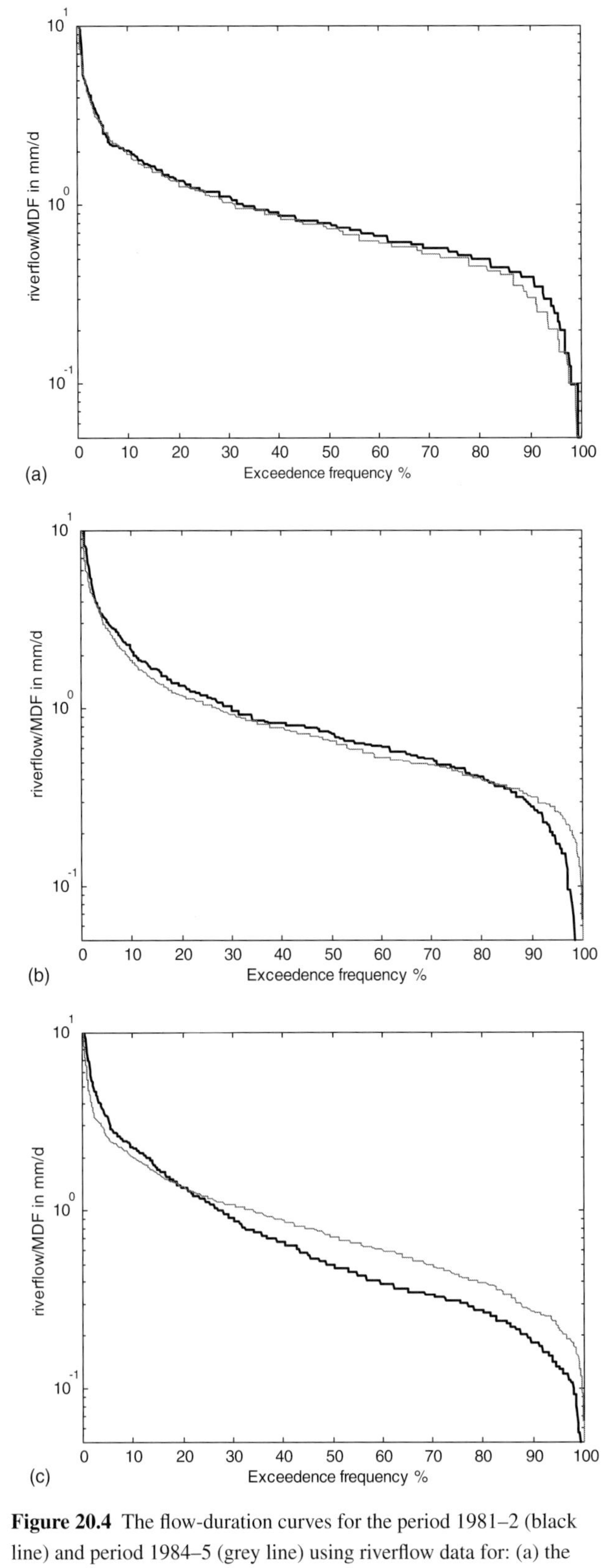

Figure 20.4 The flow-duration curves for the period 1981–2 (black line) and period 1984–5 (grey line) using riverflow data for: (a) the control catchment, (b) the conventionally-logged catchment, and (c) the RIL catchment. All Bukit Berembun data are normalised by the respective mean daily flows (MDFs).

whole range of response characteristics within a single model. Indeed, a model's parameters can be described as the model's 'Dynamic Response Characteristics' or DRCs (Jakeman *et al.*, 1993; Post and Jakeman, 1996). Any changes in these parameters following the establishment of forestry may be caused by the subsequent vegetation and/or terrain modifications. Young and Beven (1994) state that such parameter interpretation is, however, only successful with parsimonious modelling approaches (i.e. those with simple model structures, requiring only few parameters). This is because of the increasingly acknowledged problem of parameter interaction during the identification process. Data-Based-Mechanistic (DBM) modelling is one such parsimonious approach. The DBM technique combines physically-based understanding of system behaviour with model-structure identification based on linear transfer functions and objective statistical inference (Young *et al.*, 1997). Under the Wheater *et al.* (1993) classification of catchment hydrological models, the DBM model, like the IHACRES model of Jakeman and Hornberger (1993), is a type of 'hybrid metric-conceptual model'. Middleton (2000) provides a good introduction to modelling with transfer functions. Elsewhere in this volume, Schreider and Jakeman and Barnes and Bonell elaborate on IHACRES and other DBMs and their links with DRCs.

The DBM modelling approach was applied (within MATLAB) to the same daily data-series for the conventionally-logged Bukit Berembun experimental catchment (C1) for a two-year period (1981–2) just prior to the start of forestry in July 1983, and a two-year period shortly after the activity had started (1984–5). The non-linearity in the catchment behaviour was characterised using the 'Store-Surrogate Sub-Model' (SSSM; Young and Beven, 1994; Chappell *et al.*, 1999a; Young, 2001). Many different model structures were then used to attempt to describe the linear component of the relation between incoming rainfall and the outgoing riverflow. The 'best model' structure was the one that had the highest efficiency (cf. Nash and Sutcliffe, 1970), while maintaining a large negative YIC value (Young Information Criterion: Young, 2001). Small negative or positive YIC values indicate that the information content of the observed data-series is insufficient to justify the level of complexity (i.e. number of model parameters) of that particular model structure.

A simple, first-order model structure was seen to best capture the rainfall – riverflow behaviour of the C1 catchment during the first two-year period, 1981–2 (Figure 20.5a.), and explained 89% of the variance in the riverflow dynamics. In transfer function form, the optimal model structure was:

$$q(k) = \frac{P_0 - P_1 z^{-1} - P_2 z^{-2}}{1 - \mathfrak{R} z^{-1}} r_{\mathrm{eff}}(k) \qquad (20.1)$$

where $q(k)$ is the riverflow at the time index k, $\mathfrak{R}$ is the recession or lag parameter, P_0, P_1 and P_2 are the three system production or

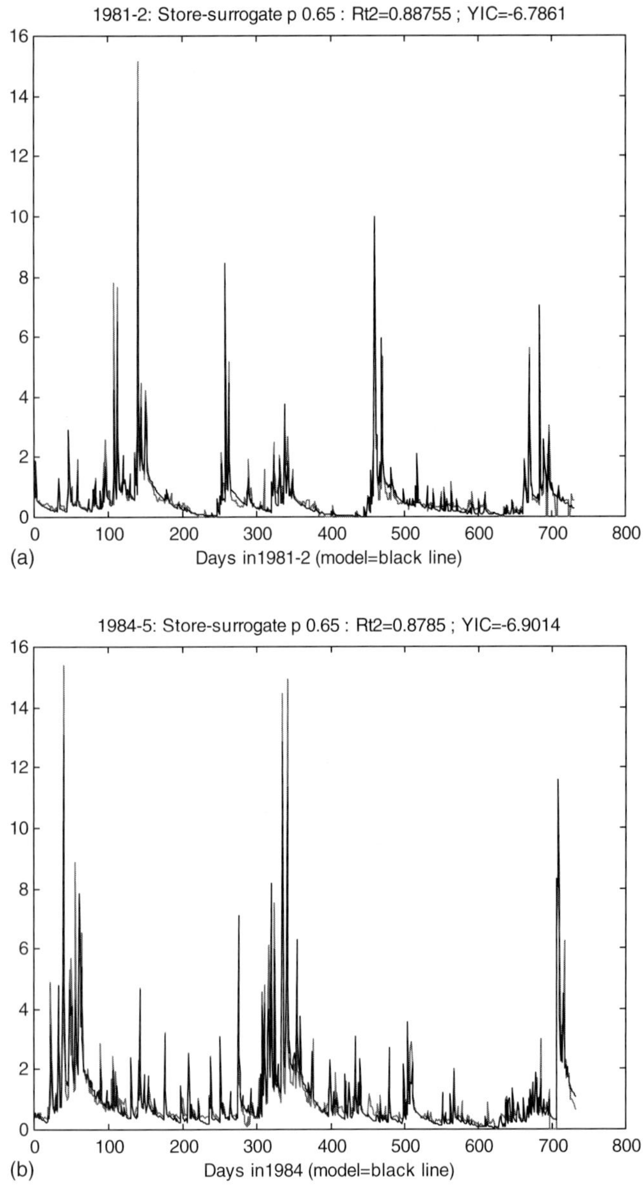

Figure 20.5 Observed (grey broken line) and simulated (black solid line) riverflow generated by the Bukit Berembun C1 catchment during: (a) 1981–2 and (b) 1984–5.

gain parameters, z^{-1} is the backward shift operator (i.e., $z^{-i}\, r(k) = r(k-i)$) which allows the expansion to higher-order models, and r_{eff} is the transformed rainfall input (Young, 1984). A pure time delay to the initial response (δ) was not necessary with the Bukit Berembun data-series. The model shown with the parameters of the linear component of the model applied to the C1 1981–82 data-series is:

$$q(k) = \frac{0.0393 - 0.0276z^{-1} - 0.0068z^{-2}}{1 - 0.9575z^{-1}} r_{\text{eff}}(k) \qquad (20.2)$$

The initial transformation on the rainfall input to describe the behavioural non-linearity (r_{eff}), was described by:

$$r_{\text{eff}}(k) = r(k)\{q(k)^{\beta}\}; \text{ where } \beta = 0.65 \qquad (20.3)$$

where r_{eff} is the transformed input, $r(k)$ is the catchment-average rainfall at time index k, $q(k)$ is riverflow at time index k, and is the estimate of the power-law exponent (after Young and Beven, 1994). To maintain the same mass balance, the transformed input was then normalised in relation to the catchment-average rainfall to give the 'normalised transformed input' (see Chappell *et al.*, 1999a). This non-linear transform, described as the 'Store-Surrogate Sub-Model', captures the non-linear effects resulting from subsurface water storage (Young and Beven, 1994; Young *et al.*, 1997), wide macro-micropore flow distributions and/or layered soils.

Two key DRCs of this overall model structure capture the catchment's responsiveness – the power-law exponent of the non-linear transform (β) and the recession or lag parameter ($\Re$). As the power-law exponent was held constant (at 0.65) for both time periods (1981–2 and 1984–5), any change in the catchment responsiveness will be compounded within the recession parameter ($\Re$). The recession parameter is usually presented in terms of a time constant (TC), where:

$$\text{TC} = \frac{-t_{\text{base}}}{\log_e(-\Re)} \qquad (20.4)$$

and t_{base} is the time-base of the data-series (i.e. daily time-steps for the Bukit Berembun data-series). This TC term can be equated either with the residence time of the water within the whole catchment (Young, 1992) or within the saturated downslope soil-rock.

The resultant TC or residence time for the C1 catchment prior to conventional selective logging was 23.059 ± 0.189 days. This is a relatively long residence time and probably relates to the percolation of a significant proportion of water deep into the underlying weathered granite (Abdul Rahim, 1990; George, 1992) before emerging in the river. For comparison, the similarly-sized Baru catchment in East Malaysia that was within relatively impermeable mudstones (Chappell *et al.*, 1998), had a residence time of only 47 minutes with a similar model structure (Chappell *et al.*, 1999a).

During the logging and immediate post-logging period, the model efficiency remained similarly high, with 88% of the riverflow variance being explained (Figure 20.5b). Critically, the residence time for the logging period remained virtually unchanged at 23.469 ± 0.167 days. Therefore, by characterising changes in the catchment's responsiveness with a single characteristic that describes the whole range of behaviour, no significant change in the river's flashiness is observed. This characteristic, and hence associated interpretations, are likely to be more robust than characteristics derived from only local parts of the flow-duration-curves described earlier. We might then ask, why do phenomena such as forestry road construction not always have a significant impact on catchment responsiveness? First, tropical rainforest slopes in Africa (e.g. Dabin, 1957), South America (e.g. Cailleux, 1959) and SE Asia (e.g. Chappell *et al.*, 1999a) usually generate only a few per cent overland flow per unit slope area. Perhaps the exceptions are those areas experiencing very intense tropical cyclones (e.g. Bonell *et al.*, 1983; Bonell, this volume). Secondly, the surface area of haulage roads or (well-used) skid trails that would increase the proportion of overland flow, occupies a relatively small area of the typical size of experimental catchment. For example, they occupy only 2.1% and 4.9% of the Baru and Bukit Berembun C1 catchments, respectively. Thirdly, visual observations during extreme storms (N. A. Chappell, *pers. observ.*) indicate that even where overland flow is generated on road surfaces, a significant proportion will drain on to surrounding slopes where it will infiltrate rather than enter a stream channel directly.

Research needs

While the impact of selective forestry on water yield seems relatively small compared with the impact of clearfelling and conversion from natural forest (cf. Bruijnzeel, 1990; 1996; 2001), understanding the relative role of changes in the transpirational losses compared to changes in the wet-canopy evaporation losses remains uncertain. New studies undertaken through the first cycle of selective logging of natural forests are needed to apportion catchment-scale evapo-transpiration accurately into the components of transpiration and wet-canopy evaporation. Given the complexity of the canopy generated by selective harvesting, these studies may need to combine *direct measurements* of: (i) catchment-wide, transpiration (cf. Sellers *et al.*, 1995), (ii) catchment-wide, wet-canopy evaporation (cf. Bidin, 2001), and (iii) lumped losses from catchment scale, precipitation-minus-riverflow (P-Q) data (cf. Hudson *et al.*, 1997). As the Bukit Berembun catchment failed to show a return to pre-logging water-yields some six years post activity, *long records post-logging are important* for this analysis (cf. Swank *et al.*, 1988).

The analysis of flow-duration-curves and a data-based numerical model did not give fully consistent results, though large, sustained impacts on river flashiness were not observed by either method. The small inconsistency in the interpretations may, at least in part, be related to uncertainty in whether there is a sufficient volume of surface flow localised on newly created road surfaces to short-circuit the catchment's subsurface system and affect the river hydrograph directly. New catchment experiments with road and trail networks fully instrumented to measure a significant proportion of all of the road-related surface flows are needed to provide more concrete evidence of road impacts on river responsiveness.

NUTRIENT FLOWS

Most hydro-chemical studies examining the impacts of temperate and tropical forestry indicate that timber harvesting results in accelerated nutrient flows along catchment rivers (e.g. Bruijnzeel, 1990; Stevens *et al.*, 1995; Swank, 1988). This has the potential to both (a) reduce the fertility of the catchment system being harvested (Attiwill and Weston, 2001), and (b) increase the likelihood of downstream eutrophication (Tundisi, 1990). Data on the effect of selective tropical forestry on catchment-scale nutrient flux in natural forests are, however, restricted to a very few studies. The most rigorous of these were the studies undertaken within the Bukit Berembun catchments in Peninsular Malaysia (Yusop, 1989) and a 6.2 ha catchment in the Mabura Hill area of Guyana (Brouwer, 1996). Pertinent data are also available from the treatments applied to the West Creek Catchment, Surinam (Poels, 1987) and the Baru Catchment, East Malaysia (Douglas *et al.*, 1992). The plot-scale work of van Dam (2001) in Guyana, and the detailed nutrient budget for the already disturbed Bukit Tarek catchment in Peninsular Malaysia (Yusop, 1996) provide further insight into the likely processes of change.

Harvesting year impacts

The Bukit Berembun study demonstrates that during the year of timber harvesting, the flux of river-dissolved macro-nutrients (nitrate, phosphate, potassium, calcium and magnesium) increase by 1.7 to 5.6-fold where conventional selective logging is practised, and by 1.2 to 2.1-fold where 'closely-supervised / RIL' techniques are practised (Table 20.3). The process studies of van Dam (2001) in selectively logged Guyanan rainforest indicate that the accelerated nutrient flux results from (i) an increase in the amount of 'percolation water', (ii) a decrease in the nutrient uptake by the vegetation remaining in canopy gaps, and (iii) an increase in the amount of decomposable nutrients contained in the crowns of felled trees.

If the change in the losses associated with the enhanced suspended-sediment flows (and other non-dissolved losses) are taken into account, then the change in the total flux of nutrients may be a little greater. Working in the Bukit Tarek Catchments also in Peninsular Malaysia, Yusop (1996) demonstrated that only 2 to 19% of the macro-nutrient load was in the form of non-dissolved material. However, it should be noted that these catchments were selectively-logged 40 years prior to his study so that the rates of

Table 20.3. *Accelerated macro-nutrient flux from the Bukit Berembun rivers following selective timber harvesting*

Catchment (harvesting method)	Harvesting year (1 July 1983–30 June 1984)	3-year 'recovery' period (1 July 1984–30 June 1987)
C1	+ 5.6-fold NO_3[b,c]	+ 2.0-fold NO_3
(commercial)[a]	+ 3.0-fold PO_4	+ 2.3-fold PO_4
	+ 2.4-fold K^+	+ 2.1-fold K^+
	+ 1.8-fold Ca^{2+}	+ 1.8-fold Ca^{2+}
	+ 1.7-fold Mg^{2+}	+ 1.8-fold Mg^{2+}
C3	+ 2.1-fold NO_3	+ 1.6-fold NO_3
(closely supervised)[a]	+ 1.2-fold PO_4	+ 1.5-fold PO_4
	+ 1.3-fold K^{2+}	+ 1.2-fold K^{2+}
	+ 1.3-fold Ca^{2+}	+ 1.2-fold Ca^{2+}
	+ 1.4-fold Mg^{2+}	+ 1.1-fold Mg^{2+}

[a] 'Commercial logging' is also known as 'conventional' or 'unsupervised' selective logging, while 'closely supervised logging' is also known as 'reduced-impact logging (RIL)'.
[b] Macro-nutrient flux is equivalent to nutrient load.
[c] Increase in the flow of dissolved macro-nutrients (nitrate, phosphate, potassium, calcium and magnesium) over the Bukit Berembun river gauging structures within C1 and C3 relative to that in C2, the control catchment. Some bias may be incorporated due to natural differences between the control and other catchments.

sediment delivery are likely to be much less than from a period of forest harvesting. An additional source of uncertainty in the rates of change of nutrient flux is the presence of subsurface nutrient flows across catchment divides defined only by surface topography (Bruijnzeel, 1991). The small size of the Bukit Berembun C1 catchment (0.133 km^2) and the possible presence of preferential flow within the weathered granite bedrock makes such inter-basin nutrient flows a possibility (cf. earlier discussion on C2 lowflows.)

The much shorter nutrient records from a 6.2 ha catchment in the Mabura Hill area of Guyana gave a doubling of potassium and nitrate flux directly after the light selective harvesting (21 m^3 ha^{-1} : Brouwer, 1996). This change is broadly consistent with those observed for the Bukit Berembun Catchments. In some contrast, however, a clear change in the phosphorus, calcium and magnesium losses could not be observed at Mabura Hill. Reported rates of acceleration of nutrient loss for the West Creek Catchment, Surinam, in the year affected by a 'refinement process' were similarly more modest than those for the harvesting year at Bukit Berembun. A 0.29-fold increase in potassium load, a 0.15-fold increase in calcium load and a 0.07-fold increase in magnesium was observed at West Creek (Poels, 1987). These data are, however, much more uncertain, given that (a) a paired catchment approach was used without calibration, (b) water-flows were assumed to be unaltered by the forestry, and (c) the disturbance process was one of 'refinement' (a process that involved the cutting of lianas and poisoning of non-commercial trees) of an already selectively-logged forest, rather than one of selective logging of virgin forest (Bruijnzeel, 1992).

Recovery

During the harvesting year in Bukit Berembun, it is the nitrate losses via riverflow that increase the most. These accelerated nitrate losses do, however, return to near natural conditions within six months of the cessation of the harvesting activities (Table 20.4 column 6) (Yusop, 1989). The other nutrients took between three and five years to return to conditions where nutrient flux was only slightly elevated above the natural condition (Table 20.4). Such rapid recovery of the nutrient losses indicates a catchment with a high 'ecosystem resilience' (Swank, 1988), with accelerating biological processes rapidly utilising the additional nutrients released to soil-water (Prof. R. Jones, pers. comm.). Further, it is interesting to note that clearfelling and burning of already selectively-logged forest in East Malaysia resulted in a similarly fast rate of recovery in the rates of catchment-scale leaching (Malmer and Grip, 1994).

The long-term inputs of macro-nutrients into the subsurface waters of a catchment derive from within (above canopy) rainfall and the weathering of soil and rock. Comparison of these 'natural rates' of input with (a) the loss of catchment nutrients in hauled timber, and (b) via the accelerated export of nutrients in riverflow (Table 20.4) demonstrates that it is the losses in timber that have the most significant impact on catchment fertility. Yet, quantification of the nutrient content of all of the timber removed from a river

Table 20.4. *Natural rates of input of macro-nutrients into subsurface water, export in harvested timber and additional mass of macro-nutrient flux from the Bukit Berembun rivers following selective timber harvesting*

Catchment (harvesting method)[a]	Input: natural rates of new nutrient delivery to sub-surface water (in rainfall and by natural weathering)[b]	Export in harvested timber[c]	Additional export in riverflow during harvesting year[d] (1 July 1983–30 June 1984)	Additional export in riverflow during 'recovery' period[d] (1 July 1984–30 June 1987)	Recovery rate[e]	C1 vs. C3 (recovery period)[f]
C1	11 kg N-total ha^{-1} yr^{-1}	–	+ 2.0 kg NO_3 ha^{-1} yr^{-1}	+ 0.4 kg NO_3 ha^{-1} yr^{-1}	0.8	+ 1.6 fold
(commercial)	0.4 kg P ha^{-1} yr^{-1}	–	+ 0.21 kg PO_4 ha^{-1} yr^{-1}	+ 0.1 kg PO_4 ha^{-1} yr^{-1}	0.5	+ 2.5 fold
	19 kg K^+ ha^{-1} yr^{-1}	200 kg K^+ ha^{-1}	+ 14.5 kg K^+ ha^{-1} yr^{-1}	+ 12.1 kg K^+ ha^{-1} yr^{-1}	0.2	+ 5.8 fold
	15 kg Ca^{2+} ha^{-1} yr^{-1}	45 kg Ca^{2+} ha^{-1}	+ 5.0 kg Ca^{2+} ha^{-1} yr^{-1}	+ 5.2 kg Ca^{2+} ha^{-1} yr^{-1}	−0.04	+ 4.3 fold
	9 kg Mg^{2+} ha^{-1} yr^{-1}	20 kg Mg^{2+} ha^{-1}	+ 1.8 kg Mg^{2+} ha^{-1} yr^{-1}	+ 2.5 kg Mg^{2+} ha^{-1} yr^{-1}	−0.4	+ 12.5 fold
C3	11 kg N-total ha^{-1} yr^{-1}	–	+ 0.53 kg $NO_3{}^1$ ha^{-1} yr^{-1}	+ 0.25 kg NO_3 ha^{-1} yr^{-1}	0.5	
(closely supervised)	0.4 kg P ha^{-1} yr^{-1}	–	+ 0.04 kg PO_4 ha^{-1} yr^{-1}	+ 0.04 kg PO_4 ha^{-1} yr^{-1}	0.0	
	19 kg K^+ ha^{-1} yr^{-1}	165 kg K^+ ha^{-1}	+ 2.8 kg K^+ ha^{-1} yr^{-1}	+ 2.5 kg K^+ ha^{-1} yr^{-1}	0.1	
	15 kg Ca^{2+} ha^{-1} yr^{-1}	37 kg Ca^{2+} ha^{-1}	+ 1.8 kg Ca^{2+} ha^{-1} yr^{-1}	+ 1.2 kg Ca^{2+} ha^{-1} yr^{-1}	0.3	
	9 kg Mg^{2+} ha^{-1} yr^{-1}	16 kg Mg^{2+} ha^{-1}	+ 1.2 kg Mg^{2+} ha^{-1} yr^{-1}	+ 0.2 kg Mg^{2+} ha^{-1} yr^{-1}	0.8	

[a] 'Commercial logging' is also known as 'conventional' or 'unsupervised' selective logging, while 'closely supervised logging' is also known as 'Reduced-Impact Logging (RIL)'.

[b] Rates from Table 20.2 in Bruijnzeel (1992), where the rate of 'natural weathering' is the long-term average rate of nutrient export in riverflow under natural conditions minus the long-term average rate of nutrient input from (above-canopy) rainfall.

[c] Masses from Bruijnzeel (1995).

[d] Rates from Yusop (pers. comm.).

[e] An index of the rate of improvement post the harvesting period relative to the additional losses in the harvesting period. Calculated from: *(additional export in riverflow during harvesting period – additional export in riverflow during 'recovery' period) / additional export in riverflow during harvesting period.* A larger positive number indicates a more rapid recovery from the impact experienced during the harvesting year, while a larger negative value indicates even greater losses during the 'recovery' period in comparison to the harvesting period.

[f] Impact in 'recovery period' of commercial (unsupervised) harvesting versus closely supervised (RIL) harvesting. Calculated from: *(additional export in riverflow during 'recovery' period in C1) / additional export in riverflow during 'recovery' period in C3.*

catchment or harvesting coupe is perhaps even more uncertain than quantifying accelerated nutrient flux in rivers (Table 20.4) (Nykvist, 1994).

Research needs

It is clear that there is a particularly acute dearth of studies on the selective forestry impacts on catchment-scale nutrient export. Further nutrient sampling programmes within existing (or new) catchments is, therefore, important (Bruijnzeel, 1989b). The rate of longer-term (i.e. 5–30 years) biochemical recovery, and hence the true ecological sustainability of the forestry practices adopted, can be estimated only roughly from short-term studies (Yusop, 1996), and would benefit from long-term sampling programmes, such as those at the Coweeta Experimental Watershed, in southeast USA (cf. Swank, 1988).

The rate of catchment-scale nutrient flux is controlled more by variations in the waterflow than in the nutrient concentration (Swank, 1988). Indeed, within undisturbed forest catchments, the annual fluctuations in the nutrient export are correlated strongly with (a) the proportion of rainfall-generating discharge, and (b) the discharge. As a result, separating relatively small but long-term accelerated rates of nutrient loss (by catchments affected by selective forestry) from the natural cycles in nutrient flux is very difficult. New studies that combine the modelling of cycles and trends in rainfall–runoff behaviour with those in nutrient behaviour (e.g. Eshleman, 2000), should more accurately quantify longer-term impacts on nutrient flux related to selective forestry operations.

SEDIMENT FLOWS

Several reviews indicate that the spatial and temporal variation in annual, net sediment flux for large tropical catchments could be as large as three to four orders of magnitude (Douglas, 1996; Milliman *et al.*, 1999; Walling and Webb, 1983). Depending on what are the dominant controlling factors, this may indicate that catchment sediment flows are sensitive to terrain disturbance during forestry or with other land-use activities. As discussed by Douglas and Guyot elsewhere in this volume, the factors controlling the rate of soil detachment, sediment flux and deposition include: (1) the local relief, properties of the soil and rock materials and the presence of tectonic activity, (2) the Intensity-Duration-Frequency (IDF) characteristics of the rainfall and the presence of marked cyclical behaviour, and (3) anthropogenic disturbance to the soil/rock and vegetation cover during forestry or other activities. Clearly, the relative impact of different selective forestry operations within natural forest need to be measured against the variations expected between different relief-geology and climatic regimes within the tropics. Additionally, the impact of each selective forestry practice may differ between contrasting geological and/or climatic settings.

Relief and geological controls

Walling and Webb (1983) reviewed sediment delivery data from almost 1 500 large catchments across the globe. These analyses indicate that the highest rates of erosion and sediment delivery in the tropics ($>1\,000\,\mathrm{t\,km^{-2}\,yr^{-1}}$) occur in (i) the Cordillera Central-Andes range of South America, (ii) Taiwan Island (China), (iii) the Hongha catchment (northern Vietnam), (iv) Java Island (Indonesia), and (v) the Aure catchment (Papua New Guinea). Steep slopes are an important factor in Cordillera Central-Andes mountains, Papua New Guinea and Java (Walling and Webb, 1983; Pickup *et al.*, 1981), and the dominance of Andosols derived from pyroclastic parent materials is also important in Java (see also Hardjowitjitro, 1981). The lowest rates of erosion and sediment delivery in the tropics ($<50\ \mathrm{t\ km^{-2}\ yr^{-1}}$), as mapped by Walling and Webb (1983), cover (i) the Congo catchment (Central Africa), and (ii) the north-western and southern headwaters of the Amazonian basin. These regions are dominated by stable Ferralsol soils and have relatively low relief (see also Douglas and Guyot, this volume).

Rainfall regime controls

On a global basis, the annual rainfall total is of lesser importance in determining the spatial patterns of erosion and sediment delivery than is the relief-geological control (Walling and Webb, 1983 p80). Rainfall is, however, critical to the understanding of the temporal variability in tropical erosion and sediment delivery, with extreme rainfall events often being responsible for most of the sediment flows (as discussed in Douglas and Guyot, Scatena *et al.*, both this volume). For example, the five largest storms during 1987–9 generated 45 and 54% of the suspended-sediment flux in the 19.9 $\mathrm{km^2}$ Batangsi and 12.5 $\mathrm{km^2}$ Chongkak catchments, respectively, in Peninsular Malaysia (Lai, 1992). Similarly, a single storm event on the 19 January 1996 mobilised 43% of the suspended-sediment flux over the period 1 July 1995 to 30 June 1996 from the 0.44 $\mathrm{km^2}$ Baru catchment in East Malaysia (Chappell *et al.*, 1999a). The impact of extreme events are also observed over significantly larger space and time scales. For example, the eight-year records for the 721 $\mathrm{km^2}$ Ulu Segama catchment, gauged close to the Baru catchment, show that extreme storms occurring on only six separate days (or 0.2% of the time-series), mobilised 25% of the suspended-sediment (Douglas *et al.*, 1999). Extreme events are important to erosion and sediment delivery within the tropics (and elsewhere) because they (i) trigger new mass movements along channels (Balamurgan, 1997) and on slopes (Chappell *et al.*, 1999a, b; Larsen and Torres-Sanchez,

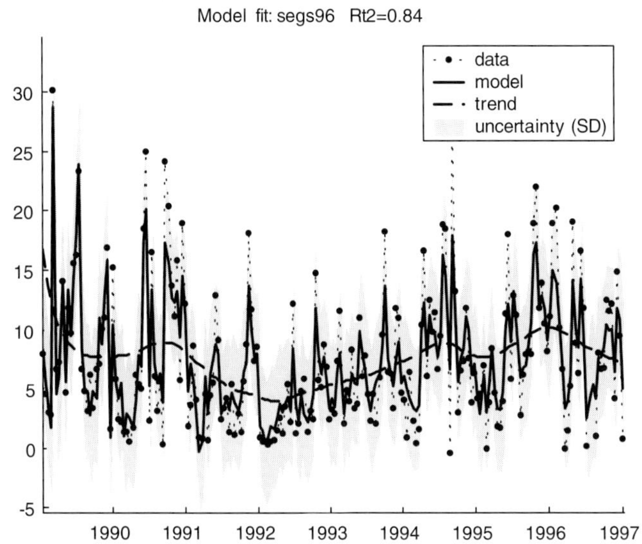

Figure 20.6 Results of the DHR modelling of the daily values of suspended-sediment concentration generated by the 721 km^2 Ulu Segama catchment against the measured data (both in rainfall equivalents).

1998), (ii) markedly expand the contributing areas of sediments mobilised by surficial erosion (Croke *et al.*, 1999), and mobilise channel-bed sediments (Swanston and Swanson, 1976).

In addition to the sediment time-series being punctuated by the effects of extreme events, the impacts of cyclicity in tropical rainfall may be observed. Given that the sediment is mobilised by rainfall, greater sediment delivery would be expected during peaks in these cycles. To illustrate this effect, the daily suspended-sediment records for the 721 km^2 Ulu Segama catchment, East Malaysia (Ian Douglas, pers. comm.) are analysed using the Dynamic Harmonic Regression (DHR) model. This model is a recursive interpolation, extrapolation and smoothing algorithm for non-stationary time-series, and identifies three components in the time-series: (i) the trend, which includes inter-annual cyclicity and longer-term drifts, (ii) the within-year cycles or 'seasonality', and (iii) white noise (see Appendix). For this analysis, the relative magnitude of the cyclicity and drift dynamics in the daily suspended-sediment concentration and flux (1989–97) are compared with those in the daily rainfall records (1986–98) monitored at the DVFC meteorological station within the Ulu Segama catchment. It is important to know whether seasonal and/or inter-annual, cyclical phenomena such as the El Niño Southern Oscillation (ENSO) (see Callaghan and Bonell; Mahé *et al.*, this volume), often observed within tropical rainfall records, are damped or magnified in the sediment records. The suspended-sediment concentration and flux data were, therefore, linearly scaled to the mean of the rainfall drift.

Application of models to the rainfall, suspended-sediment concentration and suspended-sediment flux for the Ulu Segama river produced model efficiencies (sensu. Nash and Sutcliffe, 1970) of 80%, 84% and 83%, respectively. Figure 20.6 shows the model of the suspended-sediment concentration plotted with the measured data.

The modelling indicates that the rainfall records monitored at the DVFC meteorological station exhibit marked cyclical behaviour (Figure 20.7) about the mean rainfall of 7.4 mm day^{-1}. The within-year or 'seasonal' components of this cyclicity, amounting to about 3 to 5 mm day^{-1} (Figure 20.7a), are dominated by 12 and 1 month cycles (see Bidin, 2001; Chappell *et al.*, 2001). The inter-annual cyclicity within the relatively short records for DVFC amounts to about 2 mm day^{-1} and has a periodicity of about five years (Figure 20.7b), which is coincident with the El Niño Southern Oscillation within the region (Chappell *et al.*, 2001). The variations in the longer-term drift over the 1986–98 period are not statistically significant (Figure 20.7c).

Like the rainfall, the within-year cyclicity in the records of the suspended-sediment concentration and flux are dominated by 12 and 1 month cycles, however, the strength of the cyclicity is magnified, being 5 to 20 mm day^{-1} rainfall equivalents in the case of the suspended-sediment flux (Figure 20.7d). The inter-annual cyclicity, presumably related to short-term ENSO behaviour, is also magnified in the suspended-sediment behaviour, having a magnitude of about 4 mm day^{-1} (Figure 20.7e). Figure 20.7f shows that the longer-term drift component of the suspended-sediment flux increases slightly post-1992. While the monitoring period is still too short (1989–97) to be conclusive, the increasing drift in suspended-sediment flux may be a physical phenomenon, given that a 16 km^2 downstream area of the 721 km^2 Ulu Segama catchment was selectively-logged (with a timber yield of 103 m^3 ha^{-1}) by RIL techniques in 1993 (Greer *et al.*, 1996; Pinard *et al.*, 1995). This short period of forestry activity within the catchment was some years after the selective logging of the headwaters of the Ulu Segama in the 1970s (Greer *et al.*, 1996), and so may have punctuated the recovery from the earlier activity.

Magnification of the cyclicity means that the effects of seasonality and ENSO phenomena have a much greater impact on annual sediment budgets than would be expected from the dynamics in the rainfall. As a consequence, quantification of the impact of different selective forestry operations on the sediment yield becomes strongly dependent on the season and position within the ENSO cycle that the operations take place. Road construction and harvesting conducted within the peak of the short-term ENSO cycle (La Niña period) would be expected to have a greater relative impact on the sediment delivery compared to the same operations taking place within the ENSO trough (El Niño period).

The Baru Catchment, just downstream of the gauged 721 km^2 Ulu Segama watershed, East Malaysia, was disturbed by selective forestry between August 1988 and June 1989 which was coincident with an ENSO peak. The catchment generated a 5.3-fold

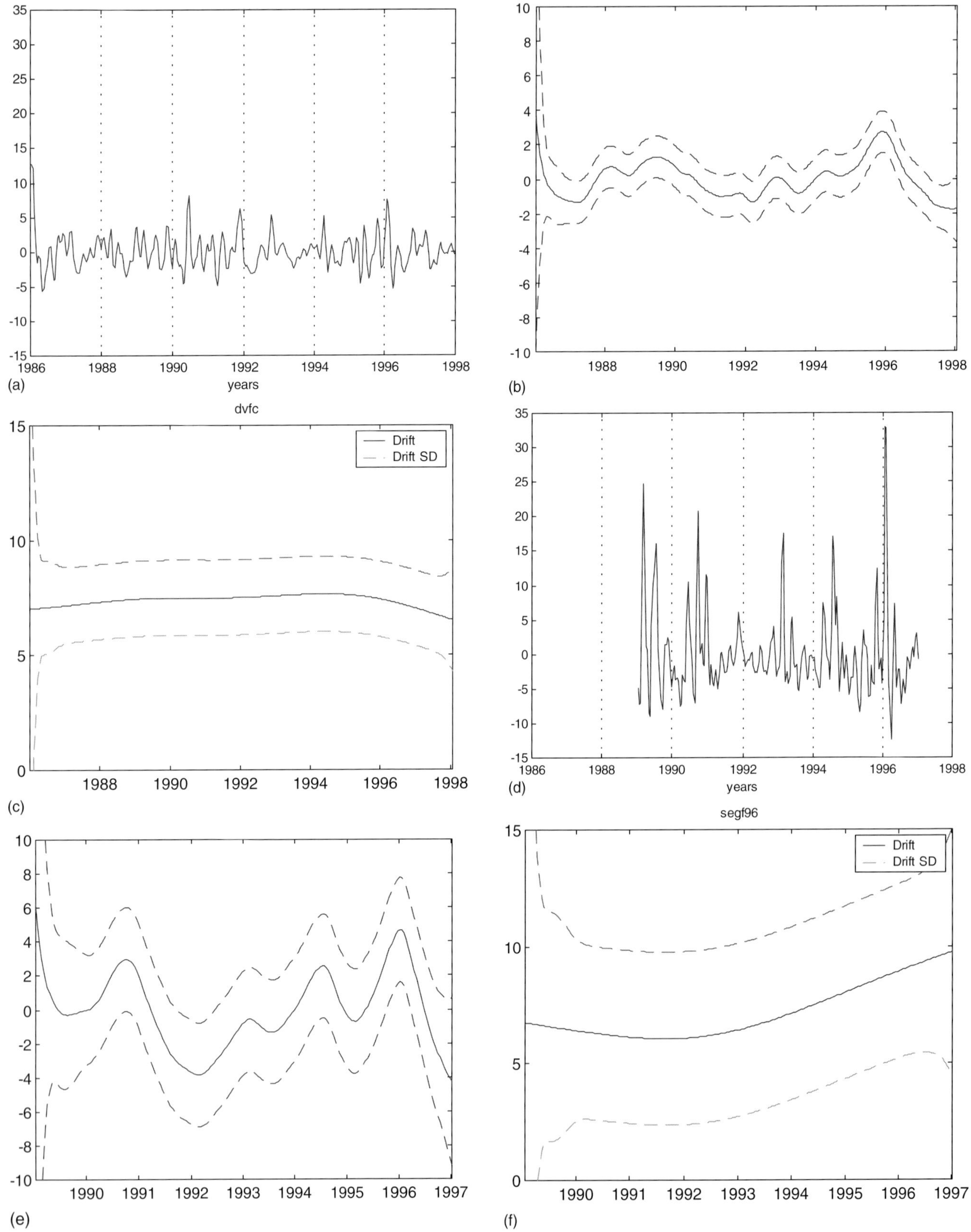

Figure 20.7 Results of the DHR modelling of rainfall and sediment trend and cycles in the Ulu Segama region of Sabah, Malaysia. (a) Seasonal cyclicity in the daily rainfalls (mm d^{-1}) at the DVFC meteorological station, Sabah, Malaysia. (b) Inter-annual cyclicity in the DVFC rainfall (mm d^{-1}). (c) Inter-annual drift in the DVFC rainfall (mm d^{-1}). (d) Seasonal cyclicity in the daily flux of suspended-sediment (in mm d^{-1} rainfall equivalents) generated by the 721 km^2 Ulu Segama catchment. (e) Inter-annual cyclicity in the daily flux of suspended-sediment (in mm d^{-1} rainfall equivalents). (f) Inter-annual drift in the daily flux of suspended-sediment (in mm d^{-1} rainfall equivalents).

Table 20.5. *Selective natural forestry impacts on catchment-scale suspended-sediment delivery*

Study number	1	2	3	4
Key reference(s)	Douglas *et al.* (1992), and Greer *et al.* (1996)	Douglas and Bidin (1994)	Lai (1992)	Lai *et al.* (1999)
Name	Baru	Jauh	Batangsi and Chongkak	Lawing
Size (km^2)	0.44	1.5	19.9 and12.7	4.7
Location	Sabah, Malaysian Borneo	Sabah, Malaysian Borneo	Peninsular Malaysia	Peninsular Malaysia
Forest type	Lowland dipterocarp	Lowland / hill dipterocarp	Hill dipterocarp	Hill dipterocarp
Relief	14–260 m, undulating	125–725 m, steep	1082 and 1265 m relief, steep	1210 m relief, steep
Slope	8°	16°	22° and 23°	24°
Geology	Siltstone / sandstone	Serpentinite	Granite and granodiorite	Granite
Rainfall regime	Equatorial	Equatorial	Equatorial	Equatorial
Practice	Conventional	Reduced-impact logging	Conventional	–
Timber yield	80 m^3 ha^{-1} (coupe 89)[a]	–	–	–
Forestry period	1988–1989	Apr–May 1992	1985– and 1944–	1993
Sediment components[b]	Suspended only	Suspended only	Suspended only	Suspended only
Forestry active period:				
Record year(s)	1988–89	May 1992 –Nov 1993	1987–9 and 1987–8	1994
Disturbed (and control) (t km^{-2} yr^{-1})	1600 (300)	431 (100)	2826 and 2476 (54)	1389 (–)
Difference	5.3-fold greater	4.3-fold greater	52- and 46-fold greater	20-fold greater
Recovery period:				
Record year(s)	1993	–	–	–
Disturbed (and control) (t km^{-2} yr^{-1})	283 (38)	–	–	–
Difference	7.5-fold greater	–	–	–

[a] Hamzah Tangki, pers. comm.

[b] The sediment rating approach of Gilmour (1977) indicated that a partial cut of the 0.183 km^2 North Creek catchment (Queensland, Australia) increased the suspended sediment flux by 2-fold.

increase in suspended-sediment delivery (1 600 t km^{-2} yr^{-1} versus 300 t km^{-2} yr^{-1}) during that forestry period in comparison with the nearby control catchment (Table 20.5) (Douglas *et al.*, 1992). Another nearby catchment, the 1.5 km^2 Jauh, was selectively harvested in the relatively dry period of 1992. This catchment generated a slightly smaller, 4.3-fold, increase in suspended-sediment delivery (431 t km^{-2} yr^{-1} versus 100 t km^{-2} yr^{-1}) (Table 20.5) during that period compared with the Rafflesia control catchment (Douglas and Bidin, 1994). While timing of the logging activities in the ENSO rainfall cycle between these two catchments may account for part of the small difference in the rate of accelerated sediment delivery, unfortunately, the catchments differ in two other important aspects. First, the Baru Catchment is within a region of highly unstable Alisol soils (Chappell *et al.*, 1999b) with landsliding being present even on undisturbed slopes. In contrast, the Jauh catchment has steep, but very thin, mountain soils which after some initial erosion of road surfaces leave resistant, Serpentinite rock surfaces (K. Bidin, pers. comm.). Secondly, while the Baru Catchment was managed by conventional, selective techniques, forestry in the Jauh catchment was by RIL methods which reduced the extent of haulage roads and skid trails, involved careful supervision of buffer zones, and gazetted large tracts of the catchment as too steep for logging (Douglas and Bidin, 1994; K. Bidin, pers. comm.; Douglas *et al.*, 1999). As a consequence, the difference in selective forestry impact may be related to the natural climatic cycles, relief-geology or selective forestry operations, or a combination of all three factors. Indeed, such a complex, equifinite situation was observed with a 'partial-clearfell' forestry study in a montane region of Puerto Rico (Larsen *et al.*, 1997).

Forestry land-use controls

Generalisation of the catchment-scale impacts of different practices of selective logging of tropical natural forests is very difficult given the dearth of experimental studies, and the effects of variations in the relief/geology and rainfall controls just described. The authors are aware of only four studies addressing catchment-scale, selective logging impacts on natural forests, and these were all undertaken within the same tropical country (Table 20.5). These studies indicate that the delivery of suspended-sediments may increase by 2–50-fold in the periods of road construction and selective timber harvesting. Douglas *et al.* (1992) and Chappell *et al.* (unpublished data) have suggested that selective forestry operations generate new sediment sources, notably road gullies, rain-splash and surficial wash on haulage roads and skidder-vehicle trails, collapses along streams (particularly at road crossings), landslides in cleared areas, landslides in road-cut materials, and soil piping under roads. The limited catchment-scale data available (Table 20.5) makes more precise quantification of the 10-fold range unrealistic, making the need for new studies to quantify which sediment sources make up most of the changes to catchment-scale sediment flux most acute (Bruijnzeel, 1996). A key issue that needs to be addressed at the catchment scale is the degree to which RIL methods lessen the physical impacts seen with conventional, unsupervised methods of selective forestry.

A further, often overlooked issue (see Douglas and Guyot, this volume), is the contribution of bedload to the total sediment budget. During the 1987–89 period, bedload accounted for an additional 1 367 and 1.6 t km^{-2} yr^{-1} of the total sediment delivery in the selectively-logged Batangsi and Chongkak catchments, respectively, and a further 125 t km^{-2} yr^{-1} in the then undisturbed Lawing catchment (Lai, 1992). Bedload data have not been collected routinely within the two other catchments (Baru and Jauh) listed in Table 20.5. Given the large difference in bedload delivery seen between the two disturbed catchments, Batangsi and Chongkak (with their similar relief, geology, vegetation and scale), exptrapolation of these results to catchments without such data would be unrealistic, yet the bedload component may be critical to the quantification of the forestry impacts on the total sediment delivery.

The annual suspended-sediment delivery for the 721 km^2 Ulu Segama catchment, East Malaysia, derived from daily riverflow and concentration data from 1989 to 1996, is 306 t km^{-2} yr^{-1}. This rate results from the history of selective-logging of over 400 km^2 of the southern headwaters primarily in the 1970s and a more recent harvesting of a further 16 km^2 in 1993, combined with the effects of natural ENSO rainfall cycles. As this scale contains annual logging coupes spanning tens of years and, thus, terrain at a range of stages from road construction to recovery with some persistent impacts, it probably provides a better estimate of sediment flux at larger time and space scales than do the results of small experimental catchments, such as those presented in Table 20.5. It should be remembered, however, that with increasing scale generally comes a reduction in mean channel slope and, therefore, an increase in the in-channel storage or residence time of the sediment (Dietrich and Dunne, 1978). This may damp the local effects associated with individual forest management coupes. Even with this scale effect and the more complex land-use histories of larger catchments, they do provide results that complement those from small-scale 'experimental catchment studies' (Singh, 1998).

CONCLUSIONS

At the 'field-scale' of say 1 hectare, a high degree of heterogeneity is observed in the natural pattern of water, nutrient and sediment flows (Bidin, 2001; van Dam, 2001; Chappell *et al.*, 1999a,b), which is then further compounded by the localised nature of forest/terrain disturbance associated with all forms of selective forest management. This means that observation of statistically meaningful changes to the rates and distribution of water processes (i.e. water, nutrient and sediment flows) need to be made only after integration over the scale of at least the 0.1–50 km^2 'experimental catchment'.

This modelling-supported review of catchment-scale data shows that all forms of selective management of natural forests in the tropics have observable, if not always large, impacts on water yield and pathway-dynamics, nutrient leaching and sediment mobilisation. The relatively modest impacts might be summarised as: (i) catchment water-yield is increased, but by less than a factor of two, (ii) the rate of migration of rainfall through the catchment system to the river and hence the 'river responsiveness' may be increased, but only slightly and will be short-lived, and (iii) nutrient flows increase by a factor of one to six in the harvesting year. Larger relative change resulting from selective forest development is seen within the suspended-sediment flux, which seems to increase by a factor of 2 to 50, though the limited set of reliable data make interpretation of this wide range unrealistic. This makes it imperative that we obtain further data-series of the impacts of the various selective forestry practices (particularly those currently considered to be 'Reduced Impact Logging' methods) on the mobilisation of sediments (and associated loss of water quality and downstream sedimentation). Such an impact not only affects aquatic ecology but also the economic and social livelihoods of all river users.

This chapter also demonstrates that precise quantification of ecohydrological change is often difficult due to large natural dynamics in the driving hydroclimatic regime, and persistence of some forestry impacts over several years. This shows that future research programmes need to observe data over perhaps 10 to 30 years, and need to use the latest modelling technologies to separate the dynamics of particular selective forestry practices (notably Reduced-Impact-Logging practices) from those of other selective

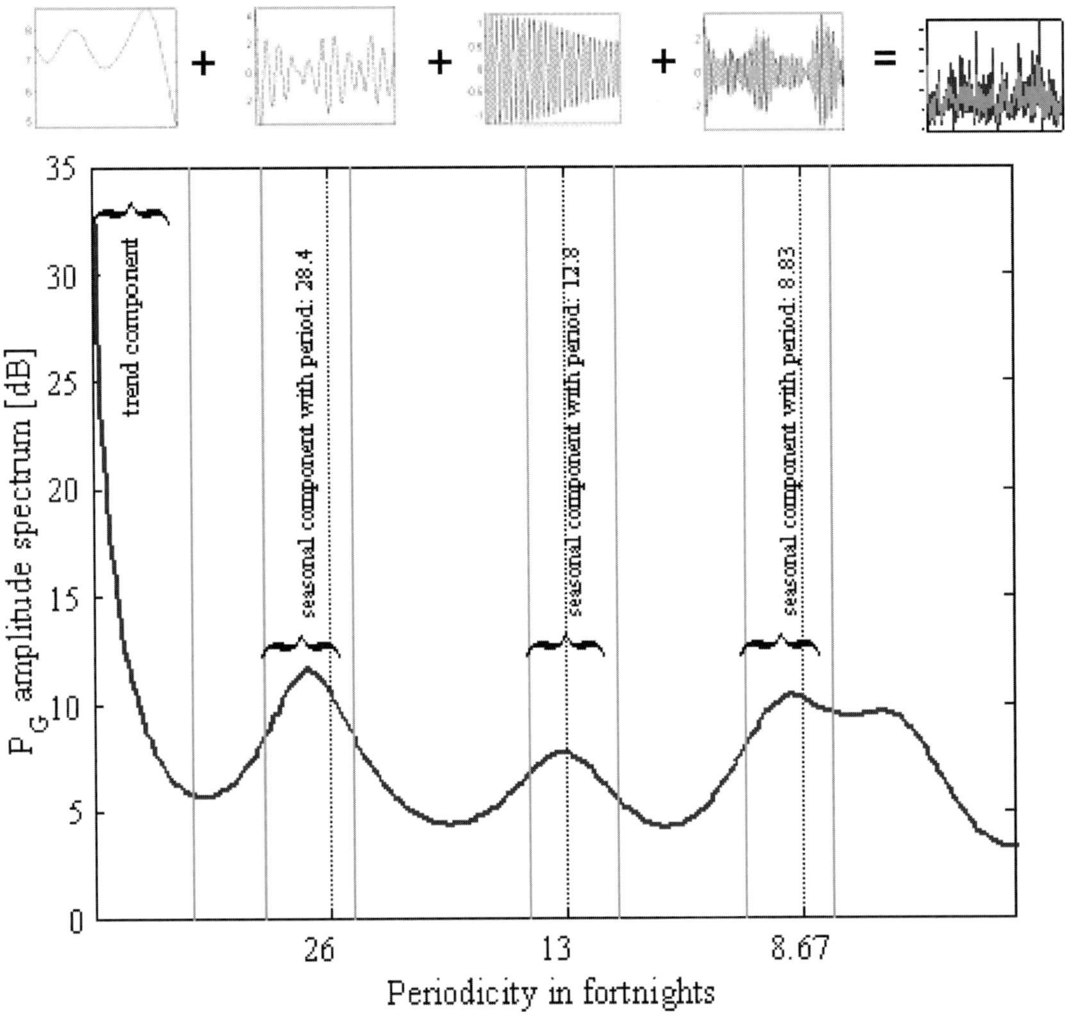

Figure 20.A1 The power spectrum and the derived trend and seasonal components of the daily rainfall time-series for the Danum Valley Field Centre, East Malaysia.

forestry activities and the spatio-temporal changes inherent in the natural system.

Maintenance of natural forests within the tropics, whether as virgin forest or selectively-managed natural forest, is important for (i) keeping drinking water abstractions free from agricultural chemicals (cf. Bolstad and Swank, 1997), and the even higher turbidity levels associated with urban construction or steepland agriculture (Douglas, 1996), (ii) the maintenance of forest flora and faunal resources, and possibly (iii) maintaining existing climatic regimes (Polcher, 1995). Many natural forests in the tropics are currently being converted or lost to agriculture, urban development and agroforestry, so that tropical rainforest research that leads to improved guidelines for the sustainable management of natural forests (as discussed in Part V of this volume) may provide support to those who wish to make the case to retain some of their natural forests.

APPENDIX 20.1
N. A. CHAPPELL
THE DYNAMIC HARMONIC REGRESSION MODEL

The Dynamic Harmonic Regression (DHR) model is a recursive interpolation, extrapolation and smoothing algorithm for non-stationary time-series, and identifies three components in the time-series: (i) the trend, which includes inter-annual cyclicity and longer-term drifts, (ii) the within-year cycles or 'seasonality', and (iii) the white noise, i.e.

$$SS_{(t)} = T_t + S_t + e_t \qquad (20.A1)$$

where $SS_{(t)}$ is the time-series of observed suspended-sediment flux, T_t is the trend (see Figure 20.A1) which includes the drift in long-term average suspended-sediment flux and the inter-annual cycles, S_t is the periodic component related to annual and intra-annual seasonality (see Figure 20.A1), and e_t is the white noise. The S_t term is further defined as:

$$S_t = \sum_{i=1}^{R} \{a_{it} \cos(\omega_i t) + b_{it} \sin(\omega_i t)\} \qquad (20.A2)$$

where $a_{i,t}$ and $b_{i,t}$ are the Time-Variable-Parameters or TVPs of the model, R is the number of seasonal components, and ω_i are the set of frequencies chosen by reference to the spectral properties of the time-series (Young, 1998; Young *et al.*, 1999). Optimisation of the TVPs is achieved by first estimating the Noise-Variance-Ratio (NVR) of the TVPs. This is achieved in the frequency domain by fitting the logarithmic pseudo-spectrum of the DHR model to the estimated logarithmic AutoRegressive (AR) spectrum of the observed rainfall series. The order of the AR model is identified via the Akaike Information Criterion. Once NVR parameters are optimised (NB these define the widths of bands of each seasonal component shown in Figure 20.A1), a single run of two recursive algorithms, the Kalman Filter and Fixed-Interval-Smoothing equations provide estimates of the various components. The estimated trend component (the first segment of the power spectrum shown in Figure 20.A1) is further split into a very slowly changing drift and the inter-annual cyclic component. Since no assumptions are made as to the periodicity of the cycle, it is unlikely that any artifacts are introduced in the procedure. Further details and examples of the DHR model are given in http://www.es.lancs.ac.uk/cres/captain.

References

Abdul Rahim, N. (1988). Water yield changes after forest conversion to agricultural landuse in Peninsular Malaysia. *Journal of Tropical Forest Science*, 1: 67–84.

Abdul Rahim, N. (1990). *The effects of selective logging methods on hydrological parameters in Peninsular Malaysia*. Unpublished PhD thesis, Bangor : University of Wales.

Abdul Rahim, N. and Harding, D. (1992). Effects of selective logging methods on water yield and streamflow parameters in Peninsular Malaysia. *Journal of Tropical Forest Science*, 5: 130–154.

Abdul Rahim, N. *et al.* (1997). *EIA Guidelines for harvesting of Natural Forests*. FRIM Technical Information Handbook No. 14. Kuala Lumpur : Forestry Research Institute of Malaysia.

Abdul Rahim, N. and Zulkifli, Y. (1994). Hydrological response to selective logging in Peninsular Malaysia and its implications on watershed management. In *Proceedings of the International Symposium on Forest Hydrology*, pp. 263–274. Tokyo, Japan, October 1994.

Adjers, G., Hadengganan, S., Kuusipalo, J., Nuryanto, K. and Vesa, L. (1995). Enrichment planting of dipterocarps in logged-over secondary forests – effect of width, direction and maintenance method of planting line on selected Shorea species. *Forest Ecology and Management*, 73: 259–270.

Asdak, C. Jarvis, P. G., van Gardingen, P. and Fraser, A. (1998). Rainfall interception loss in unlogged and logged forest areas of Central Kalimantan, Indonesia. *Journal of Hydrology*, 206: 237–244.

Attiwill, P. M. and Weston, C. J. (2001). Forest soils. In *The Forests Handbook*, ed. J. Evans, pp157–187, Oxford: Blackwell.

Baharuddin, K. (1995). *Effects of properties and soil erosion in a hill dipterocarp forest of Peninsular Malaysia*. Research Pamphlet No. 119. Kuala Lumpur: Forestry Research Institute of Malaysia.

Balamurgan, G. (1997). *Suspended solids dynamics in small forested catchments in Ulu Segama, Sabah, Malaysia*. Unpublished PhD thesis, Manchester : University of Manchester.

Becker, P. (1996). Sap flow in Bornean heath and dipterocarp forest trees during wet and dry periods. *Tree Physiology*, 16: 295–299.

Bidin, K. (2001). Spatio-temporal variability in rainfall and wet-canopy evaporation within a small catchment recovering from selective tropical forestry. Unpublished PhD thesis. Lancaster: University of Lancaster.

Bolstad, P. V. and Swank, W. T. (1997). Cumulative impacts of landuse on water quality in a Southern Appalachian watershed. *Journal of the American Water Resources Association*, 33: 519–533.

Bonell, M., Gilmour, D. A., and Cassells, D. S. (1983). Runoff generation in tropical rainforests of north-east Queensland, Australia, and its implications for land use management. In *Proc. of symp. on the hydrology of humid tropical regions, IAHS Publ. 140*, ed. R. Keller, pp. 287–297. Paris: IAHS.

Brouwer, L. C. 1996. *Nutrient cycling in pristine and logged tropical rainforest*. Unpublished PhD thesis. Utrecht : Utrecht University.

Bruijnzeel, L. A. (1989a). Nutrient dynamics of tropical forest land before and after conversion to other types of land use. In *Regional Seminar on Tropical Forest Hydrology*, 4–9 September 1989, Kuala Lumpur.

Bruijnzeel, L. A. (1989b). Nutrient cycling: the hydrological framework. In *Mineral Nutrients in Tropical Forest and Savanna Ecosystems*, ed. J. Proctor, pp. 383–415. Oxford: Blackwell Scientific Publications.

Bruijnzeel, L. A. (1990). *Hydrology of moist tropical forest and effects of conversion: a state of the art review*. Paris : UNESCO.

Bruijnzeel, L. A. (1991). Nutrient input-output budgets of tropical forest ecosystems: a review. *Journal of Tropical Forest Ecology*, 7: 1–24.

Bruijnzeel, L. A. (1992). Managing tropical forest watersheds for production: where contradictory theory and practice co-exist. In *Wise Management of Tropical Forests*, Proceedings of the Oxford Conference on Tropical Forests 1992, eds. F. R. Miller and K. L. Adam. pp. 37–75. Oxford: Oxford Forestry Institute (University of Oxford).

Bruijnzeel, L. A. (1996). Predicting the hydrological impacts of land cover transformation in the humid tropics: the need for integrated research. In *Amazonian deforestation and climate*, eds. J. H. C. Gash, C. A. Nobre, J. M. Roberts, and R. L. Victoria, pp. 15–55. Chichester: Wiley.

Bruijnzeel, L. A. (2001). Forest hydrology. In *The Forests Handbook*, ed. J. Evans, pp. 301–343, Oxford: Blackwell.

Cailleux, A. (1959). *Etudes sur l'érosion et la sédimentation en Guyane*. Mem. Serv. Carte Geol. Fr., pp . 49–73.

Chappell, N. A. and Ternan, J. L. (1992). Flow-path dimensionality and hydrologic modelling, *Hydrological Processes*, 6: 327–345.

Chappell, N. A., Franks, S. W. and Larenus, J. (1998). Multi-scale permeability estimation for a tropical catchment. *Hydrological Processes*, 12: 1507–1523.

Chappell, N. A., McKenna, P., Bidin, K., Douglas, I. and Walsh, R. P. D. (1999a). Parsimonious modelling of water and suspended-sediment flux from nested-catchments affected by selective tropical forestry. *Philosophical Transactions of the Royal Society of London Series B.*, 354: 1831–1846.

Chappell, N. A., Ternan, J. L. and Bidin, K. (1999b). Correlation of physicochemical properties and sub-erosional landforms with aggregate stability variations in a tropical Ultisol disturbed by forestry operations. *Soil and Tillage Research*, 50: 55–71.

Chappell, N. A., Bidin, K. and Tych, W. (2001). Modelling rainfall and canopy controls on net-precipitation beneath selectively-logged tropical forest. *Plant Ecology*, 153: 215–229.

Chappell, N. A., Douglas, I., Hanapi, J. M. and Tych, W. (2004). Source of suspended-sediment within a tropical catchment recovering from selective logging. *Hydrological Processes*, 18: 685–701.

Collison, A. J. C. and Anderson, M. G. (1996). Using a combined slope hydrology stability model to identify suitable conditions for landslide prevention by vegetation in the humid tropics. *Earth Surface Processes and Landforms*, 21: 737–747.

Conway, S. (1982). *Logging Practices*. San Francisco: Miller Freeman Publications.

Croke, J., Hairine, P. and Fogarty, P. 1999. Runoff generation and redistribution in logged eucalyptus forests, south-eastern Australia. *Journal of Hydrology*, 216: 56–77.

Dabin, B. (1957). *Note sur le fonctionnement des parcelles expérimentales de l'érosion a la station d'Apiopodioume (Cote d'Ivoire)*. Dakar, Decret Permanent Bureau Sols AOF. Mimeographed, pp. 16.

Davies, S. J. (1998). Photosynthesis of nine pioneer *Macaranga* species from Borneo in relation to life history. *Ecology*, 79: 2292–2308.

Dietrich, W. E. and Dunne, T. (1978). Sediment budget for a small catchment in mountainous terrain. *Zeitschrift fur Geomorphogie, Supplements*, 29: 191–206.

Dixon, J. C. (1986). Solute movement on hillslopes in the alpine environment of the Colorado Front Range. In *Hillslope Processes*, ed. A. D. Abrahams, PP. 139–159, Boston: Allen and Unwin.

Douglas, I. (1996). The impact of land-use changes, especially logging, shifting cultivation, mining and urbanisation on sediment yields in humid tropical South East Asia: a review with special reference to Borneo. In *Erosion and Sediment Yield: Global and Regional Perspectives*. pp. 463–471, IAHS publication 236, Paris: IAHS.

Douglas, I., Greer, T., Sinun, W., Anderton S., Bidin, K., Spilsbury, M., Jadda, S., and Azman, S. (1995). Geomorphology and rainforest logging practices. In *Geomorphology and Land Management in a Changing Environment*, eds. D. F. M. McGregor and D. A. Thompson. Chichester: Wiley.

Douglas, I., Spencer, T., Greer, T., Bidin, K., Sinun, W. and Wong, W. M. (1992). The impact of selective commercial logging on stream hydrology, chemistry and sediment loads in the Ulu Segama rainforest, Sabah. *Philosophical Transactions of the Royal Society of London Series B*, 335, 397–406.

Douglas, I. and Bidin, K. (1994). *Effectiveness of buffer strips in reducing the impact of logging on streamflow and stream sedimentation (January 1991–January 1994)*. R4603 Final Report, London: Overseas Development Adminstration.

Douglas, I., Bidin, K., Balamurgan, G., Chappell, N. A., R. P. D. Walsh, Greer, T. and Sinun, W. (1999). Role of extreme events in the impacts of selective tropical forestry on erosion during harvesting and recovery phases at Danum Valley, Sabah. *Philosophical Transactions of the Royal Society of London Series.*, 354, 1749–1761.

Eschenbach, C., Glauner, R., Kleine, M. and Kappen, L. (1998). Photosynthesis rates of selected tree species in lowland dipterocarp rainforest in Sabah, Malaysia. *Trees-Structure and Function*, 12: 356–365.

Eshleman, K. N. (2000). A linear model of the effects of disturbance on dissolved nitrogen leakage from forested watersheds. *Water Resources Research*, 36: 3325–3335.

George, R. J. (1992). Hydraulic-properties of groundwater systems in the saprolite and sediments of The Wheat-Belt, Western Australia. *Journal of Hydrology*, 130: 251–278.

Gilmour, D. A. (1977). Effect of rainforest lgging and clearing on water yield and quality in a high rainfall zone of North-east Queensland. In *Symposium on The Hydrology of Northern Australia, Brisbane*. pp. 156–160. Old National Conference Publication No. 7715, Institute of Engineers (Australia).

Greer, T., Sinun, W., Douglas, I. and Bidin, K. (1996). Long term natural forest management and land-use change in a developing tropical catchment, Sabah, Malaysia. In *Erosion and Sediment Yield: Global and Regional Perspectives*. pp. 453–461, IAHS publication No. 236. Paris: IAHS.

Gregory, K. J. and Walling, D. E. (1973). *Drainage basin form and process*. London: Arnold.

Hafkenscheid, R. (2000). *Hydrology and biogeochemistry of tropical montane rainforests of contrasting stature in the Blue Mountains, Jamaica*. Unpublished PhD thesis. Amsterdam: Free University.

Hardjowitjitro, H. (1981). Soil erosion as a result of traditional cultivation in Java Island. In *Proc. South East Asian Regional Symposium on Problems of Soil Erosion and Sedimentation*, eds, T. Tingsanchali and H. Eggers, pp 173–179. Bangkok: Asian Institute of Technology.

Hudson, J. A., Crane, S. B. and Blackie, J. R. (1997). The Plynlimon water balance 1969–1995: the impact of forest and moorland vegetation on evaporation and streamflow in upland catchments. *Hydrology and Earth System Sciences*, 1: 409–428.

Iremonger, S., Ravilious, C. and Quinton, T. (1997). A statistical analysis of global forest conservation. In *A global overview of forest conservation*, eds. S. Iremonger, C. Ravilious and and T. Quinton. Cambridge: CIFOR/WCMC.

Jakeman, A. J., Chen, T.-H., Post, D. A., Hornberger, G. M., Littlewood, I. G. and Whitehead, P. G. (1993). Assessing uncertainties in hydrological response to climate at large scale. In *Macro-scale Modelling of the Hydrosphere*, ed. W. B. Wilkinson, pp. 37–47. IAHS publication 214. Paris: IAHS.

Jakeman, A. J. and Hornberger, G. M. (1993). How much complexity is warranted in a rainfall-runoff model? *Water Resources Research*, 29: 2637–2649.

Jetten, V. G. 1994. Modelling the effects of logging on the water balance of a tropical rainforest: A study in Guyana. Unpublished PhD thesis. Utrecht: University of Utrecht.

Jetten, V. G., Riezebos, H. Th., Hoefsloot, F. and van Rossum, J. 1993. Spatial variability of infiltration and related properties of tropical soils. *Earth Surface Processes and Landforms*, 18: 477–488.

Jones, J. A. A. (1990). Piping effects in humid lands. In *Groundwater Geomorphology*, eds. C. G. Higgins and D. R. Coates, pp. 111–138, Special Paper 252. Boulder: Geological Society of America.

Kobayashi, S., Turnbull., J. W. and Cossalter, C. 2001. Rehabilitation of degraded tropical forest ecosystems project. In *Rehabilitation of degraded tropical forest ecosystems*, eds. S. Kobayashi, J. W. Turnbull, T. Toma, T. Mori, and N. M. N. A. Majid, pp. 1–16, Workshop proceedings, 2–4 November 1999, Bogor, Indonesia. Bogor: Center for International Forestry Research.

Lai, F. S. (1992). *Sediment and solute yields from logged, steep upland catchments in Peninsular Malaysia*, Unpublished PhD thesis. Manchester: University of Manchester.

Lai, F. S. Geoffery, J. G. Hamdan, A. G. and M. Jafri, H. (1999). Post-logging sediment yeild of the Sungai Lawing Basin, Selango, Malaysia. In *Water: Forestry and Landuse Perspectives*, Proceedings of the CFFPR conference, Forestry Research Institute of Malaysia (FRIM), 31 March to 1 April 1999. Kuala Lumpur: FRIM.

Larsen, M. C. and Torres-Sanchez, A. J. (1998). The frequency and distribution of recent landslides in three montane tropical regions of Puerto Rico. *Geomorphology*, 24: 309–331.

Larsen, M. C., Torres-Sanchez, A. J. and Concepcion, I. M. (1997). Watershed sensitivity to land use and lithology: a comparison of fluvial sediment yield in four small montane tropical catchments, Puerto Rico. *EOS Transactions American Geophysical Union*, 78:.

MacDonald L. H., Sampson R. W., Anderson D. M. . 2001 Runoff and road erosion at the plot and road segment scales, St John, US Virgin Islands. *Earth Surface Processes and Landforms*, 26: 251–272.

Malmer, A. and Grip, H. (1990). Soil disturbance and loss of infiltrability caused by mechanised and manual extraction of tropical rainforest in Sabah, Malaysia, *Forest Ecology and Management*, 38: 1–12.

Malmer, A. and Grip, H. (1994). Converting tropical rain-forest to forest plantation in Sabah, Malaysia. 2. Effects on nutrient dynamics and net losses in streamwater. *Hydrological Processes*, 8: 195–209.

Martin-Smith K. 1998 Biodiversity patterns of tropical freshwater fish following selective timber extraction: a case study from Sabah, Malaysia. *Italian Journal of Zoology (Suppl. S)*, 65: 363–368.

Middleton, G. V. (2000). *Data analysis in the earth sciences using MATLAB*. pp. 260. Upper Saddle River: Prentice-Hall.

Milliman, J. D., Farnsworth, K. L. and Albertin, C. S. (1999). Flux and fate of fluvial sediment sleaving large islands in the East Indies. *Journal of Sea Research*, 41: 97–107.

Nash, J. E. and Sutcliffe, J. V. (1970). River flow forecasting through conceptual models, 1: A discussion of principles. *Journal of Hydrology*, 10: 282–290.

Nykvist, N. (1994). Removal of plant nutrient in logs when clear-felling tropical rainforests – a literature review. In *Proc. Symposium on Harvesting and Silviculture for Sustainable Forestry in the Tropics*, eds. Wan Razali, W. M., Shamsuddin I., S. Appanah and M. Farid, A. R., pp. 192–196, Kuala Lumpur: Forest Research Institute of Malaysia.

Oyebande, L. (1988). Effects of tropical forest on water yield. In *Forest, Climate and Hydrology: Regional Impacts*, eds. E. R. C. Reynolds and F. B. Reynolds, pp 16–50, Japan: The United Nations University.

Pickup, H., Higgins, R. J. and Warner, R. H. (1981). Erosion and sediment yield in Fly River drainage basins. In *Erosion and Sediment Transport in Pacific Rim Steeplands*, eds. T. R. H. Davies and A. J. Pearce, pp. 438–456, IAHS publication No. 132. Paris: IAHS.

Pinard, M. A., Putz, F. E., Tay, J. and Sullivan, T. E. (1995). Creating timber harvest guidelines for a reduced-impact logging project in Malaysia. *Journal of Forestry*, 93: 41–45.

Poels, R. L. H. (1987). *Soils, water and nutrients in a forest ecosystem in Suriname*. Unpublished PhD thesis. Wageningen: Agricultural University.

Polcher, J. (1995). Sensitivity of tropical convestion to land surface processes. *Journal of Atmospheric Sciences*, 52: 3143–3161.

Post, D. A. and Jakeman, A. J. (1996). Relationships between catchment attributes and hydrological response characteristics in small Australian mountain ash catchments. *Hydrological Processes*, 10: 877–892.

Rahman, Y. (1993). Soil of Singapore. In *Physical Adjustments in a Changing Landscape*, eds. A. Gupta and J. Pitts, pp. 144–189. Singapore: Singapore University Press.

Restom, T. G. and Nepstad, D. C. (2001). Contribution of vines to the evapotranspiration of a secondary forest in eastern Amazonia, *Plant and Soil*, 236: 155–163.

Roberts, J., Cabral, O. M. R., Fish, G., Molion, L. C. B., Moore, C. J., and Shuttleworth, W. J. (1993). Transpiration from an Amazonian rainforest calculated from stomatal conductance measurements. *Agricultural and Forest Meteorology*, 65: 175–196.

Searcy, J. K. (1959). *Flow-duration curves*. Wat. Sup. Pap. 1542-A. Washington D.C.: USGS.

Searcy, J. K. and Hardison, C. H. (1960). *Double-mass curves*. Water Supply Paper 1541-B. Washington D.C.: USGS

Sellers P. J., Heiser M. D., Hall F. G., Goetz S. J., Strebel D. E., Verma S. B., Desjardins R. L., Schuepp P. M., MacPherson, J. I. (1995). Effects of spatial variability in topography, vegetation cover and soil moisture on area-averaged surface fluxes: A case study using the FIFE 1989 data. *Journal of Geophysical Research – Atmospheres*, 100(D12): 25607–25629.

Sheffield A. T., Healy T. R., McGlone M. S. 1995 Infilling rates of steepland catchment esturary, Whangamata, New Zealand. *Journal of Coastal Research*, 11: 1294–1308.

Singh, R. B. (1998). Land use cover changes, extreme events and ecohydrological responses in the Himalayan region. *Hydrological Processes*, 12: 2043–2055.

Spaans, E. J. A., Bouma, J., Lansu, A. and Wielemaker, W. G. 1990. Measuring soil hydraulic properties after clearing of tropical rainforest in a Costa Rican soil. *Trop. Agric. (Trinidad)*, 67: 61–65.

Stevens, P. A., Norris, D. A., Williams, T. G., Hughs, S., Durrant, D. W. H., Anderson, M. A., Weatherley, N. A., Hornung, M. and Woods, C. (1995). Nutrient losses after clearfelling in Beddgerlert Forest – a comparison of the effects of conventional and whole-tree harvest on soil-water chemistry. *Forestry*, 68: 115–131.

Subba Rao, B. K., Ramola, B. C. and Sharda, V. N. (1985). Hydrologic response of a forested mountain watershed after thinning. *Indian Forester*, 111: 418–431.

Swank, W. T. (1988). Stream chemistry responses to disturbance. In *Forest Hydrology and Ecology at Coweeta*, eds. W. T. Swank and D. A. Crossley Jr., pp. 339–357, Ecological Studies, Vol. 66. New York: Springer-Verlag.

Swank, W. T., Swift, L. W. and Douglass, J. (1988). Streamflow changes associated with forest cutting, species conversions, and natural disturbances. In *Forest Hydrology and Ecology at Coweeta*, eds. W. T. Swank and D. A. Crossley Jr., pp. 297–312, Ecological Studies, Vol. 66. New York: Springer-Verlag.

Swanston, D. N., and Swanson, F. J. 1976. Timber harvesting, mass erosion, and steepland forest geomorphology in the Pacific Northwest. In *Geomorphology and Engineering*, ed. D. R. Coates, New York: Halsted Press.

Tundisi, J. G. (1990). Perspectives for ecological modeling of tropical and subtropical reservoirs in South America. *Ecological Modelling*, 52: 7–20.

van der Plas, M. C. and Bruijnzeel, L. A. (1993). Impact of mechanized selective logging of rainforest on topsoil infiltrability in the Upper Segama areas, Sabah, Malaysia. In *Hydrology of Warm Humid Regions*. pp. 203–211, IAHS publication 216. Paris: IAHS.

van Dam, O. (2001). *Forest Filled with Gaps: Effects of Gap Size on Water and Nutrient Cycling in Tropical Rainforest*, Unpublished PhD thesis. Utrecht: Utrecht University.

van der Hout, P. (1999). *Reduced impact logging in the tropical rainforest of Guyana: Ecological, economic and silvicultural consequences*. Unpublished PhD thesis. Utrecht: Utrecht University.

Walling, D. E. and Webb, B. W. (1983). Erosion and sediment yield. In *Background to Palaeohydrology*, ed. K. J. Gregory, pp 69–100. Chichester: Wiley.

Ward, R. C. (1981). River systems and river regimes. In *British Rivers*, ed. J. Lewin, pp. 1–33. London: Allen and Unwin.

Wheater, H. S., Jakeman, A. J. and Beven, K. J. (1993). Progress and directions in rainfall-runoff modelling. In *Modelling Change in Environmental Systems*, eds. A. J. Jakeman, M. B. Beck and M. J. McAleer, Chichester: Wiley.

Whitmore, T. C. and Burnham, C. P. (1969). The altitudinal sequence of forests and soils on granite near Kuala Lumpur. *Malayan Nature Journal*, 22: 99–118.

Young, P. C. (1984). *Recursive estimation and time series analysis*. Berlin: Springer.

Young, P. C. (1992). Parallel processes in hydrology and water quality: a unified time-series approach. *Journal of the Institute of Water and Environmental Management*, 6: 598–612.

Young, P. C. 2001. Data-based mechanistic modelling and validation of rainfall-flow processes. In *Model validation: perspectives in hydrological sciences*, eds. M. G. Anderson and P. D. Bates, Chichester: Wiley.

Young, P. C. and Beven, K. (1994). Data-based mechanistic modelling and the rainfall-flow non-linearity. *Environmetrics*, 5: 335–363.

Young, P. C., Jakeman, A. J. and Post, D. A. (1997). Recent advances in the data-based modelling and analysis of hydrological systems. *Water Science and Technology*, 36: 99–116.

Young, P. C. (1998). Data-based mechanistic modelling of environmental, ecological, economic and engineering systems. *Environmental Modelling and Software*, 13: 105–122

Young, P. C., Pedregal, D. J. and Tych, W. (1999). Dynamic harmonic regression. *Journal of Forecasting*, 18: 369–394.

Yusop, Z. (1989). Effects of selective logging methods on dissolved nutrient exports in Berembun Watershed, Peninisular Malaysia. In *Regional Seminar on Tropical Forest Hydrology*, 4–9 September 1989, Kuala Lumpur.

Yusop, Z. (1996). *Nutrient cycling in secondary rainforest catchments of Peninsular Malaysia*. Unpublished PhD thesis, Manchester: University of Manchester.

21 Effects of shifting cultivation and forest fire

Anders Malmer
Swedish University of Agricultural Science, Umeå, Sweden
M. van Noordwijk
International Centre for Research in Agroforestry, Bogor, Indonesia
L. A. Bruijnzeel
Vrije Universiteit, Amsterdam, The Netherlands

INTRODUCTION

Fire has always been apparent to some extent in humid tropical forest as an agent of disturbance leading to forest renewal through succession and even to long-term changes in the biome (Flenley, 1979; 1992; 1998). Under climatic conditions of occasional drought there is an element of natural forest fires occurring without human interference (Goldammer, 1992) although this is difficult to establish because the use of fire also links back to the earliest forms of agriculture (Boserup, 1965; Steensberg, 1993). Today however, the role of man is more evident than ever before in understanding the dynamics of fire, humans and vegetation ecology (Uhl, 1998).

Perceptions by lowlanders of a loss of 'forest catchment functions' due to 'upland shifting cultivators' are often strong but these may not be based on a clear understanding of the cause-effect chains involved. For example, most major and capital cities in South East Asia have been built on floodplains at the mouths of rivers, i.e. in areas where occasional flooding is to be expected regardless of the forest cover of the uplands (Hamilton and King, 1983). When floods do occur, however, land use change in the uplands provides an easy scapegoat, especially if the uplanders have a different ethnic and cultural background, as for example in Northern Thailand. These conflicts over land use change in the uplands have reached such an intensity in some areas that basic research findings are not likely to modify the perceptions and standpoints of different stakeholders in the conflict (Van Noordwijk, *pers. obs.*). Clearly, in the longer run progress can only be expected if negotiations between stakeholder groups are based on realistic expectations of the consequences for water and sediment flows of changes in the pattern of vegetation cover of hillslopes and riparian zones, i.e. how the various filters function that together translate rainfall into streamflow.

To a large extent, the catchment hydrological budget offers a convenient framework when discussing many of the on-site and off-site effects of fire. Bruijnzeel (1990 and 1998) presented comprehensive reviews of the hydrological effects of various human impacts on moist tropical forests, including erosion and nutrient aspects. In view of the increased occurrence and extent of fires in the humid tropical forest biome, knowledge of the hydrological processes influencing topsoil fertility maintenance and losses of nutrients to streamflow is crucial. This chapter reviews the current knowledge of the immediate hydrological and hydrochemical effects of forest and wild fires, and of the establishment and cropping phase of shifting cultivation in the humid tropics. The water and nutrient dynamics of the subsequent secondary growth is discussed by Hölscher *et al.* (this volume). In general, the volume of scientific work on processes and quantification of effects of fires in the humid tropics is limited, especially when it comes to forest and wild fires. Here, we discuss the ecosystem processes related to fire and their inter-linkages, and the implications for management and further research. With the limited amount of literature available, we consider that the aim of providing general quantifications of the water and nutrient budgets associated with different fire impacts in different environments is not possible at this stage.

Natural and wild fires in tropical rainforest

Fire in the tropical biome continues to stir the headlines of international media. The fires in South East Asia of late 1997, which gave smoke problems over the entire region, provide a case in point. Most of the problem was linked to particles emanating from burning peat soils (G. Applegate, pers. comm.), in which this region is rich (Laumonier and Legg, 1998). Even though the extent was probably the largest in modern time, the smoke problem is not new

Forests, Water and People in the Humid Tropics, ed. M. Bonell and L. A. Bruijnzeel. Published by Cambridge University Press.

to the region nor to the tropics as a whole. Massive burning in various states of the Brazilian Amazon under development schemes have closed down airports in that region since the eighties (e.g. Schulze, 1998) and similar problems have also occurred in Southern Africa (Douguedroit and Bart, 1997). The smoke created serious problems for air traffic and human health (Fang *et al.*, 1999). The term 'forest fire' will be used here to denote forest fires by natural causes in little disturbed forests whereas 'wild fire' will be used for uncontrolled fires in anthropogenically-fragmented landscapes with strong links to human use of fire. It will be apparent from the following that the difference between the two is not always entirely clear but the distinction helps to facilitate the discussion when it comes to causes and effects.

Charcoal fragments dated by ^{14}C give evidence for the occurrence of fires in humid tropical forests in South America (Sanford *et al.*, 1985; Fölster, 1992) and South East Asia (Goldammer and Seibert, 1989; 1990) dating back to as much as 6000 and 17 500 years BP, respectively. Preserved pollen records also indicate past vegetation changes that were partially induced by fire (Flenley 1992). However, it is difficult to establish whether these fires were indeed 'natural' forest fires and not caused by early human activities. Sanford *et al.* (1985) argued that old charcoal fragments can also be found in places where human activity is not known or must be regarded as highly unlikely. Charles-Dominique *et al.* (1998) presented a similar case for French Guyana.

The perception of the stable humid tropical forest with its perennially humid climate preventing fire (as used for instance to burn dried residues in clearings) from entering into the forest, has been the general picture until rather recently (e.g. Richards, 1952; 1969). Since then, the evidence for early forest disturbance has led to a more modern and dynamic view of disturbance in humid tropical forest (cf. Whitmore, 1978; Spencer and Douglas, 1985; Goldammer, 1992; Whitmore and Burslem, 1998). Examples of major disturbing agents other than fire include periodic droughts (often connected to the El Niño Southern Oscillation (ENSO) phenomenon; Baillie, 1976; Nicholls, 1993; Walsh, 1996; Walsh and Newbery, 1999), high-magnitude, low-frequency storms (Whitmore, 1974; Spencer, *et al.*, 1990; Malmer, 1996a), tectonic activity (Pain and Bowler, 1973) and, on a smaller scale, treefalls creating gaps (Brown, 1993), pits and mounds (Putz, 1983). The chapter by Scatena *et al.* (this volume) discusses such disturbances in some detail.

Walsh (1996) used long-term meteorological records and historical descriptions to evaluate the degree of periodicity of drought on Borneo and Java. In north-eastern Borneo periods longer than three consecutive months with less than 100 mm of rain per month occur several times within 20-year periods. From this perspective the occurrence of forest fires is plausible, for example through ignition by lightning (Middleton *et al.*, 1997). However, it has been argued that lightning is not very common during dry periods (Wibowo *et al.*, 1997) and that there is very little, if any, evidence of forest fires by natural causes in wet tropical forests today (Tutin *et al.*, 1996). Whatever the cause of forest fires in earlier times or at present, today there are strong links between human activities and the occurrence and extent of wild fires in the humid tropics (Goldammer, 1988; Uhl, 1998). During the major droughts connected to the ENSO events of 1982/83, 1987 and 1997/98, fires in the three major tropical biomes were more severe and widespread than those previously experienced during the last century. On Borneo alone more than 4 million hectares of primary and secondary forest burned in 1982/83 (Malingreau *et al.*, 1985). Cochrane and Schulze (1998) and Cossalter and Cauvin (1998) have given accounts of the fires occurring in 1997/98 in the Brazilian Amazon and in South East Asia, respectively. Naturally, the increasing pressure from slash and burn agriculture and other human uses of fire described above very much increase the risk of ignition and occurrence of wild fires during dry spells. Furthermore, shortened fallows and the presence of large portions of forests that have been logged selectively have changed dramatically the structure of the vegetation of many tropical forest areas (Woods, 1989). Beaman *et al.* (1985) noted that during the 1982/83 fires in northern Borneo mostly secondary forests and logged-over forest burned and that often the fire stopped upon reaching the primary forest. The same observation was made during the fires of 1997 (Laumonier and Legg, 1998). Nevertheless, in 1982/83 some 1.35 million hectares of primary forest were estimated to have burned in Borneo alone (Whitmore, 1998). Goldammer and Seibert (1990) have also shown a positive relation between fire intensity and the degree of damage from previous selective logging (see also Bruijnzeel (1992) and Chappell, Tych *et al.* (this volume) for an in-depth discussion of the various hydrological impacts of selective logging.).

The increasing human pressure on humid tropical forest land and the global change of climate will indeed continue to make the effects of fire increasingly important to understand. Goldammer and Price (1998) have used global climate modelling to investigate the possible impacts of climate change on fire regimes in the tropics. They predicted the most dramatic increases in dry season length for evergreen forests in South East Asia and South America. For the Amazon, Cochrane *et al.* (1999) estimated that wild fires caused more deforestation than intentional clearing in recent years.

Shifting cultivation

Shifting cultivation is the term for non-sedentary agriculture where patches of forest are cleared and residues on the ground burned to enable cultivation. Without input of fertilisers, only one to a few crops are sustained and the cultivators then move on to a new patch (Whitmore, 1990). Other terms used in the literature are 'swidden agriculture' and 'slash and burn agriculture' (Peters and Neuenschwander, 1988). In some cases the term 'slash and burn' has been used for these forms of cultivation where the duration of the fallow period is not sufficient to allow adequate nutrient accumulation in the vegetation and soil, like in the sustainable form of 'traditional shifting cultivation', or simply more loosely to any felling and burning of forest (Tinker *et al.*, 1996).

Wibowo *et al.* (1997) presented a conceptual model of the various causes and effects of fire, including a distinction between four different types of farmers practising shifting cultivation:

- Traditional communities with strong local institutions restricting the actions of individual farmers.
- Communities that are more fully integrated in the market economy, often combined with a loosening of traditional restrictions.
- Spontaneous settlers/migrants.
- Government-sponsored (trans) migrants.

In the following the term 'shifting cultivation' will be used for all of these forms of cultivation where conversion to permanent agriculture is not aimed for, whereas 'traditional shifting cultivation' is used for the first of the four categories listed above. However, it should be noted that what is going on in most tropical landscapes is often a mix of many of these different modes of shifting cultivation in combination with the use of fire in plantation forestry, in sedentary agriculture and the 'wild fires' sparked off by human beings in secondary vegetation. New roads for the development of sparsely populated regions or for forestry operations often create a front of intensive cultivation, including burning (Riswan and Hartanti, 1995; Nepstad *et al.*, 1999). The shortened fallows associated with the higher pressure on the land in areas with high population, often combined with grazing of livestock, leads to a vegetation succession where trees are out-competed by xerophytic and pyrophytic plants like grasses (e.g. Goldammer and Price, 1998). Such secondary vegetation types will have shorter cycles of fires due to the fact that they dry out more easily (Uhl and Kauffman, 1990) and due to intensified human use of fire. The fires and cropping over time (Uhl, 1987) eventually lead to soils and ecosystems that are degenerated in terms of organic material and nutrients (e.g. Tiessen *et al.*, 1994; Hölscher *et al.*, this volume).

Jackson (1983) estimated that 150 million people are sustained by shifting cultivation whereas more recently Sanchez (1996) even assumed this to be as many as 300–500 million. Within the perspective of these developments, there has been much effort put into both describing and preserving sustainable traditional forms of shifting cultivation (Schmidt-Vogt, 1998) and to introduce improvements in cultivation systems, ranging from improved shifting cultivation and alternatives to shifting cultivation (Alegre and Cassel, 1996; Van Noordwijk *et al.*, 1998a; Kato *et al.*, 1999) to a host of agroforestry systems (Young, 1989, 1997; Wallace *et al.*, this volume).

EFFECTS OF FIRE ON HYDROLOGY: FROM RAINFALL TO STREAM

Climate, especially rainfall

Physical reasoning and modelling indicate that large-scale deforestation may cause a decrease in rainfall (Salati and Vose, 1984; Lean *et al.*, 1996; Costa, this volume), but reliable verification of long-term trends in rainfall over large areas is extremely difficult (Van Rompaey, 1995; Tinker *et al.*, 1996). The opening of the forest canopy by fire or felling with subsequent burning alters the energy and water balance of the site. Several factors such as increased albedo, changed rainfall interception, aerodynamic roughness and active rooting depths, interact to create different effects at different sites (Tinker *et al.*, 1996; Pielke *et al.*, 1998). In continental settings like Amazonia where locally derived convective rainfall is relatively important, the strongest case for lowered rainfall due to deforestation can be made (Salati *et al.*, 1986; Shuttleworth, 1988; Costa, this volume). Polcher and Laval (1993) confirmed the importance of large land masses when modelling energy balances with and without a forest cover. They derived a 10% reduction in evapotranspiration for Amazonia, 8% for Central Africa and only 3% for South East Asia, 'the maritime continent'. The large uncertainty associated with this kind of large-scale modelling was demonstrated by Henderson-Sellers (1993) who predicted a much larger decrease in evapotranspiration, ET (30% during the wet season and 10% during the dry season) for the Amazon basin after complete conversion to pasture. Lean *et al.* (1996) arrived at a reduction in Amazonian rainfall of less than 10% after a similar exercise. Gedney and Valdes (2000) recently presented a model that predicted effects of complete Amazonian forest conversion on rainfall over geographically separate areas far away from the region of forest removal, including changes in winter rainfall over the north-east Atlantic. It should be noted that these simulation attempts all dealt with completely deforested landscapes. However, in the case of fire in shifting cultivation, or after natural and wild fires, a fragmented landscape is created consisting of a mosaic of many different vegetation types. As such, any effects on climate can be expected to be correspondingly smaller. The albedo and energy partition of regenerating tropical vegetation resembles that of old-growth forest within one to three decades (Giambelluca, 1996, 2002; Giambelluca *et al.*, 1999, 2000; see also Hölscher, Mackensen and Roberts, this volume). In addition, there are indications that the water use (E_T) of 5–25-year-old secondary vegetation exceeds that of old-growth forest (Hölscher *et al.*, 1997; Giambelluca *et al.*, 2000). This implies significant lateral inputs of advected sensible heat from surrounding clearings and very young regenerating vegetation (Giambelluca, 2002). It is also thought that the mosaic of secondary vegetation in various stages of regrowth, remnant patches of old-growth forest, and recently cleared fields leads to greater aerodynamic roughness at the landscape scale due to the associated spatial contrasts in canopy height. This should enhance atmospheric turbulence and thus evaporation (Veen *et al.*, 1991). Indeed, Giambelluca (2002) relates how sapflow measurements in the trees of an isolated forest patch in northern Vietnam revealed enhanced transpiration rates which were attributed to the warmer and drier conditions prevailing in surrounding clearings.

There is a relative crudeness in the current climatic models and their difficulty to show statistically significant changes in rainfall patterns to be the result of forest removal (see review by Bruijnzeel, 2004). This confirms the impossibility to claim that small-scale forest clearing for traditional shifting cultivation and minor natural forest fires would change rainfall significantly (Bastable *et al.*, 1993). However, Tangtham and Sutthipibul (1989) reported a significant negative correlation between 10-year-moving averages of annual rainfall and remaining forest area in Northern Thailand for the period 1951–1984 whilst a positive correlation was found between forest area and the number of rainy days. At the same time, the authors were quick to point out that the effect of deforestation, if any, was still within one standard error of the means for the respective time series. Indeed, Wilk *et al.* (2001) were also unable to detect any widespread changes in rainfall totals or patterns in the 12 100 km^2 Nam Pong basin in Northeast Thailand between 1957 and 1995, despite a reduction in the area classified as forest from 80% to 27% during the last three decades. Nevertheless, rainfall in the whole of Thailand shows a remarkable decreasing trend since the 1950s during the month of September, i.e. when the southwest monsoon current is weakening. In July and August, when the monsoon is still strong, no such decrease is noted (Yasunari, 2002). Interestingly, a related atmospheric modelling study by Kanae *et al.* (2001) suggested that this decrease in September rainfall may indeed, at least partly, be related to changes in surface albedo and roughness due to deforestation. Also, there is a weak consensus that large-scale degradation to fire climax grasslands by uncontrolled wildfires and deliberate burning will have an effect in the long run (Tinker *et al.*, 1996).

Carbon losses to the atmosphere and particles in smoke from fires can also contribute to climate change on a global scale (Esser, 1992). Fires that are followed by fixation of carbon during forest succession will not make a net contribution of carbon to the atmosphere but large-scale, repeated wild fires that degrade biomass amounts will do so (Tinker *et al.*, 1996). Particles in smoke will affect energy balances both at smaller and larger scales by adding reflection and absorption of radiation. At the local scale and for specific fires, the effect is short-lived and probably of little

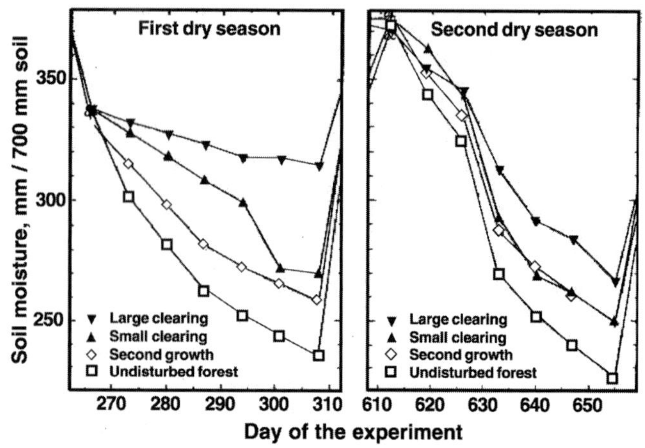

Figure 21.1 Soil water contents in undisturbed rainforest, in 6-year-old regrowth and in narrow (10 m × 50 m) and large (50 m × 50 m) clearings during the dry season, Costa Rica. (After Parker, 1985.)

significance. But regional periodical haze has an unquestionable influence on climate (Douguedroit and Bart, 1997). For a further review of these large-scale effects of fire, see e.g. Levine (1996).

Evapotranspiration

The felling of forests in humid tropical environments leads to decreased evapotranspiration and increased streamflow totals temporarily (Bruijnzeel, 1990; 1996), as has also been found in humid temperate environments (Bosch and Hewlett, 1982). This is mainly due to the reductions in transpiration and interception which exceed the increase in evaporation from the soil surface associated with the higher soil temperatures after clearing (Lal, 1987). As the forest vegetation recovers, so does evapotranspiration (e.g. Hibbert, 1967; Malmer, 1992; Hölscher *et al.*, this volume). As for the additional effect of fire, the killing of most remaining vegetation further reduces evapotranspiration. For example Malmer (1992 and 1996b) showed that burning of residual slash and vegetation for the establishment of a forest plantation in Sabah, Malaysia, added runoff increases during the first three years to 1008 mm compared with 447 mm for a non-burning practice.

Starting with the effects of small clearings for traditional shifting cultivation, these range in size between those of natural tree fall gaps and up to one or two hectares (cf. Scatena *et al.*, this volume). There will be temporarily reduced evapotranspiration even for small openings (Parker, 1985; Williams and Melack, 1997; Brouwer and Riezebos, 1998; Klinge *et al.*, 2001). It could be argued that the reduction in water use would be smaller close to the edges and in most of a small gap because of roots from the surrounding forest extending into the clearing (Figure 21.1). However, Brouwer and Riezebos (1998) did not find such edge effects for soil moisture across a 730 m^2 clearing in Guyana. Leaving a few large trees standing (which is practised sometimes with traditional shifting cultivation in tall climax forests) could also be expected to increase water use over that of crops only. The presence of only a few standing trees in clear-fellings in Sweden was also shown to increase atmospheric turbulence down to the soil surface (Granberg *et al.*, 1993). Furthermore, such trees also contribute to the dynamics of evapotranspiration due to shading and competition. Published studies on such inter-clearing dynamics have been lacking, but the subject is to some extent covered by recent work by Van Dam (2001), also in Guyana. In this respect, there is much to be learned from agroforestry situations (see Wallace *et al.*, this volume).

There are few descriptions of forest fires in tall humid tropical forests, but water use can be expected to vary with the degree of tree survival (cf. Goldammer and Seibert, 1990). As many dead trees will remain standing for a period of time, the loss of canopy closure and interception storage due to fire can be expected to be not as large as for felling and removal of the trees. For obvious reasons, there are few field data on water budget effects of forest fires (Brown, 1972; O'Loughlin *et al.*, 1982). In April 1998, a forest fire hit the paired catchment experiment of Sabah Forest Industries and the Swedish University of Agricultural Sciences at Mendolong, Sabah, Malaysia. Regardless of vegetation type, which ranged from natural dipterocarp forest via secondary vegetation to exotic tree plantations, the response of evapotranspiration reflected the number of still standing and living trees and other vegetation (Malmer, *unpubl. data*).

Moving up in scale, again there are very few studies that have actually assessed changes in the water budget after large wild fires in logged-over forests, secondary vegetation, forest plantations and grasslands and larger areas burned for shifting cultivation or permanent cultivation. We know of none from the humid tropics. Immediate effects on evapotranspiration can be assumed to be similar to conversions including felling and burning (cf. Grip *et al.*, this volume). However, the type and thus the standing biomass of the vegetation that is burnt (primary or secondary rainforest or monsoonal forest) and the amount of fuel close to the ground will determine the survival of vegetation and the spatial variability of transpiration and evaporation. Furthermore, topography will also be of some importance as the lower parts of slopes are often wetter and winds during fires tend to make fire in upper slope positions more intense (Pyne *et al.*, 1996).

Soil physical properties, infiltration and surface runoff

The physical properties of the soil surface and the hydraulic properties of the subsoil determine the fate of the water arriving at the soil surface (Dunne, 1978). Few undisturbed humid tropical forest soils have been reported to have insufficient steady-state infiltration capacity to avoid regular surface runoff on forested slopes under normal rainfall regimes. Bonell and Gilmour (1978) described a case from Queensland, Australia, where a combination of high rainfall intensity and unfavourable soil hydraulic properties resulted in saturation overland flow as the dominant water pathway to the stream during storms (Bonell *et al.*, 1981). Similarly, under less extreme rainfall conditions but in an area with highly clayey soils in the western Amazon, Elsenbeer and Cassel (1990) found overland flow (both of the infiltration-excess and saturation types) to be rampant. Both soil infiltrability and hydraulic properties as well as rainfall, throughfall and stemflow delivery showed strong spatial variability (Elsenbeer *et al.*, 1992; Lloyd and Marques Filho, 1988). This may produce local surface runoff (Herwitz, 1986) or concentrate macropore flow in the topsoil of hillside depressions (Brammer and McDonnell, 1996), but in most cases surface runoff under undisturbed forested conditions has been shown to be negligible, e.g. in Peninsular Malaysia (Peh, 1981), Brazil (Northcliff *et al.*, 1990), West Africa (Roose, 1977) and Sabah, Malaysia (Malmer 1996b). For an early and comprehensive review of surface runoff (and erosion) in traditional tree-based ecosystems, see Wiersum (1984); cf. Bonell (this volume).

There is a great deal of scientific work on the effects of shifting cultivation on soil properties (see Lal, 1987, and Ramakrishnan, 1992, for reviews). The felling of the trees itself does not disturb topsoils too much, nor does it create much compaction unless heavily mechanised methods are used, which is not the case for shifting cultivation (Dias and Northcliff, 1985; Fritsch, 1993). However, the increased amounts of water in the soil due to reduced water use may lead to an increased risk of saturation overland flow, especially in the lower parts of slopes (Dunne, 1978; Bonell, 1993). As for the impact of fire on soil physical properties, this depends on the maximum temperatures reached, as well as the texture and mineralogical properties of the soil. Zinke *et al.* (1978) recorded temperatures at the soil surface of 400–600 °C for fires with a 'heavy' fuel load in reburn piles, and 200 °C for 'moderate' fuel loads in northern Thailand. Measurements by Ewel *et al.* (1981) in Costa Rica, Mackensen *et al.* (1996) in north-eastern Amazonia and Ketterings *et al.* (1999) in Sumatra were in the same range, with temperature levels again related to fuel amount. The depth of penetration of the temperature pulse into the soil depends on the soil's moisture content. Just below 100 °C soil water evaporates rapidly, arresting the rise in soil temperature until nearly all water has gone (Hartford and Frandsen, 1992). The thermal conductivity of the soil, on the other hand, decreases strongly when the soil dries out and dry topsoil can reduce penetration of the temperature pulse. Zinke *et al.* (1978) recorded a temperature of 70 °C at 3 cm depth for primary burns with a heavy fuel load, and temperatures in the 150–300 °C range at the same depth under reburn piles (in probably drier soil) for the same or slightly lower temperatures at the soil surface.

The impact of heat on soil physical properties relates to a loss of (specific fractions) of soil organic matter and resulting changes in pore distribution (Neary *et al.*, 1999). Most field measurements show little effect of slash-and-burn fires as such on soil bulk density but subsequent field activities like planting and trampling in the ashes may lead to some soil compaction, as the soil has lost part of the fabric that maintained its pore structure. At temperatures of 600 °C and above, the texture of a soil can change drastically as clay particles turn into small bricks, and certain mineralogical changes may start at temperatures of about 300 °C (Ketterings *et al.*, 2000). Land clearing by slash-and-burn usually involves piling up the debris for secondary burns. These piled-up fires are normally more intense (with intensity defined as the product of duration and temperature (Hartford and Frandsen, 1992)).

Hydrophobicity may occur with certain types of organic material and within certain temperature limits in topsoil. Neary *et al.* (1999) describe hydrophobicity in terms of a coating of soil aggregates or soil minerals in a discrete layer in the topsoil. This coating is formed above 176 °C but destroyed above 288 °C. Leitch *et al.* (1983) reported severely decreased infiltrability and subsequent increase in surface runoff with water-repellent soil properties remaining for several months after the occurrence of a wild fire in south-eastern Australia. Malmer and Grip (1990) on the other hand observed increased infiltrability of more than 100% compared with previous forest topsoil on otherwise physically undisturbed Ultisols/Acrisols right after fire in Sabah, Malaysia. They hypothesised that calcium released from the ash might have improved the soil structure initially. However, the same hillslopes later experienced a period of increased surface runoff (see below).

Subsequent to the immediate effects of fire on soil physical properties, soil organic matter can be expected to decline due to reduced inputs via litterfall and enhanced decomposition (higher temperature and soil moisture, Van Dam, 2001). Changes in soil organic matter during cropping and initial fallow periods have also been reported by, *inter alia*, Wadsworth *et al.* (1988) and Eden *et al.* (1991). As for effects of soil moisture, excess wetness of the soil after clearing can also be limiting for soil respiration (Ilstedt *et al.*, 2000) whereas desiccation of bare soil reduces structural development (Neary *et al.*, 1999).

Due to the variable effects of fire on soil physical properties (depending on fire intensity and its spatial variation) it is difficult to give typical values for the changes in surface runoff associated with fire. However, most field studies report moderate to large increases in overland flow (e.g. Mishra and Ramakrishnan, 1983; Collinet, 1984; Wiersum, 1984). Despite initial indications of higher infiltrability right after fire in Sabah, Malaysia (Malmer and Grip, 1990), Malmer (1996a and 1996b) reported increased surface runoff at a later stage but before the establishment of ground vegetation. Possible reasons include the development of hydrophobic properties in parts of slopes and gradual clogging and sealing of topsoil pores by fine ash particles and rain-splash. Including six months of observations on the same slope surface prior to soil disturbance, surface runoff after clearing and burning during a total period of nine months was still lower than that for an undisturbed control. This was thought to be due to the presence of drier topsoils after clearing, due to enhanced insolation, which tended to increase soil sorptivity and infiltration during rain events (Malmer, 1996b). Bruijnzeel (1990) discussed several examples of high rates of surface runoff during the second year of cropping in a shifting cultivation context which was ascribed to a gradual deterioration of soil structure under intensified use (cf. Mishra and Ramakrishnan, 1983; Lal, 1987). Bons (1990) reported a similar effect for field access trails that became gradually more compacted during *taungya* cultivation in upland Java.

So far in this review of the effects of fire on soil physical properties, infiltrability and surface runoff, we have concentrated on shifting cultivation, with some additional examples of other uses of fire. However, in relation to forest or wild fires in the humid tropics there appear to be no studies with respect to their effect on topsoil physical properties. Any inferences in this respect would have to rely on what has been stated above. As fires in primary or lightly logged stands can be expected to be light, the effects on soil organic matter will be correspondingly mild. As the intensity of wild fire increases in forests and vegetation types that are more open to drying and have more fuel lying about on the ground (e.g. logged-over forests; cf. Chappell, Tych *et al.*, this volume), so will the effects on soil physical properties and surface runoff increase. As in the case with high-intensity logging operations, during which up to 100 m^3 of timber may be removed per hectare of forest, up to one-third may be opened up (Bruijnzeel,1992). So too is the forest opened up when many large trees die after a forest fire and shed their leaves. The simultaneous addition of large volumes of extra litter and exposure of the forest floor to radiation creates conditions that will promote fire hazard (e.g. Phillips, 1987).

Streamflow

Because detailed studies of the immediate effects of shifting cultivation and forest fires on streamflow are scarce (Bailly *et al.*, 1974; Fritsch, 1993), this section necessarily bases its discussion mainly on the similarities in processes that can be drawn from findings concerning other land uses that include forest clearing and fire. Due to the reduction in evapotranspiration after fire, the normal effect on streamflow in a humid tropical environment is a distinct increase in total streamflow. Bruijnzeel (1990 and 1996) gave a general range of 125–800 mm y^{-1} for streamflow increase after complete removal of forest cover during the first years, with site rainfall being a major determinant. However, not only mean annual rainfall should be considered in this aspect but also inter-annual

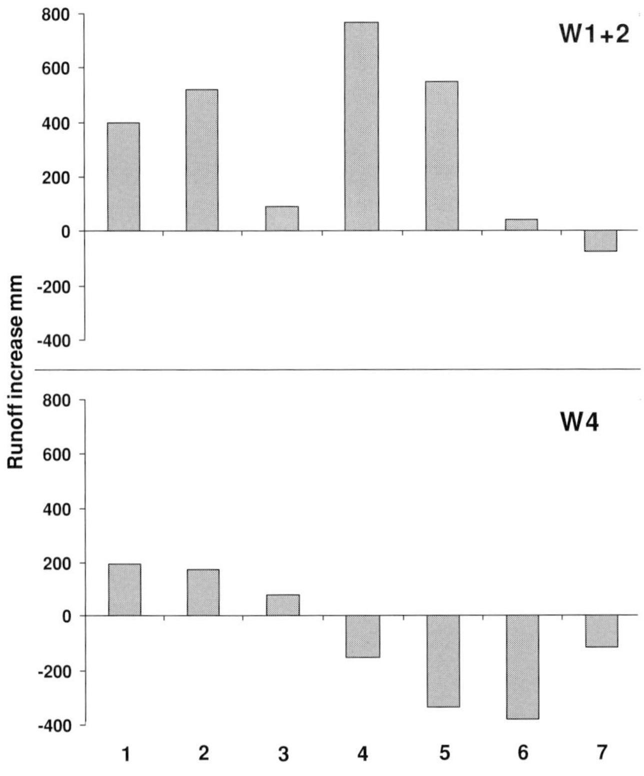

Figure 21.2 Annual runoff increase (mm, negative = decrease) as effect of establishment of *Acacia mangium* in Sabah, Malaysia. W1+2 was clearing of 4-year-old secondary vegetation after forest fire before planting, resulting in poor growth. W4 was plantation after clear-felling and wood extraction of lightly selectively logged rainforest where soil disturbance and burning was avoided before planting, resulting in relatively high production of plantation trees. See text for elaboration of relations between vegetation development and water use. Data for years 1–3 from Malmer (1992) and for later years previously unpublished.

variation in precipitation (Malmer, 1992), because precipitation is the most variable element of tropical climates (McGregor and Nieuwolt, 1998). The effect on streamflow will depend further on the portion of the catchment affected by the fire, but hardly at all on the spatial position of the latter (i.e. close to or further away from the stream). The return of streamflow to pre-burn levels will depend mainly on the water use by the regenerating vegetation plus the depth of rootable soil (Trimble *et al.*, 1963; Hibbert, 1967; Hornbeck *et al.*, 1993; see also Hölscher *et al.*, this volume). Figure 21.1 illustrates the rapid return to pre-clearing soil moisture levels during forest regrowth in Costa Rica (Parker, 1985).

Figure 21.2 depicts the declines in initial streamflow increases as observed during seven years after forest clearing using different methods in Sabah, Malaysia (Malmer, 1992 and unpublished data). Despite the fact that the biomass of the cleared secondary vegetation (which had experienced a forest fire four years earlier) in catchment W1+2 (cleared and burned before planting) was much smaller compared to the natural forest of catchment W4 (clear-felled but not burned before planting), the effect of burning on total streamflow amounts during the first year was distinctly greater (Figure 21.2). This relates to the fact that burning left the soil completely bare, while in the non-burned case there was still some vegetation which continued to use water after the clear-felling (Malmer, 1996b). In the following years the effects on streamflow depended on the water use by the newly planted vegetation. In catchment W4 the planted *Acacia mangium* performed well and developed a homogenous stand accumulating about 20 m^3 yr^{-1} (Nykvist *et al.*, 1996 and unpublished data). In catchment W1+2 on the other hand the *Acacia* trees performed much worse, despite intense weeding. The correspondingly reduced water use of the vegetation (Cienciala *et al.*, 2000) is reflected in the large and more prolonged streamflow increases (Figure 21.2). The more varied response with time of streamflow from catchment W1+2 can be interpreted as an effect of the vegetation succession after the fire. In the third year after planting, the above-ground biomass of the *Acacia* had increased to 20 t ha^{-1}, but in the fifth year it had again decreased to 16 t ha^{-1}. During the same period, the mass of such fire-promoted pioneer plants like *Neprolepis* fern, *Saccarum* grass and *Mikania* vines had increased by 1.1, 0.8 and 0.7 t ha^{-1} respectively (Sim and Nykvist, 1991; Nykvist *et al.*, 1996). Die-back of *Acacia* trees because of competition with especially *Mikania cordata* is probably the reason for the repeated increase in streamflow amounts in the fourth year (Figure 21.2).

As for effects of fire on stormflows, the 'variable source area concept' (Hewlett, 1961) stresses the relative importance of wet areas close to the stream and those with shallow soil (smaller storage) for streamflow generation. The two major potential off-site effects of fire on streamflow regime concern changes in stormflow and risk of flooding as well as the possibility of reduced delayed flows and water shortages during the dry season.

Bonell with Balek (1993) reviewed hill slope hydrological processes in relation to stormflow generation. They emphasised the interaction between high rainfall intensities, especially in tropical cyclonic climates as opposed to more continental or equatorial climates (cf. Callaghan and Bonell, this volume), in combination with patterns of sub-soil hydraulic conductivity. Bruijnzeel (1996) suggested that the most drastic changes in stormflow after fire (and soil disturbance) were recorded where the treatment induced a shift from subsurface pathways to overland flow as the dominant supplier of stormflow (e.g. Bailly *et al.*, 1974; Fritsch, 1992; Malmer, 1992, 1996b). In environments where runoff generation in natural forest is governed already to a high degree by superficial sources, the change in stormflows can be expected to be less dramatic (cf. Gilmour, 1977). Bruijnzeel (1990) discussed the results of more than ten studies of the effects of changes in land cover (including burning) on storm- and peak flows. He concluded that

Table 21.1. *Effects of (changes in) land cover on stormflow volumes and peak flows from selected studies including the use of fire*

Land use/conversion	Effect on stormflow volume			Effect on peakflow			Location and reference
	1974	1975		1974			
(1) Poor scrubland subjected to:							
• annual burning	+26%	+218%		+225%			Chandigarh, India; Gupta *et al.* (1974 and 1975)
• logging + overgrazing	+178%	+236%		+52%			
• overgrazing	+31%	+552%		+47%			
• forestation + trenches	−75%	−29%		−73%			
	Annual average			Mean annual maximum			
(2) Natural grassland to:							
• burned grassland	Increase of 100 mm y^{-1}			66% increase			Manankazo, Madagascar; Bailly *et al.* (1974)[a]
• improved cropping	reduction of 75 mm y^{-1}			53% decrease			
• pine plantation	reduction of 80 mm y^{-1}			69% decrease			
	1st year	2nd year	3rd year	1st year	2nd year	3rd year	
(3) Forest fire succession to:							
• poor *Acacia* plantation (W1+2; see Figure 21.2)	+35%	+83%	+76%	+68%	+111%	+58%	Sabah, Malaysia; Malmer (1992)
(4) Natural forest to:							
• *Acacia* plantation (W5)	ns	−30%	ns	ns	ns	ns	Sabah, Malaysia; Malmer (1992)
(5) Natural forest to:							
• shifting cultivation		+25%	+30%				French Guyana; Fritsch (1993)
• *Digitaria* grassland	+112%	+59%	+62%				
• (natural) forest	+147%	+42%	+37%				
• *Pinus* plantation	+79%	+58%	+32%				
• *Eucalypt* plantation	+58%	+42%	+11%				

[a] Annual means for 11 years.
Source: After Bruijnzeel (1990), Malmer (1992) and Fritsch (1993).

the most pronounced effects were observed after regular burning and reduction in ground cover by overgrazing. Table 21.1 is an extract from Bruijnzeel's 1990 table with subsequent published data added from Malmer (1992) and Fritsch (1993), and highlighting the few studies that included the use of fire. The scarcity of studies of the effects of fire in shifting cultivation, and the almost complete absence of data on the effects of natural/wild fire, is striking. In addition, most of the cited studies of changes in streamflow regimes following forest clearing and burning include some effects of mechanical soil disturbance, which interferes with the effect of the fire (Table 21.1). Although the fourth study in Table 21.1 represents an example of heavy soil disturbance, the relatively rapid development of a new vegetation cover (both *Acacia* trees and understorey) resulted in mostly insignificant increases in stormflow and peakflows. The reduction in the former during the second year was ascribed by Malmer (1992) to detention of surface flow in tractor tracks away from the stream. As erosion on the tracks created new, more efficient channels to the stream, in the third year stormflows and peakflows increased again (though not significantly so; Table 21.1). However, despite such soil disturbance effects, the effect of highly transpiring and soil covering vegetation may be more important to the reduction of storm and peakflows as indicated by the Indian case study with contour trenches and forestation and several other studies in Table 21.1. Finally, the critical role of differences in rainfall intensities between studies is rarely considered when discussing effects on peakflows (Fritsch, 1992; cf. Bonell with Balek, 1993).

In traditional shifting cultivation the mosaic of open, newly burned fields and patches of secondary vegetation in various stages of regeneration may be expected not to have dramatic effects on stormflows, except in small catchments with a spatial dominance of new fields (Qian, 1983). As discussed in previous sections, for natural fires in primary forests the effect on streamflow regimes may be less pronounced than that of shifting cultivation, but empirical evidence for this contention is lacking. Based on the field studies listed in Table 21.1, and the general discussion of the effects of fire on soil physical properties and surface runoff given above, it is plausible that large regional wildfires will have a more marked effect on stormflows than small-scale shifting cultivation. In such cases the effect might also be noticeable in at least medium sized rivers (with catchments up to about 100 km^2 in size). However, it is notoriously difficult to distinguish the various causes of increased peak flows for large basins due to their inherent heterogeneity in land use, relief, soil water storage and, above all, spatial variability in rainfall (Hewlett, 1982; Richey *et al.*, 1989; Bruijnzeel with Bremmer, 1989; Wilk, 2000). Despite such complications, Bruijnzeel (1990) stressed that cumulative effects may not be excluded, citing widened riverbeds as an example of deteriorated runoff regimes in extensive degraded areas in the Himalayas (Pereira, 1989) and Nigeria (Odemerho, 1984).

Local people in many parts of the tropics have claimed that clearing forest has resulted in reduced dry season flows. This may seem contradictory to the common notion by researchers that forest clearing increases the total amount of streamflow (Hamilton and King, 1983). However, the empirical observation of reduced baseflows and groundwater levels can in many cases be explained by a reduction in groundwater recharge due to lowered infiltration capacity by soil deterioration after clearing (Bruijnzeel, 1989; Sandström, 1995). Hence, in such cases any increases in baseflow that would be expected because of the reduction in evapotranspiration after clearing are exceeded by the amounts of water that are lost in increased stormflows (Bruijnzeel, 1989). As such, the eventual effect on dry season flow depends on a delicate balance between changes in vegetation water use and infiltration opportunities after the fire. The scale of traditional shifting cultivation is generally not large enough for such a marked reduction in dry season flows. Wilk (2000) has shown how the retention of heterogeneous land use can buffer the effects on streamflow in larger monsoonal watersheds in India and Thailand. Qian (1983) came to a similar conclusion for smaller catchments subject to shifting cultivation in Hainan, P.R.C. However, where fire assumes larger spatial dominance, or occurs more frequently, soil properties will risk serious deterioration over larger areas and the effect will gradually manifest itself, e.g. in the form of widened stream beds (Odemerho, 1984) and, presumably, also reduced low flows (Bruijnzeel, 1990).

EFFECTS OF FIRE ON EROSION AND SEDIMENTATION

Depending on the rate of particle detachment by rain drop impact and the spatial pattern of surface runoff, erosion on slopes may vary widely. Soil cover by understorey vegetation, herbs and litter greatly reduces drop impact (Wiersum, 1985), even though raindrop size is known to vary with the type of stand (Calder *et al.*, 1993). Concentrated surface runoff may also create severe erosion, even under an apparently lush vegetation cover in natural forest (Ruxton, 1967; Herwitz, 1986) as well as in forest plantations (Malmer *et al.*, 1998). Erosion by sheet wash or concentrated rills further depends on slope morphology (length, steepness and form). On the flatter parts of slopes (depressions, footslopes, alluvial fans), or in places with dense ground vegetation that tends to act as a filter, there will be deposition of eroded sediment (Douglas and Spencer, 1985). Sediment deposited on slopes or behind log dams in streams, however, should be seen as being stored temporarily, as large storms with long return period may reactivate such sediments (Spencer *et al.*, 1990; Malmer, 1996a; cf. Chappell, Tych *et al.*, and Douglas and Guyot, both this volume).

As stated earlier, fires as used in the context of shifting cultivation can be expected to have a more pronounced impact on the soil than the usually less intense fires that occur occasionally in primary forest. This is because the fuel load on the ground is larger and surface temperatures can thus reach higher values in shifting cultivation. Usually, the entire surface litter and root mat on top of the mineral soil is burnt, but charred remnants of woody roots may remain, as well as tree stems and thicker branches lying about. These may provide additional surface roughness and act as traps for sediment on the slope. The orientation of the felled trees has some influence on their subsequent trapping efficiency. For instance, Sabhasri (1978) documented how Lua' traditional shifting cultivators in northern Thailand make deliberate attempts to avoid rill and gully erosion by positioning logs across steep slopes in fields after burning, and by constructing log frameworks on small hillside drainage paths. During a recent study in Jambi (Sumatra, Indonesia), however, no preferential orientation of tree stems and branches remaining on the soil after burning could be detected: all angles from parallel to perpendicular to the slope occurred at equal frequencies, but with few exceptions the orientation was downhill (Prayogo, 2000). Professional foresters are trained to apply directional felling, as it helps to reduce damage to the surrounding trees (Dykstra and Heinrich, 1996; see also Cassells and Bruijnzeel, this volume), but this level of sophistication is rarely seen in shifting cultivation. Part of the tree stumps and their woody proximal roots may burn, leaving holes in the ground that may act as further sediment traps (Prayogo, 2000).

Fire normally destroys the litter layer and the seedbank it contains, but seeds of trees and pioneer species may survive in the soil, depending on their depth distribution and temperature tolerance (Whitmore, 1998). Vegetation recovery after fire, and its impacts on water balance and soil movement, thus depend on the intensity of the fire. Seeds are more likely to survive and germinate in relatively wet micro-patches, where they may later create a vegetative filter contributing to reduced sediment transport

(Van Noordwijk *et al.*, 1998a). Wiersum (1984) reviewed the results of numerous field trials on surface erosion in various tropical forest and tree based cropping systems. Although variability is high, both the median (0.3 t ha^{-1}) and range (0.04–7 t ha^{-1}) of reported erosion rates in natural forests and shifting cultivation fallows are roughly the same. As expected, maximum erosion rates during the cropping phase of shifting cultivation can be substantial (up to 70 t ha^{-1}, versus a median of 2.8 t ha^{-1}). However, at the lower end of the range erosion values are not too different from those encountered in natural forest (0.4 versus 0.04 t ha^{-1}, respectively). The high range in values obtained for erosion during the cropping phase after burning reflects both variability in natural settings and operations as well as indicating the potential for erosion reductions by wise management (cf. chapters by Critchley, Hamilton and Cassells and Bruijnzeel, all this volume).

EFFECTS OF FIRE ON HYDROCHEMISTRY

This section describes the changes in processes governing nutrient losses associated with the various types of fire. To cover all relevant nutrient losses, atmospheric losses during burning are discussed as well, even though this does not represent a true hydrological process.

Leaching losses of dissolved elements

There are several factors that interact to increase the risk of water-borne losses of dissolved elements to the stream after clearing (Jordan, 1985; Bruijnzeel, 1992; 1998; Malmer, 1996c). Firstly, the temporarily reduced water use of the vegetation increases the amounts of water passing through the soil column (Figures 21.1 and 21.2). Secondly, the most nutrient-rich parts of the felled biomass (leaves, etc.) will quickly start to decompose and be leached by rains, together with the ashes left by the fire. Thirdly, the inactivation of the root system of the former vegetation reduces drastically the possibilities for nutrient uptake by plants. Fourthly, increased soil temperatures (due to enhanced insolation and absorption of this energy by the blackened surface) may increase decomposition of soil organic matter and weathering (in younger soils) to further increase nutrient release. Fifthly, the inactivation of the deep roots of dead or felled trees may lead to larger quantities of the nutrients released by deep weathering becoming lost to the ecosystem, at least until new deep roots develop. And finally, there is the immediate nutrient loss to the atmosphere by volatilisation during burning.

Reduced soil biological activity directly after fire may however temporarily counteract increases in surficial decomposition because most of the soil fauna and flora on the forest floor and in the upper topsoil are killed (e.g. Andriesse, 1987). Mycorrhizal infections have been shown to be significantly reduced for a longer time after burning, which further reduces effective uptake of nutrients during the most sensitive period after burning. For example, Högberg and Wester (1998) observed 18% less mycorrhizal infection in cleared and burned soil compared to cleared and unburned soil one year after fire in Sabah, Malaysia. In the forest plantations replacing the original forest in that study, the burned and unburned sites showed more similar levels of mycorrhizal infection at the end of the rotation period, i.e. nine years after the fire (Lindquist, 1999). The role of deep tree roots in taking up mineral nutrients released from weathering in deeper soil horizons is difficult to investigate (cf. Brouwer, 1996; see also Proctor, this volume), but Andriesse and Schelhaas (1987a) emphasised the importance of living trees during the cropping phase of shifting cultivation to counteract initial enhanced nutrient leaching. Malmer *et al.* (1994) hypothesised that the prolonged occurrence of elevated concentrations of certain mineral nutrients (K, Ca, Mg, Si) in streamwater throughout a 2.5 year period after forest clearing (and fire) in Sabah could be the result of weathering in deeper soil horizons that had not yet been reached by new deep tree roots (cf. Malmer, 2004). By contrast, nitrogen (an element which is derived mainly from organic activity in litter and top-soil) experienced elevated concentrations during one year only after clear-felling and burning in this particular study.

Table 21.2 presents published data on enhanced solute leaching losses following the deliberate use of fire in tropical forests. In an earlier review, Bruijnzeel (1998) compared the losses associated with various kinds of forest disturbance (selective logging, clearing and burning) with nutrient gains via bulk precipitation and weathering to assess approximate ecosystem recovery times. The data collected in Table 21.2 clearly illustrate the increases in amounts of leached nutrients following fire as well as a trend of larger losses with increasing runoff (percolation) volumes.

The effect of shifting cultivation and light forest fires on nutrient leaching can be expected to be determined primarily by the size of the clearing and the amount of remaining living vegetation. Parker (1985) and Uhl *et al.* (1988) showed that soil water nutrient concentrations in small artificial tree fall gaps in Costa Rica and Venezuela did not differ from those observed below undisturbed forest. However, as the size of the clearing was increased to 500 m^2, soil water concentrations of nitrate, calcium and magnesium showed a response and remained at least twice as high during more than a year (Parker, 1985). Others have confirmed such effects for large clearings: e.g. for a 5000 m^2 clearing in Venezuela (including burning, Uhl, 1982; Uhl and Jordan, 1984) and a range of gap sizes up to 3440 m^2 in Guyana (Brouwer and Riezebos, 1998; Van Dam, 2001). In the latter study, soil water nutrient losses from larger gaps were substantially higher

Table 21.2. *Nutrient losses via leaching (kg ha^{-1} yr^{-1}) after tropical forest clearing and burning compared to the losses from forested control systems (in parentheses) for several sites experiencing different rainfall totals (annual control forest runoff, mm). Also indicated is the estimated time (years, second row) required to compensate leaching losses by nutrient inputs via bulk precipitation*

Treatment	Location	Annual runoff (mm)	N	P	K
			(kg ha^{-1} yr^{-1})		
Burning clear-felled forest for forest plantation establishment[b]	Sabah, Malaysia	1950	18 (6)[a] (3)	0.6 (0.4) (–)	59 (12) (34)
Burning 5-year-old secondary vegetation for forest plantation establishment[c]	Sabah, Malaysia	1950	10 (5)[a] (2)	0.9 (0.5) (–)	68 (20) (39)
Burning forest for cultivation[d]	Central Amazon, Brazil	1650	6.4 (3.6) (1)	0.3 (0.05) (3)	4.8 (0.4) (6)
Burning fallow for cultivation[e]	Belém, Brazil	1230	12.4 (0.0) (5)	0.4 (0.3) (<1)	3.9 (0.6) (2)

[a] Values of N rounded to whole kg.

[b, c] Malmer *et al.* (1994), catchments with 100% treatment, yearly losses are means over 33 months during and after treatments; controls are: (*b*) mean of two catchments with somewhat different geology (treatments on a mix of these geologies); (*c*) catchment with similar secondary vegetation, soil disturbance by crawler tractors in *b*). P not detected in bulk precipitation.

[d] Williams and Melack (1997), partially deforested catchment undergoing forest burning for cultivation for the last 5 years; control is the same catchment in forested state in 1984 (data from Lesack, 1993). Bulk precipitation input data from Lesack and Melack (1996).

[e] Hölscher *et al.* (1997), plots in fields using vacuum tube cup lysimeters and meteorological data; control is a nearby 10 year-old fallow. Annual runoff in forested state estimated by Klinge *et al.* (2001). Bulk precipitation input data from Hölscher *et al.* (1998).

than corresponding ones in the smaller clearings (Figure 21.3). Brouwer and Riezebos (1998) also compared soil water contents and quality in different patches of their clearings. They did not find any significant differences between zones located beneath the crowns of felled trees and other surfaces in their clearings. One reason for this may be that most of the crowns landed at the edges of the clearing. Hence, roots from the surrounding forest may have reduced amounts of water percolating through the soil (Brouwer and Riezebos, 1998). In a follow-up study, however, Van Dam (2001) did find increases in soil water concentrations of nitrate away from the gap edge in the largest gaps. The action by roots from the side in small gaps has been inferred also for smaller gap-sized clearings elsewhere in the tropics (Parker, 1985; Vitousek and Denslow, 1986; Uhl *et al.*, 1988). Although soil physical and chemical properties will influence the various processes to some extent, the studies cited were carried out on a range of soil textures and fertility levels, suggesting enhanced leaching following disturbance to be a general phenomenon.

As will be clear from the above examples, the size of a shifting cultivation clearing is most probably decisive for the risk of leaching. Even though the threshold size for detectable leaching is probably rather small (somewhere between 200 and 500 m^2) (Bruijnzeel, 1998; Van Dam, 2001), edge effects and the 'filtering' effect of vegetation surrounding the clearing will determine to what extent these on-site losses are lost in a downslope direction or into the nearest stream (Jordan, 1985; Van Noordwijk *et al.*, 1998b). There is clear evidence of the efficiency of intense biotic activity in the riparian zone as a control of nutrient losses to streams under forested conditions (McDowell *et al.*, 1992; Williams *et al.*, 1997; see also Proctor, this volume). However, until recently there have been no studies of the effects of riparian buffer zones on nutrient leaching to streams in connection with land use change in the humid tropics, even though such buffer zones are generally proposed as a conservation measure (e.g. Clinnick, 1985; Bruijnzeel, 1992; Bren, 1993). Williams *et al.* (1997) reported on nutrient concentrations in rainfall, overland flow, shallow throughflow, groundwater, stream bank seepage and streamwater in a catchment that was partially cleared for shifting cultivation in central Amazonia. They observed that solute transfers from leaching after clearing and burning were effectively diminished during passage of the water through a 15 m wide riparian zone, which was not cleared (Figure 21.4). In larger scale

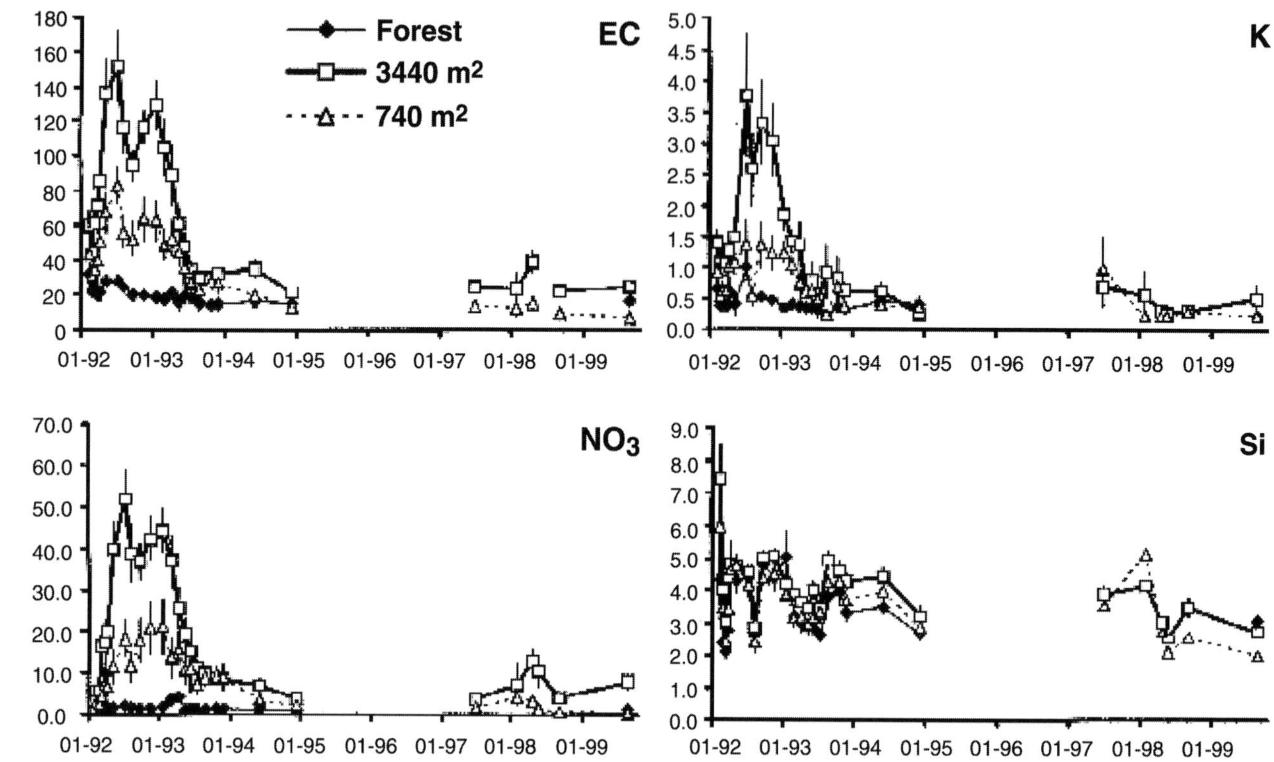

Figure 21.3 Soil water electrical conductivity (μS cm^{-1}) and concentrations (mg l^{-1}) of nitrate, potassium and silica under a rainforest gaps of different size in Guyana during seven years after logging. (After van Dam, 2001.)

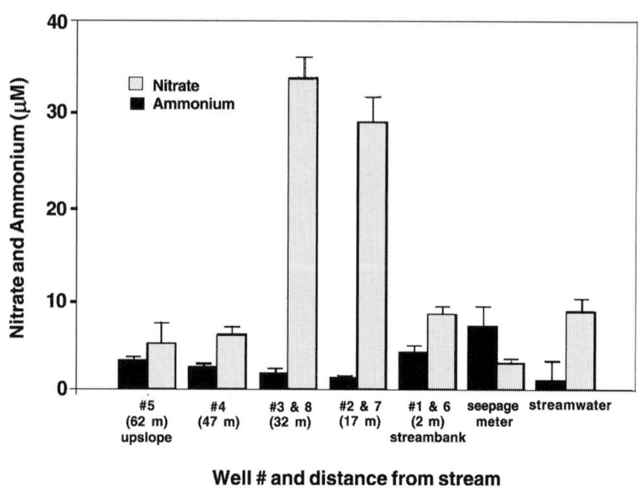

Figure 21.4 Comparison of NO_3^- and NH_3^+ along a transect from an upslope well (#5) to the stream in an Amazonian forest. Error bars represent standard errors. (After Williams *et al.* (1997) reproduced with kind permission from Kluwer Academic Publishers.)

operations and at higher densities of shifting cultivation, the dense drainage patterns prevailing in many humid tropical environments can be a problem when trying to implement and maintain riparian buffer zones of the required width throughout the landscape (cf. Bren, 1993; 1995; Thang and Chappell, this volume). For example, at high rainfall sites such as Mendolong, Sabah (Malmer, 1996b), the distance between first-order perennial streams is typically less than 100 m. The problem is compounded further by the observation of Pearce *et al.* (1980) that riparian buffer strips needed to be extended to zero-order gullies to be effective at retaining sediments and solutes carried in surface wash after clear-felling and burning a forest in New Zealand.

One clear difference between wildfires and forest fires on the one hand, and shifting cultivation on the other hand, is that the small-scale spatial heterogeneity in fire intensity that is typical of the former is normally less pronounced in the latter. Most shifting cultivators put a lot of energy into making the burn more even over the surface, so as not to leave anything unburned except for large woody debris. This is achieved by careful lopping, i.e. the cutting of residues into smaller pieces to make the fuel bed more homogenous. However, sometimes piling of residues is also practised to achieve a more effective burn of woody parts or parts that are still thought to be somewhat moist. Andriesse and Schelhaas (1987b) noted that the ash from such piles has very high alkalinity, which they feared would lead to increased leaching of any residual nitrate left after the combustion of the biomass.

As the size of most shifting cultivation fields is too small to cover a catchment of any reasonable size, the few studies that have attempted to measure leaching losses in the field during clearing and into the cropping phase have mostly used soil water modelling in combination with vacuum-tube lysimetry to sample soil water *in situ* in the deeper soil horizons (e.g. Cerri *et al.*, 1991; Hölscher *et al.*, 1997). Hölscher *et al.* (1997) reported leaching losses associated with shifting cultivation in north-eastern Brazil to be equivalent to only 1–5 years of nutrient inputs in precipitation (Table 21.2). One reason for these comparatively low leaching losses could be that the topography in the area is flat, combined with relatively low percolation (runoff). Another reason may relate to the effort put into homogenising the fuel bed as discussed above. The latter factor tends to make the effects of fire for clearing purposes other than shifting cultivation more heterogeneous, including the creation of micro-sites with higher concentrations of nutrients (cf. Gillman *et al.*, 1985). It would be an interesting exercise to compare relative differences in spatial variability in soil water chemistry for different clearing practices.

An alternative explanation for the low rate of nutrient leaching under shifting cultivation (Table 21.2) may be of a more methodological nature. The experiments relied on soil water nutrient concentrations down to a limited depth, thereby possibly not including nutrients released by weathering in deeper soil layers and not taken up by tree roots during the tree-less period (Malmer, unpublished data; cf. discussion above). This problem is not easily overcome as most shifting cultivation treatments are too small to cover a full catchment, whereas if the catchment also includes other land uses and additional landscape elements the effect measured at the outlet becomes difficult to interpret as well. One way to address this would be to install soil water samplers all the way down to the weathering front. Indeed, Eernisse (1993) in Brouwer (1996), observed slightly enhanced concentrations of Ca and Mg in the soil water just above the bedrock interface at 4.3 m in Guyana, although the data were too few to allow statistical testing. Schellekens *et al.* (2004) determined nutrient concentrations in soil water mixtures for soil samples taken at 50 cm intervals down to a depth of almost 7 m in Puerto Rico and observed a distinct increase in concentrations at depths of 5.5–6 m. Because geophysical soundings had indicated that the fresh bedrock was only found at depths of 20 m or more, the increase in concentrations were thought to mark a transition from highly leached rotten rock (saprolite) to less weathered material (Shellekens *et al.*, 2004).

In the case of light to medium intensity forest fires, once again, the amount of living vegetation after the fire will be an important determinant of the severity of leaching, but for obvious reasons hardly any literature exists on the subject. In April 1998 the catchments of the long-term forest conversion study in Sabah, Malaysia, referred to previously (Malmer, 1996b) were struck almost entirely by wildfires raging through the region at the end of a four-month dry period. This proved a rare occasion for the evaluation

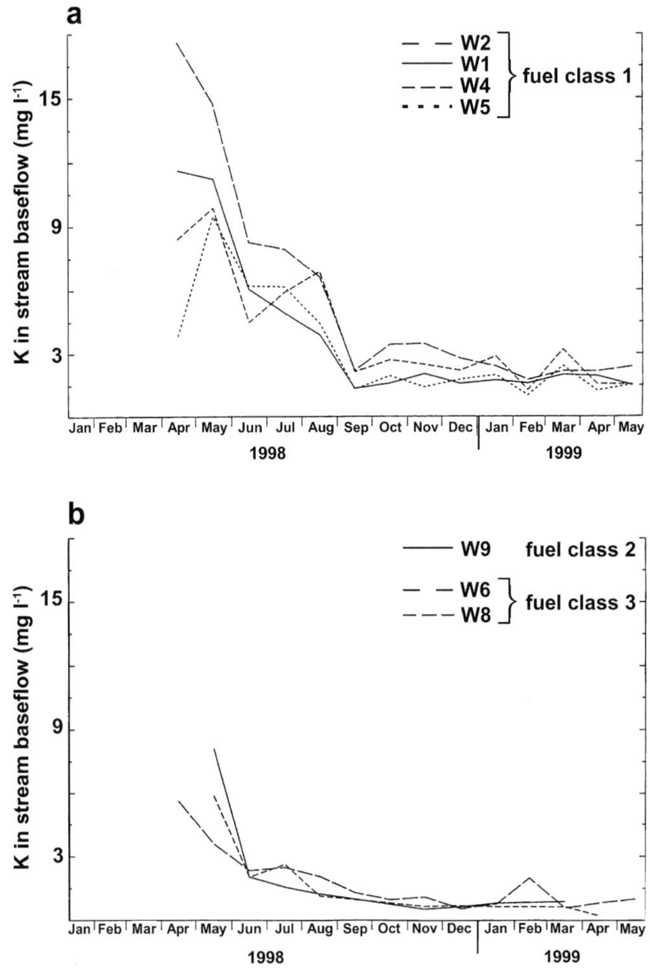

Figure 21.5 Concentrations (mg l^{-1}) of potassium in stream baseflow during one year after a wild fire in different vegetation types in April 1998 in Sabah, Malaysia (Malmer, 2004) (Reproduced with kind permission from Wiley).

of the effects of forest fire / wild fire on soil and water quality in different vegetation types. The 11 catchments under study at that time included control dipterocarp forest (lightly logged 17 years earlier), 2.5-year-old secondary vegetation and newly felled (but not entirely harvested) mature *Acacia mangium* plantations. Figure 21.5 shows some typical potassium concentrations in streamwater after the fires (Malmer, 2004). The patterns shown in Figure 21.5 not only provide insight into some of the aspects discussed above (e.g. nutrient retention) but they also illustrate the importance of the amount of fuel available for burning and its effect on the amount of nutrients released. In the closed dipterocarp forest the organic material on the forest floor was dry, but as there was little fuel, the fire was swift. Even though most large trees died from burning and thus nutrient uptake could be expected to be small, the small amounts of biomass that were burned on the forest floor resulted in the comparatively low concentrations observed in the streams. The young secondary vegetation still had more fuel from woody residues from previous clearing for shifting cultivation a few years before. In addition, most of the biomass consisted of dense and low, mostly herbaceous, vegetation with a few pioneer trees and low total aboveground biomass. Consequently, this young secondary vegetation burned more easily, which was reflected in the higher initial (but more short-lived) enhanced potassium concentrations in the streamwater (Figure 21.5b). Finally, the newly felled (10-year-old) forest plantation had not yet been fully harvested and large amounts of biomass were still lying on the ground awaiting to be extracted. In such places the fire was devastatingly hard (combusting most topsoil organic matter as well) and this resulted in the highest potassium concentrations in streamflow measured in the area so far throughout 15 years of measuring effects of various treatments and recovery (Figure 21.5a; cf. Malmer *et al.*, 1994, 1996b).

Particulate nutrient losses

Apart from dissolved nutrient losses, nutrients may also be removed in eroded particles carried by surface runoff. Arguably, focusing on particulate nutrient losses is especially important after fire, because ash is mobile and rich in nutrients (Pearce *et al.*, 1980; Andriesse and Schellhaas, 1987; Malmer, 1996d). To distinguish them from dissolved losses, particulate losses are defined here as mineral and organic materials with sizes exceeding 0.5μm. There is some methodological uncertainty in this respect because many studies of soil and streamwater losses do not state explicitly whether the water samples were filtered and if so at what filter retention capacity. With unfiltered water samples there is also an uncertainty as to the meaning of the reported concentrations. For example, analysis of dissolved elements with ion specific electrodes is not very sensitive to the presence of particles in suspension unless at high concentrations, whereas results obtained with atomic absorption spectroscopy or induced coupled plasma spectroscopy will partly include elements released from the particles during the analysis (Emteryd, 1989). In plot studies of surface runoff and erosion it is more common to analyse water and particles in suspension separately for their respective nutrient contents (e.g. Toky and Ramakrishnan, 1981; Hudson *et al.*, 1983; Maass *et al.*, 1988). However, this is still quite uncommon in the case of streamflow studies even though there is a wealth of tropical catchment sediment yield studies (Douglas, 1996; Bruijnzeel, 2004; cf. Douglas and Guyot, this volume).

For most studies conducted on small humid tropical forest streams, total dissolved solids (TDS) transport of elements dominates over elements carried as total suspended solids (TSS) for most of the time. This is contrary to the more general picture for most rivers of the world (Gregory and Walling, 1973), and can be explained by the intensive chemical weathering in the humid tropics and by the fact that the major output of sediment from undisturbed forest systems is connected to rare events of high meteorological and hydrological magnitude (Spencer and Douglas, 1985; Spencer *et al.*, 1990). However, there is considerable variation in the ratio between solute and particulate (nutrient) exports between different regions, depending on geomorphological and biological processes (Lewis *et al.*, 1995). From a single study investigating ratios of TDS and TSS in streams draining different geologies in Sabah, Malmer (1996a) noted that a stream draining less easily weathered sandstones had a DS/SS ratio of only 0.45 (considering only mineral elements and particles) versus a value of 2.2 for a nearby stream draining an area underlain by more easily weathered shales. Steep topography enhances the competence of rivers to generate and carry TSS whereas a more easily weathered and/or younger geology enhances TDS (Lewis *et al.*, 1995). Bedload transport is also part of the total transport of sediment by streams. However, bedload is notoriously difficult to estimate accurately (Task Committee, 1971; Douglas and Guyot, this volume), whilst estimates of its contribution to total nutrient export is practically lacking. Furthermore, in most cases, bedload contributions to total sediment transport make up less than 10% except in the steepest landscapes, where it can become dominant because of intensive mass wasting (Gregory and Walling, 1973). Furthermore, when considering nutrients, bedload material generally represents subsoil material of low nutrient content. Therefore, it should be of little importance from the perspective of ecosystem productivity.

Some studies have included quality analysis of eroded material after the use of fire in shifting cultivation (e.g. Toky and Ramakrishnan, 1981; Maass *et al.*, 1988; Brand and Wilfred, 1997 in Brand and Pfund, 1998; Williams *et al.*, 1997; Hoang-Fagerström, 2000) or in forest management (Hudson *et al.*, 1983) but there are none after a natural forest fire. The general picture is that eroded ash and soil particles show increased nutrient content and decreased carbon contents compared to topsoil quality. However, not all particles will end up in the stream but may settle on the lower parts of slopes or be filtered by ground vegetation (Malmer, 1996d; Williams *et al.*, 1997; Hoang-Fagerström, 2000). Malmer (1996d) studied concentrations of phosphorus in particles carried in hillslope surface runoff and streamflow as well as dissolved in streamwater in Sabah. This was done both for a forest control situation and for burned and unburned clear-felled catchments without riparian buffer zone vegetation. On burned slopes there was an increase in amounts of eroded particles. In addition, these particles were mostly of mineral origin. After clearing and burning there was a sharp increase in amounts of eroded particles which also had a distinctly lower organic matter

Table 21.3. *Relative nutrient losses (per cent) due to volatilisation during the burning of residual biomass in selected tropical forest areas as a function of fuel mass (t ha^{-1}) and mass reduction (per cent)*

Location	Fuel mass (t ha^{-1})	Mass reduction (%)	N	P	K	Ca	Mg	S	Reference
			(%)						
Turrialba, Costa Rica	38	83	23	negligible?				44	Ewel *et al.* (1981)
Belem, site 1	33	90	95	26	16	9	17	67	Mackensen *et al.* (1996)
Brazil, site 2	95	96	98	33	31	25	43	68	
Viti Levu, Fiji	40	86	84	52	79	78	60	–	Waterloo (1994)
IBGE, Brazilia D. F., Brazil (Cerrado)	10	84	45	43	49	38	–	66	Kauffman *et al.* (1994)

Source: After Bruijnzeel (1998) and Kauffman *et al.* (1994).

content and much higher phosphorus concentrations compared to eroded material after forest clearing without fire. This confirms observations from many reports that ash is easily eroded downslope (Ramakrishnan, 1992) and, depending on the situation, possibly into streams also (Pearce *et al.*, 1980). In the stream there were dramatically higher concentrations of dissolved P in stormflow after burning. However, particulate P only added 15–30% to the already enhanced dissolved amounts after clearing and burning (Malmer *et al.*, 1994). This was hypothesised to depend on the rate of dissolution of the ash as well as the amount of readily available P to bind to particles. Williams *et al.* (1997) reported similar observations following slash and burn in central Amazonian forest. Here, more than half of the total P export in the stream was in a dissolved form, even though streamflow was dominated by baseflow (see Bonell and Elsenbeer, this volume). The Malaysian stream had a much higher stormflow percentage (Malmer, 1992). It should be kept in mind, however, that a large portion of the particles carried in suspension by a stream does not necessarily derive from slope erosion but may also be supplied by stream bank erosion due to the increase in stormflows after clearing (e.g. Malmer 1996a). Rijsdijk and Bruijnzeel (1990) also reported a case with rates of bank erosion in cleared areas in East Java that exceeded overall sediment yields from forested headwater catchments.

Water quality

Apart from on-site aspects of nutrient losses and sustained soil fertility, additions of dissolved elements and particles to surface water or groundwater present an important environmental off-site effect of fire (DeBano *et al.*, 1996). Such effects relate to household use of water, fisheries, as well as to stream and coastal ecology and biodiversity. In Sabah, Malmer *et al.* (1994) noted that nitrate (NO_3-N) levels in stormflows after burning during forest plantation establishment were sometimes close to, but never exceeded, the recommended WHO limit of 10.3 mg l^{-1} for drinking water. However, after the intense wild fires of 1998, this limit was exceeded during some storm flow events on the catchments with the largest amounts of organic fuels. Soil water concentrations of NO_3-N may be expected to have been even higher than those in streamflow (Williams *et al.*, 1997; cf. Figure 21.4). The abovementioned limit has been reported to have been exceeded momentarily in clearings even without the use of fire in both Costa Rica (Parker, 1985) and Guyana (Brouwer and Riezebos, 1998; van Dam, 2001; Figure 21.3). Levels of Fe above the recommended WHO limits for potable use have also been observed in streamwater after heavy selective logging in Peninsular Malaysia (Zulkifli and Abdul Rahim, 1991).

Accelerated erosion after fire can certainly contribute to siltation of streams. The degree of siltation is clearly dependent on the degree of soil exposure, topographical and climatic settings, distances and obstacles between fields and streams, etc. (see also Douglas and Guyot, this volume). However, as a rule, the dominant cause of stream siltation after clearing is usually topsoil disturbance (Bruijnzeel, 1990; Fritsch and Sarrailh, 1996; Douglas, 1996). Both for specific types of soil disturbance, such as roads and foot paths, and on a more general note, there are many examples of improved practices to reduce adverse soil and water impacts during and after forest disturbance or conversion (see the respective chapters by Cassells and Bruijnzeel, Hamilton and Critchley, all this volume).

FIRE EFFECTS ON NUTRIENT LOSSES TO THE ATMOSPHERE

Most studies of nutrient losses to the atmosphere by biomass burning in the humid tropics have compared nutrient contents in biomass (plus ashes and topsoil) before and after the fire. Bruijnzeel (1998) reviewed several studies of atmospheric losses from burning. The reported percentage mass reduction of above-ground biomass by shifting cultivation fires ranges from 83% (Ewel *et al.*,

1981) to 97% (subhumid shrubland and grassland, Kauffman *et al.*, 1994) (Table 21.3). The main elements that are volatilised almost entirely within the normally occurring range of temperatures include carbon, nitrogen and sulphur, while losses of the other chief mineral nutrients occur mainly in the form of aerosols in smoke driven by thermally induced winds (Mackensen *et al.*, 1996; Tinker *et al.*, 1996). Right after the fire there is also a high risk of continued wind erosion of light 'white ash' which is mostly devoid of carbon but rich in mineral elements (Ewel *et al.*, 1981; Toky and Ramakrishnan, 1981; Peters and Neuenschwander, 1988).

Table 21.3 summarises the most important studies available to date. Nitrogen losses are high in most cases (23–98% of pre-fire contents of above- ground biomass) and relate to the high portion of biomass combusted, regardless of the temperature and intensity of the fire. On the other hand, atmospheric losses of mineral nutrients like phosphorus, potassium, calcium and magnesium will be more dependent on fire behaviour and temperature (Mackensen *et al.*, 1996). This is because a hotter fire will increase: (1) thermal lift; (2) production of carbon-devoid small particles for further transport as aerosols and white ash; as well as (3) the risk of true volatilisation. Hence, reported amounts of atmospheric losses of these elements are highly variable. In tropical grasslands relative losses of up to 52, 79, 78 and 60% have been reported for P, K, Ca and Mg, respectively (Kauffman *et al.*, 1994; Waterloo, 1994). Somewhat lower values were observed in central Amazonia after burning cleared forest debris: 33, 31, 25 and 43% for P, K, Ca and Mg respectively (Mackensen *et al.*, 1996), while burning of 7-year-old secondary forest elsewhere in (subhumid) Brazil induced losses that were intermediate between those for older forest and values for grassland and savannah (Table 21.3). This gradient of gradually smaller losses when going from grass through secondary vegetation to older forest must reflect the increasing portions of woody biomass that are combusted less efficiently. On the other hand, the presence of more wood could also lead to more intense fires and higher temperatures if sufficiently dried and with sufficient oxygen supply. Consequently, more 'woody' fires may be expected to show larger spatial variation. This is confirmed by the much higher standard errors associated with percentage losses of N and P for Cerrado vegetation compared to other vegetation types of lower stature in Brazil by Kauffman *et al.* (1994). Even though, relatively speaking, atmospheric nutrient losses are smaller in the case of burning tall forest (Table 21.3), such situations are possibly more problematic in terms of future maintenance of site productivity because the absolute losses are much larger. Furthermore, grasslands and savannahs generally have larger portions of their total ecosystem nutrient contents stored below ground (Kauffman *et al.*, 1994).

Nutrients contained in aerosols and smoke can be expected to be redeposited at some stage. Depending on local and regional weather conditions these (potentially large) transfers of nutrients may move over relatively short distances (e.g. trapped by local forest edges acting as filters), or as far as other continents (e.g. Prospero, 1999).

CONCEPTUALISATION OF PROCESSES AND EFFECTS OVER TIME

Clearly, from the previous process-oriented description of the hydrological and hydrochemical effects of fire, there appear to be intricate links and feedbacks with vegetational successions and land cover. These relationships vary in strength, in space and in time and describe relatively complex multi-process systems, which are not easily conceptualised without major simplification. However, the need to improve our understanding of the processes governing degradation and rehabilitation of tropical forest ecosystems under increasingly changing conditions, may justify reasonable simplifications in the conceptualisation of the various hypotheses offered in the previous sections, even though they may not always be proven yet. This section aims to conceptualise how the vegetation of the landscape elements (here: parts of a slope) may change after the repeated occurrence of fire and with it the hydrological and hydrochemical effects of fire. Figures 21.6 to 21.9 represent simplified conceptualised images of successional stages of vegetation over time after disturbance by various kinds of fire. Atmospheric and hydrological losses of nutrients are indicated as well as the general effects of fire on surface runoff and stormflow by variations in the size of the arrows.

Traditional shifting cultivation

Figures 21.6 A–D illustrate the various phases of traditional shifting cultivation which offers good possibilities for forest regeneration during a sufficiently long fallow period (Andriesse, 1987). Forest recovery in this context refers primarily to the restoration of pre-burn nutrient contents in biomass and soil. Nutrient contents in biomass during regrowth develop rather quickly (Uhl and Jordan, 1984; Andriesse, 1987; Nykvist *et al.*, 1994) whereas the full multi-layer structure and maximum height of a forest canopy takes much longer to develop (Saldarriaga *et al.*, 1987). Atmospheric losses of nutrients associated with the burning of biomass can be high (Table 21.3), but the water-borne losses may be reduced in various ways. Damage-limiting circumstances include: (1) the establishment of a successful fast growing crop capable of fully exploiting available nutrients, (2) limiting the size of the clearing and (3) ensuring the 'filtering' of nutrients in runoff and soil water on their way to the nearest stream by a buffer zone of forest (Figures 21.6B and 21.6C). If the number and sites of clearings per hillside are kept small, the effect on streamflow should be minimal. Assuming that the total atmospheric and hydrological losses of nutrients can be kept small in this way, thus maintaining reasonable soil quality in terms of organic matter and physical characteristics, then this should provide good conditions for fast and efficient secondary successional vegetation (Figure 21.6D).

Forest fire

Figures 21.7A and 21.7B conceptualise the occurrence and consequences of a forest fire in primary forest. Such a fire may occur after a severe drought (cf. chapter by Scatena *et al.*, this volume). As there is comparatively little fuel available for combustion under closed-canopy rainforest in the form of litter, such a fire will be generally swift and atmospheric losses relatively low (Figure 21.7A). Most trees may die, but will remain standing for some time, during which they may afford some soil protection by shedding their leaves and possibly even provide some degree of rainfall interception. Furthermore, the fire may be less intense on the lower, wetter parts of the slope where surviving trees may provide additional nutrient filtering and water use. There will be temporarily increased lateral flow and water-borne losses of nutrients to the stream, but they are likely to be modest compared to situations with more intense or more widespread fires. As in the situation portrayed in Figure 21.6D, site qualities associated with Figure 21.7B should be relatively favourable for efficient secondary succession, which ultimately goes back to the stage of Figure 21.6A.

Intensified shifting cultivation

A regenerating forest like that of Figure 21.6D may be used for intensified shifting cultivation (Figure 21.8A). In this case, the young vegetation is cleared before full rehabilitation of site qualities has been attained. Furthermore, a larger area is cleared for cultivation compared to the typical size of field in traditional shifting cultivation (Figure 21.8B vs. Figures 21.6B and 21.6C). Because the clearing is conducted before full site rehabilitation in terms of nutrients has been reached, subsequent crop yields can be expected

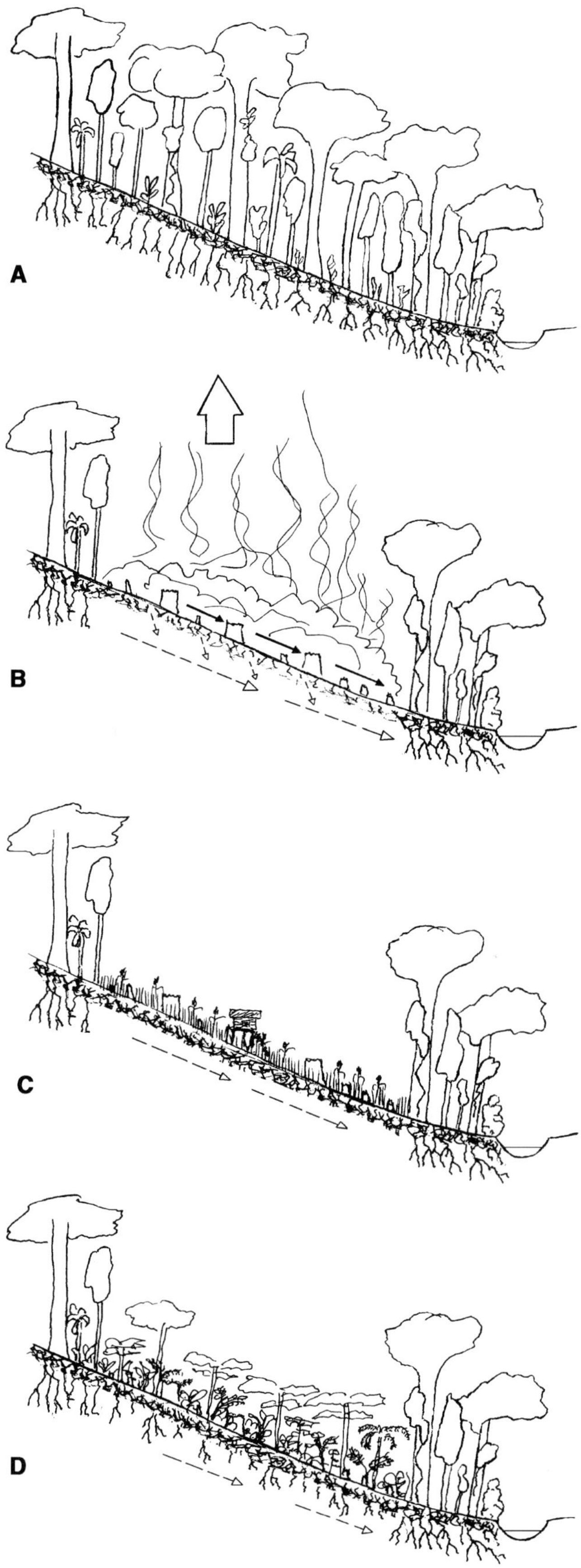

Figure 21.6 Conceptualised and generalised figures of hydrological pathways of nutrient losses through atmospheric losses, surface runoff and erosion and soil water leaching along a slope. (A) primary forest, (B) burning of cleared primary forest for traditional shifting cultivation, (C) cultivation phase of shifting cultivation and (D) secondary regrowth during the fallow. For descriptions of the processes at work and the size of the arrows in each figure, refer to the text. (Drawings by A. Malmer.)

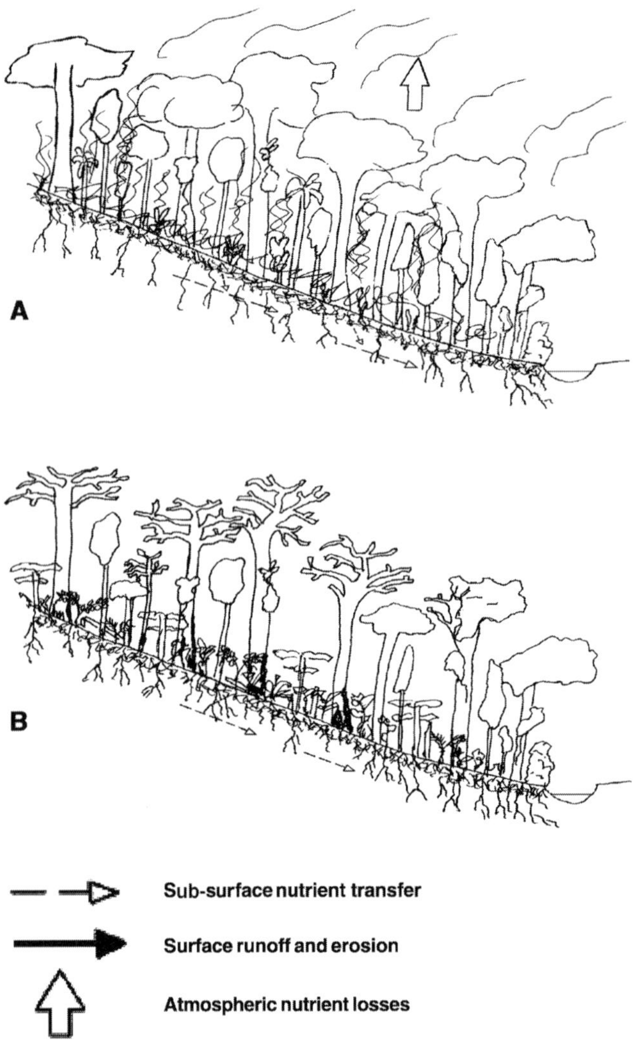

Figure 21.7 Conceptualised and generalised figures of hydrological pathways of nutrient losses through atmospheric losses, surface runoff and erosion and soil water leaching along a slope. (A) Forest fire in primary forest and (B) secondary regrowth after forest fire in primary forest. (Drawings by A. Malmer.)

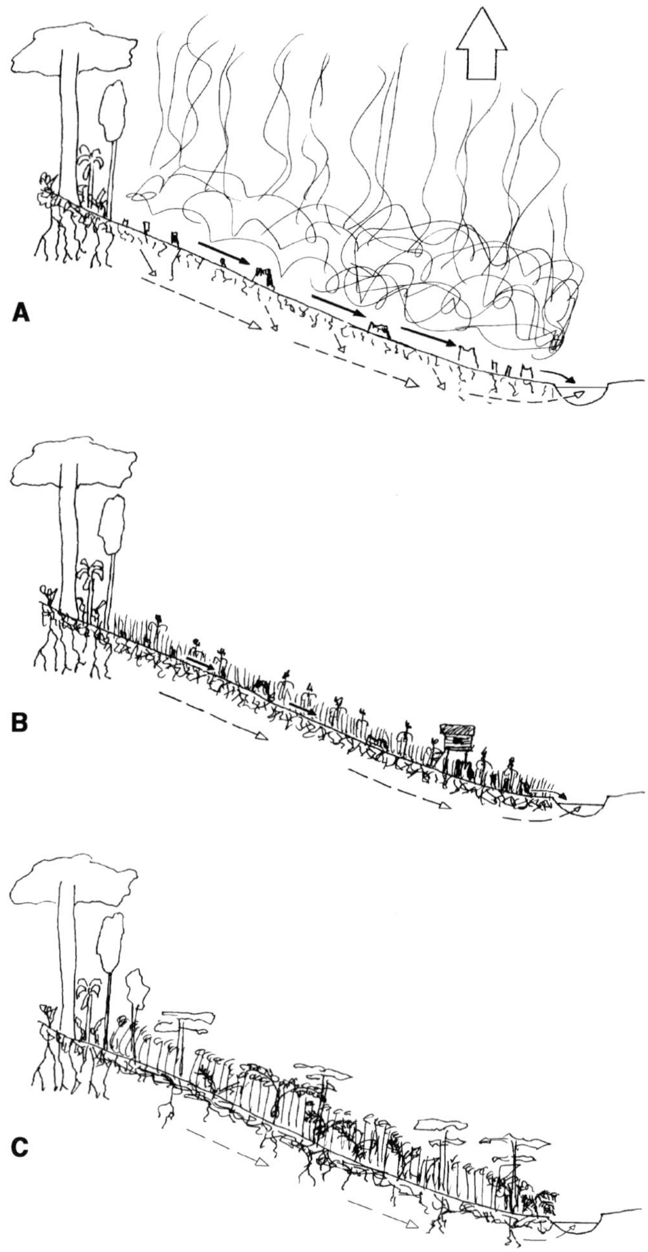

Figure 21.8 Conceptualised and generalised figures of hydrological pathways of nutrient losses through atmospheric losses, surface runoff and erosion and soil water leaching along a slope. (A) Clearing a larger area for cultivation in the secondary fallow situation of Figure 21.6D, (B) cultivation phase of (A), and (C) secondary regrowth after (B). (Drawings by A. Malmer.)

to be lower per unit area. This fact in itself is often one of the incentives for the creation of larger clearings (e.g. Wibowo *et al.*, 1997). With the fields being larger, the filter functions of remaining riparian forest can be expected to be weaker (Figure 21.8B) and as the relative area occupied by fields increases, there will be a corresponding effect on streamflow. In this case, the starting point is a situation with somewhat lower ecosystem qualities in terms of nutrient reserves and soil organic matter. In addition to that, the total losses associated with intensified shifting cultivation are likely to be larger than for traditional shifting cultivation (Figure 21.8 vs. Figure 21.6). Hence, in this case, secondary succession may be less efficient in terms of restoring site qualities over time (Figure 21.8C vs. Figure 21.6D).

Wild fires in logged forests and secondary forests

Figure 21.9 portrays how wild fires of varying frequency or intensity (e.g. associated with ignition by humans, possibly in combination with seasonal or long term drought) may affect vegetation recovery and hydrological processes. Taking a selectively

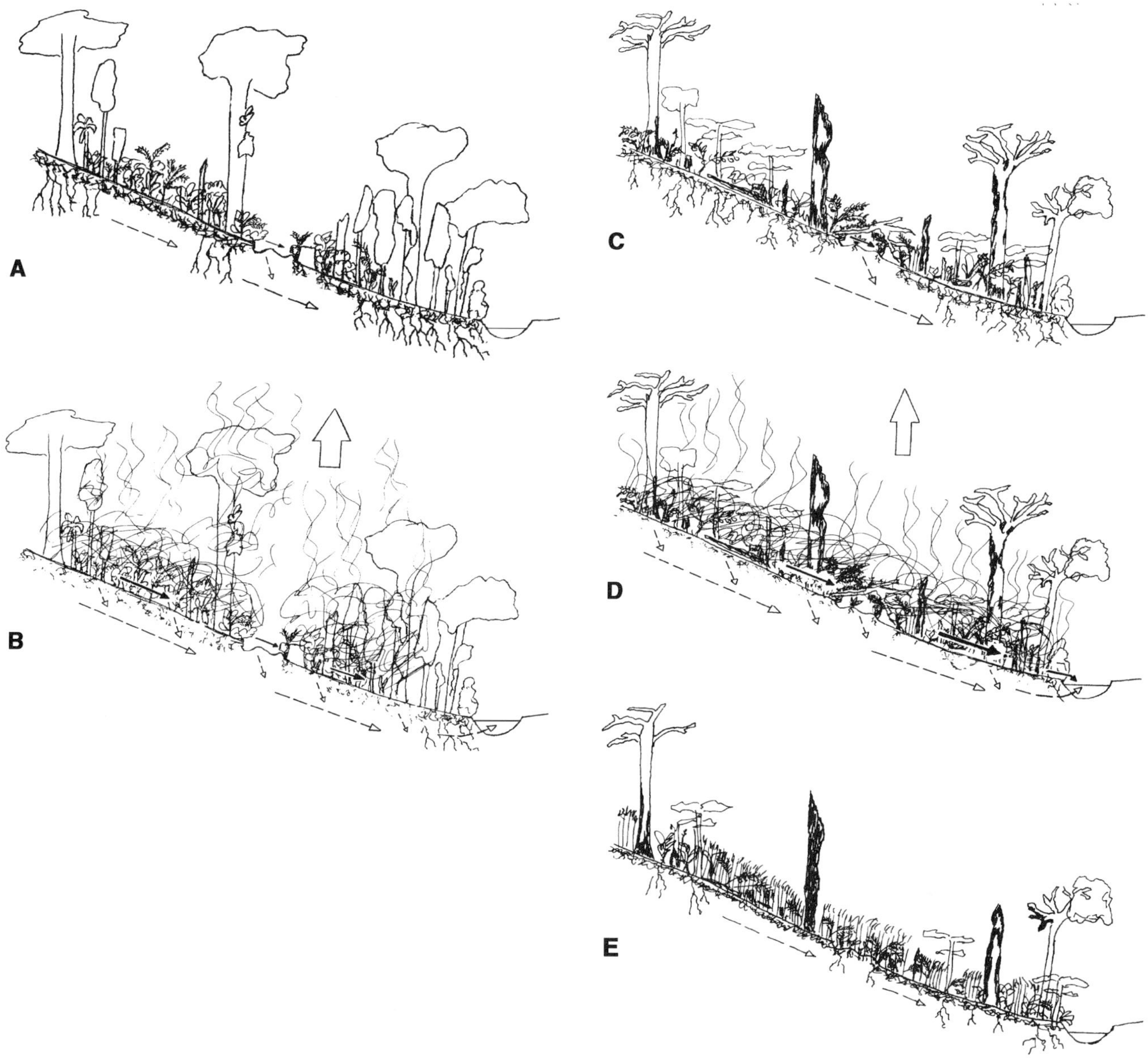

Figure 21.9 Conceptualised and generalised figures of hydrological pathways of nutrient losses through atmospheric losses, surface runoff and erosion and soil water leaching along a slope. (A) Selectively logged primary forest with some soil disturbance by tractors, (B) wild fire in selectively logged forest, (C) secondary regrowth after wild fire in selectively logged forest, (D) wild fire in secondary regrowth after previous logging and fire (C) and (E) grass-dominated secondary regrowth after wild fire in (D). (Drawings by A. Malmer.)

logged forest as the starting point (Figure 21.9A), it will be clear that compared with the undisturbed forest of Figure 21.6A, the logged forest contains numerous openings with secondary vegetation or heavily disturbed soil (former tractor tracks or roads). The nutrient status of the disturbed parts of the forest will be highly variable, depending on the degree of disturbance (Gilman *et al.*, 1985; Bruijnzeel, 1992; van Dam, 2001). The secondary vegetation in the openings easily dries during extended dry periods and abandoned logging debris further adds to the amount of fuel on the forest floor. Hence, when a fire occurs in logged-over forest it will be more intense than in the undisturbed situation. There is also a higher risk of developing into a crown fire, with accordingly larger losses to the atmosphere (Figure 21.9B). Circumstances for rehabilitation of site quality are not very good because of the combined effects of the previous logging and the fire. However, because the spatial variability of these cumulative effects will be very high, their prediction becomes rather uncertain (Figure 21.9C). As described earlier for the situation of Figure 21.8C (regrowth following intensified shifting cultivation) the efficiency of site recovery is likely to be relatively low due to the combined constraints imposed by lower nutrient retention in poorly regenerating vegetation and low soil quality (particularly at micro-sites disturbed by logging). Increases in streamflow are likely to be significant, especially when the affected portion of the catchment is large and will be added to any increases already prevalent since the previous logging.

Because the degree of openness and amounts of fuel potentially available for combustion (e.g. in dead wood and litter on the forest floor) will be particularly large now, any subsequent fire occurring in this low vegetation can be expected to be highly intense. The associated atmospheric losses may be substantial, even though the total amount of nutrients accumulated in the biomass is far less than that of a natural forest (Figure 21.9D). Consequently, nutrient reserves after this second fire can be expected to be lowered considerably, as will be the efficiency of subsequent site recovery (Figure 21.9E). This situation is often called the 'vicious circle of fire', in which a site becomes progressively poorer in nutrients, seeds, etc. to the extent that finally a spontaneous forest succession cannot be supported any more and only fire climax grassland remains (Quimio, 1996) (Figure 21.9E).

All kinds of young secondary vegetation must be considered to be more vulnerable to drying during droughts and thus more susceptible to wild fire (Goldammer, 1988). One important aspect of this increased risk of wild fires (and forest fires) relates to increased human activity (e.g. Uhl, 1998), which changes the kind of vegetation and increases the spatial dominance of young vegetation, but also increases the intensity of the use of fire, both in space and time. Hence, spatial distribution of vegetation types and land uses at the landscape level becomes an important factor in determining effects of fire.

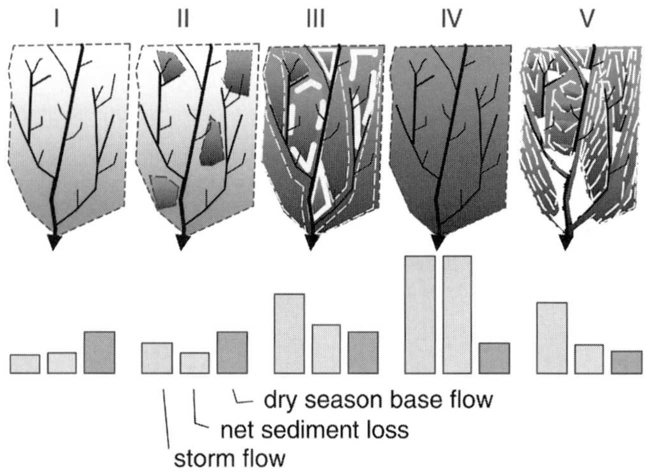

Figure 21.10 Schematic development of the landscape in a subwatershed and its effects on storm flow, net sediment loss and dry-season base flow. I: Original forest cover. II: Patches of forest opened for shifting cultivation. III: Intensification of land use has brought most land into cultivation, except for riparian zones and hedges along paths. IV: Reclamation of all 'wastelands' has removed all filter strips causing a disproportional rise in net sediment loss. V: Restored agroforestry landscape with permanently vegetated contour strips and riparian woodlands. (After Van Noordwijk *et al.* (1998b), reproduced with kind permission from CAB International.)

EFFECTS OF FIRE AT THE LANDSCAPE LEVEL

So far, this chapter has dealt mainly with on-site and off-site impacts of fire for more or less homogenous and unilateral changes in land cover (e.g. from forest to field). However, real-world landscapes usually consist of complex mosaics of forests in different phases of regeneration and various land-uses with different fire regimes and hydrology. With respect to erosion, the net loss of sediment per unit area depends strongly on the scale of measurement, and attempts to scale up from small plots to catchments may easily be wrong by an order of magnitude (Walling, 1983; Bruijnzeel, 1990). In the present context, where shifting cultivation fields include concave sections at the bottom end of a slope, these may act as a sink in which much of the eroded sediment (and the nutrients it contains) is captured (Zinke *et al.*, 1978). Recent measurements in Jambi, Sumatra, demonstrated hardly any soil loss to the stream from a field cleared by slash-and-burn, despite high within-field rates of sediment movement; an upland rice crop growing on the footslopes provided an effective sediment trap during the first few months after clearing, until weeds on the middle and higher sections of the slopes started to develop (Prayogo, 2000). Such 'filter effects' may increase crop yields downslope from a site where yields are temporarily lowered by erosion and so maintain both overall hillslope productivity and reduce off-site effects in streams. This holds for particulate erosion and sedimentation as well as for leaching and retention/uptake of nutrients. It is thus important in this respect to not only see the one-dimensional interaction but also the more complex two- or even three-dimensional interactions in a landscape. In other words, lateral transfers of matter need to be taken into account (Figure 21.10). As a consequence, erosion and sedimentation processes will affect the heterogeneity of soil fertility within a landscape (Van Noordwijk *et al.*, 1998b; Van Noordwijk and Lusiana, 1999). Despite the complexity of real-world landscapes, scaling up from an understanding of processes in homogenous land use experiments to real landscapes under change can help to develop more realistic models for land management.

Van Noordwijk *et al.* (1998b) provided a tentative classification of landscapes ranging from completely forested, via landscapes with recently opened plots with still intact riparian or mid-slope

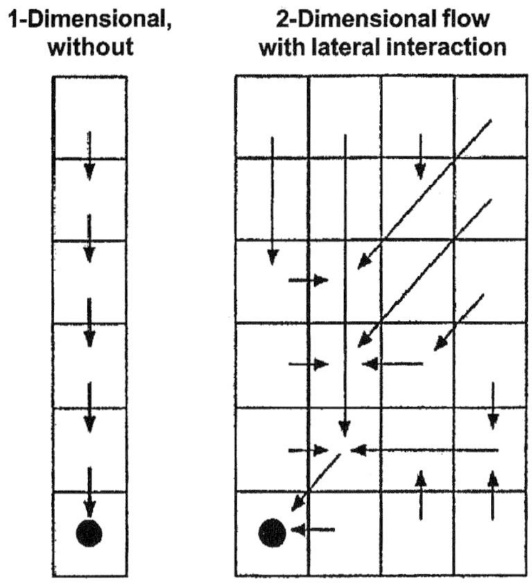

Figure 21.11 Lateral interactions in a 1-dimensional runoff model for a series of grid cells and a 2-dimensional frame where each outflow from a grid cell goes to the lowest of its eight neighbouring cells. (After Van Noordwijk *et al.* (1998b), reproduced with kind permission from CAB International.)

filter zones, to landscapes that are fully utilised for agriculture and lack vegetative filter functions or landscapes where such filter functions are deliberately restored (Figure 21.11). Patch-level erosion rates can be high in the second type of landscape, without necessarily having much impact on the sediment load of streams and rivers, as long as the riparian filter zones remain effective (Figure 21.11b). However, small incidents in the riparian strips, or gullies, footpaths and roads cutting through the filter zones, can have a substantial impact by impairing their efficiency (Pearce *et al.*, 1980; Bons, 1990). Landscape level impact assessment of land use change thus cannot be based on simple one-dimensional cause and effect statements, but has to consider the combined impacts of an increased demand for (and reduced supply of) filter functions compared to the forested state (Van Noordwijk *et al.*, 1998b).

Land reclamation patterns may differ between farmers of different cultural and ethnic backgrounds. For example, in the Sumber Jaya benchmark area (Lampung, Sumatra, Indonesia) studied by ICRAF and partners, a shifting cultivation-derived system of coffee-fallow cycles by local Sumendo farmers contrasts with the more intensive cultivation style of migrant farmers from West and Central Java. They first leave riparian strips intact; they do not weed intensively in the young coffee gardens and allow a bush fallow vegetation to take over after the main harvest period of the coffee (3–5 years after planting). The second group of farmers buy old coffee gardens or fallow land from the first, rejuvenate the coffee, weed intensively and transform the riparian strips and valley vegetation into rice paddies. The rice paddies may provide some filtering effect (cf. Purwanto, 1999; Prayogo, 2000), but possibly less so than the spontaneous vegetation in the valley bottoms used to do before conversion to agriculture. The overall impact of land use change on water and sediment flows at the landscape level is thus complex. However, to be successful, interventions aimed at reducing negative environmental impacts should at least include the complexity of land use decisions that affect plot-level land use (cf. the chapter by A. Hall, this volume), the presence and effectiveness of vegetative filters, and the channelling effects of paths, roads and gullies (zero-order streams).

CONCLUDING REMARKS

In the introduction of this chapter it was stated that fire in the humid tropics can lead to increased forest loss and undesirable off-site effects. At the same time it is a natural disturbing agent and a tool that is often used with great care in traditional cultivation systems to adapt to biogeochemical processes for optimum sustained productivity and desired ecosystem functions. The undesirable effects of fire are often caused by interactions between natural variations in climate and human-induced changes in ecosystem use and structure. Human-induced changes in global climate, in combination with intensified land use involving the use of fire, may further speed up these problem complexes over the next few decades. From this perspective a better, process-based understanding of hydrological processes and the associated pathways for plant nutrient transfers is absolutely crucial. Below, various suggestions are offered for research priorities to help achieve this. Furthermore, to be useful to policy makers involved in land management issues, our current process understanding must be applied more to develop improved agricultural practices and other uses of the land. This creates a demand for closer links between basic systems ecological and hydrological research, applied land use trials and distributed, landscape-scale modelling.

Towards integrated fire policy

This review has focussed to a high degree on how processes interact and how they are affected by fire. However, it is important to stress once again the importance of the geological and pedological setting as basic variables determining a landscape's vulnerability to changes in vegetation cover. For example, the Ultisols / Acrisols and Inceptisols / Cambisols associated with younger and steeper landscapes often have a lighter textured topsoil and thus are often more susceptible to erosion. By contrast the Oxisols /

Ferralsols of older and less steep landscapes will be more sensitive to additional nutrient losses upon disturbance because they are more weathered and nutrient poor (Baillie; 1989; Van Wambeke, 1998).

On the basis of the quantification of nutrient losses via hydrological and atmospheric pathways listed in Tables 21.2 and 21.3, it can be noted that atmospheric nutrient losses from burning in general seem to be higher than the losses by leaching (Bruijnzeel, 1998). This is the case not only for nitrogen, but often also for mineral nutrients, even though the latter are highly variable between sites (Table 21.3). These high atmospheric losses have sparked recommendations for the reduction of burning and the implementation of fire-free practices when clearing land (e.g. Mackensen *et al.*, 1996; Hamilton, this volume). Burning also increases leaching losses compared to non-burning practices (Malmer *et al.*, 1994). The traditional and modern use and misuse of fire are not easily changed though. For example, research results linking exclusion of fire and minimum soil disturbance during plantation forest establishment with subsequently doubled production of wood and strongly reduced environmental impacts in East Malaysia (Nykvist *et al.*, 1994) have not automatically paved the way for a reduced use of fire in that region. The large amounts of felling debris produced upon clearing tall tropical forest make the preparation of ground surface for subsequent planting and cultivation difficult and labour intensive. Practical alternatives for mechanical windrowing and burning of slash that could be adopted easily by the plantation industry or farmers are still lacking. Researchers need better knowledge about process linkages within landscapes and the tools (models) to evaluate the consequences of improved vs. old practices in a spatial and temporal dynamic context (see chapters on distributed modelling and remote sensing application by Chappell, Bidin *et al.* and by Held and Rodrigues, both this volume). Furthermore, such efforts should be conducted in co-operation with both farmers and policy makers. Nevertheless, for the vast number of farmers that still will not be able to afford fertilisers in the years to come, fire will continue to be an efficient tool for cultivation. Therefore, the evaluation of regionally suitable fallow periods within traditional or improved systems that include filter functions to check erosion and leaching losses will have to be promoted. To help prevent forest fires and wild fires, recommendations have also been made for improved fire protection practices in forest operations (e.g. Pinard *et al.*, 1999). However, fire prevention efforts and causative knowledge of ecological and biogeochemical linkages are not enough to deal efficiently with the problem. Socio-economic factors also have to be involved to develop comprehensive policies aiming to reduce the use and impacts of fire. It is not only that the growing numbers of shifting cultivators need to be sustained with food. They also need to be given a decent income as they are entering market economies, where abandoned old functional production systems need to be exchanged with new functional ones (Tomich *et al.*, 1998).

Recommendations for biogeochemical research

New biogeochemical research on the effects of fire should focus more on linkages between processes of water use, runoff generation and lateral nutrient transfers, especially below ground. In this context, lateral filter functions like retention of leached nutrients and eroded sediments in riparian forests appear crucial (Van Noordwijk *et al.*, 1998b). Such nested approaches at landscape level are difficult to follow in any system, perhaps even more so in humid tropical ecosystems where variability in space and time may be higher than elsewhere and the number of datasets fewer (Bonell with Balek, 1993; Bruijnzeel, 1996). At present, too little is known, not only of the retention of eroded soil particles in the landscape, but also of nutrient uptake by deep roots and mycorrhiza or the (biological) retention of nutrients in riparian zones to make firm recommendations on the best practices in the spatial and temporal complexities of a landscape (cf. Proctor, this volume). Only recently, mycorrhizal infection of fine roots at great depths have been shown from Asia (Boström, 2000) and from the Amazon (Nepstad *et al.*, 2001). Improved soil water sampling (i.e. reducing soil disturbance upon sampling) and modelling of soil water use and drainage at short time resolutions using recording tensiometers seems a promising way to explore small-scale spatial variations in nutrient transfer (Klinge *et al.*, 2001; van Dam, 2001). Furthermore, the division of nutrient transport in particulate and dissolved forms needs to be better covered by future analyses of nutrient transport dynamics at various levels of scale (cf. Malmer, 1996d; Williams *et al.*, 1997). The study of root activity in deeper soil horizons should preferably also be coupled with studies of weathering inputs in gradients throughout deep soil profiles (Burnham, 1989; Brouwer, 1996; Bruijnzeel, 1998; Schellekens, 2004). As to atmospheric losses by fire, initial attempts to quantify filtering effects of aerosols by surrounding or local vegetation would be desirable. Particles have been shown to cross oceans (e.g. Prospero, 1999) but, depending on local weather, some might also be retained in nearby tree crowns, hence there is a need for more integrative, multi-disciplinary studies. We also need better information on the relationships between fuel load in different vegetation types, soil moisture, burning intensity and subsequent nutrient retention.

Finally, it is important that humid tropical biogeochemical research is not only descriptive but also takes part in and explores the usefulness of new findings from other regions. For example, the recent finding that mycorrhizal hyphae may actively contribute to weathering by penetration of minerals (Van Breemen *et al.*, 2000), or the finding that some plants are able to take up organic nitrogen, and not only nitrate and ammonium (Näsholm *et al.*, 2000), may change our perception on nutrient use and retention in tropical forest ecosystems. The time is ripe for a concerted effort in the field of environmental hydrology and biogeochemistry in the tropics (Bruijnzeel, 1996; 1998). The increasing frequency and severity of fires in the humid tropics call for such efforts, not least in secondary and degraded forests. Once again, a nested approach to water and nutrient sources, pathways and outputs in the landscape and the landscape elements that includes the latest in spatial data collection and distributed modelling, is called for.

References

Adejuwon, J. O., Balogun, E. E. and Adejuwon, S. A., 1990 On the annual and seasonal patterns of rainfall fluctuations in sub-Saharan West Africa. *International Journal of Climatology* 10:839–848.

Alegre, J. C. and Cassel, D. K., 1996. Dynamics of soil physical properties under alternative systems to slash-and-burn. *Agriculture, Ecosystems and Environment* 58:39–48.

Andriesse, J. P., 1987 Monitoring project of nutrient cycling in soils used for shifting cultivation under various climatic conditions in Asia. Final report, Research report, Royal Tropical Institute, Amsterdam, 141 pp.

Andriesse, J. P. and Schelhaas, R. M., 1987a A monitoring study of nutrient cycles in soils used for shifting cultivation under various climatic conditions in tropical Asia. II. Nutrient stores in biomass and soil – results of baseline studies. *Agriculture, Ecosystems and Environment* 19:285–310.

Andriesse, J. P. and Schelhaas, R. M., 1987b A monitoring study of nutrient cycles in soils used for shifting cultivation under various climatic conditions in tropical Asia. III. The effects of land clearing through burning on fertility level. *Agriculture Ecosystems and Environment* 19:311–332.

Applegate, G., pers. comm. *Research Division, CIFOR, P.O. Box 6595, JKPWB, Jakarta 10065, Indonesia.*

Bailly, C., Benoit de Cognac, G., Malvos, C., Ningre, J. M. and Sarrailh, J. M., 1974 Etude de l'influence du couvert naturel et de ses modifications à Madagascar; expérimentations en bassins versants élémentaires. *Cahiers Scientifiques du Centre Technique Forestier Tropical* 4:1–114.

Baillie, I. C., 1976 Further studies on drought in Sarawak, East Malaysia. *Journal of Tropical Geography* 43:20–29.

Baillie, I. C., 1989 Soil characteristics and classification in relation to the mineral nutrition of tropical wooded ecosystems. In: J. Proctor (ed.) *Mineral Nutrients in tropical forest and savanna ecosystems.* Blackwell Scientific Publishers, Oxford, 15–26.

Bandara, C. M. and Kuruppuarachchi, T. A., 1988 Land-use change and hydrological trends in the upper Mahaweli basin. Paper presented at the Workshop on Hydrology of Natural and Man-made Forests in the Hill Country of Sri Lanka, Kandy, October 1988, 18 pp.

Beaman, R. S., Beaman, R. H., Marsh, C. V. and Woods, P. V., 1985 Drought and forest fires in Sabah in 1983. *Sabah Society Journal* 8:10–30.

Bastable, H. G., Shuttleworth, W. J., Dallarosa, R. L. G., Fisch, G. and Nobre, C. A., 1993 Observations of climate, albedo and surface radiation over cleared and undisturbed Amazonian forest. *International Journal of Climatology* 13:783–796.

Bonell, M., 1993 Progress in the understanding of runoff generation dynamics in forests, *Journal of Hydrology* 150:217–275.

Bonell, M., 1998 Possible impacts of climate variability and change on tropical forest Hydrology. *Climatic Change* 39:215–272.

Bonell, M. and Gilmour, D. A., 1978 The development of overland flow in a tropical rainforest catchment. *Journal of Hydrology* 39:365–382.

Bonell, M., Gilmour, D. A. and Sinclair, D. F., 1981 Soil hydraulic properties and their effect on surface and subsurface water transfer in a tropical rainforest catchment. *Hydrological Sciences Bulletin* 26:1–18.

Bonell, M. with Balek, J., 1993 Recent scientific developments and research needs in hydrological processes of the humid tropics. In: M. Bonell, M. M. Hufschmidt and J. S. Gladwell (eds.) *Hydrology and water management in the humid tropics. Hydrological research issues and strategies for water management.* UNESCO, Paris and Cambridge University Press, Camridge, pp. 167–260.

Bons, C. A. 1990 Accelerated erosion due to clearcutting of plantation forest and subsequent taungya cultivation in upland West Java, Indonesia. In: R. R. Ziemer, C. L. O'Loughlin and L. S. Hamilton (eds.) *Research needs and applications to reduce erosion and sedimentation in tropical steeplands.* IAHS Publications No. 192:279–288.

Bosch, J. M. and Hewlett, J. D., 1982 A review of catchment experiments to determine the effect of vegetation changes on water yield and evapotranspiration. *Journal of Hydrology* 55:3–23.

Boserup, E., 1965 The conditions of agricultural growth. George Allen and Unwin, London, 124 pp.

Boström, B., 2000 Abundance of roots and mycorrhiza in deep soil horizons under a humid tropical forest. *Stencilserie* 61, Swedish University of Agricultural Science, Department of Forest Ecology, Umeå, Sweden, 23 pp.

Brammer, D. D. and McDonnell, J. J., 1996 An evolving perceptual model of hillslope flow at the Maimai catchmnet. In: M. G. Anderson and S. M. Brooks (eds.) *Advances in Hillslope Processes.* John Wiley and Sons, Chichester, 961–973.

Brand, J. and Pfund, J. L., 1998 Site- and watershed-level assessment of nutrient dynamics under shifting cultivation in eastern Madagascar. *Agriculture, Ecosystems and Environment* 71:169–183.

Bren, L. J., 1993 Riparian zone, stream, and floodplain issues: a review, *Journal of Hydrology* 150:277–299.

Bren, L. J., 1995 Aspects of the geometry of riparian buffer strips and its significance to forestry operations. *Forest Ecology and Management* 75:1–10.

Brouwer, L. C., 1996. Nutrient cycling in pristine and logged tropical rainforest. A study in Guyana. PhD thesis, Department of Physical Geography, Utrecht University, the Netherlands. *Tropenbos-Guyana Series* No. 1, Tropenbos-Guyana Programme, Georgetown, Guyana.

Brouwer, L. C. and Riezebos, H. Th., 1998 Nutrient dynamics in intact and logged tropical rainforest in Guyana. In: A. Schulte and D. Ruhiyat (eds.) *Soils of tropical forest ecosystems: characteristics, ecology and management.* Springer Verlag, Berlin, 73–86.

Brown, J. A. H., 1972 Hydrologic effects of a bush fire in a catchment in South-eastern New South Wales. *Journal of Hydrology* 15:77–96.

Brown, N., 1993 The implications of climate and gap microclimate for seedling growth conditions in a Bornean lowland rainforest. *Journal of Tropical Ecology* 9(2):153–168.

Bruijnzeel, L. A., 1989 (De)forestation and dry season flow in the tropics: a closer look. *Journal of Tropical Forest Science* 1:145–161.

Bruijnzeel, L. A., 1990 Hydrology of moist tropical forests and effects of conversion: A state of knowledge review. UNESCO – IHP, Humid Tropics Programme, Paris, 224 pp.

Bruijnzeel, L. A., 1992 Managing tropical forest watersheds for production: where contradictory theory and practice co-exist. In: F. R. Miller and K. L., Adam (eds.) *Wise Management of Tropical Forests.* Oxford Forestry Institute, Oxford, 37–75.

Bruijnzeel, L. A., 1996 Predicting the hydrological impacts of land cover transformation in the humid tropics: the need for integrated research. In: J. H. C. Gash, C. A. Nobre, J. M. Roberts and R. L. Victoria (eds.) *Amazonian deforestation and climate.* Institute of Hydrology, pp. 15–55.

Bruijnzeel, L. A., 1997 Hydrology of forest plantations in the humid tropics. In S. E. K. Nambiar and A. G. Brown (eds.) *Management of soil, nutrients and water in tropical plantation forests.* ACIAR, Canberra, Australia, 125–167.

Bruijnzeel, L. A., 1998 Soil chemical changes after tropical forest disturbance and conversion: The hydrological perspective. In: A. Schulte and D. Ruhiyat (eds.) *Soils of tropical forest ecosystems: characteristics, ecology and management.* Springer Verlag, Berlin, 45–61.

Bruijnzeel, L. A., 2004 Hydrological functions of tropical forests: not seeing the soil for the trees. *Agriculture, Ecosystems and Environment*, doi:10.1016/j.agee.2004.01.015.

Bruijnzeel, L. A. with Bremmer, C. N., 1989 Highland-lowland interactions in the Ganges Brahmaputra river basin: a review of published literature. *ICIMOD Occasional Paper*, 11, International Centre for Integrated Mountain Development, Kathmandu, Nepal.

Burnham, C. P., 1989 Pedological processes and nutrient supply from parent material in tropical soils. In: J. Proctor (ed.) *Mineral Nutrients in tropical forest and savanna ecosystems*. Blackwell Scientific Publishers, Oxford, 27–41.

Calder, I. A., Hall, R. L. and Prasanna, K. T., 1993 Hydrological impact of *Eucalyptus* plantation in India. *Journal of Hydrology* 150:635–648.

Cerri, C. C., Volkoff, B. and Andreaux, F., 1991 Nature and behaviour of organic matter in soils under natural forest, and after deforestation, burning and cultivation, near Manaus. *Forest Ecology and Management* 38:247–257.

Charles-Dominique, P., Blanck, P., Larpin, D., Ledru, M-P., Riéra, B., Sarthou, C., Servant, M. and Tardy, C., 1998 Forest pertubations and biodiversity during the last ten thousand years in French Guiana. *Acta Oecologica* 19(3):295–302.

Cienciala, E., Kucera, J. and Malmer, A., 2000 Tree sap flow and stand transpiration of two *Acacia mangium* plantations in Sabah, Borneo. *Journal of Hydrology*, 236:109–120.

Clinnick, P. F., 1985 Buffer strip management in forest operations: a review. *Australian Forestry* 48:34–45.

Cochrane, M. A. and Schulze, M. D., 1998 Forest fires in the Brazilian Amazon. *Conservation Biology* 12(5):948–950.

Cochrane, M. A., Alencar, A., Schulze, D., Souza Jr., C. M., Nepstad, D. C., Lefebvre, P. and Davidson, E. A., 1999 Positive feedbacks in the fire dynamic of closed canopy tropical forests. *Science* 284:1832–1835.

Collinet, J., 1984 Hydrodynamique superficielle et érosion comparées de quelques sols ferrallitiques sur défriches forestières traditionelles (Cote d'Ivoire). *Challenges in African hydrology and water resources*, IAHS Publication No. 144 499–516.

Cossalter, C. and Cauvin, B., 1998 L'Indonesie et ses feux de foret. *Informations Foret, Afocel*, 566, Association Foret Cellulose (AFOCEL), Paris, 6 pp.

DeBano, L.-F., Ffolliott,-P. F. and Baker,-M. B.-Jr., 1996 Fire severity effects on water resources. US Dept. of Agriculture, Forest Service, Rocky Mountain Forest and Range Experiment Station, *General technical report RM* GTR-289:77–84.

Dias, A. C. C. P. and Nortcliff, S., 1985 Effect of tractor passes on the physical properties of an oxisol in the brazilian Amazon. *Tropical Agriculture* (Trinidad) 62:137–141.

Douglas, I., 1996 The impact of land-use changes, especially logging, shifting cultivation, mining and urbanization on sediment yields in humid tropical South East Asia: a review with special reference to Borneo. *IAHS Publications* 236:463–471.

Douglas, I and Spencer, T., 1985 Present-day processes as a key to the effects of environmental change. In: I. Douglas and T. Spencer, *Environmental change and tropical geomorphology*, George Allen and Unwin, Boston, 39–73.

Douguedroit, A., and Bart, F., 1997 Winter dust haze in tropical Southern Africa. *International Journal of Climatology* 17:725–744.

Dunne, T., 1978 Field studies of hillslope flow processes. In: M. J. Kirby (ed.) *Hillslope Hydrology*. John Wiley, Chichester, pp 227–293.

Dykstra, D. P. and Heinrich, R., 1996 *FAO Model Code of Forest Harvesting Practice*. FAO, Rome, 85 pp.

Eden, M. J., Furley, P. A., McGregor, D. F. M., Milliken, W. and Ratter, J. A., 1991 Effects of forest clearance and burning on soil properties in Northern Roraima, Brazil. *Forest Ecology and Management* 38:283–290.

Elsenbeer, H. and Cassel, D. K., 1990 Surficial processes in the rainforest of western Amazonia. In: R. R. Ziemer, C. L. O'Loughlin and L. S. Hamilton (eds.) *Research needs and applications to reduce erosion and sedimentation in tropical steeplands*. IAHS Publications 192:289–297.

Elsenbeer, H., Cassel, D. K. and Castro, J., 1992 Spatial analysis of soil hydraulic conductivity in a tropical rainforest catchment. *Water resources Research* 28:3201–3214.

Emteryd, O., 1989 Chemical and physical analysis of inorganic nutrients in plant, soil, water and air. Swedish University of Agricultural Science, Department of Forest Site Research, *Stencil*, No. 10, Umeå, Sweden, 181 pp.

Esser, G., 1992 The role of the tropics in the global carbon budget: Impacts and possible developments. In: J. G. Goldammer (ed.) *Tropical forests in transition*. Birkhäuser, Basel, 241–252.

Ewel, J., Berish, C. and Brown, B., 1981 Slash and burn impacts on a Costa Rican wet forest site. Ecology 62(3):816–829.

Fang, M., Zheng, M., Wang, F., To, K. L., Jaafar, A. B. and Tong, S. L., 1999 The solvent-extractable organic compounds in the Indonesian biomass burning aerosols – characterization studies. *Atmospheric Environment* 33(5):783–795.

Flenley, J. R., 1979 The equatorial rainforest: a geological history. Butterworths, London,162 pp.

Flenley, J. R., 1992 Palynogogical evidence relating to disturbance and other ecological phenomena of rainforests. In: J. G. Goldammer (ed.) *Tropical forests in transition*. Birkhäuser, Basel, 17–24.

Flenley, J. R., 1998 Tropical forests under the climate of the last 30,000 years. *Climatic Change* 39:177–197.

Fleming, T. H., 1986 Secular changes in Costa Rican rainfall: correlation with elevation. *Journal of Tropical Ecology* 2:87–91.

Fritsch, J. M., 1992 Les effects du défrichement de la foret Amazonienne et de la mise en culture sur l'hydrologie des petits bassins versants. *ORSTOM Editions*, Paris.

Fritsch, J. M., 1993 The hydrological effects of clearing tropical rainforest and of the implementation of alternative land uses. *IAHS Publication* 216: 53–66.

Fritsch, J. M. and Sarrailh, J. M., 1986 Les transports solides dans l'écosystème forestier tropical humide en Guyane: les effets du défrichement et de l'installation de paturages. *Cah. ORSTOM, Série Pédol.* 22: 93–106.

Fölster, H., 1992 Holocene autochthonous forest degeneration in Southeast Venezuela. In: J. G. Goldammer (ed.) *Tropical forests in transition*. Birkhäuser, Basel, 25–44.

Gedney, N. and Valdes, P. J., 2000 The effect of Amazonian deforestation on the northern hemisphere circulation and climate. *Geophysical Research Letters* 27:3053–3056.

Giambelluca, T. W., 1996 Tropical land cover change: characterizing the post-forest land surface. In: Giambelluca, T. W. and Henderson-Sellers, A. (eds.) *Climate Change: Developing Southern Hemisphere Perspectives*. J. Wiley, Chichester, 293–318.

Giambelluca, T. W., 2002 The hydrology of altered tropical forest. Invited Commentary. *Hydrological Processes*, 16:1665–1669.

Giambelluca, T. W., Fox, J., Yarnasarn, S., Onibutr, P., and Nullet, M. A., 1999. Dry-season radiation balance of land covers replacing forest in northern Thailand. *Agric. For. Meteor.* 95:53–65.

Giambelluca, T. W., Nullet, M. A., Ziegler, A. D. and Tran, L., 2000 Latent and sensible energy flux over deforested land surfaces in the Eastern Amazon and northern Thailand. *Singapore Journal of Tropical Geography*, 21:107–130.

Gillman, G. P., Sinclair, D. F., Knowlton, R. and Keys, M. G., 1985 The effect on some chemical properties of the selective logging of a north Queensland rainforest. *Forest Ecology and Management* 12:195–214.

Gilmour, D. A., 1977 Effect of rainforest logging and clearing on water yield and quality in a high rainfall zone of Northeast Queensland. In: *Proceedings of the Brisbane Hydrology Symposium 1977*, Institution of Engineers Australia, Canberra, 156–160.

Goldammer, J. G., 1988 Rural land-use and wildland fires in the tropics. *Agroforestry Systems* 6(3):235–252.

Goldammer, J. G., 1992 Tropical forests in transition. Ecology of natural and antropogenic disturbance processes – An introduction. In: J. G. Goldammer (ed.) *Tropical forests in transition*. Birkhäuser, Basel, 1–16.

Goldammer, J. G. and Seibert, B., 1989. Natural rainforest fires in Eastern Borneo during the Pleistocene and Holocene. Naturwissenschaften 76:518–520.

Goldammer, J. G and Seibert, B. 1990 The impact of drought and forest fires on tropical lowland rainforest of East Kalimantan. *Ecological studies* 84:11–31.

Goldammer, J. G. and Price, C., 1998 Potential impact of climate change on fire regimes in the tropics based on MAGICC and a GIS GCM-derived lightning model. *Climatic Change* 39:273–296.

Granberg, H., Ottosson Löfvenius, M. and Odin, H., 1993 Radiative and aerodynamic effects of an open pine shelterwood on calm clear nights. *Agricultural and Forest Meteorology* 63:171–188.

Gregory, K. J. and Walling, D. E., 1973 *Drainage basin form and process A geomorphological approach*. Arnold, London.

Gupta, R. K., Mishra, P. R., Shankar, P., Kaushal, R. C. and Sajwan, S. S., 1974 *Studies on the effect of different land treatments on water yield in the*

Siwalik – Chandigarh. Annual report 1974, Central Soil and Water Conservation Research and Training Institute, Dehradun, India, pp. 18–21.

Gupta, R. K., Mishra, P. R., Mittal, S. P. and Singh, K., 1975 *Effect of different land treatments on water yield in the Siwalik - Chandigarh*. Annual report 1975, Central Soil and Water Conservation Research and Training Institute, Dehradun, India, pp. 15–17.

Hamilton, L. S. and King, P. N., 1983. *Tropical forested watersheds. Hydrologic and soils response to major uses or conversions*, Westview Press, Colorado.

Hartford, R. A. and Frandson, W. H., 1992 When it's hot, it's hot . . . or maybe it's not! Surface flaming may not portend extensive soil heating. *International Journal of Wildland Fire* 2(3):139–144.

Henderson-Sellers, A., 1993 Continental vegetation as dynamic component of a global climate model: a preliminary assessment. *Climatic Change* 23:337–377.

Herwitz, S. R., 1986. Infiltration-excess caused by stemflow in a cyclone-prone tropical rainforest. *Earth Surface Processes and Landforms* 11:401–412.

Hewlett, J. D., 1961 Watershed management. In: *Report for 1961 Southeastern forest experiment station*, US Forest Service, Asheville, N.C., pp 61–66.

Hewlett, J. D., 1982 Forests and floods in the light of recent investigation. In: *Hydrological Processes of Forested Areas*. Nat. Res. Council of Canada Publ. 20548, NRCC, Ottawa, 543–559.

Hibbert, A. R., 1967 Forest treatment effects on water yield. In: W. E. Sopper and H. W. Lull (eds.) *Internationa Symposium on Forest hydrology*. Pergamon Press, Oxford, 525–543.

Hoang-Fagerström, M. H., 2000 Agroforestry systems in Northern Vietnam with *Tephrosia candida* as an alternative to short-fallow crop rotations. PhD-thesis,m Swedish University of Agricultural Science, *Acta Universitatis Agriculturae Sueciae, Agraria*, 213, 45 pp.

Hornbeck, J. W., Adams, M. B., Corbett, E. S., Verry, E. S. and Lynch, J. A., 1993 Long-term impacts of forest treatments on water yield: a summary for northeastern USA. *Journal of Hydrology* 150:323–344.

Hudson, J, Kellman, M., Sanmugadas, K. and Alvardo, C., 1983 Prescribed burning of *Pinus oocarpa* in Honduras II. Effects on nutrient cycling. *Forest Ecology and Management* 5:283–300.

Högberg, P. and Wester, J., 1998 Root biomass and symbioses in *Acacia mangium* replacing tropical forest after logging. *Forest Ecology and Management* 102:333–338.

Hölscher, D., Möller, R. F., Denich, M. and Fölster, H., 1997 Nutrient input – output budget of shifting agriculture in Eastern Amazonia. *Nutrient Cycling in Agroecosystems* 47:49–57.

Hölscher, D., Sá, T. D. de A., Möller, R. F., Denich, M. and Fölster, H., 1998 Rainfall partitioning and related hydrochemical fluxes in a diverse and in a mono specific (*Phenakospermum guyannense*) secondary vegetation stand in eastern Amazonia. *Oecologia* 114:251–257.

Ilstedt, U., Nordgren, A. and Malmer, A., 2000 Optimum soil water for soil respiration before and after amendment with glucose in humid tropical Acrisols and a boreal mor layer. *Soil Biology and Biochemistry* 32:1591–1599.

Jackson, I. J., 1983 The tragedy of our tropical rainforests. *Ambio* 12:252–254.

Jordan, C. F., 1985 *Nutrient cycling in tropical ecosystems*. John Wiley, Chichester.

Kanae, S., Oki, T. and Musiake, K., 2001 Impact of deforestation on regional precipitation over the Indochina Peninsula. *Journal of Hydrometeorology* 2:51–70.

Kauffman, J. B., Cummings, D. L. and Ward, D. E., 1994. Relationships of fire, biomass and nutrient dynamics along a vegetation gradient in the Brazilian cerrado. *Journal of Ecology* 82:519–531.

Ketterings, Q. M., Wibowo, T. T., Van Noordwijk, M. N. and Penot, E., 1999 Slash-and-burn as a land clearing method for small-scale rubber producers in Sepunggur, Jambi Province, Sumatra, Indonesia. *Forest Ecology and Management* 120:157–169.

Ketterings, Q. M., Bigham, J. M. and Laperche, V., 2000 Changes in soil mineralogy and texture caused by slash-and-burn fires in Sumatra, Indonesia. *Soil science Society of America Journal* 64(3):1108–1117.

Kato, M. S. O., Kato, O. R., Denich, M. and Vlek, P. L. G., 1999 Fire-free alternatives to slas-and-burn for shifting cultivation in the Eastern Amazon region: the role of fertilizers. *Field Crops Research* 62:225–237.

Klinge, R., Schmidt, J. and Fölster, H., 2001 Simulation of water budgets of a rainforest and forest conversion plots using a soil water model. *Journal of Hydrology* 246:82–95.

Lal, R., 1987 *Tropical Ecology and Physical Edaphology*. J. Wiley, New York.

Laumonier, Y. and Legg, C., 1998 Le suivi des feux de 1997 en Indonesie. Apports d'Internet et de la téledétection dans une situation d'urgence. *Bois et Fôrets des Tropiques* 258:5–18.

Lean, J., Bunton, C., Nobre, C., and Rowntree, P. R., 1996 The simulated impact of Amazonian deforestation on climate using measured ABRACOS vegetation characteristics. In:. J. H. C. Gash, C. A. Nobre and J. M. Roberts (eds.) *Amazonian deforestation and climate*. John Wiley, Chichester, 549–576.

Leitch, C. J., Flinn, D. W. and Van de Graaff, R. H. M., 1983 Erosion and nutrient loss resulting from ash Wednesday (February 1983) wildfires: a case study. *Australian Forestry* 46:173–180.

Lesack, L. F. W., 1993 Export of nutrient and major ionic solutes from a rainforest catchment in the central Amazon basin. *Water Resources Research* 29:743–758.

Lesack, L. F. W. and Melack, J. M., 1996 Mass balance of major solutes in a rainforest catchment in the Central Amazon: Implications for nutrient budgets in tropical rainforests. *Biogeochemistry* 32:115–142.

Levine, J. S. (ed), 1996 *Biomass burning and global change*, Volume I and II. The MIT Press, Cambridge, Massachussetts.

Lewis, W. M. Jr., Hamilton, S. K. and Saunders III, J. F., 1995 Rivers of Nothern South America. In: C. E. Cushing, K. W. Cummins and G. W. Minshall (eds.) *Ecosystems of the world 22, Rivers and stream ecosystems*. Elsevier, Amsterdam.

Lindquist, L., 1999 Fineroot biomass and root symbioses after a first rotation of *Acacia mangium* stand on former tropical rainforest soil. Swedish University of Agricultural Sciences, Department of Forest Ecology, *Stencilserie*, No. 46, Umeå, Sweden, 19pp.

Lloyd, C. R. and Marques-Filho, A., 1988 Spatial variability of troughfall and stemflow measurements in Amazonian rainforest. *Forest and Agriculture Meteorology* 42:63–73.

Maass, J. M., Jordan, C. F. and Sarukhan, J., 1988 Soil erosion and nutrient losses in seasonal tropical agroecosystems under various management techniques. *Journal of Applied Ecology* 25:595–607.

Mackensen, J., Hölscher, D., Klinge, R. and Fölster, H., 1996 Nutrient transfer to the atmosphere by burning debris in Eastern Amazonia. Forest Ecology and Management 86:121–128.

Mahé, G. and Citeau, J., 1993 Interactions between the ocean, atmosphere and continent in Africa, related to the Atlantic monsoon flow. General pattern and the 1984 case study. *Vieille Climatologie Satellitaire* 44:34–54.

Malingreau, J. P., Stephens, G. and Fellows, L., 1985 Remote sensing of forest fires: Kalimantan and North Borneo in 1982–83. *Ambio* 14:314–321.

Malmer, A., 1992 Water-yield changes after clear-felling tropical rainforest and establishment of forest plantation in Sabah, Malaysia. *Journal of Hydrology* 134:77–94.

Malmer, A., 1996a Observation on slope processes in a tropical rainforest environment before and after forest plantation establishment. In: M. G. Anderson and S. M. Brooks (eds.) *Advances in Hillslope Processes, Volume 2*. John Wiley and Sons, Chichester, 961–973.

Malmer, A., 1996b Hydrological effects and nutrient losses on forest plantation establishment on tropical rainforest land in Sabah, Malaysia. *Journal of Hydrology* 174:129–148.

Malmer, A., 1996c Nutrient losses from dipterocarp forests – a case study of forest plantation establishmant in Sabah, Malaysia. In: A. Schulte and D. Schöne (eds.) *Dipterocarp forest ecosystems, towards sustainable management*. Word Scientific, Singapore, 52–73.

Malmer, A., 1996d Phosphorus loading to tropical rainforest streams after clear-felling and burning in Sabah, Malaysia. *Water Resources Research* 32(7):2213–2220.

Malmer, A., 2004 Stream water quality as affected by wild fires in natural and man made vegetation in Malaysian Borneo. *Hydrological Processes*, 18, 853–864.

Malmer, A. and Grip, H., 1990 Soil disturbance and loss of infiltrability caused by mechanical and manual extraction of tropical rainforest in Sabah, Malaysia. *Journal of Hydrology* 38:1–12.

Malmer, A., Grip. H. and Wong, F. K., 1994 Converting tropical rainforest to forest plantation in Sabah, Malaysia. II. Changes of nutrient dynamics and net losses in streams due to treatments. *Hydrological Processes* 8:195–209.

Malmer, A., Kluge, M. and Johansson, E., 1998 Natural rehabilitation of disturbed tropical rainforest soils in Sabah, Malaysia. In: A. Schulte and D. Ruhiyat (eds.) *Soils of tropical forest ecosystems:characteristics, ecology and management*. Springer Verlag, Berlin, 199–204.

McDowell, W. H., Bowden, W. B. and Asbury, C. E., 1992 Riparian nitrogen dynamics in two geomorphologically distinct tropical rainforest watersheds: subsurface solute patterns. *Biogeochemistry* 18:53–75.

McGregor, G. R. and Nieuwolt, S., 1998. *Tropical Climatology*, 2nd ed., John Wiley and Sons, Chichester, 339 pp.

Middleton, B. A., Sanchez-Rojas, E., Suedmeyer, B. and Michels, A., 1997 Fire in tropical dry forest of Central America: a natural part of the disturbance regime? *Biotropica* 29(4):515–517.

Mishra, B. K. and Ramakrishnan, P. S., 1983 Slash and burn agriculture at higher elevations in North-Eastern India. I. Sediment, water and nutrient losses. *Agriculture, Ecosystems and Environment* 9:69–82.

Neary, D. G., Klopatec, C.-C., DeBano, L. F. and Ffolliott, P. F., 1999 Fire effects on below ground sustainability: a review and synthesis. *Forest Ecology and Management* 122:51–71.

Nepstad, D. C., Moreira, A. G. and Alencar, A., 1999. *Flames in the rainforest: Origins, impacts and alternatives to Amazonian fire*. World Bank Pilot Program to Conserve the Brazilian Rain Forest, Brazilia, 190pp.

Nepstad, D. C., Moutinho, P. R. S. and Markewitz, D., 2001 The recovery of biomass, nutrient stocks and deep soil funktions in secondary forests. In: M. E. McClain, R. L. Victoria and J. E. Richey *The Biogeochemistry of the Amazon basin*. Oxford University Press, pp 137–155.

Nicholls, N., 1993 ENSO, drought and flooding rain in South East Asia. In: H. Brookfield and Y. Byron (eds.) *South East Asia's environmental future: the search for sustainability*. United nations University Press, Tokyo/Cambridge University Press, Cambridge, 154–175.

Nortcliff, S., Ross, S. M. and Thornes, J. B., 1990 Soil Moisture, runoff and sediment yield from differentially cleared tropical rainfores plots. In: J. B. Thornes (ed.) *Vegetation and erosion*. John Wiley and Sons, Chichester, 451–469.

Nykvist, N, Grip, H., Sim, B. L., Malmer, A. and Wong, F. K., 1994 Nutrient losses in forest plantations in Sabah, Malaysia. *Ambio* 23(3):210–215.

Nykvist, N., Sim, B. L. and Malmer, A., 1996 Effects of tractor logging and burning on biomass production and nutrient accumulation in *Acacia mangium* plantations in Sabah, Malaysia. *Journal of Tropical Forest Science* 9(2):161–183.

Näsholm, T., Huss-Danell, K. and Högberg, P., 2000 Uptake of organic and inorganic nitrogen by a grass and three herbs grown in the field. *Ecology* 81:1155–1161.

Odemerho, F. O., 1984 The effects of shifting cultivation on stream-channel size and hydraulic geometry in small headwater basins of southwestern Nigeria. *Geografiska Annaler* 66A:327–340

O'Loughlin, E. M., Cheney, N. P. and Burns, J., 1982 The bushrangers experiment: hydrological response of a eucalypt experiment. In: E. M. O'Loughlin and L. J. Bren (eds.) *The First national Symposium on Forest Hydrology*. National Conference Publications, no 82/6, Institute for Engineers of Australia, 132–138.

Olivry, J. C., Bricquet, J. P. and Mahe, G., 1993 Vers un appauvrissement durable des ressources en eau de l'Afrique humide? *IAHS Publications* 216:67–78.

Pain, C. F. and Bowler, J. M., 1973 Denudation following the November 1970 earthquake at Madang, Papua New Guinea. *Zeitschrift für Geomorphologie Neue Folge Supplement Band* 18:92–104.

Parker, G. G., 1985 The effect of disturbance on water and solute budgets of hillslope tropical rainforest in Northeastern Costa Rica. PhD thesis, University of Georgia, Athens, Ga, USA, 161 pp.

Pearce, A. J., Rowe, L. K. and O'Loughlin, C. L., 1980 Effects of clearfelling and slashburning on water yields and storm hydrographs in evergreen mixed forests, western New Zealand. *IAHS Publications* 130: 19–127.

Peh, C. H., 1981 The suspended and dissolved sediment load of three small forested drainage Basins in peninsular Malaysia. *Malaysian Forester*, 44(4): 438–452.

Pereira, H. C., 1989. *Policy and Practice of Water Management in Tropical Areas*. Westview Press, Boulder, Colorado, USA.

Peters, W. J. and Neuenschwander, L. F., 1988 *Slash and burn farming in the third world forest*. University of Idaho Press, Moscow, Idaho.

Phillips, C., 1987 Preliminary observations on the effects of logging on a hill forest in Sabah. In: Y. Hadi, K. Awang, N. M. Majid and S. Mohamed (eds.) *Impact of man's activities on tropical upland forest ecosystems*. Universiti Pertanian Malaysia, Serdang, Malaysia.

Pielke, R. A., Sr. Avissar, R., Raupach, M., Dolman, A. J., Zeng, X. and Denning, A. S., 1998. Interactions between the atmosphere and terrestrial ecosystems: influence on weather and climate. Global Change Biology 4:461–475.

Pinard, M. A., Putz, F. E. and Licona, J. C., 1999 Tree mortality and vine proliferation following a wildfire in a subhumid tropical forest in eastern Bolivia. *Forest Ecology and Management* 116:247–252.

Polcher, J. and Laval, K., 1993 A statistical study of the impact of deforestation on climate using the LMD-GCM, In: W. B. Wilkinson (ed.) *Macroscale modelling of the hydrosphere*. IAHS Publication, 214:113–118.

Prayogo, C., 2000 Kuantifikasi modal dan distribusi karbon sistem tebang bakar pada laham berlereng di Rantau Pandan Jambi.[Quantification of carbon stock and its distribution in a slash and burn system on sloping land in Rantau Pandan – Jambi], MSc thesis Universitas Brawijaya, Malang (Indonesia).

Prospero, J. M., 1999 Long-term measurement of the transport of African mineral dust to the southeastern United States: implications for regional air quality. *Journal of Geophysical Research* 104(13):15917–15927.

Purwanto, E., 1999 *Erosion, sediment delivery and soil conservation in an upland agricultural catchment in West Java, Indonesia*. PhD thesis, Vrije Universiteit, Amsterdam, The Netherlands.

Putz, F. E., 1983 Treefall pits and mounds, buried seeds and the importance of soil disturbance to pioneer trees on Barro Colorado Island, Panama. *Ecology* 64(5):1069–1074.

Pyne, S. J., Andrews, P. L. and Laven, D. L., 1996 *Introduction to wildland fire*. 2nd edition, Wiley, New York, 769pp.

Qian, W. C., 1983 Effects of deforestation on flood charateristics with particular reference to Hainan island, China, *IAHS Publication*, 140: 249–258.

Quencez, P., 1989 Evolution de la pluvouisité en Afrique del'Ouest. *Oléagineux* 44(4):39–41.

Quimio, J. M., 1996 Grassland vegetation in western Leyte, Philippines, *Schriftenr*. 22, Inst. Landespflege Univ. Freiburg, Freiburg, Germany, 195 pp.

Ramakrishnan, P. S., 1992 Shifting agriculture and sustainable development. *Man and the Biosphere Series*, Vol. 10, Pharthenon Publications, Cranforth, UK, 424 pp.

Richards, P. W., 1952. *The tropical rainforest*, 1st edn., Cambridge University Press, Cambridge.

Richards, P. W., 1969 Speciation in the tropical rainforest and the concept of the niche. *Biological Journal of the Linnean Society* 1:149–154.

Richey, J. E., Nobre, C. and Deser, C., 1989 Amazon river discharge and climate variability: 1903–1985. *Science* 246:101–103.

Rijsdijk, A., Bruijnzeel, L. A., 1990 *Erosion, sediment yield and land use patterns in the upper Konto watershed, East Java, Indonesia*. Konto River Project Commun. 18, 3 Vols., Konto River Project, Malang, Indonesia.

Riswan, S. and Hartanti, L., 1995 Human impacts on tropical forest dynamics. *Vegetatio* 121:41–52.

Roose, E. J., 1977 Erosion et ruissellement en Afrique de l'Ouest. *Travaux et Documents de l'ORSTOM* 78, 108 pp.

Ruxton, B. P., 1967 Slopewash under mature primary rainforest in nothern Papua. In: J. N. Jennings and J. A. Mabbutt (eds.) *Landform studies from Australia and New Guinea*. Australian National University Press, Canberra, 85–94.

Sabhasri, S., 1978 Effects of forest fallow cultivation on forest production and soil, In: P. Kunstadter, E. C. Chapman and S. Sabhasri (eds.) *Farmers in the Forest: Economic Development and Marginal Agriculture in Northern Thailand*. East-West Center, University of Hawaii Press, Honolulu, 160–184.

Salati, E. and Vose, P. B., 1984 Amazon basin: a system in equilibrium. *Science* 225:129–138.

Salati, E, Vose, P. B. and Lovejoy, T. E., 1986 Amazon rainfall, potential effects of deforestation, and plans for future research. In: G. T. Prance (ed.) *Tropical* rainforests and the world's atmosphere. Westview Press, Boulder, Colorado.

Saldarriaga, J. G., 1987 Recovery following shifting cultivation, In: C. F. Jordan (ed.) *Amazonia rainforests. Disturbance and recovery*. Springer Verlag, Heidelberg.

Sanchez, P. A., 1996 Introduction. *Agriculture, Ecosystems and Environment* 58(1):1–2.

Sandström, K., 1995 *Forests and water – friends or foes? Hydrological implications of deforestation and land degradation in semi-arid Tanzania*. PhD. Thesis, Linköping Studies in Arts and Science, 120, Linköping University, Linköping, Sweden.

Sanford, R. L., Saldarriaga, J., Clark, K. E., Uhl, C. and Herrera, R. 1985 Amazon rainforest fires. Science 227:53–455.

Schellekens, J., Scatena, F. N., Bruijnzeel, L. A., van Dijk, A. I. J. M., Groen, M. M. A. and van Hogezand, R. J. P., 2004 Stormflow generation in a small rainforest catchment in the Luquillo Experimental Forest, Puerto Rico. *Hydrological Processes*, 18:505–530.

Schulze, M. D., 1998 Forest fires in the Brazilian Amazon. *Conservation Biology* 12(5):948–950.

Schmidt-Vogt, D., 1998 Defining degradation: the impacts of swidden on forests in Northern Thailand. *Mountain Research and Development* 18(2):135–149.

Shuttleworth, W. J., 1988 Evaporation from Amazonian rainforest. *Proceedings of the Royal Society of London Ser. B.* 233:321–346.

Sim, B. L. and Nykvist, N., 1991 Impact of forest harvesting and replanting, *Journal of Tropical Forest Science* 3(3):251–284.

Spencer, T. and Douglas, I., 1985 The significance of environmental change: diversity, disturbance and tropical ecosystems. In: I Douglas and T. Spencer (eds.) *Environmental Change and Tropical Geomorphology*. Allen and Unwin, London, 39–73.

Spencer, T., Douglas, I., Greer, T. and Sinun, W., 1990 Vegetation and fluvial geomorphic processes in South East Asian tropical rainforests. In: J B. Thornes (ed.) *Vegetation and erosion*. John Wiley and Sons, Chichester, 451–469.

Steensberg, A., 1993 *Fire-clearance husbandry: traditional techniques throughout the world*. Poul Kristensen, Heming.

Stohlgren, T. J., Chase, T. N., Pielke, R. A., Sr. Kittel, T. G. F. and Baron, J. S., 1998 Evidence that local land use practices influence regional climate, vegetation and stream flow patterns in adjacent natural areas. *Global Change Biology*, 4:495–504.

Tangtham, N. and Sutthipibul, V., 1989 Effects of diminishing forest area on rainfall amount and distribution in north-eastern Thailand. Paper presented at the FRIM-IHP-UNESCO Regional Seminar on Tropical Forest Hydrology, 4–9 September 1989, Kuala Lumpur, Malaysia.

Task Committee, 1971 Sediment transportation mechanics. H. Sediment discharge formulas. Journal of Hydraulics Division, Proceedings of the American Society of Civil Engineers HY4:523–567.

Tinker, P. B., Ingram, J. S. I. and Struwe, S., 1996 Effects of slash-and-burn agriculture and deforestation on climate change. *Agriculture, Ecosystems and Environment* 58:13–22.

Tiessen, H., Cuevas, E. and Chacon, P., 1994 The role of soil organic matter in sustaining soil fertility. *Nature* 371:783–785.

Toky, O. P. and Ramakrishnan, P. S., 1981. Runoff and infiltration losses related to shifting agriculture (Jhum) in Northeastern India. *Environmental Conservation* 8:313–321.

Tomich, T. P., Van Noordwijk, M., Vosti, S. A. and Witcover, J., 1998 Agricultural development with rainforest conservation: methods for seeking best bet alternatives to slash-and-burn, with applications to Brazil and Indonesia. *Agricultural Economics* 19: 159–174.

Trimble, G. R., Reinhart, K. G. and Webster, H. H., 1963 Cutting the forest to increase water yields. *Journal of Forestry* 61:635–640.

Tutin, C. E. G., White, L. J. T. and Mackanga, A., 1996 Lightning strike burn large forest tree in the Lopé Reserve, Gabon. *Global Ecology and Biogeography Letters* 5:36–41.

Uhl, C., 1982 Tree dynamics in a species rich terra firme forest in Amazonia, Venezuela. *Acta Cientifica Venezolana* 33:72–77.

Uhl, C., 1987 Factors controlling succession following slash-and-burn agriculture in Amazonia. *Journal of Ecology* 75:377–407.

Uhl, C., 1998 Perspectives of wildfire in the humid tropics. *Conservation Biology* 12(5):942–943.

Uhl, C. and Jordan, C. F., 1984 Successions and nutrient dynamics following forest cutting and burning in Amazonia. *Ecology* 65:1476–1490.

Uhl, C., Clark, K., Dezzeo, N and Maquirino, P., 1988 Vegetation dynamics in Amazonian tree fall gaps. *Ecology* 69:751–763.

Uhl, C. and Kauffman, J. B., 1990 Deforestation, fires susceptibility and potential tree responses to fire in the Eastern Amazon. *Ecology* 71(2):437–449.

Van Breemen, N. R., Finlay, R., Lundström, U. S., Jongmans, A. G., Giesler, R., Melkerud, P.-A., 2000 Mycorrhizal weathering: A true case of mineral plant nutrition? *Biogeochemistry* 49:53–67.

Van Dam, O., 2001 Forest filled with gaps. Effects of gap size on water and nutrient cycling in tropical rainforest. A study in Guyana. *Tropenbos – Guyana Series*, 10, 208 pp.

Van Noordwijk, M., Murdiyarso, D., Hairia, K., Wasrin, U. R., Rachman, A. and Tomich, T. P., 1998a Forest soils under alternatives to slash-and-burn agriculture in Sumatra, Indonesia. In: A. Schulte and D. Ruhiyat (eds.) *Soils of tropical forest ecosystems: characteristics, ecology and management*. Springer Verlag, Berlin, 175–185.

Van Noordwijk, M., Van Roode, M., McCallie, E. L. and Lusiana, B., 1998b Erosion and sedimentation as multiscale, fractal processes: implications for models, experiments and the real world. In: F. Penning de Vries, F. Agus and J. Kerr *Managing Soil Erosion*. CAB International, Wallingford, 223–253.

Van Noordwijk, M. and Lusiana, B., 1999. WaNuLCAS, a model of water, nutrient and light capture in agroforestry systems. *Agroforestry Systems* 43:217–242.

Van Rompaey, R. S. A. R., 1995 Climate change and deforestation in West Africa: a space- time trend analysis of rainfall series from Cote d'Ivoire and Liberia. *Studies in Environmental Science* 65A:417–420.

Van Wambeke, A., 1998 Managing red and lateritic soils for sustainable agriculture. In: J. Sehgal, W. E. Blum and K. S. Gajbhiye (eds.) *Red and lateritic soils, Volume 1, Red and lateritic soils of the world*. A. A. Balkema, Rotterdam, 29–46.

Veen, A. W. L., Hutjes, R. W. A., Klaassen, W., Kruijt, B. and Lankreijer, H. J. M., 1991 Evaporative conditions across a grass–forest boundary: a comment on the strategy for regionalizing evaporation. *International Association of Hydrological Sciences Publication* 204:43–52.

Vitousek, P. M. and Denslow, J. S., 1986 Nitrogen and phosphorus availability in tree fall gaps of a lowland tropical rainforest. *Journal of Tropical Ecology* 74:1167–1178.

Wadsworth, G., Southard, R. J. and Singer, M. J., 1988 Effects of fallow length on organic carbon and soil fabric of some tropical udults. *Soil Science Scociety of America Journal* 52:1424–1430.

Walling, D. E., 1983 The sediment delivery problem. *Journal of Hydrology* 65:209–237.

Walsh, R. P. D., 1996 Drought frequency changes in Sabah and adjacent parts of northern Borneo since the late nineteenth century and possible implications for tropical rainforest dynamics. *Journal of Tropical Ecology* 12:385–407.

Walsh, R. P. D. and Newbery, D. M., 1999 The ecoclimatology of Danum, Sabah, in the context of the world's rainforest regions, with particular reference to dry periods and their impact. *Phil. Trans. R. Soc.Lond. Ser. B* 354:1869–1883.

Waterloo, M. J., 1994 *Water and nutrient dynamics of Pinus caribaea plantation forests on former grassland soils in Southwest Viti Levu, Fiji*. PhD thesis, Vrije Universiteit, Amsterdam, The Netherlands, 478 pp.

Whitmore, T. C., 1974 Change with time and the role of cyclones in tropical rainforest on Kolombangara, Solomon Islands. *Commonwealth Forest Institute Paper* No. 46.

Whitmore, T. C., 1978 Gaps in the forest canopy. In: P. B. Tomlinson and M. H. Zimmerman (eds.) *Tropical trees as living systems*. Cambridge University Press, Cambridge, 639–655.

Whitmore, T. C., 1998 *An introduction to tropical rainforests*. Clarendon Press, Oxford.

Whitmore, T. C. and Burslem, D. F. R. P., 1998 Major disturbances in tropical rainforests. In: *Dynamics of tropical communities*, 37th Symposium of the British Ecological Society, Blackwell Science, Oxford, 549–565.

Wibowo, A., Suharti, M., Sagala, A. P. S., Hibani, H. and Noordwijk, M. Van, 1997 Fire management on *Imperata* grasslands as part of agroforestry development in Indonesia. *Agroforestry Systems* 36:203–217.

Wickramagamage, P., 1998 Large-scale deforestation for plantation agriculture in the hill country of Sri Lanka and its impacts. *Hydrological Processes* 12:2015–2028.

Wiersum, K. F., 1984 Surface erosion under various tropical agroforestry systems. In: C. L. O'Loughlin, and A. J. Pearce (eds.) *Symposium on effects of forest land use on erosion and slope stability*. IUFRO Vienna and East West Center, Honolulu, Hawaii, USA, pp 231–239.

Wiersum, K. F., 1985 Effects of various vegetation layers in an *Acacia auriculiformis* forest plantation on surface erosion in Java, Indonesia. In: S. El-Swaify, W. C. Moldenhauer, and A. Lo (eds.) *Soil erosion and conservation*, Soil Conservation Society of America, Ankeny, Iowa, USA, pp 79–89.

Wilk, J., 2000 *Do forests have an impact on water availability? Assessing the effects of heterogeneous land use on streamflow in two monsoonal river basins*. PhD thesis, Linköping Studies in Arts and Science, 222, Linköping University, Linköping, Sweden.

Wilk, J., Andersson, L. and Plermkamon, V., 2001 Hydrological impacts of forest conversion to agriculture in a large river basin in northeast Thailand. *Hydrological Processes*, 15:2729–2748.

Williams, M. R. and Melack, J. M., 1997 Solute export from forested and partially deforested catchments in the central Amazon. Biogeochemistry 38:67–102.

Williams, M. R., Fisher, T. R. and Melack, J. M., 1997 Solute dynamics in soil water and groundwater in a central Amazon catchment undergoing deforestation. *Biogeochemistry* 38:303–335.

Woods, P. V., 1989 Effects of logging, drought and fire on structure and composition of tropical forests in Sabah, Malaysia. *Biotropica* 21(4):290–298.

Yasunari, T., 2002 The role of large-scale vegetation and land use in the water cycle and climate in monsoon Asia. In: Steffen, W., Jäger, J., Carson, D. J. and Bradshaw, C. (eds.) *Challenges of a Changing Earth*. Springer-Verlag, Berlin, pp. 1129–1132.

Young, A., 1989 *Agroforestry for soil conservation*. CAB International, Wallingford.

Young, A., 1997 *Agroforestry for soil management*. CAB International, Wallingford.

Zinke, P. J., Sabhasri, S. and Kunstadter, P., 1978 Soil fertility aspects of the Lua' forest fallow system, of shifting cultivation. In: P. Kunstadter, E. C. Chapman and S. Sabhasri (eds.) *Farmers in the Forest: Economic Development and Marginal Agriculture in Northern Thailand*. East-West Center, University of Hawaii Press, Honolulu.

Zulkifli, Y. and Abdul Rahim, N., 1991 *Logging and forest conversion: can we minimise their impacts*. Paper presented at ASEAN seminar on 'Land use decisions and policies: Will tropical forest survive their impact?' Penang, Malaysia, 28–30 Oct. 1991.

22 Soil and water impacts during forest conversion and stabilisation to new land use

H. Grip

Swedish University of Agricultural Sciences, Umeå, Sweden

J.-M. Fritsch

Institut de Recherche pour le Développement, Montpellier, France

L. A. Bruijnzeel

Vrije Universiteit, Amsterdam, The Netherlands

INTRODUCTION

The rainforests of the humid tropics are being converted to alternative land uses at an increasing rate (Drigo, this volume). In many cases the initial forest disturbance preceding conversion occurs during mechanised harvesting of (large volumes of) valuable timber (Chappell, Tych *et al.*, Cassells and Bruijnzeel, Thang and Chappell, all this volume). Timber extraction causes a number of disturbances to the soil surface, notably during the construction and subsequent use of haulage roads, tractor tracks and log landings, but also in the form of scars by falling trees, particularly in steep terrain (cf. Kamaruzaman, 1991; Dykstra and Heinrich, 1996). Soil impacts may be lessened in some cases by the manual skidding of logs on wooden rails (usually in relatively flat terrain) or through the use of high-lead or low-lead cable yarding (Figure 22.1) (Bruijnzeel, 1992; Bruenig, 1996). Uphill logging, in which log landings are located on ridge tops, is the preferred system nowadays as this has a number of advantages: (i) access roads can be built along the ridges which usually form the driest part of the landscape, thereby minimising the frequency of road closures during periods of high rainfall; (ii) the log landings are also situated on dry spots which facilitates vehicle operation; (iii) the timber extraction network fans out in a downhill direction, thereby reducing surface runoff accumulation (and thus erosion) on tractor tracks; and (iv) the number of stream crossings is minimised in this way (Dykstra and Heinrich, 1996; cf. Figure 35.4 in Cassells and Bruijnzeel, this volume).

The subsequent use of the cleared land may vary. Less attractive parts, such as those with shallow soils or very steep slopes, are usually abandoned to be more or less rapidly re-colonised by species from the original forest, from nearby patches of secondary vegetation, or by grasses. Better land may be used for the establishment of forest plantations, tree crops, pasture, or agricultural crops. In some cases the land use change is dramatic and irreversible (e.g. urban areas, mines, or large dams), in others the forest may return through regeneration after abandonment once productivity has declined to an unacceptable level (cf. Hölscher, Mackensen and Roberts, this volume).

Soil preparation for planting after the timber extraction phase normally includes windrowing and burning of logging residues, leaving a mosaic of debris, bare patches, nutrient-rich ash beds and incipient regrowth (Figure 22.2) (Scott, Bruijnzeel and Mackensen, this volume). In the case of forest plantations, planting is done directly in the ash and mineral soil. Manual weeding or application of herbicides around the seedlings is normally required once or twice (Sim and Nykvist, 1991). For the establishment of more intensively managed operations, such as rubber, oil palm or cocoa plantations, additional measures are needed to facilitate access to individual trees for the harvesting of produce, controlling of surface runoff, erosion and competition by understorey regrowth, or to restore soil structure. Examples of such additional measures include the construction of low-gradient or horizontal terraces, deep ploughing to break up compacted soil, and the planting of a leguminous cover crop. Access lanes with their compacted and usually bare surface areas are an integral part of commercial tree plantations and need to have adequate drainage facilities (Critchley and Bruijnzeel, 1996).

The transition phase is followed by a maturation phase during which the soil surface is more or less fully protected again whilst the new vegetation actively accumulates biomass (Bruijnzeel,

Forests, Water and People in the Humid Tropics, ed. M. Bonell and L. A. Bruijnzeel. Published by Cambridge University Press.

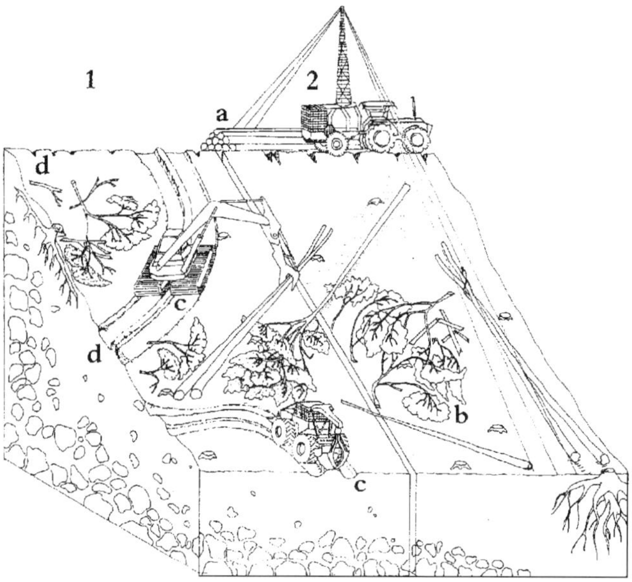

Figure 22.1 Timber harvesting in tropical forests generally uses heavy machinery such as caterpillars (1) or other tracked or wheeled vehicles; cable crane systems (2) generally have less impacts on the soil. (a) Log landing on ridge top, (b) residual slash, (c) soil surface compaction and (d) sheet and rill erosion on tractor tracks. (After Mackensen *et al.* (2003); drawing by Uta Mackensen (used with permission).)

Figure 22.2 Site preparation after logging involves burning of logging debris (1) or, less frequently, leaving the slash as a protective mulch (2). (a) Volatilisation and ash particle transport, (b) ash beds and bare mineral soil, (c) sheet, rill and incipient gully erosion, and (d) initial regrowth. (After Mackensen *et al.* (2003); drawing by Uta Mackensen (used with permission).)

1997). The soil and water impacts of tropical forest conversion have been reviewed in some detail by Bruijnzeel (1990, 1997, 1998). The present chapter complements the analyses of Bruijnzeel by taking a more explicit process-based approach. In addition, the chapter pays particular attention to the results obtained by two well-documented case studies of rainforest conversion, viz. the ECEREX (Ecology – Erosion – Experiment) project in French Guyana (Roche, 1981, 1982b; Fritsch, 1992, 1993; Fritsch and Sarrailh, 1986) and, to a lesser extent, the Mendolong project in Sabah, East Malaysia (Malmer, 1993, 1996ab; Nykvist *et al.*, 1994; Nykvist, 1997).

CHANGES IN ENERGY AND WATER BALANCES DURING FOREST CONVERSION AND STABILISATION OF NEW LAND USE

It is well known that the felling of native forest results in increases in catchment water yield under a broad range of climatic and topographic conditions. By now, there is a dataset comprising more than 100 long-term, controlled paired catchment experiments, including a number of tropical ones. The increase in flow is positively related to the proportion of biomass removed (Bosch and Hewlett, 1982; Sahin and Hall 1996; Stednick, 1996; for the humid tropics: Bruijnzeel, 1990, 1996). Reported increases in water yield during the first few years following clearfelling of tropical forest vary between 12 mm year^{-1} per 10% of biomass removal under subhumid conditions to as much as 80 mm year^{-1} per 10% forest removal under perhumid conditions (Bruijnzeel, 1990). However, the great variation in hydrological response to forest clearing can be explained only partially by differences in rainfall between locations or years (Bruijnzeel, 1996). Other factors probably include differences in elevation and distance to the coast (affecting evaporation; Roberts *et al.*, this volume); catchment steepness, soil depth and changes in permeability with depth (governing the residence time of the water, speed of baseflow recession and stormflow generation patterns; Bonell, this volume); and last, but not least, the degree of disturbance of undergrowth and topsoil by machinery or fire (determining both rainfall infiltration opportunities and rates of regrowth; Malmer, 1992; cf. Malmer, Van Noordwijk and Bruijnzeel, this volume; Hölscher *et al.*, this volume).

Because the relative importance of the respective factors varies between sites, additional process studies are usually required if the results of paired basin experiments (which essentially represent a black box approach) are to be fully understood and extrapolated to other areas (Bruijnzeel, 1990, 1996; Bonell and Balek, 1993). It is pertinent to note that the bulk of the increase in flow is usually observed in the form of baseflow rather than as greatly increased stormflows. Arguably, therefore, a better understanding may be obtained in particular by examining the changes in vegetation water use associated with forest conversion in terms of the underlying processes and their controls (cf. Roberts *et al.*, this volume).

The Penman-Monteith equation (Monteith, 1965) provides a suitable starting point for examining the factors and underlying controls governing the changes in evaporation that may accompany various changes in land cover under given climatic conditions:

$$LE_d = [\Delta(Rn - G) + \rho c_p \, \delta e \, g_a]/[\Delta + \gamma(1 + g_a/g_s)] \quad (22.1)$$

L is the latent heat of vaporisation, and E_d the actual rate of water loss from a dry vegetation surface; Δ is the rate of change in saturated vapour pressure with a concurrent change in air temperature, γ the psychrometric constant (66 Pa K^{-1}), ρ the density of dry air, c_p the specific heat of air at constant pressure and δe the vapour pressure deficit. The surface (canopy) and aerodynamic conductances to evaporation, g_s and g_a, represent the ease with which evaporated water is removed from within the leaves to the surface of the leaves, and from the latter into the atmosphere, respectively (see Roberts *et al.*, this volume for a fuller discussion and dimensions of the respective terms). The aerodynamic conductance g_a (mm s^{-1}) is frequently calculated from:

$$g_a = k^2 u/\{\ln [(z - d)/z_0]\}^2 \quad (22.2)$$

where k is von Kármán's constant (0.41) and u the wind speed at the measuring height z. The displacement height d and the roughness length z_0 represent the roughness of the vegetation and are both expressed in metres. Values of d and z_0 may be derived from wind profile measurements (Thom, 1975) but more often they are assumed to simply equal 75% of the vegetation height h_v in the case of d and 10% in the case of z_0 (Roberts *et al.*, this volume).

The surface conductance g_s at canopy level can be evaluated as the product of vegetation leaf area index L^* and stomatal conductance at the leaf level g_{sto} but more often it is derived by inverse application of Eqn. 22.1 for known values of LE_d (see Roberts *et al.*, this volume for details). An alternative approach involves calculating g_s from g_{smax} and the so-called Lohammar equation (Lohammar *et al.*, 1980; Kelliher *et al.*, 1997):

$$g_s = g_{smax} f_Q(Q) f_D(D) \quad (22.3)$$

The physiologically based function f_Q is equal to $Q / (Q + Q_{50})$, where Q_{50} is the value of the global radiation intensity for which $g_s = 0.5\ g_{smax}$. The function f_D equals $(1 + D / D_{50})^{-1}$, where D_{50} is the value of the vapour concentration deficit for which $g_s = 0.5\ g_{smax}$ (under light-saturated conditions). Reported values of g_{smax} (mm s^{-1}), Q_{50} (W m^{-2}), and D_{50} (kg m^{-3}) in the context of the Lohammar equation include: 70, 137, and 0.01 for *Salix viminalis* (Grip, Halldin and Lindroth, 1989); 20, 125, and 0.005 for *Pinus sylvestris*; 18, 125, and 0.02 for *Quercus sessiliflora* (Halldin, Saugier and Pontallier, 1984/85); and 10, 22, and 0.004, respectively, for *Larix gmelinii* (Kelliher *et al.*, 1997). For L^* values greater than 3, g_{smax} is stable for tree species and natural grasses, but it may be larger for cultivated grasses and other vegetation types with a high nitrogen status such as *Salix viminalis* (Grip *et al.*, 1989). Examining maximum conductances for global vegetation types Kelliher *et al.* (1995) found g_{smax} to be remarkably close to 18 mm s^{-1} for natural herbaceous and woody communities vs. *c.* 32 mm s^{-1} for (fertilised) agricultural crops having a much higher nitrogen status. Only a few values of g_{smax} have been reported for lowland tropical rainforests, e.g. 13 mm s^{-1} in Central Amazonia (Shuttleworth, 1988), 25 mm s^{-1} in Rondonia (Wright *et al.*, 1996), 7.5 mm s^{-1} in Panama (Meinzer *et al.*, 1993) and 20 mm s^{-1} in Peninsular Malaysia (Tani *et al.*, 2003). Wright *et al.* (1996) also gave values of g_{smax} for pastures in Amazonia, which ranged between 11.5 and 28.5 mm s^{-1}.

The above discussion has considered evaporation under conditions of ample water supply only. When soil water becomes limited, g_s is markedly reduced and transpiration (E_d) is restricted accordingly. As discussed more fully by Roberts *et al.* (this volume), information on changes in g_s under conditions of diminishing soil water reserves is scanty for tropical vegetation. A simple alternative is to multiply E_d as predicted for non-limiting conditions with a reduction coefficient whose magnitude is a function of soil water deficit. Above a certain threshold, there is no restriction to water uptake whereas below the threshold the reduction coefficient decreases from 1.0 to reach zero at the wilting point. The form of the reduction function may be linear, curved or stepwise (Calder, Harding and Rosier, 1983), but for forests with their relatively deep roots a straight line is usually adequate (Dunin, O'Loughlin and Reyenga, 1985; see also discussion in Roberts *et al.*, this volume). The KAUSHA model (Halldin and Grip, 1979; Halldin *et al.*, 1984/85) also used a linear reduction function which came into operation when two-thirds of all plant-available water had been removed from the root zone. The value of the threshold depends on the soil's unsaturated hydraulic conductivity (Gardner, 1983) as well as the prevailing evaporation rate (Denmead and Shaw, 1962; Dunin *et al.*, 1985). Using isotope tracers, Plamboeck, Grip and Nygren (1999) found for a *Pinus sylvestris* stand in Sweden that relative water uptake from different layers within of the root zone did not depend on root density distribution but rather on unsaturated hydraulic conductivity. The magnitude of the latter explained 76% of the observed variation in relative uptake. The lysimeter-based work of Dunin *et al.* (1985) on young eucalypts in south-eastern Australia showed rather elegantly how evaporation rates started to fall below potential values at a progressively earlier stage (i.e. at lower soil water deficits) as the evaporative demand increases.

When the canopy is wetted by rain, evaporation of the intercepted rainfall is dominated by the aerodynamic conductance (g_a)

as the Penman–Monteith equation simplifies to a form that omits g_s and becomes (Monteith, 1965):

$$LE_{wet} = [\Delta(Rn - G) + \rho c_p \delta_e g_a]/(\Delta + \gamma) \quad (22.4)$$

In the absence of physiological control, evaporation rates during rainfall tend to go up markedly (other climatic conditions remaining equal) and are strongly influenced by the magnitude of g_a. Values of g_a associated with tall and aerodynamically rough vegetation such as forests tend to be much higher than for grassland or crops (cf. Table 22.1 below), which is one of the reasons (apart from contrasts in leaf surface area) for the enhanced rates of wet canopy evaporation from forests compared to short vegetation types (Calder, 1979). As such, it is important to treat evaporation during times of wet and dry canopy conditions separately. As discussed more fully by Roberts *et al.* (this volume), Eqn 22.4 often appears to underestimate wet canopy evaporation rates. Because the latter are also difficult to measure directly during times of rainfall, a frequently used alternative approach is to subtract measured amounts of throughfall (*Tf*) and stemflow (*Sf*) from incident rainfall (*P*) to give total rainfall interception (E_i, i.e. cumulative wet canopy evaporation):

$$E_i = P - Tf - Sf \quad (22.5)$$

The capacity of vegetation to store intercepted rainfall is proportional to the total surface area (leaves, branches and stems) of the vegetation. As such, it may vary seasonally (e.g. in the case of agricultural crops or deciduous trees) or with time as the vegetation matures (e.g. plantations, Van Dijk and Bruijnzeel, 2001).

Evaporation from post-forest landscapes can thus be expected to differ from that of the original forest because of the associated changes in leaf area, albedo (short-wave reflection), physiological and aerodynamic characteristics, as well as rainfall intercepting capacity. More or less representative values of the respective surface characteristics associated with a number of important tropical land cover (or land use) types have been collated in Table 22.1. Because information on evaporation totals from various tropical post-forest land cover types is still scanty, Eqn 22.1 is used below to derive approximate comparative evaporation rates for a given set of climatic conditions and using the respective plant parameters listed in Table 22.1.

The albedo of most tropical rainforests is typically 0.12–0.13 (Roberts *et al.*, this volume) whereas values for pasture or fire-climax grassland (having low values of L^*) are closer to 0.17–0.20 (Culf, Fisch and Hodnett, 1995; Waterloo *et al.*, 1999). Rain-fed crops, on the other hand, exhibit still higher reflection values whereas wet soils reflect much less (Table 22.1). As such, the surface albedo associated with a mixed cropping system may vary seasonally from 0.05–0.07 shortly after planting (minimum soil cover) to 0.21–0.25 just before harvesting (maximum cover) (Van Dijk, 2002). The reflection coefficients of orchards and various tree crops are intermediate between those observed for natural forest and tall grassland (Table 22.1).

Values for g_{smax} do not differ too much between vegetation types, with the only noteworthy contrast being the higher value found for agricultural crops which was signalled earlier. Interestingly, the average values for surface conductance g_s derived for rain-fed crops on degraded soils of presumably low nitrogen status in Indonesia (Van Dijk, 2002) were as low as those for mature pine forest in Fiji on similarly poor soils (Waterloo *et al.*, 1999): viz. 12.7 vs. 12.5–17 mm^{-1}. Values associated with young and vigorously growing pines in Fiji were higher at 20–30 mm s^{-1} whereas those for oil palm in West Africa were close to 17.5 mm s^{-1} (Dufrene and Saugier, 1993). Oil palm water use under non-limiting soil water conditions can be high. However, during the dry season in West Africa, after the fraction of plant-available water in the top 0.8 m of the soil had decreased below 0.35 (i.e. the same value as used by Halldin *et al.* (1984/85) in the KAUSHA model), transpiration dropped sharply to very low values. Corresponding average values of g_s *at the leaf level* (i.e. stomatal conductance) were 6.0–6.6 mm s^{-1} and 1.5 mm s^{-1} during the wet and the dry season, respectively (Dufrene *et al.*, 1992).

The contrast in magnitude of g_a between tall and short vegetation types hinted at earlier is illustrated by the different values associated with short rain-fed crops in Indonesia (9.4 mm s^{-1}), *c.* 2 m tall fire-climax grassland in Fiji (21 mm s^{-1}), pine plantations of intermediate height in Fiji (64–70 mm s^{-1}) and lowland rainforest in Amazonia (100 mm s^{-1}) (Table 22.1).

To compare annual evaporation totals from the original tropical rainforest with those for various alternative land uses (after stabilisation), the Penman–Monteith Eqn 22.1 was applied for given climatic conditions, but using different plant parameters as appropriate (Table 22.1). Global incoming radiation was set at 190 W m^{-2}, net long-wave radiation at –40 W m^{-2}, and net energy storage at 84 W m^{-2} for dry days vs. 4 W m^{-2} for days with rain. Vapour pressure deficits (VPD) for dry and wet days were set at 1.2 and 0.38 kPa, respectively. Aerodynamic conductance g_a was calculated from stand height using Eqn 22.2 whereas surface conductance g_s was derived with the Lohammar Eqn 22.3 using $Q_{50} = 125$ W m^{-2} and $D_{50} = 0.01$ kg m^{-3}, with g_{smax} values taken from Table 22.1. Next, the year was divided into 180 dry days and 185 wet days. During a wet day, interception loss was calculated as the product of leaf area index L^* and a specific storage capacity of 0.24 mm m^{-2} (cf. Van Dijk and Bruijnzeel, 2001). After the interception capacity was evaporated, the rest of the day was treated as dry, that is evaporation took place through transpiration only. It is acknowledged that total wet canopy evaporation will be underestimated in this way because evaporation from the wet vegetation during rainfall is not included (cf. Roberts *et al.*, this volume). The resulting average annual evapotranspiration (E_T) and interception

Table 22.1. *Albedo* r, *(maximum), surface conductance* g_{smax}, *aerodynamic conductance* g_a, *vegetation height* h_v, *leaf area index* L*, *canopy storage capacity* S, *measured and calculated totals of annual water use* (E_T) *and interception evaporation* (E_i), *and evaporation ratio LE/Rn for lowland tropical rainforests and various post-forest land cover types*

Land use	r	g_a (mm s^{-1})	g_{smax} (mm s^{-1})	h_v (m)	L^* (m^2 m^{-2})	S (mm)	E_T (mm yr^{-1})	E_T (calc.) (mm yr^{-1})	E_i (mm yr^{-1})	E_i (calc.) (mm yr^{-1})	LE/Rn
Tropical rainforest											
Reserva Ducke, Amazonia	0.131[a]	100[b]	13[c]	35[c]	5.2[d]	0.74[c]	1319[c]	1382	328[c]	271	0.89[c]
Pasoh, Peninsular Malaysia	0.12[e]		20[e]	43[e]	6.25[e]			1710		289	
Janlappa, West Java		200[f]		30[f]		1.1[f]	1481[f]		595[f]		0.96[f]
Stabilisation phase											
Grassland											
Fire climax, Fiji	0.18[g]	47[h]	7[g]				746[g]	677	195[g]	138	0.46[i]
wet and dry season			4.4[i]								
Managed, Fazenda	0.163[a]	46.4[k]	8[k]	0.3[k]	1.2[k]			1040		107	
Dimona, Amazonia											
Tree crops											
Pinus caribaea, 6 yr, Fiji	0.10[g]	70[g]	38[g]	12[g]	3.5[g]	0.8[g]	1926[g]	1718	382[g]	155	1.69[i]
Pinus caribaea, 15 yr, Fiji	0.126[g]	64[g]	42[g]	18[h]	3.1[g]	1.2[g]	1717[g]	1683	371[g]	138	1.49[i]
Oil palm, Ivory Coast	0.18[l]	60[m]	17.5[n]	12[m]	5.1[m]			1302		155	
Cocoa	0.19[l]	43[o]	17[p]	7	3.5			1446		222	
Agricultural crops											
Maize/cassava, Java	0.18[h]	9.4[h]	18[h]		1.3[q]	0.09[q]	1228[h]	1098	198[q]	78	0.97[q]
Maize	0.17[h]			2.2[q]							
Cassava	0.22[h]			1.5[q]							
Open water	0.08[r]	5[r]						1471			

[a] Bastable *et al.* (1993); [b] Gash and Shuttleworth (1991); [c] Shuttleworth (1988); [d] Wright *et al.* (1996); [e] Tani *et al.* (2003); [f] Calder *et al.* (1986); [g] Waterloo *et al.* (1999); [h] Van Dijk (2002); [i] M. J. Waterloo, pers. comm. (2003); [k] Wright *et al.* (1992); [l] Ling and Robertson (1982); [m] Dufrene *et al.* (1992); [n] Dufrene and Saugier (1993); [o] Balasimha, Daniel and Bhat (1991); [p] Radersma and de Ridder (1996); [q] Van Dijk and Bruijnzeel (2001); [r] Oke (1987).

evaporation (E_i) totals are also listed in Table 22.1, together with selected values from the literature.

Calculated annual E_T values were similar to measured totals for Amazonian rainforest, fire-climax grassland and 15-year-old *Pinus caribaea* in Fiji, but smaller for 6-year-old pines in Fiji and mixed rain-fed crops (Table 22.1). The calculation of E_T for the fire-climax grassland took changes in g_{smax} during the year into account but not changes in plant development. In reality, grass height and L^* are known to vary seasonally (Wright *et al.*, 1992; Waterloo, 1994). The calculations for the mixed maize–cassava crop did take seasonal plant development into account (Van Dijk, 2002) but evaporation of intercepted rainfall, and therefore total E_T, were underestimated. As indicated previously, this must have been the case for the other stands as well because evaporation during rainfall was not accounted for. Arguably, a more representative result may be obtained by a combined approach in which wet canopy evaporation is evaluated using the analytical model of interception (Gash, 1979 and derivatives) and daily rainfall data next to the use of Eqn 22.1 for the computation of evaporation from a dry canopy. Waterloo *et al.* (1999) and Van Dijk (2002) applied this approach successfully to derive annual totals of E_T for pine plantations in Fiji and mixed crops in West Java, respectively. Calculated interception totals in Table 22.1 underestimated measured E_i by 17–63%, illustrating the shortcomings of the present approach.

From the simple calculations presented here, it may be concluded, however, that the conversion of tropical rainforest to oil palm, cocoa plantations or large water storage reservoirs would not change evaporation losses (assuming soil water is not limiting) by more than 10% (Table 22.1). Lysimeter-based observations of oil palm water use in lowland Peninsular Malaysia (Ling, 1979; Foong, Sofi and Tan, 1983) suggested higher water use (by about 350 mm year^{-1}) compared to forest when the palms were irrigated. No difference was found, however, in the absence of oil palm irrigation (Foong *et al.*, 1983). A less consistent result is obtained for cocoa considering that Imbach *et al.* (1989) derived annual E_T values of 800 and 1025 mm for agroforestry systems involving cocoa with *Erythrina poeppigiana* and *Cordia alliodora* as the respective shade trees in upland Costa Rica. Based on the latter (lysimeter-based) values for E_T one would expect a gain in water yield following conversion to cocoa of about 200–400 mm year^{-1} (cf. Bruijnzeel, 1990). It is possible that the lower water use derived by Imbach *et al.* (1989) compared to the present estimate reflects an effect of the shade trees. More work is needed to resolve the issue. Both fire-climax and managed grasslands, and rain-fed mixed cropping all use considerably smaller volumes of water compared to rainforest and a conversion of the latter to any of the former should increase catchment water yield by 280–760 mm year^{-1} (Table 22.1). Planting *Pinus caribaea* on fire-climax grassland, on the other hand, would reduce streamflows substantially (possibly by as much as 1000 mm year^{-1}; cf. discussions in Waterloo *et al.*, 1999; Scott *et al.*, this volume).

The sensitivity of computed evaporation rates to changes in the respective controlling parameters was also examined. A 10% increase in g_{smax} caused evaporation to increase by 0.6% (mixed crops) to 5.6% (oil palm); a similar increase in vapour pressure deficit produced increases in E_T ranging from 0.9% (managed pasture) to 3.9% (Pasoh rainforest), whereas increased stand height (read: aerodynamic conductance) had only modest effects (1.7% increase in E_T from 15-year-old pines and even a slight reduction (0.7%) in E_T from fire-climax grassland). As indicated previously, the effect of g_a is underestimated by the present calculations because of the exclusion of wet canopy evaporation rates during rainfall, which are strongly governed by the magnitude of g_a (cf. Eqn 22.4). Finally, a 10% increase in the Q_{50} value decreased E_T by 1.4% (oil palm) to a marginal 0.2% (rain-fed crops) whereas a similar increase in albedo gave decreases in E_T of only 0.7% (Pasoh rainforest) to 2.6% (mixed rain-fed crops). Thus, canopy conductance appears to be the most influential variable, as is also illustrated by the effects of the changes in VPD and Q_{50} which both have an effect on g_s as well (cf. Eqn 22.3).

STORMFLOW CHANGES DURING FOREST CONVERSION AND STABILISATION OF NEW LAND USE

An extensive small-scale experiment was conducted between 1977 and 1984 in French Guyana (the ECEREX project) to assess the effects on stormflow of mechanised harvesting, clearing and conversion of primary forest to various other land uses (Roche, 1982b; Fritsch, 1992, 1993). Use was made of a multiple-paired catchment approach (cf. Hewlett and Helvey, 1970; Roche, 1981). The results of this unique experiment are worthy of re-examination here in some detail, also because much of the project's more detailed literature is in French.

Context and experimental design

Ten very small headwater catchments under natural forest, each 1–1.6 ha in size and lacking perennial flow, were selected and labelled sequentially A to J, following the order from the beginning of the observations (Roche, 1981) (Table 22.2). Measurements began in January 1977 on catchments A, B and C, followed by catchments D–H at the start of 1978 and catchments I and J in December 1979. The two most distant catchments (D and H) were nearly 5 km apart. The catchments were equipped either with an H-flume (catchments F, G and H) or a 30° V-notch sharp-crested weir (remaining seven catchments). Water levels were monitored

Table 22.2. *Basic characteristics of the experimental catchments*[a] *at the ECEREX location (arranged according to percentage free-draining soils)*

Characteristics	C	I	D	E	B	J	A	G	F	H
Drainage area (ha)	1.6	1.1	1.4	1.6	1.6	1.4	1.3	1.5	1.4	1.0
Slopes% (maximum on each bank)	20–17	23–23	28–18	30–20	17–17	32–29	20–20	34–26	35–31	24–19
Vertical drainage soils area (%)	99	60	60	57	10	2	0	0	0	0
Water table extension (%) (coming to the surface during the rainy season)	0	0	0	0	0	0	0	10	4	14

[a] For details of catchments see text.

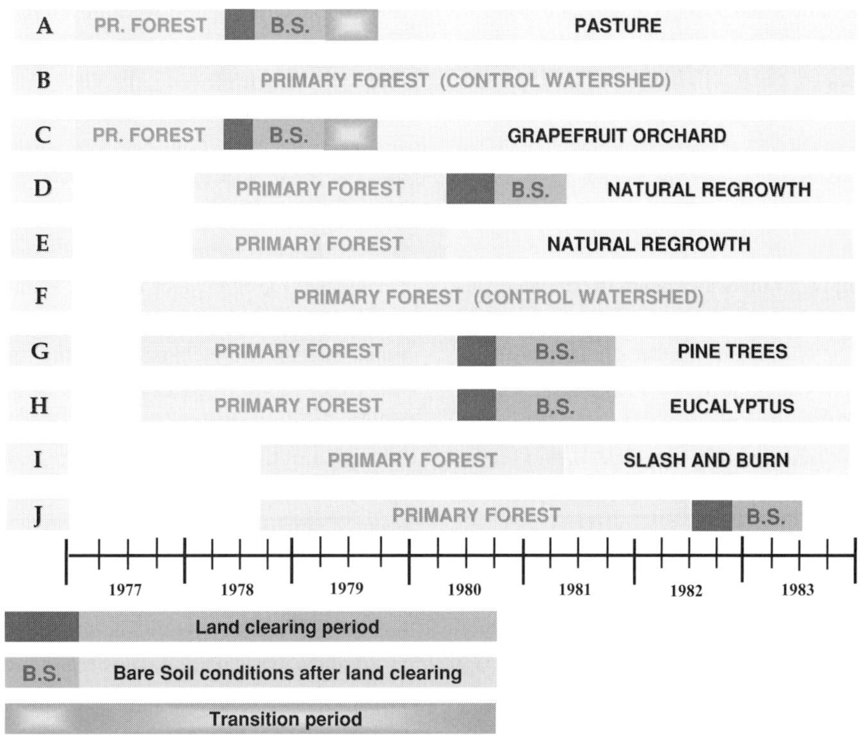

Figure 22.3 Overview of the ECEREX experiment.

by high-speed chart recorders whereas a recording rain gauge was installed in a clearing near each hydrometric station. Areal rainfall was estimated using the Thiessen polygon method. All hydrological monitoring under natural forest conditions was carried out during a minimum of two years of calibration. Catchments B and F were then assigned as controls and left in their natural state whereas various treatments were implemented on the remaining eight catchments (Figure 22.3).

The climate of the ECEREX area (5°30′N, 53°W) is perhumid tropical (type Af according to the Köppen classification). Average annual precipitation is *c.* 3325 mm (1977–1984; range 2394–3680 mm), with a rainy season between December and June, and with a relatively dry season from July to November. However, mean monthly rainfalls during the 'dry' season generally still exceed 100 mm, with the exception of September (*c.* 65 mm) and October (*c.* 95 mm). The wettest month is May (*c.* 515 mm of rain on average), but monthly totals of over 1000 mm are not uncommon and have been observed twice between 1977 and 1987. Typical 15-minute rainfall intensities amount to 90 mm h^{-1} and 130 mm h^{-1} for recurrence intervals of 2 and 10 years, respectively. Corresponding 24-hour totals are *c.* 145 mm and *c.* 195 mm, respectively (Fritsch, 1992).

The geological substrate is composed of deeply weathered mica-schists in a landscape composed of small hills less than 100 m high with short and moderately steep slopes (15%–40%). The soils belong to the red ferralitic soil family (French classification system) and classify as Ferralsols and Ultisols in the FAO and USDA classification systems. In some areas the drainage of the soil profile is good and infiltration rapid and deep, even during heavy storms, and catchments developed on these 'Vertical Drainage' (VD) soils only show significant stormflow and peak discharges after prolonged wet periods. Some hills may be composed entirely of the VD-soil type, which is found predominantly around the ridges (Boulet, 1979). In other locations, infiltration is blocked by a compact layer occurring at a depth of 20–50 cm, which causes the development of (more or less) short-lived perched water tables and pockets of stagnant water in the upper horizon. Soils with such poor vertical infiltration have been called 'Lateral Drainage' (LD) soils by Boulet (1979). Even under moderate rainfall intensities, superficial runoff is generated on these soils in the form of saturation excess overland flow (SOF) and return flow (Dunne, 1978; cf. Bonell, this volume). LD-type soils are the most common soil type found in the ECEREX area (Boulet, 1979) (Table 22.2).

The occurrence of shallow groundwater tables in flat, wide valley bottoms constitutes another major feature of the region. As soon as a catchment reaches 2–5 ha in size, the stream comes out onto flat land. Even at the very small scale considered here (<2 ha), the water table in some of the experimental catchments (generally fed by lateral subsurface drainage along the slopes), came up to the surface of the valley bottom at the height of the rainy season. The combination of these three features (VD-soils, LD-soils and contrasting groundwater table dynamics) results in surprisingly different hydrological regimes on catchments that otherwise look very similar in terms of their geology, size and shape, topography and vegetation cover (Fritsch, 1992, 1993; cf. Figure 22.4a,b below). VD-soils occupied between zero and 99% of basin area whereas three basins (F, G and H) frequently had their groundwater tables coming to the surface, with areal extensions of these saturated areas ranging from 4–14% (Table 22.2). Under such conditions, SOF is likely to become an important runoff-producing mechanism (cf. Bonell, this volume).

Hydrological regimes under natural forest conditions

Most of the runoff emerging from these very small catchments occurred in the form of quickflow. Only three of the ten basins examined (catchments F, G and H) had a weak baseflow component during dry periods but in all other cases delayed flows/baseflow usually stopped within several hours after the cessation of rainfall (Figure 22.4a). Given the very small size of the catchments under consideration, the large number of storms, and this near-complete dominance of total surface flows by stormflow, the following analysis pertains to variations in total stormflows, i.e. quickflow plus delayed flow, but excluding baseflow between storm events (Fritsch, 1993).

The variation in runoff between catchments was determined during a two-year calibration period (1978–79) when all catchments were under natural forest conditions (Table 22.3; Figure 22.4a). Despite similar rainfall conditions, stormflow volumes varied between 7.3% (catchment C) and 34.4% (catchment H) of the rainfall. As illustrated in Figure 22.4b, the magnitude of the stormflow is closely linked to the proportion of the catchment underlain by the free-draining VD-soils. As expected, the three catchments having no free-draining soils and a significant area of saturated valley bottom (cf. Table 22.2) – i.e. catchments F, G and H – exhibited the highest stormflow totals. Interestingly, although catchment A did not have any VD-soils either (Table 22.2), its water table did not come close enough to the surface to generate significant amounts of *SOF*, with a much smaller stormflow production as the result (Table 22.3; cf. Bonell, this volume). During seven years of observation, annual stormflow totals from the control catchment B varied between 300 mm and 723 mm (12 and 20% of the corresponding rainfall).

Ungauged groundwater leakage from these very small catchments can be expected to constitute an important portion of the water budget. Water balance studies on larger (320–1240 ha) watertight rainforested catchments in the same region (Roche, 1982a) suggest an average annual E_T of 1440 mm. Subtracting this value from annual rainfall gave streamflow totals that could be expected from each ECEREX catchment if no leakage would occur. Comparing the expected with actually measured streamflows suggested a ten-fold range in annual groundwater leakage, viz. from 148 mm on catchment H (no free-draining soils) to 1493 mm on catchment C (100% free-draining soils). As such, amounts of groundwater leakage exceed measured runoff totals on all but the three catchments having marshy valley bottoms (F, G and H) (Table 22.3, cf. Figure 22.4b). It cannot be excluded, therefore, that the effects of forest conversion as determined for surface flows only (see below) underestimate the overall impact of the treatment on the 'full' water balance, as some of the increases in total water outputs after forest conversion may well go unnoticed as increased groundwater leakage.

Experimental treatments

Two catchments were selected as controls. Catchment F with its high storm runoff (Figure 22.4) was assigned as a control for those catchments having shallow groundwater tables in their valley bottoms and LD-soils (G, H, and J). Catchment B was used as a control for the remaining catchments having mixed soil types (both VD- and LD-soils) and less dynamic water tables (catchments A,

Table 22.3. *Annual rainfall, stormflow and overall runoff totals for the ECEREX catchments during the 2-year calibration phase; inferred catchment leakage totals added (see text for explanation)*

Catchment	A	B	C	D	E	F	G	H	I	J
Rainfall (mm)	3423	3267	3265	3257	3350	3102	3173	3165	3285	3219
Stormflow (mm)	650	595	239	480	426	1058	947	1088	364	748
Percentage of rain	19.0	18.2	7.3	14.8	12.7	34.1	29.9	34.4	11.1	23.3
Total flow (mm)	665	615	332	511	434	1493	1370	1577	460	831
Percentage of rain	19.4	18.8	10.2	15.7	13.0	48.1	43.2	49.8	14.0	25.8
Leakage (mm)	1318	1212	1493	1306	1476	169	363	148	1385	948
Percentage of rain	38.0	37.0	46.0	40.0	44.0	5.5	11.0	4.7	42.0	29.0

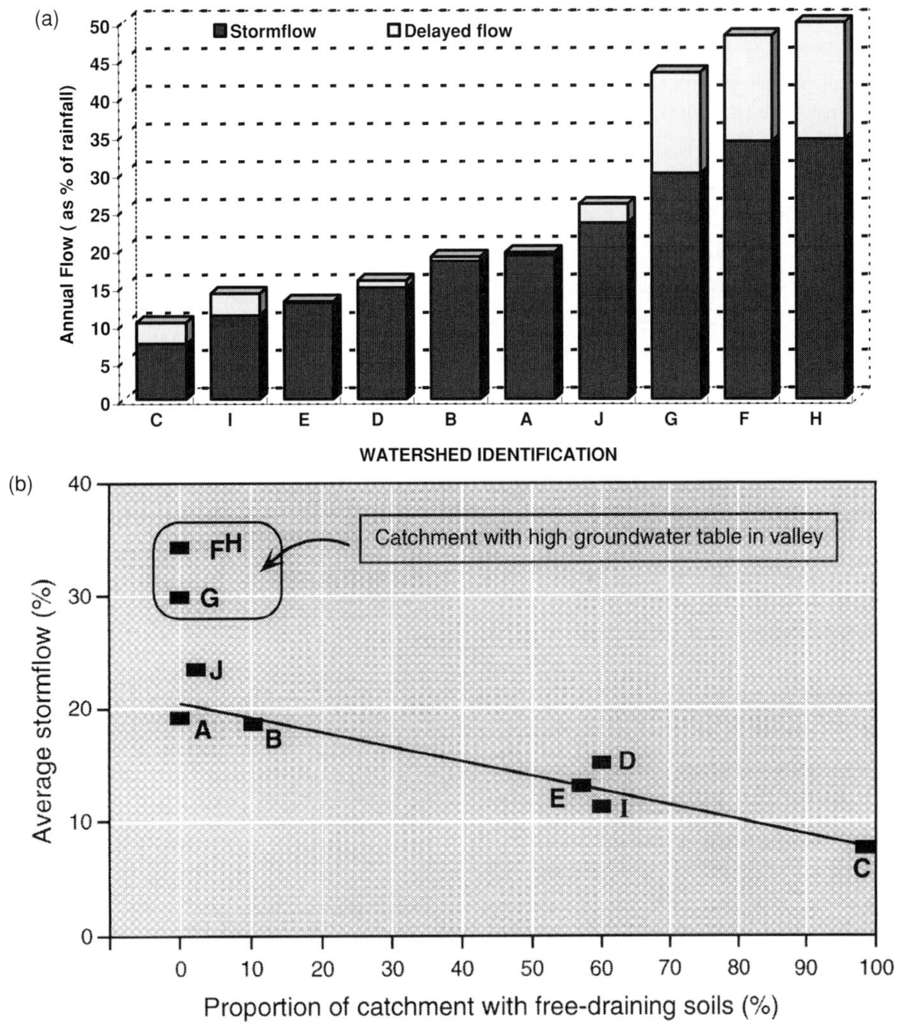

Figure 22.4 Variation in storm runoff among the *ECEREX* catchments in French Guyana under forested conditions: (a) relative amounts of stormflow (b) relative amounts of total stormflow as a function of the proportion of catchment area underlain by free-draining soils (% of rainfall) (Adapted from Fritsch, 1993.)

D, E, and I; cf. Figure 22.4a,b). Catchment C with its very low runoff coefficient was also regressed against control catchment B, although the hydrological behaviour of the two catchments was significantly different (Table 22.3; Figure 22.4a).

The applied treatments were similar to those associated with regional development projects (industrial tree plantations for paper pulp, establishment of pastures for cattle raising) with the logging of large trees constituting the first step in the conversion process. All trees over 40 cm diameter were felled using chain saws and cut into logs. A light crawler tractor (D4) with a straight blade opened tracks for subsequent skidder access (typical extension of tracks: 240 m ha^{-1}), whereas a wheeled skidder yarded the logs uphill. The smaller trees were left uncut, the stumps left in place, and the crowns and slash left on site. In a number of treatments land clearing for agricultural purposes followed the logging (Figure 22.3). Clearing was achieved with heavy crawler tractors (D8 or D9) equipped with a frontal cutting blade to push over the remaining trees, and a rear hydraulic claw system to remove the stumps. Finally, a crawler tractor with a rake blade gathered the slash in rows along the contours (windrowing) after which the slash was burnt whenever a relatively dry period occurred.

After logging the large trees, the following contrasting land cover situations were imposed (cf. Figure 22.3):[1]

(1) *Natural regrowth with no further disturbance after logging*. Two different regrowth trials were compared: natural regrowth after logging only (catchment E), and natural regrowth after logging and land clearing (catchment D).
(2) *Planting fast-growing trees after logging and land clearing*: *Pinus caribaea*, var. *hondurensis* (catchment G, 1067 trees ha^{-1}) and *Eucalyptus urophylla* prov. *Flores* (catchment H, 1300 trees ha^{-1}).
(3) *Planting fruit trees after logging and land clearing* in the catchment with the best soil conditions: grapefruit (pink pomelo) with a ground cover of *Brachyaria* grass for soil protection (catchment C, 300 trees ha^{-1}).
(4) *Planting* Digitaria swazilandensis *grass after logging and burning*: grazed by 5–10 young Zebu cattle (average weight 240 kg at the start of the experiment) per hectare (catchment A).
(5) *Traditional slash-and-burn shifting cultivation for two years* (catchment I). Although outside the scope of modern development projects, this traditional approach was used as a reference for comparison with the previous heavily mechanised treatments.

In six of the eight treated catchments the soil surface was more or less completely bare at some stage (Figure 22.3). The exceptions were catchment I, which was subjected to two years of traditional slash-and-burn cultivation, and catchment E where all large trees were logged but no subsequent land clearing took place as the contracted firm went bust (Fritsch, 1992).

Prediction of runoff from the treated catchments for forested conditions

As part of the (multiple) paired catchment technique, measured runoff totals after an experimental treatment are compared with amounts predicted for the same catchment(s) under pre-treatment (in this case, forested) conditions (Hewlett and Fortson, 1983). In the ECEREX study, various multiple regression models were tested using the data collected during the two-year calibration period. The best performing model was of the multiple linear regression type, using the runoff from the control catchment and the difference in rainfall between control and experimental catchments (see Fritsch (1992) for details). For seven out of eight catchments, these multiple regression models provided the most accurate estimates of flows under forested conditions using single storm data. For one of the catchments (C), however, a non-linear model employing a 10-day time step had to be used. A cross calibration/validation procedure of the selected models was applied to each catchment. The data collected during the catchment calibration period were divided into two samples (named periods I and II). The data from period I were first used to calibrate the correlation models, the accuracy of which was then tested using data for period II. Next, the data from period II were used for cross-calibration and the data from period I for validation. On a cumulative annual basis, the accuracy of predicted runoff under forested conditions at the 90% confidence interval ranged between 2.6% and 12%, with most values being close to or better than 5% (Table 22.4). The method was thus able to detect any modifications in runoff exceeding the respective thresholds for a catchment. It is pertinent to note that the poorest fit was obtained for catchment C which lacked a proper control catchment (i.e. having similar dominant hillslope hydrological controls). In addition, the potential confounding effect of ungauged groundwater outflows has already been commented upon.

Changes in storm runoff after land clearing (bare soil conditions)

Depending on the catchment, the bare soil period pertained to different years (Figure 22.3) but always included the main part of the rainy season (from December or January to July) (Table 22.5). The respective increases in storm runoff during the first rainy season after logging and clearing, that were attributable to the treatment according to the regression model, are listed in Table 22.5 along with measured totals.

1 Catchment J was intended originally for conversion to *Brachyaria* pasture. In June 1984, i.e. after the main experimental period, it was planted to *Terminalia ivorensis*, an African tree of moderate growth rate.

Table 22.4. *Accuracy of calculated annual runoff totals for the respective experimental catchments under forested conditions (validation at 90% confidence interval)*

Treated / control:	A / B	C / B	D / B	E / B	G / F	H / F	I / B	J / B
	7.4%	12.0%	5.3%	5.6%	2.6%	3.2%	6.4%	4.3%

Table 22.5. *Increases in storm runoff during the first rainy season following logging and land clearing (period with predominantly bare soil)*

Catchment	Period/year	Rainfall (mm)	Runoff (mm)	Increase after clearing (mm)
C	Dec 78–Jul 79	3140	682	304
D	Dec 80–Jul 81	2472	479	244
A	Dec 78–Jul 79	3400	1616	762
J	Jan 83–Jul 83	2526	1037	384
G	Jan 81–Dec 81	2888	1388	621
H	Jan 81–Dec 81	3136	1453	560

Table 22.6. *Increases in storm runoff after forest clearing (period of stabilised response during conditions of predominantly bare soil)*

Catchment	C	D	A	J	G	H
Period of stabilised response	Apr–Jul 79	Jan–Jul 81	Mar–Jul 79	Mar–Jul 83	Mar–Jul 81	
Rainfall (mm)	1448	2207	2349	2071	1445	1620
Observed storm runoff (bare phase) (mm)	342	450	1341	954	772	787
Percentage of rainfall	23.6	20.4	57.1	46.1	53.4	48.6
Predicted runoff under forest (mm)	114	181	627	483	414	475
Percentage of rainfall	7.9	8.2	26.7	23.3	28.7	29.3
Increase in runoff (mm)	228	269	714	471	358	312
Percentage of rainfall	15.7	12.2	30.4	22.7	24.8	19.3
Percentage of flow expected under forest	199	149	114	97	87	66

Increases in storm runoff ranged between 244 mm (catchment D) and 762 mm (catchment A) whereas high values (560–621 mm) were also derived for catchments G and H (Table 22.5). The sheer size of some of these increases is highlighted by the fact that they are exceeded only by the increases in *total* flows (i.e. stormflows plus baseflows) observed in a very limited number of studies. Examples include the clearing of rainforest for cocoa and oil palm in Peninsular Malaysia (706–822 mm year^{-1} during wet years; Abdul Rahim, 1988), the clear-felling of deciduous forest in Watershed No.17, Coweeta, North Carolina (662 mm year^{-1}; Swank and Douglass, 1974) or evergreen southern beech forest in southern New Zealand (650 mm year^{-1}; Pearce, Rowe and O'Loughlin, 1980). However, in the ECEREX case, a comparison of absolute increases in runoff figures for the different catchments (Table 22.6) does not provide a clear understanding of the effect of the respective treatments because of differences in the timing of the bare soil period (cf. Figure 22.3). In other words, the results are confounded by inter-annual climatic variation (including an ENSO event; cf. Mahé, Servat and Maley, this volume). Furthermore, the regressions linking flows from the treated and control catchments also lacked stability at the beginning of the treatment and it took a few weeks to get a stabilised response. The identification of this stabilised response period from the original runoff data has been described in detail by Fritsch (1992).

Amounts of stormflow associated with the respective periods of stabilised hydrological response are summarised in Table 22.6. Relative amounts of stormflow (expressed as a percentage of rainfall) during this critical period of more or less completely bare soils were typically 20–24% for catchments having at least a portion of their area underlain by free-draining soils (such as catchments C and D). By contrast, catchments without free-draining soils

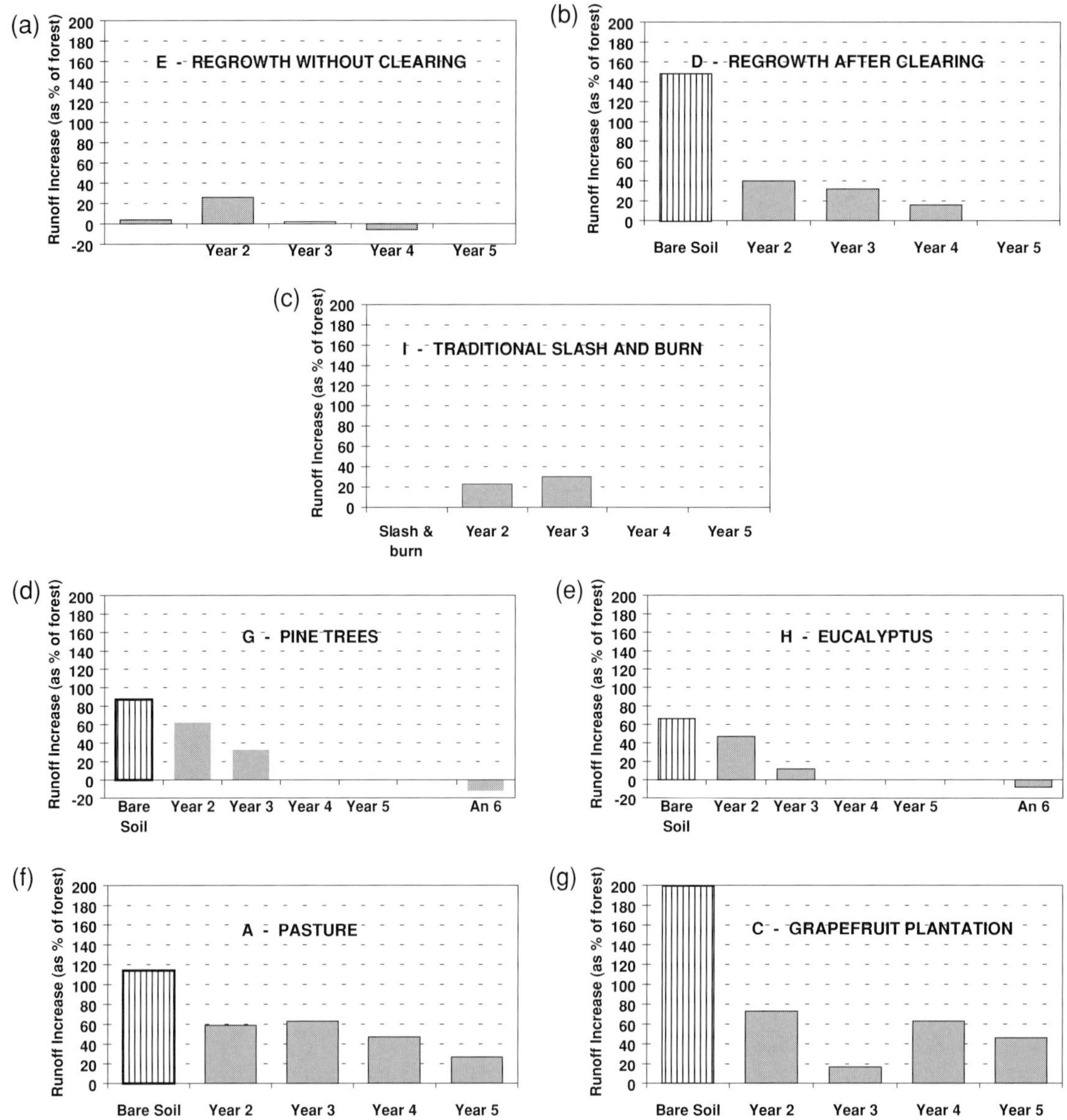

Figure 22.5 Changes in stormflow with time for the respective treatments (expressed as a percentage of corresponding runoff totals under forested conditions). (a) Regrowth following logging, (b) regrowth after logging and land clearing, (c) traditional slash and burn agriculture, (d) conversion to pine plantation, (e) conversion to eucalypt plantation, (f) conversion to pasture, and (g) conversion to grapefruit plantation. (Adapted from Fritsch, 1993.)

exhibited much higher stormflow percentages, typically 46–57% (catchments A, J, G and H). For forested conditions, the two groups of catchments would have shown storm runoff values of *c*. 8% and 23–29%, respectively (Table 22.6). Overall, relative increases in stormflow (compared to volumes expected under forested conditions) varied between 66% and 199%. It is pertinent to note that the highest relative increases (149–199%) were obtained for the catchments with predominantly free-draining soils and vice versa (cf. Bonell, this volume). As such, the relative impact of forest clearing was much more pronounced on the (agronomically) better – i.e. free-draining – soils. However, expressed in absolute terms (mm of water), the highest increases were observed for those catchments which already had high storm runoff under natural conditions in the absence of free-draining soils (Table 22.6; cf. Figure 22.4b).

Evolution of runoff with time after application of the treatments

The effects of the respective treatments on storm runoff during the years after the initial disturbance are summarised in Figure 22.5 and Table 22.7.

Table 22.7. *Absolute (mm) and relative (%) changes in stormflow with time for the respective treatments in comparison to flows which would have been observed under forest*

Catchment	Treatment	Runoff parameter	Years after treatment 2	3	4	5	6
E	Regrowth after logging only	Observed storm runoff (mm)	294	314	275		
		Change in storm runoff (mm)	+60	+6	−17		
		Relative change in storm runoff (%)	+26	+2	−6		
D	Regrowth after logging + clearing	Observed storm runoff (mm)	99[a]	525	421		
		Change in storm runoff (mm)	+29	+127	+59		
		Relative change in storm runoff (%)	+40	+32	+16		
I	Traditional slash and burn	Observed storm runoff (mm)	430	440			
		Change in storm runoff (mm)	+79	+102			
		Relative change in storm runoff (%)	+ 23	+ 30			
G	Pine trees	Observed storm runoff (mm)	1484	1195			429
		Change in storm runoff (mm)	+566	+297			−52[b]
		Relative change in storm runoff (%)	+62	+33			−12
		Observed storm runoff (mm)	1691	1159			513
H	Eucalypts	Change in storm runoff (mm)	+540	+128			−46[b]
		Relative change in storm runoff (%)	+47	+12			−9
		Observed storm runoff (mm)	755	464	739	642	
A	Pasture	Change in storm runoff (mm)	+279	+179	+235		+135
		Relative change in storm runoff (%)	+ 59	+ 63	+ 47	+ 27	
		Observed storm runoff (mm)	463	176	364	409	
C	Grapefruit	Change in storm runoff (mm)	+196	+25	+166	+128	
		Relative change in storm runoff (%)	+73	+17	+63	+46	

[a] August to December only (i.e. driest part of the year).
[b] Extremely dry year (2394 mm), hence reliability of used regression model questionable.

One general conclusion which can be drawn from Figure 22.5 and Table 22.7, is that storm runoff totals, which had increased dramatically during the first year immediately following logging and clearing, were reduced significantly in subsequent years. The respective results are discussed below in more detail.

Differences in the hydrological effect of logging-only and logging followed by mechanised land clearing can be assessed by comparing the experiments carried out on catchment E (natural regrowth after logging; Figure 22.5a) and catchment D (ditto, followed by extensive land clearing, Figure 22.5b). Because the two catchments showed very similar hydrological regimes under forested conditions (Figure 22.4a) and were deforested during the same year, it is probably safe to assume that the observed differences reflect the effect of the treatments. The intensive logging, which removed all large trees from catchment E, caused a maximum annual increase in storm runoff that was 26% higher than expected under forest (Table 22.7). However, with the additional effect of land clearing the increase became as high as 149% during the initial period when bare soil conditions prevailed (Table 22.6). Therefore, the effect of mechanised land clearing on the magnitude of storm runoff is most pronounced and considerably greater than that of high intensity timber harvesting (Figures 22.5a,b). During the fourth year after logging and land clearing, storm runoff was only 16% greater than it would have been under forest. Conversely, by then storm runoff from catchment E (logging only) was already 6% below the expected pre-disturbance value (Table 22.7). Based on the accuracy of calculated runoff values under forested conditions (estimated at 5.3–5.6% for catchments D and E, Table 22.4), this negative value is not significantly different from zero. However, whilst catchment E may be considered to have recovered in hydrological terms, this does not mean that the ecosystem as such has also regained its natural characteristics. The floristic composition, and to a lesser extent the soil conditions (see below), will remain affected for decades or longer (Bruenig, 1996).

No machinery at all was used on catchment I where traditional slash-and-burn cultivation was practiced for two years

(Figure 22.5c). Planting and harvesting were done by hand, the only mechanical input being the use of portable chain saws to fell the trees. This kind of subsistence cultivation comprises a broad variety of plants such as watermelon, corn, cassava, banana, pineapple and sweet potato. Compared to most of the mechanised trials, the hydrological impacts were rather low, with only a 23% increase in storm runoff during the first year and a 30% increase during the second year (Table 22.7). The latter may reflect the extension of the area planted to corn during the second year. At any rate, the increases in flows were light compared to those associated with mechanised clearing, in line with the findings of Lal (1987, 1996) in Nigeria. Nevertheless, if practised simultaneously over large areas, slash-and-burn cultivation may have a significant effect on basin storm runoff. For example, Odemerho (1984a,b) reported significant stream channel widening in south-west Nigeria following a widespread shift from traditional slash and burn cultivation with long fallow periods through one with short fallows to permanent cropping. Although the observations on catchment I were not extended beyond the third year (Fritsch, 1992), recovery after the cessation of cultivation can be expected to be at least as rapid as for catchment E (natural regrowth, no clearing; Figure 22.5a).

The tree plantations established in catchments G and H, which had the poorest soil conditions (impeded drainage and marshy valley bottoms) consisted of fast-growing pines and eucalypts, respectively. Although absolute and relative increases in storm runoff during the first year were rather modest compared to the other two cleared catchments (A and J; Table 22.6), the effect remained fairly pronounced during the second year at +62% (year 2) for catchment G, and +47% (year 2) for catchment H (Table 22.7). Stormflow from the eucalypt plantation became reduced significantly only by the third year after clearing (+12%) but remained at a high +33% for the pine plantation. Such findings may be explained partly by the frequent weeding applied around the young trees to prevent them from being overgrown by natural regrowth (cf. catchment D). The weeding helped to maintain soil conditions that were similar to those observed during the period shortly after the clearing. Additional runoff measurements were made in the sixth year after clearing (when the trees were five years old). Based on the regression model, annual storm runoff totals from the plantation catchments by then were less than those predicted under primary forest, viz. –12% for the pines and –9% for the eucalypts (Figure 22.5d,e; Table 22.7). Based on model performance during the calibration period (Table 22.4), these figures were significant at $F = 0.05$. However, it has to be noted that the sixth year was exceptionally dry (2394 mm). Such dry conditions had not been observed during the calibration period and the model may therefore not have been able to simulate the runoff under forest with the required accuracy for this particular year. Nevertheless, it cannot be excluded that by then the growth of the trees was such that the associated water use caused the soils to become drier, with lower runoff response to rainfall as the result (cf. Scott *et al.*, this volume). Such ambiguities in results obtained with the paired catchment technique highlight the value of additional process-based work (e.g. on tree water use; cf. Bruijnzeel, 1990, 1996; Bonell and Balek, 1993).

The grazing experiment conducted on catchment A involved creation of a *Digitaria swazilandensis* pasture grazed by 5–10 young bulls per hectare. Such a semi-intensive design is very different to the ranching system commonly in use in Amazonia, with intensities typically ranging from 0.3–1.3 animals per hectare (Fearnside, 1979; Myers, 1982). The technique used for planting the grass was propagation by cuttings, thus creating conditions for fast expansion of the forage and close vegetal cover within a few weeks after the beginning of the rainy season. Nevertheless, storm runoff levels remained high for three years, with values of around 60% above that expected under forest for the two first years after establishment and 50% during the third year. It was only four years after planting that the beginning of a decreasing trend was observed, with stormflows of about 27% above those expected for forest (Figure 22.5f; Table 22.7). As discussed more fully by Scott *et al.* (this volume), stormflow volumes from grasslands are generally higher compared with those under forest, although the effect diminishes progressively with increasing storm size. Such contrasts reflect differences in water use and thus soil water storage between forest (plantations) and grasslands, as well as differences in surface conditions (e.g. degree of compaction by cattle or degradation by repeated burning; Gilmour, Bonell and Cassells, 1987; Chandler and Walter, 1998). Although experimental work at the catchment scale from tropical Latin America seems to be lacking to confirm this, the above finding of increased stormflows from grazing land is potentially important, considering that grazing is widespread in South and Central America and indeed constitutes the dominant form of land use over extended areas (cf. Drigo and Serrao and Thompson, both this volume). In practice, much will depend on the actual status of the surface (cf. Table 25.4 in Scott *et al.*, this volume).

The grapefruit orchard with its protective grass surface cover was established on catchment C which had the best soils in terms of internal drainage and agricultural potential. Catchments with a high proportion of free-draining soils (100% for catchment C) exhibited very low storm runoff values under forested conditions (Figure 22.4b) and as a result they are potentially very sensitive to soil disturbance. Indeed, the highest increase in stormflow (199%) was observed for this particular catchment during the critical period of predominantly bare soils (Table 22.6). Although stormflows dropped to lower levels in subsequent years (from 73% above the level predicted for forest in year two to 46% in year five (Figure 22.5g), the associated volumes were by no means negligible in absolute terms (196 mm in year two, to 128 mm in year five

(Table 22.7). As indicated previously, there was no suitable control catchment with a similar stormflow regime as catchment C. As such, the accuracy of storm runoff volumes under forest conditions by the model was the poorest of the overall experimental set (12%, Table 22.4). Thus, the observed runoff increases during the second, fourth and fifth years were not statistically different from each other ($F = 0.05$). Therefore, it can be safely assumed that stormflow from the orchard catchment was about 50% higher than under forest, although the changes over time could not be assessed properly. Arguably, the observed pattern over the years somewhat resembles that found for the pasture catchment, albeit at lower absolute levels (cf. Figures 22.5g and 22.5f; Table 22.7).

Chief conclusions from the ECEREX study

Under forested conditions the hydrological response to rainfall (i.e. stormflow volumes) of the ten experimental catchments exhibited a five-fold range which largely reflected differences in soil type (free-draining vs. impeded drainage) and the presence or absence of marshy valley bottoms (Figure 22.4b). Thus, one of the most important conclusions that may be drawn from the ECEREX work is that precise knowledge of the hydrological baseline situation is required before the effects of any treatment can be assessed properly. In this case, the precise delineation of the occurrence of free-draining soils and soils with impeded drainage provided the key to such understanding. Nevertheless, additional quantitative insight into related processes such as rainfall interception and water uptake from the soil for the various post-forest land covers would have facilitated the interpretation of some of the findings (e.g. the occurrence of a declining trend in stormflows after four years for the catchments under pasture or fruit orchard, the 'negative' stormflows associated with young tree plantations in year 6) (Figure 22.5). Furthermore, given the very high degree of groundwater leakage from all but three catchments, additional groundwater observations would have been equally desirable (Table 22.3).

On six of the catchments subjected to mechanised logging and clearing, a phase with predominantly bare soils occurred during the first subsequent rainy season. After initial stabilisation of the hydrological response, increases in stormflow totals during this particularly critical period ranged from 66% to 199%. As a rule, the strongest relative impacts were observed on those catchments having the lowest storm runoff under forested conditions, and vice versa. As such, catchments with free-draining soils proved much more sensitive to disturbance than catchments with impeded drainage and marshy valley bottoms. This most probably reflects the fact that changes in the surface hydraulic properties of free-draining soils (notably compaction by machinery) caused a shift in dominant runoff pathways during rainfall, from predominantly vertical percolation to more rapid surface routes (notably infiltration-excess overland flow). By contrast, the soils with impeded drainage already exhibited rapid shallow lateral subsurface flow which, although displaced upwards towards the surface after soil disturbance, would have less overall impact on the size of the storm hydrograph (cf. Elsenbeer, 2001; Bonell, this volume).

Finally, despite the pronounced initial effect on stormflows, storm runoff decreased to original values within four to five years for all of the forestry treatments (natural regrowth after logging and/or clearing, tree plantation establishment). In the case of pasture and orchard establishment, however, increases in stormflow remained at a steady level of 30–45% throughout the first four years. Because most of the runoff occurred during and shortly after rainstorms only, the increased response to rainfall of the pasture catchment is thought to primarily reflect changes in surface conditions (notably compaction during clearing and subsequently due to grazing) rather than increased soil water contents due to a reduction in water use by grassland compared to forest (cf. Table 22.1). Concerning the persistently high increases in stormflow for the orchard catchment, it should be remembered that the runoff prediction model for catchment C was the least satisfactory of all (cf. Table 22.4). Therefore, no firm conclusion can be drawn with respect to a potentially decreasing trend from year 5 onwards (Table 22.7).

CHANGES IN FLOW PATHWAYS DURING FOREST CONVERSION AND STABILISATION OF NEW LAND USE

During dry weather, streamflow is generated by the slow discharge of groundwater to the drainage network. The deeper the groundwater level, the smaller the hydraulic gradient in the vicinity of the drains (streams) and the shorter the total length of drainage channels receiving groundwater contributions. In addition, the hydraulic conductivity of the soil (K^*) often decreases with depth below the soil surface (Elsenbeer, 2001). Together, these two characteristics act in the same direction to decrease groundwater discharge into streams as groundwater levels decrease, and this is the reason for the gradual recession of streamflow during baseflow conditions (Ward and Robinson, 2000).

Hydrological pathways during rainfall on undisturbed tropical forested hillslopes may be quite diverse (Elsenbeer, 2001; cf. Ward, 1984; Bonell, this volume). At one end of the spectrum, K^* is large enough to let rainfall of any intensity percolate vertically to the groundwater table which then continues laterally as groundwater flow. Examples include the deep Oxisols at the Reserva Ducke in Central Amazonia where hillside overland flow and shallow lateral subsurface flows are virtually absent and most, if not all, storm runoff is generated in the form of so-called saturation-excess overland flow (SOF) in flat and wide, near-saturated valley bottoms (Nortcliff and Thornes, 1984; cf. Bonell, this volume). Catchment C within the ECEREX experiment in French Guyana, which is underlain by freely draining soils throughout its area (Figure 22.4b), constitutes another example of such groundwater-dominated stormflows. At the other end of the spectrum (Elsenbeer, 2001), many Ultisols show an abrupt decrease in K^* at some depth, above which a perched water table will form if the rate of percolation exceeds conductivity, and subsurface runoff will occur laterally. Where such an impeding layer occurs at shallow depth and rainfall intensities are sufficiently high, the entire upper layer may become saturated and widespread hillside SOF will occur (Bonell and Gilmour, 1978), unless the top layer has sufficient macropores to prevent this from happening. Elsenbeer and Lack (1996) described such a case from the Peruvian Amazon in which macropores from decaying roots transmitted large volumes of percolating water downhill as shallow piped subsurface runoff. Catchments F, G and H in the ECEREX experiment represent a similar case in which stormflows are dominated by shallow subsurface flow on the hillsides feeding wet valley bottoms prone to the generation of SOF (cf. Figure 22.4b). Although catchment A in the same area also suffered impeded drainage, contributions by SOF were very limited because the groundwater table there generally did not reach the surface, except during periods of extreme rainfall (Figure 22.4b; cf. Bonell, this volume).

Generally speaking, the stormflow : rainfall ratio for a given catchment will be larger when groundwater and antecedent soil moisture levels are higher (Ward, 1984; Figure 22.4b). As such, the ratio will also be larger closer to the stream in a given hillslope section because the accumulation of groundwater travelling downslope along a streamline will require a steadily increasing transmissivity of the soil to prevent it from discharging at the soil surface. Where the accumulated groundwater flow exceeds the hydraulic gradient times the local transmissivity, the excess water will appear on the soil surface and form so-called return flow, supplemented by SOF during times of rainfall (Ward, 1984; cf. Bonell, this volume). In such a 'discharge area', the hydraulic gradient will have a component directed upwards from the groundwater surface, the soil particles will tend to separate and erosion hazard is consequently high (Vertessy *et al.*, 1990). The limit of the critical slope length (as measured from the divide) where such footslope return flow / SOF will appear (X) can be calculated from:

$$R \cdot X = T \cdot \mathrm{d}h/\mathrm{d}x \qquad (22.6)$$

where R is groundwater recharge or net precipitation, T the transmissivity of the soil profile at limit X (i.e. the product of local K^* and saturated soil depth), and $\mathrm{d}h/\mathrm{d}x$ the maximum hydraulic gradient (equal to the local slope gradient; Kirkby, 1988)). As discussed below, the hydrological implications of this in the context of land cover change are potentially profound.

Hydrological changes associated with tropical forest clearing and conversion to other land uses typically include increases in amounts of (net) rainfall reaching the ground (Bruijnzeel, 1997; Scott *et al.*, this volume) and (locally) decreased topsoil K^* due to compaction by machinery, particularly during the early post-forest stages (Malmer and Grip, 1990; Kamaruzaman, 1991). Furthermore, evaporation is (at least temporarily) reduced and overall soil water contents and groundwater levels can thus be expected to increase as well. Together, these changes will lead to increases in both baseflow and stormflow (Bruijnzeel, 1990), through increases in groundwater outflow and return flow, as well as increased contributions by overland flow (both HOF on disturbed sites and SOF in valley bottoms; cf. Table 22.6).

It follows from Eqn 22.6 that this increase in net precipitation and/or groundwater outflow will decrease the critical slope length X at which return flow and SOF can be expected. As a result, the drainage density of the cleared area will be increased through incision of new streams or headward extension of existing waterways (Vertessy *et al.*, 1990; Constantini *et al.*, 1993). A rough estimate of the magnitude of this effect may be derived as follows. Based on the difference between annual rainfall and E_T, a typical runoff figure for lowland tropical rainforest would be about 1200 mm year^{-1} (Bruijnzeel, 1990). Given an average increase in stormflow of about 480 mm year^{-1} after clearfelling (average value of first-year increases in French Guyana; Table 22.5), this would mean an average increase in streamflow of 40%. Inserting this figure into Eqn 22.6 and keeping T and $\mathrm{d}h/\mathrm{d}x$ constant would imply a corresponding decrease in mean slope length X between the water divide and the point of groundwater discharge. Although the effect of enhanced (storm) runoff after conversion will last a couple of years only in most cases (cf. Figure 22.5a–e), the drainage network would definitely have to expand more permanently to accommodate the increased production of runoff associated with treeless (Figure 22.5f) or degraded landscapes (cf. Odemerho, 1984a,b; Scott *et al.*, this volume).

It is well-established that the infiltration capacities of most undisturbed tropical rainforest soils are high to very high and therefore capable in principle of allowing complete infiltration of most short showers of high intensity. HOF (but not necessarily SOF) is therefore a relatively rare phenomenon in most undisturbed forests (Elsenbeer, 2001; Bonell, this volume). However, this may change dramatically upon forest disturbance during mechanised timber harvesting or conversion operations. Kamaruzaman (1991) demonstrated how repeated tractor passage decreased the intake capacity of an Ultisol with initially very high infiltration values (630–780 mm h^{-1}) in Peninsular Malaysia down to 40–45 mm h^{-1} in the case of tracked vehicles and to 15–20 mm h^{-1} when using rubber-tyred skidders. In addition, the effect was more pronounced when the soils were wetter (see also Figure 35.2 in Cassells and Bruijnzeel, this volume).

Malmer and Grip (1990) found topsoil infiltration capacity shortly after crawler tractor operation on a clayey Ultisol at their Mendolong site in Sabah to be reduced from an initial 154 mm h^{-1} to only 0.28 mm h^{-1}. Corresponding values for a sandy gleyic podzol in the same area were 49 and 1.3 mm h^{-1}, respectively. Similarly low values have been reported elsewhere in Sabah in the Danum area by Brooks, Richards and Nussbaum (1994). To a large extent such reductions in the soil's water intake capacity reflect the destruction of macropores, typically down to a depth of 15–20 cm (Lal, 1987; Malmer and Grip, 1990; Van der Plas and Bruijnzeel, 1993).

Naturally, infiltration-excess overland flow (HOF) will be much more frequent and intense on such compacted surfaces (Sinun *et al.*, 1992; Van der Plas and Bruijnzeel, 1993). Furthermore, the humus and litter layer has been destroyed in such places and the bare soil is therefore directly exposed to rain drop impact and further surface deterioration may follow (Lal, 1987). Another, and possibly more long-term effect of tractor tracks cross-cutting steep hillsides (cf. Figure 22.2), is that they may act as an extension of the drainage network, thereby causing a further (permanent?) increase in stormflows (Malmer, 1996a; cf. Chappell, Tych *et al.* and Chappell, Bidin *et al.*, this volume). Conditions for growth on former skidder tracks are adverse in many respects. Apart from compaction hampering the establishment of seedlings, the penetration of roots and soil faunal activity, there is the added complication of the removal of the relatively fertile top layer and the subsequent exposure of often much more infertile (as well as more erodible) subsoil material (Gillman *et al.*, 1985; cf. Proctor, this volume). As such, the recovery of the infiltration capacity on former tractor tracks can be expected to be slow. Malmer and Grip (1990) determined a value of 1.26 mm h^{-1} for former skid tracks on a clayey Ultisol at Mendolong some six years after their abandonment whereas Van der Plas and Bruijnzeel (1993) reported typical final infiltration rates of only 15–20 mm h^{-1} after 12 years on a similar soil in the Danum area. Kamaruzaman (1996) demonstrated a steady improvement in dry bulk density, total porosity, loss on ignition (LOI, a common measure of soil organic matter content) and, to a lesser extent, K^* over a period of ten years on former tractor tracks on an Ultisol in Peninsular Malaysia. Assuming a linear rate of change it was estimated that full recovery of density, porosity and LOI to pre-disturbance conditions would take approximately 22 years vs. more than 50 years for K^*. Using a similar false-time series approach at Mendolong, Malmer, Johansson and Kluge (1998) reported a near-complete recovery of average bulk density values on former tracks some 13 years after logging although topsoil infiltration rates (62 mm h^{-1} on average) were still much lower than pre-disturbance values (154 mm h^{-1}; Malmer and Grip, 1990). In addition, at only 13 mm h^{-1} the average infiltration rate on the most intensively used tracks was much lower than on tracks which had been subject to intermediate (26 mm h^1) or limited use (147 mm h^1). Malmer *et al.* (1998) suggested that the process of soil recovery may be divided into two distinct phases: an initial phase in which soil aggregation, bulk density and porosity improve in a more or less linear fashion by the addition of organic matter for LOI values up to about 10%, and a second phase during which LOI continues to increase at a slower rate while overall porosity and bulk density change little. The organic matter added during this second phase is thought to contribute primarily to the formation of macro-aggregates, thereby increasing K^* and soil aeration. Also at Mendolong, Lindquist (1998) observed a remarkable increase in fine root biomass on former skidder tracks after 10 years. Whilst Högberg and Wester (1997) had estimated a fine root biomass of only 515 ± 162 kg ha^{-1} on former tracks within a year after forest clearing (1988), 10 years later this had increased to 3635 ± 1135 kg ha^{-1} (Lindquist, 1998). Corresponding figures for relatively undisturbed soils outside the tracks were 2047 ± 416 kg ha^{-1} after one year and 4437 ± 719 kg ha^{-1} after 10 years, respectively. Thus, the former tractor tracks are seen to be recovering although judging by the large standard error, fine root growth was rather variable. Furthermore, root biomass in the top 15 cm of the main (and formerly most intensively used) tractor tracks was about 1000 and 3400 kg ha^{-1} on clayey and sandy soils, respectively (Lindquist, 1998).

Naturally, the areal extent of soil disturbance is critical to the overall hydrological effect at the catchment scale. Typically, 18–30% of the area becomes more or less seriously disturbed during mechanised logging and conversion operations, depending on local topography and timber extraction methods (Bruijnzeel, 1992) although Fritsch (1983) reported such disturbance, followed by massive erosion, over 81% of (the very small) catchment H in the ECEREX experiment. Whilst manual extraction may limit soil damage effectively (e.g. down to 4% surface area at Mendolong in Sabah vs. 24% in an adjacent catchment subjected to mechanised clearing; Malmer, 1996a), such low-impact techniques are generally not considered economical (Bruenig, 1996; cf. Cassells and Bruijnzeel, this volume).

Like tractor tracks and log landings, roads too are usually very compacted and exhibit similarly low K^* rates (e.g. 15 mm h^{-1} for a dirt road in northern Thailand; Ziegler, Sutherland, and Giambelluca, 2000; cf. Van der Plas and Bruijnzeel, 1993). In the Thai case, this represented one order of magnitude less than observed for any other land-surface type in the area (rainfed agriculture, secondary vegetation, remnant patches of disturbed old-growth forest). Ziegler and Giambelluca (1997) simulated rainfall excess occurrence (i.e. HOF) associated with the respective land-surface types and found that the hydrologically very responsive road surfaces contributed a large proportion of catchment-wide rainfall excess during frequent, small rainfall events, even though roads occupied less than 0.5% of the catchment area. However, during larger rainfall events, other land cover types became more important contributors simply because of their larger areal extent. Rijsdijk and Bruijnzeel (1990) also obtained very high annual runoff coefficients (65–70%) for various types of non-metalled rural roads in East Java, Indonesia. Clearly, the proper drainage of road and track surfaces assumes great importance if adverse off-site effects are to be avoided (cf. Bruijnzeel and Critchley, 1994; Chappell and Thang, this volume).

As indicated earlier in this section, both baseflows and stormflows are generally observed to increase after tropical forest clearing. However, where the soil's water intake capacity becomes seriously hampered over a large portion of the catchment (cf. Fritsch, 1983), it cannot be excluded that in some cases the associated shift to HOF-dominated stormflow production will impair groundwater recharge and, ultimately, reduce baseflows. Such a mechanism may explain the rather widely observed (but poorly documented) deterioration in hydrological regime associated with severely degraded tropical catchments (Bruijnzeel, 1989; see discussion in Scott *et al.*, this volume). However, it is difficult to predict when and under what hillslope hydrological circumstances such a reduction in baseflow can be expected. Within the ECEREX project the largest relative increases in stormflow (149–199%) were observed on catchments with (mostly) free-draining soils (catchments C and D; Table 22.6). Under undisturbed conditions, stormflow from these catchments was generated predominantly by vertical percolation to the groundwater table whereas this was replaced by more rapid surface routes (notably HOF) after clearing. At the other extreme, catchments with impeded drainage and wet valley bottoms (G, H) already exhibited rapid shallow lateral subsurface flow which, when displaced upwards to the soil surface after soil disturbance, would be expected to have less of an effect on the storm hydrograph (66–87%, Table 22.6). Nevertheless, in absolute terms, increases in stormflow volumes after clearing were larger for catchments G and H (312–358 mm) than for catchments C and D (228–269 mm, Table 22.6). On the other hand, the increase in stormflows for catchment A (also underlain by soils with impeded drainage but not having a saturated valley bottom like G and H) was by far the largest (714 mm), although intermediate in relative terms (114%, Table 22.6). It is difficult to decide whether this is due to the fact that soil disturbance in catchments G and H was proportionally less compared to catchment A because the marshy areas in the former were avoided by machinery (Fritsch, 1992) or whether other factors have played a role.

It is unfortunate that the ECEREX catchments were too small to sustain perennial flow and so enable an investigation of the effects of clearing on baseflow levels as well. However, compared with the 500–700 mm of reduced infiltration per year^{-1} that would be required to cause a significant negative effect on low flows (see discussion in Scott *et al.*, this volume, for details), the observed absolute increases in stormflow in years 2–5 after treatment in French Guyana (Table 22.7) suggest that there would be no such adverse effect on baseflow levels. At Mendolong, Sabah, specific baseflow levels first went up after clearing and burning (during which the soil surface was more or less seriously disturbed over one quarter of the catchment) but decreased during the second hydrological year after conversion (Malmer, 1992). Further work is necessary to resolve this important question and should explicitly include the groundwater component.

EROSION DURING FOREST CONVERSION AND STABILISATION OF NEW LAND USE

Through weathering and erosion the land surface of the Earth is continuously worn down, also under 'undisturbed' forested conditions. The eroded material is redistributed and may be deposited at the foot of slopes, in fans, alluvial plains or riverbeds, but is eventually transported to the ocean. The relative importance of the various sediment contributing mechanisms (surface erosion, gully erosion, mass wastage) varies widely between locations as a function of geological and climatic settings as well as vegetation type and cover. As a result, catchment sediment yields in the humid tropics exhibit a very large range (Douglas and Guyot, this volume). Even within the climatically reasonably homogeneous region of Malaysia, the Philippines and western Indonesia, baseline sediment yields for forested catchments have been reported to vary almost a thousand-fold (Figure 22.6, categories I–IV). At the lower end of the spectrum one finds rainforested catchments in tectonically stable areas underlain by igneous or sedimentary rocks whose soils are neither subject to significant surface erosion nor to extensive gullying or mass wasting (categories I and II, Figure 22.6). Conversely, in the tectonically or volcanically active steepland areas of the Pacific rim, yields may increase to as much as 65 t ha^{-1} yr^{-1}, especially where both surface erosion and mass wastage are rampant, as in the seasonally dry teak forests on unstable marls in Java (category IV, Figure 22.6). As such, the effects of forest clearing on catchment sediment yield can be expected to vary accordingly, with increases being much more pronounced in areas where natural rates of sediment production are low and vice versa (cf. categories IX–XI vs. I–IV in Figure 22.6). Therefore, caution is needed when comparing (absolute) increases in sediment yield after forest conversion between locations. Furthermore, the fact that a considerable portion of the material delivered to streams by (deep-seated) mass movements tends to be rather coarse and will be transported mainly as generally unmeasured bedload represents a further complication in this respect (Pickup, Higgins and Warner, 1981; Simon and Guzman-Rios, 1990).

The data base for humid tropical catchment sediment yields is growing steadily (e.g. Figure 22.6 is based on more than 60 studies from South East Asia alone). Yet studies that provide additional

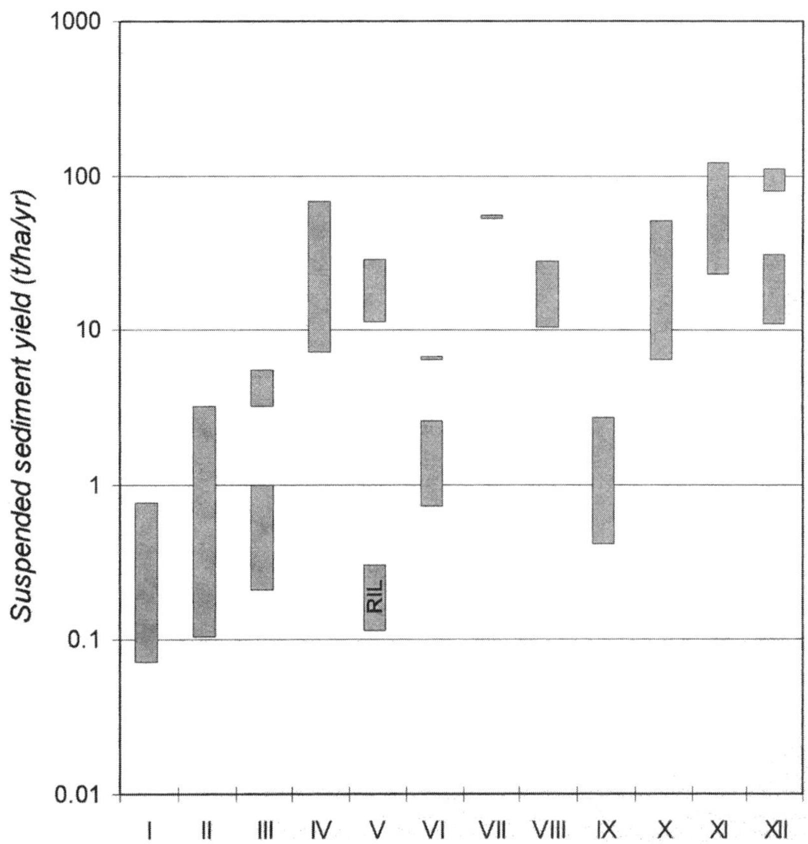

Figure 22.6 Ranges in reported catchment sediment yields in South East Asia as a function of geological substrate and land use. Categories: I, forest, granite; II, forest, sandstones/shales; III, forest, volcanics; IV, forest, marls; V, logged (RIL, reduced-impact logging); VI, cleared, sedimentary rocks (lower bar: micro-catchments); VII, cleared, volcanics; VIII, cleared, marls; IX, medium-large basins, mixed land use, granite; X, idem, volcanics; XI, idem, volcanics plus marls; XII, urbanised (lower bar), mining and road building (upper bar), mostly on granite. (After Bruijnzeel, 2004.)

insight into the underlying causes of the observed differences by explicitly quantifying the relative contributions from different sediment-generating mechanisms (e.g. mass wastage vs. surface erosion) and sources (e.g. roads and settlements vs. agricultural land) are rare (Rijsdijk and Bruijnzeel, 1991; Brown *et al.*, 1995; cf. Douglas and Guyot; Chappell, Tych *et al.*, both this volume). Based on the isotopic signatures of eroded sediment deposited in the river beds, and source materials on ridge crests and hillsides in a small forested catchment in Puerto Rico, Brown *et al.* (1995) were able to attribute about 55% of the sediment to mass wasting on the slopes whereas the remaining 45% would have been mobilised by sheet wash on the ridges and slopes. This high proportion of material supplied to mass wasting is in line with the steep nature of the terrain and the regular occurrence of hurricanes bringing large amounts of rainfall within a concentrated time period (Scatena, Planos-Gutierrez and Schellekens, this volume). Using more classical methods, Rijsdijk and Bruijnzeel (1990/1991) compared surface erosion rates for a range of land-surface types (including roads, settlements and trails) in the partly deforested 233 km^2 Konto catchment in the volcanic uplands of East Java, Indonesia. In addition, they quantified gully erosion, mass wasting intensity and volumes along roads and trails, as well as streambank retreat rates to derive an estimation of total sediment supply to the drainage network. Although roads, trails and settlements made up less than 5% of the total area, their inferred relative contribution to overall sediment production was disproportionally large at 54%. By contrast, 37% of the total sediment was estimated to have been contributed by the 20% of the catchment occupied by rain-fed agriculture whereas less than 10% was supplied by the more than 65% of forested land, mostly in the form of occasional landslides in the uppermost headwaters and bank erosion.

The above examples show that it is important to distinguish between surface erosion, gully erosion and mass movements when assessing the effects of land use change on erosion and sedimentation, because the ability of a vegetation cover to control each of these is rather different[2]. In the following, the principal changes in the frequency of occurrence and magnitude of the three main forms

2 Strictly speaking, mass wasting is not a form of erosion as there is no active agent (water, wind) involved.

of sediment-generating processes in relation to various changes in land use will be discussed (see also reviews by Hamilton and King, 1983; Wiersum, 1984; Pearce, 1986 and Bruijnzeel, 2004). In doing so, particular attention will be paid to new results about the role of roads and the importance of various forms of mass wastage in tropical landscapes.

Surface erosion: The infiltration capacity of the often clayey tropical soils depends to a large degree on their structure. As we have seen, infiltration capacities under forested conditions are generally high to very high due to the combined effect of a continuous supply of leaf litter, soil faunal activity and root decay. As a result, the potential for infiltration-excess overland flow (HOF) occurrence is low (Bonell, this volume). Naturally, processes that destroy the soil structure will decrease infiltration capacity and increase the risk of surface erosion. One such process relates to the impact of raindrops falling on the soil surface. Vis (1986) measured the drop size distribution of rainfall and throughfall (crown drip) in four rainforest ecosystems at different altitudes in Colombia and found the kinetic energy of the drip to be 20–70% higher than for rainfall in the open. After correction for interception loss the erosive power of the crown drip was still 4–30% higher. This finding is not universal, however. Hall and Calder (1993) reported that the sub-canopy drop spectra below *Pinus caribaea* and *Eucalyptus camaldulensis* trees remained essentially unchanged, whereas below *Tectona grandis* the median drop volume increased considerably. These differences were ascribed to contrasts in leaf form and density (e.g. needles vs. broad leaves) and, especially, leaf size (e.g. small eucalypt leaves vs. large teak leaves). Whether or not the higher kinetic energy of throughfall leads to splash erosion depends primarily on the degree of soil exposure. Intact understorey, litter and humus layers effectively prevent splash erosion in most forests but their destruction by grazing, burning or litter harvesting may have disastrous consequences in terms of erosion (Wiersum, 1984; Zhou *et al.*, 2001).

Similarly, disturbance of the soil during logging and forest clearing leads to strong increases in surface erosion until a new cover has become established. Important factors include the increased exposure to rain splash already mentioned and, especially, increases in overland flow occurrence and magnitude, as well as the generally much greater availability of loose soil material awaiting further transport. Although absolute levels of hillslope erosion during the vulnerable transition phase can be enormous in the case of mechanised operations, the effect is usually rather short-lived, as illustrated by the following two classical cases from Sabah and French Guyana. The results from these two studies may be considered typical for rainforested terrain not subject to significant mass wasting and therefore experiencing very low background sediment yields despite high annual rainfall (>3300 mm). In addition, under these conditions changes in catchment sediment outputs reflect changes in supply by surface erosion, modified by deposition of sediment on footslopes and in depresssions (Fritsch, 1983; cf. Bruijnzeel, 1997).

The changes in headwater sediment yield associated with the conversion of rainforest to a plantation of *Acacia mangium* at Mendolong, Sabah, can be summarised as follows (Malmer, 1990). Pre-disturbance values were very low at 0.05 and 0.2 t ha^{-1} $year^{-1}$ for two catchments, W4 and W5, respectively. Catchment W4 was cleared manually without burning of slash whilst catchment W5 underwent mechanical clearing followed by burning. Increases in sediment yield during the period of actual clearing (3.5 months) plus the next five months necessary to establish some sort of new ground vegetation, were six- and tenfold for catchments W4 and W5, respectively. Surprisingly, in view of the difference in severity of the treatments, sediment production over the next ten months was equal (1.9 t ha^{-1} yr^{-1}) in both catchments. Compared to the preceding transition period, this represented a threefold increase in the case of catchment W4 but a 50% reduction for catchment W5. Malmer (1990) interpreted the unexpected increase in sediment production from catchment W4 in terms of suppressed understorey vegetation due to the presence of large amounts of slash. In addition, the slash had been arranged in rows running down the slope rather than along the contour, which may have had a concentrating effect on surface runoff. However, the overall effect was relatively short-lived as sediment yields were virtually back to normal within 20 months after clearing at *c.* 0.1 and 0.15 t ha^{-1} yr^{-1} for catchments W4 and W5, respectively. Total sediment exports from the two catchments during the first three years after clearing were 2.3 and 4.1 t ha^{-1} $year^{-1}$ for catchments W4 and W5, respectively. By contrast, surface erosion on some of the more intensively used tractor tracks in catchment W5 were as high as 500 t ha^{-1} during the same period. Away from tractor tracks, however, surface erosion increased three- to fivefold compared to conditions under forest during the first nine months after clearing (Malmer, 1996b). The doubled increases observed at the catchment scale (compared to those on the hillslopes) reflect the extra contributions from the more intensively eroding tractor tracks although a quick calculation learns that most of the latter material must have been deposited on its way to the stream (Malmer, 1996a).

Sediment outputs from the forested ECEREX headwater catchments in French Guyana were also very low prior to disturbance and ranged between 0.05 (basin C) and 0.75 t ha^{-1} $year^{-1}$ (basin F), depending on stormflow regime and soil type (cf. Figure 22.4b; Fritsch and Sarrailh, 1986). Despite the very similar pre-clearing runoff behaviour of the two catchments converted to plantations of pine and eucalypt (catchments G and H, Figure 22.4b), increases in sediment yield during the first year after clearing differed markedly, viz. 50- and 16-fold increases to 17 and 6 t ha^{-1} $year^{-1}$, respectively. This unexpected contrast can be explained by the fact that the valley bottom of catchment H was too wet to allow access to machinery at the time of clearing. As a result, this part of the catchment became surrounded by a wall of earth and slash acting as a filter for sediment coming from upslope. An independent slope survey revealed that about 650 m^3 of earth had become displaced during the operation over an area of 8100 m^2, suggesting an average hillslope surface erosion rate of about 1200 t ha^{-1} (equivalent to a layer of soil of 84 mm depth on average; Fritsch, 1983). No information has been published on sediment yields for these catchments during subsequent years but observations for the nearby catchment A (converted to *Digitaria* grassland) suggested a rapid decline in sediment yield during the second and third year after clearing (to a stable value of about 0.5 t ha^{-1} $year^{-1}$, or three times the value expected under forested conditions), compared to an initial 34-fold increase to 12 t ha^{-1} $year^{-1}$ during the first year (Fritsch and Sarrailh, 1986).

Tractor tracks are deeper than surrounding surfaces and therefore tend to concentrate water coming from upslope or from the sides (cf. Figure 22.2). This accumulation of surface flow will soon start to erode the tracks, beginning as small rills but ultimately developing into *gullies*. Gully erosion progresses as tension cracks form during dry periods. During subsequent rainfall the cracks are filled with water and the pore-pressure increases until the edges finally collapse (Collison, 1996). Stream banks show the same process but in addition scouring at the base of the bank causes additional instability and collapse (Duijsings, 1985; Rijsdijk and Bruijnzeel, 1990).

Similarly, all types of roads and footpaths are potential places for the development of surface, rill, and gully erosion, especially on sloping ground where surface water can accumulate over longer distances and so gain erosive power (cf. Yu, this volume). In mountainous northern Thailand, trails had K^* values that were typically 80–120 mm h^{-1} lower than those determined in adjacent agricultural fields (Ziegler *et al.*, 2000). Such values were low enough to generate substantial volumes of HOF and sediment. Rijsdijk and Bruijnzeel (1990) demonstrated that the very high production of sediment associated with rural footpaths in volcanic upland Java (up to 42 kg m^{-2} $year^{-1}$) largely reflected a steady supply of loose material from the adjacent side walls. Conversely, a more compacted trail draining yards and fields produced *c.* 7 kg m^{-2} $year^{-1}$. In northern Thailand, surface runoff from dirt roads under dry antecedent soil moisture conditions reached 80% of artificial rainfall supplied at a rate of 105 mm h^{-1} during 45 min simulations. Initially, the sediment concentration exceeded 100 g l^{-1} (a very high value), but once loose surface material had been washed off concentrations decreased rapidly, because the compacted road surface underneath was resistant to detachment. Loose material was generated by various processes, including the passage of traffic, road maintenance works, and mass wasting on the steep road cuts. In a related study Ziegler *et al.* (2001) examined sediment production on roads in connection with motorcycle or truck passage during simulated rain. They obtained a three- to fourfold increase in instantaneous sediment concentration after a motorcycle passage and a sevenfold increase after a truck passage. Likewise, on the island of St John, US Virgin Islands, MacDonald, Sampson and Anderson (2001) found that relatively undisturbed, vegetated hillslopes generated runoff only during very large storm events and produced little sediment. In contrast, unpaved roads commonly produced runoff once rainfall exceeded 6 mm. Sediment yield at the plot scale was high at 10–15 kg m^{-2} $year^{-1}$. In addition, cutslopes intercepted subsurface flow from upslope, thereby increasing overall road runoff volumes even more. On a road segment with heavy traffic the erosion rate was at least 7 kg m^{-2} $year^{-1}$, more than twice that of comparable road segments with less traffic and similar to values obtained by Rijsdijk and Bruijnzeel (1990) for rural roads of variable surface condition in Java (1.7–6 kg m^{-2} $year^{-1}$). The sediment produced by such roads normally accumulates in drainage channels which are cleared rather infrequently during large storms which then suddenly deliver large volumes of sediment to the streams (MacDonald *et al.*, 2001) Chatterjea (1998) described a similar case in relation to a public park in Singapore.

Whilst the physical characteristics of some tractor tracks have been shown to recover with time, Douglas *et al.* (1993) identified continued erosion of old forest access roads and their immediate surroundings (road cuts and embankments), together with gullying on some of the more intensively used tracks, as the main source of enhanced stream sediment concentrations in logged-over terrain in Borneo. Even after annual catchment sediment yields had stabilised at about 250% of pre-logging values after two years, extreme events continued to reactivate sediment delivery to the streams occasionally through the collapse of the old logging roads and hollow-log culverts, and through the initiation of road-related landslides (Douglas *et al.*, 1999; cf. Chappell, Tych *et al.* (a), this volume). Such roads, built only for timber transport, can be expected to continue to erode, thereby delivering suspended and bedload material to streams and rivers for a long time. Grace (2000) identified roadside slopes as one of the major sources of sediment losses from managed forest systems, adding that erosion on cut and fill slopes could be typically reduced by 90% if they were to be stabilised by vegetation or protected by erosion mats.

Larsen and Parks (1997) investigated the intensity of *mass wasting* at distances perpendicular to roads along a highway system in mountainous forested terrain in Puerto Rico. Landslide density decreased sharply with distance from the roadside, and was practically down to the background value of six slides km^{-2} once a distance of 85 m from the road was reached. At smaller distances, there were about 30 landslides km^{-2}. The average surface area of the scars also decreased with distance from the roadside. Within the 85 m limit they covered about 16 300 m^2 km^{-2} vs. about 2 100 m^2 km^{-2} beyond. Also in Puerto Rico, Larsen and Torres-Sánchez (1998) analysed time series of aerial photographs to evaluate the temporal frequency of landsliding as a function of surface cover. On forested land 0.08–0.14 new landslides occurred per km^2 and year vs. *c.* 0.25 landslides km^{-2} $year^{-1}$ on deforested, agricultural land and 0.32–1.18 landslides km^{-2} $year^{-1}$ on land in the vicinity of roads and 'structures'. Clearly, human activities increase landslide frequency in this area, particularly the construction of roads in steep terrain. Sidle, Pearce and O'Loughlin (1985) listed increased weight on the hillslope from road fill material, plus hillslope oversteepening, removal of slope support by roadcuts, alteration of surface runoff paths, and enhanced runoff rates as principal causes of increased slope instability following road building. On the island of St Lucia, West Indies, Anderson (1983) noted that angle and slope plan curvature were significant factors in determining whether a road cut was likely to fail or not. In addition, elevated pore pressure resulting from heavy rainfall caused frequent slope failures along road cuts. Similarly, Tan (1984) noted that infiltration during downpours on cut and fill slopes constituted a major problem when trying to maintain slope stability in the context of highway construction, building excavations, mining and quarrying, and the construction of housing in steep terrain in Peninsular Malaysia and elsewhere in South East Asia (cf. Chatterjea, 1998). Moore, Dibb and Billing (1991) attributed major landslide occurrence in the Philippines to seismic activity and heavy tropical storms, but found that deforestation and construction of logging roads dramatically increased the extent and impact of landsliding.

Finally, conversion of tropical forest to rainfed agricultural cropping over large areas can have equally drastic impacts on the quality of drainage water, both in terms of suspended sediment load and concentrations of dissolved nutrients. A comparison of annual sediment yields for medium to large catchments in South East Asia having mixed land use (categories IX–XI, Figure 22.6) with those for forested catchments (of various sizes; categories I–IV) illustrates that yields may be quite enhanced over large areas. This, despite the moderating effects of the generally observed reductions in sediment delivery ratios with increasing basin size (Walling, 1983; Bruijnzeel, 1990). In response to these enhanced sediment yields, a range of soil conservation methods and programmes have been designed and applied, often with very limited success (see Critchley, this volume for details). Lal (1997a,b) reported on a major study of the relative impacts at the plot- and catchment-scale on amounts of runoff, sediment and nutrient losses associated with various forest clearing and tillage methods, as well as several contrasting cropping systems and soil restorative practices under subhumid conditions in Nigeria. Forest clearing methods included a comparison of: (i) manual slash and burn, (ii) mechanised clearing using a tractor equipped with a shear blade; and (iii) idem with tree-pushing and root rake facilities. These were then followed by cropping of maize applying conventional tillage or no-tillage, plus the following comparisons: (i) alley cropping with *Leucaena leucocephala* planted along the contour at 4-m intervals; (ii) restorative fallowing with mucuna (*Mucuna utilis* Lam.) on severely degraded soil; (iii) mucuna fallowing on moderately degraded soil; and (iv) rotational farming on moderately degraded soils involving a pasture phase. All treatments were imposed on small catchments of 2–4 ha each (i.e. flows were not perennial), and replicated twice although no calibration between catchments was performed. As such, the results must be considered indicative only. Summarising the multitude of results from this study, manual clearing, no-tillage practices, alley cropping, and the use of cover crops all decreased losses of sediment and dissolved nutrients in surface runoff and seepage water (see Lal, 1997a,b for details).

SOIL FERTILITY CHANGES DURING FOREST CONVERSION AND STABILISATION OF NEW LAND USE

Soil provides plants with anchorage, storage of water, exchangeable nutrients (more or less directly available for plant growth), as well as nutrients locked in minerals that may be mobilised by weathering, and oxygen for root respiration. Under humid climatic conditions, the net percolation of infiltrated precipitation pulses will tend to leach nutrients down the soil profile, and ultimately from the ecosystem via deep drainage and streamflow. The higher the fertility of the substrate, the more nutrients will be leached from the ecosystem per unit of streamflow (Bruijnzeel, 1991; Proctor, this volume). Uptake of nutrients by plant roots constitutes an important counteracting process by which nutrients on their way down the soil profile are recirculated and, in time, returned to the soil surface via litterfall (Proctor, 1987). On infertile soils, a large part of overall amounts of nutrients in the ecosystem is stored in living and dead biomass. Under such conditions, nutrient conservation becomes important. The very high concentration of fine roots at or near the soil surface found in rainforests on infertile soils is usually interpreted as an effective means through which the forests achieve more rapid access to nutrients reaching the forest floor in leaf litterfall, crown drip and stemflow (Jordan, 1989; Proctor, this volume).

In this section we discuss to what extent the properties of the major soil types of the humid tropics are altered through a transition from the original forest to other land uses. Soil physical properties that are essential for plant growth include abundant available

Table 22.8. *Averaged physical and chemical key characteristics of topsoils (0–20 cm) of dominant soil groups in the humid tropics and their major constraints for agricultural use*

Land quality	Acrisols	Ferralsols	Cambisols	Luvisols	Arenosols/Podzols	
Area in (sub)tropics (%)	31	26	5	3	7	
Soil moisture[a]	X	X		X	XXX	
Nutrient availability[a]	XXX	XXX	XX		XXX	
Nutrient retention[a]	XX	XXX	XX	X	XXX	
Aluminium toxicity[a]	XX	XX	X		X	
Mechanisation[a]				X		
Erosion hazard[a]				X		
pH_{H2O}	4.8	4.8	5.3	6.4	5.3	4.5
Organic carbon (%)	2.0	2.3	2.3	2.2	0.8	5.0
C/N ratio	14	16	11	17	16	23
Bases, $cmol_C\ kg^{-1}$	2.2	1.8	11.5	21.2	2.0	1.0
CEC_{pH7}, $cmol_C\ kg^{-1}$	9.9	8.8	19.3	22.7	6.6	20.0
Base saturation (%)	26	19	49	87	44	18
K^*, 0–0.2 m[b] ($mm\ h^{-1}$)	9.4	38.6	31,2	15.7	21.5	76.4
K^*, 0.7–1.0 m[b] ($mm\ h^{-1}$)	1.5	8.8	10.3	4.1	64.9	61.4
Plant available water[c]	128	110	34	122	150	170

[a] X, moderately limiting; XX, severely limiting; XXX, very severely limiting.
[b] Hydraulic conductivity without macropores (calculated from textural data in Kauffman *et al.* (1998) their Table 2 according to method outlined by Campbell (1985)).
[c] In mm of water down to 1 m of soil depth.
Source: After Kauffmann *et al.* (1998).

soil water, good aeration, and high penetrability for roots. In other words, a suitable pore-size distribution, good aggregation, high soil organic matter content (SOM), and a moderate bulk density. A forest soil with plenty of soil fauna mixing litter into the topsoil and creating macropores will have such properties (Pritchett, 1976). Important soil chemical properties for plant growth include a high cation exchange capacity, a high content of essential nutrients and intermediate pH values to keep aluminium in the soil solution below toxic levels (Sanchez, 1976; cf. Proctor, this volume). Apart from the importance of soil biological activity for the creation of macropores and enhancing levels of SOM, the decomposer communities (microbes and fungi) and their mycorrhizal associations have been shown to enhance weathering and nutrient uptake as well (cf. Proctor, this volume).

Kauffman, Sombroek and Mantel (1998) reviewed the major constraints for agricultural use of the dominant soil groups of the humid tropics (Table 22.8; cf. Baillie, 1996). Using the FAO classification, soil groups and their areal percentages included: Acrisols (Ultisols in the USDA classification), 31%; Ferralsols (Oxisols), 26%; Cambisols (Inceptisols), 5%; Luvisols (Alfisols), 3%; and Arenosols (Entisols) plus Podzols (Spodosols), 7%. Generally, nutrient availability and retention, and aluminium toxicity are moderately to severely limiting for all soil groups. The sum of base cations, the cation exchange capacity, and base saturation are all low to very low for Ferralsols, Acrisols, Arenosols and Podzols (Table 22.8). In other words, these soils are highly leached. Luvisols and Cambisols have moderate to high fertility but together these soils occupy only 8% of the total humid tropical land area. Therefore, about two-thirds of all humid tropical soils must be considered as being quite infertile. In addition, soil moisture can be severely limiting in Arenosols and Podzols, even under perhumid equatorial rainfall (Bruenig, 1969; Baillie, 1976).

Average topsoil pH_{H2O} varies between 6.4 (Luvisols) and 4.5 (Podzols) (Table 22.8). Organic carbon content is generally low, ranging between 5.0% for Podzols and 0.8% for Arenosols. Naturally, SOM is important as it enhances soil structure and hydraulic properties as well as nutrient exchange capacity and retention. The clay fraction in most tropical soils is dominated by kaolinite, a 1:1 lattice clay having low cation exchange capacity. Average C/N ratios vary between 23 for Podzols and 11 for Cambisols (Table 22.8). According to Kauffman *et al.* (1998) the high C/N of Podzols reflects the slow decomposition rates associated with these soils due to their low pH and, especially, their very low nitrogen status (Proctor, this volume). Cambisols on the other hand seem to have better quality SOM (Kauffmann *et al.*, 1998). A lesson from temperate forest soils is that low C/N-ratios indicate an enhanced susceptibility to nitrate leakage after fertilisation (Dise, Matzner and Forsius, 1998).

Conversion of tropical rainforest to other land uses involves the harvesting of commercially valuable timber, pushing over the remaining trees, pulling stumps and roots, and burning the slash after windrowing (Figures 22.1 and 22.2; Martin, 1970; Lal, 1987). In some cases the land is subsequently levelled or simple terraces are made along the contours. Nutrients in extracted timber are lost from the ecosystem. Repeated tractor passage leads to compaction and destruction of the valuable topsoil, which is often scraped away and pushed aside into heaps or depressions where its nutrients are of little use (Gillman *et al.*, 1985). Malmer *et al.* and Scott *et al.* (both this volume) discuss the adverse effects of burning (such as volatilisation, enhanced leaching) in more detail. Further loss of soil fertility during the transition phase is caused by soil erosion (Figures 22.1 and 22.2).

Most studies of soil fertility changes associated with tropical forest conversion consider only losses of exchangeable nutrients because contributions by weathering are often thought to be negligible (Nykvist, 2000; cf. Baillie, 1996). As shown by Hamdan, Burnham and Ruhana (2000), most base cations are lost in the primary weathering stages from the parent rock into saprolite and thus the roots will have to penetrate sufficiently deep into the saprolitic zone to reach the weathering front. Nevertheless, where rooting is deep enough, nutrient losses associated with logging and clearing may be compensated partly by deep uptake of nutrients, and their subsequent return to the soil surface via litterfall. However, little experimental evidence for this phenomenon seems to be available for the humid tropics (Proctor, this volume). Bruijnzeel (1998) estimated the time span required to compensate the nutrient losses associated with timber harvesting, burning of slash and leaching through nutrient inputs from atmospheric deposition alone or in conjunction with contributions by weathering. For a catchment in perhumid Sabah underlain by Podzols (62%) and Acrisols (38%), it would take *c.* 50 years to compensate the losses of potassium, calcium and magnesium. Taking contributions by weathering into account as well would reduce these periods to *c.* 35, 15 and 6 years, respectively. For an adjacent catchment subjected to manual clearing (but not fire) and having more clayey (66%) than sandy (34%) soils, a 55-year compensation period was estimated for calcium vs. 35–45 years for potassium and magnesium (no weathering). Inclusion of nutrients released by weathering reduced these periods again, to 20, 10 and *c.* 4 years for potassium, calcium and magnesium, respectively (see Bruijnzeel, 1998 for details). Conversely, estimated recovery times for potassium, calcium and magnesium levels in a moderately fertile Acrisol under somewhat drier conditions in Viti Levu, Fiji, were much lower at 10, 15 and 6 years (rainfall only) and 6, 1 and <1 year (including weathering), respectively. The extent of the nutrient losses in connection with forest conversion and the duration of the estimated recovery periods under wet tropical conditions (Malaysia) derived by Bruijnzeel (1998) not only illustrate the fragile nature of many tropical soils but also indicate the need for fertilisation of post-forest phase crops. With the exception of the period required to recover soil magnesium levels after logging and burning, recovery times typically exceed the rotation length of fast-growing tree plantations (<10 years). In other words, the soil will be disturbed once again upon harvesting the next tree crop before full recovery of soil nutrient levels after the previous disturbance has been achieved. Also, the observation of Nykvist (2000) that amounts of *total* calcium down to 0.5 m depth in five out of 17 examined tropical soils were lower than in harvested logs, causes one to question how sustainable forestry (or agriculture) on these calcium-poor soils really is from a nutrient perspective (cf. Sanchez, 1976; Nykvist, 1997; Mackensen *et al.*, 2003; Scott *et al.*, this volume).

Forest conversion to other land uses does not necessarily decrease the nutrient status of the soil, although it does decrease overall ecosystem nutrient contents very much (see Hölscher *et al.* and Scott *et al.*, this volume). SOM (carbon) is usually the most limiting factor for microbial growth in tropical (and most other) soils (Duah-Yentumi, Røn and Christensen (1998) and is the chief determinant of soil cation exchange capacity (Sanchez, 1976; Kauffmann *et al.*, 1998). Asio *et al.* (1998) studied the effects of forest conversion to other land uses on soil chemical characteristics (including SOM) using a 'false-time series' approach (Fölster and Khanna, 1997) on the island of Leyte, the Philippines. Their comparison included residual forest, *Imperata* grassland (fallow or grazed), and low secondary scrub on an Andisol (volcanic Inceptisol); and secondary forest, recently reforested land, former shifting cultivation fields planted to coconut plus secondary shrubs, five-year-old shifting cultivation fields, and old-growth coconut with a leguminous cover crop (kudzu, *Pueraria phaseoloides*), all on Alisols (Ultisols). On the Andisol topsoil SOM decreased in the following order: forest (14.2%), 'pasture' (11.6%), scrubland (11.0%) and grassland fallow (6.9%). Overall, the Alisol had lower SOM levels, viz. forest (7.0%), coconut + kudzu (5.3%), shifting cultivation (4.8%), coconut + shrub (4.4%), and recently reforested crop land (3.2%). Furthermore, total and available nitrogen, and exchangeable potassium, followed the trend for SOM on the Andisol whereas pH, available calcium and magnesium were increased after deforestation on both the Andisol and the Alisol. No change was found for available phosphorus, which was very low and probably the most limiting nutrient (Jahn and Asio, 1998). Phosphorus is generally the most deficient nutrient on Ferralsols and Alisols/Acrisols (Sanchez, 1976; Kauffmann *et al.*, 1998).

Ekanade (1998) compared the properties of soils beneath undisturbed rainforest with those resulting from 30 years of unfertilised mixed cropping of cocoa and oil palm on Luvisols in the Nigerian cocoa belt. In the top 15 cm of soil, the pH_{KCl} was 6.6 under forest and 6.3 under plantation, whereas exchangeable calcium levels were 11 and 6.8 meq 100 g^{-1}, respectively. Exchangeable magnesium on the other hand was much higher in the plantation soil (7.5 meq 100 g^{-1} vs. 1.5 under forest) and so was extractable phosphorus (20.4 ppm vs. 7.4 under forest). One interpretation of these results would be that the plantation trees somehow extracted these nutrients more efficiently from deeper soil layers than the original forest and returned magnesium and phosphorus to the surface soil via litterfall. Both topsoil properties (Ekanade, 1998) and crop yields (Afolami and Ajobo, 1983; Kolade, 1986) were better under the mixed cropping system, as compared with monocropping of either cocoa or oil palm.

The relatively level terraces required by tree crops such as oil palm have the disadvantage of potentially exposing saprolite ('rotten rock') where it is found close to the surface. Hamdan *et al.* (2000) compard the characteristics of exposed saprolitic material in oil palm plantations in Peninsular Malaysia as derived from granite, basalt and schists, and found a definite degrading effect on soil quality. In a basalt profile more than 95% of the base cations had been lost during the initial weathering phase from parent material to saprolite, vs. 55–90% on granite. On schists, only magnesium had been lost in appreciable quantities in this way. Where the saprolite was exposed, land suitability ratings for oil palm dropped from suitable or marginally suitable to non-suitable. The major factors contributing to the lowered rating were poor fertility, absence of aggregation, high content of rock fragments, limited possibilities for rooting, and low water availability. The absence of organic material and extensive leaching during the saprolitisation process both contributed to the poor fertility status of the saprolitic material. On a related note, Bruijnzeel and Critchley (1996) drew attention to the disturbance of subsurface hillslope flow patterns by bench terracing steep terrain in Java. Runoff volumes from terraces having saprolitic material exposed in their riser faces were larger and carried more sediment in suspension (cf. MacDonald *et al.*, 2001).

CONCLUDING REMARKS

Many of the soil and water impacts of tropical forest conversion and stabilisation to new land use are known, at least in general terms. For example, converting rainforest to oil palm, cocoa or tree plantations does not affect long-term water yields by more than 10%, whereas conversion to rain-fed cropping or pasture would typically lead to increases in flows of *c.* 300 mm $year^{-1}$ (non-seasonal conditions) to as much as 700 mm $year^{-1}$ (seasonally dry conditions). However, the humid tropics are far from being homogeneous in terms of climatic and soil conditions and results from one place may not necessarily be applicable to other places. For example, reported first-year increases in streamflow after tropical forest clearing range from 125–820 mm $year^{-1}$ (Bruijnzeel, 1996). It is therefore of great importance that observed phenomena are parameterised and conceptualised into simple general, yet robust, predictive models. One problem with this is that often the available data (e.g. on vegetation water use) are not directly comparable because the underlying calculations are quite diverse. In addition, background information is not always complete enough for a full assessment of the results, particularly in the context of assessing the effects on streamflow of a certain land use change. To make matters worse, good-quality time series of sufficiently long duration (e.g. for streamflow) are rare for the humid tropics and are, in fact, becoming increasingly rare as less funds are being made available for water resources monitoring in both tropical and temperate countries (cf. the introductory chapter to this volume).

Evaporation from vegetated surfaces can be calculated reliably using the Penman–Monteith equation. It is encouraging to note in this respect that information on the values of such important parameters as albedo, (maximum) surface (or stomatal) conductance, leaf area index development and (specific) canopy storage capacity is gradually becoming available for a range of important tropical vegetation types (Table 22.1; see also Roberts *et al.*, Hölscher *et al.*, and Scott *et al.*, all this volume). On the other hand, and as also highlighted in the chapters just cited, information on maximum rooting depths (determining, *inter alia*, maximum plant available soil water and thus plant physiological behaviour) is still badly needed for most tropical vegetation types (see also below).

Similarly, the chief variations in tropical hillslope hydrological response to rainfall under forested conditions are known in principle (Bonell, this volume) but there is comparatively little sound information on soil transmissivity and water retention characteristics, and even less on the spatial distribution of these characteristics, or on their relationship with more easily measured parameters like soil texture. This pertains especially to the situation after deforestation or during the establishment and maturation of replacement vegetation in the context of the spatial prediction of surface erosion patterns (cf. Yu, this volume). In addition, groundwater flow is seldom quantified, both under undisturbed conditions (cf. Table 22.3) and after clearing. Yet such information is vital if we are to better understand the causes underlying, for example, the decreases in baseflow during dry periods that are so widely observed (but poorly documented) following severe land degradation (cf. Scott *et al.*, this volume). Despite these shortcomings in our knowledge, it appears to be well-established that effects of tropical forest clearing and conversion on stormflow volumes, although considerable during the critical initial period of bare soils, are comparatively short-lived (2–4 years and often less). The only exception is when the post-forest land use does not lead to gradually recovering infiltration opportunities (e.g. intensive grazing, continued cultivation without proper soil conservation).

Under such conditions, stormflow volumes will remain elevated and may even continue to increase with time as the degradation process proceeds (cf. Scott *et al.*, this volume).

The fact that many tropical tree plantations are established on relatively nutrient-poor soils suggests that they may experience more or less serious nutrient deficits within only a few rotations. Better knowledge of the amounts of nutrients contributed by rock weathering, in conjunction with better information on rooting depths associated with various post-forest land uses would diminish the present uncertainties surrounding the approximate recovery times for key soil nutrients (cf. Scott *et al.*, this volume).

Although the wish list for additional information could be extended, it is important to acknowledge that a considerable body of knowledge concerning the soil and water impacts of land cover change in the humid tropics has been accumulated during the last two decades, and the database is still increasing. Nevertheless, progress could be more rapid through the implementation of strategic, coordinated pan-tropical measurement programmes focussing on a few, carefully chosen key questions (cf. Bruijnzeel, 1996). In the present context, one such key question concerns elucidation of the threshold values for the degree of soil degradation upon exceedance of which groundwater recharge becomes so seriously impaired that baseflows will start to decline. Equally important is an evaluation of the amounts of fertilisation required for environmentally and economically sustainable plantation forestry for the most widely occurring combinations of tree species and soil types (cf. Scott *et al.*, this volume).

References

Abdul Rahim, N. (1988). Water yield changes after forest conversion to agricultural land use in Peninsular Malaysia. *Journal of Tropical Forest Science*, 1, 67–84.

Afolami, C. A. and Ajobo, O. (1983). An economic evaluation of a cropping system: the case of cocoa grown in combination with oil-palm. *Café Cacao Thé*, 27, 121–125.

Anderson, M. G. (1983). Road-cut slope topography and stability relationships in St. Lucia, West Indies. *Applied Geography*, 3, 104–114.

Asio, V. B., Jahn, R., Stahr, K. and Margraf, J. (1998). Soils of the tropical forests of Leyte, Phillipines II: Impact of different land uses on status of organic matter and nutrient availability. In *Soils of Tropical Forest Ecosystems – Characteristics, Ecology and Management*, ed. A. Schulte and D. Ruhiyat, p. 37–44. Berlin: Springer Verlag.

Baillie, I. C. (1976). Further studies on drought in Sarawak, East Malaysia. *Journal of Tropical Geography*, 43, 20–29.

Baillie, I. C. (1996). Soils of the humid tropics. In *The Tropical Rain Forest*, P. W. Richards, pp. 256–286. Cambridge: Cambridge University Press.

Balasimha, D., Daniel, E. V. and Bhat, P. G. (1991). Influence of environmental factors on photosynthesis in cocoa trees. *Agricultural and Forest Meteorology*, 55, 15–21.

Bastable, H. G., Shuttleworth, W. J., Dallarosa, R. L. G., Fisch, G. and Nobre, C. A. (1993). Observations of climate, albedo, and surface radiation over cleared and undisturbed Amazonian forest. *International Journal of Climatology*, 13, 783–796.

Bonell, M. and Balek, J. (1993). Recent scientific developments and research needs in hydrological processes of the humid tropics. In *Hydrology and Water Management in the Humid Tropics – Hydrological Research Issues and Strategies for Water Management* ed. M. Bonell, M. M. Hufschmidt and J. S. Gladwell, pp. 167–260. Cambridge: Cambridge University Press.

Bonell, M. and Gilmour, D. A. (1978). The development of overland flow in a tropical rainforest catchment. *Journal of Hydrology*, 39, 365–382.

Bosch, J. M. and Hewlett, J. D. (1982). A review of catchment experiments to determine the effect of vegetation changes on water yield and evapotranspiration. *Journal of Hydrology*, 55, 3–23.

Brooks, S. M., Richards, K. S. and Nussbaum, R. (1994). Simulator experiments of the varied consequences of rainforest logging for runoff and erosion. *Geografiska Annaler*, 76A, 143–152.

Brown, E. T., Stallard, R. F., Larsen, M. C., Raisbeck, G. M. and Yiou, F. (1995). Denudation rates determined from the accumulation of in situ-produced ^{10}Be in the Luquillo Experimental Forest, Puerto Rico. *Earth and Planetary Science Letters*, 129, 193–202.

Bruenig, E. F. (1969). On the seasonality of droughts in the lowlands of Sarawak (Borneo). *Erdkunde*, 2, 127–133.

Bruenig, E. F. (1996). *Conservation and Management of Tropical Rainforests. An Integrated Approach to Sustainability*. Wallingford, UK: CAB International, 339pp.

Bruijnzeel, L. A. (1989). (De)forestation and dry season flow in the tropics: a closer look. *Journal of Tropical Forest Science*, 1, 229–243.

Bruijnzeel, L. A. (1990). *Hydrology of Moist Tropical Forests and Effects of Conversion: A State of Knowledge Review*. Paris: UNESCO-IHP, Humid Tropics Programme, 224pp.

Bruijnzeel, L. A. (1991). Nutrient inut-output budgets of tropical forest ecosystems: a review. *Journal of Tropical Ecology*, 7, 1–24.

Bruijnzeel, L. A. (1992). Managing tropical forest watersheds for production: where contradictory theory and practice co-exist. In *Wise Management of Tropical Forests 1992*, eds. F. R. Miller and K. L. Adam, pp. 37–75. Oxford: Oxford Forestry Institute.

Bruijnzeel, L. A. (1993). Land-use and hydrology in warm humid regions: where do we stand? *International Association of Hydrological Sciences Publication*, 216, 3–34.

Bruijnzeel, L. A. (1996). Predicting the hydrological effects of land cover transformation in the humid tropics: the need for integrated research. In *Amazonian Deforestation and Climate*, ed. J. H. C. Gash, C. A. Nobre, J. M. Roberts and R. L. Victoria, pp. 15–55. Chichester: J. Wiley.

Bruijnzeel, L. A. (1997). Hydrology of forest plantations in the humid tropics. In *Better Management of Soil, Nutrients, and Water in Tropical Plantation Forests*, ed. E. K. S. Nambiar and A. G. Brown, pp. 125–167. Canberra: ACIAR.

Bruijnzeel, L. A. (1998). Soil chemical changes after tropical forest disturbance and conversion: The hydrological perspective. In *Soils of Tropical Forest Ecosystems – Characteristics, Ecology and Management*, ed. A. Schulte and D. Ruhiyat, pp. 45–61. Berlin: Springer Verlag.

Bruijnzeel, L. A. (2004). Hydrological functions of tropical forests. *Agriculture, Ecosystems and Environment*. doi:10.1016/j.agee.2004.01.015.

Boulet, R. (1979). Méthodes d'analyse et représentation des couvertures pédologiques des bassins versants ECEREX. In *l'Ecosystème Forestier Guyanais, Bulletin de liaison du groupe de travail* No.1, février 1979, pp. 11–17. Cayenne, French Guyana, ORSTOM.

Calder, I. R. (1979). Do trees use more water than grass? *Water Services*, 83, 11–14.

Calder, I. R. (1998). Water use by forests, limits and controls. *Tree Physiology*, 18, 625–631.

Calder, I. R., Harding, R. J. and Rosier, P. T. W. (1983). An objective assessment of soil moisture deficit models. *Journal of Hydrology*, 60, 329–355.

Calder, I. R., Wright, I. R. and Murdiyarso, D. (1986). A study of evaporation from tropical rainforest – West Java. *Journal of Hydrology*, 89, 13–31.

Campbell, G. S. (1985). *Soil Physics with Basic. Transport Models for Soil – Plant Systems. Developments in Soil Science* No. 14. Amsterdam: Elsevier, 150pp.

Chandler, D. G. and Walter, M. F. (1998). Runoff responses among common land uses in the uplands of Matalom, Leyte, Philippines. *Transactions of the American Society of Agricultural Engineers*, 41, 1635–1641.

Chatterjea, K. (1998). The impact of tropical rainstorms on sediment and runoff generation from bare and grass-covered surfaces: A plot study from Singapore. *Land Degradation and Development*, 9, 143–157.

Coleman, D. C. and Crossley, D. A. (1996). *Fundamentals of Soil Ecology*. San Diego: Academic Press, 205pp.

Collison, A. (1996). Unsaturated strength and preferential flow as controls of gully head development. In *Advances in Hillslope Processes*, Vol. 2, ed.

M. G. Anderson and S. M. Brooks, pp. 753–769. Chichester: John Wiley and Sons.

Constantini, A., Dawes, W., O'Loughlin, E. and Vertessy, R. A. (1993). Hoop pine plantation management in Queensland: I. Gully erosion hazard prediction and watercourse classification. *Australian Journal of Soil and Water Conservation*, 6, 35–39.

Critchley, W. R. S. and Bruijnzeel, L. A. (1996). Environmental impacts of converting moist tropical forest to agricultrue and plantations. *IHP Humid Tropics Programme Series*, 10. Paris: UNESCO, 48pp.

Culf, A. D., Fisch, G. and Hodnett, M. G. (1995). The albedo of Amazonian forest and ranchland. *Journal of Climate*, 8, 1544–1554.

Denmead, O. T. and Shaw, R. H. (1962). Availability of soil water to plants as affected by soil moisture and meteorological conditions. *Agronomy Journal*, 54, 385–390.

Dise, N. B., Matzner, E. and Forsius, M. (1998). Evaluation of organic horizon C:N ratio as an indicator of nitrate leaching in conifer forests across Europe. *Environmental Pollution*, 102, S1, 453–456.

Dolman, A. J., Gash, J. H. C., Roberts, J. and James, W. (1991). Stomatal and surface conductance of tropical rainforest. *Agricultural and Forest Meteorology*, 54, 303–318.

Douglas, I., Bidin, K., Balamurugan, G., Chappell, N. A., Walsh, R. P. D., Greer, T. and Sinun, W. (1999). The role of extreme events in the impacts of selective tropical forestry on erosion during harvesting and recovery phases at Danum Valley, Sabah. *Philosophical Transactions of the Royal Society London B*, 354, 1749–1761.

Douglas, I., Greer, T., Bidin, K. and Sinun, W. (1993). Impacts of roads and compacted ground on post-logging sediment yield in a small drainage basin, Sabah, Malaysia. *International Association of Hydrological Sciences Publication*, 216, 213–218.

Duah-Yentumi, S., Rønn, R. and Christensen, S. (1998). Nutrients limiting microbial growth in a tropical forest soil of Ghana under different management. *Applied Soil Ecology*, 8, 19–24.

Dufrene, E., Dubois, B., Rey, H., Quencez, P. and Saugier, B. (1992). Changes in evapotranspiration from an oil palm stand (*Elaeis-guineensis Jacq.*) exposed to seasonal soil-water deficits. *Acta Oecologica – International Journal of Ecology*, 13, 299–314.

Dufrene, E. and Saugier, B. (1993). Gas-exchange of oil palm in relation to light, vapour-pressure deficit, temperature and leaf age. *Functional Ecology*, 7, 97–104.

Duijsings, J. J. H. M. (1985). *Streambank Contribution to the Sediment Budget of a Forest Stream*. PhD Thesis. Amsterdam: University of Amsterdam, 190pp.

Dunin, F. X., O'Loughlin, E. M. and Reyenga, W. (1985). A lysimeter characterization of evaporation by eucalypt forest and its representativeness for the local environment. In *The Forest-Atmosphere Interaction*, ed. B. A. Hutchinson and B. B. Hicks, pp. 271–291. Dordrecht: D. Reidel.

Dunne, T. (1978). Field studies of hillslope flow processes. In *Hillslope Hydrology*, ed. M. J. Kirkby, pp. 227–293. New York: John Wiley.

Dykstra, D. P. and Heinrich, R. (1996). *FAO Model Code of Forest Harvesting Practice*. Rome: Food and Agriculture Organization of the United Nations.

Ekanade, O. (1998). An edaphic evaluation of a mixed cropping system based on cocoa grown in combination with oil palm in Southwestern Nigeria. *Biological Agriculture and Horticulture*, 16, 117–128.

Elsenbeer, H. (2001). Hydrologic flowpaths in tropical rainforest soilscapes – a review. *Hydrological Processes*, 15, 1751–1759.

Elsenbeer, H. and Lack, A. (1996). Hydrological pathways and water chemistry in Amazonian rainforest. In *Advances in Hillslope Processes*, Vol. 2, ed. M. G. Anderson and S. M. Brooks, pp. 939–959. Chichester: John Wiley and Sons.

Fearnside, P. M. (1979). O desenvolvimento da Floresta Amazonica: problemas prioritarios para a formulacao de directizes. *Acta Amazonica*, 9(4) Suplemento, 123–129 (in Portuguese). [The development of the Amazonian forest: problems and priorities for the formulation of directions].

Fölster, H. and Khanna, P. K. (1997). Dynamics of nutrient supply in plantation soils. In *Management of Soil, Nutrients and Water in Tropical Plantation Forests*, ed. E. K. S. Nambiar and A. G. Brown, pp. 339–373. Canberra: ACIAR/CSIRO and Bogor: CIFOR.

Foong, S. F., Syed Sofi, S. O. and Tan, P. Y. (1983). A lysimetric simulation of leachig losses from an oil palm field. In *Proceedings of the Seminar on Fertilisers in Malaysian Agriculture*, ed. Malaysian Soil Science Society, pp. 45–68. Kuala Lumpur: Malaysian Soil Science Society.

Fritsch, J. M., 1983. Evolution des écoulements, des transports solides à l'exutoire et de l'érosion sur les versants d'un petit bassin après défrichement mécanisé de la forêt tropical humide. *International Association of Hydrological Sciences Publication*, 140, 197–214.

Fritsch, J. M. (1992). *Les effets du défrichement de la forêt amazonienne et de la mise en culture sur l'hydrologie de petits bassins versants*. Editions de l'ORSTOM, Collection Études et Thèses. Paris : ORSTOM, 392pp.

Fritsch, J. M. (1993). The hydrological effects of clearing tropical rainforest and of implementation of alternative land uses. *International Association of Hydrological Sciences Publication*, 216, 53–66.

Fritsch, J. M. and Sarrailh, J. M. (1986). Les transports solides dans l'écosystème forestier tropical humide en Guyane: les effets du défrichement et de l'installation de paturages. *Cahiers de l'ORSTOM, Série Pédologie*, 22, 93–106.

Gardner, W. R. (1983). Soil properties and efficient water use: An overview. In *Limitations to efficient water use in crop production*, ed. H. M. Taylor, W. R. Jordan and T. R. Sinclair, pp. 45–64. Madison, Wisconsin, USA: American Society of Agronomists.

Gash, J. H. C. (1979). An analytical model of rainfall interception by forests. *Quarterly Journal of the Royal Meteorological Society* 105, 43–55.

Gash, J. H. C. and Shuttleworth, W. J. (1991). Tropical deforestation: albedo and the surface-energy balance. *Climatic Change*, 19, 123–133.

Gillman, G. P., Sinclair, D. F., Knowlton, R. and Keys, M. G. (1985). The effect on some soil chemical properties of the selective logging of a north Queensland rainforest. *Forest Ecology and Management*, 12, 195–214.

Gilmour, D. A., Bonell, M. and Cassells, D. S. (1987). The effects of forestation on soil hydraulic properties in the Middle Hills of Nepal: a preliminary assessment. *Mountain Research and Development*, 7, 239–249.

Grace, J. M. (2000). Forest road sideslopes and soil conservation techniques. *Journal of Soil Water Conservation*, 55, 96–101.

Grip, H., Halldin, S. and Lindroth, A. (1989). Water use by intensively cultivated willow using estimated stomatal parameter values. *Hydrological Processes*, 3, 51–63.

Hall, R. L. and Calder, I. R. (1993). Drop size modification by forest canopies: measurements using a disdrometer. *Journal of Geophysical Research*, 98, 18465–18470.

Halldin, S. and Grip, H. (1979). Actual forest evapotranspiration by a simple model. *EOS (Transactions of the American Geophysical Union)*, 60: 580. (Abstract of paper read at the 6th Annual European Geophysical Society Meeting, Vienna; extended abstract available on request).

Halldin, S., Saugier, B. and Pontailler, J. Y. (1984/1985). Evapotranspiration of a decidous forest: Simulation using routine meteorological data. *Journal of Hydrology*, 75, 323–341.

Hamdan, J., Burnham, C. P. and Ruhana, B. (2000). Degradation effect of slope terracing on soil quality for *Elaeis guineensis* Jacq. (oil palm) cultivation. *Land Degradation and Development*, 11, 181–193.

Hamilton, L. S. and King, P. N. (1983). *Tropical Forested Watersheds. Hydrologic and Soils Response to Major Uses or Conversions*. Boulder: Westview Press, 168pp.

Hewlett, J. D. and Fortson, J. C., 1983. The paired catchment experiment. In *Forest Water Quality*, (ed.) J. D. Hewlett, pp. 11–14. Athens, Ga., ISA: School of Forest Resources, University of Georgia.

Hewlett, J. D. and Helvey, J. D. (1970). Effects of forest clear-felling on storm hydrographs. *Water Resources Research*, 6: 768–782.

Högberg, P. and Wester, J. (1997). Root biomass and symbiosis in *Acacia mangium* replacing tropical rainforest after logging. *Forest Ecology and Management*, 102, 333–338.

Imbach, A. C., Fassbender, H. W., Borel, R., Beer, J. and Bonnemann, A. (1989). Modelling agroforestry systems of cacao (*Theobroma cacaco*) with laurel (*Cordia alliodora*) and poro (*Erythrina poeppigiana*) in costa Rica. IV. Water balances, nutrient inputs and leaching. *Agroforestry Systems*, 8, 267–287.

Jahn, R. and Aiso, V. B. (1998). Soils of the tropical forests of Leyte, Phillipines I: Weathering, soil characteristics, classification and site qualities. In *Soils of Tropical Forest Ecosystems – Characteristics, Ecology and Management*, ed. A. Schulte and D. Ruhiyat, pp. 29–36. Berlin: Springer Verlag.

Jordan, C. F. (1989). *An Amazonian Rain Forest*. Man and the Biosphere Series, Vol. 2. Paris/Carnforth: UNESCO/Parthenon Publishers, 176pp.

Kamaruzaman, J., (1991). Effect of tracked and rubber-tyred logging machines on soil physical properties of the Berkelah Forest Reserve, Malaysia. *Pertanika*, 14, 1–11.

Kamaruzaman, J., (1996). Estimation of rate of recovery of disturbed soils from ground-based logging in Peninsular Malaysia. *Journal of Tropical Forest Science*, 9, 88–100.

Kauffman, S., Sombroek, W. and Mantel, S. (1998). Soils of rainforests: Characterization and major constraints of dominant forest soils in the humid tropics. In *Soils of Tropical Forest Ecosystems – Characteristics, Ecology and Management*, ed. A. Schulte and D. Ruhiyat, pp. 9–20. Berlin: Springer Verlag.

Kelliher, F. M., Hollinger, D. Y., Schulze, E.-D., Vygodskaya, N. N., Byers, J. N., Hunt, J. E., McSeveny, T. M., Milukova, I., Sogatchev, A., Varlargin, A., Ziegler, W., Arneth, A. and Bauer, G. (1997). Evaporation from an eastern Siberian larch forest. *Agricultural and Forest Meteorology*, 85, 135–147.

Kelliher, F. M., Leuning, R., Raupach, M. R. and Schulze, E.-D. (1995). Maximum conductances for evaporation from global vegetation types. *Agricultural and Forest Meteorology*, 73, 1–16.

Kirkby, M. J. (1988). Hillslope runoff processes and models. *Journal of Hydrology*, 100, 315–339.

Kolade, J. A. (1986). Influence of different densities of cocoa and oil-palm on yield performances of cocoa. *Turrialba*, 36, 345–353.

Lal, R. (1981). Deforestation of tropical rainforest and hydrological problems. In *Tropical Agricultural Hydrology*, ed. R. Lal and E. W. Russell, pp. 131–140. New York: J. Wiley.

Lal, R. (1987). *Tropical Ecology and Physical Edaphology*. New York: J. Wiley, 732pp.

Lal, R. (1997a). Deforestation and land-use effects on soil degradation and rehabilitation in western Nigeria. IV. Hydrology and water quality. *Land Degradation and Development*, 8, 95–126.

Lal, R. (1997b). Deforestation, tillage and cropping systems effects on seepage and runoff water quality from a Nigerian Alfisol. *Soil and Tillage Research*, 41, 261–284.

Larsen, M. C. and Parks, J. E. (1997). How wide is a road? The association of roads and mass-wasting disturbance in a forested montane environment. *Earth Surface Processes and Landforms*, 22, 835–848.

Larsen, M. C. and Torres-Sánchez, A. J. (1998). The frequency and distribution of recent landslides in three montane tropical regions of Puerto Rico. *Geomorphology*, 24, 309–331.

Lindquist, L. (1998). *Fine root biomass and root symbiosis after a first rotation of Acacia mangium stand on former tropical rainforest soil*. Graduate thesis, Department of Forest Ecology, Swedish University of Agricultural Sciences, Umeå, Sweden. Stencilserie No. 46, 19 p.

Ling, A. H. (1979). Some lysimetric measurements of evapotranspiration of oil palm in central Peninsular Malaysia. In *Proceedings of the Symposium of Water in Malaysian Agriulture*, ed. Malaysian Soil Science Society, pp. 89–99. Kuala Lumpur: Malaysian Soil Science Society.

Ling, A. H. and Robertsson, G. W. (1982). Reflection coefficients of some tropical vegetation covers. *Agricultural Meteorology*, 27, 141–144.

Lloyd, C. R., Gash, J. H. C. and Shuttleworth, W. J. (1988). The measurement and modelling of rainfall interception by Amazonian rainforest. *Agricultural and Forest Meteorology*, 43, 277–294.

Lohammar, T., Larsson, S., Linder, S., and Falk, S. (1980). Fast – simulation models of gaseous exchange in Scots pine. In *Structure and Function of Northern Coniferous Forests – An Ecosystem Study*, ed. T. Persson, *Ecological Bulletin (Stockholm)*, 32, 505–523.

MacDonald, L. H., Sampson, R. W. and Anderson, D. M. (2001). Runoff and road erosion at the plot and road segment scales, St John, US Virgin Islands. *Earth Surface Processes and Landforms*, 26, 251–272.

Mackensen Klinge, R., Ruhiyat, D. and Fölster, H. (2003). Assessment of management-dependent nutrient fluxes in tropical industrial tree plantations. *Ambio*, 32, 106–112.

Malmer, A. (1992). Water yield changes after clear-felling tropical rainforest and establishment of forest plantation in Sabah, Malaysia. *Journal of Hydrology*, 134, 77–94.

Malmer, A. (1993). *Dynamics of hydrology and nutrient losses as response to establishment of forest plantation. a case study on tropical rainforest land in Sabah, Malaysia*. PhD Thesis, Swedish University of Agricultural Sciences, Umeå, Sweden, ISBN 91-576-4651-1, 181pp.

Malmer, A. (1996a). Hydrological effects and nutrient losses of forest plantation establishment on tropical rainforest land in Sabah, Malaysia. *Journal of Hydrology*, 174, 129–148.

Malmer, A. (1996b). Observations on slope processes in a tropical rainforest environment before and after forest plantation establishment. In *Advances in Hillslope Processes*, Vol. 2., ed. M. G. Anderson and S. M. Brooks, pp. 961–973. Chichester: John Wiley and Sons.

Malmer, A. and Grip, H. (1990). Soil disturbance and loss of infiltrability caused by mechanized and manual extraction of tropical rainforest in Sabah, Malaysia. *Forest Ecology and Management*, 38, 1–12.

Malmer, A., Johansson, E. and Kluge, M. (1998). Natural rehabilitation of disturbed tropical rainforest soils in Sabah, Malaysia. In *Tropical Forest Soils and Their Management*, ed. A. Schulte and D. Ruhiyat, pp. 199–203. Berlin: Springer Verlag.

Martin, G. (1970). Le défrichement mécanique pour la création des palmeraies industrielles. *Oléagineux*, 25, 575–580.

Meinzer, F. C., Goldstein, G., Holbrook, N. M., Jackson, P. and Cavalier, J. (1993). Stomatal and environmental control of transpiration in a lowland tropical forest tree. *Plant, Cell and Environment*, 16: 429–436.

Miranda, R. A. C. (1994). Partitioning of rainfall in a cocoa (*Theobroma cacao* Lour.) plantation. *Hydrological Processes*, 8, 351–358.

Monteith, J. L. (1965). Evaporation and environment. The state and movement of water in living organisms. *Symposium of the Society of Experimental Biology* No. 19, Cambridge University Press, pp. 205–234.

Moore, R., Dibb, T. M. and Billing, D. W. (1991). The distribution and causes of mass movements in Aurora Province, Phillippines. In: D. H. Bell (ed.), *Landslides – Glissements de Terrain: Proceedings of the Sixth International Symposium*, Christchurch, New Zealand, Balkema, Rotterdam, 1023–1029.

Myers, N. (1982). Depletion of tropical moist forests: a comparative review of rates and causes in the three main regions. *Acta Amazonica*, 12(4), 745–758.

Nortcliff, S. and Thornes, J. B. (1984). Floodplain response of a tropical stream. In *Catchment Experiments in Fluvial Hydrology*, ed. T. P. Burt and D. E. Walling, 73–85. Norwich, UK: Geo-Books.

Nykvist, N. (1997). Uptake of nutrients in a plantation of *Acacia mangium* in relation to decrease in soil amount. *Journal of Sustainable Forestry*, 4, 131–139.

Nykvist, N. (2000). Tropical forests can suffer from a serious deficiency of calcium after logging. *Ambio*, 29, 310–313.

Nykvist, N., Grip, H., Sim, B. L., Malmer, A. and Wong, F. K. (1994). Nutrient losses in forest plantations in Sabah, Malaysia. *Ambio*, 23, 210–215.

Odemerho, F. O. (1984a). A reduced-rank prediction model of stream-channel size response to traditional agricultural land-use practices in south-western Nigeria. *Journal of Hydrology*, 70, 85–100.

Odemerho, F. O. (1984b). The effects of shifting cultivation on stream-channel size and hydraulic geometry in small headwater basins of south-western Nigeria. *Geografiska Annaler*, 66A, 327–340.

Oke, T. R. (1987). *Boundary Layer Climate*. 2nd ed. London: Routledge, 435 pp.

Pearce, A. J. (1986). *Erosion and Sedimentation*. Working Paper, Honolulu, Hawaii: Environment and Policy Institute, 18pp.

Pearce, A. J., Rowe, L. K. and O'Loughlin, C. L. (1980). Effects of clearfelling and slash-burning on water yield and storm hydrograph in evergreen mixed forests, Western New Zealand. *International Association of Hydrological Sciences Publication*, 130, 119–127.

Pickup, G. R., Higgins, R. J., Warner, R. F., 1981. Erosion and sediment yields in the Fly River drainage basins, Papua New Guinea. *International Association of Hydrological Sciences Publication* 132, 438–456.

Plamboeck, A. H., Grip, H. and Nygren, U. (1999). A hydrological tracer study of water uptake depth in a Scots pine forest under two different water regimes. *Oecologia*, 119, 452–460.

Pritchett, W. L. (1979). *Properties and Management of Forest Soils*. New York: J. Wiley, 500pp.

Proctor, J. (1987). Nutrient cycling in primary and old secondary rainforests. *Applied Geography*, 7, 135–152.

Radersma, S. and de Ridder, N. (1996). Computed evapotranspiration of annual and perennial crops at different temporal and spatial scales using published parameter values. *Agricultural Water Management*, 31, 17–34.

Rijsdijk, A. and Bruijnzeel, L. A. (1990/1991). Erosion, sediment yield and land use patterns in the upper Konto watershed, East Java, Indonesia. *Konto River Project Communication*, 18, 3 Volumes. Amersfoort, the Netherlands: DHV International Consultants.

Roche, M. A. (1981). Watershed investigations for development of forest resources of the Amazon region in French Guyana. In: *Tropical Agricultural Hydrology*, ed. R. Lal and E. W. Russell, pp. 75–82. New York: J. Wiley.

Roche, M. A. (1982a). Evapotranspiration réelle de la forêt amazonienne en Guyane. *Cahiers de l'ORSTOM, Série Hydrologie*, XIX, 37–44.

Roche, M. A. (1982b). Comportements hydrologiques comparés et érosion de l'écosystème forestier amazonien Ecerex en Guyane. *Cahiers de l'ORSTOM, Série Hydrologie*, XIX, 81–114.

Sahin, V. and Hall, M. J. (1996). The effects of afforestation and deforestation on water yields. *Journal of Hydrology*, 178, 293–309.

Sanchez, P. A. (1976). *Properties and Management of Soils in the Tropics*. New York: J. Wiley, 618 p.

Shuttleworth, W. J. (1988). Evaporation from Amazonian rainforest. *Proceedings of the Royal Society London*, B233, 321–346.

Sidle, R. C., Pearce, A. J. and O'Loughlin, C. L. (1985). *Hillslope Stability and Land Use. American Geophysical Union Water Resources Monograph Series*, 11, 140pp.

Sim, B. L. and Nykvist, N. (1991). Impact of forest harvesting and replanting. *Journal of Tropical Forest Science*, 3, 251–284.

Simon, A. and Guzman-Rios, S. (1990). Sediment discharge from a montane basin, Puerto Rico: implications of erosion processes and rates in the humid tropics. *International Association of Hydrological Sciences Publication*, 192, 35–47.

Sinun, W., Wong, W. M., Douglas, I. and Spencer, T. (1992). Throughfall, stemflow, overland flow and throughflow in the Ulu Segama rainforest, Sabah. *Philosophical Transactions of the Royal Society (London), Series B*, 335, 389–395.

Stednick, J. D. (1996). Monitoring the effects of timber harvest on annual water yield. *Journal of Hydrology*, 176, 79–95.

Swank, W. T. and Douglass, J. E. (1974). Streamflow greatly reduced by converting deciduous hardwood stands to pine. *Science*, 185: 857–859.

Tan, B. K. (1984). Landslides and their remedial measures in Malaysia. In: *Proceedings of the 4th International Symposium on Landslides*, Toronto, September 16–21, 1984, Vol 1, Toronto Univ. Press, Toronto, pp. 705–709.

Tani, M., Abdul Rahim, N., Ohtani, Y., Yasuda, Y., Sahat, M. M., Baharudin, K., Takanashi, S., Noguchi, S., Zulkifli, Y. and Watanabe, T. (2003). Characteristics of energy exchange and surface conductance of a tropical rainforest in Peninsular Malaysia. In: *Pasoh: Ecology of a Lowland Rain Forest in South East Asia*. ed. T. Okuda, N. Manokaran, Y. Matsumoto, K. Niiyama, S. C. Thomas and P. S. Ashton, pp. 73–88. Tokyo: Springer Verlag.

Thom, A. S. (1975). Momentum, mass and heat exchange. In: *Vegetation and the Atmosphere*, Vol 1, ed. J. L. Monteith, Academic Press, London, 57–109.

Van der Plas, M. C. and Bruijnzeel, L. A. (1993). Impact of mechanized selective logging of rainforest on topsoil infiltrability in the Upper Segama area, Sabah, Malaysia. *International Association of Hydrological Sciences Publication*, 216, 203–211.

Van Dijk, A. I. J. M. (2002). *Water and Sediment Dynamics in Bench-Terraced Agricultural Steeplands in West Java, Indonesia*. PhD thesis, Amsterdam: Vrije Universiteit, 363pp.

Van Dijk, A. I. J. M. and Bruijnzeel, L. A. (2001). Modelling rainfall interception by vegetation of variable density using an adapted analytical model. Part 2. Model validation for a tropical upland mixed cropping system. *Journal of Hydrology*, 247, 239–262.

Vertessy, R. A., Wilson, C. J., Silburn, D. M. Connolly, R. D. and Ciesiolka, C. A. (1990). Predicting erosion hazard areas using digital terrain analysis. *International Association of Hydrological Sciences Publication*, 192, 298–308.

Vis, M. (1986). Interception, drop size distributions and rainfall kinetic energy in four Colombian forest ecosystems. *Earth Surfaces Processes and Landforms*, 11, 591–603.

Walling, D. E. (1983). The sediment delivery problem. *Journal of Hydrology*, 65, 209–237.

Ward, R. C. (1984). On the response to precipitation of headwater streams in humid areas. *Journal of Hydrology*, 74, 171–189.

Ward, R. C., Robinson, M. (2000). *Principles of Hydrology* (4th edition). McGraw Hill, London, 450pp.

Waterloo, M. J. (1994). *Water and Nutrient Dynamics of Pinus caribaea Plantation Forests on Former Grassland Soils in Southwest Viti Levu, Fiji*. PhD Thesis. Amsterdam: Vrije Universiteit, 478pp.

Waterloo, M. J., Bruijnzeel, L. A. and Vugts, H. F. (1999). Evaporation from *Pinus caribaea* plantations on former grassland soils under maritime tropical conditions. *Water Resources Research*, 35, 2133–2144.

Wiersum, K. F. (1984). Surface erosion under various tropical agroforestry systems. In *Effects of Forest Land Use on Erosion and Slope Stability*, ed. C. L. O'Loughlin and A. J. Pearce, pp. 231–239. Vienna: IUFRO.

Wright, I. R., Gash, J. H. C., Da Rocha, H. R., Shuttleworth, W. J., Nobre, C. A., Maitelli, G. T., Zamparoni, C. A. G. P. and Carvalho, P. R. A. (1992). Dry season micrometeorology of central Amazonian ranchland. *Quarterly Journal of the Royal Meteorological Society*, 118, 1083–1099.

Zhou, G. Y., Morris, J. D., Yan, J. H., Yu, Z. Y. and Peng, S. L. (2001). Hydrological impacts of reafforestation with eucalypts and indigenous species: a case study in southern China. *Forest Ecology and Management*, 167, 209–222.

Ziegler, A. D. and Giambelluca, T. W. (1997). Importance of rural roads as source areas for runoff in mountainous areas of northern Thailand. *Journal of Hydrology*, 196, 204–229.

Ziegler, A. D., Sutherland, R. A. and Giambelluca, T. W. (2000). Runoff generation and sediment production on unpaved roads, footpaths and agricultural land surfaces in northern Thailand. *Earth Surface Processes and Landforms*, 25, 519–534.

Ziegler, A. D., Sutherland, R. A. and Giambelluca, T. W. (2001). Interstorm surface preparation and sediment detachment by vehicle traffic on unpaved mountain roads. *Earth Surface Processes and Landforms*, 26, 235–250.

23 Large-scale hydrological impacts of tropical forest conversion

M. H. Costa
Federal University of Viçosa, Brazil

INTRODUCTION

Hydrologists are facing important changes in the direction of their science. New tools, non-traditional datasets and a better understanding of the connection between land-surface hydrology and the rest of the climate system are being developed. Changes are so fundamental that a new agenda for land surface hydrology research has been proposed (Entekhabi *et al.*, 1999).

One of the major challenges is related to large-scale hydrology. Part of this challenge is to reconcile the differences between two sciences. Large-scale hydrology is being studied typically by scientists with a background in hydrology or the atmospheric sciences, and the historical approach of each discipline is clearly present. Table 23.1 shows the different definitions of the spatial scales in common use. Here, we consider large-scale to be the scale between 100 and 10 000 km, a mix of the large catchment and continental scales of hydrology and a mix of synoptic and large scales found in atmospheric sciences.

Hydrologists and atmospheric scientists advanced into the new research field of large-scale hydrology a few decades ago: hydrologists scaled up their methods, atmospheric scientists scaled down their techniques. Of course there were conflicts caused by the different assumptions, jargon, orientations and methodologies used by the two sciences because of decades of separate development. Even so, the latter is hardly the beginning of the difficulties: in one of the pioneer essays in this area, Eagleson (1986) pointed out that large-scale hydrology has two characteristics that make it unique: interdisciplinarity and the presence of strong feedbacks. Considering the biophysical aspects of the hydrological cycle only, the study of large-scale hydrology involves the knowledge of radiation physics, planetary fluid dynamics, precipitation processes, micrometeorology, plant physiology, natural and managed ecosystems, and the analysis of random fields. Furthermore, although feedback mechanisms happen at all scales, additional feedback mechanisms operate at the large-scale when hydrosphere, lithosphere, biosphere and atmosphere interact.

Classical problems studied by large-scale hydrologists include the availability of water resources under climate variability or under wide changes in land cover. Of primary interest are the hydrological impacts of the conversion of tropical forests to other kinds of land use, typically agricultural ones, such as annual cropping or pasture. This chapter is an introduction to the mechanisms associated with large-scale tropical forest conversion, the hydrological impacts of which are much more complex than the same effects at the micro-scale. To establish a connection to the other subjects covered in this book, the next section presents a short review of experimental studies on the hydrological impacts of land cover conversion, executed at the small catchment scale (<100 km^2). Subsequently, some modelling experiments are reviewed which concern the hydrological impacts of large-scale tropical conversion, with emphasis on the feedback mechanisms, whilst the concluding section discusses the role of the large-scale tropical forest in the climate system.

HYDROLOGICAL IMPACTS OF TROPICAL FOREST CONVERSION

Tropical forest conversion disrupts the hydrological cycle of a drainage basin, by altering not only the balance between rainfall and evaporation but also the runoff response of the area. The conventional approach to quantifying the hydrological impacts of changes in land cover has involved the use of experimental catchments, either singly or in pairs. In the case of single catchments, following an initial calibration period with one land cover, changes are made and measurements are continued over a comparable period. In the case of paired study areas, either two catchments that are similar in all characteristics except land cover are

Forests, Water and People in the Humid Tropics, ed. M. Bonell and L. A. Bruijnzeel. Published by Cambridge University Press.

Table 23.1. *Spatial scales in hydrology (Dooge, 1997) and atmospheric sciences*

Horizontal scale (km)	Scale: Hydrology	Scale: Atmospheric sciences
>10 000	Macro (planetary)	Planetary
1000–10 000	Macro (continental)	Large
100–1000	Macro (large catchment)	Synoptic
10–100	Meso (small catchment)	Meso
1–10	Meso (sub-catchment)	
0.1–1	Meso (module)	Micro
<0.1	Micro	

monitored simultaneously, or one of the pair is subjected to a land-cover change after the initial calibration period, while the other remains in its initial condition as a control over the following measurement period. These approaches, although expensive and time-consuming, have been applied for more than a century and extensive results have been documented.

Bosch and Hewlett (1982) reviewed the results of 94 basin experiments throughout the world and concluded that removal of forest leads to higher stream flow, and reforestation of open lands generally leads to a decline in the overall stream flow. Bruijnzeel (1990, 1996) reviewed the effect of land cover transformation in the humid tropics and concluded that: (a) carefully executed light selective harvesting of trees (up to 20% removal of biomass) has little (if any) effect on stream flow; (b) removal of the natural forest cover may result in a considerable increase in water yield (up to 800 mm yr^{-1}), depending on the amount of rainfall received and the degree of surface disturbance; and (c) there is a decline in stream flow with time associated with a reforestation. Sahin and Hall (1996) used regression analysis and data from 145 experiments to study the effects of land use change on water yields. Their analysis of the tropical forest data (only five experiments were included) suggests that runoff increased 10 mm yr^{-1} after a 10% deforestation and 213 mm yr^{-1} after a 100% deforestation. Although the reviews above show an increase in water yield after a forest conversion, in certain types of montane tropical forest the inverse can happen (see discussion in Bruijnzeel, this volume).

LARGE-SCALE HYDROLOGICAL IMPACTS OF TROPICAL FOREST CONVERSION: THE IMPORTANCE OF FEEDBACKS

Due to the costs of a large-scale experiment and the enormous environmental impacts associated, it is not feasible to use the methodology as described in the previous section to determine experimentally the large-scale hydrological impacts of forest conversion. The standard method recommended in this case is modelling. Empirically based, problem-oriented hydrological models are usually inappropriate here, and the preferred tools to evaluate the effects of forest conversion are the process-based hydrological models.[1]

Large-scale, process-based, hydrological models usually have two or three components. The first one is the vertical flux component, a Soil-Vegetation-Atmosphere Transfer model – SVAT (see detailed description in Sellers *et al.*, 1992); it uses meteorological variables as input and, according to the vegetation cover and soil type, calculates evapotranspiration, surface runoff and deep drainage. Some examples in this category include BATS (Dickinson *et al.*, 1986, 1993), SiB/SSiB/SiB2 (Sellers *et al.*, 1986; Xue *et al.*, 1991; Sellers *et al.* 1996), and LSM (Bonan *et al.*, 1996).

The second component routes the runoff produced at each cell throughout the basin, effectively calculating the river discharge at each point. Some examples in this category include WTM (Vörösmarty *et al.*, 1989), the routed VIC-2L (Lohmann *et al.*, 1998) and HYDRA (Coe, 2000).

A third component, a climate system model (see a full description in Trenberth, 1992) or a simplified version, an Atmospheric General Circulation Model (AGCM), can replace the meteorological input variables. Widely-used AGCMs include the pioneer GFDL model, the NCAR CCM/GENESIS/CSM series, the Hadley Centre and the French LMD models. A good historical perspective on the origins and development of AGCMs is found in Randall (2000).

The type of model used depends on the nature of the experiment and the computer resources available. For example, coupling large-scale surface hydrological models to atmospheric models allows scientists to study many of the feedbacks involved, although at a higher computational cost.

As discussed before, the main difference between small and large-scale hydrological phenomena is the feedbacks through the atmosphere. Feedback mechanisms in hydrological processes are present at all scales. A classical example of a non-linear feedback in hydrology is the variability of soil hydraulic conductivity with soil moisture during the infiltration process. Another example of a micro-scale feedback in the soil-vegetation-atmosphere interaction is the humidity feedback on transpiration within the planetary boundary layer (Jarvis and McNaughton, 1986; Monteith, 1995). When the local evapotranspiration decreases, the planetary boundary layer becomes drier, which increases the water vapour pressure deficit, leading to an increase in the evapotranspiration. Although such feedback cannot be estimated by a simple, semi-empirical

1 Although the term 'physically-based models' has been widely used in the past, it is appropriate to change it to 'process-based models', since modern models usually include the physical, chemical and biological processes associated with the hydrological cycle.

Table 23.2. *Changes in precipitation* (P), *evapotranspiration* (E_T) *and runoff* (R) *in two experiments of the hydrological impact of Amazonian tropical forest conversion*

	ΔP (mm day^{-1})	ΔE_T (mm day^{-1})	ΔR (mm day^{-1})	Reference
Prescribed climate study	–	–0.5	+0.5	Costa and Foley (1997)
Interactive climate study	–0.7	–0.6	–0.1	Costa and Foley (2000)

evapotranspiration equation (e.g., Penman, 1948; Priestley and Taylor, 1972), soil-vegetation-atmosphere transfer (SVAT) models of intermediate complexity can represent this phenomenon more realistically. (See also Mahe *et al.*, this volume, for examples related to Africa.)

It is unlikely that the conversion of one square kilometre of forest to agricultural cover may cause an impact in the atmospheric circulation and precipitation patterns. According to Pielke *et al.* (1998), the threshold scale above which land use changes cause changes in atmospheric circulation, cloudiness and local climate, is apparently around or greater than 100 km^2. The analysis of GOES images shows that, at the land cover conversion scale between 1000 and 10 000 km^2, feedbacks in the distribution of clouds are evident. These feedbacks have been observed on lowland tropical forest (Cutrim *et al.*, 1995), and on montane tropical forest, a biome that depends more critically on the presence of clouds (Lawton *et al.*, 2001). Mesoscale climate simulation studies (e.g. Eltahir and Bras, 1994a; Dias and Regnier, 1996; Lawton *et al.*, 2001; Bonell *et al.*, this volume) have also shown that cloud distribution and precipitation patterns change after a mesoscale deforestation. On the other hand, large-scale tropical forest conversions (>1 M km^2) may cause important changes in the large-scale atmospheric circulation, affecting significantly the precipitation patterns over the cleared land.

An interesting way to look at the hydrological impacts of large-scale forest conversion is to evaluate the feedback mechanisms themselves. To do this, one should perform two different modelling experiments, one using a soil-vegetation-atmosphere transfer model (SVAT) alone, and another experiment using the same SVAT model coupled to an atmospheric circulation model. The feedback is the difference between the effects of the coupled experiment and the effects of the offline experiment.

Table 23.2 summarises the results of a paired experiment as described above. Costa and Foley (1997) used the IBIS SVAT/Ecosystem model to evaluate the changes in land cover (from forest to pasture) in the hydrological budget of the Amazon basin using a prescribed climate, i.e. without considering any feedbacks in the atmosphere. Later, Costa and Foley (2000) used the same SVAT model coupled to the GENESIS global climate model to evaluate the changes in the surface climate, including all sorts of atmospheric circulation feedbacks.

The prescribed climate study shows no change in precipitation, since the climate is the same in both forested and deforested simulations. Removing the forest causes a decrease in the evapotranspiration and an increase in the runoff, consistent with the field experiments reviewed in the previous section. Although the simulation experiments were conducted at the regional scale, the changes in surface fluxes are comparable to changes that happen at smaller scales. However, when feedbacks through the atmosphere are allowed, the deforestation causes a decrease in the local precipitation, which induces drier soils and a further decrease in the evapotranspiration. As a result, runoff changes from an increase of 0.5 mm day^{-1} to a decrease of 0.1 mm day^{-1} (Table 23.2).

It is important to note here that the feedback through the atmosphere is of the negative kind, and the absolute changes in runoff in the presence of such feedbacks are smaller than in the absence of it. In other words, changes in runoff tend to be attenuated when atmospheric feedbacks are included (Table 23.2).

There are several modelling studies of hydrological effects of tropical forest conversion at the large-scale. Some concentrated on forest conversion within the Amazon region alone, while others studied all three tropical regions of the world. Although most of the large-scale experiments focused on a non-realistic full deforestation, a few studies (Walker *et al.*, 1995; Lean *et al.*, 1996) evaluated the effects of the present level of Amazonian tropical forest conversion.

The results from present level forest conversion (synoptic-scale forest conversion) studies are very similar to those from the full area deforestation, discussed below. The major difference is that the position of the weather/climate anomalies in the synoptic-scale experiment is located downstream of the cleared area, while the position of the anomalies in the large-scale experiments is centred over the cleared area. The difference in the position of the anomalies is a consequence of the advection of humidity and other atmospheric properties (Walker *et al.*, 1995). Since the same pattern is not observed in mesoscale simulations, it is possible that AGCMs are not suitable to simulate the effects of synoptic-scale forest conversion; a more appropriate tool in this case would be a regional (mesoscale) model nested inside a global model.

Of course the results of hydroclimatological modelling experiments like these vary according to the model used, the parameters selected and the design of the experiment. A way to validate modelling experiments is to compare similar studies that used different models. Table 23.3 summarises the results of selected full area tropical forest conversion experiments using global climate models. It is usually true that the experiments show a decrease in local evapotranspiration (E_T) and precipitation (P) in the cleared area.

The reduction in evapotranspiration is basically a consequence of three factors: the increased albedo reduces the net radiation at the surface; the reduced roughness length decreases atmospheric turbulence, weakening vertical motions; and the reduced root depth leaves less soil moisture available to plants. Additional factors that can also influence local evapotranspiration include compaction of the soil surface or sub-surface (Lean *et al.*, 1996) and reduction of leaf area index (LAI) through grazing (Eastman *et al.*, 2001).

The decrease in the simulated precipitation is a consequence of the general decreased vertical motion in the atmosphere above the deforested area. According to Eltahir (1996), the reduction in the net surface radiation after deforestation cools the upper atmosphere over the deforested area, inducing a thermally-driven high-level converging circulation that results in subsidence, less cloudiness and reduced precipitation. Although apparently this is the major process involved, at least three additional internal atmospheric interactions are present, contributing to the magnitude of the whole feedback through the atmosphere. An important attenuating interaction is introduced by the increased radiative forcing as cloud cover decreases with the weakened convection, explaining why higher sensitivities to tropical forest conversion are found in some prescribed-clouds atmospheric models (Zeng and Neelin, 1999). A second interaction is the moisture convergence feedback, which not only affects the atmospheric water budget (Lean *et al.*, 1996) but also, and more importantly, changes the atmospheric stability (Zeng and Neelin, 1999). A third interaction, of minor importance, is a positive feedback caused by the reduced contribution of surface evapotranspiration to the atmospheric water budget.

Being the difference between both quantities, the changes in runoff depend on the magnitude of the changes in P and E_T, and therefore models have not agreed on the sign and magnitude of the runoff changes. Despite the disagreement among the runoff results, it is generally accepted that – all other conditions being equal – the large-scale water yield change after forest conversion would be smaller than the equivalent micro-scale change discussed earlier.

The feedbacks described above are relatively simple, and involve only changes in the atmospheric circulation. A more complex feedback mechanism is a cyclical one, when changes in land cover cause changes in climate that, in turn, cause other changes in the land cover. For example, if the precipitation after a forest removal decreases hypothetically so much that it becomes too low to support even a grassland biome, there is a trend towards desertification, feeding back again in the atmosphere and leading to a very dry climate. Some feedbacks may even include changes in ocean circulation (Kiang and Eltahir, 1999; Delire *et al.*, 2001) or the planetary carbon cycle (Cox *et al.*, 2000). Since most of these feedbacks are still in their initial phase of study, they are not reviewed in depth here. However, the interaction between the hydrological, atmospheric and ecosystem dynamics may have a much more fundamental role in the climate system, as discussed in the next section.

HOW TROPICAL FORESTS MANIPULATE THEIR OWN CLIMATE

Tropical rainforests are an obvious consequence of the hot and humid climate in that region. However, in recent decades, evidence is accumulating that the rainforest and the hot and humid climate are strongly connected, forming a two-way interacting system that perpetuates each other. In other words, the tropical humid climate allows the existence of the rainforest which, in its turn, helps to produce the rainy climate it needs.

A rainy climate requires two necessary conditions: a humid atmosphere and sufficient ascending vertical motion to form clouds and induce precipitation. Taking again the Amazon basin as a case study, let us show the role of the Amazonian rainforest as a source of water vapour and as a driver of vertical motion. Most of the water vapour that enters the Amazon basin atmospheric column is advected from the Atlantic Ocean. On an annual average basis, around 70% of the water vapour that enters the atmospheric column is of oceanic origin and 30% is evaporated locally (Eltahir and Bras, 1994b; Costa and Foley, 1999).

The contribution of the surface evaporation is more important during the dry season. Matsuyama (1992) and Zeng (1999) determined that, during the dry season, the convergence of the oceanic water vapour is nearly zero and virtually all the water vapour that enters the Amazon basin atmospheric column is evaporated inside the basin. Measurements based on the eddy flux correlation technique (Shuttleworth, 1988) and estimates of the atmospheric water budget (Matsuyama, 1992) show that the evapotranspiration of the Amazon rainforest is relatively constant throughout the year, not showing signs of soil water stress.

The stability of the local evapotranspiration is also evident at the interannual and decadal time scale. Costa and Foley (1999) found that there was a weakening of the trade winds that transport water vapour from the tropical Atlantic into the Amazon basin during the period 1976–1996, which caused a decrease in the input of water vapour in the Amazon basin. In this case, where the main source of water vapour to the basin has decreased considerably, the Amazon basin maintained the precipitation and runoff constant

Table 23.3. *Comparison of selected recent AGCM Amazonian deforestation experiments designs and results*

	Dickinson and Kennedy (1992)	Henderson-Sellers *et al.* (1993)	Lean and Rowntree (1993)	Polcher and Laval (1994)	Sud *et al.* (1996)	Manzi and Planton (1996)	Lean and Rowntree (1997)	Hahmann and Dickinson (1997)	Costa and Foley (2000)
AGCM	CCM1	CCM1-OZ	UKMO	LMD	GLA	EMERAUDE	UKMO	RCCM2	GENESIS
Resolution	4.5° × 7.5°	4.5° × 7.5°	2.5° × 3.75°	2.0° × 5.6°	4.0° × 5.0°	2.8° × 2.8°	2.5° × 3.75°	2.8° × 2.8°	4.5° × 7.5°
Surface model	BATS (Dickinson *et al.*, 1986)	BATS (Dickinson *et al.*, 1986)	Warrilow (1986)	SECHIBA (Ducroudé *et al.*, 1993)	SSiB (Xue *et al.*, 1991)	ISBA (Noilhan and Planton, 1989)	Warrilow (1986) modified	BATS 1e (Dickinson *et al.*, 1993)	IBIS (Foley *et al.*, 1996)
Ocean	Mixed layer	Mixed layer	Prescribed SST	Prescribed SST	Prescribed SST	Prescribed SST	Prescribed SST	Mixed layer	Mixed layer
Simulation length	3 yrs	6 yrs	3 yrs	1.1 yr	3 yrs	3 yrs	10 yrs	10 yrs	15 yrs
Roughness (m)	2.00/0.05	2.00/0.20	0.80/0.04	2.30/0.06	2.65/0.077	2.00/0.026	2.10/0.026	2.00/0.05	1.51/0.05
Albedo	0.12/0.19	0.12/0.19	0.14/0.19	0.098/0.177	0.092/0.142	0.12/0.163	0.13/0.18	0.12/0.19	0.135/0.173
ΔP (mm day^{-1})	–1.4	–1.6	–0.8	+1.1	–1.5	–0.4	–0.4	–1.0	–0.7
ΔE_T (mm day^{-1})	–0.7	–0.6	–0.6	–2.7	–1.2	–0.3	–0.8	–0.4	–0.6
ΔR (mm day^{-1})	–0.7	–0.9	–0.2	+3.8	–0.3	+0.3	+0.4	–0.6	–0.1
ΔT (°C)	+0.6	+0.6	+2.1	+3.8	+2.0	+1.3	+2.3	+1.0	+1.4

by increasing the relative contribution of the local source of water vapour (regional evapotranspiration).

This lack of a seasonal cycle in evapotranspiration is a consequence of the deep roots in the Amazon basin. According to Canadell *et al.* (1996), the worldwide average maximum root depth for tropical evergreen rainforests is 7.3 m, whilst Nepstad *et al.* (1994) demonstrated that parts of the Amazon forest have access to water stored at depths greater than 10 m, with roots reaching 18 m in eastern Amazonia, where precipitation is more seasonal. Around 3000 mm of water can be stored in such a thick soil layer, although not all of it can be available to plants due to the low deep root density. The root density decreases from more than a kilogram of roots per cubic metre near the surface to a few tens of grams per cubic metre below two metres, being relatively constant below this level downwards (Nepstad, 1989, cited by Bruijnzeel, 1996). Despite the low deep root density, research done by Hodnett *et al.* (1996) near Manaus and by other groups at several other sites has demonstrated that, in many years, it is impossible to close the dry season water balance of the Amazonian rainforest without using water stored at depths greater than 2 m. Other tropical evergreen forests throughout the globe, like the Babinda experimental catchment in northeastern Australia and the Western Ghats of India, may have access to large amounts of soil moisture or groundwater (Bonell, 1998). Not every tropical forest has deep roots, however. Canadell *et al.* (1996) report that the average maximum root depth of deciduous tropical forests is only 3.7 m. In addition, the maximum root depth can be limited geologically. In part of the Guyanas, for example, roots cannot penetrate deeper than a few metres because of less deeply weathered rocks (Brouwer, 1996, p. 22). Further details about the importance of deep roots in hydrology and ecosystem functioning are presented by Jackson (1999).

The evergreen tropical forest has yet another role in the local climate. As discussed earlier in this chapter, theoretical (Eltahir, 1996; Zeng and Neelin, 1999) and modelling studies (Dirmeyer and Shukla, 1994; Lean and Rowntree, 1997; Costa and Foley, 2000) demonstrate that the low albedo rainforest favours convection over the basin, while an increase in the surface albedo causes a subsidence anomaly over the region. Because water vapour and convection are the key contributors to precipitation, it is very likely that large-scale rainforests have some ability to maintain their own climate.

A discussion about the role of the Amazonian tropical forest in the local climate would not be complete without a discussion about the existence of 18 m deep roots in the eastern Amazon basin. Deep roots are found typically in regions with low precipitation volume, or irregular precipitation patterns, allowing the plants to use the water that was previously stored at deeper layers in rainier periods. It is puzzling why some tropical rainforests have developed roots so deep in a climate so wet. In a competitive environment, species that unnecessarily allocate a big fraction of the fixed carbon to grow roots, at the expense of leaves and branches, would be at a disadvantage when competing against species that concentrated the allocation of carbon to the aerial part.

Although 18 m deep roots may be unnecessary today, they might have been useful in the past. It has been proposed that, during the Last Glacial Maximum (21 000 years before present) and until the mid-Holocene (14 000 years BP), the trade winds were more zonal, precipitation rates in Amazonia were lower and parts of the rainforest were replaced by savannas (Haffer, 1969; van der Hammen and Absy, 1994; Kubatski and Claussen, 1998; Maslin and Burns, 2000; Mayle *et al.*, 2000). If nature selected trees with 18 m deep roots to compete for water during the Last Glacial Maximum, it is very likely that the climate then also had a strong interannual variability. Dry periods may have been long enough to require the presence of 18 m deep roots (several years), followed by long wet periods that would recharge the soil[2]. Under such a climate, deep roots may have helped the survival of the forest (Kleidon and Lorenz, 2001).

CONCLUSIONS

Large-scale impacts of tropical forest conversion differ from the micro-scale impacts because of the interdisciplinarity and the feedbacks involved. At the small scale, it is usually assumed that conversion of forest to agriculture leads to increased water yield, although in certain types of montane tropical forest the inverse can happen (see discussion in Bruijnzeel, this volume). However, at the large-scale, feedbacks through the atmosphere generally reduce the precipitation over the cleared area, reducing the magnitude of the runoff increase, or even decreasing it.

There is enough evidence, at the seasonal, interannual and millennial time scale, to argue that the tropical forests and the humid tropical climate are intimately connected. An anthropogenic disturbance of this climatic equilibrium can lead to important changes in the local climate and in the biogeography of the tropical forests.

References

Bonan, G. B. (1996). A land surface model (LSM, version 1.0) for ecological, hydrological, and atmospheric studies: technical description and user's guide, NCAR Technical Note 417, 150 pp.

Bonell, M. (1998). Possible impacts of climate variability and change on tropical forest hydrology. *Climatic Change*, 39, 215–272.

Bosch, J. M. and Hewlett, J. D. (1982). A review of catchment experiments to determine the effect of vegetation changes on water yield and evapotranspiration. *Journal of Hydrology*, 103, 323–333.

Brouwell, L. C., 1996: *Nutrient cycling in pristine and logged tropical rainforest: a study in Guyana*. Tropenbos Guyana Series no. 1, Tropenbos, Georgetown, Guyana, 224 p.

2 For a detailed review of groundwater in runoff hydrology in selected parts of Amazonia, see Bonell, this volume.

Bruijnzeel, L. A. (1990). Hydrology of moist forests and the effects of conversion: a state of knowledge review. UNESCO – Int. Inst. for Aerospace Survey and Earth Science. Publication of the Humid Tropics Programme. Free University, Amsterdam, 224 p.

Bruijnzeel, L. A. (1996). Predicting the hydrological impacts of land cover transformation in the humid tropics: the need for integrated research. In: *Amazonian deforestation and climate*. J. H. C. Gash, C. A. Nobre, J. M. Roberts and R. L. Victoria, Eds. John Wiley and Sons, 1st ed. p. 15–56.

Canadell, J., Jackson, R. B., Ehleringer, J. R., Mooney, H. A., Sala, O. E. and Schulze, E.-D., (1996). Maximum rooting depth of vegetation types at the global scale. *Oecologia*, 583–595.

Coe, M. T., 2000: Modeling terrestrial hydrologic systems at the continental scale: Testing the accuracy of an atmospheric GCM. *Journal of Climate* 13, 686–704.

Costa, M. H. and J. A. Foley, 1997: The water balance of the Amazon basin: Dependence on vegetation cover and canopy conductance. *Journal of Geophysical Research-Atmospheres*, 102, 23,973–23,989.

Costa, M. H. and J. A. Foley, 1999: Trends in the hydrological cycle of the Amazon basin. *Journal of Geophysical Research – Atmospheres*, 104, 14,189–14,198.

Costa, M. H. and J. A. Foley, 2000: Combined effects of deforestation and double atmospheric CO_2 concentrations on the climate of Amazonia. *Journal of Climate*, 12, 18–35.

Cox, P. M., R. A. Betts, C. D. Jones, S. A. Spall, I. J. Totterdell, 2000: Acceleration of global warming due to carbon-cycle feedbacks in a coupled climate model. *Nature*, 408, 184–187.

Cutrim, E., D. W. Martin, R. Rabin, 1995: Enhancement of cumulus clouds over deforested lands in Amazonia. *Bulletin of the American Meteorological Society*, 76, 1801–1805.

Delire, C., P. Behling, M. T. Coe, J. A. Foley, R. Jacob, J. Kutzbach, Z. Y. Liu, S. Vavrus, 2001: Simulated response of the atmosphere-ocean system to deforestation in the Indonesian Archipelago. *Geophysical Research Letters*, 28, 2081–2084.

Dias, M. A. F. S., P. Regnier, 1996: Simulation of mesoscale circulations in a deforested area of Rondônia in the dry season. In: *Amazonian deforestation and climate*. J. H. C. Gash, C. A. Nobre, J. M. Roberts and R. L. Victoria, Eds. John Wiley and Sons, 1st ed. p. 531–547.

Dickinson, R. E., A. Henderson-Sellers, P. J. Kennedy and M. F. Wilson, 1986: Biosphere-Atmosphere Transfer Scheme (BATS) for the NCAR CCM, NCAR/TN-275-STR, National Center for Atmospheric Research, Boulder, Colorado.

Dickinson, R. E., A. Henderson-Sellers, P. J. Kennedy, 1993: Biosphere-Atmosphere Transfer Scheme (BATS), version 1e as coupled to the NCAR Community Model. NCAR Technical Note NCAR/TN-387+STR, 72 pp.

Dickinson, R. E., P. Kennedy, 1992: Impacts on regional climate of Amazon deforestation. *Geophysical Research Letters*, 19, 1947–1950.

Dirmeyer, P. A., Shukla, 1994: Albedo as a modulator of climate response to tropical deforestation. *Journal of Geophysical Research*, 99D, 20863–20877.

Dooge, J. C. I., 1997: Scale problems in hydrology. In: *Reflections on hydrology – Science and practice*. N. Buras, ed. American Geophysical Union, 1st ed., p. 84–145.

Ducroudé, N., K. Laval and A. Perrier, 1993: A new set of parameterizations of the hydrologic exchanges at the land/atmosphere interface within the LMD atmospheric general circulation model. *Journal of Climate*, 6, 248–273.

Eagleson, P. S., 1986: The emergence of global-scale hydrology. *Water Resources Research*, 22, 6S–14S.

Eastman, J. L., M. B. Coughenour, R. A. Pielke, 2001: Does grazing affect regional climate? *Journal of Hydrometeorology*, 2, 243–253.

Eltahir, E. A. B., 1996: Role of vegetation in sustaining large-scale atmospheric circulations in the tropics. *Journal of Geophysical Research*, 101, 4255–4268.

Eltahir, E. A. B. and R. L. Bras, 1994a: Sensitivity of regional climate to deforestation in the Amazon basin. *Advances in Water Resources*, 17, 101–115.

Eltahir, E. A. B. and R. L. Bras, 1994b: Precipitation recycling in the Amazon basin. *Quarterly Journal of the Royal Meteorological Society*, 120, 861–880.

Entekhabi, D., G. R. Asrar, A. K. Betts, K. J. Beven, R. L. Bras, C. J. Duffy, T. Dunne, R. D. Koster, D. P. Lettenmaier, D. B. McLaughlin, W. J. Shuttleworth, M. T. van Genuchten, M. Y. Wei, E. F. Wood, 1999: An agenda for land surface hydrology research and a call for the second international hydrological decade. *Bulletin of the American Meteorological Society*, 80, 2043–2058.

Foley, J. A., I. C. Prentice, N. Ramankutty, S. Levis, D. Pollard, S. Sitch, A. Haxeltine, 1996: An integrated biosphere model of land surface processes, terrestrial carbon balance, and vegetation dynamics. *Global Biogeochemical Cycles*, 10, 603–628.

Haffer, J. 1969: Speciation in Amazonian forest birds. *Science*, 165, 131–137.

Hahmann, A. N. and R. E. Dickinson, 1997: RCCM2-BATS model over tropical South America: applications to tropical deforestation. *Journal of Climate*, 10, 1944–1964.

Henderson–Sellers, A., R E. Dickinson, T. B. Durbidge, P. J. Kennedy, K. McGuffie, A. J. Pitman, 1993: Tropical deforestation: modeling local to regional scale climate change. *Journal of Geophysical Research*, 98, 7289–7315.

Hodnett, M. G., J. Tomasella, A. de O. Marques Filho, M. D. Oyama, 1996: Deep soil uptake by forest and pasture in central Amazonia: predictions from long-term daily rainfall data using a simple water balance model. In: *Amazonian deforestation and climate*. J. H. C. Gash, C. A. Nobre, J. M. Roberts and R. L. Victoria, Eds. John Wiley and Sons, 1st ed. p. 79–100.

Jarvis, P. G. and K. G. McNaughton, 1986: Stomatal control of transpiration: scaling up from leaf to region, *Advances in Ecological Research*, 15, 1–49.

Jackson, R. B., 1999: The importance of root distributions for hydrology, biogeochemistry, and ecosystem functioning. In: *Integrating hydrology, ecosystem dynamics, and biogeochemistry in complex landscapes*. J. Tenhunen, P. Kabat, Eds. John Wiley and Sons, 1st ed. p. 219–240.

Kiang, J. E. and E. A. B. Eltahir, 1999: Role of ecosystem dynamics in biosphere-atmosphere interaction over the coastal region of West Africa. *Journal of Geophysical Research-Atmospheres*, 104, 31,173–31,189.

Kleidon, A., S. Lorenz, 2001: Deep roots sustain Amazonian rainforest in climate model simulations of the last ice age. *Geophysical Research Letters*, 28, 2425–2428.

Kubatzki, C. and M. Claussen, 1998: Simulation of the global bio-geophysical interactions during the Last Glacial Maximum, *Climate Dynamics*, 14, 461–471.

Lawton, R. O., U. S. Nair, R. A. Pielke, Sr., R. M. Welch, 2001: Climatic impact of tropical lowland deforestation on nearby montane cloud forests. *Science* 294: 584–587.

Lean, J., P. R. Rowntree, 1993: A GCM simulation of the impact of Amazonian deforestation on climate using an improved canopy representation. *Quarterly Journal of the Royal Meteorological Society*, 119, 509–530.

Lean, J., C. R. Bunton, C. A. Nobre, P. R. Rowntree, 1996: The simulated impact of Amazonian deforestation on climate using measured ABRACOS vegetation characteristics. In: *Amazonian deforestation and climate*. J. H. C. Gash, C. A. Nobre, J. M. Roberts and R. L. Victoria, Eds. John Wiley and Sons, 1st ed. p. 549–576.

Lean, J. and P. R. Rowntree, 1997: Understanding the sensitivity of a GCM simulation of amazonian deforestation to the specification of vegetation and soil characteristics. *Journal of Climate*, 10, 1216–1235.

Lohmann D, E. Raschke, B. Nijssen B., D. P. Lettenmaier, 1998: Regional scale hydrology: I. Formulation of the VIC-2L model coupled to a routing model. *Hydrological Sciences Journal*, 43, 131–141.

Manzi, O. and S. Planton, 1996: Calibration of a GCM using ABRACOS and ARME data and simulation of Amazonian deforestation. In: *Amazonian deforestation and climate*. J. H. C. Gash, C. A. Nobre, J. M. Roberts and R. L. Victoria, Eds. John Wiley and Sons, 1st ed. p. 505–530.

Maslin, M. A. and S. J. Burns, 2000: Reconstruction of the Amazon basin effective moisture availability over the past 14,000 years. *Science*, 290, 2285–2287.

Matsuyama, H., 1992: The water budget in the Amazon river basin during the FGGE period. *Journal of the Meteorological Society of Japan*, 70, 1071–1083.

Mayle, F. E.; R. Burbridge, T. J. Killeen, 2000: Millennial-scale dynamics of southern Amazonian rainforests. *Science*, 290, 2291–2294.

Monteith, J. L., 1995: Accommodation between transpiring vegetation and the convective boundary layer, *Journal of Hydrology*, 166, 251–263.

Nepstad, D. C., C. R. de Carvalho, E. A. Davidson, P. A. Lefebvre, G. H. Negreiros, E. D. da Silva, Thomas A. Stone, S. E. Trumbore, S. Vieira, 1994: The role of deep roots in the hydrological and carbon cycles of Amazonian forests and pastures. *Nature*, 372, 666–669.

Noilhan, J. and S. Planton, 1989: A simple parameterization of land surface processes for meteorological models. *Monthly Weather Review*, 117, 536–549.

Penman, H. L., 1948: Natural evaporation from open water, bare soil and grass. *Proceedings of the Royal Society of London*, A198, 116–140.

Pielke, R. A., R. Avissar, M. Raupach, A J. Dolman, X. B. Zeng, A. S. Denning, 1998: Interactions between the atmosphere and terrestrial ecosystems: influence on weather and climate. *Global Change Biology*, 4, 461–475.

Polcher, J. and Laval, K., 1994: The impact of African and Amazonian deforestation on tropical climate. *Journal of Hydrology*, 155, 389–405.

Priestley, C. H. B. and R. J. Taylor, 1972: On the assessment of surface heat flux and evaporation using large-scale parameters. *Monthly Weather Review*, 100, 81–92.

Randall, D. A., Ed., 2000: General Circulation Model Development: Past, Present, and Future. Academic Press, 1st ed., 807 p.

Sahin, V. and M. J. Hall, 1996: The effects of afforestation and deforestation on water yields. *Journal of Hydrology*, 178, 293–309.

Sellers, P. J., 1992: Biophysical models of land surface processes. In: *Climate system modelling*, K. E. Trenberth, ed. Cambridge University Press, 1st ed. p. 451–490.

Sellers, P. J., Y. Mintz, Y. C. Sud, and A. Dalcher, 1986: A simple biosphere model (SiB) for use within general circulation models, *Journal of Atmospheric Science*, 43, 505–531.

Sellers, P. J., S. O. Los, C. J. Tucker, C. O. Justice, D. A. Dazlich, G. J. Collatz and D. A. Randall, 1996: A revised land surface parameterization (SiB2) for atmospheric GCMs. Part I: Model formulation *Journal of Climate*, 9, 676–705.

Shuttleworth, W. J., 1988: Evaporation from Amazonian rainforest. *Proc. R. Soc. London, B* 233, 321–346.

Sud, Y. C., G. K. Walker, J.-H. Kim, G E. Liston, P. J. Sellers, W. K.-M. Lau, 1996: Biogeophysical consequences of a tropical deforestation scenario: a GCM simulation study. *Journal of Climate*, 9, 3225–3247.

Trenberth, K. (Ed.), 1992: *Climate system modeling*. Cambridge University Press, 1st ed. 788 p.

van der Hammen, T. and M. L. Absy, 1994: Amazonia during the last glacial. *Palaeogeography, Palaeoclimatology, Palaeoecology*, 109, 247–261.

Vörösmarty, C. J., B. Moore III, A. L. Grace, M. P. Gildea, 1989: Continental scale models of water balance and fluvial transport: an application to South America. *Global Biogeochemical Cycles*, 3, 241–265.

Walker, G. K., Y. C. Sud, R. Atlas, 1995: Impact of the ongoing Amazonian deforestation on local precipitation: A GCM simulation study. *Bulletin of the American Meteorological Society*, 76, 346–361.

Warrilow, D. A., A. B. Sangster, A. Slingo, 1986: Modelling of land surface processes and their influence on European climate. UK Meteorological Office Met O 20 Dynamical Climatology Tech. Note DCTN 38. 92 p.

Xue, Y. K., P. J. Sellers, J. L. Kinter III, J. Shukla, 1991: A simplified biosphere model for global climate studies. *Journal of Climate*, 4, 345–364.

Zeng, N., (1999). Seasonal cycle and interannual variability in the Amazon hydrologic cycle *Journal of Geophysical Research-Atmospheres*, 104, 9097–9106.

Zeng, N. and Neelin, J. D. (1999). A land-atmosphere interaction theory for the tropical deforestation problem. *Journal of Climate*, 12, 857–872.

24 Forest recovery in the humid tropics: changes in vegetation structure, nutrient pools and the hydrological cycle

D. Hölscher
University of Göttingen, Germany
J. Mackensen
United Nations Environmental Programme (UNEP), Nairobi, Kenya
J. M. Roberts
Centre for Ecology and Hydrology, Wallingford, UK

INTRODUCTION

Many of the agricultural land use systems replacing tropical forest can be regarded as only semi-permanent. In shifting agriculture cultivated fields are abandoned within two to three years while pastures in Amazonia may have a productive period of only six to twelve years (Sanchez and Hailu, 1996; Uhl *et al.*, 1988). Abandoned fields are usually left for secondary succession with trees and shrubs establishing in these areas initiating the forest recovery. In some tropical regions secondary forests already form the most important land cover and are likely to become more important as long as forest conversion proceeds and sustainable management systems for the deforested areas are lacking. Fearnside (1996) estimated that 30% of the deforested area of the Brazilian Amazon was covered by secondary vegetation in 1990. Similarly, in mainland South East Asia, a region with a long history of shifting cultivation, secondary forest comprises about one third in representative areas in northern Thailand (Fox *et al.*, 1995), northern Vietnam (Fox *et al.*, 2000) and southern China (Xu *et al.*, 1999).

The conversion from forest to agricultural land use causes manifold changes in water and nutrient cycles and has been studied for different site conditions and degrees of disturbance (Malmer *et al.*, Grip *et al.*, this volume). This chapter reviews the changes in vegetation structure, dynamics of nutrient pools in soil and phytomass, and subsequent alterations of the hydrological cycle during the process of forest recovery on abandoned pastures and in shifting cultivation systems. A guiding question is whether the effects of conversion from old-growth forest to agricultural land are reversed during forest recovery, and if so to what extent. In the case where a restoration can be observed, it seems important to estimate at which age we can expect secondary forests to regain the function of old-growth forests with respect to water and nutrient cycles.

For the description of structural and functional characteristics of the different forest types we follow Clark (1996) who proposed to substitute classifications like 'primary' or 'undisturbed' by old-growth forest because establishing whether or not a forest has ever been affected by human influences is problematic. In this context secondary forest is defined as forest regenerating after complete clearing (Corlett, 1994). Forest recovery is a long-term process which takes decades at least but may take centuries if species composition within the forests is considered. Thus, studies on forest recovery are often restricted to chronosequences in which the spatial differences in forest age at a given time represent different stages of succession (Saldarriaga *et al.*, 1988; Uhl *et al.*, 1988; Kappelle *et al.*, 1996). A problem in studying chronosequences in this way is the comparability of the environmental site conditions and previous land management. This, however, can be reduced by a sufficient number of replications and careful site selection. As water and nutrient cycles are sensitive to variations in weather conditions, results should also be interpreted with caution when only short term data are available.

CHANGES IN VEGETATION STRUCTURE DURING FOREST RECOVERY

Structural characteristics of vegetation such as species diversity, species composition, leaf area, canopy height and rooting pattern

Forests, Water and People in the Humid Tropics, ed. M. Bonell and L. A. Bruijnzeel. Published by Cambridge University Press.

influence the water and nutrient cycles and change during forest recovery. Forest succession begins at significantly different starting points on abandoned pastures and shifting cultivation fields.

Pastures are usually composed of a small number of grass species and colonised by herbs and woody species. Species composition and diversity in abandoned pastures in the Paragominas area, Amazonia, depend largely on previous use intensity (Uhl *et al.*, 1988). Eight years after abandonment tree species richness was comparatively high (about 23 tree species per 100 m^2) on pastures subjected to light use. Almost all of those species also occurred in native forest. On pastures subjected to moderate use tree species richness was lower (about 17 tree species per 100 m^2) and there were fewer forest tree species which were again reduced on heavily used pastures (about 1 tree species per 100 m^2). The seed bank and potential for vegetative regeneration are largely destroyed on highly disturbed sites and generative propagation from surrounding areas is thus the main means for tree succession. The majority of native forest species at Paragominas have fruits that are dispersed by animals of which many do not move into large clearings. Seed dispersal is a severe impediment for tree succession at these sites (Nepstad *et al.*, 1991). A study on the limitations of forest regeneration on abandoned pastures in southern Costa Rica by Holl (1999) suggested that seed dispersal by animals was the most critical factor limiting forest recovery. In Puerto Rico, Aide *et al.* (1996) found that by 40 years after abandonment forest stands on former pastures could not be distinguished from undisturbed sites in terms of stem density, basal area and species number. However, species composition was still very different.

Woody vegetation on abandoned fields following shifting cultivation systems usually exhibits a higher species diversity than vegetation on abandoned pastures. In the Rio Negro region, Saldarriaga *et al.* (1988) found that on lightly used sites the number of species increased during the early stages of succession and stands 20 to 40 years old had a species richness similar to that found in old-growth forests. On shifting cultivation fields in eastern Amazonia, Baar (1997) found 60 to 80 woody species per 60 m^2 (1 to 10 years after abandonment) although the fields had been slashed and burned repeatedly. However, the phytodiversity in the fallow vegetation was greatly reduced when compared to 10 000 m^2 plots in old-growth Amazonian forests (Baar, 1997). For the Neotropics, Gentry (1988) documented Shannon indices (H') for old-growth forests to range between 5.4 and 7.7, whereas a comparable analysis of the woody fallow vegetation gave an average H' of 1.8. The woody fallow in eastern Amazonia contained only 5 tree species that were also found in native forests of the region. This situation may be caused by the long-term use of the shifting cultivation system in the study region (Baar, 1997). In western Amazonia Fujisaka *et al.* (1998) found that secondary forests support only slightly fewer species than old-growth forests. Again the secondary forests contained a large proportion of species not found in native forests. Vegetative regeneration was common after forest cutting and burning in northern Amazonia but burning and weeding reduced abundance of sprouting and on-site regeneration from the seed bank (Uhl and Jordan, 1984; Uhl, 1987). In eastern Amazonia, Clausing (1994) estimated that 75% of the woody species on fallow fields regenerated vegetatively.

In an upper montane *Quercus* forest in Costa Rica, Kappelle *et al.* (1996) did not find significant changes in tree species diversity indices along a chronosequence including early secondary forest, late secondary forest and old-growth forest. However, dominant tree species, stem density, canopy height and basal area changed significantly between the successional stages. The herb layer contained larger species diversity in secondary than in old-growth forests. The process of forest recovery is slower at montane sites when compared to the lowlands (Ewel, 1980; Kappelle *et al.*, 1996).

We conclude that species diversity and composition in recovering forests depend largely on the use intensity during the agricultural period. The impact of moderate and heavy pasture use is more severe than shifting cultivation. It may take centuries for recovering forests to regain the species diversity and composition of old-growth forests. Differences may even remain irreversible.

Information on the development of root systems during forest recovery is rare. A comparison of root biomass under pasture and old-growth forest in Paragominas, eastern Amazonia, revealed large differences (pasture 9.7 t ha^{-1}; forest 35 t ha^{-1}) and demonstrated the importance of root penetration in deep soil layers especially under forest (Nepstad, 1989, Nepstad *et al.*, 1994). However, root length density (km m^{-3}) integrated to 10 m depth was 2.4 times greater in the grass vegetation than in the forest. Under secondary vegetation in eastern Amazonia, Sommer *et al.* (2000) found a well-developed root system down to the studied depth of 6 m. Shallower rooting patterns were evident under oil palm (4.5 m) and passion fruit (2.5 m). A significant increase in root mass during forest recovery was observed by Saldarriaga and Uhl (1982) in the perhumid Rio Negro region. In the 0–50 cm depth the mass increased from 20 t ha^{-1} in 18 year old secondary forest to over 42 (60 year old secondary forest) to 58 t ha^{-1} in old-growth forest (Table 24.1). Increase in root mass under younger stands of eastern Amazonia was less distinct (Wiesenmüller, 1999). In a chronosequence in upper montane forest of Costa Rica, D. Hertel *et al.* (2003) observed clear changes in the root system. The percentage of ectomycorrhizal root tips in the humus layer increased significantly from early (16%) to late secondary (66%) to old-growth forest (78%). A comparable pattern was observed in the mineral soil. While most of the fine root mass in the early secondary forest was found in the mineral soil (related to a depth of 10 cm), the old-growth forest showed far higher masses in the humus layer with the late secondary forest at an intermediate

Table 24.1. *Root biomass of agricultural, secondary forest and old-growth forest sites in Latin America*

	Age (years)	Region	Country	Humus layer root mass ($t\ ha^{-1}$)	Mineral Soil depth (cm)	Root mass <2 mm ($t\ ha^{-1}$)	Root mass >2 mm ($t\ ha^{-1}$)	Root mass total ($t\ ha^{-1}$)	Reference
Maize		Turrialba	Costa Rica		50	1.5	8.2	9.7	Berish and Ewel, 1988
Secondary vegetation	5	Turrialba	Costa Rica		85	1.9	8.0	9.9	Berish and Ewel, 1988
Pasture		Paragominas	Brazil		1000		9.7		Nepstad *et al.*, 1994
Old-growth forest		Paragominas	Brazil		1000		35.4		Nepstad *et al.*, 1994
Secondary vegetation	1	Turrialba	Costa Rica		85	4.9	11.5	16.4	Berish, 1982
Secondary vegetation	8	Turrialba	Costa Rica		85	3.1	9.8	12.9	Berish, 1982
Secondary forest	70				85	3.4	12.1	15.6	Berish, 1982
Secondary forest	18	Rio Negro	Venezuela		50		20.0		Saldarriaga and Uhl, 1990
Secondary forest	60	Rio Negro	Venezuela		50		31.0		Saldarriaga and Uhl, 1990
Secondary forest	80	Rio Negro	Venezuela		50		42.0		Saldarriaga and Uhl, 1990
Secondary forest	200	Rio Negro	Venezuela		50		58.0		Saldarriaga and Uhl, 1990
Secondary vegetation	1	Rio Negro	Venezuela		50	2.8	2.5	5.3	Sanford, 1985
Secondary vegetation	3	Rio Negro	Venezuela		50	2.1	1.6	3.7	Sanford, 1985
Secondary vegetation	7	Rio Negro	Venezuela		50	5.5	2.4	7.9	Sanford, 1985
Old-growth forest, terra firme		Rio Negro	Venezuela		50	12.7	33.4	46.1	Sanford, 1985
Old-growth forest, terra firme		Rio Negro	Venezuela		50	11	48.4	54.8	Sanford, 1985
Secondary vegetation	1	Eastern Amazonia	Brazil		50	3.9	11.9	15.8	Wiesenmüller, 1999
Secondary vegetation	4	Eastern Amazonia	Brazil		50	5.5	13.4	18.9	Wiesenmüller, 1999
Secondary vegetation	9	Eastern Amazonia	Brazil		50	8.2	21.7	29.9	Wiesenmüller, 1999
Secondary forest	10	Andes	Colombia	2.5	10	2.5		5.0	Cavelier and Estevez, 1996
Secondary forest	20	Andes	Colombia	7.0	10	2.3		9.3	Cavelier and Estevez, 1996
Old-growth forest		Andes	Colombia	4.8	10	2.7		7.5	Cavelier and Estevez, 1996
Early secondary forest	8	Talamanca	Costa Rica	0.3	10	1.6		1.9	Hertel *et al.*, 2003
Late secondary forest	40	Talamanca	Costa Rica	3.4	10	3.0		6.4	Hertel *et al.*, 2003
Old-growth forest				11.3	10	1.6		12.9	Hertel *et al.*, 2003

Table 24.2. *Leaf area index (L) and height (H) of agricultural, secondary forest and old-growth forest sites in Latin America*

	Age (years)	Region	Country	L (m^2 m^{-2})	H (m)	Reference
Maize (young to mature)		Turrialba	Costa Rica	1.0–2.6	1.2–2.0	Ewel *et al.*, 1982
Sweet potato		Turrialba	Costa Rica	2.9	0.5	
Pasture		Amazonia, Ji-Paraná	Brazil	1.6–3.9	0.5	Roberts *et al.*, 1996
Pasture		Amazonia, Manaus	Brazil	1.2–1.9	0.5	Roberts *et al.*, 1996
Pasture		Amazonia, Marabá	Brazil	0.5–1.6	1.0	Roberts *et al.*, 1996
Secondary forest	1	Rio Negro	Venezuela	4.6	2.0	Uhl, 1987
Secondary forest	5	Rio Negro	Venezuela	4.0	10	Uhl, 1987
Secondary forest	1	Turrialba	Costa Rica	5.1	5.0	Ewel *et al.*, 1991
Secondary forest	1	Western Amazonia	Peru	5.0		Szott *et al.*, 1994
Secondary forest	3	Eastern Amazonia	Brazil	4.6	2.0	Hölscher, 1995
Secondary forest	5	Eastern Amazonia	Brazil	5.0	3.0	Denich, 1990
Old-growth forest		Amazonia, Ji-Paraná	Brazil	4.4	30–35	Roberts *et al.*, 1996
Old-growth forest		Amazonia, Manaus	Brazil	5.7	30–35	McWilliam *et al.*, 1993
Old-growth forest		Amazonia, Marabá	Brazil	5.4	40–45	Roberts *et al.*, 1996
Old-growth forest		Rio Negro	Venezuela	5.2	25–30	Jordan and Uhl, 1978

stage. The living fine root mass in the upper mineral soil and the humus layer was 1.9, 6.4 and 12.9 t ha^{-1} in early secondary, late secondary and old-growth forest, respectively.

Leaf area and canopy height of a vegetation may influence the water cycle significantly. Agricultural crops like maize and pastures show a high temporal variability in their leaf area (Table 24.2). Young maize was reported to have a leaf area index (L) of 1 m^2 m^{-2} (Ewel *et al.*, 1982). Amazonian pastures had a L between 0.5 and 3.9 with the minimum in the dry season and a maximum in the wet season (Roberts *et al.*, 1996). Young secondary vegetation one year after slash and burn agriculture showed a L of 5.0 with a substantial part contributed by grasses and herbs (Uhl, 1987). During the following years the proportion of herbs and grasses decreased and leaf area of shrubs and trees increased. Five years after abandonment Uhl (1987) measured a L of 4.0. Other studies in eastern Amazonia documented a L of 4.6–5.0 in young secondary forest stands (Denich, 1989; Hölscher, 1995). Leaf area indices obtained in young secondary stands did not differ significantly from Amazonian old-growth forests for which Roberts *et al.* (1996) reported values between 4.4 and 5.7. However, information collated by Cannell (1982) shows that the values from Amazonia are substantially lower than those reported for example for old-growth forests sites in Thailand and Malaysia. In a montane area of Costa Rica Hölscher *et al.* (2003) compared the foliage area indices of three forest stands differing in their successional stage. A significantly higher value was obtained in a late secondary forest when compared with a young secondary and an old-growth forest (Figure 24.1). McWilliam *et al.* (1996) found the highest leaf area density of an Amazonian old-growth forest in the range between 12 to 18 m tree height. In a 3 year old secondary vegetation the highest leaf area density was in the 1 to 1.5 m height range (Hölscher, 1995). Thus the leaf area of recovering forests can reach values of old-growth forests within a few years but vertical distribution remains different for some decades.

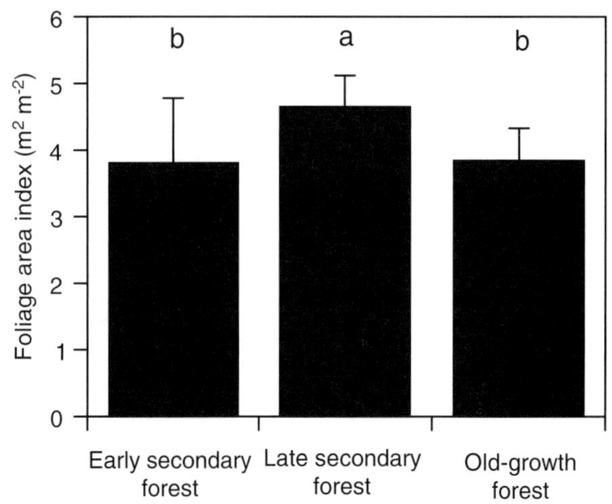

Figure 24.1 Foliage area index measured with an optical canopy analysis system (LAI 2000, $n = 30$) at three different successional stages in a montane Quercus forest, Costa Rica. (From Hölscher *et al.*, 2003.)

NUTRIENT DYNAMICS DURING FOREST RECOVERY

Phytomass accumulation, soil fertility and previous management intensity

Nutrient dynamics during forest recovery are closely related to the rate of above- and below-ground phytomass accumulation. The accumulation rate depends on factors such as previous site management and soil fertility, both of which are closely interrelated.

According to Brown and Lugo (1990a) phytomass accumulation on tropical sites is generally characterised by high to very high increment rates (up to 100 t ha^{-1}, above-ground phytomass) in the first 15 years or so regardless of specific climatic conditions within the lowlands of the humid tropics. During the first years of succession, phytomass accumulation appears to be almost linear (Uhl and Jordan, 1984). A differentiation in the phytomass production rate between sites appears mainly during subsequent stages of development depending on different abiotic factors such as climate, water availability and soil fertility (e.g. Andriesse and Schelhaas, 1987). Few exceptions were found to this assumption (Bartholomew *et al.*, 1953; Scott, 1977). Generally, few stands accumulated more than 200 t ha^{-1} by age 80 years (Brown and Lugo, 1990a). On relatively fertile vitric Andosols in Mexico, Hughes *et al.* (1999) recorded up to 290 t ha^{-1} in 50-year old secondary forests. The estimated recovery of forest phytomass and basal area for Tierra-firme sites in Colombia and Venezuela on Oxisols and Ultisols may take between 190 to 250 years (Saldarriaga *et al.*, 1988), while for sites on Andosols about 70 years were estimated (Hughes *et al.*, 1999).

Nutrient availability and thus soil fertility influence phytomass recovery rate significantly (Uhl, 1982, 1987). Applying a principal component analysis Aweto (1981) found a strong positive relationship of topsoil nutrient content and organic matter status of an Oxisol in Nigeria with tree size and vegetation cover in secondary forests. Among a variety of soil parameters topsoil nutrient content and organic matter status explained 33% of total variance. The water-holding capacity was found also among the parameters of the first component, however, it explained less than 3% of total variance. In Sarawak Ewel *et al.* (1983) found three times higher phytomass accumulation of 4.5 years old secondary forest on nutrient-rich alluvial sites (5.4 kg m^{-2}) compared to more deficient Oxisol/Ultisol slope sites (2.1 kg m^{-2}). Unfortunately, specific soil parameters, to which this correlation could be linked were not analysed.

The importance of P nutrition was confirmed by Mackensen *et al.* (2000) who detected a strong correlation ($R^2 = 0.76$) between the total P storage of weathered soils (e.g. Oxisols, Ultisols) and above ground phytomass of tropical old-growth forests. Similarly, the recovery rate of secondary forests could be controlled by the P availability.

In comparison to studies on nutrient-poor soils elsewhere in the tropics Hughes *et al.* (1999), studying forest recovery on fertile Andosols in Mexico, found high growth rates continuing for more than 30 years. However, Harcombe (1977) showed that growth rates on relatively fertile soils do not necessarily respond to fertiliser input; it was found that fertilisation (NPK, 70–166 kg ha^{-1}) on andesitic Inceptisols in Costa Rica did not enhance phytomass production significantly but forb growth was stimulated.

Uhl and Jordan (1984) and Uhl (1987) noticed that pioneer species (e.g. *Vismia* spp.) exhibit high growth rates even on very poor sites (Oxisols) in the Amazon Territory of Venezuela. High growth rates in secondary forests do not necessarily indicate good soil conditions. They may be caused by specific traits of pioneer species. In southern Venezuela a relatively fast growth of secondary vegetation was accompanied by low foliar nutrient concentrations, which was interpreted as a relatively effective nutrient use (Uhl and Jordan, 1984). Another feature which permits pioneer-species to survive under conditions of restricted nutrient availability is the high root:shoot ratios (0.14–0.41:1) compared to crop species such as cassava, *Manihot esculenta* (0.06:1; Uhl, 1987).

The previous management intensity determines the site nutrient status and thus the forest recovery rate. On abandoned pastures in eastern Amazonia, Uhl *et al.*, 1988 found that above-ground biomass accumulation of secondary vegetation averaged 10 t ha^{-1} yr^{-1}, while moderately and intensively used pastures amounted to 5 t ha^{-1} yr^{-1} and 0.6 t ha^{-1} yr^{-1} respectively. Abandoned pastures that were only lightly used also showed higher nutrient stocks in biomass than abandoned pastures under former medium and heavy use (Buschbacher *et al.*, 1988). These pastures showed no differences with respect to soil nutrient stocks. According to the study by Hughes *et al.* (1999), mean annual above-ground phytomass accumulation was inversely correlated to the duration of prior land use in Mexico. Aide *et al.* (1996) found forest recovery on abandoned pastures in Puerto Rico was significantly delayed compared to areas of other natural or anthropogenic disturbance. Forest recovery on degraded pastures in the Brazilian Amazon was slower in terms of phytomass and carbon accumulation than sites under former shifting cultivation (Fearnside and Guimaraes, 1996). Other studies (e.g. Brown and Lugo, 1990b) found that pastures may contain as much or more C as the preceding forests, depending on their establishment and management history. In the case of shifting cultivation, management intensity is mainly defined through the frequency of slash burning, the duration of the cropping cycle and the length of the fallow period. Nutrient export through harvest, slash burning and subsequent leaching losses on shifting cultivation sites deteriorate site nutrient stock significantly (e.g. Hölscher *et al.*, 1997b). The stripping of all phytomass and soil surface organic matter results in prolonged nutrient depletion of successional plots (Uhl, 1987). Also, management practices as related to different crops, can have a significant impact on soil fertility. Weaver *et al.* (1987) found that soil organic matter of secondary forests preceded by coffee had less organic matter than those preceded by pasture. In addition, intensive or regular slash burning prior to field abandonment is likely to reduce the soil seed pool and coppicing potential (Herrera *et al.*, 1981; Uhl *et al.*, 1981) and together with other common practices (e.g. weeding) can clearly decrease site regeneration potential.

Buschbacher *et al.* (1988) showed also that the impact of management intensity is not so much reflected by the soil nutrient status but rather by the total nutrient stock of the system including phytomass, litter and slash. The effect is worse in soils with an already critical nutrient status (e.g. Spodosols, sandy Ulti- and Oxisols, see Guggenberger and Zech, 1999).

Nutrient accumulation in phytomass

The rate at which vegetation is accumulating nutrients is especially high during the first successional stages. Herbs, shrubs and young trees store relatively more nutrients than older trees, which have a higher proportion of structural tissue (e.g. heart-wood) with low nutrient concentrations (e.g. Williams-Linera, 1983; Uhl, 1987). The studies by Fölster *et al.* (1976), Uhl and Jordan (1984) and Hughes *et al.* (1999) allow for a direct comparison of nutrient storage in old-growth and regrowth forests. The relative recovery rate of P and base cations in forest regrowth is 2–5 times higher than phytomass recovery, while N-recovery generally follows phytomass recovery (Table 24.3). This can be illustrated by an example: compared to the control (old-growth forest), above-ground phytomass and related N-stock in a 20-yr-old regrowth stand account for 22% and 27% respectively, while the P-stock in the same stand already represents 85% of the P-stock in the control. Beside these constraints, some early successional species are also known to be specific nutrient accumulators and might thus cause a relatively faster recovery of nutrients (Stark, 1971; Tanner, 1977; Juo and Lal, 1977; Swamy and Ramakrishnan, 1987).

Although the relative accumulation rate of nutrients slows down as succession proceeds, total nutrient stock increases with accumulating phytomass (e.g. Snedaker, 1980; Swamy and Ramakrishnan, 1987; Hughes *et al.*, 1999). Nutrients in leaves and twigs have a high turn-over rate whereas nutrients in woody plant components are immobilised for a longer time. An exception to the above was found in north-east India by Rao and Ramakrishnan (1989) who showed that bamboo had high growth rates which were sustained over the first 15 years, continued high rates of nutrient uptake and relatively constant turn-over times.

Changes in soil chemical properties during forest recovery

Secondary succession is generally accompanied by an increase in soil organic matter due to enhanced litter fall and root production. The first years after field abandonment, however, may exhibit a soil carbon decrease which was found to reach a minimum after 3–7 years (Popenoe, 1957; Zinke *et al.*, 1978; Street, 1980; Aweto, 1981; cf. Palm and Szott, 1984; Uhl, 1987; Juo *et al.*, 1995; Szott and Palm, 1996). At the time of field abandonment enhanced decomposition of easily decomposable material is accompanied by high leaching losses of nitrate and cations. The bulk of organic material (e.g. trunks) will take more time to decompose and also litterfall is temporarily reduced during this very early stage of succession. As a result of the lack of organic matter input and enhanced decomposition, soil carbon storage decreases temporarily.

On very wet tropical sites in Venezuela, Uhl (1987) found that 75% of above-ground phytomass was decomposed after 3–4 years and that after 5 years soil organic matter content was higher on secondary succession sites than on control plots. The difference can amount to 40 t ha^{-1} in the first 20 cm of topsoil (Uhl, 1987). Juo *et al.* (1995) quoted a recovery time of 13 yr for soil C-concentration in kaolinitic Alfisols in Nigeria. Recovery of soil C-storage during succession in moist, wet and dry subtropical climate zones in Puerto Rico and US Virgin Islands was estimated to take 40–50 years (Brown and Lugo, 1990b). The same time scale was given by Wadsworth *et al.* (1990) for recovery of soil C-concentration in Ultisols in humid areas of Mexico. The N-dynamics in several studies were found to mirror C-dynamics (e.g. Wadsworth *et al.*, 1990; Palm and Szott, 1984), but were generally not as pronounced (e.g. Jordan *et al.*, 1983; Szott and Palm, 1996). In the study by Brown and Lugo (1990a) recovery of N-storage appeared to be faster (15–20 years), while Wadsworth *et al.* (1990) reported similarly long recovery periods for soil N-concentration as for soil C. According to Uhl (1987) N-concentrations had already reached or exceeded initial levels under moist conditions after 5 years in the Venezuelan Amazon. Direct comparisons between studies, however, are difficult because of the different stand and management histories (Bruijnzeel, 1991). Unfortunately, the absence of bulk density data means that nutrient stocks per soil depth often cannot be calculated. However, C-losses after conversion seem to be larger in moist than in dry climate zones, while recovery rates again seem to be much faster in moist areas.

In comparison to other elements P is only leached at low rates from tropical weathered soils (e.g. Klinge, 1998). Changes in soil P-content during forest recovery are minimal and obscured by spatial heterogeneity of soil characteristics (e.g. Wadsworth *et al.*, 1990). Considering also the large percentage of immobilised P in weathered soils in relation to a generally small 'plant-available' P-pool, may help to explain why distinct patterns in total P-storage are hard to find. Referring to P of organic origin, the dynamics are similar to those found for C and N (e.g. Aweto, 1981; Szott and Palm, 1996).

In relatively fertile soils such as the vitric Andosols in Mexico studied by Hughes *et al.* (1999) where 58–91% of C and 94–99% of N of total ecosystem pools are stored in the soil (related to 100 cm soil depth), no changes in soil C- and N-storage during

Table 24.3. *Nutrient recovery in above-ground phytomass of secondary forests in Latin America*

Forest type	Stand age (years)	Country	Site history[a]	Soil type	Total aboveground phytomass (t ha^{-1})	C (t ha^{-1})	N (kg ha^{-1})	P (kg ha^{-1})	K (kg ha^{-1})	Ca (kg ha^{-1})	Mg (kg ha^{-1})	Reference
Secondary forest	0.5	Mexico	30	Vitric Andosol	5	2	75	5				Hughes *et al.*,1999
Secondary forest	4	Mexico	1	Vitric Andosol	60	28	379	46				Hughes *et al.*, 1999
Secondary forest	8	Mexico	27	Vitric Andosol	36	17	195	27				Hughes *et al.*, 1999
Secondary forest	8	Mexico	7	Vitric Andosol	109	51	474	71				Hughes *et al.*, 1999
Secondary forest	10	Mexico	30	Vitric Andosol	23	11	241	24				Hughes *et al.*, 1999
Secondary forest	16	Mexico	5	Vitric Andosol	293	137	1177	121				Hughes *et al.*, 1999
Secondary forest	20	Mexico	13	Vitric Andosol	88	41	462	89				Hughes *et al.*, 1999
Secondary forest	20	Mexico	1	Vitric Andosol	270	129	1081	158				Hughes *et al.*, 1999
Secondary forest	26	Mexico	2	Vitric Andosol	217	104	1113	53				Hughes *et al.*, 1999
Secondary forest	30	Mexico	15	Vitric Andosol	254	122	976	92				Hughes*et al.*, 1999
Secondary forest	50	Mexico	–	Vitric Andosol	287	136	1167	147				Hughes *et al.*, 1999
Old-growth forest	–	Mexico	–	Vitric Andosol	403	195	1705	105				Hughes *et al.*, 1999
Secondary forest	2	Colombia		Oxisol	19	23	276	23	132	153	37	Fölster *et al.*, 1976
Secondary forest	5	Colombia		Oxisol	69	40	609	30	394	309	65	Fölster *et al.*, 1976
Secondary forest	16	Colombia		Oxisol	203	113	1045	67	517	664	182	Fölster *et al.*, 1976
Old-growth forest	–	Colombia		Oxisol	185	100	1413	45	305	526	161	Fölster *et al.*, 1976
Old-growth forest	–	Colombia		Oxisol	326	155	1651	54	419	899	250	Fölster *et al.*, 1976
Secondary forest	5	Venezuela		Oxisol	45		237	10	86	127	41	Uhl and Jordan, 1984
Old-growth forest	–	Venezuela		Oxisol	269		1054	32	253	209	57	Uhl and Jordan, 1984
Secondary forest	8	Brazil	Light	Oxisol / Ultisol	108			29	511	934	172	Buschbacher *et al.*, 1988
Secondary forest	8	Brazil	Medium	Oxisol / Ultisol	52			20	253	922	87	Buschbacher *et al.*, 1988
Secondary forest	8	Brazil	Heavy	Oxisol / Ultisol	10			8	62	182	32	Buschbacher *et al.*, 1988

[a] Site history: duration of prior land use (years) or former land use intensity.

the 50-year-long course of forest succession could be detected. This example clearly indicates that the vulnerability of a given soil depends on the initial soil nutrient status. Unlike poor, deeply weathered soils, soils well supplied with nutrients are not expected to exhibit critical changes in soil nutrient status in reaction to e.g. forest conversion or field abandonment. On very acid soils and in montane areas old-growth forests have a significant humus layer (Table 24.4) forming an important nutrient capital that recovers only slowly during secondary succession.

In the case of primary succession on volcanic ashes in Hawaii the reciprocal influence of plant community and soils becomes apparent (Matson, 1990). Different native and exotic pioneer species proved to have a distinct impact on soil properties compared to bare soil, resulting in higher C, N, Ca-, K- and Mg-concentrations in the topsoil (0–15 cm). No significant change was found in P-concentrations under successional vegetation and bare soil. The study by Matson (1990) also showed that N-fixing species (for example *Myrica faya*) generally account for higher soil N-concentrations than non-fixers. Shifting cultivation fallows on loamy Ultisols in the Peruvian Amazon which were enriched with N-fixing species (e.g. *Inga edulis, Cajanus cajan*) generally showed improved soil fertility status (N, P and K) compared to natural fallows (Szott and Palm, 1996).

As a consequence of soil organic matter accumulation, soil effective exchange capacity for cations (ECEC), water infiltration and retention capacity increase and bulk density decreases. In the study by Aweto (1981) water retention capacity, total porosity and bulk density of an Oxisol in Nigeria returned to control levels within 10 years. More intensive and longer lasting disturbance may occur where soil was heavily compacted by machinery (Van der Plas and Bruijnzeel, 1993). The extent and rapidity of changes in soil physical properties depend also on soil texture, aggregation, site history and management intensity (Popenoe, 1957; Sanchez, 1976; Juo and Lal, 1977; Lal and Cummings, 1979; cf. Palm and Szott, 1984) as well as soil organic matter content.

Interpretations of changes in soil base cations during forest recovery depend largely on the starting point. The initial flush of base cations in seepage water (leaching) following forest clearing and burning is well documented (e.g. Klinge, 1998; Malmer *et al.*, this volume). In addition, 20–40% of base cations stored in slash are lost through volatilisation and ash particle export through slash burning (e.g. Ewel *et al.*, 1981; Mackensen *et al.*, 1996). As a consequence of forest conversion and agricultural land use with low inputs, nutrient stocks of the total system drop to lower levels. However, base cations which remain in the system accumulate in the topsoil and lead to an increase in soil pH (Stromgaard, 1984; Khanna *et al.*, 1994; Hölscher *et al.*, 1997a). Along with the remaining N and P this stock of base cations serves as the major nutrient source for the recovering forest. Additional sources of nutrients are bulk precipitation and weathering. The persistence of the fertilisation effect may vary among different ecosystems depending on soil type and vigour of growth. In pasture chronosequences on Ultisols in Amazonia, for example, de Moraes *et al.* (1996) found that a higher soil pH as a result of the forest conversion lasted for 20 years in one case, but for 81 years in a second case.

Remaining nutrient stocks in the soil, however, are depleted through nutrient accumulation in the vegetation during forest development (e.g. Scott, 1987). Buschbacher *et al.* (1988) estimated the net nutrient uptake from soil through recovering vegetation on abandoned pastures with vigorously growing vegetation. They found that especially on pasture with former light use the soil cation stock can be depleted in the top 50 cm by the vegetation, in excess of the estimated nutrient inputs in bulk precipitation and organic litter. Szott and Palm (1996) compared nutrient accumulation in vegetation with changes in soil nutrient stocks over 53 months in differently managed and unmanaged fallows on previously cultivated Ultisols in the Peruvian Amazon. The decreases in soil P-, K-, Ca- and Mg-stocks exceeded the respective increases in the vegetation, illustrating the extent of leaching.

Where leaching and intensive nutrient export through harvesting led to base cation losses it will take time for base cation stocks to recuperate to original levels. Aweto (1981; Oxisol, 1500 mm yr^{-1}) reported lower-than-control base cation concentrations after 10 years. For a chronosequence study on Andepts in perhumid Costa Rica, Werner (1984) found low Ca- and Mg-concentrations compared to controls even after 31 years of succession. Build-up of nutrient stocks in topsoil is either based on the nutrient pump function of roots transporting nutrients from greater depths to topsoil via litterfall or through input by bulk precipitation. For former shifting cultivation sites in the Amazonian sector of Peru Scott (1987) found that N, P and base cation stocks did not recover within the first 25–30 years.

Nutrient accumulation by successional species, especially those having a rapidly expanding root system, has also been observed to increase the base cation level in the topsoil through litterfall (e.g. Ewel, 1976; Aweto, 1981; Scott, 1987). This may become apparent especially where former land use did not lead to base cation accumulation in topsoil (e.g. lightly used, old pastures). The overall process of topsoil nutrient accumulation during succession may occur stepwise. Swamy and Ramakrishnan (1987) were able to correlate the disappearance and decomposition of the herbaceous plant layer after 4 years in successional plots with a simultaneous increase in soil N, P and base cations in northeast India. Base cation enrichment in the topsoil during the course of succession leads to higher fine root allocation near the soil surface (Berish and Ewel, 1988), which is thought to demonstrate a switch from abiotic to biotic nutrient sources (e.g. Vitousek and Reiners, 1975).

Table 24.4. *Dry weight of humus layer (including the litter layer) of different secondary and old-growth forests in Latin America*

Forest type	Age (years)	Region	Country	Altitude (m a s l)	Rainfall (mm yr^{-1})	Soil pH (salt)	Humus layer (t ha^{-1})	Reference
Early secondary forest	0–10	Los Tuxtlas	Mexico	200	>4000	6.3 H_2O	4.3	Hughes *et al.*, 1999
Late secondary forest	10–50	Los Tuxtlas	Mexico				6	Hughes *et al.*, 1999
Old-growth forest		Los Tuxtlas	Mexico				6	Hughes *et al.*, 1999
Early secondary forest	2	Magdalena Valley	Colombia	50	3152	3.8	7	Fölster *et al.*, 1976
Early secondary forest	5	Magdalena Valley	Colombia			3.6	16	Fölster *et al.*, 1976
Late secondary forest	16	Magdalena Valley	Colombia			3.7	24	Fölster *et al.*, 1976
Old-growth forest		Magdalena Valley	Colombia			3.8	53	Fölster *et al.*, 1976
Early secondary forest	5	Eastern Amazonia	Brazil	39	2442	4.4	7.8	Denich, 1989
Old-growth forest		Eastern Amazonia	Brazil	24	2762	3.5	14.2	Klinge, 1998
Early secondary forest	8	Talamanca	Costa Rica	2900	2700	4.9	24	Hölscher, unpublished
Late secondary forest	50	Talamanca	Costa Rica			3.5	72	Hölscher, unpublished
Old-growth forest		Talamanca	Costa Rica			3.4	147	Hölscher, unpublished
Old-growth forest (high)		Gran Sabana	Venezuela	1200	2063	4.25	48.3	Priess *et al.*, 1999
Old-growth (cloud) forest		Andes	Venezuela	2300	1500	3.5	38	Steinhardt, 1979

There are few data on nutrient outputs via leaching from secondary forests. For young fallows in eastern Amazonia, Hölscher *et al.* (1997a) demonstrated that nitrate leaching was already negligible two years after field abandonment. Leaching losses were assumed to be low because of a high root uptake. In southern Venezuela large increases in leachate concentrations of K, Mg and nitrate occurred during the first two years following forest conversion (Uhl and Jordan 1984). After 5 years concentrations were not significantly different from those for an undisturbed control.

In summary, the rate of aboveground biomass accumulation is closely related to former land use intensity and soil fertility. Accumulation of nutrients in phytomass is relatively rapid during the initial stages of succession and slows down as succession proceeds. The recovery of the total nutrient stock in soil and vegetation depends on the rate of external inputs to the ecosystem via bulk precipitation, weathering and nitrogen fixation and may require a comparable time frame as the phytomass recovery. Detailed data on nutrient fluxes during forest recovery in the tropics are rare and only a few studies have surveyed total changes in nutrient stocks in relation to the intensity of former land uses.

HYDROLOGICAL CHANGES DURING FOREST RECOVERY

Rainfall partitioning

In vegetation canopies, precipitation is divided into throughfall, stemflow and interception loss. The importance of these water fluxes is significantly influenced by the canopy water storage capacity. In vegetation types with a high canopy water storage the rain water is more likely to be retained and evaporated back into the atmosphere whereas in vegetation types with a low water storage stemflow and throughfall tend to increase. Due to their large area leaves contribute substantially to the canopy water storage capacity. The maximum thickness of the water film per unit of leaf area is influenced by the leaf inclination and surface structure. Leaf morphological characteristics of tree species can change during forest recovery (Kappelle and Leal, 1996) but changes in water storage per unit leaf area during succession are not yet documented. As the leaf area of young secondary forests and old-growth forests does not differ significantly in many successions one must assume that water storage on leaves does not change very much during succession after canopy closure. The bark structure and area can also influence the storage capacity. Herwitz (1985) showed that species with large woody surface areas and small projected crown areas are capable of storing the greatest depth equivalents of rainwater under heavy rainfall conditions. The significance of bark water storage capacity increases under turbulent conditions. Herwitz (1985) measured 2.2 to 8.3 mm interception storage capacity per unit projected crown area of different tree species in tropical Australia with the height being dependent on the bark structure. Some species, such as oaks, are known to produce a rougher bark surface with age that would imply an increasing storage capacity. Epiphytes in the canopy can also contribute significantly to the water storage capacity. Mosses that are abundant in montane cloud forest can take up two to five fold their own dry weight (Pócs, 1980; Nadkarni, 1984; Frahm, 1990). Pócs (1980) estimated the canopy water storage capacity of a moss-rich primary forest in Tanzania at 5 mm. The same value was found by Veneklaas and van Ek (1990) in a mossy old-growth upper montane cloud forest in Colombia. For a detailed analysis of the hydrological implications, see Bruijnzeel, this volume. The succession of epiphytes in recovering forests in relation to species composition and biomass is largely unknown. In the cloud forest of Monteverde, Costa Rica, Nadkarni (2000) found that the rate of recovery of the epiphytes was extremely slow even on experimentally stripped branches within the intact canopies of mature trees. Summing up the contribution of leaves, bark and epiphytes to the total canopy water storage capacity we can safely assume that the capacity increases during succession.

Only a few studies are available on rainfall partitioning in successional forests. In two stands in eastern Amazonia throughfall averaged only 65% of bulk precipitation in a diverse secondary vegetation dominated by woody species and 38% in a monospecific *Phenakospermum guyannense* (Strelitziaceae) stand (Hölscher *et al.*, 1998). The values obtained in Amazonian old-growth forests range between 82% (Tobón Marin, 1999) and 91% (Lloyd and Marques Filho, 1988) (Table 24.5). The main reason for the differences was the importance of stemflow in the early successional stages. Stemflow in the young vegetation stands was estimated to account for as much as 23 and 41% of gross rainfall. For Amazonian old-growth forests much lower values were reported for the stemflow fraction (1–8%; Ubarana, 1996; Jordan and Heuveldop, 1982). Relatively high stemflow in Amazonian old-growth forests was measured for individuals having low diameters (Jordan and Heuveldop, 1981; Lloyd and Marques Filho, 1988). In Costa Rican secondary vegetation Raich (1983) also measured a high stemflow rate on plants belonging to the Heliconiaceae, a family characterised by species with banana-like leaves (like *Phenakospermum guyannense*) which is a typical component of early stages of forest succession.

As pastures often have a low leaf area, inclined leaves, no bark and virtually no epiphytes, the canopy storage capacity is assumed to be low. McNaughton and Jarvis (1983) conclude from studies in the temperate zone that the rate of evaporation of intercepted water is usually somewhat higher from forests than from grassland and arable crops. In a typical secondary vegetation stand in eastern Amazonia interception was estimated to amount to 227 mm yr^{-1} which corresponds to 12% of gross rainfall (Hölscher *et al.*,

Table 24.5. *Rainfall partitioning in Amazonian vegetation types*

Vegetation type	Region	Country	Period (years)	Rainfall (mm)	Throughfall (%)	Stemflow (%)	Interception (%)	Reference
Old-growth forest	Rio Negro	Venezuela	2	7328	87	8	5	Jordan and Heuveldop, 1981
Old-growth forest	Manaus	Brazil	1.1	2721	91	2	7	Lloyd and Marques, 1988
Old-growth forest	Marabá	Brazil	1.2	1650	86	1	13	Ubarana, 1996
Old-growth forest	Ji-Paraná	Brazil	1.7	3564	87	1	12	Ubarana, 1996
Old-growth forest	Caquetá	Colombia	1.3	3274	87	1	12	Tobón Marin, 1999
Old-growth forest	Caquetá	Colombia	1.3	3293	87	1	12	Tobón Marin, 1999
Old-growth forest	Caquetá	Colombia	1.3	3158	86	1	13	Tobón Marin, 1999
Old-growth forest	Caquetá	Colombia	1.3	3121	82	1	17	Tobón Marin, 1999
Secondary vegetation (diverse)	East Amazonia	Brazil	1	1956	65	23	12	Hölscher *et al.*, 1998
Secondary vegetation (monospecific)	East Amazonia	Brazil	1	1956	38	41	21	Hölscher *et al.*, 1998

1998). In Amazonian lowland old-growth forests values between 5% (184 mm yr^{-1}) and 17% (530 mm yr^{-1}) were reported (Tobón Marin, 1999). In a cloud free, lower montane region of Panama, Cavelier *et al.* (1997) estimated the interception to be 37% (1299 mm yr^{-1}) of gross rainfall for the rainforest that had an abundance of epiphytes. The results of rainfall partitioning studies in tropical montane forests were recently summarised by Bruijnzeel (2002). Evident in that review was that little information is available for recovering forests except for that provided by a single study in a lower montane cloud forest in Costa Rica (Fallas, 1996).

Elsewhere in Amazonia Godsey and Elsenbeer (2002) found similarly reduced *Ks* values for secondary vegetation in the earliest stages of regrowth on an abandoned pasture. The effect of soil compaction by the former grazing was found at a depth of 12.5 cm but not at 20 cm. Based upon a comparison of *Ks* values with median rainfall intensities, the authors predicted the frequent occurrence of shallow lateral subsurface flow and saturation overland flow at this particular location. Giambelluca (2002) demonstrated how values of topsoil *Ks* under secondary vegetation in northern Vietnam were initially low (*c*. 20 mm h^{-1}) but increased rapidly to about 60 mm h^{-1} within 8 years after abandonment. This was followed by a more gradual increase to values similar to those associated with isolated patches of old-growth forest (80–90 mm h^{-1}) within 25–30 years of regeneration.

Studies involving models of the canopy water balance including interception are available for a series of lowland old-growth forest sites (Calder *et al.*, 1986; Lloyd *et al.*, 1988; Hutjes *et al.*, 1990; Ubarana, 1996; Jetten, 1996, Tobón Marin, 1999). It has been shown that models of the Rutter (Rutter *et al.*, 1971) and Gash type (Gash, 1979) can give reasonable results under 'continental' climatic conditions (Lloyd *et al.*, 1988). Even so, Calder *et al.* (1986) and Schellekens *et al.* (1999) found the Rutter type models gave unsatisfactory results under maritime tropical climatic conditions. Jetten (1996) introduced a multi-layer approach based on the Rutter model. For these models an accurate estimation of the storage capacity of the canopy at saturation is required. The stochastic model (Calder, 1986) that was also successfully applied under tropical conditions requires values for raindrop volumes as explicit input variables (Calder *et al.*, 1986; Hall *et al.*, 1996). Modelling studies for tropical montane forests are scarce (Juvik and Nullet, 1995; Hafkenscheid, 2000) despite the high importance of interception in this forest type (Veneklaas and van Ek, 1990; Bruijnzeel and Proctor, 1995; Bruijnzeel, 2002). Aboal *et al.* (1999) gave an example for laurel forests on Tenerife, Canary Islands, that are also exposed to orographic clouds with low radiation and nearly saturated conditions occurring during post-saturation rainfall periods. Modelling studies for tropical pastures and secondary forests are missing. Waterloo (1994) gave an example for fire climax

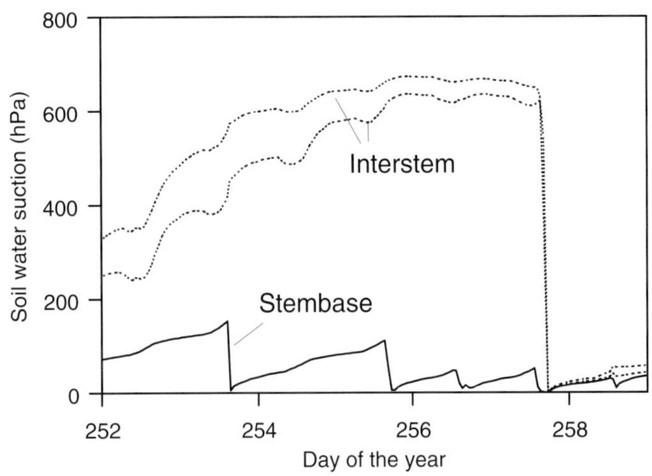

Figure 24.2 Soil water suction (10 cm depth) close to the stembase of a species with high stemflow (*Phenakospermum guyannense*) and in the interstem area in a 3 year old secondary vegetation, eastern Amazonia

grassland on Fiji where he obtained 4% interception for tall grass and a further 7% from dead grass. Due to a supposed increase in water storage capacity and a closer vegetation-atmosphere-coupling we assume that re-evaporation of intercepted water increases during forest succession but we do not know after what time period the interception returns to the old-growth forest value.

Soil water dynamics

The use of heavy machinery during land preparation and animal trampling on pastures cause soil compaction and reduced water infiltrability. Martins *et al.* (1991) observed that physical soil degradation through forest clearance and subsequent cultivation was partly reversed under woody fallow vegetation. Giambelluca (1996) studied the soil saturated conductivity (K_s) of forests, pastures, crops and successional forests of different ages in northern Thailand and eastern Amazonia *in situ*. He found that K_s was significantly reduced on agricultural soils but did not differ under successional forests compared to exploited old-growth forests. On the measured sites saturated conductivity was rarely exceeded by gross rainfall intensities.

However, the funnelling of rainwater via stemflow can induce large spatial differences in water fluxes reaching the forest floor and in soil water content. Hölscher (1995) measured the soil matric potential at the stem base of a species with high stemflow (*Phenakospermum guyannense*) and in the area between stems (Figure 24.2). Soil at the stem base remained wet and even small rainfalls were channelled to the stem base. The area between stems dried out quickly and was only wetted by larger rainfall events. High soil water content that may be induced by

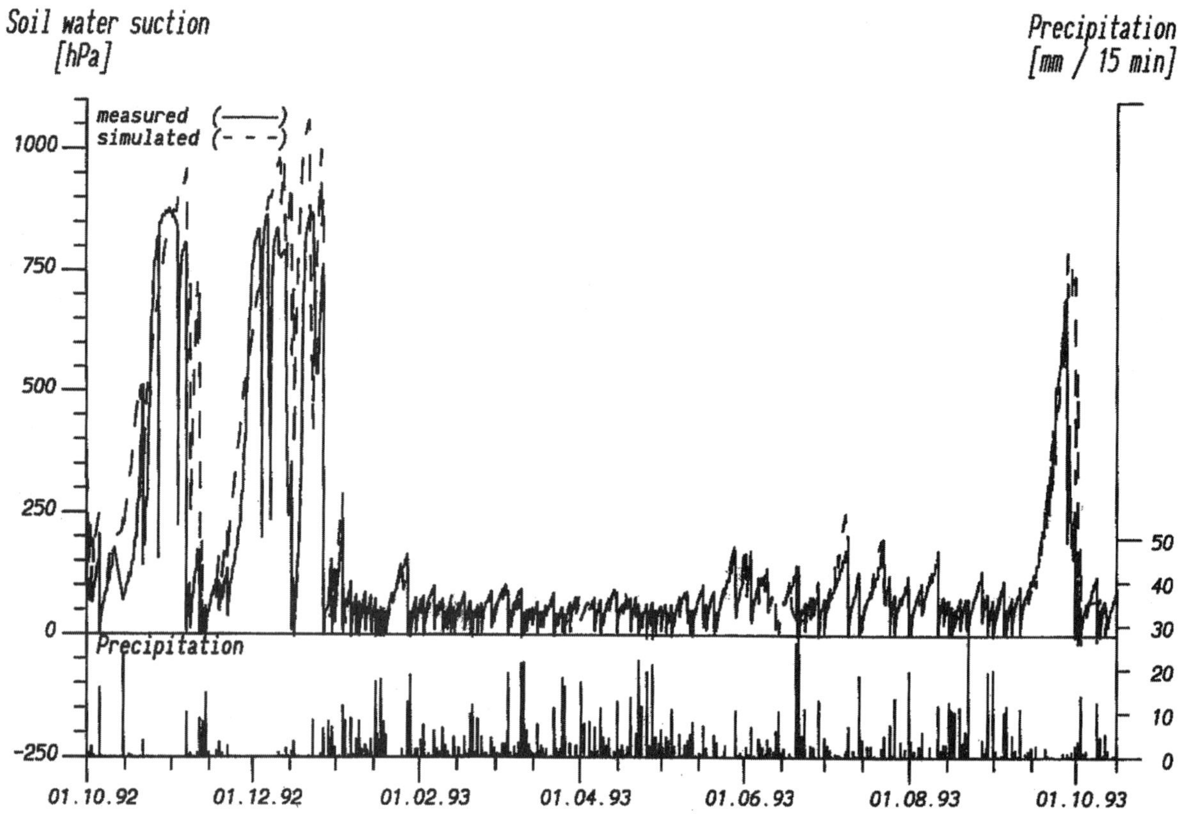

Figure 24.3 Measured and simulated soil water suction (10 cm depth) of an old-growth forest, eastern Amazonia. (After Klinge, 1998).

stemflow would favour rapid water flux in macropores. Herwitz (1986) showed that even overland flow can be induced by stemflow under conditions of high rainfall intensities. Elsewhere in northern Queensland, Bonell (1993) reported that subsurface stormflow travelled preferentially along root channels. Another reason for preferential water movement might be the hydrophobic state of soils caused by frequent burns in pastures and regrowing forests. As studies are not available, the importance of the changes in stemflow, root channels and burn frequency during forest succession for the generation of preferential flow can hardly be estimated (cf. Malmer *et al.*, this volume, and Bonell *et al.*, both this volume).

Soil water modelling studies based on Darcy's law are available for old-growth forest sites in Guyana (Jetten, 1994), a sequence of forest cutting, burning and forest plantation establishment in eastern Amazonia (Klinge, 1998), and four different old-growth forest sites in western Amazonia (Tobón Marin, 1999). It was demonstrated that soil water contents and matric potentials could be predicted adequately by the models (Figure 24.3) and thus information on the soil water balance including drainage can be acquired. However, the Darcy approach requires high conductivity close to water saturation to include rapid movement in macropores. A way to overcome this problem could be modelling with dual permeability approaches that differentiate between preferential flow paths and soil matric flow (Gerke and van Genuchten, 1993). These types of soil water modelling studies are not yet available for tropical sites.

Thick humus layers as may occur in late successional stages on sites low in soil pH and nutrient availability or at high elevations exhibit high water storage capacities. Tobón Marin (1999) estimated that in a forest on a Tertiary sedimentary plain in Amazonia with a litter layer up to 35 cm thick, 35% of the total uptake was obtained from the humus layer. The importance of the humus layer is confirmed by a study in a temperate beech-oak forest where 37% of the yearly total stand water uptake was derived from an 11 cm thick humus layer (Leuschner, 1998). This important water reservoir is absent in the first few years of forest succession following agricultural land use as sufficient time must elapse before surface humus accumulates.

Using a soil water balance approach Nepstad *et al.* (1994) showed the importance of water uptake by deep roots in an old-growth forest and pastures in eastern Amazonia. In both ecosystem types >75% of the water extracted from the soil during the dry period was from below 2 m depth. More water was extracted from the forest soil. For three pairs of Amazonian pastures and old-growth forests, soil water studies consistently showed larger seasonal variation in soil water content beneath forest compared to pasture which was attributed to a higher water use by the forest (Hodnett *et al.*, 1995; 1996). An estimate of the water balance using micrometeorological techniques revealed that water uptake from deeper soil layers can also be important under a young secondary forest in eastern Amazonia (Hölscher *et al.*, 1997c). This finding was supported by measurements on soil water dynamics at these locations (Sommer *et al.*, 1998). Direct methods for the measurement of fine root water uptake are now available (Senock and Leuschner, 1999) and could give useful information on the plant water uptake from different soil depths.

Temporal and spatial patterns of below-ground resource utilisation still remain largely unknown in old-growth and secondary forests. Excavations to study rooting depth are limited by the intricate nature of root systems and by high species diversity and density in old-growth and also secondary forests. In any case, root distribution studies provide little information about water utilisation rates or patterns of uptake because the presence of roots alone is not a reliable indicator of root activity. Stable isotopes provide an alternative means of evaluating integrated water uptake by forest plants (Goldstein *et al.*, 1996). Differences in stable isotope concentration which occur through the soil profile have been used to indicate different patterns of exploitation by evergreen and deciduous species in Panama and, also, to show that light-demanding gap species with higher transpiration rates exploit deeper soil water than shade-tolerant shrub species (Jackson *et al.*, 1995). The different isotopic signature, and predominantly surface distribution of soil water supplied as fog drip, was used by Dawson (1998) to conclude that in the coastal Redwood forest of California, compared to the larger trees the transpiration of the smaller trees is derived to a much greater extent from surface soil water resources.

Controls of transpiration

A useful framework for the consideration of the controls of transpiration and the changes in those controls as tropical croplands return to forest cover is the Monteith version of the Penman equation (Eqn 24.1) (Monteith, 1965).

$$E = \frac{\Delta(R_n - G) + \rho c_p D g_a}{\lambda[\Delta + \nu(g_a/g_a + g_c)]} \qquad (24.1)$$

where

- E is rate of water loss (kg m^{-2} s^{-1})
- Δ is rate of change of saturated vapour pressure with temperature (mb °C^{-1})
- R_n is net radiation (W m^{-2})
- G is soil heat flux (W m^{-2})
- ρ is air density (kg m^{-3})
- c_p is specific heat of air at constant pressure (1005 J kg^{-1})
- D is vapour pressure deficit (mb)
- λ is latent heat of vaporisation (2465000 J kg^{-1})
- v is the psychrometric constant (0.66)
- g_c is surface conductance (m s^{-1})
- g_a is aerodynamic conductance (m s^{-1}).

The equation allows us to focus on key features of vegetation which can significantly influence transpiration (or evaporation) over and above that caused by fluctuations in weather variables such as radiation, temperature and air humidity deficit. The solar reflection coefficient, *albedo*, has a strong control on the net radiation available at the vegetation surface to provide energy to drive evaporation. With windspeed, the *aerodynamic roughness* of the vegetation, represented in the Penman-Monteith equation by the aerodynamic conductance, g_a, determines the turbulent interchange of heat, water vapour and momentum of the vegetation with the atmosphere in both wet and dry conditions. *Surface conductance* (g_c) describes the physiological control of transpiration for dry canopy conditions. For portions or all of the canopy that have been wetted by rain the surface conductance will be infinitely high and will not limit evaporation losses. The surface conductance, g_c, can be equated to the leaf stomatal conductance (g_s) of all foliage summed by multiplying the average g_s by the leaf area index (L) or the average leaf conductance of different categories of foliage by their fraction of the total L and summing the total (Roberts *et al.*, this volume).

Albedo

The important aspect of vegetation which influences its albedo or reflection coefficient is the amount of foliage and its vertical distribution. Both of these features combine to capture radiation within the vegetation canopy. As vegetation becomes taller there is more likelihood of radiation being trapped and not reflected back into the atmosphere. Albedo can range from about 25% for vegetation below 0.5 m high to 10% in vegetation taller than 10 m (Monteith and Unsworth, 1990).

Table 24.6 summarises much of the available information on albedo of tropical land surface covers. There is now substantial information about the albedo properties of old-growth tropical forest and also for pastures which have replaced them. There is also some information available for albedo of many other crops established on cleared tropical forest areas and how albedo changes when abandoned pastures and slash and burn agriculture revert to secondary forest. Nevertheless more information on albedo changes of recovering forests and the relationship with leaf area index would be very valuable. Information to enable the comparison of old-growth tropical rainforest and pasture comes from the ABRACOS project (Culf *et al.*, 1995). In this study the albedo of three old-growth tropical forests located across the Brazilian Amazon (Ji-Paraná, Rondônia; Manaus, Amazonia and Marabá, Pará) was compared with that from adjacent pasture areas. The

Table 24.6. *Mean albedo values for tropical land covers*

Vegetation		Country	Albedo	Reference
Rainforest				
Lowland tropical rainforest		Brazil	0.134	Culf *et al.*, 1995
Lowland tropical rainforest		Nigeria	0.120	Oguntoyinbo, 1970
Lowland tropical rainforest		Thailand	0.130	Pinker *et al.*, 1980
Soil				
Bare soil		Brazil	0.170	Giambelluca *et al.*, 1997
Bare soil		Nigeria	0.140	Oguntoyinbo, 1970
Irrigated bare soil		Thailand	0.085	Giambelluca *et al.*, 1999
Agriculture				
Pasture		Brazil	0.180	Culf *et al.*, 1995
Burned pasture		Brazil	0.080	Fisch *et al.*, 1994
Pasture		Brazil	0.175	Giambelluca *et al.*, 1997
Short grass		Nigeria	0.210	Oguntoyinbo, 1970
Tall grass		Thailand	0.150	Pinker, 1982
Mature cassava		Brazil	0.176	Giambelluca *et al.*, 1997
Fallow rice		Thailand	0.141	Giambelluca *et al.*, 1999
Harvested barley		Thailand	0.163	Giambelluca *et al.*, 1999
Harvested corn		Thailand	0.167	Giambelluca *et al.*, 1999
Irrigated beans		Thailand	0.125	Giambelluca *et al.*, 1999
Secondary vegetation				
Slashed vegetation		Brazil	0.142	Giambelluca *et al.*, 1997
Burned slash		Brazil	0.097	Giambelluca *et al.*, 1997
Secondary vegetation		Nigeria	0.150	Oguntoyinbo, 1970
Secondary vegetation	(0.5 year)	Brazil	0.173	Giambelluca *et al.*, 1997
Secondary vegetation	(1 year)	Brazil	0.161	Giambelluca *et al.*, 1997
Secondary vegetation	(1 year)	Brazil	0.157	Giambelluca *et al.*, 1997
Secondary vegetation	(2 year)	Brazil	0.162	Giambelluca *et al.*, 1997
Secondary vegetation	(2 year)	Brazil	0.172	Giambelluca *et al.*, 1997
Secondary vegetation	(2 year)	Thailand	0.134	Giambelluca *et al.*, 1999
Secondary vegetation	(3 year)	Thailand	0.171	Giambelluca *et al.*, 1999
Secondary vegetation	(8 year)	Thailand	0.115	Giambelluca *et al.*, 1999
Secondary vegetation	(10 year)	Brazil	0.135	Giambelluca *et al.*, 1997
Secondary vegetation	(25 year)	Thailand	0.135	Giambelluca *et al.*, 1999

mean forest albedo was 0.134 while that of the pasture was 0.180 (Table 24.6). Culf *et al.* observed a seasonal pattern of albedo for the forest that was related to seasonal changes in soil moisture rather than seasonal patterns in cloudiness or sun angle. The relationship showed that albedo is highest when soil moisture is lowest. As no bare soil will be visible in dense tropical forest, the relationship between soil moisture and albedo must be indicative of a response of the vegetation to the soil moisture content. The relationship is probably due to changes in leaf water status. Studies on individual leaves have shown that in general, leaf dehydration leads to an increased reflectance (e.g. Carlson *et al.*, 1971; Mooney *et al.*, 1977). Other factors such as leaf angle may also play a part. Burning will have a major effect on albedo as will the irrigation of bare soils (Table 24.6).

Giambelluca *et al.* (1997) made albedo measurements over a range of agricultural and forest regeneration sites in the eastern Amazon. They showed that the albedo of 10 year old secondary vegetation was similar to that measured for old-growth tropical forest in other studies e.g. Culf *et al.* (1995). The albedo of relatively new secondary growth (0–2 years old) is about 0.03 higher than older secondary vegetation. Sites typical of agricultural land in the region have albedoes in the range 0.170 to 0.176. Giambelluca *et al.* (1997) also compared surface temperature (T_s) to air temperature (T_a) in a range of the same vegetation types. Typical values of temperature difference of actively transpiring forest are in the order of 1 K (Bastable *et al.*, 1993). For deforested land covers in the eastern Amazon, Giambelluca *et al.* (1997) showed large values of T_s-T_a to occur, generally declining with increasing vegetation age and height. Further studies of this nature have been made by Giambelluca *et al.* (1999) in northern Thailand and showed that although albedo of irrigated exposed soil was lowest (0.085), other agricultural crops were lower and higher than values published for old-growth rainforest (range 0.125–0.167). Secondary vegetation aged around 25 years had an albedo of 0.135, i.e. very similar to old-growth forest.

From their measurements, Giambelluca *et al.* (1999) showed that the ratio of net radiation to incoming solar radiation, R_{net}/R_{sh}, for intermediate and advanced secondary vegetation lies within the range of old-growth forest values. The ratio generally increases with the age of the secondary vegetation. For deforested land covers in northern Thailand the difference between surface temperature and air temperature (T_s-T_a) is quite high for agricultural sites, especially those with dry soil surfaces. Surface temperatures are lower for intermediate and advanced secondary vegetation than for early secondary forest growth and agricultural crops due to more effective turbulent exchange of heat and water vapour of the taller and presumably deeper rooted vegetation.

Recently, however, Giambelluca (2002) reported increased sapflow rate in trees of an isolated forest patch in northern Vietnam. He attributed this to the warmer and drier conditions prevailing in surrounding clearings. Similarly, Giambelluca (2002) interpreted the very high (i.e. exceeding values for old-growth forest) evaporation rates found for young secondary vegetation in eastern Amazonia and northern Thailand (Giambelluca *et al.*, 2000; cf. Sommer *et al.*, 2002) in terms of positive heat advection from neighbouring cleared areas (see also below).

In some cases forest recovery may be occurring in areas in which short vegetation with shallow roots predominate, e.g. tropical pastures or crops. In dry seasons when soil water supply has become limited it is possible for advected energy from dry pasture areas to be used to augment the direct radiant energy on to the forest to enhance evaporation. The enhancement of evaporation from irrigated crops with advected energy has been reported commonly from arid and semi-arid regions. It has also been reported for forests in temperate regions (e.g. Lindroth and Iritz, 1993; Hall *et al.*, 1998; Calder, 1998). In contrast, there is little information to determine the importance of advection in heterogeneous land covers in the humid tropics.

Aerodynamic roughness

Aerodynamic roughness of vegetation affects the turbulent transfer of water vapour, sensible heat and momentum transfer. The roughness of the vegetation has a strong influence on the evaporation rates of water held on vegetation canopies during and after rainfall, and exerts some degree of control over transpiration and also the surface temperature of the vegetation which will influence the net radiation balance of the vegetation.

Aerodynamic conductance (g_a) in the Penman–Monteith equation incorporates a laminar boundary layer conductance for a 'big leaf' and a conductance that accounts for turbulent transport from the leaf to a reference point where windspeed and air saturation deficit are measured. The size of g_a is determined by windspeed and canopy characteristics such as plant density and height, which determine the zero plane displacement, d, and the roughness length, z_0 (Eqn 24.2).

$$g_a = \frac{k^2 u}{\{\ln[(z-d)/z_0]\}^2} \quad \mathrm{m\,s^{-1}} \qquad (24.2)$$

where

d is the zero plane displacement (m)
z_0 is the roughness length (m)
k is von Kármán's constant (0.41)
u is windspeed (m s^{-1})
z is the height at which windspeed is measured (m).

Roughness length is a measure of the aerodynamic roughness of the vegetation surface and is defined by an equation for the logarithmic wind profile, such that for conditions of neutral atmospheric thermal stability, the windspeed equals zero at height z_0, the roughness length. The magnitude of z_0 depends on the size and the spacing of elements making up the vegetation surface. For vegetation, z_0 is normally about one-tenth of the average vegetation height.

Unfortunately the majority of measurements made of aerodynamic properties of vegetation are from temperate regions where detailed micro-meteorological studies have been made in different vegetation types. Kelliher *et al.* (1993) compared the range of values of g_a of grass in many temperate regions at a windspeed of 3 m s^{-1}. The highest value observed is for a tussock grass in New Zealand and is 70 mm s^{-1} (1 mm s^{-1} $\approx$ 40 mmol m^{-2} s^{-1}); the particular roughness of this type of grass was thought to contribute to its high g_a. For short vegetation such as grassland g_a is generally less than 30 mm s^{-1} over the range of windspeeds normally observed. Two studies are particularly relevant for tropical grasslands in the Amazon region. From the data presented by Wright *et al.* (1992) from the Fazenda Dimona site near Manaus in which the grass was around 0.3 m tall, g_a at a windspeed of 3 m s^{-1} would be expected to be about 27 mm s^{-1}. Grace *et al.* (1998) made measurements over pasture at Ji-Parana, Rondonia and at a windspeed of 3 m s^{-1}, g_a was around 19 mm s^{-1}. Studies of the aerodynamic properties of tropical grassland dominated by mission grass (*Pennisetum polystachyon*) have been made in wet and dry seasons in Fiji (M. J. Waterloo, pers. comm. 2000). The leaf area index in the wet season was 2.0 and the height was 1.6 m. In the dry season the height and leaf area index were less. Daily average g_a was 21 and 25 mm s^{-1} for the wet and dry seasons respectively. Wind speeds varied diurnally with the

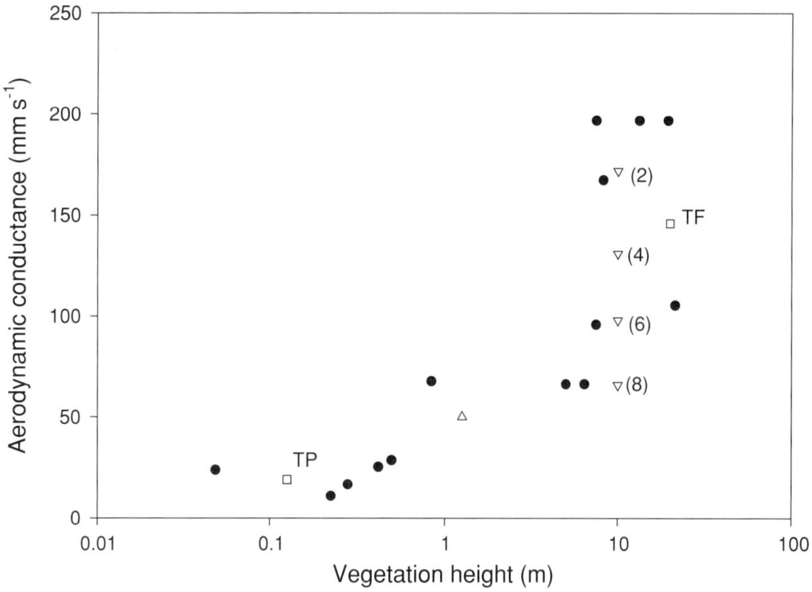

Figure 24.4 Variation of aerodynamic conductance with vegetation height for a range of grass and forest types (•, after Kelliher *et al.*, 1993; TP, tropical pasture, after Grace *et al.*, 1998; □, tropical grassland, Fiji (M. J. Waterloo, unpublished data); TF, tropical forest, after Shuttleworth, 1989; ∇ Spruce trees at various spacing (m) shown in brackets, after Teklehaimanot *et al.*, 1991).

highest values occurring in the middle of the day, consequently g_a was highest around midday reaching values in the region of 50 mm s^{-1}.

The aerodynamic conductance of tall forest is substantially more than for short vegetation. There are a number of direct measurements of the roughness parameters of old-growth forest from which aerodynamic conductances can be calculated. Shuttleworth (1989) presents data for the Reserva Ducke at Manaus in the central Amazon. The roughness length, z_0, was 2.2 m and the zero plane displacement, *d*, was 30.9 m. Using these values and for a windspeed of 3 m s^{-1}, g_a would be estimated as 146 mm s^{-1}.

There are no reliable values for the aerodynamic conductance of the Marabá pasture used in the ABRACOS studies. This is unfortunate because, particularly in the dry season, this pasture had a marked tussocky appearance. This clumping was accentuated by invading shrubs and palms.

There have been no direct observations made in previously-cropped tropical forest lands in which secondary forest is invading from which the roughness parameters can be derived and from which aerodynamic conductance can be calculated. As an approximation, it would not be unreasonable to use the values of z_0 and *d* reported by Shuttleworth (1989) with assumed vegetation heights of a developing secondary forest to estimate g_a. Such a calculation shows that g_a increases rapidly from 40 mm s^{-1} when the vegetation is 1 m tall to 3 times that value when the secondary forest reaches a height of 10 m. What is not at all simple to predict is the influence on g_a of the greater patchiness or clumping that might occur in some secondary forests. Veen *et al.* (1991) suggested that aerodynamic roughness, and therefore atmospheric turbulence and surface exchange, could be much enhanced in heterogeneous landscapes due to the spatial contrasts in vegetation height (cf. Klaassen, 1992). The studies of Teklehaimanot *et al.* (1991) showed that the boundary-layer conductance, g_a of a stand of trees depends on the tree density. g_a increased from 66 mm s^{-1} with 156 trees ha^{-1} to 172 mm s^{-1} when there were 3000 trees ha^{-1}. Figure 24.4 shows the increase in aerodynamic conductance with increasing vegetation height for grass and forests (Kelliher *et al.*, 1993). Added to the diagram are values for short grass and tropical forest derived from studies in Brazil (Shuttleworth, 1989; Grace *et al.*, 1998) and for tropical grass (1.6 m high) in Fiji (M. J. Waterloo, pers. comm., 2000). Examination of Figure 24.4 suggests that values of g_a for secondary forest of around 10 m height would be of the order 60–75 mm s^{-1} but may vary if spacing of trees becomes large.

Surface conductance

With respect to water use by young secondary vegetation, the results obtained by Kuraiji and Paul (1994, cited by Bruijnzeel, 1996) for 3–4-year-old regrowth in Sabah suggest that streamflow from such areas may be about 250 mm yr^{-1} higher than that for catchments with mature forest. Measurements made of throughfall at the same time in the respective forests showed that the differences in total evaporation could be accounted for entirely by differences in rainfall interception (Paul and Kuraiji, 1993, cited by Bruijnzeel, 1996). In contrast, Hölscher *et al.*, (1997c) used Bowen ratio measurements to estimate evaporation from secondary forest vegetation in the eastern Amazon and produced results similar to those from old-growth tropical forests in the central Amazon (Shuttleworth, 1989). Studies reported by Jipp *et al.* (1998) showed very similar amounts and patterns of soil water abstraction by undisturbed forest and 15-year old secondary forest, also in the eastern Amazon.

The similarity of water use by old-growth forest and regrowth forest illustrated by the results of Jipp *et al.* (1998), Hölscher *et al.* (1997c), Giambelluca *et al.* (2000) and Sommer *et al.* (2002) contrasts markedly with interpretations of changes in streamflow associated with regrowth after logging or fires reported from Australia and the USA which have shown marked decreases in catchment streamflow associated with forest regeneration after the initial increases in streamflow *immediately* following logging or burning (e.g. Cornish, 1993; Hornbeck *et al.*, 1997). A number of factors are thought to be responsible for the increased evaporation of regenerating forest. In forests in south-east Australia regeneration tends to be with the same species as the old-growth and the principal factors thought to be responsible for increased evaporation are increased leaf area index and increased stomatal conductance (Watson *et al.*, 1999). It is suspected that the trees comprising old-growth forests can be regarded as somewhat moribund physiologically compared to young trees of the same species which constitute the regenerating forest. In studies in the USA the situation is thought to be somewhat different. Pioneer regenerating species with higher stomatal conductance (g_s) than the logged forest are considered to be responsible for the enhanced evaporation of regrowth after logging (Hornbeck *et al.*, 1997). Likewise, Giambelluca *et al.* (2000) and Giambelluca (2002) found the evaporative fraction of net radiation for 8–25 year-old regrowth in eastern Amazonia and northern Thailand to be higher even than that for old-growth forest (cf. Sommer *et al.*, 2002). They suggested advected heat and enhanced aerodynamic roughness as potential causes.

Malmer (1992) studied rainfall and streamflow after clear-felling of lowland Dipterocarp forest in Sabah, Malaysia, followed by planting with fast-growing *Acacia mangium*. In a catchment with clear-felling, manual log extraction but no burning, runoff increase was nearly 200 mm but this fell to only 80 mm by the third year after treatment. In a catchment that was also burned after logging, the increase in runoff was sustained for up to three years. The timber from the burnt catchment was removed with tractors rather than manually as in the unburnt catchment. Therefore it is not obvious which particular factor is responsible for differences in yield response between burnt and unburnt catchments. Malmer (1992) also reported large increases (up to 522 mm yr^{-1}) in yield between a control catchment and ones with secondary forest that had been cut and burnt. In this comparison the differences in yield were maintained for two years but declined sharply to 89 mm in the third year. In French Guyana, Fritsch (1993) compared stormflow from undisturbed forest with that from areas which had been logged and allowed to regrow. Stormflow on the logged catchments increased considerably to over 200 per cent greater than the controls after treatment. However this difference in stormflow diminished to just a few percent greater than the controls catchments in only two to three years after the treatments were imposed. It would be expected that there would be a different catchment response to re-establishment of vegetation depending on whether re-growth comes from sprouts or from seedling establishment. Unfortunately information is only available for comparison from studies in temperate forests. Kochenderfer and Wendel, 1983; Kochenderfer *et al.*, 1990 reported studies from Fernow Experimental Forest, West Virginia, USA that compared catchment responses (compared to controls) following clear-cutting and subsequent recovery from sprouts or by establishment of seedlings. The effect of clear-cutting was to increase catchment yield by more than 250 mm yr^{-1}. In the case of the cleared catchment on which sprouting was allowed to occur, the difference in streamflow compared to the control fell to between 50 and 100 mm in the following year. In the catchment with seedling establishment providing the new vegetation cover the return to a water yield difference of 100 mm took over 5 years and the decline to a yield difference between 50 and 100 mm took more than 10 years. Based on data from the Fernow Experimental Forest in West Virginia, Trimble *et al.* (1963) discuss the importance of soil depth in determining the time for recovery of streamflow to pre-clearing or pre-thinning levels after forest logging. Soil depth influences the length of time that increases in streamflow persist after cutting. The deeper the soil the longer it takes for roots of new growth or the expanding root system of trees left after thinning to occupy the soil volume. When the soil volume is reoccupied, transpiration reaches maximum levels again, and increases in streamflow disappear. Trimble *et al.* (1963) consider that in a regime of abundant rainfall, a soil depth of 30 cm might be very nearly occupied by roots in one growing season after a heavy cutting; on a soil which is 1.5 to 2.0 m deep, the process might take several years.

Although there is a pressing need for many more studies of the changes in hydrology and contributing processes as secondary forests replace and mature on former agricultural land, it is constructive to reflect on the differences in L and g_s in the sequence from the initial invasion of clearings by seedlings or sprouts to the stage of a mature secondary forest. Table 24.2 shows that L of secondary forest is somewhat less than that of old-growth forest in the same region. However, given that L is reasonably high i.e. greater than ~4 in the case of secondary forest and old-growth forest, any differences are unlikely to contribute to differences in transpiration. Roberts *et al.* (1993) have shown that the foliage at the base of an old-growth forest at Manaus contributes little to the total forest transpiration. It is far more difficult to be categorical

Table 24.7. *Maximum stomatal conductances (g_s) of young trees differing in successional status, growing in open conditions*

Species	Stomatal conductance (mmol m^{-2} s^{-1})	Country	Status	Reference
Solanum crinitum	80	Brazil	Gap invader	Dias-Filho and Dawson, 1995
Vismia guianensis	98	Brazil	Gap invader	Dias-Filho and Dawson, 1995
Macaranga hypoleuca	593	Malaysia	Pioneer	Barker *et al.*, 1997
Cecropia obtusifolia	986	Costa Rica	Pioneer	Poorter and Oberbauer, 1993
Dipteryx panamensis	315	Costa Rica	Climax	Poorter and Oberbauer, 1993
Jacaranda copaia	393	French Guiana	Pioneer	Huc *et al.*, 1994
Goupia glabra	170	French Guiana	Pioneer	Huc *et al.*, 1994
Carapa guianensis	238	French Guiana	Pioneer	Huc *et al.*, 1994
Dicorynia guianensis	230	French Guiana	Late stage	Huc *et al.*, 1994
Eperua falcata	153	French Guiana	Late stage	Huc *et al.*, 1994
Hymenaea coubaril	70	Brazil	Emergent	Langenheim *et al.*, 1984
Hymenaea parviflora	51	Brazil	Emergent	Langenheim *et al.*, 1984

about changes in stomatal conductance (g_s) of the vegetation of abandoned agricultural lands in tropical regions as the cover progresses from scattered invading plants to a dense cover of mature secondary forest. The main reason for this is that no detailed studies which followed closely the changes in stomatal conductance of a substantial number of species comprising the many stages towards the development of a mature secondary forest canopy are available. The picture emerging so far is very incomplete and comes mainly from studies of the physiological functioning of tropical forest tree seedlings in canopy gaps and clearings. Some consistent patterns do emerge, however, but the impression is that there is considerable variability in the magnitude of g_s and its response to environmental variables. Much of this variability is associated with the successional status of the species of the young plants that have been studied i.e. pioneer, climax, etc, over and above variation in g_s due to environmental factors such as radiation, air humidity deficit, temperature and soil moisture status. Therefore the average g_s of an area of developing secondary forest will depend on the successional status and on the proportions of species of different successional categories comprising the vegetation at any one time as well as environmental conditions.

On a diurnal basis, the factors which have the major influence on g_s of young rainforest trees invading previously cleared areas are photon flux density and atmospheric humidity deficit. In this sense the behaviour is similar to that observed in the well illuminated foliage of the largest trees of old-growth forest. There is a tendency for stomata to open rapidly in the morning daylight period, reach their maximum conductance before noon and show a decline in conductance throughout the afternoon as the air vapour pressure deficit increases. The literature is now extensive on the physiological behaviour, e.g. net photosynthesis and stomatal conductance (g_s), of rainforest seedlings growing in understoreys, canopy gaps and in clearings. Table 24.7 presents a selection from this literature giving maximum g_s from a number of species growing in full daylight conditions which might then represent situations where trees are invading abandoned agricultural land. The species listed in Table 24.7 cover a range of successional species from gap-invaders and pioneers to trees which occur typically as adults in the rainforest canopy. The range in g_s is very large, with the highest values coming from pioneer species e.g. *Macaranga* and *Cecropia* while in contrast some of the seedlings of canopy emergents exhibit low g_s. High g_s has also been found in other pioneers of the tropical rainforest (Chiarello *et al.*, 1987; Alexandre, 1991). Roberts *et al.* (1990) found in foliage in old-growth forest in the Reserva Ducke in Manaus that species with the highest initial g_s values in the day showed the largest degree of reduction in g_s as air vapour pressure deficit increased. A similar relationship has been observed in rainforest tree seedlings growing in open areas (e.g. Clearwater *et al.* (1999)).

Huc *et al.* (1994) compared maximum g_s of young trees in the wet and dry seasons in French Guiana and showed that the largest change in g_s was in the pioneer species which had the highest initial values of g_s. g_s of the pioneer species declined by around 40% on average whereas the trees of climax rainforest species showed a reduction in g_s of only 11%. Sá *et al.* (1996) found that two woody species and the palm, Babaçu (*Orbygnia martiana*), invading a pasture (*Panicum maximum*) at Marabá, Brazil, had very high g_s in the wet season that declined by over 50% in the dry season. Huc *et al.* (1994) thought that the modest decline in g_s of non-pioneer species in the dry season in French Guiana was linked to a deeper rooting pattern than that of pioneer species. In the same region, Alexandre (1991) found a deeper rooting system in the late successional stage species *Eperua falcata* compared to two pioneer species *Goupia glabra* and *Trema micrantha.*

Studies reported earlier in this chapter exploited chronosequences of secondary forests to examine the dynamics of biomass, leaf area index and nutrient accumulation. Unfortunately, studies of a similar type that examine fluctuations in evaporation and its components and the detailed ecophysiology of a range of species over a range of representative ages of secondary forest still remain to be made. Without such studies we are unable to provide the insight into how controls on fluxes of water by secondary vegetation influence its nutrient and biomass accumulation.

CONCLUSIONS

Forest recovery after the abandonment of agricultural land use is largely influenced by the previous land use intensity. The analysed changes in vegetation structure, nutrient pools and the hydrological cycle on sites subjected to light use approach values of old-growth forests at different time scales:

- Differences in plant species composition may persist for centuries or even remain irreversible.
- Above-ground biomass recovery was estimated between 70 years on fertile soils and 200 years on poor soils.
- Leaf area index and canopy height reach the levels of old-growth forests within years and decades, respectively.
- Albedo and aerodynamic conductance of recovering forests can approach reference values within decades.
- Stomatal conductance is highly plant species specific and is likely to change significantly with the replacement of pioneer tree species by late stage species. Due to a lack of reliable studies the recovery time can hardly be estimated.
- Topsoil hydraulic conductivities appear to approach values associated with old-growth forest within three decades after moderate soil disturbance upon clearing.

We conclude that the effects of old-growth forest conversion on water and nutrient cycles are less severe where secondary forests can establish and develop. However, the direct evidence on which to base this conclusion is very limited and much further work is needed.

References

Aboal, J. R., Jimenéz, M. S., Morales, D. and Hernández, M. (1999) Rainfall interception in laurel forest in the Canary Islands. *Agricultural and Forest Meteorology* 97, 73–86.

Aide, T. M., Zimmerman, J. K., Herrera, L., Rosario, M. and Serrano, M. (1995) Forest recovery in abandoned tropical pastures in Puerto Rico. *Forest Ecology and Management* 77, 77–86.

Aide, T. M., Zimmermann, J. K., Rosario, M. and Marcano, H. (1996) Forest recovery in abandoned cattle pastures along an elevational gradient in northeast Puerto Rico. *Biotropica* 28, 537–548.

Alexandre, D. Y. (1991) Comportement hydrique au cours de la saison sèche et place dans la succession de trois arbres guyanais: *Trema micrantha, Goupia glabra* et *Eperua grandiflora*. *Annales des Sciences Forestieres* 48, 101–112.

Andriesse, J. P and Schelhaas, R. M. (1987) A monitoring study of nutrient cycles in soils used for shifting cultivation under various climatic conditions in tropical Asia. II. Nutrient stores in biomass and soil – Results of baseline studies. *Agriculture, Ecosystems and Environment* 19, 285–310.

Aweto, A. O. (1981) Secondary succession and soil fertility restoration in south-western Nigeria. II. Soil fertility restoration. *Journal of Ecology* 69, 609–614.

Baar, R. (1997) Vegetationskundliche und -ökologische Untersuchungen der Buschbrache der Feldumlagewirtschaft im östlichen Amazonasgebiet. Goltze, Göttingen. 202 pp.

Barker, M. G., Press, M. C. and Brown, N. D. (1997) Photosynthetic characteristics of dipterocarp seedlings in three tropical rainforest light environments: a basis for niche partitioning? *Oecologia* 112, 453–463.

Bartholomow, W. J., Meyer, J. and Laudelot, H. (1953) Mineral nutrient immobilization under forest and grass fallow in the Yamgambi (Belgian Congo) region. *Pubs. Inst. Natn. Etude Agron. Congo Belge. Serie Scientifique* 57, 1–27.

Bastable, H. G., Shuttleworth, W. J., Dallarosa, R. G., Fisch, G. and Nobre, C. A. (1993). Observations of climate, albedo and surface radiation over cleared and undisturbed Amazonian Forest. International Journal of Climatology 13, 783–796.

Berish, C. W. (1982) Root biomass and surface area in three successional tropical forests. *Canadian Journal of Forest Research* 12, 699–704.

Berish, C. W. and Ewel, J. J. (1988) Root development in simple and complex tropical successional ecosystems. *Plant and Soil* 106, 73–84.

Bonell, M. (1993) Progress in the understanding of runoff generation dynamics in forests. *Journal of Hydrology* 150, 217–275.

Brown, S. and Lugo, A. E. (1990a) Tropical secondary forests. *Journal of Tropical Ecology* 6, 1–32.

Brown, S. and Lugo, A. E. (1990b) Effects of forest clearing and succession on the carbon and nitrogen content of soils in Puerto Rico and US Virgin Islands. *Plant and Soil* 124, 53–64.

Bruijnzeel, L. A. (1991) Nutrient input-output budgets of tropical forest ecosystems. A review. *Journal of Tropical Ecology* 7, 1–24.

Bruijnzeel, L. A. (1996) Predicting the hydrological impacts of land cover transformation in the humid tropics: the need for integrated research. In: *Amazonian deforestation and climate*. Edited by J. H. C. Gash, C. A. Nobre, J. M. Roberts and R. L. Victoria. John Wiley, Chichester, UK. Pp. 15–55.

Bruijnzeel, L. A. (2002). Hydrology of tropical montane cloud forests: a reassessment. In Hydrology and Water Management of the Humid Tropics, ed. J. S. Gladwell, pp. 353–383. *UNESCO IHP Technical Document in Hydrology* No 52. Paris: UNESCO.

Bruijnzeel, L. A. and Proctor, J. (1995) Hydrology and biochemistry of tropical montane cloud forests: What do we really know? In: L. S. Hamilton, J. O. Juvik and F. N. Scatena (eds): *Tropical Montane Cloud Forests*, p. 38–78. Ecological Studies 110. Springer, Berlin.

Buschbacher, R., Uhl, C. and Serrão, E. A. S. (1988) Abandoned pastures in eastern Amazonia. II. Nutrient stocks in the soil and vegetation. *Journal of Ecology* 76, 682–699.

Calder, I. R. (1986) A stochastic model of rainfall interception. *Journal of Hydrology* 89, 65–71.

Calder, I. R. (1998) Water use by forests, limits and controls. *Tree Physiology* 18, 625–631.

Calder, I. R., Wright, I. R. and Murdiyarso, D. (1986) A study of evaporation from tropical rainforest - West Java. *Journal of Hydrology* 89, 13–31.

Cannell, M. G. R. (1982) *World Forest Biomass and Primary Production Data*. Academic Press, London. 391 pp.

Carlson, R. E., Yarger, D. N. and Shaw, R. H. (1971) Factors affecting the spectral properties of leaves with special emphasis on leaf water status. *Agronomy Journal* 63, 186–189.

Cavelier, J., Jaramillo, M., Solis, D. and de León, D. (1997) Water balance and nutrient inputs in bulk precipitation in tropical montane cloud forest in Panama. *Journal of Hydrology* 193, 83–96.

Cavelier, J. and Estevez, J. (1996) Fine root biomass in three successsional stages of an Andean cloud forest in Colombia. *Biotropica* 28, 728–736.

Chiarello, N. R., Field, C. B. and Mooney, H. A. (1987). Midday wilting in a tropical pioneer tree. *Functional Ecology* 1, 3–11.

Clark, D. B., (1996). Abolishing virginity. *Journal of Tropical Ecology* 12, 735–739.

Clausing, G. (1994) Frühe Regeneration und Wiederbesiedlung auf Kulturflächen der Wald-Feld-Wechselwirtschaft im östlichen Amazonasgebiet. Goltze, Göttingen. 151 pp.

Clearwater, M. J., Susilawaty, R., Effendi, R. and van Gardingen, P. R. (1999) Rapid photosynthetic acclimation of *Shorea johorensis* seedlings after logging disturbance in central Kalimantan. *Oecologia* 121, 478–488.

Corlett, R. T. (1994). What is secondary forest? *Journal of Tropical Ecology* 10, 445–447.

Cornish, P. M. (1993). The effects of logging and forest regeneration on water yields in a moist eucalypt forest in New South Wales, Australia. *Journal of Hydrology* 150, 301–32.

Culf, A. D., Fisch, G. and Hodnett, M. G. (1995) The albedo of Amazonian forest and ranchland. *Journal of Climate* 8, 1544–1554.

De Moraes, J. F. L., Volkoff, B., Cerri, C. C. and Bernoux, M. (1996) Soil properties under Amazon forest and changes due to pasture installation in Rondonia, Brazil. *Geoderma* 70, 63–81.

Denich, M. (1989) Untersuchungen zur Bedeutung junger Sekundärvegetation für die Nutzungssystemproduktivität im östlichen Amazonasgebiet, Brasilien. Göttinger Beiträge zur Land- und Forstwirtschaft in den Tropen und Subtropen 46, Goltze, Göttingen. 265 pp.

Dias-Filho, M. B. and Dawson, T. E. (1995) Physiological responses to soil moisture stress in two Amazonian gap-invader species. *Functional Ecology* 9, 213–221.

Dawson, T. E. (1998) Fog in the California redwood forest: ecosystem inputs and use by plants. *Oecologia*, 117: 476–485.

Ewel, J. J. (1976) Litter fall and leaf decomposition in a tropical forest succession in eastern Guatemala. *Journal of Ecology* 64, 293–308.

Ewel, J. J. (1980) Tropical succession: manifold routes to maturity. Special Issue: Tropical succession. *Biotropica* 12, 2–7.

Ewel, J., Benedict, F., Berish, C. and Brown, S. (1982) Leaf area, light transmission, roots and leaf damage in nine tropical plant communities. *Agro-Ecosystems* 7, 305–326.

Ewel, J., Berish, C. and Brown, B. (1981) Slash and burn impacts on a Costa Rican wet forest site. *Ecology* 62, 816–829.

Ewel, J. J., Chai, P. and Tsai, L. M. (1983) Biomass and floristics of three young second-growth forests in Sarawak. *The Malaysian Forester* 46, 347–364.

Fallas, J. (1996) Cuantificacion de la intercepcion en un bosque nuboso, Mte. De los Olivos, Cuenca del Rio Chiquito, Guanacaste, Costa Rica. CREED Technical Note 6, CREED Costa Rica, Tropical Science Center, San José, Costa Rica. 37 pp.

Fearnside, P. M. (1996) Amazonian deforestation and global warming: Carbon stocks in vegetation replacing Brazil's Amazon forest. *Forest Ecology and Management* 80, 21–34.

Fearnside, P. M. and Guimaraes, W. M. (1996) Carbon uptake by secondary forests in Brazilian Amazonia. *Forest Ecology and Management* 80, 35–46.

Fisch, G., Wright, I. R. and Bastable, H. G. (1994) Albedo of tropical grass: A case study of pre- and post-burning. *International Journal of Climatology* 14, 103–107.

Fölster, H., de las Salas, G. and Khanna, P. K. (1976) A tropical evergreen forest site with perched water table, Magdalena valley, Columbia: Biomass and bioelement inventory of primary and secondary vegetation. *Oecologia Plantarum* 11, 297–320.

Fox, J., Krummel, J., Yarnasarn, S., Ekasingh, M. and Podger, N. (1995). Land use and landscape dynamics in northern Thailand: assessing change in three upland watersheds. *Ambio* 24, 328–334.

Fox, J., Truong, D. M., Rambo, A. T., Tuyen, N. P., Cuc, L. T. and Leisz, S. (2000). Shifting cultivation: an old paradigm for managing tropical forests. Bio Science 50, 521–528.

Frahm, J.-P. (1990) The ecology of epiphytic bryophytes on Mt. Kinabalu, Sabah (Malaysia). *Nova Hedwigia* 51, 121–132.

Fritsch, J.-M. (1993) The hydrological effects of clearing tropical rainforest and of the implementation of alternative land uses. *International Association of Hydrological Sciences Publication* 216, 53–66.

Fujisaka, S., Escobar, G. and Veneklaas, G. E. (1998) Plant community diversity relative to human land uses in an Amazon forest colony. *Biodiversity and Conservation* 7, 41–57.

Gash, J. H. C. (1979) An analytical model of rainfall interception of forests. *Quarterly Journal of Royal Meteorological Society* 105, 43–55.

Gentry, A. H. (1988) Changes in plant community diversity and floristic composition on environmental and geographical gradients. *Annals of the Missouri Botanical Garden* 75, 1–34.

Gerke, H. H. and van Genuchten, M. T. (1993) A dual-porosity model for simulating the preferential movement of water and solutes in structured porous media. *Water Resources Research* 29, 305–319.

Giambelluca, T. W. (2002). The hydrology of altered tropical forest. Invited Commentary. *Hydrological Processes* 16, 1665–1669.

Giambelluca, T. W. (1996) Tropical land cover change: Characterising the post-forest land surface. In: T. W. Giambelluca and A. Henderson-Sellers. *Developing Southern Hemisphere Perspectives*, p. 293–318. John Wiley, New York.

Giambelluca, T. W., Fox, J., Yarnasarn, S., Onobutr, P. and Nullet, M. A. (1999) Dry-season radiation balance of land covers replacing forest in northern Thailand. *Agricultural and Forest Meteorology* 95, 53–65.

Giambelluca, T. W., D. Hölscher, Bastos, T. X., Frazão, R. R., Nullet, M. A. and Ziegler, A. D. (1997) Observations of albedo and radiation balance over post-forest land surfaces in Eastern Amazon Basin. *Journal of Climate* 10, 919–928.

Giambelluca, T. W., Nullet, M. A., Ziegler, A. D. and Tran, L. T. (2000). Latent and sensible heat energy flux over deforested land surfaces in the eastern Amazon and northern Thailand. *Singapore Journal of Tropical Geography* 21, 107–130.

Godsey, S. and Elsenbeer, H. (2002). The soil hydrologic response to forest regrowth: a case study from southwestern Amazonia. *Hydrological Processes* 16, 1519–1522.

Goldstein, G., Meinzer, F. C., Sternberg, L. D. S. L., Jackson, P., Cavalier, J. and Holbrook, N. M. (1996) Evaluating aspects of water economy and photosynthetic performance with stable isotopes from water and organic matter. In S. S. Mulkey, R. L. Chazdon and A. P. Smith (eds), *Tropical Forest Plant Ecophysiology*, Chapman and Hall, London. Pp. 244–267.

Grace, J., Lloyd, J., Miranda, A. C., Miranda, H. and Gash, J. H. C. (1998) Fluxes of carbon dioxide and water vapour over a C_4 pasture in south-western Amazonia (Brazil). *Australian Journal of Plant Physiology* 25, 519–530.

Guggenberger, G. and Zech, W. (1999) Soil organic matter composition under primary forest, pasture, and secondary forest succession, Region Huetar Norte, Costa Rica. *Forest Ecology and Management* 124, 93–104.

Hafkenscheid, R. L. L. J. (2000) Hydrology and biogeochemistry of tropical montane rainforests of contrasting stature in the Blue Mountains, Jamaica. PhD Thesis, Vrije Universiteit, Amsterdam.

Hall, R. L., Allen, S. J., Rosier, P. T. W. and Hopkins, R. L. (1998) Transpiration from coppiced poplar and willow measured with sap-flow methods. *Agricultural and Forest Meteorology* 90, 275–290.

Hall, R. L., Calder, I. R., Gunawardena, E. R. N. and Rosier, P. T. W. (1996) Dependence of rainfall interception on drop size: 3. Implementation and comparative performance of the stochastic model using data from a tropical site in Sri Lanka. *Journal of Hydrology* 185, 389–407.

Harcombe, P. A. (1977) The influence of fertilization on some aspects of succession in a humid tropical forest. *Ecology* 58, 1375–1383.

Herrera, R., Jordan, C. F., Medina, E. and Klinge, H. (1981) How human activities disturb the nutrient cycles of a tropical rainforest in Amazonia. *Ambio*, 10, 109–114.

Hertel, D., Leuschner, C. and Hölscher, D. (2003) Size and structure of fine root systems in old-growth and secondary tropical montane forests (Costa Rica). *Biotropica* 35, 143–153.

Herwitz, S. R. (1985) Interception storage capacities of tropical rainforest canopy trees. *Journal of Hydrology* 77, 237–252.

Herwitz, S. R. (1986) Infiltration-excess caused by stemflow in a cyclone-prone tropical rainforest. *Earth Surface Processes and Landforms* 11, 401–412

Hodnett, M. G., Pimental da Silva, L., da Rocha, H. R. and Senna, R. C. (1995) Seasonal soil water storage changes beneath central Amazonian rainforest and pasture. *Journal of Hydrology* 170, 233–254.

Hodnett, M. G., Oyama, M. D., Tomasella, J. and Marques Filho, A. de O. (1996) Comparisons of long-term water storage behaviour under pasture and forest in three areas of Amazonia. In: J. H. C. Gash, C. A. Nobre, J. M. Roberts and R. L. Victoria (eds), *Amazonian Deforestation and Climate*. John Wiley, New York.

Hölscher, D. (1995) Wasser- und Stoffhaushalt eines Agrarökosystems mit Waldbrache im östlichen Amazonasgebiet. Göttinger Beiträge zur Land- und Forstwirtschaft in den Tropen und Subtropen, Goltze, Göttingen. 134 pp.

Hölscher, D., Köhler, L., Leuschner, C. and Kappelle, M. (2003) Nutrient fluxes in stemflow and throughfall in three successional stages of an upper montane rain forest in Costa Rica. *Journal of Tropical Ecology* 19, 557–565.

Hölscher, D., Ludwig, B., Möller, R. F. and Fölster, H. (1997a) Dynamic of soil chemical parameters in shifting agriculture in the Eastern Amazon. *Agriculture, Ecosystems and Environment* 66, 153–163.

Hölscher, D., Möller, R. F., Denich, M. and Fölster, H. (1997b) Nutrient input-output budget of shifting agriculture in Eastern Amazonia. *Nutrient Cycling in Agroecosystems* 47, 49–57.

Hölscher, D., Sá, T. D. de A., Bastos, T. X., Denich, M. and Fölster, H. (1997c) Evaporation from young secondary vegetation in Eastern Amazonia. *Journal of Hydrology* 193, 293–305.

Hölscher, D., Sá, T. D. de A., Möller, R. F., Denich, M. and Fölster, H. (1998) Rainfall partitioning and related hydrochemical fluxes in a diverse and a mono specific (*Phenakospermum guyannense*) secondary vegetation stand in eastern Amazonia. *Oecologia* 114, 251–257.

Holl, K. D. (1999) Factors limiting tropical rainforest regeneration in abandoned pasture: Seed rain, seed germination microclimate and soil. *Biotropica* 3, 229–242.

Hornbeck, J. W., Martin, C. W. and Eagar, C. (1997) Summary of water yield experiments at Hubbard Brook Experimental Forest, New Hampshire. *Canadian Journal of Forest Research* 27, 2043–2052.

Huc, R., Ferhi, A. and Guehl, J. M. (1994) Pioneer and late stage tropical rainforest species (French Guiana) growing under common conditions differ in leaf gas exchange regulation, carbon isotope discrimination and leaf water potential. *Oecologia* 99, 297–305.

Hughes, R. F., Kauffman, J. B. and Jaramillo, V. J. (1999) Biomass, carbon, and nutrient dynamics of secondary forests in a humid tropical region of México. *Ecology* 80, 1892–1907.

Hutjes, R. W. A., Wierda, A. and Veen, A. W. L. (1990) Rainfall interception in the Tai Forest, Ivory-Coast – Application of 2 simulation-models to a humid tropical system. *Journal of Hydrology* 114, 259–275.

Jackson, P., Cavalier, J., Goldstein, G. and Meinzer, F. C. (1995) Partitioning of water resources among plants of a lowland tropical forest. *Oecologia*, 101, 197–203.

Jetten, V. G. (1994) Modelling the effects of logging on the water balance of a tropical rainforest. A study in Guyana. Tropenbos series 6, Wageningen.

Jetten, V. G. (1996) Interception of tropical rainforest: Performance of a canopy water balance model. *Hydrological Processes* 10, 671–685.

Jipp, P. H., Nepstad, D. C., Cassel, D. K. and Carvalho, C. R. de (1998) Deep soil moisture storage and transpiration in forests and pastures of seasonally dry Amazonia. *Climatic Change* 39, 395–412.

Jordan, C. F. and Uhl, C. (1978) Biomass of a 'tierra firme' forest of the Amazon Basin. *Oecologia Plantarum* 13, 387–400.

Jordan, C. F. and Heuveldop, J. (1981) The water budget of an Amazonian rainforest. *Acta Amazonica* 11, 87–92

Jordan, C. F., Caskey, W., Escalante, G., Herrera, R., Montagnini, F., Todd, R. and Uhl, C. (1983) Nitrogen dynamics during conversion of primary Amazonian rainforest to slash and burn agriculture. *Oikos* 40, 131–139.

Juo, A. S. R., Franzluebbers, K., Dabiri, A. and Ikhile, B. (1995) Changes in soil properties during long-term fallow and continuous cultivation after forest clearing in Nigeria. *Agriculture, Ecosystems and Environment* 56, 9–18.

Juo, A. S. R. and Lal, R. (1977) The effect of fallow and continuous cultivation on the chemical and physical properties of an Alfisol in western Nigeria. *Plant and Soil* 47, 567–584.

Juvik, J. O., and Nullet, D. (1995) A proposed standard fog collector for use in high-elevation regions - comment. *Journal of Applied Meteorology* 34, 2108–2110.

Kappelle, M., Geuze, T., Leal, M. E. and Cleef, A. M. (1996) Successional age and forest structure in a Costa Rican upper montane *Quercus* forest. *Journal of Tropical Ecology* 12, 681–689.

Kappelle, M. and Leal, M. E. (1996) Changes in leaf morphology and foliar nutrient status along a successional gradient in a Costa Rican upper montane Quercus forest. *Biotropica* 28, 331–344.

Kelliher, F. M., Leuning, R. and Schulze, E.-D. (1993) Evaporation and canopy characteristics of coniferous forests and grasslands. *Oecologia* 95, 153–163.

Khanna, P. K., Raison, R. J., and Falkiner, R. A. (1994) Chemical properties of ash derived from Eucalyptus litter and its effects on forest soils. *Forest Ecology and Management* 66, 107–125.

Klaassen, W. (1992). Average fluxes from heterognous vegetated regions. *Boundary-Layer Meteorology* 58, 329–354.

Klinge, R. (1998) Wasser- und Nährstoffdynamik im Boden und Bestand beim Aufbau einer Holzplantage im östlichen Amazonasgebiet. Göttinger Beiträge zur Land- und Forstwirtschaft in den Tropen und Subtropen, Goltze, Göttingen. 260 pp.

Kochenderfer, J. N. and Wendel, G. W. (1983) Plant succession and hydrologic recovery on a deforested and herbicided watershed. *Forest Science* 29, 545–558.

Kochenderfer, J. N., Edwards, P. J. and Helvey, J. D., 1990. Land management and water yield in the Appalachians. In: *Watershed Planning and Analysis in Action*, Symposium Proceedings, Durango, Colorado, American Society of Civil Engineers, New York. Pp. 523–532.

Kuraiji, K. and Paul, L. L. (1994) Effects of rainfall interception on water balance in two tropical rainforest catchments, Sabah, Malaysia. In: *Proceedings of the International Symposium on Forest Hydrology*, Tokyo, Japan, 291–298.

Lal, R. and Cummings, J. D. (1979) Changes in properties of an Alfisol produced by various crop covers. *Soil Science* 127, 377–382.

Langenheim, J. H., Osmond, C. B., Brooks, A. and Ferrar, P. J. (1984) Photosyntheticc responses to light in seedlings of selected Amazonian and Australian rainforest tree species. *Oecologia* 63, 215–224.

Leuschner, Ch. (1998) Water extraction by tree fine roots in the forest floor of a temperate *Fagus-Quercus* forest. *Annales Sciences Forestières* 55, 141–157.

Lindroth, A. and Iritz, Z. (1993) Surface energy budget dynamics of short rotation willow forest. *Theoretical and Applied Climatology* 47, 175–185.

Lloyd, C. R., Gash, J. H. C., Shuttleworth, W. J. and Marques Filho, A. de O. (1988) The measurement and modelling of rainfall interception by Amazonian rainforest. - *Agricultural and Forest Meteorology* 43, 277–294.

Lloyd, C. R. and Marques Filho, A. de O. (1988) Spatial variability of throughfall and stemflow measurements in Amazonian rainforest. *Agricultural and Forest Meteorology* 42, 63–73.

Mackensen, J., Hölscher, D., Klinge, R. and Fölster, H. (1996) Nutrient transfer to the atmosphere due to burning of debris in East-Amazonia. *Forest Ecology and Management* 86, 121–128.

Mackensen, J., Tillery-Stevens, M., Klinge, R. and Fölster, H. (2000) Site parameters, species composition, phytomass structure and element stores of a Terra-firme forest in East-Amazonia, Brazil. *Plant Ecology* 151(2), 101–119.

Malmer, A. (1992) Water-yield changes after clear-felling tropical rainforest and establishment of forest plantation in Sabah, Malaysia. *Journal of Hydrology* 134, 77–94.

Martins, S. P., Cerri, C. C., Volkoff, B., Andreaux, F. and Chauvel, A. (1991) Consequences of clearing and tillage on the soil of a natural Amazonian ecosystem. *Forest Ecology and Management* 38, 273–281.

Matson, P. (1990) Plant-soil interactions in primary succession at Hawaii Volcanoes National Park. *Oecologia* 85, 241–246.

McNaughton, K. G. and Jarvis, P. G. (1983) Predicting effects of vegetation changes on transpiration and evaporation. In: T. T. Kozlowski (ed.) Water Deficits and Plant Growth. Vol. 7, Academic Press, New York.

McWilliam, A.-L. C., Roberts, J. M., Cabral, O. M. R., Leitao, M. V. B. R., de Costa, A. C. L., Maitelli, G. T. and Zamparoni, C. A. G. P. (1993) Leaf area index and aboveground biomass of terra firme rainforest and adjacent clearings in Amazonia. *Functional Ecology* 7, 310–317.

Monteith, J. L. (1965) Evaporation and environment. In: *The state and movement of water in living organisms*, 19th Symposium of the Society for Experimental Biology, (Edited by G. E. Fogg), Cambridge University Press, London. Pp. 215–234

Monteith, J. L. and Unsworth, M. H., 1990. *Principles of Environmental Physics*, Edward Arnold, London. 291 pp.

Mooney, H. A., Ehleringer, J. and Bjorkman, O. (1977) The energy balance of the leaves of the evergreen shrub *Atriplex hymenelytra*. *Oecologia* 29, 301–310.

Nadkarni, N. M. (1984) Epiphyte biomass and nutrient capital of a neotropic elfin forest. *Biotropica* 16, 249–256.

Nadkarni, N. M. (2000) Colonization of stripped branch surfaces by epiphytes in a lower montane cloud forest, Monteverde, Costa Rica. *Biotropica* 16, 249–256.

Nepstad, D. C. (1989) Forest regrowth in abandoned pastures of eastern Amazonia: Limitations to tree seedling survival and growth. PhD Thesis, Yale University. 226 pp.

Nepstad, D. C., de Carvalho, C. R., Davidson, E. A., Jipp, P. H., Lefebvre, P. A., Negreiros, G. H., da Silva, E. D., Stone, T. A., Trumbore, S. E. and Vieira, S. (1994) The role of deep roots in the hydrological and carbon cycles of Amazonian forests and pastures. *Nature*, 372, 666–669.

Nepstad, D. C., Uhl, C., and Serrão, E. A. S. (1991) Recuperation of a degraded Amazonian landscape: forest recovery and agricultural restoration. *Ambio* 20, 248–255.

Oguntoyinbo, J. S. (1970) Reflection coefficient of natural vegetation, crops and urban surfaces in Nigeria. *Quarterly Journal of the Royal Meteorological Society* 95, 430–441.

Palm, C. A. and Szott, L. T. (1984) Unpublished draft on soil and vegetation dynamics during fallows. Proyecto Suelos Tropicales, INIPA-NCSU, Yurimaguas, Peru.

Paul, L. L. and Kuraiji, K. (1993) *Transition Report of Hydrological Study on Forested Catchments in Sabah, 1990–1992*. Research and Development Division, Sabah Forestry Department, Sandakan, Sabah, Malaysia.

Pinker, R. T. (1982) The diurnal asymmetry in the albedo of tropical forest vegetation. *Forest Science* 28, 297–304.

Pinker, R. T., Thompson, O. E. and Eck, T. F. (1980) The albedo of a tropical evergreen forest. *Quarterly Journal of the Royal Meteorological Society* 106, 551–558.

Pócs, T. (1980) The epiphytic biomass and its effect on the waterbalance of two rainforest types in the Uluguru Mountains (Tanzania, East Africa). *Acta Botanica Hungarica* 26, 143–167.

Popenoe, H. (1957) The influence of the shifting cultivation cycle on soil properties in Central America. Proc. 9th Pacific Science Congress, p. 72–77.

Poorter, L. and Oberbauer, S. F. (1993) Photosynthetic induction responses of two rainforest tree species in relation to light environment. *Oecologia* 96, 193–199.

Priess, J., Then, C. and Fölster, H. (1999) Litter and fine-root production in three types of tropical premontane rainforest in Venezuela. *Plant Ecology* 143, 171–187.

Raich, J. W. (1983) Throughfall and stemflow in mature and year-old wet tropical forest. *Tropical Ecology* 24, 234–243.

Rao, K. S. and Ramakrishnan, P. S. (1989) Role of bamboos in nutrient conservation during secondary succession following slash and burn agriculture (jhum) in north-east India. *Journal of Applied Ecology* 26, 625–633.

Roberts, J. M., Cabral, O. M., De Aguiar, L. F. (1990) Stomatal and boundary-layer conductances measured in a terra firme rainforest, Manaus, Brasil. *Journal of Applied Ecology* 27, 336–353.

Roberts, J., Cabral, O. M. R., Fisch, G., Molion, L. C. B., Moore, C. J., Shuttleworth, W. J. (1993) Transpiration from an Amazonian rainforest calculated from stomatal conductance measurements. *Agricultural and Forest Meteorology* 65, 175–196.

Roberts, J., Cabral, O. M. R., McWilliam, A-L. C., Da Costa, J. De P., Sá, T. D. De A. (1996) An overview of the leaf area index and physiological measurements during ABRACOS. In: *Amazonian Deforestation and Climate*. J. H. C. Gash, C. A. Nobre, J. M. Roberts and R. L. Victoria (eds) John Wiley, Chichester. pp. 287–306.

Rutter, A. J., Kershaw, K. A., Robins, P. C. and Morton, A. J. (1971) A predictive model of rainfall interception in forests: I. Derivation of the model from observations in a plantation of Corsican pine. *Agricultural Meteorology* 9, 367–384.

Sá, T. D. De A., Costa, J De P. Da, and Roberts, J. M. (1996) Forest and pasture conductances in southern Pará, Amazonia. In: *Amazonian Deforestation and Climate*. Edited by J. H. C. Gash, C. A. Nobre, J. M. Roberts and R. L. Victoria. John Wiley, Chichester, UK. pp. 241–264.

Saldarriaga, J., West, D. C., Tharp, M. and Uhl, C. (1988) Long-term chronosequence of forest succession in the upper Río Negro of Colombia and Venezuela. *Journal of Ecology* 76, 938–958.

Saldarriaga, J. G. and Uhl, C. (1982) Recovery of forest vegetation following slash-and-burn agriculture in the upper Rio Negro. In: A. Gomez-Pompa, T. C. Whitmore and M. Hadley (eds). *Rain Forest Regeneration and Management*, UNESCO, Paris. Pp. 303–313.

Sanchez, P. A. (1976) Properties and management of soils in the tropics. New York, Wiley. 618 pp.

Sanchez, P. A. and Hailu, M. (1996) (eds) Alternatives to slash-and-burn agriculture. *Agriculture, Ecosystems and Environment*, Special Issue.

Sanford, R. L. (1985) Root ecology of mature and successional Amazon forests. Dissertation, University of Berkeley, California.

Schellekens, J., Scatena, F. N., Bruijnzeel, L. A. and Wickel, A. J. (1999) Modelling rainfall interception by a lowland tropical rainforest in northeastern Puerto Rico. *Journal of Hydrology* 225, 168–184.

Scott, G. (1977) The importance of old-field succession biomass increments to shifting cultivation. *Great Plains-Rocky Mountain Geographical Journal* 6, 318–327.

Scott, G. (1987) Shifting Cultivation where land is limited – Case study No. 3: Campa Indian agriculture in the Gran Pajonal of Peru. In: Jordan, C. F. (ed.) *Amazonian Rain Forests – Ecosystems Disturbance and Recovery*. Springer, Berlin, pp. 34–45.

Senock, R. and Leuschner, Ch. (1999) Water flux dynamics in small diameter roots of a fast growing tropical tree. *Plant and Soil* 208, 57–71.

Shuttleworth, W. J. (1989) Micrometeorology of temperate and tropical forest. *Philosophical Transactions of the Royal Society, London*, B324, 299–334.

Snedaker, S. C. (1980) Successional immobilization of nutrients and biologally mediated recycling in tropical forests. *Biotropica* 12 (Suppl.), 16–22.

Sommer, R., Denich, M., Kanashiro, M. and Vlek, P. L. G. (2000) Carbon storage and root penetration in deep soils under small-farmer land-use systems in the Eastern Amazon region, Brazil. *Plant and Soil* 219, 231–241.

Sommer, R., Carvalho, E. J. M., Vlek, P. L. G. and Fölster, H. (1998) Soil water extraction through a secondary vegetation during the dry season in 1997 in eastern Amazonia. Proceedings of the Third SHIFT-Workshop Manaus, March 15–19, 1998, BMBF, Bonn.

Sommer, R., de Abreu Sá, T. D., Vielhauer, K., Carioca de Araújo, A., Fölster, H. and Vlek, P. L. G. (2002). Transpiration and canopy conductance of secondary vegetation in the eastern Amazon. *Agricultural and Forest Meteorology* 112, 103–121.

Stark, N. (1971) Nutrient cycling II: Nutrient distribution in Amazonian vegetation. *Journal of Tropical Ecology* 12, 177–201.

Steinhardt, U. (1978) Untersuchungen über den Wasser- und Nährstoffhaushalt eines andine Wolkenwaldes in Venezuela. Göttinger Bodenkundliche Berichte 56. 185 pp.

Street, I. M. (1980) Changes in the carbon inventories in live biomass and detritus as a result of the practice of shifting agriculture and the conversion of forest to pasture: Case studies in Peru, New Guinea and Hawaii. International Seminar on Geography and the Third World, Kuala Lumpur, Malaysia.

Stromgaard, P. (1984) The immediate effect of burning and ash-fertilization. *Plant and Soil* 80, 307–320.

Swamy, P. S. and Ramakrishnan, P. S. (1987) Contribution of *Mikania micrantha* during secondary succession following slash-and-burn agriculture (Jhum) in North-East India II: Nutient cycling. *Forest Ecology and Management* 22, 239–249.

Szott, L. T. and Palm, C. A. (1996) Nutrient stocks in managed and natural humid tropical fallows. *Plant and Soil* 186, 293–309.

Szott, L. T., Palm, C. A. and Davey, C. B. (1994) Biomass and litter accumulation under managed and natural tropical fallows. *Forest Ecology and Management* 67, 177–190.

Tanner, E. V. J. (1977) Four montane rainforests of Jamaica: a quantitative characterization of the floristics, the soils, and the foliar mineral levels, and a discussions of the interrelations. *Journal of Ecology* 65, 883–918.

Teklehaimanot, Z., Jarvis, P. G. and Ledger, D. C., (1991) Rainfall interception and boundary layer conductance in relation to tree spacing. *Journal of Hydrology* 123, 261–278.

Tobón Marin, C. (1999) Monitoring and modelling hydrological fluxes in support of nutrient cycling studies in Amazonian rainforest ecosystems. Tropenbos Series No 17, Wageningen. 162 pp.

Trimble, G. R., Reinhart, K. G. and Webster, H. H. (1963) Cutting the forest to increase water yields. *Journal of Forestry* 61, 635–640.

Ubarana, V. N. (1996) Observations and modelling of rainfall interception at two experimental sites in Amazonia. In: *Amazonian Deforestation and Climate*. J. H. C. Gash, C. A. Nobre, J. M. Roberts and R. L. Victoria (eds) John Wiley, Chichester. Pp. 151–162.

Uhl, C. (1982) Tree dynamics in species rich tierra firme forest in Amazonia, Venezuela. *Acta Cientifica Venezolana* 33, 72–77.

Uhl, C. (1987) Factors controlling succession following slash-and-burn agriculture in Amazonia. *Journal of Ecology* 75, 377–407.

Uhl, C., Clark, K., Clark, H. and Murphy, P. (1981) Early plant succession after cutting and burning in the upper Rio Negro region of the Amazon Basin. *Journal of Ecology* 69, 631–649.

Uhl, C., Buschbacher, R. and Serrão, E. A. S. (1988) Abandoned pastures in eastern Amazonia. I. Patterns of plant succession. *Journal of Ecology* 75, 663–681.

Uhl, C., and Jordan, C. F. (1984) Succession and nutrient dynamics following forest cutting and burning in Amazonia. *Ecology* 65, 1476–1490.

Van der Plas, M. C. and Bruijnzeel, L. A. (1993) The impact of mechanized selective logging of lowland rainforest on topsoil infiltrability in the upper Segan area, Sabah, Malaysia. *International Association of Hydrological Sciences Publication* 216, 203–211.

Veen, A. W. L., Hutjes, R. W. A., Klaassen, W., Kruijt, B. and Lankreijer, H. J. M. (1991). Evaporative conditions across a grass-forest boundary: a comment on the strategy for regionalizing evaporation. International Association of Hydrological Sciences Publication 204, 43–52.

Veneklaas, E. and van Ek, R. (1990) Rainfall interception in two tropical montane rainforests, Colombia. *Hydrological Processes* 4, 311–326.

Vitousek, P. M. and Reiners, W. A. (1975) Ecosystem succession and nutrient retention: a hypothesis. *Bioscience* 25, 376–381.

Wadsworth, G., Reisenauer, H. M., Gordon, D. R. and Singer, M. J. (1990) Effects of length of forest fallow on fertility dynamics in a Mexican ultisol. *Plant and Soil* 122, 151–156.

Waterloo, M. J. (1994) Water and nutrient dynamics of *Pinus carribea* plantation forests on former grassland soils in southwest Viti Levu, Fiji. PhD Thesis, Vrije Universiteit, Amsterdam.

Watson, F. G. R., Vertessy, R. A., and Grayson, R. B. (1999) Large-scale modelling of forest hydrological processes and their long-term effect on water yield. *Hydrological Processes* 13, 689–700.

Weaver, P. L., Birdsey, R. A. and Lugo, A. E. (1987) Soil organic matter in secondary forests of Puerto Rico. *Biotropica* 19, 17–23.

Werner, P. (1984) Changes in soil properties during tropical wet forest succession in Costa Rica. *Biotropica* 16(1), 43–50.

Wiesenmüller, J. (1999) Einfluß landwirtschaftlicher Flächenvorbereitung auf die Dynamik des Wurzelsystems und die oberirdische Regenerartion der Sekundärvegetation Ostamazoniens, Pará, Brasilien. Dissertation, University of Göttingen.

Williams-Linera, G. (1983) Biomass and nutrient content in two successional stages of tropical wet forest in Uxpanapa, Mexico. *Biotropica* 15, 275–284.

Wright, I. R., Gash, J. H. C., da Rocha, H. R., Shuttleworth, W. J., Nobre, C. A., Maitelli, G. T., Zamparoni, C. A. G. P. and Carvalho, P. R. A. (1992) Dry season micrometeorology of central Amazonian ranchland. *Quarterly Journal of the Royal Meteorological Society* 118, 1083–1099.

Xu, J., Fox, J., Xing, L., Podger, N., Leisz, S. and Xihui, A. (1999) Effects of swidden cultivation, population growth, and state policies on land cover in Yunnan, China. *Mountain Research and Development* 19, 123–132.

Zinke, P. J., Sabhasri, S. and Kunstadter, P. (1978) Soil fertility aspects of the Lua forest fallow system of shifting cultivation. In: P. Kunstadter, E. C. Chapman and S. Sabhasri (eds). *Farmers in the Forest: Economic Development and Marginal Agriculture in northern Thailand*. East-West Population Institute University Press of Hawaii, Honolulu. p. 134–159.

25 The hydrological and soil impacts of forestation in the tropics[1]

D. F. Scott

Okanagan University College, Kelowna, Canada

L. A. Bruijnzeel

Vrije Universiteit, Amsterdam, The Netherlands

J. Mackensen

United Nations Environment Programme, Nairobi, Kenya

INTRODUCTION

In response to the continuing degradation and disappearance of the world's tropical forests (Drigo, this volume) the establishment of plantation forest on degraded and previously forested sites as well as into (sub)tropical grasslands is becoming increasingly common (Evans, 1999). The hydrological effects of this practice and the potential of forestation to improve or restore the hydrological behaviour of degraded catchments constitute the prime focus of this chapter, expanding and updating an earlier review of the subject by Bruijnzeel (1997). Three aspects are highlighted in particular, namely: (i) the effects of tree plantations on annual and seasonal streamflow totals; (ii) the associated impacts on stormflow and sediment production; and (iii) concurrent changes in soil chemical characteristics (fertility). Because the hydrological changes associated with forest clearing and the establishment of a new vegetation cover during the first few years are discussed at length in the chapter by Grip, Fritsch and Bruijnzeel, much of what follows pertains to the post-canopy closure phase of plantations.

EXTENT, DEVELOPMENT AND IMPORTANCE OF TROPICAL TREE PLANTATIONS

The establishment of timber plantations is a notable and accelerating land-use development of the last half-century. It has been estimated that there are now some 40 to 50 million ha of forest plantations in the tropics and warmer subtropics, trees being planted nowadays at a rate of *c.* 2 million ha yr^{-1} compared to *c.* 1 million ha yr^{-1} a decade ago (Evans, 1999). One of the most widely used types of tree is the eucalypt (*Eucalyptus* spp.). Eucalypt plantations cover more than 17 million ha world-wide (Table 25.1) (FAO, 2001). More than 90% of these have been established since 1955 and roughly 50% during the last decade (Turnbull, 1999).

Despite the increase in rates of tropical plantation establishment in recent years, the rate of closed forest destruction far exceeds that of plantation establishment. Evans (1986) estimated a ratio of 11:1 between these two early on, whereas the data presented by Drigo (this volume) suggest even higher ratios (18–24:1). However, the latter estimates are based on remote sensing surveys in which plantations with poor establishment and low cover caused by weak growth are often missed (R. Drigo, pers. comm.). Based on pan-tropical remote sensing surveys, between 1980 and 1990 most (visible) new plantations were raised on sites that originally supported natural forest whereas between 1990 and 2000 more than half of the new plantations were established on other land cover types, including grasslands (degraded or otherwise), and, to a lesser extent, savannahs (Drigo, this volume). A distinction is sometimes made between plantations in the dry tropics (mean annual precipitation <800 mm; rainy season less than 4.5 months) and the more humid parts of the tropics and subtropics having rainy seasons of variable length (cf. Chang and Lau, 1993). Similarly, a distinction can be made between low-lying and high altitude areas, of which those in humid

1 Strictly speaking, the term 'afforestation' should be used for the planting of trees in areas where forest is naturally absent, as opposed to 'reforestation' for previously forested, and subsequently degraded areas. To avoid semantic problems, the term 'forestation' (Wiersum, 1984a) is used mostly throughout this chapter.

Forests, Water and People in the Humid Tropics, ed. M. Bonell and L. A. Bruijnzeel. Published by Cambridge University Press.

Table 25.1. *The area of plantations around the globe by region and type*

Region	Area: 10^3 ha	Area: Percent of total	Plantation areas by genus groups (10^3 ha): *Acacia*	*Eucalyptus*	*Hevea*	*Tectona*	Other broadleaf	*Pinus*	Other conifer	Unspecified
Africa	8 036	4.3	345	1 799	573	207	902	1 648	578	1 985
Asia	115 847	61.9	7 964	10 994	9 058	5 409	31 556	15 532	19 968	15 365
Europe	32 015	17.1	–	–	–	–	15	–	–	32 000
North and Central America	17 533	9.4	–	198	52	76	383	15 440	88	1 297
Oceania	3 201	1.7	8	33	20	7	101	73	10	2 948
South America	10 455	5.6	–	4 836	183	18	599	4 699	98	23
World total	**187 086**	**100.0**	**8 317**	**17 860**	**9 885**	**5 716**	**33 556**	**37 391**	**20 743**	**53 618**

Source: FAO (2001).

tropical uplands are commonly thought to be the most productive (Evans, 1999).

Clearly, the various types of plantations fulfil significant economic and ecological functions. Usually, a distinction is made between industrial and non-industrial forest plantations (FAO, 1995). Non-industrial plantations supply fuelwood and construction timber, charcoal, fodder, shelter and non-wood products for local consumers on a small scale, while industrial plantations are established for the larger-scale production of pulpwood, sawlogs and veneer. The area of industrial plantations is often small compared to the area of native forest, but their share in wood production is usually highly significant. For example, plantations in Brazil, Zimbabwe or Zambia each represent <1.3% of the national forest area but they contribute 50–60% of the wood production (FAO, 1995). This is because industrial plantations generally exhibit high productivity and are readily manageable. Especially fast-growing tree plantations using eucalypts, pines or acacias in short rotations of between 10–30 years are favoured in this regard (Brown *et al.*, 1997). In South Africa, by way of example, roughly 95% of all the country's timber needs are provided by productive and intensively managed plantations of pine, eucalypts and wattle (mainly *Acacia mearnsii*) covering just 1.5 million ha (1.2% of the land area), while wood exports contributed a further US$1.01 billion in foreign exchange earnings in 1998 (R. Godsmark, pers. comm.). Besides their role for wood production, plantations are also valued for environmental benefits such as soil protection against erosion and rehabilitation of degraded land, for diverting pressure off natural forest and, in some cases, the enhancement of biodiversity (Evans, 1999).

It is difficult to find reliable figures on the areas of plantation established on previously degraded land as opposed to landscapes in good hydrological condition, but it can be assumed that the largest growth is found in the area of essentially commercial plantations, where there is a strong profit motive, rather than in the establishment of plantations for broader social or environmental reasons. As productivity is related to the quality of the site, most industrial plantations can be expected to be on sites and soils that were not degraded. In South Africa, for example, large areas of the humid grassland regions are degraded due to over-grazing, and these lands are specifically avoided by private industrial forestry companies because of their low productivity.

HYDROLOGICAL IMPACTS OF FORESTATION

To predict the hydrological effects of a land cover change, it is important to understand what the treatment involves: is it just a change in vegetation cover, or are there associated changes to important soil properties as well? For example, deforestation followed by heavy grazing or intensive cropping accompanied by marked changes in soil physical properties (notably bulk density and infiltration capacity; Lal, 1987; Godsey and Elsenbeer, 2002) should be viewed differently from deforestation followed by the development of secondary forest with comparatively little change in vegetation leaf biomass and soil structure (Giambelluca, 2002; Hölscher, Mackensen and Roberts, this volume). Similarly, the establishment of forest in catchments in good hydrological condition (i.e. where soil infiltration capacities have been maintained) is fundamentally different from forestation of truly degraded catchments. Before reviewing the effects of forestation on precipitation, annual and seasonal water yields, storm runoff and sediment production, the water use of tropical tree plantations will be compared with those of the rainforests, grasslands and agricultural crops they may be replacing. Effects of forestation of severely degraded catchments will be treated separately in view of their different hydrological behaviour.

Water use of tropical tree plantations

Compared to short vegetation types like grass, scrub or agricultural crops, forests have greater leaf area and canopy height. Because of this their aerodynamic roughness and capacity to intercept rainfall are enhanced; in addition, forests absorb more radiation (typically by 5–10%) and have deeper root systems. The latter enables them to maintain transpiration during dry spells when more shallow-rooted plants are likely to experience water stress and thus are forced to reduce their water uptake (cf. Roberts *et al.*, this volume). The contrast is particularly pronounced in the seasonal (sub)tropics where grasslands and sclerophyllous scrub become dormant during the extended dry season (Waterloo *et al.*, 1999; Smith and Scott, 1992) and crop cover is much reduced or fields lie bare (Harding *et al.*, 1992; Van Dijk, 2002). As such, establishing tree plantations on grass-, scrub- or crop lands whose soils have not become degraded can be expected to lead to more or less serious reductions in soil moisture and, ultimately, groundwater recharge and water yield (Hamilton and King, 1983; Bruijnzeel, 1990).

Total evapotranspiration (E_T) of a forest consists of evaporation from a wet canopy (rainfall interception, E_i), evaporation from a dry canopy (transpiration, E_t), and that from the forest floor (soil or litter evaporation, E_s). E_s is generally considered to be small under fully closed canopies, although it cannot be neglected under more open conditions (Waterloo *et al.*, 1999; Putuhena and Cordery, 1995; Ashby, 1999). Bruijnzeel (1997) reviewed the results of some 20 rainfall interception studies in humid tropical tree plantations for which comparatively good data are available. Eucalypts generally exhibit low values (*c.* 12%), broad-leaved hardwood species such as teak and mahogany typically intercept about 20%, whereas pines and other conifers (*Araucaria, Cupressus*) mostly fall in the range of 20–25%, with the higher values usually found in upland situations. Well-developed stands of the particularly fast-growing *Acacia mangium*, on the other hand, exhibited values of 20–40%. Rainfall interception by another fast grower, *Paraserianthes falcataria* (a.k.a. *Albizzia*), which is increasingly planted in South East Asia for soil rehabilitation purposes (Van Dijk and Bruijnzeel, 2003), typically reaches 18–20%, reflecting its much lighter canopy compared to *A. mangium*. Typical values for the rainforests replaced by these plantations range from 10 to 20% in most lowland situations to 20–35% in montane forests (Bruijnzeel, 1989a; Bruijnzeel and Proctor, 1995).

The database for *annual* estimates of tropical tree plantation water use (E_T, E_t) is very limited. Table 25.2 (modified and updated from Bruijnzeel, 1997) summarises some key results. To facilitate comparisons between species and locations, values for ET in Table 25.2 have been normalised by dividing them by the corresponding open water evaporation (Penman's *Eo*) wherever such information was available. Corresponding estimates for natural forest or grassland in some of the locations have been added to illustrate the contrast or similarity in water use by the respective vegetation types.

The data in Table 25.2 are of variable quality (see footnotes). In many cases E_T has been estimated from the catchment water budget and may have been overestimated somewhat because of potential catchment leakage (cf. Roberts *et al.,* this volume). Similarly, in several cases E_t has been derived by subtracting interception losses from E_T, thereby introducing additional uncertainty, whereas in others annual totals were estimated by simple extrapolation of short-term measurements (e.g. for *A. mangium*). Despite these uncertainties, the data permit a few tentative conclusions:

(1) Water use by mature tropical plantations of average stocking resembles that of old-growth forest in the same area (East Malaysia, Indonesia, Jamaica, Kenya) and possibly exceeds it in the case of particularly vigorous growth (*A. mangium* in East Malaysia, *Pinus caribaea* in Fiji).
(2) Interception losses constitute a significant portion of overall E_T, particularly in humid tropical upland areas.
(3) Very high transpiration totals are found in maritime tropical situations (Fiji, Jamaica, Atlantic lowlands of Costa Rica).
(4) None of the forests listed in Table 25.2 seem to be subject to serious soil water stress.
(5) Plantation water use invariably exceeds that of rain-fed crops, pasture or fire-climax grassland in the same area, particularly under more seasonal conditions (Fiji, Tanzania).

Additional modelling-supported process-based work along the lines of Waterloo *et al.* (1999), Ciencala, Kuczera and Malmer (2000) or Bigelow (2001) is needed to obtain more reliable estimates and test some of these conclusions. However, whilst plausible estimates of E_t appear feasible from short-term measurements of stomatal behaviour made under representative climatic conditions and using the Penman-Monteith model of evaporation, a similar approach for wet canopy evaporation seems to produce less reliable results. For example, the interception estimates derived from short-term observations and subsequent modelling for various young stands in Costa Rica by Bigelow (2001) seem distinctly low (Table 25.2), possibly because advective effects were not incorporated. Waterloo *et al.* (1999) reported similar discrepancies between measured and modelled E_i for pine stands in Fiji (cf. discussion in Roberts *et al.*, this volume).

In view of its widespread use, the almost complete absence of reliable estimates of annual water use by *Eucalyptus sp.* under humid tropical conditions is striking, although such data are available for more subhumid tropical conditions, as well as from subtropical and warm-temperate areas. Examples include Australia (reviewed by Vertessy *et al.,* 2003), south-east Brazil (Lima *et al.,* 1990), South Africa (Scott and Smith, 1997; see also

Table 25.2. *Estimates of annual evapotranspiration* (E_T)*, interception* (E_i) *and transpiration* (E_t) *for tree plantations in the humid tropics after canopy closure (values rounded off to the nearest 5 mm).*

Species	Location	Elevation (m)	Mean annual precipitation (mm)	Age (yr)	E_T (mm)	E_T/E_0	E_i (mm)	E_t (mm)
Acacia mangium[a]	Sabah	700	3280	9–10	1495	>1.0	655	840
Agathis dammara[b]	Java	600	4770	11–35	1070	0.79	665	405
Cedrela odorata[c]	Costa Rica	100	4210	4–5	1320	–	75	1245
Cordia alliodora[c]	Costa Rica	100		4–5	1385	–	280	1110
Eucalyptus robusta[d]	Madagascar	1010	2000	>50	1505	–	–	–
Montane rainforest[d]	Madagascar	1010			1295	–	–	–
Hyeronyima alch.[c]	Costa Rica	100	4210	4–5	1510	–	375	1135
Pinus caribaea[e]	Fiji	80	1800	6	1925	1.13	540	1385
	Fiji	230		15	1715	1.01	495	1220
Fire-climax grassland[e,f]	Fiji	90			750	0.47	195	555
P. caribaea[g]	Jamaica	1020	3745	19	1850	–	635	1215
P. elliottii[h]	Queensland	60	1260	>35	1082	0.76[+]	–	–
P. merkusii[i]	Java	1300	2100	31	900	0.84	555	445
P. patula[k]	Kenya	2400	2305	>10	1160	0.77	560	600
Montane rainforest[k]	Kenya	2440			1155	0.78	465	690
Crops and pine seedlings[k]	Kenya	2400		<3	1030	0.69	–	–
Crops and grassland[l]	Tanzania	2500	1925		970	0.65	–	–
Lowland rainforest[m]	Manaus, Brazil		2075		1315	0.77	335	980
Pasture[n]	Manaus, Brazil		2065		915	0.54	125[o]	790[o]
Lowland rainforest[p]	Java		2850		1480	0.90	595	885

[a] Cienciala *et al.* (2000): approximate annual E_t derived by extrapolation of short-term sapflow measurements in stand of average basal area; E_i based on pers. comm. by A. Malmer; E_T of rainforest *c.*1835 mm (Malmer, 1992).
[b] Bruijnzeel (1988): catchment water balance (CWB), E_t by subtracting E_i from E_T and hence approximate only.
[c] Bigelow (2001); E_t modelled with Penman–Monteith equation, E_i derived with Rutter model and possibly seriously underestimated.
[d] Bailly *et al.*, 1974: CWB, some leakage possible; E_T for lower montane rainforest (LMRF) *c.*1295 mm.
[e] Waterloo *et al.* (1999): E_t via micro-meteorological methods, E_i including c.160 mm evaporated from litter layer.
[f] Waterloo *et al.* (in press): methods as in ([e]).
[g] Richardson (1982): CWB, leakage possible; E_T for LMRF *c.*2000 mm.
[h] Bubb and Croton (2001): CWB, [+]pan evaporation.
[i] C. A. Bons, pers. comm.
[k] Blackie (1979): CWB, E_i based on Pereira (1952).
[l] Edwards (1979): CWB, long dry season.
[m] Shuttleworth (1988): methods as in ([c, e]).
[n] Kabat *et al.* (1999).
[o] Ashby (1999): as in ([c, e]).
[p] Calder *et al.* (1986): as in ([c]).

below), south India (Harding *et al.*, 1992; Calder *et al.*, 1997), and Pakistan (Morris *et al.*, 1999). The planting of eucalypts in seasonal climates has met particularly vigorous opposition in the popular environmental literature, mainly because they are thought to be 'voracious consumers of water' (e.g. Vandana Shiva and Bandyopadhyay, 1983). Indeed, young plantations of *Eucalyptus camaldulensis* and *E. tereticornis* in south India exhibited such behaviour with transpiration rates of up to 6 mm day^{-1} when unrestricted by soil water deficits at the end of the monsoon, although values fell to 1 mm day^{-1} during the subsequent long dry season (Roberts and Rosier, 1993). Because of this stomatal regulation, the annual water use of the plantations on soils of intermediate depth (*c.* 3 m) was not significantly different from that of indigenous dry deciduous forest (Calder *et al.*, 1992; Harding *et al.*, 1992). However, on much deeper soils (>8 m) eucalypt water use exceeded annual rainfall considerably, suggesting 'mining' of soil water reserves that had accumulated previously in deeper layers during years of above-average rainfall. Moreover, the rate of root penetration was shown to be at least as rapid as 2.5 m yr^{-1} and roughly equalled above-ground increases in height (Calder *et al.*, 1997). Similar observations have been made for *E. grandis* in South Africa (Dye, 1996; see also below). It is probably pertinent in this respect that Viswanatham *et al.* (1982) observed strong decreases in streamflow after coppicing of *E. camaldulensis* in northern India. A recent catchment experiment involving *E. globulus* in south India (Sharda *et al.*, 1998) confirmed the enhancing effect of coppicing: the reduction in water yield during the second rotation of 10 years (first generation coppice) was substantially higher (by 156%) compared to that observed during the first rotation although the roots did not reach the groundwater table (Samraj *et al.*, 1988). All of the above findings confirm the original fears of Vandana Shiva and Bandyopadhyay (1983) although Collopy *et al.* (2000) recently reported surprisingly conservative water use by a highly productive *E. urophylla* plantation on the Leizhou Peninsula in China under conditions of low vapour pressure deficit. In summary, planting of eucalypts, particularly in sub-humid climates, should be based on judicious planning, i.e. away from water courses and depressions or where the roots would have rapid access to groundwater reserves.

Similar caution may have to be applied in the case of *Acacia mangium*, another tree capable of very high growth rates (Lim, 1988). Although the extrapolated annual E_T and E_i values for an *A. mangium* stand of average growth and stocking in East Malaysia were not excessive (Table 25.2), average transpiration rates in a denser and more productive stand were 170% higher and showed no consistent stomatal control in response to changes in atmospheric vapour pressure or soil water deficits even though the measurements were made during a relatively dry part of the year (Cienciala *et al.*, 2000). Since the rainfall interception fraction of the more productive stand was also higher (28% vs. 20% in the average stand; A. Malmer, pers. comm.) it cannot be excluded that total annual water use for these vigorously growing trees exceeds that of the forests they are replacing by at least several hundreds of mm (cf. Malmer, 1992). Further work is needed to test this contention.

Effect of forestation on precipitation

Because tree plantations share many of the vegetation characteristics considered to be responsible for the high evaporation from tropical rainforests (such as high leaf area index, low albedo, high aerodynamic roughness and deep roots; Roberts *et al.*, this volume), the establishment of extensive fast-growing plantations over large areas with degraded vegetation may be expected to exert a positive influence on the moisture content of the atmospheric boundary layer (ABL), and (by implication) on cloud formation and rainfall (Dolman *et al.*, 2004). Direct observational evidence to test this contention appears to be lacking for plantation forest, but both physical reasoning and model simulations indicate that large-scale forest conversion to pasture (i.e. the reverse of forestation) may cause moderate decreases in rainfall (as reviewed by Costa, this volume). Again, verification of corresponding long-term changes in rainfall in areas subject to major land use change has proved difficult (Van Rompaey, 1993; Wilk, Andersson and Plermkamon, 2001). This is because of natural climate variability, uncertainty about the relative importance of atmosphere-ocean interactions and regional surface characteristics, phase lags between rainfall and vegetation development, and (not least) uncertainties in rainfall and vegetation data (Zeng *et al.*, 1999; Bruijnzeel, 2004). Nevertheless, there is observational evidence that forest conversion to pasture over areas between 1000 and 10 000 km^2 has caused changes in the timing and distribution of clouds in parts of Amazonia (Cutrim *et al.*, 1995) and Costa Rica (Lawton *et al.*, 2001).

To explore the potential effects of large plantations of *Eucalyptus camaldulensis* on the regional climate of southern India (Karnataka), Harding (1992) used a simple one-dimensional ABL model. The simulations were limited to the post-monsoon season (December) for which previous work had shown very low evaporation rates for crop land (<1 mm day^{-1}) but much higher values for the eucalypts (3–4 mm day^{-1}). Despite its simplicity, the model was able to reproduce the daily maximum height of the ABL to within the resolution of measured radiosonde data. In addition, the model reproduced the broad features of the observed daily variation in surface temperature and humidity. Next, the model was used to predict the temperature and moisture content of the ABL associated with the two land cover types. The ABL forming above the eucalypt plantation was predicted to be 3.7 °C cooler and 0.4 g kg^{-1} moister than that above the crops, implying a specific humidity deficit above the forest that was reduced to 57% of the value predicted for bare soil.

Furthermore, the model was used to estimate the minimum size of a forest that would have a measurable effect on atmospheric humidity. A 10 km stretch of forest was shown to have a small effect on humidity which was measurable but not large enough to have a significant effect in terms of feedback to atmospheric evaporation demands downwind of the forest. Conversely, forests in excess of 50 km were predicted to have a significant effect on downwind evaporation rates. It should be noted, however, that the one-dimensional nature of the model used by Harding is likely to have overestimated the effect of changes in the vegetation because three-dimensional circulations, which tend to increase horizontal mixing of the air, were not incorporated (cf. Costa, this volume). In addition, only the feedback between evaporation and evaporative demand of the atmosphere could be addressed by the model, but not the effect on rainfall. This would require the use of a full three-dimensional meso-scale circulation model (such as RAMS; Pielke *et al.*, 1992) and realistic surface parameterisation (Lawton *et al.*, 2001; Van der Molen, 2002; Dolman *et al.*, 2004). The information on vegetation characteristics required for such simulations is now gradually becoming available for tropical grasslands (Wright *et al.*, 1996; Waterloo *et al.*, 1999, in press), various types of tree plantations (Beadle, 1997; Waterloo *et al.*, 1999; Bigelow, 2001) as well as rain-fed upland crops (Van Dijk and Bruijnzeel, 2001; Van Dijk, 2002). However, similar information on concurrent changes in soil physical properties (needed for the adequate representation of soil water feedbacks on tree water uptake in the model) is still scarce (Lal, 1987; see also section on forestation of degraded land below).

Effects of forestation on water yield

In view of the contrasts in water use between forests and shorter vegetation types discussed in the previous sections (Table 25.2) it is not surprising that the establishment of timber plantations into grassland or scrub leads to an increase in evaporative losses and a resultant decrease in *annual* streamflow totals. Although no stringent (paired) catchment studies have been conducted to actually demonstrate such reductions in flow upon forestation in the humid tropics proper, there is overwhelming evidence to this extent from the subhumid tropics (Samraj *et al.*, 1988; Sikka *et al.*, 2003), the subtropics (Scott and Smith, 1997; Scott *et al.*, 1999) and the temperate zone (Bosch and Hewlett, 1982; Trimble *et al.*, 1987; Fahey and Jackson, 1997). By analogy, there is a dataset comprising more than 100 long-term, controlled paired catchment experiments showing unequivocally that the felling of native forest or timber plantations (i.e. the reverse of forestation) under a broad range of climatic and topographic conditions results in *increases* in catchment water yield. The increase is positively related to the proportion of the catchment that is affected or the proportion of biomass removed (Bosch and Hewlett, 1982; Sahin and Hall 1996; Stednick, 1996; for the tropics: Bruijnzeel, 1990, 1996).

Reported initial increases in water yield following the clear-felling of forest, in both tropical and temperate regions, generally range between 25–60 mm yr^{-1} per 10% of catchment affected. However, maximum increases of 80–90 mm yr^{-1} per 10% forest removal have been recorded in some tropical studies (Bruijnzeel, 1990; Grip *et al.*, this volume). The observed variation in initial hydrological response to clearing is considerable and can be explained only partially by differences in rainfall between locations or years (Bruijnzeel, 1996). Other factors include differences in elevation and distance to the coast (affecting evaporation; Roberts *et al.*, this volume); catchment steepness, soil depth and changes in permeability with depth (governing the residence time of the water, speed of baseflow recession and stormflow generation patterns; Bonell, this volume); and, above all, the degree of disturbance of undergrowth and topsoil by machinery or fire (determining both infiltration and rate of regrowth; Grip *et al.*, this volume; Hölscher *et al.*, this volume). Because the relative importance of the respective factors varies between sites, additional process studies are usually required if the results of paired basin experiments (which essentially represent a black box approach) are to be fully understood and extrapolated to other areas (Bruijnzeel, 1990, 1996; Bonell and Balek, 1993). It is pertinent to note that the bulk of the increase in flow is usually observed in the form of baseflow rather than as greatly increased stormflows. This appears to be in contrast to the deterioration in flow regime that is so often observed in practice following tropical forest conversion to grazing or cropping (Pereira, 1989). Indeed, it may reflect a lack of realism on the part of most experimental studies in which either soil disturbance upon clearance remained limited or post-forest land use did not last long enough to degrade the soil sufficiently (Bruijnzeel, 1996; Sandström, 1998).

Whether the increases in water yield following forest clearing are temporary or permanent depends on the type of the new vegetation. In the case of natural regeneration, streamflows can be expected to revert to pre-clearance values within a decade (Giambelluca, 2002; Hölscher *et al.*, this volume), whereas forest conversion to pasture or annual cropping produces permanent increases (Grip *et al.*, this volume) (Table 25.2). Similarly, in one of the few examples where tropical tree plantations replaced natural forest, water yield returned to original levels after 6–10 years in the case of *Pinus patula* in a montane area with deep volcanic soils in Kenya (Blackie, 1979). An analysis of the changes in flow during the first seven years after converting logged-over rainforest to vigorously growing *Acacia mangium* in East Malaysia revealed a drop below pre-clearing streamflow levels from the fourth year onwards (cf. Figure 21.2 in Malmer *et al.*, this volume). Such rapid reductions undoubtedly reflect the high transpiration and interception of *A. mangium* cited earlier (cf. Ciencala *et al.*, 2000).

Waterloo *et al.* (1999) investigated the hydrological effects of reforesting fire-climax grassland with vigorously growing stands of *P. caribaea* in Fiji (MAP 2000 mm, dry season of 5 months) following earlier reports by Kammer and Raj (1979) of strongly diminished low flows some years after planting. Combining micro-meteorological and soil water balance techniques with modelling, Waterloo *et al.* (1999) derived (very high) annual water use totals for six- to 15-year-old pine stands (1717–1926 mm) (Table 25.2) which far exceeded evapotranspiration from the seasonally dormant grassland (*c.* 748 mm yr^{-1}). Because infiltration capacities of the (ungrazed) grassland soils had remained high (Waterloo, 1994), no change in the generation of overland flow after forestation was expected and any changes in streamflow would thus largely reflect contrasts in water use between the two vegetation types. Despite the massive difference in annual E_T between pines and grassland at the site level (Table 25.2), Waterloo *et al.* (1999) suggested that actual reductions in annual streamflow would be rather in the order of 500–700 mm. Reasons for this include the fact that the pine stands had a much poorer stocking at the catchment scale due to repeated disturbance by hurricanes and the fact that indigenous forest still lined the riparian zones of grassland catchments. This riparian forest was expected to transpire at potential, if not higher rates because of positive heat advection from the warmer and drier surrounding grasslands (cf. Giambelluca, 2002). In fact, it cannot be excluded that the very high E_T totals derived for the pines themselves may be partly due to a similar mechanism (M. J. Waterloo, pers. comm.).

The planting of *Eucalyptus globulus* over 59% of a montane grassland catchment with swampy forested valley bottoms in subhumid south India (MAP 1380 mm) did not produce a noticeable decline in streamflow during the first three years but from then onwards annual water yield declined to 120 mm below original values until the trees were coppiced at 10 years of age. This represented a reduction in flow by 21% between age 3–10 and 16% over the entire rotation (Samraj *et al.*, 1988). During a second, 10-year rotation (first-rotation coppice), flows were reduced from the beginning because the root network had remained intact. As a result, flows were reduced even more during the second rotation (by 25.4% overall; Sharda *et al.*, 1998). The increased water use during the coppice phase corresponded with increased production of woody biomass (14 t ha^{-1} yr^{-1} vs. 10 t ha^{-1} yr^{-1} during the first rotation) and a slight increase in root penetration depth (from an average 2.8 m at the end of the first rotation to 3.2 m at the end of the second; Sikka *et al.*, 2003). Such reductions in flow may seem modest compared to the 500–700 mm yr^{-1} inferred earlier by Waterloo *et al.* (1999). However, not only does the Indian work represent a partial conversion over less than 60% of the catchment but also both rainfall and evaporation in the Nilgiris are lower than in lowland Fiji.

Case study: effects of afforestation of subtropical grasslands in South Africa on water yield

In addition to these few tropical examples, much can be learned from a particularly comprehensive series of long-term (since the 1930s) paired catchment experiments of the hydrological effects of afforesting natural grasslands, sclerophyllous shrub and evergreen, broad-leaved forest in subtropical South Africa (summarised at various stages by Bosch, 1982; Scott and Smith, 1997; Scott *et al.*, 1999). Of particular relevance here are four experiments where grasslands in good condition were afforested, one with *Eucalyptus grandis* and three with *Pinus patula*, plus one experiment where native evergreen forest was replaced with *Eucalyptus grandis*. The research sites are all in the high rainfall zone of South Africa (MAP between 1100–1600 mm). Experimental control was provided by catchments kept under native vegetation. The results of the planting experiments are summarised in Table 25.3, and illustrate the range and variability of the key criteria used to express the effect of forestation. To facilitate comparisons the results have been standardised to a 10% level of planting by assuming a linear relationship to the area that is treated. It is important to note that these experimental catchments were all in good condition, with no or little soil erosion. Although generally steep, they have deep, well-drained soils and very low storm response ratios (Hewlett and Bosch, 1984). Therefore the experimental comparison is between the two vegetation covers (i.e. differences in total evaporation only), and does not involve the effects of concurrent changes in soil condition as would be expected upon forestation of degraded catchments (see below).

The resulting streamflow reductions over time after planting follow a sigmoidal pattern comparable to a growth curve (Figure 25.1). There are clear differences between the effects of eucalypts and pines, but there is also a large amount of variation from year to year within a single experiment and between different experiments, even in comparable catchments in one locality (see discussions in Bosch (1982), Smith and Scott (1997) and Scott *et al.* (1999) for details). The highest flow reductions occur while the tree crop is mature and range, for a 10% level of planting, from 17.3 mm or 10% yr^{-1} in a drier catchment to 67.1 mm and 6.6% yr^{-1} in wetter catchments (Table 25.3). The former are similar to the results obtained in South India (*c.* 20 mm per 10% forest yr^{-1}) whereas the latter rather resemble the results obtained in Fiji (50–60 mm per 10% yr^{-1}).

The timing of the first significant reductions in flow after planting varies quite widely depending on the rate at which catchments are dominated by the plantation crop. The pine plantations in the high altitude grasslands at Cathedral Peak usually took several years to have a clear impact on streamflow (up to 8 years) (Table 25.3, column 5; Figure 25.1).

However, the same species of pine had an earlier effect on streamflow (within 3 years) under the drier conditions prevailing in the Mokobulaan B catchment at Mpumalanga. Other conditions remaining the same, eucalypts have a slightly earlier impact on streamflows than pines, normally within 2–3 years (Table 25.3, column 5; Figure 25.1). Like in the case of deforesting humid tropical (Bruijnzeel, 1996) and temperate (Hewlett and Bosch, 1982) catchments, the presently found changes in streamflow following the establishment of exotic timber species into native vegetation are related to the proportion of the catchment affected, the contrast in biomass, rainfall depth (water availability) and growth rate (vigour) of the forest (Smith and Scott, 1992; cf. Sikka *et al.*, 2003).

The high values of plantation water use (E_T) inferred from the afforested South African catchment experiments are confirmed by process studies that have measured transpiration totals of 1300 mm yr^{-1} and total E_T of 1500 mm yr^{-1} (Dye, 1996; Burger *et al.*, 1999). Such values are comparable to some of the higher estimates for tropical plantations listed in

Table 25.3. *Summary of annual flow reductions in South African afforestation experiments, showing the age from which reductions were first statistically significant (column 5), the peak reductions over a 5-year interval (columns 6 and 7), and the decline in flow reductions towards the end of the rotation or period of measurement (column 8)*

1. Catchment and native vegetation	2. Area and trees planted and plant date (all at 1360 sph)	3. Latitude and mid-elevation (m)		4. Median annual rainfall (mm)	5. Initiation of reductions (age, yrs)	6. Peak reductions, absolute (5-yr mean) (mm/10%) (age, yrs)		7. Peak reductions, relative (5-yr mean) (%/10%) (age, yrs)		8. Declining effect in last 5 years[a] (mm/10%; %/10%)	
Cathedral Peak II: montane	75%; *Pinus patula* 1951	29° 0′ S	2150	1399	7	67.1	19	7.9	26	2	0.3
Cathedral Peak III: montane grassland	86% *Pinus patula* 1959	29° 0′ S	2100	1399	8	51.7	19	6.6	23	4.1	0.8
Mokobulaan A: seasonal grassland	97% *Eucalyptus grandis* 1969	25° 17′ S	1360	1189	2	41	7	10	14–18	0	0
Mokobulaan B: seasonal grassland	95% *Pinus patula* 1971	25° 17′ S	1360	1197	3	17.3	18	10	15–20+	0	0
Westfalia D: evergreen afromontane forest	98% *Eucalyptus grandis* 1983	23° 44′ S	1280	1182	3	37.3	5	9.8	10	13.6	3.4

[a] Declining effect is calculated as the streamflow reduction over the last 5 years of record divided by the maximum streamflow reduction recorded over a 5-year interval during the rotation.
Source: (Modified from Scott *et al.*, 1999).

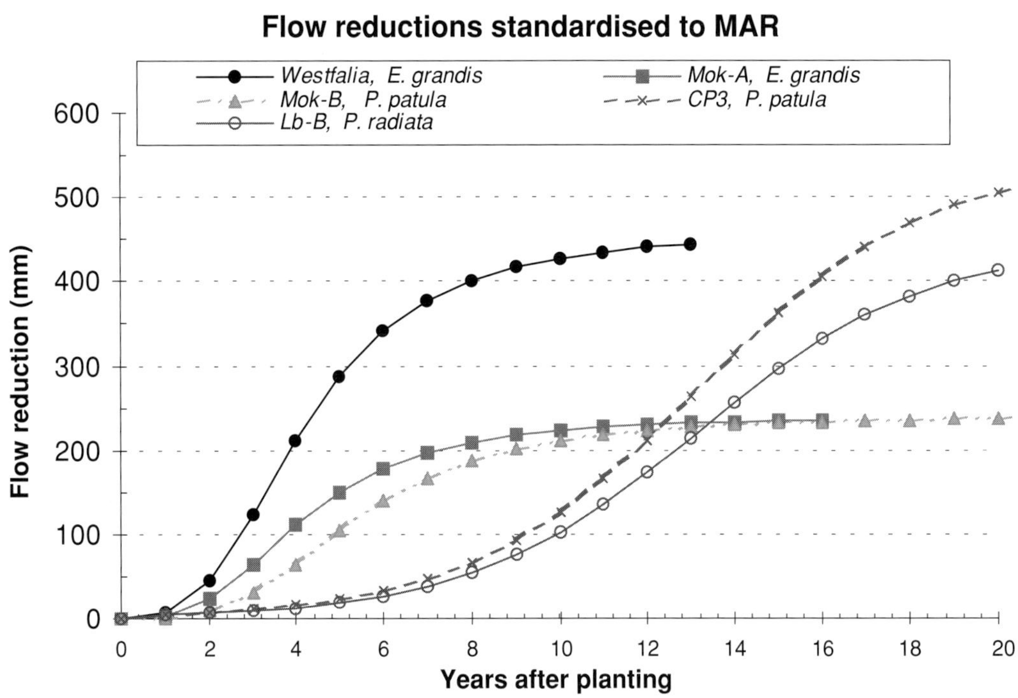

Figure 25.1 Smoothed streamflow reduction curves measured in five South African afforestation experiments where pines and eucalypts replaced grassland, scrub (Lb-B) or native forest (Westfalia). These curves were generated by plotting the product of measured percentage flow reduction (Scott and Smith, 1997) and mean annual runoff (MAR) of the experimental catchment.

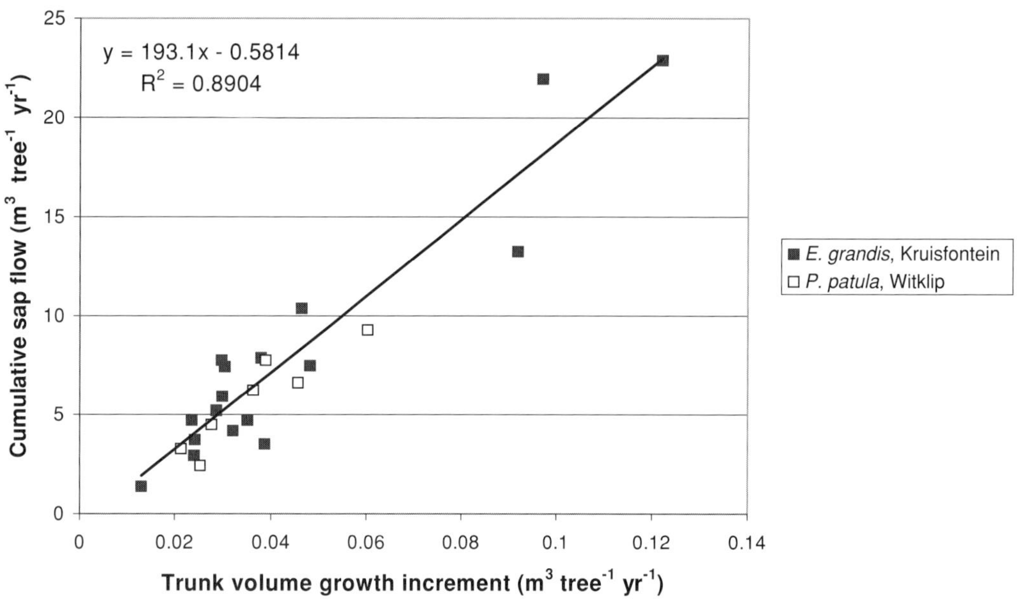

Figure 25.2 Experimental results showing the relationship between trunk volume growth increment and annual cumulative sap flow recorded in sample trees of *Eucalyptus grandis* and *Pinus patula* in the Mpumalanga Province of South Africa. (Dye *et al.*, 2001).

Table 25.2. Therefore, it is feared that in sub-humid catchments (MAP < 1200 mm) in the region all streamflow will cease if entire catchments are planted, particularly during the dry season (cf. Figure 25.5 below).

The link between productivity and water use

The generalised curves of Scott and Smith (1997) describing the changes in streamflow reduction over time (Figure 25.1) in the South African experiments closely resemble growth curves for these species (Scott and Lesch, 1997; Dye *et al.*, 2001). The rapid desiccation of the catchments is attributed more to the increased transpiration of the tree crop than to an increase in interception losses (Dye, 1996; Scott and Lesch, 1997). The exotic trees are evergreen and have deeper root systems, which allow them to sustain transpiration year-round, while native vegetation would have closed its stomata in response to dry season stress (Dye, 1996; cf. Waterloo, 1994). Recent work by Dye *et al.* (2001) has shown the strong link between the growth rate of these plantation trees and their transpiration (Figure 25.2). Their findings also support a contention raised in the early review by Bosch and Hewlett (1982) that the magnitude of the responses in water yield to forest clearing or establishment was positively related to the productivity of the forest. Other examples of the close connection between stand productivity (using leaf area index as a proxy) and water use include the studies of Dunin *et al.* (1985) on eucalypts in south-east Australia, Granier *et al.* (1992) on young plantations of two indigenous species in French Guyana, and Waterloo *et al.* (1999) on grassland and *Pinus caribaea* in Fiji.

A new finding from the up-dated analysis of the South African afforestation experiments (Scott *et al.*, 1999, 2000), is that flow reductions are definitely diminished towards the end of longer timber rotations, and this applies both to pines and at least one eucalypt experiment (Table 25.3, column 8). Obviously this trend is clearest in the longer-term experiments. The diminution of final flow reductions (mean values over the last 5 years measured) compared to the highest 5 year average reduction at any one time during the rotation ranges from zero (no change over time, usually in short-term experiments) to 60% and 50% less, for absolute and relative measures, respectively (Table 25.3).

The finding of streamflow increasing again with plantation age during the post-maturation phase of the trees agrees with the Australian experience in regenerating mountain ash (*Eucalyptus regnans*) forests near Melbourne, Victoria, following the 1939 wild fires (Langford, 1976; Kuczera, 1987; Jayasuriya *et al.*, 1993). Water yields from the old-growth (>150 years) forest were nearly 60 mm yr^{-1} per 10% forest cover higher that those from the young, vigorously regrowing forest that developed after fire (Figure 25.3). Whilst the mountain ash forest results are in apparent contradiction to the generalisation that streamflows decrease in proportion to increasing forest cover, the observed water yield declines in the first 25–30 years during the regrowth of the mountain ash forest in fact parallel the flow reductions observed after the establishment of plantations in South Africa (cf. Figures 25.1 and 25.3). However, the accent in the Australian work has always been on the (very) mature forest stages, which are accompanied by flow increases with forest age that justify the maintenance of a

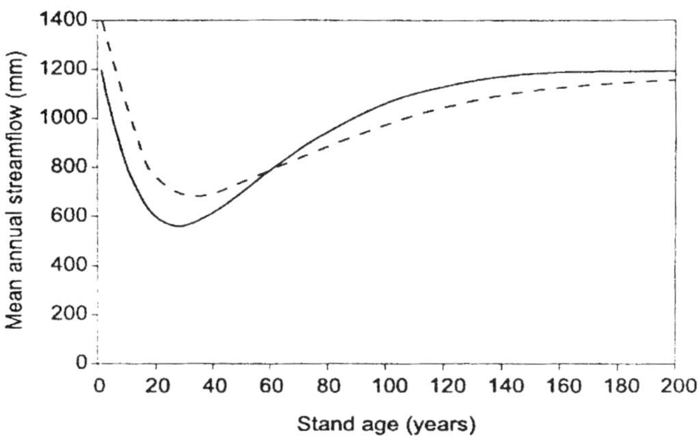

Figure 25.3 Two models of the response pattern of streamflow over time after fire in mountain ash (*Eucalyptus regnans*) forest north of Melbourne, Australia. (After Watson *et al.*, 1999.)

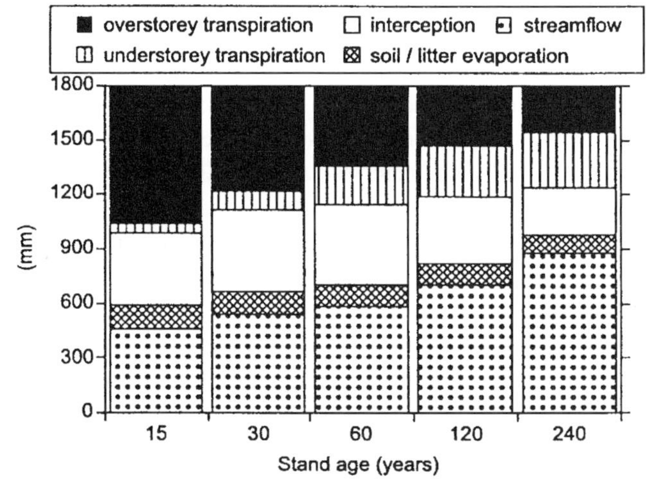

Figure 25.4 Changing contributions of various evaporation components with age in mountain ash forest, Melbourne, Victoria, Australia. (After Vertessy *et al.*, 1998a.)

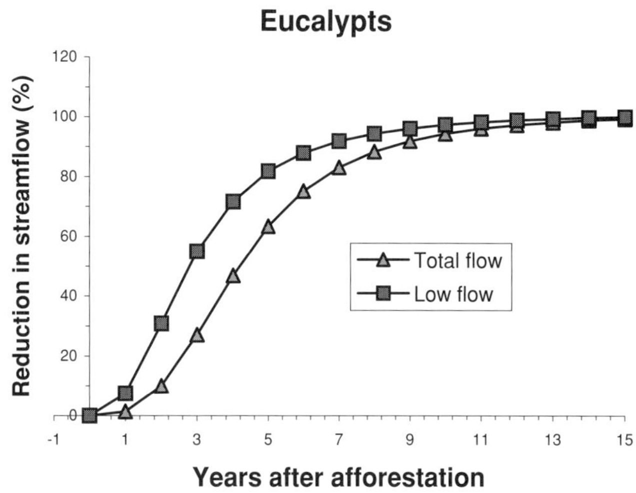

Figure 25.5 Smoothed curves fitted to observed streamflow reductions measured in two experimental catchments planted to *Eucalyptus grandis* in South Africa, showing the larger and earlier effects of afforestation on the low flow component. (After Scott and Smith, 1997.)

forest cover in these catchments whose main purpose is the supply of water to greater Melbourne.

Process studies by Haydon *et al.* (1996) and Dye (1996) show that increases in eucalypt age are associated with declines in leaf area index, sapwood conducting area and transpiration per unit of leaf area. Modelling of the various evaporation components in the mountain ash forests over time further shows that transpiration by the overstorey dominates the water balance during the initial 15–30 years, but declines gradually over time to end up being less than interception losses in over-mature forest (Vertessy *et al.*, 1998a, b) (Figure 25.4).

The implication of this is that in highly productive or vigorously growing forests, transpiration can be expected to dominate the water balance, total water use will be at its highest, and streamflow reductions at a maximum. As the forest matures, so transpiration will decline, though it might be compensated for by increases in interception losses and understorey transpiration (Figure 25.4). Overall, though, mature forest of lower vigour can be expected to have higher streamflow yields than when it was younger.

Two other well-known phenomena are also accounted for by this hypothesis. In Australia, pine plantations are known to have a larger impact on streamflows than the native eucalypts in a similar climate (Vertessy, Zhang and Dawes, 2003). This, it has been suggested, is because the productive pine plantations are being compared to the natural (old-growth) eucalypt forests, which may have a higher biomass and greater structural complexity (and hence a higher interception loss) but are less vigorous (productive) and thus transpire less (cf. Vertessy *et al.*, 1998a, b). However, the productivity of the native eucalypt forests of Australia cannot be compared to that of the disease-free, mono-specific eucalypt plantations of South Africa, Brazil or other tropical and subtropical sites, where growth rates in excess of 30 m^3 ha^{-1} yr^{-1} have been recorded (Lima, 1993; Dye, 1996). Interestingly, a similar change in water use with age is now beginning to be revealed for regenerating tropical rainforest vegetation as well. Here too, there are indications that water use by very young and vigorously growing secondary vegetation is similar to, or even higher than that of old-growth forest in the same area (see discussion in Hölscher *et al.*, this volume; Giambelluca, 2002).

Effect of forestation on low flows

As pointed out earlier, the bulk of the streamflow increases resulting from experimental forest clearing are manifested as baseflows rather than in the form of greatly increased stormflows. Figure 25.5

illustrates the similarity in the pattern of reductions in total and dry season (low) flows found for two catchments afforested with eucalypts in South Africa, though low flows are decreased more than are total flows at the same age. The effect of forestation on low flows has two supposed sources. First, these and other exotic plantations, in contrast to the native vegetation they replace, do not go dormant in the dry season (Smith and Scott, 1992; cf. Waterloo *et al.*, 1999). The second component, though less easily quantified, is that of steadily reducing soil water stores through several years, as found under eucalypts in India (Calder *et al.*, 1997; Sharda *et al.*, 1998) and in South Africa (Scott and Lesch, 1997). Low flows are a reflection of the amounts of soil water and groundwater stored in the catchment and as these are steadily depleted by tree water uptake so will low flows diminish accordingly. It is clear from the South African experiments that total water use by the tree crop can exceed annual rainfall in many years (Dye, 1996; Burger *et al.*, 1999; Dye *et al.*, 2001), and that, once summer streamflow has ceased altogether (Figure 25.5), the occurrence of rainstorms may not easily cause the streams to flow again. Once again, the changes in flows observed in South Africa reflect differences in vegetation water use only, not in soil infiltration capacity. The possibility of improved groundwater recharge through enhanced infiltration afforded by tree planting on degraded soils will be discussed below.

The greater effect of forestation on low flows observed in South Africa has also been reported for plantations of *Pinus caribaea* in seasonal grasslands in Fiji (Kammer and Raj, 1979) as well as under *Eucalyptus globulus* in South India (Sharda *et al.*, 1988), particularly after coppicing (Sharda *et al.*, 1998; Sikka *et al.*, 2003). It was also recorded after afforestation of an extensive upland area in northern Malawi where between 1966 and 1976 almost 93% of a 13.3 km^2 grassland catchment was progressively planted with a mixture of *Pinus patula, P. kesiya* and *Eucalyptus saligna*. Streamflow records for nine pre-planting years and 13 years following planting, showed no significant change in peak discharges, but the 10 and 30-day low flow totals showed a significant reduction over this period (Mwendera, 1994). *Thus, the finding of strongly reduced baseflows after forestation of* non-degraded *grasslands can be expected to be generally applicable and should be heeded when planning industrial plantations* (Calder, 1999).

Effects of forestation on storm flows

Forest hydrological research has shown that the influence of vegetation cover or type on stormflows is inversely related to the size of the rainfall event that generates the flows. Hewlett and Doss (1984) showed this to be the case for a broad range of humid catchments of the eastern USA, whereas Hewlett and Bosch (1984) demonstrated a similar independence of stormflows on vegetation cover

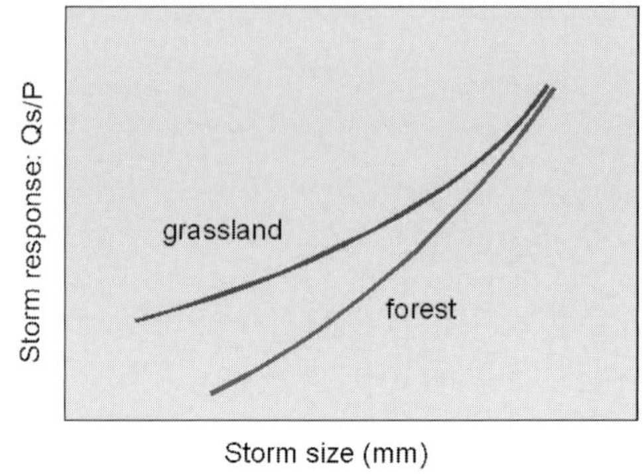

Figure 25.6 Postulated relationship between size of a stormflow generating event (P) and the resultant stormflow (Qs) and how these are affected by vegetation type under non-degraded topsoil conditions.

in afforested and control catchments in South Africa. Likewise, peak flows in the lower ranges were affected by the establishment of a vigorously growing eucalypt plantation in a montane grassland catchment in south India but the effect was almost negligible in the higher ranges (Sharda *et al.*, 1988; Sikka *et al.*, 2003). Soil infiltration capacities were not affected appreciably by the land-cover change in any of these studies, however, nor were groundwater levels in the swampy valley bottoms that produced most of the storm runoff in the Indian study (Samraj *et al.*, 1988).

In small storm events the combined storage capacity of vegetation canopies, ground-covering litter, surface micro-topography and the soil mantle can be substantial in proportion to the size of the storm depth. Of these the soil mantle is potentially the largest water store, but its capacity to accommodate additional rain varies as a function of soil wetness. Where previous uptake by the trees has depleted soil water reserves, storage capacities will be relatively high but once the soil has become thoroughly wetted by frequent rains (typically at the height of the wet season), opportunities to absorb large additional amounts of rain will be limited even under fully forested conditions. Furthermore, as precipitation events increase in size, so does the relatively fixed maximum storage capacity of the soil become less influential in determining the size of the stormflows that are generated (Figure 25.6). In other words, under conditions of extreme rainfall and soil wetness, large stormflows may also emerge from forested areas (Hewlett, 1982; Hamilton and King, 1983; cf. Bonell, this volume).

It can be seen, therefore, that the influence of vegetation on the magnitude of stormflows will be determined by its proportional contribution to the total storage capacity of the catchment, but in large storm events this contribution may easily become unimportant (Hewlett, 1982; Bruijnzeel and Bremmer, 1989). However, as discussed more fully in the next section, where degradation of a catchment has produced strong reductions in canopy and ground cover (including litter), and above all in infiltration capacity and soil depth through continued erosion (and thus overall soil water storage opportunity), reforestation could clearly lead to an improvement of most or all these factors over time.

FORESTATION OF DEGRADED LANDS: PROSPECTS FOR IMPROVED FLOW REGIME

The term 'degraded' refers to areas where the imposed land use has caused vegetation to be strongly simplified in terms of diversity and structure and where such changes are associated with soil compaction or crusting and erosion, such that surface hydrological processes are negatively affected (Lal, 1987). The expected hydrological changes would include a more or less pronounced reduction in infiltration capacity and thus increased contributions by infiltration-excess overland flow to stormflows, usually an increase in total water yield but with a decreased baseflow component compared to the pre-degradation situation because of the deteriorated infiltration and soil water recharge opportunities (Pereira, 1989; Sandström, 1998). Physical soil degradation is a widespread phenomenon, particularly in the dry tropics (Eswaran, Lal and Reich, 2001), but also under humid tropical conditions. For example, about 45% of the total land area of South and South East Asia was considered to be affected by some form of human-induced soil degradation (mostly surface erosion) in the mid 1990s. On 10–15% of the affected area, degradation was considered to have a strong to very strong impact on plant productivity (and by implication, on overland flow occurrence), whereas on 22–28% the impact was considered moderate. Impacts were light on 40–48% of the land and considered negligible on the remaining 12–23% of the land (Van Lynden and Oldeman, 1997).

Reforestation is often recommended as a means of reducing the enhanced surface water losses associated with soil degradation, returning water to a subsurface route through the soil profile, thereby ultimately restoring baseflows (e.g. Bartarya, 1989; Negi, Joshi and Kumar, 1998). The question is whether this process really occurs, in which circumstances might it occur, and would it also occur if forestation is by means of a productive timber plantation of high water uptake? The answer to this question is that baseflow is only likely to be improved if gains through enhanced infiltration after forestation exceed the associated increase in evaporative losses (Bruijnzeel, 1989b). The information on excess plantation water use over that of the grassland or rain-fed crops they replace (Tables 25.2 and 25.3) suggests that infiltration volumes would need to be increased by at least 175–200 mm yr^{-1} under subhumid tropical conditions (MAP 1200–1400 mm) and by as much as 450–700 mm yr^{-1} under wetter conditions (MAP > 2000 mm). Whether this can be attained depends on the size of the overland flow component, which, in turn, is largely decided by the prevailing climate (notably rainfall intensity and duration), the degree of soil degradation, and the nature (texture, erodibility, depth) of the soils. Together with the ability of the new vegetation to create additional soil water storage opportunity through enhanced water uptake, these climatic and soil factors will determine the extent to which stormflow components (overland flow, subsurface stormflow) can be reduced by a change in vegetation cover (Bailly *et al.*, 1974; Chandler and Walter, 1998; cf. Bonell, this volume). It is important to note that it is not necessarily the vegetation cover per se that will be responsible for a change in near-surface hydrological behaviour, but rather the changes (or not) in associated soil characteristics such as topsoil bulk density and infiltrability, soil depth, as well as surface detention and retention storage. The following three examples from East Africa illustrate this.

Working under semi-arid (MAP *c.* 800 mm) tropical conditions in northern Tanzania, Sandström (1998) concluded (mostly on the basis of model simulations) that where soils were fine-textured and liable to crusting upon exposure to rainfall, the creation and maintenance of macropores by trees within a setting of closed woodlands was crucial for effective groundwater recharge. By contrast, in a degraded nearby catchment deforested more than 40 years ago, overland flow formed a large component of the water balance, and overall groundwater recharge was strongly diminished despite the fact that water use by the non-forest vegetation must be much reduced. This led to perceived desiccation of the deforested catchment although the soils were deep and potentially had a high capacity to store water. Thus, surface characteristics rather than the contrast in vegetation water use proved to be the dominant factor here. A notable aspect of this study is the long time between the clearing of forest and the measurements. It is likely that the degradation of the soil hydraulic properties of the cleared catchment represents a gradual process that might not have been apparent closer to the time of the original conversion (Sandström, 1998). Indeed, at a somewhat wetter (MAP 1960 mm) site elsewhere in Tanzania, Edwards (1979) did not observe significant surface soil degradation nor diminished annual or low flows throughout an 11-year period of subsistence cropping. In fact, both total and summer flows were strongly increased after clearing (Table 25.2). He attributed this to a fortuitous combination of the low rainfall intensities prevailing at this montane location and the low erodibility of the deep volcanic soils. Similarly, in a broader scale study of rainfall partitioning and groundwater recharge in nearby central Uganda, Taylor and Howard (1996) estimated that of an annual rainfall of 1400 mm, around 200 mm became recharge by infiltration of rain water. Moreover, using a combination of isotope tracing and modelling, the authors were able to determine that recharge had doubled over a period of 30 years as a result of 26% of the catchment area having been converted from forest to mixed agricultural use. In this and the previous instance, the surface properties of the affected land at the catchment scale

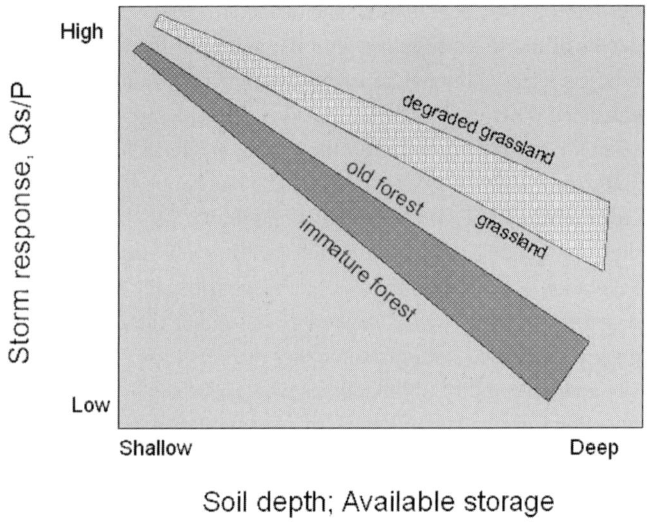

Figure 25.7 A postulated relationship between catchment storage capacity and stormflow response, and the likely role of vegetation cover.

apparently remained suitable for water to move deeply through the soil profile rather than over the surface, thereby allowing the reduction in vegetation water use to become manifest as increased groundwater recharge and baseflows. Interestingly, the magnitude of the changes in recharge or flow at the two locations was very similar at 40–45 mm yr^{-1} per 10% forest removal (Taylor and Howard, 1996; Edwards, 1979).

The degree of surface runoff reduction required to affect dry season flows under humid tropical conditions (MAP 2500 mm) was examined through modelling by Van der Weert (1994). He simulated the relative contributions to annual water yield by three streamflow components, viz. fast surface runoff (overland flow), delayed subsurface flows (called 'interflow' by Van der Weert), and deep groundwater outflow (baseflow) for a large river basin in West Java, Indonesia, under fully forested and cleared conditions as well as for gradually increased surface runoff coefficients. The simulations showed that baseflow levels would be little affected by land use change as long as the overland flow coefficient remained below 15% of the rainfall (equivalent to *c.* 350 mm yr^{-1}). However, should surface runoff become as high as 40% (close to 1000 mm yr^{-1}), baseflow (dry season flow) would be roughly halved. These estimates are almost double the amounts inferred from contrasts in vegetation water use only (Tables 25.2 and 25.3). A second modelling exercise by the same author suggested that dry season flows would diminish more rapidly following severe surface disturbance of deep soils with a large storage capacity than in the case of more shallow soils having little capacity to store water anyway (Van der Weert, 1994; cf. Sandström, 1998). Figure 25.7 is a conceptual attempt to represent the combined effect of soil depth (storage) and land cover (infiltration) on catchment stormflow response (cf. Table 25.4).

There is very little direct information on the hydrological behaviour of degraded catchments in the humid tropics, at least in the 'official' literature. Some idea may be gained, however, from the stormflow coefficients reported in the grey literature for 'typical' actively eroding ('semi-degraded') agricultural catchments in the volcanic uplands of Java, viz. 30–40% of incident rainfall vs. 5–10% for nearby forested catchments (Pramono Hadi, 1989; Rijsdijk and Bruijnzeel, 1991; Jongewaard and Overmars, 1994). Likewise, at *c.* 35–50 t ha^{-1} yr^{-1} the sediment yields from such deforested catchments are 10–50 times higher than those typically associated with similarly sized forested catchments (cf. Figure 22.6 in Grip *et al.*, this volume). Such increases largely reflect the enhanced contributions of HOF from degraded agricultural fields and compacted surfaces like trails, roads and settlements which all exhibit very low infiltration capacities (Grip *et al.*, this volume). However, whilst storm runoff in the most extreme cases can be increased to as much as 70% of the rainfall (e.g. where the underlying rock has become exposed), the associated sediment yields are relatively low because most of the erodible material has already been lost (Flatfjord, 1976). Such examples illustrate the very considerable (local) potential for highly increased storm runoff from severely degraded catchments.

There are even fewer studies of how, and to what extent, a change in land use may improve low flows (Bruijnzeel, 2004). Therefore, the remainder of this section explores to what extent infiltration and soil water recharge may be boosted (or stormflows reduced) after forestation of degraded land. These amounts may then be compared with the expected increases in water use (Tables 25.2 and 25.3) in a first attempt to derive when the net result will be positive (i.e. increased low flows) or negative (i.e. further decrease in low flows).

Table 25.4 summarises the available evidence with respect to the reductions in stormflow volumes observed after forestation or soil conservation works in a series of degraded (sub)tropical catchments that were mostly too small to sustain perennial flow. As such, the corresponding effects on low flows cannot be evaluated directly. The evidence collated in Table 25.4 does show that major relative (and sometimes also absolute) reductions in stormflows may be achieved by forestation and various soil conservation measures (and *vice versa* through adverse practices like overgrazing or repeated burning). However, almost none of these reductions seem to be large enough in absolute terms to overcome the typical increases in vegetation water use after forestation inferred earlier (175–200 and 450–700 mm yr^{-1} in subhumid and humid areas, respectively (Table 25.2)), let alone the values implied by Van der Weert (1994) in his simulations (up to 1000 mm yr^{-1}). It is probably pertinent that the only exception (Leyte, Philippines; see below) represents a high rainfall site (MAP 2200 mm). Usually, as for example in southern China (MAP 1500 mm), 'direct' stormflows are reduced after forestation compared to severely degraded

Table 25.4. *Effects of forestation or soil conservation works on stormflow volumes (and their components) in selected small degraded (sub)tropical headwater catchments*

Location	Vegetation and treatment	Area (ha)	Actual stormflow HOF (mm yr^{-1})	Actual stormflow SSF (mm yr^{-1})	Actual stormflow Total (mm yr^{-1})	Change in total storm flow vs. baseline (%)	Reference and notes
Leyte, Philippines	Degraded pasture	0.13	1518	154	1672	Baseline	Chandler and Walter (1998): direct comparison of hillslope plots with contrasting cover, shallow soils, MAP *c.* 2200 mm
	Pasture + hedgerows	0.23	396	286	682	−59	
	Conservation cropping	0.25	242	132	374	−78	
	Conservation cropping (mulched)	0.13	66	220	286	−83	
	Regrowth (10–20 yrs)	0.20	<22	44	<66	−96	
Luzon, Philippines	Burned grassland	1.6	–	–	1244/905	Baseline	Daño (1990): paired catchment, sequential treatment on experimental catchment, firstly 4 years of no-burn, secondly reforestation (8 years), MAP *c.* 3800 mm, catchments highly leaky (Bruijnzeel, 1990).
	Protected grassland	1.0	–	–	1359	+9	
	6-yr *Gmelina* forest	1.0	–	–	1002	+11	
Guangdong, China	Badland	3.7	151	589	740	Baseline	Zhou *et al.* (2001): direct comparisons, total stormflow equated to measured non-perennial flow totals (Q_t) and will include some delayed flow; HOF equated to direct runoff (Q_q) based on traditional hydrograph separation into surface runoff, throughflow and baseflow by Zhou *et al.*; SSF = $Q_t - Q_q$; deep soils, MAP *c.* 1455 mm.
	Eucalyptus	3.8	90	241	330	−55	
	Eucalyptus plus other trees	6.4	<0.5	41	42	−94	
Madagascar	Burned grassland	3.25	–	–	221	Baseline	Bailly *et al.* (1974): direct comparisons, results probably influenced by differences in leakage, MAP 1717 mm
	Protected grassland	3.18	–	–	121	−45	
	Conservation cropping	4.77	–	–	48	−78	
	Pinus patula (<10 yr)	3.90	–	–	42	−81	
North India	Degraded scrub	1.27	–	–	37/70	Baseline	Gupta *et al.* (1974, 1975): paired catchments, coefficients of determination for calibration and treatment equations unknown; runoff totals pertain to rainy seasons of 1974/75 with rainfall totals of 623 and 718 mm, respectively.
	Annual burning	4.15	–	–	66/57	+218/26	
	Overgrazing	0.74	–	–	73/117	+552/30	
	Forestation/trenching	1.93	–	–	17/12	−29/−75	

land (in this case by 60 and 150 mm yr^{-1} after forestation with eucalypts and mixed species, respectively) but both delayed and total (ephemeral) flows were decreased as well (Table 25.4). As such, no improvement in baseflows was obtained. Most probably, this reflects the higher water use of the mixed forest although measurements of soil water (to a depth of 200 cm) and groundwater levels suggested both to be lower under the pure eucalypt plantation (Zhou *et al.*, 2001). The importance of a well-developed litter layer and understorey vegetation for minimising infiltration-excess overland flow (HOF) and surface erosion (cf. Wiersum, 1985) was also demonstrated by the Chinese study. A patch with very little litter and shrub cover was protected from litter harvesting (litter is collected routinely for use as fuel in this region). After two years, litter biomass already attained 90% of its value after eight years whilst understorey cover continued to increase exponentially throughout this time. HOF was reduced from 31% of incident rainfall at the start of the experiment to less than 9% after eight years. The more completely developed forest floor in the mixed plantation exhibited negligible overland flow and surface erosion (Zhou *et al.*, 2001). Thus, this long-term study (>10 years of measurements) demonstrates the effectiveness of using eucalypts as a pioneer crop in restoring at least some hydrological functions in degraded catchments in this part of China, particularly the hydrological benefits of allowing litter accumulation on site and of encouraging succession to a mixed forest ecosystem. Once again, no improvement in low flows was achieved, however.

The finding of Daño (1990) of *increased* runoff after protecting and subsequently reforesting a very small (and very leaky) annually-burned grassland catchment in the Philippines is difficult to explain and is at odds with the previous results as well as those obtained by Bailly *et al.* (1974) and Gupta *et al.* (1974, 1975) for similar (but drier) situations in Madagascar and northern India, respectively (Table 25.4). One would have expected the higher water use of protected grassland and fast-growing trees to result in both improved infiltration and higher water use, and thus into less overall water yield. Such results illustrate the inherent limitation of the paired catchment approach which, although statistically sound, is essentially a black box (Bruijnzeel, 1996). Elsewhere in the Philippines, and at the other end of the range, Chandler and Walter (1998) reported massive reductions in stormflow (HOF plus subsurface stormflow, SSF) from degraded hillslope-sized plots after being subjected to various conservation practices whereas the smallest volumes were observed under 10–20-year-old secondary forest. However, the size of the inferred reductions (which are based on direct comparisons between plots rather than previous intercalibrations) hinges very much on the very high runoff total established for the most degraded plot (1670 mm year^{-1} or 76% of incident rainfall) (Table 25.4). Also, it is not clear whether the soil of the forested plot had been equally degraded before abandonment. Nevertheless, taking the runoff reductions observed by Chandler and Walter (1998) at face value, at 990–1605 mm yr^{-1} they should be more than sufficient to compensate any increases in evaporative losses compared to degraded grassland (presumably <750 mm yr^{-1}) (Table 25.2). It is interesting to note that the large reductions in HOF when going from severely degraded grassland to various types of conservation cropping were not matched by concurrent changes in SSF and delayed throughflow although both HOF and throughflow were lowest under forest (Table 25.4). As such, one may expect a significant portion of the increased infiltration to contribute to the recharging of groundwater reserves (and thus baseflow) rather than to be lost from the soil profile via rapid near-surface flows, although neither groundwater levels nor soil physical characteristics, rainfall intensities or tree water uptake rates were given by Chandler and Walter (1998). Once again, additional process-based work would be needed to obtain more comprehensive answers.

An alternative (non-catchment) approach bases predicted changes in occurrence of overland flow and SSF on a comparison of changes in soil infiltration capacity and saturated hydraulic conductivity (K^*) profiles with depth after forestation vs. rainfall intensity – duration distributions. Although the dataset is still small and covers only a small portion of the pan-tropical spectrum of soil, tree species and rainfall intensity combinations, a few tentative observations seem possible.

Firstly, the recovery of surface K^* following forestation with teak (*Tectona grandis*, usually planted on calcareous heavy clay soils in seasonally dry areas) is generally very limited and HOF (as well as surface erosion) remains rampant (Bell, 1973; Wolterson, 1979; Mapa, 1995). Development of surface organic matter in these forests is often hampered by various forms of disturbance, including grazing, litter collection and, above all, fire (Wiersum, 1984b)[2]. Secondly, where initial values of surface K^* are particularly low (<5–10 mm h^{-1}?), modest increases of *c.* 3–5 mm h^{-1} per year since planting or abandonment seem to be typical, regardless whether the trees are planted or regenerate naturally. For example, Giambelluca (2002) reported increases of up to *c.* 80 mm h^{-1} over 25 years of forest regeneration after shifting cultivation on red latosols in northern Vietnam whereas under more seasonal climatic conditions and on lateritic soils in Karnataka, India, B. K. Purundara and collaborators found log mean values of K^* to increase from 5.75 mm h^{-1} under degraded grassland to 22.5 mm h^{-1} in a five-year-old plantation of *Acacia auriculiformis*, which further increased to 44.0–56.8 mm h^{-1} after 10–12 years. Since K^* under natural forest in the area is about 250 mm h^{-1} it may take 50 years for the surface infiltration characteristics

2 It is potentially pertinent that Elsenbeer *et al.* (1999) observed rather high values for surface K^* in a non-disturbed teak plantation of unspecified age (but possibly 8–10 years old; Godsey and Elsenbeer, 2002) under rather wet conditions in Rondonia, Brazil, although the relatively low clay content of the latter soil may also be responsible.

to fully recover (M. Bonell, pers. comm.). Thirdly, under similar climatic but somewhat more favourable initial soil conditions (log mean surface K^* *c.* 40 mm h^{-1}), much larger increases in K^* (up to 140 mm h^{-1}) have been observed either through direct measurement (Gilmour *et al.*, 1987) or may be inferred from reductions in HOF volumes (Zhou *et al.*, 2001) within 8–12 years after reforesting degraded grass- and scrubland in Nepal and China with pines and eucalypts, respectively. In the Nepalese study, log mean surface K^* under more or less undisturbed forest ranged from 370 to 525 mm h^{-1}. On the basis of the rapid (apparent) increases in K^* attained between 5 and 12 years after tree planting (from *c.* 51 to 183 mm h^{-1}), it would take as little as 20–30 years to reach pre-disturbance values although the authors suggested that it would 'undoubtedly be many decades' (Gilmour *et al.*, 1987). Fourthly, under favourable initial conditions, natural regeneration may induce very rapid increases in surface K^*. Lal (1996a), for example, reported a tenfold increase within five years of regeneration (from 190 mm h^{-1} to 1930 mm h^{-1}) in Nigeria. Lal's observations would seem to be supported by the much reduced occurrence of HOF under fairly young secondary forest compared to degraded grassland in the Philippines referred to earlier (Chandler and Walter, 1998) (Table 25.4) although Giambelluca (2002) found more modest increases in K^* during forest regeneration in Vietnam. Summarising, a tentative pattern of increased rates of improvement of K^* over time for more favourable initial conditions seems to be discernible from the data although more information is needed to substantiate this and the cited rates of change.

However, the one-dimensional approach outlined above has its limitations for the prediction of changes in hillslope runoff response to rainfall associated with forestation. Naturally, any positive effects of increased surface infiltration capacity on the frequency of occurrence of HOF also depend on prevailing rainfall characteristics. For example, comparing surface K^* values with rainfall intensity-duration distributions, Gilmour *et al.* (1987) concluded that the major increase in surface K^* (by 140 mm h^{-1}) observed 12 years after reforesting a degraded pasture site in the Middle Hills of Nepal with *Pinus roxburghii* would make almost no difference to the frequency of *HOF* because of the prevailing low rainfall intensities. Furthermore, the method has come under scrutiny of late as it tends to overestimate the actual infiltration opportunities during rainfall (Loague and Kiriakidis, 1997; Yu, 1999). A tropical case in point was provided by Lal (1996a, b) in Nigeria who determined very high values of surface K^* but still observed substantial HOF, presumably because of surface sealing by raindrop impact (Lal, 1987). Additional complications arise from the notoriously high spatial variability of K^* and the inability of saturated permeability values to represent effects on infiltration exerted by spatio-temporal variations in soil water contents related to differences in vegetation water use and exposure to radiation and rain. A recently developed spatially variable infiltration model (SVIM) has been shown to represent temporal overland flow patterns from hillslope plots and small catchments much better (Yu *et al.*, 1997; Van Dijk, 2002; cf. Yu, this volume). Therefore, a 'dual-constraint' approach using on-site values of K^* and rainfall intensities on the one hand, and measured (plot-based) volumes of infiltration, HOF and SSF at the hillslope scale should potentially yield more useful results. The authors are not aware of any such integrated studies within the context of evaluating the hydrological consequences of reforesting degraded land in the tropics although the valiant attempts of Chandler and Walter (1998) and Zhou *et al.* (2001) go some way. Clearly, much more work is needed.

Arguably the most thoroughly documented case of the hydrological effects of reforestation of a degraded catchment is from outside the tropics in the White Hollow Catchment, Tennessee, USA (Tennessee Valley Authority, 1961). Both physical and vegetation restoration works were implemented and streamflows monitored for the next 22 years. However, neither total water yield nor dry season flows increased after catchment rehabilitation, and it was concluded that the additional water use of the recovering vegetation cover balanced improved infiltration (Tennessee Valley Authority, 1961). Restoration of streamflow would only seem to be possible where the water use of the forest does not deviate too much from that of the vegetation it has replaced, and where a large increase in infiltration is effected. In a further two documented cases cited by Bruijnzeel (2004) in Slovenia and the Southeastern USA, reforestation of previously degraded catchments was associated with overall reductions in streamflow, particularly during dry years and during the northern summer. This plus the large increases in water use that have been measured in productive plantations of pines and eucalypts that have replaced grasslands in South Africa, Fiji, New Zealand, India and elsewhere (Tables 25.2 and 25.3), strongly suggest that the likelihood of restoring streamflows in tropical catchments by means of establishing timber plantations, or even some other form of forestation, is most unlikely. *The balance of probability is that the already diminished dry season flows in most degraded catchments are likely to be reduced even further by forestation or the establishment of plantations* (Bruijnzeel, 2004), *possibly with the exception of very severely disturbed surface conditions* (Sandström, 1998; Chandler and Walter, 1998) (Table 25.4).

In view of the extent of the 'low-flow' problem (Bartarya, 1989; Pereira, 1989; Sandström, 1998), the testing of alternative ways of increasing water retention in tropical catchments without the excessive water use normally associated with exotic tree plantations should receive high priority. One could think in this respect of (a combination of) physical conservation measures (e.g. bench terracing with grassed risers, contour trenches, runoff collection wells in settlement areas: Negi *et al.*, 1998; Purwanto, 1999; Van Dijk, 2002; cf. Critchley, this volume), vegetative 'filter' strips at strategic points in the landscape (Dillaha *et al.*, 1989; Van Noordwijk *et al.*, 1998), and the use of indigenous species with potentially lower water use (Negi *et al.*, 1998; cf. Bigelow, 2001), possibly in an agroforestry context in which the trees may also assist in enhancing slope stability (Young, 1989, 1997; O'Loughlin, 1984). Calder (1999) suggested rotational land use, in which periods with forest alternate with periods of agricultural cropping, as a potential way of reducing long-term mining of soil water reserves by the trees. Needless to say, for such an approach to be successful, soil degradation during the cropping phases should be avoided as much as possible. Furthermore, there is the intrinsic problem that for optimum soil improvement large amounts of easily decomposable leaf litter need to be produced. Trees that are capable of doing this (e.g. various leguminous trees such as *Albizzia*) are usually also very fast-growing and can be expected to exhibit equally high water uptake rates although information is scarce. Arguably, designing optimal land use strategies that minimise water use and maximise agricultural productivity and streamflow at the catchment scale, constitutes a prime research challenge in the years to come.

FORESTATION EFFECTS ON EROSION AND SEDIMENT YIELDS

Deforestation is typically associated with general disturbance of vegetation, ground cover and soil, leading to an increase in soil erosion and sedimentation (Lal, 1987; Bruijnzeel, 1990; Grip *et al.*, this volume). Generally, the degree of surface disturbance is largely controllable, being determined by local physical conditions and the quality of management (cf. chapters by Thang and Chappell, and Cassells and Bruijnzeel, both this volume). However, the increased streamflow associated with the removal of transpiring and intercepting vegetation cover may also increase sediment yields, particularly from the channel and the riparian zone. Similarly, apart from the reduction in stream transporting power (Globevnik, 1998), forestation should also lead to a gradual reduction in surface erosion and sediment yields to the extent that it improves surface cover, restores soil hydraulic properties and allows for recovery of disturbance scars (including skidder tracks, temporary roads, and shallow landslides). Arguably, the most pronounced effects of forestation may be expected in the case of severely eroding land (as opposed to severely *eroded* land where most of the erodible soil material has been removed already and yields are low again; Flatfjord, 1976). Sediment production under natural, forested conditions may vary widely, however, depending on the relative importance of the respective contributing mechanisms (overland flow, gullying, and mass wasting: Pearce, 1986; Douglas and Guyot, this volume). By the same reasoning, forestation of degraded land can be expected to produce a range of responses in terms of reducing sediment production, ranging from minor to substantial. Sediment yields for humid tropical forested headwater catchments typically increase in the sequence: granite/metamorphic rocks < young volcanic deposits < marls / claystones (Douglas and Guyot, this volume; Bruijnzeel, 2004; cf. Figure 22.6 in Grip *et al.*, this volume). The highest values are usually associated with tectonically active steepland areas prone to mass wasting or where landsliding combines with widespread overland flow, such as in marly areas under a seasonal rainfall regime. In such geomorphologically active terrain, it is unlikely that forestation will have much of an effect in terms of reducing sediment yields (Bell, 1973; Bruijnzeel and Bremmer, 1989). However, where landscapes are relatively stable the restoration of forest may make a relatively large difference on sediment yields (Tennessee Valley Authority, 1961; Lal, 1987) (Figure 25.8). Therefore, when dealing with the effects of changes in land use on erosion and sedimentation, it is helpful to distinguish between surface erosion, gully erosion, and mass movements, because the ability of a vegetation cover to control these various forms of erosion is rather different.

Surface erosion

This form of erosion is rarely significant in areas where the soil surface is protected against the direct impact of the rain, be it through a litter layer maintained by some sort of vegetation or through the application of a mulching layer in an agricultural context (Wiersum, 1984b). Erosion rates increase somewhat upon removal of the understorey but rise dramatically only when the litter layer is removed or destroyed. The initial effect is rather small due to the effect of residual organic matter on soil aggregate stability and infiltration capacity (Wiersum, 1985) but becomes considerable upon repeated disturbance of the soil by burning, frequent weeding or overgrazing, which all tend to make the soil compacted or crusted, with impaired infiltration and accelerated erosion as a result (Wiersum, 1984b; Zhou *et al.*, 2001). Forestation has been shown to be able to reverse these processes over time, through the restoration of plant cover and with it an accumulation of litter to cover the soil surface and the return of biotic activity (Lal, 1987; Zhou *et al.*, 2001). Although erosion on grasslands in good condition (Fritsch and Sarrailh, 1986) or in agricultural fields with appropriate soil conservation measures on otherwise stable slopes (Paningbatan *et al.*, 1995; Young, 1989; Critchley, this volume) is usually low, Smiet (1987) made the pertinent observation that forests provide greater latitude with respect to protection of the soil surface against erosion as compared to grazing or annual cropping. Whilst the degraded natural and plantation forests of many tropical uplands are still able to fulfil a protective role because gaps are usually rapidly colonised by pioneer species, grazing lands are often prone to fire, overgrazing and landsliding.

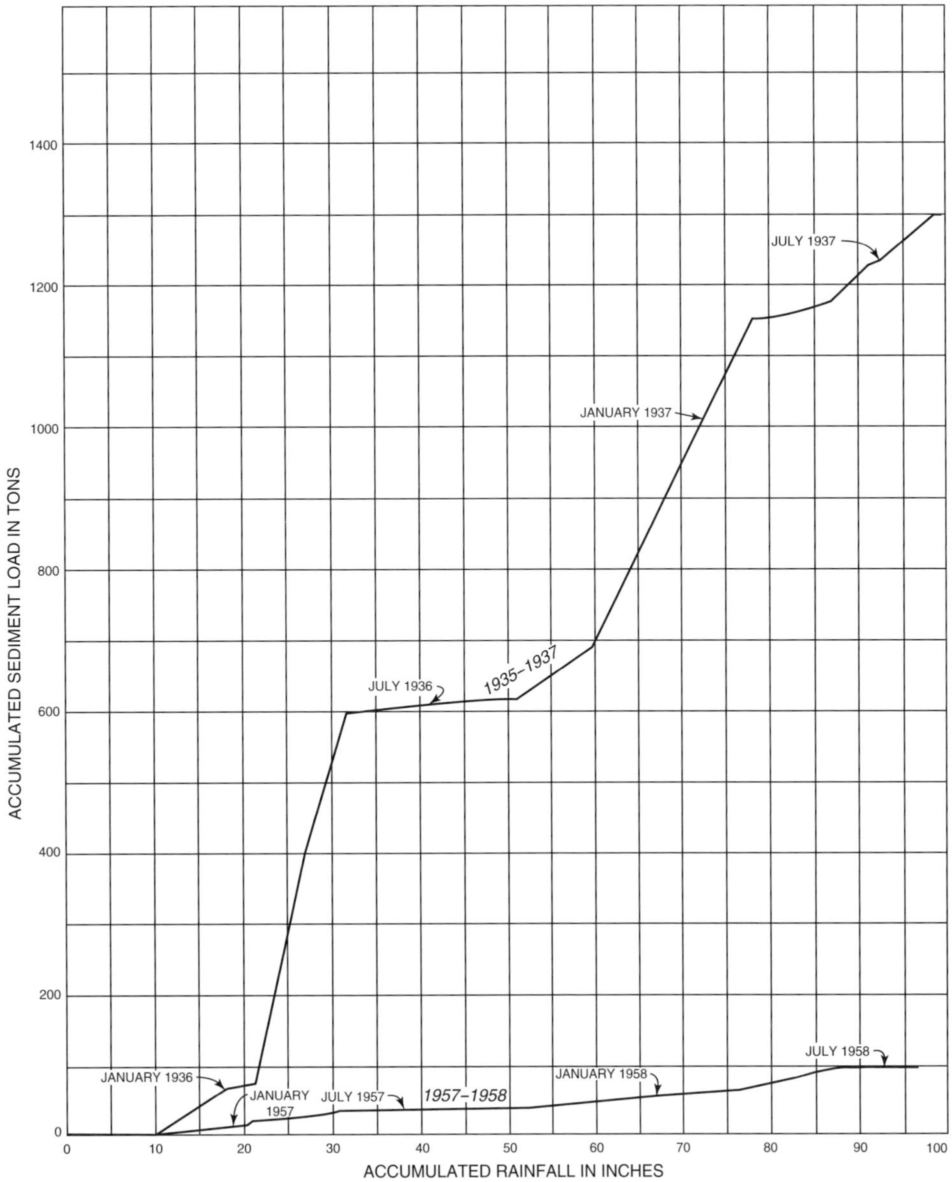

Figure 25.8 Cumulative sediment yield versus cumulative rainfall in the White Hollow catchment, Tennessee, USA before and after reforestation and other restorative measures. (After Tennessee Valley Authority, 1961.)

There is increasing evidence that erosion rates on and around such compacted surfaces as skidder tracks and log landings, roads, footpaths and settlements can be very high (35–500 t ha^{-1} yr^{-1}; see Bruijnzeel (2004) for details). In addition, the very considerable volumes of runoff generated by such impervious surfaces may promote downslope gully formation and mass wastage. Therefore, contributions of runoff and sediment to the stream network from such areas may be disproportionately large for their relatively small surface area. The lesson for forestation projects is clear, therefore: whilst the overall potential to reduce sediment production is good, frequent disturbance of ground cover, removal of surface litter, and careless harvesting, road-building or road maintenance, can reverse much of the positive effects of a forest cover. This is particularly important where forestation is part of a conversion to intensively managed timber plantations. During establishment of the new plantation road-building and site preparation, including burning, will tend to increase disturbance and thereby increase the risk of soil erosion and sediment production. Productive plantations can be harvested on a short rotation (as short as 7–10 years for pulp crops in the better eucalypt plantations in South Africa and Brazil; Gonçalves *et al.*, 1997), introducing the potential for regular disturbance associated with harvesting and road works. During the active growth phase of the plantation, there is a strong potential for desiccation of the catchments and improved ground cover, and as a result, lower overland flow risk and quickflow volumes, and consequently lower sediment yields.

Gully erosion

This is a relatively rare phenomenon in most tropical rainforests but may be triggered during extreme rainfall when the soil becomes exposed through treefall or landslips. In other cases, gullies may form by the collapse of subsurface soil pipes (Douglas and Guyot, this volume). Active gullying in formerly forested areas is often related to compaction of the soil by over-grazing or the improper discharging of runoff from roads, trails and settlements. Alternatively, gullies may develop in the centre of landslides where surface flow tends to concentrate, or when topsoil is removed during forestry operations by heavy machinery leaving more erodible subsoil material exposed (Bruijnzeel, 1997). If gullies are not treated at an early stage, they may reach a point where restoration becomes difficult and expensive. The moderating effect of forestation on actively eroding gullies is limited, however, and additional mechanical measures such as check dams, retaining walls and diversion ditches will be needed (Blaisdell, 1981; FAO, 1985, 1986).

Mass wasting

Mass wasting in the form of deep-seated (>3m) landslides is not influenced appreciably by the presence or absence of a well-developed forest cover. Geological (degree of fracturing, seismicity), topographical (slope steepness and shape) and climatic factors (notably rainfall) are the dominant controls (Ramsay, 1987a, b). However, the presence of a forest cover is generally considered important in the prevention of shallow (<1m) slides, the chief factor being mechanical reinforcement of the soil by the tree root network (O'Loughlin, 1984). Bruijnzeel and Bremmer (1989) cite unpublished observations by I. R. Manandhar and N. R. Khanal on the occurrence of shallow landslides in an area underlain by limestones and phyllites the Middle Hills of Nepal. Most of the 650 slips that were recorded between 1972 and 1986 had been triggered on steep (>33°) deforested slopes during a single cloudburst whereas only a few landslides had occurred in the thickly vegetated headwater area. However, under certain extreme conditions, such as the passage of a hurricane, the presence of a tall tree cover may become a liability in that trees at exposed locations may be particularly prone to becoming uprooted, whereas, in addition, the weight of the trees may become a decisive factor once the soils reach saturation. Scatena and Larsen (1991) reported that out of 285 landslides associated with the passage of Hurricane Hugo over eastern Puerto Rico, 77% occurred on forest-covered slopes and ridges. More than half of these mostly shallow landslips were on concave slopes that had received at least 200 mm of rain. Brunsden *et al.* (1981) described a similar case in eastern Nepal where mass wasting on steep forested slopes was much more intensive than in more gently sloping cultivated areas. Although often occurring in large numbers, such small and shallow slope failures usually become quickly revegetated and, because of their predominant occurrence on the higher and central portions of the slopes, contribute relatively little to overall stream sediment loads, in contrast to their more deep-seated counterparts (Ramsay, 1987a).

Summarising, the effect of forestation on catchment sediment yield depends on a large number of factors, most notably the relative contributions by surface erosion, gullying and mass wasting. Whilst degraded areas with widespread gullying or massive landsliding will continue to produce significant amounts of sediment following forestation, effects can be expected to be rapid and beneficial in the more typical case of rampant surface erosion and occasional gullying. The restoration of the White Hollow catchment in Tennessee, USA, provides a case in point. Despite serious surface erosion and (some) gullying and bank erosion, sediment yields from this degraded semi-forested and overgrazed catchment responded to reforestation and gully stabilisation works within two years and stabilised at a very low level ever since (Figure 25.8).

CHANGES IN SOIL CHEMICAL CHARACTERISTICS WITH LAND COVER CHANGE

The establishment of tropical tree plantations, like any other form of land use replacing old-growth forest or (degraded) grassland, leads to major changes in nutrient fluxes within and out of the ecosystem, and therefore in soil nutrient reserves (Fölster and Khanna, 1997). Whether these changes degrade or improve soil fertility depends on the original land use, the new land cover, the management practices applied as well as the type of soil and its inherent fertility level. As discussed by Bruijnzeel (1991) and Proctor (this volume), nutrient losses via drainage from undisturbed old-growth rainforests depend on the fertility of the substrate, with larger losses being associated with more fertile substrates. Site fertility is potentially threatened upon forest conversion by the bulk removal of nutrients in harvested timber, through surface erosion due to soil disturbance, and via enhanced leaching. The latter is caused by a combination of (i) the larger volumes of water percolating through the soil after opening up of the canopy (due to reductions in rainfall interception and water uptake), (ii) a temporarily reduced capacity of damaged or sparse vegetation cover to take up nutrients, and (iii) the sudden addition of large amounts of fresh organic debris to the forest floor that are either left to decompose or be burned to facilitate future access for planting (Bruijnzeel, 1998). As such, the establishment of a tree plantation on rainforest land can be expected to set in motion a series of processes that will result ultimately in overall decreased soil fertility (Ruhiyat, 1989; Spangenberg *et al.*, 1996; Mackensen *et al.*, 2003; cf. Grip *et al.*, this volume).

Moreover, plantations are usually established to harvest wood for timber or pulp and repeated harvesting of the trees inevitably diminishes soil nutrient reserves through the export of nutrients with biomass removal. The associated losses depend on a number of factors, including: site management history, tree species, rate of tree growth, timber nutrient concentrations, rotation length, harvesting method, as well as nutrient additions from atmospheric sources and weathering (Fölster and Khanna, 1997, Mackensen *et al.*, 2003). Naturally, the need to replace nutrients lost in this way assumes increased importance as more fast-growing species are being planted on low-fertility tropical soils (Brown *et al.*, 1997; Mackensen, 1998). On the other hand, reforestation of degraded grasslands by appropriate management practices may result in a general improvement of soil fertility. One mechanism through which this may be attained is by deep roots taking up nutrients from the subsoil that were previously out of reach but now added to the surface soil via leaf fall and subsequent decomposition. However, direct information on soil improvement through tree planting under humid tropical conditions is exceedingly scarce (Young, 1997).

Fölster and Khanna (1997) and Bruijnzeel (1990, 1998) have drawn attention to the methodological difficulties encountered when trying to evaluate soil fertility response to land use change. Basically, two approaches have been followed, viz. (i) monitoring the changes in (top)soil nutrient concentrations over time in one and the same area (e.g. Amir *et al.*, 1990; Gillman *et al.*, 1985) and (ii) the use of so-called 'false time series' in which soil data from a series of stands of different ages are compared (Hase and Fölster, 1983; Buschbacher, 1984; Bruijnzeel and Wiersum, 1985). Each method has its specific difficulties and limitations. The direct sampling approach not only suffers from problems related to high spatial and temporal variability in soil nutrient concentrations (cf. Proctor, this volume) but also tends to become impractical, particularly in the case of longer-term rotations. Chronosequence studies avoid the latter problem but must assume that all sites under consideration were initially comparable if the results are to be meaningful. Needless to say, the high spatial variability that is typically associated with forest soils easily confounds the results obtained with false time series (Hase and Fölster, 1983; Bruijnzeel, 1990; Waterloo, 1994). Even if based on proper sampling, detailed soil chemical information for a single site may give a limited or even erroneous idea of the magnitude of nutrient reserves or the rates of the various processes acting upon them for the forest as a whole (Van Dam, 2001). Fölster and Khanna (1997) and Bruijnzeel (1998) therefore advocated a nutrient budget approach, which compares the balance between nutrient input and output fluxes over a rotation period with adequately characterised soil nutrient reserves. No single study has measured all the respective inputs and outputs over a full rotation period but work conducted on *Pinus caribaea* on former grassland soils in Fiji (Waterloo, 1994) and *Acacia mangium* replacing logged-over rainforest in Sabah (Malmer and Grip, 1994; Malmer, 1996; Nykvist *et al.*, 1994) come close. Mackensen *et al.* (2003) applied the budget approach to evaluate net nutrient losses from stands of *A. mangium*, *Eucalyptus deglupta* and *Paraserianthes falcataria* grown in eight-year rotations in East Kalimantan, Indonesia. Estimates of losses associated with leaching, burning and erosion were based on literature data, whereas elemental concentrations in harvested material and the soils were measured.

This section discusses the changes in soil chemical characteristics occurring during plantation establishment and development in a general manner. Most published studies have focused on the changes associated with the conversion from natural forest to plantation forest, rather than the establishment of forest on degraded tropical grasslands (Waterloo, 1994). The soils of the latter are often so impoverished, however, that fertilisation is required to assure establishment of the trees (Otsamo, 1998). In the following the relevant processes affecting soil nutrient levels during land clearing and plantation establishment, maturation and harvesting are briefly described.

Processes affecting soil nutrient levels during land clearing and plantation establishment

Enhanced mineralisation

Clear-felled vegetation (logging debris) is decomposed on the site and leaching of certain nutrients (notably potassium, sodium and phosphorus) from the decomposing material has been shown to be significant (Ewel *et al.*, 1981; Mackensen *et al.*, 1996; Figure 25.9). Mineralisation rates of leaf litter and fine roots are usually enhanced upon forest cutting because of higher nutrient availability (Malmer and Grip, 1994; Silver *et al.*, 1996), increased soil temperatures (Lal, 1987; Palm, Swift and Woomer, 1996) and higher soil moisture levels (Klinge *et al.*, 1998). Harcombe (1977) reported increases in carbon losses of 7–54% associated with the mineralisation of litter including fine roots upon forest clearing in Central America whilst Sanchez, Villachica and Bandy (1983) found increases in mineralisation rates of 25% for carbon and 17% for nitrogen upon forest cutting in the Peruvian Amazon. In contrast to the above findings, *reduced* mineralisation rates have been observed where the litter was rather dry, particularly where the litter layer had also become reduced, scattered or largely destroyed, e.g. during forestry operations or slash and burn cultivation (Ewel, 1976; Brouwer, 1996). Van Dam (2001) found net-nitrogen mineralisation in logging gaps of more than 200 m^2 to be four times lower compared to those in surrounding undisturbed forest on sandy soils in Guyana. On the other hand, Klinge (1998) measured 7–10 times higher nitrate concentrations in percolating soil water at 25 cm depth in eastern Amazonia soon after forest cutting (and before burning). These high values were attributed to rapid initial mineralisation of slash and, especially, decaying fine roots in the topsoil. Evidence for the latter comes from the fact that

Table 25.5. *Relative nutrient losses due to volatilisation during burning of residual slash as a function of fuel mass and mass reduction*

	Slash (t ha^{-1})	Reduction (%)[a]	Percentage nutrient losses N	P	K	Ca	Mg	Reference
Costa Rica	38.5	83	23					Ewel *et al.* (1981)
Australia	11.4	64	64	44	55	33	37	Raison *et al.* (1985)
Brazil	6.3	90	95	51	44	52	42	Pivello and Coutinho (1992)
Fiji	40	86	84	52	79	78	60	Waterloo (1994)
Brazil	31.2	94	98	47	48	35	40	Mackensen *et al.* (1996)
	33.5	91	94	27	16	9	17	
	95.2	96	98	33	31	24	43	
Average		86	82	42	46	39	40	

[a] Relative amounts of slash reduction added for comparison.

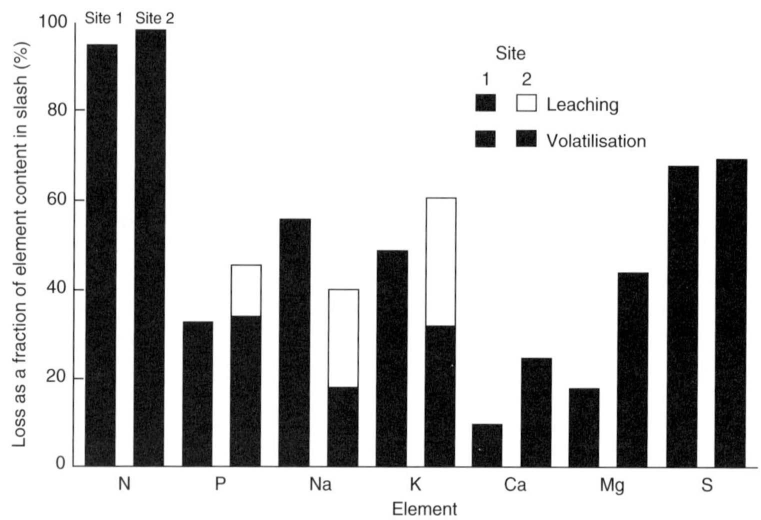

Figure 25.9 Nutrient losses upon burning of harvest slash (left histograms 33 t ha^{-1} and right 90 t ha^{-1}) in a secondary forest in eastern Amazonia. Lower part of the bars indicates losses due to volatilisation during burning, the upper part represents losses due to leaching from the slash before burning. All values given as percentage of the element content of the slash. (After Mackensen *et al.*, 1996.)

concentrations of calcium and magnesium in topsoil moisture were elevated after felling (and before burning) although these elements were not leached in detectable quantities from the residual slash (Klinge, 1998) (Figure 25.9).

Nutrient losses through slash burning

Residual phytomass is often burned on the site to facilitate access for planting. Volatilisation and ash particle transport during and shortly after burning can contribute significantly to overall losses of certain nutrients following site conversion to plantations (Figure 25.9, Table 25.5).

Reported atmospheric nutrient losses associated with burning range from 9–98%, the wide range reflecting differences in the fire intensity and volume of burned slash (Table 25.5). In the majority of studies the loss is closely related to the relative weight reduction of the slash which is, in turn, a function of the intensity of the fire (Mackensen *et al.*, 1996). Losses of phosphorus (P), potassium (K), calcium (Ca) and magnesium (Mg) are generally comparable and average 39–46% whilst the average loss of the more volatile nitrogen (N) is 82% (Table 25.5).

The transformation of organic matter into mineral ash through slash burning has a most profound and usually rather immediate impact on soil fertility since mineral ash is readily soluble and easily washed into the soil (Khanna, Raison and Falkiner, 1994). Again, the more material that is burnt and the larger the amount of mineral ash left on the site, the higher the impact on soil fertility and the higher the potential for losses through subsequent leaching (Malmer and Grip, 1994; see also the next section). A comparison of soil nutrient characteristics before and after burning generally shows increases in soil pH and base saturation levels, possibly even an increase in cation exchange capacity as well (Tomkins *et al.*, 1991). The extent of the rises in topsoil pH and base saturation is governed by initial soil acidity and the amount of ash. The duration of the effect differs strongly between sites as a function of rainfall regime (and thus leaching potential) and edaphic characteristics (notably soil texture and clay mineralogy; Sanchez, 1976). For example, Nye and Greenland (1964) reported an initial increase in pH from 5.2 to 8.2 in the top layer of an Alfisol/Luvisol in Ghana (moderate rainfall) after burning 50-year-old secondary growth. About two years later, the pH still amounted to 7. Conversely, only a modest increase in pH (from 3.8 to 4.5) was noticed after burning on a highly depleted Oxisol/Ferrasol under a more intense rainfall regime in Central Amazonia. In addition, the increase disappeared within four months (Brinkmann and Nascimento, 1973). Examples of changes in soil chemical conditions in the context of tropical plantation establishment include the plot-scale studies by Klinge (1998) in eastern Amazonia (Oxisols, covering one year) and, at the plantation scale, those by Ruhiyat (1989) and Mackensen (1998) in East Kalimantan, Indonesia (both covering one rotation or 10 years; Ultisols/Acrisols).

Nutrient leaching

Clearfelling and burning of old-growth forest results in (at least temporarily) decreased evapotranspiration (Table 25.2) and thus in increased drainage through the soil profile and water yield from a catchment (Klinge *et al.*, 1998; Malmer, 1992; Grip *et al.*, this volume). This is accompanied by enhanced nutrient leaching from the site as illustrated by a small but growing number of studies (Uhl and Jordan, 1984; Malmer and Grip, 1994; Waterloo, 1994; Klinge, 1998).

Klinge (1998) used vacuum-tube lysimeters and recording tensiometers to determine leaching losses at 25, 40, 60 and 110 cm depths during the conversion of old-growth forest to a eucalypt plantation in eastern Amazonia. Concentrations of calcium, magnesium and nitrate (NO_3-N) in the percolating water showed a significant increase at 25 cm depth within two weeks after felling, whereas concentrations of potassium were raised only upon slash-burning. Although phosphorus was found to be leached from the residual phytomass (Figure 25.9), no increase in PO_4-P concentrations in topsoil moisture was found, probably because of immediate phosphorus fixation in the topsoil (Uehara and Gillman, 1981). The observed increase in amounts of phosphorus stored in the topsoil after site conversion supports this idea (Klinge, 1998). Nutrient transfers to the subsoil were greatly enhanced upon the start of the rainy season when nutrients were rapidly leached beyond a depth of 1.1 m, and peaked about five months after forest conversion and burning (Klinge, 1998).

Brouwer (1996) used a similar (though non-automated) approach and observed comparable patterns of nutrient leaching in gaps of different sizes created by logging on sandy soils in Guyana. Although no slash was burned, nutrient concentrations in soil water at 1.2 m depth peaked after about three months, reflecting the lower water and nutrient retention capacities of these sandy soils compared to the soils studied by Klinge (1998). In addition, the magnitude of leaching was shown to depend on the size of the canopy opening, with the largest losses being associated with the largest gap which also experienced the greatest soil disturbance (Brouwer, 1996). At a somewhat larger scale (headwater catchments of 3.4–9.7 ha underlain by a mixture of clayey and sandy soils) in Sabah, Malaysia, Malmer and Grip (1994) observed an almost immediate response of potassium concentrations in streamflow during storms after logging only and after logging followed by burning, with the greatest increases in the latter case. Concentrations of NO_3-N took four weeks to respond. Much of the increased leaching losses occurred during stormflows in the form of shallow throughflow (SSF) and valley bottom saturation overland flow (SOF; Malmer and Grip, 1994; cf. Bonell, this volume). Despite the methodological differences between catchment and plot-based studies, differences in observed response times and process dynamics may be explained in terms of contrasts in precipitation and soil water (hillslope runoff) regimes and the soil exchange complex (clay and organic matter).

The interaction between percolating nutrients and the soil exchange complex is an important determinant of leaching patterns. For example, soils with 1:1-type clay minerals (kaolinite) and low organic matter content such as the Oxisols studied by Klinge (1998), retain less base cations than do Ultisols with their 2:1-type clay minerals. Under certain conditions, the 1:1-type clays also develop stronger bonds to potassium than to calcium (Levy *et al.*, 1988; Udo, 1978). This particular interaction probably accounts for the delayed peak in potassium concentration in the soil solution observed by Klinge (1998). During the period that concentrations of calcium in the soil solution peak, aluminium (and to a lesser extent hydrogen) will be replaced on the soil exchange complex and this leads to enhanced concentrations of Al^{3+} and H^+ in the soil solution (Klinge, 1998; Brouwer, 1996). This 'free' Al^{3+} in the soil solution reacts with H_2O to produce $Al(OH)_3$ and thus contributes to a further increase in H^+ concentrations. Together with the enhanced rates of mineralisation and nitrification signalled earlier, this explains the temporary decrease of the pH in the soil solution down to a depth of 0.6 m at the Amazonian site of Klinge (1998). The pH in soil solution rises again after the calcium peak subsides. Similar drops in soil water pH were observed below gaps created by logging in the absence of fire in Guyana (Brouwer, 1996) and were also described by Sollins *et al.*, 1994.

The *duration* of enhanced nutrient leaching depends on site management (including soil disturbance), the above- and below-ground nutrient pools represented by the cleared vegetation, soil texture (retention capacity) and the rate of growth of the new vegetation. In the eastern Amazonian example, increased nutrient concentrations in the deeper soil solution lasted for 6 to 9 months, with the exception of K and SO_4-S whose concentrations remained elevated down to depths of 110 cm for at least 12–15 months depending on the initial amount of slash that was burnt (Klinge, 1998). In East Malaysia, overall nutrient concentrations in baseflow remained elevated compared to a control stream for almost two years after logging or logging and burning, while the extra effect caused by the use of fire was detectable for about one year (Malmer and Grip, 1994). In Guyana, the accumulation of slowly decomposing slash in the largest gap (3440 m^2) resulted in a comparatively small nutrient flux increase which remained significantly elevated for about 15 months, although concentrations of Ca, Mg, SO_4 and NO_3 remained slightly elevated for as long as 7 years after gap creation; Van Dam, 2001, cf. Figure 21.3 in Malmer *et al.*, this volume. The substantial surface disturbance by heavy machinery retarded the regeneration of the vegetation in the large gaps in this particular case but where regrowth is rapid and vigorous, nutrient losses will be reduced by accumulation in the phytomass. Brouwer (1996) observed reduced concentrations of K in percolating soil water compared to adjacent old-growth forest after about 34 months, indicating that by then the build-up of nutrient stocks may have begun. Uhl and Jordan (1984) reported a similar phenomenon for Ca and NO_3 during forest regrowth in the Venezuelan Amazon for a site with clayey soils about two years after felling and burning. Fast-growing tree species and dense undergrowth are equally helpful in this respect. Undergrowth in particular can play an important role in the short- to medium-term storage of nutrients (Smethurst and Nambiar, 1995). Cutting of undergrowth led to increased nutrient losses through leaching in Amazonia (Klinge, 1998) whereas a similar measure during the first years of plantation development in Indonesia resulted in cumulative nutrient losses which were quantitatively comparable to losses initiated by the initial site conversion (see Mackensen, 1998, for details). Summarising, enhanced leaching is a rather temporary process and although the associated nutrient losses can be substantial, they are usually much smaller than those associated with timber removal (Bruijnzeel, 1998; Mackensen, 1998; Figure 25.10a–d).

Erosion

Even in undisturbed forest ecosystems and tree plantations with a well-developed litter layer some erosion may occur. Reported rates in the early literature (summarised by Wiersum, 1984b) range from 0.02 to a maximum of 6.2 t ha^{-1} yr^{-1} (median values of 0.3 and 0.6 t ha^{-1} yr^{-1} for forests and plantations, respectively; $n = 47$). During the establishment phase, erosion is likely to increase significantly although the effect is usually short-lived (Grip *et al.*, this volume; cf. Zhou *et al.*, 2001) (Figure 25.8). Wiersum (1984b) derived a median rate of 51 t ha^{-1} yr^{-1} for clean-weeded and otherwise disturbed and burned tree crops (range 1.2–183 t ha^{-1} yr^{-1}; $n = 24$). The associated nutrient losses can be significant for the overall nutrient budget of plantation sites, especially on low-fertility soils and for stands grown in short rotations where such disturbances will occur every 7–10 years (Gonçalves *et al.*, 1997; Mackensen, 1998) (Figure 25.10a–d). However, nutrient losses due to surface erosion may also be important in plantations grown over longer rotations and on more fertile soils. In various densely populated countries where land pressure for food production is high (e.g. Indonesia), new plantations are often established through a *taungya* system in which local farmers are allowed to grow their crops between the newly planted trees in exchange for labour during plantation establishment. The cropping phase officially lasts three years, after which the trees and undergrowth take over and the farmers move to new fields within the forest estate. However, erosion during the *taungya* phase can be substantial (Wiersum, 1984b) and in one of the few assessments in the context of a plantation nutrient budget the associated nutrient losses were considered to be at least as high as the corresponding losses via timber harvesting, particularly when considering total rather than exchangeable nutrients contained in the eroded soil material (Bruijnzeel, 1992).

Declining soil nutrient reserves in intensively managed plantations

The nutrient enrichment of topsoils after rainforest conversion to pasture or cropping is a passing phenomenon (Hölscher *et al.*, this volume) and should not lead one to assume that the system as a whole is also enriched in nutrients. Rather, because of the high management-induced nutrient losses during plantation establishment described in the previous section (volatilisation, erosion, leaching), forest conversion is likely to contribute significantly to long-term depletion of fertility in the plantation system (Spangenberg *et al.*, 1996; Fölster and Khanna, 1997). As shown in Figure 25.10a–d, tropical forest conversion and intensive plantation management result in distinct negative nutrient fluxes. As a result, the newly established nutrient equilibrium will be at a lower level than the previous one under natural forest and continue to decrease with each subsequent rotation (Jordan, 1985). Therefore, continued nutrient exports through repeated biomass removal in subsequent rotations may well endanger sustainable plantation productivity, especially on poor sites (Oxisols/Ferrasols, Ultisols/Acrisols, Spodosols/Podsols) with the critical nutrient differing between situations (Spangenberg *et al.*, 1996; Fölster and Khanna, 1997; Mackensen, 1998; Nykvist *et al.*, 2000; Mackensen *et al.*, 2003). Sufficient compensation of nutrient losses through management-independent input fluxes such as precipitation or

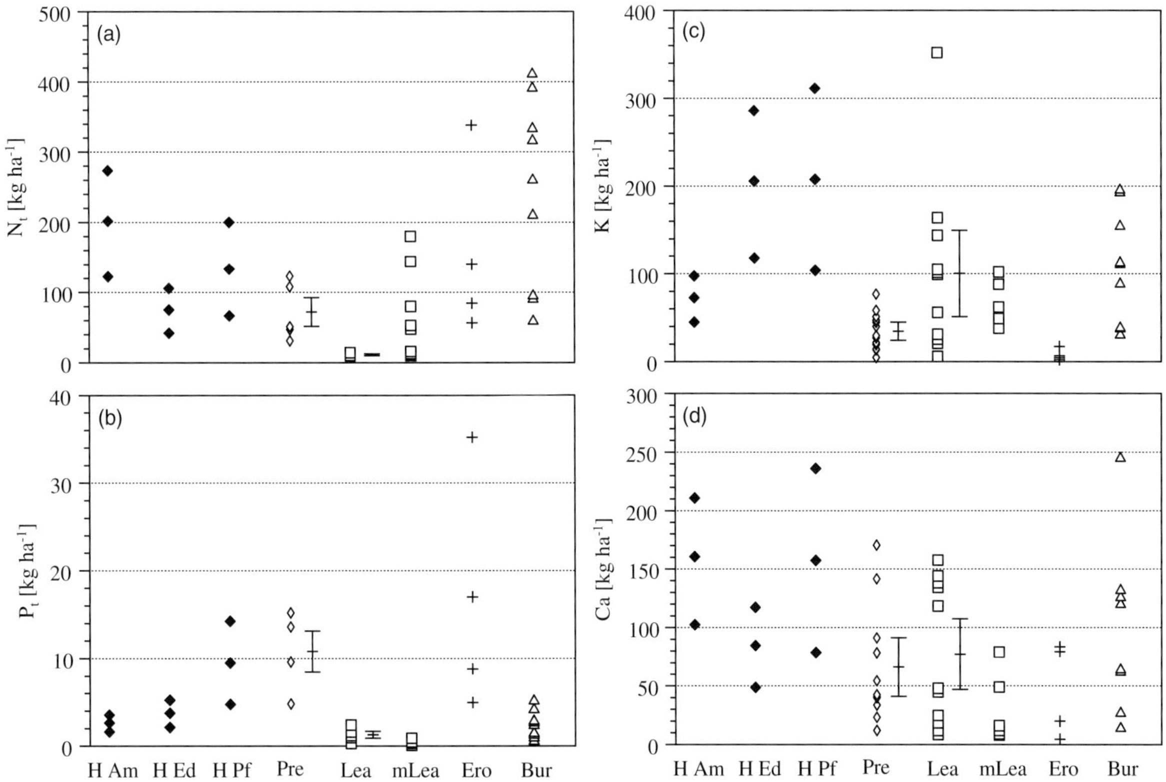

Figure 25.10 Comparison of nutrient gains and losses throughout a rotation of eight years for plantations of *Acacia mangium* (Am), *Eucalyptus deglupta* (Ed) and *Paraserianthes falcataria* (Pf) on Ultisols in East Kalimantan, Indonesia: (a) nitrogen, (b) phosphorus, (c) potassium, and (d) calcium.Explanation of remaining symbols: H Am, H Ed and H Pf – Nutrient losses associated with the harvesting of timber for the three different tree species and at harvesting volumes of 100, 200 and 300 m^3 ha^{-1} respectively; Pre – Nutrient gains through precipitation (regional literature data); Lea – Baseline nutrient losses via leaching from natural forest (regional literature data); mLea – Management-dependent nutrient losses through leaching following stand conversion and preparation; Ero – Nutrient losses through erosion (assumed losses of topsoil of 50 and 200 tonnes ha^{-1}); Bur – Nutrient losses through burning of residual slash for harvest volume of 300 m^3 ha^{-1}. (Modified from Mackensen, 1998.)

weathering is unlikely, particularly where soils are highly weathered. Klinge (1998) calculated that nutrient inputs through bulk precipitation would take between 10 and 230 years to compensate for nutrient losses triggered by plantation establishment in eastern Amazonia. Similarly, periods of 35–55 years would be needed to cover losses of Ca, Mg and K associated with the clearance of rainforest to make way for *Acacia mangium* plantations in Malaysia on the basis of atmospheric nutrient inputs only. These were reduced to 4–20 years after inclusion of potential contributions by weathering (see Bruijnzeel, 1998, for details). It should be noted, however, that these calculations do not yet include any losses associated with the removal of timber at the end of the first rotation. At any rate, therefore, growing *A. mangium* on a ten-year rotational basis is bound to deplete the system beyond naturally occurring nutrient gains.

Because of the steady depletion of overall nutrient reserves under intensive tropical tree plantation schemes fertilisation would seem inevitable. However, as discussed in detail by Fölster and Khanna (1997) and Gonçalves *et al.* (1997), fertiliser application is not without problems either. Apart from economic considerations (Mackensen, Fölster and Ruhiyat, 2000) there are dangers of acidification and exhaustion of (micro-)nutrients not covered by the application. Therefore, Fölster and Khanna (1997) advocate the harvesting of stem wood only and leaving bark material on-site where possible (cf. Crane and Raison, 1980; Bruijnzeel and Wiersum, 1985).

Conversion of grasslands into plantations

Studies of nutrient budgets following forestation of tropical grasslands are extremely rare. To make matters worse, the results of one of the very few 'false time series' studies available (Waterloo, 1994) are confounded by problems of different initial soil fertility levels between plots. Generally, the lower overall nutrient storage in degraded grassland ecosystems as compared to rainforests on similar soil types can be expected to lead to much lower volatilisation and leaching losses upon plantation establishment (cf. Malmer and Grip, 1994). Similarly, such losses might also be compensated faster by nutrient inputs from bulk precipitation or (where applicable) weathering of soil parent material. Under subhumid conditions in the Southern Congo, savanna vegetation on low fertility sands has been converted to eucalypt plantations. Both the plantation and the trees in the savanna were found to be highly efficient in retaining nutrients from atmospheric inputs, with very little loss of plant-utilisable nutrients being leached out of the rooting zone (Laclau, Boillet and Ranger, 2000).

As is the case with natural forest recovery (Hölscher *et al.*, this volume), overall nutrient availability will determine the growth rate of forestation schemes in grasslands. Equally important is the intensity and duration of previous management (e.g. grazing on pastures) as these prove to be strong determinants of site nutrient status upon forestation. For example, in abandoned pastures in eastern Amazonia, Uhl, Buschbacher and Serrão (1988) observed above-ground biomass accumulation of secondary vegetation to be considerably higher on moderately used pastures than on intensively used pastures. Similar findings were reported by Buschbacher, Uhl and Serrão (1988), Hughes, Kauffman and Jaramillo (1999), Aide *et al.* (1995), and Fearnside and Guimaraes (1996). Furthermore, the extent of nutrient losses through leaching will depend also on the management techniques applied during tree planting. Strip-planting, which opens only a narrow strip of grass for tree planting is likely to result in smaller nutrient losses than the complete destruction of the grass cover through ploughing, slash burning or herbicide application (Otsamo *et al.*, 1994).

In the long run reforestation of degraded grassland sites may have positive effects on soil nutrient status. Over time plantation stands accumulate more nutrients and organic matter in the system than the previous grassland would and nutrient build-up during long rotations (>30 years) may match those of natural forest types on similar soils. As indicated previously, enhanced nutrient and organic matter cycling in the topsoil, nutrient pumping from deeper soils layers that were out of reach for the grass roots, and improved utilisation of nutrient input by precipitation (cf. Laclau *et al.*, 2000) may all contribute to improved fertility of reforested grasslands although hard evidence for this seems to be lacking at present (Waterloo, 1994; Young, 1997). Ironically, it is this same deep-rooting nature of trees that is partly responsible for their increased water use compared to grasslands or annual crops.

CONCLUSIONS AND RECOMMENDATIONS

Based on a review of a rather limited body of applicable research and partly extrapolating from hydrological studies in other parts of the world, we conclude that the prospects for enhanced rainfall and augmented base flows as a consequence of forestation in the humid tropics are generally poor, but dependent on site-specific factors.

In terms of being able to predict the hydrological effects of land use changes it is especially important to understand the local hydrology, at least in terms of the dominant processes governing runoff generation. Erosion and sediment production from degraded landscapes may be readily moderated and stabilised at very low levels by forestation, provided they are not primarily determined by geological and topographic factors. In the longer term, the frequency of disturbance and the quality of forest management, particularly as it relates to roads and harvesting of timber and litter, will determine the on-going erosion risk and sediment yield of reforested catchments. Where rainfalls are particularly intense, and where soils are particularly clayey or degraded physically, there is greater potential for overland flow and near-surface throughflow to contribute to stormflows. In these situations there is the greatest opportunity for degraded catchments to be restored to improved hydrological function through forestation. Where soils are deep and porous and comparatively little disturbed, the effect of forestation on stormflows will be modest and more pronounced through lowered baseflows.

The establishment of plantations leads to major changes in the characteristics and nutrient fluxes of the ecosystem. Whether these changes degrade or improve soil fertility depends on the original land cover (forest or degraded grassland) and the associated management practices. Forestation of degraded land by appropriate management practices may result in a general improvement of soil fertility (although hard evidence from the humid tropics seems to be lacking), while the replacement of native forest by plantation generally initiates a range of processes resulting ultimately in a decline in soil fertility. The nutrient enrichment of topsoils after forest conversion is only very temporary and does not represent higher levels for the ecosystem as a whole. Rather, the high management-induced losses of nutrients during plantation establishment and harvesting (leaching, volatilisation, erosion) contribute significantly to overall nutrient depletion. The newly established nutrient equilibrium will be on a lower level than the previous one with each subsequent rotation.

There are very large areas of land being planted to trees in the humid tropics and given the relative paucity of direct research on the hydrological effects of this change in land use, there is a high priority for studies in this particular field. The following aspects are considered to be particularly important:

- Studies need to measure factors other than just those directly related to the vegetation change, such as effects on soil hydraulic properties which may be as important as the changes in water use and soil moisture levels imposed by forestation.

- The hydrology of degraded catchments needs to be described in terms of real measurements of actually occurring processes as too much depends on assumptions in the existing literature.
- It is suggested as a working hypothesis that in the conversion from degraded grasslands to timber plantations, the increases in infiltration that can be attributed to a forest cover are likely to be exceeded by the increase in the evaporation component of the water balance of the new forest. Research should determine under which specific situations this hypothesis does or does not apply.
- There is a need to include groundwater in the study of changed catchment water balances associated with forestation, especially as the socially important baseflows and dry season flows are normally generated from groundwater stores.
- It is expected that critical variables in determining the response of (degraded) catchments to forestation will be the nature of the rainfall and the soil properties. These aspects ought to be thoroughly measured and documented.
- All the direct and indirect hydrological effects of a major land use change will not be visible in the short term, and there is therefore a need for properly controlled long-term studies, such as paired catchment experiments backed up by process-based research.
- Knowledge of plantation nutrient dynamics and fertility over entire forest rotations requires holistic studies that are broad enough in both time and space, as well as in terms of ecological scope.

References

Aide, T. M., Zimmerman, J. K., Herrera, L., Rosario, M. and Serrano, M. (1995) Forest recovery in abandoned tropical pastures in Puerto Rico. *Forest Ecology and Management*, 77, 77–86.

Amir, H. M. S., Zakaria, M., Ghazali Hasan, M. and Ahmad, R. (1990). Nutrient dynamics of Tekam Forest Reserve, Peninsular Malaysia, under different logging phases. *Journal of Tropical Forest Science*, 2, 71–80.

Ashby, M. (1999). Modelling the water and energy balances of Amazonian rainforest and pasture using Anglo-Brazilian Amazonian climate observation study data. *Agricultural and Forest Meteorology*, 94, 79–101.

Bailly, C., Benoit de Cognac, G., Malvos, C., Ningre, J. M. and Sarrailh, J. M. (1974). Etude de l'influence du couvert naturel et de ses modifications à Madagascar; expérimentations en bassins versants élémentaires. *Cahiers Scientifiques du Centre Technique Forestier Tropical*, 4, 1–114.

Bartarya, S. K. (1989). Hydrogeology, geo-environmental problems and watershed management strategies in a central Himalayan river basin, Kumaun, India. In *Headwater Control*, ed. J. Kreek and M. J. Haigh, pp. 308–318. Plzen, Czechoslovakia: IUF RO/WASWC/CSVIS.

Beadle, C. L. (1997). Dynamics of leaf and canopy development. In *Management of Soil, Nutrients and Water in Tropical Plantation Forests*, ed. E. K. S. Nambiar and A. G. Brown, pp. 169–212. Canberra: ACIAR/CSIRO and Bogor: CIFOR.

Bell, T. I. W. (1973). Erosion in the Trinidad teak plantations. *Commonwealth Forestry Review*, 52, 223–233.

Bigelow, S. (2001). Evapotranspiration modelled from stands of three broad-leaved tropical trees in Costa Ric. *Hydrological Processes*, 15, 2779–2796.

Blackie, J. R. (1979). The water balance of the Kimakia catchments. *East African Agricultural and Forestry Journal*, 43, 155–174.

Blaisdell, F. W. (1981). Engineering structures for erosion control. In *Tropical Agricultural Hydrology*, ed. R. Lal and E. W. Russell, pp. 325–355. New York: J. Wiley.

Bonell, M. and Balek, J. (1993). Recent scientific developments and research needs in hydrological processes of the humid tropics. In *Hydrology and Water Management in the Humid Tropics – Hydrological Research Issues and Strategies for Water Management* ed. M. Bonell, M. M. Hufschmidt and J. S. Gladwell, pp. 167–260. Cambridge: Cambridge University Press.

Bosch, J. M. (1982). Streamflow response to catchment management in South Afric. In *International Symposium on Hydrological Research Basins and their Use in Water Resource Planning*, Vol. 2, pp. 279–289. Bern, Switzerland, Landesamt.

Bosch, J. M. and Hewlett, J. D. (1982). A review of catchment experiments to determine the effect of vegetation changes on water yield and evapotranspiration. *Journal of Hydrology*, 55, 3–23.

Brinkmann, W. L. F. and Nascimento, J. C. de (1973). The effect of slash and burn agriculture on plant nutrients in the Tertiary region of central Amazonia. *Turrialba*, 23, 284–290.

Brouwer, L. C. (1996). Nutrient cycling in pristine and logged tropical rainforest. A study in Guyana. *Tropenbos Guyana Series* no. 1. Georgetown, Guyana: Tropenbos Guyana, 224 p.

Brown, A. G., Nambiar, E. K. S. and Cossalter, C. (1997). Plantations for the tropics – their role, extent and nature. In *Management of Soil, Nutrients and Water in Tropical Plantation Forests*, ed. E. K. S. Nambiar and A. G. Brown, pp. 1–23. Canberra: ACIAR/CSIRO and Bogor: CIFOR.

Bruijnzeel, L. A. (1988). Estimates of evaporation in plantations of *Agathis dammara* Warb. in south-central Java, Indonesia. *Journal of Tropical Forest Science*, 1, 145–161.

Bruijnzeel, L. A. (1989a). Nutrient cycling in moist tropical forests: the hydrological framework. In *Mineral Nutrients in Tropical Forest and Savanna Ecosystems*, ed. J. Proctor, pp. 383–415. Oxford: Blackwell Scientific.

Bruijnzeel, L. A. (1989b). (De)forestation and dry season flow in the tropics: a closer look. *Journal of Tropical Forest Science*, 1, 229–243.

Bruijnzeel, L. A. (1990). *Hydrology of Moist Tropical Forest and Effects of Conversion: a State of Knowledge Review*. Paris: UNESCO and Amsterdam: Vrije Universiteit, 226 p.

Bruijnzeel, L. A. (1991). Nutrient input-output budgets of tropical forest ecosystems, a review. *Journal of Tropical Ecology*, 7, 1–24.

Bruijnzeel, L. A. (1992). Sustainability of fast-growing plantation forests in the humid tropics, with special reference to nutrition. In *Taungya: Forest Plantations with Agriculture in South East Asia*, ed. C. F. Jordan, J. Gajaseni and H. Watanabe, pp. 51–67. Wallingford: CABI.

Bruijnzeel, L. A. (1996). Predicting the hydrological effects of land cover transformation in the humid tropics: the need for integrated research. In *Amazonian Deforestation and Climate*, ed. J. H. C. Gash, C. A. Nobre, J. M. Roberts and R. L. Victoria, pp. 15–55. Chichester: J. Wiley.

Bruijnzeel, L. A. (1997). Hydrology of forest plantations in the tropics. In *Management of Soil, Nutrients and Water in Tropical Plantation Forests*, ed. E. K. S. Nambiar and A. G. Brown, pp. 125–167. Canberra: ACIAR/CSIRO and Bogor: CIFOR.

Bruijnzeel, L. A. (1998). Soil chemical responses to tropical forest disturbance and conversion: the hydrological connection. In *Tropical Forest Soils and Their Management*, ed. A. Schulte and D. Ruhiyat, pp. 45–61. Berlin: Springer Verlag.

Bruijnzeel, L. A. (2004). Hydrological functions of tropical forests. *Agriculture, Ecosystems and Environment*. doi:10.1016/j.agee.2004.01.015

Bruijnzeel, L. A. and Bremmer, C. N. (1989). Highland-lowland interactions in the Ganges Brahmaputra River Basin: a review of published literature. *ICIMOD Occasional Paper* no. 11, Kathmandu, Nepal: International Centre for Integrated Mountain Development, 136 p.

Bruijnzeel, L. A. and Proctor, J. (1995). Hydrology and biogeochemistry of tropical montane cloud forests: what do we really know? In *Tropical Montane Cloud Forests*, ed. L. S. Hamilton, J. O. Juvik and F. N. Scatena, pp. 38–78. Berlin: *Springer Ecological Studies*, 110.

Bruijnzeel, L. A. and Wiersum, K. F. (1985). A nutrient budget sheet for *Agathis dammara* Warb. plantation forest under various management conditions. *Forest Ecology and Management*, 10, 195–208.

Brunsden, D, Jones, D. K. C., Martin, R. P. and Doornkamp, J. C. (1981). The geomorphological character of part of the Low Himalaya of Eastern Nepal. *Zeitschrift für Geomorphologie Neue Folge Supplement Band*, 37, 25–72.

Bubb, K. A. and Croton, J. T. (2002). Effects on catchment water balance from the management of Pinus plantations on the coastal lowlands of south-east Queensland, Australia. *Hydrological Processes*, 16, 105–117.

Burger, C., Everson, C. S. and Savage, M. J. (1999). Comparative evaporation measurements above commercial forestry and sugarcane canopies in the KwaZulu-Natal Midlands. In *Proceedings of the Ninth South African National Hydrology Symposium*, University of the Western Cape, Bellville, 29–30 November 1999.

Buschbacher, R. J. (1984). *Changes in productivity and nutrient cycling following conversion of Amazonian rainforest to pasture*. PhD Thesis, Athens, Ga.: University of Georgia, 193 p.

Buschbacher, R., Uhl, C. and Serrão, E. A. S. (1988) Abandoned pastures in eastern Amazonia. II. Nutrient stocks in the soil and vegetation. *Journal of Ecology*, 76, 682–699.

Calder, I. R. (1999). *The Blue Revolution*. London: Earthscan Publications, 192 p.

Calder, I. R., Wright, I. R. and Murdiyarso, D. (1986). A study of evaporation from tropical rainforest – West Java. *Journal of Hydrology*, 89, 13–31.

Calder, I. R., Swaminath, M. H., Kariyappa, G. S., Srinivasalu, N. V., Srinivasa Murty, K. V. and Mumtaz, J. (1992). Deuterium tracing for the estimation of transpiration from trees. Part 3. Measurements of transpiration from Eucalyptus plantation, India. *Journal of Hydrology*, 130, 37–48.

Calder, I. R., Rosier, P. T. W., Prasanna, K. T. and Parameswarappa, S. (1997). Eucalyptus water use greater than rainfall input – a possible explanation from southern India. *Hydrology and Earth System Science*, 1, 249–256.

Chandler, D. G. and Walter, M. F. (1998). Runoff responses among common land uses in the uplands of Matalom, Leyte, Philippines. *Transactions of the American Society of Agricultural Engineers*, 41, 1635–1641.

Chang, J-H. and Lau, L. S. (1993). A definition of the humid tropics. In: *Hydrology and Water Management in the Humid Tropics – Hydrological Research Issues and Strategies for Water Management*, eds. M. Bonell, M. M. Hufschmidt and J. S. Gladwell, pp. 571–574. Paris – Cambridge: UNESCO – Cambridge University Press.

Cienciala, E., Kucera, J. and Malmer, A. (2000). Tree sap flow and stand transpiration of two Acacia mangium plantations in Sabah, Borneo. *Journal of Hydrology*, 236, 109–120.

Collopy, J., Bai Jiayu, Zhou Guoyi and Morris, J. (2000). Water use by fast growing eucalyptus plantations in Southern China. Poster paper presented at the XXI IUFRO World Congress, Kuala Lumpur, Malaysia, August 2000.

Crane, W. J. B. and Raison, R. J. (1980). Removal of phosphorus in logs when harvesting *Eucalyptus elegatensis* and *Pinus radiata* on short and long rotations. *Australian Forestry*, 43, 252–260.

Cutrim, E., Martin, D. W. and Rabin, R. (1995). Enhancement of cumulus clouds over deforested lands in Amazonia. *Bulletin of the American Meteorological Society*, 76, 1801–1805.

Daño, A. M. (1990). Effect of burning and reforestation on grassland watersheds in the Philippines. *International Association of Hydrological Sciences Publication*, 192, 53–61.

Dillaha, T. A., Reneau, R. B., Mostaghimi, S. and Lee, D. (1989). Vegetative filter strips for agricultural nonpoint source pollution. *Transactions of the American Society of Agricultural Engineers*, 32, 491–496.

Dolman, A. J., Van der Molen, M. K., Ter Maat, H. W. and Hutjes, R. W. A. (2004). The effects of forests on mesoscale atmospheric processes. In *Forests at the Land–Atmosphere Interface*, eds. M. Mencucini, J. Grace, J. Moncrieff and K. McMaughan, pp. 51–72. Wallingford, UK: CAB International.

Dunin, F. X., O'Loughlin, E. M. and Reyenga, W. (1985). A lysimeter characterization of evaporation by eucalypt forest and its representativeness for the local environment. In *The Forest-Atmosphere Interaction*, ed. B. A. Hutchinson and B. B. Hicks, pp. 271–291. Dordrecht: D. Reidel.

Dye, P. J. (1996). Climate, forest and streamflow relationships in South African afforested catchments. *Commonwealth Forestry Review*, 75, 31–38.

Dye, P. J., Vilakazi, P., Gush, M., Ndlela, R. and Royappen, M. (2001). The feasibility of using trunk growth increments to estimate water use of *Eucalyptus grandis* plantations. Water Research Commission Research Report no. 809/1/01. Pretoria, South Africa: Water Research Commission.

Edwards, K. A. (1979). The water balance of the Mbeya experimental catchments. *East African Agricultural and Forestry Journal*, 43, 231–247.

Elsenbeer, H., Newton, B. E., Dunne, T. and De Moraes, J. M. (1999). Soil hydraulic conductivities of latosols under pasture, forest and teak in Rondonia, Brazil. *Hydrological Processes*, 13, 1417–1422.

Eswaran, H., Lal, R. and Reich, P. F. (2001). Land degradation: an overview. In *Response to Land Degradation*, E. M. Bridges, I. D. Hannam, L. R. Oldeman, F. W. T. Penning de Vries, S. J. Scherr and S. Sombatpanit (eds.), pp. 20–35. Enfield (NH), USA: Science Publishers Inc.

Evans, J. C. (1986). Plantation forestry in the tropics – trends and prospects. *The International Tree Crops Journal*, 4, 3–15.

Evans, J. C. (1999). Planted forests of the wet and dry tropics: their variety, nature and significance. *New Forests*, 17, 25–36.

Ewel, J. J. (1976). Litterfall and leaf decomposition in a tropical forest succession in eastern Guatemala. *Journal of Ecology*, 64, 297–307.

Ewel, J. J., Berish, C., Brown, B., Price, N. and Raich, J. (1981). Slash and burn impacts on a Costa Rican wet forest site. *Ecology*, 62, 816–829.

Fahey, B. J. and Jackson, R. J. (1997). Hydrological impacts of converting native forests and grasslands to pine plantations, South Island, New Zealand. *Agricultural and Forest Meteorology*, 84, 69–82.

FAO (1985/1986). *FAO watershed management field manual*. FAO Conservation Guide no. 13, 2 Vols. Rome: U.N. Food and Agriculture Organization.

FAO (1995). Forest Resources Assessment 1990. Tropical countries. *FAO Forestry Paper* no. 112. Rome: U.N. Food and Agriculture Organization.

FAO (2001). Global Forest Resources assessment 2000. Main Report. *FAO Forestry Paper* no. 140. Rome: U.N. Food and Agriculture Organization.

Fearnside, P. M. and Guimaraes, W. M. (1996) Carbon uptake by secondary forests in Brazilian Amazonia. *Forest Ecology and Management*, 80, 35–46.

Flatfjord, S. (1976). Runoff and erosion measurements from an afforested and a cultivated watershed. Pidekso I and Pidekso II watersheds, 1974–1976. Upper Solo Watershed Management and Upland Development Project INS/72/006. Appendix III to Project Termination Document. Rome, FAO.

Fölster, H. and Khanna, P. K. (1997). Dynamics of nutrient supply in plantation soils. In *Management of Soil, Nutrients and Water in Tropical Plantation Forests*, ed. E. K. S. Nambiar and A. G. Brown, pp. 339–373. Canberra: ACIAR/CSIRO and Bogor: CIFOR.

Fritsch, J. M. and Sarrailh, J. M. (1986). Les transports solides dans l'écosystème forestier tropical humide en Guyane: les effets du défrichement et de l'installation de paturages. *Cahiers de l'ORSTOM, Série Pédologie*, 22, 93–106.

Giambelluca, T. W. (2002). The hydrology of altered tropical forest. Invited Commentary. *Hydrological Processes*, 16, 1665–1669.

Gillman, G. P., Sinclair, D. F., Knowlton, R. and Keys, M. G. (1985). The effect on some soil chemical properties of the selective logging of a north Queensland rainforest. *Forest Ecology and Management*, 12, 195–214.

Gilmour, D. A., Bonell, M. and Cassells, D. S. (1987). The effects of forestation on soil hydraulic properties in the Middle Hills of Nepal: a preliminary assessment. *Mountain Research and Development*, 7, 239–249.

Globevnik, L. (1998). Analysis of river morphological and environmental changes with the integration of historical data and image processing. *International Association of Hydrological Sciences Publication*, 249, 279–285.

Godsey, S. and Elsenbeer, H. (2002). The soil hydrologic response to forest regrowth: a case study from southwestern Amazonia. *Hydrological Processes*, 16, 1519–1522.

Gonçalves, J. L. M., Barros, N. F., Nambiar, E. K. S. and Novais, R. F. (1997). Soil and stand management for short-rotation plantations. In *Management of Soil, Nutrients and Water in Tropical Plantation Forests*, ed. E. K. S. Nambiar and A. G. Brown, pp. 379–417. Canberra: ACIAR/CSIRO and Bogor: CIFOR.

Granier, A., Huc, R. and Colin, F. (1992). Transpiration and stomatal conductance of two rainforest species growing in plantations (Simarouba amara and Goupia glabra) in French Guiana. *Annales des Sciences Forestières*, 49, 17–24.

Gupta, R. K., Mishra, P. R., Shankar, P., Kaushal, R. C. and Sajwan, S. S. (1974). Studies on the effect of different land treatments on water yield in the siwalik-Chandigarh. In *Annual Report 1974*, pp. 18–21. Dehradun, India:Central Soil and Water Conservation Research and Training Institute.

Gupta, R. K., Mishra, P. R., Mittal, S. P. and Singh, K. (1975). Effect of different land treatments on water yield in the siwalik-Chandigarh. In *Annual Report*

1975, pp. 15–17. Dehradun, India:Central Soil and Water Conservation Research and Training Institute.
Hamilton, L. S. and King, P. N. (1983). *Tropical Forested Watersheds. Hydrologic and Soils Response to Major Uses or Conversions*. Boulder: Westview Press, 168 p.
Harcombe, P. A. (1977). Nutrient accumulation by vegetation during the first year of recovery of a tropical forest ecosystem. In *Recovery and Restoration of Damaged Ecosystems*, ed. J. Cairns, K. L. Dickson and E. E. Herricks, pp. 347–378. Charlottesville, VA: University Press of Virginia.
Harding, R. J. (1992). The modification of climate by forests. In *Growth and Water Use of Forest Plantations*, eds. I. R. Calder, R. L. Hall and P. G. Adlard, pp. 332–346. Chichester: Wiley and Sons.
Harding, R. J., Hall, R. J., Swaminath, M. H. and Srinivasa Murthy, K. V. (1992). The soil moisture regimes beneath forest and an agricultural crop in Southern India – Measurements and modelling. In *Growth and Water Use of Forest Plantations*, ed. I. R. Calder, R. L. Hall and P. G. Adlard, pp. 244–269. Chichester: Wiley and Sons.
Hase, H. and Fölster (1983). Impact of plantation forestry with teak (*Tectona grandis*) on the nutrient status of young alluvial soil in West Venezuela. *Forest Ecology and Management*, 6, 33–57.
Haydon, S. R., Benyon, R. G. and Lewis, R. (1996). Variation in sapwood area and throughfall with forest age in mountain ash (*Eucalyptus regnans* F. Muell.). *Journal of Hydrology*, 187, 351–366.
Hewlett, J. D. (1982). Forests and floods in the light of recent investigation. In *Hydrological Processes of Forested Areas*, pp. 543–559. *National Research Council of Canada Publication* no. 20548, Ottawa, NRCC.
Hewlett, J. D. and Bosch, J. M. (1984). The dependence of storm flows on rainfall intensity and vegetal cover in South Africa. *Journal of Hydrology*, 75, 365–381.
Hughes, R. F., Kauffman, J. B. and Jaramillo, V. J. (1999) Biomass, carbon, and nutrient dynamics of secondary forests in a humid tropical region of México. *Ecology*, 80, 1892–1907.
Jayasurya, M. D. A., Dunn, G., Benyon, R. and O'Shaughnessy, P. J. (1993). Some factors affecting water yield from mountain ash (*Eucalyptus regnans*) dominated forests in south-east Australia. *Journal of Hydrology*, 150, 345–367.
Jongewaard, Th. and Overmars, M. (1994). Comparison of erosion and sediment delivery in two small catchments with different lithologies in the Gunung Kidul District, Java. MSc Thesis. Amsterdam: Faculty of Earth Sciences, Vrije Universiteit, 106 p.
Jordan, C. F. (1985). *Nutrient Cycling in Tropical Forest Ecosystems*. New York: J. Wiley, 190 p.
Kabat, P., Dolman, A. J., Ashby, M., Gash, J. H. C., Wright, I. R., Culf, A., Calvet, J. C., Delire, C., Noilhan, J., Jochum, A., da Silva Dias, M. A. F., Santos Alvala, R. C., Nobre, C. A., Prince, S. D. and Steininger, M. (1999). Use of integrated modelling for experimental design. Final Report. DLO *Winand Staring Centre Report* no. 145. Wageningen: DLO WSC, 120 p.
Kammer, D. and Raj, R. (1979). Preliminary estimates of minimum flows in the Varaciva Creek, Ba Province, and the effect of afforestation on water resources. *Technical Note* 79/1. Suva, Fiji: Public Works Department.
Khanna, P. K., Raison, R. J., and Falkiner, R. A. (1994) Chemical properties of ash derived from Eucalyptus litter and its effects on forest soils. *Forest Ecology and Management*, 66, 107–125.
Klinge, R. (1998) Wasser- und Nährstoffdynamik im Boden und Bestand beim Aufbau einer Holzplantage im östlichen Amazonasgebiet. *Göttinger Beiträge zur Land- und Forstwirtschaft in den Tropen und Subtropen*, 122. Göttingen: University of Göttingen.
Klinge, R., Schmidt, J. and Fölster, H. (2001). Simulation of water drainage from a rainforest and forest conversion plots using a soil water model. *Journal of Hydrology*, 246, 82–95.
Kuczera, G. (1987). Prediction of water yield reductions following bushfire in ash-mixed species eucalypt forest. *Journal of Hydrology*, 94, 215–236.
Laclau, J. P., Bouillet, J. P. and Ranger, J. (2000). Canopy and soil modification of precipitation chemistry in a clonal eucalypt stand in Congo: comparison with an adjacent savanna ecosystem. *Proceedings XXI IUFRO World Congress*, 7–12 August 2000, Kuala Lumpur, Malaysia. Sub-Plenary sessions, Volume 1, pp. 706–718.
Lal, R. (1987). *Tropical Ecology and Physical Edaphology*. New York: J. Wiley, 732 p.
Lal, R. (1996a). Deforestation and land-use effects on soil degradation and rehabilitation in western Nigeria. I. Soil physical and hydrological properties. *Land Degradation and Development*, 7, 19–45.
Lal, R. (1996b). Deforestation and land-use effects on soil degradation and rehabilitation in western Nigeria. III. Runoff, soil erosion and nutrient loss. *Land Degradation and Development*, 7, 99–120.
Langford, K. J. (1976). Change in yield of water following a bushfire in a forest of *Eucalyptus regnans*. *Journal of Hydrology*, 29, 87–114.
Lawton, R. O., Nair, US, Pielke, US and Welch, R. M. (2001). Climatic impact of tropical lowland deforestation on nearby montane cloud forests. *Science*, 294, 584–587.
Lesch, W. and Scott, D. F. (1997). The response in water yield to the thinning of *Pinus radiata, Pinus patula* and *Eucalyptus grandis* plantations. *Forest Ecology and Management*, 99, 295–307.
Levy, G. J., Watt, H. V., Shainburg, I. and Plessis, H. M. D. (1988). Potassium-calcium and sodium-calcium exchange on kaolinite and kaolinitic soils. *Soil Science Society of America Journal*, 52, 1259–1264.
Lim, M. T. (1988). Studies on *Acacia mangium* in Kemasul Forest, Malaysia. I. Biomass and productivity. *Journal of Tropical Ecology* 4, 293–302.
Lima, W. P. (1993). *Impacto Ambiental do Eucalipto*. Sao Palo, Brazil: University of Sao Paulo, 301 p.
Lima, W. P., Zakia, M. J. B., Libardi, P. L. and Sousa Filho, A. P. (1990). Comparative evapotranspiration of eucalyptus, pine and cerrado vegetation measured by the soil water balance method. *IPEF International Piracicaba*, 1: 5–11.
Loague, K. and Kyriakidis, P. C. (1997). Spatial and temporal variability in the r-5 infiltration data set: Déja-vu and rainfall-runoff simulations. *Water Resources Research*, 33, 2883–2895.
Mackensen, J. (1998). Untersuchungen zur nachhaltigen Nährstoffversorgung in schnellwachsenden Plantagensystemen in Ost-Kalimantan, Indonesien – ökologische und ökonomische Implikationen. *Göttinger Beiträge zur Land- und Forstwirtschaft in den Tropen und Subtropen*, 127. Göttingen: University of Göttingen, 209 p.
Mackensen, J., Hölscher, D., Klinge, R. and Fölster, H. (1996) Nutrient transfer to the atmosphere due to burning of debris in East-Amazonia. *Forest Ecology and Management*, 86, 121–128.
Mackensen, J., Fölster, H. and Ruhiyat, D. (2000). Cost analysis for suitable management of fast-growing tree plantations in East Kalimantan, Indonesia. *Forest Ecology and Management*, 131, 239–253.
Mackensen, Klinge, R., Ruhiyat, D. and Fölster, H. (2003). Assessment of management-dependent nutrient fluxes in tropical industrial tree plantations. *Ambio*, 32, 106–112.
Malmer, A. (1992). Water yield changes after clear-felling tropical rainforest and establishment of forest plantation in Sabah, Malaysia. *Journal of Hydrology*, 134, 77–94.
Malmer, A. (1996). Hydrological effects and nutrient losses of forest plantation establishment on tropical rainforest land in Sabah, Malaysia. *Journal of Hydrology*, 174, 129–148.
Malmer, A. and Grip, H. (1994). Converting tropical rainforest to forest plantation in Sabah, Malaysia. II. Effects on nutrient dynamics and net losses in streamwater. *Hydrological Processes*, 8, 195–209.
Mapa, R. B. (1995). Effect of reforestation using *Tectona grandis* on infiltration and soil water retention. *Forest Ecology and Management*, 77, 119–125.
Morris, J., Collopy, J. and Mahmood, K. (1999). Canopy conductance and water use in Eucalyptus plantations. Poster paper presented at the *IUFRO Workshop on Canopy Dynamics and Forest Management – A Missing Link*? Vindeln, Sweden, August 1999.
Mwendera, E. J. (1994). Effect on the water yield of the Luchelemu catchment in Malawi of replacing indigenous grasses with timber plantations. *Forest Ecology and Management*, 65, 75–80.
Negi, G. C. S., Joshi, V. and Kumar, K. (1998). Spring sanctuary development to meet household water demand in the mountains: A call for action. In *Research for Mountain Development: Some Initiatives and Accomplishments*, pp. 25–48. Nainital, India: Gyanodya Prakashan.
Nye, P. H. and Greenland, D. J. (1964). Change sin the soil after clearing tropical forest. *Plant and Soil*, 21, 101–112.
Nykvist, N., Grip, H., Sim, B. L., Malmer, A. and Wong, F. K. (1994). Nutrient losses in forest plantations in Sabah, Malaysia. *Ambio*, 23, 210–215.
Nykvist, N. (2000). Tropical forests can suffer from a serious deficiency of calcium after logging. *Ambio*, 29, 310–313.

O'Loughlin, C. L. (1984). Effectiveness of introduced forest vegetation for protection against landslides and erosion in New Zealand's steeplands. In *Effects of Forest Land Use on Erosion and Slope Stability*, ed. C. L. O'Loughlin and A. J. Pearce, A. J., pp. 275–280. Vienna: IUFRO.

Otsamo, A. (1998). Effect of fertilizing on establishment and early growth of tree plantations on *Imperata cylindrica* grasslands. In *Tropical Forest Soils and Their Management*, ed. A. Schulte and D. Ruhiyat, pp. 137–143. Berlin: Springer Verlag.

Otsamo, A., Adjers, G., Hadi, T. S. and Kuusipalo, J. (1994). Effect of site preparation and initial fertilizing on the growth of *Acacia mangium, Gmelina arborea, Paraserianthes falcataria* and *Swietenia macrophylla* on *Imperata cylindrica* dominated grassland. *Technical Report ENSO Forest Development Ltd.*, 44. ENSO Forest Development Ltd., Finland.

Palm, C. A., Swift, M. J. and Woomer, P. L. (1996). Soil biological dynamics in slash-and-burn agriculture. *Agriculture Ecosystems and Environment*, 58, 61–74.

Paningbatan, E. P., Ciesiolka, C. A., Coughlan, K. J. and Rose, C. W. (1995). Alley cropping for managing soil erosion of hilly lands in the Philippines. *Soil Technology*, 8, 193–204.

Pearce, A. J. (1986). *Erosion and Sedimentation*. Working Paper. Honolulu, Hawaii: Environment and Policy Institute.

Pereira, H. C. (1952). Interception of rainfall by cypress plantations. *East African Agricultural Journal Kenya*, 18, 73–76.

Pereira, H. C. (1989). *Policy and Practice of Water Management in Tropical Areas*. Boulder, Colorado: Westview Press, 236 p.

Pielke, R. A., Cotton, W. R., Walko, R. L., Tremback, C. J., Lyons, W. A., Grasso, L. D., Nichlis, M. E., Moran, M. D., Wesley, D. A., Lee, T. J., Copeland, J. H., 1992. A comprehensive meteorological modelling system – RAMS. *Meteorology and Atmospheric Physics*, 49, 69–91.

Pivello, V. R. and Coutinho, C. M. (1992). Transfer of macro-nutrients to the atmosphere during experimental burnings in an open Cerrado (Brazilian Savanna). *Journal of Tropical Ecology*, 8, 487–497.

Pramono Hadi (1989). Hydrological study of the upper Bengawan Solo catchment, Indonesia. MSc Thesis. Enschede, The Netherlands: International Institute for Aerospace Survey and Earth Sciences, 110 p.

Purwanto, E. (1999). Erosion, Sediment Delivery and Soil Conservation in an Upland Agricultural Catchment in West Java, Indonesia. PhD thesis. Amsterdam: Vrije Universiteit, 218 p.

Putuhena, W. and Cordery, I. (1995). Estimation of interception capacity of the forest floor. *Journal of Hydrology*, 180, 283–299.

Raison, R. J., Khanna, P. K. and Woods, P. V. (1985a). Mechanisms of element transfer to the atmosphere during vegetation fires. *Canadian Journal of Forestry Research*, 15, 132–140.

Raison, R. J., Khanna, P. K. and Woods, P. V. (1985). Transfer of elements to the atmosphere during low intensity prescribed fires in three Australian subalpine Eucalypt forests. *Canadian Journal of Forestry Research*, 15, 657–664.

Ramsay, W. J. H. (1987a). Sediment production and transport in the Phewa valley, Nepal. *International Association of Hydrological Sciences Publication*, 165, 461–472.

Ramsay, W. J. H. (1987b). Deforestation and erosion in the Nepalese Himalaya: is the link myth or reality? *International Association of Hydrological Sciences Publication*, 167, 239–250.

Rijsdijk, A. and Bruijnzeel, L. A. (1991). Erosion, sediment yield and land use patterns in the upper Konto watershed, East Java, Indonesia. *Konto River Project Communication*, 18, Volume III. Amersfoort, the Netherlands: DHV International Consultants, 59 p.

Richardson, J. (1982). Some implications of tropical forest replacement in Jamaica. *Zeitschrift für Geomorphologie Neue Folge, Supplement Band*, 44, 107–118.

Roberts, J. and Rosier, P. T. W. (1993). Physiological studies in young Eucalyptus stands in southern India and derived estimates of forest transpiration. *Agricultural Water Management*, 24, 103–118.

Ruhiyat, D. (1989). Die Entwicklung der standörtlichen Nährstoffvorräte bei naturnaher Waldbewirtschaftung und im Plantagenbetrieb, Ostkalimantan, Indonesien. *Göttinger Beiträge zur Land- und Forstwirtschaft in den Tropen und Subtropen*, 35. Göttingen: University of Göttingen, 206 p.

Sahin, V. and Hall, M. J. (1996). The effects of afforestation and deforestation on water yields. *Journal of Hydrology*, 178, 293–309.

Samraj, P., Sharda, V. N., Chinnamani, S., Lakshmanan, V. and Haldorai, B. (1988). Hydrological behaviour of the Nilgiri sub-watersheds as affected by bluegum plantations, Part I. The annual water balance. *Journal of Hydrology*, 103, 335–345.

Sanchez, P. A. (1976). *Properties and Management of Soils in the Tropics*. New York: J. Wiley, 618 p.

Sanchez, P. A., Villachia, J. H. and Bandy, D. E. (1983). Soil fertility dynamics after clearing a tropical rainforest in Peru. *Soil Science Society of America Journal*, 47, 1171–1178.

Sandström, K. (1998). Can forests 'provide' water: widespread myth or scientific reality? *Ambio*, 27, 132–138.

Scatena, F. N. and Larsen, M. C. (1991). Physical aspect of hurricane Hugo. *Biotropica*, 23, 317–323.

Scott, D. F. and Lesch, W. (1996). The effects of riparian clearing and clearfelling of an indigenous forest on streamflow, stormflow and water quality. *S.A. Forestry Journal*, 175, 1–14.

Scott, D. F. and Lesch, W. (1997). Streamflow responses to afforestation with *Eucalyptus grandis* and *Pinus patula* and to felling in the Mokobulaan experimental catchments, Mpumalanga Province, South Africa. *Journal of Hydrology*, 199, 360–377.

Scott, D. F. and Smith, R. E. (1997). Preliminary empirical models to predict reductions in annual and low flows resulting from afforestation. *Water SA*, 23, 135–140.

Scott, D. F., Prinsloo, F. W. and Moses, G. (1999). Results of the afforested catchment experiments: range and variability of effects and the controlling variables. Paper presented at the *Ninth South African National Hydrology Symposium*, University of the Western Cape, Cape Town, November 1999.

Scott, D. F., Prinsloo, F. W., Moses, G., Mehlomakulu, M. and Simmers, A. D. A. (2000). *A re-analysis of the South African catchment afforestation experimental data*. WRC Report no. 810/1/00, Pretoria: Water Research Commission, 138 pp.

Sharda, V. N., Samraj, P., Chinnamani, S. and Lakshmanan, V. (1988). Hydrological behaviour of the Nilgiri sub-watersheds as affected by bluegum plantations, Part II. Monthly water balances at different rainfall and runoff probabilities. *Journal of Hydrology*, 103, 347–355.

Sharda, V. N., Samraj, P., Samra, J. S. and Lakshmanan, V. (1998). Hydrological behaviour of first generation coppiced bluegum plantations in the Nilgiris sub-watersheds. *Journal of Hydrology*, 211, 50–60.

Shuttleworth, W. J. (1988). Evaporation from Amazonan rainforest. *Philosophical Transactions of the Royal Society (London), Series B*, 333, 321–346.

Sikka, A. K., Samra, J. S., Sharda, V. N., Samraj, P. and Lakshmanan, V. (2003). Low flow and high flow responses to converting natural grassland into bluegum (*Eucalyptus globulus*) in Nilgiris watersheds of South India. *Journal of Hydrology*, 270, 12–26.

Silver, W. L., Scatena, F. N., Johnson, A. H., Siccama, T. G. and Watt, F. (1996). At what temporal scales does disturbance affect belowground nutrient pools? *Biotropica*, 28, 441–457.

Smethurst, P. J. and Nambiar, E. K. S. (1995). Changes in soil carbon and nitrogen during the establishment of a second crop of *Pinus radiata*. *Forest Ecology and Management*, 73, 145–155.

Smiet, A. C. (1987). Tropical watershed forestry under attack. *Ambio*, 16, 156–158.

Smith, R. E. and Scott, D. F. (1992). The effects of afforestation on low flows in various regions of South Africa. *Water SA*, 18, 185–194.

Sollins, P., Sancho, F., Mata, R. C. and Sanford, R. L. J. (1994). Soils and soil process research. In *La Selva, Ecology and Natural History of a Neotropical Rain Forest*, ed. L. A. McDade, K. S. Bawa, H. A. Hespenheide and G. S. Hartshorn, pp. 34–53. Chicago: The University of Chicago Press.

Spangenberg, A., Grimm, U., da Silva, J. R. S. and Fölster, H. (1996). Nutrient store and export rates of *Eucalyptus eurograndis* plantations in eastern Amazonia (Jari). *Forest Ecology and Management*, 80, 225–234.

Stednick, J. D. (1996). Monitoring the effects of timber harvest on annual water yield. *Journal of Hydrology*, 176, 79–95.

Taylor, R. G. and Howard, K. W. F. (1996). Groundwater recharge in the Victoria Nile basin of East Africa: support for the soil moisture balance approach using stable isotope tracers and flow modelling. *Journal of Hydrology*, 180, 31–53.

Tennessee Valley Authority, 1961. *Forest cover improvement influences upon hydrologic characteristics of White Hollow watershed, 1935–1958*. Report no. 0-5163A. Knoxville, Tennessee, USA: Tennessee Valley Authority, 112 p.

Tomkins, I. B., Kellas, J. D., Tolhurst, K. G. and Oswin, D. A. (1991). Effects of fire intensity on soil chemistry in a Eucalypt forest. *Australian Journal of Soil Research*, 19, 25–47.

Trimble, S. W., Weirich, F. H. and Hoag, B. L. (1987). Reforestation and the reduction of water yield on the southern Piedmont since circa 1940. *Water Resources Research*, 23, 425–437.

Turnbull, J. W. (1999). Eucalypt plantations. *New Forests*, 17: 37–52.

Udo, E. J. (1978). Thermodynamics of potassium-calcium and magnesium-calcium exchange reactions on a kaolinitic soil clay. *Soil Science Society of America Journal*, 42, 556–560.

Uehara, G. and Gillman, G. (1981). The mineralogy, chemistry and physics of tropical soils with variable charge clays. *Westview Tropical Agriculture Series*, 4. Boulder, Co: Westview Press, 70 p.

Uhl, C. and Jordan, C. F. (1984). Succession and nutrient dynamics following forest cutting and burning in Amazonia. *Ecolology*, 65, 1476–1490.

Uhl, C., Buschbacher, R. and Serrão, E. A. S. (1988) Abandoned pastures in eastern Amazonia. I. Patterns of plant succession. *Journal of Ecology*, 75, 663–681.

Vandana Shiva and Bandyopadhyay, J. (1983). Eucalyptus – a disastrous tree for India. *The Ecologist*, 13, 184–187.

Van Dam, O. (2001). Forest filled with gaps: Effects of gap size on water and nutrient cycling in tropical rainforest. A study in Guyana. *Tropenbos Guyana Series*, 10. Georgetown: Tropenbos Guyana, 208 p.

Van der Molen, M. K. (2002). *Meteorological Impacts of Land Use Change in the Maritime Tropics*. PhD thesis, Amsterdam: Vrije Universiteit, 262 p.

Van der Weert, R. (1994). *Hydrological Conditions in Indonesia*. Jakarta: Delft Hydraulics.

Van Dijk, A. I. J. M. (2002). *Water and Sediment Dynamics in Bench-Terraced Agricultural Steeplands in West Java, Indonesia*. PhD thesis, Amsterdam: Vrije Universiteit, 363 p.

Van Dijk, A. I. J. M. and Bruijnzeel, L. A. (2001). Modelling rainfall interception by vegetation of variable density using an adapted analytical model. Part 2. Model validation for a tropical upland mixed cropping system. *Journal of Hydrology*, 247, 239–262.

Van Dijk, A. I. J. M. and Bruijnzeel, L. A. (2003). Environmental benefits of planting fast-growing trees on eroded cropland in Java, Indonesia. *Bois et Fôret des Tropiques*, 279, 73–79.

Van Lynden, G. W. J. and Oldeman, L. R. (1997). *The Assessment of the Status of Human-induced Soil Degradation in South and South East Asia*. Nairobi/Rome/Wageningen: UNEP/FAO/ISRIC, 35 p. + 2 maps.

Van Noordwijk, M., Van Roode, M., McCallie, E. L., Lusiana, B., 1998. Erosion and sedimentation as multiscale, fractal processes: Implications for models, experiments and the real world. In *Soil Erosion at Multiple Scales*, ed. F. W. T. Penning de Vries, F. Agus and J. Kerr, pp. 223–253. Wallingford: CABI Publishing.

Van Rompaey, R. S. A. R., 1993. *Forest Gradients in West Africa, a Spatial Gradient Analysis*. P h D Thesis, Wageningen: Wageningen Agricultural University.

Vertessy, R, Zhang, L and Dawes, WR, 2003. Plantations, river flows and river salinity. *Australian Forestry*, 66: 901–907.

Vertessy, R. A., Hatton, T. J., O'Shaughnessy, P. J. and Jayasuriya, M. D. A. (1993). Predicting water yields from a mountain ash forest using a terrain analysis based catchment model. *Journal of Hydrology*, 150, 665–700.

Vertessy, R. A., Watson, F. G. R., O'Sullivan, S. K., Davis, S., Campbell, R., Benyon, R. G. and Haydon, S. R. (1998a). *Predicting water yield from mountain ash forest catchments*. Industry Report 98/4. Clayton, Victoria, Australia: Cooperative Research Centre for Catchment Hydrology.

Vertessy, R. A., Hatton, T. J., Reece, P., O'Sullivan, S. K. and Benyon, R. G. (1998b). Estimating stand water use of large mountain ash trees and validation of the sap flow measurement technique. *Tree Physiology*, 17, 747–756.

Viswanatham, N. K., Joshie, P. and Ram Babu (1982). Influence of forest on soil erosion control – Dehradun. In *Annual Report 1982*, pp. 40–43. Dehradun, India: Central Soil and Water Conservation Research and Training Institute.

Waterloo, M. J. (1994). *Water and Nutrient Dynamics of Pinus caribaea Plantation Forests on Former Grassland Soils in Southwest Viti Levu, Fiji*. PhD Thesis, Amsterdam: Vrije Universiteit, 478 p.

Waterloo, M. J., Bruijnzeel, L. A., Vugts, H. F. and Rawaqa, T. T. (1999). Evaporation from *Pinus caribaea* plantations on former grassland soils under maritime tropical conditions. *Water Resources Research*, 35, 2133–2144.

Waterloo, M. J., Bruijnzeel, L. A., Vugts, H. F., Bink, N. J. and Rawaqa, T. T. (in press). Evaporation from a *Pennisetum polystachyon* grassland under maritime tropical conditions. *Agricultural and Forest Meteorology* (in press).

Watson, F. G., Vertessy, R. A. and Grayson, R. B. (1999) Large-scale modelling of forest eco-hydrological processes and their long term effect on water yield. *Hydrological Processes*, 13, 689–700.

Wiersum, K. F. (1984a). *Strategies and Designs for Afforestation, Reforestation and Tree Planting*. Wageningen: Pudoc, 432 p.

Wiersum, K. F. (1984b). Surface erosion under various tropical agroforestry systems. In *Effects of Forest Land Use on Erosion and Slope Stability*, ed. C. L. O'Loughlin and A. J. Pearce, pp. 231–239. Vienna: IUFRO.

Wiersum, K. F. (1985). Effects of various vegetation layers in an *Acacia auriculiformis* forest plantation on surface erosion in Java, Indonesia. In *Soil Erosion and Conservation*, ed. S. A. El-Swaify, W. C. Moldenhauer and A. Lo, pp. 79–89. Ankeny, Iowa, USA: Soil Conservation Society of Americ.

Wilk, J., Andersson, L. and Plermkamon, V. (2001). Hydrological impacts of forest conversion to agriculture in a large river basin in northeast Thailand. *Hydrological Processes*, 15, 2729–2748.

Wolterson, J. F. (1979). *Soil erosion in the teak forests of Java*. Report, 197. Wageningen: Forest Research Institute 'De Dorschkamp', 40pp.

Wright, I. R., Nobre, C. A., Tomassella, J., da Rocha, H. R., Roberts, J. M., Vertamatti, E., Culf, A. D., Alvalá, R. C. S., Hodnett, M. G. and Ubarana, V. N. (1996). Towards a GCM surface parameterization of amazonia. In. In *Amazonian Deforestation and Climate*, ed. J. H. C. Gash, C. A. Nobre, J. M. Roberts and R. L. Victoria, pp. 473–504. Chichester: J. Wiley.

Young, A. (1989). *Agroforestry for Soil Conservation*. Wallingford: CABI Publishers, 276 p.

Young, A. (1997). *Agroforestry for Soil Management*. Wallingford: CABI Publishers, 320 p.

Yu, B. F. (1999). A comparison of the Green-ampt and a spatially variable infiltration model for natural storm events. *Transactions of the American Society of Agricultural Engineers*, 42, 89–97.

Yu, B. F., Rose, C. W., Coughlan, K. J. and Fentie, B. (1997). Plot-scale rainfall-runoff characteristics and modeling at six sites in Australia and South East Asia. *Transactions of the American Society of Agricultural Engineers*, 40, 1295–1303.

Zeng, N., Neelin, J. D., Lau, K. M. and Tucker, C. J. (1999). Enhancement of interdecadal climate variability in the Sahel by vegetation interaction. *Science*, 286, 1537–1540.

Zhou, G. Y., Morris, J. D., Yan, J. H., Yu, Z. Y. and Peng, S. L. (2001). Hydrological impacts of reafforestation with eucalypts and indigenous species: a case study in southern China. *Forest Ecology and Management*, 167, 209–222.

26 The potential of agroforestry for sustainable land and water management

J. S. Wallace
CSIRO Land and Water, Townsville, Australia

A. Young
University of East Anglia, Norwich, UK

C. K. Ong
International Center for Research in Agroforestry, Nairobi, Kenya

INTRODUCTION

Improving the efficiency of land and water resource use

Much of the future increase in food and wood production in the humid tropics (and elsewhere), necessary to meet the needs of increasing populations and to reduce hunger and poverty, will have to be achieved from land and water resources already in use. Field observation shows that the extent of the 'land balance' – land that could be used for productive purposes but is not currently in use – is very limited. Estimates by FAO and associated organisations appear to show substantial areas which are cultivable but not presently cultivated (Alexandratos, 1995; Bot *et al.*, 2000). However, the validity of these estimates has recently been challenged, suggesting that the 'land balance' may be 50% or less of that in the official estimates (Young, 1998; 2000). Moreover, a large proportion of the 'land balance' is under forest, for example in Brazil, Congo Democratic Republic (formerly Zaire), Indonesia, Peru and Venezuela, clearance of which is strongly opposed for reasons of environment and biodiversity (Alexandratos, 1995).

The above 'land balance' issue focuses the associated research agenda on the challenge of improving the *efficiency* with which *existing* land and water resources are used. Over the past half-century, great progress has been achieved in this respect. In agriculture, this has been through the advances generally referred to as the green revolution; in forestry, it has been brought about through a variety of improvements in forest management systems, including fast-growing, high-yielding plantations, and by means of genetic improvement. In the early stages of the green revolution, research was directed mainly at plant breeding, fertiliser use and plant protection. However, the pace of advances by these means is slowing; the annual increase in cereal yields in developing countries, which over 1967–82 was 2.9%, has fallen to close to 1%.

As a consequence, more attention has been directed recently at greater efficiency in the use of land and water resources, for example through nutrient recycling and water conservation. A further powerful incentive in this direction has come from considerations of sustainability. Applied to land and water, sustainability means meeting the production needs of present land users whilst conserving for future generations the resources on which that production depends.

Agroforestry as a management option

Agroforestry offers one promising option for efficient and sustainable use of land and water. Agroforestry is defined as a collective name for land use systems in which woody perennials (trees, shrubs) are grown in association with herbaceous plants (crops, pastures) or livestock, in a spatial arrangement, a rotation, or both, and in which there are usually both ecological and economic interactions between the trees and other components of the system. In simplified terms, agroforestry means combining the management of trees with productive agricultural activities. Agroforestry has been practised by farmers for centuries but its recognition as a scientific discipline, and hence a basis for research, dates from 1977 (Young, 1997a).

The term 'forest conversion' is sometimes used as a euphemism for forest clearance. However, agroforestry provides opportunities for forest conversion in the true sense of the term, replacement of natural forest with other tree-based land-use systems

Forests, Water and People in the Humid Tropics, ed. M. Bonell and L. A. Bruijnzeel. Published by Cambridge University Press.

(see Hamilton, this volume, for further comment on the terms 'forest conversion' and 'conversion'). There are also opportunities to use agroforestry for the prevention or reversal of land degradation in the humid tropics (Cooper *et al.*, 1996).

There are approximately 20 major agroforestry systems, or ways of arranging trees and other components in time and space, a classification that has stood the test of time (Nair, 1989; Young, 1989, 1997; ICRAF, 1993). The oldest system is a rotational one, shifting cultivation, ecologically effective under low population densities but almost invariably non-sustainable in the modern world. Apart from this, the systems that are particularly suited to the humid tropics are:

(1) *Multistrata systems (also called forest gardens or agroforests)* Trees, shrubs and herbaceous plants are grown together in a dense, intimate, irregular spatial mixture, often with a large number of plant species; includes home gardens.
(2) *Perennial crop combinations* One or more tree or shrub crops ('plantation crops') are grown in association with herbaceous crops or timber trees, in a dense, regular or irregular spatial pattern. Combinations of perennial crops (e.g. coffee-cacao-banana) may also be treated as agroforestry systems.
(3) *Managed tree fallows* Rotations of fast-growing trees with crops. After a period of pure cropping, trees are planted and remain for one to five years. When cleared for the next cropping period, the leafy matter is generally retained on the soil.
(4) *Biomass transfer* In some indigenous systems (e.g. Nepal, China, Vietnam), tree leaf litter is collected from natural forests and carried on to the cropped land, sometimes after composting as bedding material for livestock. A modern version is to plant blocks of trees and transfer the litter.
(5) *Contour hedgerows* Hedges are planted along the contour on sloping land, with the primary objective of soil and water conservation.
(6) *Reclamation agroforestry* The name applied to systems intended to reclaim degraded land, in which after an initial period of reclamation forestry, the land is restored to agricultural use whilst retaining a component of trees.

Other systems with more limited potential include *taungya*, *trees on pastures* (common in South America), *aquaforestry* (trees employed to fertilise fishponds; mangrove systems), and *entomoforestry* (trees with insects, e.g. for honey, shellac). Taungya systems, in which farmers are permitted to grow crops on forest land, interplanted with young trees which they are required to plant and tend, whilst not uncommon in the humid tropics, have been of limited success (Jordan *et al.*, 1992). It may be noted that *hedgerow intercropping (alley cropping)*, the focus of much early agroforestry research, has rarely been practised spontaneously by farmers in the humid tropics.

There is a wide range of potential benefits which agroforestry systems can achieve, ranging from diversification of production to improved natural resource utilisation. The key benefits in terms of natural resource use are:

(1) Soil conservation in terms of protection against erosion.
(2) Improvement or maintenance of soil fertility.
(3) Water conservation, and more efficient use of water.
(4) Providing environmental functions required for sustainability.

The following sections discuss each of these issues in more detail.

SOIL CONSERVATION: PROTECTION AGAINST EROSION

In its wider sense, soil conservation refers to maintenance of the productive capacity of soil, and hence includes both protection against erosion and maintenance of fertility. In the narrower but more widely used sense, however, it is regarded as the physical conservation of the soil, or protection against erosion. Summaries of the evidence for erosion control through agroforestry systems are given in Young (1993; 1997a, b).

Complete clearance of rainforest commonly results in rapid soil loss on steep slopes, often attaining rates of the order of 50 t ha^{-1} per year. Partial clearance, however, can be followed by very much lower rates, not much exceeding those under natural forest (Ross *et al.*, 1990). An elegant experiment based on selective removal of different strata demonstrated that it is the ground surface cover of litter, not the canopy nor understorey, which provides the main protection (Wiersum, 1985). A litter cover does not have to be complete; a cover of 60% during the period of erosive rains will reduce erosion to 10% of its value on bare soil (Rose and Freebairn, 1985). These findings have implications for land use where forest clearance cannot be avoided. If some proportion of the tree cover can be retained, or replaced by planted trees, then the losses of soil material, organic matter and nutrients is greatly reduced. Agroforestry systems provide opportunities for such inclusion of trees in productive land use systems.

There is little experimental evidence for erosion control under multistrata systems and plantation crop combinations, for the reason that the existence of this control is so evident! The density of plants in these systems results in an almost complete ground-surface cover of leaf litter which, coupled with the dense root system, is highly effective in checking erosion. Observation in many countries, for example Indonesia (Sumatra), Sri Lanka, the Philippines, shows that these systems provide a means for sustainable use of steeply-sloping land in the humid tropics (Wiersum, 1984; Young, 1997a).

In systems of managed tree fallows and biomass transfer, the tree component is separated from the crops in space or time. In tree fallow systems, the degree of erosion depends on conservation under the cropping element. In biomass transfer systems, there is an intrinsic element of conservation arising from the litter cover applied to the soil under crops; cover is greatest at the start of the cropping season, generally the period of most highly erosive rains, but declines as the litter decays (Kiepe, 1995a; Young, 1997a).

The contour hedgerow system, originating on Flores Island, Indonesia, and subsequently developed in the Philippines, initially appeared to have much promise as a means of soil conservation. A series of studies have demonstrated that under controlled experimental conditions, soil loss is reduced to rates of the order of 10% of those without hedgerows, in some cases as low as 2% (Young, 1997a, 1997b). This remarkable result was attributed at first to a barrier effect; it was supposed that plant litter, washed against the stems of the hedgerows, checked the rate of overland flow and so led to soil accumulation above each hedgerow, a mechanism which can lead to the development of micro-terraces (Figure 26.1). If this were the case, one would expect the reduction in runoff to be small. However, studies at Machakos, Kenya, using *Senna siamea* hedgerows, showed that even in heavy storms, the presence of hedgerows reduced runoff by an order of 30–50%, a result which must be due to increased infiltration. This hypothesis was confirmed by observations using a drip infiltrometer, small enough to be placed within a hedgerow; the steady infiltration rate averaged 69 mm h^{-1} under the hedgerows, compared with 8–11 mm h^{-1} on adjacent cropped alleys (Kiepe, 1995a,b; Young, 1997a). Over time, this type of agroforestry system can produce almost level terraces between the hedgerows, improving the infiltration of rainfall, and thus the water supply to crops grown between the hedgerows.

By 1990 it had been demonstrated that, at least on gentle to moderate slopes (<30%), contour hedgerows (spaced at distances of the order of 4–8 m) provide a technically viable alternative to conventional methods of soil conservation based on earth structures. On steep slopes, common in the humid tropics, their effectiveness is more questionable. Based on empirical formulae used to design soil conservation by means of earth structures (e.g. Hudson, 1995; Young, 1997a), to achieve the necessary control of soil loss would require an inter-hedge spacing of about 2 m, which limits the area under crops. A variant developed in Malawi, with very closely spaced hedges, pruned low, requires further investigation before it can be validated (Banda *et al.*, 1994).

Farmers do not always accept contour hedgerows. Attempts in the Philippines to promote the system by extension have met with variable success (Queblatin, 1985; Garrity, 1996), and adoption in other countries is no more than sporadic. In projects that promote this system, the hedgerows are abandoned by some farmers once aid is discontinued. Reasons are usually given as the labour

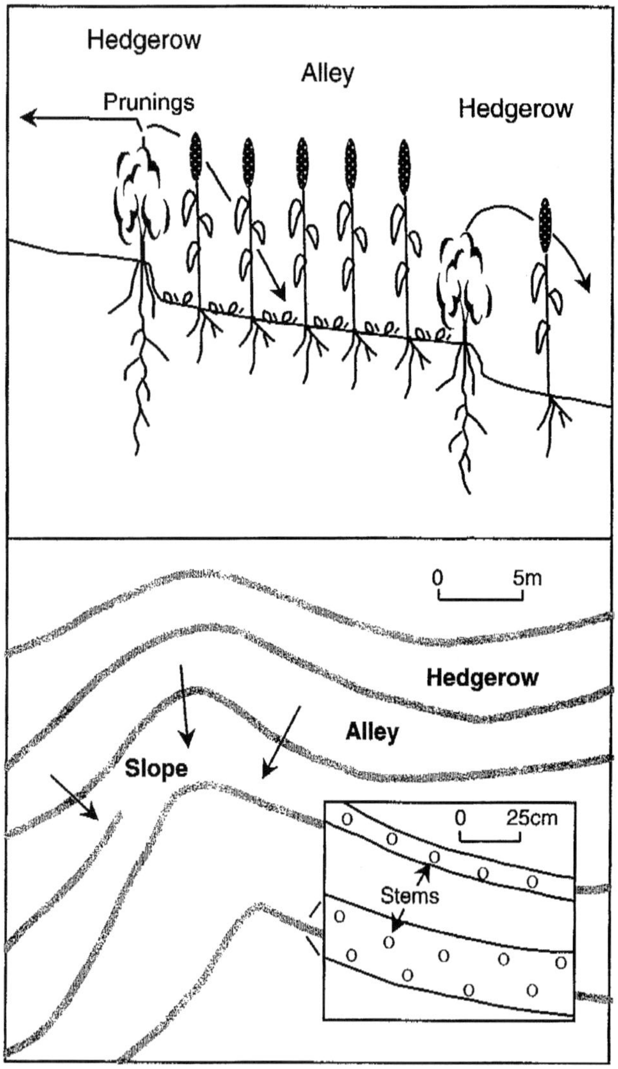

Figure 26.1 The contour hedgerow system. (Reproduced from Young, 1997a.)

requirement for hedgerow maintenance, competition with adjacent crop rows (primarily for water rather than nutrients) and, more fundamentally, the fact that farmers are reluctant to lose the space taken by the hedgerows which could otherwise be occupied by crops. However, set against such failures there are also success stories (e.g. Fujisaka, 1993; Young, 1997a), and at very least, the system is no less acceptable to farmers than conservation by means of bunds or terraces (Sajjapongse and Leslie, 1998).

Finally, reclamation agroforestry, that is the use of trees for restoring degraded land to production and subsequently conserving its fertility, has become a major branch of agroforestry, not least in the humid tropics. Thus, for example, in South East Asia the system has been employed for restoration of degraded *Imperata cylindrical* grassland (Kuusipalo *et al.*, 1995; Menz *et al.*, 1999). Other examples are cited in Young (1997a) and continue to make up a substantial section of *Agroforestry Abstracts*. The key feature in reclamation agroforestry, distinguishing it from many forms of reclamation forestry, is the inclusion of an element of production, from the start or at an early stage.

SOIL FERTILITY MAINTENANCE AND IMPROVEMENT

The general soil–agroforestry hypothesis

Why should it be supposed that the inclusion of trees in land use systems tends towards the improvement of soil fertility? Four basic reasons can be given:

(1) Soils found under natural forest and woodland are fertile, a fact well known to farmers.
(2) The practice of shifting cultivation has demonstrated the capacity of trees to restore fertility lost during cropping.
(3) Ecosystem studies have demonstrated that the cycles of nutrients under many forest ecosystems are relatively closed (as outlined in Proctor, this volume), in contrast to the more open cycles, with higher inputs (if fertilised) and higher outputs under annual cropping.
(4) There is abundant experience from reclamation forestry of the ability of trees to build up fertility on degraded lands (see Scott *et al.*, this volume).

These observations give rise to the general hypothesis on agroforestry and soil fertility, that *appropriate and well-managed agroforestry systems have a potential to maintain soil organic matter and physical properties, and to promote nutrient cycling and efficient nutrient use*. A consequence is that the higher the ratio of trees to herbaceous plants in a land use system, in terms of area covered or biomass production, the greater is the potential for maintenance of fertility. This intrinsically sets limits to the potential of agroforestry systems where the trees are widely-spaced or confined to specific sites, such as hedgerow intercropping, as compared with those where the tree cover is dense and ubiquitous, as in multistrata systems.

The processes by which trees in agroforestry systems assist in maintenance of soil fertility fall into three groups; (1) maintenance of soil organic matter, together with associated soil biological activity and physical properties; (2) nitrogen fixation; (3) nutrient recycling. Space allows for the citation of only a small selection of experimental data. A more extended review of evidence, with sources, is given in Young (1997a).

Soil organic matter and associated properties

Biomass inputs to soils under agroforestry systems come from the decay of leaf litter or prunings and, a source often neglected, the continuous decay of fine root systems. To maintain topsoil carbon at a typical value for the humid tropical land use systems of 2%, a dry matter input of some 14 000 kg ha^{-1} per year is needed. Inputs of above-ground dry matter from agroforestry systems in the humid tropics of the order of 5000–12 000 kg ha^{-1} per year have been measured, to which should be added an input of the order of 40% from decay of roots (Young, 1990; 1997a). Thus the orders of magnitude involved are broadly sufficient. This is confirmed by observation of high soil organic matter levels under multistrata systems and plantation crop combinations. Under systems with discontinuous trees, such as hedgerow intercropping, soil carbon levels frequently do not differ significantly from soils under pure cropping. However, significant differences are found between experimental treatments of hedgerow prunings removed versus retained; for example, on a degraded lixisol (alfisol) in Nigeria, soil carbon after two years was 0.7% when prunings were retained compared with 0.15% with prunings removed (Yamoah *et al.*, 1986).

Associated with organic matter levels, soil biological activity and physical properties can be improved under agroforestry. Increases in earthworm activity and populations of soil microinvertebrates have been reported under tree litter. Tree-soil transects, in which soil properties are measured radially from trees, or across hedgerows, invariably show low bulk density and substantially higher hydraulic conductivity under and adjacent to trees (see Young, 1997a, for references).

NITROGEN FIXATION

Input of nitrogen by the so-called fast-growing nitrogen-fixing trees (FGNFTs) was one of the first proven benefits of agroforestry (for reviews, see Bowen *et al.*, 1990; Danso *et al.*, 1992; Young, 1997a). A large number of the tree species commonly employed in agroforestry systems are nitrogen-fixing, including among legumes, species of *Leucaena, Calliandra, Erythrina, Gliricidia, Sesbania, Inga, Prosopis* and *Acacia*, together with *Faidherbia albida*, and, among non-legumes, *Casuarina* spp.. Quantities fixed vary from 50 to 200 kg N ha^{-1} per year, occasionally higher. Transfer of nitrogen to crops takes place primarily through decomposition of litter and prunings. Nitrogen cycling studies have demonstrated that 10–30% of nitrogen released by decomposition of prunings reaches associated crops (e.g. Sanginga *et al.*, 1995).

Comparisons of nitrogen fixation from trees with that from herbaceous nitrogen-fixing plants have shown no intrinsic differences, either in quantities of nitrogen fixed (per unit area of ground surface) or in their transfer to crops (Danso *et al.*, 1992). Thus intercropping with legumes can be treated on an equal basis with the use of nitrogen-fixing trees in the design of cropping systems.

NUTRIENT RETRIEVAL

The retrieval of nutrients from soil horizons below the reach of crop roots by the deep root systems of trees, including from weathering rock, was for long an unproved hypothesis, uptake from weathering rock being hard to detect experimentally. Evidence has come recently from measurements of soil nitrogen at different depths under a tree fallow system. Nitrate nitrogen at 50–200 cm depth decreased during the growing season under both natural fallow and planted *Sesbania sesban*, but not under maize, indicating greater uptake by the trees (Buresh and Tian, 1997). These studies concur with the evidence for abstraction of water from depth in the soil profile discussed in later in the section on drainage.

NUTRIENT RECYCLING

The earliest review of soils and agroforestry proposed what has since become known as the agroforestry nutrient-cycling hypothesis (Nair, 1984). This states that agroforestry systems can lead to more closed nutrient cycling than agriculture, and hence to more efficient use of nutrients. This is illustrated in Figure 26.2. Under mature natural forest, the uptake of nutrients from the soil by plants is balanced by returns to the soil from the decay of plant litter, whilst inputs to and outputs from the plant-soil system are relatively small, leaching being checked by the dense tree root system. Under agriculture (annual cropping), recycling is less, leaching losses greater, and there is a further nutrient loss in harvest; unless these higher losses are balanced by fertiliser inputs, there will be nutrient losses and hence soil nutrient decline (Stoorvogel and Smaling, 1990).

The hypothesis is that by the inclusion of a tree component, agroforestry can attain a position intermediate between these extremes, through reducing leaching losses and increasing recycling. It is applicable to the major, secondary and micronutrients. Let the nutrient recycling ratio (NRR),

$$\text{NRR} = 100^{*}(U + R)/(G + L + U + R) \qquad (26.1)$$

applicable to any nutrient, be defined as:
where uptake (U) and return (R) are soil-plant and plant-soil transfers respectively, and gains (G) and losses (L) refer to the plant-soil system (Young, 1997a). Gains (G) are the sum of atmospheric additions (rain and dust), organic additions (imported compost, manure), fertiliser, rock weathering, and symbiotic and non-symbiotic fixation of nitrogen. Losses (L) comprise erosion, leaching, harvest, removals other than harvest (e.g. burning), and gaseous losses of nitrogen (denitrification and volatilisation). Uptake (U) refers to uptake by plant roots from the soil. Return (R) from plants to the soil includes transfers by mineralisation and humication of litter, throughfall and stemflow, and ash from burning.

The former supposition that all nutrient cycles under tropical forest are relatively closed has been questioned. More recent reviews show that a wide variation exists, dependent on soil type; thus, nutrient losses in drainage water are low for

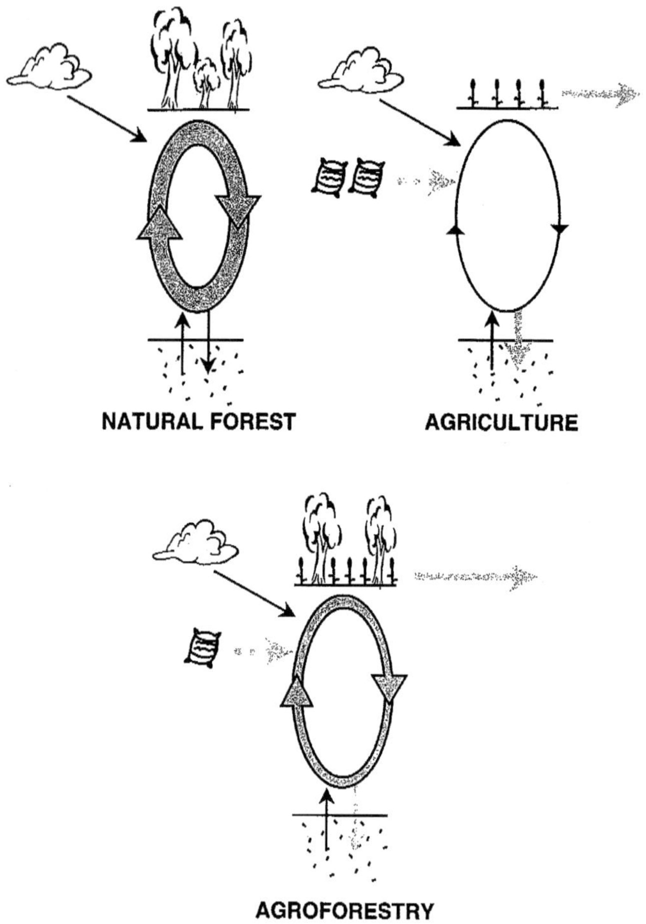

Figure 26.2 The agroforestry nutrient-cycling hypothesis. (Reproduced from Young, 1997a.)

highly-weathered spodosols and oxisols but can be substantial for forests on younger soils such as millisols and inceptisols (Bruijnzeel, 1989, 1991; Baillie, 1989; Whitmore, 1989). However, there are examples of natural forest ecosystems with NRR values as high as 80–90% (e.g. Golley *et al.*, 1975), whereas under cereal crops, they are typically of the order of 40–50%. It can be demonstrated by modelling that agroforestry systems raise the NRR to 60–80%, dependent on the number of trees in the system (Young, 1997a, 2000/2001; Young *et al.*, 1998).

There are two nutrient recycling paths, short-term and longer-term. The short-term path is via mineralisation of plant residues and uptake from the soil solution, taking place within a cropping season. The longer-term path is via fixation of nutrients in soil organic matter and their subsequent release by its mineralisation, a process extending over many years. A major advantage arising from maintenance of soil organic matter, is that its ongoing, steady mineralisation provides a balanced nutrient supply, including not only N, P, K and the secondary nutrients but also micronutrients, thus helping to avoid the micronutrient deficiencies which commonly arise if there is over-dependence on fertilisers.

The quality of tree leaf litter is defined as the ratio of nitrogen to lignin and polyphenols (Swift *et al.*, 1979; Cadisch and Giller, 1996). Most agroforestry trees produce litter of relatively high quality, decomposing rapidly to give short-term peaks in nutrient release. Systems that employ pruning have an intrinsic advantage over those based on litter fall, since trees translocate nutrients out of their leaves prior to leaf fall.

Because of the speed with which nutrients are lost from the soil solution by leaching and (for nitrogen) volatilisation, it is desirable that release should take place during the growing season of crops. The synchrony hypothesis, formulated by the Tropical Soil Biology and Fertility programme (Woomer and Swift, 1994), states that the release of nutrients from decomposition of plant residues can be synchronised with the requirements for nutrient uptake of associated crops. Systems based on pruning offer opportunities to achieve synchrony, through selection of tree species with differing rates of decomposition, timing of pruning, and manner of addition of the residues to the soil. High-quality residues, applied to the soil at about the time of crop planting, show a peak in nutrient release at about the time of maximum crop growth requirement.

Also of relevance to land use on cleared forest areas is the finding, based on studies in the Amazonian zone, of a two-phase system of water movement in forest soils, highlighted by Proctor elsewhere in this volume. Macropore water flows rapidly through the system, with short residence times and consequently low solute concentrations. By contrast, water held in meso and micropores has a longer residence time and probably higher solute concentrations; more of this water is taken up by trees, thus remaining within the plant-soil system (Nortcliff and Thornes, 1989). An implication is that it is important to maintain the soil structure following clearance, particularly pores in the diameter range 1–50 μm. There are clear indications that this is most likely if the cleared forest can be replaced by systems which include trees.

Soil fertility under agroforestry systems

Experimental data for *perennial crop combinations* is available from systems of coffee or cacao with species of *Erythrina, Inga* and *Cordia* (often called 'shade trees', although farmers are aware of their fertility-improving and other functions (Beer, 1987)). Long-term experiments have shown returns to the soil, in leaf litter and prunings, of (kg nutrient ha^{-1} per year): N 100–300, P 10–30, K 50–160, Ca 120–330 (Beer *et al.*, 1990). In some cases of fertilised systems, the nutrient content of decomposing leaf litter exceeded the amounts added in fertiliser. Despite the high annual returns of nutrients to the soil in these systems, inferred nutrient losses in drainage water are surprisingly low. This has been interpreted in terms of active uptake by the deep-rooting trees and the high nutrient-retention capacity of young volcanic soils (Imbach *et al.*, 1989). Nutrient cycling measurements are rarely made for *multistrata systems* since it is so clear, from the large quantity of litter fall and high, apparently sustained, productivity, that these achieve a substantial degree of recycling.

Large increases in crop yields are observed following *managed tree fallows*. This result is found almost invariably, and with statistical significance. There is a residual effect in the second year after ending of the fallow. An example comes from a lixisol in eastern Zambia, fallowed with *Sesbania sesban* and fertilised at four levels (Kwesiga and Coe, 1994). With no fertiliser, after 0, 1, 2 and 3 years of tree fallow, maize yields were respectively 1.6, 3.5, 5.3 and 6.0 t ha^{-1}; with the addition of 37 kg N ha^{-1}, the respective yields were 3.5, 4.6, 5.8 and 7.0 t ha^{-1}. This raises the question of whether, over the period of rotation, the yield loss during years under trees is compensated by the higher yields during cropping. Analysis of the results quoted, extended to a hypothetical rotation of six years, suggests that for nitrogen fertiliser inputs of zero, 37 and 74 kg ha^{-1} per year, one or two-year fallows give a higher total crop yield over the rotation than continuous cultivation (Young, 1997a).

In a number of countries (e.g. Zambia) it has been found that managed tree fallow systems are acceptable to farmers, which was rarely the case with hedgerow intercropping. The planting of trees as a rotational fallow appears to be accepted by farmers as a natural extension of the (less efficient) process of shifting cultivation (Buresh and Cooper, 1999). This experience has led to a major shift in the focus of agroforestry research.

Biomass transfer systems also show large increases in crop yield where tree leaf litter is added. There may be residual effects after the first year. Not all the favourable effects of tree litter are due to added nutrients. There may also be benefits to soil temperature and physical conditions such as soil moisture. Advantages of biomass transfer over fallow systems are the greater permanence of the trees, giving opportunity for their management on forestry principles, coupled with the potential to site the tree component on poorer (sloping or rocky) parts of farms. Biomass transfer by the carrying of leaf litter from natural forest onto farmed land is found in some indigenous systems, for example in Nepal.

WATER CONSERVATION AND MORE EFFICIENT USE OF WATER

Successful plant mixtures appear to be those which make 'better' use of resources by using more of a resource, by using it more efficiently, or both. In terms of the water use of an agroforestry system a central question is, therefore, does intercropping woody and non-woody plants increase total harvestable produce by making more effective use of rainfall? It is possible, at least theoretically, that

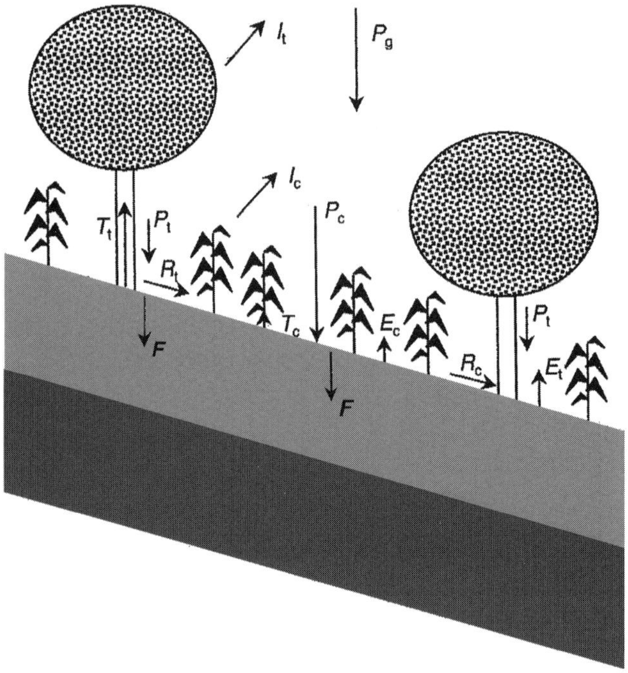

Figure 26.3 A schematic representation of the water balance of an agroforestry system on a hillslope. Gross precipitation P_g is intercepted by the tree and crop canopies, giving rise to interception losses from the trees, I_t, and crop I_c. Rainfall input to the ground beneath the trees, P_t, may be different from that beneath the crop, P_c. Infiltration rates below the trees, F_t, and crop F_c, may produce different rates of surface runoff, R_t and R_c. Evaporation from the soil surface proceeds at rates E_t and E_c beneath the trees and crop respectively. The water contents of the soil zones beneath the trees, θ_t, and the crop, θ_c, due to the different surface inputs and transpiration rates, T_t and T_c, may lead to different drainage rates, D_t and D_c. There may also be some lateral sub-surface water movement, R_s.

a mixture of trees and crops may improve the overall rainfall use efficiency either directly, by more rain being used as transpiration, or indirectly by increasing the water use ratio (or transpiration efficiency), i.e. more dry matter produced per unit of water transpired. This theoretical possibility requires systematic study of the water balance of agroforestry systems such as that carried out by Ong. *et al.* (2000) for an agroforestry system in a sub-humid part of Kenya.

The water balance of an agroforestry system

The complexity of the water balance of an agroforestry system is shown schematically in Figure 26.3. Some of the gross rainfall, P_g, is intercepted by the tree and crop canopies, giving rise to interception losses from the trees, I_t, and crop, I_c. The input of rainfall to the ground beneath the trees, P_t, can be different from that beneath the crop, P_c. P_t and P_c include both throughfall and stemflow. However, in widely spaced trees stemflow is usually a very small fraction of P_g, for example, ~0.7% in the *Grevillea*/maize agroforestry system studied by Jackson (2000). In crops such as maize, with much higher planting densities, stemflow can be between 2 and 4% of P_g (van Dijk and Bruijnzeel, 2001b). Rainfall reaching the ground may infiltrate at different rates below the trees (F_t) and crop (F_c), producing different rates of surface runoff, R_t and R_c respectively. In some circumstances F_t may be sufficiently high not only to absorb P_t, but also any runoff (R_c) from the cropped area. In this case the total runoff from the entire plot would be negligible, as observed in agroforestry studies of runoff on sloping land at Machakos, Kenya, where runoff was <2% of annual rainfall (Kiepe and Rao, 1994).

Water will evaporate directly from the soil surface at potentially different rates E_t and E_c and the trees and crop may also have different transpiration rates T_t and T_c. The water content of the soil zones beneath the trees, θ_t, and the crop, θ_c, is therefore likely to be different, due to the different surface inputs, soil evaporation and transpiration rates. This in turn may lead to different drainage rates, D_t and D_c. Where the agroforestry system is grown on a hillslope, there may also be some lateral sub-surface water movement, R_s, particularly in high rainfall areas where the soil may saturate for significant amounts of time or where perched water tables occur above soil horizons of low permeability (as outlined in Bonell, this volume). The total amount of water transpired by the tree and crop mixture is therefore,

$$T_t + T_c = P_g - I_t - I_c - E_t - E_c - D_t - D_c - R_t - R_c - \delta\theta_t - \delta\theta_c \quad (26.2)$$

In a monoculture (tree or crop) on the same area of land, the equivalent total transpiration is

$$T_c^m = P_g - I_c^m - E_c^m - D_c^m - R_c^m - \delta\theta_c^m \quad (26.3)$$

where the superscript m indicates that these terms can be different in the absence of a companion species. The hypothesis that agroforestry systems can use rainfall more effectively can therefore be expressed mathematically as

$$T_t + T_c > T_c^m \quad (26.4)$$

This defines the way in which the different water balance components need to be managed in order to benefit from the mixture of trees and crops. For example, soil evaporation, runoff and drainage should be minimised in the agroforestry system and this could be achieved via the utilisation of as much as possible of the locally 'non-productive' components of the monoculture water balance (e.g. E_c^m, R_c^m and D_c^m). Where agroforestry is used in the upper reaches of a catchment, the effects of reducing runoff and drainage on the water supply 'downstream' may need to be taken into account. In the following section we will look in more detail at how agroforestry systems can change each of the water balance components.

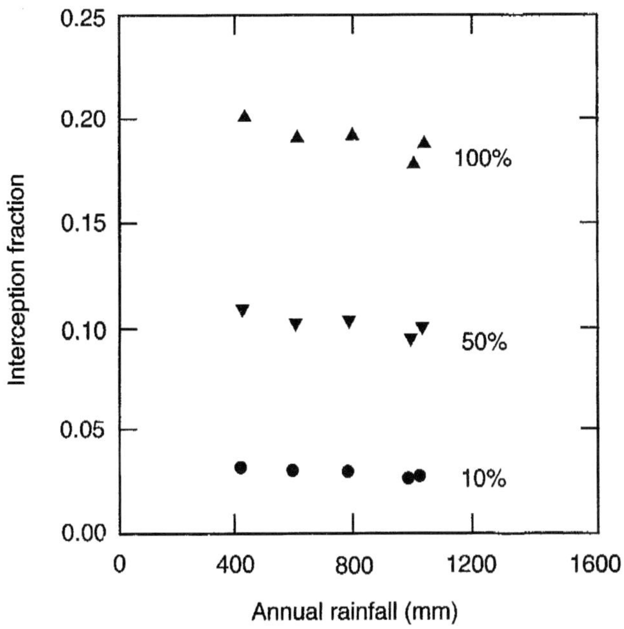

Figure 26.4 Estimates of the annual fraction of rainfall lost as interception (I_t/P_g) made using the Gash *et al.* (1995) sparse forest model with rainfall data for 1984 to 1988 from Machakos, Kenya. Different degrees of cover are input to the model to simulate dense (100%, Δ), intermediate (50%, ∇) and sparse (10%, •) canopies.

INTERCEPTION

The interception process in agroforestry systems differs from that in other forests in two main ways. Firstly, many agroforestry systems tend to have relatively sparse tree densities and secondly, additional complexity is introduced by the crop component of the system with its rapidly varying canopy cover. Van Dijk and Bruijnzeel (2001a) have developed an interception model that deals explicitly with variable crop cover. The sparse nature of the tree component of agroforests affects two key factors that influence the interception, i.e. the amount of water stored on the tree canopy and the rate of evaporation from the tree canopy. For a given species, the storage of water on an individual tree is broadly independent of tree density (Teklehaimanot and Jarvis, 1991; Gash *et al.*, 1995). However, it is the storage of water per unit area of ground that affects interception and this is directly related to canopy cover (Gash *et al.*, 1995) and tends to zero as cover decreases. The rate of evaporation from the wet canopy is primarily determined by the weather conditions during rainfall, but the tree density can affect this rate with more efficient exchange in sparser, more well ventilated canopies (Teklehaimanot and Jarvis, 1991).

Figure 26.4 shows calculations of the ratio of annual interception loss to annual rainfall (the interception ratio) for the trees in a *Grevillea robusta* agroforestry system in Kenya, made using the sparse canopy interception model described by Gash *et al.* (1995). The model assumes a canopy storage per unit area of cover of 0.8 mm, a mean evaporation rate during rainfall of 0.2 mm h^{-1} and a mean rainfall rate of 2.3 mm h^{-1}, consistent with the values estimated for this site by Jackson (2000). Figure 26.4 shows that with complete canopy cover the annual interception loss is ~20% of rainfall and this fraction decreases slightly as rainfall increases. This interception loss is intermediate between values reported for (dense) forests in temperate climates (e.g. ~40%, see Calder and Newson, 1979) and tropical climates (e.g. ~10%, see Lloyd *et al.*, 1988 and Shuttleworth, 1988). This is a consequence of the mean rainfall rate at the Kenyan site being higher than that in temperate areas and lower than that in the humid tropics. The tendency for the annual interception fraction to decrease as rainfall increases from ~400 to 1000 mm has also been reported in temperate climates by Calder and Newson (1979).

Figure 26.4 also shows the modelled interception ratios (I_t/P_g) for sparse canopies, more typical of those in semi-arid agroforestry systems (i.e. 10 to 50% cover). In this case the annual interception loss is between 3 and 10% of rainfall. The 10% interception loss for a cover of 50% is similar to the observations of interception made in the same agroforestry system by Jackson (2000). Other sparse and multi-strata forest stands also have interception losses between 3 and 10% (e.g. Schroth *et al.*, 1999; Valente *et al.*, 1997). Higher interception losses have been reported in the much denser multi-storey agroforestry systems in Costa Rica (>30%) by Imbach *et al.* (1989), where the rainfall is higher and more intense than at the Kenyan site described here. Furthermore, high interception losses have also been reported for montane forests in humid tropical regions, e.g. ~35% by Hopkins (1960) and as much as 50% by Schellekens *et al.* (1999). The main reason put forward for these high forest interception losses in humid regions is the advection of energy from nearby oceans. This point is discussed in more detail by Roberts *et al.* (this volume). Further information and discussion of interception by crops and crop mixtures is given by van Dijk and Bruijnzeel (2001a, b).

SOIL EVAPORATION

Significant quantities of water can be lost as evaporation from the soil surface, particularly in tropical regions with frequent rainfall, high radiation and sparse ground cover. In agroforestry systems the presence of a tree canopy decreases the radiation intensity at the ground thereby reducing soil evaporation. This is because soil evaporation theory and several field studies of soil evaporation suggest that total soil evaporation is determined (at least in part) by the radiant energy reaching the soil surface. This 'first phase' of soil evaporation could therefore be reduced by canopy shade. However, after the first phase is over, soil evaporation rates are determined by the soil hydraulic properties and should therefore be independent of shade. Total evaporative loss from the soil is the sum of losses in the energy-limited first phase and

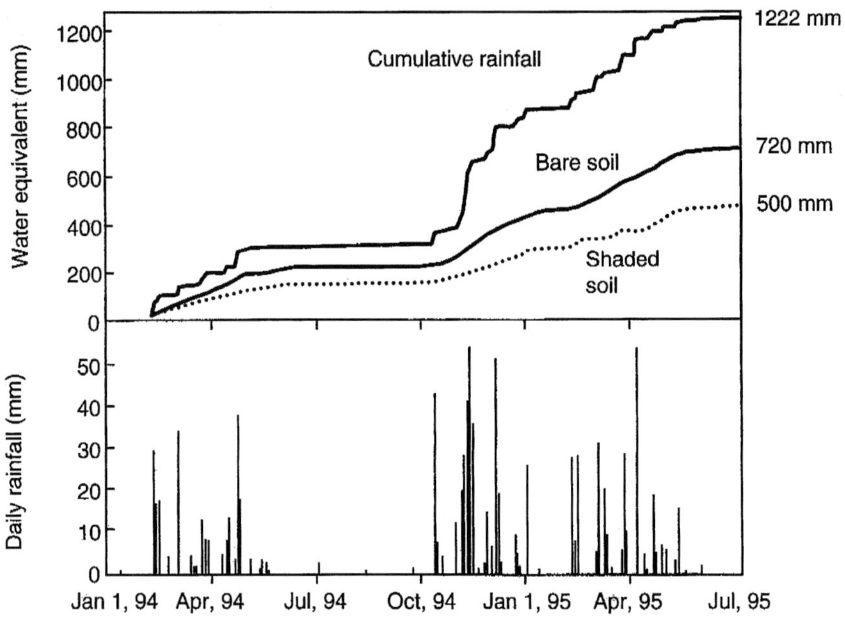

Figure 26.5 Cumulative evaporation from bare soil and soil beneath a tree canopy calculated using a model based on the Ritchie (1972) approach. Rainfall and cumulative rainfall are also shown.

the hydraulically-limited second phase. The net effect of shade on cumulative soil evaporation over periods of several weeks or more will therefore depend on the total amount of time the soil spends in first and second stage drying. This will be a function of soil type and the frequency with which the surface is re-wetted by rainfall.

The effect of shade on soil evaporation has been studied in an agroforestry system in Kenya by Jackson and Wallace (1999) and Wallace *et al.* (1999). Direct measurements of soil evaporation made using mini-lysimeters showed large reduction, up to 30%, in soil evaporation due to the presence of the tree canopy. The data obtained were used to calibrate a simple soil evaporation model, which was then used to calculate shaded and bare soil evaporation over a period of 18 months (Figure 26.5). During this period soil evaporation was reduced from 59% of rainfall in completely bare soil to 41% of rainfall for soil directly beneath the tree canopy. The mean annual reduction in soil evaporation due to full canopy shade, 157 mm or 21% of rainfall, was therefore very significant. The reduction in soil evaporation is smaller in sparser tree canopies, 15% of rainfall when cover is ~0.5 and 6% of rainfall when cover is ~0.2 (Wallace *et al.*, 1999). This analysis is for the tree component of the agroforestry system: however, during the part of the season when the crop canopy is present, there will be additional shading of the soil which may decrease soil evaporation further.

Clearly the reductions in soil evaporation produced by tree canopy shade can help offset the losses of water associated with the tree canopy interception. This is illustrated in Figure 26.6, where the annual saving in soil evaporation and the annual interception loss are plotted as a function of annual rainfall. The rainfall data used in this figure are from two sites in Kenya (Machakos and Kimakia) and both soil evaporation and interception were calculated assuming a tree cover of 10%, 50% and 100%. This analysis indicates that when the annual rainfall is low the saving in soil evaporation due to canopy shade may be greater than the interception loss. However, once rainfall exceeds ~700 mm per annum, the reverse is true, with interception losses exceeding the saving in soil evaporation. The exact point at which the two effects cross over will depend mainly on rainfall intensity and soil type.

RUNOFF

When rainfall reaches the soil surface some of it will infiltrate into the soil. If the rainfall rate is greater than the infiltration rate the excess water starts to collect at the surface and when the surface storage is exceeded, runoff will occur. Infiltration is therefore a dynamic process which changes during the course of a rainstorm depending on the soil characteristics, slope of the land and the rainfall intensity (as outlined in Bonell, this volume). Where the intercropping of woody and non-woody plants alters any of these factors, then the infiltration and runoff may be affected (Kiepe, 1995a).

Soil characteristics that affect infiltration are surface crusting, surface storage, saturated hydraulic conductivity and the presence or absence of plant residues. Vegetation cover generally increases infiltration and reduces runoff by altering one or more of these factors. For example, in Senegal runoff decreased from 456 mm

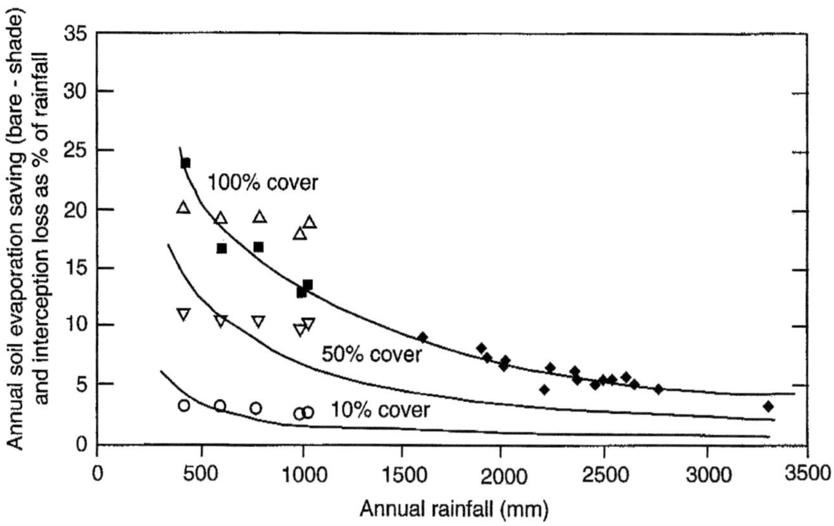

Figure 26.6 Annual saving in soil evaporation (*Es*) at Machakos (■) and Kimakia (◆) in Kenya compared to annual interception loss (open symbols). Calculations have been made at 100% cover (Δ), 50% cover (∇) and 10% cover (O).

in bare soil to 264 mm in cultivated land and further to 200 mm in fallow land containing a mixture of shrubs and herbs (Lal, 1991). Vegetation cover can affect surface infiltration via the canopy modification of the rainfall kinetic energy (e.g. see Wallace, 1996) which may alter soil particle detachment and crust formation. Another effect is via a reduction in surface crusting and improved soil hydraulic conductivity as a result of the incorporation of plant residues into the soil (Kiepe and Rao, 1994). This effect of plant residues was found to be dominant in contour hedgerow systems (see earlier section). Mulching is widely used in the tropics for conserving soil water and reducing soil erosion (e.g. see Stigter, 1984) and the distribution of plant residues in one form or another is usually a major feature of many agroforestry systems. An additional beneficial effect of mulching is through increased activity of soil fauna, further improving the soil structure and water holding capacity (Lavelle *et al.*, 1992).

There are a number of agroforestry practices which are designed to conserve water and reduce runoff by the direct effect trees can have on soil slope. Planting of trees or hedgerows on the contours of sloping land can have the effect of forming natural terraces as water and soil are collected on the up-slope side of the hedgerow (see Figure 26.1). The barrier effect of the hedgerow not only reduces soil loss but also runoff, commonly to the order of one-third of its value without hedges.

DRAINAGE

Drainage is the component of the water balance which is most difficult to measure directly. Most deductions about drainage are therefore made from observations of soil water content. This is illustrated in Figure 26.7 which shows soil moisture content data below a *Grevillea*/maize agroforestry system in Kenya and compares this with equivalent data from a sole maize crop. In December 1994 the soil moisture deficit over the entire depth of the soil profile on this site (*c.* 1.6 m) was low and similar in both the crop and agroforestry plots. As the season progressed, the soil moisture deficit under the agroforestry system developed more rapidly than in the sole crop and by April 1995 was ~100 mm greater in the tree/crop mixture. Figure 26.7(b) and (c) shows how soil moisture content varied throughout the soil profile towards the beginning and end of this time series. These observations show that the soil water contents decreased by a much greater amount in the tree/crop combination, especially at depth. Further details are given by Jackson *et al.* (2000) and they have concluded that drainage from the tree/crop mixture was much less than in the sole crop.

Another way in which trees can affect soil moisture is via the possibility of 'hydraulic lift', in which water taken up by plant roots from moist zones of soil is transported through the root system and released into drier soil (Dawson, 1993). Rainfall captured through stem flow, especially by a woody canopy, can be stored deep in the soil for later use when it is returned to the topsoil beneath the canopy by hydraulic lift. Recently, the opposite of hydraulic lift has been reported in Machakos and elsewhere, in which water is taken from the topsoil and transported by roots into the subsoil (Burgess *et al.*, 1998; Smith *et al.*, 1999a). This mechanism, termed 'downward siphoning' by Smith *et al.*, would lead to the opposite effect of hydraulic lift and would enhance the competitiveness of deep-rooted trees and shrubs.

The likely effect on each of the water balance components of the combination of trees with a crop compared to growing the crop alone are summarised in Table 26.1.

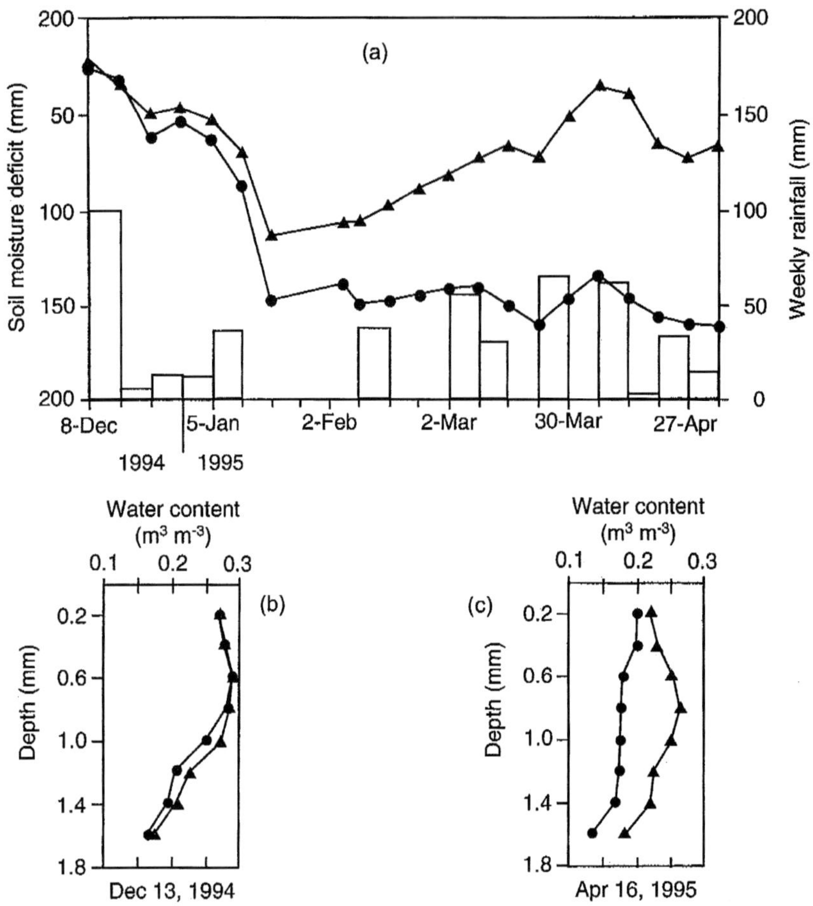

Figure 26.7 (a) Changes in total profile soil moisture deficit under maize (▲) and maize plus *Grevillea robusta* trees (●) at Machakos, Kenya. The water contents changes with depth in this profile are also given for the beginning (b) and end (c) of the period shown.

Table 26.1. *The change in water balance component between an agroforestry system with 50% tree cover and a monocrop*

Water balance component	Change (% of rainfall)	
	Semi-arid climate	Humid tropical climate
Interception loss	+10%	+10–50%
Runoff	Decrease	Decrease
Soil moisture	Decrease	Decrease
Soil evaporation	−10%	−5%
Transpiration	Increase	Increase
Drainage	Decrease	Decrease

Interception losses are around 10% in semi-arid areas, but can be between 10 and 50% in humid tropical climates, depending on whether the location is continental, montane or coastal. This loss will be compensated for completely by a decrease in soil evaporation in a semi-arid climate, but only partially so in a humid tropical climate. Runoff, soil moisture and drainage are all likely to decrease in an agroforest in either climatic regime, with the amount varying according to soil type, slope and species. The extra canopy and the ability of tree roots to exploit water at depth in the soil will lead to a general increase in transpiration in the agroforestry system.

Water use efficiency in tree/crop mixtures

The water use efficiency of any crop or tree/crop mixture can be improved by increasing the water use ratio, e_w (i.e. the amount of carbon fixed per unit of water transpired). This is inversely proportional to the mean saturation deficit of the atmosphere, d (Monteith, 1986),

$$e_w = k/d \tag{26.5}$$

where k is a physiological characteristic specific to a given species. Total dry matter production (W, per unit area in a given time) is simply the product of T_c or T_t and e_w, where T_c and T_t are the crop and tree transpiration respectively. Theoretical considerations and experimental studies have shown that (at least under fairly idealised conditions) the product $e_w\ d$ is quite conservative among species groups (Azam-Ali, 1983; Ong *et al.*, 1996). For example, in C_3 species (e.g. rice, beans and trees) $e_w\ d$ is

~4 kg mm^{-1} kPa and about twice this (8 kg mm^{-1} kPa) in C_4 species such as maize (Squire, 1990). The net effect of atmospheric humidity on any given species is therefore one of the most important factors affecting productivity, since dry matter production per unit of water transpired decreases by a factor of two as saturation deficit increases from ~2 kPa in moist temperate climates to ~4 kPa in semi-arid areas (Squire, 1990). For example, experiments in India under similar mean saturation deficits (2.0–2.5 kPa) provided season-long values of 3.9 and 4.6 g kg^{-1} for millet, compared to 1.5–2.0 g kg^{-1} for groundnut (Ong *et al.*, 1987). However, e_w is not always higher in C_4 species, since similar values have been reported for drought-tolerant C_3 species such as cowpea and cotton and relatively drought-sensitive cultivars of the C_4 species, sorghum and maize.

Equation (26.5) shows that there are two ways that overall production could be increased by increasing e_w. The first is by increasing k, the physiological characteristic which depends on the biochemistry controlling the photosynthetic processes in plant cells. This may be achieved by plant selection (e.g. C_3 or C_4 species), or by breeding or genetically engineering crops with a higher value of k. The second way to increase e_w is to reduce d, either by manipulating the micro-climate, or growing plants in a more suitable macro-climate. This means that agroforests growing in humid tropical regions, where the air is more humid (i.e. low d), will have higher water use ratios. Clearly biomass production is potentially more efficient in the humid tropics than in semi-arid areas.

In theory, the potential of agroforestry to improve e_w is limited compared to intercropping, as the understorey crops are usually C_4 species and the overstorey trees are invariably C_3 species. Improvement in e_w is most likely if the understorey crop is a C_3 species, which are usually light saturated in the open, so partial shade may have little effect on their assimilation. However, the shade will reduce transpiration with the result that e_w increases. This may explain why cotton yield in the Sahel is not reduced by the heavy shading of karite (*Vitellaria paradoxa*) and nere (*Parkia biglobosa*) in parklands, while yields of millet and sorghum were reduced by 60% under the same trees (Kater *et al.*, 1992). The same reason may explain the observation in the South and Central American savannahs that C_3 grasses are found only under trees and never grow in open grassland dominated by C_4 grasses.

There is also the potential for micro-climate modification in agroforestry systems due to the presence of an elevated tree canopy. This may alter not only the radiation but also the humidity and temperature around an understorey crop. Some evidence for this has been found where crops have been grown using trees as shelterbelts, and decreases in d have been reported for several crops (Brenner, 1996). Data from an agroforestry trial in Kenya also show that the air around a maize crop growing beneath a *Grevillea robusta* stand is more humid than the free atmosphere above the trees (Wallace *et al.*, 1995).

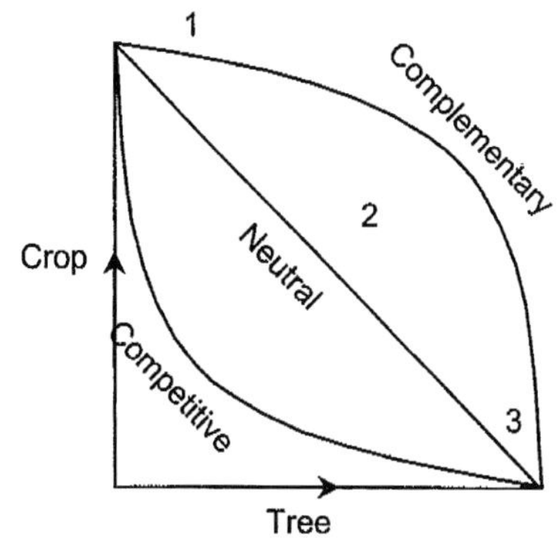

Figure 26.8 Resource capture by tree and crop showing complementarity, competitive and neutral interactions, (1) parkland or savannah, (2) boundary planting, and (3) alley cropping.

In Kenya, Belsky and colleagues (Belsky and Amundson, 1997; Rhoades, 1997) observed improved microclimate due to the presence of trees, along with higher soil biotic activity and N mineralisation, higher infiltration rate and greater beneficial effects in more xeric environments. Plants grown in the open sites were more nutrient-limited than under the tree canopy but artificial shade generated smaller increases in understorey vegetation. However, Belsky was unable to conclude decisively whether micro-climate changes or nutrient enrichment were more important in increasing understorey productivity. It is significant to note that the positive tree effects on understorey vegetation are limited to certain sites and species combinations, including both nitrogen and non-nitrogen fixing trees. It is also difficult to determine precisely whether the tree–grass interactions in tropical savannahs are typical of the competitive category (Figure 26.8) as the literature evidence is primarily based on plot level analysis. If the improvement in soil fertility is due to redistribution in the landscape, this would be an example of the neutral category in Figure 26.8.

Wallace and Verhoef (2000) have developed a multi-species interaction model (ERIN) that can be used to quantify the effect of tree cover on the water use ratio (carbon fixed per unit of water transpired) of an understorey crop. Since fluxes of heat and water vapour from the understorey crop (and soil) can also affect the micro-climate around the overstorey trees, this may alter the water use ratio of the trees. Figure 26.9 shows the results of an ERIN model simulation of both these effects in an agroforesty system with tall (4 m) C_3 trees over a 1 m high C_3 crop. The model predicts that *both* the crop and tree water use ratios will increase by ~25% as the tree cover increases from 0 to 1. The total system water use ratio includes the evaporation from the soil, so it is lower than the water use ratios of the component species. However, this also increases by ~25% in this simulation. Clearly, this improved micro-climate is only of benefit to the crop as long as there is adequate light for crop growth and water in the soil to meet both the tree and crop requirements. This highlights the need to identify

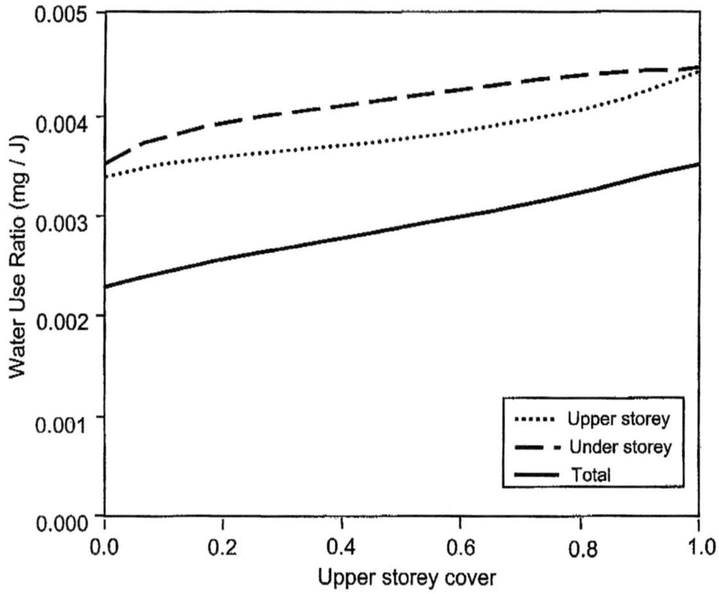

Figure 26.9 Variation of water use ratio with tree fractional ground cover in an agroforestry system with a dominant tree species. (From Wallace and Verhoef, 2000.)

the tree/crop mixtures and soil and climate combinations within which this may be the case.

Evidence from a series of shade cloth trials on maize and beans at Machakos confirmed the small but beneficial effects of shading on crop temperature and crop production when rainfall is inadequate for crop production (Ong *et al.*, 2000) but, unlike the savannah situations, the crops failed because below ground competition consistently outweighed the benefit of shade. In contrast, Rhoades (1997) reported increased soil water (4 to 53% greater than in the open) in the crop root zone beneath *Faidherbia albida* canopies in Malawi. In theory, trees can increase soil water content underneath their canopies if the water 'saved' by their shade effect on reducing soil evaporation and rainfall redistribution e.g. funnelling of intercepted rainfall as stem flow, exceeds that removed by the root systems beneath tree canopies (Ong and Leakey, 1999). At high tree densities, the proportion of rainfall 'lost' as interception by tree canopies and used for tree transpiration would exceed that 'saved' by shading and stem flow, resulting in drier soil below the tree canopy. Van Noordwijk and Ong (1999) expressed this as the amount of water used per unit of shade. This may be one of the most important factors for the observed difference between savannah and alley cropping systems.

ENVIRONMENTAL AND PHYSIOLOGICAL ASPECTS OF AGROFORESTRY

It is often assumed that appropriate agroforestry systems can provide the environmental functions needed to ensure sustainability and maintain micro-climatic and other favourable influences, and that such benefits may outweigh their complexity (Sanchez, 1995). Second, it is also assumed that agroforestry might be a practical way to mimic the structure and function of natural ecosystems, since components of the latter result from natural selection towards sustainability and the ability to adjust to perturbations (Van Noordwijk and Ong, 1999). It is this opportunity for agroforestry to mimic the interactions between trees and other plants in natural ecosystems that led to the recent redefinition of agroforestry, in which different agroforestry practices are viewed as stages in the development of an agro-ecological succession akin to the dynamics of natural ecosystems (Ong and Leakey, 1999). Recent reviews of agroforestry findings have, however, highlighted several unexpected but substantial differences between intensive agroforestry systems and their natural counterparts, which would limit their adoption for solving some of the critical land use problems in the tropics (Rhoades, 1997; Ong and Leakey, 1999; van Noordwijk and Ong, 1999). The most intractable problems for agroforestry appear to be in the semi-arid tropics. In this section, we describe recent insights into the physiological mechanisms between trees and crops in agroforestry systems and how they might be employed to reduce the trade-offs between environmental functions and crop productivity i.e. retain the positive effects of trees observed in natural ecosystems.

Resource capture: complementarity or competition?

The principles of resource capture have been used to examine the influence of agroforestry on ecosystem function, i.e. the capture of light, water and nutrients (Ong and Black, 1994) and to better understand the ecological basis of sustainability of tropical forests (Ewel, 1986). The concept of complementary resource use is not new in ecological studies and it was proposed by de Wit (1960) and others that mixtures of species may have greater capacity to exploit growth resources and hence be more productive than monocultures (Fukai and Trenbath 1993; Sinoquet and Cruz, 1995). However, the recent re-interpretations of published results indicate that increased yield of combinations of annual crops was not always associated with greater resource capture or utilisation. Nevertheless, current ideas on agroforestry interactions continue to be rooted in the complementary resource use concept. For example, Cannell *et al.* (1996) proposed that successful agroforestry systems depend on trees capturing resources that crops cannot. The capture of growth resources by trees and crops can be grouped into three broad categories to show competitive, neutral or complementary interactions (Figure 26.8). In the neutral or trade-off category, trees and crops exploit the same pool of resources so that increases in capture by one species result in a proportional decrease in capture by the associated species. If trees were able to tap resources unavailable to crops, then the overall capture would be increased as shown by the convex curve i.e. complementary use of resources. In the third category, negative interactions between the associated species could result in serious reduction in the ability of one or both species to capture growth resources (concave curve). It is important to bear in mind that tree-crop interactions may change from one category to another depending on the age, size and population of the dominant species as well as the supply and accessibility of the limiting growth resources.

Such ideas on capture of deep water and nutrients coupled with recent innovations in instrumentation (mini-rhizotrons, sap flow gauges) have stimulated a resurgence in root research (Van Noordwijk and Purnomosidhi, 1995; Khan and Ong, 1996) and increased attention on spatial complementarity in rooting distribution and the potential beneficial effects of deep rooting. Agroforestry is also considered to be critical for maintaining ecosystem functioning in parts of Australia, where deep-rooted perennial vegetation has been removed and replaced by annual crops and pastures, leading to a profound change in the pattern of energy capture by vegetation, rising water-tables and associated salinity (Lefroy and Stirzaker, 1999). The Australian example showed that compared to the natural ecosystem it replaced, the agricultural system is 'leaky' in terms of resource capture. Recent investigations in West Africa suggest that a similar magnitude of 'leakiness' is possible when native bush vegetation or woodland, which provide little runoff or groundwater recharge (Culf *et al.*, 1993), is converted into millet fields. The expectation is that agroforestry systems will be able to reduce this leakiness because of extensive tree root systems. Earlier research on South African savannahs has shown that tree roots extend into the open grassland, providing a 'safety net' for recycling water and nutrients and accounting for 60% of the total below-ground biomass (Huntley and Walker, 1982).

One of the earliest detailed studies of resource capture in agroforestry systems was that described by Monteith *et al.* (1991) and Corlett *et al.* (1992a, b) in semi-arid India (Hyderabad) for a C_4 crop, millet (*Pennisetum americanum*) and C_3 tree, *Leucaena leucocephala*. Total intercepted radiation during the rainy season was 40% greater in the alley crop than in sole millet, primarily because the presence of leucaena increased fractional interception during the early stages of the growing season. The sole leucaena and alley leuceana intercepted twice as much radiation again during the following long dry season when cropping was not possible. The evidence from this study shows that the main advantage of alley cropping was in extending the growing period into the dry season and increasing the annual light interception. However, interception by the more efficient C_4 crop was reduced to only half that of the sole millet. This system falls within the lower end of the complementary curve (Figure 26.8).

The alley crop produced 7 t ha^{-1} biomass compared to 4.7 t ha^{-1} of sole millet despite the high amount of light interception because of the low photosynthetic rate or conversion coefficient of a C_3 species. The conversion coefficient (e_r) is defined here as the ratio of biomass production to intercepted light per unit area and provides a measure of the 'efficiency' with which the captured light is used to produce new biomass; the alternative term radiation use efficiency is also commonly used. This and other studies by Ong and co-workers showed that the less efficient C_3 overstorey (tree) component dominated the total light interception while the increased e_r of the understorey (crop) component was insufficient to compensate for the reduced light interception. These results are typical of many alley cropping studies where the tree populations were so high that reduction in crop yield was inevitable since the trees captured most of the resources at the expense of the crops. Although crop yields were seriously reduced, these are examples of complementary interactions which are often misinterpreted as competitive since the sole tree controls are not available.

As for light, agroforestry offers substantial scope for spatial and temporal complementarity of water use resulting from improved exploitation of available water. However, the opportunity for significant complementarity is likely to be limited unless the species involved differ appreciably in their rooting patterns or duration. Recent findings for a range of tree species at Machakos, Kenya, showed that when rainfall was low (250 mm) maize yield is linearly and negatively related to the amount of water used by the trees. This relationship breaks down when rainfall exceeds 650 mm. This example illustrates that the trees were using water from the same soil profile as the maize, i.e. neutral response.

Competition for below ground resources: root structure and function

Early studies of spatial complementarity in agroforestry began by examining the rooting architecture of trees and crops grown as pure stands. For example, Jonsson *et al.* (1988) described the vertical distribution of five tree species at Morogoro, Tanzania, and concluded that the tree root distribution and maize were similar except for *Eucalyptus camaldulensis*, which had uniform distribution to 1 metre. Thus they concluded that there is little prospect of spatial complementarity if these trees and crops were grown in combination. Recent reviews of the rooting systems of agroforestry systems by Gregory (1996) and Ong *et al.* (1999) essentially supported the earlier conclusion of Jonsson *et al.* (1988).

What is the extent of spatial complementarity in water use when there is such a considerable overlap of the two rooting systems? Results at Machakos, Kenya, showed consistently that there was no advantage in water uptake when there was little water recharge below the crop root zone (McIntyre *et al.*, 1996; Smith *et al.*, 1999b; Jackson *et al.*, 2000). However, when recharge occurred following heavy rainfall, tree roots were still able to exploit more moisture below the rooting zone of the crops, even when there was a complete overlap of the root systems of trees and crops. This is an example of temporal complementarity, which demonstrates that soil water distribution, rather than root distribution, is the controlling variable.

Direct measurement of tree function was facilitated by the availability of robust sap flow gauges, which offer a unique opportunity for quantifying the amount of water extracted from the crop rooting zone and hence for assessing spatial complementarity. Experiments in which the lateral roots were progressively severed or excavated indicated that three-year old trees were capable of extracting up to 80% of their water requirements from beneath the crop rooting zone (Lott *et al.*, 1996; Howard *et al.*, 1997). This demonstrates the *potential* for taproots to extract large amounts of water from deep in the soil; however, this does not mean that such large amounts of deep water abstraction will occur when the lateral roots are present. As the trees grew larger in Lott and Howard's experiments, they depleted the soil water and became more and more dependent on current rainfall and severe shoot pruning was necessary to improve infiltration of soil water and below ground complementarity (Jackson *et al.*, 2000).

The lack of spatial complementarity in alley cropping was highlighted by Van Noordwijk and Purnomosidhi (1995), who observed that repeated prunings of trees in alley cropping had the danger of enhancing below-ground competition by promoting the proportion of superficial roots. They imposed three pruning heights (50, 75, 100 cm) on five tree species (*Paraserianthes falcataria, Gliricidia sepium, Peltophorum dasyrachis, Senna siamea, and Calliandra calothyrsus*) at a sub-humid site at Lampung, Sumatra, Indonesia. Recent measurements of the long term (three years) effects of pruning on two species, *S. spectabilis* and *G.sepium*, at Machakos, Kenya, confirmed the findings in Lampung and showed that rooting depths of both pruned trees and crops were almost identical in the alley cropping treatment. The evidence so far suggests that below ground competition is inevitable in alley cropping systems where water is limiting.

The most remarkable example of temporal complementarity in water use is the unusual phenology of the Sahelian tree, *Faidherbia albida*, which retains its leaf shedding habit in the rainy season even when planted in the Deccan plateau of India, where the water-table is too deep for tree roots. One of the few deliberate experiments in which *F. albida* is compared with a tree with 'conventional' leaf phenology is that reported by Ong *et al.* (1996) at Hyderabad, India. Comparison of sap flow rates of *F. albida* and a local Indian tree, *Albizia lebbek*, show that transpiration of *F. albida* begins in August when the understorey crops have developed a full canopy. In contrast, *A. lebbeck* produces a full canopy in May, well before the onset of the rains and sheds its leaves when the *F. albida* starts to develop its canopy. Crop yields beneath both trees were about the same suggesting that both tree species were utilising water from the same soil profile. This is clearly an example of a neutral category (Figure 26.8). Thus, phenology on its own is not adequate for complementary use of resources.

Where groundwater is accessible to tree roots there is clear evidence for spatial complementarity. For instance, measurements of stable isotopes of oxygen in plant sap, groundwater and water in the soil profile of windbreaks in the Majjia valley in Niger showed that neem trees, *Azadirachta indica*, obtained a large portion of their water from the surface layers of the soil only after rain when water was abundant, but during the dry season tree roots extracted groundwater (6 m depth) or deep reserves of soil water (Smith *et al.*, 1997). In contrast, at a site near Niamey, West Africa, where groundwater was at a depth of 35m, they found that both the trees and millet obtained water from the same 2–3 m of the soil throughout the year.

CONCLUDING REMARKS

The understanding of the hydrological, ecological and physiological processes in alley cropping and other simultaneous agroforestry systems has advanced considerably during the last few years. Although much remains to be studied, we conclude that sufficient is now known to make broad recommendations for what types of agroforestry systems are suited to which climatic and soil conditions.

In the humid tropics, agroforestry systems offer opportunities for conversion of forested land to productive use, whilst retaining many of the beneficial effects of a tree cover. Multistrata systems (forest gardens, agroforests) and perennial crop combinations appear to be the most appropriate agroforestry systems for sustainable land use in the humid tropics, including on sloping land; these systems are commonly found acceptable by farmers. Managed tree fallows and biomass transfer systems provide opportunities for the retention of a tree component in land use systems directed primarily at annual cropping. Contour hedgerow systems offer a technically viable alternative for soil conservation; their acceptability to farmers is variable, although no worse than for conventional methods of conservation. Reclamation agroforestry offers opportunities for restoring degraded lands to productive use.

Agroforestry systems also offer viable alternatives for soil conservation. In the design of such systems, the primary objective should be the maintenance of ground cover. The major mechanisms for improvement of soil fertility through agroforestry arise from the potential of trees for maintaining soil organic matter and associated physical properties, nitrogen fixation, and nutrient recycling. A primary mechanism for this is via the addition to the soil of tree leaf litter or prunings which leads to large and consistent increases in yield of associated crops, arising from both short-term and longer term litter decay and nutrient release mechanisms.

In semi-arid conditions the beneficial effects of savanna and parkland trees on soil properties are linked to trees with a high proportion of woody above-ground structures. It would therefore take a long time (20–40 years) before the beneficial effects could be realised, since investment in woody structure slows tree growth. Such long time scales are well beyond the planning horizon of many farmers for the relatively small benefit in crop productivity and may well explain why farmers rarely plant these trees, even though they are well aware of tree species which are soil 'improvers'. Instead, it may be more worthwhile to focus attention on selection of trees that provide more direct and immediate benefits to farmers (rather than selection for soil enrichment), with minimum loss of crop productivity. It is perhaps not surprising that farmers are already beginning to experiment with such systems. For example, in the drylands of eastern Kenya farmers have recently developed an intensive parkland system using a fast-growing indigenous species, *Melia volkensii* (Meliaceae), which provides high value timber in five to eight years and fodder during the dry season without an apparent loss in crop productivity (Stewart and Bromley, 1994).

There are two main ways of increasing the efficiency with which water is used in agroforestry. The first is to convert more of the rainfall input into transpiration and the second is to increase the transpiration water use ratio. The former may be achieved using a range of physical engineering, hydrological and agronomic techniques which reduce soil evaporation, runoff and drainage. The latter can be achieved either by plant selection or climate modification. In semi-arid climates, significant improvements (~10 to 25%) in the (generally low) transpiration water use ratio appear to be possible, whereas there is less scope for this in the humid tropics where the transpiration ratio is higher. Savings in soil evaporation due to canopy shade may compensate significantly for losses due to interception of rainfall by the canopy. In humid climates interception losses will normally exceed soil evaporation savings, whereas in semi-arid climates it is possible that savings in soil evaporation may exceed losses due to interception. Agroforestry systems can also 'convert' runoff and drainage into transpiration, thereby increasing rainfall use efficiency at the scale of the agroforestry plot. However, if the runoff and/or drainage used previously contributed significantly to downstream water resources, this needs to be taken into account in an overall assessment of water use efficiency at a catchment scale.

The importance of obtaining more information using a catchment wide approach is underlined by pointing out that current understanding of resource capture by agroforestry systems is based on well-managed small plots, often in research stations, in which about 30–45% of the rainfall is used for transpiration. Such level of rainfall utilisation is rarely achieved in subsistence agriculture or on a basin scale and there are still ample opportunities for increasing water use by incorporating trees in the landscape. For example, Rockstrom (1997) reported that only 6 to 16% of the total rainfall in a catchment in Niger is utilised by pearl millet for transpiration and the remainder is lost by soil evaporation (40%) or by deep drainage (33 to 40%). In contrast, plot level studies at Machakos by McIntyre *et al.* (1996) reported a rainfall utilisation of 40–45% by maize and cowpea for transpiration and the rest was lost as soil evaporation, thereby limiting the opportunity for agroforestry. Thus, future opportunities for simultaneous agroforestry systems should be explored within the landscape as well as on under-utilised niches within and around the farms, such as boundary plantings.

Finally, although there is clearly great potential for agroforestry systems to conserve and improve resource use, it is by no means suggested that agroforestry automatically brings about all of the above benefits. In order to do so, an agroforestry system must be appropriate for the environment (climate and soil), practicable (within the local and on-farm constraints), economically viable, and acceptable to the farmer. Finally, as with any system of agriculture or forestry, to achieve the potential benefits an agroforestry system needs to be well managed. Provided that these conditions are fulfilled, there is considerable potential for agroforestry to combine production with conservation, and thus to achieve sustainable land use.

References

Alexandratos, N. (1995) *World agriculture: towards 2010.* An FAO study. Chichester, UK: Wiley, for FAO.

Azam-Ali, S. N. (1983) Seasonal estimates of transpiration from a millet crop using a porometer. *Agricultural and Forest Meteorology* 30: 13–24.

Baillie, I. C. (1989) Soil characteristics and classification in relation to the mineral nutrition of of tropical wooded ecosystems. In *Proctor* (1989, q.v.), 15–26.

Banda, A. Z., Maghembe, J. A., Ngugi, D. N., and Chome, V. A. (1994) Effect of intercropping maize and closely spaced *Leucaena* hedgerows on soil conservation and maize yield on a steep slope at Ntcheu, Malawi. *Agroforestry Systems* 27, 17–22.

Beer, J. (1987) Advantages, disadvantages and desirable characteristics of shade trees for coffee, cacao and tea. *Agroforestry Systems* 5, 3–13.

Beer, J., Bonnemann, A., Chavez, W., *et al.* (1990) Modelling agroforestry systems of cacao (*Theobroma cacao*) with laurel (*Cordia alliodora*) or poro (*Erythrina poeppigiana*) in Costa Rica. V. Productivity indices, organic material models and sustainability over 10 years. *Agroforestry Systems* 12, 229–249.

Belsky, A. J. and Amundson, R. G. (1997) Influence of savannah trees and shrubs on understorey grasses and soils: new directions in research. In: *Carbon and Nutrient Dynamics in Natural and Agricultural Tropical Ecosystems* (Eds. Bergstrom, L. and Kirchmann, H.) pp 153–171, CAB International, Wallingford, UK.

Bot, A. J., Nachtergaele, F. O. and Young, A. (2000) Land resource potential and constraints at regional and country levels. *World Soil Resources Report 90*. Rome: FAO.

Bowen, G. D., Sanginga, N., and Danso, S. K. A. (1990) Biological nitrogen fixation in agroforestry – an overview. *Transactions of the 14th International Congress of Soil Science*, *3*, 170–175.

Brenner, A. J., 1996. Micro-climatic modifications in Agroforestry. In: *Tree-crop interactions – a physiological approach* (Eds. Ong, C. K., and Huxley, P. A.) CAB International, Wallingford, UK, pp 159–187.

Bruijnzeel, L. A. (1989) Nutrient cycling in moist tropical forests: the hydrological framework. In *Proctor* (1989, q.v.), 383–415.

Bruijnzeel, L. A. (1991) Nutrient input-output budgets of tropical forest ecosystems. *Journal of Tropical Ecology* 7, 1–24.

Buresh, R. J. and Cooper, P. J. M. (eds.) 1999. The science and practice of short-term improved fallows. *Agroforestry Systems* 47, 1–358.

Buresh, R. J. and Tian, G. (1997) Soil improvement by trees in sub-Saharan Africa. *Agroforestry Systems* 38, 51–76.

Burgess, S. S. O., Adams, M. A., Turner, N. C. and Ong, C. K. (1998) Redistribution of water within plant root systems. *Oceologia* 115, 306–311.

Cadish, G. and Giller, K. R. (eds.) (1996) *Driven by nature: plant litter quality and decomposition*. Wallingford, UK: CAB International.

Calder, I. R. and Newson, M. D. 1979. Land-use and upland water resources in Britain-a strategic look. *Water Resources Bulletin*, 15, 1628–1639.

Cannell, M. G. R., Van Noordwijk, M. and Ong, C. K. (1996) The central agroforestry hypothesis: the trees must acquire resources that the crop would not otherwise acquire. *Agroforestry Systems*, 34, 27–31.

Cooper, P. J. M., Leakey, R. R. B., Rao, M. R., and Reynolds, L. (1996) Agroforestry and the mitigation of land degradation in the humid and sub-humid tropics of Africa. *Experimental Agriculture* 32, 235–290.

Corlett, J. E., Black, C. R., Ong, C. K. and Monteith, J. L. (1992a) Above and below-ground interactions in a leucaena/millet alley cropping system. 2. Light interception and dry matter production. *Agricultural and Forest Meteorology* 60, 73–91.

Corlett, J. E., Ong, C. K., Black, C. R. and Monteith, J. L. (1992b) Above and below-ground interactions in a leucaena/millet alley cropping system. 1. Experimental design, instrumentation and diurnal trends. *Agricultural and Forest Meteorology* 60, 53–72.

Culf, A. D., Allen, S. J., Gash, J. H. C., Lloyd, C. R. and Wallace, J. S. (1993) Energy and water budgets of an area of patterned woodland in the Sahel. *Agricultural and Forest Meteorology* 66, 65–80.

Danso, S. K. A., Bowen, G. D. and Sanginga, N. (1992) Biological nitrogen fixation in trees in agro-ecosystems. *Plant and Soil* 141, 177–196.

Dawson, T. E. (1993) Hydraulic lift and water use by plants: implications for water balance, performance and plant-plant interactions. *Oceologia* 95, 565–574.

de Wit, C. (1960) On competition. *Verslagen van Landbouwkunige Onderzoekingen* 66, 1–82.

Ewel, J. J. (1986) Designing agricultural ecosystems for the humid tropics. *Ann Rev Ecol and Syst*, 17, 245–271.

Fujisaka, S. (1993) A case of farmer adaptation and adoption of contour hedgerows for soil conservation. *Experimental Agriculture* 29, 97–105.

Fukai, S. and Trenbath, B. R. (1993) Processes involving intercrop productivity and yields of component crops. *Field Crops Res* 34, 247–271.

Garrity, D. P. (1996) Tree-crop-soil interactions on slopes. In *Tree-crop interactions: a physiological approach* (ed. C. K. Ong and P. Huxley, CAB International, Wallingford, UK), 299–318.

Gash, J. H. C., Lloyd, C. R. and Lachaud, G. (1995) Estimating sparse forest rainfall interception with an analytical model. *Journal of Hydrology*, 170, 79–86.

Gregory, P. J. (1996) Approaches to modelling the uptake of water and nutrients in agroforestry systems. *Agroforestry Systems* 34, 51–65.

Golley, F. B., McGinnis, J. T., Clements, R. G., Child, G. I. and Duever, M. J. (1975). *Mineral cycling in a tropical moist forest ecosystem*. University Georgia Press, Athens. 248 pp.

Hopkins, B. (1960) Rainfall interception by a tropical rainforest in Uganda. *East African Agricultural Journal* 25, 255–258.

Howard, S. B., Ong, C. K., Black, C. R. and Khan, A. A. H. (1997) Using sap flow gauges to quantify water uptake by tree roots from beneath the crop rooting zone in agroforestry systems. *Agroforestry Systems* 35, 15–29.

Hudson, N. W. (1995) *Soil conservation*. Batsford, London.

Huntley, B. J. and Walker, B. H. (ed) (1982) *Ecology of Tropical Savannas*. Springer Verlag, Berlin, pp 669.

ICRAF (1993) *International Centre for Research in Agroforestry. Annual Report* 1993. Nairobi: ICRAF.

Imbach, A. C., Fassbender, H. W., Borel, R., Beer, J. and Bonnemann, A. (1989) Modelling agroforestry systems of cacao (*Theobroma cacao*) with laurel (*Cordia alliodora*) and cacao with poro (*Erthrina poeppigiana*) in Costa Rica IV. Water balances. *Agroforestry Systems* 8, 267–287.

Jackson, N. A. and Wallace, J. S. (1999) Soil evaporation measurements in an agroforestry system in Kenya. *Agricultural and Forest Meteorology* 94, 203–215.

Jackson, N. A. (2000) Measured and modelled rainfall interception loss from an agroforestry system in Kenya. *Agricultural and Forest Meteorology* 100, 323–336.

Jackson, N. A., Wallace, J. S. and Ong, C. K. (2000) Tree pruning as a means of controlling water use in an agroforestry system in Kenya. *Forest Ecology and Management* 126, 133–148.

Jonsson, K., Fidgeland, L., Maghembe, J. A. and Hogberg, P. (1988) The vertical root distribution of fine roots of five tree species and maize in Morogoro, Tanzania. *Agroforestry Systems* 6, 63–69.

Jordan, C. F., Gajeseni, J. and Watanabe, H. (eds.) (1992) *Taungya: forest plantations in South East Asia*. Wallingford, UK: CAB International.

Kater, L. J. M., Kante, S. and Budelman, A. (1992) Karite (*Vitellaria paradoxa*) and nere (*Parkia biglobosa*) associated with crops in South Mali. *Agroforestry Systems* 18, 89–105.

Keipe, P. and Rao, M. R. (1994) Management of agroforestry for the conservation and utilization of land and water resources. *Outlook on Agriculture* 23, 17–25.

Kiepe, P. (1995a) No runoff, no soil loss: soil and water conservation in hedgerow barrier systems. *Tropical Resource Management Papers* 10. Wageningen, Netherlands: Agricultural University.

Kiepe, P. (1995b) Effect of *Cassia siamea* hedgerow barriers on soil physical properties. *Geoderma* 66, 113–120.

Khan, A. A. H. and Ong, C. K. (1996) A low cost heat pulse method for measuring tree root water uptake. *Agroforestry For* 7(2):19–22.

Kuusipalo, J., Adjers, G., Jafarsidik, Y. *et al.* (1995) Restoration of natural vegetation in degraded *Imperata cylindrical* grassland: understorey development in forest plantations. *Journal of Vegetation Science* 6, 205–210.

Kwesiga, F. and Coe, R. (1994) The effect of short rotation *Sesbania sesban* planted fallows on maize yield. *Forestry Ecology and Management* 64, 199–208.

Lal, R. (1991) Current research on crop water balance and implications for the future. In: *Soil Water Balance in the Sudano-Sahelian Zone* (Eds. M. V. K. Sivakumar, J. S. Wallace, C. Renard and C. Giroux) Proc. Int. Workshop, Niamey, Niger, February 1991) *IAHS Publ. no.* 199, 31–44. IAHS Press. Institute of Hydrology, Wallingford, UK.

Lavelle, P., Spain, A. V., Blanchart, E., Martin, A. and Martin, S. (1992) The impact of soil fauna on the properties of soils in the humid tropics. In: *Myths and Science of Soils of the Tropics. Soil Science Society of America Special Publication* 29, 157–185.

Lefroy, E. C. and Stirzaker, R. J. (1999) Agroforestry for water management in southern Australia. *Agroforestry Systems* 45, 277–302.

Lloyd, C. R., Gash, J. H. C., Shuttleworth, W. J. and Marques, F. A.de O. (1988) The measurement and modelling of rainfall interception by Amazonian rainforest. *Agricultural and Forest Meteorology* 43, 277–294.

Lott, J. E., Khan, A. A. H., Ong, C. K. and Black, C. K. (1996) Sap flow measurements of lateral roots in agroforestry systems. *Tree Physiol* 16, 995–1001.

McIntyre, B. D., Riha, S. J. and Ong, C. K. (1996) Light interception and evapotranspiration in hedgerow agroforestry systems. *Agricultural and Forest Meteorology* 81, 31–40.

Menz, K., Magcale-Macandog, D. and Rusastra, I. W. (eds.) (1999) Improving smallholder farming systems in *Imperata* areas of South East Asia: alternatives to shifting cultivation. *ACIAR Monographs* 52. ACIAR, Canberra.

Monteith, J. L., (1986) How do crops manipulate water supply and demand? *Phil Trans Roy Soc London, B.*, 316, 245–259.

Monteith, J. L., Ong, C. K. and Corlett, J. E. (1991) Microclimatic interactions in agroforestry systems. *For Ecol Manage* 45, 31–44.

Nair, P. K. R. (1984) Soil productivity aspects of agroforestry. Nairobi: ICRAF.

Nair, P. K. R. (1989) *Agroforestry systems in the tropics*. Dordrecht, the Netherlands: Kluwer.

Nortcliff, S. and Thornes, J. B. (1989) Variations in soil nutrients in relation to soil moisture status in a tropical forested ecosystem. In *Mineral nutrients in tropical forest and savanna ecosystems* (ed. J. Proctor, Blackwell, Oxford, UK), 43–54.

Ong, C. K., Simmonds, L. P. and Matthews, R. B. (1987) Responses to saturation deficit in a stand of groundnut (*Arachis hypogaea* L.) 2. Growth and development. *Ann Bot* 59, 121–128.

Ong, C. K. and Black, C. R. (1994) Complementarity in resource use in intercropping and agroforestry systems. In: *Resource Capture by Crops* (Eds. Monteith J. L., Scott R. K. and Unsworth M. H.) Nottingham University Press, pp 255–278 Loughborough, UK.

Ong, C. K., Black, C. R., Marshall, F. M., and Corlett, J. E., (1996) Principles of resources capture and utilisation of light and water. In: *Tree-crop interactions – a physiological approach* (Eds. Ong, C. K., and Huxley, P. A.) CAB International, Wallingford, UK, pp. 73–158.

Ong, C. K. and Leakey, R. R. B. (1999) Why tree crop interactions in agroforestry appear at odds with tree-grass interactions in tropical savannahs. *Agroforestry Systems* 45, 109–129.

Ong, C. K., Deans, J. D., Wilson, J., Mutua, J., Khan, A. A. H. and Lawson, E. M. (1999) Exploring below ground complementarity in agroforestry using sap flow and root fractal techniques. *Agroforestry Systems* 44, 87–103.

Ong, C. K., Black, C. R., Wallace, J. S., Khan, A. A. H., Lott, J. E., Jackson, N. A., Howard, S. B. and Smith, D. M. (2000) Productivity, microclimate and water use in Grevillea robusta based agroforestry systems on hillslopes in semi-arid Kenya. *Agriculture, Ecosystems and Environment* 80, 121–141.

Proctor, J. (ed.) (1989) *Mineral nutrients in tropical forest and savanna ecosystems*. Blackwell, Oxford.

Queblatin, E. (1985) Upland agricultural development: the Central Visayas regional project-I experience. In Soil erosion management. *ACIAR Proceedings Series* 6 (Eds. Craswell, E. T., Remenyi, J. V. and Nallana, L. G, Canberra, ACIAR), 71–76.

Rhoades, C. C. (1997) Single-tree influences on soil properties in agroforestry: lessons from natural forest and savanna ecosystems. *Agroforestry Systems* 35, 71–94.

Ritchie, J. T. (1972) Model for predicting evaporation from a row crop with incomplete cover. *Water Resources Research 8*, 1204–1213.

Rockstrom, J. (1997) On – farm agrohydrological analysis of the Sahelian yield crisis: rainfall partitioning, soil nutrients and water use efficiency of pearl millet. Ph.D thesis, University of Stockholm.

Rose, C. W. and Freebairn, D. M. (1985) A new mathematical model of soil erosion and deposition processes with applications to field data. In *Soil erosion and conservation* (ed. S. A. El-Swaify, W. C. Moldenhauer, and A. Lo, Ankeny, Iowa, USA: Soil Conservation Society of North America), 549–557.

Ross, S. M., Thornes, J. B., and Nortcliff, S. (1990) Soil hydrology, nutrient and erosional response to the clearance of *terra firme* forest, Maracá Island, Roraima, northern Brazil. *Geographical Journal* 156, 267–282.

Sajjapongse, A. and Leslie, R. N. (eds.) (1998) Farmers' acceptance of soil-conservation technologies. *IBSRAM Proceedings* 17. Bangkok: IBSRAM.

Sanchez, P. (1995) Science in agroforestry. *Agroforestry Systems* 30, 5–55.

Sanginga, N., van Lauwe, B., and Danso, S. K. A. (1995) Management of biological N_2 fixation in alley cropping systems – estimation and contribution to N balance. *Plant and Soil* 174, 119–141.

Sinoquet, H. and Cruz, P. (1995) *Ecophysiology of Tropical Intercropping*. INRA. Paris. 483 pp.

Schellekens, J., Scatena, F. N., Bruijnzeel, L. A. and Wickel, A. J. (1999) Modelling rainfall interception by a lowland tropical rainforest in northeastern Puerto Rico. *Journal of Hydrology* 225, 168–184.

Schroth, G., Ferreira da Silva, L., Wolf, M-A., Teixeira, W. G. and Zech, W. (1999) Distribution of throughfall and stemflow in multi-strata agroforestry, perennial monoculture, fallow and primary forest in central Amazonia, Brazil. *Hydrological Processes* 13, 1423–1436.

Shuttleworth, W. J. (1988) Evaporation from Amazonian rainforest. *Proceedings of the Royal Society of London Series B*, 233, 321–346.

Smith, D. M., Jarvis, P. G. and Odongo, J. C. W. (1997) Sources of water used by trees and millet in Sahelian windbreak systems. *J Hydrol* 198: 140–153.

Smith, D. M., Jackson, N. A., Roberts, J. M. and Ong, C. K. (1999a) Reverse flow in tree roots and downward siphoning of water by *Grevillea robusta*. *Funct Ecol* 13(2), 256–264.

Smith, D. M., Jackson, N. A., Roberts, J. M. and Ong, C. K., (1999b) Root distributions in a *Grevillea robusta* – maize agroforestry system in semi-arid Kenya. *Plant and Soil* 211, 191–205.

Squire, G. R., (1990) *The Physiology of Tropical Crop Production*. CAB International, Wallingford, UK.

Stewart, M. and Bromley, T. (1994) Use of Melia volkensii in a semi-arid agroforestry systems in Kenya. *Comm For Rev* 73(2), 128–131.

Stigter, C. J. (1984) Mulching as a traditional method of microclimate management. *Arch Me Geophys Bioclim Ser B* 35, 147–154.

Stoorvogel, J. J. and Smaling, E. M. A. (1990) Assessment of soil nutrient depletion in Sub-saharan Africa: 1983–2000. *Staring Centre Report* 28. Wageningen, the Netherlands: Staring Centre.

Swift, M. J., Heal, J. W., and Anderson, J. M. (1979) *Decomposition in terrestrial ecosystems*. Oxford, UK: Blackwell.

Teklehaimanot, Z. and Jarvis, P. G. (1991) Direct measurement of evaporation of intercepted water from forest canopies. *Journal of Applied Ecology* 28, 603–618.

Valente, F., David, J. S. and Gash, J. H. C. (1997) Modelling interception loss for two sparse eucalypt and pine forests in central Portugal using reformulated Rutter and Gash models. *Journal of Hydrology* 190, 141–162.

Van Dijk, A. I. J. M. and Bruijnzeel, L. A. (2001) Modelling rainfall interception by vegetation of variable density using an adapted analytical model. Part 1. Model description. *Journal of Hydrology* 247, 230–238.

Van Dijk, A. I. J. M. and Bruijnzeel, L. A. (2001) Modelling rainfall interception by vegetation of variable density using an adapted analytical model. Part 2. Model validation for a tropical upland mixed cropping system. *Journal of Hydrology* 247, 239–262.

Van Noordwijk, M. and Purnomosidhi, P. (1995) Root architecture in relation to tree-soil-crop interactions and shoot pruning in agroforestry. *Agroforestry Systems* 30, 161–173.

Van Noordwijk, M. and Ong, C. K. (1999) Can the ecosystem mimic hypotheses be applied to farms in African savannahs? *Agroforestry Systems* 45, 131–158.

Wallace, J. S. (1996) The water balance of mixed tree-crop systems. In: *Tree-crop interactions – a physiological approach* (Eds. Ong, C. K., and Huxley, P. A.) CAB International, Wallingford, UK, pp. 189–233.

Wallace, J. S., Jackson, N. A. and Ong, C. K. (1995) Water balance of agroforestry systems on hillslopes – phase I. *Final Report to the forestry Research Programme*, UK Department for International Development pp 40, Centre for Ecology and Hydrology Wallingford.

Wallace, J. S., Jackson, N. A., and Ong, C. K., (1999) Modelling soil evaporation in an agroforestry system in Kenya. *Agricultural and Forest Meteorology* 94, 189–202.

Wallace, J. S., and Verhoef, A., (2000) Modelling interactions in mixed-plant communities: light, water and carbon dioxide. In: *Leaf Development and Canopy Growth* (Eds. Marshall, B., and Roberts, J. A.) Sheffield Academic Press, Sheffield, pp 204–250.

Whitmore, T. C. (1989) Tropical forest nutrients: where do we stand? A *tour de horizon*. In *Proctor* (1989, q.v.), 1–14.

Wiersum, K. F. (1984) Surface erosion under various tropical agroforestry systems. In *Symposium on effects of forest land use on erosion and slope stability* (ed. C. L. O'Loughlin and Pearce, A. J., Honolulu, USA, East-West Center), 231–239.

Wiersum, K. F. (1985) Effects of various vegetation layers in an *Acacia auriculiformis* forest plantation on surface erosion in Java, Indonesia. In *Soil erosion and conservation* (ed. S. A. El-Swaify, W. C. Moldenhauer, and A. Lo, Ankeny, Iowa, USA: Soil Conservation Society of North America), 79–89.

Woomer, P. L. and Swift, M. J. (1994) *The biological management of tropical soil fertility*. Chichester, UK: Wiley.

Yamoah, C. F., Agboola, A. A. and Mulongoy, K. (1986) Soil properties as affected by the use of leguminous shrubs for alley cropping with maize. *Agriculture, Ecosystems and Environment* 18, 167–177.

Young, A. (1989) *Agroforestry for soil conservation*. Nairobi: ICRAF, 276pp.

Young, A. (1990) Agroforestry for the management of soil organic matter. In *Organic-matter management and tillage in humid and subhumid Africa*. IBSRAM Proceedings 10 (ed. E. Pushparajah, M. Latham and C. R. Elliott, IBSRAM, Bangkok), 285–303.

Young, A. (1993) Agroforestry as a viable alternative for soil conservation. *Entwicklung und Ländlicher Raum* 27(5), 3–7.

Young, A. (1997a) *Agroforestry for soil management*. Wallingford, UK: CAB International.

Young, A. (1997b) The effectiveness of contour hedgerows for soil and water conservation. *Agroforestry Forum* 8(4), 2–4.

Young, A. (1998) *Land resources: now and for the future*. Cambridge, UK: Cambridge University Press.

Young, A. (2000/2001) How much spare land exists? *Bulletin of the International Union of Soil Sciences* 97, 51–55.

Young, A., Menz, K., Muraya, P. and Smith, C. (1998) SCUAF Version 4. A model to estimate soil changes under agriculture, agroforestry and forestry. *ACIAR Technical Reports* 41. Canberra: ACIAR.

Part IV
New methods for evaluating effects of land-use change

SUMMARY

This part contains eight chapters dealing with new methods to detect, evaluate or predict the hydrological effects of land-use change. They range from remotely sensed observations of terrain and vegetation characteristics, through statistical analysis of trends in streamflow data, to various model approaches simulating hydrological and related processes such as erosion and deposition at the hillslope to catchment scale. Also examined is the usefulness of isotope tracers as a tool to enhance hydrological process understanding under humid tropical conditions. The final chapter explores the possibility of using aquatic organisms as an indicator of water quality.

Held and Rodriguez present an overview of *new remote sensing technologies for the derivation of key forest and terrain attributes*. These can be divided into three groups: (i) those relevant to assessment (vegetation cover, forest type and structure, age of regenerating forest, and fire history); (ii) indicators of stress; and (iii) those more directly relevant to hydrology (terrain attributes, soil characteristics, photosynthesis and transpiration). Airborne systems (balloons, planes) allow for higher spatial resolution image collection than satellite systems and contain sensors (e.g. laser detection and ranging, LDR) capable of collecting data under cloud cover, which would make them particularly useful in (tropical) areas experiencing frequent cloud. Satellite systems provide a more stable platform for operational data collection and have the advantage of covering larger areas more rapidly at a lower cost per unit area. Data on various forest attributes (e.g. leaf area index) have traditionally been inferred from broad-band reflectances (measured in the red and near-infrared spectrum of wavelengths) using the so-called normalised difference vegetation index (NDVI). With the availability of new airborne and satellite sensors (such as ERS-1 and Radsat), there are now opportunities for their use in connection with attributes such as forest height, foliar chemistry, forest age, type, condition and function (including degree of stress). Of particular interest to hydrologists is the use of remotely sensed forest structural information in the calculation of aerodynamic roughness. The latter is required for the computation of evaporation, for example as part of soil vegetation atmosphere transfer schemes (SVATs; see Costa, Chapter 23). Furthermore, vegetation water content can be derived indirectly from the imaging spectrometers and passive microwave sensors used in the detection of thermally emitted radiation for deriving the sensible heat flux. Extrapolation of transpiration estimates based on thermal data and vegetation indices are now possible but as yet not used extensively because of the need for field calibration against flux measurements. However, there is now a growing network of flux towers associated with several large-scale field experiments, e.g. in the Amazon and South East Asia, which provide validation opportunities and thus will enable routine estimates of total evaporation to be made over large areas of rainforest. Held and Rodriguez are less optimistic about the possibilities for detecting soil moisture using radar or passive microwave remote sensing over large areas with dense forest, however. Key obstacles include a lack of detailed topographical information beneath dense forest (from which relative catchment wetness patterns can be inferred) as well as the above-ground biomass interfering with the signal. Greater success has been achieved at smaller scales (*c.* 10 ha) in more open, agricultural areas where the spatial distribution of soil moisture can be mapped based on its strong dependence on the dielectric constant of the soil. Developments are occurring with airborne laser scanning systems, however, which in the future may be able to generate digital elevation maps of the 'average' forest floor with a minimum of 2–5% of the total forest area occupied by new forest gaps.

Next, Kundzewicz and Robson discuss various *statistical methods for detecting changes in trends within river flow series arising from land-use change, climatic variability or climate change*. Given the relative data scarcity in the humid tropics (especially with regard to long, unbroken time series for rainfall or discharge), the possibilities for application of these methods will be somewhat restricted there. Nonetheless, the descriptions of the main stages within the process of obtaining and preparing data sets and subsequent exploratory data analyses (EDA) are highly relevant to any statistical analysis. The authors emphasise EDA as a powerful graphical technique for detecting hidden patterns as well as exploring and understanding the data better. No statistical applications should be undertaken until after EDA which also assists in selecting the appropriate test statistics. Different test statistics tend to detect different aspects of change, e.g. a steady gradual change (trend) vs. an abrupt jump in the data (step change). Distribution-free methods are preferred because they do not make assumptions about the underlying probability distribution of the data. Particular attention is also given to the problem of serial correlation, especially where there is a fine temporal resolution of data points. Solutions are offered such as decreasing the frequency of the data series (e.g. aggregation to monthly or annual averages), the use of the so-called modified seasonal Kendal test for data with seasonality, and block permutation and boot strapping (resampling methods). However, even after the data have been tested in this way, and after a trend has been detected, Kundzewicz and Robson still underline the need for further assumption checking by removing the trend from the data, and checking the residuals for auto-correlation and constancy of distribution. The test results should also be linked with historical information (e.g. data collection methods, land-use records, introduction of river engineering works), climatic variables (notably temperature and rainfall) and proxy data to ensure that the interpretation is undertaken with care. The authors conclude by stressing that the search for weak changes in time series of hydrological data is not a straightforward task. Even where good quality data and long-term records are available from higher latitudes, a consistent 'signature' of climate change in river flows has yet to be found. In the meantime, high priority in the humid tropics should be given to improving data collection and extending monitoring networks. In addition, concurrent historical information on land-cover change and related river engineering works should be compiled for tropical river basins.

After compiling and checking their data, hydrologists are faced with the question of *how best to assess which model to use* to describe the behaviour of their catchments over time. This question assumes particular relevance under the generally data-scarce conditions prevailing over much of the humid tropics. Hence, Barnes and Bonell focus on a 'universal model *approach*' rather than on the search for a 'universal process model'. In doing so they focus in particular on 'lumped' conceptual (as opposed to distributed physically based) models because these are simpler to understand and less data demanding. Particular emphasis is given to the concept of *emergent properties*. Basically, this concept implies that for complex systems made up of large numbers of smaller components with only local interactions (as is the case for catchments and river basins), the system may exhibit a collective behaviour that simply cannot be represented by the up-scaling (aggregation) of small-scale descriptions to yield large-scale models. Instead, 'effective' values are derived for selected physical parameters at these larger scales even though they are not directly measurable (e.g. hydraulic conductivity at the hillslope scale). One reason for the mixed success of more complex deterministic distributed models has been the lack of measurement techniques to represent hydraulic potentials, control volumes (e.g. groundwater storage) and fluxes (e.g. preferential flow at the soil to hillslope scale) at larger scales. Thus, there has been a tendency recently to move away from the use of distributed models towards so-called parametrically parsimonious conceptual models (PPCMs). Such models treat entire catchments as single spatial units (i.e. they are spatially lumped), but at the same time they try to maintain some representation of the underlying physical process. Moreover, whilst PPCMs may not be physically-based at the catchment scale, by using 'effective' values for key parameters they avoid the uncertainty associated with physical models as to what extent the assumed physically-based algorithms (which are mostly based on point measurements anyway) are indeed appropriate at larger scales of application. A further advantage of PPCMs is the low data demands (more suited to the humid tropics) using as few as six or seven parameters.

In recognition of the deficiency of hydrological data in the humid tropics referred to already, Schreider and Jakeman outline a *method for modelling stream flow from gauged and ungauged sub-catchments within large river basins*, taking the Mae Chaem basin in northern Thailand as an example. Their approach involves two main steps. First a PPCM model (the IHACRES rainfall–runoff model) was calibrated against stream flow recorded at the outlet of the Mae Chaem basin using basic climatic data (precipitation and temperature) as inputs. The criterion for model parameter optimisation was a best fit of predicted to measured discharge values. Because irrigation diverted some of the base flow, IHACRES was applied in conjunction with a standard irrigation consumption module based on information on the surface areas under different crops and their associated water use per unit area. The latter was added to measured discharge to obtain an estimate of the expected natural (unregulated) stream discharge. The second step uses TOPMODEL to disaggregate (down-scale) natural stream flow within elements of the drainage basin to derive estimates of natural flows in ungauged sub-basins. This procedure is based on the assumption that the contribution from each part of the drainage basin to the total water yield is proportional to the topographic index given by TOPMODEL (as outlined in Chappell *et al.*, see below). The stream flow disaggregation procedure was tested in two sub-basins for which stream flow records were available, although they were treated as being 'ungauged'. The stream flow disaggregation algorithm was seen to yield a relative error of 13–17% for monthly flows which must be considered a very promising result.

The difficulties of representing field-saturated hydraulic conductivity at the appropriate scale, highlighted earlier, are further elaborated by Chappell, Bidin *et al.*, who present a *critical assessment of how key soil parameters like soil-rock permeability are parameterised best within dynamic rainfall-runoff models*. The central issue is that point-scale measurements, even after upscaling, may not be meaningful any more for physically-based modelling of runoff at the catchment scale. Once again, the focus is on limiting the complexity of models (i.e. enhance their parsimony) by reducing the number of model parameters as well as improving the transparency of model component structures. Particular attention is paid to topographically-based, dynamic rainfall-runoff models, such as TOPMODEL and (variants of) TOPOG, which provide a spatial perspective of catchment wetness through the use of either a topographic or a wetness index. Chappell *et al.* note that comparatively little attention has been given to the model component within both TOPMODEL and TOPOG that contains the terms associated with the spatial distribution of permeability. It is the poor parameterisation of the latter variable which was partly responsible for the recent unsatisfactory results whilst testing TOPOG-DYNAMIC in a very small zero-order catchment in the Peruvian Amazon, despite the availability of numerous point observations of soil hydraulic conductivity (see Bonell, Part II). The authors present an equation to derive estimates of lateral 'block' permeability through inverse modelling. Estimated permeabilities obtained in this way are compared with first, point measurements using ring permeametry and second, a new field technique known as the hillslope pulse-wave experiment There was a promising consistency in the magnitude of lateral permeability estimates derived by model inversion vis-à-vis those obtained with the field pulse-wave experiment but both estimates proved larger than (up-scaled) core-based estimates of permeability, probably because large conductive pathways (such as soil pipes) are not included in core-scale measurements. Thus a principle message is that the use of whole-hillslope tests of lateral permeability linked with inversion of a simple catchment model points the way towards a more realistic parameterisation at the appropriate scale of topographically-based, dynamic rainfall-runoff models. Chappell *et al.* conclude by calling for additional critical assessments of the vexed question of whether complex physics-based approaches are indeed better than parsimonious models (i.e. PPCMs) in predicting the effects of land-cover change on the runoff generation process.

Buttle and McDonnell extend the account on runoff generation given earlier by Bonell (Part II) by discussing the accomplishments of isotope tracing studies, conducted mainly in humid temperate environments, to stimulate future application in the humid tropics. Stable environmental isotopes (such as oxygen-18, deuterium) are ideal tracers since they are inherently conservative and thus do not undergo chemical reactions when in contact with soil or weathered rock. Consequently they can be used as a tool for quantifying the age, origin and hillslope pathway followed earlier by the water of headwater streams. Other important applications of isotopes in the context of tropical forest hydrology and ecology have been discussed in the chapters by Proctor, Bruijnzeel, and Hölscher *et al.* First, Buttle and McDonnell critically review the methods and assumptions connected with isotopic hydrograph separation (IHS) techniques as a means of determining the proportion of the storm hydrograph contributed by pre-event or 'old' water (following subsurface routes to the stream) and event or 'new' water (following mostly rapid, near-surface pathways). The authors present convincing evidence that IHS can provide some valuable insights into hillslope hydrological processes, even when one or more key assumptions have been violated. Yet, they stress the need for complementary hydrometric as well as geological (e.g. bedrock 'topography') and geochemical information to allow realistic interpretation of IHS results. Buttle and McDonnell also note that stormflow in many environments (usually those that are little or undisturbed) is dominated by pre-event water, especially under conditions of high antecedent catchment wetness. Despite more than three decades of IHS studies reporting this dominance of storm hydrographs by pre-event water, the authors remark that this finding has not yet made a significant impact on process-based hydrological modelling. In disturbed (e.g. urban) environments the proportion of event water increases in line with the increase in compacted surface promoting overland flow. A similar result (yet to be demonstrated) would be expected for cleared tropical catchments where soils have become compacted by grazing or crusted by exposure to intense rainfall. As for the effect of catchment size on relative contributions by event / pre-event water, there does not seem to be a universal trend. It is further expected that the greater diversity of stormflow pathways occurring in humid tropical forests, including the greater occurrence of overland flow, will provide contrasting IHS results. The almost complete absence of hillslope hydrological studies that also address the role of runoff processes occurring within the riparian zone and within the stream channel itself is noted. Any exchange of surface- and groundwater in riparian areas could have a major impact on the incoming isotopic signal of hillslope runoff contributions and this needs investigation. Arguably, such research could be usefully combined with the work on the efficiency of riparian buffer zones in trapping eroded soil and nutrients on their way to the stream, as called for by Cassells and Bruijnzeel and Thang and Chappell (Part V). Finally, another potentially important application of IHS is in separating the origins of storm runoff within road drainage ditches and natural drainage networks that could be affected by forest harvesting operations. 'New' water would originate mostly from infiltration-excess overland flow from the compacted surfaces of the roads themselves whereas 'old' water would represent subsurface stormflow from upslope that is intercepted and channelled by the cutbank of the road. For example, a sudden increase in 'new' water within natural channels downslope of roads would indicate the need to re-adjust the forest logging plans (and related management guidelines on drainage facilities) to ensure that cross drains divert the overland flow before meeting natural channels and so encourage higher rates of off-road infiltration on slopes.

Next, recent advances in physically-based erosion prediction modelling are reviewed by Yu. These models are being considered as a potential replacement in the longer term of the simple, traditional factor-based universal soil loss equation (USLE). The latter has been widely used in the tropics to estimate annual soil losses as a function of rainfall erosivity, soil erodibility, slope steepness and length, as well as cropping patterns and conservation measures. Three representatives of the new generation of process-orientated models are discussed and compared: the Australian GUEST model, the North American WEPP model and its European counterpart EUROSEM. The new models are trying to capture the physics of erosion on an individual runoff event basis to estimate the associated soil loss using rainfall intensity data of high temporal resolution (ideally 1-minute intervals). The models rely heavily on improved descriptions of soil erosion, transport and deposition processes connected with particle detachment and entrainment by rainfall and overland flow, as well as re-entrainment after deposition. Moreover, the erodibility of different soil types, both temporally (even within storm events) and spatially, has to be represented. Considerable parameterisation is thus required to apply these equations, while the quantification of several key variables require laboratory support and controlled flume experiments. As a result, the costs associated with data collection and developing a data base for the humid tropics are likely to limit the routine application of these physically-based erosion models for a considerable time to come. Nonetheless, these models are ideal tools for erosion research (notably GUEST) and allow exploration of various sub-processes in erosion and deposition and their interactions.

The contribution of Connolly and Pearson is particularly important because it is the only one in this book devoted to a discussion of the *impacts of forest conversion on the ecology of tropical streams*. Stream ecology requires a multi-disciplinary and multi-scale approach because streams and rivers act as veins interconnecting aquatic and terrestrial ecosystems by way of riparian zones in which soil particles, leaves and woody matter; and the organic matter and nutrients these contain, are transported and recycled (see Proctor). In addition, riparian zones provide a conduit for closed-forest species to move about within otherwise undisturbed landscapes. Consequently, considerable attention is given to the ecological and hydrological functioning of riparian zones. However, even basic information on the taxonomy and ecological functioning of streams in tropical forest areas is still very sparse and there is as yet no descriptive model of how a tropical stream 'works'. As Connolly and Pearson note, the lack of long-term stream ecological studies and the corresponding paucity of researchers in the tropics make it hard to be certain of the stream ecological impacts associated with different kinds of anthropogenic influences. Current efforts are confined to only a few areas (Hong Kong, Australia and Costa Rica). Whilst it is thus difficult to generalise about the impacts of tropical forest conversion on aquatic ecosystems, Connolly and Pearson draw upon a wide range of sources (temperate as well as tropical) to evaluate the effects of changes in flow regime, the light and insolation regimes, suspended sediment and bed loads, availability of nutrients and thus the impacts of water quality on aquatic ecology in general. The authors also provide a general outline of the detrimental effects of pollutants on the flora and fauna of streams. In particular, the adverse impacts of enhanced acidity, toxicity and deoxygenation (e.g. in relation to mining) of stream water are highlighted. In conclusion, it is evident that the present lack of detailed understanding of even basic ecological patterns and processes in tropical streams means that the complex physical, chemical and biological interactions associated with pollution are even less understood. Much more work is needed.

27 Remote sensing tools in tropical forest hydrology: new sensors

A. A. Held
CSIRO, Canberra, Australia
E. Rodriguez
Jet Propulsion Laboratory, National Aeronautics and Space Administration, Pasedena, USA

INTRODUCTION

Increasingly, the monitoring of forest condition and productivity requires better spatial context, more detail and higher temporal resolution, if we are to be able to visualise key environmental or biodiversity indicators and their changes across space and time. For the accurate prediction of how changes in climate, forest condition or land-use can impact large-scale hydrological processes, researchers and water managers are also faced with the need to extrapolate patch-scale measurements to whole catchments or river basin scales. This is especially challenging across large, often inaccessible areas or where the ground information available is sparse and where the costs and effort required in setting up a dedicated, sufficiently dense network of ground-based sampling sites are very high.

Although not always measuring the required hydrological variables directly, remote sensing techniques can provide valuable information about the spatial variability of key surface characteristics across large scales. For this reason it is often used as a cost-effective mechanism for extrapolation of point-based measurements, or as a constraint for hydrological model outputs. The key to the effective use of this type of data, in conjunction with hydrological models, lies in the recognition of the characteristics and trade-offs found in the different forms of remotely sensed derived variables, and how they can be used to complement ground-based hydrological measurements effectively.

Up to now, mostly broad-band, low resolution remotely sensed data have been used as input and constraints for regional/continental hydrology and productivity models (e.g. Running and Coughlan, 1988; Schultz and Engman, 2000). This chapter will discuss not only such applications but also cover the use of new sensors such as high resolution laser profilers, imaging radars, hyperspectral imagers and new analysis methodologies, which begin to provide more absolute, spatially-dense measurements of variables previously only measurable from the ground. In addition to the improvements in vegetation measurement, recent advances in remote sensing technologies offer more precise measurements of other environmental indicators, such as water quality and sedimentation in rivers, associated with forest harvesting operations.

After a brief description of the fundamentals of remote sensing and a description of the new emerging technologies, this chapter will specifically address key forest variables such as leaf area/cover, height and biomass, which are now measurable synoptically with remote sensing technologies and in a more direct way.

Basics of remote sensing

The dramatic changes in the extent and condition of tropical rainforests world-wide have been made apparent largely through remote sensing observations over the last 20 years (see Drigo, this volume). Such 'non-contact' sensors use electromagnetic radiation as a carrier of information, to deduce the properties and spatial relationships of chemical compounds in rock, soil, vegetation, water and atmosphere. Images derived from remote sensing instruments were used initially as analogue prints from photographic film, but today most commercial remote sensing instruments produce images in digital format, which are visualised and manipulated on computers. Digital remote sensors are usually mounted on board satellites or aircraft and detect the reflected or emitted radiation in a number of specific, well-defined wavelengths of the electromagnetic radiation spectrum (EMS), as shown in Figure 27.1.

A number of commonly used satellite sensor systems like the Landsat MSS (Multispectral Scanner) and TM (Thematic

Forests, Water and People in the Humid Tropics, ed. M. Bonell and L. A. Bruijnzeel. Published by Cambridge University Press.

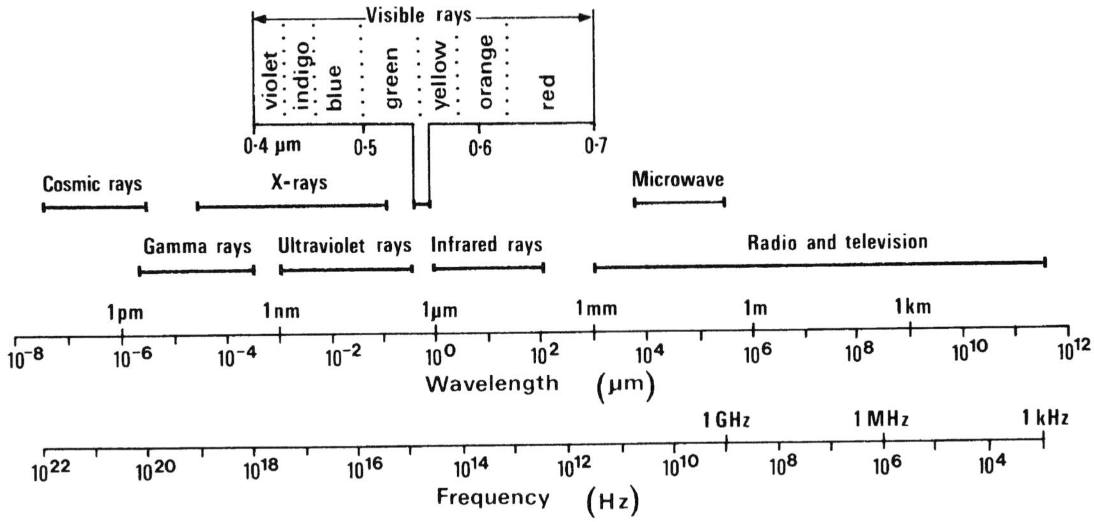

Figure 27.1 The electromagnetic spectrum. (After Harrison and Jupp, 1989.)

Mapper), Spot HRV (Systeme Probatoire d'Observation de la Terre – High Resolution Visible), 'sense' light in the visible (400–700 nm), the near infra-red (700–1300 nm) and, in the case of Landsat, also in the shortwave infra-red (1300–3000 nm) part of the EMS (c.f. Manual of Remote Sensing, Volume 1, 1983). While most sensors use the sun's reflected radiation as the light source for their measurements, some also detect radiation emitted from the ground in the thermal infra-red portion of the electromagnetic spectrum. The generic term for such sensors are passive detectors. Other sensors, called active sensors, generate their own energy, which is transmitted towards the ground and its interaction with objects or surfaces on Earth is then measured and interpreted. Active systems which operate in the microwave part of the spectrum are called 'Radio Detection and Ranging – RADAR' sensors, while those which operate in optical frequencies and use mostly concentrated laser pulses are called 'Laser Detection and Ranging – LIDAR' systems. By virtue of the EMS frequencies involved, most optical–thermal passive systems provide information on the chemical nature of the surfaces they originate from, while the radar and laser systems provide information primarily on the three-dimensional nature and structure of the objects they encounter (Schultz and Engman, 2000).

OPTICAL SENSORS

The general principle behind the use of a number of the optical sensors has its foundation in a laboratory technique for material identification and quantification known as 'spectroscopy'. Here, materials are illuminated with a source of light of known spectral characteristics and brightness, and the radiation absorbed, transmitted or reflected by the material is analysed for its identification. The absorption of radiation by materials is mainly due to the effects of photons on chemical bonds in molecules. This technique has been used for laboratory analysis and for identifying the makeup of stars in astronomy for over a century; it is now a common quality-control measurement technique in areas of industrial manufacturing, food production and medicine (Wendlandt and Hecht, 1966).

Early spectrometers on satellite and airborne platforms, designed mainly for geological exploration and weather monitoring purposes, consisted of a single line of adjacent spectrometers with broad spectral sensitivity in a number of spectral regions, also called spectral bands. The forward motion of the platform would then allow for generation of a two-dimensional spectral measurement of the Earth's surface (Figure 27.2).

It did not take very long to increase the sensitivity of these instruments and also to increase the number of spectral bands. Satellite sensors in use today, such as those on the Landsat TM and SPOT HRV satellites, are essentially of this nature, and collect two-dimensional images of the Earth's surface in up to seven bands. Each two-dimensional digital image produced by these sensors is composed of thousands of picture elements (pixels), each containing digital information on the radiance measured at the different wavelengths. When extrapolated to the ground, the pixel size or spatial resolution in most commonly used satellites ranges from 30 m to 1000 m, while in airborne systems, the pixel size can be as small as 10 or 20 cm, depending on the altitude of the aircraft and the field of view of the instrument. Satellite data of the Earth's surface have been collected since the mid–1970s, hence a good archive is now available for investigation of recent changes in rainforest cover and land-use.

With the evolution of electronics, integrated circuits and miniaturisation of detectors, systems composed of thousands of

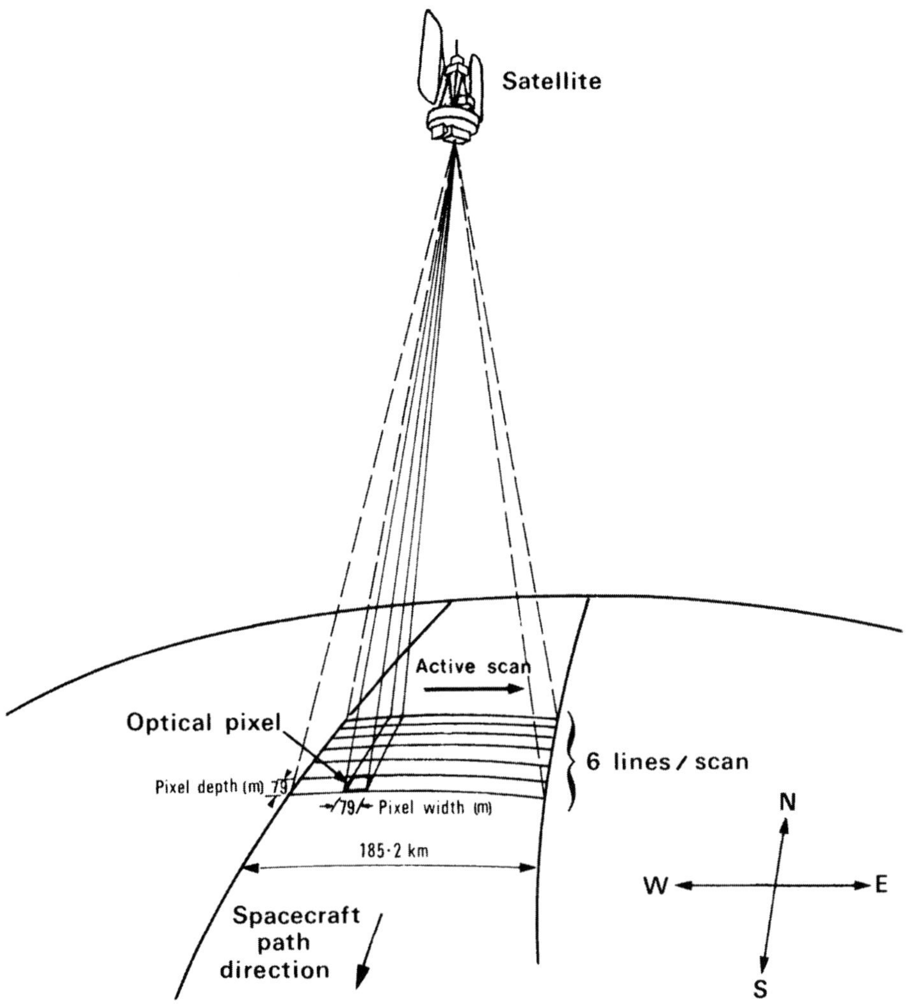

Figure 27.2 Diagram of a typical earth observing satellite (Landsat Multispectral Scanner). (After Harrison and Jupp, 1989.)

detectors, arranged as linear or two-dimensional arrays, are being used with the required focusing fore-optics, to create two-dimensional imaging spectrometer systems. These instruments collect reflected radiation in over 100 contiguous, spectral bands, forming a full reflected radiation spectrum, and are now becoming much more accessible for operational applications. The advantages of some of these so-called 'hyperspectral' instruments, as opposed to the 'multispectral' sensors with only 4–7 broad spectral bands, is that the true principles of spectroscopy can be applied to distinguish the small chemical differences between objects (Curran, 1994; Goetz, 1992). With multispectral systems, the effects of atmospheric absorption of the reflected radiation are difficult to quantify and subtle differences between materials are more difficult to differentiate. Naturally, work with hyperspectral imaging systems and their larger amounts of information collected requires more computer memory and more powerful systems (Plate 9).

Although widely used for mineral exploration and geological mapping (e.g. Hunt and Ashley, 1979), the use of hyperspectral sensors for studies of vegetation dynamics (e.g. Miller *et al.*, 1991; Ustin *et al.*, 1993), vegetation biochemical composition (e.g. Wessman, 1989), stress detection (Merton 1999; Rock *et al.*, 1986) and for plant species discrimination (e.g. Clark *et al.*, 1995; Held *et al.*, 2001; Roberts *et al.*, 1999), is still an active area of research. A number of optical sensors, some of which were launched in late 2000, are listed in Table 27.1, along with their general specifications.

IMAGE ANALYSIS

Once the digital image data are downloaded from the airborne/satellite sensors or purchased from commercial vendors, they require what is called 'base processing' before they can be used, for instance, for final analyisis and forest classification purposes. Base processing of image data usually entails some radiometric

Plate 9 Three-dimensional representation of a 126-band hyperspectral image acquired by the 'Hymap' airborne sensor over a 1.5 km × 1.5 km area surrounding Cape Tribulation in Australia. In this representation, each 3 m resolution image pixel contains the geographic, spatial dimensions in the *x*- and *y*-dimension, while the spectral reflectance information, from 400 to 2500 nm, is shown in the third, *z*-dimension. The brighter colours represent higher reflectance and dark colours represent low reflectance areas in the specific wavelegths. Notice the sharp changes in reflectance between land-based and water-based areas. (After Held, unpublished data.) (See also colour plate section.)

correction for atmospheric and sensor calibration factors, masking of cloud-affected areas and geo-rectification to the geographic coordinates of interest. Some of these processes, in some cases including atmospheric correction, are increasingly now carried out by the data suppliers. During the atmospheric correction process, the data are converted from 'at-sensor radiance' to 'target-leaving' radiances. At this point, the data can either be used for physically-based modelling and estimation of material concentrations or for classification into the different materials or vegetation types visible in the data. When divided (normalised) by the incident irradiance at the time of the overpass, the data are transformed into reflectance units and can then be used for classification purposes. This commonly entails comparing the observed reflectance values in the imagery to ground-based measurements made of known, typical species or often pure materials. Unfortunately, image pixels even as small as 1 × 1 m are seldom composed of pure materials or single plant species but represent a mix of 'spectral signatures'. For this reason new techniques of analysis of image data have been developed to characterise the mixed nature and composition of the pixels. Traditionally, image classification has been carried out by using statistical procedures, assuming that pixels are composed of pure materials. This approach produces images where pixels are 'binned' into a range of typical categories or classes of land-cover, to represent the spatial distribution of these vegetation types (Manual of Remote Sensing, volume II, 1983; Wickland, 1989). More recently the concept of 'sub-pixel spectral unmixing' has been introduced (e.g. Goetz, *et al.*, 1985; Kruse, 1999), which disaggregates each pixel's characteristic spectral signature into a set of possible pure 'endmember' signatures, provided these are available as part of an input 'spectral library' for the analysis. In the case of vegetation, spectral libraries can be composed of reflectances of sample sunlit and shaded leaf, wood and litter material for key species expected in the geographical region of interest (e.g. Roberts *et al.*, 1999).

Table 27.1. *Characteristics of some airborne and spaceborne optical sensors in current use*

Sensor name	Airborne or satellite	Country or agency and URL	Spectral range wavelength (nm)	Number of bands	Ground resolution (m)
AVHRR	Satellite	NOAA, USA http://edc.usgs.gov/glis/hyper/guide/avhrr	580–12 500 nm	5	1 100
Landsat 7	Satellite	NASA–USGS, USA http://landsat.gsfc.nasa.gov/	450–12 500	7	30 (60 m, thermal)
SPOT	Satellite	France Spot Image http://www.spotimage.fr/	500–1 750	4	20
AVIRIS	Airborne	NASA, USA http://aviris.jpl.nasa.gov/	410–2 450	224	5–20
Casi	Airborne	Itres, Canada http://www.itres.com	410–925	228/19	0.8–5
Hymap	Airborne	Integrated Spectronics, Australia http://www.intspec.com/	400–2 543	59	3–5
Ikonos	Satellite	Space Imaging, USA	500–700	1	1
		http://www.spaceimaging.com	400–1 000	4	4
MODIS	Satellite	NASA, USA	405–14 385	36	250–1 000
		http://modis.gsfc.nasa.gov	400–1 000	4	4
DAIS	Airborne	DLR, Germany	400–2 500	72	5
		http://www.op.dlr.de/dais/dais-scr.htm	400–1 000	4	4
Hyperion	Satellite	NASA, USA http://eo1.gsfc.nasa.gov/Technology/Hyperion.html	400–2 500	220	30

RADAR

On a par with better, more sensitive optical sensor evolution, radars have also evolved to become important components of the Earth's constellation of monitoring sensors. This technique has its origins in wartime applications early in the 20th century, where radio waves (now called microwaves) were generated and their return to the detectors was timed to measure accurately the distance and size of approaching objects. Unlike optical sensors, which use primarily natural illumination from the sun, radars generate their own energy in different wavelengths and polarisations, and measure the type and strength of the return which is then used for identification of the objects. Since a radar provides its own power, it can be used at night, and very accurate calibration of the intrinsic backscattering properties of the objects is possible, independent of solar illumination or time of day. The great advantage of radar, as far as the tropics is concerned, is that at the wavelengths used, clouds are practically transparent, making these instruments very suitable for mapping of tropical rainforest areas frequently affected by cloud cover. Due to the wavelength differences between optical systems and microwave systems (cf. Figure 27.1), the former are mostly sensitive to the molecular composition of the forest canopy while the latter, covering typically wavelengths between 1 cm and 1 m, are sensitive to differences in structure and water content of the canopy. A number of commonly used wavelength regions in radar have been termed X-,C-.,L-, and P-band, according to their wavelength (2.4–3.75 cm, 3.75–7.5 cm, 15–30 cm and 30–100 cm, respectively)(e.g. Curlis *et al.*, 1986; Elachi, 1980; Moore, 1983; Schultz and Engman, 2000). Research is also being carried out using even lower frequencies (longer wavelengths) typically in the Very High Frequency range (VHF) with wavelengths in excess of several metres (Imhoff, 1998).

The attributes of the radar return signal (called 'backscatter') are a result of many factors, including the wavelength and polarisation of the transmitted signal, as well as the topography and the water content and geometric properties of the vegetation canopy. Hence, radar sensors respond well to the structure of vegetation depending upon their dielectric properties and the size of the canopy components relative to the wavelength used. High dielectric constants, related primarily to high water content in vegetation, make for good radar reflection, and the strength of the return for a given wavelength is generally in proportion to the size of the canopy components.

In addition to the use of different wavelengths, electromagnetic waves can also be produced and received by radars in different polarisations, where the microwaves are made to oscillate in specific spatial planes (Moore, 1983). Hence, additional information can be obtained from analysis of the interactions between the radar waves, at specific polarisations, and the forest. The strength of these interactions between the incident microwaves and components of the forest canopy depends strongly on the relative

Table 27.2. *Characteristics of some current and planned airborne and space-borne radar sensors*

Sensor name	Country or agency and URL	Frequency/wavelength	Polarimetry	Interferometry
ERS 1 and 2	ESA http://earth.esa.int/ers/	C-band (5.6 cm)	No	Repeat pass
Radarsat	Canada RSI http://www.rsi.ca/home.htm	C-Band (5.6 cm)	No	Repeat pass
JERS	Japan NASDA http://www.eoc.nasda.go.jp/guide/satellite/satdata/jers_e.html	L-Band (23 cm)	No	No
ASAR	Europe http://envisat.estec.esa.nl/instruments/asar/	C-Band (5.6 cm)	Yes	Repeat pass
SIR-C/X–SAR	USA/Germany http://southport.jpl.nasa.gov/desc/SIRCdesc.html	L-Band (23 cm) C-Band (5.8 cm) X-Band (3.1 cm)	Yes	Repeat pass
PALSAR	Japan http://alos.nasda.go.jp/index-e.html	L-Band (23 cm)	Yes	Repeat pass
SRTM	USA NASA http://www.jpl.nasa.gov/srtm/	C-Band (5.8 cm)	No	Yes
AIRSAR	USA NASA http://airsar.jpl.nasa.gov/	P-Band (80 cm) L-Band (23 cm) C-Band (5.8 cm)	Yes	Yes

orientation of the incident wave polarisation and canopy branches and, to a lesser extent, the leaves. This feature of radar has been used extensively and with good results to map different vegetation types (van Zyl, 1993). As an example, a branch oriented at a 45-degree angle relative to the incident polarisation (e.g. H), will cause the scattered wave to have components in both the incident (H) and the cross-polarised (V) component, which are of similar magnitude. This can be contrasted with interactions with bare soil, where a low cross-polarised component will be present in the scattered field. In order to fully exploit polarisation information, radar systems must be capable of receiving the return wave in a fully polarimetric mode (e.g., SIR-C or AIRSAR). An introduction to radar polarimetry can be found in Ulaby and Elachi (1990).

The use of radar specifically for mapping forest canopy properties has been investigated extensively in the last 15 years, and continues to be an area of active research (e.g. Saatchi *et al.*, 2000; Ulaby and Elachi 1990,). So-called 'Synthetic Aperture Radar (SAR)' systems now allow for detailed forest mapping at spatial resolutions that can vary from a few metres to a hundred metres, depending on the radar sensor. In addition to differences in spatial resolution, radar sensors can also be divided by the transmit wavelengths (or frequencies) and whether they offer polarimetric or interferometric capabilities (see below). Table 27.2 summarises the characteristics of some recent and future airborne and space-borne radar sensors that have provided useful data for the study of forests.

In general, the principles of radar image analysis and interpretation are guided by simple interactions between the microwaves and objects on the ground (see Ulaby and Elachi, 1990 or Ulaby *et al.*, 1986, for an introductory review). These interactions can be separated into the following components, described in a schematic form in Figure 27.3: (1) single scattering return from the leaves at varying incidence angles; (2) single scattering from the branches or trunks; (3) double bounce scattering from ground-to-trunk to sensor or ground-to-leaf canopy interactions; and (4) other multiple scattering within the canopy.

As mentioned above, the relative amount of these different backscattering mechanisms for particular vegetation types depends on the structure of the canopy and the radar wavelength. In broad terms, lower wavelengths (e.g. X-, or C-Band) will interact more strongly with the upper canopy leaves or branches, and can be attenuated substantially before reaching the bottom of the canopy. These radar systems are highly effective for mapping discontinuities in canopy coverage, forested versus cleared areas and general land-use patterns. In contrast, longer wavelengths (e.g. P-band) will penetrate further through the tree canopy and have a scattering mechanism which is dominated by either ground-trunk double bounce scattering (for smooth terrain), or by interactions with the large branches and the trunk. Hence, such systems are better suited for mapping aspects of forest structure and biomass (see below). Intermediate wavelengths (e.g. L-Band) can show a combination of both of these characteristics, depending on the canopy type and canopy closure.

Another technique, being actively explored for geological and environmental applications, is radar interferometry, also called IFSAR (Interferometric SAR). Here the traditional radar

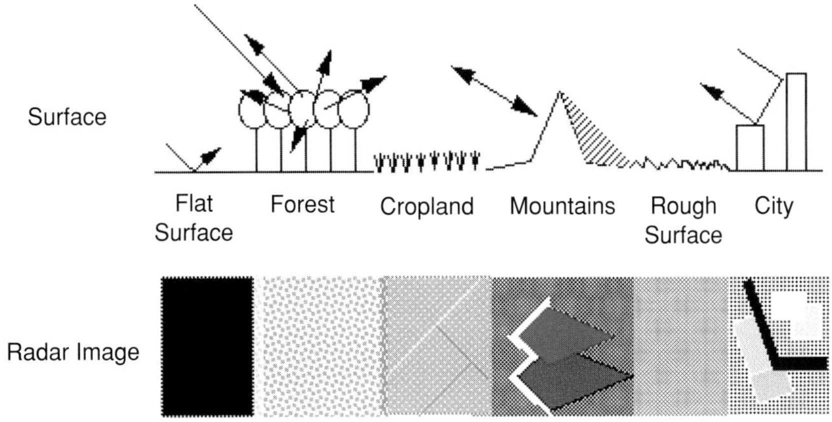

Figure 27.3 Schematic graph depiciting different interactions between the radar signal and the ground surface. (After T. Freeman, NASA – JPL.)

system is enhanced with an additional antenna, adding the ability to resolve height differences in a somewhat analogous fashion to stereo photography (for details on the technique see recent review by Hanssen, 2001). Digital elevation models (DEMs) created with radar interferometry can be used not only as terrain inputs to hydrological models but the interferometric data are also used for very detailed canopy structure mapping (Treuhaft *et al.*, 1996).

AIRBORNE VERSUS SATELLITE SYSTEMS

Once the sensor type has been selected for a specific application, a number of trade-offs need to be considered before choosing between the use of airborne or satellite systems. Airborne systems allow for higher spatial resolution image collection and provide better flexibility of image acquisition after special events or natural disasters. Airborne systems often contain more advanced sensors, provide image data at higher spatial resolution and are also capable of data collection under uniform clouds. On the other hand, satellite systems are more stable platforms, provide generally lower spatial resolution, but can cover larger areas more rapidly and at a lower cost per unit area. These systems also have fixed overpass times, capable of routine and multi-temporal data collection over larger areas.

Box 27.1 Cost and trade-offs

Satellite data

- Low cost for low to medium spatial resolution data.
- Low information content and radiometric calibration per image pixel.
- Often discount for historical time-series of data over same area.
- Data processing costs initially low, due to low data volume.
- Advanced processing and image classification can often be expensive due to problems with cloud contamination and low radiometric quality.

Airborne data

- On a per unit area basis, often considerably higher cost than satellite data.
- Information content per unit area is much higher, however, due to higher resolution and often more spectral bands.
- Aircraft and pilot costs are about the same, regardless of sensor on-board.
- Costs of acquisition of advanced sensor data (e.g. hyperspectral) becoming competitive with air photography.
- Data processing, analysis and ‘value-adding’ varies depending on levels of processing, atmospheric correction, data volume and level of spatial accuracy (3 metre rms, or better).
- When compared to standard field sampling, combined with air photo interpretation, hyperspectral scanner data has been shown to be more cost-effective in some coastal habitat mapping applications (Mumby *et al.*, 1999).

SPECTRAL REFLECTANCE OF FORESTS

Before discussing the unique advantages of new high spectral resolution (hyperspectral) sensors in remote sensing of vegetation, it is important to consider first the key features of general vegetation reflectance ‘signatures’ in more detail (see Curran, 1994, for general discussion).

As light strikes a leaf, some of this light is absorbed, some is transmitted through the leaf and some is reflected back. It is mostly the reflected signal which is analysed to provide information on foliar chemistry and physiological condition. While the characteristic sharp difference between red and near-IR reflectance, found in green plants, is detectable from imaging sensors such as the

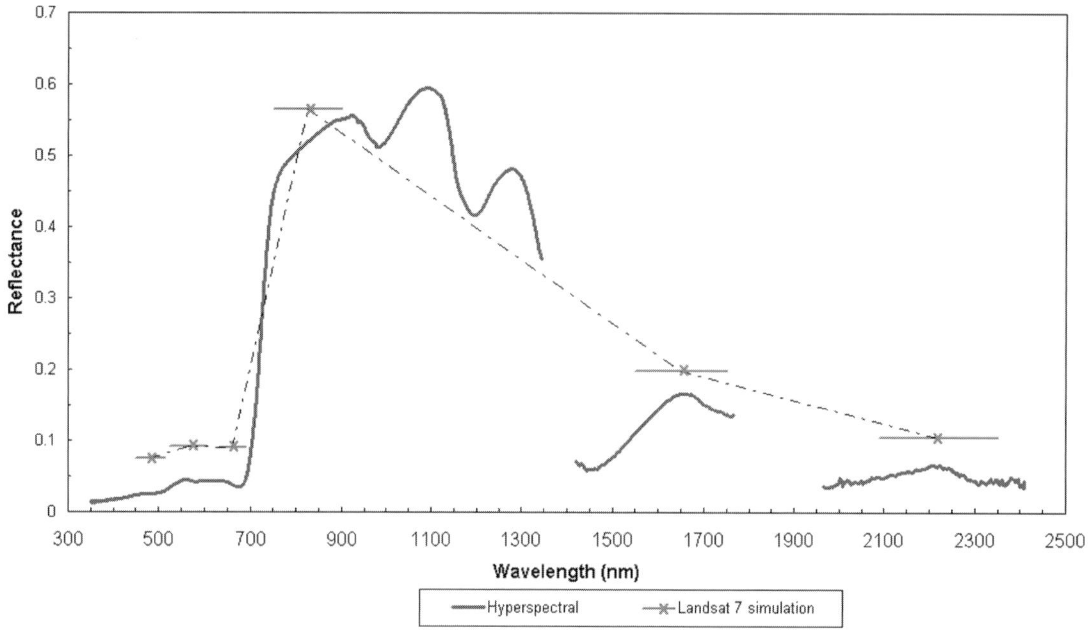

Figure 27.4 Typical reflectance spectrums of green vegetation, in this case a *Eucalyptus minifera* tree, measured outdoors with a full-range (400–2500 nm) spectroradiometer. Discontinuities in the spectral signatures are caused by strong atmospheric water vapour absorption, and hence undetectable light levels in these spectral areas. For comparison, the corresponding Landsat 7-bandset signature (displaced upwards by 0.05 reflectance units) is displayed as horizontal lines (After Held, unpublished data.)

Landsat TM or SPOT HRV satellites (Figure 27.4), there are a number of important features which cannot be seen at this spectral resolution but which can be measured explicitly with high spectral resolution (hyperspectral) sensors (see below).

Leaf reflectance is quite low ($<10\%$) in the visible spectrum (400–700 nm), due to the absorption of photons by photosynthetic pigments (mainly chlorophylls and carotenoids) (Figure 27.5). The prominent rise in reflectance between 680 and 750 nm that is typical for green vegetation, is commonly termed the 'red edge' and is caused mainly by the combination of strong chlorophyll pigment absorption and leaf internal light scattering properties. This sharp increase in reflectance has been a key factor in the development of 'greenness indices' such as the 'Normalised Difference Vegetation Index (NDVI) or the Simple Ratio (SR). They are composed of reflectances measured in the red (670–690 nm) and near infra red (750–800 nm) where the contrast caused by chlorophyll absorption in the 550–690 nm range is very high (cf. Figure 27.4). Several studies have also used changes in the inflection point of this red-edge as indicators of plant stress (e.g. Merton 1999; Rock *et al.*, 1996). In wavelengths above 700 nm, reflectance is dominated mostly by internal light scattering and light absorption by water, cellulose, lignin and leaf proteins (Figure 27.4).

At the whole forest canopy level, characteristic leaf reflectance features are still prominent but, depending on canopy density, are also influenced by tree architecture, sun illumination geometry, as well as branch, understorey vegetation, background soil and litter spectral features. For this reason, special care is required when making comparisons between sites or collecting multi-temporal data over a single site with different sensor-target-sun illumination geometries, as the reflected radiation and its spectral characteristics will be different under different conditions (Liang *et al.*, 2000). At this canopy level, spectral indices such as NDVI or the SR exhibit a curvilinear relationship with increasing leaf area due to the overlapping nature of leaves in canopies (e.g. Sellers *et al.*, 1996). Although dependent on leaf angle distribution and chlorophyll concentration, the NDVI for most temperate forest types tends to saturate above a leaf area index of 4–5, and can reach values of 6–7 before saturating in tropical rainforests (Guillevic and Gastellu-Etchegorry, 1999; Ustin *et al.*, 1993). This poses a problem in detection of early stress, separation of late regrowth stages or changes in phenology, in particular in the tropics, where leaf area index (LAI) can reach values as high as 10. Under these circumstances, broad-band NDVI or SR values would only show significant changes once the forest canopy has lost considerable leaf area due to the stress. However, with the advent of hyperspectral imaging systems, a number of case studies has shown that it is now possible to measure changes in concentrations of some pigments and other leaf chemicals more directly (see below), suggesting that this technology might be of use in detection of early signs of stress or phenological changes, even before changes in broad-band greenness and leaf area are observed remotely.

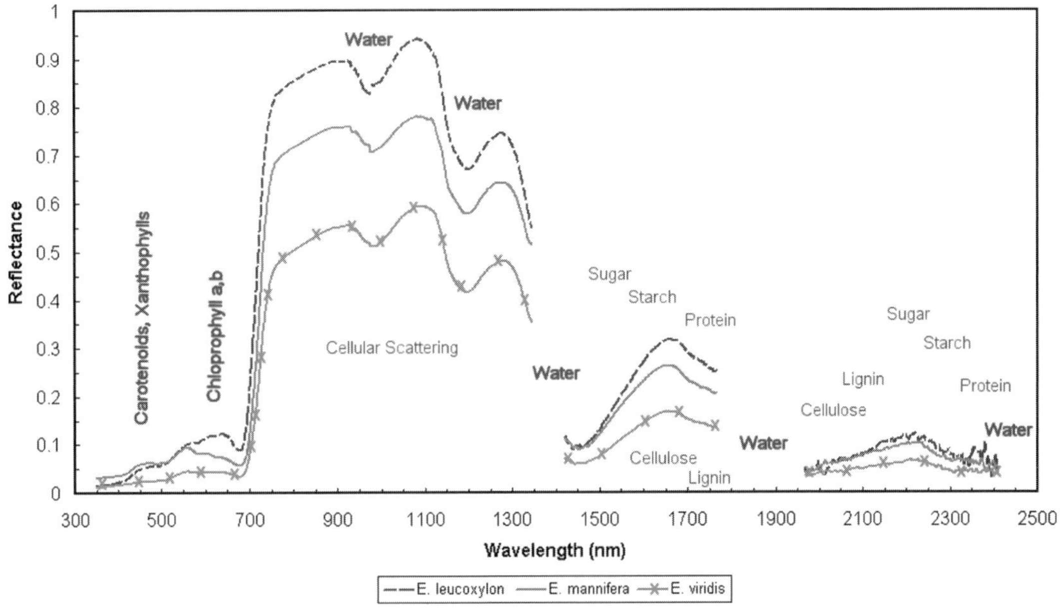

Figure 27.5 Spectral reflectance curves for three *Eucalyptus* tree canopies. Important foliar chemical absorption regions are shown. Discontinuities in the spectral signatures are caused by strong atmospheric water vapor absorption. (After Held, unpublished data.)

REMOTE MEASUREMENT OF KEY BIOPHYSICAL ATTRIBUTES

Forest cover

Canopy cover is considered to be one of the main indicators of forest condition, and which has a strong influence on forest productivity and hydrology. Due to the strong inverse correlation between levels of vegetation cover and direct runoff and erosion, an accurate picture of vegetation cover for modelled catchments is a key element in accurate estimation of water yields, peak flows and erosion potential (Gorte, 2000). A number of spatially explicit forest productivity and hydrological models in use today employ remotely sensed data to derive some key input variables. Models such as Forest-BGC, RHESSyS, TOPOG, 3-PGS, SCAM, Macaque (Running and Coughlan 1988; Band *et al.*, 1993; Vertessy *et al.*, 1993; Coops *et al.*, 1998; Raupach, 2002; Watson *et al.*, 1999, respectively), use estimates of forest leaf area, derived from spatially explicit ground measurements or from satellite data. Most of these models exhibit common features and known relationships (Figure 27.6), and most conceptualise vegetation cover as a 'big leaf' distributed over the landscape.

The inclusion of remotely sensed data in these forest hydrological or productivity models has traditionally taken the form of broad-band reflectances, measured in the red and near-infrared spectral regions, and transformed to spectral indices which are then correlated to the green cover of vegetation (Tucker, 1979). One such index, the normalised difference vegetation index (NDVI), is often used, and correlated empirically to field measurements of stand leaf area index (LAI), canopy conductance or the fraction of absorbed photosynthetically active radiation (fAPAR). The derived information is then used in the models either as a single, uniform layer across the landscape, or as a spatially and seasonally varying input, to help provide regional estimates of carbon and water fluxes (e.g. Running and Nemani, 1988; Raupach, 2002; Sellers, 1987).

A major assumption in most of these models is that the leaf area of a specific plant species at a given location is the result of an integrated optimisation process, where the amount of active leaf area is in equilibrium with the available resources, at a level that minimises stresses such as drought or low nutrients (Chapin, 1991; Eagleson 1982; Field *et al.*, 1995; Running and Coughlan, 1988).

While measurements of greenness, and derivation of green vegetation cover, can reveal broad vegetation responses to changing conditions, these methods and assumptions sacrifice some accuracy in the estimate of short-term changes in ecosystem productivity caused by short-term environmental factors or stresses. By its very nature, this integrated approach can miss subtle changes in canopy chemical composition, often associated with short-term variations in physiology, due to climatic variability or short-term stress agents (e.g. soil water deficit), that have no noticeable effects on the LAI and therefore the NDVI. In addition, many plant types, notably woody perennials, exhibit strong responses in stomatal conductance and physiological activity on time scales of minutes to days, rather than changes in leaf area, in response to changes in resource or stress levels (e.g. Farquhar *et al.*, 1980; Grantz

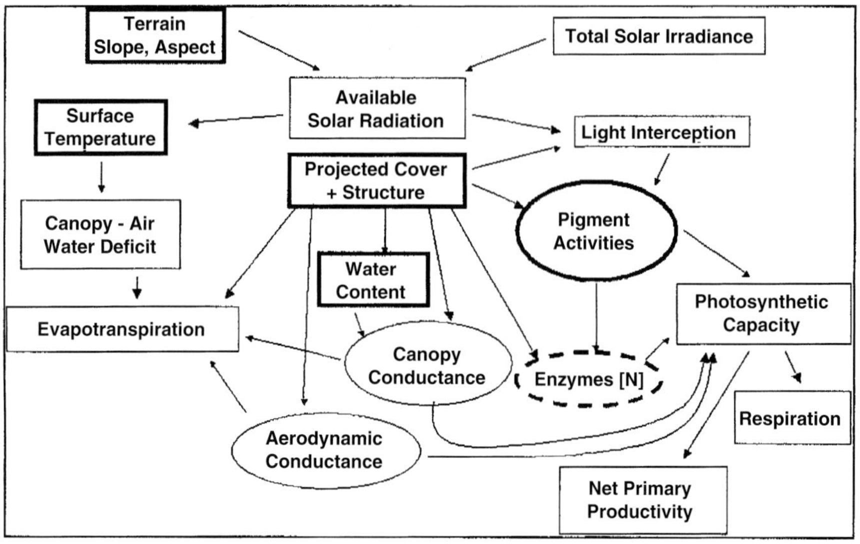

Figure 27.6 Conceptual model of forest productivity and water use. Variables inside boxes are measurable by traditional and new remote sensing means.

and Meinzer, 1990; Rawson *et al.*, 1977; Schulze and Hall, 1982; Tenhunen *et al.* 1987;). Temperate forest ecosystems, for instance, can exhibit major changes in forest age distribution, tree density, gap frequency and plant composition, all of which can have large impacts on net primary productivity, without showing significant change in the NDVI (Ustin *et al.*, 1993).

Similar and perhaps even more marked responses to short-term environmental variability such as drought, may also occur in dense tropical forests, where root systems are often shallow, but where sensitivity of the NDVI or SR-type indices to subtle changes in leaf area or changes in leaf reflectance is still very low. Factors such as these occurring at the leaf-, and ecosystem-level may cause significant errors in the estimates of ecosystem net primary productivity when the productivity models rely only on measurements of remotely detected greenness and leaf area. Very little information is available on synoptic vegetation-stress detection in tropical areas. For this reason, improved ways of using remote sensing tools, capable of detecting subtle changes in canopy reflectance are being investigated, which can be associated more directly with changes in plant function due to short-term stress or natural events.

For example, high spectral-resolution imaging systems, with the capacity to detect changes in very narrow parts of the leaf reflectance spectrum (associated with pigment levels or water content) (Figure 27.4), have been proposed for the remote and synoptic detection of direct changes in physiological activity (Gamon *et al.*, 1990; Ustin *et al.*, 1993). With more spatially explicit measurements derived from a number of such sensors, broad-scale models can also become more sensitive to smaller-scale events such as localised forest decline, logging, fires or disease (Coops, 1999). Tighter control of model simulations is also now possible through the use of multi-temporal remote sensing inputs, which allow for closer boundary conditions in terms of maximum and minimum levels of cover throughout the seasons (Landsberg and Coops 1999). Such improvements in the temporal and spatial detail of model inputs will also allow for more effective model output validation.

With the availability of new airborne and satellite sensors, some of which are detailed below, the opportunity exists to assess and describe the forest ecosystem in terms of its terrain, forest height, canopy chemistry, age, function and temporal dynamics, across the landscape in a more mechanistic manner, thereby reducing the need for pre-defined forest types and functional classes as baseline model inputs.

Terrain attributes

Topographical data such as slope, aspect and elevation, provide key variables used in distributed hydrological and carbon transfer models as they contribute to the effective mapping of catchment geometry, solar radiation distribution and in delineation of drainage networks (e.g. Vertessy *et al.*, 1993). In addition, this information is of great value in day-to-day forest management, road design and sustainable utilisation of timber resources. Digital terrain models (DTMs) used in forested catchment hydrological models have been derived traditionally from digitised topographical maps, air photography or directly from costly ground surveys. These often suffer from discontinuities between

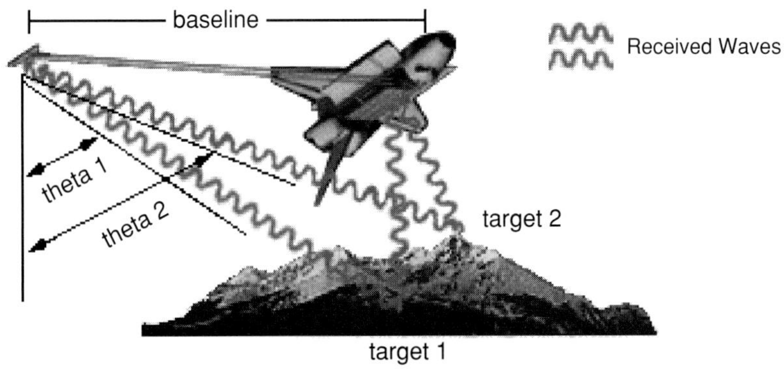

Figure 27.7 Diagram depicting the use of a second antenna on the Space Shuttle, which was used in early 2000, for interferometric mode data collection during the Shuttle Radar Topography Mission. (From: NASA–JPL.)

neighbouring orthophotos, non-real pits, and in some cases lack of true topographical delineation below dense forests, which occlude the ground below. A number of relatively coarse (250 m–1000 m resolution) global datasets are now available online (via internet) or for purchase from national mapping agencies. These general datasets however, may well be insufficient for the design of 'minimum-impact' timber extraction roads or for high resolution modelling of energy, water and carbon exchanges within catchments. A number of remote sensing technologies including airborne laser and radar technologies now allow for cost effective and more direct terrain mapping as well as forest structure estimation. New airborne LIDAR systems available for topographical mapping from commercial operators permit very detailed DTM generation over thousands of hectares per day (http://www. airbornelasermapping.com/). These systems are based on rapidly pulsating laser systems (1 laser pulse per nanosecond or more), scanning like brooms across the landscape as the aircraft passes over. Fast digital detectors and accurate aircraft movement and position measurement systems allow for creation of DTMs of good accuracy and spatial resolution (e.g. sub-metre height measurement at 1–5 m spacings across a 2 km wide swath). By collecting data as a series of parallel strips across the terrain, large areas (up to 1000 sq. km per day) can be covered with these airborne systems (Gutelius *et al.*, 1998; Kraus and Pfeifer, 1998, Wehr and Lohr, 1999).

Digital elevation information can also be derived from radar data, specifically 'radar interferometry' gathered from aircraft and now from space platforms. This technique uses radar systems in a fashion analogous to stereo photography, through a combination of images taken at slightly different view angles, thereby generating three-dimensional depictions and accurate elevation measurements of the ground below (e.g. review by Hanssen 2001). The 'all-weather' nature of radar makes this technique very appealing in wet tropical areas where cloud cover is often a problem for optical systems. Radar interferometry from aircraft systems such as the airborne NASA TopSAR system can reach height accuracies of 5 metres or better. Satellite-based radar interferometry is also available and either collected by the same satellite, imaging the Earth from two slightly different view angles over neighbouring orbits (ERS-1 and Radarsat) or with two antenna systems capable of collecting information of the returning pulses simultaneously from two different viewing angles (e.g. SRTM) (Figure 27.7).

This landmark mission, called the Shuttle Radar Topography Mission, carried out by NASA in early 2000, has collected interferometric radar data around the world with a coverage of about 80% of the Earth's surface The spatial resolution of this publicly available data is expected to be 90 metres or better, with expected height accuracies in the vicinity of 20 metres or better (http://www.jpl.nasa.gov/srtm/). This dataset is the first global, uniform and contiguous digital terrain dataset of such high quality (e.g. Plate 10).

Soil characteristics

The spatial distribution of soil types and their associated hydraulic properties are still difficult to measure by remote sensing means, in part due to the cover of the soil by vegetative matter or litter on the forest floor. A set of studies has used terrain characterisation, aspects of vegetation condition and canopy openness, all measurable by high resolution remote sensing, to provide indirect descriptions of soil characteristics under forests (e.g. Coops *et al.*, 1998; Skidmore *et al.*, 1997). Others, using environmental correlation, geostatistical and pedogenesis models, have incorporated airborne gamma-ray spectroscopy, also called 'gamma radiometrics' (Minty, 1997; Wilford *et al.*, 1997), to map key soil characteristics and physical properties through dense forest vegetation (McKenzie and Ryan, 1999; Ryan *et al*, 2000). With this technique, the relative proportions of naturally emitted gamma rays, originating from uranium, thorium and potassium elements

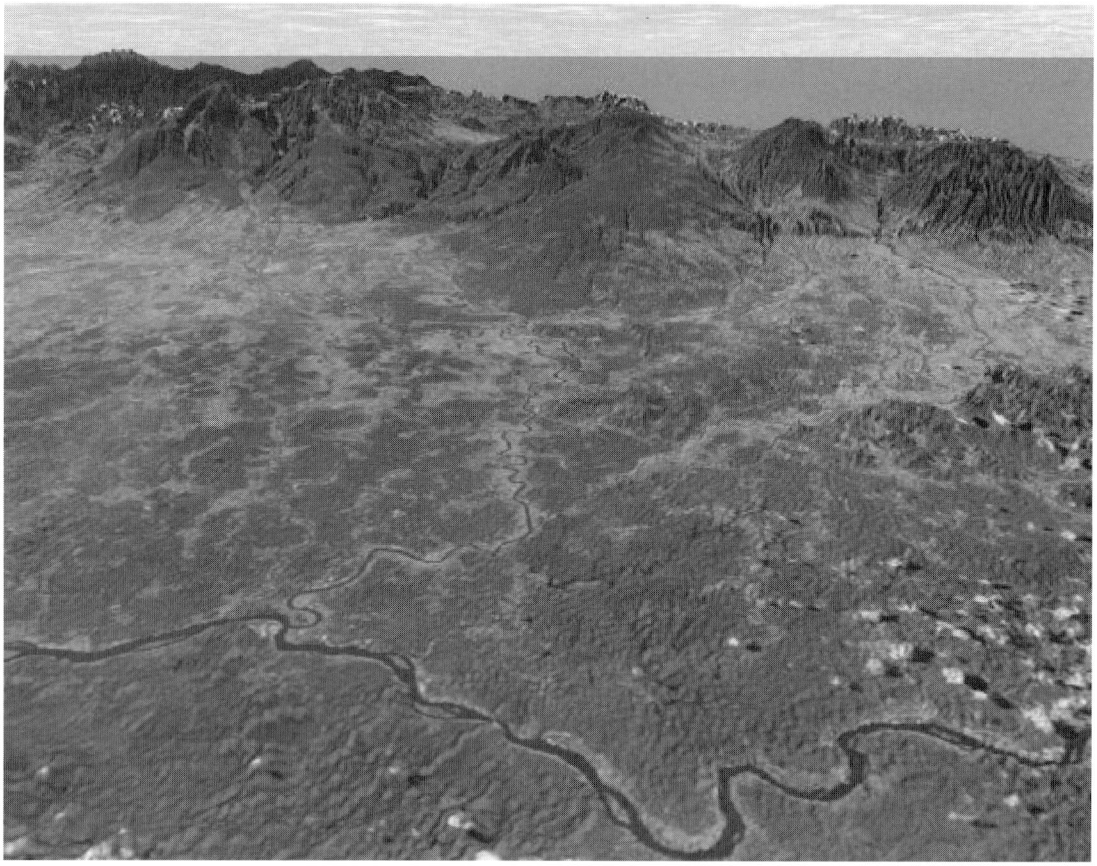

Plate 10 Perspective view of the northern coastal plain of Costa Rica and the Rio San Juan, combining a Landsat TM image with SRTM-determined topography, exaggerated about six times vertically. Both datasets were acquired in February, 2000. (From NASA/JPL/NIMA/USGS.) (See also colour plate section.)

present in most soils, are measured from a low flying aircraft (Cook *et al.*, 1996; Minty, 1997). Once interpolated in two dimensions and adequately interpreted, gamma radiometrics images can provide information on the basic geochemical origin of the soils. This then can lead to better spatial extrapolation of characteristics such as nutrient status (e.g. phosphorus) in the top 30–50 cm soil layer. When combined with additional terrain information, such as a DTM and sound geomorphological experience, this information has been effectively used to derive soil profile depth, total phosphorus and total soil carbon content, accounting for 42%, 78% and 54% of the variance, respectively, found in validation samples (McKenzie and Ryan, 1999). Water content was also estimated using similar approaches in a 50 000-ha catchment, with an r^2 of 0.67 (Ryan *et al.*, 2000). Airborne gamma radiometrics analysis techniques, combined with DTM analysis, has also been used to describe historical sediment movement across the landscape (e.g. Pickup and Marks, 2000).

Radar has also shown potential in its use to map the spatial distribution of soil moisture, mostly in small scale (~10 ha) agricultural systems, due to its strong dependence on the dielectric constant (related to water content, texture and salinity) of the top soil layers (Dobson and Ulaby, 1998; Schmugge, 1983, 1985; Wang and Schmugge, 1980). For larger areas and most forested systems, however, lack of detailed information about the topographical effects and the above-ground biomass makes this an area of still active and challenging research (Walker *et al.*, 2000). Passive microwave remote sensing has now also become an area of active research as it has shown some promise for soil moisture mapping (e.g. Owe *et al.*, 2001).

Forest type, clearing and regrowth stage

Measurements of forest type and condition have been collected with good results at the regional and continental scale using traditional satellite data (see review by Lucas *et al.*, 2002). The separation into major forest types such as hardwoods, softwoods, mixed woods, wetlands and clear-cuts, is done quite routinely in several countries and used effectively for inventory of forest resources and as an aid to regional or national forest management policies (e.g. Leckie, 1990).

While satellite data are used mainly for measurement of vegetation cover or baseline vegetation mapping, the data are not yet used widely for long-term monitoring, or as input to dynamic model simulations of the effects of forest-use/regrowth dynamics on water use or carbon uptake. Due to the importance of forest clearing and regrowth on the global carbon cycle, a number of national and international mapping programmes are under way for characterisation of forest areas and of land-use using mostly satellite data like Landsat TM, SPOT and more recently from the JERS-1 (e.g. FAO, 1990, 1996; Malingreau *et al.*, 1995; Rosenqvist *et al.*, 2000; Skole and Tucker, 1993). Assessments based on optical satellite data are especially difficult in tropical areas, where frequent cloud cover reduces the amount of data available for such analysis. Nonetheless, innovative uses of multi-temporal satellite data such as production of monthly image mosaics, composed of cloud-free subsets of multi-temporal images, have been developed for the mapping of the various tropical forest types, including regrowth stages after clearing, using optical Landsat and Advanced Very High Resolution Radiometer (AVHRR) data (e.g. Plate 11; Lucas *et al.*, 2001).

Radar data have also been used for forest type and land-use classification. Single-band, single-polarisation satellite radar systems such as the Japanese Earth Resources Satellite (JERS-1), the European Resources Satellites (ERS 1 and 2) and the Canadian Radarsat systems (c.f Table 27.2), have now been used for mapping of forested areas around the world (e.g. Rosenqvist *et al.*, 2000, or Saatchi *et al.*, 2000). Plate 12 presents an example of terrain classification in the Amazon River basin using multi-season JERS data (Siqueira *et al.*, 1997). This example highlights one of the advantages of radar data for rainforest mapping: since radar is able to see through clouds and does not require scene-to-scene calibration, large scale mosaics of even perennially cloudy regions can be obtained. Data from these imaging radar systems have permitted mapping of major land-use types, clear-cuts and uniform forested areas (e.g. Saatchi *et al.*, 1997).

Airborne systems such as the NASA AIRSAR on the other hand, provide multi-band, multi-polarisation radar data and higher resolution for even further detailed extraction, segmentation and terrain classification of different vegetation types. Ulaby and Elachi (1990) discuss polarimetric classification algorithms, and Rosen *et al.* (2000) summarise research using interferometric data. Demonstrations carried out with this system over rainforests in Guyana and Colombia have shown good delineation of major forest types, classification of regrowth stages and estimation of biomass (Hoekman and Quiñones, 2000; Van der Sanden and Hoekman, 1999) (Plate 13). With such detailed information at hand, much more accurate estimation of biomass and carbon stores and sinks is possible, for use in ecohydrological models.

SPECIES DIVERSITY MAPPING

Species-level maps of forests provide valuable information for resource inventories, sustainable forest use and biodiversity mapping. With the advent of calibrated imaging spectrometers, detailed species mapping can now be improved through the use of spectral reflectance libraries for specific plant types, where spectral differentiation (unmixing) is performed on the basis of the linkages between spectral reflectance and the vegetation's chemical composition (Martin *et al.*, 1998; Roberts *et al.*, 1999). It is important to state, however, that since all vegetation types are composed basically of the same limited types of chemicals (e.g. chlorophyll, water, accessory pigments, cellulose, proteins), it is only when a specific mix of these chemicals and their three-dimensional display can be associated uniquely with a specific species, that it can be detected and identified as such with this technology. So, in contrast to most geological applications, where the presence of different minerals on the earth's surface is 'detected' among a 'library' of thousands of different chemicals, imaging spectroscopy of vegetation is focused much more on the quantitative measurement of the absolute levels of only a few chemicals, which are present in almost all plants, but assembled in three-dimensional space (Peterson, 2000).

A number of examples shows that species identification is even possible with Landsat TM data, where certain species with unique spectral signatures can be detected (e.g. Lucas *et al.*, 1993, Roberts *et al.*, 1998). It has also been proposed that spectral libraries developed for the purpose of species identification should be restricted to regional libraries due to the inherent variability in spectral signature for the same species occurring in different regions (Roberts *et al.*, 1999). Such 'regionally-specific libraries' should incorporate seasonal changes in reflectance of the different species as best as possible, so that any possible deviations can be taken into account.

In addition to the increased spectral information afforded by hyperspectral scanners, the ability to collect multi-band imagery at resolutions of 1 metre or less can provide even better opportunities for individual tree mapping in highly mixed forests by combining object recognition and spectral analysis methodologies (see below).

FOREST STRUCTURE

Forest structure and type has traditionally been inferred at large regional scales and coarse resolution from seasonal trends in NDVI, together with basic ground-based information on vegetation types (e.g. Roderick *et al.*, 1999; Stone *et al.*, 1994). Due to its linkage with aerodynamic resistance, this information has been an important input to soil-plant-atmosphere- transfer models (i.e. Raupach and Finnigan, 1995; Raupach 1998). In 1999, the race for high resolution civilian space imagers (1 metre resolution or better) began, with the launch of the IKONOS sensor, thereby

Plate 11 Age map of a 275 km^2 area of regenerating rainforest near Manaus, Brazil, as derived from time-series of 30-m resolution Landsat sensor data. (After Lucas *et al.*, 2001.) (See also colour plate section.)

fulfilling the promise of high resolution data over large areas and often at a lower cost per unit area than airborne systems. At the high precision forest mapping level, the combination of high-resolution digital imagery (<1m pixels), with new object recognition software (e.g. Gougeon, 1995; Held and Billings, 1996; Ticehurst *et al.*, 2001), permits delineation and characterisation of individual tree crowns visible in the imagery (Plate 14). Airborne spectral scanner systems now allow for collection of ten or more spectral bands at sub-metre pixel resolutions and with geopositional accuracies of ± 3 metres or better, while high resolution commercial satellite imagers such as IKONOS collect panchromatic images (greyscale) at 1 metre resolution and multispectral data at 4 metre resolution.

Such information can yield detailed quantitative information on the density of the forest, spectral diversity, crown-size distribution and their spatial distribution over the landscape (Plate 15), and

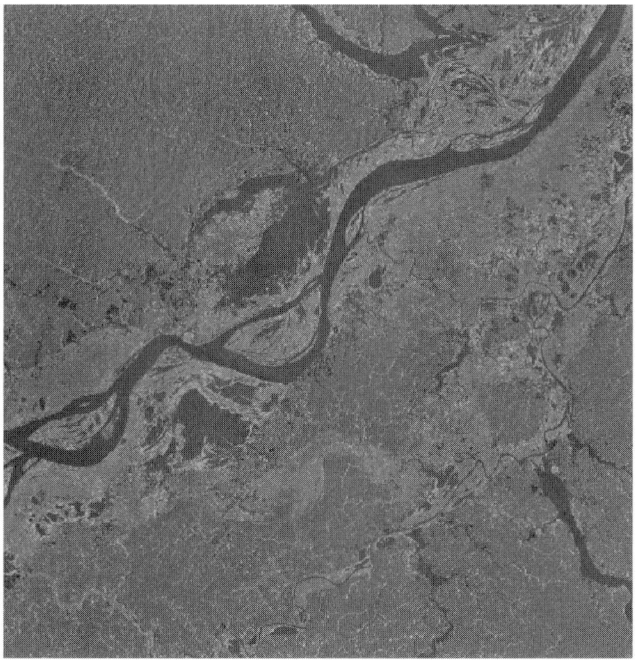

Plate 12 Sample terrain classification of a 1,140 km^2 rainforest area in the Amazon using JERS-1 (satellite radar) data processed to a 90 m pixel resolution. Water and flood areas are classified, bare soil and sand areas are yellow, whereas forest is green. (After Siqueria *et al.*, 1997.) (See also colour plate section.)

is of direct use in forest resource management as a tree-counting tool, especially in inaccessible areas. In addition, such detailed data are likely to be useful for the calculation of aerodynamic roughness terms or resistances, which are often required as input to soil-plant-atmosphere transport models (Raupach and Finnigan, 1995; Raupach, 1998). The multi-spectral nature of these new high resolution data also creates the potential for spectral discrimination of individual tree crowns in forests of high biological diversity.

Technologies such as RADAR and LIDAR, combined with new analysis methodologies, have also been used to derive estimates of forest structure and height. In the case of radar, recent investigations have shown that interferometric data can also be used for the estimation of canopy height and canopy type (Rodriguez *et al.*, 1996), which in turn can be related to the aerodynamic resistance and biomass of the forest for the purposes of hydrological modelling. A summary of radar interferometry applications for canopy mapping can be found in the review paper by Rosen *et al.* (2000). Plate 16 shows radar-based (AIRSAR) estimates of forest canopy heights in Northern Queensland, Australia. The scene shows mangroves in the Endeavour River estuary, with surrounding rainforest and mixed *Eucalyptus* and *Melaleuca* woodlands.

Plate 13 (a) AirSAR Total Power image, (b) land cover map and (c) Biomass map, for a 4 × 7 km^2 section of a 40 × 8 km^2 area mapped in Guaviare, Colombia. In the biomass map the areas of recently cut forest are masked (black). (After Hoekman and Quiñones, 2000.) (See also colour plate section.)

Plate 14 (Left) High spatial resolution (1m pixel) image of a tropical canopy collected with the casi airborne imager (cf. Table 27.1) over rainforests in Queensland, Australia. (Right) Image analysis product in the form of polygons, corresponding to delineated crowns. Different colour polygons (reproduced in greys) represent crowns of different spectral characteristics (After Held *et al.*, 2001.) (See also colour plate section.)

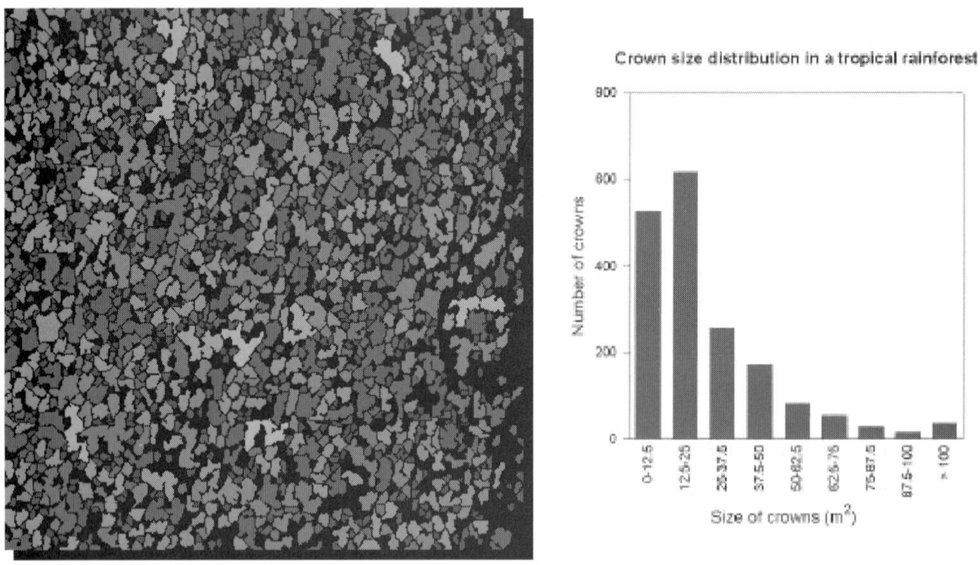

Plate 15 Automatic, computer assisted crown mapping image and crown-size distribution, derived from a high resolution forest canopy image. (After Held and Billings, 1998.) (See also colour plate section.)

Full polarimetric RADAR data such as those provided by the AIRSAR system can also be inverted to provide other canopy characteristics such as the density and orientation of tree branches. These inversions are typically done in conjunction with vegetation scattering models (e.g. Saatchi and Moghaddam, 2000; Ulaby and Elachi, 1990). An X-band (1.5 cm) SAR system was also shown to provide information on crown size characteristics when tested in interferometric mode over forests in Indonesia (Varekamp and Hoekman, 1998).

Built on the basis of the airborne laser systems designed for topographical mapping, an experimental airborne LIDAR system developed by NASA, called the 'Laser Vegetation Imaging Sensor' (LVIS) (Blair *et al.*, 1999), has been designed to collect information on the returning signal with very high levels of digitisation, hence allowing for collection of more detailed vertical information on forest structure. When collecting data over forested areas, such LIDAR systems typically detect a number of returning signals after the initial laser pulse emission. In some

Plate 16 Estimated canopy height in the Endeavour River estuary, Queensland, Australia using JPL/NASA C-band interferometric radar data. (After Rodriguez *et al.*, 1996.) (See also colour plate section.)

cases, the returning signal may be coming from the top of the trees, while in other cases it can be bouncing off the understorey or the ground, e.g. when a forest gap is present. With careful post-processing of such data, some of these systems can estimate the height and vertical structure of the forest above the underlying terrain (Aber, 1979; Lefsky *et al.*, 1999; Nilsson, 1996; Nelson *et al.*, 1997; Weishampel *et al.*, 2000). The accuracy of these estimates naturally depends on the density of the forest, the diameter of the laser pulse on the ground, topography and the electronics of these systems (see Blair *et al.*, 1999; Nilsson, 1996 for details). Figure 27.7 depicts a horizontal two-dimensional LIDAR profile taken while flying over forested terrain, where a vertical canopy profile can be derived for each laser pulse, and hence separate the different forest layers in great detail. When flown in a grid pattern or along parallel lines, a three-dimensional image of forest structure can be produced over large areas and at comparatively low cost in this way.

BIOMASS AND CARBON STORES

Remote sensing technologies are well positioned for large-scale, operational assessment of canopy above-ground biomass. While forest biomass and carbon stores have been derived from classified optical satellite images up to now (e.g. Foody *et al.*, 1996), both RADAR and LIDAR technologies, by their very nature, are poised to provide more direct assessments of the spatial variations in forest density and structure, and hence biomass and carbon stocks.

Due to its strong response to forest structure, RADAR data can readily be inverted to provide estimates of standing biomass and information on forest structure, because one of the primary determinants of the strength of the interaction between the microwaves and the canopy structure is the relative size of the electromagnetic wavelength and the size of the forest components. This correlation has stimulated investigation into the estimation of biomass from information obtained on the strength of the radar return signal at different radar wavelengths (e.g. Imhoff, 1995a; Imhoff *et al.*, 1998, Luckman *et al.*, 1997). As the wavelength of the radar pulses increases from the X-band (>1 cm) to the P-band (60+ cm), the sensitivity to changes in biomass in dense forests improves. In the early stages of regrowth, C-band (5.8 cm) radar data show a strong correlation between biomass and radar backscatter. However, as the forest biomass increases above 20 tons ha^{-1}, the C-band intensity saturates and loses sensitivity to higher levels of biomass. While L-band (23 cm) data can also show significant correlation, it may saturate for some canopy types. The best sensitivity to dense, above-ground biomass (>100 tons ha^{-1}) is obtained using P-band data, with very good correlation for boreal forests and some pine forests (Imhoff, 1995b). A substantial amount of work has been done for boreal forests (see Saatchi and Moghaddam, 2000), pine forests (Harrell *et al.*, 1997), and regenerating tropical forests (Foody *et al.*, 1997; Luckman *et al.*, 1997). Developed recently are several new SAR systems designed specifically for biomass measurement of heavy forest stands. Operating in the VHF range, these systems show no sign of saturation at above ground dry biomass densities in excess of 290 tons ha^{-1} (Plate 17) (see Imhoff *et al.*, 1998 and 2001, for BioSAR; Fransson *et al.* 2001 and Smith *et al*, 2001 for CARABAS).

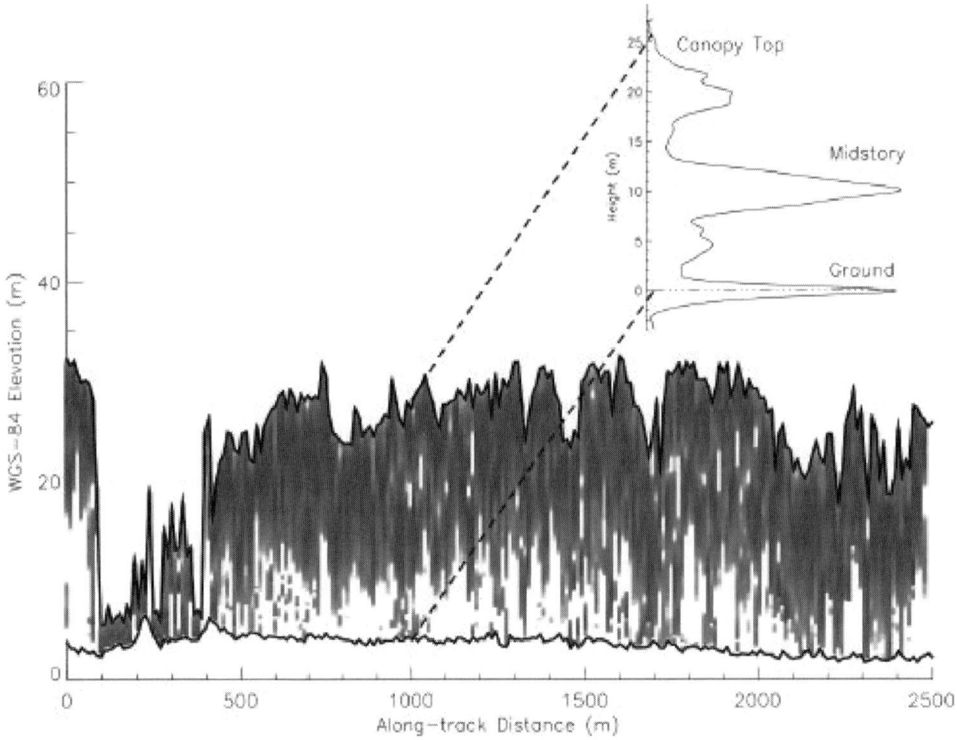

Figure 27.8 Laser scanner-derived forest height and three-dimensional structure information. (After Blair *et al.*, 1999.)

As LIDAR systems become more sophisticated and provide the opportunity to measure the underlying topography and height of various forest elements separately, estimates of standing biomass can also be derived from laser scanner data, and used for direct forest resource assessment and accurate determination of forest stratification (e.g. Aber, 1979; Blair *et al.*, 1999). Once such systems become more accessible from aircraft or spaceborne equipment, data on the three-dimensional structure of forest and its biomass may be available for most forest types around the world. A satellite laser system termed 'Vegetation Canopy Lidar – VCL', based on the airborne prototypes SLICER and LVIS, has been proposed for space deployment (Dubayah *et al.*, 1997). Such a system would provide near-global coverage of canopy structure at a 25-metre spot-size resolution. When available, such information would be of immense value for repetitive monitoring and better estimation of global biomass and carbon stores, better validation of global productivity models, as well as for natural and plantation forest resource management.

Although not yet used in tandem over the same areas, RADAR and LIDAR may be used synergistically in the future. Whereas RADAR can be used to provide information on total standing biomass and size of the major structures, LIDAR can provide additional information on the stratification of this biomass within the overstorey and understorey vegetation in the forests.

Forest condition and function

Baseline mapping and assessment of forest condition has been carried out for the last 10–15 years using broad-band satellite data, air photography and, more recently, airborne digital data. As discussed above, derivation of general vegetation condition has commonly involved analysis of time-series of greenness indices, such as the simple ratio (SR) or the normalised difference vegetation index (NDVI). These have been used for the derivation of landscape behaviour via empirical or mechanistic models that link reflected radiation to condition/productivity (e.g. Running and Nemani, 1989; Sellers, 1987). Forest condition, as defined more specifically, relates directly to the biophysical, biochemical and physiological status of the forest, and tends to be an integrated state of the vegetation in response to slower-changing conditions. Vegetation function, on the other hand, is generally considered to be an instantaneous response of vegetation to environmental changes, or a range of stress agents, that may be more dynamic in nature.

In order to understand the effects of stress agents on vegetation spectral reflectance, we require knowledge and understanding of key biochemical and functional plant processes such as photosynthesis, transpiration and light absorption by leaf pigments. At the leaf level, a number of these processes are well known and simulation models of reflectance as related to leaf chemistry and

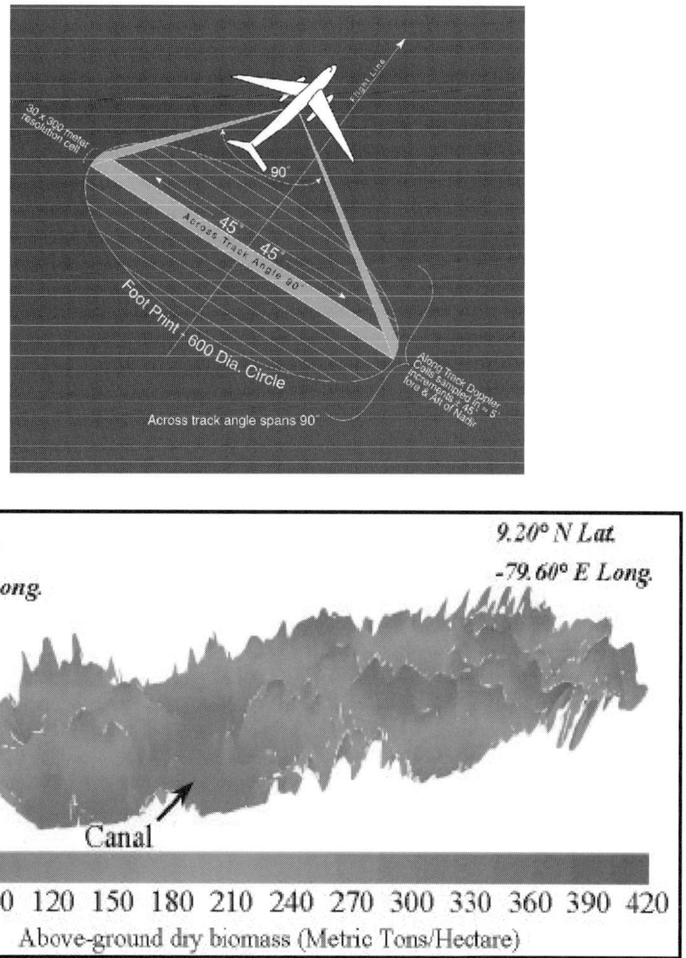

Figure 27.9 (a) Artist's impression of the aircraft and the mapping footprint of the BioSAR™ system, built for specific measurement of biomass in high density forests. (b) Biomass map produced by BioSAR™ over an area surrounding the Panama Canal and Barro Colorado island in Panama. (After Imhoff *et al.*, 2001.)

canopy architecture are now being used as inversion tools for the derivation of chemical composition of whole forests or ecosystems (e.g. Martin and Aber 1996; Ollinger *et al.*, 2001). Mechanistic models are now being used to understand better the effects of various leaf constituents directly on leaf reflectance (e.g. Jaquemoud and Baret, 1990). When combined with models of the interaction of light with the forest's three-dimensional nature, these models provide a powerful tool for detailed measurement and modelling of forest chemical composition across the landscape (e.g. Ganapol *et al.*, 1999, Jaquemoud, *et al.*, 2001; Zarco-Tejada *et al.*, 2001).

Direct detection of the chemical signature of vegetation and its correlation to vegetation condition and composition, has become increasingly feasible with the advent of high sensitivity imaging spectrometer systems (Green *et al.* 1998). Since the early 1990s, a number of case studies using the 224-spectral band Airborne Visible Infra-Red Imaging Spectrometer (AVIRIS) from NASA or other imaging spectrometers, have paved the way for use of these systems in routine forest biophysical or condition assessments (e.g. Martin and Aber, 1997; Nichol, *et al.*, 2000; Zarco-Tejada *et al.*, 2001). Contiguous spectral signatures arising from mixed or pure plant surfaces can now be 'dissected' and correlated to subtle changes in foliar chemistry such as proteins, nitrogen, chlorophyll, lignin, cellulose, water and starch (e.g. Green and Roberts, 1995; Martin *et al.*, 1998; Wessman, 1989, 1994; Zarco-Tejada *et al.*, 2001). Whole forested ecosystems have been characterised in terms of their carbon:nitrogen balance using imaging spectroscopy and related to landscape-level net primary productivity (e.g. Hallett *et al.*, 1997; Martin and Aber, 1996; Smith *et al.*, 2001).

In addition to studies where the full dimensionality of the spectral data in each pixel is used for spectroscopic and chemical analyses, several studies have been conducted to investigate the simplification of these data by using narrow-band image transformations

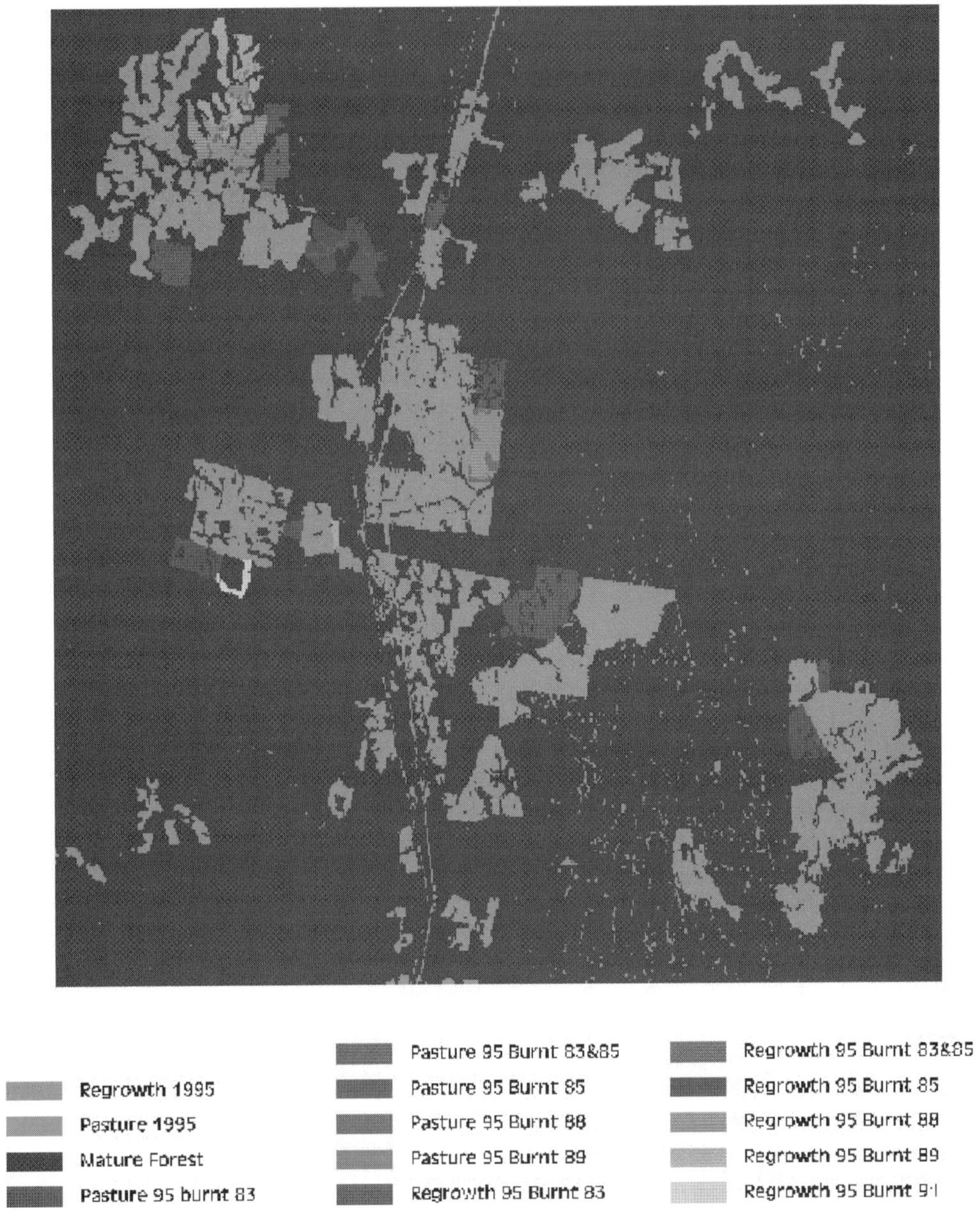

Plate 17 The fire history of a 275 km^2 area of rainforest near Manaus, Brazil, as derived from time-series comparisons of mapped fire scars within 80 m resolution Landsat MSS and 30 m Landsat TM data. (After Lucas *et al.*, 2001.) (See also colour plate section.)

or discrete-band indices, which in turn can be related to levels of specific pigments, water and nitrogen content in plants. The use of 'spectral indices' also has the advantage that brightness changes often observed across the images, and which can be caused by topographical factors, cloud shadows or haze, are reduced considerably. This is of special value in many tropical areas with steep terrain and frequent cloud shadowing. Based on laboratory and field studies of senescing leaves, Blackburn (1998, 1999) has developed a range of indices which are directly correlated to carotenoid, chlorophyll-a and chlorophyll-b pigment concentrations in the leaves. Leaf water content has also been shown to be measurable directly through imaging spectroscopy, thereby permitting generation of two-dimensional maps of liquid water content across the landscape (Green *et al.*, 1998; Gamon and Qiu, 1999; Roberts *et al.*, 1999; Ustin *et al.*, 1998). These methodologies derive estimates of vegetation liquid water content from the 970 nm, 1000 nm and 2000 nm water absorption features normally seen in spectral reflectance curves of all green leaves (cf. Figure 27.4). In turn, atmospheric water vapour absorption features extractable from hyperspectral images are also used now for direct atmospheric correction of these data in an iterative process (Green *et al.*, 1998).

A number of studies has shown that stress or phenological changes have an effect on the relative concentrations of chlorophyll-a, chlorophyll-b, anthocyanins and carotenoids in leaves before noticeable changes are detected with the standard greenness indices (e.g. Blackburn, 1998; Carter 1998; Rock *et al.*, 1986). A number of non-deciduous *Eucalyptus* species in Australia, when exposed to cold temperatures and high light levels, produce a range of accessory pigments (e.g. anthocyanins) in current-year or emerging leaves. These pigments are considered to be involved in protection of the light-harvesting systems from excess photochemical energy (Close *et al.*, 2000; M. Ball, pers. comm.). In addition, it has been observed that as chlorophyll concentrations are reduced due to early stress or phenological change, a shift of the red edge towards the shorter, blue spectral region occurs (e.g. Blackburn, 1998; Horler *et al.*, 1983; Merton, 1999).

Leaf chlorophyll concentration can also be inferred from spectral reflectance through measurement of the inflection point position of the red edge (Curran, *et al.*, 1997; Gittelson *et al.*, 1996). Others have found high correlations between chlorophyll concentration and narrow-band reflectance indices, mostly relating the reflectance observed in the 680–700 nm spectral range and normalised by a reference band located near 750 nm (e.g. Carter, 1998; Gittelson *et al.*, 1996; Peñuelas and Filella, 1998). As chlorophyll concentration increases, the red-edge feature in plants tends to deepen and broaden, thereby moving the actual red edge towards the longer wavelengths. Some types of stress, diseases or natural leaf senescence also cause changes in the proportions of accessory pigments (carotenoids, anthocyanins) relative to chlorophyll. Such changes in canopy colour are clear candidates for early detection by specially designed narrow-band indices such as SIPI (Peñuelas and Filella, 1998) which is sensitive to relative changes in reflectance due to carotenoids (445 nm) and chlorophyll-a (680 nm) (cf. Figure 27.4).

Remote, spatially explicit measurements of foliar chemicals, such as those described above, will also help constrain ecosystem process models by adding new biophysical and biochemical properties, which are linked more directly to physiological function (Field *et al.*, 1995; Gamon and Qiu, 1999). Once operational global-coverage imaging spectrometers of high sensitivity make it into space in the early 2000s, multi-temporal change detection of forest foliar chemistry will be possible at continental scales.

FOREST AGE AND FIRE HISTORY

Forest age and fire history can be considered as a representation of a specific chemical and biological status of the forest. As such, they are key aspects of forest ecological condition, and although not explicitly included in most hydrological and carbon balance models, they can have profound effects on the hydrology and carbon balance of the forest (e.g. Hoelscher *et al.*, this volume; Malmer *et al.*, this volume; Paw U *et al.*, 2000; Vertessy *et al.*, 2001; Watson *et al.*, 2001). Most remote sensing studies, where forest age mapping has been attempted, have used correlations of forest age with ensuing image variations in broad band brightness, as caused by structural effects on the 'graininess' and, hence, the spatial statistics of the imagery (e.g. Collins and Woodcock, 1999; Lambin and Strahler, 1994). Using multi-temporal satellite imagery, Lucas *et al.* (2001) have produced maps of fire history for areas of Brazilian Amazon (Plate 17). Since 2001, the US Forest Service has also been using data from the 36-band Moderate Resolution Imaging Spectroradiometer (MODIS) for rapid, daily-progression mapping and monitoring of fires across the whole continental US (http://www.fs.fed.us/eng/rsac/fire_maps.html), with a methodology proposed for this sensor by Kaufman *et al.* (1998). There are only a few examples where combined mapping of forest age and fire history have been attempted via imaging spectroscopy. Recent work using NASA AVIRIS imaging spectrometer data collected over forested areas in Washington State showed that the relative proportions of characteristic reflectance signatures for green vegetation tissue (chlorophyll), non-photosynthetic tissue (e.g. lignin content) and a background shade/soil component correlated to time after fire and regrowth stage (Sabol *et al.*, 1995). This simple approach has shown that these proportions have a strong temporal signal as the forest progresses from young to mature and fire-affected forest.

The potential therefore exists to use spectroscopy from satellites to disaggregate the spectral features of different forest age classes routinely into the main fundamentally detectable compounds like chlorophyll, accessory pigments, lignin, protein, cellulose, water and soil minerals. It is important to state, though, that in mixed forest systems, the 'chemical signal' detectable from imaging spectroscopy may also be affected by different dominants and understorey elements appearing at different forest age classes. Then again, this fact may be also viewed as an opportunity to refine the discrimination potential of this technology further.

VEGETATION FUNCTION MONITORING

Remote sensing-derived measurements of forest water use and productivity have so far used measurements of seasonal changes in the NDVI and surface temperature, which in turn have been linked through mechanistic models to changes in leaf area index, surface resistance, biomass and evapotranspiration (e.g. Nemani *et al.*, 1993; Running and Coughlan, 1998, Running *et al.*, 2000; Sellers, 1987). For example, global, or regional-scale estimates of net primary production are now being derived from 8-daily MODIS data (http://edcdaac.usgs.gov/modis/dataprod.html), using estimates of the fraction 'absorbed photosynthetically active radiation (fAPAR)', derived green cover estimates and a light-use efficiency factor derived for different vegetation functional types (Monteith, 1972; Running *et al.*, 2000). All these LAI-based methods for estimation of carbon gain and transpiration in vegetation suffer from the complication of non-linear, asymptotic relationships with primary productivity and water loss in forests, where LAI exceeds values of 4 or 5 and self-shading dramatically reduces light interception efficiencies. Across such ecosystems with high LAI, it is often observed that differences in productivity and water use still occur. These are most likely caused by variations in canopy photosynthetic activity and transpiration, which in turn can be caused by differences in microclimate or soil edaphic conditions (Reich *et al.*, 1992). Recent studies have exploited observed variations in the ratio of NDVI and surface temperature, measurable from orbiting satellites, and changes in the vegetation's resistance to water transport, to map regional progressions of drought in the tropics (e.g. McVicar and Bierwirth, 2001).

More direct methods for estimation of photosynthesis and transpiration using remotely sensed data have also been sought, which can be used to either scale flux-tower measurements or to constrain soil-plant-atmosphere exchange simulations. One such approach has been the use of imaging spectroscopy to provide maps of ecosystem nitrogen concentration, which in turn have then been correlated to photosynthetic capacity, forest productivity and soil C:N ratios (e.g. Martin and Aber, 1997; Ollinger *et al.*, 2001). This provides a unique tool for comparison of ecosystem productivity and carbon budgeting across various climatic zones or other environmental gradients. In addition, it also provides more realistic inputs of maximum photosynthesis or tighter boundary conditions to further constrain forest growth and hydrology models.

For direct stress detection, or for model validation where the simulation of the spatial distribution of a more instantaneous photosynthesis rate requires validation, other remote sensing products are required which are more responsive to short-term changes in leaf chemistry and function. This has now become possible with the advent of high-sensitivity spectrometers, available as portable field instruments or on board aircraft, and now satellites. Changes in canopy reflectance measured with a spectroradiometer a few metres above the canopy have recently been linked to fluctuations in the light-use efficiency and photosynthesis of crops (Gamon *et al.*, 1990, 1992). In these studies, it was found that changes in components of the xanthophyll pigment cycle were responsible for changes in canopy reflectance at 531 nm (Demming-Adams and Adams, 1992; Thayer and Björkman, 1990). These pigments are generally considered as accessory pigments to the chlorophylls taking part in the light-harvesting activities within the tylacoid membranes of chloroplasts. On a seasonal time-scale, it was also observed that reductions in photosynthetic efficiency caused by prolonged water and nutrient stress affected the reflectance at 531 nm (Gamon *et al.*, 1992). Hence, a narrow-band reflectance index, termed the 'photochemical reflectance index' (PRI), which is based on the reflectance at 531±10 nm and normalised by a reference band, has been proposed as a practical index for the specific detection of short-term changes in photosynthetic capacity (Gamon *et al.*, 1992). Oscillations in the PRI in response to moisture stress, and a close linkage to photosynthetic light use efficiency, have now been documented for a range of crops, native shrubs and boreal forests (Gamon *et al.*, 1997; Nichol *et al.*, 2000).

Spectral indices such as the PRI or direct measurements of surface temperature (T_s) are most suitable for the detection of relatively short-term (sub-daily) changes in stomatal aperture, leaf temperature or water content. At slightly longer time scales (days), indices such as movements of the red-edge inflection point, canopy water content or the 'Red-edge Vegetation Stress Index – RVSI (Merton, 1999), could be of use for monitoring plant stress responses that begin to induce changes in leaf chemical composition such as a drop in chlorophyll concentration or water content (Table 27.3).

TRANSPIRATION AND WATER BALANCE

For the purposes of drought monitoring or for water-balance derivations, measurements of canopy water content have been derived recently using imaging spectrometers (e.g. Gamon and Qiu, 1999; Green *et al.*, 1998; Roberts *et al.*, 1999; Ustin *et al.*, 1998;), as well as from passive microwave remote sensors (Becker and Choudury, 1988; Calvet *et al.*, 1994; Matzler, 1994, Njoku and Li, 1999). While remote sensing is not able to measure the water balance directly, the spatial context of thermally emitted radiation has been used for derivation of the sensible heat balance, which in turn forms part of regional hydrology models. In most cases, both estimates of green cover and surface temperature (Ts) have been used to provide information on the spatial distribution of transpiring area and of the driving force of evaporation, respectively (Goetz, 1992; Menenti, 2000).

Extrapolation of evapotranspiration (ET) measurements using thermal data and vegetation indices has been widely tested with varying levels of success and accuracy (e.g. Held *et al.*, 1995; Jupp and Kalma, 1989; Kustas, *et al.*, 1994; Price, 1982; Sun and Mahrt, 1994; Seguin *et al.*, 1994). Validation of the temporal dynamics of such spatial ET estimates has proven difficult in many of such studies, however, in part because of the lack of adequate regional land cover description and the cost of installation of a distributed network of long-term ground measurement towers or lysimeters for verification of the methodology (Kite and Droogers, 2000). While the linkages between the various remotely sensed vegetation greenness indices is often well established and relatively insensitive to atmospheric factors, the accurate estimation of surface temperature requires rigorous sensor and atmospheric corrections, combined with estimates of the relative contributions of soil surface temperature and emissivity

Table 27.3. *Sample stress responses to various stress agents and possible remotely sensed indices, which can be used for detection and monitoring of these stress effects*

	Low stress	Medium stress	High stress
Sample factors	Low atmospheric humidity, high light intensity	Mild drought, high temperatures (days), air pollution, heavy metals, low nutrients	Prolonged drought (weeks-months), salinity, root anoxia, high 'free' hydrogen or aluminium.
Forest responses	Stomatal closure, lower instantaneous photosynthesis rate	Wilting, growth reduction, low photosynthetic capacity, build-up of accessory pigments	Leaf drop, stunted growth
Remotely detectable signals	Reflectance change in xanthophyll signal	Drop in leaf-water absorption, red-edge shift	Drop in red–near-IR difference, drop in percent cover
Indices	PRI, T_s	WBI, RVSI, SIPI, NDTI, [N]	NDVI, SR, T_s/NDVI

in mixed pixels, and information on surface roughness or aerodynamic coupling of the vegetation with its surrounding atmosphere (Prata *et al.*, 1995; Price 1990). At the regional scale, innovative, and somewhat simpler approaches for use of thermal and optical reflective data as measured by airborne or satellite systems, are being used for mapping of moisture availability or to monitor changes in surface resistance caused by moisture stress (e.g. Lambin, 2000; McVicar and Jupp, 2000; Nemani *et al.*, 1993; Price, 1990). In some of these studies, changes in the slope of the NDVI and T_s relationship, as derived from remotely sensed data, are used in conjunction with ground-based meteorological data and soil-vegetation-atmosphere transfer (SVAT) models for estimation of seasonal patterns in surface resistance and soil moisture status.

Operational and routine uses of remotely sensed data as part of regional hydrology measurements are still not widespread, despite the high expense in providing equivalent spatially dense estimates of E_T over space and time, based on ground measured data alone. A growing world-wide network of flux towers (Euroflux, Fluxnet, Ozflux, etc.), as well as large-scale regional experiments such as FIFE over grasslands in the US, BOREAS over the boreal forests in Canada, LBA over the tropical rainforest in South America and OASIS over crop lands in Australia, have been designed specifically to provide the necessary applications-development and remote sensing validation opportunities for techniques which can be used to provide routine estimates of evapotranspiration and ecosystem water balance (Running *et al.*, 1999).

CONCLUSION

In this chapter a number of new technologies and potential uses for remotely sensed data have been discussed. These can provide unprecedented insight into the spatial extent and patterns of key indicators of forest condition and function. Many of the new technologies discussed here, such as LIDAR, imaging radar and imaging spectroscopy also have a use in the direct provision of detailed inputs to hydrology/productivity simulation models. Several of the current models still use only very basic inputs derived from remotely sensed data, such as broad-band 'vegetation cover indices', which are then related to the leaf area index. However, as demonstrated in this chapter, the possible array of remote sensing data products for use as inputs to these models is far larger than that. As image analysis methodologies become more efficient, and as data delivery occurs closer to 'real-time' and on a routine basis, provision of the possibly large range of remotely sensed data types as inputs to simulation models will be more timely and cost-effective. Access to such data may lead to new generations of hydrological/production models, based primarily around the different remotely measurable input data.

References

Aber, J. D. (1979). A method for estimating foliage-height profiles in broad-leaved forests. Journal of Ecology 67: 35–40.

Anger, C. D., Babey, S. K., and Adamson, R. A. (1990). A new approach to imaging spectroscopy. Proceedings of SPIE, 72:72–86.

Bailey, J. (1990). The potential value of remotely sensed data in the assessment of evapotranspiration and evaporation. *In:* Satellite Remote Sensing for

Operational Hydrology. Ed. Francois Becker. Remote Sensing Reviews Vol 4. Harwood Academic Publishers. New York, p. 349–378.

Band, L. E., Patterson, P., Nemani., R., Running, S. (1993). Forest ecosystem processes at the watershed scale: incorporating hillslope hydrology. Agricultural and Forest. Meteorology, 63:93–126.

Becker, F. and Choudhury, B. J. (1988). Relative Sensitivity of Normalized Difference Vegetation Index (NDVI) and Microwave Polarization Difference Index (MPDI) for Vegetation and Desertification monitoring. Remote Sensing of Environment, 24: 297–311.

Benson, M. L., Landsberg, J. J. and Borough, C. J. (1992). The biology of forest growth experiment: An introduction. Forest Ecology and Management, 52:1–311.

Blackburn, G. A. (1999). Relatonships between spectral reflectance and pigment concentrations in stacks of deciduous broadleaves. Remote Sensing of Environment 70:224–237.

Blackburn, G. A. (1998). Spectral indices for estimating photosynthetic pigment concentrations: a test using senescent tree leaves. International Journal of Remote Sensing 19:657–675.

Blair, J. B., Rabine, D. L., Hofton, M. A. (1999) The Laser Vegetation Imaging Sensor (LVIS): A medium-altitude, digitisation-only, airborne laser altimeter for mapping vegetation and topography. ISPRS Journal of Photogrammetry and Remote Sensing, 54:115–122.

Calvet, J. C., Wigneron, J. P., Mougin, E., Kerr, Y., and Brito, J. S. (1994). Plant water content and tempeatrue of the Amazon forest from satellite microwave radiometry. IEEE Transactions on Geoscience and Remote Sensing. 32:397–408

Carter, G. A. (1998). Reflectance wavebands and inices for remote estimation of photosynthesis and stomatal conductance in pine canopies. Remote Sensing of Environment. 63:61–72.

Carter, G., (1994). Ratios of leaf reflectance in narrow wavebands as indicators of plant stress. International Journal of Remote Sensing. 15:697–703.

Chapin, F. S. III, (1991). Effects of multiple environmental stresses on nutrient avaiability and use. *In*: Response of Plants to Multiple Stresses. Academic Press Inc. pp 67–88.

Choudhury, B. J., Idso, S. B. and Reginato, R. J. (1987). An analysis of infrared temperature observations over wheat and calculation of latent heat flux. Agricultural and Forest Meteorology. 37:75–88.

Clark, R. N., Trude, V. V., King, V. V., ager, C., Swayze, G. A. (1995). Initial vegetation species and senescence/stress indicator mapping in the San luis Valley, Colorado, using imaging spectrometer data. Pp. 35–38. *In*:. Proc. Fifth Airborne/Visible/Infrared Imaging Spectrometer (AVIRIS) Workshop. January, 1995.

Close, D. C., C. L. Beadle, P. Brown., and G. K. Holz. (2000). Cold-induced photoinhibition affects establishment of *Eucalyptus nitens* (Deane and Maiden) and *Eucalyptus globulus* (Labill). Trees: Structure and Function. 15: 32–41.

Collins, J. B., Woodcock, C. E. (1999). Geostatistical estimation of resolution-dependednt variance in remotely-sensed images. Photogram. Eng. Remote Sens. 65:41–50.

Cook, S. E., Corner, R. J., Groves, P. R., Grealish, G. J., (1996). Use of airborne gamma radiometric data for soil mapping. Aust. J. Soil Res. 34, 183–194.

Coops, N. C. (1999). Improvement in predicting stand growth of *Pinus radiata* (D.Don) across landscape using Noaa AVHRR and Landsat MSS imagery combined with a forest growth process model (3-PGS). Photogram. Eng. Remote Sens. 65:1149–1156.

Coops, N. C., Ryan, P. J., Bishop, A. P. (1998). Investigating CASI responses to soil properties and disturbance across an Australian eucalypt forest. Can. J. Rem. Sens. 24:153–168.

Coops, N. C., Waring, R. H. and Landsberg, J. J. (1998). Assessing forest productivity in Australia and New Zealand using a physiologically-based model driven with average monthly weather data and satellite derived estimates of canopy photosynthetic capacity. Forest Ecology and Management 104, 113–127.

Curlis, J. D., Frost, V. S., and Dellwig, L. F. (1986) Geological mapping potential of computer-enhanced images from the shuttle imaging radar: Lisbon Valley Anticline, Utah. Photogram. Eng. And Remote Sensing 52:525–532.

Curran, P. J., Kupiec, J. A., Smith, G. M. (1997) Remote sensing the biochemical composition of a slash pine canopy. IEEE Transactions on Geoscience and Remote Sensing. 35:415–420.

Curran, P. J. (1994). Imaging spectrometry – its present and future role in environmental research. *In Imaging Spectrometry – a Tool for Environmental Observations*. Joachim Hill and Jacques Megier (eds.). Kluwer Academic Publishers, Dotrecht. EU., pp. 1–24.

Demming-Adams, B., and Adams, W. W., III. (1992). Photoprotection and other responses of plants to high light stress. Ann. Rev. Plant Physiol. Plant Molec. Biol. 43:599–626.

Dobson, M. C., Ulaby, F. T. (1998). Mapping soil moisture distribution with imaging radar. In: Principles and applications of imaging radar. Manual of Remote Sensing, 3rd Edition. (R. A. Ryerson ED). John Wiley and Sons. New York.

Dubayah, R., Blair, J. B., Bufton, J. L., Clark, D. B., JaJa, J., Knox, R. G., Luthcke, S. B., Prince, S., and Weishampel, J. F. (1997). The Vegetation Canopy Lidar mission. Pages 100–112. *Land Satellite Information in the Next Decade II: Sources and Applications*. American Society for Photogrammetry and Remote Sensing, Bethesda, MD, USA.

Eagleson, P. S. (1982). Ecological optimality in water-limited natural soil-vegetation systems, I. Theory and hypothesis. Water Res. Research 18:325–340.

Elachi, C. (1980), Spaceborne imaging radar: geologic and oceanographic applications, Science, 209, 1073–1082.

FAO (Food and Agriculture Organisation of the United Nations) (1996). Forest Resources Assessment. Survey of tropical forest cover and study of change processes. FAO Forestry Paper 112, Food and Agricultural Organisation of the United Nations, Rome

FAO (Food and Agriculture Organisation of the United Nations) (1990). Forest Resources Assessment: Global Synthesis, FAO Forestry papers 124, Rome, Italy.

Farquhar, G. D., Schulze, E.-D., and Küppers. (1980). Responses to humidity by stomata of *Nicotiana glauca* L. and *Corylus avellana* L. are consistent with the optimisation of carbon dioxide uptake with respect to water loss. Aust. J. Plant Physiol., 7:315–327.

Field, C. B., Randerson, J. T., and Malmström, M. (1995). Global net primary production: Combining ecology and remote sensing. Remote Sensing of Environment 51:74–88.

Field, C. B. (1991). 2. Ecological scaling of carbon gain to multiple stress. *In*: Responses of Plants to Multiple Stresses. H. A. Mooney, W. E. Winner and E. J. Pell (eds.) Academic Press. Pp. 35–65.

Foody GM, Green RM, Lucas RM, (1997). 'Observations on the relationship between SIR-C radar backscatter and the biomass of regenerating tropical forests', International Journal of Remote Sensing 18: (3) 687–694.

Foody, G. M., Palubinskas, G., Lucas, R. M., Curran, P. J., and Honzak, M. (1996), Identifying terrestrial carbon sinks: Classification of successional stages in regenerating tropical forest from Landsat TM data. Remote Sensing of Environment 55: 205–216.

Fransson, J. E. S., Walter, F, and Ulander, L. M. H. (2001) Estimation of Forest Parameters Using CARABAS-II VHF SAR Data, IEEE Transactions on Geoscience and Remote Sensing, Vol. 38, No. 02, p. 720.

Freemantle, J. R, Pu, R. and Miller, J. R. (1992). Calibration of imaging spectrometer data to reflectance using peudo-invariant features. Proc. 15th Canadian Symposium on Remote Sensing, Toronto, Ontario, June 1992.

Gamon, J. A., Qiu, H.-L. (1999). Ecological applications of remote sensing at multiple scales. Pp 805–845. In: Handbook of Functional Ecology. Pugnaire, F. I. and Valladares, F. (eds). Marcel Dekker. New York.

Gamon, J. A., Serrano, L., Surfus, J. S. (1997). The photochemical reflectance index: an optical indicator of photosynthetic radiation use efficiency across species, functional types, and nutrient levels. Oecologia, 112:492–501.

Gamon, J. A., Peñuelas, J., and C. B. Field, (1992). A narrow-waveband spectral index that tracks diurnal changes in photosynthetic efficiency. Remote Sensing of Environment. 41:35–44.

Gamon, J. A., Field, C. B., Bilger, W., Bjôrkman, O., Fredeen, A. L. and Peñuelas, J., (1990). Remote sensing of the xanthophyll cycle and chlorophyll fluorescence in sunflower leaves and canopies. Oecologia 85:1–7.

Ganapol, B. D., Johnson, L. F., Hlavka, C. A., Peterson, D. L., Bond, B. (1999) LCM2: A coupled leaf/canopy transfer model. Remote Sensing of Environment. 1999.

Gilmore, A. M., Yamamoto, H. Y. (1991). Resolution of lutein and zeaxanthin using a non-endcapped, lightly carbon-loaded C_{18} high-performance liquid chromatographic column. J Chromatography 543:137–145

Gittelson., A. A., Merzlyak, M. N., Lichtenhaler, K. (1996). Detection of red edge and chlorophyll content by reflectance measurements near 700 nm. J. Plant Physiol. 148:501–508.

Goetz, A. F. H., Vane, G., Solomon, J. E., and Rock, B. N., (1985). Imaging spectrometers for earth remote sensing, Science, vol. 228, no. 4704, pp. 1147–1153.

Goetz, S. J. (1992). Multi-sensor analysis of NDVI, surface temperature and biophysical variables at a mixed grassland ite. International Journal of Remote Sensing. 18:71–94.

Gorte, B. G. H. (2000). Chapter 7: Land-use and catchment characteristics. *In: Schultz, G. A. and Engman, T. E. (eds.), 2000. Remote sensing in hydrology and water management. P. 133.* Springer Verlag, New York.

Goetz, A. F. H. (1992). Imaging spectrometry for earth remote sensing. In: Imaging Spectroscopy: Fundamentals and Prospective Applications. F. Toselli and J. Bodechtel (eds.) pp. 1–19. ECSC, EEC, EAEC, Brussels and Louxenbourg.

Gougeon, F. A. (1995). A crown-following approach to the automatic delineation of individual tree crowns in high spatial resolution aerial images. Can. J. Rem. Sens. V.21 No. 3.

Grantz, D. A., and Meinzer, F. C. (1990). Stomatal response to humidity in a sugarcane field: Simultaneous porometric and micrometeorological measurements. Plant, Cell and Environ., 13:27–37.

Green, R. O.; Eastwood, M. L.; Sarture, C. M.; Chrien, T. G.; Aronsson, M.; Chippendale, B. J.; Faust, J. A.; Pavri, B. E.; Chovit, C. J.; Solis, M.; Olah, M. R.; and Williams, O. (1998). Imaging Spectroscopy and the Airborne/Infrared Imaging Spectrometer (AVIRIS). Remote Sensing of Environment., 65:227–248.

Green, R. O. and Roberts, D. A. (1995). Vegetation species copmposition and canopy architecture information expressed in leaf water absorption measured in the 1000 nm and 2200 nm spectral region by an imaging spectrometers. Proceedings, 5th Annual JPL Airborne Earth science Workshop, Pasadena, California. NASA JPL Publication 95-1, vol 1.

Guillevic, P., and Gastellu-Etchegorry, J. P. (1999) Modelling BRDF and radiation regimes of tropical and boreal forests, part II:PAR regime. Remote Sensing of Environment, 68:317–340.

Gutelius, G., W. E. Carter, R. L. Shrestha, E. Medvedev, R. Gutierez, and J. G. Gibeaut (1998) 'Engineering Applications of Airborne Scanning Lasers: Reports from the Field,' Journal of American Society for Photogrammetry and Remote Sensing, Vol. LXIV, No. 4, pp. 246–253, 1998.

Hallett, R. A., Martin, M. E., Hornbeck, J. W. (1997). Predicting elements in white pine and red oak foliage with near infrared reflectance spectroscopy. J. Near Infrared Spect. 5:77–82.

Harrell P. A., Kasischke E. S., BourgeauChavez L. L., (1997), 'Evaluation of approaches to estimating aboveground biomass in southern pine forests using SIR-C data', Remote Sensing of Environment 59: 223–233,

Harrison, B. A. and Jupp, D. L. B. (1998). Introduction to Remotely Sensed data. Part One Resource manual. CSIRO Publications, Melbourne, Australia.

Hanssen, R. F. (2001). *Radar Interferometry: Data Interpretation and Error Analysis*. Kluwer Academic Publishers. Dotrecht, The Netherlans.

Held, A., Phinn, S., Scarth, P., Stanford, M. Ticehurst, C., Hartini, S. and Lymburnber, L. (2001). Hyperspectral Mapping of Rainforests and Mangroves. In: Proceedings of the International Geosciences and Remote Sensing Symposium July 9–13, 2001, Sydney, CD-ROM Proceedings, IEEE-Piscataway NY, USA.

Held, A. A. and Billings, S. (1996). Automatic Tree Crown Recognition and Counting from high resolution casi data. Proceedings, 7th Australasian Remote Sensing Conference, Sydney – Australia. July 1996.

Held, A. A., D. Pidsley, T. J. Hatton and D. L. B. Jupp. (1995). Use of Landsat TM data to scale measurements of evapotranspiration in the Northern Territory of Australia. Proceedings of the 2nd North Australian Remote Sensing and GIS Forum. Darwin, July 1995, Darwin, Australia. AURISA Monograph No. 11.

Hoekman, D. H. and Quiñones, M. J. (2000). Land cover type and biomass classification using AirSAR data for evaluation of monitoring scenarious in the Colombian Amazon. IEEE Transactions on Geoscience and Remote Sensing. March 2000.

Horler, D. N. H., Dockray, M., Barber, J. (1983) The red-edge of plant leaf reflectance. International Journal of Remote Sensing, 4:273–289.

Hunt, G. R. and Ashley, R. P., (1979). Spectra of altered rocks in the visible and near-IR. Economic Geology 74:1613-1629.

Imhoff, M. L., Johnson, P., Carson, S., Lawrence, W., Condit, R., Stutzer, D. and Wright, J. (2001). VHF radar mapping of forest biomass in Panama. In: Proceedings of the International Geosciences and Remote Sensing Symposium July 9–13, 2001, Sydney, CD-ROM Proceedings, IEEE-Piscataway NY, USA.

Imhoff, M. (1998). A low frequency radar experiment for vegetation biomass measurement. IEEE Transactions on Geoscience and Remote Sensing, 36:1988–1991.

Imhoff, M. L. (1995a). A theoretical analysis of the effect of forest structure on synthetic aperture radar backscatter and the remote sensing of biomass. IEEE Transactions on Geoscience and Remote Sensing 33:341–351.

Imhoff, M. L. (1995b). Radar backscatter and biomass saturation: Ramifications for global biomass inventory. IEEE Transactions on Geoscience and Remote Sensing. 33:511–518.

Jacquemoud, S., Bacour, C., Poilvé, H., Frangi, J.-P. (2001) Comparison of four radiative transfer models to simulate plant canopies reflectance – Direct and inverse mode. Remote Sensing of Environment. (in press)

Jacquemoud, S., and Baret, F, (1990). PROSPECT: a model of leaf optical properties spectra. Remote Sensing of Environment, 34:75–91.

Jupp, D. L. B. and Kalma, J. D. (1989). Distributing evapotranspiration in a catchment using airborne remote sensing. Asian-Pacific Remote Sensing Journal. 2:13–25.

Kaufman, Y. J., Justice, C. O., Flynn, L. P., Kendall, J. D., Prins, E. M., Giglio, L., Ward, D. E., Menzel, W. P., and Setzer, A. W., (1998), Potential global fire monitoring from EOS-MODIS. Journal of Geophysical Research, 103:32215–32238.

Kite, G. W. and Droogers, P. (2000). Comparing evapotranspiration estimates from satellites, hydrological models and field data.

Kraus, K., Pfeifer, N. (1998). Determination Of Terrain Models In Wooded Areas With Airborne Laser Scanning Data. ISPRS J. Photogramm. Remote Sensing 53 (4), 193–203.

Kruse, F. A., (1999), Visible/Infrared Sensors and Case Studies, Chapter 11, in Rencz, A., (ed.), Remote Sensing for the Earth Sciences, Manual of Remote Sensing, Volume 3, p. 567 -606.

Kustas, W. P., Perry, E. M., Doraiswamy, P. C., Moran, M. S. (1994). Using satellite remote sensing to extrapolate evapotranspiration estimates in ime and space over a semiarid rangeland basin. Remote Sensing of Environment., 49:275–286.

Lambin, E. F., Strahler, A. H. (1994). Indicators of land cover change for change-vector analysis in multitemporal space at coarse spatial scales. International Journal of Remote Sensing. 15:2099–2119.

Lambin, E. (2000). Land cover assessment and monitoring. In Encyclopedia of analytical chemistry. J. Wiley. Sept. 2000. P. 2311.

Landsberg, J. J. and Coops, N. C. (1999) Modelling forest productivity across large areas and long periods. Nat. Res. Model. 12:1–28.

Leckie, D. G. (1990) Advances in remote sensing technologies for forest surveys and management. Can. J. For. Res. 20:464–483.

Lefsky, M., Cohen, W., Acker, S., Parker, G., Spies, T. and Harding, D., (1999). Lidar remote sensing of biophysical properties and canopy structure of forests of Douglas-fir and western hemlock. Remote Sensing of Environment 70, pp. 339–361.

Liang, S., A. H. Strahler, M. J. Barnsley, C. C. Borel, S. A. W. Gerstl, D. J. Diner, A. J. Prata, and C. L. Walthall, (2000), Multiangle Remote Sensing: Past, Present and Future, Remote Sens. Rev., 18, 83–102.

Lucas, R., Held, A., Phinn, S. (2002) 'Remote sensing of tropical rainforests,' *In: Manual of Remote Sensing Volume 4 – Remote Sensing of Natural Resources*, Editor S. Ustin, American Society for Photogrammetry and Remote Sensing, Bethesda, Maryland, United States, 90p., (in press).

Lucas, R. M., Honzak, M., Curran, P. J., Foody, G. M., Mline, R., Brown, T. and Amaral, S. (2001). The regeneration of tropical forests within the Legal Amazon. *International Journal of Remote Sensing*.

Lucas, R. M., Honzak, M., Foody, G. M., Curran, P. J., Corves, C. (1993). Characterising tropical secondary forests using multi-temporal landsat sensor imagery. International Journal of Remote Sensing. 14:3061–3067.

Luckman A, Baker J, Kuplich TM, (1997). A study of the relationship between radar backscatter and regenerating tropical forest biomass for spaceborne SAR instruments, Remote Sensing of Environment 60: (1) 1–13.

Malingreau, J. P., Achard, F., D'Souza, G., Stibig, H. J., DSouza, J., Estreguil, C., Eva, H. (1995). AVHRR for global tropical forest monitoring: The lessons of the TREES project. Remote Sens., Review. 12:29–40.

Manual of Remote Sensing. Second Edition (1983). Volume 1: Theory, Instruments and Techniques. David Simonett (ed.). American Society of Photogrammetry and Remote Sensing. Bethesda, Maryland, USA.

Manual of Remote Sensing. Second Edition (1983). Volume 2: Interpretation and Applications. John E. Estes (ed.). American Society of Photogrammetry and Remote Sensing. Bethesda, Maryland, USA.

Martin, M. E., S. D. Newman, J. D. Aber, and Congalton, R. G. (1998) Determining forest species composition using high spectral resolution remote sensing data. Remote Sensing of Environment. 65:249–254.

Martin, M. E., and Aber, J. D. (1997). Estimation of forest canopy lignin and nitrogen concentration and ecosystem processes by high spectral resolution remote sensing. Ecol. Appl. 7:431–443.

Martin, ME and JD Aber (1996). Estimating canopy characteristics as inputs for models of forest carbon exchange by high spectral resolution remote sensing. *In* The use of remote sensing in the modeling of forest productivity. eds. Gholz, HG, K Nakane and H Shimoda. Kluwer Academic, Dordrecht, The Netherlands, pp 61–72.

McKenzie, N. J. and Ryan, P. J. (1999). Spatial prediticons of soil properties using environmental correlation. Geoderma 89:67–94.

McVicar, T. R., and Bierwith, P. (2001). Rapidly assessing the 1997 drought in Papua New Guinea using composite AVHRR imagery. International Journal of Remote Sensing, 22:2109–2128.

McVicar, T. R. and Jupp, D. L. B. (2000) Using covariates to spatially interpolate moisture availability in the Murray-Darling Basin: a novel use of remotely sensed data. Remote Sensing of Environment. (in press).

Menenti, M. (2000). Chapter 8: Evaporation. *In: Schultz, G. A. and Engman, T. E. (eds.), Remote sensing in hydrology and water management. P. 157.* Springer Verlag, New York.

Merton, R. (1999). Monitoring community hysteresis using spectral shift analysis and the red-edge vegetation stress index. Proceedings, 8th AVIRIS Earth Science Workshop, JPL, Pasadena, California.

Miller, J. R., Jiyou Wu, Boyer, M. G., Belanger, M. and Hare, E. W. (1991). Seasonal patterns in leaf reflectance red-edge characteristics. International Journal of Remote Sensing. 12:1509–1523.

Minty, B. R. S. (1997). Fundamentals of gamma ray spectroscopy. Journal of Australian Geology and Geophysics. 17:39–38.

Monteith, J. L., (1972). Solar radiation and productivity in tropical ecosystems. J. Appl. Ecology, 9, 747–766

Moore, R. K. (1983). Chapter 10. Imaging Radar Systems. *In: Manual of Remote Sensing. Second Edition Volume 1: Theory, Instruments and Techniques.* David Simonett (ed.). American Socienty of Photogrammetry and Remote Sensing. Bethesda, Maryland, USA.

Moran, M. S., Jackson, R. D., Raymond, L. H., Gay, L. W. and Slater, P. N. (1989). Mapping surface energy balance components by combining Landsat thematic mapper and ground-based meteorological data. Remote Sensing of Environment. 30:77–87.

Mumby, P. J., Green, E. P., Edwards, A. J. and Clark, C. D. (1999). The cost-effectiveness of remote sensing for tropical coastal resources assessment and management. Journal of Environmental Management 55:157–166.

Nelson, R. F., R. G. Oderwald, and T. G. Gregoire, (1997). Separating the ground and airborne sampling phases to estimate tropical forest basal area, volume, and biomass, Remote Sensing of Environment, 60 (3), 311–326.

Nemani, R. R., Pierce, L., Running, S. W., Goward, S. (1993). Developing satellite-derived estimates of soil moisture status. J. Appl. Meteorol. 32:548–557.

Nemani, R. R. and Running, S. W. (1989). Estimation of regional resistance to evapotranspiraion from NDVI and thermal–IR AVHRR data. J. Appl. Meteorol. 28:276–284.

Nichol., C. J., Huemmrich, K. F., Black, T. A., Jarvis, P. G., Walthall, C. L., Grace, J., Hall, F. G. (2000). Remote sensing of photosynthetic-light-use efficiency of boreal forest. Agric. For. Meteorol., 191:131–142.

Nilsson, M. (1996). Estimation of tree heights and stand volume using an airborne lidar system. Remote Sensing of Environment, 56, 1–7.

Njoku, E. G. and Li, L. (1999) Retrieval of Land Surface Parameters Using Passive Microwave Measurements at 6–18 GHz. IEEE Transactions on Geoscience and Remote Sensing. 17:79–93.

Ollinger, S. V., Smith, M. L., Martin, M. E., Hallett, R. A. Aber, J. D., Goodale, C. L. (2001). Regional variation in foliar chemistry and soil nitrogen status among forests of diverse history and composition. Ecology, *in press.*

Owe, M., De Jeu, R. A. M., and Walker, J. (2001) A Methodology for Surface Soil Moisture and Vegetation Optical Depth Retrieval Using the Microwave Polarization Index. IEEE Transactions on Geoscience and Remote sensing August (In Press).

PawU, K. T., Suchanek, T. H., Ustin, S. L., Chen, J., Hsiao, T., Shaw. R., Falk, M., King, T., Pyles, R. D., Matista., A. (2000). A new view of carbon flux in old-growth forests, Science (submitted).

Peñuelas, J. and Filella, I. (1998). Visible and near-infrared reflectance techniques for diagnosing plant physiological status. Trends in Plant Science 3:151–156.

Peterson, D. L. (2000) NIR spectroscopy in space? NIR News. 11:10–12.

Pickup, G., Marks, A. (2000). Identifying large-scale erosion and deposition processes from airborne gamma radiometrics and digital elevation models in a weathered landscape. Earth Surface Processes and Landforms. 25:535–557.

Prata, A. J., Caselles, V., Coll, C., Ottle, C., and Sobrino, J. (1995) Thermal remote sensing of land surface temperature from satellites: current status and future prospects. Remote Sensing Reviews, 12:175–224.

Price, J. C. (1990). Using spatial context in satellite data to infer regional scale evapotranspiration. IEEE Transactions Geoscience and Remote Sensing 28:940–948.

Price, J. C. (1982). On the use of satellite data to infer surface fluxes at meteorological scales. J. Appl. Meteorol. 21:1111–1122.

Raupach, M. R. and Barret, D. (2002) *In*: Vertessy, R. A. and Elsenbeer, H. (eds.) The State of the Art in Forest Hydrology. Union of Forestry Research Organizations (IUFRO).

Raupach, M. R. (1998). Influences of local feedbacks on land-air exchanges of energy and carbon. Global Change Biology. 4, 477–494.

Raupach, M. R. and Finnigan, J. J., (1995). Scale issues in boundary layer meteorology: surface energy balances in heterogeneous terrain. Hydrological Processes 9, 589–612.

Rawson, H. H., Begg, J. E., and Woodward, R. G. (1977). The effect of atmospheric humidity on photosynthesis, transpiration and water use efficiency of leaves of several plant species. Planta, 134:5–10.

Reich, P., Walters, M., Ellsworth, D. (1992). Leaf life-span in relation to leaf, plant and stand characteristics among diverse ecosystems. Ecological Monographs 62:365–392.

Roberts, D. A., Dennison, P., Ustin, S., Reith, E., Morais, M. (1999). Development of a regionally specific library for the santa monica mountains using high resolution AVIRIS data. Proceedings, 8th AVIRIS Earth Science Workshop, JPL, Pasadena, California.

Roberts, D. A., Batista, G., Pereira, J., Waller, E., Nelson, B. (1998). Change identification using multitemporal spectral mixture analysis: Applications in eastern amazonia, Chapter 9 in Remote Sensing Change Detection: Environmental Monitoring Applications and Methods. Elvidge., C and Lunetta, R., (EDS.), Ann Arbor Press, Ann Arbor MI, 137–161.

Roberts, D. A., Brown, K. J., Green, K. J., Ustin, S., Hinckley, T. (1998). Investigating the relationship between liquid water and leaf area in clonal populus. Proceedings 7th JPL Airborne Earth Science Workshop. January 1998, JPL – Publication 97–21, Pasadena, California.

Rock, B. N., Vogelman, J. E., Williams, D. L., Vogelman, A. F., Hoshizaki, T. (1996). Remote detection of forest damage. BioScience 36:439–445.

Roderick, M. L., Noble, I. R., Cridland, S. W. (1999). Estimating woody and herbaceous vegetation cover from time series satellite observations. Global Ecology and Biogeography 8:501–508.

Rosen, P. A., Hensley, S., Joughin, I. R., (2000). Synthetic aperture radar interferometry – Invited paper', Proceedings of the IEEE 88: (3) 333–382.

Rosenqvist, A., Shimada, M., Chapman, B., (2000). The Global Rain Forest Mapping project – a review', International Journal of Remote Sensing 21: (6–7) 1375–1387.

Running, S. W., P. E. Thornton, R. R. Nemani, J. M. Glassy. (2000) Global Terrestrial Gross and Net Primary Productivity from the Earth Observing System. In: Methods in Ecosystem Science, O. Sala, R. Jackson, and H. Mooney Eds. Springer-Verlag New York.

Running, S. W., Baldocchi, D. D., Turner, D. P., Gower, S. T., Bakwin, P. S., Hibbard, K. A. (1999). A global terrestrial monitoring network integrating tower fluxes, flask sampling, ecosystems modelling and EOS satellite data. Remote Sensing of Environment. 70:108–127.

Running, S. R. and Nemani, R. R. (1988). Relating seasonal patterns of the AVHRR vegetation index to simulated photosynthesis and transpiration of forests in different climates. Remote Sensing of Environment. 24:347–367.

Running, S. W. and Coughlan, J. C. (1988). A general model of forest ecosystem processes for regional applications. 1. Hydrologic balance, canopy gas exchange and primary production processes. Ecological Modelling, 42, 125–154.

Ryan, P. J., McKenzie, N. J., O'Connell, D., Lounghead, A. N., Leppen, P. M., Jacquier, D., Ashton, L. (2000). Integrating forest soils information across scales: spatial prediction of soil properties under Australian forests. Forest Ecology and Management 138:139–157.

Saatchi SS, Nelson B, Podest E, *et al.* (2000), Mapping land cover types in the Amazon Basin using 1 km JERS-1 mosaic, International Journal of Remote Sensing 21: (6–7) 1201–1234.

Saatchi, S. S., and Moghaddam, M., (2000). Estimation of crown and stem water content and biomass of boreal forest using polarimetric SAR imagery, IEEE Transactions on Geoscience and Remote Sensing 38: (2) 697–709 Part 1.

Sabol, D., E., Smith, M. O., Adams, J. B., Zukin, J. H., Tucker, C. J., Roberts, D. A., Gillespie, A., R. (1995). Proceedings, 5th Annual JPL Airborne Earth science Workshop, Pasadena, California. NASA JPL Publication 95–1, vol 1.

Schmugge, T. J. (1983). Remote sensing of soil moisture: Recent Adavnces. IEEE Transactions on Geoscience and Remote Sensing. GE–21(3):33–344.

Schmugge, T. J. (1985) Remote sensing of soil moisture, *In* Hydrological Forecasting, M. G. Anderson and T. P. Burt (Eds), John Wiley, New York.

Schultz, G. A. and Engman, T. E. (2000). Remote sensing in hydrology and water management. Springer Verlag, New York.

Schulze, E.-D. and Hall, A. E. (1982). Chapter 7. Stomatal responses, water loss and CO_2 assimilation rates of plants in contrasting environments. in: Encyclop. of Plant Physiol 12B. Physiological Plant Ecology II. O. L. Lange, P. S. Nobel, C. B. Osmond and Zeigler, H. (eds.) Springer Verlag, Berlin.

Sellers, P. J., Los, S. O., Tucker, C. J., Justice, C. O., Dazlich, D. A., Collatz, G. J., and Randall, D. A. (1996). A revised land surface parameterisation (SiB-2) for atmospheric GCMs. Part 2: The generation of global fields of terrestrial biophysical parameters from satellite data. Journal of Climate, 9:706–737.

Sellers, P. J., (1987). Canopy reflectance, photosynthesis, and transpiration. II. The role of biophysics in the linearity of their interdependence. Remote Sensing of Environment. 21:143–183.

Seguin, B., Courault, D., Guerif, Martine. (1994). Surface temperature and evapotranspiration: Application of local scale methods to regional scales using satellite data. Remote Sensing of Environment. 49:287–295.

Siqueira, P., Chapman, B., Saatchi, S., Freeman, A., (1997). Amazon rainforest visualization/classification by orbiting radar, enabled by supercomputers (ARVORES). In: Proceedings of the International Geosciences and Remote Sensing Symposium, Singapore. July, 1997, IEEE-Piscataway NY, USA.

Skidmore, A. K., Varekamp C,. Wilson, L., Knowles, E., Delaney, J., (1997). Remote sensing of soils in a eucalypt forest environment. International Journal of Remote Sensing 18:39–56.

Skole, D. L. and Tucker, C. J. (1993). Tropical deforestation and habitat fragmentation in the Amazon: satellite data from 1978 to 1988. Science, 260, 1905–10.

Smith, M. L., Ollinger, S. V., Martin, M. E., Aber, J. D., Hallett, R. A., Goodale, C. L. (2001). Direct prediction of aboveground forest productivity by hyperspectral remote sensing of canopy nitrogen. Oecologia (in review).

Snowdon, P., and Benson, M. L. (1992). Effects of combinations of irrigation and fertilization on the growth and above-ground biomass production of *Pinus radiata*. Forest Ecology and Management 52:87–116.

Stone, T. A., Schlesinger, P., Houghton, R. A. and Woodwell, G. M., (1994). A map of the vegetation of South America based on satellite imagery. Photogrammetric Engineering and Remote Sensing, Vol 60, 541–551

Sun, J., Mahrt, L. (1994). Spatial distribution of surface fluxes estimated from remotely sensed variables. J. Appl. Meteorology 33:1341–1353.

Tenhunen, J. D., Pearcy, R. W., Lange, O. L. (1987) Diurnal variations in leaf conductance and gas-exchange in natural environments, pp. 323–351. In: Zeiger, E., farquhar, G. D. and Cowan, I. R. (eds.) Stomatal Function. Stanford University Press. Stanford California.

Thayer,S.S. and Björkman, O. (1990). Leaf xanthophyll content and composition in sun and shade determined by HPLC. Photosynthesis Research 23:331–343.

Ticehurst, C., Lymburner, L., Held, A., Palylyk, C., Martindale, D., Sarosa, W., Phinn, S., and Stanford, M. (2001). Mapping tree crowns using hyperspectral and high spatial resolution imagery. Proceedings Third International Conference on Geospatial Information in Agriculture and Forestry, Denver, Colorado, 5–7 November 2001.

Treuhaft, R. N., Madsen, S. N., Moghaddam, M., van Zyl, J. (1996). Vegetation characteristics and underlying topography from interferometric radar. Radio Science 31:1449–1485.

Tucker, C. J. (1979). Red and photographic infrared linear combinations monitoring vegetation. Remote Sensing of Environment, 8:127–150.

Ulaby, F. T., and Elachi, C., (eds). (1990). Radar Polarimetry for Geoscience Applications, Artech House.

Ulaby, F. T., R. K. Moore, and A. K. Fung. (1986). Microwave Remote Sensing, Volume III, Artech House.

Ustin, S. L., Roberts, D. A., Pinzon, J. E., Jacquemoud, S., Scheer, G., Castaneda, C. M., and Palacios, A. (1998). Estimating canopy water content of chaparral shrubs using optical methods. Remote Sensing of Environment. 65:280–291.

Ustin, S. L., Smith, M. O. and Adams, J. B. (1993). Remote sensing of ecological processes: A strategy for developing and testing ecological models using spectral mixture analysis. *In*: Scaling Physiological Processes: Leaf to Globe. Academic Press Inc.

Van der Sanden, J. J. and Hoekman, D. H. (1999) Potential of Airborne radar to support the assessment of land cover in a tropical rainforest environment. Remote Sensing of Environment. 68:26–40.

Van Zyl, J. (1993). The effect of topography on radar scattering from vegetated areas. IEEE Transactions on Geoscience and Remote Sensing. 31:153–160.

Varekamp, C., Hoekman, D. H., (1998). An inversion algorithm for automatic retrieval of tree crown characteristics from high-resolution interferometric SAR data. Proc. 2nd Intl. Workshop on Retrieval of bio- and Go-physical parameters from SAR data for land applications. 21–23 October, 1998. ESTEC, Noordwik, The Netherlands.

Vertessy, R. A., Watson, F. G. R. and O'Sullivan, S. K. (2001). Factors determining relations between stand age and catchment water yield in mountain ash forests. Forest Ecology and Management, in press.

Vertessy, R. A., Hatton, T. J., O'Shaughnessy, P. J., Jayasuriya, M. D. A., (1993). Predicting water yield from a mountain ash forest catchment using a terrain analysis based catchment model. Journal of Hydrology. 150:665–700.

Walker, J. P., Kurera, D. F., and Houser, P. R. (2000). A field evaluation of ERS-2 synthetic aperture radar data for soil moisture measurement, EOS, Transactions on the American Geophysical Union, 81:S207.

Wang, J. R., Schmugge, T. J. (1980). An empirical model for the complex dielectric permittivity of soils as a function of water content. IEEE Transactions on Geoscience and Remote Sensing, 18:288–295.

Watson, F. G. R., Vertessy, R. A., McMahon, T. A., Rhodes, B. and Watson, I., (2001). Improved methods to assess water yield changes from paired catchment studies: Application to the Maroondah catchments. Forest Ecology and Management, in press.

Watson, F. G. R., Vertessy, R. A., Grayson, R. B. (1999) large-scale modelling of forest hydrological processes and their long-term effects on water yield. Hydrological Processes 13:689–700.

Wehr, A. and U. Lohr, (1999). Airborne laser scanning – an introduction and overview, ISPRS Journal of Photogrammetry and Remote Sensing, 54 (2–3), 68–82.

Weishampel, J. F., Blair, J. B., Knox., R. G., Dubayah, R., Clark, D. B. (2000) Volumetric lidar return patterns from and old-growth tropical rainforest canopy. International Journal of Remote Sensing 21:409–415.
Wendlandt, W. W., and H. G. Hecht, (1966). Reflectance Spectroscopy, Interscience Publisher, New York, 298p.
Wessman, C. A. (1994) Estimating canopy biochemistry through imaging spectrometry. *In*. Imaging Spectrometry – a Tool for Environmental Observations, pp. 57–70 J. Hill and J. Megier (eds.) Kulwer Academic Publishers, Dotrecht.
Wessman, C. A. (1989). Evaluation of Canopy Biochemistry. *In*: Remote Sensing of Biosphere Functioning. R. J. Hobbs and H. A. Mooney (Eds.). Springer Verlag, New York. pp. 65–86.
Wickland, D., (1989). Future directions for remote sensing in terrestrial ecological research. In Theory and Applications of Optical Remote Sensing. G. Asrar (Ed.), Wiley – New York, 691–724.
Wilford, J., Bierworth, P. N., Craig, M. A. (1997). Application of airborne gamma-ray spectrometry in soil/regolith mapping. Australian Geological Survey Organization Journal of Australian Geology and Geophysics 17:201–216.
Zarco-Tejjada, P. J., Miller, J. R., Noland, T. L., Mohammed, G. H., Sampson, P. H. (2001). Scaling-up and model inversion methods with narrowband optical indices for chlorophyll content estimation in closed forest canopies with hyperspectral data. IEEE Transactions on Geoscience and Remote Sensing. 39:1491–1507.

28 Detecting change in river flow series

Z. W. Kundzewicz
Research Centre for Agricultural and Forest Environment, Polish Academy of Sciences, Poznań, Poland

A. J. Robson
Centre for Ecology and Hydrology, Wallingford, UK

INTRODUCTION

Detection of trends in long time series of hydrological data is of paramount scientific and practical importance. Water resources systems are typically designed and operated based on the assumption of stationary hydrology (in particular, an assumption of stationarity of the stochastic proces of river stage or discharge). If this assumption is incorrect then existing procedures for example in the design of levees, dams and reservoirs will have to be revised. Without revision, the systems can be over- or underdesigned and either not serve their purpose adequately or be overly costly.

Studies of change are also of importance because of our need to understand the impact that man is having on the 'natural' world. Changes caused directly by man (deforestation, land-use changes, changes in agricultural practices, drainage systems, dam construction, water abstraction, river regulation, urbanisation, etc.) or indirectly via emissions of greenhouse gases, are just a few examples of anthropogenic activities that may be altering important aspects of the hydrological cycle. In addition, natural catchment changes, such as to the channel morphology, can also occur.

The search for climate change signatures in hydrological data has been of much interest recently, driven by the possibility of climate change causing more frequent and severe floods in the future. There are several non-climate mechanisms which may contribute to this effect. Some of them relate to the anthropogenic pressures such as reduction in water resources storage capacity, acceleration of flow in water courses, plus those arising from increasing populations and wealth accumulated in endangered areas. There are also several climate change related mechanisms which seem plausible, such as more frequent intense precipitation events, more frequent mid/high-latitude winter wet spells, more intense mid-latitude storms, higher peak wind intensities, higher mean/peak rainfall intensities and more frequent tropical cyclones, more El Niño-like mean state of ENSO and more intense El Niños (IPCC, 2001). Confrontation of prognostications for the future, increasingly warmer and wetter world, where an accelerated hydrological cycle will be observed – and the message given by the data already observed – is a challenging study area.

Significant changes in river flow in the humid tropics areas may be attributed to changes in land use, such as vegetation change, logging, clearfelling, deforestation and reforestation, plantation, agroforestry (e.g. rubber tree and oil palm plantations), urbanisation, road construction, river regulation, dam construction, flow diversions, flow augmentation, groundwater pumping, natural (and/or man-influenced) disturbances – wildfire, water abundance (floods). Part III of this volume includes several contributions which outline the impacts of forest disturbance, conversion and restoration on streamflow; mostly, but not all, at headwater scales.

The main stages in an analysis of change in long time series of hydrological data are as follows:

- Acquisition and preparation of a suitable dataset
- Exploratory analysis of the data
- Application of statistical tests
- Interpretation of the results

This chapter outlines these components, discussing and explaining in general terms the various activities involved to help the reader understand the arsenal of modern tools available for detecting changes in long time series of river flow data, with recommendations on how to approach detection of change.

The analysis of change in the context of humid tropics hydrology is also considered: in particular, the problems associated with detecting climate change and the ongoing problems with data availability and data quality.

Forests, Water and People in the Humid Tropics, ed. M. Bonell and L. A. Bruijnzeel. Published by Cambridge University Press.

DATA

There is a huge variety of possible types of hydrological data. These can be collected at a range of temporal intervals: continuously, hourly, daily, monthly, annually, or irregularly. Data records may contain either instantaneous values (e.g. flow, stage) or totals, averages or some other summary measure for a time interval such as annual precipitation. Data may also pertain to different spatial scales, from point or experimental plot to large areas (including the Globe). All such data can be analysed for existence of change.

A checklist of issues to be taken into account when considering what data to use includes:

(1) *Quality of data.* Data should be quality controlled *before* commencing an analysis of change. However, even with good quality control, some problems may be missed and it is helpful to be open-minded at any stage of an analysis. Possible sources of heterogeneity due to changes in instruments and observation techniques, for example, should be identified.
(2) *Length of record.* Data series should be as long as possible. For investigation of climate change, a minimum of 50 years of record is suggested. Even 50 years may be not be sufficient if the climate change signal is weak and is set against a background of strong natural variability – as is typical for hydrological data.
(3) *Missing values and gaps.* Missing values and gaps in records are complicating factors. Whether or not to fill missing values and gaps, and if so, in what way, is a complex issue. Particular care is needed if the gaps are non-random as when equipment is damaged during an exceptional event.
(4) *Frequency of data.* Processing very frequent data is computationally intensive. Furthermore, such data may be highly correlated from one time step to the next. In many instances it can be worth simplifying the record by reducing the frequency (aggregating) or using summary measures.
(5) *Use of summary measures.* It may be worth using summary measures of the data, e.g. maxima, averages, variability over a period.
(6) *Use of transformation.* Use of a data transformation can be useful to compensate for undesirable properties such as where data are highly skewed or non-normal. Commonly-used transformations include logarithms, ranks and normal scores (see below).

EXPLORATORY DATA ANALYSIS

Exploratory data analysis (EDA) is a very powerful graphical technique (cf. Cleveland, 1993, 1994) that should be a key component of any analysis of change in hydrological records. Its essence is to allow the raw data to speak for themselves. EDA is a technique that enables the analyst to display obvious features of the data, to detect hidden patterns and to explore and understand the data better. It is an iterative process, which should be used at several stages of analysis. Examining the raw data may identify interesting aspects, such as seasonality, which in turn invite further investigation. Or, having fitted a trend to the data, EDA can be used to examine the residuals to check for independence. Finally, EDA can provide a very valuable means of presenting both the data and the results in a way that maximises understanding and impact. An exposition of how EDA may be used for change detection in hydrological data is given in Grubb and Robson (2000) who presented examples of time-series plots (simple and multiple), scatter plots and spatial plots. They also dealt with seasonal variation (smoothing seasonal data, deciphering seasonal pattern, seasonal decomposition, and seasonal sub-series) and residual analysis and checking test assumptions (about the distribution and about independence).

APPLYING STATISTICAL TESTS

This section provides a brief summary of ways of selecting and applying statistical tests. The section is deliberately concise, and is a summary of the material presented in Robson *et al.* (2000). Further details of various aspects of statistical analyses are provided in Appendices 28.1–28.4. An overview of some of the important statistical principles used in change detection is given in Appendix 28.1.

It is helpful to view the choice of a statistical test as being composed of two parts: (1) selecting the test statistic, and (2) selecting a method for determining the significance level of the test statistic. By viewing the process in these two parts it becomes possible to separate out the issue of how to select a test statistic from that of how to evaluate the significance level.

Selecting the test statistic

There are many procedures that can be used to test for change. For most studies, it is recommended that more than one of these tests should be used. Different test statistics tend to detect different aspects of change; for example, there are those that look for an abrupt jump in the data (step-change) and those that detect steady, gradual change (trend). The main criteria to be aware of when selecting test statistics are

- type of change that is of interest (e.g. trend, step-change)
- power – more powerful tests are to be preferred
- including a variety of types of test statistic
- whether the test is for a known or unknown change-point time.

A short review of common useful tests is given in Appendix 28.2. Any of the test statistics applied in these procedures may be used. Additionally, if resampling techniques are to be used (see below) then it is also possible to construct new test statistics to evaluate any specific type of change.

Selecting the method for determining significance level

Hydrological series are rarely normally distributed and are often highly skewed. It is therefore recommended that distribution-free methods, i.e. those that do not make assumptions about the underlying distribution of the data, are used for evaluating significance levels (see also under *understanding assumptions* below).

The main types of distribution-free approaches include:

- rank based tests
- normal scores
- resampling methods (permutation and bootstrapping).

Further details on these approaches are given in Appendices 28.3 and 28.4. Of these methods, the first two are the most straightforward to apply but are less flexible than resampling methods. In particular, resampling methods allow significance levels to be estimated for any choice of test statistic. This means that traditional statistical tests can be adapted for application to hydrological series by extracting the test statistic but using resampling methods to determine significance.

The main steps in obtaining the significance level for a selected test statistic using resampling methods are as follows:

- calculate the test statistic for the observed data
- resample the data many times (e.g. 1000) to generate new data series
- for each generated data series, recalculate the test statistic
- estimate the significance level by comparing the observed test statistic with the generated test statistics.

Further details of resampling methods are given in Appendix 28.4.

It is recommended that resampling methods be used more broadly for testing hydrological data and are also particularly recommended for where there is serial dependency (autocorrelation – see next section) in the data. In this case, block resampling methods can be used (see Appendix 28.4).

UNDERSTANDING ASSUMPTIONS

Three types of assumption are commonly made when carrying out statistical tests. It is important to check that these assumptions hold, and where necessary, to choose testing methods that avoid making particular assumptions. Test assumptions are usually needed for the estimation of significance level. Use of resampling approaches for estimating significance levels usually involves making fewer assumptions than many other testing methods. The most common assumptions used in statistical testing are:

(1) *The form of the distribution.* Many classical statistical tests assume that the data being tested are normally distributed. This assumption can be avoided by using distribution-free methods.

(2) *Constancy of the distribution (i.e. all data points have an identical distribution).* Most basic statistical tests assume, under the null hypothesis, that the distribution of the data does not change. This assumption is violated if there are seasonal variations or any other cycles in the data, or if there is an alteration over time in the variance or any other feature of the data that is not part of the test.

If there are seasonal cycles in the data, then the options are either to (a) deseasonalise the data, i.e. estimate the seasonal structure and remove this from the data series, or (b) to use a testing approach that allows for seasonality (e.g. block resampling – see Appendix 28.4).

(3) *Independence.* Data values can be said to be independent if they are completely unrelated to one another. For many hydrological series, this is not the case: knowing the flow in the river today is a guide as to what tomorrow's flow is likely to be. Such series show *autocorrelation* (correlation from one time value to the next: this is also referred to as *serial correlation* or *temporal correlation*). The more frequent the data points, the more likely it is that there will be important serial correlation in the data. As a rough guideline, daily hydrological series are usually strongly correlated, annual series are often approximately independent and monthly values are intermediate.

Most common statistical tests do not allow for serial correlation in the data. If serial correlation is present then possible options are

- use block permutation or block bootstrap methods (see Appendix 28.4)
- decrease the frequency of the data series (e.g. by calculating monthly or annual averages)
- use methods that build in serial correlation (cf. Robson *et al.*, 2000)
- for data with seasonality, consider the modified seasonal Kendall test (Hirsch and Slack, 1984).

Methods for checking assumptions

Assumption checking may need to be carried out both prior to and after application of tests. For example, if a trend is detected, then the trend should be estimated and removed from the data and the residuals checked for autocorrelation and for constancy of distribution.

The principal method for assumption checking is to use visual techniques (see also Grubb and Robson, 2000). For example:

- histograms and normal probability plots – to examine distribution
- time series plots – to spot time dependent patterns or possibly changes in variance
- autocorrelation plots.

If visual methods for assumption checking are not sufficient, then formal tests are also available for checking some assumptions, e.g. tests for normality of data and tests for data independence (cf. Robson *et al.*, 2000).

INTERPRETATION OF THE RESULTS

Great care is often needed when interpreting results. The final interpretation brings together the information gained out of all stages of the analysis. Thus it combines information about how the data were obtained, historical information about the catchment or region, graphical information gained from exploratory data analysis and the statistical test results. It also builds in knowledge gained from other related variables such as rainfall or observations from nearby catchments.

If test results suggest that there is a significant change in a data series, then it is always important to try to understand the cause. Although the investigator may be interested in detecting climate change, there may be many competing (and better) explanations.

Understanding test results

It is important to remember that no statistical test is perfect, even if all test assumptions are met. A 5% significance level means that we are likely to be in error 5% of the time: i.e. if the null hypothesis was true then 1 in 20 test results will lead to incorrect conclusion.

Often, there are many test results to be examined. If the type of trend is not known a priori, one should apply a selection of tests, e.g. a few tests for trend and a few tests for step-change. Interpretation of multiple test results can be complex. In some cases the results from these tests will generally be in agreement, but in other cases there will be differences. Some suggestions for interpretation of such results are (Robson *et al.*, 2000):

- Use visualisation methods. Where there are multiple test results per site, it is easier to interpret tables of results if the tables show symbols rather than figures. If there are multiple sites, then plotting significance levels or size of change on a geographical map is helpful.
- Take care in interpreting significance levels. The presence of a single significant test result may be only weak evidence of change – even if this test is highly significant. If many of the tests are significant then this provides stronger evidence of change. However, if tests are very similar then multiple significant values are not extra proof of change.
- Examine the test results alongside graphs of the data, and with as much historical knowledge about the data as possible. For example, if both step-change and trend results are significant, and historical investigations reveal that a dam was built during the period, and this is consistent with the time series plot, then a reasonable conclusion is that the dam caused a step change.
- Look out for patterns in the results that may indicate further structure e.g. regional patterns in trends. These may suggest the need for and the direction of further investigation.

Gathering additional information

The best way to improve understanding of change is to gather as much information as possible. Examples of additional, and useful, information include (Robson *et al.*, 2000):

- Historical information about changes in the catchment, river engineering, land-use change, etc.
- Historical information about data collection methods, stage-discharge relationship, etc.
- Data from nearby sites – if data from other nearby sites show similar patterns then the cause is probably widespread (e.g. linked to climate or to extensive land-use change).
- Related variables, e.g. information on temperature and rainfall, can help determine whether changes in flow can be explained by climatic factors.
- Proxy data that extend record lengths. A primary problem with many hydrological records is that they are too short. If related data can be obtained that extend to a longer period then this may be of assistance.

DETECTING CLIMATE CHANGE

Climate change is probably the most difficult type of change to detect, yet it is also of great interest that any hydrological effects of climate change are detected and understood. When searching for a climate change signal, it is very important to understand the difference between climate variability and climate change. The former is the natural variation in the climate from one period to the next. The latter refers to a long term alteration in the climate. Climate variability appears to have a very marked effect on many hydrological series. This has two important effects for trend detection:

(1) *Climate variability can cause apparent trend.* Climate variability can easily give rise to apparent trend when records are short – these are trends which would be expected to disappear once more data has been collected. Because of climate variability, records of 30 years or less are almost certainly too short for detection of climate change. It is suggested that *at least 50 years of record is necessary* for climate change detection.
(2) *Climate variability obscures other changes.* Because climate variability is typically large, it can effectively obscure any underlying changes due either to climate change or to anthropogenic causes such as urbanisation.

Choice of stations

The issue of detecting a climate change signature in river flow data is particularly complex because the process of river flow is the integrated result of many variables – precipitation inputs, catchment storage, evaporation losses. Furthermore, climate change signals may well be weak relative to the background natural variability. These factors mean that care is needed in selecting data and sites for use in studying climate change.

To study climate change signature in river flow records, ideally data should be taken from pristine/baseline rivers and should be of high quality and should extend over a long period. Where pristine sites are not available, it may be possible to eliminate other influences or reconstruct natural flows, or to use conceptual flow naturalisation. Detailed suggestions on how to select a network of stations for climate change detection are given in Pilon (2000).

Example studies

There have been a number of studies in different countries, searching for climate change signatures in river flow records. Summary results from two such studies are presented here.

Lins and Slack (1999) studied streamflow trends in the United States, using daily discharge data from 395 climate-sensitive stream gauging stations in the conterminous US, furnishing continuous records over at least a 50-year period (1944–1993). The data used are an update of the Hydro-Climatic Data Network (HCDN), cf. Slack and Landwehr (1992). The methodology used by Lins and Slack (1999) for trend analysis was the non-parametric Mann-Kendall test (cf. Appendix 28.2), examining whether changes are monotonic in time. They found that trends were most prevalent for annual minima to medium flows, and least prevalent in annual maxima. These results were summarised as 'getting wetter, but less extreme' (Lins and Slack, 1999), since streamflow has increased except for the highest quantiles.

To evaluate interdecadal streamflow variability, quantile trends were calculated for 30-, 40-, 50-, 60-, 70- and 80-year periods, all ending in 1993. The number of stations with the longest 80-year data series was 34. Decreases were found only in parts of the Pacific Northwest and the Southeast, see Table 28.1 and Figure 28.1 (from Lins and Slack, 1999). Table 28.1 shows the aggregate statistics illustrating changes of selected quantiles of streamflow. Figure 28.1 presents results of spatial studies of change in flow data, with maps showing (a) trends in annual maximum daily, (b) annual median daily, and (c) annual minimum daily.

Robson and Reed (1996) conducted a comprehensive study of flood records in the UK. They used a database consisting of *c.* 600 stream gauges with long data series (from 15 to over 100 years). These included both rural and urban sites with relatively natural flows. The peak-over-threshold (POT) and the annual maxima approaches were used for testing for step change and trend. Several statistical tests were applied to annual maxima, POT magnitude, POT intervals and POT annual flood count. A range of methods was used including linear regression (and 'normal scores' linear regression), Spearman's rank, distribution-free CUSUM test (cf. Appendix 28.3) and permutation tests (cf. Appendix 28.4). Figure 28.2 (taken from Robson and Reed, 1996) shows a summary measure (trend significance level and direction of change) plotted at the geographical location at each site. There are more incidences of increased flooding than decreasing flooding, particularly in Scotland and in south-east of England. However, the study concluded that many of the changes detected were either not linked to climate (e.g. were due to urbanisation), or were probably due to climate variability rather than climate change and resulted from insufficiently long records.

Overall, most attempts to find a general climate change signature in river flow records of Europe and North America have not led to convincing evidence of climate change. This is despite the availability of relatively long and high quality datasets.

ADVANCING CHANGE DETECTION IN RIVER FLOW DATA FOR THE HUMID TROPICS

The problem of change detection is of much relevance to the humid tropics. There are many possible causes of change either from direct human impacts (see Drigo, this volume), from natural effects or from possible climate change effects.

The main problem in detecting such changes is the relative data scarcity in the humid tropics and in particular the dearth of long time series. In many countries data collection programmes are

Table 28.1. *Aggregate statistics illustrating changes of selected quantiles of streamflow (Mann–Kendall test results)*

	Years of record					
	30	40	50	60	70	80
Beginning year	1964	1954	1944	1934	1924	1914
Ending year	1993	1993	1993	1993	1993	1993
Stations tested	395	395	395	193	70	34
Annual minimum (daily mean) discharge						
No. of significant trends ($p \leq 0.05$)	112	177	163	85	34	13
Percent with significant trends	28	45	41	44	49	38
No. with increasing trend	74	145	127	76	32	10
No. with decreasing trend	38	32	36	9	2	3
Annual 10th percentile of daily discharge						
No. of significant trends ($p \leq 0.05$)	101	155	143	89	32	12
Percent with significant trends	26	39	36	46	46	35
No. with increasing trend	63	132	117	80	30	8
No. with decreasing trend	38	23	26	9	2	4
Annual 30th percentile of daily discharge						
No. of significant trends ($p \leq 0.05$)	109	160	135	81	28	9
Percent with significant trends	28	41	34	42	40	26
No. with increasing trend	76	148	125	79	27	8
No. with decreasing trend	33	12	10	2	1	1
Annual 50th percentile of daily discharge						
No. of significant trends ($p \leq 0.05$)	98	174	116	82	27	10
Percent with significant trends	25	44	29	42	39	29
No. with increasing trend	76	167	113	81	26	9
No. with decreasing trend	22	7	3	1	1	1
Annual 70th percentile of daily discharge						
No. of significant trends ($p \leq 0.05$)	59	130	64	58	19	6
Percent with significant trends	15	33	16	30	27	18
No. with increasing trend	55	124	61	58	19	6
No. with decreasing trend	4	6	3	0	0	0
Annual 90th percentile of daily discharge						
No. of significant trends ($p \leq 0.05$)	43	94	40	20	10	5
Percent with significant trends	11	24	10	10	14	15
No. with increasing trend	18	76	23	19	10	5
No. with decreasing trend	25	18	17	1	0	0
Annual maximum (daily mean) discharge						
No. of significant trends ($p \leq 0.05$)	37	53	35	20	9	4
Percent with significant trends	9	13	9	10	13	12
No. with increasing trend	12	31	14	11	5	2
No. with decreasing trend	25	22	21	9	4	2

Source: Lins and Slack (1999).

weak, decreasing, or even non-existent. Even in those countries where data are being collected, there is a reluctance to share hydrological data. For change detection studies in the humid tropics to advance, it is important that improvements be made to both the collection and ease of access to data.

Improvements in data collection

To obtain more long-term high quality datasets there is a need for more strategic collection of hydrological data and for climate change detection in particular, there needs to be long time series

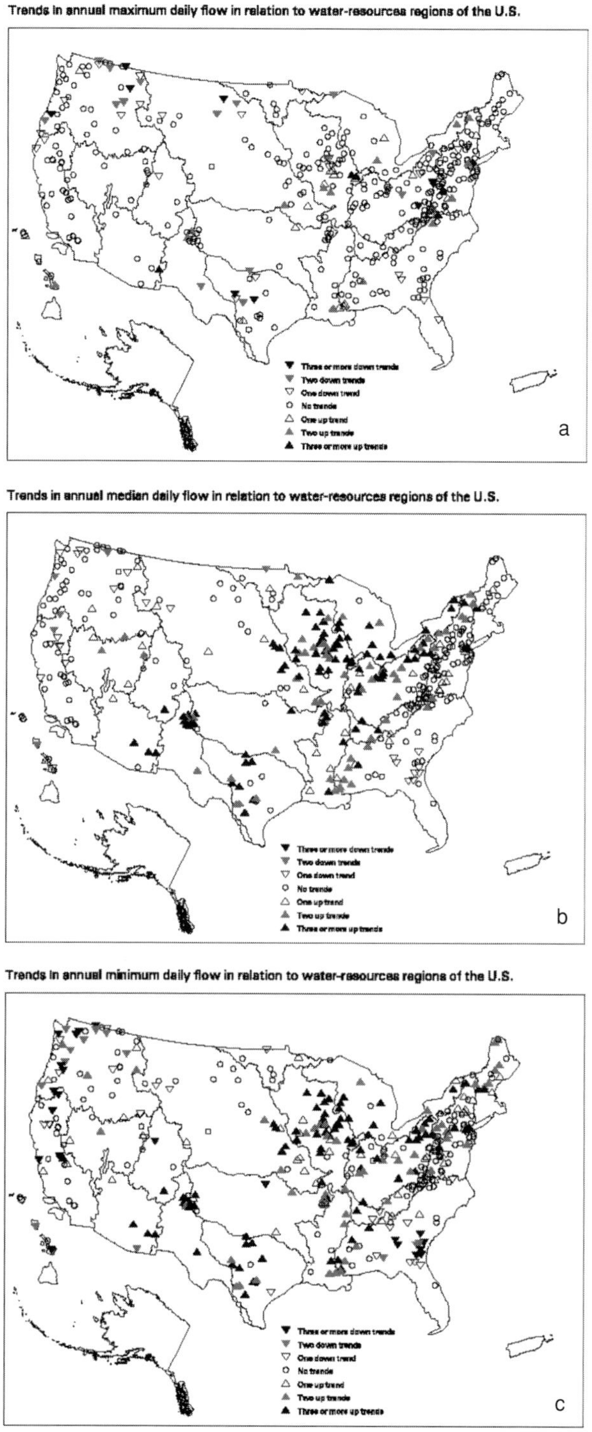

Figure 28.1 Results of spatial studies of change in river flow data (from Lins and Slack, 1999). Upward-pointing triangles indicate increasing flow, downward-pointing decreasing. Open triangles, grey-shaded triangles and solid triangles refer to trends in 1, 2, and 3 or more time periods, respectively. Open circle denotes no trends in any time period. (a) Annual maxima daily, (b) annual median daily, (c) annual minimum daily.

of quality controlled data from little disturbed areas. This latter condition may be difficult to meet because of human pressures and intensive population growth.

Improvements in data access

Even when data are available, access can be a problem. This can arise either because of the strategic importance of hydrological data for some regions, or because of a desire to raise income from the data. Below, we describe three inititatives that aim to get round these problems. Each is worthy of praise and deserves suitable attention and encouragement.

WMO GLOBAL RUNOFF DATA CENTRE

The Global Runoff Data Centre (GRDC), was set up in the German Federal Institute of Hydrology (postal address: Global Runoff Research Center, Federal Institute for Hydrology, Am Mainzer Tor 1, 56068 Koblenz, Germany), as an implementation of a World Climate Programme – Water (WCP-Water) Project A.5 – Collection of Global Runoff Data. Its aim is to help alleviate the data availability problems and to promote the idea of free and unrestricted exchange of hydrological data and products, consistent with the policy of the World Meteorological Organization (WMO). The Centre collects and stores sets of daily and monthly river flow data at the global scale, and generates a range of data products. The Centre has a considerable collection of river flow data related to the humid tropics. Requests for data, and data products, can be made in writing to the GRDC. The charges requested cover the costs of handling, diskettes, packing and postage but could be waived if the requesting body was a contributor of data to GRDC. Similar initiatives for collecting hydrological datasets and making them available for research purposes has been undertaken within the FRIEND (Flow Regimes from International and Experimental Network Data.) Project of UNESCO.

REGIONAL HUMID TROPICS HYDROLOGY AND WATER RESOURCES CENTRE

The Regional Humid Tropics Hydrology and Water Resources Centre for South East Asia and the Pacific was established in Kuala Lumpur (Malaysia) in 1996 with the objective 'to promote a conducive atmosphere for collaboration among countries in the region . . . through technology and information exchange, education and science' and 'to increase scientific and technological knowledge about the hydrological cycle, thus increasing our capacity to better manage and develop our water resources in a holistic manner'. It is a partner to another UNESCO Centre: Centro de Aqua del Tropico Humeda para America Latina y El Caribe (CATHALAC).

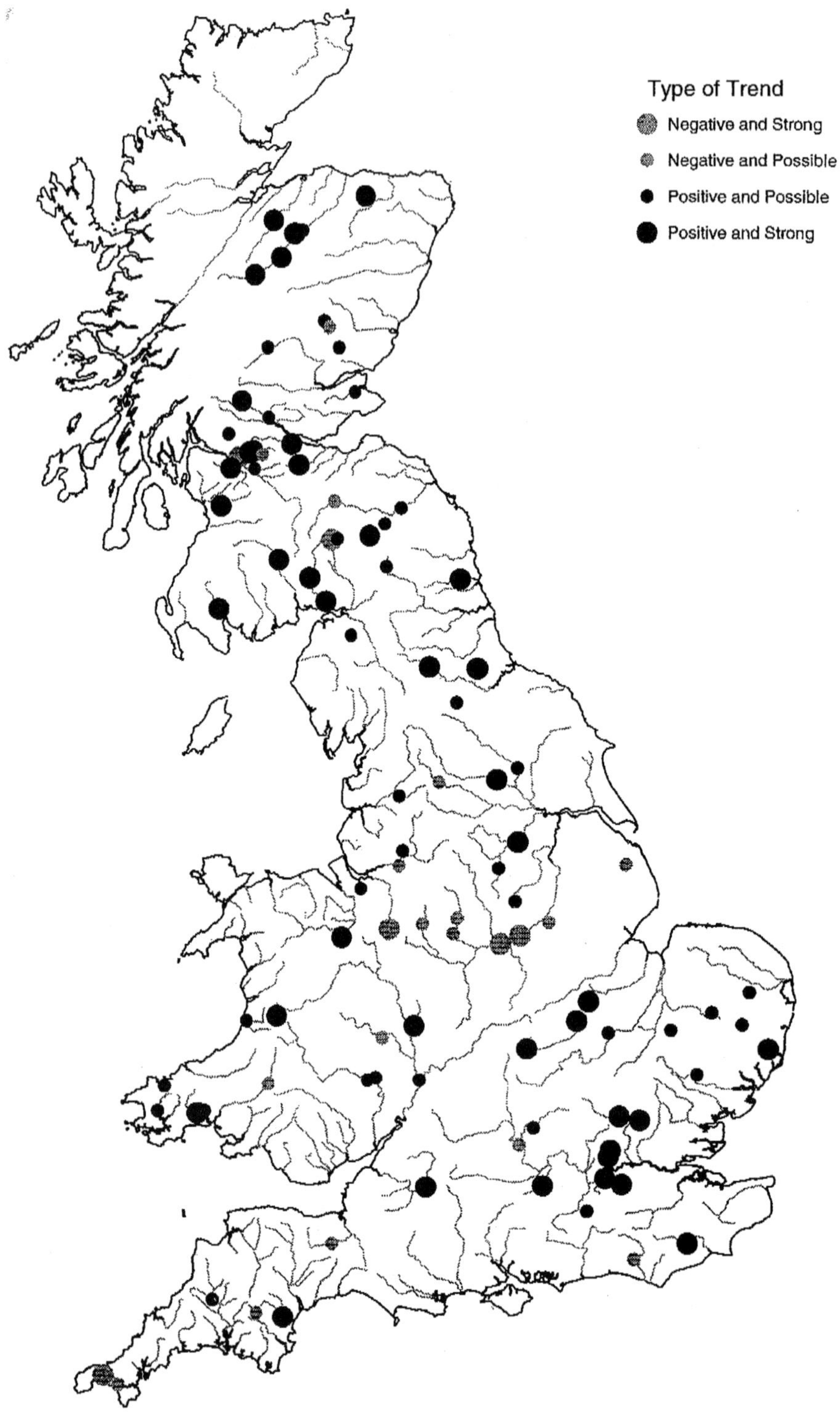

Figure 28.2 Non-stationarity in British flow records (changes in annual maxima or any peak-over-threshold variable). (From Robson and Reed, 1996.)

CATALOGUE OF RIVERS OF SOUTH EAST ASIA AND THE PACIFIC

The Catalogue of Rivers of South East Asia and the Pacific (Takeuchi *et al.*, 1995, Jayawardena *et al.*, 1997) is aimed at improving understanding of changing hydrology in the humid tropics. It is produced within the framework of UNESCO's FRIEND Project. Among the objectives of these serial publications is promotion of intra-national information exchange between different organisations in each country and promotion of an international data exchange and collaborative data network in the region. As mentioned by Takeuchi *et al.* (1995), 'We are not mutually well acquainted with nor informed of rivers in neighboring countries . . . In any country, . . . it is often difficult to gather information belonging to different organisations in order to draw a comprehensive picture of a basin'. This latter statement is an illustration of difficulties in communication between the many different institutions collecting hydrological data at the national level.

Figure 28.3 shows samples of the types of information contained in the first two volumes of the Catalogue (Takeuchi *et al.*, 1995, Jayawardena *et al.*, 1997). They illustrate the features of long time series of river flow data in humid tropics – strong natural variability, possible periodic components, changes in flow variability caused by dam construction, and the difficulties encountered through interruption of monitoring programmes.

CONCLUDING REMARKS

It has been demonstrated that the search for weak changes in time series of hydrological data is not an easy task and that even where much good quality data is available for a country or a region, a consistent climate change signature in river flow data has yet to be found. For the humid tropics, the current priority is the broader collection and dissemination of suitable hydrological series. Simulataneously, the available material should be carefully examined and tested for change. Only then will it become possible to document and understand thoroughly the hydrological changes that are occurring in these areas.

This chapter gives guidance on methods for change detection in time series of river flow records and is limited to fundamental ideas only rather than presenting individual tests in great detail. Readers may be interested therefore to consult a much broader report (Kundzewicz and Robson, 2000), which is available free of charge upon request from the World Meteorological Organization, Hydrology and Water Resources Department, 7 bis av. de la Paix, Case Postale No. 2300, CH 1211 Geneva, Switzerland. An update on 'Detecting change in the hydrological data' was presented in a special Issue of the IAHS *Hydrological Sciences Journal* (Kundzewicz, 2004).

APPENDIX 28.1 ELEMENTS OF STATISTICAL TESTING FOR CHANGE DETECTION: A PRIMER

This Appendix (after Robson *et al.*, 2000) contains an overview of some of the most important statistical principles that apply when testing for change.

Hypotheses

The starting point for a statistical test is to define the *null* and *alternative hypotheses* – statements that describe what the test is investigating. The null and alternative hypotheses are usually framed in terms of the types of change described above. For example, to test for trend in the mean of a series the *null hypothesis* (H_0) would be that there is no change in the mean of a series, while the *alternative hypothesis* (H_1) would be that the mean is either increasing or decreasing over time. To test for step-change in the mean of a series, the null hypothesis would again be that the mean of the series remains constant, but the alternative hypothesis would be that the mean of the series has suddenly changed.

The starting point for statistical testing is to assume that the null hypothesis is true, and then to check whether the observed data is consistent with this hypothesis. The null hypothesis is rejected if the data are not consistent. Otherwise, there is no reason to reject the null hypothesis (even so this is not a proof that the null hypothesis holds, only that there is no reason to reject it).

Test statistics

The *test statistic*, a numerical value calculated from the data series undergoing testing, allows a comparison of the null and alternative hypotheses. A good test statistic is designed so that it highlights the difference between the two hypotheses. A simple example of a test statistic is the linear regression gradient, which can be used to test for a trend in the mean. If there is no trend (the null hypothesis) then the regression gradient should have a value near to zero. If there is a large trend in the mean (the alternative hypothesis) then the value of the regression gradient would be very different from zero. In order to carry out a statistical test it is necessary to compare the observed test statistic with the expected distribution of the test statistic under the null hypothesis. The *significance level* of a test statistic expresses this concept more formally.

Significance levels

The *significance level* is a means of measuring whether a test statistic is very different from values that would typically occur under the null hypothesis, H_0. It is the probability that a value is as extreme as, or more extreme than the observed value, assuming the null hypothesis of no change (i.e. probability that a test detects trend when none is present). A possible interpretation of the significance level might be:

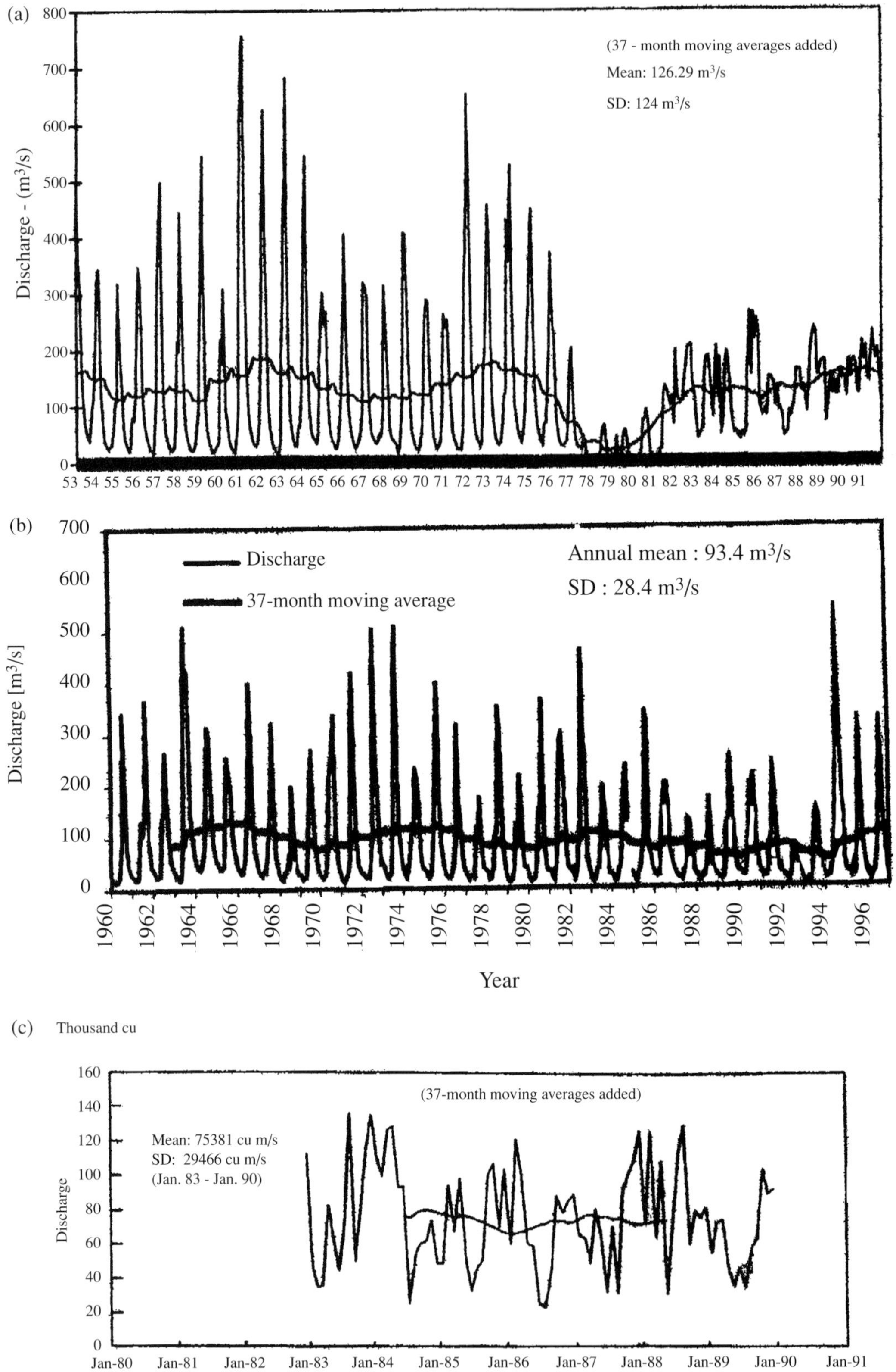

Figure 28.3 Long time series of flow records in South East Asia (from *Catalogue of Rivers* – Takeuchi *et al.*, 1995, Jayawardena *et al.*, 1997). (a) Mae Klong River, Thailand (evident change in variability caused by reservoirs) (from Takeuchi *et al.*, 1995, p. 245); (b) Nam Khane River, Lao People's Democratic Republic (strong seasonal behaviour) (from Jayawardena *et al.*, 1997, p. 185); (c) Rajang River, Malaysia (long time series of records became interrupted. After 1990 the presented record is not complete) (from Takeuchi *et al.*, 1995, p. 197); (d) Brantas River, Indonesia (decreasing variability) (from Takeuchi *et al.*, 1995, p. 108); (e) Solo River, Indonesia (strong seasonality) (from Takeuchi *et al.*, 1995, p. 97).

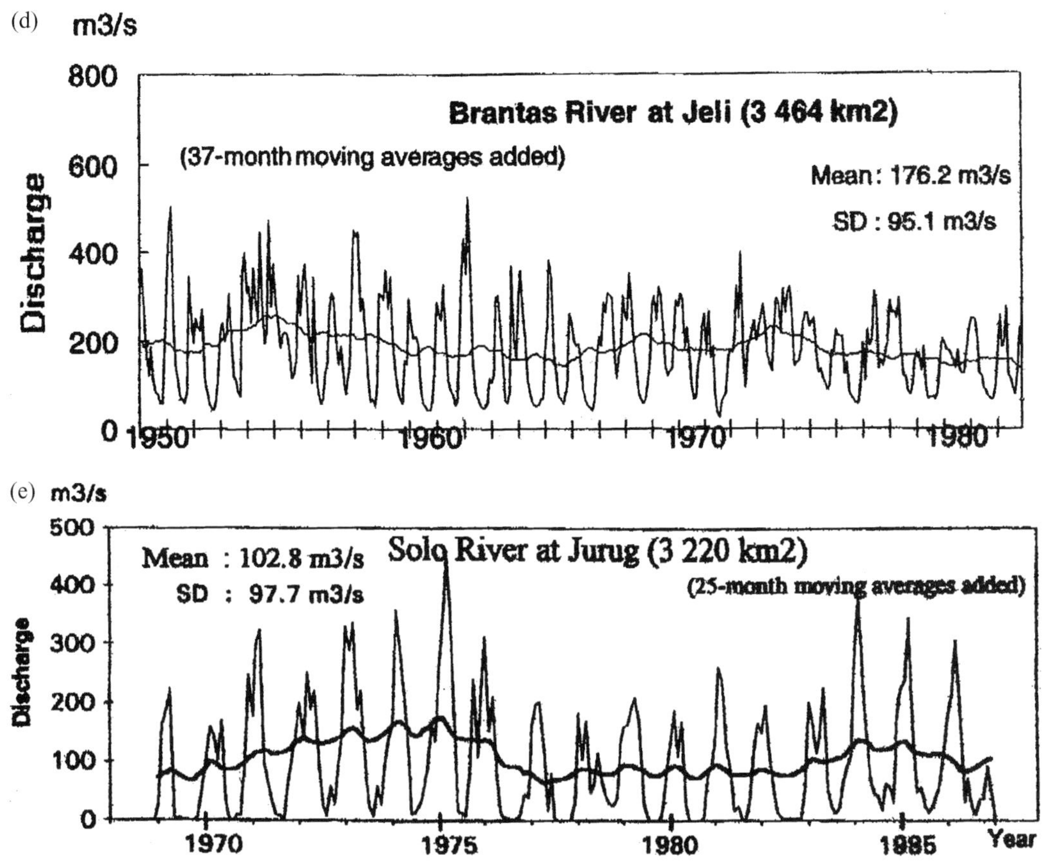

Figure 28.3 *(cont.)*

Significance level

- >10% – very little evidence against the null hypothesis, H_0
- 5% to 10% – possible moderate evidence against H_0
- 1% to 5% – strong evidence against H_0
- below 1% – very strong evidence against H_0.

Note that when reporting results the actual significance levels should normally be quoted (e.g. a significance level of 3.7%).

For many traditional statistical methods, significance levels can be looked up in reference tables or calculated from simple formulae, provided that the required test assumptions apply. In general, the significance level can be found if the distribution of the test statistic under the null hypothesis (i.e. assuming the null hypothesis is true) is known or can be estimated. One case where this distribution is usually easy to determine is where the data are independent and normally distributed. Resampling methods (Appendix 28.4) provide an alternative approach for estimating test statistic distributions. These methods are robust, require minimal assumptions to be made and are very general.

Power and errors

There are two possible types of error that can occur in a test result. The first is that the null hypothesis is incorrectly rejected (type I error). The significance level expresses the probability of this error. If the null hypothesis is accepted when the alternative hypothesis is true, a type II error is made. A test which has low type II error probability is said to be powerful. The power of the test is the probability of correctly detecting a trend when one is present.

Misconceptions

In order to warn the readers to avoid misconceptions, the following cautionary statements are offered (Robson *et al.*, 2000):

Poorly understood data gives poor results

Quality control and exploratory analysis should have been carried out. The 'Garbage In : Garbage Out' principle applies in statistical testing.

Inappropriate test assumptions are dangerous

If the assumptions made in a statistical test are not fulfilled by the data then test results can be meaningless. It is very important to understand what restrictions apply to a particular statistical test, and in what situations it is valid to apply the test.

A statistical test provides evidence not proof

Statistical test results express probability and not certainty. There is always a chance that the null hypothesis was true when a test result suggests it should be rejected. Similarly, if the null hypothesis is accepted, then this result says only that the available evidence does not contradict the null hypothesis, it is not a proof that the null hypothesis is true.

Each statistical test frames only a very specific question
A test result that shows no conclusive evidence of a trend in the mean does not establish that the variance of the same series is unchanged, or that frequency and magnitude of the extremes are unchanged.

Significance is not the same as importance
A test result may be highly significant (i.e. provide strong evidence against the null hypothesis) but the size of the observed change may be so small that it is of no importance. Conversely, an important level of change might not be statistically significant, either because of a short length of the time series of observed data or because noise in the data means it cannot be statistically distinguished from the null hypothesis.

APPENDIX 28.2 REVIEW OF STATISTICAL TESTS FOR DIFFERENT TYPES OF CHANGE

The following list contains some of the most commonly used tests for detecting step change and trend. Note that the different tests make different assumptions about the data. The tests may be used in their original form, providing test assumptions are met. Alternatively, where test assumptions cannot be met, it is recommended that only the test statistic from each test is used, and that the significance level is evaluated using resampling methods (see Appendix 28.4).

Tests for step change

Median change point test / Pettitt's test for change
Powerful rank-based test for a change in the median of a series with the exact time of change unknown, considered to be robust to changes in distributional form.

Wilcoxon–Mann–Whitney test / Mann–Whitney test / Mann test / Rank-sum test
Rank-based test that looks for differences between two independent sample groups, based on the Mann–Kendall test statistic.

Distribution-free CUSUM test
Rank–based test in which successive observations are compared with the median of the series with the maximum cumulative sum (CUSUM) of the signs of the difference from the median as the test statistic.

The Kruskal–Wallis test
Rank-based test for equality of sub-period means.

Cumulative deviations and other CUSUM tests
E.g. rescaled cumulative sums of the deviations from the mean.

Student's t-test
Standard parametric test for testing whether two samples have different means, assumes normally distributed data and a known change-point time.

The Worsley likelihood ratio test
Similar to Student's t-test but suitable for use when the change-point time is unknown; assumes normality.

Tests for trend

Spearman's rho
Rank-based test for correlation between time and the ranks series.

Kendall's tau / Mann–Kendall test
Another rank-based test, similar to Spearman's rho but using a different measure of correlation which has no parametric analogue. NB: There exists a seasonal Kendall test that allows for seasonality in the data, and a modified seasonal Kendall test that additionally allows for some autocorrelation in the data.

Linear regression
One of the most common tests for trend – it uses the regression gradient as a test statistic and assumes that data are normally distributed.

APPENDIX 28.3 DISTRIBUTION-FREE APPROACHES

The majority of hydrological series are non-normally distributed and it therefore makes sense to use distribution-free testing methods. Distribution-free methods are ones in which no assumption about the underlying distribution of the data need be made. Among commonly used distribution-free approaches are:

Rank-based tests
Rank-based tests are tests that use the ranks of the data values (not the actual data values). A data point has rank r if it is the rth largest value in a data set. There are a number of widely used and useful rank-based tests (Kundzewicz and Robson, 2000). Most rank-based tests assume that data are independent and identically distributed. Rank based tests have the advantage that they are robust and usually simple to use. They are usually less powerful than tests that are directly based on the data.

Tests using a normal scores transformation
Many standard tests for change rely on assumptions of normality. When data are non-normally distributed, as is often the case for hydrological data, these tests can still be used if the data are first transformed.

The normal scores transformation results in a dataset that has a normal distribution. It is similar to using the ranks of a data series, but instead of replacing the data value by its rank, r, the data value is replaced by the typical value that the *rth* largest value from a sample of normal data would have (the *rth* normal score). The advantages of using normal scores are that the original data need not follow a normal distribution, and the test is relatively robust to extreme values. Normal scores tests are likely to give slightly improved power for detection of change relative to equivalent rank-based tests.

Resampling methods
Resampling methods use the data to help determine significance levels. This avoids the need to make any assumptions about the underlying distribution of the data. Resampling methods can be applied to almost any test statistic and provide an alternative way of obtaining significance levels. The advantages of resampling methods are that they are flexible and robust and that when used with parametric test statistics they also allow the degree of change to be measured. Resampling tests are relatively powerful, e.g. for large samples, permutation tests can be shown to be as powerful as the most powerful parametric tests. Furthermore, resampling methods can be adapted to test data which are not independent (see Appendix 28.4).

APPENDIX 28.4 MORE ON RESAMPLING METHODS

Resampling methods are a very powerful, general and flexible approach to estimating significance levels. In essence, they use the data to determine significance levels – which means that minimal assumptions about the data need to be made. Re-sampling methods are based on changing the order of data points and comparing test statistics calculated on these generated series with the test statistic for the original data series.

There are two main types of resampling: permutation testing and bootstrapping methods.

Permutation tests

A *permutation test* works by shuffling the data very many times. Consider a time series of data with a possible trend. One measure of the trend is the regression gradient: an example of a possible *test statistic*. Suppose first that there is no underlying trend in the data. If that is true, then it should not matter very much if data are re-ordered – the regression gradient should not change very much. Each time the data are shuffled as part of the permutation test, the selected test statistic (in this case the regression gradient) is re-calculated. At the end of all the shuffling, we have generated a distribution of possible values of the test statistic under permutation, the *permutation distribution*. The permutation distribution usually depends on the data and must be recalculated for each dataset. If there is no trend, then we would expect that the observed test statistic (regression gradient) for the original data is not very different to any of the generated test statistic values, i.e. it is somewhere in the middle of the permutation distribution. So to test for trend, the observed test statistic (regression gradient) is compared with the permutation distribution. If the gradient is larger (or smaller) than almost all the values in the permutation distribution, we conclude that a trend is present. Conversely, if the original gradient is somewhere in the middle of the permutation distribution, we conclude that there is no evidence of trend.

Bootstrapping

Bootstrapping approaches are similar to permutation techniques. The main difference is that instead of reordering the data, the new data series are generated by sampling with replacement. For example, for a series of 50 values, a bootstrap sample would take 50 values at random from the original series: the resulting series might perhaps include three lots of the original first value, but no instances at all of the last value.

For both permutation and bootstrap methods, the generated series has the same distribution as the empirical (i.e. observed) distribution of the data. The bootstrap is generally, but not always, less powerful than a permutation test. However, bootstrap methods are often to be preferred where a test is looking for change in variance. Further, permutation tests cannot be applied with test statistics that do not change when the data are permuted, e.g. tests for which the test statistic is the mean or median. The tests given here can be used with either method. In general, bootstrap methods are more flexible than permutation methods and can be used in a wider range of circumstances.

Summary of method for resampling

The basic method for carrying out a permutation or bootstrap test, once the test statistic has been selected, is as follows:

- calculate the test statistic for the observed data
- resample the data series many times (e.g. 1000) to generate new data series
- recalculate the test statistic for each of these series
- estimate the significance levels.

To estimate the significance level, the data are resampled, by either permutation or bootstrapping, a large number of times, S. For each of these generated series, the test statistic, T, is calculated to give S artificial values of T. These are then ordered as

$$T_1 \leq T_2 \leq \cdots . \leq T_S$$

If the original test statistic is T_0 and

$$T_k \leq T_0 \leq \cdots . \leq T_{k+1}$$

then the probability of the test statistic being less than or equal to T_0

under the null hypothesis is approximately:

$$p = k/S \tag{28.1}$$

p may also be estimated as

$$p = (k + 0.5)/(S + 1) \tag{28.2}$$

or even

$$p = (k + 1)/(S + 2) \tag{28.3}$$

Assuming that large values of T indicate departure from the null hypothesis, the significance level for this test is then

$$100 * min(p, 1 - p)\% \tag{28.4}$$

(assuming a two-sided test, i.e. a test in which the direction of change is assumed unknown).

Further details on the application of resampling methods can be found in Robson *et al.* (2000).

Block resampling: resampling when data are not independent

The basic resampling methods, as described above, avoid any distributional assumptions but they still assume that data values are independent of one another. Frequently measured hydrological data (daily or hourly data in particular) are typically not independent: they show serial dependency (autocorrelation).

Block resampling methods are a way of testing for change when there is dependency in the data. For this, the data are permuted or bootstrapped in blocks (e.g. all values within a year are kept together). With this approach the dependency structure within each block is built into the test and independence assumptions are thus no longer violated (see Kundzewicz and Robson, 2000).

References

Cleveland, W. S. (1993) *Visualizing Data*, Hobart Press, New Jersey, USA.

Cleveland, W. S. (1994) *The Elements of Graphing Data*, Hobart Press, New Jersey, USA.

Grubb, H. and Robson, A. (2000) Exploratory / visual analysis. In: Kundzewicz, Z. W. and Robson, A. (ed.) *Detecting Trend and Other Changes in Hydrological Data*. World Climate Programme – Water, World Climate Programme Data and Monitoring, WCDMP-45, WMO/TD – No. 1013, Geneva, May 2000, 19–47.

Hirsch, R. M. and Slack, J. R. (1984) A non parametric test for seasonal data with seasonal dependence, *Water Resour. Res.*, **20**: 727–732.

IPCC (Intergovernmental Panel on Climate Change) (2001) *Climate Change 2001: The Scientific Basis* (edited by Houghton, J. T., Ding, Y., Griggs, D. J., Nouger; M., van der Linden, P. J., Dai, X., Maskell, K. and Johnson, C. A.). Contribution of the Working Group I to the Third Assessment Report of the Intergovernmental Panel on Climate Change, Cambridge University Press, Cambridge, UK.

IPCC (Intergovernmental Panel on Climate Change) (2001a) *Climate Change 2001: Impacts, Adaptation and Vulnerability* (edited by Mc Carthy, J. J., Canziani, O. F., Leary, N. A., Dokken, D. J. and White, K. S.). Contribution of the Working Group II to the Third Assessment Report of the Intergovernmental Panel on Climate Change, Cambridge University Press, Cambridge, UK.

Jayawardena, A. W., Takeuchi, K. and Machbub, B. (1997) *Catalogue of Rivers for South East Asia and the Pacific* – Volume II, Publications of the UNESCO-IHP Regional Steering Committee for South East Asia and the Pacific.

Kendall, M. and Ort, J. K. (1990) *Time Series*. Edward Arnold, London, 3rd ed.

Kundewicz, Z. W. (ed.) (2004). Detecting change in hydrological data. Special Issue, *Hydrol. Sci. J.*, **49**: 3–12.

Kundzewicz, Z. W. and Robson, A. (ed.) (2000) *Detecting Trend and Other Changes in Hydrological Data*. World Climate Programme – Water, World Climate Programme Data and Monitoring, WCDMP-45, WMO/TD – No. 1013, Geneva, May 2000, 157 pp.

Lins, H. F. and Slack, J. R. (1999) Streamflow trends in the United States, *Geoph. Res. Letters* **26**(2): 227–230.

Pilon, P. (2000) Criteria for the selection of stations in climate change detection network. In: Kundzewicz, Z. W. and Robson, A. (ed.) *Detecting Trend and Other Changes in Hydrological Data*. World Climate Programme – Water, World Climate Programme Data and Monitoring, WCDMP-45, WMO/TD – No. 1013, Geneva, May 2000, 121–131.

Robson, A., Bardossy, A., Jones, D. and Kundzewicz, Z. W. (2000) Statistical methods for testing for change. Chapter 5 in: Kundzewicz, Z. W. and Robson, A. (red.) *Detecting Trend and Other Changes in Hydrological Data*. World Climate Programme – Water, World Climate Programme Data and Monitoring, WCDMP-45, WMO/TD – No. 1013, Geneva, May 2000, 49–85.

Robson, A. J. and Reed, D. W. (1996) Non-stationarity in UK flood records. Flood Estimation Handbook Note 25, Institute of Hydrology, Centre for Ecology and Hydrology, Wallingford, UK, October 1996.

Slack, J. R. and Landwehr, J. M. (1992) Hydro-climatic data network: a US Geological Survey streamflow data set for the United States for the study of climate variations, 1874–1988, US Geol. Surv. Open-File Rept. 92–129.

Takeuchi, K., Jayawardena, A. W. and Takahasi, Y. (1995) *Catalogue of Rivers for South East Asia and the Pacific* – Volume I, Publications of the UNESCO-IHP Regional Steering Committee for South East Asia and the Pacific.

29 How to choose an appropriate catchment model

C. Barnes
Bureau of Rural Sciences, Canberra, Australia
M. Bonell
UNESCO, Paris, France

INTRODUCTION

The existence of a large number of catchment hydrology models, evident from even a cursory glance at the literature, is likely to cause trepidation or confusion for even expert modellers, let alone practitioners of hydrology who merely require something 'off the shelf' which they can use with confidence. Typically, available catchment models often come with exaggerated claims for the breadth of their applicability, and little or no in-depth discussion of their inherent assumptions and consequent limitations. Two questions will therefore be addressed in this chapter:

(1) How can an appropriate model for my catchment be chosen, given an intended application?
and
(2) How can an appropriate model be constructed (or an existing model be modified) if none exists at present?

There would appear to be several reasons for the present wide range of models, including:

- a diverse range of catchments and purposes (for example, forecasting or regulatory support) which in turn implies interest in many different kinds of processes;
- availability of different levels of information or data quantity and quality; and
- the fact that catchments are complex systems, having a huge number of potentially significant processes, and consequently 'emergent behaviour' (defined later on) which is not evidently a simple sum of the component parts.

Taken together, these three factors imply that to represent catchment behaviour efficiently, much of what is deemed to be of secondary importance must inevitably be either ignored or greatly simplified by using specific assumptions. From this point of view, differences between models of the same system are due to differences in model assumptions.

This observation suggests that models may be classified according to their inherent assumptions. The problem of model choice is then reduced to a classification problem, where the choice is made on the basis of which assumptions the modeller feels are appropriate for the particular system under consideration.

Firstly, let us look at some general characteristics of model development before exploring the different types of assumptions that are commonly encountered and the implications of this analysis for model choice. We shall then propose a possible way forward, where the concept of a *universal model process* replaces that of a *universal process model*. That is, instead of attempting to develop models with a wide range of applicability, we devise a generic process that can be applied to a wide range of problems to produce appropriate (efficient, relevant) models.

As will become clear, although no one class of model is considered to be inherently superior to any other, most of the examples considered belong to the model class sometimes termed 'spatially lumped' catchment models. This is simply an expedient, because these models are somewhat simpler to understand and can be represented diagrammatically and not because of any assumed intrinsic advantages or disadvantages.

The case for concentrating in this chapter on simple 'lumped' catchment models is persuaded by practical reality. The humid tropics are relatively data deficient (e.g. spatial density and observational frequency of national and regional networks are sparse) and less intensively studied (in terms of the number of experimental catchments) by comparison with humid temperate forests (e.g. Walker Branch Watershed, USA, reviewed in Mulholland, 1993; the Maimai catchment, New Zealand, reviewed in Seibert and McDonnell, 2002). At best, only rainfall-runoff records are

Forests, Water and People in the Humid Tropics, ed. M. Bonell and L. A. Bruijnzeel. Published by Cambridge University Press.

generally available for tropical forest experimental catchments; hence the need to consider relatively simple models.

Alternative deterministic (spatially distributed) models are structurally more complex and require large volumes of data to encompass the hydrometeorological drivers (e.g. fine temporal resolutions for precipitation, temperature and solar radiation), the physical nature of the catchment (e.g. geology, topography, soils, vegetation), as well as good understanding of hydrological processes and their parameterisation (e.g. soil-weathered rock hydraulic properties) (Littlewood, 2001).

A recent evaluation of the status of distributed modelling in *Hydrological Processes*, **16**, *2002* raises even more fundamental issues. Beven and Feyen (2002) in the opening Preface observed that over the last decade there has been much greater progress in computing and modelling developments than in field measurement techniques. They suggested that a principal barrier to more rapid progress is the lack of new measurement techniques to represent hydraulic potentials, storages and fluxes within hillslopes *at larger scales* (Beven, 2002).

The current generation of distributed models extends back to the review paper of Freeze and Harlen (1969) (known as the FH69 blue print, Beven, 2002) which provided the equations and boundary conditions for a 'physically-based' digitally-simulated hydrological response model. A principal constraint of the FH69 blueprint (and its later variants) is that it is based on Darcian theory which is applicable at small scales but 'certainly not applicable at large scales due to the effects of the non linearity of the unsaturated Darcy equation, the heterogeneity of soil (and rock) properties and preferential flow of different types' (Beven, 2002, p. 192). The current generation of field measurements, which depend on 'point' measurements, do not characterise the fundamental soil (and rock) hydraulic properties at the scale of interest in distributed models, that is, from hillslope up to catchment scale (Beven, 1989, 1993; Wheater *et al.*, 1993). Within the context of the well-known issue of soil heterogeneity (e.g. Rogowski, 1972; Nielsen *et al.*, 1973; Baker, 1978), soil macropores and pipes notoriously present formidable problems both in their measurement and representation in deterministic models (Bonell, 1998) because potential flow theory using Darcy's Law and Richard's equation does not physically describe water flow in macropore soils (Germann, 1990).

Both environmental and artificial tracer studies (e.g. Kendall and McDonnell, 1998, ed.; Bronswijk *et al.*, 1995; Lange *et al.*, 1996; see Bonell this volume) have highlighted the critical role of preferential (by-pass) flow and correspondingly, only a small proportion of total porosity is active in the rapid transfer of 'mobile' water during storm events.

It is for the above reasons that Vertessy and Elsenbeer (1999) concerning the La Cuenca (Peru), and later Schellekens (2000) referring to the Bisley II experimental catchment in Puerto Rico, had comparatively limited success in the modelling of storm events using the more heavily parameterised TOPOG-DYNAMIC model (see Bonell, this volume). Because of the above difficulties in representing the complexity of surface and subsurface stormflows both hydrometrically and hydrochemically (age, origin, pathway) (Bonell, 1998), there has been a tendency away from the use of distributed models (e.g. Barnes and Bonell, 1996; Seibert and McDonnell, 2002) towards what Littlewood (2001) termed parametrically parsimonious[1] conceptual models (PPCMs).

Such models, rather than attempting to integrate point-scale physical processes explicitly, attempt to discern new types of physical relationships between quantities which are appropriate at the scale of application. These conceptual models may treat the whole catchment as a single spatial unit (i.e. they are spatially lumped) but also maintain a connection with the underlying physical processes at smaller spatial scales. This is similar to the attempts at a smaller scale to understand the behaviour of the hydraulic conductivity, arising from Darcy's law, in terms of the soil micro-porosity (cf. the work of G. de Josselin de Jong, quoted in Bear, 1972 and Bear and Verruijt, 1987). For example, see the Swedish HBV conceptual model, which has been applied to a wide range of scales with minimal modification to its structure following its introduction in the 1970s (Bergström, 1995; Lindström *et al.*, 1997; Bergström *et al.*, 2002[2]).

It is important to recognise that parameters in conceptual models may not be directly measurable independently of the rainfall-runoff relationship itself (this applies at all scales, cf. the hydraulic conductivity). On the other hand, they may provide a means of obtaining effective values for smaller scale parameters at a much larger scale than the common measurement scale (usually the point scale), dependent on the validity of assumptions used to integrate between scales. (This process unfortunately can result in 'effective' values that lie outside the physically acceptable range – an indication that the integration assumptions are inappropriate).

An advantage of PPCMs is that they depend on as few as five parameters, in contrast to the more deterministic, distributed models that generally implicitly require upwards of a dozen parameters. Such low data demands make them attractive for use in more data-deficient environments (such as the humid tropics) where detailed within-catchment hydrology understanding is lacking. This has led to a debate as to whether some 'physically-based' rainfall-runoff models are over-parameterised, highlighting one of the major perceived advantages of PPCMs (Jakeman and Hornberger, 1993), which was later supported by the comparative

1 Parsimony is defined as: no more causes or forces should be assumed than are necessary to account for the facts.

2 HBV originates from the model developed by the Hydrologiska Byråns Vattenbalansavdelning (The Water budget section of the Hydrological Department) within SMHI (Swedish Meteorological and Hydrological Institute) (A. Rodhe, pers. comm., 2002).

testing of several models across 429 catchments by Perrin *et al.* (2001).

One PPCM model which has received considerable attention is the IHACRES (**I**dentification of unit **H**ydrographs **A**nd **C**omponent flows from **R**ainfall, **E**vaporation and **S**treamflow data) by Jakeman and co-workers (reviewed by Littlewood, 2001) which is a conceptually-lumped unit hydrograph model developed from earlier work. Being a lumped parameter model, IHACRES is not able to provide a spatial perspective of parameters such as catchment wetness (cf. TOPMODEL, TOPOG; see Bonell and Chappell, Bidin *et al.*, this volume). Rather, IHACRES is more properly applied to quantify the impacts of climate or catchment management variations (land use change, for example) on streamflow generation at the catchment scale.

The structure of IHACRES consists of a non-linear loss module based on rainfall and temperature (to estimate effective rainfall), coupled with a linear module for routing streamflow. The preferred routing option is two linear storages in parallel, representing the 'quick' and 'slow' flow contributions to total streamflow. One version of IHACRES estimates four *dynamic response characteristics* (known as DRCs) in the non-linear module and, of relevance to this chapter, there are three additional DRCs that are determined from the routing module. These latter three coefficients represent a time constant for each linear reservoir, and the proportional volumetric contribution of quick flow to total streamflow. A more detailed description of IHACRES is given elsewhere in this volume by Schreider and Jakeman in connection with a northern Thailand study.

Barnes and Bonell (1996) adapted IHACRES to the Wyvuri, South Creek tropical rainforest catchment to model a monsoon event. During the midst of the summer monsoon season when maximal catchment wetness persists, and for specific rain events of high intensity, the non-linear effective rainfall module of IHACRES was replaced with a linear proportional loss model, where streamflow generation is derived simply as a fixed proportion of rainfall (runoff coefficient = 0.88). Separate analyses showed that during the monsoon season, losses due to total evaporation (wet canopy plus dry canopy losses) represented only a small fraction of total rainfall (runoff coefficient 0.85–0.90). As with IHACRES, the rain (and associated isotopic concentration) was routed through two parallel stores despite the observed complexity of the runoff generation process at a plot scale (Bonell *et al.*, 1998). These stores were modified by the inclusion of a simplified quasi-linear storage element following the Boughton model (Boughton, 1984) for the respective surficial layer (dominated by saturation overland flow, SOF, and shallow subsurface stormflow, SSF) and a deeper, permanent passive groundwater store. As observed by Beven (2002), neither storage volumes nor fluxes can generally be measured directly at larger scales. On the other hand, the use of this PPCM provides an indirect means of detecting, as well as estimating the volume of, the deeper groundwater storage in South Creek.

Two other aspects need mention. Previous hydrometric hillslope hydrology studies and subsequent combined hydrometric-hydrochemistry campaigns provide a physical basis for interpretation of the respective upper and lower conceptual stores (as reviewed in Bonell, this volume). Furthermore the surface-permanent groundwater is more strongly coupled than previously thought (c.f. Bonell *et al.*, 1981; 1998) to the storm runoff generation process. Such interactions are not presently taken into account in TOPMODEL and TOPOG (see Chappell, Bidin *et al.* this volume) for example, so their application cannot represent observed reality in this particular catchment study. The use of a PPCM provides an alternative model that is consistent with the information available for this study. We will later outline further improvements to this conceptual model.

A more sophisticated adaptation of a box model (Seibert and McDonnell, 2002) is the representation of the hillslope hydrology of the Maimai-8 catchment in New Zealand based on a qualitative perceptual model, which represents the culmination of work by several research teams since the 1970s. Boxes represent: the hillslope; topographic hollows (which hillslope SSF converges into); and a low riparian zone; so that there is coupling downslope as well as a coupled formulation of the saturated and unsaturated zone storage (Seibert *et al.*, 2002).

CHARACTERISTICS OF MODEL DEVELOPMENT

The basis for model choice

When faced with a situation where the use of a model appears desirable, the busy practitioner is confronted by a bewildering array of models and variants, each potentially appropriate to his needs. In attempting to cover the field as widely as possible, many of these models are actually quite complex, requiring significant training and experience to use them competently. Small wonder then that there is a tendency amongst these practitioners for the model choice to be made based on familiarity, rather than on appropriateness for the problem under consideration. (If the only tool you have is a hammer, every fastener looks like a nail). In this section, we examine the process of model development, highlighting aspects that may lead to a more practical and rational methodology for model choice.

The most critical determinant of model form is the *purpose* of the model (including the required output). As an example of this, we have argued elsewhere that management models must have significantly different forms to those intended for research, because their purpose or function is entirely different (Barnes *et al.*, 1997a).

Whereas, for research purposes, models are primarily for explanation of observed behaviour (hypothesis testing), management models are required for prediction and manipulation of system outcomes, within a specified degree of uncertainty. These differences in function often lead to the choice of significantly different assumptions in representing what are effectively equivalent systems. Hence, the choice of a model requires an alignment between the purpose for which the model was constructed and its intended use.

One significant consequence of the effect of model purpose on model form is the way in which models can be used. Whereas it is sometimes desirable for the model to mimic system behaviour as closely as possible, regardless of catchment complexity (e.g. some regression models), for explanatory purposes it is often preferable to limit model complexity. In these circumstances, one of the major uses of a system model is to determine to what extent the observed system behaviour departs from some ideal system, defined by particularly simple assumptions. This ideal then becomes a baseline, displaying variation due to well-known causes so that more subtle variations can be better examined. In this mode, it is unnecessary to try for verisimilitude, and clarity of structure is paramount.

Another constraint on model choice is the availability and quality of system information. To be representative of a particular system, a model will generally require input data or information with which to determine parameter values. Lack of information will not generally prevent the development of a model, but will require additional assumptions to take the place of the missing information. (Hopefully, these assumptions will be tested in the course of model development, but even so, they will remain a source of uncertainty in model outputs). Unless the information requirements of the chosen model matches closely what is readily available, it will be difficult to implement the model satisfactorily; while data quality will determine the accuracy of the model outputs. Similar considerations also apply to the resources necessary to implement the chosen model. A model that requires a full-scale DEM (Digital Elevation Model) and GIS to be implemented may not be affordable even though technically feasible.

Catchments as complex systems

The most important facet of catchment systems for model development, and hence for model choice, is the recognition that they are *complex systems*, necessitating a quite distinctive modelling approach (see Barnes *et al.*, 1997a). Here, we define a complex system as one in which the behaviour of the whole is not simply the sum of that of the parts. What we exclude with this definition are those systems that are merely detailed, like information stored in a telephone book. This viewpoint has certain important consequences for modelling.

Catchment systems can be described in terms of processes at many levels or scales; for example ranging from chemical interactions at the molecular level, to chemographs (chemical hydrographs) at the whole catchment scale. In addition, flow processes for real catchments are affected by heterogeneity that is either unknown (sub-surface processes) or is impractical to measure (micro-relief). Several contributions in Part II of this book refer to the hydrological impacts from the spatial and temporal variability of rainfall and soil hydraulic properties, for example. When an attempt is made to represent catchment behaviour in terms of a model, therefore, simplifications of the system inevitably have to be made, and the vast majority of potential information or knowledge usually has to be excluded. The choice of which information to include, and in what detail, will depend on which aspect of system behaviour is the current focus, and represents the simplifying assumptions of the model.

One of the results that have come out of the recent increased focus on complex systems is the concept of *emergent properties* of such a system. This concept implies that, for complex systems made up of large numbers of smaller components with only local interactions, the system may exhibit collective behaviour that is not simply related to the properties of the individual components, but is a property of the whole system. A familiar illustration of this is the sand pile, formed by dropping sand grains one at a time, which eventually exhibits avalanches of all sizes up to a significant fraction of the whole pile. At the aggregated or system level, new (emergent) processes (such as avalanches) may arise for which no clear analogue exists at the lower level. The implications of this observation is that even if it were possible to correctly perform the averaging implicit in upscaling lower-scale process descriptions, it may be *far more efficient to look for a description of system behaviour in terms of 'new' processes defined at the system scale*.

Given the validity of this concept of emergent behaviour for catchment systems, profound consequences arise for modelling endeavours such as up-scaling (the processes of aggregating small-scale descriptions to yield large-scale models, meaning in this context small or large distance- or time-scales). It may be that upscaling is not possible or not useful in such systems. An example of an emergent property at another scale is the macroscopic description of water flow through soils. Although microscopically this process can be well described in terms of the properties of water and the pore structure of soil, the macroscopic description is usually in terms of Darcy's law, where water moves through porous material in a diffusive manner in response to a potential gradient. The response function, the hydraulic conductivity, is not usually calculated from measurements of the microscopic soil structure on which it depends, but is either measured directly, or obtained from empirical data based on relating many years of measurements of soil physical and hydraulic properties in connection with soil classification. A hydrological example of an emergent property may be the more or less universally observed exponential behaviour of recession hydrographs, which does not seem to be easily derived from the properties of an assemblage of simple hillslopes.

The process of model development

Initially, on encountering a new type of system, the modeller may begin with a relatively simplistic model of low complexity (parameterisation) that explains the broad features of observed system behaviour. At this stage little information is generally available, and it is appropriate to make many broad assumptions of unknown validity in order to cover missing information. A detailed model is usually not appropriate, as it is likely that any gains in accuracy will be swamped by errors introduced by barely appropriate assumptions. As further information is obtained, model refinement proceeds, guided by discrepancies between model output and observed behaviour, with a consequent increase in model complexity.

Typically, a point is reached where a given increase in model complexity (number of model parameters) leads to only small gains in the accuracy with which system behaviour can be represented. Worse, whereas initially it is easy to associate model parameters with particular aspects of observed system behaviour, with a large number of model parameters this is often difficult, and parameter estimates become highly correlated. At this stage we are only able to say that the system behaviour is fairly well described in terms of a combination of processes represented by this (relatively large) parameter set.

From this point on, the focus of model refinement must be on *reducing* model complexity, while maintaining as far as possible the previous explanatory power of the model. The struggle now is to increasingly relax model assumptions without losing significant accuracy. Conceptually, what we are saying now is that system behaviour is effectively determined (to a given accuracy) only by a few processes, *and is independent of all other processes*; a much stronger statement than the one made at the end of the previous paragraph. This represents the most intellectually difficult and labour intensive stage of modelling. The final product may be a relatively simple model, of complexity not much greater than the initial simplistic models, but with vastly greater utility (see Figure 29.1). In the (hopefully exaggerated) words of Oliver Wendell Holmes Jr., 'I don't give a fig for simplicity this side of complexity, but I would give my life for simplicity the other side of complexity'!

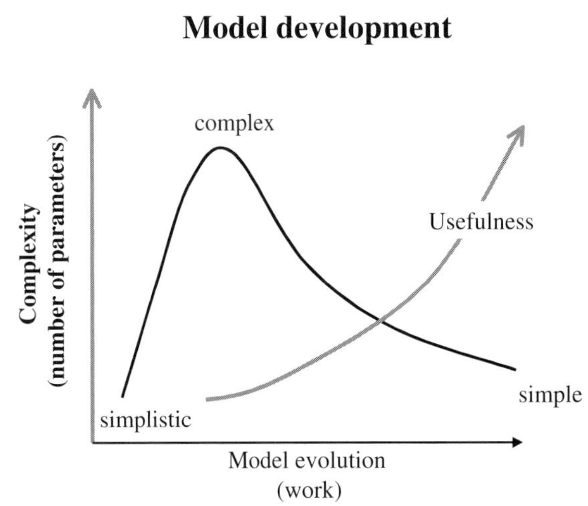

Figure 29.1 Diagram representing the process of model development in terms of model complexity and model utility.

Classification of models according to assumptions

In the previous section, we argued that differences between models of the same catchment system could arise because of differences in purpose or in information or resources available. These differences can be considered as the consequences of different assumptions made during model development, such as scale assumptions, or assumptions about dominant processes, and so on. In this way, we seek to arrive at an objective classification of catchment models in terms of their inherent assumptions. Although the variety of possible assumptions is effectively limitless, for any group of models it is possible to produce an assumption tree, showing significant relationships between the models. The structure of such a tree will depend on which assumptions are regarded as the more fundamental.

As a starting point, we propose to classify assumptions into the following major types:

1. Purpose (system monitoring and data warehousing, process mapping or system definition, forecasting and scenario exploration, communication . . .)
2. Availability of relevant information (quantity and quality-determines allowed complexity);
3. Scale (temporal, spatial);
4. Dominant processes (physical, chemical, biological . . .).
5. Other assumptions

There is an implied hierarchy in this list of assumption classes. For instance, what information is relevant will depend on the purpose of the model and not vice versa; whereas the scale of the model will be influenced both by the purpose and available information.

Scale assumptions

As an illustration of these ideas let us consider the classification of models in terms of their scale assumptions alone. Scale assumptions can be either temporal or spatial.

SPATIAL ASSUMPTIONS

One classification of catchment models that is often used is the division into so-called 'lumped' or 'distributed' models. In the simplest form, the former does not recognise the spatial dimension at all (e.g. IHACRES and HBV discussed earlier and below), whereas the latter allows system properties to vary over space (e.g. TOPOG and TOPMODEL discussed elsewhere in this volume by Chappell, Bidin *et al.*, and Bonell).

For example, the continuity or mass balance equation for solute concentration in a catchment may be written in differential form as

$$\partial \theta c/\partial t = -\nabla \cdot \mathbf{q}c \tag{29.1}$$

where $\mathbf{q}$ is the water flux, c is the solute concentration, and θ is the (volumetric) water content. This is the relationship from which most of the spatially explicit catchment water- and solute-balance models are derived, in conjunction with Darcy's law. On the other hand, if we integrate over the system volume, we obtain the equivalent expression for the mass balance of the whole system:

$$\partial \Theta c/\partial t = \iint_B \mathbf{q}c.d\mathbf{n} \tag{29.2}$$

where Θ is the total system water content, c is the average system concentration of solute and the integral is over the boundary of the system ($\mathbf{n}$ is the outward normal vector to the boundary B).

These two equations are equivalent mathematical statements of the concept of conservation of mass (or continuity equations); to be useful both require a constitutional relationship between water flux and water storage. Typical constitutional relationships are that water flux is proportional to potential gradient (Darcy's law) for differential models, or discharge is proportional to storage (linear reservoir model) for lumped system models.

On the one hand, the differential formalism is essential if the model output is required to give spatially explicit information on the internal states of the system, but requires spatially detailed input information and numerically 'smooth' data (which may not be possible if macropores or other preferred pathways have a significant effect). On the other hand, when the model output does not have to be spatially explicit, use of the integral formulation has the advantage of much less stringent conditions on data smoothness, and requires no knowledge of (and can give no information on) the internal state of the system.

Clearly, neither lumped nor spatially explicit models are implicitly superior under all circumstances; the utility of one or other models depends on the availability of information, and the purpose for which the model is required. Even in the absence of required information, a lumped model can be transformed into a spatially explicit one through the use of assumptions, such as the spatial uniformity of system properties. Similarly, spatially explicit formulations can in principle be translated into lumped models through the use of appropriate averaging, though this may not be an easy task. In general, a lumped model of a system will be less complex than a spatially explicit one, because it incorporates less system information.

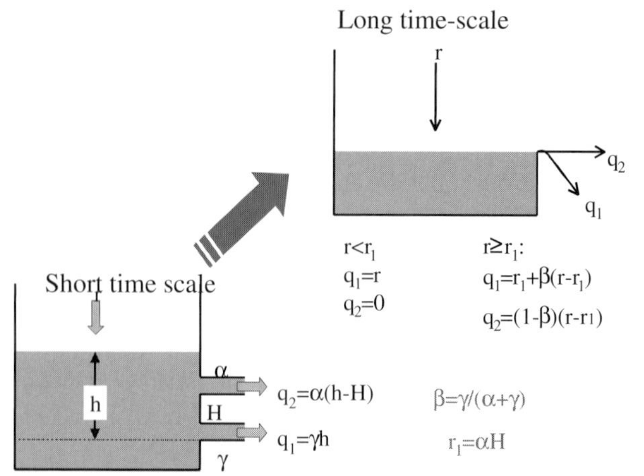

Figure 29.2 The relationship between a short time-scale and long time-scale model of the same catchment. The inverses of the short time-scale parameters, α^{-1} and γ^{-1}, are the recession time-constants for the different flow pathways, while H is a threshold storage below which the upper pathway does not flow. In the long time-scale representation, r_1 is the threshold rainfall intensity, below which the upper pathway is inactive, and β is the proportionality constant that determines the relative flows when rainfall intensity is above the threshold. Notice that the complexity of the long time-scale model is reduced.

TEMPORAL ASSUMPTIONS

Similar considerations also apply to temporal scales. Figure 29.2 shows an example of a lumped parameter model of a catchment in which runoff pathways are dependent on rainfall intensity. On a short time scale, it is assumed that the rainfall-runoff relationship can be represented in terms of a simple development of a linear reservoir (cf. Boughton, 1984). The conceptual model is that of a catchment where flow pathways are dependent on rainfall intensity. At low intensity, the catchment water storage acts as a simple linear reservoir, with flow proportional to storage (time constant α^{-1}, flux q_1 in Figure 29.2). For high intensity precipitation, this pathway is unable to cope with the increased volume of water, and a second, rapid, pathway is invoked (time constant γ^{-1}, flux q_2), Upscaling of the short time-scale model results in a model of reduced complexity because information on the absolute response times is lost; only their ratio is preserved.

Even when short-term information exists, it may be desirable to use the long-time behaviour for parameter estimation. For instance, it may be possible to estimate the two long-time parameters of our example with considerably greater accuracy than is possible for any of the short-time parameters (for instance, if the threshold, H, is not well determined independently of α by the data). Another way in which it is possible to focus on certain aspects of the long-time behaviour of a catchment system is to use a cumulative representation.

Cumulative models

In the same way that spatially explicit models are sometimes preferred to lumped parameter models, so-called 'cumulative models'[3] are often regarded as poor relatives of 'event-based' models. Cumulative models are obtained from the more familiar instantaneous discharge models (for example, a chemograph), by accumulating the mass of a particular species discharged during each time-step. Cumulative models have similar properties to up-scaled models, and act like a low-pass filter on the data, but without the loss of information inherent in upscaling (see Barnes, 2000).

It is often objected that the high R^2 values customarily associated with cumulative model fits are spurious. For instance, if the rainfall time series can be regarded as stationary, a cumulative model assuming average stream discharge will produce a near perfect fit in the long term. However this apparent anomaly can be avoided by comparing model variance to the total variance relative to the constant discharge model, or some similar modification that removes the effect of a linear trend from the total sum of squares. This procedure, termed the adjusted goodness of fit, is the one we follow in examples considered here.

It is interesting to observe that, unlike the previous example, there can be no loss of information associated with cumulative models relative to the associated discharge model, as all the original data are explicitly recoverable. However, if parameter estimation is carried out through the minimisation of an objective function[4] (such as the residual sum of squares), parameter estimates depend on the particular form of the chosen objective function, and may differ markedly for different choices. In particular, significantly different parameter values may be obtained from model fits to equivalent objective functions, one based on the instantaneous discharge and the other on the cumulative discharge, with the mean trend removed.[5]

Figure 29.3 shows a hydrograph of a stream, Broken River, at Crediton in the subtropics of southeast Queensland, Australia, in both daily discharge and cumulative form, compared to rainfall (data from the Australian 'benchmark' catchments of Chiew and McMahon, 1988). The catchment outlet is at 21°10'E and 148°31'S, and drains an area of 41 km^2, with average annual precipitation of 2100 mm.

Assuming that rainfall-runoff data are the only relevant information available, a spatially lumped parameter model is appropriate. Figure 29.3 also shows the results of fitting a unit-hydrograph model[6], with a total of five parameters, to hydrograph data using the objective function:

$$\mathrm{RSS} = \sum_{j=1}^{N}\left(\sum_{i=1}^{j} q_i - f_j\right)^2 \tag{29.3}$$

equivalent to minimising model deviations from the cumulative hydrograph (q_n and f_n are the measured and fitted discharge, respectively, at time n).

Optimal parameter values obtained by minimising the residual sum of squares

$$\mathrm{RSS} = \sum_{j=1}^{N}(q_j - f_j)^2 \tag{29.4}$$

for the discharge model gives a goodness of fit (R^2) value of 0.644 for discharge, and 0.673 for the adjusted goodness of fit (as described above) for the cumulative model. Conversely, using the objective function based on the cumulative model, Eqn 29.3, gives corresponding values of 0.541 and 0.994, respectively.

The optimal 'unit hydrograph' model based on the discharge residual sum of squares is evidently not a particularly good choice for the hydrograph, showing the usual (unit hydrograph) propensity for under-fitting the high peaks and over-fitting the smaller peaks; resulting in only moderate goodness of fit. The use of 'unit hydrograph' in this chapter simply denotes the expression of discharge as a weighted sum of delayed 'effective rainfall', which differs somewhat from Sherman's original usage (Sherman, 1932). The use of a unit hydrograph to represent water routing within a catchment can be shown to be exact if the velocity distribution within the catchment does not change with the size of the event. Because larger events generally lead to higher water velocities, and least squares optimisation favours the fitting of peaks, small peaks are over-fitted and large peaks are under-fitted.

The resulting cumulative hydrograph fit is significantly worse, because effective rainfall is consistently overestimated. On the

3 If q = f(t) is the instantaneous mass flux of some quantity (i.e., not a concentration), then the associated cumulative model, Q = F(t) is defined (in the continuous case) by

$$F(t) = \int^{t} f(t)dt$$

4 An objective function is a representation of the 'goodness of fit' as a single number, which is optimised (usually minimised) by varying parameter values. The optimised parameter values will depend on the objective function chosen. Different functions give different values. The most common objective function in this context is the residual sum of squares based on the hydrograph.

5 This is independent of the particular objective function chosen, provided that effectively the same objective function is applied to each. We are not concerned here with arguments about the most appropriate objective function.

6 The fitted model is a 2-reservoir linear routing model coupled with a constant loss-proportional loss model for effective rain (that part of rainfall that eventually becomes discharge); see below.

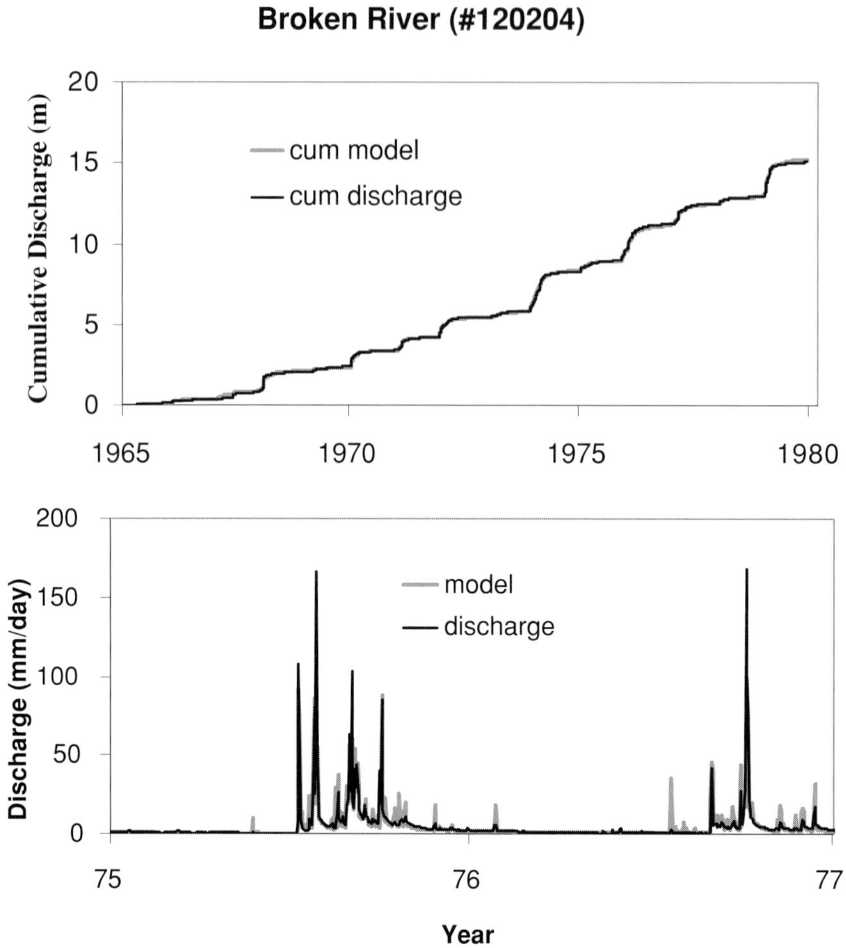

Figure 29.3 Cumulative discharge and daily discharge data (black lines); and fitted model (grey lines) for Broken River at Crediton, in Queensland, Australia.

other hand, the model optimised on the cumulative discharge objective function does barely worse on the discharge, and is near perfect on the cumulative discharge. The two models are based on two different representations of the same conceptual model, and exactly the same information.

What can we conclude from these observations? Firstly, we note that whereas the discharge model is biased towards fitting the peaks of the hydrograph, the cumulative version de-emphasises the high-frequency components in favour of the total runoff amount. This implies that whereas the unit hydrograph assumptions are not particularly appropriate for representing rapid discharge for this catchment, the simple assumptions leading to estimated effective rainfall and the slow drainage component give a very good representation of observed rainfall generation behaviour.

Adding a third linear reservoir of intermediate time constant to better define the high frequency response results in only marginal improvement to the discharge hydrograph fit ($R^2 = 0.65$). This suggests that unit hydrograph assumptions are not adequate for explaining the rapid flow component, or that there is simply too little information to accurately characterise the hydrograph, or both. (It is clearly not possible to resolve the major peaks for this relatively small catchment with a daily sampling period).

On the other hand, the runoff generation parameters (loss parameters and initial storage deficit) and slow flow component are quite well defined by the available information. We have also demonstrated that, for a given set of model assumptions, the choice of objective function or optimisation technique has a strong effect on parameter estimates. It is necessary therefore, in choosing a model, to also make a choice of objective function (see for example Seibert (1999, Table 1) for listing of types of objective function) that is in accordance with the overall purpose of the model. For instance, the choice of an objective function

Table 29.1. *Critical comparison between agricultural non-point-source model (AGNPS) and catchment management for sediment and solutes (CMSS)*

Class/model assumption	AGNPS	CMSS
Purpose	Management for water quality (nitrogen, phosphorous)	
Scale	Spatially distributed, event based	Partially spatial disaggregation, time averaged
Critical processes	Solute and particulate generation and routing	Solute and particulate generation
Information	Topographic, soils, climate, chemical data; geographical information systems (GIS)	Chemical, land use
Remarks	Inconsistent, inexplicit and unjustified assumptions and parameterisation; unknown uncertainty; heavy data requirements; complex	Largely consistent assumptions and structure; explicit uncertainty estimates; moderate data requirements; very limited application – semi-quantitative only

based on the cumulative representation of a model is appropriate for determining the parameters of streamflow generation such as average evapo-transpiration, E_T; whereas for modelling peak flow characteristics, the objective function should be based on an instantaneous discharge representation.

Critical process assumptions

For a given model purpose, and the consequent implications for appropriate scales, it is also necessary to recognise the implications for process representation. For instance, a model describing water quality and quantity for a large basin will emphasise the morphology of the stream network, and seek to avoid explicit representation of hillslope processes (equivalent to assuming that the time taken for precipitation to reach the stream network, from where it falls on the hillslope, is negligible). The role of the connected groundwater system also becomes more important in large systems, and more attention has to be paid to the non-uniform distribution of rainfall in both time and space (see Bonell, Callaghan and Connor, this volume). Similarly, different assumptions on the relative significance of different processes (e.g. generation processes versus routing) will result in differences in forms between two models (see below).

Thus, depending on the intended scale (and ultimate purpose) of the model, different processes may be de-emphasised or even ignored. Even at the same scale, essentially the same process may have different representations in different models. For example, in the second comparison given below, the effect of topography on hillslope drainage is represented in two quite distinct ways, due to different assumptions on the relationships between water storage and local discharge.

Model comparisons

To further illustrate the effectiveness of model classification by assumptions, we compare distinct models developed for similar purposes. The first comparison is between two models developed for the declared purpose of water quality management; the second compares models designed to represent stream generation from rainfall.

AGNPS VS. CMSS

The first pair of models chosen for comparison are: the model of Young *et al.* (1989), called AGNPS (*AG*ricultural *N*on *P*oint-*S*ource model), designed to evaluate the contribution of nitrogen and phosphorus from agricultural lands to pollution of watersheds; and the model of Davis and Farley (1997), called CMSS (Catchment Management for Sediment and Solutes), designed for essentially the same purpose. Table 29.1 gives a comparison of the respective models in terms of their inherent assumptions. The primary differences between the two models are scale assumptions, both spatial and temporal (see Barnes *et al.*, 1997b for a fuller discussion).

CMSS is a semi-lumped parameter model, assuming the availability of annually averaged data, and low-resolution spatially averaged data, dependent on a small number of distinct land use classes. Consequently, no routing information is required, and the emphasis is on nitrate and phosphorus generation, through 'export coefficients' which are constant for each land use class. This simple structure allows very flexible use of available information, including the use of qualitative information. It also allows at least crude estimates of uncertainty in the model outputs.

Table 29.2. *Comparison of five streamflow generation models. Note that all of these models have undergone significant development in diverse directions since initial publications. This classification most closely follows the original development of the separate models, rather than the later embellishments*

Assumption class/model	THALES TAPES TOPOG	TOPMODEL	IHACRES	HBV
Purpose	Streamflow generation			
Scale	Spatially explicit, event based	Pseudo-spatially explicit, event based	Spatially averaged, event-based	
Data	Topographic DEM, rainfall, soils data	Summary topographic and soils data, rainfall	Rainfall, (temperature), streamflow calibration data	
Critical process assumptions	Effect of topography on water routing (drainage capacity vs supply); water loss; quasi- steady-state; Darcy's law discharge with soil water potential parallel to topographic surface	As for TAPES, but with empirical relationship between local storage and drainage	Water loss occurs only during rainfall; stationary routing distribution; linear dependence of system discharge on storage	Water lost from upper soil reservoir; routing depends on rainfall intensity; linear discharge/ storage relationship
Remarks	Virtually identical conceptual models and major assumptions; differences in analytical methods; can be quite complex-difficult to analyse uncertainties; most complex models	Similar, but inherently simpler model than TAPES, TOPOG; intermediate complexity	Spatial structure subsumed into statistical routing model (cf. unit hydrograph) allows focus on temporal variations; least complex models	

AGNPS, on the other hand, is an event-based, spatially distributed model. It therefore requires spatially distributed information on catchment characteristics, and high-frequency solute generation and routing information. Because the model is supposed to be for water-quality management purposes, recognition is made of the low quality of the available data by effectively downscaling lumped parameter sub models, an exercise that is fraught with difficulties. The result is a complex model that is quite unsatisfactory, especially from the point of view of the ad hoc manner in which sometimes-conflicting assumptions are introduced. In particular, it is virtually impossible to estimate the degree of uncertainty associated with any forecasts of the model, a critical requirement of a management model. In fact, rather than being a practical management tool, the model is far more suited to research purposes, for which purpose some of the shortcomings of AGNPS are less significant.

COMPARISON OF STREAMFLOW GENERATION MODELS

Table 29.2 compares five models of streamflow generation in terms of their most significant inherent assumptions. THALES, a dynamic model based on the element network created by TAPES (**T**errain **A**nalysis **P**rograms for the **E**nvironmental **S**ciences, Moore *et al.*, 1988; Moore and Grayson, 1991; Moore, 1992) and TOPOG[7] (O'Loughlin, 1981; 1986) are essentially different expressions of the same broad conceptual model. THALES is a combination of separate surface and subsurface hydrological models based on the TAPES-C contour analysis and is named after Thales of Miletos, a Greek philosopher who recognised the influence of topography on water movement (Grayson *et al.*, 1995). TOPOG relies on the insight that, for soils of finite hydraulic depth, the water status at any point is a trade-off between the soil columns' capacity to transmit water laterally (depending on gradient and conductivity), and the up-slope contributing drainage area. TOPMODEL (a TOPography based hydrological MODEL) (Kirkby, 1975; Beven and Kirkby, 1979; Beven *et al.*, 1984) was independently developed at about the same time, based on a similar insight.

The IHACRES model was developed principally by Littlewood and Jakeman (see Jakeman *et al.*, 1991; Schreider and Jakeman, this volume) as an extension of the instantaneous unit hydrograph

7 TOPOG is used in the generic sense to include all software that is used to model hydrological response. It is the 'root' (known as the kernal) module developed to calculate the topographic attributes essential for modelling the impact of land use change in natural terrain (O'Loughlin *et al.*, 1989) with various supplementary process hydrology modules (see Vertessy *et al.*, 1996; Vertessy and Elsenbeer, 1999).

(IUH) tradition (see Sherman, 1932). Its success depends on the statistical result that any distribution function can be decomposed into a series and parallel combination of exponential distributions. As we outlined earlier, since each exponential distribution can be associated with a linear reservoir, this gives an objective way of representing the stationary part of the linear response of any catchment. The assumptions of stationarity, independent of antecedent conditions, and linear routing for the fast response part of the hydrograph, are probably the most limiting ones for this technique. In fact, a comparative study of the relationship between lumped versions of the above topographically explicit models and IHACRES may lead to useful insights into the most significant effects of topography and geometry on variations in catchment response.

The HBV model, in its original form, is similar to and predates IHACRES, and was developed from about 1970 by Bergström, and subsequently used extensively, particularly in northern Europe. It uses a lumped catchment (spatially implicit) approach, where streamflow generation is determined also using a non-linear module (identified with the soil-water and snow-pack store), which is then routed through two quasi-linear reservoirs effectively in parallel. The output from these two reservoirs is then combined and further routed through a fourth linear reservoir connected in series, representing the flow delay caused by the stream channel itself.

This version of HBV was used for catchments in Sweden and Germany in the comparative study of models by Seibert (1999). Lindström *et al.* (1997) modified this structure by introducing an alternative parameterisation for the response function in the upper zone, while the lower zone representation remained unchanged. Later Lindström (2000) added a by-pass flow component to the upper soil-moisture storage zone that gave improved results for the modelling of the stream discharge and associated isotope (^{18}O) chemograph at times of high groundwater levels.

A comparison of the HBV model structure shown in Figure 29.4, with that of Figure 29.2 (see also Figures 29.7 and 29.9), demonstrates the close relationship between HBV on the one hand, and derivatives of IHACRES (incorporating quasi-linear reservoirs developed by Boughton, 1984) on the other. Apart from the channel routing through a serial reservoir, appropriate to large catchments (a modification of Muskingum routing) the structures of the conceptual model units are very similar.

MODEL DEVELOPMENT AS A PROCESS, RATHER THAN A PRODUCT

We have examined the way in which the development of models of catchment systems is influenced by differences in purpose

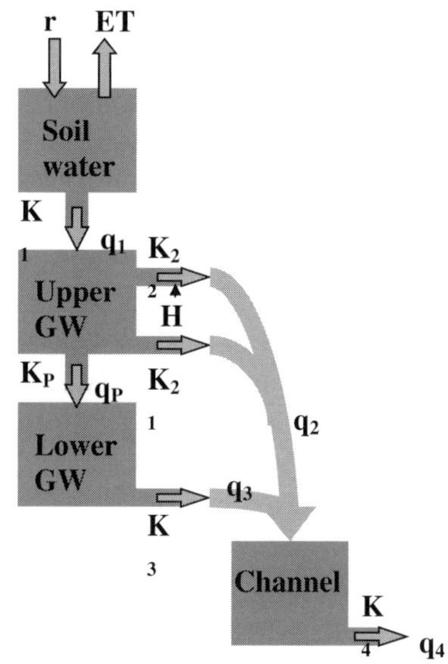

Figure 29.4 Schematic of the HBV model. Note the combination of a non-linear streamflow generation module with a series and parallel combination of linear or quasi-linear reservoirs. The quasi-linear reservoir involves a threshold H, representing an additional pathway at high rainfall intensities (cf. Figure 29.2).

and available information, and how this results in differences in assumptions. We have also shown that there may be many representations of the same system, or even the same process, depending on different assumptions which arise from considerations such as model purpose, available information, or which processes are assumed dominant. We have suggested that a classification of available models in terms of these underlying assumptions is a possible way by which an appropriate selection from available models may be made.

Two questions now arise:

(1) Is there a sufficient number of readily available models for any purpose that is likely to arise?
and
(2) If not, what modelling options are open to a hydrological practitioner?

We shall explore these questions by examining two contrasting approaches.

The universal process model

We have already remarked on the large number of published models, each with its own very wide claims of applicability. However, we also mentioned the relatively complex nature of the models, as a consequence of this desire for a wide range of application. This complexity makes it very hard for the hydrological practitioner to either understand whether a particular model is appropriate for his/her system/purpose/information, and to develop the necessary expertise to be able to use more than a very few models competently.

Furthermore, use of a complex model produces inherent problems in parameter identification, cost of information, uncertainty estimates, interpretation and transparency of the output results. In attempting to provide the necessary information to run a general model, one is often forced to rely on surrogate data using assumptions that may conflict with those of the model itself; and to consider processes of little or no relevance to the desired output.

The desire for a model with wide capabilities to cope with a large range of purposes has been called the search for the 'universal' model (Barnes *et al.*, 1997b). In addition, we have suggested that such a model is in fact inferior to more specific models, which implicitly make a much stronger statement about the relevant system. Hence, the more specific and less complex models may paradoxically require a better understanding of the system. In the next section we begin to outline a possible alternative approach which, while retaining important advantages of the universal model approach, avoids many of the difficulties enumerated above.

The universal model process: AMP

From the last section we see that the Holy Grail of modelling would be a simple, efficient model with a very wide range of applicability – an apparent impossibility! What we want to do here is to take a step back and, rather than look for a single universal model, to attempt to define a *process* which will yield near-optimal models which are efficient for each situation. The methodology borrows from theory developed principally in the USA and Japan in the 1980s for the manufacturing industry, adapted to apply to environmental systems. The specific formulation is adapted from a methodology termed Integrated Process Management (IPM) (Slater, 1991).

Figure 29.5 shows a schematic outline of the adaptive modelling process (AMP) approach, which may be termed a *five step, double-feedback loop* method. Several nearly equivalent approaches have been developed in parallel with IPM, but we will not attempt to compare alternative approaches here, except to say that the IPM approach appears to lend itself particularly well to the current purpose.

We note that this *meta*-modelling process is iterative, hierarchical and adaptive. The hierarchical nature of the process allows the resultant model to identify and concentrate on the critical areas or 'hot spots' of system behaviour (those areas which are important, or to which model output is particularly sensitive). The adaptive and iterative properties produce models that are close to optimal, while being capable of responding to changing circumstances.

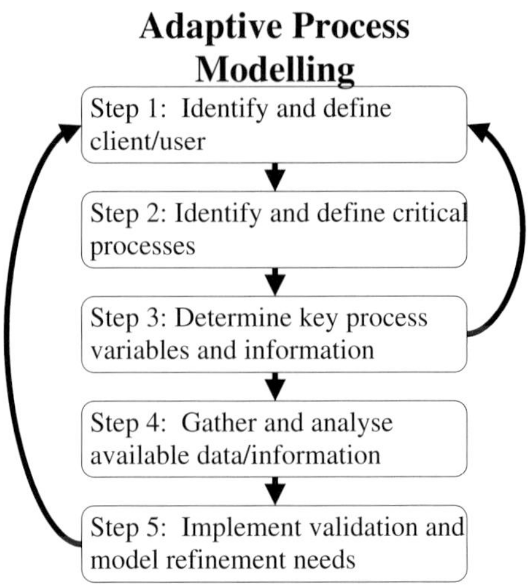

Figure 29.5 Schematic of the adaptive process modelling (APM) methodology. Note the feedback loops associated with steps 3 and 5, which make adaptation possible.

STEP 1: IDENTIFY AND DEFINE CLIENT NEEDS

Notice first of all the emphasis on the needs of the client. This is not just a token effort undertaken at the beginning of the modelling process, but an integral partnership of the modeller with the client, which is continued throughout the modelling process. The importance of the iterative (feedback) part of the procedure is the recognition that initial communications will not be perfect, either because of perspective or language (technical) not shared, or because the client may need to change his mind on reflection about some aspects.

The modeller and client together must identify:

- The purpose, the need, context and intended use of the model (engineering, regulatory, socio-legal, ethical . . .).
- Available resources (time, man-power, finance . . .)
- The scales (temporal, spatial)
- Output precision and accuracy
- Required output form (platform, graphics, publication of results . . .).

STEP 2: IDENTIFY AND DEFINE CRITICAL PROCESSES

This is the stage at which the conceptual model is defined and refined. It is the most intellectually demanding stage, during which the major part of the real intellectual property is generated. It is almost always easier to start with a model which is too simple, only adding to the complexity when required by Occam's razor, given the data available for validation (steps 3 and 5). On subsequent iterations from step 5, refine the model and assumptions by adding a representation of the most significant process not yet modelled. The stages under this part of the process are:

- Define/revise the system boundaries (and hence the extent of the system) through consideration of what input and output information is available, and client needs.
- Transform (revised) client needs into key model characteristics. Define the output quantities and formats, and the range of scales relevant to these needs.
- Identify and prioritise the relevant/revised sub-processes and inputs, and the significant interactions between processes and sub-systems.
- Define (revised) ***critical*** sub-processes and dependencies (model structure, structural assumptions) at the client's scales of interest by pruning the sub-process/ parameter list to the bare minimum.
- Begin with a simple, whole-system and long time-scale behaviour to attempt to characterise the observed gross behaviour of the system.

STEP 3: DETERMINE KEY PROCESS VARIABLES AND INFORMATION

This is the stage at which the model assumptions are spelled out, input data is identified and model structure is decided. There are four points:

- For each of the critical sub-processes, identify what type of information is required;
- Determine what type of input data and information is available;
- Hence, determine what assumptions are necessary (to compensate for missing information). This is the stage at which the use of surrogate data (data which are used in place of unavailable data ostensibly required by the model, such as turbidity data in place of sediment concentration) would be considered;
- Determine what data are available to validate model output against observed system behaviour. In particular, will the validation data support the model complexity?

At this point (Figure 29.5) the process loops back to the first stage with further consultation with the client, until it converges and both client and modeller are content that what has been arrived at truly reflects the clients needs as effectively as possible.

STEP 4: GATHER AND ANALYSE AVAILABLE DATA/INFORMATION

Here the conceptual model is represented quantitatively, and an attempt is made at validation using existing system data.

- Gather the available data identified in the previous step.
- Assess data integrity to assess and improve the information content. Data integrity involves such things as consistency, transcription errors, and tampering (i.e. lack of objectivity in obtaining information).
- Analyse data/information for parameter values and uncertainty. Can all parameters be identified from the validation data alone?

STEP 5: IMPLEMENT VALIDATION AND MODEL REFINEMENT

Finally:

- Model output is compared with observed system behaviour, using historical or other data.
- Model residual variance is checked to determine if it meets client requirements for precision. Where possible, assumptions are also checked against model outputs and observed system behaviour to verify that they are met.
- Estimate the uncertainty in the model output due to uncertainties in parameter estimation, simplifying assumptions, etc. Does this meet the client's needs?

The process then loops back to the first step, and the model is continually revised until the process converges, and the model represents the bare minimum of processes required to satisfy client requirements. If we begin with the simplest possible conceptual model in step 2, identifying the single most important sub-process at each major iteration (that which gives the best fit or smallest variance), the process then yields a hierarchy of increasingly complex models, of increasing explanatory power. The process stops either when the available information on system behaviour will not support further model refinement, or when the client's precision requirements have been met (whichever is first).

In practice it is necessary to balance the requirement to converge rapidly by making optimal refinements to the model, with the need to 'go round the loop' many times to make simple refinements one at a time. The latter strategy also allows near-optimal simulation of non-stationary systems, as long as the period of iteration is short compared to the rate of change of the system.

The problem of validation

It is relatively easy to construct a plausible model representation of a catchment system. Given some exploratory data analysis, and knowledge of the sort of processes that are likely to be of interest, standard representations of sub-processes can be strung together like pieces of Lego. Such a model is often (but not always) quite capable of reproducing observed system behaviour to the desired degree of accuracy; but the same would also be true of a purely empirical (polynomial) model with a similar number of degrees of freedom. The argument that models like this are justified, because the included processes have been validated at the sub-system scale, is itself invalid for several reasons.

Firstly, as has been already indicated, the phenomenon of emergent processes introduces constraints into the simple process of summing the behaviour of component parts to obtain collective behaviour. Secondly, as complexity is increased (involving the introduction of extra parameters) the number of potential ways in which a model can be extended increases exponentially. This may lead to adoption of a plausible but erroneous model unless considerable caution is exercised.

For instance, consider a simple linear catchment routing model, represented by two linear reservoirs in parallel, as in Figure 29.6a. (The quantity r* represents 'effective rain', which is divided (proportionately constant β) between a fast and slow response linear reservoir with time constants α_f^{-1} and α_s^{-1} respectively). However, exactly the same hydrograph is produced by the alternative structure shown in Figure 29.6b. Whereas in the conceptual model of 29.6a, the proportion of 'fast' flow is β, in Figure 29.6b (given the parameter identifications shown in the figure) the proportion is increased to

$$\beta + {}^{\alpha_s}/_{\alpha_f}(1-\beta) \qquad (29.5)$$

with a corresponding decrease in the estimated 'slow' flow proportion. In other words, hydrograph decomposition depends on the class of model structures considered, and ***determining the 'slow' – and 'quick' – flow proportions of a hydrograph is not a well-posed problem***.[8] Residence time distributions[9] of water in the two representations are also not the same. There are therefore two distinct but plausible representations of the same hydrograph with quite different basic properties, which cannot be distinguished based on hydrograph data alone, independent of the effects of measurement error.

What we have shown with the above simple example is that the problem of hydrograph separation from the hydrograph alone cannot be solved uniquely, unless further (a priori arbitrary) restrictions are placed on the class of models considered. If additional independent information is available to further constrain the models (e.g. tracer data), it may then be possible to distinguish unequivocally between the two systems (or exclude them both). This result is clearly not an artefact of errors in the data, but is an inherent limitation due to the lack of information contained in the output of a hydrograph resulting from more than one process. Merely being able to accurately reproduce observed behaviour is insufficient to fully validate the structure of this, or any other, model (Beven, 2002; Seibert and McDonnell, 2002; Buttle and McDonnell, this volume).

Seibert and McDonnell (2002) for example showed that a R^2 of 0.93 was determined when rainfall-runoff was simulated using runoff alone (hard data) for the Maimai Catchment 8

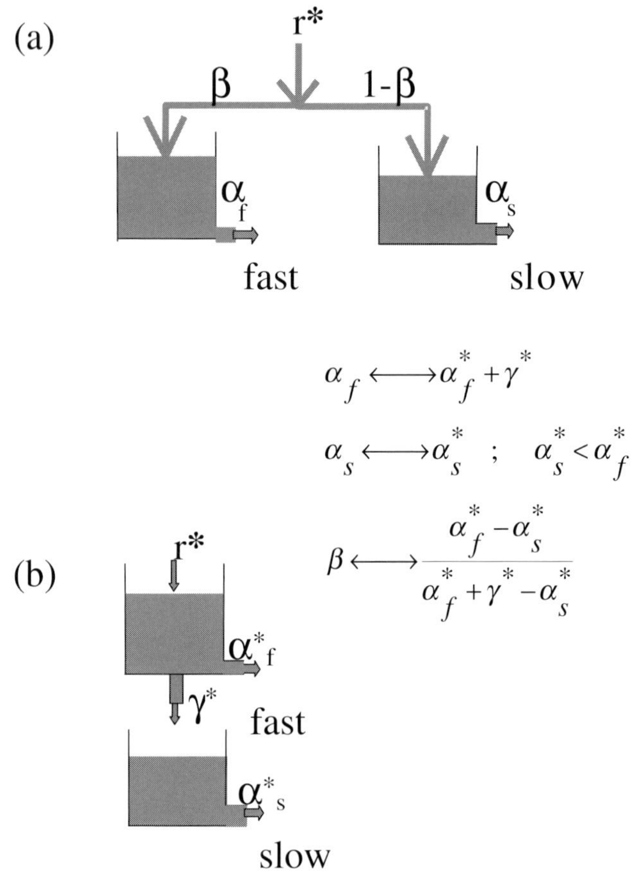

$$\alpha_f \longleftrightarrow \alpha_f^* + \gamma^*$$

$$\alpha_s \longleftrightarrow \alpha_s^* \quad ; \quad \alpha_s^* < \alpha_f^*$$

$$\beta \longleftrightarrow \frac{\alpha_f^* - \alpha_s^*}{\alpha_f^* + \gamma^* - \alpha_s^*}$$

Figure 29.6 Conceptual unit hydrograph model consisting of two linear reservoirs. (a) The two reservoirs are arranged in parallel, with effective rainfall r being partitioned proportionately (constant β) between the fast and slow response reservoirs. (b) The reservoirs are arranged vertically. There is a one-to-one correspondence between the two representations, which produce mathematically identical unit hydrographs. However, the hydrograph separations (and residence time distributions) are not the same.

(New Zealand). On the other hand, simulations of new water contributions to peak runoff, and separately simulated groundwater levels for each of the three boxes of their hillslope-model using other 'fuzzy' evaluation criteria showed in general a poor goodness of fit in comparison to the preceding runoff-only calibration.

Thus despite the high model efficiency for runoff, the within hillslope stormwater transfer as indicated by saturation status (groundwater levels) and storage across the respective hillslope,

8 A problem is said to be well-posed if the solution is unambiguous, i.e. there is a single solution which solves the problem within the class of solutions under consideration.

9 The residence time distribution of a chemical species in a system at any given time, is defined as the distribution of times which currently exiting molecules of that species have remained in the system. For stationary systems, the residence time distribution does not change with time.

hollow and riparian zone boxes were poorly represented. Subsequent inclusion of 'soft-data' criteria in the model calibration process resulted in lower R^2 values (0.84) but led to a better representation of hillslope runoff generation dynamics based on experimental hydrology (Seibert and McDonnell, 2002). Buttle and McDonnell (this volume) provide more details on the inclusion of 'soft-data' criteria in the model calibration process.

What can we conclude, and what can we do, if even the exact reproduction of system behaviour is insufficient to fully validate a catchment model? Firstly, it must be stressed that all models are only representations of the systems they describe, within the ambit of their inherent assumptions. If it turns out that a model represents system behaviour outside the region expected from the assumptions incorporated, that must be considered fortuitous, and unless and until the reasons for this extended region of validity are understood, it remains an unreliable validation.

Accepted scientific methodology involves Occam's razor,[10] whereby, in the face of uncertainty, the simplest hypothesis is chosen unless it can be invalidated. This allows us to semi-objectively choose a particular representation at each step of the modelling process, and only to discard the model (the null hypothesis) for another (more complex) representation when the first model can be shown to be inadequate, based on observed data. From what has been said earlier, validation data must be at the scale at which it is intended to apply the model. Confidence in the adequacy of the model can be inferred only for the type of data for which it has been validated. For instance, it is not justifiable to use a catchment representation, validated for hydrograph prediction, to predict solute discharge, though it may be instructive and suggest further hypotheses to be tested.

In practice, the number of degrees of freedom implicit in most environmental time-series records that are related to processes of interest is quite limited (usually ≤ 5), in the sense that this is the maximum number of parameters that can be reliably and independently determined for any reasonable model. Extraneous 'noise' makes additional parameters difficult to determine with any degree of certainty. This property of time series measurements places severe constraints on the complexity of any catchment-scale model that can be properly validated.

Numerous schemes for model validation that have been proposed attempt to minimise the problem of the lack of information in environmental time series, including various forms of split sample techniques (see review of Seibert, 1999). The problem becomes particularly acute when a large range of models is implicitly considered to explain a limited set of data. Because of the potentially huge range of processes that could be significant in explaining the rainfall-runoff relationship for a complex catchment, the difficulty of artificially enhanced apparent performance of models is important (the phenomenon of 'artificial skill'). Whereas in a well-studied catchment a new model may appear to explain much of the variance of discharge, for example, a proper validation must recognise the dependence of the new model on known previous attempts to represent the relationships, which may greatly increase the effective degrees of freedom represented by the model (see Michaelsen, 1987; Ganeshanandam and Krzanowski, 1989, for example).

One popular method for maximising the amount of data on which to validate new models is to use the complete (implicit) fitting process on all data points bar one, which is excluded from the process. The predicted value for the one point is then compared with the measured one; and the process is repeated for each point. The resulting optimised fit (utilising the predicted points) is sometimes called the 'miss-one-out' cross-validation method; it is useful because it focuses attention on the dangers of ignoring this sort of bias, or 'artificial skill' as it is sometimes called, and avoids various forms of bias which can occur with other split sample techniques. Although a number of techniques for estimating uncertainties in non-linear model parameters have recently been discussed in the literature, little attention appears to have been paid to this form of bias outside the climate modelling literature (see Michaelson, 1987).

AN EXAMPLE OF MODEL DEVELOPMENT: SOUTH CREEK, BABINDA, (WYVURI, HOLDING), NORTHEAST QUEENSLAND

To exemplify the somewhat abstract argument of the previous section, we consider the development of model representations of a catchment system. The common purpose of these models was:

(1) To characterise streamflow generation in this humid tropical catchment as accurately as possible from the data and
(2) To establish the significance of groundwater pathways relative to the total flow, and the time-scale of the delayed flow.

This is essentially a research motivation, as we are looking for the best possible explanation of catchment behaviour, rather than to manage the system within a range of outcomes. The model development follows a logical trail, looking at each step to explain the maximum observed streamflow variance with the minimum of parameters.

10 Occam's razor is the principle, attributed to the 14th century figure William of Occam, that of two possible explanations of the same facts, one should accept the simpler.

Catchment description and experimental design

The Wyvuri catchment, located near the coastal town of Babinda in the wet tropics of north-east Queensland, consists of two sub-catchments, called North Creek (18.3 ha) and South Creek (25.7 ha), and has been extensively described elsewhere (see Bonell, this volume; Bonell *et al.*, 1998). Annual rainfall is approximately 4000 mm, falling principally in the wet season (December–March) because of monsoonal depressions or cyclonic activity. Peak rainfall intensities greater than 100 mm h^{-1}, or 500 mm d^{-1}, are relatively common in this catchment.

Extensive data collection occurred in the 1990 and 1991 wet seasons (December–May), extending for about six months in each case. The purpose of this study was to develop an understanding of runoff generation processes, and particularly the influence of groundwater on streamflow, in these fairly steep and heterogeneous headwater basins. Because this was essentially a research model, we were our own clients in this case, so that client-modeller communications were not an issue.

In fact, data collection involved over 100 time series of tensiometer, piezometer, suction lysimeter, streamflow, rainfall and isotopic tracer data, collected from instruments distributed at key points over both sub-catchments (see Bonell *et al.*, 1998). Frequency of collection was event-based, but was also governed by physical limitations in the case of isotopic samples. Previous work had indicated that rapid response times of less than 10 minutes occurred with major events (Bonell *et al.*, 1981). With such a complex system of instrumentation in a relatively remote and hostile environment, equipment failures were inevitable, so that many of the time series were incomplete. Additionally, detailed soil, topographic, geomorphic and geological data were available, as were background data from previous investigations extending back for more than a decade (see Bonell *et al.*, 1981; 1987; 1998) that provided the physical basis for interpreting this modelling exercise.

The intention was to provide sufficient information so that detailed runoff generation and routing mechanisms could be distinguished from alternatives. The storm runoff-generation mechanisms are also given in some detail elsewhere in this volume (Bonell, Chapter 11). The purpose here is to attempt to understand the information available at the whole-catchment scale from measurements of rainfall, streamflow and tracer concentrations by following the above APM methodology. Although the modelling is restricted to information available at this scale, we use insights gathered from the more detailed measurements to comment on and interpret the results obtained.

Model development

Initially, a long-time view of the data was sought to attempt to understand the water balance. Further refinements are equivalent to focusing down on shorter time-scales, given that the longer time-scale behaviour is now explained. Although climate data (e.g. temperature, humidity) were available, as a first pass it was assumed that evaporation was a fixed proportion of rainfall throughout the season.[11] Data from each time-series were aggregated to three-hourly intervals to cut down the analysis time for this exploratory stage of data analysis, implying that the rapid-flow components may not be fully resolved.

ANNUAL TIME-SCALE

At long time-scales (of the order of a year or more), only the effects due to the major loss processes are evident in the cumulative hydrograph, and a catchment model in the form shown in Figure 29.7 is appropriate. The only parameters that can be determined at this time-scale are those associated with average evapotranspiration loss, and an additional one representing the initial state of the catchment. A cumulative plot of rainfall and streamflow for the undisturbed rainforest-covered South Creek is shown in Figure 29.8. The relative magnitude of the two scales suggests that about 70% of rainfall emerged as runoff during this period, so that about 30%, or 7 mm/day, was lost to E_T on average.

In the earlier water balance study of Gilmour (1975), the E_t/E_o ratios determined over short periods were in the range 0.9–1.0 when volumetric soil water contents in the 0–3.0 m profile were in excess of 0.45. (E_t was calculated as the difference between 0–3.0 m profile soil moisture volumes for consecutive measurements; E_o were Penman estimates of potential evaporation demand). E_t/E_o ratios in excess of 1.0 (up to 1.5) were also found, when substantial rain (in excess of 100 mm) had fallen during the previous measurement period, and little or no rain occurred during the current period. Apart from E_t, Gilmour (1975) attributed such high rates to protracted lateral (as well as vertical) drainage which causes the hydrograph to have attenuated recessions. Thus, bearing in mind the preceding constraints, the modelled 7 mm day^{-1} here is within the range of Penman E_o estimates of Gilmour (1975) during periods of maximal catchment wetness in the summer wet season.

For Australian catchments which undergo long dry spells, rainfall-runoff cumulative data over decadal time-spans are often well fitted (Barnes, unpublished) by a combination proportional loss/constant loss model of the form

$$\sum_{i=1}^{N} q_i = S_0 + R \sum_{i=1}^{N} r_i - N.T \qquad (29.6)$$

11 An alternative assumption of a constant rate of E_T was also considered. Results for models based on this assumption gave similar results for a given degree of parameterisation, but were generally not as good as for the proportional loss assumption. However, the more complex the model became, the smaller the difference between the two assumptions became.

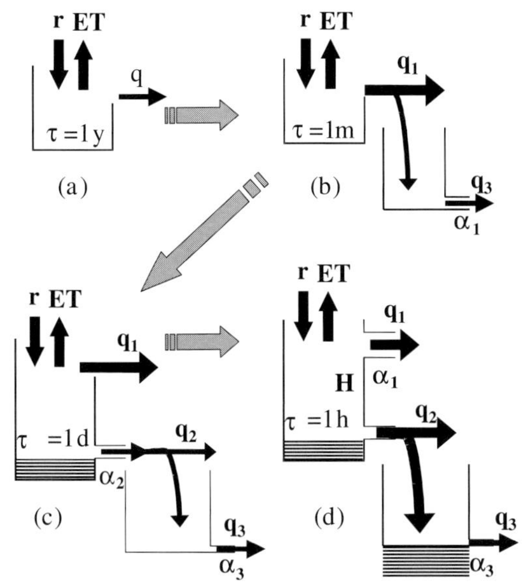

Figure 29.7 Development of a hydrograph model for Babinda South Creek, for decreasing time-scale τ. (a) Long time-scale view (τ = 1y), where only average losses are evident (2-parameter model); (b) τ = 1 m, delays due to deep groundwater are apparent; (c) τ = 1 d, delays due to shallow subsurface flow become significant; (d) τ = 1 h, delays can be discerned even in the rapid flow component. Note that the structure of the detailed model of Figure 29.7d is essentially the same as that of the HBV model, Figure 29.4. The areas of horizontal grey lines represent storage volumes that do not actively participate in discharge dynamics, but represent non-linear runoff behaviour during dry conditions. The lowest storage in (d) has no effect on discharge dynamics, but is included to explain differences between the runoff response and the response of tracers.

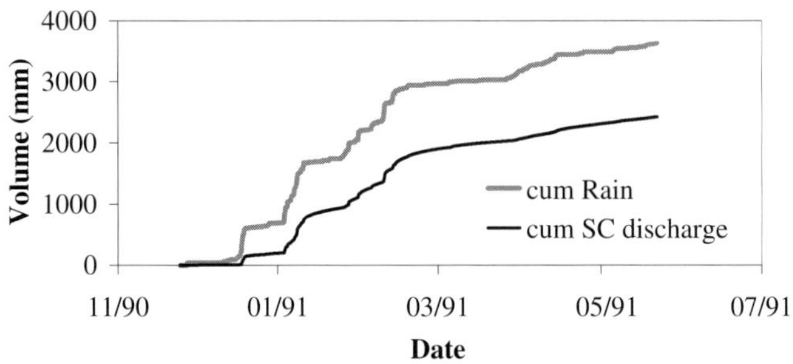

Figure 29.8 Cumulative rainfall and South Creek cumulative discharge data for 1991 wet season.

where q_i and r_i are the discharge and rainfall for the ith period, R ($0 \leq R \leq 1$) is the proportional loss coefficient, T represents (effectively constant) transpiration by deep-rooted vegetation, and S_0 is an initial storage deficit, which allows for initial variations from average conditions. This loss model can be restricted to either a constant loss model ($R = 0$, $T = E_T$, the evapotranspiration rate) or a proportional loss model ($T = 0$, $R = RO$, the runoff coefficient).

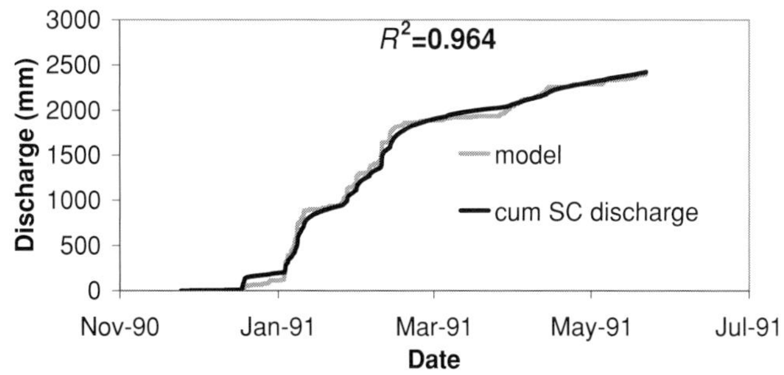

Figure 29.9 South Creek cumulative discharge data fitted by two-parameter proportional loss model. The overall shape is good, but the data is considerably smoother than the model.

Alternatively, in place of these linear loss models, a compatible non-linear loss module from other models, such as IHACRES or HBV, could be substituted. Figure 29.9 shows the results of fitting the two-parameter (proportional loss model (Eqn (29.6), $T = 0$; see Figure 29.7a) to the discharge and rainfall data, with the result that over 96% of the adjusted cumulative variance is explained (parameter values given in Table 29.3, first line).

THE SEASONAL TIME-SCALE

As the data available for the Babinda catchments are only seasonal in extent, the largest relevant time-scale for representing this catchment is of the order of months, represented by Figure 29.7b. Despite the fact that these catchments are extremely responsive, the relatively smooth nature of cumulative discharge relative to cumulative rainfall at the end of the season (Figure 29.8) suggests a significant component of delayed groundwater.

It would seem that the catchment is acting like a low-pass filter, smoothing out the high-frequency components of rainfall. Accordingly, a revised conceptual model was developed which directed a proportion of the instantaneous flow from the loss model (Eqn 29.6) through a second, linear, reservoir as delayed flow (Figure 29.7b). This view of catchment discharge, as the sum of an effectively instantaneous component plus a delayed component, is the next step as time-scales in focus are progressively reduced. Figure 29.10 shows the resulting four-parameter model fit, which gives a particularly good representation of the cumulative discharge in terms of cumulative rainfall, explaining over 99.5% of the adjusted variance.

In fact, if the parameter estimates are optimised using a modified objective function (obtained by adding the logarithms of the variance for both the cumulative and instantaneous hydrograph, the resulting four-parameter model (two loss parameters as above, a time constant and a coefficient which partitions between instantaneous and delayed flow) also explains over 78% of the hydrograph variance. The optimal constant loss model explains the same amount of variance for the cumulative model but only 55% of

Table 29.3. *Parameters and statistics for models fitted to South Creek discharge data, assuming a proportional loss model*

Parameters[a]	Units	2 parameters	4 parameters	5 parameters	6 parameters	7 parameters
S_0	mm	−416	−255	−250	−263	−252
RO		0.78	0.75	0.75	0.75	0.75
α^{-1}	days		10.1	10.1	11.4	12.0
β			0.47	0.653	0.09	0.61
α^{*-1}	days				0.15	0.48
β^*					0.51	0.58
r_1	mm h^{-1}			5.8		5.5
R^2_{cum}		0.964	0.9953	0.9967	0.9955	0.9965
R^2_{inst}			0.762	0.775	0.820	0.877

[a] Parameters obtained using a constant loss model were quite different.

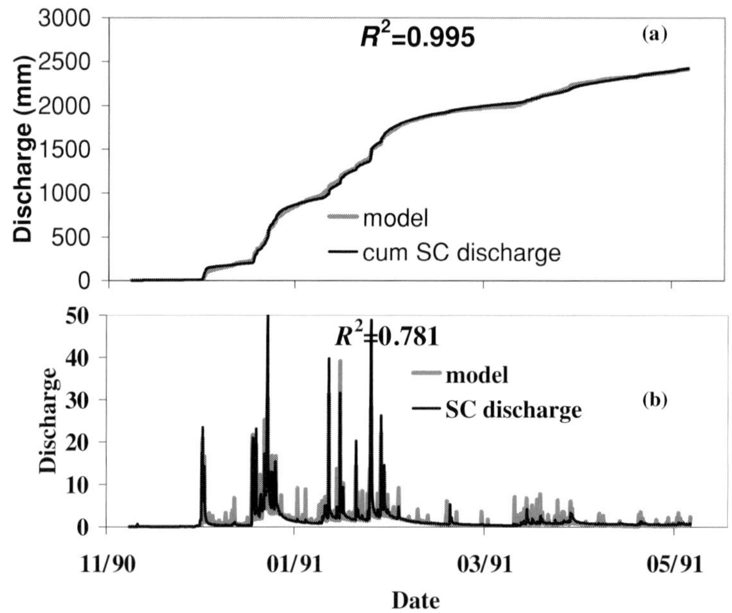

Figure 29.10 A four-parameter model fitted to South Creek data. (a) Cumulative plot, now a good fit to the data; (b) Hydrograph plot. Note the high frequency components predicted by the model, but not present in the data.

the instantaneous hydrograph variance. The combined constant loss/proportional loss model (see above) barely improved the fit over the constant loss assumption, and, being more complex, was discarded.

THE DAILY TIME-SCALE

When the residuals of the hydrograph fit are examined at a daily time-scale, a number of systematic deficiencies in the ability of this four-parameter model to represent observable behaviour are evident.

Firstly, the assumption of instantaneous discharge for the fast flow component is not adequate even at a daily time-scale, as some delays in this component are evident when the hydrograph is viewed at this resolution. The attenuated response of the catchment from the cessation of rain could be represented, in a way consistent with the conceptual development so far, by the introduction of a second linear reservoir with a smaller time-constant (of the order of a day), through which part of the fast flow component is routed (Figure 29.7c). Physically, from earlier work, this pathway can be associated with shallow sub-surface flow contributions to streamflow that persist for a day or so.

Secondly, this model predicts a rapid response to the numerous smaller rainfall events that is not evident in the measured data.

It appears that small rainfall events, particularly after a period without rain, do not cause a rapid rise in the hydrograph, whereas events of high intensity cause disproportional amounts of rapid flow. In process terms, it had previously been observed that only high intensity events on a wet catchment cause saturation overland flow, whereas lower intensity events were discharged through the upper soil matrix as subsurface stormflow (SSF) and did not appear to cause saturation overland flow (SOF) (cf. Bonell *et al.*, 1981). The simplest way to represent this intensity-dependent pathway in the current context is to introduce a reservoir of the form shown in Figure 29.2, representing the surface/upper soil water pathway (cf. the upper groundwater store of the HBV conceptual model of Figure 29.4). Under low intensity rainfall, the storage drains sufficiently quickly that the threshold for rapid drainage (representing saturation overland flow) is never reached. This threshold is identified with the constant H in Figure 29.7c and d. High intensity events quickly saturate the upper soil layer, and the flow may become dominated by rapid flow, which disappears abruptly after rainfall ceases.

Elsewhere, a detailed assessment of the temporal variability of hydraulic potentials of piezometers and soil-water pressure transducers, and the corresponding temporal variability in the isotopic concentrations (i.e. deuterium) in soil water, groundwater and streamflow for the event shown in Figures 29.12 and 13 (below), all suggested the possibility of an additional third store in between the upper and lower store of Barnes and Bonell (1996), as shown in Figure 7c and d. This intermediate store represents contributions from deeper SSF in the unsaturated zone (i.e. unsaturated in-between storms) during storm events via preferential flow (see detailed discussion in Bonell *et al.*, 1998, summarised elsewhere in this volume by Bonell).

The proposed conceptual model representation appropriate to a daily time-scale, and allowing for both a third reservoir and an intensity threshold, is shown in Figure 29.7c. At this scale, it may be appropriate to consider that, for a period of a few days, we may consider the slowest flow component contribution to be approximately constant, but we have chosen to keep the variation explicit in order to be able to compare cumulative fits.

These modifications to the conceptual model, taken one at a time, resulted in only modest improvement in R^2 values, but including both together (Figure 29.7d), a value of $R^2 = 0.877$ was obtained through the introduction of three extra parameters (a threshold, time constant, and proportionality constant), effectively halving the residual variance. Parameter estimates for each of the models evaluated are shown in Table 29.3.

Also from Table 29.3 it is apparent that whereas some parameters appear quite stable and independent (cf. α^{-1}, RO), others can change dramatically if the model is altered in some minor way (cf. β). Further, whereas the total water loss must be almost identical between models to conserve the water balance, there are large differences in both time constants and proportions of water flowing through the different pathways, depending on whether a constant or proportional loss model is assumed. For instance, the time constant α^{-1} varies by a factor of four between four-parameter models based on the two assumptions. This suggests that the discharge data alone do not contain sufficient information to be able to determine these parameters accurately, and either further data are required (cf. the three box model of Seibert and McDonnell, 2002 and similar research campaigns) to distinguish between the loss models or a re-parameterisation of the model is required.

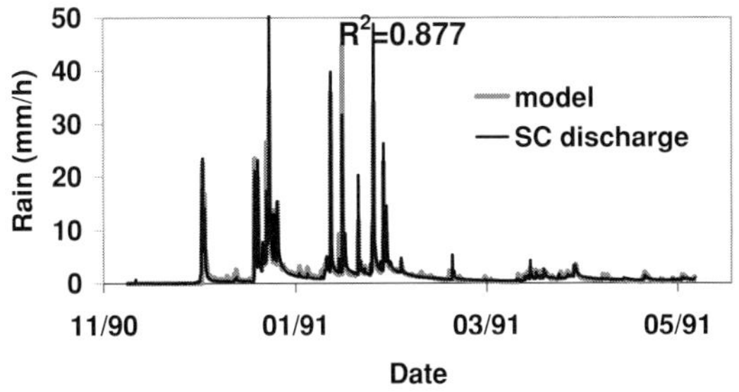

Figure 29.11 A seven-parameter fit to the hydrograph. The high frequency hydrograph structure is better represented, and the spurious small peaks are much reduced.

Another feature of the analysis is that as the complexity of the models, and the number of parameters increases, the parameter surface that needs to be searched in order to minimise the chosen objective function becomes increasingly complex itself. For example, the objective function for the seven-parameter model often has multiple local minima in close proximity, whereas for the four-parameter model the objective function usually has distinct and pronounced minima (Figure 29.11). For the more complex models, it is also possible to find widely separated minima for small changes in initial conditions or model structure, so that it is difficult to have confidence in model outcomes.

THE HOURLY TIME-SCALE

Finally, although the results are not presented here, it is possible to resolve the 'instantaneous' discharge component of Figure 29.7c by going to a sub-hourly perspective, represented by Figure 29.7d (see Barnes and Bonell, 1996). This requires the introduction of another time constant representing the initial recession of the large peaks, which then exhausts the temporal resolution of this data set, and further improves the representation of the hydrograph peaks in this model. From this (event) perspective, the slow component of streamflow, and possibly the intermediate component, appear constant (baseflow), but will vary from event to event.

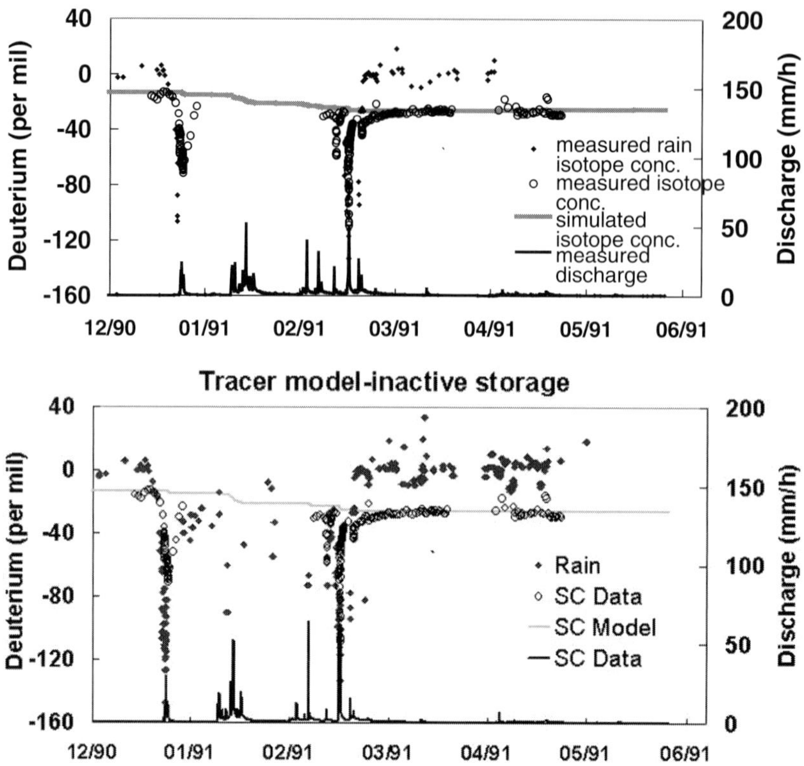

Figure 29.12 Plot of measured stable isotope data for rainfall and streamflow. Also shown is the model streamflow deuterium concentration in the slow flow reservoir, predicted from rainfall data. (a) The four-parameter rainfall-runoff model, discussed above is assumed to apply to tracer concentrations in discharge. The hydrograph response times are evidently too small for tracer response. (b) Augmenting the model with a large inactive mixing volume (*c.* 6 m storage) appears to produce a satisfactory result.

The resulting eight-parameter model may be considered to give an excellent fit to this high-quality data set, especially considering the relative crudeness of some of the assumptions (e.g. unit-hydrograph assumptions). We emphasise the relative independence of parameter estimates, as a result of the hierarchical method employed, where each new group of parameters (one-three at a time) were introduced in the context of a fairly precise prior determination of the earlier parameters.

Incorporating tracer data into the model

Tracer data give a somewhat independent picture of the discharge data, being much more sensitive to proportions of flow through different pathways and less directly dependent on time constants. We used the natural variations in stable isotopic components (deuterium and oxygen-18) of rainfall as tracers. Figure 29.12 shows isotopic concentrations in rainfall and streamflow samples collected during the monitoring period.

Assuming validity of the linear reservoir streamflow model, established from the discharge data alone, leads to an isotopic model that fails to reproduce the observed stream isotopic concentrations (Figure 29.12a) (see also Barnes and Bonell, 1996; Barnes *et al.*, 1997b). In particular, the low-flow concentrations show little variation over time, whereas the model (Figure 29.7d) without active storage in the slow reservoir suggests quite rapid variations should be observed. Also, exploration of the parameter space of this combined model (now with eight free parameters and two initial values) suggests that at least 20% of discharge volume should be rapid flow in order to reproduce the magnitude of high frequency isotopic behaviour.

A simple modification of the hydrograph model by extending the slow flow reservoir to have a large, well-mixed, but hydraulically passive, volume allows reasonable simulation of the long-time isotopic behaviour (see Figure 29.12b). Incorporating this feature into the more detailed conceptual model (cf. Figure 29.7d), gives a plausible qualitative representation of the observed isotopic behaviour (Figure 29.13). Comparing observed and modelled deuterium concentrations, it appears that failure of the model to reproduce the observed isotopic minimums can be associated with failure of the hydrograph model to adequately represent peak response, rather than the effect of a passive storage (i.e. not affecting the hydrograph) in the quick flow reservoir, similar to that of the slow flow reservoir. The quality of the tracer data (particularly missing values) causes some problems in model evaluation (Figure 29.13), so that although the direction of necessary modifications to the model is clear, parameter evaluation is only semi-quantitative

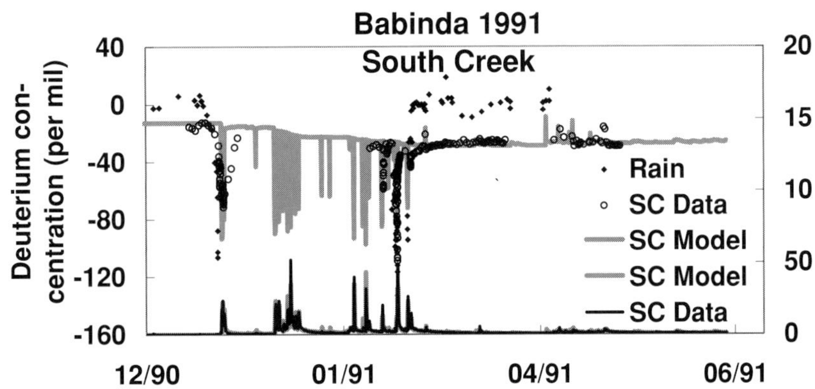

Figure 29.13 Isotope hydrograph predicted using the seven-parameter proportional loss hydrograph model, plus two passive storage volumes associated with the intermediate and slow groundwater reservoirs. There are still a number of features of the isotopic data unexplained.

at best. Minimisation of an objective function therefore was not considered appropriate with these data. The size of the passive component of the slow storage was adjusted by trial and error to optimise both the long-term isotopic behaviour, and the magnitude and dynamics of the more rapid response.[12]

From this analysis we conclude that the groundwater system appears to consist of two semi-independent bodies of water, having mean residence times of the order of a year or more (compared with maximum hydrograph response times of 10–40 days; see Table 29.3). The combined size of these reservoirs, implied from tracer data and model analysis, appears to be at least 2500–3000 mm, a magnitude that is completely unexpected from previous investigations. This probably explains why the constant loss model appears to give a good fit to the long-term data and also why the background deuterium concentration of streamflow remained virtually unchanged in between storms, despite in excess of 3000 mm over the wet season (Bonell *et al.*, 1998; Bonell, this volume). In addition, it appears that rainfall routing is intensity dependent, with overland flow only initiated when the intensity exceeds a certain level (about 6 mm h^{-1}). These conclusions can now be fed back into the modelling process for another iteration.

We mentioned earlier a much greater range of data than we have used here, including within-catchment observations such as tensiometer, piezometer, topographic and tracer data. The conclusions of the last paragraph are generally supported by this additional information, but the overall picture that emerges is of a hydrologically heterogeneous catchment as summarised elsewhere in this volume by Bonell. On the one hand, very fast response times suggest rapid streamflow generation from parts of the catchment. On the other hand, the rainfall intensity dependent response and the complex interaction of ground- and surface-water, make any simple description in terms of the effect of topography, e.g. TOPOG; TOPMODEL, of doubtful value. In fact, analysis of point hydrometric and isotopic measurements do support the existence of each of the mechanisms adduced above (Bonell *et al.*, 1998), but the same measurements are incapable of defining the overall significance to catchment response. These results of more detailed investigations have a secondary role in suggesting mechanisms that may be important at the catchment scale, rather than a primary role.

Finally, some remarks about the temporal variability of deuterium in rainfall (as shown in Figures 29.12 and 29.13) are needed. The available rain isotopic data show considerable differences in the range of deuterium concentrations between large monsoon storm events during the middle of the wet season as opposed to the dominant 'stream shower' activity found during the opening and closing stages of the wet season. The much lower deuterium concentrations are, in part, associated with cells of deep convection, with 'cold' cloud top temperatures as part of Mesoscale Cloud Clusters (MCSs, see Bonell, Callaghan and Connor, this volume). In contrast, the much more shallow 'stream shower' clouds provide much heavier (higher) deuterium concentrations arising from 'warm' rain. The differences in rain-producing processes within clouds are important here (Riehl, 1954; 1979).

An alternative response function

Recently, Kirchner *et al.* (2000, 2001, 2004) has correctly pointed out that a single linear reservoir (tank or box model) cannot even qualitatively represent the observed (stationary) response function for some tracers (e.g. chloride) in experimental catchments. They found that an apparently superior model was to use a response function $G(t)$ of the form:

$$G(t) = (\alpha t)^{\beta-1} \exp(-\alpha t)/\,\Gamma(\beta) \qquad (29.7)$$

where α^{-1} is a characteristic reponse time, β is a shape factor and $\Gamma(\beta)$ is the generalised factorial, or gamma, function. They found

12 It is not being claimed here that this is a particularly good model for isotopic response of this catchment; just that for the quality of the data, and assuming an extended unit-hydrograph type model, the methodology has developed a hierarchical set of this class of models which adequately describe the data.

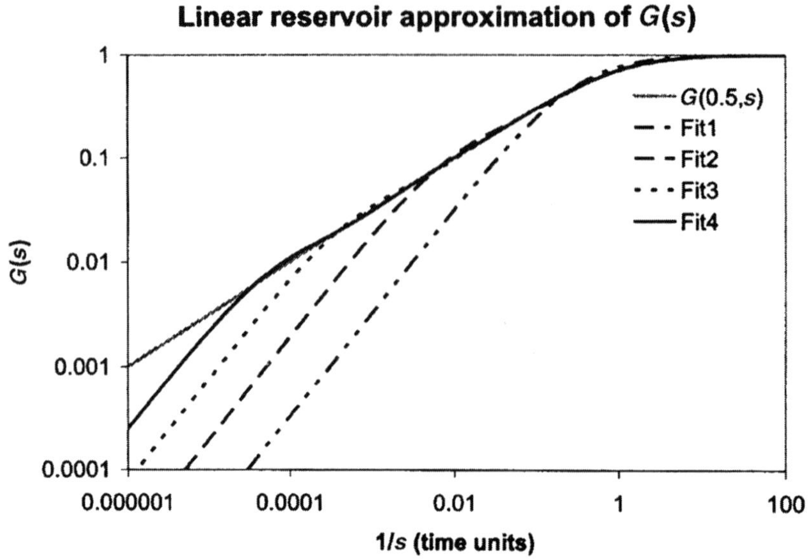

Figure 29.14 Approximation of a response function as a sum of linear reservoirs in Laplace transform space. Note that this is a log–log plot, in order to demonstrate the different (linear) slopes for the two types of response functions: Kirchner *et al.*'s function has a slope of β (= 0.5), whereas linear reservoir models have a corresponding slope of 1.

that values of β between 0 and 1 (e.g. $\beta = 0.5$) gave the best fit to the observed power spectrum.

In order to consider Kirchner *et al.*'s results, we first re-interpret them in the perspective of the current work. Rather than use the implicit Fourier analysis of Kirchner *et al.*, it seems more natural for these causal (impulse-response) systems to use Laplace transforms to analyse the problem. In Laplace transform space, response functions can be quite general, providing they satisfy the following constraints:

As a function of the transform variable s (units of 1/time), the transform of G, $G(s)$ must be monotonic decreasing$^+$, and satisfy the conditions

$$\lim_{s\to 0} G(s) = 1 \qquad \textit{(conservation of mass)}$$

and

$$\lim_{s\to\infty} G(s) = 0 \qquad \textit{(causality)}$$

The transform of the above response function used by Kirchner *et al.* is

$$G(s) = (1 + s/\alpha)^{-\beta} \tag{29.8}$$

It is easy to verify that $G(0) = 1$, and that for $\beta > 0$, $G(s)$ approaches zero monotonically as s increases, so that the above constraints are obeyed. We note that for $\beta = 1$, Kirchner's response function reduces to that of a simple linear reservoir. Elsewhere we have shown that any response function G can be represented to arbitrary accuracy by a finite combination of series and parallel linear reservoirs,[13] although for some response functions such a representation may not be particularly efficient. At least at larger times, the work of Jakeman and co-workers (reviewed in Littlewood, 2001) indicates that for real catchments this form of representation is usually efficient. For a greater range of timescales (from minutes to a year), this chapter has shown that a good response representation can be obtained with only three characteristic times in the particular catchment we considered.

In this case, the main difference between the two types of response function is the way in which they approach zero for large s. For combinations of linear reservoirs, the associated transform of the response function approached zero linearly as $1/s$, whereas the transform of Kirchner's function approaches zero as $1/\sqrt{s}$. Figure 29.14 indicates the best fit representation of the above function (for $\beta = 0.5$) as a parallel sum of 1–4 linear reservoirs, assuming a unit time constant ($\alpha = 1$). As more reservoirs are used in the approximation, a good fit (in transform space) is achieved for larger and larger values of s (smaller values of $1/s$, units of time). For 1–4 reservoirs (1–7 fitted parameters) significant departures only occur for response time of 0.125, 0.0030, 0.00025 and 0.000020 time units, respectively. This is despite the fact that ultimately, as $s\to\infty$, the Kirchner response function approaches zero as $s^{-1/2}$, whereas sums of linear reservoirs approach zero as s^{-1}.

As we note elsewhere, the unit hydrograph (response function) approach is deficient in that it underestimates the large peaks in the hydrograph, due primarily to faster stream etc. velocities at high flows. The use of Kirchner's function generally allows greater weight to be estimated for the rapid response part of

13 In mathematical terms, the space spanned by sums of linear reservoirs in series and parallel is dense in the space of all response functions.

the hydrograph, and hence may allow a partial compensation of this nonlinearity of the response, compared to linear reservoir representations.

CONCLUSION

It has been demonstrated that the classification of models in terms of their implicit assumptions can be useful for guiding appropriate model choice. We have argued that the use or construction of a model without appropriate validation is of little worth (and actually inimicable to progress), and that this constraint favours relatively simple models in practice. Furthermore, we suggested that, far from being simplistic, simple models of complex systems may contain greater catchment understanding (i.e. the relative insignificance of omitted terms) and be evolutionarily more advanced than models that are more complex.

In recent years, the advent of such tools as GIS has greatly facilitated the use and storage of large and complex data sets. Consequently, there is a tendency for models to evolve which use such data almost indiscriminately, without regard to the purpose and appropriateness of the model. While not seeking to deny the enormous utility of such data management tools, it appears that understanding and knowledge are sometimes sacrificed for completeness of information. In particular, the ability to disaggregate spatial information using GIS tools predisposes a viewpoint of the land or catchment system as a simple sum of its parts, which as we have discussed above is not necessarily valid or useful. The modelling approach we suggest here may be a useful corrective to this tendency, stressing the need to consider the system as a unit before yielding to the temptation to disaggregate.

The example given to show that the problem of hydrograph separation is not well-posed has wider ramifications. It is necessary to have a much higher degree of scepticism than is normally evident over the meaning of the 'validity' of any system model. A model is a (very simplified) representation of a real system, and can be usefully thought of as a method of testing the consequences of assumptions. It can only (at best) preserve information that is incorporated into it, and usually will in fact degrade information, due to the effect of simplifying assumptions. Therefore, although it is appropriate to talk about a model as being validated, it is necessary to recognise the limitations of any validation process, and be aware of potential applications that unreasonably extend model use. In short, it is important to come to terms with the quite subjective nature of modelling, without denigrating the very valid advantages of such a formal inferencing procedure.

We have provided here examples of simple model options for application in the less data-rich humid tropics. It is still necessary however to have an adequate appreciation of the dominant storm pathways to confidently apply the models outlined here. In the absence of detailed process studies (the S3 stage of Elsenbeer and Vertessy, 2000), the emergence of a conceptual framework of hillslope hydrology responses (Elsenbeer, 2001) (both of which are summarised in Bonell, this volume) supplemented by preliminary surveys at the S1 stage would provide some indications of the hillslope hydrology to assist modelling.

During the course of developing the Wyvuri South Creek example, it has become evident that the approach taken here is somewhat less sophisticated than the three-box model of Seibert and McDonnell (2002) where a prolonged series of research campaigns in the Maimai catchment have built a much more comprehensive experimental hydrological understanding of the hillslope hydrology across a range of topographic features. Rather, in the South Creek study, the approach has been to disaggregate the stream hydrograph using simple conceptual models (i.e. PPCMs; Littlewood, 2001) to detect the contributions from reservoirs of subsurface water (i.e. an intermediate unsaturated zone store, and groundwater store) at larger scales (i.e. the catchment scale) than available from point measurements.

Unlike Seibert and McDonnell (2002), available 'soft data' were not directly imported into the modelling, but a conceptual framework for the understanding of the runoff generation process (Bonell *et al.*, 1987; 1998) was already available for interpretation of the modelling results. A key contribution towards this understanding was also the integration and analysis of environmental isotope and hydrometric data. The environmental tracer data provided an independent means of detecting the contributing catchment storages and dominant storm pathways.

Finally, it is interesting that the basics of the spatially-lumped conceptual models outlined in this chapter were already being established from the 1950s through to the 1970s (e.g. Nash, 1957; Dooge, 1959; Chow, 1964; Dawdy and O' Donnell, 1965; see review of Shaw, 1988). Further, the DISPRIN subcatchment simulation model (**D**ee **I**nvestigation **S**imulation **P**rogram for **R**egulating **N**etworks) (Jamieson and Wilkinson, 1972; reviewed in Shaw, 1988) had already coupled 'tank' models connecting three hydrological zones, i.e. the uplands, the hillslopes and lower valley areas, designated as bottomslopes, prior to the more recent adaptations of conceptual models using 'soft data' and hillslope hydrology research in general (e.g. Seibert and McDonnell, 2002; Seibert *et al.*, 2002). Their recent revival within runoff process hydrology arises because of the aforementioned challenges associated with distributed modelling (Beven, 2002) and also because field research undertaken in the intervening period can now provide plausible, physical interpretations for these 'tank' models.

References

Baker, F. G. (1978). Variability of hydraulic conductivity within and between nine Wisconsin soil series. *Wat. Resour. Res.* 14, 103–108.

Barnes, C. J. (1995). The art of catchment modelling: What is a good model? *Environment International* **21**, 747–751.

Barnes, C. J., Walker, D. and Short, D. L. (1997a). *Models for Integrated Catchment Management*, *MODSIM97*: International Conference on Modelling and Simulation Proceedings, Hobart, Tasmania (Eds. McDonald and McAleer); Vol 4, 1647–1652.

Barnes, C. J. and Bonell, M. (1996). Application of unit hydrograph techniques to solute transport in catchments. *Hydrol. Proc.*, **10**: 793–802.

Barnes, C. J., Short, D. L. and Bonell, M. (1997b). Modelling water, nutrient and sediment fluxes using catchment scale parameters. *Hydrochemistry* (Proceedings of the Rabat Symposium, April 1997) IAHS Publ. No. **244**, 195–205.

Barnes, C. J. (2000). The use of cumulative data for characterising hydrological systems. In: *Proceedings of Runoff Generation and Implications for River Basin Modelling*, Leibundgut, C., Uhlenbrook, S. and McDonnell, J. (Eds.), Freiburger Schriften zur Hydrologie. 82–89.

Bear, J. (1972). *Dynamics of Fluids in Porous Media*. American Elsevier, New York.

Bear, J. and Verruijt, A. (1987). *Modelling Groundwater Flow and Pollution*, Theory and Applications of Transport in Porous media, Reidel, Dordrecht.

Bergström, S. (1995). The HBV model. In: *Computer Models of Watershed Hydrology*, Singh, V. P. (Editor). Water Resource Publications. Highlands Ranch, Colorado, USA. 443–476.

Bergström, S., Lindström, G. and Pattersson, (2002). Multi-variable parameter estimation to increase confidence in hydrological modelling. *Hydrol. Process*. 16, 413–421.

Beven, K. J. (1989). Interflow. *In*: Unsaturated Flow in Hydrologic Modelling, Morel-Seytoux, H. J. (ed.), Kluwer. 191–219.

Beven, K. J. (1993). Prophecy, reality and uncertainty in distributed hydrological modelling. *Advances in W. Resour*. 16: 41–51.

Beven, K. J. (2002). Towards an alternative blueprint for a physically-based digitally simulated hydrologic response modelling system. *Hydrol. Process*., 16: 189–206.

Beven, K. J. and Kirkby, M. J. (1979). 'A physically-based variable contributing area model of basin hydrology', *Hydrol. Sci. Bull*., **24**: 43–69.

Beven, K.J and Feyen, J. (2002). The future of distributed modelling. *Hydrol. Process*., 16, 169–172.

Beven, K. J., Kirkby, M. J., Schofield, N. and Tagg, A. F. (1984). Testing a physically-based flood forecasting model (TOPMODEL) for three UK catchments. *J. Hydrol*. **69**: 119–143.

Bonell, M. (1998). Selected challenges in runoff generation research in forests from the hillslope to headwater drainage basin scale. *J. Amer. Water Res. Assoc*., **34**: 765–785.

Bonell, M., Gilmour, D. A. and Sinclair, D. F. (1981). Soil hydraulic properties and their effect on surface and subsurface water transfer in a tropical rainforest catchment. *Hydrol. Sci. Bull*. **26**: 1–18.

Bonell, M., Cassells, D. S. and Gilmour, D. A. (1987). 'Spatial variations in soil hydraulic properties under tropical rainforest in north-eastern Australia' in Fok Yu-Si (Ed), *Proc. Int. Conf. on Infiltration Development and Application*. Wat. Resour. Res. Center, Univ. of Hawaii at Manoa, Jan. 1987. 153–165.

Bonell, M., Barnes, C. J., Grant, C. R., Howard, A. and Burns, J. (1998). High rainfall response-dominated catchments: A comparative study of experiments in tropical north-east Queensland with temperate New Zealand. *In*: Isotope Tracers in Catchment Hydrology, Kendall, C. and McDonnell, J. J. (Editors), Elsevier, Chapter 11, 347–390.

Boughton, W. C. (1984). A simple model for estimating the water yield of ungauged catchments. *Civ. Engg. Trans., I. E. Aust., CE* 26(2), 83–88.

Bronswijk, J. J. B., Hamminga, W. and Oostindie, K. (1995). Field-scale solute transport in a heavy clay soil, *W. Resour. Res*., 31: 517–526.

Chiew, F. H. S. and McMahon, T. A. 1992. Complete set of daily rainfall potential evapotranspiration and streamflow data for 28 unregulated catchments, Department of Civil and Agricultural Engineering, The University of Melbourne (Unpubl.)

Chow, Ven Te (1964) (ed.). *Handbook of Applied Hydrology*, McGraw-Hill, New York.

Davis, J. R. and T. F. N. Farley. (1997). CMSS: Policy analysis software for Catchment Managers. *Environmental Software and Modelling*, 12, 197–210.

Dawdey, D. R. and O'Donnell, T. (1965). Mathematical models of catchment behaviour. *Proc. ASCE HY4*, **91**: 132–137.

Dooge, J. C. I. (1959). A general theory of the unit hydrograph technique. *J. Geophys. Res*. **64**: 241–256.

Elsenbeer, H. (2001). Hydrologic flowpaths in tropical rainforest soilscapes-a review. *Hydrol. Process*. **15**: 1751–1759.

Elsenbeer, H. and Vertessy, R. A. (2000). Stormflow generation and flowpath characteristics in an Amazonian rainforest catchment. *Hydrol. Process*. **14**: 2367–2381.

Freeze, R. A. and Harlen, R. L. (1969). Blue print for a physically-based, digitally-simulated hydrologic response model. *J. Hydrol*. **9**: 237–258.

Ganeshanandam, S. and Krzanowski, W. J. 1989. On selecting variables and assessing their performance in linear discriminant analysis. Austral. J. Stat., **31**: 433–447.

Germann, P. F. (1990a). Macropores and hydrologic hillslope processes. In: Anderson, M. G. and Burt, T. P. (eds.), *Process Studies in Hillslope Hydrology*. John Wiley and Sons, Chichester. 327–363.

Germann, P. F. (1990b). Preferential Flow and the Generation of Runoff. 1. Boundary Layer Flow Theory. *Water Resour. Res*., **26**: 3055–3063.

Gilmour, D. A. (1975). *Catchment water balance studies on the wet tropical coast of North Queensland*. Unpublished PhD Thesis, Dept. of Geography, James Cook University of North Queensland, Townsville, Australia.

Grayson, R. B., Bloschl, G. and Moore, I. D. (1995). Distributed parameter modelling using vector elevation data: THALES and TAPES-C. In: *Computer Models of Watershed Hydrology*, Singh, V. P. (Ed.), Water Resource Publications, Highlands Ranch, Colorado, USA, pp. 669–696.

Jakeman, A. J. and Hornberger G. M. (1993). How much complexity is warranted in a rainfall-runoff model? *Water Resour. Res*. **29**: 2637–2649.

Jakeman, A. J., Littlewood, I. G. and Symons, H. D. (1991). Features and applications of IHACRES: A PC program for the identification of unit hydrographs and component flows from rainfall, evapotranspiration and streamflow data. In: *Proceedings of the 13th IMACS World Congress on Computation and Applied Mathematics*, vol. 4, edited by R. Viehnevetsky and J. J. H. Miller, pp. 1963–1967, Criterion, Dublin, Ireland.

Jamieson, D. G. and Wilkinson, J. C. (1972). River Dee research program. 3. A short-term control strategy for multi-purpose reservoir systems. *Water Resour. Res*. **8**: 911–920.

Kendall, C. and McDonnell, J. J. (1998). *Isotope Tracers in Catchment Hydrology*, Elsevier, Amsterdam, 839 pp.

Kirchner, J. W., Feng, X. and Neal, C. (2000). Fractal stream chemistry and its implications for contaminant transport in catchments. *Nature*, **403**: 524–527.

Kirchner, J. W., Feng, X. and Neal, C. (2001). Catchment-scale advection and dispersion as a mechanism for fractal scaling in stream tracer concentrations. *J. Hydrol*., **254**: 82–101.

Kirchner, J. W., Feng, X., Neal, C. and Robson, A. J. (2004). The fine structure of water-quality dynamics: the (high-frequency) wave of the future. *Hydrol. Process*., **18**(7): 1353–1359.

Kirkby, M. J. (1975). Hydrograph modelling strategies. In : *Processes in Physical and Human Geography*, Peel, R., Chisholm, M. and Hoggett, P. (eds). 69–90.

Lange, H., Lischeid, G., Hoch, R. and Hauhs, M. (1996). Water flow paths and residence times in a small headwater catchment at Gardsjon, Sweden, during steady state storm flow conditions. *W. Resour. Res*. **32**: 1689–1698.

Lindström, G. (2000). HBV model simulations of oxygen-18 flow in small forested basins in Sweden. In : *Nordic Hydrological Conference*, 2000, Nilsson, T, editor, Nordic Association for Hydrology, Nordic Hydrological Programme, NHP-Report No 46, Swedish Hydrological Council, Uppsala, 2000, 374–379.

Lindström, G., Johansson, B., Persson, M., Gardelin, M. and Bergström, S. (1997). Development and test of the distributed HBV-96 hydrological model. *J. Hydrol*., **201**: 272–288.

Littlewood, I. (2001). Practical aspects of calibrating and selecting unit hydrograph-based models for continuous river flow simulation. *Hydrol. Sci. J*., **46**: 795–811.

Michaelsen, J. 1987. Cross-validation in statistical climate forecast models. J. Climate Appl. Meteor., **26**: 1589–1600.

Moore, I. D. (1992). *Terrain Analysis Programs for the Environmental Sciences – TAPES*, Vol. 4 No 2, Agricultural Systems and Information Technology, Bureau of Resource Science, Canberra, ACT, pp. 37–39.

Moore, I. D. and Grayson, R. B. (1991). Terrain-based catchment partitioning and runoff prediction using vector elevation data. *Wat. Resour. Res*., **27**: 1177–1191.

Moore, I. D., O'Loughlin, E. M. and Burch, G. J. (1988). A contour-based topographic model for hydrological and ecological applications. *Earth Surf. Proc. and Landforms*, **13**: 305–320.

Mulholland, P. J. (1993). Hydrometric and stream chemistry evidence of three stream flowpaths in Walker Branch Watershed. *J. Hydrol.*, **151**: 291–316.

Nash, J. E. (1957). The form of the instantaneous unit hydrograph. *IAHS Pub. No. 45*, 114–121.

Nielsen, D. R., Biggar, J. W. and Erth, K. T. (1973). Spatial variability of field measured soil water properties. *Hilgardia*, **42**: 215–259.

O'Loughlin, E. M. (1981). Saturation regions in catchments and their relations to soil and topographic properties, *J. Hydrol*, **53**: 229–246.

O'Loughlin, E. M. (1986). Prediction of surface saturation zones in natural catchments by topographic analysis. *Wat. Resour. Res.*, **22**: 794–804.

O'Loughlin, E. M., Short, D. L. and Dawes, W. R. (1989). Modelling the hydrological response of catchment to land use change. In: *Hydrology and Water Resources Symposium, Comparison in Austral Hydrology*. Inst. Engrs., Canberra, 28–30 Nov. 1989, Christchurch New Zealand, 335–340.

Perrin, C., Michel, C. and Andreassian, V. (2001). Does a large number of parameters enhance model performance? Comparative assessment of common catchment model structures on 429 catchments. *J. Hydrol.* **242**: 275–301.

Rogowski, A. S. (1972) Watershed physics: Soil variability criteria. *Wat. Resour. Res.* **8**: 1015–1023.

Schellekens, J. (2000). Hydrological Processes in a Humid Rain Forest: A Combined Experimental and Modelling Approach. *Vrije Universiteit*, Amsterdam, The Netherlands, 158 pp.

Seibert, J. (1999). *Conceptual Runoff Models- Fiction or Representation of Reality*? Acta Universitatis Upsaliensis, Comprehensive Summaries of Uppsala Dissertations from the Faculty of Science and Technology 436. 52 pp+ 6 papers, Uppsala, Sweden.

Seibert, J. and McDonnell, J. J. (2002). On the dialogue between experimentalist and modeller in catchment hydrology: Use of soft data for multi-criteria model calibration. *Wat. Resour. Res.* **38**(11): 23.11–23.64.

Seibert, J., Rodhe, A. and Bishop, K. (2002). Simulating interactions between saturated and unsaturated storage in a conceptual runoff model. *Hydrol. Proc.* (in press). Revised manuscript from the *Proc. Int. Workshop on Runoff Generation and Implications for River Basin Modelling* (Leibundgut, C., Uhlenbrook, S. and McDonnell, J., eds.), 9–12 October, Univ. Freiberg, Germany, Institut für Hydrologie der Universitat Freiburg i. Br., 118–126.

Shaw, E. M. (1983). *Hydrology in Practice*, Van Nostrend Reinhold (International), London (in 2nd Edition 1988, Chapter 14, 322–348).

Sherman (1932). Streamflow from rainfall by the unitgraph method. *Eng. News Record*, **108**: 501–505.

Slater, R. (1991). Integrated Process Management: A Quality Model. McGraw Hill, NY.

Vertessy, R. A. and Elsenbeer, H. (1999). Distributed modeling of storm flow generation in an Amazonian rainforest catchment: Effects of model parameterization, *W. Resour. Res.* **35**: 2173–2187.

Vertessy, R. A., Dawes, W. R., Zhang, L., Hatton, T. J. and Walker, J. (1996). Catchment scale Hydroloic modelling to assess the water and salt balance behaviour of eucalypt plantations. CSIRO Division of Water, *Technical Memorandum 96.2*, February 1996, Resources, 23 pp.

Wheater, H. S., Jakeman, A. J. and Beven, K. J. (1993). Progress and directions in rainfall-runoff modelling. In: *Modelling Change in Environmental Systems* (Jakeman, A. J., Beck, M. B. and McAleer, editors), Wiley, Chichester, West Sussex, UK. 101–132.

Young, P. and Beven, K. J. (1991). Computation of the instantaneous unit hydrograph and identifiable component flows with application to two small upland catchments – Comment. *J. Hydrol.* **129**: 389–396.

Young, R. A., C. A. Onstad, D. D. Bosch and W. P. Anderson. (1989). AGNPS: A nonpoint-source pollution model for evaluating agricultural watersheds. *J. Soil and Water Conservation*, **44**: 168–173.

30 The disaggregation of monthly streamflow for ungauged sub-catchments of a gauged irrigated catchment in northern Thailand

S. Y. Schreider
Royal Melbourne University of Technology, Melbourne, Australia
A. J. Jakeman
The Australian National University, Canberra, Australia

INTRODUCTION

Background

Accurate water resource assessment in northern Thailand is of crucial importance for sustainable agricultural development as well as for resolving upland-downstream conflicts related to allegedly excessive water use by highland rural communities. Fair water resource allocation is especially important during the dry season because in regions with a monsoon climate, such as Northern Thailand, the availability of water for irrigation during that season offers a prospect of growing a second crop in one year.

The major problem for accurate streamflow modelling in Thailand as well as in many developing countries is that catchments are often poorly instrumented. For instance, the Mae Chaem catchment at Kong Kan (2157 km^2) has only three gauging stations: at Kong Kan (Station number 04061302), the 1180 km^2 catchment of the Mae Chaem River at Huai Phung (04061201) and the 68.5 km^2 Mae Mu River subcatchment at Ban Mae Mu (04061202). A map of the Mae Chaem basin, showing the subcatchments under consideration and gauging station locations is presented in Figure 30.1.

Climate and physiography

Thailand, including the Mae Chaem catchment of the Ping Basin in its northern part, has a monsoonal climate. The climate is characterised by a wet period from the end of May to September and a very dry period in the remainder of the year. In some years no rain falls for three months or more. The heterogeneity of rainfall intensities within the catchment is very high: in the lowest sites of the valley mean annual precipitation is about 1000 mm, whereas on the top of the surrounding ridges it is about 1700 mm (average value for the period of 1989–1994 for the Mount Doi Intanon meteorological station located 2500 m ASL). The mean monthly rainfall in August (the wettest month of the year) in the central part of the catchment is about 240 mm (Mae Mu meteorological station). The mean rainfall for the driest month (January) is close to zero. The mean monthly rainfalls for two sites in the Mae Chaem catchment are demonstrated in Figure 30.2. However, because of their substantial base flow, many rivers in the Mae Chaem catchment are perennial and could be used as a source of irrigation for dry season cropping. The availability of water for irrigation during the dry season and, therefore, the possibility to produce dry season crops, is an important aspect for development of highland rural communities in this region.

A global survey of soils in the Mae Chaem basin was implemented by the Department of Land Development of Thailand (DLD, 1979). Generally speaking, upland soils in Northern Thailand are dominated by light to medium textured soils (Chaiwanakupt and Chiangprai, 1990). The most common in this area are deep and shallow loam and gravel soils with 2–10% rock outcrops. Geology of the Mae Chaem Basin is very heterogeneous. The dominant rocks are plutonic of Mesozoic and Paleozoic ages, which are characterised by granite and granodiorite. Sedimentary rocks comprising red sandstone, shale and conglomerate are also common in the study area (NRC, 1997).

Aims of the study

The major aim is to develop a method for predicting streamflow in ungauged or poorly instrumented subcatchments of interest. The approach taken here is one possible method for linking terrain

Forests, Water and People in the Humid Tropics, ed. M. Bonell and L. A. Bruijnzeel. Published by Cambridge University Press.

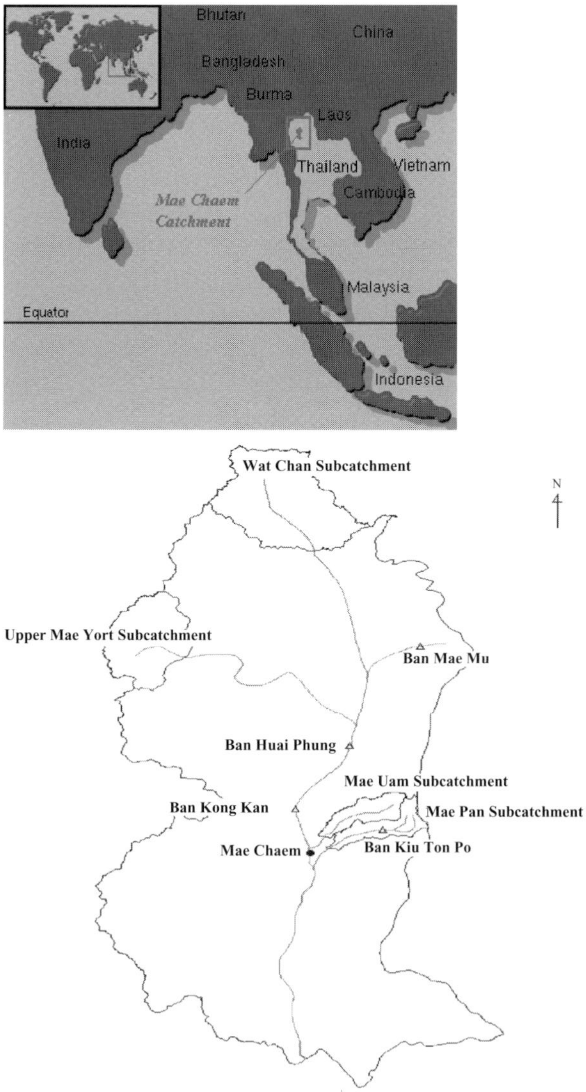

Figure 30.1 Map of the Mae Chaem catchment and locations of gauging stations.

attributes with the parameters of a conceptual hydrological model such as IHACRES (Jakeman *et al.*, 1990; Jakeman and Hornberger, 1993). Post and Jakeman (1996) explored relationships of the IHACRES model parameters with terrain attributes using regression analysis for a series of small (with areas less than 100 km^2) catchments in Victoria, Australia. The limitation of this approach is that the regression analysis requires a large number of gauged catchments with a long history of streamflow observations for calibrating the relationship between model parameters and catchment attributes. An alternative method is suggested here.

The algorithm proposed below is based on a spatial redistribution of streamflow water yield within a catchment according to the generalised topographic index introduced by Beven and Kirkby (1979) and developed in a series of further works (e.g. see Beven, 1997). The method is tested for the gauged subcatchments of Huai Phung and Mae Mu by comparing the modelled and observed streamflow series. The suggested algorithm allows one to predict natural streamflow (here we use this term to define streamflow prior to irrigation diversion) for each point in the Mae Chaem catchment if terrain characteristics above this point are available. The regulated flow available for each village in the area can then be calculated as the difference between natural flow and irrigation diversion above this village.

The proposed streamflow modelling algorithm will be applied to predict surface runoff in several subcatchments, ungauged or treated as ungauged, in the Mae Chaem catchment within the framework of the Integrated Water Resource Assessment and Management (IWRAM) Project. Objectives and general methodology of the IWRAM (Jakeman *et al.*, 1997; Scoccimarro *et al.*, 1999) project are described below. This project is developing a Decision Support System for management of water resources in rural catchments in Northern Thailand (Merritt *et al.*, 2004; Letcher *et al.*, 2002). Four highland subcatchments of the Mae Chaem catchment (Mae Pan, Mae Yort, Mae Uam and Wat Chan) and an area in the vicinity of Mae Chaem town were selected as case studies (see Figure 30.1). All these subcatchments are ungauged except Mae Pan where the quality of recorded streamflow is very poor and unsuitable for use in the present work.

The IWRAM Program seeks to provide resource managers at national, provincial and local levels with analytically robust and user-friendly methodologies for the assessment and management of land and water resources. One of the main challenges for resource managers is to weigh up the various socio-economic and environmental trade-offs that are associated with resource management decisions. The IWRAM framework provides them with tools that they can use in this complex assessment process. At the heart of the IWRAM framework is a computer-based Decision Support System that allows resource managers to develop and model scenarios so they can explore the expected *and* unexpected impacts of resource management decisions. Model outputs can be displayed and manipulated in ways that best suit the needs of user groups. The resource pressures emerging in Mae Chaem are typical of many catchments in South East Asia and the IWRAM methodology has been developed to be site-independent.

The IWRAM Project maintains collaboration between a number of Thai and Australian organisations. The lead agency in Thailand is the Thai Royal Project Foundation who has worked closely with the Department of Land Development, Kasetsart University, Mae Jo University and Chiang Mai University. Recently, other agencies have become involved, including the Royal Forest Department, the Royal Irrigation Department, Department of Agriculture and Office of the National Water Resources Committee. In Australia the lead organisation is The Australian National University with

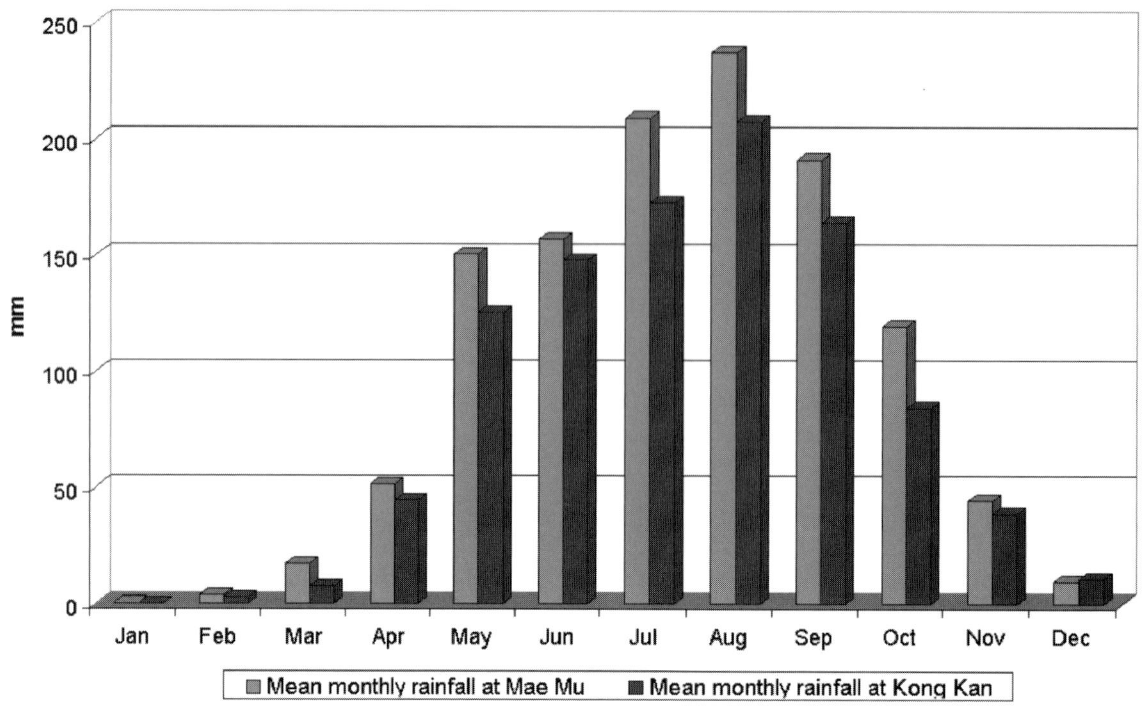

Figure 30.2 Mean monthly rainfall for two meteorological stations in the Mae Chaem catchment.

substantial funding support from the Australian Centre for International Agricultural Research.

The IWRAM DSS does not attempt to achieve precise modelling of all catchment processes. Rather, it seeks to analyse some of the key relationships within the catchment and to simulate how some of these key relationships may respond to various resource management decisions by enabling users to explore some of the key relationships that are relevant to the various environmental and socioeconomic trade-offs in catchment management. In this sense the DSS seeks to provide a structured, replicable and transparent framework for analysing and assessing the alternative scenarios that may be discussed and debated during resource planning processes.

The biophysical component includes the modelling of surface hydrology, soil erosion and crop yield in relation to climatic forcing and land use in the subcatchments considered. This chapter describes a general structure for the surface hydrology modelling used in the biophysical component of the IWRAM Project. A major challenge was to derive an algorithm for flow disaggregation, which works better than simple scaling according to relative catchment area and the variation in climatic factors between catchments. The modelling for ungauged catchments described here was implemented to address the very practical needs of the IWRAM Project. With the limited data at our disposal, it was not possible to derive and test a sophisticated regionalisation procedure. It should also be stressed that although the procedure uses the basic TOPMODEL topographic index as weighting, it does not purport to invoke all the assumptions of that model.

METHODOLOGY

The proposed surface runoff modelling algorithm has two major steps. Firstly, a rainfall-runoff model is calibrated at the catchment outlet against the streamflow recorded at the Kong Kan station. Secondly, the modelled streamflow is calculated at each point within this catchment by disaggregation of the streamflow discharge at the catchment outlet using a topographic index weighting (Beven and Kirkby, 1979).

As will be described in more detail later, the first step implies an application of some model allowing one to compute the streamflow using basic climatic data (precipitation and temperature) as input. The criterion for the model parameter optimisation is a best fit of predicted discharge to measured values. The IHACRES rainfall-runoff model (Jakeman *et al.*, 1990; Jakeman and Hornberger, 1993) in conjunction with an irrigation consumption module, will be used for this purpose. In a predominantly agricultural catchment, such as Mae Chaem at Kong Kan, the major source of water consumption is irrigation supply for the crops. Such irrigation can be calculated using information on the surface areas under different crops and the consumptive use of the respective crops per unit area. At the catchment outlet, the expected natural flow can be calculated as the sum of measured discharge and irrigation diversion:

$$F(k) = Q(k) + D(k), \tag{30.1}$$

where $F(k)$ is natural (unregulated) streamflow, $Q(k)$ is a measured streamflow (regulated discharge) and $D(k)$ is irrigation diversion at the k-th time step. The IHACRES model will be calibrated against the natural flow values $F(k)$ as predicted by Eqn 30.1.

The second step is based on the approach related to one proposed for the TOPMODEL hydrological model (Beven and Kirkby, 1979; Beven *et al.*, 1984 and Quinn *et al.*, 1995). This approach involves the assumption that the contribution from each part of a catchment to the total water yield is proportional to a topographic index. It allows one to predict the natural flow $f(k)$ for each part of a catchment using just a terrain characteristic of this area within the catchment considered. Therefore, if crop irrigation data for the selected subcatchment are available, the simulated regulated discharge $q(k)$ at its outlet can be expressed as a difference between natural flow $f(k)$ and irrigation diversion $d(k)$:

$$q(k) = f(k) - d(k) \tag{30.2}$$

Thus the irrigation diversion module is used for model calibration at the outlet of the gauged catchments (Eqn 30.1). It is also employed for restoring natural flow using the recorded discharge and the total diverted amount of water, and for the simulation of discharge in the ungauged catchments, where the regulated streamflow will be calculated as a difference between modelled natural flow and diversions as in Eqn 30.2.

The major components of the methodology proposed can therefore be outlined as follows:

(1) Modelling of irrigation consumption and prediction of the natural (unregulated) streamflow in the gauged catchment using Eqn 30.1,
(2) Modelling of the natural flow in the gauged catchment using the IHACRES rainfall-runoff model,
(3) Modelling of natural streamflow in ungauged subcatchments of this catchment using the streamflow disaggregation procedure. The regulated streamflow values can then be computed in these subcatchments using the irrigation consumption module, and
(4) Testing the model performance in smaller gauged catchments.

IRRIGATION CONSUMPTION MODULE

A water consumption module is constructed for river discharge prediction in subcatchments with considerable agricultural activities and, therefore, a significant artificial component in streamflow regime. It is based on the assumption that water consumption is defined solely by irrigation use. Thus, water consumption is calculated as a sum of products of areas under each irrigated crop and the irrigation consumption of each crop per unit of area.

Let N be the number of crop types cultivated in the catchment. Crops that are grown in wet and dry seasons are classified into two different types. Let $v_n{}^m$ be the irrigation consumption for crop n ($n = 1, 2, \ldots, N$) for the month m (only monthly irrigation consumption data were obtained for this region from the Royal Irrigation Department of Thailand, hence $m = 1, 2, \ldots, 12$) and $A_n{}^i$ be area under the nth crop in the ith year. The yearly differences in $A_n{}^i$ values reflect the changes in agricultural practice and may be used to represent future crop/land use development in this area. Therefore, monthly water diversion for the mth month in year number i can be calculated as:

$$d_m^i = r \sum_{n=1}^{N} v_n^m A_n^i \tag{30.3}$$

where r is a unit correction constant.

The regulated flow $q(k)$ can then be calculated for each monthly time step as

$$q(k) = f(k) - e^{-1}\left[d_m^i(k) + P + W\right], \tag{30.4}$$

where $f(k)$ is natural streamflow (flow value without diversion) and e is an irrigation efficiency coefficient. This coefficient is defined as the ratio of the amount of water used for irrigation to the amount of water diverted. The Royal Irrigation Department of Thailand has estimated this efficiency coefficient as 0.60 for the wet season and 0.85 for the dry season. The parameters P and W in Eqn 30.4 characterise the amount of water lost due to deep percolation (P) and for land preparation in irrigated paddies (W). According to the information provided by the Royal Irrigation Department, deep percolation can be estimated roughly as 1 mm day^{-1} for the wet season and 1.5 mm day^{-1} for the dry season. Land preparation in paddy fields needs 250 mm for the wet and 300 mm for the dry season (Karn Trisophon, Office of Highland Development, Department of Land Development, Chiang Mai; pers. comm.). The method for calculation of the crop water consumption here is based on the monthly crop coefficient approach, which is a lumped parameter taking into account relevant climatic factors for the month, including temperature and solar radiation.

MODELLING OF UNREGULATED (NATURAL) FLOW IN GAUGED CATCHMENTS

Several modelling approaches are used in hydrology for predicting surface runoff in instrumented catchments. Following the model

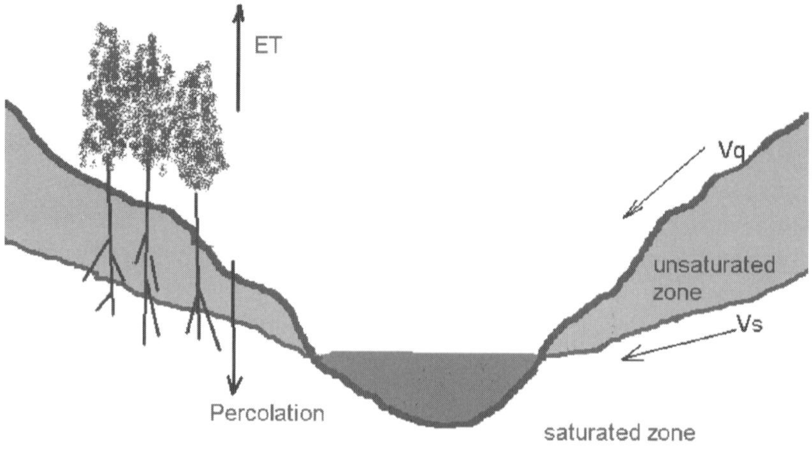

Figure 30.3 Schematic separation of streamflow components: quick – upland runoff and subsurface storm flow, and slow – ground water horizontal flow in saturated zone.

classification suggested in Wheater *et al.* (1993), three major model classes can be identified: empirical, physically-based and conceptual. The class of conceptual models is most appropriate in the present context, because of their potential for relative simplicity and relatively low data requirements (see also discussion in Barnes and Bonell, this volume). The conceptual rainfall-runoff model IHACRES, based on the Instantaneous Unit Hydrograph technique, is particularly suitable for streamflow modelling within the framework outlined above in the 'Methodology' Section. This model was described in Jakeman *et al.*, 1990; Jakeman and Hornberger, 1993 (see also the Appendix to this Chapter). It is based on the concept that effective rainfall passes through two parallel 'reservoirs' corresponding to the quick flow and slow flow. These 'reservoirs' can be interpreted as a quick surface runoff with volume V_q ('quick reservoir') and a slow (often horizontal movement of subsurface waters) component of volume V_s ('slow reservoir'). Schematically, this concept is presented in Figure 30.3. In catchments with monsoonal climates quick recession occurs predominantly during the wet season whereas the slow flow component dominates during the dry season. Certainly, the slow flow component is superimposed upon the quick flow component in the wet season, but the latter provides a significantly higher contribution to the total catchment yield during this period. The history of application of the IHACRES model in the Mae Chaem catchment and additional justification of the selected approach is described in the series of research publications (Schreider and Jakeman, 1999; Schreider *et al.*, 1999; Scoccimarro *et al.*, 1999).

The IHACRES model has two modules. A non-linear loss module which at each time step k (daily and monthly time steps are proposed in this work) transforms measured rainfall $r(k)$ into effective rainfall $u(k)$ using temperature or pan evaporation data $t(k)$. The non-linear loss module is used to account for the effect of antecedent weather conditions on the current status $s(k)$ of soil moisture and vegetation conditions, and for evapotranspiration effects. Here the effective rainfall $u(k)$ is calculated from the measured rainfall $r(k)$ and temperature $t(k)$ in the catchment area by the formulae:

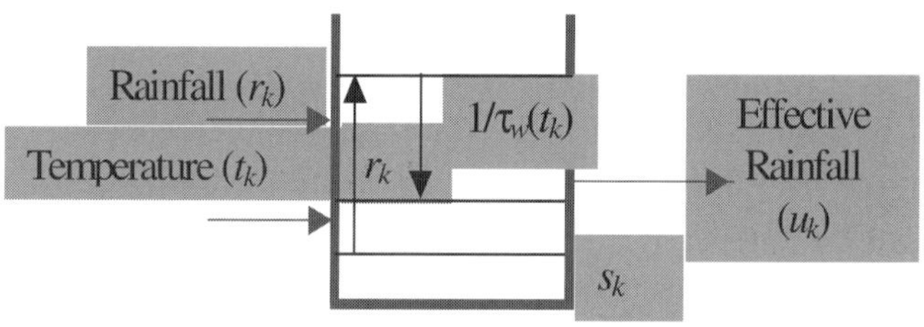

Figure 30.4 Non-linear module of the IHACRES model representing catchment soil moisture store as an evaporating bucket.

$$u(k) = r(k)[s(k) + s(k-1)]/2 \qquad (30.5a)$$

$$s(k) = cr(k) + 1 - 1/\tau_w(t(k))\,s(k-1) \qquad (30.5b)$$

$$\tau_w(t(k)) = \tau_w \exp\{t_{max}[20 - t(k)]\} \qquad (30.5c)$$

Schematically, this loss module can be represented as an evaporating bucket (storage) with rainfall input (Figure 30.4). The constant c is optimised so that the volume of effective rainfall is equal to the total streamflow for the calibration period. Importantly, it reflects the amount of potential storage in the catchment. The quantities τ_w and t_{max} are parameters to be optimised. The parameter τ_w

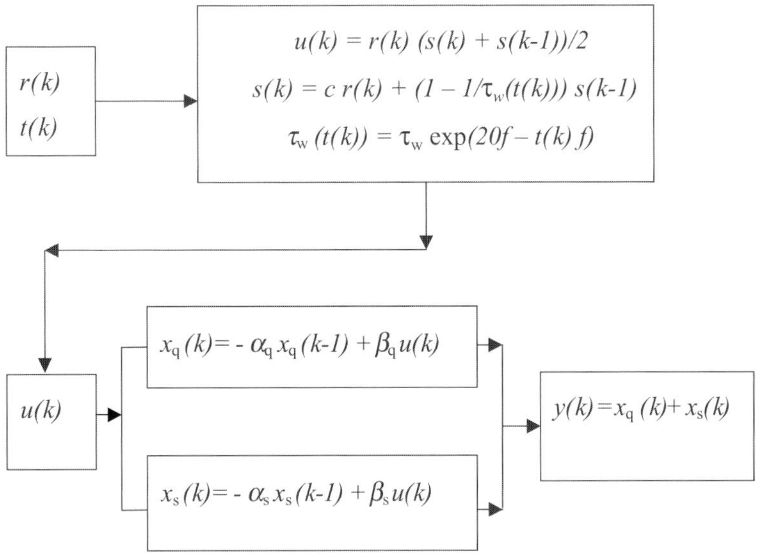

Figure 30.5 Chart diagram of the IHACRES model.

reflects the rate of drying of the catchment at 20 °C and t_{max} is a factor which modulates this rate as temperature varies.

A linear module then describes the travel of effective rainfall to streamflow $y(k)$ on the basis of a total unit hydrograph approximation. This module invokes a recursive relation at time step k for modelled streamflow $y(k)$, computed as a linear combination of its past values and current and past effective rainfall. The model's conceptual structure implies that the effective rainfall is considered to travel through two parallel stores (or reservoirs). This means that the recession of streamflow is a superposition of two exponential decay functions, one of them being responsible for quick recession (recession of high flow events) and the other for recession of the slow flow component. In other words, the modelled streamflow $y(k)$ can be represented as a superposition of the quick and slow components $x_q(k)$ and $x_s(k)$:

$$y(k) = x_q(k) + x_s(k), \qquad (30.6a)$$

where x_q and x_s decay in an exponential fashion according to:

$$x_q(k) = -\alpha_q x_q(k-1) + \beta_q u(k) \qquad (30.6b)$$

$$x_s(k) = -\alpha_s x_s(k-1) + \beta_s u(k) \qquad (30.6c)$$

In regions with a monsoon climate, such as northern Thailand, the constants α_q and α_s could be interpreted roughly as the rate of wet and dry season flow recessions, respectively, whereas β_q and β_s represent the peak hydrographs for these seasons (see also Barnes and Bonell, this volume). The flow chart diagram of the IHACRES model is shown in Figure 30.5.

STREAMFLOW DISAGGREGATION PROCEDURE

Beven and Kirkby (1979) and Beven *et al.* (1984) argue that the patterns of hydrological response within a catchment are linked to a topographic wetness index (Chappell, Bidin *et al.*, this volume, provide a detailed review of TOPMODEL, including the topographic index). This topographic index ω_j is defined for each grid cell in the catchment, considered as:

$$\omega_j = \ln[(A_j/l_j)/\tan(\phi_j)], \qquad (30.7)$$

where A_j is the drainage area above the grid cell with width l_j of the contour of the grid cell and the ϕ_j value is the average slope of this grid cell. The topographic index coefficient over each subcatchment ω is computed as the mean arithmetic value of these indices calculated for each of n grid cells of the subcatchment considered:

$$\omega = \frac{1}{n}\sum_{j=1}^{n} \omega_j \qquad (30.8)$$

This index is normally used for predicting spatial patterns within small catchments (Quinn *et al.*, 1995). Here we postulate that the relative streamflow contributions from different subcatchments within a larger catchment are also proportional to the mean topographic wetness index within each subcatchment. Hence, if the modelled streamflow volume F (which is modelled daily streamflow $y(k)$ aggregated to some larger time interval, that is, monthly here) is known at the catchment outlet, the natural (unregulated) streamflow f for each subcatchment can be computed using the assumption that f is proportional to F according to the proportionality coefficient ω/Ω, where ω is the topographic index in this subcatchment and Ω is the topographic index calculated for the entire catchment. As part of the use of the IHACRES model, this scaling was implemented for the volumetric constant c of the non-linear loss module of this model (see Eqn 30.5b).

Several researchers have questioned use of the logarithmic function in the calculation of the topographic index (Eqn 30.7). The use of the logarithmic function is based on the assumption that the downslope transmissivity profile of the catchment considered is approximated by an exponential function. In the case of zero recharge the assumptions of an exponential transmissivity profile and constant hydraulic gradient lead to hyperbolic recession curves or, in other words, to the use of a power function in computing the topographic index (Beven, 1997; Iorgulescu and Musy, 1997). Ambroise *et al.* (1996) demonstrated that the topographic index approach could be extended to the case of a linear transmissivity profile. Kirkby (1997) also mentioned that one of the strategies in the TOPMODEL development was replacement of the exponential equation for the transmissivity profile by the power law.

In the present work, a linear approximation of the soil transmissivity profile was selected. The lack of detailed information on soil types and geology in the area inhibits the appropriate selection on physical grounds. On this basis it was decided to select the simplest approximation, that is a linear one.

TESTS OF MODEL CALIBRATION AND SIMULATION

Calibration at the gauged catchment outlet

The calibration procedure is implemented in two steps:

- The natural (unregulated) streamflow discharge F_i is restored for the gauged catchment (Mae Chaem at Kong Kan here) using Eqn 30.1. The irrigation diversion component of the water balance is computed as stated above in the 'Irrigation Consumption Module' Section (Eqns 30.3 and 30.4).
- The hydrological model (IHACRES) is calibrated against the restored values of natural flow.

Simulation for ungauged subcatchments

The simulation procedure for simulating natural and regulated flow in subcatchments is as follows:

- For each selected site in the catchment the natural streamflow is calculated by disaggregating the natural flow of the entire catchment according to the topographic index weighting.
- The actual amount of water (regulated discharge) at each site is calculated according to Eqn 30.2, using the data on areas under different irrigated crops and irrigation consumption for each crop, as supplied by the irrigation consumption module.

Model testing

Two different ways to test the model were implemented:

- Streamflow data for the catchment outlet (Mae Chaem at Kong Kan) were simulated for a period other than that for which the model was calibrated. This means that the model was run with the same parameter values, which were established during the calibration, in order to predict streamflow outside the calibration period.
- The second test is a comparison of the simulated flow series obtained through the disaggregation procedure with the measured flow values available for some subcatchments (i.e. the Mae Mu and Huai Phung subcatchments in the case of the Mae Chaem catchment).

DISAGGREGATION

The 'first pass approach' developed here is based on the major assumption that a topographic index can be computed using the lumped terrain characteristics of these catchments. Such an approach avoids calculating the characteristics of interest for a set of grid cells constituting this subcatchment. This approach was employed despite some concerns about the spatial resolution at which the TOPMODEL disaggregation method can be sensibly applied, as expressed in Quinn *et al.* (1995). The match of the modelled monthly values to the observed discharge was selected as an ultimate criterion to test the applicability of the suggested methodology. The accuracy of the modelling results is evaluated using the testing procedures described in the previous Section. Note that the 'ungauged' catchments selected here (Huai Phung and Mae Mu) are actually gauged so that the approach can be tested.

The disaggregation by topographic index is applied to the volumetric coefficient c in the non-linear loss module of the IHACRES model:

$$C\frac{a}{\tan(\phi)} = c\frac{A}{\tan(\Phi)} \tag{30.9}$$

Here c (see Eqn 30.5b) represents the value of this volumetric coefficient in the subcatchment, a is the area of this subcatchment and ϕ is its mean slope. A, C and Φ are the values of these parameters for the entire catchment (the Mae Chaem catchment at Kong Kan here). As previously indicated, we chose to avoid the logarithmic transformation used in the original TOPMODEL wetness index as this element of the index arose from an assumption of exponential decline in soil transmissivity with depth (Beven

Table 30.1. *Average monthly irrigation water demand (in mm) used in the irrigation consumption module*

Month	Jan	Feb	Mar	Apr	May	Jun	Jul	Aug	Sep	Oct	Nov	Dec
Wet season rice (wr)	0	0	0	0	250[a]	300	350	150	50	50	0	0
Dry season rice (dr)	250	200	200	0	0	0	0	0	0	0	300[a]	500
Cash crops (cc)	150	150	100	0	0	0	0	0	0	0	300[a]	100

[a] water requirement for paddy preparation.

and Kirkby, 1979). Instead, we assumed a linear approximation of the soil transmissivity profile (Ambroise *et al.*, 1996; Beven, 1997).

IRRIGATION DIVERSION

Two significant simplifications applied here are:

- Dry season irrigated crops are assumed to be grown only on the furrow-irrigated paddy fields, which means ignoring sprinkler irrigation of upland fields, and
- The data provided by the Royal Irrigation Department on long-term average values for irrigation demands, calculated for the central part of Thailand, are relevant for the Mae Chaem area located in northern Thailand.

If a monthly time step is selected for modelling then the natural flows can be calculated using monthly consumption data provided by various agricultural institutions. When a daily time step is selected some additional assumptions are needed, for instance that irrigation diversion is uniform during each day of the month.

Land use data for the entire Mae Chaem basin are available in GIS format for three time slices: 1985, 1990 and 1995 (NRC, 1997). These data allow one to calculate the areas under rice paddies and upland fields. With respect to irrigation water use, the only functional information obtained from the GIS data is the area under paddy fields because no particular crops growing on these paddies are specified.

Table 30.1 presents the irrigation consumption data for all type of crops, obtained from the Royal Irrigation Department, where the consumption values are rounded off to the nearest 50 mm. Onion, garlic, soybean, tobacco, barley, melon, groundnuts, cabbage and other vegetables, which all have similar irrigation demands, are treated as a single class of cash crops (cc).

The monthly irrigation diversion in month *i* for the 'first pass approach' in the selected year can be calculated during the wet season as (see 'Irrigation Consumption Module' section):

$$d_i = \lambda wr_i Sp(i = \text{June, July}, \ldots, \text{Oct}), \quad (30.10)$$

where λ $(0 \leq \lambda \leq 1)$ is the proportion of total paddy area (Sp) covered by wet rice with irrigation demand (wr_i) as shown in Table 30.1. During the dry season monthly irrigation diversion is calculated as:

$$d_i = (\mu_1 dr_i + \mu_2 cc_i)Sp\,(i = \text{Nov, Dec}, \ldots, \text{Mar}), \quad (30.11)$$

where μ_1 and μ_2 $(0 \leq \mu_1 + \mu_2 \leq 1)$ are the proportions of total paddy area (Sp) covered by dry season rice, with monthly irrigation demand dr_i, and cash crops, with monthly irrigation demand cc_i, respectively.

Table 30.2. *The characteristics of subcatchments used in the streamflow modelling*

Subcatchment	Area (km^2)	Mean slope (°)	Total area under irrigation (km^2)		
			1985	1990	1995
Mae Chaem at Kong Kan	2157	19.0	11.8	18.6	20.1
Mae Chaem at Huai Phung	1180	18.5	8.3	12.5	13.5
Mae Mu at Mae Mu	68.5	14.3	0.04	0.10	0.10

The GIS data required for application of this streamflow modelling algorithm are summarised in Table 30.2. Despite the total irrigated area in the Mae Chaem catchment reaching only 1% of total catchment size, the irrigation consumption in the catchment can be very significant. The 'maximum consumption' scenario corresponds to rice cultivation on 100% of the paddy area during both the wet and dry seasons ($\lambda = 1$, $\mu_1 = 1$ and $\mu_2 = 0$). The 'medium consumption' scenario represents rice grown on 50% of the paddy area during the wet season and cash crops grown on 50% of the paddy area during the dry season ($\lambda = 0.5$, $\mu_1 = 0$ and $\mu_2 = 0.5$).

CALIBRATION AND TESTING

The model calibration was performed for five one-year periods between 1990 and 1994. The parameter values derived for the 1994 model run were selected for model simulation in the Mae Chaem catchment at Kong Kan over the whole period when streamflow

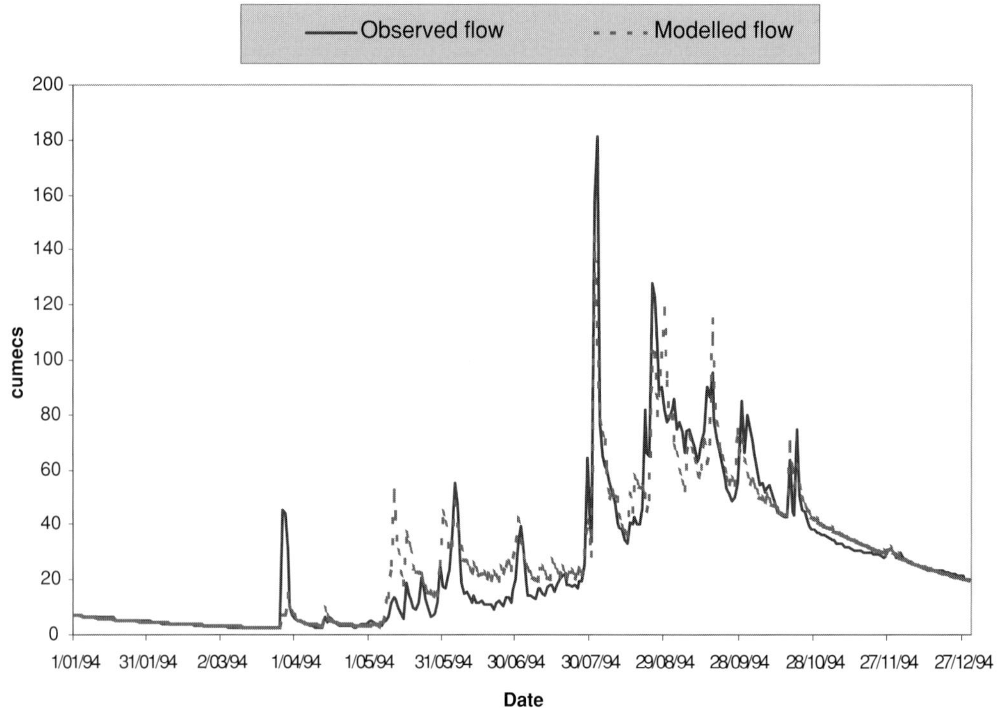

Figure 30.6 Calibration results for daily streamflow in the Mae Chaem catchment at Kong Kan ('medium' irrigation diversion scenario).

data were available (Type 1 model test). Model tests of Type 2 (disaggregation) were performed for two gauged subcatchments in the Mae Chaem catchment: Mae Mu and Huai Phung. The results of model simulation performed with and without the irrigation consumption module were compared. The simulations of natural streamflow for ungauged areas were performed for two nodes of the Mae Uam subcatchment (Figure 30.1) and then this information was employed for the integrated water resource management. However, these streamflow simulation results are not presented here because this would focus on verification of the algorithm suggested and measured streamflow is not available in the Mae Uam subcatchment.

The results of model calibration for the entire Mae Chaem catchment at Kong Kan are presented in Table 30.3. It is shown in Table 30.2 that the area under paddy fields remained similar during the 1990–95 period. The calibration was implemented for five one-year periods in 1990–94; the year 1995 was not considered as temperature data were not available in this period. Therefore Table 30.3 illustrates the calibration results for five calibration periods. The predictive capacity of the model calibration for each year was estimated using the Nash-Sutcliffe (1970) efficiency parameter R^2 and bias (mean daily error). Figure 30.6 illustrates graphically the model calibration results for 1994. As indicated earlier, the values of model parameters optimised for this calibration period were used throughout.

Table 30.3. *Calibration results for the Mae Chaem catchment at Kong Kan for 'no irrigation' and 'medium irrigation' diversion scenarios*

Year of calibration	No irrigation diversion		Medium irrigation diversion	
	R^2	Bias (cumecs)	R^2	Bias (cumecs)
1990	0.655	−2.32	0.680	−2.05
1991	0.763	−0.14	0.759	0.03
1992	0.609	−1.43	0.614	−0.71
1993	0.619	−1.73	0.675	−1.46
1994	0.877	−1.31	0.882	−1.13

The model parameters calibrated for 1994 were used for modelling streamflow for the six-year period of 1989–94 when the area under irrigated crops, and hence the water consumption for irrigation, can be assumed to remain reasonably constant (see Table 30.2). The quality of the monthly flow simulation was evaluated using the mean monthly absolute and mean relative errors. Monthly streamflow values for the model calibration in 1994 for 'no irrigation' and 'medium irrigation' scenarios were calculated. The mean monthly absolute errors for these scenarios were 9500 ML/month and 8600 ML/month, respectively, whereas the corresponding values for the mean relative errors were 24% and

Table 30.4. *Results of the flow disaggregation procedure using the scaling of the volumetric coefficient c for two subcatchments within the Mae Chaem catchment at Kong Kan (model test 2)*

Subcatchment	Irrigation diversion scenario	Mean monthly relative error (%)	Mean annual relative error (%)	Mean monthly absolute error (ML)
Mae Chaem at Huai Phung	No irrigation	16	10	4800
	Medium irrigation	13	15	4600
Mae Mu at Ban Mae Mu	No irrigation	18	5	330
	Medium irrigation	17	3	320

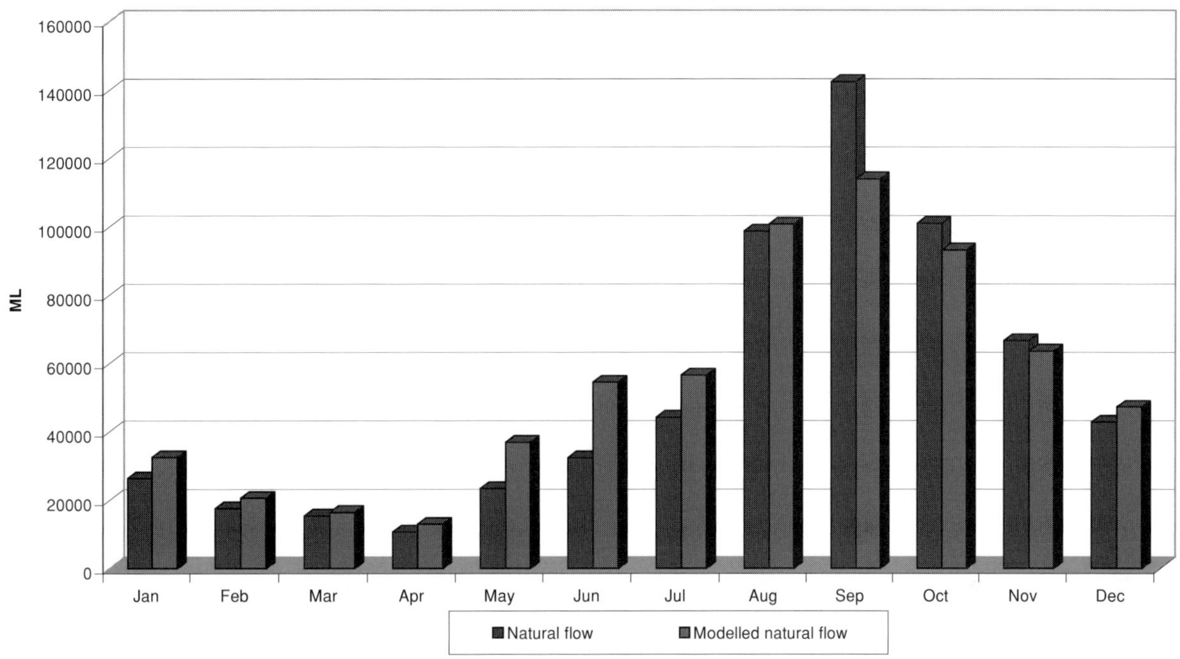

Figure 30.7 Model testing results for monthly flow from the Mae Chaem in Kong Kan. The model is applied for the period 1989–94 using the 'medium' irrigation diversion.

23%. The Nash-Sutcliffe statistic is 0.73 calculated on a monthly basis for the 'no irrigation' diversion scenario, and 0.74 for the 'medium irrigation' diversion case. Mean relative errors reached values of 30% and 22%, respectively. Mean monthly absolute errors, calculated for these model tests, were 10 300 ML/month ('no irrigation' diversion) and 8800 ML/month ('medium irrigation' diversion). The mean annual relative errors are low, at 8% and 5% for these two scenarios, respectively. This model test for medium irrigation diversion is illustrated in Figure 30.7.

DISAGGREGATION RESULTS

The streamflow disaggregation procedure was employed for streamflow modelling in two instrumented subcatchments: Mae Mu and Huai Phung. This test was also implemented for the 'no irrigation' and 'medium irrigation' cases. The mean relative errors and monthly residuals for this model test are summarised in Table 30.4. The mean monthly absolute errors for the Huai Phung subcatchment indicated in Table 30.4 are higher because absolute values for mean monthly discharges in the Huai Phung subcatchment are more than tenfold larger than those in Mae Mu. Figures 30.8 and 30.9 show the mean monthly observed and modelled discharges for the Huai Phung and Mae Mu subcatchments simulated using the model parameters estimated in the Mae Chaem catchment at Kong Kan for the 'medium irrigation' scenario.

An interesting comparison to consider is how these results compare with those based on the simplistic idea that streamflow discharge is contributed uniformly over the catchment area. The ratio of areas of the Huai Phung subcatchment to that of the Mae Chaem

Table 30.5. *Comparative results of streamflow simulation in the Mae Mu subcatchment using the ratio of the catchment topographic indices and ratio of catchment areas multiplied by the precipitation coefficient*

		Modelled flow using the ratio of the catchment topographic indices		Modelled flow using the areas' ratio multiplied by the precipitation coefficient 1.2 (ML)	
Month	Observed flow (ML)	Flow (ML)	Relative error (%)	Flow (ML)	Relative error (%)
Jan	1 732	1 394	20	1 232	29
Feb	1 261	885	30	782	38
Mar	1 229	707	42	625	49
Apr	1 058	561	47	496	53
May	1 622	1 588	0.2	1 403	13
Jun	2 012	2 348	17	2 075	3
Jul	2 531	2 433	4	2 150	15
Aug	3 810	4 334	14	3 830	0.5
Sep	4 348	4 904	13	4 334	0.3
Oct	4 013	4 010	0.1	3 544	12
Nov	3 208	2 739	15	2 421	25
Dec	2 101	2 035	3	1 798	14
Annual	28 925	27 938	3.4	24 689	15

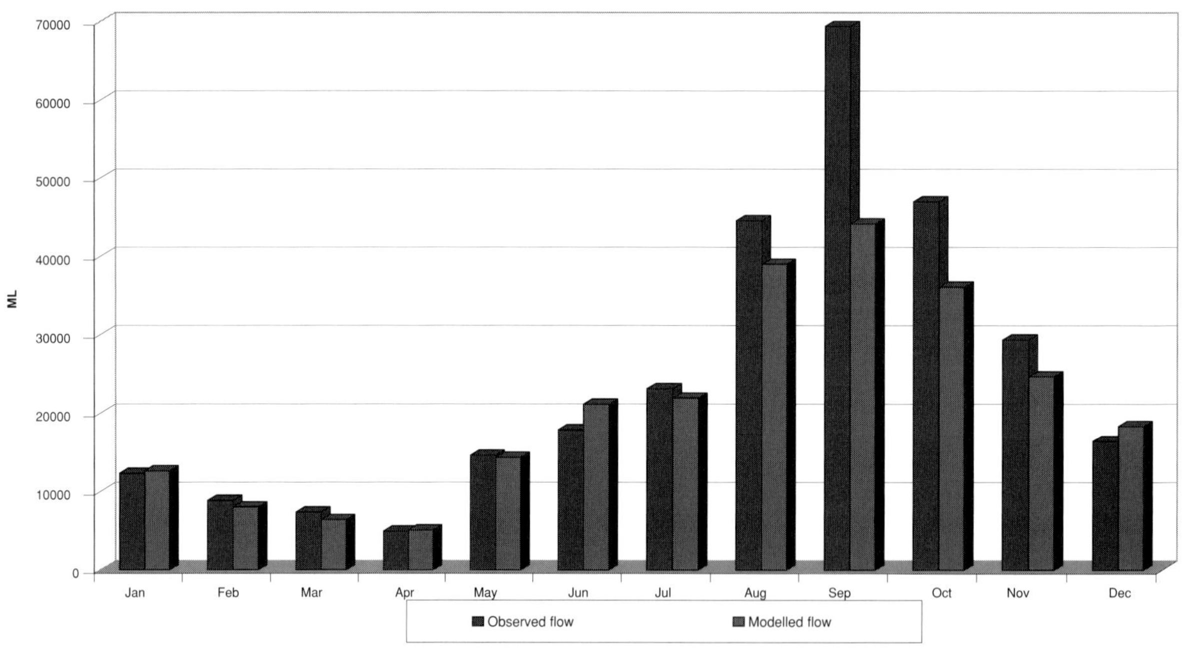

Figure 30.8 Results of streamflow modelling based on the disaggregation procedure for the Huai Phung subcatchment and observed average monthly discharge for the period of 1989–94 ('medium' irrigation diversion scenario).

catchment at Kong Kan is very close to the ratio of their respective topographic indices because these catchments have very similar slopes (Table 30.2). However, the difference in average slopes of the Mae Mu and Kong Kan catchments (19° and 14.3°, respectively) makes the values of the ratios of topographic indices (0.043) and catchment areas (0.032) quite different. The streamflow for the Mae Mu subcatchment was simulated using the *c* value scaled according to the catchment area for the 'medium' irrigation diversion scenario. Comparison of streamflow simulated in this way with observed values (Type 2 model test) provided a mean monthly relative error of 33% and a mean annual relative error of 28%. These results are significantly worse than those obtained using

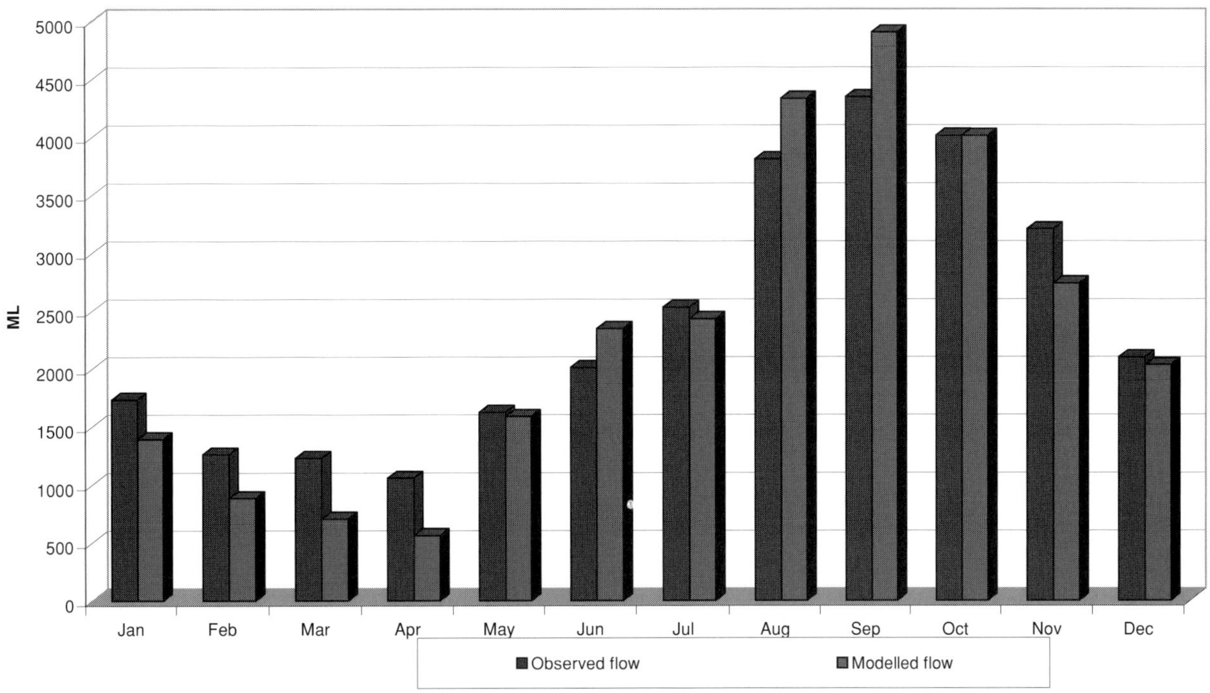

Figure 30.9 Results of streamflow modelling based on the disaggregation procedure for the Mae Mu subcatchment and observed average monthly discharge for the period of 1989–94 ('medium' irrigation diversion scenario).

the c value scaled according to the topographic index (17% for the monthly relative error and 3% for annual relative error, see Table 30.4).

Another test was implemented to check the following assumption: that the substantial difference between observed monthly flow and modelled flow, when scaling the IHACRES c coefficient according to the ratio of catchments sizes, is explained only by different climatic conditions for the Mae Mu subcatchment versus the entire Mae Chaem catchment at Kong Kan. The mean annual rainfall calculated over the last decade in the Mae Mu subcatchment is 1200 mm, compared with only 1000 mm for the Kong Kan station, which was used for the model calibration. The model was re-run therefore for the Mae Mu subcatchment with the c value recalculated from C calibrated for the Mae Chaem at Kong Kan catchment by the formula:

$$c = 1.2\,C\frac{a}{A}, \qquad (30.12)$$

where a is the area of the Mae Mu subcatchment and A the area of the Mae Chaem catchment at Kong Kan. The results of this exercise are presented in Table 30.5.

Table 30.5 shows that the results obtained by scaling the volumetric coefficient (c) according to the topographic indices are usually better than those obtained by scaling c according to the catchment areas and precipitation coefficient. The first algorithm provides consistently improved results for all months of the dry season, when water supply is crucially important. For only three months of the wet season (June, August and September) the discharge is modelled better using the latter case. However, the relative error for mean annual discharge is significantly lower for the first algorithm (3% compared to 15%, see Table 30.5).

DISCUSSION AND CONCLUSIONS

The methodology of surface runoff modelling in ungauged catchments proposed here has been applied and tested in the Mae Chaem catchment at Kong Kan, northern Thailand. The algorithm of catchment discharge disaggregation to subcatchment level is based on the assumption that the streamflow water yield in each subcatchment is proportional to the generalised topographic index calculated for this subcatchment using the linear downslope transmissivity profile. It is very important to note that the disaggregation algorithm utilised is not a direct application of TOPMODEL, as was implemented for instance by Quinn *et al.* (1991) and Molicôva *et al.* (1997) for tropical catchments in Africa and South America. It is just that a version of the easily estimated terrain attribute, similar to that used in TOPMODEL, was selected as a basis for streamflow disaggregation. Below, the data requirements and limitations of the proposed algorithm will be discussed.

The major limitation of the algorithm is that the scaling is restricted to the non-linear effective rainfall module of the IHACRES model. The rates of recession of the quick and slow flow components, α_q and α_s, in the linear component of the IHACRES model (Eqns 30.6b and 30.6c) are quite different for different subcatchments but have been assumed to be the same in this work. Subcatchments in the higher part of the catchment have flatter (slower) recession rates during the dry season than catchments with lower mean elevation. As a result, the dry season streamflow in the Mae Mu subcatchment (Figure 30.9) is consistently underestimated because the recession of dry season streamflow in the Mae Chaem catchment at Kong Kan (which was scaled down for modelling of the much smaller Mae Mu subcatchment) is steeper than that in the Mae Mu subcatchment.

Nevertheless, the results of the first-pass approach to streamflow modelling described here are encouraging. The relative errors for monthly streamflow modelled in the 'ungauged' subcatchments, estimated in the Mae Mu and Huai Phung subcatchments, range between 13% and 17%. The algorithm allows one to predict the natural streamflow and the remaining discharge after irrigation diversion. The input information required for this modelling is restricted to subcatchment area and slope for natural flow modelling, and to areas under different crops grown in the subcatchment for estimating irrigation diversions. Both are readily available from maps and data supplied by the Royal Irrigation Department and Department of Land Development of Thailand.

An important component of the algorithm is an irrigation consumption module. The results of the first-pass approach demonstrate that the application of this module provides consistent improvement in the modelling results for both model calibration and tests (see Tables 30.3 and 30.4).

There is considerable scope for further research. A major future development would be to invoke the application of crop models, used for calculation of irrigation demand, to calculate the partitioning of surface runoff versus deep drainage under different land covers (Croke *et al.*, 2004). Such new modelling work allows the introduction of demand for a wider spectrum of crop types and consideration of an 'irrigation flow return' component in the general water balance. Modelling the impacts of forest cover changes to the water availability for irrigation is another important task of the IWRAM Project (Jakeman and Letcher, 2003). Another important direction of the future modelling work is separate disaggregation of the slow flow (baseflow) and quick flow (runoff) components according to land cover changes. Some initial results obtained for the above developments of disaggregation are presented in Merritt *et al.* (2001).

APPENDIX: THE IHACRES MODEL

General information

IHACRES (**I**dentification of unit **H**ydrograph **A**nd **C**omponent flows from **R**ainfall, **E**vaporation and **S**treamflow data is a conceptual rainfall runoff model allowing one to predict surface streamflow in catchments using precipitation and temperature (or evaporation) time-series data.

Structure

IHACRES has two components

- a nonlinear loss module which produces effective rainfall from precipitation and temperature data
- a linear module which converts effective rainfall to modelled streamflow

Parameters

IHACRES has from 4 to 8 parameters to be optimised depending on the assumed and identified structure of its linear and non-linear modules (Croke and Jakeman, 2004).

Parameters are related to land use, landscape and vegetative conditions, not climate. Parameters also characterise the lumped response of catchment discharge e.g. quick flow and slow flow recession rates of discharge, slow flow index volume, potential storage capacity of the catchment.

Time step

IHACRES performs successfully for a range of time steps: from minutes to monthly. A daily time step is most commonly used.

Data requirements

IHACRES requires continuous time series of streamflow discharge, precipitation and temperature (or evaporation) data for model calibration, and just the latter two for simulation runs.

Applicability

IHACRES has been successfully applied in several regions worldwide for catchments of different sizes and under different climate conditions. It has been tested in catchments ranging from 500 m^2 (small experimental catchments) to 100 000 km^2 (the Thames River in the UK and the Avon River catchment in Western Australia). It was successfully tested for various hydroclimatologies, including alpine catchments in northern Scotland and the Australian highlands, temperate catchments in many countries, semi-arid catchments in Australia and tropical catchments in Thailand.

Abilities

IHACRES can be used for a range of purposes related to water resource assessment and management.

IHACRES is a reliable tool for estimating the climatic influence on streamflow. It allows one to predict streamflow for different climatic conditions related to climate variability or to global climate change.

IHACRES can be employed for estimating the impacts of land and water use change on streamflow. In particular it can be used for indicating the effect of deforestation and forest growth on a catchment's water resources.

Assumptions and limitations

IHACRES is a lumped parameter model. Its parameters reflect the integral, lumped properties of the catchment under consideration. However, subcatchments can be modelled separately and their output discharge linked.

IHACRES performance is limited by the quality of input climatic data (precipitation and temperature) and the quality of measured streamflow used for the model calibration.

IHACRES models surface runoff only. It is not applicable for assessing ground water resources. However, in calibration of the relationship between rainfall and measured streamflow discharge, it captures the effects of permanent losses, such as to groundwater, springs and evapotranspiration, and of delay in rainfall eventually reaching the stream, such as temporary detention in subsurface systems.

Irrigation off-takes can be inferred by the model by calibrating it on a non-irrigation period and simulating on an irrigation period. The difference between the simulation and observed stream discharge is the irrigation amount. Alternatively, irrigation can be estimated independently and added to the observed stream discharge to yield the 'natural' discharge. IHACRES can then be calibrated on the rainfall, temperature and natural discharge, so that the latter can be simulated as well as the effects of different irrigation practices on streamflow.

IHACRES is marketed by the Centre for Ecology and Hydrology, Wallingford, UK, and has been sold to many European water management agencies and researchers.

References

Ambroise, B., Beven, K. J. and Freer, J. (1996) Towards a Generalization of the TOPMODEL Concept: Topographic Indices of Hydrologic Similarity, Water Resources Research, 32, 2135–2145.

Beven, K. J. and Kirkby, M. J. (1979) A Physically Based Variable Contributing Area Model of Basin Hydrology, *Hydrological Sciences Bulletin*, 24(1), 43–69.

Beven, K. J., Kirkby, M. J., Schofield N. and Tagg, A. F. (1984) Testing a Physically-Based Flood Forecasting Model (TOPMODEL) for Three UK Catchments, *Journal of Hydrology*, 69 119–143.

Beven, K. J. (1997) TOPMODEL: a Critique, *Hydrological Processes*, 11, 1069–1085.

Chaiwanakupt, S. and Chiangprai, C. (1990) Resources and Problems Associated with Sustainable Development of Upland in Thailand, Proceedings of the seminar 'Technologies for Sustainable Agriculture on Marginal Uplands in South East Asia', Ternate, Cavite, Philippines, december 10–14, 1990.

Croke, B. F. W. and Jakeman, A. J. (2004) A catchment moisture deficit module for the IHACRES rainfall-runoff model, *Environmental Modelling and Software*, 19, 1–5.

Croke, B. F. W., Merritt, W. S. and Jakeman, A. J. (2004) A dynamic model for predicting hydrologic response to land cover changes in gauged and ungauged catchments. *J. Hydrology*, 291, 115–131.

DLD (1979) Department of Land Development of Thailand, Soil Map of Chiang Mai Province, scale 1:50000.

Iorgulescu, I. and Musy, A. (1997) Generalization of TOPMODEL for a Power Low Transmissivity Profile, *Hydrological Processes*, 11, 1353–1355.

Jakeman, A. J., Chaithawat Saowapon, Attachi Jintrawet, Karn Trisophon, Evans, J. P. and Wong, F. (1997) Biophysical Component of an Integrated Water Resources Assessment Project in the Upper Chao Phraya Headwaters, Northern Thailand, *International Congress on Modelling and Simulation (MODSIM97)*, Hobart, 8–11 December 1997, 2, 687–691.

Jakeman, A. J., Littlewood, I. G. and Whitehead, P. G. (1990) Computation of the Instantaneous Unit Hydrograph and Identifiable Component Flows with Application to Two Small Upland Catchments, *Journal of Hydrology*, 117, 275–300.

Jakeman, A. J. and Hornberger, G. M. (1993) How Much Complexity is Warranted in a Rainfall-Runoff Model?, *Water Resources Research*, 29(8), 2637–2649.

Jakeman, A. J. and Letcher, R. A. (2003) Integrated assessment and modelling: features, principles and examples for catchment management. *Environmental Modelling and Software*, 18, 491–501.

Kirkby, M. J. (1997) TOPMODEL: a personal view. Hydrological Processes, 11, 1087–1089.

Letcher, R. A., Merritt, W. S., Croke, B. F. W., Jakeman, A. J. and Buller, C. (2002) Integrated Water Resources Assessment and Management (IWRAM) Project: Integrated Toolbox, iCAM Working Paper 2002/02, Canberra, The Australian National University, pp. 23.

Merritt, W. S., Cropke, B. F. C., Jakeman, A. J. and Perez, P. (2001) Predicting Flow in Ungauged Catchments and Catchments Subject to Forest Cover Changes, Proceedings: *International Congress on Modelling and Simulation MODSIM01*, Vol. 1, Ghassemi, F., Post, D., Sivapalan, M., and Vertessy, R. (eds.), Canberra, Australia, 59–64, 10–13th December 2001.

Merritt, W. S., Croke, B. F. W., Jakeman, A. J., Letcher, R. A., and Perez, P. (2004) A biophysical toolbox for assessment and management of land and water resources in rural catchments in Northern Thailand. *Ecological Modelling*, 17, 279–300.

Moličova, H., Grimaldi, M., Bonell, M. and Hubert, P. (1997) Using TOPMODEL towards Identifying and Modelling the Hydrological Patterns within a Headwater, Humid, Tropical Catchment, *Hydrological Processes*, 11, 1169–1196.

Nash, J. E. and Sutcliffe, J. V. (1970) River Flow Forecasting Through Conceptual Models. Part I – A Discussion of Principles, *Journal of Hydrology*, 10, 282–290.

NRC (1997) Thailand Land Use and Land Cover Change Case Study. Report Produced for the IGBP-STRAT Initiative for South East Asia. Bangkok, National Research Council of Thailand.

Quinn, P. F., Beven, K. J., Chevallier, P. and Planchon, O. (1991) The Prediction of Hilslope Flow Pats for Distributed Hydrological Modelling Using Digital terrain Models, *Hydrological Processes*, 5, 59–79.

Quinn, P. F., Beven, K. J. and Lamb, R. (1995) The ln(a/tanß) Index: How to Calculate it and How to Use It Within the TOPOMODEL Framework, *Hydrological Processes*, 9, 161–182.

Post, D. A. and Jakeman, A. J. (1996) Relationships Between Catchment Attributes and Hydrological Response Characteristics in Small Australian Mountain Ash Catchments, *Hydrological Processes*, 10, 877–892.

Schreider, S. Yu. and Jakeman, A. J. (1999) Surface Runoff Modelling in Ungauged Subcatchments of the Mae Chaem Catchment, Northern Thailand: Part I, Methodology, Proc. *International Congress on Modelling and Simulation MODSIM99*, Hamilton, Oxley, L. and Scrimgeour, F. (eds.), New Zealand, 6–9th December 1999, Vol. 1, pp. 81–86.

Schreider, S. Yu., Gallant, J., Jakeman, A. J. and Merritt, W. S. (1999) Surface Runoff Modelling in Ungauged Subcatchments of the Mae Chaem Catchment, Northern Thailand: Part II, First Pass Approach, Proc. *International Congress on Modelling and Simulation MODSIM99*, Oxley, L. and Scrimgeour, F. (eds.), Hamilton, New Zealand, 6–9th December 1999, Vol. 1, pp. 87–92.

Scoccimarro, M., Walker, A., Dietrich, C. R., Schreider, S. Yu., Jakeman, A. J. and Ross, A. H. (1999) A Framework for Integrated Catchment Assessment in Northern Thailand, *Environmental Modelling and Software Journal*, 4, 567–577.

Wheater, H. S., Jakeman, A. J. and Beven, K. J. (1993) Progress and Directions in Rainfall-Runoff Modelling, Chapter 5 in: *Modelling Change in Environmental Systems*, A. J. Jakeman, M. B. Beck. and M. J. McAleer, eds., John Wiley and Sons, Southampton, UK, 101–132.

31 Parsimonious spatial representation of tropical soils within dynamic rainfall–runoff models

N. A. Chappell
Lancaster University, UK
K. Bidin
Universiti Malaysia Sabah, Malaysia
M. D. Sherlock
Lancaster University, UK
J. W. Lancaster
Arup Water, UK

INTRODUCTION

Models are used increasingly to simulate hydrological processes within tropical regions. There is now a wealth of publications addressing evaporation modelling (particularly wet-canopy evaporation) of local areas of tropical forest in, for example, Niger (Gash *et al.*, 1997), Guyana (Jetten, 1996), Puerto Rico (Schellekens *et al.*, 1999), Columbia (Marin *et al.*, 2000) and Indonesia (Asdak *et al.*, 1999; van Dijk and Bruijnzeel, 2001). Elsewhere in this volume, Roberts *et al.* provide an overview of evaporation processes and modelling. Other modelling studies have addressed the impact of such tropical evaporation on regional climates and global circulation (e.g. Polcher and Laval, 1994; Zeng, 1999; Zeng and Neelin, 1999; Zhang *et al.*, 2001). New studies using time-series models are highlighting the effects of cycles in the rainfall, such as the El Niño Southern Oscillation (ENSO) on tropical evaporation, riverflow and water quality (e.g. Zeng, 1999; Chappell *et al.*, 2001; Krishnaswamy *et al.*, 2001; Whitaker *et al.*, 2001; Chappell, Tych *et al.*, this volume). Similarly, models that simulate the generation of riverflow from the rainfall received by a tropical catchment are also beginning to be applied more frequently. These models include:

(1) Metric-conceptual models of waterflow, based upon transfer functions.[1]* For example, application of the DBM modelling approach to a nested catchment system in Malaysian Borneo (Chappell *et al.*, 1999a) and the application of IHACRES to a large Thai basin (Scoccimarro *et al.*, 1999).

(2) Conceptual models of waterflow based upon stores and predetermined empirical relationships. For example, application of the Nash model to Kenyan catchments (Onyando and Sharma, 1995), the Modhydrolog model to a tropical catchment (Chiew *et al.*, 1996), the Reservoir-Water-Balance-Simulation model to Namibian catchments (Hughes and Metzler, 1998), and the HBV-96 model (discussed in Barnes and Bonell, this volume) to catchments in Zimbabwe, Tanzania and Bolivia (Liden and Harlin, 2000).

(3) Conceptual models of waterflow incorporating spatially distributed, topographic information. For example, application of TOPMODEL in Cote d'Ivoire (Quinn *et al.*, 1991), French Guiana (Molicova *et al.*, 1997) and Malaysian Borneo (Chappell *et al.*, 1998) and TOPOG in Peru (Vertessy and Elsenbeer, 1999) and Puerto Rico (Schellekens, 2000).

(4) Hydrochemical mixing models for water-path identification. For example, analysis of natural chemical signals within catchment waters of Queensland, Australia (Elsenbeer *et al.*, 1995) and Tanzania (Sandstrom, 1996) and environmental isotopes also in Queensland, Australia (Barnes and Bonell, 1996).

(5) Hydrological models based on Geographic Information System (GIS) mapping. For example, use of remote sensing and other GIS data in runoff prediction in West Africa (Schultz,

1 Technical words with asterisks are detailed in Appendix 31.1 at the end of the chapter.

Forests, Water and People in the Humid Tropics, ed. M. Bonell and L. A. Bruijnzeel. Published by Cambridge University Press.

1994) and the ANSWERS model in Queensland, Australia (Connolly *et al.*, 1997).

(6) Process-based catchment models solving the Richards Equation (Richards, 1931). For example, application of the Système Hydrologique Européen (SHE) model in India (Refsgaard *et al.*, 1992; Singh *et al.*, 1999) and Zimbabwe (Refsgaard and Knudsen, 1996).

One of the key questions to be addressed when deciding which of these modelling approaches should be applied at a new location, is how complex does the model structure need to be to describe adequately the rainfall-runoff behaviour of a tropical catchment (and possibly also the behaviour of certain internal characteristics, such as saturation extent)? This becomes critical when it is appreciated that very little information is contained within most time-series of riverflow, so that *very simple models are often sufficient to forecast the rainfall-runoff behaviour of catchments* (Kirkby, 1975; Jakeman and Hornberger, 1993; Beven, 2001a; Kokkonen and Jakeman, 2001; Young, 2001; see also Barnes and Bonell, this volume), whether in tropical or temperate regions.

This can be illustrated with the application of a Data-Based-Mechanistic (DBM*) model to the rainfall-runoff behaviour of the 0.44 km^2 Baru Experimental Catchment in Sabah, Malaysian Borneo (cf. Chappell *et al.*, 1999a). The structure of the model can be described in transfer-function* form:

$$q(k) = \frac{P}{1 - \Re z^{-1}} r_{\text{eff}}(k - \delta) \tag{31.1}$$

where

$$r_{\text{eff}}(k) = r(k)\left[\theta_{k-1} + \frac{1}{\tau_\theta}\{r - \theta_{k-1}\}\right] \tag{31.2}$$

and *q(k)* is the riverflow at the time index *k*, $\Re$ is the recession or lag parameter, *P* is the system production or gain parameter, z^{-1} is the backward shift operator (i.e., $z^{-i}\, r(k) = r(k-i)$) which allows expansion to higher-order models, δ is the pure time delay to the initial response, and *r* is the catchment-average rainfall input, r_{eff} is the transformed input or the waterflow after the catchment non-linearity has been characterised, term θ_{k-1} is the linear component of the internal flow at the previous time-step, and τ_θ is the time constant (or residence time) of the non-linear component of the catchment behaviour (Young, 1984; Whitehead *et al.*, 1979). Further explanation of the DBM approach is given in Chappell *et al.* (in this volume).

Where the pure time delay is zero, then State Dependent Parameter (SDP) identification (Young, 2001) may identify as few as three catchment behaviour parameters that capture the rainfall-runoff behaviour of a catchment. These parameters are: the basin lag or recession, the production or water-balance term, and the term describing the form of the catchment non-linearity (i.e., $\Re$, *P*, and τ_θ). To illustrate this point, the DBM approach is applied to a one-year record of hourly rainfall and riverflow data for the Baru

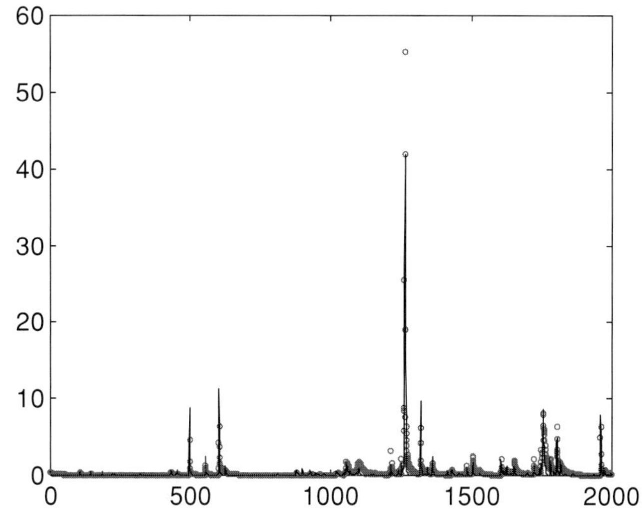

Figure 31.1 The results of a DBM model, incorporating the Bedford-Ouse Sub-Model to characterise the catchment non-linearity, applied to 1 year of hourly rainfall and riverflow data for the Baru Experimental Catchment, Sabah, Malaysian Borneo. The model has a Nash and Sutcliffe (1970) efficiency of 0.876 and a YIC of −9.93. The abscissa in time in hours, and the y-ordinate is riverflow in mm hr^{-1}. Observed riverflow is shown with a dotted line, and the modelled riverflow a solid line.

Experimental Catchment. A model efficiency (i.e. one minus the ratio of variance of the model errors to the variance of the observed data, expressed as a per cent) of 88% was achieved (Figure 31.1), indicating that the model captured most of the key dynamics inherent in the relationship between the incoming rainfall and outgoing riverflow. Such a three-parameter model indicates that one pathway dominates the catchment behaviour. DBM models do, however, allow for more complicated structures, i.e. multiple flow pathways. An objective statistic known as the 'Young Information Criterion' (YIC) can be used to examine if multiple pathways are observable within the dynamics of the catchment under study (Young, 2001). In the case of the year-long Baru Catchment dataset, however, the YIC did not support the use of more than one *dominant pathway* to route rainfall to the river.

Other studies have shown that complex (often physics-based) models do little better at forecasting riverflow time-series (in a split-sample validation) when compared with models with simple structures (and hence requiring few parameter values, i.e. parsimonious) e.g. Loague and Freeze, 1985; Franchini and Pacciani, 1991; Michaud and Sorooshian, 1994). In the tropical context, for example, the study of Refsgaard and Knudsen (1996) demonstrated that there was little additional forecasting benefit from applying the complex, physics-based MIKE-SHE model when compared to the NAM and WATBAL 'bucket models' to rainfall-runoff data from catchments in tropical Zimbabwe.

Potential value of physics-based modelling

Physics-based catchment models such as MIKE-SHE usually divide the model catchment into approximately 100 to 1000 landscape units. Over this distribution of landscape units, they then solve: (1) the momentum equation for waterflow (e.g. Darcy–Buckingham Equation for the subsurface flow: Darcy, 1856; Buckingham, 1907), and (2) the continuity equation.

These process-descriptions (or physical theories) incorporate soil-topographic parameters that (in principle) can be measured independently of the catchment's rainfall-runoff behaviour. As it may be possible to see the impact of a particular land-use activity on each of these soil-topographic parameters (e.g. soil permeability, porosity), it is then assumed that the impacts of land-use change on rainfall-runoff behaviour can be forecast via modifications to the model's parameters. This would suggest that such models have a significant advantage over 'bucket' and transfer-function models in their ability to predict the effects of internal catchment (e.g. land-use) changes on rainfall-riverflow behaviour. Physics-based models would, therefore, seem to have considerable value in the assessment of forestry, agricultural or urban impacts on the hydrological behaviour of tropical catchments.

Limitations to the testing and hence reliability of physics-based models

Currently, the physical algorithms on which many of these catchment-scale models are based were derived from observations of small-scale (10^{-1} m) phenomena and not on theory developed at the scale of the terrain-elements into which the modelled catchment has been divided (i.e. the model grid-scale*). These model grid elements are essentially the size of whole hillslopes, and tracer studies that strongly support Richards' formulations of waterflow in variably-saturated soil at this scale arguably do not exist (see e.g. Sherlock *et al.*, 1995; Sherlock, 1997). As the rainfall-runoff behaviour of a catchment or indeed a single hillslope is a non-linear system, one cannot assume that a Constant of Proportionality such as soil permeability which represents the lumped behaviour of a small core, will be appropriate at the scale of a highly heterogeneous hillslope (Beven, 2001b). Validation of the model's assumed 'grid-scale physics' (and hence the parameter estimates indicated by model calibration) is clearly necessary. This testing process has, however, been hampered primarily by the lack of available techniques to measure the 'same' lumped grid/hillslope scale parameters independently in the field (Beven, 2001a). This problem is then compounded by:

(1) The perceived need to represent all of the profile and catenal changes in soil-rock permeability within a catchment (Chappell and Ternan, 1992) which has resulted in the development of physics-based models that require, for example, the specification of numerous permeability values over the modelled catchment.
(2) At the small-scale, some of the most sensitive* soil and topographic parameters (e.g. soil permeability), have a very high degree of unstructured spatial variability (Bonell *et al.*, 1983; Elsenbeer and Cassel., 1993; Bonell with Balek, 1993). This has made it easy for those undertaking the modelling to not question almost any estimate of grid-scale* parameters identified by model calibration as being 'within the measured range'.
(3) The difficulty of identifying model parameters by calibration (i.e. parameter inversion process*), is magnified by the relatively recent observation that *many 'sets of parameters', each with very different values of each parameter, will give acceptable simulations of the same rainfall-runoff time-series* (Freedman *et al.*, 1998; Beven, 2001a; Thiemann *et al.*, 2001). The greater the model complexity, the more interaction between modelled parameters can take place during the calibration, so the more different the resultant parameter sets become, and finally the wider (or more uncertain) is the range of each model parameter.

There is, therefore, considerable merit in attempting to *constrain the complexity of catchment models to reduce the number of parameter sets that give acceptable model predictions, and thereby constrain the likely range of each model-calibrated parameter*. This then allows a more realistic comparison of grid-scale model-parameters estimated by inversion with those derived from field measurements (and appropriately up-scaled). Modelling where the *objective is to limit complexity* so that each parameter can be more narrowly defined, is called parsimonious. The group of catchment models, known as 'topographically-based hydrological models' (e.g. TOPMODEL and TOPOG_SBM) have at their core a simple/parsimonious structure, though they too are sometimes extended to become complex, parameter-demanding models. Clearly, in the tropics where there is a dearth of soil data, models which require fewer parameters and less complex spatial distributions of each parameter could be helpful.

TOPOGRAPHICALLY BASED, DYNAMIC RAINFALL-RUNOFF MODELS

The two most widely used topographically-based, dynamic catchment models are TOPMODEL (Beven and Kirkby, 1979; Beven, 1997), and TOPOG_SBM / TOPOG_DYNAMIC (Vertessy *et al.*, 1994; Vertessy and Elsenbeer, 1999). The TOPOG variants were derived originally from the WETZONE model of O'Loughlin (1986). These models might be seen to be based on two key structural components:

(1) an index describing the degree of profile-saturation distributed in plan throughout the catchment (known as the 'topographic index' in TOPMODEL and the 'wetness index' in TOPOG), and
(2) a momentum equation capable of capturing the non-linear relation between rainfall and the river discharge generated.

(1) SPATIAL MAPPING OF SOIL SATURATION

The topographic index (λ) within TOPMODEL is normally (cf. Ambroise *et al.*, 1996) defined as:

$$\lambda = \ln\left(\frac{a}{\tan\beta}\right) \tag{31.3}$$

where a is the upslope contributing area (in plan) to a given profile (m^2), and $\tan\beta$ is the local slope angle at that profile. A greater λ indicates a greater likelihood of saturation and hence surface-flow generation. The topographic index components indicate first, that as the subsurface contributing area to a local soil-rock profile in the catchment increases, so does the likelihood of increasing relative saturation. Second, the index incorporates the Darcian assumption of subsurface flow being proportional to the hydraulic gradient, and this can be approximated by the tangent of the ground-surface slope ($\tan\beta$). Thus, a steeper topographic slope is expected to give a greater hydraulic gradient and, therefore, increase the 'drainage potential' of a local soil-rock profile and thereby reduce its level of saturation. The topographic index mapped over the 0.44 km^2 Baru catchment in equatorial Borneo (using a 20 × 20 m Digital Terrain Model, or DTM) is given in Figure 31.2. Like the 'wetness index', λ is an index of 'hydrological similarity', which means that *elements of the terrain with the same index value are expected to behave in a similar hydrological manner.*

Over the last 20 years numerous studies have attempted to test this index against field observations of the dynamic spatial patterns of:

(a) the extent of surface saturation (e.g. O'Loughlin, 1981; Barling *et al.*, 1994),
(b) soil moisture content (e.g. Burt and Butcher, 1983; Chappell and Franks, 1996; Sulebak *et al.*, 2000),
(c) capillary potential (Molicova *et al.*, 1997), and
(d) water-table level within boreholes (e.g. Troch *et al.*, 1993; Jordan, 1994; Moore and Thompson, 1996).

The level of agreement between the observed and predicted saturated extent has been mixed, though Chappell and Franks (1996) have demonstrated that saturated extent may be predicted better on some slopes within a single catchment than others, and this may indicate spatial differences in hydrological complexity of the studied catchment. Some improvements to saturated area estimation have been made by:

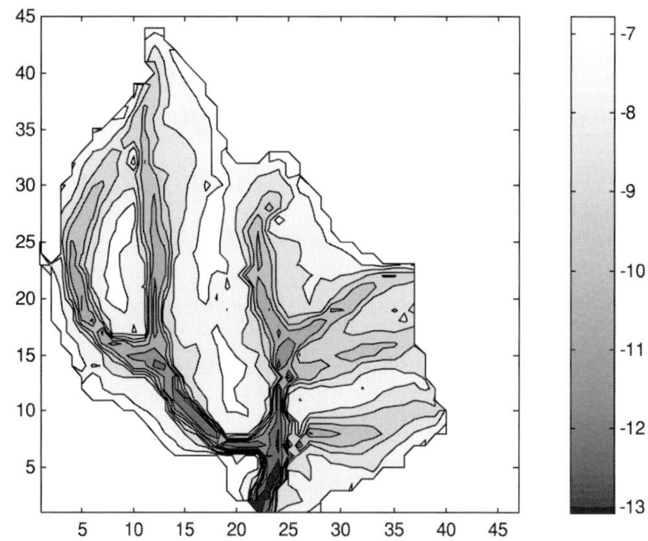

Figure 31.2 The map of the topographic index (λ) for the 0.44 km^2 Baru Experimental Catchment, Sabah, Malaysian Borneo. The index ranges from 5.99 to 12.99. The abscissa and y-ordinate axes show the number of grid cells. North is to the top of the figure.

(a) varying the model's transmissivity profile where borehole data are available (Lamb *et al.*, 1997),
(b) development of Dynamic-TOPMODEL to reproduce varying subsurface contributing areas by allowing the index to vary (Beven and Freer, 2001), though allowing spatial variations in the transmissivity and index may add significantly to model complexity,
(c) use of the bedrock-surface, rather than ground-surface, to derive the topographic index (Freer *et al.*, 1997), and
(d) evaluation of the effect of methods of analysing topography, and also the effects of different topographic grid sizes* on the values of transmissivity identified by TOPMODEL (Franchini *et al.*, 1996; Saulnier *et al.*, 1997; Brassington and Richards, 1998). Typically, these studies have shown that above a certain grid-size for the DTM (often 50 × 50 m), the effect on the catchment-average topographic index is to increase the T_0 term (i.e. lateral transmissivity, when the soil and weathered rock profile is saturated to the ground surface) within TOPMODEL calibrations. Clearly, this has implications for the interpretation of model-derived transmissivity and permeability values where coarse grids and parameter calibration routines are used.

While much effort has been directed towards the analysis of the topographic index, *comparatively little effort has been directed towards the second component of TOPMODEL (or indeed TOPOG), which contains the terms associated with the spatial distribution of soil permeability* (i.e. T_0 and m, where m is the exponential decay rate of lateral transmissivity with depth). Yet, soil permeability is invariably seen as the most important soil–rock property specified within complex physics-based models (Freeze, 1980; Rogers *et al.*, 1985; Sherlock *et al.*, 2000) and also in topographically-based models (Franchini *et al.*, 1996).

(2) LATERAL PERMEABILITY DISTRIBUTION

Within TOPMODEL the river discharge (q_i) is generated by:

$$q_i = a_i r = T_0 e^{-\bar{S}_i/m} \tan\beta \qquad (31.4)$$

where a_i is the upslope contributing area to a given point i per unit contour length (m^2), r is the catchment-average rainfall (m hr^{-1}), T_0 is the lateral transmissivity, when the soil and weathered rock profile is saturated to the ground-surface (m^2 hr^{-1}), $\bar{S}$ is the catchment-average, subsurface storage deficit (m), m is the exponential decay rate of lateral transmissivity with depth, and $\tan\beta$ is the tangent of the local slope angle at point i (m m^{-1}) (Quinn *et al.*, 1991). The lateral transmissivity for varying saturated depths, T (m^2 hr^{-1}) is then:

$$T = T_0 e^{-\bar{S}/m} \qquad (31.5)$$

where the catchment-average saturation deficit is:

$$\bar{S} = -\frac{m}{A}\int \sum \zeta + \ln r(dA) \qquad (31.6)$$

and A is the total catchment area (m^2) and ζ is the combined soil-topographic index. This can then be related to a catchment-average, water-table head using:

$$H = D_r - \left(\frac{\bar{S}}{\eta_{\mathrm{eff}}}\right) \qquad (31.7)$$

where H is the water-table head above the solid rock (m), D_r is the depth to the solid rock (m), and η_{eff} is the effective porosity (m^3 m^{-3}) (Chappell *et al.*, 1998). The lateral block permeability over the saturated part of the profile (K_{SM}, m hr^{-1}) is then:

$$K_{\mathrm{SM}} = \frac{T}{H} \qquad (31.8)$$

It is clear that the vertical distribution of lateral permeability is described by a single exponential function (Eqn 31.5). The rate of decline of the exponential function is governed largely by the m parameter, which can be determined from the average hydrograph recession (i.e. Master Recession Curve) independent of rainfall-runoff calibration. Indeed, Lamb and Beven (1997) have developed an algorithm, 'MRC tool', to automate this task. Many river hydrographs, in both tropical and temperate catchments, can be described by an exponential recession component. So the application of a topographic model that assumes an exponential recession is a good starting point. What is perhaps 'fortuitous' is that many tropical catchments have soils where the permeability declines exponentially with depth (Figure 31.3; Beven, 1982; Sherlock, 1997). Often this is the result of a combination of the argillation process in the upper profile, combined with the effect of uninterrupted *in situ* chemical weathering of the lower layers (Fitzpatrick, 1971). A fuller description of the theoretical basis, assumptions and limitations of TOPMODEL is given in Kirkby (1975), Beven and Kirkby (1979), and Beven (1997), and on the internet site: http://www.es.lancs.ac.uk/hfdg/topmodel.html. Freely available model code and executables are also available on this internet site.

Clearly, soil permeability need not follow a monotonic exponential decline with depth but follow other monotonic relationships such as a power (Lancaster, 2000) or linear decline. In soils developed on deposits such as head and glacio-lacustrine drift in temperate environments or river terrace alluvium and volcanic tuff in temperate and tropical environments, then simple monotonic reductions of permeability with depth might not be observed (Chappell and Ternan, 1992, L. A. Bruijnzeel, pers. comm.). Non-monotonic changes of permeability with depth (e.g. where a C or R soil horizon is more permeable than the overlying B soil horizon) would be difficult to equate with TOPMODEL assumptions and also to reproduce within the parameterisation. Where soils exhibit non-monotonic declines in permeability, not only is model structure more difficult to define but so is the up-scaling of the point-scale, field measurements so that they are representative of the lateral permeabilities of whole soil profiles.

The point-scale measurements of soil permeability obtained with field tests such as ring permeametry (Bonell *et al.*, 1983; Chappell and Ternan, 1997) cannot be compared directly to estimates of lateral permeability representative of the catchment-average soil profile or even the hillslope-average soil profile derived from model inversion* (Chappell *et al.*, 1998). *The point-scale measurements of soil permeability need to be first 'up-scaled' to give estimates of lateral permeability of the catchment-average soil profile.* The permeabilities derived from up-scaling* are known as block permeabilities.* If we attempt to up-scale point-permeabilities for comparison with TOPMODEL-derived values, the resultant 'lateral block permeabilities' should be equal to something between an arithmetic and a harmonic mean of the core-scale values (Cardwell and Parsons, 1945). This uncertainty relates to the uncertain geometry of the flow pathways. It is often difficult to define a priori the up-scaling method more precisely (Wen and Gómez-Hernández, 1996).

To illustrate a methodology for evaluating soil representation within rainfall-runoff models and also highlight some of the difficulties associated with such a test, we will now summarise an approach described more fully in Chappell *et al.* (1998). This study examined a small tropical catchment, the Baru Catchment in Malaysian Borneo, comprised of Acrisol-Alisol (i.e.

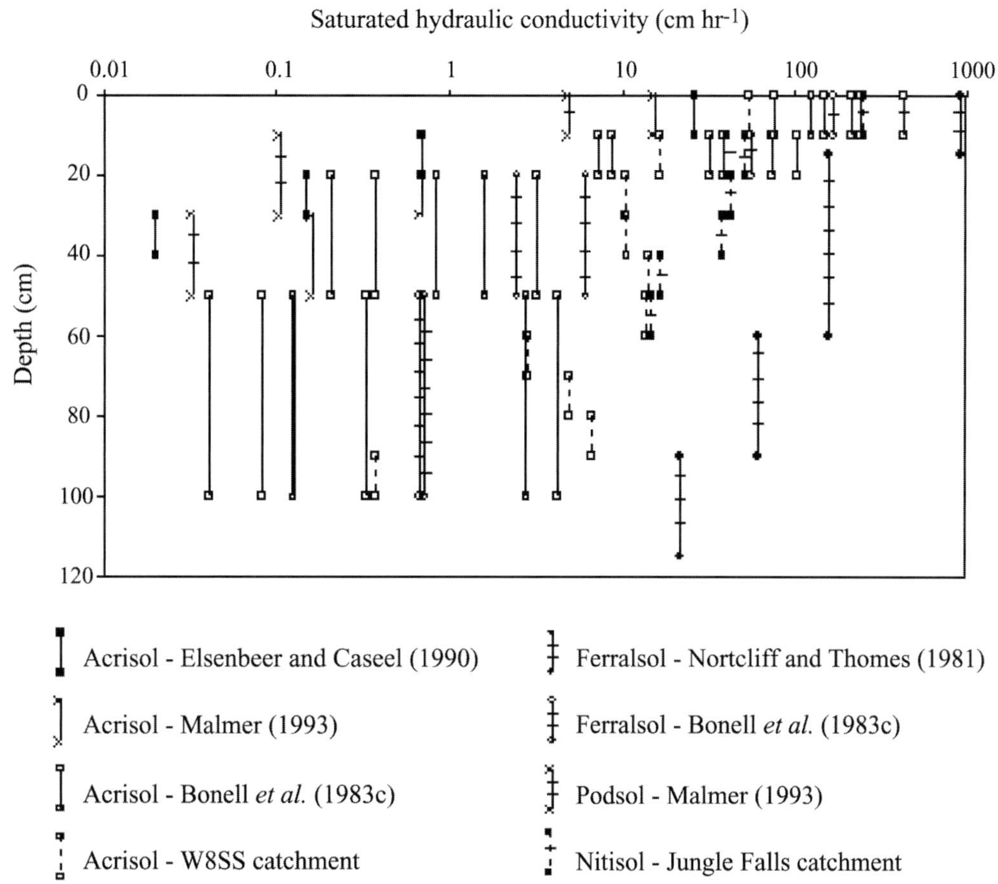

Figure 31.3 Profiles of saturated hydraulic conductivity (or 'soil permeability') from selected tropical soils. The vertical range that each published result represents is shown. (After Sherlock, 1997.)

USDA-Ultisol) soils (Chappell *et al.*, 1999b). This soil group is widespread throughout the tropics of SE Asia, West Africa and Amazonia (Bridges *et al.*, 1998).

EQUATING FIELD AND MODEL PERMEABILITY: A CASE STUDY OF A TROPICAL CATCHMENT

The Chappell *et al.* (1998) study was undertaken in natural forest (managed and disturbed) close to the Danum Valley Field Centre, Malaysian Borneo (5°01′N, 117°48.75′E), and had five key steps. These were: (1) field measurement of permeability, using relatively large soil cores, (2) up-scaling the core-based measurements to give lateral permeabilities representative of the whole hillslope soil profile, (3) deriving the catchment-average, lateral permeability profile – a procedure which is only realistic when a parsimonious model such as TOPMODEL is used, (4) comparison of up-scaled, field-derived permeabilities with model-derived permeabilities, and (5) derivation and preliminary testing of a new method for measuring the lateral permeability of the whole hillslope soil profiles.

(1) Point measurement of permeability

Ring permeametry (Bonell *et al.*, 1983; Chappell and Ternan, 1997) was used to measure the spatial distribution of soil permeability within a 12 km^2 region of only Acrisol-Alisol soils containing the Danum Valley Field Centre (Malaysian Borneo) and the catchment to be modelled. A total of 70 such permeability measurements on undisturbed, 30 cm diameter, 10 cm deep cores were taken from depths ranging from 0.05 to 3 m at crest-slope, side-slope and valley locations (Sinun, 1991; Chappell and Binley, 1992; Bidin, 1995; Sherlock, 1997; Chappell *et al.*, 1998). These data indicated that the spatial (vertical) variability in the permeability followed a simple monotonic decline with depth. The exponential function fitted to median permeability estimates for each depth, followed:

$$K_{S(med)} = 42e^{-2.188D}(R^2 = 0.92) \qquad (31.9)$$

where $K_{S(med)}$ is the core-scale, median permeability (m s^{-1}) at measurement depth D (m). This function fitted almost as well to the geometric mean permeability for each depth, i.e.

$$K_{S(geo)} = 52e^{-2.2907D}(R^2 = 0.88) \qquad (31.10)$$

where $K_{S(geo)}$ is the core-scale, geometric mean permeability (m s^{-1}) at measurement depth D (m). Clearly, these two models give some permeability at depths of many metres, but the local value will be several orders of magnitude less than that in the uppermost 3 metres of the soil-rock profile. Below 3 m, the soil-rock profile can, therefore, be assumed to have effectively no permeability.

(2) Up-scaling point measures of permeability

Within the study, an uncertain range of lateral block permeabilities were then estimated by using both arithmetic and harmonic averaging of the core-scale values. The calculated estimates of the lateral block permeability per the saturated part of the profile (K_{SC}) are presented in Figure 31.4. These estimates can be multiplied by the saturated depth to give the lateral transmissivity distribution. These data, unlike the original field measurements, could then be compared with results from the model inversion undertaken using catchment-scale data.

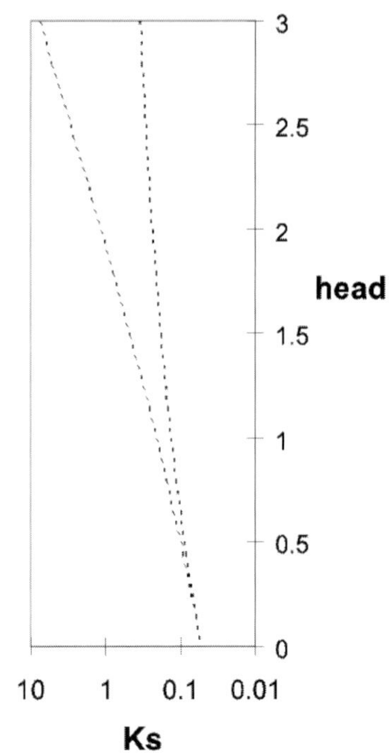

Figure 31.4 Lateral block permeability distribution derived by up-scaling core-based measurements from a 12 km^2 region around the Danum Valley Field Centre (Sabah, Malaysian Borneo) using a harmonic mean (right-hand curve) and an arithmetic mean (left-hand curve). The abscissa shows the lateral block permeability per saturated part of the soil profile (K_{SC}) in units of $\times$ 10^{-6} m s^{-1}. The *y*-ordinate shows the water table depth in metres, where zero is the rock-head (i.e. surface of the impermeable rock) and 3 m is the ground surface.

(3) Permeability estimation by catchment-model inversion

The 12 km^2 region in which the point permeability measurements were undertaken contains the 0.44 km^2 Baru Experimental Catchment. This catchment contains a distribution of data-logged raingauges and river gauging structures (Chappell *et al.*, 1999a). The Chappell *et al.* (1998) study used hourly rainfall and riverflow data lumped over the whole catchment for the period 6 May to 8 November 1995. Within this study a simple form of TOPMODEL requiring only four parameters (T_0, *m*, *Td* and SRMAX) was used to reduce the number of parameter sets that give acceptable efficiencies*, and hence help to constrain the range of individual parameters. The *Td* is the unsaturated zone time delay and SRMAX is the maximum root zone storage. Even with only four parameters, the range in parameters within physics-based and quasi-physics-based models can be so large as to include any field measurements. The Chappell *et al.* (1998) study, therefore, followed the approach of Franks *et al.* (1997). They demonstrated that where additional data are available for the catchment, such as a knowledge of the maximum and minimum extent of saturated areas, then this information could be used to reject model simulations that gave very unrealistic saturated extents. They further demonstrated that a large reduction in the uncertainty in the critical T_0 parameter could be obtained using this approach. More recently, Wooldridge *et al.* (2001) have similarly suggested that rejection of some parameter sets on the basis of inconsistency of simulations with additional internal state data (i.e. non-behavioural sets) constrains parameter uncertainty significantly.

As a result of the acknowledged effects of overly coarse DTM resolutions on the topographic index and thence the T_0 parameter (Franchini *et al.*, 1996; Saulnier *et al.*, 1997; Brassington and Richards, 1998), a fine resolution DTM (i.e. 10 $\times$ 10 m) was used. Ten thousand Monte Carlo simulations of TOPMODEL were run to identify the 'better' parameter sets. Field observations of the spatial extent of saturated soil profiles (Sinun, 1991; Sherlock, 1997; Chappell *et al.*, 1999b; P. Kukuon, pers. comm.) are were well within the range of 2 to 10% of the catchment during the hydrometeorological conditions experienced in 1995. The Monte Carlo results were, therefore, conditioned further by rejecting those model parameter sets that produced saturated areas either less than 2% or greater than 10% saturated area (i.e. parameter estimates associated with non-behavioural simulations

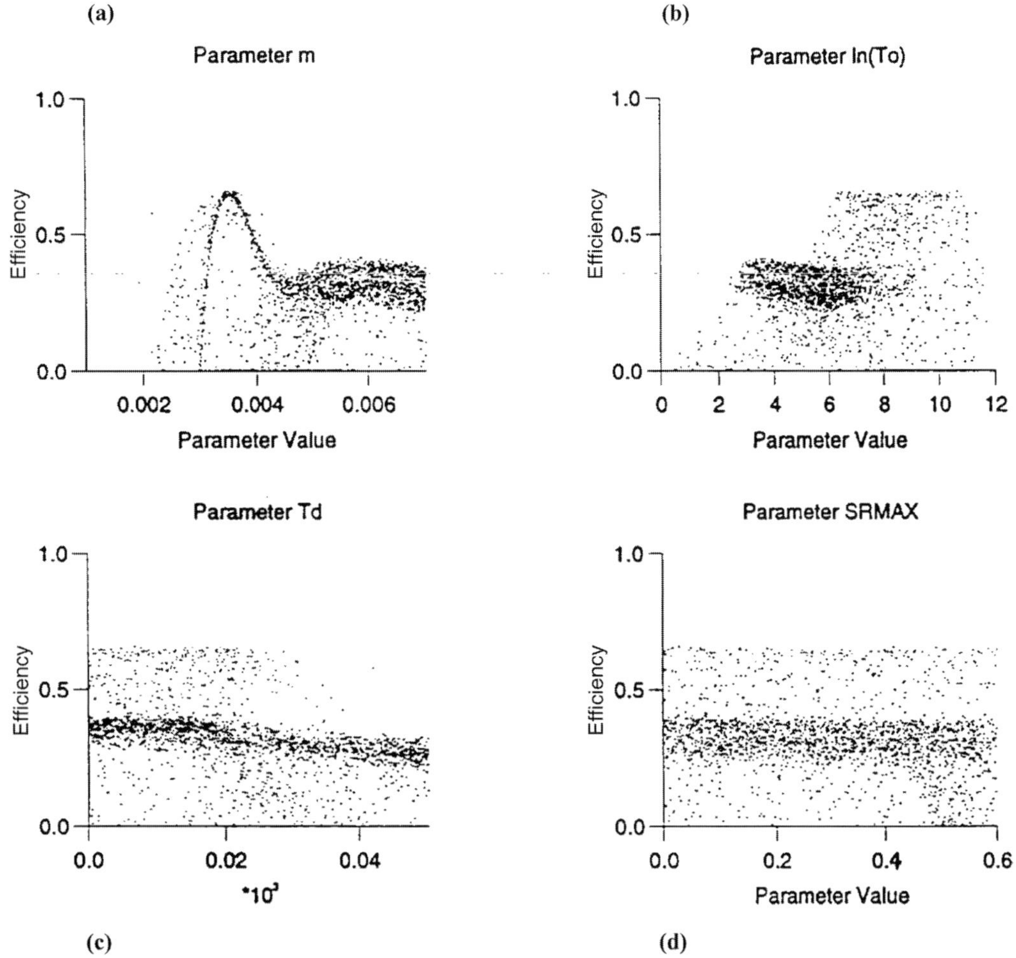

Figure 31.5 Scattergrams of model efficiency against (a) m (metres), (b) T_0 ($\ln[m^2\ s^{-1}]$), (c) Td ($\times\ 10^{-5}\ s\ m^{-1}$), and (d) SRMAX (metres) parameters after comparison with discharge predictions that produce between 2% and 10% saturated area. (After Chappell *et al.*, 1998.)

were discarded). The result of the conditioning by saturated area extent was indeed seen to constrain the T_0 parameter (Figure 31.5), giving a plateau in the efficiency surface ranging from 0.189 to 4.910 $m^2\ s^{-1}$. This range, combined with the peak value for m of approximately 3.5×10^{-3} m, was then used to calculate an uncertain range of lateral block permeability. This was presented over the range of water-tables that the model yielded, given the maximum and minimum T_0 values chosen (Figure 31.6). It is unrealistic to present lateral block permeabilities (or transmissivities) outside of the range of behavioural simulations (hence at the extremes of the exponential distribution). The predicted water-tables were indeed consistent with the patterns observed throughout the 12 km^2 Danum Valley study region (Bidin, 1995; Sherlock, 1997; P. Kukuon, pers. comm.). The mean predicted water-table head (above the effective rock-head of 3 m) during the simulation period was 1.24 m (Chappell *et al.*, 1998) and the lateral block permeability at this mean head was between 0.527×10^{-6} and 13.7×10^{-6} m s^{-1} from the modelling (K_{SM}).

(4) Comparison of model-derived permeabilities with up-scaled field values

Two key differences between the up-scaled, core-based permeability data and model-derived permeability data are observed in Figure 31.6:

(a) The range from the up-scaled core data (K_{SC}) of 0.158×10^{-6} to 0.311×10^{-6} m s^{-1} for the Baru Catchment soils is clearly much less than the model-derived range (Figure 31.6). This may indicate that: (i) some of the key conductive pathways (e.g., natural soil pipes*, percolines* or fractures) are not characterised adequately by core-scale measurements, or (ii) that the rainfall-riverflow response is not best characterised by permeabilities and hydraulic gradients but by more rapid response mechanisms such as the migration of a solitary wave across a hillslope water-table surface (Chappell *et al.*, 1998). A range of physics-based modelling studies (e.g., Sloan *et al.*, 1983; Blain and Milly, 1991) have similarly

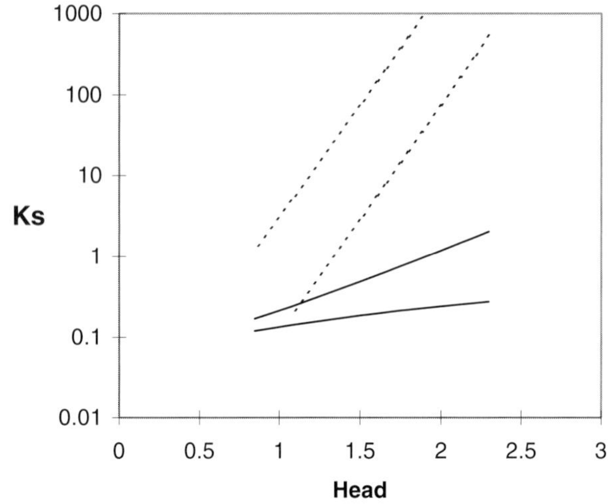

Figure 31.6 The uncertain band of lateral block permeability derived from the TOPMODEL inversion (between the broken lines, K_{SM}) and that derived by up-scaling core-based measurements (between the solid lines, K_{SC}). Lateral block permeability has units of $\times\ 10^{-6}$ m s^{-1} and water-table head is in metres. (After Chappell *et al.*, 1998.)

indicated that effective permeabilities* derived by hillslope and catchment-scale inversion give higher estimates that those derived from point-scale, field measurements. Vertessy and Elsenbeer (1999) noted that their model-derived permeability estimates were much larger than their point estimates measured in an Amazonian catchment (and presumably larger than the lateral block permeabilities, if they had derived them). Bonell (1998), assessing current challenges of research on river generation processes, reported similar differences between field- and model- derived permeability estimates. The suggestion that models such as TOPMODEL may generate artificially high effective permeabilities as a result of its steady-state assumption (Wigmosta and Lettenmaier, 1999) does, however, need further investigation.

(b) The slope of the exponential distribution of the permeability band for up-scaled core measurements and model-derived measurements are significantly different (Figure 31.6), which may indicate that there is a greater non-linearity in the catchment response (here related largely to the subsurface system) than can be obtained from the core-scale measurements. Application of TOPOG_SBM to the generation of infiltration-excess overland flow within a small catchment in the headwaters of the Amazon similarly demonstrated that the measured permeability profile could not reproduce the degree of non-linearity observed in the rainfall-runoff response (Vertessy and Elsenbeer, 1999). Similar problems in trying to parameterise TOPOG_SBM and TOPOG_DYNAMIC with field data were found by Schellekens (2000) studying a small catchment in Puerto Rico (see Bonell, this volume). New versions of TOPMODEL using linear and power declines of permeability with depth rather than an exponential decline, have been developed recently (Ambroise *et al.*, 1996; Duan and Miller, 1997; Iorgulescu and Musy, 1997). These models have been applied successfully to catchments that do not have simple exponential recession curves (e.g. Lancaster, 2000). Given that the average rate of loss of permeability with depth is orders of magnitude greater with a catchment-scale inversion than that observed with point-scale measurements, it could be argued, therefore, that the poor correspondence of the field and model values are not attributable simply to the type of permeability function.

Given both of the differences highlighted, perhaps there is value in approaches that yield accurate hillslope-scale estimates of a 'mean' permeability, as these data might be compared with: (i) point-scale measures, up-scaled to the hillslope unit, and (ii) catchment-average permeability, derived from model inversion.

(5) Hillslope-scale permeability estimation

Chappell *et al.* (1998) developed a simplified methodology for estimating whole-hillslope permeability for an Acrisol-Alisol soil.

It is important to remember that the permeability of a soil and/or weathered rock is the 'rate of water-flow through a unit area of saturated media under a unit hydraulic gradient'. Therefore, if the propagation rate of a water pulse through a hillslope is monitored and then corrected for the gradient (and other effects such as lateral dispersion of a local source), then a very approximate estimate of the lateral block permeability can be obtained. Three further key details are needed:

(a) the distance between the stream and half way to the local catchment divide is assumed to be the approximate average travel distance of rainwater migrating through the subsurface system (i.e. pathline* length), at least for shallow groundwater catchments,

(b) by applying a steady-state pulse of water to the slope at this mid-point and tracing the pulse to the stream (using tensiometers and/or boreholes), we can obtain a distribution of propagation velocities (V) for travel to any down-slope location, and

(c) if we assume that the hydraulic gradient between the mid-slope and monitoring location can be approximated by the sine of the slope angle (i.e. the difference in gravitational potential over the slope distance) then lateral block permeability estimates can be derived from the propagation velocities.

Given the approximations within the method and the need for parametric simplicity, Chappell *et al.* (1998) estimated only two propagation velocities to each sampling point. The first velocity, V_S (m s^{-1}), is the ratio of: (a) the time from water injection to a steady-state response at a sampling location, and (b) the length of the pathline.* The second velocity, V_C (m s^{-1}), is the ratio of: (a) the centroid time between injection and steady-state response and (b) the length of the pathline. These values, assumed to be pore-water velocities, were multiplied by the effective porosity (η_{eff}) to estimate block permeabilities (discharges per unit area). An estimate of the η_{eff} (i.e., 0.025 m^3 m^{-3}) was derived from the difference between the total porosity and the moisture content at −1.5 kPa capillary potential on moisture release curves determined by Sherlock (1997).

During the pulse-wave tests of Sherlock (1997), Chappell *et al.* (1998) and Lancaster (2000), water is normally applied over a 1 m width normal to the slope. The width of the resultant plume is observed to expand with distance from the injection point. The propagation of the water, therefore, needs to be corrected for this dispersion by calculation of a dispersion factor, τ, which is the across-slope width of plume at the monitoring point, normalised by the width at the injection site.

The complete calculation of the estimate of the lateral block permeability derived from, for example V_S, is therefore

$$K_{SVS} = \frac{V_S \eta_{eff}}{\tau \sin \beta} \qquad (31.11)$$

where K_{SVS} is the lateral block permeability between the mid-slope and down-slope monitoring point, as derived from V_s (m s^{-1}). Chappell *et al.* (1998) state that their aim was to present a tractable solution rather than one that includes all terms (after O'Loughlin, 1990).

Chappell *et al.* (1998) presented preliminary results of the application of this technique to the same tropical soil as that to which the catchment-scale model inversion was applied. The estimates of lateral block permeability derived from a single hillslope pulse-wave experiment (Reference name: TIKO) were K_{SVS} of 8.9 × 10^{-6} m s^{-1} and K_{SVS} 14.8 × 10^{-6} m s^{-1}. These results are not inconsistent with those derived from their model inversion (i.e. 0.527 × 10^{-6} and 13.7 × 10^{-6} m s^{-1}) (Figure 31.6). Further, the hillslope-derived estimates are clearly considerably larger than any of the up-scaled, core-based values (Figure 31.6). These results do not falsify the idea that the water plume applied to the experimental hillslope migrates under the influence of the natural soil pipes* and is less affected by the permeability distribution indicated by core-scale measurements. This same conclusion was reached by Lancaster (2000), who was able to undertake the same pulse-wave experiments, but replicated over several hillslopes within a 0.1 km^2 catchment in northwest England (UK). These preliminary results, therefore, support the idea that pulse-wave experiments may be a better method of deriving 'best estimates' of whole-hillslope permeabilities, and that they could be used to evaluate the estimates derived from catchment-scale modelling.

Clearly, derivation of a spatial distribution of whole-hillslope permeabilities do not give sufficient information on the form of the rainfall-runoff non-linearity to become the sole method of parameterising the soil-rock component of a catchment model. However, it is hoped that this may be a first step, with more conceptual work at the whole-hillslope scale being required. The central issue that arises from this case study is that core-scale measurements of key tropical soil parameters (notably soil-rock permeability), even after up-scaling, may not be meaningful for modelling catchment rainfall–runoff.

CONCLUSIONS

The objective of this chapter was to present a critical assessment of how soil-rock permeability is parameterised within dynamic rainfall-runoff models. As a result of the problem of 'parameter uncertainty'* arising from too complex a model structure, only a structurally-simple (i.e. parsimonious) model was thought capable of evaluation.

Four key conclusions arise that we believe are important to further developments in tropical catchment modelling:

(1) Further rainfall–runoff modelling in the tropics (and indeed elsewhere) that *challenges established model structures and data collection approaches* is required (cf. Refsgaard and Knudsen, 1996; Beven, 2001a,b; Young, 2001).
(2) More *tests of the internal-consistency of model predictions* (e.g. permeability distributions, water-table elevation, saturated area extent) would be helpful. Indeed, validation is greatly enhanced when several such data-series are available, as part of what has been described as multi-response validation (Mroczkowski *et al.*, 1997; Kokkonen and Jakeman, 2001).
(3) Hydrological processes and/or water *pathways that dominate at the hillslope scale* (so-called 'effective hillslope-scale processes') need to be identified more robustly (Bonell, 1998). Possible methodologies that may help are replicated, whole-hillslope tests (e.g. Lancaster, 2000) and models that explicitly address the source of rainfall–runoff non-linearity (Young, 2001).
(4) We need more assessments of whether complex physics-based approaches are better than *parsimonious, transfer-function approaches (e.g. DBM-model) based on 'good experience' (i.e. a range of case studies) at predicting the effects of land-use change* on rainfall–runoff processes.

Clearly, addressing the issue of how best to characterise those tropical soil parameters that regulate catchment rainfall–runoff behaviour is an important precursor to the reliable prediction of how land-use change might alter tropical soil parameters and thence streamflow generation and nutrient transport.

APPENDIX 31.1 GLOSSARY OF KEY MODELLING TERMS

Block permeability	An estimate of the soil or rock permeability typically derived by statistical manipulation of measured values to give estimates representative of a much larger scale (i.e. 'up-scaling'; see Wen and Gómez-Hernández, 1996).
Catchment parameters	Properties of a catchment that are largely unchanging with time (e.g. soil-rock permeability, porosity), but may be spatially variable.
DBM approach	An approach to modelling that incorporates transfer identification, with objective statistical evaluation and physical interpretation (see Young, 2001)
Effective parameters	Normally, values of parameters that give acceptable model simulations, and can be different to those from values measured in the field.
Grid-scale	In order to perform distributed (or semi-distributed) simulations with a physics-based model, the catchment is often divided into 100s or 1000s of elements in plan (Chappell and Ternan, 1992). The area of one of these elements is the grid-scale.
Impulse response function	The form of the output given a unit input (cf. Young, 1984).
Index of hydrological similarity	Catchment elements with the same index are assumed to have the same hydrological behaviour (cf. Beven, 2001a).
Model efficiency	How well the model output fits some statistical measure of the measured system output (i.e. the 'objective function'). Here the Nash and Sutcliffe (1970) efficiency measure is used (i.e. the one minus the ratio of the variance of the errors to the variance of the observed data, but then sometimes presented as a per cent).
Model parameterisation	The process by which parameters, such as soil parameters, are associated with each component of the model structure.
Model structure	The number of parameters in a model and their functional relationships with each other and the input-output variables.
Monte Carlo simulation	This is a process whereby 100s to 10 000s of simulations are undertaken with the same model structure, but with values of each parameter randomly selected from a normal or uniform distribution of values within an observed or 'realistic' range.
Natural soil pipes	Natural conduits or tunnels in the soil, formed by the action of water. Examples ranging in size from 0.05 m to 2 m diameter can be found within the humid tropics (Jones, 1990).
Parameter inversion	In the case of rainfall–runoff modelling, the identification of values of a particular parameter that is consistent with the model structure chosen and a good model efficiency.
Parameter uncertainty	Where values of a particular model parameter are derived by the 'parameter inversion' process, one set of parameters may give the same lumped output (e.g. riverflow) as a very different set, due to interaction between the model parameters. This problem and the subsequent uncertainty becomes worse as model complexity increases.
Pathline	The route taken by a fluid particle through the subsurface system.
Percolines	Zones of the soil, much broader than individual pores, natural soil pipes or fractures, where lateral water movement is much higher than that in the surrounding soil (after Bunting, 1961).
Permeability	The term is commonly used instead of the Coeficient of Permeability or more strictly the Saturated Hydraulic Conductivity. It is defined as the velocity of subsurface water through a unit area of saturated media under a unit hydraulic gradient (Darcy, 1856). It is distinct from (i) the term Unsaturated Permeability, which is the velocity of subsurface water through a unit area of unsaturated media under a unit hydraulic gradient, and (ii) the term Intrinsic Permeability, which is the intrinsic characteristic of the media, independent of the fluid (e.g., fresh water, salt water, oil) passing through it (Hubbert, 1940).
Preferential flow	The movement of one component of subsurface flow at a much greater velocity than the surrounding soil, and includes phenomena such as pipeflow, fissure flow and flows in percolines.

Sensitive model parameter	A model parameter, that if varied within the observed range of spatial variability would result in a large change in the model output (e.g., riverflow).
Transfer function	Basic z-domain representation of a linear digital filter between input(s) and output(s), expressing the filter as a ratio of two polynomials. See Middleton (2000) for an introduction to the topic of transfer function identification.
Up-scaling	Derivation of larger-scale estimates of a parameter (e.g., permeability) through rigorous statistical or numerical modelling of small-scale observations (see Wen and Gómez-Hernández, 1996).

References

Ambroise, B., Beven, K., and Freer, J. (1996). Toward a generalization of the TOPMODEL concepts – topographic indexes of hydrological similarity. *Water Resources Research*, 32, 2135–2145.

Asdak, C., Jarvis, P. G. and Gardingen, P. V. (1998). Modelling rainfall interception in unlogged and logged forest areas of Central Kalimantan, Indonesia. *Hydrology and Earth System Sciences*, 2: 211–220.

Barling, R. D., Moore, I. D. and Grayson, R. B. (1994). A quasi-dynamic wetness index for characterizing the spatial-distribution and soil-water content. *Water Resources Research*, 30, 1029–1044.

Barnes, C. J. and Bonell, M. (1996). Application of unit hydrograph techniques to solute transport in catchments. *Hydrological Processes*, 10: 793–802.

Beven, K. (1982). On subsurface stormflow – an analysis of response times. *Hydrological Sciences Journal*, 27, 505–521.

Beven, K. (1997). TOPMODEL: A critique. *Hydrological Processes*, 11, 1069–1085.

Beven, K. J. (2001a) *Rainfall-runoff modelling: The primer*. Chichester: Wiley.

Beven, K. J. (2001b) How far can we go in distributed hydrological modelling?. *Hydrology and Earth System Sciences*, 5: 1–12.

Beven, K. J. and Freer, J. (2001). A dynamic TOPMODEL. *Hydrological Processes*, 15: 1993–2011.

Beven, K. J., and Kirkby, M. J. (1979). A physically based variable contributing area model of basin hydrology. *Hydrol. Sci. Bull.*, 24, 43–69.

Bidin, K, 1995. *Subsurface flow controls of runoff in a Bornean natural rainforest*. Unpublished MSc. thesis, Manchester: University of Manchester.

Blain, C. A. and Milly, P. C. D. (1991). Development and application of a hillslope hydrologic model. *Advances in Water Resources*, 14: 168–174.

Bonell, M. (1998). Selected challenges in runoff generation research in forests from the hillslope to headwater drainage basin scale. *Journal of the American Water Resources Association*, 34: 765–785.

Bonell, M., Gilmour, D. A., and Cassells, D. S. (1983). A preliminary survey of the hydraulic properties of rainforest soils in tropical north-east Queensland and the implications for the runoff processes. In *Rainfall simulation, runoff, and soil erosion*, Catena Suppl. 4, ed. J. De Ploey. pp. 3–24. Cremlingen-Destedt: Catena Verlag.

Bonell, M with Balek, J. (1993). Recent scientific developments and research needs in hydrological processes of the humid tropics. In *Hydrology and Water Management in the Humid Tropics*, eds. M. Bonell, M. M. Hufschmidt, and J. S. Gladwell, pp. 167–260. Cambridge: Cambridge University Press.

Brassington, J. and Richards, K. (1998). Interactions between model predictions, parameters and DTM scales for TOPMODEL. *Computers and Geosciences*, 24, 299–314.

Bridges, E. M., Batjes, N. H. and Nachtergate, F. O. (1998) *World reference base for soil resources: Atlas*. Leuven: Acco.

Buckingham, E. (1907). *Studies on the movement of soil moisture*. Publication 38. USDA Bureu of Soils.

Bunting, B. T. (1961). The role of seepage moisture in soil formation, slope development, and stream initiation. *American Journal of Soil Science*, 259, 503–518.

Burt, T. P. and Butcher, D. P. (1985). Patterns of soil-moisture and throughflow in relation to defined tropographic indexes – some preliminary results. *Journal of the Geological Society of London*, 140: 322.

Cardwell, W. T. and Parsons, R. L. (1945) Averaging permeability of heterogeneous oil sands. *Transactions of the American Institute for Mining, Metallurgical and Petroleum Engineers*, 160: 34–42.

Chappell, N. A. and Binley, A. M. (1992). The impact of rainforest disturbance upon near-surface groundwater flow: modelling of hillslope flow experiments. *Annales Geophysicae*, 10 (II): c330.

Chappell, N. A. and Ternan, J. L. (1992). Flow-path dimensionality and hydrologic modelling, *Hydrological Processes*, 6: 327–345.

Chappell, N. A., and Franks, S. W. (1996). Property distributions and flow structure in the Slapton Wood Catchment. *Field Studies*, 8: 559–5758.

Chappell, N. and Ternan, L. (1997). Ring permeametry: design, operation and error analysis. *Earth Surface Processes and Landforms*, 22: 1197–1205.

Chappell, N. A., Franks, S. W. and Larenus, J. (1998). Multi-scale permeability estimation for a tropical catchment. *Hydrological Processes*, 12: 1507–1523.

Chappell, N. A., McKenna, P., Bidin, K., Douglas, I. and Walsh, R. P. D. (1999a). Parsimonious modelling of water and suspended-sediment flux from nested-catchments affected by selective tropical forestry. *Philosophical Transactions of the Royal Society of London Series B*, 354: 1831–1846.

Chappell, N. A., Ternan, J. L. and Bidin, K. (1999b). Correlation of physicochemical properties and sub-erosional landforms with aggregate stability variations in a tropical Ultisol disturbed by forestry operations. *Soil and Tillage Research*, 50: 55–71.

Chappell, N. A., Bidin, K. and Tych, W. (2001). Modelling rainfall and canopy controls on net-precipitation beneath selectively-logged tropical forest. *Plant Ecology*, 153: 215–229.

Chappell, N. A. Yusop, Z., Nik, A. R. H., Tych, W. and Kasran, B. (this volume). Spatially-significant effects of selective tropical forestry on water, nutrient and sediment flows: a modelling-supported review.

Chiew, F. H. S., Pitman, A. J. and McMahon, T. A. (1996). Conceptual catchment scale rainfall-runoff models and AGCM land-surface parameterisation schemes, *Journal of Hydrology*, 179: 137–157.

Connolly, R. D., Silburn, D. M. and Ciesiolka, C. A. A. (1997). Distributed parameter hydrology model (ANSWERS) applied to a range of catchment scales using rainfall simulator data. 3. Application to a spatially complex catchment. *Journal of Hydrology*, 193: 183–203.

Darcy, H. (1856). *Les fontaines publiques de la ville de Dijon*. Paris: Victor Dalmont.

Duan, J. F. and Miller, N. L. (1997). A generalised power function for the subsurface transmissivity profile in TOPMODEL. *Water Resources Research*, 33: 2559–2562.

Elsenbeer, H., and Cassel, D. K. (1993). Surficial processes in the rain-forest of western Amazonia. In *Research needs and applications to reduce erosion and sedimentation in tropical steeplands, IAHS Publication 192*, eds. R. R. Ziemer, C. L. O'Loughlin, and L. S. Hamilton, pp 289–297. Paris: IAHS.

Elsenbeer, H., Lorieri, D. and Bonell, M. (1995). Mixing model approaches to estimate storm flow sources in an overland flow-dominated tropical rain-forest catchment, *Water Resources Research*, 31: 2267–2278.

Fitzpatrick, E. A. (1971). *Pedology: A Systematic Approach to Soil Science*. Edinburgh: Oliver and Boyd.

Franchini, M. Wendling, J., Obled, C. and Todini, E. (1996). Physical interpretation and sensitivity analysis of the TOPMODEL. *Journal of Hydrology*, 175: 293–338

Franchini, M. and Pacciani, M. (1991). Comparative analysis of several conceptual rainfall-runoff models. *Journal of Hydrology*, 122: 161–219.

Franks, S. W., Beven, K. J., Chappell, N. A. and Gineste, P. (1997). The utility of multi-objective conditioning of a distributed hydrological model using uncertain estimates of saturated areas. In *Proceedings of the international congress on modelling and simulation: MODSIM '97: volume 1*, eds. A. D. McDonald and M. McAleer, pp 335–340. Canberra: Modelling and Simulation Society of Australia.

Freedman, V. L., Lopes, V. L., and Hernandez, M. (1998). Parameter identifiability for catchment-scale erosion modelling: a comparison of optimization algorithms. *Journal of Hydrology*, 207: 83–97.

Freer, J., McDonnell, J., Beven, K. J., Brammer, D., Burns, D., Hooper, R. P. and Kendal, C. (1997). Topographic controls on subsurface storm flow at the hillslope scale for two hydrologically distinct small catchments. *Hydrological Processes*, 11: 1347–1352.

Freeze, R. A. (1980). A stochastic-conceptual analysis of rainfall-runoff processes on a hillslope. *Water Resources Research*, 16: 391–408.

Gash, J. H. C., Kabat, P., Monteny, B. A., Amadou, M., Bessemoulin, P., Billing, H., Blyth, E. M., deBruin, H. A. R., Elbers, J. A., Friborg, T., Harrison, G., Holwill, C. J., Lloyd, C. R., Lhomme, J. P., Moncrieff, J. B., Puech, D., Soegaard, H., Taupin, J. D., Tuzet, A. and Verhoef, A. (1997). The variability of evaporation during the HAPEX-Sahel intensive observation period. *Journal of Hydrology*, 189: 385–399.

Hubbert, M. K. (1940). The theory of groundwater flow. *Journal of Geology*, 48: 785–944.

Hughs, D. A. and Metzler, W. (1998). Assessment of three monthly rainfall-runoff models for estimating the water resource yield of semi-arid catchments in Namibia. *Hydrological Sciences Journal*, 43: 283–297.

Iorgulescu, I. and Musy, A. (1997). Generalisation of TOPMODEL for a power law transmissivity profile. *Hydrological Processes*, 11: 1353–1355.

Jakeman, A. J. and Hornberger, G. M. (1993). How much complexity is warranted in a rainfall-runoff model? *Water Resources Research*, 29: 2637–2649.

Jetten, V. G. (1996). Interception of tropical rainforest: Performance of a canopy water balance model. *Hydrological Processes*, 10: 671–685.

Jones, J. A. A. (1990). Piping effects in humid lands. In *Groundwater Geomorphology*, Geological Society of America Special Paper 252, eds. Higgins, C. G., and Coates, D. R., pp. 111–138, Boulder: Geological Society of America.

Jordan, J.-P. (1994). Spatial and temporal variability of stormflow generation processes on a Swiss catchment. *Journal of Hydrology*, 153: 357–382.

Kirkby, M. J. (1975). Hydrograph modelling strategies. In *Processes in Physical and Human Geography*. eds. Peel, R. Chisholm, R. and Haggett, P., pp. 69–90, Oxford: Heinemann.

Kokkonen, T. S. and Jakeman, A. J. (2001). A comparison of metric and conceptual approaches in rainfall-runoff modeling and its implications. *Water Resources Research*, 37: 2345–2352.

Krishnaswamy, J., Halpin, P. N. and Richter, D. D. (2001). Dynamics of sediment discharge in relation to land-use and hydro-climatology in a humid tropical watershed in Costa Rica. *Journal of Hydrology*, 253: 91–109.

Lamb, R. and Beven, K J. (1997). Using interactive recession curve analysis to specify a general catchment storage model. *Hydrology and Earth System Sciences*, 1: 101–113.

Lamb, R., Beven, K. J. and MyrabØ, S. (1997) A generalised topographic-soils hydrological index. In *Landform Monitoring, Modelling and Analysis*, eds. Lane, S. N., Richards, K. S. and Chandler, J. H. Chichester: Wiley.

Lancaster, J. (2000). *Multi-scale estimation of effective permeability within the Greenholes Beck catchment.* Unpublished PhD thesis. Lancaster: Lancaster University.

Liden, R. and Harlin, J. (2000). Analysis of conceptual rainfall-runoff modelling performance in different climates. *Journal of Hydrology*, 238: 231–247.

Loague, K. M. and Freeze, R. A. (1985). Comparison of rainfall-runoff modelling techniques on small upland catchments. *Water Resources Research*, 21, 229–248.

Malmer, A. (1993). *Dynamics of hydrology and nutrient losses as response to establishment of forest plantation. A case study of a rainforest in Sabah, Malaysia.* Unpublished PhD thesis. Upsala: Swedish University of Agricultural Sciences.

Marin, C. T., Bouten, W. and Sevink, J. (2000). Gross rainfall and its partitioning into throughfall, stemflow and evaporation of intercepted water in four forest ecosystems in western Amazonia. *Journal of Hydrology*, 237: 40–57.

Michaud, M. J. and Sorooshian, S. (1994). Comparison of simple versus complex distributed runoff models on a midsized semiarid watershed. *Water Resources Research*, 30: 593–605.

Moore, R. D. and Thompson, J. C. (1996) Are water table variations in a shallow forest soil consistent with the TOPMODEL concept? *Water Resources Research*, 32: 663–669.

Mroczkowski, M., Raper, G. P. and Kuczera, G. (1997). The quest for more powerful validation of conceptual catchment models. *Water Resources Research*, 33: 2325–2335.

Molicova H, Grimaldi M, Bonell M, and Hubert P. (1997). Using TOPMODEL towards identifying and modelling the hydrological patterns within a headwater, humid, tropical catchment. *Hydrological Processes*, 11: 1169–1196.

Nash, J. E. and Sutcliffe, J. V. (1970) River flow forecasting through conceptual models, 1: A discussion of principles. *Journal of Hydrology*, 10: 282–290.

Northcliff, S. and Thornes, J. (1981). Seasonal variations in the hydrology of a small forested catchment near Manaus, Amazonia, and the implications for its management. In *Tropical Agricultural Hydrology*, eds, R. Lal and E. W. Russell, Chichester: Wiley.

O'Loughlin, E. M. (1981). Saturation regions in catchments and their relations to soil and topographic properties. *Journal of Hydrology*, 53: 229–246.

O'Loughlin, E. M. (1986). Prediction of surface saturation zones in natural catchments by topographic analysis. *Water Resources Reearch,*, 22: 794–804.

O'Loughlin, E. M. (1990). Perspectives on hillslope research. In *Process Studies in Hillslope Hydrology*, eds. Anderson, M. G. and Burt, T. P., pp 501–516. Chichester: Wiley.

Onyando, J. O. and Sharma, T. C. (1995). Simulation of direct runoff volumes and peak rates for rural catchments in Kenya, East-Africa. *Hydrological Sciences Journal*, 40: 367–380.

Polcher, J. and Laval, K. (1994). The impact of African and Amazonian deforestation on tropical climate. *Journal of Hydrology*, 155: 389–405.

Quinn, P., Beven, K., Chevallier, P. and Planchono, O. (1991). The prediction of hillslope flow paths for distributed hydrological modeling using digital terrain models. *Hydrological Processes*, 5: 59–79.

Refsgaard, C., Seth, S. M., Bathurst, J. C., Erlich, M., Storm, B., Jorgensen, G. H. and Chandra, S. (1992). Application of the SHE to catchments in India. 1. General results. *Journal of Hydrology*, 140: 1–23.

Refsgaard, C. and Knudsen, J. (1996). Operational validation and intercomparison of different types of hydrological models. *Water Resources Research*, 32: 2189–2202.

Richards, L. A. (1931). Capillary conduction of liquids through porous media. *Physics*, 1: 318–333.

Rogers, C. C. M., Beven, K., Morris, E. M. and Anderson, M. G. (1985). Sensitivity analysis, calibration and predictive uncertainty of the Institute of Hydrology Distributed Model. *Journal of Hydrology*, 81: 179–191.

Sandstrom, K. (1996). Hydrochemical deciphering of streamflow generation in semi-arid East Africa, *Hydrological Processes*, 10: 703–720.

Schellekens, J. (2000). *Hydrological processes in a humid tropical rainforest: A combined experimental and modelling approach*, Unpublished PhD thesis. Amsterdam: Free University.

Schellekens, J. Scatena, F. N., Bruijnzeel, L. A. and Wickel, A. J. (1999). Modelling rainfall interception by a lowland tropical rainforest in northeastern Puerto Rico. *Journal of Hydrology*, 225: 168–184.

Scoccimarro, M., Walker, A., Dietrich, C., Schreider, S., Jakeman, T. and Ross H. (1999). A framework for integrated catchment assessment in northern Thailand. *Environmental Modelling and Software*, 14: 567–577.

Saulnier, G-M., Obled, C., and Beven, K. (1997). Analytical compensation between DTM grid resolution and effective values of saturated hydraulic conductivity within a TOPMODEL framework, *Hydrological Processes*, 11: 1331–1346.

Schultz, G. A. (1994). Mesoscale modelling of runoff and water balances using remote-sensing and other GIS data. *Hydrological Sciences Journal*, 39: 121–142.

Sherlock, M. D. (1997). *Plot-scale hydrometric and tracer characterisation of soil water flow in two tropical rainforest catchments in South East Asia.* Unpublished PhD thesis, Lancaster: Lancaster University.

Sherlock, M. D., Chappell, N. A., and Greer, T. (1995). Tracer and Darcy-based identification of subsurface flow, Bukit Timah forest, Singapore. *Singapore Journal of Tropical Geography*, 16: 197–215.

Sherlock, M. D. Chappell, N. A. and McDonald, J. J. (2000). Effects of experimental uncertainty on the calculation of hillslope flow paths. *Hydrological Processes*, 14: 2457–2471.

Singh, R., Subramanian, K. and Refsgaard, J. C. (1999). Hydrological modelling of a small watershed using MIKE SHE for irrigation planning. *Agricultural Water Management*, 41: 149–166.

Sinun, W. (1991). *Hillslope hydrology, hydrogeomorphology and hydrochemistry of an equatorial lowland rainforest, Danum Valley, Sabah, Malaysia*. Unpublished M.Sc. thesis. Manchester: University of Manchester.

Sloan, P. G., Moore, I. D., Coltharp, G. B. and Eigel, J. D. (1983). *Modelling surface and subsurface stormflow on steeply-sloping forest watersheds*. Research Paper 142. Lexington: Water Resources Institute.

Sulebak, J. R., Tallaksen, L. M. and Erichsen, B. (2000). Estimation of aeral soil moisture by use of terrain data. *Geografiska Annaler Series A – Physical Geography*, 82A: 89–105.

Thiemann, M., Trosset, M., Gupta, H., and Sorooshian, S. (2001). Bayesian recursive parameter estimation for hydrologic models. *Water Resources Research*, 37: 2521–2535.

Troch, P. A., Mancini, M., Paniconi, C. and Wood, E. F. (1993). Evaluation of a distributed catchment scale water-balance model. *Water Resources Research*, 29: 1805–1817.

van Dijk, A. I. J. M. and Bruijnzeel, L. A. (2001). Modelling rainfall interception by vegetation of variable density using an adapted analytical model. Part 2. Model validation for a tropical upland mixed cropping system. *Journal of Hydrology*, 247: 239–262.

Vertessy, R., O'Loughlin, E., Beverly, E. and Butt, T. (1994). Australian experiences with the CSIRO Topog model in land and water resources management. In *Proceedings of UNESCO International Symposium on Water Resources Planning in a Changing World*, pp. III-135–144. Paris: UNESCO.

Vertessy, R. A. and Elsenbeer, H. (1999). Distributed modeling of stormflow generation in an Amazonian rainforest catchment: Effects of model parameterization. *Water Resources Research*, 35: 2173–2187

Wen, X. and Gómez-Hernández, J. J. (1996). Up-scaling hydraulic conductivities in heterogeneous media: a review. *Journal of Hydrology*, 183: 9–32.

Whitaker, D. W., Wasimi, S. A. and Islam, S. (2001). The El Niño-Southern Oscillation and long-range forecasting of flows in the Ganges. *International Journal of Climatology*, 21: 77–87.

Whitehead, P. G., Young, P. C. and Hornberger, G. M. (1979). A systems model of stream flow and water quality in the Bedford-Ouse River: I stream flow modelling. *Water Research*, 13: 1155–1169.

Wigmosta, M. S. and Lettenmaier, D. P. (1999). A comparison of simplifed methods of routing topographically driven subsurface flow. *Water Resources Research*, 35: 255–264.

Wooldridge, S., Kalma, J., and Kuczera, G. (2001). Parameterisation of a simple semi-distributed model for assessing the impact of land-use on hydrologic response. *Journal of Hydrology*, 254: 16–32.

Young, P. C. (1984). *Recursive estimation and time series analysis*. Berlin: Springer.

Young, P. C. (2001). Data-based mechanistic modelling and validation of rainfall-flow processes. In *Model Validation in Hydrological Science*. ed. Anderson, M. G. pp 117–161. Chichester: Wiley

Zeng, N. (1998). Understanding climate sensitivity to tropical deforestation in a mechanistic model. *Journal of Climate*, 11: 1969–1975.

Zeng, N. (1999). Seasonal cycle and interannual variability in the Amazon hydrologic cycle. *Journal of Geophysical Research-Atmospheres*, 104(D8): 9097–9106.

Zeng, N. and Neelin, J. D. (1999). A land-atmosphere interaction theory for the tropical deforestation problem. *Journal of Climate*, 12: 857–872.

Zheng, H., Henderson-Sellers, A. and McGuffie, K. (2001). The compounding effects of tropical deforestation and greenhouse warming on climate. *Climate Change*, 49: 309–338.

32 Isotope tracers in catchment hydrology in the humid tropics

J. M. Buttle
Trent University, Peterborough, Canada
J. J. McDonnell
Oregon State University, Corvallis, USA

INTRODUCTION

Isotope tracers are an important tool for quantifying the age, origin and pathway of water to streams in headwater catchments. While used regularly in temperate and high latitude areas, applications of isotope tracer techniques in the humid tropics have been minimal to date. In the developing world, finances and logistics often preclude the use of traditional hydrometric measures of streamflow, groundwater dynamics and soil water recharge. Thus, isotopic tracers and isotopic hydrograph separations (IHSs) can serve as a valuable tool for extending our understanding of streamflow generation in poorly gauged areas (Shuttleworth, 2002), particularly in humid low latitude regions. They can also provide complementary information on water sources and pathways that are often required in order to draw conclusions about streamflow generation (reviewed by Bonell, this volume), effects of disturbance on tropical forest ecosystems (Bruijnzeel, 1990), or the closure of nutrient cycles in tropical forest ecosystems (Elsenbeer *et al.*, 1995).

The aim of this chapter is to provide an overview of what has been accomplished in isotope tracing studies (mainly in mid-latitude environments to date) as an impetus for those working in the humid tropics to consider the merits of this approach. While we do not advocate use of this technique exclusively in research settings, we argue that it can be a valuable approach in the 'toolkit' of land managers and catchment scientists working on hydrological processes related to land use change in humid tropical areas. This chapter presents a number of specific uses of isotope tracers in disturbed tropical systems, including general land use change detections, quantification of logging road runoff contributions to streamflow, as a tool in the design of model structures and model calibration for these systems, and as a means of quantifying a system's memory of disturbance through use of water age dating techniques. Although this chapter will focus on isotopic tracers, we will highlight the use of geochemical tracers (both conservative and non-conservative) and geochemical hydrograph separations (GHSs) as a means of complementing the insights into hydrological processes that can be obtained from isotopic tracers.

ISOTOPE HYDROGRAPH SEPARATION BASICS

The use of isotope tracers in hydrological research problems has grown rapidly. Aggarwal (2002) reports that a Georef search for the period 1965–1970 shows 650 papers using isotope tracers, while a search for the period 1995–2000 shows over 6500 papers using isotope tracers in groundwater studies alone. The main use of environmental isotopes in catchment hydrology to date has been in hydrograph separation (Burns, 2002). The use of these isotopes in hydrograph separation has been reviewed by Genereux and Hooper (1998) and Rodhe (1998). The reader is referred to Kendall and Caldwell (1998) for an excellent discussion of the geochemistry of environmental isotopes. Hence only the main issues pertaining to use of this tool in catchment hydrology will be covered here.

Isotopic hydrograph separations have generally been conducted using tritium (^{3}H or T), oxygen-18 (^{18}O) and deuterium (^{2}H or D) (Buttle, 1994). Oxygen-18 and D have been used in the majority of IHSs (Sklash, 1990) and, unlike T, they are stable and do not undergo radioactive decay. The ratios of ^{18}O and D to their more common counterparts in the hydrosphere (^{16}O and H) are 1:500 and 1:6700 (Drever, 1988). Determination of the abundance of stable isotopes in a water sample is based on isotopic ratios (e.g. $^{18}O/^{16}O$ and D/H). Abundance is reported as δ values in parts per

Forests, Water and People in the Humid Tropics, ed. M. Bonell and L. A. Bruijnzeel. Published by Cambridge University Press.

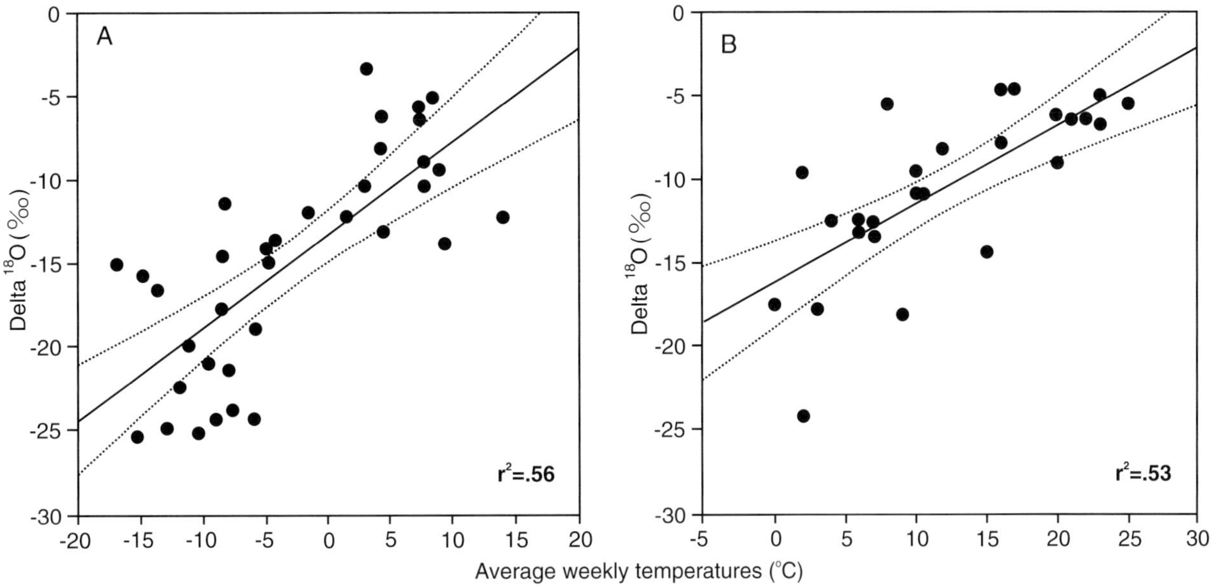

Figure 32.1 $\delta^{18}O$ vs. average weekly temperature (T_a) for precipitation collected at the Niwot Ridge, Colorado site (A) and at the North Platte, Nebraska site (B). The regression equation for the Niwot Ridge site was $\delta^{18}O = 0.55\ T_a - 13.3$ and for the North Platte site was $\delta^{18}O = 0.46\ T_a - 16.1$. (From Welker, 2000.)

thousand (‰ or per mil) and δ values are calculated by:

$$\delta^{18}O or \delta D = (R_{\text{sample}}/R_{\text{VSMOW}} - 1) \bullet 1000 \qquad (32.1)$$

where R_{sample} is the ratio of the heavy to light isotope in the sample and R_{VSMOW} is the reference standard, which is Vienna Standard Mean Ocean Water (VSMOW) for ^{18}O and D (Kendall and Caldwell, 1998).

Isotope tracers have a number of unique virtues as water tracers in catchment studies:

1. They are applied naturally over entire catchments, thus avoiding problems of realistic application rates and extent of application associated with artificial tracers (Sklash, 1990).
2. They do not undergo chemical reactions during contact with soil/regolith at temperatures encountered at or near the Earth's surface (Drever, 1988).
3. They undergo fractionation during evaporation and condensation. During evaporation, water vapour is relatively depleted in the heavy isotopes while the remaining liquid water becomes progressively enriched in D and ^{18}O. Conversely, there is preferential movement of molecules containing the heavy isotopes to the liquid phase during condensation, leaving the vapour relatively depleted. Thus, meteoric water has negative δ values which have been found to decrease with surface air temperature (Figure 32.1), increasing latitude (Figure 32.2), increasing altitude, increasing distance of vapour transport (Figure 32.3), and increasing amounts of precipitation (Dansgaard, 1964; Ingraham, 1998).
4. Variations in the isotopic signature of precipitation are often dampened as water transits the unsaturated zone to the water table (Ingraham and Taylor, 1991), such that groundwater δ values may approach uniformity in time and space, and are changed only by mixing with waters of different isotopic contents (Sklash, 1990). This means that there is frequently a difference between the δ of water input to the catchment's surface and water stored in the catchment before the event.

This difference between the isotopic signature of incoming water (event or 'new' water) and water stored in the catchment before the event (pre-event or 'old' water) often permits the separation of a stormflow hydrograph into its event (new) and pre-event (old) components:

$$Q_t = Q_p + Q_e \qquad (32.2)$$

$$C_t Q_t = C_p Q_p + C_e Q_e \qquad (32.3)$$

$$X = (C_t - C_p)/(C_p - C_e) \qquad (32.4)$$

where Q_t is streamflow; Q_p and Q_e are contributions from pre-event and event water; C_t, C_p and C_e are δ values in streamflow, pre-event and event waters, respectively; and X is the pre-event fraction of streamflow (Figure 32.4). The IHS in Figure 32.4 illustrates the partitioning of event and pre-event water in streamflow; it also demonstrates the consequences of δ in streamflow falling outside of the bounds set by C_p and C_e – namely, physically-unrealistic estimates of X that exceed 1 or are less than 0. In addition to the constraint that C_t values must fall between C_p and

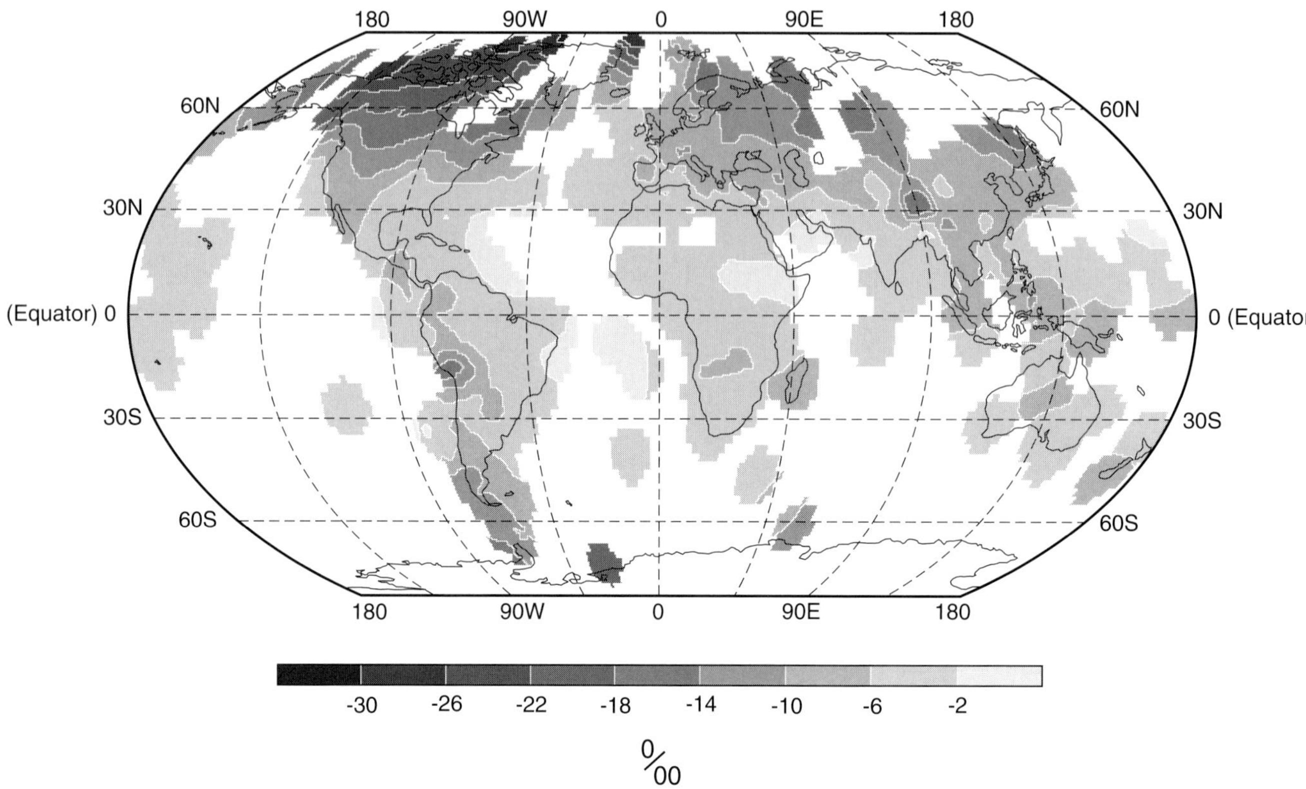

Figure 32.2 Weighted annual $\delta^{18}O$ in global precipitation. The latitudinal effect of decreasing $\delta^{18}O$ values with increasing latitude can be seen particularly clearly over North America. (From International Atomic Energy Agency, 2001.)

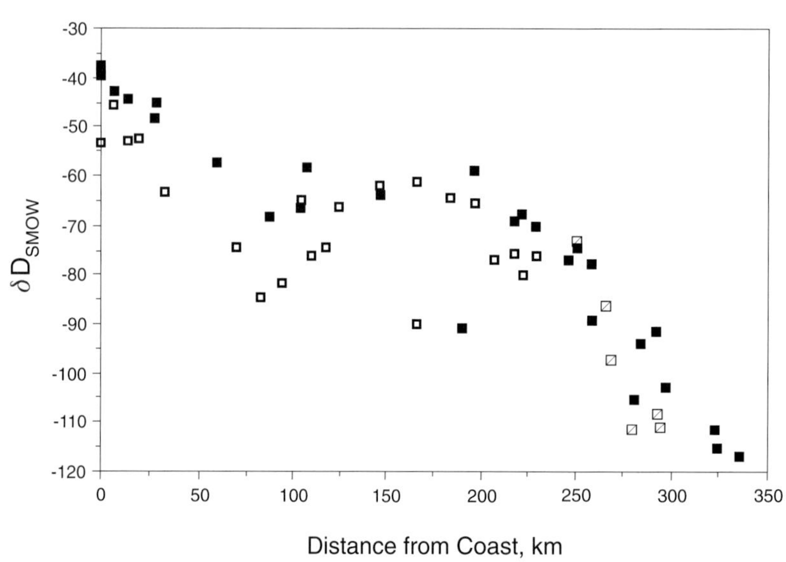

Figure 32.3 Weighted average δD of rain (open squares), snow (divided squares), and surface water and shallow groundwater (solid squares) vs. distance along a transect extending from the Pacific Ocean through central California and Nevada, USA. (From Ingraham and Taylor, 1991.)

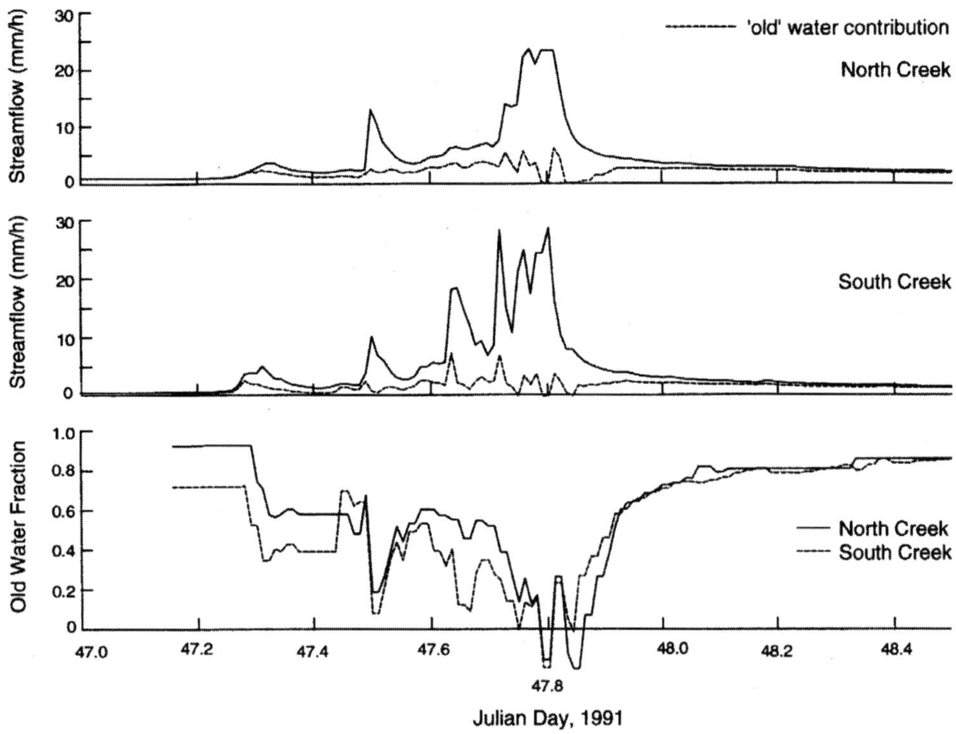

Figure 32.4 Isotopic hydrograph separation results from the Babinda catchments, Queensland, Australia. (From Bonell *et al.*, 1998.)

C_e to separate the streamflow hydrograph into event and pre-event components, the use of IHS is based on several key assumptions.

ASSUMPTIONS IMPLICIT IN THE TECHNIQUE

The assumptions that underlie the use of Eqns 32.2–32.4 to solve for the event and pre-event fractions of streamflow include:

1. There is a significant difference between the isotopic content of the event and pre-event components.
2. The isotopic signature of event water is constant in space and time, or any variations can be accounted for.
3. The isotopic signature of pre-event water is constant in space and time, or any variations can be accounted for.
4. Contributions of water from the vadose zone must be negligible, or the isotopic content of soil water must be similar to that of groundwater.
5. Contributions to streamflow from surface storage are negligible.

Rodhe (1987), Sklash (1990) and Buttle (1994), among others, have reviewed these assumptions and their implications for the reliability of IHS. The one assumption that appears to have been met in most if not all IHS studies is assumption (1). Nevertheless, there are aspects of the annual cycle of δ in precipitation in lower latitudes (examined below) that mean that this assumption may become particularly important in IHS studies in the humid tropics. Buttle (1994) has shown that most studies have been conducted in mid-to-high latitude environments. These often experience pronounced annual oscillations in the isotopic signature of precipitation, which increases the possibility that for any given event there will be a significant difference between δ for the precipitation event and that in pre-event water, the mean δ of which approximates the mean δ of annual precipitation (Clark and Fritz, 1997; Gremillion and Wanielista, 2000).

Several early studies that employed IHS did not pay excessive attention to whether the remaining assumptions had been violated. The isotopic signature of event water was often represented by samples from a single rain gauge, snowcore or snowmelt sample that may not have been obtained from within the catchment (Unnikrishna *et al.*, 2002). Bulk samples were employed in many cases, necessitating the frequently-untested assumption that temporal variations in event water δ were insignificant. Precipitation in many IHS studies is measured and sampled at open sites. This presents a problem when using IHS in forested catchments, where interception has been shown to result in isotopic enrichment of precipitation (Saxena, 1986; DeWalle and Swistock, 1994; Brodersen *et al.*, 2000) which may reduce the difference between the isotopic content of the event and pre-event components to such a degree as

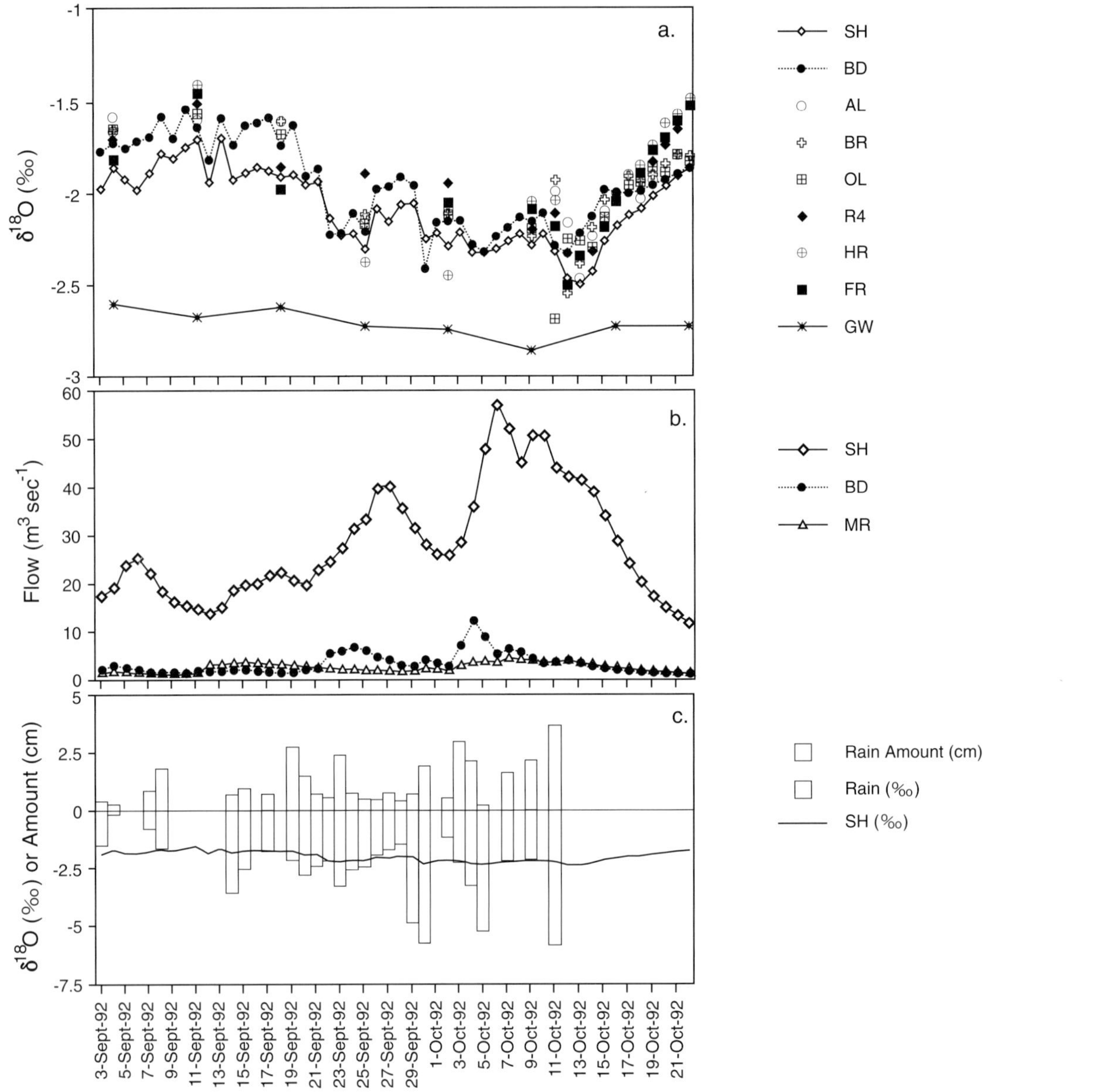

Figure 32.5 Weekly mean groundwater (GW) $\delta^{18}O$ and the $\delta^{18}O$ in river water sampled at various stations in the Econolockhatchee River catchment, central Florida, USA. (From Gremillion and Wanielista, 2000.)

to preclude hydrograph separation. There has been considerable work documenting spatial and temporal variations in event water δ at the scale of small catchments (e.g. McDonnell *et al.*, 1990; DeWalle and Swistock, 1994; Bariac *et al.*, 1995). Many initial IHS studies tested the hypothesis that pre-event water δ was constant in time and space. Sklash (1990) argued that baseflow δ is the best index of C_p on the grounds that baseflow integrates the δ of near-stream groundwater that is likely to reach the stream during an event. This assumption has been supported by the close correspondence between groundwater and baseflow δ observed in some studies (e.g. Hooper and Shoemaker, 1986; Hill and Waddington, 1993). Conversely, this argument has been called into question by documentation of substantial variability in baseflow δ along the stream channel (Bishop, 1991; Unnikrishna *et al.*, in press) and by significant differences between δ in near-stream groundwater and in the baseflow that presumably integrates these groundwater δ values (Buttle *et al.*, 1995; Bonell *et al.*, 1998; Burns and McDonnell, 1998; Gremillion and Wanielista, 2000); Figure 32.5). Bonell *et al.* (1998) attributed these differences to variations in groundwater residence times in different parts of the catchment arising from geological complexity. Gremillion and Wanielista (2000) noted the importance of evaporative enrichment of river water in central Florida in producing differences between the isotopic signature of baseflow and groundwater. Such a process might be expected to be of greater significance to IHS studies in the tropics relative to humid mid-latitude environments.

Invocation of assumption (4) allowed some early workers (e.g. Sklash and Farvolden, 1979) to assume that X represented both the pre-event and groundwater fraction of total stormflow. However, this also necessitated the assumption that C_p has a constant value, such that streamflow δ returns to its pre-storm value once discharge declines to baseflow values (e.g. Bonell *et al.*, 1990). Several studies have noted a shift in the δ of baseflow or groundwater in response to inputs during snowmelt or rainfall (e.g. Hooper and Shoemaker, 1986; Buttle *et al.*, 1995; McDonnell *et al.*, 1991a). Under these circumstances, C_p must reflect these temporal variations in baseflow or groundwater δ if X is to be interpreted as the groundwater fraction of discharge at the time of streamflow sampling.

A number of studies have documented substantial differences between δ in soil water and in groundwater or baseflow (e.g. Kennedy *et al.*, 1986; DeWalle *et al.*, 1988; Peters *et al.*, 1995). The critical question is whether such water contributes to stormflow in significant quantities. There have been two basic approaches to addressing this question. In the first case, a soil water component of stormflow is inferred based on inadequate explanation of runoff sources using the standard two-component model (e.g. DeWalle *et al.*, 1988; Ogunkoya and Jenkins, 1993; Hinton *et al.*, 1994). A second approach uses hydrometric measurements to quantify soil water contributions to catchment streamflow (e.g. Buttle and Peters, 1997). In situations where significant contributions to stormflow from one or more additional flow components (e.g. soil water) have been identified, the standard mixing equations have been modified:

$$Q_t = Q_1 + Q_2 + Q_3 + \ldots + Q_n \qquad (32.5)$$

$$C_tQ_t = C_1Q_1 + C_2Q_2 + C_3Q_3 + \ldots + C_nQ_n \qquad (32.6)$$

where Q_t = streamflow

Q_n = discharge of a particular runoff component
C_t = tracer concentrations in streamflow
C_n = tracer concentration of a particular runoff component

These equations can be solved using matrix algebra. The important point to note is that solution of the expanded mixing equations requires additional constraints and increases output uncertainty. In the case of solution for three flow components (e.g. event water, soil water and groundwater), either a second tracer or a physical measurement of flow from one component is required (Genereux and Hooper, 1998). A number of studies used a stable isotope tracer in conjunction with a geochemical hydrograph separation (GHS) to identify contributions from three flow components (e.g. Wels *et al.*, 1991; McDonnell *et al.*, 1991b; DeWalle and Pionke, 1994; Hinton *et al.*, 1994). Unlike isotopic tracers, geochemical tracers provide information about hydrological flowpaths, provided the kinetics of tracer reactivity in the subsurface are known (Burns and McDonnell, 1998; Burns *et al.*, 2001). A few studies have tried to quantify flow for a particular component – the best example being that of DeWalle *et al.* (1988), who estimated the rate of channel precipitation (direct supply of event water to streamflow) as the product of throughfall and stream surface area. The latter was estimated using a regression relationship derived from measured stream surface areas and corresponding streamflow rate. In the event that such hydrometric measurements are not available, n tracers are required to separate stormflow into $n + 1$ components.

Many initial IHS studies explicitly avoided catchments that possessed appreciable amounts of surface storage (e.g. lakes, ponds, wetlands). This appears to have been well advised, since several studies (e.g. Buttle and Sami, 1992; Hill and Waddington, 1993; Metcalfe and Buttle, 2001) have shown that mixing within wetlands can complicate interpretation of results from traditional two-component IHS. Although wetlands sometimes serve as a complicating factor in IHS, the enrichment of wetland water resulting from surface water evaporation may serve as an additional tracer through a hydrological system. Burns and McDonnell (1998) used departure from a local meteoric water line (i.e. the linear relation between ^{18}O and D) as a way to quantify wetland influences on water at the catchment outlet (Figure 32.6). They found that evaporation from a beaver pond caused a seasonal decrease in the slope of the meteoric water line for streamflow that was absent in a nearby catchment that did not contain any wetlands. This evaporative enrichment would be accompanied by an increase in solute concentrations in water held in the beaver pond, and such an evolution in isotopic and geochemical signals has implications for both IHS and GHS, especially in terms of assumptions (2) and (3) noted earlier.

The increased awareness among practitioners of IHS of the need for clear evaluation of the degree to which the underlying assumptions have been satisfied has been accompanied by explicit error analysis of IHS results (beginning with Rodhe (1987)). Genereux (1998) provided a first order uncertainty propagation approach to such an analysis, and showed that decreasing difference between C_e and C_p results in a marked increase in the uncertainty associated with resolving the contributions of these components to C_t. Both this approach and the Monte Carlo method of Bazemore *et al.* (1994) assume that the variability in the flow components signatures can be represented as normal distributions. Joerin *et al.* (2002) avoided this restriction in their analysis of statistical uncertainty in hydrograph separations (due to isotopic

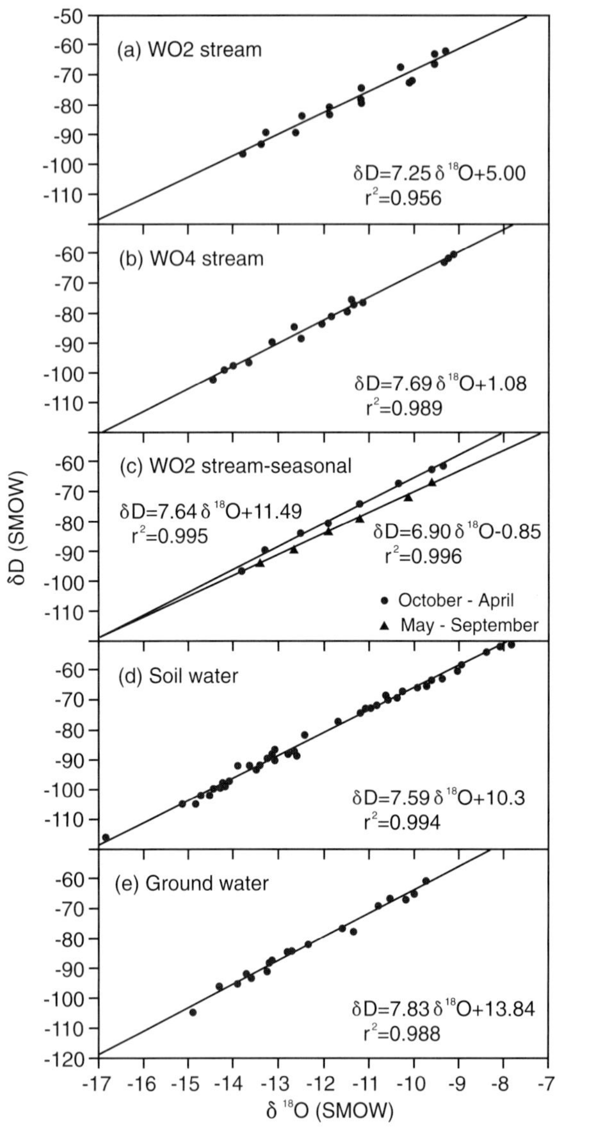

Figure 32.6 Meteoric water lines of δD vs. δ^{18}O in study catchments in northern New York, USA: (a) aggregate data for the catchment containing a beaver pond (WO2); (b) aggregate data for the catchment without a beaver pond (WO4); (c) seasonal data for the catchment containing a beaver pond (WO2); (d) soil water; (e) groundwater. (From Burns and McDonnell, 1998.)

and chemical variability of flow components) by combining a Monte Carlo approach with component frequency distributions determined directly from field samples. They also distinguished between statistical uncertainty and 'model uncertainty', which is affected by model assumptions such as temporal uniformity of flow component signatures.

All of these studies showed that spatial and temporal variability in isotopic end-member signatures can introduce substantial uncertainty in the estimated flow components. It is critical to take such uncertainty into account when interpreting hydrograph separation results in terms of physical processes operating in the catchment. An example would be the attempts to estimate event water fluxes via subsurface flowpaths through the simultaneous use of isotopic and geochemical (e.g. silica) tracers (e.g. Wels *et al.*, 1991). Here, D was used in an IHS to separate event and pre-event water in a mid-latitude forested catchment during snowmelt. A concurrent GHS using silica as a tracer was used to separate the hydrograph into surface and subsurface flow components. The subsurface flow fraction was found to exceed the pre-event water component, and was interpreted as evidence for the movement of event water via subsurface pathways. However, a complete error analysis of IHS and GHS results in the same catchment during a spring rainstorm indicated substantial overlap between the estimated pre-event and subsurface flow components, suggesting that there were no grounds for inferring such event water fluxes *solely on the basis of the tracer results* (Buttle and Peters, 1997). This illustrates the need to constrain our inferences of processes using hydrometric or other data.

Although we take a critical view of the hydrograph separation assumptions and review some situations in which they have been shown to be unsupported, we do not seek to dissuade scientists from the use of IHS. Quite the opposite – IHS provides valuable insights into hydrological processes in catchments even in situations where one or more of the key assumptions has been shown to be violated. The true usefulness of IHS is in using hydrograph separations to the level of accuracy warranted by the approach, and not to read more precision into the results of IHS than is realistic. Current as well as potential users of the IHS approach would be well-advised to bear Fretwell's Law in mind:

> 'Warning! Stable isotope data may cause severe and contagious stomach upset if taken alone. To prevent upsetting reviewers' stomachs and your own, take stable isotope data with a healthy dose of other hydrologic, geologic, and geochemical information. Then, you will find stable isotope data very beneficial' (Kendall and Caldwell, 1998, p 52).

FINDINGS IN SMALL CATCHMENTS TO DATE

What we know

Perhaps the most important overall outcome from work in small catchments to date is the general finding that stormflow in many environments is dominated by pre-event water. Buttle (1994) provided an initial review of literature results from IHS studies, and this was updated by Richey *et al.* (1998). There is a great range in pre-event water fractions (X) of stormflow for various catchment sizes, land uses and event types (Shanley *et al.*, 2002). Despite this variability in X, two salient points have emerged. The first is that X is generally smaller for catchments with land uses (e.g. urban – Halldin *et al.*, 1990; Buttle *et al.*, 1995; Gremillion *et al.*, 2000) or surface types (e.g. permafrost – Cooper *et al.*, 1991; Metcalfe and Buttle, 2001) that promote the contribution of surface runoff to stormflow production through such mechanisms as Horton overland flow. These areas would be likely to generate a greater event water fraction of stormflow as a result of these mechanisms. Secondly, X is generally smaller in forest catchments in humid temperate climates during spring snowmelt as a result of a greater tendency for surface runoff from frozen soils and maximum extents of saturated near-stream areas. This finding would initially seem to be of little or no relevance to conditions in the humid tropics. Nevertheless, the observation that X tends to decrease with catchment wetness is important, such that humid tropic catchments that experience an annual cycle in precipitation amount might also produce intra-annual variations in X.

What we think we know

In terms of scale effects on new/old water partitioning, the results are equivocal. While some have found increases in new water percentages with increasing catchment size (Shanley *et al.*, 2002; McDonnell *et al.*, 1999) others have found the opposite (Brown *et al.*, 1999). McGlynn *et al.* (2002) used tritium (T) to define the mean residence time (MRT) of water in the catchment, and tested the hypothesis that baseflow MRT increases with increasing absolute catchment size. The Maimai catchments where they worked are, relative to many catchments around the world, simple hydrological systems with uniformly wet climatic conditions, little seasonality in temperature and precipitation, uniform and nearly impermeable bedrock, steep short hillslopes, shallow soils, and well-characterised hillslope and catchment hydrology. As a result, this was a relatively simple system and an ideal location for new MRT-related hypothesis testing. While hydrologists have used T to estimate water age since the 1960s nuclear testing spike, atmospheric T levels have now approached near-background levels and are often complicated by contamination from the nuclear industry. McGlynn *et al.* (2002) were able to use results for T sampled from the Maimai catchments in nuclear industry-free New Zealand. Because of high precision analysis, near natural atmospheric T levels and well-characterised rainfall T inputs, they were able to estimate the age of young (i.e. less than three years old) waters. Their results showed no correlation between MRT and catchment size. However, MRT was correlated to the median sub-catchment size of the sampled catchments, as shown by landscape analysis of catchment area accumulation patterns of McGlynn and Seibert (2003). Their preliminary findings suggest that landscape organisation, rather than total area, is a first-order control on MRT and points the way forward for more detailed analysis of how landscape organisation affects catchment runoff characteristics.

Some hydrologists have argued that the general observation that pre-event water often makes a significant contribution to stormflow has promoted a paradigm shift in hydrological thought. We question if this is in fact the case. There is abundant evidence from the hydrological modelling literature to suggest that IHS results have not been incorporated into many current catchment-scale hydrological models. The view that one's model captures the real-world processes correctly if one 'fits' the hydrograph correctly still persists. Some hydrologists have apparently forgotten, or never learned, the point that was well captured by Hooper (2001, p 2040): 'Agreement between observations and predictions is only a necessary, not a sufficient, condition for the hypothesis to be correct'. Seibert and McDonnell (2002) have argued that the experimentalist often has a highly detailed, yet highly qualitative, understanding of dominant runoff processes – and thus there is often much more information content on the catchment than we use for calibration of a model. While modellers often appreciate the need for 'hard data' for the model calibration process, there has been little thought given as to how modellers might access this 'soft' or process knowledge, especially that derived from isotope tracer studies. Seibert and McDonnell (2002) presented a new method whereby soft data (i.e. qualitative knowledge from the experimentalist that cannot be used directly as exact numbers) are made useful through fuzzy measures of model-simulation and parameter-value acceptability. They developed a three-box lumped conceptual model for the Maimai catchment in New Zealand, where the boxes represent the key hydrological reservoirs that are known to have distinct groundwater dynamics, isotopic composition and solute chemistry. The model was calibrated against hard data (runoff and groundwater-levels) as well as a number of criteria derived from the soft data (e.g. percent new water). They achieved very good fits for the three-box model when optimising the parameter values with only runoff ($R_{\mathrm{eff}} = 0.93$). However, parameter sets obtained in this way showed in general a poor goodness-of-fit for other criteria such as the simulated new-water contributions to peak runoff. Inclusion of soft-data criteria in the model calibration process resulted in lower R_{eff}-values (around 0.84 when including all criteria) but led to better overall performance, as interpreted by the experimentalist's view of catchment runoff dynamics. The model performance with respect to soft data (like, for instance, the new water ratio) increased significantly and parameter uncertainty was reduced by 60% on average with the introduction of the soft data multi-criteria calibration. This work suggests that hydrograph separation information may have new applications in model calibration, where accepting lower model efficiencies for runoff is 'worth it' if one can develop a more 'real' model of catchment behaviour based on the information content of the isotope approach.

WHAT WE NEED TO KNOW: HUMID TROPICS

Perhaps the paradigm shift in conceptualising the rainfall-runoff process from one of event water dominance to pre-event water dominance is limited in geographical scope. The pre-event water paradigm seems to hold in environments where infiltration excess overland flow is relatively rare. However, the intense rainfalls that may occur in some parts of the humid tropics may mean that saturation (and even infiltration) excess overland flow may be more prevalent in forests than in the western European and North American studies that have used environmental isotopic tracers (Elsenbeer, 2001).

Work with tracers in the humid tropics to date has been limited and has focused on attempts to link flow components estimated using environmental isotopic tracers to causal hydrological processes. Bonell *et al.* (1998) used a relatively-extensive hydrometric record in the Babinda catchments in northeastern Australia to estimate seasonal changes in catchment storage. This assisted in the interpretation of catchment responses to tracers at various times of the year. Many IHS studies employ a moving average of the isotopic signature of rainfall composition to serve as the event water signature. 'Such a moving average is equivalent to assuming a well mixed store for event rainfall, with a size that is large compared to runoff, and hence mean residence times that are large compared to the length of the event' (Bonell *et al.*, 1998, p 360). Based on hydrometric data demonstrating the exceptionally-rapid response times of the Babinda catchments, Bonell *et al.* argued against the use of such a moving average, proposing instead that event water δ should correspond to the composition of current rainfall, lagged by a constant amount. They documented the role of rainfall intensity in controlling a shift from a 'first-in – first-out' routing to a 'last-in – first-out' routing as slower intensity pathways are short-circuited (Figure 32.7).

Bariac *et al.* (1995) employed a combined isotopic, geochemical and hydrometric approach to streamflow generation in two small catchments in French Guiana. They observed highly-variable $\delta^{18}O$ in rainfall during short time intervals that exhibited a temporal variability that was similar in magnitude to spatial variability in soil water δ. These soil water δ profiles indicated rapid infiltration in the upper soil, followed by slower infiltration and homogenisation of input δ signatures deeper in the soil. The authors went on to identify streamflow contributions from these various soil layers.

A further example of relevant work in the humid tropics is that of Elsenbeer *et al.* (1995) and Elsenbeer and Lack (1996) in western Amazonia. This work did not employ environmental isotopes as tracers; nevertheless, its use of geochemical tracers and the conclusions that were drawn are of relevance to IHS studies in tropical catchments. The study showed the benefit of reconnaissance studies in guiding subsequent hydrometric and hydrochemical studies. At the same time, it identified the importance of overland flow as a distinct end-member in any hydrograph separations that would be conducted in this landscape (see Bonell, this volume).

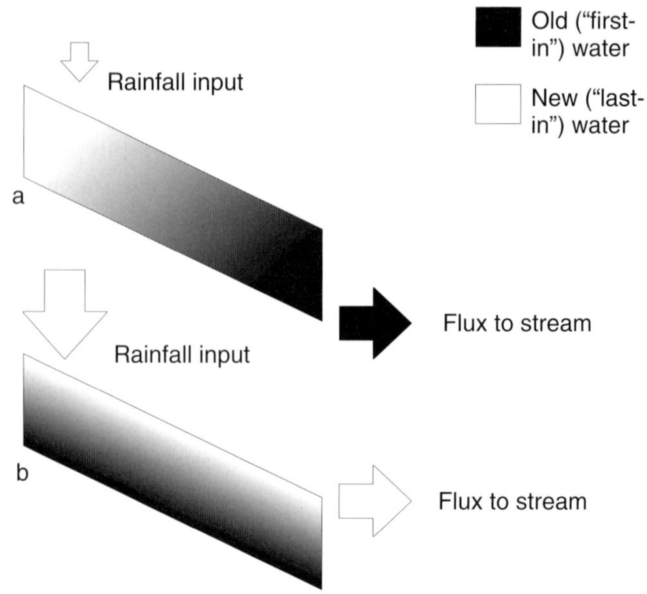

Figure 32.7 Schematic representation of the short-circuiting mechanism proposed by Bonell *et al.* (1998) for the Babinda, Queensland catchments. Under low-intensity rainfall inputs, slope water fluxes to the stream channel are dominated by old ('first-in') water moving via relatively deep subsurface slope pathways (a). This process has been referred to as translatory flow by Hewlett and Hibbert (1967). Under high-intensity rainfall inputs, new ('last-in') water moves quickly to the stream channel via more-conductive near-surface pathways, and dominates slope water fluxes to the channel (b). This effectively 'short-circuits' the translatory flow process, allowing new water inputs to bypass earlier water inputs stored deeper in the soil profile on the slope.

A consensus?

Beyond the general importance of pre-event contributions to stormflow, what can we agree on with regard to the results of IHS studies? First, we must recognise that the issue of equifinality (the same outcome can be generated by a range of alternative processes) applies to the interpretation of IHS results. Buttle (1994) has reviewed the various mechanisms that may be responsible for the rapid delivery of significant quantities of pre-event water to the stream channel, and concluded that the isotopic response of a catchment may be the result of several hydrological processes.

These can vary in their degree of importance both spatially and temporally. The prospect of equifinality undermines (perhaps fatally?) attempts to infer intra-catchment processes based on the isotopic responses at the catchment outlet. It also provides a compelling case for integrating isotopic tracers with other tracers as well as hydrometric techniques to constrain a process interpretation more fully. Second, we are not really advancing hydrological science appreciably by using IHS *by itself* in humid temperate environments (Bonell *et al.*, 1998; Rice and Hornberger, 1998; Burns, 2002). We suspect that the same comment would apply if we were to use IHS alone in the humid tropics. Third, we cannot infer the correct hydrological pathways from the stormflow hydrochemical (or isotopic) signal *alone* (Elsenbeer *et al.*, 1995); we must combine isotopic and geochemical tracers with hydrometric measurements. Fourth, IHS and GHS studies are essentially black-box approaches that assume that flowpaths and other hydrological properties are distributed homogeneously and that input waters have uniform isotope and chemical compositions (Kendall *et al.*, 2001). As Kendall *et al.* (2001) note, these assumptions 'are often adequate for general characterisation of catchment response to bulk storms, but separations made using them do not have sufficient resolution to help answer questions about intrastorm changes in flowpaths and water sources, and processes occurring along the various flowpaths' (p 1878).

RESEARCH AND TECHNICAL ISSUES THAT REMAIN TO BE ADDRESSED

There are a number of conceptual and methodological challenges facing the linkage of isotopic tracers, geochemical tracers and hydrometric evidence (Uhlenbrook *et al.*, 2003). These challenges apply to the use of the IHS approach in both temperate and tropical environments. They require us to obtain answers to the following questions:

How important are processes acting parallel or sub-parallel to the stream channel in controlling its isotopic response?

Many of our conceptual models of streamflow generation envisage that the delivery of water via various pathways from the hillslope to the stream channel occurs normal to the channel margin, and often fail to consider the role of processes operating in the channel itself. Nevertheless, it is increasingly recognised that complex exchanges of water between the stream and its bed and banks may occur as hyporheic exchange during flow along the channel (Bencala, 2000), and such exchanges might be expected to alter the isotopic signal of slope runoff from point of entry to the channel to the sampling point at the mouth of the catchment. This presents both a challenge and an opportunity. The challenge is determining to what extent mixing processes in the hyporheic zone encourage divergence between flood wave and water particle travel times. This relates to the issue of hydrological linkages between landscape elements that is addressed in a subsequent section of this chapter. The opportunity is the potential to use isotopic tracers to help us distinguish between the residence times of water on hillslopes and in the hyporheic zone. Knowledge of the latter would be particularly valuable in studies of stream ecology, given the dependence of such key ecological metrics as DOC and dissolved oxygen on water residence time.

Examples of studies that have addressed the implications of channel processes for IHS results include Bonell *et al.*'s (1998) observation of an initial rise in pre-event water at the start of each hydrograph pulse in the Babinda catchments in NE Australia. They discounted the role of groundwater ridging close to the main stream based on soil water content – matric potential data that did not support the presence of a tension-saturated capillary fringe. Instead, they opted for the mechanism suggested by Nolan and Hill (1990), whereby sudden upstream inputs of new water set up a flood wave composed of pre-event channel water which reaches downstream locations in advance of the translation of the subsequent event water. This process is distinct from the evaporative enrichment of streamflow during passage along the stream channel that has been observed by Sklash *et al.* (1976) and Gremillion and Wanielista (2000). Such enrichment would result in the overestimation of X in streamflow by shifting the δ in streamwater towards the groundwater isotopic signature, and should be considered when using IHS in situations where water residence times in the stream channel are appreciable (see below).

What is the most appropriate way to incorporate temporal variations in event water δ in IHS studies?

McDonnell *et al.* (1990) studied various approaches to treating temporal variations in the event water isotopic signature and their influence on IHS results, including the use of volume weighted means, incremental means and incremental input intensity. Nevertheless, each approach assumes that the δ of input water early in the event still exerts an influence on the stream water signature by the end of the event. This may be a realistic assumption when dealing with relatively short-lived storms of a few hours or days in duration. However, during long duration events (e.g. entire snowmelt periods), these initial water inputs could be exported from the catchment before the end of the event, and their δ should not have any influence on the stream water signature at that time. Joerin *et al.* (2002) used the unit hydrograph concept to approximate the progressive decrease in the influence of rainfall on the event water δ signal with time. Conversely, Laudon *et al.* (2002) proposed a 'runoff corrected event water' approach that bases the estimated input water δ at a given time on the amount of event water discharged from the catchment prior to that time. Nevertheless, both approaches are consistent with Bonell *et al.*'s (1998) call for the use of event water δ that corresponds to the lagged composition of current water inputs. We need to test these and other approaches over a greater range of basin and water input characteristics.

Does all the water that falls on saturated areas retain the signature of event water?

As Kendall *et al.* (2001) noted: 'In theory, rainfall that flows over the soil surface (as infiltration excess overland flow or saturation excess overland flow) or has been transported to the stream via preferential flow in the soil (as vertical bypass flow and/or lateral pipeflow) should be chemically 'new' (event) water' (p 1878). However, recent work questions the assumption that the near-stream saturated area is connected to the stream channel so that direct precipitation onto saturated areas (DPSA) can actually reach the channel unaltered. Crayosky *et al.* (1999) found that most overland flow moves only a few metres before infiltrating, such that the re-infiltration and re-exfiltration of event water inputs as DPSA may account for tracer data (both isotopic and geochemical) that shows a difference between the input signature and that of overland flow. Bonell *et al.* (1998) concluded that regions of lateral interflow and exfiltration in catchments will inevitably promote mixing of event and pre-event water within saturation overland flow, thus modifying the isotopic content of this flow. Observations in the Sleepers River catchment in Vermont, USA, suggest that the degree of mixing in the near-stream saturated zone varies down-valley, depending on the local rates of groundwater exfiltration into this zone (McGlynn *et al.*, 1999). This mixing appears to change through the hydrological year.

What water is actually being sampled from standard soil water samplers?

DeWalle *et al.* (1988) found no statistically-significant difference in δ in samples collected from soil water samplers (assumed to favour matrix water) and from pan lysimeters (assumed to favour macropore water). Conversely, Leaney *et al.* (1993) argued that suction lysimetry preferentially removes soil water from the larger capillaries, and such samples should therefore be biased towards the δ of soil water moving through the profile via relatively rapid pathways. The question does not become any clearer when attempting to characterise the isotopic signature of hillslope runoff using soil water samplers. Thus, Buttle and McDonald (2002) showed that the chemistry of soil water sampled at the base of the soil profile using suction samplers on forested slopes with thin soil cover differs significantly from slope runoff moving in a thin layer above the soil-bedrock interface. Burns *et al.* (2001) suggested that this debate can be avoided completely by sampling soil water for IHS and GHS using throughflow trenches. This ensures the sampling of mobile hillslope water that actually has the potential to participate in stormflow generation. Whatever the approach, the effect of sampling on how end-members are defined and described is an important point to note in any study employing IHS methods. The humid tropics offer great potential for samples to be altered by evaporation before they can be extracted from a lysimeter or trench. Thus, extreme care must be taken to ensure that samples are isolated from evaporative enrichment in any sampling device.

Is the inclusion of additional runoff components warranted, or does it simply reflect a mathematical correction to apparently-erroneous IHS results?

Failure of the standard two-component IHS to describe flow components in a realistic manner may suggest that contributions from one or more additional flow components (e.g. soil water) be included. This call for the consideration of more than two flow components has been made in numerous IHS studies reported in the literature. However, these components need to be supported by independent observations of the hillslope processes; otherwise they simply reflect a mathematical correction to apparently-erroneous IHS results (Bonell *et al.*, 1998). The key issue is our ability to define a priori what these additional components might be and to sample them adequately. Uhlenbrook and Leibundgut (2002) developed a conceptual catchment model by defining three components contributing to streamflow in the Brugga basin in Germany. They found that direct runoff (with a MRT of a few months), shallow groundwater (32 months MRT), and deep groundwater (MRT of 7.1 yr) could be blended to give the combined stream signal. Uhlenbrook and Leibundgut (2002) validated the model output with silica concentration data where each of the three components could be characterised by uniquely different silica end-member concentrations. It appears from their work that the combination of conceptual model development and runoff component characterisation may be a way forward for identifying what minimum set of components define any given hydrological system.

How do temporal and spatial variations in hydrological linkages between landscape units (slopes – riparian zone – stream) affect a catchment's isotopic and chemical response?

As Welsch *et al.* (2001) observed, we need to be able to quantify the processes that affect the spatial distribution of solute concentrations in source water throughout catchments if we are to predict the hydrochemical response to such perturbations as forest harvesting and climate change. Are we seeing biogeochemical resetting of solute signatures as hillslope runoff transits the riparian zone to

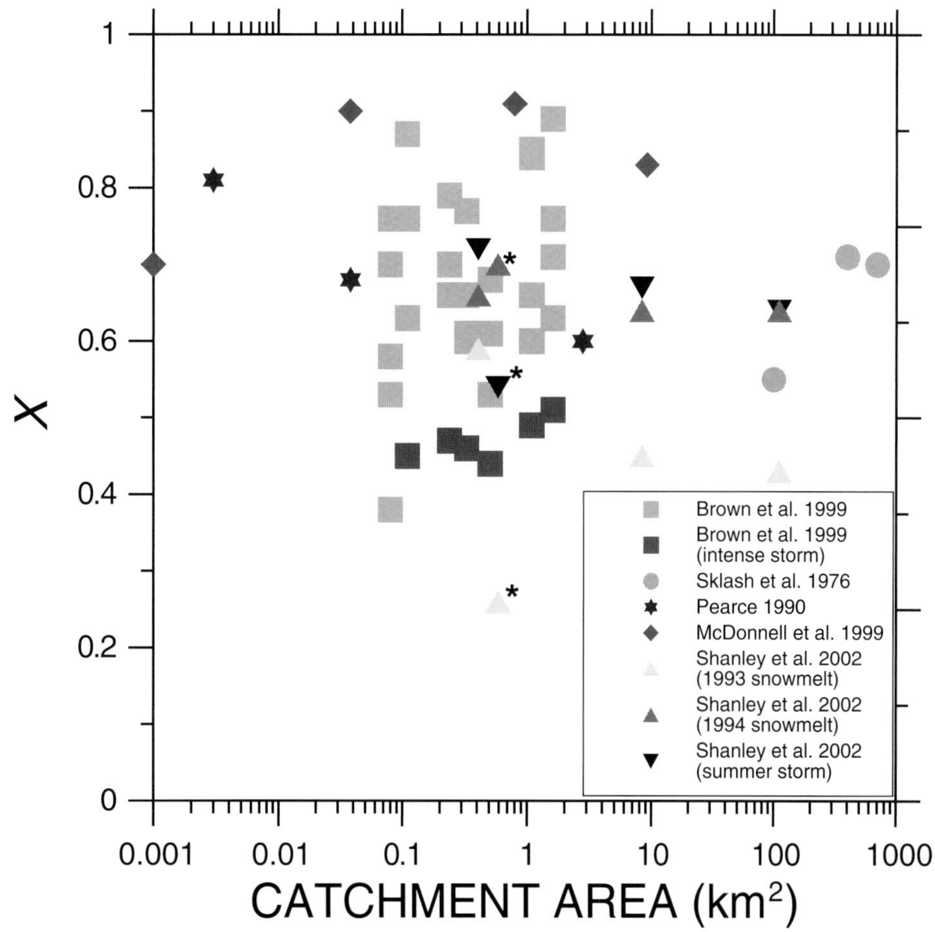

Figure 32.8 Reported pre-event water fractions in streamflow (*X*) vs. catchment area. The * associated with the Shanley *et al.* (2002) results indicates data from a small catchment largely in pasture. The other catchments from Shanley *et al.* (2002) were forested.

the stream (e.g. Robson *et al.*, 1992; Hill, 1993) or simple mixing processes where a small volume of hillslope runoff is diluted by a larger volume of riparian storage during water transit (e.g. Burns *et al.*, 2001)? We require further study of mixing and geochemical interactions in the riparian zone. Such interactions have important implications for our understanding of riparian dynamics, particularly in the context of the use of forested riparian buffer zones in attempts to mitigate the impacts of forest harvesting on aquatic ecosystems.

McGlynn and McDonnell (2003) examined the spatial sources and delivery mechanisms of DOC to streams. They examined the relationship between storm DOC dynamics, catchment landscape units and catchment scale to elucidate controls on DOC export dynamics. They focused on the controls on the characteristic hysteresis in DOC export dynamics (i.e. larger DOC concentrations on the rising relative to the falling limb of the discharge hydrograph), previously ascribed to a flushing mechanism. McGlynn and McDonnell (2003) found that the proportion of riparian runoff during the storm event was larger on the hydrograph's rising rather than falling limb, while the proportion of hillslope runoff was larger on the falling limb. The delayed response of hillslope runoff resulted in a disconnection between hillslope and riparian areas early in the event and thus greater DOC concentrations on the rising limb of the event water hydrograph. Later in the event, hillslope and riparian areas became connected once the hillslope soil moisture deficits were satisfied. They suggested that the relative timing of riparian and hillslope source contributions, and the connections and disconnections of dominant runoff contributing areas, are the first-order catchment controls on stream DOC concentrations and mass export.

How and why do IHS results vary with catchment scale?

As noted earlier, there is no agreement on how the partitioning between event and pre-event water in streamflow changes with catchment scale (Figure 32.8). Sklash *et al.* (1976) showed that *X* increased with catchment size for three catchments in southern Ontario. A similar result was found by Brown *et al.* (1999) for an intense rainstorm over six nested catchments in the Catskill Mountains of New York state. They attributed this to increased flux

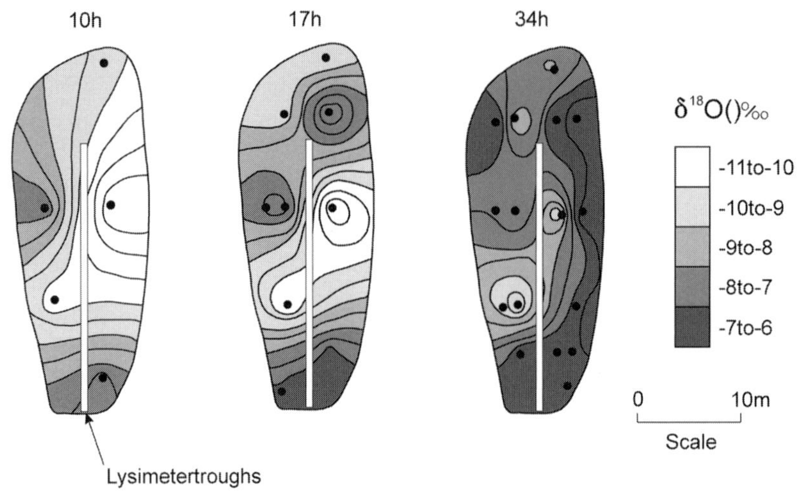

Figure 32.9 Inferred patterns of $\delta^{18}O$ in groundwater sampled from the Hydrohill experimental catchment, China, during a rainfall event that began on 5 July 1989. The dots indicate the locations of wells where $\delta^{18}O$ values were available at the indicated times since the start of 5 July. (From Kendall *et al.*, 2001.)

of shallow perched pre-event groundwater in larger catchments. Conversely, Pearce (1990) suggested that larger catchment size at Maimai was associated with an increase in the relative size of saturated floodplains, which would enhance the event water contribution to stormflow. McDonnell *et al.*'s (1999) work at Maimai demonstrated greater pre-event water contributions as one moved from the plot to the small catchment scale, but a decrease in X as catchment scale increased from ~1 to ~10 km^2. Shanley *et al.* (2002) also found a decrease in X with increasing scale for forested catchments in Vermont USA for snowmelt and rainfall events, the exception to the general trend being a small catchment largely in pasture. Thus, further work is required to determine if there is a relationship between catchment morphology and the relative partitioning of event and pre-event water in stormflow, and the degree to which inter-catchment differences in land use, pedological and geological characteristics might influence any scale-dependence of IHS results.

FUTURE RESEARCH NEEDS

There are a number of avenues of study with considerable promise for the more effective use of environmental isotope tracers to help address the outstanding research issues raised above, including:

Controlled experiments that incorporate the use of environmental isotope tracers

These studies allow us to explore the role of specific processes and controlling factors by enabling manipulation of input rates and isotopic signatures. This lets us avoid the complication of marked temporal variations in the δ of precipitation inputs that often occur in natural events, and that may result in distinct event and pre-event water signatures that preclude IHS (Turton *et al.*, 1995; Collins *et al.*, 2000). Such studies vary widely in the degree of experimental control that has been employed. The Coos Bay experiment (Anderson *et al.*, 1997; Montgomery *et al.*, 1997; Torres *et al.*, 1998; Anderson and Dietrich, 2001) involved artificial irrigation of a deforested hillslope on the west coast of the Oregon Coast Range in Oregon USA, and the examination of unsaturated and saturated zone processes and their implications for landscape evolution. Somewhat greater experimental control was exerted in the Gårdsjön covered catchment experiments in southwestern Sweden (Nyberg, 1995; Rodhe *et al.* 1996; Lange *et al.*, 1996). Here a small catchment was roofed over to exclude natural precipitation inputs, and artificial precipitation of known intensity and composition was applied to the catchment surface. Finally, the Hydrohill experiment in China probably represents the extreme in controlled hydrological experiments (Kendall *et al.*, 2001). A 490 m^2 artificial catchment was constructed, containing a detailed array of groundwater wells, runoff collectors and neutron probes. Kendall *et al.* (2001) were able to map changes in groundwater $\delta^{18}O$ during the course of the event (Figure 32.9). They obtained evidence that geochemical tracers (specifically, Cl^- and SO_4^{2-}) did not behave conservatively, that there were differences in the relative mobility of pre-event water within the catchment, and that agreement between GHS and hydrometric results was largely fortuitous. The results also suggested shifts from bypass flow to matrix flow during storms depending on rain intensity and amount of water stored in the soil zone. Kendall *et al.* (2001) argued that assessment of the impact of this process shift on IHS results requires more information on isotopic exchange rates in pore waters.

Estimation of water residence times at the point, slope and catchment scales using environmental isotopic tracers

There is abundant evidence from the literature demonstrating the important control that water residence time exerts on soil water, groundwater and streamflow chemistry. Isotopic tracers provide a valuable means of estimating MRTs at various scales in catchments. The benefits of such data are wide-ranging, and include DeWalle *et al.*'s (1997) suggestion that we can use these MRTs to estimate the length of time needed to observe catchment response to treatment or disturbance in the design of hydroecological monitoring programmes. Williard *et al.* (2001) state that we need to know the residence time of precipitation to assign a growing or dormant season $\delta^{18}O$ when estimating the proportion of atmospheric NO_3 deposition in streamflow samples. Bonell *et al.* (1998) go further, and contend that unambiguous IHS requires estimates of the travel time distribution for rainfall of a particular isotopic composition. McDonnell *et al.* (1999) attempted to implement such an approach whereby the age spectra of the new water were computed through the event. The approach of Rodhe *et al.* (1996) to estimating these distributions during controlled experiments at the Gårdsjön covered catchment is particularly promising, but may be difficult to apply in uncontrolled field conditions where marked short-term oscillations in input δ are the norm. These oscillations were partly responsible for the wide ranges in MRTs that provided significant fits to observed $\delta^{18}O$ time series in groundwater (Buttle *et al.*, 2001) and soil water (Murray, 2003) during snowmelt in forested and clearcut landscapes in central Ontario, Canada.

The link between water residence times and IHS can be examined from another perspective. The standard IHS uses Eqns 32.2–32.4 as a steady state model that assumes negligible temporal changes in the volume and isotopic signature of channel storage. Gremillion *et al.* (2000) compared IHS results using steady state (SS) and non-steady state (NSS) solutions to Eqns 32.2–32.4 and found little difference in predicted pre-event water fractions from a catchment in central Florida. However, modelling studies showed increasing divergence between predicted pre-event water fractions with increasing water residence times in the stream channel. This issue needs to be considered when applying IHS to large catchments where flow time on hillslopes is small relative to the residence time of water routed along stream channels (Bras, 1990).

Integration of more advanced hydrometric techniques

Promising examples of this include work by Zollweg (1996) and Srinivasan *et al.* (2002), who have deployed arrays of surface saturation sensors in their studies of runoff generation on agricultural slopes in Pennsylvania, USA. By comparing estimated saturation overland flow to runoff recorded at flumes, Zollweg (1996) was able to estimate the amount of Horton overland flow and to gain insight into infiltration and exfiltration processes as overland flow moved downslope. Srinivasan *et al.* (2002) extended this work by documenting the temporal and spatial dynamics of surface saturation areas and surface runoff source areas (generating infiltration excess and saturation excess overland flow) in relation to water table dynamics and slope runoff. The complexity of stormflow generation revealed at the slope scale presents a disturbing challenge to the simplistic views of stormflow generation that are founded on isotopic, geochemical and hydrometric observations made at a few locations within a catchment and at the catchment outfall. There is a need to couple arrays of this type with tracer studies to address some of the challenges noted above.

Use of environmental isotopic tracers to identify process thresholds

Partitioning between event and pre-event water in slope runoff and catchment stormflow may have important implications for the transport of reactive substances such as dissolved organic matter (DOM) and atmospheric inputs of nitrogen from slopes to receiving water bodies. Repeated IHS studies in the same catchment under a variety of event and antecedent conditions have often noted changes in the partitioning of event and pre-event water, and several studies have attempted to account for such changes (e.g. Kendall *et al.*, 2001; Buttle *et al.* 2001). In the case of the latter study, the proportion of pre-event water in runoff from a forested slope was observed to increase with antecedent soil water content, consistent with the hypothesised increase in translatory flow contributions to slope runoff during wet conditions (Hewlett and Hibbert, 1967). Conversely, transport of reactive ammonium (NH_4^+) applied to the slope during controlled irrigations was found to be greatest when event water supplied a significant fraction of slope runoff. This event water was inferred to travel via vertical and lateral preferential flowpaths, but was overwhelmed by pre-event water contributions to slope runoff when antecedent soil wetness was maximised. Greater documentation and understanding of the controls on shifts in the dominant processes operating at the slope and catchment scales is important to our ability to monitor and model catchment hydrochemistry.

Integration of isotopic and geochemical tracers and hydrometric techniques with greater consideration of topographic properties

We need to take advantage of the increased availability of digital topographic information to estimate where most of the hydrological/ hydrochemical action is going to take place in a catchment (e.g. Kendall *et al.*, 1999). For example, topographic indices such as the $\ln(\alpha/\tan\beta)$ index of Beven and Kirkby (1979) has been used to estimate depth to groundwater (Moore and Thompson, 1996;

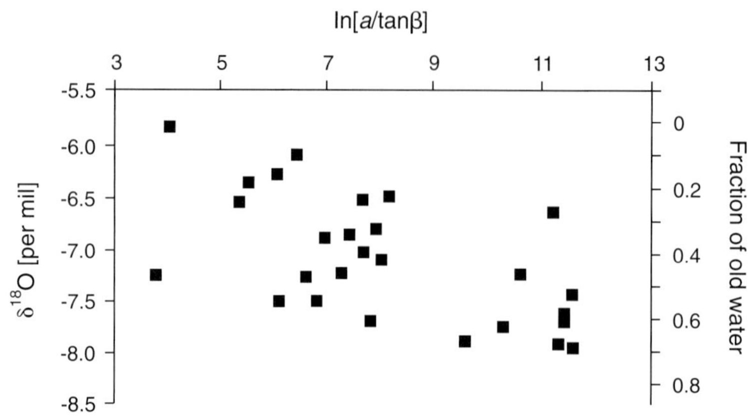

Figure 32.10 The δ^{18}O of all groundwater samples taken on 23 September 1991 from the Gårdsjön covered catchment experiments in Sweden vs. the Beven and Kirkby (1978) ln(α/tanβ) topographic index, where α is upslope contributing area to a groundwater well location and tanβ is the local gradient at that location. Corresponding fractions of groundwater consisting of old water are also indicated. (from Rodhe *et al.*, 1996.)

Seibert *et al.*, 1997) and to interpret spatial variations in the δ^{18}O of groundwater (Rodhe *et al.*, 1996) (Figure 32.10). This information can be used to design field experiments to ensure that hydrologists get the greatest information return on their investment of time, effort and money. Concurrent with this work, we need to explore the use of different types of topographic data (e.g. surface topography vs. bedrock topography – e.g. Freer *et al.*, 1997; 2002) and topographic indices (e.g. Burch *et al.*, 1987; Barling *et al.*, 1994; Chaplot *et al.*, 2000).

Explicit integration of models into our study designs

The call for greater integration between field and modelling studies is an old one, but is still being made by hydrologists. A recent example of this is Hooper's (2001) point that we should adopt sampling strategies that might permit the generation of data that could then be used to test a range of models. Another view is to use tracer information and hydrograph separation results in the model calibration process. Seibert and McDonnell (2002) used the peak new water percentage as 'soft data' in a multicriteria model calibration exercise. This and other process knowledge helped to improve model simulations, where usually only hard data such as the continuous runoff signal are used to calibrate the model.

APPLICATION POTENTIAL OF IHS FOR LAND USE CHANGE STUDIES IN THE HUMID TROPICS

Potential problems facing application of IHS in the humid tropics

Figure 32.11 shows δ^{18}O in precipitation for Ottawa, Darwin, Manaus and Harare. Ottawa (mid-latitude) values vary during the course of the year by up to ~15‰. Conversely, low-elevation sites in tropical latitudes have an annual range in δ^{18}O in precipitation of ~10‰, which would reduce the potential of observing a significant difference between δ for the precipitation event and that in pre-event water. However, Gremillion and Wanielista (2000) found that a range in δ^{18}O for precipitation in central Florida of between −6.64 and −0.17‰ provided sufficient variability between event and pre-event water signatures to permit IHS. In addition, the amplitude of the annual δ^{18}O cycle increases with altitude in the tropics, such that δ^{18}O in precipitation for Harare may be only slightly less than that in mid-latitude environments. This point is supported by work on the Table Mountain of South Africa funded through the International Atomic Energy Agency (IAEA) that indicates elevation increases from the coast to Table Mountain (3567 m) produce a clear 'signal' greater than might be expected based solely on latitudinal position. Seasonal changes in air mass type in the outer tropics may also result in promoting significant differences between event and pre-event δ signatures. For example, Barnes and Bonell (this volume) show a seasonal trend from generally lighter D values associated with the deeper (and colder) convection of monsoon disturbances towards heavier D signatures of 'warm rain' identified with clouds of the southeast trade winds that have higher temperatures at the condensation level. The low cloud temperatures associated with the high intensity, deep convection monsoonal disturbances resulted in an event water D signature that was much lighter than that of pre-event streamflow in Queensland, Australia (Bonell *et al.*, 1998), thus permitting a successful IHS.

Changes in water flowpaths

Hydrograph separation and solution of Eqns 32.2–32.4 provides a basic description of water sources contributing to the stream.

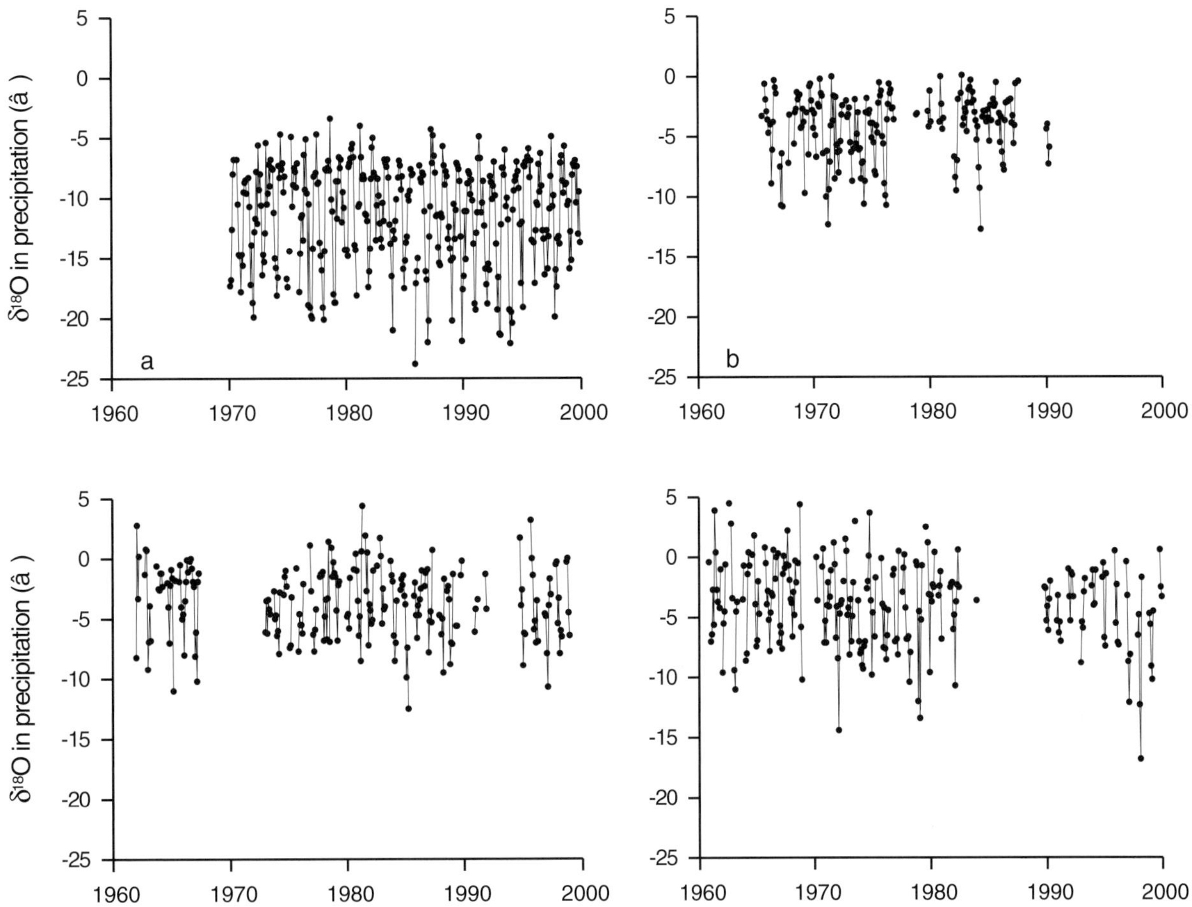

Figure 32.11 Time series of $\delta^{18}O$ in precipitation sampled at (a) Ottawa, Canada (114 m asl; 45.32°N latitude); (b) Manaus, Brazil (72 m asl; 3.12°S latitude); (c) Darwin, Australia (26 m asl; 12.43°S latitude); and (d) Harare, Zimbabwe (1471 m asl; 17.83°S latitude). Data from International Atomic Energy Agency/World Meteorological Organization (2001).

Furthermore, the relative proportion of old and new water before and after disturbance or land use change may be a useful change detection tool. While studies to date have been very few (but see Gremillion *et al.* 2000), the tool offers the potential to quantify changes in hydrograph *composition* after disturbance. In particular, the juxtaposition of labile nutrient tracers of flow path and stable isotope tracers of flow source can be a powerful tool for resolving water flow dynamics at the catchment scale. In some land-use change studies, alteration of surface conditions due to compaction, downed woody debris, etc., may force water to move laterally at shallower depths than it did prior to disturbance. Murray (2003) compared vertical profiles of mean residence times estimated from soil water $\delta^{18}O$ time series during snowmelt at forest and clearcut sites in central Ontario. This suggested a short-circuiting mechanism in clearcuts that restricted deep infiltration of inputs and diverted a portion of incoming event water laterally downslope, with important consequences for the quantity and quality of slope runoff reaching receiving waters (see also Bonell, this volume). The potential for forest harvesting to induce changes in water flowpaths was also suggested by Bariac *et al.* (1995), who used a combined IHS – GHS approach to examine differences in water flowpaths in forested and deforested (pasture) catchments in French Guiana. Peak flow from the forest catchment was largely via subsurface flow, whereas flow through the superficial soil layers dominated peak streamflow in the catchment with pasture land use. Brown *et al.* (1999) used analysis of DOC – $\delta^{18}O$ variations to quantify a shallow flow pathway through the organic layer on steep forested slopes in the Catskill Mountains of New York State. They showed that the combination of high DOC and rainwater-like $\delta^{18}O$ signatures could be used to determine that rain followed a shallow flowpath during intense summer thunderstorms. Similarly, Peters *et al.* (1995) used IHS to confirm that initial slope runoff over the forest organic mat during a spring rainstorm in central Ontario was predominantly event water. This flow was attributed by Buttle and Turcotte (1999) to the hydrophobicity of dry organic matter (Burch *et al.*, 1989, Wilson *et al.*, 1990) which promoted runoff over and through the organic 'thatch'. Buttle and Turcotte (1999) demonstrated that this overland flow decreased in quantity with antecedent soil wetness, hypothesising that a wetter litter layer would increasingly redirect event water inputs vertically into the underlying mineral soil.

Detection of forest road interception effects

Forest roads can affect the stormflow response of a catchment in a variety of ways (Ziegler and Giambelluca, 1997): (i) increased overland flow production and flow velocities on compacted road surfaces and disturbed roadside margins; (ii) interception of subsurface flow at cutbanks and re-routing via overland flow; (iii) capture and channelling of surface and subsurface flow by ditches and culverts; and (iv) capture and re-routing of surface water by erosion gullies initiated by the initial disturbance caused by the road. The standard two-component mass balance approach in Eqn 32.1 is well suited to applied problems such as the effects of forest road construction on water re-routing at the catchment scale. Resource managers often need to determine the possible increase in peak flow associated with forest harvesting and the presence of forest roads. Ditch flow can be separated using Eqn 32.1 into direct road runoff (event water) and intercepted subsurface flow from the cut back (pre-event water). This separation can be a valuable tool for quantifying these relative inputs at specific cross-drain and road culvert sections. Ziegler *et al.* (2001) were amongst the first to demonstrate the use of this approach in studies of forest road runoff in the humid tropics of Thailand, and Luce (2002) has called for increased use of hydrograph separation in land-use change studies involving roads.

Changing runoff composition as a result of suburban development

Wolman (1967) characterised the cycle of land use change in eastern North America following European settlement as a transition from forest to agriculture to woods and grazing to suburban and urban development that encompassed ~150 years. However, the growth of urban and suburban areas in the humid tropics is faster than anywhere else on the planet, and forested areas adjacent to such cities as Kuala Lumpur, Bangkok and Manila are converted directly to residential and industrial uses without passing through phases of agriculture and reversion to forest cover prior to development. There are a variety of mitigation measures (e.g. infiltration swales, detention ponds) that can be used to ensure that suburban development does not result in significant changes to the pre-development hydrograph form. However, development may result in a shift in stormflow pathways from largely subsurface during the pre-development phase to overland and channelised flow after development (Buttle, 1990). This shift can have major implications for the coupling between the riparian zone and the stream channel, since subsurface flow often must transit the riparian zone before reaching the stream. This in turn has consequences for alterations in the retention, transformation and mobilisation of substances in the riparian zone following development, as well as the response of riparian hydroperiod and overall wetland health to changes in subsurface flow contributions (Gremillion *et al.*, 2000). IHS provides a useful tool for identifying water sources in urban and suburban catchments, particularly if the link between event water generation and infiltration excess overland flow from modified surface cover can be made (cf. Halldin *et al.*, 1990; Buttle *et al.*, 1995). For example, Gremillion *et al.* (2000) performed IHS for rural and suburban sub-catchments of the Econlockhatchee River in central Florida. They noted greater event water contributions to stormflow from the suburban sub-catchment, which was attributed to an increased proportion of surface runoff in the storm hydrograph. This change in water flow paths to the river may alter groundwater flow through riparian zones, with implications for river water quality and riparian zone ecology (Gremillion *et al.*, 2000). Information on event and pre-event water partitioning of stormflow can also assist in interpreting and modelling the export of surface-applied chemicals from catchments, such as radionuclides deposited in fallout from the Chernobyl accident (Halldin *et al.*, 1990) and de-icing salts (Buttle *et al.*, 1992). Therefore IHS is another tool that can be used by hydrologists to assess the overall hydro-ecological impacts of suburban development.

Quantifying where mixing occurs in the landscape

While much of the work reviewed in this chapter has focused on the stream signal as an integrated measure of catchment-wide mixing, more needs to be done to define where this mixing occurs, how riparian zones modulate runoff and solute load from hillslopes and how these discrete units mix from the headwaters to the catchment outlet. The potential of riparian zones to buffer hillslope runoff depends partly on the size of the riparian zone relative to the adjacent upland area. McGlynn and Seibert (2002) presented a simple approach for quantifying the local contributions of hillslope and riparian areas along a stream network based on digital elevation data, and computed such catchment characteristics as the distribution of riparian and hillslope inputs to the network, the variation of riparian area percentage along the network, and sub-catchment area distributions. They found that sub-catchments with areas <20 ha comprised 85% of the total catchment area contributing to streams near Maimai in New Zealand, while only 28% of the catchment's total riparian area was found along these small streams. In addition, the median riparian-to-hillslope-area ratio along these tributaries was only 0.06, indicating that the 'effective' riparian-to-hillslope-area ratio would be significantly overestimated by the average value of 0.14 for the entire 280 ha Maimai research area. This landscape analysis and discretisation approach may be highly effective in land use change issues in the humid tropics where terrain-based measures of sensitivity can be used to develop hypotheses to then be tested with isotope tracer approaches.

CONCLUSIONS

Isotope hydrograph separation studies have gone through two stages – unbridled use and enthusiasm for the technique, followed by careful reflection and consideration of the assumptions and limitations. Given that we are presumably passing through the second stage, we are now well poised for applying isotopic tracing techniques in new environments (like the humid tropics), especially for detecting quantitative shifts in hydrological processes in the context of land-use change. This chapter has outlined the basic principles surrounding the use and implementation of the IHS technique and the various assumptions and limitations associated with its use. The reader is advised to reflect on how these new approaches can be applied in the context of what is known about the runoff process in the humid tropics from their reading of the chapter by Bonell. We hope that new students and researchers will consider using isotope tracer tools as they seek to define a robust quantitative description of how their humid tropical catchments work.

References

Aggarwal P. 2002. Isotope hydrology at the International Atomic Energy Agency. Hydrological Processes 16: 2257–2259.

Anderson SP, Dietrich WE. 2001. Chemical weathering and runoff chemistry in a steep headwater catchment. Hydrological Processes 15: 1791–1815.

Anderson SP, Dietrich WE, Montgomery DR, Torres R, Conrad ME, Loague K. 1997. Subsurface flow paths in a steep, unchanneled catchment. Water Resources Research 33: 2637–2653.

Bariac T, Millet A, Ladouche B, Mathieu R, Grimaldi C, Grimaldi M, Hubert P, Molicova H, Bruckler L, Bertuzzi P, Boulègue J, Brunet Y, Tournebize R, Granier A. 1995. Stream hydrograph separation on two small Guianese catchments. Tracer Technologies for Hydrological Systems. IAHS Publication 229: International Association of Hydrological Sciences: Wallingford; 193–209.

Barling RD, Moore ID, Grayson RB. 1994. A quasi-dynamic wetness index for characterizing the spatial distribution of zones of surface saturation and soil water content. Water Resources Research 30: 1029–1044.

Bazemore DE, Eshleman KN, Hollenbeck KJ. 1994. The role of soil water in stormflow generation in a forested headwater catchment: synthesis of natural tracer and hydrometric evidence. Journal of Hydrology 162: 47–75.

Bencala KE. 2000. Hyporheic zone hydrological processes. Hydrological Processes 14: 2797–2708.

Beven KJ, Kirkby MJ. 1979. A physically based, variable contributing area model of basin hydrology. Hydrological Sciences Bulletin 24: 43–69.

Bishop KH. 1991. Episodic increases in stream acidity, catchment flow pathways and hydrograph separation. Doctoral dissertation. Department of Geography, University of Cambridge, 246 p.

Bonell M, Pearce AJ, Stewart MK. 1990. The identification of runoff-production mechanisms using environmental isotopes in a tussock grassland catchment, eastern Otago, New Zealand. Hydrological Processes 4: 15–34.

Bonell M, Barnes CJ, Grant CR, Howard A, Burns J. 1998. High rainfall, response-dominated catchments: a comparative study of experiments in tropical northeast Queensland with temperate New Zealand. In Isotope Tracers in Catchment Hydrology, Kendall C, McDonnell JJ (eds). Elsevier, Amsterdam, 347–390.

Bras RL. 1990. Hydrology: An Introduction to Hydrologic Science. Addison Wesley, Reading, MA, 643 p.

Brodersen C, Pohl S, Lindenlaub M, Leibundgut C, v Wilpert K. 2000. Influence of vegetation structure on isotope content of throughfall and soil water. Hydrological Processes 14: 1439–1448.

Brown VA, McDonnell JJ, Burns DA, Kendall C. 1999. The role of event water, a rapid shallow flow component, and catchment size in summer stormflow. Journal of Hydrology 217: 171–190.

Bruijnzeel S. 1990. Hydrology of moist tropical forests and effects of conversion: A state of knowledge review. UNESCO International Hydrological Program, 224 pp.

Burch GJ, Moore ID, Burns J. 1987. Soil hydrophobic effects on infiltration and catchment runoff. Hydrological Processes 3: 211–222.

Burns DA. 2002. Stormflow-hydrograph separation based on isotopes: the thrill is gone – what's next? Hydrological Processes 16: 1515–1517.

Burns DA, McDonnell JJ. 1998. Effects of a beaver pond on runoff processes: comparison of two headwater catchments. Journal of Hydrology 205: 248–264.

Burns DA, McDonnell JJ, Hooper RP, Peters NE, Freer JE, Kendall C, Beven K. 2001. Quantifying contributions to storm runoff through end-member mixing analysis and hydrologic measurements at the Panola Mountain Research Watershed (Georgia, USA). Hydrological Processes 15: 1903–1924.

Buttle JM. 1990. Effects of suburbanization upon snowmelt runoff. Hydrological Sciences Journal 35: 285–302.

Buttle JM. 1994. Isotope hydrograph separations and rapid delivery of pre-event water from drainage basins. Progress in Physical Geography 18: 16–41.

Buttle, JM, McDonald DJ. 2002. Coupled vertical and lateral preferential flow on a forested slope. Water Resources Research 38: 10.1029/2001WR 000773.

Buttle JM, Peters DL. 1997. Inferring hydrological processes in a temperate basin using isotopic and geochemical hydrograph separation: a re-evaluation. Hydrological Processes 11: 557–573.

Buttle JM, Sami K. 1992. Testing the groundwater ridging hypothesis of streamflow generation during snowmelt. Journal of Hydrology 135: 53–72.

Buttle JM, Turcotte DS. 1999. Runoff processes on a forested slope on the Canadian Shield. Nordic Hydrology 30: 1–20.

Buttle JM, Taylor CH, Vonk AM. 1992. Environmental isotope hydrograph separation during snowmelt in a suburban catchment. Proceedings of the Eastern Snow Conference 49: 1–12.

Buttle JM, Vonk AM, Taylor CH. 1995. Applicability of isotope hydrograph separation in a suburban basin during snowmelt. Hydrological Processes 9: 197–211.

Buttle JM, Lister SW, Hill AR. 2001. Controls on runoff components on a forested slope and implications for N transport. Hydrological Processes 15: 1065–1070.

Buttle JM, Hazlett PW, Murray CD, Creed IF, Jeffries DS, Semkin R. 2001. Prediction of groundwater characteristics in forested and harvested basins during spring snowmelt using a topographic index. Hydrological Processes 15: 3389–3407.

Chaplot V, Walter C, Curmi P. 2000. Improving soil hydromorphy prediction according to DEM resolution and available pedological data. Geoderma 97: 405–422.

Clark ID, Fritz P. 1997. Environmental Isotopes in Hydrogeology. CRC Press, Boca Raton, FL, 328 p.

Collins R, Jenkins A, Harrow M. 2000. The contribution of old and new water to a storm hydrograph determined by tracer addition to a whole catchment. Hydrological Processes 14: 701–711.

Cooper LW, Olsen CR, Solomon DK, Larsen IL, Cook RB, Grebmeier JM. 1991. Stable isotopes of oxygen and natural and fallout radionuclides used for tracing runoff during snowmelt in an Arctic watershed. Water Resources Research 27: 2171–2179.

Crayosky TW, DeWalle DR, Seybert TA, Johnson TE. 1999. Channel precipitation dynamics in a forested Pennsylvania headwater catchment (USA). Hydrological Processes 13: 1303–1314.

Dansgaard W. 1964. Stable isotopes in precipitation. Tellus 16: 436–468.

DeWalle DR, Pionke HB. 1994. Streamflow generation on a small agricultural catchment during autumn recharge: II. Stormflow periods. Journal of Hydrology 163: 23–42.

DeWalle DR, Swistock BR. 1994. Differences in oxygen-18 content of throughfall and rainfall in hardwood and coniferous forests. Hydrological Processes 8: 75–82.

DeWalle DR, Swistock BR, Sharpe WE. 1988. Three component tracer model for stormflow on a small Appalachian forested catchment. Journal of Hydrology 104: 301–310.

DeWalle DR, Edwards PJ, Swistock BR, Aravena R, Drimmie RJ. 1997. Seasonal isotope hydrology of three Appalachian forest catchments. Hydrological Processes 11: 1895–1906.

Drever JI. 1988. The Geochemistry of Natural Waters. Prentice-Hall, Englewood Cliffs, NJ, 437 p.

Elsenbeer H. 2001. Hydrologic flowpaths in tropical rainforest soilscapes – a review. Hydrological Processes 15: 1751–1759.

Elsenbeer H, Lack A. 1996. Hydrometric and hydrochemical evidence for fast flowpaths at LaCuenca, Western Amazonia. Journal of Hydrology 180: 237–250.

Elsenbeer H, Lorieri D, Bonell M. 1995. Mixing model approaches to estimate stormflow in an overland flow dominated rainforest catchment. Water Resources Research 31: 2267–2278.

Freer J, McDonnell J, Beven KJ, Brammer D, Burns, D, Hooper RP, Kendall C. 1997. Topographic controls on subsurface storm flow at the hillslope scale for two hydrologically distinct small catchments. Hydrological Processes 11: 1347–1352.

Freer J, McDonnell, JJ, Beven KJ, Peters NE, Burns DA, Hooper RP, Aulenbach B, Kendall C. 2002. The role of bedrock topography on subsurface storm flow. *Water Resour. Res.*, 38(12), 1269. doi:10.1029/2001 WR000872, 2002.

Genereux DP. 1998. Quantifying uncertainty in tracer-based hydrograph separations. Water Resources Research 34: 915–919.

Genereux DP, Hooper RP. 1998. Oxygen and hydrogen isotopes in rainfall-runoff studies. In Isotope Tracers in Catchment Hydrology, Kendall C, McDonnell JJ (eds). Elsevier, Amsterdam, 319–346.

Gremillion P, Wanielista M. 2000. Effects of evaporative enrichment on the stable isotope hydrology of a central Florida (USA) river. Hydrological Processes 14: 1465–1484.

Gremillion P, Gonyeau A, Wanielista M. 2000. Application of alternative hydrograph separation models to detect changes in flow paths in a watershed undergoing urban development. Hydrological Processes 14: 1485–1501.

Halldin S, Rodhe A, Bjurman B. 1990. Urban storm water transport and wash-off of caesium-137 after the Chernobyl accident. Water, Air, and Soil Pollution 49: 139–158.

Hewlett JD, Hibbert AR. 1967. Factors affecting the response of small watersheds to precipitation in humid areas. In International Symposium on Forest Hydrology, Sopper WE, Lull HW (eds). Pergamon, New York, NY, 275–290.

Hill AR. 1993. Base cation chemistry of storm runoff in a forested headwater wetland. Water Resources Research 29: 2663–2673.

Hill AR, Waddington JM. 1993. Analysis of storm run-off sources using oxygen-18 in a headwater swamp. Hydrological Processes 7: 305–316.

Hinton MJ, Schiff, SL, English MC. 1994. Examining the contributions of glacial till water to storm runoff using two- and three-component hydrograph separations. Water Resources Research 30: 983–993.

Hooper RP. 2001. Applying the scientific method to small catchment studies: a review of the Panola Mountain experience. Hydrological Processes 15: 2039–2050.

Hooper RP, Shoemaker CA. 1986. A comparison of chemical and isotopic hydrograph separation. Water Resources Research 22: 1444–1454.

Ingraham NL. 1998. Isotopic variations in precipitation. In Isotope Tracers in Catchment Hydrology, Kendall C, McDonnell JJ (eds). Elsevier, Amsterdam, 87–118.

Ingraham NL, Taylor BE. 1991. Light stable isotope systematics of large-scale hydrologic regimes in California and Nevada. Water Resources Research 27: 77–90.

International Atomic Energy Agency. 2001. GNIP maps and animations. International Atomic Energy Agency, Vienna, Austria. Accessible at http://isohis.iaea.org

International Atomic Energy Agency/World Meteorological Organization. 2001. Global Network of Isotopes in Precipitation. The GNIP Database. Accessible at http://isohis.iaea.org

Joerin C, Beven KJ, Iorgulescu I., Musy A. 2002. Uncertainty in hydrograph separations based on geochemical mixing models. Journal of Hydrology 255: 90–106.

Kendall C, Caldwell EA. 1998. Fundamentals of isotope geochemistry. In Isotope Tracers in Catchment Hydrology, Kendall C, McDonnell JJ (eds). Elsevier, Amsterdam, 51–86.

Kendall KA, Shanley JB, McDonnell JJ. 1999. A hydrometric and geochemical approach to test the transmissivity feedback hypothesis during snowmelt. Journal of Hydrology 219: 188–215.

Kendall C, McDonnell JJ, Gu W. 2001. A look inside 'black box' hydrograph separation models: a study at the Hydrohill catchment. Hydrological Processes 15: 1877–1902.

Kennedy VC, Kendall C, Zellweger GW, Wyerman TA, Avanzino RJ. 1986. Determination of the components of stormflow using water chemistry and environmental isotopes, Mattole River basin, California. Journal of Hydrology 84: 107–140.

Lange H, Lisheid G, Hoch R, Hauhs M. 1996. Water flow paths and residence times in a small headwater catchment at Gårdsjön, Sweden, during steady state storm flow conditions. Water Resources Research 32: 1689–1698.

Laudon H, Hemond HF, Krouse R, Bishop KH. 2002. Oxygen 18 fractionation during snowmelt: Implications for spring flood hydrograph separation. Water Resources Research. 38(11), 1258, doi:10.1029/2002 WR001510.

Leaney FW, Smettem KRJ, Chittleborough DJ. 1993. Estimating the contribution of preferential flow to subsurface runoff from a hillslope using deuterium and chloride. Journal of Hydrology 147: 83–103.

Luce, C. H. (2002). Hydrological processes and pathways affected by forest roads: what do we still need to learn? Hydrological Processes, 16: 2901–2904.

McDonnell JJ. 1990. A rationale for old water discharge through macropores in a steep, humid catchment. Water Resources Research 26: 2821–2832.

McDonnell JJ, Bonell M, Stewart MK, Pearce AJ. 1990. Deuterium variations in storm rainfall: implications for hydrograph separation. Water Resources Research 26: 455–458.

McDonnell JJ, Stewart MK, Owens IF. 1991a. Effect of catchment-scale mixing on stream isotopic response. Water Resources Research 27: 3065–3073.

McDonnell JJ, Owens IF, Stewart MK. 1991b. A case study of shallow flow paths in a steep zero-order basin: A physical-chemical-isotopic analysis. Water Resources Bulletin, 27: 679–685.

McDonnell JJ, Rowe L, Stewart M. 1999. A combined tracer-hydrometric approach to assessing the effects of catchment scale on water flowpaths, source and age. International Association of Hydrological Sciences, Publication 258: 265–274.

McGlynn B, McDonnell JJ. 2003. The role of discrete landscape units in controlling catchment dissolved organic carbon dynamics. Water Resources Research. 39(4), 1090, doi:10.1029/2002 WR0001525, 2003.

McGlynn B, Seibert J. 2003. Distributed assessment of contributing area and riparian buffering along stream networks. Water Resources Research. 39(4), 1082, doi:1029/2002 WR0001521, 2003.

McGlynn BL, McDonnell JJ, Shanley JB, Kendall C. 1999. Riparian zone flowpath dynamics during snowmelt in a small headwater catchment. Journal of Hydrology 222: 75–92.

McGlynn B, McDonnell JJ, Brammer D. 2002. An evolving perceptual model of hillslope flow in a steep forested humid catchment: A review of the Maimai catchment. Journal of Hydrology 257: 1–26.

McGlynn B, McDonnell JJ, Stewart M, Seibert J. 2003. On the relationships between catchment scale and streamwater mean residence time. Hydrological Processes. 17(1): 175–182.

Metcalfe RA, Buttle JM. 2001. Soil partitioning and surface store controls on spring runoff from a boreal forest peatland basin in north-central Manitoba, Canada. Hydrological Processes 15: 2305–2324.

Montgomery DR, Dietrich WE, Torres R, Anderson SP, Heffner JT, Loague K. 1997. Hydrologic response of a steep, unchanneled valley to natural and applied rainfall. Water Resources Research 33: 91–109.

Moore RD, Thompson JC. 1996. Are water table variations in a shallow forest soil consistent with the TOPMODEL concept? Water Research Research 32: 663–669.

Mosley MP. 1979. Streamflow generation in a forested watershed, New Zealand. Water Resources Research 15: 795–806.

Mosley MP. 1982. Subsurface flow velocities through selected forest soils, South Island, New Zealand. Journal of Hydrology 55: 65–92.

Murray CD. 2003. Snow accumulation, melt and infiltration on forested and clearcut slopes during spring snowmelt, Turkey Lakes Watershed, central Ontario. MSc thesis, Trent University, 76 pp.

Nolan KM, Hill BR. 1990. Storm-runoff generation in the Permanente Creek drainage basin, west central California – an example of flood-wave effects on runoff composition. Journal of Hydrology 113: 343–367.

Nyberg L. 1995. Water flow path interactions with soil hydraulic properties in till soil at Gårdsjön, Sweden. Journal of Hydrology 170: 255–275.

Ogunkoya OO, Jenkins A. 1993. Analysis of storm hydrograph and flow pathways using a three-component hydrograph separation model. Journal of Hydrology 142: 71–88.

Pearce AJ. 1990. Streamflow generation processes: an Austral view. Water Resources Research 26: 3037–3047.

Pearce AJ, Stewart MK, Sklash MG. 1986. Storm runoff generation in humid headwater catchments. 1. Where does the water come from? Water Resources Research 22: 1263–1272.

Peters DL, Buttle JM, Taylor CH, LaZerte BD. 1995. Runoff production in a forested, shallow soil, Canadian Shield basin. Water Resources Research 31: 1291–1304.

Rice K, Hornberger G. 1998. Comparison of hydrochemical tracers to estimate source contributions to peak flow in a small, forested headwater catchment. Water Resources Research 34: 1755–1766.

Richey DG, McDonnell JJ, Erbe M, Hurd T. 1998. A critical appraisal of published chemical and isotopic hydrograph separations from New Zealand, North America and Europe. Journal of Hydrology (New Zealand) 37: 95–111.

Robson A, Beven K, Neal C. 1992. Towards identifying sources of subsurface flow: a comparison of components identified by a physically based runoff model and those determined by chemical mixing techniques. Hydrological Processes 6: 199–214.

Rodhe A. 1998. Snowmelt-dominated systems. In Isotope Tracers in Catchment Hydrology, Kendall C, McDonnell JJ (eds)., Elsevier, Amsterdam, 391–433.

Rodhe A. 1987. The origin of streamwater traced by oxygen-18. PhD. Thesis, Department of Physical Geography, Division of Hydrology, Report Series A 41, Uppsala University, 260 p + appendices.

Rodhe A, Nyberg L, Bishop K. 1996. Transit times for water in a small till catchment from a step shift in the oxygen 18 content of the water input. Water Resources Research 32: 3497–3511.

Saxena RK. 1986. Estimation of canopy reservoir capacity and oxygen-18 fractionation in throughfall in a pine forest. Nordic Hydrology 17: 251–260.

Seibert J. and McDonnell JJ. 2002. On the dialog between experimentalist and modeller in catchment hydrology: Use of soft data for multi-criteria model calibration. Water Resources Research. 38(11): 1241, doi:10.1029/2001 WR000978, 2002.

Seibert J, Bishop KH, Nyberg L. 1997. A test of TOPMODEL's ability to predict spatially distributed groundwater levels. Hydrological Processes 11: 1131–1144.

Shanley J, Kendall C, Smith T, Wolock D, McDonnell JJ. 2002. Controls on old and new water contributions to stream flow at some nested catchments in Vermont, USA. Hydrological Processes 16: 589–610.

Sklash MG. 1990. Environmental isotope studies of storm and snowmelt runoff generation. In Process Studies in Hillslope Hydrology, Anderson MG, Burt TP (eds). Wiley, Chichester, UK, 401–435.

Sklash MG, Farvolden RN. 1979. The role of groundwater in storm runoff. Journal of Hydrology 43: 45–65.

Sklash MG, Farvolden RN, Fritz P. 1976. A conceptual model of watershed response to rainfall, developed through the use of oxygen-18 as a natural tracer. Canadian Journal of Earth Sciences 13:271–283.

Sklash MG, Stewart MK, Pearce AJ. 1986. Storm runoff generation in humid headwater catchments, 2. A case study of hillslope and low-order stream response. Water Resources Research 22: 1273–1282.

Shuttleworth J. 2002. Some thoughts on remote sensing versus extrapolation. Hydrological Processes 16: 2037–2040.

Srinivasan MS, Gburek WJ, Hamlett JM. 2002. Dynamics of stormflow generation on a hillslope – a hillslope-scale field study in east-central Pennsylvania, USA. Hydrological Processes 16: 649–665.

Torres R, Dietrich WE, Montgomery DR, Anderson SP, Loague K. 1998. Unsaturated zone processes and the hydrologic response of a steep, unchanneled catchment. Water Resources Research 34: 1865–1879.

Turton DJ, Barnes DR, de Jesus Navar J. 1995. Old and new water in subsurface flow from a forest soil block. Journal of Environmental Quality 24: 139–146.

Uhlenbrook S. 1999. Examination and modeling of runoff generation in a mesoscaled basin. PhD thesis, Institute of Hydrology, University of Freiburg.

Uhlenbrook S, Leibundgut C. 2002. Process-oriented catchment modelling and multiple-response validation. Hydrological Processes 16: 423–440.

Uhlenbrook S, McDonnell JJ, Leibundgut C. 2002. Runoff generation and implications for river basin modelling. *Hydrological Processes* 17(2) (Special Issue): 197–493.

Unnikrishna PV, McDonnell JJ, and Kendall C. (2002). Isotope variations in a Sierra Nevada snowpack and their relation to meltwater. Journal of Hydrology 260(1–4): 38–57.

Unnikrishna PV, McDonnell JJ, Tarboton D, Kendall C. In press. Snowmelt runoff response of a small semi-arid catchment. Journal of Hydrology.

Welker JM. 2000. Isotopic ($\delta^{18}O$) characteristics of weekly precipitation collected across the USA: an initial analysis with application to water source studies. Hydrological Processes 14: 1449–1464.

Wels C, Cornett RJ, LaZerte BD. 1991. Hydrograph separation: a comparison of geochemical and isotopic tracers. Journal of Hydrology 122: 253–274.

Welsch DL, Kroll CN, McDonnell JJ, Burns DA. 2001. Topographic controls on the chemistry of subsurface stormflow. Hydrological Processes 15: 1925–1938.

Williard KWJ, DeWalle DR, Edwards PJ, Sharpe WE. 2001. ^{18}O isotopic separation of stream nitrate sources in mid-Appalachian forested watersheds. Journal of Hydrology 252: 174–188.

Wilson GV, Jardine PM, Luxmoore RJ, Jones JR. 1990. Hydrology of a forested hillslope during storm events. Geoderma 46: 119–138.

Wolman MG. 1967. A cycle of sedimentation and erosion in urban river channels. Geografiska Annaler 49A: 385–395.

Ziegler AD, Giambelluca TW. 1997. Importance of rural roads as source areas for runoff in mountainous areas of northern Thailand. Journal of Hydrology 196: 204–229.

Ziegler A, Giambelluca T, Sutherland RA, Vana T, Nullet M. 2001. Horton overland flow contribution to runoff on unpaved mountain roads: A case study in Northern Thailand. Hydrological Processes 15: 3203–3208.

Zollweg JA. 1996. Field study to support hydrologic modeling and analysis of watershed function at the microscale. In Proceedings of the AWRA Annual Symposium, Watershed Restoration Management: Physical, Chemical, and Biological Considerations, McDonnell JJ, Stribling JB, Neville LR, Leopold DJ (eds). American Water Resources Association, Herndon, VA, 129–134.

33 Process-based erosion modelling: promises and progress

B. Yu
Griffith University, Nathan, Australia

INTRODUCTION

Accelerated soil and nutrient losses and their off-site impacts are of great concern for sustainable agriculture and, more generally, for sustainable land management. Prediction of the rate of soil erosion over a range of temporal and spatial scales is important for land use planning, erosion risk assessment, and for evaluating the effects of land use change (Penning de Vries *et al.*, 1998). With urbanisation and population growth in traditionally rural areas, farming on steeplands has continued to increase in recent years, especially in the developing regions of the world (Oldeman *et al.*, 1991; Fisher and Heilig, 1997). In plantation forestry, logging and site preparation activities on steep slopes also expose forested areas temporarily to high risk of soil erosion and nutrient losses (cf. the chapter by Grip *et al.*, this volume).

The most widely used soil erosion prediction technology is the Universal Soil Loss Equation (USLE) (Wischmeier and Smith, 1978) and its successor the Revised USLE (Renard *et al.*, 1997). In recent years, however, a new generation of physically based models such as WEPP (Nearing *et al.*, 1989; Flanagan and Nearing, 1995; Laflen *et al.*, 1997), LISEM (De Roo *et al.*, 1996a), EUROSEM (Morgan *et al.*, 1998) and GUEST (Misra and Rose, 1996; Rose *et al.*, 1997) has been developed to describe and quantify soil erosion processes. These models are particularly suitable for adaptation across a range of scales in the landscape because physical principles and physically meaningful parameters are involved. In contrast to conceptual runoff and soil erosion models, there are at least three strong arguments in support of approaches based on physical principles and our understanding of the underlying runoff and sediment generation processes. First, physically-based models are more likely to succeed outside the environment in which they were developed, hence more likely to achieve broad applicability. Secondly, model parameters, having clearer physical meaning, may be estimated from measurable variables such as soil hydraulic properties, plant or litter biomass, and various land-use attributes and management practices. Finally, to calibrate conceptual models, long-term experimentation is often required. For physically-based models, the period of experimentation, and hence the cost, can be reduced considerably. For example, rainfall simulators can be used effectively to measure runoff, and sediment and nutrient loads during brief intensive campaigns in order to derive the values of key model parameters that may then be used for longer term predictions.

There have been a number of reviews of erosion modelling. Rose (1985) reviewed developments in soil erosion and deposition models with a particular emphasis on differences in the conceptualisation of water-induced erosion and deposition processes. Bingner (1990) discussed the methods used by such semi-empirical erosion models as CREAMS, SWRRB, EPIC, ANSWERS, and AGNPS. Lane *et al.* (1992) reviewed the erosion prediction technology as developed within the US Department of Agriculture with particular emphasis on the factor-based USLE/RUSLE, the hybrid approach of the CREAMS model and the process-based WEPP model. Rose (1993) gave a detailed description of the theory underlying GUEST at the time and called attention for the need to better address the issue of erosion prediction at the catchment scale. Rose (1997) further reviewed progress in water and wind erosion modelling. Finally, Coughlan and Rose (1997), and Rose and Yu (1998) summarised research findings of a multiple-country project on soil erosion and sustainable production in tropical steeplands with an emphasis on South East Asia and northeastern Australia.

The objectives of this chapter are to define and highlight new developments in erosion modelling techniques that could be considered or used by resource managers to assess erosion risks and evaluate alternative conservation technologies.

AN OVERVIEW OF EROSION MODELLING

It is useful to classify erosion models according to the conceptual framework in which the model was developed. The first group

Forests, Water and People in the Humid Tropics, ed. M. Bonell and L. A. Bruijnzeel. Published by Cambridge University Press.

Table 33.1. *Erosion rates reported at selected scales for Australia, South East Asia and China*

Country	Scale	Area (km^2)	Resolution (m^2)	Model	Erosion rate [t/(ha.yr)]	References[a]
Australia	National	7.6×10^8	275×275 (9″)	RUSLE	6.3	Lu *et al.*, 2001
China	Regional	2.54×10^6	–	–	21	Wen, 1993
Indonesia	Catchment	1.35×10^3	?	USLE	41	Agus *et al.*, 1997
Laos	Catchment	0.67	–	?	95?	Phanthasith and Chanthavongsa
Malaysia	Regional	8.2×10^3	30×30	USLE	13	Lok and Li
Myanmar	Catchment	10^3–10^4?	?	Fournier	?	Htut and Myint
Philippines	Catchment	2.81×10^2	100×100?	'Modified' USLE	13	Collado and Baloloy
Thailand	National	5.13×10^5	250×250	USLE/RUSLE	12.5	Funnpheng and Tawanron
Vietnam	National	3.31×10^5	1000×1000?	USLE/nonstandard application	~10	Yu and Lang
The World	–	1.5×10^{10}	?	?	6.3	Myers, 1993

[a] Undated authors are participants at the Regional Workshop on Erosion Risk Assessment, Kuala Lumpur, October 2001.

of models includes those of an essentially empirical nature. They rely on a comprehensive database to determine the model structure and relevant parameter values. The USLE is a prime example. The second group begins to describe the erosion processes, but still relies on some of the USLE factors to facilitate application in the field. Models in this group are best regarded as hybrid models, i.e. those in a transitional phase leading to full-blown process-oriented erosion models. The third and last group represents a new generation of erosion models that breaks away from the factor-based USLE approach. Therefore, physically based erosion models can be simply defined as those that do not contain nor rely on USLE factors. These process-oriented erosion models including GUEST, WEPP and EUROSEM/LISEM are the main focus of this chapter.

It is quite fitting and tempting to begin a review of erosion modelling with the USLE (Wischmeier and Smith, 1965; 1978). Although the USLE may not be the first soil erosion prediction model ever developed (see Renard *et al.*, 1997, for a brief history of erosion prediction equations), it is certainly the best-known and most widely used model, both in the US and elsewhere. In the USLE, the following equation is used to predict mean annual soil loss:

$$A = RKLSCP \tag{33.1}$$

where A is the predicted mean annual net soil loss per unit area, R is the rainfall and runoff erosivity factor, K the erodibility factor, L and S are slope length and slope steepness factors, C is the cover and management factor, and P is the support practice factor. Each of these USLE factors was meant to represent and quantify important processes in relation to soil erosion. Because of this particular framework, much of the work to apply and evaluate the USLE was to evaluate these individual factors in different biophysical environments and for different management practices.

The USLE has been used widely in the tropics. The work in West Africa (Roose, 1977) and in Hawaii (El-Swaify and Dangler, 1977) has been widely cited. Lal (1990) provides a good summary of USLE-based erosion research in the tropics. Almost all erosion assessments at regional scales were carried out in the USLE framework. Table 33.1 is a recent compilation of estimates of average rates of soil erosion, mostly at the regional scale and in southeast Asian countries. The table shows that the USLE/RUSLE has been used widely in this region for large-scale erosion assessment. The table also shows, interestingly, that at the large scale a typical rate of soil loss is of the order 10–20 t ha^{-1} yr^{-1} for all case studies compiled.

Given that the soil loss is expressed as a product of several independent factors, the USLE is particularly amenable to being used with a geographical information system (GIS) at the catchment and regional scale. The hybrid AGNPS model (Young *et al.*, 1989) is a prime example, while de Roo (1998) and Dickinson and Collins (1998) also presented a number of case studies of this nature. More recently, the RUSLE has been applied with GIS to the entire Australian continent to assess the magnitude and seasonal distribution of rill and sheet erosion (Lu *et al.*, 2001).

This particular factor-based framework has for many years had a profound influence on agricultural research sites around the world where experimental work at the plot and hillslope scale is aimed at demonstrating the effect of certain conservation technologies. Typically, a number of (mostly bounded) runoff plots with similar soil, slope and slope length are established at a site.

Runoff, soil and nutrient losses, and sometimes crop yields, from these plots are then contrasted to quantify the effectiveness of various treatments imposed at the site (Mutchler *et al.*, 1994). USLE lends considerable support for this type of 'one-factor analysis' of conservation technology in comparison to farmer's conventional practices.

Explicit attempts have been made to modify the USLE for small catchments by taking streamflow variables into account (Williams, 1975; Onstad and Foster, 1975; Kinnell, 1997, 1998). The factors considered in the USLE, especially K and C, have found their way into quite a few erosion models in which some of the erosion processes are considered. Notable in this group are CREAMS/GLEAMS (Knisel, 1980; 1991), EPIC (Williams *et al.*, 1983), ANSWERS (Beasley *et al.*, 1980), ANSWERS-2000 (Bouraoui and Dillaha, 1996), and CASC2D-SED (Johnson *et al.*, 2000).

As distinct from the factor-based USLE or similar approaches, mean annual soil loss in physically-based erosion models is evaluated as the sum of the soil losses from individual runoff events, M_i (kg). The event soil loss is the product of sediment concentration, $c(t)$ (kg m^{-3}) and runoff rate, $Q(t)$ (m^3 s^{-1}) integrated over its duration of T_i (s). Thus:

$$A = \frac{1}{N}\sum M_i \tag{33.2}$$

$$M_i = \int_0^{T_i} c(t)Q(t)dt \tag{33.3}$$

where N is the number of years to which the mean annual value refers. While Eqns 33.1 and 33.2 are intended to predict the same long-term average soil loss, the conceptual framework for Eqns 33.2 and 33.3 is fundamentally different from that for the USLE. Equation 33.3 shows that prediction of the runoff rate has become critical not only because the runoff rate, $Q(t)$, is required explicitly, but because the sediment concentration, $c(t)$, also depends on $Q(t)$ in addition to other variables, such as rainfall intensity, slope and cover. Arguably one of the most distinguishing features between the USLE and the more recently developed process-oriented models is that the USLE assumes or asserts that rainfall characteristics capture all of the climatic and most of the hydrological influences on soil erosion. The rainfall-runoff erosivity factor was intended to quantify the effect of raindrop impact and must also reflect the amount and rate of runoff likely to be associated with the rain (Renard *et al.*, 1997). In contrast, prediction of runoff amount and runoff rate for individual storm events has become crucial and indispensable to the latest physically-based prediction methods. Runoff generation processes and their prediction are highly relevant to erosion modelling (see also the chapter on runoff generation by Bonell, this volume). Figure 33.1 presents a conceptual framework in which the role of runoff is highlighted

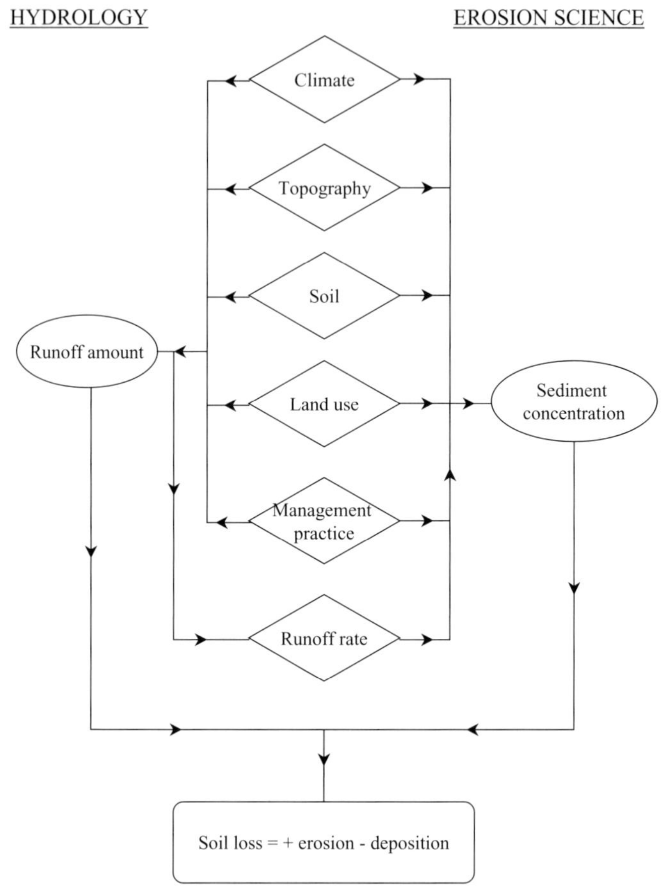

Figure 33.1 A conceptual framework for soil erosion prediction. The role of runoff in erosion prediction is highlighted.

in erosion modelling. It is no exaggeration to say that prediction of water-induced soil erosion is mostly about runoff generation and, to a lesser extent, erosion science. Once this framework for the process-based erosion prediction is established, the description and comparison of a variety of process-based erosion models become much easier because the models differ only in the way in which runoff rate and sediment concentration are predicted.

Other landscape erosion models have also been developed on the basis of physical principles. For instance, TOPOG was used to identify erosional hazards in the landscape under steady–state hydrologic conditions (Vertessy *et al.*, 1990; Dietrich *et al.*, 1992; Constantini *et al.*, 1993), while SIBERIA was used to evaluate long-term landscape evolution (Willgoose *et al.*, 1991; Hancock *et al.*, 2002). These models are mostly concerned with the topographical control on runoff generation, and the potential for erosion and sediment transport. In this chapter, these landscape erosion models were excluded deliberately because they do not explicitly address the impacts of land use change (especially the change in management practices) on the rate of erosion and deposition across the landscape.

In the following sections, the theory behind, and application of, four different physically-based soil erosion models are discussed, namely GUEST, WEPP, EUROSEM, and LISEM.

GUEST

The GUEST erosion model (Griffith University Erosion System Template) was developed originally as a tool to investigate soil erosion processes with measured runoff rate, and event-based average sediment concentration (Rose *et al.*, 1983a; 1983b). The theory behind GUEST has evolved over a period of years and the theory when overland flow is the dominant cause of erosion was described by Hairsine and Rose (1992a and 1992b). There have since been some minor changes, including the introduction of a soil erodibility parameter β as a surrogate variable for the original soil erodibility parameter J (Rose, 1993). More recently, the concept of saltation stress (Bagnold, 1977), which can become important when the sediment concentration is high, and the effect of sediment concentration on stream power have been taken into consideration. The theory and parameter sensitivity for both rainfall- and runoff driven processes were presented in Misra and Rose (1996), although the latest version of GUEST contains only the module on runoff-driven processes, as these are considered to be dominant on steeplands (Fentie *et al.*, 1997; Rose *et al.*, 1997; Yu and Rose, 1997). Below, the theory underlying GUEST is outlined to show how sediment concentration at the transport limit can be calculated for a given runoff rate, soil type and topography. Although the focus here is on the methodological aspects of GUEST and how the theory can be implemented, it is useful to consider the development of soil erosion and deposition theory over the years in general because GUEST is best regarded as a particular application of that theory.

Theoretical framework for GUEST

The theory on which GUEST is based begins with a governing equation for soil erosion, transport and deposition:

$$\frac{\partial(c_i D)}{\partial t} + \frac{\partial(c_i q)}{\partial x} = e_i + e_{ri} + r_i + r_{ri} - d_i \qquad (33.4)$$

where D is water depth (m), q is unit discharge, i.e. flow rate per unit flow width ($m^2\ s^{-1}$), c_i is the sediment concentration for particle size class i (kg m^{-3}), e_i and e_{ri} are rates of rainfall detachment and re-detachment (kg $m^{-2}\ s^{-1}$), r_i and r_{ri} are rates of flow entrainment and re-entrainment (kg $m^{-2}\ s^{-1}$), and d_i is the rate of deposition (kg $m^{-2}\ s^{-1}$). The detachment and entrainment terms are related to the process of dislodging primary particles and aggregates from the original soil. The dislocated particles and aggregates are continuously being returned to the soil surface under gravity. The processes when these loose, deposited materials are detached once again by the rain or entrained by the flow are called re-detachment and re-entrainment, respectively. Equation 33.4 is based on a mass balance for individual particle size classes. It is necessary to separate sediment into different classes according to their particle size because the associated settling velocity, which characterises the rate of deposition, is closely related to particle size. While there are obvious interactions between rainfall-driven and runoff-driven erosion processes (Moss, 1988; Rose, 1993), in practice relevant theories were developed and tested separately for various scenarios in which either rainfall or runoff dominates the erosion process. For example, Hairsine and Rose (1991) and Proffitt *et al.* (1991) considered rainfall detachment and deposition in the absence of flow-driven processes, whereas Hairsine and Rose (1992a; 1992b) and Proffitt *et al.* (1993) developed the theory for soil erosion by overland flow in the absence of rainfall. These studies provided the equilibrium solutions whereby the left hand side of Eqn 33.4 was assumed to vanish. Sander *et al.* (1996) considered an unsteady situation with sediment concentration varying in time, but assumed that sediment concentration does not vary in space (i.e. $\partial c_i/\partial x = 0$). More recently, Eqn 33.4 was solved under steady state conditions (i.e. sediment concentration does not vary with time) to model sediment deposition as a result of an abrupt reduction in slope gradient (Beuselinck *et al.*, 2002a; 2002b; Hairsine *et al.*, 2002; Sander *et al.*, 2002). Sediment deposition has also been modelled resulting from a reduction in velocity due to an abrupt increase in flow resistance (Rose *et al.*, 2002). Table 33.2 summarises these theoretical developments in a chronological order. It can be seen from Table 33.2 that Eqn 33.4 has provided a rich framework to stimulate active research in this area, and allowed examination of various combinations of processes and simplified scenarios.

GUEST methodology

For practical application of the GUEST theory to tropical steeplands, it is assumed that runoff dominates the erosion process and therefore the first two terms on the right-hand side of Eqn 33.4 (i.e. detachment and re-detachment by rainfall) are ignored. Furthermore, under equilibrium conditions when sediment concentration does not vary in time and space, the flow entrainment term (r_i) also vanishes because there is no net erosion once equilibrium is reached. This is equivalent to assuming the re-entrainment term (r_{ri}) equal to the rate of deposition. Equation 33.4 thus shows that the settling term due to gravity (d_i) is balanced by the re-entrainment term (r_{ri}) under these conditions. It is further assumed that at equilibrium, a certain fraction of the stream power, F, is involved in the re-entrainment of sediment, thus maintaining the sediment in suspension. Stream power is defined here as energy expenditure per unit area and has the unit of W m^{-2}. Without the

Table 33.2. *Historical development of the erosion and deposition theory from which GUEST was originated*

Scenarios	Comments	References
Left Hand Side (LHS) = 0 $r_i = r_{ri} = 0$	Equilibrium condition Rainfall-driven	Hairsine and Rose (1991) and Proffitt *et al.* (1991)
LHS = 0 $e_i = e_{ri} = 0$	Equilibrium condition Flow-driven in the absence of rain	Hairsine and Rose (1992a and 1992b) and Proffitt *et al.* (1993)
$\partial (cD)/\partial x = 0$ $r_i = r_{ri} = 0$	Unsteady, but uniform Rainfall-driven	Sander *et al.* (1996)
$r_i = r_{ri} = 0$	Rainfall-driven	Hairsine *et al.* (1999)
$\partial (cD)/\partial t = 0$ $e_i = e_{ri} = 0$ slope varying q constant	Steady and non-uniform Flow-driven in the absence of rain Abrupt change in slope gradient	Beuselinck *et al.* (2002b), Hairsine *et al.* (2002), and Sander *et al.* (2002)
$\partial (cD)/\partial t = 0$ $e_i = r_i = 0$ q constant	Steady and non-uniform Combined effects of rain and flow on net deposition	Beuselinck *et al.* (2002a)
$\partial (cD)/\partial t = 0$ $e_i = e_{ri} = 0$ slope constant velocity varying	Steady and non-uniform Flow-driven in the absence of rain back water effects	Rose *et al.* (2002)
$r_i = r_{ri} = 0$	Unsteady, non-uniform Numerical solution rainfall-driven	Hogarth *et al.* (2002)

stream power of overland flow above some threshold value, Ω_0, all sediment would eventually have settled out of the overland flow. In short, at equilibrium, the capacity to maintain sediment in suspension is balanced by the downward flux of sediments due to gravity. Thus, it may be shown that:

$$\frac{\sigma - \rho}{\sigma} c_t g D\phi = F(\Omega - \Omega_0)b \qquad (33.5)$$

where c_t is the sediment concentration under equilibrium condition (kg m^{-3}), or that at the transport limit (Hairsine and Rose 1992a; 1992b); ϕ is the average settling velocity of all classes when the sediment is divided into size classes of equal mass (Lisle *et al.*, 1996) and is also known as depositability (m s^{-1}), σ and ρ are sediment and water density (kg m^{-3}), respectively, g the acceleration due to gravity (m s^{-2}). Ω and Ω_0 are stream power and threshold stream power, respectively. In Eqn 33.5, b is a shape factor depending on rill geometry (Yu and Rose, 1999) (Table 33.3). Stream power in Eqn 33.5 is given by:

$$\Omega = \rho g R_h V S \qquad (33.6)$$

where R_h is the hydraulic radius (m) and V mean flow velocity (m s^{-1}). The term $F(\Omega - \Omega_0)$ can be interpreted as the fraction of the excess stream power that is effective in sediment transport. Eqn 33.5, as given in Hairsine and Rose (1992a), was used to determine the sediment concentration in overland flow at the transport limit up until the mid 1990s at various steepland sites in South East Asia and Australia (Soil Technology, 1995). Complications arise when low runoff occurs in combination with steep slope. Use of Eqn 33.5 would then lead to a physically unrealistic (high) sediment concentration at the transport limit. At small water depth, the D term in Eqn 33.5 can be of the same order of magnitude as the diameter of larger soil particles (aggregates). Thus, when runoff rate, and hence water depth, is low, it is reasonable to expect only the finer soil particles to be fully immersed in the flow and actively involved in the erosion and deposition processes. The effective settling velocity, or the effective depositability (ϕ_e) and the fraction of the eroded soil fully immersed in the flow (C) would be low when the water depth is low. When the effects of a variable water depth and sediment concentration are taken into account, the sediment concentration

Table 33.3. *Shape factor b for GUEST*

Rill geometry	Rill shape	Shape factor b
Rectangular	$z = 0$, $W_t = W_b$	R_h/D
Trapezoidal rills with steep side slopes	$z < z_0$	R_h/D
Trapezoidal rills with gentle side slopes	$z < z_0$	R_h/D
Plane (no rills)	$z \to \infty$, $R_h = D$	1

Source: Yu and Rose (1999).

at the transport limit can be written (Rose *et al.*, 1997) as:

$$c_t = \frac{Fb\frac{\sigma}{\sigma-\rho}(1-C)(\rho g RSV - \Omega_0)}{gD\phi_e(1+K_1-K_2)} \quad (33.7)$$

where the dimensionless term K_1 is associated with the effect of saltation stress and is given by:

$$K_1 = \frac{FbV^2(1-C)}{gD} \quad (33.8)$$

and the other dimensionless term K_2 is associated with the consequences of fluid density enhancement by sediment (ρ_e), and is given by:

$$K_2 = \frac{FbR_hSV(1-C)}{D\phi_e}\frac{\sigma}{\sigma-\rho}\frac{\rho_e-\rho}{\rho} \quad (33.9)$$

Equation 33.7, with Eqns 33.8 and 33.9, while not hugely different from Eqn 33.5 under most circumstances, has replaced Eqn 33.5 as a more general expression for the sediment concentration at the transport limit since the mid-1990s at various sites in South East Asia (Coughlan and Rose, 1997; Rose and Yu, 1998; Yu and Rose 1999; Yu *et al.*, 1999).

The sediment concentration at the transport limit (Eqn 33.5 or 33.7) is achieved only when the supply of sediments into the flowing water is unlimited. This occurs when the soil surface is free from any contact cover of vegetation, residue or rock fragments, and when the erosion takes place over soils of minimum cohesion, such as freshly tilled soil or newly deposited sediments. The actual sediment concentration from areas without cover, c_b, can be related to the sediment concentration at the transport limit by a soil erodibility parameter, β (Rose, 1993):

$$c_b = c_t^{\beta} \quad (33.10)$$

The higher the erodibility, the higher the sediment concentration for unprotected areas. The theoretical limit is reached when $\beta = 1$. The erodibility parameter is broadly related to a more fundamental erodibility parameter, *J*, in terms of the amount of energy required to erode a unit mass of soil (Rose, 1993), and experimental evidence suggests that the soil erodibility parameter β depends on soil strength to some extent (Misra and Rose, 1995). Although GUEST allows calculation of the instantaneous sediment concentration if the erodibility parameter is known, measurements of the instantaneous sediment concentration within a runoff event are rarely taken. Usually, only the average sediment concentration on an event basis is available. Instantaneous runoff rates are more readily measured and these can be used to determine the average soil erodibility parameter for the event as a whole.

The effect of surface contact cover on soil loss is modelled in GUEST methodology by an exponential function of the form:

$$c = c_b\, e^{-\kappa C_g} \quad (33.11)$$

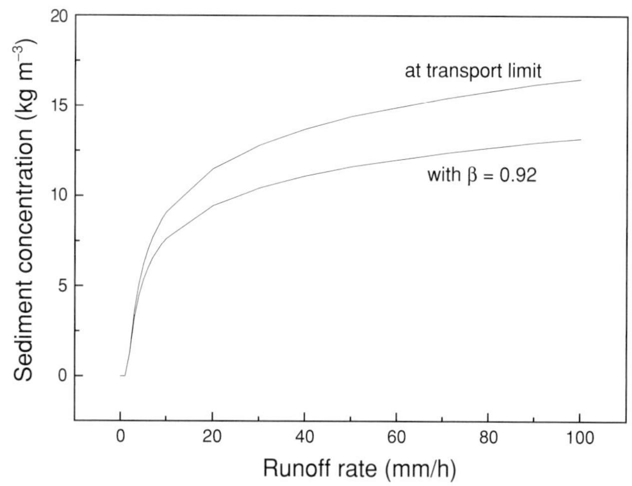

Figure 33.2 The sediment concentration at the transport limit as a function of the runoff rate predicted with Eqn 33.7. Other relevant parameter values are: slope = 5%; slope length = 36 m; Manning's $n = 0.03$; average settling velocity = 0.149 m s^{-1}; surface contact cover = 10%.

where C_g is the surface contact cover and κ an empirically determined coefficient having a value of approximately 0.1 when C_g is measured in percent. For steep slopes, the direct effect of rainfall on soil erosion has been ignored in several recent studies (Soil Technology, 1995). Since sediment concentration due to rainfall detachment is modelled as being linearly proportional to rainfall intensity, the rainfall term can be included if needed. Yu *et al.* (1999) used the following equation to estimate soil loss while taking into account the effect of both rainfall and runoff:

$$M = 0.01\,Q_{tot}\left(\frac{aP}{3.6\times10^6\phi} + \lambda c_t^{\beta}\right) \quad (33.12)$$

where *M* is event soil loss in tonnes ha^{-1}, Q_{tot} is event runoff in mm, *a* is soil detachability in kg m^{-3}, *P* the 10-min peak rainfall intensity in mm h^{-1}, whereas λ is a binary variable assuming a value of 0 or 1 depending on whether or not the direct effect of rainfall is considered. For erosion driven by flow only, $a = 0$ and $\lambda = 1$. For erosion driven by rainfall only, $\lambda = 0$. Use of λ introduces additional flexibility in Eqn 33.12.

To compute the sediment concentration at the transport limit with Eqn 33.7, *F* (= 0.1, Proffitt *et al.*, 1993), σ, ρ, g (= 9.81 m s^{-2}), and *S* are input parameters; *V*, *D*, *R* are computed by solving numerically the flow continuity equation, assuming Manning's equation. ϕ_e and $(1 - C)$ can be calculated using particle size distribution data (Lisle *et al.*, 1996) for given water depth *D*. The rill shape factor *b* depends on a critical side slope, z_0 (Yu and Rose, 1997). When rill side slope is less than z_0, the banks of the rill would be so steep that deposited sediments would not stay on them. Re-entrainment of the eroded material would therefore only take place from the bottom of the rill. In the current version of GUEST (Fentie *et al.*, 1997), a critical side slope $z_0 = 1.0$ is assumed. Figure 33.2 shows a plot of the sediment concentration at the transport limit as a function of the runoff rate to illustrate Eqn 33.7.

Input requirements for Eqn 33.12 fall in three groups: (1) soil properties, (2) topography, (3) hydrology and hydraulics. The minimum data requirement on *soil properties* includes particle size distribution (both wet-sieved and that resulting from mechanical analysis). The program GUDPRO can be used to determine the fraction of soil immersed $(1 - C)$, and the effective sediment depositability, ϕ_e, as a function of water depth from either wet-sieved size distribution, or from direct measurement of the distribution of settling velocities (Lisle *et al.*, 1996). Soil samples should be taken from the surface layer from which erosion is most likely to occur. The sand fraction (0.02 mm) can be used to estimate the wet density of the eroded soil particles (Loch and Rosewell, 1992):

$$\sigma = 1460 + 48^{*}1.0326^{x} \tag{33.13}$$

where σ and x are wet density and the percentage of sand grains (0.02 mm) in the soil, respectively.

Topographical variables in GUEST include slope length, plot width and slope steepness. Plot geometry is characterised by length (m), width (m) and slope (%). If there are rills, information on rill geometry is also needed. The dimension of a rill can be specified by its depth, D_r, top and bottom width, W_t and W_b, and the inter-rill spacing, W_r. The side slope can then be determined by:

$$z = \frac{W_t - W_b}{2D_r} \tag{33.14}$$

Hydrological and hydraulic variables are runoff rate, Q, and Manning's roughness coefficient, n. These are used to compute the mean flow velocity, the mean water depth and the hydraulic radius.

Hydrological drivers for GUEST

It can be seen from the above that runoff rate $Q(t)$ and total runoff amount Q_{tot} are critical variables in determining soil loss in the GUEST methodology. Although rainfall and runoff rates may vary greatly within a storm event, the temporal variation of sediment concentration, $c(t)$, is not modelled directly. This is partly because generally there are no measurements of sediment concentration within a runoff event, and partly because sediment concentration as a function of time is not needed when the aims are to predict event and long-term soil losses. Instead, an average sediment concentration is sought such that soil loss is the product of this sediment concentration and total runoff amount. The average concentration is termed the flow-weighted average because it is given by:

$$\bar{c} = \frac{\int_0^{T_i} c(t)Q(t)dt}{\int_0^{T_i} Q(t)dt} \tag{33.15}$$

To predict this average sediment concentration for individual events, one naturally seeks a single, 'effective' runoff rate that is logically related to this concentration. In GUEST, this effective runoff rate is given by (Ciesiolka *et al.*, 1995):

$$Q_e = \left[\frac{\int_0^{T_i} Q^{1.4}(t)dt}{\int_0^{T_i} Q(t)dt}\right]^{2.5} \tag{33.16}$$

The effective runoff rate is defined in this way so that the relationship between sediment concentration and runoff rate remains essentially invariant, regardless of whether we use steady-state sediment concentration and runoff rate, or the average concentration and effective runoff rate (Yu *et al.*, 1997a). Note that calculation of this effective runoff rate requires flow data for the entire hydrograph, typically at 1-minute intervals (Ciesiolka *et al.*, 1995).

At several experimental sites in Australia and several South East Asian countries, tipping bucket technology was used to measure runoff rate during storm events to meet GUEST data requirements (Ciesiolka *et al.*, 1995; Ciesiolka and Rose, 1997). Where detailed runoff rates are not available, a number of techniques have been developed and tested to estimate the effective runoff rate, depending on the kind of data available. For example, if data on runoff amount and peak rainfall intensity are available, a scaling technique is recommended (Yu *et al.*, 1997a). The effective runoff rate is given by:

$$Q_e = \alpha\left(\frac{Q_{tot}}{P_{tot}}\right)P_{\Delta t} \tag{33.17}$$

where $P_{\Delta t}$ is the peak intensity at a particular time interval, Δt, and α is a scaling coefficient depending on the time interval used. To examine the validity of Eqn 33.17, Yu *et al.* (1999) used rainfall and runoff data for a total of 180 site-events to estimate the scaling factor α for each of six sites in Australia and South East Asia. The 30 largest events in terms of rainfall total were selected. The same data set was used to develop and validate a rainfall-runoff model for application at small time scales (Yu *et al.*, 1997b, 1998). The mean annual rainfall for the above sites varies from 1200 mm to 3500 mm, and long-term runoff coefficients vary from 0.1 to 0.8 (Coughlan and Rose, 1997). Given the considerable range in runoff coefficient, good relationships between the effective runoff rate and the product of the gross

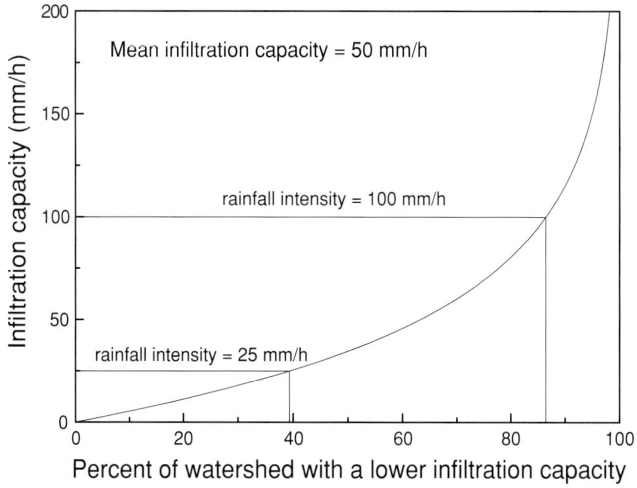

Figure 33.3 Exponential distribution of infiltration capacity with an average of 50 mm h^{-1}.

runoff coefficient and peak rainfall intensity were obtained with r^2 values of around 0.83–0.84 for time intervals between 5 to 10 minutes (Yu and Rose, 1999).

At the plot scale, the excess rainfall rate at any point in time can be regarded as the difference between the actual rates of rainfall intensity and infiltration. Yu *et al.* (1997b) proposed and evaluated a simple infiltration model in which the actual rate of infiltration, f (mm h^{-1}), is given as:

$$f = I_m[1 - \exp(-P/I_m)] \tag{33.18}$$

where P (mm h^{-1}) is the rainfall intensity and I_m (mm h^{-1}), a parameter representing the spatially averaged maximum rate of infiltration (Yu *et al.*, 1997b). This simple model for actual infiltration shows that at a given rainfall intensity, runoff occurs from a fraction of the area only (Figure 33.3). As the intensity increases, both the runoff area and the runoff rate increase. Actual infiltration also increases with rainfall intensity (Figure 33.3, Eqn 33.18). When the total runoff amount, Q_{tot} (mm), is known, which is given by the sum of the excess rainfall rates over the entire runoff event, then the parameter I_m can be readily determined by solving the non-linear equation:

$$Q_{tot} = \sum_{i=1}^{N}\{P_i - I_m[1 - \exp(-P_i/I_m)]\} \tag{33.19}$$

where N is the number of time intervals for the storm event. One of the distinct advantages of Eqn 33.18 is that the model parameter I_m can be determined uniquely from runoff amount, thus removing the need for parameter selection as in the more traditional approaches to modelling infiltration that are used in other models (cf. Table 33.7).

Yu *et al.* (1998) compared this model with two other infiltration models (constant runoff coefficient and constant runoff rate) in terms of the hydrographs generated for 30 events at each of the six sites referred to earlier using rainfall and runoff data at time intervals of 1, 6, and 15 minutes and found that Eqn 33.18 yielded the best results using the 1-min data, with model efficiency values (Nash and Sutcliffe, 1970) for predicted peak and 'effective' runoff rates of 0.83 and 0.89, respectively. Fentie *et al.* (2002) compared eight different methods to estimate peak and effective runoff rates for a pasture catchment in central Queensland and they provided further support for the scaling technique and the spatially variable infiltration model as efficient means of predicting runoff rates for erosion predictions (Fentie *et al.*, 2002). Van Dijk (2002) and van Dijk and Bruijnzeel (2001) applied the spatially variable infiltration model (Eqn 33.18) to experiments on bench terraces in West Java, Indonesia, for prediction of both runoff hydrograph and runoff amount. The model efficiency was typically in the range from 0.72 to 0.82, suggesting good model performance.

Once the hydrograph is generated, the erosion model within GUEST can be used in the same way as if the runoff rates had been actually measured. Alternatively, if runoff rates have been measured, an effective runoff rate can be computed for each event using Eqn 33.16 before the erosion model is used to determine the sediment concentration at the transport limit.

Application of GUEST to tropical steeplands

To predict soil loss, information on soil erodibility, i.e. a value for β, is required (cf. Eqn 33.10). To evaluate β, data on runoff rates at small time intervals are needed. Rainfall and runoff rates at 1-minute intervals are measured routinely at a series of ACIAR-funded research sites to evaluate soil erodibility parameters (Coughlan and Rose, 1997) (Table 33.4). For sites where data on runoff rate are not available (typical of USLE type of experimentation), the program GOSH (Yu, 1997) can be used first to estimate runoff rates from total runoff amount and rainfall rates before estimating soil erodibility parameters. Therefore, the minimum requirements for hydrological data are rainfall rate at small ($\leq$ 30 minutes) intervals in addition to total runoff amount, Q_{tot}, for each storm event.

Application and validation of GUEST therefore typically involve the following four steps:

- Use GUDPRO (Lisle *et al.*, 1996) to process data on particle size distribution to determine the effective depositability, ϕ_e, and the fraction of soil immersed, $1 - C$, as a function of the water depth, D; wet density, ρ_s is also required;
- Use GOSH (Yu, 1997) to process data on rainfall rates and event runoff amount to determine runoff rates as a function of time;
- Assemble information on plot and/or rill geometry before using the programme GUEPS (Yu and Rose, 1997) to estimate the soil erodibility parameter, β, for the event.

Table 33.4. *Sites in Australia and South East Asian countries where GUEST technology has been used in soil erosion research*

	Location	Mean annual rainfall (mm)	Plot size (m^2)	Slope (%)	Soil texture	Period	Mean annual runoff (mm/year)	Mean annual soil loss [tonnes/(ha.year)]	Treatment	GUEST erodibility
Goomboorian, Australia	26° 04′ S, 152° 48′ E	1 200	110	5.5	Sand	1992–1995	261	199	Bare	1.049
Imbil, Australia	26° 26′ S, 152° 41′ E	1 200	35.2	33	Loam	1989–1991	390	116	22 m Up–down slope cultivation, pineapple	0.674
Kemaman, Malaysia	4° 18′ N, 103° 19′ E	3 230	20	17	Sandy loam	1993–1996	1 880	340	Bare	0.319
Kuala Dal, Malaysia	4° 46′ N, 100° 54′ E	2 040	250	15	Clayey	1993–1996	498	39.5	Rubber tree with minimum ground cover	0.421
Los Banos, the Philippines	14° 6′ N, 121° 12′ E	2 040	72	18	Clay	1989–1994	393	184	Bare	0.879
ViSCA, the Philippines	10° 45′ N, 124° 49′ E	2 800	71.4	50	Clay	1989–1991	111	137	Bare	0.890
Kohn Kaen, Thailand	16° 30′ N, 102° 50′ E	1 100	50	4.0	Sand	1990–1991	372	48.4	Bare	0.928
Chiang Mai, Thailand	19° 37′ N, 99° 04′ E	1 800	360	28	Silty or clay loam	1989–1993	218	18.6	Bare	0.248
Chiang Rai, Thailand	20° 19′ N, 99° 50′ E	1 400	360	30	Silty clay or clay loam	1992–1994	314	397	Bare	0.962
Luodian, P. R. China	25° 26′ N, 106° 46′ E	1 200	200	40	Silty clay loam	1992–1994	390	177	Bare	0.421
Malangbong, West Java, Indonesia	7° 03′ S, 108° 04′ E	2 600	3.7–2.2[A]	17–59[A]	Silty clay	1998–2000	494–572[A]	94–200[a]	Bare	0.99

[a] The data range represents the average for terrace beds and risers, respectively (van Dijk, 2002).

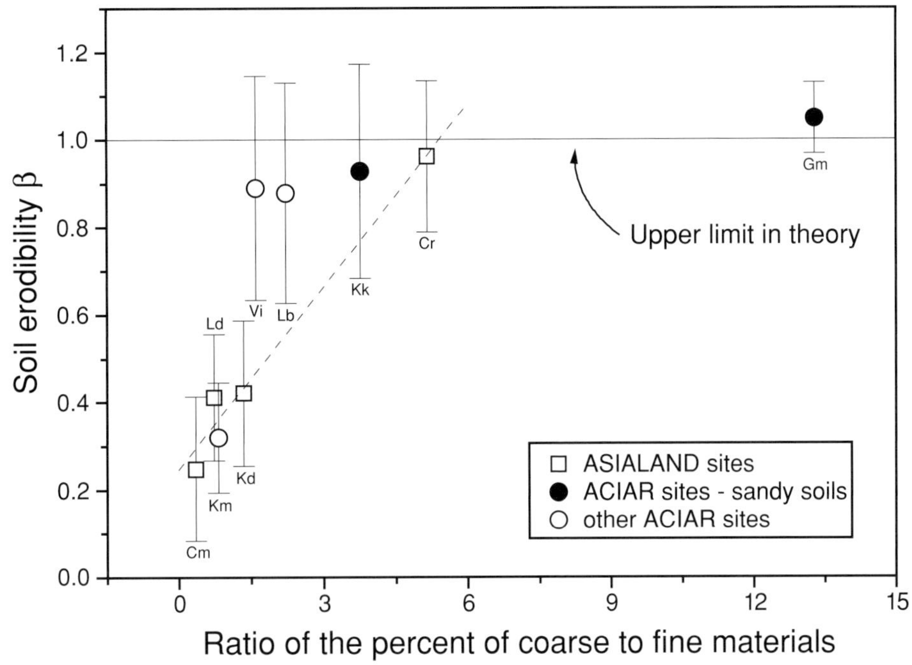

Figure 33.4 The relationship between relative abundance of coarse materials in the soil and soil erodibility β. (After Figure 9 In Yu *et al.* 1999b.)

- The model can be tested by estimating the amount of soil loss for other events at the same site using the estimated soil erodibility. A stronger test of model performance would involve applying the model and estimated soil erodibility parameter values to adjacent sites with similar soils, and crops/land use.

GUEST, as briefly described in this chapter, has been applied to a number of sites with slope up to 70% in Australia and various South East Asian countries. Table 33.4 compiles information on these sites including an average erodibility of the soil. When GUEST was used to predict event soil loss using estimated soil erodibility parameters, an average model efficiency of 0.68 was achieved for four sites in China, Malaysia and Thailand (Yu *et al.*, 1999). Predicted soil loss is sensitive to the erodibility value for the event, and the calculated β value at the ACIAR and ASIALAND sites in Table 33.4 showed considerable variation from event to event (Soil Technology, 1995; Coughlan and Rose, 1997). Although the average of these event-based β values was broadly related to measurable soil properties (Figure 33.4), this relationship has not been tested against observed soil loss in a predictive sense.

Recently, van Dijk and Bruijnzeel (2001) and van Dijk (2002) have developed the GUEST model further for use in bench-terraced terrain in Indonesia by including splash erosion from short, steep slopes as an additional term in Eqn 33.12. Total storm energy when rainfall intensity exceeded 10 mm h^{-1} was closely related to the observed rate of splash erosion in a volcanic steepland in West Java. Van Dijk *et al.* (2002) have presented a reformulation of transport by rain splash based on exponential distribution theory. Average model efficiency of the modified GUEST, known as TEST, ranged from 0.47 for the terrace bed and riser, to 0.64 for entire terrace units (van Dijk and Bruijnzeel, 2001).

More recent developments using GUEST include integration into a GIS framework to generate spatial erosion and deposition patterns at the catchment scale (Fentie, 2001) and using the erodibility parameter β as a switch from erosion to deposition down a complex hillslope (Siepel *et al.*, 2002). It is, however, also worthwhile to make a number of critical observations. First, the theory on the interactions between rainfall-driven and runoff-driven processes has not been rigorously developed, with most solutions appropriate for either rain-only or flow-only scenarios. Second, the approach to solving the governing equation has been dominated by analytical techniques, which limits the application to all but very simple configurations. Only recently were numerical solutions attempted to examine more complex scenarios (Rose *et al.*, 2002; Hogarth *et al.*, 2004). Finally, all the support for the theory relies on well-controlled flume experiments largely using disturbed soils. Even for these controlled experiments, some of the model parameters were found to vary widely. For instance, the fraction of stream power effective in soil entrainment was found to be approximately 0.1 by Proffitt *et al.* (1993), while a more recent study showed that the fraction ranges from 0.0013 to 0.0075 (Beuselinck *et al.*, 2002a; Hairsine *et al.*, 2002), and there was no obvious explanation of this order-of-magnitude discrepancy.

WEPP

WEPP (Water Erosion Prediction Project) is a process-based model for runoff and soil erosion prediction developed largely by USDA-Agricultural Research Services with an intention to replace the traditional USLE (Nearing *et al.*, 1989, Laflen *et al.*, 1997). WEPP attempts to model the fundamental erosion processes and to achieve broad applicability in a range of erosion environments such as rangeland, forests and urbanized areas, in addition to the traditional farm fields. WEPP requires none of the factors in USLE/RUSLE. By design, WEPP has all the capabilities of USLE/RUSLE, but with considerable added functionality. In particular, WEPP can handle complex hillslope profiles with ease and is able to address the effect of intrinsic or externally imposed climate variability on daily runoff, soil erosion and sediment yield at the hillslope and catchment scale. The Technical Documentation and User Summary (Flanagan and Nearing, 1995; Flanagan and Livingston, 1995) constitute the primary reference materials on WEPP. The WEPP program and its documentation are readily available at http://topsoil.nserl.purdue.edu/.

WEPP can be used to predict soil erosion and net soil loss for individual events or on a continuous daily basis. Normally, ten daily weather variables are used by WEPP to predict the hydrological variables required for soil erosion prediction. The most important of these include daily precipitation amount, storm duration and peak intensity. In addition, WEPP requires information on soil properties, land use and management practices. Of particular importance are the effective hydraulic conductivity, rill and interrill erodibility, and the critical shear stress of the soil. Given their strong spatial and temporal variability, these parameters are best calibrated using field data, although empirical relationships are available to estimate these parameters from measurable soil properties such as particle size distribution, organic matter content and cation exchange capacity (Alberts *et al.*, 1995).

The Green-Ampt infiltration equation under conditions of unsteady rain (Chu, 1978) and kinematic wave theory are used in WEPP to determine runoff amount and peak runoff rate (Stone *et al.*, 1995). The erosion component of WEPP is based on a steady-state continuity equation of the form (Foster *et al.*, 1995):

$$\frac{\mathrm{d}(cq)}{\mathrm{d}x} = D_\mathrm{i} + D_\mathrm{f} \tag{33.20}$$

where c and q are sediment concentration (kg m^{-3}) and unit discharge (m^2 s^{-1}), respectively, D_i the interrill erosion rate, and D_f rill erosion rate. The term steady state implies that these variables are assumed not to vary in time within each runoff event, but can vary in the downslope direction (x) only. The interrill erosion rate is modelled as being proportional to the product of rainfall intensity and runoff rate:

$$D_\mathrm{i} = K_\mathrm{iadj} I_\mathrm{e} Q S_\mathrm{d} F_\mathrm{n}(W/Wr) \tag{33.21}$$

where K_iadj is the adjusted interrill erodibility (kg s m^{-3}), I_e effective rainfall intensity (m s^{-1}), Q interrill runoff rate (m s^{-1}), S_d the interrill sediment delivery ratio, and F_n accounts for irrigation nozzle impact energy variation. W and Wr are rill spacing and rill width, respectively. Effective rainfall intensity is defined in WEPP as the average intensity for the duration when rainfall rate exceeds infiltration rate. Interrill sediment delivery ratio, S_d, depends on interrill roughness factor, settling velocity and fractions of five size classes considered in WEPP (Flanagan and Nearing, 2000). The rill erosion rate is given as:

$$D_\mathrm{f} = K_\mathrm{r}(\tau - \tau_\mathrm{c})(1 - c/c_\mathrm{t}) \tag{33.22}$$

where K_r is the rill erodibility parameter (s m^{-1}), τ flow shear stress (Pa) and τ_c is the critical flow shear stress (Pa). When the actual sediment concentration c exceeds that at transport capacity, c_t, deposition occurs and D_f becomes negative and is given by:

$$D_\mathrm{f} = \phi(c_\mathrm{t} - c) \tag{33.23}$$

where ϕ is the average settling velocity (m s^{-1}).

The concentration at the transport limit in WEPP is determined using Yalin's equation (Yalin, 1963) as formulated by Foster (1982).

For each runoff event, the peak runoff rate is considered the 'representative', steady-state runoff rate while effective runoff duration is the runoff amount divided by this representative runoff rate.

To evaluate the predictive capabilities of WEPP, predicted runoff and soil loss have been compared with measurements at a number of sites in the United States. Ghidey and Alberts (1996) compared measured and predicted runoff and soil loss from cultivated fallow, corn and soybean cropping systems for a site in Missouri, and found that WEPP tends to over-predict runoff and soil loss in dry years, but overall the model's performance was satisfactory. Zhang *et al.* (1996) and Liu *et al.* (1997) validated WEPP at the hillslope and catchment scale, respectively. Zhang *et al.* (1996) used data for 4124 events from 556 plot years and eight locations to evaluate WEPP runoff and soil loss predictions. Runoff and soil loss data used in this study were all collected from the standard USLE plots. They noted that WEPP over-predicted runoff for small events and under-predicted for large events. Using internally estimated hydraulic conductivity values, they were able to obtain an r^2 value of 0.67 for runoff for 491 selected events, and 0.40 for event soil loss. Tiwari *et al.* (2000) extended the work by Zhang *et al.* (1996). They used more than 1600 plot-years data from 20 locations and compared WEPP with the USLE and RUSLE in terms of soil loss prediction. For annual soil loss, the r^2 value for WEPP was 0.43, which was lower than for the USLE (0.58) and RUSLE (0.62). On a mean annual basis, the r^2 for WEPP was 0.72. This again was lower than that for the USLE (0.80) and RUSLE (0.75). Overall, WEPP did not perform as well as the USLE for the sites included. Liu *et al.* (1997) used data from 15 small catchments (0.35–5.14 ha) and obtained r^2 values of 0.86 for total runoff and 0.91 for total sediment for these catchments. On a runoff event basis, the r^2 values ranged from 0.01 to 0.8 for runoff and 0.02 to 0.90 for sediment yield. Bjorneberg *et al.* (1999) tested WEPP for furrow irrigation from three field sites in Idaho. They calibrated WEPP using data from the upper quarter of two fields and used the calibrated parameter values for effective hydraulic conductivity, critical shear stress and rill erodibility to predict runoff and soil loss at field ends. They noted that the calibrated rill erodibility was some 100 times smaller than the recommended value for the soil, and the predicted infiltration remained essentially constant (around 50 mm), and not nearly as variable as the observed infiltration (about 35–90 mm). At the field ends, WEPP over-predicted soil loss and under-predicted runoff. The r^2 values ranged from 0.08 to 0.44 for soil loss and from 0.02 to 0.86 for runoff. Kincaid (2002) further demonstrated WEPP's capability to predict runoff and soil erosion under sprinkler irrigation and found that prediction variability for individual furrows was high. Such findings are expected given the inherent spatial variability in the rate of infiltration at the field scale. It was recommended that WEPP is best used to indicate when serious runoff is likely to occur under sprinkler irrigation for different soils and crop management practices.

Elsewhere in the world, several attempts have also been made to test, validate, and use WEPP for runoff and soil loss predictions. Povilaitis *et al.* (1995) applied WEPP to five runoff and erosion plots in Lithuania and found that WEPP performed fairly well. Klik *et al.* (1997) found that the interrill erodibility needed re-calibration for two sites in Austria: runoff was better predicted ($R^2 = 0.98$) than soil loss. Soto and Diaz-Fierros (1998) presented an interesting case study of the effects of bush fire on runoff and soil loss and subsequent forest regeneration in Spain. They demonstrated the applicability of WEPP in addressing some of the management issues as being much wider than what can be dealt with using conventional erosion prediction technology such as the USLE/RUSLE. In Australia, Yu *et al.* (2000) applied WEPP to a pineapple farm in south-east Queensland to determine its predictive potential in a hitherto untested biophysical environment. The average coefficient of efficiency (Nash and Sutcliffe, 1970) for runoff and soil loss prediction was –0.02 using soil property based parameter values, and 0.66 using calibrated parameter values. Yu and Rosewell (2001) validated WEPP using runoff and soil loss data for bare fallow and winter wheat plots in southeastern Australia (New South Wales). Model parameter values were derived directly from soil properties rather than from calibration. For the latter site, in contrast to the site in southeast Queensland which had a sandy soil, WEPP worked quite well. Model efficiency was 0.97, both for predicted runoff and soil loss from bare fallow plots, in contrast to an average of –0.02 for the pineapple site in southeast Queensland. WEPP was also able to reproduce the effect of slope length on sediment concentration at the NSW site. This study provided additional information to establish the relationships between WEPP parameters and specific soil properties without which WEPP can never be used in a truly predictive sense. In New Zealand, Su *et al.* (1999) found that for a well-structured clayey soil (clay content 66%), the baseline hydraulic conductivity was under-estimated using soil properties resulting in an over-prediction of runoff from a bare plot near Auckland (slope 14%, area 13.1 m × 3.1 m). More recently, use of WEPP to simulate runoff and soil loss was attempted at Perieni station, in eastern Romania (Popa, 2002). Considerable difficulties were encountered, especially with preparing climate inputs for WEPP and in the end, CLIGEN-generated climate data for one site in Nebraska, USA, with similar average precipitation and temperatures, were used. WEPP tended to over-predict runoff for fallow, corn, winter wheat and bean crops, and under-predict soil loss for the same treatments. The quality of the simulated runoff and soil loss was difficult to assess for this Romanian site because of the problems with input preparation for WEPP. Given the inconsistent model performance, there is a need to continue to validate various components of WEPP for a range of climates, topography, soils, land uses and management practices, particularly in the humid tropics.

Recently, a trimmed-down version of WEPP, known as HEM (Hillslope Erosion Model) has been developed and adapted for rangeland (Lane *et al.*, 1995; 2002). On-line application of HEM is available at http://eisnr.tucson.ars.ag.gov/hillslopeerosionmodel/. Cogle *et al.* (2001) tested HEM at one site each in India, Australia and New Zealand. The calibrated erodibility values for tropical sites in Australia and India were much lower (by a factor of 6.1–15) than the default values for texturally similar soils. Unvalidated HEM and WEPP applications using GIS (ArcInfo/ArcView) have been performed by Wilson *et al.* (2001) and Flanagan *et al.* (2001). Another noticeable development is the possibility to run WEPP live on the Web, either as a general application or as a customised application in a forested environment (http://octagon.nserl.purdue.edu/weppV1/; http://forest.moscowfsl.wsu.edu/fswepp/).

EUROSEM AND LISEM

LISEM (LImburg Soil Erosion Model) was developed at Utrecht University in the Netherlands (de Roo *et al.*, 1996a, b), while EUROSEM (EUROpean Soil Erosion Model) is primarily a result of the collaborative efforts from a number of countries in the European Community (Morgan *et al.*, 1998), and the USDA. These two models are discussed together because they show considerable similarity in terms of the approach adopted, the science used and applicability intended.

Both are single-event runoff and erosion simulation models. Rainfall, topographical and soil data along with a number of model parameters are used to simulate time series of runoff rate and sediment concentration during a storm event. Hydrographs and sedigraphs, as well as maps showing areas of net erosion and net deposition are standard model outputs.

Local net erosion rate, e, is modelled in both models by:

$$e = e_r + e_f \tag{33.24}$$

where e_r is a rainfall detachment term and always positive, and e_f is the flow detachment term, which can be either positive for erosion, or negative for deposition when the sediment concentration exceeds that at the transport capacity. The rainfall detachment term is modelled mainly as a function of rainfall kinetic energy, but reduced as water depth increases. The flow detachment term is modelled as a linear function of the difference between sediment concentration at the transport limit, c_t, and actual sediment concentration, c, i.e.:

$$e_f \propto \phi(c_t - c) \tag{33.25}$$

where ϕ is the average settling velocity (essentially depositability in the context of GUEST). Both EUROSEM and LISEM use the set of equations for c_t as a function of the unit stream power developed by Govers (1990). Some minor differences in erosion modelling between the two erosion models include an additional option in EUROSEM of using Everaert (1991)'s equation for inter-rill transport capacity.

Both models use similar methods to simulate rainfall interception by vegetation and surface depression storage and both use Manning's equation and a kinematic wave approximation to determine surface runoff hydrographs. However, the two models differ in the infiltration models used to calculate the rainfall excess. EUROSEM uses the infiltration equation developed by Smith and Parlange (1978) as in KINEROS/KINEROS2 (Woolhiser *et al.*, 1990; Smith *et al.*, 1995), while LISEM has a number of options including the Holtan (1961) equation as used previously in ANSWERS (Beasley *et al.*, 1980), as well as the well-known Green-Ampt and Richards equations.

The major difference between the two models is that EUROSEM describes the flow and sediment routing capabilities within a catchment using flow elements which are either planes or channels following the method used in KINEROS and KINEROS2 (Woolhiser *et al.*, 1990; Smith *et al.*, 1995). LISEM, like ANSWERS, uses regular grid cells to represent the catchment. In addition, LISEM is tightly coupled with PCRaster. PCRaster is a raster-based spatial modelling system that has extensive cartographical, dynamic and geo-statistical modelling capabilities (Wesseling *et al.*, 1996; van Deursen, 1995). It has been used not only for hydrological and erosion modelling but also for flood simulation in large basins (de Roo *et al.*, 2000).

EUROSEM and LISEM have been calibrated and validated mostly in the UK and the Netherlands at plot to catchment scales (Morgan *et al.*, 1998; Folly *et al.*, 1999, de Roo *et al.*, 1996b; de Roo and Jetten, 1999). In addition, Quinton and Morgan (1998) tested EUROSEM for the C-3 catchment, Oklahoma, whereas Albaledejo *et al.* (1994) applied EUROSEM in Spain at the plot scale (75 m^2). De Roo and Jetten (1999) also presented the validation results using LISEM for a 0.69 km^2 catchment in South Africa. Lately, Hessel (2002) attempted to apply LISEM to a 3.5 km^2, deeply dissected catchment on the loess plateau in China and he argued for the need to modify a number of components of LISEM to simulate runoff, soil erosion and transport better in this environment, known for its high erosion rates. While these modifications are well justified on theoretical grounds, both the existing and modified versions of LISEM require calibration. Interestingly, there is no noticeable improvement in terms of the calibrated hydrographs and sedi-graphs when the modified version of LISEM was used (Hessel, 2002). Stolte *et al.* (2002) reported an attempt to calibrate LISEM for a 9 ha catchment near Chengdu, Sichuan Province, China, and then apply LISEM to quantify runoff and soil loss for a 7 km^2 catchment in which the catchment for calibration is located. The simulated hydrograph for calibration was highly fluctuating and much more responsive to rainfall than the measured hydrograph for the 9 ha catchment. Calibration results for soil loss were not reported for this catchment (Stolte *et al.*, 2002).

Of the three small catchments instrumented for the LISEM project in the southern hilly region of the Netherlands, the Catsop catchment (0.45 km^2) was used for calibration and validation of a number of erosion prediction models (Catena, 1999). Five events were used for model calibration and five additional events were reserved for validation purposes. Table 33.5 shows the percentage differences between simulated and measured values for five validation events using EUROSEM and LISEM for this catchment. It can be seen that EUROSEM tends to over-estimate runoff and soil loss, a trend which is consistent with the findings of Albaledejo *et al.* (1994) in Spain, while LISEM tends to under-estimate them

Table 33.5. *Validation results in terms of the percentage difference from the observed (Obs.) values for the Catsop catchment, the Netherlands, using EUROSEM (E) and LISEM (L)*

	Total runoff			Peak runoff rate			Soil loss[a]			Sediment concentration[a]		
Date	Obs. (mm)	E(%)	L(%)	Obs. (mm h^{-1})	E(%)	L(%)	Obs. (t ha^{-1})	E(%)	L(%)	Obs. (kg m^{-3})	E(%)	L(%)
13 May 87	1.45	−44	−64	0.69	17	−49	0.014	61	−69	0.96	187	−15
15 Dec 89	0.92	156	−94	0.23	770	−96	0	−H	−H	0	−H	−H
22 Jan 93	0.96	73	−14	0.76	199	25	0.065	662	−69	6.73	339	−64
30 May 93	0.84	91	−25	2.75	−44	81	0.652	−74	10	77.6	−87	48
14 Oct 93	1.44	55	−94	1.23	37	−97	0.201	182	−100	14.0	81	−99
Overall[b]	4.48	55	−63	1.13	46	−26	0.932	203	−88	16.6	96	−67

[a] H, observed soil loss was zero, while simulated soil loss was positive. Percentage difference was indeterminate for this event.

[b] For total discharge, soil loss, the sum for the five events was used. For peak discharge, the average was used. For sediment concentration, the flow-weighted average was used.

Source: de Roo and Jetten (1999) and Folly *et al.* (1999).

Table 33.6. *A comparison of two alternative process-based approaches to water erosion and deposition modelling*

Model	GUEST	WEPP
Equivalent conceptions	Rainfall-dominated processes	Interrill erosion
	Lateral sediment input	Interrill sediment delivery
	Runoff-dominated processes	Rill erosion and deposition
Concentration at transport limit, c_t	$\frac{\sigma b F(\Omega - \Omega_0)}{(\sigma - \rho) g D \phi}$	T_c/q with T_c being a calculable quantity from flow and soil characteristics
Flow entrainment/detachment	Proportional to excess stream power	Proportional to excess shear stress
Parameters for flow entrainment/detachment	$F\,J^{-1}$	K_r
Sedimentation	Proportional to c_t	Proportional to c_t
Parameters for sedimentation	ϕ	ϕ
Rain detachment	Proportional to rainfall intensity	Proportional to rainfall intensity
Parameters for rain detachment	$H\,a_d\,\phi^{-1}$	K_i

for the same events. More importantly, the percentage differences between measured and predicted values are quite large for both models. Favis-Mortlock *et al.* (1996) and Jetten *et al.* (1999) addressed a number of issues associated with model validation at field and catchment scales.

Takken *et al.* (1999) reported a most interesting and unique evaluation study of LISEM for an extreme storm event on a 2.96 km^2 catchment in Belgium. They surveyed erosion and deposition areas after the storm and calibrated the Green-Ampt infiltration parameters in LISEM so that the simulated total sediment yield from the catchment equalled the observed sediment yield. Spatially uniform rainfall at a rather coarse temporal resolution (0.5 h) was used to drive the erosion model for the catchment. Saturated hydraulic conductivity, as well as initial and saturated soil moisture contents were assumed to be homogeneous over the entire catchment. The simulated erosion rates for individual land units were poor. While the model reproduced the overall sediment delivery for the catchment (by necessity because the model was calibrated this way), a comparison of the simulated and observed spatial distributions of erosion and deposition within the catchment showed that the model over-estimated considerably the spatial extent of deposition and under-estimated the depth of deposition where deposition actually occurred. This study highlights the need to compare the spatial distributions of erosion and deposition rates, and the inadequacy of model validation using data collected at the catchment outlet alone (cf. van Dijk, 2002).

A COMPARISON OF EROSION PREDICTION MODELS

Recent work has shown that the basic equation of GUEST (Eq. 33.4) is essentially equivalent to the erosion and deposition equations (30–33) used in WEPP under steady-state conditions (Yu, 2003). In particular, the rainfall-driven terms (i.e. rainfall detachment e_i and re-detachment e_{di}) in Eqn 33.4 can be interpreted as representing the lateral sediment input from the interrill area in WEPP, while the runoff-driven processes in rills or preferred flow pathways in the context of GUEST (i.e. r_i, r_{ri} and d_i in Eqn 33.4) are equivalent to the rill erosion and deposition in WEPP (equations 33.22 and 33.23). There are also considerable similarities between GUEST and WEPP in terms of how the individual processes are parameterised (Table 33.6). For instance, the rill erodibility in WEPP, K_r, is broadly related to F/J in GUEST, where J is the specific energy of entrainment (Hairsine and Rose, 1992a; Rose 1993). The similarity between GUEST and WEPP is not a coincidence, since the governing equations for both models are based on mass balances. The implication of this similarity is that choosing a particular model may be not nearly as important as how individual erosion processes are formulated and whether the parameter values are widely available to promote model use.

In practice, WEPP offers considerable applicability in terms of representing complex topography, climate variability, or various land use and management scenarios. WEPP can be quite useful for resource managers when a database for input parameter values has been prepared. Overall, WEPP is currently too complex, and hence too rigid, to be ideal for research purposes, while at the same time not simple enough in terms of the demand for data to be adopted widely for routine applications in less developed regions of the world. In contrast, GUEST and the theoretical framework from which GUEST was derived are ideal tools for erosion research. They allow exploration of various sub-processes in erosion and deposition and their interactions. However, because there is a general lack of methods for the determination of parameter values from measurable soil properties and land use practices for GUEST, it will require considerable effort to use GUEST in a predictive sense without prior calibration against field measurements.

Schröder (2000) compared WEPP, EUROSEM and Erosion-2D (Schmidt, 1996) in terms of their model components. Sensitivity analysis strongly suggested that predictions are particularly sensitive to infiltration-related parameters for all models. Runoff and soil loss data collected under a rainfall simulator at two sites in Germany were used to evaluate the three models without calibration. In terms of model efficiency (Nash and Sutcliffe, 1970), WEPP performed better for runoff prediction for a silt loam while EUROSEM was better for a sandy loam. WEPP over-predicted sediment concentration by a factor of 2. For EUROSEM, the predicted concentration depends on which transport equation is used and whether rills are assumed to be present. Better results can be achieved using Govers' (1990) equation while assuming that no rills exist, although the equation was originally developed for rills. The large difference in predicted and measured runoff has precluded any meaningful comparison in terms of soil loss (Schröder, 2000). Kandel *et al.* (2001) selected several equations from GUEST, WEPP and EUROSEM to fit the plot data from Nepal, and showed that for these physically-based models, high-resolution (2-min) rainfall and runoff were needed. Using daily data would lead to a noticeable deterioration in model performance.

A comparison of all the erosion models discussed in this chapter is presented in Table 33.7 including those that combine the USLE factors with sub-models for various hydrological and erosion processes.

A CRITIQUE OF PHYSICALLY BASED EROSION MODELLING

In this section, we attempt to identify the challenges facing all those working with physically based models for soil erosion prediction. These challenges also serve to indicate the areas in which improvements in the current suite of physically based erosion models are likely to occur.

Different mechanisms

Different mechanisms of runoff generation have not received sufficient attention in physically based erosion models (Table 33.7). As argued throughout this chapter, runoff generation plays a crucial role in determining the magnitude and location of erosion and deposition in the landscape. It can be said that the ability to predict runoff rate determines and limits the potential of all erosion prediction models. Various aspects of runoff generation in the humid tropics are addressed by Bonell in this volume. It is, however, worthwhile to highlight a common trend in erosion models to over-emphasise the importance of Hortonian overland flow as evidenced by the various classic infiltration equations that are used (Table 33.7). Infiltration excess is often the only process considered in erosion modelling. This is at least in part because most fundamental erosion research has been carried out at the plot scale with a minimum of vegetation cover. Under such conditions surface runoff is indeed often generated by the Hortonian mechanism in which the infiltration capacity is exceeded by rainfall intensity. At the catchment scale, and in rainforest environments especially, however, for most of the catchment and for most of the time, runoff occurs only where an area is locally saturated. It is widely documented that in forested and humid environments, Hortonian flow in fact rarely occurs (Bonell, this volume). However, in deforested tropical uplands where valley bottoms are often narrow and slopes straight to convex, opportunities for the generation of saturated overland flow (SOF) are usually limited (cf. Dunne, 1978; Bruijnzeel, 1990). Under such conditions, stormflow at the headwater catchment scale represents a mixture of rapid subsurface stormflow ('throughflow') and, depending on the degree of surface disturbance, Horton overland flow generated on impervious surfaces (trails, roads, settlements) and agricultural fields (Rijsdijk and Bruijnzeel, 1990; Ziegler *et al.*, 1997; Purwanto, 1999). Although SOF and the associated erosion processes have not been accommodated at all in any of the physically based erosion models discussed in the previous section, the controlled experimental results of Huang *et al.* (1999) have shown that much higher erosion rates occurred from discharge areas generating SOF compared to recharge areas. It is imperative to incorporate necessary switches between the two different mechanisms for runoff generation in response to changes in topography, vegetation cover, land use and management practices (cf. Vertessy *et al.*, 1990).

Overland flows

Overland flow shows highly complex and irregular hydraulic characteristics (Abrahams and Parsons *et al.*, 1994). With respect to erosion modelling, there are at least two important issues that have not yet been resolved satisfactorily. The first is the issue of flow concentration over the landscape. Natural hillslopes do not have planes. Flows always occur in irregular rills or depressions. The density and dimension of these flow pathways determine the unit discharge which in turn is crucial to prediction of sediment concentration no matter which soil erosion model is considered. For erosion prediction using the concept of stream power (as in GUEST, LISEM and EUROSEM) or shear stress (WEPP), velocity and water depth are required, implying that a flow resistance equation, such as Manning's equation, must be used. Overland flow hydraulics is still an area of active research partly because the flow is neither laminar nor fully turbulent due to the presence of a bewildering array of resistance elements (Takken and Govers, 2000; Giménez and Govers, 2000).

Table 33.7. *Sediment generation models*

	USLE-derived, e.g. ANSWERS, AGNPS, CREAMS/GLEAMS, CASC2D-SED	WEPP	LISEM	EUROSEM	GUEST/GUEPS
Mode of operation	Event, or continuous	Event, or continuous	Event	Event	Event, continuous – to be developed
Climatic data requirements	*R*-factor, event EI_{30}, pluviograph, etc.	Rainfall amount; storm duration; peak intensity; time to peak	Pluviograph (break-point)	Pluviograph (break-point)	Rainfall amount Peak rainfall intensity
Hydrological data requirements	Peak flow rate, unit discharge, etc.	Runoff amount; peak runoff rate	Runoff hydrograph	Runoff hydrograph	Runoff amount and effective runoff rate
Runoff generation	Curve number, Green–Ampt	Green–Ampt (Curve number derived)	Holtan; Green–Ampt; Richards	Smith–Parlange	Water balance model coupled with a spatially variable infiltration model
Main drivers for soil erosion	Unit and/or peak discharge	Rainfall intensity Peak runoff rate; sheer stress	Kinetic energy of rainfall Stream power	Kinetic energy of rainfall Stream power	Rainfall intensity Stream power
Size distribution	Up to three classes?	Five classes	Single class	Single class	Up to 20 classes
Catchment representation	Often RASTER-based	Hillslope/channel	Plane/channel	RASTER-based	RASTER-based
Comments	Many in this category. Rigorous comparison can be difficult and probably unnecessary	Essentially does what RUSLE can do with added functionality	Critically dependent on PCRaster, a standing-alone software; suitable for small catchments	Integrated with KINEROS, similar in many aspects to LISEM, appropriate for small catchments	On sound physical basis, but much yet to be developed to provide sensible parameter values for predictive use of the model

Soil erodibility

Soil erodibility is widely understood to represent the inherent susceptibility of the soil to erosional agents such as rainfall and runoff. The concept no doubt is sound. Measures of soil erodibility, however, depend on a particular conceptual framework in which soil erodibility is presented. In practice, soil erodibility usually represents what is unexplained and inexplicable. Consequently, the numerical value of the erodibility depends not only on the properties and the state of the soil, but also on the way in which soil erodibility is defined in a model. The net result is that the soil erodibility parameters for various models are likely to be incomparable. For example, Laflen *et al.* (1994) show that there was essentially no meaningful relationship between the USLE's K factor and the interrill and rill erodibility parameters of WEPP as determined using rainfall simulators. Furthermore, although the erosion equations themselves are physically based, the relationship describing the temporal variation of the erodibility parameters embedded in these equations is highly empirical because a complex sequence of land disturbances and the varying nature of these disturbances are inherently involved in changing the state of the soil in time. Bryan (2002) argued strongly that full understanding of the spatially variable and temporally dynamic (including variability within storm events) soil properties are essential to erosion predictions.

Unsteady rain

Modelling of erosion, transport and deposition under unsteady rain is clearly needed. At present, EUROSEM and LISEM consider unsteady rain but only use a single representative particle size. GUEST and WEPP, while taking into account sediment sorting due to contrasting settling velocities, are essentially steady-state models. Realistic simulation of the dynamic processes of erosion, transport and deposition for a wide range of particle sizes is called for to represent the varied erosive forces in relation to a wide range of settling velocity characteristics during natural storm events.

Although in this critique a number of issues in relation to erosion modelling are highlighted, these are not meant to discourage the continued effort in developing, testing and validating physically based erosion models. Apart from its usefulness for erosion assessment and conservation planning, modelling provides an important link between theoretical work and practical outcomes. Allowing some delay in time, many important concepts in erosion science will have themselves expressed in erosion models as our understanding of the processes and our ability to describe them quantitatively increase. In addition, modelling provides a framework to learn about the relative importance of different erosion processes (e.g. splash vs. wash or rill erosion; cf. van Dijk, 2002) and test various formulations of these processes. Erosion theory and modelling also provide a framework for experimental design and environmental monitoring programmes. For instance, for development and use of GUEST methodology, accurate measurement of runoff rate in addition to rainfall rate was regarded as essential because such data are required to determine soil erodibility parameters.

CONCLUSION

Compared to the factor-based USLE/RUSLE, which has long been established and widely applied for soil erosion prediction, physically-based erosion models with a rather different approach to soil erosion modelling are now available to simulate various hydrological and erosion processes at a range of spatial scales. Common to most of these models is the considerable increase in their functionality in terms of dealing with complex topography and land use, sequences of management practices, realistic storm events and continuous weather sequences. More importantly, the physically-based models are capable of simulating spatial and temporal patterns of erosion as well as deposition. The latter is of considerable significance at the catchment scale. The ability to evaluate the effects of climate and land use change, as well as the changes in management practices, is the main attraction of this new generation of erosion prediction technology. The accuracy of the predictions by the physically-based models, even at the hillslope scale, is however yet to be fully demonstrated. In the humid tropics, use of these physically-based models certainly requires further calibration. However, routine application of physically-based erosion prediction models in the tropics is unlikely in the near future, if only because of the cost involved. The cost of data collection for calibration, of developing a useful database for predictive use of models, and of training with respect to both using the model and interpreting the results will be prohibitive in most cases. Physically-based models are also inevitably limited by the empirical relationships that describe the variation of model parameters in space and time. Testing and validating these process-based models in diverse biophysical environments, notably in the humid tropics, are crucial to their long-term acceptance as a viable alternative to predict the rate of soil erosion routinely and operationally and to replace, if needed, factor-based models such as the USLE/RUSLE.

References

Abrahams, A. D. and Parsons, A. J. (1994) hydraulics of interrill overland flow on stone-covered desert surfaces. Catena 23:111–140.

Agus, F., Gintings, A. N., Kurnia, U., Abdurachman, A. and van der Poel, P. (1997) Soil erosion research in Indonesia: Past experience and future direction. *In*: F. W. T. Penning de Vries, F. Agus and J Kerr (eds). Soil Erosion at Multiple Scales, CABI Publsihing, pp. 255–267.

Alberts, E. E., Nearing, M. A., Weltz, M. A., Risse, L. M., Pierson, F. B., Zhang, X. C., Laflen, J. M. and Simanton, J. R. (1995). Ch. 7 Soil component. *In* D. C. Flanagan and M. A. Nearing (ed.) USDA-Water Erosion Prediction

Project: Hillslope Profile and Watershed Model Documentation. NSERL Report No. 10. West Lafayette, Ind.: USDA-ARS Nat. Soil Erosion Research Laboratory.

Bagnold, R. A. (1977) Bedload transport by natural rivers. Water Resour. Res. 13:303–11.

Beasley, D. B., Huggins, L. F. and Monke, E. J. (1980) ANSWERS: A model for watershed planning. Trans ASAE, 23(4), 938–944.

Beuselinck L., Govers, G., Hairsine, P. B., Sander, G. C. and Breynaert, M. (2002a) The influence of rainfall impact on sediment transport by overland flow over areas of net deposition, J. Hydrol. 257:145–163.

Beuselinck, L., Hairsine, P. B. Sander, G. C. and Govers, G. (2002b) Evaluating a multi-class net deposition equation across a range of conditions. Water Resour. Res. 38(7), doi: 10.1029/2001 WR000250, 2002.

Bingner, R. L. (1990) Comparison of the components used in several sediment yield models. Trans ASAE 33:1229–1238.

Bjorneberg, D. L., Trout, T. J., Sojka, R. E. and Aase, J. K. (1999) Evaluating WEPP-predicted infiltration, runoff, and soil erosion for furrow irrigation. Trans. the ASAE 42:1733–1741.

Bouraoui, F. and Dillaha, T. A. (1996) ANSWERS-2000: Runoff and sediment transport model. J. Env. Engg., ASCE, 122:493–502.

Bruijnzeel, L. A. (1990) Hydrology of moist tropical forests and effects of conversion. A state-of-knowledge review. UNESCO International Hydrological Programme, Paris, 224 pp.

Catena (1999) Vol. 36, Issue 3–4, Special issue on 'Modelling soil erosion by water at the catchment scale.

Chu, S. T. (1978) Infiltration during unsteady rain. Water Resour. Res. 14:461–466.

Ciesiolka, C. A. A., Coughlan, K. J., Rose, C. W., Escalante, M. C., Hasim, G. M., Paningbatin, E. P., Jr. and Sombatpanit. S. (1995). Methodology for a multi-country study of soil erosion management. Soil Tech. 8: 179–192.

Cogle, A. L., Lane, L. J. and Basher, L. (2001) Testing the Hillslope erosion model for Application in India, New Zealand and Australia. In MODSIM 2001, 10–13 December 2001, The Australian National University, Canberra, Australia, pp. 173–178.

Coughlan, K. J. and Rose, C. W. (Eds). (1997) 'A New Soil Conservation Methodology and Application to Cropping Systems in Tropical Steeplands'. ACIAR Technical Report No. 40. (ACIAR: Canberra.)

De Roo, A. P. J. (1998) Modelling runoff and sediment transport in catchment using GIS. Hydrol. Process: 12, 905–922.

De Roo, A. P. J. and Jetten, V. G. (1999) Calibrating and validating the LISEM model for two data sets from the Netherlands and South Africa. Catena 37: 477–493.

De Roo, A. P. J., Offermans, R. J. E. and Cremers, N. D. T. (1996b) LISEM: A single-event physically based hydrological and soil erosion model for drainage basins. II. Sensitivity analysis, validation and application. Hydrol. Processes, 10:1119–1126.

De Roo, A. P. J., Wesseling, C. G. and Ritsema, C. J. (1996a) LISEM: A single-event physically based hydrological and soil erosion model for drainage basins. I. Theory, input and output. Hydrol. Processes, 10:1107–1117.

De Roo, A. P. J., Wesseling, C. G. and Van Deursen, W. P. A. (2000) Physically based river basin modelling within a GIS: the LISFLOOD model. Hydrol. Proc. 14: 1981–1992.

Dickinson, A. and Collins, R. (1998) Predicting erosion and sediment yield at the catchment scale. *In*: F. W. T. Penning de Vries, F. Agus and J Kerr (eds). Soil Erosion at Multiple Scales, 317–342.

Dietrich, W. E., Wilson, C. J., Montgomery, D. R., McKean, J. and Bauer, R. (1992) Erosion thresholds and land surface morphology. Geology 20: 675–679.

Dunne, T. (1978) Field studies of hillslope flow process. *In*: M. J. Kirkby (ed.) *Hillslope Hydrology*. Wiley, Chichester, 227–293.

El-Swaify, S. A. and Dangler, E. W. (1977) Erodibility of selected tropical soils in relation to structural and hydrologic parameters. In Soil Erosion: Prediction and Control, Soil and Water Conservation Society, Ankeny, Iowa, pp. 105–114.

Everaert, W. (1991) Empirical relations for the sediment transport capacity of interrill flows. Earth Surface Processes and Landforms 16:513–532.

Favis-Mortlock, D. T., Quinton, J. N. and Dickinson, W. T. (1996) The GCTE validation of soil erosion models for global change studies. J. Soil and Water Conservation 51:397–403.

Fentie, B. (2001) Sustainability of the resource base in grazing lands, Queensland: Modelling erosion-productivity pf pasture lands. Unpubl. Ph.D. thesis, University of Queensland.

Fentie, B., Coughlan, K. J. and Rose, C. W. (1997) Manual for Use of Program GUEST30. Faculty of Environmental Sciences. (Griffith University: Brisbane.)

Fentie, B. Yu, B., Silburn, M. D. and Ciesiolka, C. A. A. (2002) Evaluation of eight different methods to predict hillslope runoff rates for a grazing catchment in Australia. J. Hydrol. 261:102–114.

Fisher, G. and Heilig, G. K. (1997) Population momentum and the demand for on land and water resources. Philosophical Transactions of the Royal Society of London, B352:862–869.

Flanagan, D. C. and Nearing, M. A. (Ed.) (1995) WEPP Technical Documentation. NSERL Report No. 10. USDA-ARS-MWA: West Lafayette, IN.

Flanagan, D. C. and Livingston, S. J. (Ed.) (1995) WEPP User Summary. NSERL Report No. 11. USDA-ARS-MWA: West Lafayette, IN.

Flanagan, D. C. and Nearing, M. A. (2000) Sediment particle sorting on hillslope profiles in the WEPP model. Trans. ASAE, 43: 573–583.

Flanagan, D. C., Renschler, C. S., Frankenberger, J. R., Cochrane, T. A., and Angel, B. A. (2001) Enhanced WEPP model applicability for improved erosion prediction. In MODSIM 2001, 10–13 December 2001, The Australian National University, Canberra, Australia, pp. 1817–1822.

Folly, A., Quinton, J. N. and Smith, R. E. (1999) Evaluation of the EUROSEM model using data from the Catsop watershed, the Netherlands. Catena 37:507–519.

Foster, G. R. (1982) Modelling the erosion process. Chapter 8 in C. T. Haan (ed). Hydrologic Modelling of Small Watershed. ASAE Monograph No. 5, pp. 297–360.

Ghidey F. and Alberts, E. E. (1996) Comparison of measured and WEPP predicted runoff and soil loss for Midwest claypan soil. Trans. ASAE 39:1395–1402.

Giménez, R. and Govers, G. (2001) Interaction between bed roughness and flow hydraulics in eroding rills. Water Resour. Res. 37:791–799.

Govers, G. (1990) empirical relationships for the transport capacity of overland flow. In erosion Transport and depositiion Processes, Proceedings of Jarusalem Workshop, 187. IAHS Publi No 189.

Hairsine, P. B. and Rose, C. W. (1991) Rainfall detachment and deposition: Sediment transport in the absence of flow-driven processes. Soil. Sci. Soc. Am. J. 55:320–324.

Hairsine, P. B. and Rose, C. W. (1992a) Modelling water erosion due to overland flow using physical principles: I. Uniform flow. Water Resour. Res. 28:237–44.

Hairsine, P. B. and Rose, C. W. (1992b) Modelling water erosion due to overland flow using physical principles: II. Rill flow. Water Resour. Res. 28:245–50.

Hairsine, P. B., Sander, G. C., Rose, C. W., Parlange, J.-Y., Hogarth, W. L., Lisle, I. and Rouhipour, H. (1999) Unsteady soil erosion due to rainfall impact: a model of sediment sorting on the hillslope. J. Hydrol. 220:115–128.

Hairsine, P. B., G. C. Sander, and L. Beuselinck. 2002. Sediment transport through an area of net deposition. Water Resour. Res. 38(6), 10.1029/2001 WR000265, 2002.

Hancock, G. R., Willgoose, G. R. and Evans, K. G. (2002) Testing of the SIBERIA landscape evolution model using the Tin Camp Creek, Northern Territory, Australia, field catchment. Earth Surface Processes and Landforms 27:125–143.

Hessel, R. (2002). Adapting the LISEM model to Loess Plateau conditions. ISCO 2002, 26–31, May, Beijing, China, pp. 110–115.

Hogarth, W. L., Rose, C. W., Parlange, J. Y., Sander, G. C. and Carey, G. (2004) Soil erosion due to rainfall impact with no inflow: a numerical solution with spatial and temporal effects of sediment settling velocity characteristics. J. Hydrol. 294(4): 229.

Holtan, H. N. (1961) A concept for infiltration estimates in watershed engineering. ARS-4-51, Agricultural Research Service, USDA, 25 p.

Huang, C., Wells, L. K. and Norton, L. D. (1999) Sediment transport capacity and erosion processes: Model concepts and reality. Earth Surface Process and Landform 24:503–516.

Jetten V., de Roo, A. and Favis-Mortlock, D. (1999) Evaluation of field-scale and catchment-scale soil erosion model. Catena 37:521–541.

Johnson, B. E., Julien, P. Y., Molnar, D. K. and Watson, C. C. (2000). The two–dimensional upland erosion model CASC2D-SED. J. Amer. Water Resourc. Assoc. 36:31–42.

Kandel, D. D., Western, A. W., Grayson, R. B. and Turral, H. N. (2001) Testing current generation soil erosion models at two–minute and daily scales against plot scale data from mid-hill catchment of Nepal. In MODSIM 2001, 10–13 December 2001, The Australian National University, Canberra, Australia, pp. 173–178.

Klik A., Hebel B., Zartl A. and Rosner J. (1997) Measured vs. WEPP simulated runoff and erosion from differently tilled plots. In 'ASAE Annual International Meeting'. 10 pp. (Minneapolis, MN).

Kincaid, D. C. (2002) The WEPP model for runoff and erosion prediction under sprinkler irrigation. Trans ASAE 45:67–72.

Kinnell, P. I. A. (1997) Runoff ratio as a factor in the empirical modelling of soil erosion by individual rainstorms. Aust. J. Soil Res. 35:1–13.

Kinnell, P. I. A. (1998) Converting USLE soil erodibilities for use with $Q_R EI_{30}$ index. Soil and Tillage Res. 45:349–357.

Knisel, W. G. (1980) CREAMS: A field-scale model for chemicals, runoff and erosion from agricultural management systems. USDA Conservation Research Report No. 26. (USDA-ARS: Washington, DC)

Knisel, W. G. (1991) CREAMS/GLEAMS: A development overview. In Beasley, D. B., Knisel, W. G. Rice, A. P. Eds. Proc. of the CREAMS/GLEAMS Symposium, Publ. 4, Agric Engineering, U. Georgia, Athens, p. 9–17.

Laflen, J. M., Elliot, W. J., Flanagan, D. C., Meyer, C. R. and Nearing, M. A. (1997) WEPP-Predicting water erosion using a process-based model. J. Soil and Water Conserv 52:96–102.

Lal, R. (1990) Soil erosion in the tropics: Principles and management. McGraw Hill, New York.

Lane L. J., Renard K. G., Foster G. R. and Laflen, J. M. (1992) Development and application of modern soil erosion prediction technology – The USDA experience. *Aust. J. Soil Res.* 30:893–912.

Lane, L. J., Nichols, H. M. and Paige, G. B. (1995) Modelling erosion on hillslopes: Concepts, theory and data. In Proceddings of the International Congress on Modeliing and Simulation (MOSIM'1995), University of Newcastle, NSW, Australia, 1–7.

Lisle, I. G., Coughlan, K. J., and Rose, C. W. (1996) 'GUPRO 3.1: A Program for Calculating Particle Size and Settling Characteristics'. ENS Working Paper, 7/96. Faculty of Environmental Sciences. (Griffith University: Brisbane.)

Liu, B. Y., Nearing, M. A., Baffaut, C. and Ascough, II J. C. (1997) The WEPP watershed model: III Comparison to measured data from small watersheds. Trans. the ASAE 40: 945–952.

Loch, R. J. and Rosewell, C. J. (1992) Laboratory methods for measurement of soil erodibilities (K-factor) for the Universal Soil Loss Equation. Aust. J. Soil Res. 30:231–48.

Lu, H., Gallant, J., Prosser, I. P., Moran, C. and Priestley G. (2001) Prediction of sheet and rill erosion over the Australian continent: Incorporating monthly soil loss distribution. Technical report 13/01, CSIRO Land and Water, Canberra, Australia.

Myers, N. (1993) *Gaia: An Atlas of Planet Management*. Anchor and Doubleday, Garden City, NY.

Misra, R. K. and Rose, C. W. (1995) An examination of the relationship between erodibility parameters and soil strength. Aust. J. Soil Res. 33:715–32.

Misra, R. K. and Rose, C. W. (1996) Application and sensitivity analysis of process-based erosion model GUEST. European Journal of Soil Science 47:593–604.

Morgan, R. P. C., Quenton, J. N., Smith, R. E., Govers, G., Poesen, J. W. A., Auerswald, K., Chisci, G., Torri, D. and Styczen, M. E. (1998) The European soil erosion model (EUROSEM): A dynamic approach for predicting sediment transport from fields and small catchments. Earth Surface Processes and Landforms 23:527–544.

Moss, A. J. (1988) The effects of flow-velocity variations on rain-driven transportation and the role of raindrop impact in the movement of solids. Aust. J Soil Res. 26:443–450.

Mutchler, C. K., Murphree, C. E. and McGregor, K. C. (1994) Laboratory and field plots for erosion research. In R. Lal (ed.) Soil Erosion Research Methods, Soil and Water Conservation Society and St Lucie Press.

Nash, J. E. and Sutcliffe. J. V. (1970) River flow forecasting through conceptual models. Part 1: A discussion of principles. J. Hydrol. 10:282–90.

Nearing, M. A., Foster, G. R., Lane, L. J. and Finkner, S. C. (1989) A process based erosion model for USDA water erosion prediction project technology. Trans ASAE 32:158–1593.

Oldeman, L. R., Hakkerling, R. T. A. and Sombroek, W. G. (1991) 'World Map of the Status of Human-induced Soil Degradation: An Explanatory Note.' (International Soil Reference and Information Centre:Wageningen.)

Onstad, C. A., and Foster, G. R. (1975). Erosion modelling on a watershed. Trans. ASAE 18:288–92.

Penning de Vries, F. W. T., F. Agus and J Kerr (eds) (1998) Soil Erosion at Multiple Scales. CABi Publishing, Wallingford, UK.

Popa, N. (2002). Soil erosion models used in Romania. In ISCO 2002, 26–31, May 2002, Beijing, China, pp. 222–226.

Povilaitis, A., Robichaud, P., Dumbrauskas, A., Tumas, R., Ruseckas, J. and Heatwole, C. (1995) Erosion prediction in the Republic of Lithuania with the WEPP model. In 'Water quality modelling: Proceedings of the International Symposium'. pp. 333–340. (Orlando, FL)

Proffitt, A. P. B., Hairsine, P. B. and Rose, C. W. (1993) Modelling soil erosion by overland flow: application over a range of hydraulic conditions. Trans ASAE, 36:1743–1753.

Proffitt, A. P. B., Rose, C. W. and Hairsine, P. B. (1991) Rainfall detachment and deposition: experiments with low slopes and significant water depths. Soil Sci. Soc. Am. J. 55:325–332.

Quinton, J. N. and Morgan, R. P. C. (1998) EUROSEM: an evaluation with single event data from the C5 watershed, Oklahoma, USA. In: Proceedings of the NATO Advanced Research Workshop 'Global change: modelling soil erosion by water', J. Boardman and D. Favis-Mortlock (ed.), 65–74.

Renard, K. G., Foster, G. A., Weesies, G. A., McCool, D. K., Yoder, D. (1997) Predicting Soil Erosion by Water: A Guide to Conservation Planning with the Revised Universal Soil Loss Equation (RUSLE), USDA Agriculture Handbook No. 703, (US Gov. Print. Office: Washington DC.)

Renard, K. G., Foster, G. R., Weesies, G. A., McCool, D. K. and Yoder, D. C. coordinators. (1997). Predicting Soil Erosion by Water: A Guide to Conservation Planning with the Revised Universal Soil Loss Equation (RUSLE), US Department of Agriculture, Agriculture Handbook No. 703, 404 pp.

Roose, E. (1977) Use of the Universal Soil Loss Equation to reduce erosion in West Africa. In Proceedings of the National Conference on Soil Erosion. Soil Conservation Society of America, Ankeny, Iowa.

Rose, C. W. (1997) Modelling erosion by water and wind. In Advances in Soil Science, Eds R. Lal, W. H. Blum, C. Valentine, B. A. Stewart. 57–88.

Rose, C. W. (1993) Erosion and sedimentation. *In* 'Hydrology and Water Management in the Humid Tropics – Hydrological Research Issues and Strategies for Water Management'. (Eds M. Bonnell, M. M. Hufschmidt and J. S. Gladwell) pp. 301–43. (Cambridge University Press: Cambridge)

Rose, C. W., Coughlan, K. J., Ciesiolka, C. A. A. and Fentie, B. (1997). Program GUEST (Griffith University Erosion System Template). *In* 'A New Soil Conservation Methodology and Application to Cropping Systems in Tropical Steeplands'. (Eds K. J. Coughlan and C. W. Rose) ACIAR Technical Report, No. 40, pp. 34–58. (ACIAR: Canberra.)

Rose, C. W., Williams, J. R., Sander, G. C. and Barry D. A. (1983a) A mathematical model of soil erosion and deposition processes. I. Theory for a plane land element. Soil Sci. Soc. Am. J. 47:991–995.

Rose, C. W., Williams, J. R., Sander, G. C. and Barry, D. A. (1983b) A mathematical model of soil erosion and deposition processes. II. Application to data from an arid-zone catchment. Soil Sci. Soc. Am. J. 47:996–1000.

Rose, C. W. and Yu, B. (1998) Dynamic process modelling of hydrology and soil erosion. In: F. W. T. Penning de Vries, F. Agus, and J. Kerr eds. Soil Erosion at Multiple Scales. CAB Internationals, 269–286.

Rose, C. W., Yu, B., Hogarth, W. L., Okom, A. E. A., and Ghadiri, H. (2002) Theoretical interpretation of the spatial and size distribution of sediment deposited by buffer strips from flow at modest land slopes. J Hydrol. In review.

Sander, G. C., Hairsine, P. B., Rose, C. W., Cassidy, D., Parlange, J. Y., Hogarth, W. L., Lisle, I. (1996) Soil erosion due to rainfall impact during the early stages of a rainstorm. J. Hydrol. 178:351–367.

Sander, G. C., P. B. Hairsine, L. Beuselinck, and G. Govers. 2002. Multiclass net deposition solutions. Water Resour. Res. 38(6), 10.1029/2001 WR000323, 2002.

Schmidt, J. (1996). Entwicklung und Anwendung eines physikalisch begründeten Simulationsmodells für die Erosion geneigter, landwirtschaftlicher Nutzflächen. Berliner Geographische Abhandlungen 61. Department of Geography of the Free University, Berlin.

Schröder, A. (2000) WEPP, EuROSEM, E-2D: Results of applicatios at the plot scale. 199–250. In J Schmidt (ed.) Soil Erosion, Springer-Verlag, Berlin.

Siepel, A. C., Steenhuis, T. S., Rose, C. W., Parlange, J.-Y. and McIsaac, G. F. (2002) A simplified hillslope erosion model with vegetation elements for practical applications. J. Hydrol. 258:111–121.

Smith, R. E., Goodrich, D. and Quinton, J. N. (1995) Dynamic distributed simulation of water shed erosion: the KINEROS2 and EUROSEM models. J. Soil Water Conserv. 50:517–520.

Smith, R. E. and Parlange, J. (1978) A parameter-efficient hydrologic infiltration model. Water Resour. Res. 14:533–538.

Soil Technology (1995). Special Issue: Soil Erosion and Conservation. Soil Tech. 8:177–258.

Soto, B. and Diaz-Fierros, F. (1998) Runoff and soil erosion from areas of burnt scrub: comparison of experimental results with those predicted by the WEPP model. Catena 31:257–270.

Stolte, J., Chen, Y., Ritsema, C. J., van den Elsen, E., Reijnders, J. and Thompson, J. (2002) Calibration and up-scaling of an erosion model as a tool for farmers involvement to define alternative land use. ISCO 2002, 26–31, May, Beijing, China, pp. 218–221.

Stone, J. J., Lane, L. J., Shirley, E. D. and Hernandez, M. (1995). Ch. 4 hillslope surface hydrology. *In* D. C. Flanagan and M. A. Nearing (ed.) USDA-Water Erosion Prediction Project: Hillslope Profile and Watershed Model Documentation. NSERL Report No. 10. West Lafayette, Ind.: USDA-ARS Nat. Soil Erosion Research Laboratory.

Su, N., Basher, L., Barringer, J. and Doscher, C. (1999). Reconstructing the patterns of sediment transport and related hydrological processes using the WEPP model. In MODSIM 1999, 6–9 December 1999, University of Waikato, New Zealand, pp. 209–213.

Takken, I., Beuselinck, L., Nachtergaele, J., Govers, G., Poesen, J. and Degraer, G. (1999) Spatial evaluation of a physically-based distributed erosion model LISEM. Catena, 37:431–447.

Takken, I. and Govers. G. (2000) Hydraulics of interrill overland flow on rough, bare soil surfaces. Earth Surface Processes and Landforms 25:1387–1402.

Tiwari, A. K., Risse, L. M., Nearing, M. A. (2000) Evaluation of WEPP and its comparison with USLE and WEPP. Trans ASAE 43:1129–1135.

Van Deursen, W. P. A. (1995) Geographical information systems and dynamic models. Netherlands Geogr. Studies, 190, 198 pp.

van Dijk, A. I. J. M. (2002) Water and sediment dynamics in bench-terraced agricultural steeplands in West-Java, Indonesia. Unpublished Ph.D. thesis. Free University, Amsterdam, The Netherland.

van Dijk, A. I. J. M. and Bruijinzeel, L. A. (2001) Modelling runoff and soil loss from bench terraced hillslope in the volcanic uplands of west Java, Indonesia. MODSIM 2001, 233–238.

Van Dijk, A. I. J. M., Bruijnzeel, L. A. and Meesters, A. G. C. H. (2002) Exponential distribution theory applied to splash detachment and transport experiments. Soil Sci. Soc. Amer. J., in press.

Vertessy, R. A., Wilson, C. J., Silburn, D. M., Connoy, R. D. and Ciesiolka, C. A. A. (1990) Predicting erosion hazard areas using digital terrain analysis. IASH Proc. Int Symposium on Research Needs and Applications to Reduce Erosion and Sedimentation in Tropical Steeplands. Suva, Fiji. Pp. 298–308.

Wen, D. (1993) Soil erosion and conservation in China. In: World Erosion and Conservation, D. Pimentel (ed.), Cambridge University Press. pp. 63–107.

Wesseling, C. G., Karssenberg, D., P. A. Burrough, P. A. and Van Deursen, W. P. A. (1996). Integrating dynamic environmental models in GIS: The development of a dynamic modelling language. Transactions in GIS 1:40–48.

Willgoose, G. R., R. L. Bras, and I. Rodriguez-Iturbe (1991). A physically based coupled network growth and hillscope evolution model: 1. Theory. Water Resour. Res. 27:1671–1684.

Williams, J. R. (1975) Sediment-yield prediction with universal equation using runoff energy factor. *In* 'Present and Prospective Technology for Predicting Sediment Yields and Sources'. pp. 244–52. (USDA-ARS-S-40)

Williams, J. R., Renard, K. G. and Dyke, P. T. (1983) A new method for assessing the effect of erosion on productivity – The Epic model. J. Soil Water Conserv. 38: 381–383.

Wilson, C. J., Carey, J. W., Beeson, P. C., Gard, M. O. and Lane, L. J. (2001). A GIS-based hillslope erosion and sediment delivery model and its application in the Cerro Grande burn area. Hydrol. Processes 15:2995–3010.

Wischmeier, W. H. and Smith, D. D. (1965) Predicting Rainfall-Erosion Losses from Cropland east of the Rocky Mountains: Guide for Selection of practices for Soil and Water Conservation. USDA Agriculture Handbook No. 282. US Gov. Print. Office: Washington DC.

Wischmeier, W. H. and Smith, D. D. (1978) Predicting Rainfall Erosion Losses: A Guide to Conservation Planning. USDA Agriculture Handbook No. 537. US Gov. Print. Office: Washington DC.

Woolhiser, D. A., Smith, R. E. and Goodrich, D. C., (1990) KINEROS: A kinematic runoff and erosion model: documentation and user manual, USDA Agricultural Service, ARS-77.

Yalin, M. S. (1963) An expression for bed-load transportation. J. Hydraulic Div., ASCE, 98(HY3):221–250.

Young, R. A., Onstad, C. A., Bosch, D. D. and Anderson, W. P. (1989) AGNPS: a nonpoint-source pollution model for evaluating agricultural watersheds. J. Soil Water Cons. 44:168–173.

Yu, B. (1997) GOSH: A Program for Calculating Runoff Rates Given Rainfall Rates and Runoff Amount, User Guide and Reference Manual. ENS Working Paper, 2/97. (Faculty of Environmental Sciences, Griffith University: Brisbane.)

Yu, B. (2003) A unified framework for water erosion and deposition equations. Soil Science Society of America Journal, 67: 251–257.

Yu, B., Cakurs U. and Rose, C. W. (1998) An assessment of methods for estimating runoff rates at the plot scale. Trans. ASAE 41:653–61.

Yu, B, Ciesiolka, C. A. A., Rose, C. W. and Coughlan, K. J. (2000) A validation test of WEPP to predict runoff and soil loss from a pineapple farm on a sandy soil in subtropical Queensland, Australia. Aust. J. of Soil Res. 38:537–554.

Yu, B. and Rose, C. W. (1997) GUEPS: A Program for Calculating the Soil Erodibility Parameter β and Predicting the Amount of Soil Loss Using GUEST Methodology. ENS Working Paper 3/97. (Faculty of Environmental Sciences, Griffith University: Brisbane.

Yu, B. and Rose, C. W. (1999) Application of a physically based soil erosion model in the absence of data on runoff rates: I. Theory and methodology. Aust. J. Soil Res. 37:1–11.

Yu, B., Rose, C. W., Ciesiolka, C. C. A., Coughlan, K. J. and Fentie, B. (1997a). Toward a framework for runoff and soil loss prediction using GUEST technology, Aust. J. Soil Res., 35:1191–1212.

Yu, B., Rose, C. W., Coughlan, K. J. and Fentie, B. (1997b). Plot-scale rainfall-runoff characteristics and modelling at six sites in Australia and South East Asia, Trans. ASAE 40:1295–1303.

Yu, B. and Rosewell, C. J. (2001) Evaluation of WEPP for runoff and soil loss prediction at Gunnedah, NSW, Australia. Aust. J. Soil. Res. 38: 1131–1145.

Yu, B., Sajjapongse, A., Yin, D., Eusof, Z., Anecksamphant, C., Rose, C. W. and Cakurs, U. (1999) Application of a physically based soil erosion model in the absence of data on runoff rates: II. Four case studies from China, Malaysia and Thailand. Aust. J. Soil Res. 37:13–31.

Zhang X. C., Nearing M. A., Risse L. M. and McGregor, K. C. (1996) Evaluation of WEPP runoff and soil loss predictions using natural runoff plot data. Trans. ASAE 39:855–863.

34 Impacts of forest conversion on the ecology of streams in the humid tropics

N. M. Connolly and R. G. Pearson
Australian Centre for Tropical Freshwater Research, Townsville, Australia

INTRODUCTION

To understand the ecology of streams it is vital to appreciate the links within and between aquatic habitats and the landscape of which they are part. Many of the problems of poor land management stem from ignorance of the interactive nature of ecosystems and from a lack of appreciation of the scales at which such interactions operate. Human-induced changes can disrupt or override natural processes. The goal of contemporary land and water resource managers is to manage landscapes in an ecologically sustainable way. However, in the past, management has focussed most on the need to provide raw materials, food and fibre, as well as land and water for urban and industrial development. Socio-economic pressures towards development have resulted in widespread perturbations to aquatic ecosystems and irreversible degradation in many of them. The continuing broad-scale conversion of natural vegetation and landscape into forestry, agriculture, water storage or other land uses has dramatic effects on the ecology of streams. Impacts vary with the extent of the conversion relative to the size of the catchment, the nature of the new land use, land management under the new land use, the climate, and the extent of riparian and contiguous buffer zones. Generalisations regarding these effects are difficult because of this diversity, which creates a complex array of environmental conditions.

The increasing awareness of the considerable threats to natural aquatic environments and of their importance to human life and livelihood has shifted attention to include not only water supplies but also to ensure the maintenance of good water quality, river health and ecological processes. This is a complex issue that demands a multidisciplinary approach, with input from diverse fields such as hydrology, geomorphology, chemistry, microbiology, botany and zoology, and can be viewed from the perspectives of human well-being or nature conservation.

The concerns over global environmental change have brought the impact of anthropogenic activities on hydrology, geomorphology and ecology into sharper focus, especially in tropical regions (Gupta, 1993). Expanding human population pressures, and intensive agricultural and extractive industries, have increased the pressure in many tropical countries to exploit tropical forests in recent decades, often very destructively and with little regard for the long-term sustainability of these activities. Tropical forests contribute to global goods and services in ways that are not yet fully appreciated, but include important influences on the climate and the maintenance of the highest levels of global biodiversity. Stream systems, too, are major contributors to ecosystem goods and services (Table 34.1). The conversion of humid tropical forest is occurring very rapidly over extensive areas, leaving little chance for animals and plants to accommodate the changes. The humid tropical regions contain the largest and most productive river systems in the world but many, if not most, of them are in poor condition because of the lack of adequate protection and conservation (Dudgeon, 1992b, 1999, Dudgeon and Lam, 1994). Poor management is exacerbated in the tropics, as tropical environments are especially vulnerable to careless land-use practices because of the extreme climatic events that occur in these regions.

Tropical ecosystems are very different from their counterparts in higher latitudes. They have different geological and evolutionary histories, and different climatic extremes and dynamics. It is in the tropics that the world's biodiversity reaches its zenith. The number of interacting species is typically much higher in tropical ecosystems, including streams (Pearson *et al.*, 1986; Lake *et al.*, 1994, Jacobsen, 1997), and interactions are often more complex (Begon *et al.*, 1990). Unfortunately, information about tropical streams is relatively sparse and very scattered. For example, while there have been many reviews of the topic of pollution in fresh waters (e.g. Hynes, 1960; Hellawell, 1986; Chapman, 1992), none deals with tropical systems; and it is particularly noteworthy that, of approximately 1400 references included in a bibliography on the effects of deforestation (Blackie *et al.*, 1980), only about 0.5% referred to the tropics. However, Jackson and Sweeney (1995), in a synopsis of tropical stream research, found that significant progress had been made in the understanding of the ecology

Forests, Water and People in the Humid Tropics, ed. M. Bonell and L. A. Bruijnzeel. Published by Cambridge University Press.

Table 34.1. *Ecosystem goods and services provided by streams and rivers*

Goods and services	Requirements for sustainable use
Habitat	Maintenance of natural flow regime Maintenance of natural water quality Maintenance of riparian zone Maintenance of in-stream and riparian structure and vegetation Maintenance of natural inputs
Biodiversity	Maintenance of habitat
Water resources	Reliable supply Appropriate quality
Fisheries (freshwater and estuarine)	Maintenance of habitat
Drainage	Maintain natural channel form and riparian vegetation
Water quality enhancement	Maintain natural flows and habitats
Recreation/tourism	Maintain natural attributes of stream and contiguous habitats
Transport of wastes	Maintain volume for adequate dilution
Shipping	Maintain channel
Wetland replenishment	Maintenance of natural flow and flooding regime

of tropical streams and rivers and that the number of pertinent papers had increased markedly over the last few decades. Many subjects had been studied including taxonomy, life-history traits, production, trophic structure, transport of materials and organisms, nutrient dynamics and ecological interactions. Nevertheless, they concluded that although most subjects had received at least some attention, they remained relatively unknown at most sites, mainly because of there being fewer freshwater ecologists in tropical regions than in Europe or North America. Consequently, it is difficult to develop general theories to describe the function of tropical streams. For example, there are no studies in the tropics to match the long-term research at Hubbard Brook in the north-eastern United States, where multi-scale and multi-disciplinary investigations have led to substantial understanding of the many processes that take place in temperate streams (Likens, 1999), although current efforts may address this problem in a few places, such as Hong Kong, Australia and Costa Rica. Table 34.2 summarises the scope of ecological research on streams undertaken at some geographical foci in the tropics. In these places, while studies on biogeography and stream ecology are increasing, research on ecological processes has mostly been at the scale of individual sites. Extensive reviews of studies of tropical freshwater ecology (e.g. Bishop, 1973a; Pearson, 1994; Dudgeon, 1999; Pearson, 2002) provide many small pieces of a massive jigsaw, which is giving only inklings of its overall pattern: that is, as yet, we have no descriptive model of how a tropical stream works.

The lack of generalised models, major gaps in knowledge of taxonomy, basic biology and ecology, and the relative paucity of researchers (and research institutes) in the tropics mean that there is no capacity to record change adequately, making it difficult to be sure of the effects and extent of anthropogenic influences. However, change is taking place so rapidly and in some places so extensively that ecological impacts are clear (e.g. Pearson and Penridge, 1987; Jacobsen, 1998; Dudgeon, 1994). Unfortunately, ecological research needed for adequate conservation in the tropics is perpetually trying to keep pace with the rate of development and change.

In this chapter we aim to provide an introduction to the multi-scale and multi-disciplinary nature of streams and rivers and to stress that the biotic community is complex and is maintained by a diverse array of physical and biotic processes that interact at various temporal and spatial scales. Despite this complexity, it is feasible to understand the system by teasing apart the different processes so that appropriate stream management can be applied to sustain or restore water quality, habitat integrity and conservation values. We distinguish between factors that act directly on plants and animals and those that act indirectly by altering habitats and ecological processes. We provide more than the simple physico-chemical view of water quality by taking a broad view of habitats, of which water quality is a part. We attach ecological significance to biotic processes that are often overlooked when assessing impacts, thereby providing an ecological basis for remedial actions, to assist land managers in maintaining ecological processes as well as the physical environment. This approach is especially important given that change is inevitable in many landscapes, such that a goal to sustain pristine conditions is rarely set or achievable. A more generally acceptable goal for systems that are not in reserves might be to sustain ecological integrity in a productive environment.

THE ROLE OF STREAMS AND RIVERS IN THE LANDSCAPE

To define what a stream or river is requires a multi-disciplinary and multi-scale approach. In general, streams may be regarded as the upstream end and rivers the downstream end of a running water continuum. Here we use the term 'stream' generically to include all parts of the continuum. A stream is an integrated functional system, which includes both abiotic components (non-living physical properties) and biotic components (living organisms including bacteria, fungi, plants and animals). Stream ecosystems are obviously structured by physical and chemical properties and

Table 34.2. *Selected references indicating scope of ecological studies in streams in several tropical geographical regions*

Topic	Locality	Reference
Organic matter – inputs, processing	Hong Kong	Dudgeon 1982a, 1982b, 1984, 1994, 1995b, Dudgeon and Bretschko 1995a
	Malaysia	Bishop 1973a
	Queensland Wet Tropics	Pearson *et al.* 1989
		Nolen and Pearson 1993
		Benson and Pearson 1993
	Tropical America	Covich, 1987
Primary production	Hong Kong	Dudgeon and Bretschko 1995a
	Costa Rica	Pringle *et al.* 1986
Invertebrate ecology	Sri Lanka	Fernando 1980
		Starmuhlner 1984a, 1984b
		Benzie 1984
	Malaysia	Bishop 1973a, 1973b
	Hong Kong	Dudgeon 1984, 1985, 1988, 1992a,1989a,1989b, 1989c, 1995c
	Costa Rica	Stout 1980
		Pringle and Blake 1994,
		Pringle and Ramirez 1998,
		Pringle *et al.* 1993
	Queensland Wet Tropics	Benson and Pearson 1988,
		Pearson *et al.* 1986
		Lake *et al.* 1994
	Bouganville	Yule, 1996
Fish ecology	Sri Lanka	Wikramanayake 1990
		Wikramanayake and Moyle 1989
		Kortmulder *et al.* 1990
		Indrasena 1970
	Malaysia	Bishop 1973a
	Hong Kong	Dudgeon 1987b
	Queensland Wet Tropics	Pusey and Kennard 1996,
		Pusey *et al.* 1995a, 1995b, 1997, 2000
Ecological processes (disturbance etc.)	Hong Kong	Dudgeon 1992a, 1993
		Dudgeon and Bretschko 1995a
	Queensland Wet Tropics	Smith and Pearson 1987
		Hearnden and Pearson 1991
		Rosser and Pearson 1995
		Pearson and Connolly 2000
		Bunn *et al.* 1997, 1998
Drift and migration	Hong Kong	Dudgeon 1983, 1990
	Malaysia	Bishop 1973b
	Queensland Wet Tropics	Benson and Pearson 1987a, 1987b
Life history patterns	Hong Kong	Dudgeon 1985, 1992a, 1995b
	Malaysia	Bishop 1973a
	Queensland Wet Tropics	Nolen and Pearson 1992
Human impacts	Malaysia	Bishop 1973a
	Hong Kong	Dudgeon 1988, 1992b, 1995a
	Queensland Wet Tropics	Pearson and Penridge 1987
		Hortle and Pearson 1990
	Northern Australia	Humphrey *et al.* 1990, 1995

processes of the environment, including channel structure, landform, soils, flow variability, fluid dynamics, dissolution, chemical reactions, etc. Consequently, the study of stream ecology crosses several disciplines and many scales. Running waters function as a crucial part of the hydrological cycle, as they drain water from the land, and supply lakes, wetlands, estuaries and the sea with water, sediments and nutrients. In doing so, streams and rivers act as veins for terrestrial systems. They perform a major role in transporting and cycling materials and thereby interconnect aquatic and terrestrial systems, providing an important link in global biogeochemical cycling.

There are major links between terrestrial environments and streams in the movement of energy and materials, such that most upland forest streams are dependent on terrestrial inputs to fuel their food webs. And while floodplain systems to a great extent drive their own food webs through aquatic plant production, this is also to a large extent dependent on nutrients derived from the land and upstream.

Streams provide special habitats, support rich biotic communities and provide important biological linkages. Microbes, plants and animals are abundant and species-rich. Most major groups of animals occur in streams, and many of them are quite spectacular (including dragonflies, crayfish, fish, frogs, snakes, water dragons, tortoises, crocodiles, kingfishers, herons, rodents, various carnivorous mammals and, in Australia, the platypus). Many of the species that occur in streams are either semi-terrestrial, or have distinct terrestrial phases in their life cycles. For example, most insects that live in the water as larvae emerge to breed on land. Emerging insects can represent an important food source for terrestrial predators, especially the many species of insectivorous birds. Other essentially terrestrial animals, such as kingfishers, are important predators on animals within the stream. Amphibians (frogs, etc.) have important aquatic and terrestrial phases in their life cycles. Many fish species migrate along river and wetland networks. Eels navigate the full length of river systems from the ocean to their headwaters as tiny elvers, and then years later, when fully grown, migrate back to the sea again to reproduce and complete their life cycles. Many other fish migrate between the sea and streams, as well as between rivers and their floodplains, when sufficient inundation allows. Several crustaceans (e.g. *Macrobrachium* shrimps) also undertake migrations between estuaries and fresh waters. Some undergo quite complex movements: for example, the atyid shrimp *Australatya striolata* releases its larvae in the wet season in the upland streams where it lives. The larvae drift down to the estuary where they undergo several moults before becoming juvenile shrimps (they require saline water to moult, and they can live for a month in fresh water; so they have a month to get from the headwaters to the estuary). The shrimps then migrate back upstream, eventually arriving as young males, 20–25 mm in length. They remain and breed as males, then one or two years later grow quite rapidly to double their size, and change sex. They then breed as females, completing the cycle. The furthest upstream that they have been recorded is 160 km (Smith, 1987). Such complexities are common amongst the many species that make up the stream community. This shrimp species demonstrates how the system is a continuum, not only physically but also biologically, and indicates why it is important to regard river systems holistically.

Stream systems also provide connectivity for terrestrial organisms in many landscapes. Riparian vegetation is usually quite distinct from the vegetation of the surrounding catchment, because of the greater availability of water. In dry tropical regions, the contrast is very distinct, with riparian ribbons of closed forest following the stream through more sparsely vegetated woodlands. The effect is also obvious where forest has been cleared for agriculture, leaving riparian strips of natural vegetation. These ribbons of vegetation not only provide habitat for riparian organisms, but also provide vital connections in the landscape for closed forest species, which will move freely along the streams but not across alien woodland, grassland or agricultural landscapes.

IMPACTS OF FOREST CONVERSION

While it is difficult to generalise about the impacts of forest conversion on aquatic ecosystems because of the complex and interactive connection between streams and their catchments, it is expected that a significant change of any type in the catchment will result in a change in the stream ecosystem. The type and magnitude of such changes will depend on physical and chemical characteristics of the catchment as well as the extent and duration of the new land use relative to the size of the catchment. Changes in the catchment alter hydrological and/or geomorphologic processes that determine the character of the stream, such as the flow regime, the particle sizes and distribution of stream substrata, and the availability of nutrients and energy that support food webs. Consequently, they determine the habitat in which the biota (the microbes, plants and animals) must grow, reproduce and interact. In a pristine stream that has existed for millions of years, the biota has adapted to and evolved with the habitat which, therefore, strongly influences the structure and composition of the biotic community.

When forest is cleared, hydrological processes are affected by changes in soil infiltration, evapotranspiration by trees and possibly by capture of atmospheric water (P. Riddell, pers. comm.), potentially altering the quantity of water delivered to the stream channel. The pathway the water takes to the stream, and its interactions with physical, chemical and biological processes along the way, all affect its quality by changing the concentrations and types of dissolved and suspended materials it carries. Geomorphologic

Table 34.3. *Activities that may occur during and after forest conversion and some of the resultant agents that cause environmental impacts to rivers and streams*

Conversion to	Effect
Cropping	Increased sediments
	Increased nutrients
	Change in run-off and stream flow
	Introduction of pesticides
	Change in type of carbon inputs
	Flows (water harvesting, etc.) decreased through dams etc. for irrigation; increased in irrigation delivery and drainage network
Grazing	Erosion
	Nutrients from faeces and urine
	Increased sediments
	Increased nutrients
	Bank habitat damage
	Increased insolation and light inputs
	Loss of recruitment of native species
	Weed introduction
	Resuspension of streambed sediments and nutrients
Mining	Increased sediments
	Toxic contaminants
	Flows reduced by harvesting; increased by dewatering of aquifers
Water storage	Increased and/or decreased total flows
	Alteration of flow regime (seasonal pattern and extent and number of floods)
	Nutrient inputs
	Poor water quality (released from dam)
	Barrier to dispersal and migration
Urbanisation	Increased storm flows
	Reduced base flows
	Increased sediments
	Increased nutrients
	Other contaminants
	Riparian damage
Clearing/logging	Erosion
	Increased sediments
	Increased weeds
	Increased runoff
	Altered flows

processes, such as surface or bank erosion, may be altered, affecting the amount of suspended material transported to the stream channel (Parts II and III of this book provide details on all these processes in baseline and disturbed environments). The removal of vegetation may alter the light and insolation regime in the channel, changing temperature dynamics and promoting growth of primary producers (plants). Thus, changes in ecosystem components, such as increases or decreases in quantities of water, sediments, nutrients, or light, cause changes in the stream habitats. Wc have termed these components the 'causal agents' of change – that is, the products of various processes that act and interact to determine habitats. Table 34.3 gives examples of anthropogenic activities that may occur during and after forest conversion and the essential agents that cause environmental impact.

Many of these agents can be referred to as contaminants, but it is important to make a distinction between toxic contaminants, such as pesticides and heavy metals, and non-toxic substances or processes that change habitats and so influence the associated biota. Habitat effects tend to be understated in studies of impacted stream ecosystems due to the preoccupation with water quality rather than habitat quality. Agents that affect habitats often originate from non-point sources. They may include sediments from erosion, nutrients from agricultural runoff, and invasive weeds (promoted by increased nutrients and light levels). They may also include changes in the dynamics of water entering the stream, affecting in-stream flow and other attributes.

It is also important to recognise that many of these agents are normal components of aquatic systems. However, their concentration, extent and duration can be altered dramatically by forest conversion. Thus, large quantities of sediments may enter streams in undisturbed catchments during storm events, often as a result of landslides or bank collapses, but these events are normally infrequent and spatially isolated. Conversely, sedimentation due to inappropriate land-use tends to be prolonged and is usually extensive over a reach or a catchment. Similarly, nutrients such as nitrogen and phosphorus occur in the stream naturally but the quantity and delivery are often altered greatly by human activities. The altered distribution and abundance of these agents affect the outcome of many biotic interactions and thus alter the community composition, with knock-on effects on ecological processes. Below we discuss several of the major influencing agents in more detail.

Water flow to the stream

Changing the quantity and timing of water entering and flowing in stream channels is one of the most fundamental ways in which the ecology of streams can be altered. The biota is adapted to living in particular flow environments and there are distinct differences between the animals and plants living in flowing and still waters. Living in a flowing environment can have benefits, such as the supply of resources to the organisms and the transport of the organisms themselves, but also has risks, such as being swept away from suitable habitat. The current speed influences the substratum particle size and the delivery of food and nutrients. It also presents a direct physical force that organisms must be adapted to deal with and that to a large extent determines their distributions. Some animals and plants exhibit morphological and behavioural adaptations to live in the current and to avoid being swept away. Running water animals frequently have streamlined shapes, low vertical profiles and/or attachment devices. However, the complexity of the dynamics of water passing over heterogeneous substrata means that the relationships between flow and the biota in streams are also complex. This situation is further complicated by behaviours that allow organisms to use or avoid currents while feeding, dispersing or seeking shelter from predators or disturbance.

Flow, which describes the combination of discharge and current velocity, is not constant in any natural stream and varies both temporally and spatially at a variety of scales. In unregulated streams, flow varies inter-annually, seasonally, weekly, daily and sometimes from hour to hour. Very large floods, often produced by unpredictable climatic events such as cyclones (hurricanes), affect entire river systems, including upper catchments, stream channels, floodplain, estuary and even offshore environments. Seasonal floods that occur during the annual wet season or monsoon are more predictable but are also spatially extensive. Localised precipitation may increase the flow for a short period in a tributary but its effects alone will be negligible in terms of the whole system. Within a stream reach, current velocities vary considerably according to the channel gradient and width and the turbulence caused by the roughness of the substratum. These factors play a large role in determining the local distributions of more mobile substrata and the plants and animals that are associated with them.

The type of substratum in a stream bed is dependent on parent rock, stream cross-section and longitudinal profile, sediment inputs to the stream, the presence of riparian and in-stream plants, disturbance by domestic stock or vehicles, and so on. Distribution and extent of different substrata within the stream are dictated by variations in flow due to weather, slope and in-stream obstacles, resulting in a constantly changing dynamic mosaic on the stream bed. The distributions and relative abundances of different types, sizes and shapes of particles have a very large influence on the plants and animals living on them (Minshall, 1984). Many invertebrates are adapted to particular components of this mosaic and have very specific habitat requirements. For example, sedentary insects, such as the larvae of the blackfly, *Simulium*, and the midge, *Rheotanytarsus*, require attachment sites in high flow areas for feeding, growth and pupation. Some larval insects that live in fast flowing water have suction discs that require smooth surfaces to function – examples are some mayfly nymphs such as *Kirrara* species (family Leptophlebiidae) and fly larvae from the family Blephariceridae. Some animals are restricted to living in other particular habitats: for example, some moth larvae (family Pyralidae) live and feed on lily leaves; some mayflies such as *Jappa* species (family Leptophlebiidae) are active burrowers; and some beetle larvae (family Elmidae) are wood borers. Case-building caddis larvae (order Trichoptera) require particular materials for case construction (for example, different species use sand grains, hollow sticks, leaf fragments and even pieces of living bryophytes – A. Cairns, pers. comm.). Detrital substrata, which consist of decaying organic matter, vary with the type of organic material, the degree of conditioning by bacteria and fungi, the degree of physical fragmentation, and position in the water column – that is, floating, submerged or rooted in the stream bed. Associated with detritus is an abundant and diverse fauna that feeds on the detritus and its associated microbes (Figure 34.1).

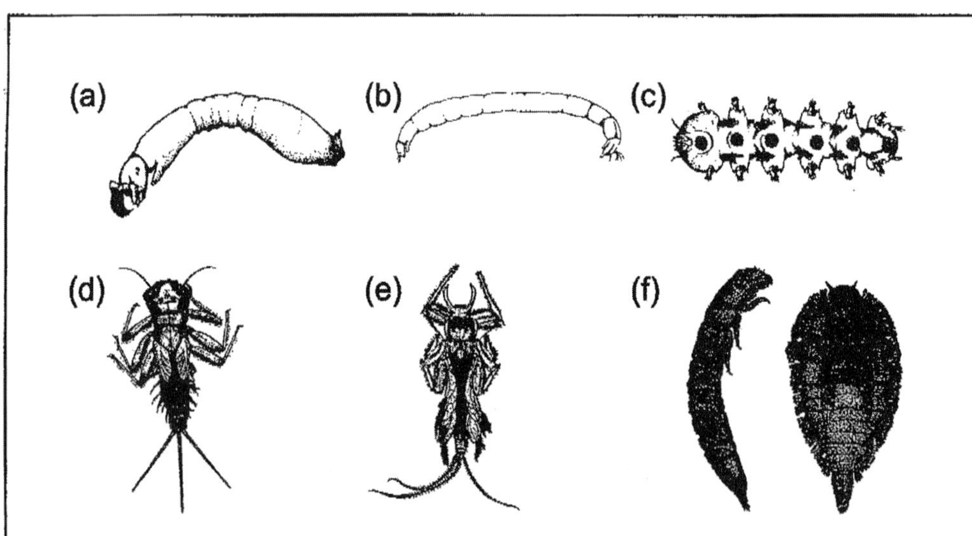

Figure 34.1 Examples of stream invertebrates and their adaptations: (a) *Simulium* sp. (Simulidae), (b) *Rheotanytarsus* (Chironomidae), (c) *Kirrara* sp. (Leptophlebiidae), (d) Blephariceridae, (e) *Jappa* sp. (Leptophlebiidae) and (d) Elmidae.

It is therefore necessary to understand how changes to the flow regime of a river alter the structure and distributions of habitats at relatively small scales to understand how flow affects the ecology of stream communities. High levels of hydrological activity in the tropics may exaggerate these processes, increasing the movement of sediments, and the temporal and spatial patchiness of substrata. Particle distribution is dependent largely on the hydrodynamic forces, particularly those occurring during high flow events. However, because substratum composition depends on flow but the substratum influences the current around it, differentiating the effects of substratum and current can prove difficult. Generally, larger particles (gravel and cobbles) are associated with faster currents, and smaller particles (sand, silt and clay) are associated with slower currents (Gordon *et al.*, 1992) although the relationship is not linear (Hjulstrom, 1939). Therefore, altering the flow conditions of a stream has a structuring effect on substratum and habitat dynamics with major consequences for the biota. Consideration of the biotic community of a stream thus needs to take account of the substratum composition, including particle types, sizes, shapes, and textures, as well as degree of packing and embeddedness. This heterogeneity and complexity is very important in providing a variety of microhabitats that enhance the biodiversity of the system. It has long been recognised that total numbers and biomass of animals often increase with increasing particle size, for example from sand through to stones, but decrease when substratum size increases to boulders or bedrock (Pennak and Van Gerpen, 1947, Ward, 1975). It is clear, therefore, that a major aim in stream conservation and flow management should be to maintain normal particle size distribution of the substratum.

Natural flows maintain stream form, habitats and biotic communities, but it is only in recent years that it has been recognised at a management level that it is the full range of flow intensities and flood frequencies that determines the character of a river system. While the effects of very large floods are obvious, as they can change the channel form and the distribution of substrata, lesser flows also have important, if more subtle, effects on sediment distribution, plant growth, etc. Increasing human demand for water has had a profound impact on the flow regime of most river systems in the developed world, through impoundment, diversion and water harvesting directly from streams. The construction of dams and weirs alters a stream system dramatically, both upstream and downstream of the impoundment. The many effects of dams and other forms of river regulation have long been documented (e.g., Baxter, 1977; Petts, 1984; Ward and Stanford, 1987; Littlehammer and Saltveit, 1984; Craig and Kemper, 1987; Gore and Petts, 1989). The impact on the stream system depends on the impoundment size, its intended purpose and its operation (for example, water quality may be seriously affected, depending on the timing of releases to the stream and whether these releases come from deep or surface waters – see below). Such structures cause fundamental changes affecting ecosystem function and the biotic community structure and break the upstream-downstream connectivity that is a normal feature of stream.

Upstream of a dam or weir the river section is transformed into a still-water environment, radically altering the ecology of that section of the river. Downstream there are changes to the flow, temperature regimes, suspended solids loads and water chemistry. Diversions and evaporative losses result in an overall reduction in discharge and current downstream. The stream's natural periodicity of flow is often altered, and its normal flow variability is changed considerably (Petts, 1984). Variability may be increased or decreased at a variety of temporal scales depending on the way in which the water release is managed. In some cases, seasonal patterns can be eliminated or even reversed, and daily fluctuations can become extreme (e.g. in hydroelectric schemes). Maintenance of a natural flow regime, including floods, is (or should be) a goal of contemporary water managers. The normal flow regime not only sustains normal channel morphology, but also provides habitat diversity, stimulus for migration and spawning of fish and other animals, weed removal and sediment transport. While flood flows have clear functions, the need to retain normal low flows is often overlooked. Supplementation of flow in dry times to deliver water downstream makes maintenance of normal low flows a difficult management objective, particularly because the lower the natural flow, the greater is the demand for water downstream. Unseasonably high flows in the channel have several important effects, including extending the period of water-logging of riparian vegetation, and subsequent die-back, encouraging weed growth on sand banks, removing remnant pool or slow-flowing habitats which are used by opportunistic and specialist species, and giving false cues for spawning movements by fish and other animals.

Transport of suspended particles and the distribution of fine sediments are greatly affected by dams, which cause much of the sediment (especially coarser fractions) to settle in the reservoir. These sediments would normally supply the river, estuary and coastal environments downstream and their removal leads to erosion downstream of the dam, where sediments are removed but not replaced and, in some cases erosion of beaches beyond the river system. Channel morphology is largely flow-dependent and therefore the channel form is modified by the altered flow regime. Dams that release very high discharges variously cause scouring of the fine sediments, armouring of the streambed, channel and bank erosion, and incision of the streambed. Where flushing flows are eliminated, on the other hand, sediments accumulate further downstream or below unregulated tributaries; this may lead to smothering of normal substrata, reduced water quality and weed invasion. River temperature regimes can be altered by impoundments, in which water is typically stratified into upper warm and lower cool layers. Releases from the bottom of the dam lead to major changes in temperature in the river, sometimes for many kilometres

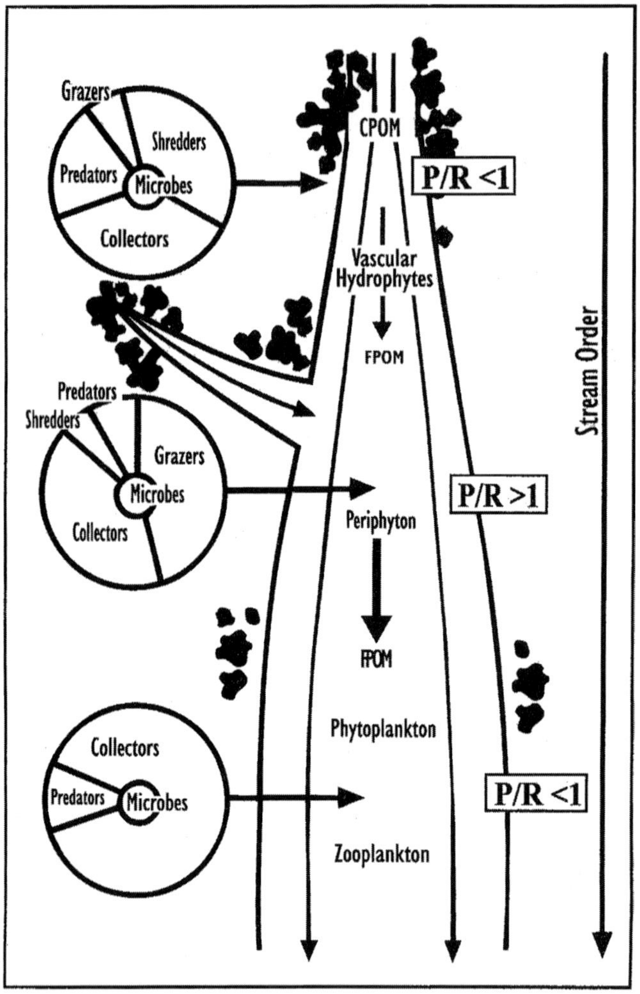

Figure 34.2 The River Continuum Concept, showing changes in primary energy input and the importance of functional feeding groups of invertebrates with stream size (after Vannote *et al.*, 1980). CPOM, coarse particulate organic matter; FPOM, fine particulate organic matter; P/R, Primary productivity to respiration ratio.

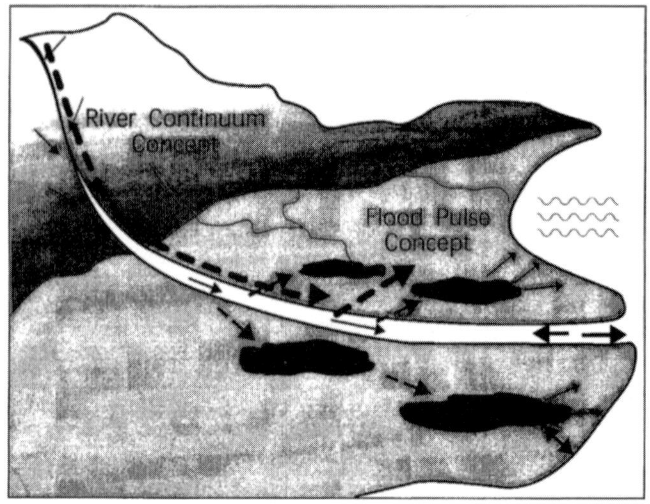

Figure 34.3 Schematic showing the dominant river system models in a large river with extensive floodplain. In the lower reaches the Flood Pulse Concept model (Junk *et al.* 1989) essentially overrides the patterns predicted by the River Continuum Concept. (Vannote *et al.* 1980).

downstream (Ward, 1975). Stratification leads to major changes in water quality (e.g. reduction in dissolved oxygen and increase in nutrients and other contaminants such as iron and manganese sulphides), with sometimes devastating impacts, measurable for many kilometres downstream (Boulton and Brock, 1999).

Dams and weirs are major barriers to dispersal of organisms. Many fish, crustaceans and other animals are migratory, moving between river tributaries, estuaries, floodplain wetlands and the sea during their life cycles. Barriers to any part of the cycle expel those animals from the system. Even alterations to the flow regime can decrease connectivity along the river and between the river and its floodplain. Connectivity and predictable change along a river is the basis of the River Continuum Concept (RCC) (Vannote *et al.*, 1980; Minshall *et al.*, 1983), which attempts to explain observed predictable patterns of gradients of physical features, community composition and ecological processes along rivers (Figure 34.2). Although the RCC was developed for snow-melt-fed streams of the northern hemisphere, and while it may not apply in detail to systems that are subject to very different flow regimes and unpredictable disturbance (e.g. in parts of Australia – Lake, 2000), or to systems with extensive floodplains (Davis and Walker, 1986; Welcomme *et al.*, 1989; Dudgeon and Bretschko, 1995a), the principles that underlie it are very useful in considering rivers in their entirety. The continuity and gradual change along rivers are of particular importance, and they are severely interrupted by dams. In fact, so bad are the effects of impoundment along rivers in the USA that the so-called Serial Discontinuity Concept was developed to describe processes in a system with multiple dams (Ward and Stanford, 1979). Serial discontinuity in the broad sense may also apply to those seasonal tropical streams that cease flowing in the dry season, becoming a series of disconnected lagoons.

Not only does river regulation interrupt connectivity along a river but it can also break links between the river and its floodplain. The major environmental factor in large river-floodplain systems is the alternation of flooding and draining or drying of the floodplain and the lateral transport of nutrients and organic material due to pulses in river discharge. This transport of organic matter from the floodplain to the river, in greater quantities than from upstream sources, essentially overrides the patterns predicted by the RCC for the lower reaches of these systems and productivity is strongly dependent on this seasonal cycle (Figure 34.3). This description of floodplain river dynamics has been expressed as the Flood-Pulse Concept (Junk *et al.*, 1989; Osborne, 2000). Periodic inundation and oscillation between terrestrial and aquatic phases drive floodplain ecology. Floodwaters convert the floodplain into an aquatic system, promoting rapid growth of associated aquatic plants and animals. As the water levels retreat and drying occurs, the aquatic plants die and decompose, releasing nutrients to the water column. Some of this vegetation may be transported to the river channel, contributing significant amounts of organic material to the riverine food webs. Many fish and invertebrate species migrate on to the floodplain with the flooding water to breed and exploit the rich food resources. The floodplain provides important feeding areas, spawning sites and nursery areas for juvenile organisms. Rivers with floodplain systems are thus much more productive than rivers that lack them (Osborne, 2000), even after water drains back into the main river channel. As the floodplain dries, the silts carried by the river and decomposing aquatic plants provide a source of nutrients to the terrestrial plants that grow during the dry season. This ensures that high rates of production are maintained through the dry period and that nutrients are retained on the floodplain. Junk (1997) discusses in detail the ecology of a pulsing system, describing the structure and function of the Central Amazon floodplain, one of the largest floodplains in the tropics.

Dams can dampen the flood flows and limit the connection between the river and floodplain, thereby reducing productivity in both environments and a loss of ecological processes vital to the biotic communities, including productive fisheries (Glaister, 1978). However, the relationship between the ecology and the flow regime of a river system is so complex that we have, at present, very poor ability to predict what changes will result in the biological community when a particular river reach becomes regulated. This is particularly true in lowland tropical systems where much of the basic ecology is poorly known.

Sedimentation

Erosion, transport and deposition are natural processes in streams and are vital to the maintenance of instream habitats. However, forest conversion and subsequent changes in land use accelerate natural erosion processes and result in an increase in the transport of fine sediment and its deposition in river channels. Changes to the flow regime of rivers also strongly influence erosion and transport and sorting of sediments, thereby modifying the instream environment. Rather than being a short-term, infrequent perturbation followed by rapid recovery, sediment deposition can become an almost continuous disturbance. Forest conversion can lead to the mobilisation of large volumes of sediment into streams by activities such as the construction of roads (Extence, 1978; Reid and Dunne, 1984), extractive forestry (Scrivener and Brownlee, 1989), mining (Davies-Colley *et al.*, 1992) and agriculture (Walling, 1990; Richards *et al.*, 1993) and also discussed in this volume in chapters by Chappell, Tych *et al.*; Grip *et al.*; Cassells and Bruijnzeel; Thang and Chappell. On the other hand, dams and weirs can trap sediments, modifying the geomorphology of the river downstream where bottom sediments continue to be eroded but are not replaced. Thus, changes to the sedimentation dynamics (i.e. mobilisation, transport and storage of sediments) may cause major changes to the physical environment and consequently to the biota in rivers.

The geomorphological processes that mobilise and transport sediment are complex and depend on the type of sediment and the hydraulic forces involved. At the same time, geochemical processes and sediments interact in their effects on water quality and toxicity to the biota. Geochemical processes (such as flocculation) may also affect rates of deposition and the type of material deposited, and there are numerous associations between fine sediments and chemical contaminants, such as transport and storage of heavy metals by clay particles (Thoms, 1987), with important implications for the stream biota (Wood and Armitage, 1997).

Larger sediments are sourced largely from the bedrock in the stream channel. Fine sediments are derived mainly from the bed and banks of the stream and its tributaries, and from erosion and transport of soils external to the channel. The nature of the river and its surrounding environment has a strong influence on the type and volume of sediments generated and transported to the channel and the degree of sedimentation and storage within it. Within the stream channel, banks and bars are subject to erosion from the shear forces exerted by the water, so the degree of in-channel erosion is strongly related to current velocity, the stability of the channel bed and banks, the duration of exposure to these forces, position in the channel and the presence of abrasive materials in the water (such as sand). Bank stability is also a function of the material it is made of (e.g. sand or rock) and the presence of vegetation, whose root masses can help bind the bank. Bank stability is often greatly diminished by the loss of riparian vegetation and damage by cattle and other domestic animals. Other sources of fine sediment include the bed interstices, backwaters where it accumulates during base flow conditions, and stands of aquatic plants. Much of this material is derived from non-channel sources and stored in the channel only temporarily.

Anthropogenically-induced changes in river flow and the supply of sediment have widespread effects on the spatial distribution of bed sediments and the channel form, with major consequences for the biotic communities. Dams and weirs have a negative downstream impact by changing the supply, erosion rates, conveyance and storage of sediments within the river channel. Excessive levels of sediment affect the biotic communities in streams by increasing the turbidity and abrasive action while it is in suspension and by changing the benthic habitats when it is deposited. Duration of the input of sediment is a major determinant of the degree of impact. At the community level, response to sedimentation can be limited in the short term, with surprising resistance exhibited by the fauna (Connolly and Pearson, 1998). Such resistance is

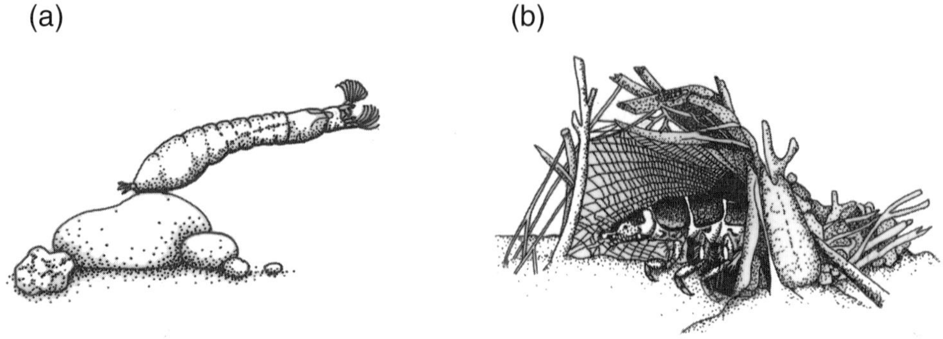

Figure 34.4 Examples of filter-feeding invertebrates: (a) Simuliidae and (b) Hydropsychidae.

probably a result of adaptation to spates of short duration. But prolonged high sediment levels reduce diversity and abundance of stream biota such as fishes (Hortle and Pearson, 1990). Many of the activities that occur during and after forest conversion have long-term and widespread effects on stream geomorphology and ecology that give the biota no opportunity to seek refugia from which to recolonise and recover.

Suspended sediments are abrasive and affect the turbidity in the water column. Sediments can abrade animals, plants, microbes and organic detritus from substrata within the stream, thereby reducing population levels and available food in the food web. A prolonged increase in turbidity reduces light penetration into the water column and so reduces photosynthesis, and the diversity and abundance of aquatic plants, including single-celled diatoms that live on stable surfaces (rocks, wood, living plants) and are a major producer of organic material in the food web. This reduction in plant biomass reduces available habitat and food for animals, with substantial effects through the food web. Changes in abundance of large plants also alter flow, nutrient, pH and dissolved oxygen dynamics, and other physico-chemical processes. High turbidity may have a short duration following storm events, or may be of extremely long duration – for example, the Burdekin River in northern Queensland mobilises very fine sediments during storms, which remain in suspension for many months in the Burdekin Dam and downstream. Water from the dam is used to supply a major irrigation scheme 100 km downstream. Consequently, for 100 km downstream from the dam the river is constantly very turbid whereas its natural state is very clear, except during storm flows.

Prolonged turbidity has an important influence on the feeding capabilities of those animals that feed by sight, such as many fish species, and insectivorous and piscivorous birds. Suspended fine sediments also hamper the feeding activities of filter-feeding animals. Many of the invertebrate animals in streams feed by filtering fine organic particles out of the water column using a variety of techniques, such as brush-like appendages (e.g. blackfly larvae – Simuliidae) and fine silken nets (e.g., net-spinning caddis larvae – Hydropsychidae) (Figure 34.4). These animals attach to solid substrata in high currents to maximise the filtration rate of organic material. Large quantities of suspended inorganic particles can reduce the efficiency of filtering activities by diluting the organic content and hence the nutritional value of the material that is filtered, and eventually by clogging the filters, rendering them non-functional (Aldridge *et al.*, 1987). Clogging of the gills of fish and invertebrates is also a major effect of suspended sediments (Bruton, 1985), which in extreme cases leads to abrasion of respiratory surfaces, introduction of pathogens, and consequent deterioration of the animals' condition.

Fine sediments deposited on the stream bed block up interstices, often penetrating into the deeper sediment (Richards and Bacon, 1994), thereby reducing diversity of habitats and their associated plants and animals. Fine sediment reduces the abundance of the biota, including eggs of invertebrates and fish species that deposit their eggs on the substratum. The sediments dilute the organic detritus, and reduce growth of microorganisms, reducing the food-gathering efficiency of detritivores that collect and scrape material from the stream bed (Cline *et al.*, 1982; Graham, 1990). Fine sediment impedes the attachment of organisms that require a hard substratum, and may favour colonisation by soft-bottom organisms such as some midge larvae (Chironomidae) and worms (Oligochaeta) (Armitage, 1995). A typical response of invertebrates to high levels of sediment is to attempt to move away, usually by drifting downstream in much higher numbers than normal (Culp *et al.*, 1985). The presence of deposited fine sediments can have detrimental effects on the embryonic development of salmonid fish because of the reduced sediment permeability and consequent reduced supply of dissolved oxygen and removal of waste metabolites. Although similar experiments have not been carried out on tropical fish and invertebrate species, it is likely that many species will be similarly affected. Again there is a need to understand specific faunal habitat requirements at micro-scales to appreciate the response to fine sediment deposition on the biotic community as a whole.

Vegetation modification

Modification of the vegetation in a catchment can have major consequences not only for the quantity of materials delivered to the stream, but also for the composition of those materials and for the nature of the energy base of the food web (Gregory *et al.*, 1991). Replacement of terrestrial plant species leads to changes in the type of leaf material entering the stream, both qualitatively and quantitatively. This effect is particularly pronounced when the native riparian vegetation is replaced by exotic species. For example, in much of southern Australia the riparian vegetation has been converted from native species to European willows. Animals that normally eat organic detritus are not able to deal with exotic species such as willows, which therefore break down only very slowly, thus depriving the stream of the organic matter essential to fuel its biotic communities. Inevitably, similar effects occur when weeds replace native riparian vegetation or when mixed forest is replaced by exotic or even native monocultures, such as pines.

Conversion of forest means a drastic change in terrestrial habitat, and associated loss of plants and animals. Where such loss encroaches on the riparian zone, there may be substantial impact on the stream. The riparian zone provides habitat for partly aquatic species. For example, emerging insects require cover to avoid predators, and in many cases they require feeding and breeding sites provided by the vegetation. Reptiles require basking sites that fallen timber provides, and birds that feed in streams require perches from which to dive in pursuit of prey (e.g. kingfishers), or on which to roost or nest (e.g. herons, cormorants). Loss of the native riparian vegetation and replacement by weedy species, especially non-woody species, removes these functions and thereby affects stream ecology by removing a pool of top predators, with the potential for cascading effects through the biotic community.

When land is converted from forest to other uses, some of the impact on streams can be minimised by retaining a strip of native vegetation along the stream banks. This riparian strip can maintain most of its normal functions, including habitat provision, shade, organic input to the stream, protection of the stream from inputs of materials from the surrounding land, and so on. The appropriate width of such strips has not been determined adequately for most situations, but is known to vary with vegetation type, size of stream and particular function. Thus, while a narrow riparian strip might maintain an adequate shade regime, it may be inadequate as habitat for forest species.

Loss of the riparian vegetation leads to loss of normal organic input and, therefore, loss of normal stream productivity. This may be partly compensated for by an increase in plant production in the stream as a consequence of increased light and temperature resulting from the removal of shade. Primary production in forest streams is severely restricted by shade, with plant life restricted to slow-growing bryophytes (liverworts and mosses) and sparse microscopic algae (diatoms etc). Increased light levels allow establishment of larger plants, which are fast growing, as well as much larger populations of diatoms. Larger plants include rooted forms that can form dense beds where sediments are soft enough for their establishment, and large filamentous algae that attach to any stable substratum, including sand beds and rocks. Mineral substrata and the larger plants themselves are then colonised by smaller algae. Together, these plants boost primary production substantially.

However, the nature of the stream community is altered to a large extent by increased plant production. Plants overtake fine substrata, and algae may cover rocky substrata. The predominance of leaf shredders and detrital collectors is replaced by grazers, which scrape the microalgae off the surfaces of larger plants. Plant beds alter flow regimes and allow colonisation by slow-water species, replacing the normal flowing-water fauna. Altered flow regimes may lead to increased sediment deposition, further encouraging plant growth. These changes in instream vegetation and sediment deposition may affect drainage severely and exacerbate flooding and bank erosion.

Ecosystem energetics

All ecosystems are based on transfer of energy through food webs. Mostly this energy is derived from the sun, and harnessed by plants during photosynthesis. Light is therefore an important resource and its relative abundance dictates how productive a system is. In streams there are several sources of photosynthetically derived organic material. Most obvious, but not necessarily most important, are the large plants ('macrophytes') that grow in streams. In open slow-flowing streams there can be extensive growth of plants, including submerged 'water weeds', emergent reeds and rushes, and floating forms such as water lilies. In lowland tropical systems, the strong light, the high temperatures and the generally adequate levels of nutrients, stimulate abundant and diverse growth. Interestingly, however, these plants tend not to be eaten by aquatic animals until they decay and enter a detrital food web. Only a few specialist aquatic herbivores (certain birds and moth larvae, for example) ingest material from larger plants in streams (Newman, 1991), although many macrophytes are eaten by large mammals, including domestic cattle.

Less obvious are the microscopic plants that grow on solid surfaces, including the surfaces of macrophytes. These are commonly grouped as micro-algae, and include filamentous and unicellular forms. They are grazed by a host of invertebrate animals, and probably form the most important part of the food web in these systems. In some cases the filamentous algae form dense mats such that they can be regarded as macrophytes. These algae may be fed upon by several fish species.

In smaller streams that are shaded by riparian vegetation, the lack of light inhibits the growth of plants, such that diatoms and the like are not very abundant and do not grow very fast. In these situations, grazing animals are scarce. Many shaded upland streams, however, are characterised by extensive growth of bryophytes which might suggest that plant productivity is quite high. However, growth of bryophytes is apparently very slow and, in any case, few animals eat them. In these small dark streams, the main photosynthetic input is derived from the forest: just as litter falls to the forest floor and is eaten by a host of soil animals, so litter falling into the stream is the basis of an important food web. In the stream there are many animals known as 'shredders' which chew up and fragment leaves, making them available for other animals to eat. At the same time this rich source of organic material is colonised by bacteria and fungi, which break down the organic material and make it more nutritious for animals (Arsuffi and Suberkropp, 1985; Pearson and Connolly, 2000).

Detrital material on the forest floor and in the stream is leached by water percolating through the soil, and flowing downstream. Leaching provides another source of food – the dissolved organic material – that is fed upon by bacteria which live on stones and on large plants. These bacteria are in turn scraped from surfaces by invertebrates and tadpoles.

The overall picture of energy input in a stream system, therefore, involves a number of processes, dependent on autochthonous (instream) organic production and allochthonous inputs from the surrounding catchment. The relative importance of these inputs depends on several factors. Light is very important, but so is habitat. For example, large macrophytes do not normally grow in small upland streams because firstly, there is insufficient light and, secondly, the substratum is unsuitable for their roots to establish. Always there is interplay between stream flow, substratum, light availability and nutrient concentrations. And in all cases, whether or not aquatic plants are an important source of organic material, detritus derived from the land or from upstream is a major component of the food web.

Different food webs depend equally on the type of organic input and the available habitat. Grazers are only abundant where there are suitable plants to graze, while some animals will only occur if there is a suitable flow environment and substrate. Therefore, the interaction between organic input and habitat determines how the food web works – that is, which species are part of the web, and what the linkages are between them. Figure 34.5 shows simplified food webs for small streams and larger rivers. The different components of webs in different sections of streams are the basis of the River Continuum Concept, described above (Figure 34.2).

Alteration of riparian conditions therefore has a very marked effect on the food webs and on the biodiversity of the stream community. Loss of riparian vegetation in the uplands results in a loss of the main source of organic input for the detrital food web, but may stimulate growth of encrusting algae. It may also stimulate the growth of exotic weeds on the stream bank leading to completely alien inputs that the fauna cannot process. Downstream, where the waterway broadens and becomes lighter, absence of riparian vegetation leads to growth of reeds and other shallow-water emergents, which thereby change not only the make-up of the normal organic input, but also the littoral and bank habitats.

Changes to organic inputs can be more subtle. For example, exotic weeds such as para grass and crops such as sugar-cane can contribute substantial quantities of organic material to streams, but this material is apparently not assimilated by the resident fauna (Bunn *et al.*, 1997). It is not clear what happens to that material nor what the outcome is for the community structure as it is difficult to separate the different effects of organic detritus composition and habitat changes. It has long been known that addition of organic material can stimulate population expansion of the resident fauna – for example, downstream of sugar mills, which can produce concentrated organic effluents, there can be substantial growth of populations of pollution-tolerant organisms such as midge larvae and burrowing worms. But at the same time, the fauna becomes dominated by these tolerant species, with the overall biodiversity reduced as the intolerant species are eliminated (Pearson and Penridge, 1987).

Nutrients

Probably the most ubiquitous ecological change that is occurring in aquatic systems throughout the world is the increased input of nutrients and consequent fertilisation and eutrophication of fresh waters. It is generally accepted that deforestation increases the amounts of particulate and dissolved material present in stream water in temperate (Likens *et al.*, 1970; Webster *et al.*, 1990) and tropical catchments (Williams *et al.*, 1997; Chappell, Tych *et al.*; Grip *et al.*, both this volume). Similarly, the concentrations of dissolved nutrients in stream water correlates with the proportion of agricultural landuse in the catchments (e.g. Dillon and Kirchner, 1974, 1975; Smart *et al.*, 1995; Jordan *et al.*, 1997). In many disturbed streams the effects of increased nutrients are exacerbated because the clearing of riparian vegetation increases light levels and encourages vigorous growth of invasive weeds and the instream growth of algae and larger plants.

The term nutrient refers to compounds that are necessary for the building of tissue for growth, reproduction and repair, and provision of energy for metabolic activities (see Proctor, this volume). Nutrients are essential for life and when in short supply can limit metabolic processes and productivity in all ecosystems. In fresh waters, carbon, nitrogen, phosphorus and silicon are the most heavily utilised elements. For aquatic plants carbon is normally readily available in carbon dioxide dissolved in the water,

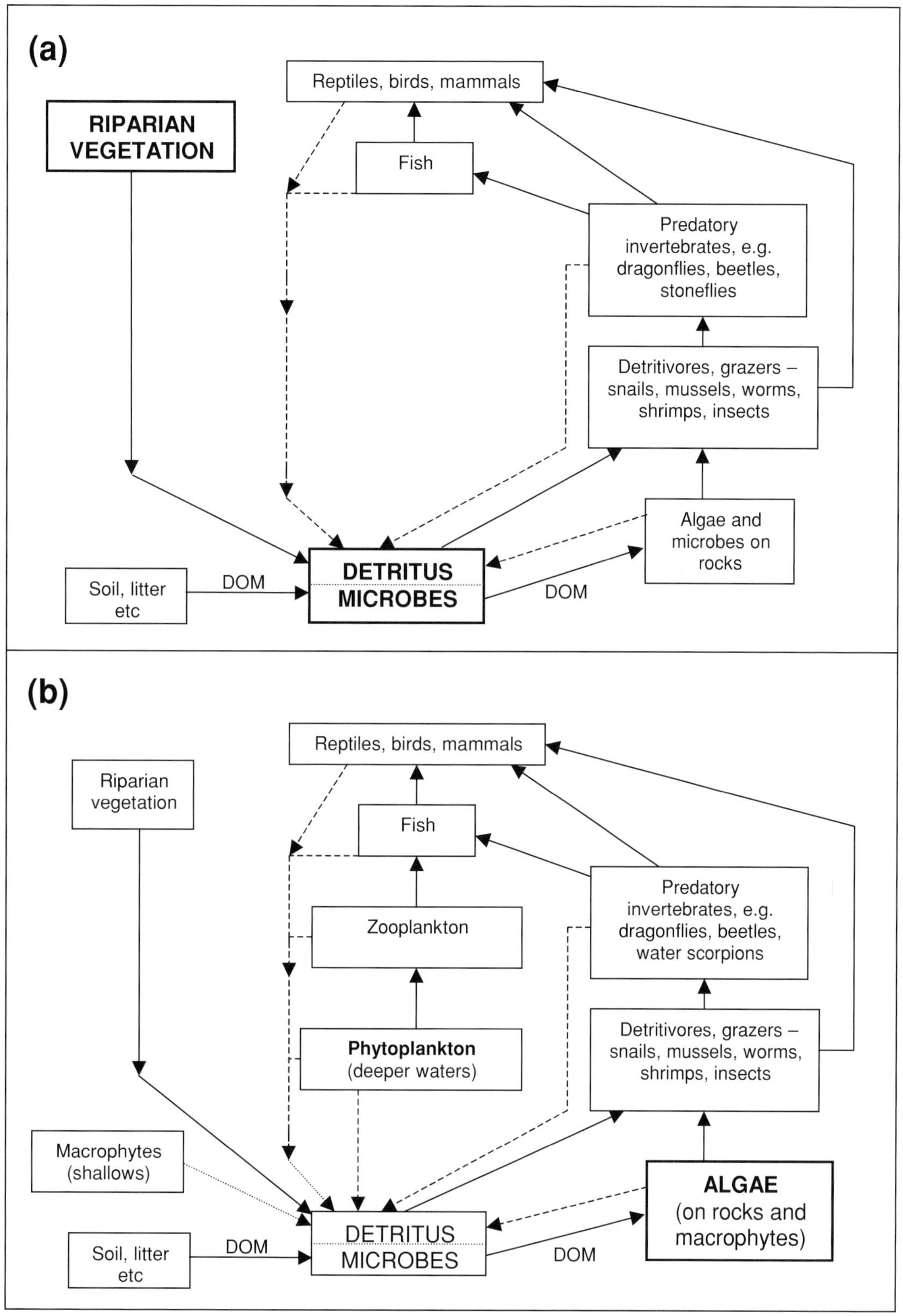

Figure 34.5 Simplified food webs: (a) small shaded stream, and (b) large, open, shallow flowing stream. DOM, dissolved organic matter.

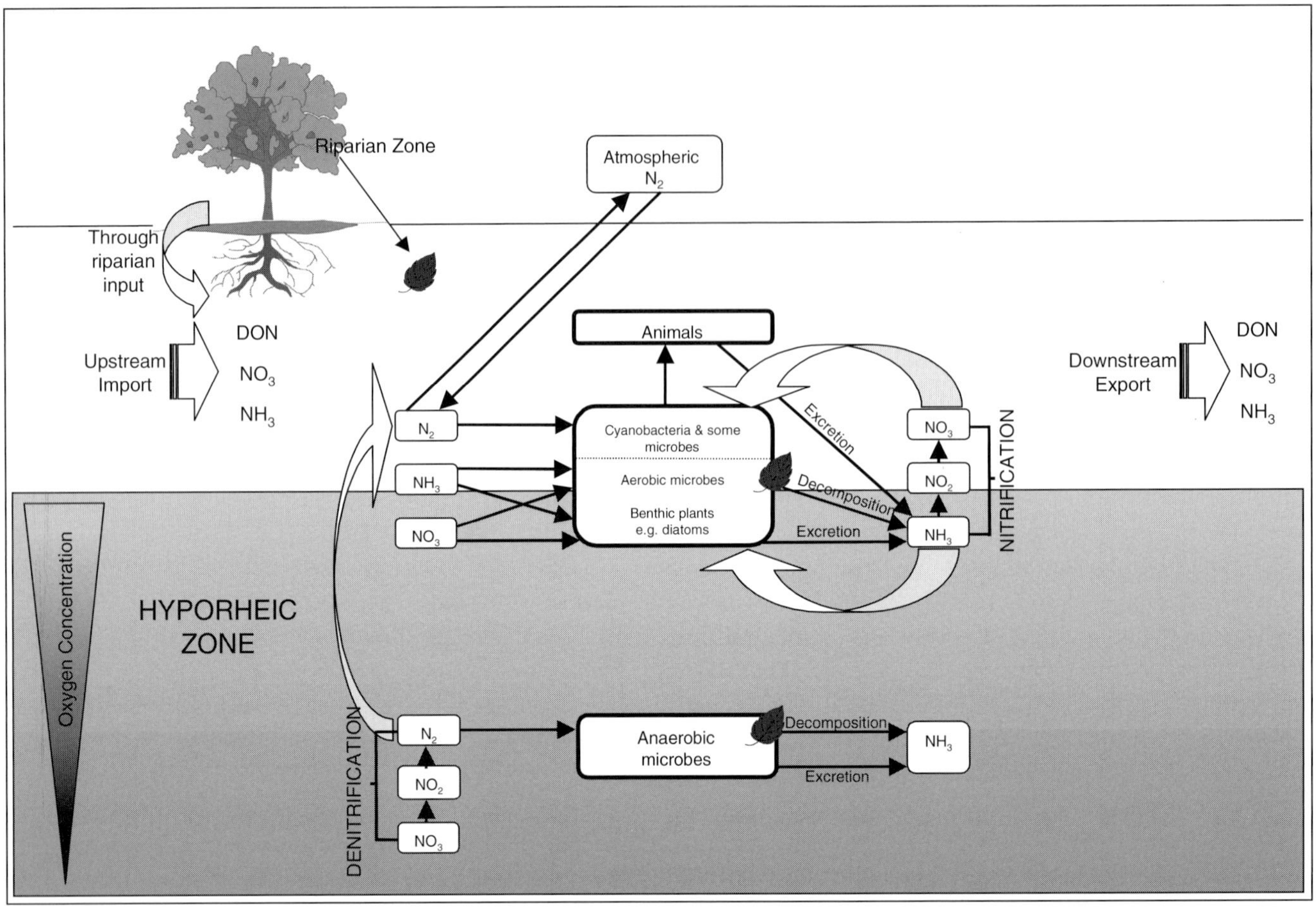

Figure 34.6 The processes of nitrogen cycling in a typical stream reach.

so it is usual to consider nitrogen, phosphorus and silicon (the so-called macronutrients) to be the elements of primary importance. Plants require many other materials in smaller quantities. These are referred to as micronutrients and include the elements sulphur, potassium, magnesium, sodium, calcium, iron, manganese, zinc, copper, boron and vanadium, and the vitamins thiamine, cyanocobalamin and biotin (Hutchinson, 1967). Animals in turn obtain carbon and other nutrients through the consumption of plant material either directly or passed through a food web. Many aquatic animals, particularly invertebrates, will also obtain much of their nutrition through the consumption of micro-organisms such as bacteria or fungi that colonise detritus particles (Pearson and Connolly, 2000). Micro-organisms have the ability to obtain the essential nutrients they need from the decomposition of dead organic material but can also utilise nutrients dissolved in solution.

Most nutrients are transported in dissolved form, but phosphorus, trace metals and hydrophobic organics are transported mainly as particles (Allan, 1995). Nutrients can occur in various chemical forms in water, as ions or dissolved gases, and their dynamics can be very complex. Processes that affect nutrients in solution include transformations from one chemical species to another, physical changes such as adsorption (the attachment of ions on to inorganic surfaces), desorption and chemical precipitation, and biological uptake, transformation and eventual release. The passage of a particular element will depend on the supply and demand for that element which may change with time and location. These processes occur on the streambed and banks, on the floodplain and in subsurface water. They cycle essential nutrients and slow down their downstream movement, so that particular atoms or molecules pass repeatedly through cyclical processes, essentially spiralling as they progress downstream (Webster and Patten, 1979; Newbold *et al.*, 1983; Newbold, 1992). An example of nutrient cycling is shown in Figure 34.6, which describes schematically how nitrogen is cycled in a typical stream reach. The principal biotic processes are assimilation, excretion and decomposition. However, nitrogen cycling is also complicated by additional transformations carried out by various nitrogen-fixing bacteria (and legumes).

In natural ecosystems many of the nutrients are not freely available in large quantities. They are quickly taken up and assimilated by micro-organisms and plants. Nutrients are required in particular ratios and the production of these organisms is thus frequently limited by the availability of a particular nutrient. If environmental factors, such as temperature, light and oxygen supply are adequate, then primary productivity is limited by whichever nutrient is present in least supply relative to demand. If a limiting nutrient becomes available in excess amounts then production will increase until another nutrient becomes limiting. It is widely held that phosphorus is the most limiting nutrient in many fresh waters (Allan, 1995). The addition of phosphorus will thus increase microbial and plant growth substantially to where only the availability of light or space prevents further growth. Unfortunately many of the activities associated with forest conversion increase input of phosphorus, as well as other nutrients such as nitrate and ammonia, into fresh waters. The process by which increased nutrient inputs, caused by anthropogenic influences, induce high levels of biological production is referred to as cultural eutrophication.

The availability of nutrients is a major factor that determines the level of primary production (plant growth etc), and so is a controlling factor in all ecosystems. Nutrient concentrations (along with light and other factors) control the productivity of instream macrophytes and algae, and plants in the riparian zone. They also influence production of bacteria and fungi, which in turn affect decomposition of dead organic material; these processes then influence the nutrition of animals, and the level of secondary productivity (i.e. size or turnover rate of animal populations) (Pearson and Connolly, 2000). Most of the nutrients that support the growth of plants, microbes and animals in streams are derived from the surrounding terrestrial environment. The flux of nutrients entering streams is highly variable and depends on the type of the nutrient, its source and climate variables. Natural sources of nutrients include: weathered rocks and soils; wind-eroded terrestrial dust and salt spray; decomposition products of organic material; animal excreta; suspended particulate and dissolved material carried in overland flow and through soil; dissolution of materials in groundwater; and materials released when bank and channel bed sediments are eroded. In a particular system, the level and type of productivity and all subsequent processes (animal-plant and animal-animal interactions and foraging mechanisms) are determined in part by the supply and pathways of nutrients. Major changes in nutrient supply lead to shifts in the composition and structure of the biotic community: organisms may flourish, be unaffected or decline, with knock-on effects through the food web.

The input of nutrients to streams typically increases with changes in land use in the catchment, through such factors as increased erosion resulting from disturbances to vegetation and soil and the presence of cattle and other stock, urban and agricultural runoff, and the disposal of waste effluent such as sewage. In undisturbed forest, concentrations of dissolved and particulate materials in rivers and streams are closely linked to soil biogeochemical processes. The clearing of forest and subsequent adoption of other land uses has major consequences for these processes. Even in the absence of waste effluent or the runoff from the application of fertilisers, changes in land-use can alter the chemical environment of streams substantially, changing the quantity, timing and chemical form of the materials delivered from the catchment (Likens *et al.*, 1970; Williams *et al.*, 1997). Neill *et al.* (2001) describe how the clearing of moist, lowland tropical forest for cattle pasture has consequences for soil biogeochemical cycles and the concentrations of dissolved and particulate materials in streams in the Brazilian Amazon. Such changes have important implications for biotic production that controls aquatic trophic dynamics. In the account of Meill *et al.* (2001), it is suggested that there is a switch from phosphorus limitation to nitrogen limitation in streams where the catchment has been converted to pasture.

Malmer and Grip (1994) monitored the water chemistry in six streams in a paired catchment experiment for five years in Sabah, Malaysia, comparing the effects of different forestry practices. Their treatments involved: (i) clearing of secondary vegetation, burning and planting; (ii) clear-felling, crawler-tractor extraction, burning before planting; and (iii) clear-felling, manual extraction, with no burning before planting. They found that concentrations of major plant nutrients (N, P and K) became correlated to stream flow during treatments. Leaching was higher from clear-felled residues with a strong effect of burning. Base-flow concentrations of these nutrients were found to be significantly elevated for a year after treatments but still detectable after three years. They observed an initial large pulse of leaching associated with mineralisation after tree felling and particularly burning. The more long-term elevated concentrations were attributed to the input of weathering products. Overall, Mulmer and Grip concluded that the amounts of nutrients entering the stream from the surrounding catchments depended on the amount of vegetation killed and the extent of soil disturbance. Consequently, forestry practices using tractor extraction and burning before planting created the largest leaching losses. They also observed lowered leaching for several elements after repeated forest fire and pre-planting fire, indicating the potential for nutrient deficiencies and exhaustion if these practices are continued at a site for a long time.

This example raises a major issue concerning humid tropical forest being cleared for agriculture, namely the frequently low fertility of the soil and its inability to support crop production in the long term. There has been a long-standing paradigm that undisturbed tropical rainforests exist on nutrient-poor soils with little or no leakage, and that most of the nutrients are contained in above-ground biomass, maintained by virtually closed nutrient cycles. This view may be an over-generalisation (Proctor, 1987 and also this volume; Whitmore, 1989), but is a concept that can explain why crops fail on many sites cleared of lush tropical forest. While not all tropical forests occur on nutrient-poor soils, there is a disproportionate area of the moist lowland tropics situated on moderate to very low fertility soils (Bruijnzeel, 1991). An example is the central Amazon forest, analysed by Klinge (1976). There is also evidence that process rates are higher in the tropics for comparable moisture regimes and soil fertility (Jordan, 1989). The amount of 'leakage' or output of nutrients from the forest to the stream is dependent on the nutrient status and biogeochemical processes in the surrounding catchment. Unfortunately, the dynamics of nutrient input-output budgets of tropical forest ecosystems are difficult to measure and are still poorly known (see review by Bruijnzeel, 1991), although a recent report of water quality in pristine South American streams indicates that most nutrients are tightly retained in the forest, and emerge in an organic form only (dissolved or suspended) (Perakis and Hedin, 2002). It is expected that in catchments with low fertility soils and where there is little leakage and considerable retention within the forest, the streams themselves will have low levels of nutrients. This is the case, for example, in parts of northern Australia (Pearson *et al.*, 1986), although Bruijnzeel (1990) notes that streams draining areas with infertile soils may still carry considerable amounts of nutrients if the stream has incised itself sufficiently into the weathering mantel to expose the fresh bedrock.

Another difficulty in predicting how catchment changes will affect the nutrient dynamics and ecology in the stream is caused by the variety of processes that occur in the riparian and hyporheic (within the substratum) zones, essentially the interfaces between the stream and its surroundings (Tabacchi *et al.*, 2000). Nutrients enter the stream largely through these pathways, which are areas of rapid and extensive nutrient transformation (Bonell, this volume). The processes that occur within these zones, such as mineralisation, nitrification, plant and microbial uptake and denitrification, are still poorly described, but without doubt play a very significant role in determining the final input of nutrients into the stream. This topic is particularly important as riparian zones are often disturbed yet have the capacity to buffer or moderate the effects of landuse practices on stream ecology. Chestnut and McDowell (2000), in their studies of tropical streams in Puerto Rico, have shown that these zones immobilise significant amounts of dissolved N and C from adjacent hill slopes and groundwater as water moves towards the stream. Their results show remarkably constant concentrations of dissolved N and dissolved C in stream water despite variable effluent groundwater, suggesting significant N and C retention or loss in the riparian and/or hyporheic zones. Differences in both speciation and concentration of nutrients within the riparian zone groundwater appear to be determined by the geology and hydrogeology (McDowell *et al.*, 1992). It is therefore important to understand the processes that are occurring in these zones before the impact of nearby landuse on stream nutrient fluxes and ecological processes can be assessed fully. However, riparian biogeochemistry is poorly understood because of variability in soil type, slope, rainfall and complex biotic interactions, especially involving micro-organisms. Nevertheless, it is clear that disturbance of these zones increases nutrient fluxes into the stream and interferes with their buffering effects. While it is possible that grasses and other weeds my take over these functions in some cases, they cannot replace all other functions of the natural riparian vegetation simultaneously.

Absolute values of nutrient concentrations may not be indicative of their ecological effects and may be less important than timing and other factors such as temperature and light intensity. For example, most of the nutrients that are transported through a stream pass through during short duration but large storm events, in terms of total rainfall, when flow volumes are high. Most of these nutrients do not remain at a site within the stream long enough to affect the productivity greatly. On the other hand, although base flows and small storms are responsible for transporting smaller amounts of nutrients, they will be retained within a reach of a stream for a longer period and consequently may strongly influence ecological processes. Even then, the effects will depend on factors such as temperature and light levels which are affected by season, reach aspect and clearing of the riparian vegetation. Our knowledge of nutrient fluxes between forest and stream is largely derived from the use of stream chemistry as an indicator of an area's nutrient status before and after a change in land use (Likens *et al.*, 1977). Although this is useful as a relative measure, it is difficult to construct accurate nutrient budgets from it. Water samples from the water column will not include nutrients that have already been incorporated into organisms. Abundant aquatic macrophytes and algae that may occur when light is abundant can strip nutrients from the water column so efficiently that water samples may be misleading and suggest low productivity potential when in fact considerable productivity is occurring. Hence it is necessary to understand where nutrients are utilised and stored in order to interpret the significance of nutrient concentrations in the water (a point that many routine monitoring programmes miss).

Pearson and Penridge (1987) found high abundances of macroinvertebrates below the outfall of a sugar mill in tropical northern Australia (Figure 34.7). They associated the increase in macroinvertebrate production with high levels of nutrients and organic matter in mill effluent. In experiments using artificial stream channels on the bank of a first order rainforest stream Pearson and Connolly (2000) were able to increase macroinvertebrate abundance by 75%. However, despite this almost doubling of macroinvertebrate abundance, the composition of the community remained remarkably unchanged. In contrast, the faunal composition below the sugar mill outfall was changed markedly. Despite the increase in abundance, Pearson and Penridge (1987) observed a decrease in the number of macroinvertebrate species,

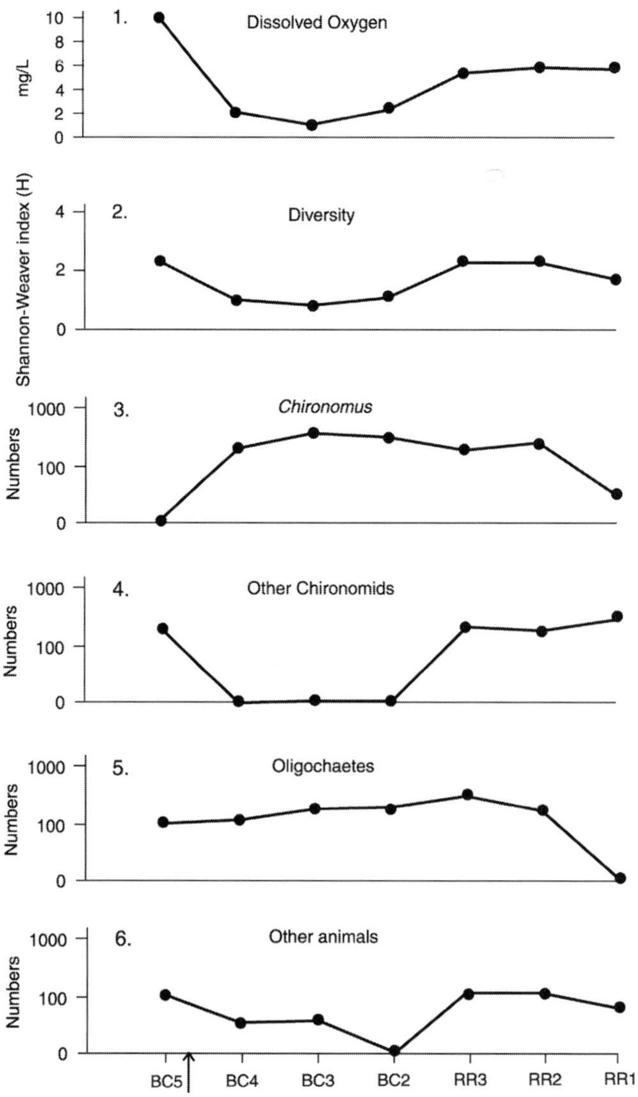

Figure 34.7 Effects of organic enrichment (sugar mill effluent) on a tropical stream. Sites are arranged from upstream (BC5) to downstream (RR1). The arrow indicates the site of the effluent outfall. 1. Dissolved oxygen decreased, caused by bacterial respiration, with recovery taking place downstream. 2. The diversity (H) of the invertebrate fauna declines, tracking the oxygen curve and recovering downstream. 3. The pollution-tolerant midge larvae, *Chironomus*, increases in abundance through the polluted section. 4. Other non-tolerant midge larvae (Chironomidae) decline in the polluted section. 5. Pollution tolerant Oligochaetes increase in abundance with the organic effluent. 6. Other animals decline in the polluted section.

the community being dominated by a few taxa, notably a pollution-tolerant species of chironomid midges (Figure 34.7) as a result of depressed dissolved oxygen levels (this effect of organic effluent on stream ecology occurs world-wide – Hynes, 1960). These two examples demonstrate how the relationships between nutrients, biotic production and community composition are not simple and depend not only on the level of nutrient input, but also on the characteristics of the particular site, and levels of other inputs, as explained below.

The relationship of increased biological production resulting from increasing the availability of a limiting nutrient, although apparently straightforward, actually involves several complex interactions. The increase in production may induce physicochemical changes – for example in pH and dissolved oxygen – and ecological processes will play a role in the final outcome. The effects depend on characteristics of the particular water body, such as depth, flow and shade. Increases in plant growth affect the distribution and proportions of different habitats for benthic invertebrates and fish. Increases in microbial populations affect detritus breakdown and the nutritional value of this detritus to consumers. In the above artificial stream experiments (Pearson and Connolly, 2000) the site was heavily shaded by rainforest, and pH and dissolved oxygen were not affected by nutrient enhancement because of the cool, fast flowing and well mixed water in the channels. Although the experiment ran for seven months, there was no detectable increase in primary production and the increase in secondary production was explained by increased fungal and bacterial growth on leaf detritus, which is the basis of the food web in this system. Nutrient enhancement thus improved the quality of the detritus, enabling it to support greater numbers of macroinvertebrates even though the quantity had not changed. Because the community composition did not change, despite the large increase in abundance, it was concluded that the community structure in this system was controlled by other factors. In the study on the effects of sugar mill effluent, Pearson and Penridge (1987) found that the physico-chemistry of the water and the structure of benthic habitat had been altered significantly by the effluent and increased levels of microbial growth. The physico-chemical conditions had changed outside the tolerances of many species and the habitat had changed due to waste sediments displacing other benthic habitats. The environmental conditions had changed to suit fewer species and the community had consequently shifted to be dominated by these taxa. The effluent supplied nutrients and organic material, which in turn supported very high abundances of these more tolerant species. Consequently sites below the sugar mill effluent outfall had very high abundances but fewer species than would be expected in the absence of this effluent. These are typical effects of cultural eutrophication in freshwater environments (Hynes, 1960).

Nutrient dynamics within a stream are clearly complex and variable, depending on prevailing environmental conditions and the biotic community present. To predict how an increase in nutrients may affect a biotic community requires the understanding that small increases may result in very different effects in comparison with the impacts on biota by large nutrient increases. It is also probable that effects of short-term changes will differ substantially from those that occur over the long term as ecological processes have time to shape the community.

Dissolved oxygen

The proliferation of microbial and plant growth resulting from nutrient enhancement has several consequences, the most biologically important of which in many situations is the change in the dynamics of dissolved oxygen, particularly its diel (24 hour) pattern, as suggested above. The availability of oxygen influences nearly all chemical and biological processes in aquatic ecosystems. Although a few aquatic organisms can obtain oxygen from the air (e.g. some fish can gulp air; some beetles take bubbles from the surface; some aquatic plants are emergent), and some do not need oxygen (anaerobic bacteria), the majority of aquatic organisms rely on the relatively small quantities of oxygen dissolved in the water. Dissolved oxygen levels are determined by a number of factors, including: the interplay between oxygen consumption (respiration) by animals, plants and aerobic microbes; photosynthetic oxygen production by plants during daylight hours; losses and gains of oxygen from the overlying air or from other parts of the water body through flow and mixing (enhanced by turbulence through riffles and waterfalls); salinity, atmospheric pressure and temperature; and groundwater flow (which is often low in dissolved oxygen concentration). Hence, dissolved oxygen concentrations can fluctuate greatly over space and time, either sporadically (e.g. in response to changes in flow rate, turbulence or wind-induced aeration and mixing) or systematically (e.g. seasonally or daily, induced by temperature and biological activity).

In shallow, swift streams, dissolved oxygen concentrations are maintained at, or close to, 100% saturation by re-aeration from the air and mixing due to turbulent flow. In small, well-shaded, low order streams, with low ambient nutrient concentrations, re-aeration is greater than oxygen consumption. Dissolved oxygen concentrations are maintained in these streams even though respiration is generally greater than photosynthetic oxygen production. Plant growth is relatively sparse in the stream because of high levels of shade. Organic input is mainly terrestrial plant material and complex dissolved humic substances that are not rapidly degraded by microbes. Continuous flow reduces opportunity for the accumulation of fine organic material and so microbial activity (and oxygen depletion) are low. Consequently, biological oxygen demand (BOD) is generally low and oxygen saturation is maintained at high levels.

Changes to these conditions, whether natural or human-induced, have substantial bearing on the dissolved oxygen status of the water. For example, reduction in shade leads to increased temperature and plant growth; nutrient run-off encourages plant growth; organic effluents, such as sewage or mill waste, add nutrients and organic material, on which microbes thrive; organic matter accumulates where stream flow is reduced (for example, by extraction for irrigation and by dams and weirs). Furthermore, dams and weirs change shallow flowing systems into deep lakes, possibly leading to thermal stratification, and promoting growth of phytoplankton and floating macrophytes. All of these changes have substantial effects not only on habitats but also on physico-chemical properties of the water, especially the dissolved oxygen dynamics and its relationship with the organisms that live in the system.

The principal consumers of oxygen in aquatic systems are microbes and plants, including microscopic algae. Plants consume oxygen constantly when it is available, day and night. During daylight, when they are exposed to adequate light levels they photosynthesise and in the process produce much larger quantities of oxygen than they consume. However, at night, they only respire, consuming oxygen from the water column. Accordingly, a high biomass of aquatic plants can oxygenate a water body very efficiently during daylight hours, but deoxygenate it during the night, especially if the plant biomass is high and re-aeration from other sources is not sufficient to compensate. This process is responsible for the typical pattern of dissolved oxygen concentrations observed in aquatic systems, in which oxygen levels increase to a peak during daylight, then decline to a minimum during the night in a continual diel cycle. During the day, dissolved oxygen levels typically reach 100% saturation or greater. Cloud cover can affect this pattern to the extent that on cloudy days hypoxia (low levels of dissolved oxygen) can develop if there is a large aquatic plant biomass present. The minimum concentration typically occurs just prior to dawn and is critical to the aquatic animals. Generally, the greater the plant biomass, the lower is the minimum concentration of oxygen due to the higher amount of nocturnal respiration.

Knowledge of the tolerance of individual species of tropical animals to reduced levels of dissolved oxygen is very limited, despite frequently reported cases of anthropogenically induced hypoxia in freshwater systems. Most of the published work on this topic has been carried out on northern-hemisphere temperate species (e.g. Hall, 1969; Nebeker, 1972; Nebeker *et al.*, 1992; 1996; Lowell and Culp, 1999). Our research group is currently addressing this situation for some tropical Australian species. For species tested so far, there appear to be sharp thresholds in survival. We have shown this to be the case for quite different organisms, such as fish and benthic insects in short-term tolerance tests. Longer-term sublethal effects on growth and reproduction have yet to be investigated, although insect emergence (to the terrestrial adult stage) was observed to be inhibited long before lethal levels were reached. (unpublished data).

Stream current has an important influence on the dissolved oxygen requirements of aquatic fauna: for example, Jaag and Ambuhl (1964) showed that some mayflies tolerated low levels of dissolved oxygen concentrations except in slow currents, while

Knight and Gaufin (1963) demonstrated that a stonefly experienced 100% mortality in low dissolved oxygen concentrations at a current velocity of 1.5 cm s^{-1}, but no mortality at 7.6 cm s^{-1}. The interplay between current and dissolved oxygen concentrations can also induce behavioural responses: for example, Wiley and Kohler (1980) observed mayfly nymphs utilising positions that were more or less exposed to the current according to their tolerance of low dissolved oxygen concentrations. These examples suggest that the consequences of organic pollution and the proliferation of aquatic plants and microbes are more acute where flow has been reduced by the installation of dams and weirs and river regulation.

Slow flowing, deep and poorly mixed water bodies become thermally stratified as warm water floats on cooler water. Microbes in the sediments consume organic material and oxygen which, because of the stratification, is not replenished from the surface; neither is it replenished by plants, which cannot grow in the dark deep water. The lower strata thus become oxygen deficient (hypoxic, or anoxic). In large lakes, wind mixing can extend to a substantial depth and prevent stratification. But, in small, sheltered water bodies, such as are typically retained behind weirs, surface mixing is limited. The relationship between water density is not linear – differences in density are relatively greater at higher temperatures. Hence stratification can occur more readily in the tropics than it does in cooler climates (Serruya and Pollingher, 1983). In the tropics, if flow is slow, even shallow water bodies can be prone to diurnal stratification, developing a shallow warm surface layer during the day but cooling down and mixing at night. In addition to these effects, an anoxic zone at the sediment-water interface can allow the interchange of many noxious or undesirable substances between the water and sediments.

In slow-flowing and still waters, floating macrophytes, which may normally be restricted to off-channel lagoons in the floodplain, can proliferate. Like emergent macrophytes, these exchange gases with the air and so do not contribute to dissolved oxygen cycling in the water. However, they shade the water column, preventing submerged plants from photosynthesising, and if growing in high densities can impede wind-induced re-aeration of the water column. This typically results in conditions of low dissolved oxygen concentrations beneath dense growths of floating macrophytes, such as the water hyacinth, *Eichhornia*. These plants release a large amount of readily biodegradable organic matter into the water column and their substantial root mass provides a large surface area that supports bacterial growth imposing a high biohemical oxygen demand.

Aquatic organisms have developed some tolerance to short periods of low dissolved oxygen concentrations so as to cope with diel fluctuations and other natural events. However, like many of the factors discussed previously, anthropogenic influences have increased the frequency, duration and intensity of hypoxia in many aquatic systems, which, along with other changes, have resulted in a loss of biodiversity and major changes in productivity in these systems. It is likely that the stress imposed by low oxygen conditions will leave these organisms vulnerable to other pressures and vice versa. The importance of dissolved oxygen has thus been long recognised and dissolved oxygen concentration has been used as a measure of pollution, particularly that induced by organic matter and nutrients. Unfortunately, its monitoring has been poorly executed in many cases, with oxygen values typically being measured during the day, at a single depth, giving a misleading impression of the condition within the water body as a whole. Clearly, from the discussion above, the critical time to measure dissolved oxygen is during the trough of the oxygen sag towards dawn, at the end of the non-photosynthetic period; and the critical concentration can only be determined by measuring the oxygen-depth profile at each location.

pH

As with dissolved oxygen, the pH dynamics of water bodies fluctuate according to physical and biological influences, especially plant growth that may be stimulated by clearing of riparian vegetation. The measured pH and the pH-buffering capacity are vital properties of a solution because they affect all chemical and biological reactions. Changes in pH may affect the uptake of nutrients by plants (Mengal and Kirby, 1978); respiration (Omerod *et al.*, 1987) and increased mortality and failure of eggs (Carrick, 1979) in fish; drift behaviour in aquatic macroinvertebrates (Hall *et al.*, 1980, 1987); and the decomposition of organic matter by microbial organisms (Hildrew *et al.*, 1984). Even small changes in pH can alter greatly the toxicity of many contaminants, including substances that can occur naturally in significant concentrations, such as aluminium (Hall *et al.*, 1987). Extreme pH values can be lethal to aquatic organisms but effects on the toxicity of other substances can be so strong that it is often difficult to discriminate direct effects.

Headwater streams that receive large amounts of rainfall are often poorly buffered and mildly acidic. However, most watercourses, especially in lowland sections, receive sufficient quantities of bicarbonate to keep pH values above about 6.4. Particular situations, such as drainage from some wetland vegetation and acid sulphate soil deposits may reduce this value (Butler *et al.*, 2002). Weak organic acids (humic and fulvic, such as produced from forest soil litter) can convert toxic substances (e.g. aluminium and zinc) into innocuous chemical forms, enabling aquatic organisms to tolerate lower pH levels than might be expected, as was found, for example, by Winterbourne and Collier (1987) in a study of acid brown-water streams in New Zealand. However, runoff or leachate from acid sulphate soils can contain sufficient sulphuric acid to create extremely low pH values. Sulphuric acid does not have the same capacity as humic acids to inhibit the toxicity of metals such as aluminium and tends instead to dissolve large quantities of these metals from the soil.

Although the pH of a water body depends initially on the underlying geology and the soils, it is also affected by organisms in the water body, which in some circumstances can have an overwhelming effect. The effects are greatest when flow rates are low and plant, animal and microbial biomass are high. As aquatic plants respire they consume dissolved oxygen and release carbon dioxide into the water column. However, during daylight hours they photosynthesise and consume far more carbon dioxide than they produce. Carbon dioxide readily dissolves in water to produce carbonic acid, which then dissociates into bicarbonate ions and hydrogen ions and so increases pH. When photosynthetic activity is very high and/or the water column is relatively stagnant (preventing rapid delivery of CO_2 from the air) carbon dioxide levels fall during the daytime and then rise again during the night. This produces a cyclical pattern in pH levels.

When the aquatic biomass is very dense, as occurs with nutrient pollution, large quantities of free carbon dioxide are consumed from the water column and pH values rise to a maximum at which point the water is so alkaline that it can no longer hold free carbon dioxide in solution. When carbon dioxide is not available most aquatic macrophytes can utilise bicarbonate instead. This further increases pH values as bicarbonate concentrations are reduced. However, in situations where phytoplankton dominates primary productivity, such as in deep, still water bodies, only cyanobacteria (blue-green algae) can obtain carbon dioxide from bicarbonate. Therefore, if phytoplankton production is very high it can increase pH levels to where blue-green algae will dominate. If this occurs, blue-green algal production will increase pH values even further, often to extreme values. This effect, coupled with toxicity and deoxygenation of the water can have severe impacts on other aquatic organisms, resulting in a reduction in the numbers of species and individuals and altered ecosystem processes. Thus, like dissolved oxygen, pH is a complex issue that is very poorly understood, especially by agencies charged with monitoring it. Clearly, the cyclical nature of pH means that occasional measurements, not taken according to a strict temporal protocol, are of little use.

Other contaminants

Any material dissolved or suspended in water may be considered a contaminant, in that it reduces the purity of the water, and innumerable contaminants can have impacts on the stream biota when in sufficient concentration. Direct impacts include, especially, the effects of toxic substances. Those not considered already that are most commonly implicated in changes in the plant and animal communities are metals and solvents associated with mining, organic pesticides associated with agriculture and urban weed control, oils and their derivatives from industry and roads, and various chemicals from industrial plants and waste-disposal facilities. Each of these groups of chemicals and their effects could be given a chapter to themselves; here we simply draw attention to them for the sake of completeness. In general in the tropics, the greatest impacts of land-use change result from the effects of contaminants discussed previously, largely because of the sheer extent of the area of land (and therefore volume of drainage) affected or the length of waterway impacted. However, at local scales and occasionally at broader scales these other factors can have major impact. In particular, urban development can lead to a combination of many or all of these groups, with resultant additive or synergistic effects, such that waterways in tropical cities are typically smelly polluted channels that support little or none of the original community, and create barriers to passage between any cleaner zones that may exist.

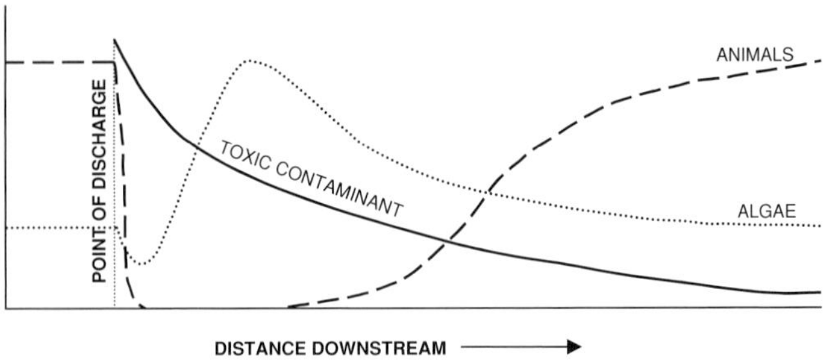

Figure 34.8 Typical pattern of contamination concentration and biological responses in stream. (Modified after Hynes, 1960.)

As with any contamination, the source may be diffuse or from particular points, and the impacts are greatest close to the source or a short distance downstream. If there is no additional contamination downstream, contaminants become diluted by tributaries entering the stream, by adsorption on to sediments and by chemical or biological processing. The resultant pattern in contaminant concentration and abundance and diversity of plants and animals has been long recognised (Hynes, 1960) (Figure 34.8).

The impacts of contaminants on plants and animals vary with the species. As Hynes (1960) pointed out, individuals of each plant or animal species can cope with a particular level of poison under particular conditions. A higher level causes deleterious effects such as reduced growth, reduced reproductive ability and death. Organisms can tolerate lower levels of poison for longer, but eventually poisons may accumulate to cause these same effects, or may simply act to a lesser degree. It is possible to find organisms tolerating apparently quite severe conditions, but the presence of those organisms may belie longer-term or more subtle effects that reduce viability of a population.

It should be restated here that presence or absence of components of the biota may relate to particular contaminants, but that a combination of factors is often responsible (for example, zinc and aluminium become much more toxic to fish at lower levels of pH), and these factors include not only water quality variables but also presence of suitable habitat and food resources. As Hynes (1960) put it, without the food, shelter and breeding sites they need, organisms (fishes in this case) cannot thrive, even though the water may not be poisonous.

Another category of agents that affect stream ecology, and which might loosely be regarded as contaminants, is introduced species – exotic weeds that replace riparian vegetation, that cover water surfaces in slow-flowing reaches and which lead to reduced flows, deoxygenation and extreme habitat modification (e.g. para grass and water hyacinth) – and exotic animals, especially fishes, that thrive particularly in modified systems. Many such species invade through careless or misguided behaviour of aquarists and agriculturalists or misguided attempts to 'improve' fish production. The example of the devastating effects of the Nile perch (*Lates nilotica*) introduced into lakes in Africa, where it did not occur naturally and where it has caused the extinction of a large number of endemic fish species, should be sufficient to forestall other attempts at introduction of exotic species.

Metals are found dissolved naturally in water in varying concentrations, and can be responsible for substantially altering the ecology of streams. For example, the concentration of calcium in natural waters seems to have a major role in determining biological productivity (Sutcliffe and Carrick, 1973). Metals can act as direct poisons, so that streams downstream of mine drainage or leaking tailings ponds may be impoverished with regard to plants, invertebrates and fish. The offending contaminant may not be the target mineral at a mine site – for example, arsenic, which is highly toxic, may be associated with tin, which is not toxic.

As long ago as the 1920s, impacts of mine drainage were recorded in streams, and included decline of various organisms (Carpenter, 1924). Since then, numerous field studies and toxicity tests in the laboratory have demonstrated effects of different metals, particular with regard to biomonitoring of natural waters. For example, in Australia, effluent from mining operations carried several metal contaminants that led to declines in the abundance and richness of invertebrates in the stream, and their effects were detectable up to 80 km downstream (Norris *et al.*, 1982). In the tropics, less work has been done, but in some cases sophisticated biomonitoring systems have been developed to indicate contamination (e.g. by uranium – see Humphrey *et al.*, 1990, 1995). Metals introduced into tropical streams are likely to have similar effects to those recorded for temperate zones, except that higher temperatures are likely to exaggerate them.

There are many pesticides available and their effects are probably equally diverse. Given that they are designed to control plant or animal pests, it is not surprising that in many cases they cause collateral damage. Largest impacts are probably sourced from extensive application in broad-scale agriculture, for example for such crops as cotton (Arthington, 1996) and sugar-cane (Russell *et al.*, 1996). Acute effects may be felt from limited application of, for example, herbicides used to control grasses and other weeds in drainage channels, and exotic plants that choke waterways, such as water hyacinth (*Eichhornia*) (Kevan and Pearson, 1993). Unfortunately, knowledge of the acute and chronic impacts of pesticides in tropical waters is very limited. It should be noted that pesticides that are withdrawn from use in western countries, because of environmental or health concerns, often continue to be used in developing countries which cannot afford the more modern substitutes. Hellawell (1986) discusses effects of many pesticides on freshwater organisms.

CONCLUDING REMARKS

The foregoing discussion is an outline of some important issues that relate to the impacts of forest conversion on stream ecology. It skims only the surface of the selected issues and virtually ignores others. The intent is to provide some insight into the complexity of systems, to ensure water quality, ecosystem function and biodiversity are not given short shrift. In particular, what should be abundantly clear is that expertise is required to deal with these issues: geomorphological problems need geomorphological expertise to address them, just as ecological problems require ecological expertise. But when examining ecosystems, no single line of expertise is enough – ecologists need geomorphologists, chemists, and so on, to be able to design appropriate research to investigate ecosystem function. And environmental chemists need the expertise of ecologists to help sort out the biological influences on water chemistry. That is, none of the disciplines can safely operate in isolation, because the links between them are so strong and complex.

The issues summarised above are regarded as the major effects of forest conversion in humid tropical zones. Not surprisingly, similar issues would apply in other climatic zones, but it must be stressed that the nature of tropical systems lends them substantial differences from the temperate northern hemisphere systems on which so many ecological paradigms are based. Regional differences have been long recognised (e.g. Lake *et al.*, 1994), but none is so pronounced as the differences between the tropics and elsewhere, as discussed above. In particular, the contrast in climate (equable temperatures, extremes of rainfall), biological activity (continuing apace, all year-round) and the interactions between them, separate tropical ecosystems conceptually from others. It is possible that these features make tropical systems so much more vulnerable to change because things can happen so fast – invasion by exotic 'weedy' species, for example, can happen very quickly, long before remedial measures can be put in place.

While we know enough about the tropics to be sure about such differences, detailed understanding of even basic patterns and processes in tropical stream systems is very sparse. The tradition of studies of natural history of organisms in the northern hemisphere over two centuries that has provided essential background to ecological understanding is largely missing in the tropics. Therefore, current work in the tropics has to be set in this Spartan context, and generalisations need to be made with extreme caution. More, and more detailed, case studies are required, and while they may not individually advance science dramatically, they are necessary to help address some major gaps. The need is for a combination of long-term detailed studies of ecological processes at targeted sites, and broad-scale studies across many sites.

The discussion presented in this chapter highlights the importance of multidisciplinary approaches to research and management, because stream (and other) ecological systems are the result of interacting physical, chemical and biological processes. It is sometimes assumed that the physico-chemical environment provides a template for biological communities; however, organisms themselves contribute to these templates, and to apparently fundamental properties of the environment, such as water quality. The very term BOD (Biochemical Oxygen Demand) – a frequently measured variable in fresh waters – captures this idea, as does DOM (dissolved organic matter). Both affect water quality to a greater or lesser extent. A final example is the effect of microbial breakdown of organic material in the sediments of lakes, which causes deoxygenation of the water and dissolution of otherwise rather insoluble substances, including compounds of iron and manganese. These compounds can be major contaminants of water supplied from dams.

Not only is it necessary to consider multiple disciplines to understand ecosystem function, it is also necessary to work at multiple scales. Managers tend to make decisions that apply at a broad scale, and while many ecological processes can also be described at comparable scales, it is at the scale of the individual or population that most processes operate. Thus, apparently minor changes to overall sediment dynamics in a catchment might have disproportionately large impacts in particular reaches (e.g. clogging gravel beds used for egg-laying by some fish species). Ecological and physiological studies of individual animals and plants are necessary to understand fundamental needs for day-to-day survival, growth and reproduction of at least key species, so that modelling and predicting changes to result from land-use modification have hard data attached to them.

It is a major challenge to attempt such work in much of the tropics where human pressure on production so often outweighs any consideration for conservation. Perhaps the main goal is to convince managers that natural ecosystems can function in parallel with productive systems, and that these natural systems are vital to human health. Nowhere is this more evident than in streams, which when managed properly can continue to drain catchments, strip contaminants and supply water for human needs, as well as providing habitat for a wealth of diverse aquatic and riparian organisms.

References

Aldridge, D. W., Payne, B. S. and Miller, A. C. (1987) The effects of intermittent exposure to suspended solids and turbulence on three species of freshwater mussel. *Environmental Pollution* **45**: 17–28

Allan, J. D. (1995) *Stream Ecology: Structure and function of running waters*. Chapman and Hall, London, UK.

Armitage, P. D. (1995) Faunal community change in response to flow manipulation. **In** Harper, D. M. and Ferguson, A. J. D. (eds.) *The ecological basis for river management*. Plenum Press, New York, pp 59–78.

Arsuffi, T. L. and Suberkropp, K. (1985) Selective feeding by stream caddisfly (Trichoptera) detritivores on leaves with fungal-colonized patches. *Oikos* **45**:50–58

Arthington, A. H. (1996) The effects of agricultural land use and cotton production on tributaries of the Darling River, Australia. *GeoJournal* **40**: 115–125.

Baxter, R. M. (1977) Environmental effects of dams and impoundments. *Ann. Rev. Ecol. Syst*. **8**: 255–284.

Begon, M., Harper, J. L. and Townsend, C. R. (1990) *Ecology: Individuals, Populations and Communities*. Blackwell Scientific, Oxford.

Benson, L. J. and Pearson, R. G. (1987a) Drift and upstream movements of macro- invertebrates in a tropical Australian stream. *Hydrobiologia* 153: 225–239.

Benson, L. J. and Pearson, R. G. (1987b) The role of drift and effect of season in colonization of implanted substrata in a tropical Australian stream. *Freshwater Biology* 18: 109–116.

Benson, L. J. and Pearson, R. G. (1988) Diversity and seasonality of adult Trichoptera captured in a light-trap at Yuccabine Creek, a tropical Australian rainforest stream. *Aust. J. Ecol*. 13: 337–344.

Benson, L. J. and Pearson, R. G. (1993) Litter inputs to a tropical Australian upland rainforest stream. *Aus. J. Ecol*. 18:377–383.

Benzie, J. A. H. (1984) The colonization mechanisms of stream benthos in a tropical river (Menik Ganga: Sri Lanka). *Hydrobiologia* 111: 171–179.

Bishop, J. E. (1973a) *Limnology of a small Malayan river Sungai Gombak*. Dr. W. Junk B. V., Publishers, The Hague, Netherlands.

Bishop, J. E. (1973b) Observations on the vertical distribution of the benthos in a Malaysian stream. *Freshwat. Biol*. **3**: 147–156.

Blackie J. R., Ford E. D., Horne J. E. M., Kinsman D. J. J., Last F. T. and Moorhouse P. (1980) Environmental Effects of Deforestation. An annotated bibliography. Freshwater Biological Association Occasional Publication No. 10. FBA, Ambleside, UK.

Boulton, A. J. and Brock, M. A. (1999) *Australian Freshwater Ecology Processes and Management*. Gleneagles Publishing Australia. 300 pp.

Bruijnzeel, L. A. (1990) *Hydrology of moist tropical forest and effects of conversion: A state of knowledge review*. International Hydrological Programme, UNESCO.

Bruijnzeel, L. A. (1991) Nutrient input-output budgets of tropical forest ecosystems: a review. *Journal of Tropical Ecology* **7**: 1–24

Bruton, M. N. (1985) The effects of suspensoids on fish. *Hydrobiologia* **125**: 221–241

Bunn, S. E., Davies, P. M. and Kellaway, D. M. (1997) Contributions of sugar cane invasive pasture grass to the aquatic food web of a tropical lowland stream. *Mar. Freshwater Res*. **48**: 173–179

Bunn, S. E., Davies, P. M., Kellaway, D. M. and Prosser, I. P. (1998) Influence of invasive macrophytes on channel morphology and hydrology in an open tropical lowland stream, and potential control by riparian shading. *Freshwater Biology* **39**: 171–178

Butler, B., Crossland, M. and Pearson, R. G. (2002) Freshwater ecosystems of the Herbert River catchment: Relationships between contaminants, water quality and biological integrity. Australian Centre for Tropical Freshwater Research Report.

Carpenter K. E. (1924) A study of the fauna of rivers polluted by lead mining in the Aberystwyth district of Cardiganshire. *Annals of Applied Biology* **9**: 1–23.

Carrick, T. R. (1979) The effect of acid water on the hatching of salmonid eggs. *J. Fish* Biol. **14**: 165–172

Chang, H. H. (1988) *Fluvial processes in river engineering*. Krieger Publishing, Malabar, Florida, 432pp.

Chapman, D. (ed) (1992) *Water Quality Assessments. A guide to the use of biota, sediments and water in environmental monitoring (2nd ed.).* E and FN SPON (Chapman and Hall), London, UK.

Chestnut, T. J. and McDowell, W. H. (2000) C and N dynamics in the riparian and hyporheic zones of a tropical stream, Luquillo Mountains, Puerto Rico. *J. N. Am. Benthol. Soc.* **19**: 199–214

Cline, L. D., Short, R. A. and Ward, J. V. (1982) The influence of highway construction on the macroinvertebrates and epilithic algae of a mountain stream. *Hydrobiologia* **96**: 149–159

Connolly, N. M. and Pearson, R. G. (1998) The effect of sedimentation on benthic invertebrates in tropical streams. **In** Pearson, R. G. and Connolly, N. M. (eds.) Macroinvertebrates as monitors of river health in the tropics. Australian Centre for Tropical Freshwater Research Report No. 98/20 James Cooke University, Townsville, Australia.

Covich, A. P. (1987) Atyid shrimps in the headwaters of the Luquillo Mountains, Puerto Rico: filter feeding in natural and artificial streams. *Verh. Internat. Verein. Limnol.* **23**: 2108–2113

Culp, J. M., Wrona, F. J. and Davies, R. W. (1985) Response of stream benthos and drift to fine sediment deposition versus transport. *Can. J. Zool.* **64**: 1345–1351

Craig, J. F. and Kemper, J. B. eds. (1987) *Regulated Streams – Advances in Ecology*. Plenum, New York.

Davies-Colley, R. J., Hickey, J. M., Quinn, J. M. and Ryan, P. A. (1992) Effects of clay discharges on streams: 1. Optical properties and epilithon. *Hydrobiologia* **248**: 215–234

Davis, B. R. and Walker, K. F. (1986) *The Ecology of River Systems*. Monographiae Biologicae, Vol. 60, Dr. W. Junk Publishing, The Hague, Netherlands.

Dillon, P. J. and Kirchner, W. B. (1974) The effects of geology and land use on the export of phosphorus from watersheds. *Water Resources Research* **9**: 135–148

Dillon, P. J. and Kirchner, W. B. (1975) Reply. *Water Resources Research* **11**: 1035–1036

Dudgeon, D. (1982a) Aspects of the hydrology of Tai Po Kau Forest Stream, New Territories, Hong Kong. *Arch Hydrobiol., Suppl.* **64**: 1–35.

Dudgeon, D. (1982b) Spatial and temporal changes in the sediment characteristics of Tai Po Kau Forest Stream, New Territories, Hong Kong, with some preliminary observations upon within-reach variations in current velocity. *Arch. Hydrobiol., Suppl.* **64**: 36–64.

Dudgeon, D. (1983) An investigation of the drift of aquatic insects in Tai Po Kau Forest Stream, New Territories, Hong Kong. *Arch. Hydrobiol.* **96**: 434–447.

Dudgeon, D. (1984) Seasonal and long-term changes in the hydrobiology of the Lam Tsuen River, New Territories, Hong Kong, with special reference to benthic macroinvertebrate distribution and abundance. *Arch. Hydrobiol., Suppl.* **69**: 55–129.

Dudgeon, D. (1985) The population dynamics of some freshwater carideans (Crustacea: Decapoda) in Hong Kong, with special reference to *Neocaridina serrata* (Atyidae). *Hydrobiologia* **120**: 141–149.

Dudgeon, D. (1987) The ecology of a forest stream in Hong Kong. *Arch. Hydrobiol. Beih., Ergebn. Limnol.* **28**: 449–454.

Dudgeon, D. (1988) The influence of riparian vegetation on macroinvertebrate community structure in four Hong Kong streams. *J. Zool., Lond.* **216**: 609–627.

Dudgeon, D. (1989a) The influence of riparian vegetation on the functional organization of four Hong Kong stream communities. *Hydrobiologia* **179**: 183–194.

Dudgeon, D. (1989b) Life cycle, production, microdistribution and diet of the damselfly *Euphaea decorata* (Odonata: Euphaeidae) in a Hong Kong forest stream. *J. Zool., Lond.* **217**: 57–72.

Dudgeon, D. (1989c) Resource partitioning among Odonata (Insecta: Anisoptera and Zygoptera) larvae in a Hong Kong forest stream. *J. Zool., Lond.* **217**: 381–402.

Dudgeon, D. (1990) Seasonal dynamics of invertebrate drift in a Hong Kong stream. *J. Zool., Lond.* **222**: 187–196.

Dudgeon, D. (1992a) *Patterns and Processes in Stream Ecology: A Synoptic Review of Hong Kong Running Waters.* Schweizerbart'sche Verlagsbuchhandlung, Stuttgart: **147** pp.

Dudgeon, D. (1992b) Endangered ecosystems: a review of the conservation status of tropical Asian rivers. *Hydrobiologia* **248**: 167–191.

Dudgeon, D. (1993) The effects of spate-induced disturbance, predation and environmental complexity on macroinvertebrates in a tropical stream. *Freshwater. Biol.* **30**: 189–197.

Dudgeon, D. (1994) Functional assessment of the effects of increased sediment loads resulting from riparian-zone modification in a Hong Kong stream. *Verh. Internat. Verien. Limnol.* **25**: 1790–1792.

Dudgeon, D. (1995a) Environmental impacts of increased sediment loads caused by channelization: a case study of biomonitoring in a small Hong Kong river. *Asian J. Environ. Manag.* **3**: 69–77.

Dudgeon, D. (1995b) Life histories, secondary production and microdistribution of Psephenidae (Coleoptera: Insecta) from a tropical forest stream. *J. Zool., Lond.* **236**: 465–481.

Dudgeon, D. (1999) *Tropical Asian streams: zoobenthos, ecology and conservation.* Hong Kong University Press, Hong Kong.

Dudgeon, D. and Bretschko, G. (1995a) Land-water interactions and stream ecology: comparison of tropical Asia and temperate Europe. In:Timotius, K. H. and Goltenboth, F. (Eds), *Tropical Limnology Volume 1. Present Status and Challenges.* Satya Wacana University Press, Salatiga, Indonesia: 69–108.

Dudgeon, D. and Lam, P. K. S. (1994) Inland waters of tropical Asia and Australia: Conservation and management. *Mitt. Internat. Verein. Limnol.* **24**: 1–3

Extence, C. A. (1978) The effects of motorway construction on an urban stream. *Environmental Pollution* **17**: 245–252

Fernando, C. H. (1980) The freshwater invertebrates fauna of Sri Lanka. *Spolia Zeylanica* **35**: 15–42.

Glaister, G. P. (1978) The impact of river discharge on distribution and production of the school prawn *Metapenaeus macleayi* (Haswell) (Crustacea: Penaeidae) in the Clarence River region, northern New South Wales. *Aust. J. Mar. Freshwater Res.* 29:311–323

Gordon, N. D., McMahon, T. A. and Brian L. Finlayson (1992) *Stream Hydrology. An introduction for ecologists.* John Wiley and Sons, UK.

Gore, J. A. and Petts, G. E. (1989) *Alternatives in Regulated River management.* CRC Press, Boca, Raton, FL.

Graham, A. A. (1990) Siltation of stone-surface periphyton in rivers by clay-sized particles from low concentrations in suspension. *Hydrobiologia* **199**: 107–115

Gregory, S. V., Swanson, F. J., McKee, W. A. and Cummins, K. W. (1991) An ecosystem perspective of riparian zones. *Bioscience* **41**: 540–551

Gupta, A. (1993) The changing geomorphology of the humid tropics. *Geomorphology* **7**: 165–186

Hall, C. A. S. (1969) Mortality of the mayfly nymph, *Ephemerella rotunda* (Morgan), at various low dissolved-oxygen concentrations. *The Journal of the Elisha Mitchell Scientific Society* **85**: 34–39

Hall, R. J., Driscoll, C. T. and Likens, G. E. (1987) Importance of hydrogen ions and aluminium in regulating the structure and function of stream ecosystems: an experimental test. *Freshwater Biology* **18**: 17–43.

Hall, R. J., Likens, G. E., Fiance, S. B. and Hendrey, G. R. (1980) Experimental acidification of a stream in the Hubbard Brook Experimental Forest, New Hampshire. *Ecology* **61**: 976–989

Hearnden, M. R. and Pearson, R. G. (1991) Habitat partitioning among mayflies (Insecta: Ephemeroptera) in an Australian tropical stream. *Oecologia* **87**: 91–101.

Hellawell, J. M. (1986) *Biological Indicators of Freshwater Pollution and Environmental Management.* Elsevier Applied Science Publishers, London, UK.

Hildrew, A. G., Townsend, C. R., Francis, J. and Finch, K. (1984) Cellulolytic decomposition in streams of contrasting pH and its relationship with invertebrate community structure. *Freshwater Biology* **14**: 323–328.

Hjulstrom, F. (1939) Transportation of detritus by moving water. In *Recent Marine Sediments, a Symposium* (ed. P. D. Trask), American Association of Petroleum Geologists, Tulsa, Oklahoma, pp 5–31.

Hortle, K. R. and Pearson, R. G. (1990) The fauna of the Annan River, north Queensland, with reference to the effects of tin mining. I. Fishes. *Aust. J. Mar. Freshwat. Res.* **41**: 677–694.

Humphrey C. L., Bishop K. L. and Brown V. M. (1990) Use of biological monitoring in the assessment of effects of mining wastes on aquatic ecosystems of the Alligator Rivers Region, tropical northern Australia. *Environmental Monitoring and Assessment* **14**: 139–181.

Humphrey C. L., Faith D. P. and Dostine P. L. (1995) Baseline requirements for the assessing of mining impact using biological monitoring. *Australian Journal of Ecology* **20**: 150–166.

Hutchinson, G. E. (1967) *A treatise on Limnology. Volume 2. Introduction to Lake Biology and the Limnoplankton*. John Wiley and Sons, New York.

Hynes, H. B. N. (1960) *The biology of polluted waters*. Liverpool University Press, Liverpool, UK.

Indrasena, H. H. A. (1970) Limnological and freshwater fisheries work in Ceylon. In: *Regional Meeting of Inland Water Biologists in South East Asia. Proceedings*. Unesco Field Science Office for South East Asia, Djakarta: 45–47.

Jaag, O. and Ambuhl, H. (1964) The effect of the current on the composition of biocoenoses in flowing water streams. *Adv. Water Pollut. Res*. **1**: 31–49

Jackson, J. K. and Sweeney, B. W. (1995) Present status and future directions of tropical stream research. *J. Nth Am. Benthol. Soc*. **14**: 5–11

Jacobsen, D. (1997) Structure and diversity of stream invertebrate assemblages: the influence of temperature with altitude and latitude. *Freshwater Biology* **38**: 247–261

Jacobsen, D. (1998) The effect of organic pollution on the macroinvertebrate fauna of Ecuadorian highland streams. *Arch. Hydrobiol*. **143**: 179–195

Jordan, T. E., Correll, D. E. and Weller, D. E. (1997) Effects of agriculture on discharge of nutrients from coastal plain watersheds of Chesapeake Bay. *Journal of Environmental Quality* **26**: 836–848

Jordan, C. F. (1989) Global patterns of process rates. **In** Proctor, J. (ed.) *Mineral Nutrients in Tropical Forest and Savanna Ecosystems*. Blackwell Scientific Publications, London, UK.

Junk, W. J. (ed.) (1997) *The Central Amazon Floodplain: Ecology of a pulsing system*. Ecological Series 126. Springer, New York.

Junk, W. J., Bayley, P. B. and Sparks, R. E. (1989) The flood pulse concept in river-floodplain systems. *Canadian Fisheries and Aquatic sciences Special Publication* **106**: 110–127.

Kevan, S. D. and Pearson, R. G. (1993) Toxicity of Diquat pulse exposure to the tropical freshwater shrimp *Caridina nilotica* (Atyidae). *Bulletin of Environmental Contamination and Toxicology* **51**: 564–567.

Klinge, H. (1976) Bilanzierung von Hauptnahrstoffen im Okosystem tropischer regenwald (Manaus) – vorlaufige daten. *Biogeographica* 7: 59–76

Knight, A. W. and Gaufin, A. R. (1963) The effect of water flow, temperature, and oxygen concentration on the Plecoptera nymph *Acroneuria pacifica* Banks. *Proc. Utah Acad. Sci., Arts, Lett.*, **40**:175–184

Kortmulder, K., Padmanabhan, C. and de Silva, S. S. (1990) Patterns of distribution and endemism in some cyprinid fishes as determined by the geomorphology of South-west Sri Lanka and South Kerala (India). *Ichthyol. Explor. Freshwaters* **1**: 97–112.

Lake, P. S. (2000) Disturbance, patchiness, and diversity in streams. *J. Nth. Am. Benthol. Soc*. **19**: 573–592

Lake, P. S., Schreiber, E. S. G. Milne B. J. and Pearson, R. G. (1994) Species richness in streams: patterns over time, with stream size and with latitude. *Verhandlungen der Internationalen Vereinigung fur Theoretische und Angewandte Limnologie* **25**:1822–1826.

Likens, G. E. (1999) The science of nature, the nature of science. Long-term ecological studies at Hubbard Brook. *Proceedings of the American Philosophical Society* **143**: 558–572.

Likens, G. E., Bormann, F. H., Johnson, N. M., Fisher, D. W. and Pierce, R. S. (1970) Effects of forest cutting and herbicide treatment on nutrient budgets in the Hubbard Brook watershed-ecosystem. *Ecological Monographs* **40**: 23–47

Likens, G. E., Bormann, F. H., Pierce, R. S., Eaton, J. S. and Johnson, N. M. (1977) *Biogeochemistry of a Forested Ecosystem*. Springer-Verlag, New York.

Littlehammer, A. and Saltveit, S. J. (1984) *Regulated Rivers*. Universittsforlaget, AS, Oslo, Norway.

Lowell, R. B. and Culp, J. M. (1999) Cumulative effects of multiple effluent and low dissolved oxygen stressors on mayflies at cold temperatures. *Can. J. Fish. Aquat. Sci*. **56**: 1624–1630

Malmer, A. and Grip, H. (1994) Converting tropical rainforest to forest plantation in Sabah, Malaysia. Part II. Effects on nutrient dynamics and net losses in stream water. *Hydrological Processes* **8**: 195–209

McDowell, W. H., Bowden, W. B. and Asbury, C. E. (1992) Riparian nitrogen dynamics in two geomorphologically distinct tropical rainforest watersheds: subsurface solute patterns. *Biogeochemistry* **18**: 53–75

Mengal, K. and Kirby, E. A. (1978) *Principles of Plant Nutrition*. International Potash Institute, Worblaufen-Bern, Switzerland.

Minshall, G. W. (1984) Aquatic insect-substratum relationships. ***In*** Resh V. H. and Rosenberg D. M. *The ecology of Aquatic insects*. Praeger Publishers, New York.

Minshall, G. W., Petersen, R. C., Cummins, K. W., *et al*. (1983) Interbiome comparisons of stream ecosystem dynamics. *Ecol. Monogr*. **53**: 1–25

Nebeker, A. V. (1972) Effect of low oxygen concnetration on survival and emergence of aquatic insects. *Trans. Amer. Fish. Soc*. No.4: 675–679

Nebeker, A. V., Onjukka, S. T., Stevens, D. G., Chapman, G. A. and Dominguez, S. E. (1992) Effects of low dissolved oxygen on survival, growth and reproduction of *Daphnia, Hyalella, and Gammarus. Environmental Toxicology and Chemistry* **11**: 373–379

Nebeker, A. V., Onjukka, S. T., Stevens, D. G., and Chapman, G. A. (1996) Efect of low dissolved oxygen on aquatic life stages of the caddisfly *Clistoronia magnifica* (Limnephilidae). *Arch. Environ. Contam. Toxicol*. **31**: 453–458

Neill, C., Deegan, L. A., Thomas, S. M. and Cerri, C. C. (2001) Deforestation for pasture alters nitrogen and phosphorus in small Amazonian streams. *Ecological Applications* **11**: 1817–1828

Newbold, J. D. (1992) Cycles and spirals of nutrients. **In** Calow, P. and Petts, G. E. (eds.) *The Rivers Handbook – Volume 1: Hydrological and Ecological Principles*. Blackwell Scientific Publications, Oxford, UK, pp. 379–408.

Newbold, J. D., Elwood, J. W., O'Neill, R. V. and Sheldon, A. L. (1983) Phosphorus dynamics in a woodland stream ecosystem: A study of nutrient spiralling. *Ecology* **64**: 1249–1265

Newman, R. M. (1991) Herbivory and detritivory on freshwater macrophytes by invertebrates: a review. *J. N. Am. Benthol. Soc*. **10**: 89–114

Nolen, J. A. and Pearson, R. G. (1992) Life history studies of *Anisocentropus kirramus* Neboiss (Trichoptera: Calamoceratidae) in a tropical Australian rainforest stream. Aquatic Insects **14**: 213–221.

Nolen, J. A. and Pearson, R. G. (1993) Processing of litter from an Australian tropical stream by *Anisocentropus kirramus* Neboiss (Trichoptera: Calamoceratidae). *Freshwater Biology* 29: 469–479.

Norris R. H., Lake P. S. and Swain R. (1982) Ecological effects of mine effluents on the South Esk River, north-eastern Tasmania. III. Benthic macroinvertebrates. *Australian Journal of Marine and Freshwater Research* **32**: 165–173.

Omerod, S. J., Boole, P., McCahon, C. P. *et al*. (1987) Short-term experimental acidification of a Welsh stream: comparing the biological effects of hydrogen ions and aluminium. *Freshwater Biology* **17**: 341–356

Osborne, P. L. (2000) *Tropical Ecosystems and Ecological Concepts*. Cambridge University Press, UK.

Pearson, R. G. (2002) Biodiversity of the freshwater invertebrates of the Wet Tropics region of north-eastern Australia: patterns and possible determinants. In B. Bermingham and C. Moritz *Rainforests: Past Present and Future*. University of Chicago Press.

Pearson, R. G. (1994) Limnology in the north-eastern tropics of Australia, the wettest part of the driest continent. *Mitt. Internat. Verein. Limnol*. **24**: 155–163.

Pearson R. G. and Connolly, N. (2000) Nutrient enhancement, food quality and community dynamics in a tropical rainforest stream. *Freshwater Biology* **43**: 31–42.

Pearson R. G. and Penridge, L. K. (1987) The effects of pollution by organic sugar mill effluent on the macro-invertebrates of a stream in tropical Queensland, Australia. *J. Environmental Management* 24: 205–215.

Pearson R. G., Benson L. J. and Smith, R. E. W. (1986) Diversity and abundance of the fauna in Yuccabine Creek, a tropical rainforest stream. In P. de Deckker and W. D. Williams (eds) *Limnology in Australia*. CSIRO, Melbourne. 329–342.

Pearson, R. G., Tobin, R. K. Benson L. J. and Smith, R. E. W. (1989) Standing crop and processing of rainforest litter in a tropical Australian stream. *Arch. Hydrobiol*. **115**: 481–498.

Pennack, R. W. and Van Gerpen, E. D. (1947) Bottom fauna production and physical nature of the substrate in a northern Colorado trout stream. *Ecology* **28**: 42–48.

Perakis, S. S. and Hedin, L. O. (2002) Nitrogen loss from unpolluted South American forests mainly via dissolved organic compounds. *Nature* **415**: 416–419.

Petts, G. E. (1984) *Impounded Rivers*. John Wiley, Chichester, UK.

Pringle, C. M. and Blake, G. A. (1994) Quantitative effects of atyid shrimp (Decapoda: Atyidae) on the depositional environment in a tropical stream: use of electricity for experimental exclusion. *Can. J. Fish. Aquat. Sci.* **51**: 1443–1450.

Pringle, C. M., Blake, G. A., Covich, A. P., Buzby, K. M. and Finley (1993) Effects of omnivorous shrimp in a montane tropical stream: sediment removal, disturbance of sessile invertebrates and enhancement of understorey algal biomass. *Oecologia* **93**: 1–11.

Pringle, C. M., Paaby-Hansen, P., Vaux, P. D. and Goldman, C. R. (1986) *In situ* nutrient assays of periphyton growth in a lowland Costa Rican stream. *Hydrobiologia* **134**: 207–213.

Pringle, C. M. and Ramirez, A. (1998) Use of both benthic and drift sampling techniques to assess tropical stream invertebrate communities along an altitudinal gradient, Costa Rica. *Freshwater Biology* **39**: 359–373

Proctor, J. (1987) Nutrient cycling in primary and old secondary rainforests. *Applied Geography* **7**: 135–152

Pusey, B. J., Arthington, A. H. and Read, M. G. (1995a) Species richness and spatial variation in fish assemblage structure in two rivers of the wet tropics of northern Queensland, Australia. *Env. Biol. Fish.* **42**: 181–199

Pusey, B. J., Read, M. G. and Arthington, A. H. (1995b) The feeding ecology of freshwater fishes in two rivers of the Australian wet tropics. *Env. Biol. Fish.* **43**: 85–103

Pusey, B. J. and Kennard, M. J. (1996) Species richness and geographical variation in assemblage structure of the freshwater fish fauna of the wet tropics region of northern Queensland. *Mar. and Freshwater Res.* **47**: 563–573

Pusey, B. J., Bird, J., Kennard, M. J. and Arthington, A. H. (1997) Distribution of the Lake Eacham Rainbowfish in the wet tropics region, North Queensland. *Aust. J. Zool.* **45**: 75–84

Pusey, B. J., Kennard, M. J. and Arthington, A. H. (2000) Discharge variability and the development of predictive models relating stream fish assemblage structure to habitat in northern Australia. *Ecology of Freshwater Fish* **9**: 30–50

Reid, L. M. and Dunne, T. (1984) Sediment production from forest road surfaces. *Water Resour. Res.* **20**: 1753–1761

Richards, C., Host, G. H. and Arthur, J. W. (1993) Identification of predominant environmental factors structuring stream macroinvertebrate communities within a large agricultural catchment. *Freshwater Biology* **29**: 285–294

Richards, C. and Bacon, K. L. (1994) Influence of fine sediment on macroinvertebrate colonisation of surface and hyporheic stream substrates. *Great Basin Naturalist* **54**: 106–113

Rosser, Z. and Pearson, R. G. (1995) Responses of rock fauna to physical disturbance in two Australian tropical rainforest streams. *J. N. Am. Benth. Soc.* **14**: 183–196.

Russell D. J., Hales P. W. and Helmke S. A. (1996) Pesticide residues in aquatic biota from north-east Queensland coastal streams. *Proceedings of the Royal Society of Queensland*

Scrivener, J. C. and Brownlee, M. J. (1989) Effects of forest harvesting on spawning gravel and incubation survival of chum (*Oncorhynchus keta*) and coho salmon (*O. kisutch*) in Carnarvon Creek, British Columbia. *Can. J. Fish. Aquat. Sci.* **46**: 681–696

Serruya, C. and Pollingher, U. (1983) *Lakes of the warm belt*. Cambridge University Press, Cambridge, UK.

Smart, M. M., Jones, J. R. and Sebaugh, J. L. (1995) Stream-watershed relations in the Missouri Ozark plateau province. *Journal of Environmental Quality* **14**: 77–82

Smith, R. E. W. (1987) Ecology of the shrimp *Australatya styriolata*. PhD thesis, James Cook University.

Smith, R. E. W. and Pearson, R. G. (1987) The macro-invertebrate communities of temporary pools in an intermittent stream in tropical Queensland. *Hydrobiologia* **150**: 45–61.

Starmuhlner, F. (1984a) Mountain stream fauna, with special reference to Mollusca. In: Fernando, C. H. (Editor), *Ecology and Biogeography in Sir Lanaka*. Dr. W. Junk Publishers, The Hague: 215–255.

Starmuhlner, F. (1984b) Checklist and longitudinal distribution of the meso- and macrofauna of mountain streams of Sri Lanka (Ceylon). *Arch. Hydrobiol.* **101**: 303–325.

Stout, J. (1980) Leaf decomposition rates in Costa Rican lowland tropical rainforest streams. *Biotropica* **12**: 264–272.

Sutcliffe, D. W. and Carrick, T. R. (1973) Studies on mountain streams in the English Lake District. I. pH, calcium and the distribution of invertebrates in the River Duddon. *Freshwater Biology* **3**: 437–462

Tabacchi, E.; Lambs, L.; Guilloy, H.; Planty-Tabacchi, A; Muller, E. and Decamps, H. (2000) Impacts of riparian vegetation on hydrological processes. *Hydrological Processes* **14**: 2959–2976.

Thoms, M. C. (1987) Channel sedimentation within the urbanized River Tame, UK. *Regulated Rivers: Research and Management* **1**: 229–246

Vannote, R. L., Minshall, G. W., Cummins, K. W., Sedell, J. R., Cushing, C. E. (1980) The river continuum concept. *Can. J. Fish. Aquat. Sci.* **37**: 130–137

Walling, D. E. (1990) Linking the field to the river: Sediment delivery from agricultural land. **In** Broadman, J., Foster, I. D. L. and Dearing, J. A. (eds.) *Soil erosion on agricultural land*. John Wiley, Chichester, UK., pp. 129–152.

Ward, J. V. (1974) A temperature-stressed stream ecosystem below a hypolimnial release mountain reservoir. *Arch. Hydrobiol.* **74**: 247–275

Ward, J. V. (1975) Bottom fauna-substrate relationship in a Northern Colorado stream: 1945 and 1974. *Ecology* **56**: 1429–1434

Ward, J. V. and Stanford, J. A. (1987) The ecology of regulated streams: past accomplishments and directions for future research. **In** Craig, J. F. and Kemper, J. B. *Regulated Streams – Advances in Ecology*. Plenum, New York, pp. 391–401

Ward, J. V. and Stanford, J. A. (1979) The serial discontinuity concept of lotic ecosystems. In: Fontaine, T. D. and Bartell, S. M. (Eds.) *Dynamics of Lotic Ecosystems*. Ann Arbor Science publishers, Michigan, pp. 29–42.

Webster, J. R., Golladay, S. W., Benfield, E. F., D'Angelo, D. J. and Peters, G. T. (1990) Effects of forest disturbance on particulate organic matter budgets of small streams. *J. N. Am. Benthol. Soc.* **9**: 120–140

Webster, J. R. and Patten, B. C. (1979) Effects of water-shed perturbation on stream potassium and calcium dynamics. *Ecol. Monogr.*, **49**:51–72.

Welcomme, R. L., Ryder, R. A. and Sedell, J. A. (1989) Dynamics of fish assemblages in river systems – a synthesis. Special Publication *Can. J. Fish. Aquat. Sci.* **106**: 569–577

Whitmore, T. C. (1989) Tropical forest nutrients, where do we stand? A *tour de horizon*. **In** Proctor, J. (ed.) *Mineral Nutrients in Tropical Forest and Savanna Ecosystems*. Blackwell Scientific Publications, London, UK.

Wikramanayake, E. D. and Moyle, P. B. (1989) Ecological structure of tropical fish assemblages in wet-zone streams of Sri Lanka. *J. Zool., Lond.* **218**: 503–526.

Wikramanayake, E. D. (1990) Ecomorphology and biogeography of a tropical stream fish assemblage: evolution of assemblage structure. *Ecology* **71**: 1756–1764.

Wiley, M. J. and Kohler, S. L. (1980) Positioning changes of mayfly nymphs due to behavioural regulation of oxygen consumption. *Can. J. Zool.* **58**: 618–622

Williams, M. R., and Melack, J. M. (1997) Solute export from forested and partially deforested catchments in the central Amazon. *Biogeochemistry* **38**: 67–102

Williams, M. R., Fisher, T. R. and Melack, J. M. (1997) Solute dynamics in soil water and groundwater in a central Amazon catchment undergoing deforestation. *Biogeochemistry* **38**: 303–335

Winterbourne, M. J. and Collier, K. J. (1987) Distribution of benthic invertebrates in acid, brown water streams in the South Island of New Zealand. *Hydrobiologia* **153**: 277–286

Wood, P. J. and Armitage, P. D. (1997) Biological effects of fine sediment in the lotic environment. *Environmental Management* 21(2):203–217.

Yule, C. M. (1996) Trophic relationships and food webs of the benthic invertebrate fauna of two aseasonal tropical streams on Bougainville Island, Papua New Guinea. *Journal of Tropical Ecology* **12**: 517–534

Part V
Critical appraisals of best management practices

SUMMARY

This section contains four chapters providing a critical assessment of existing guidelines (also called best management practices) that have been designed to minimise adverse effects on soils and streams during timber harvesting and land clearing operations in areas with tropical forest, as well as under rainfed cropping in deforested tropical steeplands. In addition, the main factors hampering the widespread application of these best management practices are examined in some detail and areas indicated requiring additional research.

Cassells and Bruijnzeel provide an overview of *guidelines aiming to minimise adverse impacts on the residual vegetation, soils and streams during tropical timber harvesting operations.* Research and land management experience over many decades have demonstrated that poorly planned or managed logging operations generally have deleterious environmental impacts. Most commonly, logging and the associated road construction lead to problems with erosion and sedimentation, and thus to a reduction in the quality of streamwater and aquatic habitat. At the same time, there is considerable evidence indicating that, provided forest managers and planners respect broad land capability limits, appropriately managed logging operations can be compatible with the maintenance of good quality water supplies. In this regard, experience has shown that particular attention needs to be given to the careful location of roads, extraction trails and stream crossings; minimising ground disturbance and maintaining effective ground cover; as well as maintaining undisturbed buffer strips (as stressed by Connolly and Pearson; Hamilton) around key streams and waterways. Further, while additional site specific (regional) guidelines for operations on different soil types and under different climatic regimes are desirable, practitioners are well-advised to follow the effective guidelines developed for extreme climatic (cyclone-prone) and hydropedological conditions (including the occurrence of widespread saturation overland flow on hillslopes, as discussed by Bonell) in the coastal zone of northeast Queensland. However, despite being known and field tested in a variety of forested environments for a number of decades now, these reduced-impact logging (RIL) technologies and general 'watershed specifications' have been ignored or inadequately practised in most tropical countries, causing unnecessary environmental and ecological damage. Ongoing research on the implementation of RIL techniques is clarifying many of the key issues associated with the continuation of inadequate management practices in commercial timber harvesting. In addition, changes in market demand and the emergence of independent certification of forest management systems are providing powerful new incentives for improved practice. However, widespread implementation of RIL techniques still depends on a number of factors including:

- translation of 'watershed specifications' into simplified locally applicable field guides that can easily be understood by key field operators;
- targeted, ongoing training for key field operatives and supervisors;
- a commitment to outcome monitoring and evaluation (both physical and economic); and
- introducing incentives and penalties to increase local community participation and encourage internalisation of environmental costs by logging companies and their employees.

Furthermore, Cassells and Bruijnzeel call for the establishment of regional demonstration forests where the costs and benefits of sustainable forest management can be rigorously monitored and ownership by local communities be fostered. Finally, although the authors recognise that significant improvements are already possible *now* by applying available knowledge, there is above all a need to address prevailing institutional, political and socio-cultural behavioural patterns, from governments down to forest enterprises and their employees. Ultimately, such aspects represent a bigger obstruction to implementing improved practices than the need for additional research results.

Thang and Chappell elaborate on the issues set forth by Cassells and Bruijnzeel by elaborating on measures to *minimise the hydrological impacts of high-intensity timber harvesting operations in Malaysia.* They indicate that Malaysia is probably one of the most advanced tropical countries in terms of progressively adopting international guidelines for sustainable forest management, including the application of an independent timber certification scheme in a number of forest management units in recent years. The Malaysian Criteria and Indicators used to assess the sustainability of timber harvesting practices are discussed in some detail, paying particular attention to those guidelines designed to minimise off-site sedimentation. Whilst the current set of guidelines is impressive, Thang and Chappell concede that there is room for improving their scientific base. Many guidelines reflect the field experience of foresters and civil engineers, plus the application of general hydrological concepts, but they are rarely based on robust hydrological process studies. The sensitive question whether to also apply streamside buffer zones around the channels of seasonal and ephemeral streams (which are likely to generate considerable amounts of water and sediment during times of peak rainfall) and not just along perennial streams provides a case in point. Thang and Chappell also encourage the use of topographically-based, process hydrology models (e.g. TOPMODEL, TOPOG) for providing a more scientific basis for buffer zone demarcation rather than depending solely on empirical criteria. The design of effective stream buffer zonation thus constitutes a key area for further academic and applied research. In addition, Thang and Chappell argue for more hydrological field studies to evaluate the landscape-scale hydrological benefits of specific skidder trail / feeder road densities, and optimum spacings of cross drains on forest roads under different physiographic and climatic conditions. As such, they recommend the instalment of quantitative river water quality monitoring in areas undergoing logging to help ascertain the general efficiency of reduced-impact logging and other guidelines aiming to reduce stream sediment loads at the landscape scale. With such information, the environmental benefits of applying specific measures can be compared more accurately with the economic considerations that are also part of the sustainability equation within the timber certification process.

Next, Hamilton complements the preceding chapters by Cassells and Bruijnzeel, Thang and Chappell (and others) on best management practices for timber harvesting operations by providing an overview of *guidelines for forest clearing operations*. In situations where clearing is inadvisable, Hamilton raises 'red flags' as guidelines where soil and hydrological impairment would be unacceptably severe. Examples include: unstable, slip-prone hillslopes, (Bonell *et al.*; Douglas and Guyot; Scatena *et al.*), riparian buffer zones (Bonell; Proctor; Connolly and Pearson; Chappell, Bidin *et al.*, significant freshwater wetlands (Drigo; Hooijer), mangroves, high-quality water supply headwater areas (including montane cloud forests) (Drigo; Aylward; Kaimowitz; Chappell, Tych *et al.*; Connolly and Pearson; Bruijnzeel), flood-prone areas (Douglas and Guyot; Hooijer; Connolly and Pearson) and areas with serious limitations such as very shallow or peat soils (Hooijer). An additional argument for Hamilton's raising 'red flags' for areas with montane cloud forest is their high biodiversity and endemism, in addition to these areas having an usually high water yield (see chapter by Bruijnzeel, Part III). Given the important soil and nutrient conservation values of riparian zones, Hamilton proposes that forested buffer (protection) strips adjacent to streams should err on the wide side. When concerning safe, high-quality water, Hamilton goes even as far as to suggest that entire catchments (not just their headwaters) should be retained under forest cover to prevent turbidity (sediment) and other pollutants from impairing water quality. Note, however, that Thang and Chappell stress the economic consequences of such rigorous protective measures. However, Hamilton concludes with the firm message that if the 'red flag' warnings are ignored then severe environmental consequences will be the unavoidable result. Prevention of degradation is much cheaper in ecological, economical and social terms than the complex, difficult and lengthy process of restoring severely degraded areas.

Finally, Critchley expands on the message that social acceptability is the key to the successful implementation *of soil and water conservation measures* for rainfed farming in tropical steeplands. He observes a paradigm shift in soil conservation thinking from being based solely on 'top-down' engineering schemes to one increasingly focused on 'farmers and communities' rather than 'watersheds'; and on 'sustained production' rather than 'erosion prevention and runoff control'. This more pragmatic view reflects acceptance of the fact that increasingly steep hillslides are being brought into cultivation due to increasing population pressure. 'Damage limitation' is thus the key to better management. Considerable attention has been given to 'bio-terraces' (as opposed to the classic bench terraces which require more labour to construct and maintain) as a recommended structure for the cultivation of tropical steeplands. These are formed naturally above narrow vegetative barriers which impede soil movement further downslope. Such contour barrier hedgerows can reduce annual erosion rates by an order of magnitude for slopes less than 20% but for steeper gradients their erosion effectiveness is more doubtful. Also, there is considerable resistance among farmers to their adoption due to maintenance requirements (regular pruning and weeding). The recent, emerging emphasis on within-field practices next to support structures is welcomed by Critchley. The goal of this 'land husbandry' is the imitation of forest floor conditions (by way of inter-cropping, crop rotation, cover crops–green manures, mulching, manure, compost) to induce higher crop productivity and thus surface protection. Such objectives are closely aligned to those connected with agroforestry (Wallace *et al.*, Part III). Particularly for steep slopes, Critchley emphasises in-field mixtures of trees and crops as one of the better options for building up organic matter and controlling erosion. Finally, Critchley examines the changing external factors and new priority areas in relation to land degradation and its remedies. In doing so, his arguments do not always follow the conventional perspectives. For example, he presents evidence from Africa that increasing population pressure does not necessarily promote soil erosion, as long as high value crops are adopted. Population pressure also results in plot subdivision and reduction in size and where hedges or structural boundaries are used, these can be an enhanced soil conservation measure. There are unlikely to be many new technical solutions: the best available are already in place. Rather, there is a need to improve the societal acceptability of conservation schemes which fulfil the criteria of low investment, low maintenance and increased crop productivity. To stimulate interest and investment in more expensive conservation schemes, these need to be associated with the use of high value crops. Poignantly, Critchley concludes that far more is known about the technical design and performance of conservation schemes *vis à vis* erosion reduction than on how farmers evaluate them or their reasons for adoption or not.

35 Guidelines for controlling vegetation, soil and water impacts of timber harvesting in the humid tropics

D. S. Cassells
The World Bank Environment Department, Washington DC, USA
L. A. Bruijnzeel
Vrije Universiteit, Amsterdam, The Netherlands

INTRODUCTION

Research and land management experience over many decades has demonstrated that poorly planned or managed logging operations in the tropics generally have a deleterious impact on the ecological and hydrological system (Burgess, 1971; Lal, 1987; Bruijnzeel, 1992; Bruenig, 1996). However, there is also a long history of research and management experience that indicates that, provided forest managers and planners respect broad land capability limits, appropriately managed logging operations can be compatible with the maintenance of hydrological values and high quality water supplies (Gilmour, 1977a, b; Poels, 1987; Baharuddin, 1988; Abdul Rahim, 1990; Bruijnzeel, 1992). On the other hand, despite being known and field-tested in a variety of tropical forested environments for a number of decades now (e.g. Cameron and Henderson (1979), Shepherd and Richter (1985) and Cassells *et al.* (1984) in north-east Australia; Marn and Jonkers (1981) and Pinard *et al.* (1995) in Borneo; De Graaf (1986), Jonkers (1987) and Hendrison (1990) in Surinam), available reduced impact logging (RIL) technologies for timber harvesting have often been ignored or inadequately practised in many countries, leading to the generation of unnecessarily high damage to remaining stands, soils and the hydrological system at large (Bruijnzeel, 1992; Bruenig, 1996). Research now being undertaken on the implementation of RIL techniques (e.g. Barreto *et al.*, 1998; Hammond *et al.*, 2000; Pinard *et al.*, 2002) is clarifying many of the key issues associated with the continuation of inadequate management practices in commercial timber harvesting. In addition, changes in market demand and the emergence of independent certification of forest management systems are providing powerful new incentives for improved practices (cf. Thang and Chappell, this volume).

The aim of the present chapter is to critically review these trends to identify specific actions that might lead to better practices in tropical rainforests during industrial timber harvesting activities. In particular, the chapter will provide:

- A brief overview of the nature of timber harvesting in (lowland) tropical rainforests;
- A brief review of the principal soil and water impacts associated with timber harvesting in these forests;
- An outline of RIL techniques;
- An outline of recent research results and management developments, and
- Some reflections on the requirements for the widespread application of reduced impact harvesting techniques, including some reflections on the continued disjunct between theory and field practice in tropical forest catchment management.

THE NATURE OF TROPICAL TIMBER HARVESTING

The nature of tropical timber harvesting can vary considerably from region to region. However, tropical timber harvesting has focused traditionally on the use of selective logging and harvesting systems that range from the crude high grading of a relatively limited number of commercially valuable species to sophisticated silvicultural systems designed to sustain the resource base and maintain and improve the composition of valuable species (Dawkins and Philip, 1998). Key to the development of such systems is the notion that forests are a potentially renewable source of timber as long as the inherent resilience of the forest ecosystem is not exceeded. In other words, timber harvesting should not simply be a mining operation but provide the manipulation of the canopy needed to favour the regeneration and growth of valuable species,

Forests, Water and People in the Humid Tropics, ed. M. Bonell and L. A. Bruijnzeel. Published by Cambridge University Press.

some of which are light-demanding and others more shade-loving. This is achieved by the creation of gaps whose size is determined by the light requirements of the species whose growth is to be enhanced (Bruenig, 1996; Whitmore, 1998).

The silvicultural systems that have been developed for use in tropical rainforests are either polycyclic or monocyclic. The former are based on the repeated removal of selected trees in a continuing series of felling cycles, whose duration is less than the time needed for individual trees to grow from seedlings to maturity (called the rotation age). The idea is to remove trees before their growth begins to stagnate and to make room for others already growing in the understorey to increase the future yield. Because of the high species density of most rainforests, and the relatively small number of species considered commercially viable, timber harvesting in polycyclic systems tends to create comparatively few scattered gaps in the forest canopy. By contrast, monocyclic systems skip the increment already accumulated in these 'adolescent' trees and rely almost exclusively on seedlings to produce the next crop (Whitmore, 1998). All saleable trees are removed during a single operation and the length of the cycle more or less equals the rotation age of the individual trees involved. Typical rotation periods for polycyclic systems range from 20 years (e.g. the CELOS system in Surinam; De Graaf, 1986) to 40–50 years (e.g. in Queensland; Shepherd and Richter, 1985) vs. 60–80 years in the monocyclic systems historically practised in the lowland forests of Malaysia (Meijer, 1970; Bruenig, 1996).

Compared to wholesale forest clearance, selection logging may be considered an intermediate form of forest disturbance (Bruijnzeel, 1992). However, the intensity of timber harvesting can vary considerably. It ranges from a few to 20 m^3 ha^{-1} in polycyclic systems in Africa (Jonkers and van Leersum, 2000) and Latin America (De Graaf, 1996; Van der Hout, 1999) to as much as 120–150 m^3 ha^{-1} in the more intensive harvesting systems practised in South East Asia where the forests are particularly well-stocked with commercial species (Bruenig, 1996). As the intensity of the harvest rises, the impacts of the logging on standing vegetation and soils become more severe and the resulting disturbance level resembles that of the earlier monocyclic harvesting systems.

Historically, tropical timber extraction methods relied on animal traction using bullocks or elephants for log removal and these systems usually lead to relatively low levels of soil disturbance and negligible hydrological impacts. The same can be said of the so-called *kuda-kuda* system practised in Borneo in which groups of 10–15 men pushed and pulled the logs on sledges along wooden 'rails' (Malmer and Grip, 1990), although Bruenig (1996) drew attention to the high incidence of work-related accidents with this method. However, to keep up with the dramatic expansion in the logging of tropical hardwoods from the 1960s onwards, tropical timber harvesting has relied almost exclusively on the use of mechanised logging and extraction techniques and this has increased significantly the levels of disturbance to the remaining forest, soils and the hydrological system, not least because of the construction of tractor tracks and a road network (cf. Grip, Fritsch and Bruijnzeel, this volume). The economic necessity to achieve an adequate return on the substantial amounts of capital invested in equipment, vehicles, roads, sawmills and plywood mills, has made it increasingly desirable to harvest as many marketable logs as possible during a single felling cycle, thus favouring a shift to more monocyclic-like systems in many places in South East Asia (Whitmore, 1998; Bruenig, 1996). Not surprisingly, damage to the forest and soil is much more drastic with monocyclic systems than under a polycyclic cutting regime. On the other hand, the frequency of disturbance is higher for polycyclic systems and this is why they have sometimes been considered to cause too much damage to be of much silvicultural use (Dawkins and Philip, 1998).

Quantitative data on the extent of damage to trees and soil associated with conventional tropical forest logging have been reviewed by, *inter alia*, Bruijnzeel (1992), Pinard *et al.* (1995), Bruenig (1996) and Whitmore (1998). Especially for South East Asia, where harvesting intensities tend to be very high, the emerging picture is rather gloomy. Regardless of whether logging was effected by means of wheeled skidders or tractors, winch lorry, or high-lead yarding, damage to the vegetation is generally considerable. The reported range suggests that, in the case of standard tractor / skidder logging, often 40–60% of all trees with a diameter at breast height (dbh) of >10 cm are killed or damaged (on top of those already harvested), with the fraction of casualties among saplings (<10 cm dbh) being of similar magnitude or higher (Figure 35.1). As a rule of thumb, for every harvested tree, at least one tree is killed and another damaged beyond recovery during conventional mechanised logging (Whitmore, 1998).

In addition, 10–30% of the soil surface is usually laid bare in the form of roads, tractor tracks and log landings. Furthermore, the use of heavy (ground-based) equipment to extract the timber (wheeled skidders or crawler tractors) generally involves considerable disturbance of the topsoil, causing both compaction of the top 15–20 cm and strong reductions in infiltration capacity (Malmer and Grip, 1990; Van der Plas and Bruijnzeel, 1993; see also Grip *et al.*, this volume), especially when wheeled tractors are used under wet conditions (Figure 35.2) (Kamaruzaman, 1991).

'High-lead yarding' (timber extraction using suspended cables to keep the logs above the ground during transport to a central 'spar') has the potential to be less damaging to the soil and residual vegetation than conventional skidder/tractor logging but this is usually achieved only by highly skilled crews (Bruenig, 1996; Dykstra and Heinrich, 1996). The few comparative 'real-world'

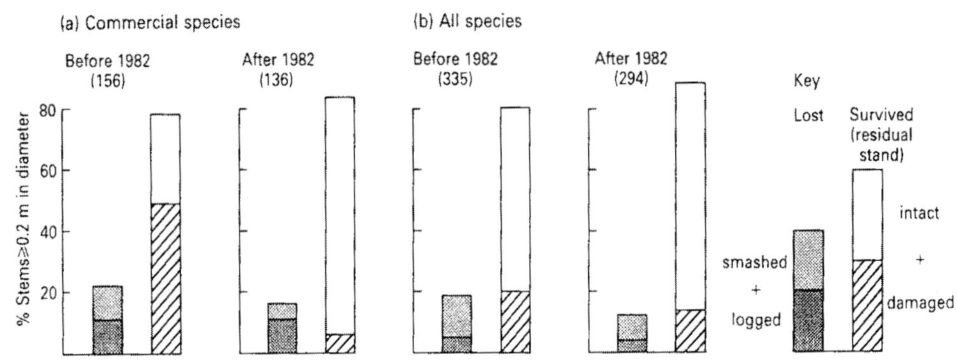

Figure 35.1 Effects of selective logging, Queensland on (a) commercial species and (b) all species, before and after the introduction of strict silvicultural rules in 1982. Total stems ha^{-1} in parentheses. (After Whitmore, 1998.)

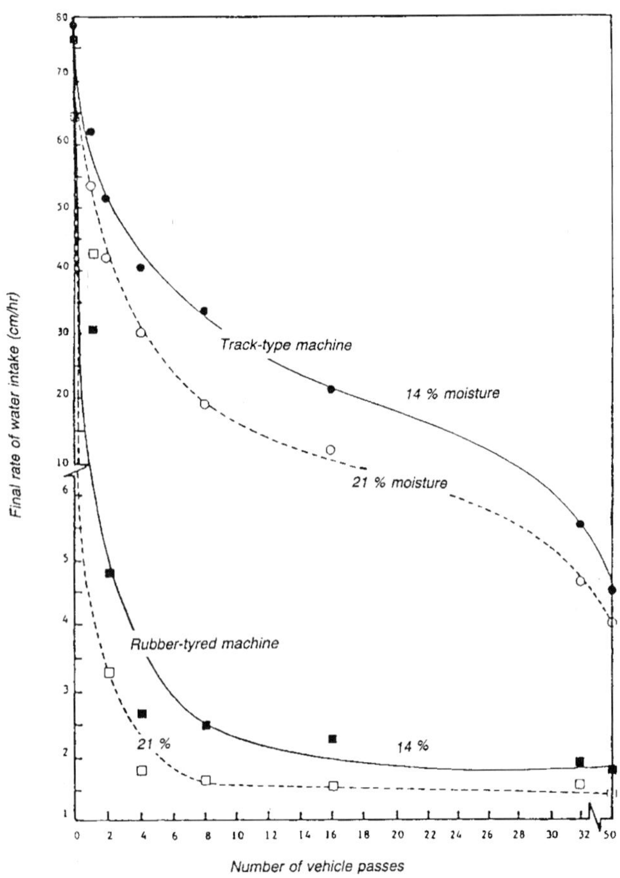

Figure 35.2 Reduction of topsoil hydraulic conductivity (cm hr^{-1}) as a function of machine type, number of machine passes and moisture content. (Redrawn after Kamaruzaman, 1991.)

studies available (e.g. Marcello and Tagudar, 1956; Borhan, Johari and Quah, 1987) not only suggest similar levels of ground disturbance on an areal basis but also very high damage to the vegetation along the path of the cables. In addition, the soil around the central spar trees (typically an area of 0.5 ha per pole) normally becomes completely devastated, with ensuing problems of erosion and regeneration (Marcello and Tagudar, 1956). On a related note, timber extraction by helicopter in remote steep terrain has been shown to be capable of achieving reductions in damage to the residual stand by up to 90% but the costs are typically 35–100% higher than for conventional ground-based logging. This makes the helicopter approach feasible under very specific conditions only (Chua 1996; Bruenig, 1996).

The type of disturbance to achieve effective regeneration will vary with the species concerned. The regeneration dynamics of some valuable hardwoods like mahogany (*Swietenia macrophylla*) and cedro (*Cedrella odorata*) are adapted to catastrophic events such as regular hurricanes and flooding and, as a consequence, require considerable canopy and, in some instances, soil disturbance to regenerate (Hammond *et al.*, 2000). Other more shade tolerant species such as greenheart (*Chlorocardium rodiei*) are adapted to low levels of canopy fragmentation where competition from fast growers is reduced. Conventional selective logging often creates gaps far greater than what is normally experienced by the forest under natural disturbance regimes. This has important consequences for the light regime experienced by seedlings which, in turn, affects the composition of the regenerating vegetation (Van Dam, 2001; Rose, 2000).

From the perspective of both maintaining sustainable yields and minimising deleterious on-site and off-site effects on soil and water resources, it is evident that more care needs to be taken than is usually the case (cf. Cassells, 1992; Pinard *et al.*, 1995). Bruenig (1996) gave a telling example from Sabah in which a well-stocked and highly productive rainforest was essentially reduced to a tangle of lianas and climbing bamboos with only a few scattered tall relict trees, some small original trees and patchy pioneer species, because of successive premature re-entries. Natural recovery of the growing stock was considered to take at least two centuries and a return to pre-logging levels of biodiversity much longer than that (cf. Hölscher *et al.*, this volume).

SOME CATCHMENT MANAGEMENT CONCERNS IN TROPICAL RAINFORESTS

The hydrological changes associated with selective forest harvesting in the humid tropics have been reviewed in detail by Bruijnzeel (1992), Douglas (1999) and Chappell, Tych *et al.* (this volume). Changes in water yield as a result of reductions in biomass and leaf area (and thus in evapotranspiration) proved negligible for extraction intensities up to 20 m^3 ha^{-1} but increased above this threshold with the amount of timber removed. One may wonder, therefore, to what extent the hydrological changes associated with the more intensive harvesting practised in South East Asia would approach those observed after clearfelling. In the absence of actual data, there is no straight answer to this question because normally there is the added soil disturbance by windrowing and burning of logging debris during clearing operations (cf. Grip *et al.*, this volume). Malmer (1992) compared changes in water yield for small catchments in East Malaysia which were cleared for plantation development, either manually without burning (i.e. minimum soil disturbance) or mechanically with burning (i.e. maximum soil disturbance). Increases in total water yield were 447 mm (+8.5%) and 1190 mm (+22%) for the first 32.5 months after clearance, respectively (or *c.* 165 and 440 mm $year^{-1}$). As the degree of soil disturbance associated with a conventional harvesting operation will be intermediate between these two cases, the resulting change in water yield may be expected to lie in between the cited extremes under a comparable rainfall regime[1].

Changes in stormflow volumes and peak flows after (low to medium-intensity) selective logging are usually minor or insignificant (see examples in Bruijnzeel, 1992). Also, amounts of nutrients lost in harvested timber during selective logging and through temporarily enhanced leaching after logging, appear to be balanced roughly by inputs via bulk precipitation (rain and dust) and, in the case of soils of medium to high fertility or freshly weathered rock within reach of the root system, by contributions from soil and rock weathering as well. Depending on the intensity of the harvest and thus the amounts of nutrients removed, the time required for the approximate recovery of nutrient reserves varies between 30 and 60 years (also depending on the nutrient under consideration; Bruijnzeel, 1992; 1998; cf. Proctor, this volume).

In most cases, therefore, the most immediate catchment management concern associated with logging is the impact of the extraction network (tractor tracks, roads, log landings) on erosion and sedimentation, especially in areas with deeply weathered and highly erodible subsoils (see the detailed discussions in Douglas (1999); Douglas and Guyot, this volume; and Chappell, Tych *et al.*, both this volume).

Finally, a major concern relates to the potential of logging operations to act as a catalyst for colonisation and more intensive land use change. In some cases, the roading network associated with forest harvesting has facilitated the encroachment of hunters, farmers and miners into previously forested areas (Wyatt-Smith, 1987; cf. chapters by Schweithelm and A. Hall, both this volume). This, in turn, has precipitated forest clearance and other forms of land use. In recent times, there has also been an increasing trend to convert logged-over forests to forest plantations in a number of countries (cf. Drigo, this volume). These areas are likely to be subject to periodic clearfelling and its associated land clearing implications (cf. Bruijnzeel, 1990; Grip *et al.*, this volume).

REDUCED-IMPACT LOGGING (RIL) TECHNIQUES

The potential of relatively simple precautionary measures to reduce the adverse impacts of logging on soils and streamflow quality, even in intense tropical environments, has long been established (Gilmour, 1971, 1977a; Cassells *et al.*, 1984). The development of catchment management control measures in humid temperate forest areas was based in part on the understanding that under such conditions stormflow and peak flows are produced not so much by infiltration-excess overland flow on the hillsides (as thought previously), but rather by so-called saturation-excess overland flow (*SOF*) occurring in valley bottoms, hillside concavities and other depressions in the landscape that are fed by more or less rapid subsurface flows from upslope (see Bonell, this volume for an exhaustive discussion of runoff generation under tropical conditions). In other words, it became of prime importance to minimise disturbance to such runoff source areas and thus keep the operations away from valleys and water courses (both perennial and ephemeral).

These principles were subsequently applied during the development of improved harvesting guidelines for more hydrologically active wet tropical environments, such as northern Queensland (Cassells *et al.*, 1984; 1985). Empirical monitoring of the effectiveness of these measures was limited at the time but did indicate substantial reductions in vegetation damage (Figure 35.1), soil exposure (Gillman *et al.*, 1985), as well as erosion and stream sedimentation (Gilmour, 1977a, b). Bruijnzeel (1990, 1992) suggested that, whilst additional site specific (regional) guidelines for operations on different soil types and under different climatic regimes were desirable, practitioners elsewhere in

1 Harvesting 40% of the commercial stocking at Bukit Berembun in much drier Peninsular Malaysia caused an average increase in water yield of *c.* 65% (*c.* 160 mm^{-1}) during the first four years (Abdul Rahim, 1990).

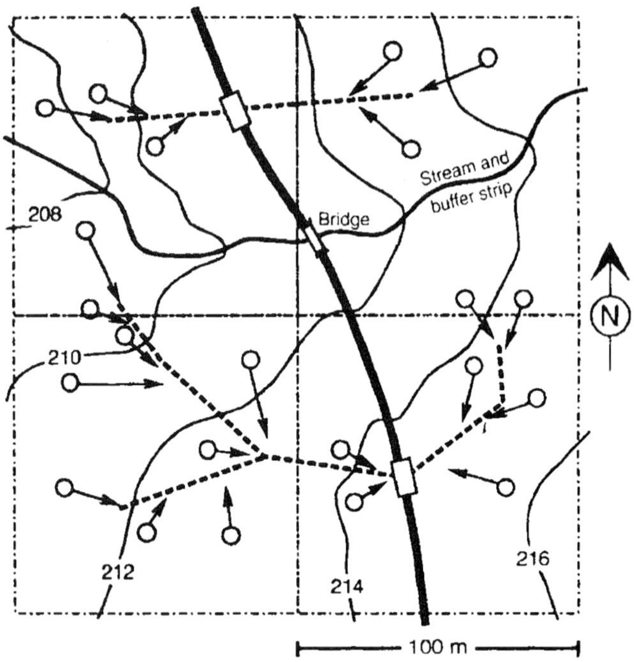

Figure 35.3 Part of a 1:5000 tactical planning map for a harvesting operation in a tropical forest. Circles indicate trees to be felled, and arrows the planned direction of felling. Heavy dashed lines represent skidder trails, the two open rectangles are log landings. Note the riparian buffer strip. (Redrawn after Dykstra and Heinrich, 1996.)

the tropics would be well advised to follow the (effective) guidelines developed for the extreme climatic and pedological conditions (including widespread *SOF* on hillsides; Bonell and Gilmour, 1978) in the coastal zone of Queensland. In general, existing prescriptions could be adhered to much more strictly (see Hendrison, 1990; Cassells, 1992; Pinard *et al.*, 1995; and, especially, Bruenig, 1996).

Key elements of the Queensland control measures to minimise on-site and off-site damage to soils and the hydrological system, and many of the subsequent measures adopted under reduced-impact logging (RIL) techniques in other tropical environments (Pinard *et al.*, 1995; Nussbaum, Anderson and Spencer, 1995; Dykstra and Heinrich, 1996; Van der Hout, 1999; Chappell and Thang, this volume) include the following:

(1) *Pre-logging planning*:
- Definition of vegetation retention areas (e.g. particularly vulnerable parts in the landscape, such as wet spots and steep slopes prone to mass wastage or gully erosion upon disturbance (the 'red flag' areas of Hamilton, this volume);
- Controls on the location of haulage roads, skidder tracks and log landings so as to stay away from streams and steep slope sections as much as possible and minimise the number of vehicle passes (Figure 35.3); also, promote uphill yarding where possible (Figure 35.4);
- Design and location of stream crossings (to minimise their number) (Figures 35.3 and 35.4);
- Design and frequency of drainage structures on roads and tracks as a function of slope and soil type / geological parent material;
- Pre-harvest inventory and tree-marking for directional felling (where possible into existing gaps so as to both limit damage to surrounding vegetation and minimise the number of tractor passes) (Figure 35.3);

(2) *Retention of streamside buffer strips* to trap sediment from upslope and provide shade to the aquatic ecosystem, bearing in mind the following aspects:
- Maintain buffer strips along all perennial streams;
- Likewise, for specified catchment areas (commonly 60–100 ha);
- Relate width of buffer strips to stream dimensions;
- Add strips along those seasonal or ephemeral channels where the flow convergence is likely to increase the risk of erosion and downstream sedimentation (cf. Thang and Chappell, this volume).

(3) *Timing of operations*:
- Seasonal logging bans, e.g. during very wet periods to avoid excess soil compaction (cf. Figure 35.2);
- Road construction conducted during period of least rainfall;
- Definition of shutdown conditions, based on cumulative precipitation or soil moisture wetness indices for different soil types;

(4) *Equipment specifications*:
- Use of integrated logging arches; raise the leading end of the logs to prevent them from ploughing into the soil;
- Use of narrow tractor blades to reduce unintended vegetation damage;

(5) *Post-logging treatment of tractor tracks and log landings*:
- Removal of temporary stream crossings;
- Construction of cross drains or bunds on critical tracts;
- Breaking up, fertilising and replanting log landings;
- Road maintenance and reseeding of roadside earthworks.

RESEARCH DEVELOPMENTS OVER THE LAST DECADE IN RELATION TO CATCHMENT MANAGEMENT IN HUMID TROPICAL FOREST AREAS

A survey of the literature on tropical forest hydrological research in relation to (selective) logging during the last decade or so reveals that most of the work has been concentrated in two areas: Sabah (Danum Valley) and Guyana. In the geologically and geomorphologically more varied Danum area, research has focused primarily

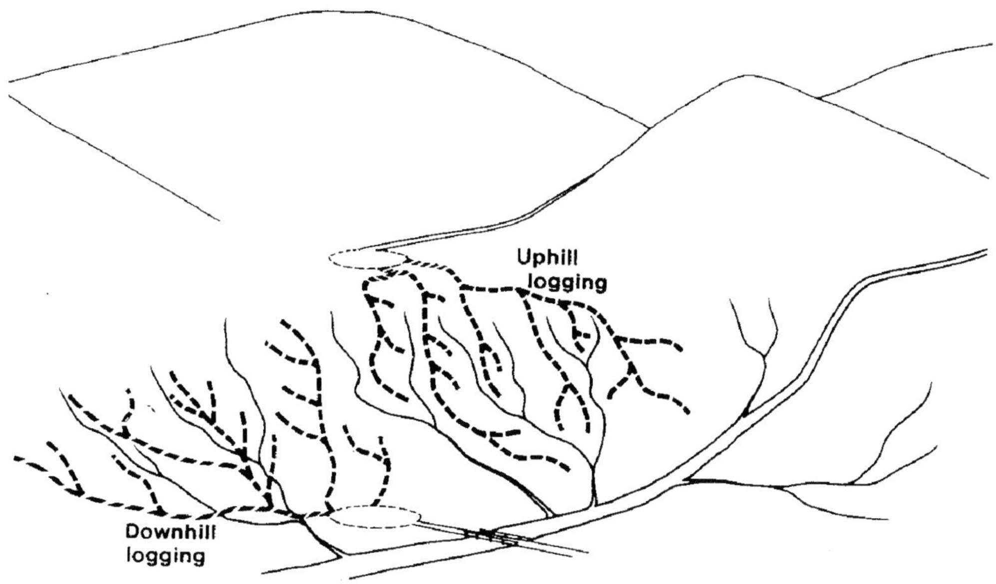

Figure 35.4 Skidder track networks associated with uphill and downhill log extraction. Uphill extraction tends to divert runoff and sediment away from streams, in contrast to downhill extraction. (After Gilmour, 1977a.)

on the movement of sediment through the landscape before and after intensive logging (reviewed in detail by Douglas (1999), Douglas and Guyot; Chappell, Tych *et al.*, both this volume), although rainfall interception as a function of the degree of canopy disturbance was also studied (Chappell *et al.*, 2001; Bidin, 2001).

In Guyana the topography is much less pronounced and the extremely infertile soils are often rather sandy and physically not very vulnerable to disturbance (Jetten, 1994). As a result, stream sedimentation is not so much an issue in the area but the loss of soil fertility after timber harvesting or clear-felling is potentially important and might eventually impair the regeneration of the more desirable species. Hence, a series of hydrological and nutrient cycling studies were conducted in close association with forest ecological (Zagt, 1997; Rose, 2000) and silvicultural (Van der Hout, 1999) research as part of a larger programme seeking, *inter alia*, to develop an ecologically and economically sustainable management system for forest relatively rich in the prized greenheart (*Chlorocardium rodiei*). Effects of (low-intensity) logging on water yield and leaching losses at the small catchment scale were shown to be negligible (Jetten, 1994; Brouwer, 1996). Brouwer (1996) also initiated a study of the effect of gap creation on nutrient leaching which was subsequently expanded by Van Dam (2001). Key results of the gap study include: (i) nutrient losses via leaching are increased dramatically in gaps exceeding 400 m^2 but do not differ from those experienced below closed forest in gaps smaller than 200 m^2; (ii) topsoil moisture contents do not differ much between differently sized gaps due to various compensatory effects but the shape of the gap (round vs. elongated or irregular) greatly affects radiation levels and thus growing conditions for seedlings of the more commercially attractive shade-loving species (Van Dam, 2001; Rose, 2000).

Needless to say, such findings are particularly relevant to the development of improved silvicultural methods. They also demonstrate that progress is being made to address the contention of some critics of industrial forestry that many forest management specifications are based on little more than intuition and educated guesswork that have yet to be rigorously tested in different management situations (e.g. Norton and May, 1993). Moreover, the ecological and hydrological advantages of better managed harvesting operations can hardly be denied (cf. Figures 35.1 and 35.2; Tables 35.1 and 35.2 below).

Many studies have also confirmed the broader ecological importance of such measures as buffer strip protection for riparian areas, both because they frequently support relatively high levels of species diversity in their own right and also because of the linkages they provide between the forest and various aquatic ecosystems (Bury, Corn and Aubry, 1991; Franklin, 1992; Naiman, Decamps and Pollock, 1993; cf. Proctor; Thang and Chappell, both this volume). In tropical forests, the function of riparian buffer strips has been studied thus far primarily in relation to the chemical composition of passing soil water and groundwater in both lowland (McClain, Richey and Pimentel, 1994; Williams, Fisher and Melack, 1997) and montane (McDowell, Bowden and Asbury, 1992) situations (cf. Proctor, this volume). For example, in Central Amazonia, concentrations of nitrate in groundwater travelling downslope have been found to decrease dramatically at the hillslope – riparian interface. This was ascribed by McClain *et al.* (1994) to a combination of denitrification in anaerobic riparian groundwater, active uptake by riparian vegetation, and possibly dilution by groundwater with low nitrate concentrations following deeper flow paths and discharging in the valley bottom. However, the fact that the phenomenon was observed both before and after the hillsides (i.e. the non-riparian areas) were deforested (Williams *et al.*, 1997) illustrates the profound influence that intact riparian buffer zones may exert on the chemical composition of subsurface flows entering streams.

Some recent studies on buffer strip management and efficiency in subtropical and temperate forests (mostly in Australia) which have led to a refinement of existing guidelines and which can be considered relevant to tropical catchment management practice as well include:

- Efforts to identify and protect areas of flow convergence (concavities and other depressions in the landscape) using comparatively simple topographical terrain models (Costantini *et al.*, 1993; Bren, 2000);
- Efforts to test the effectiveness of different buffer strip specifications in terms of sediment deposition (Prosser *et al.*, 1999; Lock *et al.*, 1999);
- Efforts to understand the generation and movement of sediment off forest roads towards streams to give more detailed guidance on road location, design and management (Costantini, *et al.*, 1999; Croke *et al.*, 1999; cf. Douglas *et al.*, 1999);
- Recognition that extensive buffer strip protection has real costs in terms of lost income from forgone harvesting and that blanket specifications may lead to over-protection of divergent stream bank areas and under-protection of upslope convergence areas with high erosion risks (Bren, 1997; 2000);

Another important research development during the last 10–15 years has been the greatly increased ability to model the temporal, and to a lesser extent spatial, dynamics of hydrological processes. Although the high data requirements of these physically-based process models cannot generally be used for operational planning in most humid tropical areas, arguably their greatest usefulness at present lies in helping researchers to understand the complex feedback mechanisms in forested catchments, identify key controlling variables, and allow scenario modelling. The already cited work by Van Dam (2001) on the effects of gap size and shape on micro-climate, soil water and leaching provides a case in point.

In the absence of detailed input data needed for the more sophisticated process models, there is scope for the use of topography-driven distributed hydrological models like TOPOG for several important on-site applications that require little more than a good contour map and basic soils information. Examples include the prediction of saturated zones in the landscape (important for the delineation of vegetation retention areas; O'Loughlin, 1986), the most likely spatial distribution of sheet erosion and gully erosion occurrence (Vertessy *et al.*, 1990; Constantini *et al.*, 1993), and landslide hazard (Dietrich *et al.*, 1992). The practical relevance of these applications to the planning of timber harvesting operations is self-evident (Figures 35.3 and 35.4).

OTHER RELEVANT DEVELOPMENTS

During the 1990s, considerable efforts were made through both intergovernmental frameworks, such as the International Tropical Timber Organization (ITTO), and voluntary frameworks to develop general guidelines and criteria and indicators for sustainable tropical forest management ahead of subsequent efforts for temperate and boreal forests (cf. Thang and Chappell, this volume). These developments included:

- The development of the ITTO Sustainable Tropical Forest Management guidelines, criteria and indicators, year 2000 objective (ITTO, 1990; 1992; 1997);
- A similar development of regional criteria and indicators for various 'intergovernmental processes' (e.g. the Tarapoto process for Amazon cooperation treaty countries, the African Timber Organization Sustainable Forest Management principles);
- Emergence of the Forest Stewardship Council (FSC) principles, criteria and independent performance-based certification systems[2] (Nussbaum, Jennings and Garforth, 2002; cf. Thang and Chappell, this volume); and
- Development of national certification systems (e.g. Indonesia, Malaysia).

While the field application of these various systems has varied (see Thang and Chappell, this volume, for successful examples from Peninsular Malaysia), their development is consistent with the recognition of the importance of indicators and performance monitoring in management systems in general to allow for adaptive management that builds on successes and learns from mistakes (Norton and May, 1993; Walker and Reuter, 1996).

Another potentially important development that may indirectly enhance the wider application of RIL techniques in tropical rainforests relates to the worldwide concern about rising atmospheric concentrations of greenhouse gases. This has prompted the search for methods to reduce emissions and increase carbon sequestration in plant biomass. The predominant approach recommended for actively offsetting CO_2 emissions is the planting of trees, though the scale at which this could be undertaken is unlikely to have a significant impact on atmospheric carbon levels (Bruenig, 1996). However, an ecologically sound and potentially cost-effective way

2 A forest certification system puts in place a process whereby a forest area is inspected by an independent certification body to determine whether its management meets clearly defined criteria and performance standards.

Table 35.1. *Estimated carbon stocks (C) and biomass (M) (both in t ha^{-1}) in mixed dipterocarp forest in Sabah, Malaysia, prior to logging and 2 years after logging by conventional and reduced-impact logging (RIL) methods*

			Post-logging			
	Pre-logging		Conventional		RIL	
Component	C	M	C	M	C	M
Above-ground biomass (living) 200	200	400	93	186	128	256
Root biomass (living)	50	100	23	46	32	64
Soil organic matter (0–30 cm)	70	146	63	131	67	134
Other necromass[a]	28	56	52	104	40	80
Total carbon/biomass	348	702	231	467	267	534
Difference in living biomass						
to pre-logging situation				−268		−180
to conventional logging						88

[a] Includes coarse and fine litter, dead roots, coarse woody material and standing dead trees.
Source: After Putz and Pinard (1993) to which reference is made for details on calculations and data sources.

to help reduce carbon emissions is to control the intensity of timber harvesting. In this way damage to the residual stand is reduced and with it the release of carbon to the atmosphere from decomposition of logging debris and destroyed trees, as well as from disturbed topsoils exposed to direct insolation (Putz and Pinard, 1993). Table 35.1 presents preliminary estimates of the carbon stocks present in living and dead vegetation and in the soil for a dipterocarp-dominated rainforest in Sabah, Malaysia, prior to (intensive) logging and two years after conventional or reduced-impact harvesting. In this particular example, non-conventional RIL left about 90 t ha^{-1} (36%) more living biomass on the site than did conventional logging (equivalent to about 44 t ha^{-1} of carbon). In addition, RIL produced about 24 t ha^{-1} (23%) less logging debris compared to conventional harvesting (Putz and Pinard, 1993).

EXPERIENCE WITH REDUCED-IMPACT LOGGING

The growing interest in reduced-impact logging (RIL) in tropical forests in the 1990s has been seen both as a step to more sustainable forest management (SFM) or to better protection of particular forest values such as maintaining the hydrological function or carbon sequestration and storage (Table 35.1). This interest grew in part because a large number of studies suggested 'win-win' outcomes (less damage to forest and soil and therefore better future harvesting prospects, and maintenance of the forest's hydrological functioning) were feasible and could be introduced in a cost-neutral or even cost-saving manner in practice. Key examples of such work include the early studies in Sarawak by FAO-UNDP (Marn and Jonkers, 1981), the introduction of rigorous 'watershed management controls' in northern Queensland (Cassells *et al.*, 1984, 1985; Figures 35.1 and 35.4), the CELOS harvesting and forest improvement system developed in Surinam (De Graaf, 1986; Jonkers, 1987; Hendrison, 1990), the Tropical Forest Foundation studies at Belem, Amazonia (Holmes *et al.*, 1999), and the work of the TROPENBOS Foundation in Guyana (Van der Hout, 1999).

It could be argued that many of the above examples concern low to medium harvesting intensities in terrain of low to moderate relief, in which it is comparatively easier to limit damage to soil and residual vegetation by applying only moderate levels of care during extraction. Nevertheless, the findings of Marn and Jonkers (1981) for mixed dipterocarp forest in hilly (but not excessively steep) terrain in Sarawak are pertinent. They found that conventional harvesting at an intensity of 50 m^3 ha^{-1} damaged, at least temporarily, more than 40% of the residual forest. In addition, almost half of the young growing stock (which is to supply future yield) was killed. As a result, it proved impossible to achieve a subsequent harvest after 25–30 years. On the other hand, extracting a similar volume of timber as part of a well-planned and supervised operation reduced the number of trees killed by 33% and limited temporary damage to the residual stand by about 40% (down to 23%). More importantly, this was achieved against significantly reduced (by 19%) operational costs, mainly due to the savings in machine operation time associated with a planned extraction network as opposed to non-planned ('random') extraction routes (cf. Figure 35.3). Marn and Jonkers (1981) concluded that, although damage to the residual stand was by no means negligible, it was at least reduced to the extent that the undamaged growing stock remained sufficiently large to provide a further crop after 30 years.

Pinard *et al.* (1995) have given figures on the relative damage to vegetation and soil caused by conventional and RIL methods in similar forest during very high intensity logging (100–150 m^3 ha^{-1}) operations in much steeper terrain elsewhere in Borneo. Together, the applied harvesting intensity and steepness of the land present an extreme case. Their results are summarised in Table 35.2.

Although very considerable reductions in damage to residual vegetation and soil (and thus to future growth) have been achieved with RIL methods, even under the extreme topographic conditions studied by Pinard *et al.* (1995) (Table 35.2), later work by the same group (Pinard, Putz and Tay, 2002) showed how the introduction of RIL in such steep terrain brought about increases in harvesting costs of *c.* \$135 per hectare. Indeed, despite the

Table 35.2. *Timber volumes extracted and associated damage to residual vegetation and soil associated with conventional and reduced-impact logging of mixed dipterocarp forest in very steep terrain in Sabah, Malaysia*

		Logging method			
		Conventional		RIL	
		Mean	SD	Mean	SD
Number of trees harvested per ha (rounded off)	14	3	9	4	
Volume of timber extracted (m^3 ha^{-1})		152	23	103	54
Percentage of area with soil disturbance	17	2	7	3	
Density of skid trails (m ha^{-1})		199	36	67	26
Percentage of trees killed (5–60 cm DBH[a])		41	11	15	7
Density of undamaged growing stock (5–20 cm DBH[a])	49	24	104	62	

[a] DBH, diameter at breast height.
Source: After Pinard *et al.* (1995).

various promising examples given above, the practice of RIL/SFM remains the exception rather than the norm, even where cost-neutral or cost-saving outcomes are likely. Hammond *et al.* (2000) have reviewed the experience obtained with RIL up to now and identified a series of benefits, bottlenecks and uncertainties related to the pan-tropical application of RIL techniques. These include:

- RIL is only part of good forest management: sustainability will not be achieved through RIL alone when other aspects of forest management such as securing appropriate regeneration and the maintenance of wildlife values are not considered within the same framework;
- The potential to limit damage to the residual forest is highly sensitive to felling intensity: where felling intensities are high, effective stand retention may require reduction in logging intensities and involve trade-offs between present and future timber yields and associated cash flows (cf. Van der Hout, 1999);
- The potential to reduce biomass damage is significant under the high volume logging intensities typical in South East Asia but considerably less at the lower logging intensities prevailing in Africa and Latin America;
- Logging must respect land capability limits: even the best harvesting techniques will lead to considerable acceleration of erosion and sedimentation on lands such as steep slopes not suitable for logging (cf. Hamilton, this volume). The benefits of corrective action will vary with background conditions, e.g. waterways which are not normally subject to large influxes of sediment will undergo more drastic biogeochemical change after poorly managed logging than streams which, through their proximity to steep, unstable slopes, are continuously fed large volumes of sediment even with controlled forest harvesting;
- RIL techniques depend on well-trained, motivated and satisfactorily compensated field crews;
- Monitoring and verification are vital components of RIL but costs have not always been adequately quantified, notably off-site costs like increased damages by enhanced peak flows, increased stream and estuarine sedimentation, etc. (cf. Easter, Dixon and Hufschmidt, 1986; Aylward, this volume); and
- The lack of (science-based) information targeted specifically at the needs of the timber industry is a significant obstacle to good forest management in many countries.

KEY OBSTACLES TO SUSTAINABLE FOREST MANAGEMENT IN TROPICAL FORESTS

In the early months of the year 2000, the ITTO commissioned a number of studies with regard to progress of Member Governments with the Organization's objective of sourcing international trade in tropical timber from sustainably managed forests by the year 2000. In a general review of progress with the Year 2000 target, Poore and Thang (2000) observed that considerable shifts in policy, rhetoric and, in some cases, legal mandate had occurred in the 1990s in response to ITTO's guidelines and the Year 2000 target but they could not ascertain in the context of their study by how much this had translated into actually improved practice at the operational level. Earlier, both Bruenig (1996) and Whitmore (1998) were rather skeptical in this respect and stressed the need for training and close supervision.

In a parallel review, Cassells and Hall (2000) found that while the production guidelines by ITTO and others had influenced government policies, they were still not widely applied in the field in many countries. Their discussions with forest managers, concession owners and others also identified some key obstacles to the widespread practice of sustainable management in tropical forests (see also Bruenig, 1996; Whitmore, 1998 for earlier discussion of these issues), including:

- The lack of public education, understanding and awareness of the issues associated with SFM and the consequent lack of commitment to SFM by political leaders and policy makers;
- Inappropriate national policies and the frequent absence of modern legal and planning frameworks to support SFM;
- The lack of transparency in forest resource allocation processes, allowing the development of corrupt practices;

- The difficulties experienced by forest planners, administrators and managers in understanding the needs, requests and levels of decision-making in local communities;
- The plethora of guidelines, criteria and indicators and other related instruments developed and published by the various international agencies (notably ITTO) as well as various more regionally operating intergovernmental 'forest processes' (e.g. the Amazon Treaty referred to earlier);
- Lack of skilled personnel in all areas of forest management, administration and utilisation; the high mobility of field labourers which is one of the reasons for the continuation of poor practices;
- Inadequate *local* forest research and development programmes (as opposed to research results in distant continents, e.g. South East Asia or Queensland vs. Amazonia or Central Africa);
- Outdated and inappropriate industry equipment that causes excessive collateral damage to the residual stand; and finally:
- Lack of targeted financial resources for affordable investment in improved forest management.

Cassells and Hall's review (2000) also identified a number of key priorities for accelerating progress towards achieving SFM in tropical forests. These included:

- Developing simple and practical *local* guidelines with a focus on applicability at the forest management unit level, as opposed to guidelines at a more generic or national code of practice level;
- Developing guidelines on the economic aspects of SFM, highlighting effective incentive structures for the private sector that will ensure more sustainable management and utilisation of forest resources;
- Developing regional demonstration forests where the costs and benefits of SFM can be rigorously monitored, while targeted training in key aspects of SFM such as RIL can be provided;
- Providing best practice guidelines in areas such as RIL and community participation in forest management (cf. Schweithelm, this volume);
- Promoting cooperation between international agencies with responsibilities for tropical forests (e.g. ITO, FAO) to reduce confusion and to foster ownership at the national level.

CONCLUDING REMARKS

Our knowledge of the hydrological functioning of logged-over tropical forests may still be rudimentary but, in most areas, significant improvements in forest practice and environmental outcomes are possible by the application of knowledge already existing. Institutional, political and cultural factors in government regulatory agencies and forest enterprises are as important (or more important) influences on forestry sector behaviour than technical knowledge and understanding. To make significant progress, understanding incentive structures and communicative structures is crucially important. The recent emergence of independent certification of sustainably produced timber is an important development that offers the potential to reward responsible stewardship with enhanced market access or, in some cases, early price premiums.

One of the key challenges is the continuing disjunct between the costs and benefits of both forest conservation and forest exploitation. Those who exploit forest rarely pay the full costs of their exploitation (i.e. including off-site costs like stream sedimentation) while those who bear the costs of conservation activities are often those who benefit the least (cf. Dixon and Easter, 1986; Aylward, this volume). To overcome these problems, and build on developments such as independent forest certification, there is a critical need for accessible regional demonstration forests where the costs and benefits of good practice can be rigorously monitored and training in key aspects of Sustainable Forest Management can be provided. Such demonstration forests should become powerful demonstration tools for local industry leaders, policy makers and political decision-makers. They should also logically become a focus for training and capacity building for improved forest management by providing access to good practice experienced *in conditions similar to those where particular forest managers or companies are operating.*

References

Abdul Rahim, N. (1990). *Effects of Selective Logging Methods on Hydrological Parameters in Peninsular Malaysia.* PhD thesis. Bangor: University College of North Wales.

Baharuddin, K. (1988). Effect of logging on sediment yield in a hill dipterocarp forest in Peninsular Malaysia. *Journal of Tropical Forest Science*, 1, 56–66.

Barreto, P., Amaral, P., Vidal, E. and Uhl, C. (1998). Costs and benefits of forest management for timber production in eastern Amazonia. *Forest Ecology and Management*, 108, 9–26.

Bidin, K. (2001). *Spatio-temporal Variability in Rainfall and Wet-Canopy Evaporation Within a Small Catchment Recovering from Selective Tropical Forestry.* PhD thesis. Lancaster, UK: University of Lancaster.

Bonell, M. and Gilmour, D. A. (1978). The development of overland flow in a tropical rainforest catchment. *Journal of Hydrology*, 39, 365–382.

Borhan, M., Johari, B. and Quah, E. S. (1987). Studies on logging damage due to different methods and intensities of harvesting in a Hill Dipterocarp forest of Peninsular Malaysia. *The Malaysian Forester*, 50, 135–147.

Bren, L. J. (1997). The effects of stream buffer width on available resources. *Australian Forestry*, 60(4), 260–263.

Bren, L. J. (2000). A case study in the sue of threshold measures of hydrologic loading in the design of stream buffer strips. *Forest Ecology and Management*, 132, 243–257.

Brouwer, L. C. (1996). *Nutrient Cycling in Pristine and Logged Tropical Rain Forest. A Study in Guyana.* Tropenbos – Guyana Series 1. Georgetown: Tropenbos-Guyana Programme.

Bruenig, E. F. (1996). *Conservation and Management of Tropical Rainforests. An Integrated Approach to Sustainability.* Wallingford, UK: CAB International.

Bruijnzeel, L. A. (1990). *Hydrology of Moist Tropical Forests and Effects of Conversion: A State of Knowledge Review*. Paris: UNESCO International Hydrological Programme and Amsterdam: Vrije Universiteit.

Bruijnzeel, L. A. (1992). Managing tropical forest watersheds for production: where contradictory theory and practice co-exist. In *Wise Management of Tropical Forests 1992*, eds. F. R. Miller and K. L. Adam, pp. 37–75. Oxford: Oxford Forestry Institute.

Bruijnzeel, L. A. (1998). Soil chemical changes after tropical forest disturbance and conversion: The hydrological perspective. In *Soils of Tropical Forest Ecosystems. Characteristics, Ecology and Management*, eds. A. Schulte and D. Ruhyat, pp. 45–61. Berlin: Springer Verlag.

Burgess, P. F. (1971). The effect of logging on hill dipterocarp forests. *Malayan Nature Journal*, 24, 231–237.

Bury, R. B., Corn, P. S. and Aubry, K. B. (1991). Regional patterns of terrestrial amphibian communities in Oregon and Washington. In *Wildlife and Vegetation of Unmanaged Douglas-Fir Forests*. USDA Forest Service General Technical Report PNW-GTR-285, pp. 341–352.

Cameron, A. L. and Henderson, L. E. (1979). *Environmental Considerations for Forest Harvesting*. Canberra: CSIRO Harvesting Research Group.

Cassells, D. S. (1992). Forested watershed controls in North-east Australia as an interim model for other humid tropical forest environments. *ITTO Tropical Forest Management Update*, 2, 6–8.

Cassells, D. S., Gilmour, D. A. and Bonell, M. (1984). Watershed forest management practices in the tropical rainforests of North-eastern Australia. In *Effects of Land Use on Erosion and Slope Stability*, ed. C. L. O'Loughlin and A. J. Pearce, pp. 289–298. Vienna: International Union of Forest Research Organisations.

Cassells, D. S., Gilmour, D. A. and Bonell, M. (1985). Catchment response and watershed management in the tropical rainforests in north-eastern Australia. *Forest Ecology and Management*, 10, 155–175.

Cassells, D. S. and Hall, C. (2000). *Proposals for the Development of a Comprehensive Framework and Practical Working Manuals on All Relevant Aspects of Sustainable Forest Management*. Repros prepared for the International Tropical Timber Organization. Georgetown, Guyana: Iwokrama International Centre for Rain Forest Conservation and Development.

Chappell, N. A., Bidin, K. and Tych, W. (2001). Modelling rainfall and canopy controls on net precipitation beneath selectively-logged tropical forest. *Plant Ecology*, 153, 215–229.

Chua, D. K. H. (1996). Helicopter logging lifts off in Sarawak. *ITTO Tropical Forest Management Update*, 6, 2–4.

Constantini, A., Dawes, W., O'Loughlin, E. M. and Vertessy, R. A. (1993). Hoop pine plantation management in Queensland: I. Gully erosion hazard prediction and water course classification. *Australian Journal of Soil and Water Conservation*, 6, 35–39.

Constantini, A., Loch R. J., Connolly, R. D. and Garthe R. (1999). Sediment generation from forest roads: bed and eroded sediment size distributions and runoff management strategies. *Australian Journal of Soil Research*, 37, 947–964.

Croke, J., Hairsine, P. and Fogarty, P. (1999). Runoff generation and redistribution in logged eucalyptus forests, south-eastern australia. *Journal of Hydrology*, 216, 56–77.

Dawkins, H. C. and Philip, M. S. (1998). *Tropical Moist Silviculture and Management: A History of Success and Failure*. Wallingford, UK: CAB International.

De Graaf, N. R. (1986). *A Silvicultural System for Natural Regeneration of Tropical Rainforest in Suriname*. PhD thesis, Wageningen, The Netherlands: Wageningen Agricultural University.

Dietrich, W. E., Wilson, C. J., Montgomery, D. R., McKean, J. and Bauer, R. (1992). Erosion thresholds and land surface morphology. *Geology*, 20, 675–679.

Douglas, I. (1999). Hydrological investigations of forest disturbance and land cover impacts in South East Asia: a review. *Philosophical Transactions of the Royal Society (London), Series B*, 354, 1725–1738.

Douglas, I., Bidin, K., Balamurugan, G., Chappell, N. A., Walsh, R. P. D., Greer, T. and sinun, W. (1999). Role of extreme events in the impacts of selective tropical forestry on erosion during harvesting and recovery phases at Danum Valley, Sabah. *Philosophical Transactions of the Royal Society (London), Series B*, 354, 1749–1761.

Dykstra, D. P. and Heinrich, R. (1996). *FAO Model Code of Forest* Harvesting Practice. Rome: Food and Agriculture Organization of the United Nations.

Easter, K. W., Dixon, J. A. and Hufschmidt, M. M. (1986). *Watershed Resources Management: An Integrated framework with Studies from Asia and the Pacific*. Boulder, Colorado: Westview Press.

Franklin, J. F. (1992). Scientific basis for new perspectives in forests and streams. In *Watershed Management: Balancing Sustainability and Environmental Change*, ed. R. J. Naiman, pp. 25–72. New York: Springer-Verlag.

Gillman, G. P., Sinclair, D. F., Knowlton, R. and Keys, M. G. (1985). The effect on some soil chemical properties of the selective logging of a north Queensland rainforest. *Forest Ecology and Management*, 12, 195–214.

Gilmour, D. A. (1971). The effects of logging on streamflow and sedimentation in a north Queensland rainforest catchment. *Commonwealth Forestry Review*, 50, 39–48.

Gilmour, D. A. (1977a). Logging and the environment, with particular reference to soil and stream protection in tropical rainforest situations. In *FAO Conservation Guide* 1, ed. S. H. Kunkle, pp. 223–235. Rome: Food and Agriculture Organization of the United Nations.

Gilmour, D. A. (1977b). Effect of rainforest logging and clearing on water yield and quality in a high rainfall zone of north-east Queensland. In *Proceedings of the First National Symposium on Forest Hydrology*, eds. E. M. O'Loughlin and L. J. Bren, pp. 156–160. Melbourne: Institution of Engineers Australia.

Hammond, D. H., van der Hout, P., Zagt, R. J., Marshall, G. Evans, J. and Cassells, D. S. (2000). Benefits, bottlenecks and uncertainties in the pantropical implementation of reduced impact logging operations. *International Forest Review*, 2, 45–53.

Hendrison, J. (1990). *Damage-Controlled Logging in Managed Tropical Rain Forest in Suriname*. PhD thesis. Wageningen, The Netherlands: Wageningen Agricultural University.

Holmes, T. P., Blate, G. M., Zweede, J. C., Pereira, R. and Boltz, F. (1999). *Financial Costs and Benefits of Reduced-Impact Logging Relative to Conventional Logging in the Eastern Amazon*. Forestry Private Enterprise Initiative Working Paper. Research Triangle Park, North Carolina, USA: Tropical Forest Founation / Fundação Floresta Tropical / USDA Forest Service. Southeastern Center for Forest Economics Research.

ITTO (1990). *Guidelines for the Sustainable Mangement of Natural Tropical Forests. ITTO Policy Development Series*, No. 1. Yokohama: International Timber Trade Organization.

ITTO (1992). *Criteria for the Measurement of Sustainable Tropical Forest Management. ITTO* Policy Development Series, No. 2. Yokohama: International Timber Trade Organization.

ITTO, 1997. *Criteria and Indicators for Sustainable Management of Natural Tropical Forests. ITTO* Policy Development Series, No. 7. Yokohama: International Timber Trade Organization.

Jetten, V. G. (1994). *Modelling the Effects of Logging on the Water Balance of a Tropical Rain Forest. A Study in Guyana*. Tropenbos Series 6. Wageningen, the Netherlands: Tropenbos Foundation.

Jonkers, W. B. J. (1987). *Vegetation Structure, Logging Damage and Silviculture in a Tropical Rain Forest*. PhD thesis, Wageningen, The Netherlands: Wageningen Agricultural University.

Jonkers, W. B. J. and van Leersum, G. J. M. (2000). Logging in South Cameroon: current methods and opportunities for improvement. *The International Forestry Review*, 2, 11–16.

Kamaruzaman, J. (1991). Effect of tracked and rubber-tyred logging machines on soil physical properties of the Berkelah Forest Reserve, Malaysia. *Pertanika*, 14, 1–11.

Lal, R. (1987). *Tropical Ecology and Physical Edaphology*. New York: J. Wiley.

Lock, R. J., Espigares, T., Costantini A., Garthe, R. and Bubb, K. (1999). Vegetative filter strips to control sediment movement in forest plantations: validation of a simple model using field data. *Australian Journal of Soil Research*, 37, 929–946.

Malmer, A. (1992). Water yield changes after clear-felling tropical rainforest and establishment of forest plantation in Sabah, Malaysia. *Journal of Hydrology*, 134, 77–94.

Malmer, A. and Grip, H. (1990). Soil disturbance and loss of infiltrability caused by mechanized and manual extraction of tropical rainforest in Sabah, Malaysia. *Forest Ecology and Management*, 38, 1–12.

Marcello, H. B. and Tagudar, E. T. (1956). Residual stands in selective high-lead logging. *The Philippines Journal of Forestry*, 12, 101–116.

Marn, H. M. and Jonkers, W. B. (1981). *Logging Damage in Tropical High Forest*. Working Paper 5. Kuching: FAO/UNDP Forestry Development Project Sarawak.

McClain, M. E., Richey, J. E. and Pimentel, T. P. (1994). Groundwater nitrogen dynamics at the terrestrial-lotic interface of a small catchment in the central Amazon basin. *Biogeochemistry*, 27, 113–127.

McDowell, W. H., Bowden, W. H. and Asbury, C. E. (1992). Riparian nitrogen dynamics in two geomorphologically distinct tropical rainforest watersheds: subsurface solute patterns. *Biogeochemistry*, 18, 53–75.

Meijer, W. (1970). Regeneration of tropical lowland forest in Sabah, Malaysia, forty years after logging. *The Malaysian Forester*, 32, 204–229.

Naiman, R. J., Decamps, H. and Pollock, M. (1993). The role of riparian corridors in maintaining regional biodiversity. *Ecological Applications*, 3, 209–212.

Norton, T. W. and May, S. A. (1993). *Integrated Forestry Harvesting in Eastern Australia: Ecological Impacts and Priorities for Conservation*. Working Paper 1993/5. Canberra: Centre for Resource and Environmental Studies, Australian National University.

Nussbaum, R., Anderson, J. and Spencer, T. (1995). Factors limiting the growth of indigenous tree seedlings planted on degraded rainforest soils in Sabah, Malaysia. *Forest Ecology and Management*, 74, 149–159.

Nussbaum, R., Jennings, S. and Garforth, M. (2002). *Assessing Forest Certification Schemes: a Practical Guide*. Oxford: Proforest.

O'Loughlin, E. M. (1986). Prediction of surface saturation zones in natural catchments by topographic analysis. *Water Resources Research*, 22, 794–804.

Pinard, M. A., Putz, F. E., Toy, J. and Sullivan, T. E. (1995). Creating timber harvest guidelines for a reduced-impact logging project in Malaysia. *Journal of Forestry*, 93, 41–45.

Pinard, M. A., Putz, F. E. and Tay, J. (2002). Lessons learned from the implementation of reduced impact logging in hilly terrain in Sabah, Malaysia. *The International Forestry Review*, 2, 33–39.

Poels, R. L. H. (1987). *Soils, Water and Nutrients in a Forest Ecosystem in Suriname*. PhD thesis. Wageningen, The Netherlands: Wageningen Agricultural University.

Poore, D. and Thang, H. C. (2000). *Review of Progress towards the Year 2000 Objective*, International Tropical Timber Council Document XXVIII/9/Rev.2, Yokohama: International Tropical Timber Organization.

Prosser, I., Bunn, S., Mosisch, T., Ogden, R. and Karssies, L. (1999). The delivery of sediment and nutrients to streams. In *Riparian Land Management Technical Guidelines. Volume One: Principles of Sound Management*, eds. S. Lovett and P. Price, pp. 39–60. Canberra: Land and Water Research and Development Corporation.

Putz, F. E. and Pinard, M. A. (1993). Reduced-impact logging as a carbon-offset method. *Conservation Biology*, 7, 755–757.

Rose, S. A. (2000). *Seeds, Seedlings and Gaps – Size Matters. A Study in the Tropical Rainforest of Guyana*. Tropenbos – Guyana Series 9. Georgetown: Tropenbos – Guyana Programme.

Shepherd, K. R. and Richter, H. V. (1985). *Managing the Tropical Forest*. Canberra: Australian National University.

Van Dam, O. (2001). *Forest Filled with Gaps. Effects of Gap Size on Water and Nutrient Cycling in Tropical Rain Forest. A Study in Guyana*. Tropenbos – Guyana Series 10. Georgetown: Tropenbos – Guyana Programme.

Van der Hout, P. 1999. *Reduced Impact Logging in the Tropical Rain Forest of Guyana. Ecological, Economic and Silvicultural Consequences*. Tropenbos – Guyana Series 6. Georgetown: Tropenbos – Guyana Programme.

Van der Plas, M. C. and Bruijnzeel, L. A. (1993). Impact of mechanized selective logging of rainforest on topsoil infiltrability in the Upper Segama area, Sabah, Malaysia. *International Association of Hydrological Sciences Publication*, 216, 203–211.

Vertessy, R. A., Wilson, C. J., Silburn, D. M., Connolly, R. D. and Ciesiolka, C. A. (1990). Predicting erosion hazard areas using digital terrain analysis. *International Association of Hydrological Sciences Publication*, 192, 298–308.

Walker, J. and Reuter, D. J. (1996). Key indicators to assess farm and catchment health. In: *Indicators of Catchment Health: A Technical Perspective*, eds. J. Walker and D. J. Reuter, pp. 21–33. Melbourne: CSIRO.

Whitmore, T. C. (1998). *An Introduction to Tropical Rain Forests*, 2nd ed. Oxford: Clarendon Press.

Williams, M. R., Fisher, T. R. and Melack, J. M. (1997). Solute dynamics in soil water and groundwater in a central Amazon catchment undergoing deforestation. *Biogeochemistry*, 38, 303–335.

Wyatt-Smith, J. (1987). Problems and prospects for natural management of tropical moist forests. In) *Natural management of Tropical Moist Forest*, eds. F. Mergen and J. R. Vincent, pp. 5–22. New Haven, Connecticut: Yale University.

Zagt, R. J. (1997). *Tree Demography of the Tropical Rain Forest of Guyana*. Tropenbos – Guyana Series 3. Georgetown: Tropenbos – Guyana Programme.

36 Minimising the hydrological impact of forest harvesting in Malaysia's rainforests

H. C. Thang
Forestry Department Peninsular Malaysia, Malaysia
N. A. Chappell
Lancaster University, UK

INTRODUCTION

Malaysia has a long experience of forestry management, with the Forestry Department of Peninsular Malaysia being established in 1901. Measures to improve the prospects of natural forest regeneration after the first cut were first applied in 1910, improved in 1927, and again in 1950 with the introduction of the Malayan Uniform System (MUS)[1]* and latterly with the application of the Selective Management System (SMS)* since the late 1970s (Thang, 1987; Jusoff and Mustafa, 1996; Dawkins and Philip, 1998, p. 154). The guidelines associated with these management systems aim to enhance regeneration in part by limiting collateral damage to trees remaining after the cut. This has the indirect effect of: (a) restricting damage to natural canopies and hence reducing changes to transpiration and wet-canopy evaporation (Asdak, *et al.*, 1998, Chappell, *et al.*, 2001, van Dam, 2001), and (b) requiring more careful skidding* or yarding* systems, thereby reducing ground damage (Pinard *et al.*, 2000) and hence soil erosion (Douglas *et al.*, 1993; Chappell *et al.*, 1999).

Methods of commercial forestry in Malaysia developed greatly from the mid-1950s onwards, with Malaysian companies now expanding their operations to other tropical countries. Inevitably, with the advancement of forestry technology has come a greater degree of mechanisation, and the potential for greater impact on the physical and biological environment (Wyatt-Smith *et al.*, 1964). Greater mechanisation within agriculture or urban development has magnified the detrimental impacts on the environment, as well as magnified the positive social effects (Pereira, 1973; Trewin *et al.*, 1998), so there should be no reason to believe that the intensified terrain and vegetation manipulation associated with more mechanised forestry should be immune from such impacts. Traditionally, Malaysian timber was cut and then hauled manually ('kuda-kuda') to landing areas* where it was then transported by river or rail (Sabah Forestry Department, 1989; Jusoff and Mustafa, 1996). Compared with the use of tracked vehicles ('skidders') and haulage lorries, such extraction is slow and costly, and would not give the state or nation the revenue to develop schools, hospitals and other social necessities (FAO, 1997; Sabah Forestry Department, 1989). In 2001, the revenue generated from the export of timber and timber products from Malaysia amounted to US$3770 million.

As forestry methods changed, guidelines have had to be modified to reflect the changing effects on forest regeneration (Dawkins and Philip, 1998). With the intensification of the operations, concerns have extended beyond the need to sustain timber production to consider off-site hydrological impacts such as turbid water supplies (Mohamed, 1987) and enhanced flood risk from increased channel sedimentation (Sheffield *et al.*, 1995). As an acknowledgement of this, state forestry policies were developed to mitigate impacts other than those impinging directly on timber production, including those related to the soil and water environment (Wyatt-Smith *et al.*, 1964).

During the 1990s the International Tropical Timber Organisation (ITTO), the Forest Stewardship Council (FSC) and several other organisations, notably Rainforest Alliance, Scientific Certification Systems (SCS) and Société Générale de Surveillance S.A. (SGS), developed guidelines allowing assessors to judge the sustainability of Natural Forest* management. In response, key organisations in Malaysia, notably the Forestry Department Headquarters (Peninsular Malaysia), the State Forestry Departments, the Forest Research Institute of Malaysia (FRIM), the Malaysian Timber Council (MTC) and the Malaysian Timber Certification Council (MTCC), restructured the existing guidelines to make them compatible with the ITTO criteria* and indicators*

1 Technical words with asterisks are detailed in Appendix 36.1 at the end of the chapter.

Forests, Water and People in the Humid Tropics, ed. M. Bonell and L. A. Bruijnzeel. Published by Cambridge University Press.

of sustainable forestry (ITTO, 1999). These are formally known as the 'Malaysian Criteria, Indicators, Activities and Standards of Performance for Forest Management Certification' or 'MC&I' (Thang, 1996). Further revisions of the MC&I have taken place (MTCC, 2001) in response to the major revision of the ITTO guidelines in 1998/9 (ITTO, 1999), and remain ongoing with, for example, the recent adoption of a set of 'Malaysian Criteria and Indicators for Forest Management Certification' in October 2002 which is technically compatible with the 'FSC Principles and Criteria' (Forest Stewardship Council, 2000; Thang, 2003).

Furthermore, Malaysia has been at the forefront of research on the hydrological, ecological and botanical impacts of forestry operations at field stations throughout Malaysia. These include Pasoh (Leigh, 1982; Condit *et al.*, 1999), Bukit Berembun (Abdul Rahim and Yusop, 1994; Abdul Rahim *et al.*, 1997), Bukit Tarek (Yusop, 1996), Hulu Langat and Sg. Lalang Forest Reserves (Lai, 1992) in the States of Peninsular Malaysia, the MFMA area of central Sarawak State (Holmes, 1995; Chua, 1996, 2001), and Sipitang (Malmer and Grip, 1994), Deramakot (Huth and Ditzer, 2001), and Ulu Segama / Danum Valley (Douglas *et al.*, 1993, 1995, 1999; van der Plas and Bruijnzeel, 1993; Pinard *et al.*, 1995, Chappell *et al.*, 1999) in Sabah State. Findings from this ongoing research are furthering the developing forestry guidelines.

THE MC&I SOIL AND WATER CRITERION OF SUSTAINABLE FORESTRY MANAGEMENT

ITTO (1999) defined seven criteria of sustainable forestry applicable at both the national and Forest Management Unit (FMU) levels. An FMU is a clearly defined forest area, managed to a set of explicit objectives and according to a long-term management plan (ITTO, 1999). Within Malaysia, the FMU can be one of the forest blocks under the management of a District Forest Officer, e.g. 'FMU 19' in Sabah State which consists of the 55 083 ha Deramakot Forest Reserve (FR) and the 57 240 ha Segaliod-Lokan FR, or it could be a whole state, e.g. the Selangor FMU which includes 234 644 ha of Permanent Forest Estate (PFE). PFE is land, whether public or private, secured by law and kept under permanent forest cover (ITTO, 1999, MTCC, 2001). Within Malaysia, 14.45 million ha (or 72% of all forest) is designated as PFE (under the National Forestry Policy, 1978, revised 1992). Some 3.81 million ha (or 26%) of the PFE exists as large blocks of 'Protection Forest', leaving 10.64 million ha (or 75%) of PFE as 'Production Forest' (Table 36.1). Forest Reserves gazetted using the National Foresty Act (1984, Section 10:1) as 'PFE Protection Forest' comprise the formal classes of Soil Protection Forest, Soil Reclamation Forest, Flood Control Forest, Water Catchment Forest, Forest Sanctuary for Wildlife, Virgin Jungle Reserved Forest, Amenity Forest, Education Forest and Research Forest. Commercial felling of trees is prohibited within all areas gazetted as PFE Protection Forest. Commercial forestry is undertaken with areas gazetted as PFE Production Forest (also called Class II Forest Reserve in Sabah State), though harvesting may be prohibited in some parts of this PFE Production Forest (e.g. stream buffer zones*, local areas > 1000 m a.s.l.). The respective forestry departments aim for all forestry with the PFE Production Forest to be undertaken sustainably (Thang, 2003), though the forestry operations within a significant proportion of this forest have yet to be certified as 'sustainably managed' by third party, international assessors.

Table 36.1. *Permanent Forest Estate (PFE) within Malaysia (million ha in 2001)*

Region	Protection Forest	Production Forest	Total land under PFE
Peninsular Malaysia	1.90	2.95	4.85
Sabah	0.91	2.69	3.60
Sarawak	1.00	5.00	6.00
All Malaysia	3.81	10.64	14.45

Except for a few thousand hectares of plantation forests, the PFE is Natural Forest*, which is forest land composed of indigenous trees, not planted by man (ITTO, 1999). Certification of sustainable management practices within Malaysia's Natural Forest (or rainforest) is currently undertaken at the scale of the FMU. For this certification* process the Malaysian Timber Certification Council and the associated independent, third parties (e.g. SGS (Malaysia) Sdn. Bhd.) are currently using six criteria of sustainability. These are: Criterion 1: Enabling conditions for sustainable forest management, Criterion 2: Forest resources security, Criterion 3: Flow of forest produce, Criterion 4: Biological diversity, Criterion 5: Soil and water, Criterion 6: Economic, social and cultural aspects (MTCC, 2001; SGS, 2002). These criteria names are as those within the most recent ITTO guidelines (ITTO, 1999), except that ITTO criterion 'Forest ecosystem health and condition' (which requires details of any natural or anthropogenic damage to the PFE) has been omitted because of perceived duplication. Some forest reserves in Malaysia have been certified according to slightly different criteria, e.g. the Deramakot FR certificate issued in 1997 (SGS, 1997), was based on compliance with the pre-1999 MC&I (Thang, 1996) and the FSC Principles and Criteria (Forest Stewardship Council, 2000), and the certificate for the Perak Integrated Timber Complex (ITC) concession of the Temengor FR issued in 2002 (SCS, 2002), was based on compliance with FSC Principles and Criteria.

Indicators of the MC&I Soil and Water Criterion

The criterion that explicitly addresses practices that protect the hydrological environment is the MC&I Criterion 5: Soil and Water. Of the nine ITTO-defined indicators, six are currently utilised for MC&I certification (MTCC, 2001; Thang, 2002), viz.

(5.1) Extent and percentage of total forest area managed primarily for the protection of soil and water.
(5.2) Extent and percentage of area to be harvested which has been defined as environmentally sensitive (e.g. very steep or erodible) and protected before harvesting.
(5.3) Percentage of length of edges of watercourses, water bodies, mangroves and other wetlands protected by adequate buffer strips.
(5.4) Existence and implementation of procedures to identify and demarcate sensitive areas for the protection of soil and water.
(5.5) Availability and implementation of guidelines for forest road lay-out, including drainage requirements and conservation of buffer strips along streams and rivers.
(5.6) Availability and implementation of harvesting procedures: (a) to protect the soil from compaction by harvesting machinery, and (b) to protect the soil from erosion during harvesting operations (MTCC, 2001).

Performance standards associated with the indictors of the MC&I Soil and Water Criterion

Associated with each of the indicators are a series of activities* and a series of very specific standards of performance or SOPs*. Within Malaysia, the same SOPs are applied to those states that form the Forestry Department Peninsular Malaysia (i.e. Perlis, Kedah, Penang, Kelantan, Perak, Terengganu, Pahang, Selangor, Negeri Sembilan / Melaka and Johore), though different SOPs are currently applied within the East Malaysian states of Sabah and Sarawak. This chapter will focus primarily on the SOPs used within Peninsular Malaysia, as these are currently the most clearly defined (MTCC, 2001). The SOPs associated with the six indicators listed above are discussed briefly.

(5.1) EXTENT AND PERCENTAGE OF TOTAL FOREST AREA MANAGED PRIMARILY FOR THE PROTECTION OF SOIL AND WATER

Within Peninsular Malaysia, the SOP for this indicator comprises the extent of PFE Protection Forest gazetted as 'Soil Protection Forest' and 'Water Catchment Forest', and the extent of PFE Production Forest that is excluded from logging because it is more than 1000 m a.s.l., within areas of >40° slope, or water catchment areas not formally gazetted under the National Forestry Act (1984). The SOP requires that felling is prohibited within all of these areas (MTCC, 2001; SGS, 2002)

(5.2) EXTENT AND PERCENTAGE OF AREA TO BE HARVESTED WHICH HAS BEEN DEFINED AS ENVIRONMENTALLY SENSITIVE (E.G. VERY STEEP OR ERODIBLE) AND PROTECTED BEFORE HARVESTING

The statistics under this indicator involve some overlap with those under indicator 5.1. Within Peninsular Malaysia these SOPs require that: (a) areas in the PFE Production Forest with an elevation above 1000 m cannot be felled, (b) areas in the PFE Production Forest with an elevation less than 1000 m but having slopes generally greater than or equal to 40° cannot be felled, (c) trees equal to and those above cutting limits for environmentally sensitive spots within the PFE Production Forest are marked and protected, and (d) areas in the PFE Protection Forest are not felled.

(5.3) PERCENTAGE OF LENGTH OF EDGES OF WATERCOURSES, WATER BODIES, MANGROVES AND OTHER WETLANDS PROTECTED BY ADEQUATE BUFFER STRIPS

The three principal forest types classified within current Malaysian FMU forest certification documents (e.g. SGS, 2002) are 'Inland / Dipterocarp Forest', 'Peat Swamp Forest' and 'Mangrove Forest'. Within Peninsular Malaysia these SOPs require that: (a) buffer strips along permanent watercourses (called Stream Buffer Zones, or SBZs*) in Inland / Dipterocarp Forest and Peat Swamp Forest of at least 5 m wide on either side of the watercourse, (b) buffer strips of permanent watercourses in Mangrove Forest of at least 3 m wide on either side of the watercourse, and (c) buffer strips of at least 50 m on the seaward side of Mangrove Forest, be protected from felling of trees.

(5.4) EXISTENCE AND IMPLEMENTATION OF PROCEDURES TO IDENTIFY AND DEMARCATE SENSITIVE AREAS FOR THE PROTECTION OF SOIL AND WATER

Within Peninsular Malaysia, the SOPs under this indicator are exactly the same as those under indicator 5.2. In some contrast, within Sabah the SOPs under this indicator are not simply listed (MTCC, 2001) but are stated as being contained within the documents: (a) Handbook on Forest Management, (b) Guideline on Forest Management Planning, (c) Forest Management Plan, (d) Annual Work Plan, (e) Comprehensive Harvesting Plan (e.g. Sabah Forestry Department, 1998a), (f) technical specifications for reduced-impact logging (RIL)* given in the Schedule F of the long-term Sustainable Forest Management License Agreement (SFMLA) for each FMU, (g) RIL Operation Guidebook

Specifically for Tracked Skidder Use (Sabah Forestry Department, 1998b), (h) Cable Logging Techniques, (i) Reference Manual for Timber Harvesting Operations in Commercial Class II Forest Reserves in Sabah, and (j) the report on the measures undertaken to implement these procedures (MTCC, 2001).

(5.5) AVAILABILITY AND IMPLEMENTATION OF GUIDELINES FOR FOREST ROAD LAY-OUT, INCLUDING DRAINAGE REQUIREMENTS AND CONSERVATION OF BUFFER STRIPS ALONG STREAMS AND RIVERS

Within Malaysia, forest roads are defined as main roads, secondary roads, feeder roads and skid trails (Forestry Department Peninsular Malaysia, 1999). Particular consideration is given to the construction of: (1) feeder roads* which are the temporary roads used to transport timber by lorry out of an annual coupe*, and (2) skid trails* which are the routes made by skidders and tractors to transport timber from the stump to the log landing areas*. Within Peninsular Malaysia the SOPs required under indicator 5.5 are:

- density of feeder roads of less than or equal to 40 m ha^{-1}
- density of skid trails of less than or equal to 300 m ha^{-1}
- right of way (corridor width) for feeder roads less than or equal to 15 m
- gradient of feeder roads of less than or equal to 20%, but will follow natural benches and topographic features when using existing roads or when newly specified by the Forest Department's (FD) Forest Engineer
- road camber (i.e. surface curvature) of feeder roads of at least 5%
- cross-fall (i.e., average transverse slope) for feeder roads of at least 3%
- carriageway of feeder roads (single lane) with width of at least 4 m, except at corners and lay-bys
- construction of V-shaped, earth side-drains along feeder roads
- adequate culverts (made from hollow logs, concrete, metal or high density polyethylene) located at stream or river crossings, where required or as specified by the FD Forest Engineer
- bridges (made from timber, concrete box culvert, or steel) of at least 3.5 m in width at stream or river crossings, where required
- silt traps in erosion-prone areas along feeder roads, as specified by the FD Forest Engineer
- buffer strips for permanent streams and rivers in Inland / Dipterocarp Forest and Peat Swamp Forest of at least 5 m in width on either side of the stream or river
- buffer strips for all perennial streams are marked and felling of trees is prohibited (MTCC, 2001).

Within Sabah, the SOPs under this indicator are again quite different, requiring: (a) following the technical specification for reduced impact logging in Schedule F of SFMLA, (b) following the guidelines in the Reference Manual for Timber Harvesting Operations in Commercial Class II Forest Reserves in Sabah., (c) road density of feeder roads not exceeding 20 m ha^{-1} for tractor skidding area and not exceeding 5 m ha^{-1} for skyline yarding* area, (d) road gradient of feeder roads must not exceed 15% (or 20% for sections of less than 150 m) but will follow natural benches and topographic features when using existing roads or when newly specified by the FD Forest Engineer, (e) total area occupied by skid trails not exceeding 6% of the total net logged area, (f) gradient of skid trails not exceeding 47%, (g) right-of-way of feeder roads of less than or equal to 15 m, (h) width of skid trails not exceeding 4.5 m on slopes up to 20° and 5 m on slopes $>20°$, (i) carriage way of feeder roads (single lane) with width of 5–6 m, and (j) cross drains* are constructed on roads and skid trails after timber harvesting has ceased, and are 0.5 m in height, approximately 45° to the road alignment, and have drain intervals of 20–30 m for gradient of 5–15° and <20 m for gradient of $>15°$ (MTCC, 2001).

(5.6) AVAILABILITY AND IMPLEMENTATION OF HARVESTING PROCEDURES: (A) TO PROTECT THE SOIL FROM COMPACTION BY HARVESTING MACHINERY, AND (B) TO PROTECT THE SOIL FROM EROSION DURING HARVESTING OPERATIONS

Within Peninsular Malaysia, the SOPs under this indicator are: (a) density of feeder roads of less than or equal to 40 m ha^{-1}, (b) density of skid trails of less than or equal to 300 m ha^{-1}, (c) gradient of feeder roads of less than or equal to 20%, but will follow natural benches and features when using existing roads or newly specified by the Forest Engineer, and (d) skid trails regenerated with appropriate species, mainly indigenous tree species (MTCC, 2001).

AN EXAMPLE MC&I CERTIFICATION ASSESSMENT

Three FMUs in Peninsular Malaysia have recently (December 2001) achieved certification* under the 1999 MC&I, and 'Certificates for Forest Management' have been awarded by the MTCC. These certificates apply to the Selangor FMU, Terengganu FMU and Pahang FMU. The sustainability of the forestry practices undertaken within these FMUs was assessed by SGS (Malaysia) Sdn. Bhd., a subsidiary of Société Générale de Surveillance S.A. of Geneva. The audit of the certified Selangor FMU will be examined as an example (SGS, 2002).

The main assessment of the Selangor state FMU was undertaken over 11–14 July 2000. Several areas of non-conformance were identified, and major Corrective Action Requests (CARs)* were issued which needed to be addressed and closed prior to eligibility for certification against the MC&I. All major CARs were subsequently addressed by the forest managers and then closed during follow-up visits by SGS in November 2000, making the Selangor FMU eligible for certification against the MC&I. The few minor infringements of the SOPs (and associated minor CARs) did not preclude the award of the certificate.

The certification process requires that audits of certified forests are undertaken. During the period 26th March to 5th April 2002, the Selangor FMU was re-assessed (audited), and a final report produced on the 3rd June 2002 (SGS, 2002). How well the forestry practices of the Selangor FMU met the standards of performance for Criterion 5 (Soil and Water) within this audit will be discussed below.

Re-assessment of MC&I Criterion 5 (Soil and Water) within the Selangor FMU

As part of indicator 5.1, SGS noted that the forest officers had properly identified the 22.0% (51,716 ha) of the forest area in the 234 644 ha Selangor PFE managed primarily for the protection of soil and water. This comprised areas gazetted as 'Soil Production Forest' and 'Water Catchment Forest' (PFE Protection Forest) and areas of PFE Production Forest excluded from logging due to height above 1000 m a.s.l., slopes greater than 40° and water supply catchments not legally gazetted (SGS, 2002). Outside of the certification requirement, it can be noted that a further 22 025 ha (9.4%) of the Selangor PFE is gazetted as other types of PFE Protection Forest, and thus protected from harvesting activities.

SGS noted that all areas having in excess of 40° slope showed no evidence of harvesting activities. Indeed the whole of Compartment* 75 in the 13 132 ha Ulu Langat FR area was excluded from logging due to the large areas with a slope of more than 40°.

Under indicator 5.3, SGS noted that the forest officers had correctly quantified the proportion of the permanent watercourse in the PFE (as identified on topographic maps). The forest department recorded that 96.45% of the permanent watercourse was protected during forest harvesting* with a buffer strip. SGS further noted that the minimum width of 5 m on either side of the channels, as prescribed in the MC&I, was generally observed, that buffer zone boundary trees were well marked, and no trees were felled within the buffer zone. Only in one location (Compartment 71, Gadong FR), was it noticed that earth had been pushed into a permanent or perennial stream* and trees felled into the buffer zone.

Under indicators 5.3 and 5.4, the SGS assessors noted that 'smaller streams or seasonal streams' were not protected by buffer zones (SGS, 2002). Seasonal or intermittent streams*, like ephemeral streams* are not required to be protected under current MC&I (MTCC, 2001), and should not have been mentioned in the report. There is some ambiguity in the terminology used in the report, so it is possible that some very small perennial streams, perhaps draining only a few hectares, were not properly protected. This issue will be addressed latter within the chapter.

With regard to the roads (under indicator 5.5), two minor CARs were identified during the re-assessment. Some road-side drains channelled water directly into perennial streams rather than divert it across the surrounding slopes outside the buffer zone (minor CAR SEL-012: SGS, 2002). Secondly, some roads (including skid trails) used for the current harvesting activities utilised old roads which were built too close to streams (minor CAR SEL-013). The high quality of the bridge and culvert construction was, however, recorded.

Under indicator 5.6, SGS noted that cross drains* were not added to all sections of skid trials after cessation of harvesting operations (minor CAR SEL-012); however, this is not a required SOP for Peninsular Malaysia under the current MC&I. Contractors are, however, sometimes required by the FD Forest Engineer to construct cross drains on particular sections of feeder roads where erosion problems have occurred. Also under indicator 5.6, SGS noted that in the 5645 ha Bukit Lagong FR and 3399.3 ha Batang Kali FR, planting of Leguminosae had been undertaken on road shoulders, skid trails and log landing areas as a soil conservation measure, and that the SOP of feeder road and skid trail density had kept within the MC&I prescriptions of 40 and 300 m ha^{-1}, respectively (SGS, 2002).

Re-assessment of other MC&I criteria pertinent to hydrological impacts within the Selangor FMU

In addition to Criterion 5 (Soil and Water), the protection of the hydrological system is also indirectly covered in Criterion 1 (Enabling Conditions for Sustainable Forest Management) and Criterion 3 (Flow of Forest Produce). Within the re-assessment of the Selangor FMU, national and state laws, policies and regulations were available to provide the framework for the sustainable management of the FMU, including the hydrological aspects (criterion 1: indicator 1.1). Notable legislation pertinent to water-related aspects in the Selangor FMU includes the National Forestry Policy (1978), the National Forestry Act (1984), Forest Rules (1988) for Selangor, the Land Conservation Act (1960) and the Environmental Quality Act (1974). For example, the prescription of an Environmental Impact Assessment (EIA) for sustainable forestry operations is prescribed in the Environmental Quality Act (1974).

While Criterion 3 is aimed at maintaining sustainable timber production, the restrictions on the volume of timber that can be harvested will directly affect: (a) the degree of canopy disturbance and hence the rates of wet-canopy evaporation and transpiration and resultant availability of surface- and soil-water, as well as (b) the intensity of skid trails or cable yarding systems and hence degree of ground disturbance and erosion (see Abdul Rahim and Yusop, 1994; Chappell, Tych *et al.*, this volume). In one area (Compartment 18 in the 5,645 ha Bukit Lagong FR) the actual production of 70 m^3 ha^{-1} did exceed the SOP limit of 61 m^3 ha^{-1} for Inland / Dipterocarp Forest with a 30-year cutting cycle (prescribed under indicator 3.4), and so a minor CAR (SEL-008) was issued. SGS acknowledged that guidelines for Reduced Impact Logging (RIL)* were available for the Selangor FMU (under indicator 3.7), notably the 'Field Manual for the Selective Management System – Volume 4' (Forestry Department Peninsular Malaysia, 1997), and a draft 'Guidelines for Reduced Impact Logging (January 2002)'. This later document was an amalgamation of SOPs stipulated throughout the existing MC&I, and SGS stated that this document should be formalised, and as a result issued a minor CAR (SEL-011). Lastly, it was noted that one of the new RIL* methods (involving skidding and rail haulage to the main log landing area) was difficult to apply in Compartment 25 of the 36 161 ha Raja Musa Peat Swamp Forest (SGS, 2002).

ITTO Indicators of the Soil and Water Criterion not currently applied in MC&I certification

The Timber Certification Scheme of MTCC was only launched in October 2001 and was done so using a 'phased approach' (Thang, 2002) in which those MC&I and associated SOPs utilised for certification would be refined with experience. Three of the indicators of the current ITTO Soil and Water Criterion have not been included within the first applications of the revised MC&I (Thang, 2003). These are:

(a) ITTO indicator 6.2: Extent and percentage of area to be harvested for which off-site catchment values have been defined, documented and protected before harvesting,
(b) ITTO indicator 6.4: Extent and percentage of area to be harvested for which drainage systems have been demarcated or clearly defined and protected before harvesting, and
(c) ITTO indicator 6.9: Existence and implementation of procedures for assessing changes in the water quality of streams emerging from production forests as compared with streams emerging from the same forest type kept free from human intervention.

These indicators will be used for certification and be included in the prevailing MC&I once field testing of their application has been completed (Thang, 2002). The SGS re-assessment of the Selangor FMU against the current MC&I also addressed changes needed in the SOPs that would be required for certification of FMUs under the Hallmark Programme of the Keurhout Foundation in the Netherlands. With respect to the omitted ITTO indicator 6.9, trials of qualitative monitoring of stream water quality at the logging block scale are being undertaken by FD staff, and SGS have suggested that this may need to be formally incorporated within the MC&I used for certification under the Hallmark Programme (SGS, 2002).

The successful award of certificates of sustainable forestry management* for the Selangor FMU and the other two state FMUs in Peninsular Malaysia, indicates that the current MC&I and associated SOPs in Peninsular Malaysia are sufficiently unambiguous and readily assessed to allow their use in the forest management certification process. The Forestry Department of Peninsular Malaysia does, however, suggest that this process could be improved if there were to be greater supervision of field operations by FD staff, and if contractors and FD field staff were to be given more training. The more open, transparent and ever improving process of forest certification facilitates the continual re-evaluation of the scientific basis of the criteria, indicators and standards of performance. In light of the review of selective forestry impacts on tropical hydrological systems detailed by Chappell, Tych *et al.* (this volume), how consistent are the current MC&I used for certification of Malaysia's rainforests?

CONSISTENCY WITH CURRENT HYDROLOGICAL SCIENCE

Chappell, Tych *et al.* (this volume), demonstrated that the greatest relative impact of selective harvesting of tropical forest on the hydrological system, is potentially the acceleration of sediment flows. This relates to a combination of increased surficial erosion and mass movements in disturbed forest areas. The resultant input of sediment into rivers leads to damage to fish populations (Martin-Smith, 1998; see also Connolly and Pearson, this volume), reduced quality of water supplies, reductions in channel capacity affecting flood risk and boat traffic (Sheffield *et al.*, 1995), and the inundation of offshore corals (MacDonald *et al.*, 2001). Impacts on water-supply availability, losses of vapour to the atmosphere, river flashiness, and nutrient losses through enhanced leaching from soil and litter, while observable, had a smaller magnitude of relative change (Chappell *et al.*, this volume).

There are several features associated with tropical, selective forest harvesting operations where the hydrological impacts could arise and where protective measures are currently incorporated or should be considered for incorporation. These are: (a) main and secondary roads, (b) feeder roads, (c) skid trails, where present, (d) stream buffer zones (SBZs), and (e) areas of disturbed forest canopy.

Main and secondary forest roads

Under the 'Forest Road Specification for Peninsular Malaysia' (Forestry Department Peninsular Malaysia, 1999), a main road is a road in/out of a forest area that is designed to accommodate a high number of vehicles (notably timber haulage lorries) to transport logs to the sawmill. A secondary road* is a road connecting a feeder road and main road for the purpose of log transport, rehabilitation and monitoring works (Forestry Department Peninsular Malaysia, 1999). Main roads are normally dual lane (with a formation width of 9–12 m), while secondary roads are single lane (with a formation width of 8–10 m). The main and secondary forest roads are permanent roads that normally have an all-weather running surface (e.g. gravel). For the forest management company, the most costly enterprise is the construction and maintenance of the permanent main and secondary forest roads (Putz, 1994; W. Sinun, Forestry Division of the Sabah Foundation, pers. comm.). It is in the company's interest to build permanent roads that are less likely to become un-trafficable due to surficial gullying, failure by mass movements or bridge collapse. The standards for main and secondary forest roads are normally high (Lim, 1999), though not as high as those of asphaltic, public highways (see e.g. Transport Research Centre, 1993) as the cost of the road construction would be higher than the revenue generated from sustainable timber production (D. H. K. Chua, Sarawak Forest Department, pers. comm.). Thus some enhanced risk of sediments entering rivers due to gradual road surface deterioration and periodic slope failure is likely. The gravel surfacing of main and secondary roads should offer some protection from erosion of fine sediments into watercourses. Soil exposure and disturbance around and below bridges or culverts could be significant point inputs of sediment into rivers, however, even within properly managed / certified forests (see the related work of Madej, 2001).

Forest feeder roads

Feeder roads classified by the Forestry Department Peninsular Malaysia (1999), in contrast to the main and secondary forest roads, are temporary roads that are used to transport logs out from a harvest site. These roads are typically unsurfaced and used to transport timber from the annual coupe and used only during timber extraction. From a commercial perspective, feeder roads need to be built to a sufficient standard to allow timber extraction during the logging of a single harvesting coupe. Deterioration of the road after this time only becomes a commercial issue after some 30–60 years when a further cut can be taken. Regulations governing the construction and maintenance of feeder roads are, therefore, required to lessen the short- and medium-term environmental impacts of road erosion and mass movement. Such regulations are contained in the standards of performance of the MC&I Soil and Water Criterion, particularly those under Indicator 5.5 on road lay-out. Within Peninsular Malaysia, these SOPs aim to: (a) minimise the risk of erosive flows developing on the feeder road (refer to earlier sections: 5.5d–h), (b) minimise the risk of culvert and bridge failure which would lead to accelerated erosion and mass movement (5.5ij), and (c) capture eroded sediments before they reach the stream (5.5l). Within the report of the audit of the Selangor FMU, SGS noted that some roads not originally built to current MC&I specifications for slope and proximity to streams had been re-opened (SGS, 2002). When considering this issue, it should be noted that abandoned feeder roads can continue to generate sediment as a result of log culvert collapses and slope failure (Chappell *et al.*, 1999). Thus the combined effect of sediment generated from an abandoned feeder road and from a newly constructed road (built to current MC&I and specification of the Forestry Department Peninsular Malaysia, 1999) may, in certain circumstances, generate more sediment than re-opening a feeder road having a lower specification. Careful consideration of the specific site details needs to be made in such cases. The restriction of feeder road density and size (5.5a–c), will restrict the extent of slope cutting, stream crossings and soil exposure within an area and so restrict the magnitude of erosion and mass movement. It also restricts the amount of canopy disturbance and resultant reductions in wet-canopy evaporation and transpiration (Asdak, *et al.*, 1998; Chappell, Tych *et al.*, this volume; van Dam, 2001), and thus increases in overland flow and/or subsurface water.

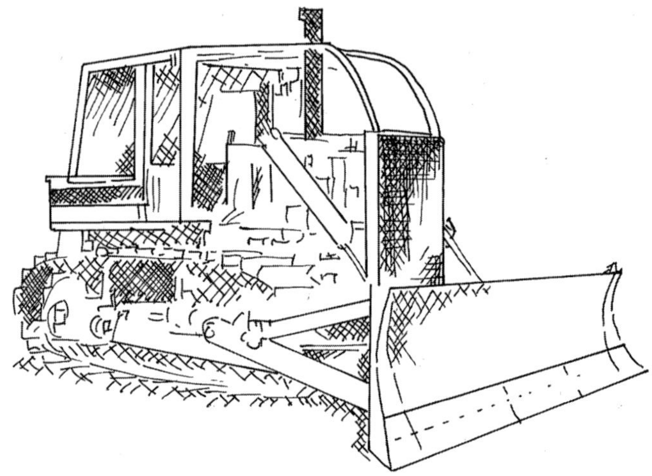

Figure 36.1 A common type of ground skidder.

Skid trails

Skid trails are the routes from the main, secondary or feeder roads made by wheeled skidders, crawler skidders or crawler tractors (Sist *et al.*, 1998) Figure 36.1). Example skid trail networks are shown in Figure 36.2 (from Dykstra and Heinrich, 1996) and in Chappell, Tych *et al.* (this volume). In Malaysia, ground skidding* is the principal method for moving logs from where a tree is felled

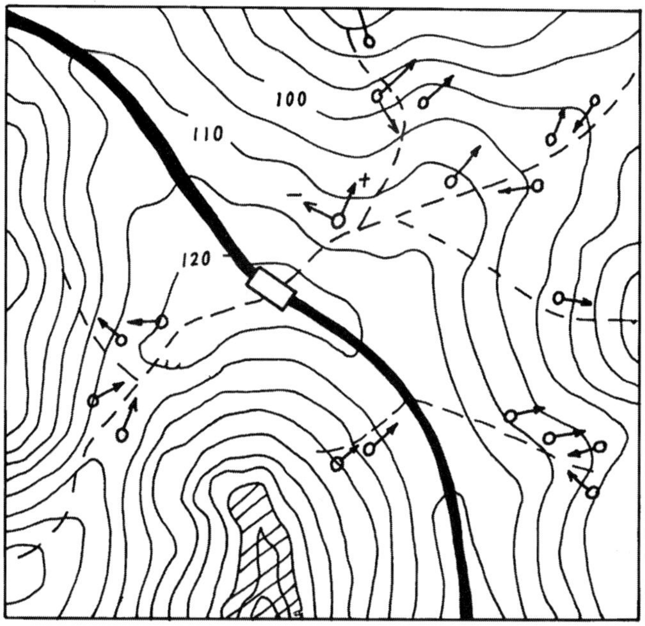

Figure 36.2 Example of a tactical logging map (scale 1:2000, 5 m contour lines) issued after the pre-harvesting inventory. Heavy solid line indicates a main, secondary or feeder road and dashed lines present skid trails*. The open rectangle is a log landing area*. Circles indicate trees to be felled, with the arrows showing the planned direction of felling. The hatched area is a stream buffer zone.* (After Sist *et al.*, 1998.)

to the landing area* (Figure 36.2) where the timber is loaded onto haulage lorries (Jusoff and Mustafa, 1996). Logs are dragged behind the skidder or tractor using a cable and winch.

Where there is a high local timber density and where slopes are relatively steep, the more expensive skyline yarding* system can be used to complement ground skidding in other parts of the logging coupe (Nydegger, 1998). With this method, each log is normally lifted above the forest canopy at the felling location to be lowered only on reaching a landing area by a road (Figure 36.3). Aerial logging*, where the cut timber is lifted by large payload helicopters (e.g. the Sikorsky S-64E skycrane) to the landing area on the roadside, has been carried out by the Sarawak Forestry Department (Chua, 1996; 2001). Aerial logging and skyline yarding with their more modest ground disturbance (Dykstra and Heinrich, 1996) have, however, not replaced ground skidding due to the significantly greater costs of their application (Conway, 1982; Dykstra and Heinrich, 1996). The use of skyline yarding in combination with ground skidding has been shown to be profitable within the Deramakot area of Sabah (Mannan and Awang, 1997).

Un-regulated skidder use, where new trails are cut to fetch individual logs, slopes that are extensively cut with the skidder's front blade, and where timber is skidded down very steep slopes, would lead to considerable opening of the forest canopy (Bruijnzeel, 1992; Dykstra and Heinrich, 1996; Pinard and Putz, 1996) and hence reductions in wet-canopy evaporation and transpiration,

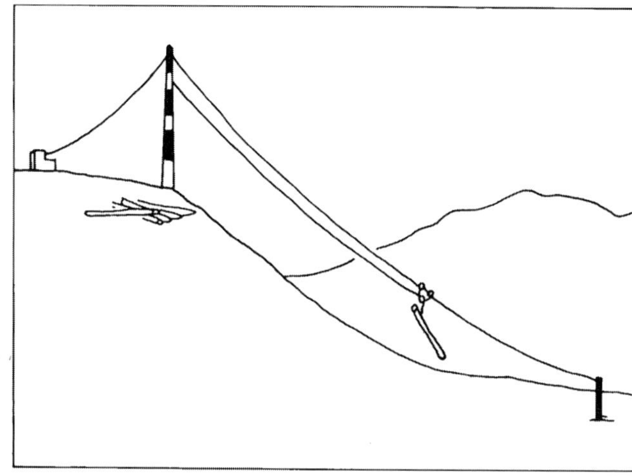

Figure 36.3 An example of a skyline with fully suspended logs.

giving greater amounts of overland flow and subsurface water. Poor management where skidders cross small rivers without the use of bridges or culverts would greatly disturb soils and sediments within and by streams, and would, therefore, greatly enhance stream sediment loads (see related studies by Grayson *et al.*, 1993; Brown, 1994). Ground skidding activities on steep slopes within Malaysia have been shown to give greater soil disturbance in comparison with those on shallower slopes (Jusoff, 1990). This observation justifies the prohibition of such activities on slopes greater than 40° under indicator 5.2 and 5.4 of the MC&I (MTCC, 2001). The research indicates that significant reductions in soil erosion could be gained if the upper slope limit were to be reduced below 40°. Further reductions in soil erosion may be obtained if MC&I required that roads were only built near ridges, so that logs were only skidded uphill (Bruijnzeel, 1992).

As skid trails are used to bring logs to a central point (i.e. the landing area) for subsequent bulk transport with lorries, a greater density of skid trails relative to main, secondary and feeder roads is expected, thus giving a greater potential for canopy disturbance. Compliance with Peninsular Malaysia's MC&I means that these potential impacts on the canopy can be checked by restricting the density of the skid trails to less than or equal to 300 m ha^{-1} (Indicators 5.5 and 5.6), by restricting harvesting* operations to below 1000 m and 40° slopes (indicators 5.2 and 5.4), by planting indigenous trees on skid trails after use (indicator 5.6), and by preventing skidder access (or cutting) near to small and large rivers (i.e. Stream Buffer Zones or SBZs; indicators 5.3 and 5.5). Direct input of sediments from skid trails into small and large rivers with permanent flows is similarly checked by the designation of SBZs.

Stream buffer zones

For forest management certification within Malaysia, stream buffer zones (SBZs) must be defined on permanent watercourses (i.e. streams and rivers). Within the stream buffer zones, skidder use is barred and tree cutting is prohibited (MTCC, 2001). Only bridges for main, secondary or feeder roads are permitted. In addition to the ecological benefits of maintaining riverine microclimate, habitat and seed pools, limiting canopy and terrain disturbance, SBZ delineation will limit stream bed and bank erosion of the protected channel (see review of Wenger, 1999).

Within Peninsular Malaysia, all reaches with permanent or perennial flows* should be buffered, though comments made by SGS during their certification of the Selangor FMU (SGS, 2002) may, however, indicate that very small (e.g. <1 m wide) permanent / perennial streams may not always be protected. Outside Peninsular Malaysia, e.g. in Sabah, streams with a permanent flow but with a width of less than 5 m, are not required to be buffered under current regulations.

Within the Danum Valley research area of Sabah State, perennial streamflow can be observed to be generated by areas as small as 1 ha (Chappell *et al.*, 1999), with streams having channels of about 5 m width being generated by areas of between 3 to 10 km^2 (300–1000 ha). First, second and perhaps third order headwater streams (Strahler, 1957) would, therefore, not be protected by buffer zones in this state. Similar channel-width based limits to buffer zone designation are present within the forestry guidelines of other countries having tropical forest (e.g. Cassells *et al.*, 1984; Sist *et al.*, 1998). Using the example of the Danum Valley area, if all permanent / perennial streams were to be protected with buffer zones with 10 m either side of the channel (20 m total width), timber harvesting would not be permitted within 13% of the area (assuming a drainage density of 6.5 km km^{-2} as observed for the 44 ha Baru catchment in the Danum area: Chappell *et al.*, 1999). Chappell *et al.* (1999) and Chappell *et al.* (2004) show high rates of sediment delivery from some perennial streams draining areas as small as 1–5 ha (i.e. a range from 14 to 1,467 t km^{-2} yr^{-1}). Similarly, Pearce *et al.* (1980), observing the effectiveness of buffer strips placed only on larger streams in the Tawhai State Forest (New Zealand), showed that un-buffered streams draining areas ranging from 1.6 to 8.3 ha can add significant quantities of sediment to the buffered main channels. As current MC&I already require the protection of all permanent streams with buffer zones, perhaps greater assurance that this is done could be achieved if FD field staff and indeed third party assessors of forest certification were to be given training by forest hydrologists in the (field and map) identification of perennial streams. Greater erosion protection would be achieved in the Malaysian state of Sabah, and indeed within several other tropical countries, if the channel-width limit for those perennial streams to be protected were to be removed.

The SGS assessors of the Selangor State FMU implied that even intermittent* and ephemeral* streams should be protected by buffer zones (SGS, 2002). Others (e.g. Durst, 1999) have similarly suggested that all channels should be protected. If all ephemeral, intermittent and perennial channels within the Danum area had 20 m total width SBZs, then timber harvesting would not be permitted within 40% of the region purely on the basis of SBZs (assuming a total drainage density of 20 km km^{-2}: Walsh and Bidin, 1995). This would also make the areas outside the SBZ largely inaccessible. This, combined with the reduced timber volumes available, would make harvesting within this region uneconomic. Given the high drainage density within most equatorial forest regions (Gregory, 1976; Morgan, 1976; Walsh, 1996) such a policy may make commercial forestry uneconomic within these regions also, and thus inconsistent with the fundamental 'People-Planet-Profit' principles of sustainable management*. Indeed, suggesting any environmental protection measures that might have a significant impact on the viability of commercial forestry should be assessed for their economic impacts (see review of RIL costs in Tay, 1999; Hammond *et al.*, 2000; Putz *et al.*, 2000; Bull *et al.*, 2001) prior to their imposition, particularly by non-producer countries. Chappell *et al.* (1993) have shown, via the monitoring of 14 nested contributory areas within the Baru catchment, that greater sediment yields (per unit area) are generated by perennial channels (1st to 3rd order basins), than by the channels with intermittent or ephemeral flows. This may imply that there is less hydrological justification for buffering intermittent and ephemeral channels, in comparison to all reaches of perennial channel. In addition to the greater training required for foresters in the identification of perennial channels, there is a need to generate simple methods for this procedure. Such channels could be identified by:

(a) the visual observation of the presence of water in the channels during inter-storm periods, or
(b) a minimum contributory area (e.g. 1 ha), or
(c) an index of minimum contributory area and slope angle.

Option (b) could be identified using Geographical Information System (GIS) methods applied to the digital Harvest Plan Maps currently used for managing certified / RIL forests (e.g. Sabah Forestry Department, 1997). An even better predictor of the start of a perennial stream would be gained by combining a minimum contributory area with the effect of the slope characteristics within the contributory areas (Option c). Topographic or wetness indices (see Barnes and Bonell, this volume; Chappell, Bidin *et al.*, this volume) could be readily incorporated within the same GIS framework as in Option (b).

Although some of the channels supporting intermittent or ephemeral flows within Malaysia's headwaters PFEs are already protected by being within areas that are either above 1000 m or on slopes $>40°$, it must be concluded that under the present regulations many zero- to second-order streams are not protected. Indeed, Bren (2000) has shown that the use of fixed buffer widths leads to underprotection of streamhead hollows and other places where flows tend to converge (places with high 'hydrological loading'), and to overprotection of slopes further downstream that tend to produce more divergent patterns of flow (places with low hydrological loading). Bren (2000) explored the usefulness of two measures of flow convergence for buffer zone delineation, viz. specific area and a slope index (both derived from digital elevation models) and their respective threshold values. Although areas of flow convergence received (more than) adequate protection in this way, downstream areas with low hydrological loading often did not get any protection at all. In addition, the results were not always predictable and showed large spatial variability, rendering the method less suitable from an administrative point of view. Moreover, despite the fact that areas with high hydrological loading were identified, the method by itself did not meet the criteria for buffer zone designation. Finally, the requirements for detailed topographic information as well as computational skills and hardware are such that convergence-based buffer zone delineation must be considered feasible only for very well-equipped and staffed organisations. As such, prospects for application in remote tropical terrain are very limited for the time being (cf. Cassells and Bruijnzeel, this volume).

Areas of disturbed forest canopy

The greater the disturbance to the canopy of Natural Forest, the greater is the likelihood of changes to the overall forest albedo, aerodynamic roughness and Leaf Area Index (LAI), and hence the greater likelihood of changes in wet-canopy evaporation and transpiration losses (van Dam, 2001). Research outside the tropics (e.g. Swanson and Dyrness, 1975; Sidle *et al.*, 1985) suggests that reducing the components of evapotranspiration with vegetation removal, increases subsurface water levels and hence the risk of slope failure. Where surficial erosion is an issue, the greater is the area of soil exposed along skid trails and feeder roads during timber cutting and extraction, and hence the greater potential for surficial erosion processes. Larger rates of canopy disturbance are also likely to give greater losses of nutrients via greater leaching (Brouwer, 1996; Yusop, 1996). Adherence to the MC&I specifications reduces the degree of canopy disturbance not only by limiting the density and size of skid trails and feeder roads (indicators 5.5 and 5.6), prohibiting felling and skidding activities in areas over 1000 m, >40° slopes (indicator 5.2 and 5.4) and within stream buffer zones (indicator 5.3), but also by limiting the volume of the annual cut (indicator 3.4). The work of Pinard *et al.* (1995) and Pinard and Putz (1996) in the RIL trial areas of the Sabah Foundation concession in Sabah state, clearly demonstrates the value of such guidelines for the quality of the forest canopy remaining after harvesting. Similar studies are identified in the review of Putz *et al.* (2000).

CONCLUSIONS AND RECOMMENDATIONS

The MC&I currently used for assessing the sustainability of forestry practices used within Malaysia's rainforest address directly the protection of the hydrological system within MC&I Criterion 5: Soil and Water. The associated indicators, activities and standards of performance clearly address the problem of accelerated losses of sediment into rivers. This focus is important, given that Chappell *et al.* (this volume) identify accelerated erosion and mass movement as having the greatest potential for relative change to the hydrological system as a result of forestry operations within areas of tropical rainforest maintained for sustainable timber production.

In the case of Peninsular Malaysia, the standards of performance that the State Forest Officers, timber concessionaires and logging companies need to follow and that are used by the national or international certification organisations to judge the (physical) environmental sustainability of these operations are very detailed and largely unambiguous. Similarly clear and simple lists of SOPs (MTCC, 2001) should be stated for the East Malaysian states of Sabah and Sarawak. The current debate between international assessors and local foresters over what is and what should be the SOPs (e.g. SGS, 2002), is valuable as guidelines may need to differ slightly with local site conditions (Nussbaum *et al.*, 2002), and should be continually refined (Thang, 2002). In this respect, we believe that the definition of stream buffer zones (indicators 5.3 and 5.5), is a key area for discussion and further academic and applied research. We suggest that such research would lead to the clearest results if landscape-scale (at least 0.5–50 km^2) impacts on sediment delivery (derived from stream monitoring: see Abdul Rahim *et al.*, 1997) were to be combined with statistically meaningful sets of plot erosion/mass movement studies within riverside areas (Chappell, Tych *et al.*, this volume).

The standards of performance pertinent to the hydrological system that are used within Malaysia, or within tropical rainforests elsewhere (e.g. Dykstra and Heinrich, 1996; Armatage, 1996; Sist *et al.*, 1998) are based largely on an assimilation of preexisting guidelines (Dykstra, 1999); the field experience of FD Forest Engineers; civil engineering practice covering road design; general hydrological concepts found within the literature (e.g. Hewlett, 1982); a few processes or applied hydrology case studies from other regions of the world (e.g. Gilmour, 1977; Swanson and Dyrness, 1975; Bonell *et al.*, 1983) and critical reviews of scientific data (Bruijnzeel, 1990, 1992). Unfortunately, however, there are very few robust scientific studies undertaken within similar physiographic locations using similar forestry manipulations (Chappell *et al.*, this volume). More hydrological field studies are needed within specific physiographic-climatic-vegetation regions (e.g. the Maritime Continent of Malaysia-Indonesia-Philippines with its relatively high relief, unstable soils and sensitivity to monsoon and ENSO cycles) that produce clear results of the landscape-scale, hydrological benefits of specific RIL practices. We suggest that such research should address the hydrological value and benefits of:

(a) perennial stream identification for improved stream buffer zone definition,
(b) maximum permitted skid trail / feeder road densities (Pinard *et al.*, 2000),
(c) maximum allowable slope for ground skidding operations (Yusop, 1990),
(d) benefits of uphill versus downhill skidding (Bruijnzeel, 1992), and the
(e) required use of cross drains* at regular intervals along forest roads (Sabah Forestry Department, 1998b; Sist *et al.*, 1998).

With such data, the (physical) environmental benefits of applying specific SOPs can be judged more accurately against the economic considerations that are an equal part of sustainable forestry management (ITTO, 1999; Tay, 1999; Hammond *et al.*, 2000; Putz *et al.*, 2000; Bull *et al.*, 2001; Thang, 2003).

Given that so little landscape-scale data (i.e. catchment water-balances, river regimes, river sediment and nutrient loads) are available to evaluate the impacts of specific forestry practices, the incorporation of quantitative monitoring of river water quality (ITTO indicator 6.9) into the MC&I used for certification would help forest hydrologists to define the SOPs and associated RIL systems. Such incorporation would have real benefits on the landscape-scale, hydrological system of Malaysia and other tropical regions with similar characteristics.

It is clear that the guidelines for reducing the impacts of logging operations within Malaysia's rainforests have developed considerably, particularly in recent years. The openness and transparency generated by the forest management certification process gives the forest hydrologist the opportunity to add substantially to the science that underpins the ever-improving forestry guidelines.

APPENDIX 36.1 GLOSSARY OF FORESTRY TERMS USED

Activity An action to be taken for compiling quantitative or qualitative information required to achieve the desired result of the Indicator* (MTCC, 2001).

Aerial logging Logs are moved from the stump to the landing area* by being lifted by helicopter or balloon (Conway, 1982).

Annual coupe Area of forest to be harvested within a period of one year (Sist *et al.*, 1998).

Certification This is the process of establishing whether or not the forestry System Standard and or Standard of Performance* has been met (Nussbaum *et al.*, 2002).

Compartment A fragment of homogeneous forest as regards its basic natural features, and requiring uniform silvicultural management, delineated in accordance with the principles of forest management.

Corrective Action Requests (CARs) A statement of non-compliance with SOPs that must be corrected before certification (major CAR) or within a few months after (minor CAR).

Criterion An aspect that is considered important by which sustainable forest management may be assessed. A criterion is accompanied by a set of related Indicators* (ITTO, 1999).

Cross drain In tropical areas, this is a trench and/or bund constructed across feeder roads* and/or skid trials* after timber harvesting has ceased, to divert surface water from the road surface or side drain.

Ephemeral stream A stream that flows only during storm events (Dingman, 1994). Compare with the definition for a perennial stream* and intermittent stream*.

Feeder road A temporary road that is used to transport logs out from a harvest site (Forestry Department Peninsular Malaysia, 1999).

Forest Management Unit (FMU) A clearly defined forest area, managed to a set of explicit objectives and according to a long-term management plan (ITTO, 1999).

Harvesting Timber harvesting operations include the construction of forest roads, tree cutting, ground skidding, cable yarding and timber haulage.

Indicator A quantitative, qualitative or descriptive attribute that, when periodically measured or monitored, indicates the direction of change (ITTO, 1999).

Intermittent stream A stream that flows only during the wet season (Dingman, 1994). Such stream types may not be common in equatorial rainforests when rainfall seasonality is small. Compare with the definition for permanent stream* and ephemeral stream*.

Landing area A roadside area used for the temporary storage of logs before loading on to haulage lorries, rail or water transportation. These areas are also called 'log landing areas', 'loading areas', 'logyards' and 'mantau' (in Malaysia).

Main road A road in/out of a forest area that is designed to accommodate a high number of vehicles (notably timber haulage lorries), normally having dual carriageway, to transport logs to timber processing mills, especially sawmills (Forestry Department Peninsular Malaysia, 1999).

Malayan Uniform System (MUS) A 'monocyclic' harvesting system in which the next cut is from seeds, seedlings or saplings (see Wyatt-Smith *et al.*, 1964; Dawkins and Philip, 1998; Putz *et al.*, 2000).

Natural Forest Forest land composed of indigenous trees, not planted by man, which is further classified using the criteria of forest formation (or type), degree of human disturbance or modification, and human interference (ITTO, 1999).

Perennial stream A stream that flows all year (Dingman, 1994); also called a permanent stream. Compare with the definition for an intermittent stream* and ephemeral stream*.

Permanent Forest Estate (PFE) Land, whether public or private, secured by law and kept under permanent forest cover. This includes land for the production of timber and other forest products, for the protection of soil and water, and for the conservation of biological diversity, as well as land intended to fulfill a combination of these functions (ITTO, 1999).

Reduced Impact Logging (RIL) The intensively planned and carefully controlled implementation of harvesting operations to minimise the impact on forest stands and soils, usually in individual tree selection cutting (Bull *et al.*, 2001).

Secondary road A road connecting a feeder road and main road for the purpose of log transport, rehabilitation and monitoring works (Forestry Department Peninsular Malaysia, 1999), normally having only a single carriageway.

Selective Management System (SMS) A 'polycyclic' harvesting system in which the next cut is from trees of intermediate size (e.g. 20–40 cm dbh) at the time of first cutting, (Dawkins and Philip, 1998; Putz *et al.*, 2000).

Skidding Logs are moved from the stump to the landing area* by being dragged by wheeled skidders, crawler skidders or crawler tractors (Conway, 1982). Also called 'ground skidding'.

Skid trail A route from the main, secondary or feeder roads made by wheeled skidders, crawler skidders or crawler tractors (Conway, 1982; Sist *et al.*, 1998). These routes are also called 'skid tracks' and 'snig tracks'.

Skyline yarding Logs are moved from the stump to the landing area* by being lifted on a wire rope suspended between two or more points (after Conway, 1982). To minimise ground and canopy disturbance, skylines that fully suspend the logs are preferable (Dykstra and Heinrich, 1996).

Standard of Performance (SOP) A set of requirements used as benchmarks to measure the attainment of a particular Indicator* (MTCC, 2001).

Stream Buffer Zone (SBZ) An area adjoining permanent streams where harvesting activities are restricted and/or prohibited.

Sustainable forest management Sustainable forest management is the process of managing forest to achieve one or more clearly specified objectives of management with regard to the production of a continuous flow of desired forest products and services without undue reduction of its inherent values and future productivity and without undue undesirable effects on the physical and social environment (ITTO, 1999). Thus the three elements of sustainable management of People-Planet-Profit (PPP) are explicitly included.

Yarding Logs are moved from the stump to the landing area* by being lifted on a wire rope either by highlead, skyline* or shotgun methods (after Conway, 1982).

References

Abdul Rahim, N. and Yusop, Z. (1994). Hydrological response to selective logging in Peninsular Malaysia and its implications on watershed management. In *Proceedings of the International Symposium on Forest Hydrology*, pp. 263–274. Tokyo, Japan, October 1994.

Abdul Rahim, N., Wan Mohd, W. R., Yusop, Z., Chan, H. T., Lim, H. F., Ratnam, L., Appanah, S., Mohd Sharif, A. H., Kasran, B., Manokaran, N., Yahya, N. A., and Yaacob, A. (1997). *EIA Guideline for Harvesting of Natural Forests*. FRIM Technical Information Handbook No. 14.

Armitage, I. (1998). *Guidelines for the management of tropical forests 1. The production of wood*. FAO forestry paper No. 135. Rome: Food and Agriculture Organisation.

Asdak, C. Jarvis, P. G., van Gardingen, P. and Fraser, A. (1998). Rainfall interception loss in unlogged and logged forest areas of Central Kalimantan, Indonesia. *Journal of Hydrology*, 206: 237–244.

Bidin, K. (2001). *Spatio-temporal variability in rainfall and wet-canopy evaporation within a small catchment recovering from selective tropical forestry*. Unpublished PhD thesis. Lancaster: University of Lancaster.

Bonell, M., Gilmour, D. A., and Cassells, D. S. (1983). Runoff generation in tropical rainforests of north-east Queensland, Australia, and its implications for land use management. In *Proceedings of the symposium on the hydrology of humid tropical regions, IAHS Publ. 140*, ed. R. Keller, pp 287–297. Paris: IAHS

Bren, L. J. (2000). A case study in the use of threshold measures of hydrologic loading in the design of stram buffer strips. *Forest Ecology and Management*, 132, 243–257.

Brouwer, L. C. 1996. *Nutrient cycling in pristine and logged tropical rainforest*. Unpublished PhD thesis. Utrecht: Utrecht University.

Brown, K. J. (1994). Riverbed sedimentation caused by off-road vehicles at river fords in the Victorian Highlands, Australia. *Water Resources Bulletin*, 30: 239–250.

Bruijnzeel, L. A. (1990). *Hydrology of moist tropical forest and effects of conversion: a state of the art review*. Paris: UNESCO.

Bruijnzeel, L. A. (1992). Managing tropical forest watersheds for production: where contradictory theory and practice co-exist. In *Wise Management of Tropical Forests*, Proceedings of the Oxford Conference on Tropical Forests 1992, eds. F. R. Miller and K. L. Adam. pp. 37–75. Oxford : Oxford Forestry Institute (University of Oxford).

Bull, G. Q., Pulkki, R., Killmann, W., and Schwab, O. (2001). Does it cost or does it pay?: An investigation of the costs and benefits of reduced impact logging. *Tropical Forest Update*, 11(2). Yokohama: International Tropical Timber Organisation.

Cassells, D. S., Gilmour, D. A., and Bonell, M. (1984). Watershed forest management practices in the tropical rainforests of north-eastern Australia. 289–298. In *Effects of Land Use on Erosion and Slope Stability*, C. L. O'loughlin and A. J. Pearce (eds). Vienna: IUFRO.

Chappell, N. A., McKenna, P., Bidin, K., Douglas, I. and Walsh, R. P. D. (1999). Parsimonious modelling of water and suspended-sediment flux from nested-catchments affected by selective tropical forestry. *Philosophical Transactions of the Royal Society of London Series B.*, 354: 1831–1846.

Chappell, N. A., Bidin, K. and Tych, W. (2001). Modelling rainfall and canopy controls on net-precipitation beneath selectively-logged tropical forest. *Plant Ecology*, 153: 215–229.

Chappell, N. A., Douglas, I., Hanapi, J. M., and Tych, W. (2004). Source of suspended-sediment within a tropical catchment recovering from selective logging. *Hydrological Processes*, 18: 685–701.

Chua, D. K. H. (1996). Helicopter logging lifts off in Sarawak. *Tropical Forest Update*, 6(3). Yokohama: International Tropical Timber Organisation.

Chua, D. K. H. (2001). Helicopter harvesting in the hill mixed dipterocarp forests of Sarawak. Paper presented at the *International Conference on Application of Reduced Impact Logging to Advance Sustainable Forest Management – Constraints, Challenges and Opportunities*, 26th February – 1st Match 2001, Kuching.

Condit, R., Ashton, P. S., Manokaran, N., LaFrankie, J. V., Hubbell, S. P., and Foster, R. B. (1999). Dynamics of the forest communities at Pasoh and

Barro Colorado: comparing two 50-ha plots. *Philosophical Transactions of the Royal Society of London Series B.* 354: 1739–1748.

Conway, S. (1982). *Logging Practices.* San Francisco : Miller Freeman Publications.

Dawkins, H. C., and Philip, M. S. (1998). *Tropical moist forest silviculture and management: A history of success and failure.* Wallingford: CAB International.

Dingman, S. L. (1994). *Physical Hydrology.* Englewood Cliffs: Prentice Hall.

Douglas, I., Greer, T. Bidin, K., Sinun, W., and Wong, W. M. (1993). Impact of roads and compacted ground on post-logging sediment yield in a small drainage basin. In *Hydrology of Warm Humid Regions*, 213–218, J. S. Gladwell (ed), Publication 216. Wallingford: IAHS Press.

Douglas, I., Greer, T., Sinun, W., Anderton S., Bidin, K., Spilsbury, M., Jadda, S., and Azman, S. (1995). Geomorphology and rainforest logging practices. In *Geomorphology and Land Management in a Changing Environment*, eds. D. F. M. McGregor and D. A. Thompson. Chichester: Wiley.

Douglas, I. Bidin, K., Balamurgan, G., Chappell, N. A., Walsh, R. P. D., Greer, T. and Sinun, W. (1999). Role of extreme events in the impacts of selective tropical forestry on erosion during harvesting and recovery phases at Danum Valley, Sabah. *Philosophical Transactions of the Royal Society of London Series B.*, 354: 1749–1761.

Durst, P. B. (1999). *Code of Practice for Forest Harvesting in Asia-Pacific.* RAP Publication 1999/12. Rome: Food and Agriculture Organisation.

Dykstra, D. P. (1999). Codes of Forest Practice: Historical background and conceptual framework of the FAO Model Code of Forest Harvesting Practice. In *Research on environmentally sound forest practices to sustain tropical forests.* Rome: Food and Agriculture Organisation.

Dykstra, D. P. and Heinrich, R. (1996). *FAO model code of forest harvesting practice.* Rome: Food and Agriculture Organisation.

FAO. (1997). *Asia-Pacific Forestry Sector Outlook Study: Country Report – Malaysia.* Working Paper APFSOS/WP/07. Rome: Food and Agriculture Organisation.

Forest Stewardship Council. (2000). *FSC Principles and Criteria.* Document 1.2. Oaxaca: Forest Stewardship Council.

Forestry Department Peninsular Malaysia. (1997). *Field Manual for Selective Management System – Volume 4.* Kuala Lumpur: Forestry Department Peninsular Malaysia.

Forestry Department Peninsular Malaysia. (1999). *Specification of Forest Roads for Peninsular Malaysia.* Kuala Lumpur: Forestry Department Peninsular Malaysia.

Gilmour, D. A. (1977). Effect of rainforest logging and clearing on water yield and quality in a high rainfall zone of North-east Queensland. In *Symposium on The Hydrology of Northern Australia, Brisbane.* pp. 156–160. Old National Conference Publication No. 7715, Institute of Engineers (Australia).

Grayson, R. B., Haydon, S. R., Jayasuriya, M. D. A., and Finlayson, B. L. (1993). Water-quality in mountain ash forests – separating the impacts of roads from those of logging operations. *Journal of Hydrology*, 150: 459–480.

Gregory, K. J. (1976). Drainage networks and climate. 289–315. In *Geomorphology and Climate.* E. Derbyshire (ed). London: Wiley.

Hammond, D. S., Van der Hout, P., Zagt, R. J., Marshall, G., Evans, J., and Cassells, D. S. (2000). Benefits, bottlenecks and uncertainties in the pantropical implementation of reduced impact logging techniques. *International Forestry Review*, 2: 45–53.

Hewlett, J. D. (1992). *Principles of Forest Hydrology.* Athens: University of Georgia Press.

Holmes, A. (1995). Baseline hydrological survey and hydrometric network. In Proceedings of the *Workshop on Topography, Soil, Hydrology and Forest Reservation*, 9 May 1995, Kuching. Kuching: Sarawak Forest Department.

Huth, A., and Ditzer, T. (2001). Long-term impacts of logging in a tropical rainforest – a simulation study. *Forest Ecology and Management*, 142: 33–51.

ITTO. (1999). *Manual for the Application of Criteria and Indicators for Sustainable Management of Natural Tropical Forests – Part B / Forest Management Unit Indicators.* Policy Development Series No. 10. Yokohama: International Tropical Timber Organisation.

Jusoff, K. (1990). A survey of soil disturbance from tractor logging in a hill forest of Peninsular Malaysia, 16–21, In *Proceedings of the Conference on Malaysian Forestry and Forest Products Research*, ed. S. Appanah, F. S. P. Ng, and R. Ismail, 3–4 October 1990. Kepong: Forest Research Institute of Malaysia.

Jusoff, K. and Mustafa, N. M. S. N. (1996). Guidelines on logging practices for the hill forest of Peninsular Malaysia. In *Forest Codes and Practice.* FAO Forestry Paper No. 133. Rome: FAO.

Lai, F. S. (1992). *Sediment and solute yields from logged, steep upland catchments in Peninsular Malaysia*, Unpublished PhD thesis. Manchester: University of Manchester.

Leigh, C. H. (1982). Sediment transport by surface wash and throughflow at the Pasoh Forest Reserve, Negri Sembilan, Peninsular Malaysia. *Geografiska Annaler A*, 64, 171–179.

Lim, K. Y. (1999). The role of road infrastructures in the tropical forests of Malaysia. In *Road Infrastructures in Tropical Forests.* Rome: Food and Agriculture Organisation.

MacDonald, L. H., Sampson, R. W., and Anderson, D. M. (2001). Runoff and road erosion at the plot and road segment scales, St John, US Virgin Islands. *Earth Surface Processes and Landforms*, 26: 251–272.

Madej, MA. (2001). Erosion and sediment delivery following removal of forest roads. *Earth Surface Processes and Landforms*, 26: 175–190.

Malmer, A. and Grip, H. (1994). Converting tropical rain-forest to forest plantation in Sabah, Malaysia. 2. Effects on nutrient dynamics and net losses in streamwater. *Hydrological Processes*, 8: 195–209.

Mannan, S. and Awang, Y. (1997). Sustainable forest management in Sabah. Presented at the *Seminar on Sustainable Forest Management*, 22nd November 1997, Kota Kinabalu, Sabah, Malaysia.

Martin-Smith K. (1998). Biodiversity patterns of tropical freshwater fish following selective timber extraction: a case study from Sabah, Malaysia. *Italian Journal of Zoology (Suppl. S)*, 65: 363–368.

Mohamed, M. (1987). Water quality studies and water quality problems in Sabah. 41–54. In *Environmental Conservation in Sabah: Issues and Strategies*, ed. S. Sani. Kota Kinabalu: Institute of Development Studies.

Morgan, R. P. C. (1976). Role of climate in the denudation system: a case study from West Malaysia. 317–343. In *Geomorphology and Climate.* E. Derbyshire (ed). London: Wiley.

MTCC. (2001). *Malaysian Criteria, Indicators, Activities and Standards of Performance for Forest Management Certification (MC&I).* Kuala Lumpur: Malaysian Timber Certification Council.

Nussbaum, R., Jennings, S., and Garforth, M. (2002). *Assessing forest certification schemes: a practical guide.* Oxford: Proforest.

Nydegger, E. (1998). *Skyline Yarding Systems for Low-impact Harvesting in Tropical Natural Forests.* Sandakan: Sabah Forestry Department.

Transport Research Centre. (1993). *A guide to the structural design of bitumen-surfaced roads in tropical and sub-tropical countries.* Overseas Research Note 31. Crowthorne: Transport Research Centre.

Pereira, H. C. (1973). *Land use and water resources in temperate and tropical climates.* Cambridge: Cambridge University Press.

Pinard, M. A., Putz, F. E., Tay, J. and Sullivan, T. E. (1995). Creating timber harvest guidelines for a reduced-impact logging project in Malaysia. *Journal of Forestry*, 93: 41–45.

Pinard, M. A., Barker, M. G., and Tay, J. (2000). Soil disturbance and post-logging forest recovery on bulldozer paths in Sabah, Malaysia. *Forest Ecology and Management*, 130, 213–225.

Pinard, M. A., and Putz, F. E. (1996). Retaining forest biomass by reducing logging damage. *Biotropica*, 28: 278–295.

Putz, F. E. (1994). Working towards a sustainable forest. *Asian Timber*, 9: 28–32.

Putz, F. E., Redford, K. H., Robinson, J. G., Fimbel, R., and Blate, G. M. (2000). *Biodiversity Conservation in the Context of Tropical Forest Management.* Biodiversity series – impact studies, Paper No. 75. Washington: World Bank.

Roberts, J., Cabral, O. M. R., Fish, G., Molion, L. C. B., Moore, C. J., and Shuttleworth, W. J. (1993). Transpiration from an Amazonian rain-forest calculated from stomatal conductance measurements. *Agricultural and Forest Meteorology*, 65: 175–196.

Sabah Forestry Department. (1989). *Forestry in Sabah.* Sandakan: Sabah Forestry Department.

Sabah Forestry Department. (1997). *SFM-Committee Work Plan.* Malaysian-German Sustainable Forest Management Project. Sandakan: Sabah Forestry Department.

Sabah Forestry Department. (1998a) *Final Harvesting Plan: Compartment 43, Deramakot FR, FMU 19, Sandakan district.* Sandakan: Sabah Forestry Department.

Sabah Forestry Department. (1998b). *RIL Operation Guidebook Specifically for Tracked Skidder Use*. Sandakan: Sabah Forestry Department.

SCS. (2002). *Forest Management Certification: Evaluation on the Forest Concession Area of Perak Integrated Timber Complex*. SCS-FM/COC-00046N. Oakland: Scientific Certification Systems.

SGS. (1997). *Certificate of Forest Management*. CFM-97/001. Kuala Lumpur: SGS (Malaysia) Sdn. Bhd.

SGS. (2002). *Selangor FMU Re-assessment Report*. Final Report, 3 June 2002. Kuala Lumpur: SGS (Malaysia) Sdn. Bhd.

Sheffield, A. T., Healy, T. R., and McGlone, M. S. (1995). Infilling rates of steepland catchment estuary, Whangamata, New Zealand. *Journal of Coastal Research*, 11: 1294–1308.

Sidle, R. C., Pearce, A. J., and O'Loughlin, C. L. (1985). *Hillslope stability and land use*. Washington: American Geophysical Union.

Sist, P., Dykstra, D, and Fimbel, R. (1998). *Reduced-Impact Logging Guidelines for Lowland and Hill Dipterocarp Forests in Indonesia*. Occasional Paper No. 15. Bogor: Center for International Forestry Research.

Strahler, A. N. (1957). Quantitative analysis of watershed geomorphology, *Transactions of the American Geophysical Union*, 28: 913–920.

Swanson, F. J., and Dyrness, C. T. (1975). Impact of clear-cutting and road construction on soil erosion by landslides in the Western Cascade Range, Oregon. *Geology*, 3: 393–396.

Tay, J. (1999). *Economic assessment of reduced impact logging in Sabah*. Unpublished PhD thesis, University of Wales, Bangor, UK.

Thang, H. C. (1987). Forest management-systems for tropical high forest, with special reference to Peninsular Malaysia. *Forest Ecology and Management*, 21: 3–20.

Thang, H. C. (1996). Formulation and implementation of criteria and indicators for sustainable forest management in Malaysia. Paper presented at the *Workshop on Forest Management Certification*, Forest Research Institute Malaysia, Kuala Lumpur. 67pp.

Thang, H. C. (2002). Development and application of criteria and indicators for sustainable forest management, including forest management certification and products labeling in Malaysia – a case study. Paper presented at the *Workshop on the Development of Brunei Darussalam Criteria and Indicators for Sustainable Forest Management*, Bandar Seri Begawan, Brunei Darunssalam, 25–28 March 2002.

Thang, H. C. (2003). Malaysia's experience in applying criteria and indicators for sustainable forest management, including forest management certification. Paper presented at the *International Conference on Criteria and Indicators for Sustainable Forest Management – The Contribution of Criteria and Indicators to Sustainable Forest Management: The Way Forward*. 3–7 February 2003, Guatemala City, Guatemala.

Trewin, R., Menz, K., and Grist, P. (1998). *Estimates of local and economy-wide costs and benefits of land uses in the Indonesian uplands through linked biological and economic models*. ACIAR Indonesian Research Project – Working Paper 98.01, Adelaide: University of Adelaide.

van Dam, O. (2001). *Forest Filled with Gaps: Effects of Gap Size on Water and Nutrient Cycling in Tropical Rain Forest*, Unpublished PhD thesis. Utrecht: Utrecht University.

van der Plas, M. C. and Bruijnzeel, L. A. (1993). Impact of mechanized selective logging of rainforest on topsoil infiltrability in the Upper Segama area, Sabah, Malaysia. In *Hydrology of Warm Humid Regions*. pp. 203–211, IAHS publication 216. Paris: IAHS.

Walsh, R. P. D. (1996). Drainage density and network evolution in the humid tropics: evidence from the Seychelles and the Windward Islands. *Z. fur Geom. N.F. Suppl. Bd* 103: 1–23.

Walsh, R. P. D., and Bidin, K. (1995). Channel head erosion in primary and logged forest in Sabah. In *Abstracts of the International Association of Geomorphologists, South East Asia Conference*, Singapore, 18–23 June 1995, 79.

Wenger, S. (1999). *A Review of the Scientific Literature on Riparian Buffer Width, Extent and Vegetation*. Office of Public Service and Outreach, Institute of Ecology. Athens: University of Georgia.

Wyatt-Smith, J., Panton, W. P., and Mitchell, B. A. (1964). *Manual of Malayan silviculture for inland forest*. Malayan Forest Record No. 23. Kuala Lumpur.

Yusop, Z. (1996). *Nutrient cycling in secondary rainforest catchments of Peninsular Malaysia*. Unpublished PhD thesis, Manchester: University of Manchester.

37 Red flags of warning in land clearing

L. S. Hamilton
Cornell University (Emeritus) Ithaca, USA

INTRODUCTION

Land clearance in the humid tropics needs to be considered very carefully. This chapter is an attempt to put the brakes on, or at least to insert a filter of conditions, so that any clearing does not have adverse, or only minimally adverse, effects on soil and water resources and processes. It follows on from the many precautionary statements, some even labelled 'guidelines' that have appeared since the early 1970s. As a matter of forest land policy, setting the framework for planning, there should be a thorough examination of alternatives to clearing more land. Poore and Sayer (1987) put it succinctly:

> 'Before deciding to modify or transform untouched areas, every consideration should be given to alternatives. This may include adapting areas that have already been changed, to more productive uses, for example, using savannas for pine plantations. Alternatively, it may involve intensification of existing uses, or using areas for more than one purpose, if these are compatible.'

The high cost (in both ecological and economic terms) of any new land development dictates that the starting point in planning should be a detailed analysis of alternatives including possibilities of **intensification** on land already converted (Hansen and Dickensen, 1987). Similarly, there is an early need for an assessment of the capability and suitability of an area under planning scrutiny for the proposed use if alternatives are not available. Such an assessment, using well-developed procedures (e.g. Carpenter, 1981; FAO, 1974), if rigorously carried out may abort a proposed conversion.

Global interest began to focus on the loss of tropical forests, particularly tropical rainforests, shortly after the 1972 Stockholm Conference on the Environment. One of the early efforts to slow what became known as deforestation[1], and put more ecological rationality into land clearing or other developments, was a set of regional meetings sponsored by the International Union for the Conservation of Nature and Natural Resources (IUCN, now called The World Conservation Union) and the young United Nations Environment Programme, and co-sponsored or supported by a host of other international organisations. From the 1974 Caracas, Venezuela, meeting came a landmark publication: 'The Use of Ecological Guidelines for Development in the American Humid Tropics' (IUCN, 1975). The guidelines were quite general and it was intended that individual countries would take these and make them specific to their own culture, economy and political realities, thus taking ownership. To my knowledge this never happened in any of the three regions. The guidelines stand the test of time, however, because they do reflect general principles. Most of those relative to the land clearing question have been extracted from the summary conclusions from the Caracas meeting by Poore (1975), and are presented in Box 37.1.

Subsequently, a number of other 'guidelines', manuals and books were produced. Most notable among these for their scope and practicality were: *Assessing Tropical Forest Lands: Their Suitability for Sustainable Use* (Carpenter, 1981); *Land Use, Watersheds and Planning in the Asia-Pacific Region* (Pearce and Hamilton, 1986); *Ecological Development in the Humid Tropics: Guidelines for Planners* (Lugo *et al.*, 1987); *The Management of Tropical Moist Forest Lands: Ecological Guidelines* (Poore and Sayer, 1987). Outstanding for its research-based focus on land clearing and subsequent agricultural use was the book by Lal *et al.* (1986) entitled *Land Clearing and Development in the Tropics*, based on a famous Seminar in 1982 on this topic at the International Institute of Tropical Agriculture in Ibadan, Nigeria. It contained many highly relevant papers on soil and water impacts such

1 Deforestation is a very ambiguous term, a slippery word that at one time or another has been applied to: forest clearing for conversion to annual cropping, shifting cultivation, fuelwood cutting, commercial logging, burning, flooding with reservoirs, gathering medicinal plants and even killing of wildlife (Hamilton, 1988). Each of these, and other conversions, have vastly different effects, yet have often been all called deforestation.

Forests, Water and People in the Humid Tropics, ed. M. Bonell and L. A. Bruijnzeel. Published by Cambridge University Press.

Box 37.1 Some general early guidelines with respect to land clearing in the humid tropics

1. If land use policies are to be the greatest possible benefit, each activity should be planned to take place in the areas most suitable for it, i.e. agriculture on the most fertile soils; the conservation of fauna and flora in a sample of those areas where they are richest and most characteristic or in areas where they are unique. If there is a reason to believe constraints should be imposed because of the nature of the site, e.g. liability to erosion, very low fertility, those constraints should be observed.
2. The full social costs and benefits of proposed actions should be assessed, especially indirect or hidden costs such as adverse changes in water flow, the siltation of a reservoir, or slow erosion or deterioration of the soil.
3. If natural areas are affected by development they lose irreversibly some of their intrinsic value. Natural areas for preservation as such should be chosen as an integral part of land use planning and, if unique, then conservation should as far as possible have priority over other uses.
4. One general goal of regional development should be the maintenance of the productivity of rivers, lake systems and estuaries, and especially of species valuable for food. Any proposal to change the water regime and the land use in a catchment should be assessed for its effect on the productivity of these systems.
5. Particular care is necessary in making decisions to modify or to transform forest or other land, because these processes can only be reversed with difficulty, if at all.
6. Decisions to modify or transform should only be taken when it has been clearly demonstrated that it is in the general public interest to do so. Unmodified land retains its potential for all uses and should be kept in reserve for future need.
7. The demand to modify or transform new areas should be reduced as far as possible by:
   (a) adapting areas which have already been changed to more productive uses, (e.g. savannas for pine plantations);
   (b) intensification of existing uses; and
   (c) using areas for more than one purpose if these are compatible.

(Adapted from Poore, 1975.)

as by Lal (1986), Fölster (1986), and Opari-Nadi and Lal (1986). It concluded with a set of 'DOs and DON'Ts'.

A subsequent compendium of experience on land clearing for agriculture was organised in an Indonesian Workshop by the International Board for Soil Research and Management and partners (Lal *et al.*, 1987), though not expressed as guidelines. Ross and Donovan (1986) summarised in a very readable way for lay persons, donors and government officials an excellent set of non-technical guidelines in booklet format. Very little has appeared since this flurry of activity in the early and mid-1980s, except in narrowly focused areas dealing with particular subsequent land uses following conversion. The most recent presentations of soil and water impacts of clearing and subsequent use of the land by Ghuman *et al.* (1991) and Lal (1996, 1997) were still based on the relatively early and fine work in the early 1980s at IITA, Ibadan.

This chapter aims to synthesise and update this excellent work and to present 'guidelines' in a slightly different way. What follows will consider initially the criteria and guidelines that deal with forest areas that should **not** be cleared or where any proposed clearing should trigger a thorough environmental impact assessment (EIA). It must also be constantly borne in mind that only soil and water considerations are dealt with in this paper. There are many other criteria which should be considered as preventing or inhibiting the clearing of a particular site. These may be cultural (a sacred site); legal (it belongs to someone or some group, not the clearing proposer: this would include lands traditionally occupied by land-users who might not have a legal title); legal/institutional (it may be designated as a national park or some other kind of protected area); economic (there may be no reasonable market for the proposed alternative use e.g. a golf course too far from a source of golfers); ethical (it may be the only remaining habitat for an endangered plant or animal – and also it could well be illegal for this reason to destroy this habitat).

The second part of the guidelines will relate to soil and water conservation, given that land clearing is about to take place, (hopefully having passed through filters such as indicated above).

SOIL AND WATER SITUATIONS WHERE CLEARING IS INADVISABLE: THE 'RED FLAG' AREAS

The aim here is raise to metaphorical 'red flags' where soil and hydrological impairment can be particularly severe for land clearing in a selected number of situations. This suggests that any proposed land clearing in these situations come under thorough scrutiny and impact assessment of a rigorous nature. Such 'red flag' areas include: tropical montane cloud forests, unstable slip-prone areas, riparian buffer zones, significant freshwater wetlands, mangroves, high-quality water supply headwaters, flood-prone areas, and areas where serious soils limitations preclude sustainable alternative use. In the writer's opinion these areas should not be cleared.

Tropical montane cloud forests

Aside from the rich biological diversity (especially endemism) of plants and animals in these ecosystems, their hydrological values puts them in a special category. In spite of the fact that they are often on acidic, wet and miserable sites in persistent or recurring cloud, with high levels of aluminium, and having other limitations as described by Bruijnzeel and Proctor (1995), they are being converted to other uses. These uses include: temperate vegetable production (Mount Kinabalu in Sabah, Mount Data in Philippines and in Sri Lanka); coffee or cardamom (Sri Lanka, Cameroon, Philippines); flowers (El Ávila in Venezuela); berries (Dominican Republic); opium or coca (Colombia and Thailand); kava/sakau (in several Pacific Islands); golf courses and resorts (Cameron and Genting Highlands, Malaysia, and Mount Kinabalu, Sabah) (Hamilton *et al.*, 1995). The high rate of loss of tropical montane cloud forest to clearing and other factors has begun finally to arouse global concern (Hamilton, 1995; Aldrich, 1998). Drigo's account at the beginning of this volume indicates that these high loss rates are still being maintained.

While the additional water capture from horizontal precipitation varies widely, and though the evapotranspiration 'use' of TMCF also varies, there is usually a net gain of water, above that from vertical precipitation, that is lost if these forests are cleared (Zadroga, 1981). The extent of this loss and under what climatic and soil conditions it takes place still needs research in key locations (Bruijnzeel, 2000a; Bonell, 2004). Additions of moisture for true cloud forests may reach hundreds of millimetres per year, with typical values ranging between 5 and 20% of vertical rainfall (Bruijnzeel and Proctor, 1995; Bruijnzeel, 2002). In seasonally dry or dry areas where clouds are stripped by forest or scattered trees, this horizontal precipitation value may go many times higher (over 100%), and in some areas, almost all of the moisture reaching the ground is captured by such trees. Reported net precipitation figures for tall montane forest not subjected to frequent low cloud and fog range from 55 to 80%. The corresponding figures for lower TMCF were 80–110%. Even higher values were found for the upper TMCF, ranging from 85 to 180% (Bruijnzeel, 2000a).

These moss- and bryophyte-rich cloud forests include what have been called upper montane forest, sub-alpine forest and in some cases part of the lower montane forest (Bruijnzeel and Hamilton, 2000). They are usually found at elevations of 1200 m and above on equatorial inland mountains, but may also occur as low as 300 m on small islands away from the Equator (Hamilton *et al.*, 1995).

Though the water capture value of these forests having frequent and particularly wind-driven cloud has not been fully quantified, the dry season flow reductions following clearing that have been reported would indicate that the precautionary principle should prevail. These are 'red flag' forests, and when their soil erosion protection value and high biodiversity value are factored in, this 'red flag' is a very bright red. They are disappearing along with other montane forests, and for cloud forests this loss can be considered virtually irreversible due to lack of knowledge and high cost of ecological restoration (Hamilton, 1995; Bruijnzeel and Hamilton, 2000).

Guideline
Where persistent cloud cover or regular wind-driven cloud occurs on tropical mountains, producing a characteristic tropical montane cloud forest, land clearing should be prohibited except under unusual circumstances. Their biodiversity, hydrological and erosion control values, plus general unsuitability for other sustainable use, argues strongly for their retention. Moreover, their uniqueness and complexity in terms of vegetation-soil-climate interaction would seem to make their loss irreversible. There are compelling reasons for having these as protected areas of some kind, including as National Parks and Watershed Reserves.

Unstable slip-prone areas

On 16 and 17 December 1999, on the coastal side of Venezuela's Cordillera de la Costa, hundreds of landslides drowned or buried alive one-tenth of the 500 000 people living on the slopes and at the base of this range (Myers, 2000). Forty thousand homes were destroyed and most roads, with damage estimated in the billions of dollars. It is considered to be the worst disaster of the Western Hemisphere in the past 500 years, with a death toll more than five times that resulting from the catastrophic Hurricane Mitch in 1998 which triggered thousands of landslips in Central America. This catastrophe was triggered by 900 mm rainfall over three days, preceded by twice the normal rainfall in the previous 16 days (Scatena *et al.*, this volume). A large number of internationally-noteworthy erosion events occurred at the end of the 1980s and initiated a substantial amount of research: Southern Thailand in 1988 after torrential rainfall (Rao, 1988), Cyclone Bola in 1988 in New Zealand (Trustrum and Page, 1992), Puerto Rico during the 1989 Hurricane Hugo (Scatena, 1990), and in the Philippines during intense hurricane-associated rainfall in 1990 (Hamilton, 1992). It is of interest to note that the devastation caused in Southern Thailand and the Philippines were both blamed on logging, whereas the failures were mostly on land cleared for crops. These slope failures of a widespread and massive nature have occurred periodically and generated much discourse over the role of forest cover in reducing the incidence of these mass movements.

Sorting out 'human-caused' slope failures from 'natural' is difficult, as reported in a review of these in the Ganges-Brahmaputra River Basin by Bruijnzeel and Bremmer (1989). There is a common misconception that slope failures do not occur on undisturbed forest lands. There is also a popular confusion about the types of mass movements that can be influenced by vegetation. From a study in Tanzania, a classification was developed that seems to be applicable in many places. Rapp *et al.* (1972) proposed:

1. Numerous, small, 1–2 m deep, 5–20 m wide earthslides and mudflows, triggered by heavy rainstorms, occurring with an interval of about 10–20 years (in the Uluguru Mountains).

2. Occasional large earthslides cutting many metres down into the weathered bedrock and below the anchoring effect of tree roots. They occur with much longer intervals in time.

Something very close to this is now generally accepted by the scientific and professional community.

There has been a substantial scientific spotlight thrown on to the role of forest cover in reducing the incidence or severity of shallow landslips (Class 1 above). Unfortunately, most research has been done in the temperate countries, especially Japan (protection forests), western United States (associated with logging), New Zealand (pasture land and logging) and Taiwan (protection forests).[2]

Large portions of major scientific symposia have been devoted to this topic, for instance: Erosion and Sediment Transport in Pacific Rim Steeplands (Davies and Pearce, 1981), Effects of Forest Land Use on Slope Stability (O'Loughlin and Pearce, 1984), Erosion and Sedimentation in the Pacific Rim (Beschta *et al.*, 1987), Research Needs and Applications to Reduce Erosion and Sedimentation in Tropical Steeplands (Ziemer *et al.*, 1990). It has seldom been presented more succinctly than by Rapp (1997) who studied land degradation in the Uluguru Mountains of Tanzania. Referring to numerous, small (1–2 m deep, 5–20 m wide) earthslides and mudflows in a 75 km^2 area after more than 100 mm of rain fell in less than three hours, he found that of 840 landslides, only three started on slopes under forest cover – the remainder were in cultivated or grazed areas on similar steep slopes. The importance of tree roots in providing shear resistance in slip-prone soils has been demonstrated in the landmark studies by O'Loughlin (1974), O'Loughlin and Ziemer (1982) and Ziemer (1981). Basically, it is this work and subsequent confirmation that allows us to produce guidelines in this arena. All steep lands benefit from being retained under forest cover, especially those in seismically active areas, but the pressure to clear land occurs on lands where slopes are intermediate but are still slip-prone, – and this is where the 'red flag' needs to be raised. Prudence in land use could avoid the catastrophic results as seen in Thailand in the 1988 storm when formerly forested land which had been cleared for rubber plantations, failed in thousands of landslips. This led to a ban on logging, but not on land clearing! (Hamilton, 1991).

The problem then is to identify these 'red flag' areas in advance and to retain them in forest. While we know the general factors that influence mass wasting as: presence of water, type of rock or mineral and state of weathering, number and density of natural fracture planes, and structure and inclination of slope (Carson, 1989), some practical field guides are needed for identification of slip-prone areas where forest retention is desirable. No one has done this better than in a 1985 guideline by Megahan and King (1985). They point out that in areas of high erosivity: the major hazards are on slopes greater than 45–55%, with a maximum frequency at about 70%; concave slopes which concentrate water are more slip-prone; soils which are low in cohesion are more slip-prone; shallow soils over bedrock or with pronounced discontinuity in texture/structure may become saturated, buoyant and slip-prone. They discuss the rainfall erosivity question, and the difficulty of obtaining good data in the tropics. Hudson (1995) suggests various ways in which erosivity can be estimated and in some places, e.g. Malaysia (Morgan, 1974), there are iso-erosivity maps. On a global scale, Blaschke *et al.* (2000) have mapped the approximate extent of areas where mass movement erosion affects land productivity and provided a table giving, by country or region, the land use, rainfall, landform, area affected, duration of event and soil loss rate, based on reported research worldwide. Good data are few in number and scattered; studies such as that by Humphreys and Brookfield (1991) in Papua New Guinea report that forms of shallow slope failures are by far the most common erosion forms in cultivated steeplands. The result is not only reduced productivity but also increased sediment loads in watercourses.

Surface erosion can of course result also in reduced productivity and serious sedimentation following clearing. We can but hope for adequate soil and water conservation methods to reduce the problem substantially. What is being 'flagged' here are those slip-prone areas that will fail even under terracing and good cropping or grazing practices. Trees offer the safest land use.

Guideline

In summary, where high intensity or prolonged rainfall is characteristic of an area with slopes around 70% (but even as low as 45%) and having spoon-shaped concavity or shallow planar surfaces, the risk of slope failures if tree roots are gone, is high. Clearing is inadvisable (especially for roads that cut across such slopes) and should only be sanctioned if it can be certain that there will be: (a) quick re-establishment of tree crops such as rubber or tree plantations (though there will be a long period of vulnerability until new tree roots become effective); (b) immediate terracing for crop production plus trees in an agroforestry system (see also Critchley, this volume), and assurance of adequate terrace maintenance and ability to repair quickly minor landslip damage. (Note: this latter alternative is difficult to assure, and is only mentioned because I have seen it effectively done in Nepal and the Philippines, in well-drained soils on forward sloping terraces with cover-stabilised risers.)

Riparian buffer zones

Maintenance of natural forest vegetation along perennial streams has many values that render these green buffers important. They

2 The names associated with these landmark studies are well-known and legion. Among those whom I recall with fondness and gratitude are: Tsukamoto, O'Loughlin, Pearce, Trustrum, Sidle, Ziemer, Swanson, Omura, Fahey, Megahan, Lin and Hsia.

are often critical areas for conserving biodiversity since they provide cover, access to water and migration corridors for both animals and birds; and a site for rich riparian plant conservation; they keep streams cooler and more sediment-free for aquatic organisms including fisheries; they provide organic debris to the water which is a food source for aquatic organisms; they trap sediment from upslope erosion to keep it out of streams used for human domestic purposes or reservoirs; they reduce streambank erosion; and if there is agriculture upslope, they can filter and immobilise fertiliser or pesticide compounds thus reducing water pollution from chemicals (Hamilton, 1997). Connolly and Pearson (this volume) elaborate on these important functions of riparian zones. Buffer strips have been defined as protective areas adjacent to streams or lakes that shield them from the effects of land management activities (O'Laughlin and Belt, 1995). But as indicated above they are much more than that if one thinks of them as habitat for biodiversity. However, here we are considering only hydrological and soil aspects, and that definition will suffice. Riparian buffer zones form the vital link between watershed lands and stream systems. Unfortunately much of the research on quantifying the benefits of sediment, pesticide and nutrient trapping, and the contribution made to bank stability and organic debris to aquatic nourishment, has been carried out in the temperate zone. This is particularly true in the United States where control of non-point source pollution (mainly sediment) has been enshrined in laws and best management practice regulations. The 'process' research however, is applicable to the tropics to supplement the few studies made there (see Connolly and Pearson, this volume). A good review of this riparian link between land and water is that of Gregory *et al.* (1991).

In most official forest harvesting guidelines in the tropics, there have for some time been requirements for maintenance of no-logging riparian buffer strips at least 20 m each side (more as slopes steepen) e.g. for Malaysia (Mok, 1986) or tropical Queensland, Australia (Cassells and Bonell, 1986). These are largely for sediment reduction but reap a host of other benefits, as enumerated above. How much more important then it is to 'red-flag' them when extensive land clearing is being contemplated, for the subsequent use (cropland, grazing, settlement, etc.) brings with it a range of other water quality impairers that include faecal bacteria, pesticides, herbicides and fertiliser chemicals, as well as soil from erosion. Megahan and King (1985) discuss various aspects of the value of maintaining riparian buffer zones. With respect to trapping sediment from up-slope, it should be recognised that only sheet and rill erosion material are controlled or reduced. Channelised sediment, depending on size and length, will break through most buffer zones of normal width.

How wide should these buffers be? One of the best studies carried out at eight different sites on one stream, but having different soils, slopes and vegetation cover, concluded that in most cases a 15 m buffer width was adequate (Budd *et al.*, 1987). Unfortunately, it was in an area with maximum slopes of around 40%, without high intensity rainfall pattern, though high annual rainfall (US Pacific Northwest). An idea of the variation recommended comes from: 20–30 m (Bosch and Hewlett, 1980), 10–30 m (Cassells *et al.*, 1984), minimum width 25 m (Megahan and Schweithelm, 1983), up to 50 m (O'Laughlin and Belt, 1995). The width will depend on the slope and the type of material that requires trapping or buffering. Given the other values of these riparian zones, it is prudent to err on the wide side in spite of some economic loss from foregoing activity. Thang and Chappell (this volume) also take up this issue of riparian buffer zones within national forest management guidelines.

POTENTIAL FOR APPLICATION OF TOPOGRAPHIC-WETNESS MODELS

It should be recalled that these specifications of buffer strips recommended above have been based on empirical (trial and error) experience. Since the 1980s, topographically- and physically-based hydrological models, based on digital terrain models, have been in the process of development (Moore *et al.*, 1991; reviewed in Bonell with Balek, 1993; Chappell, Tych *et al.*, this volume). These spatial models are capable of detecting the more vulnerable zones of preferred production of storm runoff and erosion. Simple indices (e.g. wetness index, TOPOG, O'Loughlin, 1986; topographic/soil index, TOPMODEL, Beven *et al.*, 1995) are calculated which infer the spatial variabilty of specific processes occurring in the landscape such as soil water content, and in turn, the potential for runoff generation and sheet erosion. The higher the magnitude of these indices, the more vulnerable the landscape is to disturbance. The advantage of using these models is that they require, at a minimum, only topographic data for their application on the above lines. Consequently, they are suitable for testing in environments such as the humid tropics where hydrological data is commonly deficient. These models also seem to represent the redistribution of soil water in between storms within headwater basins quite well, even though their representation of the storm runoff generation process in humid tropic environments is less certain (see Bonell, this volume) Thus a logical step is to allocate a threshold index value above which no forest disturbance should be allowed during the planning of a logging operation (Bonell with Balek, 1993; Bonell, 1998).

Therefore there is a need to 'ground truth' these spatial models with hydrological measurements by which to select more optimally the appropriate width of these buffer strips (Bonell, 1998). It is likely that more flexible, buffer strip widths will be produced which are also in line with prevailing hydrological processes. In many cases it may prove that the current empirical specifications are inadequate (i.e. the buffer strips may have to be wider). Detailed descriptions of more recent developments and applications asociated with these spatial models are provided elsewhere in this volume (see Chappell, Bidin *et al.*; Bonell).

Guideline
Riparian buffer zones that shield perennial streams and lakes from effects of land management activities are 'red flag' areas, and should be maintained. It is clear that because of the water body itself, many kinds of development and hence clearing will occur. The policy is to keep as much of the zone as intact as possible, and to keep it as wide as possible, but at least 20 m in width, and wider, as slope angle increases. The field testing of spatial, topographic-wetness models is required to refine buffer strip specifications.

Significant freshwater wetlands

Wetlands have been defined as those areas that are inundated or saturated by surface water or groundwater at a frequency and duration that normally supports a prevalence of vegetation typically adapted for life in saturated soil conditions (Maltby, 1986). Wetlands generally include swamps, marshes, bogs, peatland, fens and similar areas. They are the only ecosystem type to have their own international convention, the Ramsar Convention of 1971 (Convention on Wetlands of International Importance Especially as Waterfowl Habitat). They are common in the tropical moist forests as swamps, as depressions in periodically inundated flood plains of the major rivers of the tropical world, and as other kinds of wetlands within the forest due to special conditions of soils and topography. Many tropical countries have large areas of palm swamps (see Hooijer, this volume) and coastal freshwater swamps (not mangroves, which are dealt with separately). The extent of swamps in the tropical parts of the continents is given by Balek (1989) in thousands of square kilometres as: South America 1200; Asia 350; Africa 340; Australia 2. These are valuable areas for both hydrological and biological functions, and are the source of many animals and plants used by local communities (e.g. Sago palm). Bullock (1993) has listed the hydrological and water resource values of wetlands from a literature survey.

Hydrologically, wetlands can mitigate flooding. They store potential floodwaters, at least temporarily, – a detention function that can reduce floodwater peaks locally (Maltby and McInnes, 1997). The riverine floodplain swamps, as water-spreading areas may have even a more regional influence. Valid tropical data as to flood mitigation are lacking (Howes, 1995). One landmark study by the US Army Corps of Engineers in the wetlands of the city of Boston catchment area valued the flood prevention benefit at US$13 500 per hectare per year (Hollis *et al.*, 1988). Wetlands sift dissolved and suspended materials from floodwaters, including non-lethal levels of pollutants, thus maintaining water quality (Nutter and Brinson, 1994). They are important sequesterers of carbon. Many of them are floodplain forests, associated with river systems e.g. the 'varzea' and 'igapó' of the Amazon.[3] They are important both as nutrient and energy traps but also as suppliers to the river system in which their detritus is the base for nourishing associated river fisheries and aquatic complexes. The Amazon fishery is dependent on its associated inundated forest.

In the tropics, swamps and forest-associated wetlands have been, and still are, considered fair game for clearing and converting to rice production. The vast swamp areas of Indonesia, and the Amazon and Congo floodplains, represent a conversion opportunity to many. These should be 'red-flag' areas, requiring careful ecological and long-term economical analysis as to the perceived benefits and the real costs of such conversion. Perhaps we will learn this in time, but the lesson comes only after tremendous damage has been done already. Box 37.2 brings recent news (1999) of the failure of the Indonesian Mega-Rice project that cleared almost 1 million hectares of swamp forest in Kalimantan. However, a news article in the *Indonesian Observer*, 30 March 2000, announced that as part of the transmigration policy, 'the disastrous peatland project can be continued because certain areas of the land are suitable for profitable farming.' This may be stalled however, by the demands of the original Dyak inhabitants who are claiming compensation for the wetlands taken for the scheme (Anonymous, 2000). The First International Conference on Wetlands and Development held in Kuala Lumpur in 1995 involving 64 countries stated what will be our guideline:

Box 37.2 Failure of the Mega-Rice Project in Kalimantan swamp forests

On 13 July 1999 President Habibie ended one of the most unsavoury and unsuccessful episodes in the history of land development in Indonesia. The Mega-Rice Project initiated by former President Suharto in 1995, which had been put on hold since shortly after his downfall in 1998, was consigned to the scrap heap. An ill-conceived attempt to convert 1 million hectares of peat wetland in Central Kalimantan into rice paddies was put to rest. In the course of its implementation three trillion Indonesian Rupiah were squandered and all the swamp forest within the area was removed or degraded. No productive rice has been grown and one million hectares (the size of Northern Ireland) lies devastated and useless. Its biodiversity has gone. The natural resource functions have been disrupted, probably irreparably, by more than 4500 km of drainage channels, excavated for irrigation and to prevent flooding. After only two years the main channels are losing water and silting up. People are nonetheless still using them to gain access to the interior. All remaining timber is being removed and, in the process, debris is set alight and the surface peat catches fire, generating more of the dense, unhealthy haze that has beset South East Asia in recent years.

Source: Reilly (2000).

3 Varzea forests are periodically inundated by turbid and relatively nutrient-rich waters, whereas igapó receive clear and nutrient-poor water, though often humic stained (Kling, 1983).

Guideline

That activities likely to affect wetlands should proceed only after consideration of environmental assessment and include appropriate review stages. Prior socio-economic and environmental impact analysis of individual project components should be carried out and taken into account during the implementation phase. Long-term and cumulative impacts, and the effectiveness of wetland management and restoration, should be monitored and fed back into reviews of these procedures.

Mangrove forests

Salt-tolerant mangroves are a type of swamp forest, but are separated out for 'red-flag' status because of some unique characteristics and functions. They cover some 14 million ha in the tropics and sub-tropics (Maltby, 1986). Unfortunately they are still being appropriated for non-sustainable production systems by both agriculture and aquaculture (Aksornkoae, 1993). They are also being cleared for waste dumps, oil drilling, salt production, coastal roads and coastal resorts (Hamilton *et al.*, 1990). Aksornkoae (1988) reported on an area in Thailand which had 367 900 ha of mangrove forest in 1961, but had lost 46% of this in only 26 years. They are one of the world's most threatened ecosystems, but reliable statistics are difficult to obtain (Hamilton *et al.*, 1990).

These inter-tidal ecosystems produce an under-rated variety and amount of direct products and services. A compilation of these goods and services by country is given in Saenger *et al.* (1983) and by species in Hamilton and Snedaker (1984). As well as all the wood and non-wood forest products, the services include nourishing the near-shore fisheries and shellfisheries, provision of avian, aquatic and amphibian habitat, tourism value, coastal erosion reduction, storm surge flood reduction, sediment and pollution trapping and immobilisation function (including immobilising some heavy metals) and a nursery and spawning area for fishes (Hamilton *et al.*, 1990). A Global Mangrove Database and Information System, containing information on distribution, values and threats is held by the International Society for Mangrove Ecosystems, University of the Ryukyus in Okinawa, Japan. It will be available on the web in year 2000 (ISME, 2000).

From the hydrological and soils perspective, mangrove services are extremely important, and offer sound reasons for questioning clearing. In certain coastal locations, they provide storm surge and flood protection from hurricanes (or typhoons). Hurricanes can generate storm surges up to 7.5 m high, sending destructive flooding several km inland (Maltby, 1986). In Bangladesh for instance, some 15% of the population live less than 3 m above sea level. Thousands and even hundreds of thousands of people are killed, homes destroyed and fields salted every time there is one of these major events. One of the worst was in 1970 when between 150 000 and 300 000 people were killed (Wijkman and Timberlake, 1984). The vast mangrove forests of the Sundarbans in the deltas of the Ganges and Brahmaputra serve to damp the effects of these surges, and the greatest damage has been incurred where these mangroves have been reclaimed for agriculture and aquaculture (Maltby, 1986). The Philippines has for many years now had a ban on mangrove clearing for 40 m inland along rivers and 100 m inland for areas facing bays and the sea (Saenger *et al.*, 1983). Unfortunately clearing continues in many countries in spite of this protective service. Shoreline erosion from pounding seas is also reduced by the root barriers of fringing mangroves, a function particularly important on small islands in the tropics.

As with freshwater wetlands, particularly riverine ones, riverine mangroves provide a filtering sediment-trapping function as they receive the freshwater inputs from upstream. This has important benefits to the nearshore and estuarine fisheries and to the coral reefs. Moreover, if these sediments from upstream communities contain heavy metals such as cadmium, mercury or lead, these are immobilised in the sediment accumulation. It should be noted that if the mangroves are cleared, these heavy metal accumulations can be released all at once in a storm event, and at this point may exceed toxic levels (Hamilton *et al.*, 1990). Even if not released, they may plague *in situ* use.

The main purposes globally for clearing on a large scale are for agriculture and aquaculture. Added to the loss of products and services suggested above, a major impediment relates to the soils of the mangrove areas. Iron-rich sediments combine with the sulphates from sea water and organic matter to produce potential acid sulphate soils in many areas of the world. When cleared, these soils oxidise to form acid sulphate soils which are extremely difficult to ameliorate and use. Such acid sulphate soils have been noted in large areas of the Mekong delta in Vietnam, in areas in Senegal (where pH of 2.5 has been recorded), Sierra Leone, Indonesia, Venezuela, Bangladesh and India, (Hamilton and Snedaker, 1984). Pannier (1979) reported on a now often-cited land reclamation failure in Venezuela's Orinoco delta where acid sulphate soils developed following a governmental scheme to increase arable land. The Guaraunous Indians had to abandon what had been their traditional subsistence lands and were moved to another area. Here, a series of floods from 1976 to 1980 claimed the lives of most of the population of 4000–5000 persons. There is now no Guaraunous people, only a few displaced persons.

While clearance for mariculture (both fish and shrimp farming) would seem attractive, and has indeed been carried out extensively in places such as Ecuador, Philippines, Panamá and Thailand, here also acid sulphate or even just highly acid soils present major problems. The pond waters become highly acid, with resulting drastic reduction in fish or shrimp growth and even mortality. Heavy use of lime and phosphate fertiliser or continuous dilution with fresh or seawater increases costs and may make the enterprises uneconomic. The development and abandonment of mariculture ponds is so common in many areas of the world that it merits the label 'shifting aquaculture'.

Guideline

In summary, mangroves are another 'red-flag' area because of important goods and services lost (often to traditional coastal-dwelling societies) and because of the limitations that their soils and topographic position (floodways and storm surge areas) place on subsequent use. They depend on a balance between freshwater inputs and a salinity regime. This regime is easily altered by both on-site and off-site land clearing. Any proposals for clearing should be examined carefully. Any proposals for clearing should face rigorous analysis of benefits foregone, all costs, and the long-term viability of the proposed new use. (See for example in the Philippines, Janssen and Padilla, 1999.) Any clearing or road construction on associated land may also change the balanced freshwater/saltwater regime and affect mangroves adversely, through altered drainage.

High-quality water supply headwaters

High-quality water suitable for human and animal consumption, and water of sufficient quality for irrigation, loom large on the screen of critical issues in the new millennium. Many are calling it **the most important** factor for sustainable development and peace. It would seem to make both short-term and long-term sense to raise 'red flags' on key headwater supply areas, to make certain that any changes in land use do not impair the necessary quality standards of the water emanating from them. Any such areas now in forest cover have the safest hydrological regime in terms of maintaining natural flows, and the highest quality. The importance of headwater forests is well demonstrated in the highland-lowland system of Mount Kenya, a vital water tower in a semi-arid region (Liniger, 1998). Not that grassland or agricultural cropland cannot be hydrologically benign under a regime of high standard soil and water conservation practices. Unfortunately, in the humid tropics most steepland grassland will be grazed or over-grazed with resulting soil compaction and erosion, and often with use of fire to maintain grassland. And while well-maintained terraced padi or dryland crops may minimise erosion, and husband water, there are far too many examples of unravelling, unsustainable croplands on tropical steeplands in headwater areas. Moreover, animal manure and any agricultural chemicals represent additional potential water quality impairment. Traditional shifting agriculture under low levels of population and technology, with long fallow periods and a mosaic pattern of small field clearing is not seriously damaging to land and water (Hamilton with King, 1983). Unfortunately these necessary conditions for sustainable shifting agriculture are rapidly disappearing, and it develops into 'shiftless agriculture' which is not suitable in key water supply areas.[4] Well-adapted agroforestry systems which have due regard for soil and water movement may be acceptable catchment cover, but for maximum safety of potable water, unaltered forest cover is paramount. Forest harvesting for non-wood forest products, or even fuelwood and timber, if done under high standards of harvesting practice, are an acceptable alternative to clearing, if it is important to derive direct economic income products from the catchment. Several other chapters in this book address the hydrological and soils impacts of these human activities and also the best management guidelines for timber harvesting and upland agriculture (see: fire – Malmer *et al.*, grazing and shifting cultivation – Hoelscher *et al.*, clearing for conversion – Grip *et al.*, timber harvesting – Cassells and Bruijnzeel, agriculture – Critchley, and agroforestry – Wallace *et al.*).

Any conversion of forest to other uses in critical water source areas for water supply is to be regarded skeptically because of the changes in water delivery to the streams and in water quality. In some areas, what was a high degree of lateral subsurface flow (with its benefits) can transfer the water impeding layer from depth to the surface with infiltration excess overland flow as a result (Bonell with Balek, 1993). Subsequent land uses can further reduce infiltration, and in addition promote more erosion (both surface and mass) and introduce other pollutants (fertiliser, pesticides, hydrocarbons, animal wastes, etc.).

Many cities and towns have taken steps to assure the existing quantity and quality of their water supply reservoirs by acquiring all or key parts of the catchment forest lands above their supply sources, or by instituting controls on clearance, and limitations on uses or practices in catchment areas. The creation of National Parks, State Forest Reserves and Municipal Catchment Reserves has been a characteristic act of many public water supply protection efforts in such places as Freetown (Sierra Leone), Harare (Zimbabwe), Quito (Ecuador), San Juan (Puerto Rico), and Singapore. Some of the best forest research in the municipal water supply arena over the years has been done by the Melbourne and Metropolitan Board of Works, and though not tropical, the process research is applicable to the tropics (see for example, Langford and O'Shaughnessy, 1977).

Irrigators too have an interest in reducing the rate at which their water storages fill with sediment, whether it be a relatively small pond for an individual or a large reservoir serving many. Maintaining key catchment areas in forest cover is prudent policy, for most alternative land uses following clearing – and the clearing operation itself – produce increased sediment downslope and

4 My definition of 'shiftless agriculture' is where new colonists (usually), clear forest and plant crops repeatedly until the productivity has gone through nutrient depletion, soil erosion or pest build-up, then abandon the plot and move to another forest area. I believe Australians would refer to this as 'flogging the land'.

downstream. At a minimum, a forested buffer zone around the margins of any pond or reservoir will assist greatly in reducing sediment and other pollutants from the land surface (not the feeder stream).

Bruijnzeel (2002) has given an excellent review of the influence of the presence or absence of good forest cover on total and seasonal water yield from headwaters in the humid tropics.

Guideline

Surface water users where safe, high-quality water is required, be they an individual, a small community, a city, an industry or an irrigation group, should address the problem of forest clearing in the catchment of their water supply source. In some cases entire catchments, in other cases critical areas likely to generate sediment or other pollutants, are best retained under forest cover. Clearing in these should be prohibited because of the extra sediment from the clearing operation itself, and also because alternative land uses represent actual or potential hazards to water quality and may affect water yield or timing adversely. Designation of protected areas such as Forest Watershed (catchment) Reserves, Wildlife Sanctuaries and in some cases even National Parks, is appropriate.

Flood-prone areas

Where forests now occupy historic flood plains they should be considered for retention as preserves or as forest production areas. Forests and forestry represent a land use which can most ecologically and economically be subject to either frequent innundation or major rare flood events, thus minimising damage and suffering. The famous flooded forests of the Amazon River, mentioned already in the Wetlands section, which nourishes an impressively valuable fishery and precious wildlife habitat is such a forest with a major ecological role. During the period of innundation here, each year, the human inhabitants become 'boat people', and there is little or no flood damage. They are largely foragers, hunters, gatherers and fishers, with only small cleared areas for gardens and agroforests. Ecotourists are increasingly gaining access to this forest and its wildlife by boat.

It is primarily the damage aspect of floods that hoists this 'red flag'. The clearing of flood-prone forest foretells alternative land use, all of which is more intensive than forest, with investments in human safety, in crops, animals or structures. It is the loss of life, the suffering and the financial damage which should lead us to question whether forest clearance is a wise move. It is the increased occupance and infrastructure of flood-prone areas which pushes financial damage figures ever higher, plus our river constriction actions, rather than increased river flow as a result of deforestation in the far headwaters on the Ganges and Brahmaputra (Hamilton, 1987), for instance. It is well recognised that human beings are gamblers on the flood plains of the world, and willing to accept a risk which they perceive to be far in the future, or uncertain due to supposed safety provided by flood 'control' actions upstream. We learn flood plain occupancy lessons with reluctance and with short memories. The geographic convenience of flood plains is a compelling force. Pragmatically, some uses such as hay meadows or grazing may represent a low enough risk. Aerial photographs, skillfully interpreted, can provide much information on extent and frequency of flooding, where records and experience are lacking (Burgess, 1967). Satellite-derived sensing with high resolution has also proven useful in delineating flood-prone areas and has been tested in tropical developing countries, Honduras and Paraguay, by the Organization of American States (OAS, 1984) though it is more difficult in forested areas. For an analysis and overview of flood hazard assessment see Chapter 8 'Floodplain definition and flood hazard assessment' in the *Primer on Natural Hazard Management in Integrated Regional Development Planning* (OAS/DRDE, 1991).

It is of some interest to note the small amount of forest cover within 5 km of the major rivers of the world where persistent flood catastrophes have occurred involving hundreds to thousands of deaths and millions to billions of dollars of property damage. In the publication *Watersheds of the World* (Revenga *et al.*, 1998), for example: the Limpopo, which in February and March of 2000 produced such devastation, has zero forest cover shown; the perennially disastrous Indus shows zero, the Ganges shows about 3%, and the Brahmaputra about 10%. In summary, it would be appropriate to put the brakes on land clearing in those flood-prone areas that still have forest cover.

Guideline

The historic flood plains of rivers and major perennial streams are 'red flag' areas for clearing, since forest cover is adapted to the flood history of the river regimen and is the least damageable of any land uses. Any clearing proposal should analyse rigorously the human health and safety, and economic consequences, of occupying a flood plain with more intensive use. A zoning system at various distances from the river, based on flood frequency and extent of records (where they exist over a reasonable length of time) would indicate the degree of risk associated with alternative uses. In the absence of such records, skillful airphoto interpretation and ecological reading of the forest and soil conditions (vegetation type, relief, presence of flood debris and soil type) can provide sufficient information for planting the 'red flag' in the most critical areas.

Soil limitations that will not sustain proposed use

There is a large and long-time literature documenting the need for, and the methodology for, using soils information to determine the suitability and capability of various kinds of soil types to support an array of sustainable uses. It would seem that the earliest and most widely used system of soils-based land use capability classification was articulated by the United States Soil Conservation Service in the 1940s (USDA, 1943). The suitability of given mapped soil units for an array of uses, based on limitations inherent in the soils, became a widely used, though much-modified method around the world in fairly short order. Though developed initially in the agriculture sector (where reasonably detailed soil mapping had occurred) it also became used as a method of looking at forest lands for appraising possible future use, conversion, or levels of management intensity (USDA, 1967). FAO, by 1974 was promoting land evaluation through land use suitability assessment in the tropical world (FAO, 1974), based on suitability for land units, determined (within a climatic regime) by terrain features, soil depth, soil texture, soil drainage and salinity. A state-of-knowledge review and synthesis was carried out in 1979 at the East West Center on 'Assessing tropical forest lands: their suitability for sustainable uses' and it included case studies of systems which had developed in Malaysia, Philippines, Australia, Mexico, Indonesia, New Zealand, India, Pakistan, Sri Lanka, Canada, Thailand, Papua New Guinea and Pacific Islands (Carpenter, 1981). Environmental impact assessments as to suitability, prior to forest land clearing, are much more common today, but still not a requirement in most tropical countries and indeed not in all temperate developed countries, including the USA.

The 'red flag' has been hoisted already in this chapter for certain areas which, under a land classification suitability methodology, would show up as having severe limitations to conversion to many more intensive uses (e.g. slip-prone soils, flood-prone areas). In this guideline the focus is on sites with particularly severe soil fertility limitation. The importance has long been recognised, particularly in tropical soils, of the presence of the forest for recycling and quickly 'grabbing' any loose nutrients before they are leached out of the rooting zone by high rainfall. Even in the relatively rich volcanic soils in Costa Rica, for instance, Krebs (1975) found that 'conversion of forest to field results in a decline in soil organic matter, nitrogen, pH, calcium and magnesium and an increase in aluminium.' There are, in some parts of the tropics, ancient soils of low nutrient status where the nutrient level depends even more on the presence of the forest for recycling (see Proctor, this volume, for elaboration on biogeochemical cycling). For instance, most of the soils in the Amazon basin are inherently low in nutrients and those on podzolised quartz sand are at the extreme end of the oligotrophic scale (Herrera *et al.*, 1984). Following clearing in one such soil at San Carlos, Venezuela, nutrient loss through leaching below the rooting zone was substantial, and the biomass recovery after three years following cutting and burning was only 67% of that for cutting alone and was only 6% where the forest was cleared by bulldozer. Jordan *et al.* (1980) believe that a basic cause of the repeated failures to establish sustainable agriculture in much of Amazonia is due to the failure to recognise those soils possessing limitations because of rapid loss of what nutrients they have. This situation may well prevail in other tropical regions where we are not dealing with rich volcanic or flood plain soils. Sanchez (1979) identifies spodosols, oxisols and ultisols in particular as having low nutrient status and having a high susceptibility to nutrient loss from leaching or/and erosion. Biomass removal losses and volatilisation compound the problem when forest is cleared.

Guideline

Prior to clearing and conversion to a use where the nutrient status of the soils is important, it is imperative to determine if the soils are such that erosion or leaching following forest removal will render them unsuitable for the proposed use. This is especially critical where supplies of replacement fertiliser nutrients are unavailable physically or economically. Such soils are generally old, weathered, acid or coarse-textured. Note also the acid sulphate soils identified in connection with mangroves which also merit a 'red flag'. A good manual on assessing tropical forest lands as to their suitability for sustainable uses was produced by the East West Center in 1980 (Qureshi *et al.*, 1980). Intensification on proven soils is a better alternative.

Transition

The previous eight guidelines have the objective of providing barriers, or at least filters, to inhibit forest clearing under certain situations common in the humid tropics where hydrology and soils are likely to be adversely affected (and not even considering biological diversity, tenurial, social justice or other concerns related to culture and economics). It is recommended that in those situations, rigorous environmental assessment be carried out as to consequences, even while recognising that the assessment of ecological impacts in any kind of quantifiable way is imperfect. But, if an assessment does lead to the conclusion that the benefits exceed the costs or harms (economic, ecological and social), are there guidelines for land clearance itself that will reduce adverse hydrological or soils related impacts?

IF CLEARING IS HAPPENING, WHAT GUIDELINES ARE USEFUL?

The adverse environmental (soil and water) impacts of land clearing (deforestation) in the humid tropics have mostly to do with the subsequent poor land use practices **after** the forest is removed (Hamilton, 1991; Hamilton with King, 1983; Bruijnzeel, 1990; Critchley and Bruijnzeel, 1996). In the previous section, 'red flags' were raised for certain forest situations because of values impaired by these new uses following clearing. In those situations where, in spite of ecological costs, the arguments for perceived social or economic benefits carry the day, it is possible to implement some uses with appropriate practices so that soil and hydrological values are largely maintained (though perhaps at some new, somewhat less desirable level such as slightly higher storm flows in streams). For agriculture and plantations, Critchley and Bruijnzeel (1996) have set forth the details of minimising impacts. The hydrological and soils response to these two uses and several others following clearing were also documented by Hamilton with King (1983) though often based on research results from temperate forested areas. Methods of reducing negative impacts were suggested.

Much excellent work has been done on the effects on soil and water **during** the conversion process by different methods of forest land clearing at the International Institute of Tropical Agriculture in Ibadan, Nigeria. (See for instance Lal, 1981, 1983, 1986, 1991, 1996; Opari-Nadi and Lal, 1986; and Agboola, 1987.) Most alternative uses require land clearing, whether for agriculture, reservoirs, human settlements or other conversion types. Agboola (1987) and Ross and Donovan (1986) have presented a typology of land clearing methods which are largely valid today, as shown in Box 37.3.

One significant hydrological consequence of land clearing is an increase in water yield from the area. If the area is small, this may be unimportant but in large conversion schemes, or cumulatively small clearing added to small clearing, the effect may be significant. As Bosch and Hewlett (1982), Hamilton with King (1983), Bruijnzeel (1990) and others have reported from reviews of all temperate and tropical research, removal of more than around 33% of forest cover results in significant increases in total annual stream flow from the area in question. Bruijnzeel (1990, 2000) states that for forest clearance, the increases range between 145 and 820 mm yr^{-1} for humid tropical forests. What happens after that depends on the subsequent land use which might range from concrete (permanent yield increase) to a forest plantation (declining yield increase until perhaps 7–10 years later when it may return to original level, and because of rapid growth even result in a decline of water yield). The timing of yield from the area will depend largely on the degree of compaction that has occurred during the clearing. For instance, manual clearing of forest (with no tillage for three years) resulted in only a small increase in surface runoff (1% of rainfall) compared to no clearing, but use of tree pushing and root raking equipment in clearing resulted in a 12% increase in surface runoff (Lal, 1983). This affects timing of water delivery in streams, with duration of impact depending on subsequent land use.

Increases in bulk density of soil (Lal, 1996) and decreases in infiltration rates (Ghuman *et al.*, 1991) were shown to be a product of land clearing. For example the 1-hour infiltration rate for forest was 89 cm hr^{-1} but 32 for manual clearing and 20 for shear blade clearing, three months after treatment (Ghuman *et al.*, 1991).

Box 37.3 Land clearing methods

Logging followed by clearing

In most cases commercial timber will be harvested prior to the clearing operation. This usually involves heavy equipment characteristic of commercial log extraction, but in rare circumstances, in traditional societies, may involve manual cutting and removal of the commercially marketable trees by hand and animal power.

Then, either manual or mechanical clearing may occur depending on land ownership and purpose for clearing.

Again, burning usually follows, for this not only gets rid of some physical barriers to subsequent use, but for agriculture may provide some nutrients, even though organic matter is lost.

Manual methods

1. Clearing and burning – usually secondary forest, and carried out in dry season. This is the traditional method for swidden or shifting agriculture. Axe, machete or chain saw.
2. Selective tree felling and/or girdling so that there are still a few canopy trees that are useful for fruit production or later cash realisation. It becomes in essence an agroforestry system.

Mechanical methods

1. Bulldozer blade used for felling and stump removal. Where trees are widely spaced the undergrowth is bulldozed clear first before pushing over the trees. Rakes are often attached in order to pile or windrow the slash for subsequent drying and burning. A rear-end ripper may also be employed on the dozer.
2. Heavy chain pulled between two crawler tractors may be applicable in a few cases in tropical humid forests, but the system is most useful in clearing brush or 'bush' in drier areas.
3. Tracked vehicles with cutting or shearing blade that can cut off trees at ground level.
4. Heavy tracked vehicles with specialised equipment such as tree pushers or destumpers (front or rear mounted).
5. Disc harrows pulled after the trees are removed to loosen the soil and remove remaining roots.

Debris is usually burned after all of the above.

(After Agboola (1987) and Ross and Donovan (1986).)

Soil erosion effects also vary with the method used. For instance, manual clearing on sloping land kept soil erosion as low as 0.4 t ha^{-1} yr^{-1}, but crawler tractor clearing with tree pusher/root rake was as high as 15 t ha^{-1} yr^{-1} (Couper *et al.* 1981). Any method that removes the stump and requires filling of the hole is inflicting the greatest damage (topsoil fill in cavity and additional compaction).

Burning is common to all of the methods except an unusual one observed in Costa Rica, which was essentially slash-and-trash, rather than slash-and-burn. Critchley and Bruijnzeel (1996) succinctly sum up the impacts of burning:

> 'Not only do the more volatile constituents like nitrogen and carbon go up in smoke during burning of slash. Depending on the intensity of the fire, between 25 and 80% of all the calcium, potassium and phosphorous present in the slash may be lost this way as well . . . To make matters worse, nutrients which remain in the ash are vulnerable to removal in runoff and leaching. High temperature burns may render topsoil temporarily water repellent, causing potentially dramatic increases in overland flow frequency and intensity and therefore surface erosion.'

If mechanical methods of clearing have been used prior to burning, the surface compaction adds to the overland flow. If the subsequent use permits it (tree crop establishment in some cases), non-burning reaps fertility benefits. Malmer *et al.* in this volume describe the adverse effects of burning on soil and water.

Logging of commercially valuable wood, or harvest of much of the above-ground biomass for chipping, or logs plus fuelwood, are often part of land clearing. In many of the less-fertile humid tropical forests, between 10 and 75% of all the calcium and phosphoros of the above- and below-ground biomass plus that in the root-zone soil, is in the stems of the trees (Critchley and Bruijnzeel, 1996). Similar figures have been obtained for other key nutrients such as potassium (20–80%) and magnesium (20–65%). Nutrient loss by timber removal off-site thus represents a very serious decline in fertility, and on spodosols, oxisols and ultisols is of concern. Following timber or wood harvesting then the remaining slash is often burned, further depleting the nutrient pool.

The duration of the effects of land clearing on soil and water depend on the subsequent land use and the degree to which soil and water conservation practices are instituted, whether it be for cropland or a golf course. It is important to note that even when rapid establishment of food tree, extractive tree crops, or tree plantations occurs that mimic original forest conditions somewhat, the same conversion effects will take place every time the crop needs replacement with a new one, for the standing trees again require clearing.

A comprehensive review of the effects of the various methods of land clearing on not only soil and water but also on the success of subsequent use of the land, and the social and economic consequences, have been presented qualitatively by Ross and Donovan (1986) following a thorough review of the research on this topic until this time, and readers are referred to this excellent document, 'Land Clearing in the Humid Tropics'. Lal *et al.* (1986) in their book *Land Clearing and Development in the Tropics* have presented the best of the limited amount of research on impacts of clearing for agriculture and the subsequent effects of various cropping practices. An up-to-date technical review and synthesis of the soil and water impacts of conversions to various uses is presented by Bruijnzeel (2000).

Since such a high percentage of the nutrients of a tropical forest site are in the biomass, the removal of wood by logging or burning can severely deplete nutrients. For Spodosols, Ultisols and Oxisols in particular, loss of the sparse nutrients is of concern (Sanchez, 1979). Moreover, with vegetative cover removed (reduced rainfall interception and water uptake by forest) nutrient leaching is greatly increased out of the root zone of subsequent plants. Soil compaction from whatever source increases overland flow and hence nutrient flight, and surface erosion if the site is on sloping land. Changes in microclimate and in microorganisms also accompany land clearing and this may have implications for the new use. The guidelines suggested below are attempts to mitigate the undesirable effects of these consequences.

Guidelines

1. Timing. Undertake clearing during seasons without predictable high-intensity or long-duration rainstorms. Wet soils are more easily compacted. Make provision for temporary cessation of clearing activity in the event of unexpected periods of high rainfall.
2. Speed. Minimise the length of time that soil is exposed to leaching and erosion by rapid clearing and prompt revegetation with the new cover or provide mulch/litter cover until the new use is in place.
3. Burning. Avoid burning where possible (depends on subsequent use). Though responsible for losses of nitrogen, carbon and other key nutrients (see above) it probably has less adverse effect on soil than mechanical methods of land clearing, if it can be substituted. Avoid high temperature burns by burning debris where felled. If it must be done in windrows or piles, rake out the piles to spread the ash after the burn.
4. Stumps. If it is not absolutely necessary that stumps be removed (as in conversion to grazing or agroforestry) leave them in order to minimise topsoil displacement and additional soil compaction, as well as retaining shear strength in soils until the roots decay. Tree pushing, stump pulling equipment should be avoided, if low stump cutting by chainsaws or shearing blades on tracked equipment can be used.
5. Flat land. Where erosion is not a concern, the main problem is to keep soil compaction to a minimum and minimise leaching or volatilisation of nutrients. Therefore manual methods are better than mechanical. If mechanical methods are used: avoid mechanisation overkill in size and weight of equipment; minimise number of passes over the area; avoid dragging stems, logs or raking debris for windrowing or piling in order to minimise topsoil displacement.

6. Sloping land. All of the guidelines applicable to flat land are relevant for sloping lands but in addition, because erosion is a hazard, some additional precautions are needed. Avoidance of wet soils and periods of intense or long duration rainfall are even more critical, and speed in the conversion process is likewise more urgent. If mechanical clearing is used and residues are left over the rainy season, windrows along contour lines are indicated. Leave natural drainage ways undisturbed and do not block by dumping debris into these waterways. Where clearing of slip-prone lands is involved (see 'red flag' section on Unstable Slip-Prone Areas) it is especially important to retain stumps and root systems if at all possible, but best practice involves *not* clearing these lands. Erect sediment barriers at the downslope edge of the clearing as well as maintaining a sediment trapping vegetated area downslope of the clearing. Artificial barriers of straw or pegged plastic are often used on a temporary basis, until the cleared area can be stabilised from an erosion standpoint.
7. Roads. Most clearing will involve roads on, or providing access to, the site of conversion, for logging or land clearing equipment. Guidelines for hydrologically and erosionally safe location, construction and maintenance should be implemented. Megahan (1977) produced guidelines for reducing the erosional impact of roads that have stood the test of time. A succinct up-date review was provided for humid tropical steeplands by Adams and Andrus (1990): stabilise and rehabilitate roads (and landings or staging areas) after the clearance is finished.

CONCLUDING REMARKS

In the final analysis, the choice of whether to clear, and the technology and practices for clearing land, will be based not on ecological impacts, or hydrological and erosional safeguards. It will be based on private land-clearing 'fever', government programmes of land development and settlement, spontaneous even illegal clearing by landless or land hungry peasants or urban colonists, and land developers (possibly even foreign or multi-national corporations with money as the bottom line). Nonetheless it is in the long-term global, national and local, even individual, interest to maintain the productivity of the area under scrutiny for clearing. Rather than subsequent costly inputs of fertilisation, erosion control, rehabilitation and even restoration, consideration of and adherence to the 'red flag' warnings of Section 1 and the 'clearing guidelines' of Section 2 should avoid the ecologically, socially and economically costly mistakes that have characterised so much land development in the humid tropics. And, what has been set forth deals mainly with soil and water. When one adds in the likely adverse effects on biological diversity, carbon sequestration and production of CO_2, wildlife population dynamics, and on indigenous cultures and other socio-cultural values, clearing activities should indeed have to surmount the hurdles of adequate impact assessment.

Hopefully this review of guidelines provides grist for such a procedure. Soil degradation in the humid tropics has been widespread due to mistakes in forest clearing of inappropriate sites. Some testimony is evident in the millions of hectares of *Imperata* and *Themeda* grasslands in South East Asia alone. Restoration of cleared, degraded lands is one of the major ecological and economic tasks of the future. Meanwhile, we should constantly bear in mind that prevention of degradation due to clearing unsuitable land and poor clearing practices is cheaper ecologically, economically and socially than the complex, difficult process of restoration.

References

Adams, P. W. and Andrus, C. W. (1990) Planning secondary roads to reduce erosion and sedimentation in humid tropic steeplands. In: *Research Needs and Applications to Reduce Erosion and Sedimentation in Tropical Steeplands*. R. R. Ziemer, C. L. O'Loughlin and L. S. Hamilton (eds.) IAHS Publication No 192, Washington, 318–327.

Agboola, A. A. (1987) Current programs, problems and strategies for land clearing and development in Nigeria. In: *Tropical Land Clearing for Sustainable Agriculture*, R. Lal, M. Nelson, H. W. Scharpenseel and M. Sudjadi (eds.). International Board for Soil Research and Management Proceedings No 3, Bangkok, 177–193.

Aksornkoae, S. (1988) Mangrove habitat degradation and removal in Phauguga and Ban Don Bays, Thailand. *Tropical Coastal Area Management* 3(1):16.

Aksornkoae, S. (1993) *Ecology and Management of Mangroves*. IUCN Regional Office. Bangkok.

Aldrich, M. (1998) *Tropical Montane Cloud Forests*, Planning and advisory workshop. WCMC, IUCN, WWF, DFID report. Cambridge, UK.

Anonymous (2000) Government to press ahead with peatland project. *Indonesian Observer*, 30 March 2000. Jakarta.

Balek, J. (1989) Humid warm flatlands. In: *Comparative Hydrology: an Ecological Approach to Land and Water Resources*. T. Chapman (ed.). UNESCO, Paris. 353–369.

Beschta, R. L., Blinn, T., Grant, G. E., Swanson, F. J. and Ice, G. G. (eds.) (1987) *Erosion and Sedimentation in the Pacific Rim*. IAHS Publication No 165, Washington.

Beven, K; Lamb, R., Quinn, P., Romanowicz, R. and Freer, J. 1995 TOPMODEL (Chapter 18). In: *Computer models of Watershed Hydrology*, Singh, V. P. ed., Water Resources Publications, Boulder, USA, pp. 627–668.

Blaschke, P. N., Trustrum, N. A. and Hicks, D. L. (2000) Impacts of mass movement erosion on land productivity: a review. *Progress in Physical Geography* 24(1):21–52.

Bonell, M. 1998. Challenges in runoff generation research in forests from the hillslope to headwater drainage basin scale. *J. Amer. Water Resources Association*, 34, 765–785.

Bonell, M. (1999) Selected issues in mountain hydrology of the humid tropics. Keynote address in *Water: Forestry and Land Use Perspectives*, FRIM-WWF-UNESCO IHP, Kuala Lumpur, March 31–April 1, 1999, pp. 34–56.

Bonell, M. with Balek, J. (1993) Recent Scientific Developments and Research Needs in Hydrological Processes of the Humid Tropics. In: *Hydrology and Water Management in the Humid Tropics*. M. Bonell, M. M. Hufschmidt and J. S. Gladwell (eds.). International Hydrology Series, UNESCO/Cambridge University Press, Paris and Cambridge, 167–260.

Bosch, J. M. and Hewlett, J. D. (1980) Sediment in South African forest and mountain catchments. *South African Forestry Journal* 115:50–55.

Bruijnzeel, L. A. (1990) *Hydrology of Moist Tropical Forests and Effects of Conversion: A State of Knowledge Review*. UNESCO International Hydrological Programme, Paris.

Bruijnzeel, L. A. (2000) Hydrology of tropical montane cloud forests: a reassessment. *Second International Colloquium on Hydrology and Water Management in the Humid Tropics*, 22–24 March 1999. Panama City.

Bruijnzeel, L. A. (2002) Tropical forests and environmental services: not seeing the soil for the trees? *Agriculture, Ecosystems and Environment*. Papers from Changmai Workshop on Tropical Hydrology 30 May–2 June, 1999. IHP Technical Document in Hydrology No. 52.

Bruijnzeel, L. A. and Bremmer, C. N. (1989) *Highland-lowland interactions in the Ganges Brahmaputra River Basin: a review of published literature*. International Centre for Integrated Mountain Development Occasional Paper No. 11. Kathmandu.

Bruijnzeel, L. A. and Hamilton, L. S. (2000) *Decision Time for Cloud Forests*. UNESCO International Hydrological Programme, Paris and WWF International, Gland.

Bruijnzeel, L. A. and Proctor, J. (1995) Hydrology and biochemistry of tropical montane cloud forests: What do we really know? In: *Tropical Montane Cloud Forests*. L. S. Hamilton, J. O. Juvik and F. N. Scatena (eds.). Springer-Verlag Ecological Studies 110. New York, 38–78.

Budd, W. M., Cohen, P. L., Saunders, P. R., and Steiner, F. R. (1987) Stream corridor management in the Pacific Northwest: 1. Determination of stream-corridor widths. *Environmental Management* 11(5):587–597.

Bullock, A. (1993) Perspectives on the hydrology and water resource management of natural freshwater wetlands and lakes in the humid tropics. In: *International Hydrology and Water Management in the Humid Tropics*, M. Bonell, M. M. Hufschmidt and J. S. Gladwell (eds.). International Hydrology Series, UNESCO/Cambridge University Press, Paris and Cambridge, 273–300.

Burgess, L. C. N. (1967) *Airphoto Interpretation as an Aid in Flood Susceptibility Determination*. Department of Energy, Mines and Resources, Ottawa.

Carpenter, R. A. (ed.) (1981) *Assessing Tropical Forest Lands: Their Suitability for Sustainable Uses*. Tycooly International. Dublin.

Carson, B. (1989) *Soil Conservation Strategies for Upland Areas in Indonesia*. East-West Center Occasional Paper No. 9. Honolulu.

Cassells, D. S., Gilmour, D. A. and Bonell, M. (1984) Watershed forest management practices in the tropical rainforests of north-eastern Australia. In *Symposium on Effects of Forest Land Use on Erosion and Slope Stability*, C. L. O'Loughlin and A. J. Pearce (eds.). East-West Center, Honolulu.

Cassells, D. S. and Bonell, M. (1986) Logging operations in forest watersheds: an Australian perspective. In *Land Use, Watershed and Planning in the Asia-Pacific Region*, A. J. Pearce and L. S. Hamilton (eds.), FAO Regional Office, RAPA Report 1986/3. Bangkok, 44–58.

Couper, D. C., Lal, R. and Claasen, S. L. (1981) Land clearing and development for agricultural purposes. In: *Tropical Agricultural Hydrology*, R. Lal and F. W. Russell (eds.). John Wiley, New York, 119–130.

Critchley, W. R. S. and Bruijnzeel, L. A. (1996) *Environmental Impacts of Converting Moist Tropical Forest to Agriculture and Plantations*. UNESCO International Hydrological Humid Tropics Programme Series No 10. UNESCO, Paris.

Davies, T. R. H. and Pearce, A. J. (eds.) (1981) *Erosion and Sediment Transport in Pacific Rim Steeplands*. IAHS-AISH Publication No 132, Washington.

FAO. (1974) *Approaches to Land Classification*. Soils Bulletin No. 22. FAO, Rome.

FAO (1993) *Forest Resources Assessment 1990, Tropical Countries*. Forestry Paper 112. FAO, Rome.

Fölster, H. (1986) Nutrient loss during forest clearing. In: *Land Clearing and Development in the Tropics*, R. Lal, P. A. Sanchez and R. W. Cummings, Jr. (eds.). Balkema, Rotterdam, 241–246.

Ghuman, B. S., Lal, R. and Shearer, W. (1991) Land clearing and use in the humid Nigerian tropics, 1. Soil physical properties. *Soil Science Society of America Journal* 55(1):178–188.

Gregory, S. V., Swanson, F. S., McKee, W. A. and Cummins, K. W. (1991) An ecosystem perspective of riparian zones. *BioScience* 41(8):540–551.

Hamilton, L. S. (1987) What are the impacts of Himalayan deforestation on the Ganges-Brahamaputra lowlands and delta? Assumptions and facts. *Mountain Research and Development* 7(3):256–263.

Hamilton, L. S. (1988) Semantics, definitions and deforestation. *IUCN Bulletin* 18(4–6):8–9.

Hamilton, L. S. (1991) Tropical forests: identifying and clarifying the issues. *Unasylva* 166(42):19–27.

Hamilton, L. S. (1992) The wrong villain? *Journal of Forestry* 90(2):7.

Hamilton, L. S. (1995) *A Campaign for Cloud Forests*, Unique and valuable ecosystems at risk. IUCN Focus Series. IUCN, Gland.

Hamilton, L. S. (1997) El desarrollo sostenible en tierras altas y las áreas clave de control para la conservación de montañas. In: *Desarrollo Sostenible de Ecosistemas de Montaña: Manejo de Áreas Frágiles en los Andes*. M. Liberman and C. Baied (eds.). Instituto de Ecología, La Paz, 269–276.

Hamilton, L. S., Dixon, J. A. and Miller, G. O. (1990). Mangrove forests: an undervalued resource of the land and of the sea. In *Ocean Yearbook 8*, E. M. Borgese, N. Ginsburg and J. R. Morgan (eds.). University of Chicago. Chicago.

Hamilton, L. S., Juvik, J. O. and Scatena, F. N. (1995) The Puerto Rico tropical cloud forest symposium: introduction and workshop synthesis. In: *Tropical Montane Cloud Forests*. L. S. Hamilton, J. O. Juvik and F. N. Scatena (eds.). Springer-Verlag Ecological Studies 110. New York, 1–23.

Hamilton, L. S. with King, P. N. (1983) *Tropical Forested Watersheds: Hydrologic and Soils Response to Major Uses or Conversions*. Westview Press, Boulder.

Hamilton, L. S. and Snedaker, S. C. (eds.). (1984) *Handbook for Mangrove Area Management*. East West Center/IUCN/UNESCO/UNEP, Honolulu.

Hansen, A. J. and Dickensen III, J. C. (1987) Settlement and land development. In: *Ecological Development in the Humid Tropics*, A. E. Lugo, J. R. Clark and R. D. Child (eds.). Winrock International, Morrilton, 81–109.

Herrera, R., Medina, E., Klinge, H., Jordan, C. F. and Uhl, C. (1984) Nutrient retention mechanisms in tropical forests: the Amazon caatinga, San Carlos pilot project, Venezuela. In: *Ecology in Practice: Part 1. Eco-system Management*, F. DiCastri, F. W. G. Baker and M. Hadley (eds.). Tycooly International, Dublin, 85–97.

Hollis, G. E., Holland, M. M., Maltby, E. and Larson, J. S. (1988) Wise use of wetlands. *Nature and Resources* 24(1):2–13.

Howes, J. (1995) Report on First International Conference on Wetlands and Development. *Asian Wetlands News* 8(2):3.

Hudson, N. (1995) *Soil Conservation* (Fully revised and updated third edition). Iowa State University Press, Ames.

Humphreys, G. S. and Brookfield, H. (1991) The use of unstable steeplands in the mountains of Papua New Guinea. *Mountain Research and Development* 11:295–318.

ISME 2000. *Mangroves*, International Society for Mangrove Ecosystems Newsletter No 23.

IUCN. (1975) *The Use of Ecological Guidelines for Development in the American Humid Tropics*. IUCN, Morges.

Janssen, R. and Padilla, J. E. (1999) Preservation or conversion? Valuation and evaluation of a mangrove forest in the Philippines. *Environmental and Resource Economics* 14:297–331.

Jordan, D. F., Golley, F. B., J. D. Hall and Hall, J. (1980) Nutrient scavenging of rainfall by the canopy of a tropical rainforest. *Biotropica* 12:61–66.

Klinge, H. (1983). Forest structures in Amazonia. In: *Ecological Structures and Problems of Amazonia*. IUCN Commission on Ecology Papers No. 5, Gland, pp. 13–23.

Krebs, J. E. (1975) A comparison of soils under agriculture and forests in San Carlos, Costa Rica. In: *Tropical Ecological Systems: Trends in Terrestrial and Aquatic Research*, F. B. Golley and E. Medina (eds.). Springer-Verlag, New York, 381–390.

Lal, R. (1981) Deforestation of tropical rainforest and hydrological problems. In: *Tropical Agricultural Hydrology*, R. Lal and E. W. Russell (eds.). John Wiley, Chichester, 131–140.

Lal, R. (1983) Soil erosion in the humid tropics with particular reference to agricultural land development and soil management. In: *Hydrology of Humid Tropical Regions*, R. Keller (ed.). IAHS Publication No. 140, Wallingford, 221–239.

Lal, R. (1986) Deforestation and soil erosion. In: *Land Clearing and Development in the Tropics*. R. Lal, P. A. Sanchez and R. W. Cummings Jr. (eds.). Balkema, Rotterdam, 299–316.

Lal, R. (1996) Deforestation and land-use effects on soil degradation and rehabilitation in Western Nigeria, 1. Soil physical and hydrologic properties. *Land Degradation and Development* 7:19–45.

Lal, R. (1997) Deforestation effects and soil degradation and rehabilitation in Western Nigeria, 4. Hydrology and water quality. *Land Degradation and Development* 8:95–126.

Lal, R., Nelson, M., Scharpenseel, H. W., and Sudjadi, M. (eds.) (1987) *Tropical Land Clearing for Sustainable Agriculture*. IBSRAM Proceedings No. 3. International Board for Soil Research and Management, Bangkok.

Lal, R., Sanchez, P. A. and Cummings Jr., R. W. (eds.) (1991) *Land Clearing and Development in the Tropics*. Balkema, Rotterdam.

Langford, K. J. and O'Shaughnessy, P. J. (eds.) (1977) *Water supply catchment hydrology research. First Progress Report*. Melbourne and Metropolitan Board of Works Department, Melbourne.

Liniger, H. P. (1998) Mount Kenya: a vital water tower in a semi-arid region. In: *Mountains of the World: Water Towers for the 21st Century*. Mountain Agenda (eds.). Institute of Geography, Berne, Switzerland, 10–12.

Lugo, A. E., Clark, J. R. and Child, R. D. (eds.) (1987) *Ecological Development in the Humid Tropics: Guidelines for Planners*. Winrock International, Morrilton.

Maltby, E. (1986) *Waterlogged Wealth*. International Institute for Environment and Development and Earthscan, London.

Maltby, E. and McInnes, R. J. (1997) Functions and degradation of wetlands. In *The Global Environment, Science, Technology and Management*, D. Brune, D. V. Chapman, M. D. Gwynne and J. M. Pacyna (eds.), Vol 1. Scandinavian Science Publisher. Weinheim, Germany, 165–185.

Megahan, W. F. (1977) Reducing erosional impact of roads. In: *Guidelines for Watershed Management*, S. H. Kunkle and J. L. Thames (eds.). FAO Conservation Guide 1, Rome, 237–261.

Megahan, W. F. and King, P. N. (1985) Identification of critical areas on forest lands for control of nonpoint sources of pollution. *Environmental Management* 9(1):7–18.

Megahan, W. F. and Schweithelm, J. (1983) Guidelines for reducing negative impacts of logging. In *Tropical Forested Watersheds: Hydrologic and Soils Response to major Uses or Conversions*, L. S. Hamilton with P. N. King. Westview Press, Boulder, 143–154.

Mok, S. T. (1986) Sustained use and management of forests: a Malaysian perspective. In *Land Use, Watersheds and Planning in the Asia-Pacific Region*, A. J. Pearce and L. S. Hamilton (eds.). FAO Regional Office RAPA Report 1986/3. Bangkok, 34–43.

Moore, I. D., Grayson R. B. and Ladson, A. R. (1991). Digital terrain modelling: A review of hydological, geomorphological and biological applications. *Hydrol. Processes* 5: 3–30.

Morgan, R. P. C. (1974) Estimating regional variations in soil erosion hazard in Peninsular Malaysia. *Malaysia Nature Journal* 28:94–106.

Myers, L. (2000) Students raise funds for disaster relief in Venezuela. *Cornell Chronicle* 31(21):1, 3.

Nutter, W. L. and Brinson, M. M. (1994) Application of hydrogeomorphic principals and functions to assessment of drainage intensity in forested wetlands. In *Water Management in Forested Wetlands*. Technical Publication R8-TP20, US Forest Service Southern Region. Atlanta.

OAS (1984) *Integrated Regional Development Planning: Guidelines and Case Studies from OAS Experience*. Organization of American States, Washington.

OAS/DRDE (1991) *Primer on Natural Hazard Management in Integrated Regional Development Planning*. Organization of American States/ Department of Regional Development and Environment, Washington.

O'Laughlin, J. and Belt, G. H. (1995) Functional approaches to riparian buffer strip design. *Journal of Forestry* 93(2):29–32.

O'Loughlin, C. L. (1974) The effect of timber removal on the stability of forest soils. *Hydrology* 13:121–134.

O'Loughlin, C. L. and Pearce, A. J. (eds.) (1984) *Proceedings of the International Symposium on Effects of Forest Land Use on Slope Stability*. East-West Center, Honolulu.

O'Loughlin, C. L. and Ziemer, R. R. (1982) The importance of root strength and deterioration rates upon edaphic stability in steepland forests. In: *Carbon Uptake and Allocation: a Key to Management of Subalpine Ecosystems*. R. H. Waring (ed.). Oregon State University, Corvallis, 84–91.

O'Loughlin, E. M. (1986). Prediction of surface saturation zones in natural catchments by topographic analysis. Water Resour. Res. 22: 794–804.

Opari-Nadi, O., Lal, R. and Ghuman, B. S. (1986) Effects of land clearing methods on soil physical and hydrological properties in south western Nigeria. In: *Land Clearing and Development in the Tropics*, R. Lal, P. A. Sanchez and R. W. Cummings, Jr. (eds.), Balkema, Rotterdam, 215–226.

Pannier, F. (1979) Mangroves impacted by human-induced disturbances: a case study of the Orinoco delta mangrove ecosystem (Venezuela). *Environmental Management* 3(3):205–216.

Pearce, A. J. and Hamilton, L. S. (eds.) (1986) *Land Use, Watersheds, and Planning in the Asia-Pacific Region*. FAO RAPA Report 1986/3, Bangkok.

Poore, D. (1975) Ecological guidelines for the development of the American humid tropics. In: *The Use of Ecological Guidelines for Development in the American Humid Tropics*. IUCN, Morges, 225–247.

Poore, D. and Sayer, J. (1987) *The Management of Tropical Moist Forest Lands: Ecological Guidelines*. IUCN Tropical Forest Programme, Gland.

Qureshi, A. H., Hamilton, L. S., Mueller-Dombois, D., Perera, W. R. H., and Carpenter, R. A. (1980) *Assessing Tropical Forest Lands, Their Suitability for Sustainable Uses*. East West Center, Honolulu.

Rao, Y. S. (1988) Flash floods in southern Thailand. *Tiger Paper* 15(4):1–2.

Rapp, A. (1997) Erosion and land degradation in drylands and mountains. In: *The Global Environment, Science, Technology and Management Vol. 1*. D. Brune, D. V. Chapman, M. D. Gwynne and J. M. Pacyna (eds.). Scandinavian Science Publisher, Weinheim, 207–224.

Rapp, A., Berry, L. and Temple, P. H. (eds.) (1972) Studies of soil erosion and sedimentation in Tanzania. *Geografiska Annaler*, 54A:105–379.

Reily, J. (2000) Kalimantan after the rice project. *Arborvitae*. IUCN/WWF Forest Conservation Newsletter No 13:8. Gland.

Revenga, C., Murray, S., Abramovitz, J. and Hammond, A. (1998) *Watersheds of the World: Ecological Value and Vulnerability*. World Resources Institute/Worldwatch Institute, Washington.

Ross, M. S. and D. G. Donovan (1986) *Land Clearing in the Humid Tropics*. IUCN/IIED Tropical Forest Policy Paper No. 1. IUCN Publications Services, Cambridge.

Saenger, P., Hegerl, E. J. and Davies, J. D. S. (eds.) (1983). *Global Status of Mangrove Ecosystems*. Commission on Ecology Papers, No. 3, IUCN, Gland.

Sanchez, P. A. (1979) Soil fertility and conservation considerations for agroforestry systems in the humid tropics of Latin America. In: *Soils Research in Agroforestry*, H. O. Mongi and P. A. Huxley (eds.). International Council for Research in Agroforestry, Nairobi, 79–124.

Scatena, F. N. (1990) Geomorphic impacts of Hurricane Hugo on the Luquillo Mountains of Puerto Rico: preliminary observations. In: *Research Needs and Applications to Reduce Erosion and Sedimentation in Tropical Steeplands*. R. R. Ziemer, C. L. O'Loughlin and L. S. Hamilton (eds.). IAHS Publ. No. 192. Washington, 393.

Trustrum, N. A. and Page, M. J. (1992) The long-term erosion history of Lake Tutira watershed: implications for sustainable land use management. In: *Proceedings of the International Conference on Sustainable Land Management*, P. Henriques (ed.). Landcare Research, Palmerston, 212–215.

USDA. (1943) *Classifying Land for Conservation Farming*. Farmers' Bulletin No. 1853, US Department of Agriculture, Washington.

USDA. (1967) *Developing Soil-Woodland Interpretations*. Soils Memorandum No. 26. USDA, Washington.

Wijkman, A. and Timberlake, L. (1984) *Natural Disasters – Acts of God or Acts of Man?* Earthscan, London.

Zadroga, F. (1981) The hydrological importance of a montane cloud forest area of Costa Rica. In: *Tropical Agricultural Hydrology*. R. Lal and F. W. Russell (eds.). John Wiley, New York, 59–73.

Ziemer, R. R. (1981) Roots and the stability of forested slopes. In: *Erosion and Sediment Transport in Pacific Rim Steeplands*. T. R. H. Davies and A. J. Pearce (eds.). IAHS Publication No. 132, Washington, 343–359.

Ziemer, R. R., O'Loughlin, C. L. and Hamilton, L. S. (eds.) (1990) *Research Needs and Applications to Reduce Erosion and Sedimentation in Tropical Steeplands*. IAHS Publication No. 192. Washington.

38 From nature to nurture: soil and water management for rainfed steeplands in the humid tropics

W. R. S. Critchley
Vrije Universiteit, Amsterdam, The Netherlands

INTRODUCTION

Rainfed farming continues to expand in the humid tropics as the population grows, and natural forests are being replaced by agriculture on increasingly steep hillsides. This process may be abrupt but more often occurs through a series of gradual changes in land use: from forest modification, to shifting agriculture, through settlement mosaics and on to eventual complete forest replacement (Sayer *et al*., 2000). Often, the gradients in question are above theoretical legal limits for cultivation (Moldenhauer and Hudson, 1988) (Figure 38.1). The challenge is to identify and offer practical solutions for soil and water conservation in these areas that are both technically appropriate and socially acceptable. It is increasingly realised that measures must address water, fertility and production aspects simultaneously, as well as the prevention of surface erosion. These systems should be sound in technical terms so that they control damage to the soil and to the hydrological integrity of the hillsides. They should also sustain plant production. Social acceptability is the acid test of adoption: if 'solutions' are not attractive to land users they will not be sustained and they will not spread.

Since world-wide concern with soil erosion ('the problem') and soil conservation ('the solution') spread in the 1930s and 1940s, catalysed by the great dust bowl phenomenon in the United States (Anderson, 1984), there have been significant shifts in conceptual thinking. In summary, it has been recognised that top-down schemes based on engineering techniques have largely failed to deliver. An approach that embraces farmer-friendly, production-oriented, low cost strategies is more likely to succeed, especially amongst resource-poor farmers (Hudson, 1991; IFAD, 1992; Douglas, 1994). Indigenous technologies have increasingly received recognition (Wilken, 1987; Kerr and Sanghi, 1992; Critchley *et al*., 1994, Reij *et al*., 1996) and an appreciation of local technical innovation is emerging (Critchley, 2000a). Terminology has changed also, marking this evolution in conceptualisation. Simple 'soil conservation' has progressed to 'soil and water conservation' or 'soil and water management'. Some specialists prefer the terms 'conservation farming', 'sustainable land management', 'landcare' or 'land husbandry'. Almost all the new concepts embrace the notions of judicious use of resources and sustained productivity. Table 38.1, from a recent FAO publication, shows how far the new thinking is moving.

To a certain extent there has also been a change in the perception of the problem, though soil erosion is still widely accepted to be the core concern: 'a persistent crisis' (Hurni *et al*., 1996), 'one of the principal threats to agricultural sustainability worldwide' (McDonald and Brown, 1999) and 'the most important land degradation problem in the humid tropics' (Craswell, 1998). As already noted, increasingly steep hillsides are being brought into cultivation as a result of population pressure, and damage limitation rather than complete conservation becomes key (Hudson, 1988). Furthermore, 'conservation at all costs' is a slogan long discarded by donors and governments alike. Economic realities dictate that pragmatism is paramount. Linked to this point, and to recognition of poverty amongst a growing number of hill farmers, is the realisation that sustaining on-site production is often more important, and cost-effective, than preventing the negative effects of downstream sedimentation and flooding (Doolette and Magrath, 1990). Thus conservation programmes are becoming focussed increasingly on farmers and communities rather than watersheds, and on sustained production rather than erosion prevention and runoff control. This last change in emphasis implies a tangential approach to the problem, rather than a frontal attack. It is, effectively, 'soil conservation by stealth' (Shaxson, 1988).

This chapter sets out to review soil and water management technologies and systems in the light of the foregoing. There is a wide range of publications setting out technical guidelines. These include general texts (e.g. Hudson, 1981; Morgan, 1995), as well as guides for specific agroecological zones (e.g. the humid tropics, Sheng, 1989; tropical steeplands, Shaxson, 1999), for given

Forests, Water and People in the Humid Tropics, ed. M. Bonell and L. A. Bruijnzeel. Published by Cambridge University Press.

Table 38.1. *Changes in concepts and approaches for improved tropical steepland management*

Old	New
Runoff and erosion are the primary cause of land degradation	Runoff and erosion are usually consequences of land degradation
Protection of land is achieved by structures	Protection is achieved by recreating forest floor conditions
Decline in crop yield is related to losses of soil, water, nutrients	Yield after erosion is related to the quality of soil left behind
Cross-slope conservation works are pre-requisites	Soil management ensures greater benefits
Production is sustained by barriers and fertility application	Production is sustained by continuous land husbandry over time
Land management practices should be separate for each enterprise	Land management practices should embrace whole farming system
Resource-poor, small-scale farmers are naturally conservative and irrational	Farmers make rational decisions within an envelope of constraints
Extension workers = instructors and advisors	Extension workers = promoters and facilitators
Watershed = best unit of intervention	Community = best unit of intervention

Source: Abridged from Shaxson, 1999.

Figure 38.1 Increasingly steep slopes being farmed (Uganda). (All photographs in this chapter are by the author unless otherwise stated.)

countries (e.g. India: Singh *et al.*, 1990; Java: Carson, 1989; Kenya: Thomas, 1997) and for specific practices (e.g. Sloping Agricultural Land Technology – SALT: Partap and Watson, 1994). A current initiative entitled the World Overview of Conservation Approaches and Technologies (WOCAT) is working towards documentation of the main conservation systems and methods of implementation, on a global basis. One feature of the WOCAT inventory is the harmonisation of data collection through standard, comprehensive questionnaires. The intention is to provide data bases and also analytical inventories of the most effective conservation techniques (WOCAT/FAO, 2000). In the absence of this kind of comparative data, it is only possible to provide partial analysis, which at least helps to establish what are the current 'best-bet' solutions. After an overview of technology-specific aspects, including basic design, applicability, costs and benefits, two categories of challenges and concerns are raised in this chapter. These are:

- To what extent are the existing recommendations suitable under changing demographic, economic, political, climatic and environmental circumstances?
- What are – or should be – priorities for research and development in this field?

AN OVERVIEW OF TECHNOLOGIES

Categorisation

There are a number of different approaches to classifying, or categorising, on-farm soil and water conservation techniques. While the original focus was exclusively on field engineering and its various sub-sectors, the recent trend is to throw the net more widely and include a wide range of practices related to management of soil and vegetation (e.g. Sheng, 1989; Morgan, 1995; WOCAT/FAO, 2000). The reasons for this are basically twofold. First, in-field methods help reduce damage from raindrop impact, which is increasingly recognised as the fundamental cause of surface erosion. Second, these practices have a direct relationship with sustained plant production (Shaxson, 1999). Herein, a simple categorisation will be made to distinguish cross-slope barriers/ terraces ('support structures') and those practices carried out between these barriers where they exist ('in-field practices'). Most sound conservation systems on humid tropical hillsides will incorporate elements of both. These are not alternatives: they are complementary (Hudson, 1981).

Support structures

In humid tropical steeplands this category basically comprises terraces of various descriptions, with associated discharge and protection structures (waterways and cutoff drains/diversion ditches). Contour bunds, cross-slope vegetative strips and barrier hedges can be considered as starting points for terrace development. Clearly by including certain vegetative measures here, calling support structures 'mechanical methods' (as some authors do) leads to confusion. 'Bio-terracing' (sometimes rather misleadingly called 'bio-engineering') is a new and descriptive umbrella term for those vegetative interventions that have a structural function. Thomas (1997) suggests that the name 'terrace' should be retained to apply to 'a more or less permanent change in soil (surface) profile' with a reduction in gradient of the planted zone. While pointing out that terraces tend to be dynamic, i.e. changing shape and function over time, Critchley (2000b) summarises the main types of terraces found in rainfed steeplands in Table 38.2.

BENCH TERRACES

Bench terraces can be defined as those terraces that have beds with:

(a) a slope of less than 3° in any direction;
(b) a width of 10 metres or less; and
(c) well-defined risers separating them.

They are usually found in sequence on original slopes of 6°–25°. Bench terraces may be back-sloping, level (see Figure 38.2) or forward-sloping. A slight side slope may be found where such terraces drain laterally through a toe-drain. The terrace riser may be nearly vertical and faced with stone, or it may be less acutely angled and vegetated – ideally with a perennial fodder species. Technical specifications for bench terraces are provided by several authors (Hudson, 1981, 1992; Sheng, 1989; Singh *et al.*, 1990; Morgan, 1995; Thomas, 1997).

Construction of bench terraces may be achieved in one of two ways. It can be carried out on a one-off basis through cut and fill. Alternatively terraces may develop over time through gradual levelling behind a cross-slope barrier of earth, stone or vegetation. The levelling process results both from erosive rainfall and through the effect of tillage. Where the terrace is constructed through cut and fill, costs can be very high: 1500 person-days per hectare or more on the steeper slopes as an initial cost, with recurrent annual costs of 40 days per hectare (Bernsten and Sinaga, 1983; Stocking and Abel, 1992). Table 38.3 compares amounts of labour involved in bench terrace construction (and maintenance) with the costs of other support structures.

Clearly, bench terraces need to be very effective to be justified. This is not always the case. Doolette and Magrath (1990), reviewing the literature regarding terracing in Asia, find more references to *decrease* in crop yield after terracing than to *increase*. Decreased yields commonly result from land taken out of production to create risers, or through subsoil exposure. De Graaff and Wiersum (1992) note the importance of accompanying terracing with improved agronomic packages. Returns from switching to a high-value crop can help pay for the investment in terracing. Neither are bench terraces always as effective as assumed previously in reducing erosion. This becomes increasingly apparent when erosion is monitored using a specifically tailored methodology such as the 'natural boundary erosion plot' (NBEP) (see Figure 38.3).

In the case of back-sloping bench terraces, this avoids the problems associated with locating long, conventional (USLE-type) erosion plots down a series of terrace beds/risers, and thus cutting across drainage pathways and delivering spurious results. Net seasonal erosion rates of up 190 t ha^{-1} have been measured from individual back-sloping terraces by Purwanto (1999) in West Java using NBEPs in erodible volcanic tuffs. Although this is exceptional (Bruijnzeel and Critchley (1996), measured average

Table 38.2. *Overview of major types of support structures/ terraces used in rainfed steeplands*

Type of terrace	Alternative names: Variations	Typical characteristics and design	Use and application	Profile	Reference[a]
Bench terrace (BT)	Back-sloping BT; level BT; forward-sloping BT	Excavated or developed over time; width of bench (usually) 2–10 m; lateral grade (usually) up to 2°; level, or reverse/ forward grade up to 3°; often has toe-drain at foot of riser	Slopes up to 25°; typical 'improved' rainfed terrace design		*Sheng, 1988; 1989* Hudson, 1981 Morgan, 1995
Intermittent terrace (modified form of bench terrace)	Orchard terrace; hillside ditch	Width of bench 1.75–5 m; spaces between benches (see profile) ~10 m	Slopes up to 30°; for perennial crops on steep slopes	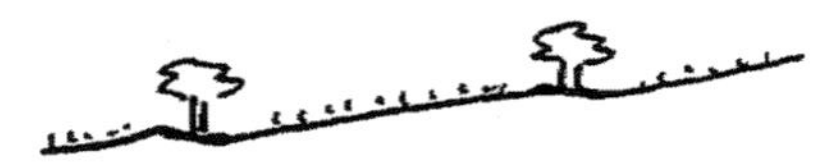	*Sheng, 1989*
Forward-sloping terrace (FST)	Outward-sloping terrace; many different forms	Bund usually earth; can also be stone; vegetation-based, etc.; forward grade of bed 3°–10° (gently sloping) forward grade of bed >10° (strongly sloping) may have toe-drain at base of riser	Wide range of slopes; common tradition in Asia and Africa		*Carson, 1989, 1992* Johnson *et al.*, 1982
Fanya juu (variation of FST)	Converse terrace; Natural terrace; Progressive BT	Width of bench (i.e. horizontal interval between structures) 8–18 m; bund 40 cm in height; trench (below bund) 60 cm × 60 cm; may evolve into BT over time	Slopes up to 12°; common recommendation in highlands of Africa		*Thomas, 1997*
Bio-terrace (variation of FST)	Vegetative-barrier based terrace; grass strip	Evolves from sedimentation behind densely planted barrier hedge of woody species (e.g. *Gliricidia*) or grass strip (e.g. *Vetiveria*); vertical interval 1–2 m; may evolve into BT over time	Slopes up to 30°; most common current technical recommendation; low-cost		*Partap and Watson, 1994*

[a] Main source is in italics.
Source: Adapted from Critchley (2000b).

Table 38.3. *Labour requirements: support structures*

Support structure	Person-days/ ha	Place	Reference
Construction/establishment			
Bench terraces (up to 26° slope)	up to 1850	Java	Bernsten and Sinaga, 1983
Bench terraces	1500	Vietnam	Stocking and Abel, 1992
Bench terraces (24° slope)	470	Jamaica	Sheng, 1988
Fanya juu (3° slope/ 11° slope)	40/135	Kenya	Thomas, 1997
Orchard terraces	110	Non-specific	Sheng, 1981
SALT[a]	350	Non-specific	Partap and Watson, 1994
Barrier hedgerows	55	Philippines	Pandey and Lapar, 1998
Natural vegetative filter strips (NVS)	4	Philippines	Pandey and Lapar, 1998
Vetiver grass (low / high slope)[b]	13/105	Non-specific	Grimshaw and Helfer, 1995
Maintenance			
Bench terraces	20–30	Non-specific	Sheng, 1989
Bench terraces	40	Non-specific	Stocking and Abel, 1992
Barrier hedgerows	14	Philippines	Pandey and Lapar, 1998
Natural vegetative filter strips (NVS)	13	Philippines	Pandey and Lapar, 1998
Vetiver grass (low/ high slope)[b]	1/8	non-specific	Grimshaw and Helfer, 1995

[a] Sloping land agricultural technology: whole field preparation: not just hedgerow establishment

[b] Figures in reference text only given for low slope (250 linear metres of grass lines/ ha) which is extrapolated here for high slope (2000 metres of grass lines/ha). Assumed that vetiver splits are available on-site.

Figure 38.2 Level bench terrace (Uganda).

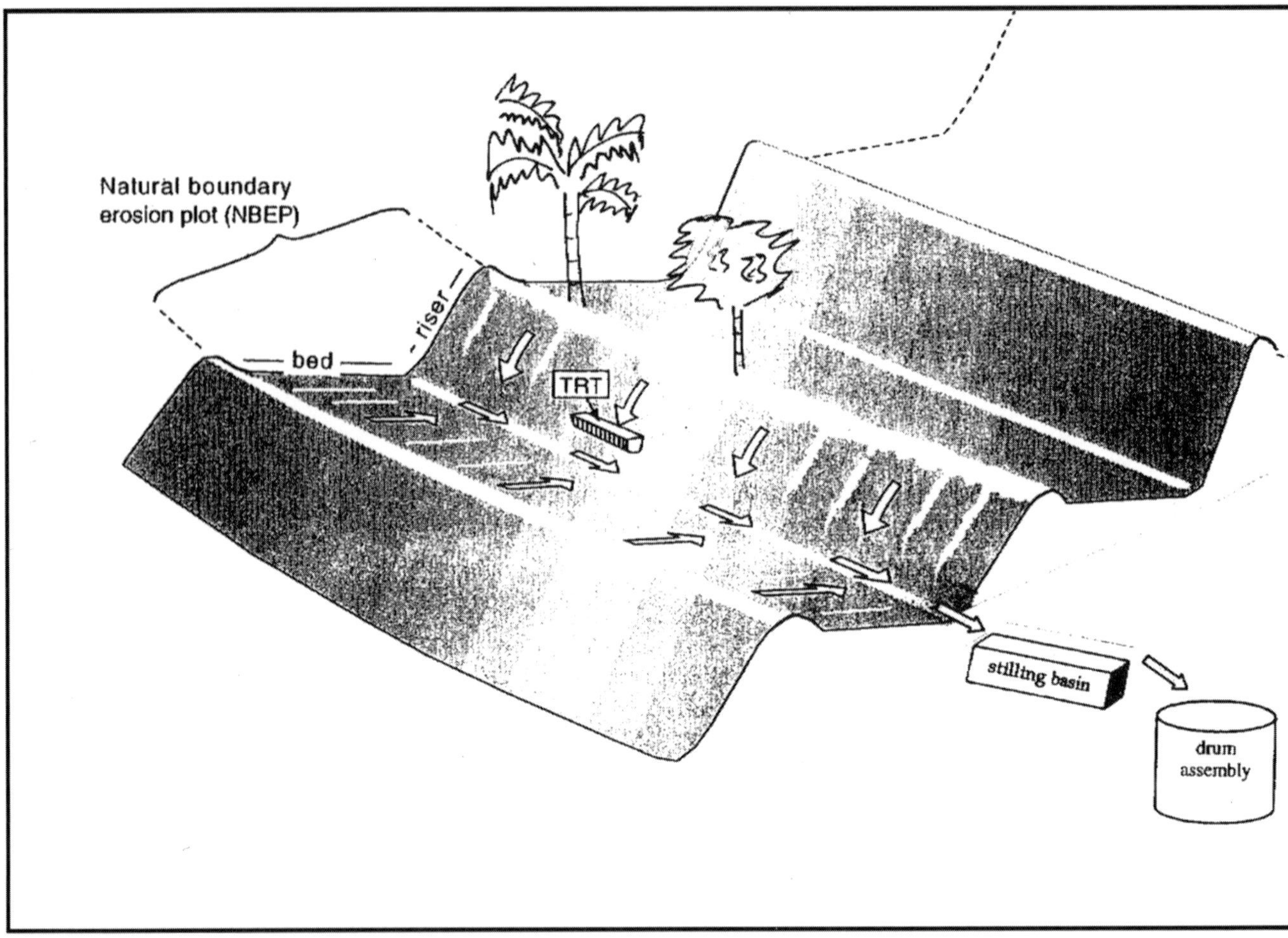

Figure 38.3 National boundary erosion plot (NBEP) – for measuring erosion from back-sloping bench terraces (note terrace riser trough, TRT).
(Source: Bruijnzeel and Critchley, 1996.)

seasonal rates of 12 t ha^{-1} in Central Java with the same methodology), there is no reason to assume that bench terracing automatically means adequate control of erosion or runoff. Table 38.4 presents erosion rates from bench terraces and other support structures. The variety of slopes, rainfall regimes and methodologies employed render direct comparison of these data unjustified. While several data sets highlight continuing erosion after terracing, it is clear nevertheless that substantial reductions in surface erosion *can* be achieved through terracing, especially good quality bench terracing.

Bench terraces are not applicable universally. There is a danger of increasing infiltration of rainfall (compared with the cultivated, non-terraced situation) leading to instability of certain soils, for example marls, and thus precipitation of landslides (Carson, 1989; Westerberg and Christiansson, 1999). Tamang (1993) shows how hill farmers in Nepal opt for forward-sloping terraces because they recognise that benches would collapse under saturation (see Figure 38.4 for example from Indonesia). Less dramatic than landslides, though much more common, is slumping of terrace risers (Euphrat, 1987; Critchley, 2000b). These, nevertheless, are usually repaired rapidly by farmers before a 'domino-effect' sequential collapse occurs. An even more important, though less conspicuous problem – in terms of downstream sedimentation at least – is that of risers yielding considerable and continuous amounts of sediment over the rainy season if unprotected by vegetation or stone. This can be measured by erosion pins, or by using specifically designed 'terrace riser troughs' (see Figure 38.3). Much of the sediment from the risers of back-sloping bench terraces is flushed out of the terraces into drainage paths and hence downstream (Critchley and Bruijnzeel, 1995; Purwanto and Bruijnzeel, 1996).

Construction of new bench terraces is going out of vogue as a technical remedy in steeplands, due not only to cost but also because of doubts about efficiency and appropriateness. The World Bank (1992), referring to terracing schemes in Asia, noted despondently that 'the effects of such structures have

Table 38.4. *Net erosion under support structures/terracing*

Location	Original slope (°)	Type structure measured[a]	Method/ scale	Net erosion (mean annual, t/ha)	Reference
Bolivia	Up to 40	(a) Stone end-bunds (b) (control)	Field estimates	38–124 (118–164)	Clark *et al.*, 1999
Burundi	26	(a) BTs : stone risers (b) (control)	Not given	5–11 (150)	Roose, 1988
China (Sichuan)	<15 (slope FST bed)	FSTs	Caesium-137-single terrace units	1–6[b]	Quine *et al.*, 1992
East Asia (five countries)	12 (average)	(a) Alley cropping (b) (control)	Not given	15 75	Sajjapongse, 1996
Ethiopia	7 and 15	(a) graded *fanya juu* (b) ('traditional')	USLE-type plot 6m × 30m	94 (252)	Grunder, 1990
India (mid-Himalayas)	Not given	(a) FSBTs (b) FSTs	2.5 m × 2.5 m plots on terrace beds	0.3–0.7 6-64	Sen *et al.* 1997
Indonesia (East Java)	27	BSBTs	USLE-type erosion plot across series of terrace units	28[c]	Rijsdijk and Bruijnzeel, 1990
Indonesia (S Central)	10 and 25	BSBTs	Natural Boundary Erosion Plots	12[d]	Bruijnzeel and Critchley, 1996
Indonesia (West Java)	8 22	BSBTs	Natural Boundary Erosion Plots	115[d,e] 190[d,e]	Purwanto, 1999
Indonesia (West Java)	8–20	BSBTs	Outlet monitoring of multiple BSBTs	95[d,e]	Purwanto, 1999
Jamaica (Smithfield)	Not given	(a) BTs (b) (control)	Not given	17 (133)	Sheng, 1988
Kenya (Mbeere)	6	(a) contour *fanya juu* (b) (control)	Bounded 'USLE-type' plots 45m long	6.0[f] (2.3)	Okoba *et al.*, 1998

(*cont.*)

Table 38.4. (*cont.*)

Location	Original slope (°)	Type structure measured[a]	Method/ scale	Net erosion (mean annual, t/ha)	Reference
Nepal (J. Khola)	Not given	FSTs	USLE-type plot across bed/riser/bed	~40[d]	Nakarmi *et al.*, 1991
Philippines	Not given	(a) SALT (b) (control)	Not given	3.4 (195)	Partap and Watson, 1994
Philippines	Not given	(a) SALT (b) (control)	Not given	[1.3%*x*] [*x*][g]	Palmer, 1996 (quoted in Ya, 1999)
Philippines	11 and 16	(a) Barrier hedge (b) (control)	Not given	20 and 63 4 and 20	Agus *et al.*, 1999
Rwanda	Not given	(a) Small earth BTs (b) (control)	Not given	12.25 (72)	Nyamulinda and Ngiruwonsanga, 1992
Taiwan	9–18	(a) BSBTs (b) level BTs (c) FSTs (d) (control)	Not given	0.4 1.4 7.6 11.6	Liao and Wu, 1987
Uganda (Kabale)	6, 14 and 24	(a) FSBTs (on bed) (b) (bare soil)	USLE-type plots 2 m × 10 m	20[h] (52)[h]	Tukahirwa, J., 1995

[a] BTs, bench terraces; BSBTs, back-sloping bench terraces; FSTs, forward-sloping terraces; SALT, sloping land agricultural technology.
[b] Erosion within upper part of terraces = 50 t/ha (majority redeposited in lower part).
[c] Calculated by extrapolation as USLE plot captured product of lowest terrace only.
[d] Over main rainfall season only.
[e] Exceptionally wet seasons cited as reason for high rates.
[f] Freshly constructed terraces (vulnerable to erosion) cited as partial reason.
[g] Comparative relationship only: SALT results expressed as a % of control (*x* t/ha)
[h] Averages of the three slopes over the two seasons of 1994.
Source: Adapted from Critchley (2000b).

Figure 38.4 Collapsed rainfed terraces (Java, Indonesia).

generally been poor' due to 'high costs, inappropriate design, inadequate drainage and exposure of subsoil'. In Central America, Bunch (1999) dismisses construction of new bench terraces as being 'seldom appropriate anywhere'. This should not be taken to mean, however, that the existing, mainly traditional, terraces that underpin farming on steep slopes of the humid tropics are without value. It merely highlights the fact that these need to be well maintained to be effective and they need to underlie profitable farming systems for this maintenance to be ensured. As an example, on mountain slopes in central Kenya bench terraces are well-maintained by smallholders because they support a profitable coffee crop. This is despite the fact that they were constructed by decree during the colonial regime (K. Mutunga, pers. comm.).

INTERMITTENT TERRACES

Intermittent terraces are a form of bench terrace. Rather than comprising a continuous series of beds separated by risers, they are narrow, back-sloping benches separated by lengths of original ground slope (Sheng, 1989; see profile in Table 38.2). These terraces are designed for very steep slopes (up to 30°) where normal bench terracing would be inappropriate due to potential instability and cost (Figure 38.5).

The alternative name for intermittent terraces, 'orchard terraces', gives a clue to their intended use, namely for fruit or other high-value tree crops. Simpler and cheaper is the construction of small platforms (metaphorically speaking 'chairs' rather than 'benches') for individual trees. While these intermittent terraces do not provide an economic basis for the production of staple food crop production on the steepest slopes, and are not a common sight outside project boundaries, they are at least one way of bringing such slopes into relatively safe, productive use.

FORWARD-SLOPING TERRACES

Forward-sloping terraces can be differentiated from forward-sloping *bench* terraces in that their forward (outward) slope is 3° or greater. Carson (1989) then differentiates between 'gently sloping' (3°–10°) and 'strongly sloping' (>10°) forward-sloping terraces. Generally, forward-sloping terraces are traditional structures, and in many areas (Nepal for example) are much more common than bench terraces under rainfed conditions. A common feature of these terraces is that a distinct fertility gradient develops behind the bund/barrier due to splash, sheet and, especially, tillage erosion within the terrace unit (Turkelboom *et al.*, 1997; Critchley, 2000b; Raussen and Siriri, 2000). Drainage may be lateral but is more usually effected through sequential overtopping of risers, thus 'cascading' downslope. One effect of cascading is to transfer fertility from the top of a sequence of terraces towards the bottom. While land users higher up may suffer, those lower down gain. There are winners and losers in this process when a landscape is viewed in its entirety (van Noordwijk *et al.*, 1998; Ives, 1999; Critchley, 2000b). Although cascading of runoff leads to a failure to maintain fertility *in situ*, it does ensure surface

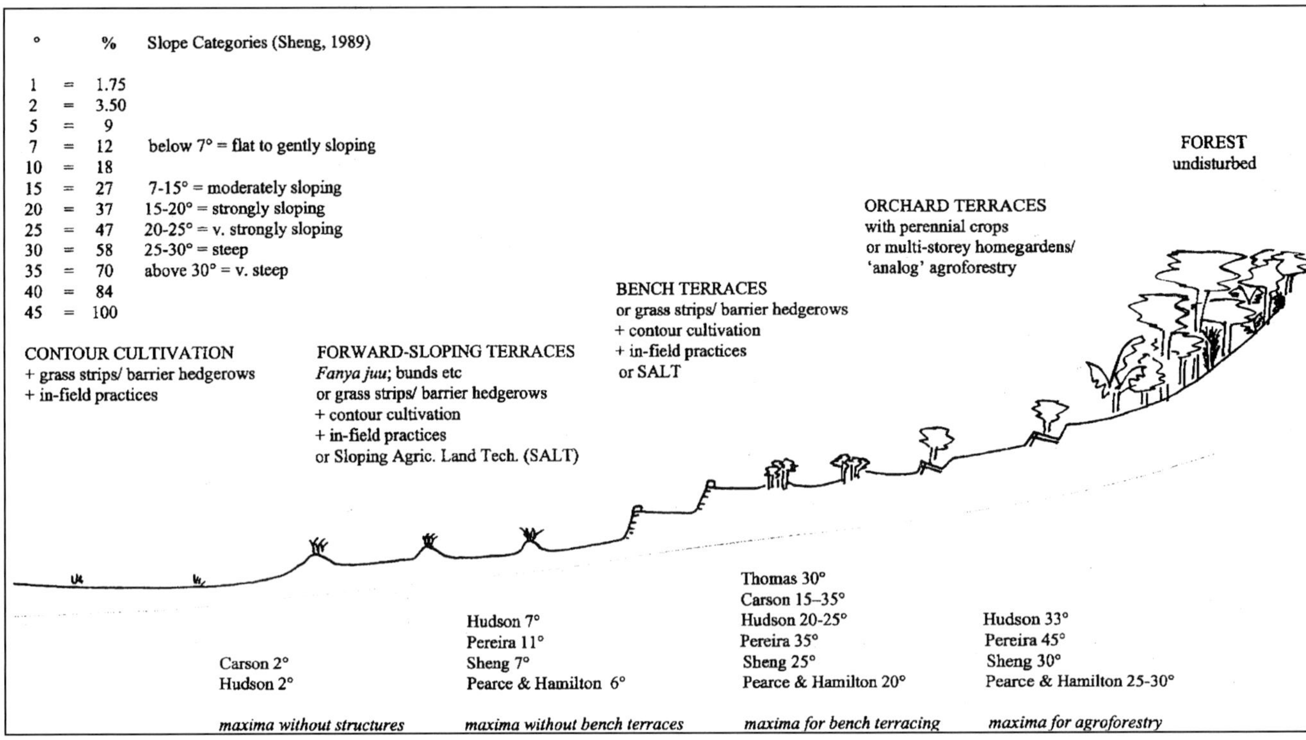

Figure 38.5 Conservation/production systems and slope limits for rainfed steeplands. Adapted from Critchley, 2000b (after Pereira, 1989); other sources: Carson, 1989; Hudson, 1981; Pearce and Hamilton, 1986; Sheng, 1989; Thomas, 1997.

drainage of rainfall that if infiltrated, may destabilise the risers or even whole hillsides. This category of terraces can be further subdivided. Before looking at two specific types (*fanya juu* and bio-terraces), it is useful to reiterate that forward-sloping terraces can actually be transitional stages on the way towards the development of benches, through the natural levelling effects referred to already.

FANYA JUU TERRACE

The *fanya juu* (sometimes called 'converse', 'progressive' or 'natural' terrace) is the mainstay of conservation in the highlands of East Africa, and in Kenya particularly. Technical specifications are given by Thomas, 1997 (see Figure 38.2 for a profile). It is a relatively cheap type of forward-sloping terrace, simple to lay out, and widely applicable in its various forms. The principle of the *fanya juu* is to throw soil upslope from a ditch to create a terrace bank/ bund immediately above. This may be excavation on a one-off basis, or soil more gradually built up on the framework of a vegetative strip or a trash line. There are three obvious advantages of this method of construction compared with the less arduous task of throwing soil downwards to form a bund. The first is that the act of throwing soil upwards counters the direction of erosion, the second is that land levelling is accelerated and the third is that runoff tends to infiltrate more evenly. In humid zones, the *fanya juu* can be graded laterally to allow for discharge of runoff into waterways. It may also require the protection of a cut-off drain above a farm. In drier zones in East Africa (less than 800 mm/annum rainfall) the terraces are often constructed along the contour and the ditch below the bund made large enough to hold as much runoff as possible (Mutunga and Mwarasomba, 1995) Figure 38.6).

BIO-TERRACES (VEGETATIVE BARRIER-BASED TERRACES)

Bio-terraces differ from the foregoing types of support structures in that they are based on narrow vegetative barriers (typically 1 metre wide) which impede soil movement downslope. Terrace beds gradually level off over time. Often these are classified under agronomic (e.g. Sheng, 1989) or vegetative (e.g. WOCAT/FAO, 2000) conservation measures. However, like the *fanya juu*, they can be considered progressive terraces but in this case the barrier is living and semi-permeable. Obviously such barriers are best suited to the more humid areas where there is adequate water supply to sustain the vegetation and moisture competition between barrier and crop is avoided. Barriers generally comprise either (a) multipurpose, leguminous trees, densely planted in double rows and pruned for fodder or mulching (e.g. *Gliricidia sepium, Calliandra calothyrsus*), (b) grass strips, or (c) combinations of the

Figure 38.6 *Fanya juu* terrace in low-rainfall zone (Kenya).

two. Grasses are of various types. They may be relatively resistant to grazing (e.g. *Vetiveria zizanioides*) which is held to be a virtue by proponents (e.g. Grimshaw and Helfer, 1995), but often not by landusers who tend to prefer a barrier species that provides a valued commodity itself, such as fodder (e.g. *Pennisetum purpureum, Setaria anceps* or *S. splendida*). *Pennisetum purpureum* (Napier grass) is particularly valuable as a cut-and-carry grass for stall-fed livestock, but at the cost of being very competitive with the adjacent crop (Thomas, 1993; Agus *et al.*, 1999). *Panicum coloratum* ssp. *makarikariensis* (Makarikari grass: Figure 38.7) and *Paspalum conjugatum* (Bahia grass) are popular grasses for barrier strips in East Africa and the Far East respectively. Both are semi-palatable. In some cases sugar cane *(Saccharum* spp.) or fruit trees may be the preferred barriers because of direct food or cash returns (Hellin and Larrea, 1998). Clearly there is a trade-off between the intrinsic value of the barrier species and its competitive effect on the crop grown in the alleys between.

In terms of technical effectiveness in reducing erosion, Young (1997) summarises results from 14 contour barrier hedgerow experiments in Asia, Africa and Latin America. An unweighted average shows 14 t h^{-1} y h^{-1} (range <1 to 65) contrasting with 96 t h^{-1} y h^{-1} (range 10 to 305) from controls. Other erosion data confirming the impact of barrier hedges are included in Table 38.4. Mercado *et al.* (1999) demonstrate that even 'imperfect' contour hedgerows of natural vegetation ('natural vegetative filter strips', NVS, apparently preferred by farmers to a labour-intensive recommendation of a tree barrier) can still achieve good erosion control, even when spaced at a vertical interval of up to 4 m. However, there is some doubt about the effectiveness of barrier hedgerows or the stability of the terraces developed from them on slopes above 20° (Sheng, 1989; Lal, 1990; Sims *et al.*, 1999). Furthermore, Garrity (1995) points out the implications of such barriers on skewing the availability of soil nutrients and moisture within the alleys, leading to fertility gradients (cf. comment under forward-sloping terraces). Scouring at the top of the crop alleys reduces production and there can also be competition at the barrier/crop interface. The same author considers it 'unrealistic' to expect sustainability of production under hedgerows without external inputs.

Bio-terraces are currently the most consistent recommendation as support structures for cultivation of tropical steeplands – and as the matrix for more complete conservation systems, such as SALT (the 'sloping agricultural land technology' system: Watson and Laquihon, 1985; Partap and Watson, 1994; Palmer *et al.*, 1999). Under SALT systems, hedgerows act as the framework between which the alleys are used for conservation-oriented cropping which incorporates rotation, intercropping, and mulching with the prunings of the hedgerow species (see Figure 38.8).

Three further SALT systems have evolved from the original SALT-I (which combines agricultural crops and hedgerows in

Figure 38.7 Contour barrier of Makarikari grass (*Panicum coloratum*) (Kenya).

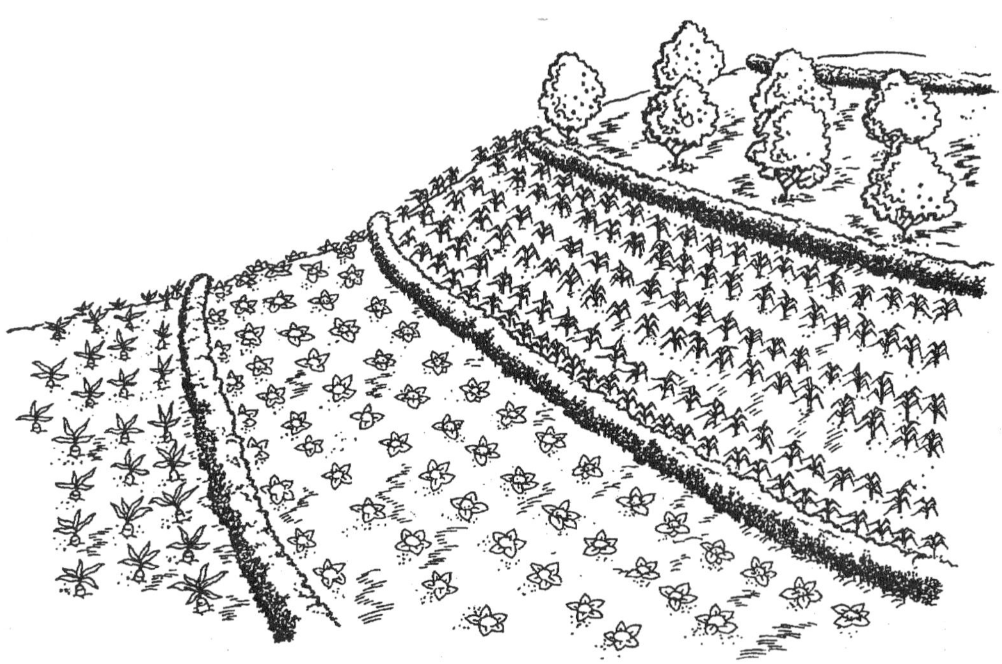

Figure 38.8 Contour barrier hedgerows forming basis for SALT (Sloping Agricultural Land Technology) farming system. Source: Ya, 1999 (reproduced with permission).

the ratio of 75:25). These are SALT-II, which integrates goat husbandry, SALT-III which is a combination of small-scale afforestation (e.g. *Albizia saman, Pterocarpus indicus, Swietenia macrophylla*) with food production, and SALT-IV which focuses on fruit trees as cash crops (Tejwani and Lai, 1992; Palmer *et al.*, 1999).

A generally assumed advantage of bio-terraces is that they can be relatively cheap to establish and are theoretically self-maintaining. Neither of these attributes should be taken for granted. Indeed, as the example of natural vegetated filter strips shows, farmers may be deterred by the (initial and recurrent) costs of recommended systems and often prefer cheap alternatives (Pandey and Lapar, 1998). Regular pruning and weeding requirements can act as a disincentive. Furthermore, where grass splits have to be planted and gapped up to establish a full hedge (e.g. vetiver grass) this can be time-consuming, and delays full conservation impact until two or more seasons have passed. Despite the enthusiastic promotion of vetiver grass by the World Bank (in particular) over the last 15 years (e.g. Grimshaw and Helfer, 1995) and its appeal to decision makers, it has been only sporadically welcomed by farmers (Thomas, 1997; Young, 1997; Hellin, 1999). Similarly, Garrity *et al.* (1998) voice concern that after two decades of recommending barrier hedges, farmer adoption of these technical packages is not widespread. Nevertheless, there may be more hope, as discussed, for farmers' modifications of these barriers where they can choose their preferred species and adapt the spacing to their requirements. Though Palmer *et al.* (1999) consider that SALT is widely adopted in Asia, they admit it is impossible to estimate how many hectares are 'SALT-ed'.

FIELD BOUNDARIES

Field and farm boundaries are rarely considered specifically as forms of conservation support structures, though this is often what they effectively become. Sometimes a field may be clearly orientated along the contour, and here a deliberately formed structure may define the upper and lower boundaries. Commonly, however, the ends of fields (or land holdings) on sloping land evolve into structures passively: they become 'default terraces' (Critchley, 2000b). This occurs due to several factors, which may act in combination. There is the phenomenon of tillage erosion. Hoeing down and away from an upper boundary creates a *de facto* riser. Stones and trash are often piled on field boundaries. Live hedges may be planted, or natural perennial vegetation may simply become established. Erosion control is thus enhanced and a potential source of organic matter established (Kuchelmeister, 1989; Briggs and Twomlow, 1998). Where the boundary marks the limits of a land holding, there is an added incentive to establish – or at least nurture – a protective barrier. This is to prevent loss of fertile soil, though erosion, to a downslope neighbour. In certain situations a field-end bund may be the only support structure farmers will tolerate (e.g. Gujarat, India: personal observation).

In-field practices

The recent, and still growing, emphasis on what happens *between* the support structures – in the cropped alleys, on the terrace beds – rather than the structures themselves has already been mentioned. The alternative names for soil conservation testify to this: 'land husbandry'; 'sustainable soil management' and so forth. Thus conservation moves away from the discipline of mechanical engineering and into the domain of 'production conservation'. Conceptually Bunch (1995) and Shaxson (1999) both envisage a goal of imitating forest floor conditions in fields: in other words recreating *natural* conditions through *nurture*. Gonsalves (1990) lists 21 desirable characteristics of a conservation-rich small farm in the humid tropics. None of these is explicitly structural. They include intensification, diversity of production for home and market, recycling of crop residues, intercropping, judicious trapping and storage of water, an emphasis on perennial crops, and integration of livestock. Metaphorically, scientists have sought to dissect conservation out of agriculture: now conservation is seen more as an organ integral to, and inseparable from, the body of farming.

The following descriptions of in-field practices attempts to highlight their actual, or potential, importance. Too often these (and support structures also) are listed in textbook form, without any clear pointers to those that are merely scientifically desirable and those that are also practical: several effective technologies are apparently more at home on research stations than in farmers' fields. In most categorisations the in-field practices detailed below will be found under headings such as 'agronomic', 'vegetative' and 'soil management' (e.g. Morgan, 1995; Sheng, 1989; Thomas, 1997; WOCAT/FAO, 2000). Here they are grouped together to distinguish them more clearly from support structures.

MIXED CROPPING; INTERCROPPING

Mixed cropping refers to the practice of growing two or more different crops together within the same field at the same time (see Figure 38.9). Intercropping is usually defined as a structured form of mixed cropping: a cereal and a legume may be grown in alternate lines, for example. The agronomic and economic advantages of these forms of crop association are several and can be important (Willey, 1979). Exploitation of different root zones is one, the ability of different species to spread various climatic and pathogenic risks is another. At the same time there are conservation benefits, realised through increased and prolonged ground cover – a fact that has been recognised scientifically for decades (e.g. Masefield, 1949). Relay cropping, meaning the planting of a second crop into a maturing stand of a primary crop, is a specific variation. Mixed cropping is a common traditional practice amongst smallholders in humid areas and – though not appropriate for all crops – is generally good for production and good for conservation too (Morgan, 1995).

Figure 38.9 Mixed cropping: including maize, groundnuts and upland rice (Java, Indonesia).

CROP ROTATION

Crop rotation technically implies a defined, temporal sequence of crops in a given field. Crop *sequencing* implies a more flexible pattern. The importance of crop rotation – as far as conservation is concerned – is that it almost always denotes the incorporation of legumes, and consequently positive implications for soil fertility and sustained soil condition (Webster and Wilson, 1966; Wrigley, 1971). Furthermore, the practice of rotation is usually an indicator of good land husbandry by association. Fallowing (resting land from production for at least one season; usually more) can be considered as a form of rotation and may be necessary to restore fertility where external inputs are not available. The potential role of 'improved fallows' has been explored recently. In this variation, fallowed land is enriched by planting leguminous shrubs or trees, which then enhance the restorative process (Young, 1989, 1997). Garrity (1995) suggests that pruned tree hedgerows (of leguminous species) could be a key component of improved fallow where fertilisers are not available. This implies protection of fallowed land from grazing. Whether smallholders will consider this to be worthwhile remains to be seen.

STRIP CROPPING

Various forms of banding of crops will be touched upon here. First, it should be noted that these bands are normally along the contour (or 'average' contour in undulating landscapes). Indeed contour cultivation in itself is well documented as a positive element in conservation, and acknowledged in the Universal Soil Loss Equation's conservation practice or *P* factor (Wischmeier and Smith, 1978; see also Yu, this volume) (Table 38.5).

Table 38.5. *P factors (estimates/ best approximations of values for application in calculations of the Universal Soil Loss Equation*[a]

	P
'Good' bench terraces	0.04
Contour bunds	0.3
'Poor' bench terraces	0.35
Contour cultivation	0.6–0.9
Up-and-down cultivation	1.0

[a] P = 1 indicates no effect on conservation; P = 0 indicates absolute control of erosion.

Sources: Hamer (1980); Hudson (1981); Morgan (1995).

Strip cropping comprises similar-width bands of different crops grown along the contour. Where one crop (for example, tobacco) is particularly prone to erosion, this can then be balanced by planting a conservation-friendly crop (for example a dense legume)

Figure 38.10 Cover crop (*Centrosema* sp.) below rubber (Java, Indonesia).

in broad bands between the tobacco. Strip cropping, however, is of most importance on large farms, on relatively low slopes (see Figure 38.5). The smaller-scale version of strip cropping, namely alley cropping ('linear hedgerow intercropping'), has proved technically proficient, but unpopular with farmers on flat land because of the high labour requirement involved in planting and pruning. Loss of space is equally important: the area taken up by the hedgerows may approach 20% of the cultivable surface (Young, 1997). *Contour* hedgerow intercropping on hillslopes is an adaptation of the latter that makes more sense to farmers (*see section on bio-terraces*).

COVER CROPS; GREEN MANURES

Cover crops and green manures are (usually) ground-covering legumes, grown primarily for their protective and/or ameliorating effect on the soil. Cover crops are generally left to grow to maturity to make the most of their protective function, while green manures are usually ploughed into the soil early in their cycle to improve soil fertility (Sheng, 1989; Thomas, 1997). Cover crops, such as *Pueraria phaseoloides, Centrosema pubescens, Mucuna pruriens* (velvet bean), *Dolichos lablab* (lablab bean) and *Canavalia ensiformis* (jackbean) are planted amongst perennial crops, or sometimes in the off-season between two annual croppings (Figure 38.10). Green manures, comprising for example *Crotalaria juncea* (sunnhemp), *Lupinus* spp. (lupins) or (once again) *Mucuna pruriens*, may be off-season crops, or relay-planted into stands of other species. While considerable recent research has confirmed the undoubted technical promise of cover crops and green manures for soil amelioration (e.g. Keatinge *et al.*, 1999; Pound, Anderson and Gundel, 1999), traditional use and new adoption of recommendations remains limited, at least amongst smallholders (Barber, 1999). Once again the main inhibiting factor is generally the relatively high supplementary labour input, combined with (often) only modest tangible benefits. In Ethiopia, a group of farmers who were questioned acknowledged the potential benefits to the soil but cited 'competition for land' and 'lack of food value or immediate benefit' as disadvantages (Amede, Belachew and Geta, 2001). Contrastingly, Bunch (1999) believes that these systems are more versatile than composting and points to their spontaneous uptake by around 200 000 farmers in Central America as being proof of their adoption potential.

MULCHING

Mulching is the practice of covering the soil surface with (normally uncomposted) organic material – comprising crop residues, weeds or other organic wastes, and/or grasses especially grown for the purpose. To maintain a mulch (to a depth of typically 10–20 cm) topping-up applications need to be made throughout the

Figure 38.11 Mulching in banana plantation (Uganda).

year. The theoretical and actual benefits of mulching are multiple. These include increasing infiltration, reducing erosion, suppressing evaporation, controlling weeds, keeping soil temperatures low, increasing soil fertility, building up soil organic matter content and encouraging soil fauna (Sheng, 1989; Njoroge, 1994; Thomas, 1997). There can be some technical demerits also. These include depression of topsoil nitrogen content (if the mulch material has a high carbon : nitrogen ratio) and mulch can introduce weed species and attract termites, as well as hosting plant pathogens and providing cover for rats and snakes. However these drawbacks do not explain why mulching is only of limited applicability. The main reason is that mulching requires regular movement of large volumes of plant residues from source to point of use. Therefore high labour and transport costs are incurred. Furthermore, mulching is often source-limited and potentially useful material may have an alternative, more directly productive use as fodder for stall-fed livestock. Another disadvantage of mulching, and one that is recognised increasingly where it is a common practice, e.g. in banana plantations in Uganda (Briggs and Twomlow, 1998; MAAIF, 1999) (Figure 38.11) is that while it helps maintain a favoured planting zone, it exacerbates the drain of nutrients from the source location. While such 'niche' farming (favouring one part of the farm at the expense of another; Hilhorst and Muchena, 2000) is a valuable subsistence strategy where inputs are limiting, it cannot sustain the fertility of the whole system without mineral fertilisers. Nevertheless, where a sustainable source of material is available, mulching can be extremely valuable, especially under high value crops.

MANURES; COMPOSTS

The virtues of manures and composts have been extolled as long as agriculture has been written about. Youatt (1852) dedicates 85 pages to the subject and points out that there are repeated references to the topic in the writings of, *inter alia*, Cato, Pliny and Virgil. Manures and composts may constitute the only supplementary plant nutrient supplies where mineral fertiliser is unavailable or unaffordable. Simultaneously they add valuable organic matter to the soil. Both (often in combination) are widely practised, have been exhaustively studied and described about (e.g. Dalzell *et al.*, 1987; Defoer, *et al.*, 2000). Improved systems continue to be promoted. This is important. There is a vital difference in end-product between good and poor practice: composting and manure preparation need care to avoid losses of nutrients through volatilisation and leaching. There are also possibilities of growing of specific plants to fortify composts: for example *Tithonia diversifolia* (Mexican sunflower) is being promoted in East Africa because of its high phosphorus (and nitrogen) content. Manures

Figure 38.12 Collecting oak leaves from the forest for fodder (India).

Figure 38.13 Fodder grass on terrace risers (Java, Indonesia).

and composts are particularly suitable for high value crops, close to the household.

Nevertheless, as is the case with mulching, when viewed in terms of a farming system as a whole, their value is necessarily limited. Reliance on supplies of decomposed organic matter from within are not enough to sustain a farming system when there are external and non-replaced losses in terms of sales, human wastes and burning (Smaling, 1993). Furthermore, as with mulch, manures (because of grazing or cut-and-carry from afar to zero-grazing units) and composts (because of collection of materials from a distance) generally imply concentration of organic matter at one point to the potential detriment of another. An example of this is the exploitation of oak and mixed forests for fodder and animal bedding – as well as firewood – in the Central Himalaya of northern India leading inevitably to their degradation (Ralhan, Negi and Singh, 1991) (Figure 38.12). However this argument can be turned on its head and work in quite a different way in other situations. Where productive stall-feeding/zero-grazing livestock systems are introduced (there are several projects dedicated to zero-grazed dairy cows in the highlands of East Africa for example), manure or manure-enriched compost is a valuable by-product. Such intensive livestock systems can help conservation through the planting of supplementary fodder grasses and legumes, which in turn can be strategically located on conservation structures in order to stabilise them (Figure 38.13). This then can stimulate a virtuous cycle of conservation in tropical steeplands. In one recent study, stall-feeding of smallstock has been demonstrated to be an essential component of intensive smallholder farming in parts of Java, where sheep are deliberately fed excess fodder to maximise manure production (Tanner *et al.*, 2001).

TILLAGE

Injudicious tillage, and especially primary cultivation, can be a major cause of soil deterioration and erosion through surface compaction, the formation of hardpans, destruction of soil structure and through accelerating the loss of organic matter. Conservation tillage is a generic term which implies reduced primary (and secondary) cultivation. Its worthy objectives are to guard against compaction, save fuel, increase moisture retention and conserve organic matter (Steiner, 1998; Biamah, Rockström and Okwach, 2000). Planting directly into crop residues without digging or ploughing is referred to as zero-tillage, and is normally associated with large scale mechanised arable farming. Under small scale farming conditions in the humid tropics, this practice is not widespread, though it characterises shifting cultivation regimes where burning of residues often precedes planting. Zero-tillage is found locally in Central America where mulching is an integral part of the system (Barber, 1999; Bunch, 2000) but this is an exception. Recent trials in Ethiopia found that small scale farmers

Figure 38.14 Home garden (Java, Indonesia).

acknowledged the reduction in labour associated with minimum tillage, but they also recognised the problems of increased weed infestation and poor crop establishment (Amede *et al*, 2001). Tillage traditions are relatively resistant to modernisation: after all it is a fundamental cultural change (in both senses of the term) for farmers to alter their ploughing and hoeing habits. At the other extreme of the cultivation spectrum, 'double digging' by hand is promoted by organisations that support organic farming methods. The concept here is to thoroughly break up compacted subsoil, thereby improving water holding capacity and root penetration, while simultaneously manuring the topsoil (Njoroge, 1994). Double digging is primarily aimed at intensive spot production of vegetables, as the very high labour requirement can only be justified by correspondingly high value outputs.

DISPERSED TREES; MULTI-STOREY CROPPING

In-field mixtures of trees and crops – ranging from a single variety of tree dispersed in a pure crop stand to complex multi-strata, multi-species 'home gardens' – are common features of traditional agriculture in the tropics (Figure 38.14). Indeed the relatively new science of agroforestry has its roots in the appreciation of such systems. Young (1997), writing about 'home' and 'forest' gardens, notes that their effectiveness in controlling erosion is clear (see also Wallace *et al.*, this volume). Smallholder cacao production in West Africa – where shade trees are used – qualifies as an example of a sustainable forest replacement system (Rice and Greenburg, 2000). Indeed these are among the best types of production/conservation combinations for the steepest slopes (up to 30° or more) (Figure 38.5), being mainly perennial and keeping the ground covered. The key to the conservation impact is the floor of the home garden, with accumulated litter and build up of organic matter. This, as already noted, is effectively 'recreating forest floor conditions' – thus mimicking nature. These systems are irregular and non-linear: their constitution and analysis is correspondingly complex. It is therefore difficult, and probably unwarranted, to make specific recommendations. The best approach may be to isolate and characterise the most promising generic groups. This is an ideal example of where land users can be involved in the processes, both helping in identification and characterisation, and describing these best-bet options to their peers. Such an approach has been developed in Sri Lanka (Senanayake, 2000) where the improved home garden system is an example of it, sometimes called 'analog forestry'. In the Embu/Meru Districts close to Mount Kenya, the multipurpose *Grevillea robusta* has been thoroughly integrated into the tea/coffee/maize/dairy zone within the last two decades – independent of project influence. This demonstrates that greater population densities *can* lead to a significant increase in tree planting (WOCAT/FAO, 2000) (Figure 38.15).

Figure 38.15 *Grevillea robusta* planted and managed by farmers (Kenya).

Yet another example is the essentially indigenous 'Quezungual system' from Honduras, which is based on pollarding naturally-regenerating trees within fields. There is growing adoption of this system both through informal farmer-to-farmer contact and by project promotion. Small-scale farmers appreciate the various direct benefits from tree products, and the improved soil condition which allows longer periods of cultivation between fallows (Hellin, Welchez and Cherret, 1999).

CHALLENGES AND CONCERNS: MYTHS AND REALITIES

Introduction

Major changes are taking place in the humid tropics and most of these are related to growing populations. Both needs and concerns are altering as a result. At the same time conventional perspectives on land degradation and its remedies are under challenge, and new approaches are evolving. Technical summaries and recommendations remain useful but can constitute merely a specialist's 'wish list' unless we see the measures from various points of view in a changing environment. This section looks briefly at some of the external factors that are on the move, and then at five priorities for a new approach, with the implicit question: how do these influence farmers' choice of technique or specialists' recommendations?

Changing external factors

INCREASE IN POPULATION

Population growth continues in the humid tropics with its inevitable consequence of increasing pressure on the land (as outlined in the rates of forest conversion, Drigo, this volume). Holdings are reduced in size, and people are squeezed onto ever-steeper slopes. There can be positive and negative implications (Blaikie and Brookfield, 1987). The mainstream view is that population increase is detrimental to the environment, because the equilibrium between people and natural resources is disturbed. If the natural resource base is inadequate to support the growth in population, exploitation or 'mining' of resources will inevitably occur. However, although cultivating steeper land (for example) brings attendant problems of how to protect and produce simultaneously on acute slopes, the very limitation of land itself can act as a stimulus to conserve and to intensify production. As already noted, people may be propelled towards stall-fed dairy production, for example, and a 'virtuous cycle' of conservation initiated. Thus population pressure can act in various ways, and not all of them are bad for conservation. The most often-quoted positive example comes from Machakos District in Eastern Kenya where the book *More People, Less Erosion* (Tiffen, Mortimore and Gichuki, 1994) documents the simultaneous growth in numbers of people and environmental recovery over the last century, confounding the Malthusian pessimists who assume that population growth inevitably leads to degradation. Nevertheless, a recent follow-up study in Africa found few such examples of degradation in reverse, and noted that good soil and water conservation practices were seldom found in the absence of high value crops (Boyd and Slaymaker, 2000). With respect to the process of transition from natural forest to domesticated forest landscapes in the humid tropics, Henkemans, Persoon and Wiersum (2000) point to the major impact of growing populations on the forest fringe. The same authors, however, conclude that the prevailing view of pioneer shifting cultivators being main culprits of environmental degradation is biased. Clearly there are two sides to the population/environment debate.

PLOT SIZE

Linked to growth in population is size of plot or field. Where continuous hereditary subdivision takes place, generation after generation, this can have a strong influence on various factors, including the very viability of holdings. In the Central Himalaya of northern India, there are examples of land being abandoned for this very reason (Negi and Joshi, 1997). But, as with growing population (and related to it) the impact on conservation, once again, is not invariably unfavourable. It is true that the smaller the plot, the less likely a farmer will be to approve methods that *cost* him/her land. If conservation structures consume precious area, then these will simply not be adopted unless they are productive and remunerative in themselves. On the other hand, smaller lots often mean more hedges and boundaries, and these themselves can act as conservation measures (Kuchelmeister, 1989). However, smaller plots (and more people) mean more footpaths *between* plots, more walking, and thus the potential for increased runoff and track erosion from the compacted surfaces.

SLOPES AND LEGAL LIMITS

The acknowledgement that cultivation was openly flouting legal slope limits (often unnecessarily prohibitively in any case) came to the fore in 1987 at an international symposium on steeplands held in Puerto Rico (Moldenhauer and Hudson, 1988). This was a benchmark conference that recognised both the increasing domination of the 'land husbandry' approach over field engineering solutions, and the futility of unrealistic conservation targets. Hudson's keynote address (*Tilting at windmills or fighting real battles?*) was a plea for realism to prevail. While not condoning farming on ever increasing slopes, it was a call to limit damage through viewing the farmer as a potential ally rather than an implacable enemy (Hudson, 1988). Accepting that there must be rules and regulations to prevent desecration of landscapes, more must be done to understand and help farmers on the steepest slopes. It may be a truism, but if the land is lost, so are their livelihoods.

LAND TENURE

It is now widely accepted that security of tenure is a prerequisite for investment in land. Investment in this sense includes conservation systems. However, security of tenure does not necessarily have to imply legal ownership. Customary systems or secure tenancy can provide sufficient security to allow investment without unease about the future (Wachter, 1996; Purwanto, 1999). Problems exist where tenancy agreements are fragile and where new land is occupied illegally and/or temporarily. These are situations where Governments can act to improve the frame conditions for conservation. An example of where this whole complex issue of the interface between land rights, production and conservation is currently being taken very seriously is in the newly democratic South Africa, where the historical distortions of apartheid are being untangled (Turner, 1998).

CHANGING INTERNATIONAL PRIORITIES

Approaches – and even technologies – are not merely determined by internal needs. There are international forces at play and it would be naïve to ignore the impact of international opinion and priorities. For example, there is a growing international disenchantment with the poor impact of soil conservation programmes, and funds are less freely dispensed than before. Further, adding more weight to this argument, a small but influential school of thought holds that the environmental problem may not be as bad as previously thought. According to this view, claims for rampant erosion, degradation and deforestation have been routinely and deliberately exaggerated. This is often achieved through 'development narratives' that seek to justify inappropriate project interventions by building up misleading and simplistic pictures of problems and solutions, based on received wisdom about the environment (Roe, 1991; Leach and Mearns, 1996; Fairhead and Leach, 1998). Simultaneously, the international conventions on climate change, biodiversity and desertification are having an increasing influence on donor funding priorities. Where conservation systems converge with the targets of the conventions there is greater hope for support. Examples are water harvesting in dry regions (*combating desertification*); multi-storey agroforestry systems (*promoting biodiversity*); and both tree planting and conservation tillage (*assisting carbon sequestration*).

Changing approaches: five new priorities

Five priority areas are suggested here to help overcome some inherent weaknesses of soil conservation/soil and water management programmes, and to address a number of the changing external factors mentioned in the previous section.

ADOPTION: SENSE AND SUSTAINABILITY

It is relatively easy to review various soil conservation manuals and to summarise the salient technical features and potential impacts of the main systems described. That has not been the main purpose of this chapter. What is more important is to look for the most appropriate options in the dynamic situation of the humid tropics, where steep slopes are *de facto* no longer out of bounds to cultivation in many countries, and where adoption rates of promoted technologies have so often been disappointing (Enters, 1996; Young, 1997). These solutions may not always be the best alternatives technically speaking. A technology cannot be judged as being 'better' than another simply on the grounds of its ability to stop erosion. Garrity *et al.* (1998) take a strong and provocative stance on this issue:

> 'Identifying practices that control erosion is not solving the problem. The technical problem is solved only when a soil erosion [control] practice is sufficiently cheap and requires so little labour and management in addition to being effective, that it is likely to increase short-term farm profitability. Spontaneous adoption is the real test of a technical innovation in conservation farming'

Much more needs to be known about what levels of adoption have actually taken place for the major techniques, what are the impacts of various systems in both technical and socio-economic terms, and what factors influence farmers' choice. This process of accountability and assessing impact is beginning (e.g. WOCAT/FAO, 2000), but there is still a long route to travel before all stakeholders – funders, development workers, researchers – are pulling in the same direction. Currently, the issue is clouded by the rhetoric of proponents of particular technologies, and by the intransigence of many conservation specialists and policy makers.

MONITORING AND EVALUATION: ASSESSING REAL IMPACT

Stocking (1988) highlights the poor standard of cost-benefit analysis of soil conservation measures. Critchley, Versfeld and Mollel (1998) point out the generally abysmal performance of monitoring and evaluation (M&E) in natural resource management programmes. Guijt (1998) talks of the 'honeymoon period' of participatory projects coming to a close, and thus the urgent need for improved M&E to gauge impact. Simultaneously, and related to the impact of conservation, is the need for continued underlying research on sources and sinks of sediment in tropical agricultural landscapes, and the separation of 'natural' and anthropogenic effects. Doolette and Magrath (1990), reviewing catchment management programmes in Asia, lament the inconsistency of monitoring methods, which renders comparisons difficult or impossible. So, factual evidence of impact and achievement under field conditions is urgently, and increasingly, required. This should cover not merely technical validation but also socio-economic and livelihood parameters. Over and above the tangible parameters of erosion control, cost-benefit and impact on soil and crops, there is the imperative to evaluate technologies and systems *together with land users*. Adoption, as has been noted, is the acid test of acceptance and durability. But where are the numbers and rankings? They seldom exist, except on local, project scale. One reason is that there has been little commitment of national conservation departments or large scale donors to follow-up impact further than direct implementational phases. They may have been deterred by the cost of such exercises, or perhaps even apprehensive about the potential findings. While WOCAT offers a common formula for recording data, there still needs to be a more concerted effort to collect that basic information, without which it is impossible to make informed judgements on current, and potential, impact.

INCENTIVES: HARMLESS STIMULANTS OR DANGEROUS ADDICTIVES?

There is only a limited amount that donor or Government-sponsored programmes can actually implement on behalf of the beneficiaries, or induce by means of incentives. While few would still agree with the first of these approaches, the second, namely the use of direct, material incentives in certain circumstances, should not always be discounted as long as these are temporary and catalytic in nature and do not act as bribes (see Sanders *et al.* (1999) for guidelines). There is a fundamental difference between overcoming a bottleneck and thus stimulating action with token support (e.g. supply of seedlings to establish barrier hedges) and on the other hand distorting priorities with wages (e.g. paying labourers to create prestige demonstration sites). In the long run, as noted already, spontaneous adoption is the only route to widespread and durable impact. If we accept that there are plenty of potentially useful conservation/production measures that have not yet been adopted by farmers, and cost is often the inhibiting factor, then a more important role for governments and donors may be through *indirect* incentives (Sanders and Cahill, 1999). It may be that the provision of enabling legislation, marketing channels and associated infrastructure brings more expensive measures into the realms of affordability and desirability.

LOCAL INNOVATION: A DYNAMIC OVERLOOKED

To some, the topic of indigenous knowledge in the developing tropics has lost its freshness of appeal. This is partly the fault of a fringe of 'indigenous knowledge fanatics' who have driven the theory forward without adequate tangible evidence of its application. This is perhaps less true of soil and water conservation, which remains largely an unexplored subject in this regard. Nevertheless, as noted in the introduction, there have been significant developments in the last decade. Looking closely at tradition in this field, there are as many instances of deteriorating traditions as of expansion (Critchley *et al.*, 1994). Looking more carefully still, the area of real importance is the *dynamic* of local innovation – where traditions *and* introduced technologies are being modified and moulded to suit changing situations. Within any population of land users, however resource-poor, there are certain individuals who are testing and trying technologies, using their initiative to adapt systems to their own situations (Critchley, 2000a). These are usually systems where conservation and production are intimately intertwined. Examples from (semi-arid) East Africa include a variety of pitting methods for concentration of nutrients and water around plants, and both vegetative and structural ways of turning gullies into fertile ribbons of land (Critchley *et al.*, 1999). These creative land users do not recognise the scientist's artificial boundaries between conservation and production – conceptual boundaries which are at last being broken down. Land user innovation is an area where specialists can spend time and energy profitably, working with local sources of inspiration.

INSTITUTIONALISATION: INFUSING SUCCESS INTO THE SYSTEM

The 'enclave' project approach to soil conservation with its narrow geographical focus, alienation from government, and with limited duration is becoming increasingly discredited. Success stories in these terms commonly prove to be transitory and isolated. Often no institutional memory is left behind, and 'achievements' on the ground soon wither when starved of project sustenance. It is surely correct that there is a move towards a more open-ended 'process approach' linked directly or indirectly to government, with an eye open continuously for the potential of scaling up success, and building capacity for initiatives to be carried forward and outward. Of course, even where institutionalisation is a stated priority of a programme, there remains a danger that there may be too much of an ideological hard-sell and the products (techniques or approaches) may not actually warrant scaling up. But, pleasingly, there is a new focus on this topic. Institutionalisation, scaling-up and promoting self-sustaining processes have been the subject of recent attention in this field – for example in India (Turton, Warner and Groom, 1998), Africa (e.g. Hagmann *et al.*, 1997) and Latin America (Bunch and Lopez, 1995).

CONCLUDING COMMENTS

This chapter has looked at specific challenges to improved land management in the humid tropical steeplands, and has identified some specific issues that need to be addressed. It has also set out the main technical options for soil and water management in these areas. Only through the use of these basic technologies, or derivations and combinations of them, will conservation be achieved. It is an illusion to expect any new technical panaceas, whether structural, vegetative or agronomic. We already have the basic, best-bet technological choices. Wholesale improvements to these are unlikely, though modifications by farmer innovation and adaptation will undoubtedly add value to these options and should be sought out actively. The main challenge is to look for the best combination of effectiveness and acceptability in a changing environment. Specialists need to be more attuned to what land users desire and appreciate. Often (but not always) these will be low-investment, low-maintenance and production-related systems with rapid payback. As a broad rule, only the introduction of high value crops will stimulate investment in more expensive conservation systems. The conservation ideal of recreating forest floor conditions is an appealing new goal but will only be obtained under certain conditions. Different 'solutions' will be better suited to, or better liked in, different locations. A further challenge is to step up efforts to monitor technical effectiveness using consistent methodologies, to evaluate systems with farmers, and to record evidence of adoption. These elements have been sorely underplayed: far more is known about technical design and research station impact upon erosion than about real life performance amongst land users. A final, hopeful conclusion is that it is in everybody's interests to sustain productivity in the steeplands. It is a common battle, which to be effectively fought must involve a joint alliance of decision-makers, scientists and land users.

References

Agus, F., Garrity, D. P., Cassel, D. K. and Mercado, A. (1999). Grain crop response to contour hedgerow systems on sloping oxisols. *Agroforestry Systems* 42, 107–120.

Amede, T., Belachew, T. and Geta, E. (2001). *Reversing the Degradation of Arable Land in the Ethiopian Highlands*. Managing Africa's Soils No. 23. London: International Institute for Environment and Development (IIED).

Anderson, D. (1984). Depression, dust bowl, demography, and drought: the colonial state and soil conservation in East Africa during the 1930s. *Africa Affairs* 83, 321–343.

Barber, R. (1999). *Land and Crop Management in the Hilly Terrains of Central America*. Soils Bulletin No. 76. Rome: Food and Agriculture Organisation.

Bernsten, R. and Sinaga, R. (1983). *'Economics'*. Technical Appendix VI, Composite report of the watershed assessment team. Jakarta: Government of Indonesia/USAID.

Biamah, E. K., Rockström, J. and Okwach, G. E. (2000). *Conservation Tillage for Dryland Farming: Technical Options and Experiences in Eastern and Southern Africa*. Workshop Report No. 3. Nairobi, London: Regional Land Management Unit (RELMA).

Blaikie, P. and Brookfield, H. (1987). *Land Degradation and Society*. London and New York: Routledge.

Boyd, C. and Slaymaker, T. (2000). Re-examining the 'More People Less Erosion' hypothesis: special case or wider trend? Natural Resources Perspectives. No. 63. London: Overseas Development Institute (ODI).

Briggs, L. and Twomlow, S. J. (1998). Organic matter flows within a smallholder farming system of the East African highlands. In *Modern Methods from Traditional Soil and Water Conservation Technologies*. eds. S. R. Briggs, J. Ellis-Jones, and S. J. Twomlow, pp. 98–113. Report IDG/98/10. Silsoe: Silsoe Research Institute.

Bruijnzeel, L. A. and Critchley, W. R. S. (1996). A new approach towards the quantification of runoff and eroded sediment from bench terraces in humid tropical steeplands and its application in south-Central Java, Indonesia. In *Advances in Hillslope Processes*, Vol. 2. eds. M. G. Anderson and S. M. Brooks, pp. 921–937. Chichester: Wiley.

Bunch, R. (1995). An odyssey of discovery. Principles of agriculture for the humid tropics. *ILEIA* 11, 18–19.

Bunch, R. (1999). Learning how to 'make the soil grow': three case studies on soil recuperation, adoption and adaptation from Honduras and Guatemala.

In *Issues and Options in the Design of Soil and Water Conservation Projects*. eds. M. McDonald and K. Brown, pp. 23–38. University of North Wales.

Bunch, R. and Lopez, G. (1995). *Soil Recuperation in Central America: Sustaining Innovation after Intervention*. London: International Institute for Environment and Development (IIED).

Carson, B. (1989). *Soil Conservation Strategies for Upland Areas of Indonesia*. Hawaii: Occasional Paper 9 of the East-West Environment and Policy Institute. East-West Centre.

Carson, B. (1992). *The Land, The Farmer and the Future. A Soil Fertility Management Strategy for Nepal*. Kathmandu, Nepal: Occasional Paper 21, International Centre for Integrated Mountain Development (ICIMOD).

Clark, R., Duron, G., Quispe, G. and Stocking, M. (1999). Boundary bunds or piles of stones? Using farmers practices in Bolivia to aid soil conservation. *Mountain Research and Development* 19, 235–240.

Craswell, E. T. (1998). Foreword. In *Soil Erosion at Multiple Scales*. eds. F. W. T. Penning de Vries, F. Agus and J. Kerr. Wallingford: CAB International.

Critchley, W. R. S. (2000a). Inquiry, initiative and inventiveness. Farmer innovators in East Africa. *Physics and Chemistry of the Earth* 25, 285–288.

Critchley, W. R. S. (2000b). *Groundtruthing. New Perspectives on Soil Erosion and Conservation in the Tropics*. PhD Thesis, Vrije Universiteit, Amsterdam.

Critchley, W. R. S. and Bruijnzeel, L. A. (1995). Terrace risers: erosion control or sediment source? In *Sustainable Reconstruction of Highland and Headwater Regions*. eds. R. B. Singh and M. J. Haigh, pp. 529–541. New Delhi: Oxford and IBH Publishing Co.

Critchley, W. R. S., Reij, C. and Willcocks, T. J. (1994). Indigenous soil and water conservation: a review of the state of knowledge and prospects for building on traditions. *Land Degradation and Rehabilitation* 5, 293–314.

Critchley, W. R. S., Versfeld, D. and Mollel, N. (1998) (eds.). *Sustainable Land Management. Some Signposts for South Africa*. Pietersberg, South Africa: University of the North Press.

Critchley, W. R. S., Cooke, R., Jallow, T., Lafleur, S., Laman, M., Njoroge, J., Nyagah, V. and Saint-Firmin, E. (1999) (eds.). *Promoting Farmer Innovation*. Nairobi: Workshop Report No. 2, Regional Land Management Unit (RELMA).

Dalzell, H. W., Riddlestone, A. J., Gray, K. R. and Thurairajan, K. (1987). *Soil Management. Compost Production and Use in Tropical and Subtropical Environments*. Soils Bulletin No. 56. Rome: Food and Agriculture Organisation.

Defoer, T. Budelman, A., Toulmin, C. and Carter, S. E. (2000). *Building Common Knowledge. Participatory Learning and Action Research (Part 1)*. In *Managing Soil Fertility in the Tropics. A Resource Guide for Participatory Learning and Action Research*. eds. T. Defoer and A. Budelman. Amsterdam, The Netherlands: Royal Tropical Institute.

Doolette, J. B. and Magrath, W. B. (1990) (eds.). *Watershed Development in Asia. Strategies and Technologies*. Technical Paper 127. Washington, D.C.: The World Bank.

Douglas, M. (1994). *Sustainable Use of Agricultural Soils. A Review of the Prerequisites for Success or Failure*. Switzerland: Group for Development and Environment, University of Berne.

Enters, T. (1996). The token line: adoption and non-adoption of soil conservation practices in the highlands of northern Thailand. In *Soil Conservation Extension. From Concepts to Adoption*. eds. S. Sombatpanit, M. A. Zobisch, D. W. Sanders and M. G. Cook, pp. 417–427. Bangkok: Soil and Water Conservation Society of Thailand.

Euphrat, F. D. (1987). *A Delicate Imbalance. Erosion and Soil Conservation in Pipal Chaur Watershed, Kabhre Palanchok District, Nepal*. MSc Thesis, University of California, Berkeley.

Fairhead, J. and Leach, M. (1998). *Reframing Deforestation. Global Analysis and Local Realities: Studies in West Africa*. London and New York: Routledge.

Garrity, D. P. (1995). Improved agroforestry techniques for conservation farming: pathways towards sustainability. In *International Workshop on Conservation Farming for Sloping Uplands in South East Asia: Challenges, Opportunities and Prospects*. pp. 145–168. Bangkok: International Board for Soil Research and Management.

Garrity, D. P., Mercado, A. Jr. and Stark, M. (1998). Building the smallholder into successful natural resource management at the watershed scale. In *Soil Erosion at Multiple Scales*. eds. F. W. T. Penning de Vries, F. Agus and J. Kerr, pp. 73–82. Wallingford: CAB International.

Gonsalves, J. F. (1990). A listing of the characteristics of an ideal regenerative agricultural system for small farmers in the humid tropics. In *Resource Book on Sustainable Agriculture for the Uplands*. eds. L. Capistrano, J. Durno and I. Moeliono, pp. 159–162. The Philippines: International Institute of Rural Reconstruction (IIRR).

de Graaff, J. and Wiersum, K. F. (1992). Rethinking erosion on Java: a reaction. *Netherlands Journal of Agricultural Science* 40, 373–379.

Grimshaw, R. G. and Helfer, L. (1995). *Vetiver Grass for Soil and Water Conservation, Land Rehabilitation, and Embankment Stabilization*. Technical Paper 273. Washington D.C.: The World Bank.

Grunder, M. (1990). Soil conservation research in Ethiopia. In *African Highlands and Mountains*. eds. B. Messerli and H. Hurni, pp. 87–105. Missouri: African Mountains Association, Walsworth Press.

Guijt, I. (1998). *Participatory Monitoring and Impact Assessment of Sustainable Agricultural Initiatives*. SARL Discussion Paper 1. London: International Institute for Environment and Development (IIED).

Hagmann, J., Chuma, E., Connolly, M. and Murwira, K. (1997). *Propelling Change From the Bottom-Up: Institutional Reform in Zimbabwe*. Gatekeeper Series 71. London: International Institute for Environment and Development (IIED).

Hamer, W. I. (1980). *Soil Conservation Consultant's Reports*. Centre for Soil Research, Bogor, Indonesia. Ministry of Agriculture, Indonesia and Food and Agriculture Organisation.

Hellin, J. (1999). Soil and Water conservation in Honduras: Addressing Whose Reality? AGREN Newsletter No. 40, pp. 12–16. London: Overseas Development Institute (ODI).

Hellin, J. and Larrea, S. (1998). Ecological and Socio-Economic Reasons for Adoption and Adaptation of Live Barriers in Guinope, Honduras. *Advances in GeoEcology* 31, 1383–1388.

Hellin, J., Welchez, L. A. and Cherrett, I. (1999). The Quezungual System: an indigenous agroforestry system from western Honduras. *Agroforestry Systems* 46, 229–237.

Henkemans, A. B., Persoon, G. A. and Wiersum, K. F. (2000). Landscape transformations of pioneer shifting cultivators at the forest fringe. In *Tropical Forest Resource Dynamics and Conservation: from Local to Global Issues*. ed. K. F. Wiersum, pp. 53–69. Netherlands: Tropical Resource Management Papers, Wageningen Agricultural University.

Hilhorst, T. and Muchena, F. (2000) (eds.). *Nutrients on the Move. Soil Fertility Dynamics in African Farming Systems*. London: International Institute for Environment and Development (IIED).

Hudson, N. W. (1981). *Soil Conservation*. London: Batsford.

Hudson, N. W. (1988). Tilting at windmills or fighting real battles? In *Conservation Farming on Steep Lands*. eds. W. C. Moldenhauer and N. W. Hudson, pp. 3–8. Ankeny. Iowa: World Association of Soil and Water Conservation.

Hudson, N. W. (1991). *A Study of the Reasons for the Success or Failure of Soil Conservation Projects*. Soils Bulletin No. 64. Rome: Food and Agriculture Organisation.

Hudson, N. W. (1992). *Land Husbandry*. London: Batsford.

Hurni, H. (*'with the assistance of an international group of contributors'*) (1996). *Precious Earth. From Soil and Water Conservation to Sustainable Land Management*. International Soil Conservation Organisation and Centre for Development and Environment, University of Berne.

IFAD. (1992). *Soil and Water Conservation in Sub-Saharan Africa. Towards Sustainable Production by the Rural Poor*. Report prepared by Centre for Development Cooperation Services, Vrije Universiteit Amsterdam. Rome: International Fund for Agricultural Development (IFAD).

Ives, J. D. (1999). The use of hillside environments for land husbandry: personal reflections. *Mountain Research and Development* 19, 173–177.

Johnson, K., Olson, E. A. and Manadhar, S. (1982). Environmental knowledge and response to natural hazards in mountainous Nepal. *Mountain Research and Development* 2, 175–188.

Keatinge, J. D. H., Wheeler, T. R., Subedi, M., Shah, P. B., Ellis, R. H. and Summerfield, R. J. (1999). Annual legume species as green manure/cover crops in low-input farming systems of Nepal. *Mountain Research and Development* 19, 325–332.

Kerr, J. and Sanghi, N. K. (1992). *Indigenous Soil and Water Conservation in India's Semi-Arid Tropics*. Gatekeeper Series No. 34. London: International Institute for Environment and Development (IIED).

Kuchelmeister, G. (1989). *Hedges for Resource-poor Land Users in Developing Countries*. Eschborn, Germany: Deutsche Gesellschaft für Technische Zusammenarbeit (GTZ).

Lal, R. (1990). *Soil Erosion in the Tropics. Principles and Management*. New York: McGraw-Hill.

Leach, M. and Mearns, R. (1996) (eds.). *The Lie of the Land. Challenging Received Wisdom on the African Environment*. Oxford: International African Institute, Heinemann/ James Curry.

Liao, M-C. and Wu, W-L. (1987). *Soil Conservation on Steeplands in Taiwan*. Taipei: Chinese Soil and Water Conservation Society.

Masefield, G. B. (1949; later edition, 1970). *A Handbook of Tropical Agriculture*. Oxford: Clarendon Press.

McDonald, M. and Brown, K. (1999) (eds.). *Issues and Options in the Design of Soil and Water Conservation Projects*. Bangor: University of North Wales.

Mercado, A. R., Garrity, D. P., Sanchez, N. and Laput, L. (1999). What is the optimum spacing of vegetative strips in tropical smallholder conservation systems? *Paper presented to the 15th Annual Scientific Conference of the Federation of Crop Science Societies of the Philippines. 10–15 April, 1999. Family Country Home, General Santos City, Philippines*.

MAAIF (Ministry of Agriculture, Animal Industry and Fisheries). (1999). *Uganda. Soil Fertility Initiative*. Entebbe, Uganda.

Morgan, R. P. C. (1986, 1995). *Soil Erosion and Conservation*. Harlow: Longman.

Moldenhauer, W. C. and Hudson, N. W. (1988) (eds.). *Conservation Farming on Steep Lands*. Ankeny, Iowa: World Association of Soil and Water Conservation.

Mutunga, K. and Mwarasomba, L. I. (1995). *Report on SWC technology development in ASAL – a survey of positive experience*. Nairobi, Kenya: Ministry of Agriculture.

Nakarmi, G., Pathak, A. R. and Schreier, H. (1991). The hydrometric, sediment and erosion monitoring program. In *Soil Fertility and Erosion Issues in the Middle Mountains of Nepal*. eds. P. B. Shah, H. Schreier, S. J. Brown and K. W. Riley, pp. 208–212. Canada: International Development Research Centre.

Negi, G. C. S and Joshi, V. (1997). Land use in a Himalayan catchment under stress: system responses. *Ambio* 26, 126–128.

Nyamulinda, V. and Ngiruwonsanga, V. (1992). Lutte anti-érosif et strategies paysannes dans les montagnes du Rwanda. *Réseau Erosion Bulletin* No 12, pp. 71–82. Montpellier: ORSTOM.

Njoroge, J. W. (1994). *Field Notes on Organic Farming*. Kenya: Kenya Institute of Organic Farming.

Okoba, B., Twomlow, S. and Mugo, C. (1998). Evaluation of indigenous soil and water conservation technologies for runoff and soil loss control in semi-arid Mbeere District, Kenya. In *Modern Methods from Traditional Soil and Water Conservation Technologies*. eds. S. R. Briggs, J. Ellis-Jones, and S. J. Twomlow, pp. 25–41. Report IDG/98/10. Silsoe: Silsoe Research Institute.

Palmer, J. (1996). *Sloping Agricultural Land Technology (SALT): Nitrogen Fixing Agroforestry for Sustainable Soil and Water Conservation*. Mindanao, Philippines: Publication of Mindanao Baptist Rural Life Centre.

Palmer, J., Guliban, E. and Tacio, H. (1999). Use and success of incentives for promotion of SALT in the Philippines. In *Incentives in Soil Conservation. From Theory to Practice*. eds. D. W. Sanders, P. C. Huszar, S. Sombatpanit, and T. Enters, pp. 309–324. Enfield, New Hampshire: World Association of Soil and Water Conservation. Science Publishers, Inc.

Pandey, S. and Lapar, M. A. L. (1998). A microeconomic analysis of adoption of contour hedgerows in the Philippine uplands. In *Soil Erosion at Multiple Scales*. eds. F. W. T. Penning de Vries, F. Agus and J. Kerr, pp 83–98. Wallingford: CAB International.

Partap, T. and Watson, H. R. (1994). *Sloping Agricultural Land Technology (SALT). A Regenerative Option for Sustainable Mountain Farming*. Occasional Paper No. 23. Kathmandu: International Centre for Integrated Mountain Development (ICIMOD).

Pearce, A. J. and Hamilton, L. S. (1986). *Water and Soil Conservation Guidelines for Land Use Planning*. Report of a seminar-workshop held at FTC Gympie, Queensland, Australia.

Pereira, H. C. (1989). *Policy and Practice in the Management of Tropical Watersheds*. Boulder and San Francisco: Westview Press.

Pound, B., Anderson, S. and Gundel, S. (1999). Species for niches. When and for whom are cover crops appropriate? *Mountain Research and Development* 19, 307–312.

Purwanto, E. (1999). *Erosion, Sediment Delivery and Soil Conservation in an Upland Agricultural Catchment in West Java, Indonesia. A Hydrological Approach in a Socio-economic Context*. PhD Thesis, Vrije Universiteit Amsterdam.

Purwanto, E. and Bruijnzeel, L. A. (1996). Soil conservation on rainfed bench terraces in upland West Java, Indonesia: towards a new paradigm. In *Towards Sustainable Land Use* (Vol 2). eds. H-P Blume, H. Eger, E. Fleischhauer, A. Hebel, C. Reij and K. G. Steiner, pp. 1267–1275. Reiskirchen, Germany: Advances in Geo-Ecology, CATENA.

Quine, T. A., Walling, D. C., Zhang, X. and Wang, Y. (1992). Investigation of soil erosion on terraced fields near Yanting, Sichuan Province, China, using Caesium-137. *Erosion, Debris Flow and Environment in Mountain Regions*. International Association of Hydrological Sciences Publication No. 209, pp. 155–168.

Ralhan, P. K., Negi, G. C. S. and Singh, S. P. (1991). Structure and function of the agroforestry system in the Pithoragarh district of Central Himalaya: an ecological viewpoint. *Agriculture, Ecosystems and Environments* 35, 285–296.

Raussen, T. and Siriri, D. (2000). Agroforestry experiences in south-west Uganda. Poster presentation at Symposium/workshop: 'Forests-Water-People in the Humid Tropics'. Kuala Lumpur, 30 July to 4 August, 2000.

Reij, C., Scoones, I. and Toulmin, C. (eds.) (1996). *Sustaining the Soil*. Earthscan Publications. London: International Institute for Environment and Development (IIED).

Rice, R. A. and Greenberg, R. (2000). Cacao cultivation and the conservation of biological diversity. *Ambio* 26, 167–173.

Rijsdijk, A. and Bruijnzeel, L. A. (1990). Erosion, sediment yield and land-use patterns in the Upper Konto watershed, East Java, Indonesia. Part II: Results of the 1987–1989 Measuring Campaigns. Project Communication No. 18, Konto River Project, Republic of Indonesia.

Roe, E. (1991). 'Development narratives' or making the best of blueprint development. *World Development* 19, 287–300.

Roose, E. (1988). Soil and water conservation lessons from steep-slope farming in French-speaking countries of Africa. In *Conservation Farming on Steep Lands*. eds. W. C. Moldenhauer and N. W. Hudson, pp. 129–139. Ankeny. Iowa: World Association of Soil and Water Conservation.

Sajjapongse, A. (1996). The sloping land network and the opportunity for technology transfer. In *Soil Conservation Extension. From Concepts to Adoption*. eds. S. Sombatpanit, M. A. Zobisch, D. W. Sanders and M. G. Cook, pp. 325–335. Bangkok: Soil and Water Conservation Society of Thailand.

Sanders, D. W. and Cahill, D. (1999). Where incentives fit in soil conservation programmes. In *Incentives in Soil Conservation. From Theory to Practice*. eds. D. W. Sanders, P. C. Huszar, S. Sombatpanit, and T. Enters, pp. 12–24. Enfield, New Hampshire: World Association of Soil and Water Conservation. Science Publishers, Inc.

Sanders, D. W., Huszar, P. C., Sombatpanit, S. and Enters, T. (1999) (eds.). *Incentives in Soil Conservation. From Theory to Practice*. Enfield, New Hampshire: World Association of Soil and Water Conservation. Science Publishers, Inc.

Sayer, J., Ishwaran, N., Thorsell, J. and Sigaty, T. (2000). Tropical biodiversity and the World Heritage Convention. *Ambio* 29, 302–309.

Sen, K. K., Rao, K. S. and Saxena, K. G. (1997). Soil erosion due to settled upland farming in the Himalaya: a case study in Pranmati watershed. *International Journal of Sustainable Development and World Ecology* 4, 65–74.

Senanayake, R. (2000). Analog Forestry: an alternative to 'clear and simplify'. LEISA Newsletter, 16, 12–13.

Shaxson, T. F. (1988). Conserving soil by stealth. In *Conservation Farming on Steep Lands*. eds. W. C. Moldenhauer and N. W. Hudson, pp. 9–17. Ankeny. Iowa: World Association of Soil and Water Conservation.

Shaxson, T. F. (1999). *New Concepts and Approaches to Land Management in the Tropics with Emphasis on Steeplands*. Soils Bulletin 75. Rome: Food and Agriculture Organisation.

Sheng, T. C. (1981). The need for soil conservation structures for steep cultivated slopes in the humid tropics. In *Tropical Agricultural Hydrology. Watershed Management and Land Use*. eds. R. Lal and E. W. Russell, pp. 357–372. Chichester: Wiley.

Sheng, T. C. (1988). Demonstrating conservation practices on steep lands in Jamaica. In *Conservation Farming on Steep Lands*. eds. W. C. Moldenhauer and N. W. Hudson, pp. 207–214. Ankeny, Iowa: World Association of Soil and Water Conservation.

Sheng, T. C. (1989). *Soil Conservation for Small Farmers in the Humid Tropics*. Soils Bulletin No. 60. Rome: Food and Agriculture Organisation.

Sims, B. G., Rodriguez, F., Eid, M. and Espinoza, T. (1999). Biophysical aspects of vegetative soil and water conservation practices in the Interandean Valleys of Bolivia. *Mountain Research and Development* 19, 282–291.

Singh, G., Venkataramanan, C., Sastry, G. and Joshi, B. P. (1990). *Manual of Soil and Water Conservation Practices*. New Delhi: Oxford and IBH.

Smaling, E. (1993). *An Agro-Ecological Framework for Integrated Nutrient Management*. PhD Thesis, Wageningen Univerisy, The Netherlands.

Steiner, K. G. (1998). *Conserving Natural Resources and Enhancing Food Security by Adopting No-Tillage*. Tropical Ecology Support Program. Eschborn, Germany: Deutsche Gesellschaft für Technische Zusammenarbeit (GTZ).

Stocking, M. A. (1988). Socioeconomics of soil conservation in developing countries. *Journal of Soil and Water Conservation* 43, 381–385.

Stocking, M. A. and Abel, N. (1992). Labour costs: a critical element in soil conservation. In *Let Farmers Judge. Experience in Assessing the Sustainability of Agriculture*. eds. W. Hiemstra *et al.*, pp. 77–85. London: Intermediate Technology Publications.

Tamang, D. (1993). *Living in a Fragile Ecosystem. Indigenous Soil Management in the Hills of Nepal*. Gatekeeper Series 41. London: International Institute for Environment and Development (IIED).

Tanner, J. C., Holden, S. J., Owen, E., Winugroho, M. and Gill, M. (2001). Livestock sustaining intensive crop production through traditional feeding practices for generating high quality manure-compost in upland Java. *Agriculture, Ecosystems and Environment* 84, 21–30.

Tejwani, K. G. and Lai, C. K. (1992). *Asia-Pacific Agroforestry Profiles*. Asia-Pacific Agroforestry Network, Bogor, Indonesia.

Thomas, D. B. (1993). Terrace riser stabilisation and fodder production. *Paper presented to the fourth National Workshop on Land and Water Management in Kenya. 15–19 February, 1993, Nairobi.*

Thomas, D. B. (1997) (ed.). *Soil and Water Conservation Manual for Kenya*. Nairobi, Kenya: Soil and Water Conservation Branch. Ministry of Agriculture, Livestock Development and Marketing.

Tiffen, M., Mortimore, M. and Gichuki, F. (1994). *More People, Less Erosion. Environmental Recovery in Kenya*. Chichester: Wiley.

Tukahirwa, J. M. (1995). *Measurement, Prediction and Social Ecology of Accelerated Soil Erosion in Kabale District, south west Uganda*. PhD Thesis, Department of Geography, Makerere University, Kampala, Uganda.

Turkelboom, F., Poesen, J., Ohler, I., van Keer, K., Ongprasert, S. and Vlassek, K. (1997). Assessment of tillage erosion rates on steep slopes in northern Thailand. *Catena* 29, 29–44.

Turner, S. D. (1998). Land reform and the environment. In *Sustainable Land Management: Some Signposts for South Africa*. eds. W. Critchley, D. Versfeld and N. Mollel, pp. 53–67. Sovenga, South Africa: University of the North Press.

Turton, C., Warner, M. and Groom, B. (1998). *Scaling Up: Participatory Watershed Development in India: A Review of the Literature*. AGREN Network Paper 86. London: Overseas Development Institute.

van Noordwijk, M., van Roode, M., McCallie, E. L. and Lusiana, B. (1998). Erosion and sedimentation as multiscale, fractal processes. Implications for models, experiments and the real world. In *Soil Erosion at Multiple Scales*. Eds. F. W. T. Penning de Vries, F. Agus. and J. Kerr, pp. 223–253. Wallingford: International Board for Soil Research and Management, CAB International.

Wachter, D. (1996). *Land Tenure and Sustainable Management of Agricultural Soils*. Development and Environmental Report 15. Berne: Centre for Development and Environment, University of Berne.

Watson, H. R. and Laquihon, W. (1985). Sloping agricultural land technology (SALT) as developed by the Mindanao Baptist Rural Life Centre. *Paper presented in a Workshop on Site Protection and Amelioration Roles of Agroforestry*. Philippines: University of the Philippines, Los Banos.

Webster, C. C. and Wilson P. N. (1966). *Agriculture in the Tropics*. London: Longmans, Green and Co.

Westerberg, L-O. and Christiansson, C. (1999). Highlands in East Africa: Unstable slopes, unstable environments? *Ambio* 28, 419–429.

Willey, R. W. (1979). Intercropping – its importance and research needs. Part 1. Competition and yield advantage. *Field Crop Abstracts, Commonwealth Bureau of Pastures and Field Crops* 32, 1–10.

Wilken, G. C. (1987). *Good Farmers: Traditional Agricultural Resource Management in Mexico and Central America*. London: University of California Press.

Wischmeier, W. H. and Smith, D. D. (1978). *Predicting Rainfall Erosion Losses. A Guide to Conservation Planning*. Washington D.C.: Agricultural Handbook no 537, United States Department of Agriculture.

WOCAT/FAO. (2000). *World Overview of Conservation Approaches and Technologies*. CD-ROM. Land and Water Digital Media Series No. 9. Rome: FAO. (www.wocat.net)

World Bank. (1992). *Strategies and Technologies for Asian Watersheds*. Washington D.C.: The World Bank.

Wrigley, G. (1971). *Tropical Agriculture*. London: Faber and Faber.

Ya, T. (1999). *Manual on Contour Hedgerow Intercropping Technology*. Kathmandu: International Centre for Integrated Mountain Development (ICIMOD).

Youatt, W. (1852). *The Complete Grazier, or Farmer's and Cattle Breeder's and Dealer's Assistant*. London: Cradock & Co.

Young, A. (1989). *Agroforestry for Soil Conservation*. International Centre for Research in Agroforestry (ICRAF). Wallingford: CAB International.

Young, A. (1997). *Agroforestry for Soil Management*. International Centre for Research in Agroforestry (ICRAF). Wallingford: CAB International.

Conclusion
Forests, water and people in the humid tropics: an emerging view

L. A. Bruijnzeel
Vrije Universiteit, Amsterdam, The Netherlands

M. Bonell
UNESCO, Paris, France

D. A. Gilmour
Private consultant, Brisbane, Australia

D. Lamb
University of Queensland, Brisbane, Australia

INTRODUCTION

Tropical forest loss: extent, patterns and causes

Land use patterns in most humid temperate countries have more or less stabilised over the past century. Changes are still taking place (e.g. urbanisation in some areas, abandonment of agricultural cropping in favour of natural regeneration in others) but they are generally rather gradual. This is not the case in most humid tropical countries where significant land use changes are still under way and often take place at an unprecedented rate. Most of these relate to changes from the original forest cover to either heavily modified (degraded) forest or non-forest land use, such as (slash and burn) agriculture, grazing or plantations of various types.

The spatial extent and trends of land cover change in the tropics have been the subject of much debate, mainly because of the different definitions for various vegetation cover types used by the respective surveys. Although it has proven difficult to obtain accurate and verifiable figures, the successive Forest Resources Assessments (FRA) of the United Nations Food and Agriculture Organization (FAO) have produced an overall picture from which a number of conclusions can be drawn. The estimated average rates of closed tropical forest loss in the developing world are about 14.6 and 14.9 million ha year^{-1} for the periods 1980–1990 and 1990–2000, respectively. As such, an area of almost 300 million ha of closed tropical forest has disappeared during the last two decades (roughly equivalent to two-thirds of continental Europe excluding Russia). A little less than 50% of this was converted to other, non-forest uses whereas the remainder represents more or less seriously degraded (open, fragmented) forest land (Drigo, this volume). Furthermore, during the last decade a shift has been noted in the rate of tropical forest loss from the 'moist deciduous forest' zone (mean annual precipitation 1000–2000 mm with a dry season of 3–5 months) towards the wetter evergreen rainforest formations (*MAP* > 2,000 mm; Drigo, this volume). Generally speaking, the main driving force behind *unplanned*, and generally more gradual, forest degradation is rural population pressure with its corresponding subsistence and energy needs. In contrast, *planned* activities result in the immediate conversion of forest to other land uses as part of government-driven programmes to stimulate resettlement, cattle ranching and permanent agriculture, as well as commercial plantations. There are important regional differences in the causes and factors influencing forest conversion. In Africa, for example, the process of forest conversion has taken the form of a stepwise degradation and conversion to other land uses rather than more immediate forest clearance. Conversely, in the Amazon Basin, the planned direct conversion of forest to large-scale cattle ranching and permanent agriculture is the chief cause of forest loss (Serrão and Thompson, this volume). The highest rates of forest degradation, as well as forest conversion, occur in Asia, however, reflecting both the effects of population pressure (slash and burn cultivation) and centrally planned conversion to agriculture (notably irrigated rice in former peat swamps) or commercial plantations (mostly oil palm, less for timber, paper pulp, rubber and cocoa), with a gradual shift towards the latter during the last decade in particular (Drigo, this volume).

Overall, therefore, there is a rapid process of ecological simplification under way in many tropical landscapes. Forests are being fragmented and the former species-rich and structurally complex forests are being replaced by plant communities that are floristically and structurally simpler (Hölscher *et al.*, this volume). Many indigenous plants and animals are disappearing from the landscape and are being replaced by exotic species (e.g. plantations, grasses).

Forests, Water and People in the Humid Tropics, ed. M. Bonell and L. A. Bruijnzeel. Published by Cambridge University Press.

These changes are being accompanied by less obvious changes in ecological processes that sustained the former forest ecosystem. For example, intricate mutualistic and very specific relationships between plants and animals (cf. Leigh *et al.*, 1983; Richards, 1996) become disrupted, patterns of nutrient cycling (and with it topsoil fertility maintenance and forest productivity; Proctor, this volume) are changed dramatically (usually for the worse; Scott *et al.*, this volume) and, of particular importance in the present context, many aspects of the hydrological cycle – notably evaporation, infiltration and storm runoff generation – may be altered profoundly after forest conversion. As an indication of scale, it has been estimated that 45% of the total land area of South and South East Asia was affected by some form of human-induced soil degradation (mostly surface erosion) in the mid 1990s. On 10–15% of the affected area, degradation was considered to have a strong to very strong impact on plant productivity (and by implication, on overland flow occurrence), whereas on 22–28% the impact was considered moderate, light on 40–48% and negligible on the remaining 12–23% of the land (Van Lynden and Oldeman, 1997).

The causes of tropical forest clearance and the often poor subsequent land management are highly complex. A prime underlying cause highlighted in many chapters of this book, however, is the inherent conflict between the desire for rapid economic development on the one hand and the need for environmental sustainability on the other. As stated by Murdiyarso (this volume), governments usually focus on the demand rather than the supply side of things and this easily causes over-use of (limited) natural resources. On a more fundamental note, C. Hall and Koo (this volume) provide several examples challenging the appropriateness of the neo-classical economic model for developing countries, arguing that its continued use in combination with an escalating population pressure is equivalent to 'an excuse to plunder' natural resources. Similarly, A. Hall (this volume) considers much of the existing poor forest-land management practices as being due to the 'arrogance of imposing the neo-classical economic model' in which the profit motive assumes priority over other, e.g. environmental or ethical, considerations. The friction between economic thought and resource conservation is illustrated further by Aylward's (this volume) attempt to translate the hydrological impacts of deforestation in economic terms. The sobering conclusion from his analysis is that, from the purely economic perspective, the production benefits of post-forest land use (agricultural produce, livestock, etc.) – and with it an area's hydrological functioning – would have to decline considerably before land rehabilitation may be considered economically justified. Ironically, therefore, severely degraded sites may be so expensive to rehabilitate that, in many cases, the costs of doing so may be prohibitive (Lamb and Gilmour, 2003).

Not one chapter in this book uses the word 'greed' explicitly but it is implied in many contributions. The point was also raised by a representative from South East Asia at the Symposium on which this book is based, in the context of obstacles towards implementing best management practices. This aspect of human nature may be observed in all layers of society. Schweithelm, for example, relates how forest dwellers suddenly exposed to the cash economy sometimes maximise their short-term financial gains by selling timber concessions to logging companies whilst putting environmental concerns aside. More often, however, large-scale corporate and export-orientated enterprises tend to view forest dwellers more as an obstacle to economic development, and the disruption of their traditional ways has resulted in mass migration to jobs in the poorest strata of society. Equally important is the notion that, once the ecosystem's carrying capacity is exceeded (e.g. because fallow periods within shifting cultivation have become progressively shorter due to increased population pressure; Malmer *et al.*, this volume), people's livelihoods will be at risk and the ensuing struggle for survival will inevitably outweigh environmental concerns (A. Hall, this volume).

Apart from the adverse effects of the blanket application of neo-classical economics and high population pressures in resource-limited environments, other important factors underlying poor land management practices in the tropics include: insecurity or lack of land ownership and resultant poverty (A. Hall, this volume); the implementation of 'top-down' soil conservation schemes without due consideration of the societal acceptability of such measures (Critchley, this volume); and, in the context of forestry operations, a lack of well-trained and satisfactorily compensated field and supervisory personnel, a lack of locally applicable practical guidelines (as opposed to guidelines at a more generic or national level), and the difficulties experienced by forestry planners and managers in understanding the needs and levels of decision-making in local communities (Cassells and Bruijnzeel, this volume).

Finally, another recurring and overarching theme in relation to poor resource management in the tropics concerns the need for adjustments in governance and improved communication between the major stakeholder groups – policy makers, resource managers (including local communities which are often the *de facto* managers) and scientists. It is clear from the examples given above and from the more than 100 'hot spots' of tropical forest land degradation identified by Drigo that the 'centralised command and control' structure generally imposed by national and state governments does not work. On the other hand, as related by Murdiyarso, a recent attempt at decentralisation by the Government of Indonesia overestimated the local human and technical capacity to manage natural resources sustainably at the District level. Furthermore, although the alternative 'bottom-up' approach – in which forest dwellers, upland farmers and other local communities are explicitly involved in land use planning, identification of key research questions and management decisions – offers a promising solution towards this much needed institutional strengthening, there is still a long way to go before the approach will be universally accepted. We come back to this important issue in the final section of this chapter.

The challenge: achieving sound environmental management

Having outlined the socio-economic setting of tropical forests and the chief processes underlying their current massive conversion to other forms of land use, what do we know of the corresponding soil and water impacts? Are we now in a position to predict the hydrological consequences of various management practices and land use changes in sufficient detail to be used by land users, resource managers or policy makers wishing to avoid adverse hydrological consequences? And is the new hydrological knowledge generated by researchers being passed on to these stakeholders in a form that is useful to them – or are there constraints limiting the extent to which this occurs? And if so, how can the communication between the scientific and applied communities be improved for the benefit of all involved? These are valid questions in a time when a common criticism from policy makers and practitioners is that most scientists seem to be insensitive to policy questions and more preoccupied with their disciplinary orientation (Murdiyarso, this volume). Although there are other, mostly economic reasons, such notions have also contributed to the general decline in funds made available for long-term hydrological research and monitoring. As such, one of the prime challenges of the present book has been to demonstrate the continued need for this kind of work if improved land and water resource management practices are to be fully realised.

The chapters in Part II have described in considerable detail the chief hydrological and related nutrient cycling and geomorphological processes in various types of old-growth ('undisturbed') tropical rainforests, including peat swamps and montane cloud forests. Such process-based knowledge is important for two reasons. Firstly, it enables a better understanding of the impacts of forest disturbance (both natural and man-made), forest conversion and recovery on the hydrology, rates of denudation (erosion), soil fertility and the aquatic ecology associated with these fragile ecosystems. Secondly, it assists in the design, critical evaluation and adjustment of guidelines for 'best management practices' (BMPs) within the context of tropical timber harvesting, land clearing and post-forest agricultural cropping and plantation management. Key indicators of sustained environmental quality after forest disturbance or conversion include a stable supply of good quality water, timber and other products needed by society without degrading soil fertility and plant productivity.

However, many BMPs have been derived from guidelines that were developed originally for use under more temperate climatic conditions. In the following, therefore, some of the factors that render the humid tropical environment different from temperate regions are highlighted before summarising our current understanding of the environmental impacts of tropical deforestation and reforestation. In the process it will be shown that, although some of the more widely held views regarding the hydrological role of tropical forests would seem mutually exclusive at first sight (such as the notions that forest removal will both lead to springs and streams drying up (i.e. reduce runoff) *and* to widespread flooding (i.e. increase runoff)), these can be reconciled by examining the underlying processes.

Next, the more pressing unresolved hydrological problems faced by tropical researchers, policy makers, resource managers, and the general public alike are highlighted. After exploring the usefulness of various new research tools for resolving these outstanding issues the chapter ends with a plea for the establishment of a pan-tropical network of demonstration forests and catchments where BMPs can be evaluated and further refined, and where carefully focused research can be conducted over sufficiently long periods to separate effects of climatic variability from those induced by land use change or management. Furthermore, particular attention is paid to ways of enhancing the interaction between the three major stakeholder groups (policy-makers, local resource managers and scientists) in terms of information exchange, formulation of the most relevant research questions, and translation of research results into locally applicable BMPs. One promising approach (advanced by Murdiyarso, this volume) is the development of decision support systems (DSS) capable of summarising in a quantitative manner the detailed components of local concerns, thereby both enhancing the negotiating position of poor local communities and reducing the potential for conflicts between the different stakeholder groups. A related approach concerns the development of payment schemes for environmental services, in which lowlanders reward upland communities for maintaining a stable supply of high quality streamflow through good land husbandry practices (Aylward, Murdiyarso, both this volume).

UNIQUE ATTRIBUTES OF THE HUMID TROPICAL (FOREST) HYDROLOGICAL CYCLE

By and large, the fundamental attributes of the hydrological cycle are similar for temperate and tropical forest ecosystems. To that extent many research findings from temperate regions should also be applicable in the tropics. At the same time, however, there are important environmental and ecological differences that have implications for land use practices in tropical catchments.

Table 39.1. *Generalised basic differences in rainfall and soil characteristics between humid tropical and humid temperate regions, and their consequences for hillslope runoff response*

Soil characteristics	Rainfall characteristics	Topographic controls
• Much deeper and infertile soils where weathering has proceeded for long time; development of surface root mats in extreme cases. • Depth of impervious horizon (Acri/Ultisols) or soil/bedrock boundary (Cambi/Inceptisols, Ferral/Oxisols) important controls of dominant runoff pathways in both regions but more strongly in tropics.	• Higher maximum short-term rainfall intensity within storm event and higher daily totals. • Principal driver in runoff generation process is the much broader range of rainfall rates linked with different rain producing systems. • Cyclonic vs. non-cyclonic areas. • Montane belts with significant extra inputs via 'occult' precipitation.	• Rate of convergent areas and slope morphology in controlling redistribution of soil water and runoff generating 'hot spots' universal to both zones. • Role of sub-surface soil-bedrock topography less clear in more deeply weathered regoliths of humid tropics.

The nature of most tropical rainfall regimes is perhaps the most obvious of these differences. Not only is the total amount of rainfall commonly higher in tropical regions than in most temperate regions at comparable elevations but also, more importantly, the intensity of this rainfall is often greater as well. This is especially true in the cyclone-prone, outer tropics (Bonell *et al.*, this volume). On a related climatic note, in many tropical montane areas one finds so-called 'cloud forests' which receive substantial additional amounts of 'occult' (i.e. not measured by conventional rain gauges) precipitation in the form of wind-driven drizzle and fog. Although fog-affected forests are also found at more temperate latitudes, mainly along the western edges of continents where advective sea fogs form over cold ocean water, their areal extent is much greater in the humid tropics where they often constitute an important part of headwater basins (Bruijnzeel, this volume). Another major difference is that the generally ample supply of soil water, together with the higher temperatures and radiation loads prevailing in the tropics and the commonly higher leaf area index of tropical plant communities, lead to much higher evaporation rates than observed for comparable vegetation types and topographic settings under temperate conditions (Roberts *et al.*, this volume). A related factor concerns the often much deeper soils in the tropics. As a result of the higher rainfall and temperatures, rock weathering has proceeded to much greater depths wherever the process has been able to continue over long periods of time, uninterrupted by volcanic or tectonic activity or climate change (Douglas and Guyot, this volume). To the extent that the roots have penetrated the entire regolith, access to the water stored in deeper layers enables tropical forests to maintain relatively high rates of evaporation even during prolonged rainless periods when vegetation on shallower soils would experience severe water stress (Roberts *et al.*, this volume). Finally, there is a much greater diversity of storm runoff pathways than in humid temperate environments. In particular, there is much more saturation-excess overland flow (*SOF*) generated on tropical hillslopes in response to the higher rainfall intensities, particularly on the widespread Acrisols (Ultisols) which often have an impeding layer at shallow depth. Based on research experience in humid temperate regions, it was a widely held view until the late 1970s that overland flow was a rare phenomenon in tropical rainforests, and forestry operations were planned and implemented on this basis. However, it is now well-established that the situation in lowland tropical rainforests subject to high intensity rainfall can be quite different (Bonell, this volume). Indeed, the importance of extreme and rare rainfall events (often related to the passage of a cyclone) in shaping the tropical landscape through widespread landsliding, the initiation of gully erosion and the deposition of sediment in flooded areas cannot be overstated (Scatena, Planos-Gutierrez and Schellekens; cf. Douglas and Guyot, both this volume).

Summarising, the much more 'intense' character of the tropical (forest) hydrological cycle has important implications for the rigour with which guidelines aiming to minimise the adverse environmental consequences of forest exploitation and conversion will need to be adhered to (Cassells and Bruijnzeel, this volume). Another compelling reason to do so relates to the generally much lower fertility of tropical soils which renders them even more vulnerable to disturbance. Once the relatively fertile topsoil has been removed by erosion or machinery, the usually much more infertile subsoil (or even saprolitic material) becomes exposed (Grip *et al.*, this volume). Once this stage is reached, plant productivity is greatly impaired and difficult to restore (Hölscher *et al.*, this volume).

Table 39.1 summarises the more pertinent differences between tropical and temperate regions in terms of rainfall, soils and hillslope runoff response to rainfall.

HYDROLOGICAL CONSEQUENCES OF DISTURBING OR CLEARING TROPICAL FOREST: THE SCIENTIFIC CONSENSUS

One of the tasks of forest and land managers has been to find ways of minimising any adverse hydrological effects arising from timber harvesting or land clearing and this was the target of much of the hydrological research initiated in tropical forests in the 1960s and 1970s. Early studies in temperate forests were usually carried out in small experimental catchments and a similar approach was subsequently used in tropical forests (Blackie *et al.*, 1979; Gilmour, 1977a, 1977b). In the following, the main findings of *c.* 40 years worth of tropical catchment research vis à vis the soil and water impacts of forest disturbance or conversion are summarised in a quantitative manner, offering the best estimates of the respective associated changes in water yield, stormflows, catchment sediment yield and soil fertility recovery times currently available.

Tropical rainforests may be subject to a range of natural disturbances (extreme rainfall, landslides, earthquakes and volcanic eruptions in some areas, droughts and fire; Scatena *et al.*, this volume) but a new closed-canopy vegetation is usually established within a few years (Hölscher *et al.*, this volume). Similarly, the manual clearing of forest in the context of shifting cultivation may produce locally increased storm runoff (typically by 25–30%) and surface erosion but, again, the effect is usually short-lived (Grip *et al.*, this volume) whereas soil fertility is not impaired significantly as long as the intervening fallow periods are not reduced below *c.* 12–15 years (Malmer *et al.*, this volume). However, the *mechanised harvesting of timber* from tropical forests introduces more severe disturbance to vegetation and soils. As a rule of thumb, for every harvested tree at least one tree is destroyed and at least one other damaged beyond recovery. The construction of haulage roads, tractor tracks and log landings causes severe soil compaction and surface erosion, the spatial extent of which varies with the intensity of harvesting. In the low-intensity selective logging operations typical for Africa and South America tractor tracks occupy *c.* 5% of the area but this figure may well increase to 20–35% in the case of the much more intensive logging practised in the richly stocked forests of South East Asia (Cassells and Bruijnzeel, this volume). The infiltration capacities associated with such compacted surfaces are very low (typically 15–20 mm h^{-1}) compared to values normally measured for undisturbed forest topsoils ($>>100$ mm^{-1}) and may reach values as low as 1 mm h^{-1} or less on the most heavily used tracks and roads. Moreover, recovery of the infiltration capacity on former tractor tracks is slow and may take at least 50 years (Grip *et al.*, this volume). Naturally, infiltration-excess overland flow (*HOF*) will be much more frequent and intense on these compacted surfaces and catchment runoff response to rainfall would therefore be expected to increase after logging. However, no statistically significant effect on stormflow volumes after selective logging has been reported, not even after extracting 40–60 m^3 ha^{-1} of timber, presumably because much of the extra overland flow still has a chance to infiltrate again before reaching the streams (Chappell, Tych, Yusop, Abdul Rahim and Kasran, this volume).

Conversely, a highly significant rise in baseflows has been observed wherever more than 20% of the standing vegetation was removed. *The joint effect of selective logging on streamflows, therefore, is an overall increase in annual water yield.* Typical increases that may be expected under 'average' rainfall conditions (mean annual precipitation *c.* 2000 mm) and harvesting intensities (removal of 33–40% of the commercial stocking) amount to 40–65% (*c.* 90–160 mm yr^{-1}) but larger increases are possible in the case of more intensive operations (see below). In addition, these increases in annual water yield have been shown to persist over a number of years. One would have expected the effect to decline again with time as vegetation water use returns to pre-disturbance values due to the re-colonisation of logging gaps and the gradual recovery of the vegetation (Hölscher *et al.*, this volume) but apparently this effect is compensated by the increased runoff contributions from roads (Chappell, Tych *et al.*, this volume). Although increases in storm runoff volumes after selective logging are modest at best, the associated amounts of sediment carried in suspension normally increase dramatically. Stream suspended sediment loads during the first few years after (intense) logging operations often increase 15–20 times (reported range: 2–50 times) although they normally stabilise within three years at a level of about two to three times the pre-disturbance value as tractor tracks become overgrown again. However, roads and their immediate surroundings (including stream crossings) continue to be sources of sediment as temporarily deposited soil material may be reactivated during large storm events (Chappell, Tych *et al.*, this volume). As such, road lay-out and drainage facilities require particular attention during construction if future stream sedimentation problems are to be avoided (Cassells and Bruijnzeel, Thang and Chappell, both this volume).

Site fertility upon forest exploitation is potentially threatened by surface erosion, the removal of nutrients contained in harvested logs, and temporarily increased leaching of nutrients from the soil. The latter is due to a combination of the larger volumes of water percolating through the soil after opening up of the canopy, the temporarily reduced uptake of nutrients by (damaged) vegetation, and the sudden addition of large volumes of fresh logging debris left to decompose on the forest floor (Scott *et al.*, this volume). Depending on the nutrient under consideration, the period required to replenish soil nutrient stocks through inputs via bulk precipitation (rainfall and dust) has been estimated at 20–60 years (Chappell, Tych *et al.*, this volume).

In most cases selective logging may be considered a disturbance of intermediate intensity because the forest may recover in time. However, large-scale mechanised *clearing* of tropical forest for the establishment of other forms of land use represents a step up in intensity and the resulting hydrological impacts are increased accordingly. Because forest conversion not only involves the clear-felling of all vegetation but also the burning of slash, the percentage of bare soil that becomes exposed to the erosive power of intensive tropical rainfall may be as high as 80–90% (with the remaining 10–20% often occupied by areas that are too wet (or too steep) to allow machine access, such as riparian zones (Grip *et al.*, this volume). Surface compaction, and with it HOF and erosion, may now occur over large parts of the catchment and both stormflows and stream sediment loads are much increased as a result.

As a rule of thumb, average stormflow volumes associated with the transition period (during which the largely bare soils are gradually covered by the newly established vegetation) are roughly doubled, with smaller events being increased relatively more than larger ones (see below). In absolute terms, the annual increase in stormflow totals typically amounts to *c.*80–100 mm under 'average' rainfall conditions (MAP *c.*2000 m) although much larger increases (250–700 mm) are possible in high rainfall areas (*MAP* >3300 mm), depending on the prevailing runoff generation mechanism under pre-disturbance conditions (Grip *et al.*, this volume). Where stormflow under forested conditions is generated mostly by deep percolation to the groundwater table (i.e. no rapid surface or near-surface lateral flows), both absolute and relative values of storm runoff tend to be small (typically 5–10% of rainfall). Upon forest clearing, contributions by *HOF* from compacted areas may easily raise the storm runoff coefficient to 20–25%, implying roughly a tripling of the runoff response to rainfall. Conversely, in catchments where rapid shallow subsurface flow and *SOF* are major contributors, storm runoff coefficients are already high under undisturbed conditions (25–30%). After clearing, the coefficient is typically raised further to 45–55% of incident rainfall, i.e. catchment response is almost doubled. As such, the greatest *relative* increases in stormflow are observed on those catchments that are the least responsive under forested conditions and vice versa (Grip *et al.*, this volume). *Such contrasting findings illustrate the need for hydrological process studies for the proper prediction of catchment runoff behaviour after tropical forest clearing* (cf. Bonell, this volume).

Depending on the land use replacing the original forest, catchment runoff response may remain elevated more or less permanently (as in the case of intensively grazed pasture: increases of +25% to +45% maintained after 5 years), return to pre-disturbance values within 4–5 years (natural regrowth following logging and burning) or even become reduced below pre-clearance values after 5–6 years (rapidly growing tree plantations creating somewhat drier soil moisture conditions) (Grip *et al.*, this volume). As such, *once a new vegetation cover becomes well-established, stormflows are no longer dramatically different compared with forested conditions*. This is true especially for the largest of storm events, i.e. floods. Indeed, the larger the event, the smaller is the relative increase in stormflow after clearing. This reflects the fact that the extra storage capacity afforded by a forest ecosystem is not infinitely large. In small storm events the combined storage capacity of vegetation canopies, ground-covering litter, surface micro-topography and the soil mantle can be substantial in proportion to the size of the storm depth. Of these the soil mantle is potentially the largest, but its capacity to accommodate additional rain varies as a function of soil wetness. Where previous uptake by the trees has depleted soil water reserves, storage capacities will be relatively high but once the soil has become thoroughly wetted by frequent rains (typically at the height of the wet season), opportunities to absorb large additional amounts of rain will be limited even under fully forested conditions. Furthermore, as precipitation events increase in size, so does the relatively fixed maximum storage capacity of the soil become less influential in determining the size of the stormflows that are generated. In other words, *under conditions of extreme rainfall and soil wetness, large stormflows may also emerge from forested areas* (Scott *et al.*; cf. Bonell, both this volume).

Similarly, although increases in catchment sediment yields are very much elevated during the transition phase (15–50 times higher), they normally return to a more or less stable level of about three times the pre-disturbance value within two to three years, provided the soil becomes protected again by new vegetation.

Although changes in stormflow volumes and sediment yields are thus seen to be limited if forest clearance is followed by a well-developed vegetation cover, the large increases in hydrological response and sediment yield observed during the transition (bare soil) period suggest that equally large changes may be possible on a more permanent basis in the case of a conversion to annual cropping, particularly if not accompanied by appropriate soil conservation measures (cf. Critchley, this volume). Typical increases in stormflow totals associated with traditional slash and burn agriculture amount to 25–30% (Grip *et al.*, this volume) but there is a surprising lack of information on the hydrological functioning of permanently cropped catchments in the humid tropics, particularly degraded ones (at least in the 'official' literature; Scott *et al.*, this volume). Some idea may be gained, however, from the stormflow coefficients reported in the grey literature for 'typical' actively eroding ('semi-degraded') agricultural catchments in the volcanic uplands of Java, viz. 30–40% of incident rainfall vs. 5–10% for nearby forested catchments. Likewise, at *c.* 35–50 t ha^{-1} yr^{-1} the sediment yields from such deforested catchments are 10–50 times higher than those typically associated with similarly sized forested catchments. Although absolute differences in sediment yield associated with different land uses vary with geological substrate, the relative increases cited appear to be widely valid (cf. Figure 22.6 in Grip *et al.*, this volume). Such increases largely reflect the enhanced contributions of *HOF* from degraded agricultural fields and compacted surfaces like trails, roads and settlements which all exhibit very low infiltration capacities (Grip *et al.*, this volume). However, whilst storm runoff in the most extreme cases can be increased to as much as 70% of the rainfall (e.g. where the underlying rock has become exposed), the associated sediment yields are relatively low because most of the erodible material has already been lost (Scott *et al.*, this volume). Such examples illustrate the very considerable (local) potential for highly increased storm runoff from severely degraded catchments (cf. Table 25.4 in Scott *et al.*, this volume).

Undisturbed lowland rainforest typically evaporates some 1400 $\pm$ 100 mm yr^{-1} (Roberts *et al.*, this volume) whereas the water use of pasture or annual crops is closer to 950 $\pm$ 50 mm yr^{-1} under moderately seasonal conditions, and *c*.750 mm yr^{-1} in areas with a well-developed dry season (Scott *et al.*, Grip *et al.*, both this volume). Such differences reflect the fact that forests tend to intercept more rainfall than shorter and aerodynamically less rough vegetation whereas the deeper rooted trees also suffer less easily from soil water deficits during dry periods than their more shallow-rooted agricultural counterparts (Roberts *et al.*, this volume). Therefore, replacing rainforest by pasture or annual cropping leads to typical *permanent* increases in annual water yield of 300–400 mm yr^{-1} under 'average' rainfall conditions. Furthermore, given the relatively modest permanent increases in stormflow totals after forest conversion to a new stable form of land use signalled earlier, it follows that the bulk of the increase in streamflows will manifest itself in the form of higher baseflows, particularly during the dry season. However, where the rainforest is replaced by fast-growing tree plantations or oil palm plantations, water yields usually return to pre-conversion values after canopy closure (5–10 years, depending on growth rate), with temporarily increased yields as long as the plantations are still young and consuming less water than the original forest. However, water yields from mature tree plantations (25–30 years) may increase again somewhat once the vigour of the trees starts to diminish (Scott *et al.*, Grip *et al.*, both this volume).

There are two important exceptions to the general rule that dry season baseflows are enhanced upon tropical forest conversion to pasture or cropping. Dry season flows in montane headwater areas whose cloud forests receive substantial amounts of 'occult' precipitation in the form of wind-driven fog captured by the trees (often 1–3 mm day^{-1}) may well be reduced if the forest is clear-felled although the evidence for this is still circumstantial only (Bruijnzeel, this volume). Furthermore, where soils are degraded to the extent that much more water is lost from the catchment in the form of surface runoff during the rainy season than is potentially gained because of reduced evapotranspiration (ET), then dry season flows will go down instead of being increased. After all, the water lost as overland flow does not infiltrate into the soil and therefore does not contribute to the groundwater reservoir feeding the baseflow. It is difficult to predict the critical threshold beyond which deterioration of the streamflow regime will occur but a first estimate supported by preliminary modelling work suggests that stormflow totals have to increase by at least 350–400 mm yr^{-1} (i.e. roughly the difference in ET) for the effect to become visible, and to *c*.1000 mm yr^{-1} for the baseflow to be halved (Scott *et al.*, this volume). We will come back to this important subject later.

A final aspect of tropical forest conversion concerns the *effect on climate*, particularly rainfall. Physical reasoning (notably the lowered return of moisture to the atmosphere via evaporation from a non-forest cover, plus decreased aerodynamic roughness creating less turbulence) and modelling, indicate that *large-scale* deforestation may cause a decrease in rainfall. In continental settings like Amazonia, where locally derived convective rainfall is relatively important, the strongest case for lowered rainfall due to deforestation can be made. Yet, the best model-based estimates predict a reduction in Amazonian rainfall of less than 10% because of various feedback mechanisms. After forest conversion, the surface energy balance becomes adjusted in favour of an increase in sensible heat (i.e. the air is warmed up) and a reduction in evaporation. The higher temperatures, however, induce an enhanced influx of moist oceanic air which partially compensates for the reduction in regional evaporation after forest conversion (Costa). In addition, these simulation attempts assume complete deforestation whereas in reality a mosaic of different vegetation types is usually created. As such, any effects of forest clearing on rainfall can be expected to be correspondingly smaller, particularly if one considers that the water use of vigorously growing young (5–25 years) secondary vegetation often exceeds that of old-growth forest (Hölscher *et al.*, this volume). Finally, model simulations for more 'maritime' tropical regions (where oceanic moisture inputs to the atmosphere override contributions by evaporation from the land) suggest that the impact of land cover on precipitation is even smaller under such conditions (Malmer *et al.*, this volume).

Although the effect of the presence or absence of a forest cover on total precipitation thus seems to be small or modest at best, there is increasing evidence (both observational and model-based) that precipitation patterns during certain (drier) periods of the year may be adversely affected by deforestation. Examples include (the end of) the dry season in Amazonia (Costa, this volume), the latter half of the summer monsoon in mainland South East Asia (Malmer *et al.*, this volume), and the opening and closing stages of the monsoon in sub-humid West Africa (Mahé *et al.*, Callaghan and Bonell, both this volume). In addition, land cover type has been shown to exert a distinct influence on long-term (decadal) rainfall patterns under more sub-humid conditions in West Africa as well (Mahé *et al.*, this volume).

With the exception of the effect on rainfall, the above review of the current scientific consensus on the soil and water impacts of tropical forest disturbance and conversion has been concerned with *effects at the local scale* (<10 km^2). However, many of the hydrological changes associated with land use alterations decrease or disappear entirely as catchment areas increase in size and the cleared or modified area becomes a smaller proportion of the whole. Large catchments can be cleared but this is usually done in a piecemeal fashion, and often over long periods of time. This makes it difficult to assess the cumulative impacts of the numerous small changes on long-term streamflow patterns. Further complications arise from the fact that larger catchments may have a range of contrasting land cover types as well as experiencing highly different amounts of rainfall, particularly during individual storms (Bonell *et al.*, this volume). Thus, the typically large changes in streamflow observed after forest removal or addition at the small headwater catchment scale are less likely to occur at larger scales.

Although a complete predictive understanding of all hydrological changes following logging or land clearing at these larger scales (>1000 km^2) has not yet been achieved, therefore, this may be less problematic than might seem at first sight. Firstly, as the database on tropical vegetation water use continues to grow (Grip *et al.*, Hölscher *et al.*, Scott *et al.*, Wallace *et al.*, all this volume), our ability to predict changes in baseflows from basic information on rainfall and land cover type (e.g. through remote sensing; Held and Rodriguez, this volume) increases correspondingly. Secondly, increases in stormflow production associated with a range of stabilised land uses have been shown not to differ dramatically from those under forested conditions, with major increases being recorded only for agricultural cropping without conservation measures, or for settlement areas and roads (Grip *et al.*, this volume). The location and areal extent of such hydrological 'hot spots' can be easily mapped. Thirdly, the confounding effect of the high spatial variability of tropical convectional rainfall is such that land use effects on event storm runoff tend to become progressively smaller with increasing basin size (Costa; cf. Bonell, both this volume). On the other hand, the prediction of changes in basin sediment yields at larger scales is more complicated because in addition to climatic and land cover variations, there is the added complexity of topographic and geological controls affecting sediment production, transport and deposition (Douglas and Guyot, this volume).

The available experimental and observational evidence suggests that the effects of (adverse) land use change on baseflows can be detected (and predicted) quite well for catchments up to *c*. 1000 km^2 (cf. Watson *et al.*, 1999) whereas the effect seems to have disappeared at a scale of 10 000 km^2 (see examples in Bruijnzeel, 2004). Although changes in stormflows per se are difficult to demonstrate for catchments larger than, say, 10 km^2 because of the confounding rainfall and land use effects listed above, strongly enhanced runoff response to rainfall is known to also occur at the intermediate basin scale (100–150 km^2) if soil degradation is sufficiently widespread (Scott *et al.*, this volume). Similarly, the presence of an intensive road network has been shown to have a peak discharge enhancing effect at this scale that increases with the size of the flood peak (La Marche and Lettenmaier, 2001). Finally, there is increasing indirect evidence in the form of increased (or earlier) wet season flows (not floods) from very large basins (4000–175 000 km^2) following a deterioration of surface intake capacities due to urbanisation or overgrazing over up to 20% of their basin area, with reported seasonal increases typically amounting to *c*.100 mm yr^{-1} (Costa, this volume; Bruijnzeel, 2004). It would be wise, therefore, not to underestimate the *cumulative* effects of adverse land use practices at these larger scales. Nevertheless, it must be repeated that the magnitude of truly large storm *events* (floods) is not influenced appreciably by land cover, because under such extreme conditions the flood generation process is governed largely by the amount, intensity, duration and spatial distribution of the rainfall (Bonell, Scott *et al.*, both this volume). In other words, the rainfall accommodating capacity of forests is not infinite (see also below).

How does the above outline of the current scientific consensus vis á vis the climatic, soil and water impacts of tropical forest logging and clearing relate to the perceptions of land managers, policy makers and the public at large? Table 39.2 holds the most commonly held perceptions against the light of the best available scientific evidence.

Table 39.2 suggests that the perceptions of the hydrological impacts of the most fundamental and widespread types of forest conversion ('deforestation') in the tropics held by practitioners, policy-makers and the general public on the one hand and researchers on the other, are arguably less different than claimed by some. Much has been made of late of these different perceptions as preventing rational land use decisions to be made (Calder, 1999; FAO, 2004; cf. Kaimowitz, this volume) but a closer inspection of Table 39.2 reveals that in many cases these contrasting views relate to differences in degree or frequency of occurrence (erosion, stormflows, even rainfall) rather than true differences in kind. As such, their reconciliation is perhaps less difficult than would appear at first sight.

Much of the confusion regarding the increase or decrease of streamflow following tropical forest clearance can be traced back to two aspects: (i) the need to distinguish between annual and seasonal water yields, and (ii) the fact that most, if not all experimental catchment studies pertain to *controlled* land use changes, the hydrological impacts of which have been monitored over relatively short periods of time only (typically up to three years,

Table 39.2. *Common perceptions about the environmental impacts of tropical deforestation*

Commonly held perceptions	Scientific experience	Qualifications
• Forests increase rainfall and their clearing results in much reduced and more irregular rainfall.	• Clearing of forests is unlikely to significantly reduce total rainfall but reduced rainfall at the onset and near the end of the rainy season is possible. • Extra inputs via occult precipitation (fog) received by montane cloud forests are lost upon full clearing.	• Forest clearing must be complete and over very large areas (>>10 000 km^2) to have an effect; proven effect <10% in the case of entire Amazon basin (continental climate) and much less under 'maritime' tropical climatic conditions.
• Forests act like sponges absorbing water during rainy season and releasing it evenly during the dry season. Cutting of forests dries up water supplies, particularly during the dry season, because the 'sponge' effect becomes lost.	• Cutting of forests increases total water yield, particularly during low flow periods. • Dry season flows much reduced if soil water intake capacity seriously impaired (as in severely degraded or urbanised catchments). • Clearing of cloud forests may well lead to reduced dry season flows and possibly total yield.	• Increased total and seasonal water yields under pasture or cropping only manifested as long as surface infiltration capacity is maintained. Fine textured soils most vulnerable to degradation.
• Cutting of forests causes floods as the 'sponge effect' is then lost.	• Cutting of forests affects stormflow volumes for small- to medium-sized events and at the local scale (<10 km^2). Little or no impact on size of extreme events (floods) at any scale but dense road network also enhances maximum peak flows. Wet season flows (but not events) from very large basins probably increase due to cumulative effect of reduced infiltration opportunities.	• Post-forest land use must afford good surface cover and road layout. Otherwise stormflows up to medium-sized events much increased (as in severely degraded catchments).
• Cutting of forests increases erosion, landsliding and stream sedimentation.	• A good vegetation cover of any kind limits or eliminates surface erosion. • Good forest cover reduces shallow landsliding but not deep-seated slides which are governed by rainfall, geology and topography.	• Soil protection afforded by undergrowth and leaf litter rather than tree canopy; grazing, fire and litter collection in forests must be strongly avoided.

occasionally longer; Grip *et al.*, this volume). As to the first, in the absence of actual streamflow measurements it is difficult to tell whether the increases in rainy season stormflows and decreases in low flows witnessed by people living in gradually degrading tropical catchments actually add up to increased total annual water yields or not. However, there is little actual difference between the layman stating that 'deforestation' leads to diminished low flows due to the loss of the 'sponge effect' of the forest and the scientist having to agree, provided that surface infiltration characteristics have been degraded sufficiently over time for this to happen. Similarly, the public view that 'floods' invariably increase after forest clearance and that of the scientist acknowledging that stormflows do increase in all but the most extreme cases is beginning to border on semantics. Therefore, it is more productive to state that stormflows are increased after forest removal *up to a certain threshold* (beyond which the effect of land cover is overridden by those of extreme rainfall and limitations in soil water holding capacity), or that low flows will decrease once a certain level of surface degradation has been reached, than to merely dismiss the 'sponge theory' as folklore or an anachronism (Calder, 1999; FAO, 2004).

Furthermore, in the heated debate on the hydrological role of tropical forests it is generally overlooked that the circumstances associated with controlled (short-term) catchment experiments may well differ from those of many 'real world' situations *in the longer term* (Bruijnzeel, 1989; 2004). The continued exposure of bare soil during agricultural cropping, the compaction of topsoil by machinery or intensive grazing, the gradual disappearance of soil faunal activity, and the increase in areas occupied by impervious surfaces such as trails, roads and settlements, all contribute to gradually reduced rainfall infiltration opportunities in cleared landscapes. No experimental catchment study has lasted long enough, however, to document the long-term effects of increasingly degraded conditions on streamflow amounts and regime. As such, *both views (diminished or increased dry season flows after clearance) must be considered correct, depending on the situation.* Where infiltration is maintained sufficiently, as under controlled experimental conditions or rational land use, the reduced water use associated with forest removal will show up as increased dry season flow (Grip *et al.*, this volume). Conversely, where infiltration and groundwater recharge become seriously impaired by surface compaction and crusting, – as is eventually the case in many real world situations – diminished dry season flows inevitably follow, despite the fact that the reduced evaporation should have produced higher baseflows. In the layman's terms: the 'sponge effect' is lost (Table 39.2).

A related aspect concerns the fact that long-term fluctuations in rainfall arising from natural climatic variability (ENSO-related, decadal and multidecadal variations; Callaghan and Bonell, Mahé *et al.*, both this volume) are not covered adequately by short-term experiments. Such fluctuations have both short- and longer-term impacts on catchment hydrology – notably the (more frequent) occurrence of peak flows or diminished dry season flows – which may be attributed erroneously to the change in land cover rather than climatic variability. This rather humbling lack of *long-term* catchment studies representing *actual* hydrological conditions experienced by countless tropical people (cf. Van Lynden and Oldeman, 1997) both calls for more modesty (some would say: less arrogance) on the part of scientists when communicating the results of (controlled) hydrological experiments to practitioners and the public at large, and for stepped-up efforts to remedy this deficiency. We will come back to this important point in the final section.

HYDROLOGICAL IMPACTS OF REFORESTING (DEGRADED) HUMID TROPICAL LANDSCAPES

Not all forest clearing leads to crops or pasture and the FAO data on tropical land cover change quoted earlier show that significant areas of cleared land are being re-vegetated by natural forest regrowth or reforested using tree plantations. The rate at which regrowth occurs after forest clearing is often quite rapid, provided a reasonable amount of topsoil remains and tree seeds or seedlings are still present or are able to colonise the site from nearby natural forest remnants. Canopy closure often occurs within the first year or two (depending on the degree of surface disturbance) and the leaf area index of the new vegetation usually increases rapidly during the first five years. Likewise, the reflection coefficient of the canopy also recovers quickly to pre-disturbance values. As a result, water use of vigorously growing secondary vegetation resembles or even exceeds that of old-growth forests after about five years for a few decades, partly also because of positive heat advection from neighbouring cleared areas (Hölscher *et al.*, this volume).

Reforestation in the form of tree plantations may result in a somewhat slower recovery of canopy cover than natural regeneration although this depends on the species planted and, again, on the degree of soil degradation experienced before planting. Whilst some plantations soon attain large leaf areas and high levels of plant productivity (notably acacias, pines and eucalypts), their structural complexity (e.g. multiple canopy levels) takes longer to develop than in the case of natural regeneration (Beadle, 1997; Hölscher *et al.*, this volume). The water use of tree plantations after full canopy closure resembles that of the old-growth rainforests they replaced but possibly exceeds it in some cases (e.g. the extremely fast-growing *Acacia mangium*). Normally, therefore, the overall effect on catchment water yield is that of a gradual return to pre-clearing levels after 5–10 years (Grip *et al.*, this volume). However, where fast-growing trees are planted on (natural or fire-climax) grasslands, degraded scrubland or abandoned agricultural fields, the effect on streamflow is invariably pronounced and lasting. Annual water yields normally start to become reduced within three years after planting, particularly under more seasonal rainfall regimes, whereas the maximum reduction in flows after full canopy closure may reach values of 200–700 mm yr^{-1}, depending on the vigour of the species, rainfall amount and seasonal distribution, and degree of positive heat advection from neighbouring non-forested areas. Although small- to medium-sized stormflows are also reduced significantly after the establishment of a good tree cover on former grasslands, the greatest decreases in flow are usually observed for baseflows and during the dry season (Scott *et al.*, this volume).

There is virtually no experimental information on the effects on low flows after reforesting severely degraded catchments in the tropics. On the one hand, the improved surface infiltration associated with forestation will (ultimately) help to increase soil moisture storage and groundwater recharge but major improvements of topsoil hydraulic characteristics take at least several decades (Scott *et al.*, this volume). On the other hand, there is the opposing force of the (much more rapidly, i.e. within a few years) increased water use by the trees which tends to reduce soil water storage, with higher rates of water use normally being associated with faster growing trees. Therefore, although it is theoretically possible that the benefits of improved rainfall infiltration override the higher evapotranspiration by the trees, *the balance of probability is that the already diminished dry season flows in most degraded catchments are likely to be reduced even further by forestation, with the possible exception of very severely disturbed surface conditions* (cf. Table 25.4 in Scott *et al.*, this volume). Because plantation water use tends to diminish somewhat after 25–30 years (when some species begin to lose

Table 39.3. *Common perceptions about the hydrological consequences of reforestation*

Commonly held perceptions	Scientific experience	Qualifications
• Reforestation increases rainfall.	• No evidence that this occurs.	• Forestation over very large areas (>10 000 km^2) may have an effect but this is yet to be demonstrated experimentally. • Planting of trees (also in hedgerows) in montane cloud belts increases net precipitation arriving at the ground due to trapping of 'occult' precipitation (fog, wind-driven rain).
• Reforestation causes rivers to flow again (particularly in the dry season). • Reforestation prevents floods.	• In the short term (*c*.30 years) reforestation decreases total water yield and low flows. • Effect for severely degraded catchments unknown. • Reforestation reduces small to medium stormflow events but cannot prevent extreme events.	• As new forests mature (>30 years) their water use declines, and water yields may increase somewhat again. • The deeper the soil, the stronger the stormflow-reducing effect of the forest.
• Species used in reforestation are all similar as to their (positive) hydrological effects. • Reforestation rapidly checks and eliminates all erosion.	• Fast-growing species tend to use more water than slow-growing species. Hydrological effects vary with proportion of area reforested and with tree density. • Gully erosion and deep-seated landslides not eliminated. • Erosion under deciduous broad-leaved forest (e.g. teak) on heavy clay soils often rampant.	• Eucalypts, acacias and pines fastest growers and among highest consumers. • High water use sometimes exacerbated by positive heat advection from surrounding grass- and scrubland. • Additional mechanical measures needed. • Undergrowth and litter layer must be able to develop for positive effect to become manifest.

their vigour) and infiltration is much improved by then, a more favourable balance may be expected during these later stages although experimental evidence for this in the context of land rehabilitation appears to be lacking (Young, 1997; Scott *et al.*, this volume).

These hydrological research findings are not always known by managers or policy makers, and certainly not by the general public, to whom reforestation simply equals the (rapid) restoration of streamflow regimes (Table 39.3). As a result, significant investment decisions continue to be made on the assumption that reforestation will cause an increase, not a decrease, in river flows. Thus, unlike the case of tropical *de*forestation (forest removal) described in the previous section (Table 39.2), there still exists a major discrepancy in the perceived hydrological benefits of *re*forestation between researchers on the one hand, and land managers (including local communities) on the other (cf. Calder, 1999; Kaimowitz, this volume). Just like researchers seem to have overlooked the importance of gradually deteriorating catchment conditions over time, so does the layman seem to forget that it will take many decades for the soil, and therefore its hydrological water accommodation and storage functions, to be restored (Bruijnzeel, 2004; Scott *et al.*, this volume). As such, there are no instant solutions. Unfortunately, despite the fundamental and societal importance of the 'low flow problem', there appear to be no experimental catchments currently operating in degraded humid tropical steeplands in which the underlying hydrological processes are studied in sufficient detail to clarify the uncertainties described above. Most published results pertain to hillslope plots or catchments that are too small to sustain perennial flow, thereby making it impossible to evaluate the effect of soil rehabilitation or reforestation on low flows (Scott *et al.*, this volume). Considering the extent of moderately to severely degraded soils signalled earlier (e.g. in South and South East Asia), further work is urgently needed (see also the next section).

In contrast to the effect of reforestation on water yield, there is usually a rapid positive effect on erosion, the occurrence and frequency of shallow landslips, and stream sedimentation *once a good vegetation cover is re-established* (usually within 2–3 years). Forestation may be less successful at checking gully erosion, and additional structural measures (such as gully plugs) may be needed (cf. Critchley, this volume). In addition, under strongly seasonal rainfall conditions, deciduous broad-leaved trees like teak planted on heavy clay soils may still experience heavy surface erosion (Scott *et al.*, this volume). Although most sedimentation problems may thus be solved by forestation, land managers would do well to remember the associated hydrological price tag (i.e. reduced streamflow) and consider alternative solutions to achieve near-complete reduction of erosion whilst limiting streamflow reductions (e.g. agroforestry; Young, 1997; Wallace *et al.*, this volume).

SOME CRITICAL UNRESOLVED PROBLEMS

One of the most notable trends in the last decades of the twentieth century was the increasing evidence of environmental damage being caused by unconstrained resource usage (forests, water, soils). Many policy makers and forest managers agreed that the logging practices then being applied in tropical forests were unsustainable and that new codes of practice were necessary if ecologically sustainable forest management was to be achieved (Cassells and Bruijnzeel, this volume). Similar concerns were felt about other common land use practices (Bridges *et al.*, 2001). Guidelines for best management practices (BMPs) have been developed in relation to timber harvesting (Thang and Chappell, this volume), land clearing (Pearce and Hamilton, 1986; Hamilton, this volume) and post-forest agricultural cropping (Critchley, this volume); but apart from socio-economic and institutional reasons preventing the full implementation of these BMPs in the tropics, there are also a number of unresolved hydrological issues needing further research. The most important of these are briefly discussed below, starting with the more pressing practical issues faced by land managers working at a local scale. These are then followed by more fundamental issues which should also concern policy makers, such as the prediction of the consequences of climate change for tropical forests, and the prediction of the hydrological impacts of land cover change at larger scales.

Chief hydrological research needs

Many questions concerning evaporation and runoff- or sediment-generating processes in old-growth tropical lowland forests are now reasonably well understood (Roberts *et al.*, Bonell, Douglas and Guyot, all this volume). However, important information is still lacking about the hydrological functioning of several specific types of rainforest, such as montane cloud forests (Bruijnzeel, this volume) and swamp forests (Hooijer, this volume). In addition, hydrological knowledge about a number of widely occurring post-clearance field situations is still far from complete, particularly with respect to managing dry season flows. Some of these problem areas are listed in Table 39.4.

Arguably, one of the more acute hydrological problems faced by tropical communities, land managers and researchers is to find ways to improve, and where possible restore, the disrupted flow regimes associated with severely degraded steeplands whilst at the same time maximising agricultural productivity (cf. Van Lynden and Oldeman, 1997; Bridges *et al.*, 2001). As indicated previously, the available evidence suggests that reforestation using fast-growing exotic tree species is likely to have an adverse effect in that low flows may well be reduced even further. The search for alternative solutions is hampered, however, by the fact that the hydrological behaviour of areas with degraded soils is poorly documented (at least in the 'official' literature), and by the absence of information on the water use efficiencies and rooting depths of many tree species and crops that could be used for land rehabilitation, e.g. in an agroforestry context (Scott *et al.*, Wallace *et al.*, both this volume). Another critical factor concerns the time required for the soil's hydraulic characteristics to recover sufficiently. Finally, particular attention needs to be given to the recharge, storage and fluxes of the groundwater feeding the baseflows. Thus far, *groundwater-surface water interactions have been barely studied in the humid tropics* (cf. Sandström, 1998). A range of strategies (e.g. block planting vs. vegetative 'filter strips' at strategic places within the catchment; natural regeneration vs. enrichment planting, mixed vs. single species plantations, exotic vs. indigenous species plantations, re-introduction of soil fauna, and mechanical measures to promote infiltration such as contour trenches) will need to be tested. Although much soil and ecosystem restoration work has been, and is being carried out already, the hydrological aspects of land rehabilitation in the tropics (other than infiltration-excess overland flow and surface erosion) seem to have received comparatively little attention so far (Young, 1997; Bridges *et al.*, 2001; Scott *et al.*, this volume). A concerted, *long-term* research effort to help solve this pressing hydrological problem is long overdue. Such work should focus not only on the general functional consequences of reforestation using some of the more widely planted tree species (pines, eucalypts, etc.) but also provide information as to where in a catchment reforestation is needed most to achieve particular hydrological benefits (Lamb and Gilmour, 2003; cf. Van Noordwijk *et al.*, 1998).

On a related note, a major outstanding research question is to what extent and over what time scales secondary forests regain their original hydrological characteristics (Hölscher *et al.*, this volume; cf. Giambelluca, 2002). Other, more locally important, questions concern the effect of montane cloud forest conversion to pasture or vegetable cropping on dry season flows, and the impacts of draining peatswamp forests. Likewise, the hydrological consequences of the seemingly more frequent fires in areas with secondary and logged-over forests (especially during El

Table 39.4. *Types of land cover or management situations in the humid tropics for which hydrological knowledge is still rather incomplete*

Land cover type or management situation	Chief outstanding research problem
Land cover type	
• Severely degraded agricultural/grazed landscapes	• Allegedly expanding but hydrological functioning poorly documented, thereby providing room for speculation as to the hydrological effects, or the best strategies for restoration.
• Montane cloud forests	• Intercept variable but poorly documented amounts of 'occult' precipitation; water use largely unknown. Hydrological significance as water producers therefore underestimated in some cases and overestimated in others (see also *Management practices* below).
• Complex secondary forests	• Areas expanding rapidly throughout tropics. Considerable variation in structure and floristic composition but hydrology only beginning to be addressed.
• Tree plantations and agroforestry systems	• Water use efficiency and rooting depths of many plantation (fast- vs. slow-growing) and agroforestry species unknown, hampering choice in a land rehabilitation context.
• Swamp forests	• Flood attenuation capacity possibly more limited than previously assumed. Virtually no information on catchment hydrological effects of their drainage.
Management practices or disturbances	
• Reduced impact logging (RIL) including maintaining riparian buffer zones	• Promoted as a means of achieving ecologically sustainable forest management, but the presumably moderating effects of RIL on runoff, erosion and stream sedimentation poorly documented. Riparian buffer strips presumably important as sediment traps and for protecting stream banks but little experimental evidence available. Controversy to what extent protection of streamhead hollows is needed/feasible.
• (Wild)fires	• Becoming more common in many parts of the tropics, especially in secondary and logged-over forests; hydrological impacts (increased peakflows?) largely unknown but overall expected to be negative.
• Clearing for agriculture or pasture	• Effect of cloud forest conversion to pasture on dry season flows unknown and possibly negative.
• Rehabilitation of land with seriously degraded soils	• Unknown whether and under what conditions low flows can be improved through soil rehabilitation / reforestation. Mixed-species approaches give better ecological results but associated hydrological effects unknown. Recovery times for soil hydraulic characteristics poorly documented.

Niño – Southern Oscillation (ENSO) periods) still await quantification (e.g. exacerbation of flooding upon the return of the rains?; Malmer *et al.*, this volume).

Finally, although much is expected from reduced-impact logging (RIL) techniques in terms of reducing the adverse soil and water impacts of timber harvesting (Chappell, Tych *et al.*, this volume), their economic viability in steep terrain is considered doubtful (Cassells and Bruijnzeel, Thang and Chappell, both this volume). Riparian buffer zones constitute an integral element of RIL methodology but their spatial lay-out and extent is still a matter of debate. Areas of flow convergence (topographic and streamhead hollows) are often 'hot spots' for runoff and sediment production (Bonell, Douglas and Guyot, both this volume) and thus warrant vegetative protection (Hamilton, this volume). Yet, riparian buffer zone protection in most internationally adopted timber harvesting guidelines is limited to channels carrying perennial streamflow only, thereby leaving the hydrologically and geomorphologically sensitive channels having only ephemeral flow unprotected. Thang and Chappell (this volume) argue that inclusion of the latter in the protected zone would make forest exploitation essentially uneconomic because of the consequential increases in protected area. Hamilton (this volume), on the other hand, calls for the complete protection of all steep headwater areas. Further work is needed to derive workable solutions that are acceptable to both planners of logging operations and those more concerned with environmental impacts. Ideally, such work should combine field experimentation to test the efficiency of strips of varying widths to trap incoming sediment and nutrients from disturbed upslope parts (Thang and Chappell; cf. Proctor, both this volume), and spatially distributed modelling of soil water conditions ('topographic wetness') to demarcate those areas most vulnerable to disturbance (Bonell, this volume). Application of such topographically driven models should introduce more flexibility to the delineation of areas in need of protection compared with the current blanket application of fixed-width buffer zones within existing forest management guidelines (Cassells and Bruijnzeel, this volume). However, as discussed in more detail by Bren (2000), application of these models is not without problems either. Resolving the buffer zone debate must be considered a prime hydrological research need for the achievement of ecologically sustainable timber harvesting in tropical rainforests.

On a more fundamental note, aside from the much publicised global warming issue and its potential consequences for tropical forest ecological functioning and plant productivity in general, the hydrological implications of the lack of stationarity in the climate of the humid tropics (Mahé *et al.*, Callaghan and Bonell, both this volume) are in need of further research. It is clear that there has been considerable paleo-climatic change as well as inherent (i.e. not human-induced) historical inter-annual, decadal and multi-decadal climatic variability. Of particular importance in this regard are the impacts of the ENSO phenomenon in the form of drought periods in some areas (including associated forest fires) (Table 39.4) and flood periods in others, e.g. in relation to shifts in the preferential tracks and frequency of tropical cyclones (Callaghan and Bonell; cf. Mahé *et al.*, both this volume). *There is a distinct need for long-term monitoring in hydrological research to incorporate all aspects of the ENSO cycle*. In particular, any inherent (cyclic) climatic variability effects on the magnitude of water and sediment yields need to be taken into account when evaluating the more immediate impacts of forest exploitation or conversion to other land uses (Chappell, Tych *et al.*, this volume).

A related aspect concerns the need for more information on the vertical distribution and function of (deep) roots in connection with sustaining evaporation (and productivity) during extended periods of drought (e.g. during ENSO periods). Better knowledge of rooting patterns is also important in the context of improved predictions of the large-scale impacts of tropical forest conversion using global circulation models through a more realistic representation of plant physiological response to soil water deficits (Roberts *et al.*, Costa, both this volume). Deep roots assume additional importance in relation to the (presumed) maintenance or restoration of soil fertility through the cycling of nutrients taken up from deeper layers and (ultimately) returned to the soil surface via leaf litterfall and decomposition although evidence for this from the humid tropics is scarce (Proctor, Scott *et al.*, Wallace *et al.*, all this volume). More work is needed.

Finally, although montane cloud forests are usually associated with more or less persistently wet and foggy conditions, there is increasing evidence that global warming threatens their ocurrence, both through a gradual lifting of cloud condensation levels and through diminished frequency and intensity of fog. Massive declines in the populations of anoline lizards and frogs have been attributed to such climatic effects. Similarly, it has been suggested on the basis of meso-scale atmospheric modelling that deforestation in coastal lowlands may affect cloud condensation levels in adjacent mountain ranges, thereby indirectly affecting the hydrological and ecological functioning of cloud forests (Bruijnzeel, this volume). In view of their importance as suppliers of good quality surface water, efforts to study the hydrology and vulnerability of cloud forests to global and regional climate change need to be stepped up.

A more general problem for managers and researchers alike is that of *scale*. Many forest and land managers work at the small sub-catchment scale (10–100 km^2). As shown in the previous sections, the hydrological consequences of many local-scale land use practices and changes therein are now reasonably well understood in a quantitative sense. On the other hand, most land use planning activities and many policy decisions are usually made over areas much larger (>1000 km^2) than those used in experimental situations. These large areas often have complex geologies, vegetation patterns and land use practices, not to mention highly variable rainfall inputs and topography. The key problem faced by land managers is how to deal with the cumulative impacts of many smaller-scale land use decisions to ensure adequate dry season flows, limit damage by stormflows and maintain overall water quality, and for the researchers to predict these quantitatively. Where do we stand in this respect? What (new) tools are at our disposal and what additional research is needed to achieve better hydrological predictions at the meso- and macro-scales?

As for the prediction of changes in *baseflows* (dry season flows) for meso-scale basins (*c.*1000 km^2), the work of Watson *et al.* (1999) has demonstrated that it is feasible to simulate long-term changes in flows associated with changing land cover, *as long as good basic information on topography, rainfall and land cover type* (e.g. through remote sensing; Held and Rodriguez, this volume) is available. Furthermore, flows from different (ungauged) parts within a river basin may be approximated by a down-scaling technique known as streamflow 'disaggregation'. The procedure is based on the assumption that the contribution from each part of the basin is proportional to a model-based 'topographic index'. Schreider and Jakeman (this volume) present the successful application of the technique to a large river basin in northern Thailand. Thus, and as indicated earlier, with the database on tropical vegetation water use continuing to grow (Grip *et al.*, Hölscher *et al.*,

Scott *et al.*, Wallace *et al.*, all this volume), our ability to predict changes in baseflows associated with land cover change can be expected to increase commensurately, even more so in view of the rapid improvements in estimating evaporation via remote sensing. Furthermore, promising developments using airborne laser scanning systems may in the near future be able to generate digital elevation maps of below-canopy topography, provided there are sufficient gaps in the canopy (2–5%) (Held and Rodriguez, this volume). As indicated earlier, good topographic information is an essential element of distributed hydrological modelling.

It should be remembered, however, that this optimism with respect to our ability to predict changes in low flows does not yet apply to river basins having a significant proportion of their area under degraded soils. Information on hillslope and catchment hydrological behaviour, and indeed vegetation water use, under such conditions is still woefully inadequate (Scott *et al.*, this volume). A further caveat concerns the possibility that, at the very large, regional scale (e.g. the whole of Amazonia) the trend in the effect of forest clearance on streamflow may be opposite to that found at the local and meso-scale because the associated absolute reduction in rainfall may exceed the reduction in evaporation. In other words, the net effect may be a (slight) decrease in streamflow instead of the normally (i.e. at smaller scales) observed increases (Costa, this volume). More research is needed to better understand the joint hydrological and climatic impacts of land cover change across a range of scales.

The situation with respect to the notoriously difficult prediction of *stormflows* over larger areas (regardless of land use change) is a lot less favourable, but to some extent this problem is universal to both the humid tropical and temperate environments. On the one hand, this reflects the already mentioned fact that land cover effects on stormflow magnitude tend to be 'diluted' over larger areas; on the other hand, there are also major data and modelling restrictions, particularly in the tropics. A key research question here is at what scales, in different environmental settings, are the effects of land cover change overridden by hydrometeorological factors, such as antecedent catchment wetness and rainfall characteristics? A major impediment to answering this question, however, is the lack of sufficient spatial and temporal resolution of rainfall inputs across large tropical river basins, especially in steep headwater areas (Bonell *et al.*, this volume). A related complication in some headwater areas (especially in the non-equatorial tropics) is that the available rainfall data may represent more or less serious underestimates as substantial amounts of wind-driven rain caught by (tall) vegetation (and reaching the soil via crown drip or stemflow to contribute to streamflow) go largely unrecorded by standard rain gauges (Bruijnzeel, this volume). This lack of precipitation data is a significant contributor to errors in hydrological predictions across a range of scales, particularly in areas where rainfall is predominantly convective. The spatial and temporal organisation of the latter is notoriously variable (Bonell *et al.*, this volume). A more integrated meteorological-hydrological approach is needed which includes the use of rain radar (and ground truthing by denser rain gauge networks) to better represent the temporal and spatial movement of rain fields across tropical river basins (Bonell, this volume). There is also an understandable global lack of data on hillslope and catchment hydrological response during extreme (flood-producing) meteorological events (such as the passage of a tropical cyclone; Bonell *et al.*, Scatena *et al.*, both this volume). Widespread occurrence of overland flow (both of the infiltration- and saturation-excess types) can be expected under such conditions (Barnes and Bonell, this volume). Yet, until hard information on the hydrological response of different land cover types during truly extreme events becomes available, there will remain room for speculation as to the flood-preventing capabilities of a good forest cover. Buttle and McDonnell (this volume) outline new hydrometric methodologies such as hillslope saturation detectors which may be used in connection with geochemical and isotope tracer-based hydrograph separation techniques to help ascertain the relative importance of contributions to overall storm runoff by 'old', pre-event water (usually subsurface flow) and 'new', event-water (usually overland flow). Other important applications of stable isotopes in this regard include the assessment of surface runoff contributions (i.e. 'new', event-water) from compacted areas (roads, settlements) or degraded fields (cf. Grip *et al.*, this volume) for different antecedent wetness conditions and rainfall intensities and durations.

In the absence of high quality information on the spatial and temporal characteristics of tropical rainfall, prospects for the application of physically-based predictive models of catchment runoff response to rainfall will remain extremely limited. Indeed, even under relatively data-rich conditions, predictions of stormflow amounts by such models are often (very) poor. One major additional source of error relates to the difficulty to adequately represent soil-rock permeabilities at the hillslope scale in these models, i.e. beyond the commonly used point scale of measurement (Bonell, this volume). The contribution by Chappell, Bidin *et al.* (this volume) presents a new and promising methodology for estimating (effective) permeability at the (sub-)hillslope scale. On a related note, there is emerging evidence for the possible occurrence of 'pressure waves' in subsurface stormflow which exhibit much higher velocities than the ordinary 'Darcyan' flow on which hydrological process models are normally based. If true, considerable error in the predictions by these models would result. More research is required to address this critical issue (Bonell, this volume).

In response to the preceding difficulties with the field parameterisation of hydrological process models, some have called for greater simplicity ('parsimony') of models whilst trying to retain some measure of physical interpretation of the model's internal functions (cf. Barnes and Bonell, this volume). These so-called Parametrically Parsimonious Conceptual Models (PPCMs) will be of increasingly practical use in both research and land-water resource management situations where only land cover and rainfall-runoff data are available, as demonstrated by the successful example given by Schreider and Jakeman (this volume). Moreover, through the application of PPCMs it is possible to derive 'emergent properties' of the catchment hydrological system (such as groundwater storage volumes and streamflow time constants reflecting effective permeability at the catchment scale) that control catchment runoff response to rainfall at larger scales. These would not be detectable using point measurements (of permeability, etc.; Barnes and Bonell, this volume). On the other hand, purely statistical approaches for the detection of trends in the streamflow records of humid tropical river basins (discussed by Kundzewich and Robson, this volume) will have restricted application because of the limited number of long-term runoff datasets available and the very considerable scatter encountered in many datasets (cf. Bruijnzeel, 2004).

The maintenance of *water quality* also remains a difficult problem for tropical land managers and there is a lack of performance indicators for assessing the overall condition of catchment or ecosystem health. One way of integrating the effect of land use changes on water quality over time might be to assess changes in aquatic biota. That is, to use these plants or animals as bio-indicators or in bioassays. There is unequivocal evidence that stream and river biota are sensitive to changes in discharge rate and sediment (as well as light and nutrient) levels but unfortunately too little of this kind of work has been done to allow predictions of the impacts of particular land use changes to be made at present. Similarly, it seems there are no generic indicators such as particular aquatic faunal groups that might be used as proxies for water quality across a variety of tropical rivers (Connolly and Pearson, this volume). Much more work is needed in this respect.

Alternatively, critical sub-catchments and other hydrological 'hot spots' within river basins may be identified through geographical information system (GIS)-based spatial analyses of such key variables as rainfall erosivity (from generalised data on intensity), slope gradient and length, soil erodibility, land cover type and percentage bare soil. At the larger scales, as well as for planning purposes, such erosion hazard predictions had best be based on the widely used empirical Universal Soil Loss Equation (USLE). However, for catchments up to a few square kilometres in size the comparatively simple steady-state application of spatially distributed topographically driven models like TOPOG can be used to predict areas of flow convergence, and thus surface and gully erosion hazards, by linking such models to relatively simple erosion models. Likewise, the most probable locations (but not the timing) of landslides can be predicted in a similar manner by linking models like TOPOG with some kind of basic slope stability function (Yu, Cassells and Bruijnzeel, both this volume). However, the current generation of physically-based erosion models are too heavily parameterised to be of practical use yet in the generally data-deficient humid tropics. In addition, their application in (small) data-rich catchments has failed to yield useful predictions of the exact spatial extent of areas with sediment deposition or of the amounts involved. Therefore, like their hydrological counterparts, these models will have to remain a research tool for the time being (Yu; cf. Barnes and Bonell, both this volume).

Outstanding economic and institutional issues

Many land use decisions involve trade-offs between environmental and economic costs and benefits. However, most standard economic evaluations (e.g. of logging or plantation operations) do not explicitly include the environmental consequences of man's activities, such as increased erosion and stream sedimentation or a negative change in streamflow regime (called 'externalities' in the economist's jargon). Indeed, there appear to be very few studies (both for tropical and temperate conditions) in which estimates of the respective environmental and economic costs and benefits of particular land use activities have been made. In addition, in the few cases where a deliberate attempt has been made to include hydrological externalities, inferences usually had to be made from the general literature rather than that actual data could be used (Aylward, this volume). A further complication relates to the contrast in time scales associated with economic policy development (usually on the basis of short-term economic information) and the assessment of environmental consequences (requiring longer term hydrological data to enable the separation of effects of climate variability from those of land cover change) (Aylward; cf. Kaimowitz, both this volume).

The friction between economic reasoning and resource conservation has already been indicated within the introduction of this chapter as being an important factor underlying the continued over-exploitation of tropical resources (C. Hall and Ko, A. Hall, this volume). Indeed, the inclusion of the true environmental costs of, for example, high intensity tropical timber harvesting in steep terrain might well render the balance between costs and benefits of such operations negative (Cassells and Bruijnzeel, this volume). As a result, certain protective measures considered necessary by the hydrologist (e.g. extending riparian buffer zones to streamhead hollows to minimise downstream sedimentation problems) are not adopted currently because the associated increase in protected area would make the logging uneconomic (Thang and Chappell, this volume). Under the prevailing economic paradigm, therefore, environmentally undesirable practices are allowed to continue. Similarly, from the economist's perspective, the production benefits from cleared forest lands would have to be reduced substantially before land rehabilitation may even be considered economically justified (Aylward, this volume). Hall and Ko (this volume) therefore propose an alternative economic model concept in which the goals of economic development are more aligned with environmental sustainability by focusing on ways to increase economic outputs against less material and energy inputs. Although the Hall and Ko analysis concentrates on the use of energy, similar changes in attitude are required in relation to the use and management of water resources (see Rosegrant *et al.* (2002) for details).

Whilst the realisation of these more visionary models may still take decades, an increasingly advocated approach towards better conservation of upland natural resources and guaranteeing a more equitable sharing of costs and benefits of good land husbandry practices in the uplands between different groups of stakeholders works within the traditional economic realm (Murdiyarso, Aylward, both this volume). A prime example concerns financial compensation by downstream communities and city dwellers to upland farmers to promote sound land use practices via so-called 'payment for environmental services' (PES) schemes (Pagiola *et al.*, 2002). In the past, the costs of, for example, leaving forests intact on steep, unstable slopes or maintaining adequate buffer zones along streams and waterways were usually borne entirely by land owners in upper catchments whereas any resulting water quality benefits were mostly restricted to downstream water users. In theory, PES provides a mechanism by which uplanders may improve their negotiating position when it comes to major land use decisions whereas before they were governed more by economic pressures (Murdiyarso, this volume). Whilst PES appears to be a promising approach, there are important obstacles. For example, it often proves difficult to assign fair monetary values to (leaving) particular forest types (in place) for the simple reason that the climatic, hydrological and biodiversity data on which to base such an assessment are lacking. Tropical montane cloud forests provide a case in point. In many areas they represent one of the last surviving sources of good quality surface water, in contrast to places downstream where the quality of the water in rivers and lakes is often degraded. In addition, cloud forests are probably the most effective water suppliers of any tropical forest type (as well as a treasure house for biodiversity) and should therefore be preserved on both accounts (Hamilton, this volume). At the same time, the hydrological functions of cloud forests are understudied (Bruijnzeel, this volume), thereby hampering their potential preservation through PES schemes. In addition, a PES system may collapse when the 'goods are not delivered', e.g. when excessive downstream flooding or sedimentation still occur due to an extreme event, despite the presence of a good forest cover in the headwater areas (cf. Kaimowitz, this volume). This requires not only much more effort on the part of Governments to enable hydrologists to supply the necessary information but also improved communication between researchers, land managers and policy makers through appropriate institutional structures (see below). Needless to say, *the importance of maintaining adequate rainfall and streamflow collecting networks cannot be overstated* in this regard. Since hydrological networks are on the decline in many tropical countries, one way of counteracting this trend could be to use some of the revenue generated through PES schemes (as is being considered in some Latin American countries) as part of a 'fine-tuning' of the size of the payments (cf. Pagiola *et al.*, 2002; Aylward, this volume).

Some unresolved problems hampering optimum land and water management in tropical river basins are *institutional* in nature. Although a considerable body of knowledge has been incorporated in various sets of guidelines for forestry and clearing operations (notably with respect to measures to prevent excessive erosion and stream sedimentation; Hamilton and Pearce, 1986; Cassells and Bruijnzeel, Thang and Chappell, Hamilton, all this volume), there is still a lack of application of existing knowledge concerning the effects of land use on baseflows (notably of reforestation; Scott *et al.*, this volume). Perhaps the most crucial problem, however, is our incomplete understanding of the social and institutional drivers leading to the degradation of hydrological functions in many upland agricultural catchment areas, including the reasons for the frequent failure of soil conservation measures to be adopted by small farmers. As hinted at in the introductory paragraphs, apart from economic and population pressures, important underlying factors relate to insecurity or lack of land ownership and the implementation of 'top-down' forestry and soil conservation policies by governments without due consideration of their societal acceptability (A. Hall, Critchley, both this volume). Inevitably this means there are often inadequate institutional structures, procedures or experience to ensure effective stakeholder involvement (including upland farmers) in the identification of key research questions, land use planning or management decision-making (Murdiyarso, this volume).

CONCLUDING REMARKS

A common criticism from policy makers and practitioners is that the agendas of researchers are too dominated by the pursuance of science for the sake of science without sufficient consideration of the practical application of research results. At the same time, policy makers lose credibility because they often lack the scientific background or inclination to incorporate new research findings in their policies. Whilst a considerable body of research results has been incorporated in guidelines for forestry and land clearing operations (Pearce and Hamilton, 1986; Cassells and Bruijnzeel, Thang and Chappell, Hamilton, all this volume), the rather serious discrepancy, highlighted earlier, between the views of policy-makers, the public at large and the scientific community regarding the effects of reforestation on streamflow aptly illustrates the insufficient level of communication between the various groups. It must be concluded, therefore, that misperceptions regarding the hydrological role of tropical forests persist despite continued attempts towards improving the communication of scientific outputs to the management and policy-making community over the last two decades. Examples include a series of publications in the early 1980s from the former Environment and Policy Institute of the East–West Center, Honolulu (e.g. Hamilton and King, 1983; Pearce, 1986; Pearce and Hamilton, 1986); a series of more popularised publications arising from UNESCO–IHP's Humid Tropics Programme (including ones on the environmental impacts of logging and forest conversion plus another on the hydrological and biodiversity values of cloud forests; Bruijnzeel and Critchley, 1994; Critchley and Bruijnzeel, 1996; Bruijnzeel and Hamilton, 2000); and the more practically orientated FAO Conservation Guides issued by the Food and Agriculture Organization's Forestry Department (e.g. FAO, 1977, 1985–1990). Although the UNESCO–IHP series in particular made a deliberate attempt at conveying the findings of hydrological research in a form that might be used by non-specialists, they may not always have been circulated in sufficient numbers (at 1500–3000 copies).[1] However, with the arrival of the internet and a new generation of (professional) resource managers and NGOs that is familiar with the worldwide web as a source of information, opportunities for the much wider as well as more accurately targeted dissemination of research results abound nowadays (cf. the UNESCO Water Portal: http://www.unesco.org/water).

Another potentially fruitful way to improve the use of research results by practitioners is through the development of interdisciplinary, more 'client'- orientated research programmes to resolve specific problem areas. Key issues once again include the managing (i.e. maintaining or boosting) of dry season flows and groundwater reserves for various land cover scenarios (Scott *et al.*, this volume), understanding the hydrological functioning of headwater cloud forests (Bruijnzeel, this volume), as well as the efficacy of riparian buffer zones of variable width to trap sediment and nutrients from upslope (Chappell, Tych *et al.*; Thang and Chappell, Proctor, all this volume), and – last but not least – the water use efficiency of different trees and crops that may be used for land rehabilitation (Scott *et al.*, Wallace *et al.*, both this volume).

Whilst an inadequate understanding of (certain elements of) land use hydrology can be an impediment to rational land use planning it is also true that there is already considerable knowledge and a number of tools that, with a little more development, might be used more effectively, including:

(1) The development of 'performance indicators' (Thang and Chappell, this volume) to assess the condition and quality of catchment outputs, ecosystem health and the adequacy of management practices (e.g. the ratio between mean monthly minimum and maximum streamflows, concentrations of suspended sediment or coliform bacteria in streamwater; Deutsch *et al.*, this volume; cf. Connolly and Pearson, this volume; Gilmour, 1977b).
(2) Better use of available stakeholder analysis techniques and participatory management tool kits to guide catchment planning and management (cf. FAO, 1986b, 1987; Anderson *et al.*, 1998; Deutsch *et al.*, this volume). Ideally, this implies the development of user-friendly, yet effective decision support systems (DSS) and extension services so that interested parties at all levels may get access to contemporary research results and soundly based guidelines for best management practices (Murdiyarso, this volume).

The latter point reiterates once more the prime importance of increasing the involvement of local communities in the entire resource management decision-making process (problem identification, planning and decision making). This 'bottom-up' approach will help to balance the previous mistakes of centralised government control in the implementation of forestry and land clearing policies (Schweithelm, A. Hall, both this volume). If a DSS is to be used, it should be capable of summarising quantitatively the detailed components of local concerns, so as to enhance the negotiating position of poor local communities and to reduce the potential for conflicts between stakeholder groups (Murdiyarso, this volume). By nature, such close cooperation with local communities is bound to be limited to the more local scale ($<10\ km^2$) although much larger areas can be addressed by using multiple stakeholder approaches to resource use planning (Anderson *et al.*, 1998; cf. Deutsch *et al.*, this volume).

Cassells and Bruijnzeel (this volume) propose the establishment of regional demonstration forests for the testing and refining of *locally applicable* guidelines for RIL, including their impacts on erosion, streamflow and sedimentation. The costs and benefits of RIL methods (including various riparian buffer zone arrangements; Thang and Chappell, this volume) can be rigorously monitored at such sites, people from the region itself can be trained in key aspects of RIL applications and ownership by local communities can be fostered through their enhanced participation in forest management (cf. Schweithelm, A. Hall, both this volume). Other catchments within such a pan-tropical network could be the subject of carefully focused investigations of the effects of various forms of land rehabilitation on low flows, or the conversion

1 The brochure on tropical montane cloud forests was circulated in larger numbers (5000 copies) through co-funding by WWF and IUCN-NL. In addition, it was translated into Spanish (1500 copies) and Indonesian (500 copies) through financial support of the UK Department for International Development (DFID) and IUCN-NL. The English and Spanish versions are linked to: http://www.cloudforestalive.org.

of montane cloud forests to grassland or annual cropping. If maintained long enough, such a network would ensure the availability of long-term hydrological reference data from which to infer the effects of climatic variability (as separate from land use impacts) on catchment water and sediment yields (Chappell, Tych *et al.*, this volume). Finally, these catchments could act as testing grounds for new research tools (cf. Buttle and McDonnell, Chappell, Bidin *et al.*, both this volume).

Arguably, of the various hydrological key issues identified earlier as requiring further research, the closely related subjects of how best to improve dry season flows from degraded landscapes and the establishment of the water use efficiencies of a range of tree and crop species, will require the most comprehensive effort. The hydrology of cloud forests is currently the subject of active research by a loosely organised network of researchers, with frequent interaction and data exchange (cf. Table 18.4 in Bruijnzeel, this volume). Similarly, long-term studies of the hydrological impacts of various logging practices are under way in Sabah, which could be extended along the lines indicated above (e.g. buffer zone studies; Chappell, Tych *et al.*, this volume). By contrast, there is virtually no experimental (process) hydrological work going on in relation to the rehabilitation of severely degraded land (cf. Table 25.4 in Scott *et al.*, this volume). The traditional, paired catchment approach of 'calibrate, cut and publish' needs to be reversed, with the 'control' catchments now representing converted land in various stages of degradation. Following a calibration phase in which the flows from the 'control' and 'experimental' or 'treatment' catchments are compared (cf. Grip *et al.*, this volume), the vegetation, soil and water impacts of various land rehabilitation strategies need to be assessed. The evaluation should include a (long-term) analysis of vegetation recuperation and the associated changes in evaporation (water use), total water yield, high and low flow characteristics (both on a seasonal and event basis), infiltration, soil- and groundwater storages, as well as the changes in erosion and sediment transport, nutrient cycling and soil fertility, and stream water quality. In other words, in-depth hydrological process studies are required to help interpret the underlying causes of the observed changes in streamflow (cf. Grip *et al.*, Scott *et al.*, both this volume).

In view of the societal importance and scale of the 'low flow' problem, the dwindling funds for hydrological in-depth research and long-term monitoring, and the general paucity of qualified researchers in the humid tropics, the pooling of available resources and experience would seem imperative. Selected sites could be part of UNESCO's HELP network of river basins (Hydrology for the Environment – Life – Policy, http://www.unesco.org/water/ihp/help) which aims to address a range of water policy issues through a tripartite stakeholder dialogue – management – scientific research agenda (UNESCO–IAEA, 2002). However, the active involvement of, *inter alia*, the International Centre for Forestry Research (CIFOR), the World Centre for Agroforestry (ICRAF), the UN Food and Agricultural Organization's (FAO) Forest Resources and Watershed Management Divisions, as well as regionally active universities, government agencies and NGOs representing local communities should also be sought to optimise financial and human resource use. Ultimately, a successful strategy depends on governments and other donors appreciating the need for a long-term vision vis-à-vis research implementation.

The bottom-up approach advocated above represents a major paradigm shift in the method of research as well as in the application of research results by meshing the concerns and efforts of scientists, local communities, professional managers and policymakers when setting a research agenda. This may seem a long way off to some but the contribution by Deutsch *et al.* (this volume) demonstrates that much can be achieved in this respect and with minimum expense. A survey of stream waterlevels, sediment concentrations and bacterial contamination conducted by members of local communities in the Manupali river basin (Mindano, Philippines) using simple but tested methods, revealed a clear west-to-east pattern of progressive land degradation that was closely associated with increasing population pressure and a corresponding change in land use from forest to fire-climax grassland and subsistence cropping. The steep environmental gradient towards progressively greater degradation while moving eastwards in the basin, of which changes were well within living memory, provided a truly dynamic appreciation of the problem by the local communities. Put simply, a person standing in the middle of the catchment could 'look west' to see how things were in the past, and 'look east' to see what was ahead. Although such community-based surveys do not have the same rigour as full scientific research (e.g. in the absence of automated recording and sampling equipment annual water and sediment yields must have been severely underestimated), the bottom-up approach does bring home the main points for environmental policy formulation, the identification of 'hot spots' in need of rehabilitation and implementation of remedial measures. It may also assist in the formulation of working hypotheses for subsequent more rigorous research efforts as described in the previous paragraphs. Most importantly, however, the involvement of local communities has been shown to lead to a much better chance that research results actually have a policy impact. As such, it is heartening to note that the approach pioneered by Deutsch *et al.* (this volume) in rural Mindanao is being adopted rapidly in other tropical uplands, not only elsewhere in the Philippines but also in mainland South East Asia, Indonesia and various countries in Latin America. Arguably, this may be interpreted as a sign that the time is ripe for the shift towards the more holistic approaches to tropical land and water management issues called for in this book.

References

Anderson, J., Clement, J. and Crowder, L. V. (1998). Accommodating conflicting interests in forestry – concepts emerging from pluralism. *Unasylva*, 194 Vol. 49(3), 3–10.

Beadle, C. L. (1997). Dynamics of leaf and canopy development. In *Management of Soil, Nutrients and Water in Tropical Plantation Forests*, eds. E. K. S. Nambiar and A. G. Brown, pp. 169–212. Canberra: ACIAR/CSIRO and Bogor: CIFOR.

Blackie, J. R., Edwards, K. A. and Clarke, R. T. (1979). Hydrological research in East Africa. *East African Agricultural and Forestry Journal*, 43, 1–313.

Bren, L. J. (2000). A case study in the use of threshold measures of hydrologic loading in the design of stream buffer strips. *Forest Ecology and Management*, 132, 243–257.

Bridges, E. M., Hannam, I. D., Oldeman, L. R., Penning de Vries, F. W. T., Scherr, S. J. and Sombatpanit, S. (2001). *Response to Land Degradation*. Science Publishers Inc., Enfield (NH), USA, 510 p.

Bruijnzeel, L. A. (1989). (De) forestation and dry season flow in the tropics: a closer look. *Journal of Tropical Forest Science*, 1, 229–243.

Bruijnzeel, L. A. (2004). Hydrological functions of tropical forests: not seeing the soil for the trees? *Agriculture, Ecosystems and Environment*, doi:10.1016/j.agee, 2004.01/015.

Bruijnzeel, L. A. and Critchley, W. R. S. (1994). Environmental impacts of logging moist tropical forests. *IHP Humid Tropics Programme Series, 7*. Paris: UNESCO, 48pp.

Bruijnzeel, L. A. and Hamilton, L. S. (2000). Decision Time for Cloud Forests. *IHP Humid Tropics Programme Series* no. 13. Paris: IHP-UNESCO, Amsterdam: IUCN-NL and Gland, Switzerland: WWF International, 44pp.

Calder, I. R. (1999). *The Blue Revolution. Land Use and Integrated Water Resources Management*. London: Earthscan Publications Ltd., 192pp.

Critchley, W. R. S. and Bruijnzeel, L. A. (1996). Environmental impacts of converting moist tropical forest to agriculture and plantations. *IHP Humid Tropics Programme Series,* 10. Paris: UNESCO, 48pp.

FAO (1977). *Guidelines for Watershed Management*. FAO Conservation Guide, 1. Rome: Food and Agriculture Organization, 293pp.

FAO (1985). *FAO Watershed Management Field Manual – Vegetation and Soil Treatment Measures*. FAO Conservation Guide, 13/1. Rome: Food and Agriculture Organization.

FAO (1986a). *FAO Watershed Management Field Manual – Gully Control*. FAO Conservation Guide, 13/2. Rome: Food and Agriculture Organization.

FAO (1986b). *Strategies, Approaches and Systems in Integrated Watershed Management*. FAO Conservation Guide, 14. Rome: Food and Agriculture Organization, 232p.

FAO (1987). *Incentives for Community Involvement in Conservation Programmes*. FAO Conservation Guide, 12. Rome: Food and Agriculture Organization.

FAO (1988a). *FAO Watershed Management Field Manual – Slope Treatment Measures and Practices*. FAO Conservation Guide, 13/3. Rome: Food and Agriculture Organization.

FAO (1988b) *FAO Watershed Management Field Manual – Landslide Prevention Measures*. FAO Conservation Guide, 13/4. Rome: Food and Agriculture Organization.

FAO (1989). *FAO Watershed Management Field Manual – Road Design and Construction*. FAO Conservation Guide, 13/5. Rome: Food and Agriculture Organization.

FAO (1990). *FAO Watershed Management Field Manual – Watershed Survey and Planning*. FAO Conservation Guide, 13/6. Rome: Food and Agriculture Organization.

FAO (2004). *Forests and Floods: Drowning in Fiction or Thriving on Fact?* Rome: United Nations Food and Agriculture Organization and Bogor, CIFOR.

Giambelluca, T. W. (2002). The hydrology of altered tropical forest. Invited Commentary. *Hydrological Processes*, 16, 1665–1669.

Gilmour, D. A. (1977a). Effect of rainforest logging and clearing on water yield and quality in a high rainfall zone of north-east Queensland. In *Proceedings of the First National Symposium on Forest Hydrology*, eds. E. M. O'Loughlin and L. J. Bren, pp. 156–160. Melbourne: Institution of Engineers of Australia.

Gilmour, D. A. (1977b). Logging and the environment, with particular reference to soil and stream protection in tropical rainforest situations. *FAO Conservation Guide*, 1, 223–235.

Hamilton, L. S. and King, P. N. (1983). *Tropical Forested Watersheds. Hydrologic and Soils Response to Major Uses or Conversions*. Boulder: Westview Press, 168pp.

La Marche, J. L. and Lettenmaier, D. P. (2001). Effects of forest roads on flood flows in the Deschutes River, Washington. *Earth Surface Processes and Landforms*, 26, 115–134.

Lamb, D. and Gilmour, D. (2003). *Rehabilitation and Restoration of Degraded Forests*. Gland, Switzerland: The World Conservation Union (*in press*).

Leigh, E. G., Rand, A. S. and Windsor, D. M. (eds.) (1983). *The Ecology of a Tropical Forest: Seasonal Rhythms and Long-Term Changes*. Oxford: Oxford University Press.

Pagiola, S., Bishop, J. and Landell-Mills, N. (editors). 2002. *Selling Forest Environmental Services: Market-based Mechanisms for Conservation*. London: Earthscan.

Pearce, A. J. (1986). *Erosion and Sedimentation*. Working Paper, Honolulu, Hawaii: Environment and Policy Institute, 18 pp.

Pearce, A. J. and Hamilton, L. S. (1986). *Water and Soil Conservation Guidelines for Land Use Planning*. Working Paper, Honolulu, Hawaii: Environment and Policy Institute, 43 pp.

Richards, P. W. (1996). *The Tropical Rain Forest*. 2nd Edition. Cambridge: Cambridge University Press, 575 pp.

Rosegrant, M. W., Cai, X. and Cline, S. A. (2002). *World Water and Food to 2025: Dealing with Scarcity*. Washington D.C.: International food Policy Research Institute, 344 pp.

Sandström, K. (1998). Can forests 'provide' water: widespread myth or scientific reality? *Ambio*, 27, 132–138.

UNESCO-IAEA (2002). *Integration and application of hydrological and ecological processes as a tool for sustainable water resources management and society development*. Expert IHP-IAEA Workshop Warsaw, 3–7 July 2002, 17 pp. Available on http://www.unesco.org/water/ihp/help.

Van Lynden, G. W. J. and Oldeman, L. R. (1997). *The Assessment of the Status of Humna-induced Soil Degradation in South and South East Asia*. Nairobi/Rome/Wageningen: UNEP/FAO/ISRIC, 35 pp. + 2 maps.

Van Noordwijk, M., Van Roode, M., McCallie, E. L. and Lusiana, B. (1998). Erosion and sedimentation as multiscale, fractal processes: Implications for models, experiments and the real world. In *Soil Erosion at Multiple Scales*, eds. F. W. T. Penning de Vries, F. Agus, and J. Kerr, pp. 223–353. Wallingford, UK: CABI Publishing.

Watson, F. G., Vertessy, R. A. and Grayson, R. B. (1999). Large-scale modelling of forest eco-hydrological processes and their long term effect on water yield. *Hydrological Processes*, 13, 689–700.

Young, A. (1997). *Agroforestry for Soil Management*. Wallingford: CABI Publishers, 320pp.